Encyclopedia of Microbiology

Second Edition

Volume 4 Q–Z

Encyclopedia of MICROBIOLOGY

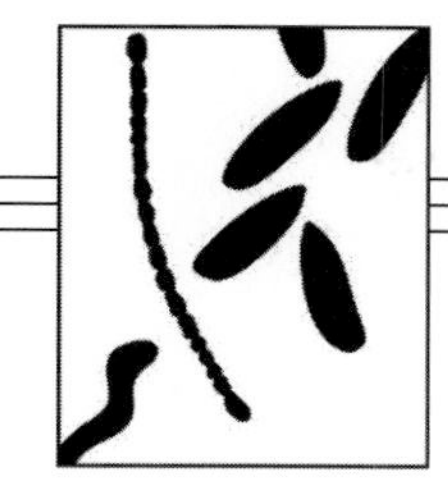

Second Edition

Volume 4 Q–Z

Editor-in-Chief

Joshua Lederberg

The Rockefeller University
New York, NY

ACADEMIC PRESS

A Harcourt Science and Technology Company

San Diego San Francisco New York Boston London Sydney Tokyo

This book is printed on acid-free paper. ♾

Academic Press
A Harcourt Science and Technology Company
525 B Street, Suite 1900, San Diego, California 92101-4495, USA
http://www.apnet.com

Academic Press
24-28 Oval Road, London NW1 7DX, UK
http://www.hbuk.co.uk/ap/

Library of Congress Catalog Card Number: 99-65283

International Standard Book Number: 0-12-226800-8 (set)
International Standard Book Number: 0-12-226801-6 Volume 1
International Standard Book Number: 0-12-226802-4 Volume 2
International Standard Book Number: 0-12-226803-2 Volume 3
International Standard Book Number: 0-12-226804-0 Volume 4

PRINTED IN THE UNITED STATES OF AMERICA
00 01 02 03 04 05 MM 9 8 7 6 5 4 3 2 1

Contents

Contents by Subject Area *xix*
Preface *xxv*
From the Preface to the First Edition *xxvii*
Guide to the Encyclopedia *xxix*
Acknowledgments *xxxi*

Q

Quorum Sensing in Gram-Negative Bacteria 1
Clay Fuqua

R

Rabies 15
William H. Wunner

Ralstonia solanacearum 32
Alan Christopher Hayward

***recA*: The Gene and Its Protein Product** 43
Robert V. Miller

Recombinant DNA, Basic Procedures 55
Judith W. Zyskind

Refrigerated Foods 65
Leora A. Shelef

Retroviruses 81
Ralph Dornburg and Roger J. Pomerantz

Rhinoviruses 97
Eurico Arruda

Rhizoctonia 109
Marc A. Cubeta and Rytas Vilgalys

Rhizosphere 117
Donald A. Klein

Ribosome Synthesis and Regulation 127
Catherine L. Squires

Rickettsiae 140
Marina E. Eremeeva and Gregory A. Dasch

RNA Splicing, Bacterial 181
David A. Shub

Rumen Fermentation 185
James B. Russell

Rust Fungi 195
Robert F. Park

S

Secondary Metabolites 213
Juan F. Martín, Santiago Gutiérrez, Jesús F. Aparicio

Selenium 238
Ronald S. Oremland

Sexually Transmitted Diseases 248
Jane A. Cecil and Thomas C. Quinn

Skin Microbiology 271
Morton N. Swartz

Smallpox 289
Donald A. Henderson

Smuts, Bunts, and Ergot 297
Denis Gaudet, Jim Menzies, Peter Burnett

Soil Dynamics and Organic Matter, Decomposition 316
Edward R. Leadbetter

Soil Microbiology 321
Kate M. Scow

SOS Response 336
Kevin W. Winterling

Space Flight, Effects on Microorganisms 344
D. L. Pierson and Saroj K. Mishra

Spirochetes 353
Lynn Margulis

Spontaneous Generation 364
James Strick

Sporulation 377
Patrick J. Piggot

Staphylococcus 387
John J. Iandolo

Starvation, Bacterial 394
A. Matin

Stock Culture Collections and Their Databases 404
Mary K. B. Berlyn

Strain Improvement 428
Sarad R. Parekh

Streptococcus pneumoniae 444
Alexander Tomasz

***Streptomyces*, Genetics** 451
Paul J. Dyson

Stringent Response 467
Michael Cashel

Sulfide-Containing Environments 478
Rutger de Wit

Sulfur Cycle 495
Piet Lens, Richard Tichý, Look Hulshoff Pol

Surveillance of Infectious Diseases 506
Robert W. Pinner, Denise Koo, Ruth L. Berkelman

Symbiotic Microorganisms in Insects 526
Angela E. Douglas

Syphilis, Historical 538
David Shumway Jones

T

Temperature Control 545
Terje Sørhaug

Tetrapyrrole Biosynthesis in Bacteria 558
Samuel I. Beale

Timber and Forest Products 571
David J. Dickinson and John F. Levy

T Lymphocytes 583
Charles A. Janeway

Tospoviruses 592
James. W. Moyer

Toxoplasmosis 598
John C. Boothroyd

Transcriptional Regulation in Prokaryotes 610
Orna Amster-Choder

Transcription, Viral 628
David S. Latchman

Transduction: Host DNA Transfer by Bacteriophages 637
Millicent Masters

Transformation, Genetic 651
Brian M. Wilkins and Peter A. Meacock

Transgenic Animal Technology 666
Simon M. Temperley, Alexander J. Kind, Angelika Schnieke, Ian Garner

Translational Control and Fidelity 690
Philip J. Farabaugh

Transposable Elements 704
Peter M. Bennett

Trypanosomes 725
Herbert B. Tanowitz, Murray Wittner, Craig Werner, Louis M. Weiss, Louis V. Kirchhoff, Cyrus Bacchi

Two-Component Systems 742
Alexander J. Ninfa and Mariette R. Atkinson

Typhoid, Historical 755
William C. Summers

Typhus Fevers and Other Rickettsial Diseases 758
Theodore E. Woodward

V

Vaccines, Bacterial 767
Susan K. Hoiseth

Vaccines, Viral 779
Ann M. Arvin

Verticillium 788
Jane Robb

Viruses 796
Sondra Schlesinger and Milton J. Schlesinger

Viruses, Emerging 811
Stephen S. Morse

Virus Infection 832
William C. Summers

Vitamins and Related Biofactors, Microbial Production 837
S. De Baets, S. Vandedrinck, Erick Jerome Vandamme

W

Wastewater Treatment, Industrial 855
Angela Bielefeldt and H. David Stensel

Wastewater Treatment, Municipal 870
Ross E. McKinney

Water-Deficient Environments 884
Bettina Kempf and Erhard Bremer

Water, Drinking 898
Paul S. Berger, Robert M. Clark, Donald J. Reasoner

Wine 914
Keith H. Steinkraus

X

Xanthomonas 921
Twng Wah Mew and Jean Swings

Xylanases 930
Pratima Bajpai

Y

Yeasts 939
Graeme M. Walker

Z

Zoonoses 955
Bruno B. Chomel

Contributors 967
Glossary 989
Index 1053

Contents of Other Volumes

VOLUME 1

Contents by Subject Area xix
Preface xxv
From the Preface to the First Edition xxvii
Guide to the Encyclopedia xxix
Acknowledgments xxxi

A

ABC Transport 1
Elie Dassa

Acetic Acid Production 13
Munir Cheryan

Acetogenesis and Acetogenic Bacteria 18
Amaresh Das and Lars G. Ljungdahl

Actinomycetes 28
Martin Krsek, Nathan Morris, Sharon Egan, Elizabeth M. H. Wellington

Adhesion, Bacterial 42
Matthew A. Mulvey and Scott James Hultgren

Aerobic Respiration: Oxidases and Globins 53
Robert K. Poole

Aerosol Infections 69
Edward A. Nardell

Agrobacterium 78
Paul J. J. Hooykaas

***Agrobacterium* and Plant Cell Transformation** 86
Peter J. Christie

AIDS, Historical 104
David Shumway Jones and Allan M. Brandt

Airborne Microorganisms and Indoor Air Quality 116
Linda D. Stetzenbach and Mark P. Buttner

Alkaline Environments 126
William D. Grant and Brian E. Jones

Amino Acid Function and Synthesis 134
Larry Reitzer

Amino Acid Production 152
Hermann Sahm and Lothar Eggeling

Aminoglycosides, Bioactive Bacterial Metabolites 162
Wolfgang Piepersberg

Amylases, Microbial 171
Claire Vieille, Alexei Savchenko, J. Gregory Zeikus

Anaerobic Respiration 180
Robert P. Gunsalus

Antibiotic Biosynthesis 189
Haibin Liu and Kevin A. Reynolds

Antibodies and B Cells 208
Ian M. Zitron

Antifungal Agents 232
Ana A. Espinel-Ingroff

Antigenic Variation 254
Luc Vanhamme and Etienne Pays

Antisense RNAs 268
Andrea D. Branch

Antiviral Agents 286
Richard J. Whitley

Arboviruses 311
Robert E. Shope

Archaea 319
Costantino Vetriani and Anna-Louise Reysenbach

Arsenic 332
Dianne K. Newman

Attenuation, Transcriptional 339
Charles Yanofsky

Autotrophic CO_2 Metabolism 349
Ki-Seok Yoon, Thomas E. Hanson, Janet L. Gibson, F. Robert Tabita

Azotobacter 359
Susan Hill and Gary Sawers

B

***Bacillus subtilis*, Genetics** 373
Kevin M. Devine

Bacteriocins 383
Rolf D. Joerger, S. F. Barefoot, K. M. Harmon, D. A. Grinstead, C. G. Nettles Cutter, Dallas G. Hoover

Bacteriophages 398
Hans-Wolfgang Ackermann

Beer/Brewing 412
Mark A. Harrison and Brian Nummer

Beet Necrotic Yellow Vein Virus 422
Renate Koenig and Dietrich-Eckhardt Lesemann

Biocatalysis for Synthesis of Chiral Pharmaceutical Intermediates 430
Ramesh N. Patel

Biocides 445
Mohammad Sondossi

Biodegradation 461
Wendy B. Bollag, Jerzy Dec, Jean-Marc Bollag

Biodeterioration: In Wood, Architecture, Art, and Other Media 472
José-Julio Ortega-Calvo

Biofilms and Biofouling 478
Karen T. Elvers and Hilary M. Lappin-Scott

Biological Control of Weeds 486
Knud Mortensen

Biological Nitrogen Fixation 492
Donald A. Phillips and Esperanza Martínez-Romero

Biological Warfare 506
James A. Poupard and Linda A. Miller

Bioluminescence, Microbial 520
J. Woodland Hastings

Biomonitors of Environmental Contamination by Microorganisms 530
Marylynn V. Yates

Biopesticides, Microbial 541
Mark A. Jackson

Biopolymers, Production and Uses of 556
William R. Finnerty

Bioreactor Monitoring and Control 567
Roland Ulber, Bernd Hitzmann, Thomas Scheper, Kenneth F. Reardon

Bioreactors 579
Larry E. Erickson

Bioremediation 587
Joseph B. Hughes, C. Nelson Neale, C. H. Ward

Biosensors 611
Yoko Nomura and Isao Karube

Biosurfactants 618
Fazilet Vardar-Sukan and Naim Kosaric

Biotransformations 636
Herbert L. Holland

C

Carbohydrate Synthesis and Metabolism 647
Robert T. Vinopal and
Antonio H. Romano

Carbon and Nitrogen Assimilation, Regulation of 669
Alexander J. Ninfa and
Mariette R. Atkinson

Careers in Microbiology 683
Alice G. Reinarz

***Caulobacter,* Genetics** 692
M. R. K. Alley

Cell Division, Prokaryotes 704
Nanne Nanninga

Cell Membrane: Structure and Function 710
Robert J. Kadner

Cellular Immunity 729
Stefan H. E. Kaufmann and
Michael S. Rolph

Cellulases 744
Pierre Béguin and Jean-Paul Aubert

Cell Walls, Bacterial 759
Joachim Volker Höltje

Chemotaxis 772
Jeffry B. Stock and
Sandra Da Re

Chlamydia 781
Jane E. Raulston and
Priscilla B. Wyrick

Cholera 789
Claudia C. Häse, Nicholas Judson,
John J. Mekalanos

Cholera, Historical 801
Christopher D. Meehan and
Howard Markel

Chromosome, Bacterial 808
Karl Drlica and Arnold J. Bendich

Chromosome Replication and Segregation 822
Alan C. Leonard and
Julia E. Grimwade

Clostridia 834
Eric A. Johnson

Coenzyme and Prosthetic Group Biosynthesis 840
Walter B. Dempsey

Conjugation, Bacterial 847
Laura S. Frost

Conservation of Cultural Heritage 863
Orio Ciferri

Continuous Culture 873
Jan C. Gottschal

Cosmetic Microbiology 887
Daniel K. Brannan

Crystalline Bacterial Cell Surface Layers 899
Uwe B. Sleytr and Paul Messner

Cyanobacteria 907
Ferran Garcia-Pichel

VOLUME 2

Contents by Subject Area *xix*
Preface *xxv*
From the Preface to the First Edition *xxvii*
Guide to the Encyclopedia *xxix*
Acknowledgments *xxxi*

D

Dairy Products 1
Mary Ellen Sanders

Detection of Bacteria in Blood: Centrifugation and Filtration 9
Mathias Bernhardt, Laurel S. Almer,
Erik R. Munson, Steven M. Callister,
Ronald F. Schell

Developmental Processes in Bacteria 15
Yves V. Brun

Diagnostic Microbiology 29
Yi-Wei Tang and David H. Persing

Dinoflagellates 42
Marie-Odile Soyer-Gobillard and Hervé Moreau

Diversity, Microbial 55
Charles R. Lovell

DNA Repair 71
Lawrence Grossman

DNA Replication 82
James A. Hejna and Robb E. Moses

DNA Restriction and Modification 91
Noreen E. Murray

DNA Sequencing and Genomics 106
Brian A. Dougherty

Downy Mildews 117
Jeremy S. C. Clark and Peter T. N. Spencer-Phillips

E

Ecology, Microbial 131
Michael J. Klug and David A. Odelson

Economic Consequences of Infectious Diseases 137
Martin I. Meltzer

Education in Microbiology 156
Ronald H. Bishop

Emerging Infections 170
David L. Heymann

Energy Transduction Processes: From Respiration to Photosynthesis 177
Stuart J. Ferguson

Enteropathogenic Bacteria 187
Farah K. Bahrani-Mougeot and Michael S. Donnenberg

Enteroviruses 201
Nora M. Chapman, Charles J. Gauntt, Steven M. Tracy

Enzymes, Extracellular 210
Fergus G. Priest

Enzymes in Biotechnology 222
Badal C. Saha and Rodney J. Bothast

***Erwinia*: Genetics of Pathogenicity Factors** 236
Arun K. Chatterjee, C. Korsi Dumenyo, Yang Liu, Asita Chatterjee

***Escherichia coli*, General Biology** 260
Moselio Schaechter

***Escherichia coli* and *Salmonella*, Genetics** 270
K. Brooks Low

Evolution, Theory and Experiments 283
Richard E. Lenski

Exobiology 299
Gerald Soffen

Exotoxins 307
Joseph T. Barbieri

Extremophiles 317
Ricardo Cavicchioli and Torsten Thomas

Eyespot 338
Paul S. Dyer

F

Fermentation 343
August Böck

Fermented Foods 350
Keith H. Steinkraus

Fimbriae, Pili 361
Matthew A. Mulvey, Karen W. Dodson, Gabriel E. Soto, Scott James Hultgren

Flagella 380
Shin-Ichi Aizawa

Food-borne Illnesses 390
David W. K. Acheson

Food Spoilage and Preservation 412
Daniel Y. C. Fung

Foods, Quality Control 421
Richard B. Smittle

Freeze-Drying of Microorganisms 431
Hiroshi Souzu

Freshwater Microbiology 438
Louis A. Kaplan and
Allan E. Konopka

Fungal Infections, Cutaneous 451
Peter G. Sohnle and
David K. Wagner

Fungal Infections, Systemic 460
Arturo Casadevall

Fungi, Filamentous 468
Joan W. Bennett

G

Gaeumannomyces graminis 479
Joan M. Henson and
Henry T. Wilkinson

Gastrointestinal Microbiology 485
T. G. Nagaraja

Genetically Modified Organisms: Guidelines and Regulations for Research 499
Sue Tolin and Anne Vidaver

Genomic Engineering of Bacterial Metabolism 510
Jeremy S. Edwards,
Christophe H. Schilling,
M. W. Covert,
S. J. Smith,
Bernhard Palsson

Germfree Animal Techniques 521
Bernard S. Wostmann

Global Burden of Infectious Diseases 529
Catherine M. Michaud

Glycogen Biosynthesis 541
Jack Preiss

Glyoxylate Bypass in *Escherichia coli* 556
David C. LaPorte, Stephen P. Miller,
Satinder K. Singh

Gram-Negative Anaerobic Pathogens 562
Arthur O. Tzianabos,
Laurie E. Comstock,
Dennis L. Kasper

Gram-Negative Cocci, Pathogenic 571
Emil C. Gotschlich

Growth Kinetics, Bacterial 584
Allen G. Marr

H

***Haemophilus influenzae*, Genetics** 591
Rosemary J. Redfield

Heat Stress 598
Christophe Herman and
Carol A. Gross

Heavy Metal Pollutants: Environmental and Biotechnological Aspects 607
Geoffrey M. Gadd

Heavy Metals, Bacterial Resistances 618
Tapan K. Misra

Helicobacter pylori 628
Sebastian Suerbaum and
Martin J. Blaser

Hepatitis Viruses 635
William S. Mason and
Allison R. Jilbert

Heterotrophic Microorganisms 651
James T. Staley

High-Pressure Habitats 664
A. Aristides Yayanos

History of Microbiology 677
William C. Summers

Horizontal Transfer of Genes between Microorganisms 698
Jack A. Heinemann

I

Identification of Bacteria, Computerized 709
Trevor N. Bryant

Industrial Biotechnology, Overview 722
Erik P. Lillehoj and Glen M. Ford

Industrial Effluents: Sources, Properties, and Treatments 738
Fazilet Vardar-Sukan and Naim Kosaric

Industrial Fermentation Processes 767
Thomas M. Anderson

Infectious Waste Management 782
Gerald A. Denys

Influenza Viruses 797
Christopher F. Basler and Peter Palese

Insecticides, Microbial 813
Allan A. Yousten, Brian A. Federici, Donald W. Roberts

Interferons 826
Bryan R. G. Williams

International Law and Infectious Disease 842
David P. Fidler

Intestinal Protozoan Infections in Humans 852
Adolfo Martínez-Palomo and Martha Espinosa-Cantellano

Iron Metabolism 860
Charles F. Earhart

VOLUME 3

Contents by Subject Area *xix*
Preface *xxv*
From the Preface to the First Edition *xxvii*
Guide to the Encyclopedia *xxix*
Acknowledgments *xxxi*

L

Lactic Acid Bacteria 1
George A. Somkuti

Lactic Acid, Microbially Produced 9
John H. Litchfield

Legionella 18
N. Cary Engleberg

Leishmania 27
Gary B. Ogden and Peter C. Melby

Lignocellulose, Lignin, Ligninases 39
Karl-Erik L. Eriksson

Lipases, Industrial Uses 49
Ching T. Hou

Lipid Biosynthesis 55
Charles O. Rock

Lipids, Microbially Produced 62
Jacek Leman

Lipopolysaccharides 71
Chris Whitfield

Low-Nutrient Environments 86
Richard Y. Morita

Low-Temperature Environments 93
Richard Y. Morita

Luteoviridae 99
Cleora J. D'Arcy and Leslie L. Domier

Lyme Disease 109
Jenifer Coburn and Robert A. Kalish

M

Malaria 131
Kostas D. Mathiopoulos

Mapping Bacterial Genomes 151
Janine Guespin-Michel and Francoise Joset

Meat and Meat Products 163
Jerry Nielsen

Mercury Cycle 171
Tamar Barkay

Metal Extraction and Ore Discovery 182
James A. Brierley

Methane Biochemistry 188
David A. Grahame and Simonida Gencic

Methane Production/Agricultural Waste Management 199
William J. Jewell

Methanogenesis 204
Kevin R. Sowers

Method, Philosophy of 227
Kenneth F. Schaffner

Methylation of Nucleic Acids and Proteins 240
Martin G. Marinus

Methylotrophy 245
J. Colin Murrell and Ian R. McDonald

Microbes and the Atmosphere 256
Ralf Conrad

Microscopy, Confocal 264
Guy A. Perkins and Terrence G. Frey

Microscopy, Electron 276
Susan F. Koval and Terrance J. Beveridge

Microscopy, Optical 288
Guy A. Perkins and Terrence G. Frey

Mutagenesis 307
Richard H. Baltz

Mycobacteria 312
John T. Belisle and Patrick J. Brennan

Mycorrhizae 328
Michael F. Allen

Mycotoxicoses 337
Stan W. Casteel and George E. Rottinghaus

Myxobacteria 349
David White

***Myxococcus*, Genetics** 363
N. Jamie Ryding and Lawrence J. Shimkets

N

Natural Selection, Bacterial 373
Daniel E. Dykhuizen

Nitrogen Cycle 379
Roger Knowles

Nitrogen Fixation 392
L. David Kuykendall, Fawzy M. Hashem, Robert B. Dadson, Gerald H. Elkan

Nodule Formation in Legumes 407
Peter H. Graham

Nucleotide Metabolism 418
Per Nygaard and Hans Henrik Saxild

Nutrition of Microorganisms 431
Thomas Egli

O

Oil Pollution 449
Joseph P. Salanitro

Oncogenic Viruses 456
Anh Ngoc Dang Do, Linda Farrell, Kitai Kim, Marie Lockstein Nguyen, Paul F. Lambert

Oral Microbiology 466
Ian R. Hamilton and George H. Bowden

Ore Leaching by Microbes 482
James A. Brierley

Origin of Life 490
William F. Loomis

Osmotic Stress 502
Douglas H. Bartlett and Mary F. Roberts

Outer Membrane, Gram-Negative Bacteria 517
Mary J. Osborn

Oxidative Stress 526
Pablo J. Pomposiello and Bruce Demple

P

Paramyxoviruses 533
Suxiang Tong, Qizhi Yao, Richard W. Compans

Patenting of Living Organisms and Natural Products 546
S. Leslie Misrock, Adriane M. Antler, Anne M. Schneiderman

Pectinases 562
Fred Stutzenberger

PEP: Carbohydrate Phosphotransferase Systems 580
Pieter Postma

Pesticide Biodegradation 594
Li-Tse Ou

Phloem-Limited Bacteria 607
Michael J. Davis

Phosphorus Cycle 614
Ronald D. Jones

Photosensory Behavior 618
Brenda G. Rushing, Ze-Yu Jiang, Howard Gest, Carl E. Bauer

pH Stress 625
Joan L. Slonczewski

Phytophthora infestans 633
William E. Fry

Phytoplasma 640
Robert E. Davis and Ing-Ming Lee

Pigments, Microbially Produced 647
Eric A. Johnson

Plague 654
Elisabeth Carniel

Plant Disease Resistance: Natural Mechanisms and Engineered Resistance 662
Karl Maramorosch and Bradley I. Hillman

Plant Pathogens 676
George N. Agrios

Plant Virology, Overview 697
Roger Hull

Plasmids, Bacterial 711
Christopher M. Thomas

Plasmids, Catabolic 730
Anthony G. Hay, Steven Ripp, Gary S. Sayler

Plasmodium 745
John E. Hyde

Polio 762
Ciro A. de Quadros

Polyketide Antibiotics 773
Annaliesa S. Anderson, Zhiqiang An, William R. Strohl

Polymerase Chain Reaction (PCR) 787
Carol J. Palmer and Christine Paszko-Kolva

Potyviruses 792
John Hammond

Powdery Mildews 801
Alison A. Hall, Ziguo Zhang, Timothy L. W. Carver, Sarah J. Gurr

Prions 809
Christine Musahl and Adriano Aguzzi

Protein Biosynthesis 824
Rosemary Jagus and Bhavesh Joshi

Protein Secretion 847
Donald Oliver and Jorge Galan

Protozoan Predation 866
Lucas A. Bouwman

Pseudomonas 876
Vinayak Kapatral, Anna Zago,
Shilpa Kamath, Sudha Chugani

Pulp and Paper 893
Philip M. Hoekstra

Contents by Subject Area

APPLIED MICROBIOLOGY: AGRICULTURE

Agrobacterium and Plant Cell Transformation
Biological Control of Weeds
Fermented Foods
Insecticides, Microbial
Methane Production/Agricultural Waste Management
Nitrogen Cycle
Nodule Formation in Legumes
Pectinases
Pesticide Biodegradation
Plant Disease Resistance: Natural Mechanisms and Engineered Resistance
Plant Pathogens
Protozoan Predation
Rumen Fermentation
Timber and Forest Products

APPLIED MICROBIOLOGY: ENVIRONMENTAL

Airborne Microorganisms and Indoor Air Quality
Alkaline Environments
Arsenic
Biocides
Biodegradation
Biodeterioration
Biofilms and Biofouling
Biomonitors of Environmental Contamination
Biopesticides, Microbial
Bioremediation
Freshwater Microbiology
Heavy Metal Pollutants: Environmental and Biotechnological Aspects
Heavy Metals, Bacterial Resistances
High-Pressure Habitats
Industrial Effluents: Sources, Properties, and Treatments
Infectious Waste Management
Low-Nutrient Environments
Low-Temperature Environments
Metal Extraction and Ore Discovery
Oil Pollution
Ore Leaching by Microbes
Selenium
Soil Microbiology
Space Flight, Effects on Microorganisms
Sulfide-Containing Environments
Wastewater Treatment, Industrial
Wastewater Treatment, Municipal
Water-Deficient Environments
Water, Drinking

APPLIED MICROBIOLOGY: FOOD

Beer/Brewing
Dairy Products
Enzymes in Biotechnology
Food Spoilage and Preservation
Foods, Quality Control
Lactic Acid Bacteria
Lactic Acid, Microbially Produced
Meat and Meat Products

Refrigerated Foods
Wine

APPLIED MICROBIOLOGY: INDUSTRIAL

Acetic Acid Production
Amino Acid Production
Aminoglycosides, Bioactive Bacterial Metabolites
Amylases, Microbial
Antibiotic Biosynthesis
Biocatalysis for Synthesis of Chiral Pharmaceutical Intermediates
Biopolymers, Production and Uses of
Bioreactor Monitoring and Control
Bioreactors
Biosensors
Biosurfactants
Biotransformations
Cellulases
Continuous Culture
Cosmetic Microbiology
Enzymes, Extracellular
Freeze-Drying of Microorganisms
Genomic Engineering of Bacterial Metabolism
Industrial Fermentation Processes
Lignocellulose, Lignin, Ligninases
Lipases, Industrial Uses
Lipids, Microbially Produced
Pigments, Microbially Produced
Plasmids, Catabolic
Polyketide Antibiotics
Pulp and Paper
Secondary Metabolites
Strain Improvement
Vitamins and Related Biofactors, Microbial Production
Xylanases

CAREERS AND EDUCATION

Careers in Microbiology
Education in Microbiology

ECOLOGY

Biological Nitrogen Fixation
Diversity, Microbial
Ecology, Microbial
Mercury Cycle
Methane Biochemistry
Microbes and the Atmosphere
Mycorrhizae
Nitrogen Fixation
Phosphorus Cycle
Rhizosphere
Soil Dynamics and Organic Matter, Decomposition
Symbiotic Microorganisms in Insects

ETHICAL AND LEGAL ISSUES

Biological Warfare
Genetically Modified Organisms: Guidelines and Regulations for Research
International Law and Infectious Disease
Patenting of Living Organisms and Natural Products

GENERAL

Conservation of Cultural Heritage
Diagnostic Microbiology
Exobiology
History of Microbiology
Industrial Biotechnology, Overview
Method, Philosophy of
Spontaneous Generation
Stock Culture Collections and Their Databases

GENETICS

Bacillus subtilis, Genetics
Caulobacter, Genetics
Chromosome, Bacterial
Chromosome Replication and Segregation
Conjugation, Bacterial
DNA Restriction and Modification
DNA Sequencing and Genomics
Escherichia coli and *Salmonella*, Genetics
Evolution, Theory and Experiments
Haemophilus influenzae, Genetics

Horizontal Transfer of Genes between Microorganisms
Mapping Bacterial Genomes
Mutagenesis
Myxococcus, Genetics
Natural Selection, Bacterial
Plasmids, Bacterial
Recombinant DNA, Basic Procedures
Streptomyces, Genetics
Transduction: Host DNA Transfer by Bacteriophages
Transformation, Genetic
Transposable Elements

HISTORICAL

AIDS, Historical
Cholera, Historical
Origin of Life
Plague
Polio
Smallpox
Syphilis, Historical
Typhoid, Historical
Typhus Fevers and Other Rickettsial Diseases

INFECTIOUS AND NONINFECTIOUS DISEASE AND PATHOGENESIS: HUMAN PATHOGENS

Arboviruses
Chlamydia
Cholera
Clostridia
Emerging Infections
Enteropathogenic Bacteria
Food-borne Illnesses
Fungal Infections, Cutaneous
Fungal Infections, Systemic
Gram-Negative Anaerobic Pathogens
Gram-Negative Cocci, Pathogenic
Helicobacter pylori
Hepatitis Viruses
Influenza Viruses
Intestinal Protozoan Infections in Humans
Legionella
Leishmania
Lyme Disease
Malaria
Mycobacteria
Myxobacteria
Oncogenic Viruses
Paramyxoviruses
Plasmodium
Prions
Pseudomonas
Rabies
Rickettsiae
Sexually Transmitted Diseases
Staphylococcus
Streptococcus pneumoniae
Toxoplasmosis
Trypanosomes
Viruses, Emerging
Zoonoses

INFECTIOUS AND NONINFECTIOUS DISEASE AND PATHOGENESIS: IMMUNOLOGY

Antibodies and B Cells
Antigenic Variation
Cellular Immunity
T Lymphocytes

INFECTIOUS AND NONINFECTIOUS DISEASE AND PATHOGENESIS: PLANT PATHOGENS, BACTERIA

Agrobacterium
Erwinia: Genetics of Pathogenicity Factors
Phloem-Limited Bacteria
Phytoplasma
Ralstonia solanacearum
Xanthomonas

INFECTIOUS AND NONINFECTIOUS DISEASE AND PATHOGENESIS: PLANT PATHOGENS, FUNGI

Downy Mildews
Eyespot

Gaeumannomyces graminis
Phytophthora infestans
Powdery Mildews
Rhizoctonia
Rust Fungi
Smuts, Bunts, and Ergot
Verticillium

INFECTIOUS AND NONINFECTIOUS DISEASE AND PATHOGENESIS: PLANT PATHOGENS, VIRUSES

Beet Necrotic Yellow Vein Virus
Luteoviridae
Plant Virology, Overview
Potyviruses
Tospoviruses

INFECTIOUS AND NONINFECTIOUS DISEASE AND PATHOGENESIS: TREATMENT

Aerosol Infections
Antifungal Agents
Antiviral Agents
Bacteriocins
Economic Consequences of Infectious Diseases
Exotoxins
Gastrointestinal Microbiology
Global Burden of Infectious Diseases
Interferons
Lipopolysaccharides
Mycotoxicoses
Oral Microbiology
Skin Microbiology
Surveillance of Infectious Diseases
Vaccines, Bacterial
Vaccines, Viral
Virus Infection

PHYSIOLOGY, METABOLISM, AND GENE EXPRESSION

ABC Transport
Adhesion, Bacterial
Aerobic Respiration
Amino Acid Function and Synthesis
Anaerobic Respiration
Antisense RNAs
Attenuation, Transcriptional
Autotrophic CO_2 Metabolism
Bioluminescence, Microbial
Carbohydrate Synthesis and Metabolism
Carbon and Nitrogen Assimilation, Regulation of
Cell Division, Prokaryotes
Chemotaxis
Coenzyme and Prosthetic Group Biosynthesis
DNA Repair
DNA Replication
Energy Transduction Processes
Fermentation
Glycogen Biosynthesis
Glyoxylate Bypass in *Escherichia coli*
Growth Kinetics, Bacterial
Heat Stress
Iron Metabolism
Lipid Biosynthesis
Methanogenesis
Methylation of Nucleic Acids and Proteins
Methylotrophy
Nucleotide Metabolism
Nutrition of Microorganisms
Osmotic Stress
Oxidative Stress
PEP: Carbohydrate Phosphotransferase Systems
Photosensory Behavior
pH Stress
Protein Biosynthesis
Protein Secretion
Quorum Sensing in Gram-Negative Bacteria
recA
Ribosome Synthesis and Regulation
RNA Splicing, Bacterial
SOS Response
Sporulation
Starvation, Bacterial
Stringent Response
Sulfur Cycle
Tetrapyrrole Biosynthesis in Bacteria
Transcriptional Regulation in Prokaryotes
Transcription, Viral
Translational Control and Fidelity
Two-Component Systems

STRUCTURE AND MORPHOGENESIS

Cell Membrane: Structure and Function
Cell Walls, Bacterial
Crystalline Bacterial Cell Surface Layers
Developmental Processes in Bacteria
Fimbriae, Pili
Flagella
Outer Membrane, Gram-Negative Bacteria

SYSTEMATICS AND PHYLOGENY

Acetogenesis and Acetogenic Bacteria
Actinomycetes
Archaea
Azotobacter
Bacteriophages
Cyanobacteria
Dinoflagellates
Enteroviruses
Escherichia coli, General Biology
Extremophiles
Fungi, Filamentous
Heterotrophic Microorganisms
Retroviruses
Rhinoviruses
Spirochetes
Viruses
Yeasts

TECHNIQUES

Detection of Bacteria in Blood: Centrifugation and Filtration
Germfree Animal Techniques
Identification of Bacteria, Computerized
Microscopy, Confocal
Microscopy, Electron
Microscopy, Optical
Polymerase Chain Reaction (PCR)
Temperature Control
Transgenic Animal Technology

Preface

The scientific literature at large is believed to double about every 12 years. Though less than a decade has elapsed since the initiation of the first edition of this encyclopedia, it is a fair bet that the microbiology literature has more than doubled in the interval, though one might also say it has fissioned in the interval, with parasitology, virology, infectious disease, and immunology assuming more and more independent stature as disciplines.

According to the *Encyclopaedia Britannica,* the encyclopedias of classic and medieval times could be expected to contain "a compendium of all available knowledge." There is still an expectation of the "essence of all that is known." With the exponential growth and accumulation of scientific knowledge, this has become an elusive goal, hardly one that could be embraced in a mere two or three thousand pages of text. The encyclopedia's function has moved to becoming the first word, the initial introduction to knowledge of a comprehensive range of subjects, with pointers on where to find more as may be needed. One can hardly think of the last word, as this is an ever-moving target at the cutting edge of novel discovery, changing literally day by day.

For the renovation of an encyclopedia, these issues have then entailed a number of pragmatic compromises, designed to maximize its utility to an audience of initial look-uppers over a range of coherently linked interests. The core remains the biology of that group of organisms we think of as microbes. Though this constitutes a rather disparate set, crossing several taxonomic kingdoms, the more important principle is the unifying role of DNA and the genetic code and the shared ensemble of primary pathways of gene expression. Also shared is access to a "world wide web" of genetic information through the traffic of plasmids and other genetic elements right across the taxa. It is pathognomonic that the American Society for Microbiology has altered the name of *Microbiological Reviews* to *Microbiology and Molecular Biology Reviews.* At academic institutions, microbiology will be practiced in any or all of a dozen different departments, and these may be located at schools of arts and sciences, medicine, agriculture, engineering, marine sciences, and others.

Much of human physiology, pathology, or genetics is now practiced with cell culture, which involves a methodology indistinguishable from microbiology: it is hard to define a boundary that would demarcate microbiology from cell biology. Nor do we spend much energy on these niceties except when we have the burden of deciding the scope of an enterprise such as this one.

Probably more important has been the explosion of the Internet and the online availability of many sources of information. Whereas we spoke last decade of CDs, now the focus is the Web, and the anticipation is that we are not many years from the general availability of the entire scientific literature via this medium. The utility of the encyclopedia is no longer so much "how do I begin to get information on Topic X" as how to filter a surfeit of claimed information with some degree of dependability. The intervention of editors and of a peer-review process (in selection of authors even more important than in overseeing their papers) is the only foreseeable solution. We have then sought in each article to provide a digest of information with perspective and provided by responsible authors who can be proud of, and will then strive to maintain, reputations for knowledge and fairmindedness.

The further reach of more detailed information is endless. When available, many specific topics are elaborated in greater depth in the ASM (American Society of Microbiology) reviews and in *Annual Review of Microbiology*. These are indexed online. Medline, Biosis, and the Science Citation Index are further online bibliographic resources, which can be focused for the recovery of review articles.

The reputation of the authors and of the particular journals can further aid readers' assessments. Citation searches can be of further assistance in locating critical discussions, the dialectic which is far more important than "authority" in establishing authenticity in science.

Then there are the open-ended resources of the Web itself. It is not a fair test for recovery on a specialized topic, but my favorite browser, google.com, returned 15,000 hits for "microbiology"; netscape.com gave 46,000; excite.com a few score structured headings. These might be most useful in identifying other Web sites with specialized resources. Google's 641 hits for "luminescent bacteria" offer a more proximate indicator of the difficulty of coping with the massive returns of unfiltered verbiage that this wonderful new medium affords: how to extract the nuggets from the slag.

A great many academic libraries and departments of microbiology have posted extensive considered listings of secondary sources. One of my favorites is maintained at San Diego State University:

http://libweb.sdsu.edu/scidiv/microbiologyblr.html

I am sure I have not begun to tap all that would be available.

The best strategy is a parallel attack: to use the encyclopedia and the major review journals as a secure starting point and then to try to filter Web-worked material for the most up-to-date or disparate detail. In many cases, direct enquiry to the experts, until they saturate, may be the best (or last) recourse. E-mail is best, and society or academic institutional directories can be found online. Some listservers will entertain questions from outsiders, if the questions are particularly difficult or challenging.

All publishers, Academic Press included, are updating their policies and practices by the week as to how they will integrate their traditional book offerings with new media. Updated information on electronic editions of this and cognate encyclopedias can be found by consulting www.academicpress.com/.

Joshua Lederberg

From the Preface to the First Edition

(Excerpted from the 1992 Edition)

For the purposes of this encyclopedia, microbiology has been understood to embrace the study of "microorganisms," including the basic science and the roles of these organisms in practical arts (agriculture and technology) and in disease (public health and medicine). Microorganisms do not constitute a well-defined taxonomic group; they include the two kingdoms of Archaebacteria and Eubacteria, as well as protozoa and those fungi and algae that are predominantly unicellular in their habit. Viruses are also an important constituent, albeit they are not quite "organisms." Whether to include the mitochondria and chloroplasts of higher eukaryotes is a matter of choice, since these organelles are believed to be descended from free-living bacteria. Cell biology is practiced extensively with tissue cells in culture, where the cells are manipulated very much as though they were autonomous microbes; however, we shall exclude this branch of research. Microbiology also is enmeshed thoroughly with biotechnology, biochemistry, and genetics, since microbes are the canonical substrates for many investigations of genes, enzymes, and metabolic pathways, as well as the technical vehicles for discovery and manufacture of new biological products, for example, recombinant human insulin. . . .

The *Encyclopedia of Microbiology* is intended to survey the entire field coherently, complementing material that would be included in an advanced undergraduate and graduate major course of university study. Particular topics should be accessible to talented high school and college students, as well as to graduates involved in teaching, research, and technical practice of microbiology.

Even these hefty volumes cannot embrace all current knowledge in the field. Each article does provide key references to the literature available at the time of writing. Acquisition of more detailed and up-to-date knowledge depends on (1) exploiting the review and monographic literature and (2) bibliographic retrieval of the preceding and current research literature. . . .

To access bibliographic materials in microbiology, the main retrieval resources are MEDLINE, sponsored by the U.S. National Library of Medicine, and the Science Citation Index of the ISI. With governmental subsidy, MEDLINE is widely available at modest cost: terminals are available at every medical school and at many other academic centers. MEDLINE provides searches of the recent literature by author, title, and key word and offers online displays of the relevant bibliographies and abstracts. Medical aspects of microbiology are covered exhaustively; general microbiology is covered in reasonable depth. The Science Citation Index must recover its costs from user fees, but is widely available at major research centers. It offers additional search capabilities, especially by citation linkage. Therefore, starting with the bibliography of a given encyclopedia article, one can quickly find (1) all articles more recently published that have cited those bibliographic reference starting points and (2) all other recent articles that share bibliographic information with the others. With luck, one of these articles may be identified as another comprehensive review that has digested more recent or broader primary material.

On a weekly basis, services such as Current Contents on Diskette (ISI) and Reference Update offer still more timely access to current literature as well as to abstracts with a variety of useful features. Under the impetus of intense competition, these services are evolving rapidly, to the great benefit of a user community desperate for electronic assistance in coping with the rapidly growing and intertwined networks of discovery. The bibliographic services of Chemical Abstracts and Biological Abstracts would also be potentially invaluable; however, their coverage of microbiology is rather limited.

In addition, major monographs have appeared from time to time—*The Bacteria, The Prokaryotes,* and many others. Your local reference library should be consulted for these volumes.

Valuable collections of reviews also include *Critical Reviews for Microbiology, Symposia of the Society for General Microbiology, Monographs of the American Society for Microbiology,* and *Proceedings of the International Congresses of Microbiology.*

The articles in this encyclopedia are intended to be accessible to a broader audience, not to take the place of review articles with comprehensive bibliographies. Citations should be sufficient to give the reader access to the latter, as may be required. We do apologize to many individuals whose contributions to the growth of microbiology could not be adequately embraced by the secondary bibliographies included here.

The organization of encyclopedic knowledge is a daunting task in any discipline; it is all the more complex in such a diversified and rapidly moving domain as microbiology. The best way to anticipate the rapid further growth that we can expect in the near future is unclear. Perhaps more specialized series in subfields of microbiology would be more appropriate. The publishers and editors would welcome readers' comments on these points, as well as on any deficiencies that may be perceived in the current effort.

My personal thanks are extended to my coeditors, Martin Alexander, David Hopwood, Barbara Iglewski, and Allen Laskin; and above all, to the many very busy scientists who took time to draft and review each of these articles.

Joshua Lederberg

Guide to the Encyclopedia

The *Encyclopedia of Microbiology, Second Edition* is a scholarly source of information on microorganisms, those life forms that are observable with a microscope rather than by the naked eye. The work consists of four volumes and includes 298 separate articles. Of these 298 articles, 171 are completely new topics commissioned for this edition, and 63 others are newly written articles on topics appearing in the first edition. In other words, approximately 80% of the content of the encyclopedia is entirely new to this edition. (The remaining 20% of the content has been carefully reviewed and revised to ensure currency.)

Each article in the encyclopedia provides a comprehensive overview of the selected topic to inform a broad spectrum of readers, from research professionals to students to the interested general public. In order that you, the reader, will derive the greatest possible benefit from your use of the *Encyclopedia of Microbiology*, we have provided this Guide. It explains how the encyclopedia is organized and how the information within it can be located.

ORGANIZATION

The *Encyclopedia of Microbiology* is organized to provide maximum ease of use. All of the articles are arranged in a single alphabetical sequence by title. Articles whose titles begin with the letters A to C are in Volume 1, articles with titles from D through K are in Volume 2, then L through P in Volume 3, and finally Q to Z in Volume 4. This last volume also includes a complete subject index for the entire work, an alphabetical list of the contributors to the encyclopedia, and a glossary of key terms used in the articles.

Article titles generally begin with the key noun or noun phrase indicating the topic, with any descriptive terms following. For example, the article title is "Bioluminescence, Microbial" rather than "Microbial Bioluminescence," and "Foods, Quality Control" is the title rather than "Quality Control of Foods."

TABLE OF CONTENTS

A complete table of contents for the *Encyclopedia of Microbiology* appears at the front of each volume. This list of article titles represents topics that have been carefully selected by the Editor-in-Chief, Dr. Joshua Lederberg, and the nine Associate Editors. The Encyclopedia provides coverage of 20 different subject areas within the overall field of microbiology. Please see p. v for the alphabetical table of contents, and p. xix for a list of topics arranged by subject area.

INDEX

The Subject Index in Volume 4 indicates the volume and page number where information on a given topic can be found. In addition, the Table of Contents by Subject Area also functions as an index, since it lists all the topics within a given area; e.g., the encyclopedia includes eight different articles dealing with historic aspects of microbiology and nine dealing with techniques of microbiology.

ARTICLE FORMAT

In order to make information easy to locate, all of the articles in the *Encyclopedia of Microbiology* are arranged in a standard format, as follows:

- Title of Article
- Author's Name and Affiliation
- Outline
- Glossary
- Defining Statement
- Body of the Article
- Cross-References
- Bibliography

OUTLINE

Each entry in the Encyclopedia begins with a topical outline that indicates the general content of the article. This outline serves two functions. First, it provides a brief preview of the article, so that the reader can get a sense of what is contained there without having to leaf through the pages. Second, it serves to highlight important subtopics that will be discussed within the article. For example, the article "Biopesticides" includes subtopics such as "Selection of Biopesticides," "Production of Biopesticides," "Biopesticide Stabilization," and "Commercialization of Biopesticides."

The outline is intended as an overview and thus it lists only the major headings of the article. In addition, extensive second-level and third-level headings will be found within the article.

GLOSSARY

The Glossary contains terms that are important to an understanding of the article and that may be unfamiliar to the reader. Each term is defined in the context of the article in which it is used. Thus the same term may appear as a glossary entry in two or more articles, with the details of the definition varying slightly from one article to another. The encyclopedia has approximately 2500 glossary entries.

In addition, Volume 4 provides a comprehensive glossary that collects all the core vocabulary of microbiology in one A–Z list. This section can be consulted for definitions of terms not found in the individual glossary for a given article.

DEFINING STATEMENT

The text of each article in the encyclopedia begins with a single introductory paragraph that defines the topic under discussion and summarizes the content of the article. For example, the article "Eyespot" begins with the following statement:

> EYESPOT is a damaging stem base disease of cereal crops and other grasses caused by fungi of the genus *Tapsia*. It occurs in temperate regions world-wide including Europe, the USSR, Japan, South Africa, North America, and Australasia. In many of these countries eyespot can be found on the majority of autumn-sown barley and wheat crops and may cause an average of 5–10% loss in yield, although low rates of infection do not generally have a significant effect. . . .

CROSS-REFERENCES

Almost all of the articles in the Encyclopedia have cross-references to other articles. These cross-references appear at the conclusion of the article text. They indicate articles that can be consulted for further information on the same topic or for information on a related topic. For example, the article "Smallpox" has references to "Biological Warfare," "Polio," "Surveillance of Infectious Diseases," and "Vaccines, Viral."

BIBLIOGRAPHY

The Bibliography is the last element in an article. The reference sources listed there are the author's recommendations of the most appropriate materials for further research on the given topic. The bibliography entries are for the benefit of the reader and do not represent a complete listing of all materials consulted by the author in preparing the article.

COMPANION WORKS

The *Encyclopedia of Microbiology* is one of a series of multivolume reference works in the life sciences published by Academic Press. Other such titles include the *Encyclopedia of Human Biology*, *Encyclopedia of Reproduction*, *Encyclopedia of Toxicology*, *Encyclopedia of Immunology*, *Encyclopedia of Virology*, *Encyclopedia of Cancer*, and *Encyclopedia of Stress*.

Acknowledgments

The Editors and the Publisher wish to thank the following people who have generously provided their time, often at short notice, to review various articles in the *Encyclopedia of Microbiology* and in other ways to assist the Editors in their efforts to make this work as scientifically accurate and complete as possible. We gratefully acknowledge their assistance:

George A. M. Cross
Laboratory of Molecular Parasitology
The Rockefeller University
New York, NY, USA

Miklós Müller
Laboratory of Biochemical Parasitology
The Rockefeller University
New York, NY, USA

A. I. Scott
Department of Chemistry
Texas A&M University
College Station, Texas, USA

Robert W. Simons
Department of Microbiology and Molecular Genetics
University of California, Los Angeles
Los Angeles, California, USA

Peter H. A. Sneath
Department of Microbiology and Immunology
University of Leicester
Leicester, England, UK

John L. Spudich
Department of Microbiology and Molecular Genetics
University of Texas Medical School
Houston, Texas, USA

Pravod K. Srivastava
Center for Immunotherapy
University of Connecticut
Farmington, Connecticut, USA

Peter Staeheli
Department of Virology
University of Freiburg
Freiburg, Germany

Ralph M. Steinman
Laboratory of Cellular Physiology and Immunology
The Rockefeller University
New York, NY, USA

Sherri O. Stuver
Department of Epidemiology
Harvard School of Public Health
Boston, Massachusetts, USA

Alice Telesnitsky
Department of Microbiology and Immunology
University of Michigan Medical School
Ann Arbor, Michigan, USA

Robert G. Webster
Chairman and Professor
Rose Marie Thomas Chair
St. Jude Children's Research Hospital
Memphis, Tennessee, USA

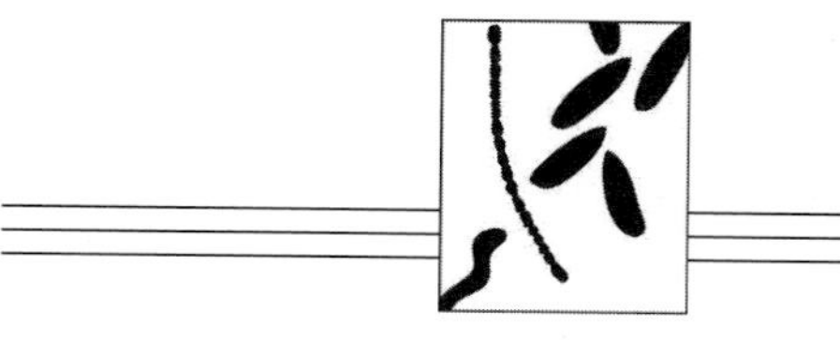

Quorum Sensing in Gram-Negative Bacteria

Clay Fuqua
Indiana University

I. Discovery: Autoinduction in Marine Vibrios
II. Quorum Sensing in Diverse Gram-Negative Bacteria
III. Biosynthesis of Acyl HSLs by LuxI-type Proteins
IV. Perception of and Response to Acyl HSLs
V. Variations on the Paradigm
VI. Regulation of Quorum Sensors
VII. Ecology and Function of Quorum Sensing
VIII. Conclusions

GLOSSARY

acyl HSLs Acylated homoserine lactones, diffusible signal molecules synthesized by a wide range of gram-negative bacteria.

biofilm A community of microorganisms associated with a surface.

bioluminescence The production of photons of visible light by living organisms.

homologous (homologs) Having a shared common ancestry, or referring to individuals that share a common ancestry; often determined by structural comparison.

Lux-type box Presumptive binding sites for LuxR-type proteins located directly upstream of quorum-regulated genes.

microcolony A microscopic aggregate of bacterial cells.

operon Genes within the same transcriptional unit under the control of a common promoter element.

regulon A set of genes under the control of a common regulator; not necessarily genetically linked on the DNA.

synthase A broad class of enzymes that catalyze biosynthetic reactions.

A GROWING NUMBER of bacteria are known to monitor their own population density via processes that are collectively described as quorum sensing. Most often, specific genes within the microbe are switched on at a defined population density, a bacterial "quorum," resulting in activation of functions under the control of the quorum sensor.

In almost all cases, the ability to sense a bacterial quorum involves the release of a signal molecule from the bacterium that accumulates proportionally with cell number. At a threshold concentration, the signal molecule interacts with a bacterial receptor that activates, either directly or indirectly, the expression of quorum-dependent genes. Quorum sensing is an example of multicellular behavior in bacteria, where individual cells communicate with each other to coordinate their efforts. Quorum sensing has been well documented in a number of systems including myxococcal fruiting body development, antibiotic production in several species of *Streptomyces,* and sporulation in *Bacillus subtilis.* In gram-positive bacteria, population density is often monitored using oligopeptide-based signal molecules that are actively secreted from cells under the appropriate conditions (for a review, see Kaiser and Losick, 1994; Dunny and Leonard, 1997). In contrast, many gram-negative bacteria produce acylated homoserine lactones (acyl HSLs), which act as cell density cues that freely permeate the cellular envelope and diffuse into the environment. These compounds were originally discovered in the bioluminescent marine vibrios, but have now been identified in a wide range of gram-negative bacteria. While the mechanism of acyl HSL quorum sensing is conserved in different bacteria, the functions under acyl HSL control are as diverse as the bacteria that produce them. This discussion will focus on the discovery and elucidation of acyl HSL-type quorum sensing in the marine vibrios, the more recent realization that these systems function in many different gram-negative bacteria, and the mechanisms by which acyl HSLs regulate bacterial behavior and influence microbial ecology.

I. DISCOVERY: AUTOINDUCTION IN MARINE VIBRIOS

A. Regulation of Bioluminescence

Acyl HSL quorum sensing, originally called autoinduction, was first discovered in the luminescent marine vibrios, *Vibrio fischeri* and *Vibrio harveyi*. *V. fischeri* is a symbiont that colonizes the light organs of several marine fishes and squids, while *V. harveyi* is an enteric bacterium that resides in the intestines and on fecal matter of certain fish. In the early 1970s, it was observed that the level of bioluminescence dropped significantly following inoculation of *V. fischeri* or *V. harveyi* into broth media, increasing again in the late stage of culture growth (Fig. 1A and 1B). However, if the conditioned cell culture fluids from dense bioluminescent cultures were added to newly inoculated cultures, bioluminescence was initiated significantly earlier. This suggested the presence of one or more inducing signals in the fluid phase of older cultures. The signal was confirmed to be bacterially derived and, hence, was described as an "autoinducer," and the phenomenon was likewise described as autoinduction. It was found that the autoinduction of bioluminescence was somewhat species-specific and that the signal was acting as an indicator of cell density, as opposed to a temporal or nutritional regulator. Initial fractionation of the cell-free culture fluids implicated a single heat-stable compound. The *V. fischeri* autoinducer was purified and chemically identified in 1981 as 3-oxo-*N*-(tetrahydro-2-oxo-3-furanyl)hexanamide, or *N*-3-(oxohexanoyl)homoserine lactone (3-oxo-C6-HSL; see Fig. 1C). Some years later, the *Vibrio harveyi* autoinducer activity was attributed to *N*-3-(hydroxybutyryl)homoserine lactone (3-OH-C4-HSL).

The nomenclature adopted for description of acyl HSLs is as follows (i) the prefix denotes the substituent at the third carbon (3-oxo for an oxygen, 3-OH for a hydroxy, and no prefix for fully reduced), (ii) C followed by the length of the acyl chain, and (iii) HSL for homoserine lactone. Unsaturated bonds in the acyl chain are indicated as Δ, followed by the first bonded position in the chain (e.g., an unsaturated double bond between the seventh and eighth carbon is Δ7). See Fig. 1C and Fig. 2.

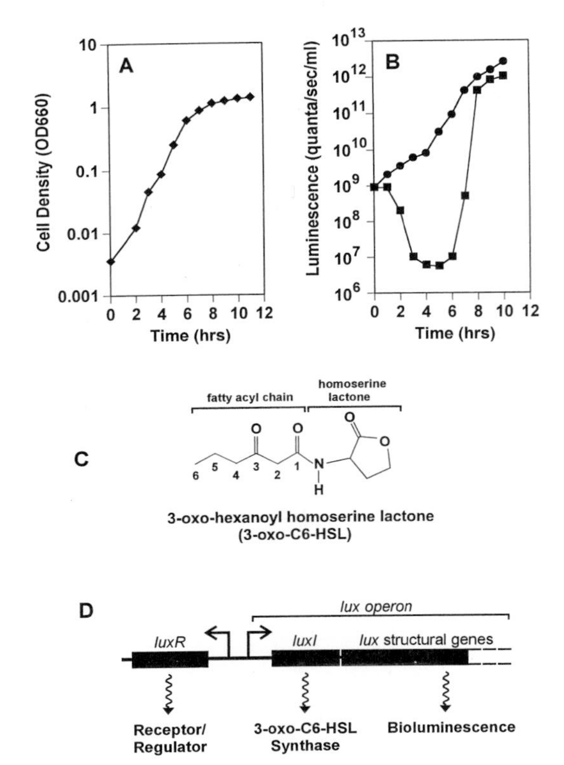

Fig. 1. Autoinduction of bioluminescence. Culture of *V. fischeri* grown in seawater complete medium. (A) Culture density (◆) measure as optical density at 660 nm. (B) Luminescence as quanta per second per ml for the unsupplemented culture (■) and for an identical culture supplemented with 3-oxo-C6-HSL, the active component in conditioned *V. fischeri* culture fluids (●). (C) Chemical structure of 3-oxo-C6-HSL. Numbers identify positions on the acyl chain. (D) Operon structure for the *V. fischeri lux* genes. A and B adapted and modified from Dunlap and Greenberg, 1991, and Dunlap and Greenberg, 1985.

B. Genetics of Autoinduction

In 1983, the *V. fischeri* bioluminescence (*lux*) genes were isolated and expressed in *E. coli*. *E. coli* expressing the cloned *lux* genes were not only bioluminescent, but exhibited cell density-dependent regulation. Molecular genetic analyses of the cloned *lux* locus revealed that two genes, designated *luxR* and *luxI*, were necessary and sufficient for autoinduction of bioluminescence (Fig. 1D). The *luxR* gene is essential for response to 3-oxo-C6-HSL and encodes a transcriptional regulator that activates expression of the *lux* operon. The *luxI* gene is required for 3-oxo-C6-HSL synthesis, but is dispensable for response to the factor. The two regulatory genes are physically linked with the *luxR* gene expressed divergently from the *lux* operon, of which *luxI* is the first followed by the structural genes responsible for bioluminescence (*luxICDABEG*).

C. A Model for Cell Density Sensing

Physiological, genetic, and biochemical analysis of *lux* gene regulation in *V. fischeri* has provided a general model for acyl HSL-type quorum sensing (also see Dunlap, 1998). At low cell densities, such as in seawater, a small amount of 3-oxo-C6-HSL is synthesized by limiting pools of LuxI. The signal molecule is rapidly depleted, due to passive diffusion across the bacterial envelope, and does not accumulate. Upon

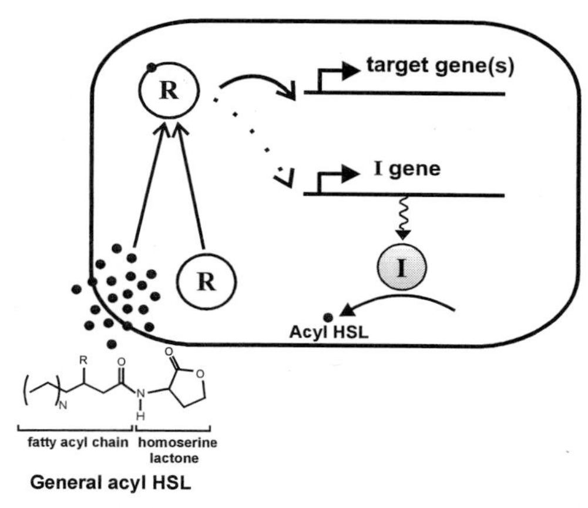

Fig. 2. General model for acyl HSL quorum sensing. LuxR-type protein and LuxI-type protein are represented as spheres labeled with R and I, respectively. The acyl HSL is represented as a filled circle. Filled arrows indicate noncovalent association and catalysis. The dashed arrow implies that activation of I gene expression is variable between different bacteria. Squiggles indicate expression and translation of the I gene. On the generalized acyl HSL structure, R can be H, OH, or O, and N equals the number of carbons 0–10.

colonization of the light organ, *V. fischeri* is provided a nutrient-rich environment. As cell number increases, the relative concentration of 3-oxo-C6-HSL also rises. At a specific threshold concentration, probably where levels inside and outside the cell are equivalent, the signal binds to the LuxR receptor protein, presumably causing a conformational change that results in LuxR-dependent activation of the *lux* operon promoter and a concomitant increase in bioluminescence. Elevated transcription of the *luxI* gene causes an increase in 3-oxo-C6-HSL synthesis, constituting a positive-feedback loop. While many of the details of this model are currently under investigation, this basic pattern of regulation continues to serve as a paradigm for acyl HSL quorum sensing.

II. QUORUM SENSING IN DIVERSE GRAM-NEGATIVE BACTERIA

A. Identification of Acyl HSL and LuxR–LuxI-type Proteins

For quite some time, the production and perception of acyl HSLs was thought to be a peculiar form of regulation restricted to the marine vibrios. Findings from several laboratories during the 1990s have radically altered this view, and it is apparent that many bacteria employ acyl HSLs as regulatory molecules (Table I, and see reviews). A variety of lines of investigation have led to discovery of acyl HSL quorum sensors.

Many of these, particularly the first few, were discovered serendipitously. Examination of elastase gene regulation in *Pseudomonas aeruginosa* led to identification of the *lasR* gene, encoding a protein homologous to LuxR. Closely linked to *lasR* was the *lasI* gene, homologous to *luxI*, that directs synthesis of 3-oxo-C12-HSL. Likewise, studies on the regulation of conjugal transfer of the *Agrobacterium tumefaciens* Ti plasmid, identified the regulatory proteins TraR and TraI, homologous to LuxR and LuxI.

Contemporaneously, purification and structural analysis of a so-called "conjugation factor" identified 3-oxo-C8-HSL as the cognate acyl HSL of the *A. tumefaciens* system. Examination of mutants repressed for elaboration of degradatory exoenzymes

TABLE I
Selected LuxR–LuxI Quorum Sensors from Gram-Negative Bacteria[a]

	Components identified[b]				
Bacterium	*Receptor*	*Synthase*	*Acyl HSL*	*Target function*	*Integrated with*[c]
Vibrio fischeri	LuxR	LuxI	3-oxo-C6	Bioluminescence (*lux*)	Glucose (CRP), Fe
Pseudomonas aeruginosa	LasR	LasI	3-oxo-C12	Virulence factors	RhlR, GacA, Vfr, LasR, RpoS regulon
	RhlR	RhlI	C4	Rhamnolipids, RpoS	
Erwinia carotovora[d]	ExpR	ExpI	3-oxo-C6	Exoenzymes	Carbon and sugar metabolism
	CarR	CarI	3-oxo-C6	Antibiotics	?
Agrobacterium tumefaciens	TraR	TraI	3-oxo-C8	Conjugal transfer	Opines
Pseudomonas aureofaciens	PhzR	PhzI	C6	Phenazine antibiotics	LemA-GacA
Pantoea stewartii	EsaR	EsaI	3-oxo-C6	Exopolysaccharide (EPS)	EPS regulation (Rcs)
Serratia liquefaciens	?	SwrI	C6, C4	Swarming motility	amino acids, cell contact
Rhodobacter sphaeroides	CerR	CerI	Δ7-C14	Cell disaggregation	?
Escherichia coli	SdiA[e]	—	?	Cell division	Division control

[a] The list of quorum sensors is not exhaustive; see Fuqua *et al.*, 1996, and other reviews for more complete listings.
[b] The LuxR homolog, the LuxI homolog, and the primary acyl HSL are indicated. Only full-length, functional proteins are included. See text for discussion of individual systems.
[c] Indicates additional regulation that affects the function of the quorum sensor.
[d] Different LuxR–LuxI homologs have been identified in different subspecies of *Erwinia carotovora,* although several lines of evidence suggest that multiple LuxR–LuxI proteins may coexist in the same strain.
[e] The *E. coli* complete genome sequence reveals no *luxI* homolog.

and antibiotic production in several subspecies of *Erwinia carotovora* led to discovery of LuxR–LuxI-type quorum sensors. As it became apparent that acyl HSLs are important regulatory molecules, several quorum sensors were identified by initial detection of the pheromone itself, using acyl HSL-responsive reporter systems, followed by genetic isolation of the presumptive *luxI* and *luxR* homologs (see Table I for a listing of bacteria that produce acyl HSLs).

B. Generalities and Mechanisms

The general features of acyl HSL quorum sensors are conserved in different species of bacteria. Most are comprised of homologs of the LuxR acyl HSL receptor and the LuxI acyl HSL synthase (Fig. 2; also see Section V for exceptions). Members of the LuxR and LuxI families of proteins share 18–25% and 28–35% identity across the length of each protein, respectively. Several bacteria, typically enteric species, synthesize 3-oxo-C6-HSL, identical to the *V. fischeri* factor. Other bacteria synthesize acyl HSLs that vary from the *V. fischeri* pheromone with respect to length and degree of saturation on the acyl chain, as well as the substituent at the third position (Fig. 1C and Fig. 2). Acyl HSLs have an acyl moiety with an even number of carbons, ranging from 4–14, associated with the homoserine lactone via an amide bond. Acyl HSLs and LuxR–LuxI-type regulators are thought to control their diverse target genes in response to population density via a mechanism analogous to that described for *V. fischeri* (Fig. 2). In environments that support only low cell density, acyl HSL, synthesized by the LuxI-type protein, dissipates by diffusion across the bacterial envelope. As conditions change to support larger numbers of cells or the flow characteristics of the environment are altered (i.e., diffusion is limited), acyl HSLs accumulate. At a specific concentration that probably varies for each microbe and each environmental niche, acyl HSLs interact with the LuxR-type receptor, elevating expression of target genes. While the *luxI* homolog is often among the target genes, as with the positive feedback described for *V. fischeri,* this is not always the case and is not essential for the function of the quorum sensor.

III. BIOSYNTHESIS OF ACYL HSLs BY LUXI-TYPE PROTEINS

A. Enzymology

It can be readily demonstrated that *luxI* and many of its homologs are required for acyl HSL synthesis. Likewise, the production of the appropriate acyl HSLs by heterologous hosts, such as *E. coli* expressing LuxI homologs, indicates that these proteins utilize common precursors and impart the biosynthetic specificity observed in the parent microbe. However, only recently has the prediction that LuxI and its homologs are acyl HSL synthases been proven. A major barrier to these findings was the identity and complexity of the substrates for biosynthesis. The general structure of acyl HSLs (Fig. 2) suggested that they were products of fatty acid and amino acid metabolism. Analysis of *V. fischeri* cell extracts using labeled precursors implicated methionine metabolic intermediates, and either biosynthetic or β-oxidative intermediates of fatty acid metabolism. More recent genetic studies with LuxI and the *A. tumefaciens* TraI protein have refined this view further, strongly implicating *S*-adenosyl methionine (AdoMet) and fatty acid biosynthetic intermediates as acyl HSL precursors.

Isolation of the purified acyl HSL synthases, either as native proteins or affinity-tagged fusion derivatives, has allowed *in vitro* analysis of acyl HSL synthesis. Under the appropriate reaction conditions, several LuxI-type proteins will synthesize acyl HSLs from AdoMet and acyl–acyl carrier protein (acyl ACP) conjugates, the primary intermediates of fatty-acid biosynthesis. A tentative reaction scheme has been formulated to describe the enzymatic reaction (Fig. 3A). First, the acyl HSL synthase binds to AdoMet and an acyl–ACP carrying the correct length acyl chain with the appropriate oxidation state at the third position (all variations found at the third position on acyl HSLs are also present at the corresponding position during fatty-acid biosynthesis). The nucleophilic amino group on the methionine moiety of AdoMet attacks the thioester bond that conjugates the fatty acyl chain to ACP. Thus forms the amide linkage found in all acyl HSLs, displacing a reduced ACP. Lactonization of the methionine on

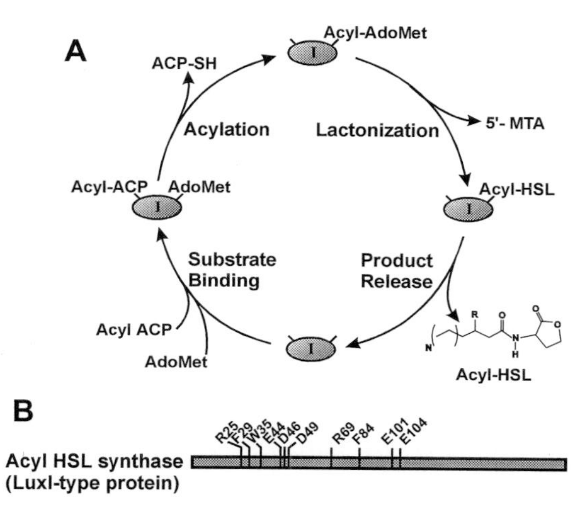

Fig. 3. (A) Tentative catalytic cycle for acyl HSL synthesis. Bars extending from LuxI indicate noncovalent interaction with substrates and products. (B) LuxI protein is represented as a bar and amino acid residues conserved throughout the LuxI family are indicated (amino acids are in single letter code: D, aspartic acid; E, glutamic acid; F, phenylalanine; R, arginine; W, tryptophan). Numbering is relative to the LuxI amino acid sequence.

AdoMet releases the acyl HSL and the reaction by-product 5′-methylthioadenosine (MTA), allowing the acyl HSL synthase to enter another catalytic cycle. While the current evidence suggests this is a plausible model, it remains to be rigorously tested.

B. Structure of Acyl HSL Synthases

Little is currently known regarding the molecular architecture of the acyl HSL synthases. They share no apparent similarity to other protein families in the sequence databases. However, the LuxI-type acyl HSL synthases now number over 15, and amino acid sequence comparisons within the family are informative. There are 10 amino acid residues, clustered in the amino terminal halves of the proteins (res. 25–104 in LuxI), conserved in all members of the family (Fig. 3B). Of these, seven have side chains that carry charges at neutral pH. The other three carry aromatic side chains. Strikingly, loss of function mutations have been isolated at each of the seven conserved, charged amino acid residues (*luxI* and *P. aeruginosa rhlI*). Mutations in the three conserved aromatic residues severely reduce, but do not abolish, the activity of acyl HSL synthases. Many acyl transferases, enzymes that catalyze the exchange acyl chains, utilize cysteine and serine residues at their active sites. At least for the acyl HSL synthases tested thus far, cysteine and serine residues do not appear to be required for the catalysis. Rather, it is likely that the conserved charged residues direct catalysis, probably via specific acid–base interactions with the substrates at the active site. There is currently no evidence that the acyl chain is transiently linked to the protein during catalysis, in contrast to some of the acyl transferases. A detailed view of acyl HSL synthesis and the architecture of the active site awaits additional biochemical and structural analyses.

IV. PERCEPTION OF AND RESPONSE TO ACYL HSLs

A. Entry into Cells and Receptor Interaction

Although the prevailing dogma is that all acyl HSLs passively diffuse across the bacterial envelope, this idea has only been tested for 3-oxo-C6-HSL. Other acyl HSLs are also exchanged very rapidly, and it is, therefore, assumed that they, too, freely cross the envelope. However, it is possible that the exchange of some acyl HSLs is facilitated by transporters. Additionally, some acyl HSLs may also substantially partition into lipid bilayers, as opposed to simply crossing them.

Once inside the cell, all evidence suggests that the acyl HSLs interact directly with LuxR-type receptor proteins. This interaction alters the conformation of the receptor and fosters transcriptional regulation of target genes. Recognition of the acyl HSL involves contacts with the homoserine lactone ring as well the acyl chain portion of the molecule. Most LuxR-type proteins can distinguish between acyl HSLs whose structure differs only by the length of the acyl chain or oxidation state at the third position. In general, LuxR–LuxI regulatory pairs have evolved so that the LuxR-type protein is most sensitive to the primary product synthesized by its corresponding acyl HSL synthase. However, LuxR-type proteins will recognize acyl HSLs other than their cognate factor, albeit with reduced efficiency. The degree of specificity varies with each quorum sensor, some demonstrating high stringency and others with a broader recognition spectrum. Interestingly, although *A. tumefaciens* strains expressing wild-type levels of the TraR protein recognize only 3-oxo-C8-HSL, elevated expression causes a dramatic loss of specificity, suggesting a role for intracellular pool sizes of the LuxR-type protein in recognition of acyl HSLs.

B. Modular Architecture of Acyl HSL Receptors

LuxR-type proteins have two distinct, physically separable activities—one is to bind acyl HSLs and the other is to bind specific DNA sequences. There is evidence that the sequences within the amino-terminal half of the LuxR protein, (amino acid residues 79–127), bind 3-oxo-C6-HSL (Fig. 4). Conversely, a protein consisting of only the carboxy-terminal 95 amino acids of LuxR binds to DNA and can also activate transcription, although it is not responsive to 3-oxo-C6-HSL. In the wild-type LuxR protein, these two functions are linked and interdependent. Interaction with acyl HSLs via the amino-

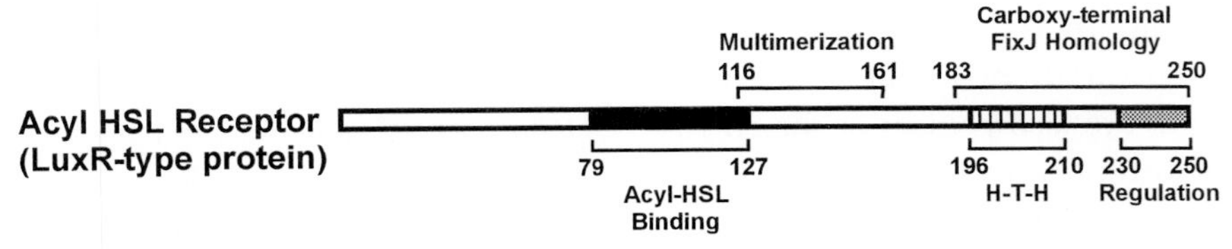

Fig. 4. Modular structure of LuxR-type proteins. Functional regions are based primarily on genetic analyses of LuxR. The numbering scheme is relative to the LuxR amino acid sequence. See text for details.

terminal region regulates binding of the LuxR-type protein to its target sites on the DNA. At least for LuxR, this appears to be a negative regulation, where the amino-terminal half of the protein inhibits the activity of the carboxy-terminal DNA-binding region in the absence of 3-oxo-C6-HSL. Binding of the ligand relieves this inhibition and allows the protein to activate transcription.

Comparisons of amino acid sequences between members of the LuxR family reflect the binary structure that has been discussed. While the overall similarity between proteins is relatively low, the similarity within the presumptive acyl HSL-binding region and the DNA-binding region is substantially higher. The acyl HSL-binding region is a signature sequence, shared only by members of the LuxR family (Fig. 4). In contrast, the carboxy terminal DNA-binding region (amino acid residues 183–250) contains a helix–turn–helix (H–T–H) motif, found in many proteins that interact with DNA. Furthermore, the HTH motif and sequences that surround it firmly place these DNA-binding modules into the much larger FixJ superfamily (Fig. 4). Most other members of this superfamily are two-component type response regulators, with amino terminal domains that are phosphorylated at conserved aspartate residues by a sensor kinase. Interestingly, for several of these FixJ-type response regulators, phosphorylation within the regulatory region relieves inhibition of an otherwise constitutive DNA-binding activity.

Genetic analysis suggests that transcriptional activation also requires multimerization of LuxR-type proteins. Nonfunctional variants of both LuxR and TraR, lacking their DNA-binding region, act as dominant negative inhibitors of their wild-type counterparts. The nonfunctional protein, in effect, poisons the wild-type protein by forming inactive complexes. On LuxR, the region between amino acid residues 116–161, overlapping with the acyl HSL ligand recognition region, is required for multimerization (Fig. 4). Formation of LuxR multimers must be integrated with the intramolecular events that occur upon ligand interaction. It is unclear whether multimerization occurs subsequently or concomitantly with the acyl HSL interaction. It is possible that binding of the acyl HSL stimulates multimerization and that this shift in contacts abolishes the inhibition of DNA-binding by the amino terminal half of the protein.

C. DNA Binding and Transcription Activation

In 1989, Devine and Shadel identified the *lux* operator, a 20 bp inverted repeat centered at −40 relative to the *lux* operon promoter. Specific mutations in this sequence caused a reduction in transcriptional activation by LuxR. As other acyl HSL-based quorum sensors were identified, it was recognized that similar sequences existed upstream of many of the genes under cell-density dependent control. These sequences not only shared the basic inverted repeat motif, and the position relative to the regulated promoter, but also primary sequence similarity with the *lux* operator. These *cis*-acting sequences are now referred to as *lux*-type boxes and they are the presumptive binding sites for LuxR-type proteins. The *lux*-type box can be 18–20 bp in length, is generally located between −40 and −44 from the transcriptional start site, and is often essential for transcriptional regulation of target genes by LuxR-type proteins. However, not all genes that are activated by LuxR-type proteins have recognizable *lux*-like boxes upstream of their promoters. Therefore, these sequences appear to be a common, although not obligatory, component of acyl HSL-regulated promoters.

The position of *lux*-type boxes relative to their associated promoters suggests that LuxR-type proteins make specific contacts with RNA polymerase, much the same as other prokaryotic transcriptional activators. While it is still unclear how transcription is stimulated, preliminary findings suggest that the carboxy terminal portion of the α subunit of RNA polymerase, a site of contact for many regulators, is required for transcriptional activation. A carboxy terminal fragment of LuxR with constitutive DNA binding activity (see Section IV.B) was shown by *in vitro* DNA-binding assays to associate with the *lux* promoter region. However, specific binding to the *lux* operator required the presence of both the LuxR carboxy terminal fragment and RNA polymerase. It is possible that this apparent synergy may be an artifact of using the truncated form of LuxR and the full-length protein may not exhibit this characteristic. Full-length LasR and TraR associate with their binding sites *in vitro* in the absence of RNA polymerase.

D. LuxR-type Proteins That Function as Repressors

The majority of LuxR-type proteins activate transcription. Consequently, null mutations in most *luxR* homologs result in low level, noninducible expression of target genes. However, mutations of the *esaR* gene from *Pantoea stewartii* (formerly, *Erwinia stewartii*) that regulates capsular polysaccharide (EPS) synthesis (Table 1), results in elevated synthesis of EPS. This and other data indicate that EsaR is a repressor and that inducing levels of 3-oxo-C6-HSL (the cognate acyl HSL produced by EsaI) cause a derepression of transcription. A simple model would have EsaR bound to regulated promoters in the absence of inducer and dissociated in its presence. If this is correct, EsaR must adopt an active conformation when not associated with ligand and convert to an inactive conformation upon binding of the acyl HSL. There are no features in the EsaR amino acid sequence that are notably different from LuxR-type activators. However, several other LuxR-type proteins are also suspected to be repressors.

E. Modulation of Responses to Acyl HSLs

Although the basic components of acyl HSL quorum sensors are LuxI- and LuxR-type regulatory proteins, there are several examples where additional factors directly impinge upon, or are necessary for, proper regulation. The clearest example of this is the *A. tumefaciens* TraR–TraI quorum sensor, where several additional regulatory proteins modulate the TraR-dependent transcriptional activation. One of these is the TraM protein, identified in several strains of *A. tumefaciens* and at least one species of *Rhizobium*. TraM acts to inhibit the function of TraR under noninducing conditions. Mutations within *traM* result in constitutive activation of TraR-dependent genes and loss of cell density-responsiveness. These results suggest that TraM is an integral component of the *A. tumefaciens* quorum sensor. In fact, the *traM* gene is itself activated by TraR and 3-oxo-C8-HSL, creating a negative feedback loop. It is unclear why the *A. tumefaciens* quorum sensor has incorporated this additional component, and the physiological role is a current topic of investigation.

There is an additional layer of regulation in at least some strains of *A.tumefaciens*. TraR is also inhibited by the activity of the Trl protein, a protein highly similar to the amino terminal half of TraR (amino acid residues 1–181) but missing the carboxy terminal DNA-binding region. Genetic analysis reveals that Trl is dominant to TraR, suggesting that it forms inactive TraR–Trl heteromultimers. The *trl* gene is itself under the control of plant-released signals, similar to signals that control expression of the *traR* gene (Table I). Apparently the balance of TraR and Trl, within the cell at any given time, also integrated with the effect of TraM, dictates whether TraR will activate target gene expression.

The LuxR protein of *V. fischeri* is also modulated, not by additional regulatory proteins, but rather by C8-HSL. This factor is synthesized by the AinS acyl HSL synthase, distinct from LuxI (see Section V). In the absence of 3-oxo-C6-HSL, the AinS-produced factor is a mild inducer of the *lux* genes. However, *V. fischeri* with mutations in *ainS*, and with a functional copy of *luxI*, activate expression of the *lux* genes at significantly lower cell densities than wild-type cells. The effect of the *ainS* mutation can be masked by addition of exogenous C8-HSL, suggesting that this factor is acting as an inhibitor of LuxR, perhaps competing with 3-oxo-C6-HSL, for access to the acyl HSL-binding site.

P. aeruginosa produces two primary acyl HSLs as well, 3-oxo-C12-HSL and C4-HSL, the products of the LasI and RhlI proteins, respectively (Table I). While each acyl HSL acts predominantly through its own LuxR type receptor (LasR and RhlR), there is evidence that 3-oxo-C12-HSL can reduce the responsiveness of RhlR to C4-HSL, possibly through direct competition for the acyl HSL-binding site. Although it is not clear how common it is for bacteria to utilize mutliple quorum sensors, as is the case in *P. aeruginosa* and *V. fischeri*, these observations suggest that acyl HSLs produced within the same cell can have physiologically relevant interactions *in vivo*.

V. VARIATIONS ON THE PARADIGM

A. Quorum Sensing in *V. harveyi*

As discussed, the original observations regarding autoinduction of bioluminescence were made with the two marine vibrios, *V. fischeri* and *V. harveyi.* Both bacteria control bioluminescence in response to cell density, by production of diffusible autoinducers, and both produce acyl HSLs, 3-oxo-C6-HSL and 3-OH-C4-HSL, respectively. For technical reasons, studies of the *V. fischeri* system proceeded more rapidly and have led to development of the LuxR–LuxI paradigm. It was assumed, quite logically, that the *V. harveyi* autoinduction mechanism would be similar. However, the reality is, in fact, strikingly more complicated (Fig. 5). While *V. harveyi* does produce an acyl HSL with which it monitors its own population density, the proteins responsible for synthesis of 3-OH-C4-HSL and its effect on *lux* gene expression share no homology to LuxR and LuxI. Production of 3-OH-C4-HSL requires the activity of two genes *luxL* and *luxM,* one or both of which comprise the acyl HSL synthase (LuxL/M). As expected, this factor is freely diffusible across the bacterial envelope and accumulates with increasing cell density. Response to 3-OH-C4-HSL by induction of *lux* genes, requires a two-component type sensor kinase, called LuxN, which interacts with a cognate response regulator called LuxO. LuxO is a repressor of *lux* genes and is, presumably, inhibited in the presence of 3-OH-C4-HSL. Inactivation of the 3-OH-C4-HSL quorum sensor, by mutating, either the *luxLM* acyl HSL synthase or the *luxN* sensor, does not, however, abolish this microbe's ability to sense its own population density. There is a second quorum sensing system that also regulates *lux* genes (Fig. 5). This apparently redundant system relies on a signal molecule(s), called AI-2, that is released from cells and accumulates with increasing cell numbers. The genes required for synthesis of AI-2 and the chemical structure(s) of this compound have not yet been elucidated. Response to AI-2 requires the activity of another sensor kinase, LuxQ, that also appears to interact with the LuxO response regulator to control *lux* gene expression. An additional transcriptional activator called LuxR (not related to the *V. fischeri* LuxR) also affects *lux* expression, but is not responsive to either 3-OH-C4-HSL or AI-2. The AI-2 based system is also dispensable for quorum sensing, provided the 3-OH-C4-HSL system is intact. Therefore, the *V. harveyi* quorum sensors appear to be functionally redundant and mechanistically distinct from the canonical LuxR–LuxI system. Recent findings suggest that 3-OH-C4-HSL may act as an intraspecies signal, while AI-2 may allow *V. harveyi* to monitor the presence of other bacteria. In support of this, AI-2-like signals have now been detected from *E. coli* and *Salmonella typhimurium,* as well as other bacteria, while 3-OH-C4-HSL has only been reported for *V. harveyi.*

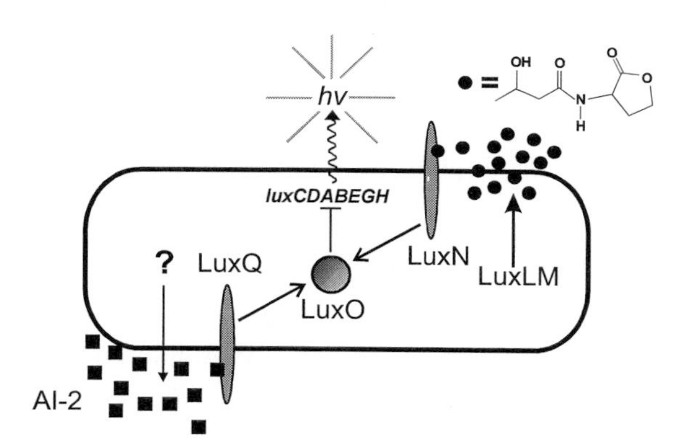

Fig. 5. Regulation of bioluminescence in *V. harveyi.* Simplified model of dual quorum sensors, adopted and modified from Bassler and Silverman, 1995. LuxQ and LuxN are transmembrane, two-component type sensor kinases that presumably phosphorylate the common response regulator, LuxO, a repressor of *lux* genes. An additional regulator of *lux* genes, the *V. harveyi* LuxR protein, has been omitted for clarity and because its role in quorum-sensing has not been established. The signal molecule produced by LuxL/M is 3-OH-C4-HSL, while the structure of AI-2 signal is not yet known. See text for additional details.

B. A Second Acyl HSL in *V. fisheri*

Homologs of the *V. harveyi* LuxM and LuxN proteins have been identified in *V. fischeri.* The AinS protein is homologous to LuxM and is required for synthesis of C8-HSL, in addition to the 3-oxo-C6-HSL synthesized by LuxI. A gene encoding a presumptive sensor kinase, AinR, homologous to LuxN, is physically linked to *ainS.* The regulatory targets of AinR have not been identified, although AinS and

C8-HSL may have a physiological role in modulating LuxR activity (see Section IV.E). The discovery of LuxL/M–LuxN in *V. harveyi* and the homologous AinS–AinR system in *V. fischeri* suggests that these proteins constitute a second family of quorum sensors, based on acyl HSLs, but mechanistically distinct from LuxR–LuxI type quorum sensors. The fact that acyl HSL quorum sensing has apparently evolved twice is remarkable and may reflect the potency of these compounds as signal molecules.

VI. REGULATION OF QUORUM SENSORS

A. Integration of Acyl HSL Quorum Sensing into Cellular Physiology

As techniques for genome-scale analysis of gene expression become available, it is more and more apparent that different aspects of cellular metabolism are linked through regulatory networks. Acyl HSL quorum-sensing regulators are often embedded in or otherwise integrated with additional control circuitry (Table I). For example, expression of the *luxR* gene is activated by the catabolite repressor protein (CRP), resulting in glucose repression of bioluminescence. In *A. tumefaciens,* expression of the *traR* gene occurs only in the presence of opines, plant-released compounds that tie the function of the quorum sensor to the rhizosphere. Most, if not all, acyl HSL regulators are positioned within a larger regulatory hierarchy. It appears that, for each microbial species, the regulatory context is different, probably reflecting the specific environmental niche the microbe occupies and the varying roles for quorum sensing (see Table I).

A well-studied case of this regulatory integration is found in *P. aeruginosa,* where there are two distinct quorum sensors, the Las and Rhl systems (Table I). LasR and 3-oxo-C12-HSL were identified as regulators of elastases and other virulence factors, while RhlR and C4-HSL were found to regulate production of rhamnolipid surfactants. LasR, in fact, controls the expression of *rhlR,* possibly facilitating a hierarchical staging of gene activation where the targets of LasR are expressed first, followed by those under RhlR control. Included within the RhlR regulon is a homolog of the *E. coli* RpoS stationary phase-specific sigma factor. Therefore, any genes under RpoS control would be superregulated by the Rhl and Las quorum sensors, perhaps integrating starvation and population density sensing in *P. aeruginosa.* Another regulatory target for Las, and possibly Rhl, are the *xcp* genes encoding an exoenzyme secretion system, modulating certain aspects of protein secretion in response to cell density. The Las–Rhl quorum sensing cascade is itself under control of at least two regulators: (i) the *P. aeruginosa* GacA protein, a two-component type response regulator, homologs of which control virulence in plant pathogenic pseudomonads, and (ii) the Vfr protein, a *P. aeruginosa* CRP homolog. The physiological significance of GacA and Vfr control of the Las–Rhl dual quorum sensors is not known.

VII. ECOLOGY AND FUNCTION OF QUORUM SENSING

A. Multiple Roles for Acyl HSL Signaling?

The diversity of bacteria known to employ acyl HSLs is expanding with the discovery of each new quorum sensor. Quorum sensing may come into play under any situation where it is beneficial for a specific process to be differentially regulated as a function of population density. Acyl HSLs often regulate genes involved in host–microbe interactions, symbiotic or pathogenic (Table I). The host environment provides conditions where high cell density responses are relevant and beneficial. However, acyl HSLs are clearly not restricted to regulation of interactions with host organisms, as they have been identified in microbes with no known host association, such as the free-living photosynthetic bacterium *Rhodobacter sphaeroides.* Acyl HSLs may also facilitate concerted and rapid responses by a population to external stimuli. For example, certain members of a population may respond to a specific, low-level stimulus, such as the presence of a host-released signal molecule, by stimulating production of an acyl HSL quorum sensor. Dissemination of the acyl HSL signal throughout the local environment might act to amplify the original host signal and enable a coordinated response by the bacterial population. Such signaling may occur between bacteria of the same species or different bacteria that employ similar acyl HSLs (see Section VII.C).

B. Microcolonies and Biofilms

Many bacteria are readily grown in laboratory culture at cell densities sufficient to trigger acyl HSL-dependent gene expression. In the natural environment, however, such luxuriant conditions are far less prevalent, and large numbers of cells at high densities are probably not the norm. The flow characteristics and physical dimensions of any given environment will also influence if and when acyl HSLs accumulate. Squid light organs colonized with *V. fischeri* contain concentrations of 3-oxo-C6-HSL well above those required to activate LuxR and affect light production. In this case, the confined (~15 nL), nutrient-rich environment of the light organ supports a high cell density (~10^{10} cfus/ml) and provides the appropriate physical characteristics to maintain inducing concentrations of the acyl HSL. However, other environmental niches occupied by quorum sensing bacteria are less tailored to provide such conditions. A number of plant pathogens control elaboration of extracellular virulence factors by acyl HSL quorum sensing and, in several cases, the appropriate timing of this attack is crucial to the success of the infection. Clearly, where active infections occur, inducing conditions have been achieved. What is the composition of the population that leads to accumulation of the acyl HSL and subsequent induction? Although the answer is not certain, plant-associated bacteria often form homogeneous or mixed microcolonies. The relative cell densities on the interior of the microcolony are likely to be quite high and, combined with the physical barriers posed by the host as well as polymeric material (e.g., EPS) produced within the microcolony, may provide conditions that lead to acyl HSL accumulation. Bacterial cohorts analogous to plant-associated microcolonies are also thought to colonize infection sites in animal hosts.

A biofilm is another bacterial structure common to a variety of environments. Biofilms are surface-associated, organized layers of bacteria, usually bound in a matrix of extracellular material. These bacterial communities can exhibit a surprising architectural complexity, with aqueous channels, chambers, and interconnected microcolonies of cells. Biofilm formation is a major concern in medical and industrial settings, where members of the biofilm demonstrate greater recalcitrance than do their planktonic counterparts to antimicrobial strategies. It is clear that biofilm communities provide environments that foster intercellular interactions between bacteria, and acyl HSL signaling may be an important aspect of existence in a biofilm. In fact, acyl HSLs have been detected in the effluent of living biofilms from several natural environments. Furthermore, *P. aeruginosa* strains unable to synthesize 3-oxo-C12-HSL form defective biofilms, with far greater sensitivity to antimicrobial treatment. This observation suggests that not only is acyl HSL signaling facilitated by conditions in the biofilm, but that these molecules may also play a role in the development of the film. It remains to be seen how universal the role of acyl HSLs is in biofilms of different bacteria.

C. Mixed Messages: Do Acyl HSLs Mediate Interspecies Communication?

Many bacteria release and respond to acyl HSLs of similar structure. If these bacteria grow in close proximity to each other, such as in a microcolony or biofilm, it is possible, perhaps even unavoidable, that one species of microbe will respond to the acyl HSL of another species. Current findings suggest that cohabiting acyl HSL-producing microbes can influence each other's behavior *in situ*. *P. aeruginosa* and *Burkholderia cepacia* are common opportunistic pathogens of patients with cystic fibrosis and may form mixed species communities within the same individual. Each bacteria produces acyl HSLs that can potentiate elaboration of virulence functions by the other bacteria, perhaps collaboratively pathogenizing the host organism. A different example has been provided by examination of artificial mixed populations of the plant biocontrol agent *Pseudomonas aureofaciens*. Coinoculation of host plants (wheat) with a mixture of wild-type *P. aureofaciens* producing C6-HSL and a mutant strain (PhzI-) responsive to, but unable to synthesize, the pheromone, resulted in strong activation of target genes in the mutant strain. Findings such as these suggest that, in many environments, opportunities for interspecies signaling are not only possible, but prevalent.

D. Host Responses to Acyl HSLs

Interactions between microbes and their hosts are often games of detection and response, with each partner evolving the capacity to perceive and manipulate the other. Animals and plants have developed elaborate mechanisms by which they defend themselves from foreign agents. Acyl HSLs provide lines of communication between infecting bacteria, and, therefore, it may be beneficial for host organisms to develop the capacity to "eavesdrop" on the coded messages of their microbial colonizers. In addition, acyl HSLs might stimulate the immune system in a manner detrimental to the host, acting directly as classic virulence factors. Several studies have shown that a subset of acyl HSLs exhibits immunomodulatory activity in animal hosts, affecting cytokine production and immune cell proliferation. While these observations are intriguing, a much more complicated question that remains to be addressed is whether this activity is relevant *in situ*? If so, how do these responses affect the interaction between microbe and host?

At least one animal host has evolved the capacity to interfere with acyl HSL signaling. The marine sponge *Delisea pulchra* produces a pair of halogenated furanones, structural analogs of acyl HSLs, that efficiently block acyl HSL-dependent induction of swarming motility in *Serratia liquefaciens* and LuxR activation of bioluminescence. It is thought that the furanones function by directly inhibiting the interaction of acyl HSLs with LuxR-type proteins. It is unclear what role(s) the furanones play in the natural setting. Are they are specific antagonists of quorum sensing or do they coincidentally inhibit the process?

VIII. CONCLUSIONS

Acyl HSL quorum sensing, for many years considered to be a specialized form of gene regulation in bioluminescent *Vibrio* species, is now known to be quite prevalent in gram-negative bacteria. The general mechanism appears to be conserved among a wide range of bacteria, although those functions regulated by the quorum sensor are extremely varied between different bacteria. As our understanding of the function and role(s) of acyl HSLs in bacterial communities improves, the prospects for utilizing these potent signal molecules in the control of bacterial behavior also improve. Likewise, studies on all forms of bacterial quorum sensing should continue to blur the distinction between unicellular and multicellular lifestyles, well as to expand our appreciation of the lines of communication that underlie complex ecosystems.

See Also the Following Articles

Bioluminescence, Bacterial • Pseudomonas • *Streptomyces*, Genetics • Transcriptional Regulation in Prokaryotes

Bibliography

Dunlap, P. V., and Greenberg, E. P. (1991). Role of intercellular communication in the *Vibrio fischeri*–monocentrid fish symbiosis. *In* "Microbial Cell–Cell Interaction" (M. Dworkin, ed.). ASM Press, Washington, DC.

Dunlap, P. V., and Greenberg, E. P. (1985). Control of *Vibrio fischeri* luminescence gene expression in *Escherichia coli* by cyclic AMP and cyclic AMP receptor protein. *J. Bacteriol.* **164**, 45–50.

Dunny, G. M., and Leonard, B. A. B. (1997). Pheromone-inducible conjugation in *Enterococcus faecalis:* Interbacterial and host–parasite chemical communication. *Annu. Rev. Microbiol.* **51**, 527–564.

Fuqua, C., and Greenberg, E. P. (1998). Self-perception in bacteria: Quorum sensing with acylated homoserine lactones. *Curr. Opin. Microbiol.* **1**, 183–189.

Fuqua, W. C., Winans, S. C., and Greenberg, E. P. (1994). Quorum sensing in bacteria: The LuxR/LuxI family of cell density-responsive transcriptional regulators. *J. Bacteriol.* **176**, 269–275.

Fuqua, C., Winans, S. C., and Greenberg, E. P. (1996). Census and consensus in bacterial ecosystems: The LuxR–LuxI family of quorum-sensing transcriptional regulators. *Annu. Rev. Microbiol.* **50**, 727–751.

Gray, K. M. (1997), Intercellular communication and group behavior in bacteria. *Trends Microbiol.* **5**, 184–188.

Kaiser, D., and Losick, R. (1994). How and why bacteria talk to each other. *Cell* **73**, 873–885.

Ruby, E. G. (1996). Lessons from a cooperative bacterial–animal association: The *Vibrio fischeri–Euprymna scolopes* light organ symbiosis. *Annu. Rev. Microbiol.* **50**, 591–624.

Swift, S., Throup, J. P., Williams, P., Salmond, G. P. C., and Stewart, G. S. A. B. (1996). Gram-negative bacterial communication by N-acyl homoserine lactones: A universal language? *Trends Biochem. Sci.* **21**, 214–219.

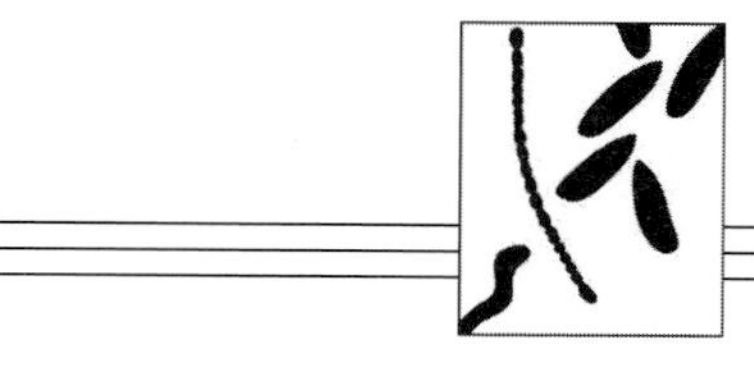

Rabies

William H. Wunner

The Wistar Institute

I. Historic Perspective of the Disease Prevalence, Virus Isolation, and Detection
II. Animal Rabies Epidemiology
III. Rabies Viruses: Serotypes, Genotypes, and Structure
IV. Genome Structure and Proteins of Rabies Virus
V. Replication of Rabies Virus
VI. Pathogenesis and Histopathology of Rabies Virus Infection
VII. Clinical Signs of Rabies
VIII. Humoral and Cell-Mediated Immune Responses to Rabies Infection and Vaccine
IX. Control and Prevention of Rabies in Domestic and Wildlife Animals

GLOSSARY

acinus Minute terminal lobule in salivary glands.

epitopes Structural sites on proteins that bind to antibodies and T-cell receptors.

fixed rabies virus Laboratory strains that grow in tissue culture at a predictable rate, with predictable biological properties, and regularly kill inoculated animals within a predictable period.

Negri body Cytoplasmic matrix composed of rabies viral nucleocapsid material.

peplomer Surface projections of homo- (or hetero-) polymers of viral glycoprotein molecules on virus particles.

street rabies virus Virus isolated from natural infection that has not been serially passaged in animals or tissue culture.

viropexis Entry of virus particles into cells by engulfment, followed by fusion of membrane-bound vesicles with lysosomes.

RABIES is an acute neurologic disease caused by virus that attacks the central nervous system of a variety of warm-blooded animals that serve as natural hosts for virus transmission. Symptomatic rabies infections of adult avians in nature are not known, but laboratory-infected avian embryonic tissue is widely used for the manufacture of antirabies vaccines. Rabies virus infection causes alternating symptoms of depression and agitation and almost invariably leads to extreme types of violent behavior and death. Rabies is enzootic in all regions of the world except Australasia (excluding Australia) and Antarctica. Several island countries, such as Britain and Japan, have remained rabies-free once rabies was eradicated, because of strict quarantine orders and vigilance. The disease is vectored by wild animals and, in most of the world, the animal reservoir of rabies is the dog. Rabies remains a sizable clinical problem for humans in those countries where canine rabies is largely uncontrolled.

Rabies virus is a well-characterized pathogen; the nucleotide sequence of the entire viral genome has been determined, and the genes for the five structural proteins have been identified. Detection of virus antigen using an immunofluorescent antibody test is generally the best single laboratory test available for the rapid diagnosis of rabies. In the pathogenesis of the disease, the virus invades the nervous system and is transported rapidly along nerve routes to the spinal cord and brain. Virus proliferation in the brain is followed by clinical signs of the disease, dissemination of virus to other body tissues, and stimulation of the host's specific and nonspecific immune responses. Because the host is unable to develop an immune defense to natural infection until late in the course of infection, the outcome of the disease is almost always fatal. Antiviral antibodies, helper and cytotoxic T lymphocytes are the major components of the specific immune response to rabies virus infection in animals and humans. Vaccine-induced virus-neutralizing antibodies are the key factor by which the level of prophylactic protection against rabies is measured. Rabies vaccines derived from nerve and brain tissue for humans, that were first developed more than a century ago, are still in use, giving protection but carrying the risk of neuroparalytic complications in humans. The more recently developed use of inactivated viruses, produced in nonneural animal cell cultures for animals and in human diploid cell culture for humans, has greatly improved the safety and potency of rabies vaccines. A third generation, live recombinant vaccinia virus containing the rabies virus glycoprotein gene, the most recently approved type of rabies vaccine, for immunization of wild animals, particularly the raccoon, against rabies, is the most recent addition to the arsenal of rabies vaccines.

I. HISTORIC PERSPECTIVE OF THE DISEASE PREVALENCE, VIRUS ISOLATION, AND DETECTION

The earliest description of rabies in literature goes back in time to the Babylonian era (ca. 2300 BC), Mesopotamian era (ca. 1000 BC), and the classical Greek and Roman periods (ca. 500 BC–AD 100), where the disease, then recognized as hydrophobia, was associated with dogs that displayed alternating symptoms of depression and agitation and extreme types of violent behavior. Fox rabies, later reported in Arabian countries (ca. 980–1037) and in Egypt (1500s), and rabies in wolves, badgers, and bears, and dogs in many parts of Europe and South America (1800s), not only indicate the global nature of rabies, they reveal that animal rabies has been a serious threat to humans as well for a long time. The reports of experiments on rabies by Zinke in 1804 and many others that followed, including Galtier and Pasteur in 1881 and 1882 and DiVestea and Zagari in 1887, provide convincing evidence that rabies is caused by virus in the saliva that is deposited in a bite wound and transmitted along nerves to the spinal cord and brain. Galtier and Pasteur studied the transmission of rabies between animals and from animals to humans and described the active role of the central nervous system (CNS) in the development of rabies. Galtier and Pasteur, and many investigators who followed them, identified the fundamental aspects of rabies pathogenesis by conducting experiments with rats, rabbits, guinea pigs, and dogs over a century ago. Yet many aspects of rabies and its spread to, within, and from the CNS are not completely understood to this day. Their studies demonstrated that when the transmissible "material" (later known to be rabies virus) was inoculated into peripheral nerves, it produced rabies. They also learned that the "rabies virus" was transported to the salivary glands from the brains of rabid animals by traveling along nerves.

Pasteur is credited with the successful development of a rabies vaccine and the first human vaccination. Pasteur had considered the possibility that persons exposed to rabies could be protected against the disease with a preparation of dried rabbit spinal cord that was "attenuated" by a process that he developed and successfully used to make dogs resistant to rabies. Pasteur's opportunity to administer the vaccine to a person came in 1885 when a 9-year-old boy, Joseph Meister, arrived in Paris from Alsace, where he had been bitten 14 times by a rabid dog. Sixty hours after being bitten, Joseph Meister was treated with a subcutaneous inoculation of the attenuated rabbit spinal cord preparation, in 12 additional subcutaneous inoculations with spinal cord preparations of increasing virulence, over a period of 10 days. Joseph Meister not only survived the rabies attack but he resisted the successive applications of highly virulent virus that were contained in the last doses of vaccine.

Pasteur's rabies virus (PAST strain), which was isolated from the brain of a cow and used to inoculate the spinal cord of rabbits, became "fixed" (a term coined by Pasteur) after a number of intracerebral passages in rabbits. The fixed Pasteur virus strain, PV-11 (derived from the original PAST strain), and the fixed challenge virus standard-11 (CVS-11) currently used in laboratory studies, which kills newborn and adult mice following intracerebral inoculation, are closely related.

Virus detection by direct fluorescent antibody (FA) staining of brain tissue is the best and most widely used diagnostic test postmortem for rabies in animals and humans. Antibodies that specifically react with nucleocapsid antigen provide a reliable and rapid indication of virus presence.

Virus isolation from humans is another common method for diagnosis of rabies. If possible, specimens of saliva, throat swabs, swabs of the nasal mucosa or eyes, and cerebrospinal fluid are taken for virus isolation during the clinical period of the disease. Specimens taken postmortem might include various parts of the CNS (brainstem, medulla, cerebellum, Ammon's horn, and cortex) and salivary glands. The classical and most frequently used method of virus detection and titration is the intracerebral inoculation of mice, first proposed by Hoyt and Jungblut in 1930. By 1966, it was established that using newborn mice (less than 3 days old) increased the sensitivity and rapidity of the mouse diagnostic test for rabies, and this is currently the procedure recommended by the World Health Organization. Because the incubation period before the onset of symptoms in the mouse will vary, depending on the virus isolate as well as the age of the animal (newborn versus adult), inoculated mice are checked daily for the onset of symptoms and clinical signs of rabies. As clinical signs of rabies in the inoculated mice are not, in themselves, always sufficient to ascertain the diagnosis of rabies, confirmation of the diagnosis by detecting viral nucleocapsid (internal) antigen in a FA test is necessary. Alternatively, tissue culture systems can be used for the isolation of rabies virus. Specimens of brain or salivary gland are homogenized to form a 20% suspension in tissue culture medium supplemented with calf serum and antibiotics. The tissue suspensions are centrifuged and individual supernatants are used to infect cells in chambers of LAB-TEK microscope slides. After incubation, the infected cells, fixed in acetone, are stained with rabies virus-specific antinucleocapsid fluorescent antibody and examined under a fluorescence microscope (the classical FA test). Providing sensitive cells, such as neuroblastoma cells, are employed, this method is as sensitive and more rapid than mouse inoculation, particularly for laboratory isolation and diagnosis of street virus. The classic virus neutralization test, using viral specific polyclonal and monoclonal antibodies (MAbs), and the identification of Negri bodies by histological examination of CNS tissue are two other tests that may be used for identification of rabies virus isolated by mouse inoculation. Negri bodies are viral-specific inclusions that can be identified in neurons of peripheral and basal ganglia of the cord, in the pons and medulla of the brainstem, in the Purkinje cells of the cerebellum, in the pyramidal cells of the Ammon's horn, and in the cortex of the brain by hematoxylin and eosin (H & E) stain and immunological or immunoenzymatic examination of paraffin-fixed or formalin-fixed tissue sections in the light microscope. Electron microscopy of tissues shows Negri bodies as compact viral nucleocapsid masses and confirms the presence of viral particles associated with cytoplasmic membranes adjacent to nearly all the Negri inclusions.

II. ANIMAL RABIES EPIDEMIOLOGY

Prior to the 1960s in the United States, the major reservoir of rabies virus was in domestic dogs (Fig. 1). Transmission of rabies virus from dogs, especially during the large outbreaks of dog rabies during the 1930s and 1940s, to other animals and particularly wildlife species, laid the foundation for the enzootic rabies disease in different geographic regions of the United States. The extensive intraspecies, as well as interspecies, rabies transmission during this period established the currently recognized terrestrial wildlife rabies virus reservoirs of the 1990s. The decline and eventual elimination of rabies in domestic dogs by the early 1970s can be attributed to the coordinated massive animal vaccination programs and stray animal control measures that were enforced by the Communicable Disease Center (CDC, now the Centers for Disease Control and Prevention) since 1947. To achieve these impressive results, major expenditures were incurred at both state and federal government levels. However, as rabies in domestic dogs was brought under control, outbreaks of rabies in wild animals began to increase, to become the dominant reservoir of animal rabies today. By the late 1950s, the striped skunk, *Mephitis mephitis,* was largely responsible for the outbreaks (epizootics) of rabies that produced the three large areas in the United States (Texas, California, and Iowa–Minnesota–South Dakota) that are endemic

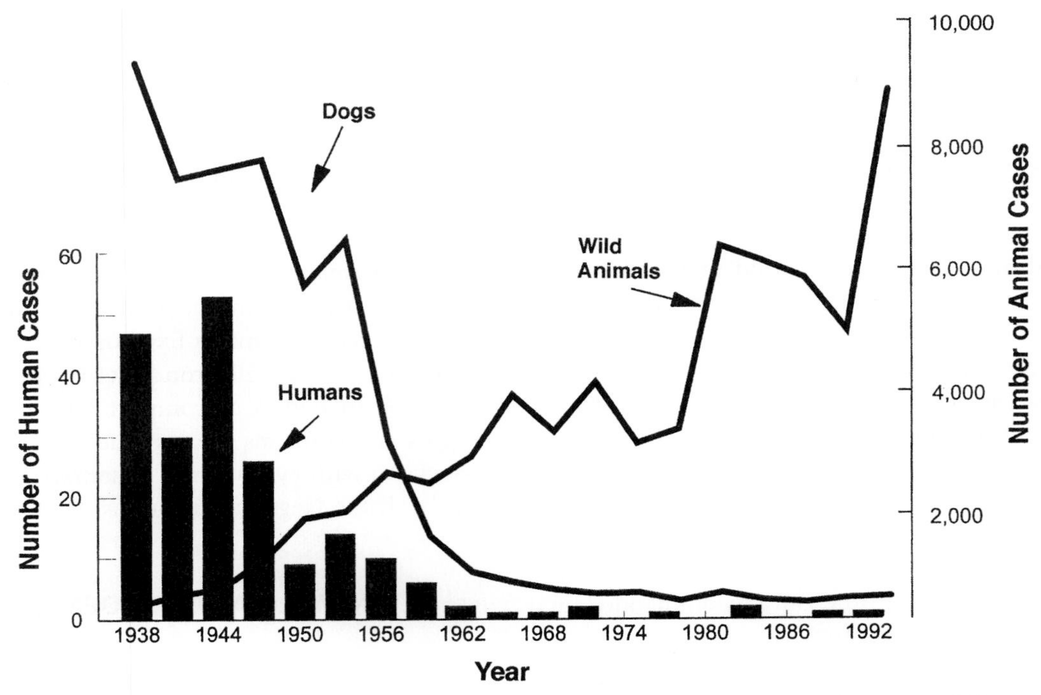

Fig. 1. Rabies cases in the United States 1938–1993. [From Smith, J. S., Orciari, L. A., and Yager, P. A. (1995). *Seminars in Virol.* **6**, 387–400.]

for skunk rabies. Three fox species are known reservoirs for rabies in four separate areas of the United States. Red foxes (*Vulpes vulpes*) in the northern counties of New York, Vermont, New Hampshire, and Maine, as well as red foxes and Arctic foxes (*Alopex lagopus*) in Alaska, and grey foxes (*Urocyon cinereoargenteus*) in Texas and Arizona, make up the present enzootic fox rabies. The coyote (*Canis latrans*) represents the newest and most rapidly expanding reservoir of animal rabies virus currently located in south Texas. Finally, the animal rabies that is most complicated epidemiologically is the nonterrestrial bat rabies virus reservoir. Different bat species across the United States transmit rabies virus strains of distinct phylogenetic lineages. There is good evidence that bats are responsible for translocating these genetically distinct rabies virus strains to terrestrial animals and that they are rapidly becoming the most frequent vector for transmission of rabies to humans. In other regions of the world, the natural reservoirs and vectors of rabies are dogs, foxes, wolves, jackals, raccoon dogs, skunks, mongooses, and vampire, frugivorous and insectivorous bats.

III. RABIES VIRUSES: SEROTYPES, GENOTYPES, AND STRUCTURE

Rabies virus and rabies-related viruses belong to the genus *Lyssavirus* of the family Rhabdoviridae. The genus name, "lyssa," derived from Greek, meaning "rage" (rabies), characterizes the relationship of these viruses to the clinical disease they produce in the infected host. Lyssaviruses have a bullet-shaped morphology, similar to other animal rhabdoviruses of the *Vesiculovirus* and *Ephemerovirus* genera that infect vertebrate and invertebrate hosts. Initially, lyssaviruses were divided into four serotypes. The classical rabies virus strains, which include most field isolates worldwide, and "fixed" laboratory strains, of which the CVS strain is the prototype, belong to serotype 1. The rabies-related viruses which were first isolated in Africa represent three distinct serotypes. Lagos bat virus (serotype 2) was first isolated in 1956 from one of seven brain pools from 42 fruit bats (*Eidolon helvum*) in Nigeria, and again, in 1974, from a fruit bat (*Micropterus pusillus*) in Central African Republic, and in 1980 from an epauleted fruit bat (*Epomophorus wahlbergi*) during an epizootic outbreak of dog rabies in South Africa. Mokola virus (serotype 3) was first isolated in 1968 from shrews in Nigeria, and again, in 1969, from a girl who recovered without sequelae, and in 1971, from a young woman who died in Nigeria, and then from wild and domestic animals in several African countries. Duvenhage virus (serotype 4) was first isolated in 1970 from a man bitten on the lip by an insectivorous bat in South Africa and, in 1981, it was again isolated from bats in South Africa. Then, unexpectedly and without any clear explanation for their presence, lyssavirus isolates were obtained on numerous occasions from cases of bat rabies in Central Europe between 1954 and 1987 that were closely related antigenically to the serotype 4 African Duvenhage virus. Initially, the Duvenhage-like European bat rabies viruses, identified as European bat lyssaviruses (EBL), were proposed to constitute serotype 5. Later, they were classified as either biotype 1 (EBL1) or biotype 2 (EBL2) and, finally, as two clearly distinct genotypes. Two other rabieslike viruses, kotonkan and Obodhiang, isolated from invertebrate hosts, have not been found in vertebrates to date and would, therefore, appear to be of no consequence to the rabies zoonosis in domestic or wildlife animal species.

The classification of lyssaviruses into the five serotypes was based on serum neutralization and MAb studies. Large panels of MAbs specific for the rabies virus glycoprotein (G) and nucleoprotein (N) revealed in patterns of MAb reactivity with the respective antigens which of the virus isolates from different species and geographic areas displayed greater similarities (were antigenically related) or showed greater antigenic diversity (unrelatedness). As a result of these studies, not only were the five serotypes established but the viruses could be identified with their particular animal host reservoirs and geographic origin. Recently, these viruses were reinvestigated by determining and comparing nucleotide sequences of part or all the N gene of representative virus isolates from the different serotypes and six genotypes were distinguished. Genotypes 1 to 4 correspond to serotypes 1 to 4, respectively, and genotypes 5 and 6 correspond to EBL1 and EBL2, respectively. A seventh genotype was recently recognized that includes the Australian bat Lyssavirus (ABL). Genetic analysis as a diagnostic technique, based on nucleotide sequences of the N gene, provides sufficient precise information for assignment of virus isolates to one of the six distinct genotypes. Genetic typing also provides data to develop phylogenetic lineages of the viruses with respect to their host species and geographic origin. By comparing phylogenetic lineages of rabies virus isolates, it is possible to determine the independent spread of rabies, trace the evolution of virus strains circulating in association with animal reservoirs, and identify traits of a rabies epizootic outbreak that indicate whether the rabies epizootic is one that represents an emerging or reemerging disease.

The commonly studied fixed laboratory strains of rabies virus (serogenotype 1) that have been used to investigate the molecular structure of rabies virus, its growth in cell culture, and its pathogenesis in animals include the CVS-11, CVS-24, Evelyn–Rokitnicki–Abelseth (ERA), Flury low egg passage (LEP), Flury high egg passage (HEP), Pitman Moore (PM), PV-11, Nishigahara (derived from the original PAST strain and used in Japan), Street Alabama Dufferin (SAD), and Vnukovo-32 (derived from the SAD strain and used in Russia) strains. Several of these fixed laboratory strains have been used for vaccine production. All genotype/serotype 1 fixed laboratory strains are antigenically similar at the viral nucleoprotein level, but antigenically distinct at the level of the viral surface glycoprotein. However, the fixed laboratory strains of rabies virus are all antigenically closer to each other than they are to the rabies-related viruses (serogenotypes 2–4 plus genotypes 5 and 6) at both the nucleoprotein and glycoprotein levels.

Rabies viruses have an average length of 180 nm (130–200 nm) and an average diameter of 75 nm (60–110 nm). Virus particles are bullet-shaped (i.e., bacilliform), hemispherical at one end, planar at the other end, and rigid. The particle contains a lipid bilayer envelope derived from the plasma membrane of the infected cell which is 7.5–10 nm thick and studded on the external surface (except at the planar end) with 10 nm-long peplomers that give the appearance of spikes projecting outward (Fig. 2). The lipid envelope surrounds a coiled (helical) ribonucleoprotein (RNP) structure 165 × 50 nm in the virus. The RNP structure, sometimes referred to as the nucleocapsid (NC) core, gives the the virus particle its cylindrical symmetry. Fully extended, the filamentous NC core of standard virus particles, as viewed in the electron microscope, ranges between 4.2 and 4.6 μm in length. These standard size NC cores are two or three times longer than the NC cores of defective interfering (DI) particles of rabies virus, which are generated during infection in cell culture.

Rabies DI particles, like DI particles of other RNA viruses, are subgenomic virions derived from the parental virus, are generally not self-replicating, and require the parental virus for propagation. DI particles are distinguishable from standard (parental) infectious virions, when viewed in the electron microscope, by their shorter particle length, which corresponds in size to the subgenomic length of the DI virion RNA. They contain the full array of viral proteins with full antigenic properties and they interfere with the replication of helper (parental) virus.

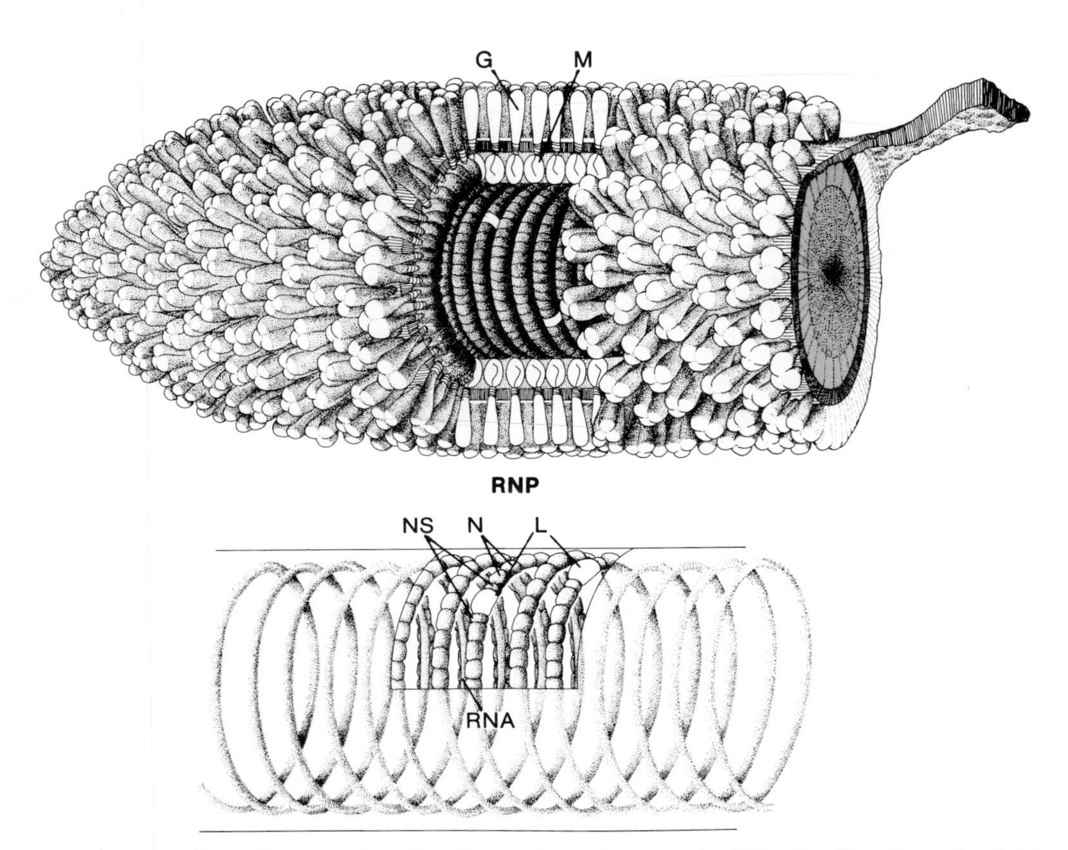

Fig. 2. A drawing of rabies virus showing surface glycoprotein (G) extending from the lipid envelope that surrounds the internal ribonucleoprotein (RNP) core. Matrix (M) protein lining the viral envelope is shown interacting with the cytoplasmic domain of the G protein. The internal helical RNP complex contains the single-strand RNA genome associated with nucleoprotein (N), phosphoprotein (NS), and transcriptase/polymerase (L). The membrane "tail" depicted in the drawing (seen in electron micrograph) represents the trailing piece of envelope that is frequently observed attached to the virus as it buds from the plasma membrane of the infected cell. [*In* "The Natural History of Rabies," 2nd ed., pp. 31–67. CRC Press, Boca Raton, FL.]

Because of their smaller size, DI particles can be physically separated from standard-size parental infectious rabies virus particles by sucrose density gradient centrifugation. Rabies DI particles have been identified in cells infected with Flury LEP, Flury HEP, ERA, CVS, and PM stains.

Although rabies DI particles appear to inhibit infection at the level of virus replication *in vitro* (in cell culture), the biological role of rabies DI particles *in vivo* is less clear. The few studies that have been performed to determine whether DI particles influence virus replication or pathogenicity *in vivo* are equivocal. Clearly, further investigation is needed in this area to substantiate and confirm the role for DI rabies particles *in vivo* because their role may provide an explanation for the periodic epizootic outbreaks of rabies in wildlife animal reservoirs.

IV. GENOME STRUCTURE AND PROTEINS OF RABIES VIRUS

A. Genome Structure

The rabies virus genome has a composition and structure that is similar, but not identical, to that of other rhabdoviruses. The helical RNP (NC core) structure contains a single-strand RNA genome of negative-sense polarity which consists of a 3′ and 5′ noncoding sequence flanking five structural genes, each encoding one of the structural proteins of the virus. The genome of the PV strain of rabies virus, which contains 11,932 nucleotides, is shown in Fig. 3. The complete nucleotide sequence consists of a noncoding leader (l) sequence at the 3′ end, followed sequentially by the genes for the nucleoprotein (N), phosphoprotein (NS or P), matrix protein (M), glycoprotein (G), and RNA transcriptase/polymerase (L), and ends with a 5′ noncoding region. The genes are separated by noncoding intergenic nucleotide sequences of varying lengths. All rhabdoviruses investigated at this level have the same genome organization as shown for rabies virus but may differ in the lengths of the coding and noncoding sequences. The rabies virus gene order was identified by transcriptional mapping studies.

The most striking differences between the genome structure of the rabies virus (genus *Lyssavirus*), apart from the nucleotide sequence itself, that distinguish it from the genome of the other rhabdoviruses, notably vesicular stomatitis–Indiana virus (VSIV) (genus *Vesiculovirus*), are the variable length and composition of their intergenic nucleotide sequences; in VSIV, the genes are all separated by two nucleotides. It is particularly interesting that, while the G to L intergenic sequence in the rabies virus genome (referred to as a pseudogene) is sufficiently long to represent a sixth gene, the longest open reading

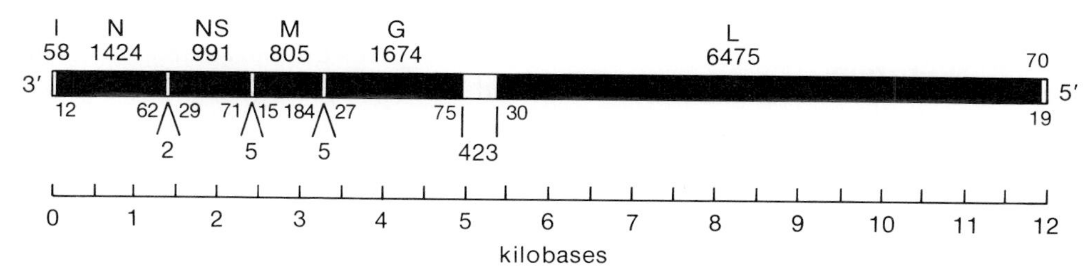

Fig. 3. Organization of the nonsegmented single-strand rabies virus RNA genome. Starting at the 3′ end, the genome (negative strand) contains a leader sequence (l), followed by the N, NS, M, G, and L genes (solid regions) and a 5′ noncoding region. The l sequence and each of the structural genes produces plus strand leader RNA and mRNA transcripts. The numbers above the line indicate the lengths of plus strand leader RNA and mRNA transcripts (including the first 7 nucleotides of the polyadenylate in each mRNA), and the 5′ noncoding sequence. The numbers immediately below the line indicate the lengths of untranslated nucleotides in each mRNA, and the numbers below each gap region indicate the lengths of the noncoding intergenic sequences. [From Wunner, W. H. *In* "Handbook of Neurovirology," pp. 455–462. Marcel Dekker, Inc., New York.]

frame within the pseudogene, beginning with the first initiation codon, contains only 17 codons. Because no mRNA transcript or short peptide that could be derived from this sequence has been detected, it may be assumed that the rabies virus genome does not encode a gene product at this site. In contrast, a sixth gene and its mRNA transcript derived from the G–L intergenic region have been reported for the rhabdovirus, infectious hematopoietic necrosis virus (IHNV), which causes an acute systemic disease in salmonid fish. The G–L intergenic nucleotide sequence in IHNV encodes a nonstructural viral protein of approximately 12,000 Da.

The structural genes of rabies virus and their complementary mRNA transcripts encode the five structural proteins of the virus. All five mRNA transcripts begin with the tetranucleotide UUGU (in the RNA genome) near the 3′ end and terminate at the polyadenylation signal of 7 Us near the 5′ end of each gene. Like the mRNA structures of other rhabdoviruses, a 5′-cap structure at the beginning of the mRNA molecules and a poly (A) tail at the end of each rabies virus-specific mRNA, is presumed to be present. A single long open reading frame, starting at the first AUG and terminating near the 3′ end with the stop codon UAA (for N, NS/P, and M) and UGA (for G and L), is present in each mRNA. The open reading frames encode proteins of 450 (N-mRNA), 297 (NS-mRNA), 202 (M-mRNA), 524 (G-mRNA), and 2142 (L-mRNA) amino acids.

B. N Protein

The nucleoprotein (N protein) is the most abundant of the viral proteins produced in rabies virus-infected cells. The relatively high level of N protein expression is consistent with the model for rhabdovirus genome transcription that proposes that the viral gene closer to the 3′ end of the RNA genome is transcribed more frequently than the gene next to it. Approximately 1800 N protein molecules encapsidate the RNA genome, making this the major protein of the RNP complex. The N protein is also the major antigen of the cytoplasmic inclusion bodies, first described by Negri as the morphologically distinct lesion of rabies found in virus-infected neurons. The amino acid sequence (primary structure) of the N protein, which is composed of 450 amino acids (50,000 Da), has a greater than 98% similarity among several of the commonly used fixed laboratory strains (e.g., PV, ERA, CVS-11, and SAD). Evidence that the N protein has a highly conserved primary structure correlates well with the comparative antigenic analysis of the fixed rabies virus strains, which shows that all but one have almost identical antigenic reactivity patterns with a panel of MAbs specific for N protein. The one exception appears to be the Flury HEP strain (obtained from Flury LEP virus after 178 to 181 passages in chicken embryos), which fails to react with 10 out of 35 anti-N protein MAbs. A somewhat lower level of conservation of antigenic determinants on the N protein is seen among street rabies viruses isolated from human cases of rabies and a variety of animal species, particularly among bat isolates. The nucleotide sequence of the Mokola virus (serogenotype 3) N gene is the most divergent of the lyssaviruses. Because of the relatively conserved nucleotide sequences intrinsic to the N gene and high degree of stability of antigenic determinants associated with the N protein, the predominant NC antigen, the N protein of fixed laboratory strains and street viruses provides the type-specificity that distinguishes rabies from rabies-related viruses and enables a definitive differentiation of viruses in epidemiologic surveys of enzootic areas.

Because protective immune responses to rabies virus antigens play a major role in preventing rabies, much effort has been made to identify the epitopes on rabies virus proteins that are recognized by MAbs and T cell receptors (TCRs). Epitopes on the N protein that specifically bind antibodies or react with T cells have been located using synthetic peptides and peptide fragments of the N protein produced by chemical (CNBr and 2-[2′-nitrophenylsulfenyl]-3′-bromo-indolenine skatole) and enzymatic (trypsin and protease V8) cleavage methods. Using peptide fragments produced by the chemical and enzymatic cleavage methods, two overlapping epitopes, recognized by MAbs 377-7 and 802-2, and another epitope, recognized by MAb 816-1, have been located between amino acid residues 374 to 390 and 313 to 337, respectively, on the N protein. All three antigenic determinants appear to be sequential (i.e., nonconformational) epitopes in the native molecule. The

MAbs appear to bind equally well to the determinants in the denatured N protein and the N protein fragments which contain them.

Studies using antigenic synthetic peptides have demonstrated, by proliferation of rabies-specific T cells *in vitro,* that the N protein is the immunodominant target antigen for rabies virus-specific T helper (T_H) lymphocytes. Prior stimulation of mouse T cells by inoculation of the animals with rabies virus or vaccine, or of humans with rabies vaccine, produced the rabies-specific T cells *in vivo.* T lymphocyte proliferation *in vitro,* measured by the uptake of [^{3}H]thymidine, can be stimulated by adding live virus, viral proteins, or synthetic peptides of specific sequences to the culture system. The mimicry of synthetic peptides has been particularly useful in these instances, since the identification of T-cell epitopes on the N protein *in vitro* can be correlated with short segments (e.g., 15 amino acids) of the primary sequence of N protein and can even be localized to individual amino acids by systematically altering the amino acid composition of the active peptide. Also, synthetic peptides bind to receptor proteins expressed by the major histocompatibility complex (MHC) and effectively present the T-cell epitope to the TCR that normally recognizes the T-cell epitope in association with MHC receptor molecules. Synthetic peptides have not only been helpful in demonstrating that the rabies virus N protein is a major target antigen for murine and human T_H cells, they have also been instrumental in identifying T-cell epitopes that cross-react with RNP and N proteins derived from different fixed rabies virus strains and from rabies-related viruses.

Although not a nominal phosphoprotein, the N protein of rabies virus is phosphorylated, unlike the N protein of VSIV, which is not phosphorylated. The serine residue at position 389 has been identified, by peptide fragment mapping and acid hydrolysis of [^{32}P]-labeled N protein, as the phosphorylation site on the rabies virus N protein.

C. NS/P Protein

The nominal nonstructural (NS) phosphoprotein (P) of rabies virus is the second most abundant protein of the viral RNP complex. The NS/P protein (originally called "M1") is also produced in abundant amounts in virus-infected cells, which, in earlier studies, gave reason to consider it a nonstructural protein. Approximately 950 molecules of NS/P protein are present in the RNP complex (clearly, a structural protein) of the rabies virus particle. It has been inferred that the proportions of N and NS proteins in the RNP complex (molar ratio of 2 : 1) are critical for the activity of the polymerase complex in virus-infected cells.

The rabies virus NS/P protein, like other rhabdovirus P proteins, is a strongly acidic protein and appears in some strains to take the form of multiple subspecies separable by electrophoresis in SDS-polyacrylamide gel. Whether more than one form of NS/P protein is required to bind to N protein in nucleocapsids, to interact with L protein, or to regulate transcription and/or replication of the RNA genome remains unclear.

The nucleotide sequence of the NS/P gene of PV, ERA, PM, CVS-11, SAD (serogenotype 1), and Mokola virus (serogenotype 3) have been reported. The deduced amino acid sequence of the 297 amino acids (33,000 Da) encoded in the rabies virus NS/P gene is 6 amino acids shorter than the NS/P protein of Mokola virus. Despite the weak overall amino acid similarity (45.9%) between the NS/P proteins of the serogenotype 1 and serogenotype 3 lyssaviruses, there are conserved regions that reveal some closer relationships (as much as 86% similarity) between the strains. The phosphorylation sites on the NS/P protein have not been determined.

D. M Protein

One of the two major membrane structural proteins, called "matrix protein," is a small protein located inside the viral envelope. The M protein in rabies virus, as in other rhabdoviruses, plays a key role in the virus assembly and budding processes. Its location, and its ability to remain attached to the viral RNP core of spikeless virus particles ("skeleton" structure) after virus particles are extracted with the detergent n-octylglucoside, suggests that it may play a role in virus assembly. It is conceivable that, as rabies virus buds from the plasma membrane of infected cells *in vitro* and *in vivo,* the M protein, which has affinity for cell membranes and nucleocapsids, is able to interact with proteins in both structures to bring them together in virus assembly.

The M protein (previously designated "M2") is the smallest and most basic of the rabies virus proteins with 202 amino acids (23,000 Da). Approximately 1500 molecules of M protein are present per virion. The overall amino acid similarity between the M proteins of fixed rabies viruses (serogenotype 1) and Mokola virus (serogenotype 3) is about 71%.

E. G Protein

The viral protein that is most crucial for the neurotropism and pathogenesis of rabies and rabies-related viruses, and in establishing the humoral and cellular immunity to these viruses, is the type 1 transmembrane glycoprotein (G protein). The G protein interacts with specific cell surface receptors to initiate infection of cells. Approximately 1800 molecules of G protein form the peplomers (spikes) located on the surface of the viral envelope of infectious virions. It is estimated on the basis of size-exclusion by gel filtration that each spike consists of three G protein molecules anchored individually in the envelope lipid bilayer. The fact that G protein monomers, stripped from virus particles by treatment of virions with nonionic detergent, are homogeneous in size and isoelectric point (pH 7) suggests that the surface spikes are indeed homopolymers of G protein. The carbohydrate in the form of N-linked complex oligosaccharides represents 6 to 12% of the G protein mass. The number and location of N-linked carbohydrate side chains found on G protein molecules varies for different rabies virus strains.

The G protein gene in all fixed virus strains examined encodes a 524-amino acid protein (58,500 Da). The first 19 amino acids (from the N-terminus) of the protein represent the signal peptide, which is cleaved after it translocates the nascent protein across the rough endoplasmic reticulum. After signal peptide cleavage, the mature G protein, which forms the spikes on the surface of standard and DI rabies virions, consists of 505 amino acids. Several distinctive features of the mature G protein have been defined: the 439-amino acid "antigenic" domain, which extends from the N terminus to the transmembrane domain; the 22-amino acid hydrophobic "transmembrane" domain, that anchors the molecule in the membrane; and the 44-amino acid "cytoplasmic tail" at the C-terminus, which, presumably, interacts with the M protein on the inner side of the viral envelope. Within the antigenic domain, three major conformational and functionally independent antigenic sites (I, II, and III) have been mapped on the G protein of the ERA and CVS-11 strain by comparative antigenic analysis, using virus-neutralizing MAbs and MAb-resistant virus mutants and amino acid sequence analysis. Two additional sites (IV and V) have been delineated, but not mapped, on the ERA strain G protein, and another site (VI) has been mapped on the CVS-11 strain by comparative antigenic and amino acid sequence analysis. Site VI is a sequential (i.e., nonconformational) determinant unaffected by protein denaturation. The epitope in site VI is both antigenic (i.e., binds MAb) and capable of inducing virus-neutralizing antibody *in vivo* in a 23-amino acid synthetic peptide, which suggests that a synthetic peptide rabies vaccine may be possible. Although preliminary, encouraging results have been obtained by immunizing mice with a synthetic peptide containing the sequential epitope identified in antigenic site VI coupled to a 15-amino acid sequence (corresponding to residues 404–418 of the N protein) that expresses a T_H-cell determinant on the N protein. Of the mice immunized with the two epitopes from individual rabies virus proteins fused into a single synthetic peptide, 75% were protected against a lethal rabies virus challenge.

The rabies virus G protein is a highly immunogenic protein, capable of producing high titers of virus-neutralizing antibodies in serum of animals immunized with inactivated virus or live recombinant virus vaccine, such as pox virus and adenovirus recombinants that express G protein. G protein molecules extracted from virions with the nonionic detergent n-octylglucoside and which spontaneously form rosette structures by the interaction of their hydrophobic membrane-anchoring domains are as capable of conferring immunity against a lethal challenge dose of rabies virus as inactivated virus vaccine. Although the virus-neutralizing antibodies directed against G protein are an important component of the immune defense against rabies, the G protein is also a target for both T_H and cytotoxic T lymphocytes (CTLs). T cells appear to recognize sequential (continuous) determinants on G protein and in synthetic peptides. Virus-neutralizing antibodies, on the other hand, recognize mostly conformational determinants, including closely arranged, as well as discontinuous, groups of epitopes on G protein.

F. Transcriptase/RNA-Dependent RNA Polymerase

The L (large) protein is thought to be a multifunctional protein located in the RNP core, which functions primarily as an RNA-dependent RNA transcriptase and polymerase. Other functions that it may carry out include capping, methylation, and polyadenylation of viral mRNAs and, possibly, protein kinase activity. The L gene is by far the largest in the rabies virus genome with its longest open reading frame coding for 2142 amino acids (244,206 Da). The number of L protein molecules per rabies virion is small (compared with VSIV), which may also account for the reduced transcriptase activity of rabies virus in comparison with VSIV. Inasmuch as clusters of amino acids in the L protein of rabies virus appear to have sequence similarity, as high as 80%, with L proteins of other single-strand, negative sense RNA viruses, it is possible that these highly conserved regions shared by L proteins represent catalytic sites and define the independent functional domains of the multifunctional L protein but appropriate research has not yet been completed to validate such presumptions.

V. REPLICATION OF RABIES VIRUS

Rabies virus attaches itself to specific receptors on the cell surface and enters the cell via the endocytic pathway (i.e., formation of coated pits that pinch off into coated vesicles within the cell). The process is known as viropexis. Several types of receptor molecules have been characterized that specifically bind rabies virus, including the nicotinic acetylcholine receptor (specifically, the neurotoxin-binding site of the α subunit) from *Torpedo californica* electric organ membranes, a high molecular weight (fibronectin-like) protein on baby hamster kidney cells, and a neural cell adhesion molecule (NCAM), found only on neurons and lymphocytes susceptible to rabies virus infection. The coated vesicles fuse with lysosomes (endosomes), from which lysosomal enzymes uncoat the virus and release nucleocapsids into the cytoplasm (Fig. 4). The viral RNA genome (negative-sense strand) is transcribed in the cytoplasm in the 3′ to 5′ direction. The leader sequence at the 3′-end of the RNA genome acts as the site of entry for the L protein. Leader RNA (complementary to the sequence of the 3′-terminus of the genome RNA, uncapped, unpolyadenylated, and untranslated) and the five monocistronic mRNAs are produced in sequential order as positive-sense strands. The mRNAs are capped at the 5′-end and polyadenylated at the 3′-end. To replicate the viral genome, a template is transcribed as a full-length, positive-strand complement of the genome RNA. Progeny genome RNA is synthesized from the antigenome RNA. The five viral proteins are translated from the individual mRNAs. The nascent N, NS/P, and L proteins encapsidate the newly synthesized progeny genome RNAs to form the helical NC cores of the newly formed virions. Excess N protein molecules aggregate into a filamentous matrix (inclusion bodies) that is associated with Negri bodies, one of the hallmarks of rabies identification upon histological examination of neurons in CNS tissue. Nascent M protein assembles progeny nucleocapsids into virions by binding to the NC cores, making the helical structure tighter, and binding the RNP-M protein complex to the cytoplasmic tail of nascent G protein that is already inserted in the endoplasmic reticulum or plasma membrane. The RNP-M protein complex is enveloped in the G protein-rich cytoplasmic membrane in a process that leads to budding and release of fully formed, infectious rabies virions.

VI. PATHOGENESIS AND HISTOPATHOLOGY OF RABIES VIRUS INFECTION

The pathogenetic process of rabies usually begins with the introduction of virus-laden saliva into a bite

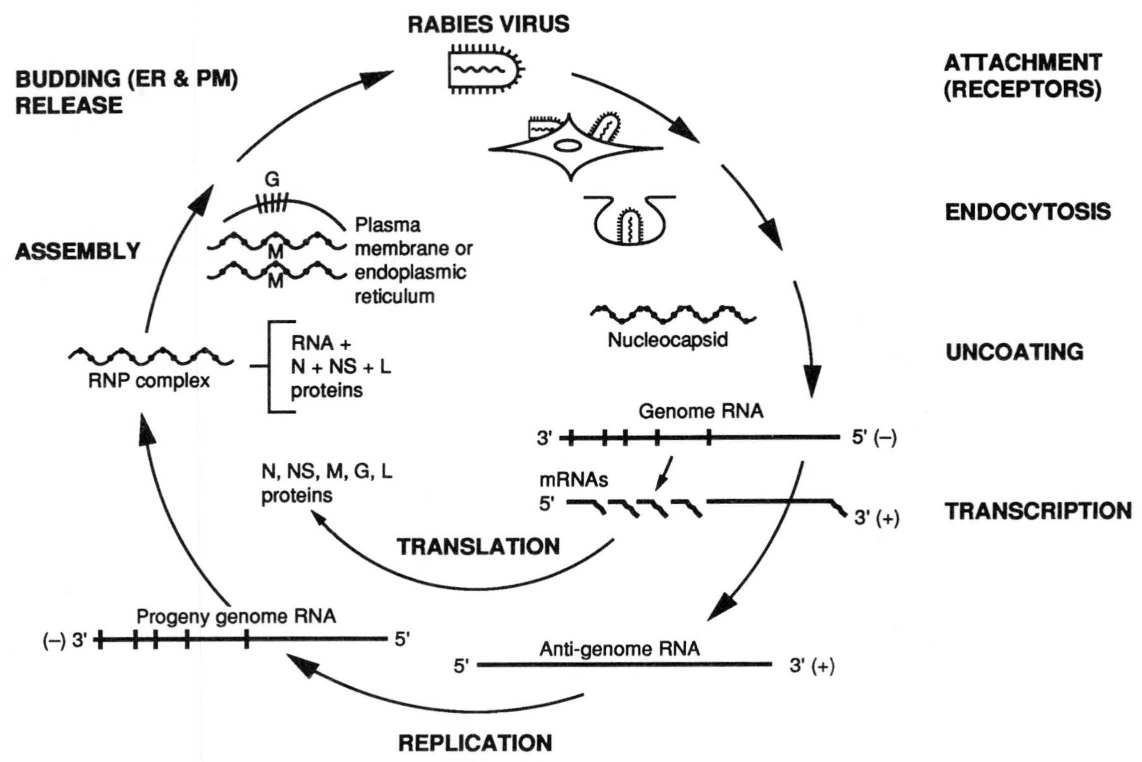

Fig. 4. Life cycle of rabies virus infection. Major steps include: virus attachment to cell surface receptors; uncoating in endosomes and release of nucleocapsid; RNA genome (− strand) transcription to produce monocistronic mRNAs and antigenome RNA (+ strands); replication of progeny genome RNA; mRNA translation into protein; progeny genome RNA encapsidation; virion assembly; virus budding and release. [From Wunner, W. H. *In* "Handbook of Neurovirology." pp. 455–462. Marcel Dekker, Inc., New York.]

or scratch wound inflicted by a rabid animal. The incubation period in warm-blooded species can be extremely variable after the bite, depending on the relative susceptibility of the species, location of the wound (to the head versus extremities), and the depth and location of virus deposit in the wound. In humans, the incubation period is generally 15 days to 1 year, but cases have been described recently in which the extremely sophisticated, highly sensitive and precise genetic (employing the polymerase chain reaction technique) and antigenic (using MAbs) diagnostic methods established that the incubation period could last several years. Other routes of infection have included intranasal (in animals that sniff) and oral routes (among bats in high-density cave dwellings) as well as corneal transplants (in human). Because the neural routes from mucosal membranes and the eye to the CNS are relatively short, the virus has almost direct access to the brain. Rabies virus in natural infections does not establish a viremia.

Perceptions of what happens to the virus at the site of entry *in vivo* and the events that precede the transit of virus toward the CNS are still vague. One puzzling aspect of preneural rabies virus infection is the sequestering of virus at the site of inoculation. Evidence of viral antigen accumulating in striated muscle in experimental animals after intramuscular inoculation of rabies virus has suggested that muscle represents a potential initial site of virus replication, although this has not been observed in humans. The "preneural phase" of virus infection in muscle is usually confined to a small number of myocytes close to the inoculation site and produces few infectious particles. Any shedding of virus into extracellular spaces around myocytes, of course, makes this virus vulnerable to the postexposure treatment with anti-rabies antibody and vaccine.

Exposed sensory nerve endings of neuromuscular or neurotendinal spindles deep in muscle are the most likely sites of virus entry into the peripheral nervous system after virus replication in muscle. The primary site of virus attachment might actually be nerve endings leading directly to neuronal infection in badly torn muscles of a bite wound. Similarly, the many sensory nerve endings of epithelial and subepithelial tissues exposed by superficial abrasions of skin and mucous membranes may also serve as primary sites of virus entry. The entry of rabies virus into peripheral nerves begins the "neuronal phase" of infection, which is characterized by the centripetal movement of virus exclusively within axons to the CNS. Schwann cells and perineural elements do not appear to be infected.

Virus migration to the brain is suspected to occur passively in the axoplasm of peripheral nerves at the estimated rate of retrograde axoplasmic flow (3 mm/hr) in peripheral nerves. Other mechanisms of virus transport may facilitate its spread (e.g., movement of virus across cell-to-cell junctions, including synaptic junctions). Observations of virus budding on axonal membranes and virus particles accumulating intra-axonally at nodes of Ranvier as the virus progresses to the brain support the possibility of other mechanisms of virus transport. The ascending paralysis seen in many infected animals and humans reflects the progressive infection of dorsal root ganglia of the CNS. The virus ascent toward the brain has been confirmed by sequential immunofluorescence studies.

Inflammatory lesions caused by perivascular infiltration of lymphocytes, macrophages, and, occasionally, polymorphonuclear cells are the most common and frequently noted histological signs of change in the spinal cord and brainstem of animals and humans infected with rabies virus. The ganglionic changes observed in rabies, although not exclusively confined to rabies, could serve as diagnostic lesions for rabies, because they are not encountered in diseases that must be considered in the differential diagnosis of rabies. The morphologically distinct lesions that are significant in rabies diagnosis are specific eosinophilic inclusion bodies (Negri bodies) within neurons of peripheral and basal ganglia of the spinal cord, in the pons and medulla of the brainstem, cerebral cortex, thalamus and Purkinje cells of cerebellum, and in ganglion cells of Ammon's horn of the hippocampus. Signs of active transport of virus in the brain include replication in the neurons, budding from the plasma membrane, and release into tissue spaces. The virus appears to infect susceptible cells by viropexis or direct cell-to-cell transmission. Once delivered in the extracellular spaces of the brain, virus can, presumably, spread by passive transport over relatively long distances through intercellular spaces large enough to pass virus by interstitial fluid movement. Virus that is transmitted directly from one nerve cell to a contiguous one could also spread and result in the same long-distance movement because of the many long, intertwined neuronal processes that connect in the brain.

Once virus begins to spread from the brain in the final phase of rabies infection, the tropism of infectious virus changes and virus is delivered to a variety of extraneural and nonneural tissues. Centrifugal spread of virus generally occurs via the same axoplasmic routes that were used for the centripetal passage of virus to the CNS. The spread of rabies virus in the absence of CNS infection is virtually nil. Among the most heavily infected tissues following propagation of virus in the CNS are the end organs in oral and nasal cavities and the head and neck. The fact that rabies antigen is occasionally detectable in skin biopsy and in corneal impressions is evidence of its nerve-mediated dissemination.

Infection of the salivary glands by dissemination of virus from the CNS is extremely important in the life cycle of the virus because it is required for effective transmission of disease. Virus titers in the salivary glands often exceed those in the brain. In some animal species, the salivary glands become infected before the signs of CNS infection become apparent and, at the height of salivary gland infection, large amounts of fluorescent viral antigen may be detected in individual acinar cells, in cell clusters, and, occasionally, in the entire acinus.

Presently, the molecular basis of rabies virus pathogenesis is linked to the surface G protein, which also plays an important role in the initiation of rabies virus infections. By testing neutralization-resistant variants of rabies virus (selected with different virus-neutralizing MAbs against G protein) for virulence, the pathogenicity of the virus appeared to correlate with the presence of an epitope on the G protein within antigenic site III. The amino acid substitution in the mutant that rendered the virus nonpathogenic for adult mice was identified at position 333 of the G protein, where arginine is normally found. Isoleucine in the ERA strain variant and glutamine or glycine in CVS strain variants replaced arginine in the respective pathogenic parent viruses. Several other fixed rabies virus strains, including attenuated viruses that were not selected by MAbs but resisted neutralization by MAbs 194-2 and 248-8, such as the Flury HEP and Kelev strains, also substituted either isoleucine, glutamine, or glycine for arginine-333. The biological relevance of arginine-333 or its influence on the pathogenic mechanism responsible for the restricted neurotropism of rabies virus infection is not completely understood. Experiments have suggested that the neural routes of entry to the CNS and the rate of virus spread may differ, depending on the amino acid at position 333, but, as yet, the complete story has not been told.

VII. CLINICAL SIGNS OF RABIES

The clinical manifestations of rabies are generally indicative of one or the other of two distinct clinical patterns of the disease that rabies assumes. The clinical patterns are known as "furious" and "paralytic" (or "dumb") rabies, although elements of both forms may occur together. Clinically, the infection is divided into three major phases: a "prodromal phase," an "acute neurologic phase," and a "comatose phase." Typically, the classic symptoms and signs of the prodromal phase, lasting several days in humans, are almost entirely nonspecific; these include general malaise, chill, fever, headache, photophobia, anorexia, nausea, sore throat, cough, and musculoskeletal pain. Complaints of abnormal sensations (itching, burning, numbness, or paresthesia), apart from pain around the contaminated site of the bite or scratch inflicted by the rabid animal, are also typical. During the acute neurologic phase, patients exhibit signs of increasing anxiety (agitation) and alternating periods of clarity and episodes of delirium. Behavior disturbances occur regularly, ranging from mild memory disturbance to severe depression, and include periods of hyperactivity, aggressiveness, intolerance to stimuli such as noise, and delirium. The patient may also experience seizures, hallucinations, and hyperventilation. The classic symptoms of hydrophobia—difficulty in swallowing and choking—are also often evident during this period. There is a steady deterioration of mental status, unless generalized convulsions suddenly precipitate coma or death. The third and final phase of the disease that progressively replaces the acute neurological phase is one of coma, lasting an average of 3–7 days and resulting in death.

In "paralytic" (or "dumb") rabies, a significant number of patients (~20%) present a neurological picture comparable to that of the Landry–Guillain–Barré syndrome, which has paralysis as the principal feature. Paralytic rabies most often develops following vampire bat bites and tends to develop after a minor bite. After a typical prodromal period, paresis often begins in the bitten extremity and then spreads either symmetrically or asymmetrically, accompanied by fasciculations that rapidly progress to flaccid paralysis. The course of paralytic rabies can be modified at any stage by the appearance of spasms, hydrophobia, and convulsions. Patients who have died unsuspectingly from paralytic rabies have, in at least four instances, been the cause of human-to-human transmission of infection and disease when the corneas from the fatal cases of dumb rabies were transplanted to susceptible recipients.

VIII. HUMORAL AND CELL-MEDIATED IMMUNE RESPONSES TO RABIES INFECTION AND VACCINE

Because rabies virus is strictly neuroinvasive and so little antigen is present in the early stages of the disease, the humoral response to viral antigens in the infected host is negligible until late in the course of infection, usually after clinical signs appear, and remains low until the terminal phase of the disease. High levels of antibody appear in serum and in cerebrospinal fluid (CSF) only at death and in cases where illness is naturally or artificially prolonged. In the rare cases of chronic and abortive rabies virus infections, virus-specific antibodies have been detected in serum and CSF and in the brains of animals. Virus-neutralizing antibodies are produced in response to the massive amount of viral antigen (G protein, in particular) that is generated through widespread infection of the CNS and made accessible to the host's reticuloendothelial system.

The critically important cell-mediated immune response to rabies virus infection is, perhaps, the most puzzling of the host's total immune response to rabies virus infection. T_H lymphocytes, which play an essential role in supporting B cell-production of antibodies, and CTLs, which function independently in cell-mediated viral clearance, are both key responses in the cell-mediated immunity derived from viral antigens. However, in spite of their importance in viral clearance, T lymphocytes appear to be suppressed in animals infected by pathogenic street viruses. As a result of virus-induced immunosuppression, the disease increases in severity and mortality rises. The strongest indications that T lymphocytes play a role in immune protection against rabies virus infection comes from studies in which athymic (nude) mice died when infected with the attenuated Flury HEP strain that normally causes nonlethal infection of the CNS. Normal mice, similarly infected, showed no signs of the disease. Likewise, when normal mice infected with Flury HEP virus were treated with the immunosuppressive agent cyclophosphamide or antithymocyte serum, the infection became as lethal as street virus (mortality in groups of mice increased to 50% and occasionally rose as high as 100%). The mice which survived the nonvirulent virus infection developed CTLs, whereas the mice treated with immunosuppressants and mice infected with virulent virus failed to develop CTLs. Further evidence for the protective role of CTLs has come from adoptive transfer experiments, in which glycoprotein-specific CTL clones were observed to protect mice against rabies virus infection. Since the function of CTLs is to destroy target cells that display virus-induced changes on their surfaces, it is clear that a strong CTL response is essential in mice that survive rabies virus infection. It is clear from these studies and others that both B and T cells play an important role in virus clearance.

Unlike the response to virulent virus infection, immunization with attenuated live and inactivated rabies virus vaccines induces humoral and cell-mediated immune responses that develop to functional levels in 7–10 days and persist for a year or more. Vaccine-induced antibody in animals and humans is regarded as a key factor in the prophylatic protection of animals and humans, and, in addition, the immunoglobulin fraction of horse or human anti-rabies serum, which provides immediate passive immunity with proven effectiveness over the initial few weeks after exposure in humans, is recommended in postexposure treatment of humans. The importance of antibody in controlling the spread of rabies virus infections is clear when one considers that antibody is capable of effectively neutralizing virus that is present in intercellular spaces or in body fluids, and it may bind to virus expressed on the cell surface, allowing complement- or antibody-dependent cellular cytotoxicity to mediate killing of infected cells.

Antibody, however, is also implicated in an immunopathologic role in regard to the "early death" phenomenon, which is associated with rabies in animals and, possibly, humans. If animals have low levels of virus- or vaccine-induced rabies-specific antibody at the time they are challenged with rabies virus, they frequently die earlier than those without antibody. In an analogous situation, where humans appear to develop clinical rabies with shorter incubation periods as a result of postexposure immunization, compared with individuals who were exposed to rabies but not immunized, the "early death" phenomenon may also apply. Inasmuch as the mechanism of early death may resemble the immune cytolysis observed in rabies virus-infected cells treated with antiserum and complement *in vitro,* the immunopathologic effect of antibody would be to accelerate the pathogenesis of the virus.

The efficacy of postexposure immunization and long-term effects of vaccine-induced prophylaxis against rabies seems also to be linked to the stimulation of a strong CTL response. Induction of CTLs and other effector T cells during infection with live attenuated rabies virus vaccine strains and in response to immunization with inactivated virus is consistent with the observation that T lymphocytes, and CTLs, in particular, are essential for protection against a lethal dose of rabies virus. There is no evidence that early death is associated with rabies-specific T cells nor does any other form of immunopathology appear to be mediated through rabies immune T cells.

For further information regarding practical approaches toward rabies prevention, including the rationale for pre-exposure and postexposure prophylaxis, treatment of wounds, vaccines and vaccine dosage, booster vaccination, and serologic testing, the reader is referred to the recommendations of the Immunization Practices Advisory Committee (ACIP) (see Bibliography).

IX. CONTROL AND PREVENTION OF RABIES IN DOMESTIC AND WILDLIFE ANIMALS

Preventing the spread of rabies to susceptible animals and humans remains a major challenge to biologists and public health officials throughout the world. To control rabies at its source is to control the disease in wildlife vector species, and this may be the most difficult task of all. A plethora of techniques have been tried; almost every species of mammal which acts as a vector for the disease requires a somewhat unique approach depending on many factors, including density of populations, restriction of movement, food sources, and functions of the species, to prevent that species from becoming infected. On the other hand, the impact that vaccines have had in dramatically reducing the incidence of canine and human rabies, clearly holds great promise for the prevention and ultimate control of rabies in wild animals.

The main transmitters of wildlife rabies throughout the world are the wild carnivores. These animals are highly susceptible to rabies, they excrete virus in the saliva, they live in high-density populations, and the long incubation period of the virus ensures maintenance of the infection. The most significant terrestrial carnivores transmitting rabies in the world today are the red fox (*Vulpes vulpes*) in central and western Europe, the red fox and raccoon dog (*Nyctereutes procyonoides*) in eastern Europe, foxes (*Vulpes vulpes* and *Urocyon cinereoargenteus*), skunks (predominantly, *Mephitis mephitis*) and racoons (*Procyon lotor*) in various parts of North America, and wolves, stray dogs, mongooses (*Herpestes auropunctatus*), meerkats, and polar foxes in other areas and countries of Africa, Asia, and Central and South America. Establishing well-organized and well-managed programs with extensive national and international cooperation among veterinary and public health personnel who share as a common goal the control of rabies in wild animals, from local eradication to inhibition of spread into uninfected areas, will undoubtedly be more difficult than controlling rabies in domestic animals. Several European and North American team efforts, nevertheless, have made significant progress evaluating the efficacy and safety of vaccines and vaccine delivery systems for oral immunization of wild carnivores. The progress is such that it can be said that the oral immunization method can be an effective and practical approach for the elimination of enzootic rabies.

Control and prevention of rabies in domestic animals and humans has been most successful in countries that advocate strong veterinary and public health programs and that have the cooperation of the people who have the responsibility for animal pets. Nationwide vaccination programs, together with rules or laws to enforce the elimination of stray dogs and cats and use of quarantines in rabies-free countries, have done the most to keep rabies from spreading. To be effective, these measures, as well as the oral immunization approach for wild animals, must reduce the number of unvaccinated animals to a level at which rabies transmission is interrupted and this level must then be maintained.

See Also the Following Articles

Cellular Immunity • DNA Sequencing and Genomics • T Lymphocytes • Vaccines, Viral • Zoonoses

Bibliography

Baer, G. M. (ed.). (1991). "The Natural History of Rabies" (2nd ed.). CRC Press, Boca Raton, FL.

Baer, G. M., Bridboard, K., Hui, F. W., Shope, R. E., and Wunner, W. H. (eds.) (1988). *Rev. Infect. Dis.* **10** (Suppl. 4).

Bourhy, H., Kissi, B., and Tordo, N. (1993). *Virology* **194**, 70–81.

Campbell, J. B., and Charlton, K. M. (eds.). (1988). "Rabies." Kluwer Academic Publishers, Boston.

Human Rabies Prevention—United States, 1999. (1999). Recommendations of the Advisory Committee on Immunization Practices (ACIP). *Morbid. Mortal. Weekly Rep.* **48**, 1–21.

Kissi, B., Tordo, N., and Bourhy, H. (1995). *Virology* **209**, 526–537.

National Association of State Public Health Veterinarians. (1991). *J. Am. Vet. Med. Assoc.* **198**, 37–41.

Rupprecht, C. E., Dietzschold, B., and Koprowski, H. (eds.) (1994). *Curr. Topics Microbiol. Immunol.* **187**.

Smith, J. S., Orciari, L. A., Yager, P. A., Seidel, H. D., and Warner, C. K. (1992). *J. Infect. Dis.* **166**, 296–307.

Smith, J. S., Orciari, L. A., and Yager, P. A. (1995). *Seminars Virol.* **6**, 387–400.

Tordo, N., Pock, O., Ermine, A., Keith, G., and Rougeon, F. (1988). *Virology* **165**, 565–576.

WHO Expert Committee on Rabies. (1984). "Technical Report Series 709." World Health Organization, Geneva, Switzerland.

Wunner, W. H. (1987). Rabies viruses—Pathogenesis and immunity. *In* "The Rhabdoviruses" (R. R. Wagner, ed.), pp. 361–426. Plenum Press, New York.

Wunner, W. H. (1988). Monoclonal antibodies against rabies virus. *In* "Clinical Applications of Monoclonal Antibodies" (R. Hubbard and V. Marks, eds.), pp. 115–137. Plenum Publishing Corporation, New York.

Wunner, W. H., and Koprowski, H. (1989). Clinical and molecular aspects of rabies virus infections of the nervous system. *In* "Clinical and Molecular Aspects of Neurotropic Virus Infection" (D. H. Gilden and H. L. Lipton, eds.), pp. 269–302. Kluwer Academic Publishers, Boston.

Ralstonia solanacearum

Alan Christopher Hayward

The University of Queensland

I. Status and Economic Importance
II. Genotypic and Phenotypic Diversity and Diagnosis
III. Host Range and Geographical Distribution
IV. Host–Parasite Relationships, Pathogenesis, and Virulence Factors
V. Epidemiology: Modes of Dissemination and Transmission
VI. Disease Control: Host Resistance and Management by Biological and Cultural Methods
VII. Summary and Prospects

GLOSSARY

***biovar* (*formerly, biotype*)** An infrasubspecific grouping (taxon) of isolates of a species differing in biochemical or physiological properties. Infrasubspecific taxa are informal and are not governed by the Code of Nomenclature.

genotype Genetic constitution of a cell, organism, or population.

hypersensitive reaction In a general phytopathological context, a violent reaction of an organism to invasion by a pathogenic organism or virus, resulting in rapid death of invaded tissue, thus preventing further spread of infection. In phytobacteriology, the hypersensitive reaction is a differential test in which, following injection of bacterial cells (ca. 10^7 cells per ml) into green tissues, usually of tobacco, a localized, rapid-onset, dry necrotic reaction may be produced.

latent infection A symptomless or inapparent infection that is chronic and in which a certain host-parasite relationship is established.

phenotype The total properties of an organism determined by the interaction between the organism and its environment.

phylogeny Evolutionary history.

RFLP Restriction fragment length polymorphism. Restriction enzymes cleave DNA at very specific base sequences (restriction sites) into fragments of different length from different sources. These cleavage sites are polymorphic and can be revealed as RFLPs by the Southern hybridization technique.

Southern blot* (*hybridization*) *technique A very sensitive method for detecting presence among restriction fragments of DNA sequences complementary to a radiolabeled DNA or RNA sequence. Restriction fragments are separated by agarose gel electrophoresis, denatured to form single-stranded chains, and then trapped in a cellulose nitrate filter onto which a probe suspension is poured. Hybridized fragments are detected by autoradiography, after washing off excess probe.

tolerant host A plant able to endure infection by a particular pathogen without showing disease symptoms and, therefore, prone to be associated with latent infections.

virulence factor An enzyme, toxin, hormone or other product associated with appearance of a specific disease symptom or with increase in agressiveness of the pathogen in the host.

SEVERAL PHYTOPATHOGENIC prokaryotes are capable of invading the xylem vessels of host plants, causing dysfunction in water transport. Among these, the most important and widespread is the bacterial wilt pathogen *Ralstonia solanacearum.* This major pathogen was first described in 1896 by the eminent pioneer of phytobacteriology Erwin F Smith; since that time, more than 2000 papers in every major language have addressed different aspects of bacterial wilt on the many hosts which are affected.

Two other gram-negative, bacterial plant pathogens which recent work has shown to be closely allied to *Ralstonia solanacearum* are considered: *Pseudomonas syzygii,* the cause of Sumatra disease of clove (*Syzygium aromaticum*) in Java and Sumatra, and "*Pseudomonas celebensis,*" the cause of blood disease of banana (BDB), a disease known only in Indonesia.

I. STATUS AND ECONOMIC IMPORTANCE

Bacterial wilt is consistently the most important disease of tomato, with an enormous impact on subsistence farmers and large-scale horticultural production systems all over the tropical and subtropical world. Indirect losses include interference with land usage, as well as discard of susceptible crops. In tomato, as in most other susceptible crops, breeding for resistance is the cornerstone of integrated disease management systems. Even resistant lines of tomato have not maintained their resistance when grown in the heat stress of the lowland humid tropics. Another major barrier in breeding for bacterial wilt resistance in tomato has been that lines with smaller fruit size often have more resistance than large-fruited types which are preferred for fresh market consumption. The presence of *R. solanacearum* in many productive soils discourages the planting of tomatoes and other solanaceous vegetables on family farms and home gardens and this represents a significant reduction in food source.

As a pathogen of quarantine importance, *R. solanacearum* has greatest impact on banana in relation to Moko disease, caused by *R. solanacearum* race 2, and, increasingly, in potato. Brown rot of potato caused by *R. solanacearum* race 3 (biovar 2, RFLP 26) has been spread worldwide on latently infected potato tubers and, in the 1990s, has been found in the Netherlands, a major producer of seed potatoes. International trade in seed potatoes assumes zero tolerance for brown rot infection. In those countries affected by brown rot, the costs of disease surveillance and eradication are considerable. Dissemination of other strains of bacterial wilt on ginger and other vegetative propagating material in Asia is a continuing problem.

II. GENOTYPIC AND PHENOTYPIC DIVERSITY AND DIAGNOSIS

A. Phylogeny, Taxonomy, and Nomenclature

For about 80 years, the bacterial wilt pathogen was a member of the genus *Pseudomonas* Migula. Phenotypic properties, and more recently, DNA-based methods, have shown that this genus is not a natural group but is exceedingly heterogeneous. In 1973, Palleroni and coworkers showed that species of *Pseudomonas* were distributed into five homology groups on the basis of rRNA : DNA hybridization. All of the fluorescent pseudomonads allied to the type species *Pseudomonas aeruginosa* were contained in homology group I and most of the nonfluorescent plant-pathogenic pseudomonads in either homology groups II or III. Later, phylogenetic relationships were based on conserved gene sequences, in particular, of the 16S rRNA genes. All of the plant pathogenic pseudomonads were contained in the alpha, beta, and gamma subdivisions of the new class Proteobacteria, with the former homology groups II and III all in the β-subdivision. The profound phylogenetic diversity within *Pseudomonas*, together with long-known differences in carbon source utilization patterns, fluorescent pigment production, differences in cellular lipid composition, and some other phenotypic properties, was the reason for the distribution of all but the fluorescent pseudomonads and their relatives into several new genera. The former pseudomonads in the β-Proteobacteria are now contained within *Acidovorax, Burkholderia, Comamonas,* and *Ralstonia.* Changes in nomenclature of the bacterial wilt pathogen are shown in Fig. 1, and the phylogenetic relationships between species of the genus *Ralstonia* and some related bacteria are shown in Fig. 2. Although *Pseudomonas syzygii* and the BDB are phenotypically distinct from *R. solanacearum* (Table I), there is no doubt of their close phylogenetic relationship with *R. solanacearum,* which is also confirmed by comparison of the 16S–23S rRNA gene intergenic spacer region sequences and partial sequences of the polygalacturonase and endoglucanase genes.

B. Infrasubspecific Classifications and Their Uses

R. solanacearum is a heterogeneous species by all of several different measures of diversity: host range and pathogenic specialization; cultural and physiological properties; and at the molecular level using DNA-based methods. Infrasubspecific classifications

Bacillus solanacearum

E.F. Smith 1896

↓

Pseudomonas solanacearum

(Smith) Smith 1914

↓

Burkholderia solanacearum

(Yabuuchi *et al.* 1992)

Validated by publication in the

IJSB[a] 43(2): 398,1993

↓

Ralstonia solanacearum

(Yabuuchi *et al.* 1995)

Validated by publication in the

IJSB 46(2): 625,1996

[a]International Journal of Systematic Bacteriology

Fig. 1. Name changes and some earlier synonyms of *Ralstonia solanacearum.*

which define and categorize this diversity are needed in targeted plant breeding, epidemiological investigations, and quarantine.

In 1962, Buddenhagen and coworkers proposed a race classification for the bacterial wilt pathogen, the first attempt to define the diversity of *R. solanacearum* in terms of pathogenic properties. Race classification is understood by plant pathologists to refer to physiological races of a parasite (particularly fungi; e.g., the wheat stem rust fungus) characterized by specialization to different cultivars of one host species, whereas, for *R. solanacearum,* races were based primarily on differences in host range and pathogenicity and, to a lesser extent, differences in colony form and pigmentation. Race 1 strains were pathogenic on solanaceous hosts, diploid bananas, and a variety of nonsolanaceous hosts; race 2 strains were restricted to triploid banana and *Heliconia* spp., and race 3 strains primarily affected potato, occasionally tomato, and a few weed hosts, such as *Solanum nigrum* and *S. dulcamara.* This race classification is widely used but was introduced long before the genetic diversity and phylogeny of this species was fully understood. Later systems of classification into biotypes, later called biovars, on the recommendation of the International Code of Nomenclature of Bacteria, to avoid the use of "type" except in the context of nomenclatural type, were based on differ-

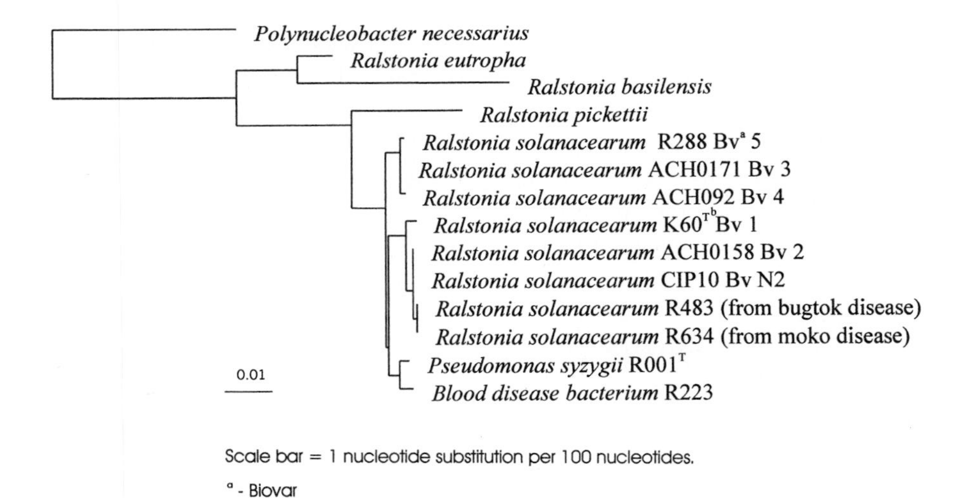

Fig. 2. Phylogenetic relationships of *Ralstonia solanacearum* and related bacteria based on 16S rDNA sequences (courtesy of Dr. Mark Fegan).

TABLE I
Comparison of *Pseudomonas syzygii*, the BDB of Banana, and *Ralstonia* spp.

	P. syzygii	*BDB*	*R. solanacearum*	*R. pickettii*
Colonies on TTC[a] medium	Tenacious, minute	viscid, <5 mm	Fluidal, >5 mm	ND[b]
Motility	–	–	v[c]	+
Growth at 37°C	–	+	+	+
Growth at 41°C	–	–	–	+
NaCl tolerance	<1%	<1.5%	<2.0%	ND
PHB inclusions[d]	ND	+	+	+
Arginine dihydrolase	–	–	–	–
Oxidase	+	+	+	+
Nitrite from nitrate	v	–	+	+
Gas from nitrate	–	–	v	+
Nutritional versatility[e]	+	++	+++	++++
Tobacco HR[f]	v	+	+[g]	ND
Plant pathogenicity and host associations	Clove	Banana	*Solanaceae*, *Musaceae*, etc.	Bacteremia in humans; intracellular growth in *Acanthamoeba* spp

[a] Triphenyl tetrazolium chloride.
[b] ND, not determined.
[c] Variable reaction between different isolates.
[d] Poly-β-hydroxybutyrate.
[e] Based on relative percentage of carbon sources utilized for growth.
[f] Hypersensitive reaction.
[g] Systemic infection with isolates from tobacco.
Based in part on Eden–Green Table 3.1, *in* "Bacterial Wilt: The Disease and Its Causative Agent, *Pseudomonas solanacearum*" (A. C. Hayward and G. L. Hartman, (eds.), p. 28. CAB International, Wallingford, UK.

ences in carbon source utilization and nitrate metabolism. A major advance occurred in 1989 when Cook and coworkers classified isolates of *R. solanacearum* into more than 40 RFLP groups (also called Multi Locus Genotypes), based on hybridization of DNA probes specifying virulence or the hypersensitive response to restriction fragments of genomic DNA using the Southern blot technique. The RFLP groups were separable into two divisions, suggesting an evolutionary divergence, which was later confirmed when the same two divisions were apparent on comparison of nucleotide sequences in the 16S rRNA genes. The metabolically more versatile biovars 3, 4, and 5 in Division 1 were separated from the metabolically less versatile biovars 1, 2, and N2 in Division 2.

The interrelationship of the different systems, which have their uses in certain contexts and limitations in others, are shown in Fig. 3. DNA-based methods provide the best measure of genetic diversity and have the level of discrimination required to detect new genotypes intercepted in quarantine, and the potential to limit further expansion of distribution on tubers and other planting material worldwide. Some genotypes have evolved in isolation for milennia; in other cases, different genotypes have coevolved in the same soil. The latter provides opportunity for lateral gene transfer, which may have occurred between biovar 1 (Division 2) and biovar 3 (Division 1).

The contribution of different approaches to classification and disease diagnosis is shown in the example of the introduction (and subsequent eradication) of Moko disease on *Heliconia* to Cairns, North Queensland, in 1989.

Because of the possibility that *Heliconia* could act as a carrier of the Moko disease pathogen, introductions into Australia are carefully examined in postentry quarantine and must be accompanied by a phytosanitary certificate. Following the discovery of *R. solanacearum* on *Heliconia* on the island of Oahu,

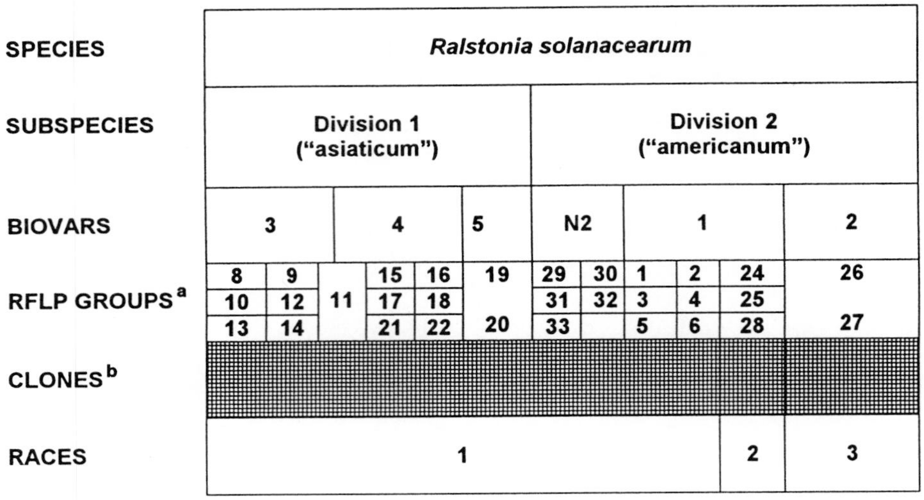

SPECIES	***Ralstonia solanacearum***											
SUBSPECIES	**Division 1 ("asiaticum")**						**Division 2 ("americanum")**					
BIOVARS	**3**		**4**			**5**	**N2**		**1**			**2**
RFLP GROUPS[a]	**8**	**9**	**11**	**15**	**16**	**19**	**29**	**30**	**1**	**2**	**24**	**26**
	10	**12**		**17**	**18**		**31**	**32**	**3**	**4**	**25**	
	13	**14**		**21**	**22**	**20**	**33**		**5**	**6**	**28**	**27**
CLONES[b]												
RACES	**1**										**2**	**3**

Fig. 3. Generalized scheme for infrasubspecific relationships within *Ralstonia solanacearum.* [a]The number of RFLP groups has since been expanded from 33 to 46. (Based on Gillings and Fahy, *ACIAR Proceedings* **45,** p. 86, 1993) [b]Isolates of the same genotype which may be distinguishable by pathogenicity or aggressiveness on a susceptible host.

Hawaii, in 1989, *Heliconia* imports into Australia were suspended and previously imported Heliconias examined for symptoms. *R. solanacearum* was isolated from five plants among a batch of 367 rhizomes imported from Oahu during April, 1989; three of the plants showed symptoms, four were still in post-entry quarantine houses, one plant had been released from quarantine. Significantly, the colonies on isolation plates were slower growing and different in appearance from those of *R. solanacearum* biovar 3, which is endemic on solanaceous and other hosts, including *Heliconia,* near Cairns, North Queensland, where the imported *Heliconia* plants were held under quarantine. Biochemical tests showed that the isolates did not utilize the hexose alcohols mannitol, sorbitol, and dulcitol or oxidize the disaccharides lactose, maltose, and cellobiose and, therefore, conformed to biovar 1, which had never previously been encountered in Australia. Observations to this point confirmed the importation of a phenotype previously unknown in Australia, but because biovar 1 does not include the Moko pathogen alone but also includes many race 1 strains, further work was necessary to identify the isolates as belonging to race 2. Later, pathogenicity to triploid bananas was confirmed. The technique of genomic fingerprinting using several restriction enzymes was applied to a panel of biovar 1 isolates known to belong to either race 1 or race 2. Race 1 isolates generated diverse genomic fingerprints, whereas the race 2 isolates produced 3 genomic fingerprint patterns corresponding to RFLP groups 24, 25, and 28 (Fig. 3) of Cook and coworkers, all associated with Moko disease in banana and plantain. The isolates from *Heliconia* imported into Australia were highly similar to RFLP 28 isolates collected from Moko disease of banana and plantain in Venezuela and Trinidad in the 1960s. Integration of information from all sources enabled a correct diagnosis to be made.

There is a phylogenetic dichotomy evident in the two divisions of *R. solanacearum* established on the basis of sequencing of the 16S rRNA genes (Fig. 2) and RFLP analysis (Fig. 3), which has led some workers to propose informally that Divisions 1 and 2 be regarded as subspecies, with names reflecting geographical location. Biovars 3, 4, and 5 are widely distributed in Asia and Australasia and unknown in most parts of the Americas. Biovar 3 occurs in northeastern Brazil, in parts of Central America and the West Indies, and southern Africa. *R. solanacearum* race 3 (biovar 2, RFLP 26), causing brown rot of potato, is believed to have evolved on potato in the Andes of Bolivia and Peru and to have been distributed worldwide on latently infected potato tubers.

III. HOST RANGE AND GEOGRAPHICAL DISTRIBUTION

The recorded host range of *R. solanacearum* is wide and diverse and is exceeded in number only by *Agrobacterium tumefaciens* among plant pathogenic bacteria. Representatives of more than 50 families of plants include one or more hosts; most are dicotyledons but banana (*Musa* spp.) and ginger (*Zingiber officinale*) and their relatives are major hosts in some parts of the world. Hosts in the *Solanaceae* include tomato, potato, tobacco, eggplant, and pepper. Leguminous hosts include groundnut (peanut), a major host in Indonesia and China, and, to a lesser extent, in the United States and some other countries. There are many solanaceous and composite weed hosts and some of these may be latently infected. Bacterial wilt has been recorded on various woody, perennial plants of economic importance in horticulture and in forestry, and most of these records have been in Asia and Australasia rather than in the Americas. The disease has been reported on custard apple (*Annona* spp.) only in Queensland, Australia, and eastern Taiwan. Cashew (*Anacardium occidentale*) is affected by bacterial wilt in Indonesia and is one of several examples of a host unknown near to its center of origin and major area of production. Cashew is not affected by bacterial wilt in Brazil though grown in parts of that country where the disease is endemic on solanaceous hosts and environmental conditions are highly conducive to the disease. The reasons for the erratic occurrence of bacterial wilt on certain hosts in some parts of the tropics and subtropics where the disease is endemic, but not in other areas where the disease is endemic, are not known. There may be differences in the environment, particularly temperature and rainfall or conducive biotic or abiotic soil factors, or the difference may be attributable to strain differences in the pathogen. There may be high soil populations of the pathogen due to wilt on a weed host closely related botanically to a crop plant. Bacterial wilt of Cassava, a member of the *Euphorbiaceae,* has been recorded only in Indonesia and was first seen in a production area where there was a high incidence of wilt on the euphorbiaceous weed *Croton hirtus.*

Bacterial wilt is of most importance and occurs on the greatest diversity of economic plants and weed hosts in the lowland humid tropics. There are few parts of the tropics and subtropics unaffected by this disease. Most records are within latitudes 40°N and 40°S with high summer temperatures and rainfall; further north and south, almost all records are of *R. solanacearum* race 3 (biovar 2, RFLP group 26), the major cause of brown rot worldwide. There has been an expansion of the distribution of this pathogen in both hemispheres due to importation of latently infected seed and, particularly in western Europe, the utilization of ware potatoes domestically and for processing. In South America, French has described how this pathogen has been spread since the early 1970s as far as 43°S, with rapid spread from southern Brazil to Uruguay, Argentina, Chile, and Bolivia in just two decades.

IV. HOST–PARASITE RELATIONSHIPS, PATHOGENESIS, AND VIRULENCE FACTORS

The early stages of pathogenesis involve host recognition and adhesion, overcoming of host defenses and production of virulence factors. In bacterial wilt, most progress has been made in understanding of the latter aspect.

R. solanacearum colonizes exudation sites at root tips or secondary root axils. Infection occurs through wounds in roots or at points of secondary root emergence; later, the intercellular spaces of the root cortex and vascular parenchyma are colonized and cell walls are disrupted, facilitating spread through the vascular system. With onset of wilt, populations can reach as high as $>10^{10}$ cells per stem.

A. Virulence Genes

Five sets of genes have been investigated to determine their role in the host–parasite interaction:

genes affecting host range, extracellular enzyme production, exopolysaccharide production, general regulatory functions, and the *hrp* gene cluster. The latter refers to *h*ypersensitive *r*esponse and *p*athogenicity genes, which are known to occur in all of the gram-negative phytopathogenic bacteria except *Agrobacterium tumefaciens*. In *R. solanacearum*, the *hrp* gene cluster spans more than 23 kb and is organized in five transcriptional units, which altogether code for 20 *hrp* genes. The *hrp* region is essential for induction of a hypersensitive reaction in tobacco leaves and for the elicitation of disease symptoms on host plants. A gene has been discovered which regulates the expression of three of the transcriptional units. The same gene also regulates the expression of *popA*, the gene product of which acts as an elicitor of the HR reaction. There are many common functions of *hrp* genes in the genera *Pseudomonas, Xanthomonas, Erwinia,* and *Ralstonia,* as well as some differences. Some are concerned with the secretion of virulence factors and proteins across the outer membrane and may be involved in the injection of proteins into plant cells.

R. solanacearum produces many extracellular proteins (EXPs) of varied function; among these, the exoenzymes degrading cell walls have been investigated in relation to pathogenesis. None appears to be absolutely required for disease; they may have a minor effect or increase aggressiveness and may be much more critical during root invasion and spread into the vascular system. Mutants in the *eep* locus, affecting export of most major proteins, greatly reduce the ability of *R. solanacearum* to infect via roots and to colonize and kill plants. Site-directed mutants deficient in endopolygalacturonase (PglA), exopolygalacturonase B (PglB), pectin methylesterase (Pme), or endoglucanase (Egl) can still cause disease, but take up to 50% longer to wilt and kill.

The major known virulence factor of *R. solanacearum* is an acidic, high molecular mass exopolysaccharide, which is released extracellularly in large quantities. EPS is a requirement for pathogenesis, but alone is insufficient for wilting and killing; its major effect is plugging of the xylem vessels rather than as a requirement for significant root invasion or multiplication *in planta*.

B. Phenotype Conversion and Regulation of Virulence

In the 1950s, Kelman showed that isolates of *R. solanacearum* typically underwent a change in colony form from fluidal to butyrous, which was correlated with coordinate loss of virulence, production of EPS, reduced endoglucanase activity, increased endopolygalacturonase activity, and motility. Later, it was shown that these normally spontaneous pleiotropic mutants could be mimicked by spontaneous inactivation of the *phcA* locus and that a functional copy of *phcA* is required by *R. solanacearum* to maintain the wild-type phenotype. The phenomenon has become known as phenotype conversion (PC) and the avirulent colonies, as PC mutants. Since the discovery of this regulatory gene, at least 12 others have been discovered, which are involved in control of distinct, but sometimes overlapping, sets of virulence genes.

V. EPIDEMIOLOGY: MODES OF DISSEMINATION AND TRANSMISSION

There are varied means of spread of bacterial wilt which are of different importance in different hosts and which may be unique to certain hosts and localities. Although bacterial wilt has been generally known as a soil-borne disease, there are instances where very little is known about the soil phase of the disease. An example is Bugtok disease of the cooking bananas saba and cardaba (ABB or BBB types) in the Southern and Central Philippines, caused by *R. solanacearum* race 2, where transmission is apparently by insects to the inflorescence. Symptoms are confined to the fruit, and there is limited penetration into the pseudostem so that daughter suckers are rarely affected and the disease is not thought to be spread in planting material. Whether there is a significant soil phase of this disease requires further investigation.

Brown rot of potato caused by *R. solanacearum* race 3 (biovar 2, RFLP 26) is believed to have been disseminated on latently infected seed potatoes worldwide from the center of origin of the potato and of this pathogen in South America. In Western Europe, introduction and dissemination of the disease may be the result of the repeated importation of potatoes from the Mediterranean region for home consumption and for processing. Thorough investigations in Sweden in the 1970s showed that the disease could be spread from river water downstream from a potato processing plant. Water used for overhead irrigation was shown to be the most probable source of infection in potato fields but an important factor in maintenance of the pathogen between seasons was the semi-aquatic weed known as bittersweet (*Solanum dulcamara*). Latent infections of the bacterial wilt pathogen occur in the adventitious roots extending from river banks. The importance of *S. dulcamara* as latent weed host has been confirmed in the European Union since emergence of brown rot in the early 1990s, in the Netherlands, Belgium, France, Italy, the United Kingdom, and several other countries. The use of overhead sprinkler irrigation from contaminated water courses and maintenance of populations in *S. dulcamara* have been common factors in several outbreaks, whereas, in other instances, outbreaks have been attributed to importation of contaminated seed potatoes. There are special features of disease biology in western Europe not applicable elsewhere; for example, in Australia, the weed *S. dulcamara* is unknown, except in Tasmania. Present evidence does not support the existence of a specifically European strain of *R. solanacearum;* rather, there appear to have been several introductions and reintroductions over the many decades in which brown rot-infested ware potatoes have been imported into western Europe. Isolates of *R. solanacearum* race 3 causing brown rot in Europe are highly similar to those found worldwide. Whole genome analysis using pulsed field gel electrophoresis does provide some evidence of different clonal lines which have been successively introduced.

The various modes of dissemination and transmission of bacterial wilt are shown in Table II. Mechanical transmission is notable in the example of *Perilla crispa* in Taiwan, where the disease is transmitted during repeated harvesting of leaves, rather than soilborne. Perilla plants rarely wilted when planted in infested soil, although roots and stems were heavily colonized with *R. solanacearum*. Symptomless carrier plants also served as an inoculum source for the spread of the pathogen to healthy *Perilla* plants during harvesting. In Japan, cucumber plants were not susceptible by artificial inoculation but wilted when grafted on to a susceptible pumpkin rootstock.

TABLE II
Dissemination and Transmission of Bacterial Wilt Caused by *Ralstonia solanacearum*

Dispersal on plant material locally and internationally
On banana corms, e.g., Central America to the West Indies; Central America to the Philippines
On ginger rhizomes, e.g., in Indonesia, Malaysia, China and elsewhere in Asia
On Heliconia rhizomes, e.g., within Central America and the West Indies; from Hawaii (USA) to Australia and to India
On potato tubers for direct consumption or processing, e.g., Mediterranean region to western Europe; on latently infected potato tubers for planting
On tomato transplants, e.g., southeastern United States to Canada
On seed of groundnut (and possible other hosts, such as tomato)
Field transmission and dispersal
Mechanical transmission by pruning knives on banana
During successive mechanical harvesting of *Perilla crispa* in Taiwan
By clipping with scissors or a rotary mower on tomato
Insect transmission on banana from diseased to healthy inflorescences
Root-to-root transmission on various hosts
Splash dispersal of leaf surface populations by wind and rain on tobacco (and possibly other hosts)
By root wounding during cultural practices
By root knot nematodes (*Meloidogyne* spp)
Spread by movement of contaminated soil, e.g., in flooding or by irrigation water
Spread by people on shoes, farm implements, and machinery

[Based on Table 1 *in* "Pathogenesis and Host Specificity in Plant Diseases" (U. S. Singh, R. P. Singh, and K. Kohmoto, eds.), p. 141. Elsevier Science, UK.]

VI. DISEASE CONTROL: HOST RESISTANCE AND MANAGEMENT BY BIOLOGICAL AND CULTURAL METHODS

Approaches to control of bacterial wilt can be best illustrated by reference to two hosts, potato and tomato. Latent infections in tolerant cultivars are a major factor in the control and dissemination of the disease in both hosts, and resistance is prone to breakdown at higher ambient temperatures. Plant breeding and selection of resistant cultivars form a major part of integrated disease management practices.

There are many components in integrated disease management, as illustrated in Table III, which shows how the weighting given to a particular strategy for control of brown rot of potato depends on the strain of the pathogen and the environment. French has emphasized that the strategy is site-specific and must take into account the socioeconomic factors that influence decision-making in the local farming community. It is acknowledged that dissemination of the disease has occurred on latently infected potato tubers and that some cultivars have been more prone than others to harbor latent infections. Breeding programs are now taking into account the need to select breeding lines which do not carry latent infections.

Control of bacterial wilt of tomato is a greater challenge than it is in potato because of the demand for cultivation in the lowland humid tropics, where the inoculum pressure from wilted weed hosts and crop plants is uninterrupted and the environmental conditions are highly conducive for disease expression. Fortunately there is no evidence that the low temperature-adapted race 3 of *R. solanacearum,* which has spread to temperate latitudes, is affecting commercial tomato production, although it has the potential to do so in the tropics and subtropics. Breeding for resistance in tomato has proceeded worldwide, but breakdown of resistance has been frequently reported in tomato cultivars grown away from the original breeding areas. Whether resistance breakdown is due to environmental causes or genetic diversity among populations of *R. solanacearum* is

TABLE III
Factors to be Weighted in Developing a Strategy for Bacterial Wilt Control with Either Race 3 (left) or Race 1 (right)

Race 3 (potato strain)	*Factors to be rated to formulate a control strategy*	*Race 1 (solanaceous strain)*
3	Resistance or tolerance	2
2	Cold climate	1
3	Healthy seed (tubers, cuttings, TPS[a])	3
7	*R. solanacearum*-free soils	7
4	Suppressive soils	2
4	Short rotation	1
3	Intercropping	2
3	Date of planting	1
2	Nematode control, resistance	4
2	Dry/heat soil	3
1	Solarization	1
4	Roguing volunteers	2
2	Roguing wilted plants	1
5	Fumigants	3
2	Control of spread in water	2
1	Minimal till	2
1	Soil amendments	1

Figures, ranging from 1 to 7, may be modified according to judgment for each location. A sum of 10 would usually be adequate for good control or even eradication.

[a] True potato seed.

[From French, Table 14.1, *in* "Bacterial Wilt: The Disease and Its Causative Agent, *Pseudomonas solanacearum*" (A. C. Hayward and G. L. Hartman eds.), p. 204. CAB International, Wallingford, UK.]

largely unknown but is being intensively investigated. Temperature is an important factor in resistance breakdown; differences of 2–3°C in average minimum and maximum soil temperatures can make a profound difference to disease incidence, but there is a differential interaction with strain of the pathogen, as with potato. Other biotic factors are important, such as the well-known breakdown of resistance due to the synergistic interaction between *R. solanacearum* and root knot nematodes. In response to these difficulties, international collaboration has increased; multilocational trials across the tropics and subtropics are in progress and are giving valuable information.

In tomato, a more analytical approach to host × pathogen and host × pathogen × environment interactions is leading to greater understanding. Resistance does not result from a physical barrier to root penetration but depends on the ability of the host plant to limit colonization and progression of the pathogen. The ability to colonize is a function of the aggressiveness of the pathogen, which is the outcome of both plant × environment and plant × pathogen interactions. An important advance by Prior, Grimault, and others in the French West Indies has been to show that, in field trials, all symptomless plants are latently infected at the collar level, and that the percentage of symptomless plants with bacteria at the midstem level was significantly correlated with the degree of resistance; the more resistant, the lower the stem colonization. Rather than basing selection of breeding lines on the presence or absence of wilt, as in the past, the use of a *Colonization Index* as a measure of bacterial invasiveness is advocated, and defined here.

$N_{wp} + (N_s \times R_s)$, where

N_{wp} = percentage of wilted plants

N_s = percentage of asymptomatic plants

R_s = percentage of those asymptomatic plants sampled from which *R. solanacearum* was recovered.

The mechanism of resistance in tomato is unknown. There is some evidence of a relationship to calcium nutrition; highly resistant cultivars were characterized by high calcium uptake and pathogen populations in stems decreased with increased calcium concentrations and degrees of resistance. However, the basis of the calcium effect is not understood.

Soil amendment with CaO (5000 kg ha^{-1}) and urea (200 kg ha^{-1}) has given encouraging results in some soils, but has not been reproduced in others. Biofumigation is a form of amendment for disease control where buried crop residues release biocidal substances, as in the example of the biocidal isothiocyanates from *Brassica* spp crop residues, which have been shown to reduce bacterial wilt populations in soil.

Biological control with antagonistic microorganisms and plants has been extensively investigated. Mutants of *R. solanacearum* which are invasive but non-wilt inducing, which compete for vascular colonization of xylem vessels, and which induce host resistance, have given promising results. French marigolds (*Tagetes patula* L) are directly antagonistic to bacterial wilt populations in soil as a result of the excretion from roots of biocidal thiophenes. Other plants which have a suppressive effect in the rhizosphere need to be identified for use in crop rotation. Crop rotation as a general control measure may be successful, depending on the strain of pathogen involved and the environment, but much more needs to be known about the role of latently infected host plants and latently infected weed hosts, and of so-called nonsusceptible hosts used in rotation, which may constitute an underestimated reservoir of inoculum from which the pathogen cannot be eradicated.

VII. SUMMARY AND PROSPECTS

DNA-based methods have brought many advances in understanding of genetic diversity. They have enabled development of highly sensitive and discriminatory approaches to identification for use in plant health management and in plant breeding. Automatic sequencing of DNA and RNA has enabled comparison of sequences of the 16S rRNA genes and shown the phylogenetic dichotomy in *R. solanacearum*. Nomenclature of the bacterial wilt pathogen is now based on a firm phylogenetic framework and is unlikely to undergo further significant change. Below species level, the establishment of subspecies is probably justified. It is likely that both *P. syzygii* and the

BDB will eventually become members of *Ralstonia.* The same molecular tools which have aided diagnosis have the potential to assist in studies on the ecology of the bacterial wilt pathogen.

In the future, progress in understanding bacterial wilt and development of control methods depends upon a balance of basic and applied research to which interacting networks of workers contribute. Because of the importance of bacterial wilt worldwide, workers in Europe and the United States have obtained continuity of support for basic research on the genetics of virulence and pathogenicity. It is much less easy to support research in the lowland humid tropics, where bacterial wilt is a major problem in the field. The greatest deficiencies in knowledge are in the ecology of bacterial wilt, particularly in soil microbiological aspects and the role of weed hosts and of nonhost crops. Latent infection and tolerance have been demonstrated in tomato and have led to the development of more discriminatory methods of selecting lines of potential value in plant breeding. More needs to be known about latently infected weed hosts and nonhost crop plants and their contribution to disease biology. Some soils and, evidently, some plants are suppressive to bacterial wilt. Much more needs to be known about the multiplicity of biotic and abiotic factors in soils, and in the environment generally, which affect bacterial wilt so that crop rotations and integrated disease management can be rationally based.

Predictive models indicate that global warming may have profound effects on the global distribution and importance of some pests and pathogens. In the light of the known breakdown of some host–plant resistance under higher ambient temperatures, bacterial wilt is likely to be one of the diseases capable of expanding its latitudinal distribution in the future.

See Also the Following Articles

Agrobacterium • Biological Control of Weeds • *Pseudomonas*

Bibliography

Boucher, C. A., Gough, C. L., and Arlat, M. (1992). Molecular genetics of pathogenicity determinants of *Pseudomonas solanacearum* with special emphasis on *hrp* genes. *Annu. Rev. Phytopathol.* **30**, 443–461.

Buddenhagen, I. W., and Kelman, A. (1964). Biological and physiological aspects of bacterial wilt caused by *Pseudomonas solanacearum. Annu. Rev. Phytopathol.* **2**, 203–230.

Denny, T. P. (1995). Involvement of bacterial polysaccharides in plant pathogenesis. *Annu. Rev. Phytopathol.* **33**, 173–197.

Hayward, A. C. (1991). Biology and epidemiology of bacterial wilt caused by *Pseudomonas solanacearum. Annu. Rev. Phytopathol.* **29**, 65–87.

Hayward, A. C., and Hartman, G. L. (eds.). (1994). "Bacterial Wilt: The Disease and its Causative Agent, *Pseudomonas solanacearum.*" CAB International, Wallingford, UK.

Hsu, S. T. (1991). Ecology and control of *Pseudomonas solanacearum* in Taiwan. *Plant Prot. Bull.* (Taiwan) **33**, 72–79.

Kelman, A. (1953). The bacterial wilt caused by *Pseudomonas solanacearum. North Carolina Agricultural Experiment Station Technical Bulletin* **99**, Raleigh, NC.

Prior, P., Allen, C., and Elphinstone, J. (eds.). (1998). "Bacterial Wilt Disease: Molecular and Ecological Aspects." Springer-Verlag, Berlin, Heidelberg, and INRA, Paris.

Schell, M. A. (1996). To be or not to be: How *Pseudomonas solanacearum* decides whether or not to express virulence genes. *Eur. J. Plant Pathol.* **102**, 459–469.

recA: The Gene and Its Protein Product

Robert V. Miller
Oklahoma State University

I. Physiology of *recA* and the RecA Protein
II. Phenotypes Associated with *recA* Mutations
III. Physical Structure of the RecA Protein
IV. Evolutionary Conservation of *recA* and the RecA Protein

GLOSSARY

conjugation Unidirectional transfer of plasmid and/or chromosomal DNA in a parasexual process requiring cell-to-cell contact between a donor and recipient bacterium; the genetic information required for conjugation is encoded in transfer-proficient (Tra^+) plasmids resident in the donor cell; strains in which the plasmid has integrated into the bacterial chromosome may transfer chromosomal DNA efficiently and are often referred to as Hfr strains.

coprotease Protein that acts to stimulate the autoproteolytic activity of another protein.

din genes Genes regulated as part of the DNA Damage Inducible Network, or SOS regulatory network, of gene regulation; usually only applied to genes of unknown function; when function is identified, the *"din"* designation is replaced by a more appropriate, function-related, name.

genetic linkage Association of genes in inheritance due to their localization on the same chromosome or other nucleic acid molecule.

LexA Product of the *lexA* gene of *Escherichia coli* and other bacterial species; responsible for repression of the SOS Regulatory Network.

lysogeny Heritable, stable potentiality of bacteria to produce and release bacteriophages, due to the presence of a prophage (viral genome) within the bacterial cell.

postreplication DNA repair DNA repair mechanism using homologous recombination to insert complementary DNA from a sister chromosome into gaps associated with inhibition of DNA replication at or near DNA damage-induced lesions.

prophage Genome of a temperate bacteriophage resident in a lysogenic bacterium in a stable, heritable manner; expression of the viral genes is repressed but, in many cases, may be induced by exposure of the host to various stress-producing agents, such as DNA-damaging ultraviolet light or mitomycin-C.

recombination Process (biochemical pathway) that leads to the interaction of elements of nucleic acid, resulting in a change of genetic linkage of genes or parts of genes.

SOS regulatory network Group of at least 15 *Escherichia coli* operons, located at different sites in the genome, that are coordinately regulated by the LexA and RecA proteins and induced in response to exposure of the cell to DNA-damaging agents; SOS-like regulons have been identified in several other bacterial species.

synapsis That portion of the recombinational process that positions homologous molecules of DNA to bring them into alignment (register).

transduction Exchange of genetic material mediated by bacteriophages; in specialized transduction, only specific genes can be transferred, whereas in generalized transduction, many genetic elements (plasmids or chromosomes) can be transferred between bacterial cells.

transformation Acquisition of DNA by a living cell from the external milieu; both plasmid and chromosomal DNA can be acquired by cells in a unique physiological state, known as competence; competence functions are chromosomally encoded and only a limited number of species are known to develop natural competence for transformation.

umuDC A genetic operon found on the *Escherichia coli* chromosome, which alters DNA Polymerase III in such a way that fidelity is relaxed and synthesis can proceed past certain types of DNA damage, such as cyclobutane dimers; this relaxation leads to an increase in mutation rate, so that the process is often referred to as "mutagenic repair;" the *umuD* gene product must be processed by the RecA protein (acting as a coprotease) for mutagenic repair to occur; many species that have *recA* (and even an SOS regulatory network) do not have functional mutagenic repair because they lack a functional *umuD* gene.

THE *recA* GENE encodes a protein, the RecA protein, which is central to the processes of general homologous recombination and DNA repair in bacteria. It is one of the most highly conserved genetic elements to be found among the eubacteria.

RecA proteins have been identified in numerous bacterial species as molecules with molecular weights ranging from 35,000 to 45,000 and carrying out several functions in common. The primary role of RecA in recombination is to direct the synapsis of homologous DNA elements leading to the reassortment of linked genetic elements. This family of proteins is unique among prokaryotic proteins in its ability to carry out these reactions. The *Escherichia coli recA* gene product is also involved in several aspects of DNA repair. It acts as both a regulatory and functional element in several of the biochemical pathways involved in the elimination of damage from the bacterial genome. As such, it is the positive effector of the SOS regulatory network of *E. coli* and many other bacteria species. In addition to gene activities directly associated with DNA repair, the SOS regulon is responsible for the control of various activities involved in virus–host interaction, cell division, bacteriocin expression, and restriction modification.

The *recA* gene probably is universally present in bacteria. A structural and functional analog of the *E. coli recA* gene has been identified in every species of eubacteria in which an intensive investigation has been carried out. To date, the *recA* gene of more than 88 bacterial species has been identified and sequenced. The high level of structural and functional conservation among the protein products of these *recA* genes from highly divergent bacterial species suggests that this gene originated very early in the evolution of the prokaryotes. A protein with synaptase activity like that of RecA was most certainly instrumental in combining nucleic acid sequences to form the first genomes and may have been required for recruitment of nucleic acid sequences into transcriptional units coding for functional proteins. Alternatively, one authority has suggested that the RecA protein originated as part of a "sloppy replication process." In this scheme, the *recA* gene product is envisioned to have acted as a single-stranded DNA (ssDNA)-binding protein, facilitating separation of parental DNA strand during replication. This role is now taken by a single-stranded binding (ssb) protein in the bacterial replication machinery. The ssb protein also acts synergistically with RecA protein in recombination, with the two proteins forming cofilaments. The gene's DNA and resultant amino acid sequences have recently been used to develop phylogenetic relationships among the eubacteria which closely mirror those identified by ribosomal RNA sequence analysis.

The role of the *recA* gene product in homologous recombination and DNA-damage repair has inspired numerous searches for *recA* analogs in a wide variety of prokaryotic and eukaryotic organisms. Several functional analogs of RecA have been found in eukaryotes. When compared to prokaryotic sequences of RecA, these proteins are not as highly conserved but many share a 20–30% amino acid identity to RecA over a conserved core region. (By contrast, prokaryotic RecAs share about 55–100% amino acid identity throughout their length.) It is interesting to note that the RecA-like protein of at least one archaeon, *Sulfolobus solfataricus*, has been identified. This protein shares significantly more amino acid identity with the eukaryotic RecA-like proteins than it does with the eubacterial RecA sequences. In addition to these RecA-like proteins, several other proteins are found in eukaryotic cells that are capable of strand transfer activities. Many of these proteins appear to operate by different molecular mechanisms, and they probably have a divergent origin from the eukaryotic RecA-like proteins.

I. PHYSIOLOGY OF *recA* AND THE RecA PROTEIN

A. Role in Generalized Recombination: Synaptase Activity of the RecA Protein

The RecA protein is central to the process of homologous recombination in eubacteria. *E. coli recA* mutants are completely devoid of generalized recombinational activity for most substrates, and the RecA protein is required for recombination of chromosomal DNA via any of the biochemical pathways of recombination known to exist in eubacteria (although some bacteriophages' genomes encode functional analogs of the RecA protein).

The first hint of the role of the *recA* gene product in this process came from biochemical studies that revealed that the *E. coli* RecA protein promotes adenosine triphosphate (ATP)-dependent annealing between ssDNA and double-stranded DNA (dsDNA) molecules that have complementary sequences of nucleotides. This process leads to the formation of a "D Loop," in which the ssDNA invades the duplexed dsDNA molecule. The ssDNA pairs with the complementary region of the dsDNA molecule and displaces the homologous strand (Fig. 1a). Other ATP-dependent reactions catalyzed by the RecA protein also highlight its role as the synaptic protein or synaptase of bacterial generalized recombination. They include (1) the annealing of two complementary ssDNA molecules to form a dsDNA (Fig. 1b), (2) the conversion of complementary circular single-stranded and linear double-stranded molecules to a linear ssDNA and a circular dsDNA (Fig. 1c), and (3) the formation of "Holiday junctions" between two double-stranded linear DNA molecules (Fig. 1d). ATP is hydrolyzed in each of these processes. Thus, the primary role of the RecA protein in pathways of genetic recombination is to promote ATP-dependent homologous pairing and strand exchange between the participating DNA molecules.

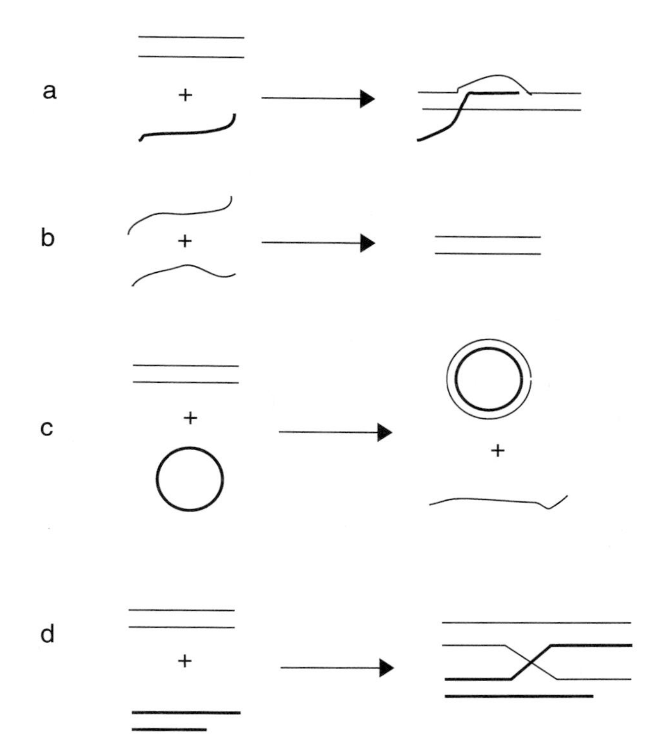

Fig. 1. Synaptic reactions of the RecA protein. The RecA protein acts as a synaptase in generalized homologous recombination. It carries out several reactions in which single-stranded DNA (ssDNA) is annealed to homologous ssDNA or double-stranded DNA (dsDNA). These reactions require adenosine triphosphate (ATP), which is hydrolyzed. In the reaction, the ssDNA becomes coated with a polymer of RecA protein. (a) Formation of "D Loop." ssDNA invades a homologous dsDNA helix, base-pairs with its complement, and displaces the identical strand from the dsDNA. (b) Annealing of two homologous ssDNA molecules. (c) Conversion of a complementary circular ssDNA and linear dsDNA to a linear ssDNA and a circular dsDNA. (d) Formation of a "Holiday junction." This is the most likely structure found in "cross-overs," which initiate the recombinational event.

The exchange reaction is initiated in a presynapsis stage that involves the binding of RecA protein to the single-stranded molecule to form a RecA-coated DNA, which is capable of participation in strand exchange. *In vitro,* the exchange rates are maximal when the reaction contains a ratio of one RecA protein monomer for every three to six nucleotides in the ssDNA. The RecA polymer formed in this process requires interaction of the RecA monomer with both the ssDNA molecule and the adjacent RecA protein molecules. When this complex is formed, conjunction occurs, which is the pairing of the RecA–ssDNA complex with the double-stranded molecule. This initial interaction does not require sequence homology but is followed by homologous alignment in the third stage of the reaction. Here, the coaggregation of the ssDNA with several double-stranded molecules acts to promote three-dimensional diffusion. This is accomplished by increasing the local concentration of DNA involved in the search process. When homologous alignment of the ssDNA with an appropriate dsDNA molecule is achieved, the RecA protein catalyzes the initial exhange of DNA strands to form a joint molecule. This junction is then extended by the branch migration process. This reaction proceeds in a polar manner ($5' \rightarrow 3'$ with respect to the incoming ssDNA). Exchange occurs at a rate of $\cong$20 nucleotides per second. Thus, a genome of the size of bacteriophage ϕX174 can be completely exchanged in 20–30 minutes.

B. Regulatory Functions of RecA: Coprotease Activity

In addition to its pivotal role in recombination, the *recA* gene product is essential to the response of bacterial cells to DNA-damaging stress encountered in their environments. Many of the genes involved in DNA repair are part of a stress-induced regulon known as the SOS regulon. This regulon contains 20 or more genes involved in DNA-damage tolerance and repair. The regulon of *E. coli* includes the *recA*, *lexA*, *umuDC*, *polB*, *recN*, *sulA*, *uvrA*, *uvrB*, and *uvrD* genes. The RecA protein is required for the elimination of repression of transcription of genes under the coordinate regulation of the SOS network in *E. coli* and many other bacteria. These genes are under the control of the LexA repressor, which binds to unique sites in their promoter regions, called "SOS boxes." Genes encoding LexA analogs have been found in several eubacteria, as have SOS box-like sequences occurring in the promoters of the *recA* and other genes of many bacterial species.

In response to exposure to various DNA-damaging agents (e.g., UVC and UVB wavelengths of light, mitomycin-C, and nalidixic acid), which alter the structure of DNA or inhibit its replication, a molecular signal is generated, which activates the RecA protein to a state in which it can act as a coprotease, accelerating the autocatalysis of the LexA repressor molecule. This molecular signal is generated when the RecA protein forms a nucleoprotein filament by binding to ssDNA gaps in the chromosome generated by the partially successful attempt to replicate damaged DNA. The activated RecA molecule, referred to as RecA*, functions to stimulate the autocatalytic breakage of a specific Ala–Gly peptide bond near the midpoint of the LexA protein. The cleaved LexA molecule cannot form dimers, required for interaction with the SOS box sequences in DNA. The resulting decrease in the cellular concentration of intact LexA proteins leads to the induction of the SOS regulon (Fig. 2). The molecular signal for the induction of the SOS regulon only occurs in cells that are incapable of handling repair of the damage to their DNA in a timely fashion and before a DNA replication fork passes the lesion. Hence, the naming of the SOS regulon is appropriate as it is only induced as a last ditch effort to "Save Our Ship" from extinction.

The stimulation of autocatalytic cleavage of protein molecules by activated RecA is not limited to the LexA repressor. A characteristic common to lysogens of temperate bacteriophages, such as coliphage λ, is the release of low levels of phage virions through the continuous, spontaneous induction of lytic functions in a small fraction of the lysogenic host populations. The frequency of this activation of expression is greatly increased by exposure of lysogens of many species of bacterial viruses to UV radiation or another DNA-damaging agent. *E. coli recA* mutants that have been lysogenized by bacteriophage λ fail to release phage virions either spontaneously or in response to UV irradiation. Early biochemical studies on RecA revealed that this phenomenon can be explained by the fact that the *recA* gene product is required to eliminate activity of the phage *cI* repressor protein. The activated RecA* protein stimulates autocatalytic cleavage of the *cI* repressor in a reaction analogous to the destruction of the LexA repressor. The carboxyl-terminal regions of the λ *cI* and LexA proteins are highly homologous and cleavage takes place at nearly identical amino acid sequences in the two proteins. Similar cleavage of lysogeny-promoting repressors by activated RecA* proteins has been documented for many other from UV-inducible prophages.

C. Role in DNA Repair

The UV sensitivity of *recA* mutants of *E. coli* and many other bacterial species is not due solely to their inability to induce the SOS network due to loss of the ability to stimulate cleavage of LexA. Besides its regulatory role in DNA repair, the RecA protein is directly involved in the repair of lesions resulting from UV irradiation of DNA. It participates in a post-replication–recombinational process, which allows repair of at least two types of UV radiation-induced lesions: (1) daughter-strand gaps, caused by skipping of a lesion in the template strand during DNA replication and (2) double-stranded breaks, resulting in the cutting of a DNA strand opposite a gap. In addition,

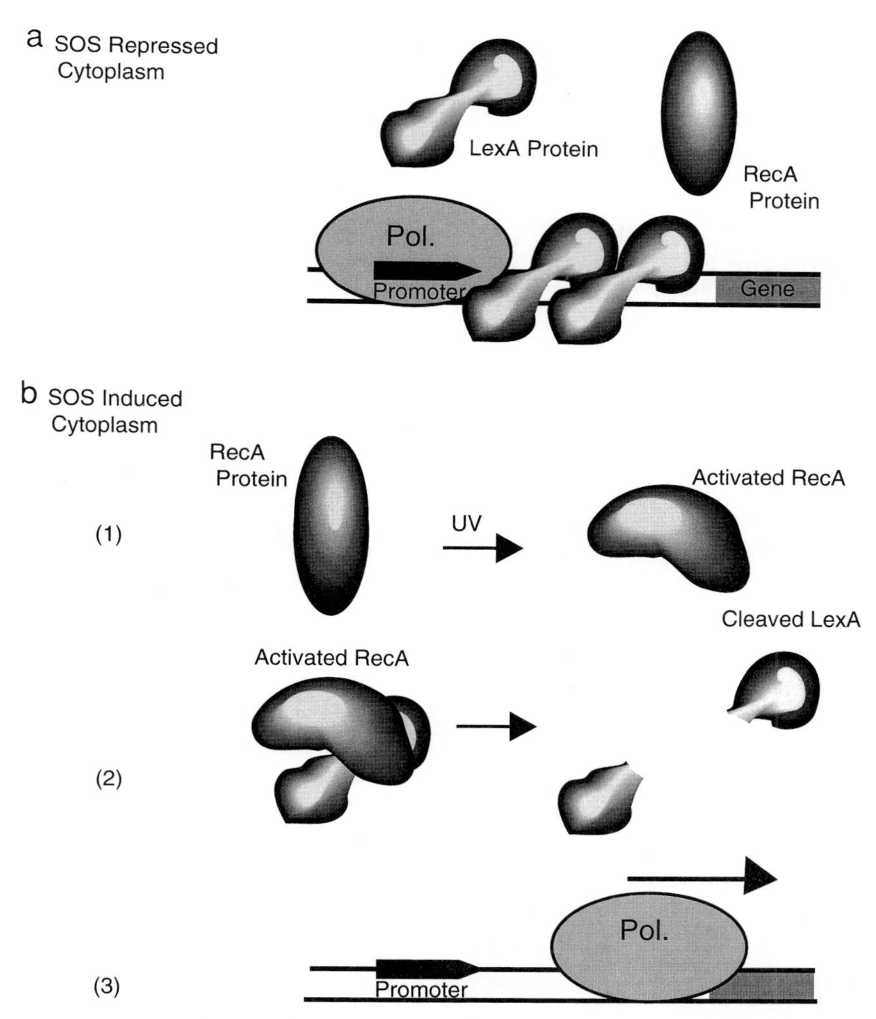

Fig. 2. RecA regulation of the SOS network. The RecA protein acts as a positive effector of the SOS network, by stimulating the inactivation of the LexA protein following exposure to DNA-damaging agents. (a) SOS-repressed cytoplasm. In the unexposed cell, the SOS network is repressed due to the binding of the LexA protein to promoters of SOS genes. Binding of the Lex dimer inhibits movement of RNA polymerase (Pol) and, thus, transcription of the gene. (b) SOS-induced cytoplasm. Exposure of the cell to a DNA-damaging agent, such as ultraviolet (UV) light, leads to the induction of the SOS network through the following events. (1) UV light exposure leads to the production of an uncharacterized molecular signal, which activates the RecA coprotease. (2) The activated protein (RecA*) stimulates cleavage of the LexA protein at a specific Ala–Gly bond in the central portion of the molecule. This cleavage inhibits LexA-dimer formation and leads to the disassociation of the LexA dimer from the SOS promoter. (3) RNA polymerase is free to transverse the SOS promoter, and transcription of the gene takes place.

recent evidence suggests that RecA protein may participate in a similar mechanism involved in the prereplication repair of excision gaps caused by excision of lesion from DNA.

D. The RecA Protein Is Required for Processing the Mutagenic Proteins of *E. coli*

In *E. coli,* mutagenic DNA repair (SOS mutagenesis) requires the products of the *umuDC* operon, which is regulated as part of the SOS network. Genetic disruption of the *umuD* or *umuC* gene eliminates most UV mutability and chemical mutability by a variety of compounds such as 4-nitroquinoline 1-oxide, methyl methanesulfonate, and neocarcinostatin. In this pathway, the coprotease activity of activated RecA protein is required in two reactions: first, for the cleavage of LexA repressor leading to induction of *umuDC* expression, and second, for activation of the UmuD protein's mutagenic potential. This activation requires posttranslational cleavage of UmuD protein by a process analogous to the cleavage of LexA and bacteriophage λ *cI* proteins. DNA sequence analysis indicates that UmuD protein is very similar to the carboxyl-terminal domain of LexA, which contains the latent autodigestive activity. Activated RecA* mediates the removal of the first 24 amino acids of the UmuD protein, to form UmuD′. UmuD′ is active in SOS mutagenesis. Cleavage of the UmuD protein is much slower than cleavage of LexA.

The various protein–protein interactions that are possible between the modified UmuD′ protein, the unmodified UmuD protein, and the UmuC protein are varied and may carry out different functions in the SOS mutagenic repair process. Interactions between UmuC and dimers of either UmuD′$_2$, UmuD–UmuD′, or UmuD$_2$ are indicated by both genetic and biochemical evidence. The UmuD–UmuD′ dimer is formed in preference to either of the homodimers. This heterodimeric interaction acts to inhibit mutagenesis, presumably by titering out the active UmuD′ protein. Interaction of the three potential dimers of UmuD and UmuD′ with UmuC have different outcomes for the mutagenic process (Fig. 3). The UmuD$_2$–UmuC complex is nonmutagenic but is involved in regulating the cell cycle of UV-damaged cells. UmuD–UmuD′–UmuC complexes are inactive, while UmuD′$_2$–UmuC complexes are required in SOS mutagenic repair.

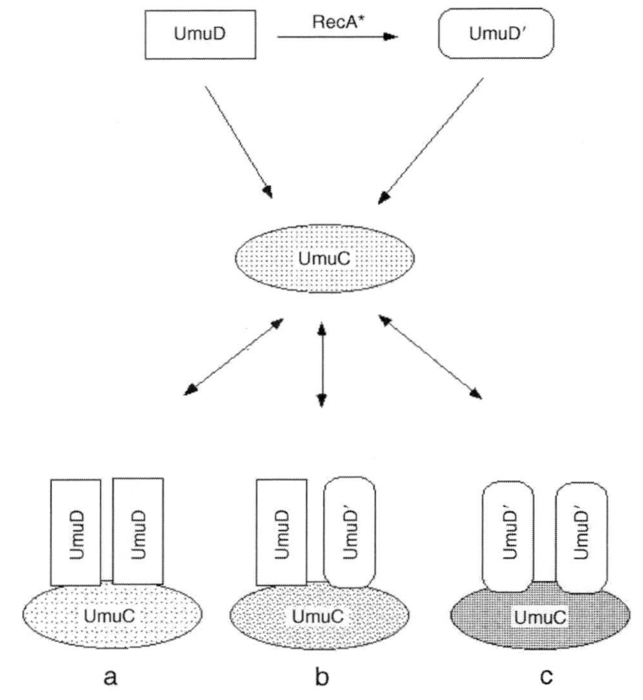

Fig. 3. Posttranslational modification of UmuD by RecA*. In addition to inducing the expression of the UmuD and UmuC proteins as part of the SOS regulon, RecA* also mediates the cleavage of UmuD to form UmuD′. UmuD and UmuD′ interact with UmuC in a variety of combinations. (a) The Umu(D)$_2$C complex seems to be involved in regulating the *E. coli* cell cycle after DNA damage. (b) UmuDD′C, does not appear to have any activity. It may play a role in shutting off SOS mutagenesis by sequestering UmuD′. (c) The Umu(D′)$_2$C complex is active in SOS mutagenesis (translesion synthesis).

The mutagenic process appears to consist of two steps: first, the incorporation of a mismatched nucleotide opposite the stress-induced lesion and, second, the continued replication of the DNA daughter strand past the lesion. This continuation of replication fixes the mismatch and establishes the mutation. It appears that the first of these steps, misincorporation of a nucleotide, is independent of UmuD′,

UmuC, and RecA. It is the second of the processes, continued replication past the lesion, that requires these three proteins.

The specific details of $UmuD'_2$–UmuC-mediated translesion synthesis are not yet understood. However, double mutants of *lexA*(Def)*recA*(Def), which have the ability to express UmuD′ and UmuC in a LexA–RecA-independent fashion, are still unable to produce UV-induced mutations. This implies a more direct role for the RecA protein in SOS mutagenesis than simply as a coprotease. This role seems to be dependent on the ability of RecA to form nucleo–protein filaments since mutants of *recA* incapable of forming these filaments are defective in SOS mutagenesis.

II. PHENOTYPES ASSOCIATED WITH *recA* MUTATIONS

Many of the biochemical and molecular roles of the RecA protein in the physiology of bacteria have been identified through the study of *recA* mutations. These studies have been most extensive in *E. coli*; however, recently some studies have been done in other species and confirm the conserved nature of the protein in function as well as structure (Table I). The *recA* gene was first discovered in 1965 by A. J. Clark and A. D. Margulies working in *E. coli*. The gene was identified by searching mutaginized F^- (*E. coli* recipient strains lacking the Tra^+ plasmid F) clones for those that showed absolutely no genetically identifiable incorporation of exogenous DNA into their endogenote (resident chromosome) following conjugation with a high frequency of recombination (Hfr) donor strain.

These mutations, which mapped to a single locus on the *E. coli* genetic map, were subsequently found to be highly pleiotropic. With our current knowledge of the physiological roles of the *recA* gene product, many of these phenotypes are explainable. Various *recA* mutations manifest themselves in one or more of these phenotypes. Genetic recombination is reduced to $<0.01\%$ of wild-type levels following transfer or acquisition of chromosomal DNA through any of the recognized mechanisms of gene transfer in bacteria (i.e., conjugation, transduction, transformation, etc.). Strains lacking a functional RecA protein are also sensitive to exposure to UV light, x-rays, mitomycin-C, methyl methanesulfonate, and other DNA-damaging agents. They are nonmutable by UV radiation and incapable of inducing the prophages of many temperate bacterial viruses including bacteriophage λ. Production of various colicins cannot be induced and degradation of UV-damaged DNA by exonuclease V cannot be regulated. In addition, pleiotropic effects of *recA* mutations in *E. coli* include lack of UV reactivation of irradiated phages, the uncoupling of DNA synthesis and cell division, and increased sensitivity to hydrogen peroxide.

Additional phenotypes have been observed to be associated with *recA* mutations in other bacterial species. For example, many *recA* mutations of *Pseudomonas aeruginosa* display a Les^- phenotype, in which temperate bacteriophages cannot establish lysogeny. Clear plaques are produced upon infection of a $RecA^-$ strain of *P. aeruginosa*.

III. PHYSICAL STRUCTURE OF THE RecA PROTEIN

The RecA protein of *E. coli* is a single polypeptide with a molecular weight of 38,742. The reactions carried out by this protein require that it can interact with several different types of molecules. As a minimum, it must (1) bind and hydrolyze ATP, (2) bind to ssDNA, (3) interact with dsDNA, (4) interact with itself to form homopolymers, (5) interact with various regulatory and enzymatic proteins, such as LexA, UmuD, and *cI* of λ. Many, if not all, of these functions have now been demonstrated in many of the eubacterial RecA proteins. The biochemical analysis of various *recA* mutants, mainly from *E. coli*, has led to the development of a rudimentary functional map of the protein (Fig. 4). Regions apparently associated with coprotease activity and regulation, repressor recognition, ATP binding and hydrolysis, self-assembly, DNA binding, and synaptase activity have been tentatively mapped. X-ray crystallographic studies have allowed the identification of functional domains

TABLE I

Characteristics of Selected *recA* Mutants from Various Bacterial Species

Species	Characteristics of recA mutants[a]	Complementation of E. coli recA mutations[b]
Acietobacter calcoaceticus	Rec^-, UV^S	?
Aeromonas caviae	UV^S	Rec^+, UV^R, $SOS\text{-}Reg^+$
Agrobacterium tumefaciens	UV^S	Rec^+, UV^R
Azospirillum brasilens	Rec^-, UV^S	UV^R
Azotobacter vinelandii	Rec^-, UV^S	Rec^+, UV^R
Bacillus subtilis[c]	Rec^-, UV^S, $\phi\text{-}I^-$, Din^-	?
Burkholderia cepacia	UV^S	Rec^+, UV^R, $SOS\text{-}Reg^+$
Erwinia carotovora	UV^S, Din^-	Rec^+, UV^R, $\lambda\text{-}S^+$, $\lambda\text{-}I^-$
Neisseria gonorrhoeae	Rec^-, UV^S, $Phas.\text{-}Var^-$	Rec^+, UV^R
Proteus mirabilis	Rec^-, UV^S, $\phi\text{-}S^-$, $\phi\text{-}I^+$	Rec^+, UV^R, UVM^+, $\lambda\text{-}S^+$, $\lambda\text{-}I^+$, $SOS\text{-}Reg^+$
Pseudomonas aeruginosa	Rec^-, UV^S, $\phi\text{-}I^-$, Din^-, Les^-	Rec^+, UV^R, UVM^+, $\lambda\text{-}S^+$, $\lambda\text{-}I^+$, SOS-Reg+
Pseudomonas putida	Rec^-, UV^S	UV^R
Pseudomoas stutzeri	Rec^-, UV^S	Rec^+, UV^R
Pseudomonas syringae	UV^S	Rec^+, UV^R
Rhizobium meliloti	UV^S	Rec^+, UV^R, $SOS\text{-}Reg^+$
Rhizobium phaseoli	Rec^-, UV^S	Rec^+, UV^R
Serratia marcescens	UV^S, Din^-	UV^R, $SOS\text{-}Reg^-$
Vibrio anguillarum	Rec^-, UV^S, $\phi\text{-}I^-$	Rec^+, UV^R, $\lambda\text{-}S^+$, $\lambda\text{-}I^+$
Vibrio cholerae	Rec^-, UV^S, $\phi\text{-}I^-$	Rec^+, UV^R, $\lambda\text{-}S^+$, $\lambda\text{-}I^+$

[a] Only phenotypes known to be affected by the *recA* mutation are listed. Rec^-, recombination eliminated; UV^S, sensitivity to UV irradiation and chemical DNA-damaging agents; $\phi\text{-}I^-$, prophage induction no longer activated by UV or chemical DNA-damaging agents; Din^-, gene expression (induction) by UV and DNA-damaging agents eliminated; Les^-, establishment of lysogeny by temperate phages reduced or eliminated.

[b] Phenotypes that have been tested are listed. Rec^+, recombination proficient; UV^R, UV and DNA-damaging agents resistant; $SOS\text{-}reg^+$, expression of the cloned RecA protein is regulated by the *E. coli* SOS system; $SOS\text{-}Reg^-$, RecA synthesis not regulated; $\lambda\text{-}S^+$, spontaneous production of λ phage from λ prophage; $\lambda\text{-}S^-$, spontaneous induction not complemented; $\lambda\text{-}I^+$, induction of λ-phage production for λ lysogens by UV and other DNA-damaging agents; $\lambda\text{-}I^-$, induction not complemented; UVM^+, UV-induced mutagenesis restored.

[c] The *B. Subtilis recA* analog is named *recE*.

within the three-dimentional structure of the protein (Fig. 5).

Three null alleles have been identified in the *E. coli recA* gene which render the protein totally defective in all biochemical functions *in vitro*. These mutations are *recA1, recA13,* and *recA56*. These sites correspond to amino acids 160, 51, and 60 in the *E. coli* RecA protein, respectively. While the *recA13* allele appears to be completely recessive, the *recA1* and *recA56* alleles show codominance with the wild-type allele for at least some reactions in merodiploids constructed by introduction of a cloned *E. coli recA* gene into the mutant cell.

X-ray crystallographic and genetic analysis emphasize the plastic nature of the RecA protein. Interaction with various effector molecules can elicit significant structural alterations throughout the protein. Because of this plastic nature, alterations in a specific amino acid may indirectly affect activities directly associated with distant regions of the primary amino acid sequence of the molecule. Thus, the creation of a precise structure–function map has required that the underlying biochemical consequences of specific structural alterations be precisely determined *in vitro* before assignment of function to unique regions of the protein can be made.

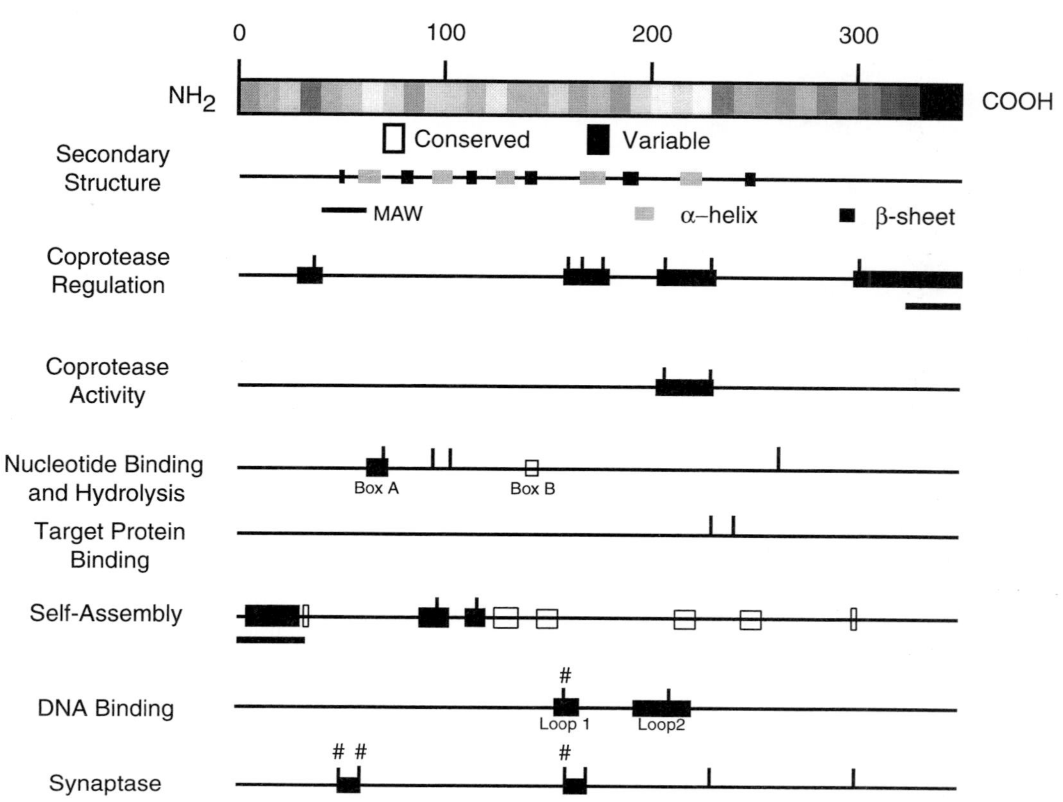

Fig. 4. A structure–function map of the RecA protein. The ever-evolving structure–function map of the RecA protein identifies regions of the molecule that are potentially required for one or more of the many activities of the protein. These regions have been identified through functional studies of mutations or from x-ray crystallography studies. The shaded map at the top of the figure shows the relative number of amino acid substitutions compared to *Escherichia coli* in 20 RecA protein sequences. The predicted secondary structure of the *E. coli* RecA protein is shown, and regions of the molecule that have been tentatively associated with various activities are indicated. Black boxes have been defined from mutational studies and vertical lines indicate point mutations, while bars below the function line indicate insights obtained from analysis of deletion mutations. Null mutations are indicated by a pound sign (#). White boxes defined functions predicted from structural analysis of the molecule. The Walker A and B boxes, the MAW sequence, and the DNA-binding loops A and B are indicated.

IV. EVOLUTIONARY CONSERVATION OF *recA* AND THE RecA PROTEIN

The DNA and amino acid sequences of the *recA* genes and their protein products from many species of bacteria have been highly conserved. Thus, both Southern (DNA) and Western (protein) analyses using probes and antisera created for the *E. coli* gene and its product react with *recA* genes and proteins from highly divergent species. This cross-reactivity has greatly simplified the search for *recA* analogs among the bacteria.

A. Phenotypic Characterization of *recA* Genes from Species Other Than *Escherichia coli*

Currently, *recA* genes have been identified in gram-negative and gram-positive bacteria, including cyanobacteria, spirochetes, streptomycetes, and my-

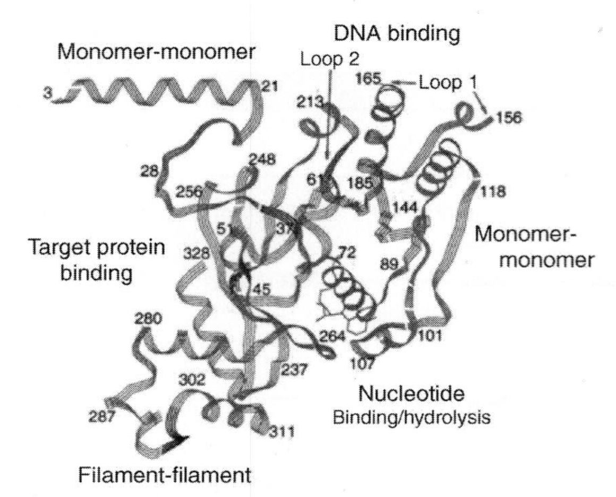

Fig. 5. Structure of RecA protein determined by X-ray crystallography. Three-dimensional structure of the RecA protein with functional domains assigned on the basis of biochemical, genetic, and structural data. See color insert. (Reprinted with permission from Kowalczykowski, S. C., Dixon, D. A., Eggleston, A. K., Lauder, S. D., and Rehrauer, W. M. (1994). *Microbiol. Rev.* **58,** 401–465.)

coplasmas. Analogs of the gene have also been found in various bacteriophages, archaea, and eukaryotic species.

Mutations inactivating the *recA* sequence of several bacterial species have been isolated and characterized (Table I). Comparison of the phenotypes associated with these mutations and the ability of these analogs to replace *E. coli recA* physiologically confirm their functional, as well as their structural relatedness.

B. DNA Sequence Similarity and the Development of Evolutionary Relationships Based on *recA* Sequence Relatedness

Comparison of the DNA and derived amino acid sequences of the known bacterial *recA* genes reveals that the DNA sequences have diverged much more than the amino acid sequences of their protein products. The divergence in DNA sequences has been mainly confined to the third position of codons, resulting in a high degree of synonymous substitutions. Evolution appears to have worked to preserve protein structure while, at the same time, altering DNA sequence to reflect the general G + C content of the species.

A comparison of the 64 bacterial RecA proteins was made by Roca and Cox in 1997. It revealed that these proteins showed from 49% (for *Mycoplasma pulmonis*) to 100% (for *Shigella flexneri*) identity to the amino acid sequence of the *E. coli* RecA. Since proteins with as low as 30% identity are often structurally related, it is likely that the crystalline structure of these bacterial homologs will closely resemble the *E. coli* RecA protein.

C. Conservation of Amino Acid Sequence and Physical Structure

Alignment of RecA sequences, such as was done by Roca and Cox, has allowed both structural comparison of the proteins and the development of phylogenetic relationships among these bacteria, based on the sequence similarity of their *recA* genes. Comparison of the derived amino acid sequences has identified 59 of the amino acid residues to be invariant and between 100 and 106 additional residues to be restricted to chemically conservative substitutions.

When conserved regions were described in reference to the crystal structure of RecA, Roca and Cox were able to identify four classes of conserved amino acid sequence. Class I residues are involved in making ADP contacts in the crystal structure and may be involved in catalyzing ATP hydrolysis. They include some of the most conserved motifs in the protein (see next paragraph). Class II residues participate in hydrophobic condensation of the RecA monomer. Class III reside at the subunit–subunit interface, while Class IV residues constitute a structurally unique region known as the MAW region (see following and Figs. 4 and 5).

Three areas of amino acid conservation are unique to RecA-like proteins. Two of these regions, the Walker Box A or P-loop motif (66-GpESsGKT-73) and Walker Box B (140-vivvD-144) and the amino acid residue Glu-96 are involved in ATP interactions. A fourth area of substantial sequence conservation occurs to the N-terminal side of the Walker A box. Unlike the other regions, it is not involved in ATP binding and/or hydrolysis. This region, known as

MAW, extends from residue 42 to residue 65. It encompasses an α-helix and a β-sheet. It is not similar to any motif found in any other class of proteins. As yet, no function has been assigned to this region.

D. Using *recA* to Develop Evolutionary Phylogenetic Trees

The highly conserved nature of the *recA* nucleotide sequence and the RecA amino acid sequence has allowed comparison of the phylogenetic relatedness of species based on the sequence characteristics of this gene. Comparisons of RecA-based evolutionary trees with those derived from 16S ribosomal RNA sequences are similar but some difference are apparent. When Lloyd and Sharp, in 1993, used the sequences of 25 bacterial *recA* genes to construct a tree of the relationships between bacterial phyla (Fig. 6), their analysis led them to conclude that the *recA* sequence alone does not carry sufficient information to resolve relationships among deep-rooted taxa. However, the relatively more rapid evolution of translated genes, such as *recA* (there is greatly reduced selection for the third base of translated codon over sequences in which the final product is an RNA molecule), is beginning to be used to allow analysis of very closely related species and for subspecies clade analysis. Such cladal structural analysis is proving useful in examining the genetic history of organisms in a variety of specific environmental habitats.

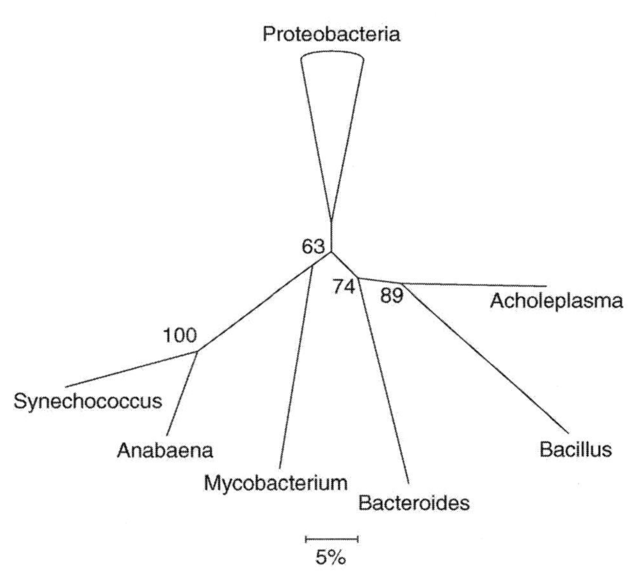

Fig. 6. **Relationship among bacterial phyla inferred from RecA protein sequences. The phylogeny was estimated by the neighbor-joining method applied to pairwise sequence differences corrected for multiple replacements. The bar is 0.05 amino acid replacements per site in length. The number at each node indicates the percentage of bootstraps which contain the branch. (Reprinted with permission from Lloyd, A. T., and Sharp, P. M. (1993). *J. Mol. Evol.* 37, 399–407.)**

Every day, new information on the structure and functions of *recA* genes from more and more diverse species becomes available. Comparative analyses of these DNA and amino acid sequences have been productive in answering questions concerning the structure–function relationships of this fascinating genetic element. The continued study of the *recA* gene and its protein product, RecA, will provide us with valuable insights into the evolution of the biochemical machinery of some of the most fundamental of DNA transactions: genetic recombination and DNA repair.

See Also the Following Articles

Conjugation, Bacterial • DNA Repair • SOS Response • Transduction • Transformation, Genetic

Bibliography

Battista, J. R. (1997). Against all odds: The survival strategies of *Deinococcus Radiodurans*. *Annu. Rev. Microbiol.* **51**, 203–224.

Brendel, V., Brocchieri, L., Sandler, S. J., Clark, A. J., and Karlin, S. (1997). Evolutionary comparisons of RecA-like proteins across all major kingdoms of living organisms. *J. Mol. Evol.* **44**, 528–541.

Camerini-Otero, R. D., and Hsieh, P. (1995). Homologous recombination proteins in prokaryotes and eukaryotes. *Annu. Rev. Genet.* **29**, 509–552.

Cox, M. M., and Lehman, I. R. (1987). Enzymes of general recombination. *Annu. Rev. Biochem.* **56**, 229–262.

Devoret, R. (1988). Molecular aspects of genetic recombination. In "The Evolution of Sex: An Examination of Current Ideas" (R. E. Michod and B. R. Levin, eds.), pp. 24–44. Sinauer Associates, Sunderland, MA.

Kowalczykowski, S. C., Dixon, D. A., Eggleston, A. K., Lauder, S. D., and Rehrauer, W. M. (1994). Biochemistry of homologous recombination in *Escherichia coli*. *Microbiol. Rev.* **58**, 401–465.

Lloyd, A. T., and Sharp, P. M. (1993). Evolution of the *recA* gene and the molecular phylogeny of bacteria. *J. Mol. Evol.* **37**, 399–407.

Miller, R. V., and Kokjohn, T. A. (1990). General microbiology of *recA*: Environmental and evolutionary significance. *Annu. Rev. Microbiol.* **44**, 365–394.

Roca, A. I., and Cox, M. M. (1997). RecA protein: Structure, function, and role in recombinational DNA repair. *Prog. Nucl. Acid Res. Mol. Biol.* **56**, 129–223.

Smith, B. T., and Walker, G. C. (1998). Mutagenesis and more: *umuDC* and the *Escherichia coli* SOS response. Genetics **148**, 1599–1610.

Recombinant DNA, Basic Procedures

Judith W. Zyskind
San Diego State University

I. Cloning Strategies
II. Cloning Vectors
III. Type II DNA Restriction Endonucleases
IV. Nucleotide Sequencing
V. Applications

GLOSSARY

***α*-complementation** Active *β*-galactosidase enzyme complex formed between peptide containing N-terminal region (~15%) of *β*-galactosidase and *β*-galactosidase peptide missing the amino terminus.

autoradiography Use of x-ray film to detect the presence of radioactive material in gels and filters.

cDNA DNA copy of mRNA synthesized by reverse transcriptase, a DNA polymerase that can use either RNA or DNA as a template.

cloning Process of inserting foreign DNA into a plasmid or bacteriophage vector and reproducing this recombinant DNA in a host cell, such as *Escherichia coli*.

cos site *cos* or cohesive end site of λ phage; the site of action of terminase, the enzyme that cuts at *cos* during packaging, leaving a 12-b single-stranded sequence at the 5′ ends.

electroporation High efficiency uptake of DNA caused by artificially inducing cell permeability with a high-voltage pulse that is applied to a suspension of cells and DNA.

endonuclease Nuclease that cleaves internal phosphodiester bonds in nucleic acids.

field inversion gel electrophoresis (FIGE) Separation of high molecular weight DNA molecules by agarose gel electrophoresis with an electrical field that pulses both forward and backward, with a pause between each pulse.

ligation Phosphodiester bond formation between two nucleic acid molecules by DNA ligase.

multiple cloning site (MCS) DNA sequence containing single copies of many sites cut by different restriction endonucleases; also called polylinker.

origin of replication Site at which DNA replication is initiated.

polymerase chain reaction (PCR) Amplification of minute amounts of DNA in a test tube using primers that flank the sequence to be amplified by a thermoresistant DNA polymerase.

replica plating Transfer of an impression of colonies, made on sterile velveteen fabric held tightly over a circular form, to a fresh agar plate, or transfer of clones from a multiwell microtiter dish containing a genomic or cDNA library, using a device with steel prongs.

restriction enzyme mapping Determining the location of restriction enzyme cleavage sites in a piece of DNA without resorting to nucleotide sequence analysis.

reverse genetics Cloning of a gene after determining a portion of the amino acid sequence of its protein product and subsequent introduction of mutations into the cloned gene that are then moved to the chromosome.

Southern blotting Procedure for transferring denatured DNA either from agarose gels or from another source, for example, colonies or plaques, to nitrocellulose or nylon membrane.

RECOMBINANT DNA TECHNOLOGY is the application of techniques currently used to isolate and analyze genes which involves inserting foreign DNA into a host cell, such as *Escherichia coli*. This technology also can include determining the nucleotide sequence of this DNA in order to characterize the functions of this and its products. The introduction of PCR technology for amplifying small amounts of DNA has provided additional tools in this field.

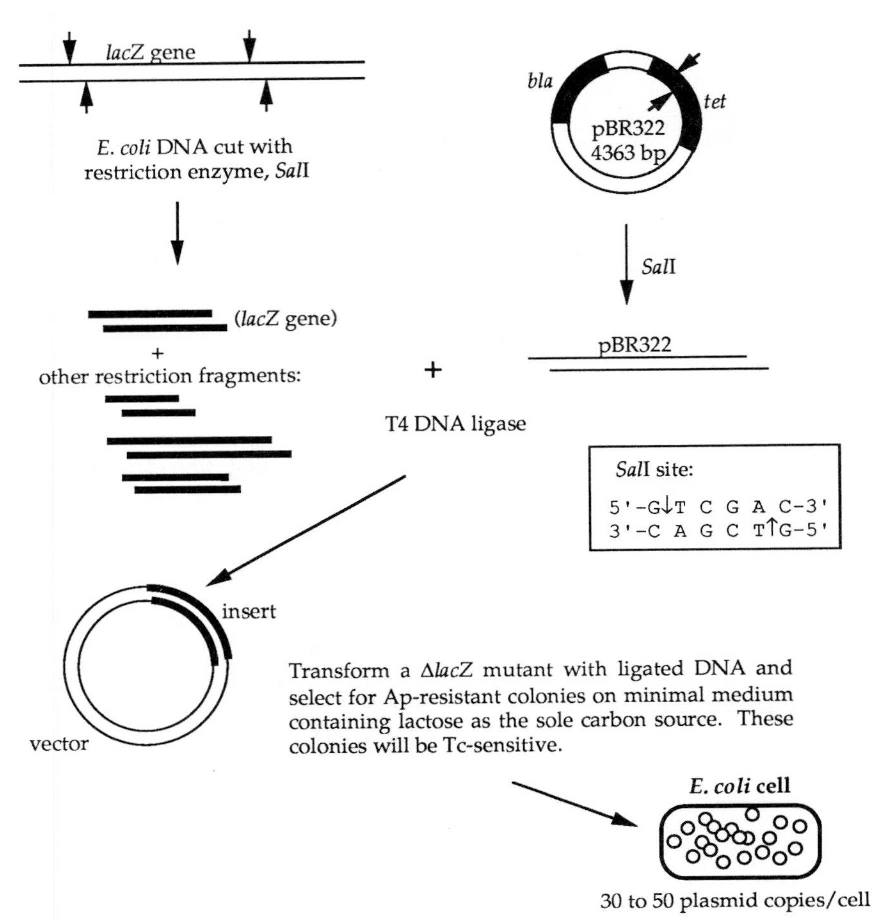

Fig. 1. Cloning the *E. coli lacZ* gene. Components include plasmid vector, pBR322, restriction enzyme that cuts once in the vector, *Sal*I, *E. coli* chromosomal DNA, DNA ligase, and a *lacZ* mutant of *E. coli*. bp, Base pairs.

I. CLONING STRATEGIES

A. Direct Selection by Complementation

Figure 1 illustrates a typical cloning experiment in which a selection method is available for the gene to be cloned. The gene in this cloning example is the *E. coli lacZ* gene encoding β-galactosidase. Plasmids containing the *lacZ* gene can be identified by complementation of a *lacZ* mutant, which is incapable of using lactose as a carbon source and, therefore, is unable to grow on minimal lactose medium. In the experiment shown in Fig. 1, the plasmid vector, pBR322, contains two genes, *bla* and *tet*, conferring resistance to ampicillin (Ap) and tetracycline (Tc), respectively. One of the genes can be used to select for the presence of the plasmid and the other gene can be used to screen for inserts by insertional inactivation. For example, DNA molecules inserted into the *tet* gene inactivate this gene. Plasmids carrying such inserts and transformed into bacteria can be selected on agar plates containing Ap and subsequently screened for inserts by the inability to grow on Tc-containing agar plates. The origin of replication of pBR322 is derived from a high copy number plasmid related to ColE1, so 30–50 copies are present in each cell. New genes introduced into pBR322 also will be present at this high copy number.

In the cloning experiment outlined in Fig. 1, chromosomal DNA is isolated from *E. coli*, then cut with the type II restriction enzyme *Sal*I, which cuts DNA at a specific sequence to give staggered or sticky ends that are complementary. The plasmid cloning vector pBR322, which has a single *Sal*I site, is also cut with *Sal*I. The two DNAs are mixed, complementary *Sal*I ends from chromosomal and vector DNA molecules hybridize to each other, and T4 DNA ligase forms phosphodiester bonds between adjacent nucleotides. The ligated DNA is introduced, either by transformation or electroporation, into a cloning host, in this case, a strain of *E. coli* deleted for the *lacZ* gene, and the cells are plated on minimal agar plates containing Ap and lactose as the sole carbon source. Colonies that develop will contain pBR322 with the *lacZ* gene inserted in the *Sal*I site.

Part of the cloning process involves plasmid replication in the host cell during colony formation. There are approximately 10^5 cells in a single colony, each with 30–50 copies of the same chimeric plasmid. This single colony and cells derived from it are frequently referred to as a clone. To analyze the cloned DNA further, plasmid DNA is isolated from clones. Depending on the scale of plasmid recovery desired, there are two main methods for isolating plasmid DNA. Miniprep procedures for plasmid isolation from small broth cultures will yield sufficient amounts of plasmid DNA to determinme the size of the insert by restriction enzyme mapping and to sequence double-stranded DNA. Large-scale plasmid isolation procedures frequently rely on cesium chloride (CsCl) density gradient centrifugation in the presence of ethidium bromide to separate linear chromosomal DNA from the plasmid molecules that are circular and supercoiled. Kits for both small- and large-scale isolation of plasmid and chromosomal DNA are available from a variety of suppliers.

Ligation of the vector ends would recreate the vector in the experiment shown in Fig. 1. To prevent self-ligation of the vector, the phosphate groups at the 5′ ends of the vector DNA can be removed by a phosphatase such as bacterial alkaline phosphatase. Self-ligation of the vector also can be prevented by cutting both the vector and insert DNA with two different restriction enzymes that produce ends that are not compatible. This highly efficient method, called forced directional cloning, yields insertion of a restriction fragment in only one orientation and is used frequently with vectors containing a multiple cloning site (MCS).

DNA can be cloned from any organism to produce chimeric DNA molecules composed of DNA from different sources. For example, in the cloning experiment shown in Fig. 1, DNA from any prokaryotic organism and cDNA from any eukaryotic organism could be substituted for *E. coli* DNA, as long as the organism produces β-galactosidase and the gene is expressed in the cloning host.

B. Screening a Library

If the gene to be cloned cannot be selected for directly, then identification becomes more difficult. A genomic library consisting of many different clones, each containing a different insert DNA in the plasmid vector, is screened for specific clones by looking for a particular physiological trait or by hybridization. For example, if a large number of Ap-resistant colonies was isolated on rich medium and screened for Tc-sensitivity in the experiment shown in Fig. 1, such a collection of Ap-resistant, Tc-sensitive clones could be considered a genomic library. The clone or clones containing the *lacZ* gene can be identified as red colonies after replica plating the genomic library onto MacConkey–lactose Ap agar plates.

The presence of the *lacZ* genes also can be identified by colony hybridization. A probe can be made from the *lacZ* gene of *E. coli* and labeled appropriately so that the probe can be detected by the presence of radioactivity, color, chemiluminescence, or fluorescence. The probe can then be used to identify the *lacZ* gene of another organism by hybridization with DNA from the genomic library that has been transferred to nitrocellulose (Southern blot). In this technique, nitrocellulose or nylon membrane cut to the size of a petri dish is placed on an agar plate, and the genomic library is transferred by replica plating, or by using toothpicks, to the surface of the membrane. After colonies form, the membrane with the colonies is moved to a lysing and denaturing solution, and the plasmid DNA is fixed to the membrane. The labeled *lacZ* probe is hybridized to the nitrocellulose or nylon membrane. Depending upon how the probe was labeled, the clone with the *lacZ* gene can be identified by detecting the presence of radioactivity, color, chemiluminescence, or fluorescence.

More commonly today, if the DNA sequence of the genome is known, the desired gene is identified by a homology search, then amplified by PCR, using primers generated from sequences flanking the gene. Restriction endonuclease sites can be included in the primers to facilitate cloning.

C. Choosing an *Escherichia coli* Host for Cloned DNA

Foreign DNA is degraded when introduced into *E. coli* cells because of the resident host restriction–modification system. Three genes are involved in the Type I *Eco*K restriction–modification system in *E. coli* K-12 strains, but only mutations in the *hsdR* gene confer the phenotype r^-m^+. Mutations in the other two genes; *hsdM* and *hsdS*, give an r^-m^- phenotype. Strains that have a mutation in the *hsdR* gene contain no *Eco*K restriction endonuclease activity but do have the *Eco*K methyltransferase. Such strains are good cloning hosts for foreign DNA.

There are at least three *E. coli* restriction endonucleases that cut methylated DNA. If the DNA to be cloned contains methylated DNA, additional mutations that eliminate these restriction endonucleases should be included in the cloning host. These mutations include *mcrA* (*m*odified *c*ytosine *r*estriction), *mrcB,* and *mar* (*m*odified *a*denine *r*estriction). If the gene to be cloned is homologous to sequences in the chromosome, then a *recA* mutation could be used to prevent homologous recombination between the cloned DNA and the chromosome. For example, in Fig. 1, a *recA* mutant host would be a preferred host for cloning the *E. coli lacZ* gene if the host strain contained a point mutation, rather than a deletion of the *lacZ gene.*

D. Reverse Genetics

The peptide product of a gene can be used to identify and clone that gene. If a partial amino acid sequence is available from an isolated protein, a mixed probe can be synthesized and labeled. A mixed probe is a set of oligonucleotides corresponding to every combination of codons for each amino acid. One of the oligonucleotides will be identical in sequence to the gene sequence and will hybridize to the gene if present in a genomic library.

Replacement mutagenesis can be carried out once a gene is cloned. A total deletion (null mutation) of the cloned gene can be engineered, then introduced into the chromosome to replace the wild-type gene. Replacement requires a double reciprocal homologous recombination event between sequences flanking the gene in the clone and on the chromosome. A *recD* mutation, which eliminates the RecBCD exonuclease V but does not destroy homologous recombination, is useful for cloning linear DNA in *E. coli* when replacing a wild-type gene on the chromosome with a mutation or deletion of that gene. The phenotype of this mutation can give valuable clues to the function of the gene and its product in the cell. If the gene being replaced is required for cell growth, the deletion of the gene cannot be recovered. For required genes, the deletion is constructed in the presence of conditional expression of the gene product where, either on a plasmid or elsewhere in the chromosome, the gene is placed under the control of an inducible promoter, such as the arabinose promoter.

E. Characterizing the Protein Products of Cloned Genes

Three methods are available to examine the sizes of proteins encoded by plasmids: use of minicells, maxicells, or a coupled *in vitro* transcription–translation system. Minicells are small chromosomeless cells that are produced by *minB* mutants of *E. coli* and can be isolated from normal cells. High copy number plasmids segregate into minicells, which are able to transcribe and translate plasmid-encoded genes. Proteins synthesized in minicells can be labeled with [^{35}S]methionine, isolated, separated by SDS–polyacrylamide gel electrophoresis, and visualized by autoradiography. Proteins encoded by cloned DNA can be correlated with sizes of open reading frames (ORFs) in the cloned DNA if the nucleotide sequence is known or with the size of a known protein that the cloned DNA is suspected to encode. The maxicell method involves UV irradiation of a host cell so double-stranded breaks occur in the large chromosome but not in all λ or plasmid cloning vectors present in the cell. Chromosomal DNA is degraded and the cloned DNA in the vector is available for *in vivo* transcription and translation. Proteins synthesized by maxicells can be characterized in the same manner as those made in minicells. Plasmid-encoded proteins also can be labeled using a cellular extract, called a Zubay extract, that contains all the necessary enzymes and components for both transcription and translation. Correlation of the size of the protein or of truncated versions encoded in cloned DNA, as determined by one of these methods, with the size of the insert will suggest where in the cloned DNA the ORF for this protein is located and how long the ORF in the nucleotide sequence is expected to be.

II. CLONING VECTORS

A. Plasmid Vectors

The first cloning vectors were constructed from plasmids with high copy numbers. DNA yields are important in cloning. As long as the gene to be cloned is not detrimental to the cell, the vector of choice is a high copy number cloning vector. When the gene product might prove lethal to *E. coli,* low copy number vectors, such as those derived from the F plasmid, are useful.

Unlike chromosomal DNA replication, most of the high copy number vectors do not require synthesis of a plasmid-encoded protein for DNA replication, so plasmid yields can be increased up to 50-fold by chloramphenicol amplification. When chloramphenicol is added to a culture in late log phase, plasmid DNA continues to replicate but chromosomal DNA is inhibited because chloramphenicol inhibits the initiation of chromosomal DNA replication.

Several properties that have been engineered into plasmid cloning vectors include (1) a selectable marker, which is almost always an antibiotic resistance gene; (2) a way to detect inserts by insertional inactivation, either by loss of antibiotic resistance or α-complementation; (3) a multiple cloning site (MCS), which is a cluster of unique restriction enzyme sites for cloning; and (4) SP6, T3, or T7 bacteriophage promoters flanking the MCS, making possible the synthesis of RNA complementary to either the coding or noncoding strand of DNA inserted into the MCS. These vectors also may contain the origin of replication from a filamentous bacteriophage, such as f1 or M13. When a helper phage is added to cells containing such vectors, called phagemids, the cloned DNA can be packaged into bacteriophage heads and isolated as single-stranded DNA that can be used as a template for the chain-termination method of sequencing DNA.

B. Bacteriophage λ Cloning Vectors

Lambda cloning vectors are useful when cloning large pieces of DNA, for example, when constructing a genomic library. DNA sequences inserted into these vectors replace λ DNA sequences not required for the phage lytic cycle. DNA of phage λ is linear, double-stranded, and 48.5 kilobase pairs (kb) long. The vector is first cut with a restriction enzyme, and the λ arms are isolated from the nonessential central region. These arms are ligated to foreign DNA with T4 DNA ligase, and the DNA is packaged with complementary extracts from two *E. coli* strains that have been infected with different λ mutants incapable of either replicating or packaging their own DNA. The requirement that DNA molecules with λ *cos* sites be separated by ~50 kb to be packaged provides a strong selection for inserts of a given size range. The advantages of producing a genomic library in phage λ are that large pieces of DNA can be cloned and that the library can be stored as 1 ml of lysate in the refrigerator.

Cosmids, plasmids containing the λ *cos* site, also can be used to clone large pieces of DNA. After ligation, cosmids containing inserted foreign DNA can be packaged in λ heads by the *in vitro* packaging extracts and transduced into an *E. coli* host. A genomic library constructed with a cosmid vector can be stored in the refrigerator after *in vitro* packaging, whereas libraries made with plasmid vectors consist of colonies stored in multiwell microtiter dishes at −70°C.

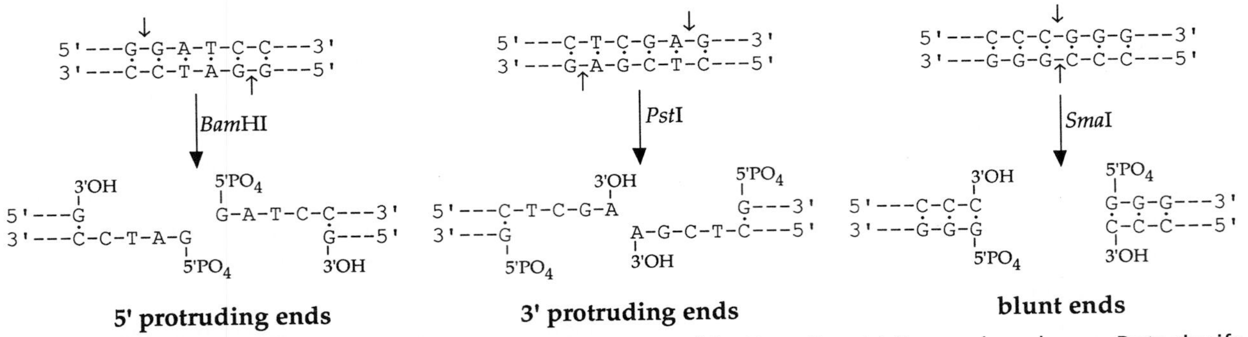

```
5'---G-G-A-T-C-C---3'        5'---C-T-C-G-A-G---3'        5'---C-C-C-G-G-G---3'
3'---C-C-T-A-G-G---5'        3'---G-A-G-C-T-C---5'        3'---G-G-G-C-C-C---5'
        BamHI                        PstI                         SmaI

3'OH         5'PO4                 3'OH        5'PO4            3'OH       5'PO4
5'---G       G-A-T-C-C---3'   5'---C-T-C-G-A        G---3'   5'---C-C-C   G-G-G---3'
3'---C-C-T-A-G       G---5'   3'---G      A-G-C-T-C---5'     3'---G-G-G   C-C-C---5'
     5'PO4       3'OH              5'PO4  3'OH                    5'PO4   3'OH
```

5' protruding ends **3' protruding ends** **blunt ends**

Fig. 2. Examples of the three types of ends that are generated by Type II restriction endonucleases. Dots signify hydrogen bonds between bases. Dashes between nucleotides indicate phosphodiester bonds.

III. TYPE II DNA RESTRICTION ENDONUCLEASES

Restriction endonucleases are enzymes that cleave DNA at recognition sites (Type II) or cleave away from recognition sites (Type I and III). The only important class of restriction endonucleases for cloning DNA are Type II because DNA cleavage occurs within or adjacent to the Type II recognition sites. It is important to be aware of Type I and Type III restriction endonucleases because their presence in the cell can interfere with cloning DNA. The Type II restriction endonucleases and methyltransferases are separate peptides and can be purchased from commercial suppliers.

Examples of Type II restriction enzyme sites are shown in Fig. 2. The name of a restriction enzyme and its corresponding methyltransferase includes the first letter of the genus name and the first two letters of the species name of the organism from which the enzyme was isolated. This abbreviation is followed by letters or numbers referring to strain or type of bacteria or bacteriophage or plasmid that encodes the restriction enzyme. The presence of "r" or "m" before the name denotes that it is the restriction endonuclease or the methyltransferase, respectively. When "r" or "m" is missing in the name, it is assumed to be the restriction endonuclease.

Restriction methyltransferases are enzymes that methylate either cytosine or adenine in sites recognized by restriction endonucleases, thereby conferring resistance of the DNA to cutting by the endonuclease. The nucleotide methylated by the m · *Eco*RI enzyme is shown in Fig. 3. Many methyltransferases can be purchased commercially and have been used in cloning experiments to protect certain sites from cutting.

Type II restriction endonucleases recognize and cleave at specific sequences, usually 4–6 base pairs (bp) long, although a few recognize sites 7–8 bp long. Most sites have dyad symmetry: the double-stranded sequence is identical to the sequence after 180° rotation. Three types of DNA ends, examples of which are shown in Fig. 2, are produced by these enzymes: 5′ protruding ends, 3′ protruding ends, and blunt ends. All cleavage reactions leave 5′ phosphate and 3′ hydroxyl ends, both of which are required substrates for ligation.

The Type II restriction enzyme site cut by r · *Eco*RI and methylated by m · *Eco*RI is shown in Fig. 3. *Eco*RI leaves 5′ protruding ends; the single-stranded 5′ overhangs are identical for all *Eco*RI fragments. These single-stranded ends, called sticky ends, are cohesive, that is, when the fragments are juxtaposed the nucleotides are complementary and can form hydrogen bonds (small dots, Fig. 3). A recombinant DNA molecule is produced when 5′ and 3′ ends from separate fragments are joined by T4 DNA ligase.

The frequency of restriction enzyme sites in DNA varies with the size of the site and the G + C content of DNA. If the G + C content is 50%, a tetranucleotide site would appear once every 256 bp or $(\frac{1}{4})^4$, whereas a hexanucleotide site would appear once every 4096 bp or $(\frac{1}{4})^6$. Restriction enzymes that recognize 4-bp sites will generate fragments that have an average length of 256 bp, whereas enzymes that rec-

```
         ↓
5'---G-A-A-T-T-C---3'
3'---C-T-T-A-A-G---5'
               ↑
        | r·EcoRI
        ▼
3'OH                    5'PO4
5'---G                  A-A-T-T-C---3'
3'---C-T-T-A-A                G---5'
          5'PO4                3'OH

        | r·EcoRI leaves staggered,
        | cohesive, "sticky" ends
        | that can reanneal
        ▼
3'OH  5'PO4
5'---G A-A-T-T-C---3'
3'---C-T-T-A-A G---5'
          5'PO4 3'OH

        | T4 DNA ligase
        ▼
5'---G-A-A-T-T-C---3'
3'---C-T-T-A-A-G---5'
phosphodiester bonds reformed


5'---G-A-A-T-T-C---3'
3'---C-T-T-A-A-G---5'
        | m·EcoRI
        ▼
         CH3
5'---G-A-A-T-T-C---3'
3'---C-T-T-A-A-G---5'
           CH3
EcoRI site modified by m·EcoRI is
no longer sensitive to r·EcoRI
```

Fig. 3. Activities of the *Eco*RI restriction endonuclease (r · *Eco*RI) and the *Eco*RI methyltransferase (m · *Eco*RI), and the recombinant DNA product produced after ligation of two *Eco*RI fragments. Dots signify hydrogen bonds between bases. Dashes between nucleotides indicate phosphodiester bonds.

ognize 6-bp sites will generate fragments that have an average length of 4096 bp, although the distribution of sizes about these averages is very high. Larger DNA fragments produced by 6-bp cutters will frequently contain the sequence of a whole gene. The average gene is about 1 kb in bacteria. Small DNA fragments generated by 4-bp cutters will rarely contain the sequence of a whole gene but are used for shotgun cloning into vectors for nucleotide sequencing projects.

To produce overlapping fragments in a genomic library, two different restriction enzymes, a 4-bp and a 6-bp cutter, are used. In this approach, chromosomal DNA is partially digested with a restriction enzyme that recognizes a 4-bp sequence, for example. *Sau*3A (↓ GATC). After size-fractionating the restriction fragments on agarose gels or sucrose gradients, the DNA is cloned into a vector that has been cut with a restriction enzyme that gives compatible ends, for example, *Bam*HI (G ↓ GATCC).

Once a fragment of DNA is cloned, a useful signature of that DNA, aside from its nucleotide sequence, is the location of restriction enzyme sites in the DNA. Restriction enzyme mapping of DNA involves cutting the DNA with restriction enzymes and determining the sizes of restriction enzyme fragments. Linear DNA molecules are separated according to size by agarose or acrylamide gel electrophoresis and visualized using the intercalating dye, ethidium bromide, which binds to DNA and fluoresces when irradiated with UV light (see Fig. 4). A restriction map can be constructed from the sizes of fragments generated with multiple enzymes, individually and in combination. Restriction enzyme mapping of large genomes, such as bacterial chromosomes, requires the use of restriction enzymes that recognize 8-bp sites [frequency is $(\frac{1}{4})^8$ or approximately every 65 kb] and field inversion gel electrophoresis (FIGE), which resolves fragments of DNA up to 2000 kb in length.

Restriction fragments carrying specific genes can be identified by Southern blotting, probe hybridization, and detection of the probe. After denaturation, DNA restriction fragments can be transferred by diffusion from agarose gels to nitrocellulose or nylon membranes; this procedure is called Southern blotting. The blot is hybridized to an appropriately labeled nucleic acid probe, then washed, and the pres-

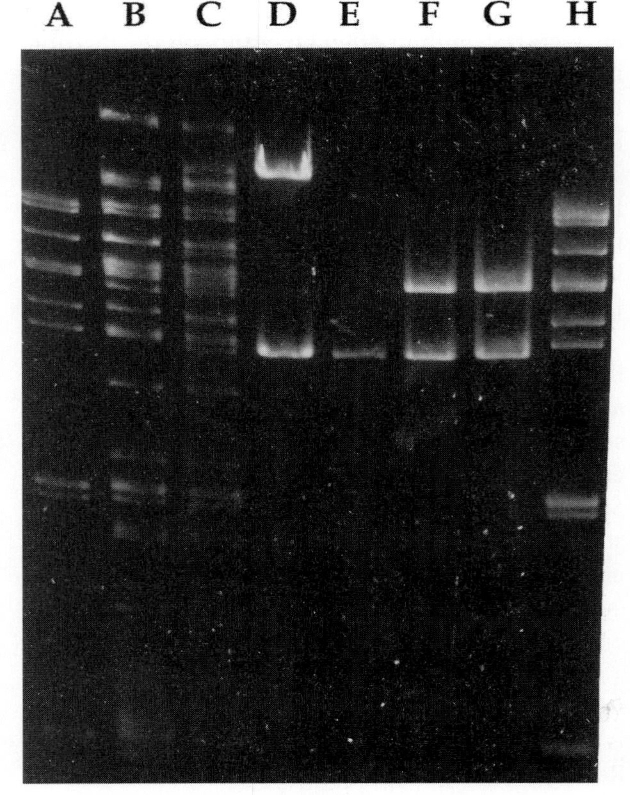

Fig. 4. Agarose gel electrophoresis of *Eco*RI-digested DNA species. Lanes A and H, F plasmid; lane B, *Salmonella typhimurium* F′ plasmid FST27-D1; lane C, *S. typhimurium* F′ plasmid FST27; lane D, pJZ1; lane E, *Eco*RI fragment carrying *kan* gene; lane F, pJZ2; lane G, pML31. Plasmid pJZ1 contains the *S. typhimurium* chromosomal origin of replication. Plasmids pJZ2 and pML31 contain the F plasmid origin of replication. [Reproduced from Zyskind, J. W., Deen, L. T., and Smith, D. W. (1979). *Proc. Natl. Acad. Sci. U.S.A.* **76**, 3097–3101.]

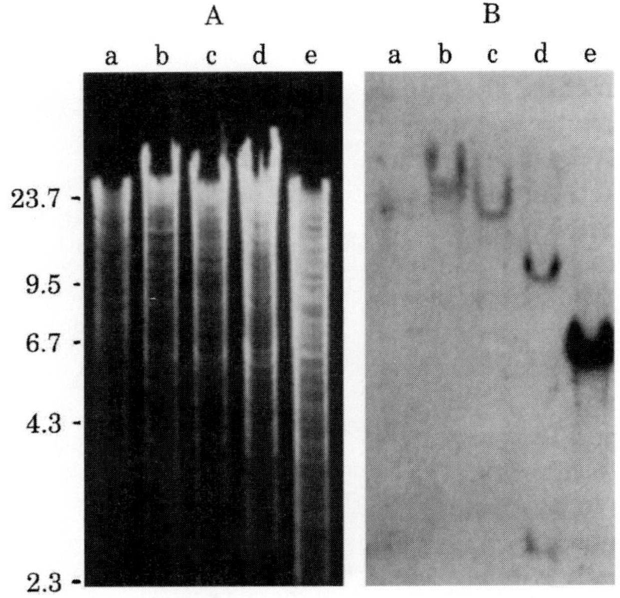

Fig. 5. Southern blot analysis of *Sal*I restriction enzyme digests of chromosomal DNA isolated from marine bacteria. Chromosomal DNA was isolated from *E. coli* (a), *Vibrio fischeri* (b), *Photobacterium leiognathi* (c), *P. phosphoreum* (d), and *V. harveyi* (e). (A) Ethidium bromide stained agarose gel. (B) Autoradiogram of a Southern blot of the DNA in (A) after the blot was hybridized with a probe containing the *V. harveyi* origin of replication. [Reproduced from Zyskind, J. W., Cleary, J. M., Brusilow, W. S. A., Harding, N. E., and Smith, D. W. (1983). *Proc. Natl. Acad. Sci. U.S.A.* **80**, 1164–1168.]

ence of the probe detected by autoradiography or chemiluminescence after exposure to x-ray film. Bands that appear on the x-ray film indicate the DNA bands that have hybridized to the radioactive probe. An example of Southern blotting is shown in Fig. 5.

IV. NUCLEOTIDE SEQUENCING

Our greatly expanded knowledge of gene structure, expression, and function, as well as of protein structure and function, is due mainly to rapid DNA sequencing methods. The two types of sequencing procedures in common use today are the chemical degradation method, developed by Maxam and Gilbert, and the dideoxynucleotide chain-termination method of Sanger and co-workers.

In the chemical degradation method, a single- or double-stranded DNA fragment is radioactively labeled at one end, the labeled DNA fragment is divided into four or five aliquots, and chemical reactions are performed so that only one or two bases are modified in each reaction. A phosphodiester bond at each modified base is cleaved with piperidine, producing a "nested set" of labeled fragments that terminates at the location of modified residues. Labeled DNA fragments from these reactions are separated by electrophoresis in adjacent lanes on a polyacrylamide gel to which the denaturant urea has been added. Autoradiography is used to detect the labeled DNA fragment ladders, and the DNA sequence can be read from the location of bands in each lane, since this corresponds to the order of the bases from the labeled end.

The dideoxynucleotide chain-termination method depends on termination of DNA synthesis after incorporation of a dideoxynucleotide. Dideoxynucleotides can be incorporated into DNA by many DNA polymerases but, because these nucleotides lack a 3′ hydroxyl group, further DNA synthesis is prevented. Template DNA can be single-stranded M13mp viral DNA, DNA amplified by asymmetric PCR, or double-stranded DNA that has been denatured. A primer is annealed to a template strand that is to be sequenced, forming hybrid molecules. The hybrid is divided evenly among four tubes, each containing a different dideoxynucleoside triphosphate and all four deoxynucleoside triphosphates, one of which is radioactively labeled. After addition of a DNA polymerase, such as modified T7 DNA polymerase (Sequenase®), the primer is elongated until incorporation of a chain-terminating dideoxynucleotide. The nested set of labeled DNA fragments in the four reactions is separated by denaturing polyacrylamide gel electrophoresis, and the sequencing ladders are detected by autoradiography (see Fig. 6). The dideoxynucleotide chain-termination method has been automated and is currently being used in commercially available DNA sequencers. In the most common variant of this method, each of the four dideoxynucleotides used to terminate DNA synthesis is covalently linked to a differently fluorescing dye. The chain extension reaction is carried out in a single reaction vessel, and the resulting products are separated by gel electrophoresis in a single lane. The characteristic fluorescence spectrum of each dye identifies the terminal base on each chain extension product (see Fig. 5).

Once the nucleotide sequence is known, it can be analyzed using computer programs that search for ORFs and for specific sites such as promoters or binding sites for DNA binding proteins. The ORFs can be translated to determine the amino acid sequences of encoded proteins. The DNA and protein sequences can be compared with sequence libraries to determine whether the gene or protein has been isolated previously. Amino acid sequence similarities with other proteins can provide clues to protein function. For example, many bacterial protein kinases involved in signal transduction have been identified after finding sequence similarities with known protein kinases.

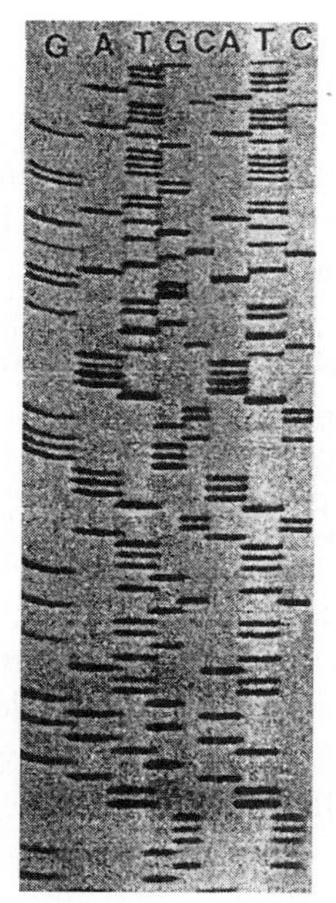

Fig. 6. Autoradiograph of dideoxynucleotide chain-termination sequencing reactions. Sequence of DNA is 5′ AGCG-CCCTTA GTAGAGTTAT GTTGCTGTTT ACCTAAAGGG CGCCTAAAAT CTGTTGGATG CTGTATTGGT TTTGTATTTC ATTTG3′. [Reproduced from Zyskind, J. W., and Bernstein, S. I. (1992). "Recombinant DNA Laboratory Manual." Academic Press, San Diego, California.

V. APPLICATIONS

Recombinant DNA methodology combined with genetic, biochemical, and immunological approaches has made possible rigorous gene analysis that was impossible 20 years ago. Promoters can be linked to reporter genes, such as *lacZ* or *chb;* these fusions can be used to determine the direction of transcription, promoter strength, and mechanisms regulating expression, including distinguishing transcriptional from posttranscriptional control. If the fusions are contained within high copy number plasmids, reporter enzyme activity may vary, due to differences in plasmid copy number or to titration of regulator proteins. For this reason, fusions often are integrated into the chromosome for single copy analysis of promoter activity using λ vectors or a simplified λ site-specific recombination system combining the λ *attP* site, the fusion with the reporter gene, and the λ *int* gene. Specific nucleotides in sequences of interest are easily changed (site-specific mutagenesis) using either (1) primer oligonucleotides that contain one base change and are complementary to the inserted sequence in M13 bacteriophage vectors or (2) PCR mutagenesis in which the base change is included in one of the primers. The development of *in vitro* mutagenesis has led to a new field called protein engineering, in which amino acid changes are introduced into a protein to change, and perhaps improve, its activity or properties. Theoretically, any chemical reaction could be catalyzed by an enzyme; perhaps in the near future it will be possible to predict the amino acid sequences of proteins with new enzymatic activity. If the amino acid sequence of a peptide is known, the corresponding gene can be chemically synthesized using codons that optimize expression in the cloning host. Expression vectors allow production of large amounts of a protein in bacterial cells by placing the appropriate gene under control of a strong but regulated promoter such as P*tac*. Currently, in the biotechnology industry, bacterial cells are used as factories for producing many important peptides, such as human insulin.

See Also the Following Articles

DNA Replication • DNA Sequencing and Genomics • Plasmids • Polymerase Chain Reaction

Bibliography

Baxevanis, A., and Ouellette, B. F. F. (eds.) (1998). "Bioinformatics: A Practical Guide to Analysis of Genes and Proteins." John Wiley & Sons, New York.

Drlica, K. (1996). "Understand DNA and Gene Cloning: A Guide for the Curious," (3rd ed.). John Wiley & Sons, New York.

Eun, H.-M. (1996). "Enzymology Primer for Recombinant DNA Technology." Academic Press, San Diego, CA.

Kalabat, D. Y., Froelich, J. M., Phuong, T. K., Forsyth, R. A., Newman, V. G., and Zyskind, J. W. (1998). Chitobiase, a new reporter enzyme. *BioTechniques* **25**, 1030–1035.

Lambrook, J., Fritsch, E. F., and Maniatis, T. (1989). "Molecular Cloning: A Laboratory Manual, Second Edition." CSH Laboratory Press, Cold Spring Harbor, NY.

Lewin, B. (1997). "Genes VI." Oxford Univ. Press, Oxford, UK.

Watson, J. D., Gilman, M., Witkowski, J., and Zoller, M. (1992). "Recombinant DNA, Second Edition," W. H. Freeman and Co., New York.

Wu, R., Simon, M. I., and Abelson, J. N. (eds.) (1995). "Recombinant DNA Methodology II (Selected Methods in Enzymology)." Academic Press, San Diego, CA.

Zyskind, J. W., and Bernstein, S. I. (1992). "Recombinant DNA Laboratory Manual," Academic Press, Inc., San Diego, CA.

Refrigerated Foods

Leora A. Shelef
Wayne State University

I. Refrigerated Storage as a Short-Term Preservation Method
II. Psychrotrophic and Psychrophilic Microorganisms
III. Adaptation of Microorganisms to Low Temperatures
IV. Food-Borne Psychrotrophic Microorganisms
V. Refrigerated Foods
VI. Modified Atmosphere Packaging

GLOSSARY

aerobe Organism able to use oxygen as an electron acceptor in metabolism.

anaerobe Organism able to grow in the absence of oxygen.

enterotoxin Microbial substance able to induce damage to the host intestine, generally leading to fluid accumulation.

facultative Able to grow in either the presence or absence of an environmental factor (e.g., "facultative anaerobe").

pathogen Disease-causing organism.

psychrophile Organism able to grow over the range of subzero to 20°C, with optimum range of 10° to 15°C.

psychrotroph Organism able to grow at 0–5°C, with optimum growth between 20° and 30°C.

REFRIGERATION maintains foods at 0–7°C. These temperatures slow down autodegradation of the food and further restrain growth of bacteria, yeast, and mold in a wide range of refrigerated meat, poultry, fish, dairy, fruit, and vegetable products. In general, bacteria that are capable of growth in refrigerated foods grow optimally at 25–30°C. Although most of these bacteria cause spoilage, certain species are food-borne pathogens capable of survival, growth, and, in some cases, elaboration of toxins at refrigerator temperatures. The growing popularity of fresh or minimally processed, refrigerated, ready-to-eat foods has increased the importance of food-borne psychrotrophic microorganisms and concerns about the safety of often preservative-free refrigerated foods. In most cases, sufficient heating before consumption is the most effective control measure to avoid illness.

I. REFRIGERATED STORAGE AS A SHORT-TERM PRESERVATION METHOD

Microorganisms have a maximum, minimum, and optimum temperature for growth. Many of those associated with foods are capable of surviving and proliferating at reduced temperatures. However, because all microbial metabolic reactions are catalyzed by enzymes and their rate depends on the temperature, the rate of metabolic reactions decreases by approximately one-half for each 10°C decrease in temperature, within a certain range (the Q_{10} rule). Consequently, reduced temperatures result in slower proliferation and metabolic changes than do higher temperatures. When foods are the substrate, restrained metabolic rates lead to an extended shelf life, because metabolic by-products often have deleterious effects on foods.

The degree to which shelf life is extended depends on a number of factors. These include the type of food and its chemical and physical characteristics, microbial flora, storage atmosphere, and temperature. Refrigeration temperatures are about 5±2°C. Chilling temperatures refer generally to the 10–15°C range, or those between refrigerator temperature and room temperature. Although lower than room temperatures, they are sufficiently different from refrigerator temperatures and, hence, influence both type of microorganisms that may proliferate in the foods and their growth rate.

Storage at low temperatures has a number of desirable effects on both fresh and processed foods. These include control of metabolic activities of fresh plant and animal tissues, chemical reactions, including those catalyzed by enzymes, and moisture loss. Moreover, most pathogens can no longer grow at the low end of the refrigerator temperatures, and growth of spoilage microorganisms is, in many instances, less efficient than at chilling and room temperatures.

Proliferation of specific microorganisms in foods depends on the combined effect of intrinsic factors (characteristics of the food) and extrinsic factors (those of the environment). Intrinsic factors include primarily the composition of the food, its pH, redox potential, water activity, and presence of antimicrobial substances. Extrinsic factors include, in addition to the temperature, the relative humidity, the gaseous atmosphere surrounding the food, and similar factors.

The composition of the foods and nutrient availability influence the type of microflora and the biochemical changes that take place in the food during microbial proliferation. The nature of the biochemical changes and the rate at which they occur will determine acceptability or rejection of the food. Changes generally result from microbial metabolism of low-molecular-weight compounds (e.g., simple sugars and amino acids). Degradation of high-molecular-weight compounds (e.g., polysaccharides and proteins) normally takes place at a later stage, by which time the food is already judged as spoiled. Odor is the most sensitive indicator of biochemical changes in foods, and changes in appearance, texture, and color follow. Aerobic spoilage of foods high in protein is described as putrefaction, resulting from production of volatile nitrogenous compounds. These same foods (e.g., fresh meats) undergo different spoilage patterns as the gaseous atmosphere changes, resulting in different odors. Foods high in carbohydrates often undergo fermentation and are described as sour, whereas spoilage of fatty foods is associated with lipolytic activity that leads to rancidity.

II. PSYCHROTROPHIC AND PSYCHROPHILIC MICROORGANISMS

The ability of microorganisms to grow at low temperature in foods was first demonstrated by Forster in 1887. The term *psychrophiles* was proposed in 1902, referring to those organisms that can grow at 0°C. Eddy, in 1960, proposed the term *psychrotrophs* for organisms able to grow at temperatures of about 5°C or less, suggesting that the term *psychrophiles* be used only for those organisms that display a low optimal growth temperature. In 1965, Ingram suggested that the term psychrophiles be used to describe organisms with a minimal generation time at 10–15°C and with generation times not exceeding the optimum values by more than a factor of 10 at 0–15°C. According to this definition, the psychrotrophs are much more widely distributed (eurythermal, or growing over a wide temperature range) than the psychrophiles (stenothermal, or growing over a narrow temperature range). Psychrotrophs include various genera of bacteria, fungi, and other organisms. Most of those associated with spoilage of foods stored at refrigerator temperatures are aerobes or facultative anaerobes; relatively few anaerobic psychrotrophs have been isolated. Selectivity is affected by the availability of O_2 in refrigerated foods.

III. ADAPTATION OF MICROORGANISMS TO LOW TEMPERATURES

Microbial adaptation to growth at low temperatures, particularly of bacteria, has been a subject of numerous studies, but the precise cause for minimal temperatures in the growth of microorganisms is not known. Adaptation to low temperatures involves overall maintenance of cellular integrity. Levels of cell proteins, RNA, cell permeability, cytoplasmic membrane lipid, and phospholipid composition are some of the characteristics that may change in response to low temperatures.

The uptake of nutrients by the prokaryotic cell is governed by the membrane, which is composed mainly of phospholipids and proteins. The phospholipids form the basic structure of the membrane and are composed of hydrophobic and hydrophilic portions, arranged in such a way that the former are directed inward and the latter toward the outside, where they associate with water. The major proteins of the membrane are hydrophobic, embedded in the phospholipid matrix. Although this structure is stabilized mainly by hydrogen and hydrophobic bonding, cations, such as magnesium and calcium, combine with some of the negative charges of the phospholipids and contribute to the stabilization of the membrane structure. The hydrophobic groups are affected by temperature, leading to dissociation or aggregation that produces changes in the membrane and influences nutrient uptake.

The lipid content of most bacteria is in the cell membrane, between 2 and 5%. A number of psychrotrophic microorganisms have been shown to respond to low temperatures by synthesizing lipids and phospholipids with increased amounts of unsaturated fatty acids at the expense of the saturated acids or fats with short-chain fatty acids. This change contributes to maintenance of membrane fluidity and its function. Studies on the composition of a psychrotrophic yeast (*Candida* sp) also showed that unsaturated fatty acids increased with decreasing growth temperatures. Because such changes decrease the melting point of the membrane lipids, thereby enabling the cell to maintain mobility at low temperatures, it can be argued that they are associated with a physiological mechanism of the cell.

Studies of change in phospholipid composition in response to temperature variations indicate that not all microorganisms undergo the same alterations in response to reduced temperatures. In a number of investigations, no difference in either fatty acid composition or phospholipid content could be demonstrated during growth at room and low temperatures. For example, psychrotrophic pseudomonads grew at a slower rate in response to lowering of temperature but exhibited no change in free fatty acid and phopholipid composition at low temperatures. It has been suggested that organisms such as psychrotrophic pseudomonads contain a sufficient proportion of unsaturated fatty acids to enable them to grow at low temperatures, whereas modification of lipid composition is necessary for psychrophiles. Experiments with a *Pseudomonas aeruginosa* strain grown at 30°C showed that the cells were susceptible to sudden lowering of temperature, whereas the same strain grown at 10°C was not susceptible, demonstrating that the growth temperature of an organism influences its ability to regulate its lipid fluidity. A reduced rate of protein synthesis is seen as temperature is decreased. The precise mechanism of this observation is also not well understood.

Psychrophiles have enzymes able to catalyze reactions more efficiently at low temperatures and, perhaps because of this, are very sensitive to elevated temperatures, undergoing rapid inactivation at 30–40°C. For example, production of extracellular dextran by *Leuconostoc* species catalyzed by dextransucrase is more effective at low temperatures; the enzyme is temperature-sensitive and is inactivated above 30°C.

Unlike thermophiles, a large number of psychrotrophs and psychrophiles, both bacteria and yeast, are pigmented. Pigment production in these microorganisms is higher under psychrotrophic conditions than under mesophilic conditions.

IV. FOOD-BORNE PSYCHROTROPHIC MICROORGANISMS

A large number of bacteria, yeast, and molds are found in fresh foods. However, most of these organisms are unable to grow during storage at refrigerator temperature; the microflora becomes more homogeneous and fewer genera and species become predominant.

A. Bacteria

Genera of psychrotrophic bacteria associated with refrigerated foods and their major properties are summarized in Table I (gram-negative) and Table II (gram-positive). A major group consists of aerobic gram-negative rods, which are oxidative rather than fermentative (Table I). From their habitat of water and soil, they get into foods such as meat, poultry, fish, shellfish, milk, dairy products, and liquid eggs. These organisms include species belonging to the genera *Acinetobacter, Alcaligens, Flavobacterium,*

TABLE I
Major Gram-Negative Bacteria Associated with Refrigerated Foods[a,b]

Family	Genus	A/An/FA	M	O	C	f/o	G + C of DNA (mol%)	Common foods
Alcaligenaceae	*Alcaligenes*	A	+	+	+	o	56–70	Raw milk, dairy products, processed meats
Enterobacteriaceae	*Enterobacter*	FA	+	−	+	f	52–59	Fresh meat, poultry, processed meats
	Erwinia	FA	±	−	+	f	53–54	Vegetables
	Escherichia (P)	FA	+	−	+	f	48–52	Raw and undercooked meat
	Hafnia	FA	±	−	+	f	48–49	Fresh and vacuum packaged meats
	Pantoea	FA	+	−	+	f	55–61	Fresh and processed meats, poultry, fish
	Proteus	A	+	−	+	f	38–42	Eggs, meat
	Salmonella (P)	A	±	−	+	f	50–53	Poultry, eggs, seafood
	Yersinia (P)	FA	±	−	+	f/o	46–50	Dairy products
Neisseriaceae	*Acinetobacter*	A	−	−	+	o	39–47	Meat, raw milk, liquid eggs
	Moraxella	A	±	+	+	o	40–46	Meat, raw milk, liquid eggs, poultry, fish
	Psychrobacter	A	−	+	+	f	44–46	Fresh and processed meats, pork, fish
Pseudomonadaceae	*Pseudomonas*	A	+	±	+	o	57–70	Raw meat, poultry, seafood, milk, liquid eggs
	Shewanella	A	+	+	+	o	43–48	Meat, poultry, seafood
Vibrionaceae	*Aeromonas* (P)	FA	+	+	+	f	57–63	Seafood, raw meat
	Vibrio (P)	FA	+	+	+	f/o	38–51	Seafood
	Flavobacterium[c]	FA	±	+	+	o	63–70 (M+) 30–42 (M−)	Fish, raw meat, poultry, raw milk, liquid eggs

[a] A/An/FA, aerobe/anaerobe/facultative anaerobe; M, motility; O, oxidase; C, catalase; f/o, fermentative/oxidative metabolism; P, contains food-borne human pathogenic species; G, guanine; C, cytosine.
[b] From Bergey's *Manual for Systematic Bacteriology* and other sources.
[c] The genus is a taxonomically heterogenous group currently under revision.

Moraxella, Pantoea, Pseudomonas, and *Shewanella*. *Pseudomonas* species are the most important bacteria in spoilage of refrigerated foods.

Several gram-positive bacteria are associated with refrigerated foods (Table II). Most of these are facultative anaerobes or anaerobes, including species belonging to the genera *Bacillus, Brochothrix, Carnobacterium, Clostridium, Lactobacillus,* and *Listeria*.

B. Food-Borne Human Pathogens

Estimates of incidence of food-borne diseases worldwide are very high, and most of these diseases are caused by bacterial pathogens that contaminate food. Although many of the bacterial pathogens are mesophilic and a substantial number of cases are caused by foods not properly cooled or not stored

TABLE II
Major Gram-Positive Bacteria Associated with Refrigerated Foods[a,b]

Family	*Genus*	*A/An/FA*	*M*	*C*	*f/o*	*G + C of DNA (mol%)*	*Common foods*
Bacillaceae	*Bacillus* (P)	A/FA	+	+	o	32–62	Fresh and processed meats, poultry, seafood
	Clostridium (P)	An	±	−	f	23–43	Fresh and processed meats, poultry, seafood
Corynebacteriaceae	*Arthrobacter*	A	−	+	o	59–70	Cheese
	Corynebacterium	A	−	+	f	51–65	Fresh and processed meats, poultry, seafood
Lactobacteriaceae	*Brochothrix*	FA	−	+	f	35	Fresh and processed vacuum-packaged meats, sausage
Lactobacillaceae	*Carnobacterium*	A/FA	±	−	f	33–37	Processed (vacuum-packaged) meat, poultry, fish
	Lactobacillus	FA/An	−	−	f	35–53	Dairy products, vegetables, processed and vacuum-packaged meats
	Lactococcus	A/FA	−	−	f	34–36	Dairy products
	Leuconostoc	FA	−	−	f	38–42	Wines, beers, fruit juice, pickles, etc.
	Pediococcus	A/FA	−	−	f	34–44	Fresh and processed meats, poultry, seafood, dairy products
Micrococcaceae	*Micrococcus*	A/FA	−	+	f/o	66–75	Fresh livers, fresh and processed meats
	Listeria (P)[c]	FA	+	+	f	36–38	Milk, soft cheese, meat, fish, vegetables, fermented products

[a] A/An/FA, aerobe/anaerobe/facultative anaerobe; M, motility; C, catalase; f/o, fermentative/oxidative metabolism; P, contains food-borne human pathogenic species; G, guanine; C, cytosine.
[b] From Bergey's *Manual for Systematic Bacteriology* and other sources.
[c] A distinct taxon within the *Clostridium–Lactobacillus–Bacillus* branch.

at refrigerator temperatures, a number of bacterial pathogens are psychrotrophs, and their contamination of refrigerated foods can result in growth to an extent sufficient to cause disease. Hence, even though refrigeration can suppress growth of many food pathogens, it is not a means of preventing food-borne illnesses. Reliance on refrigeration storage has increased in recent years, with increasing consumer preferences toward minimally preserved, ready-to-eat foods leading to growth in sales of preprepared salads, meats, poultry, and seafood. Consequently, knowledge about virulence of some of the psychrotrophic pathogens is recent, having been gathered only in the past 10–20 years, and some of the psychrotrophs have not been recognized until recently as agents of human diseases.

Pathogenic bacteria, which have been shown to cause disease in humans after consumption of refrigerated foods, belong to the genera *Aeromonas, Bacillus, Clostridium, Escherichia, Listeria, Salmonella, Vibrio,* and *Yersinia.* Of these, only *Aeromonas, Listeria,* and *Yersinia* are psychrotrophs, capable of multiplication in refrigerated foods, and *Vibrio* spp. are true psychrophiles. Some strains of *Bacillus cereus* and *Clostridium botulinum* are psychrotrophic, and pathogenic *Escherichia coli* and *Salmonella* strains can be found in refrigerated foods. *Plesiomonas shigelloides* has been recovered from fish and shellfish and from patients with diarrhea. The organism is believed to be the etiologic agent of human diarrhea and its growth has been observed at 7–10°C. These pathogens will be discussed, and major properties of *A. hydrophila, L. monocytogenes, Y. enterocolytica,* nonproteolytic *C. botulinum,* shiga toxin-producing *E. coli, B. cereus* causing diarrheal-type illness, and *V. parahaemolyticus* are summarized in Table III.

1. *Aeromonas hydrophila*

The genus *Aeromonas* consists of several species, but most studies have been conducted on *A. hydrophila.* Evidence for human pathogenicity of this gram-negative rod is indirect, based on enteropathogenicity in animal models and on isolation of the organisms from foods and stools of patients with diarrhea. Although no food-borne outbreaks have been fully confirmed so far and isolation from stools is not sufficient proof that these bacteria are the cause of the illness, because they could have invaded the intestines as a result of an altered microbial environment, *A. hydrophila* is regarded as a potential food-borne pathogen. Cause of septicemia by *Aeromonas* in immunocompromised patients is well-documented.

Aeromonas species suspected of being human pathogens are primarily aquatic bacteria, occurring in fresh and ocean waters. Isolation from municipal and hospital water systems is also documented. Feces of healthy farm animals, primarily cows, have been shown to be a source of the organism, and *Aeromonas* species have been isolated from raw offals and poultry and from foods of animal origin. Seafood, particularly raw oysters, are implicated as the possible cause of *A. hydrophila* gastroenteritis.

Most species grow in the 20–35°C temperature range, whereas a significant number are psychrotrophic and capable of growth at 5°C or below. Although growth is over a wide pH range (4–10), the strains are less tolerant to acid and salt at 5°C than at higher temperatures.

Gastroenteritis is characterized by watery stools and a mild fever in the "choleralike" illness and presence of blood and mucus in the stools in the "dysenterylike" illness. The diarrhea is usually mild but can occasionally be severe. Isolation of the organisms from diarrheal stools is most common among the young, the old, and the immunocompromised, peaking during the summer and fall. Infections spreading outside the gastrointestinal tract are also documented, including septicemia, meningitis, and wound infection. A 52-kDa polypeptide with enterotoxic, cytotoxic, and hemolytic activities is produced by virulent strains. Correlation was found between enterotoxin production and positive lysine decarboxylase reaction.

Aeromonas species are sensitive to heat and radiation and are killed at pasteurization temperatures. Proper cooking is the best method to prevent the illness. Although resistant to CO_2, the organisms grow more slowly at low temperature in the presence of the gas and will not grow in a 100% CO_2 atmosphere.

2. *Listeria monocytogenes*

Listeria monocytogenes emerged as a food pathogen of concern to humans in the 1980s. Of the six species of *Listeria* (the former *L. murrayi* was merged with *L. grayi* in 1992), only *L. monocytogenes* is recognized as a human pathogen. Although the ability of the organism to cause human listeriosis was reported in 1926, a number of large outbreaks occurred in Europe and North America since 1980, with a mortality rate of about 30%. Foods implicated in transmitting the infection included coleslaw, milk, soft cheese, and meat products. Consequently, several foods stored at refrigeration temperatures and consumed without subsequent heat treatment are considered a potential source of the infection.

The organism is widespread in the environment and has been recovered from soil, water, sewage, decaying vegetation, wild and domestic animals, animal feed, and fresh and processed foods. Isolation from raw meat and poultry samples is common.

Listeria monocytogenes strains can grow at 5°C, or even 1°C, and up to 45°C, at pH 4.1–9.6 and 10% NaCl. They survive in 16–20% NaCl and long periods of drying and freezing. All pathogenic strains lyse mammalian red blood cells. Invasiveness and cell-to-cell spread in liver have been confirmed.

All virulent strains of *L. monocytogenes* produce listeriolysin O (LLO), a 60 kDa hemolysin homologous to streptolysin O and responsible for β-hemolysis on erythrocytes. The organism is known to cause abortion, meningitis, encephalitis, septicemia, and

TABLE III
Properties of Pathogens Associated with Refrigerated Foods[a]

Property	*Aeromonas hydrophila*	*Listeria monocytogenes*	*Yersinia enterocolytica*	*Clostridium botulinum group II (type E, nonproteolytic B&F)*	*Shiga toxin-producing Escherichia coli*	*Bacillus cereus (diarrheal syndrome)*	*Vibrio parahaemolyticus*
Growth conditions							
Temperature °C, range	0–45	1–45	0–45	3.3–45	5–44	4–50	5–44
pH, range	4–10	4.1–9.6	4.8–9	4.7–8	4.4–9	4.9–9.3	4.8–11
a_w, minimum[b]	[c]	0.91–0.93	[c]	0.97	0.95	0.93	0.94
NaCl, % upper limit[b]	5	10	5–6	5–6	6	[c]	8
Known virulent factor	Cytotoxin/enterotoxin/hemolysin	Hemolysin	Heat-stable enterotoxin	Neurotoxin	Shigalike cytotoxin	Complex hemolysin BL	Thermostable hemolysin
Disease symptoms	Diarrhea, mild fever, septicemia, meningitis	Abortion, meningitis, encephalitis, septicemia, endocarditis	Enteritis, diarrhea, fever, septicemia, arthritis	Nausea, vomiting, blurred vision, paralysis, respiratory failure	Abdominal pain, watery/bloody diarrhea	Diarrhea, nausea	Diarrhea, cramps, nausea
Foods implicated	Oysters (suspected)	Milk, soft cheese, coleslaw, meat and seafood products	Pasteurized milk linked to contamination from pigs; raw pork	Seafood	Undercooked ground beef, unpasteurized milk, apple juice, cider, raw vegetables	Cereal dishes, meat products, milk	Fish and shellfish dishes
Infectious dose, cells	[c]	$<10^2$–10^6	[c]	Presence of toxin; cell numbers usually high	50–500	Presence of toxins; presence of high cell numbers (10^5–10^7)	1×10^5
Prevention	Standard pasteurization	Proper cooking or reheating	Standard pasteurization	Thermal inactivation of the toxin	Proper heating or cooking of foods	Control of spore germination & of growth of vegetative cells	Proper heating or cooking of foods

[a] Compiled from different sources.
[b] Estimates for factor considered alone in broth.
[c] Not known.

endocarditis in adults and meningitis in newborns. Immunocompromised individuals are particularly predisposed to listeriosis, whereas infected healthy individuals may experience flulike symptoms only. Susceptibility to the disease may be due to low gastric acidity.

Proper cooking of foods and reheating to greater than 74°C is required. Immunocompromised individuals are advised to avoid refrigerated ready-to-eat meat products and cross-contamination of stored foods.

3. *Yersinia enterocolitica*

Yersinia enterocolitica was identified as a human pathogen in 1939. Since the mid-1970s, a number of food-associated yersiniosis outbreaks have been reported in the United States, Japan, and the northern part of Europe, mostly during the fall and winter months. Cases in different parts of the world have been associated with different serotypes of the organism. Although 10–20% of pasteurized milk and prepared food samples surveyed contained yersinias, the potential for toxin production in refrigerated foods is not clear.

Healthy pigs appear to be a major reservoir of pathogenic *Y. enterocolitica,* and raw pork or foods contaminated with the organism are believed to be vehicles for yersiniosis cases. Although isolates have been recovered from other meat animals and seafood, these are nonpathogenic.

The organism is a psychrotroph that can grow to high numbers in raw or cooked meats at temperatures as low as 0°C and up to 45°C and over a pH range of 4.8–9.0.

During infection, the bacteria invade the intestinal mucosa and pass through the intestinal epithelium. They multiply within macrophages, and migration may give rise to systemic infection. Virulence is associated with a 44–48 kDa plasmid that encodes for several virulence-related antigens. Most clinical isolates produce a heat-stable enterotoxin. However, because strains that do not produce enterotoxin *in vitro* have been shown to cause diarrhea in mice, this enterotoxin does not appear to play a role in the infection. Yersiniosis is reported to be most common among children less than one year old. *Yersinia enterocolitica* is heat-sensitive and is destroyed in 1–3 min at 60°C and at milk pasteurization temperatures.

4. *Vibrio parahaemolyticus*

The organism was first identified as a cause of food-borne disease in 1950 by Japanese investigators. It is a psychrophile associated with food-borne outbreaks originating from seafood consumption.

Vibrio parahaemolyticus is halophilic, requiring sodium chloride for growth. It is present in marine environments and in fish and shellfish, primarily during the summer and early autumn, when water temperature is above 13–15°C. However, growth at 5°C has also been reported. Almost all strains isolated from infected patients are Kanagawa-positive, possessing a thermostable hemolysin that produces β-hemolysis in tests using human erythrocytes.

Gastroenteritis is associated with consumption of contaminated seafood, where counts per gram of 10^2–10^4 have been reported. A minimum infectious dose ranges from 10^5–10^7 cells. Heating at 60°C for 15 min kills the organism. Although refrigeration causes reduction in cell numbers, contaminated shellfish kept refrigerated for a short time and consumed raw can cause the disease.

5. *Bacillus cereus*

Growth and multiplication of vegetative cells of *Bacillus cereus* generally occur within the 10–50°C range. Contamination of raw milk appears to be the source of psychrotropic strains in pasteurized milk. Their growth at refrigerator temperatures has been reported recently in raw milk samples and in a number of other foods, but production of toxins at these temperatures has not been clearly established.

Spores and vegetative forms of the organism are common in soil, water, vegetation, and many types of foods, notably milk, where it causes spoilage referred to as "sweet curdling," meat products, cereals, and vegetables. Raw rice is frequently contaminated with *B. cereus,* and fried or boiled rice dishes are implicated in most of the emetic-type food poisoning, linked to cell numbers per gram in the foods ranging from 10^4 to 10^9. The emetic syndrome strains grow over the temperature range of 15–50°C. Cereal dishes, meat products, and milk are among foods causing the diarrheal syndrome. The diarrheal strains produce a complex hemolysin, and toxin production in milk at 6°C has been reported when cell numbers exceeded 10^7/ml.

Because cooking at temperatures of or less than 100°C does not destroy all spores, food poisoning is prevented by control of spore germination and prevention of growth of vegetative cells in cooked, ready-to-eat foods. Spore germination, demonstrated between −1° and 59°C in the laboratory, can be greatly reduced by refrigeration but will proceed when large quantities of foods that have not been cooled are refrigerated.

6. *Clostridium botulinum*

The organism is a gram-positive, spore-forming anaerobic rod. Growth of a limited number of strains of *C. botulinum* group II has been reported to occur at a temperature range of 3.3–45°C. Of the seven different types (A–G), which produce serologically distinct neurotoxins, E belongs to group II, can contaminate fish and seafood, and can produce toxin in refrigerated foods. The spores are heat-sensitive (D_{100} of <0.1 min), but the concern is with fish and seafood that are eaten without cooking and with fish products packaged in an anaerobic atmosphere (i.e., modified atmospheres), items such as smoked fish, that are eaten without further cooking. Nonproteolytic psychrotrophic types B and F are less heat-resistant than the proteolytic strains and have also been implicated in food poisoning associated with refrigerated meats.

7. *Escherichia coli*

The organism is part of the normal human intestinal microflora. Outbreaks caused by *E. coli* from consumption of cheese in 1971 and from consumption of undercooked ground beef in 1982 established the organism as a human pathogen.

Of the four groups of food-borne pathogenic *E. coli* recognized, the shiga toxin-producing *E. coli* (STEC) is the most important, having caused a series of food-borne outbreaks of hemorrhagic colitis and hemolytic–uremic syndrome (HUS) worldwide. Clinical isolates produce toxins that are indistinguishable from shiga toxin produced by *Shigella dysenteriae*, and infections have been linked primarily to consumption of undercooked ground beef and unpasteurized milk. The majority of hemorrhagic colitis cases have been associated with strain O157:H7, which is considered the most important serotype of STEC in North America. Cattle and other ruminants are major reservoirs of the organism, and principal entry to the food supply is by contamination of meat during slaughter. Several outbreaks associated with apple juice or cider have been confirmed since 1991, and manure was suspected as the source of contamination of the apples. Raw vegetables (lettuce, sprouts) have also been implicated in *E. coli* O157:H7 outbreaks in North America, Europe, and Japan. Adequate heating or cooking before consumption prevents infections.

8. *Plesiomonas shigelloides*

The organism was first isolated in 1947. Unlike other Vibrionaceae, it does not require added NaCl for growth. The organism is not considered a psychrotroph, and only a few strains grow at 8°C. *P. shigelloides* causes diarrhea in some people, and there are limited reports of food-borne enteritis associated with contaminated oysters, fish, chicken, and water in adults and children.

9. *Salmonella*

The salmonellae are widely distributed in nature and their primary habitat is the intestinal tract of animals. Food poisoning occurs when a large number of the pathogen are ingested, and eggs, poultry, meat, and meat products are the most common vehicles of human salmonellosis. Of over 2000 serovars recognized, *S. typhimurium* is the most common isolate throughout the world, while *S. enteritidis* is prevalent in poultry and eggs. Minimum growth temperatures of 4–7°C have been reported for some *Salmonella* spp. The salmonellae are heat sensitive and are killed at temperatures of milk pasteurization.

C. Yeast and Mold

A number of yeast and mold organisms are capable of growth at refrigerator temperatures. However, they do not compete well with bacteria when conditions favor bacterial growth and they often develop only after thermal processing or when conditions such as water activity and pH are not conducive to

TABLE IV
Major Yeast and Mold Commonly Found in Refrigerated Foods[a]

Genus	*Common foods affected*
Yeast	
Candida	Fresh meat and poultry, processed meats, seafood
Cryptococcus	Seafood
Rhodotorula	Fresh meat, poultry, seafood
Torulopsis	Fresh meat, poultry, seafood, processed meats, butter
Mold	
Alternaria	Fresh and processed meats, poultry, vegetables, citrus and other fruits, butter, cheese
Aspergillus	Fresh and processed meats, poultry, seafood, butter, fruit
Botrytis	Fresh and processed meats, poultry, vegetables, fruit
Cladosporium	Fresh and processed meats, poultry, eggs, fruits, potatoes, butter
Fusarium	Fresh and processed meats, vegetables
Geotrichum	Fresh and processed meats, poultry, vegetables
Mucor	Fresh and processed meats, poultry, eggs, butter, cheese, fruit
Penicillium	Fresh and processed meats, poultry, seafood, eggs, butter, cheese, fruit
Phytophthora	Vegetables, fruits
Rhizopus	Fresh and processed meats, poultry, vegetables, butter
Thamnidium	Fresh and processed meats

[a] Compiled from different sources.

bacterial growth. Table IV lists the most common yeasts and molds that are capable of growth at low temperatures, along with foods in which they may occur. In addition to meat and poultry, both yeast and mold are associated with fruits and vegetables. Yeast may reach high numbers in refrigerated fruit juices, and mold often cause spoilage of fruit and vegetables where skin damage has occurred. They are also found on low-moisture meat surfaces.

Like bacteria, yeast and mold are used to impart desirable characteristics to a number of foods. Flavor and color developments in a variety of cheeses and sausages are brought about by growth of these microorganisms.

V. REFRIGERATED FOODS

The composition and pH of typical refrigerated foods are presented in Table V. These include the highly perishable raw meat, chicken, and fish, cooked meat and sausage, dairy products (milk, butter, and cheese), eggs, two vegetables (lettuce and tomatoes), and two fruits (apples and oranges). The storage life of these foods will depend on their initial microbiological and physical quality, ranging from a few days (i.e., fish) to several months (i.e., eggs in the shell). The microbiology of these foods and their spoilage process during refrigeration are discussed later.

A. Fresh Meat, Poultry, and Seafood

1. Meat

Fresh meat microflora is heterogeneous, consisting of several genera of bacteria, yeast, and mold. Initial microbial counts vary greatly, from 10^2 to 10^5 cells/g or cm^2, depending on the type of meat and handling procedures. Minced or ground beef has the highest initial microbial load. On refrigeration, growth of most of the organisms is suppressed, and the predominant microflora under aerobic storage consists of gram-negative species of the genus *Pseudomonas*. Some members of the family Enterobacteriaceae are also common. Lactic acid bacteria and micrococci also grow occasionally. Yeast (*Candida* spp) and mold (*Geotrichum, Mucor, Penicillium, Rhizopus, Sporotrichum,* and *Thamnidium* spp) are unable to compete with the bacteria and may grow primarily on meat surfaces, where low water activity is not favorable for bacterial growth. The pH range of fresh beef and pork, 5.6–6.2, affects the organisms and their rate of growth. Total counts generally attain 10^8–10^9 cells/g or cm^2 after 7 days at 5°C, and bacterial proliferation is accompanied by pH increase. Putrid odors are easily discerned at this time. Beef or pork with a high pH (DFD, or dark, firm, dry) spoils faster than normal pH meat. Growth of lactic acid bacteria is accompanied by acid production, which lowers the pH, and production of antimicrobials, as yet not fully defined, by these organisms may influence the growth rate of gram-negative species. This phenomenon is termed lactic antagonism.

TABLE V
Approximate Composition and pH of Major Fresh Refrigerated Foods[a]

Food	pH	Water	Protein	Carbohydrate	Lipid	Ash
		Composition (%w/w)				
Beef						
Raw	5.6–6.2	66.6	20.2	0	12.3	0.9
Cooked	5.2–6.2	54.7	28.6	0	15.4	1.3
Chicken						
Raw	5.7–6.5	75.7	18.6	0	4.9	0.8
Fish						
Raw	6.6–6.8	81.2	17.6	0	0.3	1.2
Sausage						
Frankfurter, cooked	6.1–6.3	57.3	12.4	1.6	27.2	1.5
Milk						
Whole	6.3–6.5	87.4	3.5	3.5	4.9	0.7
Butter	6.1–6.4	15.5	0.6	0.4	81.0	2.5
Cheese						
Cottage	4.7–4.9	78.3	13.6	2.9	4.2	1.0
Cheddar	5.5–5.9	37.0	25.0	2.1	32.2	3.7
Eggs						
Whole	6.9–7.1	73.7	12.9	0.9	11.5	1.0
Yolk	6.2–6.8	51.1	16.0	0.6	30.6	1.7
White	7.6–9.3	87.6	10.9	0.8	0	0.7
Lettuce	5.9–6.1	95.1	1.2	2.5	0.2	1.0
Tomatoes	4.2–4.4	93.5	1.1	4.7	0.2	0.5
Apples	2.9–3.3	84.1	0.3	14.9	0.4	0.3
Oranges	3.6–4.3	88.3	0.7	10.4	0.2	0.4

[a] From Watt, B. K., and Merrill, A. L. (1963). "Composition of Foods." USDA, Washington, DC, and other sources.

2. *Poultry*

Overall, composition and microbial growth of refrigerated fresh poultry resemble that of fresh meats. The pH range is generally similar as well, although values are lower in the breast meat (~5.7) and higher in the dark meat of the leg muscle (~6.5), and these differences affect bacterial growth. For example, *Acinetobacter* species are adversely affected when the pH is 5.7 or less, but grow well in the high pH of the leg muscle. Spoilage is characterized by bacterial proliferation, appearance of off-odors, and sliminess on the outer surfaces. Aerobic counts per square centimeter can reach 10^9. The genera *Candida, Rhodotorula,* and *Torulopsis* are the most important yeasts, whereas molds grow only when bacteria are suppressed by the use of antibiotics.

3. *Fish*

Spoilage of fresh fish is essentially similar to that of fresh meat and poultry, despite differences in chemical composition. Fish contain nonprotein nitrogenous compounds, such as trimethylamine oxide, creatine, taurine, and histamine, in addition to free amino acids. Microbial growth begins on the skin, gills, and lining of the belly cavity. Initial microbial numbers are approximately 10^4/g, and numbers increase to 10^8 at overt spoilage. Classification of fish spoilage flora shows predominance (>80%) of *Pseudomonas* species and *Shewanella putrefaciens.* Utilization of the simple nitrogen-containing compounds by these organisms leads to production of various malodorous volatile compounds, several of which contain sulfur. In addition, species of *Photobacterium,* a facultative anaerobe, develop in certain fish and shellfish during storage and reduce trimethylamine oxide (TMAO) to trimethylamine (TMA).

4. *Shellfish*

Shellfish include crustaceans (shrimp, lobster, crab, crayfish), and mollusks (oysters, clams, scallops, squid). In contrast to fish that do not contain carbohydrates, crustaceans contain about 0.5% carbohydrate, and mollusks contain still higher amounts (5.6% in oysters), largely in the form of glycogen. In general, shellfish also contain higher amounts of free amino acids and nonprotein nitrogenous compounds than do fish. These differences in composition and the type of microbial flora, which depends on the quality of water and factors related to handling of the seafood, affect spoilage patterns and rates. Because of the relatively high glycogen levels, mollusks undergo fermentative spoilage, and a decrease in pH is an indication of spoilage.

B. Cooked Meat and Poultry

Yeast and lactic acid bacteria are major causes of spoilage of refrigerated cooked sausage, hot dogs, and similar products. Growth of various species may result in slime and greening on the outer surface and in souring underneath the casing. Slime consists of layers of colonies of the spoilage flora, greening is caused by reaction of bacterial (lactobacilli and *Leuconostoc*) peroxides with the cured meat pigments, and souring is caused by production of acids. Mold spoilage is seen only when conditions do not favor bacterial or yeast growth.

Bacon spoils primarily by mold growth because of its low water activity, whereas spoilage of cured refrigerated hams is accompanied by souring because solutions pumped into the hams during the curing process contain sugars that are fermented by the microorganisms.

C. Dairy Products

1. *Milk*

Milk is an ideal medium for growth of a large number of microorganisms (see Table V). Raw milk microflora is variable and depends on conditions of milking and handling. Genera typically found are *Streptococcus, Leuconostoc, Lactobacillus, Pseudomonas, Acinetobacter, Alcaligenes, Aeromonas, Bacillus,* and *Flavobacterium.* The initial numbers of about 10^4 cells/ml increase during refrigeration, and psychrotrophs become predominant, attaining numbers of up to 10^7/ml. Ropiness, a slime material caused by growth of *Alcaligenes viscolactis,* is sometimes seen in refrigerated raw milk, and gassiness may be the result of yeast growth. Microbial growth can give rise to off-odors, off-flavors, coloration, souring, and curdling. Such undesirable changes become apparent generally when the total viable count per milliliter is 10^6–10^7.

Pasteurization of milk eliminates all but thermoduric strains, primarily lactobacilli, streptococci, and sporeformers of the genera *Bacillus* and *Clostridium.* Growth of the lactic acid bacteria is accompanied by utilization of lactose and production of lactic acid that lowers the milk pH. Souring may continue even to a pH of 4.0 or less, whereas curdling takes place when the pH drops to about 4.5. Presence of molds may lead to their growth at the milk surface, utilization of the lactic acid, and an increase in milk pH. Postpasteurization contamination of refrigerated milk can lead to growth of high numbers of psychrotrophic strains belonging to the family Enterobacteriaceae. Lipolytic and proteolytic enzymes produced by psychrotrophic organisms can survive pasteurization, and their activity can lead to off-odors, change in consistency, and bitter flavors.

2. *Butter*

Microorganisms in butter are those derived from the cream used for its manufacture or from equipment. Spoilage may be of the putridity or rancidity type. Putridity is caused by growth of *Pseudomonas putrefaciens* on the butter surface. Release of certain organic acids and putrid odors may become apparent within a week in the refrigerator. Rancidity is caused by lipase from *P. fragi, P. fluorescens,* or from sources other than microorganisms and is accompanied by hydrolysis of butterfat and release of free fatty acids. Although other bacteria may cause butter spoilage, the low water content (15.5%) makes butter more susceptible to spoilage by molds. Species of the genera *Alternaria, Aspergillus, Cladosporium, Geotrichum, Mucor, Penicillium,* and *Rhizopus,* as well as the black yeasts of the genus *Torulopsis,* have all been reported to cause spoilage, often accompanied by discoloration of the butter. Contamination of pasteurized fresh uncultured cream leads to undesirable bacterial growth and off-flavors. Nowadays, uncultured cream may undergo ultrahigh temperature (UHT) treatment that extends the shelf life of butter during refrigerated storage.

3. Cheese

Selected bacterial starter cultures are used for the production of most unripened and ripened cheeses, as well as of sour cream, butter, buttermilk, yogurt, and other dairy products. Lactic starters include bacteria that convert lactose to lactic acid and consist of single or mixed strains. Although they are not psychrotrophic, a drop in pH of the products to 4.5 or less restricts spoilage by other organisms, resulting in foods that are less prone to spoilage. Salting of the curd and a prolonged ripening process decrease the water activity and further contribute to shelf stability.

Cottage cheese has a relatively short shelf life even when refrigerated, and it undergoes spoilage by bacteria, yeast, and mold. Slimy curd formation is a typical spoilage by gram-negative bacteria (e.g., *Alcaligenes, Pseudomonas, Acinetobacter* spp). Spoilage of ripened cheeses is caused primarily by molds because of their low moisture content. Anaerobic conditions support growth of *Clostridium* species (e.g., *C. tyrobutyricum, C. butyricum*) that use lactic acid with the production of CO_2 and gassiness.

D. Eggs

Freshly laid eggs are generally free of viable microorganisms, although *Salmonella enteritidis* inside the eggs has been reported. The egg yolk is an excellent growth medium but most organisms have first to penetrate the shell and membrane layers and to survive the inhibitors in the egg white, which include lysozyme, avidin, conalbumin, and a high pH of 9.0 or greater. Bacteria of the genera *Pseudomonas, Acinetobacter, Proteus, Aeromonas, Alcaligenes, Escherichia, Micrococcus, Salmonella, Serratia, Enterobacter, Flavobacterium,* and *Staphylococcus,* and mold of the genera *Mucor, Penicillium,* and others can be found on or inside eggs. Entry into the eggs is favored by high humidity. Rotting is a common bacterial spoilage, caused by *Pseudomonas* and other species. It can be colorless, green, pink, or red. Mustiness is another type of bacterial spoilage. Mold spoilage is identified by mycelial growth during candling of the eggs.

E. Vegetables and Fruits

Spoilage of vegetables and fruits is caused by preharvest and postharvest contamination. Most of it occurs after harvesting, when microbes invade bruised or damaged areas on the plants. Outbreaks of *E. coli* O157:H7 and *S. typhimurium* from consumption of apple cider are believed to have been caused by bruised apples contaminated with the pathogens. Ease of invasion is determined also by the protection provided by the skin, because a thick skin or integument acts as a barrier and natural defense against invasion. The proximate composition of representative vegetables and fruits stored at low temperatures is presented in Table V. Their nutrient content makes them susceptible to spoilage by bacteria, yeast, and molds. Storage at refrigeration temperatures is an effective method for prolonging the shelf life of a large number of vegetables and fruits. However, although microbial growth is slowed down, it is not eliminated, and optimum conditions depend on the vegetable or fruit.

1. Vegetables

The high water content, pH (generally 5–7), and the relatively high oxidation/reduction potential of vegetables favor growth of aerobes and facultative anaerobes. Microbial loads vary, from 10^3 to more than 10^6 cells/g, and mold counts tend to increase on crops harvested after rainfall. Species of the genus *Erwinia* are common causes of bacterial spoilage, referred to as bacterial soft rot. Although most of these species grow well at 37°C, they also grow well at refrigerator temperatures. A large number of vegetables are affected by the disease, giving rise to soft consistency and foul odor. Tissue softening is due to hydrolysis of pectin by pectinase. Once the plant tissue is damaged, fermentation of simple sugars takes place by a varied flora. *Pseudomonas* species are important plant disease agents, causing soft rot, blight, and leaf spot of a number of vegetables. Other diseases are caused by *Xanthomonas* and *Corynebacterium* species.

Molds are the most important group of organisms responsible for vegetable spoilage. Species of the genus *Botrytis* are the most common. Favored by high humidity and warm temperatures, *B. cinerea* causes gray mold rot on a large number of vegetables. Other examples of rot are caused by *Geotrichum candidum* (sour rot), *Rhizopus stolonifer* (soft rot), *Alternaria* species (black rot), and *Fusarium* species (brown rot).

Some fresh vegetables (e.g., cucumbers, cabbage) undergo lactic acid fermentation. The finished products (e.g., pickles, sauerkraut) have a low pH, in the range of 3.1–3.7, and they are generally pasteurized in hermetically sealed containers. Exceptions are those that undergo slow fermentation and receive no heat treatment. Such products are subject to spoilage by bacteria, yeast and mold, which cause softening, slime, and rot. Spoilage of these foods can be retarded by refrigeration but not eliminated.

2. *Fruits*

The carbohydrate content of fruits is typically higher than that of vegetables, and the water content is lower. Fruits are also characterized by a low pH (3.0–4.0), which places them in the high-acid food category. Because this pH is below the level that favors bacterial growth, bacteria are not of major importance in spoilage of most fruits. The principal spoilage microorganisms are yeast and mold, which are more aciduric than most bacteria. Infection occurs during harvesting and at different stages of marketing. Common organisms are species belonging to the genera *Penicillium, Aspergillus, Mucor, Alternaria, Cladosporium, Botrytis,* and *Monilinia,* which cause rots, spots, and other defects on the fruit. Mold growth on the infected area is favored by moist conditions. A large number of molds produce toxic metabolites, termed mycotoxins. For example, patulin is a mycotoxin produced by *Aspergillus* and *Penicillium* species. The optimum temperature for production of mycotoxins is between 24 and 28°, and they are not produced at refrigerator temperatures. A number of yeast genera are also found on fruits, where they utilize the sugars and produce alcohol and carbon dioxide.

F. Convenience Foods

A wide variety of ready-to-eat refrigerated foods is increasingly available in food stores. These are freshly prepared convenience foods, as opposed to the canned, frozen, or the chemically preserved. In addition to cooked meats, poultry, or fish discussed previously, they include various salads, baked products (e.g., pastries, quiches, pies), and pasta. The shelf life of convenience foods depends on a number of factors, including ingredients used for the preparation, their microbial quality, sanitary conditions during preparation, storage temperature history, pH, heat treatment, if any, given the food before refrigeration, and presence of any preservatives in the ingredients, which may add stability to the final product. Consequently, a large number of microorganisms may be present in the foods, and time to spoilage varies, between 7 and 10 days or even shorter. Some prepared foods require heating before consumption but most do not, raising the concern that potential contamination with pathogens will cause foodborne disease.

VI. MODIFIED ATMOSPHERE PACKAGING

The atmosphere surrounding a food can influence the type of microorganisms and their rate of growth. Alteration of the atmosphere composition is, therefore, useful for prolonging the storage life. Although respiration of fruits and vegetables is reduced at low temperatures, thereby increasing the storage life, it still contributes to loss of quality. Addition of carbon dioxide and removal of oxygen to desired concentrations are practiced to control deterioration of these foods by respiration. Alteration of atmosphere is also useful for other foods. Modified atmosphere packaging (MAP) technology was developed in the 1930s for the shipping of fresh beef stored under carbon dioxide. The growing consumer demand for fresh, preservative-free, refrigerated products in the 1980s has further increased the use of MAP. Major advantages of this technology are the pronounced shelf-life increase (from 50 to 400%), high quality of products, and decreased losses and distribution costs. Major disadvantages are the need for temperature control and for different gas formulations for different types of products, added costs associated with special equipment and trained personnel, and the concern that spoilage microorganisms, which might warn the consumers of spoilage, are inhibited, and growth of pathogens, which usually do not cause any detectable organoleptic changes, might be promoted.

In MAP, the food is packaged in an atmosphere modified to a composition other than air. Gases used for modified atmosphere include oxygen, nitrogen, and carbon dioxide. Oxygen stimulates growth of aerobic bacteria and inhibits growth of strict anaerobes. Nitrogen is an inert and tasteless gas, with little or no antimicrobial activity. Carbon dioxide is soluble in water and fat, inhibits product respiration, and is the most important from the microbiological standpoint. The reasons for its antimicrobial effects are not well understood. Among those suggested are inhibition of enzymes or their rate of activity, alteration of cell membrane function including nutrient uptake, or changes in properties of proteins, intracellular penetration, and pH drop. Inhibition is influenced by the levels of carbon dioxide used and by the solubility in the liquid phase of the food. Low storage temperature of MAP foods is required because solubility of carbon dioxide decreases with increase in temperature.

Aerobic spoilage organisms are inhibited in MAP foods. However, anaerobic and microaerophilic microorganisms, particularly anaerobic pathogens, can reach dangerous levels in the foods because of the extended shelf life of MAP products. Of major concern are the potential growth of *C. botulinum* type E and toxin production in MAP fish products and the growth of *L. monocytogenes* and *A. hydrophila* in a variety of foods.

A. Meats

Carbon dioxide, alone or in combination with oxygen, is effective in extending the shelf life of meats, particularly when the storage temperature is close to 0°C. Because preserving the fresh meat color is important, a mixture of CO_2 (10–30%) and air or O_2 is used. The latter serves to maintain myoglobin in the oxygenated form of oxymyoglobin, which is responsible for the bright red color of meat. The gram-negative meat spoilage bacteria (mainly *Pseudomonas, Moraxella,* and *Acinetobacter* spp) are generally inhibited by 20% CO_2, whereas gram-positive microorganisms (e.g., *Lactobacillus* spp, *B. thermosphacta*), which are resistant to CO_2, predominate. Spoilage is accompanied by off-odors and off-flavors arising from volatile microbial metabolites, primarily short-chain fatty acids but also other compounds produced by the lactobacilli and *B. thermosphacta*. Refrigerated MAP meats have a shelf life two to three times greater than air-packaged products.

B. Poultry

In fresh chickens, a CO_2 concentration of 60–80% causes a shift from a gram-negative to a gram-positive population and a two- to threefold increase in shelf life during refrigeration. Of the psychrotrophic pathogens, *L. monocytogenes* and *A. hydrophila* may grow to high numbers during the extended storage.

C. Fish

Although MAP extends the shelf life of fish, the concern in MAP-stored fish products is that restricted growth of normal fish spoilage bacteria may enhance growth of *C. botulinum* type E. The botulinal toxin can be detected well before any evidence of organoleptic deterioration. Moreover, even though salting adds a safety measure, botulinal spores can survive smoking. Storage below 3°C or sufficient heat treatment assures the safety of the products, but these are not always practical.

D. Vegetables

Increased carbon dioxide and decreased oxygen content have been shown to produce fungistatic effects on the common spoilage fungi in vegetables by repressing mycelial growth and spore development. However, modified atmospheres affect individual vegetables in different ways, and some of the changes in the vegetable itself tend to cancel any beneficial effects on fungal growth.

E. Miscellaneous Foods

Other foods packaged under MAP include cheese, salads, and a variety of sandwiches, where various, widely differing gas mixtures of carbon dioxide and nitrogen are employed. Aerobic counts per gram in MAP sandwiches were shown to increase on the average from about 10^5 to 10^7 at the end of the shelf life of about 30 days.

See Also the Following Articles

Dairy Products • Extremophiles • Food-borne Illnesses • Food Spoilage and Preservation • Lactic Acid Bacteria • Low-Temperature Environments • Meat and Meat Products

Bibliography

Balows, A., Truper, H. G., Dworkin, M., Harder, W., and Schleifer, K.-H. (1992), "The Prokaryotes" (2nd ed.). Springer-Verlag, New York.

"Bergey's Manual of Systematic Bacteriology," Vol. 1, N. R. Krieg, ed. (1984); Vol. 2, P. H. A. Sneath, ed. (1986), Williams and Wilkins, Baltimore, MD.

Berry, E. D., and Foegeding, P. M. (1997). *J. Food Protect.* **60**, 1583–1594.

Cousin, M. A., Jay, J. M., and Vasavada, P. C. (1992). Psychrotrophic microorganisms. *In* "Compendium of Methods for the Microbiological Examination of Foods" (C. Vanderzant and D. F. Splittstoesser, eds.), pp. 153–165. American Public Health Association, Washington, DC.

Deak, T., and Beuchat, L. R. (1996). "Handbook of Food Spoilage Yeasts." CRC Press Inc., Boca Raton, FL.

Doyle, M. P., Beuchat, L. R., and Montville, T. J. (1997). "Food Microbiology: Fundamentals and Frontiers." American Society for Microbiology Press, Washington, DC.

Farber, J. M. (1991). *J. Food Protect.* **54**, 58–70.

Gounot, A.-M. (1991). *J. Appl. Bacteriol.* **71**, 386–397.

ICMSF. (1996). "Microorganisms in Foods," 5, Microbiological Specifications of Food Pathogens". Chapman & Hall, Suffolk, UK.

Jay, J. M. (1996). "Modern Food Microbiology" (5th ed.). Chapman & Hall, New York.

Jay, J. M., and Shelef, L. A. (1991). The effect of psychrotophic bacteria on refrigerated meats. *In* "Biodeterioration and Biodegradation" (H. W. Rossmore, ed.), pp. 147–159. Elsevier, Barking, UK.

Meng, J., and Doyle, M. P. (1998). Microbiology of Shiga toxin-producing *Escherichia coli* in foods. *In* "*Escherichia coli* O157:H7 and Other Shiga Toxin-Producing *E. coli* strains" (J. B. Kaper and A. D. O'Brien, eds.), pp. 92–108. American Society for Microbiology, Washington, DC.

Retroviruses

Ralph Dornburg and Roger J. Pomerantz
Thomas Jefferson University

I. Retrovirus Genome and Life Cycle
II. Human Immunodeficiency Viruses
III. Oncogenic Retroviruses—Natural Retroviral Vectors
IV. Retroviral Vectors

GLOSSARY

AIDS Acquired immunodeficiency syndrome, a condition in which opportunistic infections cannot be cleared by the body, due to a failure of the immune system.

antigen A structure (e.g., a protein or polysaccharide) that can be recognized by an antibody.

anti-sense RNA An RNA transcript of a gene in the opposite orientation to the gene (complementary to the normal mRNA).

cDNA Complementary DNA copy transcribed from a messenger RNA template rather than a DNA template.

enhancer A DNA sequence that increases the efficiency of transcription from a promoter.

LTR Long terminal repeats; sequences located at the 5′ and 3′ ends of retroviral DNA.

oncogene A gene whose expression results in the malignant transformation of a cell.

promoter A DNA sequence that is recognized by RNA polymerases and drives gene expression.

protooncogene A cellular gene (most probably involved in differentiation and/or cell divisions), from which an oncogene evolved.

receptor A cell surface protein that binds a specific ligand, e.g., a protein or a virus.

transfection The introduction of genetic material into a cell by experimental procedures.

vector A genetically engineered DNA construct to introduce and express a gene in a target cell.

RETROVIRUSES are small enveloped viruses, which contain a single-stranded RNA genome. The RNA is reverse transcribed into a DNA intermediate, which is integrated into the host cell genome. Retroviruses are widespread in nature and can be associated with different forms of malignant tumors and other diseases. Thus, they are also termed RNA tumor viruses.

Immunodeficiency viruses (e.g., HIV-1 or SIV), which all belong to the family of lentiviruses, cause the acquired immune deficiency syndrome (AIDS) in man and monkeys. Many tumorigenic retroviruses contain a gene (oncogene) in their genome, in addition to or substituting for their replication genes. Thus, they are also termed oncogenic retroviruses. The expression of the oncogene results in the malignant transformation of the infected cells. Oncogenes are of cellular origin and were picked up by retroviruses in the course of earlier infections. Thus, retroviruses can act as vehicles to transfer cellular genes from cell to cell and from organism to organism. This ability, as well as the high efficiency and accuracy of retrovirus replication, make these viruses useful for the genetic engineering of transfer vectors to study a large variety of biological processes. Moreover, because of their properties, retroviral vectors are also being used in the first human gene therapy trials.

I. RETROVIRUS GENOME AND LIFE CYCLE

Retroviruses form a large group of viruses and have been found in all vertebrates investigated. They are divided into several subgroups, due to differences in morphology and differences in their genomic organization. However, they all use the same basic mechanism to replicate their genome. In general, the retroviral life cycle resembles that of many other animal viruses, in that it can be divided into virus attachment and entrance into cells, a synthesis period, and a period of virus assembly and release. However, the molecular mechanisms of replication are unique for this family of viruses (and a few related viruses) and follows a rather complicated, but highly efficient, pathway.

A. Virions, Genome, Taxonomy

Retrovirus virions are medium-sized particles with a diameter of approximately 100 nm, consisting of a core structure surrounded by a lipid bilayer membrane (Fig. 1). In electron micrographs, the shape of the core varies among different retroviruses, ranging from spherical (e.g., mouse retroviruses) to conelike (e.g., the human immunodeficiency viruses) forms. The core consists of viral proteins termed gag (group-specific antigen), the viral DNA polymerase and integration protein, two identical copies of the viral genomic RNA, and associated tRNA. Lentiviruses and human T-cell leukemia and related viruses also contain other regulatory proteins (for more details see the following). The lipid bilayer of the virion is derived from the host cell membrane and contains viral envelope proteins, which appear as spikes in electron microscope pictures. These proteins recognize and interact with host cell surface receptors and enable the virus to penetrate the cell. They determine the cell-type and species-specificity of the virus.

The RNA genome of all retroviruses resembles a eucaryotic mRNA, has a 5′ cap, and contains a poly(A) tail. It has the same polarity as the viral mRNAs, and therefore, it is called plus strand RNA. In simple retroviruses, it contains three genes coding for proteins (Fig. 2). The *gag* gene usually codes for four core proteins: the matrix protein (MA), which forms the junction between the core and the lipid bilayer; the capsid protein (CA), which is the major component of the core; the nucleocapsid protein (NC), which is tightly bound to the RNA; and the protease protein (PR). In all retroviruses, the core proteins are derived from a single polypeptide precursor by proteolytic cleavage via the viral protease. In some retroviruses, the protease gene is expressed as a result of a frame-shift during translation (e.g., mouse mammary tumor virus MMTV). Gag proteins are also designated with the letter p and a number which reflects the molecular weights in thousands (e.g., p20). The *pol* gene codes for the viral polymerase—an RNA-dependent DNA polymerase (i.e., reverse transcriptase, RT)—and the viral integration protein, also termed integrase (IN), which mediates the integration of the viral DNA into the host chromosome. DNA polymerase and integration proteins are first translated with the core proteins into a single polypeptide. Proteolytic cleavage of the precursor peptide by the viral protease generates individual structural core proteins and active enzymes.

The products of the *env* gene are two envelope proteins, surface protein (SU) and transmembrane protein (TM), which are found on or in the lipid bilayer membrane, respectively. They are also derived from a single precursor protein, cleaved by a cellular enzyme, and glycosylated as are cellular membrane proteins. Hence, they are also termed glycoproteins (e.g., gp70); the number reflects the molecular weight in thousands. Some retroviruses, such as the human T-cell leukemia viruses (HTLV) and the human immunodeficiency virus (HIV), carry additional genes other than gag, pol, and env. Some of these genes have essential functions in regulating viral gene expression (see following).

In addition to the protein coding genes, the retroviral RNA genome contains several regulatory sequences (also called cis-acting sequences) required for efficient viral replication. These sequences are located primarily at the ends of the genome (Fig. 2).

In modern molecular biology, the nucleotide sequence of the retroviral genome serves as the yardstick for classification. Therefore, the most commonly studied retroviruses are organized into six groups: the avian leukosis–sarcoma viruses, the avian reticuloendotheliosis viruses and mammalian leukemia–sarcoma viruses, the mouse mammary tumor viruses, the primate Type D viruses, the human T-cell leukemia-related viruses, and the lentiviruses (Table I, Fig. 2). Sequence comparisons on the nucle-

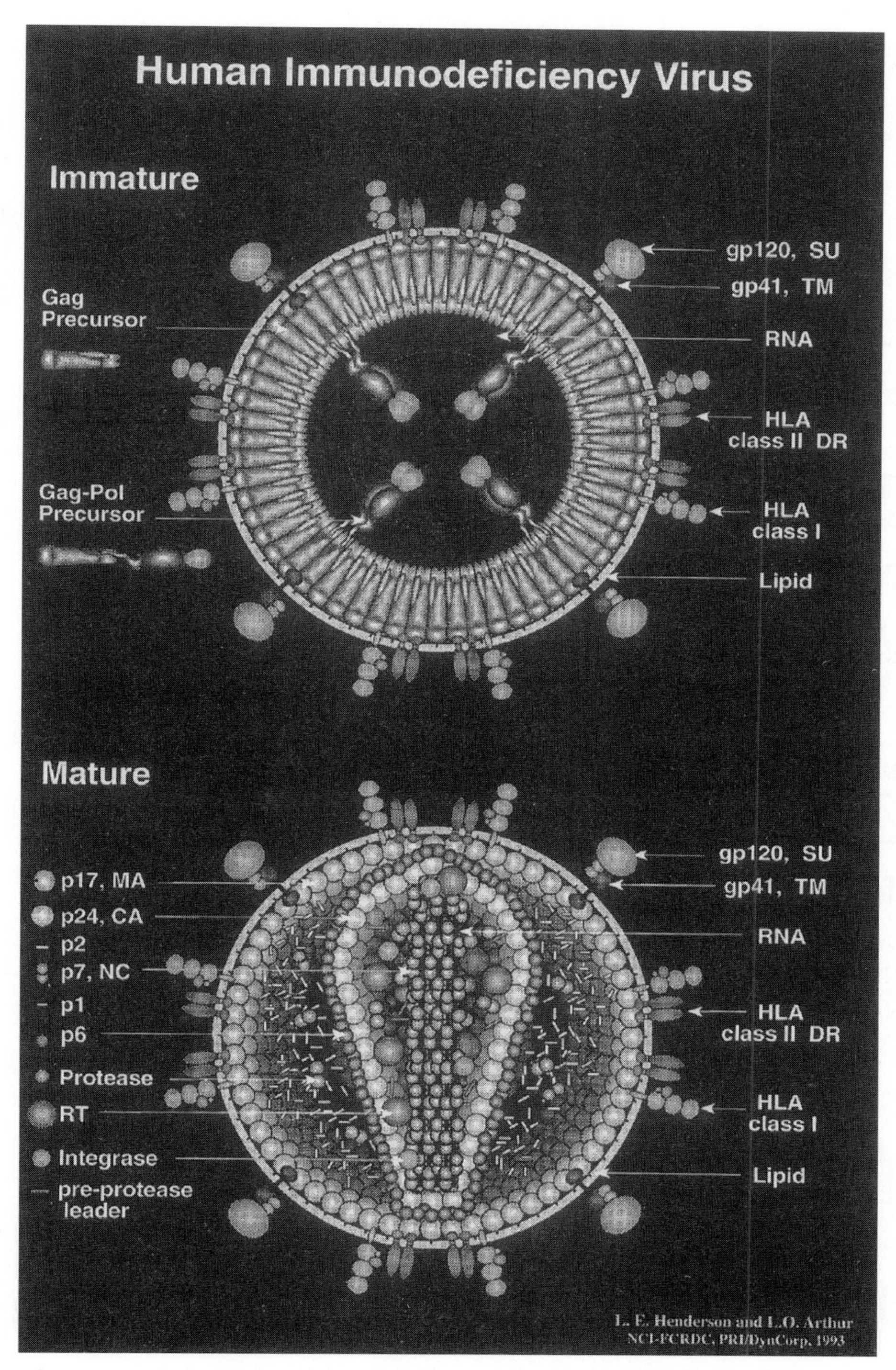

Fig. 1. Retroviral particle of the human immunodeficiency virus type I (HIV-1). During or after budding from the cell (see also Fig. 3), the particle undergoes a maturation process. Only mature particles are able to infect fresh target cells. Two genomic RNA molecules are encapsidated in the core structure, which consists of several proteins (e.g., matrix protein (MA), capsid proteins (CA), polymerase (RT), and integrase; for a detailed description, see Table II). The core is surrounded by a lipid bilayer membrane. Viral envelope proteins (gp120, gp41), as well as some cellular proteins (e.g., HLA proteins), are embedded in the bilayer membrane.

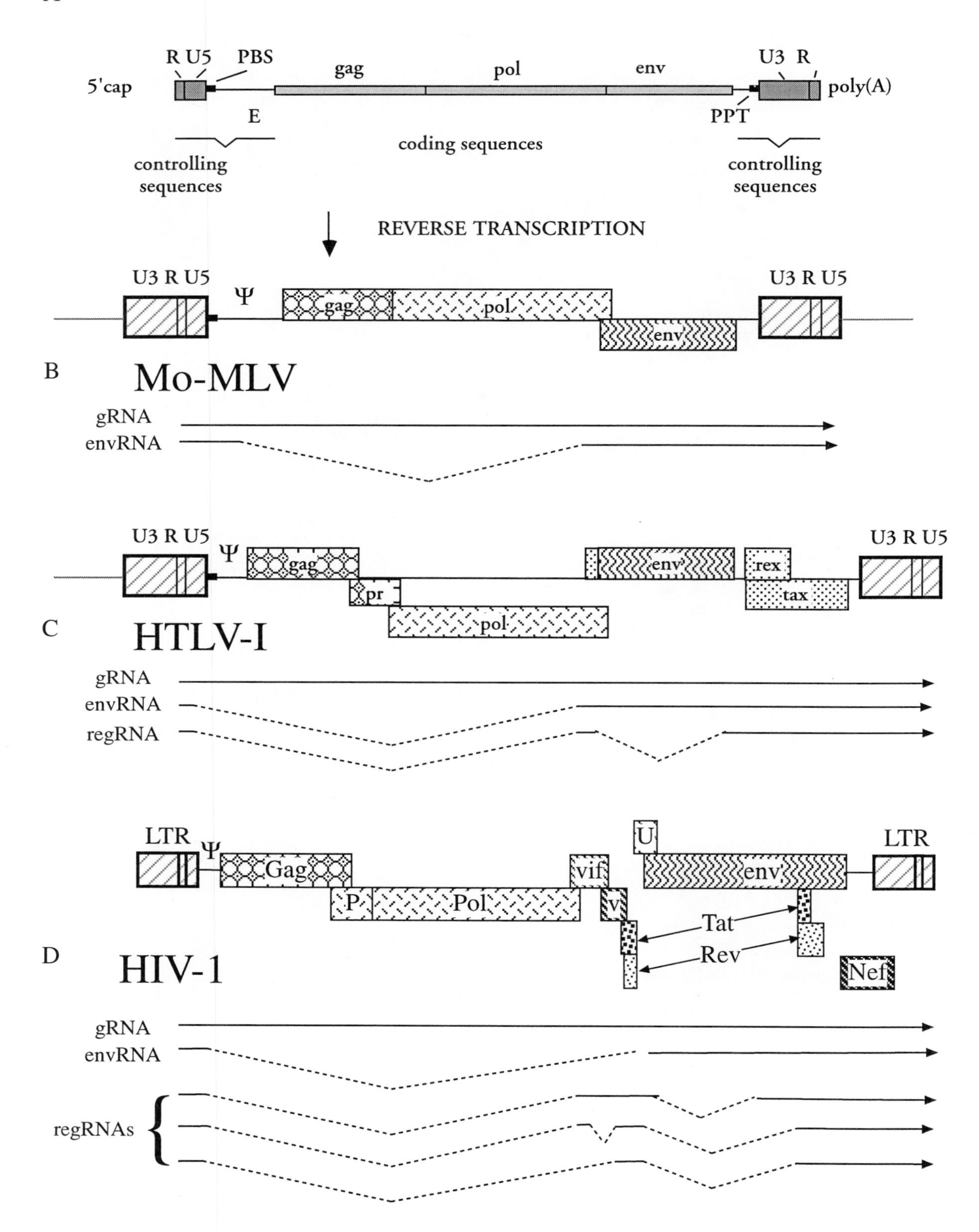
A
R U5
PBS
gag
pol
env
U3 R
5'cap
poly(A)
E
PPT
coding sequences
controlling sequences
controlling sequences
REVERSE TRANSCRIPTION
U3 R U5
Ψ
gag
pol
env
U3 R U5
B
Mo-MLV
gRNA
envRNA
U3 R U5
Ψ
gag
env
rex
U3 R U5
pr
tax
pol
C
HTLV-I
gRNA
envRNA
regRNA
LTR
Ψ
Gag
vif
U
env
LTR
P
Pol
v
Tat
Rev
Nef
D
HIV-1
gRNA
envRNA
regRNAs

TABLE I
Major Groups of Retroviruses

Virus group	Prototype	Other examples
Avian leukosis and sarcoma viruses	avian leukosis virus (ALV)	RSV, RAV-1, RAV-2
Reticuloendotheliosis viruses	spleen necrosis virus (SNV)	REV, CSV
Mammalian leukemia and sarcoma viruses	Moloney murine leukemia virus (Mo-MLV)	Mo-MSV, Ha-MSV, Fr-SFFV, FeLV, SSAV
Mammary tumor viruses	mouse mammary tumor virus (MMTV)	
Human T-cell leukemia-related viruses	human T-cell leukemia virus (HTLV-1)	HTLV-2, STLV, BLV
Immunodeficiency and lentiviruses	human immunodeficiency virus (HIV-1)	HIV-2, SIV-1, CAEV, EIAV, FIV

Note. Subgroups of retroviruses: RSV, Rous sarcoma virus; RAV, Rous-associated virus; REV, reticuloendotheliosis virus; CSV, chicken syncytial virus; Mo-MSV; Moloney-mouse sarcoma virus; Harvey mouse sarcoma virus; Fr-SFV, Friend spleen focus-forming virus; FeLV, feline leukemia virus; SSAV, simian sarcoma-associated virus; STLV, simian T-cell leukemia virus; BLV, bovine leukemia virus; SIV, simian immunodeficiency virus; CAEV, caprine arthritis-encephalitis virus; EIAV, equine infectious anemia virus.

otide and amino acid levels have led to the establishment of evolutionary trees.

B. Life Cycle

To enter a cell, the virus first has to attach to the cell membrane of the target cell (Fig. 3). This attachment is mediated by the SU protein, which recognizes a specific receptor (usually, a membrane protein) of the target cell. The interaction of the envelope with the cellular receptor is very specific and determines the host range of the virus. For example, the human immunodeficiency virus type I (HIV-1) binds preferentially to the human CD4 molecule expressed on T-cells. Other retroviruses utilize "housekeeping" receptors (e.g., Moloney murine leukemia virus (Mo-MLV) uses an amino acid transport protein) expressed on many different tissues. After attachment, retroviruses enter cells by two different mechanisms: in the case of most retroviruses, the viral and cellular membrane fuse at the cell surface and the retroviral core particle is released into the cytoplasm of the target cell (e.g., HIV). In the case of certain other retroviruses (e.g., Mo-MLV), the virus particle is absorbed by endocytosis and membrane fusion is triggered by a pH shift in the endosome. Once released into the cytoplasm, the virus has access to the cellular nucleotide triphosphates and begins to copy its RNA genome into a double-stranded DNA. In this complicated process, all enzymatic reactions are carried out by the viral reverse transcriptase (Fig. 4).

As a result of this replication process, the protein coding regions of the retroviral genome are flanked by long terminal repeats (LTRs). The LTRs carry specific sequences (short inverted repeats) at their ends, essential for the efficient integration of the provirus into the host genome: sequences at the 5′

Fig. 2. Organization of a retrovirus genome. (A) The RNA genome of a C-type retrovirus is shown on the top. It contains three protein coding genes. Regulatory sequences are located mainly at the ends. The U3 and U5 region are unique in all retroviral RNAs at the 3′ and 5′ ends, respectively; U3 contains promoter sequences for gene expression. R is a repeated region present on the ends of the RNA genome. PBS (primer binding site) is a sequence which binds a tRNA primer for first strand DNA synthesis. PPT, a purine-rich region is involved in second strand DNA synthesis. E or ψ: encapsidation sequence. As a result of retrovirus replication, the U3 and U5 region are duplicated to form long terminal repeats (LTRs; for details, see Fig. 4). (B–D) DNA genomes (proviruses) of Moloney leukemia virus (MoMLV), human T-cell leukemia virus (HTLV), and human immunodeficiency virus type I (HIV-1). RNA's transcribed and splicing products are shown below the proviruses. gRNA, genomic viral RNA; envRNA, viral RNA for expression of the envelope protein. regRNA, viral RNAs to express regulatory proteins. Dashed lines indicate the region cut out by splicing. In the case of HIV-1, only some regRNAs are shown as examples. Rex, tax, U = vpu, V = vpr, Vif, Tat, Rev, and Nef are regulatory proteins (see also Table II).

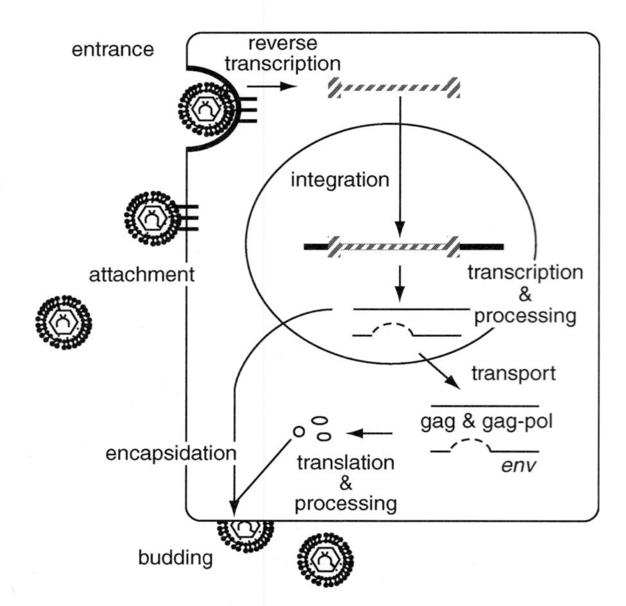

Fig. 3. Retrovirus life cycle. After entrance of the virion into the cell, the retroviral RNA is copied into a double-stranded DNA, which is integrated into the host genome. DNA transcription and RNA processing result in genomic RNA, as well as retroviral mRNAs. Virus assembly takes place at the cell membrane and results in the budding of retroviral virions.

end of the left LTR and sequences of the 3′ end of the right LTR form the attachment site (att) recognized by the viral integration protein which carries out this process. Thus, integration is sequence specific in regard to the viral DNA. However, no sequence specificity is known regarding the nucleotide sequence of the integration site in the host genome, although "hot spots" of integration have been reported. As a result of the integration, four to six nucleotides of the chromosomal DNA are duplicated, depending on the viral integration protein. The integrated DNA is called the provirus.

After integration of the provirus, RNA transcription produces genomic retroviral RNAs. RNA transcription is performed by cellular RNA polymerase II and is driven by a promoter and enhancer present in the U3 region of the LTRs. 3′-end RNA processing is regulated by viral sequences present in the (repeat) R or U3 (depending on the particular virus) and U5 regions. In either case, the RNA transcript starts at the first nucleotide of R in the left LTR and ends with the last nucleotide of R in the right LTR. As the result, RNA transcripts are identical to the original genomic viral RNA. Gag proteins and the viral enzymes are translated from genomic RNA. Expression of the envelope proteins is from spliced genomic RNAs and follows the pathway of cellular membrane proteins (Fig. 2).

To complete the viral life cycle, genomic viral RNAs are encapsidated by gag proteins to form core structures. The selective encapsidation of genomic viral RNAs is mediated by specific encapsidation sequences (called E in avian retroviruses or ψ in murine retroviruses), located 3′ of the primer binding site (PBS) on the viral RNA. Core structures interact with viral envelope proteins which are embedded the cell membrane. As a result of this interaction, virus particles bud from infected cells to yield progeny virus. Finally, proteolytic cleavage of precursor proteins takes place immediately after the budding process, resulting in virions able to infect fresh target cells.

As a result of this replication process, retroviruses often do not kill or lyse the infected cells. Instead, the cell machinery is continuously used for virus production. Moreover, after the provirus becomes a part of the cell genome, both daughter cells carry the provirus and produce retrovirus particles. Retroviral replication can be efficient: hundreds of virus particles can be produced per day from a single cell infected with one provirus.

C. Endogenous Retroviruses

Retroviruses are also able to infect germ line cells or their precursors in pre-implantation embryos. As a result, the progeny carry retroviral proviruses in all body cells. Such retroviruses are called endogenous to distinguish them from those resulting from exogenous infections. Endogenous retroviruses are found in all vertebrates (including humans) which have been investigated thoroughly. It is estimated that mice carry about 1000 endogenous (ancient) retroviral genomes. However, most are degenerated and cannot produce replication-competent virus particles. It has been hypothesized that the presence of endogenous retroviruses protects the carrier from further exogenous infections. In particular, the expression of the endogenous envelope protein appears to block the cellular virus receptor and, conse-

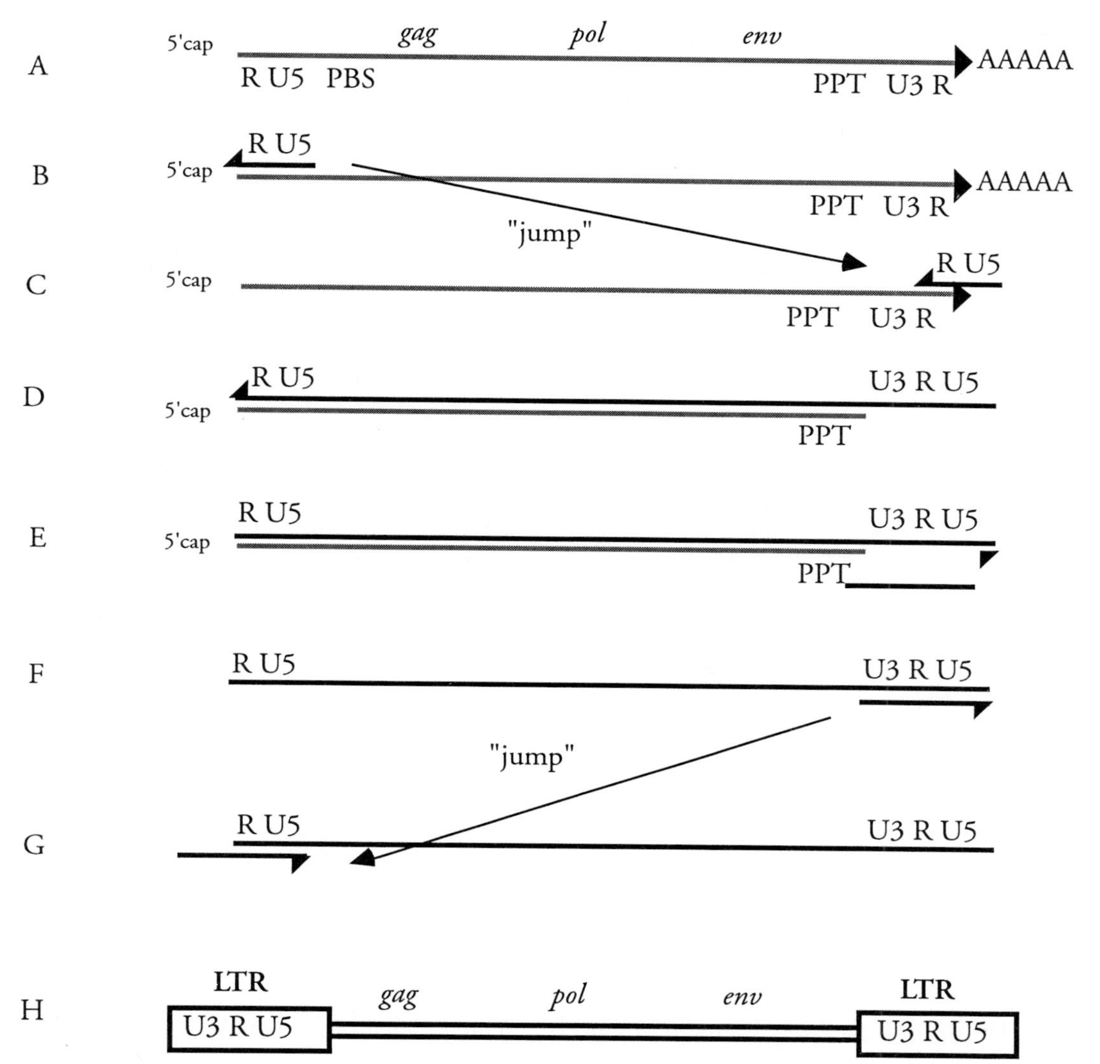

Fig. 4. Reverse transcription of the retroviral RNA. (A) Retroviral genomic RNA. (B) A short cDNA copy (solid line) of the viral RNA is synthesized using a cellular tRNA as primer which hybridizes to the primer binding site (PBS), located near the 5′ end of the viral RNA (C). Next, the RNA moiety of the resulting double-stranded DNA–RNA hybrid is removed, allowing the short single-stranded cDNA copy to "jump" to a repeated region (R) at the 3′ end of the second RNA molecule and then hybridize to it. (D) This cDNA serves as a primer for the cDNA synthesis of the complete viral genome. As a result of this "jump," a region designed U5 (because it is unique in all retroviral RNAs at the 5′ end) is now attached to the 3′ end of the single-stranded cDNA copy. Next, the RNA moiety of the DNA–RNA hybrid is removed up to a purine-rich region (PPT). (E) This polypurine track serves as a primer for the second strand DNA synthesis. (F). Then, all of the remaining viral RNA is degraded (G). The double-strand is melted, enabling the short second strand to "jump" to the other end of the first cDNA strand (H). The single-stranded regions are filled in to complete the synthesis of a double-stranded DNA molecule. In the course of this DNA synthesis, a region called U3 (unique at the 3′ end of retroviruses) is duplicated. As a result of this replication mechanism, the protein coding regions are flanked by long terminal repeats (LTR). All enzymatic reactions are carried out by the viral reverse transcriptase.

quently, infectivity of exogenous retroviruses which utilize the same receptor for virus entry.

There are also other cellular DNA sequences with partial nucleotide sequence homology to retroviruses. These and other repeated cell sequences most probably arose by the reverse transcription of RNAs. They are estimated to make up 10% of the mammalian genome.

II. HUMAN IMMUNODEFICIENCY VIRUSES

Human immunodeficiency viruses type I and II (HIV-1, HIV-2) are human retroviruses and lead to immune suppression, which is complicated by opportunistic infections and tumors. HIV-1 is the major causative agent of the acquired immune deficiency syndrome (AIDS). HIV-1 was first discovered in 1983 after the initial AIDS cases were reported in 1981. Nevertheless, this virus has been isolated in stored blood samples in Africa from the late 1950s. Over thirty million people in the world are infected with HIV-1 and HIV-2. Major epidemics exist in North and South America, Europe, sub-Sahara Africa, India, and Southeast Asia. The major growth of this pandemic is now in the developing world. In the United States, a recent change in the epidemic has shown the increase in transmission of HIV-1 in women and individuals of color.

HIV-1 and HIV-2 are the only known human lentiviruses. In general, lentiviruses lead to slowly progressive diseases in mammals. HIV-1 is closely related to the primate lentiviruses, simian immunodeficiency viruses (SIV), and to an SIV strain in chimpanzees (SIVcpz). HIV-2 is found almost solely in Western sub-Sahara Africa. It is the same as a strain of SIV from sooty mangaby monkeys. Of note, HIV-2 is more difficult to transmit sexually and is less virulent than HIV-1 in humans. Based on these findings and a number of phylogenetic trees, it has been demonstrated that HIV-1 and HIV-2 seem to have been acquired into the human population through cross-species transmission from subhuman primates.

There are a number of other related lentiviruses in mammals including visna in sheep, which leads to central nervous system and pulmonary symptoms; equine infectious anemia virus (EIAV), which leads to fevers, weight loss and anemia in horses; and caprine arthritis encephalitis (CAEV) in goats, leading to arthritis and central nervous system disease. Of note, HIV-1 is only distantly related to the other human retroviruses, human T-cell leukemia virus types 1 and 2, which lead to T-cell leukemias and lymphomas and spinal cord disease in infected individuals.

A. HIV-1 Genetics and Replication

Besides the structural proteins, which form the virus particle, lentiviruses express a number of critical regulatory genes from multiple spliced mRNAs (Fig. 2). HIV-1 contains six regulatory genes which are essential in the complex pathogenesis (Table II). First, the Tat gene is the major transcriptional transactivator of HIV-1 and is essential for the activity of the LTR promoter. The Tat protein stimulates HIV-1 transcription via an RNA intermediate called the TAR region, which is found in the R region the 5′ LTR. The product of the Rev gene rescues the unspliced viral RNA from the nucleus of infected cells by increasing transport through the nuclear pore. Other critical accessory proteins include the Vpr gene which leads to G2 arrest in the cell-cycle of infected cells, Nef which stimulates viral production and activation of infected cells, Vpu which stimulates viral release, and Vif which seems to augment viral production in either early or late steps in the viral life cycle. Of importance, the 5′ LTR has a number of motifs for binding cellular transcription factors which seem to be important in allowing the fine tuning of viral replication, based on the stimulatory state of the host cell.

HIV-1 replicates via a classic retroviral life cycle. Initially, it binds to the high affinity receptor, CD4, on CD4+ T-lymphocytes and certain monocyte/macrophage populations. Recently, it has been determined that chemokine receptors, including CXCR-4 and CCR5, serve as coreceptors and are essential for viral entry. After entry and disassembly, the viral RNA is reverse transcribed to viral DNA by reverse transcriptase (RT), like other retroviruses.

TABLE II
HIV Proteins and Their Functions

Protein	Size	Function
Gag	p25 (p24)	Capsid (CA) structural protein
	p17	Matrix (MA) protein–myristoylated
	p9	RNA binding protein (?)
	p6	RNA binding protein (?) helps virus budding
Polymerase (Pol)	p66, p51	reverse transcriptase (RT); RNase H–inside core
Protease (PR)	p10	Posttranslation processing of viral proteins
Integrase (IN)	p32	Integration of viral DNA
Envelope (Env)	gp160	Envelope precursor protein, proteolytically cleaved
	gp120	Envelope surface protein, virus binding to cell surface
	gp41 (gp36)	Envelope transmembrane protein, membrane fusion
Tat[a]	p14	Transactivation
Rev[a]	p19	Regulation of viral RNA expression
Nef[a]	p27	Pleiotropic, including virus suppression–myristoylated
Vif	p23	Increases virus infectivity and cell-to-cell transmission; helps in proviral DNA synthesis and/or virion assembly
Vpr	p15	Helps in virus replication; transactivation (?)
Vpu[a,b]	p16	Helps in virus release; disrupts gp160–CD4 complexes
Vpx (only in HIV-II)	p15	Helps in infectivity
Tev[a]	p26	Tat and Rev activities

[a] Not associated with virion.
[b] Only in HIV-1.

However, in contrast to most other retroviruses, the resulting preintegration complex is then actively transported across the nuclear membrane. Thus, HIV-1 is capable of infecting quiescent cells.

B. Pathogenesis

The *in vivo* pathogenesis of HIV-1 has been extensively studied and defined. Nevertheless, many portions of this complex pathogenetic cycle are still not fully understood. It is clear that HIV-1 leads to CD4+ T-lymphocyte depletion. Whether this occurs through direct viral infection, stimulation of programmed cell death (apoptosis), or secondary immune destruction of infected cells (or all of the above) is still not fully understood. Nevertheless, after infection of individuals by sexual and nonsexual means, there appears local replication in lymphoid tissue prior to detection of viral RNA in the peripheral bloodstream. After a primary seroconverting reaction (see following), HIV-1 infection leads to high levels of virions in the peripheral blood and in many solid tissues. At that point, the immune system, using neutralizing antibodies and cytotoxic T-lymphocytes (CTLs), seems to clear most but not all of the HIV-1 particles and infected cells (Fig. 5).

HIV-1 infected cells remain at low levels within lymphoid tissue throughout the different stages of disease and are not cleared by the immune response. Over a variable amount of time, which can be from a few months to over a decade (average, approximately 7 years), there is a slow but inexorable decrease of the CD4+ T-cell population. When CD4+ lymphocytes reach profoundly low levels (below 200 cells/mm^3 in the peripheral blood), then the opportunistic infections and tumors, the clinical syndrome of AIDS, will occur.

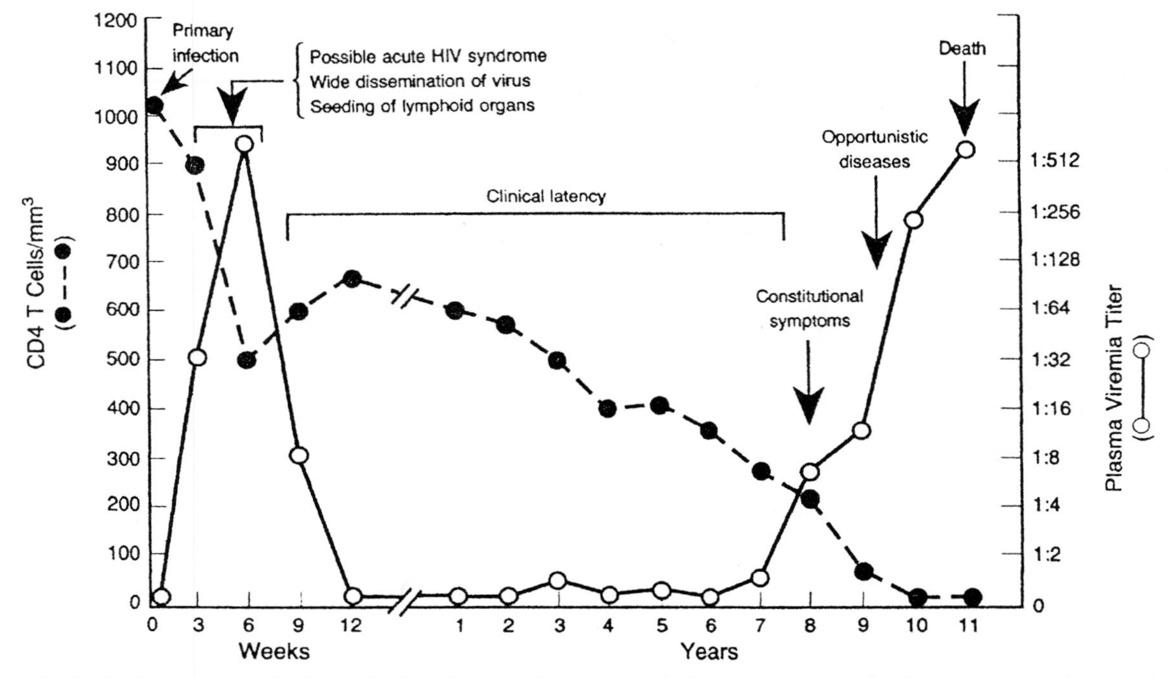

Fig. 5. Typical course of HIV-1 infection. During the early period after primary infection, there is widespread dissemination of virus and a sharp decrease of the number of CD4+ T-cells in the peripheral blood. An immune response to HIV-1 ensues, with a decrease in detectable viremia followed by a prolonged period of clinical latency. The CD4+ T-cell count continues to decrease during the following years, until it reaches a critical level, below which there is substantial risk of opportunistic diseases, which finally lead to the death of the infected person.

C. Transmission

HIV-1 is transmitted via contact with the sexual secretions of infected individuals or through direct contact with infected blood. Heterosexual or homosexual contact with infected individuals, in a nonprotected manner, remains the most common cause of transmission of HIV-1 worldwide. Transmission is more efficient via male homosexual activities, mainly receptive anal intercourse. Nevertheless, heterosexual activity is the major transmission route in the developing world and has increased as a transmission modality in the developed world. In addition, direct contact with contaminated blood, either via IV drug abuse or, still in the developing world, via infusions of infected blood remains a critical aspect in understanding the transmissibility of this agent.

Finally, vertical transmission, as compared to the horizontal modes already described, is also important, especially in the developing world. Transmission from infected mothers to infants can occur prenatally (transplacentally), perinatally, or postnatally (e.g., breast milk) (Table III).

D. Immunodeficiency

As has been known since the early 1980s, HIV-1 infection will lead to immunosuppression in between 98% and 99% of infected individuals. There is a small group of infected hosts who have been termed "long-term nonprogressors." This group of individuals seem, probably based on a variety of viral and cellular mechanisms, not to develop CD4 depletion or immunosuppression. Continued detailed research is ongoing, trying to elucidate the parameters behind these unique viral/host interactions. Of note, this group must be correctly identified, as they do not require therapy, only careful follow-up.

TABLE III
Isolation of HIV from Body Fluids

Fluid	No. with virus isolation/total no.	Estimated quantity of HIV[a]
Free virus in fluid		
Plasma	33/33	1–5,000[b]
Tears	2/5	<1
Ear secretions	1/8	5–10
Saliva	3/55	<1
Sweat	0/2	–[c]
Feces	0/2	–
Urine	1/5	<1
Vaginal and cervical fluid	5/16	<1
Semen	5/15	10–50
Milk	1/5	<1
Cerebrospinal fluid	21/40	10–10,000
Infected cells in fluid		
PBMC	89–92	0.001–1%[b]
Saliva	4/11	<0.01%
Bronchial fluid	3/24	NK[d]
Vaginal and cervical fluid	7/16	NK
Semen	11/28	0.01–5%

[a] For cell-free fluid, quantities are given as infectious particles per milliliter: for infected cells, quantities are the percentage of total cells infected.
[b] High levels associated with symptoms and advanced disease.
[c] –, no virus detected.
[d] NK, not known.

For most individuals, immunosuppression occurs years after HIV-1 infection, after a clinical "latency period." As noted, there is no true microbiological latency but only a clinical latency, in which no signs of disease are noted. During this time period, the patient will always have viral antibodies, as detected by ELISA and Western blotting. The viral RNA level is always detectable as well, utilizing the new ultrasensitive assay systems via branch-chain DNA or reverse transcriptase-(RT) PCR.

Immunosuppression usually is clinically manifest when CD4 cells are lower than 200. Since the early 1990s, the CDC has defined AIDS as occurring whenever and HIV-1-infected individual has total CD4+ T-lymphocytes in the peripheral blood below 200 cells/mm^3 (Fig. 5).

CD4+ T-lymphocyte depletion has a critical and profound affect on the total immune system of the human. As the "orchestrator of the immune system," the CD4+ T-lymphocyte has major effects on the generation of normal immune function in all wings of the system. As such, CD4+ T-lymphocyte depletion leads to alteration of macrophage activation, cytotoxic T-lymphocyte (CD8+) generation, humoral immune function via antibody generation from B cells, and natural killer (NK) cell activity. Thus, viral effects on CD4+ T-lymphocyte function and numbers lead to a generalized immune suppression, which, in turn, leads to opportunistic infections and tumors.

E. Treatment

Therapy for HIV-1 has been changed considerably in the last five years. After initial use of monotherapies to combat HIV-1 infection with only modest success, combination therapy has now led to profound improvement in the clinical well-being of large numbers of patients with HIV-1 infection. Treatment has been based on targeting two major proteins in the viral life cycle, reverse transcriptase and protease. Reverse transcriptase inhibitors are divided into two major groups: nucleoside analogs and nonnucleoside analogs. The nucleoside analogs which are presently FDA-approved include AZT, 3TC, DDI, DDC, and D4T. Nonnucleoside analogs include Nevirapine and Delavirdine. It had been determined that combining two of these inhibitors led to stronger effects on increasing CD4 T-lymphocytes and decreasing viral replication, compared to monotherapy. The addition of protease inhibitors for triple combination therapy has led to the true concept of "highly active antiretroviral therapy" (HAART). FDA-approved protease inhibitors include Indinavir, Ritonavir, Nelfinavir, and Saquinavir. It has now been demonstrated that in many cohorts 85–95% of infected individuals can be treated with HAART and have undetectable viral RNA in their peripheral blood after 2–3 months of therapy. The durability of this effect is still being followed in long-term studies but the majority of patients seem to have a durable antiviral effect for over 1–2 years after initiation of HAART. However, recent studies indicate that, even after more than 2 years of HAART treatment, all patients investigated still contain replication-competent HIV-1 virus in a small number of their immune cells.

III. ONCOGENIC RETROVIRUSES—NATURAL RETROVIRAL VECTORS

Many retroviruses (in particular, avian leukosis and mammalian C-type viruses) contain another gene (i.e., an oncogene) in their genome. Expression of the oncogene results in the malignant transformation of the infected cells. Such retroviruses are called highly oncogenic retroviruses. They arose by recombination of the viral genome with cellular protooncogenes. As a result of this process, in most oncogenic retroviruses, the viral oncogene substitutes for parts of the protein coding regions of the viral genome (Fig. 6). Thus, the viral oncogene product is often a fusion protein of virus and cell sequences. Oncogenes are always expressed from transcripts originating and terminating in the viral LTRs. Most highly oncogenic retroviruses cannot synthesize all of the proteins necessary for retroviral replication. Thus, they are replication defective.

Highly oncogenic retroviruses are natural gene transfer vehicles (retroviral vectors) which carry a

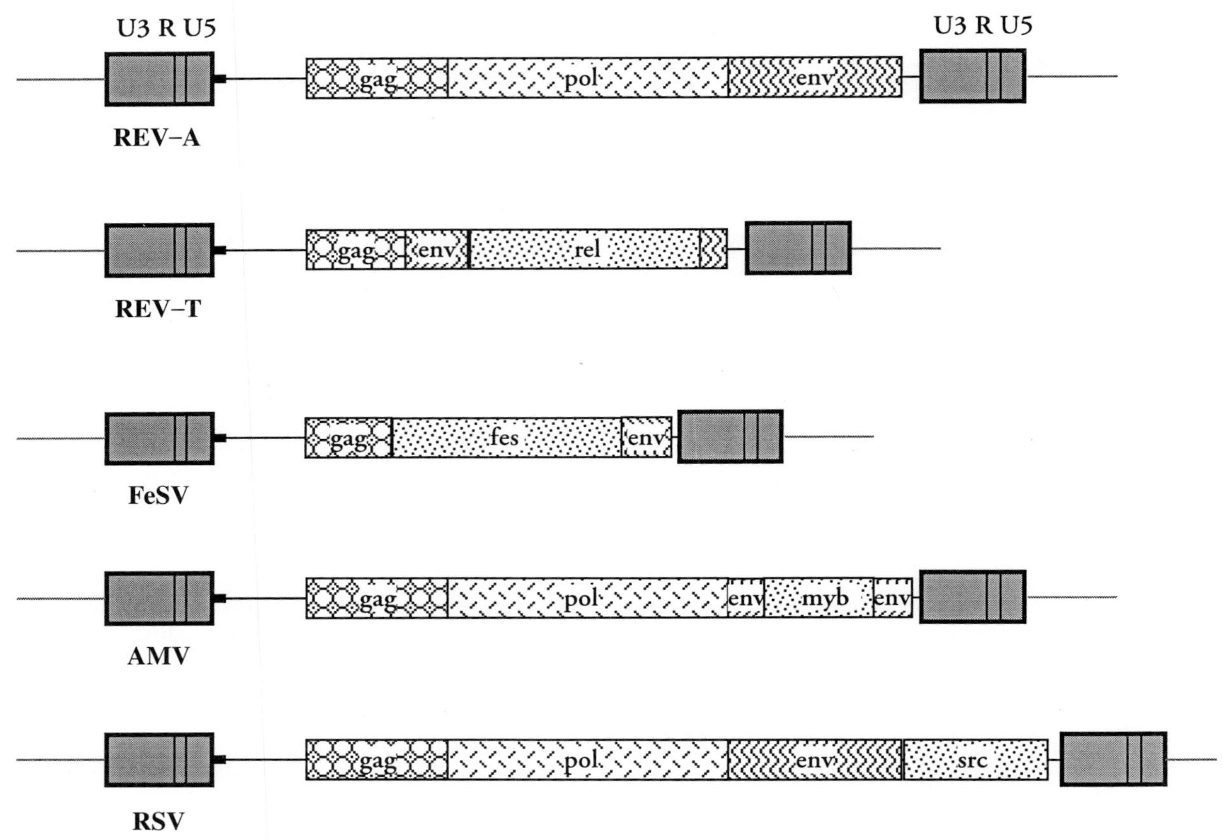

Fig. 6. Examples of highly oncogenic retroviruses. A provirus of a replication competent avian retrovirus, reticuloendotheliosis virus A (REV-A), is shown at the top. Recombination with a cellular proto-oncogene (called c-*rel*) resulted in the highly oncogenic retrovirus REV-T. In most highly oncogenic retroviruses, the oncogene substitutes for viral protein coding sequences. Retroviruses that carry an oncogene are naturally occurring gene transfer vectors. FeSV, feline sarcoma virus; AMV, avian myeloblastoma virus; RSV, Rous sarcoma virus. v-fes, v-myb, and v-src are oncogenes, the expression of which leads to the malignant transformation of the host cells. Viral oncogenes have been derived from cellular differentiation or cell-cycle regulation genes (proto-oncogenes) by recombination between the virus and the host genomes.

nonviral gene. Defective highly oncogenic viruses can only grow in fresh target cells if their host cell is also infected with a wild-type virus, called a "helper" virus. The helper virus provides those viral proteins the defective retrovirus cannot synthesize, as a result of the deletion and substitution in its genome. However, highly oncogenic retroviruses still contain all cis-acting sequences necessary for retroviral replication. Thus, their genomic retroviral RNA is encapsidated into virions (provided by the helper virus) to form infectious retroviral particles.

Retroviruses without an oncogene can also induce tumors (e.g., avian leukemia virus, ALV; murine leukemia virus, MLV; human T-cell leukemia virus, HTLV). It has been documented in several cases that insertion of a provirus near a protooncogene can result in the uncontrolled and/or increased gene expression of that oncogene, leading ultimately to the transformation of the infected cell.

IV. RETROVIRAL VECTORS

A. Vectors, Helper Cells, Experimental Design

Oncogenic retroviruses are used as a model in the construction of retroviral gene transfer systems, which usually consist of two components: the retroviral vector, which contains the gene of interest replacing retroviral protein coding sequences, and a helper cell (also termed packaging cell), which supplies the retroviral proteins for the encapsidation of the vector genome.

Retroviral vectors have been mainly derived from chicken and murine retroviruses and are constructed in several molecular cloning steps. First, a retrovirus provirus is cloned in a bacterial plasmid in which it can be amplified to obtain large quantities of DNA for genetic engineering purposes. Next, the viral protein coding genes are removed and replaced by the gene(s) of interest. Figure 7 illustrates some examples of retroviral vectors. Genes can be inserted in the same or reverse orientation to the vector. They can be expressed by the LTR promoter, additional internal (inducible) promoters, and/or spliced RNAs. Genes which are inserted in the reverse orientation must be expressed from an internal promoter. The inserted genes can contain introns. Introns of genes that are in the same orientation as the vector are lost as a consequence of retroviral replication. In many vectors, selectable marker genes (usually, resistance genes to antibiotics) have been inserted to enable the selection of gene-transduced cells *in vitro*.

First-generation helper cells carried a defective retrovirus provirus, which constitutively expressed retroviral proteins but could not encapsidate its own genomic RNAs, as a result of deletion of the encapsidation sequence in the viral nucleic acid. Modern helper cells have been created by the transfection of plasmid constructs into established cell-lines to express retroviral proteins from nonretroviral promoters. Such helper cells avoid the risk of recombination between the retroviral vector and the helper sequences.

Formation of a helper-free retroviral vector for gene transfer is outlined in Fig. 8. Helper cells are transfected with retroviral vector DNA constructs which carry appropriate encapsidation sequences. Thus, RNA transcripts of such constructs are packaged into virions provided by the helper cell. The virus produced from the helper cell is called "helper-free," since it contains no replication-competent helper virus. If the vector has a selectable gene, helper cells can be selected for the expression of the marker gene. As a result, each selected helper cell produces virions containing the RNA transcript of the transfected vector. Supernatant tissue culture medium is used to infect fresh target cells. Titers of up to 10^8 infectious virus particles per mililiter of tissue culture medium have been obtained. The efficiency of introducing a gene by other methods is several orders of magnitude lower than that by retroviral infection.

B. Tissue Culture Experiments with Retroviral Vectors

A large variety of different eukaryotic, bacterial, and viral genes have been inserted and expressed in retroviral vectors. So far, there is no report that a gene of any type could not be expressed.

Retroviral vectors have been used mainly in tissue culture experiments to investigate various aspects of

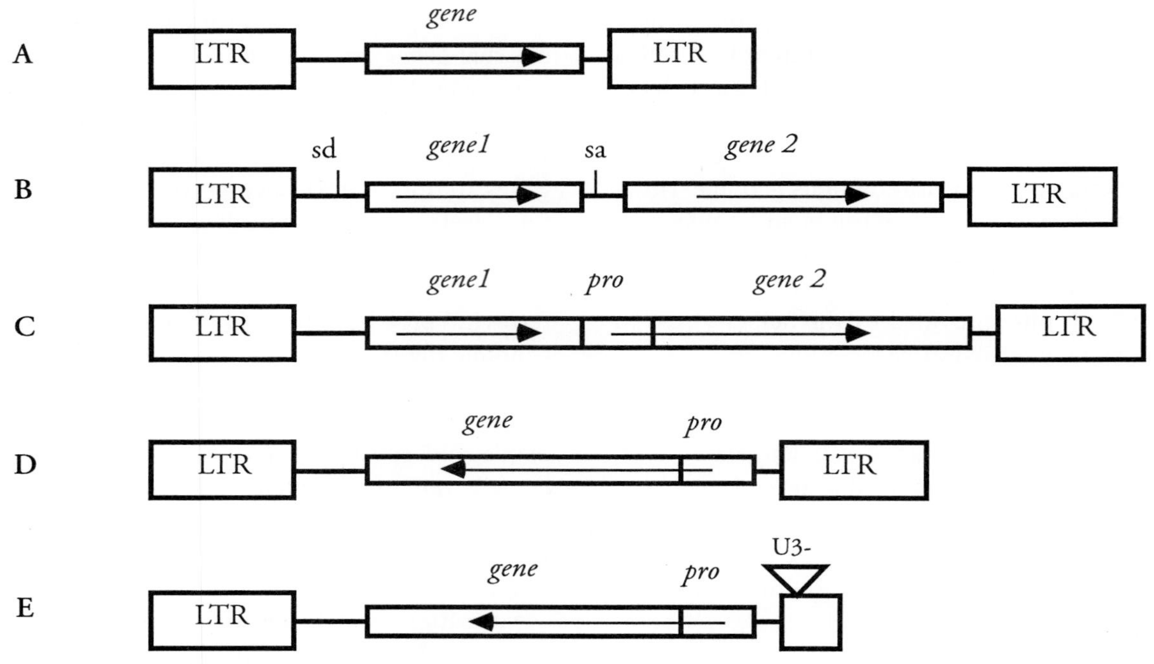

Fig. 7. Examples of retroviral vectors. All varieties of genes can be inserted into a retroviral vector in the same orientation (A–C) or in the reverse orientation to the vector (D and E). Genes can be expressed from the LTR promoter (A–C), spliced mRNAs (B), or from internal promoters (C–E). Retroviral vector constructs with a deleted U3 region in the right LTR result in a provirus without an LTR promoter after one round of replication. Gene expression is only performed from the internal promoter (E). sd, splice donor site; sa; splice acceptor site; pro, internal promoter. The vector constructs are in bacterial plasmids. The plasmid sequences which abut the LTRs are not shown.

the retroviral life cycle and to study gene expression and effects of gene products of nonretroviral genes (e.g., oncogenes or protooncogenes) in particular cells. They were also used to introduce genes for the production of anti-sense RNAs to investigate the effect of anti-sense RNAs on gene expression, particularly in cells infected with other viruses.

C. Experiments with Retroviral Vectors in Animals

Several experiments with retroviral vectors have been performed in early embryos (mainly in mice) to tag chromosomal locations of developmental genes. The idea of these experiments is to destroy genes involved in cellular differentiation by the insertion of a vector provirus. Animals in which the insertion has taken place show a characteristic developmental defect. Molecular cloning of the chromosomal sequences surrounding the integrated provirus has led to the discovery of developmental genes. Infection of murine embryos with retroviral vectors was also used for cell-lineage studies, as all cells derived from a single infected progenitor cell carry the vector provirus at the same genomic location.

In other experiments, bone marrow cells (i.e., stem cells) infected or marked with retroviral vectors have been injected in lethally irradiated mice to study hematopoiesis and the development of the immune system. The marked stem cells undergo differentiation in the injected animal. Comparison of the location of integrated vector proviruses in various fully differentiated cells from such animals led to insights into how cells of the immune system develop.

Cell transplantation experiments were also performed to study possible applications of retroviral vectors in gene therapy. For example, the gene coding for human adenosine deaminase (ADA) has been

A

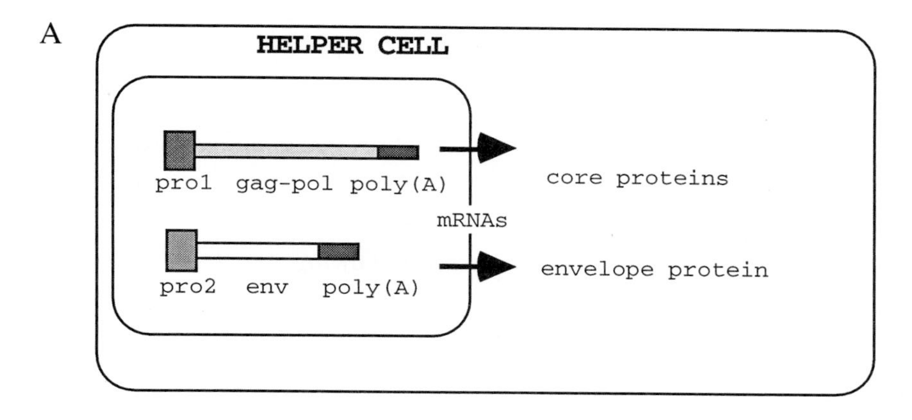

B

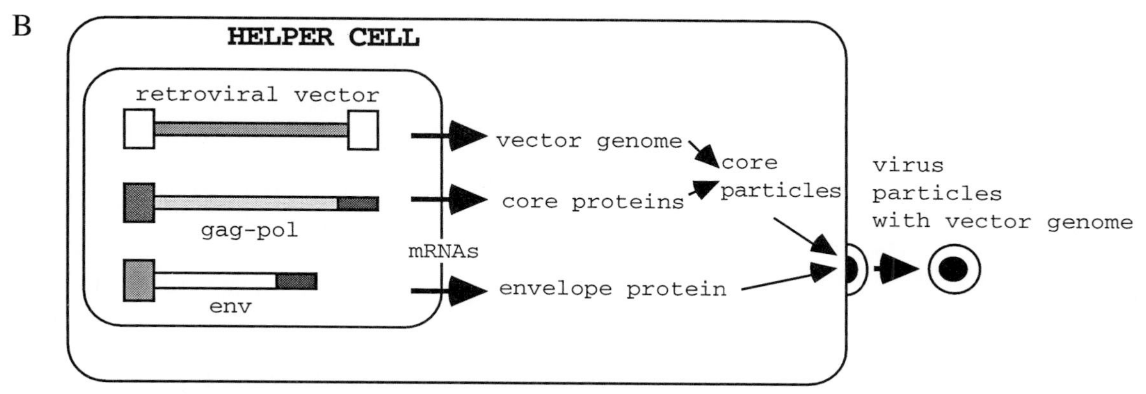

Fig. 8. Formation of a helper-free retroviral vector gene transfer. (A) In helper cells, retroviral proteins are expressed from different plasmid DNAs. These RNA transcripts do not contain encapsidation sequences. Thus, they are not encapsidated into retroviral particles. (B) Such helper cells are transfected with a retroviral vector plasmid construct. The RNA transcript of the retroviral vector contains a encapsidation sequence and, therefore, is encapsidated into virions supplied by the helper cell. Supernatant tissue culture medium is used to infect fresh target cells. pro1 and pro2, promoters to express viral protein coding sequences; poly(A), polyadenylation sequence.

introduced into mouse lymphocytes by retroviral vectors, followed by reinjection of the cells into the animals. It was found that the human gene was expressed in the animal. In other experiments, the gene coding for the human blood-clotting factor VIII was introduced by a retroviral vector into rat fibroblasts in tissue culture. Cells expressing factor VIII were transplanted in rats. Human factor VIII was detected in the blood of these rats. In addition, infection of preimplantation embryos with retroviral vectors is considered to be used in farm animals as a substitute for gene transfer by microinjections.

Retroviral vectors have also been used extensively in mice to study the effect of various therapeutic genes for the treatment of various cancers. For example, tumor cells were infected with retroviral vectors carrying interleukin genes, followed by reinjection into the animal. It was found that this treatment caused a stimulation of the immune system, which led to tumor regression in not all, but a significant percentage of, treated animals.

D. Retroviral Vectors in Human Gene Therapy

Today, approximately 3000 human diseases are known that result from a single defect gene. Some diseases are very rare; others occur with relatively high frequencies (e.g., blood clotting factor VIII deficiency, sickle cell anemia). It is estimated that 1–2%

of newborns are affected with single-gene disorders. The present therapies are mainly based on special diets, injection of proteins, and/or blood transfusions (e.g., for the supplement of clotting factors). However, in most cases, such therapies are ineffectual and unsatisfactory. Thus, the only possibility to "cure" such genetic disorders is the introduction of a functional gene into the body cells of these patients. The functional gene could be expressed even while the inherited defective gene persists.

As has been outlined, gene therapy experiments with retroviral vectors have been performed in animals (mostly mice and rats) with genes for adenosine deaminase, globin, and blood-clotting factors. The results of these experiments have been promising enough to justify the initiation of clinical trials in humans. At the end of 1991, the first human patients affected with severe combined immunodeficiency syndrome (SCID), an autosomal recessive genetic disorder caused by a deficiency of the enzyme adenosine deaminase, have been treated with retroviral vector transduced cells. These cells now contained a functional copy of the gene coding for this enzyme. The success of these first clinical trials has spurred a series of other clinical trials to introduce genes into humans to cure not only genetic diseases but mainly cancer and AIDS. Many clinical trials are now in progress and results are eagerly awaited.

In spite of the initial success of human gene therapy trials, many questions and concerns are being raised. Besides many unsolved technical problems, there are questions concerning safety, efficiency, and social effects. For instance, can the insertion of a retroviral vector into the genome lead in some cells to protooncogene gene expression and to the development of tumors? Is there a chance that infectious viruses will be produced by recombination with retroviral sequences in the helper cell, which can infect other people and/or germ line cells? What are the costs in relation to the efficiency of the treatment? A committee of the National Institutes of Health has been discussing these and related questions since 1984.

See Also the Following Articles

Antisense RNAs • Oncogenic Viruses • T Lymphocytes • Transgenic Animal Technology

Bibliography

Coffin, Hughes, & Varmus (1997). "Retroviruses," CSH Press, Cold Spring Harbor, New York.

Dornburg, R. (1995). Reticuloendotheliosis viruses and derived vectors. *Gene Therapy* **2**, 301–310.

Hunter, E., and Swanstrom, R. (1990). Retrovirus envelope glycoproteins. *Curr. Top. Microbiol. Immunol.* **157**, 187–253.

Katz, R. A., and Skalka, A. M. (1994). The retroviral enzymes. *Annu. Rev. Biochem.* **63**, 133–173.

Laughlin, M. A., and Pomerantz, R. J. (1994). "Retroviral Latency." CRC Press, Inc., Boca Raton, FL.

Levy, J. A. (1993). Pathogenesis of immunodeficiency virus infection. *Microbiol. Rev.* **57**, 183–289.

Miller, A. D. (1990). Retrovirus packaging cells. *Hum. Gene Ther.* **1**, 5–14.

Morgan, R. A., and Anderson, W. F. (1993). Human gene therapy. *Annu. Rev. Biochem.* **62**, 191–217.

Perelson, A. S., Neumann, A. U., Markowitz, M., Leonard, J. M., and Ho, D. D. (1996). HIV-1 dynamics in vivo: Virion clearance rate, infected cell lifespan and viral generation time. *Science* **271**, 1582–1586.

Varmus, H., and Brown, P. (1988). Retroviruses. *In* "Mobile DNA" (M. Howe and D. Berg, eds.). American Society of Microbiology, Washington, DC.

Weiss, R., Teich, N., Varmus, H., and Coffin, J. (eds.). (1985). "RNA Tumor Viruses: Molecular Biology of Tumor Viruses" (2nd ed.), Cold Spring Harbor Laboratory, Cold Spring Harbor, New York.

Rhinoviruses

Eurico Arruda
University of Sao Paulo School of Medicine

I. History
II. Virion Structure and Characteristics
III. Classification
IV. Replication and Propagation
V. Epidemiology
VI. Pathogenesis and Clinical Illness
VII. Diagnosis
VIII. Prevention and Treatment

GLOSSARY

common cold Clinical syndrome of acute upper respiratory symptoms (sore throat, nasal discharge, nasal obstruction, sneezing, and cough) resulting from infection by many respiratory viral pathogens.

fomite Inanimate object that harbors pathogenic microorganisms and may serve as a vehicle of transmission of infection.

icosahedral symmetry Symmetry identical to that of a polyhedron with 20 identical faces; any icosahedrally symmetric structure has 60 identical elements on its surface.

picornavirus Small virus with single-stranded (+)RNA genome in the family *Picornaviridae,* which comprises five genera: *Enterovirus, Rhinovirus, Cardiovirus, Hepatovirus,* and *Aphtovirus.*

THE MORE THAN 100 SEROTYPES OF HUMAN RHINOVIRUS (HRV) constitute the largest genus in the family *Picornaviridae,* which includes small, nonenveloped, single-stranded (+)RNA viruses. HRV is the single most frequent causative agent of common colds and probably the single most frequent etiologic agent of infectious disease in humans.

I. HISTORY

The first direct evidence of the infectious nature of colds was obtained in 1914 by W. Kruse of the Hygiene Institute of Leipzig, who transmitted colds to volunteers by inoculating them with filtered nasal secretions of patients with cold. Human rhinovirus (HRV) was isolated for the first time by Pelon and Price, independently, in 1956. Subsequently, many different HRV serotypes were identified and, in recent decades, epidemiological studies have demonstrated that they are responsible for a vast amount of acute respiratory diseases, particularly among children. Recent studies done by different research groups have resulted in considerable progress in the understanding of structure, molecular biology, and cellular effects of HRVs as well as several aspects of the pathogenesis of the illnesses they cause. It is hoped that this knowledge will eventually lead to effective means of preventing and treating HRV infections.

II. VIRION STRUCTURE AND CHARACTERISTICS

A. Structure

Rhinoviruses are approximately 300 Å in diameter with a molecular weight (MW) of approximately 8.5×10^6. The capsid of HRV consists of 60 protomers, each formed by four viral proteins (VP1–VP4), arranged in icosahedral symmetry. VP1–VP3 constitute the outer portion of the capsid, whereas the smaller polypeptide VP4 lies internally at the RNA–capsid interface. At 12 points on the capsid the convergence of five neighboring units of VP1 forms a five-fold vertex, surrounded by an ~12-Å deep canyon (Fig. 1). Neutralizing immunogenic sites correspond to hypervariable residues on the edges of the canyon and on portions of its walls. Conversely, the "footprint" of the binding domain of the cell receptor for 90% of the HRV serotypes, intercellular adhesion molecule-1 (ICAM-1), is located at the bottom and south wall of the canyon and is constituted by more conserved residues. Beneath the bottom of the canyon, within a β-barrel of VP1, there is a hydrophobic pocket which in many HRV serotypes is naturally occupied by a "pocket factor." Pocket factors found in different HRV serotypes have been identified as diverse small aliphatic molecules of cellular origin which seem to be required for capsid stabilization prior to cell entry.

Within the capsid lies the viral nucleic acid, a 7.1 or 7.2 kb single-stranded RNA of positive polarity. The viral RNA is uncapped and has a small protein

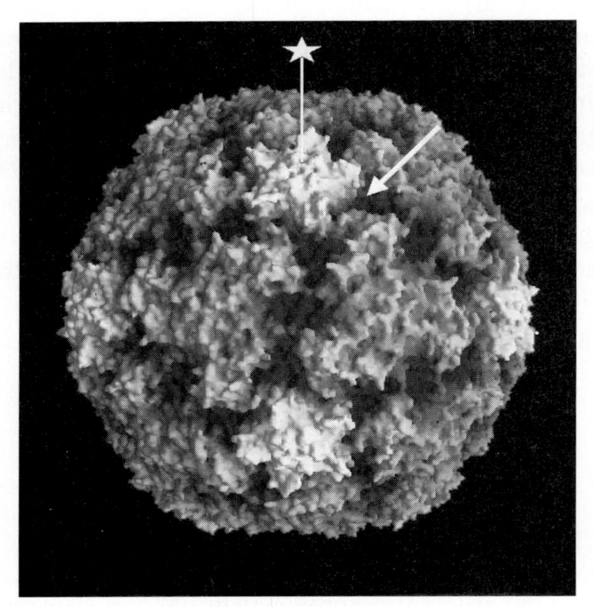

Fig. 1. Molecular graphics image of human rhinovirus 14. The star-shaped region corresponds to the pentamers formed by the convergence of five adjacent VP1 units (star symbol). ICAM-1, the receptor for the majority of HRV serotypes, binds within the canyon, off each tip of the star (arrow). See color insert. (Courtesy of Jean-Yves Sgro, University of Wisconsin–Madison, Madison, WI, who made the image based on X-ray data from Rossmann *et al.*, 1985).

called VPg (virion protein, genomic) covalently linked to its 5′ end and a poly(A) tail at the 3′ end. The first 600–700 nucleotides at the 5′ end constitute an untranslated region (UTR) with traits of highly structured RNA. Some of these tertiary structures function as an internal ribosomal entry site (IRES), thus permitting initiation of translation in a cap-independent manner. At the 3′ end, a short 40 to 50-nucleotide UTR precedes the poly(A) tail, and interaction of host cell factor(s) with tertiary RNA structure within this region apparently can modulate viral RNA replication. Translation of the viral (+)RNA produces a single polyprotein, with three major regions—P1 (coding for capsid proteins) and P2 and P3 (coding for nonstructural proteins and VPg). Proteolytic cleavage of these precursors produces 11 end products, including structural (VP1–VP4) and nonstructural (2A–2C, 3A, VPg, 3C, and 3D) proteins (Fig. 2). The proteolytic cleavages mediated by virus-coded proteases 2A and 3C occur predominantly at Gln-Gly sites. The last proteolytic cleavage (VP0 → VP2 + VP4) occurs autocatalytically after assembly of the virion at an Asn-Ser site during the final stages of virus maturation. The protein 3D is the RNA polymerase, and the remaining viral proteins 2B, 2C, 3A, and 3B (VPg) are associated with viral RNA replication.

The RNA sequences of several HRV serotypes have been determined and were found to be closely related to those of other picornaviruses. Of interest, there are several highly conserved oligonucleotide-size sequences within the genome of picornaviruses, particularly within the 5′ UTR. These sequences are involved in maintaining the conserved structure of RNA conformation within the 5′ UTR and have been used to generate primers for reverse transcription-polymerase chain reaction (RT-PCR)-based diagnostic assays for HRVs.

B. Physical, Biochemical, and Biological Characteristics

Some important properties (Table I) are useful to distinguish HRVs from other picornaviruses. The sedimentation coefficient for HRVs is 150S, whereas for foot and mouse disease virus (FMDV), another picornavirus, it is 156S–160S. Determination of

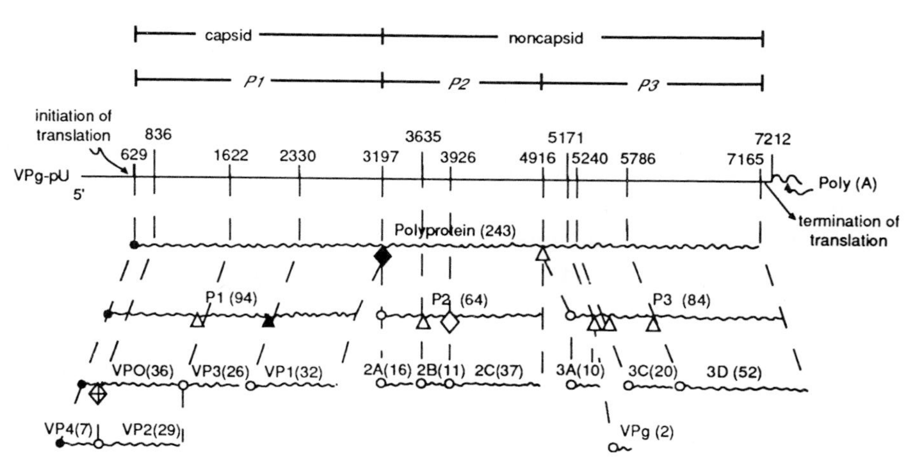

Fig. 2. Schematic representation of HRV-14 genome and gene products. The polyprotein and subsequent precursor proteins are cleaved by viral-encoded proteases 2^A and 3C, generating 11 end products. Molecular weights ($\times 10^3$) are indicated in parentheses. △, Gln-Gly; ▲, Glu-Gly; ◇, Gln-Ala; ◆, Tyr-Gly; ⊕, Asn-Ser (reproduced with permission from Cordingley *et al.,* 1990).

buoyant density in cesium chloride is one of the best ways to distinguish among picornavirus subgroups. Buoyant density in CsCl for HRV is 1.38–1.42 g/cm^3, whereas for enteroviruses and cardioviruses it is 1.34 g/cm^3 and for FMDV it is 1.43 g/cm^3. Buoyant density of HRV in metrizamide, a nonionic density medium, is 1.24 g/cm^3.

HRV is stable at pH 6.0–8.0 and loses infectivity after exposure to more acidic pH. This acid lability is the simplest and most frequently used means to distinguish HRV from enteroviruses. The optimal growth temperature for HRV is 33–35°C. At 37°C, the yield of HRV is reduced by 10–50%, and at 39°C the relative yield is only 1%. HRV is generally stable after several cycles of freezing and thawing and its infectivity is not affected by lipid solvents (such as ethanol, ether, or chloroform) or by nonionic detergents (such as Nonidet P40 and sodium deoxycholate).

TABLE I
Physicochemical Characteristics of HRV

Diameter	300 Å
Molecular weight	~8.5 × 10^6
Buoyant density (CsCl)	1.38–1.42 g/ml
pH stability	pH 6–8
Optimal growth temperature	33–55°C

III. CLASSIFICATION

A. Serotypes

Consequent to a high degree of antigenic variation, 101 numbered serotypes of HRV have been recognized, and this number should increase. Serotyping is based on neutralization of viral infectivity by specific antibody, and antisera for HRV typing may be prepared in a variety of animals. Despite the HRV serotypic diversity, serologic relationships among certain serotypes have been documented, suggesting that serogroups of HRV may exist.

B. Receptor Groups

The HRV numbered serotypes can be divided into three groups on the basis of receptor specificity. The "major" receptor group is composed of 90 serotypes that utilize ICAM-1 as receptor, the "minor" group is composed of 10 serotypes that utilize the low-density lipoprotein receptor (LDLR) as virus receptor, and HRV serotype 87 utilizes sialoproteins as receptor.

ICAM-1 (CD54) is a cell-surface protein of the immunoglobulin superfamily with broad tissue distribution whose natural ligands are integrins, such as LFA-1 and Mac-1. Interaction of ICAM-1 with integrins is important for leukocyte binding to different cell types and, consequently, for the development of immune response. ICAM-1 expression is strongly induced on epithelial cells, fibroblasts, and endothelial cells by various cytokines. The extracellular portion of ICAM-1 consists of five immunoglobulin-like domains and its most distal N-terminal domain binds to HRV capsid residues located within the canyon. ICAM-1 has one transmembrane domain and a C-terminal cytoplasmic domain.

The LDLR is a cell surface protein of approximately 120 kDa with seven consecutive complement type-A extracellular repeats (each 40-amino acids long) and a short cytoplasmic domain that contains internalization signals responsible for clustering of LDLR molecules in coated pits.

Soluble ICAM-1 and LDLR fragments can complex with major and minor HRVs, respectively, and inhibit their infectivity. Upon combining with soluble ICAM-1, different major HRVs tend to undergo uncoating and yield varying proportions of two kinds of noninfectious particles: A particles, which lack VP4 but retain intact the viral RNA, and empty capsids, which lack both VP4 and RNA.

IV. REPLICATION AND PROPAGATION

A. Replication Cycle

The replication cycle of HRV begins with the attachment of virus to the receptor. Recruitment and binding to additional receptor molecules trigger capsid structural rearrangements, which result in extrusion of VP4, virion uncoating, and release of genomic RNA into the cytoplasm (Fig. 3). Not all details of the highly organized sequence of events combining penetration and uncoating are known. Clathrin-dependent or -independent active endocytic pathways may be involved in cell penetration by different HRV serotypes.

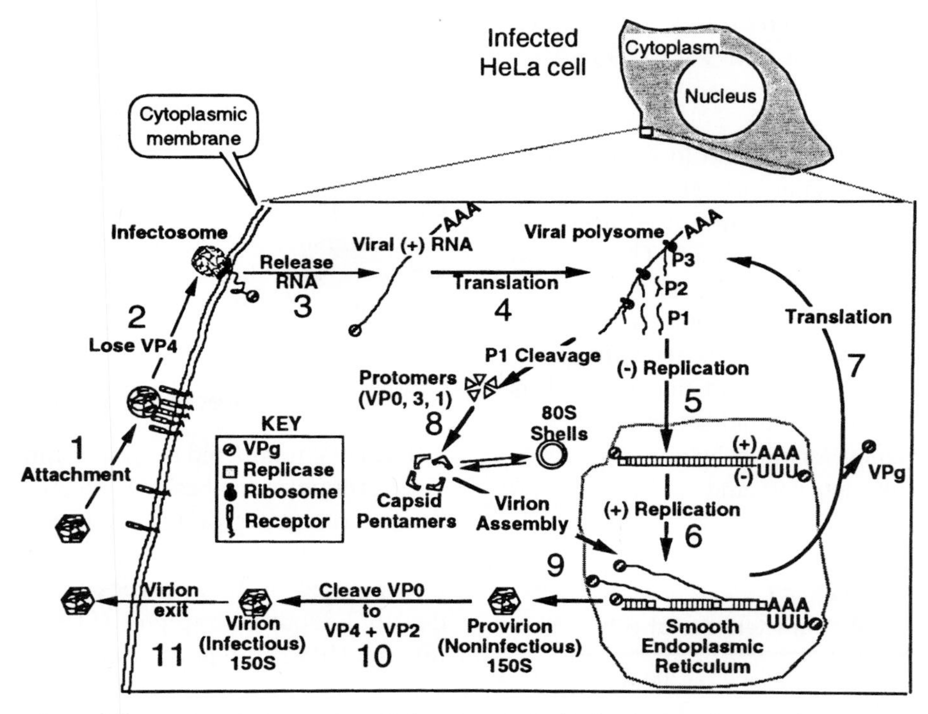

Fig. 3. Schematic representation of the picornaviral infectious cycle. Details are given in the text [reproduced with permission from R. R. Rueckert, *Picornaviridae:* The viruses and their replication. *In* "Fields Virology" (B. N. Fields, D. M. Knipe, and P. M. Howley, Eds.), pp. 609–654. Copyright 1996 by Lippincott-Raven].

Once inside the cytoplasm, the viral RNA directs the synthesis of a large protein that is cleaved during synthesis to yield three viral protein precursors (P1–P3). One of the products of the cleavage of P3 is an RNA-dependent RNA polymerase, which is required for the synthesis of a complementary RNA of negative polarity, (−)RNA, poly(U)-tailed at the 5′ end. This (−)RNA serves as a template for the generation of a rapidly expanding amount of 3′-poly(A)-tailed (+)RNA, which is ready for translation into viral proteins, although some of it will be used as template for further production of (−)RNA. The RNA replication process occurs in association with the smooth endoplasmic reticulum. With the later accumulation of viral capsid proteins, shell assembly starts to occur and part of the expanding (+)RNA pool is packaged into virions.

Prior to assembly, P1 is cleaved twice to generate VP0, VP1, and VP3, which will form immature protomers. These will later assemble into pentamers which then package (+)RNA to form noninfectious provirions. The maturation cleavage of VP0 into VP2 and VP4 occurs only after the packaging of RNA and generates mature, infectious virions that are released by disintegration of the host cell.

B. Propagation and Purification

HRV can only replicate productively in cells originating from humans or higher primates. Primary or secondary cell cultures of human embryonic kidney, human fetal tonsils, monkey kidney, human diploid embryonic fibroblast cells (such as WI-38 and MRC-5 strains), and human heteroploid cell lines (such as HeLa and Hep-2) all support HRV growth. Growth of HRV in stable cell lines provides the best way to propagate the virus to high titers, and serotypes that do not grow well initially in these cell types may be adapted to do so. HeLa cells, either in suspension or in monolayers, are generally used for large-scale preparations of HRV, and a clone of HeLa cells selected for high-level expression of ICAM-1 (HeLa-I) was found to be more sensitive than other cells for primary HRV isolation. One replicative cycle occurs every 7–10 hr and is estimated to result in production of 25,000–100,000 virions per cell, but this may vary with pH, temperature, host cell type, serotype, and multiplicity of infection.

HRV can be purified by different published procedures. Generally, a purification procedure includes a few cycles of freezing and thawing, low-speed centrifugation to clarify the supernatant from cell debris, precipitation of the virus from the supernatant by treatment with polyethylene glycol followed by centrifugation, removal of the polyethylene glycol by treatment with detergent, and banding in a density gradient. Cesium chloride, sucrose, or metrizamide gradients may be used. Although reduction of infectivity has been observed with cesium chloride gradient purification, probably as a consequence of permeation of the virus capsid, this has not been observed with metrizamide, a nonionic density medium. Infectivity to particle ratio is in the order of 1 : 24 to 1 : 240 following metrizamide gradient purification. Following virus banding in density gradient, the quantitation of particles per volume can be estimated by spectrophotometry. Optical density of 1.00 at a wavelength of 260 nm corresponds approximately to 9.4×10^{12} particles/ml.

V. EPIDEMIOLOGY

A. Incidence and Prevalence

HRV occurs in persons of all ages, on a worldwide basis, and even individuals from isolated Amazon Indian tribes have been found with serologic evidence of past infection by HRV. Multiple HRV serotypes may cocirculate in a given area within a period of time. Because of the lack of specific immunity and increased risks of exposure, rates of infection are higher in infants and children than in adults. In the United States, HRV colds occur with an approximate incidence of 0.75/person/year among adults. In children, this incidence is higher, with rates of 1.2/person/year. These incidences have been undoubtedly underestimated since they are based on HRV isolation in cell culture. School-aged children are the most important population reservoir of HRVs, and >70% of the children may become infected during school outbreaks. Children play an important role in disseminating the virus into their households, where most HRV transmission occurs. Secondary attack rates in the household range widely (25–70%) and relate closely to pre-exposure titers of specific neutralizing antibody. Similar attack rates may also occur among institutionalized groups, such as college students, military, and other groups in dormitory-type situations. Increased susceptibility to colds has been reported in smokers and individuals under psychological stress.

B. Seasonality

Although HRV infections occur year-round, for reasons not entirely understood seasonal peaks occur in populations that live in temperate climates, with highest incidence in early fall, probably related to the decrease in maximum daily temperature and the assembly of a susceptible population with the beginning of school term. During the fall peak, HRV may account for up to 80% of all colds in adults in the United States. This seasonal pattern of temperate regions has not been observed in tropical humid regions where the temperatures are more constant and the only significant climate variable may be rainfall.

C. Routes of Transmission

The transmission of HRV from person to person requires exposure to virus-contaminated respiratory secretions. HRV can be recovered from the hands of persons with natural colds and from the surfaces they touch. Under experimental conditions, transmission has been shown to occur by either hand-to-hand contamination followed by self-inoculation into the eye or nasal mucosa or by airborne spread. Small-particle aerosol and direct exchange of oral secretions as in kissing do not appear to be efficient means of transmission. Several hours of close contact between ill persons and susceptible contacts are required for efficient transmission of HRV. HRV can survive for hours on environmental surfaces, which makes possible the spread by hand–fomite–hand. Further evidence supporting hand contamination is the reduction in naturally occurring HRV infections by treatment of hands with virucidal iodine-containing lotions.

VI. PATHOGENESIS AND CLINICAL ILLNESS

A. Pathogenesis of Symptoms

Rhinoviruses infect only higher primates and produce symptoms only in humans. Experimental inoculation of susceptible human volunteers is the only well-characterized model of infection and has been used for studies of pathogenesis and immunity. Using this model, the incubation period before first symptom has been estimated to be as short as 10–12 hr. Upon inoculation in the nostril or conjunctiva, rhinovirus replication occurs in a small number of focally distributed ciliated cells of the nasal cavity and in a few nonciliated cells in the posterior nasopharynx (Fig. 4). Nasal biopsies show little or no pathologic changes. Average HRV titers of 300 $TCID_{50}$/ml are shed in nasal secretions, with a peak at 2 or 3 days and continuing for up to 3 weeks after inoculation. Optimal temperature of replication below the temperature of the body core may be one of the reasons for the localization of HRV replication in the respiratory tract.

HRV infection causes increased levels of numerous cytokines and chemokines, such as IL-1β IL-6, IL-8, IL-11, IFN-γ, TNF-α, RANTES, and GM-CSF, in respiratory secretions. Also, increased levels of kinins have been detected in nasal secretions in experimental and natural HRV colds. Symptoms of HRV colds are not caused by tissue destruction consequent to viral replication but rather by the action of cytokines, chemokines, and other inflammatory mediators, whose release is triggered by infection of low numbers of epithelial cells. These mediators cause vasodilation, tissue edema, transudation of plasma proteins, glandular secretion, and probably sensitization of neurologic reflexes associated with the commonly observed symptoms of nasal congestion, rhinorrhea, sneezing, and sore throat.

HRV infections are associated with increased bronchial responsiveness to histamine and methacholine challenges, and HRV is the respiratory virus most frequently detected in association with wheezing episodes. Allergic subjects have significantly increased responsiveness of lower airways in response to HRV infections in comparison with nonallergic ones, and

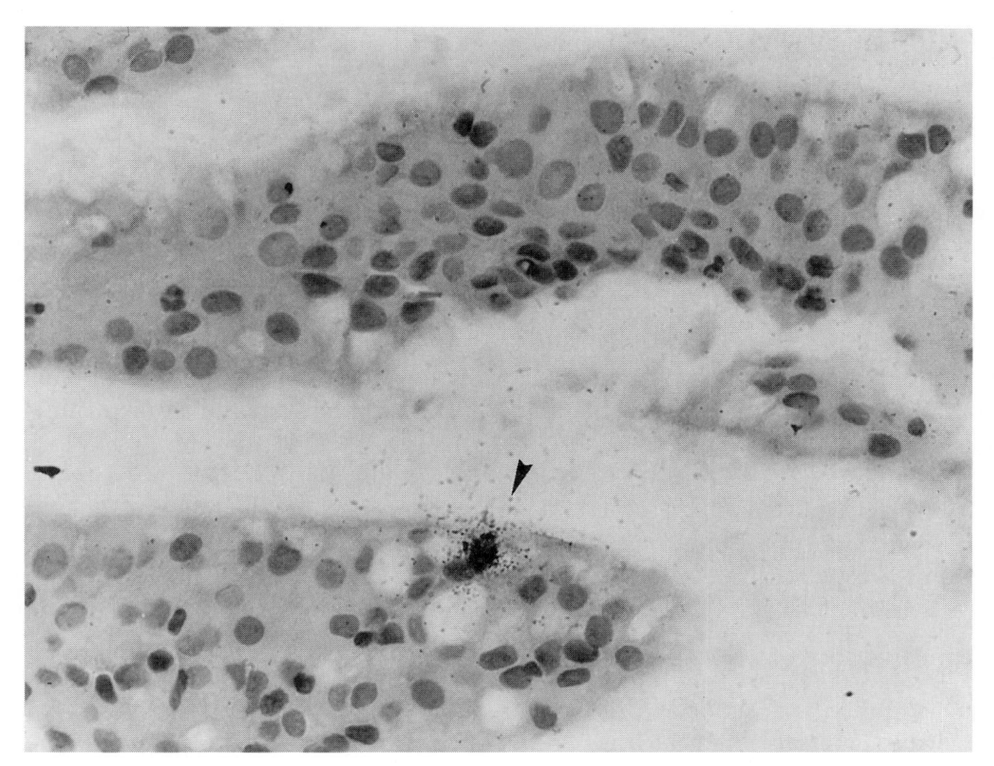

Fig. 4. Section of human nasal epithelium biopsied 3 days after experimental infection with HRV-14 and hybridized *in situ* with ^{35}S-labeled RNA probe for HRV-14. Arrowhead, discrete positive cell (reproduced from Arruda *et al.*, 1995, with permission of the University of Chicago Press).

this occurs in conjunction with airway inflammation, which is measured by increased levels of IL-8, IL-6, and eosinophil cationic protein. Furthermore, allergic individuals have an increased frequency of late allergic reactions to inhaled allergens and increased mucosal eosinophils that may persist for up to 4 weeks after HRV infection. The pathogenesis of HRV-induced bronchial hyperreactivity in asthmatics is not entirely understood, but it is likely that the cytokines and chemokines triggered by the viral infection enhance the preexisting bronchial inflammation. Increased levels of excreted soluble ICAM-1 in nasal secretions and upregulation of ICAM-1 expression on the surface of nasal epithelial cells have been found in persons with HRV colds and are likely as a consequence of the release of cytokines and chemokines triggered by HRV. Upregulated ICAM-1 can mediate adherence and recruitment of inflammatory cells to the respiratory tract and probably contributes to wheezing and asthma exacerbations associated with HRV colds. In fact, the numbers of ICAM-1-expressing epithelial cells in bronchial biopsies from asthmatic individuals with HRV colds correlate with asthma severity. In addition to the ICAM-1-mediated recruitment of inflammation, increased vagus-mediated bronchoconstriction and airway response to kinins may contribute to bronchial reactivity in HRV infection.

Recently, it has become evident that in addition to causing colds and exacerbations of asthma, HRV also predisposes to secondary illnesses such as sinusitis and otitis media. Maxillary sinus radiological abnormalities can be found in 87% of the patients with natural colds, and HRV can be detected in the sinus secretions in up to 40% of patients with clinically diagnosed acute maxillary sinusitis. Obstruction of the eustachian tube impairs drainage of the middle ear and this may contribute to the development of acute otitis media. The finding that middle ear pressure abnormalities are detected in up to 74% of subjects with HRV colds supports this view. Furthermore, HRV was detected by RT-PCR in the middle ear fluid from 25% of patients with acute otitis media and in 20% of patients suffering from otitis media with effusion who underwent tympanostomy. Whether HRV replicates in the middle ear remains to be determined.

Accumulating evidence suggests a role for HRV in diseases of the lower respiratory tract in immunocompromised adults, young children, and infants. The presence of HRV in the lower airways has been suggested by the finding of HRV RNA by RT-PCR in cell pellets from bronchoalveolar lavage, but HRV replication in the epithelium of the lower respiratory tract remains to be directly confirmed.

B. Immune Responses to Infection

HRV can be neutralized by antibodies directed against the main neutralizing immunogenic sites, and the most strongly neutralizing antibodies are those of the IgG class that cross-link adjacent capsid pentamers—an effect similar to that induced by dimers of soluble ICAM-1.

Between 7 and 21 days after HRV infection, neutralizing antibodies develop in serum and nasal secretions in the majority of infected subjects. Serum antibody against different serotypes of rhinovirus increases with age and the rates of infection and illness are low among persons with preexisting antibody. This protective effect is better early after infection and may be partially attributed to the presence of antibodies in the nasal secretions during the first few months after the infection. Serum-neutralizing antibody titers equal to or higher than 1:8 to 1:16 generally correlate with protection against infection by the homotypic virus. Under experimental conditions, however, the use of a large inoculum may induce infection in volunteers with higher titers of neutralizing antibody. The available data do not conclusively establish a role for neutralizing antibodies in nasal secretions or in the serum in the recovery of HRV infection and illness.

HRV infection stimulates a cellular immune response predominantly of the Th1 type, with production of IL-2 and IFNγ by peripheral blood mononuclear cells, and consequent increase in antiviral activities of macrophages and cytotoxic T cells. HRV infection may also increase the activity of natural killer cells and antigen-specific blastogenesis. Remarkably, it has been found that a proportion of the HRV-specific CD4 T cells can recognize serotype cross-reactive viral epitopes. It has also been noted that natural HRV infection can sometimes induce Th2 cytokines and thus augment allergic inflammation.

The finding that HRV may interfere with antigen presentation and lymphocyte activation *in vitro* by blocking ICAM-1/LFA-1 interaction is provocative. If this effect also occurs *in vivo,* it may not only favor viral replication but also contribute to the pathogenesis of secondary bacterial respiratory infections.

C. Illness

The majority of HRV infections are symptomatic and the ratio of clinically apparent to nonapparent infections is estimated to be 3 : 1. A typical HRV cold is characterized by nasal discharge, nasal obstruction, sneezing, pharyngitis, and cough. The first symptom noticed most frequently is sore throat, whereas runny nose and nasal stuffiness are frequently considered the most bothersome symptom by adults with natural HRV colds. Systemic symptoms, such as feverishness and malaise, are uncommon. Adults with naturally acquired HRV colds report significant disturbance of daily activities for 6–8 days, whereas symptoms may be present for 9–11 days on average.

HRV infection has been reported in association with lower respiratory tract manifestations, including wheezy bronchitis in children and acute exacerbations in patients with chronic bronchitis or asthma. Moreover, HRV has been associated with lower respiratory tract infection and possibly pneumonia in infants.

Upper respiratory tract complications are common after HRV infection. The frequencies of clinically diagnosed sinusitis and otitis media are estimated to be 1 in 200 and 1 in 20–50 common cold episodes, respectively.

VII. DIAGNOSIS

Until recently, diagnosis of HRV infections was based almost solely on the detection of viral cytophatic effect (CPE) in cell culture. In the past few years, serotype cross-reactive RT-PCR-based assays have become the most sensitive means of detecting HRV in clinical samples. The large number of HRV serotypes has hampered development of assays based on direct detection of viral antigens by immunofluorescence or immunoperoxidase staining.

A. Viral Isolation in Cell Culture

For optimal recovery, specimens should be obtained as early as possible in the course of illness. Specimens that cannot be inoculated fresh can be kept at 4°C for 1 or 2 days, but when longer periods of time are needed freezing specimens at −70°C is recommended. Nasal washes are the most commonly used samples, but nose and throat swabs are also acceptable. Saliva and conjunctival swabs are not useful samples and recovery from sputum is variable. Different preparations of virus transport media have been used for collection of samples for HRV isolation, and whether or not any particular type is superior is unclear.

HRV isolation from clinical samples can be accomplished in human diploid fibroblasts (MRC-5, WI-38, and fetal tonsil cells) or in HeLa cells. Selected HeLa cell clones, such as HeLa-I, may be significantly superior to other cell lines for HRV isolation and propagation. Fibroblasts are grown in minimum essential medium (MEM) containing 10% fetal bovine serum (FBS) and maintained before inoculation with MEM supplemented with 2% FBS and antibiotics. HeLa cells are usually grown in MEM with 10% FBS and should be inoculated when the monolayer is semiconfluent—in the presence of 30 m*M* $MgCl_2$. However, HeLa cells degenerate faster than fibroblasts so that it may become difficult to detect cytopathic effect without blind passage to fresh monolayers. Combinations of diploid fibroblasts and sensitive strains of HeLa cells provide the highest isolation rates. For optimal sensitivity, temperature of incubation should be 33–35°C, and the culture tube should be placed in a roller drum at 12 revolutions/hr. The tubes must be examined regularly for the appearance of typical CPE for up to 10–14 days. Typical CPE is easier to see in monolayers of diploid fibroblasts and consists of foci of pyknotic and rounded refractile cells.

B. Serologic Diagnosis

Serologic diagnosis of HRV infection is generally not a task for clinical laboratories because of the large number of serotypes and because of the lack of assays that will detect heterotypic serologic responses. Assays for detection of homotypic neutralization antibody are used for serologic confirmation of seroconversion in experimental settings. Depending on the HRV serotype, a significant increase in the neutralizing antibody titer between acute phase and convalescent serum samples can be detected in 50–80% of the infected individuals and can persist for several years. Serum-neutralization assays can be performed in tubes or in microtiter plates and are based on the inhibition of HRV CPE in cell monolayers by neutralizing antibodies present in the patient's serum. Type-specific enzyme-linked immunosorbent assays (ELISA) can detect anti-HRV antibody in serum and nasal secretions.

C. Detection of Human Rhinovirus Antigen or Nucleic Acid

In recent years, detection of HRV RNA in clinical samples by RT-PCR has become the most sensitive means of diagnosing HRV infections. Most HRV RT-PCR assays utilize primers complementary to oligonucleotide-size sequences within the 5′ UTR of the viral genome, which are conserved among the picornaviruses and thus allow for detection of virtually all known HRV serotypes. The sensitivity and specificity of the HRV RT-PCR assays can be further enhanced by confirmation of the identity of the PCR product by hybridization with oligonucleotide probes, which can be specific for HRVs or enteroviruses. Numerous studies using RT-PCR-based assays have shown that HRV colds are more frequent than previously estimated on the basis of viral isolation in cell culture. In addition, this sensitive assay has made it possible to detect HRV in secretions from which recovery of viable virus is usually hampered by low sensitivity, such as the case of middle ear secretion from children with otitis media with effusion.

An ELISA-based assay has been developed for the detection of HRV antigen in nasal secretions, but its serotype specificity limits its applicability.

VIII. PREVENTION AND TREATMENT

A. Prevention

The large number of HRV serotypes without marked serologic cross-reactivity has hampered the development of a vaccine. Antiviral chemoprophylaxis for HRV colds could have practical value during seasonal outbreaks, especially if targeted to individuals in whom HRV colds can be especially deleterious such as asthmatic patients. Interferon was the first agent to show efficacy in preventing HRV infection. Intranasal recombinant and leukocyte-derived IFN-α and -β have shown efficacy in the prophylaxis of experimental colds caused by HRV. Administration of high doses of recombinant IFN-α one day before challenge protected against infection and decreased virus shedding and rates of seroconversion. Recombinant IFN-α2b was also shown to be effective for seasonal prophylaxis of HRV colds, but the required long-term use was associated with high incidence of nasal irritation. Intranasal IFN-α2b used for postexposure prophylaxis in household contacts reduced the risk of HRV colds by 80–90% and that of acute respiratory illness by approximately 40%.

An approach to reduce the risk of HRV infection is to interrupt its spread between persons. Avoidance of self-inoculatory behavior, such as finger contact with eyes or nose, and frequent handwashing may reduce the risk of secondary infection in households.

B. Treatment

Numerous trials of antiviral drugs for HRV have been conducted, but a clinically useful specific therapy has not been identified. Lack of potency, side effects, and difficulties with drug delivery have been the main problems curbing the clinical application of antiviral therapies for HRV. Therefore, management of HRV colds remains largely symptomatic. Remarkably, it has been estimated that more than 50% of patients diagnosed with a common cold have a prescription filled for an unnecessary antibiotic, which contributes for the development of antibiotic resistance and increase in health care costs.

1. Antiviral

Several compounds have demonstrated antirhinoviral activity *in vitro,* including purine nucleoside analogs, enviroxime, capsid-binding agents such as R61837 and pirodavir, zinc salts, HRV receptor-blocking antibodies, 3C protease inhibitors, and soluble ICAM-1 (sICAM-1). Clinical trials with many of these agents yielded mostly negative results.

Intranasal IFN reduced the duration and titers of virus shedding in experimental colds but had little or no benefit in reducing symptoms of either induced or naturally occurring HRV colds. The combination of intranasal IFN, intranasal ipratropium, and oral naproxen provided greater clinical benefit than monotherapy. No consistent therapeutic benefit has been reported with intranasal enviroxime or zinc salt lozenges. Antiviral effects, but no clinical benefits, were found in studies of capsid-binding agents that bind to the hydrophobic pocket in the HRV capsid preventing receptor-mediated uncoating of the virus. The main problem with these compounds has been achieving and keeping adequate concentrations in nasal secretions.

sICAM-1 inhibits infectivity of HRV *in vitro* by blocking attachment/uncoating of the virus. Recent controlled studies have shown that intranasal tremacamra, a recombinant sICAM-1, when administered before onset of cold symptoms significantly reduced the severity of experimental HRV colds without causing adverse effects. However, clearance of medication from the nasal cavity by the mucocilliary mechanism is a problem for the clinical application of tremacamra. It also remains to be determined if this drug will be effective when given after the onset of cold symptoms.

Combinations of an antiviral with selected anti-inflammatory agents may be the best approach for treating HRV colds since they can simultaneously reduce viral replication and block inflammatory pathways. A combination of intranasal IFN, intranasal ipratropium, and oral naproxen reduced the severity of experimental HRV colds by 50%. The short duration of treatment reduced the severity of IFN side effects.

Nasal hyperthermia-inducing devices have been proposed as a means for treating HRV colds based on increasing the nasal mucosa temperature to levels nonpermissive for HRV replication (43°C). Controlled studies have provided no evidence for an antiviral effect and have demonstrated no consistent clinical benefit using this approach.

2. *Symptomatic*

Symptomatic remedies used in the treatment of colds include a large variety of nonprescription medications. Since cold symptoms result from multiple, independent inflammatory pathways, there is no single anti-inflammatory agent capable of providing relief for all symptoms. Therefore, drug combinations are required and should be started early at the initiation of the inflammatory cascade in order to provide maximum benefit.

Sympathomimetic decongestants, such as pseudoephedrine, constitute the most frequently used symptomatic treatment for common colds. Topical application of these agents is usually followed by immediate relief, but rebound vasodilation and nasal congestion occur. Oral decongestants provide less dramatic immediate relief but induce decongestion enough to be beneficial, without causing rebound vasodilation. Although generally well tolerated, systemic adverse effects include nervousness, insomnia, and increased blood pressure in predisposed persons, a finding that should restrict their use in persons with high blood pressure.

Oral antihistamines are often used for relief of sneezing based on controlled studies which have found reduction in sneezing and, to a lesser extent, rhinorrhea, with the use of these agents. First-generation antihistamines are often associated with sedation. Second-generation, nonsedating selective H1 antihistamines seem not to be beneficial for cold-related symptoms. Several trials have found that combinations of antihistamines and sympathomimetics provide symptom relief, but how much benefit results from the antihistamine component is unclear.

Anticholinergic compounds, such as ipratropium bromide, can be topically delivered to the respiratory tract without significant systemic side effects and may reduce rhinorrhea by reducing seromucous gland secretion.

Glucocorticoids, systemic or intranasal, have modest or no effect on the symptoms of HRV colds and may increase viral replication. Early treatment with nonsteroidal anti-inflammatory substances, such as naproxen, seems to be effective in reducing HRV cold symptoms.

See Also the Following Articles

Aerosol Infections • Antigenic Variation • Enteroviruses • Virus Infection

Bibliography

Arruda, E., and Hayden, F. G. (1995). Clinical studies of antiviral agents for picornaviral infections. *In* "Antiviral Chemotherapy" (D. J. Jeffries and E. De Clercq, Eds.), pp. 321–355. Wiley, Chichester, UK.

Arruda, E., Boyle, T. R., Winther, B., Pevear, D. C., Gwaltney, J. M., Jr., and Hayden, F. G. (1995). Localization of human rhinovirus replication in the upper respiratory tract by in situ hybridization. *J. Infect. Dis.* **171**, 1329–1333.

Arruda, E., Pitkäranta, A., Witek, T. J., Jr., Doyle, C. A., and Hayden, F. G. (1997). Frequency and natural history of rhinovirus infections in adults during Autumn. *J. Clin. Microbiol.* **35**, 2864–2868.

Cordingley, M. G., *et al.* (1990). Substrate requirements of human rhinovirus 3C protease for peptide cleavage *in vitro*. *J. Biol. Chem.* **265**, 9062–9065.

Gern, J. E., and Busse, W. W. (1999). Association of rhinovirus infections with asthma. *Clin. Microbiol. Rev.* **12**, 9–18.

Gern, J. E., Dick, E. C., Kelly, E. A. B., Vrtis, R., and Klein, B. (1997). Rhinovirus-specific T cells recognize both shared and serotype-restricted viral epitopes. *J. Infect. Dis.* **175**, 1108–1114.

Gwaltney, J. M., Jr., and Rueckert, R. R. (1997). Rhinovirus. *In* "Clinical Virology" (D. R. Richman, R. J. Whitley, and F. G. Hayden, Eds.), pp. 1025–1047. Churchill Livingstone, New York.

Hadfield, A. T., Lee, W.-M., Zhao, R., Oliveira, M. A., Minor, I., Rueckert, R. R., and Rossmann, M. G. (1997). The refined structure of human rhinovirus 16 at 2.15 Å resolution: Implications for the viral life cycle. *Structure* **5**, 427–441.

Pitkaranta, A., Virolainen, A., Jero, J., Arruda, E., and Hayden, F. G. (1998). Detection of rhinovirus, respiratory syncytial virus, and coronavirus infections in acute otitis media by reverse-transcriptase polymerase chain reaction. *Pediatrics* **102**, 291–295.

Rakes, G. P., Arruda, E., Ingram, J. M., Hoover, G. E., Zambrano, J. C., Hayden, F. G., Platts-Mills, T. A. E., and Heymann, P. W. (1999). Rhinovirus and respiratory syncytial virus in wheezing children requiring emergency care. *Am. J. Respir. Crit. Care Med.* **159**, 785–790.

Rossmann *et al.* (1985). *Nature* **317**, 145–153.

Rueckert, R. R. (1996). *Picornaviridae:* The viruses and their replication. *In* "Fields Virology" (B. N. Fields, D. M. Knipe, and P. M. Howley, Eds.), pp. 609–654. Lippincott-Raven, Philadelphia.

Wimalasundera, D. S., Katz, D. R., and Chain, B. M. (1997). Characterization of the T cell response to human rhinovirus in children: Implications for understanding the immunopathology of the common cold. *J. Infect. Dis.* **176**, 755–759.

Rhizoctonia

Marc A. Cubeta
North Carolina State University

Rytas Vilgalys
Duke University

I. Historical and Taxonomic Overview of *Rhizoctonia*
II. Ecology and Pathology of *Rhizoctonia solani*
III. Management of Rhizoctonia Diseases
IV. Future Perspectives

GLOSSARY

anamorph The asexual stage(s) of a fungus.

***basidium* (*pl. basidia*)** A club-shaped cell produced by fungi in the class basidiomycete during sexual reproduction.

***hymenium* (*pl. hymenia*)** The spore-bearing layer of a basidiomycete fruiting body.

***hypha* (*pl. hyphae*)** The individual vegetative filaments of a mycelium.

mycelium The vegetative body of a fungus composed of a mass of hyphae.

***sclerotium* (*pl. sclerotia*)** A survival structure consisting of a hard, frequently rounded or irregular mass of woven hyphal cells.

somatic incompatibility The inability of hyphae from two fungi to recognize, fuse, and establish a continuous network of mycelium.

***sterigma* (*pl. sterigmata*)** A finger-like structure associated with a basidium on which basidiospores are borne.

teleomorph The sexual stage(s) of a fungus.

THE GENUS *RHIZOCTONIA* includes a very large and diverse group of fungi that are frequently associated with plant roots and soil. Many *Rhizoctonia* fungi are important pathogens of agricultural, aquatic, forest, and horticultural plant species. However, not all *Rhizoctonia* fungi cause plant disease. Some species of *Rhizoctonia* can exist primarily as saprophytes on decaying organic matter, whereas others can establish beneficial relationships with certain orchids and mosses. Research studies also have shown that *Rhizoctonia* fungi can be very damaging parasites of other soil fungi. The study of *Rhizoctonia* fungi has progressed slowly because the genus *Rhizoctonia* actually represents a complex of many genetically diverse fungi with very different life histories. Although the genetic diversity among *Rhizoctonia* fungi has been recognized for many years, the biological and ecological consequences of this diversity are just beginning to be explored.

I. HISTORICAL AND TAXONOMIC OVERVIEW OF *RHIZOCTONIA*

The genus *Rhizoctonia* was first described in 1815 by the mycologist DeCandolle after observing the fungus on diseased alfalfa roots in France. Translated from Greek language, *Rhizoctonia* means "death of roots." The *Rhizoctonia* fungus originally described by DeCandolle was probably very different than the fungus recognized today as *Rhizoctonia*. Unfortunately, there are no preserved specimens of DeCandolle's *Rhizoctonia* fungus for comparison.

Rhizoctonia solani, the most widely recognized species of *Rhizoctonia*, was originally described by Julius Kühn on potato in 1858. This organism has been the subject of much research in plant pathology and will be the primary focus of this article. *Rhizoctonia solani* is a basidiomycete fungus that does not produce any asexual spores (conidia), and only occasionally will the fungus produce sexual spores (basidiospores). Unlike many basidiomycete fungi, the basidiospores are not enclosed in a fleshy, fruiting body or mushroom. The sexual fruiting structures and basidiospores (i.e., teleomorph) were first observed and described in detail by Prillieux and Delacroiz in 1891. The sexual stage of *R. solani* has undergone several name changes since 1891, but it is now known as *Thanatephorus cucumeris*.

In nature, *R. solani* reproduces asexually and exists primarily as vegetative mycelium and/or sclerotia (see Section II). The vegetative mycelium of *R. solani* and other *Rhizoctonia* fungi are colorless when young but become brown colored as they grow and mature (Fig. 1A). The mycelium consists of hyphae partitioned into individual cells by a septum containing a doughnut-shaped pore (Fig. 1B). This septal pore allows for the movement of cytoplasm, mitochondria, and nuclei from cell to cell. The hyphae often branch at a 90° angle (Fig. 1B) and usually possess more than three nuclei per hyphal cell (Fig. 1C). The anatomy of the septal pore and the number of nuclei within a hyphal cell have been used extensively by researchers to differentiate *R. solani* from other *Rhizoctonia* fungi.

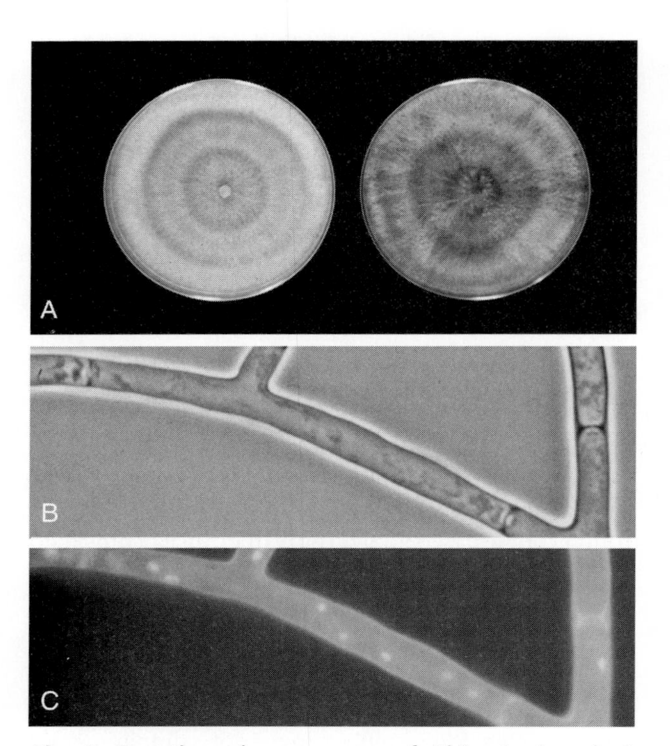

Fig. 1. Growth and appearance of *Rhizoctonia solani.* (A) Potato dextrose agar in a petri dish and on a glass slide after staining hyphae with (B) aniline blue or (C) the fluorescent, DNA-binding stain 4′, 6-diamidino-2-phenylindole. Note the septa in Fig. 1B and the seven nuclei in the hyphal cell in Fig. 1C (Figs. 1B and 1C courtesy of Bruce Martin and Eddie Echandi).

Because *R. solani* and other *Rhizoctonia* fungi do not produce conidia and only rarely produce basidiospores, the classification and identification of these fungi often have been difficult. Prior to the 1960s, researchers relied mostly on differences in morphology observed by culturing the fungus on a nutrient medium in the laboratory and/or pathogenicity on various plant species to classify *Rhizoctonia* fungi. In 1969, J. R. Parmeter and colleagues at the University of California, Berkeley, reintroduced the concept of "hyphal anastomosis" to characterize and identify *Rhizoctonia* fungi. The development of this concept was an extension of previous studies conducted in Germany and Japan in the 1930s by Schultz and Matsumoto *et al.,* respectively. The concept implies that isolates of *Rhizoctonia* that have the ability to recognize and fuse (i.e., "anastomose") with each other are genetically related, whereas isolates of *Rhizoctonia* fungi that do not have this ability are genetically unrelated. In practice, hyphal anastomosis is determined in several ways. The most commonly employed practice involves pairing two isolates of *Rhizoctonia* on a glass slide and allowing them to grow together. The area of merged hyphae is stained and examined microscopically for the resulting hyphal interaction(s). If the hyphae grow past each other and no recognition or hyphal fusion is observed (i.e., no hyphal interaction), the isolates are considered to be genetically unrelated. If the hyphae recognize, fuse, and establish a continuous network of mycelium (Fig. 2A), the isolates are considered to be genetically identical or very closely related. In this type of hyphal interaction the isolates are referred to as being "somatically compatible." If the hyphae recognize and fuse, but the fusion is followed by death of the interacting cells (Fig. 2B), the isolates are considered to be genetically related but not identical (i.e., the same genetic individual). In this type of hyphal interaction the isolates are referred to as being

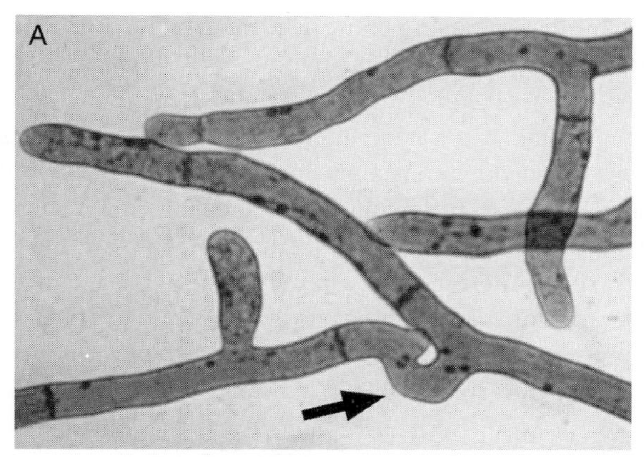

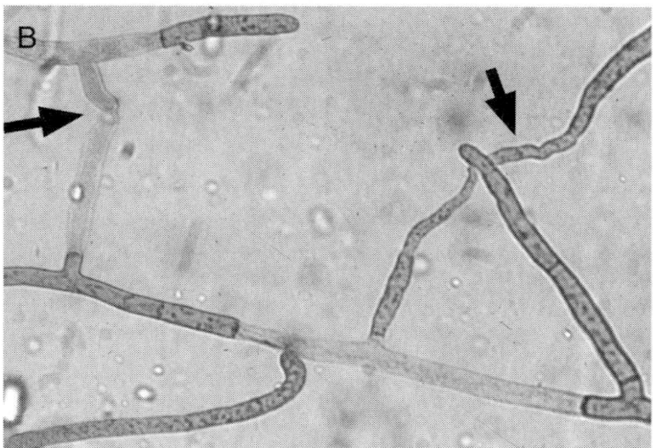

Fig. 2. Interaction of *Rhizoctonia solani* hyphae on a glass slide. (A) Compatible interaction (indicated by arrow). (B) Incompatible interactions (indicated by arrows) (Fig. 2A courtesy of Donald Carling). See color insert.

"somatically incompatible." Somatic incompatibility is a recognition system commonly found in fungi and other organisms of the same species that prevents the transfer of foreign cytoplasm, organelles, and genetic material that could disrupt cell function and metabolism. Unfortunately, very little is known about the genetic mechanisms controlling this recognition process in *Rhizoctonia.* In other filamentous fungi, somatic incompatibility is controlled by several genes with multiple alleles. For two fungal isolates to be compatible, all somatic compatibility loci must be the same.

Hyphal anastomosis criteria has been used extensively to place isolates of *Rhizoctonia* into taxonomically distinct groups called anastomosis groups. Currently, 14 anastomosis groups have been identified in *R. solani* from various geographic regions of the world. Recent protein and DNA-based studies support the separation of *R. solani* into genetically distinct groupings but have also revealed considerable genetic diversity within an anastomosis group. Hyphal anastomosis and molecular methods are currently being used to further examine the taxonomy (i.e., science of classifying organisms), ecology, and pathology of *R. solani.*

II. ECOLOGY AND PATHOLOGY OF *RHIZOCTONIA SOLANI*

Rhizoctonia solani can survive for many years by producing small (1- to 3-mm diameter), irregular-shaped, brown to black structures (sclerotia) in soil and on plant tissue (Figs. 3A and 3B). The sclerotia

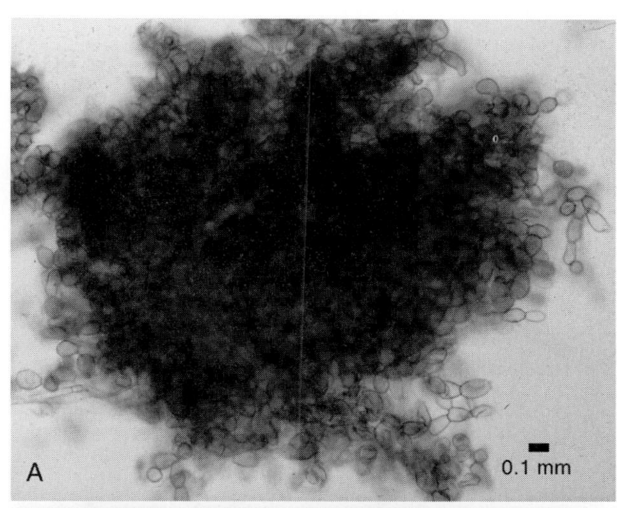

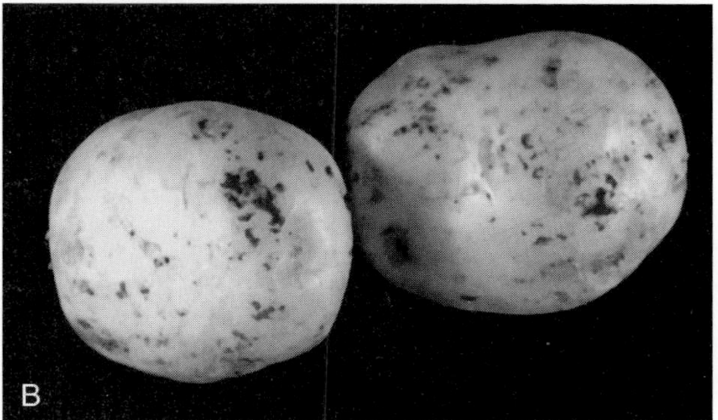

Fig. 3. Survival structures of *Rhizoctonia solani.* (A) Microscopic view of specialized cells of a sclerotium stained with Safranin O. (B) Irregular shaped, black sclerotia on potatoes.

are loose aggregates of specialized, dark-pigmented hyphal cells that are capable of surviving in soil in the absence of a host plant for many years under adverse environmental conditions. Certain *R. solani* pathogens of rice have evolved the ability to produce sclerotia with a thick outer layer that allows them to float and survive in water.

The *Rhizoctonia* fungus also survives as mycelium by colonizing soil organic matter as a saprophyte, particularly as a result of plant pathogenic activity. Sclerotia and/or mycelium present in soil and/or on plant tissue germinate to produce vegetative threads (hyphae) of the fungus that can attack a wide range of food and fiber crops when the weather favors disease development.

The fungus is attracted to the plant by chemical stimulants released by actively growing plant cells and/or decomposing plant residues. As the attraction process proceeds, fungal hyphae will come in contact with the plant and become attached to its external surface. After attachment, the fungus continues to grow on the external surface of the plant and will causes disease by producing a specialized infection structure (either an appresorium or infection cushion) that penetrates the plant cell and releases nutrients for continued fungal growth and development. The infection process is promoted by the production of many different extracellular enzymes (e.g., cellulase, cutinase, pectinase, pectin lyase, pectin methylesterase, phosphatase, and polygalacturonase) that degrade various components of plant cell walls (e.g., cellulose, cutin, and pectin). As the fungus kills the plant cells, it continues to grow and colonize dead tissue, often forming sclerotia. *R. solani* primarily

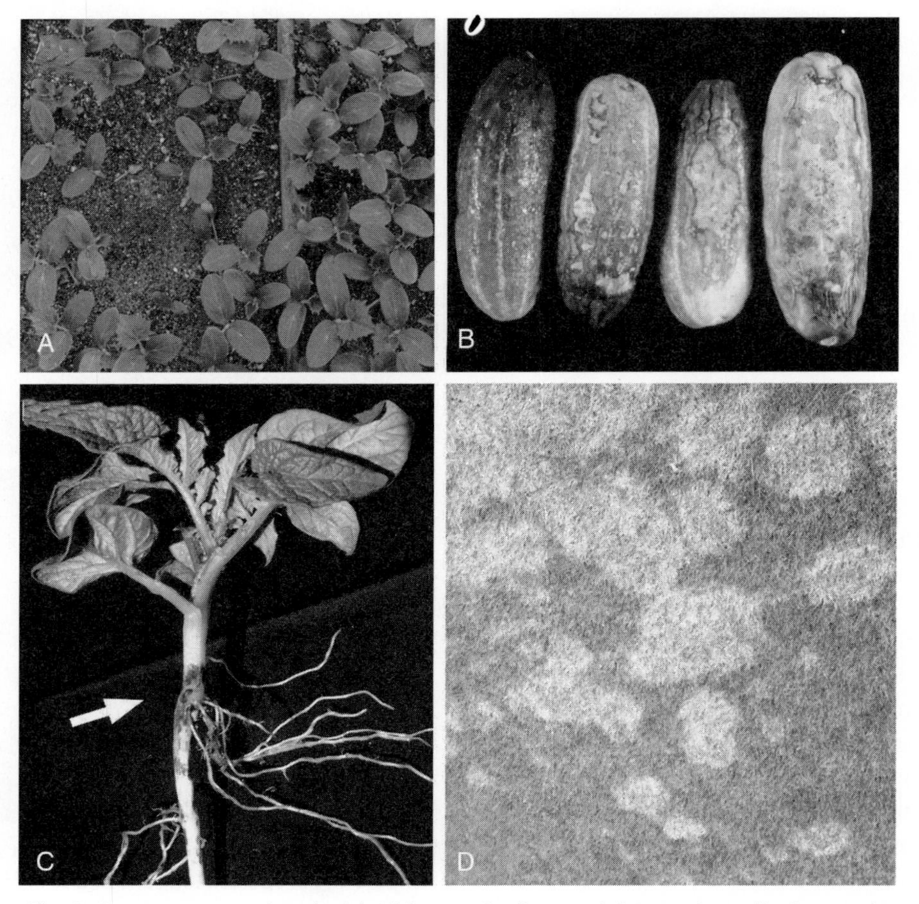

Fig. 4. Symptoms associated with Rhizoctonia disease. (A) Damping-off of cucumber seedlings. (B) Fruit rot of cucumber. (C) Stem canker of potato. (D) Brown patch of tall fescue (Fig. 4D courtesy of Lee Burpee).

attacks belowground plant parts, such as the seeds, hypocotyls, and roots, but it is also capable of infecting aboveground plant parts (e.g., fruits, leaves, and stems). Severely infected seeds usually do not germinate, whereas infected seedlings can be killed either before or after they emerge from the soil. Seedlings killed after emergence often appear to have fallen over on the soil surface as a result of excessive rainfall or overwatering. This common symptom of Rhizoctonia disease is referred to as "damping-off" (Fig. 4A). In general, seedlings are susceptible to attack during the first few weeks of their development but become progressively less susceptible as they mature and develop biochemical and physical defense mechanisms. This decreased susceptibility to *Rhizoctonia* is associated with an increased production of calcium pectate and lignin in the developing stem tissue. Infected seedlings not killed by the fungus often have reddish-brown lesions (called cankers) on stems and roots (Fig. 4C). Cankers can reduce plant productivity and yield by restricting the movement of water and nutrients within the plant. In many instances, infected plants with cankers often appear healthy. Therefore, Rhizoctonia diseases often go unrecognized because disease symptoms develop below ground and cannot be seen. Consequently, yield losses associated with Rhizoctonia diseases usually do not become evident until after harvesting. In addition to attacking below-ground plant parts, the fungus will occasionally infect fruit (Fig. 4B) and leaf tissue (Fig. 4D) located near or on the soil surface. This type of disease often occurs because the mycelium and/or sclerotia of the fungus are close to or splashed on the plant tissue.

In certain situations, *R. solani* can produce basidiospores that will cause disease. These basidiospores also serve as a source for rapid and long-distance dispersal of the fungus. The basidiospores germinate to produce hyphae that infect leaves during periods of high relative humidity and periods of extended wet weather. Under these conditions, basidiospores can often be observed on the base of stems near the soil surface or on the underside of leaves in the plant canopy (Fig. 5). Because of their sensitivity to desiccation and ultraviolet radiation, basidiospores

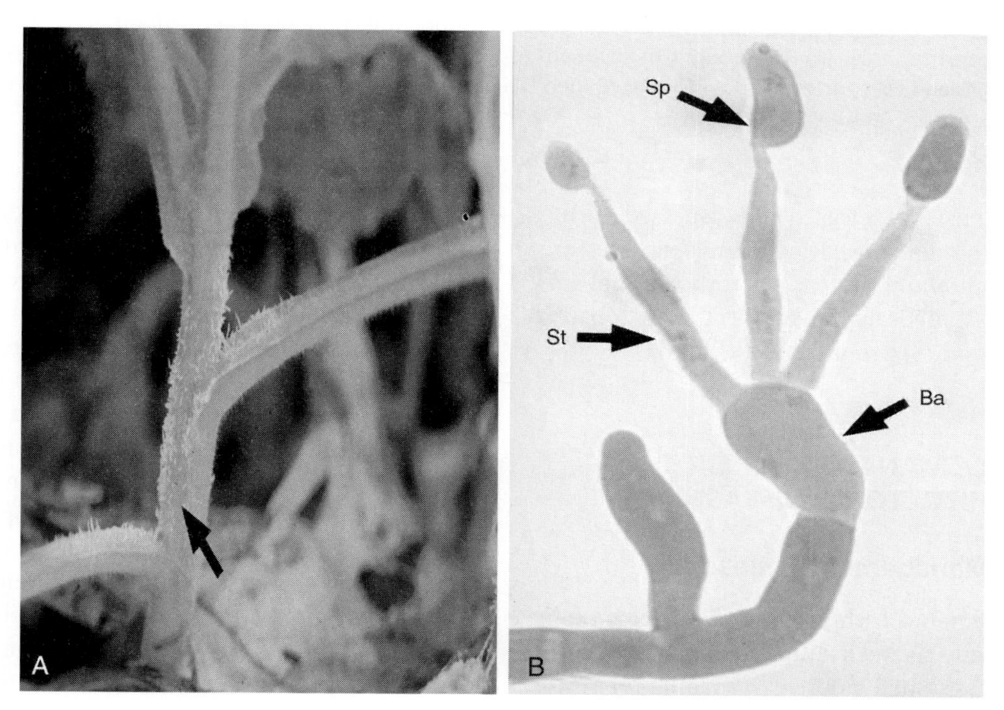

Fig. 5. Sexual fruiting structures of *Thanatephorus cucumeris* (anamorph = *Rhizoctonia solani*). (A) Hymenium of *T. cucumeris* at base of a tobacco stem. (B) Microscopic view of a fruiting structure composed of a basidium (Ba) and three finger-like sterigmata (St) bearing basidiospores (Sp) (Fig. 5B courtesy of Gerry Adams and Pam Gunnell).

Fig. 6. Symptoms of *Rhizoctonia* disease caused by basidiospore infection on leaves of different plant species. (A) Target spot of tobacco. (B) Foliar blight of soybean. (C) Foliage blight of sugar beet. (D) Web blight of snapbean (Figs. 6A, 6B, 6C, and 6D courtesy of David Shew, Shigeo Naito, and Graciela Godoy-Lutz, respectively). See color insert.

of *R. solani* are usually released and dispersed during the night. Although most Rhizoctonia diseases are initiated by mycelium and/or sclerotia, several important disease of beans, sugar beet, and tobacco result from basidiospore infection (Fig. 6).

III. MANAGEMENT OF RHIZOCTONIA DISEASES

A. Management Philosophy

The management of Rhizoctonia diseases is often challenging and has been difficult to achieve using traditional means such as breeding plants for resistance, crop rotation, and fungicides. Management decisions are often directly related to the perceived benefits (economic, environmental, reducing additional pest problems, etc.) and costs of implementing a particular management practice. In general, most successful management practices do not totally eliminate the pathogen but rather prevent or reduce infection by the pathogen. Certain practices employ the use of chemical and/or physical barriers to protect the plant from direct contact with the fungus. However, once a plant is infected, management options become limited during the current cropping season. In addition to understanding pathogen biology and ecology, the relationship between the length of time required for protection and the particular phase of the disease being targeted for management (e.g., above- versus belowground Rhizoctonia diseases) should also be thoroughly understood. For example, a 2- or 3-week period may be needed to protect plants from a damping-off disease that originated from mycelium or sclerotium in soil, whereas season-long protection may be needed for a leaf disease caused by airborne basidiospores. Therefore, an integration and implementation of multiple approaches is needed to effectively manage Rhizoctonia diseases.

B. Soil and Seed Treatment Practices

Since most Rhizoctonia diseases originate from mycelium and sclerotia in soil, a considerable amount of research has focused on the use of soil fumigants. When applied to soil covered with plastic, these materials release a toxic gas that kills or debilitates the *Rhizoctonia* fungus as well as other soil-inhabiting organisms. Soil fumigants provide very effective management of Rhizoctonia diseases, but their use is limited to high-value agricultural crops. Many soil fumigants are currently being investigated for possible damaging effects to the ozone layer, which may also limit their potential use in the future. Recent efforts to harness the sun's energy to heat soil under clear plastic in the field (a process known as solarization) have shown promising results for managing Rhizoctonia diseases. However, soil solarization may be limited to areas that receive high amounts and extended periods of soil radiation associated with a dry climate. This practice also restricts the potential use of valuable cropland during the solarization period. Treating soil or using a soil-less mix with aerated steam has provided a very effective means of managing Rhizoctonia diseases in the greenhouse and nursery industry for many years. Unfortunately, the application of aerated steam to soil in the field has not met with much success. The accidental introduction of the pathogen into recently fumigated, solarized, or steamed soil can result in more Rhizoctonia disease than before the soil was treated.

The application of fungicides to soil is not commonly practiced because it is usually not economical and their effectiveness can be limited due to microbial degradation and fixation of the fungicide to negatively charged clay minerals and organic matter. A fungicide applied to seed prior to planting may provide some relief and is relatively inexpensive compared to other production costs. However, a fungicide seed treatment will usually not be beneficial in soils highly infested with the *Rhizoctonia* fungus.

In recent years, there has been a resurgence of interest in managing Rhizoctonia diseases with bacteria (species of *Bacillus, Burkholderia,* and *Pseudomonas*) and fungi (species of *Gliocladium, Trichoderma,* and *Verticillium*). The two primary approaches for biological control of Rhizoctonia diseases have been (i) to introduce a single microorganism to soil and/or seed and (ii) to stimulate native microorganisms by adding organic amendments or green manures directly to soil. These microorganisms (i.e., biological control agents) can reduce Rhizoctonia diseases by (i) producing antibiotic-like chemicals that affect pathogen activity and survival, (ii) competing with the pathogen for space and/or nutrients, (iii) parasitizing the pathogen, and (iv) causing a biochemical change in the plant making the host less susceptible to infection (induced resistance). In many instances, several different mechanisms of biological control may be operating together to reduce Rhizoctonia disease. Although several commercial products are now available and research results appear promising for managing Rhizoctonia diseases with biological agents, their potential use in production agriculture is still relatively insignificant.

C. Cultural Practices

One of the most important initial management decisions that should be considered is whether to purchase and plant only high-quality seed. Physiologically older seed are less vigorous and more susceptible to attack by *Rhizoctonia.* Cultural conditions that promote rapid seed germination and seedling emergence, such as planting seed at a shallower depth and when soil temperature is more favorable for plant than fungal growth, are highly desirable. Since the *Rhizoctonia* fungus can persist in soil for several years, crop rotation may be of limited value; however, planting in fields with a history of severe Rhizoctonia disease should be discouraged. Deep plowing prior to planting can reduce the incidence of Rhizoctonia diseases by burying the fungus deeper in soil profile and hastening the destruction of plant residue, thus making it more difficult for the fungus to infect the plant. As with any disease, proper soil fertility and plant nutrition are important and key management components for reducing Rhizoctonia diseases.

D. Host Plant Resistance

In general, the breeding of plants for resistance to Rhizoctonia diseases through traditional approaches has not been very successful. Although certain plant species express differences in susceptibility to *R. solani,* these differences alone will not usually provide adequate protection. Therefore, the development of plants resistant to *Rhizoctonia* may lie in the realm of the molecular biologist. Recently, transgenic potato and tobacco plants have been developed which produce the antifungal proteins chitinase and glucanase. These proteins (enzymes) can break down the major structural components of *Rhizoctonia* hyphae (chitin and β-glucans) and reduce subsequent infection caused by the fungus. The continued research efforts in this area appear to offer tremendous potential for developing *Rhizoctonia*-resistant plants in the future.

IV. FUTURE PERSPECTIVES

One hundred and forty years have elapsed since the discovery of *R. solani*. During this time period, most *Rhizoctonia* research has concentrated on economically important diseases of agricultural plant species. Therefore, much of our understanding about *Rhizoctonia* fungi has been obtained from the study of *R. solani* in agricultural systems. However, very little is known about *R. solani* in nonagricultural environments or the biology and life history of other non-plant pathogenic species of *Rhizoctonia*. It seems plausible that plant pathogenic species of *R. solani* may also be beneficial on certain plant species.

Currently, *Rhizoctonia* researchers are actively examining DNA sequences from many different genes in *Rhizoctonia* to better understand the taxonomic relationships among these fungi. An understanding of their taxonomy can provide researchers with a foundation for identifying species of *Rhizoctonia* and examining evolutionary relationships and field populations of these fungi in the future. Although the primary research emphasis will undoubtedly continue to be on plant pathogenic species of *Rhizoctonia* and the development of more effective, economical, and "environmentally friendly" disease management strategies, much can be learned about fungal biology, evolution, and populations by the additional study of non-plant pathogenic species of *Rhizoctonia*. This knowledge will be attained through the integration of molecular genetics with traditional scientific disciplines, such as mycology, plant pathology, and soil microbiology.

Acknowledgments

We thank D. Michael Benson and H. David Shew for reviewing the manuscript and Mike Munster for his patience and technical assistance in preparing the figures for electronic publication.

See Also the Following Articles

Extracellular Enzymes • *Pseudomonas* • Soil Dynamics, and Organic Matter, Decomposition • Verticillium

Bibliography

Adam, G. C. (1988). *Thanatephorus cucumeris* (*Rhizoctonia solani*) a species of wide host range. Adv. *Plant Pathol.* **6**, 535–552.

Agrios, G. N. (1997). "Plant Pathology," 4th ed. Academic Press, New York.

Alexopoulos, C. J., Mims, C. W., and Blackwell, M. (1996). "Introductory Mycology," 4th ed. Wiley, New York.

Anderson, N. A. (1982). *Annu. Rev. Phytopathol.* **20**, 329–347.

Hawksworth, D. L., Kirk, P. M., Sutton, B. C., and Pegler, D. N. (1995). "Ainsworth and Bisby's Dictionary of the Fungi," 8th ed. Cambridge Univ. Press, Cambridge, UK.

Ogoshi, A. (1987). *Annu. Rev. Phytopathol.* **25**, 125–143.

Sneh, B., Jabaji-Hare, S., Neate, S., and Dijst, G. (1996). "*Rhizoctonia* species: Taxonomy, Molecular Biology, Ecology, Pathology, and Control." Kluwer, Dordrecht.

Vilgalys, R., and Cubeta, M. A. (1994). *Annu. Rev. Phytopathol.* **32**, 135–155.

The Rhizosphere

Donald A. Klein
Colorado State University

I. The World of Plant Roots and Microbes
II. Materials Released in the Root Zone by Plants
III. Microbial Responses in the Rhizosphere
IV. Rhizosphere Processes and the Plant
V. Ecology, the Environment, and the Rhizosphere
VI. Management of Rhizosphere Microbes: Biotechnological Applications
VII. Future Research Directions

GLOSSARY

associative nitrogen fixation Nitrogen fixation by free-living bacteria in the rhizosphere

biocontrol plant growth-promoting bacteria Bacteria which suppress disease by providing chemicals which inhibit pathogens or which increase plant resistance. Compare with plant growth-promoting bacteria.

cometabolism The modification of a compound not used for growth by a microbe in the presence of another compound serving as a carbon and energy source.

constructed wetlands The intentional creation of wetland areas with aquatic plants and their root-associated microbes. Water, containing contaminants, can be processed as it flows past or under the plant and its associated microbes.

denitrification Reduction of nitrate to nitrogen gas (N_2) and nitrous oxide (N_2O) during microbial respiration without oxygen present.

exudates Compounds of low molecular weight that leak from plant cells either into the intercellular spaces or directly from the epidermal cell walls into the root environment. This is a passive process.

greenhouse gases Long-lived gases, such as CO_2, CH_4, and N_2O, which absorb infrared radiation in the lower atmosphere, leading to possible warming of the earth's surface.

methanotroph An aerobic bacterium which uses methane as a source of carbon and energy.

mucigel Gelatinous material on the root surface derived from plant sources (mucilages) as well as microbial cells, soil colloids, and soil organic materials.

mucilages Gelatinous organic materials released by the plant in the root cap region derived from the Golgi apparatus, polysaccharide hydrolysis, and epidermal materials.

mycorrhizosphere The region around a mycorrhizal fungus which is deriving its carbon from the plant. Materials released from the fungus increase the microbial populations and their activities around the fungal hyphae.

phytoremediation The use of plants and their associated microorganisms to facilitate degradation, concentration, and/or extraction of soil-borne contaminants. This may involve uptake and modification by the plant, microbial modification followed by plant uptake, or direct modification by the rhizosphere microbes, possibly involving cometabolism.

plant growth-promoting bacteria Bacteria which benefit plants by means other than by suppressing deleterious microbes. Compare with biocontrol plant growth-promoting bacteria.

rhizodeposition Release of materials (gaseous, soluble, and particulate) from roots.

rhizoplane Root surface that can be colonized by microorganisms.

rhizosphere Region around the plant root where materials released from the root modify microbial populations and their activities.

siderophores Iron-binding compounds produced by microorganisms (*sidero* = iron; *phore* = carrier).

THE RHIZOSPHERE is the region in which materials released from the root, and root metabolic activities such as respiration, affect microbes. Volatile, soluble, and particulate materials are released by roots in the process of rhizodeposition. The rhizosphere microbes, after their growth on these materials and their cellular turnover, release nutrients in forms which can be utilized by plants. Plants and their rhizospheres are found in soils, in which the environment is primarily aerobic, and in many marine and freshwater environments in which oxygen is often limiting. These oxygen-limited environments include rice paddies, freshwater and saltwater marshes, and mangrove swamps. In addition, plants can be grown in contaminated soils in which rhizosphere microbes assist in phytoremediation, the treatment of soil-borne contaminants, or in artificial aquatic environments called "constructed wetlands," in which rhizosphere microbes assist in processing water-borne contaminants. With biotechnology, it is possible to modify the rhizosphere microbial community to increase plant growth and control undesirable pathogens.

I. THE WORLD OF PLANT ROOTS AND MICROBES

Vascular plants are one of the most important links which humans have to nature. For most of us, the vast majority of our food and fiber are directly derived from plants. Although it often is not evident, plant roots and their surrounding microbes (the rhizosphere) are important wherever plants are found: forests, grasslands, tundra, deserts, and wet areas such as marshes and mangrove swamps. The roots of these plants support a unique modified microbial community in an environment termed the rhizosphere, the region influenced by the root and its activities. Microbes also directly colonize root surfaces and also are found under the root surface, creating additional unique environments for microbes.

The term rhizosphere, which has been used for less than 100 years, is critical to understanding how plants interact with their environment. In essence, the microbes in the rhizosphere provide the critical link between plants, which require inorganic nutrients, and the environment, which contains the nutrients but often in organic and largely inaccessible forms, as shown in Fig. 1 for a soil.

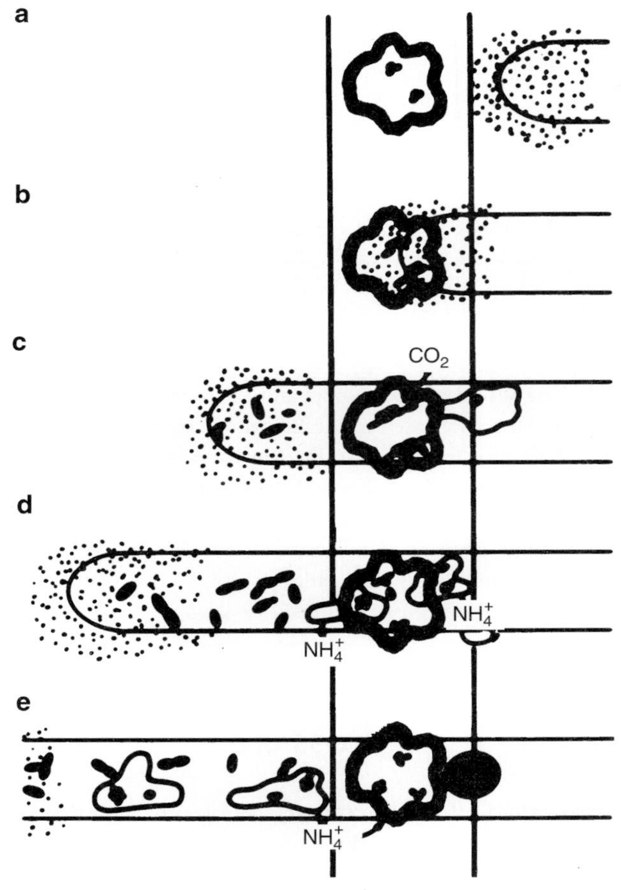

Fig. 1. Role of microbes in the plant root zone as intermediates in nutrient transfers between the soil and the plant. (a) Bacteria in soil are concentrated on the surfaces of organic matter where they exist in a low metabolic state due to restricted carbon availability. (b) When a root grows through the soil, a carbon input will take place around and just behind the tip. Fresh root-derived carbon will thus be added as a pulse when the root tip penetrates into any specific part of the soil. Bacteria will start to grow, releasing enough nitrogen from the organic matter to meet their need for growth. (c) Bacterial growth attracts protozoa. (d) When protozoa consume bacteria, one-third of the bacterial nitrogen is released as NH_4^+. (e) Part of the ammonium will be taken up by the root (reproduced with permission from Clarholm, 1985).

Microbes also colonize the plant root surface (the rhizoplane) (Fig. 2). The colonization of the rhizoplane by microbes can involve specific attachment mechanisms. For *Agrobacterium thaliana,* which forms tumors in susceptible plants, this involves a two-step process of (i) loose binding to the cell sur-

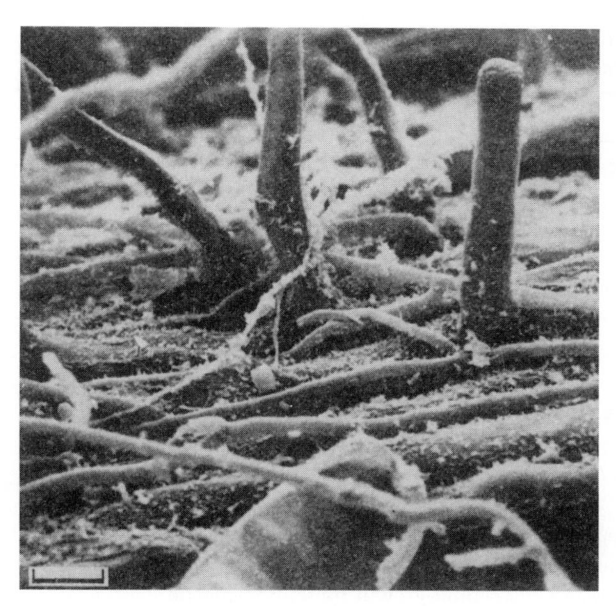

Fig. 2. Electron micrograph of microbes on the surface of a plant root showing fungi and bacteria, together with root hairs. Scale bar = 10 μM (reproduced with permission from Campbell, 1985, p. 116).

face and (ii) the synthesis of cellulose fibrils by the bacterium. This results in binding of the bacteria to the plant root surface. If mutant bacteria are used which do not have these attachment characteristics, they will not bind to the root surface.

Plants also have other microbes with which they develop unique relationships in the root environment, including the symbiotic nitrogen-fixing bacteria such as *Rhizobium,* which can form nodules on susceptible legumes, and the fixation of nitrogen by filamentous bacteria of the genus *Frankia,* an association which occurs with a wide range of shrubs and woody plants. Another important group of microbes which form direct associations with plants includes the mycorrhizae or "fungus-roots," which occur in a wide variety of plants, considered to be one of the oldest plant–microbe associations. The nitrogen-fixing bacteria and the mycorrhizae form structures within the plant root, indicating that these relationships involve longer term mutually beneficial associations. The mycorrhizal hyphal network, supported by carbon derived from the plant, also releases organic carbon. Microbes grow around the mycorrhizal hyphae in a region called the mycorrhizosphere.

II. MATERIALS RELEASED IN THE ROOT ZONE BY PLANTS

We often think of plants as consisting mostly of branches and leaves, with a minor amount of a plant's biomass consisting of roots. This impression can be misleading. For many plants the root:shoot ratio is positive—more of the plant mass is in the roots than in stems and leaves. The materials released by the plant include a wide variety of organic compounds, as noted in Table I. The types and amounts of these substances are constantly changing due to a wide range of plant and environment-related factors (Table II). These factors can include temperature and moisture stress, fertilizer additions, herbage removal (both above- and below-ground) changes in sunlight, herbicide additions, plant age, and other changes in the plant's environment. The materials lost from plant roots can be 30–40% of the carbon fixed through photosynthesis.

The fine root hairs are a critical part of the root system (Fig. 3). These can be rapidly shed when environmental conditions become less suitable for plant growth. Cortical and epidermal cells, called mucilages, and soluble metabolic products (amino acids, sugars, organic acids, etc.), described as exudates, are also released. In addition, a variety of gaseous metabolites flow from the roots. The release of these different materials is described as the process of rhizodeposition. When the mucilages combine with microbes, soil colloids, and soil organic matter, mucigels are formed which cover and protect the root tip.

III. MICROBIAL RESPONSES IN THE RHIZOSPHERE

Plant roots create new environments for microbes; due to these increased levels of nutrients, microbial populations increase, often by 1000- to 10,000-fold, and a marked change in the composition of the microbial community will also occur (Table III), as indicated by the rhizosphere:soil ratio for a soil. The numbers and types of microbes (and their activities) often increase along the root away from the tip of the plant root. The plant roots also respire (use oxygen),

TABLE I
Types of Materials Released by Plant Roots in the Process of Rhizodeposition[a]

Compound	*Exudate components*
Sugars	Glucose, fructose, sucrose, maltose, galactose, rhamnose, ribose, xylose, arabinose, raffinose, oligosaccharide
Amino compounds	Asparagine, α-alanine, glutamine, aspartic acid, leucine/isoleucine, serine, γ-aminobutyric acid, glycine, cystine/cysteine, methionine, phenylalanine, tyrosine, threonine, lysine, proline, tryptophane, β-alanine, arginine, homoserine, cystathionine
Organic acids	Tartaric, oxalic, citric, malic, acetic, propionic, butyric, succinic, fumaric, glycolic, valeric, malonic
Fatty acids and sterols	Palmitic, stearic, oleic, linoleic, linolenic acids; cholesterol, campesterol, stigmasterol, sitosterol
Growth factors	Biotin, thiamine, niacin, pantothenate, choline, inositol, pyridoxine, *p*-aminobenzoic acid, *N*-methyl nicotinic acid
Nucleotides, flavonones, and enzymes	Flavonone, adenine, guanine, uridine/cytidine, phosphatase, invertase, amylase, protease, polygalacturonase
Miscellaneous compounds	Auxins, scopoletin, fluorescent substances, hydrocyanic acid, glycosides, saponin (glucosides), organic phosphorus compounds, nematode cyst or egg-hatching factors, nematode attractants, fungal mycelium-growth stimulants, mycelium-growth inhibitors, zoospore attractants, spore and sclerotium germination stimulants and inhibitors, bacterial stimulants and inhibitors, parasitic weed germination stimulators

[a] Reproduced with permission from Curl and Truelove (1986).

which changes the environment of the rhizosphere microbes.

The microbial community which develops in this changed rhizosphere environment will face additional challenges; many of the materials released from roots do not contain sufficient nitrogen, and sometimes phosphorus, to allow rapid microbial growth. This situation limits both the plant and the associated rhizosphere microbes; the plant has an increasing demand for inorganic nutrients, which are often not available at a sufficient rate. To meet this need, the rhizosphere microbes make major contributions but at a high energetic cost for the plant.

TABLE II
Factors That Can Influence the Amount and Types of Materials Released in the Plant Root Zone (Rhizodeposition)[a]

General environmental changes	*Specific plant management/ stress-related changes*
Lower temperature	Ozone
Shorter day length	Chemicals applied to foliage
Decreased light intensity	Foliar saprophytes
Water stress	Foliar infections
Increased pH	Root saprophytes
Anaerobic conditions	Root pathogens
Ionic concentration changes	Herbage removal
Calcium, phosphorus, potassium, nitrogen	Plant age

[a] From Whipps and Lynch (1986).

The microbes grow using the available nutrient-deficient root materials as a carbon source and then develop enzymes to degrade the nutrient-rich (but plant-inaccessible) organic matter in the soil or mud surrounding the root. After the microbes have grown and assimilated the nitrogen, the microbes die and release the nitrogen and other nutrients in plant-available inorganic forms (Fig. 1). The microbes thus serve as a rapid turnover intermediate by making the inaccessible nutrients locked up in the organic matter available for plant use after the microbes die.

The filamentous fungi, including the free-living and mycorrhizal types, also play a unique role in making minerals available for the plant which cannot be provided by most unicellular bacteria. The filamentous fungi in the rhizosphere have an extensive hyphal network. With this hyphal network, they can

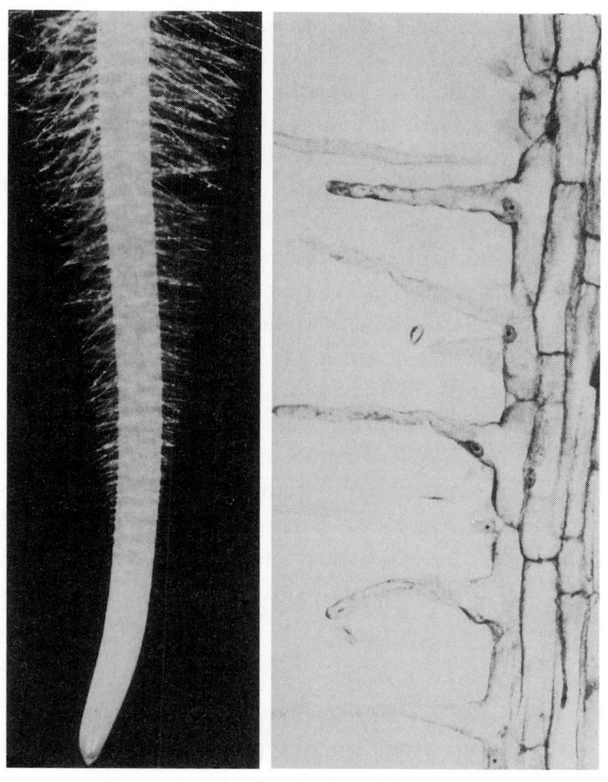

Fig. 3. Smaller root hairs that assist the plant in exploiting resources present in soils (reproduced with permission from Curl and Truelove, 1986, p. 12).

utilize carbon derived from the plant while obtaining their nitrogen and other limiting resources from outside of the immediate root zone. This means that a larger volume of the soil can be exploited for resources which limit the plant.

TABLE III
Influence of a Root on Various Groups of Organisms in a Soil Based on Cultural Procedures[a]

Organism	Microbes/g dry soil: Rizosphere soil	Microbes/g dry soil: Control soil	R:S ratio
Bacteria	1200×10^6	53×10^6	23
Actinomycetes	46×10^6	7×10^6	7
Fungi	12×10^5	1×10^5	12
Algae	5×10^3	27×10^3	0.2

[a] From Rouatt *et al.*, cited in Curl and Truelove (1986).

The rhizosphere microbes also assist the plant in acquiring needed nitrogen by a more direct means: In the presence of nitrogen-free or lower nitrogen-content substrates released from the root, free-living nitrogen-fixing bacteria can play an important role. These include bacteria of the genera *Azotobacter, Azospirillum,* and *Azoarcus*. These microbes carry out associative nitrogen fixation.

The rhizosphere community does not just consist of bacteria and fungi. The higher levels of bacteria and fungi lead to the intense development of protozoans and nematodes. These consumers feed on the nutrient-rich bacteria and fungi, leading to more rapid turnover of the microbes, which leads to an accelerated release of nutrients for plant use.

IV. RHIZOSPHERE PROCESSES AND THE PLANT

The rhizosphere is a "cloud" of microbes which literally surrounds plant roots and which is vital for the plant's survival and growth in nature. To maintain the complex rhizosphere community requires a significant portion of the plant's photosynthetic activity. From a short-term viewpoint, this energy expenditure allows the rhizosphere microbes to make minerals available at a controlled rate for use by the plant. From a longer term viewpoint, a part of the organic materials released from the plant, as well as the rhizosphere microbes, become a part of the organic matter which accumulates over time in the plant root zone. The organic matter in a "typical" soil or aquatic sediment will have an age from hundreds to thousands of years. As this organic matter gradually accumulates, carbon, originally found as carbon dioxide in the atmosphere, will accumulate together with other nutrients, including organic forms of nitrogen, sulfur, and phosphorus, which are no longer immediately available for plant use.

The microbes in the plant root zone also can affect the plant through a series of more subtle activities which can have both negative and positive effects. Many of the microbes present can produce complex organic compounds such as the gibberillins, which can alter the morphology and physiology of the plant. Other microbial metabolites include ethylene, which can be produced by rhizosphere microbes under more waterlogged conditions. Ethylene can cause the early release of flowers; this is especially critical for plants such as orchids, which must be sold within a limited time of the year.

V. ECOLOGY, THE ENVIRONMENT, AND THE RHIZOSPHERE

Plants, whether growing in soils or flooded wetland areas, have major impacts on global processes through the microbes which are found in the rhizosphere. Because of the increasing conversion of natural plant communities to intensive agriculture, whether terrestrial or aquatic environments, and the increased use of chemical control agents and fertilizers, the rhizosphere plays important roles in determining the fate and potential global-level effects of many human and natural activities.

The rhizosphere bacteria, in addition to their contributions to carbon dioxide releases, can modify the dynamics of methane releases to the atmosphere. Methane, a reduced form of carbon, is a major greenhouse gas whose concentration has been increasing at a rate of 1% per year during the past 200 years. This methane is derived from a variety of sources, including cattle and termites. Based on available estimates, between one-third and one-half of the net methane released to the atmosphere is produced by methanogens in rice paddies and wetlands.

Fortunately, methanotrophs (aerobic methane-utilizing bacteria), including those in the root zone of plants in terrestrial and aquatic environments, oxidize between 90 and 95% of the methane produced by the methanogens. This process occurs in tropical and subtropical wetlands as well as in soils from the higher latitudes. For marshland plants, many of these methane-oxidizing organisms occur in grooves in the root surface between iron oxide precipitates. The plant thus provides a unique niche for these microbes by transporting oxygen down the root into the oxygen-depleted mud. This results in two gradients: (i) organic matter released during rhizodeposition and (ii) oxygen diffusion into the anaerobic region surrounding the roots of these aquatic plants.

The rhizosphere environment is also important for the transformation of nitrogen-containing fertilizers used in agriculture. First, in an aerobic process, the ammonia present in most nitrogen fertilizers can be transformed biologically to the more water-soluble nitrate form (NO_3^-) together with traces of N_2O and NO. This process, called nitrification, which also occurs in nonrhizosphere soils, can be followed by a second process which occurs more effectively in the rhizosphere in which there is available carbon and oxygen levels are lower. This second process, denitrification, involves the use of organic matter released from the roots as a carbon and energy source, together with nitrate as a respiration alternate to oxygen, leading to the production of gaseous nitrogen and nitrous oxide (a greenhouse gas). Nitrous oxide can interact with ozone in the upper atmosphere, possibly leading to an increase in the "greenhouse effect."

VI. MANAGEMENT OF RHIZOSPHERE MICROBES: BIOTECHNOLOGICAL APPLICATIONS

Much work is currently under way to modify and manage the rhizosphere and its microbes, to improve plant health, and to minimize the occurrence of disease. This involves not only understanding nutrient "signals" but also molecular-level interactions which occur in the rhizosphere. Most of this work has been carried out using *Rhizobium* and *Agrobacterium*, both important in plant biotechnology. For example, in the presence of a potential pathogen, bean plants can alter their gene expression and synthesize products which will increase resistance to a wide variety of pathogens. These compounds, which can be produced in the presence of particular pathogens in the rhizosphere, are termed phytoalexins. These can act as antibiotics against potential invading organisms. Other molecular-level responses include "walling off" injured areas and the production of enzymes to break down the structure of potential pathogens. The plant, in defending itself against pathogens which are attempting to survive and multiply in the rhizosphere, can rapidly activate genes. This can occur in 2 or 3 min in plant cell culture systems which have been studied. The rhizosphere also harbors viruses, fungi, and nematodes which can be harmful to certain plants. Plants appear to be able to produce proteases, which will lead to altered or decreased effectiveness of these pathogens. Transgenic plants are being developed which have improved abilities to interact with and resist potential pathogens based on modified rhizosphere processes.

Often, the ability of a potential pathogen to maintain itself will depend on its ability to acquire sufficient iron to allow survival and growth. Much of the inorganic iron in soils is not available in a utilizable form. It is sequestered or "trapped" by organic chelating compounds called siderophores, which are produced by specific microbes. The phenomenon of "rhizosphere competence," by which a microbe is capable of establishing itself in the rhizosphere, may depend on the ability of a microbe to increase iron availability as it attempts to colonize the rhizosphere. To combat undesirable pathogens, it is often possible to add specific microbes to the plant rhizosphere (or to the seed surface before the seed is planted) to make iron less available to undesirable pathogens. Many of these desirable bacteria are of the genus *Pseudomonas*. An interesting aspect of plant disease control is the existence of soils, called "suppressive soils," which will retard the activities of specific plant pathogens which must establish themselves in the rhizosphere to cause disease. By modifying such soils, as with sterilization, it has been established that suppressive soils occur based on microbial sequestration of iron. In addition, bacteria can also be added to the rhizosphere which produce antibiotics effective against pathogens or which lyse the cell walls of potential invading microbes.

Major efforts are being made to modify the microbial community in the rhizosphere and to identify more effective rhizosphere-colonizing organisms. Two major groups of microbes are being added to the rhizosphere: (i) biocontrol plant growth-promoting bacteria, which suppress disease by increasing the concentration of pathogen-inhibiting chemicals or

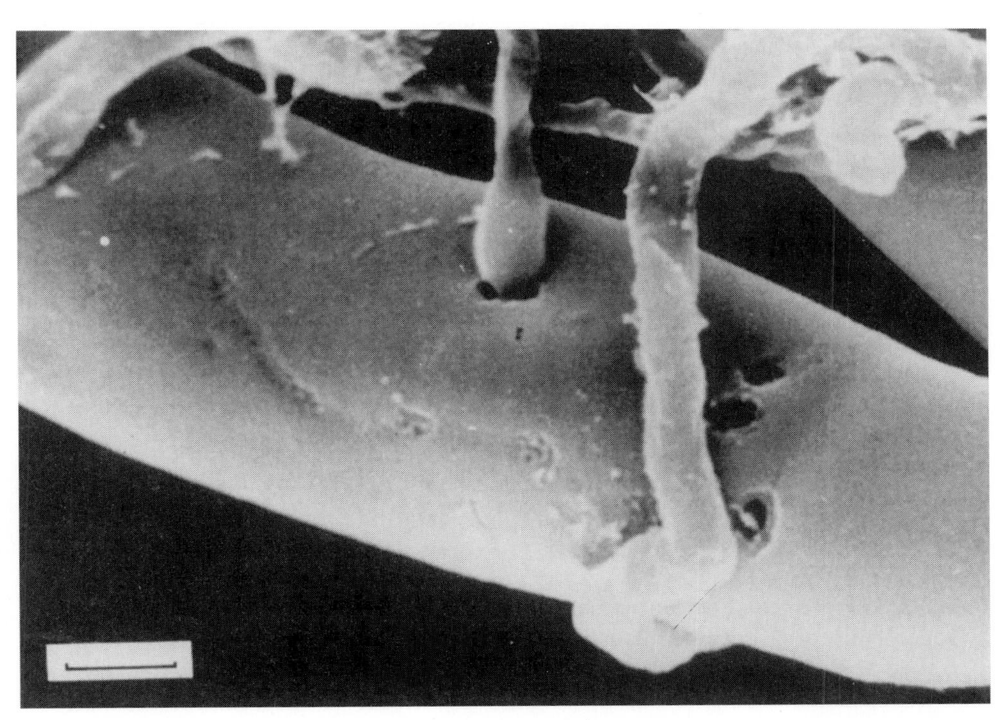

Fig. 4. Attack of a desirable *Trichoderma* against a pathogenic fungus by coiling around and parasitizing the pathogen. Scale bar = 2 μM (reproduced with permission from Lynch and Hobbie, 1988, p. 280; courtesy of C. Ridout, University of Hull and AFRC Institute of Hant Research).

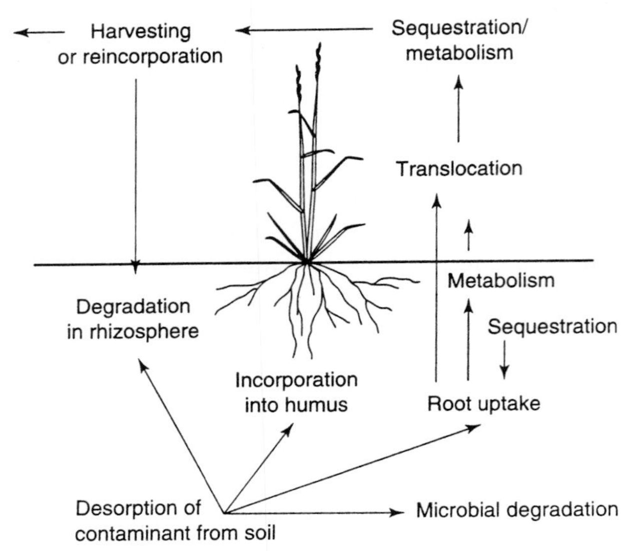

Fig. 5. Phytoremediation processes by which plants and rhizosphere microbes cooperate in degradation, modification, and phytoconcentration of chemicals from a contaminated soil (reprinted from *TIBTECH* **13,** S. D. Cunningham, W. R. Berti, and J. W. Huang, Phytoremediation of contaminated soils, 393–397, Copyright 1995, with permission from Elsevier Science).

which increase the resistance of the plant, and (ii) plant growth-promoting bacteria, which benefit plants by means other than suppressing deleterious microbes. The management of microbial processes in the rhizosphere is finding worldwide applications.

The biological control agents which can be used in the rhizosphere also include fungi. Many of these, such as *Trichoderma,* will parasitize harmful fungi and act as biocontrol agents. This organism will coil around and parasitize the potential pathogen (Fig. 4). A wide variety of additional fungi have been identified which have the ability to limit the effects of harmful pathogens in the rhizosphere, often by means of competition for nutrients or by antibiotic production.

Recently, for plant–rhizosphere interactions, it has been possible to influence the numbers and types of rhizosphere microbes by genetic variations in the plant. Ideally, such modifications can be made to increase the presence and activity of beneficial organisms, such as the nitrogen fixers and antibiotic producers, and to decrease the presence and activity of less desirable organisms which attempt to maintain themselves in the rhizosphere.

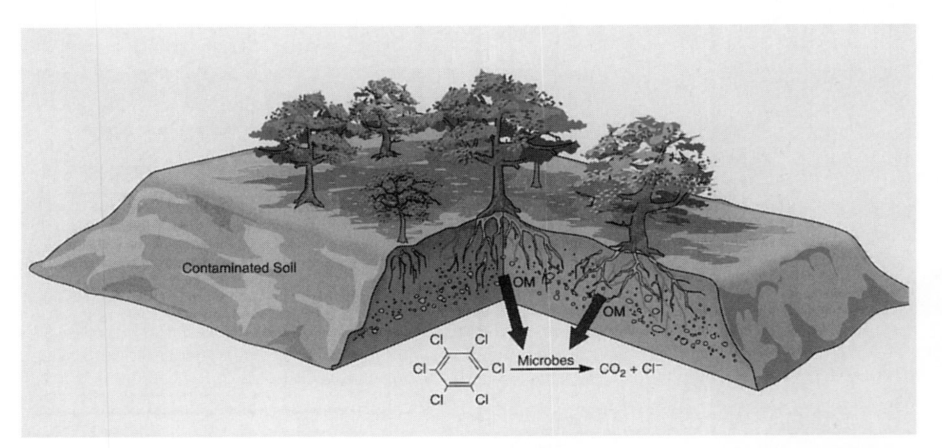

Fig. 6. Phytoremediation conceptual diagram with a cutaway view of the plant root zone. This process involves the use of the plant and the associated rhizosphere microbes: The plants can directly take up contaminating chemicals, uptake can occur after microbial modification of chemicals, and microbes can degrade the chemical using energy sources derived from the plant, possibly involving cometabolic processes. The degradation of hexachlorobenzene is shown as an example (reproduced with permission from L. Prescott, J. L. Harley, and D. A. Klein, "Microbiology," 4th ed., 1999, WCB/McGraw-Hill).

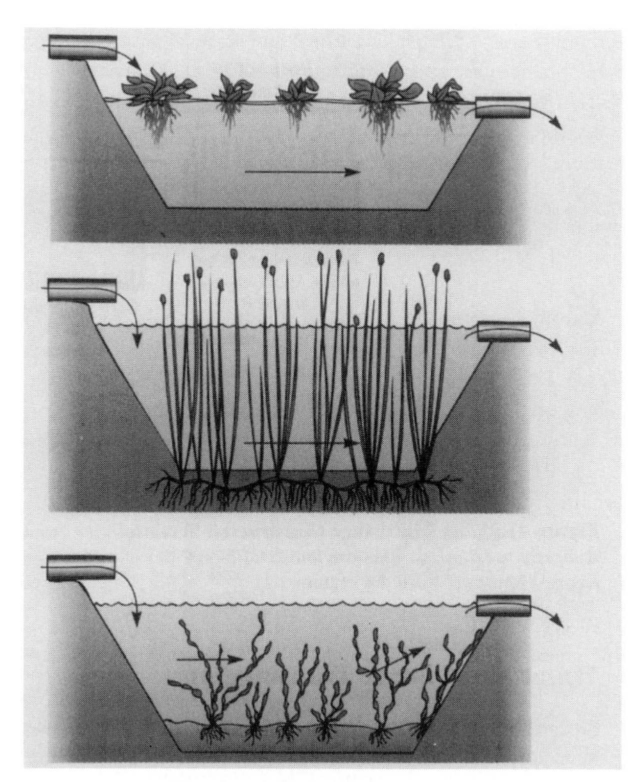

Fig. 7. Constructed wetlands for treatment of dissolved wastes. Rhizosphere microbes associated with free-floating macrophytes (top) and emergent macrophytes which will allow subsurface flow (middle) or submerged vegetation (bottom) contribute to processing of wastes (reproduced with permission from L. Prescott, J. L. Harley, and D. A. Klein, "Microbiology," 4th ed., 1999, WCB/McGraw-Hill).

A major part of recent work with plants and their rhizosphere organisms in soils involves phytoremediation (Figs. 5 and 6). A part of this process includes the plant releasing exudates, which provides energy for the rhizosphere microbes to carry out the transformation or degradation of contaminating chemicals. With the increased levels of nutrients present in the rhizosphere, the decomposition of hydrocarbons, herbicides, and insecticides can be accelerated. This involves the rhizosphere microbes, which often require secondary energy sources, derived from the plant, to carry out the cometabolic degradation of contaminants.

Another aspect of plant growth and rhizosphere processes is the increasing use of constructed wetlands in biotechnology (Fig. 7). In these wetland environments, rhizosphere microbes can contribute to toxic metal immobilization, sewage processing, and the trapping and modification of inorganic nutrients which might otherwise pollute rivers and lakes. To better understand these processes, mathematical modeling of rhizosphere processes is currently an area of major activity. Using modeling, it is possible to better understand and predict the effects of management changes on these important microbially mediated processes which occur in the rhizosphere, whether in soils or in aquatic environments.

VII. FUTURE RESEARCH DIRECTIONS

The science of microbiology is slightly more than 130–140 years old. During this time, major advances have been made in disease prevention, in improving the production of foods by use of microbiology, and in many areas of environmental management and agriculture. Despite this level of understanding of these important areas, we are only beginning to develop a meaningful understanding of the rhizosphere and its significance for plants, the environment, and for us.

As a result of rapid advances in molecular biology, our understanding of plants and how they interact with the rhizosphere microbes is increasing. A series of developing areas will be important in plant–rhizosphere research, as noted in Table IV. Studies of plant–microbe interactions in the rhizosphere should provide many new challenges and widespread benefits in the future.

TABLE IV
Developing Areas in Plant–Rhizosphere Microorganism Research

Mechanisms of recognition/binding of microbes to roots
Host–inoculant physiological and genetic studies
Phenotypic/genotypic elevation of gene products useful by microbes in the rhizosphere
Use of improved seed inoculation procedures
Selection of plants with altered rhizodeposition characteristics
Studies of rhizosphere processes in phytoremediation
Studies of rhizosphere microbes in constructed wetlands

See Also the Following Articles

Azotobacter • Biological Nitrogen Fixation • Fungi, Filamentous • Methane Biochemistry • Mycorrhizae

Bibliography

Bashan, Y., and Holguin, G. (1998). *Soil Biol. Biochem.* **30**, 1225–1228.

Calhoun, A., and King, G. M. (1997). *Appl. Environ. Microbiol.* **63**, 3051–3058.

Campbell, R. (1985). "Plant Microbiology." Edward Arnold, Baltimore.

Clarholm, M. (1985). Possible roles for roots bacteria, protozoa and fungi in supplying nitrogen to plants. *In* "Ecological Interactions in Soil" (A. H. Fitter, Ed.), pp. 355–365. Blackwell, Boston.

Cunningham, S. D., Berti, W. R., and Huang, J. W. (1995). Phytoremediation of contaminated soils. *TIBTECH* **13**, 393–397.

Curl, E. A., and Truelove, B. (1986). "The Rhizosphere." Springer-Verlag, New York.

Elliott, E. T., Coleman, D. C., Ingham, R. E., and Trofymow, J. A. (1984). Carbon and energy flow through microflora and microfauna in the soil subsystem of terrestrial ecosystems. *In* "Current Perspectives in Microbial Ecology" (M. J. Klug and C. A. Reddy, Eds.), pp. 424–433. American Society for Microbiology, Washington, DC.

Gilbert, B., Assmus, B., Hartmann, A., and Frenzel, P. (1998). *FEMS Microbiol. Ecol.* **25**, 117–128.

Kennedy, A. C. (1998). The Rhizosphere and Spermosphere. *In* "Principles and Applications of Soil Microbiology" (D. M. Sylvia, J. J. Fuhrmann, P. G. Hartel, and D. A. Zuberer, Eds.), pp. 389–407. Prentice Hall, Upper Saddle River, NJ.

Lambert, B., and Joos, H. (1989). *TIBTECH* **7**, 215–219.

Lynch, J. M. (1990). "The Rhizosphere." Wiley, New York.

Lynch, J. M. (1994). *Appl. Soil Ecol.* **1**, 193–198.

Lynch, J. M., and Hobbie, J. E. (1988). "Microorganisms in Action: Concepts and Applications in Microbial Ecology." Blackwell, Boston.

Pearce, D. A., Bazin, M. J., and Lynch, J. M. (1997). *J. Microbiol. Methods* **31**, 67–74.

Prescott, L., Harley, J. L., and Klein, D. A. (1999). "Microbiology 4th ed. WCB/McGraw-Hill, Dubuque, IA.

Rouatt, J. W., Katznelson, H., and Payne, T. M. B. (1960). Statistical evaluation of the rhizosphere effect. *Soil Sci. Am. Proc.* **24**, 271–273.

Söderberg, K. H., and Bääth, E. (1998). *Soil Biol. Biochem.* **30**, 1259–1268.

Soil Science Society of America (1984). "Microbial–Plant Interactions." Soil Science Society of America, American Society of Agronomy, Crop Science Society of America, Madison, WI.

Smart, D. R., Ritchie, K., Stark, J. M., and Bugbee, B. (1997). *Appl. Environ. Microbiol.* **63**, 4621–4624.

Whipps, J. M., and Lynch, J. M. (1986). *Adv. Microbiol. Ecol.* **9**, 187–244.

Ribosome Synthesis and Regulation

Catherine L. Squires
Tufts University

I. Regulatory Processes
II. Organization of Ribosomal RNA Transcription Units
III. Ribosome Processing and Assembly
IV. Ribosomal Protein Regulation
V. Ribosomal RNA Regulation
VI. Summary of Control Processes

GLOSSARY

antitermination A process in which RNA polymerase is modified by cellular factors such that it makes transcripts at an increased elongation rate and it becomes resistant to transcription terminators.

coupled translation When the translation of one gene requires translation of the preceding gene, they are said to be translationally coupled.

feedback control A control mechanism in which the product of a transcription unit can negatively control its own expression.

transcription unit A set of genes that are expressed from a single promoter/control region. Transcription unit and operon are the same.

UP element A sequence in the DNA of the promoter region of some transcription units that interacts directly with RNA polymerase to increase its activity at that promoter by up to 30-fold.

THE SYNTHESIS OF RIBOSOMES is one of the most fundamental of cellular processes. Present in all cells, ribosomes are the machines responsible for the production of cellular proteins.

Ribosomes have a complex structure and are composed of three RNA molecules (16S, 23S, and 5S ribosomal RNAs) and 52 ribosomal proteins. These components are assembled into two ribosomal subunits, known as the 30S (16S rRNA plus 21 ribosomal proteins) and 50S (23S rRNA, 5S rRNA, and 31 ribosomal proteins) subunits (Fig. 1). During the synthesis of a protein, one 30S and one 50S particle combine into an active protein-synthesizing particle, the 70S ribosome. Upon completion of its task, the 70S ribosome dissociates into 30S and 50S subunits, which are then ready for another cycle of protein synthesis. For microorganisms such as *Escherichia coli,* which live in environments of continually changing nutritional opportunities, the synthesis of rRNA and ribosomal protein components and their assembly into ribosomes is, of necessity, a highly regulated process. This bacterial regulatory strategy is designed to provide the cell with both ample protein synthetic capacity for times of plenty and a basic survival synthetic capacity for times of famine. Additionally, because the synthesis of ribosomes utilizes a major portion of the cell's resources, the process must be adjusted to provide the "right amount" of protein synthetic capacity for nutritional levels intermediate between feast and famine. It is a basic aspect of the regulation of ribosome synthesis in microorganisms that the number of ribosomes in the cell is closely determined by the availability of energy.

I. REGULATORY PROCESSES

Microbial cells growing in a rich nutritional environment contain as much as 40% of their dry weight as ribosomes. Thus, under such excellent growth conditions, a very large portion of the cell's synthetic capacity is devoted to making these large macromolecules. Both rRNA and ribosomal protein synthesis

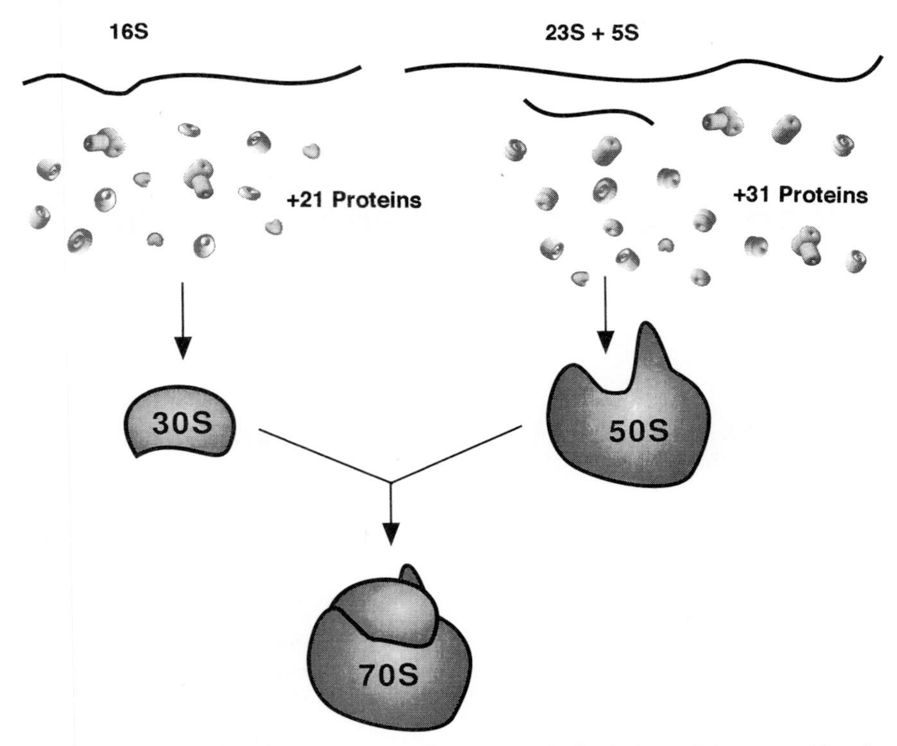

Fig. 1. The assembly of ribosomes: A diagrammatic depiction of the assembly of a 70S ribosome from the precursor molecules of rRNA, ribosomal proteins, and the 30S and 50S subunits. Wavy lines represent the 16S, 23S, and 5S rRNAs, and small tubular circles represent the rRNA proteins.

are regulated and, in turn, both control the formation of ribosomes. However, rRNA regulatory mechanisms are dominant in this process. For rRNA transcription, at least four different controls operate at the level of transcription initiation. Its transcription is also special in that it possesses a set of signals that lead to modifications of RNA polymerase and the process of transcription elongation, collectively referred to as transcriptional antitermination. Other factors that help to regulate the level of active ribosomes in the cell are the processing and assembly of precursor molecules into active ribosomes and their degradation when cells can no longer grow and divide. To guard against losing too many ribosomes, non-growing cells dimerize a portion of their 70S ribosomes into a 100S particle that is resistant to degradation. In this way, there are some ribosomes stored for future needs.

II. ORGANIZATION OF RIBOSOMAL RNA TRANSCRIPTION UNITS

To understand rRNA regulation, it is helpful to consider the physical organization of rRNA genes. rRNA genes are organized into transcription units (or operons) that have the same basic arrangement and recognizable control regions in nearly all bacterial species (Fig. 2). This arrangement consists of two promoters in an extensive regulatory region followed by a 16S gene, a spacer region containing one or two tRNA genes, a 23S gene, one or two 5S genes, sometimes a number of additional tRNA genes, and finally a transcription termination region. These units are first transcribed into a precursor rRNA, which is then processed into mature 16S, 23S, 5S, and tRNA sequences (Fig. 2). Although the basic features of this organization are preserved in most

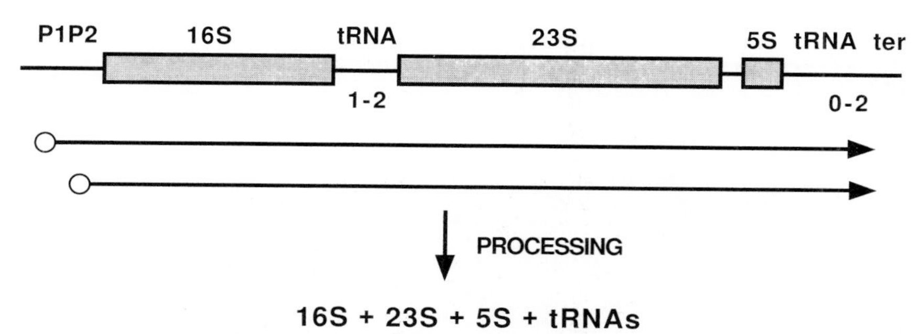

Fig. 2. Diagram of rRNA transcription units. The organization of the rRNA transcription units, the transcription products, and the mature species of RNA produced after processing are shown. P1P2 denote the two promoters; shaded boxes depict the 16S, 23S, and 5S rRNA genes. The location of tRNA-encoding genes in the spacer region between the 16S and 23S genes and after the 5S gene is also noted. The numbers below the tRNA genes indicate the number of tRNA genes found in the different transcription units. ter, the terminator for the transcription unit.

bacteria, there are exceptions in which the 16S and 23S + 5S genes reside in different transcription units (e.g., *Thermus thermophilus* and *Buchnera aphidicola*).

The number of transcription units per genome that encode rRNAs varies in different microorganisms. Some microorganisms, such as *Escherichia coli,* have 7 units; *Bacillus* and *Clostridium* have 10–13 units, *Caulobacter* and *Anacystis* have 2 transcription units, and *Mycobacterium tuberculosis* and some Archaea have only 1 rRNA transcription unit. Although it has been speculated that there might be a correlation between the number of rRNA transcription units on the genome and the ability of the microorganism to grow and divide rapidly, such a correlation has never been established. The most likely explanation for multiple rRNA transcription units in some microorganisms is the facilitation of the very rapid production of large numbers of ribosomes when environmental conditions signal good growth conditions.

An electron micrograph of a rRNA transcription unit in *E. coli* provides a striking visual glimpse of the actual transcription of these genes and their processing into the final rRNA components (Fig. 3). The RNA polymerase molecules can be identified as globular structures on the linear DNA. Their RNA products can be discerned as "ribbons" extending from the polymerase molecules. The rRNA transcription units shown in Fig. 3 are described as "double Christmas trees" because of their characteristic pattern of 16S transcription, then processing, followed by 23S transcription, and then processing again. Such photographs as that in Fig. 3 clearly show that processing events to create the individual rRNA molecules of 16S, 23S, and 5S occur while the different species of rRNA are being transcribed. The density of RNA polymerase molecules on the DNA confirms the intense transcriptional activity of the rRNA genes and the importance of turning down this high level of transcription when large numbers of ribosomes are not needed by the cell.

III. RIBOSOME PROCESSING AND ASSEMBLY

The formation of the individual, mature species of rRNAs occurs by the action of enzymes that cleave the precursor molecules at specific sites to give the final products of 16S, 23S, and 5S rRNAs. The enzyme RNaseIII makes the initial cleavages for 16S and 23S. It requires a double-stranded region for its activity, and this is provided for the 16S gene by a large based-paired region comprising the 5′ end of the transcript and part of the spacer region between the 16S and 23S genes. Likewise, the spacer region between the 16S and 23S genes and the spacer region between the 23S and 5S genes provide the long based-paired region for RNaseIII cleavage of the 23S precur-

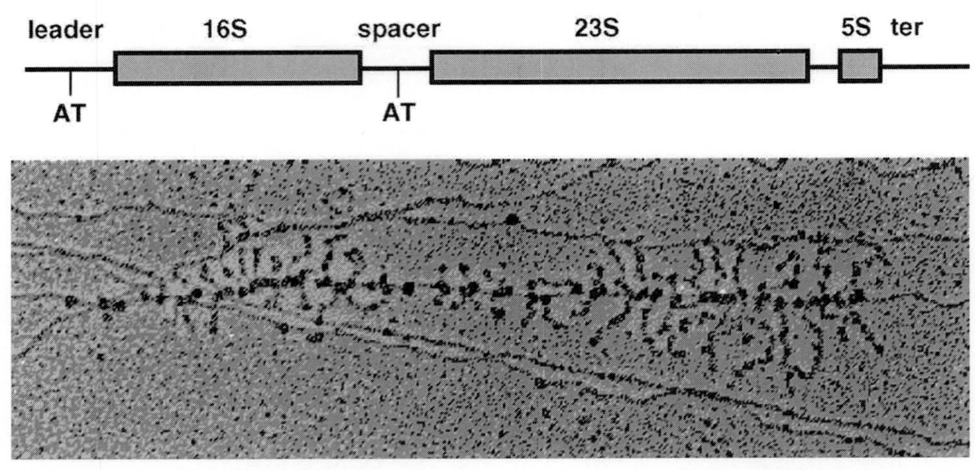

Fig. 3. Electron micrograph of a rRNA transcription unit. (Top) A diagrammatic representation of the transcription unit. Leader and spacer refer to the promoter/control region and the region between the 16S and 23S genes, respectively. AT indicates the location of the antiterminator sequences. (Bottom) The lines are DNA, and the globular balls on the DNA are RNA polymerase. Wavy extensions from the RNA polymerase molecules are the nascent rRNAs.

sor (Fig. 4). Other processing enzymes then cut the RNaseIII-cleaved molecules to their final mature 5′ and 3′ ends. It is clear from electron micrograph photographs of rRNA transcription units in *E. coli* that these processing events occur as soon as the RNaseIII processing substrates are generated, giving

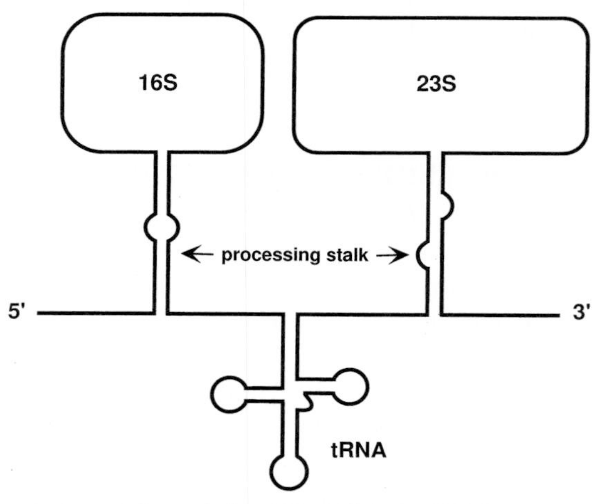

Fig. 4. Processing of the 16S and 23S rRNA precursors; a diagrammatic representation of a rRNA transcription unit with the 16S and 23S rRNA base-paired processing stalks indicated. The approximate sites of RNaseIII cleavage are indicated by arrows. Also shown is a tRNA transcript in the region between the 16S and 23S genes. The 5′ and 3′ ends of the transcript are also noted.

the characteristic double Christmas tree appearance to the operons.

The assembly of ribosomes is also thought to occur as the rRNA is being transcribed. Secondary and tertiary interactions between local and distal regions of the rRNA facilitate their proper folding. The binding and assembly of the ribosomal proteins on this backbone of RNA and the cooperative interaction between many of the ribosomal proteins complete the formation of an active 30S or 50S particle. Sometime during this process the rRNA sequences are also modified, although the precise function of these modifications is not known. The newly synthesized and assembled particles can subsequently actively participate in protein synthesis.

The assembly of ribosomes from their basic components of RNA and proteins is an extremely complicated process that is just beginning to be understood in detail. The process can be carried out *in vitro* using purified rRNA and ribosomal proteins, and current knowledge of ribosome assembly is derived from such *in vitro* studies. Assembly maps for the interaction of ribosomal proteins with their respective RNAs have been determined. These indicate an ordered assembly process that must be followed to obtain active ribosomes capable of synthesizing protein. Interestingly, this *in vitro* assembly requires incubation at fairly high salt concentrations and temperatures of approximately 40°C, suggesting that some facilitation step that occurs *in vivo* must be substituted by adding thermal energy to the *in vitro* reconstitution experiments. Other types of studies with individual ribosomal proteins have determined the specific binding sites of these proteins and worked out many protein–protein and protein–RNA interactions in the ribosome. Thus, much intricate information is available about individual interactions in the ribosome, but the overall picture of what the ribosome, with its attendant RNA and proteins, really looks like is largely unknown. Recent sophisticated biochemical and biophysical studies give better data on the structure of the assembled subunits and 70S ribosome, but a three-dimensional picture of a functional ribosome is necessary for a complete understanding of the assembly and function of these large and complex molecules.

IV. RIBOSOMAL PROTEIN REGULATION

Studies of ribosomal protein synthesis led to the discovery that their regulation is intimately tied to the level of rRNA in the cell. Several critical ribosomal proteins turn off the synthesis of operons containing multiple ribosomal protein genes. Interestingly, they accomplish this by first inhibiting translation of the genes. When levels of rRNA in the cell decrease and excess ribosomal proteins are free in the cytoplasm, they bind to their own mRNAs and inhibit translation. The fact that ribosomal protein synthesis is closely tied to the level of rRNA in the cell has led to the conclusion that the main regulation of cellular ribosome concentration is at the level of rRNA synthesis and not ribosomal protein synthesis.

A. Translational Regulation

Translational control of ribosomal protein synthesis is their major means of regulation. Many are encoded in large operons, sometimes containing up to 11 ribosomal protein genes. Interestingly, although each gene in the operon is independent, with its own translation start and stop sites, translation of the genes in the operon is not independent; it is "coupled." The basis for coupled translation is not known, but the process is known. In this process, ribosomes load onto the ribosome binding site of the mRNA at an early gene and then proceed to translate each gene through the entire operon. Translation initiation at the start of an internal gene is extremely inefficient, essentially eliminating significant independent expression.

How do key ribosomal proteins regulate coupled translation in their own operon? A few ribosomal proteins bind directly to rRNA; they are considered "primary binders" and are thought to initiate the folding and assembly of the final, active 30S or 50S ribosomal subunit. Their binding affinity to rRNA is high and thus they are usually found associated with their respective binding site on rRNA. However, when the level of rRNA in the cell decreases, perhaps due to sudden nutrient starvation conditions, new ribosomal proteins continue to be synthesized but have no substrate for binding. In this situation, select, regulatory ribosomal proteins bind to sequences in their own mRNAs that mimic their binding site on rRNA. Their affinity for this "pseudo-site" is much less than their affinity for their true binding site; thus, if rRNA is present, these ribosomal proteins will preferentially bind to it rather than to their mRNA. However, once they bind to their operon mRNA, they inhibit the binding of ribosomes to that mRNA and stop further synthesis of ribosomal proteins. This inhibition of translation has two further consequences that lower the expression of now unneeded ribosomal proteins; to expose the message to transcription termination factors so that RNA polymerase fails to synthesize more mRNA and to expose the message to nucleases which quickly degrade untranslated mRNA. In concert, transcription termination and mRNA degradation quickly stop unnecessary cellular production of ribosomal proteins.

B. Transcriptional Regulation

Although the bulk of ribosomal protein regulation is accomplished by inhibiting translation in operons in which the genes are translationally coupled, other mechanisms do occur. In one operon containing 11 genes, called S10 for the first gene in the operon, translational coupling is not sufficient to eliminate all expression from this operon; it remains leaky in its production of these 11 proteins. In this case, the regulatory protein L4 not only binds to its pseudosite on the message to inhibit translation but also is involved with another cellular factor, NusA, in terminating transcription before the first gene of the operon. Thus, the S10 operon is controlled by a combination of translational and transcriptional regulatory events. The rapid inhibition of translation followed by the cessation of transcriptional activity ensures that the cell will not invest energy and resources making these 11 ribosomal proteins when they are not required for protein synthesis.

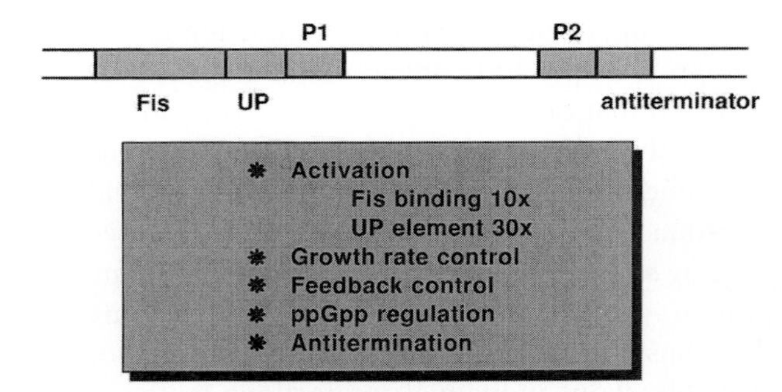

Fig. 5. Regulatory features of the rRNA transcription units. Shown is the control region containing the two operon promoters P1 and P2. Upstream of P1, the Fis-binding region and the UP element are indicated. Downstream of P2, the position of the leader region antiterminator sequences is shown.

V. RIBOSOMAL RNA REGULATION

What is involved in the complex regulation of rRNA expression? The major processes controlling the modulation of rRNA synthesis are transcription initiation, antitermination, and termination. These processes govern the formation of a transcription complex, adaptation of this complex to make it resistant to termination, and, the dissolution of the adapted (antiterminated) transcription complex at the end of the transcription unit, respectively. Only the initiation of rRNA synthesis has been clearly demonstrated, and is universally accepted, as a truly regulated process. However, antitermination and termination most likely play a more modest but nonetheless essential supporting role in the modulation of rRNA synthesis.

Currently, *E. coli* serves as the basic source of information about how rRNA gene expression is regulated in response to changes in environmental conditions. The regulation exhibited by *E. coli* is complex and occurs at multiple levels. Tightly coupled control mechanisms also ensure that the synthesis of rRNA, tRNA, and ribosomal proteins is closely coordinated in the cell.

A. Initiation

The regulatory features of a typical ribosomal operon promoter region are shown in Fig. 5. The control region contains several complex features for the coarse and fine control of rRNA synthesis. All seven *E. coli* rRNA operons contain two adjacent promoters, P1 and P2. Most regulation of transcription initiation occurs at the upstream (P1) promoter. Here, the following regulatory mechanisms intersect over a relatively short stretch of DNA: Fis-dependent activation, UP element activation, stringent regulation, and growth rate-dependent control.

1. *Activation Mechanisms*

Deletions of the activator region of the P1 promoter decrease expression from the rRNA operons by as much as 300-fold. There are two parts to the activator region; a protein-binding, upstream region that interacts with the DNA-binding protein, Fis, and an adjacent sequence, the UP element, that interacts directly with RNA polymerase. Fis binding contributes a 10-fold increase in P1 promoter expression, and the UP element is responsible for an additional 30-fold increase in expression from P1. Together, Fis activation and the UP element result in a 300-fold increase in expression from P1, which explains its exceptional strength. However, these two elements are not involved in either stringent regulation or growth rate-dependent control of rRNA synthesis.

The Fis protein is interesting in that it is a DNA-binding protein that puts a substantial bend into DNA and also has regulatory effects on a variety of other processes in *E. coli*. It has three binding sites upstream of the P1 promoter and its ability to stimulate transcription 10-fold appears to be mediated through direct contacts with RNA polymerase. The level of Fis in the cell fluctuates dramatically during different cellular growth phases. During recovery from nutrient and energy starvation, there is a sharp increase in the cellular content of Fis, with a concomitant 10-fold stimulation in rRNA synthesis from P1. However, when cell growth and division begin to slow down and the cells enter the stationery phase, Fis production is stopped, with a corresponding decrease in stimulation of the rRNA P1 promoters. Because of its dramatic effect on rRNA synthesis, there is tremendous interest in how Fis protein synthesis is regulated.

The UP element was discovered during studies designed to determine why the rRNA P1 promoters are very active in transcription. It is a sequence upstream of the well-known promoter elements called the −35 and −10 regions because of their location with respect to the start of transcription. The UP element is an additional promoter consensus sequence in the DNA that binds to a specific subunit of RNA polymerase, the α subunit. The ability of the α subunit of RNA polymerase to interact with this extra DNA sequence in the promoter region gives these promoters an extra 30-fold stimulation of transcription compared to promoters lacking this motif. The discovery of the UP element was an important contribution to understanding the exceptional activity of the rRNA promoters.

2. *Stringent Regulation*

Stringent regulation involves the complicated set of cellular adjustments made when *E. coli* cells are starved for amino acids and can no longer synthesize new proteins. In this circumstance, the cell manufactures large amounts of the alarmone, (p)ppGpp. The presence of this compound reduces expression of rRNA and tRNA operons by 90%. It has been determined that both P1 and P2 promoters of the ribosomal RNA operons are subjected to (p)ppGpp control. A short "discriminator" sequence in the promoter region between positions −10 and 1 with respect to the transcription start site is thought to be necessary but not sufficient for control by (p)ppGpp.

(p)ppGpp is synthesized by two proteins, RelA and SpoT, which are both found associated with ribosomes. RelA-mediated (p)ppGpp synthesis occurs largely in response to amino acid starvation, whereas SpoT-mediated (p)ppGpp synthesis occurs in response to carbon source deprivation and signals from the nutritional environment. As a result, the concentration of (p)ppGpp varies inversely with the growth rate. This correlation was thought to explain the growth rate-dependent control of rRNA synthesis, but it is now thought to be just one of several factors adjusting the synthesis of rRNA to properly reflect the cell's nutritional environment.

Although the role of (p)ppGpp is undisputed in the mediation of stringent regulation of rRNA expression, it is still not clear how it exerts this control. This is largely due to the difficulty in identifying exactly where in the transcription cycle this nucleotide acts and precisely what constitutes a stringent promoter. Until these problems are resolved, how (p)ppGpp inhibits the ability of RNA polymerase to transcribe rRNA genes will remain unknown.

3. *Growth Rate-Dependent Regulation*

Over a broad range of growth rates, the number of ribosomes per unit amount of cellular protein is proportional to the growth rate (μ) and the rate of ribosome synthesis per genome is proportional to μ^2. This phenomenon has been termed growth rate-dependent regulation of ribosome synthesis, and its function is to ensure a sufficient supply of ribosomes to meet the cell's demand for protein synthesis. Many tRNA genes are also regulated by growth rate, and it is assumed that the same mechanism governs both rRNA and tRNA synthesis.

Several models have been proposed to explain the growth rate-dependent regulation of rRNA and tRNA production, including a (p)ppGpp model, a ribosome feedback model, and a homeostatic model in which the availability of ATP and GTP govern rRNA synthesis. The (p)ppGpp model proposes that this alarmone functions to decrease stable RNA expression not only in stringent regulation when cells are starved for an amino acid, but also in unstarved cells growing at different rates. The ribosome feedback model, on the other hand, proposes that the rate of rRNA synthesis is governed by a feedback mechanism sensitive to the translational capacity of the cell. The homeostatic model proposes that the energy environment of the cell, namely, the availability of ATP and

GTP, governs the synthetic capacity of the rRNA and tRNA promoters.

a. (p)ppGpp Model

The inverse correlation between growth rate and the intracellular concentration of (p)ppGpp and its influence on RNA polymerase activity led to a model in which (p)ppGpp was cast as the sole effector of growth rate-dependent control of ribosome synthesis. The model suggested that (p)ppGpp directly regulated the level of rRNA synthesis either by restricting the number of RNA polymerase molecules available to initiate transcription at rRNA and tRNA promoters or by a direct interference with promotion, as in stringent regulation. The validity of this model was actively investigated for many years. The evidence leads to the conclusion that (p)ppGpp does not play a role in growth rate-dependent control of rRNA and tRNA synthesis. It is certain that (p)ppGpp is important in the regulation of rRNA expression because of its role in stringent regulation, but it does not seem to normally participate in growth rate-dependent regulation.

b. Ribosome Feedback Model

This model assumes that cells have an inherent capacity to synthesize excess amounts of all ribosomal components, but that rRNA synthesis is feedback regulated to prevent the production of more ribosomes than are needed. One effector that has been proposed is excess translational capacity (e.g., excess translating ribosomes), but it is not known whether the translation process per se activates a signal molecule or if this pathway relies on the translation of a signaling mRNA. The advantage of such a feedback regulation model is that, independently of how the growth rate is achieved, there is an appropriate concentration of ribosomes for that particular growth rate. Despite numerous efforts, however, no effector of feedback inhibition has been identified.

c. Homeostatic Model

A homeostatic feedback model has recently been proposed that explains both feedback regulation and growth rate-dependent control of rRNA synthesis. The model suggests that when cellular concentrations of ATP and GTP are high, the rRNA P1 promoters initiate efficiently and high levels of rRNA are synthesized. If levels of the initiating nucleotide triphosphates (NTPs) decrease, then the promoter-bound RNA polymerase molecules are unable to move away from the start region and transcription is essentially aborted. The model suggests that when these NTPs are consumed by translation and other energy-consuming processes in the cell, their concentrations decrease and the P1 promoters become less efficient, thereby decreasing the amount of rRNA synthesized. An important aspect of this model is that the P1 promoters are inherently unstable in the process of transcription initiation and have difficulty moving RNA polymerase from the initiation complex to an elongation mode in transcription. In this scenario, even high levels of Fis and the presence of the UP element are unable to overcome the transcription block caused by insufficient NTP levels. On the other hand, even in the absence of Fis, high levels of NTPs will stimulate transcription from P1. Indeed, in mutant strains missing the Fis protein, high levels of rRNA are still produced if the nutrient conditions are favorable. The homeostatic model ties together feedback regulation and the intricately sensitive response of rRNA synthesis to the cell's nutrient environment.

4. Location of Control Sites in the Promoter Region

To locate the sites of action of the important regulatory circuits, (p)ppGpp stringent regulation, feedback inhibition, and growth rate-dependent control, two sets of experiments were performed. In the first, precise fragments of the P1 promoter region were tested for their response to Fis activation, UP element activation, stringent regulation, and growth rate-dependent control. Using this method, the locations of the Fis binding site and the UP element were determined. These experiments also revealed that the Fis-binding region and the UP element did not participate in either stringent regulation or growth rate-dependent control. A saturation mutagenesis study of *rrnB* P1 promoter fragments missing the Fis and UP element activation regions was undertaken to locate the sites for stringent regulation and growth rate-dependent control. These experiments revealed that mutants which deviated from the wild-type promoter −35 or −10 sequences or which changed the non-consensus 16-bp spacing between these two elements lost feedback and growth rate-dependent regulation. Most of the mutants with altered growth rate-dependent control also had defective stringent regulation. However, some mutations were isolated which uncoupled the two regulatory mechanisms; that is, some of the growth rate-dependent control mutants exhibited relatively normal stringent regulation. This result implies that the two mechanisms share overlapping but nonidentical regulatory motifs in the P1 promoter area. Mutations which render the P1 promoter fragment insensitive to growth rate-dependent control are also insensitive to the cellular concentration of NTPs, supporting the homeostatic model for P1 control.

5. *Role of the rRNA P2 Promoter*

The existence of a second strong promoter in all the rRNA transcription units in *E. coli* leads to the question of what its role is in the synthesis of rRNA in the cell. Since most of the important features regulating the initiation of transcription have been located to the P1 promoter, why doesn't the P2 promoter inappropriately synthesize rRNA during times when the P1 promoter is not functioning? The P2 promoter does contain a UP element and is subject to stringent regulation, but it is neither activated by Fis nor subject to growth rate-dependent control. It is thought that during most growth conditions, transcription from the highly expressed P1 promoter blocks the ability of RNA polymerase to bind to the P2 promoter region DNA and transcription initiation is therefore blocked from P2. Nevertheless, during periods of recovery from starvation, or when there are very low levels of nutrients, it is thought that there is expression from the P2 promoter and that this expression aids the cell in a rapid adjustment to such changes in its environment. However, this aspect of rRNA expression has not been examined in great detail.

B. Antitermination

The unusually high transcription rate from rRNA promoters and the presence of an antitermination system make these transcription units unique in the cell. In normal transcription, the presence of a terminator leads to dissociation of the transcription complex and release of the RNA product. In antiterminated transcription, however, the polymerase is modified at specific sequence motifs such that it now reads through terminators. In addition to being resistant to terminators, antiterminated polymerase also elongates transcripts at an accelerated rate. Messenger RNAs are transcribed at approximately 40 nucleotides per second, whereas rRNA is transcribed at approximately 90 nucleotides per second. The current model of antitermination in the bacteriophage λ system states that the RNA from the antiterminator region stays attached to the transcription complex. This has particularly interesting implications for the ribosomal operons in terms of the processing and assembly events which must occur to generate the mature rRNA species.

Features of the ribosomal antiterminator region from *E. coli* are shown in Fig. 6. Just after P2, there is a possible stem-loop structure, boxB, followed by sequences called boxA and boxC. The boxA is a conserved sequence, TGCTCTTTAACA, whereas the stem-loop structure boxB has little apparent sequence conservation. The boxC feature contains a GT-rich sequence motif. The importance of the ribosomal boxB–boxA motif is emphasized by the extremely high level of conservation of this element in rRNA operons of both eubacteria and Archaea. These boxes derive their names from similar features in the bacteriophage λ antitermination system, which serves as the paradigm for the rRNA system. In this type of antitermination, it is thought that a modifica-

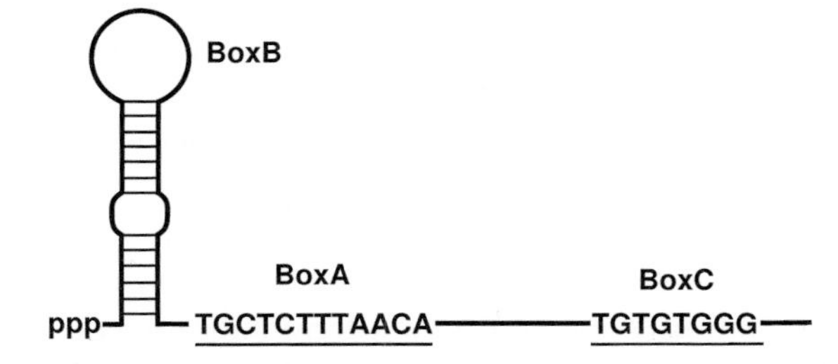

Fig. 6. Ribosomal RNA antiterminator features. Shown is the region just downstream of the P2 promoter. The stem-loop structure called boxB is followed by a conserved sequence called boxA. Slightly downstream of boxA is a GT-rich region called boxC.

tion of RNA polymerase occurs that requires the box sequences and special "factors" that are added to the polymerase such that it can no longer respond to terminators. The precise role of the antitermination system in the expression of rRNA is unknown. However, there are conserved boxA sequences in the leader and spacer regions of all seven rRNA operons in *E. coli*. Recent studies with a plasmid-borne rRNA operon in *E. coli* show that mutations in the antitermination sequences in the leader and spacer regions result in a 75% reduction in 23S rRNA synthesis. Malfunction of the rRNA antitermination system can therefore dramatically reduce the expression levels of rRNA in the cell and antitermination is apparently one of the ways in which *E. coli* maintains appropriate quantities of ribosomes under a variety of growth conditions.

The factors required for transcription antitermination in the rRNA operons have not been completely identified, but most are shared with the bacteriophage λ antitermination system. Proteins called Nus factors and ribosomal proteins are the major components needed for the rRNA antitermination system. Many of the Nus factors play a complex role in antitermination and transcription in general. Nus factors NusA, NusB, and NusG, along with ribosomal proteins S1, S4, and S10, are all likely to be associated with the rRNA antiterminated RNA polymerase complex. In addition to the boxA feature, NusA is required for increasing the transcription rate of RNA polymerase on mRNA genes from 40 nucleotides per second to 65 nucleotides per second. Interestingly, NusA is also required for a dramatic slowing of transcription in the presence of (p)ppGpp; the elongation rate decreases from 40 nucleotides per second to 19 nucleotides per second when stringent regulation is elicited and (p)ppGpp levels are elevated. This decreased rate does not occur in the absence of NusA. RNA polymerase molecules that have been modified at boxA are totally resistant to this transcription rate decrease. Another factor, NusB, is also known to be essential for transcription antitermination of *E. coli* rRNA operons. NusB and ribosomal protein S10 have been shown *in vitro* to form a complex that binds to the boxA sequence. The ribosomal protein S1 inhibits this binding to boxA RNA. This interference of S1 with the binding of NusB and S10 to the boxA RNA

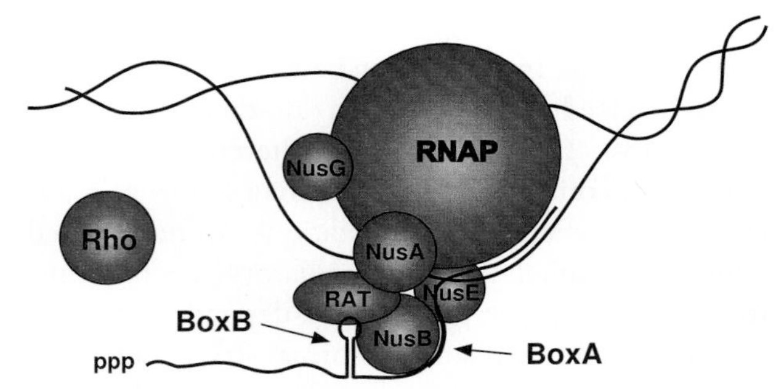

Fig. 7. Model for rRNA antitermination. Shown is RNA polymerase (RNAP) transcribing a DNA sequence containing the rRNA antiterminator region. The locations of the boxB stem-loop structure and the boxA sequence are indicated. The Nus proteins are as displayed. RAT represents other unknown cellular factors that are required for the process of transcription antitermination. Rho is a transcription termination factor that is unable to stop transcription of an antiterminated RNA polymerase. ppp indicates the 5′ starting end of the RNA transcript.

raises the possibility of an *in vivo* regulatory role for S1 in rRNA synthesis and antitermination. NusG and ribosomal protein S4 are also involved in rRNA antitermination. Understanding mechanistically how these various factors can so dramatically alter cellular transcription is one of the current challenges in RNA polymerase and rRNA antitermination studies. Figure 7 shows a model depicting an RNA polymerase molecule modified by the rRNA antitermination system.

1. *Multiple Roles of Antitermination*

Several reasons have been proposed for the presence of an antitermination system in the rRNA transcription units in addition to its role in counteracting the possibility of transcription termination within the ribosomal genes. The first of these considers the possibility that antiterminated transcription complex may aid in processing of the mature 16S and 23S rRNAs from the precursor transcript. The model borrows from the λ antitermination paradigm in which the 5′ end of the transcript is thought to be attached to RNA polymerase through the formation of the antitermination complex and remains attached throughout the transcription process. In a transcription complex in which the RNA polymerase is still attached to the beginning of the transcript, the 5′ stem of the RNaseIII processing stalk of the 16S gene would actually be brought to the newly transcribed 3′ stem so that it does not have to be found by diffusion, making the processing event much more efficient (Fig. 8). This would also explain the occurrence of another boxA sequence in the spacer region, just before the 23S gene. After the 16S RNA is cleaved from the precursor RNA as soon as its RNaseIII processing site is completed, the transcription complex no longer has an attachment to the 5′ end of the transcript. Picking up a new 5′ end in the spacer region close to the 23S gene would facilitate processing of the 23S RNA. Although it seems very likely that this method of increasing the efficiency of 16S and 23S rRNA processing occurs, there is no *in vivo* evidence suggesting that this event is physiologically important. Such a process cannot be the sole function of the antitermination mechanism since it does not explain the premature transcription termination seen in boxA mutants. It does, however, provide an interesting multifunctional aspect to the ribosomal antitermination system.

Another role proposed for the rRNA antitermination system is the establishment of the "correct" transcription elongation rate to aid in the correct folding and assembly of rRNA. It is conceivable that the rate at which defined sites within the coding sequence are made is critical in allowing newly transcribed domains to fold properly for subsequent correct assembly. An examination of the conserved domains in 16S and 23S rRNA indicates that the free energy of many intermediate structures and the final structures found in the mature ribosome is not the minimum energy state possible. If during transcription RNA polymerase fails to pause or has a prolonged pause at particular sites, incorrect secondary and tertiary interactions may result. Consistent with this idea, transcription of an *E. coli* rRNA operon by bacteriophage T7 RNA polymerase leads to largely defective ribosomes. The T7 polymerase can transcribe as much as five times faster than *E. coli* RNA polymerase. The defective ribosomes can be isolated and then totally dissociated and reassociated into functional ribosomes *in vitro*. This result indicates that misfolding is the most probable cause of the defect. Establishing the correct transcription elongation rate may also be important in preventing the type of premature termination se n in strains mutated for the Nus proteins since RNA polymerase molecules that are not antiterminated pause more frequently and are thus more likely to be terminated.

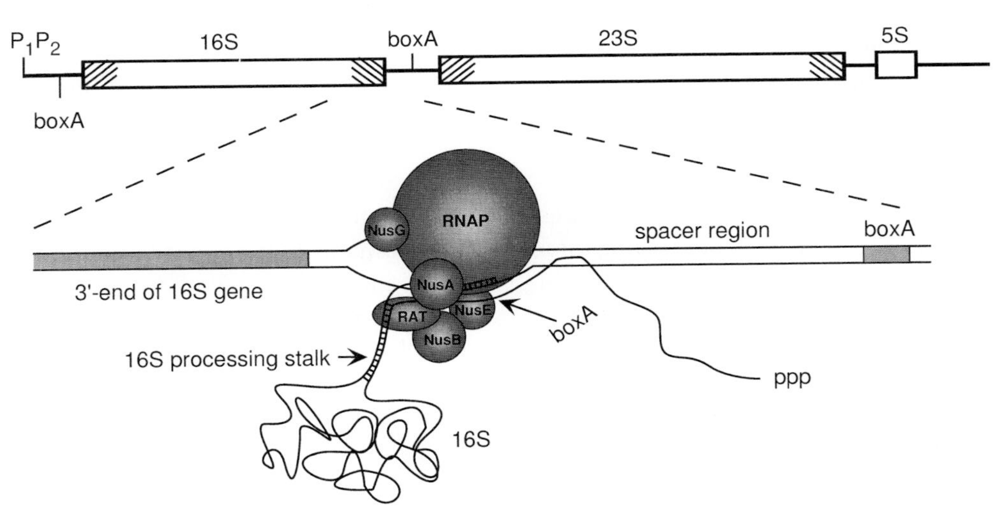

Fig. 8. Model of how transcription antitermination and rRNA processing may be coupled in the cell. Features are as defined in the legends to Figs. 2 and 7, except that the 16S precursor transcript with its processing stalk attached is shown.

2. *Is Antitermination Regulatory?*

The participation of ribosomal proteins in both the bacteriophage λ and the rRNA antitermination systems results in the following interesting question: Is there another regulatory circuit connecting the cellular level of ribosomal proteins with the vigorous synthesis of rRNA? The discovery that ribosomal proteins S1, S4, and S10 can all associate with the rRNA antitermination system resulted in the speculation that these proteins in fact play a regulatory role in rRNA synthesis. The general regulatory circuit for ribosomal proteins is that if the level of rRNA in the cell decreases and there are free ribosomal proteins in the cytoplasm, then specific, regulatory ribosomal proteins shut off the translation and then transcription of ribosomal proteins. In this circuit, rRNA determines the level of ribosomal proteins in the cell. However, if several ribosomal proteins are also involved in facilitating the synthesis of rRNA through antitermination, then these proteins would be acting as positive regulators. The result would be the increased synthesis of rRNA when there are excess ribosomal proteins in the cell and all other conditions, such as NTP levels, are conducive to rRNA synthesis. This system may complement the feedback and homeostasis regulatory circuits in which the control is in the opposite direction, from rRNA to ribosomal protein.

3. *Termination Regions and Antitermination*

rRNA transcription involves highly expressed genes transcribed by antiterminated RNA polymerase molecules. How are the numerous terminator-resistant and rapidly transcribing polymerases stopped? In lambda the antiterminated elongation complex reads through virtually all terminators in its path. Does the same thing happen in rRNA operons? The electron micrograph of a rRNA transcription unit shown in Fig. 3 clearly demonstrates that RNA polymerase molecules terminate efficiently at the end of the transcription unit since there are no RNA polymerase molecules on the DNA after the operon. Examination of many such photographs indicates that read-through from rRNA transcription units into following genes is a rare occurrence. What are the options of an antiterminated RNA polymerase at a terminator region? If it does not read through, then it might be stopped by a "super terminator." Another possibility is that the modifications leading to RNA polymerase terminator read-through ability are removed, with subsequent ease of termination. This would require an "anti-antiterminator" sequence. The sequence of the rRNA terminator regions is satisfyingly complicated, and this perhaps explains their ability to stop frequent and antiterminated transcription complexes. It is clear that the terminators function efficiently, and that the rRNA antitermination system is unable to suppress their function.

VI. SUMMARY OF CONTROL PROCESSES

The regulation of ribosome synthesis is an important aspect of overall microbial physiology. Most microbes experience constantly changing environments in which they are competing with many other microorganisms for resources. To compete well, they must be efficient in their use of the resources and able to quickly take advantage of favorable conditions. The synthesis of ribosomes reflects these fluctuating conditions; when high levels of protein synthesis are required for rapid cell growth and division, correspondingly high numbers of ribosomes are synthesized to make new proteins. However, the formation and assembly of ribosomes and their translation activity require an enormous expenditure of energy in the form of ATP and GTP; the cell can only afford to do this if environmental conditions, especially nutrient conditions, are favorable. Complex regulatory mechanisms monitor the synthesis of rRNA and adjust the output to conform to the availability of resources needed for protein synthesis; detection systems for determining levels of NTPs and protein precursors, amino acids, are specifically employed. Amino acid starvation leads to the production of (p)ppGpp with a rapid shut-down of rRNA synthesis, as does a decrease in the cellular NTP energy level. The homeostatic sensing of nucleotide levels maintains the initiation of new rRNA molecules at an appropriate level to meet the cell's protein synthesis requirements. Coordination of the levels of rRNA and ribosomal proteins is another regulatory pathway the cell uses to conserve its resources and maximize its competitive advantage.

Acknowledgments

I am very grateful to Craig Squires and the members of my laboratory for helpful comments on the manuscript and to Dmitry Zaporojets for creating the graphic illustrations.

See Also the Following Articles

TRANSCRIPTIONAL REGULATION IN PROKARYOTES • TRANSLATIONAL CONTROL AND FIDELITY

Bibliography

Condon, C, Squires, C., and Squires, C. L. (1995). Control of rRNA transcription in *Escherichia coli*. *Microbiol. Rev.* 59, 623–645.

Gaal, T., Bartlett, M. S., Ross, W., Turnbough, C. L., Jr., and Gourse, R. L. (1997). Transcription regulation by initiating NTP concentration: rRNA synthesis in bacteria. *Science* 278, 2092–2097.

Gourse, R. L., Gaal, T., Bartlett, M. S., Appleman, J. A., and Ross, W. (1996). rRNA transcription and growth rate-dependent regulation of ribosomes synthesis in *Escherichia coli*. *Annu. Rev. Microbiol.* 50, 645–677.

Keener, J., and Nomura, M. (1996). Regulation of ribosome synthesis. *In* "*Escherichia coli* and *Salmonella:* Cellular and Molecular Biology" (F. C. Neidhardt, J. L. Ingraham, E. C. Lin, K. B. Low, B. Magasanik, W. S. Reznikoff, M. Riley, M. Shaechter, and H. E. Umbarger, Eds.), pp. 1417–1431. ASM Press, Washington, DC.

Noller, H. F., and Nomura, M. (1996). Ribosomes. *In* "*Escherichia coli* and *Salmonella:* Cellular and Molecular Biology" (F. C. Neidhardt, J. L. Ingraham, E. C. Lin, K. B. Low, B. Magasanik, W. S. Reznikoff, M. Riley, M. Shaechter, and H. E. Umbarger, Eds.), pp. 167–186. ASM Press, Washington, DC.

Rickettsiae

Marina E. Eremeeva
University of Maryland at Baltimore

Gregory A. Dasch
Naval Medical Research Center

I. Introduction and Historical Perspectives
II. Classification and Ecology
III. Microenvironment, Growth, and Morphology
IV. Cultivation and Isolation Procedures
V. Genomic Characteristics
VI. Cell Wall Composition, Antigens, and Putative Virulence Factors
VII. Identification
VIII. Susceptibility to Antimicrobial Agents

GLOSSARY

phagolysosome Host cell vacuole which forms after fusion of a phagosome with lysosomal vesicles; the host cell environment of *Coxiella burnetii*.

phagosome Host cell vacuole which forms upon bacterial entry into the host cell; the host cell environment of ehrlichiae.

rickettsiae Narrowly used, any species of the family *Rickettsiaceae;* otherwise, broadly applied to fastidious organisms of similar size exhibiting intracellular growth.

S-layer Crystalline surface layer, also called the microcapsule layer, which contains the species-specific surface protein antigen(s) in *Rickettsia*.

transovarial and transstadial transmission Vertical maintenance of rickettsiae by transmission from the ovaries of infected female ticks to their eggs and ensuing larval offspring; maintenance of the rickettsiae in the molting process of the tick, larvae to nymph, and/or nymph to adult.

FOR MANY YEARS, THE TERM "RICKETTSIAE" has been loosely applied to a very wide range of gram-negative fastidious bacteria that are frequently associated with diverse arthropod vectors. Molecular approaches to rickettsial phylogeny first demonstrated that species in the genera *Rickettsia, Orientia, Ehrlichia, Anaplasma, Wolbachia pipientis, Cowdria,* and *Neorickettsia* have similar evolutionary origins and that all belong to the α subdivision of Proteobacteria. In contrast, *C. burnetii* and *Rickettsiella grylli,* which are closely related to *Legionella,* and *Francisella* (*Wolbachia*) *persica* belong to the γ subdivision of Proteobacteria. Similarly, former species of *Rochalimaea* and *Grahamella* have been renamed and combined with *Bartonella* species based on their genomic characteristics. However, they belong in a different lineage (α-2) from the rest of the rickettsiae in α-Proteobacteria. The majority of the bartonellae grow epicellularly, but some species can invade and multiply within eukaryotic cells. However, for the purpose of this article, *C. burnetii* and *Bartonella* spp. are retained. This is justified by historical aspects of the discovery of each microorganism, similarities in their metabolism, some common aspects of their ecology, and their frequent association with arthropod vectors.

I. INTRODUCTION AND HISTORICAL PERSPECTIVES

The designation "Rickettsia" honors Dr. Howard Taylor Ricketts, who isolated the etiological agent of Rocky Mountain spotted fever and established the role of the tick in its transmission. Ricketts recognized that the agents causing Rocky Mountain spotted fever and epidemic typhus are similar but distinct microorganisms. Charles Nicolle established in 1909 that typhus is transmitted by the human body louse. The specific epithet of its etiological agent, *Rickettsia prowazekii,* was named after Von Prowazek, who greatly contributed to the understanding of the epidemiology of typhus. Another louse-transmitted rickettsial disease, trench fever, is caused by *Bartonella (Rochalimaea) quintana.* It made its first explosive appearance during World War I. The Q fever agent, *Coxiella burnetii,* was first isolated by Derrick in 1935 in Australia and was characterized as a rickettsia by Burnet and Freeman. At almost the same time a filterable agent, *Rickettsia diaporica,* was isolated in Montana. It was soon found to be identical to the Q fever agent, suggesting that it had a worldwide distribution. Because of its marked biological differences from the other rickettsiae, the agent was named *Coxiella burnetii* in honor of Cox and Burnet. Historically, ehrlichiae were first discovered as animal pathogens. *Cowdria ruminantium,* which is responsible for heartwater in sheep, cattle, and goats in sub-Saharan Africa, was encountered as early as 1838. The bacterium was named after E. V. Cowdry, who discovered the rickettsial etiology of the disease and its transmission by the bont tick, *Amblyomma hebraeum.* Tick-borne fever, caused in sheep by *Ehrlichia phagocytophila,* was first described in 1932. The agent of tropical canine pancytopenia, *Ehrlichia canis,* has been known since 1935. The number of ehrlichial agents has continued to expand significantly to include several agents of human disease, particularly the monocytic agents *E. chaffeensis* and *E. sennetsu* and the agent of granulocytic ehrlichiosis which is closely related, if not identical, to *E. equi* and *E. phagocytophila.*

II. CLASSIFICATION AND ECOLOGY

Rickettsiae inhabit a range of intracellular environments, including the cytoplasm, phagosome, or phagolysosome, and may also grow epicellularly. Rickettsiae include human or animal pathogens as well as pathogens or symbionts of arthropods, plants, and fish. However, there are considerable differences in the details of the association of each species with its arthropod and vertebrate host or humans; therefore, these are discussed individually (see Tables I–III).

A. *Rickettsia*

The genus *Rickettsia* is composed of the typhus group and spotted fever group (SFG). Classically, based on the presence of a common group lipopolysaccharide (LPS), the typhus group includes two human pathogens, *R. prowazekii* and *R. typhi,* and *R. canada,* which has been isolated from ticks only, whereas the SFG contains more than 20 diverse genotypes which cause at least seven different diseases.

Rickettsia prowazekii is the etiological agent of epidemic typhus, which is transmitted by the human body louse. In the history of humankind epidemic typhus typically accompanies wars and catastrophes. Typhus epidemics are often associated with cold seasons of the year or cold climate and poor hygienic conditions suitable for body louse infestation. The louse becomes infected by ingesting the blood of humans experiencing disease. Rickettsiae multiply quickly in the gut cells of the louse and are discharged in the louse feces in large quantities. Although the rickettsial infection is deadly for the louse, infected lice can survive for about 2 weeks and transmit the infection to a healthy individual. Infection is typically acquired by scarification of skin bites with infected feces. Recovery from epidemic typhus is thought to result in nonsterile immunity, permitting persistence of *R. prowazekii* in a human reservoir for decades between epidemics. Occasionally, these individuals suffer from a relapsed or recrudescent form of typhus called Brill–Zinsser disease, which has clinical symptoms similar to those of primary epidemic typhus but is usually milder. Sylvatic cycles of *R. prowazekii* have also been described with the American flying squirrel, *Glaucomys volans.* Several human cases of flying squirrel-associated infections have been found that are typically associated with invasion of houses by squirrels during the cold seasons. Epidemic typhus rickettsiae are believed to be transmitted by the specific louse of the squirrel between animals and to humans by the fleas, which are not host-specific.

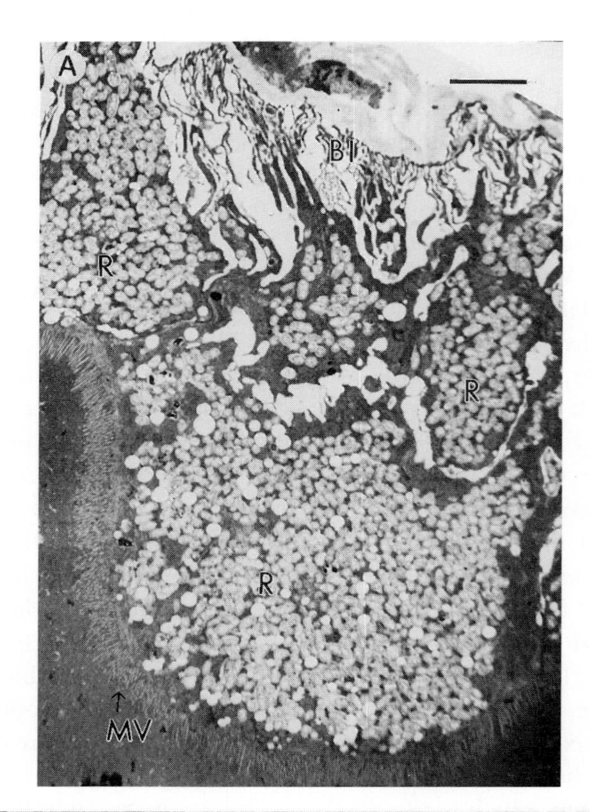
A
BI
R
R
R
MV

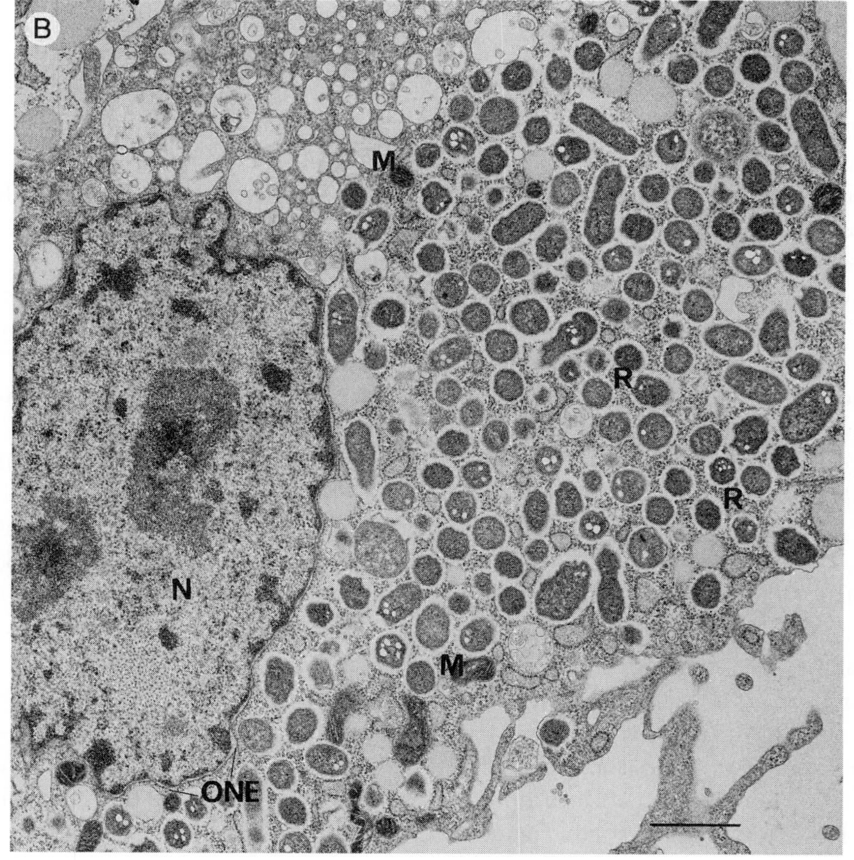
B
M
R
R
N
M
ONE

Rickettsia typhi causes endemic typhus, which is also called murine typhus. It is often a milder disease in humans than the infection caused by *R. prowazekii.* It has a worldwide distribution that corresponds to the spread of its principal reservoirs, the rats *Rattus norvegicus* and *R. rattus,* although other vertebrate hosts may be involved. *Rickettsia typhi* is transmitted by rat lice in rodent populations, but the oriental rat flea, *Xenopsylla cheopis,* is the chief transmitter of the disease to humans. *Rickettsia typhi* grows in the midgut epithelial cells of the flea (Fig. 1A) and is excreted in its feces. The life span of the flea is not affected by the rickettsiae, which are maintained by transovarial transmission in infected populations of fleas. Humans usually become infected due to scarification of infected flea tissues and feces at the flea bite site and, in rare cases, by aerosol infection with feces. *Rickettsia felis* (ELB agent) is a typhus group-like rickettsia based on its serological characteristics and flea association, but phylogenetic analysis has placed it in a distinct clade among the SFG rickettsiae. *Rickettsia felis* has been difficult to isolate and cultivate. It has been maintained in a laboratory population of cat fleas, *Ctenocephalides felis,* in which it is inherited transovarially. *Rickettsia felis* has been identified in opossum tissues and their fleas and from a patient originally thought to have murine typhus. *Rickettsia canada,* serologically and biologically another typhus-like rickettsia, has only been isolated from *Haemaphysalis* ticks from Richmond, Canada, and from Mendocino County, California. The possible role of *R. canada* as a human pathogen is not established.

The majority of SFG rickettsiae are typified by their association with specific ticks, whose ecology limits the geographic distribution of each genotype of rickettsiae. Ticks acquire rickettsiae by feeding on infected vertebrates and remain infected throughout the rest of their lives, transmitting the bacteria transstadially and transovarially to their offspring. Rocky Mountain spotted fever, caused by *R. rickettsii,* is found throughout the Western Hemisphere. It is transmitted by the wood tick, *Dermacentor andersonii,* in the western United States and the dog tick, *D. variabilis,* in the eastern United States. *Rhipicephalus sanguineus* and *Amblyomma cajennense* are thought to be vectors of *R. rickettsii* in Mexico and South America, respectively. Several other SFG rickettsiae have been identified in North America, including *R. bellii, R. parkeri, R. montana, R. amblyommii, R. peacockii* (the former East Side agent), and several other unnamed distinct serotypes. The area of distribution for some of these SFG rickettsiae overlaps with that of *R. rickettsii* and they may be transmitted by the same tick vectors. Whether they interfere with *R. rickettsii* maintenance in ticks and whether they are pathogenic for man are unknown. However, these SFG agents may cause seroconversion without apparent disease in some endemic areas.

Rickettsia conorii is responsible for boutonneuse fever, which is also known as Mediterranean spotted fever or Marseilles fever in Mediterranean countries. It also occurs in Eurasia through India and throughout Africa. Its geographical range correlates with the distribution of its principal vector, the brown dog tick, *R. sanguineus. Rickettsia sibirica* causes North Asian tick typhus, whose endemic area is western and eastern Siberia of the former Soviet Union and northern China. *Rickettsia sibirica* has been detected in more than 20 different species of ixodid ticks. Individual isolates have been found far from the limits of known foci of disease; therefore, the distribution of disease caused by *R. sibirica* may actually exist from Siberia and Pakistan through central Europe to Portugal. *Rickettsia slovaca* is primarily transmitted by *Dermacentor marginatus* in European countries, and it has been associated with sporadic cases of meningoencephalitis. *Rickettsia africae,* the agent of *Amblyomma*-transmitted tick typhus in sub-Saharan Africa, and other spotted fever agents causing the

Fig. 1. Electron microscopic observations on typhus group rickettsiae. (A) *Rickettsia typhi* infection in the midgut epithelium of *Xenopsylla cheopis* flea. BI, basal invagination of plasma membrane; MV, apical microvilli; R, rickettsia. Scale bar = 5 μm (courtesy of S. Ito, Harvard Medical School, Boston). (B) *Rickettsia prowazekii* Breinl strain grown in a primary chicken fibroblast, 72 hr after infection. Scale bar = 1.0 μm. R, rickettsia; M, mitochondria; N, nucleus, ONE, outer nuclear envelope (reproduced with permission from Eremeeva *et al.,* 2000).

TABLE I
Diseases and Habitats of Rickettsiae of the genera *Rickettsia, Orientia,* and γ Subgroup of Proteobacteria

Species	Disease or pathogenic effect	Principal arthropod vector	Principal vertebrate reservoirs[a]	Geographic distribution
Typhus group				
R. prowazekii	Epidemic typhus	Human body louse	**Human**[b]	Worldwide
	Brill–Zinsser (recrudescent typhus)	None	**Human**	Worldwide
	Sylvatic typhus	Squirrel flea, louse	**Flying squirrels**	Eastern United States
R. typhi	Endemic (murine) typhus	Rat flea, louse	**Rats, other rodents**	Worldwide
R. felis	Murine typhus-like	Cat flea	**Opossum**, rats	Texas, California
R. canada	Unknown	*Haemaphysalis* ticks	Rabbits, hare, birds	Ontario, California
Spotted fever group:				
R. akari	Rickettsial pox	Mouse mite	**Rodents**	Worldwide
R. rickettsii	Rocky Mountain SF[c]	*Dermacentor* ticks, *Amblyomma cajennense* tick	Rodents, lagomorphs, canines, birds	North America South America
R. amblyommii	Suspected mild SF rickettsiosis	*Amblyomma americanum* tick	Rodents, birds, ruminants	Southern United States
R. parkeri	Unknown	*Amblyomma maculatum* tick	Rodents, birds, ruminants	North America
R. montana	Unknown	*Dermacentor* ticks	**Rodents**, dogs	North America
R. rhipicephali	Unknown	*Rhipicephalus sanguineus* tick	Dogs	North America
R. peacockii	Unknown	*Dermacentor* tick		North America
R. conorii	Boutonneuse fever, Mediterranean SF	*Rhipicephalus* and *Haemaphysalis* ticks	**Rodents, dogs**, lagomorphs	Africa, southern Europe to India
R. helvetica	Unknown	*Ixodes ricinus* tick	Rodents, deer, cattle	Europe
R. slovaca	Suspected in meningitis	*Dermacentor* ticks	**Rodents, lagomorphs**, ruminants	Europe, Asia
R. massiliae	Unknown	*Rhipicephalus* ticks	Rodents, dogs, ruminants	Europe, Africa
R. mongolotimonae	Suspected mild SF	?	?	Europe, Asia
R. sibirica	North Asian tick typhus	*Dermacentor, Haemaphysalis* ticks	**Rodents**, canines, ruminants	Europe, Asia
R. africae	African tick bite fever	*Amblyomma* ticks	**Ruminants**, rodents	Sub-Saharan Africa
R. aeschlimanii	Unknown	*Hyalomma marginatum* tick	?	Europe, Africa
R. australis	Queensland tick typhus	*Ixodes holocyclus* tick	Rodents, marsupials	Australia
R. honei	Flinders Island SF	*Ixodes cornuatus* tick	Rodents, dogs	Australia
R. japonica	Oriental SF	*Haemaphysalis, Dermacentor* ticks	Rodents	Japan
R. helongjiangi	Unknown	*Dermacentor, Haemaphysalis* ticks	Rodents	China
R. sharonii	Israeli tick typhus	*Rhipicephalus* ticks	Dogs, rodents, hedgehog	Israel
R. sp.	Astrakhan SF	*Rhipicephalus pumilio* tick	Canines, hedgehog	Europe
R. sp. Bar 29	Unknown	*Rhipicephalus* ticks	?	Europe
Other species				
R. bellii	Unknown	Ixodid and argasid ticks	Rodents	United States
R. sp. PAR	Unknown	Pea aphid *Acyrthosiphon pisum* (Harris)	None	California
R. sp.	Papaya bunchy top disease	Leafhopper *Empoasca papayae* Oman	None	Puerto Rico, Costa Rica
R. sp. AB bacterium	Sex ratio distortion in insect host	Ladybird beetle *Adalia bipunctata*	None	Europe, Asia

continues

Continued

Species	Disease or pathogenic effect	Principal arthropod vector	Principal vertebrate reservoirs[a]	Geographic distribution
O. tsutsugamushi	Scrub typhus, tsutsugamushi fever	Trombiculid mites	**Rodents, marsupials**	Southern Asia, Australia
γ-Proteobacteria				
C. burnetii	Q fever (acute and chronic)	Ticks and other arthropods	**Domestic and wild animals**	Worldwide
R. grylli		Cricket, locust, crustacea	None	Worldwide
R. popilliae	Damage to laboratory insectary and stocks; possible control for agricultural pests	Insecta	None	Worldwide
R. chironomi		Insecta, Arachnida, midge *Chironomus*	None	Worldwide
R. phytoseiuli		Mite *Phytoseiulus*	None	Europe
Francisella (Wolbachia) persica	Cytoplasmic incompatibility, sex ratio distortion, parthenogenesis	Diverse	None	Worldwide
Piscirickettsia salmonis	Salmonid rickettsial septicemia (piscirickettsiosis)		**Fish**	Worldwide

[a] Primary vectors are indicated. Many vectors, but not all, serve as the primary rickettsial reservoir.
[b] Vertebrate hosts that have been shown to be rickettsial reservoirs are in bold print. The other vertebrates listed serve as hosts for the vectors and frequently have antibodies to rickettsiae, but rickettsiae have not been isolated from their tissues or blood.
[c] SF, spotted fever.

diseases known as Israeli tick typhus and Astrakhan spotted fever are closely related to *R. conorii. Rickettsia australis, R. honei,* and *R. japonica* are the agents of human diseases in Australia, Flinders Island, and Japan, respectively. Other tick-borne spotted fever agents are also known from Europe and Asia, but their pathogenicity for humans has not been established (Table I).

Rickettsia akari differs markedly from the other tick-transmitted SFG rickettsiae since its main vector is a mite, *Liponyssoides sanguineus,* which parasitizes domestic mice, *Mus musculus.* Mites are the main reservoir of *R. akari,* which is maintained transovarially and transstadially. Typically, mites circulate in rodent populations, but when their host animal's temperature is increased by disease the mites may leave them, infest available humans, and transmit *R. akari* to them. The distribution of *R. akari* is probably worldwide since outbreaks of rickettsialpox have occurred in the United States, Europe, and Asia.

Two plant-associated arthropods have been found to contain rickettsiae closely related to *R. bellii* and one of these causes disease in plants (Table I). Another recently discovered unusual *Rickettsia,* AB bacterium, is found in some populations of the ladybird beetle, *Adalia bipunctata.* AB agent is responsible for sex ratio distortion in its insect hosts and has been detected in beetles from England, European regions of Russia, and Japan. Although the AB bacterium is also not yet cultivable, it is more closely related to typhus and SFG rickettsiae than to the more prevalent arthropod-associated *W. pipientis* and its relatives (Table II).

B. *Orientia*

Orientia (formerly *Rickettsia*) *tsutsugamushi* is the etiological agent of scrub typhus, also known as tsutsugamushi disease. The natural foci of *O. tsutsugamushi* typically consist of transitional forms of vegeta-

TABLE II
Diseases and Habitats of Species of the Ehrlichiae

Genogroup and species[a]	*Animals affected*	*Disease or pathogenic effect*	*Invertebrate host*	*Geographic distribution*
Ehrlichiae, group I				
E. canis	Canines	Tropical canine pancytopenia	*Rhipicephalus sanguineus* tick	Worldwide in tropical and subtropical areas
E. chaffeensis	Human, deer	Human ehrlichiosis	*Amblyomma americanum* tick	United States, Europe
E. ewingii	Canines	Canine granulocytic ehrlichiosis	*Amblyomma americanum* tick?	United States
E. muris	Mouse	Unnamed	?	Japan
Cowdria ruminantium	Ruminants	Heartwater	*Amblyomma hebraeum* tick	Sub-Saharan Africa, Caribbean Islands
Ehrlichiae, group II				
E. phagocytophila	Ruminants	Tick-borne fever	*Ixodes ricinus* tick	Europe, Asia, Africa
E. equi	Horse	Equine ehrlichiosis	*Ixodes pacificus* and *I. ricinus* ticks	United States, Europe
HGE	Human, dog, horse	Human granulocytic ehrlichiosis	*Ixodes scapularis* tick	Northern United States, Europe
E. platys	Canines	Canine cyclic thrombocytopenia	Tick?	United States
"*E. bovis*"	Cattle	Bovine ehrlichiosis	*Hyalomma* ticks	Africa, Middle East
"*E. ovina*"	Sheep	Ovine ehrlichiosis	*Hyalomma* ticks	Africa, Middle East, Sri Lanka
"*E. ondiri*"	Cattle	Bovine petechial fever (Ondiri disease)	Unknown	Kenya
LGE	Llama	Llama granulocytic ehrlichiosis	*Ixodes pacificus* tick	California, USA
Anaplasma marginale	Cattle	Anaplasmosis	*Dermacentor* and *Boophilus* ticks, tabanid flies	Worldwide
Ehrlichiae, group III				
E. sennetsu	Human	Sennetsu ehrlichiosis	Fluke?	Japan, Malaysia
E. risticii	Horse	Potomac horse fever, equine monocytic ehrlichiosis	Snails Pleuroceridae (*Juga* spp.)	United States, Europe
Neorickettsia helminthoeca	Canines	Salmon poisoning, Elokomin fluke fever	Fluke (*Nanophyetus salmincola*)	Western United States
SF agent	Human, canines	Hyuganetsu disease	Fluke (*Stellantchasmus falcatus*)	Japan
Ehrlichia-like agents group IV				
Wolbachia pipientis-like symbionts:				
Group A, B	Arthropods, isopods	Cytoplasmic incompatibility, sex ratio distortion	Seven orders of insects, *Armadillium*	Europe, United States
Worm-associated group	Nematodes	Possible essential symbionts	*Dirofilaria, Onchocerca*	Worldwide

[a] Quotation marks indicate that an organism has not been officially named.

tion that are associated with changing ecological conditions. It is transmitted by the bite of trombiculid mites, particularly those of the genus *Leptotrombidium*. As with many other rickettsiae, *O. tsutsugamushi* is transmitted vertically in its invertebrate hosts; it also causes sex ratio distortion to a female bias in some species of trombiculid mites. The larval stage or chigger feeds on a wide range of vertebrate hosts, including humans, and the bite is responsible for the transmission of tsutsugamushi disease to them. Scrub typhus occurs throughout the Orient from Afghanistan and Tadzhikistan to Korea and Japan, from the maritime provinces of Russia to northern Australia and to the western islands of the Pacific Ocean.

C. *Coxiella* and Other γ-Proteobacterial Agents

Coxiella burnetii is widespread in natural environments. All known isolates belong to a single species, which causes disease in wild and domestic animals and humans. Q fever occurs in acute and chronic forms, which are caused by the same strains of *Coxiella*. The clinical manifestations of disease depend on the species, sex, age, and immunological status of infected animals. Predominant manifestations of acute Q fever in man include self-limited flu-like fever, hepatitis, and pneumonia, whose frequency varies in different countries. Chronic disease is manifest most frequently as endocarditis. *Coxiella burnetii* is very resistant to unfavorable growth conditions and is transmitted from animal to animal and to humans primarily by aerosol. Ticks are very important in maintaining natural cycles of *C. burnetii*, although infected vertebrates clearly also play a role in its transmission. Q fever is often an occupational disease which affects the workers of slaughterhouses. The bacterium is particularly abundant in the placenta of parturient domestic stock. Parturient cats also cause infection in households. Human fetal infections with *C. burnetii* have been described. Sexual transmission of *Coxiella* has been documented in animals and probably occurs in humans.

Rickettsiella currently contains four named species, *R. grylli*, *R. popilliae*, *R. chironomi*, and *R. phytoseiuli*, as well as similar unnamed organisms in other hosts, but their relatedness has not been established. *Rickettsiella* spp. have been found in insects, crustaceans, and arachnids, in which they replicate in cell vacuoles of the fat body, hepatopancreas, and other organs and undergo a cell cycle slightly resembling that of chlamydia. The potential of *Rickettsiella* as a human pathogen is not known, but they are well-known for their devastating effect on laboratory insectaries and for their potential utility as control agents for agricultural pests. Rickettsiellae are pathogenic for the larval stage of their principal invertebrate hosts and for the young and mature stages of other secondary invertebrate hosts. Rather than being maintained by transovarial transmission, infection of offspring is usually effected through ingestion of contaminated soil in which rickettsiellae may be maintained for years. Recently, *R. grylli* was shown to be closely related to *C. burnetii* based on its 16S rRNA gene sequence.

Piscirickettsia salmonis, the first characterized rickettsial pathogen of fish, was initially recognized in 1989 in southern Chile. It causes salmonid rickettsial septicemia or piscirickettsiosis, a disease affecting salmon in farmed populations. *Piscirickettsia salmonis* belongs to the γ group of Proteobacteria. Rickettsial-like microorganisms have been found among diverse species of fish from different geographic locations and aquatic environments, but their identity and relationship to *P. salmonis* are not known.

D. *Ehrlichia*

Ehrlichiae include four distinct genetic groups of intracellular microorganisms (Table II). The first group includes *E. canis*, *E. chaffeensis*, *E. muris*, *E. ewingii*, and *C. ruminantium*. *Ehrlichia canis*, which is transmitted by *R. sanguineus* ticks, is a pathogen of dogs and is found worldwide. *Amblyomma* ticks transmit *C. ruminantium*, the agent of heartwater, to ruminants in Africa and some islands in the eastern Caribbean Sea. *Ehrlichia chaffeensis*, the agent of human monocytic ehrlichiosis, most frequently occurs in the southeastern and south-central United States and is associated with *Amblyomma americanum* and white-tail deer, which are the most likely vector and reservoir for this microorganism, respectively. Natural cycles for *E. ewingii*, the agent of canine granulocytic ehrlichiosis, and *E. muris*, which was isolated from a wild mouse in Japan, are not fully understood.

The second group of ehrlichiae contains the agent of equine ehrlichiosis, *E. equi,* the agent of caprine ehrlichiosis, *E. phagocytophila,* the agent of human granulocytic ehrlichiosis, HGE, and *E. platys,* which infects platelets in dogs. *Ehrlichia bovis, E. ovina,* and *E. ondiri* have been described as agents of granulocytic ehrlichioses in domestic animals in Africa, the Middle East, and Sri Lanka, although molecular characterization of these agents is incomplete. An HGE-like agent, LGE, has been isolated from llamas, a camelid species, and from llama-associated ticks, *Ixodes pacificus,* in California. *Anaplasma marginale,* which has a tropism for erythrocytes and lacks a cell wall, is phylogenetically a member of the second group, but it resembles group I ehrlichiae in its antigenic composition and vector, *Amblyomma* ticks. The HGE agent, *E. equi,* and *E. phagocytophila* are very closely related and are transmitted by *Ixodes* ticks in the northern United States and Europe. Perinatal transmission of HGE from an infected mother to her infant has been reported.

The third group of ehrlichiae contains *E. sennetsu, E. risticii,* SF agent isolated from the fluke *Stellantchasmus falcatus,* which parasitizes fish in Japan, and *Neorickettsia helminthoeca.* Transmission of *E. sennetsu* is associated with the consumption of raw fish, but the precise mode of its transmission is unknown. Stream water snails of the family Pleuroceridae (*Juga* spp.) are the natural reservoir for *E. risticii,* consistent with the association of Potomac horse fever with pastures bordering rivers and irrigation ditches. Many recently identified isolates of *Ehrlichia* spp. exhibit significant genetic polymorphisms when compared to type strains of the same species, but their classification as separate species will require further analysis.

The fourth group of ehrlichiae contains *W. pipientis* and the many related but diverse endosymbionts of invertebrates which appear to have spread promiscuously both horizontally and vertically in nature (Table II). Clades A and B are closely related and include *W. pipientis* and its relatives which live in symbiotic relationships with arthropods and isopods. These bacteria cause sex ratio distortion and alteration of sex determination in their hosts, including parthenogenesis, female-biased sex ratios, and sterile offspring of crosses between infected males and uninfected females. Another distinct group of *Wolbachia*-like microorganisms established mutualistic interaction with several species of filarial nematodes which parasitize people and animals worldwide.

E. *Bartonella*

The number of known *Bartonella* species is rapidly increasing (Table III). Currently, *Bartonella* includes 13 species formerly classified in the genera *Bartonella, Grahamella,* and *Rochalimaea.* Four species are known to be pathogenic in humans, among which *B. bacilliformis* and *B. quintana* have the longest known histories. *Bartonella bacilliformis* is transmitted by sand flies of the genus *Lutzomyia* in the Andian regions of South America. In humans this agent causes a biphasic disease called Carrion's disease. Its acute stage, Oroya fever, is characterized by a severe life-threatening hemolytic anemia. The chronic stage, termed verruga peruana, causes unique vascular proliferative lesions of the skin. *Bartonella quintana,* formerly known as *Rochalimaea quintana,* is responsible for louse-transmitted trench fever in humans and has been isolated from patients on both sides of the Atlantic. *Bartonella quintana* is also associated with other clinical syndromes, including endocarditis, lymphadenopathy, bacillary angiomatosis, and bacteremia in patients with AIDS-associated disease. Recently, several new species of *Bartonella* have been identified as emerging pathogens. *Bartonella henselae* is responsible for many clinical syndromes, including bacillary peliosis hepatis, septicemia, and bacillary angiomatosis. It is also recognized as the etiological agent of cat scratch disease. Domestic cats and their fleas, *C. felis,* are the natural reservoir and vector of *B. henselae,* respectively. *Bartonella elizabethae* was isolated from a patient suffering from endocarditis. Genetically related to *B. elizabethae, B. tribocorum* is found in the rat, *R. norvegicus.* Six *Bartonella* species—*B. grahamii, B. taylorii, B. doshiae, B. talpae, B. peromysci,* and *B. vinsonii*—were isolated from small wild animals, but their pathogenicity for humans and animals has not been established. *Ixodes* ticks may be involved in the circulation of some *Bartonella* species in nature. *Bartonella vinsonii* subsp. *berkhoffii* was isolated from a dog suffering from endocarditis. *Bartonella clarridgeiae* has been isolated from domestic cats. Many other recently isolated *Bartonella* genotypes parasitize the erythrocytes

TABLE III
Human Diseases and Habitats of *Bartonella* Species

Species	*Reservoir or principal host*	*Major disease and clinical symptoms*	*Invertebrate host (vector or natural reservoir)*	*Geographical distribution*
B. bacilliformis	Human	Carrion's disease	*Lutzomyia*	Peruvian and Ecuadorian Andean region of South America
B. quintana	Human	Trench fever, bacillary angiomatosis, endocarditis, chronic lymphoadenopathy	Human body louse	Worldwide
B. henselae	Human, cat	Cat scratch disease, bacillary angiomatosis, peliosis hepatis, bacteremia, meningitis, endocarditis	*Ctenocephalides felis* flea?	Worldwide
B. vinsonii	Vole *Microtus pennsylvanicus*?	Unknown	Unknown	Canada
B. vinsonii, subsp. *berkhoffii*	Dogs?	Endocarditis in dogs	Unknown	United States
B. elizabethae	Human	Endocarditis	Unknown	United States
B. talpae[a]	Moles *Talpa*	Infectious for its vertebrate host	Fleas	Old World
B. peromysci[a]	Deer mouse *Peromyscus*	Infectious for its vertebrate host	Fleas	Old World
B. grahamii	Voles *Clethrionomys* and *Microtus*, mouse *Apodemus*	Unknown	Unknown	England
B. taylorii	Voles *Clethrionomys* and *Microtus*, mouse *Apodemus*	Unknown	Unknown	England
B. doshiae	Voles *Clethrionomys* and *Microtus*, mouse *Apodemus*	Unknown	Unknown	England
B. clarridgeiae	Cat	Suspected in cat scratch disease	Unknown	United States, Europe
B. tribocorum	Rat *Rattus norvegicus*	Unknown	Unknown	Eastern France

[a] *Bartonella talpae* and *B. peromysci* are two valid species, isolates of which are no longer extant.

of small rodents much as was initially described for species of *Grahamella*.

III. MICROENVIRONMENT, GROWTH, AND MORPHOLOGY

Most of the rickettsiae are obligate intracellular parasites of eukaryotic cells. Nearly all species probably enter their host cells via induced phagocytosis upon interaction with an unknown cellular receptor(s). Rickettsiae survive intracellularly by at least three different mechanisms. Species of the genera *Rickettsia* and *Orientia* rapidly escape from the phagosomal vacuole into the cytoplasm and can sometimes invade the nucleus. Ehrlichial species modify the phagosome and prevent its fusion with lysosomal vesicles, a strategy which avoids acidification and

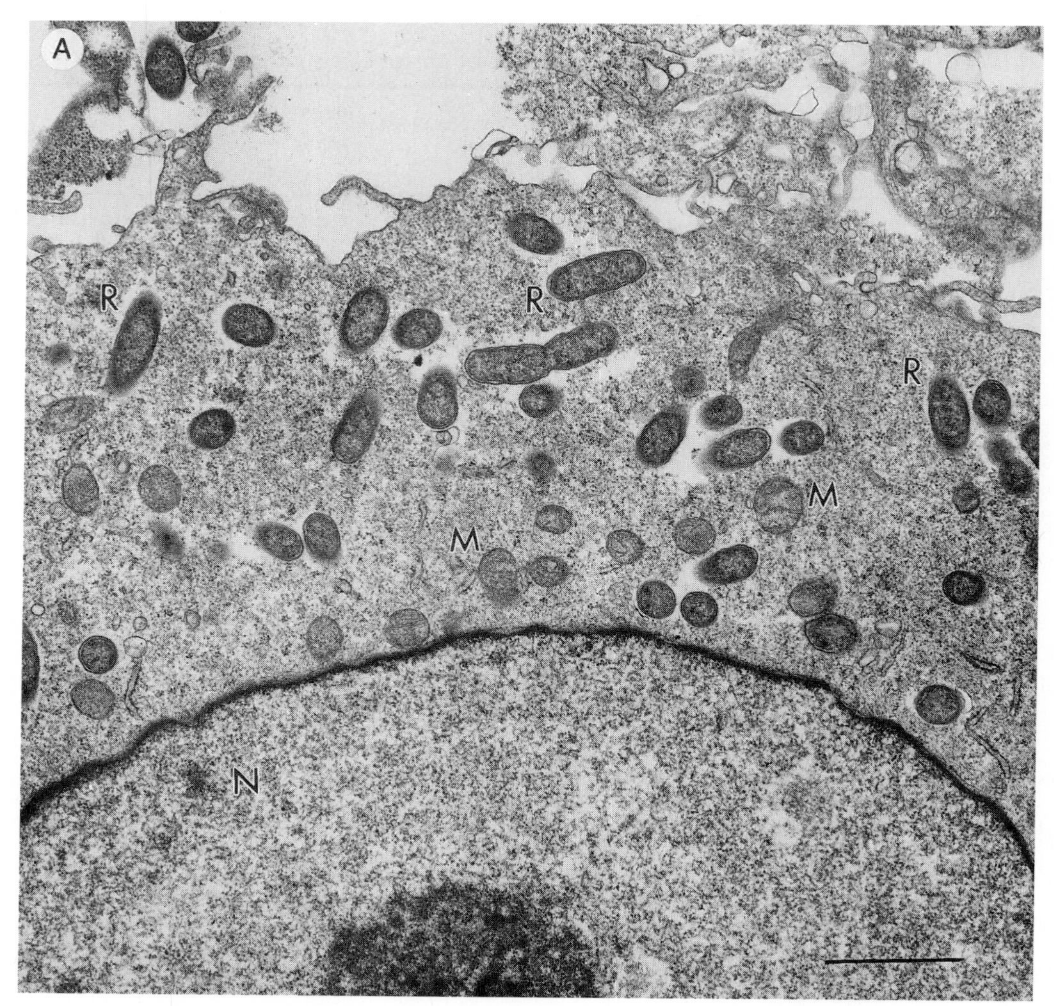

Fig. 2. Electron microscopic observations on SFG rickettsiae in the EA.hy 926 human endothelial cell line. (A) Intracytosolic multiplication of *R. sibirica* 246 strain 72 hr after infection (courtesy of D. J. Silverman, University of Maryland at Baltimore, School of Medicine). (B) Intranuclear growth of *R. sibirica* 246 strain 72 hr after infection. (C) Cytopathic effect of *R. rickettsii* Smith strain 72 hr after infection. Extensive dilatation of membranes of the endoplasmic reticulum and outer nuclear envelope are demonstrated. R, rickettsiae; M, mitochondria; ER, endoplasmic reticulum; ONE, outer nuclear envelope; N, nucleus. Scale bar = 1.5 μm. (Figs. 2B and 2C reproduced with permission from Eremeeva *et al.,* 2000).

exposure to proteases and other bactericidal factors. *Coxiella burnetii* does not avoid lysosomal fusion and is adapted to grow within an acidified phagolysosome. *Bartonella* usually grows epicellularly, often modifying the surface of erythrocytes, but some species can also invade and multiply in the cytoplasm of eukaryotic cells to varying degrees.

A. Growth in the Cytoplasm and the Nucleus

Endothelial cells of small- and middle-sized vessels are the primary sites for replication of *Rickettsia* in vertebrate hosts. *Rickettsia rickettsii* can also invade underlying smooth muscle cells. In arthropods rick-

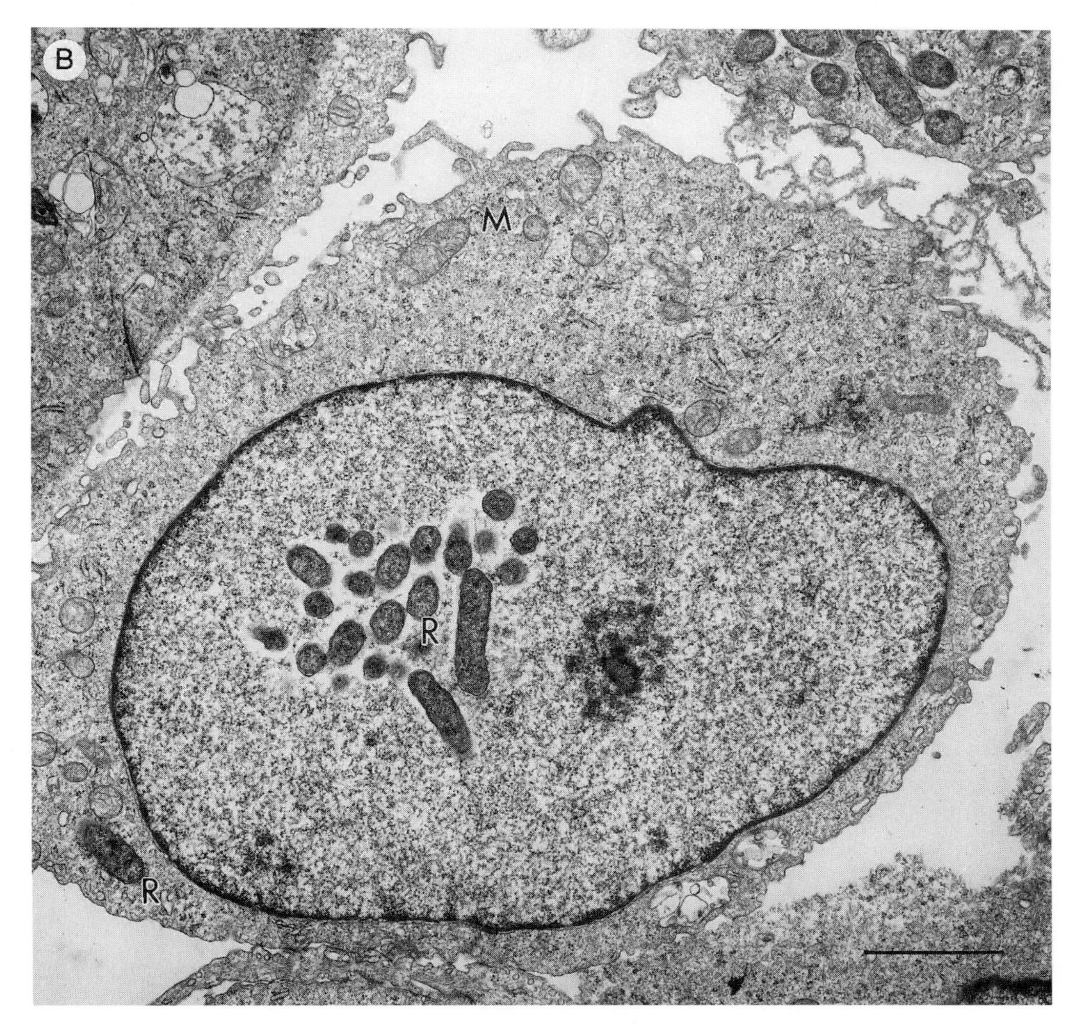

Fig. 2. Continued.

ettsiae colonize salivary gland cells, midgut cells (Fig. 1A), and intestinal and gonadal cells. Escape of *Rickettsia* from the phagosome is probably mediated by the action of rickettsial phospholipase A_2 on the phagosomal membrane. *Rickettsia prowazekii* grows in the cytoplasm to very high bacterial density and escapes by cell rupture (Fig. 1B). SFG rickettsiae typically grow to lower cell densities (Fig. 2A). *Rickettsia typhi* and SFG rickettsiae escape from their host cells without cell lysis earlier in the course of multiplication and invade other adjacent cells, resulting in a very rapid spread of infection. Occasionally, *R. canada* and SFG rickettsiae invade the nucleus and multiply there to high number (Fig. 2B). Intracellular multiplication of SFG rickettsiae, particularly *R. rickettsii*, causes very distinctive changes in cellular architecture, which are believed to be due to oxidant-mediated cell injury (Fig. 2C). Both SFG rickettsiae and *R. typhi*, to a lesser extent, catalyze the polymerization of actin tails which appear within the first few minutes after rickettsial entry into the cell (Fig. 3A). Actin polymerization causes active intracellular movement of the rickettsiae within the cell (Fig. 3B) and facilitates their exit and invasion of neighboring cells (Fig. 3C). Intracellular *R. prowazekii* does not exhibit this type of interaction with cellular actin pools (Fig. 3D).

The morphology of members of the genus *Rickettsia* is quite typical of small gram-negative bacterial rods, except they also possess microcapsular S-layers

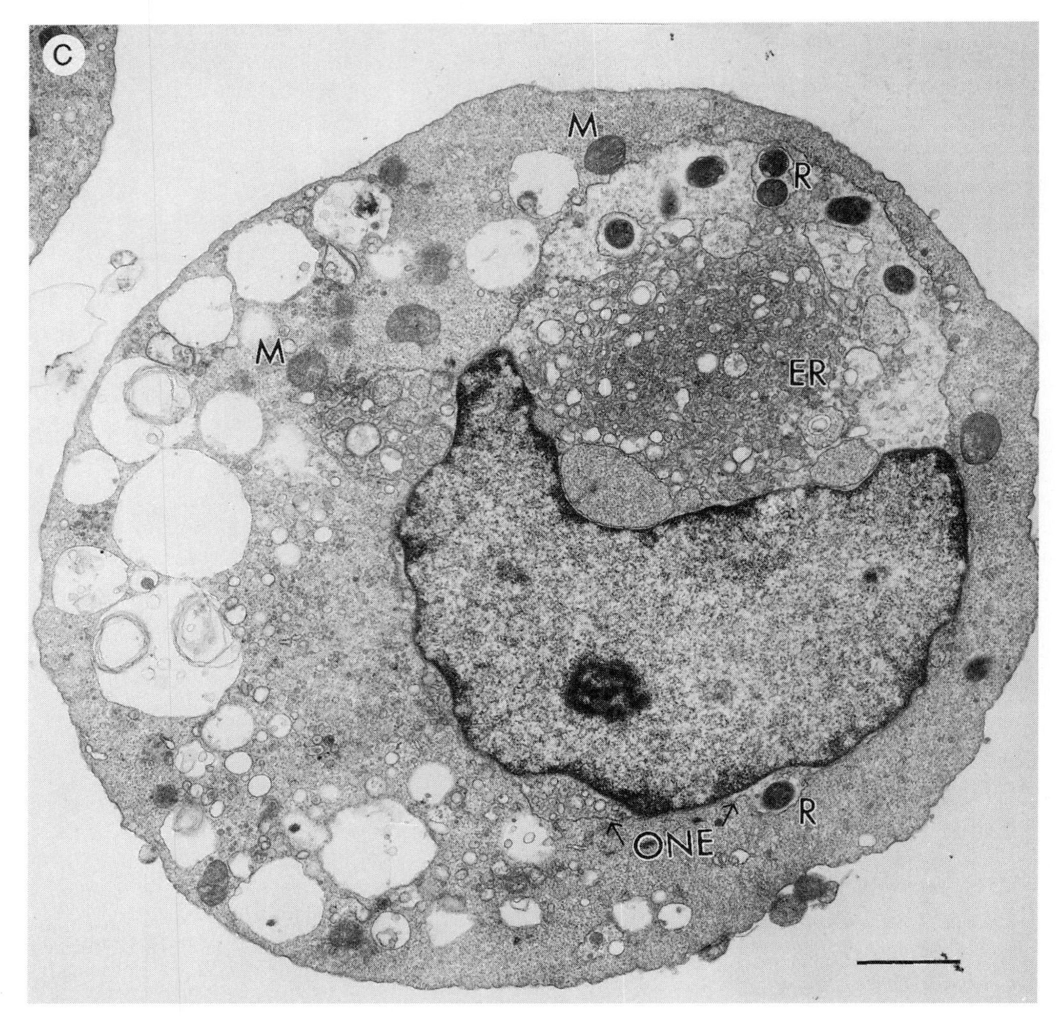

Fig. 2. Continued.

(Fig. 3D; see Section VI,A). *Rickettsia* are about 0.3–0.5 μm in diameter and 0.8–2.0 μm in length. They do not have morphologically identifiable flagella, pili, or attachment proteins.

Orientia tsutsugamushi differs from *Rickettsia* spp. in its morphology since the external layer of its outer membrane is much thicker than the inner leaflet, whereas the opposite is true of other rickettsial species. *Orientia* lacks a microcapsular S-layer. It is also approximately two times larger in diameter (1.2–2.5 μm) than species of *Rickettsia*. *Orientia tsutsugamushi* grows to high density in the cytoplasm of its infected cells and it may also occasionally invade the nucleus. To effect its release from infected cells, *Orientia* evaginates from the cell plasma membrane in a process resembling the budding process of enveloped viruses (Fig. 4). The budding rickettsiae adhere tenaciously to host cell components and accumulate to high density at the surface of infected cells. Like the other rickettsiae, *O. tsutsugamushi* grows slowly in vertebrate cells at 37°C with a doubling time between 9 and 18 hr.

B. Growth in the Phagosome

Individual ehrlichial particles are small, round bacteria with typical gram-negative trilaminar outer and inner membranes and are 0.5 μm in diameter. They stain dark blue to purple with Giemsa or Wright stains. The ehrlichiae exhibit marked differences in the type of vertebrate cells which they parasitize. Genogroup I ehrlichiae, except for *E. ewingii* and *C. ruminantium,* infect monocytes and macrophages. *Ehrlichia ewingii* infects granulocytes and *C. ruminantium* infects vascular endothelial cells and neutrophils. Genogroup II ehrlichiae, except for *E. platys* and *A. marginale,* infect granulocytes. *Anaplasma* grows in erythrocytes while *E. platys* parasitizes platelets. Genogroup III ehrlichiae infect monocytes and macrophages, but *E. risticii* also invades intestinal epithelial cells and mast cells. Upon internalization, *Ehrlichia* species prevent phagosome–lysosome fusion. In genogroups I and II, ehrlichial progeny remain within the same vacuole, which consequently enlarges as they replicate. For genogroup III ehrlichiae, the membrane of the parasitophorous vacuole adheres closely to the enclosed bacterium and divides simultaneously with it. By electron microscopy all ehrlichiae studied in cell culture exist in two morphological forms, reticulate and dense-cored cells, and both can divide by binary fusion. Dividing ehrlichial particles can form large intracellular colonies referred to as morulae for their mulberry appearance. There are group-specific differences in the structure of the morulae and in their interactions with their host cells (Fig. 5). Species in genogroup I form large morulae in monocytes with many ehrlichiae often suspended in a fibrillar matrix, and the host cell mitochondria and endoplasmic reticulum (ER) are aggregated near and in contact with the morulae, respectively (Figs. 5A and 5B). Morulae of genogroup II have no fibrillar matrix and no contacts with the host cell mitochondria, and ER does not envelope the morulae (Fig. 5C). Organisms in genogroup III usually develop in small individual vacuoles that do not fuse with each other and divide along with the ehrlichiae (Fig. 5D). *Wolbachia pipientis* is more often detected as small, round (0.4 μm) or oval (0.5 $\times$ 1.0 μm) reticulate cells growing in individual vacuoles that form during its cultivation in the Aa23 mosquito cell line. *Wolbachia*-containing vacuoles are surrounded with cisterns of the ER which are in tight contact with the vacuolar membrane and continue into the host cell cytoplasm.

C. Growth in the Phagolysosome

In contrast to the other rickettsiae, coxiellae have some attributes resembling those of gram-positive bacteria. *Coxiella burnetii* is a unique bacterium since it combines obligate intracellular parasitism and adaptation to the acidic environment of the phagosome (Fig. 6) with a developmental cycle that includes transverse binary fission and sporogenesis. Sections of *C. burnetii*-infected cells typically display a variety of bacterial forms which may be separated into two distinct classes, a small-cell variant (SCV) and a large-cell variant (LCV). Single cells are usually small rods of 0.3–1.0 μm length, whereas dividing cells are double this length, form a spore at one end, or both. The cells also contain spores in various stages of formation (Fig. 7). Upon continuous cultivation in the yolk sacs of embryonated chicken eggs, *Coxiella* strains may display phase variation which is associated with significant changes in the structure of their LPS. Phase I microorganisms are infectious for animals, whereas non-pathogenic phase II cells are unable to survive within macrophages. Metabolic and synthetic activities of *C. burnetii* are stimulated at pH 5 or lower. Expression of several enzymes, including superoxide dismutase, catalase, and acid phosphatase, is thought to help *C. burnetii* escape destruction within the phagolysosome. *Coxiella burnetii* acid phosphatase dephosphorylates tyrosine-containing peptide *in vitro,* and it may have a role in regulation of signal transduction events during phagocytosis. In particular, it may inhibit an Fc receptor-mediated signal transduction pathway which triggers the oxidative burst in macrophages.

D. *Bartonella* and Its Environment

Bartonellae are oxidase-negative, aerobic, rod-shaped bacteria. Although this genus was unified based on phylogenetic data, there are significant differences in the phenotypic characteristics of the individual species. Individual species of *Bartonella* also display differences in their metabolic characteristics

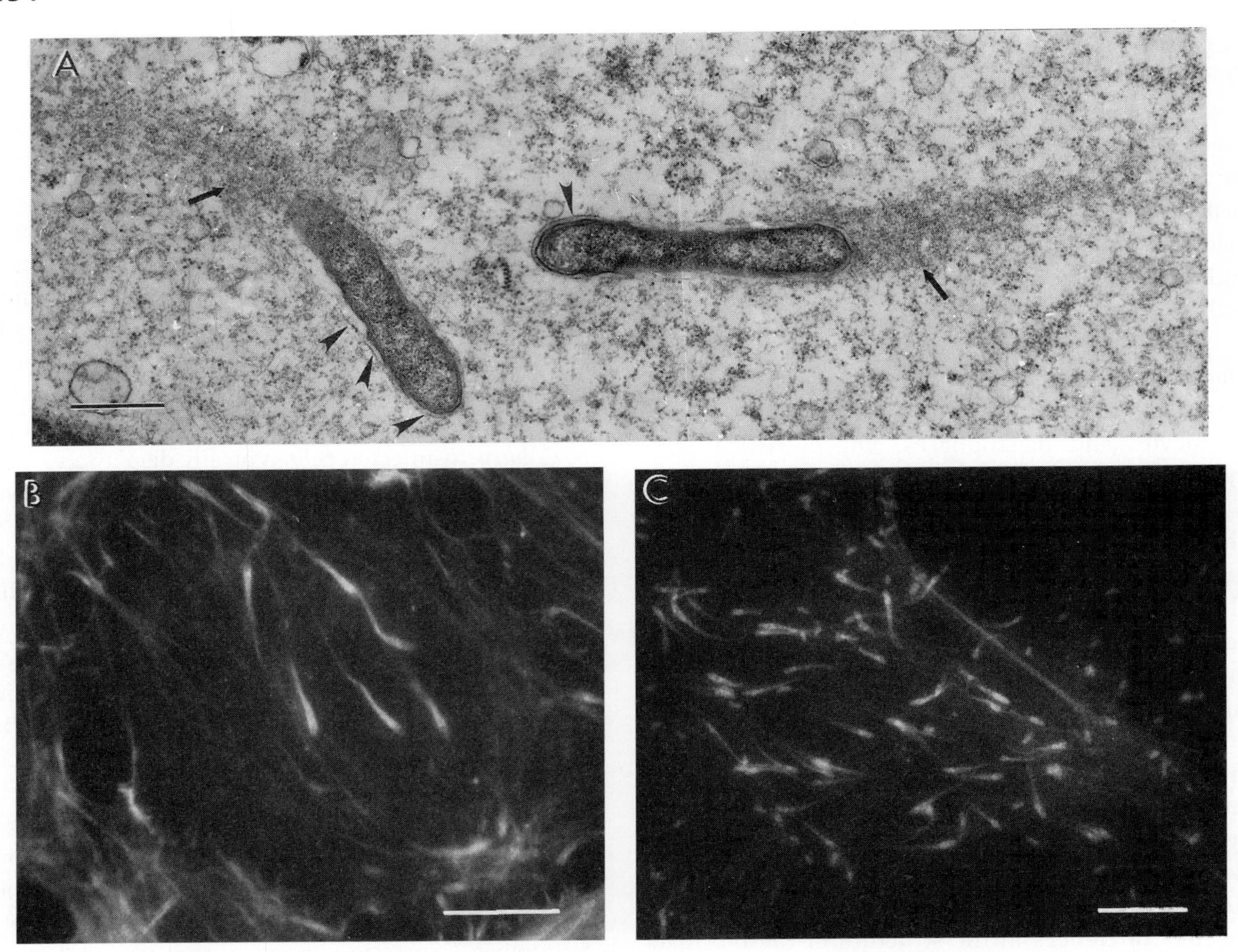

Fig. 3. Electron microscopic observations on interaction of rickettsiae with the actin cytoskeleton of human endothelial cells. (A) Formation of actin tails by *R. rickettsii* Smith strain 15 min after entry into the cytoplasm of human umbilical vein endothelial cells (HUVEC). Actin tails are indicated by arrows. Fragmentation of the phagosomal membrane is indicated by arrowheads. Scale bar = 0.5 μm. (B) Actin tail-associated mobility of *R. rickettsii* Smith strain in the cytoplasm of EA.hy 926 cells 72 hr after infection. Rickettsiae were detected with fluorescein-labeled antibody and actin was stained with rhodamin-labeled phalloidin. Scale bar = 10 μm. (C) Actin tail-associated release of *R. rickettsii* Smith strain from HUVEC 72 hr after infection. Rickettsiae were detected with Texas red-labeled antibody and actin was stained with fluorescein-labeled phalloidin. Scale bar = 10 μm. (D) *Rickettsia prowazekii* Breinl strain surrounded with actin fibers in the cytoplasm of HUVEC 72 hr after infection. Microcapsular S-layer is indicated by arrowheads. A, actin filaments. Scale bar = 1 μm (Figs. 3A–3C reproduced with permission from Eremeeva et al., 2000; Fig. 3D courtesy of D. J. Silverman, University of Maryland at Baltimore, School of Medicine).

and nutritional requirements. *Bartonella bacilliformis* is a flagellated motile bacterium which is known to parasitize erythrocytes. Intracellular multiplication of *B. bacilliformis* results in formation of trenches and excess indentations in the erythrocyte membrane. The interaction of *B. bacilliformis* with the endothelium causes the unusual lesion of verruga peruana, which is due to bacterial protein-mediated induced angiogenesis. *Bartonella quintana, B. henselae,* and *B. vinsonii* are gram-negative bacteria

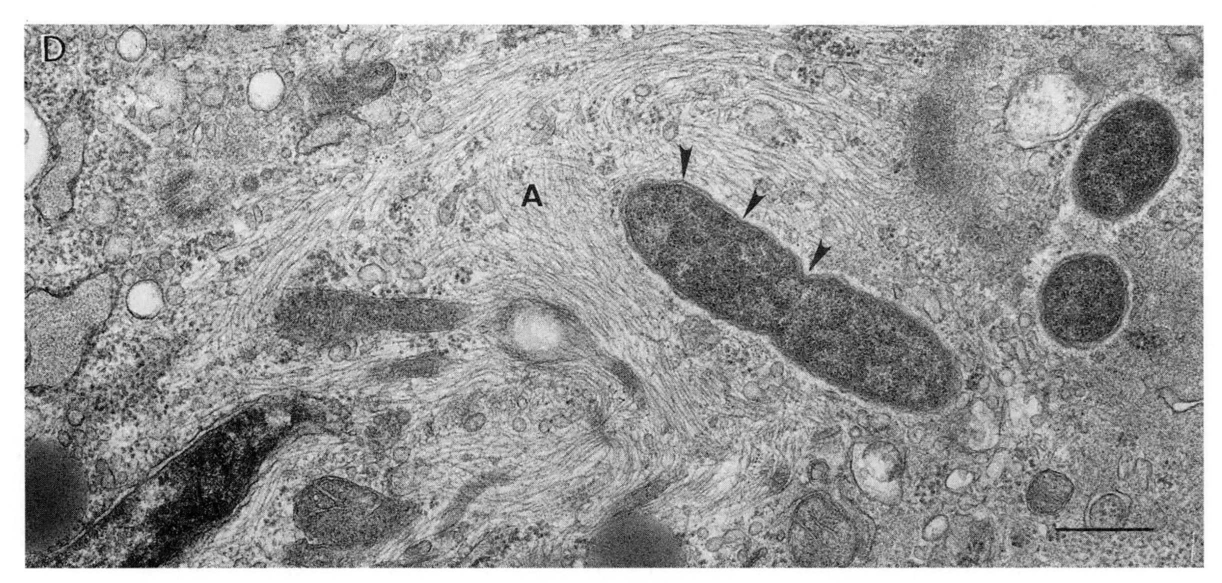

Fig. 3. Continued.

and closely resemble *Rickettsia* spp. in morphology and staining properties, but primarily they grow epicellularly (Fig. 8). However, *B. quintana* also invades cultured human endothelial and human epithelial cells and it exists intracellularly in vacuoles (Fig. 8A). *B. vinsonii* can infect L cells (Fig. 8B). *Bartonella henselae* grows in human epithelial cells and feline erythrocytes and can also invade Vero cells. Cocultivation of *B. quintana* or *B. henselae* with human umbilical vein endothelial cells (HUVEC) resulted in proliferation of HUVEC, probably due to a bacterial angiogenic factor(s). Neither *B. quintana* nor *B. henselae* possess flagella, but they produce a twitching motion when they are suspended in saline and examined microscopically under a coverslip. This motility is related to the adherence mechanism of these microorganisms that is mediated by fine fimbriae similar to type 4 pili. Adherence to, invasion, and lysis of erythrocytes by *B. bacilliformis*, and multiplication of pathogenic *Bartonella* spp. in tissues infiltrated by polynuclear and mononuclear phagocytes, are thought to expose bartonellae to a significant amount of reactive oxygen species. Expression of superoxide dismutase and glucose-6-phosphate dehydrogenase by pathogenic species of *Bartonella* may contribute to their resistance to oxidative stress and survival within the host.

IV. CULTIVATION AND ISOLATION PROCEDURES

A. Cultivation of Established Rickettsial Strains

Embryonated chicken eggs remain the medium of choice for the large-scale propagation of rickettsiae of the genera *Rickettsia, Coxiella,* and *W. persica.* Embryonated eggs, 5–7 days of age, are inoculated via the yolk sac and harvested, depending on the strain and species of rickettsia, 7–12 days after inoculation. SFG rickettsiae kill the embryo early in the course of infection but continue to grow for the next 2 days. In contrast, the maximal titers of typhus rickettsiae are achieved just before embryo death and then they decline rapidly while *Coxiella* titers remain constant. Sucrose glutamate buffer, which consists of 0.22 *M* sucrose, 0.1 *M* potassium phosphate, and 5 m*M* potassium glutamate (pH 7.0), is a medium commonly used to suspend and to freeze harvested

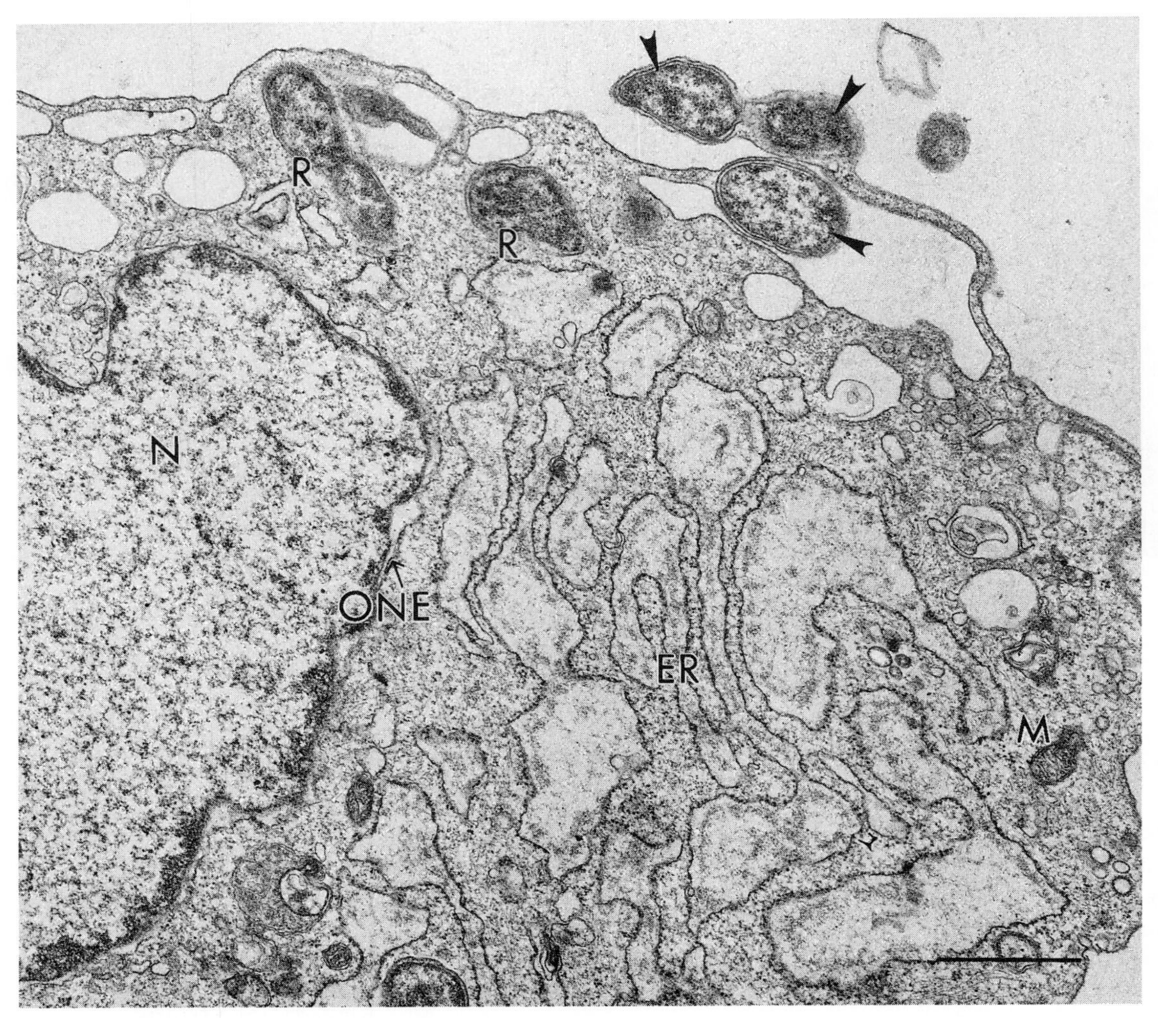

Fig. 4. Release of *O. tsutsugamushi* Karp strain from chicken embryo cells infected for 96 hr. Budding rickettsiae are surrounded by host cell membranes and are indicated by the arrowheads. R, rickettsiae; M, mitochondria; ER, endoplasmic reticulum; ONE, outer nuclear envelope; N, nucleus. Scale bar = 1.0 μm (courtesy of D. J. Silverman, University of Maryland at Baltimore, School of Medicine).

yolk sacs. Seeds frozen at −70°C in this diluent can remain stable for many decades.

1. *Rickettsia, Orientia, and Coxiella*

The growth of rickettsiae in primary chicken fibroblasts or other established permanent cell lines, including Vero green monkey kidney cells, L929 mouse fibroblasts, RAW264.6 mouse macrophage-like cells, and EA.hy 926 human endothelial cells, are common alternatives to the embryonated chicken egg system. However, cultural characteristics of many rickettsiae in cell culture cannot be maintained indefinitely *in vitro,* and occasional passages through the embryonated chicken eggs and in some cases through susceptible animals may be required. Alternatively, using master seed stocks and restricting the number of continuous cell passages are recommended. Growth of rickettsiae in cell culture can be enhanced by irradiation of the host cells prior to infection with rickettsiae, a procedure which arrests cellular division so the slowly growing rickettsiae can increase in concentration per cell. A shell vial technique, in which rickettsia-containing suspensions are inoculated onto cell monolayers for 20 min to 1 hr at room temperature by centrifugation at 1000–2000 rpm prior to cultivation, has also been used to enhance adherence of rickettsiae to the cell monolayers and thereby increase the efficiency of their uptake. Detection of growing bacteria is achieved after differential staining with acridine orange, Giemsa, or Gimenez procedures or with specific antibodies labeled with fluorophores.

Rickettsiae can also be isolated by plaquing on cell monolayers. All species of typhus group rickettsiae, despite differences in their cell culture growth characteristics, exhibit very similar plaque-forming properties and require 7–10 days to produce small plaques of 0.5- to 1.5-mm diameter. Supplementation of agarose with 2 μg/ml of emetine and 40 μg/ml NaF, or dextran sulfate results in formation of more distinct plaques and increases plaquing efficiency of typhus rickettsiae. In contrast, SFG rickettsiae typically escape the host cell and rapidly infect neighboring cells, and form 1- to 4-mm plaques in 4–7 days. *Orientia tsutsugamushi* requires 12–17 days to produce small plaques of 0.4- to 1.2-mm diameter; therefore, a second agarose overlay is typically required. Treatment of the host cells with daunomycin at 400 ng/ml also significantly increases the efficiency of plaque formation by *O. tsutsugamushi*.

Coxiella burnetii infects and multiplies efficiently in many cell cultures that are conventionally used for propagation of rickettsiae, but different cell lines vary in their susceptibility to phase I and phase II microorganisms. Plaquing of *C. burnetii* seeds on such cell monolayers is particularly valuable for the isolation of pure phase I and phase II variant forms.

2. *Ehrlichiae*

Ehrlichiae sennetsu and *E. risticii* are grown in the continuous murine macrophage cell line P388D1. *Ehrlichiae canis* and *E. chaffeensis* are propagated in a hybrid mouse–dog cell line, primary dog monocytes, the canine monocytic cell line 030, and in the dog macrophage cell line DH82. The tick cell line IDE8, derived from embryonated eggs of *Ixodes scapularis,* is highly susceptible to infection with *E. equi*. Growth conditions for HGE in HL-60 human promyelocytic leukemia cells have been established. Granulocytic differentiation of these cells, induced by retinoic acid, potentiates the growth of HGE agent, whereas monocytic differentiation, resulting from treatment with 12-*O*-tetradecanoylphorbol-13-acetate, interferon-γ (IFN-γ), or 1,25-dihydroxyvitamin D_3, correlated with resistance to growth of HGE. The continuous propagation of *E. chaffeensis* in HL-60 cells requires differentiation of the myeloid cells along the monocytic pathway toward phenotypically mature macrophages by the addition of 1,25-dihydroxyvitamin D_3 to the growth medium. *Ehrlichiae chaffeensis* has also been propagated successfully in irradiated or untreated mouse embryo, Vero, BGM, and L929 cells. The growth of *E. chaffeensis* is characterized by a rapid adaptation to each cell line, in which typical morulae can be detected as early as 5 days after inoculation. *Ehrlichiae chaffeensis* induces a characteristic cytopathic effect in mouse embryo, L929, and Vero cells, and the formation of macroscopic plaques has been detected in L929 and mouse embryo cells.

In vitro cultivation of *C. ruminantium* has been established in the calf endothelial cell line E5. *Cowdria ruminantium* organisms are also propagated in bovine pulmonary artery or aortic endothelial cell cultures. As observed with rickettsiae, irradiation of the cell monolayer prior to infection and centrifugation of the inoculum onto the cells facilitated initial growth of the organisms and allowed numerous passages *in vitro*. *In vitro* cultivation of *W. pipientis* was achieved in the continuous cell line, Aa23. The Aa23 cell line was established from eggs of a strain of the Asian tiger mosquito, *Aedes albopictus,* which is naturally infected with *W. pipientis*.

3. *Axenic Growth of Bartonella*

Bartonella spp. are cultivated on enriched agar supplemented with 5–10% rabbit, sheep, or horse blood or in endothelial cell culture systems. Growth is optimal at 35–37°C in the presence of 5% CO_2, except for *B. bacilliformis,* which has an optimal growth temperature of 25–28° C without supplemental CO_2. When growing on blood agar plates, colonies of some *Bartonella* spp. are of two morphologic types: (i) irregular, raised, whitish, rough colonies which are dry in appearance and resemble "cauliflowers" or "molar teeth" and (ii) smaller, circular, tan, and moist-appearing colonies which tend to pit and adhere to the agar. Both types are usually present in

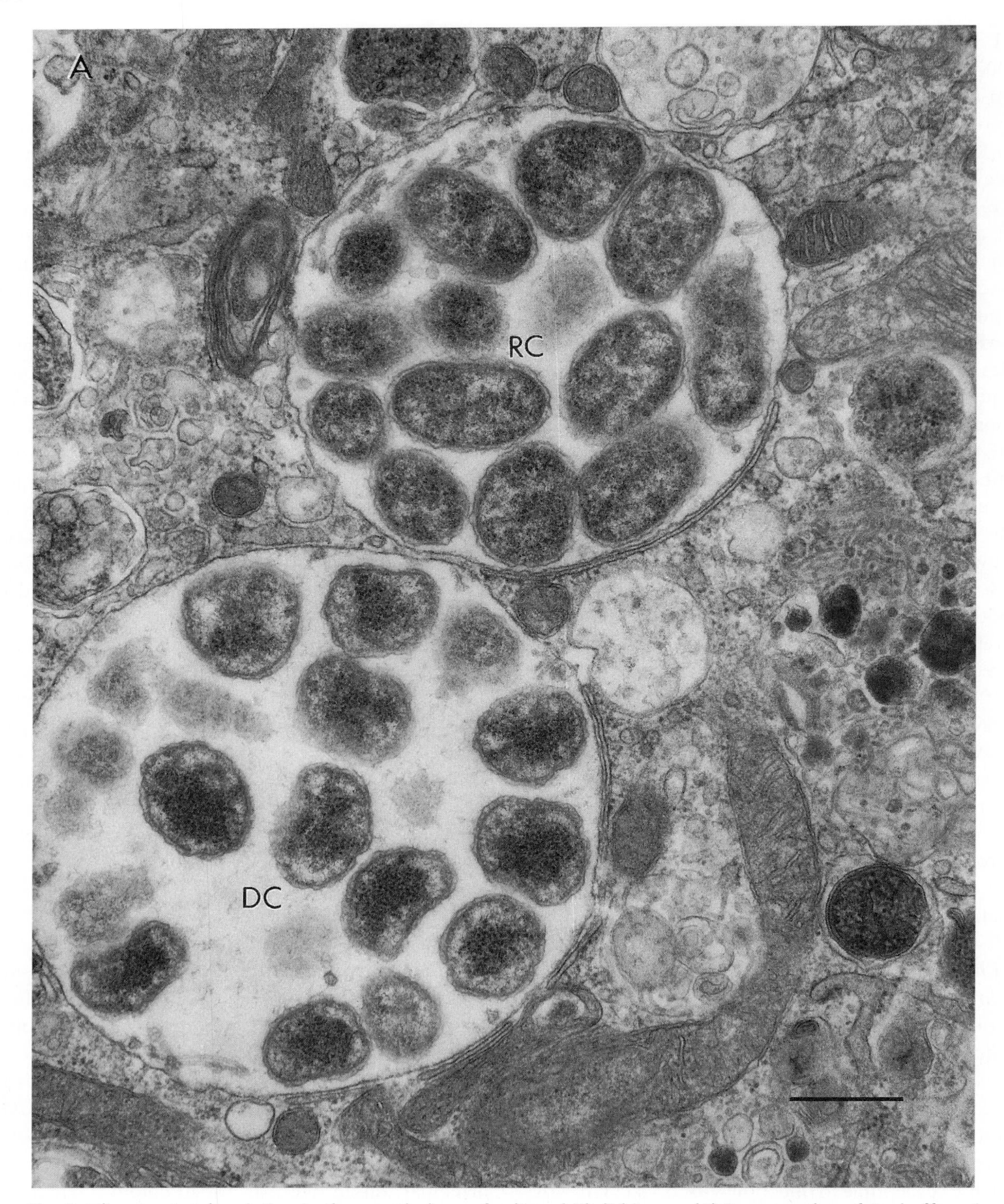

Fig. 5. Ultrastructural variation in the morphology of cultured *Ehrlichia* spp. (A) Two morulae of *E. chaffeensis* Arkansas strain with populations of either reticulate (RC) or dense-cored (DC) cells in the cytoplasm of an irradiated mouse embryo cell. Scale bar = 0.5 μm. (B) Morulae of *E. canis* Oklahoma strain in canine histiocytoma cell line DH82. Scale bar = 0.5 μm. (C) Morulae of HGE (strain No. 54) in HL60 cells. Scale bar = 1.0 μm.

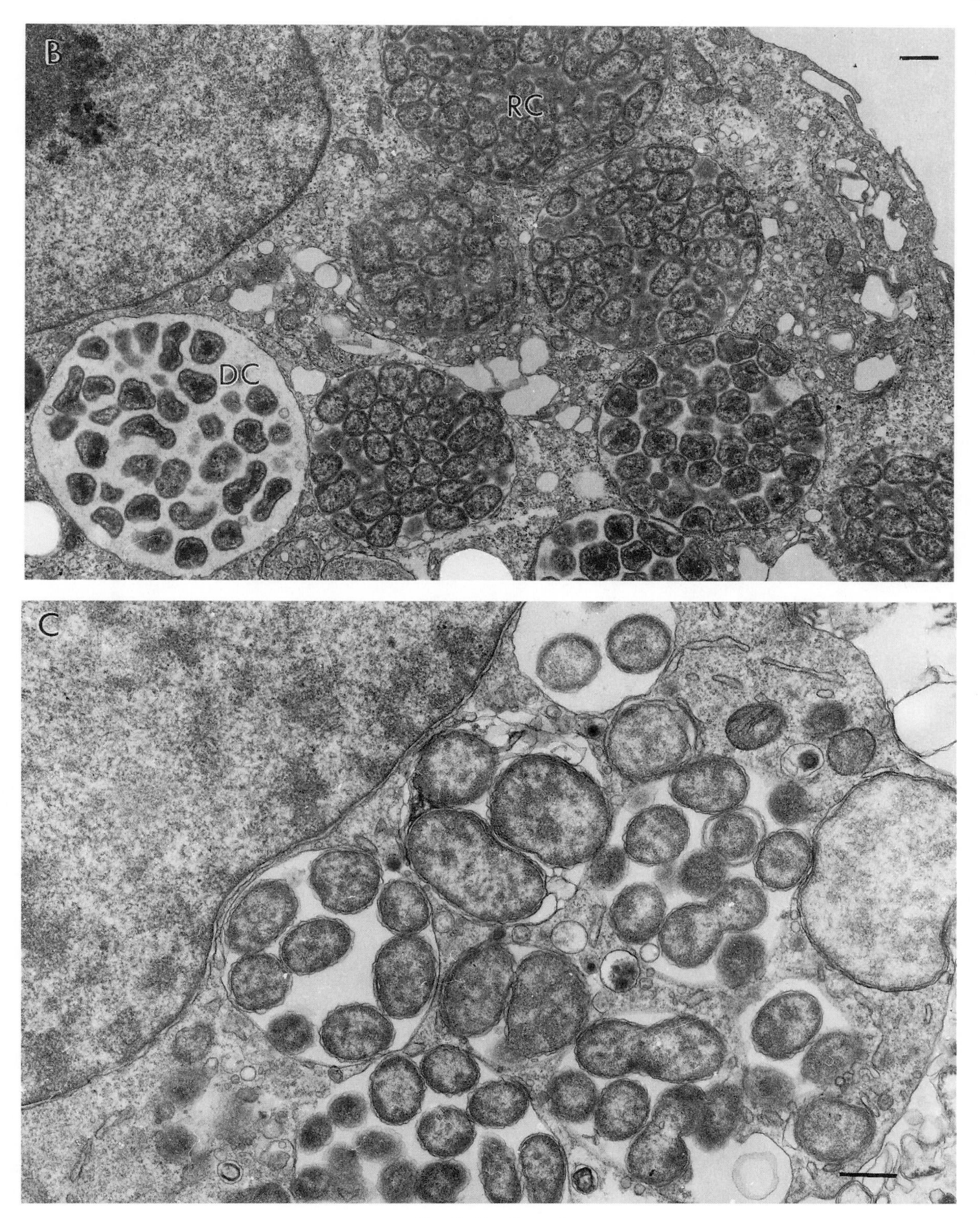

Fig. 5. Continued.

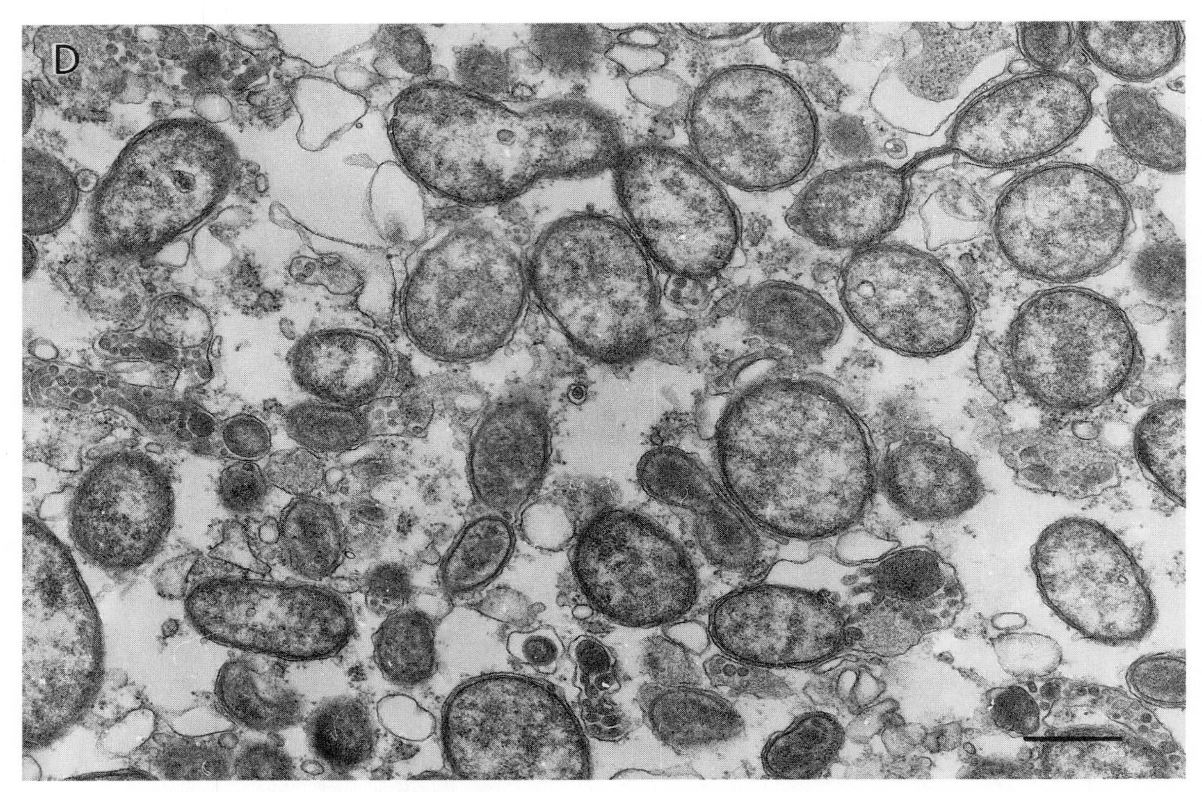

Fig. 5. *(Continued)*. (D) *Ehrlichia sennetsu* Miyayama strain in DH82 cells. Scale bar = 0.5 μm [Fig. 5A reproduced with permission from Popov, V. L., *et al.* (1995). *J. Med. Microbiol.* **43,** 411–421; Figs. 5B–5D reproduced with permission from Popov, V. L., *et al.* (1998). *J. Med. Microbiol.* **47,** 235–251].

the same culture, but their relative frequency varies with species, strain, and passage.

Separation of cultivated intracellular rickettsiae from host cell components is achieved by procedures that involve disruption of the infected host cells, followed by isolation of the rickettsiae by density gradient centrifugation. Gradients of Renografin (Verografin) and Percoll are the most commonly used. *Orientia tsutsugamushi* and *Ehrlichia* are the most difficult to purify because these agents are labile and adhere quite tenaciously to host cell components that have not been characterized.

B. Primary Clinical Isolation and Safety Conditions

Because the majority of rickettsiae are highly infectious and infection can be transmitted by aerosol, the large-scale propagation and purification of rickettsial isolates should be done only in biosafety level (BL-3) laboratories. However, preparation of samples for direct examination or for polymerase chain reaction (PCR) amplification usually does not pose significant risk of aerosol infection and is permissible with standard BL-2 procedures, except for placenta samples thought to contain *Coxiella.* Primary care must be paid to avoiding accidental needle autoinoculation or puncture with "sharps" since single organisms can cause disease following parenteral infection.

Inoculation of the yolk sacs of embryonated chicken eggs or intraperitoneal inoculation of adult male guinea pigs are two classic approaches for initial isolation of typhus and SFG rickettsiae from clinical or field samples, including patient blood, biopsy samples, or arthropod homogenates. However, both methods suffer from disadvantages such as contamination with adventitious bacteria or poor animal susceptibility to strains with low virulence. The shell

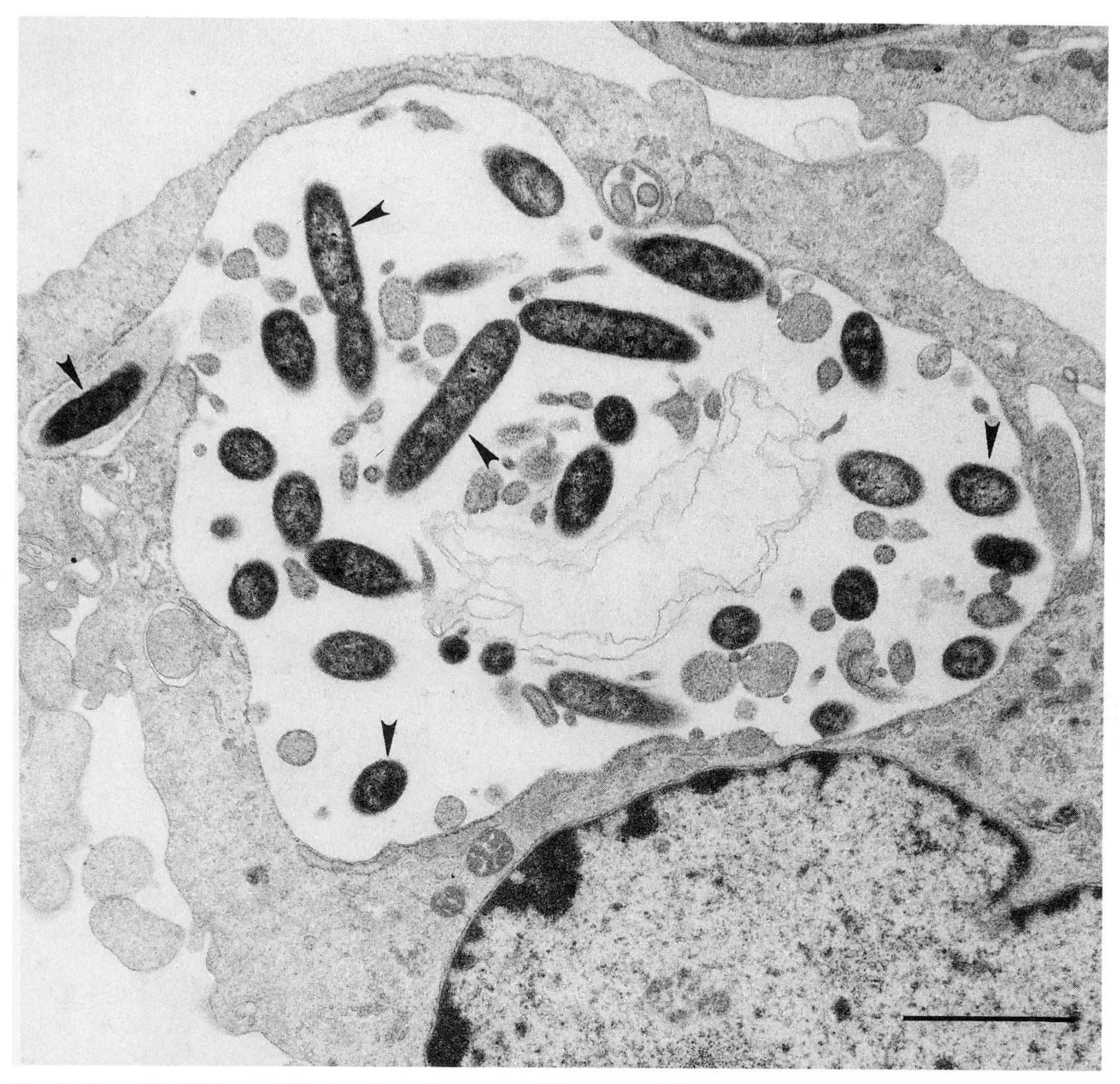

Fig. 6. Growth of *C. burnetii* in the phagolysosome of human umbilical vein endothelial cells. *Coxiella burnetii* are labeled with arrrowheads. Scale bar = 1 μm (courtesy of D. J. Silverman, University of Maryland at Baltimore, School of Medicine).

vial isolation technique is considered to be a powerful alternative since it works well for both virulent and nonvirulent isolates, although some rickettsiae may require stringent conditions for their initial cultivation *in vitro*.

Orientia tsutsugamushi isolates are most easily made from patients by inoculation of clinical samples into mice. However, strains display very different degrees of virulence for animals, and this may affect strain recovery. Alternatively, the shell vial technique with susceptible cell monolayers can be employed or direct culture of monocytes isolated from infected patients.

Primary isolation of *C. burnetii* is achieved by inoculation of specimens onto conventional cell culture monolayers or into embryonated eggs or into guinea

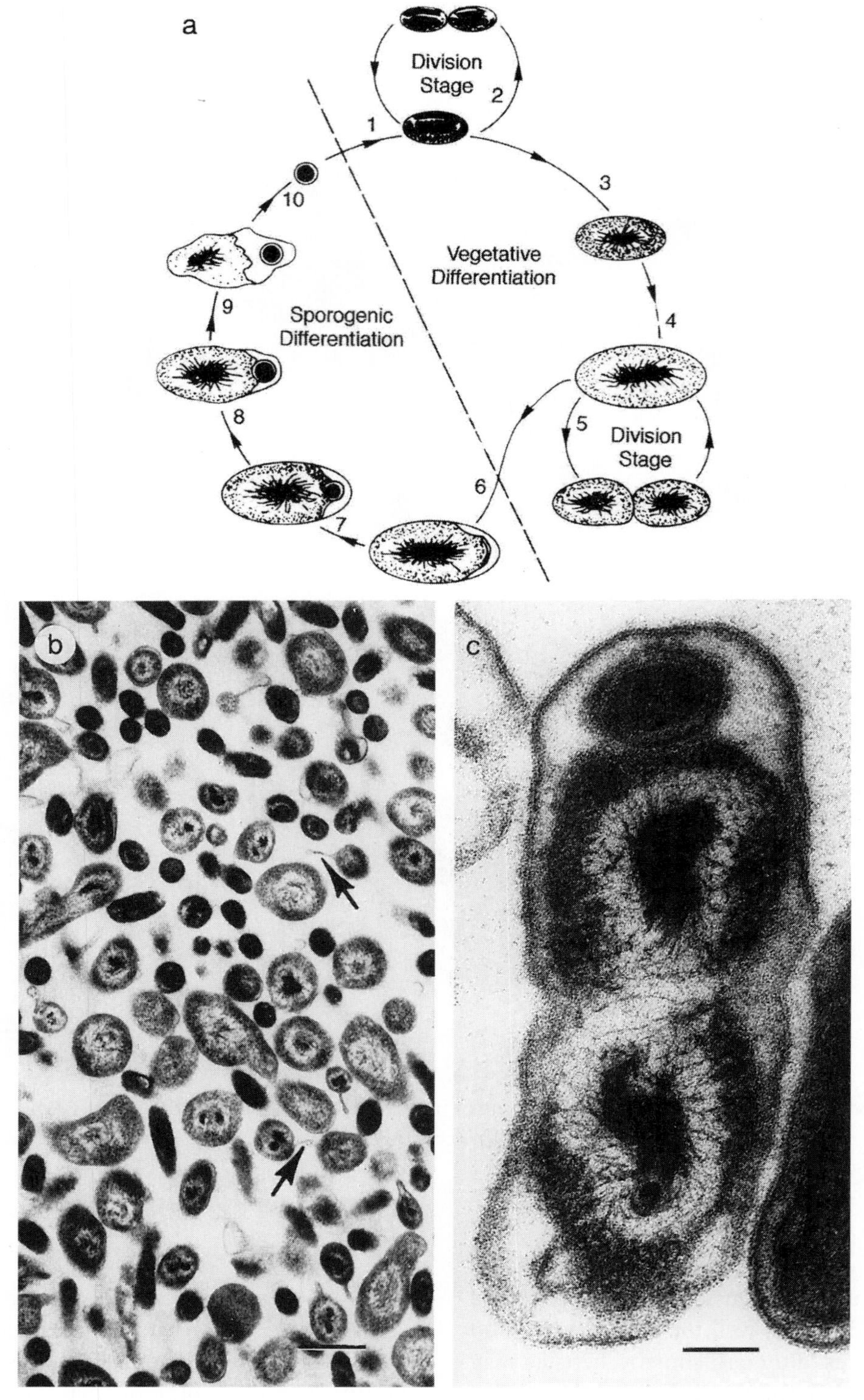

Fig. 7. Developmental cycle of *C. burnetii* within the phagolysosome of eukaryotic cells. (a) Schematic representation of the cycle. (b) Renografin-purified coxiellae cultivated in the yolk sac of chicken embryos, illustrating the diversity of forms seen in such preparations. Note that the outer membranes of some cells appear to form blebs (arrows). Scale bar = 0.6 μm. (c) Complete formation of endospore in a cell concurrently undergoing unequal division. Scale bar = 0.1 μm (reproduced with permission from McCaul and Williams, 1981).

pigs or mice. Passage through animals is used to obtain phase I free of phase II *Coxiella* microorganisms.

Isolation of ehrlichiae is accomplished by inoculation of buffy coat or mononuclear cell fractions of the peripheral blood of infected humans or animals into susceptible cell cultures. Positive isolation provides a definitive diagnosis, although negative results do not necessarily indicate the absence of infection. Moreover, the isolation procedures for ehrlichiae take from 3 days to 1 month, which is quite a long time compared to the rapid course of ehrlichial disease, so isolation is very impractical for clinical purposes.

For primary isolation of *Bartonella* spp., suitable samples include blood, lymph nodes, tissue and aspirates, and biopsy specimens of skin and other organs. The majority of isolates may require up to 3 or 4 weeks of incubation on blood plates before they can be detected. The use of lysis centrifugation or cocultivation of biopsy samples with an endothelial cell line may accelerate and increase the efficiency of isolation of *Bartonella* spp.

V. GENOMIC CHARACTERISTICS

A. Genome Sequence of *R. prowazekii* and Its Phylogenetic Analysis

The complete genome sequence has been determined for the circular chromosome of *R. prowazekii* strain Madrid E. It has 1,111,523 bp and an average G + C content of 29.1 mol%. The genome contains 834 complete open reading frames (ORFs) with an average length of 1005 bp. Protein-coding genes represent 75.4% of genome, and 0.6% of the genome encodes stable RNA. Thirty-three genes encode the 32 different isoacceptor–tRNA species. Each species of rRNA gene—16S, 23S, and 5S—is present as a single copy in the genome and each is located separately from the tRNA genes. The rRNA operons of *R. prowazekii* and the other typhus and SFG rickettsiae are unusual since the 16S rRNA gene and 23S—5S rRNA genes are not contiguous; therefore, this dislocation probably occurred before the divergence of these two groups of rickettsiae.

Functional analysis of the genome of *R. prowazekii* is based on homologies to characterized genes and proteins in other bacteria. It confirms earlier observations on the biology and physiology of rickettsiae that were derived from classical biochemical tests. Genome analysis suggests that only a very small proportion of rickettsial genes are involved in the biosynthesis of amino acids and nucleosides and their regulation when compared to free-living proteobacteria. On the other hand, identified *R. prowazekii* genes have a very great resemblance to the genes found in mitochondria. ATP production in *R. prowazekii* also resembles that of mitochondria; the identified rickettsial genes encode a complete tricarboxylic acid cycle, respiratory-chain complexes, ATP-synthase gene complexes, and the ATP/ADP translocases, whereas the genes encoding proteins which effect anaerobic glycolysis are absent. These results indicate that the functional roles of the genes which are absent in the rickettsial genome were probably replaced by homologous functions encoded by the host cell genome during the evolutionary adaptation of rickettsiae to a strict intracellular lifestyle. As a reflection of this fact, *R. prowazekii* has a reduced set of regulatory genes, particularly two-component signal transduction systems. Furthermore, approximately 25% of the genome of *R. prowazekii* is represented by noncoding sequences, which are thought to have been degraded by mutations but have not yet been lost from the genome. In this setting, essential roles may be inferred for the retained genes of *R. prowazekii* that encode a complete *sec*-dependent secretory system and the bacterial chaperones *dnaK, dnaJ, hslU, hslV, groEL, groES, htpG,* and *htrA.*

B. Comparison of the Genomes of *Rickettsia* and Other Intracellular α-Proteobacteria

DNA sequence analysis of the other typhus and SFG rickettsiae is still fragmentary and for many species has been done only for a few genes, including the 16S and 23S rRNAs, *gltA, rompA, rompB,* and the 17-kDa genus-specific protein antigen. These genes displayed very little sequence divergence between isolates, typically within the range of 0.5–2%. The phylogenetic analysis of these gene sequences confirms that all *Rickettsia* belong to a closely related cluster of microorganisms, suggesting their metabolic properties may be very similar as well.

Before sequence analysis was easily obtained, the genomic information that was garnered about *Rickettsia* reflected the limitations of extant methodology and its applicability to obligate intracellular bacteria. As determined by pulsed-field gel electrophoresis (PFGE), the genome sizes of *R. typhi* and *R. prowazekii* are about 1.1 Mbp, whereas those of 14 species of SFG rickettsiae are about 1.2 or 1.3 Mbp. *Rickettsia massiliae* and *R. helvetica* genomes are larger (1.3 or 1.4 Mbp) than those of the other SFG species, whereas the genome size of the more ancestral species, *R. bellii,* is approximately 1.66 Mbp. The mol% G + C of typhus rickettsiae and *R. bellii* DNA is 29–30 mol%, whereas the DNA of 6 species of SFG is 31–33 mol%.

DNA–DNA hybridization experiments were performed for typhus group rickettsiae and a few isolates of SFG. Seventy to 72% homology exists between *R. prowazekii* and *R. typhi,* whereas their homologies to *R. canada* and *R. rickettsii* are only 44–52 and 36–53%, respectively. Within the SFG group, *R. rickettsii* shares 91–94% DNA homology with *R. conorii,* 70–74% with *R. sibirica* and *R. montana,* 53% with *R. australis,* and 46% with *R. akari*. Homologies for other species have not been determined. Electroporation and transformation of *R. prowazekii* and *R. typhi* have been achieved recently and this will permit genetic manipulation for the first time in *Rickettsia*.

Orientia tsutsugamushi contains a chromosome of 2.4–2.7 Mbp. Whether *Orientia* has a more extensive metabolic repertoire than *Rickettsia* or its genome contains significantly more noncoding or paralogous sequences remains unknown. The G + C contents determined for six prototype isolates varied between 28 and 30.5 mol%.

The genome size of *E. risticii* and *E. canis* is 867–881 kbp and that of *E. chaffeensis* is significantly greater (1163 kbp). *Anaplasma marginale* has a 1.25-Mbp genome and 56 mol% G + C composition. *Cowdria* has a genome size of 1.9 Mbp and an extrachromosomal element migrating at 815 kbp. Two cosmid clones containing 50 kbp of *Cowdria* DNA have been partially characterized. Thirteen ORFs have been deduced, of which 6 encode genes with homologs in other bacteria but only 3 of these have defined functions, suggesting *Cowdria* has novel genes. Two of the ORFs encode putative cytoplasmic ABC binding proteins (for iron and sulfate transport), whereas the third ORF is homologous to recM or recR proteins which regulate RecA protease-mediated SOS response for DNA recombination and repair. Genomic characteristics of other *Ehrlichia* species are not known. Only a few metabolic characteristics have been evaluated and only for *E. risticii* and *E. sennetsu*. They cannot utilize glucose-

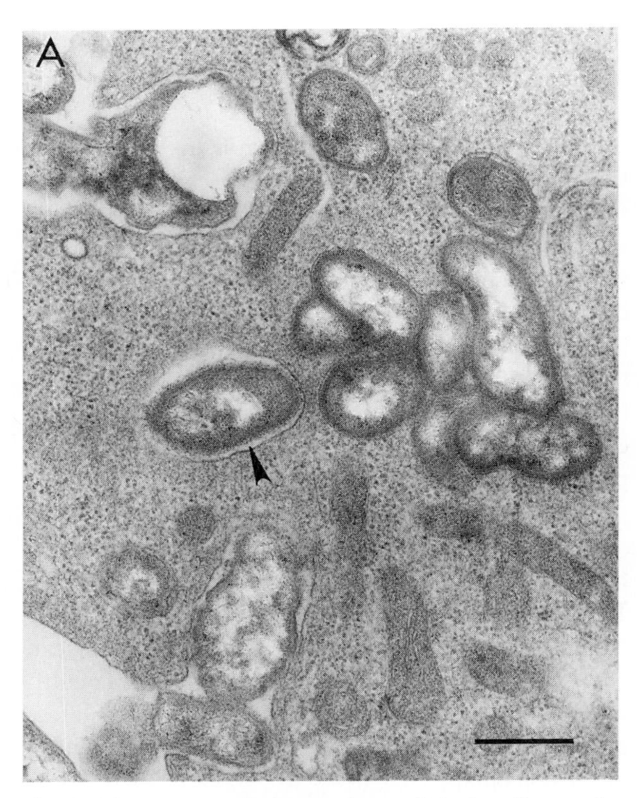

Fig. 8. Ultrastructural observations of *Bartonella* spp. interactions with eukaryotic cells. (A) Entry of *B. quintana* Oklahoma strain into ECV 304.9 endothelial cell line 1 hr after inoculation. Intracellular bacteria in the vacuole are indicated by the arrowhead. Scale bar = 0.5 μm (reproduced with permission from Brouqui and Raoult, 1996). (B) Growth of *B. vinsonii* in irradiated mouse LM_3 cell line 52 hr after infection. Epicellular and some intracellular bacteria, apparently in the vacuoles, are indicated by arrowheads and arrows, respectively (reproduced with permission from Merrell *et al.,* 1978).

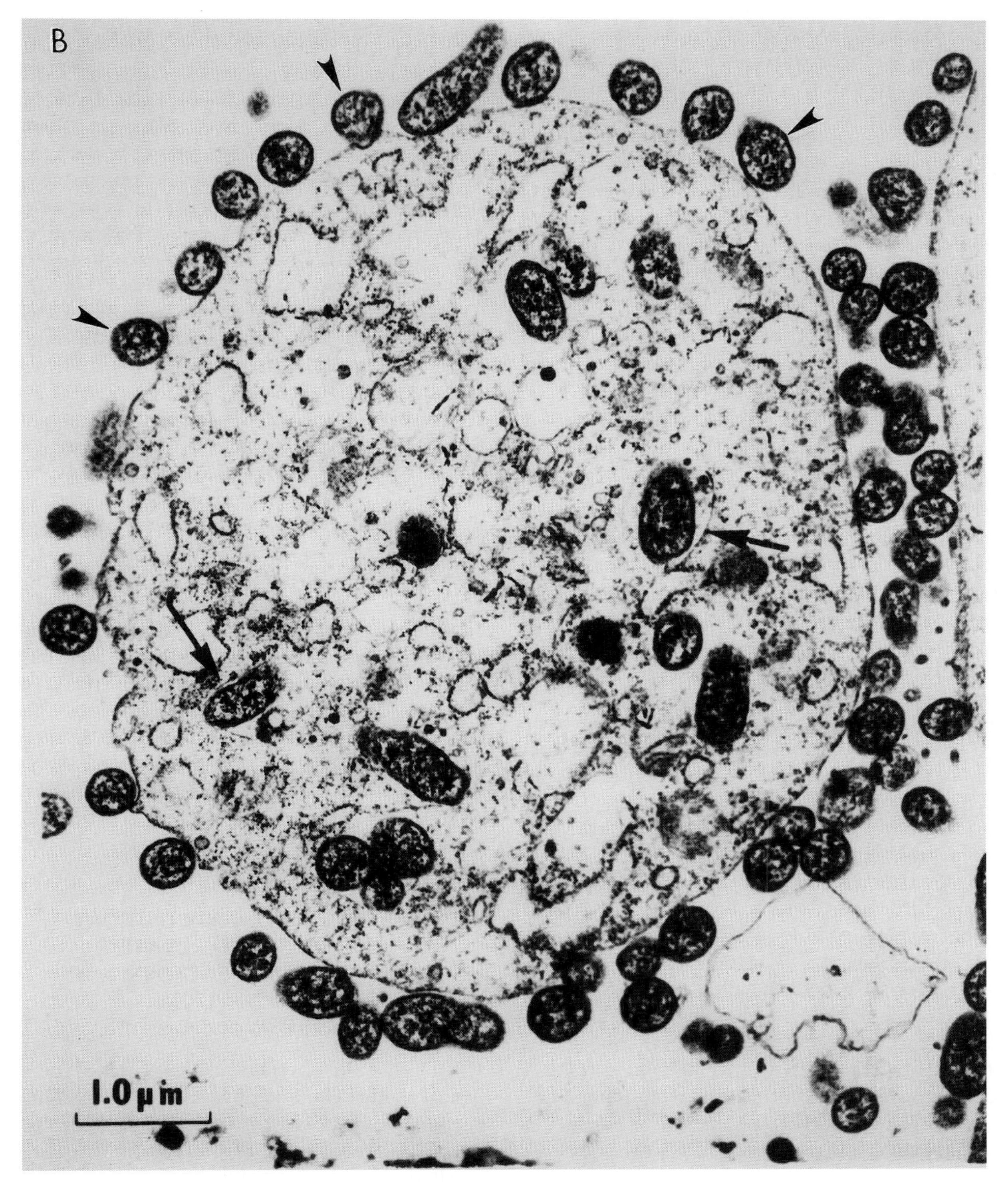

Fig. 8. Continued.

6-phosphate and glucose and rely on oxidative phosphorylation for ATP production. *Ehrlichia risticii* and *E. chaffeensis* are sensitive to desferoxamine, an intracellular iron chelator, suggesting that efficient acquisition of iron is as essential for these bacteria as it is for other pathogens. Both *E. risticii* and *E. chaffeensis* inclusions accumulate transferrin receptor and also upregulate transferrin receptor mRNA expression in their host cells, whereas HGE does not, thus suggesting that HGE uses a different strategy for acquisition of iron than utilizing iron acquired from transferrin.

C. Genome of *Coxiella*

Coxiella burnetii isolates exhibit 20 distinct restriction fragment-length polymorphism (RFLP) patterns of chromosomal DNA identified by PFGE, and their genome sizes vary from 1.4 to 2.4 Mbp. The DNA base composition of *Coxiella* is 42–43 mol% G + C. Four different types of plasmids of 33.5–56 kbp have been identified which vary with the *Coxiella* strain and comprise approximately 2% of their genomic information; some isolates contain chromosomally integrated DNA homologous to the plasmid sequences. At least eight genomic groups of *C. burnetii* have been distinguished based on plasmid type and chromosome characteristics from PFGE. A putative correlation between plasmid type and the type of disease caused by *C. burnetii* has been refuted. Nineteen copies of a repetitive element, IS*1111*, resembling a bacterial transposon have been identified within the Nine Mile strain genome. A physical linkage map of the *C. burnetii* Nine Mile isolate has been constructed and the positions for 54 genes, including *oriC*, have been identified. The mapping studies suggested that the *C. burnetii* chromosome may be linear, but additional studies are required to confirm these results and to explain the mechanism of replication. The organization of housekeeping enzyme genes in *C. burnetii* is similar to that found in *E. coli*, but only a single copy of the 16S rRNA gene was found. Genetic transformation of *C. burnetii* by electroporation has been demonstrated.

D. Genome of *Bartonella*

The *B. bacilliformis* genome is a circular 1.6-Mbp DNA molecule for which a physical map of the chromosome with the locations of characterized genes has been constructed. The genome sequence of *B. henselae* is being determined. The genome size of *B. quintana*, *B. henselae*, *B. vinsonii*, and *B. elizabethae* varies between 1700 and 2174 kbp. These four species have 49–71% DNA homology to each other but their homology to *B. bacilliformis* varied from 32 to 45%. The DNA homology of *Bartonella* species with *Rickettsia* is less than 33%. The G + C DNA content for most *Bartonella* species ranges from 38 to 41 mol%. About 98% homology among 16S rRNA gene sequences exists in *Bartonella* species, whereas values for the 5S rRNA genes were 91–93% and for citrate synthase 89–92%. Bacteriophage particles have been identified in both *B. bacilliformis* and *B. henselae* (Fig. 9). The particles consist of three major proteins and 14-kbp fragments of double-stranded DNA that are packaged in a near-random manner reminiscent of generalized transducing phage. The packaging of chromosomal DNA from bacteria and export of the resulting particles into the cell culture medium raises the possibility of genetic exchange among the species of *Bartonella*. Electroporation procedures have been developed for genetic transformation of *B. bacilliformis* and *B. quintana*. Finally, conjugal plasmid transfer from *Escherichia coli* to *B. henselae* has been reported, thus extending the range of genetic manipulation applicable to *Bartonella*.

VI. CELL WALL COMPOSITION, ANTIGENS, AND PUTATIVE VIRULENCE FACTORS

A. *Rickettsia* and *Orientia*

The cell walls of typhus and SFG rickettsiae are typical of other gram-negative bacteria. They contain peptidoglycan, LPS, a cytoplasmic membrane with numerous proteins, and an outer membrane with a much simpler protein composition and some proteins present in high concentrations. The rickettsiae also contain a microcapsular protein layer, which is

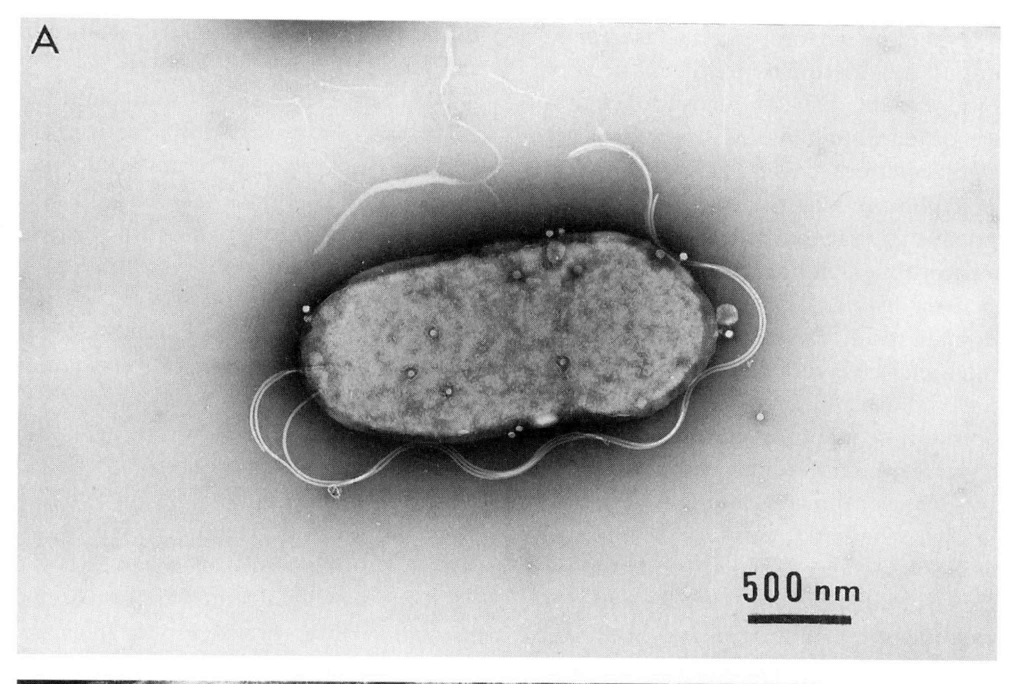

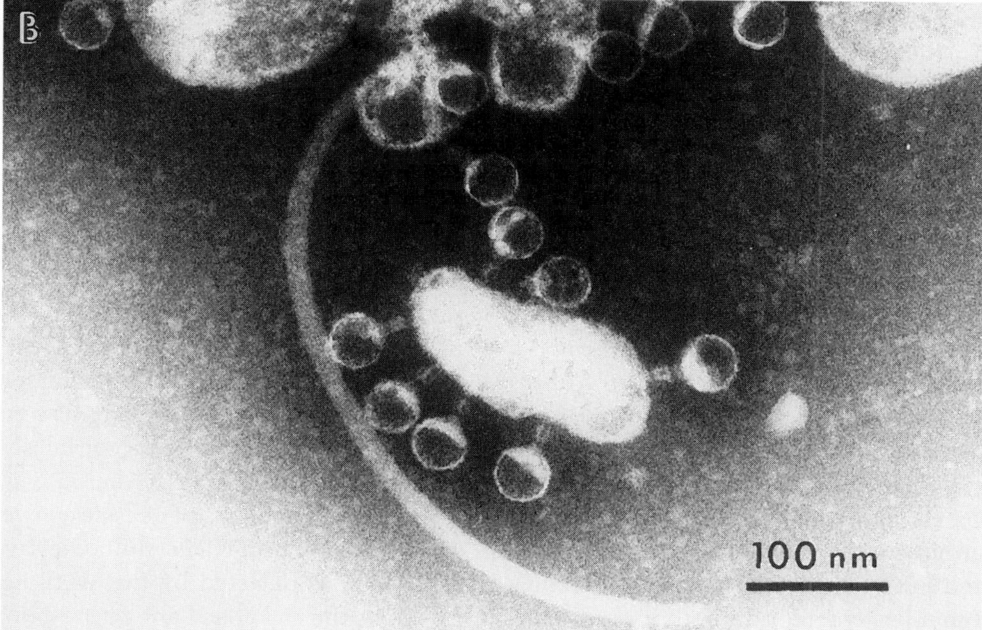

Fig. 9. Phage in *B. bacilliformis.* (A) Phage-like particles on a *Bartonella* cell. (B) Particles attached to cell fragments. Note that the majority of phage heads are empty (reproduced with permission from Umemori *et al.,* 1992).

often called a crystalline surface protein layer or S-layer. It is composed of one or two multimeric proteins, rOmpB and rOmpA, also known as the species-specific surface protein antigens (SPAs), which are arrayed in a regular periodic layer. rOmpB of *R. prowazekii* and *R. typhi* contain two and one cysteine residues, respectively, that are oxidized and form disulfide-linked multimers. rOmpB is processed from a 168-kDa precursor by cleavage of a conserved carboxypeptide domain resulting in the amino-terminal-encoded 120-kDa surface layer protein and a 32-kDa β protein which contains a membrane-anchoring domain. The β protein sequence is highly conserved among rompB genes from different species. A homologous sequence is present in some rOmpA proteins, but it is not known whether rOmpA genes also undergo carboxy cleavage. One peculiarity of rOmpA is the presence of tandemly repeated 72- and 75-amino acid sequences which vary in number and order among different SFG isolates. rOmpA and rOmpB are very rich in β sheet structure and are heat-labile antigens. rOmpA and rOmpB contain strain, species, group-specific, and group cross-reactive epitopes. Some monoclonal antibodies also recognize domains that contain methylated lysines on rOmpB of typhus group rickettsiae. These proteins are the dominant protein antigens which elicit antirickettsial responses in humans and experimental animals. They have been used as protective subunit vaccines for typhus and SFG rickettsiae.

Another well-characterized rickettsial protein is the genus-specific 17-kDa lipoprotein. *Rickettsia* also possesses several widely conserved heat shock proteins, among which DnaK and GroEL share epitopes found widely in other bacteria, whereas DnaJ, GrpE, and GroES possess more genus-specific epitopes.

LPS is the group-specific antigen which differentiates typhus and SFG rickettsiae. However, *R. bellii* LPS contains epitopes common to both groups and some additional heterogeneity exists in the LPS within each group. The LPS of both groups share epitopes with both the *Proteus* OX19 (primarily typhus rickettsiae) and *Proteus* OX2 (SFG stronger) Weil–Felix antigens and with *Legionella bozemanii* and *L. micdadei* LPS that are detected by human IgM early in the course of rickettsial infection. This IgM cross-reactivity may lead to misdiagnosis, whereas human IgG antibodies are more specific for rickettsial LPS.

Both the cell wall structure and antigens of *O. tsutsugamushi* are very different from those of typhus and SFG rickettsiae. The microcapsule, peptidoglycan, and LPS found in *Rickettsia* have not been detected in *Orientia. Orientia tsutsugamushi* lacks the constituent components of peptidoglycan and LPS, such as muramic acid, hydroxy fatty acids, and 2-keto-3-deoxyoctonic acid (KDO). Possibly reflecting the lack of peptidoglycan, *O. tsutsugamushi* is particularly osmotically fragile, it is sensitive to digestion with phospholipase A_2, and its growth is more resistant to penicillin than that of other rickettisae. Electron-lucent halo zones ("slime layer") are also not observed around growing *O. tsutsugamushi* cells as seen with *Rickettsia*. Isolates of *O. tsutsugamushi* exhibit extensive antigenic variation and the genotypes present in different geographic areas are unique. The major immunogenic surface protein, designated 56-kDa antigen, differs between serotypes of *Orientia*, exhibits considerable size variation from 52 to 62 kDa, and contains serotype-specific epitopes. Other antigenic proteins in the range of 22 to 110 kDa are also located on the cell surface. The 22- and 110-kDa *Orientia*-specific proteins are also type-specific antigens, whereas the 47-, 60-, and 70-kDa protein antigens are more conserved and are homologous to the HtrA, GroEL, and DnaK heat shock proteins, respectively. However, the DnaJ and GroES proteins of *Orientia* do not exhibit cross-reactivity with antisera against the corresponding homologs from *E. coli*. The 56-kDa proteins elicit very early IgM and IgG serologic responses but convalescent sera also contain antibodies against the 72-, 60-, 47-, and 22- to 35-kDa proteins of *O. tsutsugamushi*. Scrub typhus infection also elicits antibodies against *Proteus* OXK LPS, as detected by the Weil–Felix reaction; however, the rickettsial antigen responsible for this reaction has not been identified.

B. *Ehrlichiae*

By electron microscopy the cell envelopes of ehrlichiae appear similar to other gram-negative bacteria, but no peptidoglycan or LPS have been described for these species. Little biochemical information has

TABLE IV
Sequenced Homologous Antigen Genes and Related Protein Antigens of Ehrlichiae[a]

Genogroup and species	*70-kDa HSP*	*10-kDa HSP*	*60-kDa HSP*	*msp1α-like repeat antigen*	*map1 family 28–44 kDa*	*msp3 80–90 kDA*	*msp5 21 kDa*	*Unrelated antigen genes*
Group I								
E. canis		*groES*	*groEL*		*28 kDa*(p30)*			
E. chaffeensis		*groES*	*groEL*	*120 kDa*	*omp-1* (p28)*			
E. ewingii			GroEL		P28	MSP3		
E. muris								
C. ruminantium		*groES*	*groEL*		*map1*?*		*map2*	
Group II								
E. phagocytophila		*groES*	*groEL*		P44			
E. equi		*groES*	*groEL*		P44	MSP3		
HGE agent	*dnaK*	*groES*	*groEL*	*100 kDa, 130 kDa*	*hge-44* (p44*)*			*160 kDa* ankyrin
E. platys								
"*E. ovina*"					MAP1			
A. marginale				*msp1α* (70–105 kDa)	*msp2*, msp4**	*msp3**	*msp5*	*msp1β**
Group III								
E. sennetsu	*dnaK*	*groES*	*groEL*					
E. risticii	DnaK	*groES*	*groEL*			MSP3		*51 kDa, 80/45 kDa*
N. helminthoeca	DnaK		GroEL					
SF agent								
Group IV								
W. pipientis		*groES*	*groEL*		*wsp*			

[a] Asterisks indicate polymorphic multigene family demonstrated; HSP, heat shock protein. Genes are indicated in italics and related proteins are capitalized.

been obtained on ehrlichial antigens because only small amounts of cells can be grown and purified, host cell proteins contaminate the purified preparations, and it is difficult to characterize membrane proteins in a complex mixture of related molecules. Furthermore, differential expression of some proteins may occur during growth in the morula as part of a poorly understood developmental cycle or during cycling of ehrlichiae between invertebrate and vertebrate hosts. The presence of proteins damaged by the host immune system or degradation may also complicate antigen characterization. However, many ehrlichial genes which encode antigens present in the outer membrane have been cloned, sequenced, and overexpressed as recombinant proteins (Table IV), and their physical and immunochemical properties are becoming clearer. Sequence information has provided rationales for the observed immunological cross-reactivity between ehrlichial species and the binding of monoclonal antibodies to multiple proteins which differ in molecular weight within a single isolate. Homologous antigens are present in different species and some antigens are expressed from complex polymorphic gene families that are found in multiple copies in the genome and sometimes tandemly organized. The expression of these genes can also change, presumably in response to attacks by the immune system on specific surface antigens. Shifts in surface antigen expression have clearly been demonstrated for *A. marginale*. This type of phase variation may be a general property of ehrlichial species.

Numerous surface antigens of ehrlichiae were initially characterized by Western blotting with polyclonal sera from naturally infected or immunized animals and by radioimmunoprecipitation of iodine-labeled surface proteins. As the antigen and epitope specificities of monoclonal antibodies have been identified and monospecific sera for individual cloned protein antigens have been obtained, the complexity of cross-reactions between and within species of ehrlichiae has become more understandable but it has not been simplified.

DnaK (HSP70), GroES (HSP10), and GroEL (HSP60), the well-known molecular chaperonins and presumptive heat shock proteins, have been cloned and characterized from many ehrlichiae. Relatively small (1.2- to 2.5-fold) increases in levels of *E. sennetsu* mRNA for GroEL and DnaK can be induced by temperature upshifts of infected cells, suggesting that they may be important in ehrlichial stress responses. DnaK genes have been sequenced from both *E. sennetsu* and the HGE agent; they are most closely related to the DnaK genes of *R. prowazekii* and other α-Proteobacteria. Ehrlichial DnaK appears to be relatively conserved since both *E. sennetsu* and *E. risticii* DnaK react strongly to antibody against *E. coli* DnaK and human antibodies against *B. burgdorferi* can cross-react with the HGE agent DnaK. However, *E. sennetsu* and *E. risticii* DnaK do not react to a monoclonal antibody against *E. coli* DnaK. Immunoblots of most ehrlichial species contain bands at 70–80 kDa, but other unrelated antigens of this weight besides DnaK may also be present. Considerably more is known about the *groESL* operon of ehrlichiae since it has been sequenced from nearly all species (Table IV). The phylogenetic relationships of ehrlichiae deduced from comparison of GroES and GroEL sequences are identical to those deduced from their 16S rRNA sequences, thus confirming the reliability of these groupings. *groES* and *groEL* are contiguous genes in all ehrlichial species and appear to be present in single copies. *Ehrlichiae sennetsu* GroES reacted weakly with rabbit antibodies against *R. typhi* GroES but neither it or *E. risticii* GroES reacted with antibodies to *E. coli* GroES. Monoclonal antibodies against *E. risticii* and *E. sennetsu* GroEL did not react with either *Rickettsia* or *Bartonella* GroEL, although some broadly reactive monoclonal antibodies against the latter species react with these ehrlichiae whereas others do not. GroEL of both ehrlichiae also reacted with rabbit anti-*E. coli* GroEL. Polyclonal serum against *E. sennetsu* GroEL reacted broadly against GroEL of *Rickettsia, Orientia,* and genogroups I–III of ehrlichiae. Mice infected with HGE agent but not those infected with *B. burgdorferi* exhibit strong antibody responses against purified HGE recombinant Gst–GroEL fusion protein. Similarly, anti-GroEL antibodies from cases of human monocytic ehrlichiosis, heartwater, and Potomac horse fever are readily detected by immunoblotting against their homologous antigens. Because broadly reactive, genus- and species-specific epitopes are all present in GroEL, it is unclear whether it is a reliable antigen for serodiagnosis of ehrlichioses or whether immune responses directed against domains of ehrlichial GroEL that are conserved with host GroEL proteins may contribute to the pathogenesis of ehrlichioses.

For species in genogroup III, only surface antigens from *E. risticii* have been characterized extensively with monoclonal antibodies or by cloning and sequencing their genes. Many *E. risticii* antigens cross-react with those of *E. sennetsu* by Western blotting, but *Neorickettsia* antigens appear to be only weakly cross-reactive with these species, except for GroEL and DnaK. Eighteen protein antigens were initially identified in *E. risticii*, and 9 were prominent surface antigens. Considerable heterogeneity exists in the molecular weights of antigens from different isolates of *E. risticii*. The major 51-kDa antigen gene exhibited more variability than the *groESL* or 16S rRNA genes and it could be used to demonstrate geographic clustering of isolates. Another strain-specific antigen (SSA) varies widely in molecular weight (48–85 kDa) between isolates and contains varying numbers of diverse types of tandem repeats interspersed between eight identical common domains. Additional domains are unique to different sizes of SSA. These antigens appear to be protective antigens and their variability may account for strain-dependent vaccine failures. To date, neither the 51-kDa nor SSA antigens have sequence homology with other cloned ehrlichial antigens.

Extensive work accomplished on the antigens of the important veterinary pathogen, *A. marginale*, presaged many recent discoveries about the antigens of the other three genogroups of ehrlichial species. These discoveries suggest that all arthropod-associated ehrlichias are closely related and that other shared antigens may exist. Purified *A. marginale* initial bodies have nine major surface protein (MSP) antigens. Genes encoding MSP1–MSP5 have been characterized (Table IV).

MSP1 exists as a polymorphic dimer and consists of proteins MSP1a and MSP1b that are both disulfide and noncovalently associated. Immunization with MSP1 elicits partial protection to *Anaplasma*. In different isolates MSP1a varies greatly in size (70–105 kDa) depending on the number (two to eight) of tandem and nearly identical 28- or 29-amino acid repeat sequences that are present. A monoclonal antibody which binds to a hexapeptide sequence in the hydrophilic repeat region of MSP1a can neutralize many *Anaplasma* isolates *in vitro*. MSP1a is expressed from a single genomic copy and is a highly conserved helical transmembrane protein except for the variation in the number of repeats. Three or four nonidentical homologs of the 97- to 100-kDa MSP1b protein gene exist per genome and these differ by extensive deletions, insertions, and switching of sequences. *msp1β* is conserved in size and homology in different isolates, but these genes also differ in sequence by RFLP analysis. MSP1b contains regions of tandemly repeated sequences and glutamine-rich regions at the N and C termini, but only the repeat sequences exhibit any homology to MSP1a. MSP1a and MSP1b both lack a signal sequence, migrate at a much greater molecular weight than is predicted from their gene sequence, and function as an adhesin for bovine erythrocytes when expressed in *E. coli*.

Genes with physical similarity but insignificant sequence homology to MSP1a have been described recently in *E. chaffeensis* and the HGE agent. *Ehrlichiae chaffeensis* 120-kDa genes have been sequenced from two isolates. Sapulpa isolate lacks 65 bp of DNA found in a region of the Arkansas isolate gene which contains four near-identical 240-bp tandem repeats that comprise 60% of the gene sequence. The genes are otherwise nearly identical. Similarly, other isolates of *E. chaffeensis* have either 3 or 4 repeats by PCR analysis. A recombinant antigen composed of 2 repeats of the 120-kDa antigen has been used successfully in a dot blot assay for human monocytic ehrlichiosis. In HGE agent different tandem repeat domains were present in genes encoding 100- and 130-kDa protein antigens (3 repeats of approximately 100 amino acids and 8 repeats of 24–34 amino acids, respectively). These antigens each contain small but different domains which are each homologous to different portions of the *E. chaffeensis* repeat region, but the 100- and 120-kDa antigens exhibit little homology to each other and do not share common epitopes. Like MAP1a, the repeat regions of these proteins are highly hydrophilic and may act as surface adhesins; these antigens also exhibit substantially higher apparent molecular weights by electrophoresis than expected from their DNA sequences. HGE agent also contains a larger unrelated 160-kDa protein antigen which has eight 33-amino acid long ankyrin-like domains and both 27- and 11-amino acid repeats that increase its apparent molecular weight. Ankyrin domains are found in proteins of diverse phyla and mediate protein–protein interactions, including those between cell membrane and cytoskeletal proteins.

Anaplasma marginale MSP2 and MSP3 are expressed from polymorphic, multigene families. MSP4 is highly homologous to MSP2 and both proteins are related to immunodominant antigens found in other ehrlichial species (Table IV). MSP2 and MSP3 are major antigens detected on immunoblots with sera from infected cattle, whereas anti-MSP4 responses are variable. MSP2 and MSP3 can be resolved as multiple spots by two-dimensional gel electrophoresis at 33–41 and 80–90 kDA, respectively. Immunization with purified native MSP2 elicits protective immunity. However, MSP2 varies structurally and antigenically among isolates. Even though MSP2 expression is stable in an isolate during tissue culture passage, new types appear to be induced following passage through ticks and during persistent cyclic rickettsemia in animals. Whether these polymorphic MSP2 variants arise from differential expression of the multiple polymorphic genes or generation of mutants with insertions, deletions, and point mutations is not resolved. Native MSP2 can form disulfide-linked tetramers on the cell surface. Organisms may express more than one MSP2 gene and these may form heteromers, thus further increasing the antigenic diversity of *Anaplasma*. MSP2 variation occurs primarily in a central 120-amino acid region. Seven to 10 genomic copies of *msp2*, comprising more than 1% of the genome, are distributed throughout the chromosome. MSP4 is encoded by a single gene and is highly conserved among isolates of *A. marginale* and *A. centrale* but not *A. ovis*. Native MSP2 and native and recombinant MSP4 each induce statistically significant protection but this is variable on an individual animal basis. MSP4 is attractive as a vaccine candidate because it does not exhibit the variability of MSP2, but this monomeric protein may not be adequately immunogenic. MSP3 is a monomeric,

80- to 90-kDa protein which is also encoded by multicopy genes. Depending on the isolate, 7–15 genomic copies of *msp3*, comprising more than 3% of the genome, are distributed throughout the chromosome. MSP3 homologs are probably present in *E. risticii, E. equi,* and *E. ewingii,* but they have not been cloned. *msp3* homologs of *A. marginale* exhibited complex sequence rearrangements and MSP3 antigens can differ significantly among isolates. Surprisingly, protein variant MSP3-12 also contains a 121-amino acid fragment with homology to MSP2. *Anaplasma ovis* also has polymorphic, multigene *msp2* and *msp3* families and can cause both repetitive cycles of rickettsemia and relatively constant and persistent rickettsemia.

In genogroup I ehrlichiae, polymorphic, multigene families homologous to the Msp2/4 gene family have been identified in *Cowdria* (*map1*), *E. canis* (*p30* and *28 kDa*), and *E. chaffeensis* (*p28* and *omp-1*). Although only a fragment of a second *map1* coding region has been described in the Senegal strain, unlike *msp2* it was initially believed to exist as a single copy. The *p28* and *p30* genes can also exist in tandem and all do not appear to be expressed. MAP1 from different isolates differ by 0.6–14% and contain three variable regions. Because of the cross-reactivity of MAP1 with sera against other ehrlichiae, a recombinant fragment MAP1-B has been used for specific diagnosis of heartwater. Partial protection against *Cowdria* infection in mice was recently demonstrated with a DNA vaccine containing MAP1. At least four copies of *p30* are found in *E. canis* and two were found in tandem. The 30-kDa antigen is immunodominant and purified recombinant rP30 has been proposed as a diagnostic antigen, although it shares epitopes with *E. chaffeensis* 28-kDa antigen. In *E. chaffeensis* six copies of the *p28* (*omp-1*) gene are tandemly arrayed but additional copies are probably present elsewhere in the chromosome. Recombinant P28 accelerated the clearance of *E. chaffeensis* from mice. Twenty-eight- to 30-kDa MAP1 homologs are known from different isolates of *E. canis* and *E. chaffeensis*. They exhibit differences in their size and reactivity to monoclonal antibodies. They appear to differ in three variable regions like those found in MAP1. The immunodominant antigens of *E. ewingii, E. equi, E. phagocytophila,* and HGE agent consist of one or more proteins between 43 and 49 kDa which exhibit minimal cross-reactivity with group I and III ehrlichiae. This 44-kDa family also belongs to the MSP2 family of genes. Distinct isolates of HGE agent express remarkable pleomorphism in this major outer membrane protein antigen as well as variable reactions to monoclonal antibodies and to immune serum from patients. Multiple *p44* genes with partial sequence similarity are present both tandemly and dispersed in the chromosome. Mice could be protected against illness by passive immunization with anti-P44 monoclonal antibodies. Recombinant P44 has been used for serodiagnosis in a dot immunoblot assay. *Wolbachia pipientis* also contains a 24-kDa polymorphic outer membrane protein gene *wsp* that belongs to the *msp2* antigen family. It appears to be present in a single copy and exhibits greater variability than the 16S rRNA, *ftsZ,* and *groESL* genes used previously for strain typing and phylogenetic analysis. It has been used to devise a new classification system for these diverse isolates. Phylogenetic trees derived from ehrlichial *msp2* gene family sequences are completely in agreement with groupings deduced from the *groESL* and 16S rRNA gene sequences.

The 19-kDa antigen MSP5 is present in all species of *Anaplasma,* is encoded by a single genomic copy, and has a putative signal sequence. An ELISA based on competition of sera with a monoclonal antibody which reacts with purified recombinant MSP5 (rMSP5) exhibits excellent specificity and sensitivity for detecting *Anaplasma* infections in cattle. rMSP5 has also been very suitable for diagnosis by Western blotting. MSP5 forms a noncovalently linked complex with MSP1 and forms multimers under nonreducing conditions. Protective immunity was elicited against an outer membrane fraction that contained MSP5 and this immunity correlated with antibody titers against MSP5, MSP2, and MSP3, but purified MSP5 was not protective. Antibodies against MSP5 from protected animals recognize conformation-dependent epitopes that require internal disulfide bonding in MSP5. MAP2 is a 21-kDa protein that is conserved in all strains of *Cowdria ruminantium* and that is homologous to MSP5. Recombinant MAP2 is recognized by sera and immune $\gamma\delta$ T cells from infected animals. Although MSP5 and MAP2 homologs have not been identified in other ehrlichiae, two SCO_2 homologs from *R. prowazekii* and LipA from *Rhizobium etli* which may be involved in c-type cytochrome biogenesis, and, less closely, the genus-specific 17-kDa antigens of *Rickettsia* exhibit homology to these antigens.

C. *Coxiella*

Except for its conserved heat shock proteins, *C. burnetii* is antigenically unrelated to the other rickettsiae, as may be expected from its phylogenetic distance. In particular, its antigenic properties reflect the existence of different morphological forms during its developmental cycle, its phase variation in virulence and associated differences in LPS structure, and the limited solubility of its cell envelope components.

Differences in the chemical composition of phase I and phase II LPS of *Coxiella* are compatible with a smooth to rough transition where the O side chain polysaccharide undergoes significant truncation while the core polysaccharide region which contains KDO and lipid A remains intact. The major antigenic sugars in the O chain of *C. burnetii* LPS are the neutral sugars virenose and dihydrohydroxystreptose, which are not unique to *Coxiella.* The branched-chain fatty acid composition of LPS is unique to *Coxiella.* In nature and laboratory animals *Coxiella* exists in the phase I state. In experimentally infected animals the first antibody elicited reacts with phase II LPS antigen, whereas later antibody reacts against phase I LPS. In man, the presence of high antibody titers ($>$1:800) against phase I antigen are suggestive of chronic Q fever because phase II antibodies are predominant during acute infection. Phase I and II *C. burnetii* cells do not exhibit ultrastructural differences, although they differ in their buoyant density in cesium chloride and in their affinity for basic dyes.

Various protein antigens have been identified by immunoprecipitation of surface radioiodinated proteins and by immunoblotting. The best characterized is a 29-kDa protein P1, which is present in phase I and II cells but not in the small, dense cell type or endogenous spore. It is both immunogenic and protective for animals. P1 may be identical to the cloned 27-kDa protein Com1, or there are several antigenic proteins of similar size since a 29-kDa lipoprotein has also been described in *C. burnetii.* Macrophage infectivity potentiator (Mip) is another immunoreactive protein of *C. burnetii* which is similar to proteins involved in macrophage infection by *Legionella pneumophila* and *Chlamydia trachomatis.* Mip, which seems to be an exported protein of *C. burnetii,* may be excreted into the host phagolysosomal vacuole and perhaps into the host cytoplasm. It has peptidyl-prolyl isomerase activity that is inhibited by rapamycin. Although the function of *C. burnetii* Mip is not understood, its homologs in prokaryotic and eukaryotic cells appear to play a role in modulation of T cells resulting in the downregulation of IL-2 and IFN-γ synthesis. A 67-kDa outer membrane protein is protective in guinea pigs and mice. The conserved 62-kDa protein of *Coxiella* is homologous to the 60-kDa chaperonin GroEL. Its expression in *C. burnetii* is regulated by heat or 10% ethanol in acid-activated *Coxiella.* It is present on the surface of phase II but not phase I cells. The 62-kDa protein is immunogenic and is recognized by sera from acute and chronic Q fever patients. The protein product of *mucZ* was shown to induce capsule synthesis in association with the DnaK–chaperone system and RcsB–RcsC two-component regulatory system. Synthesis of the 20–kDa histone-like homolog Hq1 and a 30-amino acid basic peptide ScvA was demonstrated in the small-cell variant. Their degradation has been suggested to be a part of the regulatory mechanisms operative during morphogenesis from SCV to large-cell variant. An ORF for a protein involved in chromosomal and plasmid replication, Roa307, has also been identified on plasmid QpH1.

D. *Bartonella*

Antigenic proteins varying in size from 11.2 to 160 kDa have been identified in *B. bacilliformis.* Fourteen proteins ranging from 11.2 to 75.3 kDa are localized in the outer membrane, and 11 of these can be surface radiolabeled. Twelve antigens of *B. bacilliformis* ranging in size from 16 to 160 kDa are reactive with human sera from patients suffering from Carrion's disease. The major antigen named Bb65 is homologous to the GroEL family of heat shock proteins. A 67-kDa hydrophobic protein, called deformin, is associated with the vanadate- and Ca^{2+}-sensitive deformation of erythrocyte membranes; its native form probably consists of two identical subunits. A 42-kDa protein component of the flagella of *B. bacilliformis* has been identified. Antibodies raised to this protein block the association with and invasion of human erythrocytes by *B. bacilliformis*. Two other proteins, IalA and IalB with molecular masses of 21 and 18 kDa, respectively, may also mediate *B. bacilliformis* invasion into erythrocytes. These proteins are encoded by two closely linked genes within the so-called invasion-associated locus, which confers the ability to invade human erythrocytes on minimally invasive strains of *E. coli*.

The protein composition of other *Bartonella* spp. is less well characterized, although protein patterns differ by sodium dodecyl sulfate–polyacrylamide gel electrophoresis (SDS–PAGE) for *B. quintana, B. henselae, B. vinsonii,* and *B. elizabethae*. A 17-kDa protein from *B. henselae* is highly reactive with human sera from patients with cat scratch disease, but its role in the pathogenesis or immunity to infection with *B. henselae* is not known. A similar protein is also found in *B. quintana*. An immunogenic homolog of the HtrA stress response protein which confers protection against oxidative injury has also been identified; its gene has been cloned and sequenced. DnaK and heat-inducible GroEL homologs were also detected in *B. quintana, B. henselae,* and *B. vinsonii*. DnaK and GroEL from these three *Bartonella* species contain epitopes which cross-react with anti-*E. coli* DnaK and anti-*E. coli* GroEL antibodies. In contrast, GroES homologs from these *Bartonella* species share epitopes with *Rickettsia* GroES but do not react with anti-*E. coli* GroES polyclonal antiserum.

VII. IDENTIFICATION

Direct identification of rickettsiae is most often accomplished with specific differential staining procedures and by microimmunofluorescence (MIF) or immunoperoxidase assays with specific antibodies. Rickettsiae are gram-negative bacteria, but they are not stained particularly well by Gram stain. The method of Gimenez provides better and relatively specific staining of typhus and SFG rickettsiae, *Coxiella,* and *Bartonella* isolates. It is based on the selective retention of carbol fuchsin by rickettsiae, which are seen as bright red, slender coccobacillary forms against a pale greenish blue background that is produced by the counterstain, typically malachite green. *Orientia tsutsugamushi* and ehrlichiae stain more satisfactorily with Giemsa stain. Acridine orange staining visualized by ultraviolet (fluorescent) microscopy, based on the binding of this dye to RNA, provides advantages for differentiation of physiologically active microorganisms. The viable microorganisms fluoresce a bright orange to red color, whereas dead microbes appear green like the cytoplasm of their host cells.

A. *Rickettsia* and *Orientia*

PCR-RFLP analysis of the rOmpA, rOmpB, 17-kDa protein, and citrate synthase genes, RFLP analysis of chromosomal DNAs, and direct sequencing of PCR amplicons have become the primary means for rapid characterization of new isolates of typhus and SFG rickettsiae. Scrub typhus isolates are best typed by PCR-RFLP methods employing the conserved GroEL operon and the more variable 22- and 56-kDa antigen genes. PCR assays have proven to be quite sensitive for clinical use, requiring a minimum of about 10 rickettsiae in an assay sample for a positive result. Samples from animals, arthropod vectors, blood, and biopsies can be used as sources of DNA for PCR. To increase the reliability and specificity of PCR amplification with low numbers of template copies, specific nested PCR procedures have been employed. 16S rDNA sequences also can be employed for agent detection, but they have relatively low variability so they are not very useful for distinguishing among genotypes. Western blotting and SDS–PAGE analyses are often employed to distinguish closely related isolates of rickettsiae, particularly when specific polyclonal mouse typing sera or monoclonal antibodies are available. However, these procedures require cultivation of the agent and its subsequent purification for best results. Specific monoclonal antibodies are available for many typhus, SFG, and *Orientia* isolates and they have been used in blocking assays, capture assays, MIF, and Western blotting procedures to identify rickettsiae.

B. *Coxiella*

Different sets of primers are available for amplifying DNA of *C. burnetii,* including the 16S rRNA gene, the 23S rRNA gene, the 16S–23S rRNA internal transcribed sequence, plasmid sequences, the *htp*AB-associated repetitive element sequence, 27-kDa Com1 protein gene, and superoxide dismutase genes. To increase the sensitivity of conventional PCR, nested assays have also been designed. Moreover, detection of PCR-amplified product using ELISA-based technology has also permitted estimation of the number of microorganisms that are present.

Serological identification using a panel of specific monoclonal antibodies is a highly reliable technique for identifying individual *C. burnetii* isolates. Detection of antibodies is performed by indirect MIF, ELISA, Western blotting, and immunoperoxidase techniques. Immunohistochemical methods have been used to detect *C. burnetii* in paraffin-embedded tissues. *Coxiella burnetii* can also be detected in hematocytes of infected ticks after classical direct staining or following detection by immunochemical methods.

C. *Ehrlichia*

Detection of *Ehrlichia* spp. in the acute febrile phase of disease is often based on examination of peripheral blood smears or buffy coat preparations stained by Romanowsky-type techniques (Giemsa, Wright, or Diff-Quik). Microorganisms appear as round, dark purple stained dots or clusters of dots or as morulae in the cytoplasm of leukocytes. Immunohistological techniques can be performed on peripheral blood smears, formalin-fixed paraffin-embedded bone marrow biopsy specimens, or aspirated marrow. Detection is achieved using biotinylated anti-*Ehrlichia* spp. antibodies and avidin–alkaline phosphatase or avidin–horseradish peroxidase systems. PCR identification is based on primer sets derived from 16S rRNA gene sequences and subsequent sequencing or RFLP analysis of amplified fragments. Nested PCR assays have improved the sensitivity of DNA detection in clinical and entomological samples. PCR primers derived from the nucleotide sequence of the GroEL operon have provided an alternative to 16S rRNA genes for the species-specific identification of ehrlichiae. Amplification of the repeat region from the 120-kDa protein gene is particularly useful for detecting different genetic types of *E. chaffeensis* and other antigen genes have been used for studying other species (See Section VI,B). Repetitive-element PCR (rep-PCR) has also been introduced to distinguish *Ehrlichia* isolates at the species and strain levels. In this technique, intergenic spacers are recognized with tRNA-directed primers and amplified by PCR yielding species-specific fingerprint patterns of bands from 50 to 1000 bp. Serological diagnosis of ehrlichial infections is most commonly accomplished by the indirect MIF test.

D. *Bartonella*

Although identification of *Bartonella* spp. to the genus level is relatively straightforward, identification to the species level by biochemical tests is more difficult. Available identification procedures include biochemical tests, fatty acid, genotypic analyses, immunochemical staining and histopathological examination, and serological procedures. In the Microscan rapid anaerobe panel in which the test medium is supplemented with 100 μg/ml of hemin, effective differentiation of *Bartonella* spp., including *B. bacilliformis, B. quintana, B. henselae,* and *B. vinsonii,* has been obtained. Cell wall fatty acid analysis (FAA) is performed on isolates growing on blood agar. Although FAA detects differences between *Bartonella* spp., these are not very reliable markers for confident differentiation of species. Intraperitoneal inoculation of animals with whole killed *Bartonella* cells elicits strong antibody responses. The use of a panel of such polyclonal sera in immunofluorescence tests, ELISA, or immunoblotting allows for species-specific differentiation of *Bartonella* isolates. Serological procedures, primarily MIF, are the most practical means of confirming current or prior infections in animals and humans.

Many genotypic procedures are available for species-specific identification of *Bartonella* isolates. Of these, DNA–DNA hybridization has the greatest theoretical basis, but it is also the least practical since it requires access to all reference strains and is technically demanding. PCR amplification of DNA, including 16S rRNA or citrate synthase genes, and intergenic spacer regions, followed by RFLP analysis or sequencing of the amplified fragments, rapidly provides reproducible and characteristic data for each species. In particular, amplification of the 16S–23S rRNA intergenic region fragment and its digestion with *Alu*I and *Hae*III readily distinguishes *B. quintana, B. vinsonii, B. bacilliformis*, and two subtypes of *B. henselae*. Digestion of the 16S rRNA gene segment with *Dde*I and *Mnl*I, or a *gltA* segment with *Taq*I and *Aci*I, allows reliable and reproducible differentiation of the species now known to cause disease. PCR-based identification has also been used for the identification and subtyping of individual isolates of *Bartonella* and for the initial characterization of microorganisms that are refractory to culturing. *Bartonella* isolates can also be differentiated by PFGE after digestion of chromosomal DNA with the restriction enzymes *Sma*I, *Eag*I, *Hind*III, *Hae*III, or *Tag*I.

VIII. SUSCEPTIBILITY TO ANTIMICROBIAL AGENTS

A. Physical Stability and Chemical Inactivation

Rickettsiae and ehrlichiae have stringent requirements for maintaining their physiology outside of their host cells. High ionic-strength buffers or buffers supplemented with sucrose to provide osmomolarity are essential for long-term maintenance of isolated microorganisms. Even in media designed to mimic the internal host cell environment, metabolic activity or infectivity cannot be preserved longer than a few days. However, the stability of *Rickettsia* in dried arthropod feces or in properly lyophilized cultures is well-known. Chemical disinfectants and 60°C treatment are highly effective for their inactivation.

Orientia tsutsugamushi is more sensitive to chemical disinfectants and temperature changes and more labile than typhus group rickettsia under physiological conditions, whereas *R. rickettsii* and *R. conorii* can continue to divide after death of the host cell. The stability of ehrlichiae after host death may be comparable to that of SFG rickettsiae. DMSO (10%) is used for cryopreservation of viable blood stabilate vaccines used for immunization against *Anaplasma* and *Cowdria*.

Relative to all other rickettsiae, *C. burnetii* is notorious for its resistance to elevated temperatures, desiccation, osmotic shock, ultraviolet light, and chemical disinfectants. Consequently, contamination of food, buildings, and pastures with even a very small number of microorganisms (1–10) has led to human and/or animal infections. To disinfect suspensions of *C. burnetii*, 70% ethanol, 5% chloroform, and 5% EnviroChem are applied. Alcide, sodium hypochlorite, benzalkonium chloride, Lysol, and 5% formalin are not effective. Surface-applied *C. burnetii* is also difficult to inactivate under low-humidity conditions, even with paraformaldehyde gas or ethylene oxide sterilization.

B. Sensitivity to Antibiotics

The susceptibility of rickettsiae to antibiotics and their suitability for treatment of patients must be evaluated first in experimental models of infection, which include cell culture, embryonated chicken eggs, and animals. When positive effects are demonstrated with a new drug, clinical trials are necessary to further evaluate its value for prophylaxis and treatment.

Cell culture assays commonly measure the inhibition of rickettsial plaque formation by antibiotic compounds. The plaque assay allows the determination of the minimum inhibitory concentration (MIC) and also of the rate of killing of rickettsiae by a single antibiotic concentration by measuring the reduction in number of plaques with respect to time of exposure of infected cells to the antibiotic. Alternatively, a colorimetric assay in 96-well plates may be used for testing the antibiotic susceptibility of rickettsiae. In this assay, cells infected with rickettsiae for 4 days are stained with neutral red or methylthiazol tetrazolium dye, and changes in host cell cytopathology are evaluated spectrophotometrically by measuring the optical density of supernatants from treated and untreated infected cells. Finally, a variation of a shell vial procedure has been used for antibiotic susceptibility tests based on examination of treated and untreated infected cultures after direct staining or detection of rickettsiae with fluorescent antibodies. To measure the effect of antibiotics in embryonated eggs, rickettsiae are inoculated into the yolk sac with and without antibiotics and the difference in embryo mean survival time is determined. Subculture of yolk sacs from surviving infected treated embryos permits evaluation of the rickettsiacidal or rickettsiastatic activity of antibiotics. Reduction of mortality and the duration of fever in susceptible animals inoculated intraperitoneally with viable rickettsiae are taken as measures of antibiotic effectiveness. Although the animal model tests antirickettsial activity of the drug in combination with specific and nonspecific defense mechanisms, the reliability of animal models for predicting human responses is uncertain for many drugs because of differences in their physiology.

Although tetracyclines, chloramphenicol, rifampicin, and fluoroquinolones could all be used effectively for treating infections with *Rickettsia,* adverse effects may limit their administration to patients. Treatment of pregnant women and young children is the most problematic since the tetracyclines are toxic, and the use of chloramphenicol exposes these patients to the risk of aplasia.

Susceptibilities of the rickettsiae to antibiotics are summarized in Table V. *Rickettsia prowazekii* and *R. typhi* are susceptible to all macrolides tested to date. *Rickettsia canada* and SFG rickettsiae are the most susceptible to josamycin but exhibit more resistance to other macrolides. Several tick-restricted SFG rickettsiae are more resistant to rifampin compared with typhus and other SFG rickettsiae. β-Lactams, aminoglycosides, and cotrimoxazole are not effective at all.

In addition to its well-established resistance to penicillin, susceptibility of *O. tsutsugamushi* to antibiotics has not been examined extensively. Chloramphenicol and tetracyclines are active when assayed in eggs and mice and effective in treating scrub typhus in humans. Rifampicin, ciprofloxacin, doxycycline, and erythromycin are all potentially useful drugs by *in vitro* testing in cell cultures or mice. Different susceptibilities to antibiotics of distinct isolates of *O. tsutsugamushi* are documented. Possible antibiotic resistance has been described in clinical cases of scrub typhus occurring in some endemic areas of northern Thailand. Numerous genetic types of *Orientia* exist in this region, as does a previously undescribed infected chigger species, *L. chiangraiensis.*

Coxiella burnetii is resistant to treatment with chloramphenicol, erythromycin, and streptomycin in animals and chicken eggs. In eggs, rifampicin, trimethoprim, doxycycline, and oxytetracycline are bacteriostatic. In cell culture, the antibiotic susceptibility of *C. burnetii* and reported MIC may vary significantly with different isolates, the cell line employed, and whether an acute or chronic model of infection was used. In HEL cell culture, cotrimaxazole, rifampicin, doxycycline, tetracycline, monocycline, sparfloxacin, and the quinolones PD127,391 and PD131,628 are bacteriostatic. Ofloxacin, pefloxacin, chloramphenicol, fusidic acid, and erythromycin are variably noneffective. The fluoroquinolones ciprofloxacin, pefloxacin, and fleroxacin are bacteriostatic when tested in eggs and Vero cells. In the presence of lysosomotropic agents used to alkalinize the *C. burnetii*-containing phagolysosome, pefloxacin and doxycycline exerted higher bactericidal effects than rifampicin. Clarithromycin exhibited good bacteriostatic activity against four genomic types of *C. burnetii,* but its bactericidal activity was not evaluated. Since acute Q fever is generally a self-limiting disease, the use of bacteriostatic antibiotics suffices to help the patient to recover. In the case of chronic Q fever, bactericidal antibiotics are required to cure the patient, although distinct isolates from chronic disease patients display differential resistance to tetracyclines and fluoroquinolones and there is not a standard protocol for treatment of patients with distinct forms of chronic disease.

Successful clinical treatment of ehrlichioses is achieved with tetracycline. Chloramphenicol is less effective *in vivo;* furthermore, its *in vitro* minimal inhibitory concentrations for *E. chaffeensis* and HGE are beyond those safely achieved *in vivo* in humans. Rifamycins and the fluoroquinolone antibiotic trovafloxacin have *in vitro* efficacy against HGE, but *in vivo* studies have not been reported. All antibiotics tested *in vitro* were not bactericidal for *E. risticii.*

Tetracycline therapy cures pathogenic effects caused by *Wolbachia* endosymbionts. In particular, 25 mg/kg/day of tetracycline given to nematode-infected mice resulted in retardation of filarial growth and infertility—effects thought to be due to elimination of symbionts from the parasite. These observations offer new possibilities for the treatment of infec-

TABLE V
In Vitro Susceptibilities of Rickettsiae to Antibiotics

	Minimal inhibitory concentration (μg/ml)[a]								
Antibiotic	*R. prowazekii, R. typhi*[b]	*SFG rickettsiae, R. canada*[b]	*Orientia*	*Coxiella*[c]	*E. risticii,*[d] *E. sennetsu*[e]	*E. canis,*[f] *E. chaffeensis*[g]	*HGE*[h]	*Wolbachia*	*Bartonella*[i]
Doxycycline	0.06–0.125	0.06–0.25		V	≤0.01[d] 0.125[e]	2[f] 0.5[g]	0.25		0.12
Tetracycline	0.1	0.25	0.15–0.31	V	1–10[d]			4	
Minocycline					0.01–0.04[d]				
Oxytetracycline					0.01–0.04[d]				
Thiamphenicol	1–2	0.5–4							
Chloramphenicol	1	0.25–0.5	1.25–2.5	R	1–4[e]	R[g]	I		0.25
Rifampin	0.06–0.25	0.03–1[j]	0.31	R	0.45–0.9[d] 0.5[e]	2[f] 0.125[g]	0.5		0.125–0.75
Rifabutin							≤0.125		
Erythromycin	0.06–1	1–16	2.5–10	V	>10[d] 0.25–4[e]	R[g]			0.125–0.75*
Clarithromycin	0.125–1	0.5–4		S					0.015 0.03–0.07*
Josamycin	0.5–1	0.5–2							0.06
Roxithromycin	1	8–16		S					0.0625–0.25*
Pristinamycin	2–4	1–8							
Spiramycin		16–32							
Azithromycin	0.25	8–16[k]		R			R		0.0156–0.03*
Clindamycin							R		8
Ciprofloxacin	0.5–1	0.25–1		V	0.125[e]	R[g]	2	1	2
Ofloxacin	1	0.5–2		S			2		8
Pefloxacin	1	0.5–2		V					8
Trovafloxacin				S			≤0.125		
Sparfloxacin		0.125–0.25		S					0.12
PII		2.0							
PI		256							
PD 127,391		0.125–0.25		S					
PD 131,628		0.25–0.5		S					
Co-trimoxazole	2; >8	2; >8			R[e]	R[g]			1
Gentamicin	16	4–16		R	R[e]	R[g]	I	4	1–4
Amoxicillin	128	128–256		R					0.06
Ampicillin				R			R		
Penicillin G			R		R[e]	R[g]			0.06
Cefotaxime									0.25
Ticarcillin									0.25
Imipenem							R		0.5
Oxacillin									4
Cephalothin									16
Trimethoprim-Sulfamethoxazole				S			R		1–5
Fosfomycin									64
Colistin									16
Vancomycin									16
Dirithromycin									0.125–1*

[a] Unless otherwise specified, the MIC of each antibiotic is the lowest antibiotic concentration resulting in complete inhibition of bacterial growth or plaque formation. R, resistant to the antibiotic; I, antibiotic had weak inhibitory activity, but no bactericidal effect was observed; S, susceptible to the antibiotic; V, different effects were observed.

[b] MIC determined by both plaque and dye uptake assays.

[c] MICs vary significantly for different isolates, the cell line used, and acute or chronic model of infection employed.

[d] For *E. risticii*, infected cells were examined by MIF after 2–4 days of antibiotic exposure. The MIC was defined as the lowest concentration of antibiotic capable of reducing the percentage of infected cells to 50%.

[e] For *E. sennetsu*, the MIC_{90} was determined using Diff-Quick staining.

[f] MIF was used for *E. canis*.

[g] Diff-Quick staining was used for *E. chaffeensis*.

[h] MIC of each antibiotic is defined as the lowest antibiotic concentration resulting in >90% reduction of cells that are infected.

[i] Antibiotic susceptibility of bartonellae was determined on horse blood-supplemented Columbia agar plates (MIC_{90}) or in Vero cell cultures examined by MIF. *In the latter experiments, MIC was defined as the concentration which completely inhibited growth of bartonellae after 5 days of incubation.

[j] *Rickettsia massiliae, R. aeschlimanii, R. montana, R. rhipicephali,* and isolate Bar 29 exhibited some resistance to rifampin with MICs of 2–4 μg/ml.

[k] MIC for *R. akari* is 0.25 μg/ml.

tions with pathogenic filariae which contain rickettsial endobacteria.

Antibiotic susceptibilities of *Bartonella* isolates have been evaluated in axenic media and upon inoculation in Vero cell monolayers (Table V). *Bartonella quintana, B. henselae, B. vinsonii,* and *B. elizabethae* are very susceptible to β-lactams, with the greatest effects by penicillin and amoxicillin with MIC_{90}s of 0.06 μg/ml. MIC_{90}s for the aminoglycosides range from 1 to 4 μg/ml, and gentamycin is the most effective compound. Moreover, aminoglycosides are the only antibiotics which are bactericidal for *B. henselae* in axenic broth or in cocultivation with eukaryotic cell lines. The macrolides, roxithromycin, azithromycin, and clarithromycin, generally are quite active, although clindamycin is much less effective. *Bartonella* isolates also displayed a very wide range of susceptibility to different fluoroquinolone compounds.

Although many antibiotics inhibit the growth of *Bartonella in vitro,* the question of their clinical efficacy is not simple due to the very wide range of different syndromes associated with *Bartonella* infection. Treatment of bartonellosis in immunocompromised patients and in patients with chronic infections or endocarditis, which do not express universal responsiveness to traditionally applied regimens and have a tendency to relapse, presents the greatest difficulties. Since *Bartonella* is able to invade and multiply in infected tissue, the prescription of antibiotics exerting bactericidal effect is advisable. The prescription of the aminoglycoside gentamicin at 2 mg/kg intramuscularly or doxycycline, 200 mg per os, is recommended for adult patients. Erythromycin, 500 mg four times a day, is suitable for pregnant women, whereas erythromycin in a lower dose or gentamicin is prescribed for children.

Acknowledgments

We thank David J. Silverman, Vsevolod L. Popov, K. Amano, P. Brouqui, and S. Ito for supplying the photographs and Perry Comegys for reproducing them. We also thank David J. Silverman for reviewing the manuscript and for financial support from Public Health Service Grant AI-17416 from the National Institute of Allergy and Infectious Diseases.

This work was supported by the Naval Medical Research and Development Command, Research Task 61102A.001.01.BJX.1293 (G.A.D.). The opinions and statements contained herein are the private ones of the authors and are not to be construed as official or reflecting the views of the Naval Department or the Naval Service at large.

See Also the Following Articles

Crystalline Bacterial Cell Surface Layers (S Layers) • Mapping Bacterial Genomes • Typhus Fevers and Other Rickettsial Diseases

Bibliography

Anderson, B. E., and Neuman, M. A. (1997). *Bartonella* spp. as emerging human pathogens. *Clin. Microbiol. Rev.* **10**, 203–219.

Anderson, B. E., Friedman, H., and Bendinelli, M. (Eds.) (1997). "Rickettsial Infection and Immunity." Plenum, New York.

Andersson, S. G. E., Zomorodipour, A., Andersson, J. O., Sicheritz-Ponten, T., Alsmark, U. C. M., Podowski, R. M., Naslund, A. K., Eriksson, A.-S., Winkler, H. H., and Kurland, C. G. (1998). The genome sequence of *Rickettsia prowazekii* and the origin of mitochondria. *Nature* **396**, 133–140.

Brouqui, P., and Raoult, D. (1996). *Bartonella quintana* invades and multiplies within endothelial cells *in vitro* and *in vivo* and forms intracellular blebs. *Res. Microbiol.* **147**, 719–731.

Dasch, G. A., and Weiss, E. (1998). The Rickettsiae. *In* "Topley & Wilson's Microbiology and Microbial Infections" (L. Collier, A. Balows, and M. Sussman, Eds.), 9th ed., Vol. 2, pp. 853–876. Arnold/Oxford Univ. Press, London, New York.

Dumler, J. S., and Bakken, J. S. (1998). Human ehrlichioses: Newly recognized infections transmitted by ticks. *Annu. Rev. Med.* **49**, 201–213.

Eremeeva, M. E., Dasch, G. A., and Silverman, D. J. (2000). Interaction of rickettsiae with eukaryotic cells: adhesion, intracellular growth, and host cell responses. *In* "Bacterial Invasion into Eukaryotic Cells." (J. Hacker and T. A. Oelschlaeger, Eds.), pp. 479–516. Plenum, London.

Fournier, P. E., Marrie, T. J., and Raoult, D. (1998). Diagnosis of Q fever. *J. Clin. Microbiol.* **36**, 1823–1834.

Fryer, J. L., and Mauel, M. J. (1997). The rickettsia: An emerging group of pathogens in fish. *Emerg. Infect. Dis.* **3**(2), 137–144.

Ihler, G. M. (1996). *Bartonella bacilliformis:* Dangerous pathogen slowly emerging from deep background. *FEMS Microbiol. Lett.* **144**, 1–11.

Jongejan, F., Goff, W., and Camus, E. (Eds.) (1998). Tropical veterinary medicine. Molecular epidemiology, hemoparasites and their vectors, and general topics. *Ann. N.Y. Acad. Sci.* **849**, 1–500.

Marrie, T. J., and Raoult, D. (1997). Q fever—A review and issues for the next century. *Int. J. Antimicrobiol. Agents* **8**, 145–161.

Maurin, M., and Raoult, D. (1996). *Bartonella (Rochalimaea) quintana* infections. *Clin. Microbiol. Rev.* **9**, 273–292.

McCaul, T. F., and Williams, J. C. (1981). Developmental cycle of *Coxiella burnetii*: structure and morphogenesis of vegetative and sporogenic differentiations. *J. Bacteriol.* **147**, 1063–1076.

Merrell, B. R., Weiss, E., and Dasch, G. A. (1978). Morphological and cell association characteristics of *Rochalimaea quintana*: Comparison of the vole and Fuller strains. *J. Bacteriol.* **135**, 633–640.

Popov, V. L., Chen, S.-M., Feng, H.-M., and Walker, D. M. (1995). Ultrastructural variation of cultured *Ehrlichia chaffeensis*. *J. Med. Microbiol.* **43**, 411–421.

Popov, V. L., Han, V. C., Chen, S.-M., Dumler, J. S., Feng, H.-M., Andreadis, T. G., Tesh, R. B., and Walker, D. M. (1998). Ultrastructural differentiation of the genogroups in the genus *Ehrlichia*. *J. Med. Microbiol.* **47**, 235–251.

Raoult, D., and Roux, V. (1997). Rickettsioses as paradigms of new or emerging infectious diseases. *Clin. Microbiol. Rev.* **10**, 694–719.

Thompson, H. A., and Suhan, M. L. (1996). Genetics of *Coxiella burnetii*. *FEMS Microbiol. Lett.* **145**, 139–146.

Umemori, E., Sasaki, Y., Amano, K., and Amano, Y. (1992). A phage in *Bartonella bacilliformis*. *Microbiol. Immunol.* **36**, 731–736.

Walker, D. H. (1998). Tick-transmitted infectious diseases in the United States. *Annu. Rev. Public Health* **19**, 237–269.

Werren, J. H. (1997). Biology of *Wolbachia*. *Annu. Rev. Entomol.* **42**, 587–609.

RNA Splicing, Bacterial

David A. Shub
State University of New York at Albany

I. History of Splicing
II. The Splicing Pathways
III. Mechanisms of Self-Splicing
IV. Introns in Bacteriophage and Bacteria
V. The Origin of Self-Splicing Introns
VI. Concluding Remarks

GLOSSARY

exons Sequences flanking introns in precursor RNA molecules prior to splicing that are joined together in the mature RNA.

introns Intervening sequences in the coding portions of genes that are transcribed into RNA and are removed during RNA splicing.

maturases Proteins sometimes encoded within self-splicing introns that assist in the splicing reaction.

open reading frame The genetic code for the sequence of amino acids in a protein.

ribozyme An RNA molecule that can catalyze a chemical reaction.

RNA splicing An RNA processing reaction in which intervening sequences (introns) in the primary RNA gene transcript are removed and the flanking sequences (exons) are joined together.

self-splicing Describing or referring to splicing pathways in which the determination of the splicing boundaries and catalysis are properties of the intron RNA.

snurps Small *n*uclear *r*ibonucleo*p*rotein particles, made up of proteins and RNA, that provide the specificity for splice boundaries and the catalytic center for the splicing of mRNA in the eukaryotic nucleus.

spliceosome A large assembly of pre-mRNA, snurps, and other factors that constitutes the mRNA splicing machinery in nuclei.

RNA SPLICING refers to the processing of a precursor RNA molecule, in which an internal section is removed and the flanking fragments are reattached (ligated) to form the final, mature RNA. The term ''splicing'' was intended to convey an analogy to familiar editing processes for film and tape. The portion that is removed is called an intron and the flanking portions that are ligated into the mature RNA are called exons. There may be many introns in a single gene, each of which needs to be precisely excised, and the exons need to be ligated to produce the mature, functional RNA molecule.

I. HISTORY OF SPLICING

The discovery in 1977 of splicing of adenovirus mRNAs in mammalian nuclei was completely unexpected. Until then, detailed studies of gene structure (all of which had been performed on bacterial or bacteriophage genes) uniformly confirmed that the information content of DNA, RNA, and protein molecules was colinear. That is, RNA was a faithful transcript of one of the DNA strands and proteins were translated from mRNA by strict application of the triplet genetic code.

In 1981 Thomas Cech discovered an intron in the nuclear-encoded large ribosomal subunit RNA of the eukaryotic microbe *Tetrahymena thermophila.* This finding revolutionized our thinking about enzymatic mechanisms because the intron could self-splice. No additional RNA or protein cofactors were required for the splicing reaction, leading to the conclusion that RNA acted as an enzyme, a "ribozyme." The discovery of splicing and ribozymes earned Cech the

Nobel prize in chemistry in 1989. The prize was jointly awarded to Sidney Altman, who discovered, at about the same time, that the catalytic component of the bacterial enzyme RNase P is RNA.

RNA splicing, which results in elimination of large tracts of nucleotides (that were originally transcribed from DNA) from the final RNA gene products, remained an exclusive property of eukaryotic RNA molecules for almost 10 years. As a result, many textbooks cited splicing as one of the fundamental differences between eukaryotes and bacteria. Splicing of mRNAs of *Escherichia coli* bacteriophage T4 and tRNAs of archaea were discovered in the mid-1980s. Additional examples have been described recently, but RNA splicing remains much more common in eukaryotic genes. Bacterial splicing is of interest since it demonstrates the universality of this form of RNA processing. Moreover, since the self-splicing pathways found in bacteria are thought to be the earliest forms of splicing, interest has arisen in the possible occurrence of RNA splicing in the earliest common ancestor of all living things.

Within a few years of the initial discovery of RNA splicing it became apparent that, instead of a single splicing mechanism, there are four distinct splicing pathways.

II. THE SPLICING PATHWAYS

A. Spliceosomes

Spliceosomes are involved exclusively in splicing of mRNAs made in eukaryotic nuclei. This pathway relies on a large number of accessory factors, including small RNA molecules that combine with proteins into ribonucleoprotein particles (snurps) and ATP. The snurps combine with precursor mRNA into a larger ribonucleoprotein particle called the spliceosome (due to its superficial resemblance to the ribosome). The RNA components of the snurps interact by base pairing with complementary sequences within the intron and at the intron–exon boundaries to provide the specificity for removal of the introns, which can be hundreds or thousands of nucleotides in length. Although it has not been experimentally demonstrated, it is generally assumed that (because of similarities to the group II splicing pathway) the RNAs in the snurps also comprise the catalytic center of the spliceosome.

B. tRNA Splicing

This splicing pathway, also discovered in the eukaryotic nucleus, involves genes encoding transfer RNAs. In this case a single small intron (generally less than 100 nucleotides in length) per gene is always present at the same position within the anticodon loop sequence of the tRNA. Intron recognition, excision, and exon ligation are all performed by two typical protein enzymes, a ribonuclease and RNA ligase. RNA–RNA interactions are not part of this splicing pathway. A similar pathway has been described in archaea.

C. Group I and Group II Introns

Introns in genes of yeast mitochondria are similar in size to spliceosomal introns (several hundred to about 2000 nucleotides). In contrast to spliceosomal introns, which share only short conserved sequences and have no regular secondary structure in the RNA, the mitochondrial introns have extensive secondary structure elements that fold into a shared core structure. However, their conserved primary sequence elements bear no resemblance to the spliceosomal introns. The mitochondrial introns are placed into two groups based on their splicing mechanisms—shared nucleotide sequences and core secondary structure elements. Group I and group II introns are found in many other mitochondrial and chloroplast genomes in genes encoding mRNAs, tRNA, and ribosomal RNAs.

Surprisingly, these introns frequently contain complete coding sequences for proteins [open reading frames (ORFs)] that are inserted into single-stranded loops of a conserved, folded RNA core structure. In some cases the protein products of these ORFs are required for efficient splicing of the intron in the organelle (in which case they are called maturases). However, the ORFs are also frequently involved in promoting a directed gene conversion event, whereby introns are inserted into cognate intronless versions of the same gene in genetic crosses.

This ORF-promoted intron mobility may have resulted in lateral transfer of introns across species boundaries, hindering efforts to reconstruct the evolutionary history of these introns.

III. MECHANISMS OF SELF-SPLICING

The only absolute requirements for group I intron excision and exon ligation are the precursor RNA and a guanosine nucleotide. After the intron is removed it retains an active center that can catalyze other reactions in *trans*. Group II introns are also capable of self-splicing *in vitro*. Although maturases are sometimes required for efficient splicing *in vivo*, it is clear that the catalytic component of both of these splicing pathways resides in the intron.

Group I and group II introns have been shown to splice via transesterification reactions. That is, as one phosphodiester linkage is broken, another is simultaneously formed. In group I splicing the cascade of transesterification reactions is triggered by nucleophilic attack of the 3′ OH group of the guanosine nucleotide cofactor on the bond between the upstream exon and the first nucleotide of the intron. This liberates the exon for a further transesterification reaction at the other end of the intron, resulting in exon ligation, and this also results in covalent addition of the guanosine nucleotide at the 5′ end of the excised intron. Thus, if group I self-splicing reactions are performed *in vitro* with radiolabeled guanosine nucleotide, the reaction can be monitored by observing the formation of end-labeled intron molecules. This reaction has facilitated the identification of many of the group I introns that have been found in bacteria and bacteriophage genes.

In the case of group II introns, the triggering nucleophile is the 2′ OH of a ribose situated within the intron RNA backbone. This is also how the splicing reaction proceeds in the spliceosome, suggesting that nuclear mRNA splicing is derived from a group II self-splicing progenitor. It has been hypothesized that the RNA molecules in the snurps perform the same functions in *trans* that portions of the group II intron catalytic core perform in *cis*.

IV. INTRONS IN BACTERIOPHAGE AND BACTERIA

The first indication of RNA splicing in a bacterial system was obtained when the DNA sequence of the gene for thymidylate synthase (*td*) of the *E. coli* bacteriophage T4 was determined. The protein coding sequence of the *td* gene was interrupted by more than 1000 base pairs of DNA. This was subsequently shown to be an intron that conforms to the sequence and structure rules for group I introns and to be removed from precursor RNA by the group I splicing pathway. Subsequently, similar introns were found in other T4 genes and in the genes of several other bacterial viruses. Surprisingly, introns have been found only in T4 and its close relatives and not in any other viruses of gram-negative enteric bacteria, including *E. coli*. Instead, introns have been more commonly found in viruses that infect gram-positive bacteria (*Bacillus, Staphylococcus*, etc.).

The only bacterial genes that have been shown to contain group I introns are certain tRNA genes in cyanobacteria and in α and β proteobacteria. These introns are closely related in sequence and structure and are all inserted within, or immediately adjacent to, the anticodon triplet of the tRNA. The evolutionary implications of these introns are of special interest and will be discussed later.

Group II introns are found sporadically in bacterial chromosomes in which they are often inserted into transposons or cryptic prophages. The best studied bacterial group II intron is in a plasmid-borne gene required for conjugation in *Lactococcus lactis*. These introns have not been found in bacteriophages.

V. THE ORIGIN OF SELF-SPLICING INTRONS

It is likely that introns have been transferred horizontally for the following reasons: (i) Intron ORFs promote the transfer of introns into cognate genes that lack introns and (ii) most of the introns discovered in bacteria are associated with genetic elements (viruses, transposons, and conjugal plasmids) that are inherently mobile. Tracing the evolutionary history of these introns is virtually impossible. The only exceptions to this situation to date are the group I introns in bacterial tRNA genes. In this case, the genes are essential and are stably transmitted to each new generation. Interestingly, the bacteria that possess these tRNA introns (cyanobacteria and proteobacteria) share a common ancestry with organelles (chloroplasts and mitochondria, respectively) in which self-splicing introns are most commonly encountered. More remarkably, one of these introns is inserted at exactly the same genomic location (a leucine tRNA gene) in many distantly related cyanobacteria and in most chloroplasts. The available evidence is consistent with vertical inheritance of this intron from the common ancestor of chloroplasts and contemporary cyanobacteria. If substantiated, this would be the earliest example for the origin of any kind of intron.

VI. CONCLUDING REMARKS

Many questions remain unanswered about bacterial splicing, including the origin of the phenomenon and the extent to which it is present in the major bacterial groups. Evolutionary relationships between the major splicing pathways remain to be elucidated. Three kinds of introns can be found in the anticodon region of tRNA genes in a variety of genomes: the protein-spliced introns in archaeal and eukaryotic nuclear-encoded tRNAs, group I and group II introns in mitochondria and chloroplasts, and group I introns in bacteria. It will be very interesting to determine whether this indicates an evolutionary connection between these splicing mechanisms, a functional role for intron splicing in tRNA metabolism, or if this is simply a preferred target for intron insertion. The mechanisms by which the group I and group II intron ORFs function as maturases, and promote the mobility of the introns, are also under investigation. Since bacteria present advantages for genetic and biochemical manipulation compared to mitochondria or chloroplasts, bacterial splicing will continue to be a useful experimental system for the study of these questions.

Bibliography

Cech, T. R. (1993). Structure and mechanism of the large catalytic RNAs: Group I and group II introns and ribonuclease P. *In* "The RNA World" (R. F. Gesteland and J. F. Atkins, Eds.), pp. 239–269. Cold Spring Harbor Laboratory Press, Cold Spring Harbor, NY.

Lambowitz, A. M., and Belfort, M. (1993). Introns as mobile genetic elements. *Annu. Rev. Biochem.* **62**, 587–622.

Shub, D. A., Coetzee, T., Hall, D. H., and Belfort, M. (1994). The self-splicing introns of bacteriophage T4. *In* "Molecular Biology of Bacteriophage T4" (J. Karam, Ed.), pp. 186–192. ASM Press, Washington, DC.

Westaway, S. K., and Abelson, J. (1995). Splicing of tRNA precursors. *In* "tRNA: Structure, Biosynthesis, and Function" (D. Söll and U. L. RajBhandary, Eds.), pp. 79–92. ASM Press, Washington, DC.

Rumen Fermentation

James B. Russell

Agricultural Research Service, USDA, and Cornell University

I. Recognition of Ruminal Fermentation
II. Rumen as a Microbial Habitat
III. Polymer Degradation
IV. Transport and Phosphorylation
V. Crossfeeding among Ruminal Microorganisms
VI. Ruminal Fermentation Schemes
VII. Growth, Maintenance, and Energy Spilling
VIII. Effect of pH on Ruminal Fermentation
IX. Toxic Compounds
X. Effect of Ionophores on Ruminal Microorganisms
XI. Biotechnology
XII. Rumen Microbial Ecology
XIII. Ruminal Bacteria
XIV. Ruminal Protozoa
XV. Ruminal Fungi
XVI. Models of Ruminal Fermentation
XVII. Summary

GLOSSARY

abomasum The gastric stomach of a ruminant.

energy spilling The process used by microorganisms to dissipate excess energy.

eructation The process that lets fermentation gases escape from the rumen.

ionophores A class of antibiotics that alter ruminal fermentation end products.

maintenance Energy used to maintain ion gradients and turn over macromolecules.

omasum A chamber that traps large feed particles as digesta passes from the rumen to the abomasum.

reticulum A small pouch that extends from the anterior rumen. It collects either large feed particles for rumination or small particles for passage to the lower gut.

rumen The large pregastric stomach of a ruminant that acts as a fermentation chamber.

rumination The process by which a ruminant forces large feed particles from the rumen up the esophagus to the mouth to be chewed again.

RUMINAL FERMENTATION is an exergonic process that converts feedstuffs into short-chain volatile fatty acids (VFA), CO_2, CH_4, NH_3, and heat. Some of the free energy is trapped as ATP and is used to drive the growth of anaerobic ruminal microorganisms. Ruminant animals absorb VFA and digest the microbial protein to obtain energy and amino acids, respectively. CH_4 and NH_3 represent losses of energy and nitrogen to the animal. Some ruminal microorganisms can degrade cellulose, and this attribute gives ruminant animals the ability to digest fibrous materials. The rumen is inhabited by a highly diverse population of bacteria, protozoa, and fungi.

I. RECOGNITION OF RUMINAL FERMENTATION

Prehistoric man used ruminant animals to exploit the photosynthetic potential of the temperate grasslands and establish a stable food supply. The Old Testament gave special ruminants special recognition, "Whatever divides a hoof, thus making split hoofs, and chews the cud, among the animals, that may you eat" (Leviticus 11:3). Aristotle described the four compartments of the ruminant stomach, but the role of microorganisms in ruminant digestion was not recognized until 1884 when van Tappeiner added antiseptics to rumen fluid and demonstrated an inhibition of fiber digestion.

Webster's dictionary described ruminants as "any of a group of four footed, hoofed, even toed, and cud chewing mammals which have a stomach consisting of four divisions or chambers, the rumen, reticulum, omasum, and abomasum; the grass etc. that they eat is swallowed unchewed and passes into the rumen or reticulum from which it is regurgitated, chewed and mixed with saliva, again swallowed, and then passed through the reticulum and omasum into the abomasum where it is acted on by gastric juice." This definition is anatomically correct, but it lacks any mention of microorganisms or fermentation.

II. RUMEN AS A MICROBIAL HABITAT

The rumen is an ideal habitat for the growth of anaerobic microorganisms (Fig. 1). Salivary bicarbonate buffers the ruminal fluid, and VFA, arising from the fermentation, are absorbed across the rumen wall. Some O_2 enters the rumen with the feed, but the fermentation produces CO_2 and CH_4 that displace O_2 from the rumen. The remaining O_2 is consumed by a small population of facultative anaerobes. The rumen accounts for one-seventh to one-tenth of the animal's body weight and can be as large as 80 liters. The ruminal contents are forced back and forth over two pillars, and these mixing movements inoculate ingested feed with microorganisms and transfer VFA to the mucosal surface where they can be absorbed. The reticulum is a much smaller pouch that collects large feed particles so they can be forced back up the esophagus to be chewed again (ruminated). Gases also exit the rumen through the esophagus via a process known as eructation. Feed leaving the rumen passes to the omasum. Liquid and small feed particles pass through the laminae of the omasum, but large feed particles are trapped and back washed into the rumen. Feed then enters the abomasum (the gastric stomach), and from this point digestion is analogous to the process observed in simple stomached animals. Because feed enters and leaves the rumen at regular intervals, the rumen operates as a continuous culture device, but selective retention of feed creates at least

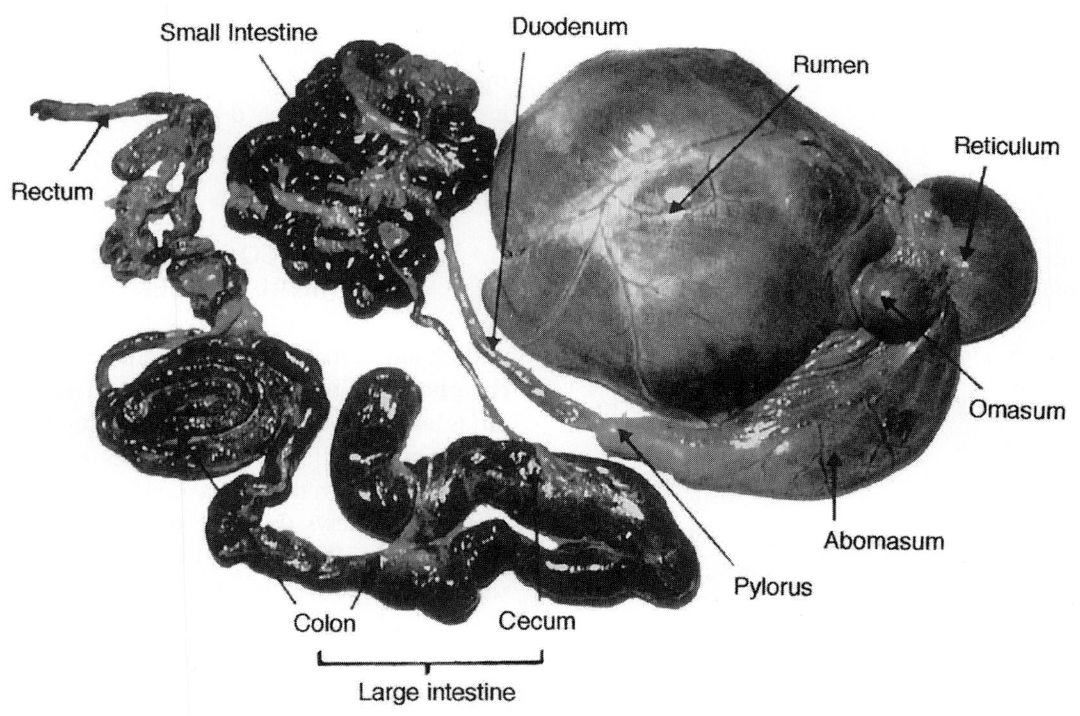

Fig. 1. The digestive tract of a sheep showing the various compartments.

two major dilution rates. Large feed particles turn over at a slower rate (2–4 times) than do the liquid and small particle fractions. Many ruminal microorganisms prolong their residence time in the rumen by attaching to feed particles with a slow dilution rate.

III. POLYMER DEGRADATION

Feedstuffs are primarily composed of large, and sometimes complex, polymers (cellulose, hemicelluloses, pectin, starch, proteins, etc.). These polymers must be degraded by extracellular enzymes before they can be utilized by ruminal microorganisms. Some enzymes are secreted into cell-free ruminal fluid, but most hydrolases remain cell associated. By sequestering enzymes to their outside surfaces, ruminal microorganisms increase their chance of receiving the end products. A variety of hydrolases have been purified and cloned, but it is still not clear if all of the important enzymes have been identified. Carboxymethylcellulose (CMC) degradation has been used as an assay for cellulases, but CMCases lack cellulose binding domains and are usually unable to degrade intact insoluble cellulose. True bacterial cellulases have not yet been purified, and they appear to be unusually sensitive to feedback inhibition by cellobiose, the hydrolytic product. Cellulase activity seems to be constitutive, but some enzymes (e.g., amylase) are repressed by glucose or other sugars. Because hydrolases can only come in contact with feed via water interactions, solubility can be a key factor regulating feed degradation. When proteins are heated, hydrophobic groups once buried in the molecule come to the surface, solubility decreases, the protein is more resistant to ruminal degradation, and more amino nitrogen passes to the lower gut. Heating also affects starch degradation, but in this case solubility and degradation are enhanced. The starch of cereal grains is covered with protein, and heat ruptures this protective coating.

IV. TRANSPORT AND PHOSPHORYLATION

Ruminal microorganisms must compete for extracellular degradation products, and most ruminal bacteria have high affinity transport systems and efficient mechanisms of phosphorylation. Some ruminal bacteria use phosphotransferase (PTS) systems to take up mono- and disaccharides, and these systems phosphorylate sugars as they pass across the cell membrane. Because at least some of the sugar is already phosphorylated, the ATP demand of a glucokinase reaction is decreased. Ruminal bacteria lacking PTS systems have an alternative strategy of conserving ATP. When disaccharides are translocated into the cell via active transport (ion or ATP driven), the disaccharides can be phosphorylated by phosphorylases. Phosphorylases conserve the free energy of the hydrolytic bond and phosphorylate one of the sugar residues. The rumen is a sodium rich environment, and many active transport systems are sodium-dependent.

V. CROSSFEEDING AMONG RUMINAL MICROORGANISMS

Cellulolytic bacteria are always outnumbered by noncellulolytic species *in vivo,* and cellulolytic and noncellulolytic species can be cocultured *in vitro* on cellulose. Wolin and colleagues hypothesized that the noncellulolytic species were living on the extracellular products of cellulose digestion, but recent work indicated that at least some cellulolytics have a reversible cellodextrin phosphorylase. Cellodextrins that are excreted by cellulolytic bacteria can be "stolen" by noncellulolytics, but the interaction is even more complicated. Cellulolytic species require branched-chain VFA, and noncellulolytic, amino acid-fermenting bacteria produce these essential nutrients. The most active amino acid-fermenting bacteria are not proteolytic and must in turn depend on other proteolytic species. Crossfeeding is also illustrated by the observation that pure cultures of ruminal bacteria can form products *in vitro* that are not detected in ruminal fluid. Formate and succinate are converted to CH_4 and propionate by methanogenic and succinate-decarboxylating species, respectively. Lactate does not accumulate in ruminal fluid until the pH declines, and this intermediate can be used by *Megasphaera elsdenii* or *Selenomonas ruminantium.*

VI. RUMINAL FERMENTATION SCHEMES

Virtually all of the ruminal hexose carbon is metabolized by the Embden–Meyerhof–Parnas (EMP) pathway, but pentose carbon can be metabolized by either the pentose pathway or a scheme involving phosphoketolase. Pyruvate arising from carbohydrate catabolism can be converted to lactate, but this method of fermentation provides only a modest amount of ATP. Acetate is the dominant end product of ruminal fermentation, and when pyruvate is converted to acetate, the NADH of the EMP must be oxidized by other methods. NADH oxidation via H_2 production is thermodynamically unfavorable, but methanogenic archaea, by scavenging H_2, can keep the partial pressure of H_2 low enough so even this reaction is possible. Interspecies hydrogen transfer promotes acetate production and increases ATP production.

Not all ruminal bacteria can produce large amounts of H_2, and some species have developed other schemes of reducing equivalent disposal. Butyrate-producing bacteria also convert pyruvate to acetyl-coenzyme A (acetyl-CoA), but two acetyl-CoA can be condensed and reduced by butyryl and butyryl-CoA dehydrogenases. Propionate can be produced by either the randomizing (through succinate) or acrylate pathway. These pathways have dehydrogenase reactions, malate dehydrogenase and fumarate reductase, or two acrylyl-CoA dehydrogenase steps. The fumarate reductase of succinate or propionate production is a cytochrome-linked, ATP-producing reaction, but schemes that use other methods of reducing equivalent disposal (alcohol, lactate, butyryl-CoA, and β-hydroxybutyryl-CoA dehydrogenases) produce less ATP than acetate production. In this regard, rumen bacteria often have to counterbalance reducing equivalent disposal with opportunities for ATP production.

Ruminal CH_4 production is a process that involves the uptake of H_2 and the stepwise reduction of CO_2, and it is a dominant mechanism of disposal of reducing equivalents in the rumen. The energetics of CH_4 production were until recently not well understood. The free energy change of the overall process is negative (-8 kJ/mol), but initial steps of CO_2 formation are low or even positive. The terminal step appears to create a proton-motive force that then drives ATP synthesis. Ruminal acetogens can also utilize H_2 and CO_2, but these bacteria have a lower affinity for H_2 than the methanogens and prefer to utilize sugar.

Some ruminal bacteria ferment amino acids, but the ATP yield of amino acid fermentation is very low. It initially appeared that carbohydrate-utilizing ruminal bacteria were responsible for all of the deamination, but their rates of amino acid fermentation could not explain all of the ruminal ammonia. The rumen also has a small population of obligate amino acid-fermenting bacteria, and these highly specialized bacteria have exceedingly high rates of amino N uptake.

Some anaerobic habitats have bacteria that can oxidize volatile and even long-chain fatty acids, but these bacteria grow so slowly that they cannot persist in the rumen. Ruminal bacteria can, however, saturate the double bonds of polyunsaturated fatty acids, and biohydrogenation is yet another mechanism of disposal of reducing equivalents. Biohydrogenation is normally complete, but recent work indicates that

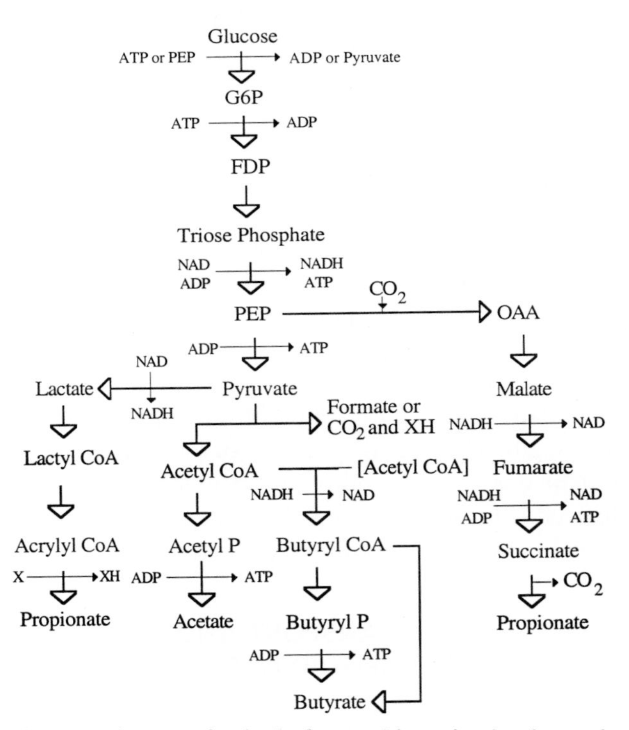

Fig. 2. Pathways of volatile fatty acid production in ruminal bacteria. "X" denotes electron carrier other than NAD.

low fiber diets and low ruminal pH can cause the accumulation of partially hydrogenated fatty acids. Trans-octadecenoate has been implicated in depressing the milk fat of lactating cattle.

VII. GROWTH, MAINTENANCE, AND ENERGY SPILLING

In its simplest sense, bacterial growth is a process that assembles simple monomers (either preformed or synthesized) into the macromolecules. Protein is usually the most abundant polymer in bacteria, and this macromolecule is relatively expensive to synthesize. Peptide bonds have less than 1 kJ/mol of free energy per mole, but protein synthesis is a four-step process that consumes approximately 4 ATP equivalents or approximately 7 kJ/mol of free energy per mole of amino acid polymerized. As much as two-thirds of the ATP consumption can be used to polymerize amino acids. Polysaccharides (e.g., glycogen) and nucleic acids account for a smaller fraction of the total mass, and the polymerization of sugars and nucleic acids requires less energy.

Bacterial growth is driven by the ATP of catabolic schemes, but observed yields are often much lower than amounts predicted by ATP availability alone (Y_{ATP}). Nongrowth ATP utilization accounts for the discrepancy between ATP production and growth, but the nature of this expenditure was, until recently, poorly defined. Nongrowth ATP can be consumed by maintenance expenditures as well as by energy-spilling reactions, and both are "turnover functions." When cells grow, protein is synthesized, but the degradation and resynthesis of protein creates an additional expenditure. Growing cells take up ions and concentrate them, but ions can also leak back across the cell membrane. The ion leakage of maintenance expenditures and energy spilling are similar processes, but their regulation is distinctly different. All cells must be "maintained," but cells only "spill energy" when it is in excess.

When cells have large amounts of all essential nutrients and growth rate is rapid, maintenance expenditures are "insignificant," but it should be realized that maintenance expenditures are in reality a "fixed overhead." When energy is restricted growth decreases, and maintenance becomes the dominant avenue of ATP utilization. It has been estimated that ruminal bacteria expend 10 to 50% of their ATP on maintenance. The rumen generally operates as an energy-limited system, but other nutrients can be limiting. When other nutrients are limiting (e.g., amino N or branched VFA), the excess energy is spilled. The utility of energy spilling to ruminal bacteria is illustrated by the observation that bacteria lacking energy spilling mechanisms can be killed by excess carbohydrate.

Streptococcus bovis is a ruminal bacterium that has very high rates of catabolism, and it has been used as a model of energy spilling. Energy spilling in *S. bovis* is regulated via changes in the concentration of intracellular fructose-1,6-diphosphate (FDP). When the glycolytic rate is fast, intracellular FDP increases, and this increase causes a decrease in intracellular inorganic phosphate (P_i). Decreases in P_i cause an increase in the ΔG of ATP hydrolysis and an increase in proton-motive force (Δp). When Δp increases, the resistance of the cell membrane to protons decreases and protons enter the cell. The inward flux of protons is counteracted by the membrane-bound ATPase, and the excess ATP is dissipated.

It has long been recognized that amino N can increase the growth efficiency of ruminal bacteria, but the mechanism of this stimulation was not explained. Because the cost of taking up an amino acid is nearly as much as the cost of synthesizing one, preformed amino acids should not have a significant impact on growth efficiency per se. These simple comparisons, however, did not address the balance of anabolic and catabolic rates. When amino acids are available, growth rate is improved, the imbalance is alleviated, and less energy is spilled.

VIII. EFFECT OF pH ON RUMINAL FERMENTATION

The rumen is well buffered by the bicarbonate of saliva, but pH can decline if the fermentation is very rapid. Acute ruminal acidosis is usually caused by lactate accumulation, but chronic acidosis can be caused by an increase in total VFA. When animals are fed grain or other nonfibrous feeds, rumination and mixing movements decrease, and VFA absorption from the rumen is impaired. Ruminal acidosis can be prevented by feeding cattle sodium bicarbonate, but this practice has little impact on the buffering capacity of the ruminal fluid per se. The rumen is one of the largest compartments in the body, and its osmotic pressure must be carefully regulated to prevent hemoconcentration or hemodilution. When the cattle are fed sodium bicarbonate, water intake increases, the fluid dilution washes VFA and unfermented starch from the rumen, and ruminal pH increases.

Ruminal bacteria differ greatly in their sensitivity to pH. Cellulolytic bacteria and methanogenic archaea are particularly sensitive to declines in pH, and even mild cases of ruminal acidosis can decrease cellulose digestion and CH_4 production. Starch-fermenting ruminal bacteria are generally pH-resistant, and low ruminal pH does not seem to decrease starch fermentation. The effect of pH on ruminal bacteria can be explained by the transmembrane pH gradient (ΔpH). The pH-sensitive ruminal bacteria often try to maintain a near neutral intracellular pH, and ΔpH causes a logarithmic accumulation of VFA anions. Acid-tolerant ruminal bacteria have adapted to tolerate a low intracellular pH, and therefore the low ΔpH does not drive so large an accumulation of VFA anions.

IX. TOXIC COMPOUNDS

Ruminal bacteria can degrade toxic plant materials (e.g., mimosine and oxalate), but these specialized bacteria may initially be present at very low numbers in the rumen. In some cases, ruminal bacteria can produce toxic compounds from relatively innocuous feed ingredients (e.g., sulfide from sulfate, nitrite from nitrate, and 3-methylindole from tryptophan).

X. EFFECTS OF IONOPHORES ON RUMINAL MICROORGANISMS

Beef cattle in the U.S. are routinely fed a class of antibiotics known as ionophores, and these compounds decrease H_2 and CH_4 production and increase propionate and energy retention. Obligate amino acid-fermenting ruminal bacteria are also sensitive to ionophores, and this inhibition decreases NH_3 production and conserves amino N. Some lactic acid-producing bacteria are inhibited by ionophores, and this activity seems to modulate ruminal pH.

Ionophores translocate ions across cell membranes. When ion gradients (e.g., potassium, sodium and protons) are dissipated, the cell must expend energy to reestablish the gradients, with the result that growth is impaired. Ruminal bacteria differ greatly in their sensitivity to ionophores, and gram-negative bacteria are generally more resistant than gram-positive species. Gram-negative bacteria have an outer membrane that prevents ionophores from reaching the cell membrane, but gram-positive bacteria lack this protective coating.

The value of many antibiotics has been diminished by the development of antibiotic resistance and by the ability of bacteria to transfer antibiotic-resistance genes on plasmids and transposons. Ionophores have been widely used for more than 20 years, but there has been little change in feed efficiency. This latter observation is consistent with the idea that the outer membrane of gram-negative bacteria is the mechanism of resistance.

XI. BIOTECHNOLOGY

In the 1980s, rumen microbiologists succeeded in cloning genes from ruminal bacteria into organisms like *Escherichia coli,* but some of these genes produced highly truncated proteins. Attempts to insert genes into ruminal bacteria were for many years unsuccessful. Many species could not take up free DNA, and vectors based on phage or bacterial conjugation were not available. By the 1990s, transformation and conjugation systems for some ruminal bacteria were available, but the impact of genetic engineering has yet to be realized. Several groups of scientists focused on potential improvements in cellulose digestion, but cellulose surface area usually seems to be a more critical factor than the amount of cellulase per se. Because ruminal cellulolytic bacteria do not digest cellulose at low pH, it is conceivable that the development of acid-resistant cellulolytic bacteria could enhance ruminal cellulose digestion when starch is added to the diet.

XII. RUMEN MICROBIAL ECOLOGY

Hungate explained the complexity of ruminal microorganisms in three ways. His first and second principles were based on the idea that feedstuffs were relatively complex materials and that "a single kind of cell cannot contain the diversity of conditions and enzymes for all the individual reactions in the combination yielding maximum cell growth." Some ruminal bacteria are, indeed, very specialized, but other have broad ranges of substrate specificity. Hungate's third hypothesis of ruminal microbial diversity was based on the supposition that even the "most-fitted" species, by its proliferation, would eventually change the habitat and create additional niches for other bacteria. Rumen microbial diversity can also be explained by generation time. Microorganisms in natural habitats generally grow very slowly, but ruminal microorganisms have been allowed to grow continuously at a relatively rapid rate (average doubling time of approximately 7 hr) for more than 70 million years. Because ruminal bacteria have relatively short generation times, there has been a much greater opportunity for adaptation, selection, and diversity.

XIII. RUMINAL BACTERIA

In 1996, Hungate listed approximately 20 species of ruminal bacteria that achieved numbers of at least 10^7 cells per gram of rumen contents (Table I). These species were classified by physiological and morphological traits typical of *Bergey's Manual,* but some were renamed in the 1970s and 1980s. Strains within a given species sometimes had DNA homologies that were as low as 20%, and 16S rRNA sequencing indicated that many strains and some species should be reclassified. Relatively few 16S rRNA genes of ruminal bacteria have been sequenced, but it is clear that the diversity of ruminal bacteria is much greater than 20 species.

Many ruminal bacteria are difficult to grow in the laboratory, and the direct microscopic count is usually 10- to 100-fold greater than the culturable count. With the advent of molecular techniques, microbiologists have been enticed by the possibility that 16S rRNA probes and the polymerase chain reaction (PCR) could be used to monitor the numbers of nonculturable bacteria in natural habitats. However, if a genetically distinct species and its closest relatives have never been cultured, how do you construct the initial probe? Some regions of the 16S rRNA are, indeed, specific enough to create species-specific probes, but the probe can be so specific that it only reacts with a particular strain and not all members of the species.

DNA "fingerprinting" has been used as a means of distinguishing people, animals, plants, and even bacteria, but this approach is only suitable when diversity in the population is low. If the diversity is very great, the gels will be nothing more than an uninterpretable smear of nucleic acid bands. Recent work indicated that polymorphisms in specific portions of bacterial DNA (e.g., terminal restriction fragment of 16S rRNA genes) could be used to assess "community structure" of natural habitats, but this approach alone does not provide species identification, evolutionary history, or physiological traits.

An ecological analysis of any habitat is dependent on quantitative measurements of activity, but the activity of pure cultures and the mixed culture population have not always been compared. For example, in the 1960s it was generally assumed that carbohydrate-fermenting ruminal bacteria were the most important NH_3-producing species, but later work indicated that these species had an NH_3 production rate 50% lower than mixed ruminal bacteria. This comparison led to the isolation of obligate amino acid-fermenting bacteria with 20-fold greater activity.

XIV. RUMINAL PROTOZOA

Ruminants can be reared without protozoa or can be defaunated with chemicals, but ruminal protozoa can account for as much as half of the microbial mass in the rumen. These eukaroytic and sometimes even sexual organisms are 20 to 100 times larger

TABLE I
Taxonomy and Physiological Traits of Predominant Ruminal Bacteria

Taxonomy		*Fermentation products*[a]	*Primary niches*
Circa 1966	*Now*		
Bacteria			
Bacteroides succinogenes	*Fibrobacter succinogenes*	S, F, A	Cellulose
Ruminococcus albus	*Ruminococcus albus*	A, F, E	Cellulose
Ruminococcus flavefaciens	*Ruminococcus flavefaciens*	S, F, A, H_2, E	Cellulose
Butyrivibrio fibrisolvens	*Butyrivibrio fibrisolvens*	B, F, L, A, H_2	Cellulose, hemicellulose, starch, pectin, sugars
Bacteroides amylophilus	*Ruminobacter amylophilus*	S, F, A, L, E	Starch
Selenomonas ruminantium	*Selenomonas ruminantium*	L, A, P, B, H_2, F	Sugars, starch, lactate
Bacteroides ruminicola	*Prevotella ruminicola*	S, A, F, P	Starch, hemicellulose, pectin, β-glucans, proteins
	Prevotella albensis	S, A, F	Starch, hemicellulose, pectin, β-glucans, proteins
	Prevotella brevis	S, A, F	Starch, hemicellulose, pectin, β-glucans, proteins
	Prevotella bryantii	S, A, F	Starch, hemicellulose, pectin, β-glucans, proteins
Succinimonas amylolytica	*Succinimas amylolytica*	S, A, P	Starch
Succinivibrio dextrinosolvens	*Succinivibrio dextrinosolvens*	S, A, F, L	Maltodextrins
Streptococcus bovis	*Streptococcus bovis*	L, A, F, E	Starch, sugars
Eubacterium ruminantium	*Eubacterium ruminantium*	A, F, B, L	Maltodextrins, sugars
Peptostreptococcus elsdenii	*Megasphaera elsdenii*	P, A, B, Br	Lactate, maltodextrins, amino acids
Lachnospira multipara	*Lachnospira multipara*	L, A, F, E, H_2	Pectin, sugars
	Peptostreptococcus anaerobius	Br, A	Peptides, amino acids
	Clostridium aminophilum	A, B	Amino acids, peptides
	Clostridium sticklandii	A, Br, B, P	Peptides, amino acids
Vibrio succinogenes	*Wolinella succinogenes*	S	Malate, fumarate
Anaerovibrio lipolytica	*Anaerovibrio lipolytica*	A, S, P	Glycerol, lactate
Methanobacterium ruminantium	*Methanobrevibacter ruminantium*	CH_4	H_2, CO_2, formate
Protozoa			
Holotrichs	*Isotricha, Dasytricha*	A, B, L	Soluble sugars
Entodinomorphs	*Entodinium, Diplodinium, Epidinium, Orphryoscolex*	A, B, H_2	Starch grains
Fungi	*Neoxallimastix, Caecomyces, Piromyces, Oprinomyces, Aneromyces*	A, L, H_2, F	Cellulose

[a] A, acetate; P, propionate; B, butyrate; F, formate; L, lactate; E, ethanol; Br, branched-chain VFA.

than bacteria, and they can engulf bacteria, feed particles, and even other protozoa. The rumen has two major classes of protozoa, the holotrichs that have cilia over their entire bodies and entodinomorphs that have cilia around the oral cavity. The holotrichs use soluble sugars as an energy source, but entodinomorphs prefer to utilize starch grains. The entodinomorphs also seem to digest cellulose, but our understanding of protozoal metabolism in the rumen has been stymied by the observation that most ruminal protozoa have an obligate requirement for bacteria. The impact of protozoa on the fermentation can be both positive and negative. Protozoal engulfment of starch grains modulates the ruminal pH and prevents acidosis. Protozoal predation of bacteria and protozoal lysis increase ruminal NH_3 and decrease amino acid availability.

XV. RUMINAL FUNGI

Early workers noted large flagellated microorganisms in the rumen, but they were mistakenly classified as protozoa. Most of these flagellates were really fungal zoospores. The ruminal fungi have a complex life cycle that is similar to phycomycetes. Zoospores give rise to a mycelium that covers feed particles, and sporangia release zoospores that can colonize fresh feed. Ruminal fungi have been classified as chytridomycetes, the most primitive group of fungi. Ruminal fungi are found in greatest numbers in animals that are consuming poor quality forage, and the fungi seem to have very active cellulases. These observations lead microbiologists to believe that fungi were better able to digest cellulose than bacteria, but there has been little direct demonstration that this hypothesis is true. The ruminal fungi grow slower than bacteria, have very low growth yields, and rarely make up more than 6% of the microbial mass in the rumen.

XVI. MODELS OF RUMINAL FERMENTATION

Computer models have been used since the 1960s to simulate complex systems, and some research groups have developed models to describe ruminal fermentation. To date, ruminal fermentation models have not provided a detailed description of the ecology and have primarily emphasized the kinetics of feed digestion and its removal from the rumen. In the 1980s, workers at Cornell University developed a model that had equations describing basic principles of microbial growth (effects of maintenance energy, pH, and amino acid availability) and substrate availability (fermentation versus passage rates) as well as different microbial pools. The microbial "ecology" was, however, still highly simplified. Ruminal bacteria were divided into only two groups (structural and nonstructural carbohydrate-fermenting types). Protozoa were simply a factor that decreased the theoretical maximum growth yield of the bacteria, and the fungi were completely ignored. Nevertheless, the impact of the Cornell model in the cattle industry was very positive. Feed savings as great as 17% were reported, and it eventually provided a basis of the 1996 National Recommendations for Beef Cattle.

XVII. SUMMARY

The rumen is one of the best studied and well-understood microbial ecosystems in nature, and it is an ideal habitat for the growth of anaerobic microorganisms. The ruminal microorganisms have been able to grow at relatively rapid growth rates for millions of years, and this long evolution has selected a diverse, and in many cases highly specialized, population of bacteria, protozoa, and fungi. The ecology is clearly interdependent, and the product of one species may be an essential nutrient for another. Organisms occupying all of the major niches have been isolated and characterized, but it is clear that not all species have been identified. A variety of genes have been cloned from ruminal bacteria, but the transfer of DNA into ruminal bacteria was initially thwarted by the lack of suitable shuttle vectors. Some ruminal bacteria use the same methods of transcription as other gram-positive and gram-negative bacteria, but some species seem to have unique promoter sequences. Chemicals and additives can modify ruminal fermentation end products, but these compounds are not highly selective. Ruminal fermentation models were developed to improve ruminant diets, but these models did not contain detailed descriptions of microbial ecology. Further work is needed to identify factors regulating ruminal fermentation.

Acknowledgments

I thank Dr. Paul Weimer, Matthew Fields, Daniel Bond, and Graeme Jarvis for valuable suggestions.

See Also the Following Articles

Diversity, Microbial • Gastrointestinal Microbiology • Growth Kinetics, Bacterial • Methanogenesis • PEP: Carbohydrate Phosphotransferase Systems • Protozoan Predation

Bibliography

Allison, M. J., Daniel, S. L., and Cornick, N. A. (1995). *In* "Calcium Oxalate in Biological Systems" (S. R. Khan, ed.), pp. 131–168. CRC Press, New York.

Bauchop, T. (1979). The rumen anaerobic fungi: Colonizers of plant fibre. *Ann. Rech. Vet.* **10**, 246–248.

Bryant, M. P., and Wolin, M. J. (1975). Rumen bacteria and their metabolic interactions. *In* "First Intersectional Congress of the International Association of Microbiology Society" (T. Hasegawa, ed.), pp. 297–306. Science Council of Japan, Tokyo.

Coleman, G. S. (1980). Rumen ciliate protozoa. *Adv. Parisitol.* **18**, 121–173.

Dehority, B. A., and Orpin, C. G. (1997). *In* "The Rumen Microbial Ecosystem" (P. N. Hobson and C. S. Stewart, eds.), pp. 196–245. Chapman & Hall, New York.

Forsberg, C. W., Cheng, K. J., and White, B. A. (1996). "Gastrointestinal Microbiology" (R. I. Mackie and B. A. White, eds.), pp. 319–379. Chapman & Hall, New York.

Hungate, R. E. (1996). "The Rumen and Its Microbes." Academic Press, New York.

Jones, R. J. (1981). Does ruminal metabolism of mimosine explain the absence of *Leucaena* in Hawaii? *Aust. Vet. J.* **57**, 55–56.

Krause, D. O., and Russell, J. B. (1996). How many ruminal bacteria are there? *J. Dairy Sci.* **79**, 1467–1475.

Russell, J. B., and Strobel, H. J. (1989). Mini-Review: The effect of ionophores on ruminal fermentation. *Appl. Environ. Microbiol.* **55**, 1–6.

Russell, J. B., and Wallace, R. J. (1997). *In* "The Rumen Microbial Ecosystem" (P. N. Hobson and C. S. Stewart, eds.), pp. 246–282. Chapman & Hall, New York.

Russell, J. B., and Wilson, D. B. (1996). Why are ruminal cellulolytic bacteria unable to digest cellulose at low pH? *J. Dairy Sci.* **79**, 1503–1509.

Russell, J. B., Strobel, H. J., and Martin, S. A. (1990). Strategies of nutrient transport by ruminal bacteria. *J. Dairy Sci.* **73**, 2996–3012.

Sauvant, D. (1997). *In* "The Rumen Microbial Ecosystem" (P. N. Hobson and C. S. Stewart, eds.), pp. 685–708. Chapman & Hall, New York.

Stewart, C. S., Flint, H. J., and Bryant, M. P. (1997). *In* "The Rumen Microbial Ecosystem" (P. N. Hobson and C. S. Stewart, eds.), pp. 10–72. Chapman & Hall, New York.

Webster's New World Dictionary of the American Language. (1960). College Edition. The World Publishing Co. Cleveland and New York.

Wolin, M. J. (1975). *In* "Digestion and Metabolism in the Ruminant" (I. W. McDonald and A. C. I. Warner, eds.), pp. 134–148. The University of New England Publishing Unit, Armidale, Australia.

Rust Fungi

Robert F. Park
University of Sydney, Australia

I. Historical Perspectives
II. Rust Diseases of Economic or Potential Importance
III. Life Cycles
IV. Phylogeny and Taxonomy
V. Cytology and Genetics
VI. Modes of Parasitism and Growth
VII. Intraspecific Variability in Rusts
VIII. Control Strategies

GLOSSARY

alternate host One or other of the different hosts infected by heteroecious rust fungi.

autoecious Requiring only one host species or group of closely related host species to complete the life cycle.

collateral hosts A group of closely related plants parasitized by a species of rust fungus. Also referred to as alternative hosts or accessory hosts.

correlated species Some species of rust fungi, which differ principally in their life cycles, that are presumed to be derived from a common ancestor because of similarities in morphology and host range.

demicyclic Lacking the uredinial state; forming only the pycnial, aecial, telial, and basidial states during the life cycle.

gene-for-gene hypothesis A concept proposed by H. H. Flor to describe the genetics of host–pathogen interactions in the flax–flax rust pathogen system; "for each gene conditioning rust reaction in the host there is a specific gene conditioning pathogenicity in the parasite."

heteroecious Requiring two different host species for completion of the life cycle.

macrocyclic Having all five spore types in the life cycle.

microcyclic Lacking the aecial and uredinial states; forming only the pycnial, telial, and basidial states during the life cycle.

obligate parasite Originally used to describe a parasite for which a parasitic phase is an essential part of its life cycle. This term later came to indicate an organism that could not be grown in axenic culture. A preferred term is obligate biotroph, which describes an organism that is dependent on a living host for its survival in nature.

physiologic race (= pathotype, strain) A group of isolates that share common pathogenic attributes. Isolates classified as a single race may or may not be genetically identical.

rust The disease that results from the interaction of a pathogenic rust fungus with a host plant. Although not strictly valid, the term is sometimes used to refer to the fungus itself.

Tranzchel's law In correlated species, telia of derived microcyclic species simulate the habit of the aecia of the parental macrocyclic species and occur on the aecial hosts of the latter.

THE RUST FUNGI are a group of phytopathogenic microfungi that comprise the order Uredinales of the phylum Basidiomycota (Basidiomycota, Teliomycetes, Uredinales). The common name "rust" derives from the characteristic rust-colored spores produced on plants by many species of the Uredinales. Rust fungi are cosmopolitan in distribution and parasitize a wide range of plants, including ferns and conifers and most families of dicotyledonous and monocotyledonous angiosperms. They are of great biological interest due to their obligate parasitism in nature and often complex life cycles. Moreover, many species are of great economic importance. Because of this, the rust fungi are one of the most intensively studied groups of fungal plant parasites, and many of the principles that developed from the work of early uredinologists have found wide applicability in studies of other fungal plant pathogens.

I. HISTORICAL PERSPECTIVES

The earliest documented accounts of rust diseases concern the cereal rusts and date from ancient times. Destructive rust epidemics (epiphytotics) in cereal crops have had a significant effect on the development of human civilization. Early accounts of such epidemics come from the Bible and from Greek and Roman literature. Other references to cereal rusts include a sacred Roman festival known as the Robigalia, dating back to about 700 B.C., in which prayers and sacrifices were made in order to placate a rust god. Observations by various scholars, including Aristotle (384–322 B.C.), Theophrastus (ca. 372–287 B.C.), and Pliny (A.D. 23?–79), associated outbreaks of rust diseases with environmental conditions such as dew and sunshine. Although not understood and disputed for some time, a relationship between rusted grain crops and barberry bushes was recognized as early as the 1600s. Barberry bushes growing near grain crops were often destroyed, and this was enforced by law in some regions (e.g., Rouen, France, 1660; Connecticut Colony, America, 1726).

Felice Fontana is regarded as the first person to recognize that rust is caused by a fungal parasite, an observation that was reported in 1767 following a severe rust epidemic in Italy. However, it was not until the latter part of the nineteenth century that the various species of rust fungi were generally recognized as being distinct and that studies of the biology of the organisms involved were initiated.

II. RUST DISEASES OF ECONOMIC OR POTENTIAL IMPORTANCE

The rust fungi are among the most destructive of plant pathogens. They have caused immense damage in many of the plants cultivated by humans, including field and vegetable crops (Table I), trees (Table II), and ornamental plants.

Because of the global importance of cereals as a food source, the cereal rust fungi are often regarded as causing not only the most important rust diseases, but also the most important plant diseases. Three rust diseases occur in wheat: stem or black rust (caused by *Puccinia graminis* Pers. f. sp. *tritici* Erik. & Henn.), stripe or yellow rust (caused by *P. striiformis* Westend. f. sp. *tritici* Erik. & Henn.), and leaf or brown rust (caused by *P. recondita* Rob. ex Desm. f. sp. *tritici* Erik. & Henn.). Stem rust has caused complete failures of wheat crops, and severe infection by the stripe and leaf rust pathogens have caused grain yield losses exceeding 50%. Epidemics of the diseases have consequently had significant adverse economic effects; an epidemic of stem and leaf rusts in southeastern Australia during 1973, for example, resulted in financial losses of about $A200 million. In cereal crops that are also grown for forage production (e.g., oats), rust infection can reduce the yield and palatability of forage in addition to reducing grain yield.

Because of the perennial nature of trees, economic damage caused by rust species in tree crops can be enormous. Estimates of worldwide annual losses in coffee crops due to the rust pathogen *Hemileia vastatrix* Berk. & Br. are of the order of $US1 to 2 billion. The cultivation of tea in Sri Lanka (Ceylon) was a consequence of severe epidemics of rust in coffee plantations in the country during the 1870s and 1880s. This lead to tea being the preferred beverage across much of the former British Empire. Damaging rust diseases also occur on important wood-producing trees such as species of *Pinus*. Fusiform stem canker of loblolly and slash pine, caused by *Cronartium quercuum* (Berk.) Miy. ex Shirai. f. sp. *fusiforme* (*C. fusiforme*), was estimated to cause annual losses of at least $US75 million in the U.S. The rust *Puccinia psidii* Wint. attacks several myrtaceous species, including guava (*Psidium guajava*) and introduced eucalypts growing in Central and South America and the Caribbean. It is now regarded as an important stem and leaf pathogen of eucalypts in this region, but has so far not been reported elsewhere (one report of a rust with identical urediniospores was made from Taiwan, but this has not been seen since). *Puccinia psidii* is a major quarantine concern for regions such as Australia, where it could cause enormous damage to eucalypts in native plant communities and plantations.

Rust diseases are also common on ornamental plants, and although not so economically important, they can cause severe damage to valued garden plants. Some examples (with causal agents given pa-

TABLE I
Examples of Rust Diseases of Economic or Potential Importance on Field and Vegetable Crops

Rust pathogen	*Disease*	*Hosts*[a]
Melampsora lini (Pers.) Lév.	Flax rust	*Linum usitatissimum* (autoecious)
Phakopsora pachyrhizi Syd.	Brown rust of soybean	*Glycine max*/unknown
Puccinia arachidis Speg.	Groundnut rust	*Arachis hypogaea*/unknown
P. coronata Cda.	Crown or leaf rust of oats	***Avena sativa***/*Rhamnus*
P. graminis Pers.[b]	Stem or black rust of barley	***Hordeum vulgare***/*Berberis, Mahonia*
P. graminis f. sp. *avenae*	Stem or black rust of oats	***Avena sativa***/*Berberis*
P. graminis f. sp. *secalis*	Stem or black rust of rye	***Secale cereale***/*Berberis, Mahonia*
P. graminis f. sp. *tritici*	Stem or black rust of wheat	***Triticum aestivum***/*Berberis, Mahonia*
P. helianthi Schw.	Sunflower rust	*Helianthus annuus* (autoecious)
P. hordei Otth.	Leaf or brown rust of barley	***Hordeum vulgare***/*Ornithogalum*
P. melanocephala Syd.	Sugarcane rust	*Saccharum* spp./unknown
P. menthae Pers.	Mint rust	*Mentha* spp. (autoecious)
P. polysora Underw.	Southern corn rust	*Zea mays*/unknown
P. purpurea Cke.	Sorghum rust	***Sorghum* spp.**/*Oxalis*?
P. recondita f. sp. *secalis*	Leaf or brown rust of rye	***Secale cereale***/*Anchusa, Lycopsis*
P. recondita Rob. ex Desm. f. sp. *tritici*	Leaf or brown rust of wheat	***Triticum aestivum***/*Thalictrum, Anchusa*
P. sorghi Schw.	Common corn rust	***Zea mays***/*Oxalis*
P. striiformis West. f. sp. *hordei*	Stripe or yellow rust of barley	*Hordeum vulgare*/unknown
P. striiformis f. sp. *tritici*	Stripe or yellow rust of wheat	*Triticum aestivum*/unknown
Uromyces appendiculatus (Pers.) Unger	Bean rust	*Phaseolus vulgaris* (autoecious)

[a] Hosts are listed as telial/aecial, with the host of greater economic importance given in bold type. Often more than one species of host is infected; examples of important host species are given where possible.
[b] Several forms of *P. graminis*, including the wheat-attacking and cereal rye-attacking forms, can parasitize barley.

TABLE II
Examples of Rust Diseases of Economic or Potential Importance on Tree Crops

Rust species	*Disease*	*Host*[a]
Cronartium coleosporioides Arth.	Stalactiform blister rust	*Melampyrum, Castilleja*/***Pinus* spp.**
C. quercuum (Berk.) Miy. ex Shirai. f. sp. *fusiforme*	Fusiform stem canker	*Quercus*/***Pinus taeda, P. elliotti* var. *elliottii***
C. ribicola Fisch. ex Raben.	White pine blister rust	*Ribes*/***Pinus* spp. (5-needle pines)**
Chrysomyxa abietis (Wallr.) Unger	Spruce needle rust	*Picea* spp. (autoecious)
Endocronartium harknessii (Moore) Hirat.	Western gall rust	*Pinus contorta* var. *latifolia* (autoecious)
Gymnosporangium juniperi-virginianae Schw.	Cedar-apple rust	***Juniperus virginiana*/*Malus sylvestris***
Hemileia vastatrix Berk. & Br.	Coffee leaf rust	***Coffea arabica***/unknown
Melampsora larici-populina Kleb.	Poplar leaf rust	***Populus* spp.**/*Larix*
M. medusae Thüm.	Poplar leaf rust	***Populus* spp.**/*Larix, Pseudotsuga*
M. pinitorqua Rostr.	Pine twisting rust	*Populus tremula*/***Pinus sylvestris***
Puccinia psidii Wint.	Guava rust	*Psidium guajava, Eucalyptus* spp. (autoecious)
Tranzschelia discolor (Fckl.) Tranz. & Litv.	Peach and plum rust	***Prunus* spp.**/*Anemone coronaria*
Uromycladium tepperianum (Sacc.) McAlp.	Acacia gall rust	*Acacia* spp. (autoecious)

[a] Hosts are listed as telial/aecial, with the host of greater economic importance given in bold type. Often more than one species of host is infected; examples of important host species are given where possible.

rentheses) include mallow and hollyhock rust (*Puccinia malvacearum* Bert. ex Mont.), rose rust (*Phragmidium* spp., particularly *P. tuberculatum* Müller and *P. mucronatum* (Pers.) Schlecht), pelargonium rust (*Puccinia pelargonii-zonalis* Doidge), snapdragon rust (*Puccinia antirrhini* Diet. & Holw.), chrysanthemum rust (*Puccinia chrysanthemi* Roze), and white rust of carnation (*Puccinia horiana* P. Henn.).

III. LIFE CYCLES

A. Hosts

One of the most fascinating features of rust fungi is the requirement of certain species for two unrelated hosts for completion of the full life cycle. This phenomenon, known as heteroecism, is rare in fungi other than rusts. *Puccinia graminis* f. sp. *tritici* is a well-known heteroecious species, requiring both wheat and species of *Mahonia, Berberis,* or certain *Mahonia* × *Berberis* hybrids to complete its full life cycle. Although either of the two different hosts of heteroecious species can be referred to as an alternate host, this term is commonly applied to the host of less economic importance. In addition to wheat, *P. graminis* f. sp. *tritici* is also able to infect certain closely related grass species, and these hosts are referred to as collateral, accessory, or alternative hosts.

In contrast to heteroecism, autoecious species complete their life cycle on one host or a group of collateral hosts.

B. Spore States

Up to five spore states may be produced during the life cycles of rust fungi. Spore states have been defined on the basis of morphological characteristics, spore ontogeny, or the nuclear cycle. For some species, spore states are the same irrespective of the scheme used, but for others, differences do occur. The cytology of many rust species has not been studied in detail, and when applied, the ontogenetic system is frequently based on morphological comparisons with species for which cytological information is available.

The five spore states are conventionally symbolized with Roman numerals as 0, I, II, III, and IV. However, caution is needed in their use since problems can arise when they are applied to ontogenetically based systems. The five spore states, as they relate to the nuclear cycle, are pycniospores, aeciospores, urediniospores, teliospores, and basidiospores.

1. *Pycniospores (0)*

Also referred to as spermatia (produced in spermogonia) and pycnidiospores (produced in pycnidia). Pycniospores are single-celled haploid male gametes produced in pycnia (singular pycnium) and exuded in a nectar that attracts insects.

2. *Aeciospores (I)*

Also referred to as aecidiospores and plasmogamospores. Aeciospores are produced in aecia (singular aecium), resulting from dikaryotization occurring after fertilization. The aecial initials that give rise to aecia are sometimes referred to as the female gametes, since they receive the pycniospores of the opposite mating type through flexuous or receptive hyphae known as trichogynes. The dikaryotic aeciospores are nonrepeating vegetative spores that germinate and produce a dikaryotic mycelium on which uredinia and/or telia are produced.

3. *Urediniospores (II)*

Also referred to as uredospores, urediospores, summer spores, and red rust spores. Urediniospores are vegetative spores produced in uredinia (singular uredinium; also referred to as uredosori, uredia), usually composed of dikaryotic mycelium. Typically, they are unicellular, are produced singly on pedicels, are pigmented, have echinulate surface ornamentation, possess two or more germ pores in various arrangements, and are capable of germinating immediately on formation. Studies of several species of rust fungi have established that urediniospores may produce compounds which act as self-inhibitors of germination. Urediniospores of bean, sunflower, corn, snapdragon, peanut, and stripe rust fungi produce methyl *cis*-3,4-dimethoxycinnamate, which inhibits germination of urediniospores at concentrations of several picograms to nanograms per milliliter. The production of self-inhibitors of germination is thought to be an adaptation to prevent spore germination in the uredinium prior to dispersal, and to prevent rapid simultaneous germination of all spores. Urediniospores are usually produced repeatedly during the main growing phase of the host and are therefore associated with the phase of maximum potential damage.

4. *Teliospores (III)*

Also referred to as teleutospores, teleutosporodesma, winter spores, resting spores, and black rust spores. Teliospores are produced in telia (singular telium; also referred to as teleutosori), and together these comprise the teleomorph. Both the spores and the sporocarps vary greatly in morphology among species. Teliospores may germinate to produce basidia immediately after formation, or may undergo a period of dormancy that is broken only by certain environmental conditions. The teliospores of species that germinate immediately are sometimes referred to as leptospores if most other species of that genus display dormancy.

5. *Basidiospores (IV)*

Also referred to as sporidia. Basidiospores are haploid unicellular spores produced on basidia (singular basidium) after meiosis. The basidia are usually four-celled and give rise to four basidiospores.

The delineation between spore types is not always distinct due to morphological similarity. For example, in some instances aeciospores resemble urediniospores (uredinioid aeciospores), and rarely, urediniospores may resemble aeciospores (aecidioid urediniospores).

Species in which all spore states are present are termed macrocyclic; those lacking the uredinial state are termed demicyclic, and those lacking both the uredinial and aecial states, microcyclic. By combining these terms with those relating to hosts required for the full life cycle, several authors have recognized five basic variations in the life cycles of rust fungi (Table III).

C. Correlated Species

In rusts, the heteromacrocyclic life cycle is generally regarded as the primitive ancestral form, from which all other life cycles were derived by the progressive loss and/or alteration of different stages. This hypothesis is supported by the observation that rust species parasitizing more "primitive" plants are often heteromacrocyclic, and those on more "advanced" hosts tend to be autoecious and microcyclic. It was also suggested that more complex life cycles were derived from certain "primitive" autoecious species occurring on hosts such as mosses and ferns, and that "modern" autoecious species were in turn derived from the complex types by reduction.

On the basis of similarities in morphology and host range, some rust species that differ principally in their life cycles are presumed to be derived from a common ancestor. Members of such groups are referred to as correlated species. A feature of corre-

TABLE III
The Five Basic Life Cycles of Rust Fungi

Life cycle	*Spore states present*[a]	*Example of fungal species (common name of disease)*
Heteromacrocyclic	(0), I–II, III, IV	*Puccinia graminis* f. sp. *tritici* (stem or black rust of wheat)
Automacrocyclic	(0), I, II, III, IV	*Melampsora lini* (Pers.) Lév. (flax rust)
Heterodemicyclic	(0), I–III, IV	*Gymnosporangium juniperi-virginiae* Schw. (cedar-apple rust)
Autodemicyclic	(0), I, III, IV	*Arthuriomyces peckianus* (Howe) Cumm. & Y. Hirat. (orange rust of *Rubus*)
Microcyclic[b]	(0), III, IV	*Puccinia malvacearum* Bert. ex Mont. (hollyhock rust)

[a] See text for details. Parentheses indicate that the state may or may not be present.
[b] Several microcyclic rust species have basidia borne on teliospores which morphologically resemble aeciospores. These are referred to endocyclic and include species in the genera *Endocronartium* Y. Hirat., *Endophyllum* Lév., and *Gymnoconia* Lagh. (=*Kunkelia* Arth.).

lated species is that the telia of the derived microcyclic species occur on the same host as the aecia of the ancestral macrocyclic type, and these telia often assume the habit of the aecia of the ancestral type. This is known as Tranzchel's law, and it appears to apply in all cases where microcyclic telia occur on the aecial host of the ancestral heteroecious species. One simple example used by several authors to illustrate this law is that of *Tranzschelia discolor*, a macrocyclic species which causes rust of peach and plum trees and which produces systemic aecia on the garden anemone, *Anemone coronaria* L. In contrast, the reduced microcyclic species *T. anemones* has systemic telia on the same host.

IV. PHYLOGENY AND TAXONOMY

The Uredinales are regarded as an ancient group of fungi, which possibly evolved from a simple ascomycete species, similar to those in the genus *Taphrina*, that parasitized ancient ferns or their ancestors during the Carboniferous period. The connection between ascomycete species and the Uredinales (Basidiomycota) is based on similarities in several morphological and cytological features, including spore ontogeny and nuclear and chromosome structure. The obligate parasitism of the Uredinales in nature has led to considerable coevolution with their hosts, making the host–pathogen relationship a particularly important biological feature of these fungi. Indeed, the host specialization exhibited by many rust fungi can greatly simplify species determination.

The Uredinales are one of the largest groups of the Basidiomycota, and although they are very distinct as a group, classification above the level of genus is still subject to considerable debate. For some time, two families were recognized, separable on the basis of teliospore morphology: the Melampsoraceae, characterized by sessile teliospores, and the Pucciniaceae, with pedicillate teliospores. Distinction of the Melampsoraceae into three further families (Melampsoraceae, Coleosporiaceae, and Cronartiaceae) was made on the basis of the arrangement of teliospores in telia. More recent appraisals of the taxonomy of the group have placed greater importance on spermogonial structure, for which 12 different morphological types were characterized. Currently, 14 families are recognized using this feature as the main criterion: Chaconiaceae, Coleosporiaceae, Cronartiaceae, Melampsoraceae, Mikronegeriaceae, Phakopsoraceae, Phragmidiaceae, Pileolariaceae, Pucciniaceae, Pucciniastraceae, Pucciniosiraceae, Raveneliaceae, Sphaerophragmiaceae, and Uropyxidaceae. However, it is generally accepted that further research is needed to clarify the taxonomy of rust fungi at the familial level. In addition to spermogonial morphology, criteria such as teliospore characters, spore ontogeny, uredinial characters, host–pathogen associations, and life cycles all need to be considered in this regard. Recently developed techniques for examining variability at the DNA level should also provide valuable insight.

Unlike other fungal groups such as the Ascomycota, the asexual or imperfect states of rust fungi generally have not been placed in the Deuteromycota, leading to difficulties with nomenclature. About 7000 species are currently accepted as valid taxa, most of which belong to 12 of the approximately 164 accepted genera (there are also 139 generic names that are regarded as either synonyms or of uncertain status). The largest genera are *Puccinia* and *Uromyces*, both within the family Pucciniaceae, and account for 3000 to 4000 and 600 to 700 species, respectively.

V. CYTOLOGY AND GENETICS

A. Nuclear Cycle

Rust fungi are either heterothallic or homothallic. The first demonstration of heterothallism in rusts was made in *P. graminis* by J. H. Craigie. These studies established the gametic function of pycniospores in this species. In a typical heteromacrocyclic species, "+" and "−" mating types arise from meiosis in the basidium. The "+" or "−" basidiospores infect a receptive host and establish a haploid monokaryotic mycelium in which pycnia are formed. Various studies have established the role of insects in aiding fertilization in rust fungi by transmitting pycniospores between mating types. Pycniospores are produced within a sticky sweetish nectar which is exuded from the pycnium and which attracts insects. In cases where a "+" pycniospore comes in contact with a flexuous hypha from a "−" pycnium, plasmogamy occurs and gives rise to dikaryotic cells which then form the aecium. The processes involved in the development of aecia, following plasmogamy, are poorly understood. Mature aecia produce dikaryotic aeciospores that infect the telial host to produce a dikaryotic mycelium within which uredinia and subsequently urediniospores are produced. The urediniospores are regarded as dikaryotic conidia, and the cycle of production and infection can lead to a rapid buildup of the pathogen population. Telia are produced on this mycelium, usually in response to physiological changes in the host such as senescence. The teliospores are initially dikaryotic, but in rusts such as *P. graminis,* they undergo nuclear fusion prior to the production of the basidium. Variations do occur with respect to the behavior of nuclei in teliospores and basidia, particularly in endocyclic and microcyclic species. Various other mechanisms for dikaryotization are also known, especially in homothallic species, which often lack pycnia (e.g., fusion of mycelia, spontaneous dikaryotization, and self-dikaryotization).

The chromosomes of rust fungi are small and difficult to resolve by light microscopy. Various attempts have been made to determine karyotypes for rust fungi, with most being based on light microscopy. A karyotype of $n = 18$ was determined for six North American isolates of *P. graminis* f. sp. *tritici* by producing three-dimensional reconstructions of teliospore pachytene nuclei based on electron micrographs of serial sections of protoplasts. This study also showed the absence of chromosomal centromeres and provided evidence that the length of individual chromosomes may vary between isolates of this species.

Genome size has been examined in various rust species using methods such as flow cytometry and light microscope photometry (to give an indication of relative size), or reassociation kinetics. The latter was used to demonstrate a genome size of 58 Mb in *P. graminis* f. sp. *tritici,* comprising 64% single copy sequences and 36% repeat sequences. Similar studies of *Puccinia sorghi* revealed a genome size of approximately 50 Mb, of which one-half comprised a moderately repetitive fraction.

B. Double-Stranded RNA in Rust Fungi

Double-stranded (ds) RNAs are found in the cytoplasm of a wide range of fungi, including the Uredinales, and occur either as encapsidated viral genomes (mycoviruses) or, less frequently, as naked or unencapsidated molecules. Most rust fungi examined harbor significant amounts of dsRNAs.

Studies have established that dsRNAs are transmitted between isolates during the sexual cycle. They are also transmitted by asexual means, possibly through anastomoses between the hyphae of different isolates of the same species, although the exact means by which dsRNAs are maintained and/or lost in isolates are unknown. It has been shown that dsRNAs present in specific isolates derived from single urediniospores remain unchanged following repeated subculturing of at least six generations. There is no evidence for transmission between species that occur on the same host (e.g., the three rust pathogens of wheat), and different species have distinctive patterns of dsRNAs, within which variation is observed. Within-species variation was particularly apparent in a study of isolates of *P. striiformis* collected from eastern Australia over a 4-year period. Although the isolates were considered to be closely related on the basis of pathogenicity, differences were found in the dsRNA present both between and within pathotypes.

The function of dsRNAs in rust fungi is unknown. Many species that contain large quantities of dsRNAs are potentially destructive pathogens, indicating that the presence of dsRNA has no observable effect on pathogenic fitness. Furthermore, controlled environment studies on *Melampsora lini* failed to show any difference between the growth of isogenic isolates differing only in the presence or absence of dsRNA.

C. Inheritance Studies

In discussing the genetics of host–pathogen interactions in rust diseases, the term pathogenicity is often used to describe the pathogen character involved in the interaction with the host. The contrasting phenotypes of pathogenicity are avirulence

TABLE IV
Major Infection Type Classes for Stem and Leaf Rusts of Cereals

Infection type[a]	*Host response*	*Symptoms*
0	Immune	No visible symptoms
;	Very resistant	Hypersensitive flecks
1	Resistant	Small uredinia with necrosis
2	Resistant to moderately resistant	Small to medium-sized uredinia with green islands and surrounded by necrosis or chlorosis
3	Moderately resistant to moderately susceptible	Medium-sized uredinia with or without chlorosis
4	Susceptible	Large uredinia without chlorosis
X	Resistant	Mesothetic, heterogeneous uredinia similarly distributed over the leaves
Y	?	Variable size with larger uredinia toward the leaf tip
Z	?	Variable size with larger uredinia toward the leaf base

[a] Variations within a class are indicated by the use of − (less than average) and + (more than average), as well as C and N to indicate more than usual degrees of chlorosis and necrosis, respectively.

and virulence. The outcome or phenotype of an interaction between a host and a pathogen is often referred to as an infection type (Table IV). This interaction can be referred to as the disease reaction or disease response, and is either incompatible (low; resistant host, avirulent pathogen) or compatible (high; susceptible host, virulent pathogen).

Genetic studies with rust pathogens are technically demanding, and for this reason relatively few have been made. Two common problems are the induction of germination in dormant teliospores and maintaining purity of the progeny from crosses. Pioneering studies of the genetic interaction between flax, *Linum usitatissimum,* and the autoecious rust pathogen *Melampsora lini* by H. H. Flor led to the formulation of the gene-for-gene hypothesis. The hypothesis was based on results of parallel experiments that examined the inheritance of pathogenicity in *M. lini* and the inheritance of disease reaction in *L. usitatissimum.* In *M. lini,* a ratio of 3 avirulent : 1 virulent was obtained when F_2 progeny of a cross between two isolates contrasting in pathogenicity were inoculated onto a flax cultivar which was resistant to one parental isolate but susceptible to the other. In *L. usitatissimum,* a ratio of 3 resistant : 1 susceptible was obtained when F_2 progeny of a cross between the same two flax cultivars contrasting in disease response were inoculated with the avirulent isolate of *M. lini* used to conduct the pathogenicity study. Flor demonstrated the presence of 26 such independent corresponding loci in the host and pathogen, and proposed that "for each gene conditioning rust reaction in the host there is a specific gene conditioning pathogenicity in the parasite." This hypothesis has found wide applicability in other plant–pathogen interactions, including fungi other than the Uredinales, bacteria, and viruses.

Studies of the inheritance of pathogenicity have been made in a variety of rust fungi, particularly those infecting cereals. Often, single loci were implicated, with avirulence being dominant to virulence; however, exceptions to dominance relationships are known. For example, a parallel genetic study of pathogenicity in *Puccinia recondita* f. sp. *tritici* and of resistance in its host, common wheat, indicated that avirulence could be dominant, incompletely dominant, or recessive, depending on the pathogen locus being considered and the genotype of the host line being used. Detailed studies of "putative" recessive avirulence in *M. lini* showed that, in fact, there was a dominant suppressor of avirulence that resulted in confounding of 1 : 3 (1 locus) and 3 : 13 (2 loci) genetic ratios. Other genetic studies of *P. recondita* f. sp. *tritici* demonstrated linkage between some genes governing pathogenicity. Reciprocal crosses of two isolates of the oat stem rust pathogen, *P. graminis* f. sp. *avenae,* indicated maternal inheritance of avirulence for the resistance gene *Pg3,* implying that the corresponding avirulence gene is not located on the nuclear genome.

Variants differing in urediniospore color are known to occur in some species. Such variants have arisen spontaneously under controlled conditions, or in the field, and have also been induced by chemical mutagenesis. The inheritance of red urediniospore color in *P. graminis* f. sp. *tritici* is governed by two independently acting dominant genes, one controlling brown pigmentation in the spore wall and the other, orange carotenoid pigmentation in the cytoplasm. Yellow mutants lacking spore wall pigmentation, gray-brown mutants lacking cytoplasmic pigmentation, and white mutants lacking both are known to exist. A study of the reversion rate of yellow to red (normal) pigmentation in *P. graminis* f. sp. *tritici* demonstrated a high rate of 0.228%, whereas mutations to yellow pigmentation occurred only rarely.

To date, few studies have examined the inheritances of biochemical (e.g., allozyme) or DNA-based markers (e.g., random amplified polymorphic DNAs, RAPDs; amplified fragment length polymorphisms, AFLPs) in rust fungi. The pattern of segregation of five isozyme loci observed in F_2 progeny derived from a cross between two North American isolates of *P. graminis* f. sp. *tritici* suggested Mendelian inheritance of nuclear genes, and provided evidence of linkage between the loci. Recently, North American workers have examined the inheritance of RAPD markers in *P. graminis* f. sp. *tritici*, using an F_2 population produced by selfing an F_1 individual. Eight dominant avirulence genes were characterized in the population, and for each of these, linked RAPD markers were identified. The ultimate aim of this work is to clone and characterize avirulence genes in this rust species.

VI. MODES OF PARASITISM AND GROWTH

A. Establishment of Rust Infections

Many histological studies have examined the processes of infection by rust pathogens in susceptible and resistant hosts. In most cases, these have focused on the establishment of the dikaryotic mycelium originating from urediniospores.

Germination of urediniospores on the leaf surface of a susceptible host requires the presence of free moisture and can be influenced by environmental conditions such as temperature and light. Typically, urediniospores produce an adhesion pad to maintain contact with the host cuticle, and a germ-tube which grows across the leaf surface and gains entry to the leaf via stomata. The germ-tubes of some rust fungi (e.g., the soybean rust fungus, *Phakopsora pachyrhizi* Syd.) are able to directly penetrate the leaf cuticle. In species that utilize stomata, thigmotropic responses to the topography of the leaf surface are thought to be involved in stomatal recognition, since germ-tubes often become oriented on the leaf surface and may grow at right angles to veins. The topography of the stomatal guard cell and/or the chemical environment around the stoma triggers the production of an appressorium from which an infection peg forms, pushing through the stomatal pore and subsequently giving rise to a vesicle in the substomatal cavity. Infection hyphae develop from the vesicle and ramify intercellularly throughout leaf tissues to produce a dikaryotic mycelium, on which small terminal branches form. In turn, the branches form haustorial mother cells, from which a penetration peg extends through the host cell wall to produce a haustorial neck and haustorium. The host plasmalemma becomes invaginated and surrounds the entire haustorium. Haustoria are generally thought to be involved in the absorption of host nutrients by the pathogen, although there is little or no direct evidence for this. Those formed on dikaryotic mycelia are referred to as D-haustoria.

The morphology of monokaryotic mycelia derived from infection by basidiospores differs substantially from that of the dikaryotic mycelium. Where studied, infection by basidiospores typically occurs by direct penetration of the host cuticle rather than via stomatal pores. In comparison with the dikaryotic mycelium, the morphology of the monokaryotic phase is less complex, and branches that form intracellular haustoria (M-haustoria) do not differ greatly from the intracellular mycelia.

In some autoecious rust fungi [e.g., *Puccinia punctiformis* (Strauss) Roehl. and *P. menthae* Pers.], dikaryotic and monokaryotic mycelia show fundamental differences in modes of infection. In the monokaryon of these species, systemic infection of vascular tissues is accompanied by alterations in host growth (see next section), whereas this does not occur in the dikaryon. Such vascular infections may become perennial. Similar differences in ability/inability to infect vascular tissues are seen between the monokaryons/dikaryons of heteroecious rusts (e.g., *Cronartium ribicola* Fisch. ex Raben. on *Pinus strobus* and *Ribes* spp., *Puccinia coronata* Cda. on *Rhamnus cartharticus* and *Avena sativa*).

Although often localized, dikaryotic mycelia can also become systemic within host tissues. Localized dikaryotic mycelia originating from single points of infection can be used to establish single spore isolates of rust fungi with reasonable accuracy by assuming that each uredinium is derived from a single urediniospore. Some species, such as *Puccinia striiformis*, establish systemic dikaryotic infections that can extend the entire length of an infected leaf.

B. Effects of Rust Infection on the Host

1. Host Physiology

The infection of plants by rust pathogens has significant effects on host physiology. A common, visibly striking effect is the formation of "green islands," where regions of host tissue in the vicinity of uredinia remain green while surrounding tissue becomes chlorotic. The formation of green islands is due to the retention of chlorophyll in infected tissue, and it is also observed with other biotrophic pathogens such as those causing mildew diseases.

Studies on the effects of rust infection on cereals have indicated that the pathogen modifies levels of hormones in infected tissues by either producing them themselves or by altering the concentrations of hormones produced by the host. Cytokinins, auxin, ethylene, and abscisic acid are all known to increase in concentration in infected leaves. Cytokinins delay senescence when applied to healthy leaves, and could therefore contribute to the retention of chlorophyll and green island formation. Both ethylene and abscisic acid may be responsible for the accelerated senescence of host tissues seen in later stages of infection.

Rust fungi also affect metabolic activity in infected tissues. Respiration increases, but the increase observed comprises not only that of the host, but also (perhaps mostly) that of the fungus. Variations in photorespiration and photosynthesis also occur. In one study, both processes declined in rust-infected bean leaves, whereas infected cereal leaves displayed increased (barley infected with *Puccinia hordei* Otth.) or unchanged (wheat infected with *P. graminis* f. sp. *tritici*) photorespiration along with a decline in photosynthesis. The decline in photosynthesis in rust-infected tissue is thought to result from the loss of chlorophyll and proteins important to the process.

The rust mycelium also affects the translocation patterns of photosynthates within infected plants. Studies with the bean rust pathogen *Uromyces phaseoli* [*U. appendiculatus* (Pers.) Unger] demonstrated that photosynthesis in uninfected trifoliate leaves of plants with rusted primary leaves was higher than that of uninfected control plants. The application of $^{14}CO_2$ to different parts of locally infected plants demonstrated that infected primary leaves imported 40 times more photosynthate from the uninfected trifoliate leaves than did those of uninfected controls. Similar results were obtained in studies with other rust pathogens and host plants. Together, the results indicate that the accumulation of photosynthates in infected leaves results from the fungus acting as a metabolic sink through increasing translocation to the infection site and by preventing translocation away from the site.

2. Host Growth

Normal growth patterns of plants can be altered radically by rust infection (Fig. 1). The changes may result from increased levels of hormones such as auxins in infected tissues or by the physical presence of hyphae resulting in enlargement of structural tissue or blocking of vascular elements.

Systemic vascular or meristematic infections established by many rust monokaryons often cause hypertrophy of host tissues (e.g., thickening of leaves, formation of galls, or changes in host plant habit, including witches'-brooms). For example, monokaryotic systemic infections by *Puccinia monoica* Arth.

Fig. 1. Symptoms of selected rust diseases. (A) Stripe or yellow rust of wheat caused by *Puccinia striiformis* f. sp. *tritici.* (B) Stem or black rust of wheat caused by *P. graminis* f. sp. *tritici.* (C) Leaf or brown rust of wheat caused by *P. recondita* f. sp. *tritici.* (D and E) Rust of *Arabis holboelii* caused by *P. monoica.* The rust transforms normal host morphology (D) by creating elevated clusters of infected leaves that mimic true flowers (E) and attract insects that fertilize the rust (photographs by B. A. Roy, Swiss Federal Institute of Technology, reprinted by permission from *Nature* **362,** 56–58). (F) Gall rust of acacia caused by *Uromycladium tepperianum.* (G) Mint rust caused by *P. menthae.* Distortion of stems occurs due to vascular infection during the aecial stage. See color insert. (Courtesy Dr. J. Edwards, University of Melbourne, Australia.)

alter the growth of *Arabis* species by invading meristematic tissues. Host growth is altered by the inhibition of flowering and the induction of pseudoflowers, which resemble flowers of unrelated co-occurring plants. The surface of the pseudoflowers is primarily composed of pycnia, receptive hyphae, and sugary pycnial fluid, and is very successful in attracting pollinating insects, which fertilize the pathogen.

C. Axenic Culture of Rust Fungi

The growth of organisms of a single species in the absence of living organisms or cells of any other species (i.e., growth on nonliving substrata) is known as axenic culture. In nature, rust fungi are obligate biotrophs, meaning that they are completely dependent on a living host for nutrition. Many unsuccessful attempts were made to grow rust fungi on artifical media in isolation from their hosts. The first documented success was obtained with the cedar-apple rust fungus, *Gymnosporangium juniperi-virginianae* Schw. Several axenic cultures were established by subculturing mycelium which grew from cultured sections of telial galls from *Juniperus* on agar medium. Success with the wheat stem rust fungus, *P. graminis* f. sp. *tritici*, was achieved by Australian scientists in 1966. In this case, the primary inoculum comprised aseptically produced urediniospores which were densely seeded onto media. Mycelia developed slowly on media containing Czapek's nutrients, 0.1% yeast extract, and 3.0% sucrose. Later studies showed that addition of 0.1% Evan's peptone resulted in increased vegetative growth and the formation of urediniospores and teliospores. The urediniospores were capable of establishing successful infections on a susceptible host under greenhouse conditions. Since this initial work, other species such as *Puccinia striiformis, P. recondita, Phragmidium violaceum* (C. F. Schultz) Wint. (causal agent of blackberry rust), *Melampsora lini, Uromyces caryophyllinus* (Schr.) Wint. (= *U. dianthi;* causal agent of carnation rust), and *Puccinia helianthi* Schw. (causal agent of sunflower rust) have been axenically cultured from urediniospores, although results are often variable. In the case of *P. graminis* f. sp. *tritici* at least, some isolates appear to be more amenable to axenic culture than others.

VII. INTRASPECIFIC VARIABILITY IN RUSTS

A. Formae Speciales

Within species of rust fungi, it is common to find variants which, although morphologically indistinguishable, are adapted to parasitizing different host species. Jakob Eriksson, a Swedish plant pathologist, was the first to coin the term formae speciales ("special forms"; f. spp.) to designate these variants. Forma specialis (f. sp.) is an informal rank that is not regulated by the International Code of Botanical Nomenclature. In some cases variants within rust species were given the designation "varietas" (var.), although this should only be used when there are distinguishable morphological characters.

Formae speciales are usually named according to the host with which the dikaryon is most commonly associated. Hence, three of the many formae speciales which occur in the stem rust pathogen *Puccinia graminis* are *P. graminis* f. sp. *tritici* (on wheat, *Triticum aestivum*), *P. graminis* f. sp. *avenae* (on oats, *Avena sativa*), and *P. graminis* f. sp. *secalis* (on cereal rye, *Secale cereale*). Despite the host specificity of these special forms, there are some host species, or genotypes of host species, that are susceptible to two or more forms, and these are referred to as common hosts. Some genotypes of wheat are common hosts for *P. graminis* f. spp. *tritici* and *secalis*. These genotypes have been used to examine the genetic basis of resistance of wheat to *P. graminis* f. sp. *secalis*. Several wheat cultivars lacking effective resistance genes to *P. graminis* f. sp. *tritici* possessed from one to four genes for resistance to *P. graminis* f. sp. *secalis*. Similarly, the resistance of two barley cultivars to *P. graminis* f. sp. *avenae* was simply inherited and possibly involved a single gene.

The monokaryons of related formae speciales generally share the same alternate host. In *P. graminis*, viable sexual crosses have been made between, *P. graminis* f. spp. *tritici* and *secalis* and, on one occasion, between *P. graminis* f. spp. *tritici* and *avenae*.

In general, the progeny from intercrosses between formae speciales have a lower range of virulence than the parental isolates, with the F_1 progeny being avirulent on the principal hosts of both parents but often virulent on a common host species.

B. Pathogenic Variability

Isolates within formae speciales may differ in ability to infect different genotypes within the host species with which they are most commonly associated. These differences were first identified in the wheat stem rust pathogen by E. C. Stakman in 1916, who referred to variants with the same pathogenicity as physiologic races (the terms strain and pathotype are now more commonly used). Pathotypes comprise a group of isolates possessing the same pathogenicity on a selection of host genotypes; these isolates may or may not be genetically identical. The host genotypes used to identify pathotypes are referred to as differential lines, and in the case of cereal rusts, each usually possesses a different genetically characterized resistance. Pathotypes are designated by a code; however, the coding systems used often differ between countries and at times even between laboratories within countries. When publishing information on pathotypic variability, it is good practice to include relevant details of the virulence/avirulence of each pathotype.

The most detailed surveys of pathogenic variability in rust fungi have involved those parasitizing cereals. In some regions, such surveys have been conducted on a continuing basis for many years, providing not only information to assist cereal breeders in developing rust-resistant cultivars, but also a long-term account of the evolution and maintenance of pathogenic variability in the pathogen populations. In Australia, which is isolated from other major cereal-growing regions of the world, and where wheat rusts do not undergo sexual recombination, ongoing pathogenicity surveys suggest that new pathotypes primarily appear via single-step mutations from avirulence to virulence. Lineages can be constructed for the three wheat rust pathogens, each comprising isolates derived from a common ancestor and which differ presumably by the addition of a single virulence (Fig. 2). Although the frequency of mutation is considered low and cereal rusts are dikaryotic (i.e., double mutations may be required to generate a virulent isolate), huge numbers of urediniospores can be produced even at relatively low levels of disease. In *P. recondita,* for example, a single uredinium can produce some 2000 spores per day for 2 weeks or more, and in *P. graminis* f. sp. *tritici,* estimates of 5000 spores per day have been made.

Another means of generating pathogenic variability in rust fungi involves somatic hybridization, a process thought to involve the fusion of dikaryotic vegetative hyphae, nuclear exchange, and, possibly, exchange of whole chromosomes between nuclei and/or parasexual recombination. Empirical studies of several cereal rust species have provided good evidence of somatic hybridization under controlled conditions. In these studies, new pathotypes were detected following the infection of a susceptible host with two pathogenically distinct "parental" pathotypes. It is extremely difficult to detect and confirm naturally occurring somatic hybrids, because of the lack of isolate- or pathotype-specific markers. Consequently, the frequency and significance of this process in nature are difficult to establish. A group of Australian *P. graminis* isolates that parasitize barley but not wheat or rye is considered to have arisen via somatic hybridization between isolates of *P. graminis* f. spp. *tritici* and *secalis.* A distinct rust isolate found on poplars in New Zealand is believed to have arisen via hybridization (most likely somatic hybridization) between the species *Melampsora medusae* and *M. larici-populina.* Two examples for which there is good evidence of somatic hybridization between pathotypes of individual formae speciales in nature involve *P. graminis* f. sp. *tritici* and *P. recondita* f. sp. *tritici.* In both cases, new pathotypes that shared pathogenic, isozymic, and RAPD features of the putative parental pathotypes were detected.

In addition to mutation and somatic hybridization, exotic cereal rust pathotypes are occasionally introduced into the Australasian region. Two pathotypes of *P. graminis* f. sp. *tritici* first detected in Western Australia in 1969 were most likely introduced from southern Africa by long-distance aerial dispersal. *Puccinia striiformis* f. sp. *tritici* was probably introduced to Australia from western Europe by humans. Following the first detection of this disease in Victoria in 1979, it spread throughout eastern cereal-

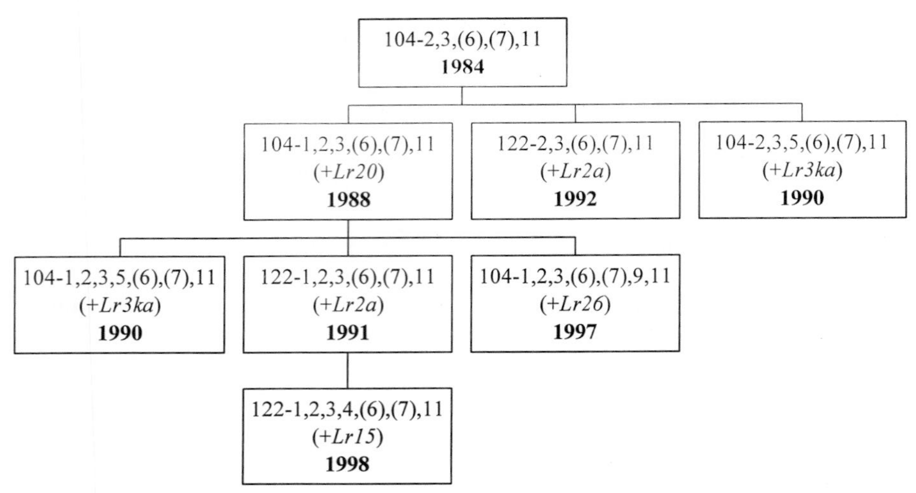

Fig. 2. Putative stepwise evolution of pathotypes in the wheat leaf rust pathogen, *Puccinia recondita* f. sp. *tritici,* in Australia, 1984–1998. Each pathotype is represented by a numerical code. The parental pathotype 104-2,3,(6),(7),11 is thought is have been accidentally introduced into the Australasian region either during or before 1984. Derivative pathotypes presumably developed via mutations at single loci governing pathogenicity; at each step the pathogen has gained virulence for specific resistance genes in the wheat host (given in parentheses); the year in which the pathotype was first detected is given is bold type.

growing regions and in the following year was detected in New Zealand, some 1600 km across the Tasman Sea. However, *P. striiformis* f. sp. *tritici* is still absent from Western Australia, illustrating the influence of westerly weather patterns on spore dispersal. There are many other well-documented cases of long-distance dispersal of rust pathogens and of migration of newly introduced species or pathotypes within regions in other parts of the world. One dramatic example is that of the southern maize rust pathogen *Puccinia polysora* Underw., which spread throughout Africa and Southeast Asia during the 1950s and 1960s following its introduction into the regions, presumably by human-mediated means.

VIII. CONTROL STRATEGIES

A. Cultural

The first attempts to control rust fungi included the removal of morning dews by drawing a rope over wheat crops, which is known to have occurred from the 1600s onward. At about the same time, barberry eradication was being practiced in regions such as Rouen, France, to control stem rust in wheat. In both cases, factors associated with epidemics (dew, barberry) were recognized well before their biological significance was fully appreciated.

Eradication or exclusion of alternate hosts of important heteroecious rust diseases still continues in many parts of the world. The most extensive eradication campaign was that of barberry in the U.S., which followed a disastrous stem rust epidemic in 1916. The early stages of this program had a dramatic effect on cereal stem rusts, and by the end of 1967, it was estimated that over 99 million barberry bushes had been destroyed. The removal of barberry did not result in the eradication of stem rust, but it was instrumental in reducing the frequency and extent of epidemics by delaying disease onset, reducing initial inoculum levels, and reducing and stabilizing variability in the pathogen population in space and in time. Cultivated and wild species of *Ribes* susceptible to white pine blister rust were eradicated in many parts of the U.S. from the early 1900s. Because of the epidemiology of this disease, eradication was not as successful in controlling this disease as was the effect of barberry eradication on cereal stem rust. In some species, it is possible to achieve complete control by the eradication of one or the other host. For example, eradication of the less important juniper host of *Gymnosporangium juniperi-virginianae* removes the possibility of apples being infected, and established infections in apple orchards do not progress since uredinia are not produced by this fungus and the aeciospores produced on the apple host can only infect junipers.

Several rust fungi have been eradicated from regions following accidental introduction. *Hemileia vastatrix* was detected in Papua New Guinea in 1965 and was eradicated by destroying infected coffee bushes. Unfortunately, the disease was reintroduced in the late 1980s and is now well established in the region.

The development of early maturing wheat cultivars by W. Farrer in Australia in the 1880s was an effective means of reducing the threat of rust epidemics. These cultivars were also better adapted to the more arid regions, which were less favorable for the rust pathogens.

Crop sanitation is important in the control of some rust diseases. This can include practices such as the destruction of out-of-season volunteer plants, which contribute to between-season survival of the pathogens, and crop rotation, deep plowing, or burning to destroy debris, which in the case of autoecious rusts may harbor dormant teliospores capable of surviving extended periods of time. The flaming of peppermint during autumn or spring, using propane gas burners, has also achieved success in controlling rust in this crop. However, repeated flaming over successive years may sometimes adversely affect plant density in crops.

B. Biological

Several fungi are known to be parasites of rust fungi, although none have been successfully deployed to control rusts on a commercial scale. The rust parasites are referred to as hyperparasites, and they include *Darluca filum* (Biv. Bern. ex Fr.), *Tuberculina maxima* Rostr., and *Verticillium lecanii* (A. Zimmerm.) Viégas.

It is possible to reduce the level of rust on plants by preexposure to related pathotypes or different rust fungi. This phenomenon is known as induced resistance or cross protection. Studies on resistance in oats to virulent pathotypes of *Puccinia coronata* induced by prior inoculation with avirulent pathotypes or with viable isolates of a nonpathogen (e.g., *P. recondita* f. sp. *tritici*) indicated that protection was due to appressoria produced over stomata by the avirulent pathotypes of *P. coronata* or by *P. recondita* f. sp. *tritici*. There was evidence that this exclusion mechanism was not purely mechanical, and that substances produced by the host and/or the appressoria may be involved. In other studies, the formation of phytoalexins by plants in response to avirulent pathotypes or nonpathogens was implicated as the mechanism of induced resistance. Phytoalexins are fungitoxic compounds produced in plants only after stimulation by microorganisms or by chemical or mechanical damage.

C. Chemical

The best known and most common means of chemical control of rust diseases has been the use of fungicides, which act directly by killing fungal structures. These compounds have provided an economical means of controlling rust fungi in various situations, and they have been used extensively in the control of rusts of cereals and other hosts as either foliar sprays or seed dressings.

Early attempts to control rusts with fungitoxic chemicals were made on cereals. These included the use of compounds such as sulfur, piric acid, and sulfonamides, all of which displayed phytotoxic properties. Several groups of organic chemicals have since been developed for the control of rusts. The dithiocarbamates (e.g., maneb, zineb) act as surface protectants and have provided economical control of cereal rusts. The development of systemic fungicides provided a more effective means of controlling plant diseases in general, since these chemicals may be translocated into new growth and provide longer protection. The carboxylic acid anilides (e.g., carboxin, oxycarboxin) are selectively toxic to basidiomycetes and interfere with mitochondrial respiration. Other systemic fungicides that have been used to control rusts in cereals include morpholine derivatives (e.g., tridemorph, fenpropemorph) and ergosterol biosynthesis inhibitors (SBIs; e.g., triadimefon, triadimenol, propiconazole, imazalil). Ergosterol is the primary sterol component in many fungal species and is essential for maintaining membrane structure and function.

The use of chemicals to control diseases carries with it the risk of resistance to the chemical developing in pathogen populations, and this has been documented in rust pathogens to chemicals such as oxycarboxin (e.g., in *Puccinia horiana*). Of additional concern is the potential for development of cross-resistance to fungicides, where resistance that develops to one compound confers resistance to others. This has occurred in some fungal plant pathogens to fungicides within the demethylation inhibiting group (DMIs) of SBIs. Evidence for moderate levels of cross-resistance to DMIs was recently obtained in western European isolates of *Puccinia recondita*.

Several chemicals capable of inducing systemic resistances in plants to rusts are known. These chemicals do not have antifungal activity, but they elicit host defense mechanisms such as increasing peroxidase or chitinase production, and as such are often referred to as abiotic elicitors. Potassium phosphate (10 m*M*) and EDTA (5 m*M*) induced systemic resistance in beans to *Uromyces viciae-fabae* (Pers.) Schröter and reduced infection by up to 75% in greenhouse tests. Similar greenhouse studies with *Puccinia sorghi* Schw. achieved reductions of up to 98% in the number of uredinia on the fifth, sixth, and seventh leaves on maize plants on which the first three leaves were treated with a single application of potassium salts (100 m*M*). Reductions in uredinial numbers of *Uromyces appendiculatus* on beans grown under field conditions have also been achieved by single applications of the chemical 2,6-dichloroisocotinic acid.

D. Genetic Resistance

The most economical means of controlling rust fungi is the development and cultivation of plants with genetic resistance. The first demonstration of Mendelian inheritance of resistance in wheat to a pathogen in a plant was made nearly a century ago by R. H. Biffin, in studies of resistance to *Puccinia striiformis* f. sp. *tritici*. The discovery was also significant because it showed the resistance was inherited independently of other traits, and demonstrated the ease with which lines breeding true for the resistance phenotype could be obtained.

A great deal of effort has subsequently gone into finding, characterizing, and utilizing resistances to rusts. In annual crops such as the cereals, this has included the genetic characterization of loci conferring resistance, to assist in their utilization. Several loci for rust resistance in wheat, for example, comprise allelic series, meaning that genes at the loci cannot be combined in a single homozygous cultivar. Other resistances comprise linked genes for resistance to two or more rust diseases, greatly simplifying their manipulation in breeding programs. Some of the genes identified are of little value since they were quickly overcome by mutations in the pathogens, but others have proved durable. Estimates of the annual value of controlling rusts in wheat primarily by resistance breeding suggest annual values of $A289 million in Australia, $C217 million in Canada, and £79.8 million in the UK.

The development of plants with resistance to rusts is contingent on the discovery of new sources of resistance. Recent advances in DNA technology have provided the tools for cloning resistance genes and transforming the cloned genes into susceptible plants, and may eventually lead to the engineering of "synthetic" resistance genes. To date, few rust resistance genes have been cloned. The first was the *L6* gene in flax, which confers resistance in flax to the rust pathogen *Melampsora lini*. Several other genes conferring resistance to *M. lini* were recently cloned, and hybrid genes have been generated by recombining regions of different genes *in vitro*. Although the new genes have not conferred new resistances following their reintroduction into susceptible flax plants, this work has shown that it is possible to recombine genes in the laboratory and signals a strategy that may lead the way for engineering new resistances.

See Also the Following Articles

Plant Disease Resistance • Rhizoctonia • Smuts, Bunts, and Ergot

Bibliography

Bushnell, W. R., and Roelfs, A. P. (1984). "The Cereal Rusts. Volume I, Origins, Specificity, Structure and Physiology." Academic Press, Orlando, Florida.

Cummins, G. B., and Hiratsuka, Y. (1983). "Illustrated Genera of Rust Fungi." The American Phytopathological Society, St. Paul, Minnesota.
Knott, D. R. (1989). "The Wheat Rusts—Breeding for Resistance." Springer-Verlag, Berlin.
Littlefield, L. J. (1981). "Biology of the Plant Rusts. An Introduction." Iowa State Univ. Press, Ames.
Littlefield, L. J., and Heath, M. C. (1979). "Ultrastructure of Rust Fungi." Academic Press, New York.
McIntosh, R. A., Wellings, C. R., and Park, R. F. (1995). "Wheat Rusts: An Atlas of Resistance Genes." CSIRO Publications, Melbourne.
Roelfs, A. P., and Bushnell, W. R. (1985). "The Cereal Rusts. Volume II, Diseases, Distribution, Epidemiology and Control." Academic Press, Orlando, Florida.
Schumann, G. L. (1991). "Plant Diseases: Their Biology and Social Impact." The American Phytopathological Society, St. Paul, Minnesota.
Scott, K. J., and Chakravorty, A. K. (1982). "The Rust Fungi." Academic Press, New York.

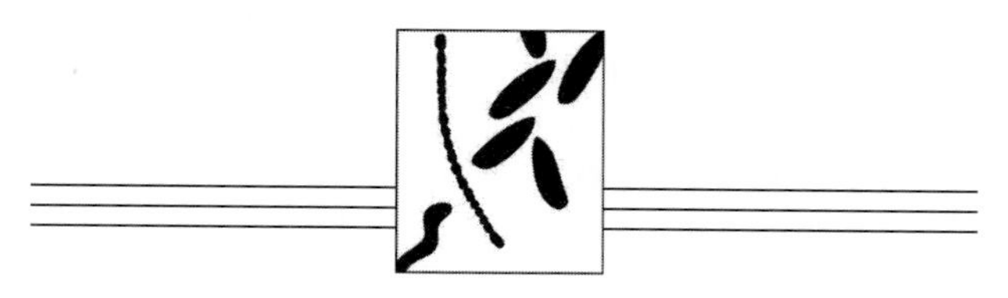

Secondary Metabolites

Juan F. Martín, Santiago Gutiérrez, and Jesús F. Aparicio

Institute of Biotechnology INBIOTEC and University of León, Spain

I. Introduction: Secondary versus Primary Metabolites
II. Specific Precursors for Secondary Metabolites
III. Conversion of Primary Metabolites to Specific Precursors for Secondary Metabolites
IV. Reactions Committing the Precursors into Specific Pathways for Secondary Metabolites
V. Polymerization and Condensation Reactions
VI. Late Modification Reactions
VII. Nonribosomal Peptides
VIII. β-Lactams
IX. Polyketides
X. Isoprenoids

GLOSSARY

adenylate-forming enzymes A large family of enzymes that activate short-chain fatty acids or amino acids as acyl-adenylates (acyl-AMP).

nonribosomal peptides Peptides that are synthesized by peptide synthetases. The sequence of amino acids in nonribosomal peptides is determined by the order of the peptide synthetase domains (the colinearity rule).

peptide synthetases Complex enzymes that catalyze a large number of enzymatic reactions (activation, condensation, elongation, etc.) leading to the formation of nonribosomal peptides.

polyketide A polyketo compound derived from acetate, propionate, or butyrate units activated in the form of acetyl-CoA (or malonyl-CoA), methylmalonyl-CoA, and ethylmalonyl-CoA, by the action of polyketide synthases.

polyketide synthases Enzymes involved in the activation and condensation of precursor units to form polyketides. Type I polyketide synthases (PKSs) are large multienzyme proteins with many catalytic centers encoded by a single large gene. Type II polyketide synthases are complexes of several polypeptides encoded by specific small genes.

prenyltransferases A family of enzymes from different organisms involved in the synthesis of isoprenoid molecules by condensation of isoprene units.

synthases and synthetases Enzymes catalyzing the condensation of precursor units to form covalent bonds. Synthetases require ATP to perform the condensation, whereas synthases do not require ATP (they use preactivated units, usually CoA thioesters).

thiotemplate mechanism A mechanism of nonribosomal peptide biosynthesis in which the amino acids are activated as thioesters and transferred from one domain to another of the peptide synthetases by using a pantetheine arm.

SECONDARY METABOLITES are microbial and plant products nonessential for growth and reproduction of the producer organisms; each secondary metabolite is produced by a relatively limited number of species and is encoded by sets of dispensable genes. The compounds are synthesized at the end of the exponential growth phase and during the stationary phase, and their formation is highly influenced by the growth conditions, especially by the composition of the culture medium.

Many secondary metabolites have chemical structures that are infrequent in the biological world. These molecules belong to many classes of organic compounds: aminocyclitols, amino sugars, quinones, coumarins, epoxides, glutarimides, glucosides, nonribosomal peptides, phenazines, polyacetylenes, polyenes, pyrroles, quinolines, terpenoids, and tetracyclines. In addition, secondary metabolites possess unusual chemical linkages, such as β-lactam rings, cyclic peptides made of normal and modified amino acids, the unsaturated bonds of polyenes, and the large rings of macrolides. As a result of the broad substrate specificity of some biosynthetic enzymes, secondary metabolites are typically produced as members of a particular family of compounds. Secondary metabolites provide a wide variety of useful pharmacological activities.

I. INTRODUCTION: SECONDARY VERSUS PRIMARY METABOLITES

The classic separation of secondary metabolites from primary metabolism now has to be reconsidered in the light of new findings about the genetics of the biosynthesis of a variety of secondary metabolites (Kleinkauf and Von Döhren, 1997). The biosynthetic pathways of the so-called secondary metabolites follow the same general rules as any other biosynthetic process in living cells, except for peculiar aspects related to the formation of precursors and polymerization reactions (Martín and Liras, 1981).

Primary metabolism is essentially the same for all living cells; primary metabolites very rarely accumulate since the primary metabolism is finely balanced. A few primary metabolites (e.g., some amino acids in corynebacteria) accumulate in nonphysiological quantities with reduced growth rates of the producing cultures and behave, therefore, as secondary metabolites. On the other hand, secondary metabolites are frequently species-specific, and the mechanism of evolution of their biosynthetic genes is intriguing. Horizontal transfer of β-lactam biosynthesis from bacteria to filamentous fungi has been proposed, and in many other cases genetic information encoding enzymes for antibiotic biosynthesis has been lost (at least in part) and silent genes, which do not result in metabolite production, remain in the genome.

The biosynthesis of secondary metabolites is closely related to the primary metabolism of the producer cells (Drew and Demain, 1977) since all secondary metabolites derive from intermediates of primary metabolism (Fig. 1). In many cases the intermediates of primary metabolism are converted to specific precursors for secondary metabolism (Table I) by specific modifying reactions (Lancini and Lorenzetti, 1993).

The number of intermediates of primary metabo-

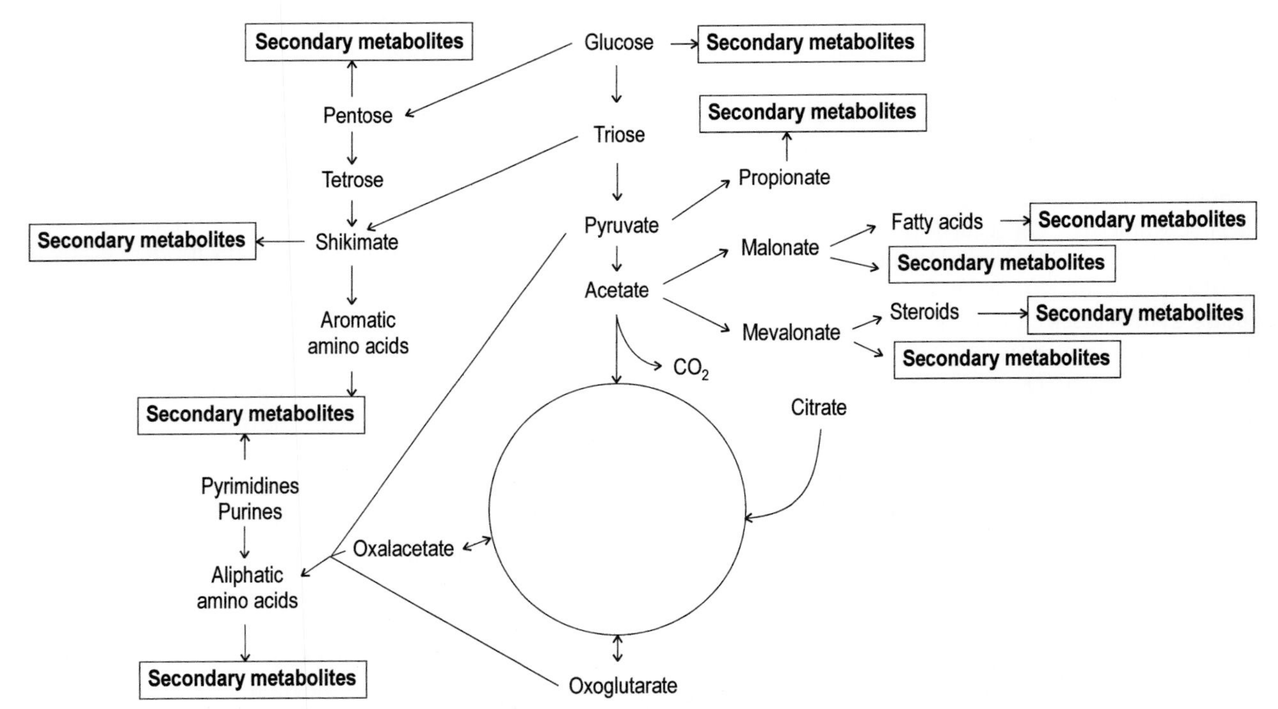

Fig. 1. Scheme of the central metabolism indicating the intermediates and reactions involved in formation of different types of secondary metabolites (boxes).

TABLE I
Examples of Intermediates That Are Precursors of Secondary Metabolites[a]

Type of compound	*Precursors*
Short-chain fatty acids	Acetate, propionate, malonate, methylmalonate, butyrate
Isoprene units	Isopentenyl pyrophosphate, mevalonate
Amino acids	Normal protein amino acids, unusual amino acids
Sugars and amino sugars	Hexoses, pentoses, tetroses, modified hexoses, modified amino sugars (e.g., *N*-methyl-L-glucosamine, mycosamine, mycaminose), branched sugars (e.g., streptose)
Cyclitols and aminocyclitols	Inositol, inosamine, streptidine, 2-deoxystreptamine, actinamine
Amidino groups	Arginine (amidino moiety)
Purine and pyrimidine bases	Adenine, guanine, cytosine, dimethyladenine, 3′-deoxypurines
Aromatic intermediates and aromatic amino acids	Shikimic acid, chorismic acid, prephenic acid, *p*-aminobenzoic acid, *p*-hydroxybenzoic acid, tyrosine, tryptophan, phenylalanine
Methyl groups (C_1 pool)	*S*-Adenosylmethionine, formyltetrahydrofolate

[a] Modified from Martín and Liras (1981).

lism used for biosynthesis of secondary metabolites is limited (Table I). These include acetate, propionate, and other short-chain fatty acids, mevalonate, amino acids, sugars and amino sugars, cyclitols and aminocyclitols, amidino groups, purine and pyrimidine bases, aromatic amino acid intermediates, the carbamoyl group, and the methyl group (C_1 pool).

It is appropriate to classify the biosynthetic reactions of secondary metabolites into four classes, namely:

Class I: Conversion of primary metabolites into specific precursors for secondary metabolites
Class II: Reactions committing the precursors to specific pathways for secondary metabolites (i.e., activation or modification)
Class III: Polymerization and condensation reactions, giving rise to polyketides, peptides, isoprenoids, etc.
Class IV: Late (a posteriori) modification reactions

II. SPECIFIC PRECURSORS FOR SECONDARY METABOLITES

Carbohydrates are converted into smaller five-, four-, three-, and two-carbon units (e.g., pentoses, tetroses, trioses, acetate, 2-oxoglutarate, oxaloacetate). Some of these intermediates may be used directly as precursors of secondary metabolites. However, very frequently specific precursors of secondary metabolites arise by modification of intermediates of glycolysis or the Krebs cycle (see Table I).

A. Acetate, Propionate, and Other Organic Acids

Many secondary metabolites are derived from acetate, propionate, butyrate, and other short-chain organic acids, through the so-called polyketide pathway (polyacetate and polypropionate). Acetate and propionate linked to coenzyme A (CoA) in the form of acetyl-CoA and propionyl-CoA act as C_2 and C_3 starter units, respectively, whereas the normal elongation units are malonyl-CoA and methylmalonyl-CoA, which act as donors of two and three carbon units, respectively. Acetyl-CoA is a key intermediate in the breakdown of glucose (or fatty acids), being formed by the action of pyruvate dehydrogenase. Malonate is formed by carboxylation of acetate via the action of acetyl-CoA carboxylase. Propionate is an important precursor of the polyketide-derived secondary metabolites produced by actinomycetes. Propionate is extensively used as a precursor in the synthesis of the lactone rings of some macrolide (e.g., erythromycin) and ansamycin (rifamycin) antibiotics. Formation of methylmalonate, the C_3 extending unit, takes place by the enzymes propionyl-CoA carboxylase and methylmalonyl-CoA carboxyltransferase.

B. Isoprene Units

A large number of fungal metabolites are formed by the condensation of five-carbon isoprene units, $CH_3-CH(CH_3)-CH=CH_2$. Isoprene units are involved in the formation of steroids and terpenes by animals, plants, and fungi. A variety of secondary metabolites from plants and fungi are derived from isoprene units. Some monoterpenes (formed from two C_5 units), sesquiterpenes (three C_5 units), or triterpenes (six C_5 units) have antibiotic activity, for example, trichodermin and fusidic acid. Others include ubiquinones, plastoquinones, vitamin K, and the ring systems of coumarins and quinolines. Some isoprene units are also involved in the biosynthesis of complex antibiotics such as novobiocin. Geosmin, which is responsible for the characteristic smell of *Streptomyces,* is also derived from isoprene units.

Isoprenoid compounds are formed by polymerizations of isopentenyl pyrophosphate, the so-called activated isoprene. Isopentenyl pyrophosphate is synthesized from acetyl-CoA through acetoacetyl-CoA and mevalonic acid. It is also formed from leucine by deamination and conversion into the intermediate 3-hydroxy-3-methylglutaryl-CoA.

C. Amino Acids

Amino acids normally found in proteins, as well as modified nonprotein amino acids, are used as building units of homopeptide antibiotics or linked to other moieties in heteropeptide antibiotics. Amino acids are synthesized from intermediates of glycolysis, the pentose phosphate cycle, and the citric acid cycle by several biosynthetic pathways, which form well-defined families of amino acids. Thus, the biosynthesis of the aromatic amino acids takes place by condensation of phosphoenolpyruvate and erythrose-4-phosphate. The branched-chain amino acids arise from pyruvate; the serine family from 3-phosphoglycerate, the glutamate family from 2-oxoglutarate, and the aspartate family from oxaloacetate.

Unusual amino acids, which play an important role in biosynthesis of secondary metabolites, are formed by modifications of normal amino acids.

D. Aromatic Intermediates

The aromatic amino acid pathway is responsible for the biosynthesis of most of the aromatic antibiotics of actinomycetes and part of the aromatic secondary metabolites of fungi and plants. These include quinic acid, the quinoline alkaloids, naphthoquinone and anthraquinone, the quinazoline and ergot alkaloids, and the phenazines. Some fungal and plant aromatic metabolites derive from the polyketide pathway.

In addition to the three aromatic amino acids, many intermediates of the so-called shikimate pathway are used as precursors of secondary metabolites. Thus, the chromophores of choramphenicol and corynecin derive from chorismic acid (through the intermediate *p*-aminophenylalanine), the C_7N unit of rifamycin from early intermediates of the shikimate pathway, the phenazine skeleton of pyocyanine from anthranilic acid, that of novobiocin and xanthocillin from tyrosine, and the aromatic rings of actinomycin, indolmycin, and pyrrolnitrin from tryptophan. The early intermediate shikimic acid is formed by condensation of erythrose-4-phosphate (an intermediate of the pentose cycle) with phosphoenylpyruvate (an intermediate of glycolysis). Shikimate is later converted to chorismic acid, which serves as the branching intermediate for the biosynthesis of anthranilate (which is transformed to tryptophan), *p*-aminobenzoate, and prephenate (which gives rise to tyrosine and phenylalanine).

E. Sugars and Amino Sugars

Both normal and unusual sugars and amino sugars occur frequently in antibiotics and other secondary metabolites (e.g., macrolides, aminocyclitols, tetracyclines, anthracyclines, and nucleoside antibiotics).

The more frequently used sugar precursors include trioses (D-glyceraldehyde, dihydroxyacetone), tetroses (D-erythrose, D-erythrulose), pentoses (L-arabinose, D-xylose, D-ribose, D-xylulose, D-ribulose, D-2-deoxyribose), hexoses (D-glucose, D-mannose, D-galactose, D-fructose, L-sorbose, L-rhamnose, L-fucose), and heptoses (D-sedoheptulose, D-mannoheptulose).

Sugars of secondary metabolites are frequently found attached by glycosidic bonds to other moieties referred to as aglycones. Attachment is generally through a hydroxyl group, giving *O*-glycosides. However, *N*-glycosides, *S*-glycosides (involving amines or thiol groups), and *C*-glycosides also occur.

Glucose serves as a precursor of most of these sugars and amino sugars; the pentoses (e.g., ribose) derive directly from the intermediates of the pentose phosphate cycle. The carbon skeleton of glucose is usually incorporated into many antibiotics without breakdown into smaller units. However, the six-carbon chain may undergo a variety of modifications.

The introduction of the amino group to form amino sugars involves transamination between the amide group of glutamine (or glutamate) and a keto-hexose.

F. Purine and Pyrimidine Bases

In addition to their role as components of the nucleic acids of all living organisms, methylated purines (e.g., caffeine, theophylline) and substituted purines having an isopentenyl group are formed as secondary metabolites in plants and microorganisms.

Several antibiotics are analogs of nucleosides. They are either modified in the ribose moiety or in the purine or pyrimidine bases. Nucleoside antibiotics containing an unmodified purine or pyrimidine base (e.g., cordycepin, psicofuranine, angustmycin) are derived directly from normal nucleosides.

G. Methyl (One-Carbon) Groups

All methylations involved in the biosynthesis of antibiotics use methionine as the methyl group donor in a transmethylation reaction catalyzed by methyltransferases. The methyl group of methionine derives in turn from the methyl group of N^5-methyltetrahydrofolate. Prior to the transfer, the methyl group of methionine is activated in the presence of magnesium and ATP through the formation of *S*-adenosylmethionine (SAM). *S*-Adenosylmethionine serves as a high energy methyl donor.

A large number of transmethylation reactions are known in primary and secondary metabolism. They involve C-, O-, and N-methylations. Occurrence of C-methylation is far rarer than N-methylation or O-methylation. Methyl groups derived from methionine occur in novobiocin, in the methylated sugar and aglycone of erythromycin, and in gentamicin, cephamycin, and many other secondary metabolites.

III. CONVERSION OF PRIMARY METABOLITES TO SPECIFIC PRECURSORS FOR SECONDARY METABOLITES

Many of the precursors used in the synthesis of secondary metabolites are not the normal monomers used for the synthesis of macromolecules in the cell. The basic intermediates of primary metabolism are often modified before being used for antibiotic synthesis of secondary metabolites. These changes involve (1) modifications of the carbon skeleton, (2) changes in the oxidation–reduction level of the molecule, and (3) cyclizations giving rise to heterocyclic rings.

Very frequently amino acids that participate in the synthesis of secondary metabolites of the peptide type are different from the 23 amino acids that occur in proteins. There are about 200 nonprotein amino acids, including D-amino acids, N- and β-methylated amino acids, dehydro- and β-amino acids, abnormal sulfur amino acids, dibasic amino acids (e.g., diaminobutyric acid), and amino acids that are intermediates of biosynthetic routes such as ornithine and α-aminoadipic acid (Table II).

Many of the sugars of secondary metabolites (e.g., those occurring in aminocyclitol and macrolide antibiotics) are rare sugars. They include deoxy sugars, branched-chain sugars, and unusual amino sugars.

Also striking is the presence in some antibiotics of rare chemical groups such as the azide group of azaserine, the pyrrolo[1,4]benzodiazepine nucleus of the pyrrolobenzodiazepine antibiotics, the cyano group of toyocamycin, the vinyl group of sarcomycin and primorcarcine, and the mitosan moiety of mitomycins and phorphyromycin, all of which have no equivalent in primary metabolism. Such abnormal precursors are synthesized by only a few microorganisms. The strain specificity for biosynthesis of such precursors is due to the fact that some pieces of

TABLE II
Some Unusual Amino Acids and Hydroxy Acids Existing in Peptide Secondary Metabolites[a]

Type	*Amino acid*	*Antibiotic*
D-Amino acids	Allo-D-hydroxyproline	Etamycin
	D-Allothreonine	Stendomycin
	D-Serine	Quinomycin
	D-Leucine	Linear gramicidin
	D-Ornithine	Bacitracin
	D-Phenylalanine	Gramicidin S, tyrocidine, polymyxin B
N-Methyl amino acids	*N*-Methyl-L-threonine	Stendomycin
	N-Methyl-L-valine	Quinomycin
	N-Methyl-L-alloisoleucine	Quinomycin B
	N,γ-Dimethyl-L-alloisoleucine	Quinomycin C
	N,β-Dimethyl-L-leucine	Etamycin, triostin C
	Sarcosine	Etamycin
	Phenylsarcosine	Etamycin
C-Methyl amino acids	3-Methyltryptophan	Telomycin
	3-Methylphenylalanine	Bottromycin A
	3-Methylvaline	Bottromycin A
	3-Methylproline	Bottromycin A
	N,β-Dimethyl-L-leucine	Etamycin, triostin C
	3-Methyl-L-lanthionine	Nisin, subtilin
β-Amino acids	β-Lysine	Streptothricin, viomycin
	L-β-Methylaspartic acid	Aspartocin
	β-Phenylalanine	Edeine
	β-Serine	Edeine
	β-Tyrosine	Edeine
Imino acids	*cis*-3-Hydroxy-L-proline	Telomycin
	trans-3-Hydroxy-L-proline	Telomycin
	4-Oxo-L-pipecolic acid	Vernamycin
S-Amino acids	L-Lanthionine	Nisin
	2-Thiazolyl-L-alanine	Bottromycin
Dehydro amino acids	Dehydrotryptophan	Telomycin
	Dehydroalanine	Nisin
	3-Ureido-dehydroalanine	Viomycin, tuberactinomycin
	Dehydroleucine	Albonoursin
	Dehydrophenylalanine	Albonoursin
	Dehydroproline	Osteogrisin A
Complex amino acids	α,β-Diaminopropionic acid	Edeine
	2,6-Diamino-7-hydroxyazaleic acid	Edeine
Basic amino acids	L-Ornithine	Gramicidin S
	L-2,4-Diaminobutyric acid	Polymyxin
Hydroxy acids	D-Allo-4-hydroxyproline	Etamycin
	2-Hydroxyisovaleric acid	Enniantin
Rare amino acids	D-2-Aminoadipic acid	Cephalosporin, cephamycin
	L-2-Aminoisobutyric acid	Alamethicin

[a] Modified from Martín (1998).

genetic information that code for the biosynthesis of such compounds are not distributed in all microorganisms.

There are many examples of conversion of amino acids or other intermediates of metabolism into specific antibiotic precursors. The molecule of streptomycin consists of three moieties attached glycosidically to one another: streptidine, streptose, and *N*-methyl-L-glucosamine. The amidino groups of streptidine and related aminocyclitol antibiotics are derived from the amidino group of L-arginine. Amidino groups are transferred from arginine by the action of the enzyme amidinotransferase.

The sequence of intermediates between *myo*-inositol and streptidine involves two aminations in which amino groups from glutamine and alanine are transferred onto keto groups by the action of aminotransferases. Two amidino groups from arginine are then transferred to the previously introduced amino groups.

IV. REACTIONS COMMITTING THE PRECURSORS INTO SPECIFIC PATHWAYS FOR SECONDARY METABOLITES

Specific pathways channeling precursors into secondary metabolites are of great interest, but there is a surprising lack of knowledge of the biosynthetic enzymes involved.

The first enzyme of such specific pathways is of great importance because it determines the amount of precursor that enters the pathway for a secondary metabolite and, therefore, the flow of intermediates and the output of the pathway. Other key enzymes of a pathway may also be limiting. Such enzymes are frequently subject to feedback or carbon, nitrogen, or phosphate regulation. Interestingly, some of them may be induced by accumulation of high intracellular levels of precursors. A few of these enzymes have been studied in some detail. They include dimethylallyltryptophan (DMAT) synthetase, the first enzyme of the biosynthesis of ergot alkaloids; phenoxazinone synthetase, which forms the phenoxazinone chromophore of actinomycin; amidinotransferase (L-arginine : inosamine phosphate amidinotransferase), which appears to be the key enzyme in the biosynthesis of the streptidine moiety of streptomycin; guanosine triphosphate-8-formylhydrolase, which forms the pyrrole ring of pyrrolopyrimidine nucleoside antibiotics; *p*-aminobenzoate synthetase, the enzyme that converts chorismic acid to *p*-aminobenzoic acid, the first specific enzyme of the biosynthesis of candicidin; and lysine-6-aminotransferase, which commits lysine to the cephamycin pathway in two different actinomycetes.

The first enzyme of the biosynthetic pathway of ergot alkaloids is induced by tryptophan (a precursor of alkaloids) and tryptophan analogs. Increased levels of DMAT synthetase appear to be due to higher intracellular pools of tryptophan in the transition from trophophase to idiophase; at this time the enzymes are induced. Similarly, lysine-6-aminotransferase is induced by diamines in *Nocardia lactamdurans*. Phenoxazinone synthetase, a key enzyme in the biosynthesis of actinomycin, is repressed by glucose. The formation of *p*-aminobenzoic acid synthetase by *Streptomyces griseus* is repressed by phosphate (Martín and Demain, 1980).

Interestingly, genes encoding enzymes that commit common intermediates into specific pathways for secondary metabolites are sometimes clustered with genes for antibiotic biosynthesis and resistance (Martín and Liras, 1989). One of the best known examples is the clustering of the *lat* gene, encoding a lysine-6-aminotransferase that converts lysine to α-aminoadipic semialdehyde, with other genes of the cephamycin pathway in *Nocardia lactamdurans* and *Streptomyces clavuligerus* (Martín, 1998).

V. POLYMERIZATION AND CONDENSATION REACTIONS

Once precursors have been synthesized, they are channeled into pathways specific for biosynthesis of secondary metabolites. In some cases, monomeric structural units are polymerized to form polymeric secondary metabolites. This is the case in the formation of polyketides, nonribosomal peptides, and isoprenoid molecules. Polymerization reactions are catalyzed by modular polyketide or polypeptide synthases with unusually high molecular weights. The products of these specific biosynthetic reactions are usually subject to late structural modifications. In some antibiotics different moieties synthesized by separate biosynthetic pathways are assembled together to form complex structures.

VI. LATE MODIFICATION REACTIONS

After the polyketide, peptide, or isoprenoid chain is synthesized, late (*a posteriori*) modifications of the secondary metabolite structure occur. These late modifications may be carried out to a different extent depending on the physiological condition of the producing cells. In the last years a few of these reactions have been studied in some detail (see the sections on nonribosomal peptides, β-lactams, polyketides, and isoprenoids).

Late modifications include glycosylation (e.g., in many polyketides and macrolides), hydroxylations, oxidation of hydroxyl to keto groups, aminations and transaminations (e.g., in the aminocyclitols), N- and O-methylations, N- and O-acetylations, dehydrations (sometimes resulting in the formation of double bonds), epoxidations (e.g., in the isoprenoids), and chlorination (e.g., in the chlortetracyclines).

The biochemistry of these late-modification reactions is only beginning to be understood. A few well-known examples are glycosylation reactions of macrolides such as erythromycin and tylosin. The two-protein system involved in the methoxylation of cephamycin in *Streptomyces clavuligerus* and *Nocardia lactamdurans* has also been elucidated (Martín, 1998). Progress in this field is facilitated by the availability of cloned genes encoding the late conversion enzymes. These genes are usually clustered with antibiotic biosynthesis and antibiotic-resistance genes (Martín and Liras, 1989).

VII. NONRIBOSOMAL PEPTIDES

Many microbial peptide antibiotics are known. Most of these antibiotics are synthesized by a nonribosomal mechanism using multienzyme complexes. In the 1990s a significant advance has been made in our understanding of the so-called thiotemplate mechanism of nonribosomal peptide biosynthesis. In addition, many genes encoding peptide synthetases have been cloned, providing strong genetic support for the previously proposed model. The extensive knowledge about nonribosomal peptide biosynthesis has allowed the rational design of peptide antibiotics by targeted replacement of bacterial and fungal domains, providing a directed way to modify the substrate specificity of peptide synthetases.

The nonribosomal peptide secondary metabolites have several particular characteristics that allow them to be distinguished from ribosomal proteins:

1. Nonribosomal peptides are generally smaller (≤3000 Da) than proteins.
2. Most nonribosomal peptides are synthesized by bacteria and fungi as families of closely related substances, as occurs also with most other secondary metabolites.
3. Nonprotein amino acids are frequently found in peptide antibiotics. These include amino acids with the D-configuration, *N*-methyl derivatives of normal amino acids, β-amino acids, imino acids, unsaturated dehydroamino acids, and hydroxyamino acids (Table II).
4. Peptide antibiotics often form cyclic structures with no free α-amino or carboxyl ends (e.g., gramicidin S). Lactones are also formed with a terminal or side-chain hydroxyl group. The presence in peptide antibiotics of proline, *N*-methylamino acids, or imino acids may preclude the formation of α-helix structures. Finally, cysteine residues may form thiazoles, whereas serine residues give rise to oxazole derivatives.
5. The presence of rare amino acids and the cyclic structures render peptide secondary metabolites largely resistant to cleavage by proteases.
6. Some peptide antibiotics are made exclusively of amino acids, whereas others contain fatty acids, hydroxy acids (so-called depsipeptides), amines, amino sugars, alcohols, and pyrimidines linked to the amino acids.

A. Biosynthesis of Peptide Antibiotics

Amino acids, imino acids, or hydroxy acids are activated, condensed, and even modified by multienzyme polypeptides or multienzyme complexes named peptide synthetases by means of the thiotemplate mechanism, which is similar to the mechanism of synthesis of fatty acids and polyketides by fatty acid or polyketide synthetases. This mechanism has four sequential steps:

1. Activation of amino acids by ATP and formation of aminoacyl-adenylates.
2. Binding of the activated amino acids to thiol groups on the enzymes with formation of thioester bonds.
3. Formation of a peptide bond between the carboxyl group of the first amino acid and the amino group of the second one, using the energy of the thioester bond.
4. Transpeptidation reactions by which the thioester of the dipeptide breaks and the carboxyl group forms a peptide bond with the amino group of the third amino acid; this step is repeated until the chain is completed. Transpeptidation requires the aid of 4′-phosphopantetheine, a well-known cofactor of fatty acid synthetases.

B. Peptide Synthetases

Peptide synthetases act as templates for the growing peptide chain, attached to the enzyme via a thioester bond. The protein templates are composed of distinctive substrate activating modules, whose order dictates the primary structure of the corresponding peptide product (the colinearity rule). According to this, a peptide synthetase forming a peptide of *n* amino acids would be encoded by a DNA sequence of *n* modules. Each module contains the catalytic functions (i) for the activation as the corresponding adenylate of an amino acid, imino acid, and hydroxy acid, respectively, (ii) for the transfer of the activated carboxyl group to the 4′-phosphopantetheine cofactor, (iii) for generating the derived thioester, (iv) for the epimerization or N-methylation of the activated intermediate, and (v) for peptide bond formation (Fig. 3).

Peptide synthetases require a complex spatial organization to direct the activation reactions and sequential polymerization of the component amino acids. Thus a peptide synthetase can be represented as a line of several domains that act in concert according to their order within this line (Figs. 2 and 3). This can be accomplished either by a very high molecular weight multienzyme polypeptide or by a set of smaller polypeptides (each carrying specific functions).

C. Carboxyl Activation

1. Adenylate Formation

The activation of amino acids, imino acids, or hydroxy acids is a key reaction in enzymatic, nonribosomal, peptide biosynthesis. The adenylate-forming domain bears the substrate recognition pocket and activates its cognate amino acid as adenylate at expense of ATP. The reaction involves binding to an ATP–Mg^{2+} (or Mn^{2+}) complex. This domain represents the nonribosomal code, since primary selection of substrates is located within this element. The protein template system is generally considered less precise than its ribosomal counterpart, and peptide isoforms are an indication of this interpretation. Nutrient conditions can affect the constitution of the peptide product, either *in vivo* by adding certain amino acids to the medium or *in vitro*.

2. Adenylate Domains

Each adenylate domain comprises about 550 amino acids and is composed of two interacting subdomains. Similar motifs have been defined for all the members of the luciferase–tyrocidine synthetase–acyl-CoA synthetase superfamily. The two subdomains are involved cooperatively in nucleotide triphosphate binding and stabilization of the (amino) acyl-adenylate intermediates.

The key event of nonribosomal peptide biosynthesis, the selection of an amino (imino, hydroxy) acid, has been tentatively assigned to the region between motifs C and D (Fig. 3).

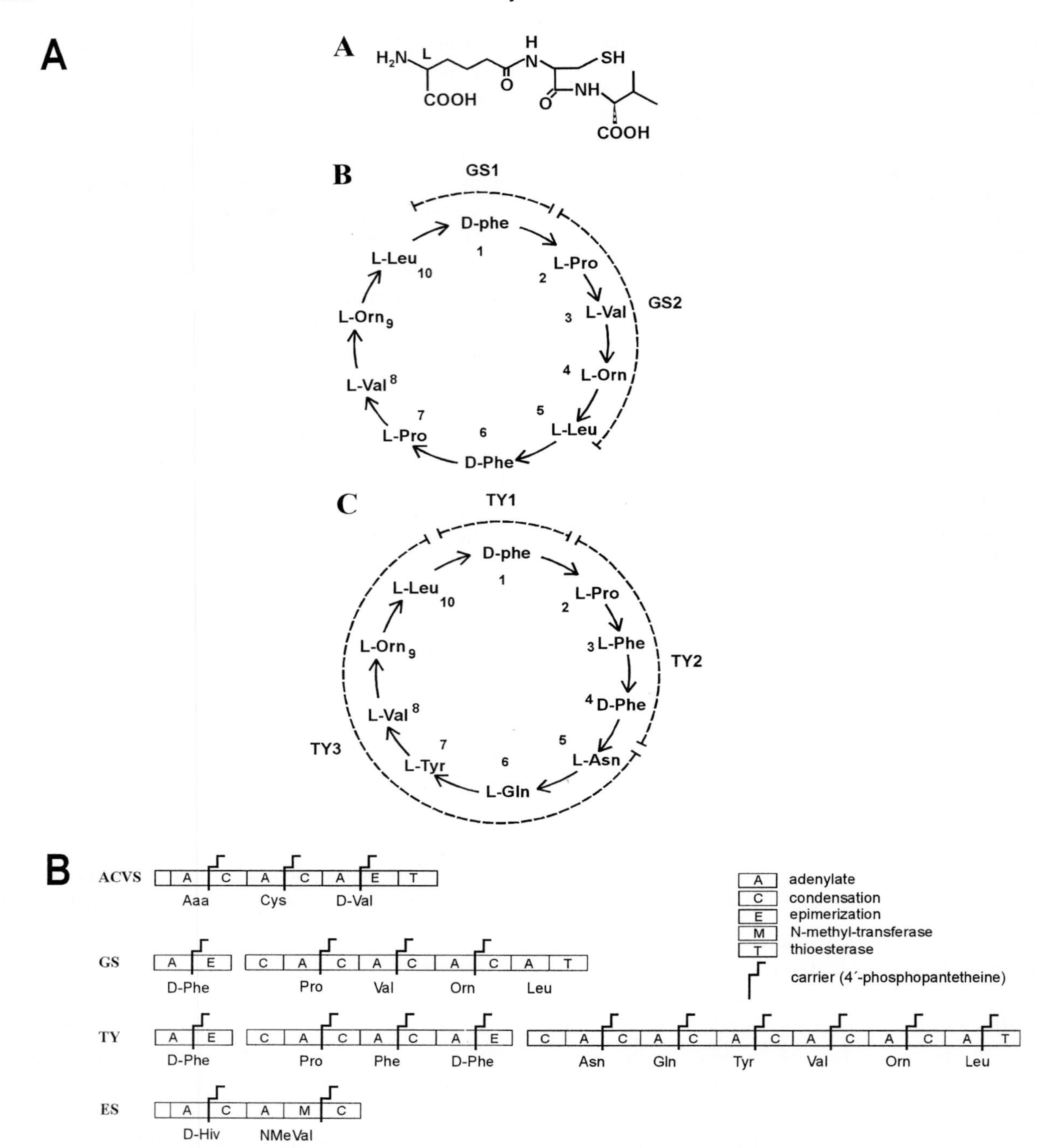

Fig. 2. Structures of some peptide secondary metabolites and modules required for their biosynthesis. (A) Chemical structures of the linear peptide α-aminoadipyl-cysteinyl-valine (ACV) and the cyclic peptides gramicidin S and tyrocidine. The component amino acids of gramicidin S are activated by gramicidin synthetases I and II (GS1 and GS2 in the figure), and the tyrocidine amino acids are activated by tyrocidine synthetases I, II, and III (TY1, TY2, and TY3). (B) Domains of the ACV synthetase (ACVS), gramicidin synthetases I and II (GS), tyrocidine synthetases I, II, and III (TY), and enniantin synthetase (ES). The symbol ┌┘ indicates the phosphopantetheine arm.

D. Aminoacylation

1. *Cofactor Binding*

The thiotemplate mechanism is characterized by the formation of a relatively stable thioester with enzyme thiol groups prior to the peptide bond constitution. These thiol groups have been identified as cysteamines of the pantetheine moiety of enzyme-bound cofactors, in complete analogy to acyl esters of carboxylic acids.

2. *Carrier Domains*

The 4′-phosphopantetheine cofactor is added in a posttranslational modification reaction to the peptide synthetase module by highly specific 4′-phosphopantetheine transferases. The attachment site is the highly conserved J motif (Fig. 3). These transferases, essential for the activity of the peptide synthetases, have been shown to be part of the biosynthetic gene clusters for enterobactin, surfactin, gramicidin S, bacitracin, iturin, and nosipeptide.

E. Modification Reactions

Modification of amino acid residues may occur (a) before entrance to the catalytic peptide-forming cycle, that is, prior to activation, (b) at the aminoacyl stage preceding elongation, (c) after elongation at the peptidyl intermediate stage, or (d) following termination as a product transformation reaction. The modifications studied in more detail have been epimerization and N-methylation.

F. Peptide Bond Formation

The peptide bond is formed on the interaction of two carrier domains with an elongation domain, allowing the contact of two thioesters with the catalytic site. The peptide bond is formed at the condensation site of the first module, and the product is generated as a peptidyl thioester at the adjacent module.

G. Termination Reactions

Release of the completed peptidyl intermediate at the final module leads to either linear or cyclic products. Most of the peptide biosynthetic clusters contain genes or domains similar to thioesterases (domain T, Fig. 3). The thioesterase of α-aminoadipyl-cysteinyl-valine (ACV) synthetase appears to catalyze the stereospecific release of the tripeptide from the peptide synthetase.

VIII. β-LACTAMS

A. Biosynthesis of Penicillins, Cephalosporins, and Cephamycins

The biosynthesis of penicillins, cephalosporins, and cephamycins has been reviewed by several authors [see Aharonowitz *et al.* (1992) and Martín (1998)]. A concise overview of the β-lactam biosynthetic pathways is presented here.

The *penam* nucleus of penicillins and the *cephem* nucleus of both cephamycins and cephalosporins are formed by condensation of the three precursor amino acids L-α-aminoadipic acid, L-cysteine, and L-valine by a mechanism designated as "nonribosomal peptide synthesis" that involves activation and condensation of the three component amino acids and epimerization of L-to D-valine to form the tripeptide δ-(L-α-aminoadipyl)-L-cysteinyl-D-valine (LLD-ACV) (Fig. 4).

B. Common Reactions in the Biosynthesis of Penicillins, Cephalosporins, and Cephamycins

1. *Formation of the ACV Tripeptide*

Pioneering studies on the biosynthesis of penicillin using cell-free systems led to the crucial observation that the cysteine-containing tripeptide δ-(L-α-aminoadipyl)-L-cysteinyl-D-valine (Fig. 2A) was the direct precursor of these antibiotics. Similar findings on the tripeptide intermediate were subsequently reported for bacterial producers of β-lactams.

The enzyme that catalyzes this function is the ACV synthetase, which requires ATP and Mg^{2+} or Mn^{2+} ions, and it has several different activities including substrate recognition, amino acid activation (adenylation), condensation, epimerization of L-valine to the D configuration, and cleavage and release of the final tripeptide.

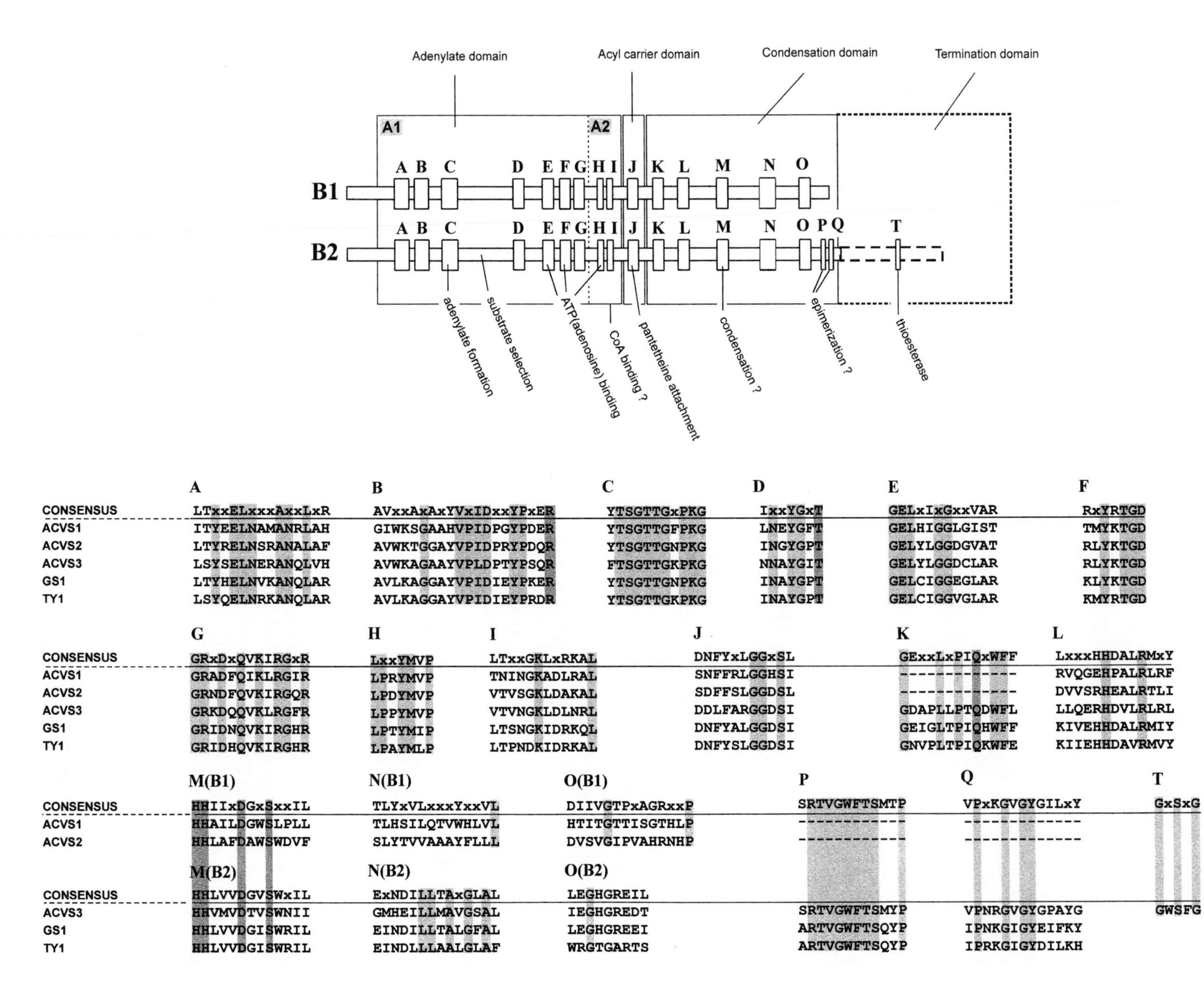

Adenylate domain
Acyl carrier domain
Condensation domain
Termination domain
A1
A2
B1
B2
A B C D E F G H I J K L M N O
A B C D E F G H I J K L M N O P Q T
adenylate formation
substrate selection
ATP(adenosine) binding
CoA binding ?
pantetheine attachment
condensation ?
epimerization ?
thioesterase
A
CONSENSUS LTxxELxxxAxxLxR
ACVS1 ITYEELNAMANRLAH
ACVS2 LTYRELNSRANALAF
ACVS3 LSYSELNERANQLVH
GS1 LTYHELNVKANQLAR
TY1 LSYQELNRKANQLAR
B
CONSENSUS AVxxAxAxYVxIDxxYPxER
ACVS1 GIWKSGAAHVPIDPGYPDER
ACVS2 AVWKTGGAYVPIDPRYPDQR
ACVS3 AVWKAGAAYVPLDPTYPSQR
GS1 AVLKAGGAYVPIDIEYPKER
TY1 AVLKAGGAYVPIDIEYPRDR
C
CONSENSUS YTSGTTGxPKG
ACVS1 YTSGTTGFPKG
ACVS2 YTSGTTGNPKG
ACVS3 FTSGTTGKPKG
GS1 YTSGTTGNPKG
TY1 YTSGTTGKPKG
D
CONSENSUS IxxYGxT
ACVS1 LNEYGFT
ACVS2 INGYGPT
ACVS3 NNAYGIT
GS1 INAYGPT
TY1 INAYGPT
E
CONSENSUS GELxIxGxxVAR
ACVS1 GELHIGGLGIST
ACVS2 GELYLGGDGVAT
ACVS3 GELYLGGDCLAR
GS1 GELCIGGEGLAR
TY1 GELCIGGVGLAR
F
CONSENSUS RxYRTGD
ACVS1 TMYKTGD
ACVS2 RLYKTGD
ACVS3 RLYKTGD
GS1 KLYKTGD
TY1 KMYRTGD
G
CONSENSUS GRxDxQVKIRGxR
ACVS1 GRADFQIKLRGIR
ACVS2 GRNDFQVKIRGQR
ACVS3 GRKDQQVKLRGFR
GS1 GRIDNQVKIRGHR
TY1 GRIDHQVKIRGHR
H
CONSENSUS LxxYMVP
ACVS1 LPRYMVP
ACVS2 LPDYMVP
ACVS3 LPPYMVP
GS1 LPTYMIP
TY1 LPAYMLP
I
CONSENSUS LTxxGKLxRKAL
ACVS1 TNINGKADLRAL
ACVS2 VTVSGKLDAKAL
ACVS3 VTVNGKLDLNRL
GS1 LTSNGKIDRKQL
TY1 LTPNDKIDRKAL
J
CONSENSUS DNFYxLGGxSL
ACVS1 SNFFRLGGHSI
ACVS2 SDFFSLGGDSL
ACVS3 DDLFARGGDSI
GS1 DNFYALGGDSI
TY1 DNFYSLGGDSI
K
CONSENSUS GExxLxPIQxWFF
ACVS1 ------------
ACVS2 ------------
ACVS3 GDAPLLPTQDWFL
GS1 GEIGLTPIQHWFF
TY1 GNVPLTPIQKWFE
L
CONSENSUS LxxxHHDALRMxY
ACVS1 RVQGEHPALRLRF
ACVS2 DVVSRHEALRTLI
ACVS3 LLQERHDVLRLRL
GS1 KIVEHHDALRMIY
TY1 KIIEHHDAVRMVY
M(B1)
CONSENSUS HHIIxDGxSxxIL
ACVS1 HHAILDGWSLPLL
ACVS2 HHLAFDAWSWDVF
N(B1)
CONSENSUS TLYxVLxxxYxxVL
ACVS1 TLHSILQTVWHLVL
ACVS2 SLYTVVAAAYFLLL
O(B1)
CONSENSUS DIIVGTPxAGRxxP
ACVS1 HTITGTTISGTHLP
ACVS2 DVSVGIPVAHRNHP
P
CONSENSUS SRTVGWFTSMTP
ACVS1 ------------
ACVS2 ------------
ACVS3 SRTVGWFTSMYP
GS1 ARTVGWFTSQYP
TY1 ARTVGWFTSQYP
Q
CONSENSUS VPxKGVGYGILxY
ACVS1 -------------
ACVS2 -------------
ACVS3 VPNRGVGYGPAYG
GS1 IPNKGIGYEIFKY
TY1 IPRKGIGYDILKH
T
CONSENSUS GxSxG
ACVS3 GWSFG
M(B2)
CONSENSUS HHLVVDGVSWxIL
ACVS3 HHVMVDTVSWNII
GS1 HHLVVDGISWRIL
TY1 HHLVVDGISWRIL
N(B2)
CONSENSUS ExNDILLTAxGLAL
ACVS3 GMHEILLMAVGSAL
GS1 EINDILLTALGFAL
TY1 EINDLLLAALGLAF
O(B2)
CONSENSUS LEGHGREIL
ACVS3 IEGHGREDT
GS1 LEGHGREEI
TY1 WRGTGARTS

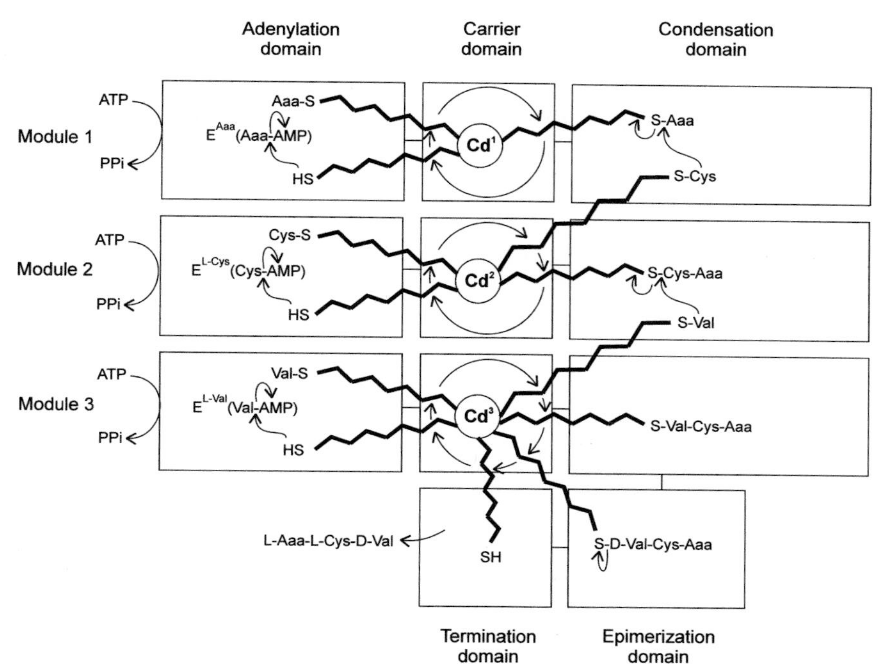

Fig. 4. Thiotemplate mechanism of activation, condensation, epimerization, and termination reactions during the formation of the LLD-ACV tripeptide by the action of ACV synthetase.

The ACV synthetase has three domains, with sequences very conserved between them, and separated by regions of low similarity. These domains, having between 500 and 1000 amino acids, seem to correspond to the activation center for each precursor amino acid, suggesting a similar organization to the adenylate-forming enzymes and to other peptide-forming multienzymes (Kleinkauf and von Döhren, 1990). The ACV synthetase contains regions similar to those defined as consensus for the phosphopantetheine binding sequence; this phosphopantetheine moiety keeps the intermediates bound to the enzyme and should translocate these intermediates to other domains (Fig. 4). The ACV synthetase also has a thioesterase consensus sequence; the thioesterase activity seems to hydrolyze the bond between the enzyme and the tripeptide.

The ACV synthetase has a molecular mass ranging from 404 to 425 kDa, depending on the microorganism, and is encoded by the *pcb*AB gene, which has been cloned from different β-lactam producers. This gene is clustered with the other genes involved in β-lactam antibiotic biosynthesis (see Fig. 6).

2. *Cyclization of the ACV Tripeptide and Formation of Isopenicillin N*

The tripeptide ACV is then cyclized to form isopenicillin N (IPN), an intermediate that contains an L-α-aminoadipyl side chain attached to the *penam* nucleus (Fig. 5) by isopenicillin N synthase (IPN synthase or cyclase). The IPN synthase has been purified from *Acremonium chrysogenum, Penicillium chrysogenum, S. clavuligerus, N. lactamdurans,* and *Flavobacterium* sp.

The enzyme of *P. chrysogenum* shows a molecular mass of 39,000 Da. It requires dithiothreitol (DTT)

Fig. 3. Conserved motifs (labeled A to T) in amino acid activating domains of peptide synthetases: ACVS1, ACVS2, and ACVS3 refer to the α-aminoadipyl-cysteinyl-valine synthetase; GS1, gramicidin synthase I; TY1, tyrocidine synthetase 1. Note that B2 domains are longer than type B1 domains due to the presence of a thioesterase domain.

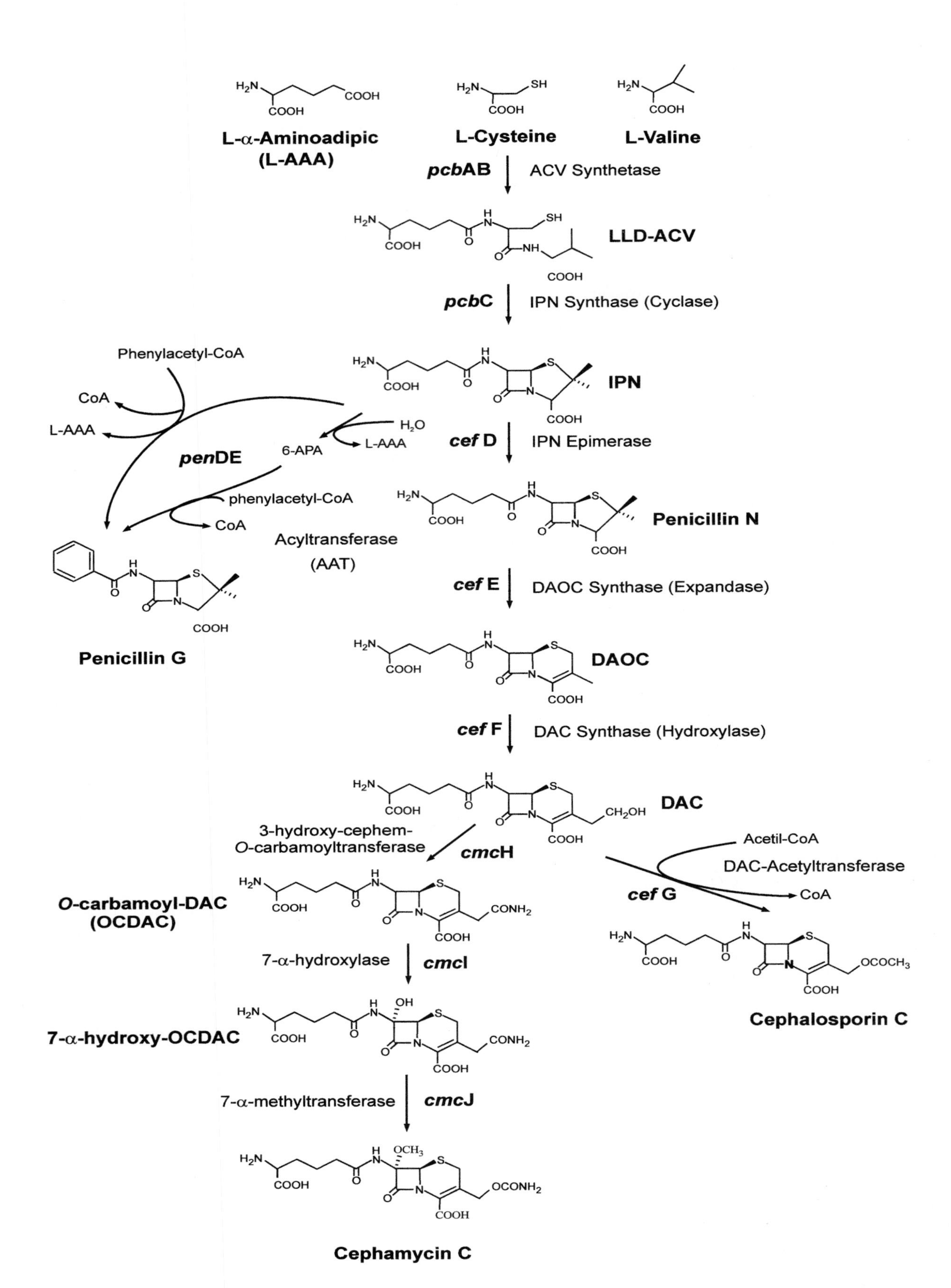
L-α-Aminoadipic
(L-AAA)
L-Cysteine
L-Valine
pcbAB
ACV Synthetase
LLD-ACV
pcbC
IPN Synthase (Cyclase)
Phenylacetyl-CoA
CoA
L-AAA
IPN
H2O
L-AAA
6-APA
penDE
cef D
IPN Epimerase
phenylacetyl-CoA
CoA
Acyltransferase
(AAT)
Penicillin N
Penicillin G
cef E
DAOC Synthase (Expandase)
DAOC
cef F
DAC Synthase (Hydroxylase)
DAC
3-hydroxy-cephem-
O-carbamoyltransferase
cmcH
Acetil-CoA
DAC-Acetyltransferase
cef G
CoA
O-carbamoyl-DAC
(OCDAC)
7-α-hydroxylase
cmcI
Cephalosporin C
7-α-hydroxy-OCDAC
7-α-methyltransferase
cmcJ
Cephamycin C

for the activity and is stimulated by ascorbate and Fe^{2+} ions (usual cofactors for oxygenases). The enzymatic reaction requires O_2 and is stimulated by the concentration of dissolved oxygen in the reaction. In fact, the cyclization occurs by the elimination, mediated by oxygen, of four hydrogen atoms from the tripeptide molecule.

The *Aspergillus nidulans* IPN synthase has been crystallized. The crystal structure revealed the active site to be unusually buried within a "jelly-roll" motif and lined by hydrophobic residues.

The IPN synthase is encoded by the *pcbC* gene which has been cloned from many microorganisms that produce β-lactam antibiotics (Fig. 6). All the *pcbC* genes show very high similarity among them, with percentages of identity from 62 to 80%, even between prokaryotic and eukaryotic β-lactam producers (Aharonowitz *et al.*, 1992).

3. The Last Step in Penicillin Biosynthesis: Conversion of Isopenicillin N to Penicillin G

The α-aminoadipyl side chain is later exchanged for phenylacetic acid in penicillin-producing fungi but not in cephalosporin producers. This reaction is catalyzed by an isopenicillin *N*-acyltransferase (IAT) found in extracts of *P. chrysogenum* and *A. nidulans* but not in *Cephalosporium acremonium* (syn. *Acremonium chrysogenum; Acremonium strictum*). The purified preparation of the acyltransferase showed, after electrophoretic analysis, three protein bands of 40, 29, and 11 kDa molecular mass. The 40-kDa protein is a heterodimer composed of two subunits of 11 and 29 kDa.

The acyltransferase protein is encoded by the *pen*DE gene, which has been cloned from *Penicillium chrysogenum* (Gutiérrez *et al.*, 1999) and *Aspergillus nidulans* (Martín *et al.*, 1997). The small subunit corresponds to the amino-terminal end of the protein, which is first synthesized as the 40-kDa protein and later is processed into the two subunits (11 and 29 kDa).

C. Specific Reactions in the Biosynthesis of Cephalosporins and Cephamycins

Isopenicillin N is converted to penicillin N in cephalosporin- and cephamycin-producing microorganisms, but not in producers of hydrophobic penicillin (i.e., penicillin G or V), by an isopenicillin N epimerase that isomerizes the L-α-aminoadipyl side chain to the D configuration. This enzyme does not require Fe^{2+}, ascorbic acid, or ATP, and the activity seems to be stimulated by pyridoxal-5-phosphate. The isopenicillin N epimerase is encoded by the *cef*D gene, which has been cloned from *S. clavuligerus* and *N. lactamdurans*. The protein deduced from these genes showed a molecular mass of approximately 44 kDa.

Penicillin N is transformed into deacetoxycephalosporin C by the deacetoxycephalosporin C synthase (the so-called ring-expanding enzyme or expandase). This enzyme converts the five-membered thiazolidine ring of penicillins into the six-membered dihydrothiazine ring of cephalosporins and cephamycins and requires Fe^{2+} and 2-ketoglutarate. Its activity is strongly stimulated by ascorbic acid and to a lesser extent by ATP, both well-known effectors of 2-ketoglutarate-dependent dioxygenases. The enzyme from *C. acremonium* has a molecular mass of 31,000 Da. It expands the β-lactam ring of penicillin N but does not accept the isomer isopenicillin N, penicillin G, or 6-aminopenicillonic acid (6-APA) as substrates (Martín *et al.*, 1997).

Deacetoxycephalosporin C is hydroxylated to deacetylcephalosporin C by another 2-ketoglutarate-dependent dioxygenase. This enzyme catalyzes the incorporation of an oxygen atom from O_2 into the exocyclic methyl group of deacetoxycephalosporin C, and is also stimulated by 2-ketoglutarate, ascorbate, dithiothreitol, and Fe^{2+}. The deacetoxycephalosporin C synthase and deacetoxycephalosporin C hydroxylase activities from *C. acremonium* are located on a single protein of approximately 33,000

Fig. 5. Biosynthetic pathways to penicillin G, cephalosporin C, and cephamycin C indicating the enzymes catalyzing each reaction and the genes (bold letters) that encode them. LLD-ACV, δ-(L-α-Aminoadipyl)-L-cysteinyl-D-valine; IPN, isopenicillin N; DAOC, deacetoxycephalosporin C; DAC, deacetylcephalosporin C.

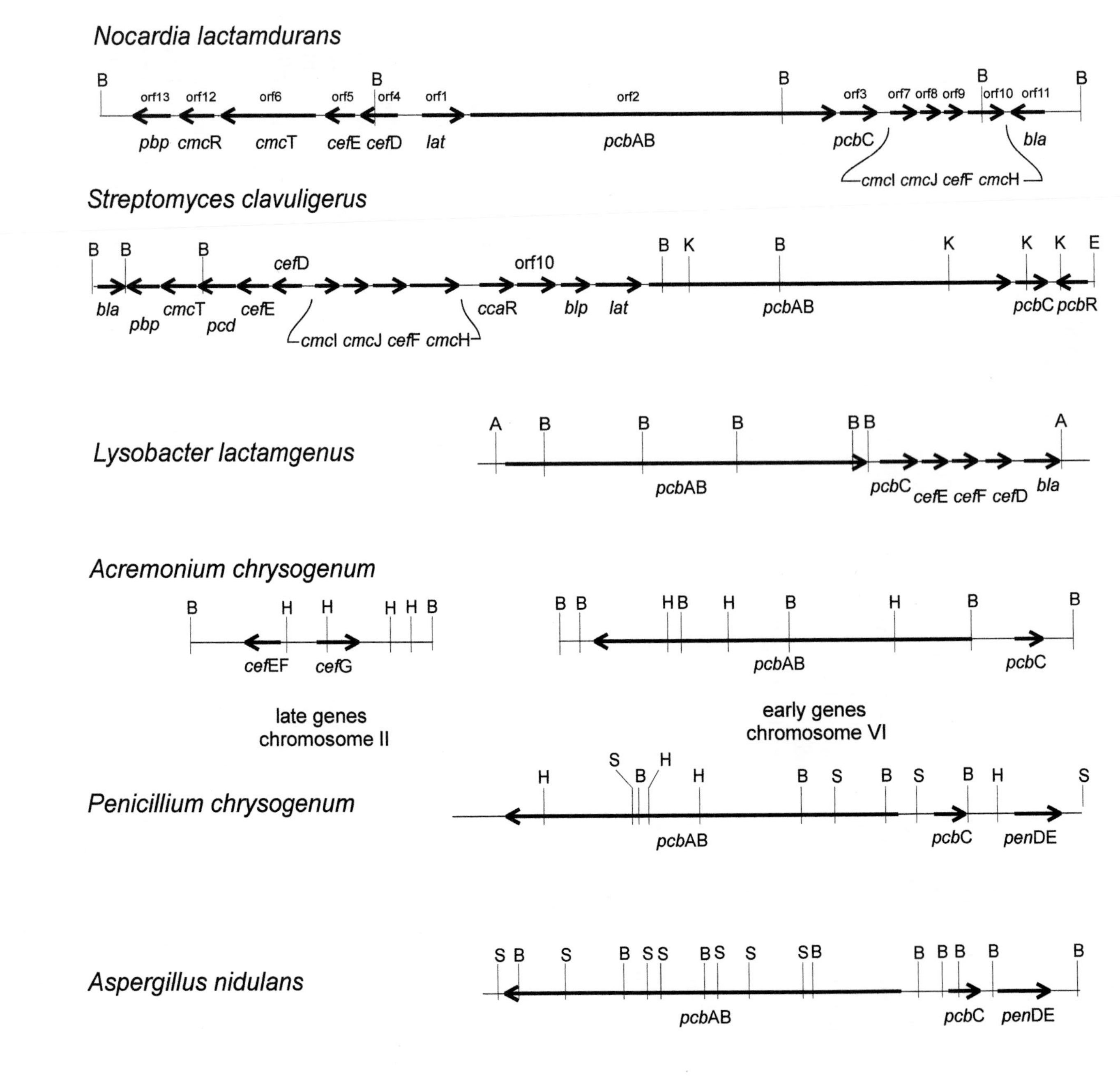

Nocardia lactamdurans
orf13 orf12 orf6 orf5 orf4 orf1 orf2 orf3 orf7 orf8 orf9 orf10 orf11
pbp cmcR cmcT cefE cefD lat pcbAB pcbC bla
cmcI cmcJ cefF cmcH
Streptomyces clavuligerus
cefD orf10
bla pbp cmcT pcd cefE ccaR blp lat pcbAB pcbC pcbR
cmcI cmcJ cefF cmcH
Lysobacter lactamgenus
pcbAB pcbC cefE cefF cefD bla
Acremonium chrysogenum
cefEF cefG
late genes
chromosome II
pcbAB pcbC
early genes
chromosome VI
Penicillium chrysogenum
pcbAB pcbC penDE
Aspergillus nidulans
pcbAB pcbC penDE
2 kb

Da, which is encoded by the *cef*EF gene. However, deacetoxycephalosporin C synthase and deacetoxycephalosporin C hydroxylase are separate enzymes in cephamycin-producing actinomycetes. These enzymes are encoded, respectively, by the *cef*E and *cef*F genes.

Acetylation of deacetylcephalosporin C (DAC) to cephalosporin C is the terminal reaction in cephalosporin-producing fungi (Fig. 5). The DAC acetyltransferase was purified to near homogeneity, and its amino-terminal end matched perfectly the amino acid sequence deduced from the gene. The purified protein has a molecular mass of 50 kDa; it is encoded by the *cef*G gene. The DAC acetyltransferase has high affinity for the hydroxymethyl group in the C-3 position of DAC and requires Mg^{2+} for activity. The cephalosporin C can be degraded in fermentation broths by the action of extracellular acetylhydrolase activities.

1. The Late Reactions in Cephamycin Biosynthesis

Further reactions are involved in the attachment of the C-7 methoxyl group and the C-3′ carbamoyl group during cephamycin C biosynthesis in the actinomycetes. The biochemistry and molecular genetics of the late reactions of cephamycins have been reviewed (Martín, 1998).

IX. POLYKETIDES

Polyketides are a large class of natural metabolites produced through the successive condensation of simple carboxylic acids. This highly diverse group of molecules are found in most groups of organisms, but they are specially abundant in actinomycetes.

These compounds have generated much interest in synthetic and biosynthetic chemistry because of their staggering structural complexity (see Fig. 7). Polyketides exhibit an impressive range of antibacterial, antifungal, anticancer, insecticidal, antiparasite, and immunosuppressant activities. Some polyketides produced by fungi and bacteria are associated with sporulation or other developmental pathways; others do not yet have a known function. Some polyketides have more than one property: the macrocyclic polyketide rapamycin, in addition to its immunosuppressant effect, has antifungal and antitumor activities; pamamycin stimulates aerial mycelium formation in *Streptomyces alboniger* and also shows antimicrobial activity; the aromatic polyketide mithramycin (aureolic acid) shows antibacterial and antitumor activity.

A. Mechanisms of Polyketide Biosynthesis

Despite their structural diversity, all polyketides share a common mechanism of biosynthesis (Hopwood and Sherman, 1990; Katz and Donadio, 1993), in a manner that is conceptually similar to the biosynthesis of the long-chain fatty acids found in all organisms. A polyketide synthase (PKS) catalyzes the formation of a polyketide through repeated decarboxylative condensations between enzyme-bound acyl carrier protein (ACP)-thioesters. Extender units are transferred from coenzyme A (CoA) to the pantetheine arm of the ACP by the acyltransferase (AT). An exception to the rule involves type III polyketide synthases, distributed in plants, that lack ACP functionalities and act directly on CoA esters of the building units rather than on ACP-linked residues. After each condensation, carried out by a β-ketoacyl ACP synthase (KS), the β-keto group of the growing chain remains unchanged or is reduced to a hydroxyl, enoyl, or methylene group by the action of a β-ketoreductase (KR), dehydratase (DH), or enoylreductase (ER), respectively (Fig. 8). Structural variations among naturally occurring polyketides arise largely from the way in which the PKS controls the number and type of starter and extender units used, and from the extent and stereochemistry of reduction at each cycle. Yet further diversity is produced by functionalization of the polyketide chain by the ac-

Fig. 6. Organization of the genes involved in the biosynthesis of β-lactam antibiotics in different eukaryotic and prokaryotic microorganisms, B, *Bam*HI; K, *Kpn*I; S, *Sal*I; A, *Apa*I; H, *Hin*dIII.

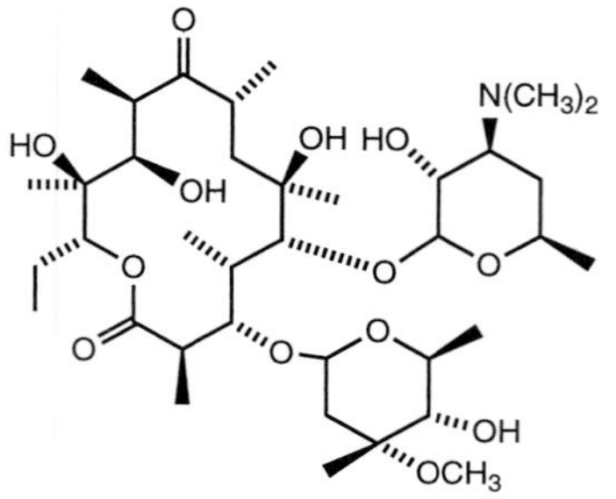

Erythromycin A

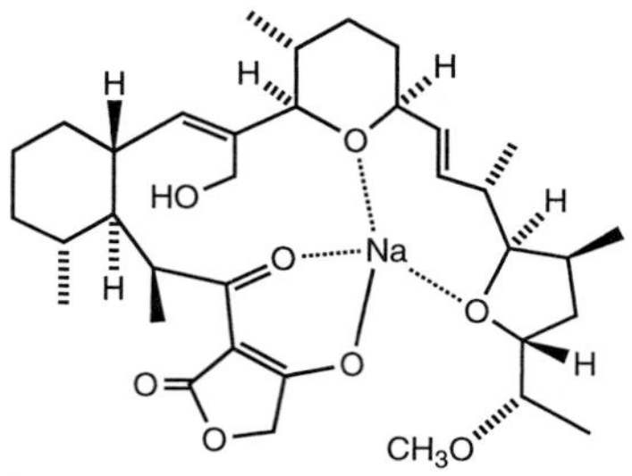

Tetronasin

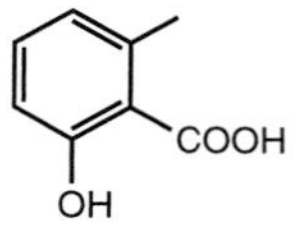

6-methylsalicylic acid

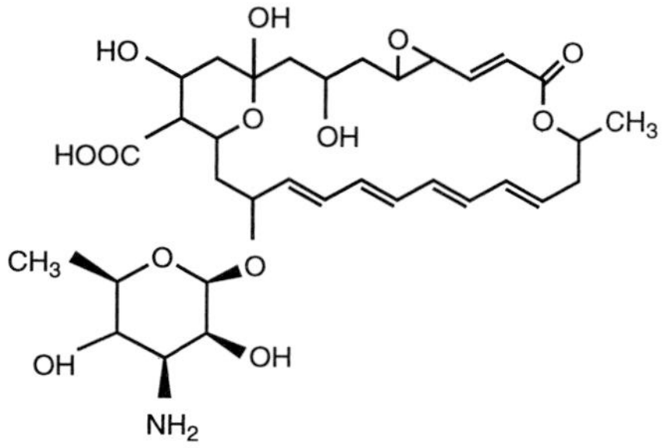

Pimaricin

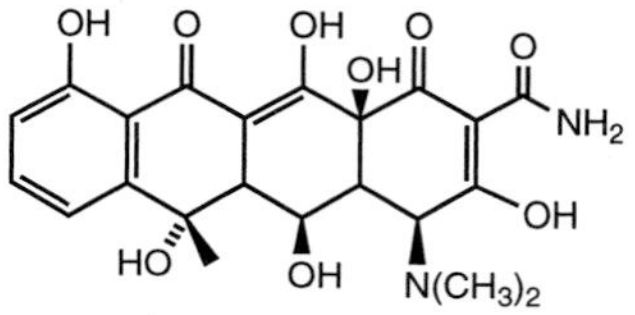

Oxytetracycline

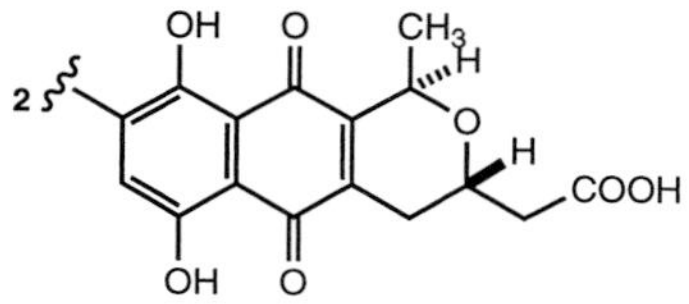

Actinorhodin

Fig. 7. Some examples of fungal and bacterial polyketide structures. Note that some of these polyketides contain aromatic rings (e.g., 6-methylsalicylic acid) formed by cyclization and aromatization reactions.

tion of glycosylases, methyltransferases, and oxidative enzymes on the product of the PKS.

B. Polyketide Precursors

The term "polyketide" (or "polyketene") was first used in 1907 with reference to the synthesis of plant aromatic metabolites, to propose, with a great insight for the time, that such molecules may be polymers of acetate or other acetate homologs. In the early 1950s, Birch and collaborators proposed that polyketides were formed through the condensation of acetate residues and indicated the relationship of polyketides with the biosynthesis of long-chain fatty acids.

Polyketides are basically formed from acetate, propionate, and more rarely butyrate units (derived from malonyl-, methylmalonyl-, or ethylmalonyl-CoA, respectively) assembled in a head-to-tail fashion. [1-, 2-, and 3-^{14}C]Propionate and [1-^{14}C]acetate are incorporated into the polyene aglycone of nys-

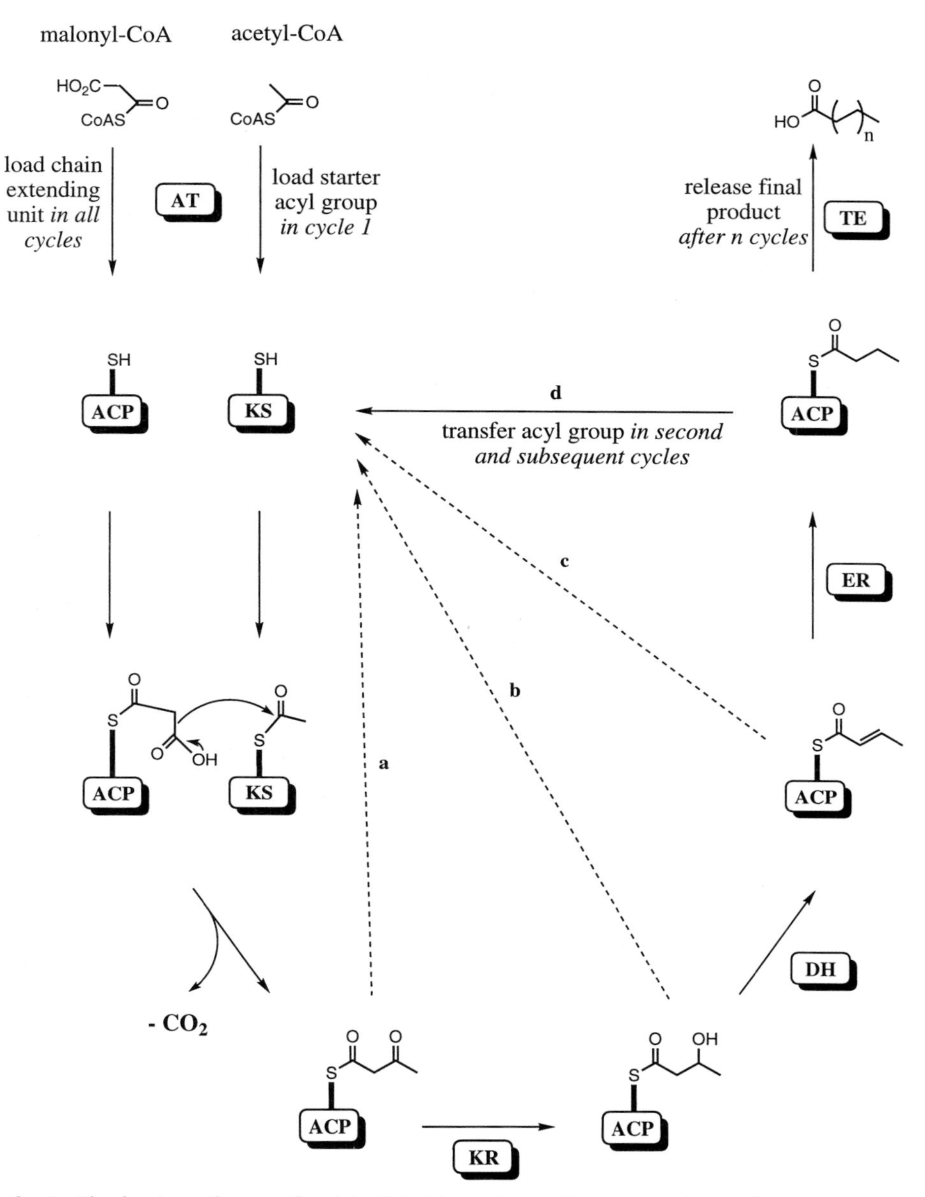

Fig. 8. The basic pathway of polyketide biosynthesis. Note that alternative starter and extender units may be accepted (see text). Dashed arrows represent shortcuts of the pathway that may occur in a given cycle to produce a (a) keto, (b) hydroxyl, (c) enoyl, or (d) methylene functionality in the polyketide backbone. The reaction steps are labeled: AT, acyltransferase; KS, ketosynthase; ACP, acyl carrier protein; KR, ketoreductase; DH, dehydratase; ER, enoylreductase; TE, thioesterase.

tatin, whereas no incorporation was observed with [*methyl*-^{14}C]methionine, indicating the absence of C_1 precursor units. Labeling studies have also established that the benzoisochromanequinone antibiotic actinorhodin (Fig. 7) is made from acetate units. Similar precursor incorporations were observed in the biosynthesis of anthracyclines, macrolides, and many other polyketides.

For type II polyketide synthases (see below) from actinomycetes whose products are typically aromatic polyketides, and for type I polyketide synthases from fungi, the extension units are derived almost exclusively from malonyl-CoA. In contrast, reduced or complex polyketides usually contain both acetate and propionate units (Hopwood and Sherman, 1990). The giant rapamycin polyketide synthase selectively incorporates seven propionate and seven acetate extender units into the polyketide chain. The basis for this selectivity has recently been uncovered. Individual acyl-CoA : ACP acyltransferase (AT) domains or enzymes make the choice of the extender unit to incorporate into the growing chain.

Besides the diversity created by the incorporation of these extender units, further variation is introduced from the choice of starter units. The starter unit is normally activated as acetyl-CoA (e.g., pimaricin) or propionyl-CoA (e.g., erythromycin; Aparicio *et al.*, 1994), but for some complex polyketides like the immunosuppressants FK506 and rapamycin, the starter unit is a cyclohexane carboxylic acid derived by reduction of shikimate. Similarly, several of the so-called aromatic polyenes use an aromatic unit as starter, either *p*-aminoacetophenone (e.g., candicidin, aureofungin) or *N*-methyl-*p*-aminoacetophenone (e.g., candimycin) (see Martín, 1977, for a review). Other examples of aromatic rings as primer units include the aromatic cinnamoyl unit present in several compounds of the chalcone or stilbene series, the *p*-coumarate in the biosynthesis of plant polyketide secondary metabolites such as flavonoids, and the C_7N aromatic moiety (3-amino-5-hydroxybenzoic acid) acting as a starter in the biosynthesis of the macrolactam antibiotic rifamycin. Some anthracyclins can also use butyrate (feudomycin) or isobutyrate (13-methyl-aclacinomycin A) as starter.

Finally, an even further source of variation becomes apparent in some macrolides containing aminoacyl or peptidyl residues, as in the peptidylmacrolactone immunosuppressants FK506, immunomycin, and rapamycin. Presumably, a peptide synthetase interacts here directly with a polyketide synthase, and the polyketide chain is transferred from a thioester linkage on the polyketide synthase to the amino group of an enzyme-bound iminoacid.

C. Organizations for Polyketide Synthesis

The genes responsible for the various steps in polyketide biosynthesis show a strong tendency to be clustered (Hopwood and Sherman, 1990). This clustering has greatly facilitated cloning and characterization of the genes involved in the biosynthesis of a given polyketide.

Polyketides fall into two structural classes: aromatic and complex. The genetic organization of the gene clusters that direct their biosynthesis is strikingly different, and so are the protein constituents that interact to build the functional PKSs. A description of these two main types of PKSs is presented below.

1. *Aromatic Polyketide Synthases*

The products of type II PKSs are typically aromatic polyketides that are synthesized by using acetyl or propionyl starter units, and only a malonyl moiety for extension (Hopwood and Sherman, 1990). Aromatic polyketide synthases are determined by four to six genes encoding mono- or bifunctional enzymes, and the polyketide synthase is used for all synthesis steps through an iterative process wherein the β-carbonyl groups formed after each condensation cycle are left largely unreduced. Fungal enzymes seem to contain all the enzymatic domains distributed over a single protein.

The basic architecture of the gene clusters encoding the sets of subunits of bacterial aromatic PKSs is very similar (Fig. 9), and so prediction of polyketide structure from gene organization is not possible. Malpartida and Hopwood first reported the cloning of the entire set of actinorhodin biosynthetic genes from *Streptomyces coelicolor*, a 26-kilobase gene cluster. Since then a number of type II gene clusters have been investigated (see Hopwood, 1997, for a review). All reported gene clusters for aromatic polyketides contain a set of three genes that encode the so-called minimal PKS. This includes a β-ketoacyl ACP synthase (KS), which also carries a putative acyl-CoA : ACP acyltransferase (AT) domain, a chain length factor (CLF), and an acyl carrier protein (ACP) (Fig. 9). ACPs are small polypeptides (less than 100 amino acids in length) that

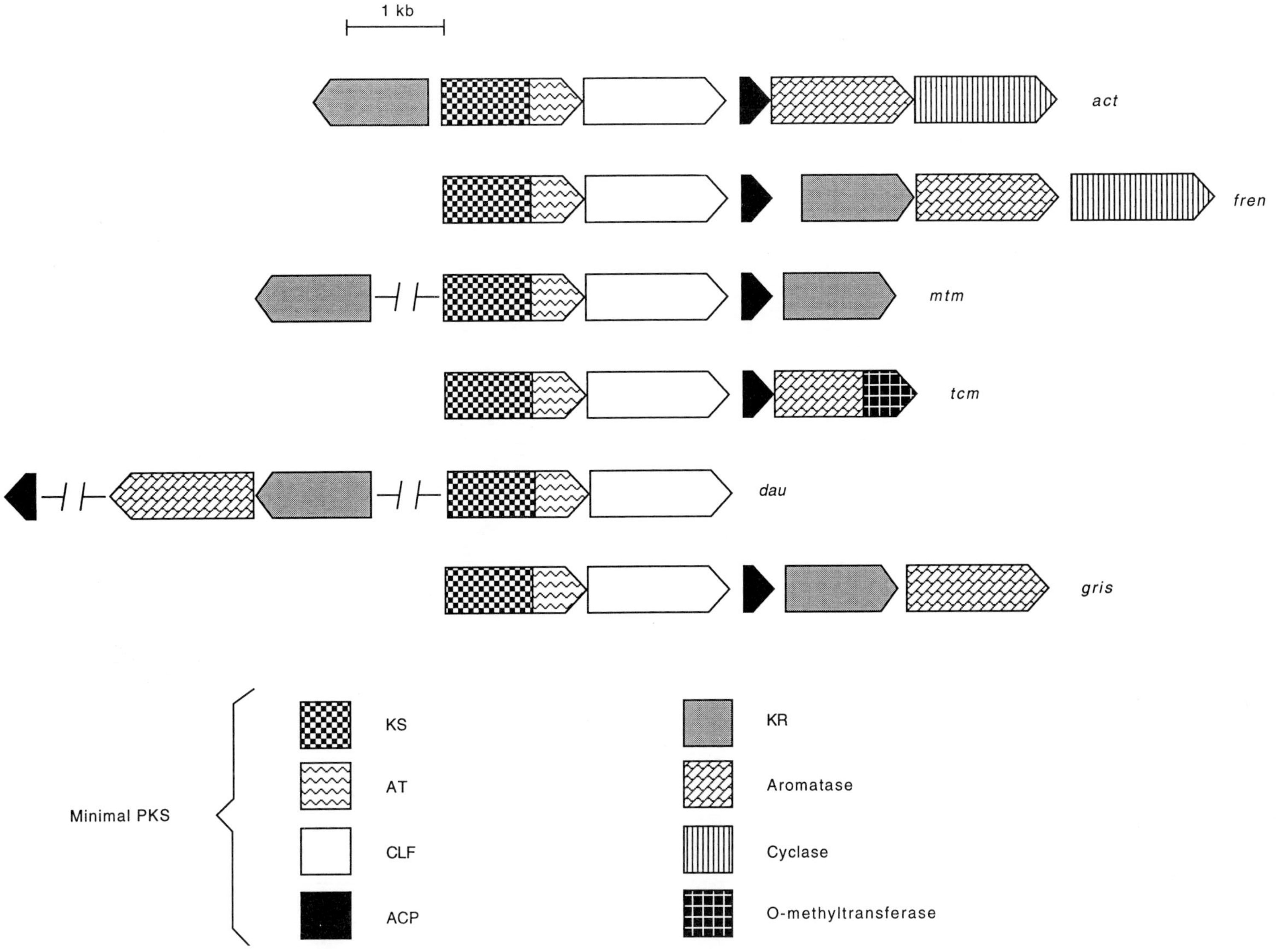

Fig. 9. Organization of gene clusters for some type II polyketide sythases in actinomycetes. Only the genes involved in carbon chain assembly, reduction, cyclization, or aromatization are shown, except for the *O*-methyltransferase domain in the tetracenomycin gene cluster. *act,* Actinorhodin PKS gene region, produced by *Streptomyces coelicolor; fren, S. roseofulvus* frenolicin gene cluster; *mtm,* mithramycin gene cluster in *S. argillaceus; tcm, S. glaucescens* tetracenomycin C PKS gene cluster; *dau, Streptomyces* sp. strain C5 daunomycin PKS gene region; *gris, S. griseus* griseusin gene cluster. Abbreviations: CLF, chain length factor; other abbreviations as in Fig. 8.

contain the highly conserved signature sequence LGXDS characteristic of the 4′-phosphopantetheine binding site, and show a good end-to-end similarity.

The remaining genes that determine PKS functions are less conserved, and their relative positions are more variable. These genes include a ketoreductase (KR) whose product catalyzes regiospecific ketoreduction in the growing polyketide chain (present, e.g., in the actinorhodin, frenolicin, griseusin, daunomycin, jadomycin, and mithramycin biosynthetic gene clusters) and an aromatase (ARO) gene whose product catalyzes the aromatization of the nascent chain (present in actinorhodin, frenolicin, griseusin, jadomycin, and mithramycin gene clusters among others). A cyclase is also present in some type II clusters. This activity is believed to reside in a bifunctional polypeptide that also carries a dehydratase domain, and is presumably responsible for the formation of the second ring in the final polyketide structure (e.g., actinorhodin, frenolicin, or dihydrogranaticin). In the tetracenomycin gene cluster the cyclase enzyme is linked to a C-terminal *O*-methyltransferase domain (Fig. 9).

Functional analysis of type II systems has involved analysis of individual components of these systems not only by gene disruption experiments (see Hopwood, 1997, for a review), but also by using biochemical approaches such as heterologous expression in *Escherichia coli*. The structure in solution of the recombinant actinorhodin ACP has now been unveiled, for the first time for any PKS component; and at least for one aromatic polyketide, the tetracenomycin F2 (a tetracenomycin C precursor), synthesis has been achieved *"in vitro"* in cell-free extracts from a *Streptomyces glaucescens* strain.

2. *Complex Polyketide Synthases*

Complex polyketide synthases are composed of several multifunctional polypeptides that contain enzymatic domains for the condensation and reduction steps; each domain is used at a unique step in the biosynthesis, and the extent of β-carbonyl processing depends on the functional domains operating at a given synthase. The result is a far more reduced product than the ones yielded by the other type of PKSs.

Furthermore, in complex PKSs the enzymatic activities are clustered into modules, so that each module contains all the different domains required for each successive round of elongation. Once the product has gone through one cycle, it moves to the next enzyme module, where it suffers a new condensation and subsequent processing of the β-carbon. The synthesis of complex polyketides is a stepwise processive mechanism in which presumably the stereochemistry of the nascent chiral centers and certainly the appropriate degree of processing of the β-carbons are fixed into the growing chain prior to subsequent elongation steps.

Leadlay and collaborators at the University of Cambridge and Katz at Abbott Laboratories first cloned and sequenced the genes from *Saccharopolyspora erythraea* responsible for synthesis of the intermediate in the biosynthesis of erythromycin. During the biosynthesis of erythromycin A, the aglycone 6-deoxyerythronolide B is first synthesized through the condensation of one propionate and six methylmalonate units. The three PKS multienzymes involved in this process are encoded by three giant genes (*ery*AI to *ery*AIII), lying at the center (spanning some 30 kb) of the cluster, containing two modules each (Fig. 10). The molecular characterization of other type I modular PKS gene clusters has since been reported.

The most remarkable examples studied so far are the PKS for the 26-membered ring polyene macrolide pimaricin (Aparicio *et al.,* 1999) and the PKS for the 31-membered immunosuppressant rapamycin from *Streptomyces hygroscopicus,* which contains 14 modules contained within only three huge multienzyme polypeptides (Aparicio *et al.,* 1996). The order of the domains within each module is KS, AT (DH, ER, KR where applicable), and ACP, exactly as for the fatty acid synthetase (FAS).

The discovery of the one-to-one correspondence between active sites and catalytic steps provides a potentially powerful model for the rational design and engineered biosynthesis of novel polyketides through genetic manipulation, particularly because the overall chain extension process is dictated by the order of the catalytic domains. This "predictibility" of the gene function characteristic of type I PKSs, together with the endless diversity of products

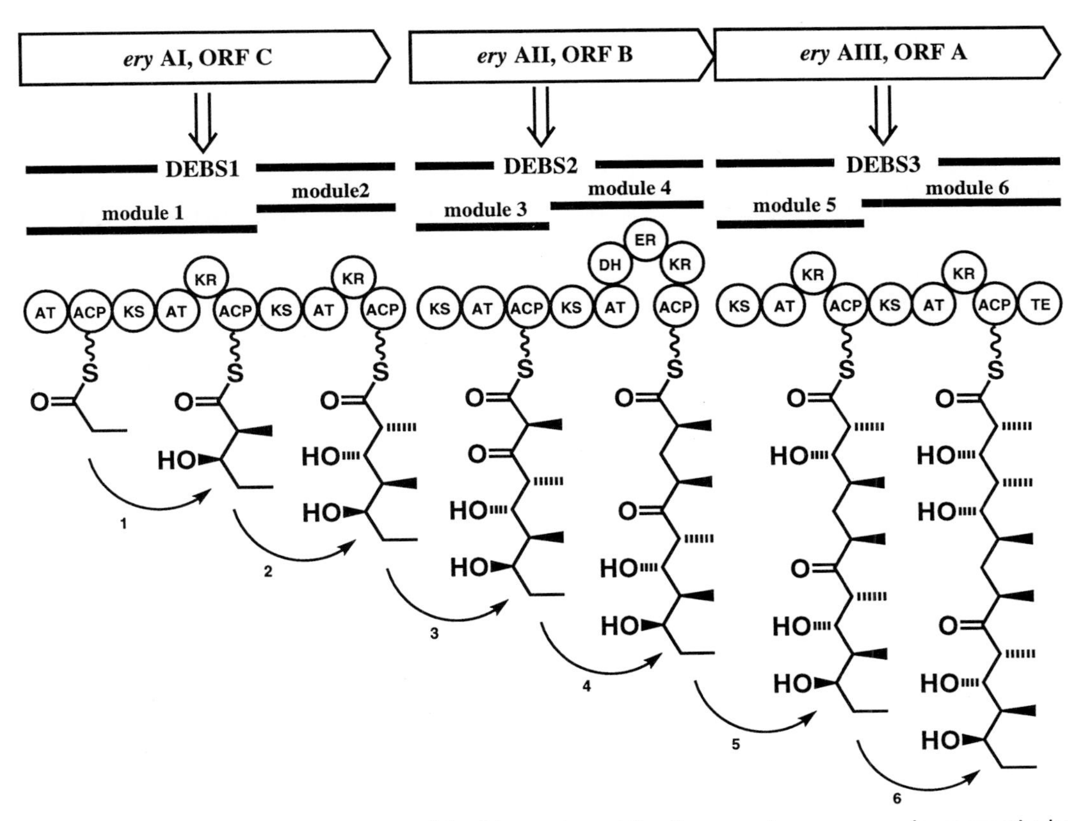

Fig. 10. The modular erythromycin polyketide synthase. The three *ery*A genes encode, respectively, DEBS1, which binds the propionyl-CoA starter unit and catalyzes the first two cycles of polyketide chain extension and reduction, DEBS2, which catalyzes the third and fourth cycles, and DEBS3, which accounts for the final two cycles and the cyclization of the macrolide ring. One model for chain growth on the DEBS proteins is also shown. Abbreviations are as in Fig. 8.

yielded by these kinds of multienzymes, has raised enormously the interest around these systems.

X. ISOPRENOIDS

A variety of natural microbial products including both primary and secondary metabolites are isoprenoids, for example, carotenoids, steroids, gibberellins, phytoalexins, and antibiotics or anticancer agents.

Isoprenoid molecules derive from the C_5 unit isopentenylpyrophosphate (IPP), which is synthesized by condensation of three acetate units (activated as acetyl-CoA) via the intermediate 3-hydroxy-3-methylglutaryl-CoA (HMG-CoA) that undergoes reduction to mevalonic acid and then activation and decarboxylation to isopentenylpyrophosphate. The central role of mevalonic acid and IPP in the biosynthesis of linear and cyclized mono-, sesqui-, and diterpenes, sterols, polycyclic triterpenes, and carotenoids is well established (Cane, 1999).

There is, however, a second biosynthetic route to IPP that involves neither acetate nor mevalonate. In the first step of this pathway, pyruvate and glyceraldehyde-3-phosphate undergo a thiamine pyrophosphate-dependent condensation to give 1-deoxy-D-xylulose-5-phosphate, which is later converted to IPP. This route, named the deoxyxylulose phosphate pathway, is the source of isoprenoids in prokaryotic microorganisms (except archaebacteria) and occur side by side with the mevalonate pathway in higher plants.

A. Isoprenoid Chain Elongation

A large number of isoprenoid compounds are derived by enzymatic chain elongation using as substrates for these transformations the acyclic isoprenoids dimethylallyl diphosphate (DMAPP), geranyl diphosphate (GPP), farnesyl diphosphate (FPP), and geranylgeranyl diphosphate (GGPP) (Fig. 11).

Isoprenoid chain elongation is catalyzed by a family of enzymes known as prenyltransferases that contain a divalent metal cofactor (Mg^{2+} or Mn^{2+}). The product of the condensation of two isoprene units may be released from the enzyme surface, or may serve as substrate for one or more additional condensations, leading to the formation of linear isoprenoids with two, three, four, or more units. The length of the isoprenoid chain is determined by each biosynthetic prenyltransferase.

B. Isomerization and Cyclization

Many natural isoprenoids are cyclic monoterpenes, sesquiterpenes, and diterpenes that are formed by cyclization of the linear geranyl, farnesyl, and geranylgeranyl diphosphate. Cyclization requires previous isomerization of the all-trans molecules, since there is a steric barrier to formation of the six-membered ring from the *trans*-allylic diphosphate precursor. Intramolecular cyclizations are mechanistically similar to the intermolecular condensations catalyzed by the prenyltransferases.

C. Biosynthesis of Steroids and Triterpenes

Biosynthesis of the more complex terpene molecules such as the triterpenes and steroids involves the head-to-head condensation of two farnesyl diphosphate molecules, which is followed by a reduction step.

D. Genetics of Isoprenoid Biosynthesis

Progress in the molecular genetics of isoprenoid biosynthesis has been slow when compared with advances in polyketide or peptide biosynthesis, but the genetic basis for understanding the formation of carotenoids and other isoprenoid molecules have been established in the last 10 years.

Carotenoid synthases show a relatively low substrate specificity, as occurs with polyketide synthases and nonribosomal peptide synthases. Many types of carotenoids are formed from a relatively small number of precursor units. The genes for carotenoid biosynthesis have been designated *crt*. Most available information derives from the carotenoid biosynthesis genes in anoxygenic photosynthetic bacteria, pigmented nonphotosynthetic bacteria, and cyanobacteria. The *crt* genes encoding biosynthetic enzymes are usually clustered (except in the cyanobacteria). Some of the genes are organized in multigenic operons that may confer an advantage for the coordinate synthesis of adequate amounts of the enzymes involved in the pathway.

Most bacterial *crt* clusters contain a *crt*E gene encoding a GGPP synthase that is specific for carotenoid biosynthesis, as *crt*E mutants lack pigment formation but are still able to form GGPP for quinone biosynthesis and other essential functions. Multiple GGPP synthases appear to occur in plants. Bacterial CrtE proteins belong to a superfamily of structurally functionally related proteins that include GGPP, FPP, and hexaprenylpyrophosphate synthases.

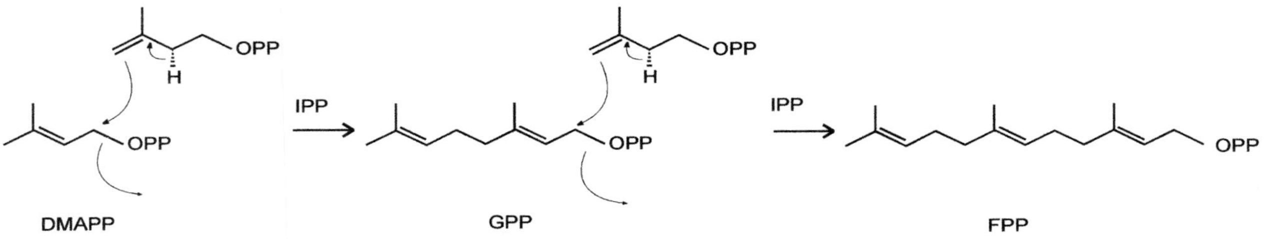

Fig. 11. Enzymatic chain elongation of isoprenoids by the prenyltransferases to form geranyl diphosphate (GPP) and farnesyl diphosphate (FPP).

Amino acid sequence comparisons revealed that members of the isoprenyl pyrophosphate synthase superfamily share five conserved stretches of amino acids.

Acknowledgments

This work was supported by grants of the European Union (BIO4-CT96-0535, BIO4-CT96-0145, FAIR-CT97-3140, BIOTECHNOLOGY BIO4-98-0051).

See Also the Following Articles

Amino Acid Function and Synthesis • Polyketide Antibiotics

Bibliography

Aharonowitz, Y., Cohen, G., and Martín, J. F. (1992). Penicillin and cephalosporin biosynthetic genes: Structure, organization, regulation and evolution. *Annu. Rev. Microbiol.* **46**, 461–496.

Aparicio, J. F., Caffrey, P., Marsden, A. F. A., Staunton, J., and Leadlay, P. F. (1994). Limited proteolysis and active-site studies of the first multienzyme component of the erythromycin-producing polyketide synthase. *J. Biol. Chem.* **269**, 8524–8528.

Aparicio, J. F., Molnár, I., Schwecke, T., König, A., Haydock, S. F., Khaw, L. E., Staunton, J., and Leadlay, P. F. (1996). Organization of the biosynthetic gene cluster of rapamycin in *Streptomyces hygroscopicus:* Analysis of the enzymatic domains in the modular polyketide synthase. *Gene* **169**, 9–16.

Aparicio, J. F., Colina, A. J., Ceballos, E., and Martín, J. F. (1999). The biosynthetic gene cluster for the 26-membered ring polyene macrolide pimaricin: A new polyketide synthase organization encoded by two subclusters separated by functionalization genes. *J. Biol. Chem* **274**, 10133–10139.

Cane, D. E. (1999). Isoprenoid biosynthesis: Overview *In* "Comprehensive Natural Products Chemistry" (D. Barton, K. Nakanishi, and O. Meth-Cohn, eds.), Vol. 2, pp. 1–13. Pergamon, Oxford.

Drew, S., and Demain, A. L. (1977). Effect of primary metabolites on secondary metabolism. *Annu. Rev. Microbiol.* **31**, 343–356.

Gutiérrez, S., Fierro, F., Casqueiro, J., and Martin, J. F. (1999). Gene organization and plasticity of the β-lactam genes in different filamentous fungi. *Antonie van Leewenhoek. Intern. J. Gen. Molec. Microbiol.* **75**, 81–94.

Hopwood, D. A. (1997). Genetic contributions to understanding polyketide synthases. *Chem. Rev.* **97**, 2465–2497.

Hopwood, D. A., and Sherman, D. H. (1990). Molecular genetics of polyketide and its comparison to fatty acid biosynthesis. *Annu. Rev. Genet* **24**, 37–66.

Katz, L., and Donadio, S. (1993). Polyketide synthesis: Prospects for hybrid antibiotics. *Annu. Rev. Microbiol.* **47**, 875–912.

Kleinkauf, H., and von Döhren, H. (1990). Non-ribosomal biosynthesis of peptide antibiotics. *Eur. J. Biochem.* **192**, 1–15.

Kleinkauf, H., and von Döhren, H. (1997). Peptide antibiotics. *In* "Biotechnology" (H. J. Rehm, G. Reed, A. Pühler, and P. Stadler, eds.), 2nd Ed., pp. 277–322. VCH, Weinheim.

Lancini, G., and Lorenzetti, R. (1993). "Biotechnology of Antibiotics and Other Bioactive Microbial Metabolites", p. 236. Plenum, New York.

Martín, J. F. (1977). Biosynthesis of polyene macrolide antibiotics. *Annu. Rev. Microbiol.* **31**, 13–38.

Martín, J. F. (1998). New aspects of genes and enzymes for β-lactam antibiotic biosynthesis. *Appl. Microbiol. Biotechnol.* **50**, 1–15.

Martín, J. F., and Demain, A. L. (1980). Control of antibiotic synthesis. *Microbiol. Rev.* **44**, 230–251.

Martín, J. F., and Liras, P. (1981). Biosynthetic pathways of secondary metabolites in industrial microorganisms. *In* Biotechnology. A Comprehensive Treatise" (H. J. Rehm and F. Reed eds.), Vol. 1, pp. 211–233. Verlag Chemie, Weinheim.

Martín, J. F., and Liras, P. (1989). Organization and expression of genes involved in the biosynthesis of antibiotics and other secondary metabolites. *Annu. Rev. Microbiol.* **43**, 173–206.

Martín, J. F., Gutiérrez, S., and Demain, A. L. (1997). β-Lactams. *In* "Fungal Biotechnology" (T. Anke, ed.), pp. 91–127. Chapman & Hall, Weinheim.

Selenium

Ronald S. Oremland
U.S. Geological Survey

I. Introduction
II. Microbial Transformations of Selenium
III. The Selenium Cycle in Nature

GLOSSARY

assimilatory reduction The process by which higher oxidation states of selenium [Se(VI) and Se(IV)] are reduced to selenide [Se(−II)] by enzymes present in microorganisms and plants for the purpose of incorporating selenium into amino acids such as selenomethionine.

bioremediation The use of plants or microorganisms to cleanse Se-contaminated waters or soils by removing Se as a solid precipitate [i.e., Se (0)] or as a volatile compound [i.e., $(CH_3)_2Se$].

demethylation A microbial process that removes the methyl groups from volatile forms of selenium such as dimethylselenide or methane selenol.

dissimilatory reduction The use of higher oxidation states of selenium by bacteria for the purpose of anaerobic respiration, whereby electrons generated from the oxidation of carbon substrates to CO_2 results in the reduction of Se(VI) or Se(IV) to lower oxidation states [e.g., Se(0)] coupled with the conservation of energy (i.e., ATP) for growth.

methylation The biological mechanism whereby inorganic forms of selenium are reduced to the Se(−II) state and receive methyl groups from a carrier protein, resulting in the formation of volatiles like dimethylselenide.

selenium cycle The flux of selenium between components of the biosphere, which includes the atmosphere, hydrosphere, and lithosphere. The chemical forms and oxidation states of selenium found in nature govern its hydrologic mobility, volatility, bioavailability, and toxicity. Important aspects of the environmental chemistry of this element are controlled by biologically mediated redox reactions, and microorganisms play a significant role in these biogeochemical processes. Because selenium is both a micronutrient required by all life forms as well as a potent toxicant when it is present at high concentrations in ecosystems, its biogeochemistry has important implications for human, animal, and plant health and nutrition, as well as for the functioning of aquatic and terrestrial ecosystems.

SELENIUM is an essential trace element for microbes, plants, and animals. Paradoxically, it is also a potent environmental toxicant when present at elevated concentrations in contaminated ecosystems. Recognition of this fact has stimulated research into the behavior of this element in nature. It is now understood that this element undergoes a complete biogeochemical cycle. Thus, oxidized forms of selenium (e.g., selenate and selenite) undergo biologically mediated reduction reactions either to achieve its incorporation into cellular organic matter or for it to serve as a terminal electron acceptor for anaerobic respiration. Conversely, reduced species of selenium (e.g., selenide and elemental selenium) are susceptible to microbial oxidation back to higher oxidation states. Other important microbial reactions include methylation of selenium, which forms volatile organoselenides, as well as demethylation reactions, which destroy these compounds.

I. INTRODUCTION

A. Origin, Properties, and Uses of Selenium

Selenium is a Group VIB element (atomic number and weight of 34 and 78.96, respectively) with chemical properties very similar to those for its neighbor sulfur, and hence it also exists in nature in four oxidation states: Se(VI) as selenate or SeO_4^{2-}, Se(IV) as selenite or SeO_3^{2-}, Se(0) as elemental or "native" selenium, and Se(−II) as selenide or Se^{2-} (e.g., FeSe). The latter two are usually found in reducing environments, while the former occur in oxidizing ones. Six stable isotopes of the element exist with natural abundances ranging from 0.87% for ^{74}Se to 49.82% for ^{80}Se. In addition, ^{75}Se is a gamma-emitting radioisotope ($t_{1/2}$ = 120 days) which has utility as a tracer in biological and chemical experiments. Native selenium [Se(0)] can occur as different allotropes (physical states): in the red or black amorphous form, as monoclinic crystals (red or transparent), or in the gray metallic state or as gray or black hexagonal crystals. The various allotropes have different physical properties.

Selenium is a rare element with an average abundance of about 0.00001% in Earth's crust. Nonetheless, it occurs in a variety of minerals with exhibited speciation ranging from selenide (e.g., ferroselite, $FeSe_2$) to selenites (e.g., challomenite, $CuSeO_3 \cdot 2H_2O$) to selenates [e.g., schmeiderite, $(Pb,Cu)_2SeO_4(OH)_2$]. Selenium is often found in association with sulfur minerals and in deposits of native sulfur. Selenium, like sulfur, is relatively depleted in abundance in igneous rocks (with the exception of aeolian ones like volcanic tuff) because it is volatilized as a gas from heated magma (e.g., H_2Se). However, soils and sediments can be highly enriched in selenium, especially those located in regions where volcanic activity either is ongoing or was once present. Hydrologic processes and the associated weathering of minerals in parent rocks also contribute to this process, resulting in the formation of highly seleniferous soils, which occur rather commonly in the western U.S. Conversely, some soils are highly depleted in selenium.

Because selenium is both a necessary trace element as well as a potent toxicant, there are fundamental agricultural and environmental problems posed by cases in which soils have either too little or too much selenium (see below). Native selenium occurs in both soils and sediments as a consequence of biological reduction of Se(IV) or Se(VI) (see below). Both Se(VI) and Se(IV) form highly soluble oxyanions and present as dissolved constituents in natural waters as well as in soils. Selenite, however, has strong adsorptive properties, and its mobility and bioavailability at neutral to acidic pH are highly constrained by its interaction with mineral phases like ferrihydrite (FeOOH). The adsorptive properties of Se(IV) decrease dramatically at elevated pH (e.g., pH > 9.0). No such adsorptive mobility constraints are associated with Se(VI). Organic forms of selenium also are present in soils, sediments, and natural waters. Selenium, like sulfur, also has an extensive organic chemistry, and organoselenium intermediates are highly suited for synthetic transformations of organic compounds.

B. Environmental Sources of Selenium

Selenium has a number of industrial applications, including its use as a semiconductor, as a chemical catalyst, as an important element in photoelectric cells, as an insecticide/fungicide, and as an ingredient in the manufacture of glass. However, none of these uses represent major sources of selenium pollution to the environment. Primary anthropogenic sources of selenium are associated with refining of petroleum, combustion of coal (fly ash is enriched in Se), mining activities, processing of metals, and from irrigation wastewater drainage derived from cultivation of naturally occurring seleniferous soils.

II. MICROBIAL TRANSFORMATIONS OF SELENIUM

A. Assimilation of Selenium

Figure 1 illustrates the overall types of microbial reactions that can occur with respect to selenium. Selenium is a micronutrient that is found in all forms of life ranging from viruses to elephants. In plants and in microorganisms, inorganic forms of selenium [e.g., Se(VI) and Se(IV)] are transported into cells by uptake and permease systems associated with the assimilation of sulfate. Plants typically contain ~0.1 mg Se/kg dry weight, although those which contain high levels of sulfur, such as cruciferous vegetables like cabbage and broccoli, can accumulate even higher levels especially when cultivated on

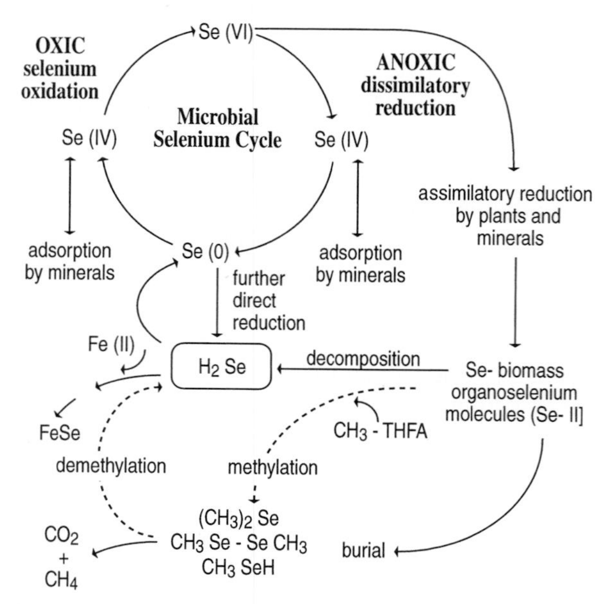

Fig. 1. Schematic representation of the biogeochemical cycle of selenium in nature.

seleniferous soils (~0.5 mg Se/kg dry weight). Certain plants (e.g., *Astragalus*) accumulate exceptionally high levels of selenium (~15 g Se/kg dry weight) and can achieve this even when grown on nonseleniferous soils. These plants have been called "locoweeds" because their ingestion by grazing animals soon leads to erratic behavioral displays which precede death.

Once selenium is present within the cells, the higher oxidation states must undergo assimilatory reduction by thiols such as glutathione with NADPH to the level of selenide, whereby they are next incorporated into amino acids and thence into proteins and enzymes. These amino acids are structural analogs of the ones which contain sulfur, primarily selenomethionine and selenocysteine, which can also be transported into cells from the surrounding milieu by transport systems designed for the sulfur-containing analogs. In microbial ecology investigations, selenium-containing amino acids have been employed as specific inhibitors of bacterial thioamino acid metabolism in anoxic sediments. A number of bacterial enzymes contain selenium, including the formate dehydrogenases, hydrogenases, and the glycine reductases of anaerobes. In the case of formate dehydrogenase, the role of selenium is associated with the coordination of the molybdenum cofactor. Selenium is also found in certain tRNAs of the Eubacteria and Archaea, which are known as seleno-tRNAs.

B. Selenium Deficiency and Toxicity

A paucity of selenium in the diet of animals leads to chronic health conditions, such as anemia and the well-known white muscle disease of cattle. In the case of microbes, a few microorganisms have an absolute nutritional requirement for specific forms of selenium, and some bacterial species show improved growth and yield characteristics when defined media are supplemented with inorganic selenium. However, inclusion of nutritionally rich supplements in complex broth media, such as yeast extract and peptone, usually provide adequate organoselenium to sustain the vast majority of cultured microorganisms. When dietary intake of selenium greatly exceeds these trace nutritional requirements, selenium toxicity can occur. In vertebrates, teratogenic effects are commonly displayed among the resident populations of waterfowl and fish inhabiting selenium-contaminated ecosystems, such as the former Kesterson Wildlife Refuge in California. Acute selenosis occurs in vertebrates when ingestion of high quantities of selenium results in death. Subacute selenosis can cause a variety of symptoms: In humans, it can induce vomiting, diarrhea, and hair loss. In microbes, high levels of Se(IV) or Se(VI) in defined medium will inhibit growth of algae and bacteria, and these effects can be ameliorated by addition of the sulfur antagonist or by provision of complex nutritional supplements (e.g., yeast extract).

C. Methylation and Demethylation of Selenium

Volatile organoselenium compounds found in nature are analogs of those which commonly occur for sulfur, the most prevalent forms being dimethylselenide [$(CH_3)_2Se$], dimethyldiselenide [$(CH_3)_2Se_2$], and methaneselenol [CH_3SeH]. A variety of fungi (e.g., various species of *Penicillium, Fusarium, Alternaria*) and bacteria (e.g., various species of *Flavobacterium, Pseudomonas, Aeromonas, Cornyebacterium, Rhodo-*

spirillum) are capable of methylating exogenously supplied inorganic forms of selenium, including Se(IV), Se(VI), and, in a few instances, Se(0). The biochemical pathway(s) by which this proceeds has not been elucidated; however, because the Se in these compounds is in the −2 oxidation state, an assimilatory Se reductase system can be inferred. Likewise, the molecule that acts as a biochemical carrier for the methyl group before it is joined to the atom has not been identified. However, because folates (e.g., tetrahydofolate or THFA) and cobalamins are common one-carbon unit carriers in many biochemical reactions, including methylation of certain metals like Hg, it is likely that they play a central role in the methylation of selenium:

$$CH_3{-}THFA + Se^{2-} + 2\,H^+ \rightarrow CH_3SeH + H{-}THFA$$
$$CH_3SeH + CH_3{-}THFA \rightarrow CH_3SeCH_3 + H{-}THFA$$

An alternative to the above mechanism is a pathway which is analogous to that for the formation of organosulfur volatiles like dimethylsulfide and methanethiol. These substances can arise from the bacterial degradation of methionine or of the osmolyte dimethylsulfoniopropionate (DMSP) (osmolytes are organic compounds cells employ to maintain an isotonic salt balance with the environment). The DMSP lyase of the marine bacteria *Pseudomonas doudoroffii* and *Alcaligenes* sp. strain M3A will also attack its selenium-containing analog, dimethylselenioproplonate (DMSeP):

$$DMSeP \rightarrow DMSe + \text{acrylate}$$

Of the various volatile organoselenides, dimethylselenide (DMSe) is the most important in nature owing to the combined effects of a slow rate of oxidation and high vapor pressure relative to the other forms. Microbial methylation of inorganic selenium contained in agricultural soils has been proposed as a mechanism to achieve their bioremediation via removal as a volatile organoselenide.

Bacterial degradation of dimethylselenide readily occurs in sediments and soils. In anoxic systems, degradation is mediated by sulfate-reducing bacteria and methanogens, and the organoselenium compounds are metabolized by enzymes employed for the attack of the corresponding organosulfur compounds. Hence, methanogens that can grow on dimethylsulfide achieve degradation of dimethylselenide as follows:

$$2\,(CH_3)_2Se + 2\,H_2O \rightarrow 3\,CH_4 + CO_2 + 2\,H_2Se$$

However, unlike the case for dimethylsulfide, methanogens cannot achieve growth on dimethylselenide, presumably because of the formation of toxic hydrogen selenide as an end product, as well as methylselenol as a toxic intermediate. Degradation of dimethylselenide under aerobic conditions has not been studied, but presumably it is oxidized to CO_2 plus selenate since methylotrophic thiobacilli can achieve the corresponding reaction for dimethylsulfide.

D. Dissimilatory and Other Microbial Reductions of Inorganic Selenium Compounds

Selenate and selenite can be used as terminal electron acceptors for the oxidation of organic matter in anoxic environments. Over the 1990s several novel species of anaerobic bacteria (both facultative and strict anaerobes) have been reported that can grow by coupling the oxidation of substrates like lactate or acetate to the reduction of millimolar quantities of Se(VI) to Se(IV) (Table I). Cell suspensions of these organisms also can readily achieve a quantitative reduction of Se(VI) to Se(0), and because this native selenium is generated as an extracellular solid, these organisms are of interest for the purpose of bioremediating Se-contaminated wastewaters. Examples of bacterial species that can grow by dissimilatory reduction of Se(VI) include *Thauera selenatis, Aeromonas hydrophila, Sulfurospirillum barnesii,* and the haloalkaliphile *Bacillus arsenicoselenatis*. Another haloalkaliphile, *Bacillus selenitireducens,* grows with the highly toxic Se(IV) as the electron acceptor, quantitatively reducing it to Se(0), but it cannot grow with Se(VI). *Sulfurospirillum barnesii, B. arsenicoselenatis,* and *B. selenitireducens* can also employ arsenate, fumarate, and nitrate as electron acceptors for growth. These species generally use simple organic molecules as electron donors (e.g., lactate, pyruvate, acetate, malate) to support selenate reduction. Growth of *S. barnesii* and *B. arsenicoselenatis* on selenate with lactate as the electron donor is highly

TABLE I
Some Characteristics of Bacteria That Can Grow by Dissimilatory Reduction of Selenium Oxyanions

Isolate	*Phylogeny*	*Electron donors*	*Electron acceptors*
Thauera selenatis[a]	β Proteobacteria, gram negative	Acetate, glucose, lactate	Se(VI), NO_3^-, O_2
Sulfurospirillum barnesii SES3[b]	ε Proteobacteria, gram negative	Lactate, pyruvate, H_2 + acetate	Se(VI), NO_3^-, NO_2^-, As(V), Fe(III), fumarate, S(0), $S_2O_3^{2-}$, low O_2
Aeromonas hydrophila[c]	γ Proteobacteria, gram negative	Glycerol	Se(VI), Fe(III), NO_3^-, Co(III), fumarate
Bacillus arsenicoselenatis[d]	Gram positive, low (G + C)	Lactate, malate	Se(VI), As(V), fumarate, Fe(III), NO_3^-
Bacillus selenitireducens[d]	Gram positive, low (G + C)	Lactate, pyruvate, (fermentative growth with sugars and starch)	Se(IV), As(V), fumarate, NO_3^-, NO_2^-, trimethylamine oxide, low O_2

[a] Macy *et al.* (1993).
[b] Oremland *et al.* (1994); Laverman *et al.* (1995).
[c] Knight and Blakemore (1998).
[d] Switzer Blum *et al.* (1998).

exergonic and is consistent with the following stoichiometry:

$$3\ \text{lactate}^- + 2\ SeO_4^{2-} + H^+ \rightarrow 3\ \text{acetate}^- + 2\ \text{Se}(0) + 3\ HCO_3^- + 2\ H_2O$$

$$\Delta G_f^\circ = -467.4\ \text{kJ/mol lactate or } -85.8\ \text{kJ/mol electrons}$$

Two new bacterial isolates (strains AK4OH1 and Ke4OH1) have been isolated that employ refractory organic molecules (e.g., halogenated aromatics) as electron donors for selenate reduction. In terms of reducing equivalents, selenate reduction is more favorable thermodynamically than is respiratory nitrate reduction to nitrite (−57.8 kJ/mol electrons). When grown on nitrate, *S. barnesii* does not express a selenate reductase that can reduce millimolar levels of Se(VI). However, a constitutive selenate reductase is present that can reduce submillimolar Se(VI) concentrations in the presence of millimolar levels of nitrate. This makes the organism attractive for the bioremediation of seleniferous agricultural wastewaters, which often contain micromolar quantities of selenate and millimolar quantities of nitrate. Another important point is that none of the above isolates are capable of reducing sulfate or sulfite. Hence, the pathway of dissimilatory reduction of selenium oxyanions in these organisms clearly does not proceed via those established for sulfate. The membrane of *S. barnesii* contains both *b*-type and *c*-type cytochromes, the former associated with selenate- or fumarate-grown cells, while the latter occurs in thiosulfate- or nitrate-grown cells. The selenate reductase of *Thauera selenatis* has been purified from the cell's periplasm and is hydrophilic. When assayed *in vitro* using reduced methyl or benzyl viologen, the enzyme is specific for selenate, showing no reactivity with nitrate, nitrite, or sulfate. It consists of three subunits, contains molybdenum, iron, and sulfur prosthetic groups, and is associated with a *b*-type cytochrome. The selenate reductase of *S. barnesii* is membrane-bound and hydrophobic, and is apparently a tetramer rather than a trimer. These early observations suggest that future research will discover many more significant differences among the various purified bacterial respiratory selenate reductases.

The reduction of selenium oxyanions to solid, elemental selenium can also be achieved by bacteria which do not respire Se(VI) or Se(IV), making them of potential interest as bioremediative agents as well. Selenite (but not selenate) readily undergoes reduction to Se(0) on encountering mild chemical reductants of microbial origin, such as sulfide or ascorbate. A number of microorganisms, including aerobic bacteria, photosynthetic bacteria, algae, and fungi, can precipitate Se(0) from highly toxic Se(IV), but they derive no apparent physiological benefit from this other than detoxification of the medium. This reduction is usually achieved by glutathione reductases. A few anaerobes, including some strains of *Wolinella succinogenes, Pseudomonas stutzeri,* and *Desulfovibrio desulfuricans,* can reduce submillimolar levels of Se(VI) to Se(0) but cannot achieve growth with Se(VI) as an electron acceptor. As can be seen from the above examples, the reduction of selenium oxyanions to the elemental state is performed in a diversity of microbes; however, Se(0) like S(0) can theoretically undergo further reduction to selenide by accepting two more electrons. Anaerobic cell suspensions of *Thiobacillus ferroxidans* are capable of reduction of Se(0) and S(0) to H_2Se and H_2S, respectively. Preliminary data suggest that cell suspensions of *S. barnesii* and *B. selenitireducens* can achieve this reduction when they are supplied with an excess of electron donor, although it is not yet known if energy is conserved by this reduction.

Cell suspensions of *Desulfovibrio desulfuricans* can reduce submillimolar levels of Se(VI) to Se(−II). Although this reduction is mediated by the sulfate-reductase system, these organisms cannot grow with Se(VI) as an electron acceptor because it destroys the cell's internal energy balance. Cells will expend ATP to "activate" selenate molecules acting as sulfate analogs via ATP sulfurylase, but will not recover additional respiratory ATPs with further sequential reduction of selenite to selenide by the reaction chain initiated by adenosine phosphosulfate (APS) reductase. This is another point of departure whereby dissimilatory selenate reduction differs from that of sulfite, namely, that the E_0' for the Se(VI)/Se(IV) couple is ~480 mV, and thus no cellular expenditure of ATP for activation of Se(VI) to achieve reduction to Se(IV) is required.

E. Oxidation of Se(−II) and Se(0)

Selenide is a highly poisonous and very reactive molecule. When present as H_2Se in suboxic environments, it will precipitate with metal ions to form selenides (e.g., FeSe) or will spontaneously autooxidize to form red Se(0) when encountering even low concentrations of oxygen. Microorganisms that oxidize sulfide or reduced metals and that typically inhabit the oxic/anoxic interface can biochemically oxidize Se(−II). For example, the photosynthetic bacterium *Chromatium vinosum* oxidizes Se(−II) to Se(0). This organism usually employs sulfide as a photosynthetic electron donor and normally accumulates intracellular native sulfur [i.e., S(0)]. However, it will also accumulate intracellular Se(0) when cultured in medium which provides selenide plus sulfide, although it does not use selenide as an electron donor to support photosynthesis. Several aerobic bacteria including *Bacillus megaterium, Thiobacillus* ASN-1, and *Leptothrix* MnB1 can slowly oxidize Se(0) to Se(VI), but the enzymatic mechanism(s) responsible for this phenomenon has not been delineated. It is also not known if selenium undergoes dismutation reactions that can occur with intermediary oxidation states of sulfur having at least two sulfur atoms (e.g., thiosulfate) which result in the oxidation of one atom and the reduction of the other.

III. THE SELENIUM CYCLE IN NATURE

A. Global Mass Balances

Figure 2 illustrates the annual balance of selenium mass transfer between major components of Earth's biosphere. The mobilization of selenium into the environment by anthropogenic means is the largest component of this element's global cycle, accounting for ~80,000 tons annually (about 40,000 tons of which enter the oceans and an equivalent value is deposited on land), with about 90% contributed to an equivalent flux entering aquatic and terrestrial systems, and only ~9% volatilized to the atmosphere. Total annual deposition (wet and dry) from the atmosphere accounts for ~15,000 tons input to terrestrial and marine systems. The amount of selenium volatilized to the atmosphere from natural processes occurring in marine and terrestrial systems is ~8000 tons annually, which when added to the anthropogenic flux to the atmosphere (~7000 tons) forms a rough balance between atmospheric input and outfall. Natural weathering of terrestrial rocks and soils results in the transfer of ~14,000 tons of particulate and dissolved Se into the marine environment, which

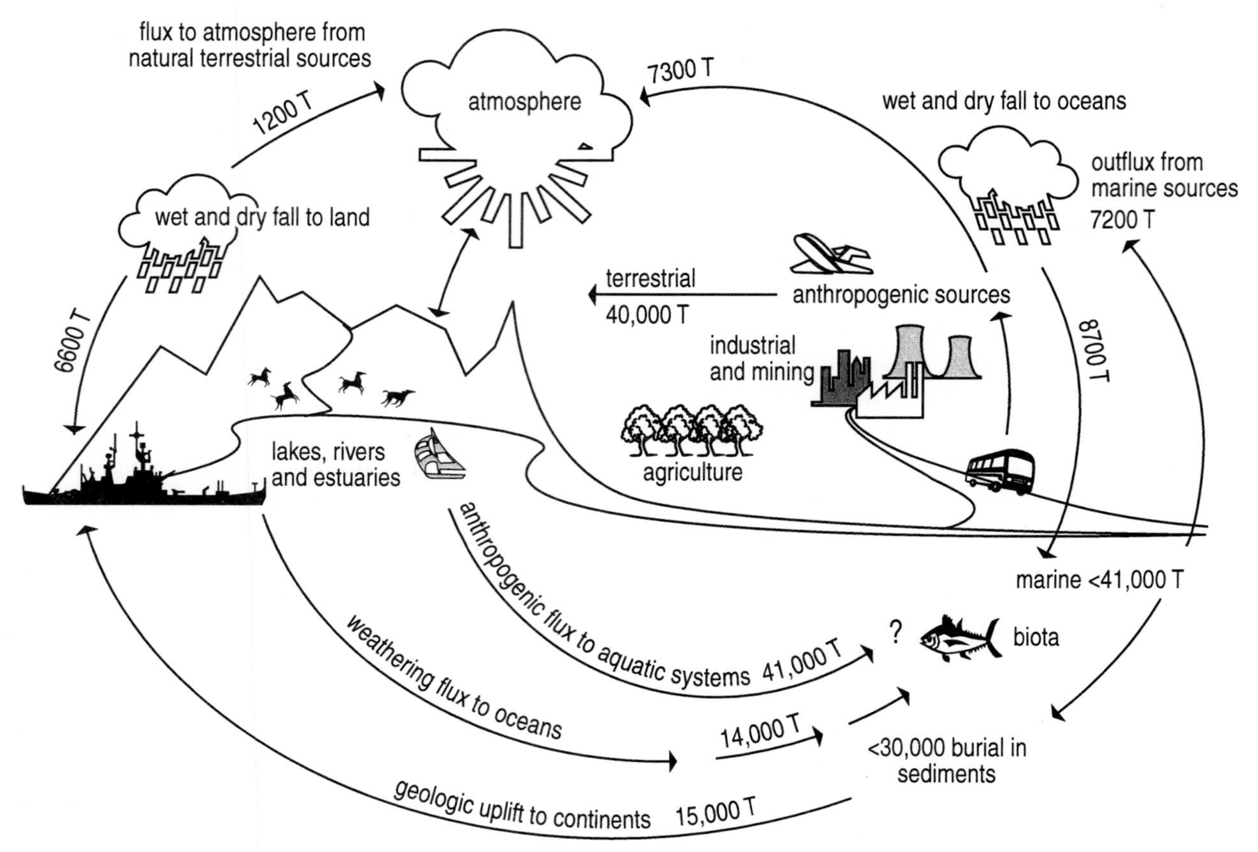

Fig. 2. Quantitative representation of the annual flux of selenium between components of the atmosphere, lithosphere, and hydrosphere. Values are given as tons and are from the review of Haygarth (1994).

when combined with anthropogenic sources totals as much as 54,000 tons. This amount does not account for Se buried into the sediments of aquatic environments on the continents, which would be subtracted from the 54,000 ton quantity. Marine biota account for ~16,000 tons of annual uptake, which when combined with volatilization to the atmosphere (~7000) leaves about 31,000 tons as ultimately deposited into marine sediments annually. It has been estimated that about one-half of this deposition value is returned to the terrestrial environment annually in the form of geologically uplifted marine rocks.

B. Biogeochemical Cycling of Selenium

All four oxidation states of selenium are present in the biosphere, and the dynamics of transfer between these chemical species are largely controlled by microorganisms. The weathering of selenium minerals in rocks and soils results in the mobilization and fluvial transport of Se(VI) and Se(IV) in groundwaters and surface waters. Selenite, being readily adsorbed to solid surfaces at neutral to acidic pH, tends to be associated with suspended particulates, while selenate accounts for most of the dissolved forms of inorganic selenium. Uptake of Se(IV) and Se(VI) by the permeases of microorganisms (including phytoplanktonic algae) or by the root systems of vascular plants results in their incorporation into organic matter when metabolized by assimilatory reductase enzymes. The occurrence of particulate and dissolved organoselenium compounds in natural waters is a consequence of the death, decay, or excretion of these organisms.

When selenium oxyanions enter anoxic environments, they serve as electron acceptors for dissimilatory reduction to Se(0). This can occur within the surface layers of sediments, in anoxic bottom waters of stratified lakes and fjords, in subsurface aquifers, in biofilms, within biodigestors that carry out anaerobic waste treatment, and within portions of the gastrointestinal tracts of selected animals (e.g., ruminant stomachs, termite hindguts, horse cecum). Biological surveys have found that many diverse sediments, including those from Se-contaminated environments as well as pristine ones, contain bacteria that readily reduce Se(VI) to Se(0). The isolation of dissimilatory Se(VI)-reducing bacteria from chemically diverse locales, including sediments from freshwater, marine, and extremely hypersaline systems, alkaline soda lakes, hot springs, landfills, as well as animal gastrointestinal tracts indicates that these organisms are truly ubiquitous in the natural environment.

Owing to the environmental problems caused by the introduction of selenium into stagnant marshes, these are the best studied ecosystems with respect to selenium biogeochemistry. They also serve as a good model to illustrate the biogeochemistry of Se in other ecosystems, such as the oceans, because the same basic processes occur in each. The rate of selenium oxyanion reduction in sediments is generally controlled by the supply of electron donor, which is in turn dependent on the amount and nature of decaying organic matter present. Vegetated marshes with abundant plant debris and porous, fine-grained sediments are an excellent locale to observe optimal conditions for Se sequestration of this kind. Turnover rate constants for Se(VI) in freshwater marsh sediments can be quite high, with values of $\sim 2\ hr^{-1}$ not uncommon. The reduction of Se(VI) to Se(0) in sediments from such environments displays Michaelis–Menten kinetics, with apparent K_m values for selenate exceeding by 1–3 orders of magnitude the ambient concentrations of dissolved selenium oxyanions. Thus, the rate of Se(VI) reduction is often concentration-dependent with respect to Se(VI). Sediments of this type are an important sink for selenium oxyanions present in the overlying waters, and they have a substantial capacity to absorb large additional quantities of Se(VI) and Se(IV) and to rapidly sequester it as insoluble Se(0).

Native selenium is the major speciated form of this element present in sediments of both freshwater and saltwater marshes. When taken together with "organoselenium," both the Se(0) plus the Se(−II) oxidation states represent most of the mass of selenium present in these reduced, organic-rich environments. However, "organoselenium" is an operational term defined by its mode of extraction rather than by actual molecular identification (e.g., "selenimethionine"), and therefore some of this fraction may also comprise various allotropes of Se(0). The two primary sinks for Se oxyanions in such environments therefore are (1) assimilatory reduction by the plant and microbial biota, which forms organoselenium compounds, and (2) dissimilatory reduction to Se(0). In the water column of oceans, lakes, estuaries, and rivers, assimilatory reduction is the major biological process. Anaerobic processes are confined to restricted locales, such as the gastrointestinal tracts of pelagic animals. A portion of the organoselenium formed by plants and microbes is channeled into the formation of volatile organics (e.g., dimethylselenide), some of which is degraded by microbes while the remainder is lost to the atmosphere by such processes as active venting through plant stomata, as trace components of gas ebullition from sediments, or as diffusive loss from the surface film layer of waters overlying the marsh. However, although Se(0) is relatively unreactive, it is still bioavailable and can be ingested by invertebrates and incorporated into their tissues.

Marshes can therefore act as biofilters in that they remove soluble Se(VI) and Se(IV) oxyanions from waters entering the marsh by reducing them to particulate Se(0) in the sediment or to particulate Se(−II) associated with organic matter of the living or decaying biota. Artificially constructed wetlands of this type have been designed to exploit these biological sinks for selenium oxyanions so that water exiting the marsh is free or greatly depleted of these dissolved species. It is for this reason that interest in the construction of artificial marshes to act as natural biological Se filters still exists even 15 years after the debacle of Kesterson Resevoir proved this approach to be unsound. Because marshes of this type represent a potentially inexpensive means to comply with regulatory requirements governing the selenium content of wastewater, this approach has continued. In actuality such a "natural" treatment process is flawed ecologically because it creates a selenium-contaminated foodchain that will serve as an attractor for mating, feeding, and nesting wildlife. The result will be a Se-poisoned resident biota, although for a time the water exiting the marsh will be in compliance with regulatory requirements. Eventually, as these systems saturate their ability to sequester Se, they can become net exporters of selenium compounds into the surrounding environment and defeat the main purpose for their construction.

Even when such "treatment" wetlands are taken off-line and are retired from use as net sinks for Se, these former marshes can become sources for the net export of selenium into the surrounding environment. Elemental selenium is subject to microbial oxidation when it comes in contact with oxygen. Oxygen contact occurs either through sediment resuspension or when hydrologic alterations are made that change the water-saturated sediments into aerated soils. In these reactions Se(0) is oxidized back to Se(IV) and Se(VI), but the reactions proceed very slowly when compared to rates associated with dissimilatory reduction. Whereas turnover of Se(VI) pools in anoxic sediments is usually measured in hours, or at most a few days, the oxidation of Se(0) in soils, or on particle surfaces, takes years. Thus, rate constants for Se(0) oxidation are three to four orders of magnitude lower than those for Se(VI) reduction. Hence, the sequestration of Se(VI) by dissimilatory reduction in anoxic environments is relatively easy to achieve because it occurs so rapidly. However, once such marsh treatment systems are drained, the immobilized Se(0) will slowly oxidize to Se(IV) and Se(VI), thereby assuring that these sites will export Se oxyanions into their surrounding milieu for many decades to come.

Another approach to bioremediation of seleniferous wastewaters is the construction of anaerobic digestors that carry out the dissimilatory reduction of Se(VI) to Se(0). The Se(0) is removed from the aqueous phase by a combination of gravitation and filtration, with the ultimate result that the exiting water is cleansed of Se compounds, as well as of other electron acceptors like nitrate. The recovered selenium is removed from the reactor and is either handled as toxic waste and buried at dump sites, or is refined and sold as commercial Se. These flow-through digestor systems are encapsulated in order to maintain anoxia, and are thus physically segregated from any surrounding biota. They cannot serve as wildlife attractors, and they do not create a Se-contaminated ecosystem to act as a filter because the recovered Se(0) is physically removed from the ecosystem by the treatment process. The disadvantage of digestors is that they are complex, and they are expensive to build and operate. Of all these obstacles, the most significant is the choice of an electron donor required to drive the reduction reactions forward, which must be both inexpensive and effective. However, considerable progress has been made in the 1990s both in the design of these systems, in isolating new species of anaerobic bacteria that show promise as bioremediative agents, and in identifying inexpensive sources of electron donor to drive this reaction forward, so there are some grounds for optimism with respect to the viability of this approach in the future.

See Also the Following Articles

ARSENIC • BIOREMEDIATION • ENERGY TRANSDUCTION PROCESSES • SULFUR CYCLE

Bibliography

Berrow, M. L., and Ure, A. M. (1989). Geological materials and soils. *In* "Occurrence and Distribution of Selenium" (M. Ihnat, ed.), pp. 213–242. CRC Press, Boca Raton, Florida.

Cutter, G. A., and Bruland, K. W. (1984). The marine biogeochemistry of selenium: A re-evaluation. *Limnol. Oceanogr.* **29**, 1179–1192.

Doran, J. W. (1982). Microorganisms and the biological cycling of selenium. *Adv. Microbial Ecol.* **6**, 17–32.

Dowdle, P. R., and Oremland, R. S. (1998). Microbial oxidation of elemental selenium in soil slurries and bacterial cultures. *Environ. Sci. Technol.* **32**, 3749–3755.

Haygarth, P. M. (1994). Global importance and global cycling of selenium. *In* "Selenium in the Environment" (W. T. Frankenberger, Jr., and S. Benson, eds.), pp. 1–28. Dekker, New York.

Heider, J., and Bock, A. (1993). Selenium metabolism in microorganisms. *Adv. Microbial Physiol.* **35**, 71–109.

Knight, V. M., and Blakemore, R. G. (1998). Reduction of diverse electron acceptors by *Aeromonas hydrophila. Arch. Microbiol.* **169**, 239–248.

Laverman, A. M., Switzer Blum, J., Schaefer, J. K., Philips, E. J. P., Lovley, D. R., and Oremland, R. S. (1995). Growth of strain SES-3 with arsenate and other diverse electron acceptors. *Appl. Environ. Microbiol.* **61**, 3556–3561.

Macy, J. M., Rech, S., Auling, G., Dorsch, M., Stackenbrandt, E., and Sly, L. I. (1993). *Thauera selenatis* gen. nov., sp. nov., a member of the beta subclass of *Proteobacteria* with a novel type of anaerobic respiration. *Int. J. System. Bacteriol.* **43**, 135–142.

Oremland, R. S., Hollibaugh, J. T., Maest, A. S., Presser, T. S., Miller, L. G., and Culbertson, C. W. (1989). Selenate reduction to elemental selenium by anaerobic bacteria in sediments and culture: Biogeochemical significance of a novel, sulfate-independent respiration. *Appl. Environ. Microbiol.* **55**, 2333–2343.

Oremland, R. S., Switzer Blum, J., Culbertson, C. W., Visscher, P. T., Miller, L. G., Dowdle, P., and Strohmaier, F. E. (1994). Isolation, growth, and metabolism of an obligately anaerobic, selenate-respiring bacterium, strain SES-3. *Appl. Environ. Microbiol.* **60**, 3011–3019.

Schröder, I., Rech, S., Krafft, T., and Macy, J. M. (1997). Purification and characterization of the selenate reductase from *Thauera selenatis*. *J. Biol. Chem.* **272**, 23765–23768.

Steinberg, N. A., and Oremland, R. S. (1990). Dissimilatory selenate reduction potentials in a diversity of sediment types. *Appl. Environ. Microbiol.* **56**, 3550–3557.

Switzer Blum, J., Burns Bindi, A., Buzzelli, J., Stolz, J. R., and Oremland, R. S. (1998). *Bacillus arsenicoselenatis* sp. nov., and *Bacillus selenitireducens,* sp. nov.: Two haloalkaliphiles from Mono Lake, California that respire oxyanions of selenium and arsenic. *Arch. Microbiol.* **171**, 19–30.

Sexually Transmitted Diseases

Jane A. Cecil and Thomas C. Quinn
The Johns Hopkins University

I. Introduction
II. Gonorrhea
III. Chlamydia
IV. Syphilis

GLOSSARY

epidemiology The study of the relationships of various factors determining the frequency and distribution of disease in a community.

incidence The rate at which new cases of a disease occur in a community in a given time period.

prevalence The total number of cases of a given disease in a community at a certain time.

THE TERM "SEXUALLY TRANSMITTED DISEASES (STDs)" refers to a number of distinct infections, which are typically transmitted through sexual contact and result in a variety of clinical manifestations. The acquisition and transmission of STDs depend primarily on sexual behavior, but some STDs such as hepatitis B and cytomegalovirus can also be acquired through nonsexual contact in childhood in areas with poor living conditions. Additionally, STDs can be transmitted from mother to child during pregnancy and childbirth. STDs are caused by a variety of organisms, including bacteria, viruses, protozoa, and ectoparasites. The clinical syndromes associated with infection by one or more of these organisms can range from asymptomatic infection to severe life-threatening illness.

I. INTRODUCTION

Over the 1980s and 1990s there have been tremendous advances in our understanding of sexually transmitted diseases (STDs). There has been an appreciation of new pathogens, new clinical syndromes, emerging antimicrobial resistance, the increasing importance of viral pathogens including the human immunodeficiency viruses (HIVs), as well as better understanding of the epidemiology of many STDs. There are approximately 50 distinct clinical syndromes associated with at least 25 different organisms. Table I summarizes some of the most important pathogens, the associated clinical syndromes, and potential complications.

Sexually transmitted diseases constitute a major public health problem worldwide, despite improvements in diagnosis, treatment, and prevention strategies. STDs are seen in virtually every society, though developing nations and underserved populations experience the greatest burden of disease. The World Health Organization (WHO) estimates that there were more than 333 million new cases worldwide of syphilis, gonorrhea, chlamydia, and trichomoniasis in adults aged 15–49 years in 1995. The global prevalence of infection with the agents of common viral STDs, such as genital herpes simplex virus, genital human papillomavirus, hepatitis B, and HIV, is estimated to be in the billions of cases, since many individuals are infected with more than one of these pathogens. In the U.S., there are approximately 12 million new cases of STDs annually, giving the U.S. the highest rate of curable STDs in the developed world. Of the top ten most frequently reported diseases to the Centers for Disease Control and Prevention (CDC) in 1995, five were STDs.

Encyclopedia of Microbiology, Volume 4
SECOND EDITION

TABLE I
Major Sexually Transmitted Agents and the Diseases They Cause

Agent	*Disease or syndrome*
Bacteria	
Neisseria gonorrhoeae	Urethritis; epididymitis; proctitis; bartholinitis; cervicitis; endometritis; salpingitis and related sequelae (infertility, ectopic pregnancy); perihepatitis; complications of pregnancy (e.g., chorioamnionitis, premature rupture of membranes, premature delivery, postpartum endometritis); conjunctivitis; disseminated gonoccal infection (DGI)
Chlamydia trachomatis	Same as *N. gonorrhoeae,* except for DGI; also, lymphogranuloma venereum,[a] Reiter's syndrome; infant pneumonia
Treponema pallidum	Syphilis[a]
Haemophilus ducreyi	Chancroid[a]
Calymmatobacterium granulomatis	Donovanosis[a]
Mycoplasma hominis	Postpartum fever; salpingitis
Ureaplasma urealyticum	Urethritis; low birth weight (?); chorioamnionitis (?)
?*Gardnerella vaginalis, Mobiluncus* sp., ?*Bacteroides* sp.	Bacterial vaginosis
?Group B β-hemolytic streptococci	Neonatal sepsis; neonatal meningitis
Viruses	
Herpes simplex virus (HSV-2, HSV-1)	Primary and recurrent genital herpes[a]
Hepatitis B virus (HBV), hepatitis C virus (HCV)	Acute, chronic, and fulminant hepatitides B and C, with associated immune complex phenomena and sequelae including cirrhosis and hepatocellular carcinoma
Cytomegalovirus (CMV)	Congenital infection: gross birth defects and infant mortality, cognitive impairment (e.g., mental retardation, 8th nerve deafness), heterophile-negative infectious mononucleosis, protean manifestations in the immunosuppressed host
Human papilloma virus (HPV)	Condyloma acuminata; laryngeal papilloma in infants; squamous epithelial neoplasias of the cervix, anus, vagina, vulva, penis
Molluscum contagiosum virus (MCV)	Genital molluscum contagiosum
Human immunodeficiency virus (HIV-1, HIV-2)	AIDS and related conditions
Human T-lymphotropic virus type 1	T-cell leukemia/lymphoma; tropical spastic paraparesis
Protozoa	
Trichomonas vaginalis	Vaginitis; urethritis (?); balanitis (?)
Fungi	
Candida albicans	Vulvovaginitis; balanitis; balanoposthitis
Ectoparasites	
Phthirus pubis	Pubic lice infestation
Sarcoptes scabiei	Scabies

[a] These infections are responsible for the syndrome known as genital ulcer disease.

Sexually transmitted diseases are becoming recognized as a source of tremendous social, health, and economic costs worldwide. The 1993 World Development Report ranked STDs, excluding AIDS, as the second leading cause of healthy life lost among women between the ages of 15 and 44 years in the developing world. There are many factors contributing to the disease burden exacted by STDs. Challenges to the control of STDs include limitations on the availability and quality of medical care, public education regarding safe sexual practices, access to condoms and other methods of prevention, convenient and affordable methods of diagnosis and treatment, the emergence of antibiotic resistance, and the social stigma associated with STDs. Furthermore, STDs are commonly asymptomatic in both women and men. All of these factors can result in a delay or complete lack of appropriate treatment.

Thus, the vast majority of disease burden from STDs is a result of the complications and sequelae that may follow initial infection. When left untreated, infections can migrate upward from the lower reproductive tract and result in pelvic inflammatory disease (PID), chronic pelvic pain, infertility, ectopic pregnancy and tubo-ovarian abscess in women, and prostatitis, epididymitis, or orchitis in men. It has been estimated that the economic cost of PID alone exceeds $2 billion per year in the U.S. Untreated infections in pregnant women may lead to fetal loss, stillbirths, low birth weight, and eye, lung, or neurologic damage in a newborn. The viral STDs are typically incurable and result in chronic disease. The most obvious example of this is HIV infection and AIDS. Another example is human papillomavirus, of which some types have been found to cause cervical cancer in women.

In general, the major STDs pose a greater threat to women's health, for a number of reasons. Women are more likely to get an STD from a man than the reverse because STDs are more efficiently transmitted from men to women. Women are more likely to be asymptomatic than men, and thus women may be less likely to seek medical attention in a timely manner. STDs are often more difficult to accurately diagnose in women than in men because symptoms are less specific in women and diagnostic tests are less sensitive. Finally, the potential complications of STDs are more common and more serious in women.

Extensive studies of the epidemiology of STDs have revealed several risk factors consistently associated with the acquisition of STDs. These risk factors appear to apply to most, if not all populations. Risk factors include multiple sexual partners, urban residence, single marital status, and young age. The risk of STDs for young adults and adolescents has not been fully appreciated. STD prevention efforts in young people have remained controversial in the U.S. despite the fact that approximately 3 million teenagers acquire an STD each year. Young adults and adolescents are at greatest risk for acquiring an STD for a number of reasons, related to patterns of sexual behavior and other behaviors associated with STDs, such as substance abuse. There is some evidence that biological factors may also play a role where young females are concerned. Thus any successful STD prevention program needs to specifically include this population. Another factor assotciated with the acquisition of an STD is a history of prior STD infection. Similarly, simulataneous infection with more than one STD is not uncommon. In fact, several studies have found that approximately 25% of individuals with gonorrhea are also infected with *Chlamydia trachomatis*. On the basis of these data, treatment recommendations include antibiotics that are effective against both organisms.

The control of STDs has become an even higher public health priority in the setting of the HIV epidemic. STDs and HIV share many behavioral risk factors, and thus efforts at educating individuals to eliminate high-risk behaviors and adopt safer sexual practices may help reduce the incidence of both STDs and HIV. In addition, many studies have clearly documented that both ulcerative and nonulcerative STDs increase the risk of HIV transmission. Thus, an individual with both HIV and another STD is more likely to transmit HIV to a sexual partner, compared to an HIV-infected individual without another STD. Similarly, an HIV-uninfected individual with another STD is more likely to get HIV from an HIV-infected sexual partner than if he or she did not have an STD at the time of sexual intercourse. This association between STDs and HIV is further strengthened by the findings of an STD treatment study in the Mwanza region of Tanzania, by Grosskurth and colleagues. In this study, communities provided with improved STD diagnosis and treatment experienced a substantial reduction in the incidence of HIV, as a result of a decrease in sexual transmission.

Fortunately, given the renewed interest in the control of STDs, there have been many recent advances in the field. Intensive research efforts have yielded improvements in the clinical recognition, microbiologic diagnosis, and treatment of STDs. Epidemiologic data have helped to identify high-risk populations, and thus prevention programs and resources can be appropriately targeted. New clinical algorithms have aided in the early recognition of PID, urethritis, cervicitis, and vaginal infections. Newer diagnostic assays are more sensitive and more specific for quickly diagnosing diseases caused by specific pathogens. The development of nucleic acid amplification tests for some STDs have enabled health care providers to easily and accurately screen large numbers of individuals, even in the absence of symptoms, with little discomfort to the individual. This has made it possible to identify and treat those individuals who may be more likely to transmit infection or to develop complications, because they would not be prompted to seek health care given an absence of symptoms. Finally, newer antibiotics ensure the possibility of effective treatment of many STDs, including those that have become resistant to older therapeutic agents.

The STDs discussed in this article have been selected because they are among the more common STDs. In addition, they are illustrative of important variations in epidemiology, diagnosis, treatment, prevention, and pathogenesis, concepts that are central to our understanding of STDs.

II. GONORRHEA

A. Epidemiology

Despite recent declines in new cases of gonorrhea, it remains one of the most commonly reported diseases in the United States. The greatest number of cases reported to the Centers for Disease Control and Prevention (CDC) was 1,013,436 in 1978. Since then there has been a significant decrease in the number of reported cases, with 392,848 reported cases in 1995. These numbers likely underestimate the true prevalence of gonorrhea in the U.S., however, given that many health care workers do not comply with reporting guidelines and many patients are treated empirically without confirmatory diagnostic testing because of highly suggestive symptoms or known exposure. In fact, the true prevalence of gonorrhea in the U.S. is estimated to be twice the number of reported cases. Estimates of disease burden attributable to gonorrhea are even less reliable in many developing countries. Available data suggest, however, that gonorrhea remains relatively common in the developing world as it does in the U.S. In contrast, gonorrhea seems to be relatively rare in Canada and much of western Europe.

As in the U.S., the incidence of gonorrhea in many developed nations has declined in recent years. The decline in cases has been felt to be due in part to the HIV epidemic and the subsequent reduction in high-risk behaviors seen in many segments of the population. Other factors believed to have contributed to the decline in gonorrhea cases include an increase in condom use, screening of asymptomatic individuals at risk for gonorrhea, and contact tracing used to identify individuals who are at risk for gonorrhea given a known exposure to an infected individual. In the U.S., the decline in gonorrhea cases has not been observed consistently throughout the population, however. The most striking decline in gonorrhea has been among homosexual men and white men and women. In contrast, smaller declines have been observed among urban, minority segments of the population, primarily African-American and Hispanic. In 1995, the reported rates of gonorrhea were 37 times higher for African-Americans than for whites. In 1996, the Centers for Disease Control reported gonorrhea rates in African-American adolescents (15 to 19 years olds) that were almost 25 times greater than for white adolescents. Similarly, greater declines have been observed among older age groups compared to the young.

Numerous factors have been identified that correlate with an increased risk of gonorrhea. Gonorrhea is clearly more common among African-Americans, although some of this disparity is a result of greater attendance of minority individuals at public health clinics, where reporting of cases is much better than in other health care settings. Young age is also a risk factor for gonorrhea. In 1995, 77% of all reported cases of gonorrhea occurred in individuals between the ages of 15 and 29 years. Furthermore, the reported incidence of gonorrhea was nearly twice as high for sexually active adolescent girls compared to sexually active women aged 20 to 24 years. Young women seem to be at particularly high risk for gonorrhea. This may be due in part to patterns of sexual behavior, but there is also a biologic basis to this observation. Young women have larger areas of cervical ectropion (columnar epithelial cells on the ectocervix) that provide a greater area of susceptible cells for infection. Low socioeconomic status, early onset of sexual activity, multiple sexual partners, unmarried marital status, and sexual activity associated with illicit drug use are also risk factors for gonorrhea. Similarly, individuals with a prior history of gonorrhea as well as prostitutes and their clients are at increased risk for gonorrhea. Although the use of oral contraceptives is not generally associated with higher rates of gonorrhea, their use may increase the risk for acquiring gonorrhea when a woman is exposed to an infected partner, when compared to her risk if she uses no contraception at all.

B. Clinical Manifestations

The term gonorrhea is often thought to refer to an illness characterized by urethral discharge in men or vaginal discharge in women. The causative organism, *Neisseria gonorrhoeae,* can, however, infect several anatomic sites including the cervix, urethra, rectum, oropharynx, and conjunctiva, and is associated with a wide spectrum of disease. An individual with gonorrhea may have no symptoms at all, may have localized symptomatic disease, may have localized complicated disease, or may be very ill with disseminated infection. Infections caused by *N. gonorrhoeae* are typically limited to superficial mucosal surfaces lined by columnar or cuboidal, nonkeratinized epithelial cells, such as those sites noted above. Infection of these mucosal surfaces is usually accompanied by a marked inflammatory response that results in the production of copious, purulent discharge. As noted, however, infection may also be asymptomatic. Of men and women who report sexual exposure to partners with gonorrhea, urethral infection in men is asymptomatic approximately 10% of the time and cervical infection in these women is asymptomatic about 40 to 50% of the time. Pharyngeal and rectal infections secondary to gonorrhea are typically asymptomatic. Of interest, individuals who develop disseminated gonococcal infection (DGI) commonly have asymptomatic mucosal infections, detected only by screening diagnostic tests. It is unknown whether strains that cause DGI are more likely to cause asymptomatic mucosal infection or if individuals who lack symptoms simply do not seek medical treatment and therefore the infection eventually spreads.

1. Gonococcal Infections in Women

The most common form of uncomplicated gonorrhea in women is infection of the uterine cervix. The vagina is usually not infected because it is lined by squamous epithelium. The urethra is colonized in 70 to 90% of infected women. In women who have undergone a hysterectomy, the urethra is the most common site of infection. The Skene glands (periurethral glands) and Bartholin glands are also commonly infected in the setting of endocervical infection with *N. gonorrhoeae.* Symptoms typically appear within 10 days of infection, although the incubation period can be quite variable. The symptoms associated with endocervical infection may include vaginal discharge, genital itching, intermenstrual bleeding, unusually heavy menstrual bleeding, or painful urination. An infected woman may have all, none, or any combination of these symptoms, and symptoms may range from mild to severe. Physical examination may be normal but typically reveals a purulent discharge from the cervix, redness and swelling of the cervix, and easily induced cervical bleeding. These findings, as well as the symptoms noted, are not specific for gonorrhea, however. Thus, evaluation of sexually active women spected of having gonorrhea must include testing for *Chlamydia trachomatis Candida albicans, Trichomonas vaginalis,* bacterial vaginosis, as well as other infections.

Pelvic inflammatory disease (PID) is the most serious complication of *N. gonorrhoeae* infection in women, affecting approximately 10 to 20% of all women with acute infection. PID is a term referring to endometritis, salpingitis, tubo-ovarian abscess, peritonitis, or any combination of these findings. Gonococcal PID results from upward spread of the organism from the initial site of infection (the endocervix) into the upper genital tract. Symptoms of PID can include fever, unilateral or bilateral lower abdominal pain, pain associated with sexual intercourse, abnormal menses or intermenstrual bleeding, or other complaints associated with an intraabdominal infection. These symptoms may be severe or mild. Findings on physical exam include abdominal or uterine tenderness, adnexal tenderness, cervical motion tenderness, and occasionally an adnexal mass or fullness associated with a tubo-ovarian abscess. PID may be caused by a number of other infections, including *Chlamydia trachomatis* as well as many other bacteria. Treatment of PID includes broad-spectrum, often intravenous, antibiotics targeting chlamydia, gonorrhea, and anaerobic bacteria. PID is a significant complication not only because of the acute manifestations, but because of the long-term sequelae, including infertility, ectopic pregnancy, and chronic pelvic pain. Approximately 10% of women will become infertile after a single episode of PID. For those women that remain fertile, there is a 10-fold increase in the risk of ectopic pregnancy.

The manifestations of gonorrhea in pregnant women are similar to those in nonpregnant women, although PID is less common. Pregnant women with genital gonorrhea are at significant risk, for spontaneous abortion, premature rupture of membranes, premature delivery, and acute chorioamnionitis, as well as for transmitting gonorrhea to their newborns during delivery. Neonates infected with gonorrhea can suffer eye and pharyngeal involvement, as well as other complications. Consequently, all newborn infants are treated prophylactically with topical antibiotics to prevent ocular infection.

2. *Gonococcal Infections in Men*

For symptomatic men, the most common manifestation of infection with *N. gonorrhoeae* is urethritis. Infected men often develop a urethral discharge that can range from scant and clear to copious and purulent. Acute infection is usually accompanied by dysuria (painful urination), which develops after the onset of discharge, or may be the sole manifestation. Symptoms typically develop 2–5 days after exposure to an infected sexual partner, although there can be a delay of over 2 weeks. Untreated acute infection can result in complications, including acute epididymitis, acute and chronic prostatitis, and rarely cellulitis, penile lymphangitis, or periurethral abscess. The most common complication is acute epididymitis, which presents with gradually worsening unilateral scrotal pain. Examination of a man with epididymitis may reveal previously unnoticed urethral discharge as well as tenderness and swelling of the posterior of the scrotum. In young men under the age of 40 years, coinfection with other STDs, particularly *Chlamydia trachomatis,* is as common as it is for their female counterparts. Thus, treatment of gonococcal urethritis and associated complications should be targeted toward both organisms. In older men, over the age of 40 years, structural abnormalities of the urinary tract predispose to infections caused by gram-negative bacteria and other urinary tract organisms.

3. *Other Gonococcal Infections*

Anorectal gonorrhea is uncommon in heterosexual men, but it is seen more frequently in homosexual men and in approximately 35 to 50% of women with gonococcal infection of the endocervix. Receptive anal intercourse is the principal mode of transmission in homosexual men, whereas rectal infections in women more commonly arise from local contamination by cervicovaginal secretions. Most men and women with anorectal gonorrhea are asymptomatic; however, some will develop symptoms of proctitis, including severe rectal pain, tenesmus, rectal discharge, and constipation. Other symptoms may include anal itching, painless purulent discharge, or a small amount of rectal bleeding.

Pharyngeal infection rarely occurs without concomitant genital infection. Estimates suggest that of patients with gonorrhea, pharyngeal infection occurs in 10 to 20% of heterosexual women, 3 to 7% of heterosexual men, and 10 to 25% of homosexually active men. The vast majority of gonococcal pharyngeal infections are asymptomatic; however, these cases may rarely be associated with acute pharyngitis, tonsillitis, fever or lymph node enlargement in the neck.

Neisseria gonorrhoeae infections of the eye occur most commonly in newborns (ophthalmia neonatorum) as a result of exposure to infected maternal secretions during delivery. These infections can result in corneal scarring or perforation of the cornea, and thus prompt diagnosis and treatment are crucial. Fortunately, the use of silver nitrate or erythromycin eye drops at birth is effective in preventing gonococcal ophthalmia neonatorum. Adults can also acquire gonococcal ocular infection through autoinoculation, and develop keratoconjunctivitis as a result.

4. Disseminated Gonococcal Infection

Disseminated gonococcal infection (DGI) occurs in approximately 1 to 2% of individuals with untreated gonorrhea as a result of hematogenous spread of the organism from typically asymptomatic mucosal infections of the pharynx, cervix, urethra, or rectum. DGI is more common in women than in men, and bacteremia probably begins 7 to 30 days after initial infection. DGI is often referred to as an arthritis–dermatitis syndrome, given the usual clinical characteristics of an asymmetric polyarthritis and associated rash. The joint manifestations in DGI may begin with arthralgias or painful tenosynovitis which progress to frank arthritis later in the course of disease. Approximately 30 to 40% of patients with DGI will present with overt arthritis, usually involving the wrist, metacarpophalangeal, ankle, or knee joints, although any joint may be involved. Joint fluid cultures are often negative except when the fluid has a high white blood cell concentration, usually more than 40,000 white blood cells per cubic millimeter. Overall, only 20 to 30% of joint fluid cultures from patients with DGI will be positive for the organism. The rash associated with DGI is composed of a small number (fewer than 30) of tender, necrotic pustules on a reddened base, usually located on the distal extremities. The skin lesions may, however, appear as macules, papules, petechiae, bullae, or ecchymoses. Despite the fact that DGI represents a bloodstream infection, patients often do not clinically appear particularly ill. Blood cultures, like joint fluid cultures, are only positive in 20 to 30% of patients with DGI. As previously noted, the mucosal infections in DGI are usually asymptomatic. Thus, the diagnosis of DGI can be difficult to make. When DGI is suspected it is very important to obtain cultures of all potential mucosal sites, in addition to blood and joint fluid, for *N. gonorrhoeae,* to help confirm the diagnosis. Rarely, DGI can be further complicated by seeding of the meninges or the heart valves, resulting in potentially life-threatening gonococcal meningitis or endocarditis.

Disseminated gonococcal infection can also occur in newborns of infected mothers, when exposed to secretions during vaginal delivery. The illness in neonates can be similar to that in adults, with associated meningitis and arthritis. Gonococcal disease in newborns usually manifests 2 to 5 days after birth.

C. Diagnosis

Gram stain and culture are the most widely used methods for detection of *N. gonorrhoeae,* although some newer nonculture techniques are proving to be more convenient and equally accurate in some settings. Specimens for both Gram stain and culture must be obtained directly from the urethra or cervix with a special swab. For detection of gonococcal cervicitis or urethritis in symptomatic or high-risk individuals, a Gram stain of an appropriately obtained specimen can provide rapid and accurate results. A Gram stain is considered positive if gram-negative diplococci are seen within polymorphonuclear leukocytes. Gram stain results are more difficult to interpret on rectal and pharyngeal specimens because of the presence of other gram-negative organisms. Isolation of *N. gonorrhoeae* by culture is the standard method for diagnosis of gonococcal infections of any site, and is necessary for antibiotic-resistance testing. Optimal recovery of *N. gonorrhoeae* requires direct inoculation of the specimen onto selective media, followed by placement into a 35°C to 37°C incubator containing 5 to 7% carbon dioxide. The most commonly used selective media include modified Thayer-Martin, Martin-Lewis, and New York City media. These media contain antimicrobial substances to suppress the overgrowth of other microorganisms that can obscure the presence of the small colonies typical of *N. gonorrhoeae. Neisseria gonorrhoeae* can survive for 5 to 7 hr in special transport media, permitting clinics without the necessary laboratory capabilities to transport clinical specimens to nearby laboratories better able to process the specimens appropriately.

Nonculture techniques detect gonococcal antigens, gonococcal enzyme activity, or gonococcal genetic material. One test uses a single-stranded DNA probe to detect gonococcal ribosomal RNA, and it is fairly sensitive and highly specific for gonorrhea. In 1996, the ligase chain reaction assay was approved in the U.S. for the diagnosis of gonorrhea. Similar assays using the polymerase chain reaction and other nucleic acid amplification techniques have also been developed. These assays are as sensitive as culture (95 to 98%) and highly specific for detecting *N. gonorrhoeae*. They have the added advantage of being equally sensitive and specific when used on first-void urine specimens, compared to swab specimens which can be impractical to obtain and uncomfortable for the patient. Currently, there are no acceptable serologic tests for the diagnosis of gonorrhea.

D. Treatment

For several decades penicillin was the drug of choice for the treatment of gonococcal infections. By the late 1980s, however, penicillin resistance had become widespread, and penicillin was no longer a reliable choice for effective therapy. Antibiotic resistance may be plasmid-mediated or chromosomally mediated, and some strains of *N. gonorrhoeae* have become resistant to a variety of antibiotics, including tetracycline, erythromycin, and quinolones. Resistance to these antibiotics is less common. There are a number of effective antibiotic regimens, which are summarized in Table II. Tetracyclines and quinolones should not be used in pregnant women or very young children. The choice of an antibiotic regimen depends on the severity of the infection, ease of administration, cost and availability, as well as effectiveness against coinfecting pathogens, such as *C. trachomatis*. This latter point is important because at least 10 to 20% of men and 30 to 40% of women with uncomplicated gonorrhea are also infected with chlamydia. Thus, in areas where the rate of coinfection is high, individuals diagnosed with gonorrhea should be treated for both gonorrhea and chlamydia. In general, individuals with one STD are at risk for having other STDs, and should be evaluated accordingly. The most common cause for treatment failure is repeated exposure to an untreated sexual partner. Given the prevalence of asymptomatic gonorrhea, treatment of all sexual partners contacted within the previous 60 days is recommended, regardless of whether they have symptoms suggestive of infection. If the patient's most recent sexual encounter was more than 60 days prior to the onset of symptoms or diagnosis, then the patient's most recent partner should be treated.

E. Organism and Pathogenesis

Neisseria gonorrhoeae belongs to the family Neisseriaceae, which includes several nonpathogenic species, as well as the well-recognized pathogen *Neisseria meningitidis*. *Neisseria gonorrhoeae* is a fastidious gram-negative diplococcus, 0.6 to 1.0 μm in diameter, with complex growth requirements described previously. *In vivo*, gonococci grow under relatively anaerobic conditions and are capable of growing under strictly anaerobic conditions *in vitro*. *Neisseria gonorrhoeae* produces high levels of catalase and produces an oxidase. Because oxidase production is easy to detect in the laboratory, oxidase-positive gram-negative diplococci that are obtained from appropriate clinical specimens, and that grow on selective media, are presumed to be *N. gonorrhoeae* in most cases. *Neisseria gonorrhoeae* can be distinguished from other *Neisseria* species on the basis of sugar fermentation patterns. *Neisseria gonorrhoeae* utilize only glucose, pyruvate, and lactate as their carbon source. Other *Neisseria* species, *Kingella dentrificans*, *Moraxella* species, and *Branhamella catarrhalis* can be mistaken for *N. gonorrhoeae*, particularly given their similar appearance on Gram stain. Such errors can have serious repercussions for patients and their health care providers given the social and medicolegal implications of being diagnosed with an STD.

The structure, molecular biology, and pathogenesis of *N. gonorrhoeae* have been studied extensively, yet many processes remain elusive. The cell wall is made up of pili, a capsule, several characteristic outer membrane proteins referred to as protein I, Opa protein (protein II), protein III (Rmp), iron regulating proteins, lipooligosaccharide (LOS), and H.8 (Lip), and other proteins that can vary with growth conditions. The pili are hairlike filamentous appendages that extend from the cell surface. Pili are an important virulence factor. Pili, in concert with protein II, enable gonococci to attach to the microvilli of host nonciliated columnar epithelial cells. Protein I is the most prominent outer membrane protein and functions as a porin, providing a hydrophilic channel

TABLE II
Therapy for Gonococcal Infections[a,b]

Uncomplicated Gonococcal Infections of the Cervix, Urethra, or Rectum

Recommended Regimens:

Ceftriaxone 125 mg intramuscularly (IM), single dose
or
Cefixime 400 mg orally (PO), single dose
or
Ofloxacin 400 mg PO, single dose
or
Ciprofloxacin 500 mg PO, single dose
and
Therapy should include a regimen effective against *Chlamydia trachomatis,* such as
Doxycycline 100 mg PO twice daily × 7 days
or
Azithromycin 1 g PO, single dose

Epididymitis

Recommended Regimens:

Ceftriaxone 250 mg IM, single dose
and
Doxycycline 100 mg PO twice daily × 7 days

Conjunctivitis

Ceftriaxone 1 g IM, single dose, and lavage the infected eye with saline solution once. Topical therapy alone is inadequate.

Disseminated Gonococcal Infection

Patients with DGI should also be treated for *Chlamydia trachomatis,* unless coinfection has been excluded.

Recommended Regimen:

Ceftriaxone 1 g IM or intravenously (IV) every 24 hr
Parenteral therapy should be continued for 24 to 48 hr following the onset of clinical improvement, after which therapy may be changed to one of the following to complete a full week of treatment:
Cefixime 400 mg PO twice daily
or
Ciprofloxacin 500 mg PO twice daily
or
Ofloxacin 400 mg PO twice daily

Pelvic Inflammatory Disease

Recommended Parenteral Regimens for Severe Disease:

Regimen A

Cefotetan 2 g IV every 12 hr
or
Cefoxitin 2 g IV every 6 hr
and
Doxycycline 100 mg PO every 12 hr × 14 days
Continue IV antibiotics until clinically improved, then complete course of therapy with PO doxycycline.
or

Regimen B

Clindamycin 900 mg IV every 8 hr
and
Gentamicin 1.5 mg/kg every 8 hr, following loading dose
Continue IV therapy until improved, then complete 14-day course with doxycycline or clindamycin 450 mg PO 4 times a day.

Recommended Oral Regimens for Milder Disease:

Ofloxacin 400 mg PO twice daily × 14 days
and
Metronidazole 500 mg PO twice daily × 14 days

[a] The regimens suggested here are for nonpregnant adults only. Recommended treatments and dosing for pregnant women and young children are different.

[b] In addition to the regimens suggested here, there are numerous alternative regimens for each of the above conditions. Please refer to the 1998 Guidelines for Treatment of Sexually Transmitted Diseases published by the CDC for more details.

through the outer membrane. Protein I triggers endocytosis of gonococci by the host mucosal cell. Following endocytosis, the membrane of the mucosal cell retracts around the organism, forming a membrane-bound vacuole which is transported to the base of the cell, where gonococci are subsequently released into the subepithelial tissue. Unlike most of the other outer membrane proteins, protein I is always expressed, and it is useful in identifying individual strains of gonococci as each strain produces a single antigenically stable protein I. Protein I also appears to inhibit neutrophil function, and thus may limit the ability of the host to respond to infection.

In addition to enhancing the adhesion of gonococci to mucosal cells, the Opa protein, or protein II, appears to mediate infection by enhancing gonococcal cell-to-cell adhesion, which may result in a more infectious unit. It specifically appears to enhance adhesion to conjunctival cells, epithelial cells, and neutrophils. Protein III (Rmp) appears to be able to block bactericidal antibodies produced by the host directed at other surface proteins acting as antigens, specifically protein I and LOS. Lipooligosaccharide (LOS) is similar to the lipopolysaccharide component characteristic of the cell wall of all gram-negative bacteria, except that it lacks long, hydrophilic, neutral polysaccharides. LOS appears to mediate most of the damage to host tissues. LOS is the primary target of host antibodies, and it regulates complement activation on the cell surface of gonococci in addition to prompting the release of enzymes, such as proteases and phospholipases. Evidence suggests that gonococcal LOS stimulates the production of tumor necrosis factor (TNF) in fallopian tube organ cultures, and inhibition of TNF has been shown to limit tissue damage. LOS also appears to be capable of molecular mimicry, and thus hinders the ability of the host to mount an effective host immune response. Pili, protein II, and LOS all also exhibit extensive intrastrain and interstrain antigenic variation, as well as phase variation where pili and protein II are concerned. Such variation likely represents a mechanism by which gonococci can escape the host immune response. The role of the other outer membrane proteins in the pathogenesis of gonococci is not well understood.

III. CHLAMYDIA

A. Epidemiology

Chlamydia, caused by *Chlamydia trachomatis,* is currently considered the most common STD in the U.S. There are an estimated 4 million cases in the U.S. each year, at a cost of $2.4 billion. In 1986 chlamydia became a reportable disease in several states, and for years thereafter the number of reported cases increased steadily, likely as a result of improved diagnostic techniques, screening practices, and reporting in more states. Chlamydia became the most common reportable disease in the U.S. in 1995, when there were 477,638 cases reported. Chlamydia is a problem internationally as well. In 1995, the WHO estimated that there were 89 million new cases worldwide. As with other STDs, data on the true incidence and prevalence of chlamydia are limited by a number of factors. First, chlamydial infections are often asymptomatic. In a screening study of women attending a family planning clinic, 70% of the women found to have genital infections secondary to chlamydia had neither clinical findings nor symptoms of infection. Thus, a significant number of cases likely go unrecognized by the patient or health care provider. Second, when symptoms are present, individuals are often treated empirically without diagnostic confirmation, and thus these cases of chlamydia are never reported. Third, some of the laboratory methods used to diagnose chlamydia are relatively insensitive and may yield false negative results, leading to an underestimate of disease burden.

Several studies in a variety of settings, utilizing different diagnostic screening strategies, have provided some estimates of the prevalence of genital chlamydial infections in the U.S. Approximately 3 to 5% of asymptomatic men and women seen in general medical settings are infected with chlamydia, whereas women and men screened in STD clinics have a prevalence of 15 to 20%. More recent studies using highly sensitive diagnostic assays reported even higher prevalence data. In one study, Burstein and colleagues demonstrated that the prevalence of chlamydial genital infections among adolescent females screened in family planning, STD, and school-based clinics in Baltimore, Maryland, was 29.1%. The majority of infected girls were asymptomatic. In a study of 13,204 new female army recruits the prevalence was 9.2%. There was wide geographic variation in the prevalence of chlamydial infection according to the state of origin of the recruit. More than 15% of women from South Carolina, Georgia, Alabama, Louisiana, and Mississippi were infected with chlamydia. In New Jersey, North Carolina, Kentucky, Texas, Oklahoma, and Arkansas the prevalence was 10 to 15%. In five states, including Washington and Oregon, the prevalence was less than 5%, most likely as a result of aggressive public health initiatives in place in those states since the late 1980s, aimed at reducing the number of chlamydial infections. In a similar study of 2245 male army recruits the prevalence of chlamydial urethritis was 5.3% overall, with similar geographic variation by state. Of the men infected with chlamydia, less than 15% reported having any symptoms to suggest infection.

Specific risk factors for chlamydial infections have been consistently identified in these studies. The single most important risk factor for chlamydial infections in women is young age. In a study of adolescent females aged 12 to 19 years, 14-year-old females had the highest age-specific chlamydia prevalence rate. In the study of female army recruits, 95% of those infected with chlamydia were less than 25 years of age. Furthermore, studies of reinfection have shown that up to 20 to 30% of female adolescents had evidence of another chlamydial infection within 6 months of the initial infection. Young age, however, does not appear to be a consistent risk factor for chlamydial infections in men. In the study of male army recruits, age did not correlate with infection. Other risk factors associated with chlamydial infection mirror those associated with other STDs and include multiple sex partners, a recent new sex partner, inconsistent condom use, and nonwhite race. Among female army recruits, African-American women had more than a threefold greater risk of infection compared to white female recruits. Another important risk factor for chlamydia is the diagnosis of gonorrhea. As noted, approximately 20% of men and 40% of women infected with gonorrhea are simultaneously infected with chlamydia. In the above study of adolescent females, however, attempts to identify the majority of those infected using such risk factors, other than young age, as a screening measure were unsuccessful. The investigators recommended that all sexually active adolescent females be screened for chlamydia every 6 months, regardless of symptoms, prior infections, condom use, or partner risks.

Chlamydia is also a problem for pregnant women and their newborns. Prevalence studies in pregnant women using relatively insensitive screening techniques yielded prevalence rates of 5 to 7%, although prevalence rates of 21 to 25% have been identified in inner city and Native American populations. Of infants born to untreated infected mothers, approximately 60 to 70% can become infected with chlamydia, a significant proportion of which go on to develop pneumonia and ocular complications as a result.

Given the results of these studies, some areas of the U.S. have instituted broad-based screening and treatment programs, and the prevalence of chlamydial infections in these areas has declined dramatically. In the Pacific Northwest of the U.S., screening for chlamydia in patients attending family planning clinics began in 1988, and in STD clinics in 1993. The prevalence of chlamydia in this region declined from 10 to 12% in the late 1980s to 4 to 5% in 1995. One challenge to these efforts is the relatively long incubation period of chlamydia of 1 to 3 weeks. Infected individuals may not develop symptoms for a prolonged period of time, if at all, and thus may remain sexually active while infectious. The efficiency with which chlamydia is transmitted between sexual partners is not well understood. Studies evaluating the sexual partners of infected individuals demonstrated an infection rate of 85% among female contacts of men with chlamydial urethritis. Traditionally researchers suspected that men were more likely to transmit chlamydia to women than the converse, but more recent data from studies using highly sensitive diagnostic techniques suggest that transmission may be equal in either direction.

B. Clinical Manifestations

There are several subtypes of *C. trachomatis,* the trachoma and LGV (lymphogranuloma venereum) subtypes being the only human pathogens. LGV subtypes are transmitted through sexual contact and cause ulceration, as well as painful lymphadenopathy and abscess formation. LGV infections are extremely rare in industrialized nations; they are less common than infections caused by the trachoma subtype and will not be discussed further here. Genital infections caused by the trachoma subtype of *C. trachomatis* have many of the same characteristics as those caused by *N. gonorrhoeae.* Both organisms preferentially infect columnar epithelial cells and do not appear to be capable of growth in squamous cells. Thus, like gonorrhea, chlamydia infects the urethra, with extension to the epididymis and possibly the prostrate gland in men, the squamocolumnar junction of the endocervix in women, with the potential for spread to the endometrium and upper genital tract, and the rectum. *Chlamydia trachomatis* can also cause conjunctivitis. In contrast to gonorrhea, however, chlamydial infections generally have a longer incubation time, are associated with a less robust inflammatory response, and are more likely to be asymptomatic. Unlike gonorrhea, chlamydia does not disseminate to cause a systemic infection like DGI. Unfortunately, there are no unique signs or symptoms to help make the diagnosis of *C. trachomatis* infection of the genital tract. The associated symptoms and physical findings, if present at all, can also result from other infections or complications, and therefore one has to have a high index of suspicion to diagnose a chlamydial infection.

1. *Chlamydial Infections in Women*

In women, *C. trachomatis* most commonly infects the cervix, although approximately 50% of infected women will have simultaneous infection of the urethra. Chlamydial infection of the cervix typically results in cervicitis with a mucopurulent discharge, present in approximately 20 to 40% of infected women. On exam, there is sometimes visible swelling and easily inducible bleeding in a zone of ectopy, an area with a relative increase in exposed columnar cells at the squamocolumnar junction. As previously noted, however, in studies in which women were screened for chlamydia, 70% or more of infected women did not have any abnormal clinical findings. Women with cervical ectopy have a higher prevalence of chlamydia than those without ectopy. Cervical ectopy is normally present in 60 to 80% of sexually active adolescent females, and declines with age. This finding may in part explain the higher prevalence of chlamydia among adolescent females. Oral contraceptives are also associated with cervical ectopy, and this may account for the increased risk of chlamydia, and gonorrhea, for women on oral contraceptives. Chlamydial urethritis has also been well documented in women and is estimated to occur without concomitant cervical infection about 25% of the time. The prevalence of isolated urethral infection increases with age. Chlamydial urethritis may be associated with painful and frequent urination, although the majority of women with urethritis do not have these symptoms. Nonetheless, *C. trachomatis*, as well as *N. gonorrhoeae*, should be considered as a possible diagnosis in sexually active women with urinary symptoms if there is evidence of coexisting cervicitis or if other STD risk factors are present.

The most serious form of *C. trachomatis* infection is upper genital tract disease. Approximately 10% of women with cervical chlamydia infection will develop symptomatic pelvic inflammatory disease (PID). There is likely an even larger proportion of infected women who develop asymptomatic upper tract disease. All of these women are at increased risk for infertility and ectopic pregnancy as a result of tubal inflammation and scarring. The clinical manifestations of PID secondary to chlamydia are similar to those of PID caused by gonorrhea, and are discussed more fully in a previous section. As with other sites of infection, chlamydial infection of the upper tract is clinically milder and more likely to be asymptomatic than infections caused by gonorrhea or anaerobic bacteria, despite ongoing tubal damage. Like gonorrhea, chlamydia can cause perihepatitis, with or following the development of PID, referred to as the Fitz-Hugh-Curtis syndrome. Women who are infected with chlamydia in the first trimester of pregnancy are at increased risk of postpartum endometritis following vaginal delivery.

2. *Chlamydial Infections in Men*

More common than gonococcal urethritis is the entity known as nongonococcal urethritis (NGU). A diagnosis of NGU is made when *N. gonorrhoeae* cannot be identified in urethral specimens from a man with urethritis. *Chlamydia trachomatis* is believed to be the causative organism in 35 to 50% of cases of NGU. Other organisms include *Ureaplasma urealyticum, Trichomonas vaginalis,* and herpes simplex virus. NGU is often characterized by the presence of inflammatory cells in the urine and a urethral discharge that is usually less purulent and relatively scant compared to the urethral discharge associated with gonorrhea. Infected men may experience painful or frequent urination. However, the symptoms are variable enough that one cannot reliably distinguish NGU from gonococcal urethritis clinically. If a Gram stain of the urethral specimen shows a large number of inflammatory cells (specifically, polymorphonuclear cells) without intracellular diplococci, then a presumptive diagnosis of NGU is made. A diagnosis of chlamydial urethritis requires specific diagnostic testing.

Men may also experience urethritis following treatment for gonorrhea, known as postgonococcal urethritis. These men have recurrent or persistent symptoms of urethritis despite appropriate treatment for gonorrhea. *Chlamydia trachomatis* is thought to account for 70 to 90% of cases of postgonococcal urethritis. Men who develop this syndrome were most likely infected with both gonorrhea and chlamydia simultaneously, but were treated with antibiotics effective against gonorrhea but not chlamydia. Because of the longer incubation period for chlamydia, symptoms of urethritis may develop even after an apparent initial response to therapy for gonococcal urethritis. Because of the high rate of coinfection, presumptive treatment for both organisms is recommended when gonorrhea is diagnosed.

Chlamydial infections of the lower genital tract in men may also become complicated. *Chlamydia trachomatis* is the most common cause of epididymitis in sexually active young men, accounting for more than 60% of cases. Epididymitis results from the spread of an untreated chlamydial urethral infection to the upper gential tract, and it is characterized by progressive unilateral scrotal pain and swelling. Chlamydial urethritis has also been associated with the subsequent development of Reiter's syndrome. Reiter's syndrome consists of urethritis, conjunctivitis, arthritis, and characteristic skin lesions. Some studies have found that 80% of men with Reiter's syndrome have evidence of prior or ongoing chlamydial infection. This syndrome is more common is men with the HLA-B27 haplotype. These patients may also develop arthritis or tenosynovitis, without the other features of Reiter's syndrome, in association with infection with *C. trachomatis*, and usually improve following treatment for chlamydia.

3. *Other Chlamydial Infections in Adults*

Both men and women can develop proctitis as a result of rectal infection with *C. trachomatis*. Rectal infections are usually asymptomatic, but if proctitis develtops patients may experience rectal pain, bleeding, and a mucous discharge. Adults can also develop an acute follicular conjunctivitis as a result of autoinoculation of the eye with infected genital secretions. Conjunctivitis can develop within 1 to 3 weeks and does not typically result in permanent damage or vision loss.

4. *Chlamydial Infections in Neonates*

Neonates exposed to *C. trachomatis* during vaginal delivery are at risk for developing conjunctivitis and pneumonia. Nasopharyngeal infection is also common. Conjunctivitis develops within 5 to 21 days and is characterized by copious purulent discharge and redness of the eye. As in adults, it is usually self-limited and does not result in permanent eye damage or vision loss. Pneumonia caused by chlamydia can be life-threatening, however. The incubation period for chlamydial pneumonia in neonates is 2 to 12 weeks. Infected infants often have a history of conjunctivitis and present with rhinitis, rapid breathing, and a characteristic cough, in the absence of fever. There is evidence to suggest that infants who have chlamydial pneumonia frequently go on to develop asthma or obstructive airway disease, even after appropriate treatment of the initial pneumonia.

C. Diagnosis

Because *C. trachomatis* is an obligate intracellular organism, cell culture has traditionally been the gold standard for the diagnosis of chlamydia. Cycloheximide-treated McCoy cells are the most commonly used cell culture system. Cell culture is often impractical, however, because of the high level of technical expertise needed, as well as the strict requirements for specimen collection and transport, and high cost. Samples of discharge or secretions containing only inflammatory cells are not adequate. Instead, carefully collected samples of cervical or urethral columnar cells are necessary. Specimens must then be refrigerated in special transport media and inoculated onto cell culture plates within 24 hr. When done correctly, cell culture has a sensitivity of 80%. Because of these limitations, several nonculture techniques have been developed, including a direct fluorescent antibody test (DFA), enzyme immunoassay (EIA), and direct DNA probes. Although these tests may be relatively inexpensive and more convenient in some settings, they still require careful specimen collection and must be processed by an experienced technician. The DFA is slightly more sensitive than the EIA when done properly, with a sensitivity of 80 to 85% and a specificity of 99% relative to culture. The EIA is only 60 to 80% sensitive when compared to culture. The performance of direct DNA probes is comparable to that of DFA.

The development of tests to detect amplified chlamydial DNA or RNA has markedly improved diagnosis. These techniques utilize the ligase chain reaction (LCR), polymerase chain reaction (PCR), or transcription-mediated amplification (TMA). Perhaps the greatest benefit of LCR and PCR tests, which detect amplified DNA, is that they can be performed on first-catch urine samples, as well as urethral and cervical specimens. In 1996, LCR was approved by the U.S. Food and Drug Administration (FDA) for the diagnosis of chlamydial genital tract infections. The urine LCR assay is both highly sensitive and specific compared to cell culture, with a sensitivity of 90 to 96% in men and 69 to 96% in women. The PCR test appears to be comparable to the LCR, and both techniques are 99–100% specific. The TMA assay detects amplified ribosomal RNA and appears to be similarly sensitive and specific.

Serologic tests are not appropriate in the diagnosis of uncomplicated genital tract infections, primarily because of the high prevalence of antibodies to *C. trachomatis* in sexually active adults. High antibody titers may be seen in patients with complicated upper tract disease, and thus support the diagnosis of chlamydial PID or epididymitis. However, sexually active patients with PID or epididymitis should be treated for chlamydia regardless of the results of serologic testing.

D. Treatment

The antibiotics typically used in the treatment of gonorrhea, such as penicillins and cephalosporins, are not effective against chlamydia. As noted, in areas where the rate of coinfection with both organisms is high, treatment regimens should include antibiotics that will eradicate both gonorrhea and chlamydia. Macrolides, tetraclyines, and quinolones, with the exception of ciprofloxacin, are all effective against chlamydia. There has not been any evidence of emerging antibiotic resistance, as has been the case for gonorrhea. Currently the treatment of choice for uncomplicated chlamydial infections of the lower gential tract is 1 gram of oral azithromycin. Alternatively, a 1-week course of doxycycline or ofloxacin can be used. Although azithromycin is more expensive than doxycycline, cost-effectiveness analyses demonstrate that this treatment option is a cost-effective choice because of the improvement in compliance with single-dose therapy. Infected patients and their partners should abstain from sexual activity until they have completed a course of antibiotic therapy, in order to prevent transmission of chlamydia prior to the infection. If single-dose therapy is given, individuals should abstain for 7 days following treatment.

Since tetracyclines and quinolones are not acceptable for use in pregnant women or infants, azithromycin or erythromycin should be used. The effectiveness of erythromycin may be reduced compared to azithromycin due to compliance issues, and therefore patients treated with erythromycin should be retested for chlamydia 2 weeks following completion of therapy (or 3 weeks if using an amplification test). Conjunctivitis should be treated with oral antibiotics, not topical therapy, because of the risk of infection at other sites. The treatment of PID should routinely include antibiotics effective against chlamydia. All sex partners within the previous 60 days should also be treated for chlamydia and gonorrhea in a timely manner. If the patient's last sexual encounter was more than 60 days prior to the onset of symptoms or diagnosis, then the patient's most recent partner should be treated. Dosing schedules and duration of treatment are summarized in Table III.

E. Organism and Pathogenesis

Chlamydia trachomatis is one of four species within the genus *Chlamydia.* Of the others, *C. pneumoniae* and *C. psittaci* are also human pathogens. The fourth, *C. pecorum,* is a pathogen of mammals but is not believed to cause disease in humans. *Chlamydia pneumoniae, C. psittaci,* and *C. pecorum* are difficult to differentiate from one another without the use of DNA hybridization and monoclonal antibody techniques. *Chlamydia trachomatis,* on the other hand, can be easily distinguished from the others using

TABLE III
Therapy for Chlamydial Infections[a]

Genital tract infection and conjunctivitis in adults and adolescents
Recommended Regimens:
Azithromycin 1 g orally (PO), single dose
or
Doxycycline 100 mg PO twice daily × 7 days
Alternative Regimens:
Erythromycin base 500 mg PO four times daily × 7 days
or
Erythromycin ethylsuccinate 800 mg PO four times daily × 7 days
or
Ofloxacin 300 mg PO twice daily × 7 days
Genital tract infection in pregnant women
Erythromycin base 500 mg PO four times daily × 7 days
or
Amoxicillin 500 mg PO three times daily × 7 days
Genital tract infection and conjunctivitis in infants and young children
Erythromycin base 50 mg/kg/day PO divided into four doses daily × 10 to 14 days
Topical therapy alone is inadequate. Follow-up cultures are necessary to ensure that treatment has been effective, as erythromycin is only effective 80% of the time.

[a] The use of tetracyclines and quinolones is not appropriate in pregnant women and young children.

two simple laboratory tests. Cells infected with *C. trachomatis* contain inclusions that are rich in glycogen and therefore stain with iodine, whereas cells infected with the other species do not. Of the *Chlamydia* species, only *C. trachomatis* is susceptible to sulfonamide antibiotics.

Within each species, there are many different strains that have been isolated from birds, humans, and other mammals. The strains within each species are genetically very similar, but there is only partial DNA homology among the four species. Recent analysis of the *C. pneumoniae* genome revealed that approximately one-fifth of the genomes are unique, when compared to the *C. trachomatis* genome. The individual strains within species are referred to as biovars or serovars, depending on the specific characteristics that distinguish them from the other strains. These individual biovars and serovars vary by preferred host species, the type of disease they cause, and their antigenic composition. The species *C. trachomatis* can be divided into the three biovars: the murine biovar, the trachoma biovar, and the LGV (lymphogranuloma venereum) biovar. Only the latter two cause disease in humans, and they can be differentiated serologically and by the type of disease they cause. The LGV biovar is the agent of lymphogranuloma venereum, an STD rarely seen in the developed world and quite distinct from syndromes associated with the trachoma biovar discussed in this chapter. The LGV biovar is much more invasive than the trachoma biovar, and it appears to preferentially infect endothelial and lymphoid cells, resulting in genital ulcers, abscess formation, and scarring of the lymphatics. The trachoma biovar infects columnar epithelial cells and is responsible for the genital tract disease and conjunctivitis discussed here, as well as for trachoma, a potentially chronic and sight-threatening infection of the eye endemic in many parts of the world where sanitation is poor.

Chlamydiae are structurally complex microorganisms. They have a cell wall and membrane that is somewhat similar to gram-negative bacteria, although it does not contain significant amounts of muramic acid. These organisms are unique from other bacteria in many ways. They are obligate intracellular bacteria and possess an unusual life cycle that distinguishes them from all other bacteria. Each of the four species shares the same developmental cycle, although the intracellular inclusions do vary in morophology between species and can help distinguish them from one another. The initial step in the life cycle is the attachment of the organism to the host cell. At this stage, the organism is in the form of an elementary body (EB). The EB is the stable, extracellular, infectious form of the organism and does not appear to be metabolically active. Attachment is partially charge dependent and appears to be mediated by heparan sulfate-like molecules that act as a bridge between receptors on the surface of the organism and the host cell. After attachment to the host cell, the organism, in the EB form, is endocytosed into the host cell, although the exact mechanism for uptake into the host cell is unclear. Following entry into the cell, the organism undergoes a morphological change within the first 6 to 8 hr, which transforms the organism from an EB into a reticulate body (RB). The RB is the intracellular, metabolically active, dividing form of the organism. The RB is not infectious and cannot survive outside the host cell. The RB rely on the host cell for precursors and energy in order to synthesize their own macromolecules such as RNA, DNA, and proteins. The RB divide by binary fission, and in the first 18 to 24 hr after entry into the host cell, some RB revert back to the EB form. The process by which chlamydiae undergo the morphological switch between the EB and RB form is not known. After the initial 24 hr, there is a progressive increase in the number and proportion of EB, and ultimately, around 48 to 72 hr after entry, the host cell ruptures, releasing the infectious EB into the extracellular environment to infect other susceptible host cells. The entire cycle takes place within the phagosome, and, interestingly, phagolysosomal fusion does not occur until cell rupture is about to take place. The inhibition of phagolysosomal fusion is attributed to a surface protein antigen on the EB that acts as an inhibitor.

The pathogenesis of *C. trachomatis* is not well understood. There is no direct *in vivo* evidence for a latent stage, where chlamydiae persist without ongoing replication. Since the life cycle ends with the destruction of the host cell, then presumably untreated *C. trachomatis* may cause progresive damage in infected individuals. The disease and associated symptoms of *C. trachomatis* infection are believed to be primarily the result of the inflammatory response of the host to both the organism and destroyed host cells. Available evidence, primarily from animal data, suggests that much of chlamydial disease manifestations stem from the host's immune reaction to the infection. This is supported by the observation that oftentimes second infections inflict greater damage. The responsible stimulus for the immune response appears to be a sensitizing antigen belonging to the 60-kDa heat-shock protein (HSP 60) family. This antigen is loosely bound to EB and is excreted by infected cells. Some women who are infertile or experience ectopic pregnancy because of tubal injury have been found to have high levels of antibodies to HSP 60.

IV. SYPHILIS

A. Epidemiology

Syphilis is a complex and fascinating disease, caused by infection with the bacterium *Treponema pallidum*. The disease is a systemic one, capable of affecting nearly every organ in the body and thus mimicking many other disease processes. Syphilis is characterized by several different stages, including primary, secondary, tertiary, and latent stages, and it can persist for decades if untreated. Syphilis is distributed worldwide and is particularly a problem in the developing world, where it is the principal cause of genital ulcer disease. Prior to the development of penicillin in the 1940s, syphilis was a huge public health problem in the U.S. At the beginning of World War II, the U.S. Public Health Service estimated that approximately 2.5% of all Americans were infected with syphilis. With the advent of penicillin, however, the prevalence of primary and secondary cases declined by about 93%. Fortunately, penicillin remains fully effective against syphilis, with no evidence of the emergence of penicillin resistance.

Nonetheless, syphilis remains a significant public health concern. Beginning in the mid 1980s there was a marked and progressive resurgence of primary and secondary (P&S) syphilis in the U.S. that peaked in 1990, with a total of 55,132 reported cases of P&S syphilis, representing a 20.2% increase over the previous year. The reasons for these trends are unclear. Of concern was the dramatic increase in cocaine use in the U.S. during the mid to late 1980s. Cocaine use has been associated with high-risk sexual behaviors. Also of concern have been declines in the resources allocated for public health programs involved in the control of syphilis over the previous two decades. During the 1980s, the majority of cases were initially observed in homosexual men, and subsequently in intravenous drug-using individuals. In the 1980s, more than 40% of infected men reported other men as their sexual contacts. With the onset of the AIDS epidemic and the associated changes observed in sexual behavior among homosexual men, the proportion of syphilis cases attributable to this segment of the population declined dramatically. Subsequently, mirroring trends in the HIV epidemic, a growing proportion of syphilis cases were seen in inner city, heterosexual, drug-using populations.

Since 1990, however, there has been a consistent decline in reported cases of syphilis in the U.S. In 1996, 11,387 cases of P&S syphilis were reported, the lowest number since 1959. The majority of these cases were from the South. As observed with many other STDs, there were marked discrepancies in the proportion of cases among specific races or ethnic groups. The rate of P&S syphilis among African-Americans was nearly 50 times that among whites. The rate of P&S syphilis among Hispanics was 3 times that seen in whites. Cases of congenital syphilis have also declined, but of the 1160 reported cases in 1996 where the race or ethnicity of the mother was known, 90% of all reported cases were either African-American or Hispanic, yet these groups comprised only 23% of the female population. Certainly, biases in reporting explain some of these findings, but cannot explain them completely. Other more fundamental risk factors for STDs that are also associated with race or ethnicity, such as socioeconomic status, access to quality health care and health education, and others, may also play an important role.

B. Clinical Manifestations

Syphilis is unique from other STDs in its ability to spread throughout the body in the bloodstream of infected individuals, resulting in involvement of multiple distant organs including the central nervous system, bones, arteries, and other sites. Because it is a systemic disease, with sometimes nonspecific signs and symptoms, syphilis can mimic many other disease processes, and patients may be easily misdiagnosed if syphilis is not considered and ruled out by appropriate diagnostic testing. Transmission usually occurs as a result of sexual contact with an infected individual who has the typical mucosal lesions associated with primary and secondary syphilis. Syphilis can also be transmitted via contact with other infectious lesions seen in the mouth or on the skin in secondary syphilis, or it may be transmitted transplacentally from mother to fetus, resulting in congenital syphilis. In addition, infected mothers can transmit syphilis to their newborns through breast-feeding.

Following contact with an infected individual, *T. pallidum* begins to multiply at the site of entry of the organism, resulting in the formation of a papule, after an incubation period of 10 to 90 days. The papule eventually converts into a painless superficial ulcer, known as a chancre, characteristic of primary syphilis. Chancres are highly contagious lesions, owing to the presence of a high concentration of *T. pallidum*. During this period, infected individuals often develop local lymph node enlargement, but often times these signs go completely unnoticed. After a period of 2 to 6 weeks, the chancre usually heals spontaneously.

Subsequent spread and multiplication of the organism in other organs results in findings characteristic of secondary syphilis. These signs and symptoms of secondary syphilis usually do not appear until after a 2- to 24-week period following resolution of the primary stage, during which the infected individual may be completely asymptomatic. Individuals with secondary syphilis typically develop a rash but often experience other symptoms such as low-grade fever, malaise, headache, sore throat, generalized lymph node enlargement, and muscle aches. Patients may also develop bone inflammation and hepatitis, both of which are usually asymptomatic. The rash associated with secondary syphilis usually begins with a mild, transient macular rash, which often goes unnoticed. This quickly evolves into a diffuse, symmetric, papular rash, which involves the entire trunk and extremities, classically including the palms of the hands and the soles of the feet. The papules are red or reddish-brown, discrete, and often scaly in appearance. Patients typically experience severe itching. Other typical mucocutaneous lesions, known as condylomata lata, are large, raised white or grayish patches that occur in warm, moist areas, such as the mouth, vagina, or rectum. These skin lesions contain a high concentration of treponemes and are thus highly contagious. Hair loss as well as hyper- or hypopigmentation of the skin can occur also. As seen in primary syphilis, the signs and symptoms of secondary syphilis usually resolve spontaneously after a 2- to 6-week period, even without treatment, as a result of the host's immune response.

About 25% of individuals with primary or secondary syphilis remain infected and may experience multiple relapses of the secondary stage, usually within the first year of infection. After that, approximately one-third of patients with primary or secondary syphilis clear the infection completely without treatment. The remainder enter a stage referred to as latent syphilis, for which there are no signs or symptoms, and a diagnosis of syphilis can only be made by serologic screening. Latent syphilis can persist for decades. This stage is arbitrarily divided into early latent and late latent stages. By definition, patients with early latent syphilis are asymptomatic and have been infected for less than 1 year. Those with late latent syphilis are asymptomatic and have been infected for more than 1 year. A diagnosis of early syphilis can be made if within the preceding year the individual has had a documented seroconversion, unequivocal symptoms of primary or secondary syphilis, or a sex partner with documented primary, secondary, or early latent syphilis. Individuals with syphilis are believed to be infectious and can thus transmit syphilis to their partners during the primary, secondary, and early latent stages. If the duration of latent syphilis infection cannot be documented for certain, then patients are assumed to have late latent syphilis for the purposes of treatment. One-half of patients who develop latent syphilis will continue to have serologic evidence of infection but will never develop late complications of the disease. The other half may remain asymptomatic for decades but will ultimately develop late manifestations of syphilis. Individuals with late latent syphilis are immune to reinfection with *T. pallidum*.

This latter group of individuals, if untreated, eventually experience disease progression and develop what is referred to as tertiary syphilis. Individuals with tertiary syphilis may have involvement of their skin, bones, central nervous system, heart, and arteries, as well as other sites. At this stage, syphilis may be characterized by the presence of destructive granulomatous lesions known as gummas. Gummas may be found in the bones and skin, but they may also be found in other organs, including rarely the heart and digestive tract. The pathogenesis of gumma formation is not known, but gummas usually result in extensive damage to surrounding tissues. Tertiary syphilis involvement of the central nervous system and cardiovascular system may also result from direct invasion by treponemes, without gumma formation. Cardiovascular involvement is thought to result from the multiplication of treponemes in the aorta and proximal coronary arteries. The associated arteritis can lead to severe aortic valve insufficiency, aneurysm formation, and myocardial infarction. The neurologic effects of tertiary syphilis take two forms. Meningovascular syphilis results from involvement of the meninges, the tissues surrounding the brain. The parenchymatous form affects the brain, referred to as general paresis, and spinal cord, known as tabes dorsalis, directly. Thus, the neurologic manifestations of tertiary syphilis may include meningitis, stroke, dementia, cranial nerve deficits, sensory disturbances, and paralysis. Approximately 80% of deaths from syphilis are attributable to cardiovascular involvement, and the remainder result from neurologic disease. Fortunately, tertiary syphilis has become extremely rare since the introduction of penicillin.

Congenital syphilis is a potentially devastating disease, resulting from transplacental transmission of syphilis from an infected mother to her fetus. Approximately 50% of infected fetuses are spontaneously aborted or stillborn. Those that survive to delivery may exhibit a wide range of signs and symptoms of syphilis. Newborns with symptoms before the age of 2 years have early congenital syphilis, and they develop skin lesions, mucous membrane involvement often with copious secretions, bone involvement, severe anemia, and enlargement of the liver and spleen. Children with late congenital syphilis may not have symptoms until after the age of 2 years, at which time they may develop interstitial keratitis with subsequent blindness, tooth deformities, deafness, neurosyphilis, cardiovascular involvement, and skeletal deformities. Pregnant women who receive prenatal care are routinely screened for syphilis given the terrible consequencs of infection for the newborn.

Individuals with syphilis and concurrent HIV infection appear to have a similar experience with regard to primary and secondary syphilis, with one notable exception. Concurrent HIV infection has been recognized to have a significant impact on neurologic involvement in individuals with syphilis. Numerous case series have documented rapid progression from early syphilis to neurosyphilis in patients infected with both HIV and syphilis. Many of these patients develop meningitis, cranial nerve deficits, including hearing loss and blindness, and even stroke, often in the context of therapeutic failure with conventional doses of penicillin for primary or secondary syphilis. Thus, evaluation and treatment recommendations vary somewhat for individuals with both HIV infection and syphilis.

C. Diagnosis

The most sensitive and specific test for the diagnosis of primary syphilis is the identification of treponemes on dark-field microscopy, when performed by an experienced observer on an adequate sample of fluid obtained from the surface of a chancre. *Treponema pallidum* has a characteristic corkscrew appearance on dark-field microscopy, and false positive results do not occur in experienced hands. Serologic tests can be useful in the diagnosis of primary syphillis, although false negative results occur 10 to 20% of the time. The RPR (rapid plasma reagin) and the VDRL (Venereal Diseases Research Laboratory) are referred to as "nontreponemal tests," because they do not measure antibodies specific to treponemal components. As a result, these assays may occasionally be positive in other disease states and are less specific than treponemal tests. These assays are also more likely than treponemal tests to be negative in primary syphilis and thus should not be used to rule out a diagnosis of primary syphilis. The RPR titer does, however, correlate with disease activity and often becomes negative following adequate treatment, although it may remain positive at a low titer throughout one's life. Thus, even when the diagnosis of syphilis is made by other means, it is useful to obtain an RPR to monitor subsequent disease activity. Tests such as the FTA-ABS (fluorescent treponemal antibody absorbed) or the MHA-TP (microhemagglutination assay for antibody to *T. pallidum*) are referred to as treponemal tests. These assays are more likely to be positive in primary syphilis than the RPR or VDRL and remain positive for life, but they are not useful in monitoring disease activity. Because these assays are specific for syphilis, but are more costly and labor intensive, they are not used for screening but rather to confirm a diagnosis of syphilis when an individual has a positive RPR or VDRL, for stages other than primary syphilis.

Secondary syphilis can be diagnosed by dark-field examination of material taken from skin lesions of individuals with suspected secondary syphilis. Serologic assays are much more reliable in diagnosing secondary syphilis compared to primary syphilis, however, and are the preferred method of diagnosis in this setting because of their relative convenience. The RPR is always positive in secondary syphilis, usually with a high titer. The treponemal tests are also always positive in secondary syphilis, and may be used to confirm that a positive RPR is in fact due to syphilis. The RPR is then used to monitor the response to therapy.

Because individuals with latent syphilis are asymptomatic, latent syphilis is usually diagnosed as a result of screening with an RPR or VDRL, followed by confirmation with a treponemal test. A diagnosis of latent syphilis is easily made if an individual recalls a recent history of a chancre or skin rash for which he or she did not seek medical care and is currently asymptomatic. Unfortunately, most asymptomatic individuals with a positive RPR are unable to provide a diagnostic medical history. For public health purposes, these individuals are assumed to have latent syphilis and are thus treated, although it may be the case that they have a positive serology despite previous adequate therapy or may have partially treated syphilis. Most local public health departments keep a record of all positive syphilis serologies and attempt to maintain records on treatment and subsequentesponse to therapy. If a current RPR titer is positive at two or more dilutions less than, or fourfold lower than, previously documented (e.g., from 1:16 to 1:4), then the individual has likely had an appropriate response to prior therapy.

A diagnosis of neurosyphilis is easily made when there are neurologic findings in the setting of positive serologies and abnormal cerebrospinal fluid (CSF). However, the presence of only two of these three criteria is usually sufficient to mandate treatment for neurosyphilis. The CSF in patients with neurosyphilis may be normal or may exhibit any combination of the following findings, including an elevated opening pressure, an elevated white blood cell count (usually 10 to 200 mononuclear cells per cubic millimeter), an elevated protein concentration, or a positive CSF VDRL. A positive CSF VDRL is diagnostic of neurosyphilis, but this test is not sensitive and is often negative in neurosyphilis. Stroke, dementia, and other neurologic abnormalities more often than not are due to etiologies other than syphilis, but may coincidentally occur in the presence of positive serology. If, however, a patient has neurologic findings possibly due to syphilis in the presence of positive RPR or treponemal test, these individuals should be treated for neurosyphilis, even if the CSF is normal, and evaluated for other disease processes as indicated (e.g., carotid artery disease in the case of stroke). CSF examination is most useful for confirming the diagnosis of neurosyphilis when the CSF is abnormal in the context of positive serology. If an individual has a positive serology and abnormal CSF, the individual should be treated for neurosyphilis even in the absence of clinical findings of neurosyphilis. This latter scenario is most common in HIV-infected individuals. In the absence of gummas, diagnosis of cardiovascular disease from tertiary syphilis is usually made on clinical grounds in the presence of positive serology or a history of syphilis and characteristic findings on angiography. Other forms of tertiary syphilis are usually diagnosed on the basis of the identification of a gumma.

D. Treatment

Parenteral penicillin G is the preferred treatment for all stages of syphilis. The exact preparation used, the dose, and the duration of therapy depend on the stage of syphilis and certain characteristics of the infected individual. HIV may alter the response to therapy, and treatment failures have been reported in HIV-infected individuals. It is thus recommended that all individuals with syphilis be tested for HIV, and in areas of high prevalence of HIV, individuals with syphilis that are negative on initial HIV testing should be retested 3 months later. Adults with primary, secondary, or early latent syphilis should be treated with a single dose of 2.4 million units of benzathine penicillin intramuscularly. Individuals with primary or secondary syphilis who have signs or symptoms of neurologic or ophthalmic involvement should undergo CSF evaluation and slit-lamp examination to rule out neurosyphilis and ocular involvement, respectively. In the absence of neurologic findings, routine lumbar puncture is not indicated, as CSF is often abnormal in primary and secondary syphilis and does not necessarily reflect true neurosyphilis. Patients should be reevaluated clinically and serologically at 6 and 12 months. If patients have persistent clinical findings, a rising RPR titer, or failure of the titer to decline fourfold within 6 months of treatment for primary or secondary syphilis, then they most likely represent treatment failures or have been reinfected. These patients should undergo lumbar puncture to rule out neurosyphilis and should be treated with three intramuscular injections of 2.4 million units each of benzathine penicillin 1 week apart, if neurosyphilis is not present. Patients with primary, secondary, or early latent syphilis who are not pregnant and are allergic to pencillin may alternatively be treated with doxycline 100 mg orally twice a day for 2 weeks. Treatment failures are more likely with doxycycline, and these patients require especially close follow-up. Pregnant patients who are allergic to penicillin should be desensitized and treated with pencillin.

About one-third of patients with primary syphilis

and two-thirds of patients with secondary syphilis develop an unusual reaction to treatment known as the Jarisch-Herxheimer reaction. This reaction typically consists of fever, chills, headache, worsening rash, and even hypotension in severe cases. The onset is usually within the first 4 hr following the initiation of therapy, and the reaction resolves within 24 hr. It is believed to result from the release of treponemal components, resulting in an endotoxin-like reaction. The Jarisch-Herxheimer reaction may be confused with an allergic reaction to penicillin but is not an indication for discontinuing therapy. Instead, treatment should continue and the patient should be treated with nonsteroidal anti-inflammatory medication, in addition to other supportive measures as indicated.

Patients with late latent syphilis, or latent syphilis of unknown duration, should be treated with a total of 7.2 million units of benzathine penicillin administered as three doses of 2.4 million units each, intramuscularly, 1 week apart. Evaluation should include a lumbar puncture to rule out neurosyphilis for all patients with latent syphilis if they have neurologic or ophthalmic signs or symptoms, evidence of active tertiary syphilis, evidence of treatment failure as described above, or HIV infection with late latent syphilis or syphilis of unknown duration. Patients should be reevaluated clinically and serologically at 6, 12, and 24 months. An alternative regimen for nonpregnant, penicillin-allergic patients is doxycycline 100 mg orally, twice daily for 4 weeks. As with early syphilis, penicillin-allergic pregnant patients should undergo desensitization, followed by treatment with penicillin. Treatment for tertiary syphilis other than neurosyphilis is the same as that for late latent syphilis, except that all patients with tertiary syphilis should undergo lumbar puncture to rule out neurosyphilis.

Neurosyphilis can occur at any stage of syphilis, and patients with any evidence of meningitis, ophthalmic or auditory symptoms, or cranial nerve palsies should undergo a lumbar puncture to evaluate for neurosyphilis. Because of an increased risk of neurosyphilis, HIV-infected individuals with late latent syphilis or syphilis of unknown duration, patients with evidence of treatment failure, and patients with active tertiary syphilis should also undergo a CSF examination prior to therapy, even in the absence of signs or symptoms of neurosyphilis. Those diagnosed with neurosyphilis should receive aqueous crystalline penicillin G 18 to 24 million units a day, administered as 3 to 4 million units intravenously every 4 hr for 10 to 14 days. This course should be followed with a dose of 2.4 million units of benzathine penicillin intramuscularly, to provide a duration of therapy comparable to that for latent syphilis. There are no adequate alternatives to penicillin in the treatment of neurosyphilis. Individuals with neurosyphilis who are allergic to penicillin should undergo desensitization, followed by the recommended course of penicillin. Many experts also recommend a follow-up lumbar puncture 6 months following the completion of therapy, to ensure an adequate response.

Partner notification and treatment are as important for cases of syphilis as for other STDs. Transmission of syphilis occurs in approximately 50% of cases where there is direct sexual contact with lesions of an individual who has primary or secondary syphilis. The Centers for Disease Control and Prevention recommend that all individuals exposed within 90 days preceding the diagnosis of primary, secondary, or early latent syphilis in a sex partner be treated presumptively, given the concern for false negative serologies shortly after exposure. For individuals exposed more than 90 days prior to the diagnosis of primary, secondary, or early latent syphilis in a sex partner, the CDC recommends presumptive treatment if serologic testing is not immediately available and follow-up cannot be guaranteed. Long-term sex partners of individuals who are diagnosed with a late form of syphilis should be evaluated for syphilis clinically and by serologic testing, then treated accordingly.

E. Organism and Pathogenesis

Treponema pallidum is one of several treponemal species associated with human disease. Pathogenic species belonging to the genus *Treponema* are responsible for yaws, endemic syphilis, pinta, periodontal disease, as well as venereal syphilis discussed here. *Treponema pallidum* is an obligate human parasite, with no animal or environmental reservoirs. It is a helical corkscrew-shaped gram-negative bacterium, 6 to 20 μm in length and 0.10 to 0.18 μm in diameter, placing it below the resolution of light microscopy. The unstained organism is best visualized by dark-field or phase-contrast microscopy. *Treponema pallidum* generally stains poorly with many dyes but can be visualized using silver impregnation techniques. Like other gram-negative bacteria, *T. pallidum* has an outer membrane, an inner membrane, and a thin cell wall consistig of peptidoglycan. The outer membrane is somewhat unique, however, in that it lacks lipopolysaccharide, thus rendering it more susceptible to damage resulting from physical disruption or detergent use during handling. The outer membrane also contains relatively few proteins, which may account for the limited antibody response seen with this pathogen. *Treponema pallidum* is also unique in that it has flagella, located in between the inner and outer membranes, which are responsible for its characteristic motility, consisting of rapid rotation along its longitudinal axis, as well as bending and flexing. Such motility is thought to play an important role in the organism's ability to invade and disseminate.

Treponema pallidum is a fastidious organism and has not been successfully cultured *in vitro,* and thus diagnosis depends on direct visualization and serologic testing. For study purposes, viable organisms can, however, be maintained for several weeks in tissue cell culture. The organism appears to be microaerophilic and is very sensitive to environmental conditions, being easily inactivated by mild temperature fluctuations, desiccation, and chemical agents including most disinfectants. These characteristics have limited study of the organism. Recently, the complete genome of *T. pallidum* was successfully sequenced. The genome consists of 1,138,006 base pairs containing 1041 predicted coding sequences. The information gained from sequencing has helped to elucidate systems for DNA replication, transcription, translation, and repair, as well as possible virulence factors and metabolic processes. Potential virulence factors include a family of 12 membrane proteins and several putative hemolysins. Catabolic and biosynthetic capabilities appear to be limited. Glucose is the main carbon and energy source for *T. pallidum,* although pyruvate appears to be an alternative.

The pathogenesis of *T. pallidum* is not well understood. Following penetration of mucosal surfaces, *T. pallidum* appears to adhere to host cells and then begin multiplication. Treponemes have been shown to adhere to a variety of cell types, perhaps through an interaction with fibronectin and host cell receptors. Following multiplication, the organisms disseminate via the circulation and invade distant organs, aided by their unique motility. Lesion resolution is thought to be the result of cell-mediated immunity, involving phagocytosis by macrophages activated by lymphokines and opsonic antibodies, released from antigen-specific sensitized T and B cells. Despite the destruction of millions of organisms, some persist and eventually cause the later stages seen in syphilis. There are several possible mechanisms by which treponemes escape the host immune response, probably the most important of which is the paucity of outer membrane proteins that would ordinarily provide ample targets for the immune response. In addition, treponemes may also have some surface components that inhibit complement-mediated killing. Secondary syphilis results from the dissemination and further multiplication of persistent treponemes. Some manifestations may also result from immune complex deposition, especally with regard to the skin and kidney involvement seen at this stage. Following a period of latency, persistent treponemes go on to invade the central nervous system, the cardiovascular system, and other organs in patients with tertiary syphilis. Here, much of the damage seen is likely due to the organism's invasive capabilities and the associated delayed hypersensitivity response of the host.

See Also the Following Articles

AIDS, Historical • Diagnostic Microbiology • Surveillance of Infectious Diseases

Bibliography

Black, C. M. (1997). Current methods of laboratory diagnosis of *Chlamydia trachomatis* infection. *Clin. Microbiol. Rev.* **10**, 160–184.

Centers for Disease Control and Prevention, Division of STD Prevention. (1997). "Sexually Transmitted Disease Surveillance, 1996." U.S. Department of Health and Human Services, Public Health Service. September.

Centers for Disease Control and Prevention. (1998). Guidelines for treatment of sexually transmitted diseases. *MMWR* 47(No. RR-1).

Cohen, M. S. (1998). Sexually transmitted diseases enhance HIV transmission: No longer a hypothesis. *Lancet* **351** (Suppl. 3), 5–7.

Drugs for sexually transmitted diseases (1995). *Med. Lett. Drugs Therap.* 37(964), 117–122. Dec22.

Erbelding, E., and Quinn, T. C. (1997). The impact of antimicrobial resistance on the treatment of sexually transmitted diseases. *Infect. Dis. Clin. North Am.* **11**(4), 889–903.

Gaydos, C. A., and Quinn, T. C. (1998). Ligase chain reaction for detecting sexually transmitted diseases. *In* "Rapid Detection of Infectious Agents" (Specter *et al.*, eds.), Plenum, New York.

Gerbase, A. C., Rowley, J. T., and Mertens, T. E. (1998). Global epidemiology of sexually transmitted diseases. *Lancet* **351**(Suppl. 3), 2–4.

Holmes, K. K., Mardh, P. A., Sparling, P. E., Wiesner, P. I., Cates, W., Lemon, S. M., and Stamm, W. E. (eds.) (1999). "Sexually Transmitted Diseases." McGraw-Hill, New York.

Hook, E. W. (1998). Is elimination of enzemic syphilis transmission a realistic goal for the USA? *Lancet* **351**(Suppl. 3), 19–21.

Jackson, S. I., and Soper, D. E. (1997). Sexually transmitted diseases in pregnancy. *Obstet. Gynecol. Clin. North Am.* **24**(3), 631–644.

Kalman, S., Mitchell, W., Marathe, R., Lammel, C., Fan, J., Hyman, R. W., Olinger, L., Grimwood, J., Davis, R. W., and Stephens, R. S. (1999). Comparative genomes of *Chlamydia pneumoniae* and *C. trachomatis*. *Nat. Genet.* **21**(4), 385–389.

Mayaud, P., Hawkes, S., and Mabey, D. (1998). Advances in the control of sexually transmitted diseases in developing countries. *Lancet* **351**(Suppl. 3), 29–32.

Mindel, A. (1998) Genital herpes—How much of a public health problem? *Lancet* **351**(Suppl. 3), 16–18.

Skin Microbiology

Morton N. Swartz
Massachusetts General Hospital and Harvard Medical School

I. Host Determinants of Colonization and Invasion
II. Bacterial Adherence as a Factor in Determining Ecological Niches
III. Resident (Normal) Skin Flora: Gram-Positive Bacteria
IV. Resident (Normal) Skin Flora: Gram-Negative Bacteria
V. Resident (Normal) Skin Flora: Fungi
VI. Transient Skin Flora: Gram-Positive Bacteria
VII. Transient Skin Flora: Gram-Negative Bacteria
VIII. Other Forms of Microbiological Involvement in Infectious Processes of the Skin

GLOSSARY

acne An inflammatory disease involving the pilosebaceous unit.

carbuncle A larger, painful, deeper, more serious, and more nodular lesion than a furuncle. It is due to *Staphylococcus aureus* and characteristically occurs on the nape of the neck, upper lip, back, or thighs and progresses to drain externally around multiple hair follicles.

cellulitis An acute edematous, suppurative, spreading inflammation of the deep subcutaneous tissues producing an area of erythema and tenderness with indistinct margins. The most common etiologic agents are group A streptococci and *Staphylococcus aureus.*

ecthyma A group A streptococcal process which begins much like impetigo at a site of minor trauma but extends through the epidermis, producing a shallow ulcer covered by a crust.

erysipelas An acute superficial form of cellulitis involving the dermal lymphatics, usually caused by group A streptococci, and characterized by a bright red, edematous, spreading process with a raised, indurated border.

furuncle A painful erythematous nodule (boil) caused by *Staphylococcus aureus* and formed by localized inflammation of the dermis and subcutaneous tissue surrounding hair follicles.

gas gangrene An acute severe infection, usually resulting from dirty penetrating wounds in which the subcutaneous tissues and muscles contain gas and a serosanguineous exudate. The process is a histotoxic infection due to *Clostridium perfringens, C. septicum,* or other *Clostridium* species. (Also known as *clostridial myonecrosis.*)

glabrous Bare, without hair.

impetigo A contagious pyoderma caused by direct inoculation of group A streptococci or *Staphylococcus aureus* into superficial abrasions of the skin. The lesion is confined to the epidermis and initially consists of a fragile vesicopustule with an erythematous halo progressing to a yellow-brown crust.

lymphangitis Inflammation of lymphatic vessels. Acute lymphangitis is evidenced by painful subcutaneous red streaks along the course of lymphatics.

necrotizing fasciitis A fulminating form of cellulitis that spreads to involve the superficial and deep fascia causing thrombosis of subcutaneous vessels and gangrene of overlying tissues. Type I necrotizing fasciitis is commonly due to a mixture of one or more anaerobes with one or more facultative species; Type II is due to group A streptococci.

paronychia An infection involving the folds of tissue surrounding the nail.

pilosebaceous Pertaining to hair follicles and sebaceous glands.

pyoderma Any purulent skin disease.

THE HUMAN SKIN SURFACE is a distinct ecosystem made up of a large number of microbial species and the chemical and physical environment with which they interact. Since the cutaneous environment varies considerably from sector to sector (glabrous as opposed to hairy regions; dry areas compared to intertriginous or moist ones), the interactions have added complexity. The normal skin flora comprises a permanent group of microbial species, the so-called resident flora, and variable types of temporary surface colonizers, the so-called transient flora. It is a set of important reciprocal effects that determines ultimately what microorganisms will persist as permanent colonizers. These include, on the human host's part, the structural integrity of the epidermis, biochemical and immunological defenses, local anatomic features (hair, pilosebaceous apparatus, etc.), and alterations produced by physiological changes (increased sebum secretion in adolescence) and environmental factors (increased surface moisture secondary to skin occulusion or heightened environmental temperature). On the part of the colonizing microorganism they include, in addition to the foregoing, important ecological determinants such as availability of nutrients, specific ligands for attachment to host keratinocytes, and capacity to survive adverse microbial interactions (competitive colonization suppression by other species by virtue of the latter's growth advantage through use of available nutrients in this particular niche or by virtue of the latter's ability to produce antibiotics that suppress the growth of another species). For primary pathogens among the transient flora, which includes microorganisms such as group A streptococci, the intrinsic virulence of the species is a major determinant of the development of skin lesions and invasive infection.

I. HOST DETERMINANTS OF COLONIZATION AND INVASION

A. Structural Integrity of the Skin

The most differentiated layer (stratum corneum) of the epidermis functions primarily as a defense against inordinate water loss from the body, but it also serves as protection against invasion by microorganisms through its relative dryness, an inhospitable environment for pathogens requiring moisture for sustained growth. Also, through continuous desquamation of the outermost keratinocytes, adherent surface microorganisms are constantly shed. In contrast to the intact stratum corneum as a strong defense against invasion by normal resident flora or by transient pathogenic colonizers, cracks produced by trauma or various primary skin diseases can provide ready ingress, particularly in moist areas, for these microorganisms.

B. Biochemical Defenses

Free fatty acids, particularly long-chain polyunsaturated ones such as linolenic and linoleic acid in the skin may have a role (not conclusively proven) in preventing permanent skin colonization by species such as *Staphylococcus aureus* and *Streptococcus pyogenes* (group A streptococcus). Removal of the skin surface lipids with solvents increases the duration of survival of *Staphylococcus aureus* on the skin, an effect reversed by replacement of lipid. It has been suggested that this inhibitory action of such unsaturated fatty acids accounts for the relatively poor survival of *S. aureus* on the skin surface vis-à-vis coagulase-negative staphylococci. Fatty acids released from the triglycerides of sebum (e.g., by hydrolysis effected by resident flora) may act bacteriostatically by lowering local pH as well. In general, these antimicrobial effects are limited to transient microbial pathogens and not to resident flora. A category of the resident flora, propionibacteria, is suppressed in growth to some extent by linolenic and linoleic acids. This effect may vary with species: *Propionibacterium acnes* is inhibited at higher concentrations of both fatty acids than is *P. granulosum*, perhaps accounting for lower population densities of the latter than of the former on the skin surfaces. Gram-negative and gram-positive bacteria have roughly similar susceptibilities to fatty acids. Thus, the presence of fatty acids on the skin is unlikely to be the major factor in accounting for the low incidence of gram-negative bacteria among the resident cutaneous flora.

C. Immunologic Defenses

Immunoglobulins, Langerhans cells of the epidermis, and cytokines (particularly interleukin 1α, IL-1α) may play roles in immunologic defense at the cutaneous surface and alter the composition and invasive capacity of some pathogens of the transient flora. In view of the very low levels of specific immunoglobulins on the normal skin surface, they are unlikely to provide any significant antibacterial action unless a primary exudative dermatosis is present. Langerhans cells, a type of dendritic cell involved in immune surveillance for bacterial and other antigens that may breach the defenses at the stratum corneum, provide for early antigen recognition that initiates an anamnestic immune response (particularly cell-mediated). The cornified epithelial cells of the skin surface contain large quantities of IL-1α that normally are not released. However, their release is stimulated by local inflammation or bacterial action, thus setting off autocrine (and paracrine) effects with production of further cytokines and attraction of inflammatory cells to the skin surface.

D. Anatomic Features

Although the skin can be considered as a single organ, its regional differences in hair, depth of cornification, numbers of sweat and sebaceous glands, and local moisture content (e.g., in contiguous surfaces) provide a variety of differing ecological niches. The majority of resident bacteria on the skin are located at two microscopic sites: within the more cornified layers of the stratum corneum and within the hair follicles.

1. *Stratum Corneum*

Cocci and bacilli, often existing in microcolonies and representing the principal genera of the resident flora, have been observed on the surface of the stratum corneum in skin scrubbing experiments, cellulose tape strippings through successive layers of the epidermis, in histologic sections of skin, and by scanning electron microscopy. Large variations occur in numbers of organisms in different anatomic areas. Abnormal keratinization and hyperproliferation, as in psoriasis, have been associated with increased colonization with *S. aureus*.

2. *Hair Follicles*

The greatest concentration of microorganisms on the skin is in the hair follicles. Members of the anaerobic genus *Propionibacterium* are distributed in the follicle closer to the skin surface than are coagulase-negative staphylococci, another component of the resident flora, which colonize the deeper portion of the follicle in larger numbers. *Propionibacterium acnes,* the most numerous anaerobic coryneform bacterium found on the human skin, is found in largest numbers in areas rich in sebaceous glands (e.g., forehead, alae nasi, scalp), and in much smaller numbers in the dry areas of the arm and leg. Abnormal keratinization in hair follicles causing occlusion, associated with abnormalities of sebum production leading to increased proliferation of resident *P. acnes* and subsequent inflammation, are pathogenic features in the development of acne lesions.

3. *Gross Regional Variations*

Resident flora vary in population densities and predominant species among various localized anatomic sites, which can be viewed as specialized ecological niches that differ from the more extensive flat, exposed surfaces. Such niches include more occluded areas such as axillae (favoring large populations of either coryneform bacteria or staphylococci), toe-web spaces (colonized commonly with large numbers of gram-negative bacilli and dermatophytic fungi), and the groin, as well as the scalp (favoring proliferation of large numbers of staphylococci, propionibacteria, and *Pityrosporum* species).

E. Physical Factors Affecting Composition of Skin Flora: Role of Hydration

Increased water content at the skin surface, when the relative humidity approaches 100%, as when the skin of the arm is occluded with a plastic wrap, leads to rapid microbial growth to high population densities. Under such conditions, counts of coagulase-negative staphylococci, lipophilic coryneforms, and gram-negative bacilli increase by four orders of magnitude in 72 hr. Similarly, densities of skin flora are highest in workers in hot, humid climates, pyodermas are more frequent during hot, humid seasons of the year, and infections by dermatophytes become more problematic under similar conditions. A normally less numerous component of the skin flora, *Acinetobacter* spp., become more numerous in the summer heat. The role of hydration in altering numbers of skin bacteria and lesional pathogenesis has been shown in the case of *Pseudo-*

monas aeruginosa. Under conditions of experimental occlusion for up to 7 days, exposure of normal skin to 10^6 *P. aeruginosa* has not produced lesions; in contrast, the addition of dressings producing hyperhydration caused development of papules and pustules. These results are congruent with the observation of the occurrence of widespread folliculitis in individuals using "hot tubs" and jacuzzis contaminated with *P. aeruginosa*.

As mentioned earlier, an occluded area such as an axilla, by virtue of its increased hydration, favors a heavy growth of staphylococci and corynebacteria. Several species of coryneform bacilli, under conditions of excessive sweating, are able to colonize hair shafts, grow extensively as visible yellow- or red-colored colonies thereon, producing the unsightly and maladorous (but unimportant) condition known as trichomycosis axillaris. The pungent axillary odor appears to result from metabolism of testosterone locally in the apocrine glands by coryneform species.

II. BACTERIAL ADHERENCE AS A FACTOR IN DETERMINING ECOLOGICAL NICHES

Bacteria tend to adhere strongly to cutaneous surfaces since vigorous washing removes less than one-half the normal flora. Thus, adhesins have been assumed to play a role, but specific components of the microorganisms adherent to corneocytes are not yet defined. It is known that *S. aureus* adhere in greater numbers to the nasal mucosa of normal nasal carriers of this species, and from the nose they may be spread as transients to the skin surface. The specific corneocyte adhesin(s) of chronic nasal carriers of *S. aureus* with which this organism's ligand interacts is not yet defined, but adherence to corneocytes can be blocked by pretreatment of these cells with staphylococcal teichoic acid. Abnormal skin, as in patients with atopic dermatitis, promotes better adherence of *S. aureus* to the cornified squamous epithelium.

Selectivity exists in adherence of bacterial cells to epithelial surfaces. *Staphylococcus epidermidis* adheres more readily to skin cells than to urinary tract cells. Strains of *S. epidermidis,* common constituents of cutaneous flora, have a capacity to adhere to vascular catheters extending through the skin surface, accounting for the prominent role of this microorganism in catheter-related bloodstream infections. Similarly, selectivity exists in regard to adherence of members of the transient skin flora. Strains of *S. pyogenes* serotypes involved in impetigo and related superficial cutaneous infections adhere better to skin cells than do strains of *S. pyogenes* isolated from the throat; likewise pharyngeal isolates adhere better to buccal epithelial cells than to skin cells.

Competition between selected strains of the same species may occur not only at the level of nutrient utilization but also at the level of adherence. Such competitive adherence was shown in the 1960s when it was noted that, during an outbreak of *S. aureus* infections, neonates whose skin had been precolonized naturally by a strain of *S. aureus* (strain 502A) acquired from a nurse were protected against the more virulent strain (type 80/81) then ambient and causing infection in the newborn. Deliberate attempts at colonization with strain 502A were subsequently shown to be effective in preventing infection with more virulent epidemic strains of *S. aureus,* but this approach was abandoned when it became evident that strain 502A could itself sometimes produce skin infections when introduced.

The skin acquires surface bacteria shortly after birth. For example, the nose is sterile in 90–100% of infants at birth. Within 72 hr, 40% of infants have become colonized with *S. aureus*. *In vitro* results parallel these: binding of *S. aureus* to nasal epithelial cells obtained during the first 4 days of life is very low, reaching a level comparable to that of adult cells on the fifth day.

A relationship between pathogenicity and adherence has been suggested by the gradation in affinity of various candidal species to corneocytes: *Candida albicans* > *C. stellatoidea* > *C. parapsilosis* > *C. tropicalis* > *C. krusei* > *C. guilliermondii*.

III. RESIDENT (NORMAL) SKIN FLORA: GRAM-POSITIVE BACTERIA

The principal components of the resident cutaneous flora consist of staphylococci and coryneform bacteria (to a lesser extent *Acinetobacter* spp. and

TABLE I
Resident (Normal) Skin Flora

Gram-positive bacteria
- Coagulase-negative staphylococci (numerous species)
- *Staphylococcus aureus*
- Coryneform bacteria
  - *Corynebacterium* spp.: *C. minutissimum, C. jeikeium,* group CLC, group D2
  - *Rhodococcus* spp.
  - *Brevibacterium* spp.
  - *Dermobacter* spp.
  - *Propionibacterium acnes*
  - *Micrococcus* spp.

Gram-negative bacilli
- *Acinetobacter* spp.: *Acinetobacter calcoaceticus–baumannii* complex, *A. johnsonii, A. lwoffii*
- *Pseudomonas aeruginosa* (localized to toe-webs)

Fungi
- *Malassezia furfur* (yeast forms formerly designated *Pityrosporum ovale* and *P. orbiculare*)

Micrococcus spp.) and lipophilic yeasts, primarily *Malassezia furfur* (Table I).

A. Coagulase-Negative Staphylococci

The Baird-Parker classification of staphylococci in the 1960s identified three species (*Staphylococcus aureus, S. epidermidis, S. saprophyticus*) and *Micrococcus*. By current taxonomy *Micrococcus* and *Staphylococcus* are considered as separate and distinct genera, and about 24 species of coagulase-negative staphylococci have been distinguished utilizing phenotype characteristics and, most importantly, DNA hybridization. These species have been isolated primarily from the skin (with the exception of *S. saprophyticus* being recovered from the urine) of humans and animals (Table II). Most of the human species have been isolated from normal human skin: *Staphylococcus capitis, S. cohnii, S. haemolyticus, S. hominis, S. simulans, S. warneri,* and *S. xylosus*. *Staphylococcus capitis* was identified on the scalp initially, and *S. auricularis* was first found in the human ear. *Staphylococcus schleiferi* and *S. lugdunensis* have caused bacteremias,

TABLE II
Coagulase-Negative Staphylococci

Human origin (colonizer, pathogen)	*Animal origin (colonizer, pathogen)*
Common pathogens	Isolates primarily from animals
Staphylococcus epidermidis	*S. arlettae*
S. saprophyticus (principally urinary tract)	*S. caprae*
Uncommon pathogens	*S. carnosus*
S. capitis	*S. caseolyticus*
S. caprae	*S. chromogenes*
S. cohnii	*S. delphini*
S. haemolyticus	*S. equorum*
S. hominis subsp. *hominis*	*S. felis*
S. hominis subsp. *novobiosepticus*	*S. gallinarum*
S. lugdunensis	*S. hyicus*
S. saccharolyticus	*S. intermedius*
S. schleiferi	*S. kloosii*
S. simulans	*S. lentus*
S. warneri	*S. muscae*
S. xylosus	*S. pasteurii*
Rare pathogens	*S. sciuri*
S. auricularis	*S. vitulus*

but attempts to define a cutaneous niche for the latter have been largely unsuccessful. Certain of the coagulase-negative staphylococcal species making up the normal skin flora share the property of novobiocin resistance with the urinary tract species *S. saprophyticus*.

Distribution studies of various coagulase-negative staphylococcal species among humans show some person-to-person variability and variability depending on anatomic areas studied; for example, striking variations have been observed in very similar populations of teenagers in the prevalence on the skin of *S. xylosus*. While *S. epidermidis* is the species of coagulase-negative staphylococci most widely distributed over the human body surface, various individual staphylococcal species predominate in individual anatomic areas, as determined in the small number of quantitative studies performed. Thus, in adults *S. epidermidis* is the dominant staphylococcal species isolated from the scalp, face, chest, and axilla, while *S. hominis* has a lesser, but still important, role in these areas. In contrast, on the dry areas of the arms and legs *S. hominis* is found about as frequently as *S. epidermidis* and roughly equals or exceeds the latter in percentage of total staphylococci found on these areas. *Staphylococcus saprophyticus* along with *S. cohnii* and *S. xylosus*, three closely related species, have been found particularly on the feet. Whereas before the age of puberty these three species account for about 5% of the staphylococcal flora on the feet of both males and females, at puberty in females they rise to account for about 45% or more of the staphyloccal flora. The more common location for *S. haemolyticus* has been the thighs.

The viable counts of individual coagulase-negative staphyloccal species vary from person to person and from area to area of the body. For example, viable counts of *S. epidermidis* in the axillae of 11 carriers averaged $3 \times 10^4/cm^2$, with a range of $8 \times 10^2/cm^2$ to $2 \times 10^5/cm^2$; on the forehead the mean count was $2 \times 10^3/cm^2$, with a range from 3×10^0 to 4×10^4 per cm^2. Great variations exist in staphylococcal counts on healthy feet. The greatest population density of staphylococci on the feet are about the toes, where counts exceed $1 \times 10^6/cm^2$.

Although the members of the "resident flora" are being considered here, their persistence in a given location may vary. Whereas 3 of 16 normal individuals were shown to carry *S. saccharolyticus* on the forehead for periods of at least 16, 27, and 38 months, another individual carried only small numbers of this microoganism and only on one occasion. In a study of staphylococcal isolates of personnel in Antarctica over a period of 42 weeks, the so-called resident staphylococcal flora was found to be made up of a mixture of "permanent" and "temporary" resident strains. Of 17 individuals studied, about one-third carried their own individual clone (defined by polyacrylamide gel electrophoresis and Western blotting) of *S. capitis* on their scalps for most or all of the prolonged period of study. To emphasize regional anatomic differences in resident floral strains (over and above distinctions at the species level), some of the individuals in this study carried a clone of *S. capitis* on their chin that was different from the one on their scalp. In contrast to the "relative permanence" of *S. capitis* on the scalps of individuals at an Antarctic base, isolates of *S. warneri, S. haemolyticus*, and *S. saprophyticus* recovered in the early stages of the study were not recovered subsequently, indicating that these were "temporary" resident flora constituents at these sites that had come from other reservoir sites. Dissemination or dispersal of members of the skin flora occurs, predominantly via squames or skin scales.

The coagulase-negative skin flora species colonize the newborn at somewhat differing rates. *Staphylococcus epidermidis, S. haemolyticus*, and *S. hominis* are present in the majority of samples obtained during the first week of life, even as early as the first day of life in some instances. Other species colonize later (10–32 weeks).

B. Coagulase-Positive Staphylococci

Coagulase-positive staphylococci have several characteristics by which these potential pathogens are recognized: coagulase, a heat stable nuclease, and protein A. Although several other coagulase-positive species are known (*S. intermedius*, a colonizer and pathogen of dogs; *S. hyicus*, a colonizer of pig and cattle skin and a pathogen of piglets; *S. delphini*, a cause of suppurative lesions in dolphins), only

S. aureus is a member of the skin flora and an important pathogen of humans.

The major habitat for *S. aureus* on healthy humans is in the anterior nares (30% of individuals), the perineum (15%), axillae (2–7%), and toe-webs (1–5%). Nasal carrier rates vary somewhat; in successive monthly random samples from normal individuals the nasal colonization rate varied from 19 to 40%. Factors affecting nasal carriage rates include local trauma to the nose, increased familial carriage rates, AIDS, and temporary colonization with specific "hospital strains" of *S. aureus*. *Staphylococcus aureus* is uncommonly found (<10%) on normal skin, but it occurs increasingly on the skin of individuals with diabetes and skin diseases such as atopic dermatitis, where colonization rates approach 100%. In the latter high densities of *S. aureus* can be found on lesions and extending to nearby uninvolved skin.

As in the case of coagulase-negative staphylococci, *S. aureus* may be dispersed to other body areas by squames or larger skin scales and to other patients by hand contact with health care personnel, particularly if there has been failure to use appropriate barrier precautions and hand washing in the care of infected wounds of another patient. Transient colonization of other skin areas by spray from the anterior nares of a colonized individual may occur, and this provides a reasonable explanation for the more frequent carriage of *S. aureus* on the cheek than on most other skin surfaces.

C. Coryneform Bacteria

In the past coryneform bacteria colonizing the skin have been poorly characterized and considered together as "diphtheroids." Subsequently, coryneform bacteria inhabiting human skin, gram-positive pleomorphic bacilli assuming V-forms and palisades in culture and including aerobic and anaerobic species, were classified on the basis of (1) forming small colonies or being lipophilic (much better growth on media containing added lipid) or lipolytic, or (2) forming large colonies or being nonlipophilic (growth not enhanced by added lipid). Currently, skin coryneform bacteria are divided among five principal genera on the basis of their cell wall composition (Table I): *Corynebacterium, Rhodococcus, Brevibacterium, Dermobacter,* and *Propionibacterium.*

1. *Corynebacterium Spp.*

The genus *Corynebacterium* encompasses a variety of species, including *C. diphtheriae* (found primarily on mucous membranes, but occasionally present as a secondary invader in underlying skin lesions) and most skin diphtheroids. The latter consist of *C. minutissimum* (a non-lipid-dependent, large-colony coryneform), *C. xerosis* (a large-colony, nonlipophilic coryneform), group CLC (cutaneous lipophilic corynebacteria), *C. jeikeium* or CDC group JK [a lipophilic coryneform which colonizes the skin most frequently of patients with malignancies or severe immunocompromise, characteristically multiantibiotic-resistant (with the exception of vancomycin), and a particular cause of infections and bacteremias taking origin in intravenous catheters], and CDC coryneform group D2 (urease positive and sometimes called *C. urealyticum,* like *C. jeikeium,* lipophilic).

a. *Sites of Colonization by Corynebacterium spp.*

Lipophilic *Corynebacterium* spp. are important components of the normal cutaneous flora. *Corynebacterium jeikeium* and group CLC strains are most commonly found in the anterior nares, axilla, perineum, and toe-web spaces. Group D2 *Corynebacterium* appear to be limited to the axillae. In their aforementioned niches these organisms reach high densities; for example, group CLC reaches 10^3 to 10^4 per cm^2 in the perineum, 10^5 to 10^6 per cm^2 in the axilla, and 10^6 per cm^2 in the toe interspace.

2. *Other Coryneform Genera*

Other coryneform genera include *Rhodococcus, Brevibacterium, Dermobacter,* and *Propionibacterium* (the one anaerobic genus). *Rhodococcus* and *Propionibacterium* are lipophilic, and *Brevibacterium* and *Dermobacter* are nonlipophilic and form large colonies.

Rhodococcus species (e.g., *R. equi*) are soil organisms that cause bronchopneumonia in horses, cattle, and sheep. They are rarely isolated from healthy humans, but almost all human infections, with or without histories of animal exposure, have occurred in patients with defects in cell-mediated immunity including AIDS. Such infections include broncho-pneumonia, lung abscess, brain abscess, and, rarely, skin infections.

Brevibacterium epidermidis, the species found on the normal human skin, is particularly localized to moist areas (perineum, toe-web spaces), where it may reach concentrations as high as 10^6/cm^2, but it is rarely found in oily areas (scalp, forehead) as is consistent with its nonlipolytic nature. *Dermobacter hominis,* a species similar to *B. epidermidis,* has been isolated in a few instances from the arms of healthy adults.

Propionibacterium spp. are anaerobic coryneforms present on human skin and include *P. acnes* (by far the most prevalent), *P. granulosum,* and *P. avidum. Propionibacterium acnes* is present in largest numbers in areas (scalp, forehead, alae nasi) rich in sebaceous glands, and *P. avidum* predominates in wet areas (axilla, perineum) where eccrine sweat glands are abundant.

3. *Coryneform Bacteria as Cutaneous Pathogens*

Coryneforms have been implicated as the etiology or as contributors to pathogenesis in several skin conditions (via extension of their normal skin carriage): erythrasma, acne, interdigitial toe-web infections, trichomycosis axillaris, and pitted keratolysis.

Erythrasma is a common superficial infection of the skin due to *C. minutissimum,* characterized by well-defined but irregular reddish brown patches, often finely scaly and finely wrinkled. It is most frequent in intertriginous sites, such as the genitocrural and axillary areas, and is commoner in the tropics than in temperate climates. The lesions show a characteristic "coral red" fluorescence (due to porphyrin production) under Wood's lamp. At present the nature of the metabolic or physical changes that convert this microorganism from its status as a normal colonizer to a pathogen producing changes in the stratum corneum are unknown. Treatment with oral erythromycin or topical clindamycin is usually effective.

Propionibacterium acnes is a contributory factor in the development of acne, a multifactorial disease. The process is initiated by an alteration in keratinization in the sebaceous follicle, whose glands in individuals with acne are larger and produce more sebum than normal sebaceous glands. The keratinous material becomes more dense, desquamates, and accumulates in the follicle, obstructing and dilating it, forming comedones ("whiteheads" or "blackheads") visible on the skin surface. During the puberty–adolescent years, a time when sebaceous glands are developing prominently under androgenic stimulation, patients with acne have very high densities ($\sim 10^5$/cm^2) of *P. acnes,* whereas individuals of the same age without acne have much lower numbers of these microorganisms in the skin. Although the fatty acids of newly formed sebum are esterified, the lipase activity of *P. acnes* on sebaceous gland triglycerides causes release of free fatty acids (nearly 50% of the lipids reaching the surface being in this form). Inflammation about sebaceous follicles, a hallmark of acne, appears to be produced in response to a variety of factors, all generated predominantly by *P. acnes,* although other species (*P. granulosum, Pityrosporum ovale*) may contribute to a lesser extent. These factors contributing to inflammation include intradermal release of free fatty acids through ruptured follicles, activation by *P. acnes* components of the classic and alternative complement pathways, release by *P. acnes* of neutrophil chemotactic factors, and release of tissue-damaging liposomal hydrolases following phagocytosis of *P. acnes* by neutrophils.

Strong additional support for the role of *P. acnes* in production of the lesions of acne comes from the suppression of *P. acnes* growth and amelioration of the lesions by systemic and/or topical therapy with antibiotics (erythromycin, clindamycin, minocycline).

The primary etiologies of interdigital toe-web infections are dermatophytic fungi, which initiate an initial superficial scaling and fissuring process. Secondary infection due to overgrowth of bacteria, commonly *Brevibacterium epidermidis* and *C. minutissimum* (but sometimes gram-negative bacteria such as *Pseudomonas* spp.), leads to a more symptomatic process characterized by malodorous maceration of toe-web spaces.

Unclassified *Corynebacterium* spp. (and also *Micrococcus sedentarius*) have been implicated as the causes of pitted keratolysis, a superficial process on the plantar surface of the foot characterized by small pits which develop in the stratum corneum and may coalesce to form large erosions. Topical antimicrobial therapy is effective.

A variety of currently unclassified coryneforms (previously inappropriately considered to be an individual species called *C. tenuis*) is responsible for trichomycosis axillaris, a malodorous process characterized by waxy, nodular coverings on axillary hair shafts that develop under conditions of poor local hygiene.

D. Micrococcus spp.

Seven *Micrococcus* species are members of the resident skin flora. They are now known on the basis of 16S RNA sequences to be more closely related to *Arthrobacter* and coryneform species than to *Staphylococcus* spp. *Micrococcus luteus* and *M. varians* are the two species of *Micrococcus* most commonly found on human skin. *Micrococcus* spp. colonize the skin of infants more slowly (at 28–32 weeks) than staphylococci, which are already present in the majority at 1–7 days. Unlike coryneform bacteria, *Micrococcus* spp. do not cause skin infection with the exception of the occasional role of *M. sedentarius* in some cases of pitted keratolysis.

IV. RESIDENT (NORMAL) SKIN FLORA: GRAM-NEGATIVE BACTERIA

Gram-negative bacilli, with the exception of *Acinetobacter* spp., are relatively rare components of the resident flora on normal skin. On the basis of DNA hydridization 17 genospecies of *Acinetobacter* have been delineated, including *A. calcoaceticus, A. baumannii, A. johnsonii,* and *A. lwoffii.* Since *A. calcoaceticus* and *A. baumannii* are closely related genotypically and phenotypically, they are often described as the *A. calcoaceticus–baumannii* complex. Carriage of *Acinetobacter* as resident flora occurs in about 25% of normal adults, primarily in axillae, groins, antecubital fossae, and toe-webs. These consist mainly of strains of *A. johnsonii, A. lwoffii,* and as yet unclassified genotypes. The carriage rate is higher in patients with eczema, particularly on lesions. Carriage of *Acinetobacter* is commoner in summer months, presumably related to increased perspiration. *Acinetobacter* skin infections, often nosocomial as are other *Acinetobacter* infections, although uncommon, include operative wound infections, cellulitis, and skin abscesses. Many of the strains of *Acinetobacter* isolated from invasive infections such as bacteremia (often vascular catheter induced), pneumonia, meningitis, and surgical wound infection belong to *A. baumannii.*

Other than *Acinetobacter* spp., gram-negative bacilli are rarely constituents on the normal skin, except as transient colonizers of exogenous origin, such as nosocomial carriage of *Klebsiella* and *Enterobacter* on the hands of health-care workers, *P. aeruginosa* colonization on use of occlusion (hyperhydration) in dermatologic therapy, and *P. aeruginosa* carriage in toe-webs. The latter probably accounts for the common involvement of the latter organism in trench foot or immersion foot of soldiers. Similarly, another pseudomonad, *P. cepacia* (now known as *Burkholderia cepacia*), has been found in the macerated web spaces of the condition known as swamp foot. Exogenously acquired *P. aeruginosa* folliculitis from use of contaminated whirlpool baths, on clearing of all lesions, may be followed, some weeks later after heavy exercise, by recurrence of the original rash, suggesting that there had been persistent carriage of *P. aeruginosa* on normal skin. *Proteus* spp. can be found normally on the nasal mucosa of about 5% of healthy individuals and in the toe-web spaces along with *Pseudomonas* spp.

V. RESIDENT (NORMAL) SKIN FLORA: FUNGI

Members of the genus *Malassezia* represent the principal resident fungal skin flora and are found on all adults, particularly around openings of sebaceous glands in the superficial layers of the stratum corneum. Although dermatophytes (*Epidermophyton* spp., *Trichophyton* spp.) may be recovered from normal-appearing skin, their role in these intances (residents or transients) is unclear as yet. *Malassezia furfur* is the name currently given for the lipophilic yeast, which assumes yeast and short hyphal forms in the superficial lesions of tinea (pityriasis) versicolor of which it is the etiology. Yeast forms of this organism, designated in the past as *Pityrosporon ovale* and *Pityrosporon orbiculare,* are those seen in carriage on normal individuals. Colonization of human skin with these fungi begins in infants, with carriage reaching a peak of 100% in early adult life. Their numbers are normally greatest on the chest, back, and scalp where sebaceous glands are plentiful, and the former two being areas where the lesions of tinea versicolor, a disease associated with overgrowth of this colonizing lipophilic yeast, is prone to develop.

VI. TRANSIENT SKIN FLORA: GRAM-POSITIVE BACTERIA

Table III lists species of transient skin flora.

A. Staphylococcus aureus

While persistent nasal carriage of *S. aureus* occurs in approximately 30% of normal individuals, resident carriage on normal skin occurs in about 6% of normal individuals and transient carriage, in another 7%.

TABLE III
Transient Skin Flora

Gram-positive bacteria
Staphylococcus aureus
Streptococcus pyogenes (group A streptococcus)
Enterococcus spp.
Streptococcus agalactiae (group B streptococcus)
Peptostreptococcus spp.
Clostridium perfringens
Erysipelothrix rhusiopathiae
Corynebacterium diphtheriae
Bacillus anthracis
Gram-negative bacteria
Pseudomonas aeruginosa
Haemophilus influenzae
Escherichia coli
Klebsiella spp.
Enterobacter spp.
Proteus spp.
Serratia spp.
Citrobacter spp.
Bacteroides spp.
Aeromonas hydrophila
Halophilic noncholera vibrios

Transient skin carriage may be generated in areas previously uncolonized by distribution of the microorganism from another site of more persistent colonization on the same individual. This may occur in chronic nasal carriers and in individuals who have resident *S. aureus* colonizing or infecting preexisting skin lesions (e.g., eczema) or in individuals with other underlying conditions such as diabetes mellitus. Transmission of *S. aureus* is likely to be carried out in the aforementioned circumstances by the hands of individuals themselves. In addition, nosocomial transmission of *S. aureus* to another patient, producing transient colonization, with or without subsequent secondary infection of skin lesions, may be effected through the hands of hospital personnel. This is of particular concern with regard to the difficult-to-treat methicillin-resistant *S. aureus* (MRSA) and is the reason why "contact precautions" are employed to prevent spread of this pathogen to vulnerable hospitalized patients.

1. *Pyodermas Due to Staphylococcus Aureus*

Resident or transient *S. aureus* are the infecting organisms in a number of superficial staphylococcal pyodermas. Impetigo is a highly contagious superficial unilocular vesicopustular process located between the stratum corneum above and the stratum granulosum below. Although group A streptococci were most commonly isolated in the past, currently *S. aureus* (of endogenous or exogenous origin) is most frequently found in the lesions, which are usually located near the openings of hair follicles. Chronologically, after appearance in the patient's nose, the *S. aureus* strain is disseminated about 11 days subsequently to normal skin and thence is spread to skin lesions (at sites of minor abrasions or insect bites) about 11 days later, producing the characteristic lesions of impetigo. A second form of impetigo, bullous impetigo, is produced by *S. aureus* strains belonging to phage Group II and is characterized by vesicles which rapidly progress to flaccid bullae. It occurs mainly in newborn and young children. *Staphylococcus aureus* folliculitis is a pyoderma located within hair follicles, producing perifollicular inflammation or superficial small dome-shaped pustules at the openings of hair follicles. A furuncle ("boil") is a deep-seated inflammatory nodule about a hair follicle (commonly in areas subject to friction and perspiration) and usually follows a superficial folliculitis. It most often occurs without evident predisposing skin lesions but may follow preexisting abrasions, scabies, or insect bites. A carbuncle is similar to a furuncle but is a serious, more extensive, deeper placed painful lesion that develops when suppuration occurs under thick inelastic skin in areas such as the nape of the neck and back. Recurrent furunculosis is not uncommon and may result from autoinoculation from previous lesions, recurrent shedding of *S. aureus* from nasal carriage, contact sports such as wrestling, and staphylococcal infections in other family members.

B. Streptococcus pyogenes (Group A Streptococcus)

1. Pyodermas Due to Group A Streptococcus

Spread of group A streptococci usually occurs by transfer of organisms from an infected individual or carrier (upper respiratory tract) through close personal contact.

a. Impetigo

Impetigo, currently, is less frequently due to group A streptococci than to *S. aureus* in Europe and the U.S., whereas formerly the reverse obtained. While streptococcal strains involved in pharyngeal disease have predominantly belonged to M serotypes 1, 3, 5, 6, or 12, those involved in group A streptococcal impetigo have newer M types such as 31, 49, 52, 53, 55–57, 60, 61, and 63. This form of impetigo occurs predominantly in preschool-age children and is highly contagious. Group A streptococci are members of the transient skin flora only during transitory carriage and while active lesions of streptococcal impetigo and other streptococcal pyodermas persist. Group A streptococci appear as transient flora on the normal skin of children approximately 10 days prior to the development of impetigo. Pharyngeal and nasal carriage of these microorganisms is not detectable until 14 to 20 days after skin colonization, indicating that spread has most likely been from skin contact with another individual with streptococcal impetigo or other streptococcal pyodermas. Minor trauma (abrasions, insect bites) prior to the acquisition of the group A streptococcal strain on otherwise normal skin predisposes to the appearance of impetigo. Nonsuppurative complications of group A streptococcal infections occur in the form of acute rheumatic fever and acute glomerulonephritis. While acute rheumatic fever may follow untreated group A streptococcal pharyngitis or tonsillitis in less than 2 to 3% of instances, it does not develop following streptococcal skin infections. Acute glomerulonephritis, on the other hand, may result from infection of either the skin or respiratory tract. Specific M serotypes are more likely to be nephritogenic: Pharyngitis-associated strains include serotype 12, 1, 4, 25; serotypes 2, 49, 55, 57, and 60 are pyoderma-associated strains. The frequency of acute glomerulonephritis following infection with a known nephritogenic strain can be as high as 10 to 15%.

b. Ecthyma

Ecthyma, also a group A streptococcal process, begins in a fashion much like impetigo with a superficial vesiculopurulent lesion, but it extends more deeply, penetrating through the epidermis and producing a shallow ulcer covered by a crust or eschar.

c. Erysipelas

Erysipelas, also almost always due to group A streptococci (uncommonly due to group C or group G streptococci; rarely *S. aureus*), is a superficial infection of the skin involving mainly the dermis and dermal lymphatics. The lesion has a bright red, edematous (peau d'orange) appearance and rapidly spreads peripherally. The process frequently begins with a very small break in the skin which has disappeared by the time of onset of the lesion. The group A streptococci have usually transiently colonized the skin of the involved lesion from a preceding upper respiratory tract infection, although when erysipelas has developed group A streptococci are often no longer isolated on throat culture. The disease is rapidly progressive, causes high fever, and requires prompt therapy with penicillin G or a β-lactamase-resistant penicillin.

d. Acute Cellulitis

Acute cellulitis is a spreading infection involving both the skin surface and, particularly, the deeper subcutaneous tissues. It is usually due to group A (sometimes group B, C, or G) streptococci and follows transient skin colonization by the microorganism at a site of a recent puncture wound, or, on a lower extremity, at the site of a stasis ulcer or spread from a minor break in the skin in an interdigital web area, commonly due to tinea pedis.

e. Acute Lymphangitis

Acute lymphangitis, most commonly caused by group A streptococci but sometimes by *S. aureus*, is an inflammatory process involving the subcutaneous lymphatics. The portal of entry is usually an acute puncture wound or abrasion that is contaminated at the time of acquisition by a group A streptococcus from the upper respiratory tract. Lymphangitis causes red linear streaks, a few millimeters to several centimeters in diameter, that progress from the local lesion toward regional lymph nodes. Prompt therapy with penicillin G or a penicillinase-resistant β-lactam drug is required to treat the frequently complicating bacteremia.

f. Streptococcal Gangrene

Streptococcal gangrene, also known as necrotizing fasciitis type II or "the flesh-eating bacterial infection" in the lay press, is a gangrenous, edematous process involving the subcutaneous tissues and fascia followed by necrosis of the overlying skin. The group A streptococci involved in the process usually have entered through a puncture wound, or laceration or surgical incision, or through a nonvisible break in the skin. Transient colonization of the skin site previously or at the time of the skin injury by group A streptococci from the upper respiratory tract or from another colonized skin surface is the source of the pathogen. Extensive surgical debridement, as well as antibiotic therapy, is necessary for treatment in view of the extensive undermining of tissue in this life-threatening process.

A similar type of process, type I necrotizing fasciitis, occurs particularly on the abdominal wall, perineum, and lower extremities and is commonly due to a mixed infection with one or more anaerobes (e.g., *Peptostreptococcus* spp., *Bacteroides* spp.) along with one or more facultative species (e.g., various non-group A streptococci, such as *Enterococcus* spp. and various Enterobacteriaceae). Crepitus (due to gas in the subcutaneous tissues) is often present. The source of this type of mixed infection is commonly from dissection of an intestinal process such as an intestinal perforation or a perirectal abscess, or spread from a decubitus ulcer or surgical wound infection. As in the case of necrotizing fasciitis due to group A streptococci, treatment consists of extensive debridement of the process, drainage of the feeding focus of infection, and, because of the mixed nature of the infecting microorganisms, broader antimicrobial coverage (based on gram-stained smears of exudate and culture results) than for type II necrotizing fasciitis.

C. Non-Group A Streptococci

As noted earlier, group C and G streptococci can occasionally produce erysipelas and cellulitis, and group B streptococci may be responsible for erysipelas (in infants) and cellulitis. *Enterococcus* spp., group B streptococci, and *Peptostreptococcus* spp. are normally members of the lower intestinal tract flora. *Enterococcus* spp. are common causes of urinary tract infections. Thus, contamination of a skin surface with fecal or urinary flora (particularly in incontinent elderly patients) can readily cause transient colonization and secondary infection of decubitus and stasis ulcers, eczematous dermatitis, and abrasions with such microorganisms.

1. Progressive Bacterial Synergistic Gangrene

A somewhat distinctive mixed infection of the skin, progressive bacterial synergistic gangrene occurs in the setting of abdominal surgery (about stay sutures or adjacent to a colostomy opening) or chronic ulceration on an extremity. The lesion consists of a central ulcer surrounded successively by a rim of gangrenous skin and an advancing zone of purplish erythema. The advancing margin contains anaerobic or microaerophilic streptococci, whereas the central ulcerated area contains *S. aureus* (rarely *Proteus* or other gram-negative bacilli); both types of organisms are necessary for this synergistic infection. Contamination (transient) of an operative wound or ulcer by such flora can lead to subsequent infection at the site of the original lesion.

2. *Streptococcus Iniae Infections*

Cellulitis and lymphangitis have occurred due to *S. iniae,* normally a fish pathogen, in individuals handling fresh aquacultured fish (tilapia) or preparing the fish for cooking in the household. Cellulitis can occur, following (or simultaneous with) transient skin colonization with the microorganism, after a percutaneous injury while handling the fresh fish.

D. Other Gram-Positive Bacteria

1. *Clostridium Perfringens*

Clostridium perfringens is a normal constituent of the bowel flora. Transient skin colonization from fecal or soil contamination, or direct wound contamination with fecal material or soil contents, can result either in anaerobic cellulitis due to this microorganism or in clostridial myonecrosis (gas gangrene), an incredibly toxemic infection primarily involving skeletal muscle. Gas gangrene has prominent cutaneous manifestations (edema of subcutaneous tissues, tense blebs containing dark brown fluid, patches of skin necrosis). Both anaerobic cellulitis and clostridial myonecrosis are characterized by gas in the tissues and thin serous exudate (containing numerous short, plump gram-positive rods without spores, but a paucity of polymorphonuclear leukocytes). Both processes develop as a result of a traumatic dirty wound with extensive muscle or soft-tissue damage or following wound contamination during surgery, usually involving bowel or gallbladder.

2. *Bacillus Anthracis*

Anthrax is primarily a disease of domestic and wild animals, but humans become involved accidentally through exposure to animals and their products. Although essentially eliminated in the U.S., anthrax still occurs in animal reservoirs in Africa, the Middle East, India, and South America. The most common site of human infection involves the skin on an exposed part of the body on which an abraded area has been inoculated with the organism from contact (usually occupational) with wool, hides, and other animal products.

Bacillus anthracis, a large gram-positive aerobic rod, forms spores in the environment and on culture but not in tissues. The microorganisms produce several toxins that can cause shock and death when the infection is extensive or disseminated. Cutaneous anthrax begins as a papule and develops into a hemorrhagic bulla which undergoes necrosis and eschar formation. The relatively painless lesion is surrounded by extensive, brawny, nonpitting edema.

3. *Erysipelothrix Rhusiopathiae*

Erysipeloid, an uncommon acute cellulitis due to *Erysipelothrix rhusiopathiae,* occurs primarily in fishermen, butchers, and housewives who handle raw fish, poultry, and meat products. After the microorganism, a thin microaerophilic gram-positive rod, is inoculated through a break in the skin, a serpiginous violaceous lesion with sharply defined borders subsequently develops.

4. *Corynebacterium Diphtheriae*

Cutaneous diphtheria still occurs in underdeveloped areas of the world, and outbreaks have occured among alcoholics on "skid row" and among Native Americans in the western U.S. Spread of infection occurs from carriage in the pharynx of the patient or contact with another carrier, but in children spread to others may occur via direct contact with skin lesions. Most cases of cutaneous diphtheria represent transient colonization and then infection by *C. diphtheriae* of a preexisting skin lesion such as a traumatic abrasion, eczema, or ecthyma, but in occasional patients cutaneous diphtheria begins as a primary pustular lesion. Whether it originates as a primary or a secondary infection, the lesion ultimately appears as an ulcer, covered partially with a grayish membrane and a purulent exudate, surrounded by a zone of edema and erythema. In 3 to 5% of patients with cutaneous diphtheria, neurologic findings develop as a result of elaboration and spread of *C. diphtheria* exotoxin.

VII. TRANSIENT SKIN FLORA: GRAM-NEGATIVE BACTERIA

A. *Pseudomonas aeruginosa*

Colonization and infection of the skin with *Pseudomonas aeruginosa* has been considered earlier under the section on Resident Flora. However, this gram-negative nonfermentative, obligately aerobic bacillus is found ubiquitously in nature, particularly on fresh vegetables and in water. Exogenous contamination and transient colonization with *P. aeruginosa* producing hot tub-associated folliculitis has already been considered. *Pseudomonas aeruginosa* colonization and secondary infection of preexisting skin lesions such as decubitus ulcers and thermal burns are not uncommon occurrences and may be complicated by bacteremia. Increased colonization of the gastrointestinal tract and skin with this microorganism occurs in patients who are granulocytopenic or immunocompromised. Secondarily infected lesions often produce a purulent exudate with a greenish color and a "fruity" odor.

Pseudomonas aeruginosa is responsible for some painful paronychias, sometimes associated with green-blue discoloration of the fingernail, in individuals who have their hands chronically in water and are thus exposed to transient colonization by this microorganism. External otitis, also known as "swimmer's ear," is frequently caused by *P. aeruginosa* which colonizes the external ear transiently from exposure to water. Subsequent mild trauma allows ingress of the microorganism into the area of the pinna, producing a characteristic macerated, swollen appearance.

B. *Haemophilus influenzae*

Most invasive *H. influenzae* infections (meningitis, bacteremia, epiglottitis, cellulitis) are caused by encapsulated type b strains and occur in children. Widespread use of the polysaccharide–protein conjugate vaccine has reduced the incidence of such infections by 95%. *Haemophilus influenzae* cellulitis characteristically occurs in younger children (aged 6–24 months) and commonly involves the face, neck, or upper extremities. Although the exact pathogenesis of the cellulitis is uncertain, in most instances it has been preceded by an upper respiratory infection. The association of otitis media, a previously common infection caused by *H. influenzae* type b, with cellulitis of the face has led to the suggestion of possible spread of infection from the ear via lymphatics in the pathogenesis of buccal cellulitis. Involvement of the face, neck, and upper extremities is suggestive of "fall-out" from the respiratory tract onto the skin of the aforementioned areas. Subsequent transient colonization and invasion locally through a small break in the skin is the likely pathogenesis of cellulitis with this striking anatomic distribution. Typically, *H. influenzae* type b cellulitis follows coryza or pharyngitis in an infant and is characterized by an increase in fever and a purple-red tender area of edema. Unlike the distinct margins of erysipelas, the border of *H. influenzae* cellulitis is indistinct.

C. Enteric Gram-Negative Bacilli

Members of the Enterobacteriaceae (*Escherichia coli, Klebsiella, Proteus, Enterobacter, Citrobacter, Serratia*) as well as *Bacteroides* spp. and other members of the anaerobic lower intestinal flora occasionally can be the etiology in acute cellulitis, particularly when it occurs in the elderly, in diabetics, or following skin trauma, surgery, or subcutaneous dissection of infection from the colon or perineum. Granulocytopenia, prior extensive antimicrobial use, and chronic illness are additional predisposing factors. Such mixed infections in the form of acute cellulitis or necrotizing fasciitis may also follow introduction of microorganisms from unhygienic skin surfaces as well as via contaminated narcotics delivered by "skin popping." Such mixed infections, in any of the foregoing settings, may exhibit subcutaneous gas formation.

D. *Aeromonas hydrophila*

Aeromonas spp. are commonly found in fresh and brackish waters associated with fish or aquatic animals. *Aeromonas* can produce diarrheal disease (most common infection), can be transiently carried in the intestinal tract, occasionally can cause soft tissue infections, and is rarely responsible for bacteremia and sepsis in immunocompromised hosts. Aeromonads are nonsporulating facultatively anaerobic bacilli. Of the over a dozen species, *A. hydrophila* is the one most commonly producing soft tissue infection. Exposure of the skin to contaminated water provides an opportunity for transient colonization. Invasion occurs through previously sustained sites of skin trauma in the colonized area or at the time of trauma coincident with the exposure to freshwater. The increasing clinical use of leeches following reimplantation or flap surgery has been an additional source of soft tissue infections due to *A. hydrophila*. This microorganism normally inhabits the foregut of leeches, and potentially contaminates the wound area. The high frequency (7–20%) occurrence of *Aeromonas* infection in patients treated with leeches has led to the prophylactic use of antibiotics with this procedure.

Cellulitis due to *A. hydrophila* may develop rapidly

TABLE IV
Infections of the Skin Caused by Direct Contact with an Individual or Environmental Source

Disease process	*Causative microorganism*
Fungal diseases	
Epidermophytosis ("athlete's foot")	*Microsporum* spp., *Trichophyton* spp., *Epidermophyton* spp.
Common yeast infections	*Candida* spp., *Torulopsis* spp.
Mycetoma	*Pseudoallescheria boydii, Madurella* spp., *Exophiala,* etc.
Sporotrichosis	*Sporothrix schenckii*
Sexually transmitted diseases	
Syphilis	*Treponema pallidum*
Endemic (nonvenereal) syphilis (Bejel)	*T. pallidum*
Yaws	*T. pertenue*
Pinta	*T. carateum*
Chancroid	*Haemophilus ducreyi*
Lymphogranuloma venereum	*Chlamydia trachomatis* (serovars L1–L3)
Granuloma inguinale	*Calymmatobacterium granulomatis*
Mycobacterial diseases	
Primary inoculation tuberculosis (susceptible host)	*Mycobacterium tuberculosis*
Tuberculosis verrucosa cutis (inoculation in person with +PPD)[a]	*M. tuberculosis*
Fishtank granuloma	*M. marinum*
Buruli ulcer	*M. ulcerans*
Pyogenic skin abscesses and ulcers	*M. chelonae* subsp. *chelonae, M. chelonae* subsp. *abscessus*
Leprosy	*M. leprae*
Bacterial diseases	
Ulceroglandular tularemia	*Francisella tularensis*
Melioidosis	*Burkholderia pseudomallei*
Glanders	*B. mallei*
Mycetoma	*Nocardia brasiliensis*
Viral diseases	
Herpes labialis	HSV type 1
Herpes progenitalis	HSV type 2
Orf	Orf parapoxvirus
Milker's nodule	Paravaccinia virus
Molluscum contagiosum	Molluscum contagiosum virus
Warts	Human papillomavirus

[a] +PPD, positive tuberculin skin test.

after an acute trauma and may be complicated by the development of bacteremia and a systemic sepsis syndrome. Another, much rarer, form of infection, *A. hydrophila* myonecrosis, can follow penetrating trauma and muscle injury occurring in a freshwater environment. It occurs within 24 to 48 hr of sustaining trauma, progresses rapidly, and is characterized by prominent pain, systemic toxicity, marked edema, and the presence of gas in fascial planes and muscle. In all these respects this process resembles clostridial gas gangrene. Bacteremia frequently is present. Extensive surgical debridement and prompt initiation of antimicrobial therapy are required. Most strains are susceptible to ciprofloxacin, trimethoprim–sulfamethoxazole, aminoglycosides (except streptomycin), third-generation cephalosporins, and carbapenems.

E. Halophilic Noncholera Vibrios and Non-O1 *Vibrio cholerae*

Four *Vibrio* species in addition to non-O1 *Vibrio cholerae* (strains not belonging to serogroup O1 and only occasionally responsible for sporadic cases and outbreaks of diarrhea) are capable of producing cellulitis or traumatic wound infections: *V. vulnificus*, *V. alginolyticus*, *V. damsela*, and *V. parahaemolyticus*. The *Vibrio* spp. are halophilic but will grow on blood agar plates, however. These microorganisms are commonly present in saltwater and estuarine sediments as well as on fish and shellfish, particularly along the coast of the Gulf of Mexico (and to a lesser extent along the Atlantic and Pacific coasts) in the summer months. *Vibrio vulnificus* is the most pathogenic of these species and is capable of causing severe wound infections and a "primary septicemia" syndrome, which may follow 24 to 48 hr after ingestion of raw oysters or other uncooked seafood. The more typical *V. vulnificus* infection occurs as cellulitis several days following a laceration sustained in seawater or brackish inland lakes. Sometimes cellulitis develops without discernible antecedent skin trauma following exposure to seawater. Here the sequence presumably has been transient skin colonization with subsequent penetration through a small break in the skin. Primary septicemia or severe cellulitis complicated by septicemia are more likely to occur in patients with underlying cirrhosis of the liver, hemochromatosis (with its high serum iron level), diabetes mellitus, renal failure, leukemia, and processes requiring corticosteroid therapy. It is important to recognize the entity of "primary septicemia" since bullous skin lesions resembling those of a primary *V. vulnificus* cellulitis may be a feature of the "primary septicemia" syndrome, usually of gastrointestinal origin.

Traumatic wound infections due to *V. vulnificus* commonly involve the lower extremity and usually consist of cellulitis with large hemorrhagic bullae, but may consist of pustular lesions with associated lymphangitis and lymphadenitis. The cellulitis may be very painful, cause high fever, and progress rapidly, producing local necrotizing vasculitis with ex-

TABLE V
Infections Involving Skin Caused by Insect Vector or Animal Bite

Disease process	*Causative microorganism*
Cat scratch disease (primary lesion)	*Bartonella henselae*
Bacillary angiomatosis	*B. henselae*
Lyme disease (erythema chronicum migrans)	*Borrelia burgdorferi*
Boutonneuse fever, South African tick-bite fever (1° lesion)	Rickettsia conorii
Rickettsialpox (1° lesion)	*Rickettsia akari*
Scrub typhus (1° lesion)	*Rickettsia tsutsugamushi*
Animal bites: cellulitis and infected wounds	*Pasteurella multocida, P. canis, Capnocytophaga canimorsus, Haemophilus felis, H. aphrophilus, Neisseria canis, N. weaveri, Weeksella zoohelcum, Prevotella melaninogenica, P. denticola, Fusobacterium nucleatum, Porphyromonas canoris, P. gingivalis, Veillonella parvula*

TABLE VI
Infections Involving the Skin Due to Hematogenous Dissemination

Bacterial infections (bacteremic)
Gram-positive bacteria
Staphylococcus aureus
Streptococcus pyogenes (group A streptococcus)
Enterococcus spp. (acute bacterial endocarditis)
Histotoxic clostridia (primarily *Clostridium septicum*)
Gram-negative bacteria
Pseudomonas aeruginosa
Burkholderia pseudomallei (melioidosis)
Salmonella typhi
Neisseria gonorrhoeae
Neisseria meningitidis
Vibrio vulnificus
Bartonella bacilliformis
Spirochaetes
Treponema pallidum (secondary syphilis)
Borrelia burgdorferi
Fungi
Histoplasma capsulatum
Blastomyces dermatitidis
Coccidioides immitis
Paracoccidioides brasiliensis
Candida spp.
Cryptococcus neoformans
Mycobacteria
Mycobacterium tuberculosis (acute miliary tuberculosis)
M. leprae (leprosy)
Rickettsia
Rickettsia rickettsii (Rocky Mountain spotted fever)
R. typhi (endemic typhus)
R. prowazekii (epidemic typhus)
Viruses
Coxsackie virus A16 (hand-foot-and-mouth disease)
Enterovirus 71 (hand-foot-and-mouth disease)
Herpes simplex type 1
Herpes simplex type 2
Varicella-zoster
Vaccinia
Variola (smallpox)
Human immunodeficiency virus (acute retroviral syndrome)

tensive skin necrosis and ulcer formation, myositis, and gangrene requiring amputation. Treatment of infected traumatic wounds due to *V. vulnificus* and similar *Vibrio* species requires debridement of necrotic lesions and antimicrobial therapy with a third-generation cephalosporin plus an aminoglycoside or with a combination of a tetracycline (e.g., doxycycline) or a fluoroquinolone plus an aminoglycoside (gentamicin or tobramycin).

Non-O1 *V. cholerae,* although primarily a gastrointestinal pathogen, may cause cellulitis and necrotizing fasciitis following skin trauma and contact with seawater in subtropical areas. As in the case of *V. vulnificus* infections, patients with chronic liver disease are particularly vulnerable, and "primary septicemia" can result in cellulitis with hemorrhagic bullae.

VIII. OTHER FORMS OF MICROBIOLOGICAL INVOLVEMENT IN INFECTIOUS PROCESSES OF THE SKIN

Other routes for ingress into the skin for infecting microorganisms not parts of the "normal resident" or "transient" flora include spread by direct contact, spread by vector inoculation, spread via bloodstream dissemination, or spread from infection of contiguous anatomic structures. Owing to space limitation examples will be cited here in tabular form with the aim to provide illustrative examples without any attempt at completeness (see Tables IV–VI).

See Also the Following Articles

ADHESION, BACTERIAL • CLOSTRIDIA • COSMETIC MICROBIOLOGY • FUNGAL INFECTIONS, CUTANEOUS • *PSEUDOMONAS* • *STAPHYLOCOCCUS*

Bibliography

Kloos, W. E., and Bannerman, T. L. (1994). Update on clinical significance of coagulase-negative staphylococci. *Clin. Microbiol. Rev.* 7, 117–140.

Noble, W. C. (1983). "Microbial Skin Disease: Its Epidemiology." Arnold, London.

Noble, W. C. (ed.) (1992). "The Skin Microflora and Microbial Skin Disease." Cambridge Univ. Press, Cambridge.

Noble, W. C. (1993). Ecology and host resistance in relation to skin disease. *In* "Dermatology in General Medicine" (T. B. Fitzpatrick, A. Z. Eisen, K. Wolff, J. M. Freedberg, and K. F. Austen, eds.), 4th Ed., Chap. 17. McGraw-Hill, New York.

Noble, W. C., and Somerville, D. A. (1974). "Microbiology of Human Skin." Saunders, Philadelphia.

Rupp, M. E., and Archer, G. L. (1994). Coagulase-negative staphylococci: Pathogens associated with medical progress. *Clin. Infect. Dis.* **19**, 231–245.

Swartz, M. N., and Weinberg, A. N. (1993). Infections due to gram-positive bacteria. *In* "Dermatology in General Medicine." (T. B. Fitzpatrick, A. Z. Eisen, K. Wolff, J. M. Freedberg, and K. F. Austen, eds.), 4th Ed., Chap. 187. McGraw-Hill, New York.

Weinberg, A. N., and Swartz, M. N. (1993). General considerations of bacterial diseases. *In* "Dermatology in General Medicine" (T. B. Fitzpatrick, A. Z. Eisen, K. Wolff, J. M. Freedberg, and K. F. Austen, eds.), 4th Ed., Chap. 186. McGraw-Hill, New York.

Weinberg, A. N., and Swartz, M. N. (1993). Gram-negative coccal and bacillary infections. *In* "Dermatology in General Medicine" (T. B. Fitzpatrick, A. Z. Eisen, K. Wolff, J. M. Freedberg, and K. F. Austen, eds.), 4th Ed., Chap. 188. McGraw-Hill, New York.

Smallpox

Donald A. Henderson
The Johns Hopkins University School of Public Health

I. History
II. Smallpox, the Disease
III. The Global Eradication Campaign
IV. Posteradication Events

GLOSSARY

case-fatality rate The number resulting from dividing the total number of deaths caused by a disease by the total number of cases of the disease.

inoculation In this context, the introduction into the skin of pustular or scab material containing smallpox virus. It is sometimes called "variolation." Vaccination, in contrast, consists of introducing into the skin vaccinia (or cowpox) virus.

variola A word used almost interchangeably with the word "smallpox." It derives from the Latin word for "pock" or "pustule" and was the name originally given to smallpox.

UNTIL ITS ERADICATION IN 1980, SMALLPOX was a disease, caused by the variola virus, that infected only humans. The more severe form of the disease, variola major, killed some 30% of its victims; there was no treatment. Smallpox was transmitted directly from person to person, and it infected persons in all parts of the world. Those who recovered from the disease were immune from acquiring a second infection. The disease was marked by high fever and the development of a characteristic pustular rash. As the disease progressed, scabs formed over the pustules and eventually separated, leaving permanent, deeply pitted scars, which were most prevalent over the face. Of all the infectious diseases, none had the capacity both to spread so widely and to inflict such a high mortality as did smallpox.

In 1798, an English physician, Edward Jenner, reported that pustular material taken from a lesion caused by the related cowpox virus could be transferred from person to person by inoculation into the skin. He showed that individuals so treated were protected from acquiring smallpox when they were later exposed. This was the world's first vaccine, and its discovery is acknowledged to be one of the most important in the history of medicine.

As vaccine use increased and more potent and heat-stable vaccines became available, smallpox steadily receded until, in 1967, with only 34 countries still reporting infections, the World Health Organization launched a global campaign to eradicate the disease. This proved to be a success. The last naturally occurring case was a Somali resident who became ill on 26 October 1977. In 1980, smallpox was declared to have been eradicated, and vaccination ceased everywhere.

I. HISTORY

Smallpox was an ancient disease and a specifically human one. Those who became ill could transmit infection only during the 2- to 3-week period of acute illness; second attacks of smallpox were rare. For the virus to survive, it was necessary for it to infect one person after another in a continuing chain of transmission. Thus, to sustain the chain of infection, a moderately large population in reasonably close contact with one another was required. For these reasons, it is assumed that the disease must have arisen less than 10,000 years ago, a time when agricultural settlements were first developing. Three Egyptian mummies from the Eighteenth to Twentieth Dynasties (1100–1570 BC), presumed victims

Fig. 1. Edward Jenner (1749–1823) of England, who introduced vaccination to Britain in 1796. (Courtesy of Wellcome Institute Library, London.)

of the disease, all show the characteristic pustules of acute smallpox.

The subsequent spread of smallpox is difficult to trace because of the paucity of historical records that describe disease symptoms and epidemics in sufficient detail to identify them. Accounts of the disease in India suggest its existence there for almost as long a period as in Egypt. From India, it appears to have spread across China (about 250 BC) and to other parts of Asia. Periodic outbreaks occurred in Europe but not until the eighth century did smallpox became firmly established throughout the continent. By the seventeenth century, smallpox was killing more than 400,000 persons in Europe every year.

The saga of smallpox in the Americas represents one of the most catastrophic, yet little known chapters in the history of mankind's struggle against disease. The central cause was the extraordinarily high case-fatality rates among the Amerindian population throughout the Americas. In 1507, soon after contact with the New World, the Spaniards brought smallpox to Hispaniola. By 1541, the island's population, estimated to have originally been between 300,000 and 1,000,000 persons, numbered only 500. Several diseases contributed to this disaster, but smallpox was the main offender. In 1520, Cortes, with an army of 500, was sent to Mexico to capture additional slaves. Smallpox traveled with him and soon spread throughout Mexico and Middle America and eventually into the vast Incan Empire of the Andean region. By the end of the sixteenth century, populations throughout Central and South America are estimated to have decreased by 80 to 90%. North American Indians fared no better.

In all parts of the world, smallpox was the most feared of all diseases, and special smallpox deities were to be found among populations in many parts of Asia and Africa. In India, however, sometime early in the Christian era, an unusual practice was discovered that served to protect against fatal smallpox. It was called inoculation. Pustular or scab material was taken from a smallpox lesion and inoculated into the skin of a susceptible person. Smallpox infection induced in this manner, rather than by the normal route of inhalation, usually resulted in a much less serious disease. Customarily, the recipient would experience fever and malaise and after 5 to 7 days would develop a pustular lesion at the inoculation site. Most persons recovered uneventfully and were thereafter immune to acquiring smallpox. The inoculated individual could transmit infection to others, however, and in some a more severe illness occurred, which was sometimes fatal.

The practice of inoculation spread to China and Southwest Asia although it is uncertain as to what proportion were so treated. In 1721, the practice was introduced into England and later into other parts of Europe as well as North America. Although it was accepted in some areas, the practice was not enthusiastically embraced everywhere, in part because of the sometimes fatal outcomes and in part because of the occurrence of new smallpox outbreaks developing as a result of spread of the virus from inoculees.

The concept of inoculation as a protective measure

effectively established this technique as a common practice. Thus, Edward Jenner utilized a form of inoculation in his now famous experiment in which he took pustular material from a cowpox lesion on the hand of a dairymaid and inoculated it into the arm of a young boy, James Phipps. The use of cowpox was suggested by local lore, which held that dairymaids did not acquire smallpox because of having had a cowpox infection. Jenner attempted to inoculate Phipps after some weeks with pustular material from a smallpox lesion and demonstrated that he could not be infected. He published this and other observations in 1798, and soon thereafter the cowpox vaccine (an orthopoxvirus, like smallpox) was shipped to countries around the world. Jenner's discovery was acclaimed at the time and is recognized even now as one of the most important of all medical discoveries.

However successful vaccination was, its widespread application proved difficult. Until late in the nineteenth century, vaccination was conducted by the arm to arm transfer of the vaccine material. When a vaccination was unsuccessful and no other material was available, fresh material had to be obtained. Cowpox, however, was found only in Europe and occurred only sporadically. Thus, in many areas, there were often long periods when no vaccine was available. When efforts were made to ship new material, it often would deteriorate en route. Eventually, it was discovered that syphilis and hepatitis B could also be transferred by inoculation. Incidents occurred, in fact, in which cutaneous lesions of syphilis were thought to be vaccinia and large-scale syphilis outbreaks resulted when the pustular material was transferred.

Late in the nineteenth century, cowpox began to be grown on the scarified flank of a calf or sheep, thus assuring a more reliable and ample supply of vaccine. In the 1920s, French and Dutch scientists showed that a dried preparation could be produced which was stable at ambient tropical temperatures. Finally, a commercial process for freeze-drying the vaccine, perfected in England in the early 1950s, opened the way for widespread vaccination throughout the world and eventually for the global eradication program.

II. SMALLPOX, THE DISEASE

A. Etiology

The causative agent of smallpox, variola virus, belongs to the genus *Orthopoxvirus*, subfamily *Chordopoxvirinae*, family *poxviridae*. Three other members of this genus can also infect humans: monkeypox, cowpox, and vaccinia. However, none of the three are readily transmitted from person to person. The poxviruses are the largest and most complex of all viruses. The virion is a brick-shaped structure with a diameter of about 200 nm. Its lipoprotein outer membrane (envelope) encloses a single linear double-stranded DNA. Replication of the poxviruses occurs in the cytoplasm. Unlike other DNA viruses, the poxviruses encode the dozens of enzymes required for transcription and replication of the viral genome. The viral core is released into the cytoplasm after fusion of the virion with the plasma membrane. Restriction endonuclease maps of the genome definitively identify the species of the *Orthopoxvirus* genus.

B. Pathogenesis

Two distinct forms of smallpox have been observed over the years: variola major with a case-fatality rate of 20 to 30% and variola minor with a case-fatality rate of 1% or less. Variola major appears to have predominated throughout the world at least though the nineteenth century. A much less virulent form of the disease, usually called variola minor or alastrim, was first definitively identified in South Africa toward the end of the nineteenth century. This form of smallpox may have been present in some areas of the world before this time, but few records are available that set forth numbers both of cases and deaths along with a reasonable description of the disease. Variola minor appeared in the U.S. at the beginning of the twentieth century and spread rapidly across the country, eventually displacing variola major. From the U.S., variola minor spread to Latin America and to Britain.

Throughout Asia, variola major was the only type of smallpox seen, whereas in Africa, other variants occurred. In Ethiopia and Somalia, variola minor was found while elsewhere in Africa, another type of smallpox predominated, called variola intermedius by some, with a case-fatality rate of between 10 and 15%.

Natural infection occurred by the implantation of variola virus on repiratory mucosa. After migration to and multiplication in regional lymph nodes, viremia developed followed by multiplication of virus in the spleen, bone marrow, and lymph nodes. A secondary viremia with fever and toxemia began about the eighth day. The virus, contained in leukocytes, then localized in small blood vessels of the dermis and mucosa. This process resulted in formation of the characteristic vesicular and pustular lesions. Hemagglutinin-inhibiting (HI) and neutralizing antibodies could be detected beginning about the sixth day of illness. Neutralizing antibodies were long lasting, whereas HI antibodies declined to low levels within 5 years. Little is known about the development of cell-mediated immunity. Vaccinia-induced antibody responses were more rapid, accounting for the fact that complete or partial protection of individuals was possible even among those vaccinated several days after exposure. Secondary bacterial infection was not common, and death, when it occurred, probably resulted from the toxemia associated with circulating immune complexes and soluble viral antigens.

As the patient recovered, the scabs separated and the characteristic pitted scarring gradually developed. The scars were most evident over the face and resulted from the destruction of sebaceous glands followed by shrinking of granulation tissue and fibrosis.

C. Symptoms

After an incubation period of 7 to 12 days, the patient experienced a 2- to 5-day period of high fever, malaise, and prostration with headache and backache. Those with variola major were usually sufficiently ill to want to stay in bed. A maculopapular rash then began to develop, at which time the patient first became contagious to others. The rash appeared first on the face and forearms and on the mucosa of the mouth and pharynx. It then spread to the trunk and legs. Within 1 to 2 days, the rash became vesicular and then pustular. The pustules were characteristically rounded, tense, and deeply embedded in the dermis. Crusts began to form about the eighth or ninth day. When they separated, they left pigment-free skin and eventually scars. The eruption was characteristically more extensive on the face and distal parts of the arms and legs, and lesions were occasionally found on the palms and soles.

Illness caused by variola major was generally more severe, with a more extensive rash, a higher fever, and a greater degree of prostration than illness caused by variola minor. A milder form of disease was also seen among those who had previously been vaccinated; the rash in such persons tended to be more scant and atypical and the evolution of lesions more rapid.

Supportive therapy, maintenance of fluids and nu-

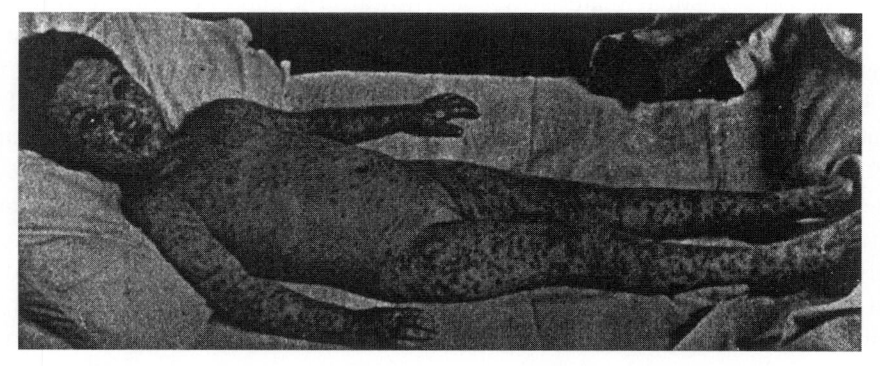

Fig. 2. Classic smallpox (pustular stage) in an unvaccinated child. [Reprinted from Horsefall and Tamm. (1965). "Viral and Rickettsial Infections of Man," Fourth ed. J. B. Lippincott Co., Philadelphia.]

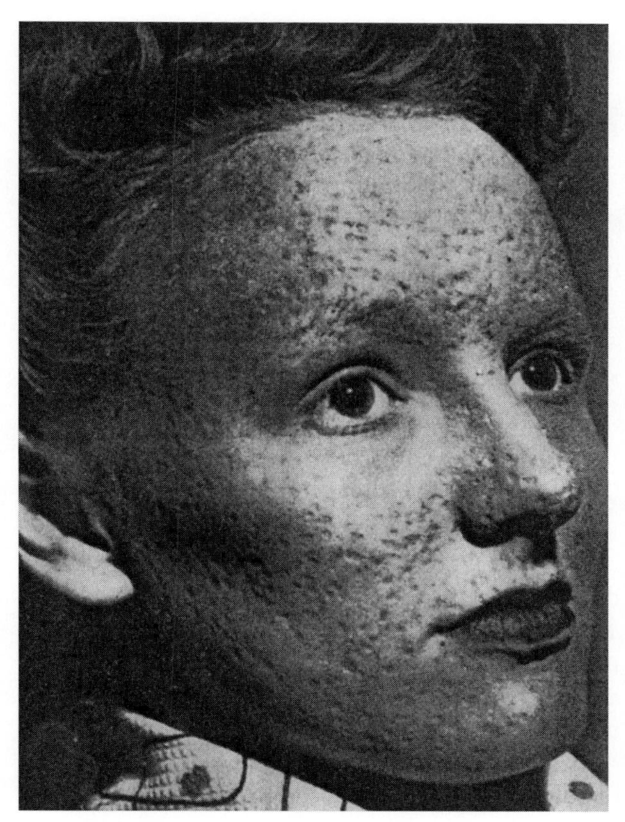

Fig. 3. Residual scarring (pockmarks) 6 months after recovery from smallpox. [Reprinted from Dixon, C. W. (1962). "Smallpox." Churchill Livingstone, Edinburgh.]

trition, and good nursing care were the best that could be offered to most patients. On the infrequent occasions when bacterial infections supervened, antibiotics were appropriate. Antiviral drugs, most notably methisazone and the arabinosides, were reported in some studies to be effective in early treatment or prophylaxis, but confirmatory studies showed otherwise.

D. Laboratory Diagnosis

Diagnosis of a proxvirus infection is confirmed rapidly by electron microscopic identification of virus particles in vesicular or pustular fluid or scabs. All orthopoxvirus virions have the same appearance. The four orthopoxviruses that infect humans grow and produce a cytoplasmic effect in cultured cells derived from many species, but they cannot usually be differentiated from each other in most preparations. For diagnostic purposes they are cutomarily grown on the chorioallantoic membrane of 10- to 12-day-old chick embryos on which they produce characteristic pocks. Each orthopoxvirus species has a distinctive DNA map demonstrable by digestion with restriction endonucleases.

III. THE GLOBAL ERADICATION CAMPAIGN

A global campaign for the eradication of smallpox was proposed by the Soviet Union in 1958 and agreed by the World Health Assembly. Over the succeeding eight years, however, little progress was made. At the time, the World Health Organization (WHO) and many of the smallpox-endemic countries were fully preoccupied with a global malaria eradication program, which had begun in 1955 and which consumed a substantial proportion of the Organization's budget, personnel, and time. As the years passed, many began to express skepticism that malaria or, for that matter, any other disease could actually be eradicated.

Soviet delegates, joined by others, expressed increasing displeasure that the Organization did not assign a higher priority to the smallpox effort and demanded, finally, that steps be taken to work out a definitive plan and cost estimates for an intensified program. Meanwhile, in recognition of International Cooperation Year, the U.S., in November 1965, announced its intention to provide support to 18 countries of western and central Africa in an effort to eradicate smallpox and control measles. At the 1966 World Health Assembly, the U.S. joined the Soviet Union in arguing for the development of a greatly intensified eradication campaign. Some countries, as well as the Director General of the WHO, opposed the concept of smallpox eradication as being technically impossible; others objected to increasing the WHO budget by $2.5 million per year to provide necessary support for the effort. Eventually, by the margin of only two votes, the decision was taken to intensify the eradication effort. Assembly delegates proposed a 10-year target.

In 1967, the year the program began, 46 countries reported smallpox, of which 33 were considered to

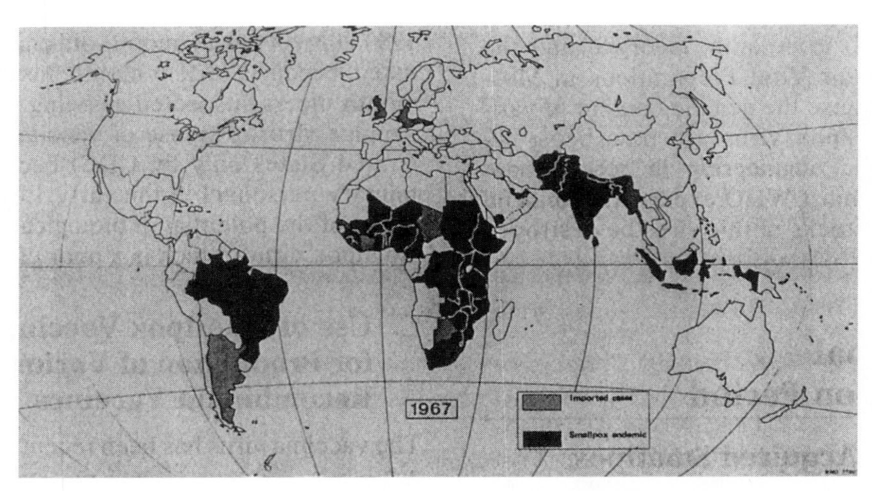

Fig. 4. Countries reporting smallpox cases in 1967. (Courtesy of the WHO, Geneva.)

have endemic disease while 13 recorded imported cases. Estimates later showed that there had occurred that year some 10 to 15 million cases and 2 million deaths.

The basic strategy for the program was twofold: population-wide vaccination and a surveillance–containment program. The vaccination program was intended to reach 80% of the population in each of the endemic countries and in neighboring countries at risk of imported cases. Those directing the program reasoned that a substantial level of vaccination immunity would sufficiently diminish the rapidity of virus spread so as to permit the surveillance–containment program to take effect. The latter program required that a reporting system be developed such that each hospital and health center would report each week as to the number of smallpox cases seen or discovered. When cases were reported, a surveillance team would go to the field, seek to find other cases, and vaccinate household and neighborhood contacts in hopes of stopping the chain of transmission of infection.

The strategy of surveillance and containment was soon discovered to be far more effective than might have been expected on the basis of the conventional wisdom of the time, which held that smallpox spread rapidly and widely. Through field epidemiology, it soon became clear that a given patient seldom infected more than 2 to 5 others, usually household members or close friends, and they normally did not develop illness until 7 to 12 days after exposure. Thus, smallpox usually spread slowly, and cases tended to cluster within cities and within geographic areas. This greatly facilitated the containment activities.

Many observations of importance gradually shaped the program. It was found that with good planning and involvement of community resources, vaccinators could regularly average 500 or more vaccinations per day, and vaccination coverage of more than 90% was to be expected. In contrast, vaccination coverage that relied solely on clinics and hospitals for dispensing vaccine seldom exceeded 60%.

Quality control was essential. As the program began, it was discovered that not much more than 10% of the vaccine then in routine use met accepted international standards. Two laboratories, one in the Netherlands and one in Canada, undertook to perform independent testing for all vaccine to be used in the program, and within 3 years, all vaccine met international requirements. In many countries independent assessment teams checked a sample of villages vaccinated by the teams to assure that both coverage and vaccine efficacy standards were met.

Research made important contributions. In 1967, a new vaccination needle was in the late experimental stages at Wyeth Laboratories in the U.S. With Wyeth approval, the WHO undertook additional studies with the needle, altered the technique for its use, and introduced it worldwide. The needle used only one-fourth as much vaccine as did the standard needles and produced a higher proportion of successful takes. Field studies soon revealed that the protection afforded by primary vaccination was not the 3 to 5 years that most textbooks asserted but, rather, showed a 90% efficacy rate at 20 years. With this discovery, it was possible to focus efforts on assuring that every person had a vaccination scar rather than assuring that everyone had been vaccinated within the preceding few years.

Rapid progress was made in the campaign so that by 1973, smallpox was largely confined to four countries of South Asia and, in Africa, to Ethiopia. In Asia, however, efforts such as had been successful in the Americas and in Africa proved far less efficacious. The problem was that with a more extensive infrastructure for road and rail travel in Asia, people traveled frequently and over considerable distances, often returning as a family to a home village when one of the family became ill. The result was that smallpox spread more rapidly than in other countries and over longer distances. In the summer of 1973, it was decided, with Indian government authorities, to assign large numbers of health personnel to undertake a 10-day search throughout all the villages, towns, and cities of the country. It was hoped that this would permit cases to be found much earlier so that an effective control vaccination effort might be mounted. The first search found tens of thousands of unreported cases, many more than had been expected, but early containment was then possible. Every few months the searches were repeated with increasing precision, eventually reaching every household in every village in India. Less than 2 years elapsed before India and all of Asia was smallpox free.

As the numbers of cases decreased, it became the policy to offer a small reward to anyone reporting a case that proved to be smallpox. As the cases became fewer, the reward was increased, so that by the end of the program, a reward of $1000 was being offered for the report of a case that could be confirmed as smallpox. This technique proved to be exceedingly valuable in the closing phases of the program in which each country had to demonstrate to an independent commission that it had a surveillance network sufficiently sensitive to detect cases if present.

The last case of smallpox occurred on 26 October 1977 in Somalia. Thereafter, 2 full years of surveillance and search for cases occurred in all previously endemic areas. An independent international team then visited each country to decide for itself whether it was satisfied that the surveillance had been sufficiently thorough as to guarantee that no possible cases of smallpox had escaped detection. Finally, on 8 May 1980, the World Health Assembly affirmed that eradication had been achieved and recommended that all countries cease vaccination. Only two further cases occurred, both in Birmingham, England, in 1978, the result of a laboratory accident.

IV. POSTERADICATION EVENTS

A. Monkeypox, a Disease in Humans Similar to Smallpox

During 1997–1998, reports of large outbreaks of human monkeypox cases in the Democratic Republic of the Congo raised questions as to whether monkeypox might be spreading so readily from person to person as to replace smallpox as a dangerous contagious disease. Monkeypox is caused by a virus that is a member of the *Orthopoxvirus* genus and so is closely related to smallpox. It resembles smallpox clinically and is associated with a case-fatality rate of 10 to 15%, similar to that of smallpox in central African countries.

Human monkeypox cases were first detected in 1970 and studied intensively by WHO teams throughout the 1980s. The WHO documented the fact that the natural monkeypox virus reservoir was actually ground squirrels and that man only infrequently became infected. However, the teams discovered that the virus could be passed from person to person, although it was much less contagious than was smallpox. Moreover, the studies indicated that it was highly unlikely that the virus could continue to spread even in populations that had never been vaccinated.

Initially, the reports in 1997 of large numbers of human monkeypox cases were widely publicized and caused alarm. As investigations progressed, however, it became clear that a large proportion of the cases were chickenpox and that there was little evidence to suggest that the virus had changed its character. However, field studies are continuing.

B. Destruction of Remaining Smallpox Virus Stocks

Following the 1980 declaration of smallpox eradication, a WHO expert committee was constituted to consider additional steps that might be taken to assure confidence in the fact of eradication and to advise regarding the continuation of a vaccine stockpile among other matters. One important issue was to decide whether the known remaining stocks of variola virus should be destroyed. As a result of WHO recommendations, 74 of 76 laboratories known to have stocks of variola in 1980 had destroyed their stocks or sent them to one of two WHO reference laboratories, located in Russia and in Atlanta, Georgia.

At meetings in 1986 and 1990, the expert committee affirmed the desirability of destroying the variola stocks but advised that this be deferred until virus strains could be appropriately mapped, cloned, and sequenced. By 1994, this work had been completed, and endorsements supporting the destruction of the virus strains had been obtained from five prominent scientific bodies including the International Union of Microbiological Societies, the Russian Academy of Medical Sciences, and the board of scientific counselors of the U.S. National Center for Infectious Diseases. At its 1996 meeting, the World Health Assembly voted to accept the recommendations of the expert committee and proposed that the virus strains at the two laboratories be destroyed in June 1999. In May 1999, it was decided to postpone destruction of the virus until 2002.

Although there are a few scientists who continue to believe that the virus should be retained for possible scientific studies at some future date, most policy makers agree that the argument for virus destruction is a compelling one, as it would make it less likely that the virus would be released either by accident or for malicious purposes.

C. Smallpox as a Terrorist Weapon

There is increasing concern that biological weapons might be deployed by terrorists acting either as a dissident group or under state sponsorship. Smallpox, until recently, was considered an unlikely agent because of the high level of population immunity, the availability of a vaccine, and the knowledge that vaccination of contacts can rapidly and effectively control epidemics. Circumstances have changed. Vaccination programs ceased in 1980, and all vaccine production has stopped. Moreover, the facilities for producing vaccine have been converted to other uses. Reserves of vaccine are presently low, and there are, as yet, no plans for replenishing them.

These considerations give special impetus to the decision to destroy all known stocks of the virus and to produce a special vaccine reserve should it be needed.

See Also the Following Articles

Biological Warfare • Polio • Surveillance of Infectious Diseases • Vaccines, Viral

Bibliography

Fenner, F., Henderson, D. A., Arita, I., Jezek, Z., and Ladnyi, I. D. (1998). "Smallpox and Its Eradication." WHO, Geneva.

Hopkins, D. R. (1983). "Princes and Peasants—Smallpox in History." Univ. of Chicago Press, Chicago.

Breman, J. G., and Henderson, D. A. (1998). Poxvirus dilemmas—Monkeypox, smallpox, and biologic terrorism. *N. Engl. J. Med.* **339**, 556–559.

Jezek, Z., and Fenner, F. (1998). Human monkeypox. "Monographs in Virology," Vol. 17. Karger, Basel.

Mahy, B. W., Almond, J. W., Berns K. I., *et al.* (1993). The remaining stocks of smallpox virus should be destroyed. *Science* **262**, 1223–1224.

Joklik, W. K., Moss, B., Fields, B. N., Bishop, D. H., and Sandakhchiev, L. S. (1993). Why the smallpox virus stocks should not be destroyed. *Science* **262**, 1225–1226.

Smuts, Bunts, and Ergot

Denis Gaudet, Jim Menzies, and Peter Burnett[1]

Agriculture and Agri-Food Canada Research Centre

I. Smuts and Bunts
II. Ergot
III. Conclusions and Future Outlook

GLOSSARY

ascigerous Tissues in which asci are produced. Generally pertains to the sexual stage in Ascomycetes.

avirulence The inability of a pathogen race or strain to cause disease on a particular strain or cultivar of the host.

biotrophic parasite Parasite that requires a living host to complete its life cycle in nature.

ergotism Poisoning in animals and humans brought about by ingestion of ergot sclerotia.

gene-for-gene concept The concept developed by H. R. R. Flor that states that for every resistance gene in the host, there is a corresponding avirulence gene in the pathogen, and for each gene for susceptibility in the host, there is a corresponding virulence gene in the pathogen.

genetic locus Position on the chromosome at which a particular gene and its alleles reside.

heterothallism A condition of fungi that produce compatible male and female gametes on two physiologically distinct mycelia. In smuts and bunts, two anatomically similar sporidia, each possessing a different mating type, fuse to form the dikaryotic infection hyphae, which is the parasitic stage of the fungus in the host plan.

hybridization The mating of two genetically distinct individuals.

hyperplasia Excessive abnormal growth in a plant due to increased cell division.

hypertrophy Excessive abnormal growth in a plant due to cell enlargement.

infection Successful establishment of the pathogen in the host plant. For the pathogen, infection involves penetrating plant tissues, obtaining nutrients for growth and development, and reproducing itself.

meristematic tissues Undifferentiated plant tissues that continually divide to produce specialized organs and tissues.

resistance Ability of the host plant to restrict or mitigate the damaging effects of the pathogen. Resistance is genetically governed in the host plant and ranges from partially to completely effective.

sclerotium Compact mass of hyphae capable of surviving environmental conditions unfavorable to the fungus.

teliospores Globose, resting-spore stage of smuts and bunts, surviving for up to 25 years in storage. Teliospores of bunt fungi contain trimethylamine, which imparts the fishy odor to the spores and functions to inhibit spore germination and permit extended survival.

virulence The ability of a strain of a pathogen to cause disease on a particular strain of host.

SMUTS, BUNTS, AND ERGOTS are plant diseases that affect the floral or seed bearing tissues of plants. These diseases, caused by plant pathogenic fungi, have had an important role in the history of agriculture. Their greatest impact has been to cause severe losses in crop production and quality throughout recorded history. Because these diseases prevent formation of the head, or replace seeds with a conspicuous fungal structure, records of their occurrence can be found in ancient Greek and Roman writings. Early philosophers believed that these diseases originated from adverse environmental conditions spawned by the supernatural as a punishment for inappropriate human behavior. Ergot of rye and wheat has had far-reaching effects on animal and human physical and psychological states throughout human history. While genetic plant resistance and fungicides currently control most smuts and bunts in Western agriculture, these diseases still

1. For the Department of Agriculture and Agri-Food, Government of Canada.

threaten yield and quality of cereals in certain parts of the world where adequate control measures are lacking. No such control for ergot exists, and this disease is currently causing a pandemic in hybrid sorghum. Owing to their destructive potential and highly variable genetic constitution, these pathogens still remain important plant pathogens worldwide. A basic understanding of the nature of pathogenicity and virulence in smuts, bunts, and ergot, and of resistance in their respective hosts, is required before the threat of these diseases can be eliminated.

I. SMUTS AND BUNTS

A. History

It is likely that the smuts and bunts were evident to prehistoric peoples when they first domesticated wheat and other cereals. Indeed, the fetid odor of the common bunt spores, resembling that of rotten fish, and the black powdery mass of spores that replaces the healthy grain (Fig. 1a–c), makes this disease difficult to ignore. The origin of the term "bunt" is unknown but is probably Latin or Greek, and is a derivation of the word "burnt". Early references to bunt or corruption of wheat, found in ancient Greek and Roman writings, attributed the cause to superstitious origins or to the environment. Theophrastus and Pliny believed adverse environmental conditions, such as rainwater lingering on the head or the burning rays of the sun, cause a corruption or degeneration of the wheat plant. A publication in 1709 by Camerarius entitled "De Ustilagine frumenti" documents a situation where the Gauls in 1550 found that their wheat crop resulted in mature heads made up of a "black foetid dust." The notions that adverse environment or superstition caused the corruption of grain persisted even beyond the Middle Ages. In his treatise, *Horse-hoeing Husbandry*, published in 1733, the agronomist Jethro Tull concluded that the condition "smuttiness of wheat" was caused

Fig. 1. Disease cycle of wheat bunt. (a) Mature wheat spike infected with common bunt. (b) Healthy seeds (left), partially bunted seeds (center), and bunt balls (right) originating from a single infected wheat head. (c) Broken bunt ball showing mass of teliospores. These become attached to healthy seeds during harvesting or spread over the soil surface. (d) Teliospores of *Tilletia controversa,* the dwarf bunt fungus, showing spore ornamentation. (e) Wheat field infected with common bunt. Infection occurs following establishment of the fungus in the apical growing point of the wheat plant within the first 3–4 weeks following seed germination. The plant remains symptomless until seed formation, when bunt balls are formed. See color insert.

by excessive soil moisture. Interestingly, he recommended seed treatment with brine, a treatment that had been discovered 70 years earlier when a shipload of wheat had sunk near Bristol, England. The following year, the entire country's wheat crop had been ravaged by bunt except the wheat originating from the salvaged shipload of wheat. The dipping of wheat seed in brine prior to sowing to control bunt became general farm practice for the next century. During the 1700s, bunted wheat was bought by gingerbread makers in England at the time because they could hide the odor and taste of the bunted wheat with ginger and treacle.

In 1755, Mathieu du Tillet of Troyes, France, published his dissertation on common bunt of wheat. Though not trained in the sciences, Tillet possessed a keen interest in science and a determination to solve practical problems in agriculture. In a series of field studies, he demonstrated that the cause of wheat bunt was the black dust associated with diseased heads. He showed that wheat grains artificially dusted with the bunt spores developed bunted heads whereas those originating from clean wheat grains were bunt-free. Tillet also noted that there were differences between the wheat bunt disease, *la carie,* and the loose smut disease, *le charbon.* Although he was unaware of the true nature of bunt dust, his early studies set the stage for Bénédict Prévost, also from France, to establish the cause of common bunt as an infectious disease in 1807. Using a microscope, Prévost described the germination of the black spores and the formation of a promycelium, and primary and secondary conidia, thereby demonstrating the fungal nature of bunt. Prévost also observed that farmers had a custom of immersing their seed in a mixture of sheep's urine and milk of lime in a copper basket, and they believed that this controlled bunt. He conducted experiments which demonstrated that copper acetate and copper sulfate at a concentration of one part per million reduced spore germination and were thus effective in controlling the disease. Prévost correctly believed that the fungus entered the germinating seedling from seed surface but was unable to prove it. Prévost's microscopical observations on teliospore germination and development were confirmed by the Tulasne brothers in 1853, who recognized that the wheat bunt fungus, which had hitherto been known as *Uredo caries,* belonged to a distinct genus of smut fungi, which they named *Tilletia,* in recognition of Tillet's earlier contributions.

B. Morphology

Most frequently, smuts and bunts attack the ovaries and anthers, which form the plant's reproductive structures. However, smuts that attack the vegetative portions of the leaf and stem are common, and a few species affect roots and underground stems. The majority have little effect on the morphology of the host plant until spore formation occurs, although a few smuts induce hyperplasia or hypertrophy resulting in galls or tumors in plant tissues. Infected plants may be stunted, and leaves of infected plants may exhibit some chlorotic flecking or streaking.

Most smuts and bunts exist inside the host plant as vegetative mycelium and produce at least two spore types: (1) teliospores, which range in color from yellowish brown to black, are generally globose, and may be surrounded by sterile cells; and (2) sporidia, which are hyaline, and elongate, globose, or subglobose in shape. Teliospores serve as the resting-spore stage, surviving for up to 25 years under dry conditions. Teliospores of bunt fungi contain trimethylamine, which imparts the fishy odor to the spores and functions to inhibit spore germination and permit extended survival. The teliospore, normally diploid, germinates to produce a promycelium that gives rise to 4–100 haploid sporidia. Meiosis likely occurs during germination, and the four haploid nuclei migrate out into the promycelium and into the developing primary sporidia. In many smuts and bunts, these unicellular sporidia grow in a yeast-like manner on artificial media.

Smuts and bunts are heterothallic, that is, two sporidia, each possessing a different mating type, must fuse to form the dikaryotic infection hyphae, which is the parasitic stage of the fungus in the host plant. In many members of *Tilletia,* sporidia fuse to form so-called H bodies, which indicates that hybridization has occurred. These fungi possess a bipolar mating type, where compatibility is governed by a single genetic locus, or a tetrapolor mating type, where compatibility is governed by two different genetic loci. Following a complex series of pheromonelike signals between the nuclei of opposite mating types, the compatible sporidia grow toward each other and eventually fuse. In some species, dikaryotic secondary sporidia may be formed. Conversely, in others, the sporidial stage may be omitted, and the dikaryotic infection hyphae develop following pairing of compatible nuclei in the promycelium. The infection hyphae grow and collectively form the vegetative mycelium that proliferates in the meristematic tissues of the susceptible host. This dikaryotic stage is often referred to the "obligate" or "biotrophic" form of the fungus since it can only develop in living plant tissues.

In general, there are four different modes of infection for the bunts and smuts: (1) seedling infection, (2) floral infection, (3) stem infection, and (4) local infection. However, there are many obscure smut fungi for which the exact nature of infection is unknown. In some smuts, the infection hyphae penetrate vegetative tissues and proliferate in the plant's meristematic tissues to form local smut infections. In others, the infection hyphae systemically invade in the plant, without obvious impact on the host plant, until it reaches the plant's reproductive tissues. Coincident with the development and maturation of the head is the rapid proliferation by the fungus in the developing anthers or ovaries. Teliospores are formed following transformation of hyphal cells into thick-walled, ornamented spores. In smuts, the spores are held together in a sorus by a persistent or semipersistent membrane of fungal origin. In bunts, the sorus wall originates from ovary wall, and thus the bunt ball often assumes the shape of the plant's seed.

C. Taxonomy

There are approximately 1200 described species of smut and bunts belonging to 33 genera, and all but a few species are parasitic on angiosperms. More than half the species attack members of the Gramineae, the grasses and cereals. There are also numerous smut species parasitic on members of the Cyperaceae, the sedges.

The classic studies conducted by the Tuslane brothers formed the foundation for taxonomy of the smuts and bunts. Their studies demonstrated that the promycelia and sporidia of the bunt, smut, and rust fungi were analogous to the basidia and basidiospores of the Tremellineae, Hymenomycetes, and Gastromycetes. These studies form the basis of the modern day taxonomy of the Basidiomycetes. The smuts and bunts are classified in the phylum Basidiomycotina, because they produce sexual basidiospores (sporidia) externally on a basidium (promycelium). They are placed in the order Ustilaginales, which contains two families. Tulasne separated the smuts and bunts into two families, the Ustilaginaceae and the Tilletiaceae, respectively, primarily on the basis of the morphology of the promycelium. The Ustilaginaceae exhibit a septate promycelium with lateral and terminal sporidia (see Fig. 2c), whereas the Tilletiaceae bear a nonseptate promycelium with terminal sporidia. Teliospores of members of the Ustilaginaceae do not contain trimethylamine and therefore are odorless. Among members of the Ustilaginaceae, teliospores develop by intercalary morphogenesis of the infection hyphae. This contrasts with acrogenous morphogenesis of teliospores among members of the Tilletiaceae.

Currently, the delimitation of species is based on spore morphology and, to some extent, the host(s) which the species attacks. The size and ornamentation of the teliospores, the germination pattern and number and size of the sporidia, and differences in mating patterns between compatible sporidia are considered relevant characteristics for species designation. Bunt and smut fungi are specific to the host species that they attack. This phenomenon, called host specialization, may also be used in species designation.

Within the Tilletiaceae, the most important genus is *Tilletia*. All but two species of *Tilletia* attack members of the Gramineae. Because of the strong tendency to form H bodies within a whorl of sporidia, these species are likely inbreeding under natural infection conditions. Two of the most important species within the genus *Tilletia* are *T. tritici* (*T. caries*) and *T. laevis* (*T. foetida*), which form the common bunt complex. *Tilletia tritici* has reticulate spore ornamentations, whereas *T. laevis* has a smooth spore. These species freely hybridize under laboratory conditions; in the field, however, hybrids between the

Fig. 2. Smut of barley and wheat (a) Covered smut of barley caused by *Ustilago hordei.* Note the compact spore masses that are liberated and spread to healthy seeds during harvesting. (b) Loose smut of wheat caused by *U. tritici.* The spores become wind-borne during heading and infect healthy flowering wheat spikes. (c) Germinating *U. hordei* teliospores on an artificial medium showing elongate promycelia and four haploid sporidia. (d) Germinating *U. tritici* teliospore on an artificial medium showing promycelia directly producing infection hyphae. See color insert.

two species are rarely observed in nature because of their inbreeding nature. *Tilletia controversa,* the causal agent of dwarf bunt, is closely related to *T. tritici,* and the two species freely hybridize under artificial conditions. Again, hybrids between these two species are rarely observed under field conditions. Sporidia of *Tilletia indica,* causal agent of Karnal bunt, or of *T. barklayana,* causal agent of rice bunt, do not fuse to form H bodies while they are attached to the promycelium. Instead, hybridization occurs on the plant surface, prior to infection. For this reason, *T. indica* and *T. barklayana* are considered outbreeding species, and are sometimes included in the genus *Neovossia.*

The taxonomy of the Ustilaginaceae is controversial. Conflicts arise when the criteria of spore morphology and host specialization are both simultaneously used for classification. For example, the loose smuts *Ustilago nuda* and *U. tritici* have identical spore morphology and disease symptoms, but *U. nuda* only attacks barley and *U. tritici* attacks only wheat. *Ustilago nigra* and *U. avenae* cause false loose smut of barley and loose smut of oats, respectively, and both have ornamented spores. *Ustilago hordei* and *U. kolleri* cause covered smut of barley and oats, respectively, and have smooth spores. The morphology of *U. nigra, U. avenae, U. hordei,* and *U. kolleri* during germination is similar. However, research has shown that all four species can hybridize in all possible combinations when inoculated to *Agropyron tsukushiense* var. *transiens*. Thus, delimiting the four species based entirely on spore morphology or host specialization is unjustified and reflects the difficulty in classifying these fungi. Molecular techniques are currently being applied to fungi, and classification schemes based on similarity of DNA among species are being compared to traditional schemes in order to better understand the relationships among the various groups of fungi.

D. Economic Impact

Bunts and smuts attack a wide variety of the world's most important food crops such as wheat, barley, oats, rye, maize, sorghum, rice, sugarcane, and millet. Before the advent of fungicides or resistant cultivars, the smut fungi caused extensive losses to crop production. Because most smuts and bunts replace healthy kernels with fungal spores, their impact on yield is directly proportional to the percentage infection. In other words, 1% infection is roughly equivalent to a 1% yield loss. Infection levels exceeding 75% have been observed. For example, in 1890, one-fourth to one-half of the wheat crop in Kansas was estimated to have been destroyed by bunt. Losses in wheat production due to bunt in the U.S. ranged from 5 million to 27 million bushels annually during 1917–1924. During the period 1920–1923, annual losses caused by smut diseases on wheat, barley, and oats in Canada were estimated at $12,831,000. Similar reports were made from other major wheat producing countries including Britain, Germany, Argentina, and Australia. Currently, losses to these fungi are negligible in countries were effective fungicides and resistant cultivars are available. For instance, in North America, losses to these fungi in cereal crops have been reduced to less than 1%, and in some instances, such as in bread wheats, the losses are less than 0.1%. Nevertheless, the potential for loss to these diseases is always present, as individual fields of barley and oats occasionally have smut as high as 10 to 25% infection.

In addition to yield losses, the bunt fungi cause quality losses. The unmistakable fishy odor associated with bunt spores, which is noticeable at concentrations of 0.5% (wt/vol) of bunt spores to healthy grain, can impart a fishy odor to the grain and to grain-based food products such as bread and biscuits. Bunted grain is therefore automatically downgraded to animal feed. An additional washing charge may be levied to bunted wheat. Although the smut diseases cause less quality problems than the bunt diseases, some quality issues can become important. Wheat and barley seed infected with loose smut exhibit reduced germination, and seedlings have a reduced and delayed emergence. In winter wheat, smut and bunt infected seedlings are more prone to winter-kill. Barley seed with high levels of loose smut infection (15–60%) is unsuitable for malting. For all cereal crops, if the teliospores of the fungus are present at high enough levels so that grain inspectors can visually detect a discoloring of the grain, the grain can be downgraded in quality, resulting in lower returns to the farmer. Furthermore, few producers or grain dealers are willing to accept grain heavily contaminated with surface-borne smuts and bunts for fear of contaminating their own grain handling systems, and this may render the crop worthless.

Smut has also been found to affect the quality of noncereal crops as well. Stem smut of sugarcane caused by *Ustilago scitaminea* also reduces the sucrose content and purity of the cane sap. Seedling survival and tillering is reduced when *Bromus catharticus* seed is infested with *Ustilago bullata.* The resulting hay crop has reduced dry matter yields and tiller numbers, although the fodder quality is only slightly affected. Infected plants also tend to develop more inflorescences but do not live as long as healthy plants. This smut fungus may thus prevent the plants from expressing their normal short-term perennial character.

Numerous "smut explosions," which followed ignition of the fine dust associated with bunt, were recorded in early part of the twentieth century, destroying harvesting equipment and causing injuries to farm workers. This dust, which is more explosive than coal dust, is readily ignited by the static electricity that builds up in the harvesting machinery. The fine spore dust created during thrashing operations has also been implicated in respiratory ailments.

Studies have been conducted to identify possible toxic effects of bunts or smuts to animals or humans, but to date these studies have been inconclusive. Wheat bunt spores have been implicated in hemorrhagic gastroenteritis and other irritations of the gastrointestinal system. Anecdotal reports have implicated *Ustilago maydis,* the causal agent of corn smut, in the human condition called "ustilaginism." The symptoms include a swelling and reddening of the hands and feet. In animals, corn smut spores have been reported to cause abortions in cattle. Conversely, numerous reports that state no harmful effects attributable to ingestion of smut or bunt spores have been published. The immature galls produced by corn smut fungus, known as *Cuitlacoche,* is considered a delicacy in Mexico and some parts of the U.S. *Ustilago esculenta,* a smut of wild rice, is also considered a delicacy in the Far East. *Cuitlacoche* is high in carbohydrates, protein, fiber, and linoleic acid, and low in fat. Smut galls generally erupt through the husk at 12 to 18 days after mid-silk stage, and this is when they are best harvested. After this time, the galls begin to break apart and can become colonized by other species of fungi such as *Fusarium, Aspergillus,* or *Penicillium.* Colonization by these secondary invaders make the galls unacceptable for market and can even pose a health risk if the galls are eaten. It is possible that reports of "ustilaginism" attributed to the corn smut fungus were due to these secondary invaders. As is the case with all fungi, smuts and bunts can induce an allergic reaction in susceptible individuals, and this feature may account for some of the reported effects in humans.

E. Infection Cycle

Seedling infection is the predominant means of infection for the common and dwarf bunt fungi. Bunt teliospores become attached to the surface of healthy seeds and become distributed on the soil surface during harvesting operations. Teliospores of most bunt species can survive at least 1 year in the soil and many years on the seed when stored under dry conditions. Dwarf bunt teliospores may survive 5–7 years in the soil. When conditions are conducive, teliospores germinate and infect the germinating seedling under the soil surface. For common bunt, the optimum temperatures for germination and infection are 14°–20°C. In contrast, teliospores of the dwarf bunt fungus, *T. controversa,* germinate best from 3° to 8°C. *Tilletia controversa* germinates and infects the dormant winter wheat plants during the winter underneath a persistent blanket of snow. Common bunt fungi develop systemically in the plant, often without any obvious symptoms. Symptoms of dwarf bunt infection may consist of flecking or streaking of the leaves in the early spring and a stunting of wheat plants and excess tillering. In the developing spike, the anthers usually fail to develop and the fungal hyphae rapidly proliferate in the developing ovary until almost all host tissues inside the ovary wall are consumed. Fully developed sori contain millions of bunt spores.

Tilletia indica, the Karnal bunt fungus, infects the developing floral tissues. Teliospores normally germinate on the soil surface following exposure to moisture, light, and temperatures of 15°–25°C. Haploid primary sporidia may undergo several mitotic divisions to form secondary sporidia that are forcibly discharged and carried by air currents to the developing grain spikes. The site of hybridization between compatible mating types is unknown but is likely to be the developing spikes. Moderate temperatures, high humidity, or rainfall during flower development favor infection. Owing to localized infection, Karnal bunt does not affect all the kernels in the spike. Kernel infection varies from a small bunted pit on a seed to a totally bunted kernel.

The disease cycle of smut species belonging to members of the Ustilaginaceae can be highly variable.

The fungus overwinters as teliospores in the soil or on the surface of the seed, or as mycelium in the seed embryo. For species that infect perennial hosts, the smut fungus can survive as dormant mycelium in dormant plant tissues. The mycelium of seedling-infecting smuts grows systemically in the host and eventually invades and proliferates in undifferentiated floral tissues. The floral tissues become replaced by masses of teliospores in sori, some of which are covered by either a delicate or tough tissue that eventually breaks open after the inflorescence of the plant emerges. The teliospores in the sori are disseminated normally by wind or rain, but occasionally by insect pollinators. Teliospores that germinate on host tissues produce haploid sporidia which fuse with compatible sporidia to form dikaryotic hyphae that infect the host. In these instances, the sporidia often fuse with sibling sporidia, thereby limiting genetic exchange between different smut isolates. Teliospores that overwinter in the soil germinate when the environment is conducive to produce sporidia, which are disseminated to host tissue before fusing with other compatible sporidia. Genetic exchange between different smut isolates occurs much more frequently in these instances.

The smuts that infect more mature plant tissues do so with binucleate mycelia that penetrate through stomata, through wounds, or directly through host cell walls. These infections can either develop locally or become systemic. In the case of sugarcane smut, teliospores germinate on host tissue in 5 to 6 hr and slowly infect the cane where the bracts attach to the stalk. The mycelium becomes systemic in the canes, and in 6 to 10 months a sporulating "whiplike" organ is produced on the cane apex. With the corn smut, it is the young, actively growing organs such as the ears, tassels, and new leaves of corn that tend to become infected.

F. Host–Parasite Interactions

Through the course of evolution in the host and plant pathogen population, members within a host species, such as wheat, have evolved specific genes that confer resistance to these pathogens. In response, many obligate biotrophs such as the smuts and bunts have evolved a series of avirulence/virulence genes to overcome the specific resistance genes in the host. This interaction between specific resistance genes in the host and specific avirulence/virulence genes in the pathogen is known as the gene-for gene concept and was developed by a U.S. Department of Agriculture pathologist named H. R. R. Flor who studied bunt and flax rust during the 1930s and 1940s in North Dakota. He determined that for every resistance gene in the host, there is a corresponding avirulence gene in the pathogen. The gene-for-gene concept is currently an area of intensive research to understand the molecular interactions between pathogenicity and virulence genes in the pathogen and resistance genes in the host. Recent research on other plant–pathogen interactions has demonstrated that the protein or enzyme products of avirulence genes in the pathogen either stimulate the production of elicitors or act as elicitors themselves. The elicitors are recognized by specific receptors in the host's cells. These resistance receptors, encoded by specific resistance genes in the host plant, initiate a series of resistance responses directed at stopping development of the pathogen. If the sequence of the avirulence protein is somehow changed in the pathogen, by mutation, for example, the host plant may fail to recognize the protein and, hence, fail to initiate the resistance responses. In this case, the pathogen is virulent, and infection and disease progression proceeds normally. Understanding the nature of variability for virulence in the pathogen and resistance in the host is necessary to develop effective, long-term strategies for controlling these diseases.

G. Control

Owing to the seed-borne nature of some smuts and bunts, quarantines can be effective in preventing disease introduction or restricting their movement on infested grain within a country or regions. Currently, dwarf bunt (*Tilletia controversa*), Karnal bunt [*Tilletia indica* (= *Neovossia indica*)], and flag smut (*Urocystis agropyri*), all of which are wheat pathogens, are subject to international quarantines, and many countries have instituted zero-tolerance standards for these fungi. This means that if a single spore is found, an entire shipload of wheat can be rejected. In 1974, China instituted a zero-tolerance policy for dwarf bunt on imported seed. This quarantine action has cost producers and consumers billions of dollars, despite the fact that the disease is considered of minor importance, worldwide. In continental North America, dwarf bunt is restricted to intermountain regions of the Pacific Northwest and a few other localized regions. Similarly, many countries have a zero-tolerance policy against Karnal bunt of wheat. Karnal bunt, originally from northwestern India, was inadvertently brought to Mexico in infested seed in the early 1970s. In 1982, the U.S. instituted a quarantine to prevent its introduction into the Southwest. Subsequently, many other countries followed suit. In 1996, Karnal bunt became established in southwestern Arizona and in California. Thus, the initial quarantine action on Karnal bunt has subsequently cost producers in both the U.S. and Mexico billions of dollars in lost sales, and shipping and cleaning costs. Although quarantines can be an effective means of preventing the introduction of smut and bunt diseases, some consider these quarantines to be an unnecessary impediment to the global trade in wheat. The implementation and continuation of these quarantines may be politically motivated and very difficult to dismantle once instituted.

The impact of most plant diseases can be reduced by adjusting the tillage practices in such a way that environmental conditions promote growth of the plant while suppressing the development of the pathogen. For example, common bunt develops best in cool soils while the wheat plant develops best in warm soils. Therefore, by delaying seeding of spring wheat until the soil has warmed, or seeding winter wheat early into warm soils, the risk of disease can be greatly reduced. In addition, the common bunt fungus will not survive more than 1 year in the soil. Therefore, rotation to a non-host crop for 1 or 2 years following a common bunt outbreak will effectively eliminate the soilborne phase. Because common bunt teliospores are known to survive passage through the intestinal track of livestock, it is not recommended that fresh manure be spread on fields that will be seeded to wheat.

Most of the smut diseases are seed-borne in nature, being transmitted either on the surface of the seed or within the embryo. This makes it difficult to control these diseases using cultural practices, with some exceptions. Some cultural practices can be employed to reduce the incidence of the smuts in corn. Common smut caused by *Ustilago maydis* is favored by dry conditions and warm temperatures. Corn receiving high levels of nitrogen fertilizer develop heavier infestations of corn smut because nitrogen promotes succulent growth. Therefore, heavy applications of N fertilizers or manures to corn is not recommended. Corn smut is favored by mechanical injury to the plants, so practices such as detasseling, buggy-whipping, or cultivation that produce plant injury can lead to an increase in disease incidence. Proper crop sanitation practices and crop rotation can aid in reducing the incidence of smut. The seed-borne smuts can also be controlled by using certified clean seed. Regulations specify that the incidence of smut infected plants cannot exceed 0.01% and 0.05% infection, depending on the seed class. Seed cleaning can also be of use when dealing with the embryo-infecting smut diseases. Generally, smut infected seed is smaller so a proper and rigorous seed cleaning may reduce the proportion of seed that is infected.

Seed treatment fungicides are chemicals that are toxic to fungi associated with the seed and developing plants. Today, fungicides are a reliable and safe means of reducing or eliminating most surface- and seed-borne smuts and bunts. In general, there are two classes of seed treatment fungicides: (a) eradicant fungicides that kill fungi situated on the seed surface and (b) systemic fungicides that are absorbed into the developing seedling and act by killing or retarding fungal growth, whether on the seed surface or within the seedling. The efficacy of any seed treatment fungicide for controlling surface and seed-borne smuts and bunts depends on whether it can prevent infection or systemic spread inside the plant.

Early seed treatments were eradicants. In 1809, Prévost demonstrated the effectiveness of copper sulfate in killing bunt spores on the seed surface and providing disease control. Although effective in reducing losses, copper sulfate was toxic to the developing seedlings, and fungicide treatment was often followed by a lime treatment to reduce the toxicity of copper sulfate to the plant. During the late 1800s, a nonfungicidal treatment for controlling bunt diseases, including those that lie dormant within the seed embryo, was developed by Jensen in Denmark. He demonstrated that soaking seed for half a day at 68°–86°F, followed by a 5-minute soak at 127°–137°F and then immersion into cold water to stop the process, resulted in control of the smut fungi. The Danish hot water treatment was quickly introduced into England and North America. This treatment had to be conducted properly because seed death could result from the high temperature treatment. However, copper sulfate (or blue stone) remained the most widely used treatment for controlling these diseases in the 1800s and early 1900s.

In the late 1800s, formaldehyde in 1 : 40 mixture with water gained popularity because it was effective in controlling all surface-borne smuts and bunts of wheat, barley, and oats, and was less expensive and less phytotoxic to the seedlings than copper sulfate. Liquid seed treatments had their problems, however. If applied to seeds that had cracks in their seed coats, some fungicides would soak into the seed and kill it. Copper carbonate and organic mercury fungicides also became popular at this time because they could be applied as a dust and were effective for controlling smuts and bunts. However, these dusts were hazardous to the health of workers applying and handling the treated seed. In 1938, a liquid mercury formulation, Panogen, was introduced as a highly effective seed treatment fungicide. However, the persistent and toxic nature of mercury-based fungicides to plants and animals was soon recognized and prompted the release of hexachlorobenzene after the Second World War. This compound was not only highly effective against seed-borne common bunt but also effective against the soilborne phase when applied at higher rates. Today, newer organic sulfur compounds known as dithiocarbamates, such as maneb, provide control for some smuts and bunts.

In the 1960s, a new class of systemic seed treatment fungicides, the oxanthiians, were discovered. These compounds, of which carboxin and oxycarboxin are the most important, inhibit succinate dehydrogenase, a mitochondrial enzyme important in respiration. These compounds, marketed under the trade names Vitavax and Plantvax, are absorbed through the seed coat into the seed, where the active ingredient is taken up by the developing seedling. They have a narrow spectrum of activity but are effective against most of the surface- and seed-borne cereal smuts and bunts. In the past 20 years, numerous triazole based fungicides have been introduced with excellent broad-spectrum activity against many fungi including the bunts and the smuts. These fungicides are effective at very low concentrations and act by inhibiting ergosterol biosynthesis in fungi. Unlike the oxanthiians, the triazoles are absorbed by both the seed and roots, which increases their efficacy. Dividend, a new and highly effective triazole-based fungicide, was recently registered in the U.S. for control of dwarf bunt. This represents a landmark for control of dwarf bunt, which up until now could not be effectively controlled by seed treatment fungicides.

In the early stages of their development, seed treatment fungicides were difficult to apply to the seed, and they were nonspecific, that is, were generally toxic to microbes, plants, and animals. Today, the seed treatment fungicides are safe, easy to apply, highly specific to fungi, and effective at very low concentrations. Some also have the advantage of controlling several seed and seedling diseases with one application. In the future, seed treatments may act by stimulating the plant's own natural defense mechanisms.

Throughout most parts of the world, there is still a heavy reliance on seed treatment fungicides to control smuts and bunts because many popular varieties are highly susceptible to these diseases. For example, many new semidwarf varieties of wheat, which are very high yielding, are also highly susceptible to common bunt. Seed treatments have not eliminated the threat of these diseases, however, because not all producers routinely employ them. Because these diseases can produce large quantities of teliospores from a single infection, they can multiply very rapidly. Even light infestations of seed, under highly conducive environmental conditions, can lead to an epidemic. Heavy reliance on seed treatment fungicides also increases the risk that the pathogen will develop resistance or tolerance to the fungicides. Resistance of fungal isolates to carboxin is already known in corn smut, *Ustilago maydis,* and barley loose smut, *Ustilago nuda.* Carboxin resistance in *Ustilago maydis* can develop following a single amino acid change in the iron–sulfur subunit of succinate dehydrogenase. In Europe, the spread of carboxin-resistant mutants of barley loose smut has been hindered through the use of other classes of fungicides that are effective against the fungus. However, this approach increases the risk that pathogen strains with resistance to several fungicides will evolve.

H. Genetic Resistance

The most economical and environmentally desirable means of controlling bunts and smuts of cereals is by deploying genetic resistance in the plant. Most commonly, this resistance is governed by the action of a single gene in the host plant, but two or more genes acting in unison may also condition resistance. In wheat, a series of approximately 15 different resistance genes (Bt-1, Bt-2, Bt-3, etc.) have been identified and catalogued for common bunt. For loose smut, up to 11 dominant genes for resistance have been reported.

Many resistance genes have been deployed during the past 100 years for control of common bunt, but the evolution of new virulences in the pathogen, in single and multiple combinations that overcome resistance, has occurred with regular frequency. Perhaps the most elegant example of the evolution of new races occurred in the U.S. Pacific Northwest. In the period from 1900 to 1930, races with a relatively simple virulence spectrum predominated. For example, race surveys during this time showed that races could only overcome the Martin gene (Bt-7). Then, in the mid-1920s, the varieties Ridit and Albit were released. Ridit carried the Bt-3 gene which conditioned resistance to 23 of the 28 known races, and Albit carried the Bt-1 resistance gene which was effective against about half of the known races. During the years from 1931 to 1947, new races predominated that could individually overcome Bt-1 and Bt-7, and new races that overcame the Bt-3 gene were detected. In the meantime, Turkey, carrying Bt-4, and Rio carrying Bt-6, were released. Both conditioned resistance to all but four of the known races. Before 1947, however, a new highly virulent race that overcame the Bt-4 resistance of Turkey was detected. During the years 1950–1957, new sources of resistance were introduced with the release of Brevor, which contained Bt-1 and -4, and Omar, which contained the Bt-1, -4, and -7 combination. During this period, races that overcame the Bt-1 and -7 predominated. Races that overcame Bt-6 and a new race that overcame the combination of Bt-1, -4, and -7 resistance genes in Omar were detected. During the years 1958–1960, the variety Columbia, which also contained the Bt-1, -4, and -7 combination, and Burt, which combined Bt-1, -6, -7 and Brevor resistance, were released. Today, there are several races that will overcome these resistance gene combinations. In fact, races that overcome the Bt-1, -2, -4, -6, and -7 combinations are widespread.

What have we learned from attempts to develop common bunt resistance in the U.S. Pacific Northwest? First, the population of common bunt present at the beginning of the 1900s was highly variable. It is likely that a small proportion of the existing races was capable of overcoming, individually, some or all of the Bt resistance genes that were introduced during the first 60 years of the twentieth century. Second, new races that were able to overcome the combinations of resistance genes gradully evolved through mutation or hybridization and recombination among existing races. Thus, new races possessing complex virulence combinations likely arose from existing races. Deploying resistance genes, one at a time, merely selects for those races that already possess virulence on those resistance genes. As a result, these races predominate because the very survival of the pathogen rests on its ability to overcome resistance genes. Combining several "defeated" resistance genes into a single variety after the pathogen is able to individually overcome the genes selects for races to develop multiple virulence combinations. This illustrates the risk in deploying resistance genes to a genetically highly variable pathogen for which virulence to the genes already exists in the population. A more balanced approach would be to pyramid two or three resistance genes, for which virulence does not exist in the pathogen population, into a single crop variety. In this scenario, the pathogen would have to simultaneously mutate from avirulence to virulence at two or three loci in order for it to overcome the resistance in the particular variety. The probability of this occurring is very small. Molecular techniques now enable the pyramiding of multiple resistance genes in a single variety.

In loose smut of wheat, there is also a form of resistance termed "morphological" or "field" resistance. In this form of resistance, cultivars that are susceptible when artificially inoculated with loose smut are resistant or nearly resistant under field conditions. The mechanism of field resistance is not known. It is effective against all races of the pathogen and cannot be lost through the evolution of new virulence genes in the pathogen population. More study is needed to determine the nature of field resistance to permit plant breeders to easily select resistant plants when developing new varieties. Some wheat and barley cultivars may also possess a different form of "incompatibility" to loose smut. In this situation, infected plants emerge from the soil and die, thus preventing further spread of the pathogen.

II. ERGOT

A. History

The term "ergot" is of French origin, derived from *argot,* meaning cock spur. It refers to the small pointed nail on the heel of a cock, which is similar in shape to the ergot body. Ergot of cereals and grasses is caused by the fungus *Claviceps purpurea.* In Latin, "claviceps" means club-headed, and "purpurea" means purple. The name pertains to the conspicuous dark purple-black ergot body or sclerotium that is visible in infected rye heads (Fig. 3a). The sclerotium can be up to 10 times longer than the normal seed. The conspicuous nature of the sclerotium probably explains the numerous references to the disease throughout recorded history. The disease has had such a dramatic impact on mankind because

Fig. 3. Ergot of wheat. (a) Wheat heads with prominent ergot sclerotia. Note distortion of the wheat head on the left. (b) Honeydew stage of ergot. The drops are high in sugar and contain millions of fungal spores. Honeydew is formed after initial infection of the developing flowers. Splashing rain and insects carry the spores to other developing spikes, which the spores infect.

the ergot body contains many compounds toxic to humans and other mammals. Prolonged ingestion of the sclerotia or of bread made from ergoty flour produces devastating physiological and psychological effects on the human body, called ergotism. The ergot body also contains psychoactive compounds similar to lysergic acid diethylamine (LSD), which only adds to the horror and pain of afflicted individuals. Many thousands died from ingestion and ergot poisoning during the many epidemics that ravaged Europe during the Middle Ages. Children were the most susceptible to the ravages of ergot. Severe chronic ergot poisoning eventually turned individuals into screaming, gibbering individuals whose fingers, toes, arms, and legs blackened and corroded away from a dry gangrene. Before the etiology of the disease was known, the only known relief from this scourge was death.

To understand the impact that ergot had on humankind, especially during the Middle Ages, it is necessary to understand the nature and importance of its primary host, rye, in bread making. The Greeks and the Romans described ergot but did not report any of its medical effects because they did not employ rye in their breads. Rye was the bread corn of the Teutons, or Germans, and was introduced into southern Europe in the early Christian era. It was most popular in France and in a large belt extending from Holland across northern Germany, Czechoslovakia, Austria, Poland, and Central Russia. It was not widely employed in England and southern Europe because of the tradition in those areas of making bread out of wheat flour. Rye was the culprit in the ergot epidemics because it is much more susceptible to ergot than wheat, barley, or oats.

The first recorded epidemic from ergot was in AD 857 in the Rhine Valley, where thousands of peasants died. The disease became known as *sacer ignis* or "holy fire" because afflicted individuals often experienced an intense burning and were thought to be cursed by God. Ergot epidemics were sporadic, striking only every 25–50 years in some regions of France and Germany, and more commonly in others. Ergot generally occurred following prolonged rainy conditions that favored development of the fungus. Ergot poisoning was most frequently observed shortly after harvest. Exhaustion of the previous year's food stores by midsummer resulted in food shortages for many peasants. Therefore, the grain that fell from the ear during harvesting was immediately milled into flour. Because the ergot sclerotia are easily detached, this early milling was rich in ergot sclerotia. Indeed, the disease was widely viewed as a disease of the poor because the wealthy could afford to eat clean grain and purchase higher quality cereals such as wheat, which are much less susceptible to ergot.

In 1039, during an ergot epidemic in France, Gaston de la Valloire built a hospital to care for afflicted individuals and dedicated it to the memory of St. Anthony. The disease was hereafter known as St. Anthony's fire, and monks eventually established the Order of St. Anthony to take care of afflicted individuals. Sufferers often improved in their care, likely because of the temporary absence of ergoty bread from their diets, but would often relapse on returning to their peasant life. The cause of St. Anthony's fire remained a mystery until the 1670s when a quiet French country physician named Thuiller made careful observations on those succumbing to St. Anthony's fire. He noted that the wealthy or city dwellers did not succumb to this disease but the country peasants were very susceptible to epidemics. He reasoned that the disease must not be infectious and had to be linked to diet. One day, his investigations brought him to a field of rye, heavily infected with ergot bodies. As a physician, he knew about the medicinal properties of ergot bodies, and of their toxicity when administered in high doses. Although he was convinced of his diagnosis, he was unable to prove it, and worse, he was unable to convince the farmers not to eat the sclerotia. The proof that eluded Thuiller remained hidden for another 200 years until Louis Tuslane, one of the Tuslane brothers, conducted microscopical and inoculation studies to determine the life cycle of the causal agent. During the 1700s, however, there was mounting evidence that the ergot sclerotia were responsible for St. Anthony's fire, and countries such as Prussia were exchanging ergoty rye for sound grain. Furthermore, rye was gradually losing importance as a staple among the peasantry as other crops such as potatoes were introduced into Europe. Other improvements in agriculture included improved drainage of cereal fields, which also contributed to the reduction of ergot severity.

Ergot poisoning still sporadically caused devastating losses to human life and, in some instances, was responsible for changing the course of history. In 1722, Peter the Great of Russia was poised to conquer Turkey and extend his empire beyond the ice-free ports on the Black Sea. On the eve of his attack on the Sultan Ahmed III's army, Peter's armies and their horses were fed on a rye originating from the Volga delta. Shortly after eating the grain, horses were immobilized with the staggers and scores of his men were writhing in pain. Over 20,000 soldiers reputedly died, and with them did Peter's dream of conquest of the Ottoman Empire. Even in the twentieth century, a major epidemic occurred in Russia 1926–1927 with more than 10,000 cases of ergot poisoning and 89 deaths reported. The last outbreak was recorded in Pont-Saint-Esprit, in Provence, France, in 1951. An unscrupulous farmer sold ergoty rye to an equally unscrupulous miller and baker who employed the grain to make bread that was consumed by the majority of the inhabitants of Pont-Saint-Esprit. At the end of the epidemic, over 200 cases of severe ergot poisoning and four deaths were recorded.

Ergot is still commonly found in temperate regions of the world. Ergot poisoning can be a threat to grazing animals in ryegrass pastures in some parts of the world, and care must be taken to prevent animals from feeding on ryegrass that has gone to seed. Today, ergot remains a "grade loss" issue to most farmers, and ergot poisoning is unknown in most developed countries. In fact, most countries have legislation governing the maximum allowable percentage of ergot in cereals used for human consumption. In the U.S., any grain containing more than 0.3% ergot sclerotia by weight is graded as ergoty and cannot be sold for human consumption. Effective sources of ergot resistance are unknown in cereals, however, and sporadic outbreaks still occur. Furthermore, a new ergot pandemic caused by *C. africana* is presently occurring in hybrid sorghum. Within three years, the disease spread from the center of origin in Africa to Australia, and throughout South, Central, and North America. Consequently, this disease has become a major threat to sorghum production.

B. Economic Impact

Ergot attacks many of the world's most important food crops such as wheat, barley, oats, rye, triticale, sorghum, and wild rice. Historically, the disease has been very important in rye because of heavy reliance on breads made with rye flour by the European peasantry, and because of the deleterious health effects of eating bread made with ergoty flour. The highly susceptible nature of rye to ergot is due to fact that rye is a cross-pollinated crop; the flowers of one spikelet must be fertilized from pollen originating from a different spikelet. To facilitate cross-pollination, the flowers stay open for an extended period of time during the summer. Cool, wet weather during flowering further extends the period that the flowers remain open. The ergot fungus requires open flowers in order to successfully infect the rye flowers. Therefore, the longer the flowers remain open, the higher are the ergot infection levels. Levels as high as 30% ergot have been recorded in rye under field conditions. Triticale, a hybrid between wheat and rye, is also quite susceptible, and high infection levels have been observed in some lines. Wheat, barley, and oats are self-pollinated plants, and the flowers do not generally remain open for long. These crops, however, are susceptible to ergot, and under conditions of high moisture, the flowers can remain open and become infected with ergot. Losses in grain yields due to ergot usually exceed the percentage of ergot bodies because sterility of the florets surrounding the developing ergot body is frequently observed. Ergoty grain must be cleaned before it can be sold, which adds to production costs.

Sorghum is the fifth most important food crop and is currently grown on over 45 million hectares, worldwide, for food, feed, and beverages. The majority of the world's sorghum production relies on hybrid seed; commercial seed companies employ male-sterile lines in hybrid seed production. Ergot is especially serious in hybrid sorghum because the flowers of male sterile sorghum lines remain open for an extended period. Low temperatures below 13°C for 2–4 weeks before anthesis can also cause male sterility in male-fertile sorghum. Male sterility, coupled with prolonged cool, wet conditions around 19°C, is highly conducive for ergot infection. Under normal environmental conditions, commerical male-fertile lines generally escape infection. However, widespread damage has been observed in these same lines following conducive conditions. In India, losses of 10–80% have been reported. In Zimbabwe, annual losses of 12–25% are observed. Ergot severity declines rapidly with increasing temperatures and is rare at 28°–30°C.

The toxic as well as beneficial properties of the numerous compounds contained in the ergot sclerotia have been elucidated since the beginning of the twentieth century, and many continue to be employed in modern medicine. Spain, Portugal, and Russia have been the traditional suppliers of ergot sclerotia. Spanish and Portuguese ergot contains the highest quantity and quality of alkaloids, which range between 0.05 and 0.3%, compared to 0.02 to 0.10% in Russian sclerotia. The total Iberian crop of ergot sclerotia between 1920 and 1929 was an estimated 725 tonnes. Ergot sclerotia from other species such ryegrass (*Lolium perenne*) are usually smaller, but the alkaloid content may be higher. Although sorghum ergot sclerotia contain some alkaloids, clinical symptoms of ergot poisoning have not been observed when ergoty grain was fed to livestock. Mild symptoms in animals include feed refusal and milk production failure. It is likely that different ergot species produce different types and quantities of alkaloids.

C. Host Range

The ergot fungi belong to the class Ascomycetes, series Pyrenomycetes, order Clavicipitales, family Clavicipitaceae. Other members of the Clavicipitaceae produce ergotlike alkaloids that are toxic to grazing animals. Species such as *Epichloe typhrina*, *Acremonium* spp., and *Balansia* spp. have recently been implicated in atypical ergotism toxicity in cattle. A large outbreak affected approximately 50,000 cattle in Louisiana in the early 1970s. These fungi belong to a group that are collectively referred to as plant endophytes and form symbiotic associations their grass hosts.

There are approximately 200 host species listed for all ergot species, worldwide, all of which are monocots. There are less than 15 species that belong to the genus *Claviceps*. The most important ergot species is *Claviceps purpurea* (Fr.) Tul., which attacks more than 150 members of the Gramineae, the family of cereals and grasses, which includes wheat, barley, and oats. In contrast, other ergot species have a very limited host range. *Claviceps gigantea* is known to cause ergot in corn only in the high humid valleys of central Mexico. Two species of ergot that attack sorghum, *Sorghum bicolor*, are *C. sorghi* and *C. africana*. *Claviceps cinerea* has only been recorded on curly-mesquite and tobosa grass. *Claviceps grohii* has been reported on *Carex* spp., members of the Cyperaceae.

Several researchers have suggested that biotypes or races of *C. purpurea* exist whose pathogenicity is limited to certain species. For example, the biotype from sweet vernal grass (*Anthoxanthum odoratum*) attacks rye and some other grasses but not barley. Wood brome (*Brachypodium sylvaticum*) and manna grass (*Glyceria fluitans*) are reported to each have their exclusive single biotypes. It is difficult to obtain a clear picture of the host range of these putative biotypes because infection studies were carried out under artificial conditions. It is likely that there is little host specificity, and the strain which attacks rye also attacks most species of grasses in addition to wheat, barley, and oats. There is evidence of some host specificity, but these reactions may be peculiar to a particular isolate rather than confirm the existence of patterns of pathogen evolution.

There is no evidence for a gene-for-gene interaction between the ergot fungus and the host. In fact, there is no evidence for immunity to ergot among commonly grown cereals. This is not surprising from an evolutionary standpoint since ergot is seldom severe enough in plants in their native habitats to direct the evolution of plant resistance.

D. Infection Cycle

Most ergot species possess three morphological stages in their life cycles, the sclerotium stage, the ascigerous (fruiting or sexual) stage, and the sphacelia (honeydew or asexual) stage. The occurrence of multiple stages led early investigators to believe that there were three different organisms involved. The pathogens overwinter as the sclerotium on the soil or near the soil surface. In rye ergot, sclerotia originate from the previous year's crop or from wild or volunteer grasses in ditches or headlands. Sclerotia can also be spread by sowing contaminated grain.

The sclerotia remain viable for about 1 year whether in soil or stored dry in grain. Sclerotia require several weeks of cold-temperature treatment near 0°C followed by several weeks at temperatures near 18°C for optimum germination. During the spring and early summer, the ergot sclerotium produces 1 to 60 flesh-colored stalked fruiting bodies 5 to 20 mm long, called stromata. These stromata possess knob-shaped heads, in which numerous saclike perithecia (the ascigerous stage) develop. Within the perithecia are borne numerous club-shaped asci, each containing 8 filiform ascospores. The ascospores are products of meiosis and form part of the sexual stage of the fungus. The ascospores are dispersed by wind and splashing rain and are carried to the flowers.

The spikes are more susceptible to ergot infection in the early stages of flowering. When the spores come in contact of the flower's stigma, the spores germinate and penetrate the ovary within 24 hr. Approximately 5 days later, a mass of mycelium known as the sphacelia develops on the ovary surface, which produces millions of conidiospores in sugar droplets. These droplets are referred to as the honeydew (Fig. 3b). The conidiospores, which are products of mitosis, may be carried to other florets or to other flowering plants by splashing rain. The honeydew attracts insects, which transport the spores to other flowering plants. The mycelium in the ovary also gives rise to the sclerotial hypha, which elongates from the base and differentiates into a hard, purple black surface layer, a fertile hyphal mass, and a central layer of storage cells. The sclerotia fall to the ground or are harvested with the grain at plant maturity.

The infection cycle of sorghum ergot is similar to that of rye ergot, but some gaps in our knowledge of the cycle remain. For example, the sclerotia of *C. sorghi* produce both the ascigerous stage and a conidiospore stage. On the basis of field observations, both stages are hypothesized to infect collateral hosts which, in turn, provide fresh honeydew for primary infection of sorghum. In *C. africana,* the ascigerous stage is not considered important in perpetuating the pathogen, and the conidiospores associated with the honeydew, sphacelia, and sclerotia are the source of primary and secondary infections in both the collateral and primary host. In Australia and the Americas, *Sorghum halepense,* which flowers continuously, is considered an important collateral host of *C. africana.* Sclerotia, sphacelia, and honeydew become mixed with healthy seed during harvest and can serve as a seed contaminant. In fact, seed stickiness due to contamination by honeydew can cause difficulties with mechanical sowing and predispose seed and seedlings to infection by soilborne pathogens.

E. Control

Quarantines can effectively prevent the spread of ergot between continents. However, the high frequency of seed exchange between countries worldwide makes quarantines difficult to enforce, particularly in developing countries that lack infrastructure and resources. *Claviceps africana* was likely transported from Africa to Brazil in contaminated seedlots in 1995. Airborne inoculum, consisting of secondary condiospores, has been implicated in the spread of *C. africana* throughout different sorghum producing regions of Africa and rapid spread throughout the Americas since 1995. Therefore, quarantines would have likely been ineffective in preventing the regional spread of this disease.

Destroying collateral hosts in the proximity of the host crop and mowing headlands and roadside ditches to prevent flowering are probably the most important means of reducing the impact of ergot. Rotation to a nonhost species can be an effective means of eliminating the sclerotia from the soil, but must be accompanied with effective weed control since many weed species can harbor ergot. Seed cleaning to remove sclerotia can be an effective means of reducing the severity and spread of ergot. In heavily infested grain, however, it is difficult and expensive to bring the percentage down to within acceptable limits. Early investigators understood that seed and sclerotia possessed different densities and the latter would float in a brine solution. A solution of 10% salt will remove up to 99% of *C. africana* sclerotia. Today, potassium chloride is more commonly employed because it is less damaging to the seed. Soaking the seed also removes honeydew and increases seed germination and seedling development.

Adjusting seeding dates to avoid ergot is a common practice in many countries. Seeding may be delayed or advanced in order that flowering coincides with the midseason dry spell. This practice is common with sorghum production in central Africa and India. By growing the crop in dry regions under irrigation, the water can be restricted at critical stages during flowering, thereby reducing ergot infection. Crop sanitation by removing infected panicles can help reduce ergot. Plowing to bury sclerotia below 4 cm of soil will prevent stromata formation and ascospore release. However, the pathogen is able to spread very quickly from a single infection source, if environmental conditions are conducive, and epiphytotics do occasionally occur despite these measures.

Fungicides are not normally employed to control ergot because of the high cost. Only in seed production operations, where there is a sufficiently high value for the seed, are fungicides employed. In addition, the fungicide must come in contact with the stigma in order to impart protection. Therefore, several sprays must be applied between anthesis and the end of flowering for effective ergot control in sorghum. Seed treatment fungicides can also reduce the severity of ergot by killing honeydew conidiospores and reducing sclerotial germination.

Resistance is the most desirable method to control ergot, but it is not widespread among commonly grown crop species. There are differences among rye and wheat cultivars in their degree of susceptibility. In less susceptible cultivars, there is a reduction in the number of florets that become infected, and in the amount of honeydew produced, compared to highly susceptible cultivars. In sorghum, resistance is related to the plant's ability to rapidly complete pollination and fertilization, thereby reducing the opportunity for infection of the stigma. Even reducing the interval of susceptibility in sorghum from 8 days in a susceptible cultivar to 5 days in a more resistant cultivar reduced the "risk period" for ergot attack by 37%. Clearly, the key to successful control of ergot involves an integrated approach. By selecting low risk locations, adjusting sowing dates, utilizing fungicides and rotations, reducing the inoculum load through sanitation, and employing the most resistant cultivars available, it is possible to effectively manage this disease in most years.

F. Ergot Poisoning

Early physicians realized that there were two distinct components of ergotism, the convulsive form and the gangrenous form. The two forms were initially considered a manifestation of the same toxic substance. However, early physicians realized that the convulsive form was especially prevalent in Germany but not in France, where the gangrenous stage was common. It is now known that the ergot sclerotium contains over 100 different groups of substances, the most important with respect to pharmacology and toxicology being the alkaloids. By 1950, 20 different alkaloids had been identified and studied. The first experiments to establish the toxic effects of ergot were conducted by Tuillier on poultry, which died shortly after being fed ergot sclerotia. In Strasbourg in 1774, Read reported that a pig fed ergoty grain lost an ear to gangrene after 17 days on a diet of ergoty grain. Even at this time, at least 100 years after publication of Tuillier's observations on St. Anthony's fire, there was still great controversy on its cause. It was not until the early 1900s, when chemists had made advances in purifying the toxins, that physiological studies shed light on the toxicology of ergot sclerotia.

It is now known that there are several toxic substances contained in the *C. purpurea* ergot body including ergosterol, ergotoxin, ergotamin, ergostetrine, ergoclarin, and lysergic acid diethylamide. Some of the compounds affect the central nervous system and cause convulsions and brain and spinal lesions, whereas others cause a constriction of the capillary blood vessels in both the smooth and skeletal muscles, as well as in other tissues. Large doses or continuous exposure for long periods can reduce blood flow to the extremities such as fingers and toes, resulting in gangrene and the eventual loss of these organs. Feeding ergot to livestock can cause abortions and prevent lactation.

Incredibly, ergot has also long been known for its beneficial effects on human health. Ergot was used by German midwives to help women in labor. The toxic substances induce a specific paralysis of the smooth muscles, particularly those associated with arteries and the uterus, which prevented excessive bleeding. In Adam Loncier's 1582 publication *Kreut-*

erbuch, he states that three sclerotia proved useful in inducing pains of the womb. This rate would be in accordance with rates used in modern medicine. Physicians were reluctant, however, to prescribe ergot because of the unpredictability of dosage and the dire effects of a single ergot overdose. Today, some are routinely prescribed for treating excessive bleeding during childbirth, high blood pressure, varicose veins, and migraine headaches.

The quantity of ergot required to cause symptoms varied widely in reports and were inconclusive. This was because the alkaloid content in ergot sclerotia varies with the ergot strain, and with the age and size of the ergot sclerotium. In the 1700s and 1800s, many felt that convulsive symptoms were caused by the lower doses and the gangrenous form was caused by higher doses. In general, 1% ergot in grain can cause acute poisoning and chronic poisoning, whereas 7% ergot can cause fatal poisoning. The diet of the peasants also influenced the severity of symptoms and may explain why the convulsive symptoms were not common in France. The convulsive form has been attributed to an ergot induced vitamin A deficiency. In France, a peasant's diet was rich in dairy products that are high in vitamin A; German peasants on the other hand, consumed very few dairy products. Consumption of dairy products likely suppressed the convulsive form, thereby permitting the afflicted individuals to eat higher quantities of ergot which, in turn, led to the expression of gangrenous form.

In 1938, Dr. Albert Hofmann, at Sandoz Laboratories in Switzerland, isolated a new ergot alkaloid, lysergic acid diethylamide (LSD). He experienced the mind altering effects of this new alkaloid on two occasions in 1943. The first was inadvertent, when he accidently absorbed some through his skin while conducting routine chemical tests. He began to experience vivid multicolored illusions and distortions that lasted for several hours. On the second occasion, he set out to determine if this new compound was responsible for his hallucinations. On publication of his results, there was intense interest in LSD, particularly by psychiatrists, who saw a possible cure for schizophrenia, depression, and even alcoholism. However, the hallucinations induced by LSD were frequently terrifying, and it became evident that the LSD was more harmful than beneficial. As chemists learned to synthesize LSD, it became more widely available. In the 1960s and 1970s, LSD was extensively employed by young adults as a recreational drug and remains so today. It is a criminal offense to possess LSD.

III. CONCLUSIONS AND FUTURE OUTLOOK

Bunts and smuts are currently controlled by a combination of genetic resistance in the host plant and efficacious seed treatment fungicides. However, these fungi are genetically highly variable, and many resistance genes in the host, and some fungicides, have been rendered ineffective due to the ability of the fungi to overcome these barriers to infection. Furthermore, new high yielding crop varieties tend to be more susceptible to bunts and smuts than the land varieties used by our ancestors. Recent developments in molecular biology offer the promise of developing an understanding of the basic mechanisms involved in pathogenicity and virulence in these pathogens, and resistance in the host crop. This knowledge will someday be employed in developing durable, effective, and sustainable control of the smuts and bunts.

With the exception of the pandemic that is occurring in sorghum, ergot is essentially a disease of the past. In the developing world, however, it may still occasionally cause problems when the seed is not sufficiently cleaned or when conducive environmental conditions prevail. Educating producers in the agronomic practices that minimize ergot severity in crops will further reduce the importance of this disease.

See Also the Following Articles

Plant Disease Resistance • Rust Fungi

Bibliography

Bandyopadhyay, R., Frederickson, D. E., McLaren, N. W., Odvody, G. N., and Ryley, M. J. (1998). Ergot: A new disease threat to sorghum in the Americas and Australia. *Plant Dis.* **82**, 356–367.

Barger, G. (1931). "Ergot and Ergotism." Gurney and Jackson, London.

Carefoot, G. L., and Sprott, E. R. (1967). "Famine on the Wind." Longmans Canada Limited.

Fischer, G. W., and Holton, C. S. (1957). "Biology and Control of the Smut Fungi." Ronald Press, New York.

Malik, V. S., and Mathre, D. E. (eds.) (1998). "Bunts and Smuts of Wheat: An International Symposium." North American Plant Protection Organization, Ottawa, Canada.

Trione, E. J. (1982). Dwarf bunt of wheat and its importance in international wheat trade. *Plant Dis.* **66**, 1083–1088.

Vánky, K. (1994). "European Smut Fungi." Gustav Fischer, Stuttgart.

Wilcoxson, R. D., and Saari, E. E. (eds.) (1996). "Bunt and Smut Diseases of Wheat: Concepts and Methods of Disease Management." CIMMYT, Mexico City.

Soil Dynamics and Organic Matter, Decomposition

Edward R. Leadbetter
University of Connecticut

I. Soil Composition
II. Origins of Organic Matter in Soils
III. Decomposition of Organic Matter in Soils
IV. Decomposition Represents Activities Associated with Growth
V. Decomposition Can Occur in the Presence or Absence of Air
VI. Soluble Organic Molecules Are Formed in Soil as a Consequence of Fermentations
VII. Anaerobic Respiration Introduces New Biomass into Soils and Releases Reduced Inorganic Ions of Nitrogen and Sulfur
VIII. Chemolithotrophic (Chemoautotrophic) Bacteria Consume Inorganic Chemicals as They Produce New Biomass
IX. Microbial Protoplasm, Too, Is Subject to Decomposition
X. Some Organic Components of Soils Are Decomposed Very Slowly and Incompletely Mineralized

GLOSSARY

aerobic respiration Use of molecular oxygen to accept electrons derived from energy-conserving oxidation reactions by cells.

anaerobic respiration Use of ions or molecules (usually inorganic) other than molecular oxygen to accept electrons derived from energy-conserving oxidation reactions by cells.

biosynthesis Synthesis of cellular components by organisms.

chemolithotrophy Derivation of energy for biosynthesis from oxidation of inorganic compounds.

fementation Energy-conserving, coupled oxidation–reduction processes in which organic molecules serve both roles in the redox reactions.

humus The organic fraction of soil, characteristically black-brown in color.

macrobes eukaryotic (usually) organisms, either single celled or multicellular, visible to the unaided eye.

microbes Prokaryotic and eukaryotic organisms.

photosynthesis Harvesting of light energy in order to synthesize cellular organic compounds and organelles.

protoplasm The inorganic and organic compounds of which cells are composed.

THE NOTION OF SOIL set out in common dictionaries is that it is a surface layer of the earth in which plants grow. A more stringent definition is not easily possible, since in no sense is soil a homogeneous material. Soils differ widely from one local area to another in terms of water content, chemical composition, acidity, size and characteristics of component particles or aggregates, and organisms (biota) present, and these differences are even more dramatic in geographical regions that differ in climates. When viewed in a vertical perspective, in contrast to horizontal ones, a given soil will likely not be identical in characteristics as one proceeds from the surface to its deeper layers. Perhaps the only general definition is that soil is a complex, natural material of the planet's surface in which a solid phase (as opposed to an aqueous one) is dominant and which is conducive to the growth and survival of different life-forms. The two principal components, apart from moisture and air, of a soil are its inorganic (mineral) and organic ones. Both of these are in a dynamic state; thus soil cannot be considered a static entity.

I. SOIL COMPOSITION

The mineral content of a soil ordinarily reflects the composition of rock that either underlies or is adjacent to the soil; for example, in a region rich in mica, soils are likely to have a lower content of sulfate and more aluminum and silica than soils in an area where, for example, gypsum abounds. The organic components of a soil have accumulated, and are accumulating, by virtue of growth and reproduction of phototrophic organisms, principally plants, although algae, cyanobacteria, and nonphototrophic bacteria (chemolithotrophs) make a smaller contribution. Apart from water, which constitutes the greater portion of an organism's protoplasm, the principal components of protoplasm (the cellular components) are the polymeric materials ("biopolymers") such as polysaccharides, proteins, and nucleic acids and nonpolymeric lipids and lower molecular weight biosynthetic precursors. All of these are a result of the growth and reproduction of the phototrophic organisms, the "primary producers"; eventually these organic materials gain entrance to and become part of soil on the death of these organisms, or when parts of them are shed (e.g., twigs, branches, leaves, flowers). These organic biopolymers undergo decay (decomposition) at variable rates, in and on soils.

An end result of the sequential decomposition processes is the conversion of the atoms (e.g., carbon, nitrogen, sulfur, oxygen, hydrogen, phosphorus) that make up the organic molecules into an inorganic form—a crucial event in maintaining functional habitats and, indeed, the totality of these—the biosphere itself. These essential recycling events, dynamic bioconversions (or biotransformations) mediated in such large part by bacteria and fungi, are described and discussed below.

II. ORIGINS OF ORGANIC MATTER IN SOILS

For all but a few environments, photosynthesis is the primary way in which the energy inherent in light is conserved in the form of chemical bond energy of the biopolymers synthesized by the primary producers. The biosyntheses carried out by the phototroph are dependent on a steady supply of carbon (in the form of carbon dioxide), nitrogen (in the form of the ammonium or nitrate ion), sulfur (often in the form of sulfate), and phosphorus (as phosphate), other minerals in smaller amounts, and, of course, water. Many of these ions are in limited supply on the planet, and were it not for the eventual decomposition and recycling of the products of photosynthesis (e.g., plant biomass), life could not continue to be reproduced (i.e., biosynthesis would cease) because of unavailability of the atoms of which protoplasm is composed.

III. DECOMPOSITION OF ORGANIC MATTER IN SOILS

Decomposition of biopolymers can be viewed as a stepwise or sequential process that begins with the breakdown of polymers into the monomeric units of which they are composed. This first step in decomposition, which, generally speaking, converts the polymer from a nonsoluble or poorly soluble state into the more soluble monomeric form(s), is carried out by a quite diverse array of bacteria and fungi. The further decomposition of the monomers into inorganic ions is brought about by different groups of these microorganisms—a succession of microbial populations is involved. Cellulose, for example, becomes solubilized as a result of its decomposition into shorter and shorter chains of glucose (the monomeric unit of cellulose). Ultimately the carbon atoms of glucose are oxidized, resulting in formation of carbon dioxide, and the hydrogen and oxygen atoms are used to form water by the bacterium or fungus that, by virtue of its growth and reproduction, is decomposing the glucose. In a similar manner the activities of yet different bacteria and fungi bring about the decomposition of the protein polymers into the some 20 different amino acids which are the monomers of proteins. Following this breakdown (or proteolysis), bacteria and fungi, often differing quite markedly in biochemical and physiological properties from the organisms involved in glucose breakdown, bring about the oxidation of the carbons of the amino acids, resulting in carbon dioxide production and the release of amino acid nitrogen as ammonia and, where present, amino acid sulfur as sulfate.

Decomposition of nucleic acids into their monomeric components—nucleotides, and ultimately purines and pyrimidines—results in the release of the organophosphate atoms of the nucleic acids as inorganic phosphate, and the conversion of the organic carbon and nitrogen atoms of the purine and pyrimidines into carbon dioxide and ammonia.

IV. DECOMPOSITION REPRESENTS ACTIVITIES ASSOCIATED WITH GROWTH

These decomposition events, and those outlined above, are carried out by the microbes whose strategy or purpose (as is the case for all living things) is to reproduce themselves; in order to do this they must have a source of energy to power the biological machinery of biosynthesis and a source of atoms such as carbon, nitrogen, hydrogen, oxygen, sulfur, and phosphorus with which to build the new polymers needed for cell duplication.

The biopolymers, then, provide both the energy source and the atoms from which the newly synthesized biopolymers of the reproducing cell are formed.

V. DECOMPOSITION CAN OCCUR IN THE PRESENCE OR ABSENCE OF AIR

When these decompositions occur in the presence of sufficient molecular oxygen (O_2), we term the cellular process respiration. Microbial life, especially that of the prokaryotes—Archaea and Bacteria—is not dependent on respiration employing O_2, however, for many of these organisms are able to grow well in the absence of the oxygen in air, and many must grow in the latter manner. Although humans often focus on their own lifestyle, and its dependence on cellular respiration processes involving air (aerobic), a great deal of the planet is anaerobic—air, and the oxygen in it, is absent, or nearly so.

Soils are no exception to this generalization, for they often are anaerobic. It is quite easy to isolate from soils an enormous variety of bacteria, and some fungi, that are able to grow by the anaerobic processes termed fermentation or anaerobic respiration. Included among those microbes that live by fermentative mechanisms are those that are able to utilize a variety of different sugars, amino acids, or purines and pyrimidines as the source of energy and the carbon, nitrogen, etc., for the biosyntheses they carry out in their anaerobic growth processes.

VI. SOLUBLE ORGANIC MOLECULES ARE FORMED IN SOIL AS A CONSEQUENCE OF FERMENTATIONS

Since the absence of molecular oxygen makes impossible the reduction of its atoms to form water (as occurs in aerobes), the necessary balancing of those oxidation reactions associated with energy provision for biosyntheses takes the form of the reduction of organic molecules derived from, as the case may be, the sugar or amino acid or purine or pyrimidine. These fermenting anaerobes thus transform, for example, sugars not only into new cell protoplasm, but also into an array of low molecular weight organic compounds; specific types of fermenting organisms are able to produce, and excrete into the soil, particular organic molecules. Some organisms produce largely lactic acid; others, formic, acetic, propionic, or succinic acids; others, any one of a variety of short-chain alcohols; and still other organisms, different low molecular weight products. The amounts of carbon dioxide produced by fermentation of a mole of glucose, for example, will be smaller than the amount produced by aerobic respiration responsible for decomposition of that amount of glucose. Carbon dioxide production from amino acids would be similarly different; in addition, amino acid sulfur is more likely to be released as a reduced end product, sulfide, rather than as the oxidized sulfate anion, as occurs in aerobic respiration. The organic molecules produced by fermenting organisms may migrate to aerobic habitats where they can undergo oxidation to support growth of aerobic bacteria, fungi, etc.

VII. ANAEROBIC RESPIRATION INTRODUCES NEW BIOMASS INTO SOILS AND RELEASES REDUCED INORGANIC IONS OF NITROGEN AND SULFUR

Another anaerobic life process—anaerobic respiration—is also the lifestyle for a variety of heterotrophic bacteria, archaea, and perhaps a few fungi and protozoa. In this form of respiration an anion such as nitrate, nitrite, sulfate, sulfite, or bicarbonate may be employed as the entity which (instead of molecular oxygen or an organic molecule) becomes reduced to balance the energy-providing oxidation reactions accompanying the decomposition of organic molecules. The reduced end products of anaerobic respiration include dinitrogen (N_2), for example, from nitrate or nitrite, sulfide (HS^-) from sulfate or sulfite, and methane or acetate from bicarbonate. Apart from the increase in the biomass of prokaryotes and eukaryotes carrying out the various decompositions of monomers (derived, as above, from polymers), the principal decomposition products of the organic molecules are similar to those in aerobic respiration.

VIII. CHEMOLITHOTROPHIC (CHEMOAUTOTROPHIC) BACTERIA CONSUME INORGANIC CHEMICALS AS THEY PRODUCE NEW BIOMASS

The dynamics of organic molecules in soils can be closely related to inorganic molecules as well. Those organisms that are able to live a chemolithotrophic lifestyle (in contrast to the heterotrophs that live at the expense of organic molecules) produce organic matter. The energy for the biosyntheses comes not from light (as in the case of the phototrophic primary producers) but rather from the aerobic oxidation of reduced inorganic molecules such as ammonia, nitrite, sulfide, thiosulfate, elemental sulfur, and hydrogen gas (H_2) (present in air but also produced by many fermenting prokaryotes). As in the case of plant photosynthesis, carbon dioxide is the form in which these chemolithotrophic organisms assimilate the carbon they incorporate into their protoplasm, and ammonium, nitrate, or sulfate are common sources of the nitrogen and sulfur in the cellular biopolymers. Although in most soils plant (and in some instances algal and cyanobacterial) photosynthesis constitutes the primary input of organic matter, it should be clear that the biopolymers making up prokaryotic microbes, and some eukaryotic ones, are also a significant part of the dynamics of soil organic chemical changes.

IX. MICROBIAL PROTOPLASM, TOO, IS SUBJECT TO DECOMPOSITION

Such microbial protoplasm is, for example, not only subject to decomposition on death of such cells in soil, but also represents an important source of nutrients for many different types of invertebrates inhabiting soils. These larger organisms (macrobes) feed on the microbial cells, often decomposing them in internal organs (e.g., intestines) and assimilating part or all of the decomposition products into new "macrobial" biomass as they reproduce. On the death of the macrobes of course, their protoplasm becomes fair game either for still larger organisms to devour, and decompose it, or for bacteria and fungi to decompose that protoplasm by the means described above in aerobic or anaerobic habitats.

X. SOME ORGANIC COMPONENTS OF SOILS ARE DECOMPOSED VERY SLOWLY AND INCOMPLETELY MINERALIZED

Although many biopolymers derived from plants (e.g., cellulose, hemicellulose, xylan) and from invertebrates (e.g., chitin) are decomposed at relatively rapid rates by microbiota resident in soils, a significant organic fraction of soil—humus—seems to be decomposed rather slowly. Humus is introduced into soils as a result of the slow and incomplete decomposition of lignins, components of a variety of plants, at the surface of a soil. The humic and fulvic acids so formed constitute a major fraction of soil organic matter and are themselves mineralized rather slowly;

this undoubtedly is a reflection of their relatively poor solubility in water as well as their aromatic chemical features. In contrast to the breakdown of polymers such as starch or protein, which may be stable for only a few days or weeks, decomposition of humus is often considered in terms of decades.

See Also the Following Articles

Anaerobic Respiration • Biotransformations • Ecology, Microbial • Nitrogen Cycle • Sulfur Cycle

Bibliography

Fenchel, T., King, G. M., and Blackburn, T. H. (1998). "Bacterial Biogeochemistry," 2nd Ed. Academic Press, San Diego.

Lengeler, J. W., Drews, G., and Schlegel, H. G. (1999). "Biology of the Prokaryotes." Thieme, Stuttgart.

Metting, F. W., Jr. (ed.) (1993). "Soil Microbial Ecology." Dekker, New York.

van Elsas, J. D., Trevers, J. T., and Wellington, E. M. H (eds.) (1997). "Modern Soil Microbiology." Dekker, New York.

Soil Microbiology

Kate M. Scow
University of California, Davis

I. Overview of the Soil Microorganisms
II. Soil Habitat and Distribution of Microorganisms
III. Factors Affecting Microbial Communities
IV. Major Microbial Processes in Soil
V. Interactions among Soil Organisms
VI. Emerging Research Areas

GLOSSARY

bulk soil The portion of the soil not under the influence of plant roots.

decomposition Breakdown of a compound into simpler compounds, often by microorganisms.

denitrification The biochemical reduction of nitrate or nitrite to gaseous nitrogen either as molecular nitrogen or as an oxide of nitrogen.

humus Dark-colored organic by-products consisting of microbial cell walls and other resistant molecules formed from free-radical reactions of sugars, amino acids, and products of lignin decomposition.

mineralization The conversion by microorganisms of an element from an organic form to an inorganic form.

nitrification The biochemical oxidation of ammonium to nitrite and nitrate by microorganisms.

rhizosphere The portion of soil in the immediate vicinity of plant roots in which the microbial communities are influenced by the presence of the roots.

soil The dynamic natural body comprising Earth's surface layer, composed of mineral and organic materials and living organisms.

SOIL MICROBIOLOGY is concerned with microorganisms, primarily bacteria and fungi, that spend at least part of their life cycles in soil. Soil is defined as the dynamic natural body comprising Earth's surface layer, composed of mineral and organic materials and living organisms. Much of soil microbiology is also relevant to microbiology in deeper subsurface environments (e.g., groundwater). The chemical and biological balance of the planet is dependent on processes carried out by soil microbial communities. Global scale processes in which soil microorganisms are active participants include energy flow, organic matter decomposition, and biogeochemical cycling. The difficulty of separating soil organisms from their environment has fostered a strong tradition in soil microbiology of studying processes, with less emphasis on studying individual organisms than is the case in other areas of microbiology.

This article is an introduction to the microorganisms found in soils, describes the soil habitat and environmental factors affecting microbial communities, considers several microbial processes in soils, discusses interactions among soil organisms (including symbioses between microorganisms and plants), and concludes with a discussion of emerging topics and research areas in soil microbiology.

I. OVERVIEW OF THE SOIL MICROORGANISMS

A. Classification

The organisms inhabiting the soil span a wide range of taxonomic and functional groups. As many as 1000 to 5000 different genotypes may be present in a gram of soil based on the reannealing kinetics of DNA extracted directly from soil and then denatured. Soil microorganisms include representatives of the three phylogenetic domains of Bacteria, Archaea, and Eukarya, domains that are differentiated by sequence patterns of the small subunit ribosomal RNA Bacteria and fungi are the most abundant groups in soil with respect to biomass and numbers. Archaea, including methanogens, extreme halophiles, and extreme thermophiles, are present, but their significance in soils that are neither flooded nor exposed to environmental extremes is not well understood. Phototrophic bacteria and green algae are abundant only when soils are flooded (e.g., in wetlands, rice paddies, poorly drained soils). Viruses of bacteria, fungi, fauna, and plants are common and exist both in and outside of their soil-dwelling hosts. Many organisms in soil are not metabolically active but appear to be in wait for an opportunity when environmental conditions favor them.

Soil microorganisms are also categorized on the basis of the functions (usually metabolic) they perform. All life-forms require a source of electrons (electron donor), an electron sink, and a source of carbon. Metabolic requirements in these three categories form the basis for a useful functional classification system that targets processes more than individual organisms. Organisms are classified as phototrophs, lithotrophs, or organotrophs according to whether they obtain energy from light, inorganic chemicals, or organic chemicals, respectively. A broad array of organic compounds, both natural and human-made, can be metabolized by organotrophs. Reduced forms of numerous inorganic elements, including nitrogen, iron, manganese, sulfur, selenium, and others, provide energy for a diverse group of lithotrophic bacteria. On the basis of whether they obtain their carbon from carbon dioxide or organic chemicals, organisms are classified as autotrophs or heterotrophs, respectively. Organisms that respire are categorized as aerobic if they use oxygen as an electron acceptor and anaerobic if they use other compounds as electron acceptors. These alternate electron acceptors include oxidized forms of nitrogen, iron, manganese, sulfur, selenium, and carbon dioxide. The terms strict and facultative are used to designate whether an organism is exclusively aerobic or anaerobic, or able to live under both conditions, respectively.

B. Major Groups

Viruses, though not technically classified as forms of life, are often considered in soil microbiology. They range in diameter from 0.02 to 0.25 μm. Although viruses are obligate parasites of the bacteria, fungi, plants, and fauna species that inhabit the soil, many can persist outside their hosts, in some cases for years. Viruses may play a role in regulating microbial populations, though parasitism, and are believed to be important vectors for the exchange of genetic material among prokaryotes.

Bacteria are a very diverse group of organisms in soil, and most major taxonomic groups are represented in most soils. As many as 13,000 bacterial species are estimated to be in soils based on analysis of DNA. Common genera found in soil include *Acinetobacter, Agrobacterium, Alcaligenes, Arthrobacter, Bacillus, Caulobacter, Cellulomonas, Clostridium, Corynebacterium, Flavobacterium, Micrococcus, Mycobacterium, Nocardia, Pseudomonas, Streptomyces,* and *Xanthomonas.* Bacteria capable of most types of metabolism (e.g., aerobes and anaerobes, organotrophs and lithotrophs, autotrophs and heterotrophs) can be found, at least at low densities, in most soils. Actinomycetes, many of which have filamentous growth forms, are common soil inhabitants. They are relatively resistant to desiccation and are abundant in desert soils. Actinomycetes are more abundant at higher or neutral pH than under acidic conditions. Previously, specific groups of bacteria that were thought to be most abundant in soil corresponded only to those organisms that could be cultured on laboratory media (e.g., gram-negative bacteria, spore-forming bacteria, and actinomycetes). More recent methods that describe microbial communities based on extraction and characterization of DNA (e.g., cloning libraries) contradict some of these previous assumptions about soil organisms and indicate that some of the most abundant species have never been described.

Bacteria range in size from approximately 0.3 to 1.0 μm in diameter and exhibit a variety of morphologies, including rods, cocci, and filamentous forms. Bacteria are usually smaller when living in soil, where nutrient conditions are poor, than when grown in nutrient-rich laboratory media. Although many soil bacteria are potentially motile, their movement is constrained by the low moisture conditions usually present in soil. Numbers of bacteria usually range from 10^8 to 10^9 cells in a gram of soil from the surface horizon, and the total biomass of bacteria usually ranges from 400 to 5000 kg in a hectare of agricultural soil (Table I). Bacteria commonly grow in small colonies of 2 to 20 cells. Many species form spores or other types of resistant bodies that permit them to endure harsh conditions in soil.

Soil fungi are a diverse group with a broad range of morphologies and life cycles. In many ecosystems, such as forests, they constitute the largest biomass (500 to 5000 kg/ha) of all the soil organisms. In fact, the largest individual organism known, inhabiting a total of 2.5 square miles, is a soil fungus inhabiting a forest soil. Fungi are far less diverse metabolically than are the bacteria; most fungi are organoheterotrophs. At least 70,000 fungal species have been identified. These include members of Oomycota, Chytridomycota, Zygomycota, Ascomycota, and Basidiomycota. The oomycetes, chytrids, and zygomycetes (including "sugar fungi") are rapid initial colonizers of organic materials added to soil. The chytrids tend to be found in very wet soils and include animal and plant parasites. The zygomycetes include the endomycorrhizal fungi. The basidiomycetes are slow growers and include the ectomycorrhizal fungi and lignin degraders such as white rot fungi.

The most common fungal growth form in soil is mycelial growth. The mycelium is defined as the filamentous network of hyphae that grows by apical extension. Filamentous organisms, particularly fungi, are uniquely suited to colonize large volumes of soil containing heterogeneously distributed food sources. Some fungi can endure harsh environmental conditions or predation through having resistant structures such as spores or because their hyphae are made up of recalcitrant organic molecules. Reproduction can be through dissemination of sexual and asexual spores. Many plant pathogens are fungi, many of which spend at least part of their life cycle

TABLE I
Relative Numbers and Biomass of Microbial and Faunal Populations in Surface Soils[a,b]

	Number		*Biomass*[c]	
Organisms	*per m²*	*per gram*	*kg/ha*	*g/m²*
Microflora				
Bacteria	10^{13}–10^{14}	10^8–10^9	400–5000	40–500
Actinomycetes	10^{12}–10^{13}	10^7–10^8	400–5000	40–500
Fungi	10^{10}–10^{11}	10^5–10^6	1000–20,000	100–2,000
Algae	10^9–10^{10}	10^4–10^5	10–500	1–50
Fauna				
Protozoa	10^9–10^{10}	10^4–10^5	20–200	2–20
Nematodes	10^6–10^7	10–10^2	10–150	1–15
Mites	10^3–10^6	1–10	5–150	0.5–1.5
Collembola	10^3–10^6	1–10	5–150	0.5–1.5
Earthworms	10–10^3		100–1700	10–170
Other fauna	10^2–10^4		10–100	1–10

[a] From THE NATURE AND PROPERTIES OF SOIL by Brady/Weil, © 1996. Reprinted by permission of Prentice-Hall, Inc., Upper Saddle River, NJ.

[b] Surface soils are generally considered 15 cm (6 in.) deep, but in some cases (e.g., earthworms) a greater depth is used.

[c] Biomass values are on a live weight basis. Dry weights are about 20–25% of these values.

in soil, outside of plants. Beneficial and, in some cases, necessary symbiotic relationships occur between mycorrhizal fungi and most plant species. Fungi are usually more abundant than bacteria in soils that are not physically disturbed, that receive large inputs of complex organic matter, and where organic materials remain on the soil surface.

Green algae and cyanobacteria, both phototrophic unicellular organisms, can be cultured out of most soils. Because these phototrophs are sensitive to desiccation and because sunlight cannot penetrate far into the soil matrix, algal and cyanobacterial population densities are usually low except in flooded or poorly drained soils. Algae and cyanobacteria may be the major primary producers in ecosystems where plant growth is limited by low temperature or moisture. In deserts, for example, blooms following rainfall events provide the main sources of carbon and nitrogen to these systems.

II. SOIL HABITAT AND DISTRIBUTION OF MICROORGANISMS

The soil habitat exhibits substantial heterogeneity which, in turn, fosters and maintains an enormous biological diversity. The spatial distribution of microorganisms is far from uniform as evidenced by the fact that microbial populations occupy only a few percent of the total mass or volume of soil. The heterogeneity in the distribution of microbial populations is manifested at the scale of the soil particle and aggregate, within the soil profile, and at the field scale.

A. Particle and Aggregate Scale

At the scale of micrometers to millimeters, soil consists of mineral particles complexed with organic material and of the voids (pores) created by their arrangements in space. The organic matter fraction of soil ranges from less than 0.1 to 80% on a volume basis, with most surface soils having organic matter contents between 1 and 7%. Microbial biomass usually makes up 3 to 5% of the total organic matter in soils. The mineral particles (clay, silt, and sand) are components of larger structures, called aggregates, in which the particles are held together by microbial polysaccharides and hyphae. Soil texture is defined by the relative proportions of clay, silt, and sand. Soil pores, between and within aggregates, contain either gas or solution, often in an approximately equal ratio. The pore size distribution depends on the relative proportions of clay, silt, and sand-sized particles, and it strongly influences which types of soil organisms are active. Numerous microenvironments, differing in redox status, pH, nutrient levels, pore size, and other factors can coexist within microns of one another.

Microbial distribution at the microscale is a function of soil pore size and access of the organisms to nutrients and electron acceptors. Figure 1 shows the distribution of microorganisms on a soil aggregate. The size of a microorganism strongly influences its

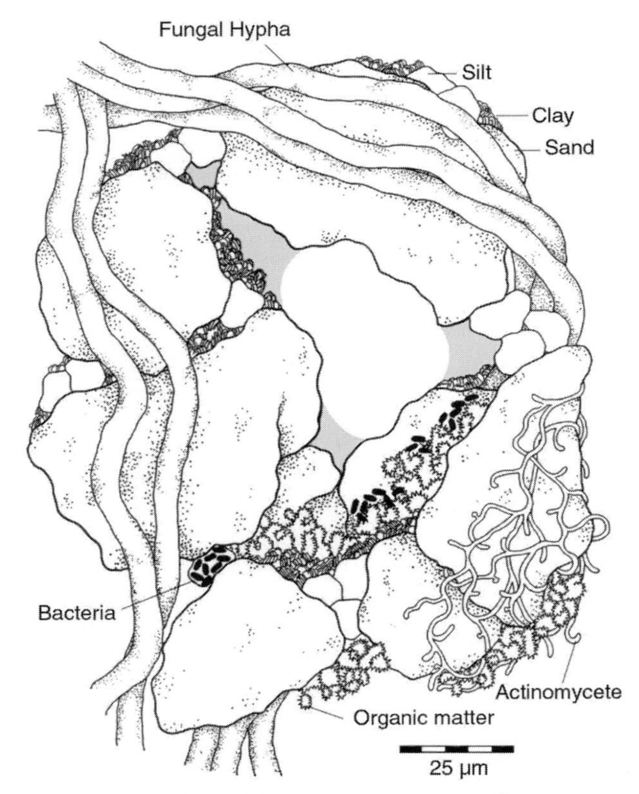

Fig. 1. Distribution of microorganisms on a soil aggregate. From PRINCIPLES AND APPLICATIONS OF SOIL MICROBIOLOGY by Sylvia, D. *et al.*, © 1991. Reprinted by permission of Prentice-Hall, Inc., Upper Saddle River, NJ.

location. Bacterial cells range from 0.3 to 1 μm in diameter and 1 to 2 μm long for nonfilamentous forms and up to 15 μm long for actinomycetes. The diameters are similar in size to large clay particles and are small enough to permit bacteria to inhabit the small pores found within soil aggregates. Fungal hyphae range in diameter from 2 to 10 μm, and their lengths vary considerably. Being too large to penetrate into most soil aggregates, fungal hyphae grow on the surfaces of and between aggregates. Fungal hyphae are important in binding small aggregates into macroaggregates. Microbial population densities decrease as one moves toward the center of an aggregate. This is due to the fact that nutrients and oxygen are consumed by organisms living at the surface of the aggregate. Organisms capable of anaerobic respiration, such as denitrifiers, are often active within the anoxic centers of soil aggregates. Most methods for measuring soil properties are for relatively large volumes of soil and lump together phenomena occurring in numerous microenvironments. Highly sensitive techniques capable of making measurements at spatial scales relevant to microorganisms are usually not possible to employ in complex systems such as soil.

B. Soil Profile Scale

At the scale of millimeters to meters, there is a horizontal pattern in a soil's composition, with a higher organic matter content in the surface, which declines with depth, and higher amounts of clay deeper in the soil profile. Figure 2 shows the distinct horizons that can be found in most soils. These horizons are referred to as the organic (O), eluvial (A), illuvial (B), and unconsolidated parent material (C) horizons. Generally the density of microbial populations declines with soil depth, primarily because of corresponding decreases in organic carbon content, and, to a lesser degree, in oxygen concentration.

There is also a lateral and horizontal pattern in the composition of soil that is created by the spatial distribution of plants. Plants, particularly their roots, are essential members of soil communities and an important source of carbon and other nutrients to microorganisms. The rhizosphere is defined as the zone of soil under the influence of plant roots

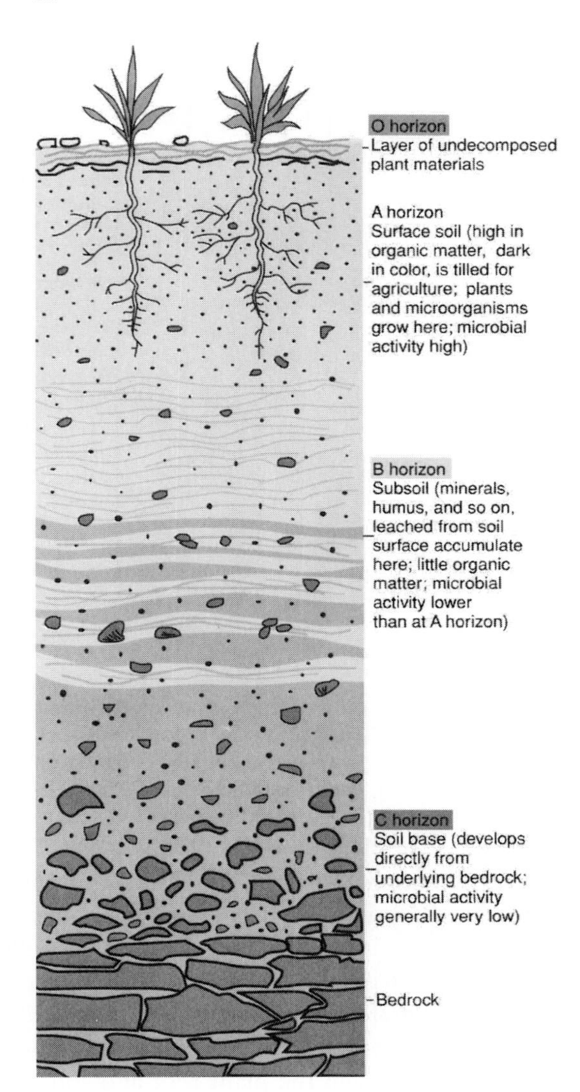

Fig. 2. Profile of a well-developed soil. From BROCK BIOLOGY OF MICROORGANISMS 8/E by Madigan/Martinko/Parker, © 1997. Reprinted by permission of Prentice-Hall, Inc., Upper Saddle River, NJ.

(Fig. 3). The soil not contained within the rhizosphere is commonly referred to as bulk soil. The rhizoplane is defined as the environment at the surface of the plant root. Rhizosphere soil often experiences greater fluctuations in water content, a higher or lower pH, and different nutrient and oxygen concentrations that does bulk soil. Plant roots release a variety of carbon compounds that support higher microbial populations and activity in the rhizosphere than in bulk soil. These compounds include low

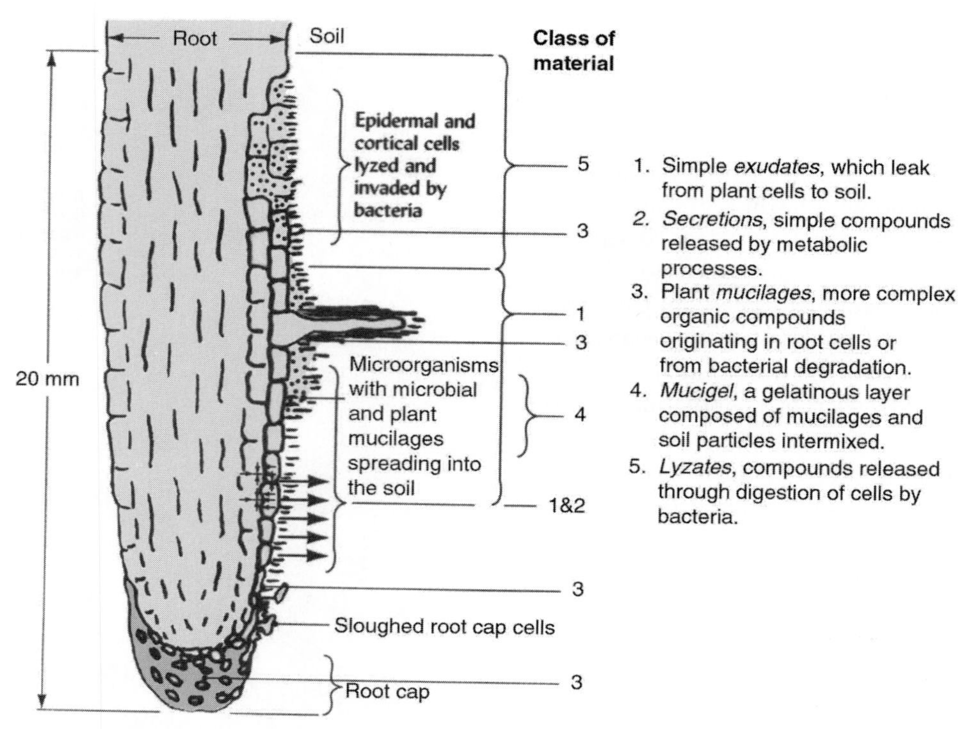

Fig. 3. Substances released by plant roots into the rhizosphere. Adapted from Rovira, Foster, and Martin (1979).

molecular weight compounds, called secretions and exudates, and larger molecular weight compounds in the form of mucilages and sloughed off plant cells (Fig. 3). Mucigel, which provides carbon to microorganisms, is a complex of plant and microbial cells, polymeric substances, and soil particles and is found at the interface between the plant root and soil.

Rhizosphere communities contain larger population densities, greater microbial biomass, and higher levels of certain types of microbial activities than do adjacent communities in the bulk soil. Gram-negative bacteria and denitrifiers are cultured in greater numbers from rhizosphere than bulk soil samples. Fungi, particularly mycorrhizal and plant pathogenic species, are also abundant in the rhizosphere. Conflicting data prevent making conclusions about whether microbial communities are more diverse in the rhizosphere or bulk soil. As tools for characterizing microbial communities continue to improve, we will have the ability to answer such questions.

Heterogeneity in soil chemistry, largely driven by microbial reactions, develops in soil when a source of reductant (e.g., organic substance) is made available or oxygen availability is greatly reduced, for example, after flooding of the soil. The redox and available electron acceptors available at a particular location determine which types of microorganisms can live there. With an increasing distance from the source of the reductant, there is a change in the dominating electron acceptors utilized, creating a spatial gradient in microbial communities. Changes in metabolism that occur along the gradient result from both changes in the dominant members of the community as well as from changes in the electron acceptors being used by the same organisms (e.g., facultative anaerobes). Soil physical structure can also create spatial gradients in electron acceptor utilization and thus microbial communities. For example, from the surface to interior of a large soil aggregate there is a change from the utilization of oxygen to that of nitrate as the major electron acceptor. Terminal electron acceptor process (TEAP) analysis is the study of the chemistry of electron acceptors and their products. This information can provide proof that biodegradation of a pollutant is occurring at a particular site.

C. Field and Landscape Scale

At the scale of meters to kilometers (the landscape scale), soil physical and chemical properties vary depending on topography, parent material, vegetation and other biota present, climate, and time since its initial formation. Figure 4 depicts the major forces at work in soil formation. Microbial activities play an important role in soil development through organic matter turnover and mineral weathering. An example of weathering is the production of carbon dioxide which is converted to carbonic acid which, in turn, dissolves limestone and other minerals. Systematic and predictable relationships between soil taxonomy and microbial community composition have not been made on a broad scale. Certain factors associated with a feature of the landscape, such as flooding in a low-lying area, have large effects on community composition. The effects of various environmental factors on microbial communities are discussed below.

III. FACTORS AFFECTING MICROBIAL COMMUNITIES

Soil microorganisms are essentially aquatic organisms and require liquid medium or at least a thin film of moisture on the surfaces of solid media. Water is held in soil by adsorption onto surfaces or as free water in the pores between soil particles. Soil water is usually measured as the work required to remove the water from the soil and is expressed in terms of suction or potential.

Soil biological activity is at a maximum when moisture is abundant, but with abundant enough water-free pores to permit adequate oxygen diffusion from the atmosphere, that is, at -10 to -100 kPa. Soil moisture also buffers temperature fluctuation and influences the diffusion, mass flow, and concentrations of nutrients and gases. At low moisture contents, the distribution of the soil's water becomes discontinuous, and the solution remaining is held in small pores. Changing the soil moisture content alters the composition of microbial communities. Fungi, actinomycetes, and spore-forming bacteria are favored when moisture is low. The filamentous growth form of fungi permits these organisms to

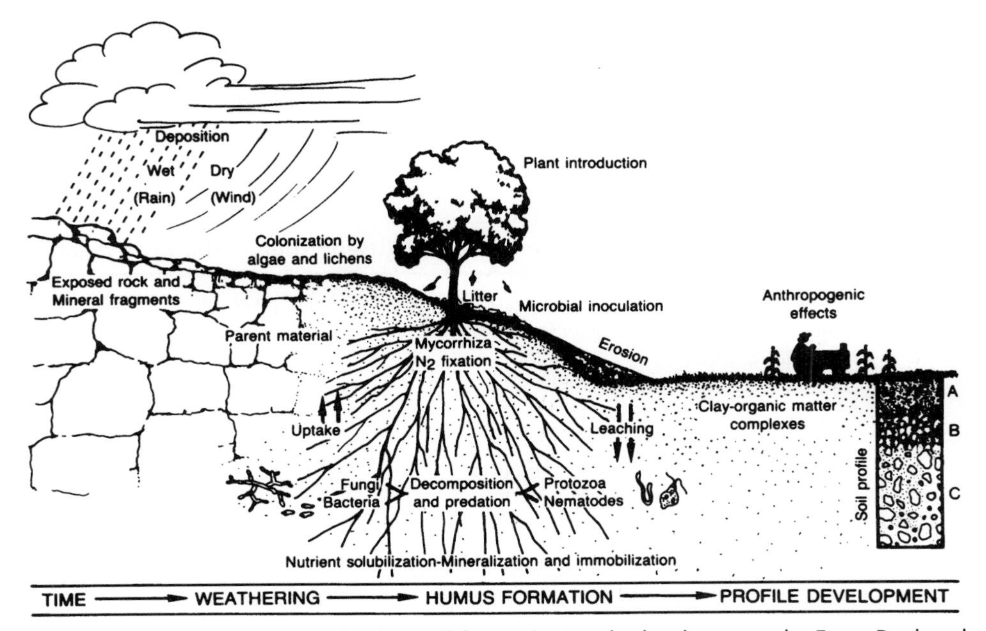

Fig. 4. The major factors involved in soil formation at the landscape scale. From Paul and Clark (1996), *Soil Microbiology and Biochemistry,* p. 21, Academic Press, San Diego.

bridge air-filled pores and thus utilize nutrients in other locations. Algae, protozoa, and facultative and anaerobic bacteria are favored by abundant moisture.

The oxygen concentration at a particular location in soil is affected by the rate of diffusion of oxygen and the biological demand for oxygen at that site, which in turn is determined by available carbon, available electron acceptors, and the abundance of organisms. The oxidation–reduction (redox) potential is a measurement of the likelihood that a substance will gain or lose electrons. When measured directly in soil, the redox potential actually represents numerous redox reactions. Table II summarizes the major categories of reactions, and their redox potentials, involved in the reduction part of redox reactions. For each of these half-reactions, there is a corresponding half-reaction in which a carbon compound is oxidized. The redox potential in soil ranges from below −240 to 820 mV at pH7. Microbial activity usually lowers the redox potential of soil as oxygen and then the other electron acceptors are consumed. Aerobic and anaerobic organisms can coexist within a soil. Anoxic environments often exist as microsites in what are largely aerobic soils; the anoxia is caused by depletion of oxygen around pockets of high carbon concentrations. Long-term flooding of soil can lead to significant changes in microbial communities. Decomposition of organic compounds is usually most rapid when coupled with the utilization of oxygen, rather than other compounds, as the electron acceptor.

Soil organisms live in an environment of constantly changing osmotic pressure, even in nonsaline soils, because of large daily and seasonal fluctuations in soil moisture content. Soil solution ranges from nearly pure water, for example, in areas of high rainfall, to high salt concentrations, such as in saline soils. Under ideal conditions, the protoplasm of cells has a slightly higher solute concentration than the surrounding soil solution, so water tends to enter cell and turgor is maintained. At low soil osmotic potentials, water molecules move into cells, which may cause expansion and rupturing. At high soil osmotic potentials, water moves out of cells and results in dehydration and shriveling. Organisms are protected from osmotic shock by rigid cell walls, and some species generate intracellular chemicals for osmotic balance.

The largest variations, both diurnal and seasonal, in soil temperature occur near the surface where soil is directly exposed to the atmosphere. Microorganisms, as a group, exist over a temperature range from 0° to 80°C, but they are most active between 20° and 40°C. Every organism has a temperature range within which it is active. Within the temperature range specific to each organism, most microbial reaction rates increase by 1.5 to 3 times for every 10°C increase in temperature. Psychrophilic, mesophilic, and thermophilic microorganisms can all be isolated from many types of soils, even in temperate climates. Temperature extremes usually only suppress microbial activity rather than kill off large portions of the commu-

TABLE II
Temporal Sequence of Terminal Electron Acceptors Used When Carbon Is Available[a]

Terminal electron acceptor and ultimate reduced product	*Environmental process*	*Redox potential at pH 7 (mV)*	*Soil biota involved*
$O_2 + e^- \rightarrow H_2O$	Aerobic respiration	+820	Plant roots, aerobic microbes, animals
$NO_3^- + e^- \rightarrow N_2$	Denitrification	+420	*Pseudomonas*
$Mn^{4+} + e^- \rightarrow Mn^{3+}$	Manganese reduction	+410	*Bacillus* etc.
Organic matter + e^- → organic acids	Fermentation	+400	*Clostridium* etc.
$Fe^{3+} + e^- \rightarrow Fe^{2+}$	Iron reduction	−180	*Pseudomonas*
$NO_3^- + e^- \rightarrow NH_4^+$	Dissimilatory nitrate reduction	−200	*Achromobacter*
$SO_4^{2-} + e^- \rightarrow H_2S$	Sulfate reduction	−220	*Desulfovibrio*
$CO_2 + e^- \rightarrow CH_4$	Methanogenesis	−240	*Methanobacterium*

[a] From Kilham (1994). Reprinted with the permission of Cambridge University Press.

nity. Storage of soils at low temperatures (e.g., at 4°C), often utilized in experiments to suppress biological activity, fails to completely shut down many microbial processes and can lead to misleading information.

Soil pH, which exhibits a broad range across soils, has both direct and indirect effects on microorganisms. Direct effects of low pH include denaturation of proteins and alteration of pH-sensitive enzyme activity. Indirect effects include impacts of pH on the availability and/or chemical forms of toxic (e.g., aluminum) or essential (e.g., phosphate) ions and of organic acids and bases. Bacteria are active across a wide range of pH values from 1 to 9; however, many have optimal activity at a neutral pH. The sulfur oxidizing bacteria are among the most acid loving of soil organisms. Fungi are competitive at pH values ranging from 2 to 7, whereas most actinomycetes and cyanobacteria are active only at pH 6 and higher. The composition of microbial communities can be modified by altering soil pH through acidification or liming, and these practices are sometimes used to control pathogens of plants. Microbial activity lowers the pH of soil through production of organic acids during fermentation and lignin degradation, through production of inorganic acids during oxidation of reduced forms of nitrogen and sulfur, and from carbon dioxide production during aerobic respiration. Microorganisms can increase soil pH via reduction reactions that consume protons, such as occur in anoxic soils. The pH of a previously acidic soil will reach neutrality within weeks of flooding due to reduction reactions.

A large fraction of many compounds in soil are not in the soil solution but instead are associated with surfaces, are partitioned into organic matter, or exist in precipitate form. Microorganisms take up undissolved compounds, if at all, at slower rates than the same compounds in solution. Therefore, a large fraction of many elements and molecules exist in forms that are not biologically available. This is the case with mineral ions that exist as precipitates or are held on exchange complexes on clays, oxides, and organic matter. Similarly, many organic compounds, both naturally occurring and human generated, exceed their water solubility limit in soil or are strongly associated with the hydrophobic portion of organic matter. The reduced bioavailability of environmental pollutants greatly decreases their rates of biodegradation or transformation, but also lowers the toxicity of these pollutants to sensitive organisms.

Soils contain numerous substances that may be toxic to some organisms. These include heavy metals, hydrogen sulfide, organic acids excreted by plant roots or generated by microorganisms (e.g., acetic, butyric, lactic acids), antibiotics, carbon dioxide (when transfer of gases is impeded), and human-made substances such as pesticides and industrial wastes. Many compounds that are toxic to one group of organisms can be utilized as carbon and energy sources, or in other ways, by another portion of the community.

IV. MAJOR MICROBIAL PROCESSES IN SOIL

A. Gas Exchange

Soil microorganisms are important regulators of Earth's atmosphere through the gases they emit and consume. Soil microorganisms are involved in the cycling of all major elements, and many of these elements have gaseous forms. Low molecular weight, volatile compounds produced by microorganisms include carbon dioxide, from respiration; methane, from methanogenic processes in anaerobic environments; and N_2, NO, and N_2O from denitrification and hydrogen sulfide from sulfate reduction in anoxic environments. Gases emitted by microorganisms are, in turn, consumed by other types of microorganisms. Thus carbon dioxide is consumed by autotrophic bacteria, methane by methanotrophic bacteria, nitrogen gas by nitrogen-fixing bacteria, and hydrogen sulfide by sulfur oxidizing bacteria. Through these activities, microorganisms affect the composition, as well as chemistry, of the atmosphere. Long-term transport of volatile compounds emitted by microorganisms can be lead to nutrient deposition into other ecosystems. Microorganisms can also produce volatile forms of heavier elements, such as selenium and mercury, through methylation reactions. These methylation reactions can lead to the transfer of toxic chemicals both form, as in the case of selenium, or into, as for mercury, environments where sensitive organisms can be exposed. Certain gases are directly, or indirectly, destructive of the ozone layer or contribute to global warming. As Earth's atmosphere is changing, there is a growing interest in better understanding soil microbial processes both to reduce the production of greenhouse gases and to enhance potential microbial sinks for greenhouse gases.

B. Elemental Cycling

The oxidation states, and in some cases the physical forms, of elements change when used by microorganisms as electron donors or acceptors. A biogeochemical cycle is the result of a series of changes in the chemistry and physical locations of a specific element; thus, biological processes are driving forces of biogeochemical cycles. In some cases, abiotic reactions are faster than biological reactions (e.g., iron oxidation at high pH) and will also contribute to the cycling of elements.

The nitrogen cycle is one of the most well-studied biogeochemical cycles (Fig. 5). Nitrogen-fixing bacteria, both free-living in the soil and symbiotic with plants (see below), are largely responsible for the transfer of nitrogen (N_2) from the atmosphere to the soil. The organic forms of nitrogen resulting from N fixation enter the soil with the death of microorganisms and plants and are eventually mineralized to ammonium by most species of bacteria and fungi. In the process called nitrification, the reduced form of N as ammonium is oxidized to nitrite and nitrate primarily by the lithoautotrophic bacterial genera *Nitrosomonas, Nitrosospira,* and *Nitrobacter.* Nitrate is also used as an electron acceptor by a variety of

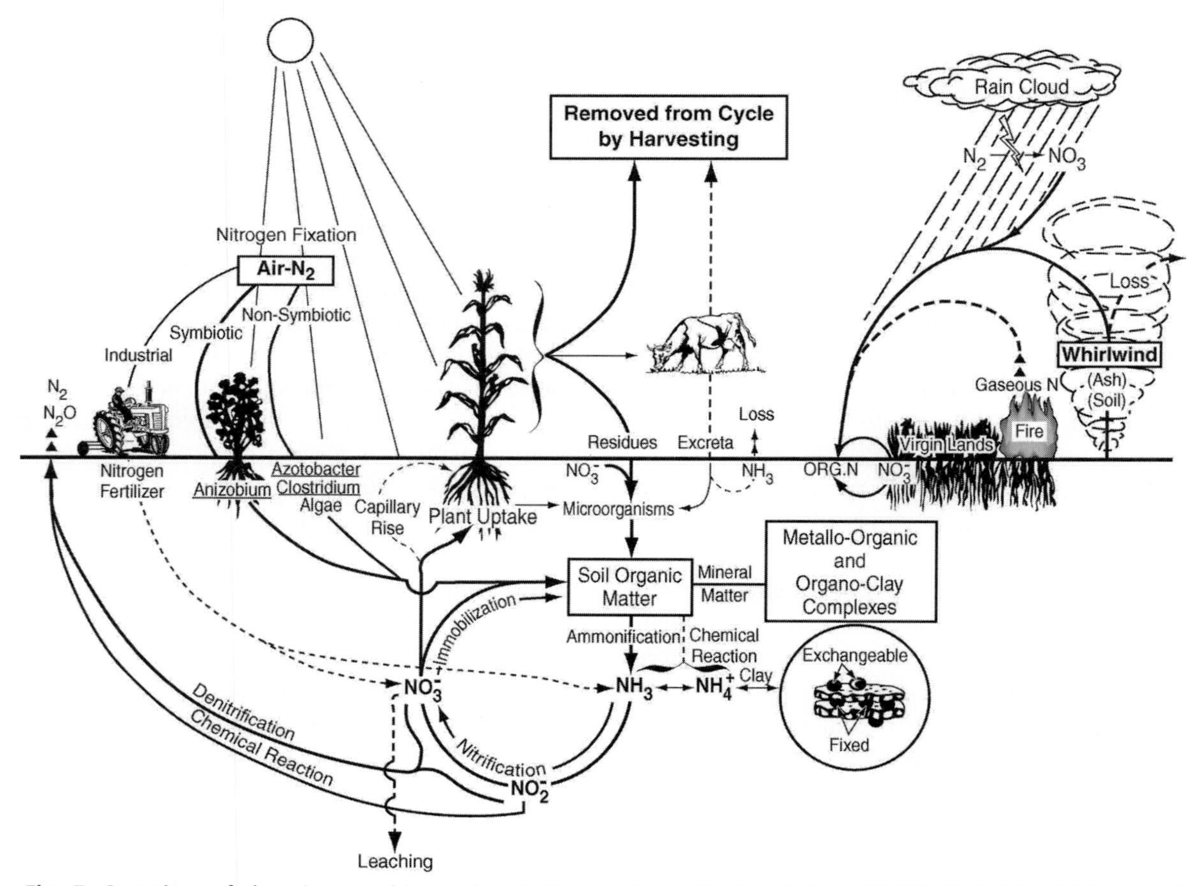

Fig. 5. Overview of the nitrogen biogeochemical cycle. From Paul and Clark (1996), *Soil Microbiology and Biochemistry,* p. 186, Academic Press, San Diego.

bacteria and transformed to dinitrogen gas and, to a lesser extent, nitric and nitrous oxides. In addition, both nitrate and ammonium are incorporated into cellular material (thus into organic form) by most microorganisms in the process of immobilization (see below). All of the above reactions form a closed loop in the cycling of nitrogen in the environment.

The significance of microbial transformations of certain elements is evident throughout the terrestrial environment. Some of the more striking colors of soil are microbial by-products. Many of the red and orange pigments in soil are from oxidized iron, some of it of microbial origin. In waterlogged soils, various shades of gray, green, and black result from reduced sulfur and iron and from their interactions with each other and other elements. Many of the characteristic odors of anoxic environments are microbial in origin. The typical rotten egg smell of anoxic soils is reduced sulfur resulting from the use of sulfate as an electron acceptor. Putrescence is associated with organic sulfur compounds resulting from decomposition or fermentation products produced by anaerobic processes. Microbial redox reactions alter the oxidation states of many inorganic pollutants (e.g., selenium and chromium) and thus can decrease (and sometimes increase) the toxicity of these elements.

C. Decomposition, Mineralization, and Immobilization

Organic compounds produced by plants via photosynthesis continually enter the soil as root exudates when the plant is living and as debris when the plant dies. These complex organic polymers are broken down into smaller organic and mineral components by the combined forces of soil fauna and microorganisms in a process called decomposition. The nutrients released by this process are thus made available for uptake and incorporation into plants and other organisms. Most carbon fixed by terrestrial photosynthesis is returned to the atmosphere within 1 year by microbial decomposition. Plant residues make up the largest fraction of carbon that enters the soil, and microorganisms and animals contribute the remainder. On entering the soil, approximately 60–70% of organic residues are decomposed in the first year, and the remainder decays much more slowly. Many hydrolytic and oxidative extracellular enzymes are essential in the early phases of decomposition because they produce smaller molecules from polymers (e.g., cellulose, hemicellulose, lignin) that are otherwise too large to be absorbed by cells. Nutrients are released from organic residues by the process of mineralization. Mineralization is defined as the conversion of organic chemicals to their inorganic constituents, primarily carbon dioxide, water, and/or ammonium.

During anabolism, soil microorganisms assimilate inorganic forms of elements into their cells in a process referred to as immobilization. Depletion of soil inorganic nutrient pools by microbial immobilization can temporarily limit plant growth. Although both mineralization and immobilization are always occurring simultaneously, the carbon to nitrogen ratio of available organic substrates determines which process will dominate. Thus net immobilization of inorganic nitrogen is expected during decomposition of compounds with high carbon to nitrogen ratios. A large portion of immobilized elements eventually becomes available for plant uptake when microorganisms are preyed on by protozoa or lysed by environmental conditions.

An important product of decomposition is soil humus. Humus is defined as dark-colored organic byproducts consisting of microbial cell walls and other resistant molecules formed from free-radical reactions of sugars, amino acids, and products of lignin decomposition. Thus, microorganisms contribute to humus formation in several ways. They form or release smaller molecules from larger organic polymers. These smaller molecules are condensed into humic substances by reactions catalyzed by primarily fungal-derived extracellular enzymes. Fungi create complex and recalcitrant molecules such as melanin and tannins to withstand predation and other stresses, which in turn become humus precursors when the fungi die. Another type of soil organic material important to soil structure are polysaccharides. Many microorganisms in soil live in a mesh of extracellular polymeric substances, usually polysaccharides. Production of these substances, formed by both bacteria and fungi, occurs under conditions of unbalanced growth (e.g., when carbon is more abundant than nitrogen) or under moisture stress.

Cross sections of bacterial colonies in soil, viewed by electron microscopy, reveal that colonies of bacteria consist of a few individual cells emeshed in a nest of polysaccharides which, in turn, are often surrounded by a coating of clay particles.

D. Biodegradation and Transformation of Pollutants

Many organic and inorganic pollutants end up in soil intentionally, through their use as pesticides, and unintentionally when they migrate from contaminated waste sites or are deposited on soil from the atmosphere. Biodegradation of organic pollutants can be considered under the general category of decomposition. In its most general form, biodegradation is defined as an alteration in the chemical composition of a molecule mediated by a biological process. Microorganisms are able to utilize many organic pollutants as sources of energy and carbon and use some of the highly chlorinated pollutants as electron acceptors under anoxic conditions. Bioremediation is defined as the decontamination of polluted environments via biological activity. Different types of bioremediation include simple monitoring of naturally occurring biological and abiotic processes leading to pollutant containment or removal (intrinsic remediation), stimulation of natural processes through amendment with nutrients and electron acceptors (biostimulation), or inoculation with microorganisms (bioaugmentation). In soil, biodegradation rates may be substantially reduced by association of pollutants with the solid phase of soil, a process which decreases the availability of the pollutant to microorganisms.

V. INTERACTIONS AMONG SOIL ORGANISMS

A single-species population of microorganisms rarely occurs in soil; instead microorganisms are members of complex communities. The soil food web describes all organisms in soil and their interrelationships. Figure 6 is a simplified depiction of a soil web. The composition of a community is governed by the biological steady state, which is a function of the associations and interactions of the members of the community. With environmental change, this steady state may be upset and may shift to a new set of relationships. These relationships can be antagonistic, positive, or neutral.

Antagonistic relationships include predation and parasitism, ammensalism, and competition. Examples of predation and parasitism among soil microorganisms include the bacterium *Bdellovibrio* infesting other bacteria and viral infections of bacteria and fungi. Examples of ammensalism include the alteration of the immediate environment through an organism's normal metabolic activities (e.g., through production of acid). Other soil microorganisms, most notably the actinomycetes, produce specific toxic compounds, such as antibiotics, that inhibit or eliminate bacteria, yeast, and other fungi. Most of the earliest antibiotics used for medical purposes were discovered to be produced by the common soil actinomycete *Streptomyces*. These antibiotics include chloramphenicol, streptomycin, neomycin, erythromycin, cycloheximide, and tetracycline. The soil microbiologist Selman Waksman was awarded the Nobel Prize for his pioneering research on antibiotics. Competition for resources, particularly carbon, is considered one of the most important controls of microbial populations in soil. Competitive organisms may have faster growth rates or higher enzyme affinities for the resource in demand than do the less competitive species. Some organisms produce specific compounds, such as siderophores, that increase their access to limiting nutrients, in this case iron.

Positive relationships include mutualism (including symbiosis), commensalism, and synergism. Mutualistic reactions are involved in nitrification and lignin decomposition, both multistep processes in which different organisms are responsible for different metabolic steps. For example, ammonium oxidizing bacteria such as *Nitrosomonas* or *Nitrosospira* produce nitrite which, in turn, is oxidized by other bacterial genera, such as *Nitrobacter*. Without *Nitrobacter*, buildup of nitrite would eventually become toxic to the ammonium oxidizers. Some of the most important symbiotic relationships involving soil mi-

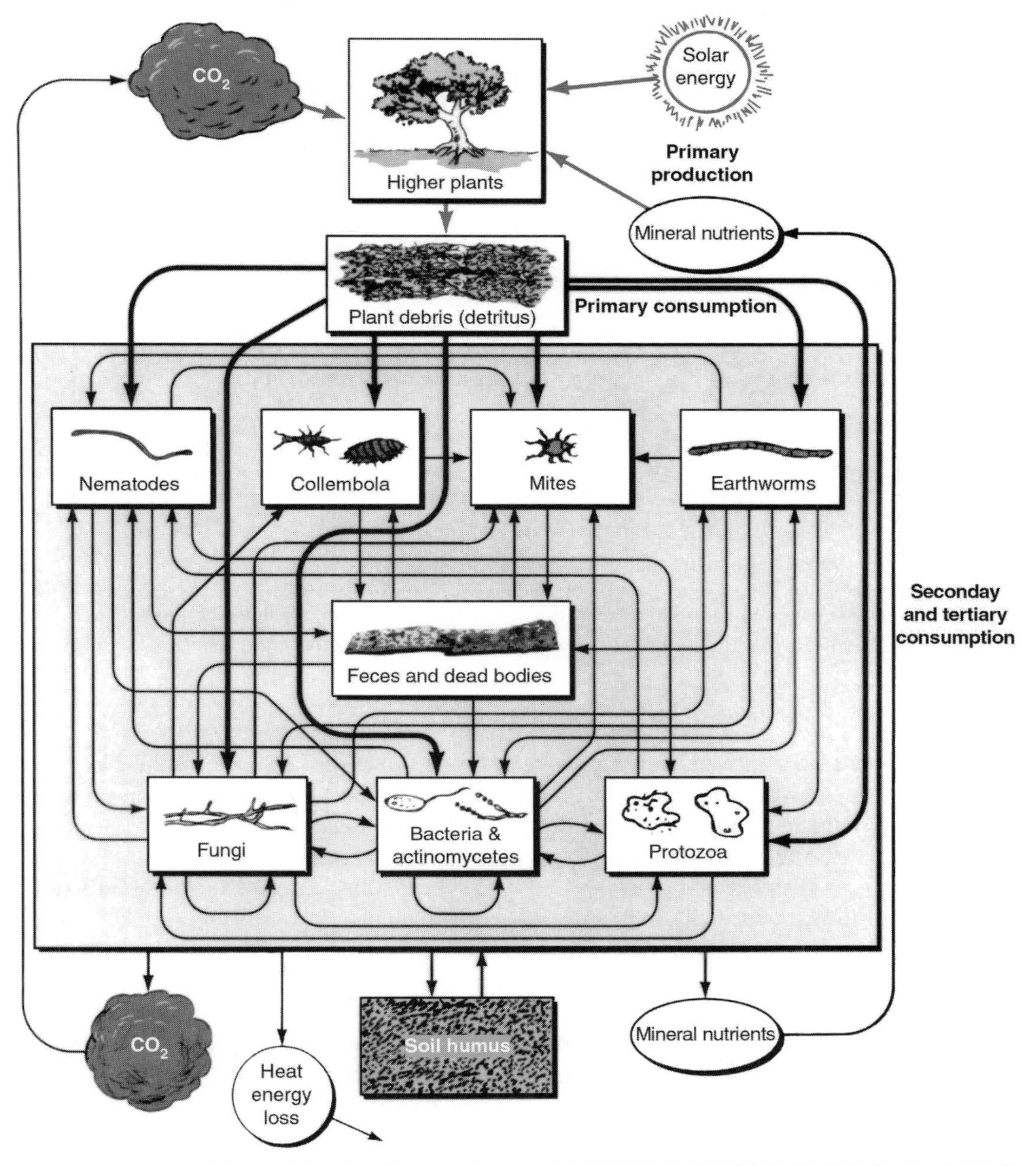

Fig. 6. Overview of the soil food web. From THE NATURE AND PROPERTIES OF SOIL by Brady/Weil, © 1996. Reprinted by permission of Prentice-Hall, Inc., Upper Saddle River, NJ.

croorganisms are relationships with plants. These symbioses between plants and microorganisms are essential to the survival or competitiveness of many plant species. One example of symbiosis is the relationship between leguminous plants (e.g., clover, alfalfa, soybeans, vetch) and root-nodule bacteria, primarily of the genera *Rhizobium* and *Bradyrhizobium* (Fig. 7). Another example is the relationship between the actinomycete *Frankia* and actinorrhizal plants, including members of the families Betulaceae, Casuarinaceae, Elaeagnaceae, Myricaceae, Rosaceae, and Rhamnaceae. In both types of mutualism, nodules

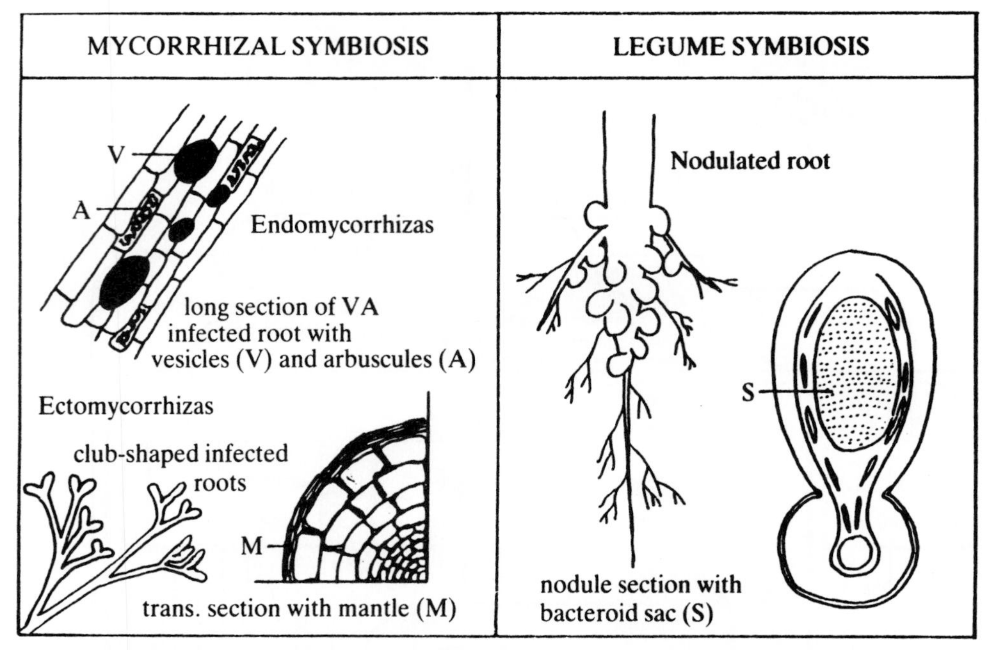

Fig. 7. Plant root/microbial relationships: the mycorrhizal and legume symbioses. From Kilham (1994). Reprinted with the permission of Cambridge University Press.

formed on the plant root provide an environment in which the bacterial symbiont can convert atmospheric nitrogen to a form that is usable by the plant. Atmospheric dinitrogen is reduced to ammonium and then transported into the plant. The plant, in turn, provides the bacteria with sugars necessary for the high energy-requiring demands of N_2 fixation. Amounts of nitrogen fixed by symbiotic bacteria far exceed amounts fixed by free-living bacteria potentially able to fix nitrogen.

The other major type of mutualism in soil involves most plant species and primarily two kinds of mycorrhizal ("fungus root") fungi: ectomycorrhizal and endomycorrhizal fungi (Fig. 7). The fungus benefits from the association by having direct access to sugars provided by the plant. The most obvious benefit to the plant is an increase in the volume of soil the plant can exploit. Mycorrhizal hyphae can extend from 5 to 15 cm into the soil from the plant root and penetrate soil pores that are too small for plant roots to enter. The fungal mycelia can increase the plant root surface area by as much as 10 times. Mycorrhizae substantially increase plant uptake of immobile nutrients, such as phosphate. Other benefits to the plant, depending on the kind of mycorrhizal fungus, include increased uptake of nitrogen and other elements, protection against plant pathogens and heavy metals, increased aggregation of soil particles, and possibly drought resistance.

The ectomycorrhizae include hundreds of species of fungi (primarily Basidiomycetes) that are associated primarily with tree species (e.g., pine, birch, oak, spruce, fir) in temperate and arid ecosystems. The plant rootlet is bound by layer of fungal material, called the mantle. The inner portion of mantle is connected to hyphae extending between plant root cells of the epidermis and the outer cortex, forming a mycelial network called the Hartig net. Many ectomycorrhizal fungi can be cultivated on laboratory media and thus have been well studied. The endomycorrhizal fungi (mostly Zygomycetes), also known as vesicular–arbuscular (VA) mycorrhizal fungi, are associated with more than 80% of plant species, including nearly all cultivated plants, forest and shade trees, shrubs, and herbaceous species. These fungi penetrate the cell wall of plant cortical cells and form highly branched hyphal structures, known as arbuscules, that are sites of nutrient exchange between plant and fungus. Because endomycorrhizal fungi cannot grow without their plant hosts, less is known about the physiology of endo- than ectomycorrhizal fungi. Plant biologists are now recognizing that if studies of plant nutrition are to be realistic, they must consider the contribution of mycorrhizae to the uptake of nutrients by plants.

VI. EMERGING RESEARCH AREAS

Soil supports an extraordinary diversity of microorganisms; however, surveys of soil indicate that a substantial number of organisms have not been identified and characterized. The development of molecular tools has permitted the recent discoveries of many new, previously unidentified bacterial genotypes. Considerable effort is directed at characterizing this diversity and particularly using methods that do not initially require traditional enrichment and isolation of organisms. The challenge of the next decade will be to link new information on the composition and structure of soil microbial communities back to the soil processes that have long been the focus of soil microbiology. In addition, the evolution of soil microbial communities; in particular the role of horizontal gene transfer in their evolution, is an area of intense study.

With the growing interest in sustainable management of agriculture and forests, and in reducing synthetic chemical inputs to these ecosystems, new research is being conducted on the role of microorganisms in soil fertility and biological control. Biotechnology is being used to manipulate microorganisms to enhance processes they already perform, or to perform new functions. The role of microorganisms in accelerating and ameliorating global climate change is also an area of active study.

See Also the Following Articles

Fungi, Filamentous • Mycorrhizae • Osmotic Stress • pH Stress • Rhizosphere • *Xanthomonas*

Bibliography

Killham, K. (1994). "Soil Ecology." Cambridge Univ. Press, Cambridge, UK.

Paul, F. A., and Clark, F. E. (1996). "Soil Microbiology and Biochemistry," 2nd Ed. Academic Press, San Diego.

Silvia, D. M., Fuhrmann, J. J., Hartel, P. G., and Zuberer, D. A. (1998). "Principles and Applications of Soil Microbiology." Prentice-Hall, Upper Saddle River, NJ.

Tate, R. L. (1994). "Soil Microbiology." Wiley, New York.

SOS Response

Kevin W. Winterling
Emory & Henry College

I. *Escherichia coli* as the Paradigm
II. Regulation of the SOS Response
III. LexA as a Repressor
IV. The SOS Inducing Signal
V. Mechanism of LexA Cleavage
VI. SOS Mutagenesis
VII. The SOS Response in *Bacillus subtilis*
VIII. The SOS System in Other Prokaryotes

GLOSSARY

constitutive Refers to the expression of a set of genes/operons in the absence of the inducer for expression.

din genes Genes that are induced in the presence of DNA damage.

lytic phase One potential pathway of bacteriophage or prophage that leads to the lysis of the host cell and the release of new phage into the surrounding medium.

prophage A bacteriophage that has had its DNA incorporated into the bacterial chromosome.

regulon A set of genes and/or operons that are all regulated by the product(s) of the same regulatory gene(s).

reporter gene A gene that is fused to another gene so that the expression of that gene may be assayed. The product of the reporter gene is typically more stable and easier to detect than the gene to which it is fused.

−10 and −35 sites Short sequences of DNA that lie approximately 10 and 35 base pairs upstream of the transcription start site of that particular gene. These two sites are necessary for the binding of RNA polymerase.

THE SOS RESPONSE refers to the expression of a global regulon in response to DNA damage. It involves the coordinated induction of a number of unlinked *d*amage *in*ducible or din genes that are involved in DNA repair, inhibition of cell division, and enhanced survival and mutagenesis of the bacterial population, as well as the induction of bacteriophage.

I. *ESCHERICHIA COLI* AS THE PARADIGM

Throughout their life cycle, all living cells will be exposed to a variety of stressful and ever changing conditions to which they must respond appropriately in order to continue to survive. One of the most significant and common traumas encountered by a living cell is the alteration of the structure of its DNA. These alterations or changes to DNA may occur as a result of exposure to a variety of physical and chemical agents or as a result of imperfect DNA replication. If left uncorrected, these changes may lead to mutations, which may ultimately be lethal. In addition, the products of DNA damage often block DNA replication and thus pose an immediate threat to the survival of the cell. Obviously, for these reasons, it would be beneficial for living cells to have a way of correcting and/or accommodating DNA damage in order to survive. Many prokaryotes have evolved such a system in which a set of unlinked genes is coordinately expressed in response to such DNA damage. This global response has come to be known as the "SOS response."

Since the SOS phenomenon was first suggested by Miroslav Radman nearly 25 years ago, the induction of the SOS response in *Escherichia coli* has been studied extensively and has become the paradigm for the prokaryotic response to DNA damage. The goal of this chapter is to provide an overview of the characteristics of the SOS response as they have come to be known through the extensive study of *E. coli*. In addition, the properties of the SOS response in the gram positive bacterium, *Bacillus subtilis* will be discussed.

II. REGULATION OF THE SOS RESPONSE

The SOS system is regulated by two proteins which are themselves expressed as part of the SOS response. One protein, RecA, has several functions (among them is the facilitation of genetic recombinational events) in addition to its role as the positive regulator of this global response. The other protein, LexA, serves as the transcriptional repressor of the system. As DNA damage accumulates in a bacterial cell, specific enzymatic or coprotease activities of RecA become activated (designated as RecA*). RecA* forms a long polymer filament by complexing with other RecA* molecules and DNA. This nucleoprotein filament, as it is called, is recognized and bound by free LexA protein. This interaction of the nucleoprotein filament and LexA leads to the cleavage of LexA, thus diminishing the pool of intact protein that is available to bind to the operators of various SOS genes and operons. This releases the SOS genes from the negative transcriptional repression of LexA and allows them to be expressed.

The first observations that a DNA damage induced survival mechanism existed in bacteria was reported by Jean Weigle. He demonstrated that when ultraviolet-irradiated bacteriophage λ was plated on *E. coli* cells that were previously irradiated with ultraviolet light, the bacteriophage survival and the rate of bacteriophage mutagenesis were greater than if the bacteriophage had been plated on unirradiated bacteria. These phenomena are referred to as W (Weigle) reactivation and W (Weigle) mutagenesis, respectively. It was not until the mid-1970s, however, that significant insights into the regulatory pathway of the SOS system in *E. coli* came from the further examination of the bacteriophage, λ. It was observed that as a consequence of exposure to DNA damage, λ cI repressor (cI controls/represses the lytic phase) was being cleaved and, subsequently, the lytic phase of λ was derepressed. Interestingly, this process of cI cleavage and subsequent λ induction was not seen in *recA* (Def) strains (strains of *E. coli* in which the *recA* gene had been mutated), suggesting that RecA was in some way regulating the induction of λ. The hypothesis that RecA was involved in λ induction was further reinforced by *in vitro* assays that showed λ cI was indeed cleaved when incubated in the presence of purified RecA protein.

About this same time, Gudas and colleagues noticed that after cells had been exposed to DNA damaging agents, there was a dramatic increase in the cellular level of a specific but unknown protein, which they called protein X. Further analysis showed that the synthesis of protein X was also similarly elevated in various *recA* and *lexA* mutant strains. Subsequent purification and characterization of the unknown protein revealed its identity to be RecA. On the basis of these observations, it was suggested that LexA repressed the expression of the *recA* gene while RecA played a role in the inactivation of the LexA protein itself. Furthermore, the connection was made that like λ repressor, LexA was most likely cleaved by RecA at the onset of SOS induction. Indeed, the purification and characterization of LexA, and the isolation of LexA (Def) mutants that allowed for the constitutive expression of the SOS system, proved this hypothesis to be correct.

Additional study of the regulation of other SOS genes has revealed that the SOS phenomenon in *E. coli* ubiquitously requires the interaction of the products of the *recA* gene and the *lexA* gene. The LexA protein is the cellular repressor for the many genes that are expressed as part of the SOS response (Table I), including *recA* and *lexA* itself. LexA functions as the cellular repressor of the SOS system by binding to a 16 base pair region of DNA that displays dyad symmetry [5′-CTGT-$(AT)_4$-ACAG-3′]. This conserved sequence of DNA is known as the "SOS box" and is located upstream of most SOS genes and SOS operons. Even in the repressed state, there is, however, a basal level of expression of genes that comprise the SOS regulon. Specifically, there is a sufficient amount of LexA protein to act as the SOS repressor and enough RecA present to fulfill the cell's need for recombinational repair and to induce the SOS response.

TABLE I
DNA Damage Inducible Genes in *Escherichia coli* That Appear to be Regulated by LexA

Gene	*Function*
dinA	DNA polymerase II
dinB	λ mutagenesis
dinD	Cold-sensitive mutant
dinF	Unknown
dinG	Helicase
dinH	Unknown
dinI	Unknown
dinJ	Unknown
dinK	Unknown
dinL	Unknown
dinM	Unknown
dinN	Unknown
dinO	Unknown
dinP	Identical to *dinB*
lexA	Transcriptional repressor of SOS
recA	Recombination, SOS regulator, SOS mutagenesis
recN	Recombinational repair
ruvAB	Recombinational repair
sbmC	Resistance to Microcin B17
ssb	Single-stranded binding protein
sulA	Inhibitor of cell division
umuDC	SOS mutagenesis
uvrA	Nucleotide excision repair
uvrB	Nucleotide excision repair
uvrD	DNA helicase

III. LexA AS A REPRESSOR

Differential expression of SOS genes is seen in *E. coli*. That is, a certain level of DNA damage does not lead to the equal expression of all of the SOS genes. Such regulation occurs, in part, because LexA binds to the operators of various SOS genes with differing affinities. As a result, certain damage inducible (din) genes are fully expressed at low concentrations of a particular inducing agent, whereas other din genes may require more time and/or an increased dosage to be significantly induced. This graded response allows cells with only minor DNA damage to induce error-free repair processes (nucleotide excision repair) without inducing other more dramatic pathways that may ultimately be error-prone (SOS mutagenesis). In fact, substantially more DNA damage must accumulate in the cell in order for the UmuDC proteins (which are required for error-prone repair of DNA and SOS mutagenesis) to be expressed. The continued accumulation of unrepaired DNA lesions ultimately leads to the arrest of cellular division (*sulA*) and the eventual induction of temperate bacteriophage that may exist as prophage in the cell. As cells begin to recover from the damage inducing conditions, the inducing signal diminishes and repression of the SOS regulon returns to preinduction levels.

As briefly mentioned above, there are minor differences among the operators of the SOS genes in *E. coli* that result in different levels of expression. The extent of repression of the individual genes of the SOS regulon depends on at least four factors. The first is operator strength: The actual sequences of individual SOS boxes will vary slightly from the consensus sequence listed above. These slight differences in individual SOS boxes may vary their relative ability to bind LexA by up to a factor of 17. Indeed, there are large differences in the experimentally determined dissociation constants (K_d) of LexA for the SOS box of one gene versus another. LexA shows the highest affinity for the *sulA* gene and a much weaker affinity for the *lexA* SOS box. LexA seems to have the weakest affinity for the SOS box of the *uvrD* gene. Intuitively, this makes sense. The prolonged presence of SulA will lead to the death of the cell. LexA must be readily available to appropriately regulate the SOS response.

The second factor influencing the extent of repression is the location of the operator with respect to the promoter: Some SOS boxes overlap the −35 site, some are located between the −35 and −10 sites, some overlap the −10 site, while still others are located downstream of the −10 site. In the case of the *uvrA* gene, the SOS box overlaps the −35 region of the promoter, which appears to allow LexA to interfere with the action of RNA polymerase at an early stage of transcription initiation. The diverse locations of other SOS boxes with respect to the promoter elements suggest that LexA may inhibit other stages of transcription as well. In those cases where LexA is bound downstream of the transcription start site (*uvrD, umuDC,* and *sulA*), it is possible that both LexA and RNA polymerase are bound simultaneously, as is the case with the classic example of the *lac* repressor. In this situation, the presence of the repressor hinders the formation of a competent transcription complex.

Promoter strength is also a factor, as evidenced by the variable levels of expression of SOS genes in *lexA* (Def) and *lexA* temperature sensitive (ts) strains. The induced level of expression of *sulA* is 110 times greater than the uninduced level of expression of *sulA*. Whereas the induction ratio for *sulA* is 110-fold, the induction ratio for another SOS gene, *uvrB*, is only 3.6-fold. If regulation of the SOS genes simply relied on LexA binding affinity, all SOS genes would presumably have similar levels of expression in *E. coli* strains that contain a nonfunctional repressor.

Finally, the presence of additional operators, upstream of certain SOS genes most likely plays a role in SOS regulation. While most SOS genes have only one SOS box associated with them, some have multiple operators (*recN* has three operators, *lexA* has two). Unfortunately, the effect of these additional SOS boxes has not yet been fully determined, but the relationship is believed to be cooperative in nature.

The fact that the SOS box displays dyad symmetry [5′-CTGT-$(AT)_4$-ACAG-3′] led many researchers in the field to suggest that LexA would bind target DNA as a dimer (two monomers of protein, one monomer to each half of the dyad). In fact, protein–DNA binding assays have shown that LexA does indeed interact with target DNA as a dimer. Many DNA binding proteins that are functionally active as dimers or higher order oligomers are readily capable of forming these structures in solution. Both CAP (dimer) and the *lac* repressor (tetramer) exhibit the ability to form stable multimers in solution, and λ repressor has been shown to form dimers in solution with an association constant (K_a) of $5.9 \times 10^{-7} M^{-1}$. Despite this precedent, researchers have shown that in all likelihood, LexA does not readily form dimers in solution. The K_a for the monomer–dimer equilibrium of LexA is a relatively low $2.1 \times 10^{-4} M^{-1}$. At the *in vivo* concentration of LexA (1300 molecules of LexA monomer per cell) one would therefore predict that the predominant portion of LexA should be in the monomeric state. Indeed, extensive sedimentation analysis revealed that LexA is predominantly a monomer in solution.

The x-ray crystallographic studies of several protein–DNA complexes show that one subunit of each protein monomer contacts each half-site of the operator. For many of these proteins, these protein–DNA complexes form by binding of preformed dimers to their target sites. Since it was evident that LexA did not form dimers in solution, it was suggested that LexA bound target DNA as a monomer, then dimerized while in contact with the target DNA. Since the amino-terminal domain was found to bind an SOS box half-site with the same affinity as the intact LexA protein, it was determined that this region, alone, is responsible for DNA binding. In similar experiments, the K_{dimer} (dimerization constant) of the carboxyl terminus was determined to be the same as the K_{dimer} for intact LexA, thus leading to the conclusion that this region alone is accountable for LexA dimerization. Finally, by demonstrating that intact LexA has a much higher affinity for the intact SOS operator than it does for a half-operator, it was concluded that LexA dimerizes on the DNA in a cooperative manner. It seems that once the first LexA monomer binds to an operator half-site, the second monomer binds much more quickly via a combination of protein–protein and protein–DNA interactions.

Like many other proteins that bind to DNA, the DNA binding domain of LexA displays a helix–turn–helix motif. The work performed by Knetgel *et al.* delineates some of the specific interactions of LexA with its binding site, the SOS box. These authors demonstrated that there are three α-helices in the amino terminus, with helices II and III constituting the primary structure of the helix–turn–helix motif. It is helix III, which comprises amino acid residues 40–52, that protrudes into the major groove of the DNA and is responsible for the affinity of LexA to DNA. Ser^{39}, Asn^{41}, and Ala^{42} contribute hydrophobic interactions in addition to Asn^{41}, Glu^{44}, and Glu^{45}, which form the direct hydrogen bonds to the CTGT half-site of the SOS box. There are many other nonspecific protein–DNA contacts that have proved important for binding. Their role is to enhance the stabilization of the LexA monomer to the operator half-site and consequently promote dimerization.

Not all of the base pairs within the consensus sequence of the SOS box are of equal importance.

On the basis of the distribution and the degree to which proteins are expressed by operator constitutive mutants, the four base pairs CTGT seem to be the most important for LexA recognition. Within these four base pairs, the two central base pairs (TG) seem to be absolutely required, since these are conserved in all known operator half-sites. The importance of the $(AT)_4$ region is very difficult to assess, owing to the high degree of variability present. This region may not be involved in LexA contact since it is not always protected in assays that methylate bases that are not specifically in contact with protein. This sequence most likely favors LexA binding indirectly by providing the proper spacing between the two half-sites of the SOS box, as has been suggested in the case of the repressor of bacteriophage 434.

IV. THE SOS INDUCING SIGNAL

As discussed above, the activated form of RecA, RecA*, is an important physiological requirement for the induction of the SOS system. It is not the DNA damage itself that leads to the activation of RecA, but most likely the accumulation of single-stranded DNA (ssDNA) that occurs when DNA replication is stalled due to bulky lesions and mutations. There is definitive evidence supporting this hypothesis. First, RecA may be activated *in vitro* by the addition of ssDNA and ATP to the reaction mixture. Additionally, infection of *E. coli* with the filamentous ssDNA phage f1 does not typically induce the SOS response. However, similar infection with an f1 mutant that is defective in the initiation of minus-strand DNA synthesis routinely does lead to the induction of the SOS response.

V. MECHANISM OF LexA CLEAVAGE

Following DNA damage, large stretches of single-stranded DNA are generated when DNA polymerase III dissociates at a lesion and then reassociates approximately 1 kb downstream from the lesion. RecA protein binds to this single-stranded DNA to form spiral nucleoprotein filaments on DNA. Free LexA recognizes this structure and binds within the deep helical groove of the RecA nucleoprotein filament. This interaction with RecA then facilitates the cleavage of LexA, at a scissile peptide bond located between residues Ala^{84} and Gly^{85}, approximately the center of the protein. This Ala^{84}–Gly^{85} bond connects the amino-terminal domain of the protein (which is responsible for DNA binding) to the carboxyl-terminal domain (responsible for dimerization). On cleavage, the level of functionally active LexA protein available to bind the SOS boxes located in the operator/promoter regions of SOS genes decreases dramatically. The end result is to release SOS genes from the negative transcriptional regulation of LexA. Once the DNA is repaired and the inducing signal subsides, RecA returns to an inactive state and the subsequent autocatalytic activity of LexA ceases. Consequently, the cellular pool of functional LexA increases, and expression of the SOS regulon returns to preinduction levels.

It was originally believed that RecA was responsible for the enzymatic cleavage of LexA via a classic protease mechanism. This hypothesis was disproved by Little and colleagues when they showed that the mechanism for LexA cleavage involves both intramolecular and intermolecular reactions that occur independently of RecA. The role of RecA appears to be that of a coprotease that increases the rate of LexA autodigestion under physiological conditions. This autocatalytic cleavage also occurs when LexA is incubated at an alkaline pH, thus proving LexA cleavage is capable of occurring in a manner independent of RecA. This type of cleavage is observed in a number of other functionally and/or structurally related proteins such as the repressors of bacteriophages λ, 434, P22, and ϕ80 and the mutagenesis proteins UmuD, MucA, and $RumA_{(R391)}$. A small number of amino acid residues have been found to be highly conserved among these proteins (the previously mentioned alanine–glycine cleavage site and appropriately spaced serine and lysine residues), and these conserved amino acids presumably play principal roles in the autocatalytic cleavage process. Earlier studies proposed that the LexA cleavage reaction was very similar to that of serine proteases or other signal peptidases, but more recent studies have shown that the reaction is more like that of the TEM1 β-lactamase. The proposed mechanism of autocatalytic cleavage suggests that the lysine residue in the carboxyl domain removes a proton from the carboxyl domain serine residue, which in turn acts as a nucleophile to attack the alanine–glycine bond. In addition, the lysine residue may donate a proton to the α-amino group when the bond is broken.

VI. SOS MUTAGENESIS

The SOS response is often thought of as being a system that is responsible for repairing DNA damage, when in reality, it is a mechanism that enhances the survival of the cell. Often, after an organism has been exposed to a DNA damaging agent such as ultraviolet light or mutagenic chemicals, all of the mutations and lesions that result cannot be repaired by the methods to which the cell has access. Typically, this DNA damage would block the replication of DNA and lead to the death of the cell. SOS mutagenesis allows the replication machinery to bypass these DNA lesions. The end result is mutated DNA, but the cell has survived. Because mutations themselves may have extremely negative effects on the viability of the cell, SOS mutagenesis is tightly regulated (discussed above).

Three of the gene products that are produced as part of the SOS response are required for SOS mutagenesis: RecA, UmuD, and UmuC. In addition to these three LexA regulated genes, DNA polymerase is also required for this process. Given the appropriately intense inducing dose, the SOS system produces the UmuD and UmuC proteins via transcription of the *umuDC* operon. In its initial transcribed state, the UmuD protein is not mutagenically active. In much the same way that the LexA protein undergoes an autocatalytic cleavage, UmuD is cleaved. UmuD is structurally related to the aforementioned proteins, LexA and λ cI. A comparison of the amino acid sequences of the three proteins reveals a conserved alanine–glycine cleavage site as well as the appropriately spaced serine and lysine residues. The cleavage of UmuD yields a modified protein that is referred to as UmuD′. UmuD′ then forms a homodimer and complexes with a UmuC monomer to form a complex ($UmuD'_2C$) that is essential for SOS mutagenesis. This complex, together with RecA, seems to allow the DNA replication enzyme, DNA polymerase III, to bypass specific types of DNA damage.

Although it is generally accepted that the primary components that are required for SOS mutagenesis are the proteins mentioned above, the mechanism by which these proteins allow DNA replication to continue past sites of DNA damage is still under intense study. The importance of SOS mutagenesis is supported by the fact that homologs of UmuD and UmuC are found in a number of other bacteria, as well as bacteriophage P1. The presence of UmuDC homologs even transcends into the eukaryotic kingdom in such organisms as the yeast *Saccharomyces cerevisiae* and the nematode *Caenorhabditis elegans*.

VII. THE SOS RESPONSE IN *BACILLUS SUBTILIS*

The SOS response has long been considered an integral part of the ability of a bacterial cell to survive environmental insults and faulty metabolic processes, but it has since become evident that a DNA repair mechanism, such as the SOS system, is capable of more than enhancing cell survival and regulating the rate of mutagenesis. DNA repair systems also play important roles in viral activation, DNA replication, genetic recombination, metabolism, and cancer. Unfortunately, *E. coli* lacks a readily identifiable developmental cycle, and therefore is not an appropriate model for studying the relationship between DNA repair systems and these other phenomena. On the other hand, the gram-positive soil bacterium *B. subtilis* appears to be an ideal paragon for studying the relationship between DNA repair mechanisms and other developmental cycles. In addition to having a defined SOS system, *B. subtilis* maintains the ability to differentially sporulate, develop motility, produce degradative enzymes, express antibiotics, and become naturally competent in response to environmental stimuli. In addition to the SOS regulon of *E. coli*, similar systems and the cognate regulatory proteins exist in many other gram-negative bacteria. The apparent importance of this stress response mechanism suggests that it is of ancient origin, and its evolution might have even preceded the divergence of the gram-negative and gram-positive eubacteria. Using a set of isogenic strains that differed only in the presence of specific mutations in genes that were thought to be involved in DNA repair and recombination, it was demonstrated that an SOS system did exist in the gram-positive, spore forming bacterium *B. subtilis*. Furthermore, the SOS systems of both *E. coli* and *B. subtilis* are induced in a RecA-dependent manner and have similar phenotypic characterizations of DNA repair, enhanced survival and mutagenesis, prophage induction, and the inhibition of cell division. A number of din genes (including *recA*) were isolated, mapped, cloned, and sequenced using reporter gene technology. This information allowed for the identification of a consensus sequence, GAAC-N_4-GTTC, that was found in the promoter region of each identified din gene. Deletion analysis of this consensus sequence allowed investigators to conclude that the sequence and/or adjacent regions are essential to the proper SOS regulation of each din gene. Since this sequence was found in the promoter regions of *recA* genes in other gram-positive organisms, it was hypothesized to be the gram-positive version of the SOS box, although there is no sequence homology present between it and the SOS box found in *E. coli*.

Owing to the sequence difference between the gram-negative SOS box and the putative gram-positive SOS box, the site in *B. subtilis* was delineated the "Cheo box." Also, since the exact role of the Cheo box had not been fully elucidated, it seemed premature to actually label it as the SOS box. At times however, editorial constraints have, in fact, led to the Cheo box being designated as the gram-positive SOS consensus sequence or SOS box.

The *B. subtilis recA* gene and its gene product have been extensively characterized; however, the identity and role of the repressor of the *B. subtilis* SOS regulon remained somewhat enigmatic. A number of studies by several researchers eventually identified the repressor as the product of a previously isolated din gene, *dinR*. The DinR protein is the same approximate size as LexA (~23 kDa) and also exhibits 34% identity and 47% similarity to LexA. These similarities include regions thought essential for autocatalytic proteolysis. The further characterization of DinR and the operator regions of SOS genes in *B. subtilis* has led to the redefinition of the SOS repressor binding site. This site is now referred to as the "DinR box" and has a consensus sequence of 5′-CGAAC-N_4-GTTCG-3′.

Unlike the *E. coli* SOS response, in which there is only one mechanism for induction, the SOS phenomena of *B. subtilis* are currently classified into four distinct types. The increased complexity of the *B. subtilis* SOS response is primarily attributable to its relationship with the development of natural competence. Again, using reporter gene technology, it was determined that a number of DNA damage inducible genes were also induced following the onset of competence. As noted above, RecA protein is activated by the presence of single-stranded regions of chromosomal DNA, which have been demonstrated to accumulate as cells become competent. It was hypothesized that activated RecA then stimulates the autocatalytic activity of a LexA homolog, and SOS genes are subsequently derepressed. Once it was noted that several *din* genes were induced by competence in *B. subtilis* strains that lacked functional RecA protein, it became evident that SOS induction is more complex in *B. subtilis* than in *E. coli*.

Four types of SOS phenomena in *B. subtilis* have been subsequently described. These types have been classified according to the nature of their mechanisms of DNA damage induction and competence induction. The type of SOS induction that appears to be the most similar to the induction of the *E. coli* SOS system is known as the Type I phenomenon. This includes the following events: expression of the genes *dinB, dinC,* and *uvrB,* induction of certain prophage (ϕ105, SPO2), error-prone repair, and W reactivation. Type I events are induced by DNA damage as well as by the development of the competent state with both being RecA dependent.

The Type II SOS phenomenon of *B. subtilis* has only one characterized phenotype, filamentous growth. Filamentation in *B. subtilis* is induced by DNA damaging agents, but not by the development of competence. Surprisingly, the phenomenon of filamentation in *B. subtilis* is a RecA-independent event.

The genes (*recA* and *dinA*) that have been classified as part of the type III phenomenon are regulated in the most complex manner yet observed. Expression of *recA* and *dinA* is induced by the presence of DNA damaging agents, but only in the presence of RecA itself. The expression of *recA* and *dinA* is also induced by the development of competence; however, under these conditions, RecA protein need not be present. It appears that damage induction of these genes occurs in the prototypical manner that requires activated RecA to stimulate the autocatalytic activity of the LexA homolog. However, the competence induction of these genes appears to occur via a separate competence specific pathway, since several genes that are specific to the competence pathway (*spo0A, spo0H, degU, comK*) are required for the competence induction of *recA*.

The Type IV phenomenon involves the induction of the prophage of the SPβ family, such as ϕ3T. There are two types of temperate bacteriophage that infect *B. subtilis:* "smart" phage and "naive" phage. The smart phage, which include SPβ and ϕ3T, are capable of differentiating between SOS induction by DNA damage and SOS induction by the development of competence. Therefore, the smart bacteriophage are induced on DNA damage only, and not by the development of competence. The induction of smart bacteriophage is a RecA-dependent event.

VIII. THE SOS SYSTEM IN OTHER PROKARYOTES

Escherichia coli has served as the paradigm for study of the SOS system in gram-negative bacteria, and *B. subtilis* has served as the model for study in gram-positive populations. However, the SOS regulon has been found to exist in many other bacteria based on a variety of observations suggesting that the two key regulatory elements, RecA and LexA, have been conserved throughout evolution.

More than 60 highly conserved homologs of RecA have been characterized in a wide variety of bacteria. In addition, there is evidence for LexA-like regulation in 30 species of gram-negative bacteria. This was demonstrated by the introduction of a plasmid containing a *recA* operator/promoter that was fused to a reporter gene (*recA-lacZ*), which was capable of being both induced and repressed in relation to the presence or absence of DNA damaging agents, respectively. More direct evidence for the existence of SOS regulatory networks in other bacteria has been provided by the successful cloning of the *lexA* genes from several organisms as well as genome sequencing projects that have identified *lexA*-like genes in other organisms.

See Also the Following Articles

Bacillus subtilis, Genetics • DNA Repair • Mutagenesis • RecA

Bibliography

Dubnau, D. (1991). *Microbiol. Rev.* 55.

Friedberg, E. C., Walker, G. C., and Siede, W. (1995). "DNA Repair and Mutagenesis," 2nd Ed. American Society for Microbiology, Washington, D.C.

Hanawalt, P. C. (1989). Concepts and Models for DNA Repair: From *E. coli* to mammalian cells. *In* "Environmental Molecular Mutagenesis," Vol. 14.

Koch, W. H., and Woodgate, R. (1998). "DNA Damage and Repair: DNA Repair in Prokaryotes and Lower Eukaryotes" (J. A. Nickoloff and M. F. Hoekstra, eds.), Humana, Totowa, New Jersey.

Yasbin, R. E., Cheo, D. L., and Bol, D. (1993). "*Bacillus subtilis* and Other Gram-Positive Bacteria, Biochemistry, Physiology, and Molecular Genetics" (J. A. Hoch, A. L. Soneshein, and R. Losick, eds.), American Society for Microbiology, Washington, D.C.

Space Flight, Effects on Microorganisms

D. L. Pierson and S. K. Mishra
NASA/Johnson Space Center

I. Microbial Survival and Growth in Space
II. Sensitivity to Antibiotics
III. Radiation, Bacteriophage Induction, and Microbial Genetics
IV. Effect of Simulated Microgravity on Microorganisms
V. Human–Microbe Interactions during Space Missions

GLOSSARY

commensalism Symbiotic relationship in which one species benefits and the other is unharmed.

endogenous Originating or produced within an organism or one of its parts.

exogenous Originating outside an organism; infections can be of exogenous or endogenous origin.

microgravity The condition of an environment in which acceleration due to gravity is approximately zero; also termed weightlessness.

pedicel Slender stalk that supports the fruiting or spore-bearing organ in some fungi.

solar particle event Sudden eruption on the surface of the sun that results in an increased flux of high-energy particles, which in turn increases the exposure of spacecraft to ionizing radiation.

Spacelab Manned laboratory developed by the European Space Agency for flights aboard the U.S. Space Shuttle; pressurized habitable modules and pallets adapted to specific missions are carried in the Shuttle's payload bay.

THE EFFECT OF SPACE FLIGHT on microbial function has been of concern to microbiologists since humans first began to explore space. Because microorganisms will be present on board manned and unmanned spacecraft, the potential exists for colonization of the vehicle itself as well as its inhabitants. The combination of the closed nature of spacecraft and the stressful nature of space flight (e.g., acceleration, weightlessness, radiation) increases the possibility of microbially induced allergic reactions and infections among space crews. Space flight is also suspected of altering human immune function as well as bringing new environmental selection pressures to bear on endogenous and exogenous microbiota. The combined effect of these processes may render normally harmless commensal or environmental microbes pathogenic to humans. Furthermore, colonization of the vehicle itself may result in system fouling, biodegradation of sealants, and perhaps the production of toxic metabolites and environmental pollutants. As the number and duration of manned flights increase, it has become imperative to characterize the effects of space flight and related factors on microbial growth, physiology, virulence, and susceptibility to antibiotic agents to protect the health of the crews and the integrity of the spacecraft.

At present, very few research data are available to address these concerns. Although many microorganisms have been used as models to study the effects of cosmic radiation, microgravity, vibration, and hypervelocity on living systems during a number of missions over the past 40 years (Tables I and II), the severe constraints involved in performing experiments in space have largely precluded exhaustive studies. The absence of gravity, which mandates the development of specialized equipment and procedures, the restrictions imposed on power, weight, and volume, and intense competition for the crew's time during space flights require that experiments be simple and easily performed with little or no crew involvement. Thus, many basic questions concerning the effects of space on microbial structure and function have yet to be

TABLE I
U.S. and Soviet Missions Carrying Microbiology Experiments

Flight	Country	Launched	Manned duration
Sputnik	USSR	1957–1961	Unmanned
Vostok	USSR	1961–1965	
Gemini	U.S.	1964–1966	10 Manned flights
Cosmos 110	USSR	Feb 1966	
Apollo	U.S.	1967–1972	
Biosatellite II	U.S.	Sept 1967	2 days
Zond 5	USSR	Sept 1968	Unmanned
Zond 7	USSR	Aug 1969	
Cosmos 368	USSR	Oct 1970	
Salyut 1	USSR	April 1971	23 days
Skylab 1	U.S.	May 1973	Unmanned
Skylab 2	U.S.	May 1973	28 days
Skylab 3	U.S.	July 1973	59 days
Soyuz 12	USSR	Sept 1973	
Skylab 4	U.S.	Nov 1973	84 days
Salyut 3	USSR	June 1974	14 days
Cosmos 690	USSR	Oct. 1974	
Salyut 4	USSR	Dec 1974	41 days
Apollo–Soyuz	U.S.–USSR	July 1975	9 days
Salyut 5	USSR	June 1976	33 days
Salyut 6	USSR	Sept 1977	1,192 days
Cosmos 1129	USSR	Sept 1979	
Salyut 7	USSR	April 1982	1,805 days
Spacelab 1	U.S.	Nov. 1983	10 days
Spacelab 3	U.S.	July 1985	7 days
Spacelab 2	U.S.	July 1985	8 days
Spacelab D-1	U.S.	Oct 1985	7 days
Mir	USSR	Feb 1986	366 days (max)

resolved. This article presents an overview of information collected to date on the effects of the space flight environment on microbial physiology and function.

I. MICROBIAL SURVIVAL AND GROWTH IN SPACE

Studies on the effect of extreme conditions on microorganismal growth and survival began as early as 1935, when high-altitude balloons were used to investigate the effects of low temperatures, decreased pressures, and increased radiation. From 1954 to 1960, experiments on nearly 30 high-altitude balloons and sounding-rocket flights revealed that *Neurospora* spores and vegetative bodies could survive direct exposure to the environment at 35–150 km above Earth's surface. In the 1960s, viable organisms from cultures of *Penicillium roquefortii* and *Bacillus subtilis* carried on the Gemini 9A and 12 missions were recovered after nearly 17 hr of direct exposure to space. Parallel attempts to detect microorganisms in the extraterrestrial environment in analyses of micrometeorites collected during the Gemini missions, and lunar samples collected during the Apollo flights, revealed no evidence of viable microorganisms nor any identifiable biological compounds. Therefore, the potential for contaminating Earth with extraterrestrial life forms, an early concern, was judged to be extremely unlikely.

Studies of microbial behavior in space performed to date, although numerous, have produced inconclusive, occasionally contradictory results (Tables III and IV). The first experiments on the unmanned Sputnik orbital satellites (1957–1961) used microorganisms to identify the gross effects of galactic radiation, weightlessness, and other related factors on biological systems. Studies of bacteriophage induction have been many (see later) and have dated back to the second Soviet satellite mission in August 1960, which included flight experiments with *Clostridium butyricum*, *Streptomyces* spp., *Aerobacteria aerogenes* 1321 bacteriophage, and T-2 coliphage. Viability and gas production by *C. butyricum* was no different after flight than that of ground-based control cultures; growth of flight cultures of *Streptomyces*, however, was accelerated sixfold on return to Earth. Growth of *Escherichia coli* cultures flown on later Soviet missions was not appreciably different than that of ground-based controls. The characteristics of an alga, *Chlorella* sp., were studied on many Soviet missions during the 1960s and 1970s, but the organism showed no change in growth characteristics, viability, proportion of photosynthetically active cells, or mutation rate; however, chloroplast volume tended to decrease.

Continuing chronologically, the first serious attempts by U.S. investigators to study the effect of space flight on microorganisms began with the launch of Biosatellite II in September 1967. The biological experiments included in this 45-hr Earth or-

TABLE II
Organisms Used in Space Flight Experiments

Prokaryotes	*Eukaryotes*
Bacteria	Protozoa
Actinomyces aureofaciens	*Euglena gracilis*
Actinomyces erythreus	*Paramecium aurelia*
Actinomyces levoris[a]	*Paramecium tetraurelia*
Actinomyces streptomycin	*Pelomyxa carolinensis*
Aerobacteria aerogenes	*Tetrahymena periformis*
Aeromonas proteolytica	*Tetrahymena pyriformis*
Bacillus brevis	Fungi
Bacillus subtilis	Molds
Bacillus thuringiensis	*Aspergillus niger*
Clostridium butyricum	*Chaetomium globosum*
Clostridium sporogenes	*Neurospora crassa*
Escherichia coli	*Penicillium roquefortii*
Hydrogenomonas eutropha	*Phycomyces blakesleeanus*
Methylobacterium organophilum	*Polyporus brumalis*
Methylomonas methanica	*Trichoderma viride*
Methylosinus sp.	*Trichophyton terrestre*
Proteus vulgaris	Yeasts
Pseudomonas aeruginosa	*Candida tropicalis*
Staphylococcus aureus	*Rhodotorula rubra*
Streptomyces levoris[b]	*Saccharomyces cerevisiae*
Bacteriophage	*Saccharomyces vivi*
Aerobacteria aerogenes phage 1321	*Zygosaccharomyces bailii*
Escherichia coli phage T1, T2, T4, T7, and λ	Slime mold
Salmonella typhimurium phage P-22	*Physarum polycephalum*
	Algae
	Chlamydomonas reinhardtii
	Chlorella ellipsoidea
	Chlorella pyrenoidosa
	Chlorella sorokiniana
	Chlorella vulgaris
	Scenedesmus obliquus

[a] Synonyms: *Actinomyces* in Russia, *Streptomyces* in the U.S.

bital flight were investigations of the effects of microgravity and radiation on the growth and life cycle of *Salmonella typhimurium, E. coli, Neurospora crassa,* and *Pelomyxa carolinensis.* Although the total cell density in cultures of *S. typhimurium* flown as nonirradiated controls on this mission increased 15% during stationary-phase growth, mean viable cell density increased by 30%. Some portion of these increases, however, was attributable to vehicular vibration and acceleration.

On Apollo 16 in April 1972, four fungal species, two yeastlike fungi, *Rhodotorula rubra* and *Sacchraomyces cerevisiae,* and two filamentous fungi, *Trichophyton terrestre* and *Chaetomium globosum,* were exposed outside the space capsule within a specially designed Microbial Ecology Evaluation Device. Both dry inocula and aqueous suspensions of vegetative yeast cells, conidia, and ascospores were exposed to the sun at a 90° angle for about 10 min in cuvettes equipped with a quartz window and bandpass filters. These cultures were subjected to many tests on their return to Earth. The two filamentous fungi showed

TABLE III
Effect of Microgravity on Microbial Growth and Sensitivity to Antibiotics

Organism	Result	Mission	Duration of experiment
Escherichia coli	Cell density slightly increased	Biosatellite II	2 days
Salmonella typhimurium	Cell density increased 20%	Biosatellite II	
Proteus vulgaris	Cell numbers increased sevenfold	Biosatellite II	
Bacillus subtilis	Biomass significantly increased	Soyuz 12	2 days
	Sporulation reduced, autolysis increased in stationary phase	Spacelab D-1	3 days
Physarum polycephalum	Growth reduced	Cosmos 1129	
Chlorella sp.	Growth not affected	Soyuz 12	
		Cosmos 573	Up to 10 days
Chlamydomonas sp.	Proliferation increased 100%	Spacelab D-1	6 days
Paramecium tetraurelia	Growth increased twofold	Salyut 6	4 days
		Spacelab D-1	5 days
Staphylococcus aureus	MIC of following increased twofold: oxacillin, chloramphenicol, erythromycin	Salyut 7	1 day
Escherichia coli	MIC of following increased two- to fourfold: colistin, kanamycin	Salyut 7	1 day

changes in the colony perimeters, growth density, protoplasmic leakage of hyphal apices damaged by ultraviolet (UV) irradiation, abnormal growth at the hyphal apex, forked hyphal branches, and irregular hyphal walls. The infectivity and ability to degrade human hair of *T. terrestre* increased after flight. The cellulolytic fungus *C. globosum* lost its ability to produce pigment, developed fewer fruiting bodies, and demonstrated variations in amylase production. Cultures of *R. rubra* also showed altered phosphoglyceride content. *Saccharomyces cerevisiae* nearly doubled its rate of phosphate uptake after space exposure, as well as being more susceptible to UV radiation and better able to survive in intradermal lesions in artificially infected mice. Finally, exposing all four strains to UV irradiation at 254, 280, and 300 nm and then maintaining them in total darkness increased their susceptibility to antibiotics.

Growth of *Proteus vulgaris* studied aboard Soyuz 12 in 1973 showed a sevenfold increase compared with ground controls; a significant increase was also observed in the growth of *B. subtilis*, with a corresponding decrease in spore production (5×10^4 spores/ml in space-grown cultures versus 8×10^5 spores/ml in ground controls). Cultures grown in space also showed increased cell lysis. Fruiting bodies of the fungus *Polyporus brumalis* grown on the satellite Cosmos 690 (Oct–Nov 1974) showed disorientation and occasional flattening, with pedicels twisted into spirals or balls; ground-control cultures,

TABLE IV
Effect of Microgravity on Microbial Genetics and Sensitivity to Radiation

Organism	Result	Mission	Duration of experiment
Escherichia coli	No effect on transduction, transformation; 40% increase in conjugation	Spacelab D-1	3 hr
Salmonella typhimurium	Increased sensitivity to P-22 phage	Biosatellite II	2 days
	Increased resistance to high doses of radiation	Biosatellite II	2 days
Escherichia coli	Increased resistance to high doses of radiation	Biosatellite II	2 days

in contrast, displayed strong negative geotropism and long pedicels.

During the joint Apollo–Soyuz mission in 1975, *Streptomyces levoris* was used to study growth rate periodicity. One of the eight cultures studied grew more quickly in the spacecraft environment than on the ground, and three of the space cultures developed double spore rings during the immediate postflight period. Of the many microbial species studied aboard the Salyut 6 mission, space cultures of the methanogenic bacteria *Methylosinus* sp. (AB-21) and *Methylomonas methanica* (AB-3) grew no differently than did ground controls; however, *Methylobacterium organophilum* (MB-67) showed a greater tendency to grow anaerobically in space. Electron microscopic analysis after landing, however, revealed no changes in cytoskeletal ultrastructure. Cultures of the protozoan *Paramecium tetraurelia* aboard Salyut 6 (1977–1981) and on Spacelab D-1 in 1985 had larger, spherical cells during the early log phase, with smaller cells during later growth stages. The growth rate of this protozoan nearly doubled in microgravity; its yield increased nearly four times after 5 days in space, and its cell protein, magnesium, and calcium concentrations had decreased relative to ground controls. In contrast, earlier cultures of another protozoan, *Pelomyxa carolinensis,* on Biosatellite II showed a possibly increased rate of cell division but no distinct effects on overall growth rate or physiological, morphological, and cytochemical variables.

A serious drawback in these early studies, both Soviet and American, was the lack of on-board controls that could be used to rule out the effect of vibration and acceleration in flight. This drawback was overcome in 1985 with the use of the European Space Agency's Biorack facility, which included an incubator-centrifuge that could be used in flight to simulate gravity. This facility was used successfully on the German Spacelab D-1 mission in 1985, during which several microbiological studies were conducted using cultures of *B. subtilis,* the alga *Chlamydomonas,* and the protozoan *Paramecium tetraurelia.* The period and phase of the photo-accumulation behavior of the two strains of *Chlamydomonas* flown on Spacelab D-1 did not change in flight, although their proliferation rates increased significantly. In contrast, growth of the slime mold *Physarum polycephalum* on Spacelab D-1 and on the Cosmos 1129 mission in 1979 was reduced relative to ground controls, and the organism also showed reduced protoplasmic movement after flight.

In sum, although many species of both prokaryotes and eukaryotes have been flown in space, the inconsistencies with regard to use of analytical equipment, procedures, and culture conditions have led in some cases to conflicting results. In addition, the relatively infrequent flight opportunities, as well as limited crew time during flight, means that the results that are available tend to be fragmentary and often lack a classic, controlled experimental context with which to interpret them. The absence of an overall structure to the study of microbial function in space is reflected throughout this article.

II. SENSITIVITY TO ANTIBIOTICS

Because microorganisms will be ubiquitous on space vehicles and in space-based habitats, the risk of infectious disease from environmental or commensal flora cannot be ruled out. Little information is available, however, as to whether exposure to microgravity affects the sensitivity of microorganisms to antibiotics, either *in vitro* or *in vivo.* Evidence suggests that some forms of bacteria isolated from astronauts after space flight may be more resistant to antibiotic agents than the same strains isolated before flight. To supplement these observations, a group of French and Soviet scientists in July 1982 studied the effects of several antibiotics on space-grown cultures of *Staphylococcus aureus* and *E. coli* as part of the Cytos 2 experiment on Salyut 7. The minimal inhibitory concentrations (MIC) of oxacillin, chloramphenicol, and erythromycin on actively growing cultures of *S. aureus* were found to be roughly twice the ground-control values of 0.16, 4.0, and 0.5 μg/ml, respectively. The MIC for colistin and kanamycin against *E. coli* were greater than 16 μg/ml, compared with 4 μg/ml in ground-control cultures. The Antibio experiment conducted 3 years later on Spacelab D-1 supported these observations. The tentative conclusion at present is that space flight may increase the required MIC for these organisms, possibly by stimulating microbial growth or by inducing physiological or biochemical changes in cell-wall structure or permeability. More data are needed to explore these speculations.

III. RADIATION, BACTERIOPHAGE INDUCTION, AND MICROBIAL GENETICS

Another environmental factor prompting concern as to whether humans and other Earth-based biological systems can survive in space is radiation, with its preponderance of high-intensity solar UV light, radionuclides, protons, electrons, decay products, and galactic cosmic rays. Cosmic rays normally reach Earth's surface at about 0.027 rem (roentgen equivalent for man) per year. By comparison, inside the U.S. Space Station, (Figs. 1 and 2), which will be located 270 nautical miles above Earth's surface, organisms will be subjected to annual cos mic radiation of about 43 rem; additional peaks up of 1400 rem can occur in free space during solar particle events such as the one that took place in August 1972. The yearly dose of cosmic radiation at the lunar surface is expected to be 30 rem; at the Martian surface, even with atmospheric shielding of 10 g/cm^2, average annual cosmic radiation has been calculated to be 12 rem, with peaks to 83 rem.

The response of *B. subtilis* to high atomic number–high energy (HZE) particles, a component of galactic cosmic radiation, was studied in the Biostack I and II experiments on board Apollo 16 and 17, and again on the Apollo–Soyuz Test Project. In the Apollo experiments, germination was not influenced by an HZE "hit," but spore outgrowth was reduced significantly. In the later project, *B. subtilis* spores were inactivated by HZE particles at distances up to 4 μm. Dose values at these distances were approximately 0.1 Gray, whereas the D37 value (the dose needed to reduce survival to 37%) for electron radiation is about 800 Gray. These investigations prompted the conclusion that the biological hazard of cosmic HZE particles, especially its high Z component, has been much underestimated.

Ionizing radiation is well known for its mutagenicity. Mutagenic doses of radiation vary among species and with the life stage of the organism. For example,

Fig. 1. Space Station with a Space Shuttle to the left.

Fig. 2. Model of the space station facility that includes the microbiological components discussed in the text.

in 1960 on the second Soviet satellite, the number of spores produced by *Actinomyces erythreus* strain 2577, which is known to be resistant to UV light, increased six times over the spores produced by the UV-sensitive strain 8594. Bacterial endospores are quite resistant to gamma radiation; a 90% kill rate can be achieved with 3000–4000 Gray, and only 10% of this amount to reach the same mortality in vegetative bacteria. Indeed, experiments with *Neurospora* spp. on Biosatellite II and Gemini XI showed that rapidly metabolizing cells displayed greater effects of radiation in the control samples on the ground than did those from the flight. For nondividing, inactive spores, there was no difference in genetic effects between control and flight samples. Similarly, *B. subtilis* and *B. thuringiensis* spores exposed to solar UV and the space vacuum on Apollo 16 survived as well as did ground controls. Later exposure of other cultures of *B. thuringiensis* to full sun in space produced a lower survival rate compared with either exposure to solar UV alone or ground controls. Also on Apollo 16, the space vacuum enhanced the lethal effect of solar UV at 254 and 280 nm in one strain of *B. subtilis* but not in a repair-defective strain. Exposure to the space vacuum alone did not alter the survival rate of dried spores. These observations suggest that other components of solar radiation in space may act as powerful biocides, either alone or in combination with UV light.

Lysogenic bacteria (those with a viral prophage integrated into the host genome) have been used frequently as biological indicators in tandem with physical methods of measuring radiation. Bacteriophage production from lysogenic bacteria can be induced by x-rays, gamma rays, protons, or neutrons, is dependent on radiation dose, and may be resistant to other flight factors. Soviet investigators have used the K-12 λ phage of *E. coli* since the second Earth-orbiting satellite as a kind of dosimeter to monitor radiation in the spacecraft interior. Phage production by these bacteria can be stimulated by as little as 0.3 rad of gamma radiation, or still smaller doses of protons or rapid neutrons; thus, information about the mutagenicity of cosmic radiation and its associated risks to the genetic structure can also be ascertained. This phage system was carried aboard most of the Sputnik missions, the manned flights of the Vostok series, the unmanned biosatellite Cosmos 110, and Zond 5 and 7, the latter of which orbited the moon before returning to Earth. In the early Soviet studies, phage production in flight did not exceed that of controls; the relationship between length of flight and number of phage produced is still not clear. Another Soviet observation that vibration increased the sensitivity of lysogenic bacteria to subsequent gamma irradiation suggests that space flight factors may have cumulative effects. Phage induction in *E. coli* cultures on the Soviet missions Vostok 4 and 6, Zond 5 and 7, and the U.S. Biosatellite II mission showed slightly higher survival rates than ground controls, with small increases in viable cell density.

Fig. 3. A Space Shuttle launch.

U.S. experiments with the lysogenic bacterium *Salmonella typhimurium* BS-5 (P-22)/P-22 on Biosatellite II found that weightlessness and gamma irradiation had significant effects on growth rate and induction of prophages: Although not different statistically, the flight cultures tended to have fewer induced bacteriophages per viable bacterium at all levels of radiation relative to ground cultures. Research with *B. subtilis* on Apollo 16 indicated that filtered solar radiation at 254 and 280 nm appeared to be less lethal to spores at these doses and induced lower mutation frequencies than in ground-control studies. The T7 bacteriophage was inactivated to a greater degree by the 254-nm solar UV than in ground studies, although the shapes of the dose-response curves were comparable.

Studies of microbial genetics with respect to radiation effects in space have been few; however, it was observed during the Apollo Microbial Ecology Evaluation Device experiments that *Trichophyton terrestre* and *Chaetomium globosom* showed greater numbers of nucleotide pairs per haploid nucleus after flight. On the Spacelab D-1 mission, space flight was found to have no effect on transduction and transformation in *E. coli;* however, conjugation was enhanced by up to 40%. Far more data are needed to separate and characterize the multiple effects of space flight factors on microbial genetics at the molecular level.

IV. EFFECT OF SIMULATED MICROGRAVITY ON MICROORGANISMS

Because actual space flight experiments are severely constrained by limited availability of electrical power, stowage space, and crew time and a general lack of sophisticated analytical equipment and expertise aboard the spacecraft, attempts are underway to study microbial structure and functions under simulated microgravity conditions in Earth-based laboratories. A high-aspect-rotating-vessel (HARV) type bioreactor developed by NASA/Johnson Space Center and Synthecon, Inc. (Houston, TX) is believed to simulate microgravity conditions by rotating slowly so that the axis of rotation is perpendicular to the gravity vector. The HARVs were originally developed to create a low-turbulence, low-shear environment that would allow human cells to grow and assemble into three-dimensional constructs. These vessels are rotated horizontally so that the cell sedimentation associated with gravitational forces is balanced by the centrifugal forces caused by rotation. Our studies with the baker's yeast, *Saccharomyces cerevisiae,* and with the opportunistic yeastlike fungus *Candida albicans* revealed a faster growth rate when cells were grown in the bioreactor, comparable to what has been noted in the past with other microorganisms in actual microgravity studies. The cells tended to be more spherical and slightly larger, and had a tendency to form clusters.

Demain and co-workers from the Massachusetts Institute of Technology have used HARV to study simulated microgravity effects on secondary metabolism. They observed reduced production of a β-lactam antibiotic by *Streptomyces clavuligerus* in simulated conditions, but noted that the stimulatory effects of phosphate and L-lysine were similar to those seen in normal gravity. Production of gramicidin S by a strain of *Bacillus brevis,* and of antibacterial polypeptide microcin B17 by *Escherichia coli* ZK 650, was also inhibited, although the test strains grew faster in the simulated conditions than in normal gravity conditions.

V. HUMAN–MICROBE INTERACTIONS DURING SPACE MISSIONS

We have monitored body microbiota of the space shuttle crew since the beginning of NASA's Space Transportation Program. Except for a slight decline in *Corynebacterium* and *Haemophilus* species in throat swabs, and some increase in the α-hemolytic streptococci and *Staphylococcus aureus* in nasal swabs, no remarkable change has been noticed in the resident microbiota of the upper respiratory tract. However, the postflight nasal swabs from a large number of crew members, especially those who supported 10- to 17-day missions, were positive for *Aspergillus* and *Penicillium* species. Isolation of these common molds from the nasal swabs does not suggest any infection or colonization, but probably a delayed clearance from the nasal mucosa or an increased exposure to their propagules proliferating in onboard locations such as the air ducts. Although fungi, including *Penicillium* and *Aspergillus* species, have been frequently isolated from the interiors of the U.S. space shuttles and the Russian space station Mir, space flight effects on mucosal immunity are still poorly understood. Similarly, except for a decline in anaerobic bacteria in the fecal flora, noted by Russian investigators, U.S. as well as Russian studies on long-duration (3 months to over a year) space missions do not reveal any remarkable change in the microbiota of the gastrointestinal tract. In contrast, postflight urine samples revealed a significant increase in the number of common skin contaminants as well as potentially pathogenic bacteria, such as *Escherichia coli, Enterococcus faecalis,* and *Proteus mirabilis.* In numerous instances, the bacterial counts rose from 0 colony-forming units (CFU) before flight to 10^5 CFU per milliliter in the postflight specimens. Such increases may to some extent be attributed to lack of a whole-body shower and reduced opportunities for personal hygiene. An effect of changes in host physiology, especially the hormonal milieu, cannot be ruled out.

Considering a single-family home with an internal area of about 700 to 1000 m^3 (approximately 2200 to 3000 square feet of floor space) as a comfortable living condition, the space shuttle, with its 65 m^3 of habitable area often shared by five to seven astronauts, offers highly crowded living conditions. It is, therefore, logical to believe that under those circumstances, microorganisms frequently will be exchanged among the crew members, and in due course all of them will harbor a comparable microbiota on their body surfaces. Using as a model two common commensals that also are important opportunistic pathogens, namely, *Candida albicans* and *Staphylococcus aureus*, we have compared genotypes of multiple isolates (from throat, nose, urine, and fecal specimens) from 61 crew members who supported 10 shuttle missions ranging from 7 to 10 days in duration. Surprisingly, the crew members mostly carried their own microbiota; person-to-person transfer of microorganisms was observed, but only on a few rare occasions.

See Also the Following Articles

Exobiology • Microbes and the Atmosphere

Bibliography

Cioletti, L. A., Pierson, D. L., and Mishra, S. K. (1991). Microbial growth and physiology in space: A review. Presented at the 21st International Conference on Environmental Systems, San Francisco, California, July 1991. Society for Automotive Engineers Technical Paper Series No. 911512, Warrendale, Pennsylvania.

Dickson, K. J. (1991). Summary of biological experiments with cells. *ASGSB Bull.* 4, 151–260.

Fang, A., Pierson, D. L., Mishra, S. K., Koenig, D. W., and Demain, A. L. (1997a). Gramicidin S production by *Bacillus brevis* in simulated microgravity. *Curr. Microbiol.* 34, 199–204.

Fang, A., Pierson, D. L., Mishra, S. K., Koenig, D. W., and Demain, A. L. (1997b). Secondary metabolism in simulated microgravity: β-lactam production by *Streptomyces clavuligerus*. *J. Indus. Microbiol. Biotechnol.* 18, 22–25.

Gmunder, F. K., and Cogoli, A. (1988). Cultivation of single cells in space. *Appl. Microgravity Tech.* 13, 115–122.

Mattoni, R. H. T., Ebsersold, W. T., Eiserling, F. A., Keller, E. C., and Romig, W. R. (1971). Induction of lysogenic bacteria in the space environment. *In* "The Experiments of Biosatellite II [NASA SP 204]" (J. F. Sanders, ed.), pp. 309–324. National Aeronautics and Space Administration, Washington, D.C.

Pierson, D. L., McGinnis, M. R., Mishra, S. K., and Wogan, C. F. (eds.) (1989). "Microbiology on Space Station Freedom [NASA Conference Publication 3108]." National Aeronautics and Space Administration, Washington, D.C.

Pierson, D. L., Mehta, S. K., Magee, B. B., and Mishra, S. K. (1995). Person-to-person transfer of *Candida albicans* in the spacecraft environment. *J. Med. Vet. Mycol.* 33, 145–150.

Pierson, D. L., Chadambaram, M., Heath, J. D., Mallary, L. L., Mishra, S. K., Sharma, B., and Weinstock, G. M. (1996). Epidemiology of *Staphylococcus aureus* during space flight. *FEMS Immunol. Med. Microbiol.* 16, 273–281.

Rodgers, E. (1986). "The ecology of microorganisms in a small closed system: Potential benefits for space station [NASA Technical Memorandum 86563]." National Aeronautics and Space Administration, Marshall Space Flight Center, Huntsville, Alabama.

Taylor, G. R. (1974). Space microbiology. *Annu. Rev. Microbiol.* 28, 121–137.

Taylor, G. R. (1977). Cell biology experiments conducted in space. *Bioscience* 27, 102–108.

Tixador, R., Richoilley, G., Gasset, G., Templier, J., and Bes, J. C. (1985). Study of minimal inhibitory concentration of antibiotics on bacteria cultivated *in vitro* in space (Cytos 2 Experiment). *Aviation Space Environ. Med.* 56, 748–751.

Volz, P. A. (1990). Mycology studies in space. *Mycopathologia* 109, 89–98.

Zhukov-Vererzhnikov, N. N., Volkov, M. N., Maiskii, I. N., Guberniev, M. A., Rybakov, N. I., Antipov, V. V., Kozlov, V. A., Saksonov, P. P., Parfenov, G. P., Kolobov, A. V., Rybakova, K. D., and Aniskin, E. D. (1968). Experimental genetic investigations of lysogenic bacteria during flight of the AES "Kosmos-110." *Kosmicheskie Issledovaniya* 6, 144–149. (In Russian; English translation: *Cosmic Res.* 6, 121–125.)

Zhukov-Verezhnikov, N. N., Volkov, M. N., Rybakov, N. I., Maiskii, I. N., Saksonov, P. P., Guberniev, M. A., Podoplelov, I. I., Antipov, V. V., Kozlov, V. A., Kulagin, A. N., Aniskin, E. D., Rybakova, K. D., Sharyi, N. I., Voronkova, Z. P., Parfenov, G. P., Orlovskii, V. I., and Gumenyuk, V. A. (1971). Biological effects of space flight on the lysogenic bacteria *E. coli* K-12 (lambda) and on human cells in culture. *Kosmicheskie Issledovaniya* 9, 292–299. (In Russian; English translation: *Cosmic Res.* 9, 267–273.)

Spirochetes[1]

Lynn Margulis
University of Massachusetts, Amherst

I. Definition of Spirochetes
II. Taxonomy and Diversity
III. History of Knowledge
IV. Physiology
V. Ecology
VI. Molecular Biology and Pathogenicity

GLOSSARY

axial fibrils, axial filaments Flagella, bacterial flagella that reside in the periplasm of the cell, periplasmic flagella, or endoplasmic flagella.

crenulations Ruffles in the outer cell wall layers characteristic of spirochetes of the genus *Pillotina*.

microbial mats Stratified, layered, aquatic, benthic communities of microorganisms that trap, bind, and precipitate sediment. Under environmental conditions of large quantities of carbonate they may lithify to form laminated limestone or chert rocks called stromatolites. They represent some of the oldest ecosystems on Earth, and their presence in the fossil record accounts for some of the most ancient evidence of life on the planet.

morphometrics Analysis of structure by measurement of the physical features of cells and organisms; morphometric analysis.

polar organelle Proteinaceous vesicular structure approximately 200 nm in diameter just below the poles of spirochetes and other motile bacteria in some cases shown to be composed of ATPase, cytochrome oxidase, and stalked donut-shaped particles 5–6 nm in diameter arranged in an hexagonal array.

protoplasmic cylinder Protoplast; that part of the spirochete cell bounded by the plasma (inner) membrane that contains the ribosomes, nucleoid, and the rest of the cytoplasmic materials.

sheath Outer layer of the spirochete covering that corresponds to standard features of the gram-negative cell wall of bacteria generally: the inner and outer coats of the outer membrane and the outer membrane itself.

sillon The longitudinally aligned groove on the spirochete surface where the outer membrane contacts the protoplasmic cylinder.

SPIROCHETES, corkscrew-shaped highly motile, thin, heterotrophic gram-negative eubacteria, are as elusive as they are ubiquitous. In marine muds, nutrient-rich fresh water, mammalian digestive tracts, and even stranger habitats dwell far more spirochetes than ever have been described. Unique ultrastructure correlated with rapid swimming and flexous behavior unite the cohesive group: in all spirochetes the rotary-motor flagella insert subterminally into the protoplasmic cylinder (cell body) and extend down the periplasmic space under the outer membrane. Both their distinctive 16S ribosomal RNA sequences and stages that display these internal (periplasmic) flagella permit the recognition of myriad unknown, uncultivable abundant populations of spirochetes in a surprising range of sizes, morphotypes, and communities.

I. DEFINITION OF SPIROCHETES

Spirochetes, a cohesive group of prokaryotes, have been known by direct microscopic observation since

1. This chapter is dedicated to the memory of Kari Hovind-Hougen (1934–1993), champion scholar and superb electron microscopist, for her little-known contribution to our understanding of spirochete morphology.

the inventor of the microscope, Dutchman Antonie van Leeuwenhoek (1632–1723), first saw them. Although spirochetes are cosmopolitan in distribution, easily recognized by those familiar with them, widely diversified in size, habitat, and physiology, the vast majority have eluded study because of their intractability to growth in pure culture.

Spirochetes are motile, helically shaped bacteria. The group name, *spiro* = coiled + *chaete* = hair, and their description as "wriggling hairs" are apt. In minimally viscous media such as water they tend to flex, bend, and squirm, whereas in thicker, dense, and more viscous solutions they are the speedsters of the microbial world. As chemotactic rapid swimmers in these environments, spirochetes will outdistance spirilla, peritrichously flagellated bacteria, and other highly motile bacteria in their search for bits of food.

Although occasionally confused with spirilla, spirochetes tend to be finer and more delicate. Unlike spirilla their walls are flexible and not rigid; they often bend many times in on their own bodies. Although in adverse conditions division products may remain together and form long chains of cells, all spirochetes are fundamentally single celled and all reproduce by transverse binary fission.

The cell walls of all spirochetes conform to the description of gram-negative eubacterial cells. All are organochemoheterotrophic; neither photoautotrophic nor chemoautotrophic spirochetes have ever been reported. Morphologically distinctive, they are defined by their most salient feature: their periplasmic flagella. These have been called by many names including endoplasmic flagella, axial fibrils, or axial filaments. Both chemical and morphological analysis have confirmed that these fibrils are homologous to the flagella of other bacteria. Thus the rotary motor-driven flagella of all spirochetes lie within the periplasm, that is, between the plasma or inner and outer membranes of the gram-negative cell wall. The smallest and simplest in morphology have but a single flagellum emerging at each pole of the cell, whereas the most complex may have as many as 300 flagella from each terminus. In the best preparations the polar organelle can be clearly seen associated with the flagella. It is very likely that this structure is universally present in spirochetes. However, polar organelles

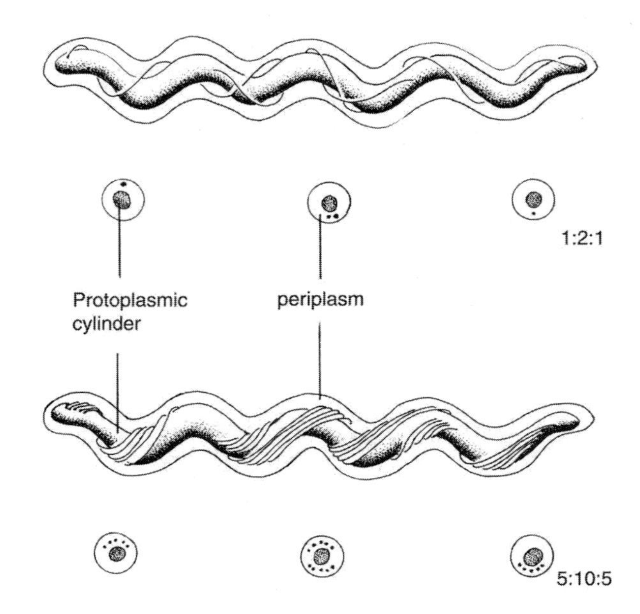

Fig. 1. Spirochete description expression of flagellar arrangement, namely, $n:2n:n$ where $n = 1$ or $n = 5$.

are impossible to demonstrate in the absence of superb high power electron micrographs, which are so far unavailable. These plate-shaped organelles were formerly called "polar membranes"; however, because they are proteinaceous structures and not lipoprotein membranes, they are now universally referred to as the polar organelle.

Because of the uniform morphology of all spirochetes with respect to their flagella, all can be described by a common general expression. If n is the number of flagella emerging from one cell terminus and 2n the number of overlapping flagella in the middle of the cell, then the characteristic array is $n:2n:n$. In those cases where the flagella do not extend sufficiently from each pole to overlap in the middle (e.g., *Treponema phagedensis* or *Leptospira*), the descriptive expression becomes $n:0:n$. This expression is illustrated in Fig. 1 for two spirochetes: a 1 : 2 : 1 (like most treponemes and many spirochaetas) and a 5 : 10 : 5 (similar to *Borrelia* and *Mobilifilum*) example.

II. TAXONOMY AND DIVERSITY

Consistent with studies of spirochete genes for the 16S ribosomal RNA sequences is the grouping of the

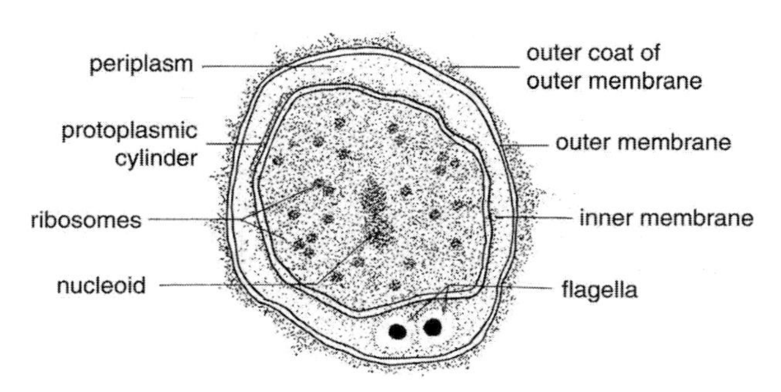

Fig. 2. Generalized structure of a 1:2:1 spirochete.

cultivable spirochetes in two families: Spirochaetaceae (which includes the treponemes, borrelias, spirochaetas, *Brachyspira,* and *Brevinema*) and the Leptospiraceae (which include *Leptonema* and *Leptospira,* aerobic spirochetes known to respire molecular oxygen). A generalized drawing of these kinds of small spirochetes with few flagella is shown in Fig. 2.

The larger spirochetes, none of which are cultivable, are so far limited to the third family, the Pillotinaceae. All of these dwell in the digestive systems of animals; they are symbionts in arthropods and mollusks.

A total of 16 genera are recognized. Each genus, the estimated number of species, and the typical habitat in which each is found are listed in Table I.

Spirochete cell diameters are constant in all genera except in *Spirosymplokos.* They vary from as tiny as 0.09 μm in some treponemes to greater than 3 μm in *Cristispira.* The larger spirochetes, those that exceed 0.5 μm in diameter, are so complex at the ultrastructural level that morphology well serves as the basis for taxonomy in all members of this group. A total of 18 morphometric characters have been used to distinguish the genera. For at least 14 of them, thin section electron microscopic ultrastructure of glutaraldehyde-fixed specimens is required to establish the features with certainty. These distinctive criteria, listed in Table II, are illustrated in Fig. 3. Many genera have conspicuous cytoplasmic tubules that vary in diameter from 6 to 24 nm in their protoplasmic cylinders. In the very best preparations, Kari Hovind-Hougen (1934–1993), the Norwegian scientist who worked in Copenhagen, was able to show that the tubules emanate from the proximal side of the rotary motor (Fig. 4, arrow). Transverse sections of the distinctive genera are illustrated in Fig. 5.

III. HISTORY OF KNOWLEDGE

Van Leeuwenhoek's single drawing (his Figure G) was presented without specific commentary to depict

TABLE I
Spirochete Genera

Genus	*Estimated species*	*Habitat*
Borrelia	20	Ticks, birds, ducks, mice, rats, guinea pigs
Brachyspira[a]	1	Humans: intestine, blood, and other tissues
Brevinema	1	Short-tailed shrew, white-footed mice
Brevispira	1	Shrews, mice
Clevelandina	3	Dry wood-eating and subterranean termite intestines
Cristispira	1	Crystalline style of bivalve mollusks
Diplocalyx	4	Dry wood-eating termite intestines
Hollandina	2	Dry wood-eating termite and cockroach intestines
Leptonema	2	Freshwater, marine water, mammalian kidney tubules
Leptospira	6	Freshwater, marine water, soil, mammalian kidney tubules
Mobilofilum	2	Intertidal microbial mats
Pillotina	4	Wood-eating termite intestines, especially reticulitermitids
Serpulina	2	Pig intestines
Spirochaeta	12	Muds, salt marshes, soil, hot springs, lacustrine sediments
Spirosymplokos	1	Intertidal microbial mats
Treponema	15	Chimpanzees/humans: genitals, mouth, skin lesions, eyes

[a] *Brachyspira* may be equivalent to *Serpulina.*

TABLE II
Morphometrics Criteria in Spirochete Analysis[a]

Criterion	Method of determination[b]	Genus[c]
1. Diameter	LM, NS, TEM, VID	—
2. Number of flagella (one terminus)	TEM	—
3. Sillon	TEM	*Pillotina*
4. Crenulations	TEM	*Pillotina*
5. Ratio: OCOM/OM	TEM	*Hollandina*
6. Ratio: ICOM/OM	TEM	*Clevelandina*
7. Ratio: OCIM/IM	TEM	*Diplocalyx*
8. Ratio: DIameter of protoplasmic cylinder/diameter	TEM	*Spirosymplokos*
9. Angle protoplasmic cylinder subtended by flagella	TEM	*Hollandina*
10. Flagellar bundle	TEM	*Cristispira*
11. Length, μm	LM, NS, VID	—
12. Amplitude, μm	LM, NS, VID	—
13. Wavelength, μm	LM, NS, VID	—
14. Tubules (in protoplasmic cylinder)	TEM	*Diplocalyx*
15. Polar organelle	TEM	*Cristispira*
16. Rosettes	TEM	*Cristispira*
17. Granulated cytoplasm	TEM	*Spirosymplokos*
18. Composite (>1 protoplasmic cylinder/periplasm)	TEM	*Spirosymplokos*

[a] See Fig. 3 for illustrative details.
[b] LM, light microscopy; NS, negative stain; TEM, transmission electron microscopy; VID, videomicroscopy.
[c] Genus in which the feature is highly conspicuous or value is large. A dash (—) means the criterion is applicable to all genera.

the "animalcules" of his own mouth. It is presumed that he drew one of today's many oral treponemes. More than a century after van Leeuwenhoek, spirochetes were next seen in marine muds. They were drawn and described as *"Vibrio serpens"* by the German naturalist C. G. Ehrenberg (1795–1876). The type species of the genus *Spirochaeta, S. plicatilis,* is a tightly and constantly coiled common mud spirochete with many flagella. The flagella were described in negative-stain preparations by E. Canole-Parola, a microbiologist well known for his pioneering work on spirochetes and for training a generation of spirochete experts. Yet *S. plicatilis* still has not been cultured, and therefore its physiology is unknown. In fact, it has not even been morphometrically analyzed because no one has succeeded in obtaining it in large enough quantity for ultrastructural, thin section electron microscopic study.

The large molluscan *Cristispira* was introduced into the literature by A. Certes in 1882. Probably because of its swimming movement and swollen "crest," Certes mistook the spirochete for a trypanosome, that is, a kinetoplastid protist. He named it *"Trypanosoma balbianni"* and drew for it a chambered body. Apparently when he thought each "chamber" to be equivalent to a single cell, he displayed his misunderstanding of both bacteria and protists.

The problems plaguing those who have studied spirochetes in the past have not disappeared today even though the quality of microscopic equipment has markedly improved. Spirochetes are still identified by their swimming behavior and appearance. The inexperienced observer frequently overlooks their presence, especially in cultures of cyanobacteria and other photosynthesizers where they often persist as contaminants. Patients with spirochetoses are misdiagnosed as spirochete-free because of the difficulty in seeing the tiny pathogens. Dark-field microscopy is the tool of choice for analysis of tissues suspected to be infected by spirochetes. Spirochetes, which of-

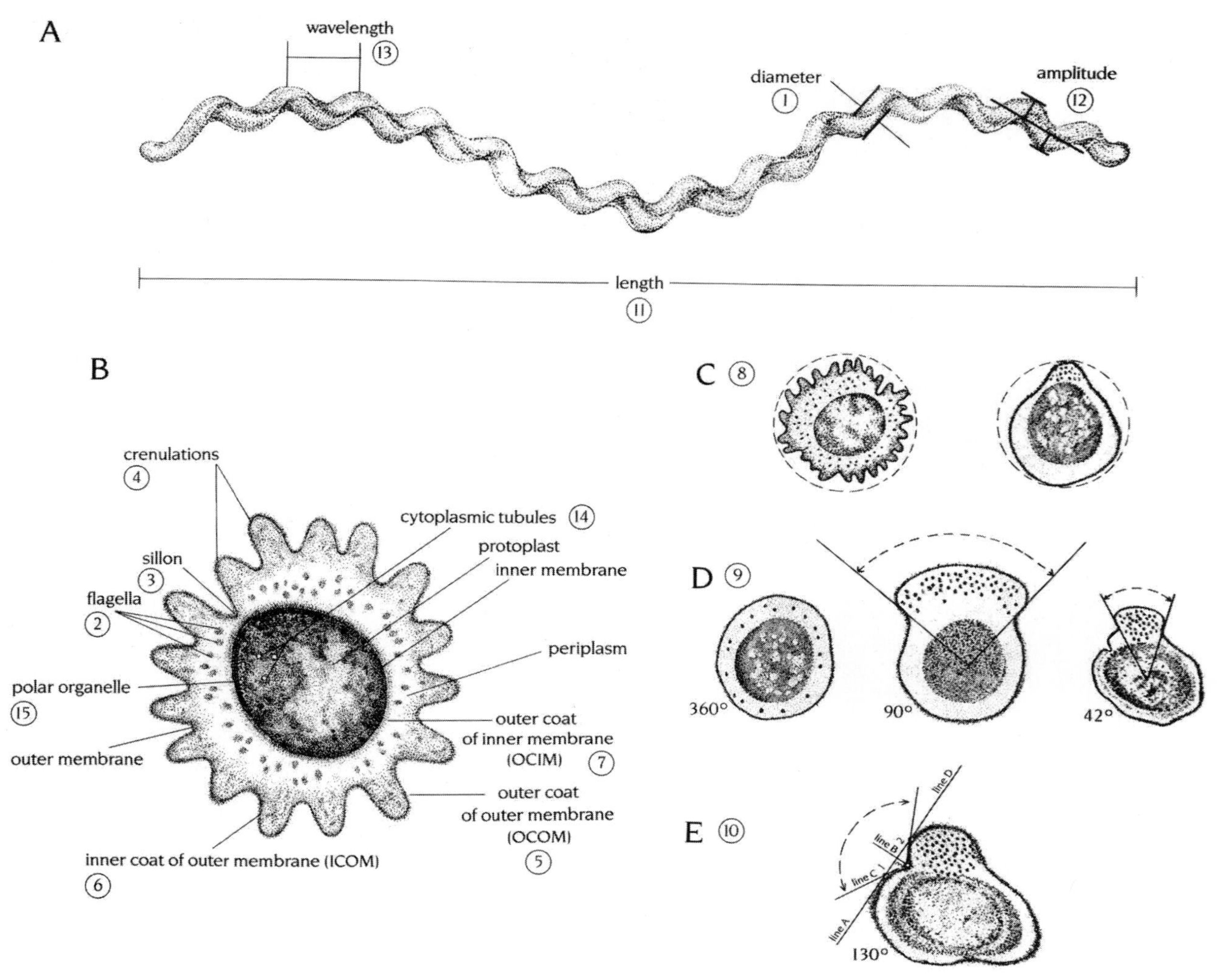

Fig. 3. Morphometric analysis. (A) Light-microscopic criteria. (B) Transverse section electron microscopic analysis. (C) Diameter. (D) Angle subtended by flagella. (E) Flagellar bundle analysis.

ten round up as environmental conditions worsen, are still misidentified as tiny protists (they resemble *Tricercomitus,* for example). They may be mistaken for tissue components or tiny sediment particles, or simply overlooked entirely.

Spirochetes became associated with certain specific diseases late in the nineteenth century. The excellent Japanese–American microbiologist Hideo Noguchi (1876–1928) recognized that the early symptoms of the veneral disease syphilis manifested itself as chancres (open sores) and that these genital sores, early in the course of the disease, were replete with tiny spirochetes. He found the spirochete in the brains of victims who suffered partial paralysis. He named them *Treponema pallidum* ("pale turning threads") because of their appearance and failure to take up stain. Noguchi, who also correctly identified the huge cristispira spirochetes of mollusks, recognized that animal-associated spirochetes were not necessarily causative of disease. He examined the digestive systems of many different marine bivalve mollusks and showed the cristispira spirochetes coursed up and down parallel to the long axis of a translucent gelatinous organ, a digestive grinding structure known as the "crystalline style." Noguchi, who recognized the essential bacterial nature of the

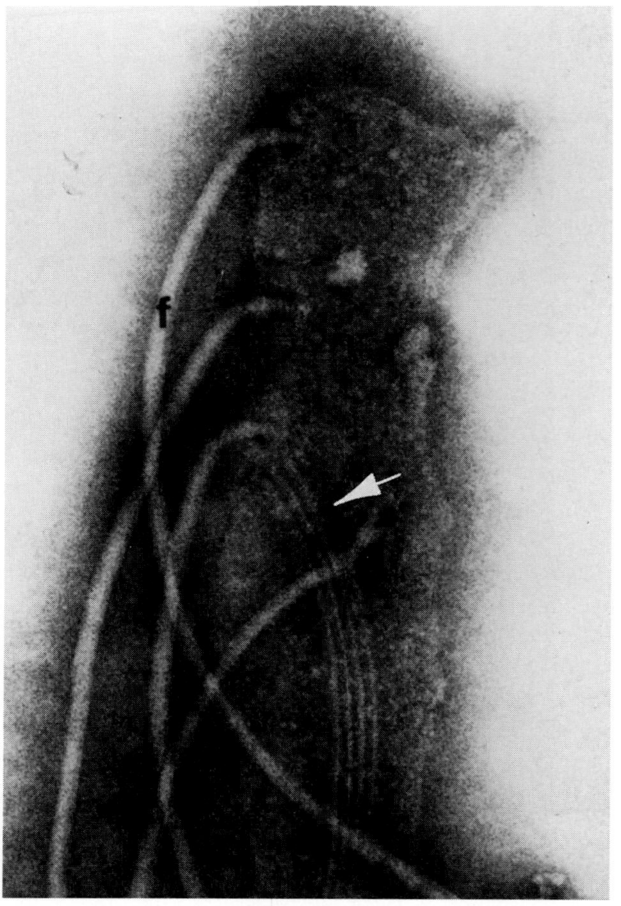

Fig. 4. Cytoplasmic tubules revealed by negative stain after deoxycholate removes the outer membrane.

large cristispires, suggested that their "crests," longitudinally aligned wavy folds, were bacterial flagella. Noguchi's interpretation was entirely verified by the electron micrographs of the French investigators A. Ryter and J. Pillot. Measurements of the internal or periplasmic flagella confirmed the homology of the spirochete structures with other bacterial flagella. Chemical studies showed the spirochete "axial fibrils" or "axial filaments" to be composed of flagellin proteins. The "crest," or "cristi," of *Cristispira* was recognized to be a flagellar bundle. With this work the concept of the unitary $n : 2n : n$ structure of these helical gram-negative bacteria became widely accepted.

A naturalist from Philadelphia, Dr. Joseph Leidy (1823–1891), in an attempt to determine the source of nutrition of the wood-dwelling subterranean termites of New Jersey, punctured the hypertrophied hindguts of workers, soliders, and other termite nest mates. The spiral shaped "wriggling threads" that Leidy found in abundance in the termites he called "*Spirillum undula*" (1850) and then later renamed them "*Vibrio termites*" (1881). Clifford Dobell (1886–1949), who recognized the termite inhabitants as spirochetes, named them *Treponema termites*.

From the beginning of studies of any spirochetes, their ubiquity and abundance in dark, organic rich habitats, including in muds and in the tissues of animals, has been realized. The consistent failure to reliably culture them *in vitro* or even grow manipulable populations of them outside their native habitats or to separate them from the tightly organized communities of microbes they inhabit has thwarted sustained attention of microbiologists. It has been the undoubted association of certain spirochete types with several loathsome human diseases that has generated all the close study of their properties (Table III). Not only has the treponeme correlation with syphilitic lesions been a stimulus to detailed investigation, but members of this genus are also associated with various infections such as yaws and pinta. In the twentieth century the leptospires, spirochetes often distinguishable in the live state by the presence of their "bent ends," have been correlated with serious systemic infections including some fatal forms of leptospirosis. Fisherman who are often subject to small skin lesions (fish hook cuts) in waters infested with rodent urine tend to be at risk for acquiring leptospiroses. Oral spirochetes are observed in elevated population numbers in patients afflicted with acute gum diseases (necrotizing ulcerative gingivitis or juvenile peridontitis).

The presence of elevated populations of spirochetes that tend to invade intact epithelium or connective tissue has been seen in HIV-positive subjects. According to spirochete expert Russell Johnson, all patients thought to have been cured of syphilis who later appeared HIV positive suffer a return of the treponema spirochete. This suggests that spirochetes are present in the tissues of HIV-positive people in an unrecognizable form. The possibility that small membranous structures form inside spirochete "round bodies" and confer some desiccation resis-

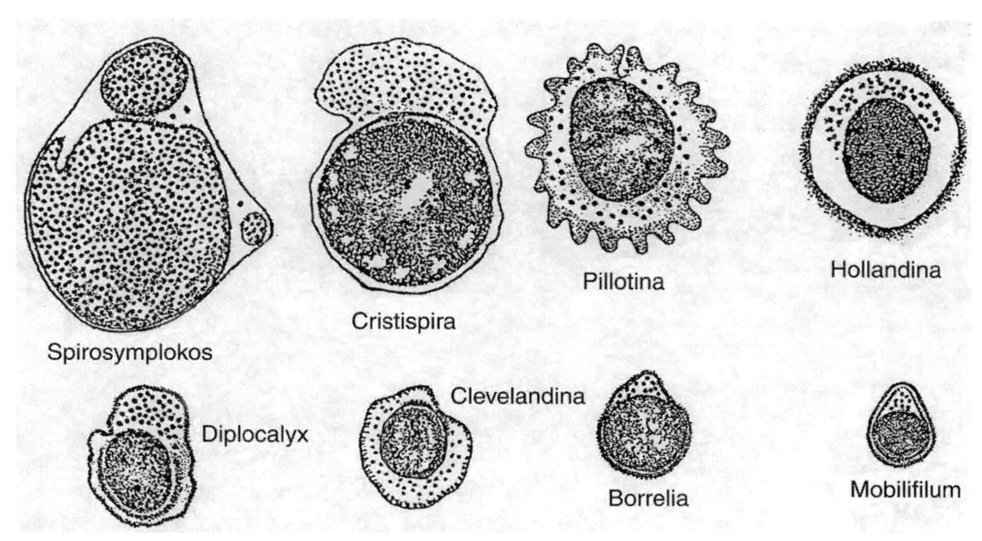

Fig. 5. Transverse sections through the larger morphometrically distinctive spirochete genera.

tance on these bacteria is high. Such small membranous structures were frequently observed in the variable-diameter microbial mat spirochete *Spirosymplokos deltaeiberi*. This organism not only forms membranous structures inside its round bodies (Fig. 6A), but, reminiscent of the baeocytes of cyanobacteria, *Spirosymplokos* is unique because it generates, probably by multiple binary fission, smaller offspring spirochetes in its cytoplasm (Fig. 6B).

Many unidentified spirochetes have been reported attached to protists or to animal tissues by elaborate, predictable attachment structures (Fig. 7). The tendency of spirochetes to form symbiotic associations to augment the rates of cell motility is well known, especially for *Mixotricha paradoxa*, a termite protist.

It is estimated that some 300,000 people per year suffer tick bites followed by the symptoms of Lyme arthritis, including erythroblastemia migrans. Not

TABLE III
Examples of Spirochetes Pathogenic for Vertebrates

Spirochete	*Disease*
Genus *Borrelia*	
B. anserma	Avian borreliosis
B. burgdorferi complex	Lyme borreliosis (humans and other animals)
B. coriaceae	Epizootic bovine abortion
B. hermsii, B. turicatae, B. parkeri	Tick borne relapsing fever (humans)
B. recurrentis	Louse borne relapsing fever (humans)
B. thelleri	Bovine borreliosis
Genus *Leptospira*	
L. interrogans (many serotypes)	Leptospirosis (humans and domestic animals)
Genus *Treponema*	
T. carateum	Pinta (humans)
T. pallidum subsp. *pallidum*	Syphilis (humans)
T. pallidum subsp. *pertenue*	Yaws (humans)
T. pallidum subsp. *endemicum*	Endemic syphilis (bejel, humans)
T. paraluescaniculi	Rabbit treponematosis

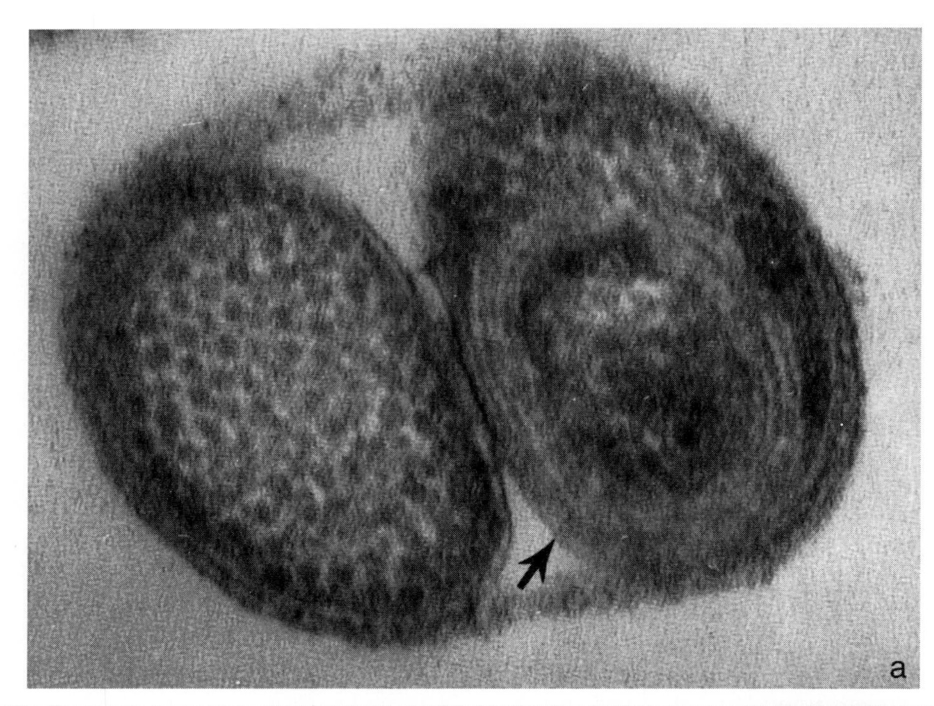

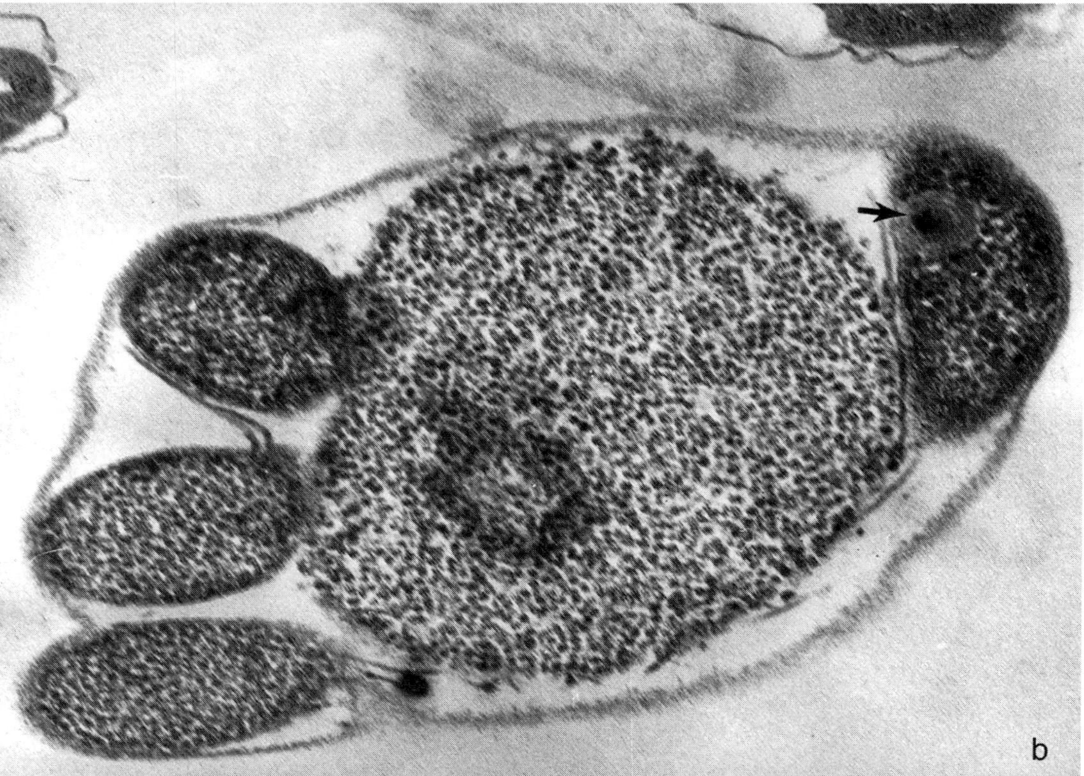

Fig. 6. (A) Membranous structure of *Spirosymplokos deltaeiberi*. (B) Composite periplasm.

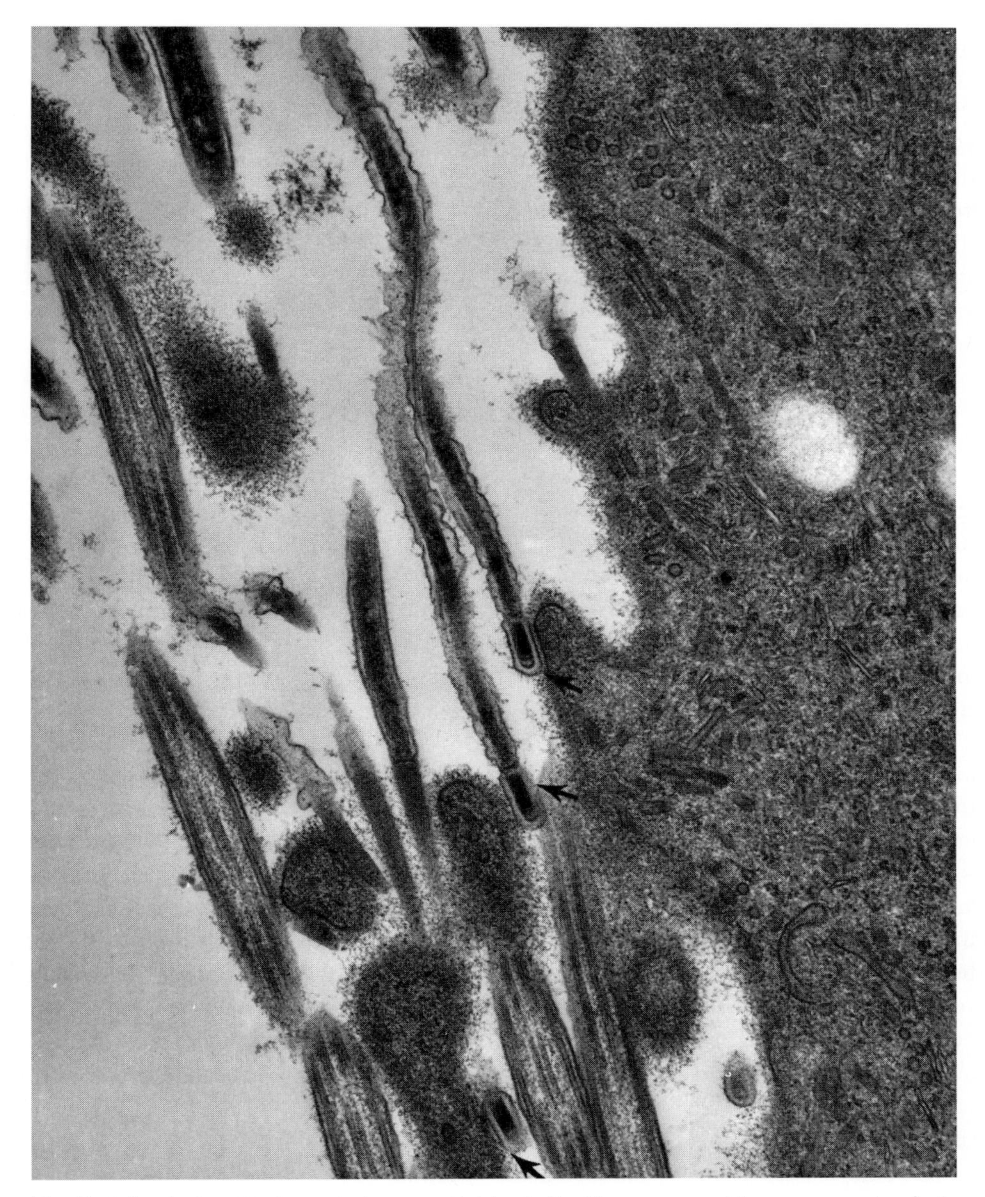

Fig. 7. Spirochete attachments to a protist in *Reticulitermes,* a subterranean spirochete. Attachment sites are at arrows.

only is the Lyme disease tick, *Ixodes dammini,* responsible for the transmission of the disease state, but pathogenic spirochetes that transmit relapsing fever are spread by another set of acarids (eight-legged arthropods). The correlation of spirochetes with this set of diseases has led to a great upsurge in the attention paid to spirochetes generally.

IV. PHYSIOLOGY

All strains of the genus *Spirochaeta* are free-living. They are either obligately or facultatively anaerobic. When grown aerobically the facultative anaerobes may produce colored pigments (red, orange, or yellow carotenoids). Many different carbohydrates are metabolized, primarily to ethanol, acetate, carbon dioxide, and hydrogen gas. *Spirochaeta zuelzerae,* isolated from freshwater muds, produced no ethanol from carbohydrates when growing anaerobically. Lactate and succinate were detected as its end products. When grown aerobically these organisms produce carbon dioxide and acetate.

All treponemes grow in association with the tissues of animals. Those that have been cultivated *in vitro* are either strict anaerobes or microaerophils with requirements of oxygen less than 5%. Some strains ferment glucose and require it or other carbohydrates as a carbon and energy source, others grow by fermentation of amino acids. Those studied require short- or long-chain fatty acids in their media as provided in rumen fluid or serum, respectively. The oral treponemes require thiamine pyrophosphate in addition to serum. Although oral spirochetes have been studied for many years, only seven types have been validly named and cultured *in vitro*. All are members of the genus *Treponema*: *T. denticola, T. pectinovorum, T. socranskii, T. maltophilum, T. medium, T. amylovorum,* and *T. vincentii*.

Unlike *Leptospira* and *Treponema,* which also occur in animal tissues but frequently go undetected, *Borrelia* stains well and clearly in blood and other tissue preparations. Those isolates of *Borrelia* that have been maintained in *in vitro* cultivation require complex media for growth, and even with serum, proteose peptone, tryptone, and the provision of other undefined substances, vigorous population growth of *Borrelia* in culture has been impossible to sustain.

Neither carbohydrates nor amino acids support growth in culture of *Leptospira*. Rather, fatty acids serve these functions: the media that support growth contain fatty acids with 15 or more carbon atoms, and at least two B vitamins, thiamine (vitamin B_1) and cyanocobalamin (vitamin B_{12}). Ammonium salts, but not amino acids, satisfy the nitrogen requirement, and purines, but not pyrimidines, will be incorporated from the medium into these spirochetes. Media are enriched by bovine serum albumin or rabbit serum. The end products of fatty acid oxidation are acetate and carbon dioxide. Cytochromes and catalases have been detected in leptospires.

All cultivated spirochetes are sensitive to penicillin and its derivatives. Many, if not all, resist rifampicin (rifampin). The ability of spirochetes to endure high concentrations of this RNA synthesis inhibitor, a drug which stops growth of so many other bacteria, has been put to good use in the isolation of new spirochetes.

None of the members of the genera of large (>0.5 μm diameter) spirochetes have ever been grown in pure culture in spite of much expended effort. Therefore virtually nothing is known about their physiology. In fact, of the myriad of insect symbiotic spirochetes only two small anaerobic treponemes (ZAS-1, ZAS-2) have been cultured (by John Breznak and colleagues) from termites (*Zootermopsis angusticollis*). Both synthesize acetate from gases: CO_2 and H_2. Acetate is utilized by the termite mitochondria. Growth with N_2 and H_2 in the gas phase led to CO_2 reduction to acetate and to formate when the relative quantities varied with conditions. Methanogenesis by spirochetes never has been documented.

V. ECOLOGY

Certain *Spirochaeta* of geological significance alter cell morphology and develop meter-long threads that retain their normal <0.5 μm diameters. The former bacterial genus *Thiodendron* was identified by G. A. Dubinina and her colleagues as a *Desulfurobacter* (sulfate-reducer)–*Spirochaeta* (carbohydrate-fermenter and sulfide-oxidizer) symbiotic association. Common in organic- and sulfide-rich environments, this spirochete–sulfate reducer association forms extensive microbial mats that prevent toxic hydrogen sulfide gas from percolating into overlying oxygenated waters.

Most intense work on the habitats of spirochetes has been limited to studies of mammalian, including human, tissue. For example, the proximal convoluted tubule of the kidney is considered the natural habitat of *Leptospira interrogans*. Yet even cursory study of organically rich–oxygen poor habitats in nature, such as termite intestines and microbial mats, makes it clear that most spirochete diversity is unknown and those communities likely to harbor many spirochetes new to science remain largely unexplored. In a study by Thomas H. Teal and colleagues, seven distinctive spirochete morphotypes, none of which exactly matched published spirochetes, were detected by electron microscopy in marine muds from Cape Cod, Massachusetts. By designing a spirochete-specific oligonucleotide probe, Bruce Paster and colleagues detected two new termite hindgut spirochetes in an African termite (*Nasutitermes lujae*). On the basis of this study he concluded that "the percentage of spirochetes in gut material of *N. lujae* was greater than 50%, which is considerably higher than the 10 to 20% estimated by phase contrast microscopy" (p. 350). Although only very few hot temperature spirochetes have been grown and characterized, there is little doubt that many more extremophiles that belong to this group await the curious investigator.

A summary of the ecological settings for an example of each of the spirochete genera is listed in Table I. Certainly this set of documented habitats grossly underestimates the extent of spirochete distribution in nature.

VI. MOLECULAR BIOLOGY AND PATHOGENICITY

The complete genome sequence is now available for two pathogenic spirochetes: *Treponema pallidum*, the spirochete associate with syphilis, and *Borrelia burgdorferi*, that involved in tick-transmitted Lyme disease. *Treponema*, with its 1.138 kilobase pairs of DNA in single circular genophore, has one of the smallest bacterial genomes known. The 36 genes encoding proteins involved in flagella structure and function most resemble those in *Borrelia burgdorferi*. *Treponema pallidum* contains only some 22 lipoproteins of the outer membrane, whereas *B. burgdorferi* has 105. Of the 100 or so open reading frames of unknown function shared by these two sequenced bacteria, nearly half seem limited to those spirochetes. In 304 open reading frames (of known and unknown function) shared by the two spirochetes, gene clusters with conserved gene order are maintained. The significance or relevance of these observations of gene organization to the pathogenicity and chronic infection caused by the spirochetes is still elusive, especially given the great differences in the genes and their sequences in the two pathogens. However, the completeness of the data augurs well for the eventual resolution of the molecular basis for pathogenicity in this group of widespread, primarily free-living, highly motile heterotrophic bacteria.

See Also the Following Articles

Flagella • Lyme Disease • Oral Microbiology • Syphilis, Historical • Timber and Forest Products

Bibliography

Bermudes, D., Chase, D., and Margulis, L. (1988). Morphology as a basis for taxonomy of large spirochetes symbiotic in wood-eating cockroaches and termites: *Pillotina* gen. nov., nom. rev.; *Pillotina calotermitidis* sp. nov., nom. rev.; *Diplocalyx* gen. nov., nom. rev.; *Diplocalyx calotermitidis* sp. nov., nom. rev.; *Hollandina* gen. nov., nom. rev.; *Hollandina pterotermitidis* sp. nov., nom. rev.; and *Clevelandina reticulitermitidis* gen. nov., sp. nov. *Int. J. System. Bacteriol.* **38**, 291–302.

Canale-Parola, E. (1984). The spirochetes. *In* "Bergey's Manual of Systematic Bacteriology," Vol. 1, pp. 38–46. Williams & Wilkins, Baltimore/London.

Dubinina, G. A., Leshcheva, N. V., and Grabovich, M. Yu. (1993). The colorless sulfur bacterium "*Thiodendron*" is actually a symbiotic association of spirochetes and sufidogens. *Mikrobiologiya* **62**, 717–732.

Dubinina, G. A., Grabovich, M, Yu., and Leshcheva, N. V. (1993). Occurrence, structure, and metabolic activity of "*Thiodendron*" sulfur mats in various salt water environments. *Mikrobiologiya* **62**, 740–750.

Leadbetter, J. R., Schmidt, T. M., Graber, J. R., and Breznak, J. A. (1999). Acetogenesis from H_2 plus CO_2 by spirochetes from termite guts. *Science* **283**, 686–689.

Margulis, L. (1993). "Symbiosis in Cell Evolution," 2nd Ed. Freeman, New York.

Margulis, L., and Hinkle, G. (1992). Large symbiotic spirochetes: *Clevelandina, Cristispira, Diplocalyx, Hollandina,* and *Pillotina*. *In* "The Prokaryotes. A Handbook on the Biology of Bacteria: Ecophysiology, Isolation, Identification, Applications" (A. Balows, H. G. Trüper, M. Dworkin, W. Harder, and K.-H. Schleifer, eds.), 2nd Ed. Vol. 4, pp. 3965–3978. Springer-Verlag, New York.

Margulis, L., Ashen, J. B., Solé, M., and Guerrero, R. (1993). Composite, large spirochetes from microbial mats: Spirochete structure review. *Proc. Natl. Acad. Sci. U.S.A.* **90**, 6966–6970.

Paster, B. J., Dewhirst, F. E., Coleman, B. C., Lau, C. N., and Ericson, R. L. (1998). Phylogenetic analysis of cultivable oral treponemes from the Smibert collection. *Int. J. System. Bacteriol.* **48**, 713–722.

Spontaneous Generation

James Strick
Arizona State University

I. Introduction
II. Early History
III. The Needham/Spallanzani Controversy
IV. Worms, Molecules, and Evolution
V. The Role of Louis Pasteur
VI. The British Debate of the 1870s: Spontaneous Generation and the Germ Theory of Disease
VII. Twentieth Century Ideas

GLOSSARY

abiogenesis Term used in 1870 by Huxley to mean life arising from a combination of inorganic starting materials. Since that time, it has come to be used more widely to indicate origin of life from any nonliving matter.

archebiosis Term coined by Bastian in 1870 to mean life arising from strictly inorganic starting materials.

biogenesis Term used in 1870 by Huxley to mean life arising only from other living things, to which he made the opposite abiogenesis. The term seems to have been first coined by Bastian earlier that year, but to mean exactly the opposite: spontaneous generation. It is Huxley's usage that became famous and remains the definition of this term today.

heterogenesis Nineteenth century term used to mean the origin of living things from organic materials, for example, from infusions of plant or animal matter, or from decaying tissue in a diseased or dying organism. It was thought to be the source of tumors, parasitic worms, and microorganisms from putrefaction, as these organized themselves from the smallest microscopically visible particles ("molecules") of organic tissues. Note that many supporters of heterogenesis might not support the more extreme position of archebiosis/abiogenesis from inorganic matter.

molecules Term used from the writings of Buffon (c. 1750) onward to indicate the tiniest (bacteria-sized) microscopically visible particles of organic matter in blood, tissue, infusions, etc. This term was used by histologists, pathologists, and cytologists until the 1870s, obviously in a way quite different from the chemists' use of the term that developed through the same time period. Important examples include the "active molecules" of Robert Brown (1828–1829), the cell-forming cytoblastema "molecules" of Theodor Schwann (1839), and the histological "molecules" of John Hughes Bennett (1840–1875). Many thought clumping together of these units was the basis for heterogenesis.

pleomorphism The doctrine that microorganisms did not come in distinct, stable Linnean species but rather were highly variable in shape and metabolic capabilities in response to changes in their environment. Many believed this to be true of the bacteria, especially in Britain and Germany through the 1880s. More extreme versions held that bacteria, yeasts, molds, and algae were all interconvertible stages in the life cycle of a single organism. Heterogenesis can be seen as the next logical step, including the transition from nonliving organic matter to the simplest bacterial forms as but one more stage in that fluid transformability.

I. INTRODUCTION

The idea that living things can originate from nonliving materials, spontaneous generation, has a long history, inseparably intertwined until about 1880 with the development of microbiology as a science. No less an authority than Aristotle claimed that cases of spontaneous generation could be observed in nature, and his support was important in establishing the idea for many centuries. Beginning around the time of the Scientific Revolution of the seventeenth century, however, the doctrine of spontaneous generation was increasingly challenged and became the subject of numerous episodes of controversy. As combatants tried to answer one another's criticisms, the new breakthroughs in technique and in experimental design that were developed served as some of the most important foundations on which a science of microbiology could be built. As just one example, the development of sterile technique as well as procedures to sterilize glassware and growth media all grew directly out of experiments trying to prove or disprove the possibility of spontaneous generation of microorganisms.

In seeking to answer a question as basic as how life originates, theory played a role as important or more so than technique or experiment. From the first, the doctrine of spontaneous generation was seen to be fraught with religious implications. If life could originate spontaneously from lifeless matter, the position of philosophical materialism, then a Creator God was irrelevant. If spontaneous generation occurred in present times, this was at odds with a single original Creation as told in the Bible. However, for those interested in a naturalistic worldview, such as supporters of Darwinian evolution, there was also a potential conflict. The doctrine of evolution was based on a profound philosophical assumption of continuity in nature, that is, that there were no sudden unbridgeable gaps between similar living forms, which would require supernatural intervention. Furthermore, Darwin's theory implied that the vast diversity of living things had come from one or at most a few original ancestral organisms, and these must have originated somehow. For many, then, to believe in evolution and a completely naturalistic worldview required the belief that no unbridgeable gap occurred between living matter and nonliving matter and that living organisms must have been capable of arising from nonlife, at least once on early Earth. Those with this view, for example, Henry Charlton Bastian, challenged hypocrisy on the part of those who supported Darwin but were unwilling to believe in the necessity of spontaneous generation.

Still another important source of disagreement, even among scientists who claimed not to care about religious implications, were fundamental epistemological assumptions about the nature of life. Some believed very deeply that all living things must reproduce by "seeds" or "germs," by analogy with the large number of organisms for which this process had been observed. So fundamental was this belief that, even in cases where microscopic life appeared in tubes of fluid boiled for hours, those scientists concluded that either (a) the tubes must have leaked after boiling or (b) there must be some kind of structures produced by microorganisms that were capable of withstanding previously unheard of temperatures, even though nobody had ever seen such structures. Their opponents, as late as the 1870s, believed that the most clearly observed and reliably established fact known about living things was that they were totally unable to survive the boiling temperature of water for more than a few minutes. From this equally tenable premise, they concluded that spontaneous generation was a less strained explanation of organisms in boiled infusions than the ad hoc invoking of "spores" that nobody had ever experimentally demonstrated with totally unprecedented heat-resisting abilities. Thus, before ever carrying out an experiment, both sides in the spontaneous generation controversies were often begging the question at issue. Clearly these philosophical issues were crucial points showing how experimenters could disagree because they were talking past one another, more than because of differences in the experiments themselves. The doctrine of spontaneous generation rose and fell in popularity repeatedly at different times in different countries in the past several centuries, and to tell the story of spontaneous generation controversies only as a series of battles about "dueling experiments" would be to misunderstand much or even most of what the controversies were really about.

II. EARLY HISTORY

From antiquity it had been maintained that frogs, eels, mice, and numerous worms and insects, especially parasitic worms living inside animals, could arise by spontaneous generation, mostly heterogenesis. With Leeuwenhoek's discovery of microorganisms, many naturalists assumed that these too could arise without parents. Indeed, as microbes seemed exceedingly simple, and bacteria simplest of all, it was believed that they were the most likely to be organizable from nonliving materials. Francesco Redi, natural philosopher to the same Tuscan court that was patron to Galileo, carried out some famous experiments in 1668 that investigated specifically the origin of insects. Redi's experiments showing that maggots come from fly eggs, not from rotting meat, usually lead off histories of spontaneous generation debates. It was commonly believed until that time that the appearance of maggots in rotting meat was a clear example of spontaneous generation. Redi placed samples of many different types of meat and fish in glass jars. One set of jars was open to the air and soon swarmed with maggots. The other set was covered with fine muslin cloth. Redi saw that, while maggots never appeared in the meat of those jars, flies crawled about on the cloth and sometimes laid eggs there. Those eggs were seen to hatch into maggots, disproving spontaneous generation as their origin. Though Redi himself continued to believe that some insects, such as gall flies, did arise by spontaneous generation, many naturalists assumed from this time onward that spontaneous generation, if it occurred, only did so among parasitic worms and microorganisms.

Of course, working at the time of the Scientific Revolution, Redi did participate in a time when experiment came to the fore as an activity ever afterward central in natural philosophy. Spontaneous generation debates, like all other natural philosophic questions, featured experiments in a more and more important place from the late sevententh century onward. Just as "natural history" or "natural philosophy" historically did not mean what we now call "science," until fairly recently, however, neither did "experiment" always mean what it means now. Many natural philosophers in the seventeenth century were interested in the power of the experimental method, but they were just as interested in public demonstrations of "experiments" as a way of convincing audiences (e.g., prospective patrons) that their enterprise was qualitatively different from the book-dominated natural philosophy of the past, and from the subjective and often bloody religious disputes that had traumatized European life for over a century. Recent studies have looked closely at Redi's career, especially his relationship to the Medici Grand Dukes, his principal patrons. One such study concludes that Redi's public demonstration of experiments was a form of court entertainment in which the final arbiter of the meaning and success of the outcome was the Grand Duke. Thus there was a world of difference between Redi's procedures and what would by the nineteenth century be called "controlled experimentation by a biologist." He set high value on repeated observation and demonstration, saying, "I do not put much faith in matters not made clear to me by experiment." Paradoxically, however, he suggested elsewhere that he experimented "in order to make myself more certain of that of which I am already most certain." For Redi, then, especially in his primary role as an early modern courtier, "experiment" meant something quite different from what we understand by that term today, particularly with regard to the role of "preconceived expectations." The paradox results only if we assume in a naively ahistorical way that what Redi called "experiment" should be assumed to have the same meaning that term came to have after "natural philosophy" had been transformed into "science."

Once we gain this richer sense of what experiment meant in the historical context in which Redi lived and worked, we may begin to question the validity of an ahistorical narrative which uncritically compares his work with experiments performed by Pasteur, Tyndall, and their antagonists two centuries later, merely because of superficial similarity in "the use of controls." Perhaps there are more similarities in the meaning of "experiment" in such different historical settings, but this is an open question that can only be answered by looking at the full context in which each investigator had to operate, including such questions as the existing system of patronage on which each depended to support work on scientific pursuits. Such points of comparison will be developed further when discussing Pasteur's work.

III. THE NEEDHAM/SPALLANZANI CONTROVERSY

Eighty years after Redi, in 1748, another series of experiments by the Irish priest John Turberville Needham was published, and this time was widely believed to support the spontaneous generation of microorganisms. Needham soon collaborated with the French aristocrat Comte de Buffon. And over the next several decades the work of Buffon and

Needham was opposed by many, including Charles Bonnet and Lazzaro Spallanzani.

Needham's first experiments found that, when mutton gravy was stoppered and sealed with mastic in glass tubes and heated to boiling for a time, the fluid afterward teemed with microorganisms (protists). Buffon and Needham later reported seeing tiny particles (in the size range of bacteria) that they called "organic molecules," which came from disintegrating organic material and moved very actively. These, they said, clumped together to form the larger animalcules (protists). They also claimed that sealing and boiling proved that the molecules did not originate from "insects or eggs floating in the atmosphere." Most naturalists considered this, if true, to be a case of spontaneous generation.

Spallanzani, an Italian cleric, carried out experiments challenging these claims and first published his results in 1765. His opposition was based on two main arguments. First, he said that with his own microscope he never saw anything like the "organic molecules" and therefore charged that Buffon and Needham were only seeing what they wanted to see, because Buffon's philosophy of nature had predicted that such particles must exist. This claim was widely repeated. Second, he said that when he boiled infusions similar to Needham's, he did not find any living microbes in them, as long as he sealed the tubes by melting the glass shut in a flame before boiling and continued the boiling for at least an hour. Thus, Spallanzani claimed that the microorganisms in Needham's infusions must indeed have gotten in, despite Needham's precautions, from the atmosphere after boiling.

Needham responded that boiling for as much as an hour would clearly be so severe a treatment as to deprive the air in the tubes of its power to generate or support life. Later experiments showing that indeed such treatment decreased or consumed the oxygen content in the tubes seemed to support Needham's claim. Thus the experimental outcome was underdetermined by the experiments themselves, so the controversy was able to continue through the 1780s.

A popular notion of experiment has it that "proper science" can only occur when the scientist approaches his experiment with no "preconceived ideas" about the outcome, and intends to "let the chips fall where they may." In many histories of the spontaneous generation debates, the "losers" are often accused of having been biased beforehand by their belief in a "vital force" or some similar overarching philosophy, while Redi, Spallanzani, Pasteur, and Tyndall epitomize the open-minded investigator. A closely related claim has been that Spallanzani, Pasteur, etc., were often able to defeat their foes by virtue of possessing better instruments, especially superior quality microscopes. Thus, as the story has been told since even the early nineteenth century, the Comte de Buffon and his collaborator Needham in the 1740s were portrayed as "armchair philosophers" who cooked up a doctrine of "organic molecules," a vital "plastic force," and spontaneous generation; and they had such an inferior microscope that they were able to interpret whatever fuzzy images they saw as supporting the fuzzy ideas they wanted the data to confirm. Victorian biologist T. H. Huxley's version established one of his most famous clichés by describing Buffon and Needham's work as an object lesson for generations of young scientists:

> Such as it was, I think it [Buffon and Needham's doctrine] will appear . . . to be a most ingenious and suggestive speculation. But the great tragedy of Science—the slaying of a beautiful hypothesis by an ugly fact—which is so constantly being enacted under the eyes of philosophers, was played almost immediately, for the benefit of Buffon and Needham.[1]

Huxley argued, perhaps on the basis of Pasteur's similar claim, that Buffon and Needham had a poor microscope and their opponent Spallanzani had a much more objective outlook and understanding of the method of controlled experiment.

Although seldom emphasized, it is very important to note that spontaneous generation was seen to be directly supportive of two other, broader doctrines: epigenesis and materialism. Epigenesis is the doctrine that all the parts of an embryo are assembled gradually, having at first been nonexistent. Epigenesis was opposed in the seventeenth and eighteenth

---

1. Huxley, T. H. (1870). Biogenesis and abiogenesis. *In* "Discourses Biological and Geological Essays," pp. 232–274. Appleton, New York, 1925, p. 239.

centuries by the doctrine of preformation, which claimed that all embryonic organs existed inside the sperm or the egg in miniature from even before fertilization and only needed to grow, unfold, and expand as gestation proceeded. This implied logically that all the generations of, for example, humanity must have been contained, each within the previous one like Russian dolls, all the way back to the eggs in Eve's ovaries or the sperm of Adam. (The overwhelming majority of naturalists writing during this period were Christian.) Since Buffon and Needham's description of the origin of protist microbes suggested their organization from originally homogeneous "molecules," this seemed a blow for preformationists. Buffon indeed ridiculed preformation theory, insisting that the microscope showed no trace of these successively enclosed generations within the sperm or eggs of animals. Spallanzani eventually came down on the side of preformation and rejected Needham's theory. We would probably admit today that, since his microscope could not have shown him little preformed homunculi inside eggs or sperm cells, Spallanzani, too, must have been willing to go some way in believing in things he could not verify and was thus also operating in an atmosphere of "philosophizing."

More than that, if Buffon and Needham's observations were correct, they showed that matter contained within itself all the properties necessary to organize into life. This was the basic tenet of philosophical materialism, a doctrine profoundly at odds with Christianity or any kind of Deism. Voltaire, among others, feared that Needham's claims would support atheism and materialism. It was thought by many that this theory implied that life could originate by a chance, random combination of substances. This was so contrary to mainstream religious beliefs, and to a natural philosophy still very much in the service of demonstrating the existence of a beneficent Creator, that it generated heated opposition. Further, this led to spontaneous generation becoming strongly associated, beginning around 1750–1760 with atheism, materialism, and political radicalism. (Since this chance combination of chemicals has become such a crucial element of modern views of the origin of life, such as the famous 1953 Miller–Urey experiment, it is interesting that Buffon and Needham are not celebrated as thinkers far ahead of the religious biases of their time. Voltaire seems to have been one of those who most actively spread the opinion that Needham and Buffon were poor scientists, though this was based on seriously misreading Needham.)

Only recently for the first time has close study been applied to the actual technical evidence on what kind of microscopes were being used by Spallanzani and by Needham. This work has shown that Buffon and Needham's results cannot be successfully explained by assuming that they worked with poor instruments, nor that their work was sloppy or biased in advance by a priori theorizing, despite the fact that these explanations are ubiquitous among previous accounts of the controversy. Indeed, a close review of the evidence on the actual experiments (Sloan, 1992) suggests that the experiments of Needham and Buffon were careful, high quality work with quite a bit of sophisticated hypothesis formation and testing.

Many of the contemporary critics of Buffon and Needham, including Spallanzani in the 1760s, had originated this claim of a priori bias, at least partly because they assumed the pair to have worked with a British Cuff compound microscope, common at that time (Fig. 1). This type of device had a maximum magnification of only about 100×, and even at that low power the image it produced suffered from severe chromatic and spherical aberration, as indeed was a problem with all compound microscopes prior to the achromatic lenses not perfected until 60 years later. By careful analysis of the original publications, however, it has been shown that the instrument used by Needham was actually a high quality single lens microscope of the Wilson screw-barrel design, capable of at least 400× magnification with outstanding resolution, that is, not plagued by the aberrations of compound scopes (Fig. 2). This was similar to the instrument with which Robert Brown discovered the cell nucleus in plant cells and Brownian movement in pollen particles in the 1820s and 1830s. In fact, Buffon and Needham's observations anticipate those that led to the discovery of Brownian movement, their "organic molecules" being what Brown called

Fig. 1. Plate from Buffon's *Histoire naturelle* depicting an eighteenth century scientist using the compound Cuff microscope. Perhaps because of the illustration, many assumed that Buffon and Needham had used this instrument to make their observations, when in fact they had used the much superior Wilson screw-barrel design (Fig. 2). Compound microscopes of that era suffered from severe chromatic and spherical aberration, and the Cuff microscope had a maximum magnification of only about 100×. Courtesy of Beinecke Rare Book and Manuscripts Library, Yale University.

"active molecules" The simple microscope employed in the famous experiments by Spallanzani was a variant of the Lyonnet aquatic microscope. Spallanzani's instrument was incapable of the short focal length, high resolution work permitted by the Wilson screw-barrel design, so that he could not even see the bacteria-sized "organic molecules" under dispute. Thus the truth of this controversy appears to be that Spallanzani was the technically handicapped participant, though exactly the opposite story has become universal. Buffon and Needham's discovery of "organic molecules" was not the work of imagination but an actual discovery of active molecules and of Brownian movement, some 80 years before Brown. This explains why Buffon and Needham refused to back down from their original observations despite accusations of atheism and decades of controversy. Furthermore, both parties seem to have had plenty of a priori philosophical commitments at stake, and nobody approached as explosive a subject as materialism with a completely open mind.

Fig. 2. Single lens Wilson screw-barrel microscope used by Buffon to make his observations in support of his interpretation of the theory of spontaneous generation. In this theory, small bacterium-size particles, which Buffon called "organic molecules," clumped together to form living "animalcules." Courtesy of Beinecke Rare Book and Manuscripts Library, Yale University.

IV. WORMS, MOLECULES, AND EVOLUTION

The appearance of parasitic worms within animals was long seen, until perhaps the 1840s, as the strongest single piece of evidence supporting spontaneous generation claims. The complex life cycles of most of these parasites involved intermediate life cycle stages that had to live in other host animals. Thus, egg-feeding experiments would not have been able to show eggs from one animal directly producing worms in another. Only in the 1840s and 1850s, with the working out of the complex life cycles and intermediate hosts, did it finally become unambiguously clear that the worms reproduced via eggs.

Despite the denunciation of Buffon and Needham, a great many pathologists, histologists, and, later, cytologists continued to observe bacteria-sized, actively moving particles in blood, tissues, and infusions. These particles continued to be called "molecules" by life scientists. They were using the term, obviously, in a way quite different from the chemists' use of the term that developed through the same time period. Important examples include the "active molecules" of Robert Brown (1828–1829), the cell-forming cytoblastema "molecules" of Theodor Schwann (1839), and the histological "molecules" of John Hughes Bennett (1840–1875). Many thought clumping together of these units was the basis for heterogenesis, and that this furthermore supported epigenesis as the correct explanation of embryonic development. Especially vocal in this belief were Lorenz Oken, Karl Burdach, and other German biologists of the *Naturphilosophie* school. (It should be emphasized that neither Brown nor Schwann themselves interpreted their "molecules" in this way, and Schwann did some significant experiments early in his career that seemed to disprove certain cases of spontaneous generation.) In Britain, both Charles Darwin and Richard Owen worked in private on the molecule theory during the 1830s, as did William Addison in the 1840s. More importantly, the British histologist Hughes Bennett was one of the first professors to teach a physiology course in a university and medical school, and he taught from 1840 until almost 1870. His theory of "molecules" as the essential units from which cells formed, explicitly embracing heterogenesis by the late 1860s, was thus widely influential among British scientific and medical circles for over a generation.

J. B. Lamarck, the early French evolutionist, had made spontaneous generation of microorganisms an integral part of his evolution theory from its beginning in 1800. Lamarckian supporters were numerous and widespread, though often associated with radical politics, as self-organization and development of matter were seen as a crucial scientific underpinning for democratic, bottom-up organizing political theories. A few such Lamarckians did rise to university teaching posts, however, and notable among them was the outspoken British comparative anatomy professor Robert Grant. Although Grant was considered not quite respectable in Victorian intellectual circles, his teaching of evolution and spontaneous generation continued from the 1820s through his death in 1874, and thus gave those ideas a presence in British thought decades before *On the Origin of Species* appeared. Charles Darwin, for instance, was one of Grant's earliest students and protégés (though Darwin later felt he had to avoid any contact with the less-than-respectable Grant when he was developing his own evolution theory in secret for 20 years). Grant and other Lamarckians had established in the public mind a strong link between evolutionary theories and spontaneous generation.

V. THE ROLE OF LOUIS PASTEUR

As Louis Pasteur's experiments on spontaneous generation, particularly his "swan-necked flasks," are among the most famous of all, the Pasteur–Pouchet debate of the 1860s in France deserves close attention. It is also worth noting that the historical introduction to Pasteur's famous 1862 memoir on the subject has served as the model on which almost all subsequent histories of the controversy were based for over a hundred years (see Bulloch, for example). This has led to the casting of the controversy as a series of what I call "dueling experiments," and has often stripped it of its crucial philosophical context. Pasteur also chose to leave out an entire chapter of the story: the argument that parasitic worms must have arisen by spontaneous generation. So dominant has Pasteur's master narrative been, that it is only with historical detective work in the early 1970s that this omission of Pasteur's was finally restored to its place of importance in the story.

Between 1860 and 1862 Pasteur carried out a lengthy series of experiments focusing exclusively on the development of microbes in previously boiled infusions, in response to the experiments in favor of heterogenesis published by Felix Pouchet. With great ingenuity, Pasteur designed numerous experiments that would allow the boiled infusions to come in contact with air that had not been itself altered by the heat, to answer Needham's objection of a century before. Air was led through dense gun cotton filters, for example, before coming in contact with the infusions. The filters were then dissolved and the solid particles trapped from the air were examined microscopically. Pasteur saw some spherical bodies that he assumed were the "germs" of microbes, since in those infusions microbial life rarely developed. Perhaps most famous of Pasteur's experiments involved a series of flasks which, after being filled with infusion, had their necks heated in a flame then drawn out into long, curved shapes, for example, like a swan's neck (Fig. 3). Air could enter through the long neck, but dust particles settled in the dip and could not come in contact with the flask's contents. Again, if these flasks were boiled and then cooled, rarely did any microbial growth ever appear in them. However, if the neck was broken off and dust allowed to enter the flask, microbial growth in the infusion soon followed. Pasteur concluded that this proved microbes or their germs were carried on dust particles, so the exclusion of dust followed by no growth meant the disproof of spontaneous generation.

As the swan-necked flasks have in our time become textbook icons for exemplary experimental practice, it is difficult for us to reconstruct the historical period when, even after Pasteur's famous public lectures on his experiments, a significant number of scientists at the time remained unconvinced. In particular Pasteur claimed to have sweepingly proven that germs must be the source of growth in previously boiled infusions, but his opponents pointed out that what he had shown could be read to be only that dust was a necessary ingredient for spontaneous generation with the yeast/sugar water infusions that he used. Recent historians have also pointed out, as did Pouchet and other supporters of spontaneous generation at the time, that Pasteur never replicated some of Pouchet's most convincing experiments: those involving boiled hay infusions. (It was only recognized a decade later that the hay bacillus, *Bacillus subtilis,* produces heat-resistant endospores.) If Pasteur had tried his famous swan-necked flask method with a hay infusion, it has been suggested that the debate might have ended quite differently on this point alone.

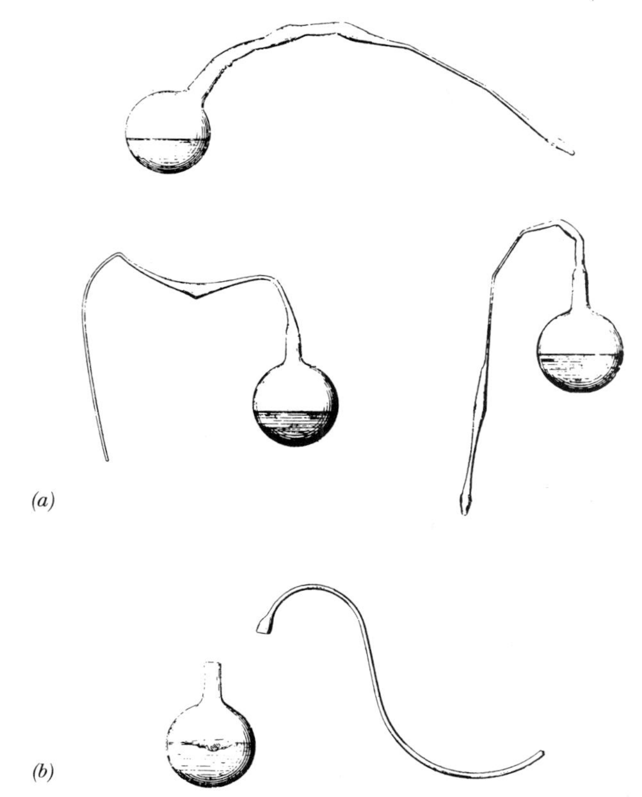

Fig. 3. Pasteur's illustrations of his "swan-necked flasks." Courtesy of the American Society for Microbiology Press.

Since even in this most famous of textbook cases, it still seems the experimental evidence at the time was not as finally conclusive as Pasteur declared it to be, why did the French Academy of Sciences award the victory to Pasteur and declare the spontaneous generation controversy settled once and for all? Spontaneous generation was still just as politically and religiously charged a subject in the French Second Empire of Louis Napoleon as it had been 100 years earlier in Buffon and Needham's time. Furthermore, although spontaneous generation had in the past tended to be associated with the doctrine of transmutation of species, the latter doctrine was currently undergoing a particularly heated wave of notoriety in the wake of the recent publication of Darwin's *On the Origin of Species*. The first French translation, by Clémence Royer, of the *Origin* had just appeared in 1862 and was prefaced by a long diatribe by Royer against the Catholic Church. The Church was, however, on closer terms than ever with the conservative government, since Louis Napoleon's coup of 1851. In this environment, since the Academy of Sciences was a state-supported institution, Darwinian evolution was regarded in France as a politico-theological doctrine allied with forces that threatened the Church and State. In addition, spontaneous generation was seen as directly undermining belief in a providential Creator, so that the outcome of the Pasteur–Pouchet debate carried implications of enormous importance to the power structure of the Second Empire. Even foreign savants observing the controversy at the time, including Britain's premier anatomist Richard Owen and Germany's Friedrich Lange, noted that opposition to spontaneous generation seemed intimately tied to support for the forces of conservatism and the established Church.

The force of Pasteur's experimental genius has been surpassed historically only by his rhetorical skill. For, while he broached all these controversial subjects in a famous 1864 Sorbonne lecture on spontaneous generation, yet he succeeded in portraying himself (and science done at its best) as able to remain entirely aloof from such matters, saying that he had approached the question in a totally unbiased manner, equally ready to accept or reject the existence that spontaneous generation, solely on the basis of the outcome of the experiments. While Pasteur was truly an experimental genius compared to Pouchet, the crucial problem with this "let the chips fall where they may" portrayal of himself is that it is not true. By this time Pasteur had been dominated for some years already by his own preconceived notions (derived from his work on crystalline asymmetry) of life as dependent on a Cosmic Asymmetric Force. He had done numerous experiments in private, trying to produce living conditions in the laboratory by application of light, electricity, and magnetism to experimental setups, and he was convinced that Pouchet was wrong, not so much because spontaneous generation was somehow impossible, but because Pouchet was not aware of the importance of asymmetric forces to the problem of life, and was not approaching the experiments by that route. Furthermore, Pasteur was fully aware that all of his important scientific patrons were entrenched in the conservative regime, and he was constantly trying to cultivate those contacts, particularly with the Emperor and Empress personally. Thus it is no surprise that during all the years of the Second Empire he kept silent about his own attempts to simulate or create life in the laboratory and concentrated in public on refuting the work of Pouchet, which he knew to be so unpalatable to the Academy and the State.

VI. THE BRITISH DEBATE OF THE 1870s: SPONTANEOUS GENERATION AND THE GERM THEORY OF DISEASE

On the British scientific scene in the 1860s and 1870s the most entrenched advocates of spontaneous generation, and those most resistant to the germ theory of disease, were among the medical community. At that time, despite Pasteur's suggestive research on silkworm diseases, British doctors strongly favored a theory of contagious disease as analogous to fermentation via chemical catalysis, a theory first promulgated by Justus Liebig in the early 1840s. This was in part due to a brief but highly publicized theory of 1849 that cholera was caused by a fungus. When the supposed cholera-fungus was soon discovered to be a common mold contaminant, many British medical men who might have had leanings toward a germ theory felt that it had been discredited. Thus, when Pasteur's claims of the mid-1860s reached Britain, most British doctors were skeptical that living microbes were the sources of contagion, and they continued to favor a chemical poison, viewing the microbes as some kind of product or concomitant of the disease process rather than its cause. Thus, when the physicist John Tyndall gave a famous lecture on "Dust and Disease" in January 1870, arguing that doctors must accept the germ theory of Pasteur in order to have any really scientific approach to disease, many British doctors took offence. Tyndall was known for his brash public persona and as a general spokesman for science in London society circles. In addition to his quite haughty tone, he was trained as a physicist and certainly had no clinical experience of disease or of the role of patients' constitutions in causing their susceptibility to be so highly varied. So it was not surprising that doctors accused Tyndall of being an interloper in biology and medicine, an area where his opinions had no weight. They thought he was totally unaware of the cholera-fungus hoax of 1849 and of other important technical reasons for medical skepticism about an oversimplified germ theory. Tyndall's version of the germ theory denied any role to the "constitution": it compared patients to so many identical tubes of infusion. If germ-laden dust particles fell into the patient, he would get sick, Tyndall said. Those in hospitals or towns with epidemics who did not get sick were those who such particles, or "germ clouds," passed by.

Chief among the medical professionals who opposed Tyndall was Henry Charlton Bastian, professor of pathological anatomy at London's University College Medical School. Bastian was an avowed supporter of Darwin's and Spencer's writings on evolution, and he did much experimental work to try to show that microorganisms could arise by spontaneous generation, or biogenesis as he at first called it. Bastian, like Alfred Russell Wallace and many others interested in natural selection, thought at that time that Darwin's theory required spontaneous generation to explain in a nonmiraculous way where the original common ancestor of all species came from. Bastian also thought that bacteria in diseased patients resulted from spontaneous generation, as by-products of the disease process. Between 1870 and 1875 he had published the results of hundreds of experiments in which he showed that bacteria could be found in tubes of various infusions boiled for periods varying from a few minutes up to an hour. Attempting to refute Bastian's work from 1875 to 1877, Tyndall devised an ingenious

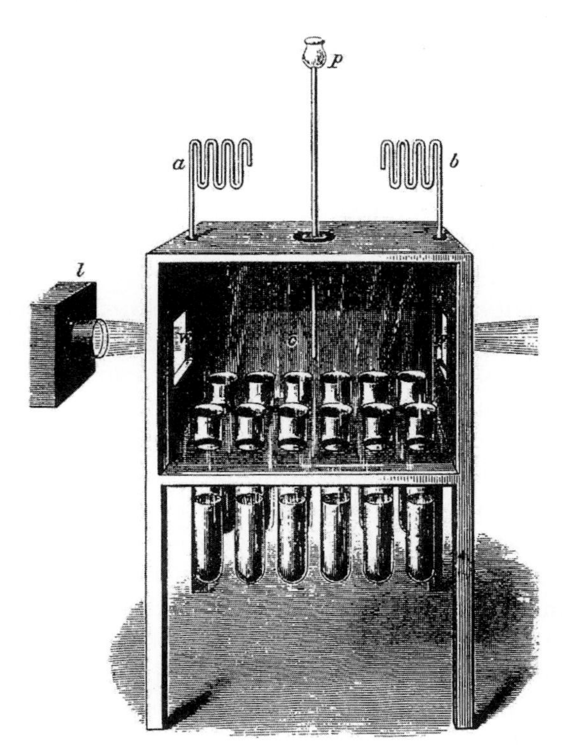

Fig. 4. Tyndall's dust-proof chamber for carrying out spontaneous generation experiments. Courtesy of the Royal Institution, London.

dust-free cabinet in which to carry out the experiments (Fig. 4).

Tyndall's close friend T. H. Huxley was at first interested in Bastian's experiments but soon concluded that the pathologist must be mistaken in his conclusions. He declared that Bastian "had gotten out of his tubes exactly what he put into them," that is, that organisms must have gotten into the tubes as contaminants. Both Huxley and Tyndall were also evolutionists, but they found it much easier to believe in living things able to survive boiling somehow (or, even more believable, that Bastian was a sloppy experimenter) than to believe that organisms as complex as protozoa could come from anything other than the "seeds" or "germs" of other like organisms. They lobbied widely among scientific colleagues to convince others that Bastian was at best a poor experimenter and at worst a fraud and a cheat. As mentioned above, Bastian, for his part, found it much more difficult to believe that life could survive the boiling temperature for more than a minute or two, than to believe (especially since he saw it as a logical necessity of continuity in nature and of evolutionary science) that the transition in stages from nonliving to living matter should be possible. A noted physiologist, John Burdon Sanderson, observed Bastian's technique carefully and reported that it was up to high scientific standards.

Huxley, in a famous address to the British Association for Advancement of Science (BAAS) meeting at Liverpool in September 1870, moved to gain the rhetorical upper hand in the debate by defining the terms. He defined "biogenesis" to mean life only from other life. The opposite belief he termed "abiogenesis" and began to argue that it very well might be possible, even probable, but only in the conditions of the primitive Earth. This convenient dualistic terminology was rapidly picked up by contemporaries and propagated in textbooks. Given this, it is more than a little ironic that Huxley had hijacked the term "biogenesis" from Bastian, who was using it up until that time to mean exactly the opposite, namely, spontaneous generation. Also important, it is from this time that all discussion of the origin of life question began to be pushed more and more into the distant past. Only after Huxley's talk, for example, did Darwin advance the argument that if spontaneous generation had occurred in Earth's distant past, it would no longer be possible after the evolution of the first heterotrophs, since they would consume any organic molecules that formed before those could assemble into a new organism.

Those opposed to both evolution and spontaneous generation had a simpler solution. The physicist William Thomson, Lord Kelvin, for example, responded to both Huxley and Bastian in his own Presidential Address to the BAAS in 1871. He suggested that the germs of life might have originally been brought to Earth from another world via a meteorite. To those demanding a completely naturalistic origin of life, of course, this merely pushed the question back a step to some other planet.

Beginning in the autumn of 1876, Tyndall began to have difficulty with the experiments carried out in his dust-free cabinet. Infusions that had been sterilized by only 5 minutes' boiling a year earlier now could not be sterilized even after hours of boiling. In 1876, the discovery by the German botanist Ferdinand Cohn that certain species of *Bacillus,* especially common in hay and in cheese (and in infusions made from these that showed microbial growth after extended boiling), were capable of producing heat-resistant endospores. This has been taken to explain why Bastian's infusions were full of microbial growth after boiling, and on being handed a copy of Cohn's article Tyndall immediately adopted this stance to explain his own recent difficulties.

Tyndall struggled for several months and finally discovered that sterility could be achieved by repeated short boilings, followed by a period of allowing the remaining spores to germinate. This process came to be known as "fractional sterilization" or "Tyndallization." This was largely superceded with the development of the autoclave in the 1880s. Equally interesting from a historical point of view is that, although this rapid about-face also showed that Tyndall and Huxley had been wrong that Bastian must be a sloppy, incompetent experimenter, that is precisely the version of Bastian (and most spontaneous generation advocates) which has usually gone into the textbooks until quite recently. The sole exception was Pasteur's student Emile Duclaux, the only contemporary writer among the winners who credited Bastian's perseverance in sticking to his experimental results with the value that it truly had, that is, that without the goad of Bastian and his popularity, Pasteur, Cohn, and Tyndall might never have discovered that they were wrong about the existence of bacteria in boiled hay infusions, and thus been led to realize that heat-resistant spores exist and require autoclaving or fractional sterilization to guarantee their being killed. Pasteur was a devout Catholic and never believed, to his dying day, the viewpoint favored by many modern origin of life hypotheses, that random or chance organization of materials could have formed life, even the very first living organisms.

VII. TWENTIETH CENTURY IDEAS

The origin of life has continued to be a subject of active interest in the twentieth century, and, despite avoidance of the term "spontaneous generation" since 1880, some modern concepts have resembled older ideas more than superficially. Beginning with Bastian (who revived experiments on spontaneous generation from 1900 to 1915), a number of investigators came to believe that the basic chemistry specific to the living was the chemistry of colloids. A burst of research in colloid chemistry from 1910 until the 1930s included many workers who believed that this avenue would lead to understanding the simplest combination of materials to cross the boundary from nonlife to living matter. Among these were biochemists Benjamin Moore and Albert Mary, bacteriologist Arthur I. Kendall, the Mexican biologist Alfonso Herrera, and medical doctor and researcher George Crile. Geneticist H. J. Muller suggested that the first chemical assembly of a gene should be considered the origin of life, though this still suggested a sudden origin of life from nonlife, perhaps the most central idea implicit in the doctrine of spontaneous generation. The Russian biochemist A. I. Oparin proposed instead a gradual chemical evolution of life, probably proceeding through the intermediate stage of coacervates. Research into life's origin had lost none of its larger associations: the issue still attracted researchers with anticlerical leanings (Herrera was one) as well as affiliations with radical, even Communist political beliefs. J. B. S. Haldane and J. D. Bernal did research in the field and were avowed Communists, as was Muller until the late 1930s. Furthermore, Oparin and Wilhelm Reich explicitly credited the philosophy of dialectical materialism as having been crucial to their origin of life research programs.

Research on viruses, particularly Wendell Stanley's crystallization of tobacco mosaic virus in 1935, lent credence for a time to the idea that a virus might be the simplest "living molecule." In addition, after the rise of molecular biology and the discovery of the Watson–Crick DNA structure, many considered Muller's idea of "the gene as the origin of life" more likely. The Miller–Urey experiment, published just 3 weeks after Watson and Crick's paper on DNA, lent further support to the notion that the building blocks of life could have assembled very rapidly. In that experiment, an electrical discharge passed through a mixture of steam, hydrogen, methane, and ammonia in a sealed container produced amino acids and other organic compounds after only a few days. Throughout the 1950s and 1960s, such views were in tension with those of Oparin and his students in the Soviet Union, that the process of chemical evolution was a more gradual one, with no single decisive "living molecule." Cold War and anti-Lysenkoist hostilities were involved, as any scientific theory openly lauding dialectical materialism was seen as suspect in the West in the aftermath of Lysenko's takeover of Soviet genetics.

Thus, "spontaneous generation" has been considered a dead-end, disproven belief since 1880 or so. But, despite the dropping of the older term, the conceptual continuities between modern and older origin of life ideas, particularly for abiogenesis in Earth's distant past, are in some ways as interesting as the discontinuities. The continued association of these ideas with larger philosophical and political concerns is also a striking feature of their history well into the twentieth century.

See Also the Following Articles

EVOLUTION, THEORY AND EXPERIMENTS • HISTORY OF MICROBIOLOGY • METHOD, PHILOSOPHY OF

Bibliography

Bulloch, W. (1938). "A History of Bacteriology." Oxford Univ. Press, Oxford.

Farley, J. (1977). "The Spontaneous Generation Controversy from Descartes to Oparin." Johns Hopkins Univ. Press, Baltimore.

Farley, J. (1978). The political and religious background to the work of Louis Pasteur. *Annu. Rev. Microbiol.* **32**, 143–154.

Findlen, P. (1993). Controlling the experiment: Rhetoric, court patronage and the experimental method of Francesco Redi. *Hist. Sci.* **31**, 35–64.

Geison, G. L. (1995). "The Private Science of Louis Pasteur." Princeton Univ. Press, Princeton, New Jersey.

Graham, L. R. (1987). Origin of life. *In* "Science, Philosophy and Human Behavior in the Soviet Union," Chap. 3. Columbia Univ. Press, New York.

Lazcano, A. (1992). Origins of life: The historical development of recent theories. *In* "Environmental Evolution: The Effects of the Origin and Evolution of Life on Planet Earth" (L. Margulis and L. Olendzenski, eds.), pp. 57–69. MIT Press, Cambridge, Massachusetts.

Pelling, M. (1978). "Cholera, Fever and English Medicine, 1825–1865." Oxford Univ. Press, Oxford.

Roe, S. (1983). John Turberville Needham and the generation of living organisms. *Isis* 74, 159–184.

Sloan, P. R. (1992). Organic molecules revisited. *In* "Buffon '88" (J. Roger, ed.), pp. 415–438, J. Vrin, Paris and Lyon.

Strick, J. E. (1999). Darwinism and the origin of life: The role of H.C. Bastian in the British spontaneous generation debates, 1868–73. *J. Hist. Biol.* **32**, 51–92.

Strick, J. E. (forthcoming). "Sparks of Life: Darwinism and the Victorian Spontaneous Generation Debates." Harvard Univ. Press, Cambridge, Massachusetts.

Sporulation

Patrick J. Piggot
Temple University School of Medicine

I. Stages of Sporulation
II. Initiation of Sporulation
III. The Sporulation Division
IV. Compartmentalization of Gene Expression, and Different σ Factors
V. Engulfment
VI. Postengulfment Development
VII. Development of Spore Resistances
VIII. Germination

GLOSSARY

cortex A peptidoglycan that is unique to spores and sporulating organisms. There are very few peptide bridges, and many of the muramic residues in the glycan chain are in the form of a lactam with no attached peptide side chain.

engulfment The process by which the developing prespore is completely surrounded by the mother cell.

mother cell One of the two cells that is formed by the sporulation division. It is required for spore formation but lyses when the spore is formed.

prespore One of the two cells that is formed by the sporulation division. It develops into the mature spore. It is sometimes called the forespore.

σ factor A protein that binds to RNA polymerase core enzyme, designated E, to form RNA polymerase holoenzyme, Eσ. The σ factor determines the specificity of the binding of the holoenzyme to promoter sequences in DNA.

vegetative cell A bacterial cell from a culture that is growing exponentially.

SPORES are a dormant form of bacteria. They are resistant to a variety of environmental stresses that would kill the vegetative (growing) form of the bacteria. The stresses include heat, desiccation, irradiation, and chemicals such as ethanol and chloroform. Sporulation is the process by which the spores are formed from vegetatively growing bacteria, and it is a response to nutrient depletion. The best studied examples of sporulation are of members of the genera *Bacillus* and *Clostridium.* The description is generally also valid for sporulation of members of related genera such as *Sporosarcina* and *Thermoactinomyces.*

Sporulation has been most extensively studied with *Bacillus subtilis.* Formation of heat-resistant spores from vegetative cells of *B. subtilis* takes about 7 hr at 37°C. This species has been favored because it has very good systems of genetic analysis. Efficient systems of genetic exchange by transformation and transduction mean that it is easy to transfer genes and mutations from one strain to another. It is also easy to identify mutants of *B. subtilis* that cannot sporulate, because, on appropriate media, the mutant colonies are poorly pigmented compared to colonies of the sporulating parental strains. The complete sequence of the *B. subtilis* genome (chromosome) has recently been determined, further facilitating studies of sporulation of this species.

I. STAGES OF SPORULATION

The morphological changes during sporulation have been characterized for a number of species by electron microscopy. The basic sequence of changes is similar for all species of *Bacillus* and *Clostridium* that have been studied. It is illustrated in Fig. 1. Identification of successive stages by Roman numerals follows the convention introduced by Ryter and now generally used. Vegetative (growing) cells are

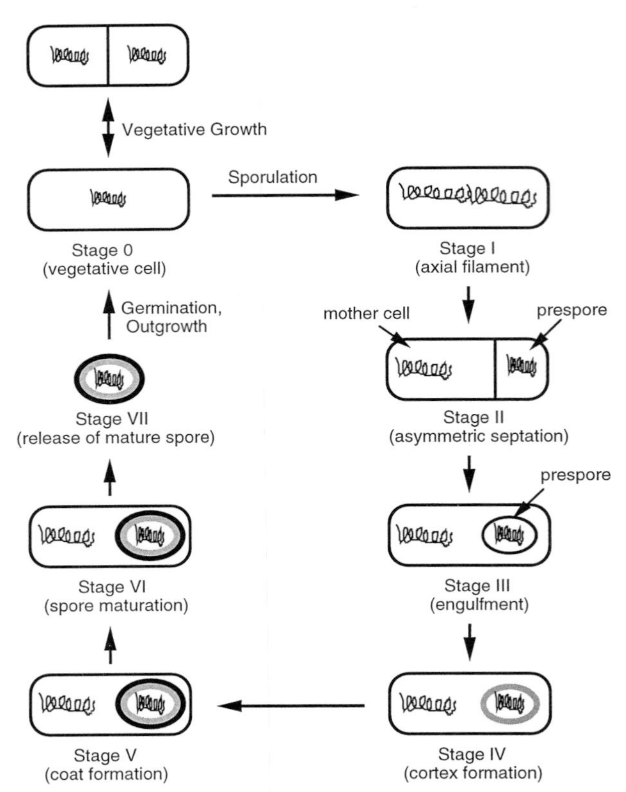

Fig. 1. Schematic representation of the stages of sporulation. (I thank Jun Yu and Edward Amaya for help in the preparation of this and the other figures.)

rod shaped. They are defined as stage 0. Stage I is the formation of an axial filament of chromatin. Stage II defines the completion of a division septum at one pole of the cell. This division produces two cell types, the smaller prespore (sometimes called the forespore) and the larger mother cell. Stage III is defined as the completion of engulfment of the prespore by the mother cell. Stage IV is the deposition of two layers of cell-wall material, the cortex and the primordial germ-cell wall, between the opposed membranes that surround the engulfed prespore. Deposition of layers of coat material around the outside of the prespore defines stage V. By Stage VI the prespore has matured into the heat-resistant spore, but the spore is still located within the mother cell. Lysis of the mother cell to release the mature spore is defined as stage VII.

It should be noted that development of the spore takes place inside the mother cell from stages III to VII. For this reason, the process is often called endospore formation. Bacteria of some less-studied genera, for example, *Thermoactinomyces* and *Sporosarcina,* also form endospores by a process very similar to that described above, although they do not have rod-shaped vegetative cells: *Thermoactinomyces* has vegetative mycelium and *Sporosarcina* cocci; interestingly, the sporulation division for *Sporosarcina* appears to be symmetrically located. Sporulation within other genera such as *Streptomyces* and *Myxococcus* does not involve endospores and is also in other ways significantly different from the process described above; it is not discussed here.

A. Spo$^-$ Mutants and *spo* Loci

Sporulation is not an obligatory part of the life cycle of endospore formers. Consistent with this, many different mutants have been described that cannot sporulate but can grow perfectly well vegetatively. Such Spo$^-$ mutants have been most extensively analyzed for *B. subtilis.* Operationally, Spo$^-$ mutants that have been tentatively identified because of the poor pigmentation of the colonies can be positively identified by phase-contrast microscopy through their failure to produce phase-bright spores. They can also be recognized by their sensitivity to heat—typically spores survive 20 min at 80°C, but Spo$^-$ mutants and vegetative cells of Spo$^+$ strains are killed by this treatment. (The heat resistance of spores is species dependent. In general, species that grow at higher temperatures give spores that resist higher temperatures.) Analysis by electron microscopy of thin sections has shown that, in general, particular mutants are blocked at a recognizable stage of sporulation and are designated accordingly. For example, SpoII mutants are blocked at stage II—the sporulation septum is formed, but engulfment is not completed; SpoIII mutants are blocked at stage III—the prespore is completely engulfed by the mother cell, but none of the protective layers, such as cortex and coat, have yet been formed.

The mutations that give rise to the Spo$^-$ mutant phenotype map, by definition, in *spo* loci. More than 40 *spo* loci have been identified in *B. subtilis.* These loci were defined initially by groups of *spo* mutations that mapped close together. In almost all cases they

TABLE I
Guide to Selected Sporulation Loci Discussed in the Text

Locus	Encoded function
cotE	Scaffold protein for spore coat assembly
cotG	Tyrosine-rich coat protein
gerA	Germination response to L-alanine
gerB	Germination response to GFAK[a]
kinA	Kinase that phosphorylates Spo0F
kinB	Membrane-associated kinase that phosphorylates Spo0F
phrA	Peptide inhibitor of RapA
rapA	Spo0F-PO_4 phosphatase
rapB	Spo0F-PO_4 phosphatase
sigK	Pro-σ^K
spo0A	Response regulator; pivotal transcription regulator
spo0B	Phosphotransferase
spo0E	Spo0A-PO_4 phosphatase
spo0F	Response regulator
spoIIA	Operon coding σ^F, SpoIIAB (anti-sigma factor), and SpoIIAA (anti-anti-sigma factor that is inactivated on phosphorylation by SpoIIAB)
spoIIE	Membrane protein; SpoIIAA-PO_4 phosphatase
spoIIG	Pro-σ^E and SpoIIGA (likely processing enzyme)
spoIIR	$E\sigma^F$-transcribed gene required for pro-σ^E processing
spoIIIE	DNA translocase
spoIIIG	σ^G
spoIVA	Scaffold protein for coat assembly
spoIVB	$E\sigma^G$-transcribed gene required for pro-σ^K processing
spoVA	DPA[b] transport to prespore
spoVF	DPA synthesis
spoVID	Scaffold protein for coat assembly

[a] GFAK, glucose, fructose, L-asparagine, and KCl.
[b] DPA, dipicolinic acid.

now been located on the completed sequence of the *B. subtilis* genome, and have been shown to correspond to individual genes or to operons, which are groups of adjacent genes expressed as a unit (Table I).

II. INITIATION OF SPORULATION

Sporulation is favored by high cell density. It can be initiated by depletion of the source of carbon, of nitrogen, or, in some conditions, of phosphorus. It can only be initiated in cells that are actively replicating their genome, and only during part of the vegetative cell-division cycle. It is known that a fall in concentration of the nucleotides GTP and GDP always occurs at the start of sporulation. Further, artificial depletion of GTP and GDP can trigger sporulation in some media that ordinarily would not support sporulation. These nucleotides, or something derived metabolically from them, are plausible candidates as mediators of the starvation signal. However, it is not known how a fall in the concentration of these nucleotides leads to the complex pattern of gene expression that is triggered after the start of sporulation and is necessary for spore formation.

The critical early event that sets in motion this complex pattern of gene expression is the activation of the transcription regulator Spo0A by phosphorylation. Phosphorylation of the Spo0A protein is achieved through a series of reactions collectively referred to as the phosphorelay (Fig. 2). In the phosphorelay a protein kinase, KinA or KinB, phosphorylates the Spo0F protein. The phosphate group is then transferred to Spo0A through the action of the Spo0B phosphotransferase. Countering this sequence that leads to Spo0A phosphorylation are several phosphatases: Spo0E specifically dephosphorylates Spo0A-PO_4; RapA and RapB specifically dephosphorylate Spo0F-PO_4. The balance of all these reactions determines if Spo0A is phosphorylated sufficiently to activate a series of genes that must be expressed for successful sporulation.

The phosphorelay is the conduit for the external and internal signals that trigger spore formation. Thus, the rarget of the cell-density signal appears to be the RapA phosphatase. This phosphatase is inhibited by a pentapeptide derived from processing of a small protein called PhrA. The PhrA protein is secreted, probably as a 19-residue peptide. It is processed and then reimported by an oligopeptide permease. At high cell density it is thought that the concentration of the 5-residue peptide derived from PhrA is sufficient to inhibit the RapA phosphatase, thus increasing the concentration of Spo0F-PO_4 and fueling the phosphorylation of Spo0A; strains deleted for the *phrA* gene or for the oligopeptide permease genes sporulate very poorly. Cell cycle, DNA replication, and tricarboxylic acid cycle signals also feed

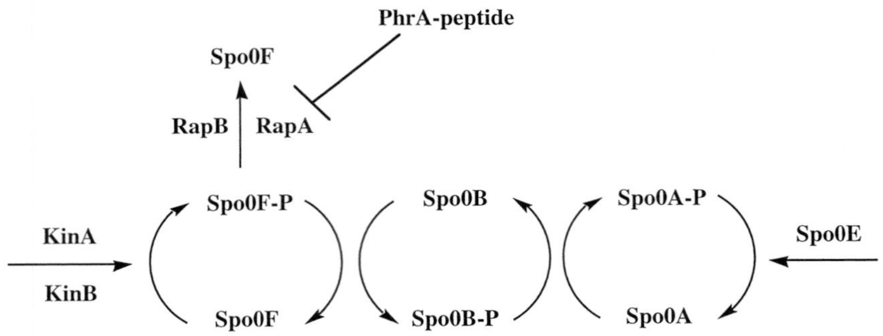

PhrA-peptide
Spo0F
RapB RapA
Spo0F-P Spo0B Spo0A-P
KinA KinB Spo0E
Spo0F Spo0B-P Spo0A

Fig. 2. Representation of the reactions of the phosphorelay (adapted from Perego and Hoch, 1996, and Piggot, 1996). Two kinases, KinA and KinB, are activated by unknown signals to phosphorylate Spo0F. In the figure, phosphorylated forms of proteins are indicated by -P, for example, Spo0F-P. The phosphate group is transferred from Spo0F to the key transcription regulator Spo0A by the phosphotransferase Spo0B. Spo0A is activated by phosphorylation. The protein-phosphatase Spo0E specifically dephosphorylates Spo0A-P. RapA and RapB are protein phosphatases that specifically dephosphorylate Spo0F-P; their action also reduces the phosphorylation of Spo0A because of the action of the phosphorelay. A pentapeptide derived from PhrA inhibits the action of RapA. It is likely that there are additional regulators of the phosphatases.

through the phosphorelay, as do nutritional-starvation signals; the molecular details of these various signal mechanisms have not been worked out.

Spo0A-PO_4 activates or represses expression of a large number of genetic loci. Crucial to subsequent events during sporulation, it activates transcription into mRNA, and hence translation into protein, of the *spoIIA, spoIIE,* and *spoIIG* loci, which are discussed below; and it is necessary for the asymmetrically located sporulation division.

III. THE SPORULATION DIVISION

The asymmetrically located sporulation division is often considered the defining early morphological event in sporulation of *B. subtilis*. The machinery for division is similar to that for vegetative division; for example, both processes require the tubulin-like FtsZ protein. However, there are several clear differences between the two types of division: (i) The sporulation division septum is much thinner than the vegetative division septum. (ii) The two cells that result from the sporulation division do not separate from each other as occurs following vegetative division. Rather, the mother cell engulfs the prespore. (iii) Autolysis of the wall material (peptidoglycan) within the sporulation septum begins in the center of the septum, and ultimately there is apparently complete loss of wall material. In contrast, autolysis of the wall material of the vegetative septum begins at the periphery of the septum and proceeds inward. Moreover, there is little loss of wall material—the split septum provides the wall for the poles of the nascent cells. (iv) The sporulation septum is formed before a chromosome is completely packaged into the prespore. The prespore initially contains only the origin-proximal one-third of a chromosome. A DNA translocase, SpoIIIE, is located in the center of the septum and is required for transfer of a complete chromosome into the prespore. (v) The septum is asymmetrically located, with respect to the cell poles, during sporulation but not vegetative growth. The *spoIIE* locus mediates this Spo0A-directed switch in septum location. (vi) Gene expression becomes compartmentalized after the sporulation division, with some genes expressed in the prespore and other genes expressed in the mother cell.

These comments about the structure of the sporulation septum apply to all species of endospore formers that have been studied, with the expception that the sporulation septum is symmetrically located in *Sporosarcina ureae*. Comparatively little is known about sporulation genes in species other than *B. subtilis*. In general, however, homologs of the *B. subtilis* genes have been found when they have been looked for. Homologs of *Spo0A* were identified in all of a wide range of endospore-forming species that were analyzed.

IV. COMPARTMENTALIZATION OF GENE EXPRESSION, AND DIFFERENT σ FACTORS

As soon as the spore septum is completed, different genes start to be expressed in the prespore and in the mother cell. This compartmetalized gene expression is a consequence of different sporulation-specific RNA polymerase σ factors becoming active, σ^F in the prespore and σ^E in the mother cell. The σ factors direct the RNA polymerase to transcribe particular genes into mRNA by recognizing DNA sequences called promoters. Different σ factors direct recognition of different promoters, and so σ^F directs transcription of a different set of genes from σ^E. These two σ factors are synthesized before the spore septum is formed, but are not active. Why they only become active on septation and why their activities are compartmentalized are questions that have not been fully answered. These questions are presently exciting much research activity. Because σ^F needs to be active in order for σ^E to be activated, and not vice versa, activation of σ^F is considered first.

The *spoIIA* locus is a three-gene operon. Its transcription is activated by Spo0A-PO_4 soon after the start of sporulation. σ^F is encoded by the third gene of the *spoIIA* operon, *spoIIAC*. The second gene, *spoIIAB*, encodes a protein, SpoIIAB, that binds to σ^F and inhibits its action. The SpoIIAB protein is thus an anti-sigma factor. The first gene of the operon, *spoIIAA*, encodes an anti-anti-sigma factor that binds to SpoIIAB and releases σ^F from σ^F–SpoIIAB complexes (Fig. 3). SpoIIAB is also a protein kinase that can specifically phosphorylate SpoIIAA; SpoIIAA-PO_4 is not able to bind to SpoIIAB. To complete the list of known participants in this system, the SpoIIE protein is a phosphatase that specifically dephosphorylates SpoIIAA-PO_4. The balance of these various reactions determines if σ^F is active or not. The critical step in activation of σ^F is thought to be the activation of SpoIIE phosphatase activity. In the sporulating cell before septation, this phosphatase is not active, SpoIIAA is completely phosphorylated, and SpoIIAB is bound to σ^F, thus blocking σ^F activity. Something about septation activates the SpoIIE phosphatase; its activity releases unphosphorylated SpoIIAA which binds SpoIIAB, thus freeing and hence activating σ^F. It is known that SpoIIE is a membrane protein that is found predominantly in the spore septum, but what activates SpoIIE is not known. σ^F becomes active on the prespore side of the sporulation septum, but not the mother cell side of the septum. It is not certain why its activity is compartmentalized, but there are data which suggest that the phosphatase domain of SpoIIE is located only on the prespore side of the septum. Of course,

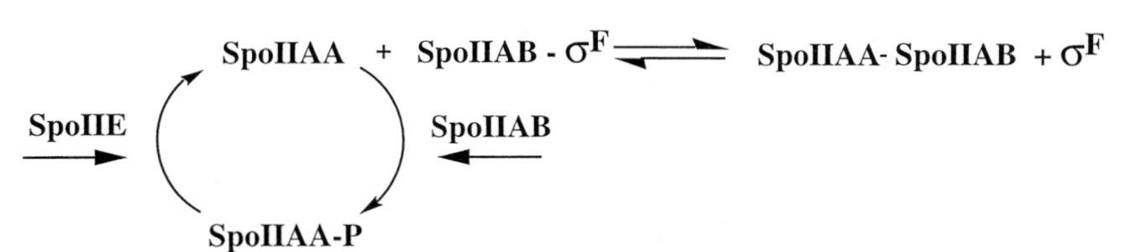

Fig. 3. Outline of the activation of σ^F during sporulation (modified from Piggot, 1996). SpoIIAB is an anti-sigma factor that binds noncovalently to σ^F and inactivates it. SpoIIAA can bind noncovalently to SpoIIAB, thereby releasing and activating σ^F. SpoIIAB also functions as a kinase to phosphorylate SpoIIAA; SpoIIAA-P cannot bind to SpoIIAB and reverse the inhibition of σ^F activity. SpoIIE is a septum-located phosphatase that specifically dephosphorylates SpoIIAA-P, thus activating SpoIIAA and hence σ^F. Activation of the SpoIIE phosphatase is thought to be the critical step leading to σ^F activation during sporulation.

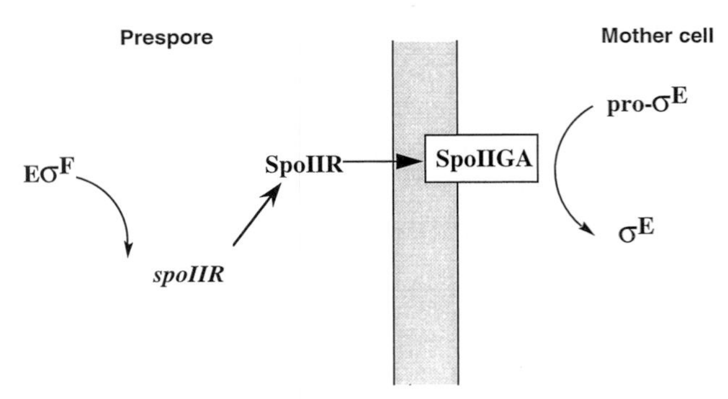

Fig. 4. Outline of the activation of σ^E during sporulation (modified from Piggot, 1996). Formation of SpoIIR in the prespore causes activation in the mother cell of pro-σ^E by its proteolysis to σ^E. This action is mediated by the septally located SpoIIGA, which is thought to be the protease that catalyzes the reaction.

this then raises the unanswered question of why SpoIIE is asymmetrically located in the septum.

The mother cell-specific σ^E is activated in a different way from σ^F. σ^E is encoded by the second gene of the *spoIIG* operon as an inactive precursor, pro-σ^E. An N-terminal pro sequence of 27 to 29 residues must be proteolytically cleaved for σ^E to become active. The likely processing enzyme is SpoIIGA, which is encoded by the first gene of the *spoIIG* operon. SpoIIGA is a membrane protein that is located in the spore septum. Activation of SpoIIGA, and hence σ^E, requires the activity of the SpoIIR protein (Fig. 4). The SpoIIR protein is produced exclusively in the prespore from a gene that is exclusively under the control of σ^F. Thus, activation of σ^F in the prespore leads very rapidly to activation of σ^E in the mother cell. These two σ factors initiate the programs of compartmentalized gene expression, but neither directs the process that leads to compartmentalization. The activation of σ^F and σ^E is required for progression to the next stage of sporulation, engulfment. Mutations in the structural genes for σ^F and σ^E and in the genes for their activation all block sporulation at stage II, septation.

V. ENGULFMENT

The process of engulfment starts soon after the sporulation division septum is formed. Autolysis of septal wall material starts from the center of the septum and proceeds to its periphery, resulting in removal of essentially all wall material. At this time the septum is very flexible. The middle of the septum appears, in electron micrographs, to bulge into the mother cell. The annulus of attachment of the septum to the cylindrical wall of the organism moves to the prespore pole and then fuses to itself. The consequence of this fusion is that the prespore is completely surrounded by membranes of opposed polarity, and is further shielded from the external medium because it is completely surrounded by the mother cell. Completion of engulfment is defined as stage III. By this stage the organism is "committed" to form spores: Addition of rich medium does not reverse the sporulation process. This contrasts with the organism at stage II, where both mother cell and prespore can revert to vegetative growth on addition of rich medium.

Sporulation mutants have been found that are blocked at different stages of the engulfment process. Some of the corresponding genes are under σ^F control, others under σ^E control. Thus, both prespore and mother cell contribute to engulfment. Still other genes are under the control of the main vegetative σ factor, σ^A. Although a number of such engulfment genes have been identified, the biochemical mechanisms involved in engulfment are not well understood.

VI. POSTENGULFMENT DEVELOPMENT

By stage III the mother cell and the prespore are very different from each other. Notably, the prespore is now completely surrounded by the mother cell and so is not in contact with the medium. The prespore has not, however, acquired any of the resistance or dormancy properties that characterize the mature spore. These are developed in the succeeding stages of sporulation.

Completion of engulfment activates two additional σ factors, σ^G in the prespore and σ^K in the mother cell. Activation of these σ factors is required for the subsequent stages of sporulation, including the development of spore resistances. Transcription of the structural gene for σ^G, *spoIIIG*, is directed by σ^F

and consequently is confined to the prespore. It also requires expression of an unidentified mother cell-specific gene (or genes). This latter control is one of several controls found throughout postseptation development, where interaction between mother cell and prespore is requried for further development. σ^G is initially inactive, and it is only activated on the completion of engulfment. Its activation also requires expression of mother cell-specific genes. The molecular mechanisms of σ^G activation have not been worked out. σ^G and σ^F have overlapping promoter specificity, and once activated, σ^G directs transcription of *spoIIIG* and hence its own synthesis. A number of prespore genes transcribed before engulfment by E-σ^F continue to be transcribed after engulfment by E-σ^G, because of this overlapping specificity. (It is suspected but not proven that σ^F is inactive after engulfment.) In addition, several new genes become active whose promoters are recognized by σ^G but not by σ^F. Of particular importance in this last category are the *ssp* genes which encode small acid-soluble proteins that are critical to ultraviolet resistance (see below).

Transcription of the structural gene for σ^K, *sigK*, is directed by σ^E, and consequently is confined to the mother cell. σ^K is initially formed as an inactive precursor, pro-σ^K. Activation of pro-σ^K, by proteolytic removal of the N-terminal pro sequence (which shows no similarity to the pro sequence for σ^E), requires the product of a particular σ^G-directed gene, *spoIVB*. Thus σ^K activation in the mother cell is tied to prior events in the prespore and, indirectly, requires completion of engulfment. In an interesting parallel to events in the prespore, σ^K and σ^E have overlapping specificity, and when activated, σ^K directs transcription of *sigK* and hence its own synthesis. Several other mother cell-specific genes are also transcribed both before and after engulfment, because of this overlapping specificity. A number of genes whose promoters are recognized exclusively by σ^K are transcribed only after engulfment; notable among the latter are several *cot* genes which encode coat proteins (see below).

In some strains of *B. subtilis* an additional, very striking control is exercised on *sigK*. In the genome of the germ line (i.e., vegetative cells, prespores, and spores) the 5′ and 3′ portions of the *sigK* gene are separated by a 48-kbp DNA sequence known as SKIN. There is a site-specific recombination specifically in the mother cell (directed by σ^E) that excises SKIN and forms an intact copy of *sigK*. In other strains of *B. subtilis* and in other species of *Bacillus*, *sigK* is intact in the germ line and there is no need for this site-specific recombination.

VII. DEVELOPMENT OF SPORE RESISTANCES

Spore resistance to heat, to irradiation, and to chemicals, such as chloroform and octanol, develops in the succeeding stages of sporulation. Several spore constituents formed during this period are thought to contribute, to varying degrees, to these resistances. Formation of two of the constituents, the cortex and the coat, is associated with stages IV and V, respectively. The possible roles of these structures are discussed below. The roles of two other major spore constituents are then considered: small acid-soluble proteins (SASPs) and dipicolinic acid (DPA). During this period the spore core (i.e., the spore cytoplasm) decreases in volume, so that the final volume of the spore core is about half that of the prespore cytoplasm immediately following engulfment.

Two types of peptidoglycan, the cortex and the primordial germ-cell wall, are assembled around the prespore following the completion of engulfment. The cortex has a very different structure from other peptidoglycans and is only found in spores and sporulating organisms. It is degraded on germination. The cortex shares with other peptidoglycans a glycan chain of alternating N-acetylglucosamine and muramic acid residues. It is unique in that approximately 50% of the muramic residues are in the form of a lactam and have no attached peptide side chain. The other muramic residues have attached to them either L-alanine or a peptide side chain. There are strikingly few peptide cross-links in the cortex peptidoglycan. These differences are thought to make the cortex a more flexible structure than vegetative cell-wall peptidoglycan. It was long thought that this distinctive structure might be critical to the dehydration that is central to spore heat resistance. However, recent results indicate that normal spore dehydration can be obtained in mutants containing no muramic acid lactam, and in mutants with a fourfold increase in cross-linking. The results suggest that these aspects of cortex structure are not important for the process of dehydration, but may still have a role in the maintenance of dehydration that has been achieved by other means. There remains a correlation between the extent of dehydration and the heat resistance of spores. Mutants with little or no muramic acid lactam, are dramatically impaired in germination; it seems likely that the muramic acid lactam is a major specificity determinant of germination-lytic enzymes.

The primordial germ-cell wall (PGCW), as its name implies, becomes the wall of the cell that is formed on germination of the spore; this contrasts with the cortex, which is degraded during germination. PGCW structure, so far as it has been determined, is that of the vegetative cell wall. Both cortex and PGCW are assembled between the opposed membranes that surround the engulfed prespore. The cortex is the outer of the two peptidoglycans. Where they have been distinguished, enzymes specific to the cortex are found to be synthesized in the mother cell and enzymes specific to the PGCW in the prespore.

The spore coat appears in electron micrographs as electron-dense layers that surround the prespore. There is an inner coat that is lamellar and lighter staining and an outer coat that is darker staining. The coat is composed of a number of proteins which together make up 50% or more of the total protein of the mature spore. Coat proteins tend to be insoluble and difficult to work with biochemically. However, for *B. subtilis* some 15 *cot* genes that encode coat proteins have been identified. All the known *cot* genes are transcribed only in the mother cell, most of them exclusively by $E\sigma^K$, although a few have promoters recognized by σ^E. Assembly of coat layers in the mother cell only loosely follows the pattern of transcription of *cot* genes. The layers are assembled on a scaffold that includes the SpoIVA, SpoVID, and CotE proteins. Each of these proteins has been shown to localize to the outer surface of the prespore. Of the three, SpoIVA is required first. *spoIVA* mutants assemble coats in whorls within the cytoplasm of the mother cell, and there are no coat layers around the prespore; in such mutants, SpoVID and CotE do not assemble around the prespore. *spoVID* and *cotE* mutants are blocked at later stages of the coat assembly process. The genes for the scaffold proteins, *spoIVA, spoVID,* and *cotE,* are all transcribed by RNA polymerase containing the earlier expressed σ^E.

The coat layers contain several types of covalent cross-linkages between protein subunits. These include disulfide bonds, ($\varepsilon-\gamma$) glutamyl–lysyl isopeptide bonds, which are found in keratins, and *o,o*-dityrosyl linkages, which are found in insect cuticle. The structures of the individual coat proteins are varied. For examples the CotG protein is tyrosine rich. There is a superoxide dismutase located in the coat, and, in combination with a peroxidase, it may help assemble CotG into the insoluble coat matrix through the formation of the dityrosyl protein–protein bridges. Other coat proteins resemble β-keratin. The effects of mutation in the various *cot* genes vary considerably. The overall picture that emerges is that the coat layers contribute to the resistance of spores to organic solvents, such as chloroform and octanol, and to the action of lysozyme; also, the coat contributes to the germination properties of spores.

Small acid-soluble proteins make up about 10% of the protein of the dormant spore. They are degraded very rapidly on germination, and the amino acids released contribute significantly to the outgrowth of the germinated spore. There are two main types of SASPs, of which one, known as the α/β type, is important to spore resistance to UV irradiation. This resistance is achieved because the α/β-type SASPs bind to DNA and change the properties of the DNA. They convert the DNA from its normal B conformation to the A conformation. UV irradiation of DNA of spores or of DNA solutions containing α/β-type SASP produces a product known as spore photoproduct. This spore photoproduct is very efficiently repaired on germination. Consequently the spores are more resistant to UV irradiation than are vegetative cells in which pyrimidine dimers (the typical product of UV irradiation of almost all living cells) are formed and are less efficiently repaired. The repair of the spore photoproduct during germination is critical to the UV resistance. The α/β-type SASPs have also been shown to confer resistance to dry heat. These last experiments indicate that the major cause of the lethality of dry heat to spores is damage to DNA. The mechanism of resistance of spores to dry heat (e.g., resistance of a freeze-dried spore preparation) is thus different from that for wet heat (e.g., resistance of an aqueous suspension of spores) where SASPs have a relatively minor role. (Papers discussing spore heat resistance usually mean resistance to wet heat unless they specifically mention dry heat; this usage is retained here.)

Dipicolinic acid (pyridine-2,6-dicarboxylic acid; DPA) is another major constituent of the spore core, accounting for as much as 10% of dry weight. Most DPA in the spore is thought to be complexed with divalent cations, predominantly Ca^{2+}. Its role in spore dormancy and possibly resistance (see below) may reflect its association with these cations, which are present in a correspondingly high concentration. DPA is synthesized from dihydrodipicolinate, an intermediate in lysine biosynthesis. Its synthesis is catalyzed by the product of the *spoVF* locus. Although DPA accumulates in the spore core, transcription of *spoVF* is directed by σ^K in the mother cell. The uptake and/or maintenance of DPA in the prespore depends, at least in part, on the σ^G-directed *spoVA* locus.

Dipicolinic acid enhances significantly the SASP-dependent formation of spore photoproduct on UV irradiation, and so sensitizes spores to UV irradiation. DPA is thought to be required for the maintenance of spore dormancy. The role of DPA in spore (wet) heat resistance has been the subject of controversy. The controversy has centered on spores of DPA$^-$ mutants: Some groups have reported that the spores have normal heat resistance, whereas other groups studying different mutants (including *spoVF* mutants of *B. subtilis*) have found the spores to be heat sensitive. It is clear that production of DPA is not sufficient to confer heat resistance. It is also clear that spores can have a greatly reduced level of DPA and still be heat resistant. However, there is a low spontaneous level of formation of DPA so that *spoVF* mutants (or other "DPA$^-$" mutants) are not completely free of DPA. This reviewer leans to the view that a low level of DPA may be required for full heat resistance. In support of this interpretation are experiments with the *spoVF* mutants of *B. subtilis* in which exogenous addition of a low amount of DPA significantly increased the heat resistance of populations of spores formed by the mutants.

It is likely that resistance to wet heat is multifactorial. As discussed above, cortex, SASP, and DPA are factors that have been shown to have or are suspected of having a role. None has been shown to be the sole determinant of heat resistance. The one strong correlation with heat resistance that has not been shaken over the years is extensive dehydration of the spore core. This dehydration suggests that a high turgor pressure must somehow be maintained by the structures of the spore. The dehydration changes the density of the developing spore, and mature spores have an unusually high density for living cells of about 1.3 g/ml. The increase in density changes the optical properties of the spore, which becomes bright when viewed by phase-contrast microscopy as opposed to the dark vegetative cell and mother cell. Operationally, phase-bright spores provide a very easy way to identify spores. For the sporulation aficionado, phase-bright spores are a joy to behold—except when a Spo$^-$ mutant is expected!

VIII. GERMINATION

Not only are mature spores highly resistant, but they are dormant. Within the limitations of the techniques used, no metabolism has been detected in dormant spores. Spores have been reported to survive for decades, centuries, and even millenia. Nevertheless, when a germinant is provided, spores germinate and return to vegetative growth. Usually this process is divided into two stages: first, germination, which is rapid (a few minutes) and results in the loss of resistances and the resumption of active metabolism; second, outgrowth, which is slower (about 2 hr) and involves the resumption of macromolecular synthesis and the development into normal, vegetatively growing bacilli.

Most known germinants are nutrients, and a wide range have been described. They include amino acids, nucleosides, and sugars. Often they are species-specific. L-Alanine probably works on the widest range of species. For *B. subtilis*, a number of *ger* loci have been defined by mutations that impair germination. In some cases the mutants are impaired in the response to particular germinants. For example, *gerA* mutants are defective in their response to L-alanine, and *gerB* mutants are defective in their response to glucose, fructose, L-asparagine, and KCl (which together function as a single germinant, GFAK). The two loci are transcribed in the prespore by $E\sigma^G$. The encoded proteins are thought to provide membrane-located receptors for their respective germinants; the mechanism of signal transmission is not known. Other *ger* loci encode proteins that are involved in the response to both L-alanine and GFAK; germination is thought to be a multistep process.

The events of germination happen very rapidly, and it is difficult to fit a precise chronology to them. This task is complicated by the heterogeneity of the response within the spore population. The heterogeneity can be reduced, but not eliminated, by first "activating" spores (a pregermination treatment) with a treatment such as mild heat. The earliest germination events include the release of DPA into the medium and the loss of heat resistance. Hydrolysis of the cortex is rapidly initiated by germination-specific lytic enzymes. All this occurs in the absence of macromolecular synthesis—thus it is not dependent on transcription. There is little ATP in the dormant spore, and energy is initially provided by phosphoglyceric acid; ATP synthesis is one of the earliest events that is initiated on addition of germinant. The germinated spore is phase-dark. During outgrowth, the germinated spore swells considerably; protein, RNA, and DNA synthesis is initiated, and the first vegetative division usually occurs about 2 hr after the initiation of germination, given that a growth medium is provided. The organism has now returned to vegetative growth.

See Also the Following Articles

Bacillus subtilis, Genetics • Clostridia

Bibliography

Driks, A., Roels, S., Beall, B., Moran, C. P., Jr., and Losick, R. (1994). Subcellular localization of proteins involved in the assembly of the spore coat of *Bacillus subtilis*. *Genes Dev.* **8**, 234–244.

Errington, J. (1993). *Bacillus subtilis* sporulation: Regulation of gene expression and control of morphogenesis. *Microbiol. Rev.* **57**, 1–33.

Foster, S. J., and Johnstone, K. (1989). The trigger mechanism of bacterial spore germination. *In* "Regulation of Procaryotic Development" (I. Smith, R. A. Slepecky, and P. Setlow, eds.), pp. 89–108. American Society for Microbiology, Washington, D.C.

Henriques, A., Melsen, L. R., and Moran, C. P., Jr (1998). Involvement of superoxide dismutase in spore coat assembly in *Bacillus subtilis*. *J. Bacteriol.* **180**, 2285–2291.

Perego, M., and Hoch, J. A. (1996). Protein aspartate phosphatases control the output of two-component signal transduction systems. *Trends Genet.* **12**, 97–101.

Piggot, P. J. (1996). Spore development in *Bacillus subtilis*. *Curr. Opin. Genet. Dev.* **6**, 531–537.

Piggot, P. J., Moran, C. P., Jr., and Youngman, P. (eds.) (1993). "Regulation of Bacterial Differentiation." American Society for Microbiology, Washington, D.C.

Popham, D. L., Helin, J., Costello, C. E., and Setlow, P. (1996). Muramic lactam in peptidoglycan of *Bacillus subtilis* spores is required for spore outgrowth but not for spore dehydration or heat resistance. *Proc. Natl. Acad. Sci. U.S.A.* **93**, 15405–15410.

Stragier, P., and Losick, R. (1996). Molecular genetics of sporulation in *Bacillus subtilis*. *Annu. Rev. Genet.* **30**, 297–341.

Staphylococcus

John J. Iandolo

University of Oklahoma Health Sciences Center

I. Cell Surface
II. Staphylococcal Genetics
III. Extracellular Proteins
IV. Exoprotein Gene Expression
V. Staphylococcal Disease

GLOSSARY

capsule Polysaccharide outermost layer of the cell. Production of the capsule imparts a viscous, slimy look to colonies.

cell wall One of the outer layers of the bacterial cell that protects the cell from osmotic perturbations and provides mechanical protection to the fragile cellular membrane.

endonuclease Class of enzymes that have specific binding sites of varying complexity at which they sever phosphodiester bonds of DNA.

plasmid An autonomously replicating small extrachromosomal DNA molecule.

superantigen Any of a number of proteins that elicit a massive T cell receptor Vβ-restricted primary response.

THE GENUS *STAPHYLOCOCCUS* is defined in *Bergey's Manual of Determinative Bacteriology* as a member of the family Micrococcaceae. There are presently 32 recognized species of staphylococci (Table I), 13 of which are indigenous to humans with the remainder associated with various nonprimate animals. Staphylococci are gram-positive cocci that occur singly and in pairs in liquid media and in clusters when grown on solid media. Over the years, they have been characterized by their variety of colonial, morphological, and biochemical activities, which has resulted in description of several biotypes of variable stability. They are aerobic or facultatively anaerobic, are catalase positive, and are capable of generating energy by respiratory and fermentative pathways. These organisms are nutritionally fastidious with complex nitrogen requirements. Most species require several amino acids, vitamins (thiamine and niacin), and uracil (to grow anaerobically) for growth. In complex, nutritionally complete growth media, the organism is able to grow at generation times $\leq$20 minutes, a rate comparable to that of *Escherichia coli.*

At the molecular level, the genus can be distinguished from other members of the Micrococcaceae by the low G + C content of its DNA (which ranges from 30 to 38%), the presence of teichoic in their cell wall, the production of catalase, and their ability to tolerate extremely low water activities. *Staphylococcus aureus* is routinely cultured at an a_w of 0.88 (15% NaCl), and cells have been reported to grow at as low as $a_w = 0.83$ (saturated NaCl solution). Anaerobiocally, tolerance to low a_w is less, with the limit at 0.90. The normal habitat of these organisms is the skin, skin glands, and mucous membranes of warm-blooded animals. However, staphylococci can be isolated from a variety of sources, including soil, dust, air, water, and food and dairy products. As a result they present a food and water public health hazard, as well as an infection risk.

I. CELL SURFACE

A. Cell Wall

The cell wall of *Staphylococcus* is a thick, electron-dense structure that provides great mechanical support to the cell. It is composed of a giant polymer consisting of peptidoglycan complexed with teichoic

TABLE I
Recognized Species of *Staphylococcus*

S. aureus	*S. epidermidis*
S. capitis	*S. warneri*
S. haemolyticus	*S. hominis*
S. saccharolyticus	*S. auricularis*
S. saprophyticus	*S. caseolyticus*
S. cohnii	*S. kloosii*
S. xylosus	*S. simulans*
S. carnosus	*S. intermedius*
S. hyicus	*S. chromogenes*
S. sciuri	*S. lentus*
S. gallinarum	*S. lugdunensis*
S. pasteuri	*S. caprae*
S. equorum	*S. arlettae*
S. felis	*S. piscifermentans*
S. schleiferi	*S. delphini*
S. muscae	*S. vitulus*

acid and other surface proteins described below. It is a heteropolymer consisting of glycan chains cross-linked through short peptides. The repeating unit in the glycan backbone is β-1,4-*N*-acetylglucosamine and *N*-acetylmuramic acid (muramic acid). About 60% of the *N*-acetylmuramic acid residues are O-acetylated. The high level of O-acetylation makes the cell wall resistant to lysozyme digestion. The carboxyl group of muramic acid is substituted by an oligopeptide that contains alternating L- and D-amino acids (L-Ala, D-Gln, L-Lys, D-Ala, D-Ala). In *S. aureus,* adjacent polypeptides are cross-linked by pentaglycine cross-bridges between the ε-amino of lysine and carboxyl-terminal D-alanine. In other species, the composition of the cross-bridge is variable.

The teichoic acid component is linked to the D-alanine component of the mucopeptide by α or β glycosidic linkage through *N*-acetyl-D-glucosamine. In *S. aureus,* the teichoic acid backbone is ribitol based, whereas in *S. epidermidis* it is glycerol based. In other species, glycerol teichoic acids are more common than their ribitol counterparts.

B. Capsule

Eleven capsular polysaccharide serotypes have been described for *S. aureus*. The most commonly occurring serotypes among clinical isolates are type 5 and type 8. The main components of the capsules are *N*-acetylaminouronic acids and *N*-acetylfucosamine. The genes for capsule production have been identified and are located in a single operon.

Encapsulated staphylococci are resistant to phagocytosis. Antibodies to capsular polysaccharides neutralize the antiphagocytic properties of the capsule and opsonize the cell for phagocytosis. Opsonization has made the capsule a prime target in the search for an effective staphylococcal vaccine.

II. STAPHYLOCOCCAL GENETICS

A. Genome

The primary strain of *S. aureus* used for genetic manipulation and gene discovery is NCTC8325 and its derivatives. Also known as PS47, this is the propagating strain for the typing bacteriophage ϕ47 and is a member of phage group III (see below). It is routinely used to generate batches of bacteriophage used for typing purposes. It is, however, lysogenized by three temperate phage, ϕ11, ϕ12 and ϕ13. A derivative (8325-4) has been cured of all demonstrable phage and is the prototype strain whose genome has been used to build a genetic map. The current map is based on genetic and physical parameters. Genetic markers have been placed on a marco restriction map generated by digestion with the restriction endonucleases *Sma*I, *Asc*I, *Csp*I, and *Sgr*A1. These enzymes recognize GC-rich restriction sites and cut the staphylococcal genome sparingly (for each enzyme there are, respectively, 16, 6, 8, and 8 sites and fragments). For more detailed information, the reader is directed to Iandolo *et al.* (1997). Presently, the genomes of NCTC8325 and a methicillin-resistant clinically relevant variant (strain COL) are being sequenced. The NCTC8325 data are freely available on the World Wide Web at www.genome.ou.edu and for strain COL at www.tigr.org.

The genome is circular and consists of 2768 kb in phage group I, 2771 kb in phage group II, 2800 kb in phage group III, and 2586 kb in phage group IV. Other than phage group III stains, little is known of the genomic organization of the phage groups, although preliminary genome maps are available for phage groups I and II. However, there is considerable information available regarding the genomic organization of phage group III strains. Genetic and sequence data indicate that in addition to the normal complement of housekeeping genes, the chromosome contains many accessory genetic elements that are not necessary for growth under laboratory conditions. There are no allelic counterparts in sensitive or negative strains for the newly described pathogenicity islands SaPI1 and SaPI2, which encode the toxic shock syndrome toxin gene. Similarly, the elements encoding staphylococcal enterotoxin B, methicillin resistance, and the converting and other temperate phage represent unique DNA. The genes and genomic organization of *S. aureus* bear a striking similarity to the genome of *Bacillus subtilis* such that the organism has been called a morphologically degenerate form of *Bacillus*.

The staphylococcal genome also contains transposable elements such as Tn*551*, insertion sequences IS256, IS257, and IS1181, and others. Tn*551* and its close relative Tn*917* have been extensively utilized to generate knock-out mutations in staphylococci, many of which have been mapped and localized near other genes.

B. Plasmids

Staphylococci are also endowed with a generous array of plasmids ranging in size from a few kilobases to 40–50 kb. Thus far, there are seven recognized incompatibility groups, but the replication characteristics of many plasmids remain undertermined. Replication of plasmids belonging to incompatibility groups I and II is stringently controlled and copy number is low, whereas that of the remaining groups is relaxed, with correspondingly higher copy numbers. Although a few of the smaller plasmids are cryptic, most encode resistance to one or more antibiotics. Among the large plasmids, the majority encode resistance to penicillin and/or heavy metals; a few that encode gentamicin resistance have been shown to carry conjugative functions and can mobilize themselves and some small plasmids. It is widely held that plasmids are exchanged among staphylococci, streptococci, and bacilli, accounting for the presence of the same or similar plasmids in each group.

C. Staphylococcal Bacteriophages

Staphylococcus aureus was among the first organisms used to demonstrate the existence of bacteriophages. In fact, intensive study of staphylococcal bacteriophages has led to the establishment of a complex method of bacteriophage typing of disease isolates identified in epidemiological investigation. A panel of 21 bacteriophages called the International Typing Series of bacteriophages is used to define strain types by specific patterns of susceptibility to these typing phages. Phage typing led to the identification of five strain groups (I–IV and miscellaneous).

Most pathogenic strains belong to groups II and III, but all phage groups are capable of causing disease. More recently, phage typing has lost favor because of the finding that the standard bacteriophage-propagating strains carry temperate bacteriophages that are distinct from the typing phage they propagate. The phage preparations derived from these strains are therefore mixtures of several bacteriophages. In fact, virtually all strains of *S. aureus* are lysogenized by at least one bacteriophage, and many carry multiple temperate phages in their genome. Some of the lysogenizing phages carry additional genes that they bring in to the recipient strain and alter its phenotype by inserting into the genome at specific attachment sites located within genes. Both negative conversion [the inactivation of expressed genes (e.g., lipase and beta-toxin)] and positive conversion [the introduction of newly expressed genes (staphylococcal enterotoxin A and staphylokinase)] are common occurrences in staphylococci. Temperate bacteriophages have been shown to alter the restriction pattern of the genome by introducing additional restriction sites that produce restriction fragment length polymorphisms (RFLPs) when analyzed by clamped homogeneous electrophoretic field (CHEF) gel electrophoresis. The creation of additional sites has caused ambiguities and difficulties in establishing clonal derivations in many epidemiological investigations of disease outbreaks.

Bacteriophages are also thought to play a major role in the transmission of virulence factors and in the occurrence of genetic rearrangements. They are responsible for the only naturally occurring means of genetic exchange in the staphylococci. The first reports of transduction in staphylococci occurred in 1960 and concerned the transmission of resistance to penicillin. Since then many detailed studies of the transductional transmission of genes have been published. Most staphylococcal bacteriophages are generalized transducers, capable of transferring about 40–45 kb of DNA (a headful) to a suitable recipient strain. DNA is packaged from specific sequences called PAC sites that are randomly dispersed around the genome.

Bacteriophages also play a role in transformation after induction of competence by Ca^{2+} treatment. Both plasmid and chromosomal DNA can be transferred by transformation, albeit at low frequency. In order for transforming DNA to be recombined, it is necessary that the recipient strain be lysogenized by a serogroup B bacteriophage (there are 12 serogroupings).

Progress in genetic mapping of the chromosome has been painstakingly slow due to the low frequency of gene transmission by either transduction or transformation. Other procedures, such as electroporation, enjoy limited effectiveness for plasmid transfer, but they are not useful for transfer of chromosomal genes.

III. EXTRACELLULAR PROTEINS

Staphylococci produce 30 or more extracellular and cell surface proteins. A large number of these are encoded on plasmids and other accessory elements. Therefore, each strain exhibits a variable array of toxins, enzymes, and other factors. A list of some of these exoproteins is presented in Table II. The extracellular proteins are subdivided into cell surface oriented and soluble. All the exoproteins are translated as precursor proteins with signal peptides that are removed at secretion. In addition, the cell surface proteins also possess a characteristic amino sequence motif (LPXTG) at the C terminus which precedes the membrane spanning region and serves as an anchor, linking the protein to the cell wall peptidoglycan. Both types of exoproteins are secreted by type II secretory mechanisms involving the SecYEG pathway. With the exception of certain bacteriocins, there is no evidence at this time that extracellular proteins are secreted by other secretory pathways.

TABLE II
Extracellular Proteins of *Staphylococcus aureus*

Protein	*Gene locus*
Hemolysins	
Alpha toxin	Chromosome
Beta toxin	Chromosome
Gamma toxin	Chromosome
Delta toxin	Chromosome
Leukocidin	Chromosome
Staphylococcal Enterotoxins (SEs)	
SEA	Bacteriophage/chromosome
SEB	Chromosome/pathogenicity island
SEC	Plasmid
SED	Plasmid
SEE	Chromosome
SEG	Chromosome
SEI	Chromosome
Enzymes and other toxins	
Lipase	Chromosome
Nuclease	Chromosome
Proteases	Chromosome
Esterase	Chromosome
Coagulase	Chromosome
Staphylokinase	Bacteriophage/chromosome
Protein A	Chromosome
Phospholipase C	Chromosome
Exfoliative toxin A	Chromosome
Exfoliative toxin B	Plasmid
Toxic shock syndrome toxin-1	Chromosome/pathogenicity island
MSCRAMMs	
Clumping factor	Chromosome
Fibronectin binding proteins A/B	Chromosome
Fibronectin binding protein	Chromosome
Collagen binding protein	Chromosome
Elastin binding protein	Chromosome
Matrix adhesin factor	Chromosome

A. Soluble Exoproteins

Soluble exoproteins include a wide array of toxins and enzymes such as food poisoning enterotoxins, exfoliative toxins, toxic shock syndrome toxin, hemolysins, coagulase proteases, lipases, and other enzymes. The exfoliative toxins are probably proteolytic and attack the epidermis of susceptible animals; the hemolysins are membrane damaging proteins whose activity is mediated through pore formation or lipolytic action. Many of the enzymes such as lipase, nuclease, and proteases are presumed to enhance invasiveness through tissue destruction.

Several of the soluble exoproteins including enterotoxins, exfoliative toxins, and toxic shock toxin are superantigens owing to their ability to stimulate mitogenic activity and cytokine production for a wide array of T-lymphocyte halplotypes. They are able to activate specific sets of T-lymphocytes by binding to major histocompatibility complex class II (MHC II) proteins. They bind to the variable region of the T cell receptor β chain. The activated cells proliferate and release cytokines/lymphokines, interferon-γ (IFN-γ), and interleukins. Because they exhibit this broad based activity, they have been called superantigens. This activity is suspected to enhance virulence by suppressing the host's response to staphylococcal antigens produced during infection.

B. Cell Surface Proteins

A major class of cell surface proteins are adhesins called MSCRAMMs (*m*icrobial *s*urface *c*omponents *r*ecognizing *a*dhesive *m*atrix *m*olecules). These molecules comprise the main adhesins of the organism and include collagen binding protein, fibronectin binding proteins, fibrinogen binding protein, elastin binding protein, clumping factor, and the matrix adhesin factor. There may be as many as 12 other surface proteins that contain membrane anchor domains and potentially qualify as MSCRAMMs. A second group of cell surface proteins includes nuclease and protein A. Staphylococcal nuclease is a thermally stable endonuclease able to withstand boiling for 30 minutes without significant loss of activity. Protein A is able to bind to the nonantigenic F_c fragment of immunoglobulin G, causing the complex to precipitate. Its role in virulence is thought to be in escape from immune surveillance.

IV. EXOPROTEIN GENE EXPRESSION

The expression of extracellular proteins is largely under the influence of a master genetic circuit called *agr* (accessory gene regulator). This signaling arm of the operon (*agrBDCA*) is activated by a quorum sensing mechanism that depends on the accumulation of an activating octameric peptide (processed from the AgrD precursor by AgrB). The peptide triggers increased expression of the entire operon via an integral signal transduction pathway (AgrC, AgrA) upregulating production of the octamer and activating a second promoter which produces a unique regulatory molecule, RNAIII. A second locus, *sar*, modulates expression of the *agr* locus by binding to the promoter of AgrBDCA. It also plays a further unknown role in exoprotein gene expression. RNAIII is the effector molecule for regulated protein expression. It is not translated, nor does it bind to the promoter regions of regulated genes. It presumably interacts with other genes, in an unknown way, as both a positive and negative regulator of exoprotein gene expression. RNAIII is required for the expression of soluble exoproteins and represses the expression of cell surface proteins. Because a threshold level of octapeptide is required for activation, RNAIII is not expressed until late in growth. Therefore, cell surface proteins are produced early, presumably to allow the organism to attach and colonize. The RNAIII-induced activation of soluble protein genes results in a necrotic effect allowing the organism to invade to deeper tissues and become bacteremic.

V. STAPHYLOCOCCAL DISEASE

A. Carriage

Staphylococci are one of the major groups of organisms that inhabit the skin of mammals. Usually several different strains are found on the same host. They may be present as transient contaminants, short-term replicating residents, or long-term colonizers that persist for long periods. The majority of species are opportunistic pathogens that become infectious when the skin or mucous membranes are compromised by trauma, inoculation by needle, or direct implantation of medical devices (foreign bodies). Staphylococci are then able to attach, colonize, and produce toxic substances that destroy host tissues.

B. Disease Types

In general, three types of diseases are usually associated with infection by *S. aureus*. These are characterized as superficial or cutaneous infections such as pimples, boils, and toxic epidermal necrolysis; systemic infections such as heart valve disease, bacteremia, and osteomyelitis; and toxinoses as food poisoning and toxic shock syndrome.

The major species associated with disease in humans are *S. aureus* and *S. epidermidis*. These species are basically identical except for the ability of *S. aureus* to produce the fibrinogen clotting substance coagulase. The coagulase-negative species are largely commensal and can be isolated in high number from all areas of human skin. It was proposed by Baird-Parker that coagulase distinguished pathogenic from nonpathogenic species, and this differentiation was used until the mid-1970s. More recent evidence has shown, however, that coagulase-negative staphylococci (CoNS) are a major cause of wound infections and infections caused by foreign bodies including intravascular catheters, peritoneal dialysis catheters, prosthetic heart valves, joint prostheses, pacemaker electrodes, and fluid shunt systems. Therefore, coagulase is no longer considered an exclusive indicator of pathogenicity.

Species of *Staphylococcus* are the leading cause of nosocomial infection. Each year about 2 million hospitalizations in the U.S. result in nosocomial infection, and approximately 50% are due to *S. aureus* and *S. epidermidis*. Community acquired infections have a similar incidence rate. A predisposing condition has been the acquisition of multiple drug resistance by *S. aureus,* which has caused the incidence of nosocomial and community acquired infection to increase steadily since the 1960s.

C. Immunity

Immunity to staphylococcal infections is poorly understood. Normal healthy humans have a high degree of innate resistance to invasive infections. Experimental infections are difficult to establish in animals and require large inocula containing millions of organisms. In humans, the organism is able to colonize mucosal and epidermal surfaces with little resistance, and so long as they remain intact, these barriers are the main source of natural immunity to infection. After invasion, however, phagocytosis by polymorphonuclear leukocytes is the main humoral defense.

Because of repeated exposure of animals to *S. aureus* and *S. epidermidis* in natural settings, antibodies to various components of the cell and its products (both cell surface and soluble) are prevalent in animals. Nevertheless, with the exception of toxic shock syndrome where antibody is an important factor in immunity, serological studies have not successfully related immunity and antibody titer. Moreover, prior infection fails to elicit immunity to reinfection. In spite of these drawbacks, vaccine research is being strongly pursued. Among the vaccines being studied, the most promising are capsular components perhaps conjugated with exoprotein or MSCRAMM antigens, which may prove effective in promoting immunity.

D. Identification

The clinical laboratory must be able to identify *S. aureus* quickly and accurately. Many types of clinical samples have been utilized for detection. Samples that are heavily contaminated with other bacterial flora (i.e., nasal, skin, or wound specimens) can be grown on solid media or liquid media containing various selective agents which exploit the resistance of staphylococci to NaCl, chemicals such as potassium tellurite and lithium chloride, or antibiotics. Suspect colonies from the primary isolation are then subjected to antibiotic sensitivity determination to provide treatment alternatives. Coagulase determination, an important consideration for treatment decisions regarding coagulase-negative strains (CoNS), is usually carried out. A variety of other phenotypic markers can be tested if further characterization is desired. These include alpha and beta hemolysins, nuclease, proteases, lipase, protein A, and the specific toxins.

E. Treatment

More than 95% of patients with *S. aureus* infections worldwide do not respond to first-line antibiotics such as penicillin and ampicillin. These strains have been routinely treated with methicillin, but resistance appears in about 30% of infections (methicillin-resistant *Staphylococcus aureus,* or MRSA). Vancomycin is the most effective treatment for multiply resistant strains, but the recent emergence of vancomycin resistance in staphylococci poses a critical problem to effective treatment of these infections. The emergence of strains that are resistant to multiple antibiotics, especially the recent appearance of vancomycin resistance, has spurred renewed interest in antibiotic discovery and vaccine therapy. New drug targets are being investigated to devise novel families of antibiotics.

See Also the Following Articles

Cell Walls, Bacterial • Mapping Bacterial Genomes • Plasmids, Bacterial • Protein Secretion • Skin Microbiology

Bibliography

Iandolo, J. J. (1989). Genetic analysis of extracellular toxins of *Staphylococcus aureus. Annu. Rev. Microbiol.* **43**, 375–402.

Iandolo, J. J., Bannantine, J. P., and Stewart, G. C. (1997). Genetic and physical map of the chromosome of *Staphylococcus aureus. In* "The Staphylococci in Human Disease" (K. B. Crosley and G. L. Archer, eds.), pp. 39–54. Churchill Livingstone, New York.

Leung, D. Y. M., Huber, B. T., and Schlievert, P. M. (1997). Historical perspective of superantigens and their biological activities. *In* "Superantigens: Molecular Biology, Immunology and Relevance to Human Disease" (D. Y. M. Leung, B. T. Huber, and P. M. Schlievert, eds.), pp. 1–14. Dekker, New York.

Novick, R. P. (1991). Genetic systems in staphylococci. *In* "Methods in Enzymology" (Bacterial Genetics) (J. Miller, ed.), Vol. 204, pp. 587–636. Academic Press, Orlando, Florida.

Starvation, Bacterial

A. Matin
Stanford University

I. Importance of Starvation
II. Appearance of Starving Bacteria
III. The Starvation Proteins
IV. How Starvation Proteins Protect
V. The Starvation Sigma Factors
VI. The Role and Regulation of σ^S
VII. The σ^S-Dependent Promoters
VIII. Starvation and Virulence
IX. Future Perspectives

GLOSSARY

ancillary factors Proteins that influence promoter recognition by an RNA polymerase.

chaperones Proteins that are required for correct folding of newly synthesized proteins. They also prevent protein denaturation during stresses, and can renature damaged proteins.

chemostat An apparatus that makes it possible to grow bacteria under steady-state conditions at submaximal growth rates due to low concentrations of an essential nutrient.

promoters Sequences upstream of the transcriptional start site of a gene (usually −10 and −35 nucleotides upstream of the start site) recognized by individual species of RNA polymerase holoenzymes.

semistarvation Conditions under which bacteria grow at a rate less than their maximal potential due to the limitation of an essential nutrient. Existence under complete or semistarvation is the norm for bacteria in nature.

sigma factors Small proteins that combine with the RNA polymerase core enzyme. The resulting RNA polymerase holoenzyme can transcribe various genes. Each species of RNA polymerase holoenzyme recognizes specific promoter sequences.

starvation An environmental condition in which bacteria do not grow at all due to the lack of an essential nutrient.

starvation promoters Promoters that are selectively switched on during starvation or semistarvation conditions. Their sequences are recognized by starvation specific RNA polymerases.

starvation proteins Proteins whose levels either go up or that are uniquely synthesized in starving bacteria.

vegetative cells Nonstarved, actively growing bacterial cells.

virulence Cellular features that enable a bacterium to cause disease.

STARVATION which is frequently experienced by bacteria, causes them to differentiate into forms that are much more resistant to killing. This results from the synthesis of special proteins, called the starvation proteins. These proteins strengthen the bacterial cell envelope, prevent damage to vital cell constituents, and enhance the cell's capacity to repair DNA and essential proteins. A secondary sigma factor, called σ^S, becomes stabilized in starving cells. Its concentration increases, leading to the formation of a new species of RNA polymerase. The latter recognizes the regulatory region of the starvation genes, increasing their transcription. Starved cells are probably also enhanced in their disease-causing ability.

I. IMPORTANCE OF STARVATION

Bacteria in nature mostly exist in a state of semi- or complete starvation because most natural environments are deficient in nutrients. The available food for bacteria in the oceans is only a fraction of a milligram per liter (6–10 μg/L in freshwater, 0.4 g/

100 g of soil), and although the quantitative aspects are not known, it is likely that disease-causing bacteria also experience nutrient deprivation while colonizing their host. In nearly all of these environments, bacteria on average grow only at a fraction of the rate of which they are genetically capable—indeed, estimated growth rates in many natural environments can range from close to zero to a generation time of hundreds of days.

Within the constraints of their genetic endowment, bacterial characteristics can very enormously depending on their growth conditions. To understand what bacterial physiology is like under natural conditions, there is increasing interest in studying bacterial characteristics under partial or complete starvation conditions. An elucidation of these characteristics is a prerequisite for purposeful manipulation of bacteria toward beneficial ends. Indeed, insights gained from such studies promise to provide radically new approaches for environmental cleanup, as well as microbial containment. Completely starving bacteria are studied in the laboratory utilizing flask (batch) bacterial cultures that enter the stationary phase due to exhaustion of an essential nutrient. For studies on semistarving bacteria, the chemostat is the instrument of choice.

II. APPEARANCE OF STARVING BACTERIA

With respect to changes in shape produced by starvation, bacteria can be divided into two major groups, those exhibiting a very marked morphological differentiation and those showing only a minimal alteration. The former group is exemplified by species of the genus *Bacillus* which, on starvation, form structures called endospores (Fig. 1a) that appear markedly different from their actively growing counterparts. Another example is provided by the myxobacteria, which form fruiting bodies (Fig. 1b) that differ strikingly from their vegetative cells. In the second group, which encompasses a majority of bacteria, starvation-induced changes are confined to diminution of cell size, protoplast shrinkage with consequent periplasm enlargement, and nucleoid condensation (Fig. 1c). In this article, I consider only the second group. Because of lack of pronounced morphological changes in the bacteria, it was recognized only relatively recently that this group, too, undergoes a profound alteration in its gene expression pattern on starvation. Studies involving this group of bacteria have concentrated mainly on *Escherichia coli*, although many other bacteria have also been examined in this respect, namely, *Vibrio*, *Salmonella*, *Pseudomonas*, and others.

III. THE STARVATION PROTEINS

Use primarily of two-dimensional gel electrophoresis showed that when bacteria such as *E. coli* enter the stationary phase, they increase the synthesis of a number of proteins, many of which are unique to the starvation state. These proteins are called starvation proteins. They are synthesized mainly in the first few hours of starvation and fall into early, middle, and late temporal groups. Their synthesis is accompanied by a progressive increase in the resistance of the cells to a variety of stresses. Examples of the stresses to which the starved cell becomes more resistant include starvation itself, hostile temperature, oxidative or osmotic state of the environment, and deleterious chemicals such as chlorine. In other words, fully starved cells of *E. coli* (i.e., those starved for about 4 hr) exhibit a marked general resistance.

Starvation is a composite stress. One of its effects is to diminish the cellular redox status, generating oxidative stress. The dearth of ATP and lowered proton-motive force undermine ion transport. Maintenance of pH homeostasis is thereby rendered more difficult, exposing the cell to acid or alkaline stress; and the inability to concentrate ions produces osmotic stress. Survival in the face of these assaults necessitates induction of chaperones (see below) to prevent cellular damage and of repair mechanisms capable of operating in energy-depleted cells. Survival during starvation thus requires mechanisms to resist along with starvation, also its constituent stresses; it is therefore logical that a starved cell exhibits enhanced resistance to multiple stresses.

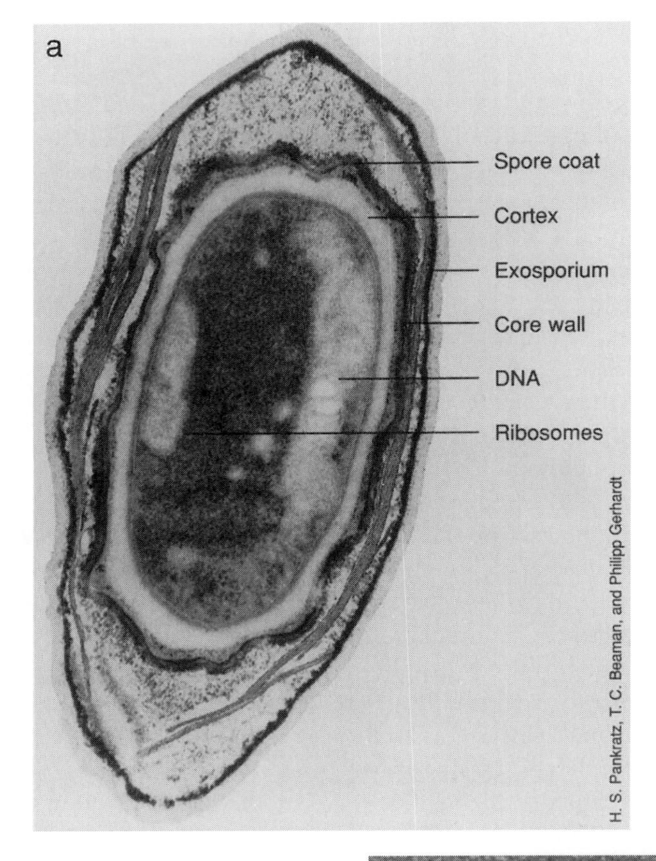

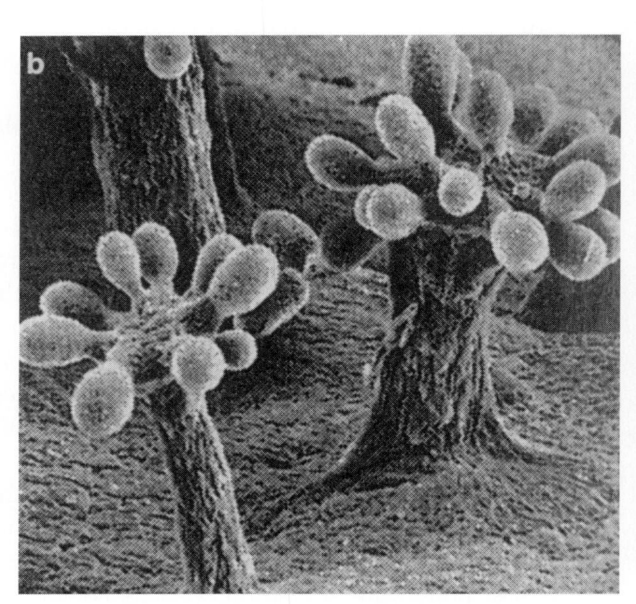

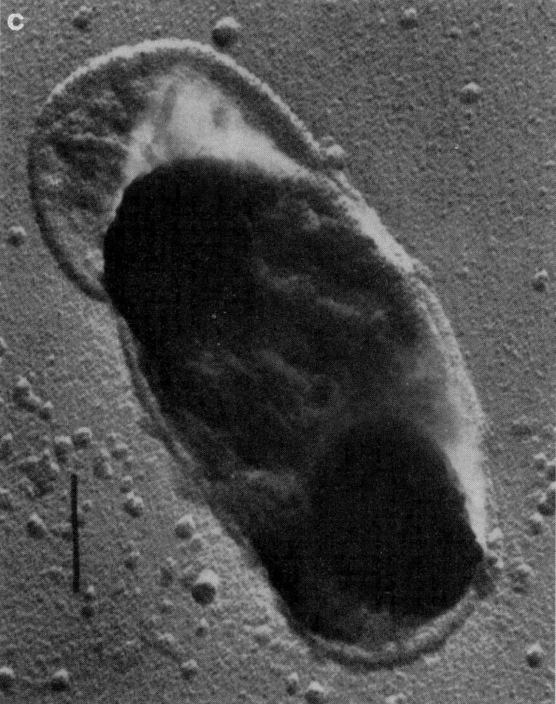

Fig. 1. (a) A bacterial endospore. Reproduced with permission from Brock *et al.* (1997). Prentice-Hall, Upper Saddle River, NJ. (b) A bacterial fruiting body. Reproduced with permission from Brock *et al.* (1997). Prentice-Hall, Upper Saddle River, NJ. (c) A starved *E. coli* cell. Reproduced with permission from Reeve *et al.* (1984). *J. Bacteriol.* 160, 1041–1046. American Society for Microbiology, Washington, D.C.

IV. HOW STARVATION PROTEINS PROTECT

A. Metabolic Amplification

The starvation proteins help the cell survive a dearth of nutrients in two broad ways. A number of these proteins are concerned with increasing the scavenging capacity of the cell for the missing nutrient. For example, under the scarcity of glycerol as carbon substrate, *E. coli* synthesizes a different pathway for its uptake and metabolism. The key enzyme of this pathway is glycerol kinase, which utilizes an ATP molecule for glycerol uptake and has a high affinity for this compound. Similarly, under phosphate starvation, a high affinity phosphate transport system is induced (PST), and when potassium is scarce the high affinity KDP system for K^+ uptake is utilized. Another strategy to escape starvation is synthesis of proteins that enable the cell to utilize substrates other than the one which became limiting. An example is the induction of the CstA protein in *E. coli* when glucose runs out. This protein appears to be concerned with the utilization of peptides, which are abundant in the gut, thereby enabling the bacterium to avoid starvation when the supply of glucose and other carbohydrates runs out (Matin, 1996).

B. Stress Resistance

Several categories of starvation proteins are known with specific protective roles in starvation.

1. *Cell Envelope Protection*

BolA is a starvation protein that regulates the expression of several genes. The product of one of these genes, DacA (D-alanine carboxypeptidase) increases the degree of peptidoglycan cross-linkage, thereby increasing the strength of the cell wall of the starved cells. The proteins encoded by the *otsBA* (*pexA*) operon promote trehalose biosynthesis; trehalose is believed to strengthen the cell membrane; it also provides protection against osmotic stress by acting as a compatible solute (Hengge-Aronis, 1996).

2. *DNA Protection*

As mentioned above, oxidative stress is likely to increase in a starved cell. Consistent with this expectation is the fact that starvation proteins include enzymes that can destroy oxidizing agents, as well as those that can repair oxidative damage. Among the starvation proteins that can destroy H_2O_2 are hydroperoxidase I and II (also referred to as KatG and KatE, respectively). The proteins concerned with repairing the effects of oxidative damage include exonuclease III (also referred to as XthA) and AidB, both involved in DNA repair. PexB (DpS) is another such protein that may be involved in inducing additional proteins concerned with protecting DNA against oxidative stress.

3. *Protein Protection*

A large number of starvation proteins are concerned with preventing denaturation of cellular proteins as well as with renaturation of damaged proteins; these are termed chaperones. As can be expected, protein denaturation is a feature of nearly all stresses. Excessive heat exposure, an altered ionic cellular milieu (which interferes with cellular pH and turgor pressure maintenance), oxidative stress, as well as starvation all share the common feature of causing protein denaturation. Increased synthesis of chaperones—the so-called heat-shock proteins—is a common feature of all these stresses. The chaperones can both minimize damage to cellular proteins as well as rescue proteins that get denatured despite their increased levels during stress.

a. Chaperones Concerned with Protein Conformation

Stress-denatured firefly luciferase can be reactivated in *E. coli* but only if the chaperones DnaK, DnaJ, and GrpE had been present during stress exposure. If human growth hormone (HGH) is overproduced in *E. coli,* it denatures, precipitating out as inclusion bodies. But if HGH production occurs in the presence of increased DnaK levels, HGH denaturation is minimized. Renaturation of damaged proteins by DnaK has been demonstrated *in vitro,* and for yeast Hsp 104 *in vivo.*

Chaperones are essential in bringing about correct folding of proteins under all conditions, and their repair function is probably an extension of this role. For the DnaJ, DnaK, and GrpE chaperones, the proposed mechanism envisages binding of DnaJ with the unfolded (or misfolded) protein, followed by interaction with the DnaK–ATP complex. ATP hydrolysis provides the energy for the protein–DnaJ–DnaK–ADP complex formation. GrpE exchanges ATP for ADP in DnaK. This exchange leads to the disintegration of the ternary complex, releasing the protein in an appropriately folded, or partially so, state. If the latter, the cycle continues until proper folding is attained. In starved cells, a close analog of DnaK, CbpA, is also induced. The latter may be more efficient at renaturation.

b. Chaperones Concerned with Covalent Damage

The chaperones discussed above deal directly with protein conformation. Another class is concerned with restoring correct conformation by repairing stress-induced covalent damage. For example, stresses can alter proline peptide bonds from their normal *trans* configuration to the *cis* form. The abnormal *cis* configuration introduces a "kink," which interferes with protein folding or refolding. Peptidyl-prolyl-*cis–trans* isomerases (PPIases) can catalyze proline isomerization to restore the normal configuration of these bonds. *Escherichia coli* has nine genes coding for PPIase homologs, which include the starvation gene *surA*. PpiA and PpiB have been shown to promote refolding of denatured proteins *in vitro*. To accomplish this, they require the HtpG chaperone. Another example is L-isoaspartyl-protein carboxylmethyltransferase (ICM), the product of the *pcm* gene. Under stress, asparaginyl residues can be converted to the isoaspartate form, which leads to kink formation and interference with correct folding. ICM can catalyze isoaspartate conversion to aspartate, thus removing the kink.

V. THE STARVATION SIGMA FACTORS

Sigma factors are relatively small proteins that combine with the RNA polymerase core enzyme. The resulting holoenzyme can transcribe genes and recognizes different promoter sequences depending on its sigma factor (Record *et al.*, 1996).

The holoenzyme $E\sigma^{70}$, which plays a major role in gene transcription in unstressed rapidly growing cells, also transcribes several starvation genes. The levels of σ^{70} do not change during starvation; an increase in the cellular cyclic AMP (cAMP) levels causes $E\sigma^{70}$ to transcribe several starvation-related genes, while the transcription of growth-related genes is drastically reduced. In *E. coli* strains missing cAMP (Δ*cya* strains), two-thirds of the starvation proteins are not expressed. These are referred to as the Cst proteins. They are concerned primarily with enabling the cell to escape starvation by the synthesis of high affinity substrate transport and capturing enzymes and with amplifying the quantitative metabolic potential of the cell, as discussed above. Their lack of expression has no effect on the cell's ability to develop starvation-mediated general stress resistance, as shown by the fact that the Δ*cya E. coli* strains are normal in developing such resistance (Matin, 1996). Some *cst* genes may have a protective role under some starvation conditions. Thus, the *cstC* gene codes for *N*-α-acetylornithine-δ-aminotransferase, which appears to be involved in amino acid catabolism in starving cells, and its absence compromises survival during nitrogen starvation.

The starvation proteins whose expression is not hampered in Δ*cya* strains of *E. coli* are termed the Pex proteins—indeed, many of these proteins show a greater induction in the Δ*cya* strains. The Pex proteins include many of the chaperones discussed above. The expression of several *pex* genes depends on RNA polymerase holoenzymes other than $E\sigma^{70}$. In *E. coli,* these holoenzymes are $E\sigma^{S}$, $E\sigma^{32}$, and possibly $E\sigma^{E}$; in *P. putida* $E\sigma^{54}$ is also involved. $E\sigma^{S}$ is the major starvation sigma factor and has received the greatest attention; it will be discussed in the next section. σ^{32} has been studied extensively in the context of the heat-shock response (Yura *et al.*, 1993). However, its levels increase also during starvation, and its dearth from a cell compromises starvation survival by preventing induction of a number of chaperones. What accounts for the increase in σ^{32} levels in starved cells is not known. A potential role for σ^{54} in starvation survival has been shown only in *Pseudomonas putida.* A carbon starvation operon in this bacterium, whose absence compromises starvation survival, is controlled by $E\sigma^{54}$; one of the genes of this operon codes for a G protein. In general Eσ^{54} is responsible for controlling nitrogen metabolism, inducing proteins that help a cell escape nitrogen starvation (Magasanik, 1996). A majority of the Pex proteins are also induced in response to starvation for nutrients other than carbon, as well as in response to other individual stresses, such as heat, hyperosmosis, and oxidation. Thus, it is Pex proteins that confer stress resistance on a starving cell.

VI. THE ROLE AND REGULATION OF σ^S

Mutants in the *rpoS* gene, which codes for σ^S, are rendered unable to induce some 30 starvation proteins, including several of the Pex proteins, and show markedly increased sensitivity to several stresses, both during starvation and rapid exponential growth. The cellular σ^S concentration increases progressively as cells are subjected to increasing degrees of starvation. This is shown in chemostat experiments where the relationship between cellular σ^S levels and different limiting glucose concentrations was directly determined (Fig. 2). What accounts for this increase?

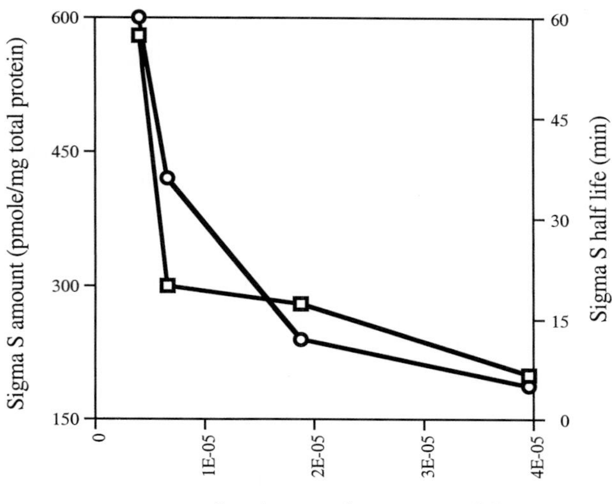

Fig. 2. Steady-state limiting glucose concentrations in growth medium in a chemostat culture versus σ^S concentration (squares) and half-life (circles). Note the exponential notation used in the abscissa; for example, 1E05 indicates a glucose concentration of 10^{-5} *M*. Modified from Zgurskaya *et al.* (1997). *Mol. Microbiol.* **24,** 643–651. Blackwell Science, Oxford.

A. σ^S Synthesis Rates

Although σ^S levels increase during starvation, its synthesis rate actually decreases. Calculations based on *rpoS* mRNA and σ^S protein levels and their half-lives reveal that both the transcription of the *rpoS* gene and the translational efficiency of the *rpoS* mRNA decrease during starvation, resulting in up to an 80% decrease in σ^S synthesis rate (from ~55 in the exponential phase to 13 pmol/mg protein/min in the stationary phase). But a concurrent increase in σ^S stability more than offsets the decreased synthesis, resulting in the observed increased levels of the sigma factor. In starved cells this protein exhibits a 7- to 16-fold greater half-life (Fig. 2) (Matin *et al.*, 1999).

B. The Biochemical Basis of σ^S Stability

The biochemical basis of the σ^S protein degradation is partly understood, but little information is available on its differential stability in exponential versus starved cells. σ^S is degraded by a protease called the ClpXP protease. The protease is made up of two different proteins, ClpP and ClpX. Neither alone has significant proteolytic activity; indeed ClpX by itself acts as a chaperone to rescue certain denatured proteins. Once combined with ClpP, however, ClpX acquires proteolytic activity against specific proteins. ClpP, incidentally, can also pair up with other chaperones, such ClpA, and ClpY to become a protease, but each combination targets different proteins (Gottesman, 1996). Thus, σ^S is degraded by ClpXP but not ClpAP protease, and many targets of the latter are not affected by the former. σ^S protein is also not affected by other proteases of *E. coli,* such as the Lon protease and very likely also the FtsH (HflB) protease. These conclusions are based on a study of *E. coli* mutants. Those devoid of either the *clpP* or *clpX* gene show a stable σ^S and increased levels of this sigma factor in all phases of growth, whereas *lon* or *clpA* mutants behave like the wild type in this respect.

The ClpXP protease levels do not change during transition to the stationary phase, nor does its activity toward another of its target proteins show any alteration during this transition. Thus, it is highly probable that σ^S protein becomes resistant to the protease in the stationary phase. Certain proteins have been found to associate with σ^S only in exponential phase cells, and it is possible that their association puts σ^S in a configuration that is vulnerable to the ClpXP-mediated proteolysis. The situation may be analogous to the postulated mechanism of σ^{32} instability in normal, unstressed cells. It is thought that the association of this sigma factor with certain chaperones (Dnak, DnaJ, and GrpE) in nonstressed cells is responsible for its sensitivity to proteases such as Lon, FtsH, and ClpQY. On heat stress, the chaperones dissociate from the sigma factor, possibly altering its configuration to a form that is resistant to proteolysis (Yura *et al.,* 1993). What proteins may be concerned with altering σ^S stability are not known; they probably do not include the key chaperones like GroEL, DnaJ, GrpE, or CbpA, as mutants devoid of these proteins show a wild-type stability pattern for σ^S. A lack of DnaK, however, is known to destabilize σ^S in the stationary phase.

The ClpXP protease appears to target amino acids 173–188 of the σ^S protein. If these amino acids are deleted from the σ^S protein, it becomes resistant to the protease. This region approximates the site at which σ^S interacts with the RNA polymerase core enzyme. This aspect may have a role in the different sensitivity of σ^S to proteolysis in different phases. Another possibility is that the 173–188 amino acid region of the sigma protein serves as the site recognized by the ClpXP protease. This protease appears to need two sites for recognizing its target protein, one of which (LDA/L) is probably common to all of its substrates, while the other is not; the 173–188 amino acid patch may be the second site needed to target σ^S.

C. Potential Regulation at the Translational Level

The 173–188 amino acid patch of σ^S (the "turnover" site) is encoded by nucleotides 519–564 of the coding region of the *rpoS* gene. Computer-assisted analysis showed that this region is complementary to the translational apparatus (i.e., the Shine–Dalgarno sequence, the start codon, and the surrounding sequences) of *rpoS* mRNA. Bonding of this "antisense element" with the translational apparatus can cause the mRNA to loop on itself, blocking or diminishing translation. Indeed, it was first thought that the major mechanism of changes in σ^S levels during the exponential to stationary phase transition is increased translational efficiency of the *rpoS* mRNA in the stationary phase due to a relaxation in this phase of the mRNA secondary structure.

As the direct measurements of the *rpoS* mRNA translational efficiency during stationary phase transition showed decreased rather than increased efficiency (see above), this is obviously not the mechanism of increased σ^S levels during glucose starvation. But it remains an intriguing possibility that this coincidence—the antisense element of the *rpoS* mRNA encoding the turnover element of the RpoS (σ^S) protein—exists to permit regulation at both the translational and posttranslational levels. σ^S has diverse roles in cell physiology, requiring a strict control of its levels. In situations where, due to conflicting requirements, σ^S sensitivity to ClpXP protease cannot be enhanced, and yet σ^S levels need to be lowered, decreased translational efficiency through the use of the antisense element could indeed be involved. The proteins H-NS and HF-1 may have a role in influencing rpoS mRNA translation. H-NS is a histonelike protein whose synthesis increases in the stationary phase, whereas HF-1 apparently has the capacity to relax helical regions around the mRNA ribosomal binding sites (Matin *et al.,* 1999; Hengge-Aronis, 1996). Similarly, the untranslated RNAs *dsrA* and *oxyS* have been postulated to influence *rpoS* mRNA translation. The stringent response regulator ppGpp also regulates σ^S synthesis, possibly at the transcriptional level.

That translational efficiency changes are responsible for the increased σ^S levels in starving cells was inferred from the behavior of protein fusions of *rpoS* to the *lacZ* gene. Unless a translational fusion contained a certain minimum coding region of the *rpoS* gene, it produced high levels of β-galactosidase in all phases of growth. Only the fusion containing the nucleotides comprising the antisense element (519–

564 nucleotides) mimicked the induction pattern of σ^S during the exponential to stationary phase transition of the wild type bacteria; fusions devoid of this element showed high levels of β-galactosidase in all growth phases. The initial interpretation of these results was that the presence of the antisense element prevented efficient translation in the exponential phase. However, the findings that the antisense element codes for the turnover element of σ^S and that the translational efficiency actually decreases in the stationary phase have led to the conclusion that the low levels of β-galactosidase in the fusions with the antisense element are due to posttranslational regulation. The hybrid RpoS–LacZ proteins formed by these fusions would contain the σ^S turnover region, leading to its degradation by the ClpXP protease. The hybrid protein generated in the fusions devoid of the antisense element would, on the other hand, not be targeted by the protease, and such fusions show high β-galactosidase levels in all phases (Fig. 3) (Matin *et al.*, 1999).

VII. THE σ^S-DEPENDENT PROMOTERS

RNA polymerase holoenzymes exercise their transcriptional selectivity by recognizing different promoter sequences. The −10 and −35 promoter sequences recognized by $E\sigma^{70}$, for example, are quite different from those recognized by $E\sigma^{32}$. Given that $E\sigma^S$ is involved in the expression of genes with a role in starvation and stresses, it was expected that the promoter sequence recognized by $E\sigma^S$ would be unique. It turns out, however, that the genes whose expression is dependent on $E\sigma^S$ possess a consensus −10 sequence that is very similar to that recognized by $E\sigma^{70}$ (CTATACT versus TATAAT); the latter holoenzyme is the major agent of gene transcription in rapidly growing cells. In agreement with this finding, it has been shown by *in vitro* experiments that several promoters can be recognized by both $E\sigma^S$ as well as $E\sigma^{70}$.

What then determines the ability of $E\sigma^S$ to affect differential gene transcription? Only tentative an-

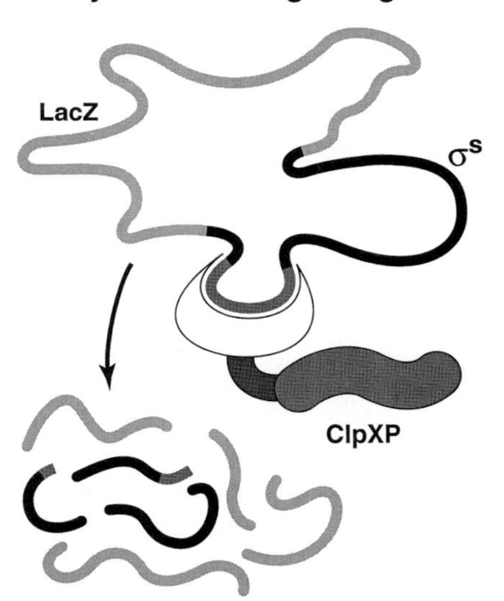

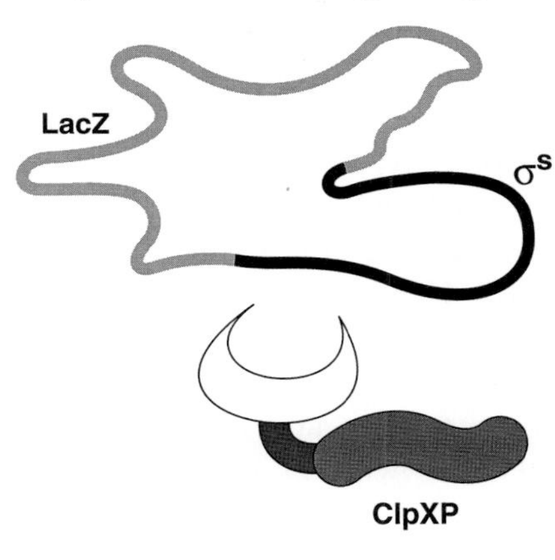

Fig. 3. In fusion strains making the RpoS–LacZ hybrid protein containing the ClpXP target region (left), the hybrid is attacked by the ClpXP protease, accounting for low levels of β-galactosidase in that fusion. Fusion strains making the RpoS–LacZ hybrids without the target region (right) show high levels of β-galactosidase because the hybrids are not attacked by the protease. Reproduced with permission from Matin *et al.* (1999). Survival strategies in the stationary phase. *In* "Microbial Ecology and Infectious Disease" (E. Rosenberg, ed.). American Society for Microbiology, Washington, D.C.

swers can be given to this question at this time. For one thing, there are some differences in the promoter region recognized by Eσ^S and Eσ^{70}. Thus, sequences upstream of the -17 position appear crucial for recognition by the Eσ^S holoenzyme. In addition, at approximately the -35 region, the Eσ^S-recognized promoters possess an AT-rich region rather than the Eσ^{70} consensus -35 sequence. Such a sequence produces curved DNA, and it appears that such a configuration facilitates promoter recognition by Eσ^S. Nor can it be discounted on the present evidence that the -35 sequence of the Eσ^S-recognized promoters is wholly unimportant. Conserved cytosines followed by a -33 guanine appear to be essential for recognition by Eσ^S. More important, however, in determining whether a gene is transcribed by Eσ^S may be factors other than the promoter sequence. These include the ionic composition of the cell (which is likely to be significantly influenced by different stresses), the presence of ancillary factors, stress-induced changes in a core component of RNA polymerase, and the proportion at a given time of the Eσ^{70} and Eσ^S holoenzymes.

The diverse stresses against which Eσ^S-transcribed genes afford protection are probably an ever present and recurrent threat in nature. But while the product of these genes is protective, it is probably inimical to growth. Rapid growth, when circumstances are conducive, is as important to survival as the ability to withstand stresses. Hence there is the need to constantly, and at short notice, shift between expression of Eσ^{70}-and Eσ^S-transcribed genes. This is probably why the two proteins, σ^S and σ^{70}, chemically resemble each other, and why the promoters their holoenzymes recognize differ not so much in their sequence, but rather in such factors as changes in cellular ionic, ancillary factor, and RNA polymerase core component composition.

VIII. STARVATION AND VIRULENCE

It is likely, though not proven, that one of the stresses encountered in the host by an invading bacterium is starvation. Inside the host, bacteria generally grow at a slower rate than their genetic potential permits, and they express many starvation genes. Both of these effects could be due to stresses other than starvation. However, the host mobilizes high affinity ligands to capture essential nutrients, and rapid exponential growth of the invader would soon deplete any environment with finite replenishment capacity, regardless of its initial plenitude. Starvation therefore is likely to be a contributing stress for a pathogenic invader.

The regulation of the expression of virulence genes in many pathogenic bacteria appears to be consistent with this premise. An essential gene cluster for *Salmonella typhimurium* virulence is the *spvABCD* operon, which is borne on a virulence plasmid. The expression of this operon is mediated by σ^S. Further, σ^S is involved in the expression of the *spvR* gene, which also regulates expression of the *spvABCD* operon. Thus, this operon is maximally expressed in starving cells of this bacterium. σ^S levels increase in *S. typhimurium* when it is phagocytosed by macrophages, and mutants deficient in σ^S are avirulent—indeed, they can serve as effective vaccines.

slyA is another gene of *S. typhimurium* that is relevant to its virulence. Transcriptional fusions to this gene show strong induction in the stationary phase as well as when phagocytosed by the macrophages. Mutants deficient in this gene synthesize fewer proteins in stationary phase and inside the macrophages, and they show reduced virulence. Similarly, the stress protein HtrA is important for the ability of *S. typhimurium* to replicate in the macrophages. HtrA is a serine protease that is believed to dispose of denatured periplasmic proteins. This protease is also important for the virulence of *Yersinia enterocolitica* and *Brucella abortus*. The starvation protein ClpC facilitates replication of *Listeria monocytogenes* in macrophages; *clpC* mutants of this bacterium are less virulent. Likewise, σ^E, which is also probably a starvation protein, controls the mucoidy phenotype of *Pseudomonas aeruginosa* that enables this bacterium to colonize the lung of cystic fibrosis patients.

IX. FUTURE PERSPECTIVES

A sharp focus on the molecular physiology of starving bacteria is only a relatively recent development. Among the questions that are presently being investigated is the nature of the growth phase-dependent sensitivity of σ^S to the ClpXP protease. This question impinges on a broader question in biology, namely, what determines the ability of different proteases to selectively target specific proteins. Whether the antisense element does have a role in translational control of σ^S is another important question. Little is known about the way in which starvation is sensed by a bacterium and linked to changes in σ^S levels. A protein called SprE or RssB, which bears homology to the response-regulator class of proteins, has been implicated in this process. However, the nature of the sensor remains unknown. Also important is further elucidation of the apparent promiscuity of the Eσ^S-regulated promoters.

Since starvation promoters are maximally induced during slow growth, they have been utilized to dissociate expression in bacteria of useful activities from a need for rapid growth. This is potentially an area of considerable importance, for instance, in improving *in situ* bioremediation and vaccine effectiveness, but awaits further research to be effectively exploited. The role of stress proteins in enhancing cellular resistance and bacterial virulence needs to be understood at a deeper biochemical and molecular levels. The resulting knowledge can provide novel means of controlling bacterial populations, both when their destruction is desired, as in disease, or when their perpetuation is the objective, as in many beneficial processes carried out by bacteria.

See Also the Following Articles

Low-Nutrient Environments • Osmotic Stress • Oxidative Stress

Bibliography

Fu, J. C., Ding, L., and Clarke, S. (1991). Purification, gene cloning and sequence analysis of an L-isoaspartyl protein carboxyl methyltransferase from *Escherichia coli*. *J. Biol. Chem.* **266**, 14452–14572.

Gottesman, S. (1996). Proteases and their targets in *Escherichia coli*. *Annu. Rev. Genet.* **30**, 465–506.

Hengge-Aronis, R. (1996). Regulation of gene expression during entry into the stationary phase. *In* "*Escherichia coli* and *Salmonella typhimurium:* Cellular and Molecular Biology" (F. C. Neidhardt, R. Curtiss III, J. L. Ingraham, E. C. C. Lin, K. B. Low, Jr., B. Magasanik, W. S. Reznikoff, M. Riley, M. Schaechter, and H. E. Umbarger, eds.), pp. 1497–1512. American Society for Microbiology, Washington, D.C.

Magasanik, B. (1996). Regulation of nitrogen utilization. *In* "*Escherichia coli* and *Salmonella typhimurium:* Cellular and Molecular Biology"(F. C. Neidhardt, R. Curtiss III, J. L. Ingraham, E. C. C. Lin, K. B. Low, Jr., B. Magasanik, W. S. Reznikoff, M. Riley, M. Schaechter, and H. E. Umbarger, eds.), pp. 1344–1356. American Society for Microbiology, Washington, D.C.

Matin, A. (1996). Role of alternate sigma factors in starvation protein synthesis—Novel mechanisms of catabolite repression. *Res. Microbiol.* **147**, 494–504.

Matin, A., Baetens, M., Pandza, S., Park, C. H., and Waggoner, S. (1999). Survival strategies in the stationary phase. *In* "Microbial Ecology and Infectious Disease" (E. Rosenberg, ed.), American Society for Microbiology, Washington, D.C.

Munster, U., and Chrost, R. J. (1990). Advanced biochemical and molecular approaches to aquatic microbial ecology. *In* "Brock/Springer Series in Contemporary Biosciences" (J. Overbeck and R. J. Crost, eds.), pp. 8–46. Springer-Verlag, New York.

Record, M. T., Jr., Reznikoff, W. S., Craig, M. L., McQuade, K. L., and Schlax, P. J. (1996). *Escherichia coli* RNA polymerase (Eσ^{70}), promoters, and the kinetics of steps of transcription initiation. *In* "*Escherichia coli* and *Salmonella typhimurium:* Cellular and Molecular Biology" (F. C. Neidhardt, R. Curtiss III, J. L. Ingraham, E. C. C. Lin, K. B. Low, Jr., B. Magasanik, W. S. Reznikoff, M. Riley, M. Schaechter, and H. E. Umbarger, eds.), pp. 792–820. American Society for Microbiology, Washington, D.C.

Yura, T., Nagai, H., and Mori, H. (1993). Regulation of heat shock response in bacteria. *Annu. Rev. Microbiol.* **47**, 321–350.

Stock Culture Collections and Their Databases

Mary K. B. Berlyn
Yale University

I. Federations Sponsoring Databases That Identify Collection Resources
II. Early Germplasm and Type Culture Collections
III. Early Genetic Stock Centers
IV. Microbial Genetic Stock Centers
V. National Collections and Consortia with Broad Holdings
VI. Organism-Specific Collections
VII. A Table of Stock Centers
VIII. Organizations Promoting Culture Collections

GLOSSARY

alkanotrophic Capable of assimilating propane, *n*-butane, and other *n*-alkane gases and liquid hydrocarbons as sole carbon and energy sources.

hyphomycetes Grouping of asexually reproducing fungi ("fungi imperfecti") in which the asexual spores (conidia) are produced on loose, cottony hyphae.

lyophilization The process of simultaneously freezing and drying materials under a vacuum.

patent depository In accordance with patent laws of most countries, novel microorganisms involved in a patent must be deposited in a recognized patent depository. Signatories to an International Budapest Treaty agreed that deposits in any approved International Depository Authority will be acceptable for meeting their patent office deposition requirements. (See, for example, *http://www.atcc.org/app_sci/patents.html#budapest.*)

A STOCK CULTURE COLLECTION is a repository for strains or varieties or species of organisms developed for the purpose of preservation and distribution of a useful range of the organisms. A distinction can be made between genetic stock collections, which are primarily mutant derivatives of one or more founder strains, along with related nonmutant strains, and germplasm and type culture collections, which have more heterogeneous holdings. This article is limited to microbial stock centers, although a few nonmicrobial centers are cited in recounting the history of genetic stock collections. The descriptions of early stock centers are presented in approximately chronological order, followed by genetic stock centers and by the later, national collections and consortia, which are grouped by country. These and more specialized collections are also listed under type of organism. Descriptions and access information are presented in a Table of Stock Centers.

I. FEDERATIONS SPONSORING DATABASES THAT IDENTIFY COLLECTION RESOURCES

The World Federation for Culture Collections (WFCC) is a Commission of the International Union of Biological Sciences (IUBS) and a federation within the International Union of Microbiological Societies (IUMS) (*http://wdcm.nig.ac.jp/wfcc/AboutWFCC.html*). Its purpose is to support and promote establishment of culture collections. In collaboration with the Microbial Resources Centres (MIRCEN) and other organizations described below it sponsors a database that is a directory of culture collections, the World Data Centre for Microorganisms, WDCM. This database began as a registry of worldwide culture collections initiated by V. B. D. Skerman at the University of Queensland in the 1960s. The registry was published in 1972 as the World Directory of Collections of Cultures of Microorganisms (WDCCM) by the WDCM, as a result of a proposal by the Japanese Federation of Culture Collections. The Directory was supported by the United Nations

Educational, Scientific, and Cultural Organization (UNESCO), the World Health Organization (WHO), the Australian Commonwealth Scientific and Industrial Research Organization (CSIRO), and the Canadian Research Council and its second edition in 1982 also by UNEP, the United Nations Environmental Program. Its fourth edition was published in 1993. Currently, the WDCM provides an on-line database of the collections and also of databases on microbes and cell lines. The server is located at the National Institute of Genetics in Japan (*http://wdcm.nig.ac.jp/DOC/menu3.html* and *http://wdcm.nig.ac.jp/wfcc/cir_iccc9.html*). It includes large and small collections, numbering nearly 500, located in public and private institutions in 55 countries. Particularly useful is an index of species and serovars held by culture collections with links to the culture collections holding the selected strain of interest. Many of these collections are cited in later sections of this article. The species list is found at *http://wdcm.nig.ac.jp/wfcc/species_list.html* and the index at *http://wdcm.nig.ac.jp/index.html*. The WDCM also allows querying by species and key word (e.g., *http://wdcm.nig.ac.jp/cgi-bin/ALGAE.pl* for algae) and presents a list of consulting services provided by the various stock centers (*http://wdcm.nig.ac.jp/wfcc/services_list.html*).

The Microbial Resources Centres (MIRCEN) is a UNESCO-sponsored network of academic and research institutes to foster collaboration in environmental, applied microbiological, and biotechnological research in cooperation with the National Commissions of Member States of the United Nations (*http://www.unesco.org/general/eng/programmes/science/life/mircen.htm*). MIRCEN is also a sponsor of the WDCM database. This is acknowledged by the official designation of the database as the WFCC-MIRCEN WDCM. *Rhizobium* is a particular focus for MIRCEN, and other areas include biotechnology, aquaculture, and bioinformatics. Historically UNESCO's role in preserving microbial diversity dates to 1946. (See DaSilva in Bibliography.) Currently, the largest members of the culture collection network include the Belgian Coordinated Collections (BCCM), the German Collection (DSMZ), Braunschweig, The Culture Collection of Hungary, the National Collection of Type Cultures (NCTC) in London, C.A.B. International Mycological Institute (IMI), and the American Type Culture Collection (ATCC), all of which are described in sections that follow. However, the network also supports collecting and coordination of collections at MIRCEN centers and groups throughout the world, including Bangkok, Dakar, Hawaii, Nairobi, and sites in African, Latin American, European, Arab, and Asian countries too numerous to mention. MIRCEN also sponsors a server located in Cairo for searching microbial collections (*http://202.41.70.55/cgi-bin/asearch/msdn/cairo/read*). In addition to the WDCM, the bioinformatics area of MIRCEN includes BITES, the Biotechnological Information Exchange System in Slovenia, and Bioinformatics MIRCEN at the Karolinska Institute in Stockholm (*http://www.unesco.org/general/eng/programmes/science/life/mircen.htm*).

Other organizational sponsors of the WDCM are the National Institute of Genetics Center for Information Biology of Japan (CIB), the Japan Science and Technology Corporation (JST), the Committee on Data for Science and Technology (CODATA), UNESCO, and UNEP.

MSDN, the Microbial Strain Data Network, is a nonprofit organization that provides links to databases of interest to microbiologists and biotechnologists. It is sponsored by UNEP. The secretariat is in Sheffield, UK, and it is managed by an international committee. The European node was established in conjunction with CABRI, the Common Access to Biotechnological Resources and Information project. The Institute of Microbiology of China is a collaborator and has established an MSDN node in China as part of the Microbial Information Network of China, and there are nodes in India and Brazil as well. Databases for small, specialized collections as well as the large heterogeneous national and international collections are included in the MSDN Web site. It allows searches of databases in the Czech Republic, Siberia, India, Japan, Russia, Slovenia, Argentina, Bulgaria, and the UK (*http://www.im.ac.cn/msdn.html*, *http://www.bdt.org.br/bdt/msdn*, *http://202.41.70.55/www/msdn.html*, and *http://panizzi.shef.ac.uk/msdn*).

The CABRI consortium provides the MSDN node in the UK and for Europe, providing links to MSDN, several of the large European culture collections, and cell culture and biological materials sources (*http://www.cabri.org/partners.html*).

The Microbial Germplasm Database (MGD) at Oregon State University emphasizes strain collections available from individual university, industry, government, and National Science Foundation (NSF)-supported laboratories and research stations, in contrast to public collections. It includes algae, bacteria, fungi, and viruses, and provides contact information for the laboratory collections. It also publishes an on-line newsletter. The MGD is supported by the U.S. Department of Agriculture (USDA) and NSF (*http://mgd.nacse.org/cgi-bin/mgd*). Bergey's Manual Trust (responsible for publishing new editions of the classic authority on bacterial taxonomy, *Bergey's Manual of Determinative Bacteriology*) maintains on its Web site a list of major collections and consortia (*http://www.cme.msu.edu/bergeys/culture.html*).

There are also directories with more general purposes which include information about stock collections, for example, *http://www.seaweed.ie*, *http://www.keil.ukans.edu/~fungi/fcollect.html*, and MycDB. MycDB is sponsored by the WHO and Follereau Foundation, originally at the Royal Institute of Technology in Stockholm and now at the Pasteur Institute in Paris, and is primarily a source of molecular biology information on mycobacteria, but it does provide information on availability of stocks of mycobacterial species in stock centers (*http://www.pasteur.fr/mycdb*). Databases that link or search multiple collections are also presented at the end of Table I (Section VII).

II. EARLY GERMPLASM AND TYPE CULTURE COLLECTIONS

The Collection of the Bacterial Strains of Institut Pasteur (CIP) traces its origins to Dr. Binot, who began to collect strains in 1891. It is a private, nonprofit collection that has become enlarged primarily by collaboration with the research laboratories of the Pasteur Institute. It maintains, preserves (by lyophilization since 1952), distributes, and provides information about strains. The CIP joined the World Federation of Culture Collections (WFCC) and the European Culture Collections Organization (ECCO, a collaborative organization for curators founded in 1982), and it is part of the Microbial Information Network Europe (MINE). Querying for strains can be performed at *http://www.pasteur.fr/CIP*. The individual records include links to other collections, as well as descriptions and literature references.

The Mycological Collection of the Catholic University of Louvain [now part of the Belgian Coordinated Collections of Microorganisms (BCCM)—see National Collections and Consortia in Table I] was founded in 1892–1894 by Prof. Philibert Biourge, and in 1901 Biourge's *Penicillium* collection was augmented with Prof. F. Dierckx's *Penicillium* and *Aspergillus* strains. During World War I, however, much of the collection disappeared, with accessions increasing again after the war. It was managed in the 1930s through the 1960s by Prof. P. Simonart, then by Prof. G. L. Hennebert as it merged with his fungus collection, which had begun in 1956. Other significant mergers brought in hyphomycetes from Prof. J. A. Meyer and the yeast collection of Biourge. The current holdings are described in Table I as part of the BCCM consortium (*http://www.belspo.be/bccm*).

The Uppsala University Culture Collection (Mykoteket or UPSC) has its origins intermingled with that of the Botanical Museum at Uppsala University and its extensive herbarium of plants, algae, and fungi, which date back to 1785 and C. Thunberg, a disciple of Linnaeus, and more specifically to the E. Fries fungal herbarium of the mid-1800s. The current fungal culture collection, the UPSC or Mycoteket, holds over 3000 strains. Its on-line database is found at *http://ups.fyto.uu.se/mykotek/index.html*.

The American Museum of Natural History/Society of American Bacteriologists Collection originated with an 1899 recommendation by the Society of American Bacteriologists (now the American Society for Microbiology) calling for a central repository for bacterial cultures. Such a collection was begun at the American Museum of Natural History in New York by MIT professor C.-E. A. Winslow in 1911.

Similarly a proposal by the Association Internationale des Botanistes led to the establishment of the Centraal Bureau voor Schimmelcultures (CBS) of the Netherlands in 1903 as a central collection of fungi, with 80 cultures available in 1906. In 1922 it moved to Delft Technical University. It was initially supported privately, but it became an institute of the Royal Netherlands Academy of Arts and Sciences, financially supported by the Dutch government, in 1968. The CBS includes as a separate unit the Phabagen Collection (see Microbial Genetic Stock Centers) and the Bacterial Collection of Kluyver Laboratory of Microbiology (LMD), which are merging to become the Netherlands Culture Collection of Bacteria (NCCB). See the section on national collections in Table I. A goal of the CBS is to preserve representatives of virtually all fungal groups that can be cultured, and the cultures currently number 35,000 with annual increases of approximately 1000 (*http://www.cbs.knaw.nl/www/collection.htm* and *http://www.cbs.knaw.nl/www/cbshome.html*).

The USDA Culture Collection also had its origins in the early 1900s. Although plant germplasm is the best-known of the collections supported and administered by the U.S. Department of Agriculture, the early history of U.S. microbial germplasm and genetic stock collections finds the USDA in a central role. Drs. Charles Thom and Margaret Church established a USDA collection in Washington, D.C., in 1913 that had begun with Dr. Thom's mold cultures in Connecticut and developed into the source of strains for both the USDA/Agricultural Research Service (ARS) Culture Collection and the American Type Culture Collection.

The USDA/ARS Culture Collection developed from this early USDA collection. In 1940, the Northern Regional Research Laboratory (NRRL) opened in Peoria, and Dr. Kenneth Raper formally established the ARS Collection there. (The laboratory is now named the National Center for Agricultural Utilization Research, NCAUR.) The NRRL collection included approximately 2000 Thom and Church cultures of molds, deposits of citric acid-producing aspergilli, bacterial strains, and L. J. Wickerham's yeasts. Large-scale use of lyophilization (freeze-drying) for preserving cultures was pioneered at the NCAUR for these cultures. Collections contributed to the ARS collection from later research programs include Mucorales, Taphrina, filamentous fungi, yeasts, aerobic spore-forming bacteria, rhizobia, and actinomycetes. The collection accepts strains that are nonpathogenic to humans and animals and are not fastidious in their growth requirements. Plant and animal cell lines are not accepted. Active research programs of curators and other scientists at the center, as well as outside researchers, utilize and enhance the collections and their applications. A notable, historic contribution was the development of an industrial process for producing penicillin during World War II. The collections include a patent culture collection as well as the "open" public collection, totaling over 80,000 strains (*http://nrrl.ncaur.usda.gov/the_collection3.htm* and *http://nrrl.ncaur.usda.gov*). On-line databases for the Actinomycetales, bacterial, and filamentous fungal collections can be searched at *http://nrrl.ncaur.usda.gov/searchcc.htm*.

The USDA/ARS also maintains a Collection of Entomopathogenic Fungal Cultures (ARSEF), begun in the 1970s at the University of Maine and moved to the Plant Protection Research Unit on the Cornell campus in Ithaca, New York. A fungal herbarium is also associated with this collection (*http://www.ppru.cornell.edu/Mycology/ARSEF_Culture_Collection.html*).

A third USDA/ARS collection is the National *Rhizobium* Germplasm Resource Center. The first cultures were isolated from northern Virginia in 1913 and increased by USDA scientists in response to research demands, particularly in the 1930s and 1940s. Because of its international agricultural significance, it was initially funded by the U.S. Agency for International Development, and in the 1980s it was designated a UNESCO Microbiological Resource Center (see MIRCEN above). Since 1990, the Agricultural Research Service has supported the collection (*http://bldg6.arsusda.gov/pberkum/Public/cc1a.html*). A searchable index can be found at the Web site for the Germplasm Resources Information Network of the USDA/ARS (*http://www.ars-grin.gov/cgi-bin/nmgp/rhy/search.pl*).

An algal collection developed by Prof. E. G. Pringsheim and colleagues at the University of Prague (Charles University) formed the basis for both the CCAP, the Culture Collection of Algae and Protozoa, now in the UK, and the CCALA, the Culture Collection of Autotrophic Organisms in the Czech Republic. The collection developed by Pringsheim, V. Czurda, and F. Mainx moved with Prof. Pringsheim to England and was later expanded and directed by Prof. E. A. George at Cambridge University, with support from the UK Natural Environment Research Council. Its two current components, the Institute of Freshwater Ecology (IFE) collection and Dunstaffnage Marine Laboratory collection, are cited in the algal collections section (see Table I). The algal collection of Profs. V. Uhlir and Pringsheim in Prague was merged with the algal collection at the Institute of Microbiology, Czech Academy of Sciences at Trebon, to form the CCALA, also cited with algal collections in Table I.

The American Type Culture Collection (ATCC) was founded in 1925 by representatives from the National Research Council, the Society of American Bacteriologists, the American Phytopathological Society, the American Zoological Society, and the McCormick Institute for Infectious Diseases. It was located at first at the McCormick Institute in Chicago, moving in 1937 to the Georgetown University School of Medicine and to houses in Washington, D.C., in 1947 and 1956. It was incorporated as an independent, nonprofit institution in 1947. It then included microorganisms, animal and plant viruses, and cell lines. In 1964 it moved into a facility in Rockville, Maryland, and in 1998 to Manassas, Virginia. It has been supported by many grants from the U.S. NSF and National Institutes of Health (NIH) directed toward the various specialized collections and activities—hybridomas, bacterial collections, databases, yeasts, protozoa, probes, clones, etc.—as well as by users' fees. The holdings are extensive, with over 14,000 bacterial strains, 500 viral strains, and 26,000 strains of 6500 yeast and fungal species. It is also an official U.S. patent repository and conducts workshops and education sessions. It is the distribution source for microbial genome clones for completely sequenced prokaryotic genomes from the Institute for Genomic Research (TIGR) in Maryland and for many cDNA clones (*http://www.atcc.org*).

The International Mycological Institute (IMI) Culture Collections are part of the institute founded in 1920 and is currently part of CAB (originally called Commonwealth Agricultural Bureaux) International, a nonprofit organization supported by 32 governments. The current collection has over 16,500 strains of filamentous fungi, yeasts, and bacteria. It is a patent depository and offers identification, preservation, testing, consultation, and training services and undertakes applied research. A database may be searched at *http://wdcm.nig.ac.jp/cgi-bin/MSDN/IMI.pl* or at *http://www.cabi.org/BIOSCIENCE/grc.html*.

III. EARLY GENETIC STOCK CENTERS

In most cases, genetic stock centers originated when early geneticists working with the species recognized that the stocks were valuable for future research and made accommodations for preserving and disseminating these stocks. The realization of the need for the stocks was often accompanied by the realization that a means for disseminating new scientific results in a rapid and somewhat informal way was also required for advancing scientific progress with the organism. For most stock centers the information function has progressed from species-specific newsletters and strain lists to comprehensive online databases.

Among the earliest stock centers in the United States are the *Drosophila* Stock Center and the Maize Cooperation Genetics Stock center. The *Drosophila* center grew from the mutant collection originating with T. H. Morgan and his students at Columbia, beginning in 1913. Their stocks, maintained by Bridges, were provided to anyone requesting them. In 1928, when Morgan, Bridges, Sturtevant, and the collection moved to the California Institute of Technology, that became the site of the *Drosophila* stock center and of the stock list published in the *Drosophila* Information Service. This stock center is currently at the University of Indiana and includes approximately 6000 stocks (T. Kaufman and K. A. Matthews, *http://flybase.bio.indiana.edu/bloom-home.html*).

The Maize Genetics Cooperation Stock center had its origins in the 1928 Winter Science meetings in New York, when a group of maize geneticists discussed work on the maize linkage maps and the idea of an organized Maize Genetics Cooperation. A formalized proposal in 1932 included provision for the Maize Genetics Cooperation News Letter and the Maize Genetics Cooperation Stock Center. Marcus M. Rhoades served as first secretary of the Newsletter and first director of the stock center. Currently, the Maize Cooperation Stock Center is at the University of Illinois, USDA/ARS (M. Sachs), and the Maize Genetics Cooperation Newsletter is produced at ARS/USDA, University of Missouri (E. Coe). The collection includes nearly 80,000 pedigreed samples, including alleles of several hundred genes, combinations of such alleles, chromosome aberrations, ploidy variants, and other variations. Details about the collection, its history, available stocks, and request forms can be found at *http://w3.ag.uiuc.edu/maize-coop/mgc-info.html*. (The newsletter can be found on the Maize Genome Database site at *http://www.agron.missouri.edu/mnl*.)

IV. MICROBIAL GENETIC STOCK CENTERS

A. Bacterial Genetic Stock Centers

Sexuality and the ability to make genetic crosses between bacteria was discovered in *E. coli* in the 1930s. The use of microorganisms to explore the relationship between genes and biochemical pathways also became established in that decade. Individual laboratories accumulated large numbers of *E. coli* mutants, primarily derivatives of the wild-type isolate *E. coli* K-12. In the 1960s, it became apparent that a national repository would greatly aid the free exchange of strains and the advance of molecular genetics.

At that time, support from the U.S. National Science Foundation was a critical element in developing such a resource, and the NSF supported a proposal to begin with E. A. Adelberg's Yale University collection of stocks and add important strains and sets of strains from laboratories worldwide. This became the *E. coli* Genetic Stock Center (CGSC) at Yale, curated for 25 years (until 1993) by B. Bachmann. The CGSC holds over 7500 strains and a plasmid library of cloned segments covering nearly all of the *E. coli* genome. Unlike the previous two centers, a newsletter was not associated with the founding of the stock center, but it soon assumed the functions of registering gene names and allele numbers, as set forth in the widely accepted guidelines for bacterial nomenclature by Demerec *et al.* (1966), and meeting subsequent needs for registry to avoid duplication of designations for deletions, insertions, and F-primes. It also took on responsibility for periodic publishing of the linkage map for *E. coli* K-12, and with that came a contingent role of "authenticating" canonical gene symbols. These information roles provided a natural progression to an on-line database covering gene names, functions, map locations, strain genotypes, mutation information, and supporting documentation, as well as links to information from other bacterial, genetic, and molecular biology databases. The development and maintenance of the database, as well as the Stock Center itself, have been supported by the NSF (*http://cgsc.biology.yale.edu*).

Phabagen, the Phage and Bacterial Genetics Collection, includes 5000 mutant bacterial strains (*E. coli* and *Agrobacterium tumefaciens*), 450 cloning vectors, 800 other plasmids, two plasmid-containing gene banks of *E. coli*, and over 100 phages. It was established in the early 1960s with deposits of bacterial mutants from researchers of the Working Community Phabagen. It has, since 1990, been part of CBS (see national consortia in Table I) and is merging with the Kluyver Institute's LMD Collection to form the NCCB, the Netherlands Culture Collections of Bacteria (*http://www.cbs.knaw.nl/nccb/database.htm*).

The National Institute of Genetics of Japan (NIG) established a Genetic Stock Center in 1976 and developed an extensive collection of *E. coli* strains. In 1997 it was reorganized as the Genetic Strains Research Center, Microbial Genetics Center, and the Center for Genetic Resource Information. The microbial collections include approximately 4000 strains of *E. coli* genetic derivatives and 400 cloning vectors (*http://www.nig.ac.jp/labs/MicBio/home.html, http://www.shigen.nig.ac.jp/ecoli, http://www.shigen.nig.ac.jp/cvector/cvector.html*).

The *Salmonella* Genetic Stock Center (SGSC) at the University of Calgary, British Columbia, Canada, had its origins in the laboratory of M. Demerec at Cold Spring Harbor Laboratory and Brookhaven National Laboratory, Long Island, New York, in the 1950s and 1960s as genetic derivatives primarily of *Salmonella typhimurium* (or *Salmonella enterica,* subsp. *enterica,* serovar Typhimurium) strain LT2. It remained with the Demerec laboratory during his long career, and then it was moved and expanded at the University of Calgary by K. E. Sanderson. It currently has several thousand strains, cosmid and phage libraries, and a set of cloned genes. Many of the mutant strains are organized into special purpose kits, useful for specific genetic techniques or analyses. In addition to the mutants, it has the *Salmonella* Reference Collection (SARC) representing all subgenera of *Salmonella.* The SGSC is supported by the Natural Sciences and Engineering Research Council of Canada (*http://www.acs.ucalgary.ca/~kesander*).

Bacillus subtilis has been studied as the model spore-forming bacterium, and many mutant strains have been isolated. In 1980 the *Bacillus* Genetic Stock Center was founded at Ohio State University. It includes over 1000 genetically characterized *B. subtilis* strains, over 200 strains of other *Bacillus* species, and *E. coli* strains bearing shuttle plasmids or cloned

Bacillus DNA. It publishes a newsletter and a genetic map for *Bacillus subtilis*. The National Science Foundation supports this stock center (*http://bacillus.biosci.ohio-state.edu*).

B. Fungal and Algal Genetic Stock Centers

The Fungal Genetic Stock Center (FGSC) was organized as a result of recommendations by the Genetics Society of America in 1960. It was originally located at Dartmouth College, directed by R. Barratt, then moved to California State University in Humboldt, and in 1985 to the University of Kansas Medical Center. Its holdings include nearly 9000 strains of filamentous fungi, mostly genetic derivatives of *Neurospora crassa, Aspergillus nidulans,* and *Aspergillus niger,* but include also *Neurospora tetrasperma* and isolates of other *Neurospora* and *Aspergillus* species. The collection also contains *Fusarium* species and mutants, and *Nectria* and *Sordaria* mutants and species. The stock center publishes, mails, and makes available on its Web site the Fungal Genetics Newsletter, meeting abstracts and announcements, and a bibliography. The Web site has information on genes, alleles, and maps, and the center supplies plasmids, clones, and gene libraries for *N. crassa* and *A. nidulans,* as well as the strains. The FGSC is supported by the National Science Foundation (*http://www.fgsc.net*).

The Yeast Genetic Stock Center (YGSC) originated at the University of California at Berkeley in 1960, founded and directed by R. Mortimer for genetic derivatives of *Saccharomyces cerevisiae,* primarily originating from "founder" stocks of C. Lindegren at Southern Illinois University in Carbondale. The Mortimer laboratory also published linkage maps and linkage data summaries. Maps and linkage data, but not stock descriptions, are currently provided by the *Saccharomyces* Genome Database (*http://genome-www.stanford.edu/Saccharomyces*). The 1200 genetic stocks in the collection will be moving to the ATCC, and stock information will be available from there (*http://phage.atcc.org/searchengine/ygsc.html*).

The Peterhof Genetic Collection of Yeasts (PGC) is located in the Biotechnology Center at St. Petersburg State University in Russia. It has over 1000 genetically marked yeast strains, mainly lines originating from a diploid cell of an inbred strain of *Saccharomyces cerevisiae* (the Peterhof genetic lines, as distinguished from the Carbondale origin of the strains in the YGSC), other genetically marked *S. cerevisiae* strains, and segregants of crosses between the latter and the Peterhof lines. The collection is searchable via the MSDN, at *http://panizzi.shef.ac.uk/msdn/peter*.

The *Chlamydomonas* Genetics Center (CGC), at Duke University, collects, describes, and distributes nuclear and cytoplasmic mutant strains, and genomic and cDNA clones, of *Chlamydomonas reinhardtii,* which has served as a model organism for algae and for photosynthetic organisms. It is supported by the National Science Foundation. The Web and gopher sites provide, in addition, information on genetic and molecular maps of *Chlamydomonas reinhardtii,* plasmids, sequences, and bibliographic citations (*http://www.botany.duke.edu/chlamy*).

C. Information Management at Genetic Stock Centers

The dissemination of information was often an equal partner with the dissemination of stocks in the call by the scientific community for founding of the stock centers. Some stock centers distributed printed catalogs, and others published periodic announcements summarizing their holdings. Many stock center curators or directors took responsibility for publishing frequently updated versions of the genetic map for the major organism in their collection. In 1988, a workshop on computerization of databases for genetic stock collections was held in conjunction with the 16th International Congress of Genetics, August 1988, Toronto. This meeting acknowledged the information functions of the stock centers in describing and distributing the organisms and also in producing genetic maps, standardizing allele designations, and documenting gene function and nomenclature, and it discussed and made recommendations for facilitating database development and database construction. In the U.S., the National Science Foundation and other funding agencies played a prominent role in encouraging stock centers to recognize the need for modernization and computerization of their valuable records.

Development of the Internet and the World Wide Web has altered and enhanced the performance of these information functions and the way that users interact with the stock centers. It has allowed information formatted in many different ways at the many stock collections sites to be searched and presented to the user and allows links among stock centers and between stock centers and databases with sequence and other types of information. It allows the centers to provide on-line direct access to the information about stocks, gene characterization, and the genetic map updates. Citations of Web addresses for all microbial stock centers and ancillary information associated with them have been given throughout this article and are included as part of the entries in Table I.

V. NATIONAL COLLECTIONS AND CONSORTIA WITH BROAD HOLDINGS

During the past three decades, many countries have moved to consolidate the administration of their stock centers into a single administrative and information unit or federation. Examples of some of these consortia are presented in Table I by country. Particularly extensive consortia include those in the United Kingdom, France, Belgium, and the Netherlands, as well as the USDA collections. These and other national collection consortia in the Czech Republic, Poland, Russia, the Ukraine, China, Japan, Indonesia, Canada, and Brazil are described under the heading National consortium, comprehensive in Table I. Other countries have long had centralized national collections rather than individual specialized or general collections. Such collections are listed under National, comprehensive type collections in Table I. The DSMZ in Germany is one of the largest, and others are located in Belarus, Bulgaria, Hungary, India, Thailand, Taiwan, New Zealand, and Mexico. Still other comprehensive collections serve national and international communities but are not administered or entirely supported by governmental agencies. For example, the ATCC (American Type Culture Collection) in the U.S. is only partially supported by federal grants and is administered as an independent, nonprofit institution. Such collections are type classified as Other—comprehensive in Table I. The U.S. collections are not organized under any single coordinating administrative body, although the USDA/ARS collections are coordinated through the USDA and the ATCC represents a consolidation of many different types of collections. There is a U.S. Federation of Culture Collections to promote cooperation and information sharing among the diverse stock centers. It cooperates with the International World Federation of Culture Collections in sponsoring international meetings among directors and curators of stock collections. The International Mycological Institute is internationally supported (CAB International), with a secretariat in the UK, and this collection and university collections in Sweden, Finland, and Australia, and Japan's Institute for Fermentation (IFO) Collection in Osaka are classified as Other—comprehensive in Table I. National collections that are specialized rather than comprehensive are presented under organism-specific headings in Table I.

VI. ORGANISM-SPECIFIC COLLECTIONS

The organism-specific collections are presented in Table I under the following broad headings: bacteria; fungi, including yeasts; and algae, protozoa, and cyanobacteria. Type headings provide a more specific indication of the holdings. Previously described collections, in either the national consortia part of the table or in narrative sections, are cited again in Table I in order to facilitate locating the collection and its Web URL.

VII. A TABLE OF STOCK COLLECTIONS

To facilitate look-up of specific kinds of collections and especially to facilitate locating the Web address of a collection, a tabular rather than text format is used to present collections by type and country, to provide brief descriptions of the collections, and to cite the Web addresses. Clearly a totally comprehensive listing of such diverse and widespread resources is not possible, and regrettable oversights have no doubt occurred. Widely used and Web-accessible sites have been emphasized. See legend for information on searching for collections not listed in the table.

TABLE I
Representative Stock Collections: Location, Description, and Access

Many of the descriptions briefly cite special functions, history, and holdings, if this information has not been previously cited. Preservation and distribution functions are assumed in all cases and are not cited specifically. The table first presents comprehensive collections, with country presented in uppercase, then organism-specific collections (Bacteria; Fungi,; Algae, protozoa, and cyanobacteria), and finally databases that cover multiple collections. Web URLs may change at any time, and searches may be necessary to update those shown in the table. No claim can be made that this is an exhaustive compilation of all the world's stock centers. Comprehensive database searches and links given at the end of the table can be used to search for collections not listed in the table. For example, *http://wdcm.nig.ac.jp/cgi-bin/CCINFO.pl?countryname* will retrieve all WDCM collections in that country, as shown in the last entry of the table.

Type	*Country*	*Designation*	*Description*	*Web URL*
COMPREHENSIVE				
National consortium, comprehensive	UNITED KINGDOM	UK National Collections include the following:	See individual collection descriptions below	Bioguide, http://dtiinfol.dti.gov.uk/bioguide/culture.htm, and http://www.ukncc.co.uk
		NCTC, National Collection of Type Cultures, London	Has about 5000 cataloged strains of medical and veterinary significance and 13,000 uncataloged strains. A patent and safe depository. Jointly with the DSM (Germany) is the Resource Centre for plasmid-bearing bacteria for Europe. Supported by the Public Health Laboratory Service.	http://www.ukncc.co.uk/html/organisation/nctc/nctc_info.htm, http://dtiinfol.dti.gov.uk/bioguide/culture.htm#type
		NCIMB, National Collection of Industrial and Marine Bacteria Ltd., Torry Res. St., Aberdeen, Scotland	Includes 3800 accessions of general industrial and scientific significance, (actinomycetes and other bacteria, plasmids, and phages) and 1500 marine bacteria. Originally funded by Ministry of Agriculture, Fisheries and Food. Patent depository. Identification, service, and contract work.	http://www.ncimb.co.uk
		NCFP, National Collection of Food Bacteria, Aberdeen, Scotland	About 2000 bacteria of significance to food and dairying. A patent and safe depository. Originally funded by Ministry of Agriculture, Fisheries and Food, Institute of Food Research and is now affiliated with NCIMB, Ltd.	http://dtiinfol.dti.gov.uk/bioguide/culture.htm#food_bacteria
		NCYC, National Collection of Yeast Cultures, Institute of Food Research, Norwich	Includes brewing yeast strains, genetically defined strains of *Saccharomyces cerevisiae* and *Schizosaccharomyces pombe,* and general yeast strains, totaling over 2700 nonpathogenic yeasts. A patent and safe depository. Performs yeast identification services. Member of the European Culture Collections Organization (ECCO) and the WFCC.	http://www.ifrn.bbsrc.ac.uk/NCYC
		NCPF, National Collection of Pathogenic Fungi, London	Has 1500 human and animal pathogenic fungi and performs identification and contract research. Supported by the Public Health Laboratory Service.	http://www.ukncc.co.uk/FrameIndex4.htm,select ncpf.
		NCWRF, National Collection of Wood Rotting Fungi in Watford	Over 500 cultures of ~300 species of wood-rotting basidiomycetes. Funded by the Department of the Environment.	As above, select ncwrf.
		NCPPB, National Collection of Plant Pathogenic Bacteria, Harpenden	Has 3000 phytopathogenic bacteria, representing a wide range of hosts and locations for species causing most of the known bacterial plant diseases, and has some bacteriophages as well. Funded by the Ministry of Agriculture, Fisheries and Food. Provides information, identification, and contract services.	As above, select ncppb.
		See also, IMI, RCR, and IACR collections	Classified in this table as Other—comprehensive, fungal—*Rhizobium,* and plant pathogens, respectively.	

continues

Continued

Type	Country	Designation	Description	Web URL
		CCAP, Culture Collection of Algae and Protozoa	Had its origins with the early culture collections, moving from Prague to England in the 1920s (see early collections section in text), undergoing development at Cambridge University and later moving to two sites—the freshwater algae and all protozoa to the Institute of Freshwater Ecology (IFE) Windermere Laboratory at Ambleside and the marine algae to Dunstaffnage Marine Laboratory (DML) in Scotland. Approximately 2000 strains of algae and protozoa. A patent depository.	http://www.ife.ac.uk/ccap and http://www.ukncc.co.uk/ FrameIndex4.htm, select ccap(f) and ccap(m)
National consortium, comprehensive	FRANCE	CNCM, National Collection of Cultures of Microorganisms	Established in 1976 by French Ministerial authorities. Housed in the Pasteur Institute. A patent and safe depository. Coordinates activities of the specialized collections of the Pasteur Institute. Member collections include the CIP (see early collections section in text and bacterial type in this table), which incorporates the *Pasteurella* laboratory collection, the Pasteur Institute collections of cyanobacteria (see PCC under collections of algae, protozoa, and cyanobacteria), *Rhodococcus, Gordonia, Lactobacillus,* coryneforms, and *Pasteurella,* as well as fungal, viral, and animal cell collections. Member of the World Federation for Culture Collections and the European Culture Collections Organization.	http://www.pasteur.fr/Bio/RAR96/ Colloq.html
National consortium, comprehensive	BELGIUM	BCCM, Belgian Coordinated Collections of Microorganisms. Includes the following:	Consortium of four research-based collections, containing 50,000 documented strains of bacteria, filamentous fungi, and yeasts, and over 1500 plasmids. Supported by the Belgian Federal Office for Scientific, Technical, and Cultural Affairs. Patent and safe deposit services as well as contract research and training.	http://www.belspo.be/bccm, search at http://www.belspo.be/bccm/ db/index.htm
		BCCM/MUCL, the Mycological Collection of the Catholic University of Louvain	Originated with the earliest collections (see early collections section of text). Holds 25,000 strains of fungi and yeast, including representatives of over 3300 species of zygomycetes, ascomycetes, hyphomycetes, and basidiomycetes. Focuses on agro-industrial cultures. Also holds a mycological herbarium of 40,000 fungal specimens. Undertakes research in systematic and applied mycology and provides software for yeast identification.	http://www.belspo.be/bccm/ mucl.htm#main
		BCCM/LMBP, the Plasmid and cDNA Collection, University of Ghent	Includes plasmids that replicate in prokaryotic (*E. coli* and *Lactobacillus*) and eukaryotic (yeast, animal cell) hosts and 10 cDNA libraries.	http://www.belspo.be/bccm/ lmbp.htm#main
		The BCCM/LMG Bacterial Collection, University of Ghent	Holds over 16,000 strains representative of 1300 species, subspecies, or pathovars, with emphasis on phytopathogenic bacteria and species of medical and veterinary significance. Also has some marine *Vibrio* species and strains of biotechnological interest.	http://www.belspo.be/bccm/ lmg.htm#main

continues

Continued

Type	Country	Designation	Description	Web URL
		The BCCM/IHEM Collection of Biomedical Fungi and Yeasts, Brussels	A collection of filamentous and yeast fungi of public health and environmental interest, containing over 6500 strains of human and animal pathogenic and allergenic species. These include specific disease isolates, e.g., 500 *Aspergillus fumigatus* and 300 *Candida albicans* isolates. Located at the Scientific Institute of Public Health—Louis Pasteur in Brussels	http://www.belspo.be/bccm/ihem.htm#main
National consortium, comprehensive	THE NETHERLANDS	CBS, The Centraal Bureau voor Schimmelcultures	The CBS collection and NCCB, described below, are separate units of the organizational structure of the CBS, an institute of the Royal Netherlands Academy of Arts and Sciences.	http://www.cbs.knaw.nl
		CBS, the Centraal Bureau voor Schimmelcultures Collection, Baarn and Delft	See early collections section of text. Currently includes separate collections and databases for filamentous fungi in Baarn (over 28,000 strains) and yeasts in Delft (4500 strains), with the IGC collection of 1250 yeast strains in Oeiras, Portugal, also included in the database, and a database for taxonomic and nomenclatural data for *Fusarium* and the Aphyllophorales. One of the largest collections of fungi in the world, totaling over 35,000 strains, its goal is to have representatives of all fungal groups that can be cultured. Plans to integrate yeasts and fungi into one institute. Serves as the Dutch node of MINE, the Microbial Information Network of Europe, an EC-sponsored network of European collections. Performs identification and patent deposit services.	http://www.cbs.knaw.nl, http://www.cbs.knaw.nl/www/search_fdb.html, http://www.cbs.knaw.nl/www/search_ydb.html, http://www.cbs.knaw.nl/fusarium/database.html, and http://www.cbs.knaw.nl/aphyllo/database.html
		NCCB, Netherlands Culture Collections of Bacteria, Utrecht	Merger of the Bacterial Collection of Kluyver Laboratory of Microbiology at the Technical University of Delft (Kluyver Institute of Biotechnology), the LMD, containing over 4500 strains, and the Phabagen Collection (see Microbial Genetic Stock Centers) at the University of Utrecht. Participant in CABRI. Patent and safe depository, taxonomic research, identification, and consultation services.	http://www.cbs.knaw.nl/nccb, http://www.cbs.knaw.nl/nccb/about.htm
National consortium, comprehensive	GERMANY	DSMZ, Deutsche Sammlung von Mikroorganismen und Zellkulturen, Braunschweig	Founded in 1969 as the national culture collection of Germany, part of the Gesellschaft für Strahlenforschung in Munich. Moved to the Gesellschaft für Biotechnologische Forschung in Braunschweig, in 1988 became a private company, and in 1996 became an independent nonprofit organization supported by the Federal Ministry of Research and Technology and the State Ministeries. Holds over 8700 bacteria and archaea, 300 plasmids, 100 phages, 2300 filamentous fungi, 500 yeasts, 700 plant viruses, as well as plant, human, and animal cell lines. Serves as a safe and patent respository, and performs teaching, service, and identification functions. On-line searchable databases for microorganisms, cell lines, and plasmids, as well as a bacterial nomenclature database.	http://www.dsmz.de, http://www.dsmz.de/bactnom/bactname.htm

continues

Continued

Type	Country	Designation	Description	Web URL
National consortium, comprehensive	CZECH AND SLOVAK REPUBLICS	Czech Federation of Culture Collections	Includes a number of microbial collections, the larger collections listed below. Member of the WFC and ECCO.	http://www.natur.cuni.cz/fccm
		The Czech Collection of Microorganisms, Brno	Established at the J. E. Purkyne University in 1963, based on the bacterial collection of the university begun in 1957 and merged with the Culture Collection of the Institute of Microbiology, of the Czechoslovak Academy of Science in Prague. Includes over 2000 strains (representing more than 170 genera and 600 species) of bacteria, over 300 strains of fungi, mycoplasmas, and viruses, particularly animal pathogens and organisms of interest to food microbiology, medicine, education, agriculture, and industry. Performs identification and consultation services and serves as a safe repository. Member of the WFCC.	http://panizzi.shef.ac.uk/msdn/ccm, http://www.sci.muni.cz/ccm/ccmang.htm
		The Czech Culture Collection of Fungi (CCF) at Charles University, Prague	Combined the collections of the Biological Institute of the Czechoslovak Academy of Sciences and the Charles University Botany Department research collection in 1964–1965. Approximately 1600 strains of zygomycetes and ascomycetes.	http://panizzi.shef.ac.uk/msdn/ccf, http://www.natur.cuni.cz/fccm/#ccf
		CCBAS, the Culture Collection of Basidiomycetes	Founded in 1959 and contains 630 strains and 253 species, including some rare species.	http://www.biomed.cas.cz/ccbas/fungi.htm
National consortium, comprehensive	POLAND	PCM, Polish Collections of Microorganisms	Combines several individual collections at the Hirszfeld Institute of the Polish Academy of Sciences, Wroclaw. These include the DMVB and the CRS, Collection of *Rhizobium* Strains, in Pulawy, the Industrial Microorganism collection in Lodz and Warsaw, the Dairy Cultures collection in Olsztyn, the National *Salmonella* Centre in Gdynia-Redlowo, and the Research and Development Center for Biotechnology in Warsaw. Represents a wide range of bacteria, yeasts, and filamentous fungi, which are searched for and described from a single Web site.	http://immuno.pan.wroc.pl:80/pcm
National consortium, comprehensive	RUSSIA	VKM, the All-Russian Collection of Microorganisms, Institute of Biochemistry and Physiology of Microorganisms	Merger in 1980 of the Institute of Microbiology's collection of yeast cultures of V. Kudryavtsev begun in the 1930s and the Institute of Biochemistry and Physiology's collections of microorganisms begun in 1968. Over 9000 strains, encompassing 3000 species of 650 genera, of bacteria, actinomycetes, filamentous fungi, and yeast. Includes genetically modified strains of *E. coli* and yeast as well as a few genetic derivatives of *Agrobacterium tumefasciens, Bradyrhizobium japonicum,* and *Streptomyces galilaeus*. Russian Academy of Sciences, Pushchino, Moscow region.	Searches and information via WDCM at http://panizzi.shef.ac.uk/msdn/vkmst.
		VKIM, All-Russian Collection of Industrial Microorganisms	Institute of Genetics and Selection of Microorganisms, Moscow.	

continues

Continued

Type	*Country*	*Designation*	*Description*	*Web URL*
National, comprehensive	UKRAINE	IMV, Zabolotny Institute of Microbiology and Virology Collection	Sponsored by the Academy of Sciences in Kiev.	See MSDN
National, comprehensive	BELARUS	INMIB Collection of the Institute of Microbiology	Academy of Science of Belarus, Minsk.	http://www.ac.by/organizations/institutes/inobio.html#off3578
National, comprehensive	BULGARIA	NBIMCC, National Bank for Industrial Microorganisms and Cell Cultures, Sofia	Founded in 1983, encompasses the Bulgarian Type Culture Collection, and includes fungi and yeasts, bacteria and actinomycetes, plasmids, viruses, and cell lines.	http://panizzi.shef.ac.uk/msdn/nbimec
National, comprehensive	HUNGARY	NCIM, National Collection of Agricultural and Industrial Microorganisms, Budapest	At the University of Horticulture and Food Industry. Includes over 1600 strains of bacteria, yeasts, and fungi. It was established in 1974 and is a patent and safe depository and performs purification, identification, consultation, and training services. Member of the WFCC and ECCO.	http://ncaim.kee.hu. For searches by keyword, http://panizza.shef.ac.uk/msdn/ncim
National, comprehensive	INDIA	NCIM, National Collection of Industrial Microorganisms, Pune	A collection of over 100 bacteria, nearly 2000 fungi and yeasts, and a few algae and protozoa. National Chemical Laboratory in Pune, sponsored by the Council for Scientific and Industrial Research.	http://www.nic.in:80/snt/ari.htm
National, comprehensive	INDIA	ITCC, the Indian Type Culture Collection, New Delhi	Division of Mycology and Plant Pathology, Indian Agricultural Research Institute.	http://panizzi.shef.ac.uk/msdn/mushroom/itcc.html
National, comprehensive	INDIA	MTCC, Microbial Type Culture Collection, Chandigash	Established 1986 at the Institute of Microbial Technology. Actinomycetes, bacteria, yeast, fungi, plasmids	http://imtech/ernet.in/mtcc See also http://wdcm.nig.ac.jp/cgi-bin/CCINFO.pl?india
National consortium, comprehensive	CHINA	CCCCM, the China Committee for Culture Collections of Microorganisms	Provides search capability by name or keyword for databases encompassing the following three collections as well as others. It also provides a national node of the MSDN that allows querying of some general and many specialized collections in many countries (CCCCM). Oversees the following three collections:	See Micro-Net of China http://www.im.ac.cn. Search collections at http://www.im.ac.cn/database/catalogs.html. MSDN node: http://www.im.ac.cn/msdn.html
		CCCM, the Center for Culture Collection of Microorganisms, Beijing	Institute of Microbiology, Chinese Academy of Science. Includes 1900 yeasts, 5500 other fungi, 2200 bacteria, and 1400 actinomycetes; also included are the type culture collection of China and patent strain repository.	http://www.im.ac.cn/imcas/junbao.html
		CCVCC, the China Center for Virus Culture Collection	Collects, preserves, classifies, and studies viruses, now numbering 600, including insect, plant, bacterial, and animal viruses. Chinese Academy of Sciences, Wuhan Institute of Virology, established in 1979.	http://www.im.ac.cn/institutes/ccvcc/ccvcc.html
		ACCC, the Agricultural Culture Collection of China, Beijing	The ACCC holds more than 2000 strains of bacteria, actinomycetes, fungi, rhizobia, and edible fungi. Chinese Academy of Agricultural Sciences, Beijing, established in 1980.	http://www.im.ac.cn/institutes/accc/accc.html
National, comprehensive	THAILAND	TISTR, the Thailand Institute of Scientific and Technological Research Collection	Linked to WDCM and can be queried at the Web site shown. The Tropical Database of Thailand is also cited by WDCM as a TISTR project. Supported by the Ministry of Science, Technology and Environment of Thailand.	http://www.jcm.riken.go.jp/TISTR

continues

Continued

Type	*Country*	*Designation*	*Description*	*Web URL*
National consortium, comprehensive	JAPAN	JFCC, Japan Federation of Culture Collections	Maintains a search capability for the various collection databases for bacteria, fungi, viruses, and algae.	http://wdcm.nig.ac.jp/wdcm/JFCC.html
		IAM Culture Collection, Institute of Applied Microbiology, University of Tokyo	Founded in 1953 and in 1993 reorganized in the IMCB (Institute of Molecular and Cellular Biosciences) to broaden coverage to higher organisms and their cultured cells. The microbial collection contains over 3400 strains of bacteria, yeasts, filamentous fungi, and algae.	http://www.iam.u-tokyo.ac.jp/misyst/ColleBOX/IAMcollection.html
		The Japan Collection of Microorganisms (JCM) in the Institute of Physical and Chemical Research (RIKEN)	Has since 1980 distributed microorganisms from a collection that includes over 3500 strains of bacteria, including actinomycetes, 2000 strains of fungi, including yeasts, and a hundred strains of archaea. Collections such as the KCC Culture Collection of Actinomycetes, Kaken Chemical Co., Tokyo, have transferred their holdings to the JCM. On-line catalog allows searching by scientific names or key words and contains information about strain history, taxonomy, reference citations, and cultivation conditions. Database also can be used to search for strains of bacteria, fungi, and yeasts, bacteriophages and other viruses, algae, and protozoa in Japan and elsewhere.	http://www.jcm.riken.go.jp/JCM/aboutJCM.html, http://wdcm.nig.ac.jp/wdcm/JFCC.html
		National Institute of Genetics Cloning Vector and *E. coli* Collections	Provides vectors of *E. coli* that can be stored stably as purified DNA. It also has the *E. coli* Genetic Resources collection—see bacterial collections in this table.	http://www.shigen.nig.ac.jp/cvector/cvector.html
National, comprehensive	TAIWAN	The CCRC, Culture Collection and Research Center	Founded by the Food Industry Research and Development Institute in Hsinchu, Taiwan, in 1982. Began with food microbiology and expanded to serve the interests of industry, agriculture, medicine, environmental research, education, and biotechnology. Includes bacteria, yeasts, filamentous fungi, bacteriophages, and plasmids and phage vectors.	http://wdcm.nig.ac.jp/database/CCINFO/43.html
National consortium, comprehensive	INDONESIA	Various collections, mostly governmental and university	Collections in Indonesia are listed at the WDCM Web site shown. Include yeasts, algae, lichens, protozoa, fungi, bacteria, and viruses.	http://wdcm.nig.ac.jp/indonesia.html
Other comprehensive	AUSTRALIA	UNSWCC, the University of New South Wales Culture Collection	Established in 1969. Provides catalogs of algae and protozoa, fungi and yeast, and bacteria.	http://www.unsw.edu.au/cult.html
		The Division of Food Research, CSIRO, in New South Wales	Has a collection of 3000 fungi and 100 yeasts.	http://wdcm.nig.ac.jp/database/CCINFO/11.html
		The University of Queensland Microbial Culture Collection in Brisbane	Has over 3500 cultures of bacteria and also has algae, fungi, yeasts, and viruses. It provides identification and training and patent deposit services.	http://wdcm.nig.ac.jp/database/CCINFO/7.html
		Other collections	Fifty other culture collections in Australia are described at the Web address shown.	http://wdcm.nig.ac.jp/cgi-bin/CCINFO.pl?australia

continues

Continued

Type	Country	Designation	Description	Web URL
National, comprehensive	NEW ZEALAND	The International Collection of Microorganisms from Plants, Auckland	Includes bacteria and fungi, with primary emphasis on plant pathogens.	http://wdcm.nig.ac.jp/database/CCINFO/298.html
		The New Zealand Reference Culture Collection of Microorganisms	Sponsored by the New Zealand Dairy Research Institute.	http://wdcm.nig.ac.jp/database/CCINFO/165.html
		Other reference culture collections	Including those in the Communicable Disease Centre in Porirua, the Animal Research Centre in Wellington, the New Zealand Forest Research Institute Culture Collection in Rotorua, the International Collection of Microorganisms from Plants, DSIR, Auckland, and the Institute of Environmental Science and Research, Medical Section, in Porirua. These and other New Zealand collections are described in records accessible from the WDCM.	http://wdcm.nig.ac.jp/cgi-bin/CCINFO_pl?Zealand
National consortium, comprehensive	BRAZIL	Base de Dados Tropical	Although culture collections remain separate, the Base de Dados Tropical presents Brazilian National Catalogs of Strains for Algae, Bacteria, Filamentous Fungi and Yeasts, Protozoa, and Viruses, which allows querying across collections varying from one (for viruses) to eighteen (for bacteria) individual collections in Brazil. Maintains a separate database for Colecao de Culturas Tropical Organisms and services and for an Entomopathogenic Fungi Databank and List of Bacterial Cultures of the Biological Institute in Sao Paulo, IBSBF.	http://www.bdt.org.br/colecoes/microrganismo
National collection	MEXICO	CENACUMI, National Center for Microbial Cultures, UNAM, Mexico City	Agricultural, industrial, medical, and marine organisms for biotechnology, food, and environmental sciences. Bacteria, fungi, yeasts, cell lines, and viruses. Patent depository	http://osuno.fsciences.unam.mx/hojos/biointeres.html, http://wdcm.nig.ac.jp/cgi-bin/CCINFO.pl?CENACUMI
National consortium	CANADA	CMGRIS, the Canadian Microbial Genetic Resources Information System	The goal is to develop a comprehensive information system for all microbial genetic resources, bacteria, fungi, yeasts, viruses, algae, and protozoa. The first of the collections to be presented by this information system was the Canadian Collection of Fungal Cultures, CCFC. See below.	http://res.agr.ca/brd/ccc/cccmicro.html
		CCFC, the Canadian Collection of Fungal Cultures	Has over 10,000 strains representing 2500 species and is the primary repository for fungi in the Agriculture and Agri-Food Canada Research Branch and a patent repository.	http://res.agr.ca/brd/ccc
		UAMH, the University of Alberta Microfungus Collection and Herbarium	Has over 9400 strains. Established in 1960 and supported by the Natural Sciences and Engineering Research Council of Canada and the University of Alberta Devonian Botanic Garden. A Canadian repository for fungus cultures of scientific or industrial importance, including all kinds of fungi, but specializing in ascomycetous and hyphomycetous microfungi.	The collection can be searched at http://www.devonian.ualberta.ca/uamh/index.html-ssi
		Other Canadian collections	Information on many other Canadian collections, including university, Forest Service, and public laboratory collections, can be found at the Web address shown.	http://wdcm.nig.ac.jp/cgi-bin/CCINFO.pl?canada

continues

Continued

Type	Country	Designation	Description	Web URL
National consortium, comprehensive	U.S.	USDA/ARS collections, including NCAUR (NRRL), ARSEF, National *Rhizobium* Germplasm Resource Center	See text section on early collections for descriptions. Also in this table, see section on Fungi. Over 80,000 strains in collections of filamentous fungi, yeasts, bacteria, nonpathogenic to animals or humans. The ARSEF and National *Rhizobium* Center have entomopathogenic and rhizobium collections. Patent depository, research programs, identification services.	http://nrrl.ncaur.usda.gov, http://www.ars-grin.gov/nmg, http://www.ppru.cornell.edu/Mycology/ARSEF_Culture_Collection.html
Other—comprehensive	U.S.	ATCC, American Type Culture	See text section on early collections. Holds over 14,000 bacterial strains, 500 viral strains, 26,000 strains of 6500 yeast and fungal species, as well as protozoa, hybridomas, probes, and clones.	http://www.atcc.org
Other—comprehensive	UK location, Sponsored by 34 CAB countries	IMI, International Mycological Institute Culture Collections, Surrey	Described in the early collections section of the text. The bacterial collection and the fungal/yeast collection of 16,500 strains can be searched at the Web site shown. See also entry under Filamentous fungi and yeast section of this table.	http://www.cabi.org/BIOSCIENCE/grc.htm
Other—comprehensive, primarily bacterial	SWEDEN	CCUG, Culture Collection, University of Goteborg	Founded in 1968, holds 40,000 strains of bacteria, also some yeast and fungi. In 1981, the CCUG was a joint founder of the ECCO, European Culture Collections Organization, and in 1984 it became part of the WDCM. Since 1991, a participant in the MINE project, Microbial Information Network Europe. Includes the genera *Helicobacter, Fusobacterium, Bilophila, Haemophilus, Neisseria, Campylobacter, Streptococcus, Actinomyces, Aeromonas, Vibrio, Bacillus, Enterococcus, Staphylococcus,* various members of Enterobacteriaceae, and others. Performs identification of bacterial species and maintains a database of bacterial names.	http://www.ccug.gu.se
Other—comprehensive	FINLAND	HAMBI, the Culture Collection, University of Helsinki	Strains of bacteria, filamentous fungi, and yeasts, ~2400. Department of Applied Chemistry and Microbiology.	http://honeybee.helsinki.fi/MMKEM/hambi
Other—comprehensive	JAPAN	IFO, Culture Collection of the Institute for Fermentation, Osaka	A large collection of over 4400 bacteria, 7700 fungi, and 3000 yeasts as well as animal cell lines and viruses. Nonprofit organization.	http://www.nacsis.ac.jp/ifo/microorg/microorg.htm, http://wdcm.nig.ac.jp/cgi-bin/CCINFO.pl?wdcm191
BACTERIAL				
Bacterial	RUSSIA	IBSO Siberian Culture Collection of Luminous Bacteria	Institute of Biophysics, Krasnoyarsk. Dates to 1970s. Expeditions in ocean waters and other waters and subsequent studies of luminescent bacteria with benthic organisms and in particular with coral reef inhabitants.	http://panizzi.shef.ac.uk/msdn/ibso
Bacterial	RUSSIA	KMMGU Russia Database of Collections in Lomonosov State University, Moscow	Department of Microbiology, Lomonosov State University.	http://panizzi.shef.ac.uk/msdn/kmmgu
Bacterial	U.S., U.K., others	Various consortia and national collections.	See Comprehensive section of this table.	See URLs in Comprehensive section, e.g., NCAUR, ATCC, NCTC, NCIMB, NCFP
Bacterial	RUSSIA	VKM	See Comprehensive national collections section of this table.	See URLs in Comprehensive section

continues

Continued

Type	Country	Designation	Description	Web URL
Bacterial	FRANCE	CIP, The Collection of the Bacterial Strains of Institut Pasteur	See National collections, CNCM. Incorporates the *Pasteurella* laboratory collection, the Pasteur Institute collections of cyanobacteria (see also PCC under Algae, protozoa, and cyanobacteria section of this table), *Rhodococcus, Gordonia, Lactobacillus,* coryneforms, and *Pasteurella*. Collecting began in 1891–1901. See text section on early collections.	http://www.pasteur.fr/applications/CIP
Bacterial	BELGIUM	BCCM/LMG Bacterial	See National consortium, comprehensive section of this table.	http://www.belspo.be/bccm/lmg.htm#main
Primarily Bacterial, some fungal and yeast	ARGENTINA	CCM-A, Collection of Microbiological Cultures, University of Buenos Aires	Has about 900 bacteria and about 50 fungal and yeast cultures and serves as a patent and safe depository and center for identification, training, and consultation.	http://wdcm.nig.ac.jp/cgi-bin/CCINFO.pl?wdcm29
Primarily bacterial	CUBA	Colleccion de Cultivos Finlay and CGEB Culture Collections of Microorganisms	Instituto Finlay, Ciudad de la Habara and Center of Genetics, Engineering and Biotechnology, Havana	http://wdcm.nig.ac.jp/database/CCINFO/471.html and 472.html
Bacterial and fungal	ARGENTINA	The National University of Cordoba, Laboratory of Microbiology	Maintains a specialized collection of about 200 strains of bacteria and fungi, which can be searched via MSDN.	http://panizzi.shef.ac.uk/msdn/uncor
Bacterial—methanogens	U.S.	OCM, the Oregon Collection of Methanogens	Approximately 150 strictly anaerobic methanogens, primarily archaebacteria, and 50 nonmethanogens. At the Oregon Graduate Institute Department of Environmental Science and Engineering. Department also maintains the Subsurface Microbial Culture Collection. Offers strains and cell masses or DNA of certain strains.	http://caddis.esr.pdx.edu//OCM/intro.html
Bacterial—lactic acid and other	JAPAN	NRIC, the Nodai Research Institute Culture Collections, Tokyo University of Agriculture	See also entry under Fungi in this table. Food-fermentation microorganisms. 2500 lactic acid bacteria (also 400 other bacteria, 1400 yeast cultures) Member Japan Society for Culture Collections	http://wdcm.nig.ac.jp/cgi-bin/CCINFO.pl?wdcm747,http://wdcm.nig.ac.jp/wdcm/JFCC.html
Bacterial—Antarctic	AUSTRALIA	ACAM, the Australian Collection of Antarctic Microorganisms	University of Tasmania, Heterotrophic bacteria collected from Antarctica and subantarctic islands and Southern Ocean. Established 1986.	http://www.antcrc.utas.edu/antcrc/micropro/acaminfo.html
Bacterial—*Escherichia* and *Klebsiella*	DENMARK	Statens Serum Institute, WHO International *Escherichia* Centre	Sponsored by the World Health Organization of the UN, maintains ~400 serovars. Collaborating Centre for Reference and Research on *Escherichia* and *Klebsiella*. Performs serotype identification services as well as providing cultures. Copenhagen.	Artellerivej 5,300 Copenhagen S, Denmark Fax: 4532 683868
Bacterial—*E. coli*	U.S.	ECOR, the *E. coli* Reference Collection and DEC	The ECOR was established by H. Ochman and R. Selander (1984) from human and animal hosts to represent variation in O and H serotypes. The DEC collection is a second such collection. Both are available from Pennsylvania State University.	http://www.bio.psu.edu/People/Faculty/Whittam/Lab/ecor and /deca
Bacterial—*Bacillus*, genetic	U.S.	BGSC, *Bacillus* Genetic Stock Center, Ohio State University	See bacterial genetic stock centers section of text. Publishes genetic map. *B. subtilis* and other *Bacillus* species in addition to strain distribution.	http://bacillus.biosci.ohio-state.edu

continues

Continued

Type	Country	Designation	Description	Web URL
Bacterial—*E. coli,* genetic	U.S.	CGC, *E. coli* Genetic Stock Center, Yale University	See bacterial genetic stock centers section of text. Publishes genetic map, registers gene symbols and allele numbers in addition to strain distribution.	http://cgsc.biology.yale.edu
Bacterial—*E. coli*	JAPAN	NIG, National Institute of Genetics	See Bacterial Genetic Stock Centers section of text. Approximately 4000 strains and over 400 cloning vectors	http://www.shigen.nig.jp/ecoli, http://www.shigen.nig.jp/cvector/cvector.html
Bacterial—*Salmonella,* genetic	CANADA	SGSC, *Salmonella* Genetic Stock Centre, University of Calgary	See bacterial genetic stock centers section of text. Publishes genetic map, registers gene symbols and allele numbers in addition to strain distribution.	http://www.acs.ucalgary.ca/~kesander
Bacterial—*Salmonella*	POLAND	National *Salmonella* Center	See National, comprehensive section of this table, PCM	http://immuno.pan.wroc.pl:80/pcm
Bacterial—*Rhizobium*	POLAND	CRS Collection of *Rhizobium* strains, Pulawy	See National, comprehensive section of the table, PCM	http://immuno.pan.wroc.pl:80/pcm
Bacterial—*Rhizobium*	UK	RCR Rothamstead Collection of *Rhizobium,* Harpenden	Collection of *Rhizobium* of the Soil Microbiology Department.	http://www.iacr.bbsrc.ac.uk/res/depts/plantpath/links/pplinks/cultures
Bacterial—*Rhizobium*	U.S.	USDA/ARS National Collection of *Rhizobium*	Part of the USDA/ARS National *Rhizobium* Germplasm Research Center, combining USDA/ARS collections with the Lipha-Tech, Boyce Thompson Institute, University of Minnesota ARS Collection, the University of São Paulo, Brazil Collection, and the Centro Internacional de Agricultura Tropical (CIAT) collection in Cali, Colombia.	http://bldg6.arsusda.gov/pberkum/Public/ccla.html
Bacterial—alkanotrophic	RUSSIA	IEGM, Institute of Ecology and Genetics of Microorganisms, Moscow	Regional specialized collection of alkanotrophs (which assimilate *n*-alkanes gases and liquid hydrocarbons as carbon and energy source) with 400 species, 1000 strains. Includes mycobacteria and actinomycetes—*Rhodococcus, Gordona, Micrococcus, Brachybacteria* species. Collecting began in 1975 for studies of bioindicators of gas and oil fields and pollution. Russian Academy of Science, Urals Branch.	http://wdcm.nig.ac.jp/wdcm/IEGM_readme.html
FUNGI, INCLUDING YEASTS				
Plant pathogens	UK	The Institute of Arable Crops—Rothamstead (IACR) Plant Pathology Culture Collections	Specialized collection that holds over 600 soilborne fungi from around the world, with particular emphasis on cereal root pathogens of the Gaeumannomyces–Phialophora complex of fungi. The GPDATA records include collecting information and pathogenicity tests, and queries based on name or origin of culture can be entered from the Web site.	http://www.iacr.bbsrc.ac.uk/res/depts/plantpath/links/pplinks/cultures
Fungi	CZECH REPUBLIC	CCF and CCBAS	See National, comprehensive section of this table.	http://www.biomed.cas.cz/ccbas/fungi.htm, and http://panizzi.shef.ac.uk/msdn/ccf
Fungi	CANADA	CCFC and UAMH	See National consortium section of this table.	http://res.agr.ca/brd/ccc/cccintro.html and http://www.devonian.ualberta.ca/uamh/index.html-ssi

continues

Continued

Type	Country	Designation	Description	Web URL
Filamentous fungi and yeasts	BELGIUM	The Mycological Collection of the Catholic University of Louvain, MUCL, and IHEM	*Penicillium, Aspergillus,* hyphomycetes and other fungi, and yeasts. See National Consortium section of this table, Belgium, BCCM/MUCL and BCCM/IHEM.	http://www.belspo.be/bccm/
Filamentous fungi and yeasts	THE NETHERLANDS	See National, comprehensive section of table, CBS	Note particularly filamentous fungi in Baarn, yeasts in Delft and Oeiras, Portugal, and taxonomic and nomenclature databases for *Fusarium* and Aphyollophorales.	http://www.cbs.knaw.nl/www/search_fdb.html and search_ydb.html
Filamentous fungi and yeasts	UK	See National consortium section, NCPF, N, NCYC, NCWRF, IACR	Note particularly NCPF for pathogenic fungi, NCYC for yeast cultures, NCWRF for wood-rotting fungi, and the NCPPB and IACR for plant pathogens.	http://dtiinfo1.dti.gov.uk/bioguide/culture.htm
Filamentous fungi and yeasts	Multinational, in UK	IMI, International Mycological Institute	See early collections section of text. Holds 16,500 strains. Part of CAB International, supported by 32 governments.	http://www.cabi.org/BIOSCIENCE/grc.htm or http://wdcm.nig.ac.jp/cgi-bin/MSDN/IMI.pl
Filamentous fungi and yeasts	SWEDEN	UPSC, Mycoteket, Uppsala University	Over 3000 fungal cultures. See early collections section of text.	http://ups.fyto.uu.se/mykotek/index.html
Fungi—mycorrhizal	U.S.	The International Culture Collection of Vesicular–Arbuscular Mycorrhizal Fungi (INVAM) at West Virginia University	Holds and distributes strains representing more than 60 of the 154 known mycorrhizal species and has other strains yet to be identified. NSF-supported collection of vesicular and vesicular–arbuscular mycorrhizal fungi, originated with N. Schenck at the University of Florida in 1985, moved to West Virginia University in 1990, combining with J. Morton's collection. Over 1100 cultures, including species in the genera *Acaulospora, Entrophospora, Gigaspora, Glomus,* and *Scutellospora.* Web site information on species, nomenclature, methods, and services; querying by name or site characteristic.	http://invam.caf.wvu.edu/invam.htm. Links to general mycorrhizal information are also maintained (http://mycorrhiza.ag.utk.edu and http://invam.caf.wvu.edu/articles.htm)
Fungi—mycorrhizal (arbuscular)	EUROPEAN	BEG, the European Bank of the Glomales	European "Stock Center without Walls," with isolates representing temperate, tropical, and polar representatives of Glomales species. Database of these European endomycorrhizal collections can be queried by genus, species, continent, country, biome, or associated plant.	http://wwwbio.ukc.ac.uk/beg/
Fungi—entomopathogenic	U.S.	ARSEF, USDA-ARS Collections of Entomopathogenic fungi, Ithaca, New York	See early collections section of the text and National collection section of this table.	http://128.253.66.86/Mycology/ARSEF_Culture_Collection.html
Fungi	U.S.	USDA/ARS Systematic Mycology Laboratory Culture Collection, Beltsville, Maryland	A fungal culture collection as well as database of cultures within USDA/ARS system.	http://nt.ars-grin.gov/sbmlweb/cultures/cultureframe.htm
Fungi (including yeast)	U.S.	USDA/ARS/NCAUR Peoria, Illinois	Extensive holdings. See early collections section in text. Largest U.S. yeast collection (including Antarctic Marine Yeast Collection, described below). Also *Aspergillus* and *Penicillium* collection and collection of other filamentous fungi. Research, identification services, and patent depository.	http://nrrl.ncaur.usda.gov, http://nrrl.ncaur.usda.gov/the_collection3.htm

continues

Continued

Type	*Country*	*Designation*	*Description*	*Web URL*
Marine yeasts		The Antarctic Marine Yeasts Collection	Established in 1989 to preserve 3000 strains of yeast isolated and studied during the 1960s from the Antarctic, South Pacific, and Indian oceans. Contributed by J. Fell of the University of Miami to NCAUR Yeast Collection.	See http://nrrl.ncaur.usda.gov
Yeast	U.S.	The Phaff Collection of Yeasts and Yeast-like Microorganisms at University of California, Davis	Specialized collection emphasizing microbial diversity in natural habitats. Dedicated in 1996, but its history traces back to wine- and food-related yeasts at the College of Agriculture and Experiment Station and the Food Science Department. In 1939, H. J. Phaff began to develop the collection into a yeast taxonomy and ecology collection. Includes over 6000 strains from North and South America, Asia, Australia, and Europe.	On-line searches of the collection database can be made from the Web site http://www.defenseweb.ccm/phaff.
Filamentous fungi and yeasts, primarily	SLOVENIA	MZKI, Microbial Culture Collection of the National Institute of Chemistry	Formerly the Filamentous Fungi Collection, Kernijski Institute in Ljubljana, now at the National Institute of Chemistry, Hajdrihova. Storage (including bacteria), identification, patent depository, and consultation services.	http://wdcm.nig.ac.jp/cgi-bin/CCINFO.pl?Slovenia
Filamentous fungi and yeasts	AUSTRALIA	See National collection section of table	Comprehensive section of table—National consortium, especially CSIRO collection, New South Wales.	http://wdcm.nig.ac.jp/CCINFO.pl?australia
Filamentous fungi and yeasts, primarily	BRAZIL	ESAP, Escola Superior de Agricultura collection in São Paulo	Approximately 1000 cultures of fungi and yeasts and 70 bacteria.	http://wdcm.nig.ac.jp/cgi-bin/CCINFO.pl?wdcm294
Filamentous fungi and yeasts	BRAZIL	URM, Federal University of Pernambuco	Approximately 2000 fungi and yeasts.	http://wdcm.nig.ac.jp/cgi-bin/CCINFO.pl?wdcm604
Filamentous fungi and yeasts	BRAZIL	IOC, Culture Collection of the Instituto Oswaldo Cruz in Rio de Janeiro	Approximately 1400 fungal cultures. Training and consultation services.	http://wdcm.nig.ac.jp/cgi-bin/CCINFO.pl?IOC
Fungi	JAPAN	NRIC, the Nodai Research Institute Culture Collection, Tokyo	Tokyo University of Agriculture. Primarily food fermentation microorganisms. Over 1400 yeast cultures and 400 molds. See also bacterial entry in table.	http://wdcm.nig.ac.jp/cgi-bin/CCINFO.pl?wdcm747
Fungi	INDIA	DUM, Delhi University Mycological Herbarium collection	Department of Botany Mycological Herbarium. 200 cultures of fungi. Performs identification services.	http://wdcm.nig.ac.jp/cgi-bin/CCINFO.pl?wdcm40
Fungi—Basidiomycetes	INDIA	ITCC, India Type Culture Collection Mushroom Cultures	Indian Agricultural Institute Department of Pathology. Has *Agaricus, Auricularia, Lentinus,* and *Pleurotus* species.	http://panizzi.shef.ac.uk/msdn/mushroom/itcc.html
Fungi	RUSSIA	RIAM, Research Institute of Applied Microbiology, Obolensk	Moscow region, Russian Fungal Collection.	http://panizzi.shef.ac.uk/msdn/riam
Fungi—Basidiomycetes	RUSSIA	Le(BIN) Basidiomycete Collection, St. Petersburg	Komarov Botanical Institute of the Russian Academy of Sciences	http://panizzi.shef.ac.uk/msdn/lebin
Yeast	RUSSIA	MSU, Moscow State University Yeast Database	Department of Soil Microbiology. See also yeast collection within VKM National collection, Comprehensive section of this table.	http://panizzi.shef.ac.uk/msdn/msu

continues

Continued

Type	Country	Designation	Description	Web URL
Yeast	GERMANY	IMET National Kultурensammlung von Microorganismen, Jena	Zentralinstitut fur Mikrobiologies und Experimentelle Therapie, ZIMET. Also yeasts within DSMZ National Collection.	Available at http://www.dsmz.de
Yeast (also see Comprehensive—other)	JAPAN	IFO, Culture Collection of the Institute for Fermentation, Osaka	A large collection of over 4400 bacteria, 7700 fungi, and 3000 yeasts as well as animal cell lines and viruses.	http://wdcm.nig.ac.jp/cgi-bin/CCINFO.pl?wdcm191, http://wwwsoc.nacsis.ac.jp/ifo/microorg/microorg.htm
Fungi and yeast	U.S.	Fungal Genetics Stock Center and Yeast Genetics Stock Center	Described in the text section on early genetic stock centers.	http://www.fgsc.net and http://phage.atcc.org/searchengine/ygsc.htm
Fungi and yeast	BRAZIL	Individual Collections	See National consortium section of this table.	
Yeast	RUSSIA	Peterhof Genetic Collection of Yeasts	Described in the text section on early genetic stock centers.	http://panizzi.shef.ac.uk/msdn/peter
ALGAE, PROTOZOA, AND CYANOBACTERIA				
Algae, protozoa, and cyanobacteria	U.S.	UTEX, the Culture Collection of Algae at the University of Texas	Over 2100 strains of primarily freshwater algae. Began at Indiana University in 1954 and moved with Prof. R. C. Starr to Austin in 1976. Supported by the NSF and the University of Texas Organized Research Units. Lists of UTEX cultures and genera can be accessed from the Web site, along with information on culture media and links to other collections.	http://www.botany.utexas.edu/infores/utex
Algae, protozoa, and cyanobacteria	U.K.	CCAP, the Culture Collection of Algae and Protozoa	See early collections section of text and National Consortium—UK entries in this table. Approximately 2000 strains of algae and protozoa at IFE and DML. Patent depository.	http://www.ife.ac.uk/ccap
Algae, protozoa, and cyanobacteria	CZECH REPUBLIC	CCALA, the Culture Collection of Autotrophic Organisms	Began with early algal collection of Uhlir and Pringsheim in Prague, 1913, and merged in 1979 with the collection of the Institute of Microbiology, Czech Academy of Sciences at Trebon, founded in 1960. See early collections section of text. Over 200 strains of cyanobacteria and over 300 algae, and also has mosses, liverworts, ferns, and duckweed. Searchable via the MSDN.	http://panizzi.shef.ac.uk/msdn/ccala
Algae, protozoa, and cyanobacteria—national center	U.S.	CCMP, Provasoli-Guillard National Center for Culture of Marine Phytoplankton, Boothbay Harbor, Maine	Originated with the collections of Dr. Luigi Provasoli at Yale University and Dr. R. Guillard at Woods Hole Oceanographic Institution and became a national center as a result of recommendations of a 1980 workshop. Originally located at Woods Hole, but moved to the Bigelow Laboratory at Boothbay Harbor in 1981. Approximately 1450 cultures of primarily well-identified organism in pure culture. Special services include culture purification, safe deposit of private cultures, supplying mass cultures, DNA extractions, and occasional courses on culturing phytoplankton. Supported by the National Science Foundation. The culture database contains taxonomic and collection information, and the Web site also presents information on culture techniques.	http://ccmp.bigelow.org

continues

Continued

Type	*Country*	*Designation*	*Description*	*Web URL*
Algae	GERMANY	SVCK, the Culture Collection of Conjugatophyceae	Began at the Free University in Berlin and in 1959 moved to Marburg. Established at the Institute of General Botany of the University of Hamburg in the 1960s. Approximately 500 strains, collected world-wide. Education, training, and research functions. The Web site lists culture and collection information by family.	http://www.rrz.uni-hamburg.de/biologie/b_online/d44_1/44_1.htm
Cyanobacteria	FRANCE	PCC, Pasteur Culture Collection of Cyanobacteria	Axenic cyanobacterial strains, one of the specialized collections of the Institut Pasteur in Paris and the Collection Nationale de Cultures de Microorganismes (CNCM) of France. See National collection section of table. Member of the WFCC and ECCO. Information includes accession number, brief description, history of isolation, cross-references with other collections with same species, medium for growth, properties, synonymous names for the species, and references. Keyword searches or specific, designated property searches are provided on their Web search pages.	http://www.pasteur.fr/Bio/PCC/General.htm, http://www.pasteur.fr/recherche/banques/PCC
Algae and cyanobacteria	RUSSIA	IPPAS, the Culture Collection of Microalgae, Moscow	Institute of Plant Physiology, Russian Academy of Sciences. Established in 1958 and contains 600 cultures of cyanobacteria and algae. Patent depository for countries of the former USSR. Member of the ECCO. On-line searches can be conducted via the MSDN.	http://panizzi.shef.ac.uk/msdn/ippas.html
Algae	CANADA	NEPCC, North East Pacific Culture Collection of Marine Microalgae at the University of British Columbia	Founded by F. J. R. Taylor in the late 1960s. Supported by the Department of Botany of the university. Focuses on species of the North Pacific, but includes algae from other areas, and has many dinoflagellates. The Web site lists the algae in the collection by family and gives collection and culturing information.	http://beluga.ocgy.ubc.ca/projects/nepcc
Algae and cyanobacteria	CANADA	UTCC, University of Toronto Culture	Maintains and distributes cultures of over 350 isolates of primarily freshwater algae and cyanobacteria. Established in 1987 from several collections at the University of Toronto, with support from the Ontario Ministry of Colleges and Universities and currently from the Canadian Natural Sciences and Engineering Council and the university Department of Botany. Services include safe deposit, custom isolation of algae, identification, training, and specialized media preparation.	http://www.botany.utoronto.ca/utcc/index.stm
Algae	JAPAN	Marine Biotechnology Institute (MBI) of Japan		See http://seaweed.ucg.ie/cultures/CultureCollections.html
Algae—*Chlamydomonas*	U.S.	*Chlamydomonas* Genetics Center (CGC), Duke University	See genetic stock centers section of text. Mutants, cloned DNA, information on mapping and sequences.	http://www.botany.duke.edu/chlamy
Diatoms	U.S.	The Loras College Freshwater Diatom Culture Collection, Dubuque, Iowa	1200 freshwater diatoms, 64 genera, 350 species and varieties.	http://www.bgsu.edu/Departments/biology/algae/html/DiatomCulture.html

continues

Continued

Type	Country	Designation	Description	Web URL
Algae	AUSTRALIA	Murdoch University's algal collection, Perth	Approximately 400 strains of algae of commercial importance or potential. Includes *Dunaliella, Haematococcus, Spirulina, Chlorella,* and others. Also serves as a safe culture deposit for companies.	http://seaweed.ucg.ie/cultures/CultureCollections.html#Murdoch
Algae	FRANCE	Thallia Pharmaceuticals, SA, Ecole Centrale de Paris	About 250 strains of microalgae.	http://seaweed.ucg.ie/cultures/CultureCollections.html#anchor480299
DATABASE LINKS				
Comprehensive database of collections	Internationally administered	MSDN, Microbial Strain Data Network	Access to large and small collections, private and public. MSDN is a nonprofit company with secretariat in Sheffield and nodes in China, India, and Brazil.	http://panizzi.shef.ac.uk/msdn (Europe), http://202.20.55/www/msdn.html (India), http://www.in.ac.cn/msdn.html (China), http://www.bdt.org.br/bdt (Brazil) (also http://202.41.70.55/cgi-bin/asearch/msdn/cairo/read for MIRCEN—Cairo)
Comprehensive database of collections	International, server in JAPAN	WDCM, World Data Center for Microorganisms	Access to large and small collections, private and public. Broad searches across multiple databases and cross-linking with collection databases.	http://wdcm.nig.ac.jp, http://wdcm.nig.ac.jp/DOC/menu3.html, http://wdcm.nig.ac.jp/cgi-bin/CCINFO.pl search interface, Directory of Culture Collections
Comprehensive database of collections	JAPAN—WDCM	WDCM—AHMII Agent for Hunting Microbial Information across the Internet	Queries across multiple databases.	http://wdcm.nig.ac.jp/AHMII/ahmii.html
Listing of collections	U.S.	Bergey's Manual Trust	Maintains on its Web site a list of major collections and consortia.	http://www.cme.msu.edu/bergeys/culture.html
Listing of collections	International	MIRCENS—Microbial Resources Centres	In addition to the on-line databases that MIRCEN sponsors, it publishes the World Directory of Collections of Cultures of Microorganisms, the World Catalogue of *Rhizobium* Collections, and the World Catalog of Algae and supports collecting and coordination of collections in MIRCEN centers and networks throughout the world. See Section I of the text.	http://www.unesco.org/science/life/life1/mircen.htm and the books cited in preceding column
Laboratory collections	U.S.	MGD, Microbial Germplasm Database	Allows querying over numerous specialized and laboratory collections.	http://mgd.nacse.org/cgi-bin/mgd
Fungal database	U.S.	University of Kansas Database of Mycological Resources	Part of the University of Kansas BioDiversity and Biological Collections Web server.	http://biodiversity.bio.uno.edu/~fungi/fcollect.html
Algal database	IRELAND	Seaweed	National University of Ireland, Galway.	http://www.seaweed.ie
Algal database	JAPAN—WDCM	WDCM—Algae	World Catalog of Algae search engine.	http://wdcm.nig.ac.jp/cgi-bin/ALGAE.pl
Any WDCM collection	Search by country	—	Use query interface shown.	http://wdcm.nig.ac.jp/cgi-bin/CCINFO.pl?countryname

VIII. ORGANIZATIONS PROMOTING CULTURE COLLECTIONS

Organizations fostering stock center activity include the previously cited ECCO, European Culture Collections Organization, national federations such as the U.S. Federation of Culture Collections and the Japan Federation of Culture Collections (which sponsor annual meetings as well as international meetings), UNESCO's MIRCEN, and the WFCC, World Federation of Culture Collections. An International Congress for Culture Collections, sponsored by the international federations, is held every 4 years.

See Also the Following Articles

Patenting of Living Organisms and Natural Products • Strain Improvement

Bibliography

DaSilva, E. J. (1997). Microbial Resources Centres: Springboard for networking in the 21st century. *http://wdcm.nig.ac.jp/a_MIRCEN.html.*

Demerec, M, Adelberg, E. A., Clark, A. J., and Hartman, P. E. (1966). A proposal for a uniform nomenclature in bacterial genetics. *Genetics* 54: 61–76.

Hawksworth, D. L. (1990). "WFCC Guidelines for the Establishment and Operation of Collections of Cultures of Microorganisms." Simworth Press, Richmond, Surrey, UK, and *http://www.wdcm.nig.ac.jp/wfcc/wfcc_guidelines.html.* Revised, *http://www.wdcm.nig.ac/wfcc/GuideFinal.html.*

Hunter-Cevera, J. C., (1998). The importance of culture collections in industrial microbiology and biotechnology. *http://www.bdt.org.br/bdt/oeaproj/cevera.*

Hunter-Cevera, J. C., and Belt, A. (1996). "The Importance of Culture Collections in Preservation and Maintenance of Microorganisms Used in Biotechnology." Academic Press, San Diego.

Jong, S.-C. (1989). Microbial germplasm. *In* "Beltsville Symposia in Agricultural Research 13: Biotic Diversity and Germplasm Preservation, Global Imperatives" (L. Knutson and A. K. Stoner, eds.), pp. 241–273. Kluwer Academic Publishers, Boston.

Kurtzman, C. P. (1986). The ARS culture collection: Present status and new directions. *Enzyme Microb. Technol.* **8**, 328–333.

Nierlich, D. P., Benson, D., Berlyn, M. K. B., Blaine, L., Karp, P., and Sanderson, K. (1999). World Wide Web Resources for Microbiologists. *Biotechniques* **6**, 70–78.

Ochman, H., and Selander, R. (1984). Standard reference strains of *Escherichia coli* from natural populations. *J. Bacteriol.* **157**, 690–693.

WFCC Executive Board. (1999). WFCC Guidelines for the Establishment and Operation and Collections of Cultures of Microorganisms. 2nd edition. WFCC Newsletter No. 30. *http://www.wdcm.nig.ac.jp/wfcc/GuideFinal.html.*

WFCC Publications: see *http://wdcm.nig.ac.jp/wfcc/publications.html.*

WFCC Workshop on the Economic Value of Microbial Genetic Resources at the Eighth International Symposium on Microbial Ecology. (1998). Halifax, Canada. *http://wdcm.nig.ac.jp/wfcc/Halifax98.html#t1.*

Strain Improvement

Sarad Parekh
Dow AgroSciences

I. Introduction
II. Attributes of Improved Strains
III. Significance, Impacts, and Benefits
IV. Strategies for Strain Development
V. Improved Strain Performance through Engineering Optimization
VI. Summary

GLOSSARY

DNA recombination A laboratory method in which DNA segments from different sources are combined into a single unit and manipulated to create a new sequence of DNA.

fermentation A metabolic process whereby microbes gain energy from the breakdown and assimilation of organic and inorganic nutrients.

gene Physical unit of heredity. Structural genes, which make up the majority, consist of DNA segments that determine the sequence of amino acids in specific polypeptides. Other kinds of genes exist. Regulatory genes code for synthesis of proteins that control expression of the structural genes, turning them off and on according to circumstances within the microbe.

gene cloning Procedure employed where specific segments of DNA (genes) are isolated and replicated in another organism.

genetic code The linear sequence of the DNA bases (adenine, thymine, guanine, and cytosine) that ultimately determines the sequence of amino acids in proteins. The genetic code is first "transcribed" into complementary base sequences in the messenger RNA molecule, which in turn is "translated" by the ribosomes during protein biosynthesis.

genetic recombination When two different DNA molecules are paired, those regions having homologous nucleotide sequences can exchange genetic information by a process of natural crossover to generate a new DNA molecule with a new nucleotide sequence.

interspecific protoplast fusion Method for recombining genetic information from closely related but nonmating cultures by removing the walls from the cells.

metabolic engineering A scientific discipline that integrates the principles of biochemistry, chemical engineering, and physiology to enhance the activity of a particular metabolic pathway.

mutation Genetic lesion or aberration in DNA sequence that results in permanent inheritable changes in the organism. The strains that acquire these alterations are called mutant strains.

plasmid An autonomous DNA molecule capable of replicating itself independently from the rest of the genetic information.

primary metabolites Simple molecules and precursor compounds such as amino acids and organic acids that are involved in pathways that are essential for life processes and the reproduction of cells.

secondary metabolites Complex molecules derived from primary metabolites and assembled in a coordinated fashion. Secondary metabolites are usually not essential for the organism's growth.

THE SCIENCE AND TECHNOLOGY of designing, breeding, manipulating, and continuously improving the performance of microbial strains for biotechnological applications is referred to as "strain improvement." The science behind developing improved cultures has been enhanced recently by a greater understanding of microbial biochemistry and physiology, coupled with advances in fermentation reactor technology and genetic engineering. In addition, the availability and application of user-friendly analytical equipment such as high pressure liquid chromatography (HPLC) and mass spectroscopy, which raised the detection limits of metabolites, have also played a critical role in screening improved strains.

I. INTRODUCTION

The use of microbes for industrial processes is not new. Improving the commercial and technical capability of microbial strains has been practiced for centuries through selective breeding of microbes. In making specialty foods and fermented beverages (such as alcohol, sake, beer, wine, vinegar, bread, tofu, yogurt, and cheese), specific strains of bacteria and fungi isolated by chance have been employed to obtain desirable and palatable characteristics. Now, with integrated knowledge of biochemistry, chemical engineering, and physiology, microbiologists have taken a more scientific approach to the identification of microbial strains with desired traits.

Later spectacular successes observed in improvement of the industrial strains by mutation and genetic manipulations in the production of penicillin and other antibiotics led to strain development as a driving force in the manufacture of pharmaceuticals and biochemicals. Microbes are now routinely used in large-scale processes for the production of lactic acid, ethanol fuel, acetone–butanol, and riboflavin as well as for the commercial production of enzymes such as amylases, proteases, and invertase. Efforts were also made by chemical engineers to improve fermenter designs on the basis of understanding the importance of culture media components, sterile operations, aeration, and agitation. Today, production of hormones, steroids, vaccines, monoclonal antibodies, amino acids, and antibiotics are testimonies to the important role of strain improvement in the pharmaceutical industry.

The intent of this article is to briefly describe strategies employed in strain improvement, the practical aspects of screening procedures, and the overall impact that strain improvement has on the economics of fermentation processes. Readers are also urged to review the additional articles described in the bibliography, especially the basic concepts as well as the theoretical basis of genetic mutations and screening improved strains.

II. ATTRIBUTES OF IMPROVED STRAINS

Microbial strain improvement cannot be defined simply in terms of modifying the strain for overproduction of bioactive compounds. Strain improvement should also be viewed as making the fermentation process more cost-effective. Some of the traits unique to fermentation process that make a strain "improved" are the ability to (a) assimilate inexpensive and complex raw materials efficiently; (b) alter product ratios and eliminate impurities or by-products problematic in downstream processing; (c) reduce demand on utilities during fermentation (air, cooling water, or power draws); (d) excrete the product to facilitate product recovery; (e) provide cellular morphology in a form suitable for product separation; (f) create tolerance to high product concentrations; (g) shorten fermentation times; and (h) overproduce natural products or bioactive molecules not synthesized naturally, for example, insulin.

A. Need for Strain Improvement

Microbes (fungi, bacteria, actinomyces) that live freely in soil or water and produce novel compounds of commercial interest, when isolated from their natural surroundings, are not ideal for industrial use. In general, wild strains cannot make the product of commercial interest at high enough yields to be economically viable. In nature, metabolism is carefully controlled to avoid wasteful expenditure of energy and the accumulation of intermediates and enzymes needed for their biosynthesis. This tight metabolic and genetic regulation, and synthesis of biologically active compounds, is ultimately controlled by the sequence of genes in the DNA that program the biological activity. To improve microbial strains, the sequence of these genes must be altered and manipulated. In essence, microbial strain improvement requires alteration and reprogramming of the DNA (or the genes) in a desired fashion to shift or bypass the regulatory controls and checkpoints. Such DNA alterations enable the microbe to devote its metabolic machinery to producing the key biosynthetic enzymes and increasing product yields. In some cases, simple alteration in DNA can also lead to structural changes in a specific enzyme that increases its ability to bind to the substrate, enhance its catabolic activity, or make itself less sensitive to the inhibitory effects of a metabolite. On the other hand, when the changes are made in the regulatory region of the gene (such as the promoter site), it can lead to deregulation of gene expression and overproduction of the metabolite. A typical example is overproduction of the enzyme amylase, where specific constitutive mutants have been developed that produce the enzyme even in the absence of the starch inducer (Elander and Vournakis, 1986).

Knowledge of the functions of enzymes, rate-limiting steps in pathways, and environmental factors controlling synthesis further help in designing screening strategies. The outcome of the strain selection, however, depends primarily on the kind of improvement desired from the microbe. For instance, increased product yield that involves the activity of one or more genes, such as enzyme production, may be enhanced simply by increasing the gene dosage. Molecules such as secondary metabolites and antibiotics that are complex in structure and require a coordinated as well as highly regulated biosynthetic process, however, may require a variety of alterations in the genome to derive a high yielding strain. Apart from modifying the strains genetically, the success of a strain improvement program also depends on developing and combining more efficient ways of screening, testing, and confirming the improved and high yielding status of the mutants against a background of nonimproved strains.

III. SIGNIFICANCE, IMPACT, AND BENEFITS

Strain improvement is the cornerstone of any commercial fermentation process. In most cases, it determines the overall economics. Fermentation economics is predominantly determined by manufacturing cost per unit of product made (e.g., "dollars per pound") and the cost associated with plant construction and start-up. Although lower fermentation manufacturing and capital costs can be anticipated from fermenter engineering design, improvement of microbial strains offers the greatest opportunity for cost reduction. Great efforts are therefore expended to develop industrial strains that have an increased ability to produce the compound of interest at a faster rate. Enhanced productivity of the fermentation process through strain improvement (more product/vessel/unit time, e.g., grams per liter per hour) is one factor that makes the most impact. It can determine the ability of a manufacturing process to meet additional demands without adding more fermenters. Furthermore, the strain that can synthesize a higher proportion of the product using the same amount of raw material can also reduce material and manufacturing costs significantly. For example, strains that utilize low-cost materials such as starch or corn syrup, or spent products like molasses (instead of refined glucose), can reduce fermentation costs significantly.

Improvement of industrial strains is clearly justified when one takes into account the additional anticipated capacity and extra fermenters (capital cost) required to meet the throughput in the absence of titer gains and strain development efforts. Through strain improvement, one can free up fermenters and facilitate the launching of other fermentation products in the pipeline. Also of great importance is the use of genetically engineered microbes that manufacture nonmicrobial products such as insulin, interferon, human growth hormone, and viral vaccines that cannot be produced efficiently by other manufacturing processes.

IV. STRATEGIES FOR STRAIN DEVELOPMENT

Several procedures are employed to improve microbial strains. All bring about changes in DNA sequence. These changes are achieved by mutation, genetic recombination, or the modern DNA splicing techniques of "genetic engineering." Each procedure has distinct advantages. In some cases, a combination of one or more techniques is employed to attain maximum strain improvement.

A. Mutation

Microbes, generation after generation, generally inherit identical characteristics as their parents.

However, when changes occur in the DNA, they too are passed on to daughter cells and inherited in future generations. This permanent alteration of one or more nucleotides at a specific site along the DNA strand is called a genetic mutation. The strain that harbors the mutation is called a mutant strain. Although a gene consists of hundreds or even thousands of base pairs, a change in just one of these bases can have a significant change in the function, operation, and expression of the gene or in the function of its protein product. A mutant by definition will differ from its parent strain in genotype, the precise nucleotide sequence in the DNA of the genome. In addition, the visible property of the mutant, its phenotype, may also be altered relative to the parent strain.

A point mutation may be associated with a change in a single nucleotide, through substitution of one purine for another purine or substitution of one pyrimidine for another pyrimidine (transition), or through substitution of one purine by a pyrimidine or vice versa (transversion). Mutations may also result from deletion of one or more base pairs, insertion of base pairs, or rearrangement of the chromosome due to breakage and faulty reunion of the DNA. These changes in base pair arrangements can alter the "reading frame" of the gene (frameshift mutations), and during the transcription and translation process also changes the amino acid sequence in the resulting protein. Most mutations occur on a chromosome structure at a specific site or locus (gene mutations).

Genetic mutations do occur spontaneously, at low frequency at any point along the gene (10^{-5} to 10^{-10} per generation). Some mutations are the result of integration or excision of insertion sequence elements and result in subtle modification of the genetic sequence. In many cases mutations are harmful, but certain mutations occur that make the organism better adapted to its environment and improve its performance. The potential for a microbe to mutate is an important property of DNA since it creates new variation in the gene pool.

Modification of the strain through mutation can also be induced at will, by subjecting the genetic material to reaction with a variety of physical and chemical agents called mutagens. Examples of some known mutagenic agents are listed in Table I. Each agent includes DNA alterations in a specific manner, and in some cases, an agent may induce more than one type of lesion. Most cause some damage to the DNA through deletion, addition, transversion, or substitution of bases or breakage of DNA strands (Table I). Although microbes have systems to repair the damaged or altered DNA and return it to its original form, the repairing and editing mechanisms are not errorproof. Thus, when DNA reacts with mutagenic agents for longer periods of time, the damage in the DNA cannot be repaired to the correct genetic sequence with the same rapidity and accuracy as in normal circumstances. The microbes (progeny) that survive the changes in their genetic DNA sequence usually acquire an altered genetic code for "reprogrammed" metabolic and biosynthetic activity. The frequency of bacterial mutation for a particular trait is low: 1 per 10^4 to 10^{10} cells per generation.

In addition to mutation, alteration in DNA can also occur by genetic recombination. Here information from two similar but different DNA molecules is brought together and recombined by crossing-over, resulting in novel DNAs of different lengths so that new combinations of mutations are produced. This allows the circumvention of slow leaps to obtain new combinations of desired characteristics in microbes. Genetic engineering is usually employed to create targeted mutations on the genes, unlike other methods of mutation that are random. It must be emphasized that this technology is not a way of constructing new forms of life. Even the genetic materials of the simplest organisms are highly complex, and insertion of a few genes from an unrelated organism will not create a new microbe.

On the basis of the method of screening and selection chosen, there are basically two methods of improving microbial strains through random mutation: (1) random selection and (2) rationalized selection.

1. Random Selection

Random mutagenesis and selection is also referred to as the classic approach or nonrecombinant strain improvement procedure. Improved mutants are normally identified by screening a large population of mutated organisms, since the mutant phenotype may

TABLE I
Mutagens Employed for Strain Development

Mutagen	Mutation induced	Impact on DNA	Relative effect
Radiation			
Ionizing radiation			
1. X rays, gamma rays	Single- or double-strand breakage of DNA	Deletions, structural changes	High
Short wavelengths			
2. Ultraviolet rays	Pyrimidine dimerization and cross-links in DNA	Transversion, deletion, frameshift, GC → AT transitions	Medium
Chemicals			
Base analogs			
3. 5-Chlorouracil, 5-bromouracil	Faulty base pairing	AT → GC, GC → AT transition	Low
4. 2-Aminopurine	Errors in DNA replication	—	Low
Deaminating agents			
5. Hydroxylamine (NH_2OH)	Deamination of cytosine	GC → AT transition	Low
6. Nitrous acid (HNO_2)	Deamination of A, C, and G	Bidirectional translation, deletion, AT → GC and/or GC → AT transition	Medium
Alkylating agents			
7. NTG (*N*-methyl-*N*′-nitro-*N*-nitrosoguanidine)	Methylation, high pH	GC → AT transition	High
8. EMS (ethyl methanesulfonate)	Alkylation of C and A	GC → AT transition	High
9. Mustards, di-(2-chloroethyl) sulfide	Alkylation of C and A	GC → AT transition	High
Intercalating agents			
10. Ethidium bromide, acridine dyes	Intercalation between two base pairs	Frameshift, loss of plasmids, microdeletions	Low
Biological			
11. Phage, plasmid, DNA transposons	Base substitution, breakage	Deletion, duplication, insertion	High

not be easy to recognize against a large background. After inducing mutations in the culture, the survivors from the population are randomly picked and tested for their ability to produce the metabolite of interest. This approach has the advantage of being simple and reliable. Moreover, it offers a significant advantage over the genetic engineering route alone by yielding gains with minimal start-up and sustaining such gains over years despite a lack of scientific knowledge of the biosynthetic pathway, physiology, or genetics of the producing microbe (Lein, 1986). This empirical approach has been widely adapted by the fermentation industry, following the successful improvement in penicillin titers since World War II. One drawback to the random selection approach is that it relies on nontargeted, nonspecific gene mutations, so many strains need to be screened to isolate the improved mutant in the mixed population. In addition when the culture is mutagenized, multiple mutations may be introduced in the strain. This may result in the enfeeblement of the organism lacking the properties of interest.

This process of strain improvement involves repeated applications of three basic principles: (1) mutagenesis of the population to induce genetic variability, (2) random selection and screening from the surviving population of improved strains by small-scale model fermentation, and (3) assaying of fermentation broth/agar for products and scoring for improved strains. It must be emphasized that the action of the mutagenic agent on DNA not only can cause genetic alteration but can also induce cell death, owing to irreversible damage to the DNA or formation of lethal mutations. Hence, after the mutagenic treatment, mutants are sought among the surviving population with the anticipation that each of the surviving cells harbors one or several mutations. Each time an improved strain is derived through mutation, it is used again as the parent strain in a new cycle of mutation, screening by fermentation (liquid or solid), and assay (Fig. 1). This random procedure of mutant selection is continued until a strain is derived that is statistically superior in performance compared to the control strain prior to the mutagenic treatment. The objective of mutagenesis is to maximize the frequency of desired mutations in a population while minimizing the lethality of the treatment. For this purpose, nitrosoguanidine (NTG) has been the mutagen of choice because it offers the highest possible frequency of mutants per survivor (Baltz, 1986). The efficiency of the random selection process is dependent on several factors: the type of culture used (such as spores or conidia), mutagen dose and exposure time, the type and damage to DNA, conditions of treatment and posttreatment, frequency of mutagen treatment, and the extent of yield increase detectable.

In addition to the mutation conditions, the test or quantitative and analytical screening procedures employed (bioassays, radioimmunoassays, chromatography, HPLC) also play a critical role in successful isolation of superior mutants. The ability to detect a gain mutant among the randomly selected mutants is greatly influenced by the process as well as the variability within the process and the actual titer differences between the improved strain and the control (Rowlands, 1984). The screening procedure therefore is usually designed to maximize the precision and selectivity of improved cultures (gain per sample tested) and to minimize the variability (measured as the coefficient of variation) when treating the unmutagenized control and reference samples. All the strains tested, including the control, are normally worked up all the way head-to-head from the initial cell clone stage to the final screening stage. This is essentially the test of significance. As such, if all the treatment conditions were the same, a successful test should show a statistical difference between the means of the control and improved cultures.

Putative mutants isolated after a primary run are subjected to secondary and tertiary confirmations (replications and repetitions) to raise the level of confidence and observe the anticipated titer differences on data collections (Fig. 1). A desired improvement in the strain is typically obtained with less testing if the selection system is less susceptible to variability and if the coefficient of variation is lower. Considerable efforts are therefore directed at troubleshooting important handling procedures to identify and eliminate key contributors to the errors in the process. Furthermore, the screening strategy is carefully chosen so that the medium and fermentation parameters mimic large-scale production. This increases the possibility that the improved performance of the mutant will be achieved at scale-up. The random approach of strain selection relies on delivering small incremental improvements in culture performance. Although the procedure is repetitive and labor intensive, this empirical approach has a long history of success and has given dramatic increases in titer improvement, as best exemplified by the improvements achieved for penicillin production in which titers over 50 g/liter are reported—a 4000-fold improvement over the original parent strain (Crueger and Crueger, 1984). These authors have also cited certain actinomyces or fungal strains capable of overproducing metabolites in quantities as high as 80 g/liter. Not surprisingly, therefore, pharmaceutical and other fermentation industries typically adopt this technique for selection of improved mutants for many of their processes. The historic successes (e.g., production of antibiotics and other secondary metabolites, enzymes, and amino acids) bear testimony to creating superior strains through this procedure (Queener and Lively, 1986).

The procedure of mutation followed by random selection is laborious and requires screening a large number of strains to obtain desired mutants. This is because in random screening procedures a high percentage of mutants examined will be carried over as survivors from the mutagenesis and will exhibit the same or lower yields than the parent strain. Factors that impact the success of the random program and accelerate strain improvement are the following: the extent of yield improvement, the frequency of induced mutations, the amount of time for turnaround of the mutation selection cycle, and the testing capacity. In addition, the success of a strain improvement program also depends on resource allocation. The key labor-intensive steps in classic strain improvement programs include the isolation of individual mutated cultures, preparation and distribution of sterile media, transfer of clones and their inoculum to initiate fermentation, assays of fermentation broth, and repeated confirmations. Furthermore, the more complex the regulation and biosynthesis of a desired compound, the greater is the number of strains that need to be evaluated and replicated. As a rule, mutants with very high yields are rarer than those with subtle improvements. So, to increase the odds, a larger number of improved strains are examined, raising the probability of detecting improved strains. Thus, if the strain improve-

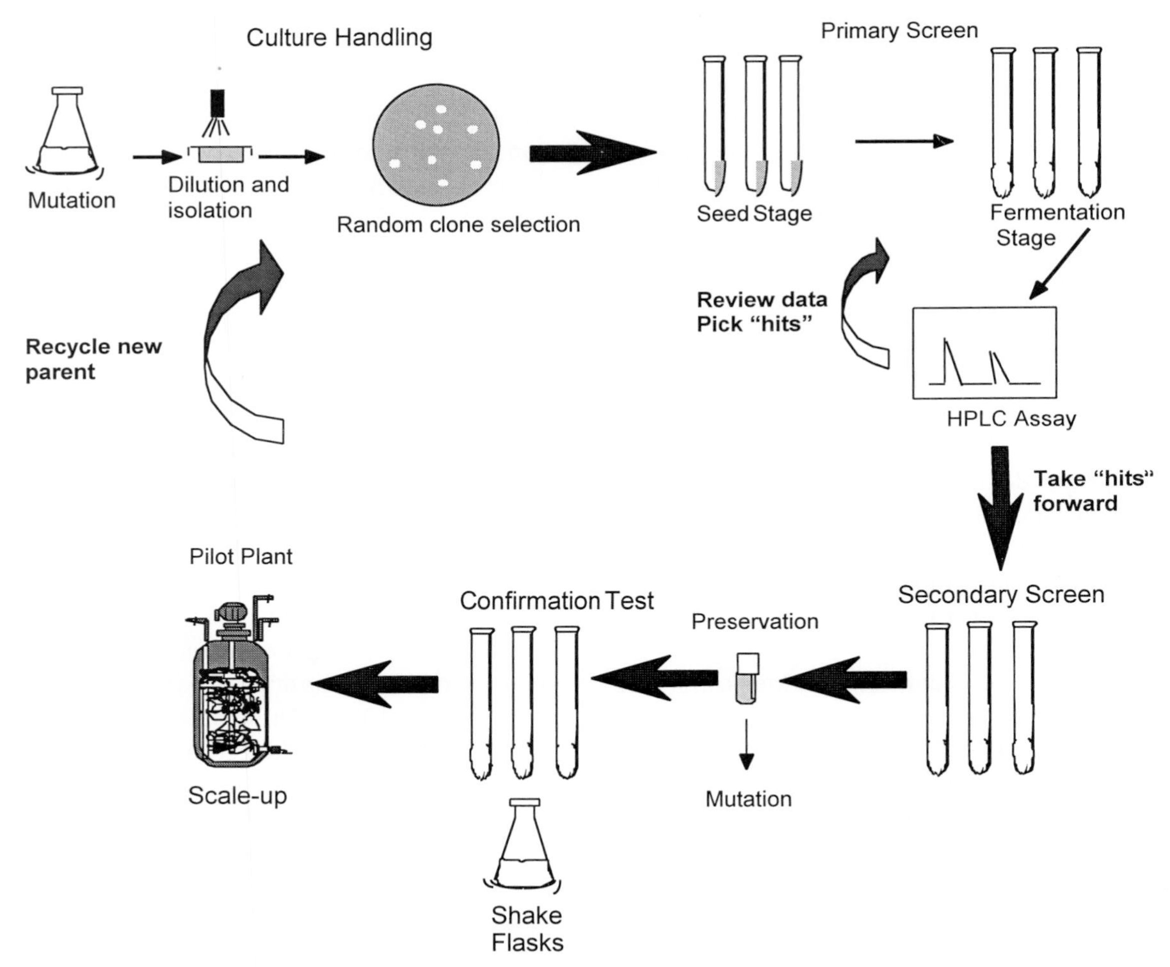

Fig. 1. Typical steps in mutation and random strain selection process.

ment program is operated manually, successful improvements detected will roughly be proportional to the number of personnel allocated (Rowlands, 1984). The advantage of manual screening is in the trained eyes of the microbiologists. They can visually detect the alteration in the morphology, pigmentation, and growth characteristics of mutants during the selection process. In fact, isolation of pelleted strains of filamentous organisms is commonly based on morphological criteria.

To increase the efficiency of random selection, ways by which the key steps in the process can drive the throughput higher without adding labor are typically sought. In some instances high throughput screens have been automated with robotics technology. This allows screening of large populations with minimum resources by miniaturizing wherever possible the equipment required and constructing an automated integrated system. In an industrial system, sterile media are robotically dispensed in custom designed sterilizable and cleanable/disposable modules, each having over 100 tubes or bottles. Individual clones are detected by an optical system and plugged from an agar based medium into liquid seed medium. The inoculations of seed stage culture to fermentation vials are also accomplished by robots. The extraction and HPLC analysis of the fermentation broth is also automated to match the throughput of the screening stage. The advantage of such automation is that it facilitates the capture and downloading of process data and allows statistical process control approaches to be implemented where refinement of the process is required. The success of automated programs requires skillful microbiologists, and constant monitoring and evaluation of the screening system to ensure that all aspects of the automation are functioning efficiently without introducing variability. The significant disadvantage of robotic systems is the initial high capital investment and continued maintenance of equipment and software (Nolan, 1986).

Although other sophisticated techniques are being developed to generate improved strains, random high throughput selection and mutation will continue to be an integral part of any strain improvement program. The random approach is least useful for microbes that are less susceptible to mutagenesis, such as some fungi (owing to their diploid or polyploid genome structure) and bacteria with very efficient repair systems. In a typical manually operated strain improvement project, the expected frequency of gain could be of the order 1 in 10,000, where about 10,000 mutants may pass through the primary screen before a higher producing candidate is identified. However, as the titer increases, depending on the pathway, the organism, the product, and the history of the production strain, significantly larger gains are required to detect an improved mutant. The use of prescreening and rational selection allows for a significant improvement in the efficiency of the selection process.

2. *Rationalized Selection*

An alternate approach to random screening requires a basic understanding of product formation and the fermentation pathway; this can be acquired through radioisotope feeding studies and isolation of mutants blocked in various pathways (Queener *et al.*, 1978). These observations can shed light on the metabolic checkpoints, and suggest ways to isolate specific mutants. For example, environmental conditions (pH, temperature, aeration) can be manipulated, or chemicals can be incorporated in the culture media to select mutants with desired traits. This approach is used in many instances as a prescreen, since selecting for a particular mutant is unlikely to guarantee a hyperyielding mutant. In some instances, by adding toxic substances to the media, the sensitive parent strains are prevented from growing, and only the resistant mutant clones propagate. Such an enrichment procedure has been used to isolate mutants with increased biosynthetic capacity through a change in a regulatory mechanism (leading to either an enzyme resistant to inhibition or an enzyme that is expressed constitutively) or mutants that are modified in the transport or degradation of compounds, which ultimately leads to higher product formation. This rationalized technique is powerful, and when the logic behind the mutant selection criteria is sound, the effectiveness of mutant gains is much greater than with random selection.

Rationalized selection for strain improvement does not generally require a sophisticated understanding of molecular biology to manipulate environmental or cultural conditions. It does, however, require some understanding of cellular metabolism and product synthesis to design the right media or environmental conditions. The procedure is useful in selecting strains overproducing metabolites, antibiotics, simple molecules, amino acids, or enzymes (Queener and Lively, 1986). Some of the mutants derived through rationalized selections are described below.

a. Auxotrophic Mutants

Many metabolic processes have branched pathways, and isolating mutants blocked in one branch of the metabolic pathway can cause accumulation of simple products such as amino acids, nucleotides, and vitamins made by other branches. Auxotrophic strains are blocked at some point in a pathway vital for growth, and unless the specific nutrients or products of the pathway are supplied in the media, the auxotrophs do not survive. Auxotrophs are primarily isolated by plating the mutagenized population on a complete medium that has all the nutrients needed for growth. The clones are then replica plated to minimal medium lacking some specific nutrients, and auxotrophs that fail to grow on minimal media are identified. Most auxotrophic strains give poor antibiotic yields; however, some prototrophic revertants have demonstrated improved antibiotic production such as in tetracycline production (Queener *et al.*, 1978).

b. Regulatory Mutants

Since anabolism and catabolism in any organism are tightly regulated, selection and screening of microbes with less efficient regulation, and optimizing culture conditions, can lead to relaxed regulation and overproduction of microbial products. A broad understanding of metabolic pathway bottlenecks is necessary for a rational approach to developing improved regulatory mutants. Isolating strains relaxed in regulation can usually be accomplished by selecting strains desensitized to feedback inhibition (enzyme activities) or feedback repression (enzyme synthesis) involved in the pathway. One difficulty in applying analog-resistant mutants to strain improvement is that many analogs of primary metabolites need to be tested and some either do not inhibit growth or inhibit growth only at very high concentrations.

(1) *Mutants Resistant to Feedback Inhibition* In many microbes, the end products of metabolism, when accumulated in the microbial cell, inhibit the enzyme activities of many pathways. The end product causes conformational changes by binding to a specific (allosteric) site on the enzyme, and inhibits activity. The binding is usually noncompetitive. Mutation in the structural gene, however, can alter the enzyme binding site and prevent these inhibitory effects. By studying the interaction of various analogs of end products and their resistance, improved strains can be selected that lack feedback inhibition and thus overproduce metabolites of interest. For example, some analogs (acting through these regulatory controls) prevent the synthesis of compounds required for growth and thus cause cell death. Supplementing the screening medium with these analogs selects only mutants with altered enzyme structure and desensitized to inhibition effects to grow. Such procedures have led to development of superior mutant strains of *Arthrobacter, Bacillus, Streptomyces, Aspergillus,* and *Corynebacterium* that overproduce amino acids, nucleotides, and vitamins (Demain, 1983). In some cases, the rational selective agent is biological rather than chemical. Resistance to actinophage has been used to isolate superior vancomycin-producing strains of *Streptomyces orientalis* (Soviet Union Patent 235-244-A, 1969).

(2) *Mutants Resistant to Repression* Here intermediates, products of catabolism (derived from breakdown of compounds containing carbon, nitrogen, or phosphorus), or end products regulate the amount of biosynthetic enzymes synthesized and, therefore, the amount of final product formed. However, mutations at the operator site or other regulatory sites on the gene relieve such end-product repression and allow overproduction of the biosynthetic enzyme. For example, it is well known that antibiotics inhibit their own biosynthesis (e.g., penicillin, chloroamphenicol, puromycin, streptomycin), where key enzymes required for the architecture of these complex molecules are repressed. Mutant strains less sensitive to antibiotic production are

therefore isolated to provide higher yields (Elander and Voumakis, 1986). In a similar context, constitutive mutants have been selected that form enzymes (amylase, glucoamylase, lipase, protease) independent of cultural conditions or the presence of inducing compounds.

Resistance to an antimetabolite is not the sole means of selecting product-excreting mutants resulting from a desensitized enzyme system. Removing enzymes sensitive to feedback repression or the end product that causes inhibition during fermentation also accelerates productivity. Elimination of end-product inhibition or repression effects have been demonstrated by adding chemicals during the fermentation process to trap the end product or the inhibitor. *In situ* end-product extraction, or adding a mechanical device during fermentation (such as specific membrane modules with a particular molecular weight cutoff), allows the percolation of the final product from accumulating in the broth. Increasing permeability of the cell membrane is another method of controlling intracellular product accumulation, enhancing the extracellular metabolite flux. This approach has been exploited to improve the titers of monosodium glutamate (MSG) from *Corynebacterium, Micrococcus,* and *Brevibacterium* (Demain, 1983).

Last, ways of stabilizing the activity of enzymes involved in the assembly of molecules have been reported to augment product formation and strain performance. For example, in gramicidin biosynthesis amino acids are added to stabilize *in vivo* gramicidin S synthetase enzymes and prolong the longevity of biosynthetic activity (Demain, 1983).

c. Other Procedures

Mutant strains suspected of metabolic impairment can also be assayed visually for the presence or absence of specific enzyme activities by plating and spraying on the "diagnostic" solid or liquid culture medium with selective reagents, dyes, or an indicator organism. For example, the agar plug method has been used to detect production of an antibiotic by measuring the extent of growth inhibition of an organism sensitive to the antibiotic. The diameter of the resulting zone of inhibition serves as a measure of antibiotic production. Other procedures rely on the use of chromogenic agents, which are normally converted to a visible product by a specific biochemical reaction or reorganization of the redox level in the media. This leads to visual detection from the large background population of a specific strain having the biochemical activity of interest. Examples of these detection substrates are phenol red for acid–base reactions, 2-nitrophenyl-β-D-galactopyranoside (ONPG) for galactosidase, 6-nitro-3-phenylacetamidobenzoic acid (NIPAB) for penicillin amidase, 4-nitrophenylphosphate (PNPP) for phosphatase, nitrocefin for β-lactamase inhibition, and azocasein for proteinases (Elander and Vournakis, 1986; Queener and Lively, 1986). Tetrazolium and methylene blue (EMB, eosin methylene blue agar for *Escherichia coli* and other coliforms) are commonly employed for detection of oxidation–reduction reaction complexes exhibited by strains of interest. These reactions can be coupled into high throughput screening systems, giving the possibility of targeting whole cells or isolated enzymes for strain selection. For example, carotenoids have been demonstrated to protect *Phaffia rhodozyma* against singlet oxygen damage. A combination of Rose Bengal and thymol in visible light has been employed to select carotenoid-overproducing strains (Schroeder and Johnson, 1995). Enrichment with a singlet oxygen system led to development of mutants with increases in certain carotenoids but a decrease in astaxanthin.

B. Genetic Recombination

In addition to the manipulation of microorganisms by mutation, the techniques of genetic recombination can be employed to get new strains containing novel combinations of mutations and superior microbial strains. Generally, genetic recombination methods include those techniques that combine two DNA molecules having similar sequences (homologs). Through the special event of crossing-over, they are reunited to give a new series of nucleotide sequences along the DNA that are stable, expressible genetic traits. This mechanism of gene alteration and strain modification is called genetic recombination. This definition includes the techniques of protoplast fusion, transformation, and conjugation. Most recently, recombinant DNA technology has been employed to assemble new combinations of DNA *in vitro,* which are then reinserted into the genome of the microbe, creating new varieties of microbe not attainable through traditional mutation and rationalized selection approaches. This approach overlaps the other methods to some extent in that it involves transformation of microbes with laboratory-engineered specific recombinant molecules via plasmid or phage vectors (Hamer, 1980).

1. *Protoplast Fusion*

Fusing two closely related protoplasts (cells whose walls are removed by enzyme treatment) is a versatile technique that combines the entire genetic material from two cells to generate recombinants with desired traits that cannot be obtained through a single mutation. The technique has the advantage of producing hybrids from cells that are sexually incompatible. The procedure of forced mating allows mingling of DNA that is not dependent on appropriate sex factors and is not influenced by barriers of genetic incompatibility.

The procedure relies on stripping the cell wall of the microbes with lytic enzymes, stabilizing the fragile protoplasts with osmotic stabilizing agents, and using a chemical agent or an electric pulse (electrofusion) to induce membrane fusion and to form a transient hybrid. In the hybrid, the genes align at homologous regions, and crossing-over of genes creates recombination within the fused cells. After recombination, the protoplasts are propagated under specific conditions that favor regeneration of cell walls. The unwanted parents are discriminated against by incorporating selective markers in the screening process (e.g., auxotrophy, extracellular enzymes, morphological differences, levels of antibiotic produced) so that only recombinants grow and form viable cells. The efficiency of the technique is influenced by the fragility of cells, the types of genetic markers, the fusing agent used, and the protoplast regeneration capability.

The use of protoplast fusion has been reported (Matsushima and Baltz, 1986) to improve a wide range of industrial strains of bacteria and fungi including *Streptomyces, Nocardia, Penicillium, Aspergillus,* and *Saccharomyces.* This technique is frequently employed in the brewing industry for improving yield and incorporating traits such as flocculation to aid beer filtering, efficient utilization of starch, contamination control, and minimizing off flavors during brewing. Many of these traits are not easily achievable through simple mutation. One advantage of protoplast fusion is the high frequencies at which recombinants are produced under nonselective conditions in the absence of sex factors and without need of specific mating types. In *Streptomyces coelicolor* frequencies as high as 20% have been observed. Another interesting feature is that more than two strains can be combined in one fusion. In some instances, four strains have been fused to yield recombinants containing genes derived from all four parents. This approach can be extremely useful in accelerating strain development. However, because of the absence of control over the amount of genetic material from any one strain retained in the recombinant, protoplast fusion may not improve the strain in the desired fashion. The big disadvantage to this approach has been the genetic instability of the fused strains and the lack of control over which genetic alterations occur. A detailed description of protocols and considerations for the application of protoplast fusion to a variety of industrial microbes is available (Matsushima and Baltz, 1986).

2. *Transformation*

Transformation is the process involving the direct uptake of purified, exogenously supplied DNA by recipient cells or protoplasts. When this occurs, the donor DNA may either combine with the recipient DNA or exist independently in the cell. This leads to changes in the amount and organization of the recipient microbe DNA, hopefully improving it with some of the characteristics coded by the donor DNA. Transformation can be mediated by total genomic DNA or cloned sequences in plasmid or phage DNA. Essentially, the cultures to be transformed are cultivated in a specific physiological manner to develop the competency to make them readily accept foreign DNA. Having selectable markers on the donor DNA allows easy identification of transformants. This procedure allows the transfer of genetic material between unrelated organisms. Certain microorganisms have a well-established gene cloning system that provides a great potential for improving strains by transformation. Transformation methods for strain improvement pertaining to primary or secondary metabolites have been demonstrated in *Streptomyces, Bacillus, Saccharomyces, Neurospora,* and *Aspergillus* (Elander and Vournakis, 1986).

3. Conjugation

Conjugation introduces mutational changes in microbes through unidirectional transfer of genetic material from one strain to the other; it is mediated by plasmid sex factors. Conjugation requires cell-to-cell contact and DNA replication. This mode of genetic exchange can achieve transfer of chromosomal DNA or plasmid DNA. Several strains have been modified by this procedure to make them resistant to specific antibiotics and microbial contamination.

The application of conjugate plasmid recombination technology is employed for strain improvement of *Lactococcus* starter cultures in the dairy industry, which is often plagued by problems with phage. Phage infection can lead to slow acid production, which can economically impact a cheese factory. Furthermore, owing to the nonaseptic nature of the dairy starter culture (open vat in cheese making, the presence of mixed flora in milk), phage-resistant strains are desired as starter cultures for fermentation. Various naturally occurring phage-resistant strains have a number of resistance mechanisms (e.g., abortive infection, restriction/modifications, and absorption inhibitions) that in many instances are carried on conjugative plasmids. Several lactic acid bacteria have therefore been modified by conjugation and transformation procedures to acquire phage resistance in dairy starter cultures. Typical methods for conjugative transfer involve mating by donor and recipient cells on milk agar, followed by harvesting of cells and further isolation on selective medium. This procedure has been applied successfully to construct nisin-producing *Lactococcus* strains (Broadbent and Kondo, 1991).

C. Cloning and Genetic Engineering

1. In Vitro Recombinant DNA Technology

By employing restriction endonucleases and ligases, investigators can cut and splice DNA at specific sites. Some endonucleases have the ability to cut precisely and generate what are known as "sticky ends." When different DNA molecules are cut by the same restriction enzyme, they possess similar sticky ends. Through a form of biological "cut and paste" processes, the lower parts of the one DNA is made to stick well onto the upper part of another DNA. These DNA molecules are later ligated to make hybrid molecules. The ability to cut and paste the DNA molecule is the basis of "genetic engineering." A useful aspect of this cut and paste process involves the use of plasmid, phage, and other small fragments of DNA (vectors) that are capable of carrying genetic material and inserting it into a host microbe such that the foreign DNA is replicated and expressed in the host. A wide array of techniques can now be combined to isolate, sequence, synthesize, modify, and join fragments of DNA. It is therefore possible to obtain nearly any combination of DNA sequence. The challenges lie in designing sequences that will be functional and useful (Hamer, 1980).

The protocol to modify and improve strains involves the following steps:

a. Isolate the desired gene (DNA fragment) from the donor cells.
b. Isolate the vector (a plasmid or a phage).
c. Cleave the vector, align the donor DNA with the vector, and insert the gene into the vector.
d. Introduce the new plasmid into the host cell by transformation or, if a viral vector is used, by infection.
e. Select the new recombinant strains that express the desired characteristics.

For successful transfer of a plasmid/phage vector, it must contain at least three elements: (1) an origin of replication conferring the ability to replicate in the host cell, (2) a promoter site recognized by the host DNA polymerase, and (3) a functional gene that can serve as a genetic marker. A great deal of literature exists on the theoretical overviews, and laboratory manuals on the use of recombinant DNA for strain modification and improvements are available.

2. *Site-Directed Mutagenesis for Strain Improvement*

So far the mutations and the modifications of the strains discussed have been randomly directed at the level of the genome of the culture. The application of recombinant technology and the use of synthetic DNA now make it possible to induce specific mutations in specific genes. This procedure of carrying out mutagenesis at a targeted site in the genome is called site-directed mutagenesis. It involves the isolation of the DNA of the specific gene and the determination of the DNA sequence. It is then possible to construct a modified version of this gene in which specific bases or a series of bases are changed. The modified DNA can now be reinserted into the recipient cells and the mutants selected. Site-directed mutagenesis has found valuable application in improving strains (Crueger and Crueger, 1984), by enhancing the catalytic activity and stability of commercial enzymes, for example, penicillin G amidase.

Since the mid-1970s the synergistic use of classic techniques along with rational selection and recombinant DNA has made a significant impact in developing improved strains. Fermentation processes for products as diverse as human proteins and antibiotics and other therapeutic agents (chymosin, lactoferrin) have benefited from these combinatorial approaches. Transcription, translation, and protein secretion, activation, and folding are one or more of the rate-limiting steps critical for the overproduction of such therapeutic proteins. Achieving overproduction of active therapeutic proteins in bacterial or fungal heterologous gene expression systems has been made amenable due to the mix of classic and rational selection procedures. Genetic engineering along with classic methods has been used on numerous occasions to improve the performance of yeast and bacteria in alcohol fermentation, expand the substrate range, enhance the efficiency of the fermentation process, lower by-product formation, design yeast immune to contamination, and develop novel microbes that detoxify industrial effluents. Cost-effective production by fermentation of alcohols (ethanol, butanol) that can be used as substitutes for fossil fuels has been aided by this technology. Bacterial manufacturing of large quantities of hormones, antibodies, interferons, antigens, amino acids, enzymes, and other therapeutic agents to combat diseases has also become possible by recombinant DNA technology and strain improvement programs. Through increased gene dosage, improved efficiencies of antibiotic production have been achieved to relieve one or more rate-limiting steps. Novel and hybrid antibiotics and bioactive compounds have also been produced by combining different biosynthetic pathways in one organism that would have been difficult or impossible to manufacture through synthetic chemistry (e.g., *Cephalosporium acremonium* and *Claviceps purpurea*). Moreover, using recombinant DNA techniques, entire sets of genes for antibiotic biosynthesis have been cloned into a heterologous host in a single step. By cloning portions of the biosynthetic genes from one producer to another strain, hybrid compounds have also been synthesized, with novel spectra of activities and pharmacological applications. An example of this is the production by *Streptomyces peucetius Subsp. caesius* of adriamycin (14-hydroxydaunomycin), an antitumor antibiotic (Crueger and Crueger, 1984).

Occasionally it has been found that certain improved mutants produce extremely high levels of a specific enzyme. When analyzed, these mutants had multiple copies of a structural gene coding for the specific enzyme of interest. Increasing the number of gene copies in the cell (through gene cloning) has therefore been employed to overproduce enzyme precursors and their end product. In addition, mutations at the promoter or regulatory site have been demonstrated to alter secondary metabolite productivity (Hamer, 1980). For example, in *Saccharopolyspora erythraea*, specific mutations at a ribosomal RNA operon terminator site altered the transcription and expression of the erythromycin gene cluster, and strains harboring these mutations overproduced enzymes involved in the later steps of erythromycin biosynthesis (Queener *et al.*, 1978).

Once an improved strain is confirmed through bench-work studies, additional efforts are necessary to validate its performance. It is normally purified by reisolation, and the reisolates are verified for strain variability, homogeneity, and performance. They are preserved in large lots for examination under pilot plant conditions before being introduced for large-scale production.

V. IMPROVED STRAIN PERFORMANCE THROUGH ENGINEERING OPTIMIZATION

Major improvements in fermentation are no doubt attributed to superior strains created through mutation or genetic alterations. Further improvement in culture performance can also be achieved by giving a strain the optimum environmental and physical conditions. During the strain improvement process, it is important to keep in mind that the ultimate success also depends on optimization of fermentation design factors. The use of batch or fed batch, continuous or draw-and-fill operation, the extent of shear, broth rheological properties, and oxygen and heat transfer characteristics all contribute to improvement in strain performance. The application of biochemical engineering principles can be used to design environmental parameters that shift kinetics of metabolic routes toward the desired product.

A. Improving Strain Performance through Optimizing Nutritional Needs

The environment in which the altered strain is grown is known to influence higher product yield and get the best performance out of the culture. Since the media commonly used for production are different from the ones in which mutants are screened, media optimization is requisite to achieving the best response from the improved strains when scaled up to production. The media for production are reformulated so that they meet all growth requirements and supply the required energy for growth and product synthesis. Early bench work is typically performed with biochemically defined media to elucidate metabolite flux and regulation (inhibition or repression) by specific nutrients and physical variables. Later research is done to develop complex media that are more cost-effective to support cultural conditions of improved strains and maximize product synthesis without producing additional impurities that may impact isolation of the product. Additional issues, such as inoculum media and transfer criteria, media sterilization, pH, cultivation variables, and the sensitivity of the culture to different batches of raw material, are addressed during media optimization.

Statistical computer-based methods and response surface modeling are available for the study of many variables at the same time. A full search is normally made of every possible combination of independent variables to determine appropriate levels that give the optimum response in strain performance. Success in this area can be enhanced if additional physiological data are available, such as the role of precursors, the steps in the biosynthetic pathway, carbon flux through the pathway, and the regulation of primary and secondary metabolism by carbon and nitrogen. Controlling the levels of metabolites and precursors during fermentation aids in controlling lag and repression or toxicity effects. The removal of inhibiting products has been practiced where increasing the concentration up to economical levels demonstrates poor process kinetics. Adding chelating agents has been beneficial if the fermentation is found to be sensitive to substrate-specific repression. Further improvement in strain performance and productivity gain has been observed when the key enzymes participating in product formation are stabilized. For example, biosynthesis of the antibiotic gramicidin has been improved greatly by adding precursor amino acids that are substrates for the key enzymes (Demain, 1983).

B. Influence of Bioengineering in Improving Strain Performance

The ultimate destination of an improved strain is a large fermenter in which the desired product is made for commercial use. Conversion of laboratory processes to an industrial operation is called scale-up. It is not a straightforward process, requiring the use of methods of chemical engineering, physiology, and microbiology for success. The goal of the scale-up team is to cultivate improved strains under optimum production conditions. Open communication, data feedback, and synergy between engineers and scientists are vital to facilitate successful launch and scaling up of new and improved strains. Factors such as media sterilization for culture seed and production, methods of aeration and agitation, power input, control of viscosity, and evaporation rates are considered when moving new strains into production. In addition, sterility factors, heat transfer, impeller types, baffle types and positioning, the geometry and symmetry of the fermenter, mixing times, oxygen transfer rates, respiratory quotient and metabolic flux, disengagement of gases, and culture stability are all important in bringing the improved mutants from the laboratory to industrial scale production. Furthermore, metabolic feeds and the impact of the addition of feeds subsurface or surface, as well as timing of additions, are also optimized for directing ways toward the desired product. In some cases, process control and parameter optimization are facilitated using near-infrared spectroscopy and Fourier transform infrared spectroscopy when integrated in the fermentation process. This allows fermentation broth analysis and the ability to assay *in situ,* avoiding sample preparation and permitting timely adjustment of environmental parameters. However, on-line mass spectroscopy analysis is helpful mostly in a fermentation that is less sensitive to specific variables, as it cuts down routine assay work, sample preparation time, and the need for expensive equipment.

After the successful introduction of the improved strain in fermenters, production processes are validated and designed to run automatically for comparative and consistent operations. In some cases mathematical modeling of the physiological state and microbiological process are elucidated for maximizing strain performance. Typically this is done in three stages: (a) qualitatively analyzing relationships among growth, substrate consumption, and production (usually based on the assumption of metabolic pathway and biogenesis of product); (b) establishing mathematical formulations and kinetic equations of the model, emphasizing the role of operator functions and technical operation associated with overproduction; and (c) estimation of parameters and simulation of the model on the basis of experimental data. During these scale-up and modeling studies, emphasis is also placed on the capital and operating costs as well as the reliability of the process (Hemker, 1972; Flynn, 1983).

Among the various strains of microbes that have been scaled up, a few problems have invariably been noticed when scaling up filamentous organisms. The viscous nature of the culture creates heterogeneity, uneven mixing and distribution of bubbles, and failure to disperse micelle and floc formation. Several of these factors can be addressed up front during strain selection. The development of morphological mutants with short mycelia (higher surface area per unit volume) has been beneficial. This change can also influence the release of heat and spent gas without causing gradients in the fermenters. In addition, methods of mixing, nutrient feeding, and pH control also play critical roles in successful scale-up of improved mycelial strains. Finally, data on performance of the broth in pilot and large-scale purifications are also crucial for approval of improved strains for market production.

C. Metabolic Engineering for Strain Development

In the broadest sense, metabolic engineering is a new technology in strain improvement that optimizes, in a coordinated fashion, the biochemical network and metabolic flux within the fermenters, with inputs from chemical engineering, cell physiology, biochemistry, and genetics. By systematically analyzing individual enzymatic reactions and pathways (their kinetics and regulation), methods are designed to eliminate bottlenecks in the flow of precursors and to balance stoichiometrically the distribution of metabolites for optimum product formation. Nuclear magnetic resonance studies of metabolic flux analysis and kinetic measurements are further combined with thermodynamic analysis of the biological process to predict better strain performance. The principles governing a biosynthetic pathway, including genetic controls, interaction with complex raw material sources, and bioreactor operations and mathematical modeling strategy, are exploited to exceed the microbe's capability and improve its productivity. Metabolic engineering applications in strain improvement have found a special niche as a result of their previously observed successes in the production of amino acids and biopolymers from strains of *Brevibacterium, Corynebacterium,* and *Xanthomonas* (Rowlands, 1984).

VI. SUMMARY

Strain development has been the icon of the fermentation industries. Discoveries in mutation, protoplast fusion, genetic manipulations, and recombinant DNA technology and the experience gained from modern reactor design and operation of fermenters have revolutionized the concept of microbial strain development. Although greater improvements in overproduction of metabolites and antibiotics of specific microbes have resulted from essentially random empirical approaches to mutation and strain development, future strain development technology will be supplemented by more knowledge-based scientific methods. With the advances in understanding biosynthetic pathways, elucidation of regulatory mechanisms related to induction and repression of genes, as well as bioengineering design, it will be possible to apply new strategies and limitless combinations for isolating improved strains. Furthermore, tailoring genes through the avenue of *in vitro* DNA recombination techniques in both bacteria and fungi has been shown to be feasible. Perhaps these areas will facilitate new strategies and have higher impact on industrial strain improvement.

See Also the Following Articles

Genomic Engineering of Bacterial Metabolism • Industrial Fermentation Processes • Mutagenesis • Recombinant DNA, Basic Procedures • Transformation, Genetic

Bibliography

Baltz, R. H. (1986). Mutagenesis in *Streptomyces* spp. *In* "Manual of Industrial Microbiology and Biotechnology" (A. Demain and N. A. Solomon, eds.), pp. 184–190. American Society of Microbiology, Washington, D.C.

Broadbent, J. F., and Kondo, J. K. (1991). Genetic construction of nisin-producing *Lactococcus cremoris* and analysis of a rapid method for conjugation. *Appl. Environ. Microbiol.* 57, 517–524.

Crueger, W., and Crueger, A. (1984). Antibiotics. *In* "Biotechnology: A Textbook of Industrial Microbiology," pp. 197–233. Akademische Verlagsgesellschaft, Wiesbaden, Germany. English translation copyright 1984 by Science Tech, Madison, WI.

Demain, A. L. (1983). New applications of microbial products. *Science* **219**, 709–714.

Elander, R., and Vournakis, J. (1986). Genetics aspects of overproduction of antibiotics and other secondary metabolites. *In* "Overproduction of Microbial Metabolites" (Z. Vanek and Z. Hostalek, eds.), pp. 63–82. Butterworth, London.

Flynn, D. (1983). Instrumentation for fermentation processes. *In* "IFAC Workshop, 1st, Helsinki, Finland, 1982. Modeling and Control of Biotechnical Processes: Proceedings," pp. 5–6. Pergamon, Oxford.

Hamer, D. H. (1980). DNA cloning in mammalian cells with SV40 vectors. *In* "Genetic Engineering, Principles and Methods" (J. K. Setlow and A. Hollander, eds.), Vol. 2, pp. 83–102. Plenum, New York and London.

Hemker, P. W. (1972). *In* "Analysis and Simulation of Biochemical Systems," Proc. 8th FEBS Meeting, pp. 59–80. Elsevier North-Holland, Amsterdam.

Lein, J. (1986). Random thoughts on strain development. *SIM News* **36**, 8–9.

Matsushima, P., and Baltz, R. (1986). Protoplast fusion. *In* "Manual of Industrial Microbiology and Biotechnology" (A. Demain and N. Solomon, eds.), pp. 170–183. American Society of Microbiology, Washington, D.C.

Nolan, R. (1986). Automation system in strain improvement. *In* "Overproduction of Microbial Metabolites" (Z. Vanek and Z. Hostalek, eds.), pp. 215–230. Butterworth, London.

Queener, S., and Lively, D. (1986). Screening and selection for strain improvement. *In* "Manual of Industrial Microbiology and Biotechnology" (A. Demain and N. Solomon, eds.), pp. 155–169. American Society of Microbiology, Washington, D.C.

Queener, S., Sebek, K., and Veznia, C. (1978). Mutants blocked in antibiotic synthesis. *Annu. Rev. Microbiol.* **32**, 593–636.

Rowlands, R. T. (1984). Industrial strain improvement: Mutagenesis and random screening procedures. *Enzyme Microbial Technol.* **6**, 3–10.

Schroeder, W., and Johnson, E. (1995). Carotenoids protect *Phaffia rhodozyma* against singlet oxygen damage. *J. Indust. Microbiol.* **14**, 502–507.

Streptococcus pneumoniae

Alexander Tomasz
The Rockefeller University

I. History
II. Structure
III. Plasma Membrane and Intracellular Membranes
IV. Epidemiology of Antibiotic Resistance
V. Guides to the *S. pneumoniae* Literature

SINCE THE FIRST EXPERIMENTAL DEMONSTRATION OF THE CAPACITY OF *STREPTOCOCCUS PNEUMONIAE* to cause disease in the 1880s, this bacterium has become recognized as one of the major causative agents of human community-acquired diseases.

I. HISTORY

Recent estimates from the United States indicate that the frequency of pneumococcal pneumonia approaches approximately 500,000 cases per year, accompanied by approximately 40,000–50,000 cases of bloodstream infections, 4000–5000 cases of meningitis, and several million cases of otitis media annually. Although mortality of these diseases has decreased substantially in the antibiotic era, introduction of antimicrobial agents did not seem to substantially reduce the attack rates of pneumococcal disease which is age related, with peaks in the early years of life and in populations of 65 years of age and older. A new peak of frequency of pneumococcal disease has appeared in HIV-positive human populations with an average age in the early 30s. Throughout the history of the microbiology of *Streptococcus pneumoniae,* efforts to understand the mechanism of the often fatal diseases caused by this bacterium have led to some landmark achievements of molecular biology, including the identification of DNA as the genetic material, chemical characterization of the first polysaccharide antigen, identification of the first bacterial autolytic enzyme and the first bacterial quorum-sensing polypeptide, and the first use of bacterial enrichment culture for the production of an enzyme (that was capable of degrading the type 3 polysaccharide). Recent important studies have led to the identification of the genetic elements involved with the control of the competent state in which pneumococci can take up and integrate exogenous DNA molecules and to the elucidation of the mechanism of uptake of DNA and recombination in the process of genetic transformation. Studies with *S. pneumoniae* also made contributions to the understanding of the mode of action of penicillin, the mechanism of penicillin resistance, and penicillin tolerance. The first report on the sequencing of the complete genome of the laboratory strain R6 of *S. pneumoniae* appeared recently. Through these studies, *S. pneumoniae* has emerged as one of the major model organisms for studies on the molecular and cell biology of gram-positive bacteria.

II. STRUCTURE

Most clinical isolates of *S. pneumoniae* appear as typical "football"-shaped cocci with diameters of 1.0–1.5 nm from tip to tip and with a "waistline" of 0.5–0.8 μm. Most clinical isolates of pneumococci express a capsular polysaccharide which covers the outermost surface of this bacterium and the function of which is to provide protection against phagocytosis. Currently, there are as many as 90 chemically distinct capsular polysaccharides identified, and in a fraction of these the complete sequence of the capsular determinants has also been determined. Despite the large number of capsular types, the overwhelming majority of human disease is associated with a smaller fraction of capsular types, most of which are included in the 23-valent antipneumococcal vaccine. Capsular polysaccharides are major virulence determinants and a subgroup of these, often referred to as the "pediatric serotypes" (primarily serogroups/types 6, 9, 14, 19, 23), have come to prominence for two reasons: (i) They represent the capsular chemistries against which the immune system of infants has difficulty mounting an antibody response, and (ii) the same serotypes also represent the capsular types associated with the overwhelming majority of antibiotic-resistant strains of *S. pneumoniae*. Major efforts are currently in progress regarding conjugate vaccines in which polysaccharide antigens are linked to protein carriers to induce production of protective antibodies in children of a young age. The rank of serotype of *S. pneumoniae* isolates varies with the geographic area and time of isolation. Reasons for these differences and fluctuations in frequency are not understood. A unique mechanism—capsular switch—for the change of polysaccharide capsule was identified through the use of molecular fingerprinting techniques (pulsed-field gel electrophoresis [PFGE] of chromosomal restriction digests). In capsular switch, a resident capsular locus is replaced by the capsular determinants of another pneumococcus. Such capsular switch events occur spontaneously in the natural environment of pneumococci, possibly via the process of genetic transformation. Capsular switching events were repeatedly detected with multidrug-resistant clones of pneumococci, and in at least one recently described case acquisition of a new capsule (switch from serotype 23F to 3) was accompanied by massive increase in virulence of the strain as determined in the mouse peritoneal assay.

An electron microscopic fine structure studied with the *S. pneumoniae* strain R6 has identified the most important morphological features of this bacterium. Strain R6 is a derivative of strain R36A, the parental strain of most pneumococcal laboratory isolates with which the overwhelming majority of laboratory studies have been carried out during the past decades. R36A is a nonencapsulated strain derived from strain D39, a serotype 2 pneumococcus, through serial passage in medium containing anticapsular type antibody. Strain R36A carries a 7510-bp deletion involving five of the genetic determinants of the capsular type 2 genetic locus.

The cell wall of *S. pneumoniae* appears in electron microscopic thin sections as a trilaminar layer directly underneath the somewhat amorphous polysaccharide capsule. The cell wall is composed in approximately equal proportions (weight for weight) of a peptidoglycan and a teichoic acid of an unusual chemical composition which includes galactosamine phosphate, 2,4,6-trideoxy-2,4-diamino hexose, ribitol phosphate, and phosphocholine. Recent high-resolution chemical analysis of the peptidoglycan performed on many clinical isolates has demonstrated that the complex, multicomponent structure of these cell wall polymers is specific for the species. Naturally occurring peptidoglycan variants were also detected, however, and these were invariably resistant to penicillin. The nature of association between the penicillin-resistance trait and the abnormal muropeptide composition of the cell walls of these bacteria is not clear. The species-specific peptidoglycan of pneumococci is built with linear muropeptide components which are directly interlinked but which also include, as a minority, muropeptide species that carry short dipeptide side chains and which are cross-linked through these branched peptides. A characteristic feature of the peptidoglycan of penicillin-resistant strains is the distortion of this composition such that the branched and indirectly cross-linked muropeptides become the majority components of the cell wall. The choline component of the wall teichoic acid, which is also present in a membrane teichoic acid of similar primary structure, appears to perform major and multiple roles both in the physiology of the pneumococcus and as a host interactive component of the pneumococcal surface. Except for some recently isolated laboratory constructs, *S. pneumoniae* strains have a nutritional requirement for choline which may be replaced by other amino alcohols such as ethanolamine. Biosynthetic replacement of the choline residues in the wall teichoic acid results in multiple and drastic changes in physiology. First, the bacteria lose their characteristic sensitivity to autolysis in the stationary phase of growth, during treatment with deoxycholate or other detergents, and they no longer lyse during treatment with penicillin and other cell wall active antibiotics. Second, ethanolamine-grown cells also have a defect in cell separation at the end of cell division so that they produce infinitely long chains of pneumococci. Finally, the capacity of pneumococci to react with the competence-inducing factor to undergo genetic transformation is also inhibited in the ethanolamine-grown cells. The autolytic defects appear to be related to at least two biochemical abnormalities: the inability of the pneumococcal autolytic enzyme (the product of the *lytA* gene) to attach and to hydrolyze cell walls that contain ethanolamine instead of choline in the teichoic acid. In addition, the autolytic enzyme produced in ethanolamine-grown cells needs to interact with choline-containing ligands before acquiring catalytic activity. The choline residues in the bacterial surface are also part of the receptors for several pneumococcal bacteriophages, and a recent study suggests that prophage carriage may be very frequent among clinical isolates of *S. pneumoniae.* Choline residues serve as attachment sites for the C-reactive protein and for a large family of myeloma proteins. Variation in the number of choline residues per cell has been proposed as one of the biochemical correlates of phase variation in pneumococci, which is linked to the colonization versus the invasive mode of interaction between *S. pneumoniae* and the human host. Recent studies identified a surprising number of choline-binding proteins with postulated roles in the interaction of the bacterium with the host. In addition to the capsular polysaccharides, many proteins also play roles in the virulence of *S. pneumoniae,* and these include the intracellular pneumolysin and the surface-located pspA, psaA, neuraminidase, and IGA protease. Purified pneumococcal cell walls and its subcomponents were also shown to have powerful inflammatory activity in several animal models, and cell walls were shown to induce production of cytokines from human peripheral blood cells and also when injected in the cisterna magna in the experimental model of meningitis.

The pneumococcal cell wall, like the cell wall of other streptococci, enlarges through incorporation of new material at a single, equatorially located growth zone. New cell wall material does not seem to incorporate elsewhere into the pneumococcal surface, and as a result the pneumococcal surface is inherited in a conservative fashion during growth: "Old" hemispheres of the cell surface assembled during the previous cell division are passed on intact to the daughter cells in each subsequent cell generation.

III. PLASMA MEMBRANE AND INTRACELLULAR MEMBRANES

On the inner side of the cell wall layer, electron microscopic thin sections show a second trilaminal structure which has the dimensions typical of biological unit membranes. Numerous intracellular membranes or "mesosomes" have also been identified in thin sections of the pneumococcus, and these structures show a peculiar geometric arrangement most frequently at the tips of the ingrowing septa of the bacteria and at the two poles of the cells, suggesting that they may represent organelles involved with the equatorial growth of the cell wall and/or segregation of the bacterial chromosome during cell division.

Important components of the plasma membrane are the penicillin binding proteins (PBPs), which are the primary targets of the β-lactam antibiotics and the natural function of which is related to terminal stages in bacterial cell wall synthesis. Six pneumococcal PBPs have been identified by the capacity of these proteins to covalently bind radiolabeled penicillin, and these are referred to in the literature, in order of decreasing molecular sizes, as PBPs 1a, 1b, 2x, 2a, 2b, and 3; also, the DNA sequences of PBP genes have been determined. Although the precise enzymatic functions of these proteins are not clear, PBP3 was identified as a D,D-carbyoxpeptidase. The mechanism of penicillin resistance (both in laboratory mutants and in clinical isolates) involves modification of the high-molecular-weight PBPs in such a manner that their binding capacity ("affinity") for the penicillin molecule is reduced. In genetic transformation of penicillin resistance, high-level drug resistance is introduced in a multistep process which involves the stepwise modification of high-molecular-weight PBPs in a sequence beginning with PBP2x. Penicillin-resistant clinical isolates contain *pbp* genes that show extensive sequence diversity, indicating that they are the products of heterologous recombinational events in which *pbp* determinants of extra species origin are taken up by pneumococci and incorporated into their *pbp* genes, and these "mosaic" genes appear to include the sequence alterations essential for production of a low-affinity PBP protein. The mosaic sequences of these altered *pbp* genes are conserved and may be used to detect the spread of penicillin resistance genes among strains of pneumococci, a process that appears to be one of the driving forces of spread of β-lactam resistance in this species. A β-lactam resistance mechanism based on β-lactamases has not been detected in pneumococci.

IV. EPIDEMIOLOGY OF ANTIBIOTIC RESISTANCE

The first penicillin-resistant strain of *S. pneumoniae* that appeared in the clinical environment and that invoked comment in the infectious diseases literature was the strain isolated from the throat of a healthy child in the mid-1960s in a remote village, Agunganak, in Papua New Guinea. Although initially the possibility of geographic spread of the penicillin-resistant pneumococcus was judged to be remote, this prediction was soon contradicted by the massive outbreak of pneumococcal disease in South African hospitals in 1977 which was caused by multidrug-resistant strains of this bacterium. Between the early 1980s and the late 1990s, reports on the detection and increase, both in frequency and in antibiotic resistance level, of drug-resistant strains of *S. pneumoniae* have appeared in increasing numbers and currently the antibiotic-resistant pneumococcus has become a global phenomenon.

Introduction of molecular typing techniques such as PFGE, multilocus enzyme analysis, and multilocus sequencing has clearly shown that among the very large number of genetic lineages of antibiotic-resistant pneumococci, a handful of highly epidemic clones have emerged that achieved massive and often pandemic spread. The most outstanding of these are the "Spanish/USA clone," usually expressing serotype 23F and carrying resistance to penicillin, tetracycline, and chloramphenicol and often to erythromycin and sulfamethoxazole-trimethoprim. This clone has been identified in numerous national and international surveillance studies both as a powerful colonizer and as a strain capable of causing the entire spectrum of pneumococcal diseases, among both adults and children. An apparent intercontinental transfer of this clone from southern Europe to the United States was demonstrated. Similar "importation" of a multidrug-resistant pneumococcal clone, presumably from southern Europe to Iceland, was demonstrated by molecular fingerprinting techniques in the early 1990s. This so-called "Icelandic clone," expressing serotype 6B and resistant to penicillin, tetracycline, chloramphenicol, erythromycin, sulfamethoxazole-trimethoprim, and occasionally to third-generation cephalosporins, quickly spread in Iceland and within 3 years of its detection was shown to be responsible for approximately 20% of all pneumococcal infections in that country. A third genetic lineage, often referred to as the "French/Spanish clone," carries resistance to penicillin and tetracycline and, occasionally, to sulfamethoxazole-trimethoprim, exists in two capsular types—either 9V or 14—and the clone has been shown to spread to virtually every country in Europe, Latin America, and the United States. The incidence of penicillin-resistant strains varies widely from country to country and from one geographic site to another. Particularly high frequencies of antibiotic-resistant pneumococci were identified in the nasopharyngeal flora of children attending day care centers, and the incidence of pneumococcal disease among these children was also shown to be high. The institution of day care has emerged as a unique epidemiological entity in which many children of preschool age are cohorted. The high carriage rate of respiratory pathogens, immunological status, and behavior, together with the frequent occurrence of viral respiratory diseases and the extensive and often imprudent use of antimicrobial agents among this age group of children, are the most likely reasons why attendance at day care centers has become a risk factor for both the carriage and the infection by antibiotic-resistant strains of pneumococci.

The impact of antibiotic resistance on the chemotherapy of pneumococcal disease varies with the particular infection. Even low-level resistance to β-lactam antibiotics requires change in chemotherapy in meningitis because of the low penetration of the cerebrospinal fluid by this class of antibiotics and because of the need for bactericidal concentration. On the other hand, penicillin therapy was shown to remain effective in pneumococcal pneumonia caused by penicillin-resistant strains, with penicillin values as high as 1 or 2 μg/ml. There is little, if any, experience concerning appropriate therapy of infections by strains that express unusually high resistance levels to β-lactam antibiotics. Such strains have been identified both in the United States and in Eastern Europe.

V. GUIDES TO THE *S. PNEUMONIAE LITERATURE*

The number of publications on various aspects of the pneumococcus—its epidemiology and molecular biology—has increased exponentially since the mid-1980s. Guides to this large body of original literature—both the older and the recent publications—may be found in three sources. The book by White (1979) provides a good source to early data. A publication in *Reviews of Infectious Diseases* in 1981 summarizes data on the mechanism of pathogenesis, resistance, and cell biology. A recent publication on the molecular biology of *S. pneumoniae* and its diseases is the book *Streptococcus pneumoniae: Molecular Biology and Mechanisms of Disease—Update for the 1990s* (1999).

See Also the Following Articles

Cell Membrane: Structure and Function • Microscopy, Electron • Staphylococcus • Transformation, Genetic

Bibliography

Appelbaum, P. C., Koornhof, H. J., Jacobs, M., Robins-Browne, R., Isaacson, M., Gilliland, J., and Austrian, R. (1977). *Streptococcus pneumoniae* resistant to penicillin and chloramphenicol. *Lancet* **2**, 995–997.

Austrian, R., Douglas, M. R., Schiffman, G., Coetzee, A. M., Koornhof, H. J., Hayden-Smith, S., and Reid, R. D. W. (1976). Prevention of pneumococcal pneumonia by vaccination. *Trans. Assoc. Am. Physicians* **89**, 184–194.

Avery, O. T., MacLeod, C. M., and McCarty, M. (1944). Studies on the chemical nature of the substance inducing transformation of pneumococcal types. Induction of transformation by a deoxyribonucleic acid fraction isolated from pneumococcus type III. *J. Exp. Med.* **79**, 137–157.

Baltz, R. H., Norris, F. H., Matsushima, P., DeHoff, B. S., Rockey, P., Porter, G., Burgett, S., Peery, R., Hoskins, J., Braverman, L., Jenkins, I., Solenberg, P., Young, M., McHenney, M. A., Skatrud, P. L., and Rosteck, P. R., Jr. (1998). DNA sequence sampling of the *Streptococcus pneumoniae* genome to identify novel targets for antibiotic development. *Microb. Drug Resistance* **4**, 1–9.

Briles, E. B., and Tomasz, A. (1970). Radioautographic evidence for equatorial wall growth in a gram positive bacterium: Segregation of choline ^{3}H-labeled teichoic acid. *J. Cell Biol.* **47**, 786–790.

Claverys, J. P., Dintilhac, A., Mortier-Barrière, I., Martin, B., and Alloing, G. (1999). Regulation of competence for genetic transformation in *Streptococcus pneumoniae*. *J. Appl. Bacteriol. Symp. Ser.*, in press.

Coffey, T. J., Berrón, S., Daniels, M., Garcia-Leoni, M. E., Cercenado, E., Bouza, E., Fenoll, A., and Spratt, B. G. (1996). Multiply antibiotic-resistant *Streptococcus pneumoniae* recovered from Spanish hospitals (1988–1994): Novel major clones of serotypes 14, 19F and 15F. *Microbiology* **142**, 2747–2757.

Corso, A., Severina, E. P., Petruk, V. F., Mauriz, Y. R., and Tomasz, A. (1998). Molecular characterization of penicillin resistant *Streptococcus pneumoniae* isolates causing respiratory disease in the United States. *Microb. Drug Resistance* **4**, 325–337.

Dowson, C. G., Hutchison, A., and Spratt, B. G. (1989). Extensive remodelling of the transpeptidase domain of penicillin-binding protein 2B of a penicillin-resistant South African isolate of *Streptococcus pneumoniae*. *Mol. Microbiol.* **3**, 95–102.

Dubos, R. J. (1937). Mechanism of the lysis of pneumococci by freezing and thawing, bile, and other agents. *J. Exp. Med.* **66**, 101–112.

Dubos, R., and Avery, O. T. (1931). Decomposition of the capsular polysaccharide of pneumococcus type III by a bacterial enzyme. *J. Exp. Med.* **54**, 51–71.

Enright, M. D., and Spratt, B. G. (1998). A multilocus sequence typing scheme for *Streptococcus pneumoniae:* Identification of clones associated with serious invasive disease. *Microbiology* **144**, 3049–3060.

Fenoll, A., Martin Bourgon, C., Munoz, R., Vicioso, D., and Casal, J. (1991). Serotype distribution and antimicrobial resistance of *Streptococcus pneumoniae* isolates causing systemic infections in Spain, 1979–1989. *Rev. Infect. Dis.* **13**, 56–60.

Garcia, E., Garcia, J.-L., Ronda, C., Garcia, P., and Lopez,

R. (1985). Cloning and expression of the pneumococcal autolysin gene in *Escherichia coli*. *Mol. Gen. Genet.* **201**, 225–230.

Garcia, P., Martin, A. C., and Lopez, R. (1997). Bacteriophages of *Streptococcus pneumoniae:* A molecular approach. *Microb. Drug Resistance* **3**, 165–176.

Goebel, W. F., and Adams, M. H. (1943). The immunological properties of the heterophile antigen and somatic polysaccharide of pneumococcus. *J. Exp. Med.* 77, 435–449.

Hakenbeck, R., and Kohiyama, M. (1982). Purification of penicillin-binding protein 3 from *Streptococcus pneumoniae*. *Eur. J. Biochem.* **127**, 231–236.

Hakenbeck, R., Konig, A., Kern, I., van der Linden, M., Keck, W., Billot-Klein, D., Legrand, R., Schoot, B., and Gutmann, L. (1998). Acquisition of five high-M_r penicillin-binding protein variants during transfer of high-level β-lactam resistance from *Streptococcus mitis* to *Streptococcus pneumoniae*. *J. Bacteriol.* **180**, 1831–1840.

Handwerger, S., and Tomasz, A. (1986). Alterations in kinetic properties of penicillin-binding proteins of penicillin-resistant *Streptococcus pneumoniae*. *Antimicrob. Agents Chemother.* **30**, 57–63.

Havarstein, L. S., Coomaraswamy, G., and Morrison, D. A. (1995). An unmodified heptadecapeptide pheromone induces competence for genetic transformation in *Streptococcus pneumoniae*. *Proc. Natl. Acad. Sci. USA* **92**, 11140–11144.

Iannelli, F., Pearce, B. J., and Pozzi, G. (1999). The type 2 capsule locus of *Streptococcus pneumoniae*. *J. Bacteriol.* **181**, 2652–2654.

Jennings, H. J., Lugowski, C., and Young, N. M. (1980). Structure of the complex polysaccharide C-substance from *Streptococcus pneumoniae* type 1. *Biochemistry* **19**, 4712–4719.

Lacks, S. (1962). Molecular fate of DNA in genetic transformation of pneumococcus. *J. Mol. Biol.* **5**, 119–131.

Lacks, S., and Greenberg, B. (1976). Single-strand breakage on binding of DNA to cells in the genetic transformation of *Diplococcus pneumoniae*. *J. Mol. Biol.* **101**, 255–275.

Laible, G., Spratt, B. G., and Hakenbeck, R. (1991). Interspecies recombinational events during the evolution of altered PBP2X genes in penicillin-resistant clinical isolates of *S. pneumoniae*. *Mol. Microbiol.* **5**, 1993–2002.

McDaniel, L. S., Sheffield, J. S., Swiatlo, E., Yother, J., Crain, M. J., and Briles, D. E. (1992). Molecular localization of variable and conserved regions of pspA, and identification of additional pspA homologous sequences in *Streptococcus pneumoniae*. *Microbial Pathogen.* **13**, 261–269.

Nesin, M., Ramirez, M., and Tomasz, A. (1998). Capsular transformation of a multidrug-resistant *Streptococcus pneumoniae in vivo*. *J. Infect. Dis.* **177**, 707–713.

Pallares, R., Gudiol, F., Linares, J., Ariza, J., Ruffi, G., Murgui, L., Dorca, J., and Viladrich, P. F. (1987). Risk factors and response to antibiotic therapy in adults with bacteremic pneumonia caused by penicillin-resistant pneumococci. *N. Engl. J. Med.* **317**, 18–22.

Paton, J. C., Andrew, P. W., Boulnois, G. J., and Mitchell, T. J. (1993). Molecular analysis of the pathogenicity of *Streptococcus pneumoniae:* The role of pneumococcal proteins. *Annu. Rev. Microbiol.* **47**, 89–115.

Ramirez, M., Severina, E., and Tomasz, A. (1999). A high incidence of prophage carriage among natural isolates of *Streptococcus pneumoniae*. *J. Bacteriol.* **181**(12).

Reviews of Infectious Diseases (1981). 3(2), 183–371.

Ring, A., Weiser, J. N., and Tuomanen, E. I. (1998). Pneumococcal trafficking across the blood–brain barrier. Molecular analysis of a novel bidirectional pathway. *J. Clin. Invest.* **102**, 347–360.

Rosenow, C., Ryan, P., Weiser, J. N., Johnson, S., Fontan, P., Ortqvist, A., and Masure, H. R. (1997). Contribution of novel choline-binding proteins to adherence, colonization and immunogenicity of *Streptococcus pneumoniae*. *Mol. Microbiol.* **25**, 819–829.

Severin, A., and Tomasz, A. (1996). Naturally occurring peptidoglycan variants of *Streptococcus pneumoniae*. *J. Bacteriol.* **178**, 168–174.

Soares, S., Kristinsson, K. G., Musser, J. M., and Tomasz, A. (1993). Evidence for the introduction of a multiresistant clone of serotype 6B *Streptococcus pneumoniae* from Spain to Iceland in the late 1980s. *J. Infect. Dis.* **168**, 158–163.

Sternberg, G. M. (1881). A fatal form of septicemia in the rabbit produced by subcutaneous injection of human saliva: An experimental research. *Natl. Board Health Bull.* **2**, 781–783.

Tomasz, A. (1965). Control of the competent state in pneumococcus by a hormone-like cell product: An example for a new type of regulatory mechanism in bacteria. *Nature 208*, 155–159.

Tomasz, A. (1968). Biological consequences of the replacement of choline by ethanolamine in the cell wall of pneumococcus: Chain formation, loss of transformability, and loss of autolysis. *Proc. Natl. Acad. Sci. USA* **59**, 86–93.

Tomasz, A. (1995). Editorial: The Pneumococcus at the gates. *N. Engl. J. Med.* **333**, 514–515.

Tomasz, A. (Ed.) (1999). "*Streptococcus pneumoniae:* Molecular Biology and Mechanisms of Disease—Update for the 1990s." Liebert.

Tomasz, A., Jamieson, J. D., and Ottolenghi, E. (1964). The fine structure of *Diplococcus pneumoniae*. *J. Cell Biol.* **22**, 453–467.

Tomasz, A., Albino, A., and Zanati, E. (1970). Multiple antibiotic resistance in a bacterium with suppressed autolytic system. *Nature* **227**, 138–140.

Weiser, J. N., Austrian, R., Sreenivasan, P. K., and Masure, H. R. (1994). Phase variation in pneumococcal opacity: Relationship between colonial morphology and nasopharyngeal colonization. *Infect. Immun.* **62**, 2582–2589.

White, B. (1979). "The Biology of Pneumococcus." Harvard Univ. Press, Cambridge, MA.

Zighelboim, S., and Tomasz, A. (1980). Penicillin-binding proteins of the multiply antibiotic-resistant South African strains of *Streptococcus pneumoniae*. *Antimicrob. Agents Chemother.* **17**, 434–442.

Streptomyces, Genetics

Paul Dyson
University of Wales

I. Genome Composition
II. The *Streptomyces* Chromosome
III. Transposable Elements
IV. Plasmids
V. Bacteriophages
VI. The Genetics of Antibiotic Production and Morphological Development
VII. Artificial Genetic Manipulation

GLOSSARY

bacteriophage A virus that infects a bacterium.

codon A group of three consecutive nucleotides in nucleic acid that codes for a single amino acid.

genome The entire genetic material of an organism.

insertion sequence A discrete DNA sequence that carries only the genes that enable its own transposition.

open reading frame (ORF) A nucleic acid sequence containing a series of codons without any stop codons, and which potentially encodes a protein.

plasmid A nonessential extrachromosomal genetic element capable of autonomous replication.

recombination The generation of new DNA molecules by breaking and rejoining the original molecules.

replication Synthesis of a copy of a DNA molecule, using the original as a template.

sigma (σ) factor Polypeptide that associates with RNA polymerase core enzyme to determine promoter specificity.

transposition Process whereby a DNA element can insert itself genome independent of the host cell recombination machinery.

transposon A discrete DNA element capable of transposition and carrying recognizable genes, for example, antibiotic-resistance determinants.

THE UNUSUAL ASPECTS OF THE GENETICS OF ANTIBIOTIC-PRODUCING STREPTOMYCES

are undoubtedly a reflection of the morphology and life cycle of these complex prokaryotes. After spore germination, they grow to form aseptate multinucleate branching hyphae which can form a ramifying network to exploit a nutritive substrate. After depletion of the resource, they subsequently erect aerial mycelia on which spores are elaborated prior to dispersal. It is perhaps not surprising that special genetic systems go hand in hand with maintaining features of such a complex *mode de vie,* although we have still much to learn about this relationship. A driving force behind investigating their genetics has been the desire to exploit the rich and varied streptomycete secondary metabolism responsible for the vast range of commercially important antibiotics, pharmacologically active drugs, veterinary compounds, and pesticides that they synthesize. This article reviews our current understanding of the area and is being written at a time when the complete genome sequence of the model species, *Streptomyces coelicolor,* is being assembled. Fulfilment of this project is likely to stimulate many new ideas concerning genome structure and function, but in the meantime readers can obtain a progress report at the Web site *http://www.sanger.ac.uk/Projects/S_coelicolor/.* It should also be noted that the genus is quite diverse, and that our present knowledge is based on genetic analysis of only a few of the several hundred cataloged species.

I. GENOME COMPOSITION

In describing a bacterial genome, it is customary to make an arbitrary division between the chromosome and other genetic elements such as plasmids, phages, and transposons, with the assumption that the latter are not essential for cell viability. Such a facile distinction can both be misleading in emphasis and conceal the dynamism of the genome as a whole, whereby each of the constituent elements may interact with each other, giving rise to associations of varying permanence. That being said, in line with convention, this article will describe each of these components individually, but allude to what we know about their interactions, and, in each case emphasize distinctive features which contrast to the *Escherichia coli* model. One of these differences, which affects the total genome, is in DNA base composition. For *Streptomyces* spp. this is about 73% G+C, compared to 50% in *E. coli*. Quite why streptomycete DNA is at the upper end of the spectrum for base composition, and close to the theoretical limit for coding conventional proteins, is unclear. One idea is that it may confer enhanced UV resistance important in protecting the genomes of the aerial mycelia and spores. One obvious implication is a different codon usage, and one example where this has been exploited is to ensure correct developmental timing of translation: Genes necessary for vegetative growth lack the rare TTA leucine codon. Mutation of the cognate tRNA gene, *bldA*, prevents progress from the vegetative stage of the life cycle and the ability to produce aerial mycelia and antibiotics, as a consequence of the inability to translate the requisite mRNAs of certain key developmental genes containing the codon.

In the context of genomic base composition, many different restriction–modification systems have been found in members of the genus, and consequently the respective recognition sequences in the relevant species are methylated at either cytosine or adenine residues. However, it is noteworthy that there is little evidence for more extensive methylation of these bases, which could result from the activities of host modification enzymes equivalent to the Dam and Dcm methylases of *E. coli*. Interestingly, at least two species, *S. lividans* and *S. avermitilis,* contain low levels of a novel and site-specific modification to guanine. The structure of the base remains to be determined, but the modifying activity, which acts postreplicatively, requires a much more complex and extensive DNA substrate compared to conventional DNA modification enzymes. It is conceivable that this modification may influence aspects of DNA metabolism such as gene expression and DNA repair in much the same way as host methylation in *E. coli*.

The high G+C content could also theoretically influence DNA topology. There is as yet no evidence for *Streptomyces* DNA adopting an alternative conformation to standard negatively supercoiled B-form DNA at any stage of the life cycle, unlike the situation in *B. subtilis* spores where the DNA is in the A-form. The area is not well researched, however, and it is entirely conceivable that local variations in DNA structure, such as cruciforms and Z-DNA, play important roles in aspects of DNA metabolism at different stages of the developmental cycle. Such conformational changes could be influenced by the action of DNA binding proteins like the histonelike protein HS1, which belongs to the same family as the *E. coli* HU protein.

II. THE *STREPTOMYCES* CHROMOSOME

A. Chromosome Architecture

Genetic mapping based on conjugal recombination in *Streptomyces* is quite different from interrupted Hfr crosses in *E. coli*. Whereas the latter would reveal a discontinuity in the chromosome, if it existed, mapping in *Streptomyces,* which usually relies on an even number of crossovers between the complete recipient chromosome and a fragment from the donor occurring over an extended time period, will suggest *a priori* chromosomal circularity. Given the precedent of circularity established in *E. coli,* it was therefore assumed to be the case for other bacteria including *Streptomyces*. A circular linkage map of the model species *S. coelicolor* A3(2) was established by the mid-1960s, revealing, unusually, that nearly all genes were located in two arcs separated by "silent" regions at opposite sides of the circular map. This symmetry, and the fact that genes of related function were mapped to positions opposite one another in the two arcs, suggested that the chromosome may have evolved by duplication.

The 1990s has brought about a minor revolution in our understanding of the organization and structure of the *Streptomyces* chromosome. Physical analysis of the genome using pulsed-field gel electrophoresis, complemented by the construction and ordering of a cosmid library (now being used as the basis of the genome sequencing project), has revealed that, instead of being circular, the chromosome of *S. coelicolor* is in fact a linear molecule of 8 Mb (Fig. 1). This makes the streptomycete chromosome among the largest of any prokaryote, more than 1.5 times larger than the *E. coli* chromosome.

Chromosomal linearity appears to be the normal architecture in all species examined, posing the question as to why this is the case. There is an inherent problem in replicating the ends of a linear molecule, owing to the ability of all known nucleic acid polymerases to proceed only in the 5′ to 3′ direction and a requirement for them to bind to a region surrounding the position at which the first base is to be incorporated. In addition, DNA polymerases can only extend by adding nucleotides one at a time to an existing 3′-OH end. Consequently, normal lagging strand synthesis proceeds initially by synthesis of short RNA primers; these are subsequently removed and the gaps filled in. The problem with chromosomal linearity is that an RNA primer cannot be synthesized opposite the extreme 3′ end of the template strand, and, even if a suitable complementary RNA primer could be synthesized from an internal template and then base-pair with the 3′-end, its subsequent removal would leave a gap which could not be filled. Circularity avoids the predicament, and consequently appears to offer a tidy solution for organisms with small genomes represented by a single chromosome.

But circularity poses another problem: after replication and prior to cell division two circular chromosomes can dimerize by any odd number of crossing-over events, with the consequence that the single large molecule will be inherited by only one daughter cell. To ensure correct segregation of monomeric chromosomes into both daughter cells, *E. coli* specifies an efficient system of site-specific recombination to resolve chromosome dimers down to "monomers." This may be a clue as to why mycelial streptomycetes have opted for chromosomal linearity. The aseptate mycelial compartments contain many chromosomal copies. Vegetative growth by extension of the apical tips, but not accompanied by regular septation, would not appear to demand active partitioning of the chromosomes. In contrast, during sporulation, the organism must ensure that each of the many spore compartments which form at the tip of an aerial mycelium receives one chromosome. If the chromosome were circular, it is possible that recombination acting on multiple copies would push the equilibrium away from "momomers" toward higher forms, even if a resolving function were operating. Linearity ensures the chromosomes remain monomeric independent of recombination. This is a testable hypothesis now that circular chromosomes have been engineered (see below).

The demands of orchestrating correct chromosomal partitioning at sporulation are likely to involve both *cis-* and *trans-*acting factors yet to be determined. It is interesting to note that the parallel process in *Bacillus* involves a conjugation-like mechanism involving DNA translocation from the mother cell to the single spore compartment mediated by a protein sharing homology with the Tra proteins encoded by *Streptomyces* fertility plasmids (see Section IV). The process in *Streptomyces* may be more complex, in that for every aerial mycelium there is not just a single spore compartment but many into which a copy of the chromosome must be partitioned, but it could potentially employ a similar mechanism.

In some respects chromosome replication in *Streptomyces* is similar to that in *E. coli.* It is initiated at a single origin of bidirectional replication, *oriC,* located at the center of the chromosome (Fig. 1). The structure and function of this origin are related to those from bacteria with circular chromosomes, and indeed it can support replication of a circular molecule. A model to explain the particular problem of how the ends of the chromosome are replicated is based on studies on replication of linear plasmids (see below). The ends of the chromosome are characterized by long terminal inverted repeats (TIRs), which vary in length between 30 and 600 kb depending on the species (Fig. 1). Terminal proteins covalently attached to the free 5′ ends are thought to prime DNA synthesis, perhaps by formation of a complex of polymerase and the terminal protein bearing a priming nucleotide, as is the case for adeno-

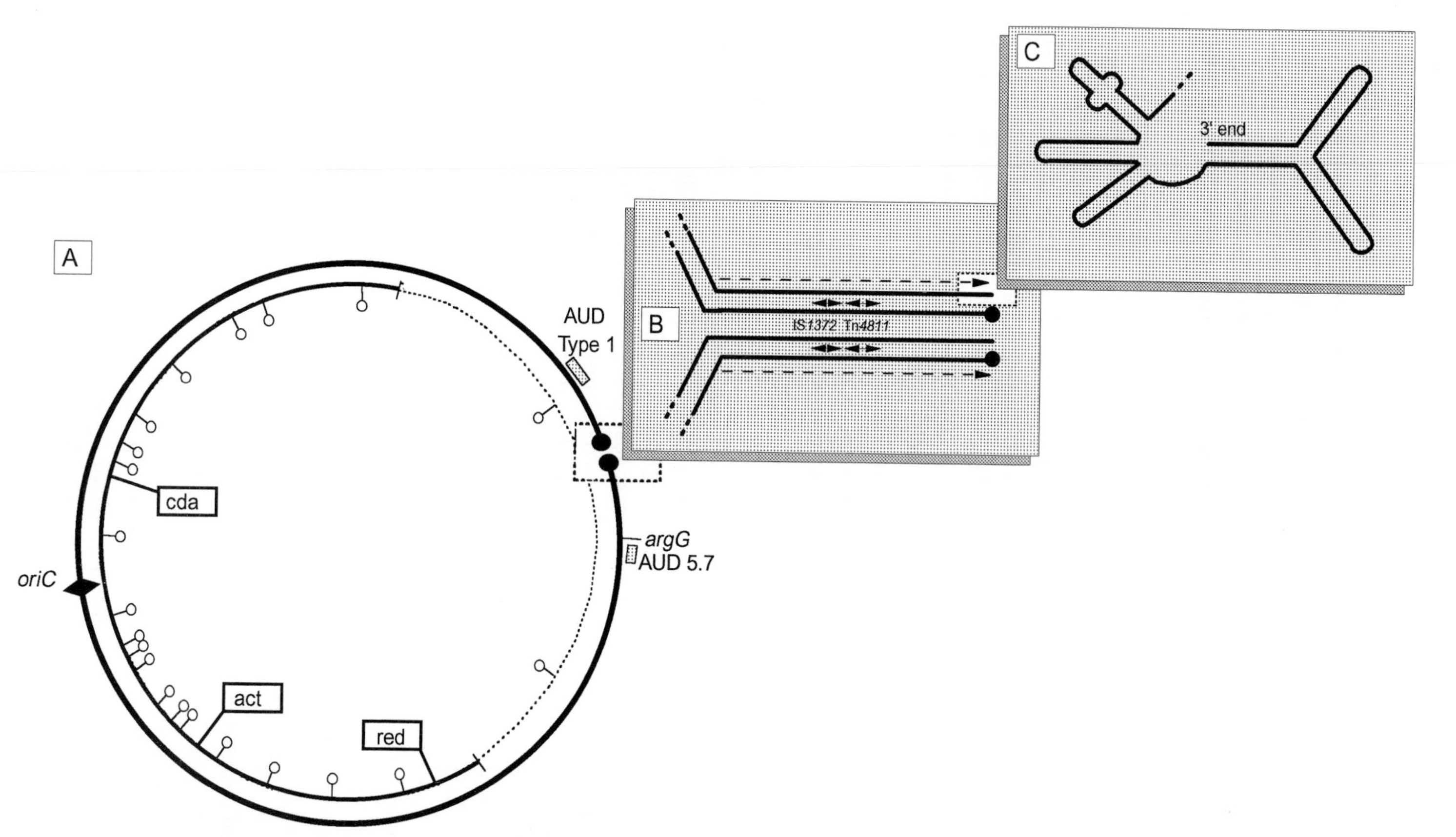

Fig. 1. The *Streptomyces* chromosome. (A) The outer arc represents the 8 Mb *S. lividans* chromosome, indicating positions of the amplifiable units of DNA (AUD). The inner circle represents the related *S. coelicolor* chromosome, the dashed arc containing nonessential or conditionally essential genes. The locations of mapped genes involved in morphological development and the control of physiological development are indicated (°), as are the three antibiotic biosynthetic gene clusters *act,* actinorhodin; *red,* undecylprodigiosin; and *cda,* calcium-dependent abtibiotic. (B) The extent of the 30 kb terminal inverted repeats of the *S. lividans* chromosome is indicated by dashed arrows, and the positions of two internal transposable elements by smaller arrows. The terminal protein bound to the 5′ ends is indicated by a solid circle. (C) The predicted secondary structure of the terminal 180 bp of 3′ single-stranded DNA.

virus replication. Terminal-protein primed synthesis is likely to be facilitated by other proteins that specifically interact with sequences within the TIRs. A comparison of TIRs from different species has revealed conserved sequences within the first 170 bp arranged as seven palindromes: This configuration may stimulate formation of DNA secondary structure important for stability of the terminal sequences, protein binding, and/or correct processing of terminal sequences during replication (Fig. 1). The significance of the extent, and variance thereof, of the TIRs is unclear as yet. The overall structure of the chromosome may resemble a "racket frame," with the TIRs forming the handle by virtue of cohesion between proteins bound to each of the inverted repeats.

At the time of writing, 1200 kb or 15% of the *S. coelicolor* genome has been sequenced, revealing 1045 potential open reading frames (ORFs) with an average length of 1.15 kb. Of these genes, 67% are orientated so as to be transcribed in the direction of replication emanating from *oriC*. Assuming an even distribution throughout the chromosome, it is assumed that the organism encodes >7000 proteins, some 20% more than the unicellular eukaryote *Saccharomyces cerevisiae*. This genetic complexity no doubt supports high metabolic versatility and the ability to endure a wide variety of physical and chemical stresses and competition with many kinds of organisms in the soil environment.

B. Genetic Instability

With the physical map of *S. coelicolor* established, it is now possible to readdress the significance of the unusual genome organization first suggested by genetic mapping. An arc can appear to be "silent" or underrepresented by markers either if there is a high concentration of crossovers in the region or if it is indeed an extent of DNA with few genes of recognizable function. One region originally identified as being silent, although now known to possess many recognizable genes, correlates with the position of *oriC*. The second silent region, on the opposite side of the circular genetic map, is coincident with the ends of the physical map, and there is evidence for an almost total absence of essential genes in the vicinity of these ends. Indeed, none of the 65 mapped genes involved in macromolecular synthesis are within 1.2 Mb of either terminus, with the exception of one potentially redundant RNA polymerase sigma factor gene, *hrdC*. Primary metabolic genes, with one exception (*argG*), are also absent from this region. The region accounts for 40% of the genome and appears to be populated only by nonessential or conditionally essential genes (Fig. 1).

The unequal distribution of essential genes (defined here as genes required for growth at least under laboratory conditions) requires some explanation, as it is reasonable to assume there might be some evolutionary significance. Indeed, the terminal regions are not simply theoretically dispensable, but are actually prone to high-frequency spontaneous deletion in the laboratory. These deletions can remove a substantial proportion of the nonessential regions on both chromosomal arms as well as the ends themselves. Initial unstable mutants commonly segregate variants with even greater extents of deletion. A processive loss of genetic information in this manner suggests that maintenance of the chromosomal termini is an error-prone process. The frequency of these events may have been the evolutionary driving force to ensure that essential genes are safeguarded by being long distances from these regions of instability.

An additional protection may be provided by amplifiable DNA sequences which commonly define the end points of deletion in the terminal regions. Up to several hundred copies of a tandemly amplified sequence can, it appears, act as a buffer to prevent further processive deletion eating into the chromosome. These amplifications can be of fixed unit length and reproducibly arise from a specific preexisting structure in the chromosome which, if it exists in a duplicated form in the wild type, is competent for reiteration (Type II amplification). Alternatively, Type I amplifications are generated which are characteristically nonreproducible, heterogeneous in size in different mutants, and originating from overlapping chromosomal regions. What both types of amplified element generally share in common is that they are maintained at high copy number in deletion mutants in the absence of an obvious selection pressure, leading to the suggestion that they may function to maintain the integrity of the remaining essential regions of the chromosome; deletions affecting the amplified DNA may be counterbalanced by new rounds of amplification. More extensive deletions of these chromosomal regions can be found in occasional mutants in which an amplified buffering DNA is lost.

These amplifiable sequences are located close to and thereby protect either chromosome end in *S. lividans* (Fig. 1), and the evolution of this architecture could be another consequence of the problems associated with maintaining the ends. Given the high frequency of instability, a cost–benefit analysis would appear to make the option of chromosomal linearity rather unsafe even with the built-in safeguards. However, simple circularization is not necessarily a viable solution either. Mutants containing spontaneous or experimentally induced circular chromosomes often exhibit very high rates of deletion and/or amplifications originating in the same unstable regions. This has been explained as being due to inefficient decatenation, lack of replication terminator sequences, or problems in partitioning at cell division. Whatever the reason or starting topology, the *Streptomyces* chromosome is inherently much less stable than that of *E. coli,* and it is worthwhile considering if this is another adaptation related to the life cycle. For example, the vegetative mycelia are programmed to undergo autolysis, allowing recycling of their macromolecules to fuel growth of the aerial mycelia. Are unstable mutants derived from mycelia which have undergone the initial steps in this programmed cell death, only to be subsequently "rescued"? Although this is speculative, it is curious that the *catA* gene encoding catalase function, a major defense against the oxidative damage which can provoke cell death, is located close to one telomeric TIR in *S. coelicolor* and hence is unstable. Perhaps the overall chromosome architecture has evolved to permit turnover of the vegetative mycelia, with genetic instability being a normal course of events resulting in deprotection from oxidative stress and ultimately leading to complete fragmentation of the genome in these terminally differentiated cells.

As yet it is not clear what types of recombination mechanisms have a role in genetic instability. RecA-mediated homologous recombination is implicated in Type II amplification, as shown in *S. lividans* where it acts on the perfect 1 kb direct repeats of the amplifiable unit of DNA, *AUD1,* present in the wild-type chromosome. A *recA* mutant containing a slightly shortened version of the gene was unable to amplify this element unless complemented with the wild-type gene. The same mutant exhibited an enhanced frequency of deletion of unstable genes, suggesting a role for recombination in maintenance of the chromosome ends. The precise role of RecA in genetic instability is difficult to define. In a very few cases, the junction sequences at which large deletions have occurred have been examined, revealing very short stretches of homology which would not normally be considered adequate substrates for homologous recombination; instead, the deletions may result from replication slippage.

III. TRANSPOSABLE ELEMENTS

Another contributory factor to genetic instability, certainly with precedence in other biological systems, could be transposition. Deletion of sequences between a "donor" and "target" site is one expected consequence of intramolecular transposition. Although there is no direct evidence to suggest a link between the movement of mobile elements and loss of terminal DNA, there are indications which are consistent with the idea. First, sequence analysis of terminal DNA has revealed the presence of transposable elements: The TIRs of *S. lividans* contain copies of Tn*4811* and IS*1372* (Fig. 1). Moreover, an insertion sequence, IS*1373,* delimits an amplifiable element in the same species. Systematic sequencing of the entire "silent" region will reveal if indeed it acts as a "sink" for transposable elements. Subsequent mobility of these elements is likely to affect the stability of the termini as shown experimentally with a heterologous insertion sequence: Transposition of the mycobacterial insertion sequence IS*6100* into one chromosome end can promote rearrangements and amplifications in *S. lividans.*

Other transposable elements have been uncovered either by direct searches involving screening disrupted genes, or having been acquired by temperate phages, sometimes rescuing the ability of an *attP*-deleted phage to lysogenize, or due to insertion stabilizing an otherwise unstable plasmid, or as a result of sequencing projects. The majority of these elements are Class I cryptic insertion sequences, but there are some noteworthy examples with atypical genetic features. An element from *S. clavuligerus*, IS*116*, neither contains terminal inverted repeats nor generates target site duplications, resembling IS*900* found in *Mycobacterium paratuberculosis*. Both these elements show similarities with IS*117*, termed the "minicircle," from *S. coelicolor*. The deduced protein encoded by the single open reading frame present in IS*116* shares 25% homology with the putative transposase product of IS*117*. The latter also does not generate target site duplications, although it does possess imperfect inverted repeats at its ends. IS*117* is notable in that it appears to transpose via a covalently closed circular intermediate generated by replication of a preexisting integrated copy of the element. Subsequent integration is quite site-specific, although there is no obvious homology between the inverted repeats which flank the cleavage site of the minicircle and the preferred chromosomal integration sites.

The only compound transposon found in *Streptomyces* to date is an amplifiable region, which includes mercury resistance genes, bounded by two copies of IS*1373* in *S. lividans*, although transposition of only a single copy of this insertion sequence has been experimentally proven. Of the two other transposons discovered, one, Tn*4556*, is a cryptic Tn*3*-like element isolated from *S. fradiae*, and the other, Tn*4811*, forming part of the TIRs of *S. lividans*, encodes a putative oxidoreductase. Although antibiotic-resistance genes are commonly associated with transposons in many other bacteria, the resistance genes of the antibiotic-producing *Streptomyces* spp. are not. On the basis of similar biochemical resistance mechanisms shared between antibiotic producers and clinical bacteria, it is entirely plausible that they originated in streptomycetes and were subsequently disseminated by horizontal transfer. However, the evidence that the resistance genes are typically chromosomally encoded and nonmobile in streptomycetes suggests indirect evolution of these transposons, possibly after acquisition of antibiotic-resistance genes by other bacterial species sharing a similar ecological niche as *Streptomyces*.

IV. PLASMIDS

A. Fertility

With the exception of the broad host range IncQ plasmid, RSF1010, typical enteric plasmids are unable to replicate in *Streptomyces*. Instead there is a considerable diversity of native plasmids differing in size, copy number, topology, and autonomy. Thus, there are small, covalently closed circular (ccc) high copy number plasmids, larger ccc low copy number plasmids, ccc plasmids which arise by reversible site-specific recombination from the chromosome, and linear plasmids which share common structural features with the chromosome; they may be considered as "minichromosomes." A property common to maybe all is their self-transmissibility and ability to mobilize chromosomal markers, making them supremely important in terms of promoting genetic exchange. As noted above, transfer functions may also have a role in partitioning of replicons at sporulation. Plasmid transfer between strains can approach frequencies of 100%, whereas chromosome mobilization can yield up to 1% recombinants. Typically, plasmid transfer is apparent due to the formation of "pocks" on a lawn of recipient bacteria. These are zones of retarded development of aerial mycelia affecting transconjugant mycelia, surrounding what is believed to be the primary site of contact between donor and recipient mycelia, and believed to be the manifestation of intramycelial spread.

The conjugation mechanism is quite unlike that of enteric bacteria as is evident from studies on small covalently closed fertility plasmids like pIJ101. A region of only 4 kb of this plasmid encodes all functions required for transfer and subsequent intramycelial spread of the plasmid in the recipient. There is no evidence for pilus formation, and the establishment of mating pairs appears to be a simpler process than in gram-negative bacteria. Despite their functional similarity, the transfer proteins of the various plasmids studied to date are divergent in sequence and size. However, they contain conserved nucleoside triphosphate (NTP)-binding motifs and, for at least two examples tested, are localized to the cytoplasmic membrane. A general model for streptomycete plasmid transfer involves cell fusion mediated either by the transfer protein(s) or possibly by chromosomally encoded functions, and subsequent mobilization of the plasmid and chromosome of the donor. The NTP-binding domains are conserved among proteins which mediate DNA transport across the membrane. Studies on pIJ101 have indicated that the transfer protein is active only during vegetative growth. A *cis*-acting sequence, *clt,* is required for plasmid transfer, and this may be functionally equivalent to the origin of transfer of enteric plasmids. Homologous sequences may be required for mobilization of the chromosome. Subsequent to intermycelial transfer, additional functions, encoded by *spdA, spdB,* and *kilB* in pIJ101, allow for spread of the plasmid throughout the mycelia of the recipient; their activities affect the size of a pock. The translated products of these three genes have strongly hydrophobic domains, suggesting they are also membrane-associated.

One naturally temperature-sensitive ccc plasmid, pSG5, although reported to be self-transmissible does not encode its own distinct transfer functions. This may indicate that it is difficult to distinguish whether a plasmid is self-transmissible or mobilizable in *Streptomyces,* with the possibility for transfer functions being provided *in trans* from chromosomally integrated elements (see below). Alternatively, the pSG5 Rep protein could be bifunctional, promoting both replication and transfer; unlike the Rep proteins of other multicopy plasmids such as pIJ101, the pSG5 Rep contains an NTP-binding domain. Interestingly, pSG5 transfer is not accompanied by pocking, consistent with there being no spread functions encoded by the plasmid.

B. Topology and Replication

With the exception of one of the larger linear plasmids, there are typically no discernible phenotypic traits other than fertility conferred by streptomycete plasmids. The modes of replication for the different types of fertility plasmids are diverse, however. The small ccc high copy number plasmids, such as pIJ101 from *S. lividans* and pSN22 from *S. nigrifaciens,* are replicated via a rolling-circle mechanism initiated from a plus origin of replication. The second-strand synthesis is subsequently initiated from a separate minus origin, *ssi* (or *sti*), deletion of which can be tolerated but results in the accumulation of single-stranded replication intermediates. This mode of replication is a feature of plasmids from several other gram-positive species. The *ssi* region is also responsible for an interesting reversal of conventional plasmid incompatibility in that two Ssi^+ or two Ssi^- plasmids are compatible, but a Ssi^+ plasmid cannot coexist with a related Ssi^- plasmid. Larger low copy number ccc plasmids, of which SCP2 of *S. coelicolor* is an example, are believed to replicate by standard theta-form replication initiated from a single bidirectional origin. The linear plasmids range in size from 12 kbp to several hundred kilobase pairs and, as for the chromosome, are characterized by possessing TIRs and protein covalently attached to the 5′ ends. The replication mechanism has been examined for two of the smaller, and hence experimentally tractable examples, the 17 kb pSLA2 of *S. rochei* and the 12 kb pSCL of *S. clavuligerus.* Both possess a centrally located origin of bidirectional replication which can support replication of either circular or linear forms, and which resembles the origins of temperate bacteriophages of the Enterobacteriacae and *Bacillus.* The replication intermediates are linear duplex molecules with recessed 5′ ends, and it is assumed that the 3′ overhangs serve as templates for nondisplacing synthesis of the lagging strand primed by the protein attached to the 5′ ends. Study of these plasmids has actually been the source of ideas about replication of the chromosome itself.

C. Interactions with the Chromosome

The 350 kb plasmid SCP1 of *S. coelicolor* is considered to be the paradigm for very large streptomycete linear plasmids. SCP1 itself is remarkable in that it contains the complete biosynthetic gene cluster and linked resistance gene for the antibiotic methylenomycin. Despite the prevalence of similar large linear plasmids in various antibiotic-producing species, however, other antibiotic biosynthetic gene clusters have been mapped to chromosomal locations. The genes for three spore-associated proteins, *sapC, D,* and *E,* are also located on SCP1, on the TIRs, although SCP1 can be lost without affecting the morphological phenotype. As is the case for several other linear plasmids, SCP1 can interact with the chromosome, and the best studied concerns an integration event which inactivated the chromosomal agarase (*dagA*) gene. This may have involved recombination between copies of the insertion sequence IS*466*. One copy is located close to the right TIR of SCP1 and another is found close to *dagA*. Sequence analysis of the recombinant junctions of one example revealed that the left junction had an almost intact terminus of SCP1, whereas the right junction was composed of IS*466*, deleting the right TIR. The same deletion may have removed *dagA*.

This indicates a possible mechanism for horizontal transfer and chromosomal integration of antibiotic gene clusters, and it is interesting that the reverse has also been reported: acquisition of biosynthetic genes by a linear plasmid. This example concerns the 387 kb linear plasmid pPZG101 of *S. rimosus;* an oxytetracycline overproduction phenotype has been attributed to the observation that the mutant contained an enlarged pPZG101 derivative of 1 Mb which included oxytetracycline biosynthetic genes. Interactions between the same plasmid and the *S. rimosus* chromosome are also reported to lead to replacement of one chromosome end with that of the plasmid. Similar types of interactions may explain the homology existing between the ends of the *S. lividans* chromosome and the coresident 50 kb linear plasmid SLP2. About 17 kb of the extreme ends of the 30 kb chromosomal TIRs are similar or identical to one end of SLP2, including one copy of Tn*4811*. Moreover, mutants in which both chromosomal ends are lost cannot support replication of the plasmid, suggesting an interdependence for functions required for replication of the plasmid linear ends that are encoded by the deleted chromosomal regions.

Together these observations suggest a fairly dynamic relationship between the chromosome and plasmids, and this is reemphasized in the case of another class of circular plasmids which undergo reproducible integration and excision. The best studied examples are SLP1, which originates from *S. coelicolor,* and pSAM2 from *S. ambofaciens.* These plasmids encode their own site-specific recombination functions similar to those of temperate phages, and recombination occurs between specific plasmid and chromosomal sequences equivalent to phage attachment sites, suggesting a possible phylogenetic relationship. The host range for integration can be effectively broader than for autonomous replication by virtue of the chromosomal attachment site overlapping a conserved tRNA gene, for example $tRNA^{Pro}$ for pSAM2. Coupled with their self-transmissibility, this makes this class of plasmids potentially very promiscuous because tRNA gene sequences are highly conserved.

V. BACTERIOPHAGES

Both virulent and temperate bacteriophages can be isolated with ease from *Streptomyces* and/or their soil habitats. Few have been characterized in any detail. In contrast to "nonparasitic" genetic elements, bacteriophages have a much greater range in base composition. For instance, the 131 kb genome of the broad host range phage FP22 has a G+C content of 46 mol%, whereas for phage SH3 it is 73 mol%. There is only one well-documented generalized transducing phage: the narrow host range ϕSV1 of *S. venezualae.* Another system of natural gene transfer is provided by ϕSF1 of *S. fradiae,* which can exist as a plasmid prophage (pUC1) and promote conjugation without chromosomal integration, thus extending the phylogenetic interrelationship between phages and plasmids in *Streptomyces.* By analogy with the integrating plasmids described above, the chromosomal *attB* site for integration of the temperate phage RP3 in *S. rimosus* is a tRNA gene, in this case $tRNA^{Arg}$. Moreover, as for the plasmid attachment sites, this *attB* site contains a region of dyad symmetry downstream of the tRNA gene that could act as a transcriptional terminator to prevent cotranscription of phage and tRNA genes.

By far the best characterized phage is ϕC31, a temperate phage whose genome organization and general biology are similar to the lambdoid phages of *E. coli.* An interesting feature of this phage is its ability to shut down host RNA synthesis at an early stage of infection by affecting transcription initiation. A second noteworthy interaction with the host concerns phage resistance and a phage growth limitation phenotype, Pgl. ϕC31 propagated on a Pgl^- strain can adsorp to and lysogenize Pgl^+ strains of *S. coelicolor* with the same efficiency as Pgl^- strains. In the Pgl^+ strain a normal lytic cycle then follows and, although the progeny phage can infect a Pgl^- host, they are attenuated in their ability to infect further Pgl^+ mycelia. In this way, the Pgl^+ host effectively limits infection to only one mycelial compartment, protecting the rest of the colony from lysis. The molecular basis for this system could be explained by an inversion of classic restriction–modification, an idea supported by the observation that *S. coelicolor* restricts methylated DNA. Thus the Pgl^+ host could modify progeny phage DNA in a manner so that it is restricted on infection of the remaining mycelia. This model remains to be substantiated, but two genes encoding proteins of unknown function which play a role in this system have been cloned.

VI. THE GENETICS OF ANTIBIOTIC PRODUCTION AND MORPHOLOGICAL DEVELOPMENT

A. Pleiotropic Control of Physiological and Morphological Development

Streptomycetes possess a large degree of promoter heterogeneity which in part reflects the occurrence of multiple σ factors, allowing RNA polymerase to discriminate between different subsets of promoter sequences. For example, *S. coelicolor* contains up to 20 different σ factors. Temporal regulation of expression of σ factor genes themselves, coordinated with the life cycle and the physiological status of the mycelia, ensures that certain sets of genes are expressed at appropriate times. This forms a framework for the regulation of both antibiotic biosynthesis and morphological development, properties which together contribute to the biological fascination and economic importance of streptomycetes. However, the emerging evidence suggests that, instead of a linear cascade model to explain this regulation, as is the case for the control of sporulation in *Bacillus*, the integration of a variety of environmental and physiological signals requires that complex regulatory networks are operating.

In that both antibiotic production and the onset of morphological development are both usually associated with the stationary phase, at least in surface-grown cultures, it is perhaps not surprising to find many genes with pleiotropic effects on these processes. Many of these genes were found in screens for mutants unable to erect aerial mycelia, the so-called bald mutants. One of these, *bldA*, encoding a rare tRNA has been referred to earlier (Section I). At least nine other classes of *bld* mutants defective in antibiotic production have been identified in *S. coelicolor*. Whereas the morphological deficiencies of these different mutants can often be suppressed nutritionally or by cross-stimulation by diffusible factors (see below), the same conditions do not generally suppress the pleiotropic defects in antibiotic production. Like *bldA*, several of the other pleiotropic *bld* loci have been cloned and their products deduced to be associated with gene expression; *bldD* encodes a transcription factor-like protein, *bldG* a protein homologous to others known to interact with transcription factors, and *bldB* a potential transcription factor.

Moreover, the timing of both sporulation and antibiotic production can be regulated by the synthesis and accumulation of small diffusible signaling compounds. The best characterized example concerns the role of a γ-butyrolactone termed A-factor in streptomycin production and sporulation in *S. griseus*. One gene, *afsA*, involved in A-factor synthesis has been cloned, and its overexpression leads to precocious antibiotic production and sporulation; addition of related γ-butyrolactones to *S. coelicolor* cultures also leads to earlier production of actinorhodin (Act) and undecylprodigiosin (Red). A-factor is freely diffusible across the cytoplasmic membrane and can bind in a 1:1 ratio with a cytoplasmic A-factor binding protein, ArpA. In *S. griseus*, unbound ArpA is believed to have a negative regulatory role on both streptomycin production and sporulation, which is alleviated by association with the ligand. Two homologous receptors, CprA and CprB, were found in *S. coelicolor*: Genetic evidence suggests that CprA is a positive regulator of Act and Red biosynthesis, and also sporulation, whereas CprB negatively regulates only Act synthesis and morphological development.

B. Organization and Specific Regulation of Antibiotic Biosynthetic Genes

Subsequent compartmentalization separating the developing aerial mycelia and the antibiotic-producing vegetative mycelia demands distinct regulation of each process, reinforced perhaps by the contrasting organization of the genes involved. Whereas the developmental genes are typically dispersed on the chromosome, the biosynthetic genes for each antibiotic are arranged in clusters in which many are cotranscribed (Fig. 1). In a given species producing more than one antibiotic, the clusters themselves are typically dispersed. Many of these clusters contain at least one, and sometimes two, pathway-specific regulatory genes, which encode transcriptional activators, and also antibiotic self-resistance genes whose expression can thereby be coordinated with antibiotic biosynthesis. Interestingly, the primary transcripts of several of these resistance genes lack untranslated leader sequences—the start points for transcription and translation are coincident; the significance of this in relation to ribosome recognition and control of translation is as yet unclear. The architecture of antibiotic biosynthetic gene clusters obviously allows for well-coordinated regulation of the various genes but is also of interest in relation to their evolution and horizontal transfer (see Section IV).

Of particular note are the genes responsible for biosynthesis of polyketides, which comprise a very large and diverse class of streptomycete secondary metabolites. *Streptomyces* have evolved two types of polyketide synthase (PKS): one which is a complex of several separate polypeptides (as for the fatty acid synthases of *E. coli* and plants) and another consisting of multifunctional polypeptides carrying the different catalytic functions as a separate domains, similar to vertebrate fatty acid synthases. However, the latter, the Type I PKSs responsible for synthesis of complex actinomycete polyketides, are notable in that for each addition to the polyketide chain there is a separate PKS module, so that the gene cluster contains multiple sets of homologous PKS genes. The evolution of this architecture is conjectured to have proceeded by successive gene duplications.

An integrated picture is yet to emerge for how different physiological conditions such as growth rate, nutrient limitation, metabolic imbalance, and cell density affect antibiotic production; there may be several routes to activation of a particular pathway and interactions between different regulatory responses. Several genes have been identified that pleiotropically affect the activation and perhaps maintenance of production of the different antibiotics synthesized by *S. coelicolor,* but which have little effect on morphological development. The pleiotropic regulators appear to act directly or indirectly to influence expression of pathway-specific activators (Fig. 2).

For example, transcription of the Act pathway-specific activator gene, *act*II-ORF4, is almost undetectable in an *afsB* mutant; the AfsB gene product (as yet uncharacterized) also influences the Red synthesis of γ-butyrolactones. Mutations in two other pleiotropic regulatory genes, *absA1* and *absB,* also reduce expression of pathway-specific activator genes and can be phenotypically suppressed by the introduction of these activator genes on high copy number plasmids to restore antibiotic biosynthesis. Similar suppression of the *absA* and *absB* mutations is observed in recombinant strains containing multiple copies of *afsR2,* consistent with a model in which the expression or activity of the product of *afsR2* depends on *absA* and *absB,* and AfsR2 in turn stimulates expression of the pathway-specific activators. The *absB* gene encodes a homolog of a double-strand specific RNase, suggesting that its control of antibiotic biosynthesis is through posttranscriptional regulation of specific target genes. In common with several other loci encoding pleiotropic regulators, the *absA* locus encodes a two-component regulatory system: The AbsA1 product is a sensor histidine kinase believed to phosphorylate AbsB2 which, in turn, can negatively regulate antibiotic biosynthesis. The mutations in *absA1* which reduce Act and Red production may lock the kinase in an active conformation; disruption of *absA1* results in early hyperproduction of both antibiotics, correlated with increased expression of the respective pathway-specific activator

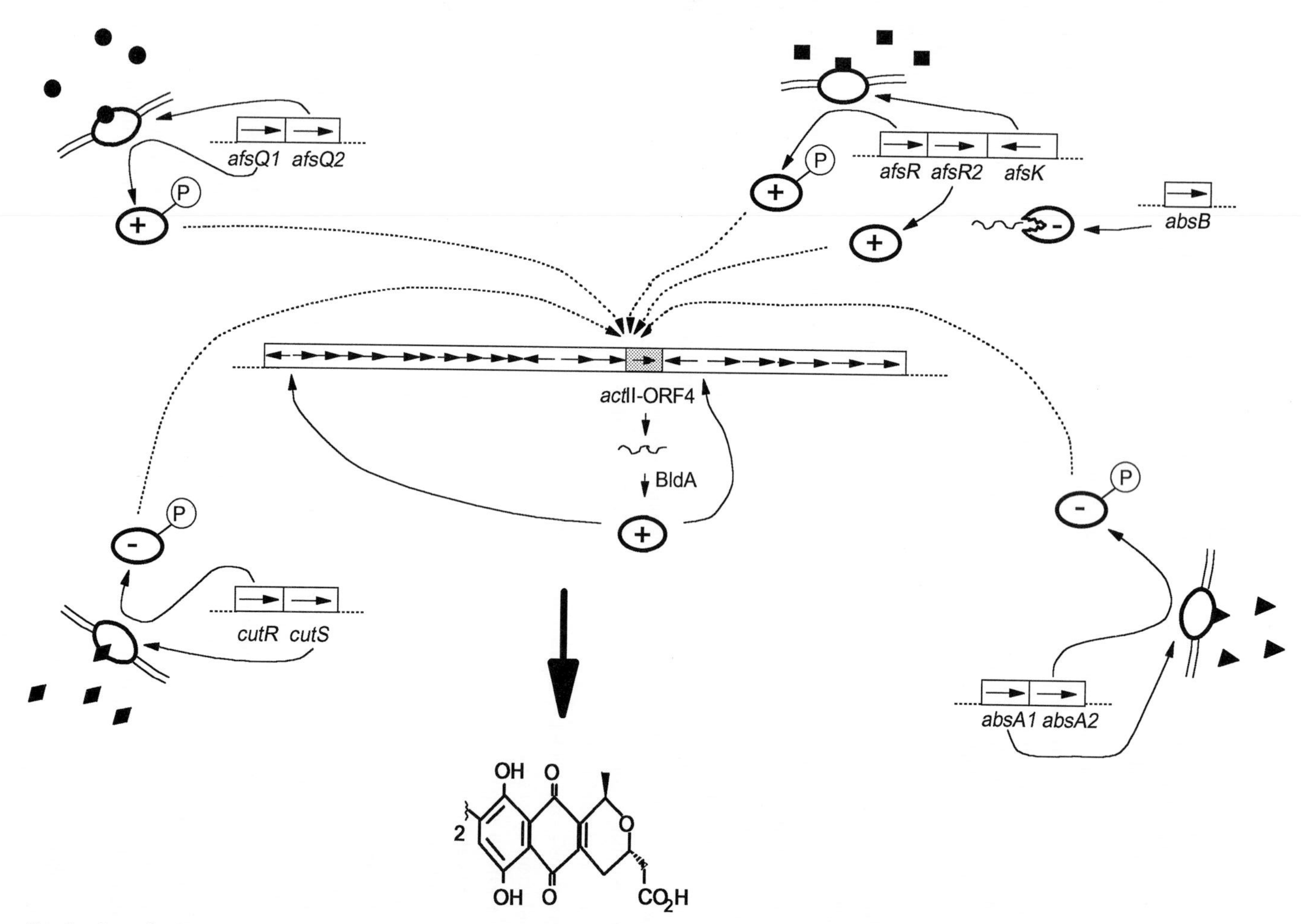

Fig. 2. Complexity of the specific regulation of antibiotic biosynthesis. The function of several two-component response regulators and their putative extracellular ligands (solid shapes) is depicted in relation to expression of the actinorhodin pathway-specific activator gene, *act*II-ORF4. Dotted arrows indicate either direct or indirect regulation of the gene, which might involve interdependencies between the different regulators. Also indicated is the AbsB RNase activity which may regulate expression of *afsR2* (see text). Expression of *act*II-ORF4 also depends on BldA. At least two of the targets for the *act*II-ORF4 encoded activator itself are intergenic regions between *act*VI-ORFs A and 1, and between *act*III and *act*I-ORF1; these regions are indicated on the map of the entire *act* cluster.

genes. In a similar manner, protein phosphorylation mediated by *cutRS* negatively regulates antibiotic production, whereas two other two-component regulatory systems encoded by the *afsK/afsR* and *afsQ1/afsQ2* loci act as positive regulators (Fig. 2).

Typically, the antibiotic production phenotypes associated with mutations in these regulatory loci are conditional on the media and growth conditions, suggesting several different routes to activation. The nature of the molecules that activate these multiple signal transduction pathways remains to be elucidated, although γ-butyrolactones, as discussed earlier, and an increase in cytoplasmic ppGpp concentration, associated with reduced growth, both have roles in activation of antibiotic biosynthesis. Moreover, activation of transcription of the pathway-specific activators is not the last level of regulation: The transcripts themselves can contain a UUA codon, as is the case for both *act*II-ORF4 and *redZ,* and consequently their translation is BldA-dependent.

C. Regulation of Morphological Development

A model for the genetic regulation of sporulation in bacteria is provided by *Bacillus subtilis*. [*See* SPORULATION.] In this case, a nongrowing, stationary phase cell undergoes septation, with one compartment maturing as a spore. A phosphorelay, triggered by the physiological status of the resting cell, initiates an orderly linear casade of sigma factors whose activities regulate successive stages in the sporulation process. Although some of these themes are utilized during streptomycete sporulation, the underlying biology is significantly different, and this is reflected in distinct systems of regulation. The onset of morphological development involves the erection and extension of aerial mycelia, and each mycelium subsequently undergoes septation to give mutliple unigenomic spore compartments (Fig. 3). The development of these mycelia is dependent on several genes, most notably the dozen or so *bld* genes referred to earlier. Many of these genes determine the nonribosomal synthesis of a 17 amino acid hydrophobic oligopeptide SapB which, at least on rich media, coats the surface of a newly developing aerial mycelium, allowing it to break surface tension at the colony–air interface and grow upward (Fig. 3). The *bld* genes concerned in SapB synthesis are believed to encode components of an extracellular signaling cascade (e.g., BldK is a membrane oligopeptide transporter) leading to expression of *sapB*, or, alternatively, they could provide biosynthetic intermediates of an end product that either is SapB itself or is a signal to activate SapB biosynthesis. SapB is not produced on defined minimal media, however, and in these conditions the phenotypes of several *bld* mutants are conditional on the carbon source: They produce aerial mycelia and sporulate on carbon sources other than glucose.

It is at present unclear how aerial mycelia are erected in these conditions. Perhaps a second surfactant protein is produced whose synthesis is dependent on the carbon source on which these *bld* mutants are grown, even though several *bld* mutants appear to be deficient in aspects of glucose catabolite repression. Moreover, the pH of the immediate environment of the colony may also influence its ability to erect aerial mycelia. To some extent this can be modified by the bacterium: During early vegetative growth, the colony secretes organic acids that lower the overall pH and, to neutralize the pH, these compounds are subsequently reabsorbed and metabolized as a prelude to morphological development. The ability to do the latter is impaired in both a *cya* mutant, deficient in cyclic AMP (cAMP) synthesis, and several *bld* mutants. It is suggested both that cAMP may serve as a signaling molecule to coordinate development and also that the regulation of carbon metabolism may have a crucial bearing on the developmental fate.

Evidence for further regulatory inputs determining a commitment to sporulation comes from studies on *afsA* mutants of *S. griseus* deficient in A-factor synthesis and also development of aerial mycelia. This defect in morphological development can be rescued by multiple copies of the cloned *amfR* gene, which encodes a response regulator-like protein. This gene, which includes a TTA codon suggesting dependence on *bldA,* is part of the *amf* locus which also contains the *amfAB* genes encoding a peptide transporter. A homologous locus, *ramABR* (although lacking in TTA codon), is found in *S. coelicolor*. The AmfR and RamR proteins may repress morphological development as a response to peptide signaling mole-

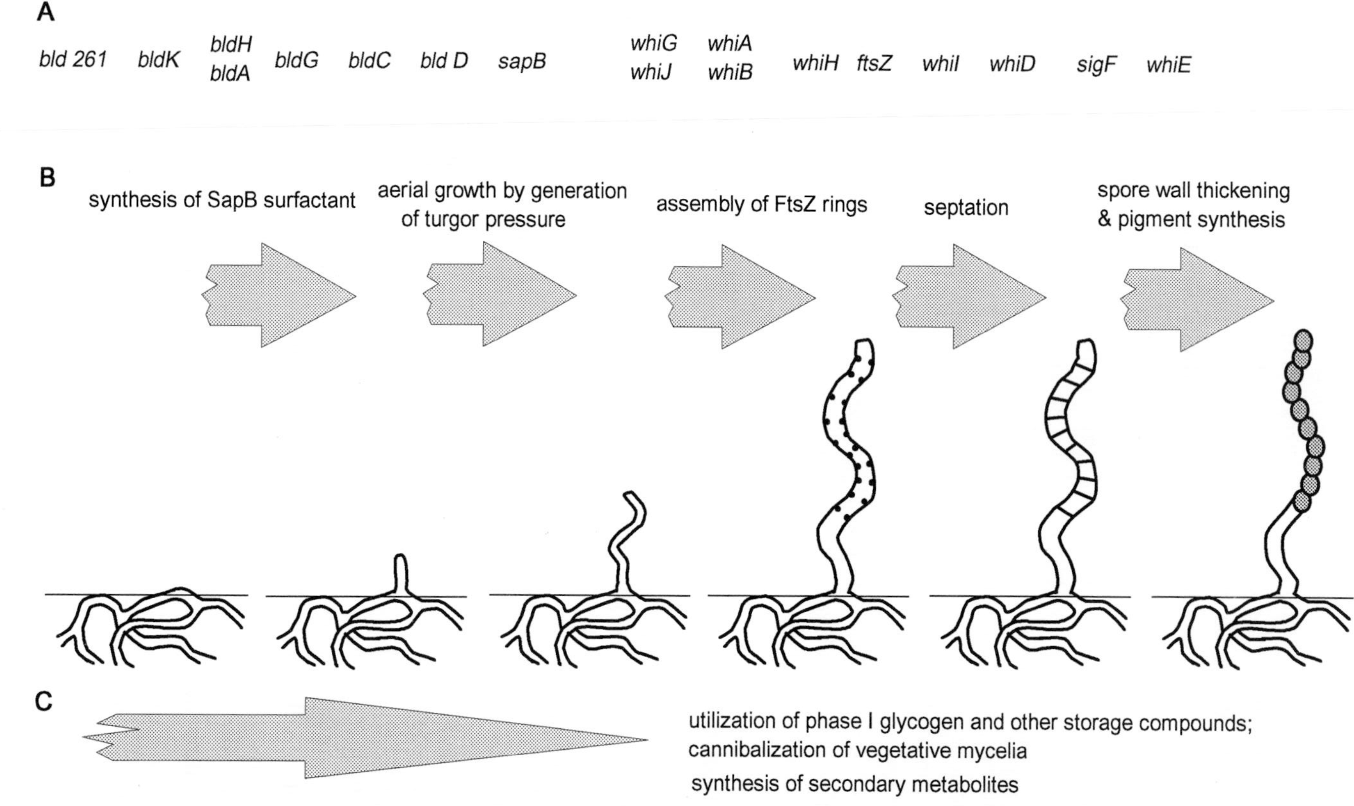

Fig. 3. Sporulation in *Streptomyces.* (A) Key genes identified as being involved in the developmental process are indicated, and the approximate order of their expression correlated to successive stages of development is depicted in (B); see text for details. (B) Fundamental physiological events associated with growth and development of the aerial mycelia. Segregation of a copy of the genome to each spore compartment immediately precedes septation, but is not depicted. (C) Physiological changes occurring in the vegetative mycelia that accompany sporulation.

cules. The *amfR* gene is itself repressed by the AdpB protein whose synthesis is part of the regulatory cascade leading to morphogenesis determined by A-factor. Yet another regulatory effect in *S. coelicolor* is exerted by a GTP-binding protein, Obg, a membrane-bound protein which, by binding GTP, can delay the onset of development.

The growth needed to extend the aerial mycelia upward is fueled, at least in part, by cannibalization of the vegetative mycelia. The required turgor pressure may be provided by degradation of osmotically inactive polymers like glycogen, oil, and polyphosphate. Indeed, glycogen deposits (phase I glycogen) are formed at the base of the aerial mycelia as a prelude to morphological development in the wild type, and this deposition is affected in *bld* mutants. Two sets of genes (*glg*) involved in glycogen synthesis have been identified: one set for phase I synthesis and the other for phase II synthesis in prespore compartments. Genetic evidence suggests that phase I glycogen synthesis and degradation are not essential to generate the turgor pressure needed for aerial growth, perhaps because metabolism of other storage compounds can fulfil the same role.

The signals which both limit extension of the aerial hyphae and control the formation of sporulation septa are unknown. Some of the genes involved were identified from studies of *whi* mutants, whose aerial mycelia fail to develop fully mature spores (the gray spore pigment is not synthesized and the aerial mycelia consequently remain white). Many of the these genes, the "early" *whi* genes, are involved in a regulatory cascade governing the formation of unigenomic prespore compartments. Thus, *whiG* encodes an RNA polymerase sigma factor whose activity is critical to the sporulation process. Overexpression due to increased gene dosage of *whiG* results in premature and/or ectopic sporulation. Transcription of *whiG* is fairly constant throughout the life cycle, suggesting posttranscriptional regulation of σ^{WhiG}, possibly by an as yet unidentified anti-sigma factor, in a manner similar to how related motility σ factors are regulated in other bacteria. One gene transcribed by the σ^{WhiG} form of RNA polymerase is *whiH*, encoding a GntR-like protein: In other bacteria, these proteins are known to repress certain genes involved in carbon metabolism, as well as repressing their own expression, the repression being alleviated by appropriate carbon metabolites. WhiH-mediated regulation of the formation of prespore compartments forms another potential link between carbon metabolism and morphological development.

Formation of prespore compartments requires the laying down of regularly spaced septa, initiated by assembly of peripheral rings of the protein FtsZ (Fig. 3). This protein resembles eukaryotic tubulins which form dynamic chains or filaments, comprising a major component of the cytoskeleton and allowing for changes in cell shape and movement of organelles. The contraction of the FtsZ rings permits in growth of the septum. Immunofluorescence techniques have revealed "ladders" of FtsZ rings in the aerial mycelia of the wild type, but FtsZ ladders are absent in *whiG* or *whiH* mutants. At least in *S. griseus*, it is evident that septation is preceded by increased expression of *ftsZ* from a sporulation-specific promoter. FtsZ ladders are also absent in *whiB S. coelicolor* mutants; the WhiB protein belongs to an expanding family of putative transcription factors containing a characteristic array of four cysteine residues and a potential basic α-helical domain which may interact with DNA. It is regulated independently of *whiG*, suggesting a second parallel pathway controlling sporulation.

The later stages of sporulation, involving spore wall thickening and spore pigment synthesis, depend on another RNA polymerase σ factor, σ^{F}. The *sigF* gene is not transcribed until the sporulation septa become detectable, and its regulation is dependent on early *whi* gene products. However, the *sigF* promoter is not recognized by $E\sigma^{\mathrm{WhiG}}$, and the exact manner in which it is regulated is not yet clear. σ^{F}, itself, is required for expression of a gene belonging to the *whiE* gene cluster. This locus specifies a type II polyketide synthase responsible for synthesis of the gray spore pigment.

VII. ARTIFICIAL GENETIC MANIPULATION

Cloning and manipulation of the streptomycete genome became feasible with the development of a method for introducing DNA by protoplast transformation. Subsequently, a wide variety of cloning vectors have been developed on the basis of small high copy number of plasmids such as pIJ101, larger low copy number replicons like SCP2, or phages. SCP2, by virtue of theta-form replication, which affords potentially greater stability to larger cloned segments, has been exploited to carry antibiotic biosynthetic gene clusters and, in a parallel to combinatorial chemical synthesis, to create novel combinations of both Type I and Type II PKS genes to permit synthesis of unnatural polyketides. The integrating plasmids like pSAM2 and the ϕC31 phase have also been adapted to allow for site-specific integration of adventitious genes.

The problems concerning structural stability of plasmids containing large cloned segments can be overcome by chromosomal integration, and this can also be part of a strategy to circumvent restriction barriers, which are commonplace in many species. Plasmid integration via generalized recombination between cloned sequences and their chromosomal homologs is enhanced if the incoming DNA is in a single-stranded form; this not only appears to stimulate recombination, possibly via activation of RecA, but also protects the DNA from restriction. Single-stranded DNA can be introduced either by conventional protoplast transformation or by intergeneric conjugation from *E. coli*. The latter method relies on plasmids containing the *oriT* of the broad host range plasmid RK2, which can be conjugated from *E. coli* strains containing plasmid RP4 integrated into the chromosome or RK2 as an autonomous replicon. So far horizontal gene transfer of this type has only been demonstrated in one direction, highlighting a potential contribution to streptomycete evolution. The efficiency of integration of the incoming DNA appears to be largely dependent on the extent and quality of DNA homology. Less than 200 bp of homology is inefficiently integrated, and genes that are 90% homologous recombine 10^5-fold less well than identical copies. The latter indicates a barrier to so-called homologous recombination; this may have a bearning on the stability of repeated Type I PKS genes in many species.

The same methodologies used to introduce adventitious genes into the chromosome by homologous recombination have been widely exploited to perform gene disruption in order to mutate specific genes. Nonreplicating plasmids, introduced by protoplast transformation or intergeneric conjugation, and *attP*$^-$ ϕC31 derivatives have been used for this purpose. Random mutagenesis using transposons appropriately modified as genetic tools has proved successful in some species, although unreliable in others. The provision of additional nonstreptomycete transposons, for example, Tn*5493* derived from the enteric transposon Tn*5*, may go a long way to solving this problem. Tools such as these will assume much importance in relating function to the many open reading frames being identified by the ongoing *S. coelicolor* genome project; at the present stage of the project, 18% of the identified ORFs are similar to genes of unknown function in other organisms, and 27% are of unknown function and dissimilar to genes from other organisms.

See Also the Following Articles

Antibiotic Biosynthesis • Mapping Bacterial Genomes • Plasmids • Sporulation • Transposable Elements

Bibliography

Baltz, R. H. (1998). Genetic manipulation of antibiotic-producing *Streptomyces. Trends Microbiol.* **6**, 76–83.

Bibb, M. (1995). The regulation of antibiotic production in *Streptomyces coelicolor* A3(2). *Microbiology* **142**, 1335–1344.

Chater, K. F. (1998). Taking a genetic scalpel to the *Streptomyces* colony. *Microbiology* **144**, 1465–1478.

Chater, K. F., and Bibb, M. J. (1997). Regulation of bacterial antibiotic production. *In* "Biotechnology, Second Edition, Products of Secondary Metabolism" (H. Kleinkauf and H. Von Döhren, eds.), pp. 57–105. VCH Press, Weinheim.

Chen, C. W. (1996). Complications and implications of linear bacterial chromosomes. *Trends Genet.* **12**, 192–196.

Hopwood, D. A. (1997). Genetic contributions to understanding polyketide synthases. *Chem. Rev.* **97**, 2465–2497.

Volff, J.-N., and Altenbuchner, J. (1998). Genetic instability of the *Streptomyces* chromosome. *Mol. Microbiol.* **27**, 239–246.

Stringent Response

Michael Cashel
National Institutes of Health

I. Stringent and Relaxed Responses
II. (p)ppGpp Is Synthesized on Ribosomes by RelA
III. RelA Is an ATP:GTP(GDP) Pyrophosphoryl Transferase
IV. The Stringent Response Is Pleiotropic
V. SpoT Is a (p)ppGppase
VI. SpoT Is Also a Weak (p)ppGpp Synthetase in *Escherichia coli*
VII. SpoT Protein Domains
VIII. Bacterial Distribution of (p)ppGpp and Rel/Spo Homologs
IX. Mechanisms of (p)ppGpp Regulation

GLOSSARY

GDP Guanosine diphosphate.

GTP Guanosine triphosphate.

hungry codon An mRNA codon present in a ribosomal A site not satiated by binding of charged tRNA.

ppGpp A GDP analog with a pyrophosphate esterified on the ribose 3′-hydroxyl; guanosine tetraphosphate; once called "magic spot I" or MSI.

pppGpp As ppGpp but the GTP analog; guanosine pentaphosphate; "magic spot II" or MSII.

(p)ppGpp Both pppGpp and ppGpp.

(p)ppGpp° A complete deficiency of both pppGpp and ppGpp.

ribosomal idling A translating ribosome whose elongation is stalled for lack of charged tRNA.

THE STRINGENT RESPONSE in *Escherichia coli* and *Salmonella typhimurium* is the complex physiological response to amino acid starvation that is dependent on the accumulation of ppGpp, a nucleotide signal of nutritional stress. Accumulation of ppGpp is often, but not invariably, accompanied by the closely related nucleotide, pppGpp. Other nutritional stress can provoke (p)ppGpp to varying extents, such as limitation for sources of carbon, nitrogen, or phosphate. It is now possible to identify (p)ppGpp-dependent regulatory events more rigorously by artificially manipulating (p)ppGpp levels without nutritional stress. This can be achieved either by altering the expression of genes involved in (p)ppGpp metabolism or by completely abolishing the capacity to synthesize (p)ppGpp. The metabolism of (p)ppGpp in *Escherichia coli* and closely related organisms is governed by three genes: *relA, spoT,* and *gpp.* RelA and SpoT are homologous proteins. The (p)ppGpp synthetic activity of the RelA protein is ribosome-associated and activated by uncharged tRNA binding to A-sites of translating ribosomes. The SpoT protein is a strong (p)ppGppase and a weak source of (p)ppGpp synthesis. The Gpp protein converts pppGpp to ppGpp. Limitation of energy source availability leads to ppGpp accumulation by inhibiting (p)ppGpp turnover rather than by stimulating synthesis. Genomic sequencing provides many examples of RelA and SpoT family homologs found exclusively in prokaryotic but not so far in Archaea or in eukaryotic organisms. Often a single gene exists in gram-positive organisms that probably provides both *relA* and *spoT* functions. Increased (p)ppGpp leads to both positive and negative regulatory effects on gene expression as well as on metabolism. From the viewpoint of cell physiology, components of the stringent response seem to curtail activities that are superfluous and favor activities that facilitate adaptation to the nutritional stress that accompanies starvation.

I. STRINGENT AND RELAXED RESPONSES

The stringent response was originally defined as a phenotypic difference between wild-type *Escherichia coli* and a *relA* mutant during amino acid starvation. It was noticed that the exponential rate of accumulation of stable RNA (tRNA and rRNA) shown by growing cells was abruptly inhibited when wild-type strains were amino acid starved. In contrast, starved *relA* mutants continued accumulating stable RNA for nearly a generation time. The wild-type dependence of RNA accumulation on the presence of amino acids was termed stringent RNA control while the mutant was said to be relaxed. This RNA control phenotype has been extended beyond *E. coli* by observations with mutants in a gene now appreciated to be a homolog of RelA (see below). Early studies using conditional aminoacyl–tRNA synthetase mutants revealed that an inability to charge tRNA could provoke the RNA control response even in the presence of abundant free amino acids. Conversely, restoring levels of charged tRNA in starved stringent cells by inhibiting consumption of charged tRNA with protein synthesis inhibitors led to relaxing RNA control. A search for the possible participation of nucleotide substrates of RNA polymerase in RNA control led to the observation that the pair of (p)ppGpp nucleotides, then called magic spots, rapidly accumulate during the stringent response (Cashel and Gallant, 1969). The absence of (p)ppGpp in *relA* mutants led to the realization that (p)ppGpp might mediate the stringent response. Such mutants could be isolated because, unlike the wild type, they failed to resume growth for several hours after dilution into fresh supplemented media.

Figure 1 is a schematic view of the (p)ppGpp cycle and summarizes the roles of enzymes encoded by *relA* and *spoT* (see below). The guanosine pentaphosphate phosphatase (*gpp*) gene shown encodes a pppGpp γphosphohydrolase (see below). Deletions of *gpp* result in higher levels of pppGpp relative to ppGpp during the stringent response but no effects on growth in the sequenced *E. coli* strain MG1655.

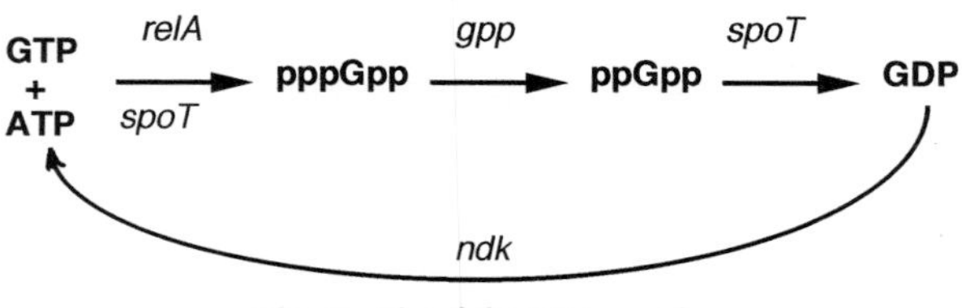

Fig. 1. The (p)ppGpp cycle.

Nucleoside diphosphate kinase (*ndk*) phosphorylates GDP to GTP, completing the cycle.

II. (p)ppGpp IS SYNTHESIZED ON RIBOSOMES BY RelA

The mechanism of (p)ppGpp synthesis on ribosomes by the RelA protein (stringent factor) was demonstrated in the classic experiments of Haseltine and Block (1973). These studies revealed that (p)ppGpp synthesis required binding of codon-specified, uncharged tRNA to ribosomal acceptor (A) sites. There was not a specific requirement for initiation or elongation factors apart from their roles in positioning the translating ribosome at a hungry codon on the mRNA to be responsive to nonenzymatic (i.e., EF-Tu independent) binding of uncharged tRNA. It was later proposed that ribosomal idling involves the RelA protein cycling on and off the ribosome, dissociating after each round of (p)ppGpp synthesis. Despite reaction requirements for ribosomes, tRNA, and mRNA, the catalytic source for (p)ppGpp synthetic activity is localized to RelA: The pure protein can be partially activated *in vitro* when exposed to methanol or mild denaturants. A simple model can be imagined in which protein synthesis comprises a means of surveying the population of mRNA codons actively translated at any given time by continuously sensing charged tRNA consumption. If nonenzymatic binding of a given species of uncharged tRNA to A-sites effectively competes with enzymatic binding of the corresponding charged tRNA, then a round of (p)ppGpp synthesis occurs, setting the stage for another cycle. One prediction of this model was verified, namely, that when protein synthesis is dominated by abundance of a message lacking codons for a particular amino acid, as during R17 phage infection, starvation for that amino acid should not give ppGpp synthesis. This model also predicts that (p)ppGpp synthesis should respond to increased ratios of uncharged to charged tRNA rather than the absolute concentration of uncharged tRNA. This prediction was verified as well. It should be noted that this simple model may not apply for certain tRNAs where unexpected charging ratios can occur during starvation.

The RelA protein has been found to be present at concentrations sufficient to bind to about 1% of ribosomes. This leads to the verified prediction that if RelA abundance were increased, (p)ppGpp levels would increase because of a higher fraction of productive hungry codon encounters. The requirement for ribosome binding also predicts ribosome mutants might exist that interfere with RelA function; such mutants were isolated, mapped to the *rplK* gene encoding the ribosomal protein L11, and termed *relC*. Although these mutants do display a relaxed phenotype, ribosome binding of RelA protein persists but activation is apparently blocked.

The response time for the appearance of (p)ppGpp after starvation occurs is a matter of seconds. The rates of synthesis in starved cells, expressed as moles per ribosome, approach rates of peptide bond formation in unstarved cells; typically, the content of (p)ppGpp increases about 50-fold above basal levels to reach concentrations equal to GTP, which drops by half as it is converted to (p)ppGpp. When the nutritional stress is reversed by supplementation, (p)ppGpp disappears with a half-life on the order to 20–30 seconds. Degradation of (p)ppGpp, mediated by SpoT (see Fig. 1) is inhibited during starvation for carbon sources. This leads to slow rates of (p)ppGpp accumulation, with the nucleotide ultimately reaching levels nearly equivalent to those seen during the stringent response to amino acid starvation with apparently similar physiological consequences.

III. RelA IS AN ATP:GTP(GDP) PYROPHOSPHORYL TRANSFERASE

The (p)ppGpp synthetic reaction by RelA is a pyrophosphoryl group transfer with the β,γ-phosphates of ATP transferred to the ribose 3′-hydroxyl of GTP (or GDP) with the γ-phosphate of ATP appearing exclusively as the 3′β-phosphate of (p)ppGpp. For the RelA protein, equivalent K_m values are found for GTP and GDP, about 0.5 m*M*, within the range of physiological concentrations of GTP, but not GDP. This is consistent with kinetic demonstrations *in vivo* that pppGpp is the precursor for ppGpp early in the stringent response. The presence of a 3′-pyrophosphate in (p)ppGpp confers two useful properties for identifying the analog. The first is that (p)ppGpp does not form complexes with borate ions, unlike guanine ribonucleotides bearing cis-hydroxyls (such as GMP, GDP and GTP). This gives a mobility difference during thin layer chromatography in borate solutions. The second property is that the 3′β-phosphate of (p)ppGpp is as labile to alkali hydrolysis as are RNA phosphodiesters. Furthermore, the stable nucleotide hydrolysis product (p)ppGp contains a phosphomonoester randomized among the ribose 2′- and 3′-hydroxyls. Mild alkali treatment (0.3 *M* KOH, 37°C for 30 min) cleaves (p)ppGpp, yielding (p)ppGp and inorganic phosphate (P_i). The RelA protein does not contain obvious sequence motifs characteristic of either G proteins, nucleotide binding proteins, or known pyrophosphoryl transferases.

It appears that domains of RelA are separable for ribosome binding and for (p)ppGpp synthesis (Schreiber *et al.*, 1991). The original *relA1* allele is an IS2 insertion between codons 85 and 86 of the 743 amino acid protein yielding two peptide fragments capable of transcomplementation when overexpressed. Subsequently, the N-terminal 455 amino acid fragment of RelA was found to display high constitutive levels of (p)ppGpp synthetic activity. Since this activity was not elevated by amino acid starvation and not lowered by a *relC* allele mutant, it was concluded that the RelA N-terminal peptide was not activated by ribosomes. On the other hand, the C-terminal RelA fragment behaves as if it possesses ribosomal binding activity but not (p)ppGpp synthetic activity. Overexpressing the RelA C-terminal peptide inhibits the activity of a single copy wild-type *relA* gene as if the truncated peptide binds to ribosomes and prevents wild-type RelA binding to ribosomes. These properties together with the catalytic inactivity of the full-length RelA protein, whether associated with ribosomes or not, suggests the C-terminal peptide inhibits the activity of the N-terminal peptide. Activation of synthetic activity by uncharged tRNA binding to ribosomal A-sites can be imagined as triggering a conformational change in RelA that interferes with the negative intramolecular interaction between the N-terminal and C-terminal portions of RelA.

Fusion of an inducible promoter to the catalytically active RelA fragment, or the full-length protein, gives a means of artificially inducing (p)ppGpp without imposing nutritional limitation. Heterologous expression of such constructs into other hosts has proven a useful means of addressing whether (p)ppGpp can complement putative *rel* mutants (see below).

IV. THE STRINGENT RESPONSE IS PLEIOTROPIC

Multiple features of the stringent response were originally uncovered by comparing wild-type and *relA* mutant behavior during amino acid starvation (Table I). Early attempts to visualize the extent of changes in gene expression on two-dimensional gels revealed that about half of the protein spots are affected, about equally divided between those whose abundance increased and those that decreased. Equivalent analyses of stable RNA species revealed that levels of most of the spots were reduced. It has come to be appreciated that the relaxed response is more complex than simply the absence of the stringent response. This is because relaxed cells show a dramatic reduction in (p)ppGpp levels existing before amino acid starvation. The basis for this behavior is unknown.

Many features listed in Table I are being examined more rigorously through the use of techniques for gratuitous manipulation of (p)ppGpp as well as taking into account the growing awareness that regulation of a given process usually involves multiple effectors. A serious qualification is that reports of sites targeted for regulation by (p)ppGpp are usually not bolstered by isolation of (p)ppGpp-resistant mutants. Gratuitously inducing intracellular levels of (p)ppGpp in otherwise growing cells does lead to severe growth inhibition. The inhibition of growth is severe enough to predict that (p)ppGpp might represent a means of reserving biosynthetic potential for adaptation to nutritional stress. Thus, before amino acid starvation renders a cell incapable of synthesizing the proteins needed to adapt to nutritional stress, superfluous and energetically expensive anabolic processes are severely, but incompletely, restricted. Steps in the biosynthesis of some, but not all, amino acids are facilitated as well as induction

TABLE I
Multiple Components of the Stringent Response[a]

Component	*Response*
rRNA synthesis (*rrnA–E, G, H*)	Inhibition
tRNA synthesis (most except *metZ, metY*)	Inhibition
RpoS	Induction
Protein turnover	Increased
Protein synthesis (*tufA, B, tsf, fusA*)	Inhibition
Nucleotide synthesis (PurA, GuaB PyrBI)	Inhibition
Amino acid synthesis (*argF, I, argECBH, gltB, glnA, gdh, hisD, B, G, ilvA, B, lysA, C, met C, F met, K, thrA, B, C*)	Inhibition
DNA binding proteins (HimA, hip dps)	Induction
Polyphosphate synthesis (Ppk)	Induction
Glycogen synthesis	Induction
Phospholipid synthesis (*plsB*)	Inhibition
Peptidoglycan synthesis	Inhibition
Transport	
α-Methylglucoside	Inhibition
Purine	Inhibition
Pyrimidine	Inhibition
Branched-chain amino acids	Induction
Phosphate	Inhibition
Antibiotic permeability	Increased
Plasmid replication	Inhibition
Mecillinam resistance	Induction
Penicillin tolerance	Induction
NaCl-induced acid sensitivity	Induction
Cold-shock (cspA)	Inhibition
Mutagenesis of amino acid biosynthesis genes (*argH, leuB*)	Increased
Transcription of retron Ec107	Induced

[a] The activities listed are an updated summary from a more extensive review of sometimes conflicting *in vivo* and *in vitro* observations (Cashel *et al.*, 1996). In addition, the Bibliography lists new reviews on regulation of lipids and phospholipids (DiRusso and Nystrom, 1998) and plasmid replication during the stringent response (Wegrzyn, 1998).

of a global regulatory protein, Lrp. In the case of histidine, (p)ppGpp has inhibitory effects on $tRNA^{His}$ transcription operating at the level of enhancing melting a supercoil-dependent promoter (see Figueroa-Bossi *et al.*, 1998) as well as attenuator-independent activation of the P_{His} promoter itself. Survival after prolonged starvation is partially dependent on induction of stationary phase sigma factor (RpoS), which is impaired in (p)ppGpp^0 strains. A more distantly related adaptive effect found by B. Wright is a mild increase in mutagenesis of amino acid biosynthetic genes (*argH, leuB*) whose transcription is increased during a prolonged stringent response.

V. SpoT IS A (p)ppGppase

SpoT of *E. coli* is the main source for degradation of (p)ppGpp due to its managanese-dependent 3′-pyrophosphohydrolase activity, as judged by slowing of (p)ppGpp decay rates for *spoT* mutants. The enzyme catalyzes a manganese-dependent release of the 3′-pyrophosphate residue from (p)ppGpp to yield GTP or GDP. Equal substrate affinities are found for both ppGpp and pppGpp.

Experimentally, decay is measured as first-order rates of disappearance of (p)ppGpp starting from high levels induced by the stringent response after chloramphenicol is added to turn off RelA synthesis. Mutants of *spoT* can slow the 20–30 second half-life of (p)ppGpp to 20 minutes or more. Less severe mutants have been isolated with intermediate effects on decay. These have been exploited as a means of systematically increasing steady-state basal levels of ppGpp over about a 10-fold range for titrating regulatory effects during the ensuing slower exponential growth. The phenotype of *spoT* mutants during the stringent response is a higher than normal accumulation of ppGpp, consistent with impaired decay rates, Curiously, *spoT* mutants accumulate ppGpp, and not pppGpp, after the first few minutes of the stringent response. Early in the response, pppGpp does accumulate and can be demonstrated to be the precursor of ppGpp. A number of explanations for this behavior have been proposed.

Carbon source starvation inhibits (p)ppGpp decay, quantitatively giving a phenocopy of a *spoT* mutant. This impairment leads to accumulation of high levels of ppGpp without appreciably increasing rates of (p)ppGpp synthesis, even in a *relA* deletion mutant during carbon source exhaustion. Carbon source starvation of a *relA*$^+$ strain does give a higher initial rate of accumulation of (p)ppGpp than for *relA* cells, presumably because of a transient amino acid deficiency. The mechanism by which energy source depletion leads to inhibition of (p)ppGppase activity remains unsolved. The accumulation of (p)ppGpp can also be provoked by interrupting aeration during growth of *E. coli* at high cell densities ($OD_{600} > 0.6$) or by turning off the lights during growth of a phototroph. Many other conditions also have been reported to lead to *spoT*-dependent accumulation of ppGpp. They include manganese chelators such as picolinic acid, 1,10-phenanthroline, and antibiotics such as tetracycline and chlortetracycline. Less readily understood conditions are polymyxin B, gramicidin, levallorphan, osmotic shock, heat shock, inhibitors of fatty acid synthesis, uncouplers of phosphorylation, and long-chain alcohols. This list is not easily understood, but might reflect membrane modifications that could activate a sensor with the ability to inhibit SpoT or affect manganese availability.

VI. SpoT IS ALSO A WEAK (p)ppGpp SYNTHETASE IN *ESCHERICHIA COLI*

The ability to accumulate ppGpp when (p)ppGppase activity was blocked in *relA* deleted strains means that a second source of (p)ppGpp synthesis exists. The search for this enzyme led to the SpoT protein itself. Simultaneous deletions of *relA* and *spoT* resulted in an apparently complete deficiency of (p)ppGpp, termed (p)ppGpp^0. The observation that (p)ppGpp^0 strains are able to grow at nearly normal rates in amino acid rich media indicates that despite widespread regulatory activity, (p)ppGpp is not essential for growth. However, (p)ppGpp^0 strains fail to grow on glucose minimal medium unless supplemented with multiple amino acids. Using mixtures of all 20 amino acids but one, requirements were established for Arg, Gly, His, Leu, Lys, Phe, Ser, Thr, and Val. The lack of (p)ppGpp also makes cells nonviable after long stationary phase exposures, salt-sensitive, and strikingly elongated. Such cells are defective in the ability to ferment a variety of sugars including glucose but, like *lrp* mutants, are able to use serine as a carbon source. Two features of the (p)ppGpp^0 phenotype were used to isolate suppressors: complete prototrophy on minimal medium and stationary phase survivors screened as prototrophs.

All such suppressors map exclusively to three (*rpoB, rpoC,* and *rpoD*) of the four RNA polymerase subunit genes. This result seems to reinforce notions that positive control exerted by (p)ppGpp operates at the level of transcription (Hernandez and Cashel, 1995).

VII. SpoT PROTEIN DOMAINS

Comparison of the SpoT and RelA protein sequences revealed that about 75% of the residues are related. Although homology extends throughout the proteins, it is less dense in the N-terminal region. This extensive homology hints that the (p)ppGpp synthetic activity of SpoT might be mechanistically similar to that of RelA. However, Western analysis revealed that the SpoT protein is not associated with ribosomes under starvation conditions. The (p)ppGpp synthetic activity of SpoT is very weak judging from slow kinetics of accumulation of ppGpp accumulation when the RelA source of synthesis is deleted and SpoT (p)ppGppase is blocked with manganese chelators. So far, the synthetic activity of SpoT is undetectable *in vitro.* However, synthetic activity can be mapped by *in vivo* tests including the ability to restore amino acid prototrophy or glycogen accumulation to an otherwise $(p)ppGpp^0$ host. Degradation activity can easily be mapped by scoring suppression of the small colony phenotype of severe *spoT* mutants.

In this way (p)ppGpp synthetic and degradation activities of SpoT were mapped to partially overlapping domains within the N-terminal half of the 702 residue SpoT protein. The (p)ppGppase mapped to the first 203 residues, whereas synthetic activity mapped from residues 85 to 375. A deletion of the first 66 amino acids abolishes (p)ppGppase, defining a necessary but not sufficient subdomain that could be imagined to contain a (p)ppGpp binding site. The peptide from residues 204 to 375 is uniquely necessary for synthesis and might comprise an ATP binding site. The shared core domain necessary for both synthesis and degradation, residues 67 to 375, could contain a catalytic center involved in pyrophosphoryl transfer from ATP (for synthesis) or from (p)ppGpp (for degradation). Overall, it appears that RelA and SpoT are homologs with almost completely separated functions, that is, they are almost complete paralogs.

VIII. BACTERIAL DISTRIBUTION OF (p)ppGpp AND Rel/Spo HOMOLOGS

Recent evidence suggests a widespread distribution of genes in the Rel/Spo family. Genomic sequencing reveals instances of genomes with a single Rel/Spo homolog, unlike the pair of genes in *E. coli* and its close relatives. Table II is a list of organisms with Rel/Spo homologs or observed to form (p)ppGpp.

A. *Streptococcus equisimilis*

Streptococcus equisimilis encodes a Rel/Spo homolog, here called Rel_{Seq}. When this protein is expressed in *E. coli,* it functions like SpoT with a strong manganese-dependent (p)ppGppase activity with equal affinities for ppGpp and pppGpp. In addition, Rel_{Seq} shows weak (p)ppGpp synthetic activity *in vitro* and *in vivo,* but this activity is not stimulated in *E. coli* by amino acid starvation. Synthesis of (p)ppGpp is occurs with essentially pure protein and is not further stimulated by adding *E. coli* ribosomes, eliminating possible RelA-like requirements for ribosomes, tRNA, or mRNA. Nevertheless, studies of patterns of pyrophosphate transfer with this enzyme are identical to those of RelA. Unlike SpoT, the bifunctional nature of the Rel_{Seq} protein is demonstrable *in vitro.*

Surprisingly, the rel_{Seq} gene seems to function like RelA in its native host. An insertion allele of rel_{Seq} substituted for the wild-type gene in the chromosome of *S. equisimilis* abolishes the accumulation of (p)ppGpp during amino acid starvation but not during glucose starvation, and (p)ppGpp accumulation was consistent with this behavior (Mechold and Malke, 1997). Thus, rel_{Seq} behaves like *relA* (but not *spoT*) in its native host but the opposite, like *spoT* (but not *relA*), when expressed in *E. coli.* This behavior is even more interesting because there is probably a single Rel/Spo gene in the genome, based on the

TABLE II
Microbial Distribution of Rel/Spo Genes, Mutants, and (p)ppGpp

Organism[a]	*Genome*[b]	*Cloned?*[c]	*Mutant?*[d]	*(p)ppGpp?*[e]
Gram-positive				
High G + C				
Corynebacteria glutamicum		1	Yes	Yes
Streptomyces coelicolor		1	Yes	Yes
Low G + C				
Bacillus subtilis	x1	1	Yes	Yes
Clostridium acetobutylicum		1		Yes
Mycoplasma pneumoniae	x1			
Streptococcus equisimilis		1	Yes	Yes
Spiroplasma citri	x1			
Gram-negative				
Alpha				
Rhodobacter sphaeroides				Yes
Beta				
Neisseria gonorrhoeae	x2			
Gamma				
Haemophilus influenzae	x2			
Vibrio sp. S14		x2	Yes	Yes
Escherichia coli	x2	x2	Yes	Yes
Salmonella typhimurium			Yes	Yes
Delta				
Myxococcus xanthus		x1	Yes	Yes
Epsilon				
Helicobacter pylori	x1			
Cyanobacteria				
Synechocystis sp. PCC6803	x1			
Thermatogales				
Thermatoga sp.	x1	Yes		

[a] Bacterial divisions are grouped according to Woese. In addition, genomic sequencing projects in progress for the following organisms show significant homology against *E. coli* RelA (Blast Score > 150 and an E value of e − 27 or lower): *Actinobacillus actinomycetemcomitan, Aquaflex aeolicus, Bordetella pertussis, Campylobacter jejuni, Chlorobium tepidum, Clostridium acetobutylicum, Deinococcus radiodurans, Enterococcus faecalis, Porphyromonas gingivalis, Pseudomonas aeruginosa, Streptococcus pneumoniae, Streptococcus pyogenes, Vibrio cholerae, Thermatoga maritima,* and *Yersinia pestis*. Analysis date, Oct. 1998.

[b] The number of Rel/Spo genes found for a completed genomic sequence; parentheses indicate results for nearly completed genomic sequence.

[c] The number of different Rel/Spo homologs found.

[d] Yes means a mutant allele exists in the chromosome.

[e] Reports of (p)ppGpp.

nearly complete sequencing of the genome of the closely related *Streptococcus pyogenes*. A key question becomes whether the (p)ppGpp accumulating during glucose starvation for the rel_{Seq} mutant arises from residual activity of the insertion allele. It has not yet been possible to obtain a complete deletion of rel_{Seq} in the *S. equisimilis* chromosome. A related question is why rel_{Seq} appears to be activated on ribosomes during the stringent response in *S. equisimilis* but not in *E. coli*.

B. *Bacillus subtilis*

Bacillus subtilis provides an example where a complete genomic sequence clearly reveals a single Rel/Spo homolog, called *relA* by Wendrich and Marahiel (1997). A chromosomal deletion of this gene gives phenotypes reminiscent of deletions of both *spoT* and *relA* in *E. coli:* undetectable (p)ppGpp, abolition of (p)ppGpp accumulation during amino acid or glucose starvation, slow growth, amino acid requirements (but strong for only valine and weak for isoleucine, leucine, and methionine), and poor survival when starved for amino acids. Comparisons of wild-type and *relA* mutant protein expression early during amino acid starvation revealed impairments of expression of protein synthesis elongation factors, amino acid biosynthetic proteins, as well as general stress proteins.

The stringent response had previously been documented in *Bacillus subtilis*. A *relA* mutant (called $relA_{BR17}$) was isolated long ago on the basis of a relaxed RNA control phenotype. Finding residual (p)ppGpp synthesis in extracts of the $relA_{BR17}$ mutant led to postulating the existence of a second synthetic enzyme, called ppGpp synthetase II (PSII), in *B. subtilis* and related bacilli. Finding (p)ppGppase activities in the mutant led to supposing a separate *spoT* gene existed, as in *E. coli*. Sequencing of $relA_{BR17}$ by Wendrich and Marahiel (1997) reveals a single lesion (G240E) affecting a conserved glycine. If the catalytic domains deduced from SpoT conform to those of *B. subtilis* RelA, then the G240E lesion would be predicted to be in the synthesis domain, leaving the (p)ppGppase domain intact. Thus, it seems a single gene exists in *Bacillus subtilis* for (p)ppGpp synthesis as well as degradation.

A caveat is that (p)ppGpp synthetic enzymes could exist without obvious homology to Rel/Spo enzymes. However, the failure to observe detectable (p)ppGpp seems to rule out a second source of (p)ppGpp. Nevertheless, there are two published examples of such enzymes with (p)ppGpp synthetic activity. One is the purine nucleotide pyrophosphotransferase (*ppk*) from *Streptomyces adephospholyticus* (alias *S. morookaensis*). The other example is purine nucleotide phosphorylase from *Streptomyces antibioticus*.

C. *Streptomyces coelicolor*

Streptomyces coelicolor is a gram-positive mycelial soil bacterium that produces a variety of secondary metabolites, including antibiotics, during slow growth and stationary phase. A hypothesis from Ochi's laboratory, arising from interfering effects of *relC*-like thiostrepton-resistant mutations, is that ppGpp accumulation induces antibiotic synthesis. Although ppGpp was found to impair ribosomal RNA synthesis in *Streptomyces* strains, there were conflicting reports of the correlation with antibiotic production. Recently, *relA* genes have been isolated from two closely related *S. coelicolor* strains, along with mutant alleles (laboratories of M. Bibb and F. Malpartida). It is interesting that genes in regions flanking Rel/Spo homologs are found to be conserved to varying extents among several gram-positive organisms (see Wendrich and Marahiel, 1997).

A chromosomal deletion allele of *relA*, but not the presumably leaky insertion allele, interferes with the appearance of the antibiotics undecylprodigiosin and actinorhodin as well as transcription of regulatory genes specifically involved in their production. The null allele also shows a relaxed RNA control response and morphological changes in aerial mycelia. Both features are complemented by either a wild-type *Streptomyces relA* gene or by heterologous expression of the *E. coli relA* gene in *Streptomyces*. The properties of *Streptomyces relA* have been assessed after expression in *E. coli* tester strains. Like the Rel_{Seq} protein, the *Streptomyces* RelA can be shown to possess both (p)ppGpp synthetic as well as (p)ppGpp degrading activity with functional domains in positions very similar to those of the SpoT homolog. Unlike the Rel_{Seq} protein, stimulation of synthetic activity by ribosomes is observed both *in vitro* and *in vivo* like *E. coli* RelA.

D. *Myxococcus xanthus*

Myxococcus xanthus has a developmental pathway for forming microspores, and there is strong evidence from D. Kaiser's laboratory that (p)ppGpp participates in the induction of this process. When cells are nutrient starved at high cell densities, they aggregate and form a multicellular fruiting body in response to a population density signal, called A factor. The starvation period has long been known to result in (p)ppGpp accumulation. The need for starvation can be bypassed by heterologous expression of the *E. coli* RelA protein in *M. xanthus* and (p)ppGpp formed in this manner activates transcription of initial steps in A factor formation. It was also shown that *E. coli RelA* could be activated with *M. xanthus* ribosomes and vice versa. Antibiotics giving specific lowering of GTP levels without (p)ppGpp accumulation did not induce transcriptional activation of A factor pathway genes, fruiting body formation, or spore formation. A mutant was isolated with an early block in the developmental cycle during the starvation period prior to aggregation. The mutant failed to form (p)ppGpp, showed relaxed RNA control, and was rescued by the *M. xanthus* RelA gene homolog. The identity of the mutant was established when a chromosomal insertion allele of the *rel/spo* gene displayed the mutant phenotype. Finally, a mutant found by Davis *et al.* (1995) in *asgC*, the *M. xanthus* RNA polymerase homolog of *rpoD,* affects expression of a putative (p)ppGpp activated gene in A factor formation. The mutant lesion is located in the same conserved region 3 as found for *E. coli rpoD* mutant suppressors of the (p)ppGpp^0 state (Hernandez and Cashel, 1995). It could be that use of (p)ppGpp as a trigger for a developmental cycle in *M. xanthus* is related to its unusual dependence on amino acids. The organism is unable to ferment carbohydrates and instead utilizes amino acids as sources of both carbon and nitrogen. The A signaling factor consists of a mixture of six amino acids secreted in micromolar concentrations.

IX. MECHANISMS OF (p)ppGpp REGULATION

With a few exceptions, conclusions regarding the ability of (p)ppGpp to regulate many processes are still qualified by the lack a detailed mechanistic verification. At the metabolic level, there are briefly studied examples of apparently simple and direct interactions between (p)ppGpp and enzyme targets such as IMP dehydrogenase and adenylosuccinate synthetase. An interesting account of the involvement of (p)ppGpp in the metabolic regulation of polyphosphate (polyP) during the stringent response comes from the work of A. Kornberg and colleagues. The conclusion that regulation operates at the metabolic level comes from observing that the activities of the two enzymes responsible for polyP synthesis (Ppk) and for polyP degradation (Ppx) were unchanged when extracted from cells undergoing the stringent response despite a 100-fold difference in the level of polyP. It was found that pppGpp was a potent competitive inhibitor of polyP degradation by Ppx ($K_i = 10\ \mu M$). The inhibition by ppGpp was weaker, but still physiologically significant ($K_i = 200\ \mu M$) because of high ppGpp levels achieved during the stringent response. The accumulation of polyP by blocking polyP degradation can be mimicked *in vitro* with reaction mixtures containing balanced amounts of Ppk and Ppx as well as (p)ppGpp. The reason why pppGpp competitively inhibits Ppx is probably because Ppx can utilize pppGpp as a substrate in pppGpp γ-phosphohydrolase catalysis with a K_m also in the 10 μM range. This activity of Ppx is precisely the catalytic activity associated with Gpp (Fig. 1). Notably Gpp and Ppx are homologs and each can degrade polyP as well as pppGpp, but with differing efficiencies. Judging from the behavior of deletion mutants, Ppx accounts for the majority of polyPase but little of the pppGppase *in vivo.* For Gpp the reverse is true; a Gpp deletion eliminates the majority of the pppGpp γ-phosphohydrolase but only a small portion of the polyPase activity. Thus, Ppx and Gpp are an example of homologs whose functions are incompletely separated: They are paralogs for their characteristic major function but orthologs for mutually shared minor functions.

Obtaining a consensus regarding the mechanism of transcription regulation by (p)ppGpp has been particularly troublesome, although there is long-standing physiological evidence that both positive and negative regulatory effects occur as well as promising physical evidence of an RNA binding site for ppGpp from Chatterji's group. Models range from no involvement of (p)ppGpp at all to effects on RNA polymerase partitioning, on transcription initiation, or on elongation, with active or passive roles. There are also highly polarized views as to whether growth rate control involves (p)ppGpp (for reviews, see Bremer and Dennis, 1996; Cashel *et al.*, 1996). Recent work from R. Gourse's laboratory argues that the activities of rRNA (and tRNA) promoters are determined by pool levels of the initiating nucleotides which are, in turn, argued to increase with growth rate without participation of (p)ppGpp. The reason for activity regulation is that for these promoters, the initiating nucleotide facilitates formation of the first phosphodiester bond, which stabilizes what would otherwise be unstable open complexes. Recently, the Gourse laboratory reported isolation of (p)ppGpp^0 suppressor mutants in *rpoB* and *rpoC* capable of altering growth rate control. An enduring and tested feature of promoters regulated during the stringent response was first noted by A. Travers. It is that negately regulated promoters possess a GC-rich "discriminator" region between positions -10 and $+1$. Conversely, an AT-rich discriminator is often noted for positively regulated promoters, and its role has been verified, particularly in work by the S. Artz group on the histidine operon promoter, which is induced by ppGpp in an attenuator-independent manner. The promoter for histidine tRNA is negatively regulated by (p)ppGpp as well as strongly supercoiling dependent. The Bibliography lists a paper (Figueroa-Bossi *et al.*, 1998) proposing a unifying view of many of the factors involved. It argues that the discriminator region promoter dependence on negative supercoiling, which is correlated with susceptibility to melting *in vivo,* leads to stabilization of open complexes by the initiating nucleotide and responsiveness to stringent control. It now appears more reasonable than ever before to hope that resolution of this problem will emerge along with a better basic understanding of promoter function.

In closing, it seems noteworthy that achieving a detailed understanding of the Rel/Spo family might be of practical value as a target for antibiotic design. The genes in this family are present almost exclusively in Eubacteria, their operation involves a variety of regulatory responses to nutritional stress, and the (p)ppGpp signaling molecule they control is a potent inhibitor of bacterial growth.

See Also the Following Articles

AMINO ACID FUNCTION AND SYNTHESIS • LOW-NUTRIENT ENVIRONMENTS • *MYXOCOCCUS*, GENETICS • RIBOSOME SYNTHESIS AND REGULATION • STARVATION, BACTERIAL • *STREPTOMYCES*, GENETICS

Bibliography

Bremer, H., and Dennis, P. P. (1996). Modulation of chemical composition and other parameters of the cell by growth rate. *In "Escherichia coli* and *Salmonella typhimurium:* Cellular and Molecular Biology" (F. C. Neidhardt, R. Curtiss III, J. L. Ingraham, E. C. C. Lin, K. B. Low, B. Magasanik, W. S. Reznikoff, M. Riley, M. Schaechter, and H. E. Umbarger, eds.), pp. 1553–1569. 2nd Ed. ASM Press, Washington, D.C.

Cashel, M., and Gallant, J. (1969). Two compounds implicated in the function of the RC gene in *Escherichia coli. Nature (London)* **221**, 838–841.

Cashel, M., Gentry, D. R., Hernandez, V. J. and Vinella, D. (1996). The stringent response. *In "Escherichia coli* and *Salmonella typhimunium:* Cellular and Molecular Biology" (F. C. Neidhardt, R. Curtiss III, J. L. Ingraham, E. C. C. Lin, K. B. Low, B. Magasanik, W. S. Reznikoff, M. Riley, M. Schaechter, and H. E. Umbarger, eds.), 2nd Ed. ASM Press, Washington, D.C.

Chakraburtty, R., and Bibb, M. (1997). The ppGpp synthetase gene (relA) of *Streptomyces coelicolor* A3(2) plays a conditional role in antibiotic production and morphological differentiation. *J. Bacteriol.* **179**, 5854–5861.

Davis, J. M., Mayor, J., and Plamann, L. (1995). A missense mutation in rpoD results in an A-signalling defect in *Myxococcus xanthus. Mol. Microbiol.* **18**, 943–952.

DiRusso, C. C., and Nystrom, T. (1998). The fats of *Escherichia coli* during infancy and old age: Regulation by global regulators, alarmones and lipid intermediates. *Mol. Microbiol.* **27**, 1–8.

Figueroa-Bossi, N., Guerin, M., Rahmouni, R., Leng, M., and Bossi, L. (1998). The supercoiling sensitivity of a bacterial tRNA promoter parallels its responsiveness to stringent control. *EMBO J.* **17**, 2359–2367.

Harris, B. Z., Kaiser, D., and Singer, M. (1998). The guanosine nucleotide (p)ppGpp initiates development and A-factor production in *Myxococcus xanthus. Genes Dev.* **12**, 1022–1033.

Haseltine, W. A., and Block, R. (1973). Synthesis of guanosine tetra- and pentaphosphate requires the presence of a codon-specific, uncharged transfer ribonucleic acid in the acceptor site of ribosomes. *Proc. Natl. Acad. Sci. U.S.A.* **70**, 1564–1568.

Hernandez, V. J., and Cashel, M. (1995). Changes in conserved region 3 of *Escherichia coli* sigma 70 mediate ppGpp-dependent functions *in vivo. J. Mol. Biol.* **252**, 536–549.

Martinez-Costa, O. H., Fernandez-Moreno, M. A., and Malpartida, F. (1998). The relA/spoT-homologous gene in *Streptomyces coelicolor* encodes both ribosome-dependent (p)ppGpp-synthesizing and -degrading activities. *J. Bacteriol.* **180**, 4123–4132.

Mechold, U., and Malke, H. (1997). Characterization of the stringent and relaxed responses in *Streptococcus equisimilis*. *J. Bacteriol.* **179**, 2658–2667.

Schreiber, G., Metzger, S., Aizenman, E., Roza, S., Cashel, M., and Glaser, G. (1991). Overexpression of the *relA* gene in *Escherichia coli*. *J. Biol. Chem.* **266**, 3760–3767.

Wegrzyn, G. (1998). Replication of plasmids during bacterial response to amino acid starvation. *Plasmid* **41**, 1–16.

Wendrich, T. M., and Marahiel, M. A. (1997). Cloning and characterization of a relA/spoT homologue from *Bactillus subtilis*. *Mol. Microbiol.* **26**, 65–79.

Sulfide-Containing Environments

Rutger de Wit

Centre National de Recherche Scientifique and Université Bordeaux 1

I. Sulfide in Metabolism
II. Overview of Sulfide-Containing Ecosystems
III. Biodiversity and Ecological Niches of Sulfide Oxidizers

GLOSSARY

anoxygenic photosynthesis Light-driven metabolic reduction of low molecular weight carbon compounds (often CO_2) using electrons derived from a compound other than water, and, consequently, oxygen is not produced; typical electron donors include H_2S, other reduced sulfur compounds, H_2, and simple organic substrates. This type of photosynthesis has been found only in the Bacteria domain.

benthos A community consisting of organisms growing on or in the sediment.

electron acceptor A compound that is reduced in a metabolic redox reaction.

electron donor A compound from which electrons are derived in a metabolic redox reaction, thus resulting in the oxidation of the electron donor.

meromixis A condition in which stratification of the water mass in a lake is maintained during the whole year, often due to a solute concentration gradient.

microbial mat A sediment ecosystem with very high population densities of microorganisms. The top millimeters comprise a clearly laminated structure: (1) the top layer is dominated by oxygen-producing phototrophs, especially cyanobacteria, (2) the anoxic bottom layer is rich in sulfide as a result of the degradation of organic matter by fermenting and sulfate-reducing bacteria, and (3) phototrophic and/or chemotrophic bacteria sandwich in between the top and bottom layers, forming a fine lamina at the oxygen–sulfide interface.

SULFIDE-CONTAINING ENVIRONMENTS are found in aquatic and sediment ecosystems and are inhabited by predominantly prokaryotic communities. In some places, geochemically formed sulfide emerges into the biosphere, but more commonly sulfide is biologically produced in anoxic environments by sulfate-reducing bacteria. Sulfide is a toxic compound but is used as a substrate for growth by certain prokaryotic organisms. These microorganisms include the following:

1. Phototrophic sulfur bacteria that oxidize sulfide in a light-driven reaction and use the electrons to reduce CO_2 for subsequent biosynthesis.
2. Chemotrophic sulfur bacteria that oxidize sulfide in a chemical reaction with oxygen or nitrate as electron acceptors and thereby obtain metabolically useful energy.

Hence, phototrophic sulfur bacteria bloom in sulfide-containing environments if sufficient light is present, whereas chemotrophic sulfur bacteria proliferate where sulfide coexists with oxygen or nitrate.

I. SULFIDE IN METABOLISM

A. Assimilatory Sulfate Reduction

Sulfur is an essential element for all living organisms. The content of sulfur in biomass averages about 1% (comparable to phosphorus), but some variation is species specific or caused by environmental conditions. Most of the sulfur in organisms occurs in proteins, namely, as a component of cysteine and methionine, and as a component in biologically important cofactors such as coenzyme A, thiamine, biotin, lipoic acid, and the ferredoxins. In these molecules, sulfur occurs in its reduced state with valence −2, or −1 as, for example, in disulfide bonds. However, in oxic environments, the most abundant sulfur source occurs in an oxidized state as sulfate (valence of sulfur = +6). All green plants, fungi, and most bacteria reduce sulfate to sulfide for biosynthetic purposes by a process called assimilatory sulfate reduction. Several bacteria highly specialized to sulfide-containing environments cannot perform assimilatory sulfate reduction, but rather take sulfide directly from the environment.

B. Sulfide Formation by Sulfate-Reducing Bacteria

In anoxic environments sulfide is formed from sulfate and sulfur by sulfate- and sulfur-reducing bacteria, respectively. Hydrogen and low molecular weight organic molecules serve as the electron donors in these dissimilatory reactions. Some sulfate reducers are versatile and can reduce both sulfur and thiosulfate in addition to sulfate. The degradation of organic matter in anoxic environments is a rather complex process in which different and mutually interacting bacteria interfere. Biopolymers are first converted by fermenting bacteria into small organic molecules and hydrogen. These products are the substrates for sulfate-reducing and methanogenic bacteria. Sulfate-reducing bacteria outcompete methanogens if sufficient sulfate is available. As a result, in sulfate-rich anoxic habitats, virtually all reducing equivalents from degradable biomass finally flow into dissimilatory sulfate reduction and appear as sulfide in the environment. Some sulfide is also formed by desulfuration of reduced organic sulfur compounds. However, this quantity is often small with respect to dissimilatory sulfide formation (e.g., 1 kg of biomass would yield 10 g of H_2S by desulfuration and potentially 570 g of H_2S by dissimilatory sulfate reduction). Dissimilatory sulfate reduction is widespread in coastal and marine sediments but restricted in many freshwater environments due to low sulfate concentrations (< 0.3 m*M*).

C. Sulfide Consumption by Phototrophic and Chemotrophic Sulfur Bacteria

Sulfide is used as an electron donor for fixation of carbon dioxide (lithoautotrophy) and in oxygen and nitrate respiration (chemotrophy). The major redox reactions performed by lithoautotrophic sulfur bacteria are listed in Table I, and the microbiological classification of nutritional types is shown in Fig. 1. The synthesis of reduced cell material from CO_2 with H_2S as the electron donor is exergonic, as demonstrated for glucose in Table I. Hence, lithoautotrophic sulfur bacteria need an energy-generating system for growth on sulfide and CO_2. In phototrophic sulfur bacteria, energy is generated from light, which is harvested by photosynthetic pigments comprising bacteriochlorophylls and carotenoids. These pigments provide these bacteria with colors that, at high population densities, are visible to the naked eye. Chemotrophic, or colorless, sulfur bacteria obtain energy from a chemical reaction, which represents

TABLE I
Standard Free Energy ($\Delta G^{\circ\prime}$) of Metabolic Reactions at pH 7

Bacterial group	*Reaction*	*kJ/reaction*	*kJ/mol O_2*
Aerobic chemoorganotrophic bacteria	Glucose + 6 $O_2 \rightarrow$ 6 CO_2 + 6 H_2O	−2865	−477
Chemolithotrophic sulfur bacteria	H_2S + 0.5 $O_2 \rightarrow S^\circ$ + H_2O	−204	−407
	S° + 1.5 O_2 + $H_2O \rightarrow SO_4^{2-}$ + 2 H^+	−583	−388
	H_2S + 2 $O_2 \rightarrow SO_4^{2-}$ + 2 H^+	−786	−393
	H_2S + 1.6 $NO_3^- \rightarrow SO_4^{2-}$ + 0.8 N_2 + 0.4 H^+	−736	
All lithoautotrophic sulfur bacteria	3 H_2S + 6 CO_2 + 6 $H_2O \rightarrow$ glucose + 3 SO_4^{2-} + 6 H^+	+506	

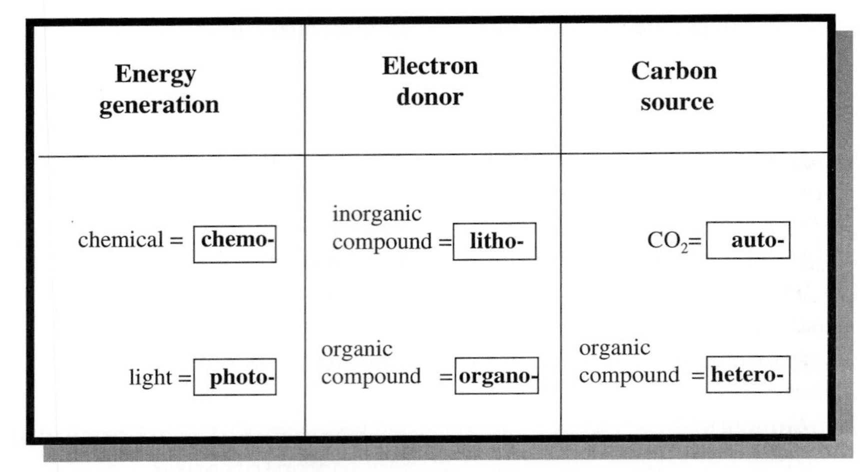

Fig. 1. General classification of nutritional types found among prokaryotes. Bacteria and Archaea encompass a much broader spectrum of nutritional types than that found among animals and green plants. Consequently, the concepts of heterotrophy and autotrophy established for plants and animals are insufficient for Bacteria and Archaea. Therefore, in microbiology, a combined triple term is used that refers to the mechanism used for the generation of metabolic energy, the electron donors, and the carbon source. Accordingly, a purple sulfur bacterium growing in the light on sulfide and CO_2 is a photolithoautotroph, and a *Thiobacillus,* which generates energy from sulfide oxidation and uses CO_2 as the only carbon source, is a chemolithoautotroph. Bacteria that respire or ferment organic compounds, such as *Escherichia coli,* are correctly designated as chemoorganoheterotrphs.

When dealing with large groups of organisms, shorter designations can be used if such groups contain different metabolic types. The major distinctive feature among the sulfide-oxidizing bacteria is the energy-generating mechanism. Accordingly, the main groups are designated as phototrophic sulfur bacteria and chemotrophic sulfur bacteria.

a respiratory type of metabolism; the electrons from sulfide are used to reduce oxygen and nitrate into water and dinitrogen (N_2) gas, respectively. Phototrophic sulfur bacteria grown in the light use all electrons from sulfide for CO_2 fixation, but, in contrast, chemotrophic sulfur bacteria shunt most of the electrons into respiration and thus obtain less biomass per amount of sulfide consumed. Accordingly, the chemotrophic yield on sulfide is 10–33% of the phototrophic yield.

Most sulfur bacteria can use low molecular weight organic compounds as a carbon source for biosynthesis. However, many of these substrates, such as acetate, are still too oxidized and need metabolic reduction prior to incorporation in biomass. Sulfide is used as an electron donor in this energy-requiring reaction. Such growth is referred to as lithoheterotrophy and is common among both phototrophic and chemotrophic sulfur bacteria.

1. Phototrophic Sulfur Bacteria

C. B. Van Niel, a Dutch microbiologist who spent a large part of his career in the U.S., highlighted the similarities between oxygenic "plant type" photosynthesis and the sulfide-dependent bacterial type of photosynthesis. His concept that CO_2 is reduced with water or hydrogen sulfide as the respective electron donors led to a breakthrough in research on photosynthesis. Photons of the appropriate wavelength are absorbed by photosynthetic pigments in the antennae complexes and are subsequently processed in the photosystems. CO_2 fixation occurs in a distinct, so-called dark reaction, which requires adenosine triphosphate (ATP) as metabolic energy and the cofactors reduced nicotinamide adenine dinucleotide (NADH) or NADPH as electron donors. Oxygen-producing phototrophs, including green plants, eukaryotic algae, cyanobacteria, and prochlorophytes, possess two photosystems (I and II). These systems are coupled in series and linked through an electron-transport chain according the so-called Z-scheme. In contrast, phototrophic sulfur bacteria and other anoxygenic phototrophs contain only one photosystem. These photosystems and the electron flow during photosynthesis is depicted in Fig. 2.

Purple sulfur bacteria contain bacteriochlorophyll *a*, or bacteriochlorophyll *b* in a few species, and carotenoids. The reaction center has a structural similarity with photosystem II of oxygenic phototrophs with quinones as the primary electron acceptor. It occurs associated with the light-harvesting antennae complexes in intracytoplasmic membranes that are continuous with the cytoplasmic membrane. The generation of metabolic energy is mediated by photosynthesis through cyclic electron transport via a cytochrome b/c_1 complex. Nicotinamide dinucleotide (NAD^+) is reduced by reduced quinone in an energy-consuming reaction (by dissipation of the proton-motive force). It is not clear whether electrons from sulfide directly reduce qui-

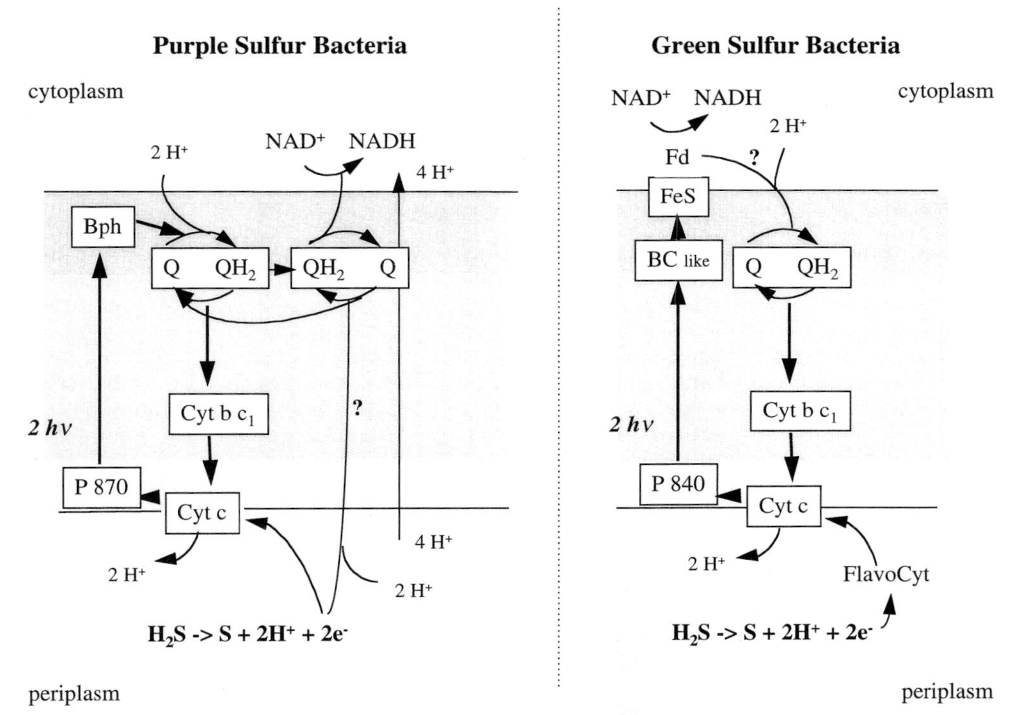

Fig. 2. Photosystems and electron flow during photosynthesis in purple sulfur bacteria (left) and green sulfur bacteria (right) and their orientation in the cytoplasmic membrane. BC like, bacteriochlorophyll *c*-like compound functioning as an electron acceptor; Bph, bacteriophaeophytin; Cyt b c_1, cytochrome bc_1 complex; Cyt c, cytochrome *c*; Fd, ferredoxin; FeS, iron–sulfide protein; FlavoCyt, flavocytochrome; NAD^+, nicotinamide adenine dinucleotide; NADH, reduced nicotinamide adenine dinucleotide; Q, quinone; QH_2, quinol, which is the reduced form of the quinone; P840, reaction center molecule of bacteriochlorophyll *a* with an absorption maximum at 840 nm; P870, reaction center molecule of bacteriochlorophyll *a* with an absorption maximum at 870 nm.

nones or whether they are shunted by a cytochrome *c* into the reaction center. Carbon fixation is mainly through the Calvin cycle as in oxygenic phototrophs.

Green sulfur bacteria contain bacteriochlorophyll *c, d,* or *e,* depending on species, and carotenoids as light-harvesting pigments. They also contain smaller amounts of bacteriochlorophyll *a.* Bacteriochlorophylls *c, d,* and *e* are unique among the chlorophylls, because rather than a single molecule, they comprise mixtures of 5–20 different allomers, which differ by their substitutions on the macrocycle and/or by their esterified alcohol. The light-harvesting antennae complexes are packed in chlorosomes, which are cell bodies surrounded by a non-unit membrane attached to the cytoplasmic membrane. Green sulfur bacteria have the highest specific, that is, expressed per unit biomass, chlorophyll contents among all photosynthetic organisms and are specially adapted to grow at extremely low light intensities. They comprise green species that have bacteriochlorophylls *c* or *d* with the carotenoid chlorobactene, and brown species that have bacteriochlorophyll *e* with the carotenoids isorenieratene. Bacteriochlorophyll *a* is found in the reaction center located in the cytoplasmic membrane; in addition, it occurs associated with the chlorosome and in the water-soluble Fenna-Mathews-Olsen protein. Green sulfur bacteria contrast with purple sulfur bacteria in several respects. First, the photosystem of green sulfur bacteria has a structural similarity with photosystem I of oxygenic phototrophs with an iron–sulfur protein as the primary electron acceptor. Second, the reduced iron–sulfide protein reduces ferredoxin and NAD^+ without additional energy input. The electrons consumed by linear transport are replenished from sulfide oxidation by a flavocytochrome. Third, CO_2 fixation in green sulfur bacteria is mainly by the reversed tricarboxylic acid cycle. This process discriminates to a lesser extent against ^{13}C than the Calvin cycle. Hence, biomass synthesized by green sulfur bacteria has an isotopic carbon signature distinct from most autotrophic organisms.

2. *Chemotrophic Sulfur Bacteria*

Chemotrophic sulfur bacteria do not contain photosynthetic pigments and are often referred to as colorless sulfur bacteria. Energy is generated in a redox reaction with oxygen as the electron acceptor, but nitrate is an alternative electron acceptor in several species. The electrons from sulfide are accepted by a cytochrome *c* and subsequently run downhill through the respiratory chain coupled to energy formation. Because NAD^+ cannot be directly reduced by reduced cytochrome *c*, reversed or uphill energy-consuming flow occurs. As in plants and purple sulfur bacteria, CO_2 fixation occurs through the Calvin cycle.

D. Sulfide Toxicity

Sulfide is a toxic compound for all living organisms; surprisingly this is also true for sulfide-producing and sulfide-consuming organisms. However, the sensitivity to sulfide is extremely variable among different species: While most aerobic organisms are markedly inhibited by 0.1 m*M* sulfide, some green sulfur bacteria grow well beyond 10 m*M* sulfide.

Sulfide strongly inhibits most aerobic respiration. For oxygenic photosynthesis it has been shown that sulfide dramatically blocks the activity of photosystem II (the water-splitting photosystem). An exception is found in some cyanobacteria from sulfide-containing habitats that possess a slightly modified photosystem II. Another feature of sulfide is its strong tendency to precipitate with metal ions forming highly insoluble salts, especially with Fe^{2+}, Zn^{2+}, and Mn^{2+}. These cations are required in many metabolic reactions as catalysts.

The noxious effects of sulfide are pH dependent. Hydrogen sulfide (H_2S) is a weak acid ($pK_{a,1} = 7.05$). The undissociated acid diffuses passively through the cellular membrane; therefore, sulfide is more toxic at lower pH values. The growth response of a sulfide-oxidizing organism is an optimum curve. At low sulfide concentrations, sulfide is limiting and growth rate increases with sulfide concentration according the Monod equation until a maximum is reached. At higher concentrations, sulfide toxicity predominates and growth rate decreases with increasing sulfide concentration; above a given sulfide concentration, growth is not possible.

II. OVERVIEW OF SULFIDE-CONTAINING ECOSYSTEMS

In some places, sulfide chemically formed in Earth's hot, deep layers emerges into the biosphere through volcanic activity. In such places, unique microbial ecosystems are found. However, most of the sulfide in the biosphere is formed by sulfate-reducing bacteria. Sulfate-reducing bacteria thrive in habitats that feature sufficient sulfate and anoxia due to a high input of organic matter and restricted oxygen input. The sulfide formed can precipitate with iron to from ferrous sulfide (FeS) and pyrite (FeS_2) or can be sequestered by organic matter. Free sulfide only appears in detectable amounts in the environment when these geochemical processes are locally saturated. In the marine environment, sulfate concentrations are normally non-limiting for sulfate reducers (e.g., the sulfate concentration of seawater of 33% salinity is 28 m*M*). In freshwater environments, however, sulfate concentrations vary by orders of magnitude dependent on geological settings and climatic conditions. Therefore, sulfide-containing environments are common in marine and coastal settings, but their occurrence in inland systems is bound by geological factors and local conditions. Sulfide-oxidizing bacteria can develop conspicuous population densities as biofilms in sediments and as planktonic plates if sulfur cycling is the dominant biogeochemical process. The term "sulfuretum" was coined by Baas-Becking to designate these types of sulfide-containing environments.

Generally, environments containing free sulfide are not closed ecosystems but, rather, are in close contact with oxidized environments. Sulfide tends to migrate into the oxidized environments, while oxidized compounds such as oxygen and nitrate migrate in the opposite direction. As a result, most sulfide-containing ecosystems feature gradients and can be divided into the anoxic and sulfide-rich habitat, the oxygen–sulfide coexistence zone, and a fully oxidized habitat. Alternatively, as observed in many bioturbated sediments, oxygen and free sulfide do not coexist, but the oxic and sulfidic zones are separated by a so-called suboxic zone dominated by denitrification and cycling of iron and manganese.

Often, oxygen and sulfide gradients run vertically, with the oxidized habitat on top and in contact with the atmosphere. These gradients run parallel with the light gradient. The different phototrophic and chemotrophic sulfide-oxidizing populations stratify along these gradients in accordance with their ecological niches. These phenomena occur both in aquatic and in sediment ecosystems, albeit, strikingly, the vertical scale is several orders of magnitude smaller in the sediment ecosystem. The idealized schemes of the gradients and the location of the different populations are shown in Fig. 3. The location of the different populations is a reflection of the gradients, but, at the same time, the metabolic activities of the populations actually determine the shape of the oxygen–sulfide profile. The effect of the populations is clearly demonstrated when comparing day and night profiles. The catabolic processes like oxygen respiration and sulfate reduction are light-independent. In contrast, all photosynthetic processes like oxygen production by primary producers and photooxidation of sulfide stop during the nighttime. As a result, the oxygen–sulfide interface shifts downward during daytime and upward during the night.

Table II gives an overview of the different natural ecosystems that feature sulfide-containing habitats and their most conspicuous sulfide oxidizers. The biodiversity and the ecology of the latter are discussed in detail in Section III. Anoxic conditions are widespread in sediments, waterlogged soils, and aquatic environments. In sediments, oxygen transport is limited to slow molecular diffusion. As a result, anoxic conditions are established only a few millimeters or centimeters below the sediment surface except in some oligotrophic deep-sea habitats. In contrast, in the water column, atmospheric and photosynthetically *in situ* produced oxygen is rapidly transported downward due to turbulent mixing. However, mixing diminishes when the water body becomes stratified due to a density gradient caused by variation of temperature or solute concentration. Thermal stratification in temperate lake is bound to seasonality, and, therefore, most of them completely mix (turn over) once or twice a year. In contrast, solute concentration gradients are normally persistent throughout the year, and, consequently, such a

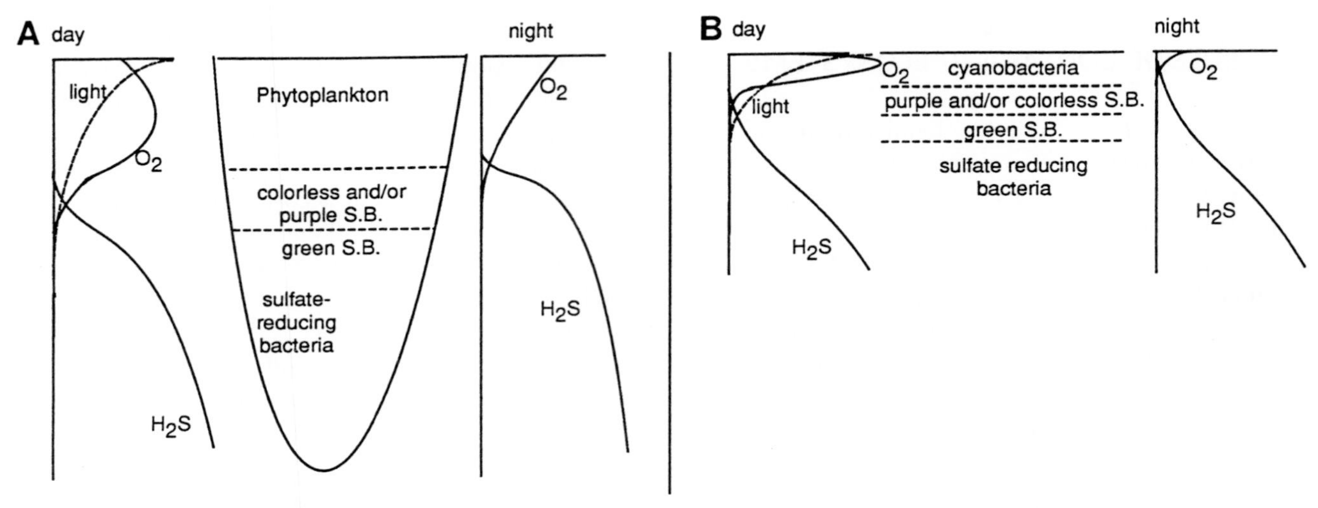

Fig. 3. (A) Stratified lake. (B) Sediment ecosystem (microbial mat). Idealized schemes show the population distributions and the physicochemical profiles found in stratified ecosystems with anoxic bottom layers containing sulfide. Striking similarities are found among aquatic and sediment ecosystems, despite large differences in the vertical scale (planktonic plates of phototrophic bacteria in lakes are 1 to several decimeters thick and may be found at great depth; microbial mats only comprise a few millimeters of the top layer of the sediment). Not all mentioned groups are always present. In fact, the occurrence of green sulfur bacteria in microbial mats is uncommon. Note the diurnal up and down shifts of the oxygen–sulfide interface. S.B., sulfur bacteria; colorless S.B., chemotrophic sulfur bacteria.

water body is stratified year-round. This phenomenon is referred to as meromixis. The hypolimnion in thermally stratified lakes and the monimolimnion in meromictic systems are the deep stagnant water layers that do not mix with the overlying water. These layers generally are or become anoxic except in very oligotrophic systems.

A. Inland Environments

Deep stratified lakes in glacier valleys contain sufficient sulfate to develop sulfide-containing habitats if gypsum is present in the parent material. Lakes with sulfide and phototrophic bacteria are found in gypsum–carbonate karstic systems. Both thermally stratified and meromictic lakes are especially common and have been extensively studied in Michigan (U.S.) and in the karstic regions in eastern Spain.

Endorheic basins are areas that do not pertain to a river catchment and are common in semiarid regions. The lower parts of these basins form permanent or ephemeral lakes that are subjected to strong evaporation and concentrate salts from runoff and groundwater flow. Albeit uncommon in temperate climates, on a global scale, inland salt lakes comprise an amount of water comparable to that in freshwater lakes. Sulfide is often present in the sediments of these lakes, where extensive microbial mats may develop. Some of the deeper lakes are meromictic and feature a sulfide-containing monimolimnion with conspicuous plates of phototrophic sulfur bacteria.

Human behavior has increased sulfide concentrations on the continents. A dramatic example of a human-made system is a sewage-treatment plant. The sulfide formed in anaerobic digestion plants is a nuisance. Trickling filters in aerobic sewage plants often develop biofilms that feature free sulfide at the bottom. The emissions of sulfur oxides resulted in continuously increasing sulfate concentrations in freshwater systems during the twentieth century, particularly important in the Northern Hemisphere. The combined effect of eutrophication and sulfur deposition are clearly reflected in the development of Lake Vechten (The Netherlands): During the last 50 years, total sulfur in this lake progressively increased, and presently moderate sulfide concentra-

TABLE II
Main Natural Ecosystems with Sulfide-Containing Environments

Ecosystem	*Geographical occurrence*	*Most conspicuous sulfide-oxidizing populations*[a]
Inland environments		
1. Stratified lakes with water rich in SO_4^{2-}	Karstic regions, glacier valleys	Planktonic plate of purple and green SB at depth
2. Inland salt lakes	Endorheic[b] basins in semiarid climates	Benthic purple SB and *Beggiatoa* (microbial mats)
3. Eutrophic lakes and sediments	Ubiquitously	Chemotrophic or purple SB, dependent on conditions
4. Biofilms as, e.g., on trickling filters	Ubiquitously	Dependent on conditions
5. Local accumulations of organic matter and rotting plant or animal tissues	Ubiquitously	Dependent on conditions
6. Volcanic lakes	Volcanic regions	Acid-tolerant thiobacilli
7. Sulfide-containing springs	Volcanic and karstic regions	Purple SB and filamentous chemotrophic SB (thermoacidophilic archaebacteria)
Coastal environments		
8. Tidal sediments; sediments of coastal lagoons	Ubiquitously on nonrocky coasts	On fine sand, laminated ecosystems with purple SB; on clay, chemotrophic SB
9. Temporal phenomenon in eutrophied lagoons	After a sudden crash of macroalgal blooms	Blooms of purple SB or chemotrophic SB in highly turbid systems
10. Stratified lagoons	Both fresh- and seawater inflow	Planktonic plate of green SB at depth
11. Evaporative hypersaline environments	Seawater inflow or seepage in (semi)-arid regions	Laminated sediment ecosystem with purple or chemotrophic SB
Marine environments		
12. Stratified water masses	Seas and fjords surrounded by land (e.g., Black Sea)	Planktonic plate of chemotrophic/green SB
13. Marine sediments in upwelling areas	Coastal plain along South America (Chili)	Chemotrophic SB, e.g., *Thioploca* spp.
14. Hydrothermal vents	Ocean floor tectonic spreading center	Chemotrophic SB in (1) the vent, (2) symbiotic association with invertebrates, and (3) the benthos

[a] SB, sulfur bacteria.
[b] Endorheic basins do not pertain to river catchment areas.

tions occur in the hypolimnion, allowing proliferation of planktonic phototrophic sulfur bacteria.

Accumulations of organic material often present sulfide-containing niches, even in sulfate-poor habitats where most of the sulfide may originate from desulfuration. On a smaller scale, sulfide-containing habitats occur in rotting and decaying tissues of animals and plants. For example, halitosis, the phenomenon of production of bad smell from a human mouth, is a cause of human discomfort. It is related to production of sulfur-containing gases including hydrogen sulfide from biofilms in the oral cavity.

Sulfide of geochemical origin is found in the volcano lakes and geysers. Acid volcano lakes are found in Japan, whereas sulfide springs occur in Yellowstone National Park (U.S.), Iceland, New

Zealand, and Russia. Some of the springs are extremely hot and/or acidic.

B. Coastal and Marine Sulfide-Containing Environments

Coastal sediments in the tidal fringe and in coastal lagoons often contain free sulfide at a shallow depth below the surface. In several coastal lagoons and estuaries fresh continental water stratifies on top of salt seawater. In the Mediterranean and tropical regions, stratified coastal lagoons with green and purple sulfur bacteria are known. Highly eutrophic coastal lagoons, particularly those which have been increasingly eutrophied by human activities, often feature abundant development of macroalgae. In summertime these blooms suddenly crash, which gives rise to sudden degradation of biomass resulting in anoxic conditions and formation of sulfide in the water column. This situation is known in the South of France as "malaigue," that is, bad smelling waters. It is often succeeded by conspicuous "eaux rouges," that is, red waters, due to a rapid colonization by purple sulfur bacteria. In highly turbid waters, the sulfide oxidation is performed by colorless sulfur bacteria, resulting in "white waters." In upwelling areas, sulfide-containing sediments cover extensive parts of the coastal plain because of high inputs of sedimented organic matter. Spectacular thick benthic mats consisting of *Thioploca ingrica* and *Thioploca chileae* occur over 3000 km along the continental shelf off southern Peru and Chile.

Stratified water masses sometimes occur in fjords and marginal seas. The best known example is the Black Sea, which is meromictic and features plates of chemotrophic or brown-colored green sulfur bacteria at the interface of both water masses (80–150 m depth).

Evaporitic hypersaline environments are found along the seacoast in semiarid and arid climates. Often they represent ephemeral coastal shallow lagoons with well-developed microbial mats that represent active sulfureta. Solar lake (Egypt) is a permanent deeper lake that has been extensively studied. This lake is stratified during winter when its hypolimnion contains 1.5 m*M* sulfide. Both benthic and planktonic phototrophic sulfur bacteria proliferate in this lake.

Extremely spectacular underwater sulfide-containing ecosystems were discovered in the mid-1970s. At the Galapagos Rift seafloor tectonic-spreading center, hydrothermal vents were found at depths of about 2500 m. Hot fluid containing geochemically formed sulfide is mixed with oxygen-containing seawater and emitted at a high flow rate into the ocean. The coexistence of sulfide and oxygen provides the habitats for chemotrophic sulfur bacteria, which occupy three different habitats: within the vent system itself, on various surfaces and sediments within the plume of the hydrothermal fluid, and in symbiotic associations with invertebrate animals. Similar symbiosis of colorless sulfur bacteria and invertebrates have also been found in the coastal areas. They often comprise less spectacular invertebrates, and although their study has been neglected for a long time, it has revived after the discovery of symbiosis in the deep sea.

III. BIODIVERSITY AND ECOLOGICAL NICHES OF SULFIDE OXIDIZERS

A. Phototrophic Sulfur Bacteria

1. *Taxonomy*

Phototrophic sulfur bacteria have characteristic morphological traits, which facilitate their microscopic observation in natural environments. Among these are the highly refractile sulfur globules, which are formed as an intermediate product during sulfide oxidation. Sulfur globules appear inside or outside the cell. Originally, several morphological (phenotypical) features were used as major criteria in taxonomy, particularly for the description of families and genera. However, recent phylogenetic studies using 16S ribosomal RNA sequencing techniques have shown that some of these criteria, such as cell shape, gas vacuoles, pigment composition, and motility, are not good indicators of evolutionary relationships. Therefore, taxonomy is continuously being reconsidered as new sequences become available, and phototrophic sulfur bacteria have been rearranged in new genera which reflect phylogenetic relationships. Interestingly, halophilic species from hypersaline environments are often phylogenetically distant from their morphologically similar counterpart species in freshwater and marine environments, indicating that specialization to hypersaline conditions occurred early in evolution. The 1998 state of the art of phototrophic sulfur bacterial taxonomy is described in Table III. An update of bacterial taxonomy can be consulted on the World Wide Web (*http://www.dsmz.de/bactnom/bactname.htm*), which is a free service provided by the German Culture Collection of Microorganisms and Cell Cultures (DMSZ).

The green sulfur bacteria constitute a phylogenetically homogeneous group within the Bacteria (formerly Eubacteria). Sulfur globules are always outside the cell. They comprise only the family Chlorobiaceae, with five genera and 14 species which have been described on the basis of morphological criteria. Pigment composition and light harvesting by chlorosomes are very similar to those of the green filamentous bacteria (Chloroflexaceae), which are, however, a phylogenetically distant group.

The purple sulfur bacteria comprise two families, namely, the Ectothiorhodospiraceae and Chromatiaceae. They belong to the γ (gamma)-branch of the proteobacteria (gram negative) within the Bacteria domain. Sulfur globules are outside the cells in the Ectothiorhodospiraceae. Ultrastructurally, sulfur in the Chromatiaceae is formed in the periplasm on the outside of the cytoplasmic membrane which invaginates in the bacterial cell. As a result, sulfur globules in the Chromatiaceae appear within the cell and are visible by optical microscopy. The Ectothiorhodospiraceae comprise two genera, namely; *Ectothiorhodospira,* the marine species, and *Halorhodospira* species adapted to grow in extremely salty and alkaline solutions (salinity $>$ 12%; pH about 9). The Chromatiaceae is the largest family with more than 30 described species, and it has been rearranged in new genera recently.

2. *Microbial Consortia-Based Sulfur Cycling*

Some green sulfur bacteria live in symbiosis with sulfur- or sulfate-reducing bacteria forming stable consortia. The number of cells involved and the shape and the topology of the consortium are often remarkably stable; consortia have been assigned a Latin name and a description if they were a single species (e.g., *Chlorochromatium, Pelochromatium, Chloroplana*). The symbiosis is based on synthrophy: The chemoorganotrophic partner produces sulfide, while this compound is recycled to sulfur by the phototrophic partner.

3. *Light Quality as an Ecological Factor*

The way light is converted to metabolic energy is described in Section I,C,1 and depicted in Fig. 3. Phototrophic sulfur bacteria grow on visible and near-infrared light, but not all wavelengths are harvested and used with the same efficiency. Light absorption is reflected by the *in vivo* absorption spectrum; such spectra are shown in Fig. 4 for different phototrophs. The *in vivo* spectra are very different from those obtained when the pigments are extracted with organic solvents. These differences occur because of intermolecular interactions in the antennae complexes and reaction centers. Several *in vivo* peaks correspond to bacteriochlorophylls: one around 400 nm and one or two (bacteriochlorophyll *a*) in the infrared region. Carotenoids have *in vivo* absorption maxima between 450 and 600 nm.

Figure 4 demonstrates that the different groups of phototrophic organisms are specialized for use of different and complementary wavelength regions. This elegant example of niche differentiation allows the coexistence of representatives from the various groups. In ecosystems light is, however, also attenuated with depth due to scattering and absorption caused by abiotic substances. While green light penetrates deepest in aquatic ecosystems, red and infrared wavelengths do so in sandy sediments. Consequently, planktonic phototrophic bacteria harvest the light mainly with their carotenoids, whereas benthic bacteria predominantly use their chlorophylls for this purpose. Not all light harvested by phototrophs is converted into new biomass, since a certain amount is needed for maintenance of cell integrity. Maintenance energy expenditure is about 10 times lower in green sulfur bacteria compared to purple sulfur bacteria. Consequently, green sulfur

TABLE III
Different Groups, Families, and Genera of Phototrophic Sulfur Bacteria

Green sulfur bacteria	
Family Chlorobiaceae (pigments: bacteriochlorophylls *c, d,* or *e* and minor amounts of bacteriochlorophyll *a* + carotenoids; chlorosomes[a]; division by binary and ternary fission; gram-negative; CO_2 fixation by reversed tricarboxylic acid cycle; sulfur globules outside the cells[b])	
Chlorobium	Straight or curved rods, nonmotile
Pelodictyon	Straight, curved, or ovoid cells, gas vacuoles[c]
Prosthecochloris	Starlike with extrusions, prosthecae
Ancalochloris	Starlike with extrusions, prosthecae, gas vacuoles
Chloroherpeton	Long gliding unicellular flexible rods, gas vacuoles
Purple sulfur bacteria	
Family Ectothiorhodospiraceae (γ-proteobacteria; pigments: bacteriochlorophyll *a* or *b*[d] and carotenoids; division by binary fission; CO_2 fixation by the Calvin cycle; sulfur globules outside the cells[b])	
Ectothiorhodospira	Rod and spirilloid cells, marine species
Halorhodospira	Spirilloid cells, halophilic and alkalophilic
Family Chromatiaceae (γ-proteobacteria; pigments: bacteriochlorophyll *a* or *b*[a] and carotenoids; division by binary fission; CO_2 fixation by the Calvin cycle; sulfur globules *inside* the cells[b])	
Chromatium[e]	Rods, large (4–6 μm), anaerobes strictly dependent on sulfide
Isochromatium[e]	Rods, anaerobes strictly dependent on sulfide
Allochromatium[e]	Rods, freshwater and halotolerant species, facultative chemoautotrophs[f]
Marichromatium[e]	Rods, marine species, facultative chemoautotrophs
Halochromatium[e]	Rods, halophilic (optimum at salinity >8‰), facultative chemoautotroph[f]
Thermochromatium[e]	Rods, thermophilic (optimum 48°–50°C), not a chemoautotroph[f]
Rhabdochromatium	Rods, marine species, motile by flagella, not a chemoautotroph[f]
Lamprobacter	Rods, motile by flagella, with gas vescicles
Thiodictyon	Rods, no flagella, with gas vescicles
Thiocapsa[g]	Spherical cells, versatile freshwater and marine species, facultative chemoautotrophs[f]
Thiohalocapsa[g]	Spherical cells, halophilic (optimum salinity 60‰), facultative chemoautotroph[f]
Amoebobacter[g]	Spherical cells, freshwater species, facultative chemoautotrophs[f]
Thiococcus[g]	Spherical cells, strict anaerobe; containing bacteriochlorophyll *b*
Thiolamprovum[g]	Spherical cells, flat shaped, freshwater, facultative chemoautotroph[f]
Thiorhodococcus	Spherical cells, nonmotile, strict anaerobe
Thiocystis	Spheres to rod shaped, motile by flagella, facultative chemoautotrophs[f]
Lamprocystis	Spherical cells, nonmotile, gas vacuoles, not a chemoautotroph[f]
Thiopedia	Spherical cells in platelets, with gas vacuoles

[a] Chlorosomes are intracellular non-unit membrane enclosed vesicles in green phototrophic bacteria that contain bacteriochlorophyll *c, d,* or *e* with minor amounts of bacteriochlorophyll *a*. These vescicles underlie the cytoplasmic membrane where the reaction centers are located.

[b] Sulfur is formed during sulfide oxidation and occurs as typical refractile globules visible with the light microscope.

[c] *Pelodictyon* cells often form microcolonies shaped as three-dimensional nets.

[d] Only a few species.

[e] Genera described in 1998 accommodating rod-shaped species previously affiliated to *Chromatium*. All species are motile by flagella.

[f] Chemolithoautotrophy, with sulfide or thiosulfate as the electron donor and oxygen as the electron acceptor.

[g] Genera emended or described in 1998 accommodating species belonging to the former complex of *Thiocapsa/Amoebobacter* consisting of spherical cells without flagella. Some species contain gas vacuoles, which is, however, not indicative of phylogenetic relationships.

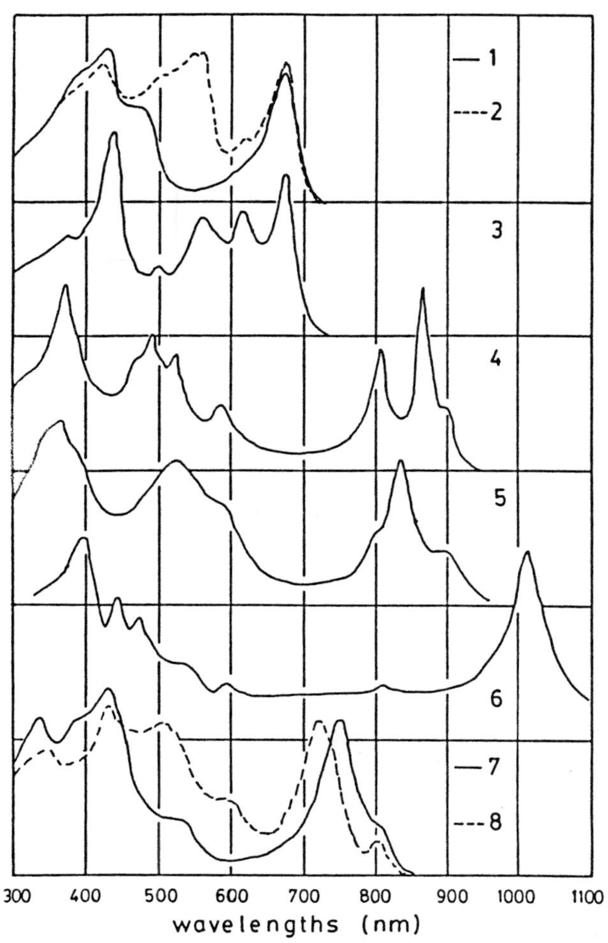

Fig. 4. *In vivo* absorption spectra of different phototrophic microorganisms. 1, green algae; 2, red algae; 3, cyanobacteria; 4, purple sulfur bacteria containing bacteriochlorophyll *a* and the carotenoid spirilloxanthine; 5, purple sulfur bacteria containing bacteriochlorophyll *a* and the carotenoid okenone; 6, purple sulfur bacteria containing bacteriochlorophyll *b*; 7, green sulfur bacteria containing bacteriochlorophyll *c* and the carotenoid chlorobactene; 8, green sulfur bacteria containing bacteriochlorophyll *e* and the carotenoid isorenieratene (actually brown). [From Caumette, P. (1989). *In* "Microbial Mats: Physiological Ecology of Benthic Microbial Communities" (Y. Cohen and E. Rosenberg, eds.), pp. 283–304. American Society for Microbiology, Washington, D.C.]

bacteria can grow and survive at lower light intensities than purple sulfur bacteria.

In lakes, green sulfur bacteria with chlorobactene (Fig. 4, spectrum 7) often occur below a layer of purple sulfur bacteria, because of their capacity to grow at lower light intensities and because of their greater sensitivity toward molecular oxygen (see below). The brown colored Chlorobiaceae (Fig. 4, spectrum 8) are specialized to grow in very deep lakes, forming plates below 8 m depth, and occur in the Black Sea even at a depth of 80 m. However, these isorenieratene-containing species are not competitive below a layer of planktonic purple sulfur bacteria, because the isorenieratene absorption maximum (approximately 510 nm) coincides with the maxima of carotenoids in purple sulfur bacteria (see Fig. 4, cf. spectrum 8 with spectra 4 and 5). Specific absorption differences in the near-infrared are also found within the bacteriochlorophyll *a*-containing purple sulfur bacteria (Fig. 4, cf. spectra 4 and 5). Thus, it was found that in evaporitic microbial mats the okenone-containing *Thiohalocapsa halophila* coexists with the spirilloxanthin-containing *Halochromatium salexigens* (see Fig. 5). *Thermochromatium tepidum* is the only known purple sulfur bacterium with an absorption peak between 900 and 1000 nm (not shown in Fig. 4). *Thiococcus pfennigii* (formerly *Thiocapsa pfennigii*) contains bacteriochlorophyll *b*, which provides an absorption peak in the far-infrared about 1030 nm (see Fig. 4, spectrum 6).

4. *Metabolic Versatility*

Hydrogen and, most often, reduced sulfur such as thiosulfate, polysulfides, and elemental sulfur are excellent electron donors for phototrophic growth of purple and green sulfur bacteria. In addition, a variety of carbon compounds is photoassimilated by some species. However, in *Chlorobium* photoassimilation only occurs with concomitant consumption of sulfide and CO_2.

Originally, purple sulfur bacteria were considered to be obligate anaerobes and obligate phototrophs. In the early 1970s, however, it was found that *Thiocapsa* species can grow in the presence of oxygen as a chemolithoautotroph resembling the colorless sulfur bacteria (sulfide as the electron donor, oxygen as the electron acceptor). Later it was demonstrated that

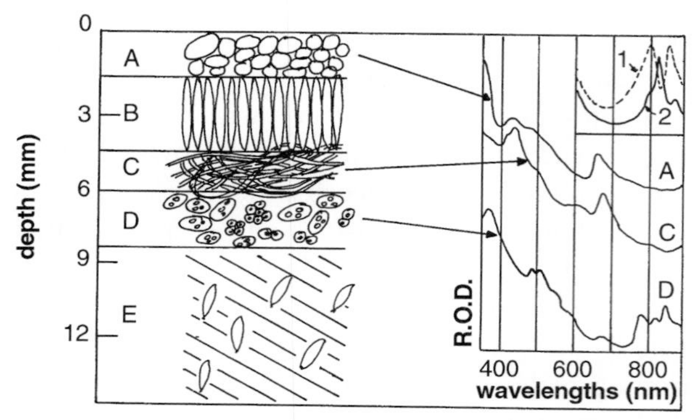

Fig. 5. Microbial mat communities that develop at 130–220‰ salinity in the Salin-de-Giraud (Camargue, France). (Left) Stratification of the community structure (A, brown layer of the cyanobacterium *Aphanothece*; B, transparent layer composed of gypsum crystals; C, green layer of the cyanobacterium *Phormidium*; D, layer of the purple sulfur bacteria; E, black FeS-containing layer. (Right) Absorption spectra of the different community layers (A, C, and D) and near-infrared absorption spectra of *Halochromatium salexigens* (1) and *Thiohalocapsa halophila* (2). [From De Wit, R., and Caumette, P. (1994). Diversity of and interactions among sulphur bacteria in microbial mats. *In* "Microbial Mats: Structure, Development and environmental Significance" (L. J. Stal and P. Caumette, eds.), pp. 377–392. NATO ASI Ser. Vol. G35 Springer-Verlag, Berlin and Heidelberg.]

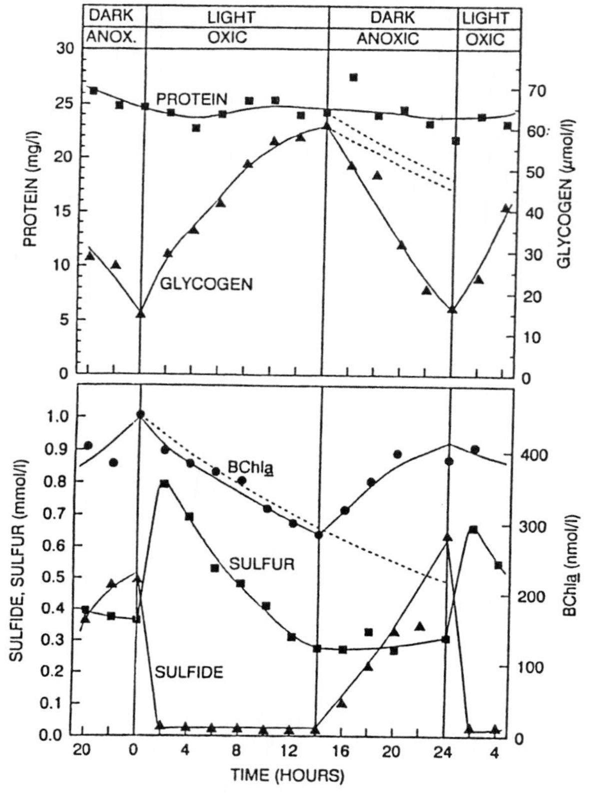

Fig. 6. Experimental simulation of the fluctuating conditions experienced by the purple sulfur bacterium *Thiocapsa roseopersicina* in its natural habitat of the microbial mat. A sulfide-limited culture (dilution rate of 0.03 hr^{-1}) was exposed to light plus oxic and dark plus anoxic alternations. Previous to the data sampling the culture was grown at this regimen for a week. Bacteriochlorophyll *a* was washed out during the day, thus indicating that no synthesis occurred. In contrast, bacteriochlorophyll *a* was synthesized at a high rate during the night with glycogen as the energy and carbon source. During the night, the sulfide concentrations rose because this compound was not consumed and was continuously fed to the culture. [From de Wit, R., and van Gemerden, H. (1990). *Arch. Microbial.* **154,** 459–464.]

this capacity of chemolithotrophic growth is widespread among small members of the Chromatiaceae (see Table III). Ectothiorhodospiraceae species only grow in the presence of oxygen with organic substrates, whereas members of the Chlorobiaceae are strict anaerobes.

Thiocapsa roseopersicina is especially common in alternating oxic–anoxic habitats below cyanobacteria in microbial mats. Typical environmental conditions in the microbial mat were simulated in continuous cultures of this species as depicted in Fig. 6. Thus, oxic periods were combined with illumination (day conditions) and anoxic periods with darkness (night conditions). During the oxic day periods, pigment synthesis is repressed by oxygen; however, during the anoxic night periods, pigment synthesis occurs at high rates. As a result, the organism remains pigmented and is able to continue photosynthesis during daytime despite the presence of oxygen. This way, it obtains a higher growth yield on sulfide rather than if it would have shifted to full chemosynthesis. This example shows that pigment synthesis and photosynthesis are not coupled.

5. Growth Kinetics and Ecological Niches

The small Chromatiaceae (e.g., *Allochromatium, Marichromatium, Thiocapsa), Ectothiorhodospira,* and *Chlorobium* all exhibit similar maximal growth rates

(μ_{max}) of about 0.1 hr^{-1}, which corresponds to a doubling time of 7 hr. These species are opportunists that are easily enriched from natural samples in appropriate nutrient-rich growth media. In contrast, large *Chromatium* species, such as *Chromatium weissii*, grow more slowly. The competitive advantage of *Chromatium weissii* with respect to smaller *Allochromatium vinosum* has been elegantly demonstrated in a series of experiments using the chemostat. *Chromatium weissii* has not only a lower μ_{max} but also a lower affinity for sulfide than the small *Allochromatium vinosum*. Thus, under constant environmental conditions, *Allochromatium vinosum* grows faster than *Chromatium weissii* at any sulfide concentrations. However, the natural environmental conditions are fluctuating: Sulfide concentrations rise at night in the absence of photosynthesis and drop in the morning. It was found that *Chromatium weissii* has a twofold greater uptake capacity for sulfide than *Allochromatium vinosum*. Sulfide taken up at high rate is only oxidized to intracellular stored sulfur and is further oxidized when sulfide concentrations are very low. Therefore, *Chromatium weissii* is better equipped to take advantage of the temporal occurrence of high sulfide concentrations.

Members of the Chlorobiaceae and Ectothiorhodospiraceae, in general, have a 5- to 10-fold higher growth affinity for sulfide than do the Chromatiaceae, which has been attributed to the fact that the former two store sulfur extracellularly. Thus the Chromatiaceae are expected to lose the competition when sulfide is growth-limiting. To test this hypothesis, it was essential to use chemostat-culture techniques. In such a culture, the losing species is drastically washed out, and, consequently, only the competitive species can maintain stable population densities. Surprisingly, when cultivating a mixture of *Chlorobium* and *Allochromatium* under sulfide limitation, the outcome was not a monoculture of *Chlorobium* but, rather, a stable coexistence of both organisms. This result indicates that, beside sulfide, another substrate is influencing the growth rates. The extracellular sulfur of *Chlorobium* appeared to be tightly bound to the cell, thus, as such, being unavailable for *Allochromatium*. However, this extracellular sulfur continuously reacts with sulfide to form soluble polysulfides. In the mixed culture the latter substrate was used exclusively by *Allochromatium*, which allowed this species to coexist with *Chlorobium*.

B. Chemotrophic Sulfur Bacteria

1. *Taxonomic Affiliation*

Chemotrophic sulfur bacteria are not a phylogentically homogeneous group; rather, their representatives occur in many lineages scattered throughout the Bacteria and Archaea domains as shown in Table IV. In the 1990s data from 16S ribosomal RNA sequencing have resulted in major taxonomic rearrangements, particularly for *Thiobacillus*-like organisms Thus, it was shown that very versatile bacteria as the former *Thiobacillus versutus* and *Thiosphaera pantotropha* belong to the α branch of the proteobacteria and have been reclassified as *Paracoccus versutus* and *Paracoccus denitrificans*, respectively. Within the β branch, *Thiomonas* is a revived name of a newly defined genus which accommodates a cluster of mixotrophic thiobacilli. Most of the species retained within the genus *Thiobacillus* are chemolithoautotrophic specialists.

The filamentous organisms *Beggiatoa*, *Thioploca*, and *Thiothrix* are closest to the phototrophic purple sulfur bacteria (γ proteobacteria). It is therefore believed that these genera evolved from a phototrophic ancestor which lost its photosynthetic capabilities.

2. *Ecological Niches*

Chemotrophic sulfur bacteria are particularly abundant in extreme environments. Extremely thermoacidophilic species (optimum temperature 85°–96°C, able to grow at pH 1–6) belonging to the Archaea have been isolated from hot sulfide-containing acidic springs. In addition, *Thermothrix* is a bacterial species adapted to growth at high temperatures up to 80°C, albeit at neutral pH. *Sulfobacillus* is a gram-positive acidophilic genus comprising mesophilic and moderately thermophilic species that have a very versatile metabolism and are capable of growth on mineral sulfides including pyrite (FeS_2).

Many *Thiobacillus* species can grow at low pH values, which, for example, explains their predominance in acid volcanic lakes. Several species grow best between pH 2 and 4. The acidity is often the

TABLE IV
Different Genera of Chemotrophic Sulfur Bacteria and Their Phylogenetic Affiliation and Major Morphological and Metabolic Characteristics

Taxon	Characteristics
Domain Bacteria	
α-Proteobacteria	
Paracoccus	Spherical to rods; very versatile mixotrophs
β-Proteobacteria	
Thiobacillus	Rods; many obligate chemolithoautotrophs
Thiomonas	Rods; mixotrophic species
Alcaligenes	Mostly heterotrophic; some species capable of sulfide oxidation
Thermothrix	Filamentous; thermophilic; capable of mixotrophy
γ-Proteobacteria	
Thiomicrospira	Spirilloid cells, motile
Beggiatoa	Single filaments; sulfur inclusions[a]; gliding
Thioploca	Filaments occur in bundles enclosed by a slimy sheath; sulfur inclusions[a]; vacuoles[b]; gliding
Thiothrix	Attached rigid filaments within a sheath that produce hormogonia[c]; sulfur inclusions[a]
Achromatium	Very large spherical to oval cells; rotating with inclusions of sulfur and $CaCO_3$
Hyphomicrobium	Mostly heterotrophic; some species capable of sulfide oxidation
Firmicutes (gram-positive bacteria) with low GC content (*Clostridium* group):	
Sulfobacillus	Rods; moderately thermophilic mineral sulfide-oxidizing bacteria
Unknown phylogenetic affiliation	
Macromonas	Straight or curved rods; motile by flagella with inclusions of sulfur and $CaCO_3$[a]
Thiovulum	Large round to ovoid cells; motile and conspicuously chemotactic; sulfur inclusions[a]
Thiodendron	Vibrio-shaped cells with stalks
Thiospira	Spirilloid; motile with polar flagella in tufts; sulfur inclusions[a]
Thiobacterium	Rods embedded in gelatinous mass; sulfur inclusions[a]
Domain Archaea	
All species in this group are extremely thermophilic and acidophilic	
Sulfolobus	Coccoid highly irregular
Acidianus	Coccoid highly irregular; also reduces sulfur to sulfide
Sulfurisphaera	Coccoid highly irregular; physiology is similar to *Acidianus*, but phylogenetically closer to *Stygiolobus*
Sulfurococcus	Coccoid

[a] Sulfur is formed during the oxidation of sulfide and occurs as characteristic refractile globules visible with the light microscope; in *Beggiatoa* and *Thiothrix* they appear to be intracellular, but in fact, they occur outside the cytoplasm within invaginations of the cytoplasmic membrane.
[b] Some species contain vacuoles where nitrate can be stored for use as an electron donor.
[c] Hormogonia are motile filaments that become detached from the parent organism and act as a means of vegetative propagation.

result of their own metabolic activity, because the formation of sulfate from sulfide results in proton release (see Table I). In contrast, in phototrophic sulfur bacteria, this pH effect is counterbalanced by the fixation of CO_2, which tends to raise the pH.

Chemotrophic sulfur bacteria are the oxygen–sulfide gradient organisms par excellence, and most species show chemotactic or chemophobic motility allowing these bacteria to locate themselves within the gradient. In addition, many species are able to use nitrate as an alternative electron acceptor under anaerobic conditions. Nitrate is formed in the oxic zone and often diffuses somewhat deeper in the sediment than oxygen; therefore, the niche of denitrifiers is generally just below the oxic–anoxic interface. In 1994, the extensive *Thioploca* mats on the Chilean coastal shelf were studied during a German–Chilean cruise. It was found that these organisms dwell in the sediment and perform vertical migrations between the lower sulfide-rich layers and the sediment surface from where they protrude into nitrate-rich water. The nitrate is taken up from the bottom water, stored in intracellular vacuoles, and used subsequently in the deeper layers as an electron acceptor for chemotrophic sulfide oxidation. By keeping sulfide spatially separated from nitrate and oxygen, these organisms are able to outcompete other colorless sulfur bacteria.

Most bacteria live embedded in a viscous water layer in both pelagic and benthic habitats. As a result, uptake of substrates is constrained by molecular diffusion. Therefore, bacteria require a high surface-to-volume ratio, which is achieved by being small in size. The chemotrophic sulfur bacteria, however, include examples of "giant microorganisms," which have been able to overcome the common bacterial constraints on surface-to-volume ratios. Giant *Beggiatoa* filaments (120 μm thick, >1 cm in length) live attached to sediments in deep-sea hydrothermal vents, exploiting a pulsed hydrodynamic flow of alternating oxygen-rich and sulfide-rich water. Similarly, the vertical migrations of the Chilean shelf *Thioploca* spp. between sulfide-rich sediments and nitrate-containing bottom water allow these organisms to overcome diffusion limitations and to evolve into a "giant microorganism" (*Thioploca araucae* filaments are 10–20 μm wide and up to 7 cm long).

Beside sulfide, other reduced sulfur compounds such as thiosulfate, sulfur, and polythionates are appropriate electron donors for many chemotrophic sulfur bacteria. *Thiobacillus ferrooxidans* can grow with Fe^{2+} as the electron donor at low pH and converts it to Fe^{3+}. Many chemotrophic sulfur bacteria are not obligatorily dependent on inorganic compounds as the electron donor and on CO_2 as the only carbon source. In fact, the colorless sulfur bacteria comprise a wide spectrum of metabolic adaptations linking them to the heterotrophic bacteria. Four nutritional groups have been identified (cf. Table I):

1. Obligate chemolithoautotrophs cannot use organic compounds.
2. Facultative chemolithoautotrophs are very versatile: Both CO_2 and organic compounds can be used as carbon sources; both sulfide and organic compounds can be used as electron donors. Often these species perform best under mixotrophic conditions.
3. Chemolithoheterotrophs cannot use CO_2, but they can obtain energy from sulfide oxidation.
4. Chemoorganotrophic species exclusively grow on organic compounds, and no energy can be obtained from sulfide oxidation. These species probably benefit from the oxidation of sulfide for detoxifying hydrogen peroxide.

Chemostat-culture experiments were essential to elucidate the ecological niches of these different nutritional groups. Specialists are competitive when one substrate is limiting. Thus, chemolithoautotrophic and chemoorganoheterotrophic bacteria perform best when sulfide or organic substrates, respectively, are the only substrates. However, versatile mixotrophic species can reach high population densities when both sulfide and organic substrates are growth-limiting. Hence, it was concluded that the ratio of sulfide with respect to organic substrates available as nutrients determines the community structure in the ecosystem.

C. Competition between Phototrophic and Chemotrophic Sulfur Bacteria

The small specialized thiobacilli like *Thiobacillus neopolitanus* and *Thiobacillus thioparus* have maximal growth rates of about 0.35 h^{-1}, which corresponds to a doubling time of 2 hr. Thus, the maximal growth rate on sulfide but, also, the affinity toward this compound highly exceed the values found for phototrophic sulfur bacteria (see Section III,A,5). Therefore, these small thiobacilli seem to be superior competitors at the oxygen–sulfide interface. However, the phototroph *Thiocapsa roseopersicina* can coexist with *Thiobacillus thioparus* in illuminated chemostats when both the oxygen and the sulfide supply are limiting. This was achieved when oxygen-to-sulfide supply ratios ranged between 0.65 and 1.6. All the sulfide was oxidized by *Thiobacillus,* but only incompletely into intermediate products like sulfur and thiosulfate, because insufficient oxygen was present for a full oxidation to sulfate. Moreover, this species depleted oxygen to extremely low levels, which allowed for pigment synthesis in *Thiocapsa roseopersicina*. This purple sulfur bacterium grew phototrophically on the intermediate products of sulfide oxidation and could, therefore, coexist with *Thiobacillus thioparus*. At higher oxygen-to-sulfide supply ratios, *Thiocapsa roseopersicina* was, however, washed out from the culture vessel and a monoculture of *Thiobacillus thioparus* was obtained. Mathematical modeling of population dynamics coupled to reaction–diffusion processes in microbal mats predicted that, depending on conditions, either colorless sulfur bacteria dominate or a coexistence occurs of colorless and purple sulfur bacteria. However, under the appropriate conditions, such as low oxygen-to-sulfide supply ratios and low light attenuation, the purple sulfur bacteria are extremely successful, as their biomass outweighs that of colorless sulfur bacteria by more than an order of magnitude.

Acknowledgments

I thank Johannes Imhoff and Pierre Caumette for discussing new insights on the taxonomy of phototrophic sulfur bacteria. Thanks are due to Hans van Gemerden for allowing me to start work on anoxygenic photosynthesis as a Ph.D. student at the University of Groningen, and for continuing our stimulating discussions on the ecology of sulfide-containing environments during many subsequent years.

See Also the Following Articles

Continuous Culture • Photosensory Behavior • Sulfur Cycle • Wastewater Treatment, Municipal

Bibliography

Brune, D. C. (1989). Sulfur oxidation by phototrophic bacteria. *Biochim. Biophys. Acta* **975**, 189–221.

Fossing, H., Gallardo, V. A., Jørgensen, B. B., Hüttel, M., Nielsen, L. P., Schulz, H., Canfield, D. E., Forster, S., Glud, R. N., Gundersen, J. K., Küver, J., Ramsing, N. B., Teske, A., Thamdrup, B., and Ulloa, U. (1995). Concentration and transport of nitrate by the mat-forming sulphur bacterium *Thioploca*. *Nature (London)* **374**, 713–715.

Imhoff, J. F., Süling, J., and Petri, R. (1998). Phylogenetic relationships among the *Chromatiaceae*, their taxonomic reclassification and description of the new genera *Allochromatium, Halochromatium, Isochromatium, Marichromatium, Thiococcus, Thiohalocapsa* and *Thermochromatium*. *Int. J. System. Bacteriol.* **48**, 1129–1143.

Madigan, M. T. (1988). Microbiology, physiology and ecology of phototrophic bacteria. *In* "Biology of Anaerobic Microorganisms" (A. J. B. Zehnder, ed.), pp. 39–111. Wiley, New York.

Pfennig, N. (1993). Reflections of a microbiologist, or how to learn from the microbes. *Annu. Rev. Microbiol.* **47**, 1–29.

Schlegel, H. G., and Bowien, B. (1989). "Autotrophic Bacteria." Science Tech Publishers, Madison, Wisconsin.

Van Gemerden, H. (1993). Microbial mats: A joint venture. *Marine Geol.* **113**, 3–25.

Van Gemerden, H., and Mas, J. (1995). Ecology of phototrophic sulfur bacteria. *In* "Anoxygenic Photosynthetic Bacteria" (R. E. Blankenship, M. T. Madigan, and C. E. Bauer, eds.), pp. 49–85. Kluwer Academic Publishers, Dordrecht, The Netherlands.

Sulfur Cycle

Piet Lens and Look Hulshoff Pol
Wageningen Agricultural University, The Netherlands

Richard Tichy
Institute of Landscape Ecology, Czech Republic

I. Planetary Sulfur Fluxes
II. The Microbial Sulfur Cycle
III. Sulfur Cycling within Ecosystems
IV. Environmental Consequences and Technological Applications

GLOSSARY

acidophiles Bacteria that prefer acidic conditions, that is, microbes with a low pH optimum, typically below pH 4.

assimilatory reduction Reduction of a compound for the purpose of introducing its building elements into cellular material.

dissimilatory reduction Reduction of a compound in an energy-yielding reaction. The element does not become incorporated into cellular material.

sulfate reducing bacteria (SRB) Name of a group of bacteria belonging to a diversity of genera that gain their metabolic energy from the reduction of sulfate to sulfide.

sulfide oxidizing bacteria Name of a group of bacteria belonging to a diversity of genera that gain their metabolic energy from the oxidation of sulfide to sulfate. Under certain conditions, the oxidation is incomplete and stops at elemental sulfur, thiosulfate, or sulfite.

sulfuretum Habitat with a complete sulfur cycle (plural: sulfureta).

sulfur reducing bacteria Name of a group of bacteria belonging to a diversity of genera that gain metabolic energy from the reduction of elemental sulfur to sulfide.

SULFUR CYCLING is a natural environmental process in which sequential transformation reactions interconvert sulfur atoms in different valence states ranging from −2 to +6. The reactions of the sulfur cycle alter the chemical, physical, and biological status of sulfur and its compounds so that sulfur cycling can occur. Many of the reactions of the sulfur cycle are mediated by microorganisms. The transformation reactions involved represent a continuous flow of sulfur-containing compounds among Earth's various compartments (soil, water, air, and biomass). Besides cycling of sulfur on a macroscale, internal sulfur cycles within one compartment also exist. These internal cycles depend on gradients of oxygen and the sulfur compounds, which can occur on a large scale (e.g., in stratified lakes) or on a micrometer scale (e.g., in laminated mats and wastewater treatment biofilms). Disrupting of the sulfur cycle can lead to several serious environmental problems. On the other hand, sulfur biotransformations are the basis of a whole set of environmental bioremediation technologies.

I. PLANETARY SULFUR FLUXES

Sulfur is the eighth most abundant element in the solar atmosphere and the fourteenth most abundant element in Earth's crust. Sulfur is present in several of the large environmental compartments present on Earth (Table I). The lithospheric compartment is the largest and contains roughly 95% of this element. The second largest compartment is the hydrosphere, and Earth's oceans contain approximately 5% of the total sulfur. Sulfate is the second most abundant anion in seawater. The other compartments in which sulfur is found together comprise <0.001% of the remaining sulfur.

Within the frame of global biogeochemical cycling (Fig. 1), sulfur is transformed with respect to its oxidation state, formation of organic and inorganic compounds, and its physical status (gas, liquid, or

TABLE I
Amount of Sulfur Contained in Various Components of Earth

Component	*Amount of sulfur (kg)*
Atmosphere	4.8×10^{9}
Lithosphere	2.4×10^{19}
Hydrosphere	
Sea	1.3×10^{18}
Freshwater	3.0×10^{12}
Marine organisms	2.4×10^{11}
Pedosphere	
Soil	2.6×10^{14}
Soil organic matter	1.0×10^{13}
Biosphere	8.0×10^{12}

solid; soluble or insoluble). In oxidizing conditions, the most stable sulfur species is sulfate. In reducing environments, sulfide is formed. Sulfide and some organic compounds containing reduced sulfur, for example, dimethyl sulfide (DMS), carbonyl sulfide (COS), and carbonyl disulfide (CS_2), are volatile and can escape to the atmosphere. From there, sulfur compounds are redeposited into the litho-, hydro-, and pedospheres, either directly or after conversion to sulfate, via the gaseous intermediate SO_2.

Sulfide has a high chemical reactivity with some metal cations, leading to the formation of poorly soluble metal sulfides. This results in an accumulation of solid-state reduced sulfur stocks within global planetary sulfur cycling, owing to the formation of insoluble metal sulfides in most anaerobic environments, such as marshes, wetlands, freshwater, and sea sediments all over the globe. The formation of solid-state reduced sulfur proceeded even in the early history of planetary biogeochemical cycling. Accumulation of sulfur in anaerobic, highly organic-rich deposits of biomass resulted in contamination of coal by metal sulfides as well as organically bound sulfur compounds (sulfur content ranging between 0.05 and 15.0%). Similarly, mineral oils and petroleum can contain substantial amounts of sulfur com-

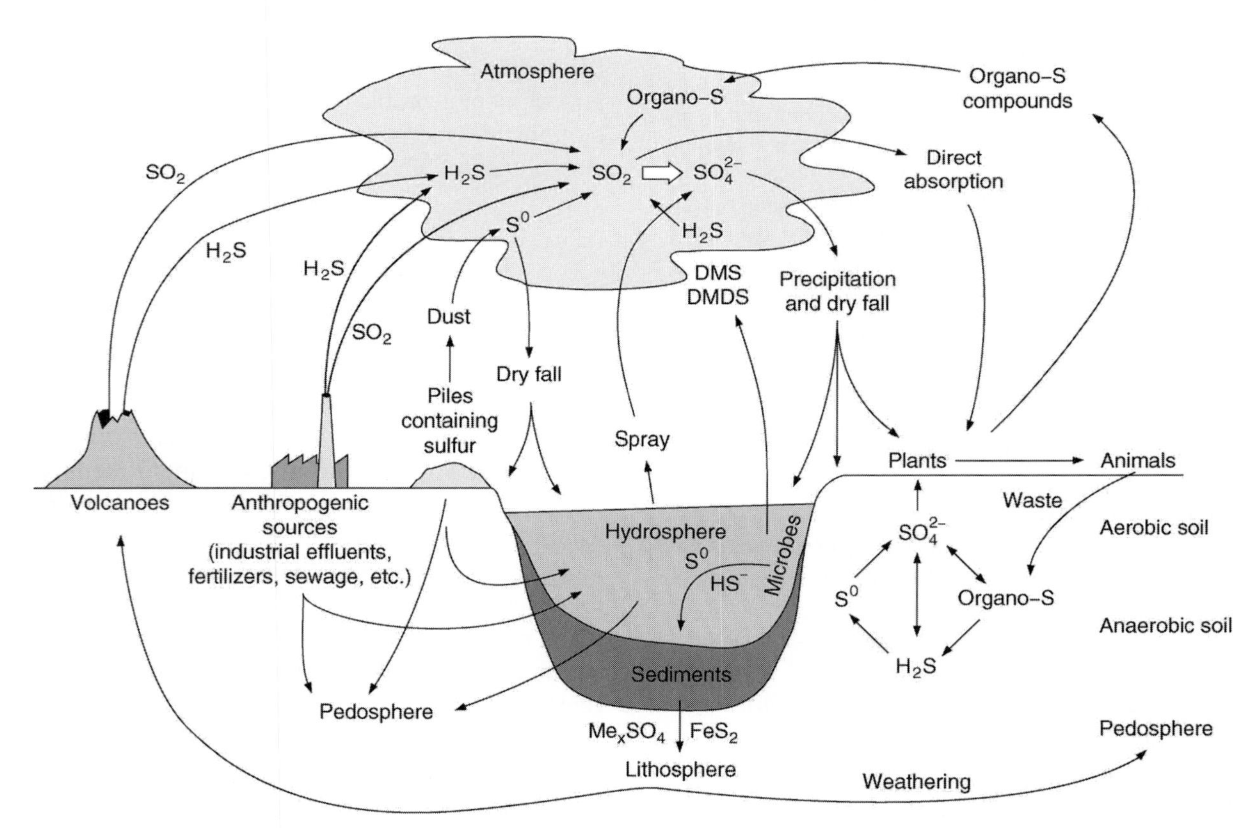

Fig. 1. Simplified version of the overall sulfur cycle in nature.

pounds (0.025–5%). Under certain conditions, large—commercially exploitable—quantities of elemental sulfur can accumulate in petroleum reservoirs (see Section III). Other stocks of accumulated solid-state reduced sulfur are the sulfidic ores, for example, pyrite (FeS_2), covellite (CuS), chalcopyrite ($CuFeS_2$), galena (PbS), and sphalerite (ZnS).

Increasing anthropogenic extraction of sulfur-containing compounds from the lithosphere considerably perturbs the global sulfur cycle. The most important anthropogenic flux is the emission of sulfur into the atmosphere (113 Tg S/year, i.e., 113×10^{12} g S/year). In the continental part of the sulfur cycle, this flux is comparable only with weathering (114.1 Tg S/year) and river runoff to world oceans (108.9 Tg S/year). The second major anthropogenic flux of sulfur is the pollution of rivers, and subsequent runoff of the river waters (104 Tg S/year). The anthropogenic sulfur inputs on the globe result in an acceleration of sulfur cycling. This is manifested in elevated levels of sulfate in runoff waters, buildup of sulfide in anaerobic environments, and, after exposure of reduced-sulfur stocks to air, acidification of the environment and leaching of toxic metals (see Section IV). At the same time, the increased anthropogenic emissions to the atmosphere bring about other adverse environmental effects, for example, acid deposition. The negative consequences of acid rain are well known and include, for example, extensive damage to forests and wildlife and detrimental effects on buildings, constructions, and artifacts such as art works, statues, etc.

II. THE MICROBIAL SULFUR CYCLE

The behavior of sulfur compounds in the environment is highly influenced by the activity of living organisms, particularly microbes. In Fig. 2, the stocks of sulfur with different oxidation status (marked by squares) are given: S^{2-} is the sulfide form, S^0 is elemental sulfur, SO_4^{2-} is sulfate, and C-SH represents the stock of organic sulfur compounds. Arrows indicate the trophic status of microbes in each process, distinguishing autotrophic (using inorganic CO_2) and heterotrophic (using organic carbon compounds, C_{Org}).

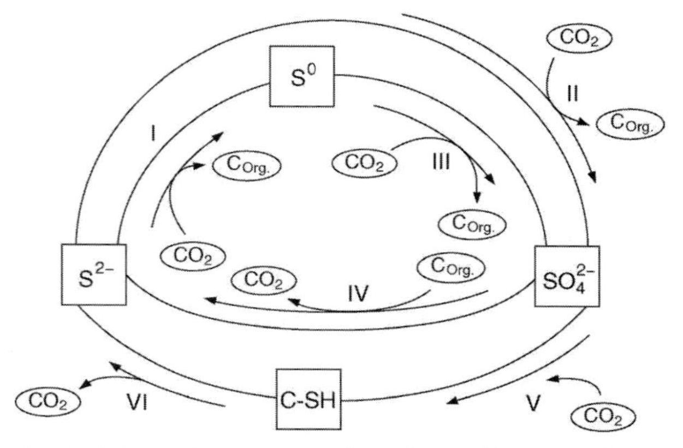

Fig. 2. Schematic representation of the different pathways occurring in the microbial sulfur cycle. After Tichý *et al.* (1998).

Since the 1980s, the ecology of bacteria with a role in the sulfur cycle has received considerable attention from different scientific fields (e.g., microbial mats and sediments, wastewater treatment biofilms, and corrosion). This involved various advanced analytical techniques, for example, quantification of reaction products at the micrometer scale using microelectrodes for oxygen, sulfide, pH, and glucose; determination of metabolic and transport processes using ^1H, ^{13}C, and ^{31}P nuclear magnetic resonance (NMR) techniques, and studies of population dynamics using 16S ribosomal RNA (rRNA) based detection methods.

A. Microbial Sulfate Reduction

Sulfate salts are the major stock of mobile sulfur compounds. They are mostly highly soluble in water, and considerable amounts can be transported in the environment. In the microbial sulfur cycle, sulfate is converted to sulfide by sulfate reducing bacteria (SRB) via dissimilatory sulfate reduction (pathway I in Fig. 2). This process of bacterial respiration occurs under strictly anaerobic conditions and uses sulfate as terminal electron acceptor. Electron donors are usually organic compounds, eventually, hydrogen:

$$8H_2 + 2SO_4^{2-} \rightarrow H_2S + HS^- + 5H_2O + 3OH^- \quad (1)$$

SRB include the traditional sulfate reducing genera *Desulfovibrio* and *Desulfotomaculum* in addition to the morphologically and physiologically different genera

Desulfobacter, Desulfobulbus, Desulfococcus, Desulfonema, and *Desulfosarcina* (Widdel, 1988). In the presence of sulfate, SRB are able to use several intermediates of the anaerobic mineralization process. Besides the direct methanogenic substrates molecular hydrogen (H_2), formate, acetate, and methanol, they can also use propionate, butyrate, higher and branched fatty acids, lactate, ethanol and higher alcohols, fumarate, succinate, malate, and aromatic compounds. In sulfidogenic breakdown of volatile fatty acids, two oxidation patterns can be distinguished. Some SRB are able to completely oxidize volatile fatty acids to CO_2 and sulfide as end products. Other SRB lack the tricarboxylic acid cycle and carry out an incomplete oxidation of volatile fatty acids, with acetate and sulfide as end products.

In addition to the reduction of sulfate, reduction of sulfite and thiosulfate is also very common among SRB. *Desulfovibrio* strains have been reported to be able to reduce di-, tri-, and tetrathionate. A unique ability of some SRB, for example, *Desulfovibrio dismutans* and *Desulfobacter curvatus,* is the dismutation of sulfite or thiosulfate:

$$4SO_3^{2-} + H^+ \rightarrow 3SO_4^{2-} + HS^- \quad (2)$$

$$S_2O_3^{2-} + H_2O \rightarrow SO_4^{2-} + HS^- + H^+ \quad (3)$$

Some SRB were found to be able to respire oxygen, despite being classified as strictly anaerobic bacteria. Thus far, however, aerobic growth of pure cultures of SRB has not been demonstrated. The ability of SRB to carry out sulfate reduction under aerobic conditions remains nevertheless intriguing and could be of significance for microscale sulfur cycles (see Section III).

In the absence of an electron acceptor, SRB are able to grow through a fermentative or acetogenic reaction (Widdel, 1988). Pyruvate, lactate, and ethanol are easily fermented by many SRB. An interesting feature of SRB is their ability to perform acetogenic oxidation in syntrophy with hydrogenotrophic methanogenic bacteria, as described for cocultures of hydrogenotrophic methanogenic bacteria with *Desulfovibrio* sp. using lactate and ethanol or with *Desulfobulbus*-like bacteria using propionate. In the presence of sulfate, however, these bacteria behave as true SRB and metabolize propionate as the electron donor for the reduction of sulfate.

B. Microbial Sulfur Oxidation

Different bacteria can oxidize various reduced sulfur compounds, for example, sulfide, elemental sulfur, or thiosulfate. Oxidation of sulfide to elemental sulfur (pathway II in Fig. 2) is performed by autotrophic bacteria. Equation (4) gives the stoichiometry of the chemoautotrophic process, which proceeds aerobically or microaerobically. In addition, photoautotrophic sulfide oxidation can also occur under anaerobic conditions. Photosynthetic sulfur bacteria are capable of photoreducing CO_2 while oxidizing H_2S to S^0 [Eq. (5)], in a striking analogy to the photosynthesis of eukaryotes [Eq. (6)].

$$2H_2S + O_2 \rightarrow 2S^0 + 2H_2O \quad (4)$$

$$2H_2S + CO_2 + h\nu \rightarrow 2S^0 + [CH_2O] + H_2O \quad (5)$$

$$2H_2O + CO_2 + h\nu \rightarrow 2O_2 + [CH_2O] \quad (6)$$

Sulfide can also be completely oxidized to sulfate. Equation (7) gives a formula for the chemoautotrophic process, although photoautotrophic oxidation of sulfide to sulfate can also occur.

$$H_2S + 2O_2 \rightarrow SO_4^{2-} + 2H^+ \quad (7)$$

This oxidation reaction, catalyzed by, for example, *Thiobacillus,* involves a series of intermediates, including sulfide, elemental sulfur, thiosulfate, tetrathionate, and sulfate (Kelly *et al.,* 1997):

$$SH^- \rightarrow S^0 \rightarrow S_2O_3^{2-} \rightarrow S_4O_6^{2-} \rightarrow SO_4^{2-} \quad (8)$$

Eventually, oxidation of sulfide may proceed in oxygen-free conditions, using nitrate as the electron acceptor [Eq. (9)]. Oxidation of sulfide in anoxic conditions is mostly carried out by bacteria from the genus *Thiobacillus,* such as *T. albertis* and *T. neapolitanus.*

$$0.422\,H_2S + 0.422\,HS^- + NO_3^- + 0.437\,CO_2 + 0.0865\,HCO_3^- + 0.0865\,NH_4^+ \rightarrow 1.114\,SO_4^- + 0.5\,N_2 + 0.0842\,C_5H_7O_2N\ (\text{biomass}) + 1.228\,H^+ \quad (9)$$

Elemental sulfur can be oxidized to sulfate via chemoautotrophic or photoautotrophic microorganisms (Fig. 2, pathway IV). The stoichiometry of the chemoautotrophic process is given in Eq. (10).

$$2\,S^0 + 3O_2 + 2H_2O \rightarrow 2SO_4^{2-} + 4\,H^+ \quad (10)$$

The biological oxidation of reduced sulfur is mediated by a diverse range of bacterial species (Table II). They can be divided into two main groups: the aerobic and microaerobic chemotrophic sulfur oxidizers (sometimes called the colorless sulfur bacteria) and the anaerobic phototrophic sulfur oxidizers (sometimes called the purple and green sulfur bacteria).

The most common microorganisms associated with sulfide oxidation are *Thiobacillus* spp. These non-spore-forming bacteria belong to the colorless sulfur bacteria. Thiobacilli are gram-negative rods about 0.3 μm in diameter and 1–3 μm long. Most *Thiobacillus* species are motile by polar flagella. All *Thiobacillus* species grow aerobically, although anaerobic growth [see Eq. (9)] has been observed for some species as well. Elemental sulfur accumulates on the cell surface, in contrast to filamentous colorless sulfur bacteria, for example, *Thiothrix* sp., *Beggiatoa* sp., and *Thioploca* sp., which accumulate elemental sulfur intracellularly. *Thiothrix* spp. are commonly found in flowing sulfidic water containing oxygen in both marine and freshwater environments such as outlets of sulfidic springs and wastewater-treatment plants. They differentiate ecologically from other filamentous colorless sulfur bacteria in that *Thiothrix* spp. prefer hard substrates to which they attach with a holdfast, whereas *Beggiatoa* spp. and *Thiploca* spp. do not attach and prefer soft bottom sediments. The diameter of the filaments is variable in all three genera and is used to define species.

The phototrophic bacteria make up the Chlorobiaceae (green sulfur bacteria), the Chromatiaceae (purple sulfur bacteria), and the filamentous thermophilic flexibacteria, examplified by *Chloroflexus aurantiacus*. Phototrophic bacteria contribute substantially to both the sulfur cycle and the primary productivity of shallow aquatic environments, but

TABLE II
Sulfide and Sulfur Oxidizing Bacteria

Genus or group	*Habitat*	*Comments*
Chemotrophs		
Thiobacillus	Soil, water, marine	Mostly lithotrophs, one Fe(II) oxidizer, some thermophiles, deposit S^0 outside cell
Sulfobacillus	Mine tips	Lithotroph, spore-former, Fe(II) oxidizer, thermophilic
Thiomicrospira	Marine	Lithotroph
Beggiatoa	Water, soil, marine	Gliding bacteria, deposit S^0 inside cell, difficult to isolate
Thiothrix	Water, soil, marine	
Thioploca	Water, soil	
Achromatium	Water	
Thiobacterium	Water	Deposit S^0 inside cell, not grown in pure cultures
Macromonas	Water	
Thiovulum	Water	
Thiospira	Water	
Sulfolobus	Geothermal springs	Archaebacterium, lithotroph, Fe(II) oxidizer, thermophilic
Phototrophs		
Chlorobiaceae (green S bacteria)	Water, marine	Lithotrophs, deposit S^0 outside cell
Chromatiaceae (purple S bacteria)	Water, marine	Mixotrophs, deposit S^0 inside cell, except for *Ectothiorhodospira* spp. which deposit S^0 outside cell
Chloroflexaceae	Geothermal springs	Mixotrophs, deposit S^0 outside cell, thermophiles
Oscillatoria (blue-green algae/ cyanobacteria)	Water	S^0 deposited outside cell under anoxic conditions

they are less significant in the mid-ocean where they are limited by light availability.

In addition to the specialized bacteria listed in Table II, a considerable number of common bacteria (e.g., *Bacillus* spp., *Pseudomonas* spp., and *Arthrobacter* spp.) and fungi (e.g., *Aspergillus* spp.) have also been shown to oxidize significant amounts of reduced sulfur compounds when grown in pure culture. Even if such transformations are incidental, these organisms are present in large numbers in soils and the aquatic environment compared to *Thiobacillus* spp., suggesting that nonlithotrophic sulfur oxidation may be quantitatively important.

Species of *Thiobacillus*, as the main representatives of the acidophiles, play a key role in the degradation of sulfidic materials (Johnson *et al.*, 1993). In addition to the oxidation of reduced sulfur compounds as their energy source [see Eqs. (7) and (10)], they can also gain energy from the conversion of ferrous (Fe^{2+}) to ferric (Fe^{3+}) iron. The best known acidophile is *Thiobacillus ferrooxidans*, which combines the ability to oxidize both sulfur compounds and ferrous iron. The bacterium is able to oxidize, at low pH values (pH 1–4), sulfidic minerals such as pyrite according to the following reaction [Eq. (11)]:

$$4FeS_2 + 15O_2 + 2H_2O \rightarrow 2Fe_2(SO_4)_3 + 2H_2SO_4 \quad (11)$$

Thiobacillus ferrooxidans was isolated from acid mine drainage water in the late 1940s, together with *T. thiooxidans*. The latter species can only oxidize sulfur and reduced sulfur compounds and lacks the ferrous iron-oxidizing capacity. In contrast to *T. ferrooxidans, T. thiooxidans* cannot attack sulfidic minerals on its own, but it can contribute to their solubilization in a syntrophic relation with ferrous iron oxidizers such as *Leptospirillum ferrooxidans*. The latter microorganism can only convert ferrous iron and has no sulfur-oxidizing capacity.

The group of acidophiles also comprises facultative autotrophs or even obligate heterotrophs (Fortin *et al.*, 1996). A representative of this subgroup is *T. acidophilus*, a facultatively autotrophic sulfur compound oxidizer that can also grow on, for example, sugars. Apart from solubilizing sulfidic minerals, facultatively autotrophic acidophiles are also crucial in removing low molecular weight organic acids, which are toxic to the obligate autotrophs even at very low concentrations (5–10 mg/liter). These toxic effects are due to the uptake of organic compounds with carboxylic groups, which are undissociated in the outer medium with low pH but are dissociated in bacterial cytoplasm with circumneutral pH.

C. Transformations of Organic Sulfur Compounds

The formation and degradation of organic sulfur compounds (C-SH) are not solely microbial processes, and numerous other organisms participate in them. Particularly, the formation of organic sulfur (Fig. 2, pathway V) is accomplished by all photosynthesizing organisms, including algae and green plants. Dimethyl sulfide (DMS) is the most common product of oceanic green algae sulfur conversions. Green plants and many microorganisms assimilate sulfate as their sole sulfur source. Therefore, they reduce sulfate via a reductive process, assimilatory sulfate reduction, in which the formed sulfide is incorporated into organic matter via a condensation reaction with serine derivatives to generate the amino acid cysteine.

Conversion of organic sulfur to sulfide occurs during the decomposition of organic matter (Fig. 2, pathway VI). Considerable environmental risks are encountered in these processes, especially regarding the volatilization of organic sulfur compounds and associated odor pollution (Smet *et al.*, 1998).

III. SULFUR CYCLING WITHIN ECOSYSTEMS

The sulfur cycle, together with the carbon and nitrogen cycle, is distinguished from most other mineral cycles (e.g., P, Fe, Si) by exhibiting transformations from gaseous to ionic (aqueous or solid) forms. Each of the transformations (see Section II) is reliant on appropriate cellular and ambient oxygen (or redox) conditions. Thus, cycling of these elements depends on the presence of oxygen (or redox) gradients. These gradients vary in size from the smallest, which can be generated over distances of only a few micrometers, for example, biofilms and sediments,

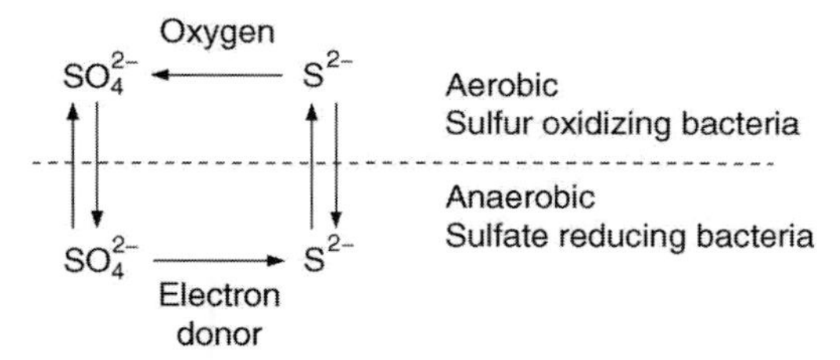

Fig. 3. Schematic representation of the cyclic reactions prevailing in a microbial sulfur cycle between sulfate reducing and sulfide oxidizing bacteria.

to larger zones (a few millimeters) such as in soil crumbs and microbial mats. In exceptional circumstances, these gradients may extend over several tens of meters, for example, in geothermal springs, stable stratified lakes, or sewer outfalls. Of particular interest for the sulfur cycle are the steep gradients present in microenvironmental or microzonal conditions, as they allow interactions between cells that normally cannot coexist, namely, between anaerobic sulfate reducers and aerobic sulfide oxidizers.

In some cases, very neat cyclic reactions are possible. The clearest example is the sulfur cycle as illustrated in Fig. 3. Habitats with a complete sulfur cycle are known as sulfureta. Both groups of bacteria may be found close to the border between the aerobic and anaerobic habitats, and neither organism can grow in the other's space. Yet, they are totally dependent on sulfur compounds that diffuse between them. Such cycles are common in marine or estuarine sediments, and they are also evident in stratified water bodies and fixed film wastewater-treatment systems (Fig. 4).

Figure 4 depicts possible blocks of the cyclic activity, for example, sulfide precipitating with heavy metals (e.g., FeS) or accumulation of elemental sulfur in sulfide oxidizing bacteria. The latter also prevails in natural, light exposed environments, as illustrated in Fig. 5, which depicts some scenarios for the biogenic deposition of elemental sulfur in a lake and in the vicinity of petroleum reservoirs. Formation of colloid/solid elemental sulfur does not necessarily imply a complete blocking of sulfur cycling activity. The sulfur can be reduced to sulfide by, for example, *Desulfovibrio desulfuricans,* thus allowing the sulfur cycle to proceed (Fig. 6).

IV. ENVIRONMENTAL CONSEQUENCES AND TECHNOLOGICAL APPLICATIONS

Disruption of sulfur cycling can also have serious environmental consequences. Below, an overview is

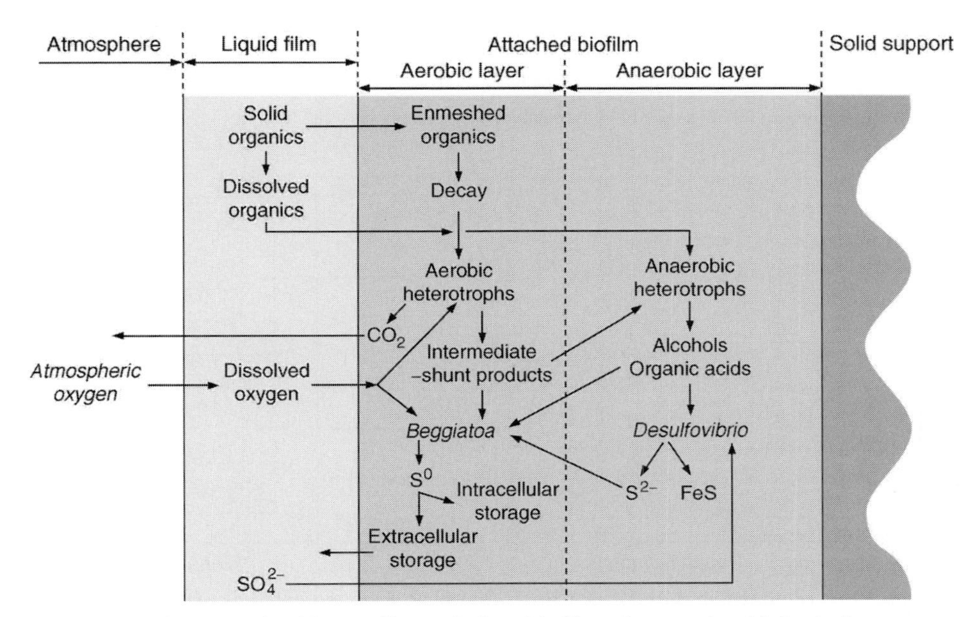

Fig. 4. Reactions involved in a sulfur cycle in a biofilm of a rotating biological contactor type wastewater-treatment system. After Alleman *et al.* (1982).

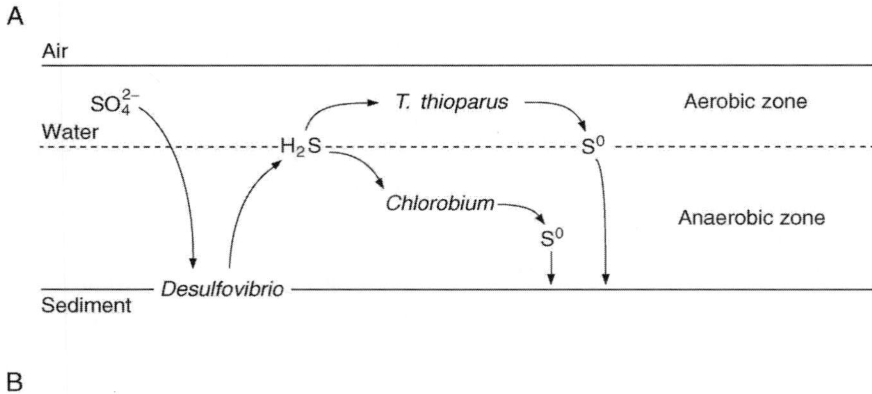

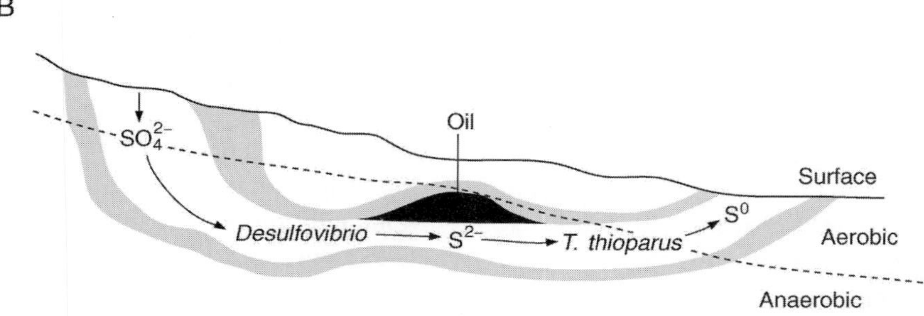

Fig. 5. The biological deposition of sulfur (A) in a lake and (B) in geological strata mediated by the disrupting of sulfur cycling activities.

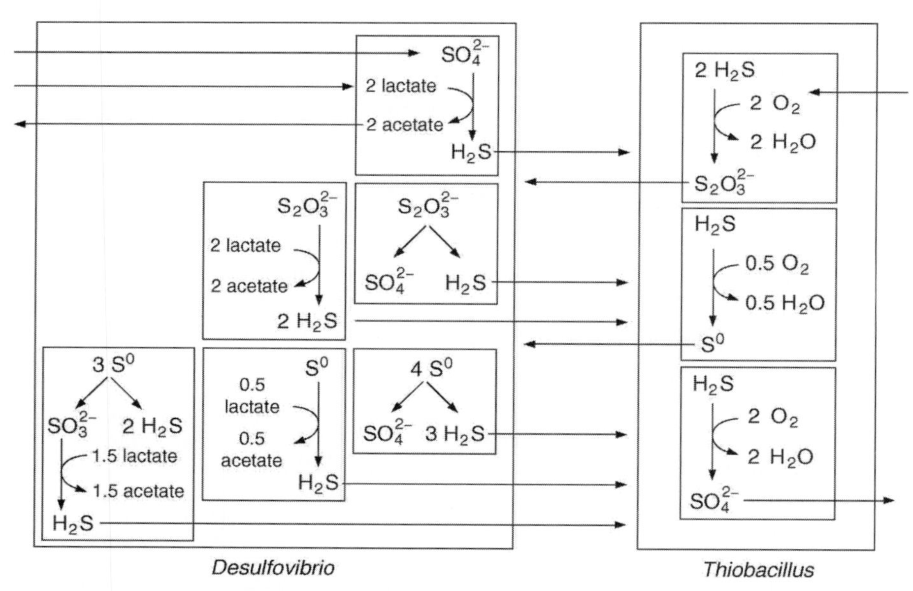

Fig. 6. Possible pathways of sulfur transformations in mixed cultures of *Desulfovibrio* and *Thiobacillus* supplied with lactate, sulfate, and oxygen. After Van den Ende *et al.* (1997).

given of the major environmental effects when the sulfur cycle is unbalanced. This imbalance can be a result of both natural and anthropogenic processes. A number of technological applications that utilize bacteria of the sulfur cycle are also presented.

Many of the volatile organic and inorganic sulfur compounds are extremely odorous. Farmyard feedlots and improperly managed rendering and composting facilities are examples of places where offensive sulfur gases are generated.

A. Environmental Consequences

Microbial transformations of solid-state reduced sulfur compounds that are used or affected by anthropogenic activities represent major environmental risks. Materials such as fossil fuels, ores, anaerobic sediments, or solid waste may undergo oxidative changes, resulting in a solubilization of sulfur from the solid phase. Thus, large amounts of sulfuric acid are formed [see Eq. (11)], which are transported off site in so-called acid mine drainage (Dvorak *et al.*, 1992). The acid inhibits plant growth and aquatic life. Its major environmental consequence is however, the solubilization of cationic heavy metals.

In tropical regions of the world, periodic flooding and draining of maritime estuarine soils leads to the accumulation of reduced sulfur and its subsequent oxidation (Begheijn *et al.*, 1978). This results in production of sulfuric acid and appearance of acid sulfate soils, sometimes called "cat clays." In many cases, treatment (neutralization) of these sites is required, for example, before farming may be allowed.

Sulfate-rich wastewater is generated by many industrial processes that use sulfuric acid or sulfate-rich feed stocks (e.g., fermentation or seafood-processing industries). Also, the use of reduced sulfur compounds in industrial processes, namely, sulfide (tanneries, Kraft pulping), sulfite (sulfite pulping), thiosulfate (fixing of photographs), or dithionite (pulp bleaching), contaminates wastewater with sulfurous compounds.

Pollution by humans has greatly increased the sulfur dioxide levels in the atmosphere. When dissolved in water in the clouds, SO_2 acidifies the rainwater. Acidic rainfall has been implicated in the reduction of growth of forests in North America and Europe. Acid rain also has a detrimental effect on buildings, constructions, and other artifacts such as ancient art works, causing etching and destroying their original beauty.

B. Technological Applications

Hydrometallurgical processes play an important role in the extraction of metals from certain low-grade ores, for example, those with $<1\%$ Cu by weight. Hydrometallurgy consists of the dissolution of metals from minerals, usually by constant percolation of a leaching solution through beds of ore (Bailey and Hansford, 1993). The process is most effective in ores that contain substantial amounts of pyrite and is based on the rapid oxidation of S^{2-} and Fe^{2+} promoted by acidophiles. In the literature, two mechanisms by which acidophiles attack the insoluble metal sulfides are proposed, namely, indirect and direct leaching (Evangelou and Zhang, 1995). In the indirect mechanism, the ferric iron acts as a chemical oxidizer of the sulfidic minerals. The ferric iron is a product of bacterial oxidation. In this mechanism, the biological sulfur-oxidizing capacity of the acidophiles is of no relevance. In the direct mechanism, sulfidic mineral solubilization involves both the biological ferrous iron and sulfur compound oxidation. Pyrite oxidation by *T. ferrooxidans* was shown to be mediated by both the biological oxidation of ferrous iron and sulfide.

Microbial leaching (bioleaching) is also proposed for decontamination of polluted solid wastes, soils, or sediments (Couillard and Mercier, 1992). Successful demonstration of the bioleaching of toxic metals from an anaerobically digested sewage sludge was demonstrated on a technological scale. Experimental attempts to use bioleaching for heavy metal removal from freshwater sediments are reported as well. The process may use the autoacidification potential of the sediment, driven by the presence of reduced sulfur and ferrous compounds. If the latter is insufficient relative to the sediment's buffering capacity, additional sulfur can be added to achieve satisfactory extraction yields.

Reduced sulfur compounds are commonly added to alkaline soils so that the sulfuric acid produced

TABLE III
Environmental Biotechnological Applications Using Processes of the Microbial Sulfur Cycle

Application	Sulfur conversion utilized	Waste stream
Wastewater treatment		
Sulfate removal	Sulfate and/or sulfite reduction + partial sulfide oxidation to S^0	Industrial wastewaters, acid mine drainage, and spent sulfuric acid
Sulfide removal	Partial sulfide oxidation to S^0	Industrial wastewaters
Heavy metal removal	Sulfate reduction	Extensive treatment in wetlands or anaerobic ponds
		High rate reactors for process water, acid mine drainage, and groundwater
Microaerobic treatment	Internal sulfur cycle in the biofim	Domestic sewage
Off-gas treatment		
Biofiltration of gases	Oxidation of sulfide and organosulfur compounds	Biogas, malodorous gases from composting and farming
Treatment of scrubbing waters	Sulfate and/or sulfite reduction + partial sulfide oxidation to S^0	Scrubbing waters of SO_2-rich gases
Solid waste treatment		
Bioleaching of metals	Sulfide oxidation	Sewage sludge, compost
Desulfurization	Oxidation	Rubber
Gypsum processing	Sulfate reduction	
Treatment of soils and sediments		
Bioleaching of metals	Sulfide oxidation	Dredged sediments and spoils
Degradation xenobiotics	Sulfate reduction	Polychlorinated bifenyl (PCB)-contaminated soil slurries

decreases the soil pH to a level acceptable for plant growth. Moreover, sulfur compounds can be used as fertilizers to improve crop production in those areas of the world where sulfur is a limiting nutrient. This applies, for example, in Western European countries, where atmospheric deposition of sulfur to farmland decreased as a result of the stringent control of sulfur dioxide emissions.

Better insight into sulfur transformations enabled the development of a whole spectrum of new biotechnological applications for the bioremediation of polluted waters, gases, soils, and solid wastes (Table III). These technologies rely on both bacterial sulfate reduction and sulfide/sulfur oxidation. They allow the removal of sulfur and organic compounds as well as heavy metals and nitrogen. Until recently, biological treatment of sulfur-polluted wastestreams was rather unpopular because of the production of hydrogen sulfide under anaerobic conditions. Gaseous and dissolved sulfides cause physiochemical (corrosion, odor, increased effluent chemical oxygen demand) or biological (toxicity) constraints, which may lead to process failure. However, anaerobic treatment of sulfate-rich wastewater can be applied successfully provided a proper treatment strategy is selected (Lens *et al.*, 1998), which depends on the aim of the treatment: (i) removal of sulfur compounds, (ii) removal of organic matter, or (iii) removal of both.

Microbial sulfur transformations can provide a unique tool to control pollution by both sulfur compounds and heavy metals. In some cases, rather extensive techniques using natural processes are applied, for example, the use of wetlands or anaerobic ponds to treat voluminous aqueous streams like acid mine drainage or surface runoff waters. These systems require low maintenance and can sustain their function for prolonged time intervals. Subsequent regeneration of these systems using bioleaching and treatment of spent extraction liquor by SRB may be a logical step. In general, technological applications using various processes of the microbial sulfur cycle may provide many beneficial effects in the future.

See Also the Following Articles

Heavy Metal Pollutants • Nitrogen Cycle • Ore Leaching by Microbes • Wastewater, Industrial

Bibliography

Alleman, J. E., Veil, J. A., and Canaday, J. T. (1982). Scanning electron microscope evaluation of rotating biological biofilm. *Wat. Res.* **16**, 543–550.

Bailey, A. D., and Hansford, G. S. (1993). Factors affecting bio-oxidation of sulfide minerals at high concentrations of solids—A review. *Biotech. Bioeng.* **42**, 1164–1174.

Begheijn, L. T., van Breemen, N., and Velthorst, E. J. (1978). Analysis of sulfur compounds in acid sulfate soils and other marine soils. *Commun. Soil Sci. Plant Anal.* **9**, 873–882.

Couillard, D., and Mercier, G. (1992). Metallurgical residue for solubilization of metals from sewage sludge. *J. Environ. Eng.* **118**, 808–813.

Dvorak, D. H., Hedin, R. S., Edenborn, H. M., and McIntire, P. E. (1992). Treatment of metal-contaminated water using bacterial sulfate reduction: Results from pilot-scale reactors. *Biotech. Bioeng.* **40**, 609–616.

Evangelou, V. P., and Zhang, Y. L. (1995). Pyrite oxidation mechanisms and acid mine drainage prevention. *Crit. Rev. Environ Sci. Technol.* **25**, 141–199.

Fortin, D., Davis, B., and Beveridge, T. J. (1996). Role of *Thiobacillus* and sulfate-reducing bacteria in iron biocycling in oxic and acidic mine tailings. *FEMS Microbiol. Ecol.* **21**, 11–24.

Johnson, D. B., McGinness, S., and Ghauri, M. A. (1993). Biogeochemical cycling of iron and sulfur in leaching environments. *FEMS Microbiol. Rev.* **11**, 63–70.

Kelly, D. P., Shergill, J. K., Lu, W.-P., and Wood, A. P. (1997). Oxidative metabolism of inorganic sulfur compounds by bacteria. *Antonie van Leeuwenhoek* **71**, 95–107.

Lens, P. N. L., Visser, A., Janssen, A. J. H., Hulshoff Pol, L. W., and Lettinga, G. (1998). Biotechnological treatment of sulfate rich wastewaters. *Crit. Rev. Environ. Sci. Technol.* **28**, 41–88.

Smet, E., Lens, P., and van Langenhove, H. (1998). Treatment of waste gases contaminated with odorous sulfur compounds. *Crit. Rev. Environ. Sci. Technol.* **28**, 89–116.

Tichý, R., Lens, P., Grotenhuis, J. T. C., and Bos, P. (1998). Solid-state reduced sulfur compounds: Environmental aspects and bioremediation. *Crit. Rev. Environ. Sci. Technol.* **28**, 1–40.

Van den Ende, F. P., Meier, J., and Van Gemerden, H. (1997). Syntrophic growth of sulfate-reducing bacteria and colorless sulfur bacteria during oxygen limitation. *FEMS Microbiol. Ecol.* **23**, 65–80.

Widdel, F. (1988). Microbiology and ecology of sulfate- and sulfur reducing bacteria. *In* "Biology of Anaerobic Microorganisms" (A. J. B. Zehnder, ed.), pp. 469–586. Wiley, New York.

Surveillance of Infectious Diseases

Robert W. Pinner, Denise Koo, and Ruth L. Berkelman
Centers for Disease Control and Prevention

I. Introduction
II. Elements of Public Health Surveillance of Infectious Diseases
III. Surveillance Systems around the World
IV. Challenges and Issues in Infectious Diseases Surveillance
V. Subtyping and the Application of Molecular Epidemiologic Methods
VI. Conclusions

GLOSSARY

active surveillance Surveillance method in which health officials or other persons conducting surveillance take measures to ensure reporting, for example, by collecting the data themselves or regularly reminding providers to report. In general, this approach has the advantage of providing more complete and better data, but it is more time-consuming and resource-intensive than passive surveillance.

case definition Specifies the event to be counted in surveillance, thereby providing uniform criteria for reporting of cases and improving ability to compare rates of disease from place to place and over time. Components of a case definition may vary depending on the characteristics of the disease or condition, and the particular purpose of surveillance.

passive surveillance Surveillance method that relies on reporting of cases from health care providers (e.g., physicians, clinical laboratories), based on a list of conditions. In general, this approach has the advantage of being simple and inexpensive to implement, but it may be of limited completeness of reporting, and variable in quality and timeliness.

population-based surveillance Surveillance in which the cases reported occur in a defined population, which permits calculation of incidence rates. Note that population-based surveillance may be passive or active.

sentinel surveillance Involves monitoring of key or "sentinel" health events in the general population or in special populations, which may be an early warning or represent the "tip of the iceberg." In general, the intent of sentinel surveillance is to provide rapid, relatively inexpensive information rather than to develop precise incidence measurements. Also used to mean a network of sentinel reporters, information from whom serves the purposes of sentinel surveillance.

PUBLIC HEALTH SURVEILLANCE is the ongoing, systematic collection, analysis, and interpretation of data essential to the planning, implementation, and evaluation of public health practice. For surveillance efforts to be complete, data should be synthesized, disseminated in a timely way, and integrated closely with decision making about prevention and control.

I. INTRODUCTION

Surveillance of infectious diseases has been a cornerstone of preventing and controlling communicable diseases of public health significance throughout the twentieth century. Antecedents of current surveillance can be found in the recording of deaths during the eighteenth and nineteenth century. In 1776, Johann Peter Frank advocated more extensive monitoring of health in Germany to support public efforts to protect the health of schoolchildren and to provide water and sewage disposal. In the mid-1800s, William Farr in Great Britain proposed using vital statistics to take effective public health action. In 1899 the United Kingdom began mandatory notification of selected infectious diseases. The United

States also began compulsory notification of infectious diseases in the late 1800s, and by 1925 all states were reporting weekly to the U.S. Public Health Service.

The dependence of the polio control program on surveillance in the 1950s focused attention on the critical role of reporting. Specifically, in 1955, several recipients of polio vaccine developed acute poliomyelitis, threatening national vaccination programs which were then in their early years. The U.S. developed an intensive national surveillance system, which assisted in demonstrating that the problem was limited to a single manufacturer of the vaccine, allowing the vaccination program to continue.

Surveillance was also the foundation for the successful global campaign to eradicate smallpox, with progressively more intensive surveillance for cases conducted as the goal of eradication was approached. The program of eradication depended on information from surveillance to enable targeting vaccinations specifically in limited areas. Reporting sources were established in schools and among agricultural extension workers and others, and towards the end of the campaign house-to-house searches were conducted in the few areas where endemic smallpox was still suspected.

More recently, surveillance data have been critical to the public health response to the AIDS epidemic, assisting with identification of modes of transmission and population groups at risk of infection. In the late 1980s, surveillance enabled researchers to detect the resurgence of measles in the United States; as a result, national immunization efforts were strengthened and recommendations for immunization were modified from a one-dose to a two-dose schedule using combined measles-mumps-rubella (MMR) vaccine (Fig. 1).

Yet, in the 1970s and 1980s, surveillance efforts lagged in the U.S. and globally as public health authorities perceived infectious diseases to be a receding threat (Berkelman *et al.*, 1994). The system of surveillance for drug-resistant tuberculosis ceased in 1986 and was not reinstituted until the early 1990s when the problem had become self-evident. Continued surveillance with earlier detection of, and response to, the increase in multi-drug-resistant tuber-

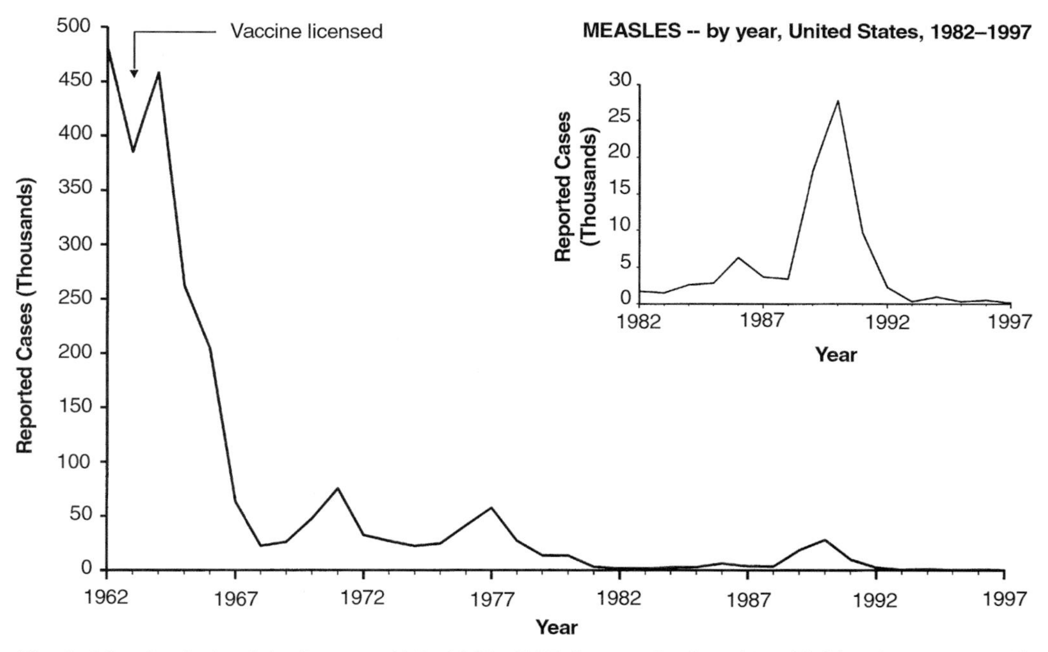

Fig. 1. Measles (rubeola)—by year, U.S., 1962–1997. Source: National Notifiable Diseases Surveillance System.

culosis would have saved considerable public health resources and human suffering. Further, even as new diseases such as *Escherichia coli* O157:H7 infection and cryptosporidiosis were recognized, surveillance was not undertaken in most states in the U.S. or in other countries during the mid-1980s and early 1990s until epidemiologic investigations revealed that outbreaks were often going unrecognized or detected late.

Recognition of the persistence and emergence of infectious diseases by the United States and globally has stimulated increased emphasis on surveillance. In the U.S., strengthening public health surveillance and response became a foundation for addressing threats: Traditional systems of notifiable diseases have been improved and new systems of early detection of infectious diseases have been initiated. Global recognition of the continued importance of public health surveillance to address infectious diseases and threats of emergence led to the enactment of a World Health Organization (WHO) resolution in 1995 to strengthen and coordinate global surveillance and control of communicable diseases, working in conjunction with its network of Collaborating Centers. During 1997, detection of cases of influenza A (H5:N1) in Hong Kong, coupled with reporting, investigation, and prompt institution of control measures, may have prevented the spread of this disease and did focus attention on the need for improved preparedness for pandemic influenza, including surveillance as an integral component.

II. ELEMENTS OF PUBLIC HEALTH SURVEILLANCE OF INFECTIOUS DISEASES

A. Purposes and Uses of Public Health Surveillance

Public health surveillance serves a variety of purposes (see Box). Although surveillance data have common uses at the local, state, and federal public health levels, different governmental levels have differing emphases. For example, individual case investigation is often more critical at the local and state levels than at the federal level, while evaluation of larger scale prevention and control measures (e.g., the impact of new vaccines) is a high priority at the federal level.

B. Sources of Data for Surveillance

Fulfilling the objectives of public health surveillance requires gathering information from numerous sources of data. Public health relies on reports from components of the health care system, e.g., health care providers and laboratories, which provide individual reports of reportable diseases such as measles, syphilis, and *E. coli* O157:H7 infection. Other sources of medical data useful for surveillance include vital statistics (birth and death certificates), hospital discharge data bases, and medical records of ambulatory patients and emergency department encounters. In

Purposes and Uses of Public Health Surveillance

- estimating the magnitude of a public health problem and predicting future trends
- determining the geographic distribution of an illness
- portraying the natural history of a disease
- detecting epidemics or defining a problem
- generating hypotheses and stimulating research
- evaluating control measures
- monitoring changes in infectious agents (e.g., development of antibiotic resistance)
- identifying individual cases for treatment of a communicable disease
- tracking contacts for prophylactic therapy
- detecting changes in health practices
- facilitating planning

addition, public health surveillance may utilize data from outside the traditional health care system. Information about the seroprevalence of infection or behavioral risk factors (such as eating undercooked beef or raw eggs) are collected most effectively from periodic surveys. Collecting serologic or disease data from animals or vectors may constitute part of surveillance for zoonotic and vector-borne diseases, such as rabies or the arboviral encephalitides. For any given disease, a complete picture of the problem may require combining information from a variety of sources.

The evolving health care system in many developed countries may impact on surveillance systems for infectious diseases. For example, anticipated decreases in laboratory testing for diarrheal illnesses may result in decreased notification of enteric illnesses, such as salmonellosis and shigellosis, which in turn may result in decreased or delayed detection of outbreaks. In addition, using specimens for rapid testing may lead to a lack of available specimens for public health or other laboratories to perform molecular typing or antimicrobial resistance testing helpful in evaluating which prevention and control measures are the most effective.

One recent example demonstrates the importance of laboratory diagnosis in detecting outbreaks. In 1994, a nationwide outbreak of salmonellosis affected nearly a quarter of a million people who ate ice cream from a nationally distributed product. This nationwide outbreak was identified after a cluster of cases was recognized in southeastern Minnesota. During the investigation, only 150 culture-confirmed cases associated with the outbreak were reported in Minnesota; 593 were reported nationwide. Nationwide, these 593 culture-confirmed cases represented only about 0.3% of the estimated 224,000 cases that occurred. Detection of even this very large outbreak rested on laboratory confirmation and serotyping of this small proportion of cases; it could easily have gone unnoticed.

Examples of useful surveys in the United States include population surveys such as the National Health Interview Survey, the National Health and Nutrition Examination Survey (NHANES), or the Behavioral Risk Factor Survey, or provider-based surveys such as the National Hospital Discharge Survey or the National Ambulatory Medical Care Survey. NHANES includes blood samples from respondents and thus can provide information about the seroprevalence of infection due to diseases such as hepatitis C.

Vital statistics such as death certificate data or other-sources of data not expressly collected for infectious diseases surveillance can nonetheless provide important information. A study of death certificate data (Armstrong *et al.*, 1999) elucidated trends in infectious diseases mortality during the twentieth century (Fig. 2). Hospital discharge data have also been used, for example, to characterize the burden of rotavirus diarrhea in the U.S.

C. Case Definitions for Public Health Surveillance

The case definition specifies the event to be counted and thus is a key element of any surveillance system. Case definitions provide uniform criteria for reporting cases to improve the ability to compare rates of disease over time and from place to place. The components of a case definition will vary depending on the characteristics of the disease or condition, the role and availability of laboratory testing, the method of reporting or data collection, and the purpose of surveillance. A surveillance case definition may be constructed from clinical or laboratory data, or both in combination. For example, isolation of *Neisseria meningitidis* from a normally sterile site (e.g., blood or cerebrospinal fluid) might define a case of meningococcal disease for a laboratory-based surveillance system; however, for surveillance during an epidemic of meningococcal disease in a developing country without laboratory resources, a clinical case definition (e.g., fever and a stiff neck) would be more appropriate after initial cases were cultured and confirmed. The case definition may also depend on the epidemiologic circumstances; for example, surveillance for childhood vaccine-preventable diseases and food-borne diseases may include criteria such as exposure to probable or confirmed cases of disease or to a point source of infection to define cases.

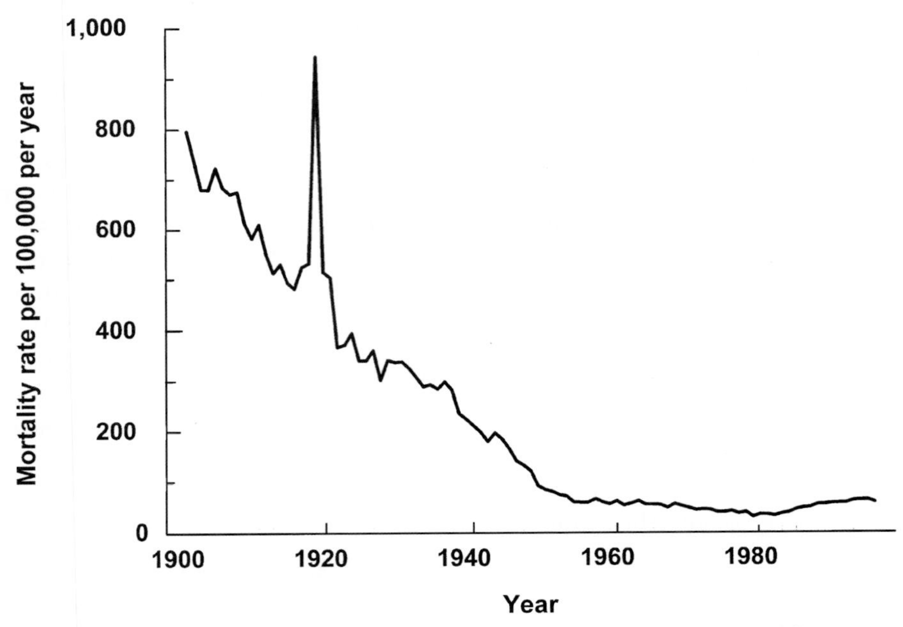

Fig. 2. Trends in infectious disease mortality in the U.S. during the twentieth century.

Case definitions are used primarily for case reporting, to permit surveillance systems to construct comparable and interpretable data bases. They are not intended to be used for establishing clinical diagnoses, determining treatments for individual patients, providing standards for reimbursement or quality assurance, or as rigid standards to dictate when public health action is warranted.

D. Approaches to Surveillance

There are a variety of methods for surveillance. The particular approach used depends on purposes of the surveillance, characteristics of the disease, sources of infomation, and available resources.

1. *Notifiable Diseases Surveillance*

A well-known approach to public health surveillance is the use of notifiable diseases systems. In notifiable diseases surveillance systems, public health authority is used to make certain diseases or conditions "reportable" to public health jurisdictions. Public health agencies then rely on clinicians or laboratories for reporting of surveillance data. Within public health systems, there is usually a chain of reporting from local to more central public health authorities.

a. National Notifiable Diseases Surveillance in the United States

One of the oldest surveillance systems in the United States, the National Notifiable Diseases Surveillance System (NNDSS), is built on a longstanding partnership between the Centers for Disease Control and Prevention (CDC) and state and local health departments. Public health officials at state health departments and the CDC collaborate in determing which diseases should be nationally notifiable. The Council of State and Territorial Epidemiologists (CSTE), with input from the CDC, makes recommendations annually for additions and deletions to the list of nationally notifiable diseases. A disease may be added to the national list as a new pathogen emerges (e.g., cryptosporidiosis and hantavirus pulmonary syndrome in 1994), and a disease may be deleted as its incidence declines (e.g., rheumatic fever in 1994). However, reporting of nationally notifiable diseases to the CDC by the states is voluntary. Reporting by providers and laboratories is mandated at the state level (i.e., by state legislation or regulation), and the list of diseases that are considered notifiable, therefore, varies slightly by state.

Because trends in many infectious diseases have the potential to shift rapidly, surveillance needs to be as timely as possible. Reports from health care providers and laboratories regarding individual cases of notifiable diseases, such as measles, hepatitis, syphilis, and *E. coli* O157:H7 infection, are essential to the function of this surveillance system at local, state, and national levels. These reports are generally sent to the local health department (by telephone, facsimile, or morbidity report form), which passes the information on to the state health department. Each state health department, in turn, reports their data electronically to the CDC on a weekly basis. Urgent matters requiring immediate public health response are reported immediately to the local, state, or federal jurisdiction (e.g., a case of botulism or meningococcal meningitis, or a multistate outbreak of food-borne disease).

For most of these diseases [acquired immunodeficiency syndrome (AIDS), in the U.S., is probably the exception], surveillance is largely a passive, provider-initiated system, in which health departments rely on health care providers or laboratories to report cases of disease. Thus, although notifiable diseases data are useful for analyzing disease trends and determining relative disease burdens, the data must be interpreted in light of reporting practices, which may vary from place to place. Some diseases that cause severe clinical illness (e.g., plague or rabies), if diagnosed by a clinician, are most likely to be reported accurately. However, persons who have diseases that are clinically mild and infrequently associated with serious consequences (e.g., salmonellosis) may not even seek medical care from a health care provider; even when these less severe diseases are diagnosed, they are less likely to be reported. The degree of completeness of reporting is also influenced by such factors as the availability and the use of diagnostic tests; control measures that are in effect; public awareness of a specific disease; the interests, resources, and priorities of government officials responsible for disease control and public health surveillance; and the quality of relationships between health care providers and public health agencies. Finally, factors such as changes in the case definitions for public health surveillance, the introduction of new diagnostic tests, or the discovery of new disease entities may cause changes in disease reporting that are independent of the true incidence of disease.

2. *Active versus Passive Surveillance*

Passive surveillance relies on the reporting of cases to public health officials. In contrast, in active surveillance, health officials or other persons conducting surveillance take measures to ensure reporting (e.g., by collecting the data themselves or by regularly reminding providers to report). Active surveillance is more time-consuming and resource-intensive than passive surveillance, but for many disease may provide more complete, better quality data. However, because of the more substantial resource requirements for active surveillance, such systems often cover a limited geographic area.

a. The Emerging Infections Program Network—Active Population-Based Surveillance

The CDC has recently collaborated with state health departments and academic institutions to develop the Emerging Infections Program (EIP) network of population-based sites to conduct active surveillance, engage in applied laboratory and epidemiologic research, and pilot test and evaluate prevention and control measures. During 1998, the EIP network included sites in California, Connecticut, Georgia, Maryland, Minnesota, New York, and Oregon. This network, with special surveillance efforts in targeted areas, complements national passive notifiable diseases surveillance. EIP activities on invasive bacterial diseases, unexplained deaths and critical illnesses, and food-borne diseases demonstrate this approach.

(1) Active Bacterial Core Surveillance (ABCs) The EIP ABCs conduct laboratory-based surveillance for invasive pneumococcal disease. Cases, defined

by the isolation of *Streptococcus pneumoniae* from normally sterile sites (usually blood or cerebrospinal fluid) from residents of the surveillance areas, are sought actively, and efforts are made to ensure comparable and complete case ascertainment from all the sites. This surveillance showed that during 1995–1996, the incidence of invasive pneumococcal disease varied substantially across the sites, from 18.8/100,000 in Minnesota to over 30/100,000 in California and Georgia (Fig. 3). Rates of pneumococcal disease vary considerably by age and race, which, along with distribution of underlying diseases like HIV that predispose to pneumococcal disease, account for this geographic variation. A convenience sample of pneumococcal isolates from assorted laboratories or passive surveillance could not have been used to characterize the incidence or demonstrate geographic variation precisely. Moreover, this surveillance involves collection of isolates as well as case reports, with testing of these isolates at a common lab using reliable methods to measure penicillin susceptibility. The project has demonstrated that the proportion of cases of pneumococcal disease caused by penicillin-resistant or intermediate isolates shows substantial geographic variation and has been essential in helping clarify the emergence of drug-resistant *S. pneumoniae* in the U.S.

Since its emergence in the 1970s, group B streptococcal (GBS) disease has been the leading bacterial infection associated with illness and death among newborns in the U.S. In May 1996, the CDC released consensus guidelines on the prevention of neonatal group B streptococcal disease, which recommended the use of either of two alternative intervention strategies involving the use of intrapartum antibiotics. In addition to conducting active surveillance for disease caused by group B streptococcus, the EIP ABC sites have been employed to monitor the implementation and impact of the prevention guidelines. For example, recent work has found an increasing number of hospitals with GBS disease prevention policies and also that geographical areas with a higher proportion of hospitals with GBS prevention policies have a lower incidence of GBS disease (Fig. 4). Continuing efforts are aimed at refining understanding of GBS prevention and monitoring for any adverse conse-

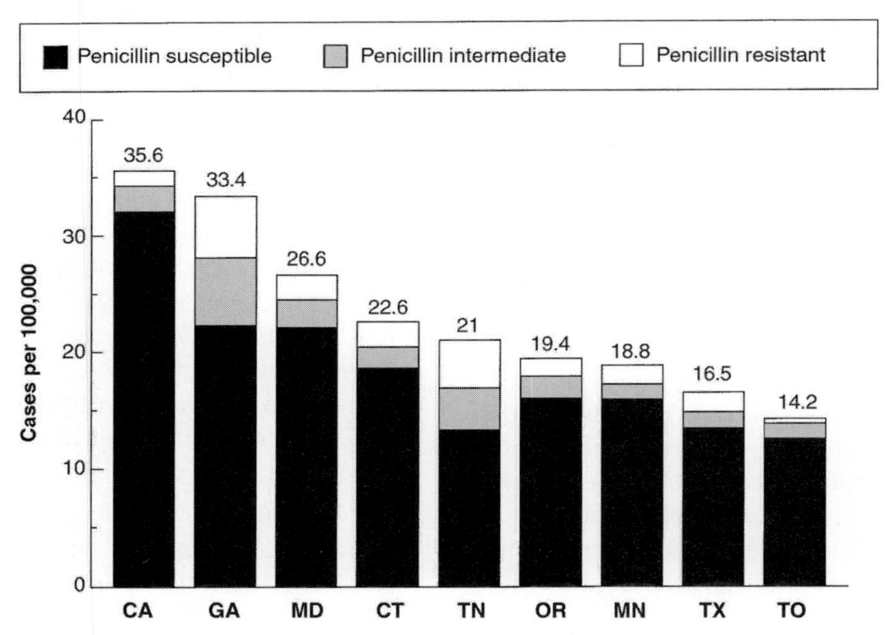

Fig. 3. Incidence of invasive *S. pneumoniae* disease and distribution of penicillin susceptibility in selected areas of the U.S., 1995–1996.

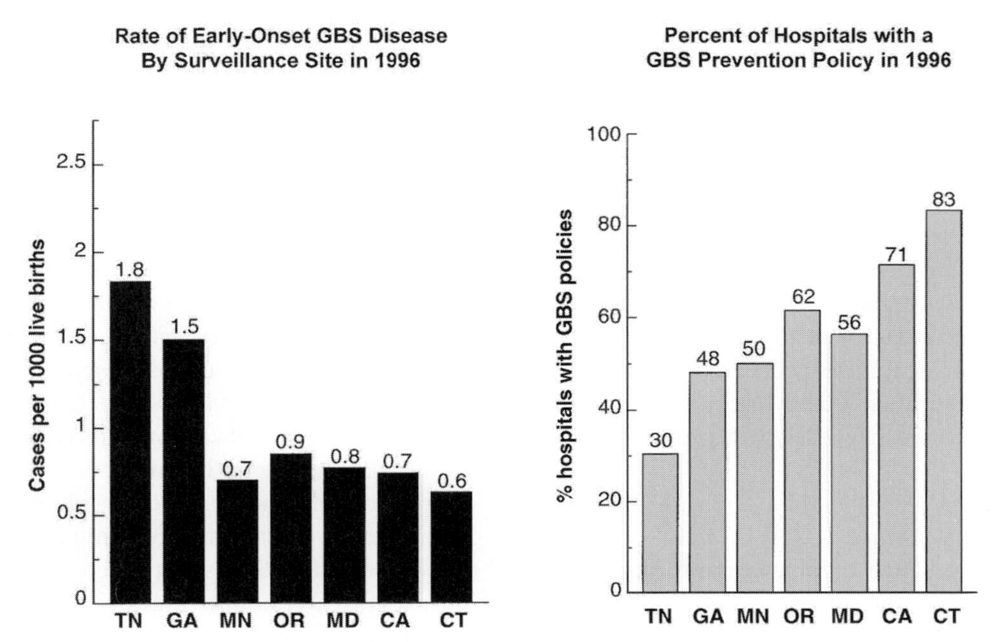

Fig. 4. Incidence of neonatal group B streptococcal (GBS) disease and proportion of hospitals with GBS prevention policy in selected areas of the U.S., 1996.

quences of the implementation of current GBS prevention strategies.

(2) *Unexplained Deaths and Critical Illnesses Surveillance* Several infectious diseases have been identified in recent years through investigation of illnesses for which a cause had not been recognized (e.g., Legionnaires' disease, toxic shock syndrome, AIDS, and hantavirus pulmonary syndrome); subsequent retrospective investigations have sometimes identified cases that occurred many years before their identification. These observations led the EIP to pilot test a surveillance project for unexplained deaths and critical illnesses due to possibly infectious causes in previously healthy persons. This surveillance involves clinician reporting of cases with syndromes that have the appearance of an infectious disease (e.g., severe illnesses with fever and abnormalities in one or more organ systems), but for which no diagnosis has been made during the course of the clinical evaluation. Information about these cases, including information about patient exposures, such as travel or contact with animals or insects, is being collected. In addition, samples of tissue specimens obtained during clinical care or diagnostic evaluation, or at autopsy in fatal cases, are being collected. By approaching surveillance for unexplained syndromes systematically, and with thorough laboratory testing, including applying modern laboratory methods such as immunohistochemistry, polymerase chain reaction (PCR) amplification, and other nucleic acid-based techniques, this activity aims to improve the chances of earlier recognition of new or previously unrecognized pathogens. In addition, this surveillance effort is identifying cases of known infections that went unrecognized because diagnostic tests were not performed or were not sufficiently sensitive; these findings may be useful in identifying areas where improved diagnostic capabilities are needed. This project is also developing a population-based bank of clinical specimens that will be valuable in future testing for newly recognized agents.

(3) *FoodNet* FoodNet is the principal food-borne disease component of the EIP and involves collaboration among the EIP sites, the CDC, the U.S. Department of Agriculture (USDA), and the Food and Drug Administration (FDA). The objectives of

FoodNet are to determine the frequency and severity of food-borne diseases; determine the proportion of common food-borne diseases that results from eating specific foods; and describe the epidemiology of new and emerging bacterial, parasitic, and viral food-borne pathogens. To address these objectives, FoodNet uses active surveillance and conducts related epidemiologic studies and surveys (Fig. 5) (*www.cdc.gov/ncidod/dbmd/foodnet*). For example, in 1997 FoodNet estimated that approximately 360 million cases of diarrheal illness occurred in the United States, and it identified *Campylobacter* as the most commonly identified bacterial pathogen in cases of diarrhea (Fig. 6).

3. *Sentinel Surveillance*

Sentinel surveillance involves monitoring of key health events, for example, in specific geographic areas or using only a sample of potential health care providers, laboratories, or other reporters. Although sentinel surveillance may not provide a complete picture, it may stimulate further investigation.

a. *National Nosocomial Infections Surveillance System*

The National Nosocomial Infections Surveillance System (NNIS) operated by the CDC is an example of a sentinel surveillance system. Approximately 245 hospitals, which voluntarily report their nosocomial infection data to NNIS, constitute this system (see *www.cdc.gov/ncidod/hip/nnis/@nnis.htm*). NNIS data have been important in quantifying the burden of nosocomial infections and evaluating the occurrence rates of a variety of noscomial infections, such as urinary catheter-associated urinary tract infections, central line-associated bloodstream infections, and ventilator-associated pneumonia. In addition, the NNIS has provided information on the emergence of antibiotic-resistant infections, including methicillin-resistant *Staphylococcus aureus* (MRSA) infections during the 1980s and vancomycin-resistant enterococcal (VRE) infections during the 1990s. The NNIS reports on infections in acute general hospital patients; however, recently changing health care delivery in the United States, with increasing emphasis on outpatient care, is blurring the distinction between inpatient and outpatient settings. In the future, surveillance through the NNIS or other approaches will need to account for healthcare-associated infections in a wider range of care settings.

4. *Application of Modern Communications and Information Technology to Surveillance*

Collection and dissemination of health information can be enhanced by recent advances in commu-

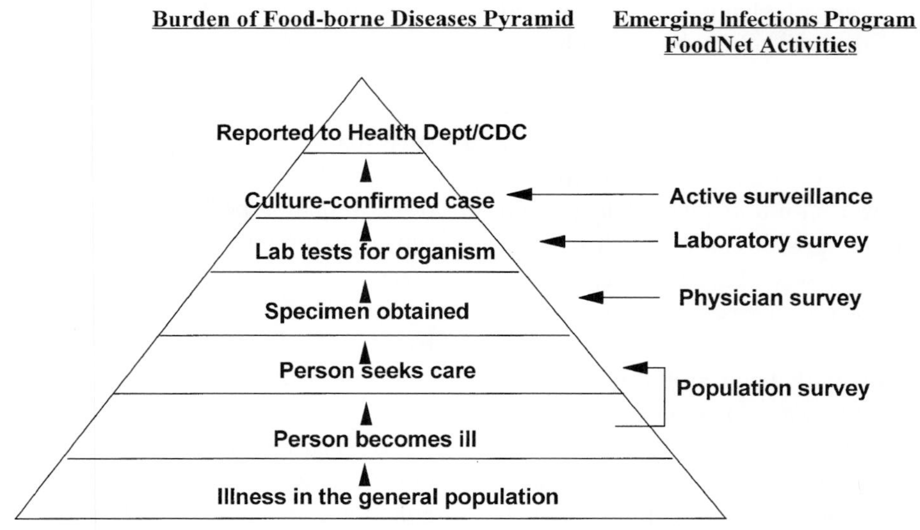

Fig. 5. Schematic of burden of food-borne diseases and corresponding FoodNet activities.

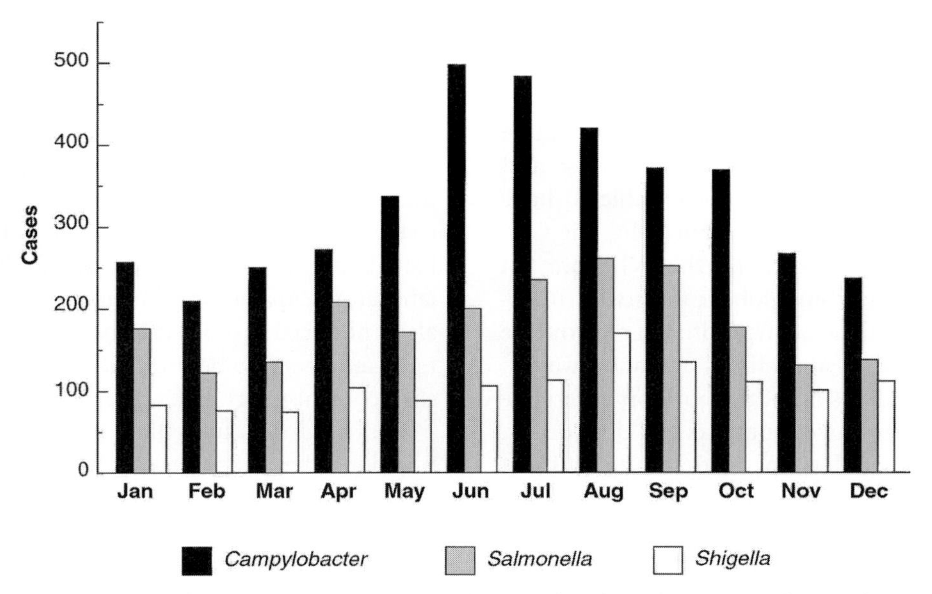

Fig. 6. Cases of disease caused by selected bacterial pathogens reported by FoodNet, by month of collection, 1997.

nications and information technology. Resources such as the Internet and other telecommunication networks have provided new opportunities for rapid communication of known or suspected outbreaks and other alerts. The ProMED Electronic Network, an initiative of the Federation of American Scientists, is a global program for monitoring emerging infectious diseases. The network establishes direct relationships and communications among interested participants around the world (generally scientists and physicians). Although ProMED is not a formal surveillance system and reports on the system cannot always be verified, the network has proved useful to a large number of professionals. The network's success has also inspired similar models such as PACNET, a Pacific Islands networking system used to support surveillance of infectious diseases in that region.

The development of technologies, standards, data-sharing agreements, and assurance of privacy and confidentiality to facilitate electronic exchange of information between health care delivery systems and public health agencies promises to reorient and improve surveillance for a number of conditions. For example, efforts are underway in several states to transmit electronic reports of clinical laboratory data to public health officials. In many countries, strong interest and substantial effort are directed toward the development of computerized patient records. Future surveillance efforts for some diseases may be better able to tap into streams of already collected and electronically stored health information (e.g., from laboratories and hospital information systems), rather than depend exclusively on the construction of freestanding public health surveillance systems to collect data.

Progress in information technology, software applications, and analytic methods are having an important impact on infectious diseases surveillance, as illustrated by these examples. Application of geographic information systems (GIS) is enabling improved display of the relationships among disease occurrence and a variety of demographic and geographic factors. Adaptation of "cumulative sums" (CUSUM), a quality control method used in manufacturing, has been used to detect outbreaks in large surveillance databases. E-mail and browser-based applications are speeding up the process of querying and response that is part of epidemiologic investigation; and improved software is facilitating the timely analysis and display of surveillance data.

III. SURVEILLANCE SYSTEMS AROUND THE WORLD

The goals of global surveillance programs include recognizing the unusual occurrence of disease and responding appropriately; it is not feasible to have all infectious diseases routinely reportable. The Collaborating Centers sponsored by the WHO are an important aspect of current global capacity for infectious disease surveillance. They often serve routine surveillance functions, and efforts are underway to enhance the capabilities of local laboratory reference centers to recognize novel outbreaks as they occur. Since 1947, the WHO has operated an international network of collaborating laboratories to monitor the emergence and spread of new epidemic and pandemic strains of influenza. The primary purpose of the network is to detect, through laboratory surveillance, the emergence and spread of antigenic variants of influenza so that the vaccine formula can be updated as appropriate to include these new antigenic variants. In recent years, use of this influenza surveillance network has resulted in good matches between the circulating strains of influenza and the vaccine strains, providing a sound basis for influenza control activities. This network now includes four international centers, and over 100 WHO national collaborating laboratories. Regional networks for influenza surveillance also contribute; in 1995, the European Influenza Surveillance Scheme was created with the participation of seven countries; this network provides clinical and virological data from the general populations and hospitals through an interactive real-time database, which can then be used for data entry, queries, and consultations. Research programs to standardize clinical data among countries have been initiated, and quality assurance is integrated into the surveillance system; a steering committee provides regular assessment.

Numerous recent reports have shown that the quality of health surveillance in most nations is suboptimal, and a number of plans by the WHO and other institutions are being developed and implemented as resources allow. Many plans are aimed at increasing the level of health surveillance within individual countries by such measures as improving laboratory capacity for diagnosing certain infections in specific areas (e.g., plague in India, influenza in China). When health information systems are limited, sentinel surveillance for specific illnesses may be established. When yellow fever emerged in Kenya in 1992, the diagnosis was not confirmed for more than 3 months; in response, surveillance for hemorrhagic fever involving 13 sentinel sites was established in the affected and surrounding areas, and laboratory capacity for diagnosis of yellow fever was also enhanced. To be sustained, however, these systems may need to be integrated into established disease control programs.

The need for surveillance that crosses international borders is increasingly evident as a consequence of such changes as increasing travel, population growth, and the increase in international food distribution networks. The signing of the Treaty of Maastricht (1992) established a basis for the European Community to take action in the field of public health, including international surveillance of infectious diseases. In Europe, a number of initiatives are underway to facilitate surveillance that crosses national boundaries, including surveillance of food-borne infections (ENTER-NET), legionellosis (European Working Group on Legionella Infections), tuberculosis (the EuroTB programme), and influenza (European Influenza Surveillance Scheme). In 1994, a laboratory-based surveillance system was established to improve the prevention and control of human salmonellosis and other food-borne infections in the countries of the European Union and the European Cooperation in Science and Technology. Epidemiologists with national surveillance responsibilities and heads of reference laboratories in more than a dozen European countries currently participate, and most are participating in an on-line network database.

In Australia, surveillance is enhanced through the Communicable Diseases Network of Australia and New Zealand. Teleconferences and other meetings occur monthly to exchange information on infectious diseases and to coordinate surveillance and control activities. Participants include a representative from New Zealand, representatives of the Australian Department of Human Services and Health, and representatives of the state and territory health authorities.

Increasingly, there is also interest in creating a functional network of laboratories within various regions of the world. For example, the Pan American Health Organization, working with the greater southern cone region of Latin America (Argentina, Bolivia, Brazil, Chile, Paraguay, and Uruguay), has recently developed a plan of surveillance for (1) disease syndromes, (2) specific pathologies, and (3) antimicrobial resistance. Influenza, hantavirus pulmonary syndrome and hantavirus disease, acute diarrhea, particularly bloody diarrhea leading to hemolytic uremic syndrome, and drug-resistant tuberculosis are four areas identified for initial surveillance by this network.

In addition to strengthening of current systems of the WHO for infectious disease surveillance, new networks are also being formed to address emergent problems. For example, the WHO has formulated a series of recommendations that emphasize enhanced surveillance of drug resistance and increased monitoring and improved usage of antimicrobial drugs.

IV. CHALLENGES AND ISSUES IN INFECTIOUS DISEASES SURVEILLANCE

The diversity of infectious diseases biology, modes of transmission, and available control measures requires a variety of surveillance approaches and means that key surveillance issues may vary by diseases. The following examples illustrate some of these key issues or features.

A. *Haemophilus influenzae*—Information from Two Surveillance Systems Combine to Demonstrate the Impact of Vaccination

Before the late 1980s, *Haemophilus influenzae* type b (Hib) was the most common cause of bacterial meningitis among children in the United States. Surveillance information has been essential in both monitoring and understanding the decline of this disease. Data from the National Notifiable Diseases Surveillance System have documented national declines in disease caused by *H. influenzae* during the mid and late 1980s, after polysaccharide vaccines and, subsequently, protein-conjugated vaccines for the prevention of Hib were licensed. However, laboratory-based surveillance information, which included serotype (not reliably included in the passive national data), was required to determine that declines were occurring specifically in type b disease and therefore were probably due to use of the vaccine. The observation drawn from the surveillance data—that the rate of Hib disease was declining at a rate faster than one might have predicted based on use of the vaccine—hinted that the protein-conjugated Hib vaccines exerted their effect both by preventing disease in the vaccinated and by reducing nasopharyngeal carriage and spread. Although it took more focused research efforts to substantiate that nasopharyngeal carriage was reduced as an effect of the conjugate Hib vaccines, surveillance data provided a clue and a context for interpreting research on the impact of the conjugated vaccines on transmission. Continued surveillance of overall rates and the distribution of disease-causing serotypes of *H. influenzae* disease is essential to monitor the effectiveness and duration of protection of immunization and to detect any emergence of disease caused by other serotypes (Fig. 7).

B. Viral Hepatitis—Advances in Diagnostic Tests Enhance Surveillance

Viral hepatitis provides an example of how advances in virology are integral for improvement in the usefulness of surveillance data and understanding of a syndrome (Fig. 8). Although experts believed, on the basis of epidemiologic and clinical observations, that the syndrome of acute hepatitis probably was caused by multiple viruses, nationally reported surveillance data on viral hepatitis could not definitely distinguish between them until serologic and virologic tests for hepatitis A and hepatitis B became available. The differing epidemiology, modes of transmission, and approaches to control of hepatitis A and B made the development of tests to distinguish between them essential for public health surveillance of hepatitis. More recently, the discovery of hepatitis C virus (HCV) has been a challenge for surveillance. Available serologic tests can detect past infection with hepatitis C but have been unable to distinguish between recent and remote infection, confounding

A

*** *Because of the low number of states reporting surveillance data during 1987–1990, rates for those years were race-adjusted using the 1990 U.S. population.*

B

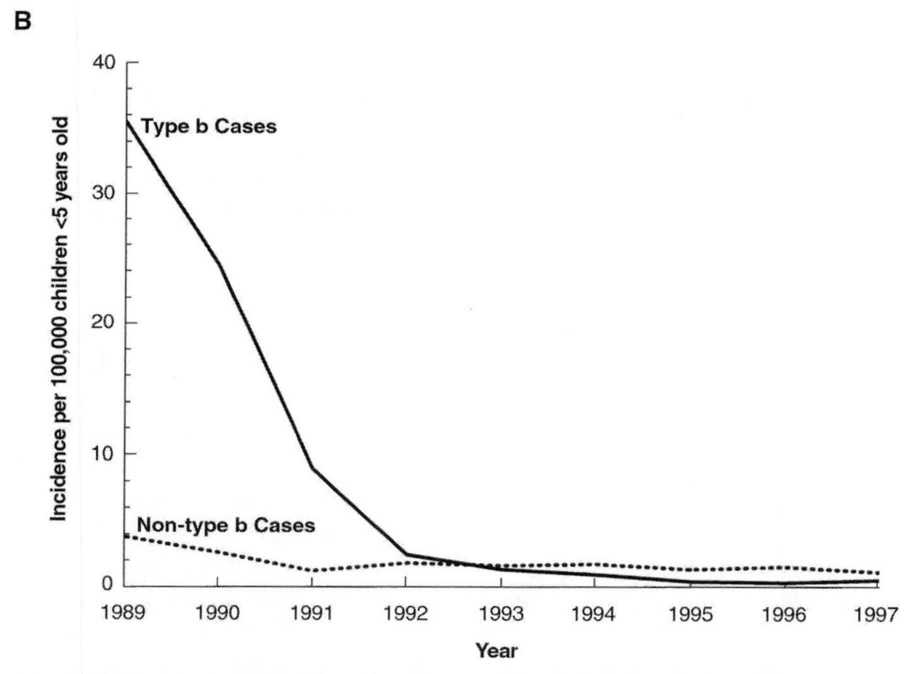

Fig. 7. (A) Incidence rate of invasive *Haemophilus influenzae,* U.S., 1987–1997, National Notifiable Diseases Surveillance System. (B) *Haemophilus influenzae* type b (Hib) and non-type b incidence rate per 100,000 population detected through laboratory-based surveillance among children aged <5 years, U.S., 1989–1997.

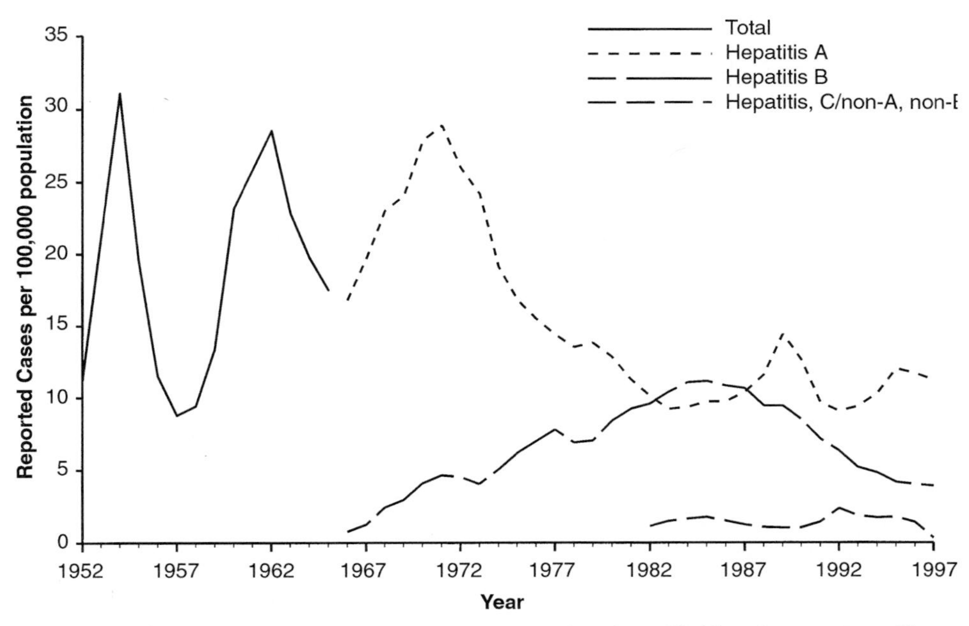

Fig. 8. Viral hepatitis by year, U.S., 1952–1997, National Notifiable Diseases Surveillance System.

efforts to determine the incidence of hepatitis C using laboratory-based surveillance data. In addition, as many as 25–30% of hepatitis C infections are asymptomatic, and chronic liver disease, which occurs in up to 70% of infected persons, occurs many years after initial infection, further complicating surveillance and resulting in an underestimate of the proportion of chronic liver disease attributable to viral hepatitis. Current methods for hepatitis C surveillance combine laboratory test-based surveillance of prevalence of HCV antibody, active surveillance and appropriate laboratory testing for the syndrome of acute hepatitis, and surveillance for causes of chronic liver disease.

C. Hantavirus Pulmonary Syndrome—A Surveillance Case Definition Improves

In late May 1993, an outbreak of unexplained respiratory failure in previously healthy young adults occurred in the southwestern part of the United States. Subsequent investigation revealed that this outbreak was due to a previously unrecognized hantavirus. Hantavirus pulmonary syndrome (HPS) demonstrates how a case definition and the role of the laboratory may evolve in surveillance of a newly recognized syndrome. The initial surveillance case definition defined a "suspected case" as unexplained respiratory disease with hypoxemia and chest x-ray evidence or autopsy findings of unexplained pulmonary edema. This case definition was not specific, which meant that cases identified using this definition would likely include background cases of other diseases as well as cases truly part of the outbreak. However, without a more definitive way to diagnose or precisely define cases of this syndrome, that was the best that could be done initially. Over the subsequent few months, laboratory methods to diagnose HPS were developed, including serologic tests, PCR, and immunohistochemical tests, and by October the case definition was revised to incorporate laboratory evidence of HPS. Using this more specific case definition in a national passive surveillance effort, 21 laboratory-confirmed cases in 11 states were identified by December 31, 1993; by August 1998, 188 cases in 29 states had been confirmed.

The virus responsible for HPS in the 1993 outbreak, since named Sin Nombre virus, is associated with a particular species of rodent host, the deer mouse (*Peromyscus maniculatus*). Subsequent surveillance and investigation of HPS in North and South America, using more vigorous case finding and current laboratory means to identify and subtype hantaviruses, have revealed that HPS is caused by several different hantaviruses, each with its own rodent host. This observation underscores how understanding the epidemiology and conducting surveillance for some infectious diseases depends on applying effective laboratory methods. It also demonstrates that a complete picture sometimes requires surveillance for infection in animal hosts and vectors, as well as for human disease.

D. AIDS—Evolution of a Surveillance Case Definition

Surveillance for the acquired immune deficiency syndrome (AIDS) has been one of the most dynamic surveillance systems in the history of public health. The increase in knowledge, the introduction of a diagnostic test for the human immunodeficiency virus (HIV), the changes in diagnostic practices for opportunistic infections, and therapeutic advances have all challenged the surveillance system for this disease. The case definitions for AIDS has been modified many times to accommodate these changes and to improve the ability to monitor trends in the diseases as well as to project future numbers of cases (Fig. 9).

As diagnostic tests improved and knowledge was gained regarding the long latency of HIV infection, systems which supplement case reporting have been developed in the U.S. and other countries, including a number of serosurveys in populations ranging from injection drug users to childbearing women. HIV reporting has been initiated in some areas, and although it is of limited usefulness in monitoring trends in infection in most populations, it does provide a framework for a variety of prevention and

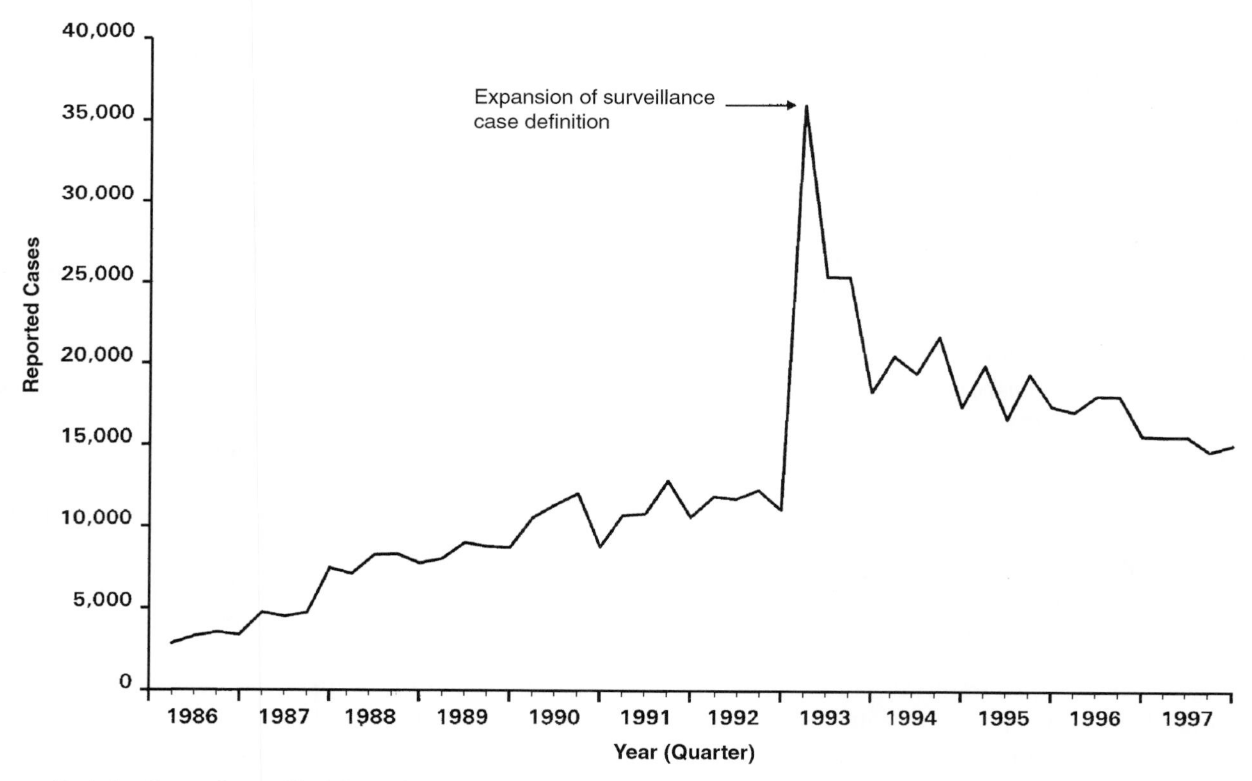

*Includes Guam, Puerto Rico, the U.S. Pacific Islands, and the U.S. Virgin Islands.

Fig. 9. Acquired immunodeficiency syndrome (AIDS) cases, reported by quarter, U.S., 1986–1997. Source: Division of HIV/AIDS Prevention—Surveillance and Epidemiology, National Center for HIV, STD, and TB Prevention.

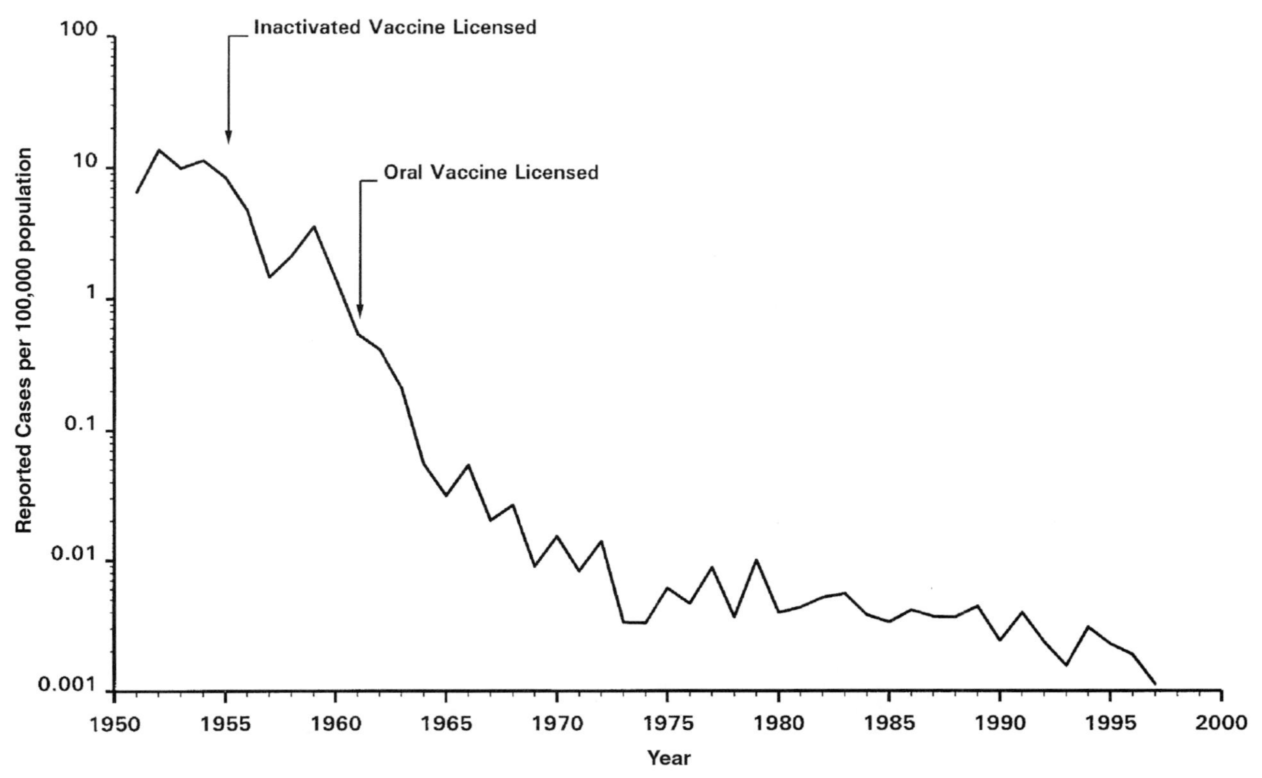

Fig. 10. Poliomyelitis (paralytic) cases, by year, U.S., 1951–1997. Source: National Notifiable Diseases Surveillance System.

health care measures. An important area for HIV will be the further development of surveillance for HIV subtypes to improve understanding of the geographic spread of the diseases, potential strain variations in pathogenicity, and antiviral drug susceptibility patterns.

E. Poliomyelitis—Increased Role of the Laboratory as Goal of Eradication Approaches

Poliomyelitis surveillance, which has been instrumental in monitoring polio's decline and will be used to certify eradication of polio, provides an example of how surveillance systems can be adapted to changing public health objectives. When disease rates were relatively high and background cases of acute flaccid paralysis were low relative to paralysis caused by poliomyelitis, a clinical surveillance case definition of acute onset of flaccid paralysis of one or more limbs without sensory or cognitive loss sufficed for monitoring trends in disease and gauging the impact of vaccination. However, as the incidence of poliomyelitis fell to low levels, and with it the predictive value for actual poliomyelitis of a clinical case definition, laboratory confirmation of cases became essential (Fig. 10). Now, as poliomyelitis has been designated for global eradication, an international laboratory surveillance network has been developed to provide laboratory confirmation of cases and to distinguish cases due to wild poliovirus from vaccine-associated cases.

V. SUBTYPING AND THE APPLICATION OF MOLECULAR EPIDEMIOLOGIC METHODS

The application of molecular epidemiologic tools has become an integral part of surveillance for a variety of infectious diseases and will continue to increase in importance, as illustrated by the following examples, as well as the above examples of strain surveillance for poliovirus and HIV.

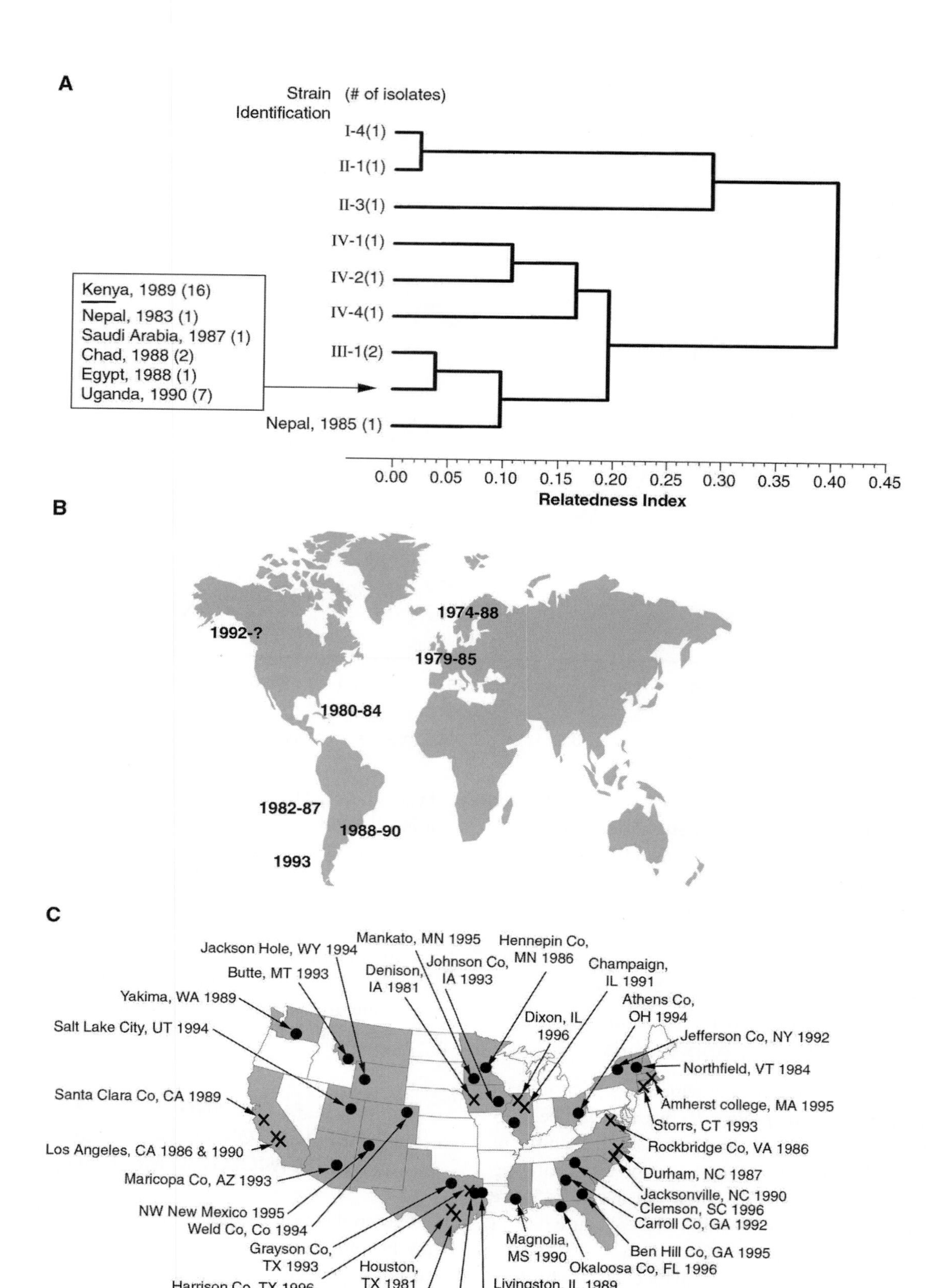
A
Strain Identification
(# of isolates)
I-4(1)
II-1(1)
II-3(1)
IV-1(1)
IV-2(1)
IV-4(1)
III-1(2)
Kenya, 1989 (16)
Nepal, 1983 (1)
Saudi Arabia, 1987 (1)
Chad, 1988 (2)
Egypt, 1988 (1)
Uganda, 1990 (7)
Nepal, 1985 (1)
0.00 0.05 0.10 0.15 0.20 0.25 0.30 0.35 0.40 0.45
Relatedness Index
B
1992-?
1974-88
1979-85
1980-84
1982-87
1988-90
1993
C
Jackson Hole, WY 1994
Mankato, MN 1995
Hennepin Co, MN 1986
Butte, MT 1993
Denison, IA 1981
Johnson Co, IA 1993
Champaign, IL 1991
Yakima, WA 1989
Athens Co, OH 1994
Dixon, IL 1996
Salt Lake City, UT 1994
Jefferson Co, NY 1992
Northfield, VT 1984
Santa Clara Co, CA 1989
Amherst college, MA 1995
Storrs, CT 1993
Rockbridge Co, VA 1986
Los Angeles, CA 1986 & 1990
Durham, NC 1987
Maricopa Co, AZ 1993
Jacksonville, NC 1990
Clemson, SC 1996
NW New Mexico 1995
Carroll Co, GA 1992
Weld Co, Co 1994
Grayson Co, TX 1993
Ben Hill Co, GA 1995
Magnolia, MS 1990
Okaloosa Co, FL 1996
Houston, TX 1981
Harrison Co, TX 1996
Livingston, IL 1989
Harris County, TX 1991
Gregg Co, TX 1994
Fort Worth, TX 1996
Community outbreak
Organization-based outbreak

A. Serotyping of *Salmonella poona*

Classic serotyping played a pivotal role in understanding an outbreak of salmonellosis. During June and July 1991, more than 400 laboratory-confirmed infections with *Salmonella poona* occurred in 23 states. Subsequent case–control studies identified cantaloupe as the vehicle, but surveillance was responsible for initial identification of the outbreak. The use of serotyping and the unusual serotype responsible permitted identification of this outbreak, and its association with a particular vehicle, in the context of many thousands of cases of salmonellosis each year.

B. Multilocus Enzyme Electrophoresis of *Neisseria meningitidis*

Multilocus enzyme electrophoresis (MEE), a subtyping method based on the differential electrophoretic mobility of constitutively expressed enzymes, has been a powerful tool in recent efforts to understand the epidemiology of meningococcal disease. It was used to demonstrate the international spread of the III-1 clone of group A *Neisseria meningitidis*. Enzyme typing profiles indicated that this clone was brought from South Asia to the Middle East by Muslims making their pilgrimage (haj) to Mecca, Saudi Arabia, in 1987; subsequently, pilgrims who became group A carriers introduced this clonal group into sub-Saharan Africa on their return from the haj. MEE was used in conjunction with active laboratory-based surveillance in the United States to demonstrate the emergence of a particular clonal group, the ET-5 complex, of group B meningococcus in Oregon. ET-5 was responsible for substantially elevated rates of disease in several European countries during the 1970s and in Cuba and several South American countries in the 1980s and 1990s. MEE has also been instrumental in showing that a recent increase in the number of clusters of group C meningococcal disease in the United States has been caused by closely related strains (Fig. 11).

C. Pulsed-Field Gel Electrophoresis of *Escherichia coli* O157:H7

Recently, the Minnesota Department of Health conducted surveillance for *E. coli* O157:H7 by performing pulsed-field gel electrophoresis (PFGE) on all available isolates, as well as by collecting case reports. Using this approach, they concluded that they were able to identify outbreaks that would have gone unrecognized by traditional methods and to differentiate when sudden increases in reported cases were due to sporadic cases or outbreaks (Bender *et al.*, 1997).

D. PulseNet

Food-borne outbreaks are now often distributed over wide geographic areas and may be associated with low level contamination of a widely distributed product. In an effort to detect and control these outbreaks earlier, CDC initiated in 1995 (and continues to coordinate) PulseNet, a national network of public health laboratories that performs PFGE subtyping on bacteria that may be food-borne. This national network employs information technology tools to transmit, analyze, and compare PFGE patterns from isolates in the participating laboratories; it also has required careful standardization of the PFGE laboratory methods. Currently, PulseNet is performing PFGE "fingerprinting" on *E. coli* O157:H7 and is developing the network to work with *Salmonella* serotype Typhimurium (www.cdc.gov/ncidod/dbmd/pulsenet/pulsenet.htm). For example, DNA "fingerprinting" has assisted surveillance and outbreak investigation of *E. coli* O157:H7 infections in several instances: (1) to connect outbreaks of *E. coli* O157:H7 in Connecticut and Illinois to a common source, mesclun lettuce; (2) to trace an outbreak in patients from four states and one Canadian province to a commercial unpasteurized apple juice; and (3) to link a cluster of cases in Colorado to a particular brand of ground beef patties. In the last example, the PFGE pattern from the patient and the ground

Fig. 11. (A) Dendrogram of group A meningococcal disease, ET-III-1 strains. (B) Global distribution of *N. meningitidis* ET-5 Complex, 1974–1995. (C) Serogroup C meningococcal disease outbreaks in the U.S., 1981–1996.

beef was transmitted to PulseNet sites and was compared with patterns from over 300 other *E. coli* O157:H7 isolates; no matching patterns were found, which provided evidence that the outbreak was not large or widely distributed nationally.

Defining more clearly the role of PFGE and other subtyping methods in infectious diseases surveillance and developing the capacity for subtyping in public health laboratories will be important efforts over the next several years. The role of nucleotide sequence-based subtyping is likely to grow in importance, especially in clarifying the relationship of epidemiologic phenomena to virulence properties of disease-causing organisms and to host susceptibility.

VI. CONCLUSIONS

The job for surveillance of infectious diseases will remain to provide high quality information on which to base and evaluate public health actions. However, modern times present considerable challenges, the most important of which is to maintain and develop surveillance systems that can keep pace with the dynamic emergence of infectious diseases. The concept of disease emergence and the factors that influence it—human demographics and behavior, technology, economic development and land use, international travel and commerce, microbial adaptation and change, and breakdown of public health measures—have been described (Lederberg *et al.*, 1992). Included in these modern challenges is the need for preparedness to detect and respond to potential acts of bioterrorism. At the same time, modern technology offers opportunities to revolutionize surveillance. For example, continued development of diagnostic and molecular epidemiologic methods, when effectively combined with epidemiologic information, can make prevention, recognition, and response to outbreaks faster and more precise. Further, developments in information technology can facilitate linking diverse sources of information, rapid analysis and synthesis of huge amounts of information, and develpment of the communications and networks needed globally for surveillance and response to infectious diseases.

See Also the Following Articles

Economic Consequences of Infectious Diseases • Emerging Infections • Hepatitis Viruses • International Law and Infectious Diseases • Polio • Smallpox

Bibliography

Armstrong, G. L., Conn, L. A., and Pinner, R. W. (1999). Trends in infectious disease mortality in the United States during the 20th century. *JAMA* **281**, 61–66.

Bender, J. B., Hedberg, C. W., Besser, J. M., Boxrud, D. J., MacDonald, K. L., and Osterholm, M. T. (1997). Surveillance by molecular subtype for *Escherichia coli* O157:H7 infections in Minnesota by molecular subtyping. *N. Engl. J. Med.* **337**, 388–394.

Berkelman, R. L., Buehler, J. W., and Dondero, T. J. (1992). Surveillance of acquired immunodeficiency syndrome (AIDS). *In* "Public Health Surveillance" (W. Halperin and E. L. Baker, eds.), pp. 108–120. Van Nostrand-Reinhold, New York.

Berkelman, R. L., Bryan, R. T., Osterholm, M. T., LeDuc, J. W., and Hughes, J. H. (1994). Infectious disease surveillance: A crumbling foundation. *Science* **264**, 368–370.

CDC (1994). Addressing emerging infectious disease threats to health: A prevention strategy for the United States. U.S. Department of Health and Human Services, Centers for Disease Control and Prevention, Atlanta, Georgia.

CDC (1996). Progress toward elimination of *Haemophilus influenzae* type b disease among infants and children—United States, 1987–1995. *MMWR* **45**, 901–906.

CDC (1997). Case definitions for infectious conditions under public health surveillance. *MMWR* **46** (No. RR-10).

CDC (1998). "Preventing Emerging Infectious Diseases: A Strategy for the 21st Century." U.S. Department of Health and Human Services, Centers for Disease Control and Prevention, Atlanta, Georgia.

Cox, N. I., and Regnery, H. L. (1996). Global influenza surveillance: Tracking a moving target in a rapidly changing world. *In* "Options for the Control of Influenza—III" (L. E. Brown, A. W. Hampson, and R. G. Webster, eds.), pp. 591–598. Elsevier Science.

Henderson, D. A. (1976). Surveillance of smallpox. *Int. J. Epidemid.* **5**, 19–28.

Hennessy, T. W., Hedberg, C. W., Slutsker, L., *et al.* (1996). A national outbreak of *Salmonella enteritidis* infections from ice cream. *N. Engl. J. Med.* **334**, 1281–1286.

Koo, D., and Wetterhall, S. (1996). History and current status of the National Notifiable Diseases Surveillance System. *J. Public Health Manag. Pract.* **2**, 4–10.

Langmuir, A. D. (1963). The surveillance of communicable diseases of national importance. *N. Engl. J. Med.* **268**, 182–192.

Langmuir, A. D. (1976). William Farr: Founder of modern concepts of surveillance. *Int. J. Epidemid.* **5**, 13–18.

Lederberg, J., Shope, R. E., and Oaks, S. C. (eds.) (1992). "Emerging Infections—Microbial Threats to Health in the United States. National Academy Press. Washington, D. C.

LeDuc, J. W., and Tikhomirov, E. (1994). Global surveillance for recognition and response to emerging diseases. *Ann. N.Y. Acad. Sci.* **740**, 341–345.

Martinez, L. J., and Heymann, D. L. (1998). A global response to a global challenge. *In* "New and Resurgent Infections" (B. Greenwood and K. DeCock, eds.), pp. 199–203, Wiley, New York.

Pan American Health Organization (1998). Meeting to establish a network of laboratories for the surveillance of emerging infectious diseases (EID) in the southern cone region—Buenos Aires, Argentina.

Perkins, B. A., Flood, J. M., Danila, R., Holman, R. C., Reingold, A. L., Klug, L. A., *et al.* (The Unexplained Deaths Working Group). (1996). Unexplained deaths due to possibly infectious causes in the United States: Defining the problem and designing surveillance and laboratory approaches. *Emerg. Infect. Dis.* **2**, 47–53.

Pinheiro, F. P., Kew, O. M., Hatch, M. H., da Silveira, C. M., and de Quadros, C. A. (1997). Eradication of wild poliovirus from the Americas: Wild poliovirus surveillance—laboratory issues. *J. Infect. Dis.* **175** (Suppl. 1), S43–S49.

Schuchat, A., Robinson, K., Wenger, J. D., *et al.* (1997). Bacterial meningitis in the United States in 1995. Active Surveillance Team. *N. Engl. J. Med.* **337**, 970–976.

Teutsch, S. M., and Churchill, R. E. (eds.) (1994). "Principles and Practice of Public Health Surveillance." Oxford Univ. Press, New York and Oxford.

Thacker, S. B., Choi, K., and Brachman, P. S. (1983). The surveillance of infectious diseases. *JAMA* **249**, 1181–1185.

Vacalis, T. D., Bartlett, C. L. R., and Shapiro, C. G. (1995). Electronic communication and the future of international public health surveillance. *Emerg. Infect. Dis.* **1**, 34.

Symbiotic Microorganisms in Insects

A. E. Douglas
University of York

I. Diversity of Symbiotic Microorganisms in Insects
II. Function of Symbiotic Microorganisms
III. Determinants of the Density of Symbiotic Microorganisms in Insects
IV. Transmission of Symbiotic Microorganisms
V. Symbiotic Microorganisms and Insect-Pest Management

GLOSSARY

bacteriocyte An insect cell harboring symbiotic microorganisms.

nitrogen recycling The metabolism of insect nitrogenous waste products (e.g., uric acid and ammonia) by symbiotic microorganisms to nitrogenous compounds (e.g., essential amino acids) that are transferred to and used by the insect.

symbiosis The intimate association between phylogenetically different organisms; often restricted to relationships from which all the organisms derive benefit.

transovarial transmission The transfer of symbiotic microorganisms to the unfertilized eggs in the ovaries of the female insect.

vertical transmission The transmission of microorganisms from a parent insect to its offspring.

SYMBIOTIC MICROORGANISMS are the components of the microbiota of an insect that contribute to insect survival, growth, or fecundity. They are borne by an estimated 10% of all insect species, and are located in the insect gut or tissues, often restricted to specialized cells called bacteriocytes. Most of the microorganisms are rare or unknown apart from the insect partner, and many have not been cultured *in vitro*. Historically, the microbiology of insects has been little studied and, until recently, most of the information available on symbiotic microorganisms has been derived from microscopic analysis of the insect regarding the morphology of the microorganisms, their location in the insect body, and their mode of transmission between insects. The advent of molecular techniques has transformed our understanding of symbiotic microorganisms, allowing the taxonomic identification of microorganisms and the elucidation of the molecular basis of their function. Because of the insects' dependence on their symbiotic microorganisms, these associations are of great potential value as a novel approach to insect-pest management.

I. DIVERSITY OF SYMBIOTIC MICROORGANISMS IN INSECTS

A. Distribution of Symbiotic Microorganisms across the Microbial Kingdoms

Symbiotic microorganisms include members of all microbial kingdoms (Table I)—various Eubacteria, methanogens (Archaea), and protists and fungi (Eukaryota). The Eubacteria, especially members of the γ-Protobacteria, are widely represented both in the guts and cells of insects, but methanogens and protists occur in the strictly anaerobic portions of the guts of certain insects. Yeasts have been reported in the gut lumen, cells, and haemocoel (body cavity) of some species. (Basidiomycete fungi are also cultivated in the nests of some insects, e.g., fungus-gardening termites and leaf-cutting ants, but these ectosymbioses are not considered here.)

TABLE I
Survey of Symbiotic Microorganisms in Insects

Insect	*Microorganism*	*Location*[a]	*Incidence*
Blattaria (cockroaches)	Flavobacteria	B in fat body	Universal
	Various bacteria	Hindgut	Universal
Isoptera (termites)	Various bacteria[b]	Hindgut	Universal
	Flagellate protists	Hindgut	Lower termites
Heteroptera	Various bacteria[b]		
Cimicidae		B in haemocoel	Universal
Coreidae		Midgut	Widespread/irregular
Lygaeidae		Midgut	Widespread/irregular
Pentatomidae		Midgut	Widespread/irregular
Pyrrhocoridae		Midgut	Widespread/irregular
Triatomidae		Midgut	Universal
Homoptera	Bacteria: including γ-Protobacteria (in aphids and whitefly) and β-Protobacteria (in mealybugs)	B in various locations; in haemocoel	Nearly universal[c]
	Pyrenomycete yeasts	Predominantly extracellular in fatbody/haemocoel	In delphacid planthoppers and hormaphidine aphids
Anoplura	Bacteria[b]	B, variable locations	Universal
Mallophaga	Bacteria[b]	B in haemocoel	Irregular
Diptera			
Glossinidae	γ3-Protobacteria	B in midgut epithelium	Universal
Diptera Pupiparia	Bacteria[b]	B in haemocoel	Universal
Coleoptera			
Anobiidae	Yeasts	B in midgut caeca	Universal
Bostrychidae	Yeasts	B in haemocoel	Universal
Cerambycidae	Bacteria	B in midgut caeca	Widespread
Chrysomelidae	Bacteria	B in midgut caeca	Irregular
Curculionidae	Bacteria	B in variable location	Widespread
Lucanidae	Bacteria	Midgut or hindgut	Universal
Formicidae (ants)			
Camponoti	γ3-Protobacteria		Universal
Formicini	Bacteria[b]		Irregular

[a] B indicates bacteriocytes.
[b] These bacteria have not been characterized by molecular techniques and their phylogenetic position is unknown.
[c] Absent from typhlocybine leafhoppers, phylloxerid aphids, and apoimorphine scale insects.

B. Symbiotic Microorganisms in Insect Guts

Most insects have a substantial gut microbiota, although there are wide differences among insect taxa and among regions of the gut. Much of the literature gives misleading estimates of the microbial diversity in insect guts because the techniques commonly used are based on culturable forms, which account for only 0.1–10% of the total microbiota. An additional complication to the study of symbiotic microorganisms in insect guts is that many or all members of the microbiota are either transient (i.e., passing through the gut with the unidirectional passage of food), or commensal (i.e., resident for extended periods, but of no discernible advantage to the insect). Detailed experimental study is required to identify which, if any, of the resident gut microbi-

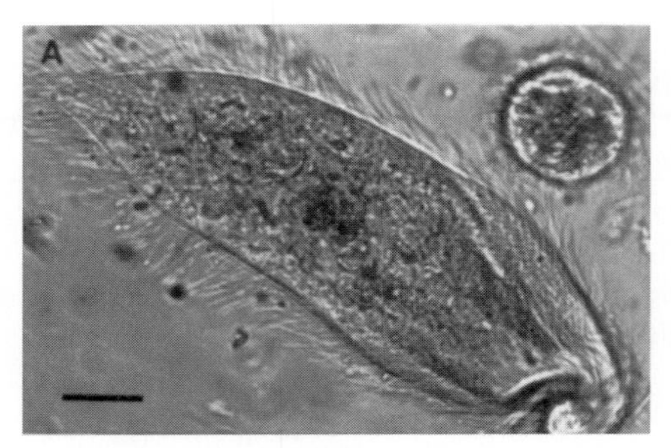

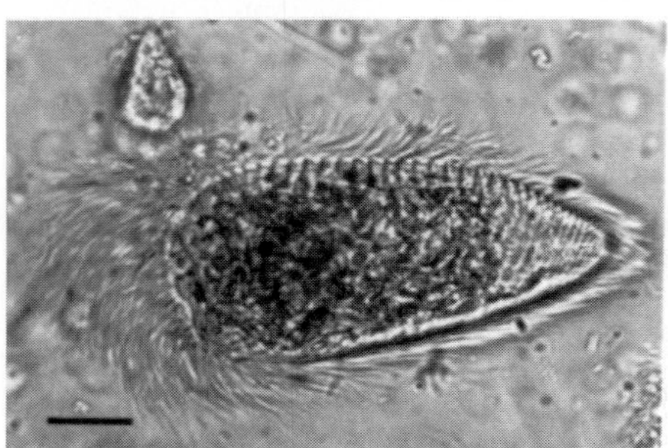

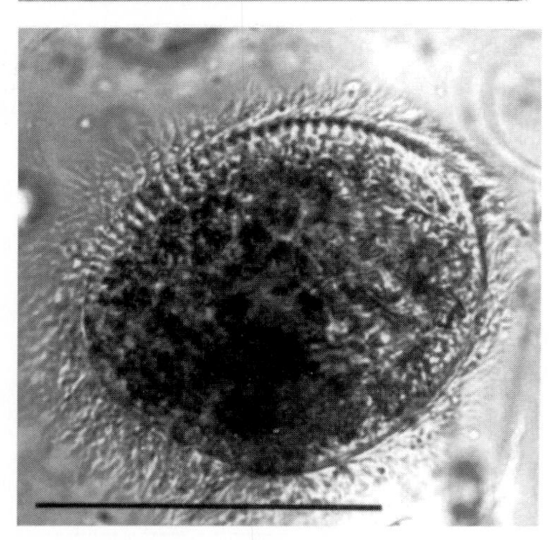

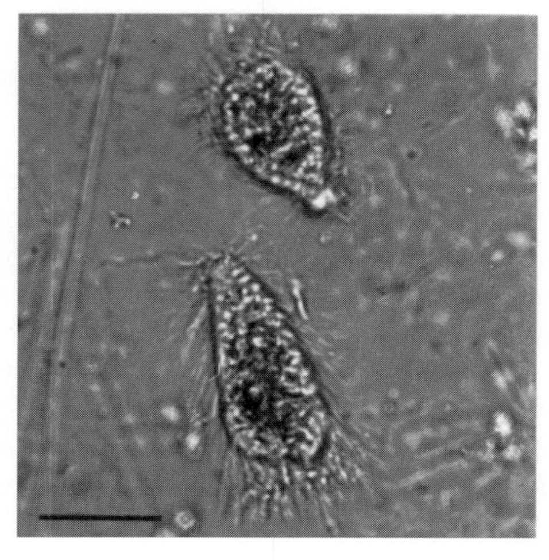

ota are symbiotic microorganisms (i.e., beneficial to the insect).

The microbiology of termite guts has been studied extensively. The greatest density of microorganisms is in the anoxic proximal portion of the hindgut, known as the paunch. In all termites, this region harbors bacteria, at 10^9–10^{10} cells/ml gut volume. All the bacteria are facultative or obligate anaerobes. They comprise methanogens, spirochaetes (pillotinas, e.g., *Hollandia, Pillotina, Diplocalyx,* and *Clevelandina,* and many other unidentified forms), and other eubacteria, including species of *Enterobacter, Bacteroides, Bacillus, Citrobacter, Streptococcus,* and *Staphylococcus*. The lower termites in addition have obligately anaerobic, flagellate protists of the orders Hypermastigida and Trichomonadida and Oxymonadida at densities of up to 10^7 cells/ml (Fig. 1). Approximately 400 species of these protists have been reported and a few, including *Trichomitopsis termopsidus* and *Trichonympha sphaerica* (both from the termite *Zootermopsis*) have been brought into axenic culture. Higher termites lack these protists.

Apart from the termites, most studies have concerned the gut microbiota of the few insect species that are routinely reared in laboratories. As examples, the American cockroach *Periplaneta americana* bears both obligately anaerobic bacteria (e.g., *Clostridium* and *Fusobacterium*) at densities of $\sim 10^{10}$ cells/ml and facultative anaerobes (e.g., *Klebsiella, Yersinia, and Bacteroides*) at $\sim 10^8$ cells/ml; the locust *Schistocerca gregaria* has exclusively facultative anaerobes, usually at considerably lower densities than the cockroach; and the blood-feeding reduviid bug *Rhodnius prolixus* has a diversity of bacteria, including *Pseudomonas, Streptococcus, Corynebacterium,* and various actinomycetes (and not a single gut symbiont, the actinomycete *Nocardia rhodnii,* as claimed in much of the early literature).

Fig. 1. Symbiotic protists in the hindgut of a lower termite *Coptotermes lacteus*. Bar = 25 μm. Photographs by R. T. Czolij and M. Slaytor, reproduced from Fig. 3a of Douglas, A. E. (1992). "Encyclopedia of Microbiology," Vol. 4, pp. 165–178.

The composition of microorganisms in the guts of insects can vary widely with environmental circumstance. The microbiota may change when insects are transferred to laboratory rearings, and specific differences in the microbiota of insects in different laboratories or under different temperature or dietary regimes have been described.

C. Symbiotic Microorganisms in Insect Tissues and Cells

Many symbiotic microorganisms are located in insect cells, where they may be well protected from the hemolymph-based defense system of the insect. In any single insect, the cells bearing intracellular symbiotic microorganisms are usually of a single morphological form (e.g., see Fig. 2) and location in the insect body (see Table I). They are called bacteriocytes or mycetocytes, and their sole function appears to be the housing of the microorganisms.

The incidence of intracellular microorganisms in insects and, where known, taxonomic information on the microorganisms are summarized in Table I. Most of the microorganisms are Eubacteria but, because they have not been brought into axenic culture, taxonomic information is available only for those forms whose 16*S* rRNA has been sequenced. Many of the bacteria are members of the γ-Protobacteria, and the microorganisms in three taxonomically disparate insect groups, aphids, tsetse flies, and ants, are particularly closely related. The bacteria in aphids and tsetse flies are known as *Buchnera* spp. and *Wigglesworthia* spp., respectively, in recognition of the research of entomologists Buchner and Wigglesworth on these systems. Other intracellular bacteria in bacteriocytes include members of the β-Protobacteria in mealybugs and flavobacteria in cockroaches. The bacteria in several insect groups, including the Anoplura and Mallophaga (the sucking and chewing lice, respectively) and various Heteroptera, have not been studied at the molecular level (see Table I).

Some insects bear microorganisms whose location (intracellular vs. extracellular) is variable. For example, delphacid planthoppers and hormaphidine aphids lack intracellular bacteria (unlike other planthoppers and aphids), and bear pyrenomycete yeasts in and between fat body cells. Some species of aphid and tsetse fly have secondary symbionts, in addition to *Buchnera* and *Wigglesworthia,* respectively. The secondary symbionts are variably located in cells and the haemocoel, and of uncertain significance to the insect.

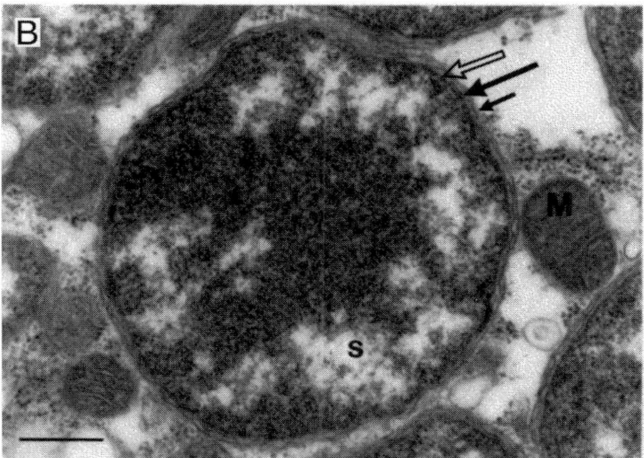

Fig. 2. *Buchnera,* the intracellular symbiotic bacterium in aphids. (A) The symbiotic bacteria occupy the greater part of the bacteriocyte cytoplasm (electron micrograph of pea aphid, *Acyrthosiphom pisum*); bar = 5 μm. Reproduced from Douglas, A. E. (1997). "Symbiotic Interactions," by permission of Oxford University Press, Oxford. (B) Each symbiotic bacterium (S) is bounded by a cell membrane (⇑), cell wall (↑), and insect membrane (↑). A mitochondrion (M) marks the insect cell cytoplasm (electron micrograph of black bean aphid, *Aphis fabae*); bar = 5 μm. Reproduced from Fig. 1a of Douglas A. E. (1992). "Encyclopedia of Microbiology", Vol. 4, pp. 165–178.

II. FUNCTION OF SYMBIOTIC MICROORGANISMS

A. Symbiotic Microorganisms as a Source of Novel Metabolic Capabilities

Symbiotic microorganisms are widely believed to contribute to the nutrition of insects. This was first deduced from the distribution of the associations among insects. In general, the microorganims are restricted to insects living on nutritionally poor or unbalanced diets. They are widespread or universal among insects feeding through the lifecycle on the phloem and xylem sap of plants, deficient in essential amino acids; vertebrate blood, deficient in B vitamins; and wood, which is composed principally of lignocellulose and is deficient in many essential nutrients for insects. The implication is that the symbiotic microorganisms variously degrade cellulose and synthesize essential amino acids and vitamins. They have also been implicated in the synthesis of sterols, which insects and other arthropods cannot synthesize *de novo*.

The symbiotic microorganisms can be considered to be a source of biochemically and genetically complex metabolic capabilities that the insect lacks. They have also been described as microbial brokers, mediating insect utilization of blood, plant sap, and wood.

B. Contribution of Symbiotic Microorganisms to the Nitrogen Nutrition of Insects

Three routes have been identified by which symbiotic microorganisms contribute to the nitrogen nutrition of insects, the fixation of N_2, nitrogen recycling, and, the synthesis of essential amino acids.

Symbiotic microorganisms in the gut of some insects fix N_2 at appreciable rates; no intracellular N_2-fixing bacteria in bacteriocytes have been described. In particular, many termite species derive significant supplementary nitrogen from N_2-fixing bacteria, (e.g., *Enterobacter agglomerans* and *Citrobacter freundii*) in the anoxic portion of their hindgut. The N_2 fixation rate varies widely among termite species, from <0.2 g N fixed/g insect weight/day in *Labiotermes* sp. and *Cubitermes* sp. to >6 g N/g/day in *Nasutitermes* species, and is also influenced by environmental conditions, including the concentration of combined nitrogen in the diet.

Nitrogen recycling refers to the microbial consumption of nitrogenous waste products of insects and the synthesis of compounds (e.g., essential amino acids) of nutritional value to the insect, which are then translocated back to the animal. Microbial utilization of insect-derived uric acid or ammonia has been demonstrated in several systems, including cockroaches, planthoppers, aphids, and termites. For example, various bacteria, including *Streptococcus, Bacteroides,* and *Citrobacter* species, in the hindgut of the termite *Reticulotermes flavipes* degrade uric acid anaerobically to ammonia, carbon dioxide, and acetic acid. Experiments using ^{14}C and ^{15}N-labeled uric acid confirmed that uric acid is degraded by the hindgut microbiota in the insect and nitrogen is subsequently assimilated by the insect tissues, in the insect.

Microbial provision of essential amino acids to the insect has been studied systematically in the symbiosis between aphids and the intracellular bacteria *Buchnera*. The core evidence is nutritional, and arises from the development of chemically defined diets consisting of sucrose, amino acids, vitamins, and minerals, on which aphids can be reared. Dietary studies in which the 20 amino acids of proteins are individually omitted have revealed that many aphids have no specific requirement for most or all the amino acids that are normally dietary essentials for animals, but that aphids experimentally deprived of *Buchnera* by antibiotic treatment require all the essential amino acids. The implication, that the insect derives essential amino acids from *Buchnera,* is supported by radiotracer studies demonstrating the synthesis *de novo* of various essential amino acids by aphids bearing *Buchnera* (Fig. 3).

The analysis of the plasmid profiles in *Buchnera* has revealed molecular support for the role of these bacteria in the amino acid nutrition of aphids. *Buchnera* in many aphids, including all members of the family Aphididae studied to date, bear two multicopy plasmids on which genes for the biosynthesis of tryptophan and leucine are amplified. Figure 4 shows the genetic organization of these plasmids in the *Buchnera* from which they were first described.

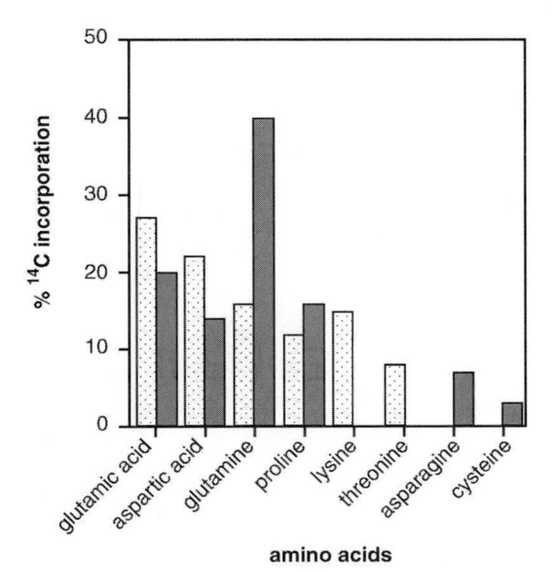

Fig. 3. Essential amino acid synthesis by pea aphid-*Buchnera* symbiosis, as demonstrated by the incorporation of ^{14}C-glutamic acid into the free amino acid pool of aphids containing *Buchnera* (open-hatched) and experimentally deprived of *Buchnera* (closed-hatched). Only the aphids with the symbiotic bacteria synthesize the essential amino acids lysine and threonine. Reproduced from Fig. 1b of Wilkinson and Douglas (1996). *Entomol. Exp. Appl.* **80,** 279–282, with kind permission from Kluwer Academic Publishers.

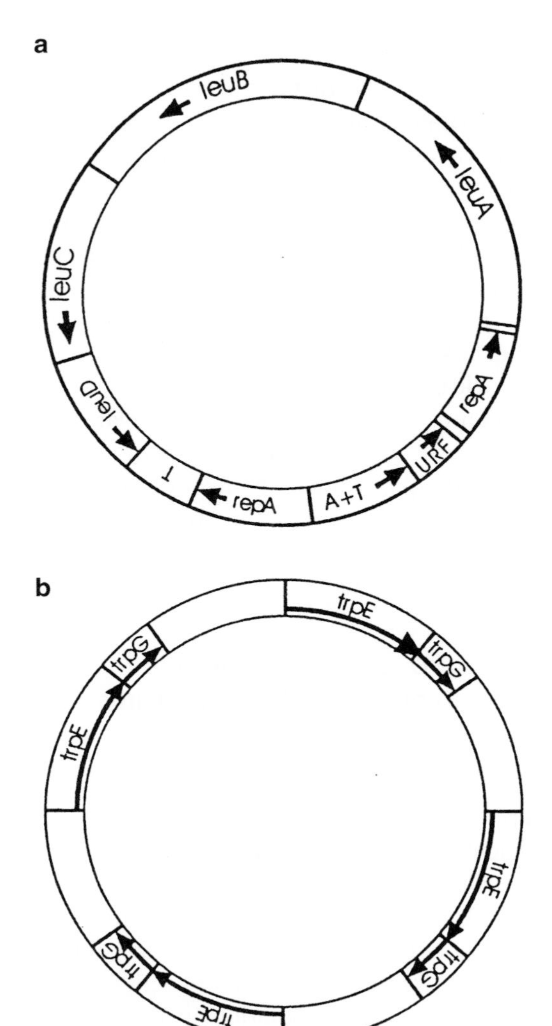

Fig. 4. The plasmids of *Buchnera* from members of the family Aphididae. (a) Gene map of the leucine plasmid pRPE of *Buchnera* from *Rhopalosiphum padi* (7.8 kb) with genes *leuA–D* coding the enzymes in the dedicated leucine biosynthetic pathway. (b) Gene map of the tryptophan plasmid pBA-trpEG of *Buchnera* from *Schizaphis graminum* (14.4 kb) with four apparently identicial tandem repeats including *trpEG,* which codes for anthranilate synthase. This enzyme is the regulatory enzyme in tryptophan biosynthesis, subject to feedback inhibitition by tryptophan. It has been argued that amplification of *trpEG* results in the overproduction of anthranilate synthase and the consequent sustained tryptophan synthesis in the presence of tryptophan. Reproduced from Figs. 2 a & b of Douglas, A. E. (1997). *FEMS Microbiol. Ecol.* **24,** 1–9.

The synthesis of essential amino acid by intracellular bacteria has not been studied systematically in any insects apart from aphids. There is, however, a strong but unproven supposition that many insects, especially phloem-feeding Homoptera (e.g., whitefly and psyllids) and cockroaches, derive these nutrients from their symbiotic microorganisms. If confirmed, the essential amino acid provisioning has evolved independently in different bacterial groups.

C. Vitamin Synthesis by Symbiotic Microorganisms

The microbial provision of B vitamins has been proposed for the diverse insect taxa that feed on vertebrate blood and bear microorganisms in either the gut or bacteriocytes. The experimental basis for this role is exclusively nutritional, based on insect performance. Triatomid bugs deprived of their gut microbiota die as larvae, but this developmental arrest is alleviated by injection of B vitamins into the insect or vitamin supplementation of the diet. Similarly, the larvae of the louse *Pediculus* lacking symbiotic bacteria suffer high mortality, unless the blood diet is supplemented with nicotinic acid, pantothenic acid, and biotin, and B vitamins have been reported to partially restore the fecundity of tsetse fly from which the bacteria are eliminated.

Other insects feeding on nutritionally poor diets, (e.g., cockroaches and timber beetles) may also derive vitamins from symbiotic microorganisms. The anobiid beetle *Stegobium paniceum* is independent of several B vitamins, riboflavin, nicotinic acid, pyridoxine, and pantothenic acid, but insects from which the yeasts are eliminated require these vitamins for normal development.

D. Sterol Synthesis by Symbiotic Microorganisms

Yeasts may contribute to the sterol nutrition of insects. Despite some claims in the literature, for example, that aphids derive sterols from their symbiotic bacteria *Buchnera*, bacterial provision of these nutrients is most improbable because the Eubacteria are not capable of substantial sterol synthesis.

Yeasts have been implicated in the sterol nutrition of some planthoppers and timber beetles. For example, when the yeast population in the planthopper *Laodelphax striatellus* is depleted, many of the insects die during the final molt to adulthood, but mortality was reduced from 94 to 40% by injecting the insects with either cholesterol or the plant sterol sistosterol.

E. Cellulose Degradation

Many insects feeding on fiber-rich plant material, especially wood, have substantial gut microbiota. For many years, these insects have been assumed to be strictly comparable to vertebrate herbivores in which microorganisms mediate cellulose degradation. This can be illustrated by microbe-mediated cellulolysis in lower termites. Protists in the hindgust of lower termites (see Section I.B) can be eliminated by incubating the insects at elevated oxygen tensions, and these protist-free insects, commonly termed defaunated termites, cannot survive on high, cellulose diets, such as filter paper (Fig. 5). The protists degrade cellulose, with carbon dioxide and short-chain fatty acids, especially acetate, as the principal products of fermentation. The acetate is absorbed across the hindgut wall and metabolized as a source of energy by the aerobic tissues of the termite. Cellulase active against crystalline cellulose has been demonstrated in the protist *Trichomitopsis* and in mixed populations of protists from *Coptotermes lacteus*.

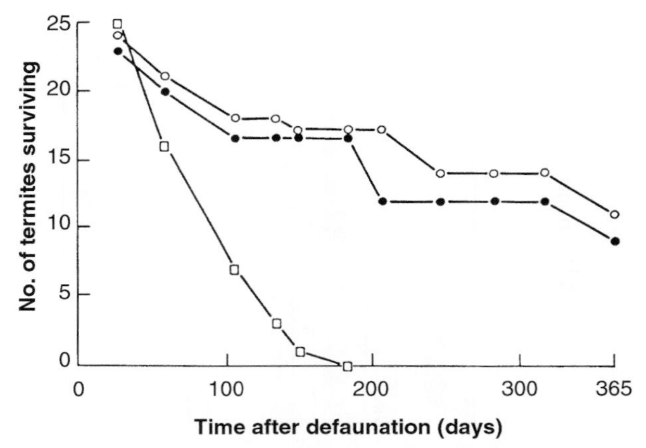

Fig. 5. Survival of the lower termite *Zootermopsis* maintained on a diet of cellulose. Twenty-five days after the removal of the symbiotic protists from the hindgut defaunation by exposure to oxygen, termites were reinfected with protists from untreated termites (23 insects, filled circles), axenic culture of *Trichomitopsis termopsidis* (24 insects, open circles), and heat-killed *T. termopsidis* (25 insects, open squares), and their survival was assayed over 1 year. Reproduced from Fig. 8.8 of Smith and Douglas (1987). "The Biology of Symbiosis." Edward Arnold.

Although the experimental data on lower termites are not in serious doubt, this system cannot be generalized to all insects feeding on high-fiber diets. It is now recognized that some insects (unlike vertebrates) have intrinsic cellulases, especially endoglucanases and β-glucosidases. For example, the higher termites (which lack protists) do not have cellulolytic microbiota, but instead possess high activities of intrinsic cellulases, especially in the midgut (Table II).

The contribution of intrinsic and microbial cellulases to cellulose breakdown has been investigated in relatively few insects other than termites. The locust *Schistocerca gregaria* has intrinsic cellulases;

TABLE II
Sites of Cellulase Activity in Termites[a]

Gut region	Cellulase activity (% of total activity)	
	Nasutitermes walkeri[b]	*Mastotermes darwiniensis*[c]
Foregut and salivary glands	5	13
Midgut	94	13
Hindgut	1	74

[a] Data from Veivers *et al.* (1982). *Insect Biochem.* **12**, 35–40 and Hogan *et al.* (1988). *J. Insect Physiol.* **34**, 891–899.
[b] A higher termite.
[c] A lower termite.

the scarab beetle *Pachnoda marginata* uses the cellulolytic capability of bacteria in its hindgut; and among the woodroaches (cockroaches that feed on wood), *Panesthia cribatus* uses intrinsic cellulases and has gut microbiota of noncellulolytic bacteria, whereas *Cryptocercus punctulatus* has up to 25 species of obligately anaerobic protists that degrade cellulose. There is no evidence that the efficiency of fiber digestion by the insect is influenced by the origin (intrinsic or microbial) of the cellulolytic enzymes.

Insects with microbial cellulolysis generally have a high population of methanogenic bacteria. These bacteria act as a sink for hydrogen produced by anaerobically respiring microorganisms, and so promote cellulose degradation.

III. DETERMINANTS OF THE DENSITY OF SYMBIOTIC MICROORGANISMS IN INSECTS

A. Gut Microorganisms

Insect guts are physically unstable environments. The gut lumen is dominated by the unidirectional bulk flow of ingested food and many microbial cells pass directly through the gut (see also Section I.B). Microbial persistence in the gut is promoted by:

1. A higher proliferation rate than the rate of passage of the food (this is probably important in insects with cellulolytic microbiota in enlarged fermentation chambers).
2. The adhesion of microorganisms to the gut wall.
3. The sequestration of microorganisms into outpocketings of the gut (e.g., midgut caeca and Malpighian tubules).

A further aspect of the instability of the gut environment is that the cuticle and contents of the foregut and hindgut are lost at each insect molt, such that the microbiota is reestablished *de novo* multiple times through an insect's lifespan. The midgut microbiota can persist through insect molts, but is commonly lost at the metamorphosis of holometabolous insects (insects with complete metamorphosis, e.g., flies and beetles). In other respects, however, the midgut of many insects is a hostile environment for microorganisms because it is the principal site of digestive enzymes.

B. Intracellular Microorganisms

The symbiotic microorganisms in the cells and tissues of insects are not generally subject to the frequent disturbance experienced by the foregut and hindgut microbiota (see Section III.A), and their populations are probably regulated by density-dependent processes, mediated by the insect.

The regulation of intracellular bacteria *Buchnera* in aphids has been well studied. In the wingless parthenogenetic morph of aphids, the bacterial population increases in parallel with aphid biomass through larval development (Fig. 6), and in adult aphids the bacteria occur at a density of $\sim 10^7$ cells/mg aphid fresh weight, equivalent to 10% of the

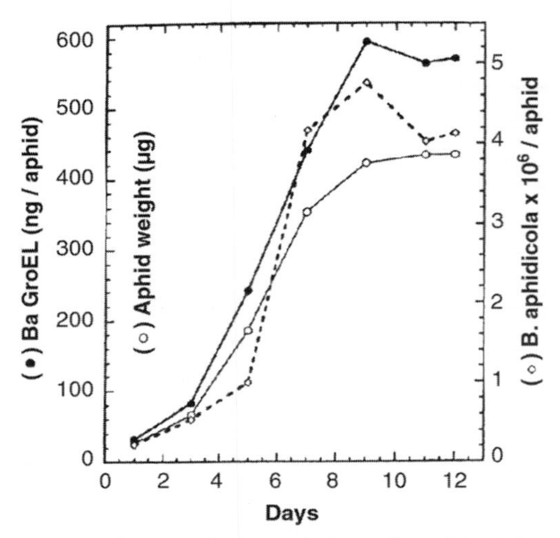

Fig. 6. Controls over the population of symbiotic bacteria *Buchnera* over larval development of the aphid *Schizaphis graminum*. The increase in the bacterial population, as quantified both directly and from the increase in the bacterial protein GroEL, broadly parallels the increase in aphid weight. Reproduced from Fig. 4 of Baumann, Baumann, and Clark (1996). *Curr. Microbiol.* **32,** 279–285, copyright notice of Springer-Verlag.

total insect volume. Regulation occurs at two levels, controls over the bacterial division rate in each bacteriocyte such that they maintain a uniform density, equivalent to 60% of the cytoplasmic volume of bacteriocytes, and controls over the number of bacteriocytes through age-dependent lysis of the bacteriocytes and all enclosed bacteria. Adults of winged female aphids and male aphids tend to have a smaller *Buchnera* population than the wingless morph and these differences are mediated by both a lower rate of bacteriocyte enlargement and higher incidence of bacteriocyte death during larval development. The possibility that these morph-specific differences are mediated by insect hormones has not been investigated. Reduced bacterial populations in males has also been demonstrated in leafhoppers, weevils, and cockroaches.

IV. TRANSMISSION OF SYMBIOTIC MICROORGANISMS

A. Gut Microbiota

All members of the gut microbiota are acquired via insect feeding. For many microorganisms, the transfer between insects is haphazard, dependent on the chance ingestion with food, but in many taxa, the insect eggs are provisioned with gut microorganisms. The transmission of some microorganisms is assured by stereotyped feeding responses of the insect. This can be illustrated by the heteropteran bug *Coptosoma scutellarum,* which bears bacteria in its midgut caeca. A capsule bearing bacteria is deposited alongside each oviposited egg. When the larva hatches, it immediately feeds on the capsule contents and acquires its complement of bacteria (Fig. 7), on which its growth and development depend.

Insect behavior is also implicated in the transmission of obligately anaerobic microorganisms, especially the cellulolytic protists in certain woodroaches and the lower termites. At each molt of the insect, the oxygen tension in the hindgut increases dramatically and all the protists are killed. In *Cryptocercus,* the protists initiate sexual reproduction just before each molt and are transformed into oxygen-resistant cysts. They are expelled from the hindgut into the environment, where they persist until ingested by the insect after molting. In contrast, the protists in lower termites rarely reproduce sexually and never encyst, and they are killed at each insect molt. The termites acquire a fresh inoculum of protists by feeding on a drop

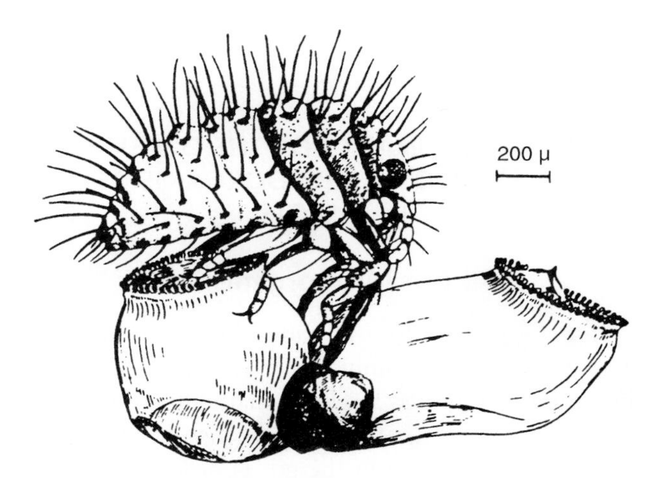

Fig. 7. Role of feeding behavior of insects in the transmission of symbiotic microorganisms. The newly hatched larva of the heteropteran bug *Coptosoma scutellatum* feeds on a capsule bearing symbiotic bacteria that the mother had deposited alongside the egg. Reproduced from Fig. 94c of Buchner (1965). "Endosymbioses of Animals with Plant Microorganisms." Wiley, London.

of hindgut contents from the anus of another colony member, a behavior known as proctodeal trophyllaxis. It has been suggested that the requirement for conspecifics as a source of protists may have been a major selection pressure for the evolution of eusociality in termites. Higher termites, which lack the protists, do not exhibit proctodeal trophyllaxis.

B. Transovarial Transmission

Microorganisms located in insect tissues and cells are generally transmitted from mother to offspring via the unfertilized egg in the female ovary. As a result, the symbiotic microorganisms are present even before fertilization and, potentially, for the entire lifespan of the insect. The timing and anatomical details of transovarial transmission vary widely among insect groups, consistent with these associations having evolved independently on multiple occasions. In some insects (e.g., aphids), the bacteriocytes are closely apposed to the ovaries and the bacteria have a fleeting extracellular stage in transit from bacteriocyte to ovaries. In other insects, the bacteria have a prolonged extracellular phase, either because they are expelled from bacteriocytes at a distance from the ovaries and migrate to the ovaries (e.g., many species of lice) or because they remain on the egg surface for extended periods (e.g., cockroaches). Varying among insect taxa, the bacteria are phagocytosed directly by the egg, usually at the time of vitellogenesis (e.g., aphids and cockroaches) or taken up by insect cells at the base of each ovariole and inoculated into the posterior pole of the egg just prior to formation of the chorion (the egg shell) and ovulation.

C. Vertical Transmission and Its Evolutionary Consequences

Transovarial transmission and the more sophisticated instances of microbial transmission by egg smearing ensure that each egg is colonized by bacteria from its mother. If vertical transmission persists

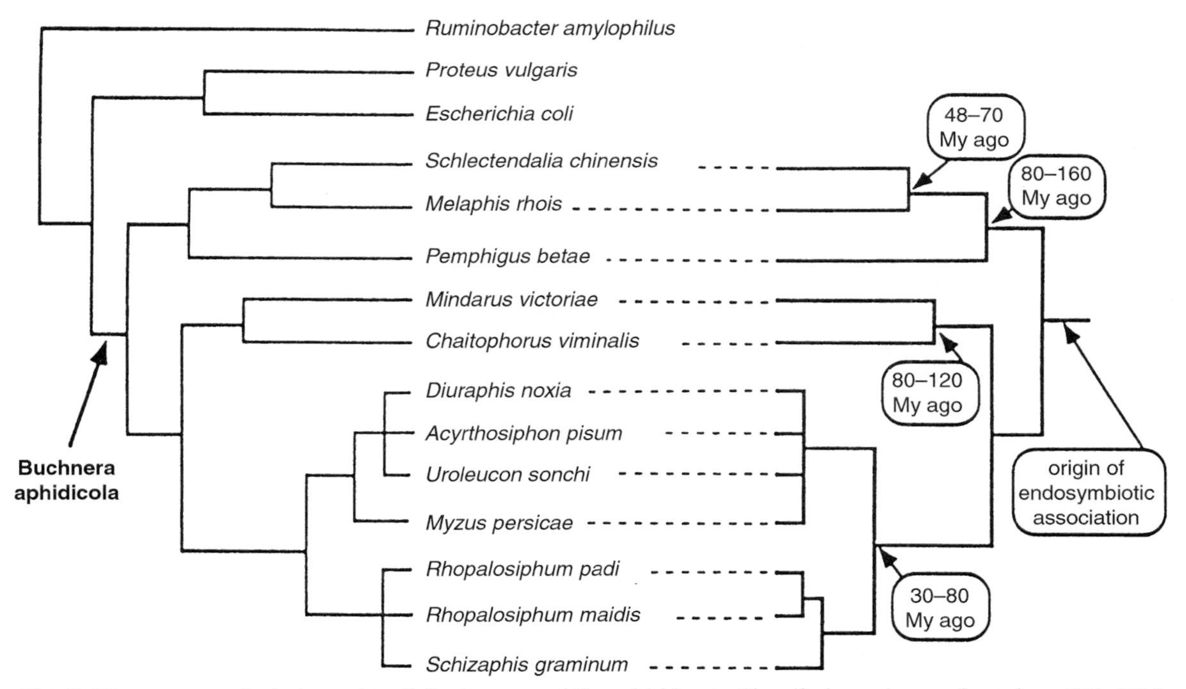

Fig. 8. The congruent phylogenies of *Buchnera* and its aphid hosts. The phylogenies are based on 16*S* rDNA sequences for *Buchnera* and morphology for aphids, and the dates are estimates from aphid fossils or biogeography. Reproduced from Fig. 1 of Moran and Baumann (1994). *Trends Ecol. Evol.* **9,** 15–20, with permission from Elsevier Science.

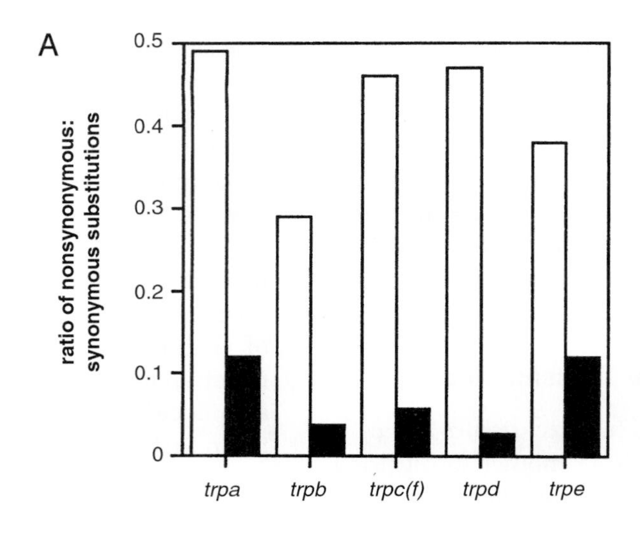

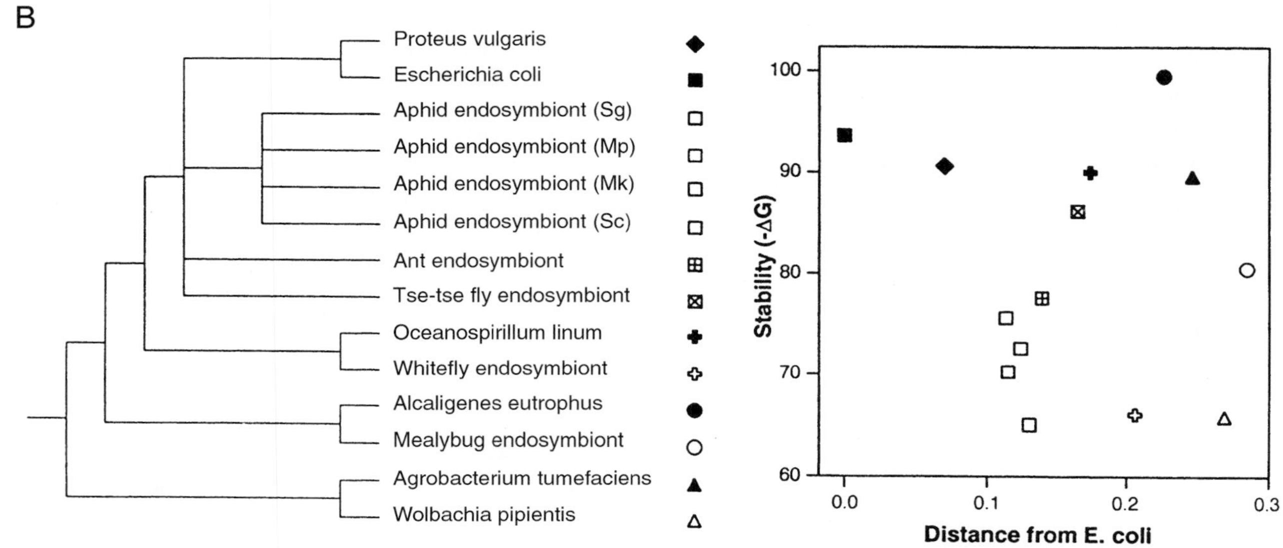

Fig. 9. The consequences of vertical transmission for the molecular evolution of *Buchnera*. (A) Nonsynonymous substitutions in protein-coding genes of *Buchnera*. The number of substitutions between *Buchnera* in the aphids *Schizaphis graminum* and *Schlechtendalia chinensis* (open bars) and between the two enteric bacteria *Escherichia coli* and *Salmonella typhimurium* (solid bars) is expressed as the ratio of nonsynonymous : synonymous substitutions per nucleotide site. Data are presented for the five genes of the *trp* operon. Reproduced with permission from Fig. 1 of Hurst and McVean (1996). *Nature* **381,** 650–651. Copyright 1996 MacMillan Magazines Limited. (B) The stability of Domain I of the 16*S* rRNA in symbiotic bacteria and free-living bacteria, showing the relationships of bacterial taxa and stabilities (-Δ*G*) summed over Domain I for each organism. Reproduced from Fig. 2 of Lambert and Moran (1998). *Proc. Natl. Acad. Sci. U.S.A.* **95,** 4458–462. Copyright (1998) National Academy of Sciences, U.S.A.

over many insect generations without cross-infection, the insect and microbial lineages evolve in parallel, and their phylogenies are congruent. Congruent phylogenies have been demonstrated for several insect–microbial symbioses, most notably the aphid–*Buchnera* association, in which the 16*S* rRNA sequence phylogeny of *Buchnera* is completely concordant with morphology-based phylogeny of aphids (Fig. 8). On the reasonable assumption that the aphids and *Buchnera* diversified in synchrony, the date of divergence between *Buchnera* and its relatives, such as *Escherichia coli,* has been estimated at 180–250 million years ago.

Vertical transmission, especially by the transovarial route, has two characteristics of significance for the evolution of symbiotic microorganisms. First, relatively small numbers of cells are transferred from parent to offspring; that is, the effective population size of the microorganisms is small. Second, the strict maternal inheritance prevents any contact between microbial populations in different insects, precluding recombination. Deleterious mutations are likely to accumulate in these small asexual populations. Supportive evidence comes from studies of sequence evolution in intracellular bacteria. First, among the *Buchnera* lineages, protein-coding genes have significantly elevated the incidence of nonsynonymous substitutions (i.e., those point mutations that alter the amino acid) and this is not paralleled by an increase in the rate of synonymous substitutions (Fig. 9A). Second, vertically transmitted intracellular bacteria have base substitutions in the 16*S* rRNA gene that tend to destablize the secondary structure of the rRNA molecule (Fig. 9B).

V. SYMBIOTIC MICROORGANISMS AND INSECT-PEST MANAGEMENT

Many insects that depend on symbiotic microorganisms for sustained growth and fecundity are pests of agricultural or medical importance. They include crop pests (e.g., aphids, whitefly, grain beetles, and timber beetles) and vectors of animal and human pathogens (e.g., tsetse fly and bedbugs). In addition, certain microbial taxa have been implicated in the vector competence of their insect hosts. The transmission of luteoviruses by aphids is promoted by binding of a protein, the chaparonin GroEL derived from the intracellular bacteria *Buchnera,* to the surface of the virus particles in the insect body, and the inhibition of trypanosome infection in the tsetse fly by an insect lectin is blocked by high *N*-acetylglucosamine titers generated by the chitinase activity of a midgut bacterium.

The selective disruption of the symbiotic microorganisms in insect pests, and the consequent depression in insect performance, would be of considerable economic value but no commercial approach has been developed. There is also interest in the genetic manipulation of symbiotic microorganisms, for example to reduce insect-vector competence. The intracellular microorganisms are generally perceived as intractable because the methods to culture and transform them and to reintroduce the transformed bacteria into insects have not been developed. Most research has been conducted on blood-feeding insects with gut microbiota. For example, a plasmid of the actinomycete *Rhodococcus rhodnii,* which inhabits the gut of *Rhodnius prolixi,* has been genetically modified to bear the plasmid replication origins for both *R. rhodnii* and *E. coli,* and sustained infections of *Rhodnius* have been achieved with the transformed bacteria. This technology is being developed as part of a strategy to reduce the *Rhodnius*-mediated transmission of the protozoan pathogen that causes Chagas's disease in humans.

See Also the Following Articles

Cellulases • Insecticides, Microbial • Nitrogen Fixation

Bibliography

Baumann, P., Lai, C-Y., Roubakhsh, D., Moran, N. A., and Clark, M. A. (1995). Genetics, physiology, and evolutionary relationships of the genus *Buchnera*—intracellular symbionts of aphids. *Annu. Rev. Microbiol.* **49**, 55–94.

Beard, C. B., O'Neill, S. L., Tesh, R. B., Richards, F. F., and Aksoy, S. (1993). Modification of arthropod vector competence via symbiotic bacteria. *Parasitol. Today* **9**, 179–183.

Breznak, J. A. (1982). Intestinal microbiota of termites and other xylophagous insects. *Annu. Rev. Microbiol.* **36**, 323–343.

Buchner, P. (1965). "Endosymbioses of Animals with Plant Microorganisms." Wiley, London.

Douglas, A. E. (1989). Mycetocyte symbiosis in insects. *Biol. Rev.* **64**, 409–434.

Douglas, A. E. (1998). Nutritional interactions in insect-microbial symbioses: aphids and their symbiotic bacteria *Buchnera. Annu. Rev. Entomol.* **43**, 17–37.

Moran, N. A., and Telang, A. (1998). Bacteriocyte-associated symbionts of insects—A variety of insect groups harbor ancient prokaryotic endosymbionts. *BioScience* **48**, 295–304.

Slaytor, M., and Chappell, D. J. (1994). Nitrogen metabolism in termites. *Comp. Biochem. Physiol.* **107B**, 1–10.

Syphilis, Historical

David Shumway Jones
Harvard Medical School

I. The Origins of Syphilis
II. The History of a Disease
III. Managing Syphilis
IV. The Limits of Biomedicine

GLOSSARY

"magic bullet" The name given to Paul Ehrlich's Salvarsan, the first specific antibiotic, representing the hope that use of a specific drug to kill the specific bacterial cause of syphilis would control the disease.

***morbus gallicus* (*the French disease*)** The most common name given to a new disease that appeared in Europe in the late fifteenth century; many physicians and historians identify morbus gallicus as syphilis.

paleopathology The technique of examining ancient skeletal remains to identify the diseases suffered by historical populations.

Treponema pallidum The causative agent of syphilis. It is extremely similar, possibly identical, to the organisms that cause endemic syphilis, yaws, and pinta.

THE HISTORY OF SYPHILIS has long been one of the most popular topics in the history of medicine. Despite 500 years of research and debate by historians, physicians, and anthropologists, many questions about the origins of the disease remain unresolved. The development of syphilis since the sixteenth century demonstrates remarkable changes in both its medical symptoms and the cultural meanings of those symptoms. The history of efforts to control syphilis reveals the inevitable moral judgments about syphilis, the balance between disease control and individual rights, and the limited ability of even powerful medical remedies to control the disease. These lessons take on new relevance as fears of HIV and AIDS motivate efforts to eradicate syphilis.

I. THE ORIGINS OF SYPHILIS

The recorded history of syphilis began in the late fifteenth century with the appearance of a new disease, widely called "morbus gallicus," the French disease. The earliest cases were described in 1495 as the mercenary army of King Charles VIII of France retreated from its siege of Naples. Victims suffered from fevers, open sores, disfiguring scars, and disabling pains; many were consumed by the disease and met gruesome deaths. As the French army disbanded, infected soldiers carried the disease throughout Europe, to Germany in 1495, and to Holland, England, and Greece by 1496. The voyage of Vasco de Gama took it to India in 1498. By 1505 it had reached Japan. Witnesses described the disease with horror. Joseph Grünbeck (1473–1532) left a typical account: "In recent times I have seen scourges, horrible sicknesses and many infirmities affect mankind from all corners of the earth. Amongst them has crept in, from the western shores of Gaul, a disease which is so cruel, so distressing, so appalling that until now nothing so horrifying, nothing more terrible or disgusting, has even been known on this earth." This dramatic appearance has puzzled observers for centuries. The 1490s were a time of great transformations in Europe—Columbus encountered the Americas, France invaded Italy, the Spanish government expelled Jews and Moors from Spain, and the Pope abolished leper hospitals. All of these factors might have contributed to the emergence of a new disease.

Spanish witnesses traced the disease to Columbus's voyages to the Americas; they heard from natives that the disease had long been endemic in Hispaniola, and they witnessed unchaste Spaniards acquiring the infection while there. Details of the timing of the early Spanish voyages suggest that syphilis could have been carried back from the New World to Spain in time to appear in Italy by 1495. This would make syphilis the counterpart to the many diseases, notably smallpox, that the Spaniards carried from Europe to America. However, definitive evidence does not exist.

The traditional alternative to this theory asserts that syphilis always existed in Europe and Asia, but was not recognized before the 1490s. Detailed analyses of Egyptian papyri, the Hebrew Bible, Hippocratic writings, and medical texts from India and China contain abundant evidence of the prevalence of sexually transmitted infections. However, these records do not provide clear evidence of the specific presence of syphilis. Ambiguous medieval descriptions of leprosy, which observers believed could be transmitted sexually, have caused particular confusion.

Several novel explanations have been suggested in more recent decades. Four distinct diseases, venereal syphilis, endemic syphilis, yaws, and pinta, are all caused by closely related subspecies of *Treponema pallidum*. The four diseases might simply be different manifestations of infection with the same microbe, with the results of infection depending on social and environmental conditions: Yaws and endemic syphilis, long present in Europe, may have transformed into venereal syphilis in the fifteenth century when changes in sanitation, personal hygiene, and sexual behavior produced new conditions favorable for the venereal transmission and manifestations of the disease. Alternatively, venereal syphilis may have emerged as a result of new and virulent mutations to existing subspecies of *Treponema pallidum*. Paleopathological evidence from thousands of bones from pre-Columbian America, Europe, Africa, and Asia suggests that yaws had long existed in Europe and America. Syphilis may have evolved by mutation from yaws in America, returned to Europe with the Spanish, thrived in its new environs, and produced the outbreak of morbus gallicus. New types of evidence continue to appear. Immunologists have sought treponemal antigens in mummies and skeletal remains, finding some in a Pleistocene bear from Indiana. The recent sequencing of the 1,138,006 bp of the treponemal genome has renewed hopes that state-of-the-art science will finally resolve this historical ambiguity.

II. THE HISTORY OF A DISEASE

The new disease had many dramatic impacts. Observers were horrified by its emergence and manifestations. Everyone blamed someone else. The Italians called it the "morbus gallicus" or the "mal francese"; the French named it the "mal de Naples." Others labeled it "scabies hispanicus," the "American disease," or, in Japan and the East Indies, the "Portuguese disease." Some names reflected its manifestations: the "Great pox," "fire-piss," "gangrene grossa," the "Neapolitan itch," and "plum-blossom sores." The term "syphilis" first appeared in the 1530 Latin poem of humanist and physician Girolamo Fracastoro (1483–1553), but this name was not widely used until the eighteenth century.

Early observers traced syphilis to astrological origins, noting the ominous conjunction in 1484 of Mars, Jupiter, and Saturn in Scorpio, the constellation most closely associated with genital affairs. By the 1520s, the disease had been connected to sexual transmission and given a new name, "lues venerea," the venereal disease. Many people saw the disease as a punishment from God for sexual debauchery (Fig. 1). Public bathhouses closed, distrust divided friends and lovers, and Platonic love emerged as a vibrant social cult.

Physicians struggled to understand the disease for centuries. Many perceived the varied symptoms, including urethral discharges, penile chancres, and skin rashes as a single phenomena, all caused by a single poison. However, in the eighteenth century, the concept of a single disease became increasingly contested; physicians argued for the existence of many different "morbi venerei." This debate drove English physician John Hunter (1728–1793) to his famous 1767 experiment. He reportedly inoculated his own penis with pus from a patient with gonorrhea; when he developed a characteristic syphilitic

¶ Tractatus de pestilentiali Scorra sive mala de Franzos. Originem. Remediaqz eiusdem continens. cōpilatus a venerabili viro Magistro Joseph Grunpeck de Burckhausen. sup Carmina quedam Sebastiani Brant utriusqz iuris professoris.

Fig. 1. Illustration from Joseph Grunpeck's *Tractatus de pestilentiali scorra, sive mala de Franzos* (1496). This image is traditionally interpreted as showing rays of divine wrath striking sinners with syphilitic lesions. It illustrates the theological interpretations of the appearance of the new epidemic.

chancre, he concluded that the existence of a single lues venerea had been proven.

Careful work in the nineteenth century settled the controversy. In 1838 Phillipe Record (1799–1889) reported the results of over 2500 experimental inoculations performed at a Paris hospital. He demonstrated that the primary, secondary, and tertiary stages of syphilis all represented a single disease, distinct from other venereal infections. Rudolf Virchow (1821–1902) and Alfred Fournier (1832–1913) extended this work, describing the characteristic pathological lesions of syphilis and the consequences of congenital infection. In 1879 Albert Neisser (1855–1916) settled lingering doubts by isolating the gonococcus, the causative agent of gonorrhea. The characterization of syphilis, in its modern form, was completed in the early twentieth century. In 1905, protozoologist Fritz Schaudinn (1871–1906) and syphilologist Erich Hoffman (1869–1959) described the slender spiral bacteria *Spriochaeta pallidum,* later renamed *Treponema pallidum.* In 1906, August Wassermann (1866–1925), working with Neisser and Carl Bruck (1879–1944) developed the Wasserman test, a complement fixation reaction that became the definitive serological test for syphilis. The disease had been defined, the agent identified, and an objective diagnostic test developed.

III. MANAGING SYPHILIS

When morbus gallicus first appeared, physicians treated its victims by purging their bodies of its poison. They recommended hothouses and extreme exercise (ball games, running, boar hunting, or farming) to induce sweating. They practiced blood-letting and prescribed purgatives. Mercury, given topically or orally to induce sweating and salivation, became popular. Its severe side effects—loss of teeth, gum ulcerations, skeletal deterioration, and gastrointestinal problems—matched the severity of the disease. Such punitive treatment seemed appropriate for a disease attributed to venery. However, patients and physicians soon became confused about which symptoms came from the disease, and which from the treatment.

In the 1510s, guaiac, a resin extracted from a West Indian tree, arrived in Europe. Although expensive and not grounded in existing medical theories, guaiac quickly became popular: rumors credited it with miraculous cures, it satisfied a popular theory that treatments for a disease ought to come from the same place (the Americas) as the disease itself, and its side effects were much less severe than those of mercury. Desperation led physicians to attempt many other treatments—massage, tobacco smoking, abstinence from pork and peas, and decoctions of vulture broth with sarsparilla. Hospitals and hospices appeared throughout Europe. Most were established as hospitals for the incurable, but as the effects of the Great Pox moderated over the century, they increasingly became places for cure and care. The Catholic Church even developed a special mass, *Missa contra morbum gallicum*.

Mercury and guaiac dominated syphilis treatment for centuries. Surgeons developed ways to excise or cauterize the sores. Medical entrepreneurs produced and marketed secret remedies. But by the late nineteenth century, some physicians had begun to suspect that their treatments provided no benefit. This suspicion was confirmed by the famous Oslo Study between 1890 and 1910, in which 2000 patients received no treatment and fared just as well (or as poorly) as those receiving mercurial ointments and other treatments. The infamous Tuskegee Syphilis Experiment sought to extend the results of this study by describing the "natural history" of syphilis among African-American men in the rural South. Public Health Service officers deceptively promised treatment, but then provided none, from 1932 until 1972. But in contrast to the Oslo Study, these physicians withheld treatments—including penicillin—that they believed to be effective.

As physicians and scientists struggled to develop effective treatments for syphilis, its prevalence steadily increased. In the nineteenth century, syphilis fueled widespread fear of cultural decay and the breakdown of social values. It was identified as a family poison that spread from profligate men to their innocent wives and children. Fears of casual transmission—through pens, pencils, drinking fountains, toilet seats, and doorknobs—proliferated. In the United States, immigrants were stigmatized as a source of infection. These fears, which reveal deep cultural anxieties about disease and sexuality, motivated many attempts at social engineering. Shocked to learn that as many as one-third of its troops were infected with syphilis, the British government passed the Contagious Disease Acts of 1864, 1866, and 1869. The laws acknowledged that sexual continence was an unrealistic goal. Syphilis could only be contained by minimizing the consequences of the inevitable transgressions: Prostitutes would be inspected and, if infected, detained until cured. These laws produced an outcry from purity reformers, who argued that prostitution needed to be criminalized and repressed, not acknowledged and regulated. The laws were repealed in 1886.

Similar calls for social hygiene dominated syphilis control in the United States from the 1890s through the 1920s. Education and moral encouragement beseeched men to be disciplined and restrained, faithful to their wives and families. During World War I, the U.S. military, which saw syphilis as a substantial threat to its effectiveness, conducted extensive campaigns against the disease. Just as they drained swamps to prevent malaria, public health officials shut down red-light districts and detained 20,000 suspected prostitutes. The Training Camp Commission distracted soldiers with organized sports and educated them with unambiguous messages: "A German bullet is cleaner than a whore." Faced with the French system of regulated prostitution, the Army refused to distribute condoms. Instead, officials established prophylaxis stations to provide treatment if exposure occurred. They hoped that the painful urethral injections would be a disincentive to sex. The Navy contributed to the campaign by removing doorknobs from battleships to prevent transmission of syphilis among sailors. Despite all of this work, syphilis rates remained high.

After the war, interest in venereal disease control diminished. Prudish censorship blocked public education in magazines or on the radio. Surgeon General Thomas Parran (1892–1968) led a campaign to reverse this. He decried the "conspiracy of silence" that surrounded venereal diseases. He argued that sexual continence remained an impossible ideal, and that victims of syphilis were victims of a disease and not criminals. His efforts led to the 1938 National Venereal Disease Control Act, which provided federal funding for diagnosis and treatment. States began requiring premarital serological tests (Fig. 2). This new pragmatic attitude dominated efforts during World War II. Moralistic campaigns against venereal diseases did continue (Fig. 3). But instead of repressing sexual behavior, officers of the Social Protection Division sought to modify it; instead of providing moral education, they distributed condoms. These efforts substantially reduced the incidence of syphilis among military personnel.

The contrast between the efforts during World War I and World War II illustrates the two extreme

Fig. 2. "Happiness Ahead—for the Healthy but not for the Diseased." During the 1930s, many states began to require premarital screening for venereal diseases. In this poster from a public health campaign, syphilis was portrayed as poison that would destroy marriages and families.

Fig. 3. "V. D.: Worst of the Three." During World War II, the U.S. military waged a campaign against syphilis that combined clearly moral messages, as seen in this poster, with condom distribution campaigns. Note that in this image, venereal disease is personified as an evil, dangerous woman, and not as a promiscuous soldier.

options of social management of syphilis. Social hygienists asserted a moral ideal and believed that abstinence was the only way to prevent infection. Pragmatists, in contrast, acknowledged that pre- and extramarital intercourse were inevitable; they sought to prevent infection by encouraging safer sexual practices and by providing unstigmatized treatment.

IV. THE LIMITS OF BIOMEDICINE

While public health officials swung between these two extremes of social management, medical researchers gradually produced powerful methods of medical management. In 1909, Paul Ehrlich (1854–1915) announced that his new drug, Salvarsan, had specific activity against syphilis. It was the first time anyone had found a specific drug that killed a specific microbe. Ehrlich called it a "magic bullet." With this drug, the promise of modern medicine seemed fulfilled: Medical scientists had established the specific cause of syphilis, they had developed a specific diagnostic test, and they had a specific treatment. The control of syphilis seemed at hand. Hope for easy control went unfulfilled. Salvarsan was toxic and difficult to administer, with some patients requiring 2 years of treatment; only 25% of patients received the full series of injections. Hope was restored in 1943 when John F. Mahoney (1889–1957) demonstrated the power of penicillin. Syphilis could be cured with a single injection: The miracle drug had been found. Incidence fell rapidly during the 1950s. Syphilis seemed vanquished.

But again syphilis defied expectations. Despite the power of penicillin, syphilis began a slow recovery during the 1960s. Many factors have been implicated—the promiscuity of the sexual revolution; the availability of oral contraceptives; and decreased funding for education, case-tracing, and other disease-control programs. Some physicians had even refused to prescribe penicillin, fearing that easy treatment only encouraged illicit sexual activity. Whatever the cause, syphilis rates began to rebound. Rising and falling in 10-year cycles, with each peak higher than the last, by the 1990s syphilis had reached the highest rates seen since the 1940s, with tremendous disparities among various populations.

How could syphilis continue to spread despite the existence of penicillin, an affordable and decisive treatment? "Magic bullets," even those as powerful as penicillin, are never panaceas. The history of syphilis shows the range of scientific, social, political, and cultural factors that contribute to the prevalence of the disease. A single intervention cannot treat the many different problems that syphilis reflects. Successful management requires concerted efforts against all of the causes of the disease.

Interest in the history of syphilis took on new urgency in the 1980s with the appearance of HIV and AIDS. Syphilis became both a significant facilitator of the transmission of HIV, and a dangerous infection in people with AIDS. In addition, public health officials and policy makers turned to the history of syphilis to guide their efforts against HIV. Many lessons were suggested. Just as penicillin had not solved syphilis, effective treatments for HIV would not end the epidemic. Educational programs, whether advocating abstinence or safe sex, would not cause people to avoid high-risk sexual behaviors. Compulsory measures, whether quarantine or serological screening, would bring few results, at high costs to civil liberties. Fear, blame, and stigmatization would shape the development of public health policy.

This new urgency, combined with effective treatment, widespread social awareness of the risks of sex, and a natural ebbing in the prevalence of syphilis, has led to calls for the eradication of the disease. Proponents of eradication realize that the most important barriers to eradication will not be scientific, but rather a series of considerable social and economic obstacles. Syphilis is still widely seen as a moral problem, reducing the willingness of victims to seek treatment. It flourishes in communities that lack basic financial and social resources; and that perceive more serious threats from lack of education, unemployment, crime, and other health issues. The shadow of the Tuskegee study leaves many afflicted communities suspicious of public health campaigns. And even if eradication of syphilis in the United States should succeed, it will only reveal the continuing international economy of disease. Reintroduction of syphilis from foreign countries will be inevitable.

The history of syphilis, therefore, demonstrates many important characteristics of diseases. Syphilis is not simply a biological phenomena. Instead, its incidence reflects political and economic structures, and cultural beliefs and practices. Its manifestations and meanings change over time, shaping efforts to control the disease. Isolated medical remedies, even those as powerful as penicillin, cannot be the only solution. Successful control requires comprehensive programs that acknowledge the cultural and social dynamics of syphilis.

See Also the Following Articles

AIDS, Historical • Chlamydia • Sexually Transmitted Diseases

Bibliography

Arrizabalaga, J. (1993). Syphilis. *In* "Cambridge World History of Human Disease" (K. F. Kiple, ed.), pp. 1025–1032. Cambridge University Press, Cambridge.

Arrizabalaga, J., Henderson, J., and French, R. (1997). "The Great Pox: The French Disease in Renaissance Europe." Yale University Press, New Haven, CT.

Baker, B. J., and Armelagos, G. J. (1988). The origin and antiquity of syphilis: Paleopathological diagnosis and interpretation. *Curr. Anthropol.* **29**, 703–737.

Brandt, A. M. (1987). "No Magic Bullet: A Social History of Venereal Disease in the United States since 1880." Oxford University Press, New York.

Brandt, A. M. (1993). Sexually transmitted diseases. *In* "Companion Encyclopedia of the History of Medicine" Routledge, New York. (W. F. Bynum and R. Porter, eds.), pp. 562–584.

Crosby, A. W. (1972). "The Columbian Exchange: Biological and Cultural Consequences of 1492." Greenwood, Westport, CI.

Fleck, L. (1979). "Genesis and Development of a Scientific Fact" (F. Bradley and T. J. Trenn., trans.) University of Chicago Press, Chicago.

Hudson, E. H. (1965). Treponematosis and man's social evolution. *Am. Anthropol.* **67**, 885–901.

Jones, J. (1981). "Bad Blood: The Tuskegee Syphilis Experiment." The Free Press, New York.

Morison, S. E. (1942). "Admiral of the Ocean Sea: A Life of Christopher Columbus." Little, Brown and Company, Boston, MA.

Quétel, C. (1990). "History of Syphilis" (J. Braddock and B. Pike, trans.) Johns Hopkins University Press, Baltimore, MD.

Rothschild, B. M., and Rothschild, C. (1996). Treponemal disease in the New World. *Curr. Anthropol.* **37**, 555–561.

St. Louis, M. E., and Wasserheit, J. N. (1998). Elimination of syphilis in the United States. *Science* **281**, 353–354.

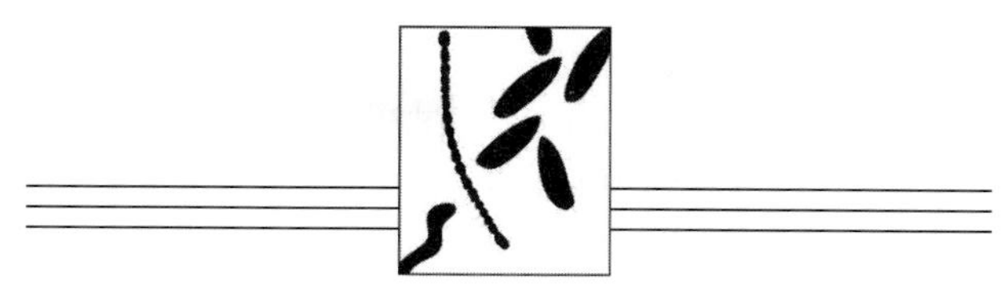

Temperature Control

Terje Sørhaug
Agricultural University of Norway

I. Temperature Ranges for Growth of Microorganisms
II. Microbial Activities Controlled by Temperature
III. Sublethal Injury at High and Low Temperatures
IV. Thermal Inactivation

GLOSSARY

***cold-shock response* (*heat-shock response*)** A rapid change in gene expression that occurs when there is a temperature shift to a low (elevated) temperature.

D value The decimal reduction time, that is, the time to kill 90% of the cells in a microbial population.

hurdle technology The intelligent use of combinations of preservation factors or techniques (hurdles) in order to achieve multitarget, mild but reliable preservation effects.

indicator organism A microorganism that when detected gives a quantitative representation of a group of organisms, often of fecal contamination or pathogens in foods.

poikilotherm An organism that is cold-blooded, with a body temperature that approximately follows that of its surroundings.

psychrotrophic microorganism A microorganism that will develop countable colonies within 10 days at 7°C.

resuscitation A procedure that restores the ability of an organism to grow and develop after a sublethal injury.

stress A condition that may inflict injury on a microorganism.

sublethal injury The temporary loss of tolerance for specific conditions in a microorganism.

thermization A mild thermal process (e.g., at 63°C for 10–20 s).

thermoduric Able to survive pasteurization.

MODERN FOOD MANUFACTURE depends heavily on temperature control to supply safe products of high quality. Thus, pathogens and spoilage microorganisms are often thermally killed or their activities are limited to acceptable levels by refrigeration or freezing. Many traditional fermentations (e.g., the production of cheeses and beer) imply strict adherence to temperature schemes that have been further refined in industry.

Thermal processing at 70–80°C for 10–20 s is sufficient to kill many vegetative microbial cells in environments of high-water activity. The destruction of some types of bacterial spores requires heating >120°C for 15–20 min. Microbes exposed to high or low temperatures may suffer sublethal injury. These stressed organisms recover and resume growth under favorable conditions, sometimes in foods. Selective, analytical procedures may not allow injured organisms to recover for detection. Thermal shock will engage adaptation mechanisms to enhance the resistance in exposed cells.

I. TEMPERATURE RANGES FOR GROWTH OF MICROORGANISMS

Microorganisms commonly grow in temperature ranges of 30–40°C. Thermophilic organisms grow >50°C, and psychrotrophs may develop at 5 to <0°C. Food-borne representatives of all groups are found and psychrotrophic bacteria, yeasts, and molds selectively grow at refrigeration temperatures.

A. Minimum, Optimum, and Maximum Temperature

The growth of microorganisms is the combined expression of a complex system of biochemical and physicochemical events. These include at least energy metabolism, biosynthesis of monomers and polymers, assembly of enzyme complexes, ribosomes, genetic material, membranes and walls, cell division, and turnover of discarded intracellular components. If any one of these processes stops, the growth of the organism will soon come to a halt.

Microorganisms often grow in temperature range of 30–40°C. Figure 1a shows an ordinary variation of growth rates with temperature for an organism. At the low temperature end for growth, the minimum temperature, membrane disfunction, and changes of hydrophobic protein interactions may be important factors in explaining the limits to growth. The near-exponential increase in growth rate following a rise in temperature is found for many microorganisms. The maximum growth rate at the top of the curve defines optimum temperature. At this point, destructive processes in the cells such as injury to nucleic acids and membranes and protein denaturation will dominate at only a small increase in temperature. The upper limit of growth is called the maximum temperature. The curve and its minimum, optimum, and maximum points, the cardinal temperatures, are characteristic for the organism, but deviations may occur if the growth conditions are changed.

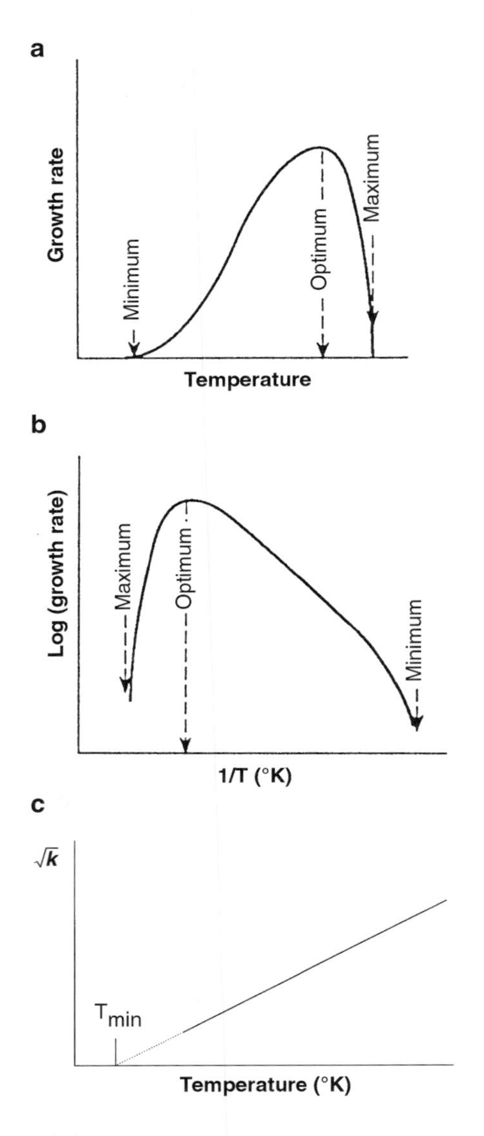

Fig. 1. Temperature characteristics for a microorganism. (a) Variation of growth rate with temperature. Cardinal temperatures: minimum, optimum, and maximum. (b) Arrhenius plot. (c) Bělehrádek–Ratkowsky square root plot.

B. Mathematical Models for Growth of Microorganisms versus Temperature

Early attempts to treat the relationship between the growth rate of microorganisms and temperature mathematically made use of the Arrhenius equation developed for chemical reactions:

$$k = Ae^{-E/RT}$$
$$\log k = -(E/2.303\mathrm{R})(1/T) + \log \mathrm{A}$$

where k = specific reaction rate constant at a given temperature; R = Boltzmann's constant (gas constant); E = energy of activation; A = frequency factor (Arrhenius factor); T = temperature in K. The validity of the equation is apparent when the plot of log k versus $1/T$, the Arrhenius plot, yields a straight line (Fig. 1b). The slope of this line is $-E/2.303\mathrm{R}$. The activation energy can thus be calculated when experimental values for log k versus $1/T$ have been obtained.

A simple linear relationship for part of the Arrhenius plot for an organism implies that one reaction or function in the cell is limiting for growth in that temperature range (Fig. 1b). Although it would be attractive to use this to explain growth kinetics, the detailed analysis of plots for many bacteria reveals that the apparent linear part of the curve is often slightly curved or multiphasic. Also the corresponding temperature range is rather narrow, which implies that the precision of the E calculated from the slope will be low. Developments of the Arrhenius equation have been made to account for nonlinearity, lag time, water activity (a_w), and pH.

The quest for useful mathematical expressions led Ratkowsky to introduce the square root relationship based on work by Bělehrádek:

$$\sqrt{k} = b(T - T_{min})$$

where k = growth rate, b = slope of the regression line, T = temperature in K, and T_{min} = the value of T where the extrapolated line in the Bělehrádek–Ratkowsky plot crosses the temperature axes (Fig. 1c). T_{min} is a conceptual value without physiological significance. It is, however, characteristic for the organism, and plots made for different pH and a_w all meet in T_{min}. The square root expression can also be developed to account for k outside the favorable ranges as well as for variable pH and a_w.

With equations for expressing the growth rates at all relevant temperatures (and a_w, pH, etc.), the tools are available for predictive microbiology. The models developed with these tools, in combination with practical laboratory testing, are becoming indispensable in safety evaluation, quality control (Hazard Analysis Critical Control Point, HACCP), product development, and education.

C. Temperature Classification: Psychrophiles, Psychrotrophs, Mesophiles, Thermophiles

The growth of microorganisms has been reported from < −10°C up to 110°C. The high-temperature extreme was claimed for bacteria inhabiting hot ocean floors near volcanic vents. In any case, such organisms are hardly of concern for food and fermentation microbiology, but among the microbes growing at <0°C, several are candidates for food spoilage in refrigerated storage. Temperature-range specialization among microorganisms has prompted the proposals of various classification schemes, both for scientific purposes and for more pragmatic reasons. The most common terms in use are psychrophiles, psychrotrophs, mesophiles, and thermophiles. Three of these definitions are based on minimum, optimum, and maximum temperature ranges (Table I).

The term psychrotroph is used mainly in food microbiology for organisms growing under refrigerated conditions. A common definition is that psychrotrophs produce colonies on agar media within 10 days at 7°C. A psychrotrophic count may thus include both psychrophiles and mesophiles.

In applied microbiology and screening work, isolates of microbes are seldom characterized strictly according to the definitions as psychrophiles, mesophiles, or thermophiles. As indicated earlier, organisms growing at low temperatures are often distinguished by a psychrotrophic count. Mesophiles are often observed by growth on agar media at 30–32°C in 2–3 days. Thermophiles are often reported as those organisms forming colonies at 55°C.

D. Critical Points on Maintaining Constant Temperature for Growth of Microorganisms

Establishing the growth rate–temperature curves requires measurements at a number of temperatures, each of which must be very exact. This is particularly evident between optimum and maximum temperatures, for which a difference of only a few degrees greatly changes growth rates. Water baths for tubes and flasks allow good control, but often practical obstacles limit the distribution of temperatures. Temperature-gradient incubators can be made, but the exactness of measurements should be carefully checked when they are in use.

Temperature fluctuations in air incubators are difficult to handle. Variations often occur in the incubator itself or may be caused by the frequent opening of doors and when large numbers of plates or flasks

TABLE I
Temperature Classification of Microorganisms

	Temperature (°C)		
	Minimum	*Optimum*	*Maximum*
Psychrophiles	<0	<15	20
Mesophiles	5–25	25–40	40–50
Thermophiles	35–45	45–65	>60

are deposited. When plates are stacked, the capacity of the incubator and the air access between the plates should be occasionally checked with accurate thermometers.

Similar considerations apply when killing microorganisms in autoclaves or otherwise in the laboratory, and also in food processing.

II. MICROBIAL ACTIVITIES CONTROLLED BY TEMPERATURE

A. How to Avoid or Select Microbial Activities

This section briefly surveys how temperature is used purposely by humans to avoid or regulate microbial activities.

Low-temperature storage of foods is now common in many parts of the world. Freezing and refrigeration, often developed into cold distribution chains, have become an indispensable means to maintain the freshness of foods while diminishing the effects of spoilage microorganisms and preventing health hazards by pathogens. The convenience of extended storage periods is important both for the industry and for the consumer.

Freezing foods to $-20°C$ or below effectively stops microbial growth because of low temperature and the low a_w; however, microorganisms may survive and develop after thawing. Important factors that influence the degree or extent of injury and death by refrigeration and freezing are the rate of temperature change during chilling and thawing, the temperature differential rather than the actual temperatures, cryoprotectants, and the growth phase of cells. Inactivation by freezing occurs in two phases. First, there may be a rather quick decrease in viable numbers of microorganisms during freezing, and, then, there may be a slower change during storage. The action of lipases from microbes has been observed in some frozen foods rich in fats.

The refrigeration of foods is used alone or in combination with a previous heat treatment. Fresh fish is preferably kept on ice; fresh meat products, and raw and pasteurized milk and cream are stored at 0–5°C. Foods that have been further processed and often packed, but without (commercial) sterility, may be kept at somewhat higher temperatures, but lower temperatures are strongly recommended to avoid potential health hazard.

Although bringing many advantages to the industry and the consumer, the refrigeration of foods has also given new and interesting challenges to the microbiologists. Although refrigeration may have been thought of by some as putting a lid on microbial activities, in reality the priorities are instead changed from mesophilic and thermophilic organisms toward psychrotrophs.

1. An Interesting Example—Handling of Milk

The cold storage of raw milk with subsequent and alternative thermal treatments, followed again by low-temperature storage, is an interesting scenario for the presentation and consideration of the advantages and problems associated with temperature control. Raw milk that has not been stored or heat treated contains a wide variety of bacteria, yeasts, and molds. The percentage ranges of the organisms vary with the total microbial load. Mesophiles account for a relatively large fraction in high-quality milk. Pathogens may occur. The milk at this stage may contain only a moderate count of fast-growing psychrotrophs. In areas with a complete cold chain, the milk is collected in farm tanks with cooling and later transported quickly to the dairy for further cold storage in silos. If this period extends to several days cold storage, the psychrotrophic gram-negative rods will outgrow other microbes until complete domination. Among this group *Pseudomonas* spp., particularly fluorescent pseudomonads, are most frequently detected, but strains of other genera are also regularly found, such as *Acinetobacter, Flavobacterium, Alcaligenes, Aeromonas,* and *Enterobacteriaceae.*

In the dairy, milk is pasteurized (e.g., at 72°C for 15 s) before distribution or further processing. Thermization, a mild heat treatment (e.g., at 63°C for 10–20 s), of the milk may also precede silo storage to safeguard against the growth of psychrotrophs; pasteurization follows later. In either case, the majority of the psychrotrophs are killed. Thermoduric bacteria are those that survive pasteurization, such as strains of *Micrococcus, Microbacterium, Alcaligenes,*

Arthrobacter, Corynebacterium, and Streptococcus, and spores of *Bacillus* and *Clostridium.*

2. *Spore-Formers*

Processed sweet milk for liquid consumption is kept in the cold chain, and among the surviving bacteria only some species of *Bacillus* have psychrotrophic strains that can develop and spoil the milk. Proteinases produced by the bacilli contribute to sweet curdling of the milk. Sweet curdling and bitter taints occur mainly in the summer and are less of a problem when refrigeration is more strictly maintained.

When pasteurized milk is used for cheese production, relatively high temperatures prevail during processing (~28–57°C) and ripening (~3–27°C), varying with the type of cheese. The higher temperatures in combination with the other conditions in certain cheeses (pH, E_h, a_w, and salt content) may allow the growth of clostridia. Thus, Swiss-type cheeses and sometimes Gouda may suffer from the development of poor flavor and excessive gas formation (late blowing) due to the butyric acid fermentation of lactate by *Clostridium tyrobutyricum.*

The spores of *Bacillus* and *Clostridium* survive the pasteurization of milk, but the heat treatment probably also contributes to the activation of the spores, encouraging later germination and outgrowth. This side effect is another good illustration of the duality that must be considered when temperature is used to control microbes.

When clostridia spoil cheese, the quantitative aspect is interesting. Fewer than 100 spores/100 ml milk is sufficient to cause problems. The detection of these small numbers is a challenge in analytical microbiology, involving the destruction of vegetative cells by heat (at 70–80°C for 10 min) with simultaneous activation of the spores, followed by most probable number or filtration methods. However, the ultimate goal is to get rid of the clostridial spores. Again, heat treatment is called for after removal by centrifugation (bactofugation) or ultrafiltration (Bacto-Catch) of >95% of the spores from the bulk milk. The fraction enriched with spores is subjected to ultra-high temperature treatment (e.g., at 145°C for 3–5 s). These are two ways to challenge the problem; others involve the use of nitrate, lysozyme, or nisin.

3. *Recontamination*

Pasteurization, as we remember, killed the psychrotrophic gram-negative rods, those organisms growing most efficiently in the cold-stored milk. Nevertheless, there are two more ways in which these bacteria may interfere with the quality of milk and dairy products.

Postpasteurization contamination (PPC) will occur. Of course, a few organisms will have access to the milk from the air, water, or equipment before the cartons and bottles are closed, and cheese production is not an aseptic process. Constant attention to minimize the risk of PPC must be part of the hygiene program in a dairy. However, if microbes, including pathogens, get into the system in rather large numbers through leakages in lines, heat exchangers, or otherwise, the situation can soon become very serious. Some pathogens can grow at < 5°C, such as strains of *Listeria monocytogenes, Yersinia enterocolitica, Aeromonas hydrophila,* enteropathogenic *Escherichia coli,* and *Clostridium botulinum* type E. Conditions in milk at 4°C may not favor the growth of several of these species, but *L. monocytogenes* has been implicated in some cases, as indicated earlier.

4. *Heat-Resistant Enzymes*

Finally proteinases, lipases, and phospholipases produced by psychrotrophic, gram-negative rods present a spoilage problem. The enzymes from *Pseudomonas* are mainly produced when the bacterial numbers reach 10^6–10^8 colony-forming units (CFU) /ml. Such dense flora are hardly compatible with high-quality milk, but, if the enzymes have been produced in milk remaining in poorly constructed or poorly washed lines or other equipment, they may later be flushed into the bulk milk. The enzyme yield in a good culture can easily reach 100,000 ng/ml, and only 1 ng/ml (proteinase) may be sufficient to spoil milk. The proteinases and lipases can be produced at low temperature, and they have reasonably high activity at refrigeration temperatures; they release bitter peptides and free fatty acids, respectively, which both contribute to quality losses in dairy products. Destabilization of casein micelles and milk fat globules may also result from the action of the proteinases and lipases or phospholipases, respectively. The unfortunate fact is that most of these enzymes are not inactivated by any of our heat-treatment processes. Very good hygiene, close control of temperature during storage, and thermization to reduce the bacterial load are important measures to fight the psychrotrophs and their enzymes.

5. *Meats, Poultry, and Fish*

Unlike milk, which is pasteurized and still considered fresh, meats, poultry, and fish are no longer "fresh" commodities after a heat treatment. While in cold storage (0–5°C), representatives of the initial psychrotrophic flora will therefore be definitive spoilers of these foods. Muscle tissues of all these foods when properly cut and handled can be obtained in nearly sterile condition. Contamination will thus be from the skin, hide, hoves, nostrils, intestinal contents, water, and processing equipment for meats and poultry; and from the skin, gills, intestinal contents, water (ice), and processing equipment for fish. The tendency in all these foods during aerobic cold storage is an increasing domination by *Pseudomonas* on meats and poultry *Pseudomonas* and of *Shewanella putrefaciens* (formerly *Pseudomonas* type III/IV, then *Alteromonas putrefaciens*) on fish, but other bacteria also occur (*Acinetobacter, Aeromonas, Alcaligenes, Moraxella, Shewanella,* and specific contaminants of the various foods).

Psychrotrophic counts from fish from temperate or cold waters are often higher than the counts at higher temperatures (e.g., at 30°C for 2 days), a situation quite different from that of meats and poultry. That many bacteria from these types of fish are relatively psychrophilic is a good example of the ecological control exerted in the vast areas of the oceans, lakes, and rivers, where the temperature often ranges from −1–5°C. Such bacteria will be natural inhabitants of the skin, gills, and intestines of the resident fish and, thus, end up in our markets. We appreciate, then, the common practice of storing fish on ice. It is impossible to completely avoid the contamination of the fish filets with the initial flora during handling, which is even worse with fish than with meats and poultry.

The contamination of tropical fish will often be characterized by gram-positive bacteria of the coryneform and micrococcus groups. Because these organisms are less psychrotrophic, if at all, fish stored in ice can have very long shelf life (20–30 days).

The aerobic spoilage of meats, poultry, and fish becomes apparent as unpleasant odors, slime formation, and discoloration from the metabolism of available low-molecular-weight compounds (e.g., carbohydrates, lactate, amino acids, nucleotides, and trimethylamine oxide (for fish)). Contrary to the situation in milk, proteinases and lipases from psychrotrophic bacteria, if they appear, are thought to contribute to a secondary stage of spoilage, with possible consequences for texture and rancidity.

B. Combination Systems–Hurdle Technology

When temperature is used to control microbial activity, other variables always influence the outcome (process) to a greater or lesser extent. This comes from the obvious fact that the microbes are part of a biological system with certain characteristics of pH, $a_w E_h$, gas atmosphere, nutrient composition, salt content, and antimicrobial content.

Often combinations of factors that regulate microbial action are taken for granted because they represent common knowledge. Food processors generally accept that microorganisms are more easily killed by thermal processing when the pH is low (<4.5), and many foods can be stored at higher temperatures when the a_w has been lowered by drying, the addition of salts or sugars, or some combination of these parameters. The combinations of parameters theoretically possible for the inhibition or inactivation of microorganisms in the processing of foods are indeed numerous. Naturally restrictive ranges (e.g., the pH of milk and meats) simplify considerably these, but still there are many alternatives to evaluate in each case.

Attempts to systematize the relationships between processes and parameters and the relative importance of parameters for a given process led to the hurdle concept, illustrated in Fig. 2. For a specific process, there are one or more primary factors or hurdles that contribute most to preventing microbial growth and other activities; additional hurdles are also common. The pasteurization of milk is based on heat as the main hurdle, but subsequent refrigerated storage is decisive for extended shelf life. Fresh meats are pre-

Parameters \ Processes	Heating	Chilling	Freezing	Freeze drying	Drying	Curing	Salting	Sugar addition	Acidification	Fermentation	Smoking	Oxygen removal	IMF (f)	Radiation
F (a)	X(c)	+	+	+	+	+	o	+	o	+	+	+	+	o
t (b)	+(d)	X	X	o	o	+	+	o	+	+	+	+	+	o
a_w	+	+	X	X	X	X	X	X	o	X	+	o	X	o
pH	+	+	o	o	+	+	+	X	X	X	+	+	+	o
E_h	+	+	+	+	o	+	+	+	+	+	+	X	+	o
Preservatives	+	+	o	o	+	X	+	+	+	X	X	+	X	o
Competitive flora	o(e)	o	o	o	o	+	o	o	+	X	o	+	+	o
Radiation	+	o	o	o	o	o	o	o	o	o	o	o	o	X

Fig. 2. Processes used in food preservation and the parameters or hurdles that they are based on: (a) high temperature, (b) low temperature, (c) main hurdle, (d) additional hurdle, (e) generally not important for this process, (f) intermediate moisture foods (IMF). (From Leistner, L. (1978). Hurdle effect and energy saving. *In* "Food Quality and Nutrition" (W. K. Downey, ed.), pp. 553–557. Applied Science Publishers, England.)

served by chilling, but the advent of modified atmosphere packaging (MAP, oxygen removal and the introduction of other gases) has contributed to an even longer shelf life. The application of preservatives is an additional hurdle of interest for meat products at low temperature (e.g., sorbate on chickens). The systematic use of hurdles can contribute to less-severe heat treatments and the reduced addition of preservatives, and the safety of a process can be increased by additional hurdles. Thus, the repertoire of hurdles has seen some development in recent times, for example, high-pressure processing, far-infrared heating, high electric field pulses, sous-vide cooking (vacuum cooking in a pouch), and the use of bacteriocins. Various factors may act additively or with synergistic effects; thus, the complexity and potential benefits of exploiting several inhibitory principles in a process must be evaluated carefully.

Food fermentations are also combination systems with temperature as one of the factors. These complex processes depend on the fine regulation of microbial growth and enzymatic conversions that imply temperature control to select some organisms and enzyme activities at the expence of others. The great majority of food fermentations have traditional roots, and the temperature during ripening was that of the storage environment (e.g., the cellar for cheeses, wine, and beer; open air for cucumbers, olives, and sausages; and silos, for forage). For some of these productions, elevated temperatures are applied during the initial processing (cheeses, beer, and sausages), which allows only the more resistant and desirable organisms (or enzymes) to survive. Modern food fermentations often demand the strict regulation of temperature to obtain the correct flavor and texture of the product. As an example, the fine balance of *Streptococcus salivarius* ssp. *thermophilus* and *Lactobacillus delbrueckii* ssp. *bulgaricus* necessary for the desirable flavor of yoghurt can only be accomplished at 42°C.

C. Mechanisms of Adjustment to Low and High Temperatures

The temperature limits for microorganisms are directly related to those of the activities and structures of the cells because, as poikilotherms, they have no shield against the surrounding temperature fluctuations. The further consequence is that any microorganism has enzymes, biochemical systems and structures more or less specialized for growth (or at least survival) over a certain (often prevailing) temperature range. Dynamic systems for functional changes or adjustments with temperature fluctuations are also operative.

Thermophilic microbes have regimes of thermoresistant proteins seldom found in psychrophiles. An upward temperature shift induces qualitative and quantitative changes, which may be quite drastic when the final temperature is above the normal growth range. The fatty acids of the membrane phospholipids are mainly saturated; thus, the lipids are in a functional liquid–crystalline state in temperature ranges for thermophiles. However, at low temperature, for thermophiles, transistion to the gel state for membrane lipids may be a growth-limiting factor.

In addition to the specialization of components and structures for the high-temperature of life of thermophiles, microorganisms in general—in fact all living cells—appear to produce large amounts of heat-shock proteins (Hsp) on exposure to temperatures above their normal growth conditions. Other inducers of the response include sulfhydryl reagents, ethanol, hydrogen peroxide, amino acid analogs, puromycin, nalidixic acid, and alkaline conditions. The regulation of the response includes polypeptide signals produced by regulatory genes (*rpoH*) that control the expression of *Hsp* genes. The induction results in the increased synthesis of a group of proteins, some of which are involved in protecting the organism at the elevated temperature. Mutants without the proteins die under similar conditions. However, some or all of the Hsps appear to have functions in the cells at all temperatures, for cellular growth, DNA replication and correct assembly of proteins, and bacteriophage.

Analogous to the heat-shock response of synthesizing Hsps, a cold-shock response has been observed in mesophiles and psychrophiles, the production of cold-induced proteins (Cip; also termed cold-shock proteins, Csp). The duration of cold shock and the extent of the temperature difference determine the magnitude of the cold-shock response. Ribosomes, apparently, function as sensors of thermal stresses.

Several bacteria may also enter the viable but nonculturable (VNC) state in times of mild but long-lasting stress, including thermal stress. VNC cells are viable, but they will not grow on regular media until an inducing event has occurred (e.g., heat shock). The composition and structures of VNC cells compared to regular vegetative cells are considered to be expressions of a dormant state with increased stability and persistence of the cells.

In addition, long-lasting and extreme thermal and other stresses promote adaptive mutations, contradicting somewhat the theory of random mutations, but acceptable as a definite survival mechanism (Archer, 1996). Thermal-shock responses as well as adaptive mutations may give pathogens a competitive edge, an increasing concern for food safety.

The limiting conditions for psychrophiles at low temperatures have received considerable attention. General evidence from proteins at low temperatures (<5°C) show a weakening of the hydrophobic bonds, which may contribute to slight conformational changes, thus interfering with enzyme activities, regulatory functions in allosteric systems, and the assembly of complexes such as ribosomes. The sensitivity of DNA and DNA replication has also been implicated. However, sufficient evidence is not available to claim that these changes alone limit growth in general. Quite to the contrary, cold-acclimatization proteins (Caps) are synthesized to a greater extent at low temperatures, and Csps are synthesized at a higher rate after a sudden decrease in temperature in mesophiles and psychrophiles, apparently to protect the organism and to compensate for reduced enzyme activities. Low temperature has not been proven to stop the energy supply in psychrophiles.

With lipids, the situation is somewhat different because there is substantial information leading us to consider that changes from a liquid–crystalline state of membrane lipids to a gel state at low temperatures influence solute (nutrient) uptake in a decisive way. Accordingly, lowering the melting point of membrane lipids should be compatible with the maintenance of functionality (homeostasis) at low temperature. The comparison of psychrophilic bacteria with mesophiles or the exposure of an organism to low temperature (0°) has revealed tendencies for changes in membrane lipid composition to occur—the desaturation of fatty acids; as well as the synthesis *de novo* of unsaturated fatty acids; the synthesis of shorter fatty acids; the increase of anteiso branched-chain fatty acids and the decrease of cyclopropane acids; rearrangements of fatty acids to the *sn*-1long–*sn*-2short configuration of *sn*-glycerol-3-phosphates that also contribute to low melting points or, rather, melting ranges. Whether changes of composition must accompany a temperature decrease has been challenged by observations of mesophiles (*Escherichia coli*) with fatty acids suitable for low-temperature growth after incubation at mesophilic temperatures.

An addition to the repertoire of melting-point depressing mechanisms is the transition of *trans*-unsaturated fatty acids into the *cis* configuration in *Vibrio* and other organisms. Both remodeling of phospholipids and the *trans–cis* alterations may be reaction systems for the rapid adaptation to low temperatures.

Other functions, although not necessarily essential, may support growth and survival at low temperatures. The optimum temperature for chemotactic response with *Pseudomonas fluorescens* has been observed below the optimum for growth, and chemotaxis was strong at 0°C. Exopolysaccharide synthesis, visible as slime, is stimulated at low temperatures. For microaerophilic organisms (e.g., *Leuconostoc mesenteroides* and certain lactococci), capsules may protect against the increased oxygen solubility at low temperatures but also against threats such as antibiotic action and dessication.

III. SUBLETHAL INJURY AT HIGH AND LOW TEMPERATURE

A. Definitions

A variety of environmental changes from conditions normal for a microorganism are stressful and may inflict injury to the organism. The injury may be lethal (irreversible), so that the cells will not be able to produce progeny under any conditions or after any resuscitation procedure known to us. In contrast, sublethal injury is reversible and can be repaired under favorable conditions. The actual damage to the cells is seldom observed, but the overall consequences of the reversible injury are shown by the disparity of viable cell numbers (colony counts) in nonselective (base) and selective (test) media. Reversible injury may thus be defined as the temporary loss of tolerance for specified conditions. This operative definition is illustrated in Fig. 3.

In the first phase in Fig. 3, a pure culture of an organism grows normally in a liquid medium, and plate counts with a test medium are close to those in a base medium. During phase two, cells are injured by exposure to a higher temperature, growth stops, and some cells die, as indicated by the decrease in counts on base-medium plate. However, although sublethally injured and uninjured cells form colonies on the base-medium plate, the test medium does not allow the injured cells to repair, thus, only the uninjured cells form colonies. The difference between the counts represent injured cells. When the stress condition is terminated (phase three) the injured cells in the culture gradually repair themselves, provided the culture medium is rich. The steep rise of the curve sometimes observed in this phase indicates that repair is taking place, not regular growth. The difference bet.veen the base medium and the test medium may, for instance, be a higher salt concentration in the test medium.

Sublethal injury to vegetative cells can be produced by a variety of stressful conditions, such as modestly higher temperatures, lower temperatures including freezing and thawing, drying and rehydration, high osmotic conditions, low pH, salts, preservatives, sanitizers, and depletion of essential ions.

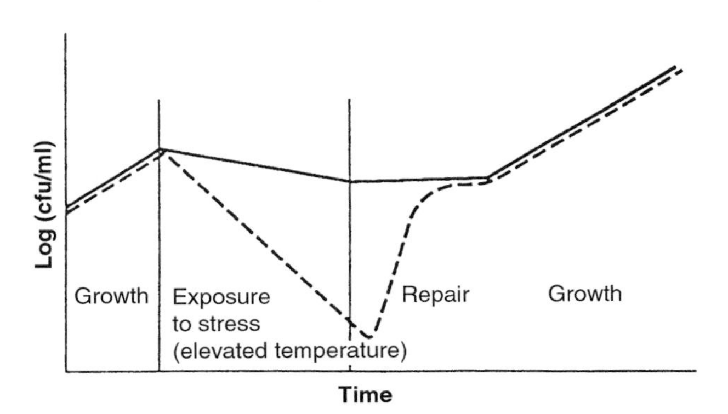

Fig. 3. Model illustrating injury and repair in a microorganism. Solid line, plate counts on base agar; dotted line, plate counts on test agar. (Adapted from Busta, F. F. (1978). *Adv. Appl. Microbiol.* 23, 195–201.)

Differential plate-count procedures to reveal that cells have been injured actually introduce additional or secondary stresses so that the cells that are reversibly injured in the first place (primary stress) do not repair and reproduce. Because reversible injury and the lack of ability to form colonies have often been related to microbiological analysis and the use of selective media, differential conditions (secondary stresses) regularly seen are increased salt concentrations; minimal media; deoxycholate; lauryl sulfate; bile salts; detergents; crystal violet, brilliant green, and other dyes; azide; and antibiotics. Secondary stress may also occur during the dilution of samples; thus Mg^{2+}-containing diluents are recommended to counteract this stress. Similarly a cold shock during dilution may kill some bacteria.

Sublethal injury has been observed in nearly all the food-borne pathogenic bacteria, in several indicator bacteria, spoilage bacteria, lactic acid bacteria, yeasts, and molds. Bacterial spores also experience injury, but generally more severe treatments are required. Drying and freezing are seldom injurious to bacterial spores. The mechanisms of injury vary in vegetative cells and spores.

B. Implications for Death or Survival of Microbes in Applied Microbiology and Research

Injury and repair in microorganisms have been studied thoroughly in food-borne pathogens and indicator organisms. Spoilage organisms and starters have also been given some attention. The importance of injury in food-borne pathogens and indicator organisms is manifested in two ways. On the one hand, injured pathogens may repair and develop to the stage of food poisoning; on the other hand, injured organisms that are able to recover in a food may not be detected in traditional microbiological analysis. Substantial efforts have been devoted to remedy this situation by developing analytical procedures that will detect both uninjured and reversibly injured organisms.

Primary stress to microorganisms in foods may occur at various stages—in the fresh raw material (from refrigeration or chlorine in water), during thermal and other processing, and during storage (from refrigeration or modified atmospheres). Injured cells may recover, or the damage may be exposed to secondary stresses in the food resulting in death. For the food processor, the synergistic effects of several stresses are desirable (cf. the hurdle concept) for better preservation and extended shelf life.

Samples for analysis are often stored transiently (frozen or refrigerated) with possible changes to the flora. Further stress may occur during the preparation of samples (from thawing, mixing, or heating) and dilutions (from type of diluent, temperature). Even pouring plates has the potential for stress, particularly if the melted agar medium has not been cooled to 45°C. For pathogens and indicator organisms, the customary selective media, although suitable for uninjured cells, are harmful for injured cells. Therefore, in microbiological analysis, great care should, in general, be exercised with samples to avoid secondary stresses, which result in accumulated injury and the subsequent death of organism, that are not necessarily experienced in the food.

The solutions to the problems with selective media focus mainly on preincubation procedures with nonselective media to allow the recovery of injured cells. Incubation on agar media are preferred over liquid incubation to avoid interference by the multiplication of uninjured cells. Thus, the diluted sample is poured (or spread) in a nonselective medium, and after a 1- to 4-hr resuscitation period, the appropriate selective medium (single or double strength) is poured on top before the final incubation. Markedly higher counts are regularly obtained in this way. Good alternatives are filter techniques with transfers between nonselective and selective media and the use of replica techniques.

Specific and important agents for injury are hydrogen peroxide and superoxide radical, which, however, are effectively destroyed by the addition of catalase or pyruvate to the media.

Vegetative bacterial cells exposed to different kinds of stress may suffer damage at several sites, such as cytoplasmic membrane injury with leakage of intracellular material; DNA, ribosome, and RNA degradation; protein denaturation; and enzyme inactivation. With heating or low temperatures, the extent of injury will vary according to the severity of the treatment, and one or rather several of the sites are influenced.

Spores with other structures have different or additional mechanisms of injury in the spore and during activation, germination, and outgrowth.

IV. THERMAL INACTIVATION

A. Classic Death Kinetics, Survivor Curve, Thermal Processing

Experiments with many different organisms, particularly with bacterial spores, in various thermal processes have shown that the number of surviving organisms decrease exponentially with time.

$$N = N_0 e^{-kt},$$

where N_0 = the initial number of viable organisms, N = the number of survivors at time t, and k = the specific reaction rate for death of the organism. In logarithms, base ten, this becomes

$$\log N/N_0 = -kt/2.303$$

A plot of log N versus time is often called the survivor curve (Fig. 4). The slope is $-k/2.303$, and the intercept on the ordinate is log N_0. The decimal reduction time D is the time for a 10-fold decrease in the number of survivors, that is, the time to kill through one log cycle (Fig. 4). A D value is given for a specific temperature (e.g., $D_{121°C}$). The D values decrease exponentially with increasing temperature; the slope of this line is called the z value (i.e., the number of degrees celsius for a 10-fold change in D value). Typical examples of D values are the following.

Psychrotrophic pseudomonads	$D_{70°C}$ = 0.001–0.5 s
Psychrotrophic sporeformers	$D_{110°C}$ = 0.4–8 min
B. coagulans, acid condition	$D_{121°C}$ = 0.05 min
B. stearothermophilus, low-acid food	$D_{121°C}$ = 4 min
C. botulinum, low-acid food	$D_{121°C}$ = 0.21 min
C. sporogenes, low-acid food	$D_{121°C}$ = 0.15–1 min

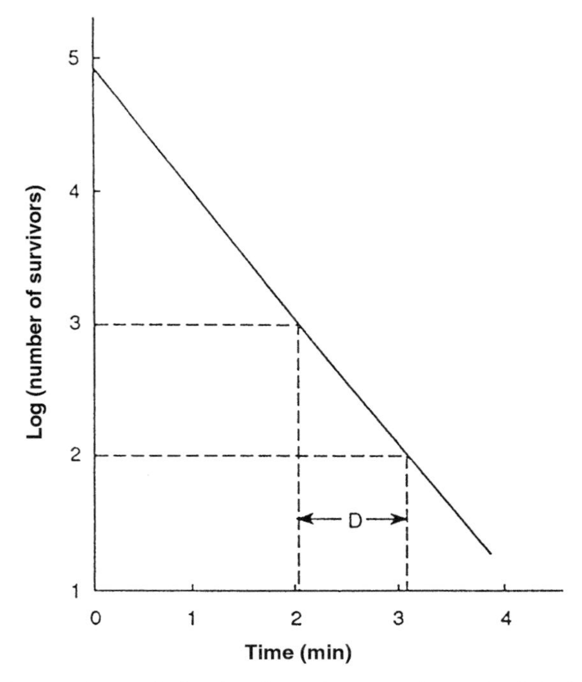

Fig. 4. Thermal death of a microorganism. Model of a survivor curve.

From these data, it is apparent that strains of *Pseudomonas*, the predominant psychrotroph in milk, will be killed by pasteurization at 72°C for 15 s, whereas psychrotrophic spore-formers will survive.

In the canning of nonacid foods, a very extended thermal treatment would be needed for the complete inactivation of *B. stearothermophilus* spores. However, spoilage by this organism is not a great concern unless the cans are cooled too slowly or stored at excessively high temperatures.

The safety requirement for a canning process with low-acid foods is to inactivate all *C. botulinum* spores because it is the most heat-resistant food-poisoning organism. This is accomplished by the 12*D* process, which kills *C. botulinum* through 12 log cycles. An initial *C. botulinum* load of 10^4 per unit will thus be reduced to 10^{-8} per unit, or one survivor in 100 million units. With sealed containers and a processing temperature of 121°C, the 12*D* process will require $12 \times 0.21 = 2.5$ min, which in practice is rounded off to 3 min. The same calculation applied to the spores of a mesophilic spoilage bacterium (e.g., *C. sporogenes*), gives a processing requirement of 12 min. However, the occasional spoilage of a can is more acceptable than food poisoning, so a less severe treatment may be chosen. The preceding calculations give the F_0 values, which are the total lethal effects expressed as minutes at 121°C. Thermal processes may, thus, be compared by calculating the integrated lethality and then transferred to the equivalent value at 121°C (250°F). *F* values include contributions from heating, holding, and cooling periods.

Spores from various bacteria vary greatly in heat resistance. Also, psychrotrophic spore-formers have spores that are much less heat-resistant than those of mesophiles, and still less than those of thermophiles. Less dramatic but still of considerable importance are the variations within strains or species caused by other factors. Lowering the pH decreases heat resistance markedly. Growth at high temperatures results in spores of increased resistance, but this may not be reflected in the cells after germination. Water activity, salts, and preservatives may variously influence heat resistance. For spores these considerations are complicated by impacts during activation, germination, and outgrowth.

Among the early regulations for food safety was the pasteurization of milk to kill *Mycobacterium tuberculosis* and *Mycobacterium bovis* at 61°C for 30 min. However, it was the heat resistance of *Coxiella burnettii,* the etiological agent of Q-fever, that led to the requirement of 63°C for 30 min that is commonly reported for the pasteurization of milk. High-temperature short-time (HTST) treatment is often referred to as 72°C for 15 s, but equivalent combinations are 89°C for 1 s, 90°C for 0.5 s, 94°C, for 0.1 s, 96°C for 0.05 s, and 100°C for 0.01 s. The thermal processing of milk is also considered in Section II.A. Somewhat higher temperatures are required for cream and ice cream mixes. Other products may not demand such severe treatments because of lower or higher pH or other factors; thus, whole eggs are pasteurized at 60°C for 3.5 min. In comparison, cooking procedures have similar requirements (e.g., the minimum internal temperature of roast beef is 63°C with a holding time of 5 min). Fruit juices of low pH may need rather severe treatments to kill yeasts and molds.

B. Deviations from Log-Linear Relationships of Survivor Curves

When D values are used, they often present simplifications of reality. The line in Fig. 4 represents the simple log-linear relationship between survivors and time, but partial deviations from linearity also occur for the death kinetics for spores. An initial increase of viable organisms may be the result of spore activation. A shoulder in the graph has been explained as a balance between activation and inactivation; however, assuming that multiple hits on sensitive sites in the spores are necessary for inactivation, a shoulder may also result from this. Heat-up time and clumping have likewise been mentioned as factors contributing to shoulders.

Concave survivor curves have mainly been associated with variations in the heat resistance of individual organisms in a population. However, other explanations have been advanced. Extended tails as well as biphasic curves have been observed.

The log-linear relationship for thermal death over time has also been assumed for vegetative cells of microorganisms. However, even greater deviations are commonly observed than for spores. Differences among cells during heating have been suggested as important factors to explain this (cf. also Section II.B).

See Also the Following Articles

Extremophiles • Food Storage and Preservation • Heat Stress • Low-Temperature Environments • Meat and Meat Products • Refrigerated Foods • Wine

Bibliography

Adams, M. R., and Moss, M. O. (1995). "Food Microbiology." The Royal Society of Chemistry, Cambridge, UK.

Andrew, M. H. E., and Russell, A. D. (eds.) (1984). "The Revival of Injured Microbes." The Society for Applied Bacteriology Symposium Series No. 12. Academic Press, London.

Archer, L. A. (1996). Preservation microbiology and safety: Evidence that stress enhances virulence and triggers adaptive mutations. *Trends Food Sci. Technol.* 7, 91–95.

Gould, G. W. (1989). Heat-induced injury and inactivation. *In* "Mechanisms of Action of Food Preservation Procedures" (G. W. Gould, ed.), pp. 11–42. Elsevier Applied Science, London.

Gounot, A.-M. (1991). Bacterial life at low temperature: Physiological aspects and biotechnological implications. *J. Appl. Bacteriol.* **71**, 386–397.

Graumann, P., and Marahiel, M. A. (1996). Some like it cold: Response of microorganisms to cold shock. *Archiv. Microbiol.* **166**, 293–300.

Knøchel, S., and Gould, G. (1995). Preservation microbiology and safety: Quo vadis. *Trends Food Sci. Technol.* **6**, 127–131.

Leistner, L., and Gorris, L. G. M. (1995). Food preservation by hurdle technology. *Trends Food Sci. Technol.* **6**, 41–46.

McDougald, D., Rice, S. A., Weichart, D., and Kjelleberg, S. (1998). Nonculturability: Adaption or debilitation? *FEMS Microbiol. Ecol.* **25**, 1–9.

Ray, B. (1989). Enumeration of injured indicator bacteria from foods. *In* "Injured Index and Pathogenic Bacteria" (B. Ray, ed.), pp. 9–54. CRC Press, Boca Raton, FL.

Roberts, T. A., Pitt, J. I., Farkas, J., and Grau, F. H. (eds.) (1998). "Microorganisms in Foods, Microbial Ecology of Food Commodities." Blackie Academic & Professional, London.

Sørhaug, T., and Stepaniak, L. (1997). Psychrotrophs and their enzymes in milk and dairy products: Quality aspects. *Trends Food Sci. Technol.* **8**, 35–41.

Whiting, R. C., and Buchanan, R. L., (1997). Predictive modeling, *In* "Food Microbiology, Fundamentals and Frontiers" (M. P. Doyle et al., eds.), pp. 728–739. ASM Press, Washington, DC.

Tetrapyrrole Biosynthesis in Bacteria

Samuel I. Beale
Brown University

I. Outline of the Tetrapyrrole Biosynthetic Pathway
II. Initial Steps from General Metabolites to Committed Tetrapyrrole Precursors
III. Formation of the First Macrocyclic Tetrapyrrole
IV. The Branch toward End Products with Reduced Methylated Pyrrole Rings
V. Shared Steps toward Hemes, Bilins, and Chlorophylls
VI. The Fe-Porphyrin Branch Leading to Hemes and Bilins
VII. The Mg-Porphyrin Branch Leading to Chlorophylls and Related Compounds
VIII. Regulation

GLOSSARY

bacteriochlorin A porphyrin with two nonadjacent reduced pyrrole rings; the tetrapyrrole macrocycle of bacteriochlorophylls.

bilin A nonmacrocyclic (i.e., open-chain) tetrapyrrole.

chlorin A porphyrin with one reduced pyrrole ring; the tetrapyrrole macrocycle of chlorophylls and heme *d*.

chlorophyll A macrocyclic tetrapyrrole that has an isocyclic ring and contains Mg as the centrally chelated metal atom.

corrin A contracted macrocyclic tetrapyrrole in which two of the pyrrole rings are connected directly at their α positions instead of by a methylene or methene bridge; the tetrapyrrole macrocycle of cofactor B_{12}.

heme A macrocyclic tetrapyrrole that contains iron as the centrally chelated metal atom.

isobacteriochlorin A porphyrin with two adjacent reduced pyrrole rings; the tetrapyrrole macrocycle of siroheme and heme d_1.

isocyclic ring A fifth ring on a tetrapyrrole formed by ligation of a bridge carbon linking two pyrrole rings to a β substituent on one of the pyrrole rings; a feature of all chlorophylls and related compounds.

macrocycle The overall ring structure formed by four pyrrole rings linked one to another through their α positions, usually via methylene or methene bridges.

porphyrin A macrocyclic tetrapyrrole that has a conjugated double-bond system, conferring aromatic characteristics.

porphyrinogen A macrocyclic tetrapyrrole that is not aromatic because at least one of the links between adjacent pyrrole rings is saturated and the double-bond conjugation system is interrupted.

tetrapyrrole A molecule containing four pyrrole rings linked one to another through their α positions, usually via methylene or methene bridges; the tetrapyrrole can have a closed (macrocycle) or open (bilin) structure.

TETRAPYRROLES are a diverse family of molecules that are nearly ubiquitous in living cells. As prosthetic groups and cofactors of enzymes and other proteins, tetrapyrroles have key roles in many biochemical processes including redox chemistry, light absorption and photochemistry, reversible binding of gaseous molecules, and carbon-bond rearrangements. A major chemical property of tetrapyrroles that contributes to their biological usefulness is the ability, in certain oxidation states, to form a macrocyclic conjugated double-bond system that provides aromaticity; that is, the tetrapyrrole becomes a rigid planar structure with delocalized bonding electrons and the binding affinity for metals is greatly increased. The ability to bind metal ions strongly within the rigid planar framework allows the metal to interact reversibly with other molecules through axial ligands that extend away from the tetrapyrrole plane. The aromatic nature of certain tetrapyrroles also enhances their effectiveness in photosynthesis in absorbing light, transferring excitation energy, and accepting and donating electrons. Among groups of organisms, prokaryotes contain the widest variety of tetrapyrroles. These include hemes, bilins, chlorophylls and bacteriochlorophylls, and coenzyme B_{12} and related corrinoids, as well as more narrowly distributed tetrapyrroles such as the Ni-binding factor F_{430} of methanogens. Owing to their often prominent colors and an early recognition of the important roles that they have in major biochemical processes such as respiration and photosynthesis, tetrapyrroles were among the first molecules whose biosynthesis was systematically studied. This article is a brief summary of the current state of knowledge of tetrapyrrole biosynthesis as it occurs in prokaryotic organisms.

I. OUTLINE OF THE TETRAPYRROLE BIOSYNTHETIC PATHWAY

The structures of the tetrapyrroles that are synthesized by prokaryotic organisms are shown in Fig. 1. These tetrapyrroles can be arranged on a biosynthetic grid as a multibranched pathway leading to the various end products (Fig. 2). Some of the end products are themselves starting points for further metabolism to other end products. There are three major branches of the pathway. One branch leads to end products with reduced methylated pyrrole rings, the second branch produces Fe-porphyrins (hemes) and their derivatives, and the third branch generates chlorophylls and related compounds.

II. INITIAL STEPS FROM GENERAL METABOLITES TO COMMITTED TETRAPYRROLE PRECURSORS

A. δ-Aminolevulinic Acid Formation from Glutamate

The universal tetrapyrrole precursor is the five-carbon compound δ-aminolevulinic acid (ALA). All of the carbon and nitrogen atoms of the tetrapyrrole nucleus are derived from ALA. There are two distinct routes of ALA formation. In the five-carbon route, glutamic acid is activated by ATP-dependent ligation to $tRNA^{Glu}$, catalyzed by glutamyl-tRNA synthetase (EC 6.1.1.17). This is the same reaction that is used to activate glutamic acid for peptide-bond formation in protein synthesis. In the next reaction, the activated glutamate is reduced to glutamic acid 1-semialdehyde by the action of the nicotinamide adenine dinucleotide-P (NADPH)-dependent enzyme glutamyl-tRNA reductase. Finally, the glutamic acid 1-semialdehyde is rearranged to form ALA by a two-step process that starts with the addition of an amino group derived from enzyme-bound pyridoxamine-P at the aldehyde carbon to form enzyme-bound diaminovaleric acid, followed by removal of the original amino group to regenerate the enzyme cofactor and yield ALA, which is isomeric with glutamic acid 1-semialdehyde. This two-step aminotransferase reaction is catalyzed by glutamic acid 1-semialdehyde aminotransferase (EC 5.4.3.8). The five-carbon route of ALA formation is used by all prokaryotic organisms except those that are in the α subgroup of proteobacteria. It is also used by plants and algae.

B. δ-Aminolevulinic Acid Formation from Glycine and Succinate

Members of α subgroup of proteobacteria, as well as all animals, fungi, and yeasts, form ALA by the glycine–succinate route. In this pathway, a molecule of glycine is condensed with the activated succinate compound, succinyl-coenzyme A, to yield ALA plus free coenzyme A and a molecule of CO_2 that is derived from the carboxyl carbon of glycine. This reaction is catalyzed by the pyridoxal-P-containing enzyme ALA synthase (EC 2.3.1.37).

The rationale for the existence of two routes to ALA is not entirely clear. Expressed heterologous ALA synthase can complement ALA auxotrophic mutants of organisms that normally use the five-carbon pathway. The five-carbon route, being much more widely distributed than the glycine–succinate route, is considered to be the evolutionarily older one. The glycine–succinate route could not have evolved until after the appearance of succinyl-coenzyme A as a metabolite, which perhaps occurred in conjunction with the rise of aerobic respiration.

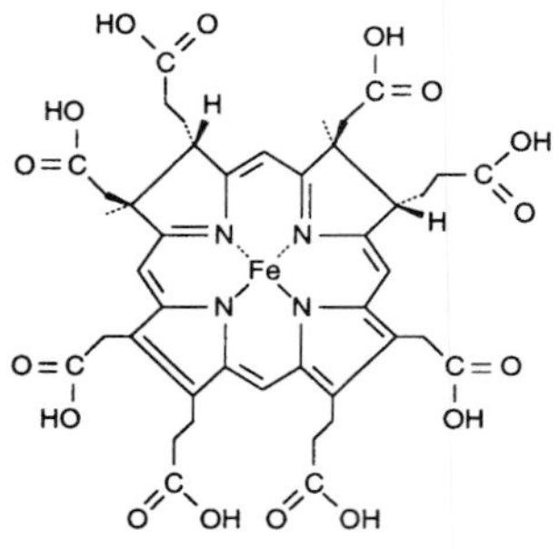

Siroheme

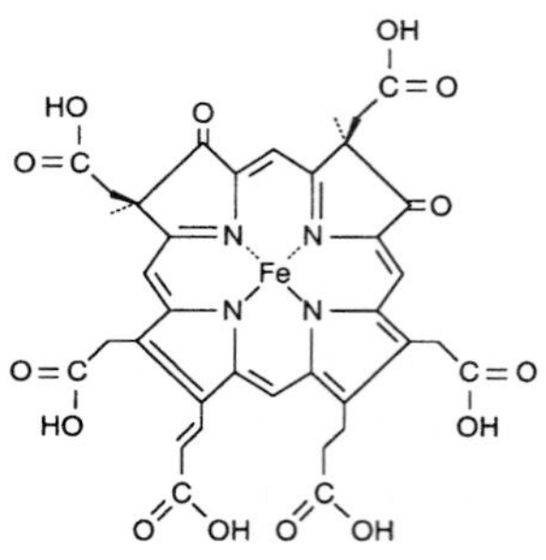

Heme d_1

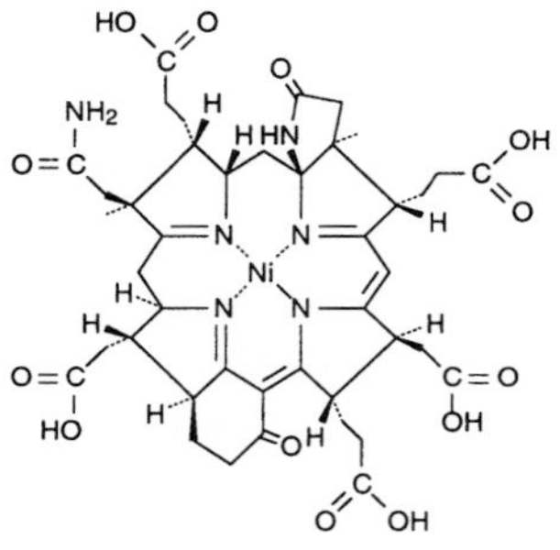

Factor F_{430}

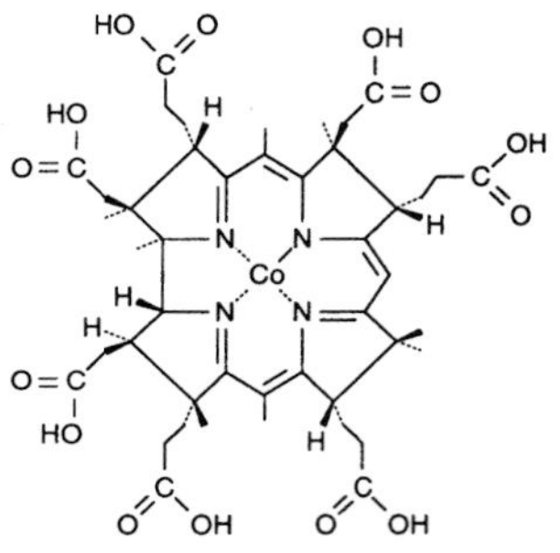

Cobyrinic Acid
(Coenzyme B_{12} nucleus)

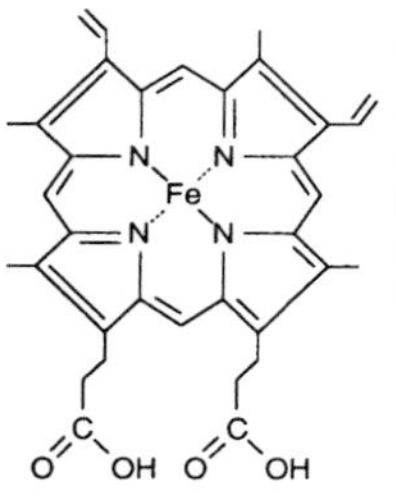

Protoheme

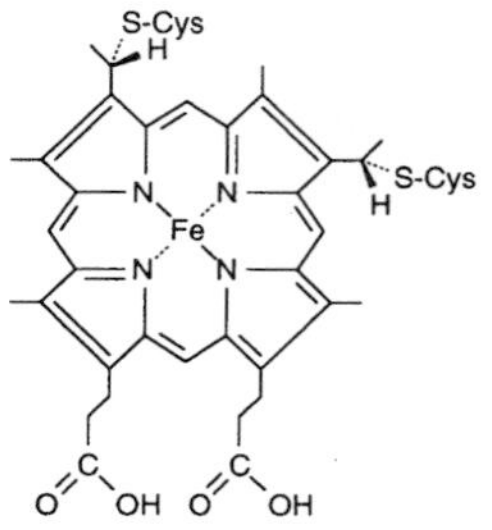

Heme *c*

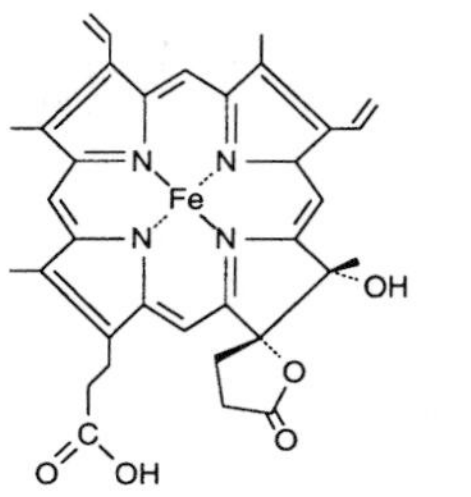

Heme *d*
(catalase type)

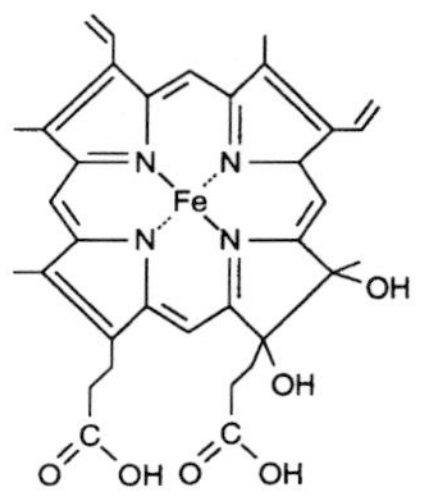

Heme *d*
(terminal oxidase type)

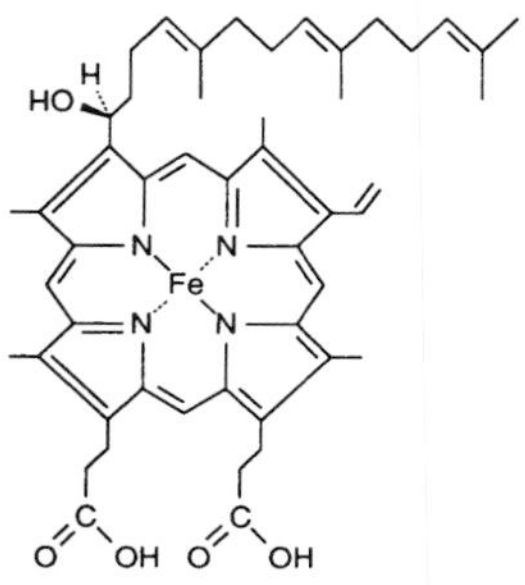

Heme *o*

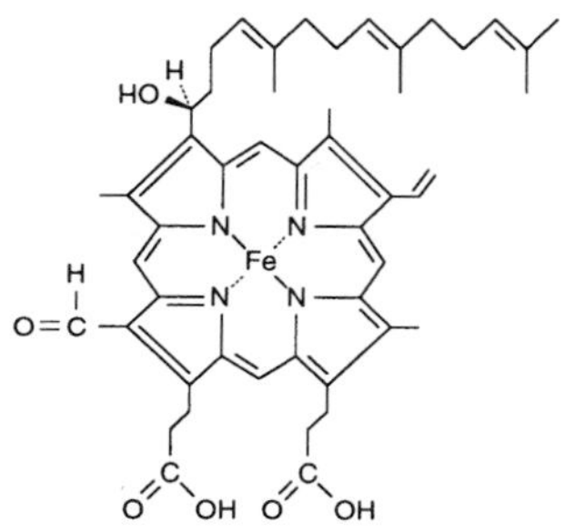

Heme *a*

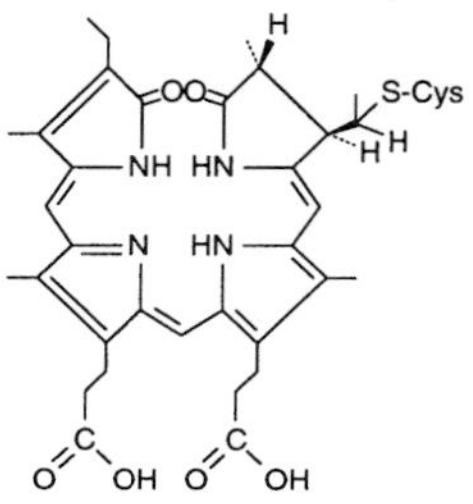

Phycocyanobilin

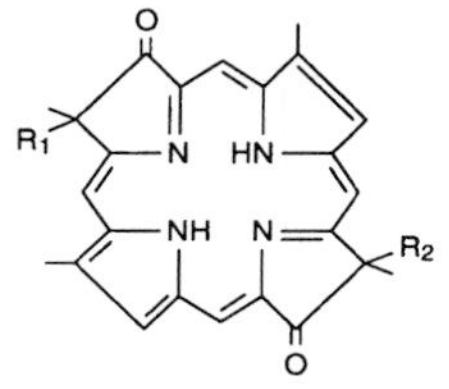

Tolyporphins
(R_1, R_2 = OH, O-acetyl,
or C-glycosyl derivatives)

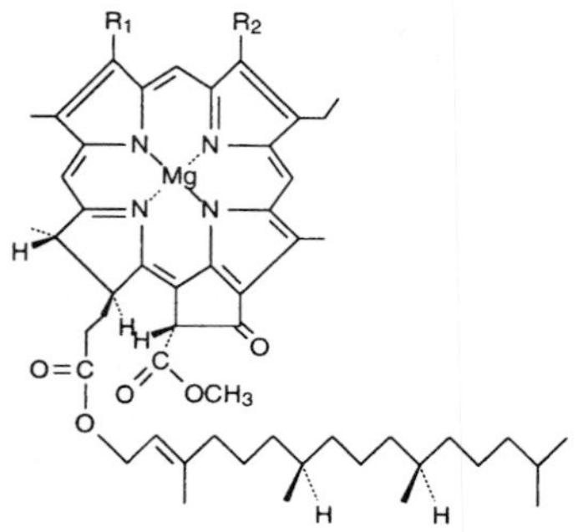

Chlorophyll *a* ($R_1 = CHCH_2$, $R_2 = CH_3$)
Chlorophyll *b* ($R_1 = CHCH_2$, $R_2 = CHO$)
Chlorophyll *d* ($R_1 = CHO$, $R_2 = CH_3$)

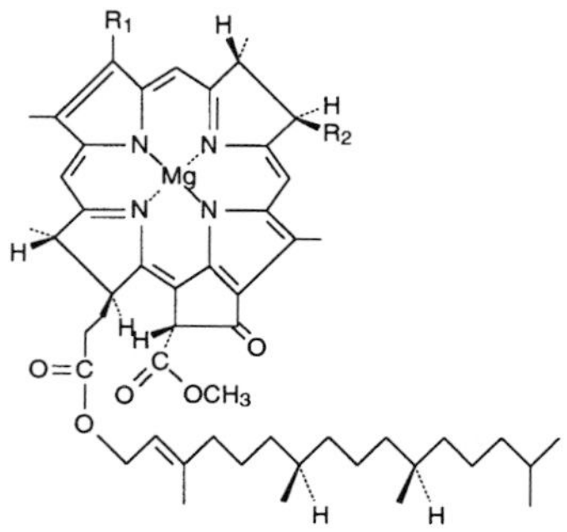

Bacteriochlorophyll *a* ($R_1 = C(O)CH_3$, $R_2 = CH_2CH_3$)
Bacteriochlorophyll *b* ($R_1 = C(O)CH_3$, $R_2 = CHCH_3$)
Bacteriochlorophyll *g* ($R_1 = CHCH_2$, $R_2 = CHCH_3$)

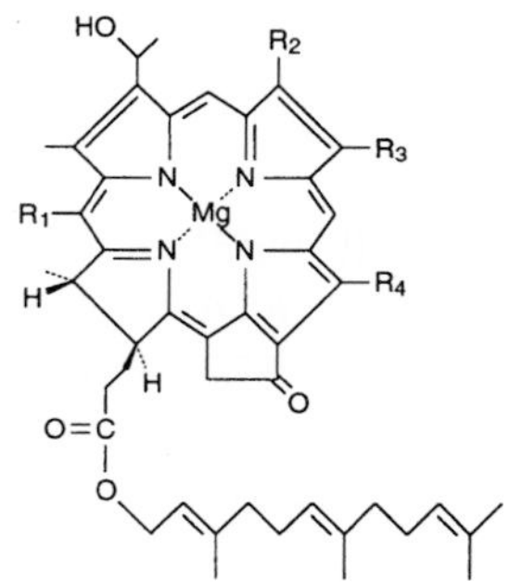

Chlorosome Chlorophylls
(R_1 = H or CH_3,
$R_2 = CH_3$ or CHO,
R_3, R_4 = C_1-C_5 alkyl groups)

III. FORMATION OF THE FIRST MACROCYCLIC TETRAPYRROLE

The first macrocyclic tetrapyrrole, from which all other biological tetrapyrroles are derived, is uroporphyrinogen III (Fig. 2). This tetrapyrrole is formed from eight molecules of ALA by the sequential action of three enzymes. First, porphobilinogen synthase (EC 4.2.1.24; also known as ALA dehydratase) catalyzes an asymmetric condensation of two ALA molecules, with the loss of two H_2O molecules, to form a pyrrole, porphobilinogen. Next, four porphobilinogen molecules are ligated together into a linear tetramer, hydroxymethylbilane, by joining the NH_2-bearing carbon atom of one porphobilinogen to the free α-position of the next porphobilinogen, with loss of the amino group. This oligomerization reaction is catalyzed by the enzyme hydroxymethylbilane synthase (EC 4.3.1.8; also known as porphobilinogen deaminase and preuroporphyrinogen synthase). The enzyme contains a bound dipyrromethane group that acts as a primer for the ligation of additional porphobilinogen molecules, and is retained by the protein upon release of hydroxymethylbilane. In the last step, uroporphyrinogen III synthase (EC 4.2.1.75; also known by the earlier name uroporphyrinogen III cosynthase) closes the macrocycle in a remarkable reaction in which one of the pyrrole rings is inverted so that its acetate and propionate substituents are in the opposite positions relative to those of the other three pyrrole rings. All biologically relevant tetrapyrroles are derived from uroporphyrinogen III. However, hydroxymethylbilane is unstable in solution and, in the absence of uroporphyrinogen III synthase, it spontaneously and irreversibly cyclizes to form uroporphyrinogen I, which does not contain an inverted pyrrole ring. Under physiological conditions, there is ample uroporphyrinogen III synthase present to prevent the formation of the incorrect uroporphyrinogen I isomer.

There has been much discussion about why all biological tetrapyrroles are based on the type III isomer. It can be calculated, and it has been experimentally verified, that random nonbiological assembly of porphobilinogen molecules into tetrapyrroles yields 50% type III isomer, 25% type I, and 12.5% each type II (two adjacent rings inverted) and type IV (alternating rings inverted). It is possible that before organisms were capable of synthesizing tetrapyrroles, they used tetrapyrroles found in the environment. These tetrapyrroles would have been formed by random processes and would therefore be predominantly type III. When organisms became capable of synthesizing tetrapyrroles, they would have been predisposed to use type III isomers.

For completeness, it should be mentioned that there is one reported case of a biologically functional molecule based on a type I tetrapyrrole. The enzyme rubredoxin: oxygen oxidoreductase, which functions as a terminal oxidase in *Desulfovibrio gigas*, contains Fe-uroporphyrin I as one of its heme groups. Nothing has been reported concerning the biosynthesis of this heme.

IV. THE BRANCH TOWARD END PRODUCTS WITH REDUCED METHYLATED PYRROLE RINGS

Several end-product tetrapyrroles of major biological importance but minor or restricted abundance are derived from uroporphyrinogen III and contain reduced methylated pyrrole rings. The first intermediate in the biosynthesis of this group of compounds is precorrin-2 (Fig. 2; also referred to as dihydrosirohydrochlorin), which is the product of S-adenosyl-L-methionine:uroporphyrinogen III C-methyltransferase (EC 2.1.1.107). There appear to be multiple enzymes that can catalyze this reaction, and the genes for these enzymes are contained in operons that carry other genes for the enzymes that catalyze the formation of the various end products.

Fig. 1. Structures of tetrapyrroles that occur in prokaryotes. The top row illustrates the end products of the reduced methylated tetrapyrrole branch; the middle two rows show porphyrin, Fe-porphyrin (heme), and bilin end products; and the bottom row shows end products of the Mg-porphyrin branch.

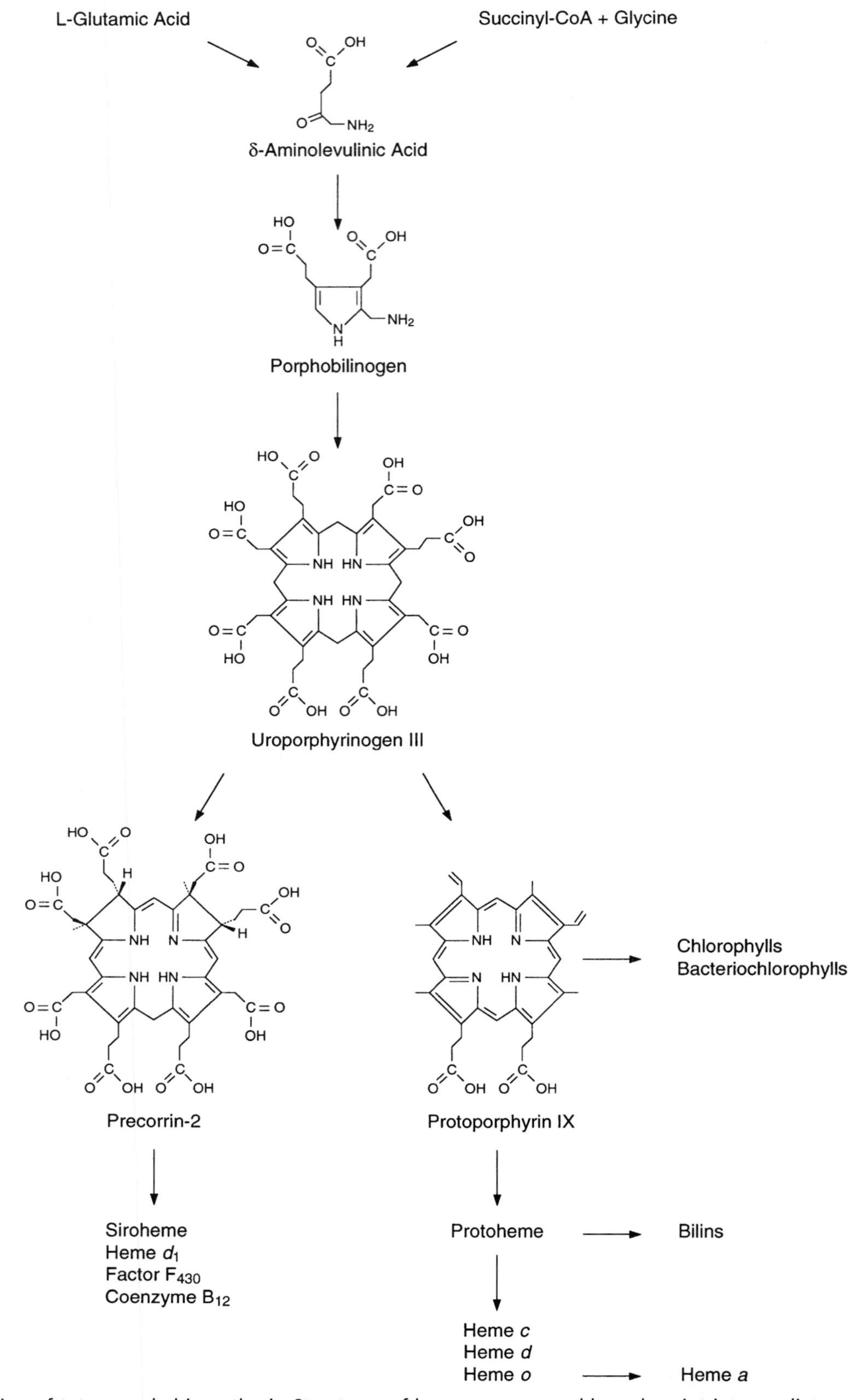

Fig. 2. Outline of tetrapyrrole biosynthesis. Structures of key precursors and branch-point intermediates are shown.

One consequence of the pyrrole ring methylations is that the two methylated pyrrole rings become locked into a reduced oxidation state, relative to their oxidation state in porphyrins.

A. Siroheme

One end product, siroheme, is the Fe-containing prosthetic group of assimilatory sulfite and nitrite reductases, and dissimilatory sulfite reductases. In some organisms in which siroheme is the sole, or major, end product of this branch, the four reactions required to convert uroporphyrinogen III to siroheme (two methylations to form precorrin-2, followed by a two-electron oxidation to form sirohydrochlorin (an isobacteriochlorin), and finally insertion of Fe into the macrocycle) are catalyzed by a single multifunctional enzyme, siroheme synthase. In the genus *Desulfovibrio,* a variant form of siroheme occurs in which one of the carboxylic acid groups is amidated. It is of interest that an amidation at the analogous position occurs on Factor F_{430} (see Section VI.C). It is not known if this similarity arises from a shared biosynthetic step.

B. Heme d_1

Heme d_1 is an Fe-containing reduced tetrapyrrole that is a prosthetic group of cytochrome cd_1, the dissimilatory (respiratory) nitrite reductase of certain *Pseudomonas* species. The methyltransferase for this product, as well as enzymes that are required for the other steps of heme d_1 formation, are encoded on genes in a *nir* operon for nitrite respiration. The mechanisms of the conversion of two propionates to oxo groups, oxidation of a propionate to an acrylate group, and the removal of two electrons from the macrocycle to form heme d_1 have not been elucidated.

Even though two adjacent pyrrole rings of siroheme and heme d_1 are reduced, there is nevertheless a conjugated double-bond system around the isobacteriochlorin macrocycle, which allows for delocalization of the eletrons and gives the macrocycle aromatic characteristics.

C. Factor F_{430}

Factor F_{430} is a reduced tetrapyrrole end product that contains Ni as the chelated metal. Factor F_{430} is a prosthetic group of methyl-coenzyme M reductase, the enzyme that produces methane in methanogenic organisms. Little about the biosynthesis of factor F_{430} has been reported except that it is derived from precorrin-2 and that the formation of the cyclohexanone ring from a propionic acid group is a late step.

D. Coenzyme B_{12}

Finally, this group of end products includes coenzyme B_{12}, which has a Co-containing cobyrinic acid nucleus. The high degree of modification of this tetrapyrrole includes the excision of one of the bridging carbon atoms and the direct linking of the adjoining pyrrole rings to form the corrin macrocycle (Fig. 1). The biosynthesis of coenzyme B_{12} is achieved through a very lengthy and complex pathway, the description of which is beyond the scope of this article.

V. SHARED STEPS TOWARD HEMES, BILINS, AND CHLOROPHYLLS

The conversion of uroporphyrinogen III to protoporphyrin IX, the last common precursor of hemes and chlorophylls (Fig. 2), is catalyzed by the sequential action of three enzymes.

A. Uroporphyrinogen III Decarboxylation

Uroporphyrinogen III decarboxylase (EC 4.1.1.37) decarboxylates all four of the acetate groups of uroporphyrinogen III to form coproporphyrinogen III. The order of the decarboxylations is clockwise around the molecule, beginning with the acetate at the 8 o'clock position, as shown in Fig. 2. A variant route from uroporphyrinogen III to coproporphyrinogen III, which does not involve uroporphyrinogen III decarboxylase but instead proceeds via precorrin-2, has been described for some members of the *Desulfovibrio* group of strict anaerobes.

B. Coproporphyrinogen III Oxidative Decarboxylation

Coproporphyrinogen III oxidative decarboxylase (EC 1.3.3.3) selectively converts two of the four propionic acid groups of coproporphyrinogen III to vinyl groups, forming protoporphyrinogen IX. Two types of coproporphyrinogen III oxidative decarboxylase have been described, one that occurs in aerobic organisms and requires O_2 as the oxidant, and the other that operates in strict anaerobes and facultative organisms under anerobiosis. The mechanism of the reactions catalyzed by these enzymes is poorly understood.

C. Protoporphyrinogen IX Oxidation

The final step in protoporphyrin IX formation is the six-electron oxidation of protoporphyrinogen IX by protoporphyrinogen IX oxidase (EC 1.3.3.4). In eukaryotic organisms, this enzyme is a well-characterized membrane-associated protein that contains a bound flavin and requires O_2 as the oxidant. A similar enzyme has been reported in *Bacillus subtilis* and a few other bacteria that are normally considered to be strict aerobes. In other prokaryotes, however, the reaction appears to be catalyzed by a different enzyme that does not require O_2. The protein components of this protoporphyrinogen IX oxidase are not well characterized, and the products of as many as three different genes may be involved. In contrast to the porphyrinogens, protoporphyrin IX has a fully conjugated double-bond system around the macrocycle, which facilitates delocalization of the electrons and gives an aromatic character to the macrocycle.

D. Protoporphyrin IX as an End Product

Although protoporphyrin IX is usually considered to be a precursor rather than an end product, it can be used as an end product with a specific biological function. For example, membrane-associated protoporphyrin IX has been shown to be the photoreceptor for light-induced carotenogenesis in *Myxococcus xanthus*.

E. Uroporphyrin III and Coproporphyrin III

There are occasional occurrences in bacteria of oxidized forms of the porphyrinogen precursors of protoporphyrin IX (uroporphyrin III and coproporphyrin III). These porphyrins are not substrates of the biosynthetic enzymes for protoporphyrin IX, which require porphyrinogens as substrates, and it is unclear whether they have a biological function. When uroporphyrin III and coproporphyrin III are present, they often have chelated metals such as Co, Cu, and Zn.

F. Tolyporphins

For completeness, mention should be made of a group of compounds called tolyporphins (Fig. 1). These molecules have a bacteriochlorin-type macrocycle and they appear from their structure to be derived from protoporphyrin IX, although no biosynthetic information is available. Tolyporphins are found in certain soil cyanobacteria, in which their function is unknown. They have been investigated for possible therapeutic use as multidrug resistance antagonists.

VI. THE Fe-PORPHYRIN BRANCH LEADING TO HEMES AND BILINS

A. Protoheme

The insertion of Fe^{2+} into protoporphyrin IX by the enzyme ferrochelatase (EC 4.99.1.1) produces protoheme. The reaction requires no energy source and ferrochelatase has no cofactor. Protoheme is the prosthetic group of *b*-type cytochromes, bacterial peroxidases, and several oxidase enzymes. Besides having these metabolic functions, protoheme serves as the precursor of all other porphyrin-based hemes and it is also the precursor of bilins.

B. Heme *c*

In addition to protoheme, four other hemes have been identified in bacteria. Heme *c* is the prosthetic group of *c*-type cytochromes. The heme is covalently attached to protein cysteine groups by fusion with the heme vinyl groups to form α-methyl thioethers (Fig. 1). Enzymes that catalyze this reaction, collectively named apocytochrome *c* heme lyase (EC 4.4.1.17), have been characterized in eukaryotic organisms, but very little is known about the enzyme(s) from prokaryotes. Many bacterial *c*-type cytochromes are periplasmic. The apo-cytochrome and protoheme are separately exported from the cytoplasm to the periplasm, the latter with the aid of a "heme chaperone," and holocytochrome assembly takes place within the periplasm.

C. Heme *o* and Heme *a*

Heme *o* and heme *a* are prosthetic groups of *o*- and *a*-type terminal oxidases, respectively. Mutagenesis and transformation experiments indicate that heme *o* is initially formed from protoheme by the attachment of a farnesyl group to one of the heme vinyl groups to form an α-hydroxyethylfarnesyl group (Fig. 1). Heme *a* is formed from heme *o* by oxidation of a methyl group to a formyl group. Some archaea and a few bacteria contain variant *a*- and *o*-type hemes that have geranylgeranyl rather than farnesyl moieties, and the α-hydroxyethyl moiety may be dehydrated to an ethenyl moiety.

D. Heme *d*

Heme *d* is the prosthetic group of bacterial catalases and cytochrome *bd*-type terminal oxidases. Heme *d* contains one reduced pyrrole ring and the macrocycle is therefore a chlorin rather than a porphyrin. The heme *d* of *Escherichia coli* catalase HPII contains vicinyl *cis*-hydroxyl groups on the reduced pyrrole ring, and one of the hydroxyl groups forms a *spiro*-lactone by condensing with the propionate group at the same position on the pyrrole (Fig. 1). This heme *d* is derived from protoheme and it is apparently formed spontaneously *in situ*, after the apo-catalase binds protoheme. The heme *d* of cytochrome *bd*-type respiratory terminal oxidases has been reported to have a somewhat different structure than the catalase-type heme *d*. The vicinyl hydroxyl groups are *trans* rather than *cis*, and a *spiro*-lactone is not present on the native heme. The absolute configuration of the *trans* vicinyl hydroxyl groups has not been determined. There have been no reports on how this heme *d* is synthesized.

E. Bilins

In animal cells, bilins are considered to be primarily heme degradation products without major physiological functions. However, in some cells that are capable of photosynthesis, including the prokaryotic cyanobacteria, certain bilins named phycobilins have important roles as the primary light-harvesting pigments for photosynthesis. These bilins occur covalently ligated to proteins known as phycobiliproteins, and the phycobiliproteins are assembled on the photosynthetic membranes into arrays that efficiently transfer the energy of light absorbed by the phycobilin chromophores to the photosynthetic reaction centers. Phycobilins are formed from protoheme by the sequential action of several enzymes. In the initial step, the macrocyclic ring of protoheme is opened by oxidative excision of the bridge carbon that connects the two vinyl-containing pyrrole groups, to form biliverdin IXα. The excised carbon is released as CO. This reaction is catalyzed by heme oxygenase (EC 1.14.99.3). Cyanobacterial heme oxygenase is soluble and requires both O_2 and reduced ferredoxin as cosubstrates. The phycobilins are derived from biliverdin IXα by a series of transformations that include isomerizations and reduced ferredoxin-dependent reduction reactions. All cyanobacteria contain the blue-colored phycocyanobilin (Fig. 1). In addition, they may contain one or more of about a dozen different phycobilins that have been characterized. Specific sets of enzymes appear to be required for ligation of individual phycobilins to specific sites on apo-phycobiliproteins to produce the functional holoproteins. Ligation is usually through one, but sometimes through two α-methyl thioether bonds between protein cysteine residues and bilin vinyl groups or their derivatives. The presence of multiple phycobilins, with their overlapping but different light-absorption spectra, allows the organisms to absorb light having a wide wavelength range and contribute the absorbed energy to the photosynthetic reaction centers.

VII. THE Mg-PORPHYRIN BRANCH LEADING TO CHLOROPHYLLS AND RELATED COMPOUNDS

A. Mg Chelation

The path toward chlorophylls begins with insertion of Mg^{2+} into protoporphyrin IX. The Mg-chelatase reaction and its enzymes are much more complex than the analogous ferrochelatase reaction for protoheme formation. Whereas ferrochelatase is a (usually) membrane-bound homodimer, Mg-chelatase has three subunits, all of which are soluble, and whereas Fe^{2+} chelation requires no energy source, Mg^{2+} chelation is dependent on ATP hydrolysis. Moreover, there are two steps that require ATP, the first being activation of the enzyme, which requires only two of the three subunit types, and the second is the chelatase reaction itself, which requires all three enzyme subunits. The stoichiometric composition of the enzyme has not been determined, nor has the number of ATP molecules consumed per Mg^{2+} chelation.

B. Isocyclic Ring Formation

After Mg^{2+} is chelated, the next steps toward chlorophyll formation involve the creation of the isocyclic ring, a feature of all chlorophylls and bacteriochlorophylls (Fig. 1). This ring is derived from one of the two propionate groups of Mg-protoporphyrin IX. The first step in this process is methylation of the propionate carboxyl group, a reaction that is catalyzed by *S*-adenosyl-L-methionine : Mg-protoporphyrin IX methyltransferase (EC 2.1.1.11). The methylation is believed to be necessary to prevent decarboxylation of the propionate at a later step where a β-keto group is introduced; β-keto acids are prone to spontaneous decarboxylation, whereas β-keto esters are not. In plants and other aerobic chlorophyll-forming organisms, the cyclization reaction begins by the introduction of an O_2-derived hydroxyl group at the β-position of the propionate ester, then the oxidation of the β-hydroxypropionate ester to a β-ketopropionate ester, and finally the ligation of the α-carbon of the β-ketopropionate ester to the adjacent tetrapyrrole bridge carbon. In anaerobic bacteriochlorophyll-forming organisms, the cyclization reaction is different in that the keto oxygen atom is derived from H_2O rather than O_2. This may occur by oxidation of the propionate ester to an acrylic ester, followed by hydration to a β-hydroxypropionate ester. The product of the cyclization process is known as Mg-divinylpheoporphyrin a_5 or divinyl protochlorophyllide.

C. Reduction of the Porphyrin to a Chlorin Macrocycle

1. *The Light-Dependent Process*

The conversion of protochlorophyllide to chlorophyllide *a* involves a reduction of one of the pyrrole rings to convert the porphyrin macrocycle to a chlorin. There are two processes to accomplish this reaction. Cyanobacteria, as well as all eukaryotic chlorophyll-synthesizing organisms, have an enzyme, NADPH : protochlorophyllide oxidoreductase (EC 1.3.1.33), which catalyzes a light-dependent transfer of two electrons, and one proton, from NADPH to protochlorophyllide (the second proton is derived from H_2O). This reaction requires the absorption of one photon by enzyme-bound protochlorophyllide for each catalytic cycle.

2. *The Light-Independent Process*

In addition to this light-dependent process, all prokaryotic chlorophyll-forming organisms, as well as all plant groups except angiosperms, have a light-independent protochlorophyllide-reducing system. Photosynthetic prokaryotes other than cyanobacteria have only the light-independent system. Three gene products have been identified as being required for the light-independent reaction. There is very little known about the enzymes or reaction mechanism for the light-independent process, but the reaction is believed to require the energy of ATP hydrolysis and to derive reducing equivalents from reduced ferredoxin.

It is not clear why cyanobacteria (as well as most plant groups) have maintained both the light-dependent and light-independent protochlorophyllide-reducing systems. Mutant cyanobacteria in which the light-dependent system has been disrupted appear to function normally at lower light intensities but cannot tolerate high light intensities. Perhaps the light-dependent system originated for some signaling role, and still functions in that capacity in organisms that have both systems, to allow the cells to gauge light intensity and respond accordingly. Its essential biosynthetic function in angiosperms may have evolved later with the loss of the light-independent system.

D. Reduction of a Vinyl to an Ethyl Group

The initial product of the isocyclic ring-forming reaction, divinyl protochlorophyllide, contains two vinyl groups, whereas chlorophyll *a* has one vinyl and one ethyl group (Fig. 1). The vinyl-to-ethyl conversion is a reduction reaction that can occur either before or after the conversion of protochlorophyllide to chlorophyllide *a*. In angiosperm plants that require light for the protochlorophyllide-to-chlorophyllide *a* conversion, a variable portion of the divinyl protochlorophyllide that accumulates in the dark may be converted to the monovinyl form. However, in prokaryotes, which do not require light for the protochlorophyllide-to-chlorophyllide *a* conversion, most or all of the divinyl protochlorophyllide that is formed by the cyclase reaction is probably converted to divinyl chlorophyllide *a* before the vinyl-to-ethyl conversion occurs. Some initial progress has been made toward describing the vinyl-to-ethyl reduction reaction in plant chloroplast membranes, but there have been no reports on the reaction in any prokaryotic organism. It is of interest that some marine *Prochlorococcus* species accumulate divinyl chlorophylls *a* and *b* in addition to, or instead of, the monovinyl forms of these pigments.

E. Phytylation to form Chlorophyll *a*

The last step in the formation of chlorophyll *a* is the esterification of the remaining propionic acid with the C_{20} polyisoprene phytol. The esterification is catalyzed by the enzyme chlorophyll synthase. Chlorophyll synthase from organisms that accumulate chlorophylls accepts chlorophyllides, but not bacteriochlorophyllides, as substrates, and the reverse is true for the enzyme from bacteriochlorophyll-accumulating organisms. The polyisoprene substrate is the pyrophosphate ester of phytol or its precursor geranylgeraniol. In greening plant tissues, geranylgeraniol can be reduced to phytol after it is ligated to chlorophyllide. Photosynthetic bacteria contain traces of bacteriochlorophylls that are esterified with geranylgeraniol and partially reduced phytol precursors, suggesting that these organisms, similarly to plants, can use geranylgeranyl-pyrophosphate as the substrate for esterification, and convert this moiety to phytol after esterification.

F. Chlorophylls *b* and *d*

Chlorophyll *a* is the only chlorophyll present in cyanobacteria. However, a group of organisms known as prochlorophytes, which are related to cyanobacteria, accumulate both chlorophyll *a* and chlorophyll *b*. The latter pigment is derived from chlorophyll *a* by oxidation of a pyrrole methyl group to a formly group (Fig. 1). Despite the ubiquity of chlorophyll *b* in plants and green algae, there has been very little progress toward understanding how it is formed from chlorophyll *a*. An attractive reaction intermediate is an hydroxymethyl group, which could be formed by oxidation of the methyl group and then further oxidized to a formyl group, but there is no direct evidence to support this hypothesis.

Chlorophyll *d* was once thought to be an artifact of extraction of pigments from certain red algae. However, it was recently found to occur as the major chlorophyll in a marine photosynthetic prokaryote, where it functions as a light-harvesting pigment for photosynthesis. Its mode of formation is unknown, but the presence of a formyl group in place of one of the vinyl groups suggests an oxidative or epoxidative reaction.

G. Bacteriochlorophylls

All photosynthetic prokaryotes except cyanobacteria and prochlorophytes) use bacteriochlorophylls instead of chlorophylls as their photosynthetic pigments. Bacteriochlorophylls differ from chlorophylls by having a second reduced pyrrole ring, the one on the opposite corner of the macrocycle from the one that is reduced in chlorophylls. This type of macrocycle is known as a bacteriochlorin. Reduction of the second pyrrole ring is catalyzed by a non-light-requiring enzyme system whose protein components are homologous to the three gene products that have been identified for light-independent protochlorophyllide reduction. The chlorophyll-to-bacteriochlorophyll conversion has not been directly observed *in vitro*.

There are three known bacteriochlorophylls, designated types *a*, *b*, and *g* (Fig. 1). Each of these can function as the primary reaction center pigment as well as a light-harvesting accessory pigment, in different groups of photosynthetic bacteria. In some organisms, geranylgeraniol rather than phytol is the esterified polyisoprene alcohol. Any given organism contains only one of these bacteriochlorophyll types, but it may also contain chlorosome chlorophylls (see Section VII.H).

H. Chlorosome Chlorophylls

Certain photosynthetic bacteria contain specialized intracellular structures called chlorosomes that absorb light and transfer the excitation energy to the photosynthetic reaction centers that are located on the cell membranes. The chlorosomes contain pigments that more closely resemble chlorophylls than bacteriochlorophylls, in that they contain only one reduced pyrrole ring (Fig. 1). These pigments are now called chlorosome chlorophylls, in preference to their former, confusing names, such as *Chlorobium* chlorophylls or bacteriochlorophylls *c*, *d*, and *e*. A common feature of all chlorosome chlorophylls is the absence of a carboxymethyl group on the isocyclic ring. The chlorosome chlorophylls have a great diversity of structural variation, brought about in part by the addition of one to several *S*-adenosyl-L-methionine-derived methyl groups to various positions on the macrocycle. Additional sources of variation are the conversion of a methyl group to a formyl group, and a change of chirality of the asymmetric hydroxyethyl group. Most chlorosome chlorophylls contain the C_{15} polyisoprene farnesol as the esterifying alcohol rather than phytol or geranylgeraniol, but some variant types have been reported to contain various other groups including fatty alcohols. The structural variation results in an extension of the effective light-absorbing wavelength range of the chlorosomes, and it may also discourage the crystallization of the densely packed pigment molecules within the chlorosomes.

I. Photosynthetic Reaction-Center Pigments

Photosynthetic reaction centers contain modified pigments including metal-free (bacterio)chlorophyll derivatives known as (bacterio)pheophytins, which function as primary electron acceptors, and molecules in which the configuration of the asymmetric carbon atom of the isocyclic ring is opposite to that of (bacterio)chlorophyll *a*, referred to as (bacterio)chlorophyll *a*′, which are primary electron donors of one type of reaction center. These pigments are thought to be derived by modification of the parent (bacterio)chlorophyll, rather than arising from an earlier precursor, because the enzymes that catalyze the late steps of (bacterio)chlorophyll formation do not accept metal-free or chirally altered pigments as substrates.

VIII. REGULATION

A. Regulation of Enzyme Activity

A single organism often forms varying amounts of two or more major tetrapyrrole end products simultaneously or at different stages of development, and this ability obviously requires a high degree of metabolic regulation. One important mechanism for regulating biosynthetic pathways is by allosteric end-product inhibition of early steps that control the entry of general metabolites into the pathway, and this mode of regulation occurs in tetrapyrrole biosynthesis. In organisms that from ALA from glycine and succinate, ALA synthase is inhibited by micromolar concentrations of protoheme, and in organisms that form ALA from glutamate, glutamyl-tRNA reductase is inhibited by micromolar concentrations of protoheme. For organisms in which hemes are the only, or major end products, allosteric inhibition of ALA formation by protoheme seems sensible. For organisms that form much larger quantities of other end products such as (bacterio)chlorophylls, more complex allosteric effects would be expected. However, in these organisms, the other end products do not appear to function as allosteric regulators of ALA formation, although protoheme does. A mechanism that allows only one end product to effectively regulate a pathway that provides precursors for more than one end product requires that certain conditions exist.

For example, Mg-chelatase has a much higher affinity for protoporphyrin IX than does ferrochelatase. This large affinity difference implies that when Mg-chelatase is active, protoporphyrin IX is preferentially directed toward (bacterio)chlorophyll formation and away from heme synthesis. Heme is consumed in the formation of hemoproteins and bilins, but its availability depends on both the rate of ALA synthesis and the rate of diversion of protoporphyrin IX to (bacterio)chlorophyll formation. When protoheme is being depleted faster than it is being formed, the decrease in free-protoheme concentration causes an increase in the rate of ALA formation because that process is allosterically inhibited by protoheme. The increased rate of ALA formation consequently causes an increase in the rate of protoporphyrin IX formation, and when enough is formed to more than satisfy the demands for (bacterio)chlorophyll formation and protoheme utilization, protoheme will again accumulate to a concentration that inhibits ALA formation. The overall result is an autoregulation of the rate of protoporphyrin IX formation to ensure that there is enough made to fulfill the needs for heme and (bacterio)chlorophyll synthesis, but not an excess that can be phototoxic. The advantage of using protoheme instead of protoporphyrin IX itself as the feedback effector is that heme is relatively inert photochemically compared to protoporphyrin IX, and the potential for photodamage that could be caused by even transient increases in protoporphyrin IX concentration is avoided.

Of the many enzymes involved in tetrapyrrole biosynthesis, the only ones that consistently exhibit allosteric inhibition by end products are those involved in ALA formation. There is very little information available about allosteric regulation of other tetrapyrrole biosynthetic enzymes, and knowledge of how the regulation is managed is clearly incomplete.

B. Regulation of Gene Expression

For certain end products, key biosynthetic enzymes are encoded on genes that are parts of operons subject to coordinated regulation. For example, the conversion of uroporphyrinogen III to precorrin-2, which is the first step in the formation of the group of end products with reduced methylated pyrrole rings, is catalyzed by several enzymes, the expression of each being controlled by factors that affect the formation not only of the tetrapyrrole, but also of other components of the biochemical process that makes use of the tetrapyrrole. In this way, precorrin-2 formation is independently induced when cells are forming the siroheme-containing sulfite–nitrite reductase system in response to nutritional needs, when they are forming the heme d_1-containing cytochrome cd_1 of the dissimilatory nitrite reductase respiratory system for anaerobic growth in the presence of nitrite, and when conditions are such that the biosynthesis of coenzyme B_{12} is induced.

A somewhat more elaborate example of coordinated regulation of gene expression occurs in some facultative photosynthetic bacteria, in which all of the enzymes that are involved in forming bacteriochlorophyll from protoporphyrin IX are part of a photosynthetic gene cluster that consists of a few operons which are coordinately regulated to ensure that all of the components needed for photosynthesis, including bacteriochlorophylls, carotenoids, pigment-binding proteins, and electron-transport components, are made in the correct proportions and only under conditions where they can be correctly assembled and effectively function in the complex photosynthetic process.

The study of the regulation of tetrapyrrole biosynthesis at the enzyme and genetic levels is in its infancy. It can be anticipated that a better understanding of how organisms control and change their tetrapyrrole content in response to metabolic, developmental, and environmental cues will follow as more is learned about the details of the interconnected and interacting control systems.

See Also the Following Articles

Coenzyme and Prosthetic Group Biosynthesis • Photosensory Behavior • Pigments, Microbially Produced

Bibliography

Beale, S. I.(1994). Biosynthesis of cyanobacterial tetrapyrrole pigments: Hemes, chlorophylls, phycobilins. *In* "The Molecular Biology of Cyanobacteria" (D. A. Bryant, ed.), pp. 519–558. Kluwer Academic, Dordrecht.

Beale, S. I. (1996). Biosynthesis of hemes. *In* "*Escherichia coli* and *Salmonella*: Cellular and Molecular Biology" (F. C. Neidhardt *et al.*, eds.), 2nd ed., pp. 731–748. ASM Press, Washington, DC.

Beale, S. I. (1999). Enzymes of chlorophyll biosynthesis. *Photosynth. Res.* **60**, 43–73.

Chadwick, D. J., and Ackrill, K. (eds.) (1994). "The Biosynthesis of Tetrapyrrole Pigments." Ciba Foundation Symposium 180. John Wiley and Sons, Chichester, UK.

Dailey, H. A. (ed.) (1990). "Biosynthesis of Heme and Chlorophylls." McGraw-Hill, New York.

Dailey, H. A. (1997). Enzymes of heme biosynthesis. *J. Biol. Inorg. Chem.* **2**, 411–417.

Jordan, P. M. (ed.) (1991). "New Comprehensive Biochemistry," Vol. 19, Biosynthesis of Tetrapyrroles. Elsevier, Amsterdam.

Suzuki, J. Y., Bollivar, D. W., and Bauer, C. E. (1997). Genetic analysis of chlorophyll biosynthesis. *Annu. Rev. Genet.* **31**, 61–89.

Timber and Forest Products

David J. Dickinson and John F. Levy
Imperial College

I. Wood Structure
II. Wood Degradation
III. Habitat
IV. From Forest to Timber in Service
V. Durability of Timber
VI. Conclusion

GLOSSARY

hardwood Timber cut from trees belonging to the botanical group of plants Angiospermae.

heartwood The inner part of the wood of the trunk or branch of a tree.

moisture content The amount of water in a piece of wood, expressed as a percentage of the dry weight of that piece of wood [% moisture content (MC)]. Equilibrium moisture content is the % MC in equilibrium with the moisture in the immediate vicinity. Fiber saturation point is the moisture content of the wood at which the cell walls are fully saturated with water while the cell lumina are empty.

sapwood The outer part of the trunk or branch of a tree between the bark and the heartwood.

softwood Timber cut from trees belonging to the botanical group of plants Gymnospermae.

substrate Growth medium on which the organisms feed.

wood cell wall Structural element in plants composed of layers: the primary wall, inside of which is a three-layered secondary wall with an outer layer (S1), a middle layer (S2), and an inner layer (S3).

DECAY AND DEGRADATION OF WOOD are part of nature's recycling process. When a tree or branch of a tree falls to the ground in a forest, the materials manufactured by the tree from the basic elements in the environment to produce its constituent parts are broken down by a range of organisms to return those elements to the biosphere. These recycling processes are vital to life as we know it. Not only do they maintain the balance of the basic elements in nature required for plant growth but also they prevent the earth's surface from becoming so littered with plant and animal remains that life as we know it would no longer be possible. It has been estimated that, without such decay processes, such littering would take little more than 20 years to achieve.

When wood is cut from a tree to be used as timber, or some other forest product, it passes through a sequence of events, including the felling of the tree, the cutting of it into logs in the forest, the removal of the logs from the forest to the saw mill, storage of the logs prior to sawing, storage of sawn timber prior to seasoning, seasoning, further processing to a finished product, and the use of the finished product in service in a variety of situations, some wet and some dry. Through all these stages, the timber is expected to remain sound, durable, stable in dimension, and free from all defects. It is expected to survive unharmed in exposed situations, such as posts and poles in ground contact, although nature has evolved processes for its destruction. This inevitably means further processing or careful selection is necessary to ensure an induced durability for a long service life.

I. WOOD STRUCTURE

Wood is a complex, anisotropic, polymeric, cellular composite. Its basic structure will depend on the tree species from which it was cut. Woods such as cedar, larch, pine, and spruce belong to the group of plants known to the botanist as gymnosperms, to the forester as conifers, and to the timber trade as softwoods. These woods are formed from two cell types (tracheids and parenchyma cells) and have some degree of similar structure among species. Woods such as beech, eucalyptus, mahogany, oak, and teak belong to the group of plants known to the botanist as angiosperms, to the forester as broadleaves, and to the timber trade as hardwoods. These woods are formed from four cell types (vessels and fibers as well as tracheids and parenchyma) and, as a result, show a wider variation in basic structure among species. Softwoods and hardwoods not only differ in the cell types from which they are formed but also in their chemical composition. Both hemicellulose and lignin constitution will depend on which of the two plant groups to which the wood belongs. When wood becomes a habitat for organisms, therefore, the structure of that habitat will depend on the tree species from which the wood was cut.

In the living tree, the wood of the stem or trunk performs three main functions: support, water conduction, and storage of food reserves. It provides mechanical strength to support the mass of branches and leaves forming the crown of the tree and contains the pathways of the water conducting system for the movement of water from the ground through the trunk and branches to the leaves. Transverse movement of water and materials between the inner and outer layers is via the rays. Here, food reserves manufactured in the leaves are stored for future use. As the tree grows older and increases in girth, the stored food reserves are moved to the outer layers. At the same time, the living parenchyma cells die and, in the process, produce materials that often color the center of the stem and are sometimes of a toxic nature. This may give the central portion of the stem a greater durability when the wood is used as timber. This part of the stem with no living cells is known as the heartwood, whereas the outer layers containing the stored food reserves in the living cells are known as the sapwood. Sapwood of all species is permeable to liquids to some degree, whereas heartwood of the same species may be impermeable.

II. WOOD DEGRADATION

A. Types of Degradation

Many factors can bring about the degradation of wood: fire; mechanical wear and tear (e.g., floors and lock gates); chemical action (e.g., dye houses and laboratory benches); weathering (a combination of chemical action and colonization by fungi and bacteria, sometimes followed by wasps stripping off the surface layers for nest building); biological activity by other plants, fungi, and bacteria; and biological activity by animals (e.g., birds, mammals, insects, termites, shipworm, and gribble).

The damage by biological factors may be casual (e.g., beaver and deer damage trees in the forest, and woodpeckers and other birds damage trees and standing poles), for protection of soft-bodied organisms (e.g., shipworm), transitory (e.g., mice in buildings), the result of a search for stored food reserves in the sapwood (e.g., insects, termites, fungi, and bacteria), because of an association between colonizing fungi and insects (e.g., ambrosia beetles and deathwatch beetles), or due to the use of the constituent materials of the wood cell walls as food (e.g., wood-degrading fungi and bacteria).

B. Biological Factors

Although all types of degradation in service are important, the greatest destruction is caused by biological agents. These agents differ in different parts of the world and at different stages in the process of converting the wood in the tree to timber in service. Termites are the major factor in the tropics; shipworm, gribble, and marine fungi are the major factors in the sea; and insects, fungi, and bacteria are widespread throughout the world. Microbial ecology is largely a matter involving the fungi and bacteria, but some mention will be made of the insects.

C. Abiotic Factors

For fungi and bacteria to colonize and decay wood, water is an essential factor. Water is present in wood as free water in the cell lumina and as bound water in the cell walls. When the cell walls are fully wet but there is no free water in the cell lumina, the wood is said to be at the fiber saturation point. This point will vary between species, but it is normally at the point at which the moisture content (MC) of the timber is approximately 30% of its oven dry weight. There must be enough water present in the wood to permit the products of extracellular enzyme activity to flow from and to the organism. If too little is present, the enzyme systems that decay the wood cannot work, whereas too much water can produce an anaerobic environment in which most wood-rotting fungi cannot grow and survive.

Although water is by far the most important factor, temperature, related to geographical location, can also be important, particularly for insects. Nutrients are also important. These are often provided during the early stages of colonization by the food reserves that had been stored in the sapwood of the tree before it was cut down. Once the wood-rotting fungi are established and have started to decay the wood, the breakdown products so formed also provide a food source for the later colonizers that are unable to degrade the cell walls.

The abiotic factors that govern the biological activity of the organisms are therefore water, nutrients, oxygen, and a suitable temperature, but the most important controlling factor is water.

III. HABITAT

A. Habitat and Substrate

It has been said that, to the succession of microorganisms that colonize it, wood simply consists of a series of conveniently oriented holes surrounded by food. The holes are a means of access for the organisms, from cell to cell through the wood, and also form a pathway for the movement of liquids and gases. The holes consist of the cell lumina, which are interconnected from cell to cell by localized thinner regions of the intervening double cell wall known as pits in which only thin membranes separate the two cell lumina. Within the complex cellular structure of wood the microbial activity that leads to its destruction takes place. The wood is thus both habitat and food substrate.

The basic structure of the habitat depends on the wood species from which it was cut and whether it is sapwood or heartwood. The suitability of the habitat for the establishment of wood-inhabiting organisms will largely depend on the moisture content. Naturally, durable heartwood may be subject to colonization without active cell wall breakdown by the usual wood-rotting fungi; however, given time, decay will occur. Any chemicals impregnated into the wood (e.g., wood preservatives) simply alter the habitat. Such treated wood is still capable of being colonized by some fungi and bacteria, which may be the start of an ecological succession capable of destroying the wood in the very long term either by being resistant to the preservative or due to the preventative being lost or broken down by other microorganisms. Therefore, all wood is degraded; naturally durable or treated wood simply takes very much longer to degrade than sapwood.

IV. FROM FOREST TO TIMBER IN SERVICE

A. Ecosystems

At each stage of the processes involved in converting wood in the standing tree to timber in service, fungi and bacteria may infect the timber. A microbial ecological succession involving many types of organism can develop. The early colonizers are scavenging for readily available nutrients, such as the stored food reserves of the sapwood. Many will be unable to penetrate cell walls or pit membranes. Some early colonizing bacteria can destroy the pit membranes, thereby opening up the structure to allow other organisms to penetrate more deeply. At the same time the destruction of the pit membranes will allow water and other liquids to flow from one cell lumen to another. It also enables gaseous diffusion of water vapor and oxygen to occur, which allows the habitat to dry and become less anaerobic. Such ecosystems may be specific to a particular stage in the conversion process and are worth considering before discussing what happens to timber in service, for which different ecosystems exist.

1. Standing Tree

Wood in the living tree consists of the living sapwood and the nonliving heartwood. In the former, the true host–parasite relationship can exist, and invading microorganisms may be controlled by the response of the living cells. In the heartwood, however, because of the absence of living tissue, there is no host response. Infection occurs through exposed sites caused by damage to the bark and outer layers of the roots, stems, or branches. Such damage is often kept in check by the host response in the sapwood, but it causes decay in the unresponsive heartwood. The most common effect is butt or heart rot, which can give rise to the "hollow" standing tree. Control is largely a matter for the forester since decayed and infected areas of logs will be rejected at the saw mill.

2. Felled Logs

Once the tree is felled and cut into logs, the host–parasite response of the sapwood was thought to cease, but it is now known that this response is only slowly lost in the log and has a direct effect on colonization of sawn logs. The wood beneath the bark is now exposed at the cross-cut ends of the logs and at the wounds where branches were cut off. Both sapwood and heartwood are open to infection from airborne spores and from organisms in the ground or those transported to it by insects or other anthropods. Climate, season, and temperature may have an effect. In climates with a cold winter and warm or hot summer, the logs are normally extracted from the forest during the cold season when the metabolic activity of the potential invading organism is low and colonization is slow. If left in the forest through the summer, activity may be high and the logs of certain species can be rendered too heavily infected to be of any use as sawn timber. Extraction has to be quick and the logs must be stacked without ground contact at the saw mill.

The logs are still metabolically active and as such are capable of a host response to the invasion of any microorganisms that enter through wounds or cut ends. The situation is parasitic rather than saprophytic, and only those fungi capable of overcoming this response will grow in the still living wood. These tend to be darkly pigmented stain fungi. As the wood dies the host response diminishes and faster growing, more competitive bacteria and mold fungi compete with the stainers for the stored food in the rays of the wood. The bacteria include species capable of destroying the pit membranes, altering the substrate and making it even more suitable to the scavenging fungi.

As the sapwood begins to dry out, the true wood-rotting fungi infect the substrate. They are of three basic types: soft rot, white rot, and brown rot. Each results in a characteristic degradation of the wood cell wall—the micromorphology of decay. The soft-rot fungi will arrive first but are very soon displaced by the white-rot fungi, at a slightly lower wood moisture content; as further drying occurs, the brown-rot fungi may be present.

In the heartwood, a similar succession of organisms will colonize. The pit membranes, following the deposition of materials during the process of heartwood formation, are no longer as easily destroyed by bacteria. This can prevent the penetration of scavenging molds beyond the surface layers; in the absence of stored food materials, the molds do not survive long. This will cause the ecological sequence of events to differ between heartwood and sapwood.

In addition to the host response, the wood moisture content in the standing tree may also differ between heartwood and sapwood. In certain pines, for example, the sapwood moisture content may be 120% of the dry weight of the wood, whereas the heartwood moisture content of the same tree may be as low as 35%. At the very high moisture content, fungi will be inhibited; therefore, logs are often stored in water or under continuous water spray to keep them safe from fungal colonization. In these circumstances, bacteria may still survive and be able to destroy the pit membranes of the sapwood. This can make the sapwood very permeable and, once dried, very susceptible to rewetting in service.

Water storage will prevent fungal activity in the log due to the high moisture content. If stored out of water, deterioration is likely to occur as drying proceeds. The logs must be rapidly converted into sawn timber or the exposed wood at the cut ends and sides of the log must be treated with a fungicide and, when necessary, an insecticide. Like extraction from the forest, the time in storage should be short.

In parts of the tropics it is very difficult to extract certain species before the sapwood is badly discolored by sapstain fungi. Only a very short time is required for this to occur. Even at the most hygenic saw mill yards, discoloration and decay can occur quickly if the climate and moisture content are suitable for fungal growth.

3. *Converted Timber Prior to Seasoning*

Once the log has been sawn into timber, all surfaces are exposed and open to infection. If it has been felled recently, the timber will have a high moisture content and will be still living and is said to be in the "green" state. After drying, often called "seasoning," it is referred to as "dry" or "seasoned" timber, with a moisture content (% MC) given or implied. "Air-dry" timber will be lower than 20% MC. Timber dried in heated drying kilns may be dried to much lower moisture contents.

Drying is necessary to bring the moisture content of the wood into equilibrium with the atmospheric moisture content of the immediate area in which the timber will be in service as a finished product. The drying process must be fast enough to prevent deterioration due to fungal and insect penetration but not so fast that it induces mechanical stresses in the wood that could cause distortion, collapse, or breakage.

If the wood must remain stacked for any length of time before heated drying or air-drying, then chemical protection may be required to reduce the rate of degradation. After drying, the wood must still be protected from wetting by rain, snow, or dew or from extreme desiccation of the surface layers by sun or wind. During such periods of storage, the wood may become a habitat for animal or insect penetration. Dry wood is free from microbial attack but may be attacked by insects.

4. *Converted Timber after Seasoning and in Service*

If seasoned timber must be dried to a known moisture content, it must be maintained in that condition. This may not be easy to achieve. Many a load of timber has arrived on a construction site at the correct moisture content only to be kept there for weeks or months. As a result, its MC increases and it is put into the building at too high an MC, with consequent problems as it dries. Depending on the end use of the timber, the prevailing moisture will vary and it is generally recognized that a range of microbial and insect hazards exist and are related to the moisture content of the timber.

B. Hazard Classes

In Europe, a series of hazard classes are recognized, and similar systems exist in other parts of the world. These vary according to local conditions, but they all tend to be related to the prevailing moisture content of the timber in service. Five European hazard classes exist covering all the uses of timber in constructions. Each hazard class has a defined moisture range and characteristic organisms associated with it. In other words, the European hazard classes are ecosystems with clearly defined environmental conditions and associated populations of organisms.

1. *Class 1: Aboveground, Covered*

In this situation, the timber is protected from exposure to rain, snow, dew, and condensation. It is therefore considered to be permanently dry, with a moisture content $<18\%$. Such a moisture content is well below the fiber saturation point, with no free water in the wood. It will therefore be free from fungal and bacterial attack but liable to attack by wood-boring beetles and termites depending on the temperature.

In Europe, the common furniture beetle (*Anobium punctatum*), the house longhorn beetle (*Hylotrupes bajalus*), and the powder post beetle (*Lyctus brunneus*) are important in different areas. Termites, both the dry-wood termites and the subterranean termites, are a very important and serious problem throughout the tropics and subtropics.

2. *Class 2: Aboveground, Covered, Risk of Occasional Wetting*

In this situation, particularly when condensation is likely to be a problem, the timber will occasionally be above 20% MC without being above the fiber saturation point. As a result, wood-boring insects and termites are likely to be more active, and quick-growing fungi can colonize when the wood is wet and, by their activity, provide a food source for the insects, termites, and other fungi. These include those wood-decaying fungi capable of growing in such intermittent wet/dry situations. The brown-rot fungi can often tolerate such conditions and are the most common wood-decay fungi present. In buildings, accidental wetting due to leaks and poor maintenance gives rise to the brown-rot fungi that cause wet rot and dry rot. Many species are involved, but the most common in Europe are *Coniophora puteana* for wet rot and *Serpula lacrymans* for true dry rot.

3. *Class 3: Aboveground, Uncovered*

In this hazard class, the uncovered timber is subject to the climate; rain, dew, or snow can result in a moisture content higher than 20% for long periods of time. Although insects and the brown rots may still be a problem, the wetter conditions will produce an ecosystem in which the other group of wood-decaying basidiomycetes, the white-rot fungi, can become established and decay such commodities as external painted joinery or fence rails. The optimum moisture conditions for their growth, when the wood is neither too wet nor too dry, will support an ecological succession of microorganisms very similar to those of logs in the forest. The white rots are often the climax organism, followed later by a colonization of microorganisms only capable of feeding on the breakdown products of the cell wall components left behind by the white rots. This occurs when the wood has remained wet for long periods of time. In this hazard class the surface of exposed timber is constantly wetted and dried and subjected to sunlight. Previously, the graying or silvering of the timber in this situation was thought to be due to ultraviolet (UV) degradation of the surface. This is in fact true but the resulting color of such action would be white due to delignification of the surface cells leaving the white cellulose exposed. However, during this process the surface is colonized by darkly pigmented fungi, typically by *Aureobasium pullulans*, a dimorphic black yeast. The combination of color results in the silver-gray appearance of weathered wood. This fungus has been shown to be able to metabolize the lignin-degradation products of the action of UV. These compounds are often toxic to other fungi allowing the *Aureobasium* to grow without competition. This fungus and other similar fungi are also capable of colonizing surface finishes on wood, on which they feed on nutrients diffusing through the paint film. This fungus also penetrates wet wood in this hazard class, causing deep-seated stain in the sapwood. It is commonly referred to as "blue stain in service" and is a major problem where maintenance of surface appearance is important.

4. *Class 4: In the Ground or Fresh Water*

Contact with the ground or with fresh water will produce a permanent moisture content in the timber higher than 20%. If the timber is a post in the ground or a lake, the bottom part of the post will take up water until the equilibrium moisture content is reached. Free water in the cell lumina will rise by capillary action until it is in the upper part of the post above the ground or water surface. Here, evaporation will occur since water will be lost to the atmosphere as that part of the post approaches an equilibrium moisture content with the relative humidity of the surrounding air. As water is lost by evaporation, more moves up to replace it; thus, the free water in the lower portion of the post is in constant movement. Any nutrients in the soil water will therefore also move upward and, unless intercepted and used by invading microorganisms, will accumulate above the groundline as the water evaporates from the post and leaves its dissolved solutes behind. Therefore, interface at the groundline or water surface will produce ideal conditions for microbial growth and development: water from below, with a supply of nutrients and oxygen from the air above, in a suitable habitat.

The other important factor is that the full range of moisture contents is present. In such a situation, a full sequence of microorganisms ranging from bacteria to the wood-rotting basidiomycetes and associated secondary molds will be found growing.

The fact that soil salts, including nitrogen, can build up in the wood, and the fact that the wood may in parts be too wet for basidiomycetes, allows ascomycetes and fungi imperfecti to be the climax organisms which can go on to cause soft-rot of the wood cell wall.

In the dryer parts the conditions are ideal for the wood-rotting basidiomycetes and rapid decay will occur. Higher up in drier regions "blue-stain in service" will be evident at the surface of the timber.

In hazard class 4, therefore, are all the microorganisms capable of growing in wood, with the soft-rot fungi being the most common group associated with the wetter regions in ground contact.

5. *Class 5: In Salt Water*

The timber is in permanent exposure to wetting by salt water and therefore permanently at a moisture content in excess of 20%. Like the post in the ground or fresh water, if part of the timber is above the high water mark, then it is in a hazard class 3 situation and subjected to the types of organisms characteristic of that hazard class.

In the sea water, there may be marine fungi capable of causing minor deterioration, but the major destruction of wood is by the group of organisms known as "marine borers." The chief of these are shipworm and gribble. The former, of which there are many species, use their bivalve shells to burrow into the wood so their long, soft body is protected by the presence of the surrounding wood. These organisms can cause severe damage, particularly when associated with fungal decay. This was particularly true of wooden ships; it has been said that, in the Napoleonic wars, the British Navy lost more ships to break up due to shipworm and rot than to enemy action. Perhaps the most famous was the loss of the H.M.S. Royal George when the bottom fell off while the ship was moored off Portsmouth Harbour.

Gribble damage is caused by organisms burrowing into the wood just below the surface. This damage is accentuated by the abrasive action of sand and gravel in the waves and surge of the sea, breaking off the thin surface layers above the gribble's burrowings. This can erode the surface to a great depth. A 12 × 12-in. cross-section of a post can be reduced to a 5 × 5-in. cross section in the region of the post between high and low water levels well within 20 years of exposure.

The five hazard classes, each with its own characteristic microbial populations, are the main types of hazard to which timber is subjected in Europe. Although other hazards are defined in other areas of the world, these ecosystems are represented worldwide.

In the tropics, the termites are the main organism responsible for the destruction to timber. Many different species are present, each with a number of basic types of life history. The "ground" or "subterranean" termites take advantage of wood decay to organize "fungal gardens" in their nests. The gardens are produced by the worker termites scavenging for timber and, when they find it, cutting off small pieces that they take back to their nests. Here, they work the wood fragments into a three-dimensional labyrinth with wet soil. Fungi grow in and on the wet wood and sporulate. The fungal hyphae and spores, and probably the breakdown products of the fungal decay, then form the basic food on which newly hatched termites are nurtured.

C. Colonization of Wood in Service: The Organisms and Their Effect on Wood

Much of the original work on colonization and the effect of microorganisms on wood was conducted in wood in ground contact (hazard class 4) in which all the organisms and decay types can be found. Much work has been carried out to establish the sequence of events that occurs at or about the groundline of a timber post, the lower portion of which is buried in the ground. It has also been shown that the sequence of events that occurs at the groundline is very similar to that which occurs in simulated window joinery exposed to the weather out of ground contact. In both cases, a succession of organisms occurs, with some being very early colonizers and others taking weeks to colonize the wood.

A great difficulty in attempting to determine what is happening inside the wood habitat is the fact that there is no technique that can be used to make direct continuous observations without altering the structure of the habitat. Which organism colonizes first? What happens when a later colonizer meets up with an earlier colonizer? If attempts are made to isolate the organisms from the wood and culture them *in vitro* for identification, what was the state of the organism at the time the identification was made? Was it a young, actively growing hypha, an old moribund hypha, or a reproductive spore of some sort? Can one be sure that every organism present in the wood at any one time has been isolated and cultured on the growth media available and, therefore, identified? Despite these difficulties, much work by many people during the past 30 years has established a sequence of events.

In wood under ideal conditions for microbial development, the early colonizers are bacteria, often within a few hours of exposure. They are quickly followed by the fast-growing molds and sapstain fungi. The former can only penetrate the wood through open pits and cell lumina because they are incapable of destroying the cell wall. The sapstain fungi, however, are able to penetrate across cell walls from one cell lumen to another by a fine constriction of the hypha and a very narrow borehole through the walls. Thus, they are not as totally dependent on other organisms' destroying the pit membranes and opening up the structure to provide a pathway into the wood habitat as are the molds. The wood-rotting fungi appear some time later and are of three main groups: the ascomycetes and fungi imperfecti, which produce the type of decay known as soft rot, and the two groups of basidiomycetes, which produce the types of decay known as white rot and brown rot. The soft-rot fungi usually appear first, followed by the white or brown rots. Once established, either basidiomycete group will suppress the soft rots and destroy the lignocellulose polymers that form the wood cell walls. As the cell walls are broken down, the final group of molds appears which use the breakdown products as a source of nutrients, although by themselves they cannot destroy the intact cell wall.

Such a complete sequence depends on the habitat having the correct moisture content and temperature. If the habitat is too wet, the sequence may not proceed beyond the bacterial colonization which, in certain circumstances, may produce bacterial decay of the wood. As the habitat dries out, normal colonization by molds, sapstain fungi, and soft-rot fungi can occur, but the moisture content may still be too high for the white and brown rots. In this case, the soft-rot fungi become the climax organisms for wood decay and will be followed by the secondary molds. This situation is often found in the timbers in water-cooling towers.

These microbial groups are composed of a range of organisms, often from widely separated genera. The grouping is independent of taxonomy and reflects the effect each group has on the structure of the wood habitat. Individual species are not necessarily confined to one group; *Philalophora fastigiata,* for example, will act as a sapstain fungus on initial colonization but can, once established, act as a soft-rot fungus and under unusual culture conditions has also been shown to cause a type of decay typical of the brown rots. Each of the groups has its own characteristic effect on the wood habitat, and the groups are considered separately.

1. Bacteria

The bacteria that colonize wood are of many types, belonging to many different genera. A high moisture content is normally required (above the fiber saturation point) so that sufficient free water is available in which the bacteria may move. Moving water in wood may well carry bacteria with it. Certainly, they have been shown to move progressively through the open pathways in the wood structure. This broad group of organisms performs the following functions:

1. Destruction of the pit membranes, thus opening up the structure of the wood to provide open pathways for the movement of liquids, gases, and microorganisms
2. Antagonism to other microorganisms, retarding their colonization
3. Synergism with other microorganisms to encourage and support their colonization
4. In specific circumstances, the destruction of the wood cell wall

The destruction of the pit membranes is the most important since it permits the entry of the molds and sapstain fungi deep into the wood. The molds are therefore enabled to scavenge in depth for stored food reserves and degraded parenchyma cell contents in the rays and to mobilize these for the use of later colonizers that have been shown capable of saprophytism or parasitism of the earlier invaders. Antagonism and synergism may be important locally but are not usually of major importance.

The wood-destroying potential of bacteria is realized if the prevailing conditions are such that the fungi are unable to grow. Originally, this was thought to be very slow in very wet situations, such as in timber piles deep in wet soil. However, failure of treated wood in some locations has led to major interest in the effect of bacteria on the cell wall. Such decay was first seen in preservative-treated softwood stakes in vineyards and farms in New Zealand. On these sites, a regime of sprays and irrigation had produced a high moisture content in the wood that combined with the presence of the wood preservative had prevented fungal colonization. This had given a range of bacteria unrestricted access to the habitat, free from competition for nutrients with the fast-growing fungi. Three basic modes of degradation have been observed and described, each with its own characteristic micromorphology. These modes are eroding, tunneling, and cavity forming. Detailed knowledge of the species involved, their life history and physiology, and the mechanisms that produce the mode of degradation of the cell wall characteristic of each group is still scanty, but in treated wood in ground contact the significance of wood-degrading bacteria is recognized mainly due to the work of Nilsson in Sweden. In the development of new preservatives for ground contact, the bacteria are now taken into account in assessing their performance by testing them in soil contact under conditions in which soft-rot fungi and bacteria will be encouraged.

2. *Primary Molds*

The fungi that are the earliest colonizers of dead wood in service are usually the fast-growing molds that scavenge for nitrogen, simple carbohydrates, and fats deposited in the rays of the sapwood. This group contains a wide variety of fungal species from many genera. Their life cycle is often short, and reproduction, particularly by asexual spores, is usually prolific. Both of these are very important criteria when intermittent wetting and drying occurs in timber in service. Entrance into the cut surface by growing mycelia, quickly followed by sporulation, allows the wood surface to dry and the mycelia to die by desiccation, leaving the spores ready to germinate when wetting recurs. The new mycelial growth can then penetrate a cell or two deeper from the surface before further desiccation occurs, followed later by spore germination which, repeated time after time, allows colonization to continue. Such mycelia and spores may provide a source of nutrients for other organisms.

These fungi, called primary molds, do not produce an enzyme system capable of breaking down the components of the wood cell wall. Their pathway into the wood must be opened for them by the action of other early colonizers, particularly the pit membrane-destroying bacteria. In freshly felled living logs, stainers will precede the molds.

3. *Sapstain Fungi*

This group is characterized by dark-colored hyphae that can penetrate intervening cell walls from the lumen of one cell to the lumen of an adjacent cell by a very small borehole in which the fungal hypha is greatly constricted. With the primary molds, they are early colonizers after the bacteria in the sapwood of dead wood in service, feeding on the stored food reserves in the sapwood rays. In living logs related species are primary colonizers that are able to resist the remaining "host response" of the logs. The dark-pigmented hyphae discolor the sapwood, which is normally a light color, turning it black, blue-black, or deep blue. This can occur quickly; timber felled in the cold season of the year must be removed from the forest very quickly and certainly well before the higher temperatures of spring and summer increase the rate of growth of the organisms. This is particularly important in the tropics and subtropics, where light-colored hardwood species such as *Antiaris* and *Pycnanthus* are very difficult to extract unstained. Again, these characteristics are not confined to a single species but are common to many genera that may vary throughout the world.

4. *Soft-Rot Fungi*

The soft-rot fungi represent a wide range of genera from the ascomycetes and fungi imperfecti. Their importance in causing decay of wood in water-cooling towers was first discovered by Savory in the 1950s. These fungi have since been shown to be widespread throughout the world, wherever the conditions in the habitat are marginal for the wood-rotting basidiomycetes because of excess moisture or treatment by wood preservatives or for other reasons. They then become the major wood-rotting fungus and have caused severe problems.

They are characteristically able to destroy the wood

cell wall, confining themselves mainly to destruction of the middle layer of the secondary cell wall, the "S2 layer." Corbett, in the early 1960s, showed that a fungal hypha lying close to the inner layer of the cell wall produces a thin side branch that penetrates the cell wall at right angles to its surface. On reaching the S2 layer, this branch changes its penetration path by turning at right angles and along the grain within the S2 layer. At the turning point, a branch of the hypha develops that grows along the grain in the opposite direction; the original branch tip forms what is known as the "T" branch. Each of the two branch tips (the "proboscis hyphae") grow parallel to the helical orientation of the cellulose microfibrils in the S2 layer for a time. Elongation stops as the hyphae increase in girth. At the same time, a cavity with characteristic pointed ends is formed around each of the proboscis hyphae. Once the cavities have been formed, proboscis hyphae grow from the distal ends of the enlarged hyphae in their cavities following the orientation of the cellulose microfibrils. After a period of growth, the elongation of the hyphae stops again, the hyphae increase in girth, and new cavities with pointed ends develop. This stop/start growth and cavity formation can continue until the whole S2 layer is destroyed. Short side branches from the hypha in a cavity can give rise to further T branches, initiating successive chains of cavities parallel to those formed by the original T branch.

Once in the S2 layer, a T branch can grow into the S2 layer of an adjacent cell wall and the whole process can continue from cell wall to cell wall without penetrating a cell lumen until the main structural component of all the cell walls in the infected timber is destroyed. The S1 and S3 layers of secondary cell wall remain, as does the primary wall–middle lamella complex.

5. *White-Rot Fungi*

The white-rot fungi comprise numerous genera belonging to the basidiomycetes and are characterized by their mode of destruction of the wood cell wall and their ability to degrade Lignin.

The hyphae characteristically penetrate the wall by enzymic action leaving boreholes through the wall. The hyphae lie on the inner surface of the cell wall and, by the action of extracellular enzymes that are believed to be contained in the mucilaginous sheath surrounding the hyphae, destroy the lignin, cellulose, and hemicelluloses in direct contact. This lysis of the wall occurs along the line of hyphal contact, forming a groove or trough with a central ridge on which the hyphae rest. As the hyphae branch, the grooves begin to coalesce, thus eroding the cell wall from the inner surface. Degradation of both lignin and the carbohydrates can result in almost complete destruction of the wood, resulting in weight losses of more than 90% of the original wood dry weight. As decay proceeds, the wood becomes bleached or white in color.

6. *Brown-Rot Fungi*

The brown-rot fungi also belong to genera of the basidiomycetes but have a different mode of destruction of the wood cell wall. They are characterized by their ability to degrade both the cellulose and the hemicelluloses (the holocellulose) and leave the lignin essentially unaltered. The hyphae penetrate the cell lumen and lie on the inner surface of the cell wall. Both the hyphae and the cell wall surface appear largely unchanged, but the layers beneath the surface are converted to a brown amorphous mass, the residual lignin, giving rise to the name brown rot to this type of decay.

7. *Secondary Molds*

The secondary molds are those fungi that appear to be incapable of attacking wood, although they possess an active cellulase enzyme system, as seen by their ability to break down ball-milled cellulose in agar culture. Like the jackals that feed on the carcass after the lion has made the kill, the secondary molds appear after the wood-rotting fungi have begun to destroy the cell wall and have broken through the lignin barrier to give access to the holocellulose and the products of cell wall breakdown. This cellulolytic food source may be a nutritional excess left behind by the wood-rotting species or may present a source of competition between the two groups of organisms. These secondary molds often produce antibiotics that restrict the activity of the basidiomycetes and have been investigated as a means of biological control of decay. However, the long-term effects have not been successful but the principle remains a tantalizing possibility.

V. DURABILITY OF TIMBER

A. Natural Durability

The sapwood of all timbers is likely to be susceptible to fungal and insect attack because of the food reserves stored in the rays. *Lyctus* and *Hylotrupes* beetles will destroy the sapwood of species in which starch is the main stored product. Most fungi and bacteria associated with wood will also scavenge for this readily available source of nutrient. However, this is not the case for the heartwood of the tree. First, the stored food reserves are moved to the sapwood during the process of heartwood formation. Second, additional materials are often deposited in the cell walls and pit membranes as heartwood is formed. These materials may have many effects on the wood habitat. Additional lignification of the pit membranes may prevent their destruction by bacteria, so the wood structure is not opened. In some species, the materials deposited in the cell wall may be toxic to the usual colonizing microorganisms and thus inhibit the destruction of the wood structure by these organisms. Such heartwoods are said to show natural durability.

Such natural durability has an interesting impact on the organisms normally colonizing wood in ground contact. Many microbial species are absent from the wood and also from the soil in immediate contact with the wood. This clearly gives rise to a situation in which the organisms found in the soil in immediate contact with the wood are able to colonize the wood freely. Given time, an ecosystem is likely to evolve in which resistant strains of wood-rotting organisms are likely to proliferate and eventually bring about the destruction of this durable substrate.

B. Induced Durability

One means of inducing durability in wood is to keep it permanently dry. When wood in service is likely to be subject to deterioration by organisms, protection can be induced by the introduction of suitable chemicals into the wood structure. Such wood preservation has long been practiced in various forms. For more than 150 years, the use of railways has necessitated the protection of the wooden sleepers or ties on which the rails are fixed. These are impregnated with chemicals toxic or inhibitory to the wood-rotting organisms. Today, a worldwide wood preservation industry attempts to keep wood in service in a sound, durable condition for its entire projected service life.

C. Wood Preservation

For effective wood preservation to occur, the toxic chemical must penetrate deeply into the wood and, in particular, into the layers of the wood cell walls. A range of chemical systems are in current use, of which creosote has the longest service record, from the early days of the railway. Waterborne salts, such as copper chrome arsenate (CCA) and boron borate mixtures, and organic solvent systems all have given long service life to wood.

Once treated with a chemical wood preservative, timber becomes simply a new habitat. An additional abiotic factor, a toxin, has been added to it. Whereas the water availability tends to be the limiting factor for decay in untreated wood, the toxic chemical becomes the overriding abiotic factor determining the types of organism that can live in wet wood.

How the toxin works is another matter. The mere presence of the toxin in the habitat will prevent infection by the majority of the normal colonists. Others may grow through the habitat without any apparent inhibition but also without any deterioration of the treated cell walls until the organism has penetrated beyond the treated zone. The wood-rotting species among them can then decay the untreated wood. In these circumstances, the chemical may be acting simply to block the sites in the cell wall at which the enzyme system of the invading fungus must bind to initiate decay. The soft-rot fungi are examples of such organisms and have been shown to colonize CCA-treated sapwood of pine and birch. The former species showed no sign of decay, whereas treated birch fibers were destroyed by soft rot. The reason for this difference is the mode of action of the soft-rot fungi and the micromorphology of the destruction of the S2 layer of the cell wall. In the pine sapwood, a permeable species, all the cell wall layers are well impregnated with the preservative, whereas the S2 layers of the fibers in the birch are largely untreated. As described earlier, once the fungus has penetrated into the S2 layer of one cell wall, it can move across the primary wall–middle lamella complex by means of a T branch into the S2 layer of the neighboring cell without penetrating a preservative-treated cell lumen. Thus, the microdistribution of the preservative through the wall structure is a very important factor in successful wood preservation.

Wet-treated timber, from an ecological point of view, must be seen as a harsh environment for microorganisms in which few can grow and develop. It is, however, an accepted concept of microbial ecology that, in due time, an organism capable of using the resource of the modified habitat will do so. If none is present among the early colonizers, others capable of doing so will eventually arrive in the habitat. Many basidiomycete species are copper tolerant and can change a copper-containing chemical into a form such as copper oxalate, which is nontoxic. The fungus can subsequently decay the wood.

The detoxification of modern organic wood preservatives by non-decay fungi or bacteria can leave the wood at risk to the other decay organisms. The industry may be forced to use biodegradable biocides which will not persist in the environment, but these need to be effective long term in the wood. Protection of the biocides in the wood, long-term effectiveness against the wood decay organisms, and biodegradability if lost to the environment or at the end of life of the commodity is the new challenge. This can only be achieved by a full understanding of the microbiol ecology of treated wood.

VI. CONCLUSION

From many years of study we now understand the role of the different groups of microorganisms in the degradation of timber. The approach of considering wood in service as a range of ecosystems with their specific prevailing conditions and characteristic populations has contributed much to understanding the problem and correctly specifying timber in service and the required type and level of wood preservation.

In the past, wood-degrading organisms have been considered as the enemies of timber. Recent environmental assessments based on "life cycle analysis" have pointed to biodegradability as a major benefit at the end of the life cycle. This allows for natural recycling of this renewable resource. In the future, perhaps we will come to view the wood-degrading organisms as allies rather than enemies, to be controlled when the wood is in service and eventually utilized at the end of life of the structure.

See Also the Following Articles

BIODEGRADATON • LIGNOCELLULOSE, LIGNIN, LIGNINASES • PULP AND PAPER • WATER-DEFICIENT ENVIRONMENTS

Bibliography

Bravery, A. F. (1971). *J. Inst. Wood Sci.* 5, 13–19.

Carey, J. K. (1983). *In* "Biodeterioration" (T. A. Oxley and S. Barr, Eds.), pp. 13–25. Wiley, New York.

Corbett, N. H. (1965). *J. Inst. Wood Sci.* 3(14), 18–29.

Dickinson, D. J. (1991). Wood preservation: The biological challenge. *In* "Chemistry of Wood Preservation" (R. Thompson, Ed.). Royal Society of Chemistry, London.

Dickinson, D. J., and Levy, J. F. (1990). The microbial ecology of timber and forest products. *In* "Methods in Microbiology, Techniques in Microbial Ecology" (R. Grigorova and J. R. Morris, Eds.), Vol. 27, pp. 479–496. Academic Press, London.

Harvey, P. J., and Palmer, J. M. (1989). *Spectrum Br. Sci. News* **217**, 8–11.

Levy, J. F. (1987). *Philos. Trans. R. Soc. London* **A321**, 423–435.

Levy, J. F. (1990). Fungal degradation of wood. *In* "Cellulose, Sources, and Exploitation" (J. F. Kennedy, G. O. Phillips, and P. A. Williams, Eds.), pp. 397–407. Horwood, London.

Nilsson, T., and Daniel, G. F. (1982). *Proceedings of the 16th Convention Deutsche Gesellschaft Holzforswch.*

T Lymphocytes

Charles A. Janeway, Jr.
Yale University School of Medicine and Howard Hughes Medical Institute

I. How T Cells Recognize Antigen
II. How T Cells Are Measured
III. How T Cells Develop in the Thymus
IV. Atypical T Cells
V. Summary

GLOSSARY

antigen-presenting cells (*APCs*) Cells that process and present antigen to T cells along with the essential costimulatory molecules required for T-cell activation.

CD4 The coreceptor on T cells specific for MHC class II molecules.

CD8 The coreceptor on T cells specific for MHC class I molecules.

cytotoxic T cells T cells that kill other cells; generally these are CD8 T cells, which recognize peptides generated in cytosol that are presented on the cell surface by MHC class I molecules.

helper T cells T cells that help B cells respond to antigen or activate macrophages to become microbicidal; generally, these are CD4 T cells that recognize peptides generated in the vesicles of the cell and presented on the cell surface by MHC class II molecules.

major histocompatibility complex (*MHC*) An important gene cluster that encodes molecules that are recognized by the antigen receptor on T cells. There are two classes of MHC-encoded proteins on the surfaces of antigen-presenting cells, called MHC class I and MHC class II proteins. These are recognized by T cells bearing two distinct coreceptor proteins, called CD8 and CD4, respectively. MHC class I and MHC class II molecules are encoded in genes that are more polymorphic than any others in the mammalian genome.

T-cell receptor (*TCR*) The antigen receptor on T cells that is encoded in a set of rearranging gene segments in the genome.

T cells T lymphocytes; they are usually called this because they arise in the thymus.

thymus The central lymphoid organ in which T lymphocytes rearrange their receptor genes and undergo selection for self-MHC specificity and for self-tolerance.

THE ROLE OF T LYMPHOCYTES in host defense against microbial infection involves the recognition of the microbe's presence in the body, the activation of specific effector responses to the microbe, and the elimination of the microbe or its products from the host. The recognition involves three variables, the microbe itself, the major histocompatibility complex (MHC)-encoded molecule that presents its peptides, and the T-cell receptor (TCR) that recognizes it. These three variables are dealt with in a unique way by the generation of the TCR repertoire, the range of T cell specificities found in any individual vertebrate. Briefly, as T cells develop in the thymus, they are selected for the ability of their receptors to interact with a particular set of MHC molecules; these are called self-MHC molecules. The MHC is highly polymorphic, so the chance of a given receptor fitting is quite low. Fortunately, the genes that encode the TCR are highly variable, so that it is estimated that between 10^{15} and 10^{18} distinct TCRs can be produced and tested for fitness during development. Only a fraction of these actually pass this test, so that there is tremendous wastage during intrathymic development. The expression of the coreceptor proteins is also dictated by the specificity of the TCR, with MHC class I-specific TCRs selecting CD8 T cells and MHC class II-specific TCRs selecting CD4 T cells. Finally, there are T cells with either conserved variable regions or different TCRs altogether, which will be dealt with briefly at the end of this chapter.

I. HOW T CELLS RECOGNIZE ANTIGEN

T cells recognize foreign antigens as complexes between the various MHC molecules of the particular individual and a foreign antigenic peptide. These peptides are short contiguous stretches of a protein that are processed inside of a cell. T cells can exist in several activation states, naive or antigen-inexperienced, activated, "armed" effector cells, memory cells, and anergic cells that do not respond to an antigen even when it is presented on a competent antigen-presenting cell.

The peptides of pathogens are handled differently depending on whether they are generated in the cell's cytosol inside the cell or in the vesicular compartment, either by intracellular vesicular infection or by receptor-mediated endocyosis. Peptides that are generated in the cytosol are transported across the endoplasmic reticulum, where they are bound to and transported as a complex with the MHC class I molecule to the cell's surface and are recognized by T cells with CD8 coreceptors. Peptides that are generated in the vesicles of the cell are transported to the cell's surface by binding to MHC class II molecules; here, they are recognized by T cells with CD4 coreceptors. These rules are true for T cells in all states of activation.

A. MHC Class I Molecules Bind to Short Cytoplasmic Peptides

MHC class I molecules have a peptide-binding site made up of their outer two domains. As the binding site is enclosed at both ends, it can only bind short sequences of proteins that show particular motifs determined by the particular allele at the MHC. The sequences of the particular MHC allele creates pockets in the peptide-binding groove, two or three of which are very restrictive as to the peptide side chains they accomodate. CD8 T cells recognize and respond to these peptides bound to MHC class I molecules. How peptides are formed from cytosolic proteins and transported into the endoplasmic reticulum is a fascinating story.

Proteins are synthesized in one of two ways, either for export or for function inside the cell. Proteins to be exported are inserted into the endoplasmic reticulum via a well-characterized translocon by means of a signal peptide. Proteins that are retained inside the cell are synthesized without a leader sequence and are not further processed. However, there is turnover of all cellular proteins at rates that differ for each protein. Most of this turnover is the function of a large multicatalytic complex called a proteasome. The proteasome cuts proteins into snippets of about 6–15 amino acids. Most are degraded futher by other mechanisms inside the cell, but some are transported into the lumen of the endoplasmic reticulum by a unique mechanism involving an ATP-binding casette (ABC) protein, called TAP for *t*ransporters associated with *a*ntigen *p*rocessing. TAP molecules are located only in the membrane of the endoplasmic reticulum, and they mediate the ATP-dependent transport of peptides into the endoplasmic reticulum. Here, the peptides meet an MHC class I molecule that is tethered to TAP by a chaperone molecule called calreticulin as well as by a *TAP-as*sociated prote*in* molecule (tapasin). These molecules bind to MHC class I molecules that have not bound peptide; the MHC class I molecules are released upon binding a peptide that fits its peptide-binding groove appropriately. The mechanism of release is not worked out, but it is believed to be due to the completion of folding of the MHC class I molecule by the antigenic peptide. Most peptides bound to MHC class I molecules are derived from normal self-proteins; only when a cell is infected with a cytosolic pathogen does the cell surface express foreign peptides that serve to alert the CD8 T cells that there is a pathogen growing in the cell. The action of CD8 T cells is to kill cells with surface-expressed foreign peptides, as we see later in this article.

B. MHC Class II Molecules Bind to Longer Peptides Generated in the Vesicular Compartments of Cells

MHC class II molecules are formed differently from MHC class I molecules, in that their outer domains form a peptide-binding groove from two chains instead of one. The peptide-binding groove of MHC class II molecules is also open at the ends, so that longer peptides can be accomodated. Finally, most peptides are bound to the MHC class II molecules by bonds formed with the peptide backbone, with much less emphasis on the precise sequence of the anchor residues, so that peptide motifs are more difficult to discern in the case of peptides that bind to MHC class II molecules.

As with MHC class I molecules, most peptides that are presented to the CD4 T cells are derived from normal self-proteins and consequently are not recognized or responded to. However, when a cell is infected with a microbe that is an obligate intracellular pathogen, such as *Mycobacterium leprae,* its peptides reach the cell surface bound to MHC class II molecules. Similarly, when a B lymphocyte takes up a toxin molecule with its antigen-specific receptor, the protein is degraded in the endocytic pathway, where the peptides bind to the MHC class II molecules.

To prevent peptides from binding to the MHC class II molecules during their synthesis in the endoplasmic reticulum, they are combined with the MHC class II invariant chain. This protein contains a region that binds tightly to the peptide-binding groove, and thus this groove is blocked from binding other peptides, either self or nonself. Later, during its journey to the surface of the cell, the invariant chain targets the MHC class II molecule to a specialized intracellular compartment called a class II vesicle or CIIV. Here, the invariant chain is progressively degraded until only a small fragment is bound in the peptide-binding groove. Then an MHC class II-like molecule called DM edits the peptide content of the MHC class II molecule by exchanging whatever peptide is available with the invariant chain peptide. Then the MHC class II molecule can complete its folding and proceed to the cell surface, where it can be recognized by a CD4 T cell.

So the secret of antigen recognition by T cells is that the antigen is not recognized directly, as in the case of B lymphocytes, but indirectly as a complex between the MHC molecules of the individual, and a peptide from the pathogen that is necessary to complete the folding of the MHC molecules. Healthy uninfected cells are covered with self-peptide–self-MHC protein complexes, whereas infected cells also express foreign peptides that make them recognizable to the passing T cells and vulnerable to attack and removal.

C. The T Cell Receptor Recognizes Foreign Peptides Bound to Self-MHC Molecules

The T cell receptor (TCR) recognizes peptides derived from pathogens only when they are bound to a particular MHC allele; this is termed MHC restriction, as it restricts the cells that can present the peptide to those that express the particular MHC allele in question. MHC restriction is an intrinsic property of a given TCR; it can be transferred to other cells by transfection of the genes encoding the variable α- and β-chains of the particular TCR. Also intrinsic to the TCR is the property known as alloreactivity, in which a foreign MHC molecule is recognized by a large number of different TCRs. This is the property that is the origin of the name of major histocompatibility complex to the MHC; it was the major set of genes that encoded cell-surface molecules that allowed or did not allow skin grafting across different mouse strains. It was only later that the antigen-presenting function of the MHC molecules became the accepted physiological function of these genes (Figs. 1–2).

The receptors of T cells are encoded in clusters of gene segments that rearrange during T cell development; we consider this later in this article.

II. HOW T CELLS ARE MEASURED

T cells are usually assayed by performing a functional assay, such as cell proliferation, cytokine production, B cell or macrophage activation, or cell killing. It has also become possible to measure the binding of antigen by T cells through making complexes of MHC molecules with specific peptides. This has been done both with MHC class I and more recently with MHC class II molecules. We briefly enumerate here the various measures of T cell function before turning to T cell development.

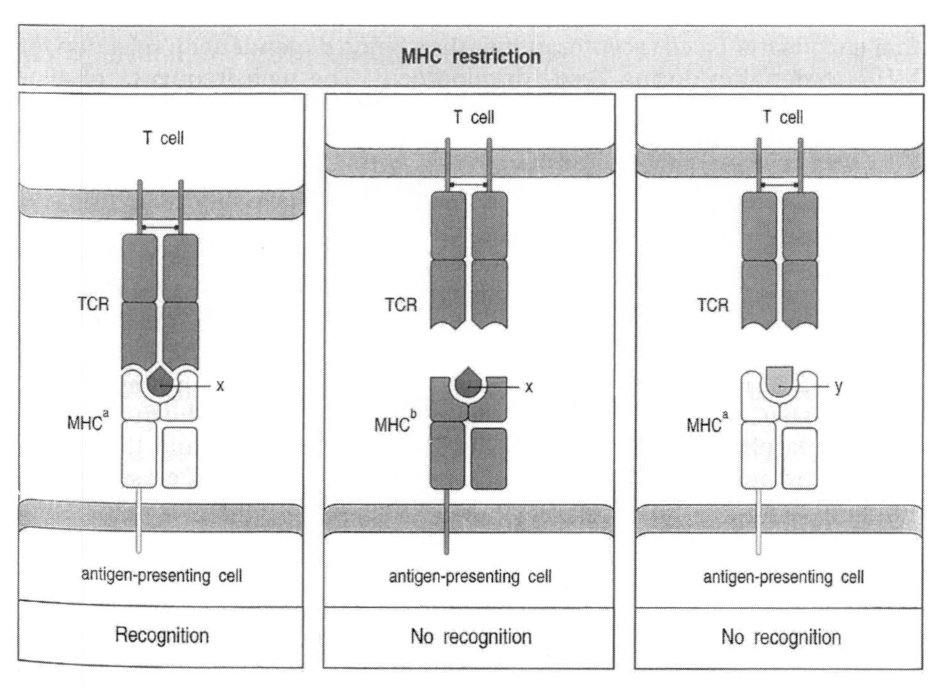

Fig. 1. The T cell receptor recognizes antigenic peptides bound to self-MHC molecules producing MHC-restricted antigen recognition. (Left) MHC^a molecule presents peptide X to a specific receptor. (Center) MHC^b molecule is not recognized by the same T cell even when it presents the same peptide. (Right) The same T cell receptor also does not recognize a different peptide presented to it by MHC^a. Thus, the T cell receptor is specific for the complex of foreign peptide bound to self-MHC. Reprinted from Janeway and Travers (1999), Fig. 4.25, by permission of Garland Publishing/Current Science.

A. T Cell Proliferation

When naive T cells are exposed to foreign peptides, they do not proliferate. However, after infection or immunization, exposure to the same foreign peptide or pathogen causes them to proliferate. Proliferation is considered a sign that a person has been exposed to a pathogen or antigen, because it is antigen-specific. Proliferation is an *in vitro* analog of clonal expansion, which is essential to host defense because T cells can only function effectively when large numbers of cells with identical specificity are produced; so, the rare cells that are specific for a particular pathogenic determinant need to expand rapidly in response to a specific antigen challenge before they can differentiate into armed effector T cells. Thus, proliferation *in vitro* is a hallmark for the preexistence of memory T cells, which only appear after an immune response has occurred; naive T cells are present in numbers that are too low to be detected by proliferation *in vitro,* although they are assumed to be able to clonally expand during a primary exposure to a pathogen.

B. Cytokine Secretion

Cytokines are small proteins that alter the behavior of the cell that makes them (autocrine), neighboring cells (paracrine), or distant cells (endocrine). Cytokines are made by many cells, especially armed effector T cells. Their action on specific cytokine receptors, again distributed widely on cells, is how these activities were discovered. However, it is possible to measure cytokine secretion quantitatively and specifically using monoclonal antibodies. Many different cytokines have been measured in this way, and distinct combinations of cytokines are known to be secreted by T cell populations with different functional activities during an immune response.

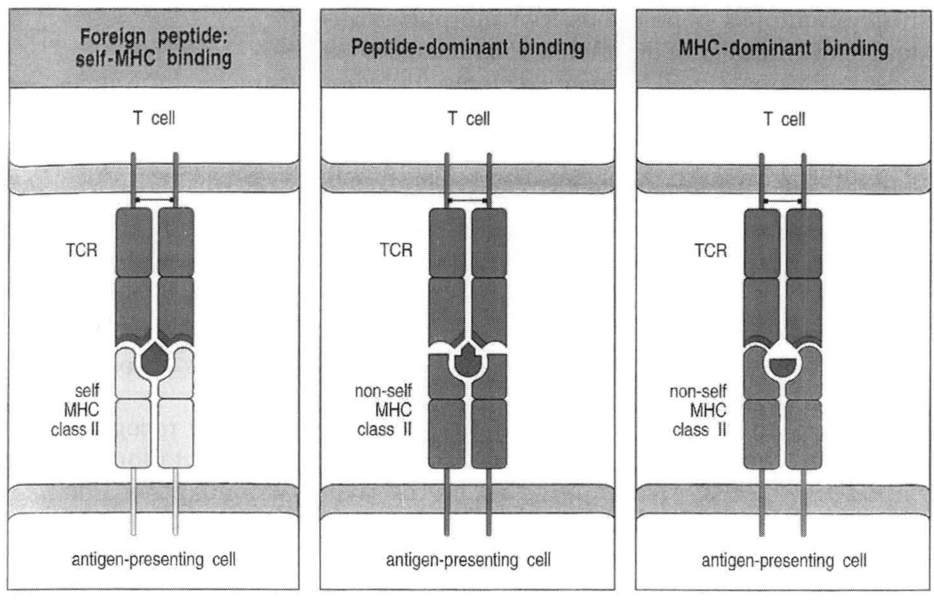

Fig. 2. All T cell receptors must recognize an MHC molecule. Therefore, MHC recognition is intrinsic to the genes encoding T cell receptors. Many TCRs that are self-MHC-restricted can recognize MHC protein of nonself, either by interacting with the peptide, the MHC molecule, or both together. (Left) TCR raised on MHC[a] recognizes a peptide presented by MHC[a], demonstrating MHC-restricted peptide recognition. (Center) and (Right) Two means of nonself recognition. Center panel shows peptide-dependent nonself-recognition; right panel shows MHC-dependent nonself-recognition. The truth appears to be somewhere between these two models. Reprinted from Janeway and Travers (1999), Fig. 4.26, by permission of Garland Publishing/Current Science.

This field remains highly active in biomedical research, so it is difficult to make generalizations. However, there are distinct patterns of cytokine secretion associated with given infections or immune responses. Cytokines that act specifically on mast cells, for instance, are prominent in allergic reactions, which are in part mediated by mast cells; likewise, cytokines that are involved in activating macrophages to kill intracellular pathogens are prominent in such infections when they are controlled, but not when they are out of control.

T cells also produce membrane-bound cytokine-like molecules. These all belong to the tumor necrosis factor (TNF) family of cytokines. These molecules are involved in cell–cell interactions that act by contact with a target cell, in which the T cell expresses a membrane-bound member of the TNF family, and the target bears the appropriate TNF family receptor (TNFR). Thus, for instance, during an immune response, most T cells express two molecules, one called Fas ligand, which is a member of the TNF family of cytokines, and the other called Fas, which is a member of the TNFR family. The interaction of Fas with its ligand, either on the same or a different cell, leads to the activation of programmed cell death that rapidly destroys most of the responding cells. This is important, as some responses involve huge numbers of specific cells, such that after a few such responses in the absence of programmed cell death there would be a massive overgrowth of T cells. That this happens in mice with mutations in Fas or its ligand proves that this mechanism is crucial to T cell homeostasis.

C. T Cells Can Kill Target Cells Infected with Viruses and Other Pathogens

A third way to measure the prevalence of T cells is to use specifically sensitized cells as targets for T-cell killing. This assay is quite rapid, as the program-

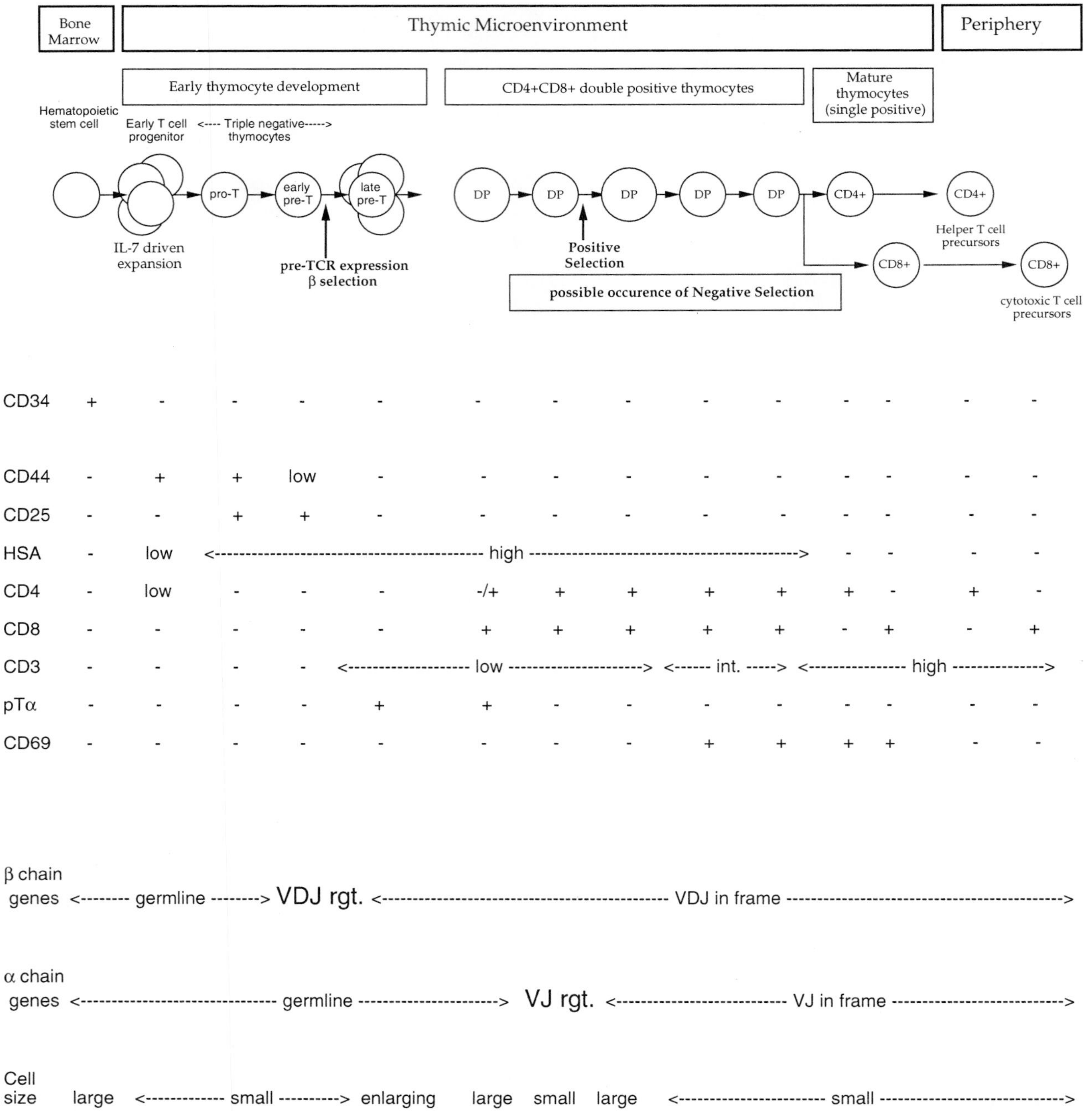

Fig. 3. Synoptic representation of intrathymic $\alpha\beta$ T-cell development. Hematopoietic stem cells from bone marrow colonize the thymus and give rise to distinct lymphoid cell types. For clarity, the emergence of NK cells, NK T cells, $\gamma\delta$ T cells, and thymic dendritic cells are not indicated. The MHC-independent phase of $\alpha\beta$ T-cell development (early thymocyte development) is defined by the expression status of the CD44 and CD25 molecules. The expression of CD117 (c-kit), which, together with the IL-7 receptor, is required for the initial survival and expansion phase, virtually correlates with the CD44 expression (not shown). The β-chain rearrangement (VDJ rgt.) occurs first at the $CD44^{low}$–$CD25^{+}$ stage. The surface expression of the β-chain protein in association with the surrogate pTα-chain (the pre-TCR) constitutes a major developmental checkpoint called β selection. At this point, the cells are fully committed to the T cell lineage. β selection requires the expression of a number of proteins, including the RAG-1 and RAG-2 proteins, the p56 lck kinase, and the

ming for lysis occurs rapidly, and the dissolution of the target cell is measurable at 3–4 hours. This assay was used in the first demonstration of MHC-restricted killing and is still commonly used. Quantitation is achieved by titrating the effector cells on a fixed number of target cells and calculating how many CD8 T cells are required to kill a certain number of targets.

D. Binding of MHC: Peptide Complexes to T Cells

In recent years, it has become possible to count T cells that are specific for a particular MHC–peptide complex by producing MHC molecules in bacteria and loading them with specific peptides. The C-terminus of the MHC molecule is modified in order to allow biotinylation; the biotinyl–MHC–peptide complex is bound to fluorscein–steptavidin to form flourescent tetramers. Such MHC–peptide tetramers bind with sufficient avidity to the low-affinity TCRs that they can be visualized and counted in a flow cytometer. In this way, it has become possible to follow the expansion and contraction of specific T cells responding to particular viruses. In the case of one virus, up to 50% of all CD8 T cells bound to a single peptide–MHC class I molecular complex. These studies show rapid expansion and loss of CD8 T cells during the course of a viral infection. Similar studies with MHC class II molecules have appeared, and it is expected that this technique will revolutionize the study of T cell dynamics during various immune responses *in vivo*.

III. HOW T CELLS DEVELOP IN THE THYMUS

T cells begin their life in the bone marrow as undifferentiated hemopoietic stem cells that produce progeny, some of which end up in the thymus. The thymus is a complex organ consisting of a stroma made up of various epithelial elements, and a major population of thymocytes or developing T cells. In the absence of a thymus, no T cells can mature. After entering the thymus, it takes about 3 weeks for a T cell to undergo full development and leave to enter the periphery. This can be split into roughly three periods of 1 week each. The first week is spent expanding in the outer cortex of the thymus, just beneath the thymic capsule; the second week is spent in rearranging first the β-chain, followed by α-chain rearrangement, which leads to TCR expression and the ability to interact with self-MHC molecules; the selected T cells then spend a week or so in the medulla before emigrating from the thymus. Most studies focus on the steps of gene rearrangement and the positive and negative selective processes that all T cells must undergo before maturing and emigrating from the thymus.

A. The Rearrangement of T Cell Receptor Genes

The genes for the receptors on most T cells consist of the TCR α-chain and the TCR β-chain genes, which are arranged as tandem sets of gene segments. The TCR β-gene segments rearrange first in thymo-

---

ε-chain of the CD3 complex. The β-locus allelic exclusion also occurs at this stage. The first α-chain rearrangement (VJ rgt.) coincides with the transition to the $CD4^{+}$–$CD8^{+}$ double-positive (DP) stage. A productive VJ rearrangement allows the surface expression of an $\alpha\beta$ TCR in which the α-chain replaces the pre-Tα. The next phases of the developmental progression are totally dependent on the interaction of the TCR with self-peptide–self-MHC complexes on thymic stromal cells. Depending on the nature of this interaction, it leads to either positive selection via a survival signal delivered to thymocytes or to clonal deletion via induction of apoptosis. Thymocytes that are positively selected become $CD4^{+}$–$CD8^{-}$ or $CD4^{-}$–$CD8^{+}$ (single positive) and populate the thymic medulla. After a week, they leave the thymic compartment and populate the peripheral lymphoid organs, where they are subjected to a phenomenon closely resembling the intrathymic positive selection–their peripheral survival requires a repeated TCR–self-peptide–self-MHC interaction on frequent contact with antigen-presenting cells, most likely with dendritic cells in the T cell area of secondary lymphoid organs. Reprinted from Viret and Janeway, *Reviews in Immunogenetics* **1**, 91–104, Fig. 1. © 1999 Munksgaard International Publishers Ltd., Copenhagen, Denmark.

cytes that are characterized as "double negative" because they lack expression of the CD4 and CD8 molecules that are used to mark different populations of developing thymocytes. When this produces an intact TCR β-chain gene that can be translated into protein, the β-chain protein pairs with a surrogate α-chain to form a pre-TCR. The expression of this pre-TCR allows the thymocyte to receive nonspecific signals that drive its proliferation, so that one rearranged β-chain can subsequently be expressed with many different rearranged α-chains. The α-chains rearrange in cells that now express both CD4 and CD8, and so are called "double positive" cells.

B. Positive Intrathymic Selection

The $\alpha\beta$ T-cell receptor is then tested for its ability to interact with the self-MHC molecules expressed on the surface of the thymic cortical epithelial cells; all of these molecules are expressed with an array of self-peptides that serve to stabilize their surface expression. As this process occurs in an environment in which the MHC molecules are by definition markers of self, only self-MHC molecules can perform this function. This process of positive intrathymic selection guarantees that the T cell will be self-MHC-restricted in its antigen recognition. Approximately 1–5% of T cells survive this process.

C. Negative Intrathymic Selection

During or after positive selection, a process known as negative intrathymic selection occurs, which serves to delete all cells with the potential to be self-reactive. This process can occur on any cell in the thymus, although the dendritic cells at the junction of the cortex and the medulla are known to be the most potent at driving negative selection. This process eliminates those positively selected cells that would be activated subsequently as peripheral T cells. It operates by activating programmed cell death in all developing thymocytes that recognize self-peptide–self-MHC complexes with sufficient avidity to trigger an effector response in the periphery. Actually, the threshold for deletion in the thymus is set considerably below the threshold for activation in the periphery, giving a margin of safety to the host that protects against inappropriate activation against self. Thus, the process of negative selection mainly guarantees that no self-reactive T cells emerge into the peripheral tissues. This is the major means of self-tolerance in T cells. Somewhere between 5 and 75% of T cells are lost at this stage of development. Thus, positive selection is numercally a far more important event than negative selection, but negative selection also plays an important role in shaping the mature TCR repertoire.

D. Onward Survival of T Cells on Self-MHC–Self-Peptide Complexes

Once T cells emigrate from the thymus, they come under the influence of survival signals that are also driven by receptor specificity and the recognition of the same self-peptide–self-MHC complexes as in the thymus. Current evidence supports the idea that these molecules are most importantly expressed by the APCs that best activate the T cells, the dendritic cells. However, the survival of T cells in the periphery has just begun to be studied, and thus there is a paucity of hard data on this subject (Fig. 3).

IV. ATYPICAL T CELLS

What has been described so far accounts for about 95–99% of peripheral T cells found in the blood, lymph nodes, and spleens of mammals. However, there are known to be at least two other sets of T cells that have similar receptors. However, in one set, the NK T cells, the α-chain is basically invariant and the β-chain is highly restricted in its use of gene segments. In the other set, known as $\gamma\delta$ T cells, the use of different genes altogether makes them quite distinct. There is indirect evidence that the $\gamma\delta$ T cells are evolutionary precursors of the $\alpha\beta$ T cells that have been retained to perform unique functions, but the true role of the $\gamma\delta$ T cells remains mysterious.

A. NK T Cells

NK T cells were originally identified as a subset of $\alpha\beta$ T cells that expressed cell-surface molecules typical of natural killer cells, and they were a mix of CD4 and double-negative T cells. They were later shown to be positively selected on the molecule CD1 in the mouse, whose human homolog is called CD1d; CD1 molecules are close cousins of MHC class I molecules, but are encoded in a distinct region of the genome, and they lack a groove for binding peptides. They are believed to be expressed at the surface of cells by binding to lipids rather than to peptides.

The receptors of NK T cells are made up of an invariant α-chain paired with one of a small number of different β-chains. Their function appears to be the immediate release of cytokines upon binding to their CD1 ligand. These cytokines in turn are thought to play a central role in determining which cytokines will be secreted by the conventional T cells that respond to specific antigen.

B. $\gamma\delta$ T Cells

The $\gamma\delta$ T cells were discovered by accident during the search for the TCR α-chain genes; a gene was cloned that was at first reported to be the α-chain, and then was shown to be a distinct gene called γ that was expressed in distinct cells; the δ-chain genes were then shown to lie within the TCR α-chain locus. Much has been learned about the genetics of $\gamma\delta$ TCR rearrangements, their timing and relationship to $\alpha\beta$ TCR gene rearrangement, and the biochemistry of $\gamma\delta$ TCRs; what is mainly missing is a coherent theory of what these atypical T cells do. This has begun to yield to experiments carried out in gene knock-out mice that lack $\gamma\delta$ cells; it appears that in some instances, $\gamma\delta$ T cells perform a regulatory function in particular diseases. How they do this is unclear.

V. SUMMARY

T lymphocytes play a central role in adaptive immunity, being responsible for most of the effective immune responses that individuals make to infection, as well as for undesirable immune responses to allergens, autoantigens, and transplanted organs. Their study is fascinating, their role in controlling the adaptive immune response is overwhelmingly important to life as we know it, and yet they have only been appreciated for the past 30 years. One hopes eventually to be able to turn on or off specific clones of T cells in order to manipulate the immune response *in vivo;* doing so represents a true challenge to the enquiring mind.

See Also the Following Articles

ABC Transport • Antibodies and B Cells • Antigenic Variation • Cellular Immunity • Interferons

Bibliography

Janeway, C. A., Jr., and Travers, P. (1999). "Immunobiology: The Immune System in Health and Disease," 4th ed., Chapters 4–8. Garland, New York.

Paul, W. E. (1999). "Fundamental Immunology," 4th ed. Raven Press, New York.

Tospoviruses

James W. Moyer
North Carolina State University

I. Taxonomy
II. Virion Morphology, Genome Structure, and Organization
III. Replication
IV. Host Range and Symptomatology
V. Transmission and Virus–Vector Interactions
VI. Control

GLOSSARY

ambisense genome Viral RNA genome that has open reading frames in both the viral and viral complementary sense on the same segment.

envelope Membrane-like structure that packages genome segments.

negative sense genome Viral RNA genome that codes for proteins in the viral complementary sense. Transcription of viral complementary mRNA is required for translation of viral proteins.

RNP Ribonucleoprotein complex consisting of the viral RNA genome segment, N protein, and a small number of RNA-dependent, RNA polymerase molecules.

virion Quasi-spherical structure bounded by a membrane-like envelope which contains the viral genome.

THE *TOSPOVIRUS* GENUS, formerly the monotypic, tomato spotted wilt virus (TSWV) group, is a collection of viral species, classified in the Bunyaviridae family of viruses. Viruses in this family typically have enveloped virions that package a tripartite, RNA genome. The genera are distinguished in part by the organization of their genomic segments. Viruses in three genera of the family have all three genomic segments with open reading frames (ORFs) in the negative sense and the other two genera have one and two segments, respectively, with an ambisense organization. The remaining segments are of negative polarity.

The *Tospovirus* genus is characterized by an L segment in the negative polarity and the M and S segments in an ambisense polarity. In addition, this is the only genus in the family that infects plants and is vectored by thrips, although viruses in other genera are vectored by other arthropods. Several viruses in this family are frequently cited on lists of "emerging viruses" due to their recent increase in frequency and the mortality rate of infected human victims. Many of the tospoviruses have similar impact in populations of infected crop plants. The incidence of tospoviruses began increasing on a worldwide basis in the mid-1980s and they continue to cause significant economic losses in many food and floral crops. The rapid increase of these viruses in crop production systems, both field and greenhouse, has been attributed to the concomitant increase in thrips populations that are vector species.

I. TAXONOMY

The tomato spotted wilt virus group of plant viruses was considered to be a monotypic (one-member) group of plant viruses. In 1989 a second member, impatiens necrotic spot virus, was distinguished based on differences in nucleic acid homology, serological relatedness, and host range. This discovery was the basis for the differentiation and classification of additional virus species in this genus. It is now widely accepted that viruses with >90% nucleotide sequence homology with the N gene of another virus should be considered a strain of that virus. Viruses with 80–90% homology may be considered strains or distinct species depending on other molecular, serological, or biological relationships. Viruses with <80% nucleotide sequence homology are considered distinct species but may be placed in a serogroup with other related viruses. Initially, each virus was placed in a serogroup in the order of discovery and designated numerically, including groups with only one member. Serogroups are currently only used for two or more serologically related viruses and are designated by a type member. Currently, there are two serogroups: tomato spotted wilt virus serogroup and the watermelon silver mottle serogroup (Table I). More than 12 species have been identified, with the greatest diversity in South America and Southeast Asia (Table I). A significant distinction is made between the serogroups in tospoviruses and the serogroups defined in other genera of the Bunyaviridae. Serogroups in tospoviruses are defined by antisera to the N protein in ELISA assays, whereas serogroups in other genera are defined by neutralization or hemagglutination assays that measure relationships between glycoproteins (GPs).

II. VIRION MORPHOLOGY, GENOME STRUCTURE, AND ORGANIZATION

A. Virion Morphology and Cytopathology

Virions have a quasi-spherical morphology with a diameter of 80–120 nm. They are defined by a membranous envelope containing two viral-coded GPs designated G1 and G2. The genome consists of three RNA molecules encapsidated by the N protein. There are a small number of molecules of the RNA-dependent, RNA polymerase (RdRp) associated with the segments and all are packaged within the envelope. Although the enveloped virion is considered to be the infectious unit for the virus, there are reports of infectious ribonucleoproteins (RNPs) that have been carefully isolated from sucrose or $CsSO_4$ gradients. Virion formation occurs within the cytoplasm. Several hypotheses for virion formation have been proposed which involve the cisternae and the Golgi apparatus. Recent evidence from semisynchronously infected plant protoplasts suggests that following infection of the cell and the accumulation of viral proteins, slightly curved, parallel membranes develop that immunolabeling experiments suggest are of Golgi origin. The GPs appear to have an affinity for these membranes as does the synthesis

TABLE I
Tospovirus Species Generated during the International Symposium on Tospoviruses and Thrips, May 1998, Wageningen, Netherlands

Serogroup	*Species*	*Abbreviation*
Tomato spotted wilt	Tomato spotted wilt virus	TSWV
	Groundnut ringspot virus	GRSV
	Tomato chlorotic spot virus	TCSV
Watermelon silver mottle	Watermelon silver mottle virus	WSMV
	Watermelon bud necrosis virus	WBNV
	Groundnut bud necrosis virus	GBNV
Ungrouped	Impatiens necrotic spot virus	INSV
	Chrysanthemum stem necrosis virus	CNSV
	Iris yellow spot virus	IYSV
	Peanut chlorotic fan-spot virus	PCFV
	Peanut yellow spot virus	PYSV
	Physalis severe mottle virus	PSMV
	Zucchini lethal chlorosis virus	ZLCV

of viral RNPs. The membranes containing GPs then envelop the RNPs, resulting in the frequently observed double-membrane particles that occur individually. The double-membrane virions then aggregate into groups of single-membrane virions that are contained within an outer membrane.

Other structures of viral origin are also observed in infected cells composed of nonstructural (NSm and NSs) proteins or structural proteins. The NSs protein accumulates in long fibrillar structures that may associate as loose bundles (e.g., TSWV) or in highly ordered paracrystalline arrays (e.g., INSV). Excess N protein is observed as electron-dense, granular material in the cytoplasm. NSm is also detected in granular electron-dense inclusions. Observation of virions and inclusions is both host and virus isolate dependent.

B. Genome Structure

The *Tospovirus* genome is distributed among three single-stranded RNA molecules designated L, M, and S in order of decreasing size (Fig. 1). Each segment is pseudocircular with a panhandle-like structure at the termini formed by base pairing of inverted complementary sequences. The length of these sequences varies between segments and viruses; however, there is an eight-nucleotide sequence (5′-AGAGCAAU-3′) at the termini of each segment of all tospoviruses. The intergenic region on the M and S segments is AU rich and computer models predict a high degree of base pairing. The intergenic regions of the S RNAs of tospoviruses range from 500 to more than 1200 nucleotides in length. The intergenic regions of the M and S segments of isolates of TSWV are also reported to be of variable length, with size differences as much as 100 nucleotides. The sequence of the intergenic regions contains highly conserved domains separated by variable sequences and deletions and insertions that appear as gaps in alignments. The deletions and insertions occur at multiple sites and may be as large as 100 nucleotides, giving rise to the variable length of the intergenic region. The occurrence of a duplicate, 33-nucleotide sequence has been associated with a loss of competitiveness in mixed infections of isolates with single or duplicate copies of the sequence.

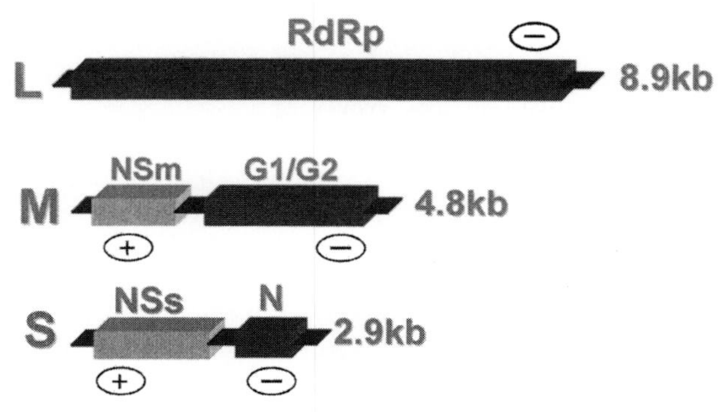

Fig. 1. Tospoviruses genome organization. Positive and negative sense ORFs are indicated by + and −, respectively.

C. Genome Organization and Protein Function

The largest *Tospovirus* genomic segment (L) is approximately 9 kb and contains a single ORF coding for a 330-kDa protein (Fig. 1). The protein is translated from the viral complementary strand and is the putative RNA-dependent RNA polymerase. It has minimal homology to L proteins in viruses of other genera of the family. However, there are motifs that suggest additional functions in RNA replication and processing. The L protein has been detected in virions, consistent with other members of the Bunyaviridae, and RdRp activity has also been detected in RNP preparations from disrupted virions.

Unlike the L RNA, both the M and S RNAs have two ORFs in an ambisense orientation. The M RNA is approximately 4.8 kb and encodes a 34-kDa protein in the viral sense, designated NSm, beginning near the 5′ terminus. This protein is the putative cell-to-cell movement protein having been associated with plasmodesmata and its ability to stimulate tubule formation in protoplasts. Both are hallmarks of viral proteins that mediate intercellular movement in plants, a function of plant viruses distinct from animal viruses. An NSm homolog is also absent in the animal infecting Bunyaviridae. A second ORF near the 3′ terminus, separated from the first by an AU-rich intergenic region, encodes a 127-kDa precursor to the two glycoproteins (GP1 and GP2) in the viral complimentary sense. GP1 and GP2 have not been individually characterized but their M_rs are 78 and 54 kDa, respectively. These two proteins are assumed to be autocatalytically cleaved coincident with synthesis based on the inability to detect the

GP1/GP2 precursor following either *in vivo* or *in vitro* synthesis. GP1 and GP2 are found in the virion envelope but have not been associated with any function in plants. There is preliminary evidence that these proteins may interact with a receptor protein found in cells lining the midgut of the thrips vector.

The S RNA is approximately 3 kb and encodes a 54-kDa protein, designated NSs, from an ORF near the 5′ end in the viral sense. Although the NSs accumulates to high concentration as fibrillar structures in cells infected with some isolates, it has no known function. The high concentration of this NSs structural protein is used as an indicator of replication because it is not found in virions. This is an especially useful assay in insect vectors since preformed virions may be acquired from infected plants, making it impossible to distinguish potential vectors which are carrying virus acquired from the host from those supporting virus replication and capable of transmission. The nucleocapsid protein (N) (29–31 kDa) is encoded by the ORF in a viral complementary sense located near the 5′ termini. The N protein encapsidates the viral RNA forming RNPs and is thought to be necessary for RNA replication based on investigations of similar viruses.

III. REPLICATION

Primary infection of a healthy plant is via a viruliferous thrips during processes coincident with thrips feeding on the potential plant host. Upon entry of the virus into a plant cell, virion RNA replication and virion assembly of tospoviruses occurs in the cytoplasm. Details of the replication process are not available; however, circumstantial evidence suggests that the RNA segments undergo replication similar to other viruses with negative and ambisense RNAs. The viral-coded RdRp contained in the virion transcribes full-length viral complementary L RNA which may serve as a template for translation of RdRp or viral sense RNA for RNP formation to be contained in mature virions. High passages in selected hosts result in synthesis of defective L RNA segments (DIs). DIs typically consist of molecules of viral L RNA with internal deletions but wild-type termini. The S and M RNA segments exhibit a distinctly different translation strategy in that subgenomic RNAs can be detected for both the ORFs in viral and viral complementary orientation. Interestingly, these subgenomic RNAs have up to 20 nucleotides as an extension of the 5′ terminus of the subgenomic RNAs. These are heterogeneous sequences of nonviral origin in single infections; however, leader sequences from alfalfa mosaic virus have been similarly detected as caps of TSWV mRNA in mixed infections of the two viruses. The viral RNA also serves as template for synthesis of full-length viral complementary RNA that, in turn, serves as template for more virion RNA. Analysis of the polarity of RNA isolated from purified virions and preparations of RNPs revealed that little or no viral complementary L RNA was detected; however, approximately 10% of the total S and M RNA segments were of viral complementary polarity. This is consistent with the accumulation of ambisense RNA segments of other viruses.

Isolation of functional replicase from native infections or expressed cDNAs remains problematic, as does the development of a system for rescue of viral cDNAs (infectious transcripts). Current research employing classical viral genetics is providing some additional insights into mechanisms of adaptation but has not contributed to our understanding of the replicative processes. Substantive advances in understanding gene function and replication will require the conception of protocols for reverse genetics in the Bunyaviridae.

IV. HOST RANGE AND SYMPTOMATOLOGY

TSWV has one of the largest documented host ranges of any virus. It is known to infect more than 900 species across several families. Single isolates have been reported to infect both dicotyledonous and monocotyledonous plants. However, most tospoviruses have moderate host ranges similar to other viruses. Some, such as iris yellow spot virus, have very narrow host ranges limited to a few species. Some of the early difficulty in isolating TSWV, especially from floral crops, can be attributed to using assay hosts and antiserum suitable for TSWV, when in fact the virus was INSV, which does not systemically infect many of the diagnostic hosts or react with antiserum routinely used for TSWV. Those TSWV-like viruses were other tospoviruses that did not share the same diagnostic hosts. During the 1980s, attempts to isolate TSWV from many floral crops were considered difficult; however, it is now known that INSV was probably the virus being isolated. INSV does not systemically infect many of the diagnostic hosts routinely used for TSWV. In general, *Nicotiana benthamiana* is now used as an assay host for most tospoviruses.

Symptoms induced by tospoviruses are highly variable and of little diagnostic value. Concentric, necrotic, or chlorotic rings often characterize symptoms of TSWV and some other tospoviruses.

Infection by TSWV of young plants of species such as tobacco, tomato, pepper, and several floral crops is often accompanied by high mortality rates. However, infection of some plants and older plants of all species may be latent, mild, or transient. Symptoms of the same virus on different plants are frequently different. Conversely, symptoms of different viruses on the same host may be quite similar. Many plants also express necrotic symptoms on foliage, leaves, and stems as well as inhibition of growth. Fruit and flowers of many plants may exhibit symptoms resulting in additional losses due to inferior quality of the crop. Furthermore, symptoms may be quite severe and mimic symptoms and injuries caused by other biotic and abiotic stresses. Sampling of plants for confirmatory serological, polymerase chain reaction, or biological diagnostic assays is often confounded by the nonuniform, systemic distribution of the virus in plants.

V. TRANSMISSION AND VIRUS–VECTOR INTERACTIONS

TSWV and the other known tospoviruses are transmitted from plant to plant in nature by eight species of thrips (Table II). They can also be spread by vegetative plant propagules but not through true seed. Tospoviruses are also transmitted experimentally in infected tissue extracts. Care must be taken to preserve the RNP complex which includes RdRp activity. As with other negative sense viruses, purified RNA preparations are not infectious. Critical factors include the following: There must be a high-titered inoculum obtained from tissue prior to necrosis formation and inoculum should be maintained below 4°C and transferred in a buffer with appropriate reducing agents. The recipient hosts must be highly susceptible, usually very young seedlings free of current or previous exposure to stress. A 24- to 48-hr pretreatment in darkness is necessary for highly efficient transmission to some species. Successful transmission may also be influenced by biological factors. Natural infections tend to exist as highly heterogeneous mixtures of isolates whose composition is sensitive to selection pressure imposed by different hosts. Thus, serial passage through an experimental host may make return inoculation to the natural host difficult or impossible.

TABLE II
Thrips Species That Have Been Demonstrated to Vector *Tospoviruses*

Frankliniella bisponsa	*Frankliniella schultzei*
Frankliniella fusca	*Thrips palmi*
Frankliniella intonsa	*Thrips setosa*
Frankliniella occidentalis	*Thrips tabaci*

One or more of eight different species of thrips in two genera transmit each *Tospovirus* in nature: *Frankliniella* and *Thrips* (Table II). There is insufficient evidence to conclude that there is a high level of virus–vector specificity among thrips species. Interestingly, only the larval stage of the thrips can acquire the virus and become viruliferous. TSWV is ingested into the gut of the thrips and invades the midgut cell of the larval stage where it initiates replication. The virus passes out of the midgut into the interior tissues of the larvae and ultimately comes in contact with the salivary glands. The virions pass across the plasmalemma into the glands, in which infectious particles are transported coincident with salivation into the recipient host plant. Once acquisition occurs, the individual insect remains viruliferous through molts for the remainder of its life. However, there is no evidence for transovarial transmission to the next generation as is typical of other arthropod-borne bunyaviruses. Replication in the midgut appears to be the barrier in adult insects that prohibits acquisition, except in *Thrips setosus* adult midgut cells, which support replication but not further invasion. There is also an emerging body of literature that suggests that the GPs bind to a receptor protein in the midgut of the larval-stage thrips. A 50- and a 94-kDa protein have been detected as putative receptor proteins.

VI. CONTROL

The diseases caused by tospoviruses result in significant economic losses due to suppressed growth and yield as well as reduced quality. Crops affected by these viruses are grown in open-field and greenhouse environments. Each presents special constraints for control tactics. Strategies focused on vector control have only been partially effective, even in controlled greenhouse conditions. Integrated approaches that target both elimination or avoidance of viral inoculum and vector management have been most successful. However, as with most vector control programs, threshold vector populations are significantly less, zero in some instances, than for control of the same insect as an insect pest. General recommendations include the utilization of planting stock certified as free of detectable tospoviruses, employment of cultural practices which reduce risk of virus spread such as planting dates and use of resistant or tolerant cultivars, and removal of cull piles and weeds that may provide sources of inoculum. Traditional insecticides have shown little promise under field conditions; however, recent insecticides that may prevent probing and feeding altogether have shown promise.

Resistance has only been observed in a few of the susceptible crops, such as peanut, pepper, tobacco, and tomato. Even in these cases, resistant varieties that survive for more than a couple of years before the native virus population adapts to the resistant cultivar have remained elusive. Recently, a resistance gene (SW-5) was bred into tomato from a wild tomato relative. It provided typical gene-for-gene resistance and has recently been isolated and characterized and shares properties of other R genes. However, natural populations of TSWV rapidly overcame this form of resistance. Genetic experiments revealed that the resistance-breaking element was located on the M RNA of the virus.

Pathogen-derived resistance has also been experimentally tested in many crops. The N gene from the S RNA segment has been used as the transgene in nearly all experiments, although the NSm gene was also shown to be effective. Recent research has demonstrated that less than full-length sequence is sufficient to impart the response when nontarget sequence is added to achieve minimal size. The search for resistant cultivars through conventional breeding and genetic engineering together with genetic analysis of the processes of adaptation are among the most pressing research areas.

See Also the Following Articles

Economic Consequences of Infectious Diseases • Plant Disease Resistance • Viruses, Emerging

Bibliography

Best, R. J. (1968). Tomato spotted wilt virus. *Adv. Virus Res.* **13**, 65–145.

Cho, J., Mau, R. F. L., Pang, S.-Z., Wang, M., Gonsalves, C., Watterson, J., Custer, D. M., Gonsalves, D. (1998). Approaches for controlling of tomato spotted wilt virus. *In* "Plant Virus Disease Control" (A. Hadidi, R. K. Khetarpal, and H. Koganezawa, Eds.), pp. 547–564. APS Press, St. Paul, MN.

Daub, M. E., Jones, R. J., and Moyer, J. W. (1996). Biotechnical approaches for virus resistance in floral crops. *In* "Biotechnology of Ornamental Plants" (R. L. Geneve, J. E. Preece, and S. A. Merkle, Eds.), pp. 335–351. CAB International, Wallingford, UK.

Daughtrey, M. L., Jones, R. K., Moyer, J. W., Daub, M. E., and Baker, J. R. (1997). Tospoviruses strike the greenhouse industry: INSV has become a major pathogen on flower crops. *Plant Dis.* **81**, 1220–1230.

De Avila, A. C., de Haan, P., Kormelink, R., Resende, de O., Goldbach, R. W., and Peters, D. (1993). Classification of tospoviruses based on phylogeny of nucleoprotein gene sequence. *J. Gen. Virol.* **74**, 153–159.

German, T. L., Ullman, D. E., and Moyer, J. W. (1992). *Tospoviruses*: Diagnosis, molecular biology, phylogeny and vector relationships. *Annu. Rev. Phytopathol.* **30**, 315–348.

Mumford, R. A., Barker, I., and Wood, K. R. (1996). The biology of Tospoviruses. *Ann. Appl. Biol.* **128**, 159–183.

Prins, M., and Goldbach, R. (1998). The emerging problem of tospovirus infection and nonconventional methods of control. *Trends Microbiol.* **6**, 31–35.

Ullman, D. E., Sherwood, J. L., and German, T. G. (1997). Thrips as vectors of plant pathogens. *In* "Thrips as Crop Pests" (T. L. Lewis, Ed.), pp. 539–565. CAB International, London.

Toxoplasmosis

John C. Boothroyd
Stanford University School of Medicine

I. Classification
II. Life Cycle
III. Clinical Aspects and Public Health
IV. Population Biology
V. Molecular Biology and Genetics
VI. Cell Biology
VII. Host Immune Response
VIII. Effect on Behavior
IX. Prospects for Future Improvements in Controlling Toxoplasmosis

GLOSSARY

apicoplast A plastid-like organelle just anterior to the nucleus, delimited by four membranes and containing its own genome. Named for its ubiquitous presence in Apicomplexa.

bradyzoite The slowly dividing, infectious, asexual stage found in tissue cysts in the intermediate host. From the Greek for slow (*brady*) little animal (*zoite*).

endodyogeny The process by which tachyzoites and bradyzoites divide, involving the creation of two daughter cells wholly within a mother cell.

micronemes Small, thin, anterior organelles that secrete their contents on the parasite's surface at the start of the invasion process. From the Greek for small (*micro*) thread (*nema*).

micropore A small, single invagination anterior to the nucleus with an apparent clathrin coat, presumed but not demonstrated to be involved in phagocytosis.

oocyst The mature, environmentally robust product of the sexual cycle, shed in its immature form in cat feces and comprising a cyst wall and two sporocysts, each of which contains four sporozoites.

rhoptries Club-shaped, apical organelles that release their contents into the nascent parasitophorous vacuole. From the Greek for club (*rhoptry*).

sporozoite The product of the sexual cycle, found as two sets of four within the oocyst.

tachyzoite The rapidly dividing, asexual stage. From the Greek for fast (*tachos*) little animal (*zoite*).

tissue cyst The protective structure containing the asexual bradyzoites in the intermediate host. Also known as the pseudocyst.

zoonosis A disease of humans that derives from a reservoir of nonhuman animal infection. From the Greek for disease (*nosos*) originating from animals (*zoo*).

TOXOPLASMOSIS is caused by the protozoan *Toxoplasma gondii,* an obligate, intracellular parasite whose definitive host is the family Felidae (cats). *T. gondii* has an exceptionally broad range of intermediate hosts, including humans, in which the asexual cycle can occur resulting in serious disease. It is found throughout the world and, in terms of the number of hosts infected, *Toxoplasma* may be the most common protozoan parasite of warm-blooded animals. The genus *Toxoplasma* falls within the phylum Apicomplexa and is a close relative of the nonhuman pathogens, *Neospora* and *Sarcocystis.* It is more distantly related to *Eimeria,* the causative agent of coccidiosis in birds and more distant still to *Plasmodium* and *Cryptosporidium,* the etiologic agents of malaria and cryptosporidiosis, respectively. The infection of healthy humans is generally of little consequence, but human disease does occur in two scenarios, in the child of a woman infected for the first time during pregnancy and in individuals who are immunocompromised. In animals, the disease is also generally self-limiting with the most serious consequences (abortion) being from acute infection during pregnancy. The relative ease of handling this haploid parasite in the laboratory and the wide range of natural hosts have made *T. gondii* a popular model for experimental study of intracellular parasitism. Through such studies, much has been learned about the way in which this common parasite interacts with its host, about its population biology, and about the molecular details behind its obligate intracellular lifestyle.

I. CLASSIFICATION

Toxoplasma gondii is a protozoan parasite within the phylum Apicomplexa, Class Sporozoasida, order Eucoccidiorida, suborder Eimeriorina, family Sarcocystidae (Levine, 1983). The genus name derives from the Greek word *toxon* meaning bow and *plasma* meaning form, and refers to the bow-shaped tachyzoite. The species name derives from the name of the North African rodent from which the parasite was first isolated in 1908 by Nicolle and Manceaux. It was simultaneously described by Splendore in Brazilian rabbits.

II. LIFE CYCLE

The complete life cycle of *T. gondii* was not established until about 1970, when four groups independently identified the cat as the definitive host in which a classic coccidian sexual cycle occurs. Figure 1 depicts the complete life cycle as consisting of two, potentially independent cycles, one asexual and one sexual. This has been done in order to emphasize that the parasite appears fully capable of propagating itself through either cycle. The degree to which these two cycles intermix (arrows 6 and 7) is not known. In practical terms, this uncertainty means that it is not clear what fraction of infection in the intermediate host (e.g., humans) is a result of ingesting meat containing tissue cysts (arrow 1) versus accidental ingestion of oocysts in the environment (arrow 8).

The tachyzoite stage is much the best studied mainly because it is easy to grow this form *in vitro*. The other asexual stage, the slow-growing, encysted bradyzoite is poorly understood. Fortunately, however, protocols have recently been developed that can cause tachyzoites to efficiently differentiate to

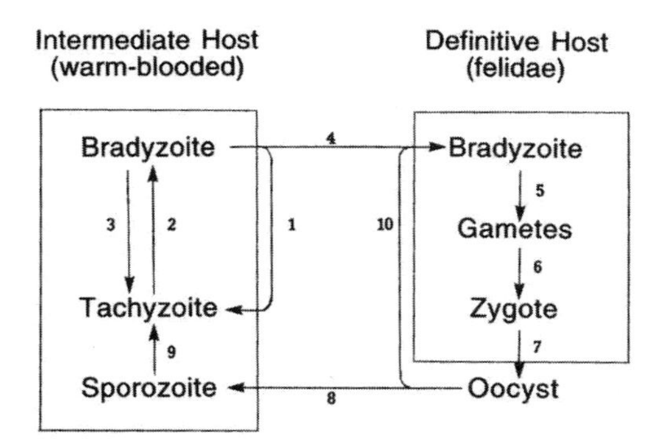

Fig. 1. The life cycles of *Toxoplasma gondii.* The asexual cycle can occur in a large number of warm-blooded animals and is shown on the left. It involves an equilibrium between the rapidly dividing tachyzoite and the more slowly dividing bradyzoite (arrows 2 and 3). Transmission in the asexual cycle is through the ingestion (carnivorism) of meat and other tissue containing the infectious encysted bradyzoites (arrow 1; e.g., rat to pig or pig to human). Scavenging allows the parasite to cycle back down the food chain (e.g., pig to rat). The sexual cycle is shown on the right and involves schizogony, gametogenesis, and fertilization (arrow 5, 6 and 7) in the gut epithelium of felines. Cat-to-cat transmission through the sexual cycle is theoretically possible by ingestion of oocysts in fecal contamination (arrow 10). Crossover between the two cycles is represented by arrows 4 and 8, but the extent of this in nature is not known.

the bradyzoite stage *in vitro*. These protocols rely on a number of stresses (certain drugs, heat, and pH) that stimulate the differentiation pathway. Whereas the differentiation may not be complete, the availability of these protocols opens up the way for more detailed studies of this critically important stage, which previously could only be obtained in small amounts from the tissue of infected animals.

The two asexual stages are similar in overall morphology, with the same complement of organelles. The major differences are in the presence of many amylopectin granules in the bradyzoite and a cyst wall that encloses the entire complement of bradyzoites in any given vacuole. The metabolic differences are not clear, although the existence of stage-specific isoforms of some glycolytic enzymes (e.g., enolase and lactate dehydrogenase) suggests the possibility of a fundamental difference in sugar metabolism. Such differences might be related to the presence of the amylopectin granules or the polysaccharide-containing cyst wall.

The sexual stages have only been described morphologically in the gut of infected cats (it has not yet been possible to grow any of these stages *in vitro*). The process of gametogenesis, which occurs in the epithelium of the small intestine, conforms to classic coccidian rules, including schizogony in which a multinucleated state is reached followed by segmentation to yield multiple uninucleated merozoites. Mating between micro- and macrogametes is followed by infection into epithelial cells and results in a zygote, which then encysts. The resulting immature oocyst is released into the environment where, after ~2 days of exposure to air, it develops into a mature oocyst comprising two sporocysts, each with four sporozoites. The genetic relationship of the eight sporozoites within a single oocyst (i.e., which are the result of mitotic and which of meiotic divisions) has not been determined.

III. CLINICAL ASPECTS AND PUBLIC HEALTH

A. Symptoms

Toxoplasma causes disease in humans primarily under two circumstances, in the developing fetus of a woman who acquires her first infection during pregnancy and in people who are immunocompromised. The disease that results varies from mild to severe and is sometimes fatal. Note, however, that in congenital infections the disease may not appear until many years later, when the child is in the second decade of life or older.

The prevalence of seropositive adults ranges from ~5–15% in the United States to as high as 60–85% in France or up to <90% in some developing countries. Historically, severe disease has not been common because of the low number of people who fall in the two disease-susceptible groups. The AIDS epidemic has obviously produced a huge increase in the number of people in the immunocompromised category. Initial results indicated that if left untreated about one-third to one-half of AIDS patients who were seropositive for *Toxoplasma* (most cases of toxoplasmosis in AIDS patients are thought to be due to the reactivation of historic quiescent infections) were liable to develop the potentially fatal condition of toxoplasmic encephalitis. This proportion has been dropping in recent years, probably as a result of three factors: 1) improved anti-HIV therapies (with a resulting improvement in general immune competence), 2) prophylactic treatment with trimethoprim and sulfamethoxazole for another opportunistic infection, *Pneumocystis*—such treatment has the added benefit of controlling *Toxoplasma* infection, and 3) deliberate prophylaxis for *Toxoplasma* in patients who are seropositive and whose CD4 counts are low.

The congenital infection rate varies by country (in proportion to the seropositivity of the adult population); in the United States, there are no exact numbers, but current estimates are that ~500–3000 congenitally infected children are born each year. The probability of transmission from an acutely infected mother to the developing fetus varies with the stage of the pregnancy; acute infections in the first trimester are the least likely to be transmitted, whereas third-trimester infections are the most likely. As for many congenital diseases, however, the outcome is worse when the infection occurs earlier in the pregnancy. First-trimester infections can lead to severe neurological disease (blindness or retardation) or even death, whereas third-trimester infections are more mild and may yield no symptoms at birth. This is not to say that no disease will occur in this latter group; it appears that the parasite can lie essentially dormant and erupt in early adulthood, leading to various symptoms, most notably sudden blindness in one eye due to severe retinal destruction. Drug treatment of acutely infected, pregnant women can reduce the probability of congenital transmission but, as discussed in Section III.C, such treatment is not without significant risk to the fetus.

B. Diagnosis

Serology is the most common method of diagnosis. Acute infection in pregnant women is usually diagnosed from a high IgM or IgA antibody titer or a rise in IgG in successive serological samples taken at multiweek intervals during the pregnancy. In countries such as France where the incidence is high, strict monitoring is in place. Women are tested for anti-*Toxoplasma* titers at the time of marriage and those who are seronegative and later become pregnant are monitored repeatedly for signs of seroconversion (equated to an acute infection). If an acute infection is detected, it is now possible to monitor for transplacental transmission to the fetus (which is the real concern) by amniocentesis and polymerase chain reaction to detect the parasite (or, rather, its DNA) directly.

C. Treatment

The current treatment for acute infection in nonpregnant adults relies on the synergistic action of pyrimethamine plus sulfadiazine, which combine to block folate metabolism through inhibition of dihydrofolate reductase and dihydropteroate synthase, respectively. Unfortunately, even this powerful combination has little effect on the chronic form of the parasite (the bradyzoite), and thus treatment is not sterilizing. As a result, the drug therapy must be maintained for long periods to prevent recrudescence of the infection. This is not a simple option, as sulfadiazine is poorly tolerated by many patients and its use often must be stopped. Other therapies, although showing some promise, have yet to prove as effective as pyrimethamine plus sulfadiazine (e.g., pyrimethamine plus clindamycin and the newer drug, atovaquone), although better formulations and treatment protocols could dramatically improve their efficacy.

Because of drug toxicity, the treatment of pregnant women is always a clinically complicated decision. In countries such as France, which has the most experience of managing such patients, an acute infection in the mother is treated with spiromycin, although the benefit of such treatment is not firmly established. If transmission to the fetus has been demonstrated, then the more potent but more toxic combination of pyrimethamine plus sulfadiazine is used because the risk of serious disease outweighs the possible harmful effects of the drugs. Therapeutic abortion is also sometimes chosen.

D. Public Health

In the absence of a vaccine and given the shortcomings of the drugs mentioned, the best approach to controlling human toxoplasmosis is avoidance. This is easier said than done, however, as evidenced by the fact that infection with this parasite is so prevalent worldwide.

Meat-eaters are especially at risk from undercooked lamb and pork, although the latter may not be such a problem because of centuries of knowledge about severe diseases that can result from pork and the resulting common knowledge that it should always be well cooked. Lamb, on the other hand, is frequently eaten extremely rare in many cultures. (Note that beef, although also often eaten rare or raw, is not such a problem because for unknown reasons, bovines do not harbor substantial numbers of tissue cysts in their muscle or other tissue.) It is thus critical that people at risk (those who are immunocompromised and seronegative women who might become pregnant) avoid eating meat that is not thoroughly cooked. Fortunately, freezing also kills tissue cysts and thus this can also be used to reduce the risk of infection.

Toxoplasma's unusual life cycle means that simple vegetarianism will not enable someone to avoid infection. The oocysts shed by cats are extremely stable in the environment and cats are particularly fond of defecating in loose soil, such as that found in vegetable gardens. It is thus fairly simple for vegetables to be contaminated with minute amounts of oocysts and so people who are at risk should take special care to eat only cooked or thoroughly washed fresh vegetables.

Finally, the stability of the oocysts also means that they can be ingested through daily activities as innocuous as gardening or playing in a dusty field or sandbox. Obviously, cleaning a pet cat's litter tray is not appropriate for people who are at risk, although because oocysts take ~48 hr to become infectious after shedding, regular daily removal of the feces can reduce the risk of infection. Even better, keeping cats strictly indoors and feeding them only tinned or dried foods can reduce the chance of their acquiring an infection.

IV. POPULATION BIOLOGY

A. Major Genotypes

One of the key deficiencies in our understanding of toxoplasmosis is what fraction of human infection comes from the ingestion of tissue cysts in undercooked meat products versus from oocysts in environmental contamination. In recent years, the population biology of *Toxoplasma gondii* has come under intense scrutiny through studies of isoenyzmes, restriction-fragment-length polymorphisms (RFLP) and direct-sequence analysis of specific loci. From these studies a very clear, consistent, but surprising picture has emerged. First, despite the existence of the well-described sexual cycle in cats, the parasite appears to be reproducing largely clonally. That is, the vast majority of strains fall into one of three genotypes (hereafter referred to as type I, II, or III) without the mixing expected for a sexual population. Because *Toxoplasma gondii* is haploid, this does not necessarily mean that the sexual cycle is not occurring. Instead, it could mean that the mating that does occur is almost entirely between self. This is not unreasonable given that a single haploid tachyzoite can give rise to a full sexual cycle in the cat (including genetically identical micro- and macrogametes) and thus the parasite does not have fixed mating types or sexes. Moreover, it is known that once a cat is infected, oocysts appear in the feces for about 1–2 weeks, but that thereafter the cat is relatively immune to further infection and oocyst shedding drops precipitously. Whether a cat infected with strain A and subsequently some weeks later with strain B will yield A/B recombinant progeny has not been directly examined, but all available data suggest the answer will be no. These factors alone could account for the rarity of recombinants.

There is a second possible explanation for the clonal propagation: Cats may play little if any role in the transmission of some strains. For example, it has been noted that type I strains grown in the lab are difficult to pass in cats. Because cats are the only known host in which the sexual cycle occurs, a loss of this function would require exclusively asexual expansion with transmission through carnivorism and scavenging, or, perhaps, vertically in some host species.

This could explain why most strains appear to be expanding clonally, but why do most fall into only three genotypes? For one of the most common protozoan parasites on Earth (at least in warm-blooded animals), it is hard to imagine that a major bottleneck arose through chance environmental factors. Instead, we must presume that the three types that have arisen are remarkably successful recombinants (which they surely are) and that there may be a small but diverse population of parasites that collectively possess tremendous genetic complexity and out of which these three particular genotypes have emerged as particularly successful for the available niches. In this context, those niches are humans, domestic animals, and livestock in Europe and North America, from which the vast majority of the strains so far examined have originated. Support for this hypothesis comes from the fact that when strains are isolated from remote regions and unusual species (e.g., South American wild monkeys), a quite different genotype can be found. These latter strains are presumably representative of the total diversity to be found when more unusual hosts and regions are eventually sampled and carefully analyzed.

B. Strain-Specific Virulence

The three major types of this parasite are not just minor or random polymorphic states. Indeed, one of the first clues that the population biology of *Toxoplasma gondii* was unusual was that the virulence of multiple different strains, when measured as LD100 in susceptible mice, was either a single parasite or more than 10 thousand without a substantial middle ground. We now know that in susceptible mice, all the type I parasites are highly virulent organisms, whereas types II and III are relatively avirulent. The genetic basis for this difference is not known.

There is very little data on how the difference in mouse virulence translates to disease outcome in humans and livestock. The reason is that the knowledge that there are a limited number of strains (essentially three) is new, and methods to distinguish them have been tedious or only recently applied. Nevertheless, some studies have been done and the data show that most human infections of AIDS patients are by type II parasites and that the congenital infections are by various types, although again type II dominates. In the absence of information on what strains are present in asymptomatic infections, however, and what the relative frequency is of different types in the various animal reservoirs of human infection, it is impossible to draw any conclusion about relative virulence. For example, it could be that type II parasites are also most prevalent in asymptomatic human infections and they cause no worse disease than other types; that is, they predominate in diseased people simply because they account for most human infections, symptomatic or otherwise. Alternatively, it could be that type II parasites are more virulent in humans and infections that are asymptomatic are the result of type I or type III strains. Discriminating between these two possibilities in human and livestock infection is an important future goal of *Toxoplasma* researchers.

V. MOLECULAR BIOLOGY AND GENETICS

A. Genome and Gene Expression

Except for the diploid zygote, all stages of *Toxoplasma* are haploid with an estimated genomic complexity of $\sim 8 \times 10^7$ bp/nucleus. A combination of physical and genetic means shows that there appear to be ~11 chromosomes in the nuclear genome.

Protein-coding genes in *Toxoplasma* are generally present in a single copy per nucleus and introns are not uncommon. Unlike some other parasitic protozoa, gene expression in *Toxoplasma* appears to be similar to that seen in model eukaryotes such as yeast or humans.

In addition to the nuclear DNA, two other genomes are believed to be present in *T. gondii*. These are the 35-kb Apicoplast genome and an incompletely characterized mitochondrial DNA. The complete sequence of the 35-kb element has been completed in *Toxoplasma* but the sequence sheds little light on the ultimate function of its organellar home.

B. Genetics

Because of its haploid nuclear genome, it is relatively easy to generate mutants in *Toxoplasma*. A large number of such have been described, including lines that are temperature-sensitive for growth, deficient in one or other surface antigen, or resistant to one or more drugs. The ease of generating such mutants is probably also due to the fact that, as already mentioned, the protein-coding genes that have been characterized to date are almost invariably single-copy.

A low-resolution genetic map for the nuclear genome has been created and used to map genes encoding important phenotypes (e.g., drug resistance). The fact that the crosses must be done in cats, however, makes mapping far from trivial.

C. Molecular Genetic Tools Available for the Study of Toxoplasma

In the past decade, there has been a substantial effort invested into the development of molecular genetic techniques for the study and manipulation of *Toxoplasma*. The large number of such techniques now available makes *Toxoplasma* one of the most easily studied and "engineered" of any protozoan parasite.

Initially, transient expression of suitably engineered DNA plasmids was achieved using electroporation. There are several selectable markers reported for stable transformation of *Toxoplasma*—bacterial chloramphenicol acetyl transferase (conferring resistance to chloramphenicol), pyrimethamine-resistant dihydrofolate reductase (*DHFR*), bacterial tryptophan synthase (*trpB*) conferring tryptophan prototrophy), bacterial *ble*, conferring bleomycin and phleomycin resistance, bacterial β-galactosidase (*lacZ*), jellyfish green fluorescent protein, and, in the appropriate null backgrounds, *Toxoplasma* hypoxanthine/xanthine/guanine phosphoribosyl transferase (HXGPRTase), uracil phosphoribosyl transferase (UPRTase), and the major tachyzoite surface antigen, *SAG1*. With the exception of the last, all have been successfully used to introduce exogenous or engineered genes into the parasite for expression.

Episomal vectors have also been developed that allow the simpler recovery of introduced DNA, multicopy suppression, and tests for essentiality using negative selectable markers on the episome.

VI. CELL BIOLOGY

A. Organelles

As a dedicated, intracellular parasite, *Toxoplasma* has many specialized organelles dedicated to this lifestyle. Some of these are shown in Fig. 2.

1. Apicoplast: This is a spherical plastid-like organelle just anterior to the nucleus. It has been recognized for decades and has been given various names (e.g., Golgi adjunct and spherical body). It is delimited by four membranes and contains its own genome, whose sequence revealed its relatedness to plastids of green algae. The data further argue that it was acquired by the ancestor of *Toxoplasma* (and other Apicomplexa) through a secondary endosymbiosis of an algal-like eukaryote. The function of this organelle is not known and its small genome has given no real clues: virtually all of the recognizable genes serve housekeeping functions in transcription and translation. The few that do not are not homologous to genes of known function in other systems.

2. Dense granules: These are small spherical secretory organelles distributed throughout the tachyzoite. They release their contents into the parasitophorous vacuole upon invasion into a host cell and are believed to help modify the vacuole to make it better suited for the growth of the parasites within. One of the most abundant constituents of these granules is a potent NTPase that may have a role in nucleoside scavenging.

3. Inner-membrane complex: This is a set of collapsed vesicles that almost completely envelope the parasite and that are closely apposed to the parasite's plasma membrane. Together with the latter, they make the parasite surface appear as if it consisted of a triple layer of membranes. It is one of the first organelles to appear *de novo* during endodyogeny and behaves as if it were the nucleating point for formation of the two daughter parasites.

4. Micronemes: These are numerous thin anterior organelles that secrete their contents onto the parasite's surface at the start of the invasion process.

5. Micropore: This small single invagination anterior to the nucleus has an apparent clathrin coat and is presumed, but not demonstrated to be involved in phagocytosis.

6. Rhoptries: These are club-shaped apical organelles that release their contents into the nascent parasitophorous vacuole early in the invasion process. The functions of their protein contents are not known, although evidence exists for at least one of the proteins being involved in recruitment of the host mitochondria into tight apposition with the parasitophorous vacuolar membrane.

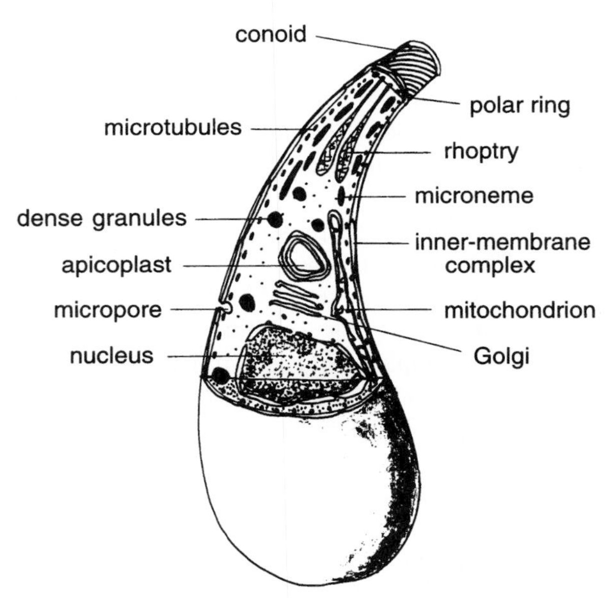

Fig. 2. Ultrastructure and morphology of the tachyzoite. This schematic shows a cut-away of the tachyzoite stage of the parasite. The overall morphology is very similar for the bradyzoite and sporozoite. Note that the microtubules gently spiral down the body, starting from the anterior polar ring and ending about two-thirds along the length of the parasite. This is why they have an oval appearance in the cross section shown.

B. The Lytic Cycle

Toxoplasma is an obligate intracellular parasite. As such, and given that it causes lysis of the host cell after going through several rounds of replication, the classic virology term, lytic cycle is perfectly suited to describing the various stages of intracellular growth. The cycle begins with attachment, followed by invasion, replication (including vacuolar modification), and egress (Fig. 3).

1. *Attachment*

Taking all the data together, the most likely model for attachment involves two steps. First, a resident surface molecule, perhaps one of the glycolipid-anchored surface antigens, may interact with unidentified host ligands. This part of the model is based on studies showing that antibodies to the major surface antigen, P30 or SAG1, can block the binding of parasites to host cells (fixed or live) and thus prevent invasion. The putative ligand is likely to be something ubiquitous on vertebrate cells because *Toxoplasma* is remarkable for its ability to invade almost any such cell it encounters, at least *in vitro*.

Following such initial interactions, a signal is sent, which results in release of higher affinity molecules, perhaps including the MIC2 protein of micronemes. MIC2 is a homolog of the better-studied thrombospondin-related anonymous protein (TRAP) molecule of *Plasmodium sp.* These two molecules have homology domains that argue for an ability to bind glycosaminoglycans (GAGs). GAGs and other proteoglycans have been implicated as a key ligand on the host side, although the connection between MIC2 and binding of GAGs has not been directly shown. Another *Plasmodium* homolog found in *Toxoplasma* is *AMA1* which, as was MIC2, was identified as part of the *Toxoplasma* expressed sequence tag (EST) sequencing effort. The ligand for this microneme protein is, as yet, completely unknown in both parasite systems.

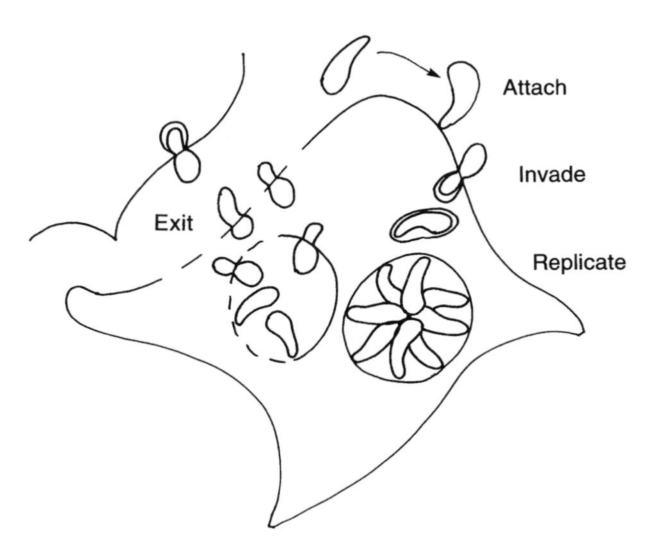

Fig. 3. The lytic cycle of asexual reproduction. In this cartoon, the stages of attachment, invasion, replication, and egress are shown. Note that the host cell may have multiple vacuoles. Egress is shown here as involving both membrane destruction and active egress through semiintact membranes. This is the least well-studied part of the lytic cycle.

Electrophysiological studies have provided further insight into this initial attachment stage. They show that there is a brief transient peak of conductance, suggesting that there is a profound perturbation of the integrity of the host-cell plasma membrane associated with the initial attachment. The significance and role of this event are not known.

2. *Invasion*

Invasion is an active process involving many parasite functions and is accompanied by the sequential discharge of several distinct intracellular compartments, including the micronemes, rhoptries and dense granules. The contents of these compartments serve many functions. Among these, it is presumed that some number of molecules facilitate the invasion process itself, perhaps providing the link between the host-cell surface and the parasite's actin–myosin-based motors that provide the driving force of invasion. Thus, molecules such as MIC2 or AMA1 might enable the parasite to literally pull itself into the host cell, creating a vacuole as it proceeds. Physiology studies have clearly demonstrated that this parasitophorous vacuole membrane (PVM) is derived from the host-cell's plasma membrane.

3. *Vacuole Modification and Parasite Replication*

The first challenge to the intracellular parasite is to create the necessary niche for optimal growth. For *Toxoplasma*, this requires modifying the vacuole created on invasion. The PVM initially appears devoid of host markers, suggesting that during its formation all resident proteins are removed through some sort of molecular sieve that must exist at the moving circular junction that migrates down the parasite surface, creating the PVM in its wake. Very quickly, however, parasite proteins begin to be detectable on the PVM. No host proteins are ever detected, suggesting that no fusion with lysosomes or other compartments of the host endocytic pathway occurs. This is consistent with ultrastructure studies on infected cells and the fact that there is no acidification of the vacuole. The source of lipid used to grow the PVM (after its initial formation) is not known and could be either host or parasite in origin.

The major features of the PVM are

1. Pores appear in the PVM that are permeable to molecules of up to about 1300 Da without regard to charge. These structures are presumably necessary to allow the parasite to feed freely on host cell nutrients. They also result in the lack of a physiological barrier between the vacuolar space and the host-cell cytosol. Thus, aspects such as pH, osmolarity, and concentration of specific ions, should be similar if not identical between these two compartments.
2. Various parasite proteins are found associated with the PVM through interactions ranging from insertion into the PVM to less well-characterized associations. The parasite molecules associated with the PVM originate from the rhoptries and dense granules. The latter are also a source of proteins that can associate with the intraparasitophorous vacuolar network (IPN), a loose meshwork of tubules of unknown composition and function. The IPN has been reported to be continuous with the PVM, suggesting it may be used to increase the effective surface area of the PVM, although there are no data to show that this is physiologically the case.
3. Host-cell organelles such as mitochondria and endoplasmic reticulum are recruited around the PVM with extraordinary efficiency and uniformity. Presumably, the parasite proteins that span the PVM are involved.

Thus, through this extraordinary vacuole, the parasite is able to create the ideal environment for its own growth, which then occurs with considerble efficiency (e.g., a doubling time of as little as 6 hours for type I strains under ideal conditions, *in vitro*). Division is not by simple binary fission, however. Instead, the parasite undergoes a complex process known as endodyogeny (Fig. 4). This term was chosen to convey the fact that two daughter parasites develop within the mother cell, dividing up the organelles between them, but at one point, some organelles are found with a complement of three (e.g., the inner-membrane complex and the rhoptries). Most of the organelles, however, are only ever present with a complement of at most two (e.g., the nucleus, the mitochondria, the Golgi, and the plastid).

4. Egress

The final stage of the lytic cycle, egress, is also the least studied. The intuitively appealing notion that the parasite's replication simply fills the host cell to the bursting point is argued against by the following data.

1. Video microscopy clearly shows that the parasites become highly motile just prior to egress, suggesting a signal is delivered that initiates the exit process.
2. Egress can be artificially stimulated by various treatments, including adding calcium ionophores,

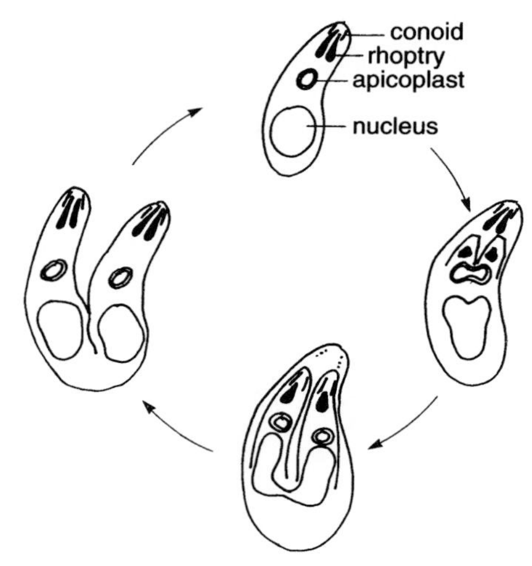

Fig. 4. Endodyogeny during asexual reproduction. Endodyogeny involves the creation of two daughter cells within a mother cell. As the two daughters develop from the anterior end backward, the apicoplast and nucleus divide, and new rhoptries, and inner-membrane and apical complexes appear. Eventually, the mother cell's organelles, including her plasma membrane, are fully co-opted by the daughters, except for the mother's inner-membrane complex, which mysteriously disappears.

such as A23187, or strong reducing agents, such as dithiothreitol, to the medium. This effect can be achieved as soon as 30 min after invasion (i.e., when there is only a single parasite per vacuole).

The simplest explanation is that in the natural cycle, as the parasite numbers grow, a critical concentration of a key molecule in the vacuolar space is reached, which then sends a signal to initiate the cascade of events leading to egress. This molecule would likely be freely diffusible (through the pores), thus explaining why the total number of parasites in a given cell is important rather than the number in a given vacuole. Hence, an infected cell harboring four vacuoles each with eight parasites will be lysed at roughly the same time as one infected with a single vacuole containing ~32 parasites.

VII. HOST IMMUNE RESPONSE

A. Nature of the Host Response

As would be expected for an intracellular parasite of nucleated cells, the major effector mechanism for host immunity is cell-mediated; humoral immunity (antibodies) can confer some protection, but this is apparently less important than the cellular response. Based on the use of antibodies that selectively deplete various T-cell subsets or mice that are deficient in these cells, it is clear that CD8+ cytotoxic T-cells are among the most important effectors. Their action is augmented by a potent Th1-type response facilitated by CD4+ cells, as evidenced by the effect of injecting infected animals with various cytokines or antibodies to various cytokines. For example, the injection of anti-interferon-γ causes the immune response to be depressed and the cerebral infection is correspondingly more severe. The cellular components of the innate immune response, (e.g., natural killer cells) also play an important role in controlling the acute stages of the infection.

In some cases, the host immune response can produce a severe pathology with lethal consequences, for example, in some inbred mice where a Th1-dependent gut pathology is seen. At the other extreme, the immune response in the brains of infected animals can be minimal, allowing for persistence of cysts in this semiprivileged site. The mechanisms by which these various responses occur are not known.

B. Genetics of Host Susceptibility

There is some information to indicate that in mice, at least, there is a genetic component that influences the outcome of infection. One key locus is Ld within the MHC cluster of genes. This is presumably affecting the presentation of a specific key antigen, which has not been identified. There are also data to suggest that MHC plays a role in human infection, at least in the context of toxoplasmosis in AIDS patients, but the effect is relatively small and the basis of the effect unknown.

C. Immunization Studies

Immunization with crude or purified antigens (e.g., SAG1) or with attenuated strains (e.g., the temperature-sensitive ts4 mutant) can elicit a protective response. Efforts have also been made to use a temperature-sensitive strain that is unable to complete the sexual cycle in cats. It generates a good immunity in the cat, not producing oocysts, and would thus be relatively safe. Neither of these approaches have been developed for commercial use.

A commercial vaccine does exist based on the S48 strain, which was serially passed 3000 times through mice. This strain will produce an infection in sheep, which, once cleared, results in an immunity that reduces the problems associated with subsequent infection during pregnancy of the ewes (i.e., abortion). The strain apparently is completely cleared from the infected animals with no tissue cysts developing. Thus, vaccination with this strain should not represent a problem if the sheep are subsequently slaughtered for consumption because no transmission to humans should be possible. Likewise, the death of the vaccinated sheep in the field should not result in the strain being transmitted to scavengers or carnivores.

VIII. EFFECT ON BEHAVIOR

Experiments in rodents have clearly established that a chronic infection, which includes the development of tissue cysts in the brain, can result in a marked change in behavior. These changes include increased activity that may make the animals more likely prey for cats, which are attracted by motion. This would have some advantage to the parasite but whether this phenomenon plays a role in natural transmission in the wild has not been established. The possibility that human behavior might be affected by the infection has not been carefully examined. Although there are no data to support it, this remains a formal possibility.

IX. PROSPECTS FOR FUTURE IMPROVEMENTS IN CONTROLLING TOXOPLASMOSIS

A. Public Health

Educating the public about this disease and ways to avoid it is an important goal for the immediate future. This effort will be helped by research to better understand the relative contribution to human disease of the two routes of transmission (eating undercooked meat containing bradyzoites and accidentally ingesting oocysts contaminating garden vegetables and hands). Such information would allow the public health message and control measures to be focused on whichever route proves to be the more serious risk.

Likewise, information about which strains cause the most serious disease will also help in the clinical management of the patient (e.g., in balancing between the use of potentially toxic drugs and the likely disease outcome if the infection is left untreated). Such improvements will depend on more refined diagnostic methods that distinguish between strains and the source of the infection.

B. Vaccination

Although there can be little doubt that a human vaccine could be developed, the relative rarity of serious disease (compared to the number infected) and the fact that such disease is largely restricted to pregnancy or people with AIDS make the testing and use of such a vaccine extremely difficult. Nevertheless, vaccines still have a tremendously important role to play because we are dealing with a zoonosis; *Toxoplasma* is not passed from human to human, but instead is acquired from animals (oocysts from cats or tissue cysts from meat animals). Thus, the successful vaccination of these animals would prevent human infection. The fact that *Toxoplasma* infection in sheep is a major cause of abortion in these animals adds a commercial incentive for the farmer, which provides further motivation for the development of an animal vaccine. This could take the form of a recombinant subunit vaccine based on heterologous expression of the genes for one or more of *Toxoplasma's* major antigens. Alternatively, the relative ease with which the parasite can be genetically engineered makes possible the creation of a specifically engineered, attenuated mutant that would give rise to a high level of immunity without being pathogenic or transmissible from one host to another.

C. Chemotherapy

Existing drugs are far from benign or are limited in their efficacy, especially against the chronic bradyzoite stage. The recent explosion in our understanding of the parasite's metabolism and cell biology, combined with concomitant improvements in the area of drug design, bode extremely well for the development of safer and more effective drugs. The fact that such drugs might also work on related Apicomplexan parasites such as *Eimeria* (the cause of chicken coccidiosis) and *Plasmodium* (the cause of human malaria) provides further incentive for efforts in this direction.

See Also the Following Articles

Cellular Immunity • Intestinal Protozoan Infections in Humans • Plasmodium • Zoonoses

Bibliography

Ajioka, J. W., Boothroyd, J. C., Brunk, B. P., Hehl, A., Hillier, L., Manger, I. D., Marra, M., Overton, G. C., Roos, D. S., Wan, K.L., Waterston, R., and Sibley, L. D. (1998). Gene discovery by EST sequencing in *Toxoplasma gondii* reveals sequences restricted to the Apicomplexa. *Genome Res.* **8**, 18–28.

Beaman, M. H., McCabe, R. E., Wong, S.-Y., and Remington,

J. S. (1995). *Toxoplasma gondii. In* "Principles and Practice of Infectious Diseases" (G. L. Mandell *et al.,* eds.), 4th ed., pp. 2455–2475. Churchill Livingston, New York.

Buxton, D. (1993). Toxoplasmosis: The first commercial vaccine. *Parasitol. Today* **9**, 335–337.

Darde, M. L., Bouteille, B., and Pestre-Alexandre, M. (1992). Isoenzyme analysis of 35 *Toxoplasma gondii* isolates and the biological and epidemiological implications. *J. Parasitol.* **78**, 786–794.

Dubey, J. P., Lindsay, D. S., and Speer, C. A. (1998). Structures of *Toxoplasma gondii* tachyzoites, bradyzoites and sporozoites and biology and development of tissue cysts. *Clin. Micro. Rev.* **11**, 267–299.

Dubremetz, J. F. (1995). *Toxoplasma gondii:* Cell biology update. *In* "Molecular Approaches to Parasitology" (J. C. Boothroyd and P. Komuniecki, eds.), pp. 345–358. Wiley-Liss, New York.

Frenkel, J. K., Pfefferkorn, E. R., Smith, D. D., and Fishback, J. L. (1991). Prospective vaccine prepared from a new mutant of *Toxoplasma gondii* for use in cats. *Am. J. Vet. Res.* **52**, 759–763.

Gross, U. (1996). *Toxoplasma gondii.* "Current Topics in Microbiology and Immunology" (U. Gross, ed.), Vol. 219. Springer, Berlin.

Holliman, R. E. (1997). Toxoplasmosis, behaviour and personality. *J. Infection* **35**, 105–110.

Howe, D. K., and Sibley, L. D. (1995). *Toxoplasma gondii* comprises three clonal lineages: Correlation of parasite genotype with human disease. *J. Infect. Dis.* **172**, 1561–1566.

Khan, I. A., Ely, K. H., and Kasper, L. H. (1991). A purified parasite antigen (p30) mediates CD8+ T-cell immunity against fatal *Toxoplasma gondii* infection in mice. *J. Immunol.* **147**, 3501–3506.

Kohler, S., Delwishe, D. F., Denny, P. W., Tilney, L. G., Webster, P., Wilson, R. J. M., Palmer, J. D., and Roos, D. S. (1997). A plastid of probably green algal origin in Apicomplexan parasites. *Science* **275**, 1485–1489.

Levine, N. D. (1983). "Veterinary Protozoology." Iowa State University Press, Ames, IA.

Luft, B., and Remington, J. S. (1992). AIDS commentary: Toxoplasmic encephalitis in AIDS. *Clin. Infect. Dis.* **15**, 211–222.

Pfefferkorn, E. R., and Pfefferkorn, L. C. (1976). *Toxoplasma gondii:* Isolation and preliminary characterization of temperature-sensitive mutants. *Experi. Parasitol.* **39**, 365–376.

Sher, A., Gazzinelli, R. T., Oswald, L. P., Clerici, M., Kullberg, M., Pearce, E. J., Berzofsky, J. A., Mossman, T. R., James, S. L., Morse, H. C., and Shearer, G. M. (1992). Role of T-cell derived cytokines in the down regulation of immune responses in parasitic and retroviral infection. *Immunol. Rev.* **127**, 183–204.

Sibley, L. D., and Boothroyd, J. C. (1992). Virulent strains of *Toxoplasma gondii* are clonal. *Nature* **359**, 82–85.

Sinai, A. P., Webster, P., and Joiner, K. A. (1997). Association of host cell mitochondria and endoplasmic reticulum with the *Toxoplasma gondii* parasitophorous vacuole. *J. Cell Sci.* **110**, 2117–2128.

Suzuki, Y., Orellana, M. A., Schreiber, R. D., and Remington, J. S. (1988). Interferon-gamma: The major mediator of resistance against *Toxoplasma gondii. Science* **24**, 516–518.

Transcriptional Regulation in Prokaryotes

Orna Amster-Choder
The Hebrew University Medical School

I. The Transcription Machinery
II. Template Recognition: Promoters
III. Transcription Initiation
IV. Transcription Elongation
V. Transcription Termination

GLOSSARY

−10 element A consensus sequence centered about 10 bp before the start point of transcription, which is involved in the initial melting of DNA by RNA polymerase.

−35 element A consensus sequence centered about 35 bp before the start point of transcription, which is involved in the initial recognition by RNA polymerase.

promoter A sequence of DNA whose function is to be recognized by RNA polymerase in order to initiate transcription. A typical *E. coli* promoter contains two conserved elements, a −10 element and a −35 element (see above).

RNA polymerase The enzyme that synthesizes RNA using a DNA template (also termed DNA-dependent RNA polymerase).

sigma (σ) factor A subunit of bacterial RNA polymerase needed for initiation. The σ factor has major influence on the selection of promoters.

start point The position on the DNA that corresponds to the first base transcribed into RNA.

terminator A DNA sequence that causes RNA polymerase to terminate transcription and to dissociate from the DNA template.

transcription The synthesis of RNA on a DNA template.

transcription unit The DNA sequence that extends from the promoter to the terminator; it may include more than one gene.

THE FIRST STEP IN GENE EXPRESSION is the transcription of the coding DNA sequences to discrete RNA molecules. Specific DNA regions, defined as promoters, are recognized by the transcribing enzyme, a DNA-dependent RNA polymerase. The RNA polymerase binds to the promoter and initiates the synthesis of the RNA transcript. The enzyme catalyzes the sequential addition of ribonucleotides to the growing RNA chain in a template-dependent manner until it comes to a termination signal (terminator). The DNA sequence between the start point and the termination point defines a transcription unit. An RNA transcript can include one gene or more. Its sequence is identical to one strand of the DNA, the coding strand, and complementary to the other, which provides the template. The base at the start point is defined as +1 and the one before that as −1. Positive numbers are increased going downstream (into the transcribed region), whereas negative numbers increase going upstream. The immediate product of transcription, termed primary transcript, which extends from the promoter to the terminator, is almost always unstable. In prokaryotes, messenger RNA (mRNA) is usually translated concomitantly with being transcribed, and is rapidly degraded when not protected by the ribosomes, whereas ribosomal RNA (rRNA) and transfer RNA (tRNA) are cleaved to give mature products.

Transcription is the principal step at which gene expression is controlled. DNA signals and regulatory proteins determine whether the polymerase will choose to transcribe a certain gene and whether the whole process of transcription will be accomplished successfully. The timing of transcription of specific genes is influenced by environmental conditions and by the growth cycle phase.

The molecular picture of how genes are transcribed and the nature of the regulatory mechanisms that control transcription are far from being complete, but lots of progress has been made. Work on relatively simple organisms, bacteriophages and bacteria, has provided new insights into the mechanisms that are involved in the regulation of gene expression, transcriptional-control mechanisms being among them. Although there are significant differences in the organization of individual genes and in the details and the complexity of the regulatory mechanisms among prokaryotes and eukaryotes, it is clear that basic principles are shared among all organisms. Due to the relative simplicity of prokaryotic biochemical pathways, their easy manipulation in the laboratory, and the advanced tools that are available for changing their genotype and for testing the resulting phenotype, it is easier to infer these basic principles by studying prokaryotes.

I. THE TRANSCRIPTION MACHINERY

A. RNA Polymerase Catalyzes Transcription

RNA synthesis is catalyzed by the enzyme RNA polymerase. The reaction involves the incorporation of ribonucleoside 5′ triphosphate precursors into an oligoribonucleotide transcript, based on their complementarity to bases on a DNA template. RNA polymerase catalyses the formation of phosphodiester bonds between the ribonucleotides. The formation of a phosphodiester bond involves a hydrophilic attack by the 3′-OH group of the last ribonucleotide in the chain on the 5′ triphosphate of the incoming ribonucleotide. The incoming ribonucleotide loses its terminal two phosphate groups, which are released in the form of pyrophosphate. In this manner, the RNA chain is synthesized from the 5′-end toward the 3′-end.

The best characterized RNA polymerases are those of eubacteria, for which *E. coli* is a prototype. Unlike eukaryotic cells, in which various types of polymerases are dedicated to the synthesis of the certain types of RNA, in eubacteria a single type of polymerase appears to be responsible for the synthesis of mRNA, rRNA, and tRNA.

The dimensions of bacterial RNA polymerase are approximately 90 × 95 × 160 Å. The molecular weight of the complete *Escherichia coli* enzyme is approximately 465 kDa. About 7000 molecules of RNA polymerase are present in an *E. coli* cell, but the number of molecules engaged in transcription at any given time varies from 2000 to 5000, depending on the growth conditions.

The DNA sequence that is being transcribed by RNA polymerase is transiently separated into its single strands, with one of the strands serving as a template for the synthesis of the RNA strand. This region is therefore defined as the *transcription bubble*. As the RNA polymerase moves along the DNA, it unwinds the duplex at the front of the bubble and rewinds the DNA at the back, so that the duplex behind the transcription bubble reforms. Thus, the bubble moves with the RNA polymerase, and the RNA chain is elongated. The length of the transcription bubble varies from 12–20 bp. The length of the transient hybrid between the DNA and the newly synthesized RNA sequence within the transcription bubble is a matter of controversy and the estimates range from 2–12 nt. Beyond the growing point, the newly synthesized RNA chain enters a high-affinity binding site within the RNA polymerase.

B. Bacterial RNA Polymerase Consists of Multiple Subunits

The RNA polymerases of certain phages consist of single polypeptide chains. These polymerases recognize a very limited number of promoters and they lack the ability to change the set of promoters from which they initiate transcription. In contrast, bacterial RNA polymerases consist of several subunits. The most studied bacterial RNA polymerase, the one in *E. coli*, exists in two forms, an holoenzyme and a core enzyme. The holoenzyme is capable of selective initiation at promoter regions, whereas the core RNA polymerase is capable of elongation and termination, but not selective initiation. The two forms of the polymerase consist of two identical α-subunits, one β-subunit and one β'-subunit, but only the holoenzyme contains an additional subunit, one of several σ proteins. The subunit composition of the holoenzyme is summarized in Fig. 1. The α-subunit plays an important role in RNA polymerase assembly, which proceeds in the pathway $\alpha \rightarrow \alpha_2 \rightarrow \alpha_2\beta\beta' \rightarrow \alpha_2\beta\beta'\sigma$. The α-subunit plays some role in promoter recognition and in the interaction of RNA polymerase with

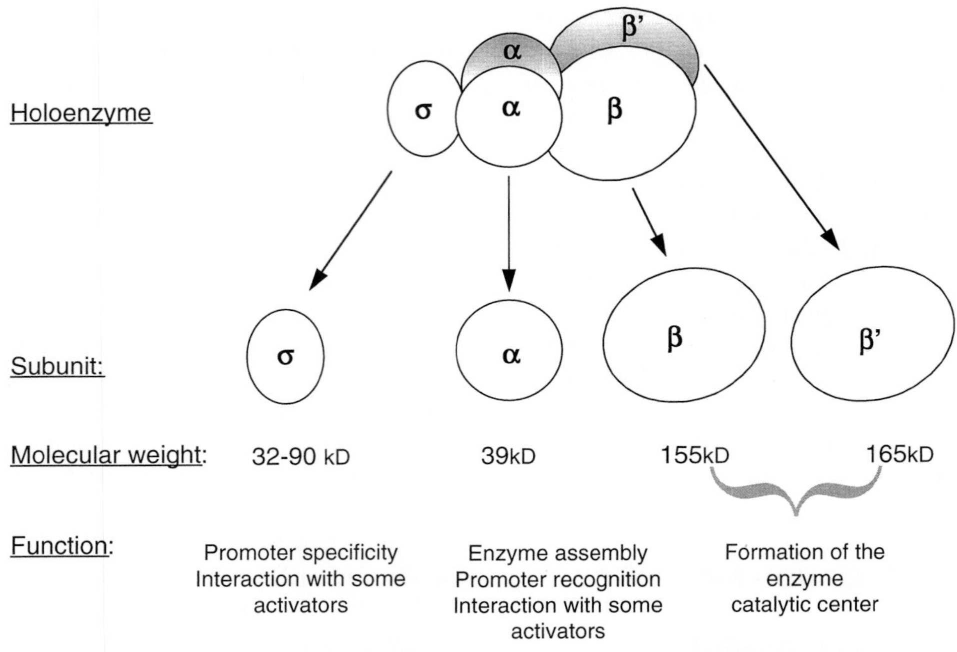

Fig. 1. *E. coli* RNA polymerase holoenzyme consists of four types of subunits.

transcriptional activators (see later). The β- and β'-subunits together constitute the catalytic center of RNA polymerase. The β-subunit was demonstrated to contact the template DNA, the newly synthesized RNA, and the substrate ribonucleotides. The β'-subunits contacts the RNA chain as well. Mutations in the genes that encode β and β' show that both subunits are involved in all stages of transcription. The sequences of β and β' show homology to the sequences of the largest subunits of eukaryotic RNA polymerases. This conservation through the evolution hints that the mechanisms by which all RNA polymerases catalyze transcription share common features. The assignment of individual functions to the different subunits of the core polymerase are only a rough estimation because most likely each subunit contributes to the activity of the enzyme as a whole.

The σ factor is involved only in transcriptional initiation. Its function is to ensure that the polymerase binds in a stable manner to DNA only at promoters, not at other sites. There are several types of σ factors in the bacterial cell (see later). The major σ factor in *E. coli*, which is required for most transcription reactions, is σ^{70}. Although σ^{70} has domains that recognize the promoter DNA sequence, as an independent protein it does not bind to DNA, perhaps because its DNA-binding domain is sequestered by another domain of the σ^{70} molecule (when σ^{70} is shortened from its N-terminus, it is able to bind to DNA, suggesting that the N-terminal region has an inhibitory effect on the ability of σ^{70} to bind to DNA). However, upon binding to the core enzyme and the formation of the holoenzyme complex, σ^{70} undergoes a conformational change and it now contacts the region upstream of the start point. The dogma is that the σ factor discharges from the core enzyme when abortive initiation is concluded and RNA synthesis is successfully initiated (although findings suggest that, at least in some cases, σ can remain associated with the polymerase at postinitiation steps). The released σ factor becomes immediately available for use by another core enzyme. *E. coli* cells contain about 3000 molecules of σ^{70}, enough to bind about one-third of the intracellular core RNA polymerase.

C. The Ability of RNA Polymerase to Selectively Initiate Transcription Is Dependent on the Presence of σ Factor

Bacterial polymerases have to recognize a wide range of promoters and to transcribe different genes on different occasions. Specificity of gene expression is in part modulated by substituting one species of σ for another, each specific for a different class of promoters. In *B. subtilis*, σ factors are implicated in the temporal regulation of sporulation. In *E. coli*, alternative σ factors are used to respond to general environmental changes. The σ factors are named either by their molecular weight (e.g., σ^{70}), or after their genes, which are usually termed *rpo* (e.g., RpoD for σ^{70}).

When cells are shifted from low to high temperature, the synthesis of a small number of proteins, the heat-shock proteins, transiently increases. The σ^{32} protein (RpoH) is responsible for the transcription of the heat-shock genes. The basic signal that induces the production of σ^{32} is the accumulation of unfolded proteins that results from the increase in temperature. The heat-shock proteins play a role in protecting the cell against environmental stress. Several of them act like chaperones, preventing the unfolding (denaturation) of proteins. The heat-shock proteins are synthesized in response to other conditions of stress, implying that the production of σ^{32} is induced by conditions other than elevation in temperature. The σ^{32} protein is unstable and it is rapidly degraded when it is not needed. Another σ factor, σ^{E}, appears to respond to more extreme temperature shifts that lead to the accumulation of unfolded proteins, which are usually found in the periplasmic space or outer membrane. Less is known about this σ factor and about the genes it controls. When *E. coli* and other enteric bacteria are nitrogen limited, the synthesis of a number of proteins is dramatically induced. The increased production of these proteins enlarges the capacity of cells to produce nitrogen-containing compounds, and to use nitrogen sources other than ammonia. The transcription of the genes that encode these proteins is dependent on σ^{54} (also known as σ^{N}). The expression of the flagellar genes under normal conditions depends on σ^{28} (also known as σ^{F}).

When *E. coli* cells are starved, they shift from the exponential-growth phase to the stationary phase. The ability of the cells to cope with starvation depends on the production of many proteins. This is enabled due to the synthesis of σ^{S} (RpoS), which transcribe the relevant genes. Unlike *E. coli*, when *Bacillus subtilis* cells are starved, they can form spores. Sporulation involves the differentiation of a vegetative bacterium into a mother cell and a forespore; the mother cell lyses and a spore is released. The process of sporulation involves a drastic change in the biosynthetic activities of the bacterium, in which many genes are involved. This complex process is concerted at the level of transcription. The principle is that in each compartment (the mother cell and the forespore) the existing σ factor is successively displaced by a new σ factor that causes the transcription of a new set of genes. Communication between the compartments occurs in order to coordinate the timing of the changes in the mother cell and the forespore.

The promoters identified by the various σ factors are organized similarly, having their important elements centered around 10 and 35 nt upstream from the start point (see later), except for σ^{54}, whose promoters have slightly different characteristics. However, the ability of different σ factors to cause RNA polymerase to initiate at different sets of promoters stems from the fact that each type of σ factor recognizes promoter elements with unique sequences. σ^{54} differs from the other σ factors also in its ability to bind to DNA independently, and by the influence of sites that are distant from the promoter on its activity. In many aspects, σ^{54} resembles eukaryotic regulators.

A comparison of the sequences of the different σ factors identifies four regions that have been conserved. Several sequences in these regions were identified with individual functions, such as interaction with core RNA polymerase or contacting the various promoter elements.

II. TEMPLATE RECOGNITION: PROMOTERS

A. Promoter Recognition Depends on Conserved Elements

Template recognition begins with the binding of RNA polymerase to the promoter. Promoter is a sequence of DNA whose function is to be recognized by RNA polymerase in order to initiate transcription. The information for promoter function is provided directly by the DNA sequence, unlike expressed regions, which require that the information be transferred into RNA or protein to exercise their function. Two main approaches were used to identify the DNA features that characterize a promoter. The first is comparative sequence analysis and the second is the identification of mutations that alter the recognition of promoters by RNA polymerase. The sequence of more than 100 *E. coli* promoters recognized by the major RNA polymerase species, σ^{70} holoenzyme, has been determined. Their comparison revealed an overall lack of extensive conservation of sequence over the 60 bp associated with RNA polymerase. Nevertheless, statistical analysis revealed some commonalities. A typical *E. coli* promoter, which is recognized by σ^{70}, contains four conserved features: the start point, the −10 region, the −35 region, and the distance between the −10 and −35 regions. The start point is usually a purine. It is often the central base in the sequence CAT, but this is not a mandatory rule. The *−10 region* is a hexanucleotide that centers approximately 10 bp before the start point, although this distance is somewhat variable. Its consensus sequence is TATAAT (in the antisense strand). The conservation is T_{80} A_{95} T_{45} A_{60} A_{50} T_{96}, where the numbers refer to the percent occurrence of the most frequently found base at each position. The *−35 region* is a hexanucleotide sequence that centers approximately 35 bp upstream of the start point. Its consensus is TTGACA and the conservation is T_{82} T_{84} G_{78} A_{65} C_{54} A_{45}. The favored spacing between the −10 and the −35 sequences is 17 bp.

For most promoters, there is a good correlation between promoter strength and the degree to which the −10 and −35 elements agree with the consensus sequences. The significance of the conserved promoter features was further emphasized by the finding that most mutations that alter promoter activity (i.e., affect the level of expression of the gene(s) under the control of this promoter) change the sequence of the particular promoter in an expected fashion. Mutations that increase the similarity to the proposed conserved −10 and −35 sequences or bring the spacing between them closer to 17 bp, usually enhance the promoter activity (*up mutations*); mutations that decrease the similarity to the conserved sequences or bring the spacing between them more distant from 17 bp, usually reduce the promoter activity (*down mutations*). The nature of down mutations in the −35 and −10 regions of various promoters led to the conclusion that the −35 region is implicated in the recognition of the promoter by RNA polymerase and the formation of a closed transcription complex, whereas the −10 region is implicated in the shift of the closed complex to the open form (see later). The fact that the −10 region is composed of AT base pairs, which require low energy for melting, makes its suitable to assist in unwinding and, thus, in converting the transcription complex to its open form.

There are several exceptions to the proposed generalized pattern. For example, some promoters lack one of the conserved sequences, the −10 or the −35 region, without a corresponding effect on promoter activity. In some cases, it was proposed that another sequence compensates for the lack of a consensus sequence. In still other cases, it was concluded that the promoter cannot be recognized by RNA polymerase alone, and the involvement of additional proteins, which overcome the deficiency in intrinsic interaction between RNA polymerase and the promoter, is required. Other exceptional promoters were discovered due to the isolation of promoter mutations that do not affect any of the conserved promoter features described so far. Rather, they are the outcome of base substitutions in other sequences in the vicinity of the conserved sequences or the start point. One explanation for these findings is that the analysis that generated the sequence characteristics of a typical promoter may have missed some sites that contribute to the transcription initiation process, possibly because it included too many promoters, both weak and strong. Alternatively, other base pairs in the promoter could be recognized by RNA polymerase, but might become significant only if canonical recognition sites are absent.

The isolation of deletions that progressively approach specific promoters from the upstream region demonstrated the involvement of specific upstream sites in the recognition of RNA polymerase in some cases. This led to the discovery that some *E. coli* promoters contain a third important element in addition to the −10 and −35 sequences. This element was named the upstream element, or *UP element,* because it is located approximately 20 bp upstream of the −35 region. Its sequence is AT-rich and it was first identified in the strong promoters of the *rrn* genes, which encode ribosomal RNA. It is believed now that promoter strength is a function of all three elements, −10, −35, and UP, with very strong promoters, such as the *rrn* promoters, having all three elements with near-consensus sequences and with weaker promoters having one, two, or three nonconsensus promoter elements. It has been found that, whereas σ^{70} is responsible for the recognition of the −10 and −35 regions, the UP element interacts with the α-subunit of RNA polymerase.

Finally, the activity of some promoters is affected by sequences downstream to the −10 region, or even downstream to the transcription start point. The sequences immediately around the start point seem to influence the initiation event. The effect of the initial transcribed region (from +1 to +30) on promoter strength is explained by the influence of this region on the rate at which RNA polymerase clears the promoter.

Unlike the promoters described so far, which are recognized by holoenzymes that contain σ^{70} or a close homolog, a minority of the cellular holoenzymes use σ^{54} and have different basal elements located at −12 and −24. Transcription initiation from these promoters relies also on enhancer-like elements that are remote (upstream) from the promoters (see later).

B. Possible Mechanisms for Promoter Recognition

How does RNA polymerase find the promoter sequences? How does it identify a stretch of ~60 bp which defines a promoter in the context of 4×10^6 bp that make up the *E. coli* genome? Three models were suggested to explain the ability of RNA polymerase to find promoters. The first model assumes that RNA polymerase moves in the cell by random diffusion. It associates and dissociates from loose binding sites on the DNA until, by chance, it encounters a promoter sequence that allows tight binding to occur. According to this model, movement of an RNA polymerase molecule from one site on the DNA to another is limited by the speed of diffusion through the medium. However, this parameter might be too low to account for the rate at which RNA polymerase finds promoters. The second model addresses this problem by assuming that once an RNA polymerase molecule binds to a DNA sequence, the bound sequence is directly displaced by another sequence; the enzyme exchanges sequences very rapidly, until it binds to a promoter that allows an open complex to form and transcription initiation to occur. According to this model, the association of the polymerase with DNA sequences and their dissociation are essentially simultaneous. Thus, the time spent on site exchange is minimal and the search process is much faster in comparison to the speed calculated based on the first model. This model fits the accepted notion that core polymerases that are not busy in transcription are stored by binding to loose sites on the DNA. The third model assumes that RNA polymerase binds to a random site on the DNA and starts sliding along the DNA molecule until it encounters a promoter. The actual mechanism by which RNA polymerase finds promoters might combine features of the various models.

III. TRANSCRIPTION INITIATION

A. Stages of the Transcription-Initiation Process

Transcription initiation is the phase during which the first nucleotides in the RNA chain are synthesized. It is a multistep process that starts when the RNA polymerase holoenzyme binds to the DNA template and ends when the core polymerase escapes from the promoter after the synthesis of approximately the first nine nucleotides.

The stages of the transcription initiation process, which are summarized in Fig. 2, can be described in terms of the types of interaction between the RNA polymerase and the nucleic acids that are involved. The first stage in transcription initiation is the formation of a complex between the holoenzyme and the DNA sequence at the promoter, which is in the form of a double-stranded DNA. This complex is termed a closed binary complex or *closed complex*. The second stage is the unwinding of a short region of DNA within the sequence that is bound to the RNA polymerase. The complex between the polymerase and the partially melted DNA is termed an open binary

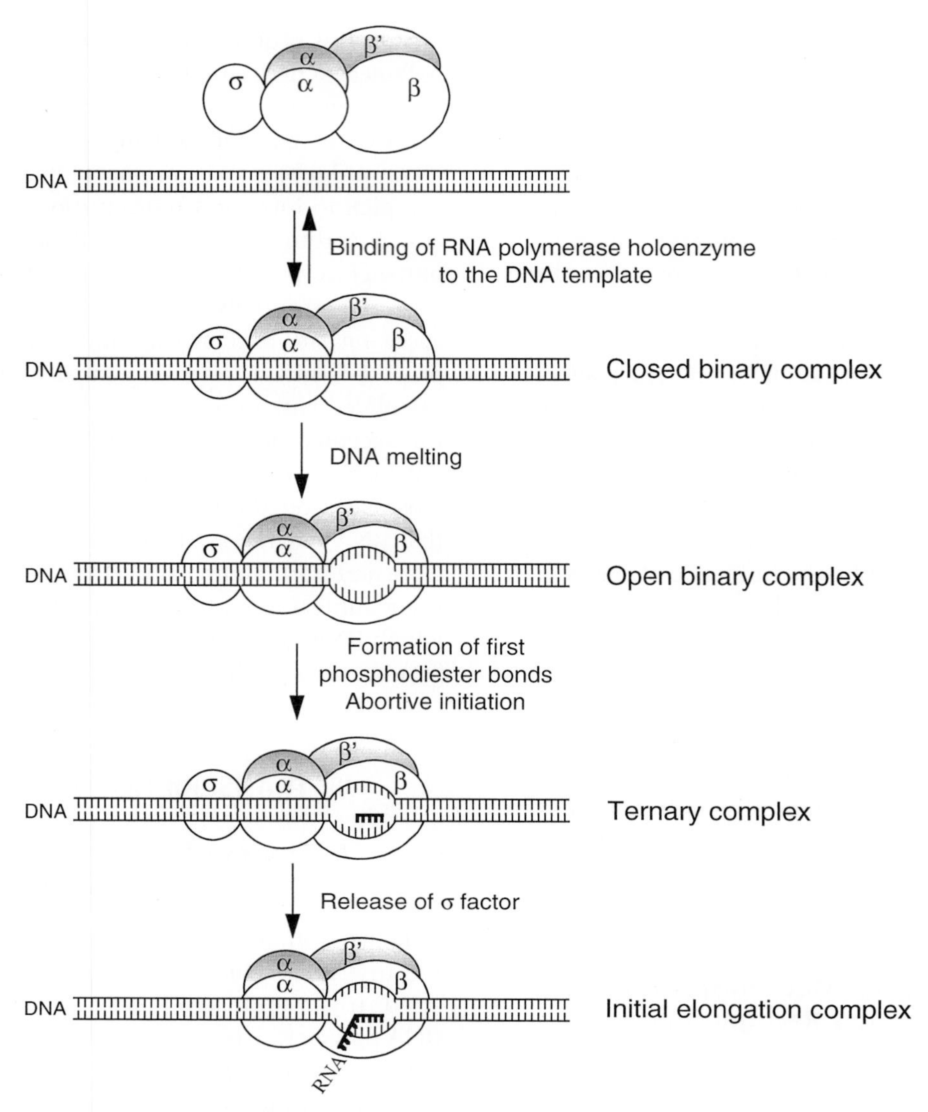

Fig. 2. Stages in the transcription initiation process.

complex or *open complex*. The conversion of the closed complex into the open complex leads to the establishment of tight binding between the RNA polymerase and the promoter sequence. For strong promoters, the conversion into an open complex is irreversible. The third stage is the incorporation of two ribonucleotides, and the formation of a phosphodiester bond between them. Because the complex at that stage contains an RNA as well as DNA, it is called an initiation ternary complex. Up to seven additional ribonucleotides can be added to the RNA chain without any movement of the polymerase. After the addition of each base, there is a certain probability that the enzyme will release the short (up to nine bases long) RNA chain. Such an unsuccessful initiation event is termed an abortive initiation. Following an abortive initiation, RNA polymerase begins again to incorporate the first base. Several rounds of abortive initiations usually occur and the result is the formation of short RNA chains that are 2–7 bases long. When initiation succeeds, that is, a nine-base-long RNA chain is formed and is not released, the last stage in transcription initiation occurs. At that stage the σ factor is released from the polymerase.

As a consequence, a complex containing core polymerase, DNA, and RNA is formed. This complex is called an elongation ternary complex. The departure of the polymerase from the promoter to resume elongation is termed promoter escape or promoter clearance.

The activities of genes are frequently regulated at the initiation step of transcription. Therefore, the initiation of transcription is a very precise event that is tightly controlled by a variety of regulatory mechanisms.

B. Repression of Transcription Initiation

Gene expression is sometimes negatively regulated by a repressor protein that, when bound to DNA, inhibits transcription initiation. The ability of the repressor to bind to DNA is in turn modulated by the binding of an effector molecule to the repressor. The regulation of the *lac* operon expression in *E. coli* is a paradigm for this type of transcriptional control. LacI is a repressor protein that blocks the initiation of transcription from the promoter of the *lac* operon. LacI binds to a site on the DNA, termed an *operator,* that overlaps with the promoter. Because of this overlap, the binding of the LacI repressor and of RNA polymerase are competitive events. That is, RNA polymerase cannot bind to the promoter until the repressor is removed from the operator. The binding of one of several β-galactoside compounds to the repressor destablizes the repressor–operator complex and allows RNA polymerase to bind to the promoter to initiate transcription. Interestingly, *lac* has two additional binding sites for LacI, an upstream site and a downstream site located in the first gene of the operon. Compared to the operator, the additional sites have a lower affinity for the repressor protein and it was suggested that they do not directly participate in the inhibition of transcription initiation. Rather, the secondary binding sites seem to stabilize the repressor–operator complex.

There are important exceptions to the *lac* repression paradigm. An example is the repression of the *gal*-operon transcription initiation by a different and less understood mechanism. The *gal* operon contains two repressor binding sites, both required for maximum efficacy of the *gal* repressor, yet neither of these operators overlap with the promoter sequences. Thus, the binding of the *gal* repressor to its operators does not seem to compete directly with binding of RNA polymerase to the *gal* promoter. It was suggested that the *gal* repressor, when bound to both binding sites, holds the DNA in a conformation that is unfavorable for binding to RNA polymerase.

C. Activation of Transcription Initiation

The frequency of transcription initiation from many promotrs is enhanced by activator proteins. In most cases, these proteins bind within or upstream from the promoter and seem to act by making a direct contact with RNA polymerase. The activators that interact with the most abundant form of RNA polymerase involved in transcription initiation in *E. coli,* the σ^{70} holoenzyme, can be roughly divided into two groups, those that interact with the σ-subunit of RNA polymerase and those that interact with the σ^{70} subunit.

The best characterized activator from the first group is the catabolite gene-activator protein (CAP). The CAP target site on the DNA was determined in a number of systems. Comparative sequence analyses led to the definition of a consensus sequence for CAP binding. CAP-binding sites are found at various location relative to the transcription start point in different systems. The most studied case is the activation of transcription initiation from the *lac* promoter by CAP. The regions that are required for the activation on both CAP and the α-subunit of RNA polymerase were defined. CAP acts as a dimer, and although the activating region is present in both subunits of the CAP dimer, transcription activation at the *lac* promoter requires only the activating region of the promoter-proximal subunit. CAP interacts with the C-terminal domain of the α-subunit (αCTD). The αCTD constitutes an independently folded domain, which is connected to the remainder of α by a flexible linker. This allows αCTD to make different interactions in different promoters. The simplest model for transcription activation by CAP is that CAP binds to the DNA and recruits the αCTD, and thus the RNA polymerase holoenzyme, to the promoter.

The best characterized activator from the second group is the cI protein of bacteriophage λ (λcI). λcI binds to a site on the DNA that overlaps the -35 element of the λP_{RM} promoter. The activating region in λcI was defined and was demonstrated to directly contact a specific region in σ^{70}.

The existence of at least two groups of activators that bind to separate targets on the DNA and to different components of RNA polymerase raised the possibility that, at some promoters, RNA polymerase might be contacted simultaneously by two or more activators. This was demonstrated in an artificial system that was engineered to contain both CAP and cI binding sites upstream from the λP_{RM} promoter. Both activators were shown to make contact with RNA polymerase, most likely with CAP contacting αCTD and with λcI contacting σ^{70}, and their effect on transcription activation was synergistic. Transcription activation by two activators was demonstrated also in natural promoters. One conclusion from studies with various types of activators is that many activators seem to function by helping recruit DNA-binding domains of RNA polymerase to DNA, thus supplementing suboptimal RNA polymerase–DNA interactions with protein–RNA polymerase interactions.

In contrast to the copious σ^{70} promoters, the rare σ^{54} promoters, which contain -12 and -24 basal elements instead of the well-known -10 and -35 elements, seem to be regulated solely by activation rather than by repression. σ^{54} activators (the most studied is NtrC) bind to enhancer-like sites on the DNA; that is, the sites are remote from the promoters (upstream) and their precise location is not critical for transcriptional activation. In fact, these sites can be moved kilobases in *cis* and retain their residual function. Thus, unlike σ^{70} activators that bind to sites that enable direct communication with RNA polymerase, σ^{54} activators, once bound to their DNA target sites, cannot touch the polymerase without looping out the intervening DNA. This seems to be the reason why σ^{54} promoters frequently require the help of integration host factor (IHF), which enhances the bending of the DNA, as a cofactor. It is accepted that σ^{54} polymerase can bind to its promoters to form a closed complex. However, this polymerase cannot transcribe because it cannot melt the DNA. Once the upstream activator binds to its target site upstream of the promoter, it loops out of the sequence between its binding site and the promoter and touches the complex. This interaction triggers the melting of DNA (with the help of a helicase activity) and the creation of a transcription bubble. Thus, σ^{54} activators catalyze the conversion of the polymerase–promoter complex from a closed state to a transcription-ready open state, rather then tethering the RNA polymerase to the promoter.

D. Regulation of Transcription Initiation via Changes in DNA Topology

The template for transcription is a negatively supercoiled DNA. Because the formation of an open transcription complex requires DNA melting, and because the degree of superhelicity affects the energy needed for the melting, it was anticipated that the superhelical character of a template would affect the properties of this template. Indeed, the efficiency of some promoters is influenced by the degree of supercoiling. Most of these promoters are stimulated by negative supercoiling, although few are inhibited. The effects of superhelicity on the process of transcription initiation have been shown *in vitro* in numerous studies and *in vivo* by the use of inhibitors of gyrase, which introduces negative supercoils. The reason why some promoters are sensitive to the degree of supercoiling, whereas others are not, might have to do with the fact that the sequence of some promoters is easier to melt and is therefore less dependent on supercoiling. Alternatively, because various regions on the bacterial chromosome are believed to have different degrees of supercoiling, the location of the promoter might determine whether it is sensitive to changes in superhelicity.

IV. TRANSCRIPTION ELONGATION

The initiation phase ends when RNA polymerase succeeds in extending the RNA chain beyond the first nine nucleotides and escapes from the promoter. At that stage the elongation process begins and the enzyme starts moving along the DNA, extending the growing RNA chain. During the transition from initiation to elongation, the size and shape of the RNA polymerase undergoes successive changes. The first change is the loss of the σ factor. Whereas the holoenzyme covers approximately 75 bp (from -55 to $+20$), after the loss of σ, the polymerase covers approximately 55 bp (from -35 to $+20$). At that stage the polymerase is displaced from the promoter (promoter clearance or escape) and undergoes a further transition to form the elongation complex, which covers only 35–40 bp, depending on the stage during elongation. The polymerase now becomes tightly bound to both the nascent transcript and the DNA template, making it very stable.

The average rate of transcript elogation by the various RNA polymerases is 40 nt/s. However, this rate varies dramatically among RNA polymerase and loosely correlates with the subunit complexity of the enzyme. Thus, the simple single-subunit bacteriophage RNA polymerases are the most rapid of all DNA-dependent RNA polymerases (several hundred nt/s), bacterial RNA polymerases transcribe at an intermediate rate (50–100 nt/s), and eukaryotic polymerases, although diversified, appear to be the slowest (20–30 nt/s).

RNA polymerases are not as accurate as DNA polymerases. The reason for the difference in fidelity between RNA and DNA polymerases might be that RNA polymerases do not have the robust proofreading mechanism that characterizes DNA polymerases. Of course, the fidelity of DNA replication is of greater importance than that of transcription because unlike replication errors, misincorporation during transcription does not result in permanent and inherited genetic changes.

A. Blocks to Transcription Elongation

Transcript elongation does not occur at a constant rate. Throughout the elongation phase, RNA polymerase can be paused, arrested, or terminated. During a *transcriptional pause,* the polymerase temporarily stops RNA synthesis for a certain amount of time, after which it can resume the elongation process. Thus, pausing can be described as transcriptional hesitation. In contrast, during a *transcriptional arrest* the polymerase stops RNA synthesis and cannot resume it without the aid of accessory proteins. Throughout both pauses and arrests, RNA polymerase remains stably bound to the DNA template and to the nascent transcript. These features distinguish paused and arrested polymerases from the those that have terminated and thus detached from the DNA. Pausing and termination are sometimes related because pausing is a prerequisite for termination, at least in the case of ρ-dependent termination (transcription termination is discussed in Section V). However, not all pauses are termination precursors. The time it takes a stalled polymerase to resume elongation varies among pause sites from very short periods of time, which cannot be accurately measured, to several minutes. The fraction of the polymerases that respond to an elongation block is also variable because ternary complexes differ in their ability to recognize pausing signals.

Transcriptional pause and arrest signals can be intrinsic; that is, sequences in the nascent transcript or in the DNA template whose interaction with RNA polymerase can inhibit the progression of the ternary complex, such as RNA regions, which have the propensity to form a stable secondary structure. In addition, extrinsic factors may obstruct the progress of RNA polymerase during transript elongation. There are numerous examples of RNA polymerases from various organisms being physically blocked by DNA-binding proteins during RNA synthesis in natural and artificial systems. An example is the purine repressor, which binds well downstream from the *purB* operon transcriptional start point and blocks the polymerase during elongation (it should be noted that in many cases RNA polymerase is able to transcribe beyond the DNA-binding proteins in its path by either displacing them or bypassing them). In addition to DNA-binding proteins, which are the most obvious obstacles for RNA polymerase, factors that perturb the structure of DNA can also inhibit the progression of RNA polymerase and thus interfere with transcript elongation, for instance, extreme positive or negative supercoiling, unusual DNA structures such as Z-DNA, and DNA lesions. The efficiency of such potential impediments to block RNA polymerase from elongating depends on various local factors and on the type of the RNA polymerase. For example, T7 RNA polymerase can efficiently bypass gaps in the DNA template strand that are 1–5 nt, and less efficiently gaps as large as 24 nt.

Transcriptional pausing is involved in various regulatory mechanisms. An example is the attenuation of amino acid biosynthetic operons, which depend on a transcriptional pause that leads to the precise coordination of the position of RNA polymerase with the ribosome translating the nascent RNA. Although transcriptional arrest has been characterized *in vitro* and the evidence for its occurrence in the cell is only circumstantial, it is also believed to be implicated in the regulation of many genes. These predictions are based on the recognition that if an arrest occurs within the coding region of a gene, the arrested complex would block subsequently initiated RNA polymerases, thereby effectively repressing RNA synthesis from the affected gene.

B. Transcript Cleavage during Elongation

RNA polymerases in ternary complexes are able to endonucleolytically cleave the nascent transcript near the 3′-end. The resulting 5′-fragment remains stably bound to the RNA polymerase, whereas the 3′-fragment is released from the enzyme. Transcript cleavage does not result in transcript termination, but rather in the creation of a new 3′-terminus for chain elongation. Following the cleavage, RNA polymerase can correctly resynthesize the discarded RNA segment and continue with the elongation. The size of the released 3′-fragment varies. There are reports of cleavage products ranging from 1–17 nt. The cleavage rate also varies, depending on the particular ternary complex. The mechanisms that cause the variations in both cleavage-product size and cleavage rate are poorly understood. It is likely that the conformational changes that occur within the ternary complex as it moves along the DNA template play a crucial role in determining the rate of the cleavage reaction and the size of the cleaved fragment.

The actual cleavage is carried out by the catalytic site of RNA polymerase, which catalyzes polymerization. However, the cleavage reaction seems to involve accessory proteins in addition to RNA polymerase. In *E. coli,* the GreA and GreB proteins stimulate the polymerase to cleave and release the 3′-fragment and to resume transcription elongation. Thus, GreA and GreB can be defined as cleavage-stimulatory factors. GreA and GreB also affect the size of the released 3′-fragment. Other factors, such as NusA, can regulate the cleavage properties induced by GreA and GreB.

The physiological role of transcript cleavage reaction catalyzed by RNA polymerase has not been determined. However, one accepted hypothesis is that transcript cleavage serves to rescue RNA polymerases that are arrested during elongation. It has been shown that arrested RNA polymerase complexes can be reactivated by transcript cleavage *in vitro.* The cleavage-stimulatory factors, GreA and GreB, were shown to release the RNA polymerase from the elongation arrest. GreA can suppress the formation of arrested complexes *in vitro;* it can act only if it is present before the polymerase arrests. GreB can act after the polymerase has stopped; its action is triggered by the appearance of an arrested complex. It has been suggested that GreA and GreB stimulate promoter escape at some promoters *in vivo.*

Another role that has been presumed for transcript cleavage is to increase the fidelity of transcription by facilitating the removal of misincorporated nucleotides. To accomplish transcript cleavage and elongation resumption, the RNA polymerase, after it has stalled, is assumed to either move backwards along the DNA template or undergo structural changes so that the catalytic site is repositioned to the new 3′-end to which bases should be added. The ability of the polymerase to withdraw, to cleave, and discard the 3′-end of the nascent transcript; to replace the removed sequence with a newly synthesized one; and to continue with the elongation in a template-directed manner is what makes this process so suitable for proofreading.

C. The Inchworm Model for Transcription Elongation

The old-fashion view of the elongation process as a smooth forward motion, during which RNA polymerase moves 1 bp along the DNA template for every base added to the newly synthesized RNA chain, might still hold for some regions of DNA.

However, evidence has accumulated for a different type of movement of the polymerase during elongation. Thus, a new model for RNA polymerase translocation has evolved: the *inchworm model*. This model, which is illustrated in Fig. 3, describes the movement of RNA polymerase on the DNA template in a discontinuous inchworm-like, fashion. The model predicts that the process of RNA chain elongation is a cyclic process that consists of discrete translocation cycles. Each cycle involves the steady compression of the RNA polymerase on the DNA template followed by a sudden expansion. According to this model, the upstream (back) boundary of the enzyme moves steadily during elongation, as the RNA chain is extended. However, the downstream (front) boundary of the enzyme does not move while several nucleotides are added; it then "jumps", that it, it moves 7–8 bp along the DNA. At the beginning of each cycle, RNA polymerase stretches across ~35 nt of template DNA; it gradually compresses from the back end till it covers only ~28 nt; it then releases from the front end and stretches back to cover ~35 nt. As the RNA polymerase compresses, the nascent RNA chain the complex becomes longer and the single-stranded transcription bubble enlarges as well. The internal tension that these changes probably create in the enzyme is released when the front end expands discontinuously.

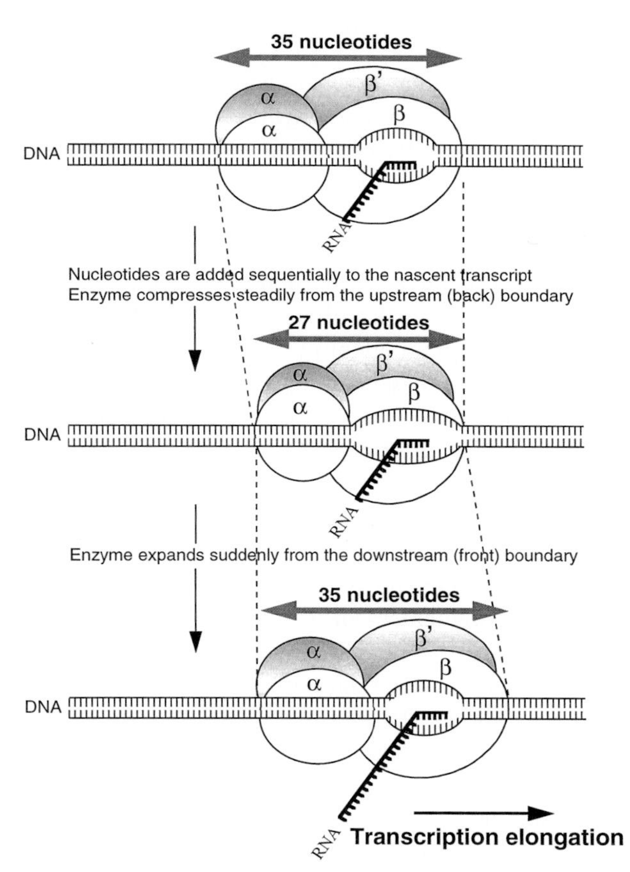

Fig. 3. Discontinuous inchworm-like movement of RNA polymerase during transcription elongation.

The inchworm model for transcription elongation postulates that RNA polymerase binds to the DNA template at two separate sites, one downstream in the direction of transcription and one upstream, that can move independently of each other. This permits the polymerase to move in an inchworm-like manner, so one DNA binding site on the polymerase remains fixed to the DNA, whereas the other moves along the DNA. There is now both direct and indirect evidence that validate this assumption. The downstream DNA binding site in the *E. coli* polymerase was found to be double-strand-specific, whereas the upstream is single-strand-specific and interacts with the template strand. The model also assumes that the catalytic site of RNA polymerase is linked to the movement of the upstream DNA site, but can move independently of the downstream DNA-binding site. The inchworm model also makes predictions about the existence of more than one RNA-binding site on the ternary complex. It has been shown that nascent RNAs interact with at least three sites on the *E. coli* polymerase, two on the β-subunit and one on the β'-subunit. It is believed that together the DNA and RNA sites account for the remarkable stability and flexibility of ternary complexes. However, the precise size, placement, and strand specificity of these nucleic acid-binding sites are currently being elucidated.

D. Transcriptional Slippage

RNA polymerase usually synthesizes RNA transcripts that are precisely complementary to the DNA template. However, in rare circumstances, RNA polymerase can undergo transcriptional slippage that results in the synthesis of a transcript that is either longer or shorter than the sequence encoded by the

DNA template. Such a slippage appears to occur only when the polymerase transcribes homopolymeric runs. It has been proposed that the generation of transcripts that are shorter than the encoding template is due to translocation of RNA polymerase without the incorporation of nucleotides, whereas the longer products are due to RNA polymerase incorporating nucleotides without translocation. Transcriptional slippage can occur during both the initiation and the elongation phases. However, the minimal length of the consecutive template nucleotides that can promote slippage in the two phases in different. During initiation, homopolymeric runs as short as 2 or 3 nt can be reiteratively transcribed by RNA polymerase. During elongation, RNA polymerase tends to slip only on longer runs, but the precise requirements have not been elucidated. In one case, slippage by the *E. coli* RNA polymerase during elongation was reported to require runs of at least 10 dA or dT nucleotides, whereas runs of dG at the same length did not result in slippage. In some cases, the ability to slip seems to require a transcriptional pause in addition to the homopolymeric run. Transcriptional slippage is sometimes an important means of regulating transcription. For example, slippage has been reported to play an important role in the regulation of transcription initiation at several bacterial operons, including *pyrBI*.

E. Implications of DNA Topology on Transcription Elongation

Because DNA has a helical secondary structure, a rotation about its axis is necessary to accomplish transcription elongation. This requires that either the entire transcription complex rotates about the DNA or that the DNA itself rotates about its helical axis. Under conditions in which the RNA polymerase rotation is constrained, for example, due to the presence of ribosomes attached to the nascent RNA chain (which is often the case in bacteria), the DNA will rotate througthe enzyme. Consequently, the process of transcription will tend to generate positive supercoils in the DNA ahead of the advancing RNA polymerase and negative supercoils behind it. Excessive torsional stress in the DNA will arise if the DNA is anchored at various pcints (as is the case for circular DNA, such as the bacterial chromosome), or from the movement of RNA polymerase in opposite directions along the DNA. DNA topoisomerases are the natural candidates to remove this tension. It was suggested that gyrase, which can relieve positive supercoils, and topoisomerase I, which removes negative supercoils, amend the situation in front of and behind the RNA polymerase, respectively. This model is supported by the finding that when the activities of gyrase and topoisomerase I are inhibited or otherwise defective, transcription causes major changes in DNA supercoiling. A possible implication of this is that transcription, in addition to having a significant effect on the local structure of DNA, is responsible for generating a significant proportion of supercoiling that occurs in the cell.

F. Implications of DNA Replication on Transcription

Transcription regulation is carefully coordinated with DNA replication and chromosome segregation. In *E. coli* and in other bacteria and bacteriophages, heavily transcribed genes are oriented such that replication and transcription occur in the same direction. Despite this arrangement, because DNA replication occurs 10–20 times faster than transcription, RNA polymerase and DNA polymerases do collide. The outcome of such an encounter is hard to predict. In *E. coli* there is evidence suggesting that the replication fork can displace the elongation complex. However, in bacteriophage T4, the movement of the replication apparatus does not seem to disrupt the elongation complexes, regardless of the direction of their motion relative to the replication fork. Interestingly, when direct collisions occur between the DNA and RNA polymerases of T4, the RNA polymerase switches from the original DNA template strand to the newly synthesized daughter strand. The mechanism that allows the strand exchange without the dissociation of the elongation complex is not known, but probably relies on the various contacts with the DNA and RNA. Whatever the mechanism, the cell needs to coordinate the replication and the transcription processes carefully.

V. TRANSCRIPTION TERMINATION

Transcriptional elongation is highly processive and can lead to the production of RNA transcripts that are thousands of nucleotides long. The processivity is due to the high stability of the complex between the RNA polymerase and the nucleic acids during elongation. It is this stability that necessitates the involvement of specific signals and factors to implement the termination of transcription. To enable efficient termination, the termination signals or factors should cause drastic alterations of the interactions that are responsible for the stable elongation. At termination, RNA polymerase stops adding nucleotides to the RNA chain, all the hydrogen bonds that hold the RNA–DNA hybrid together break leading to the release of the transcript, the DNA duplex reforms, and the enzyme dissociates from the DNA template. The sequence of these events is still not clear because attempts to determine whether the release of the RNA polymerase is simultaneous with the transcript release or occurs subsequently have given ambiguous results. Once the transcript is released from the complex, it is unable to reattach in a way that allows transcriptional elongation to resume. Therefore, the transcript release is the commitment step that makes the termination process irreversible. On one hand, this mechanism ensures the termination at the end of genes and prevents the expression of adjacent distinct genetic units; on the other hand, this mechanism provides an opportunity to control gene expression.

The exact point at which termination of an RNA molecule occurs in the living cell is difficult to define. The 3′-end of an RNA transcript looks the same whether it is generated by termination or by cleavage of the primary transcript. Therefore, the best identification of termination sites is provided by systems in which RNA polymerase terminates *in vitro*. An authentic 3′-end can be identified when the same end is generated *in vitro* and *in vivo*. In *E. coli,* two types of terminators were discovered, intrinsic terminators that do not require ancillary proteins, and terminators that require the involvement of termination factors.

A. Intrinsic Terminators

Intrinsic terminators are sites at which core polymerase can terminate transcription *in vitro* in the absence of any other factor. The best characterized intrinsic terminators are the ones recognized by *E. coli* RNA polymerase. Intrinsic terminators are characterized by a GC-rich sequence with an interrupted dyad symmetry followed by a run of about 6–8 dA residues on the template strand.

The transcription of the GC-rich sequence with the interrupted inverted repeats will give rise to an RNA segment that has the potential to fold into a stable stem-and-loop secondary structure (sometimes described as a hairpin structure). There is much indirect evidence that this structure is indeed formed in the nascent RNA. For example, mutations that interrupt the pairing in the stem part decrease the efficiency of termination, and compensatory mutations that restore the pairing recover the efficiency. There is also a strong correlation between the predicted stability of the structure and the termination efficiency. DNA oligonucleotides that are complementary to one arm of the stem in the stem-loop structure effectively reduce the efficiency of termination, presumably by annealing to the RNA sequence, and thus prevent the RNA from folding into the stem-loop structure. The sequence of the loop in the stem-loop structure also influences the stability of the RNA secondary structure, but the rules for contributing to loop stability have not been fully elucidated. How does the stem-loop structure contribute to termination? It is suggested that the formation of this structure in the newly synthesized RNA sequence, which is still in contact with the polymerase, causes the polymerase to pause and destabilizes the ternary complex.

The other structural feature of an intrinsic terminator, the run of the dA residues (which is sometimes interrupted) in the template strand, is located at the very end of the transcription unit. The transcription of this sequence will generate a run of rU residues at the 3′-end of the RNA transcript. The hybrid between the dA and the rU residues is significantly less stable than most other hybrids, due to weak base-pairing, and it thus requires the least energy to break the association between the strands. This poor base-pairing is assumed to unwind the DNA–RNA hybrid and destablize the interaction of the nucleic acids with the paused polymerase. The importance of the dA run has been established by mutational analysis. The importance of the length of the dA stretch was confirmed by introducing deletions that shortened this element; although the polymerase could still pause at the stem-loop, it no longer terminated. Interestingly, the actual termination can occur at any one of several positions towards the end of the dA run.

The DNA sequence within 30 bp downstream to the transcription stop point, which does not reveal an obvious consensus sequence, is also important for termination in certain cases. For example, changes in the sequence 3–5 bp downstream to the stop point of T7 early-gene terminator can reduce the efficiency of termination from 65 to 10%. Although these sequences are not transcribed, they are near or within the contact point between the RNA polymerase and the DNA in the transcription complex. The way these sequences can affect transcription is by influencing the unwinding of the DNA or the progression of the polymerase along the DNA. Alternatively, the stability of the binding of the polymerase could vary depending on the sequence at the contact points.

B. Rho-dependent Termination

The best characterized termination factor is the bacterial ρ (rho) protein. ρ is a classic termination factor in the sense that it provides a mechanism for dissociating nascent transcripts at sites that lack intrinsic terminators. The sequences that form a ρ-dependent terminator extend from at least 60 bp upstream to about 20 bp downstream of the actual stop point. A sequence comparison of several ρ-dependent terminators did not reveal a consensus. ρ binds to the nascent RNA chain and a common feature of the RNA sequence to which it binds is a relatively high C and low G content. In addition, ρ has a strong preference for sufficiently long segments of unstructured RNA (lacking base-pairing).

ρ causes RNA polymerase to terminate preferentially at points that are natural pause sites. There is no evidence that ρ affects the elongation-pausing specificity of the polymerase. In addition to ρ being an RNA-binding protein, ρ contains an RNA–DNA helicase activity; it hydrolyzes ATP to energize the separation of an RNA–DNA hybrid. Thus, ρ is acting primarily as an RNA-release factor. The current model for ρ action is that it binds to the RNA transcript at sites that are unstructured and rich in C residues; it then translocates along the RNA until it catches up with the polymerase at sites where the enzyme pauses; ρ unwinds the RNA–DNA hybrid in the transcription bubble; and termination is completed by the release of ρ and RNA polymerase from the nucleic acids. Some ρ mutations can be suppressed by mutations in the genes that encode the β- and β'-subunits of RNA polymerase. This implies that in addition to interacting with the nascent RNA chain, ρ also interacts with the polymerase.

The lack of stringent sequence requirements for a ρ-dependent transcripton terminator raises the possibility that such terminators might be fairly frequent in DNA sequences, not only at the ends of operons but also within genes. What prevents ρ from terminating within genes? Because transcription and translation are coupled in prokaryotes, the mRNA chain that emerges from the transcription complex is protected by ribosomes, probably preventing ρ from gaining access to the RNA. The phenomenon of polarity (a nonsense mutation in one gene prevents the expression of subsequent genes in the operon) can be explained by the release of ribosomes from the transcript at the nonsense-mutation site, so that ρ is free to attach to and move along the mRNA; when it catches up with RNA polymerase, it terminates transcription, thus preventing the expression of distal parts of the transcription unit. What prevents ρ from acting on transcripts that are not translated, such as rRNAs and tRNAs? One reason seems to be the lack of ρ-binding sites on these RNAs because they are highly structured. Ribosomal RNA molecules are further protected by the binding of ribosomal proteins. Another mechanism that protects rRNA operons against ρ-dependent termination relies on sequences near the start of the rRNA genes that dictate antitermination. It was suggested that this mechanism increases the rate of transcriptional elongation of rRNA operons by preventing pausing.

The Psu protein encoded by bacteriophage T4 antagonizes ρ-dependent transcription termination. Although the mechanism that enables Psu to oppose ρ is unknown, its general anti-ρ action suggests that it acts directly on ρ, either by binding to it or by modifying it.

C. Auxiliary Termination Factors

Although some polymerases can spontaneously terminate transcription at intrinsic terminators, the efficiency of termination *in vitro* is often enhanced significantly by the presence of additional factors. It therefore seems that the DNA signals that characterize intrinsic terminators are necessary, but sometimes not sufficient. The best characterized of these auxiliary termination proteins is NusA. A less-studied factor named τ (tau) is known to enhance and modify recognition of some strong intrinsic terminators for *E. coli* RNA polymerase. ρ-dependent termination can also be enhanced by an auxiliary factor, the NusG protein.

The ability of NusA to increase the efficiency of termination at some intrinsic terminators might be attributed to its capability to increase the rate of pausing at certain sites. Enhanced pausing would allow more time for the conformational change that leads to the release of the nascent transcript. Some intrinsic terminators predicted to form not a very stable secondary structure, such as the one in the ribosomal protein S10 operon leader, depend on NusA for their operation, and can therefore be defined as NusA-dependent terminators. In the case of the S10 leader terminator, the effect of NusA is enhanced substantially by the product of one of the genes in this operon, the ribosomal protein L4. Because L4 is an RNA-binding protein, one assumption is that it binds to the S10 leader RNA and enhances its folding into a stem-loop structure. Alternatively, L4 bound to the nascent RNA can affect the elongation properties of RNA polymerase. In the unusual case of the intrinsic terminator in the attenuator region preceding the gene for the β-subunit of *E coli* RNA polymerase, NusA, rather than enhancing termination, reduces termination efficiency. The efficiency of termination at ρ-dependent sites is not increased by NusA.

NusG has a minor effect on termination at some intrinsic terminators. However, NusG plays a significant role in the function of some ρ-dependent terminators. This role was deduced from the strong effect of NusG loss on the activity of some ρ-dependent terminators *in vivo*. NusG might be acting directly to enhance ρ activity. Alternatively, it might be acting indirectly by slowing the dissociation of ribosomes from mRNA, thus preventing ρ from binding to the transcript.

Interestingly, the presence of the σ factor in excess significantly increases the rate of RNA polymerase recycling. Hence, the σ factor can also be considered a termination factor. This activity supports the notion that σ can remain associated with the polymerase at postinitiation steps.

D. Antitermination

Antitermination is used as a control mechanism in phages to regulate the progression from one stage of gene expression to the next, and in bacteria to regulate expression of some operons. Antitermination occurs when RNA polymerase reads through a terminator into the genes lying beyond. The terminators that are bypassed can therefore be defined as conditional terminators. Antitermination is not a general mechanism that can occur in all terminators, but is, rather, dependent on the recognition of specific sites in the nucleic acids. Many of the antitermination mechanisms rely on the modification of the polymerase that makes it bypass certain terminators or on changes in the structure of the transcript caused by RNA-binding proteins.

The N protein of bacteriophage λ mediates antitermination necessary to allow RNA polymerase to read through the terminators located at the end of the immediate early genes in order to express the delayed early genes. The recognition site needed for antitermination by N, called *nut* (for N utilization), lies upstream from the terminator at which the action is eventually accomplished. The *nut* site consists of two sequence elements, a conserved 9-nt sequence called boxA, which is also a part of the bacterial *rrn* operon antiterminator signals, and a 15-nt sequence called boxB, which encodes an RNA that would form a short stem structure with an A-rich loop. A number of host proteins, including NusA, NusG, ribosomal protein S10 (NusE), and NusB, participate in the N-mediated antitermination process (Nus stands for N utilization substance). According to the current model for N-mediated antiterminatiomn at ρ-dependent terminators, N recognizes and binds to the boxB stem-loop structure formed on the nascent transcript, whereas NusB and S10 bind to the boxA sequence on the RNA. These proteins are held together through interactions with core RNA polymerase that are stabilized by NusA and NusG. Hence, a ribonucleoprotein complex is formed at the *nut* site and stays attached to t elongating RNA polymerase. This complex prevents RNA polymerase from pausing, thus denying the ρ factor the opportunity to cause termination, and the polymerase continues past the terminator. N also suppresses termination at intrinsic terminators; however, NusA suffices for N to prevent termination at these sites. Other phages related to λ have different N proteins and different antitermination specificities. Each phage has a characteristic *nut* site recognized specifically by its N-like protein. All these N-like proteins seem to have the same general ability to interact with the transcription apparatus in an antitermination capacity.

The Q protein is required later in bacteriophage λ infection. It allows RNA polymerase to read through the terminators located at the end of the immediate early genes, in order to express the late genes of bacteriophage λ. Q has a different mode of action than N. It recognizes and binds to a site on the DNA, called *qut*. The upstream part of *qut* lies within the λ-promoter P_R, whereas the downstream part lies at the beginning of the transcribed region. Thus, Q antitermination is specific for RNA polymerase molecules that have initiated at the P_R promoter. The part of *qut* that lies within the transcribed region includes a signal that causes RNA polymerase to pause just after initiation. This pause apparently allows Q to interact with the polymerase. Once bound, the Q-modified enzyme is released from the pause and is able to read through most transcription terminators, both intrinsic and ρ-dependent. It seems that the modification of the polymerase by Q increases the overall rate of transcription elongation and permits the polymerase to hurry past the terminators. Interestingly, the pause of the polymerase early in the transcription unit, which is a prerequisite for Q-mediated antitermination, involves the binding of the σ subunit of the RNA polymerase holoenzyme to the nontemplate stand of DNA in the transcription bubble up to 15 nt downstream from the start point of transcription. Thus, an initiation factor acts in concert with a DNA-binding termination factor to modify the elongation properties of RNA polymerase. Once again, it is shown σ can remain associated with the polymerase and play a role in postinitiation steps.

RNA polymerase molecules that are engaged in transcribing the ribosomal RNA (*rrn*) operons are modified in a way that makes them bypass certain terminators within the rRNA genes. The modification is established by the recognition of a sequence signal that is nearly identical to the boxA sequence involved in N-mediated antitermination. It has been shown that a heterodimer of NusB and S10 protein binds to the boxA sequence on the RNA of one of the *rrn* operon. It was therefore proposed that the mechanism of transcriptional antitermination in the *rrn* operons, similar to the mechanism mediated by N, involves formation of a ribonucleoprotein complex on the boxA complex that is carried along with the elongating polymerase. The probable purpose of this mechanism is to ensure that transcription of the rRNAs is immune from ρ action.

Transcription of the *bgl* operon in *E. coli,* which codes for proteins required for the use of β-glucoside, is also controlled by antitermination. One of the operon products, the BglG protein, prevents the termination of transcription at two intrinsic terminators. The first terminator is in the 5′ untranslated leader of the transcript and the second is in the intergenic region between the first and second genes of the operon. BglG is an RNA-binding protein that recognizes and binds to a specific sequence partially overlapping the sequence of both terminators. By binding to its RNA target site, BglG stabilizes a secondary structure, which is an alternative to the terminator structure. Thus, BglG binding to the *bgl* transcript prevents the formation of the terminators and the polymerase can read through them. BglG exerts its effect as a transcriptional antiterminator only when the expression of the operon is required, that is, when β-glucosides are present in the growth medium. In the absence of β-glucosides, BglG is kept as an inactive monomer due to its phosphorylation by BglF, the β-glucoside phosphotransferase. Upon addition of β-glucosides, BglF dephosphorylates BglG, allowing BglG to dimerize and act as an antiterminator. The effect of BglG is reminiscent of the effect of the ribosomes in the *E. coli trp* operon transcriptional attenuation.

See Also the Following Articles

Attenuation, Transcriptional • Bacteriophages • Sporulation • Starvation, Bacterial

Bibliography

Busby, S., and Ebright, R. H. (1994). *Cell* **79**, 743–746.

Chamberlin, M. J., and Hsu L. M. (1996). *In* "Regulation of Gene Expression in *Escherichia coli*" (E. C. C. Lin and A. S. Lynch, eds.), pp. 7–25. Landes, Austin, TX.

Choy, H., and Adhya, S. (1996). *In* "*Escherichia coli* and *Salmonella*: Cellular and Molecular Biology" (F. C. Neidhardt *et al.*, eds.), 2nd ed., pp. 1287–1299. American Society for Microbiology, Washington, DC.

Gralla, J. D. (1991). *Cell* **66**, 415–418.

Gralla, J. D., and Collado-Vides, J. (1996). *In* "*Escherichia coli* and *Salmonella:* Cellular and Molecular Biology" (F. C. Neidhardt *et al.*, eds.), 2nd ed., pp. 1232–1245. American Society for Microbiology, Washington, DC.

Landick, R., Turnbough, C. L., Jr., and Yanofsky, C. (1996). *In* "*Escherichia coli* and *Salmonella:* Cellular and Molecular Biology," (F. C. Neidhardt *et al.*, eds.), 2nd ed., pp. 1263–1286. American Society for Microbiology, Washington, DC.

Reznikoff, W. S., Siegele, D. A., Cowing, D. W., and Gross, C. A. (1985). *Annu. Rev. Genet.* **19**, 355–387.

Richardson, J. P. (1993). *Crit. Rev. Biochem. Mol. Biol.* **28**, 1–30.

Roberts, J. W. (1996). *In* "Regulation of Gene Expression in *Escherichia coli*" (E. C. C. Lin and A. S. Lynch, eds.), pp. 27–45. Landes, Austin, TX.

Uptain, S. M., Kane, C. M., and Chamberlin, M. J. (1997). *Annu. Rev. Biochem.* **66**, 117–172.

Transcription, Viral

David S. Latchman
University College, London

I. Small DNA Viruses
II. Large DNA Viruses
III. RNA Viruses
IV. Summary

GLOSSARY

enhancer A region of DNA that, although incapable of directing transcription itself, can enhance the ability of a promoter to do so.

genome A DNA or RNA molecule of the virus that enters the infected cell.

long terminal repeat An element found at either end of the retroviral genome that contains the promoter or enhancer element driving viral transcription.

promoter A region of DNA essential for directing the transcription of a particular gene.

reverse transcription The process whereby DNA is synthesized using an RNA template by the enzyme reverse transcriptase.

splicing The production of a mature messenger RNA molecule from the initial RNA transcript by the removal of intervening sequences (introns).

transcription The process whereby RNA is synthesized using a DNA template.

TRANSCRIPTIONAL REGULATION OF EUKARYOTIC VIRUSES is the process whereby these viruses control the expression of their proteins by regulating the transcription of the genes that encode them. Following the infection of eukaryotic cells with these viruses, the virus must sequentially express different sets of proteins—first, the regulatory proteins that control the expression of other viral genes; second, the proteins required for genome replication and metabolism: and finally, structural proteins. This will allow the virus to replicate its genetic material, package it appropriately, and leave the cell. This sequential production of specific proteins is controlled by the sequential transcription of the genes encoding them. In turn, this sequential transcription depends on the interplay between cellular transcription factors present in the cell prior to viral infection and virally encoded factors that are either synthesized in the infected cells or enter with the incoming virion. The regulatory processes that occur in the small DNA viruses, the large DNA viruses, and the RNA viruses will be discussed.

I. SMALL DNA VIRUSES

A. Simian Virus 40

Among the small DNA viruses (genome size 5–10,000 bases), transcriptional regulation is best understood for simian virus 40 (SV40). This virus, which naturally infects monkey cells, has a genome 5243 bp in size and encodes five proteins. Two of these, the large T and small t antigens, are produced immediately after the entry of the virus into the cell, whereas the others encoding the coat proteins VP1, 2 and 3 are expressed subsequently. The arrangement of the genes encoding these proteins in the circular viral genome is illustrated in Fig. 1. The large T and small t antigens play a vital role both in the replication of the viral DNA and in the activation of the genes encoding the coat proteins. Therefore, they must be produced early in infection, allowing DNA replication, coat-protein synthesis, and virion production to occur subsequently (Fig. 2).

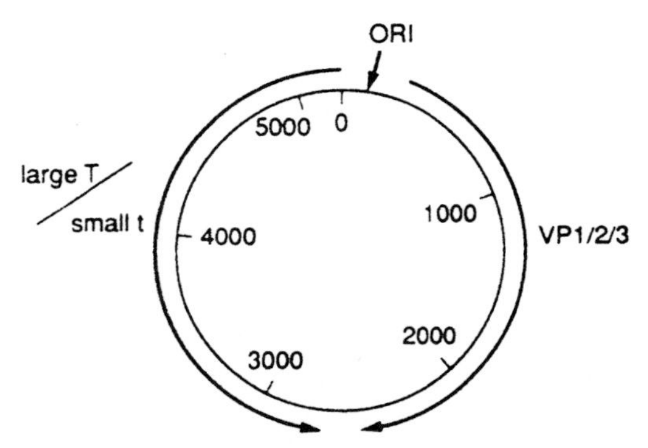

Fig. 1. The SV40 genome, showing the origin of DNA replication (OR1), and the transcription units encoding the large T and small t antigens and the coat proteins VP1, 2, and 3. The numbers indicate number of DNA base pairs.

The large T and small t antigens are encoded by a single gene in the virus, with the critical RNA product of this gene being spliced in two ways to produce the messenger RNAs (mRNAs) encoding the two proteins. Therefore, this gene must be transcribed immediately following the entry of the virus into the cell so that the large T and small t antigens can be made and fulfill their function of activating the transcription of the other genes and DNA replication.

This early transcription of the large T–small t gene depends on a promoter element located immediately upstream of the start site for transcription and an enhancer element located upstream of the promoter. The region of the viral genome containing these control elements also contains the origin of DNA replication and the late promoter driving the transcription of the coat-protein genes, which are transcribed late in infection, in the opposite direction from the gene encoding the large T and small t antigens (Fig. 1).

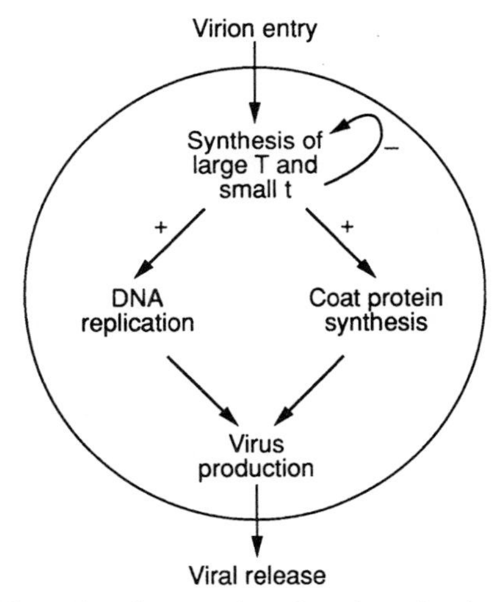

Fig. 2. Life cycle of SV40, showing the stimulatory effect of the early antigens on DNA replication and coat-protein production, and the inhibitory effect on their own synthesis.

The promoter and enhancer driving large T–small t expression contain binding sites for a number of cellular transcription factors, which are present in the uninfected cell. In particular, the SV40 promoter, in common with many cellular and viral genes, contains a TATA box sequence approximately 30 bases upstream of the transcription start site. This TATA box will bind the cellular factor TFIID, which consists of the TATA box-binding protein (TBP) and various TBP-associated factors (TAFs). Following the binding of TFIID, several other factors such as TFIIB and the RNA polymerase, itself, bind to it, forming a basal transcription complex and allowing transcription to occur. The rate of such transcription will be greatly increased by the binding of other cellular factors to their appropriate binding sites upstream of the TATA box in the promoter and enhancer. These factors will interact with TFIID, activating it and greatly increasing transcription. The binding sites for the various factors in the promoter and enhancer are illustrated in Fig. 3.

Following the entry of the virus into the cell, these multiple cellular transcription factors will bind to the viral promoter and enhancer, thereby activating transcription of the large T and small t antigen gene. Thus, the expression of these genes occurs early in infection without the need for viral protein synthesis to allow them to activate the later stage of the lytic cycle.

Once the large T antigen has been synthesized, it then binds to three binding sites (T1–T3) within the promoter region illustrated in Fig. 3. This results in the inhibition of large T–small t gene transcription because the bound large T antigen lies across the start site for early RNA production and, hence, interferes with the initiation of transcription. Moreover, the large T antigen also stimulates the replication of the viral DNA, which begins at the site where it binds, as well as the transcription of the viral coat-protein genes, which occurs from the same control

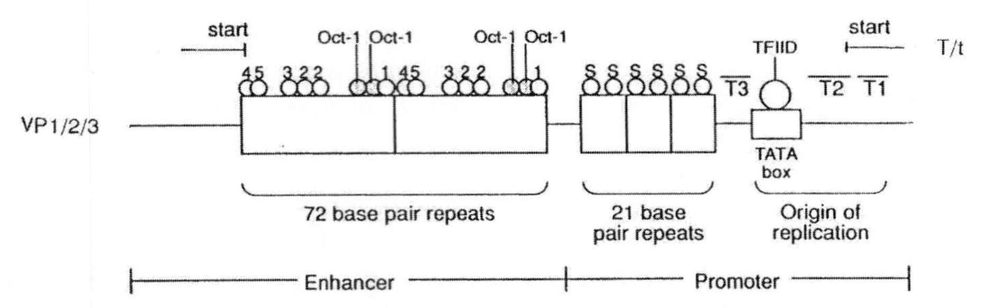

Fig. 3. The SV40 promoter–enhancer region. The binding sites of the viral large T antigen (T1, T2, T3); the cellular factors TFIID, Sp1 (S), and Oct-1; and the activating transcription factors, AP 1-5(1-5), are indicated.

region, but in the opposite direction to early gene transcription (Fig. 3).

Hence, the transcription of the SV40 genome occurs in two precisely defined stages with cellular factors stimulating the early transcription of the large T–small t antigen gene, and the proteins encoded by these genes then stimulating late gene transcription of the coat-protein genes and DNA replication.

B. Other Small DNA Viruses

A similar two-stage pattern of gene transcription is also observed in the other small DNA viruses, In the mouse virus polymoma, for example, a single gene encoding the three regulatory proteins large T, middle T, and small t is transcribed first following infection, with the subsequent transcription of the genes encoding the viral coat proteins. As in the case of SV40, the enhancer element driving early gene transcription contains binding sites for many cellular transcription factors, several of which are identical to those that bind the SV40 enhancer.

SV40 and polyoma early gene transcription can occur in most cell types, paralleling the ubiquitous expression of the cellular transcription factors such as Oct-1, Sp 1, and AP 1, which bind to the promoter–enhancer region. In contrast, however, the related human papovavirus, JC, which is the cause of progressive multifocal leukoencephalopathy (PML) infects only glial cells. This is because its promoter–enhancer region is active only in the presence of a cellular transcription factor known as Oct-6/SCIP/Tst-1, which is expressed only in glial cells. Hence, in this case, the cellular tropism of the virus is controlled by the expression pattern of a cellular transcription factor on which it depends for its expression.

As in the papovaviruses, in the papillomaviruses such as human papillomaviruses types 16 and 18 (HPV 16 and 18), the transcription of the early regulatory genes is controlled by cellular transcription factors that bind to a region of the genome known as the long control region (LCR), which also contains the origin of DNA replication. The subsequent transcription of the late genes encoding viral structural proteins depends, as before, on the prior expression of the early regulatory proteins.

As in the JC viruses, the regulatory elements of HPV 16 and 18 display a strong cell-type specificity in their activity. Thus, the LCR is considerably more active in driving early gene expression in epithelial cells than in any other cell types, whereas the late genes are only expressed in terminally differentiated keratinocytes in the upper layers of stratified epithelia. It has been shown that the ubiquitously expressed cellular transcription factor Oct-1 can bind to a site in the LCR and repress transcription by preventing the stronger cellular transcriptional activator NF1 from binding to its adjacent or overlapping binding site. Hence, in this case, unlike that of the JC virus, a cellular transcription factor represses viral transcription in most cell types. In cervical cells, this effect is likely to be overcome by cervical-specific transcription factors that activate the LCR; one such activating factor, which is expressed at high levels in cervical cancer cells and can bind to the same sites as Oct-1, has recently been identified.

II. LARGE DNA VIRUSES

A. Herpes Simplex Virus

Herpes simplex virus (HSV) types 1 and 2 are large DNA viruses with a genome of approximately 150,000 bases that can infect human cells. The large size of these viruses is paralleled by their ability to produce over 70 proteins, far more than the number encoded by the small DNA viruses. This increased number of proteins is paralleled by a more complex temporal pattern of protein production. Thus, viral proteins are synthesized following infection in three temporal phases. The first of these results in the production of the five viral immediate-early proteins, which have a regulatory role. In particular, these immediate-early proteins induce the expression of the viral early proteins, which include the viral DNA polymerase and other enzymes of DNA replication. Subsequently, the action of the immediate-early proteins also results in the expression of the late proteins, which encode viral structural proteins and are only expressed following DNA replication (Fig. 4).

As in the large T–small t gene of SV40, the genes encoding the immediate-early proteins must be expressed immediately after the viral genome enters the cell. As in the SV40 case, the promoters of the genes encoding the immediate-early proteins contain multiple binding sites for cellular transcription factors. In fact, the cellular factors that have been shown to bind to the HSV promoters such as TFIID, Spl, and Oct-1 also bind to the SV40 promoter or enhancer. The binding sites for cellular transcription factors in the major immediate-early gene IE3 encoding Vmw175 or ICP4 are illustrated in Fig. 5.

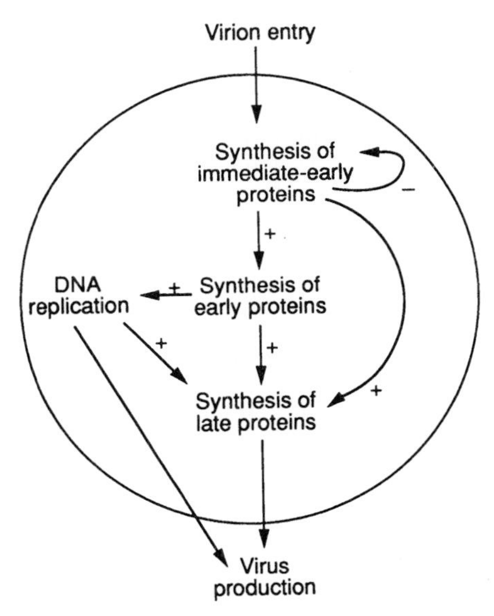

Fig. 4. Life cycle of herpes simplex virus.

Unlike SV40, however, HSV employs a unique mechanism for modulating the activity of one of these factors. Thus, the incoming HSV virion contains a protein known as VP16 (Vmw65), which, in addition to being a structural component of the virion, also forms a complex with the cellular transcription factors Oct-1 and host cell factor (HCF). When Oct-1 binds to the TAATGARAT (R = purine) binding site in the viral immediate-early promoters, it undergoes a conformational change that does not occur when it binds to its normal binding site in cellular promoters (ATGCAAATNA), which is only distantly related to the TAATGARAT motif. This conformational change allows Oct-1 to be recognized by the viral VP16 protein, which is a strong transactivator of gene expression. Its binding therefore results in the strong activation of viral immediate-early gene expression (Fig. 6). Thus, in this case, Oct-1 plays a key role in activating viral gene expression by recruiting the VP16 transactivator, as opposed to its inhibitory role in the papillomaviruses where it prevents the binding of the strong cellular transactivator NF1. Hence, in HSV, the problem of transcribing the first set of viral genes in the absence of viral proteins made in the infected cells is solved in a more complex manner than in SV40 by using both cellular transcription factors and a viral protein contained in the incoming virion.

Interestingly, as in the papilloma viruses, the transcription of the immediate-early genes of HSV is cell-type-specific. Thus, although high-level transcription occurs following infection of most cell types, such transcription is not observed following infection of neuronal cells. Hence, these cells do not support the viral lytic cycle illustrated in Fig. 4: instead, the virus establishes long-lived latent infections in which viral proteins are not synthesized.

In lytic infection with HSV, as with SV40, once the regulatory proteins are synthesized, they play multiple roles in the next stage of the lytic cycle.

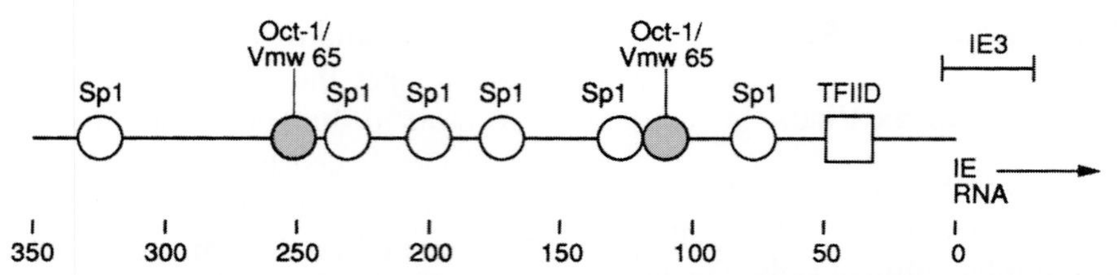

Fig. 5. The herpes simplex virus immediate-early three-gene promoter. The binding sites for the IE3 protein itself, the cellular factors TFIID and Sp1 and the complex of the viral factor VP16 (Vmw65), and the cellular factor Oct-1 are indicated.

In particular, the protein encoded by the HSV immediate early IE3 gene (Vmw175 or ICP4), as does T antigen, binds to its own gene promoter, covering the start site of transcription and, hence, inhibiting its own transcription (Fig. 5). Moreover, this protein also plays an essential, positive role in stimulating the transcription of the genes encoding the viral early and late genes, because mutant viruses that lack a functional ICP4 protein fail to transcribe either the early or late genes.

Unlike the repression of its own synthesis, however, gene activation by ICP4 apparently does not require direct binding of the protein to the early- or late-gene promoters. Hence, the activation of the viral early and late genes by ICP4 appears to be mediated via its interaction with cellular transcription factors, increasing the rate of gene transcription.

Interestingly, although both early and late promoters respond to ICP4, they can be distinguished by the binding sites they possess for these cellular transcription factors. Thus, whereas early promoters contain, in addition to the TATA box, binding sites for factors such as Spl or the CCAAT box-binding factor, late promoters appear to contain only the TATA box, and it is through this binding site that the response to ICP4 is mediated. This relatively simple structure of the late promoters is in contrast, however, to their complex pattern of regulation. Unlike the early promoters, they require DNA replication in addition to ICP4 for full activation.

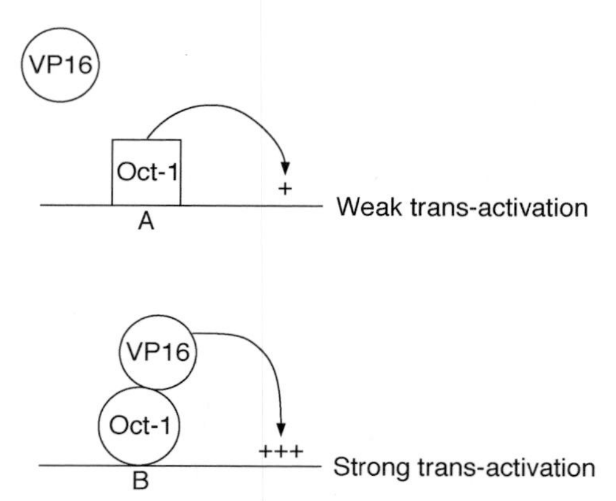

Fig. 6. Oct-1 binds to most sites (A) in a configuration that is not recognized by VP16, resulting in the weak activation characteristic of Oct-1 alone. In contrast, following binding to the TAATGARAT sites in the HSV promoters (B), it undergoes a conformational change that allows it to be recognized by the strong transactivator VP16.

Thus, the three-phase transcription of the HSV genome is controlled by the various control elements in the promoters of the different gene classes. In addition to binding sites for cellular factors, the immediate early promoters contain the TAATGARAT motif, which allows them to respond to the virion protein VP16. The early promoters contain binding sites for cellular transcription factors, allowing a response to ICP4. The late promoters have a much simpler structure, which leads to a requirement for an increased copy number following DNA replication for full activity to occur. The promoter structure of the different HSV gene promoters is summarized in Fig. 7.

B. Adenovirus

The action of the HSV ICP4 protein in activating subsequent stages of viral gene expression is paralleled in the adenoviruses by the immediate-early E1A protein, whose mechanism of action is understood in great detail. This protein, which is made in the cell immediately following infection, transactivates a number of other adenovirus genes, including the

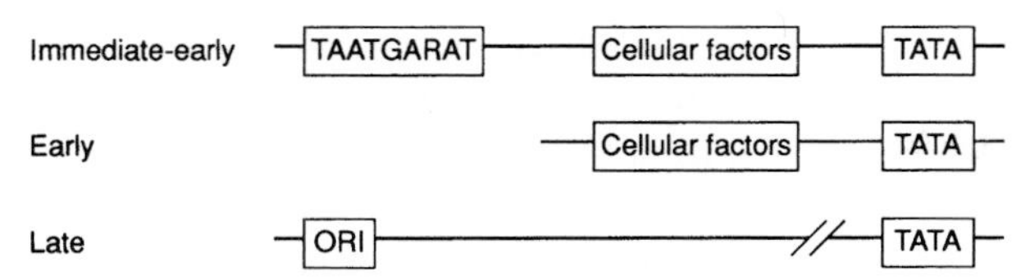

Fig. 7. Structure of different classes of herpes simplex virus promoters. Note that although only the immediate-early promoters contain the TAATGARAT binding sites for the Oct-1–VP16 complex, both the immediate-early and early promoters have binding sites for other cellular factors and a TATA box. In contrast, the late promoters have only a TATA box and depend on DNA replication initiating at a replication origin (ORI) for their expression.

early genes E1B, E2, E3, and E4, as well as the major late promoter.

This transactivation does not depend on binding of E1A to these promoters. Thus, as with ICP4, there are no specific sequences in the induced promoters that confer responsiveness to E1A; rather the promoter sequences necessary for such a response are the same as those required for basal transcription in the absence of E1A and are sites for the binding of cellular transcription factors. Hence, as with the interaction of ICP4 with the HSV early and late promoters, the E1A protein stimulates gene expression by interacting with cellular factors bound to the promoter.

Thus the cellular factor ATF-2 binds to sites in the promoters of E1A responsive genes, and the E1A protein then binds to it. The E1A protein, which is bound to the promoter indirectly in this way, then interacts with other components of the transcription machinery (such as TFIID and the RNA polymerase itself) to stimulate transcription (Fig. 8).

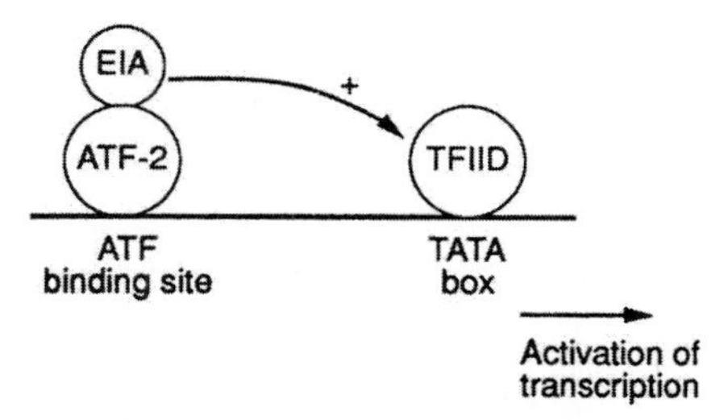

Fig. 8. Activation of transcription by the adenovirus E1A protein, which binds to the cellular activation transcription factor-2 (ATF-2) and then stimulates the TFIID factor bound to the TATA box.

In addition to this direct effect on gene transcription, E1A can also activate the adenovirus gene expression indirectly. Thus, in uninfected cells, the cellular transcription factor E2F is associated with the antioncogene protein Rb, which inhibits its activity. E1A binds to Rb and dissociates the complex, thereby releasing E2F, which can then bind to the adenovirus E4 protein and activate the viral E2 promoter (Fig. 9).

Hence, as in HSV, the interaction of cellular and viral transcription factors results in the appropriate coordinate expression of the viral genome.

III. RNA VIRUSES

Among the RNA viruses, the regulation of transcription is best understood in the retroviruses. Despite their RNA genome, these viruses face the same problems in the regulation of gene transcription as

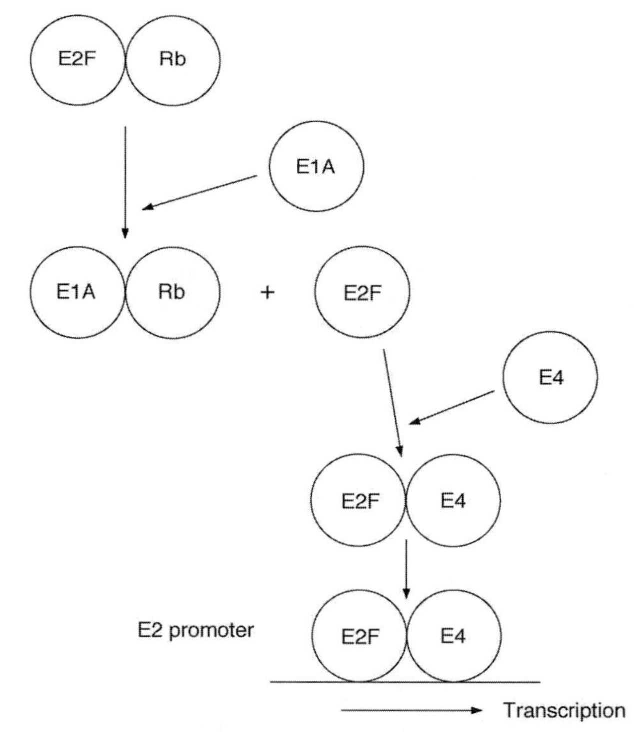

Fig. 9. E1A disrupts the complex of the cellular E2F and Rb factors, releasing free E2F. The E2F can then interact with the adenovirus E4 factor and activate the transcription of the adenovirus E2 promoter.

the DNA viruses. Thus, following entry into the cell, the viral genome is copied into its DNA equivalent by the process of reverse-transcription, and this DNA molecule then integrates into the host-cell genome. Therefore, to produce the viral proteins, this DNA genome must be transcribed into RNA, exactly as occurs for the DNA viruses. The various retroviruses that have been studied can be divided into two groups, the simpler viruses, which do not encode any transcriptional regulatory proteins of their own, and the more complex viruses, which do encode such regulatory proteins.

A. Simple Retroviruses

The simpler retroviruses are typified by the intensively studied Moloney murine leukemia virus (MoMuLV), which encodes only three proteins: the virion structural proteins gag and env, and the pol protein, which reverse-transcribes the RNA genome. The single RNA molecule encoding these proteins is transcribed from a promoter–enhancer element contained within the long terminal repeat (LTR) sequence found at either end of the viral genome (Fig. 10).

The promoter–enhancer is highly active in most cell types and evidently does not depend on the activity of a viral regulatory protein, because no such protein is encoded in the viral genome. Rather, as in the SV40 promoter, the LTR contains binding sites for a number of cellular transcription factors that are present in all cell types and whose binding activates transcription.

A similar pattern of multiple binding sites for cellular factors is also seen in the LTR of the chicken

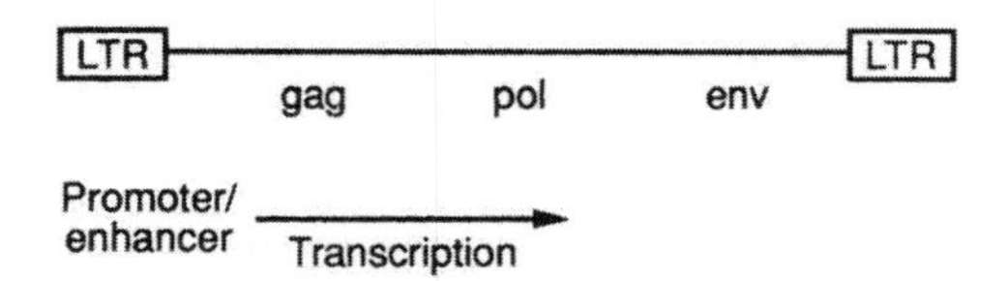

Fig. 10. The Moloney murine leukemia virus genome, showing the long terminal repeats (LTR) that contain the promoter–enhancer element, and the positions of the genes encoding the structural proteins (gag and env) and of the gene encoding the reverse transcriptase (pol).

virus Rous sarcoma virus (RSV), which, like MoMuLV does not encode its own regulatory proteins. The factors that bind to the RSV promoter–enhancer are different from those that bind to that of MoMuLV, however, and include proteins that also bind to the enhancers of SV40 or polyoma.

In both RSV and MoMuLV, therefore, the binding of multiple cellular-transcription factors to the LTR allows it to direct a high level of transcription in the absence of any viral protein. Because these cellular factors are present in all cell types, this high rate of transcription is observed in virtually all cells into which the LTR is introduced. Interestingly, however, the MoMuLV LTR cannot direct transcription in early embryonic cells, which, therefore are not permissive for viral infection. This failure of viral transcription does not arise, however, from the lack of one of the positively acting cellular factors mentioned earlier. Rather, it is due to the presence of an inhibitory protein that is found only in early embryonic cells. This protein binds to a sequence in the MoMuLV LTR known as the CCAAT box and not only fails to activate transcription itself, but also prevents the binding to this site of a positively acting factor, which is essential for the activation of transcription.

Hence, the regulation of the simple retroviruses is controlled by the interplay of positively and negatively acting cellular factors without any involvement of virally encoded regulatory proteins.

B. Complex Retroviruses

The complex retroviruses that do encode their own regulatory proteins are typified by the human immunodeficiency virus (HIV), which has been intensively studied due to its role as the causative agent of AIDS. As is the case for MoMuLV and RSV, the LTR of HIV contains binding sites for several cellular transcription factors. Several of these factors, such as Sp1 (which also binds to SV40 and HSV promoters), are present in cell types, whereas others such as NFκB and NFAT-1 are present in an active form only in T cells that have been stimulated with antigen or cytokines (Fig. 11). The binding of these factors together is sufficient to produce a low level of tran-

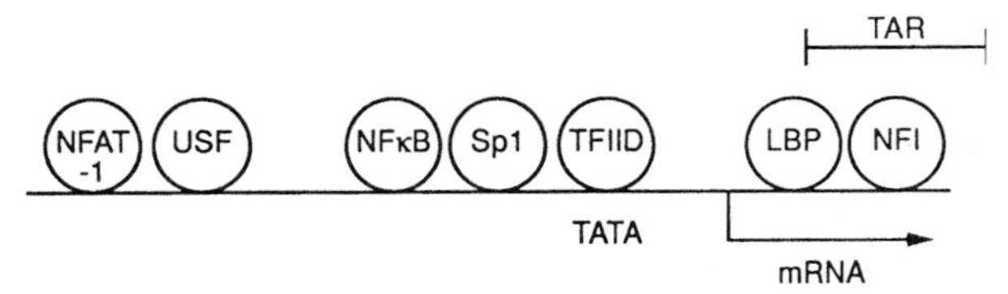

Fig. 11. The human immunodeficiency virus long terminal repeat (LTR). The binding sites for the cellular factors nuclear factor 1 (NF1), leader-binding protein (LBP), TFIID, Sp1, NFκB, upstream stimulatory factor (USF), and NFAT-1 are indicated, together with the target site for activation by the viral Tat protein (TAR), mRNA, messenger RNA.

scription from the HIV LTR, which is enhanced in activated T cells.

This low level of transcription results in the production of an mRNA, which is then multiply spliced to produce small RNAs encoding the HIV regulatory proteins Tat, Nef, and Rev. The Tat protein is a strong activator of HIV LTR-driven transcription, and its presence results in a great increase in the amount of HIV RNA that is made. Simultaneously, the Rev protein promotes the transport to the cytoplasm of the unspliced RNA encoding the viral structural proteins rather than the fully spliced RNA, which encodes Tat, Rev, and Nef (Fig. 12). Hence the effect of Rev is that its own production and that of Tat and Nef is inhibited, and the production of RNA encoding the viral structural proteins is favoured. Hence, in this case, unlike SV40 or HSV, the viral regulatory proteins made in the early phase do not promote the use of a different promoter late in infection. Rather, they stimulate the same promoter that is used early in infection, but alter the pattern of RNA transport so that new proteins are made (Fig. 12).

Interestingly, Tat increases transcription by binding not to DNA but to newly made nascent HIV RNA in a region at the 5′-end of the RNA molecule, known as Tar (see Fig. 11). When bound in this position, the Tat protein can interact with cellular transcription factors bound to the DNA upstream of the transcription start site and stimulate transcription. The action of Tat results not only in an enhanced rate of transcriptional initiation, but also in an increase in the level of full-length viral transcripts, indicationg that Tat stimulates not only viral transcription initiation, but also transcriptional elongation to produce full-length transcripts (Fig. 12).

HIV is not unique among retroviruses in encoding viral regulatory proteins that interact with cellular transcription factors to modulate LTR-driven transcription. Thus, the human T-cell leukemia virus HTLV-1 encodes the Tax protein, which greatly increases transcription from its LTR. In contrast to Tat, however, Tax does not bind to RNA, but rather interacts with the NFκB protein, which binds to an upstream region of the HTLV-1 LTR, thereby increasing LTR-driven transcription. Hence, in this case, NFκB serves not only as a means of stimulating LTR activity in response to T-cell activation (as occurs in HIV) but also as the response element for a viral transactivator protein.

Retroviruses such as HIV and HTLV-1, therefore, have a more complex pattern of transcriptional regulation than RSV or MoMuLV, involving both virally

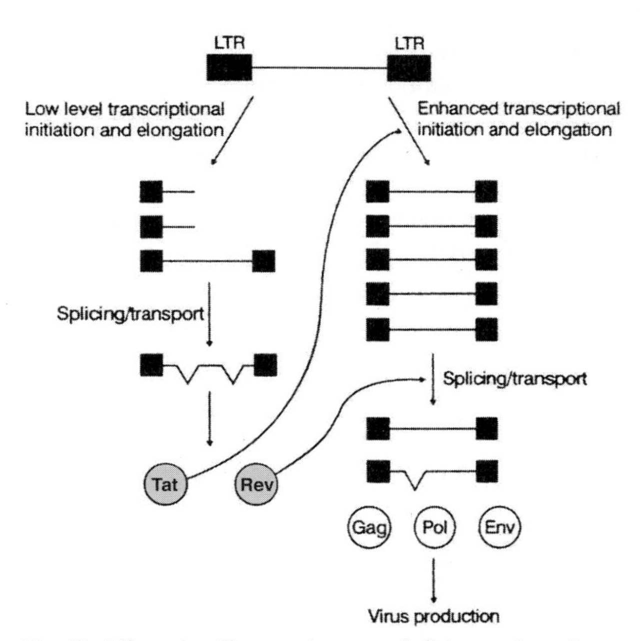

Fig. 12. Life cycle of human immunodeficiency virus, showing the early phase of viral infection, which yields RNA species encoding the regulatory proteins Tat, Nef, and Rev, and the subsequent late phase in which viral transcription initiation and elongation is stimulated by Tat while a change in RNA transport mediated by the Rev protein produces RNA encoding the viral structural proteins.

encoded transactivators and the response to cellular signals such as T-cell activation. These complex gene-regulation strategies are probably related to the ability of these viruses to produce long-lived latent infections as well as a full replicative cycle and are, therefore, central to our understanding of virally induced disease.

IV. SUMMARY

In a wide variety of viruses, the initial phase of gene transcription following entry into the cell depends on the binding of cellular factors to the viral regulatory sequences. Even in the case of HSV, in which transcription is stimulated by a virally encoded transactivator protein contained in the incoming virion, such activation occurs via a cellular transcription factor with which the virus protein forms a complex.

Similarly, although the later phases of viral transcription in the infected cell depend, in most cases, on virally encoded transactivator proteins made in the first phase of infection, these factors often act by binding to or stimulating cellular factors that can bind to sites in the viral control elements. Such a requirement for cellular factors in all phases of viral transcription exemplifies the parasitic role of viruses and their dependence on the cell for their gene expression and replication.

See Also the Following Articles

DNA Replication • Retroviruses • Virus Infection

Bibliography

Cullen, B. R. (1991). Regulation of HIV-1 gtene expression. *FASEB J.* 5, 2361–2368.

Everett, R. D. (1987). The regulation of transcription of viral and cellular genes by herpes virus immediate-early gbe products. *Anticancer Res.* 7, 589–604.

Goodwin, G. H., Partington, G. A., and Perkins, N. D. (1990). Sequence specific DNA binding proteins involved in gene transcription. *In* "Chromosomes; Eukaryotic, Prokaryotic and Viral" (K. W. Adolph, ed.) Vol. 1, pp. 31–85. ed., CRC Press, Boca Raton, FL.

Jones, N. C., Rigby, P. W. J., and Ziff, E. B. (1988). *Trans*-acting protein factors and the regulation of transcription: lessons from studies on DNA tumour viruses. *Genes Dev.* 2, 267–281.

Latchman, D. S. (1998). "Gene Regulation—A Eukaryotic Perspective," 3rd ed., Stanley Thornes, Cheltenham, UK.

Latchman, D. S. (1998). "Eukaryotic Transcription Factors," 3rd ed. Academic Press, London.

Latchman, D. S. (1999). Regulation of DNA virus transcription by cellular POU family transcription factors. *Rev. Med. Virol.* 9, 31–38.

Tooze, J. (1980). "Molecular Biology of Tumour Viruses: DNA Tumour Viruses," 2nd ed. Cold Spring Harbor Laboratory, New York.

Wegner, M, Drolet, D. W., and Rosenfeld, M. G. (1993). POU domain proteins: structure and function of developmental regulators. *Curr. Opin. Cell Biol.* 5, 488–498.

Transduction: Host DNA Transfer by Bacteriophages

Millicent Masters
Edinburgh University

I. History and Concepts
II. Specialized Transduction
III. Generalized Transduction: Basic Facts
IV. Generalized Transduction: How Transducing Particles Are Formed
V. Generalized Transduction: Transduced DNA in the Recipient Cell
VI. Other Transducing Systems
VII. Laboratory Uses of Transduction
VIII. Transduction outside the Laboratory

GLOSSARY

abortive, transductant A bacterium that has received and expresses a transduced DNA fragment in generalized transduction, but that fails to integrate it into its genome; abortively transduced DNA is not replicated, but can be unilinearly inherited.

attachment site A site on a phage (attP) or chromosome (e.g., attB, attλ, or attphi80) at which a site-specific integrase can join-circular phage DNA and host chromosome.

bacteriophage capsid The protein coat that protects and transfers the phage chromosome.

concatemer A single molecule of DNA consisting of more than two phage genomes joined head to tail.

cotransduction The delivery of two genetic markers to a recipient cell by a single phage; cotransducible chromosomal markers must normally be close enough to be included on a single piece of DNA small enough to fit in a phage head.

headful packaging Incorporating DNA into a maturing bacteriophage particle using a single site-specific cut that initiates the entry of DNA into the phage head and, when the head is filled, cutting the DNA again to create a chromosome, with a terminal repeat, that is longer than a genome.

induction The destruction of phage-replication repressor in a lysogen, leading to initiation of the phage lytic cycle.

lysogeny A quiescent state in which a temperate bacteriophage is replicated as part of the host genome, either passively, if integrated into the chromosome, or by maintenance as a plasmid.

package To actively incorporate DNA into a maturing bacteriophage particle; the initiation of phage DNA packaging is site-specific and starts from a *pac* (P1, P22) or *cos* (λ) site.

site-specific packaging Incorporating DNA into a maturing bacteriophage particle by cutting specific sites; the length of DNA packaged is determined solely by the distance between the specific cutting sites.

stable, transductant A bacterium that has received DNA via a viral vector and incorporated that DNA, or a part of it, into its genome.

temperate Being able either to infect and lyse a host with the production of further phages in an infectious cycle, or to enter the lysogenic state and be stably inherited as part of the host genome.

terminal redundancy The DNA sequence repeated at the ends of a phage chromosome; generally indicative of headful packaging.

virulent Not being able to lysogenize and, on infection, being able to enter only the lytic cycle.

GENETIC TRANSDUCTION is the transfer from a donor to a recipient cell of nonviral genetic material in a viral coat. Transduction is one of the three ways by which genetic material can be moved from one bacterium to another. The others are transformation, in which naked DNA is actively or passively taken up by recipient cells, and conjugation, in which cell–cell contact is initiated via structures encoded principally by plasmids. DNA is either transferred through these structures or protected by them during transfer. Both transformation and conjugation are evolved systems. Naturally tranformable bacteria, such as *Bacillus subtilis, Haemophilus influenzae,* and *Streptococcus pneumoniae* have sizable sets of genes concerned with the transformation process. Conjugational systems are essential to the plasmid way of life, allowing plasmids to move between hosts, not necessarily of the same species, more or less promiscuously. Transduction, on the other hand, may well be simply an accidental concomitant of the way in which bacteriophages multiply and package their DNA.

I. HISTORY AND CONCEPTS

A. Discovery of Transduction in Bacteria

Transduction was discovered by Zinder and Lederberg in the early 1950s, at a time when the existence of "the gene" was not yet universally accepted. Conjugation had been discovered in *Escherichia coli* by Lederberg and Tatum in 1946 and it was conjugational transfer of DNA between two differently marked strains of *Salmonella typhimurium* that Zinder and Lederberg were expecting to demonstrate when they discovered tranduction. Conjugation had been shown to require cell–cell contact because it failed to occur when donor and prospective recipient were grown in the same vessel, but separated by a filter that allowed macromolecules and medium, but not cells, to pass through. On the contrary, Zinder and Lederberg were to find that a filter did not impede gene transfer between *Salmonella* strains.

Because it was expected that conjugational gene transfer would allow the coinheritance of any pair of donor genes, however far apart, the strains tested were all doubly marked (i.e., contained mutations in two different genes) to avoid mistaking revertants or contaminants for recombinants. Although many strain pairings were tried, transduction was discovered only when a particular pair of differently marked strains of *S. typhimurium* was tested. This single success was the consequence of two happy accidents. We know now that in transduction only linked markers can be coinherited. By good fortune, one of the strains used, LT-22, was both marked with two closely linked genetic markers and was lysogenic for a bacteriophage, P22, that is able to transduce. Phages produced by spontaneous lysis of LT-22 crossed the filter, infected the other, sensitive, strain and carried transducing DNA home. Transformation was excluded as a mechanism by showing that transducing activity was resistant to DNase treatment and further experimentation showed a phage to be responsible for the gene transfer. A few years later, bacteriophage P1, a lysogenizing phage of *E. coli*, was also shown to mediate transduction. Because any single genetic marker can be transferred by P1 or P22, the process is called generalized transduction. Transduction by these two phages continues to be an important genetic tool. Figure 1 shows an overview of generalized transduction based on the behavior of the *E. coli*-P1 system.

In 1956, Morse, Lederberg, and Lederberg described another sort of transductional system. They found that the bacteriophage λ was able to transduce the galactose (*gal*) genes, which are located next to the site on the chromosome at which the lysogenized phage is integrated. Because they could find no other genes that λ would transduce, they named the process specialized transduction. It was later shown that the biotin (*bio*) genes, which with the galactose genes flank the λ integration site, can also be transduced. Specialized transduction in a modern guise has since achieved new importance because genetic engineering techniques permit the artifical construction of specialized transducing particles containing any desired gene.

B. Specialized and Generalized Transduction Compared

Specialized and generalized transduction are processes that are quite different in detail, but which have in common the fact that a bacteriophage acts as a vector to transfer host DNA from donor to recipient. The two processes are compared in Table I. They are described separately in the sections that follow.

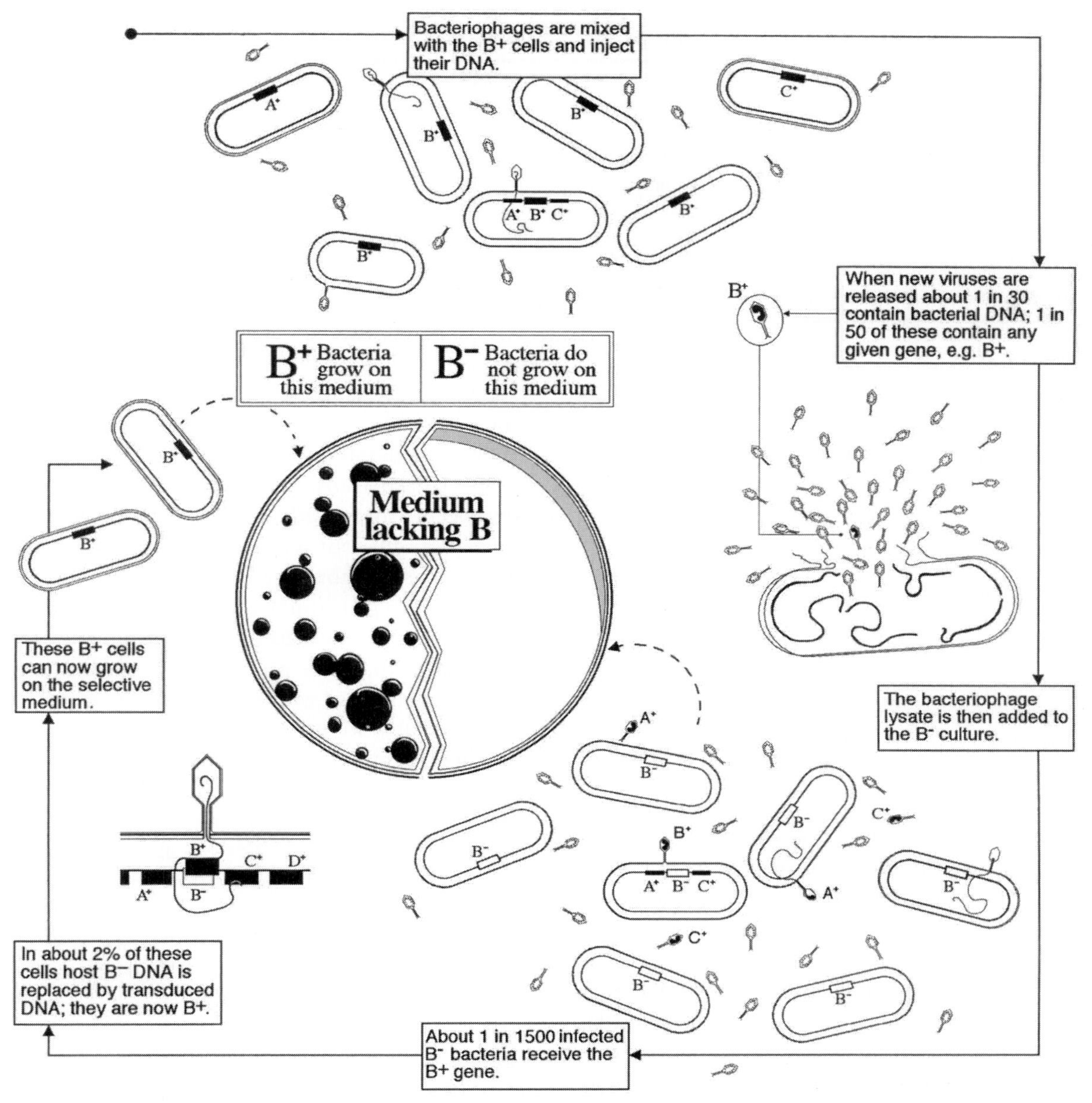
Bacteriophages are mixed with the B+ cells and inject their DNA.
When new viruses are released about 1 in 30 contain bacterial DNA; 1 in 50 of these contain any given gene, e.g. B+.
The bacteriophage lysate is then added to the B- culture.
About 1 in 1500 infected B- bacteria receive the B+ gene.
In about 2% of these cells host B− DNA is replaced by transduced DNA; they are now B+.
These B+ cells can now grow on the selective medium.
B+ Bacteria grow on this medium
B− Bacteria do not grow on this medium
Medium lacking B

Fig. 1. Overview of the generalized transduction process (from Masters, 1996, with permission). The picture, which proceeds clockwise from the upper left, applies in outline to all generalized transducing systems, but the numerical information is based on the *E. coli*-P1 system. It shows the production of a transducing lysate from a prototrophic strain and the transduction of a gene, B, from donor to recipient. The donation of B^+ to the B^- recipient confers a selectable phenotype (growth on B-deficient medium). Although A^+ and C^+ are also donated and genetic exchange can occur, these exchanges will not lead to an alteration in recipient genotype because the recipient is already A^+C^+.

TABLE I
Comparison of Specialized and Generalized Transduction

	Specialized	*Generalized*
Genes transduced	Adjacent to chromosomal insertion site	Any
Process creating transducing lysates	Induction of lysogen	Induction or infection
Transducing particles	Contain both phage and host DNA covalently linked on a single molecule	Host DNA only
Process creating transducing particles	Aberrant excision of lysogenic phage DNA	Mischoice of packaging substrate
Transduced progeny	Unstable, partially diploid lysogens	Stable, haploid recombinants
Principal enterobacterial transducing phages	λ and related phages	P1, P22, T4

II. SPECIALIZED TRANSDUCTION

A. Phage λ and the Generation of Low-Frequency Transducing Lysates

Specialized transduction by coliphage λ follows naturally from its temperate lifestyle. λ particles contain linear phage DNA with cohesive ends that result from staggered cuts within *cos* sites, made during phage DNA packaging. These cohesive (or sticky) ends permit the DNA to circularize after infection. After circularization, a choice is made between two possible outcomes of the infection. In the lytic cycle, the λ circle replicates rapidly, initiating a process that will result in the lysis of the host and release of progeny phage particles about an hour later. An intermediate in lytic growth is a DNA concatemer consisting of several full-length genomes arranged head to tail. This is the packaging substrate, which a phage enzyme cleaves at successive *cos* sites to create the cohesive-ended, genome-length fragments of fixed sequence that are packaged into phage heads.

In the lysogenic response, the repression of replication is established and a chain of events is initiated that lead, via a site-specific recombinational event between *attP* on the phage and *attB* on the host DNA (*attB* is located between the *gal* and *bio* genes), to the integration of the phage into the continuity of the bacterial chromosome. Once integrated, it is passively replicated, once per division cycle, by the host's replicative machinery. It remains quiescent, except for the continual production of a repressor that prevents entry into the lytic cycle. If repression fails or is interfered with deliberately by induction, the phage is cut out of the chromosome to reverse the process that occurred at lysogenization. If all goes smoothly, excision will be accurate and a normal phage is generated.

Specialized transducing phages arise when excision occurs inaccurately to form a circle of about the correct length, but with the wrong DNA. Provided that the replication origin is retained, such a DNA molecule can go through the replication cycle including concatemer formation, *cos* cutting, and packaging. Phages that have acquired *gal* (λd*gal*, where "d" signifies defective) will have lost phage capsid genes and be unable to undergo further lytic growth without help. However, because the cell in which the particle originates probably also contains complete phage genomes (excised perhaps from a sister chromosome), the requisite enzymes and structural components will be available to allow the assembly and liberation of defective transducing particles. λ*bio* phages, in contrast, contain replication and capsid genes and so are not defective. They will, however, lack some of the genes needed to attain lysogeny. Aberrant excisions that lead to recoverable transducing particles are rare; only 10^{-5}–10^{-6} phages are of this type. Figure 2 illustrates the events that lead to the formation of a specialized transducing particle.

B. Transduction by Specialized Transducing Phages

If a lysate such as the one described is prepared on a *gal*$^+$ donor and used to transduce a *gal*$^-$ recipient, *gal*$^+$ progeny will be found at low frequency. These progeny will frequently be diploid for the *gal*

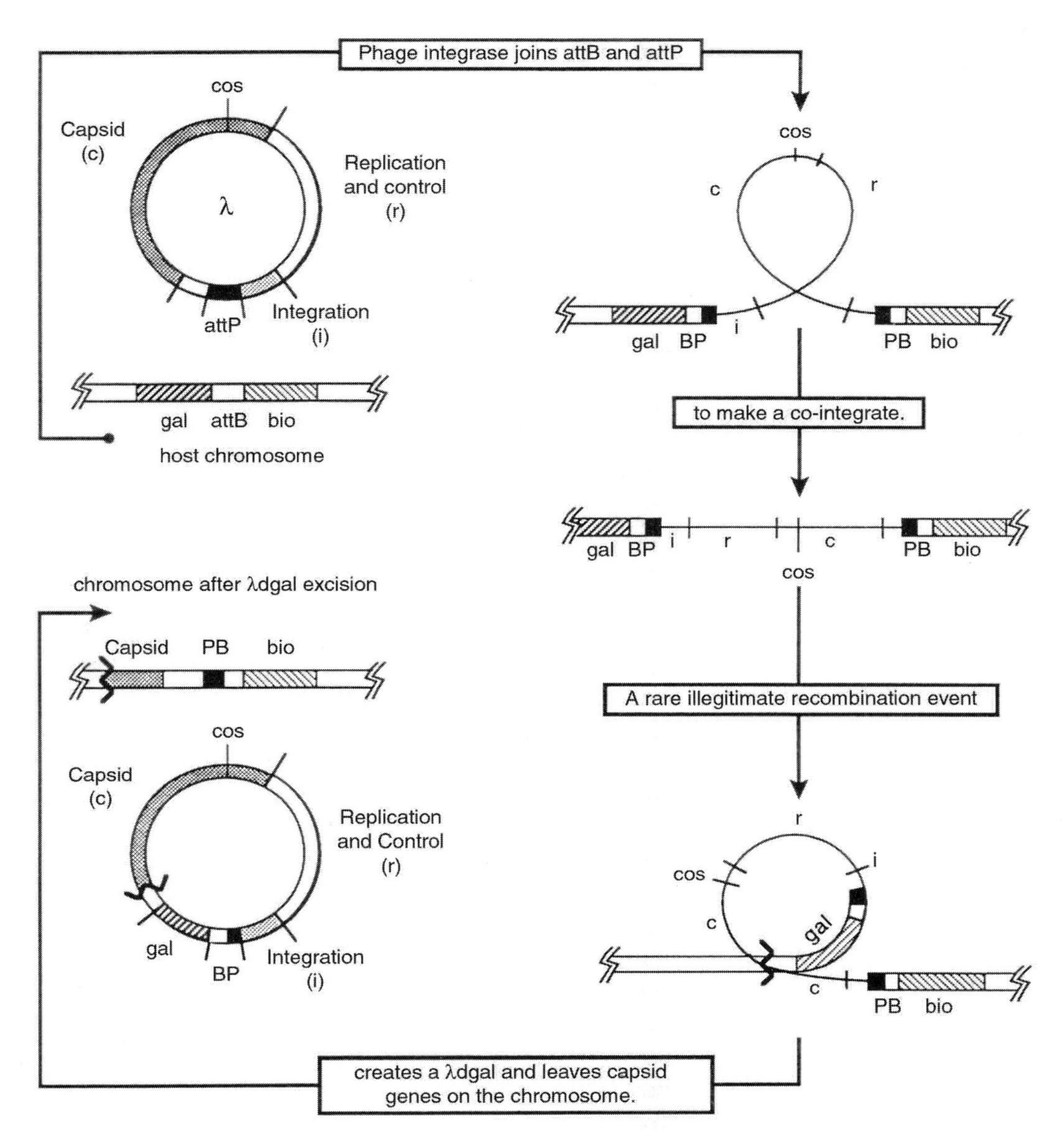

Fig. 2. The creation of a specialized transducing phage. The figure reads clockwise from the upper left; phage λ is used as an example. Site-specific recombination between *attP* and *attB* leads to the formation of a cointegrate between λ and the chromosome to create a normal λ lysogen. After induction, a further site-specific event (not shown) reverses the process in most cases. Rarely, however, an illegitimate recombination event takes place that leaves phage DNA on the chromosome and a similar amount of chromosomal DNA (*gal* in this case) as part of a defective phage. The bold zig-zags mark the site of illegitimate recombination and the consequent boundary between phage and chromosomal DNA.

genes as a result of inheriting *gal* DNA from both phage and host, although haploid progeny can arise by the replacement of gal^- by gal^+ with loss of the phage. If the ratio of phage to recipient is low, cells receiving λd*gal* will not be coninfected with an intact phage. Diploid gal^+ progeny will either be λd*gal* lysogens with the phage integrated at the attachment site, or, more often, cointegrates of phage and chromosome generated by a single homologous recombination event between *gal* sequences. These cointegrated Gal^+ progeny will be unstable because the integration is readily reversible. Both types of lysogen will be immune to λ infection, but will not produce any phage on induction because they lack coat protein genes. Stable, haploid, gal^+, λ-sensitive progeny can also arise by the resolution of the cointegrate with allele exchange and the subsequent loss of phage DNA.

If the ratio of phage to recipient is high, cells receiving λd*gal* are likely to also be infected with a normal phage. In this case, dilysogens can form with the defective and normal phages in tandem array. The Gal^+ phenotype of such progeny will be unstable because homologous recombination between the tandemly repeated genomes will lead to frequent phage loss. Dilysogens are inducible because the normal phage encodes all necessary proteins for phage maturation; the product of induction will be a mixed lysate that can contain equal numbers of transducing and infectious particles. These high-transducing lysates are not only convenient for further gene transfer, but provided a precloning, pre-PCR method for obtaining purified *gal* sequences. In *E. coli* strains from which *attB* is missing, λ can lysogenize at low frequency at alternative sites. This property was used to generate transducing phages for a variety of markers around the chromosome before it became possible to use restriction endonucleases to create genomic libraries *in vitro*.

C. Other Specialized Transducing Systems

Any temperate bacteriophage that integrates into the chromosome of its host should be capable of specialized transduction. This supposition has not been extensively tested, but high-transducing lines of the lambdoid phages P22 (*Salmonella*) and phi80 (*E. coli*), the corynephage γ, and *Bacillus* phages SPβ and H2 have been reported (for references see Weisberg, 1996).

In vitro packaging allows specialized transducing phages to be made for any gene *in vitro*, and specially constructed λ vectors are widely used for making genomic libraries or for cloning particular genes. A λ cloning vector is a phage that has been modified so as to provide unique restriction endonuclease cutting sites at the designated cloning site. Many vectors have had inessential λ DNA deleted to provide more "space" for cloned DNA and have lost the DNA needed to lysogenize. However, lysogenizing vectors are available in which DNA fragments several genes in length can be cloned, allowing such genes to be stably maintained in single copy, as part of a phage, at the phage-attachment site. Special-purpose transducing phages of this sort have been made that permit the controlled expression of foreign genes in *E. coli*, such as, for example, phage T7 RNA polymerase, used to amplify the expression of selected proteins engineered to be transcribed by the polymerase. Other phages allow the linking of gene promoters or translation–initiation regions to reporter genes, so that expression can be quantified. Because λ lysogens contain single copies of cloned genes, problems encountered when gene products are overexpressed, as they usually are from plasmid clones, are avoided. The use of specialized transducing phages as mutagens is mentioned in Section VII.A. For a fuller discussion of cloning in λ see the chapter by N. Murray in "Bacterial Genetic Systems" (Miller, 1991).

III. GENERALIZED TRANSDUCTION: BASIC FACTS

The most extensively studied and used generalized transducing systems are the *Salmonella*-P22 and *E. coli*-P1 systems. In both systems, transduction is a rare event with perhaps $1/10^5$ phage particles able to transduce a particular selected marker. Because transduction is infrequent, no single recipient cell will receive more than one transduced DNA fragment. Therefore, cotransduction will occur only if the markers are close enough together to be included on the same piece of DNA. A P1 head contains a linear DNA molecule about 100 kb, or 2% of the chromosome, in length. P22 is smaller and holds about 44 kb of linear DNA, about 1% of the chromosome. These lengths define the maximum distance between cotransducible markers. Because transduced DNA is linear, it can become incorporated into the recipient chromosome only via a double crossover event. Because homologous pairing must be achieved, the requisite crossovers seldom occur very close to the ends of the transduced fragment and, in practice, the closer two markers are, the more likely they are to be cotransduced. Figure 3 shows the frequency of cotransduction of two markers as a function of the distance between them. Cotransductional mapping has been extremely valuable in placing closely linked markers, not separable by conjugational mapping, in the correct relative positions on genetic maps.

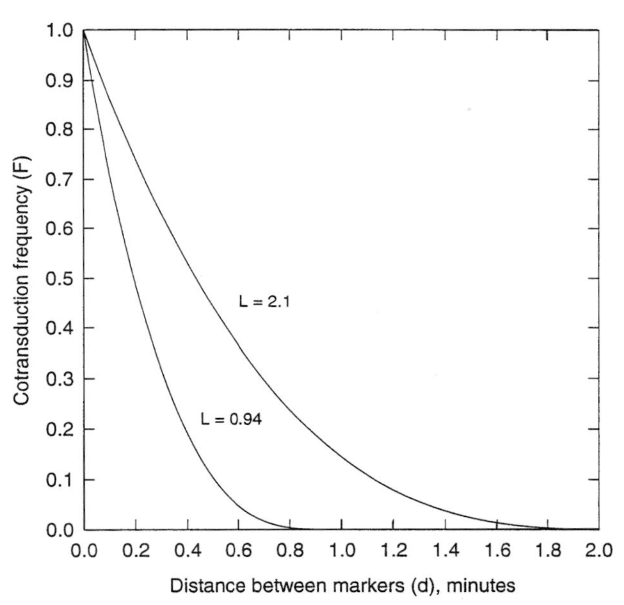

Fig. 3. Mapping by cotransduction (from Masters, 1996, with permission). Cotransduction frequencies (F) are plotted as a function of the distance between markers (d) according to the formula $F = (1 - d/L)^3$ · L is the length, in map units, of the packaged DNA, and is 2.1 for P1 and 0.94 for P22. The formula assumes that all genes are equally likely to be transduced and so is more applicable to transduction with P1 or with high-transducing mutants of P22 than to transduction with wild-type P22, which packages DNA selectively.

IV. GENERALIZED TRANSDUCTION: HOW TRANSDUCING PARTICLES ARE FORMED

A. How Generalized Transducing Phages Package Their DNA

Generalized transducing particles are formed when the phage DNA-packaging mechanism seizes on host DNA, instead of phage DNA, as a packaging substrate. In order to understand how transducing particles are formed, it is thus necessary to understand how the phages that transduce package their own DNA. P1, P22, and most other naturally transducing phages that have been studied package their DNA from concatemers by a headful packaging mechanism. Concatemers are polymers of phage genomes that arise during DNA replication. In cells infected with P1 or P22, a phage-specific packaging endonuclease recognizes a particular site on the concatemeric substrate, termed the *pac* site. Cutting at this site creates an end that then enters a waiting phage head. DNA packaging proceeds until the head is full and the DNA is then cut again (this time not at a specific site). An empty phage head replaces the full one and the process is repeated. An important point is that the second and subsequent cuts are not site-specific; the terminal cut occurs when the phage head is filled. Each phage head holds more than a phage genome's worth of DNA, leading to what is termed terminal redundancy, which describes the fact that each new phage DNA molecule has a rather long sequence repeated at both ends. Successive daughter DNA molecules cut from a single concatemer will begin and end with different sequences (Fig. 4). This permuted terminal redundancy is crucial for replication during the next cycle of infection, as it permits the replication of linear molecules without the loss of unique genetic material and provides a means for subsequent concatemer formation.

B. Packaging Host DNA

How does this packaging mechanism foster the packaging of host DNA? Both P1 and P22 *pac* sites have been added to host chromosomes experimentally. In each case, the transduction of markers to

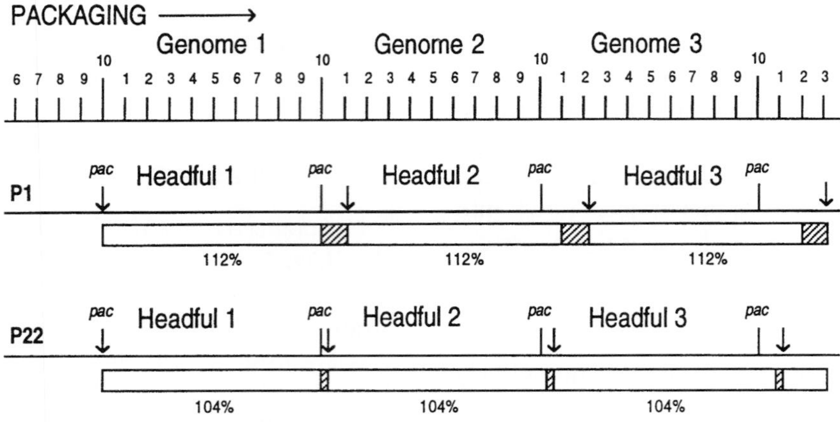

Fig. 4. Headful packaging of phage DNA results in terminal redundancy. The figure shows the packaging of phage DNA, startng at a *pac* site, from a notional concatemer in which each genome is divided into 10 units. A P1 head holds 1.12 genome equivalents, leading to a terminal redundancy of 12%. The capacity of a P22 head is only slightly larger than a genome equivalent; as a result, the terminal redundancy is only 4%. Terminally redundant DNA is shown as cross-hatched, whereas vertical arrows indicate the sites from which successive encapsidations begin.

one side of the introduced site is increased dramatically, by as much as several orders of magnitude. The DNA that is packaged with increased frequency is not only that which contains the *pac* site, but also that of subsequent "headfuls." In a *Salmonella* system in which *in situ* replication of P22, which proceeds into the neighboring chromosome, can be induced, even the 12th headful was transduced at 1000 times the normal frequency. One hundredfold increases in P1 transduction were observed when P1-*pac* was inserted into the chromosome. These observations suggest that the chromosomes of *E. coli* and *Salmonella* lack exact matches to P1-*pac*, or P22-*pac*. However, packaging of transduced DNA might none the less proceed from sites sufficiently similar to *pac* to allow the initiation of packaging at low efficiency. It appears that this is so for the *Salmonella*-P22 but not for the *E. coli*-P1 system.

1. P22 Packages Chromosomal DNA from pac-like Sites

If the chromosome had *pac*-like sites, their numbers would be unlikely to be great, and markers closer to and downstream from *pac* might be expected to be packaged with higher frequencies. Consistent with this idea, different genes are transduced with quite different frequencies, varying over a range of ~1000-fold. A further expectation, if *pac*-like sites are few, is that processive head-filling would mean that packaged fragments would tend to be in register, and cotransduction frequencies would be skewed to reflect this. A test confirmed this expectation. Making a deletion of chromosomal material altered the cotransduction frequencies of downstream markers in a way that suggested that the packaging register had shifted. Finally, P22 mutants able to transduce at greatly increased frequency (up to 1000-fold) were isolated and shown to have a mutated packaging endonuclease. The range in transduction frequencies was greatly reduced in these mutants, consistent with a loss in packaging specificity.

2. P1 Packaging of Host DNA Is Unlikely to Proceed from pac-like Sites

The range of frequencies with which P1 transduces different markers is not very great; about 20-fold is the maximum. The measurement of the relative frequencies of host genes in transducing lysates by quantitative hybridization showed that they varied over a quite narrow range, about three fold. This suggests that the packaging of transducing DNA by P1 is initiated non-specifically, either from a large variety of possible sequences or from accessible structures, such as free DNA ends. Despite this, each transducing particle does not appear to originate from a separate cell; rather cells that produce one are much more likely to produce a second, carrying a different host DNA fragment, than do infected sister cells. This suggests that the packaging of transducing DNA, once initiated, may well be processive (continue along the same molecule). The hybridization studies described allowed the fraction of host DNA in the lysate to be estimated at 5%. Because each transducing phage carries 2% of the chromosome the frequency of transduction for an individual marker would be $\sim 10^{-3}$ if all transducing particles yielded recombinants, rather than the 10^{-5} observed. Clearly, the delivery to a recipient cell is not sufficient to ensure stable inheritence (see Section V.B).

V. GENERALIZED TRANSDUCTION: TRANSDUCED DNA IN THE RECIPIENT CELL

Once packaged within phage particles, transducing DNA is delivered to recipient cells in the same manner as is phage DNA. Linear DNA molecules are injected into the cell through the wall and cytoplasmic membranes. Three possible fates await them there—degradation by nucleases, recombination with homologous recipient DNA to form a stable haploid transductant, or persistence within the cytoplasm in a form refractory both to recombination and degradation. For both the P1 and P22 systems, only about 1% of the transferred DNA becomes part of recipient chromosome.

A. Some Transduced DNA Is Degraded

E. coli contains several exonucleases able to destroy unprotected incoming linear DNA. Thus successful transformation of *E. coli* with linear DNA requires the use of nuclease-deficient strains, and, even so, the recovery of transformants is poor. Transduced DNA, in contrast, resists degradation; successful transduction does not require nuclease-deficient hosts. If the fate of radioactively labelled donor DNA is monitored, the fraction of DNA that is degraded and recycled can be measured. In recipients unable to undergo genetic exchange, no more than 15% of the label entering cells can be recovered from the host chromosomal DNA. This material is not in continuous stretches, consistent with the expectation that it likely to consist of recycled nucleotides. Surprisingly, however, recoverable chromosomal label was not increased much in recombination-proficient cells. Thus, there appears to be significant recycling of nucleotides, but very little physically detectable recombination. Most transduced DNA must have some other fate.

B. Abortive Transduction

Most of the transduced DNA, up to 90% of label in some cases, remains in the cytoplasm in a stable form. It neither replicates nor is degraded, and can be physically recovered for at least 5 hr after infection. This DNA is referred to as abortively transduced DNA, and the cells that harbor it are termed abortive transductants. Abortive transduction was first noted by Lederberg in the mid-1950s and was proposed to explain a curious phenomenon that accompanied the transduction of nonmotile *Salmonella* to motility. Motility is scored on semisolid agar, on which motile bacteria swim and form diffuse patches rather than discrete colonies. After transduction, transduced progeny were observed instead to form trails, consisting of strings of colonies of nonmotile cells, each apparently arising from a single nonmotile descendent of a motile cell. This could be understood if the parent motile cell were an abortive transductant harboring a stable nonreplicating, but transcriptionally active fragment of DNA encoding the synthesis of flagella. At division, only one daughter would inherit this DNA; this cell would continue to make flagella and would swim off. Its sessile sister would found a colony. The abortively transduced piece of

DNA would continue to be unilinearly inherited as long as the original recipient continued to divide. Later, abortive transductants for nutritional markers were found; these are barely visible colonies, each containing about 10^5 cells. Each such colony contains one cell able to found a similar colony. If that cell alone were able to make the enzyme encoded by the transduced gene, the colony would grow linearly, each daughter inheriting enough gene product to grow for a bit, but only the original cell continuing to divide to produce a visible microcolony.

Thus it appears likely that the bulk of P1- and P22-transduced DNA assumes the abortive configuration in the recipient cell. DNA in this configuration is neither degraded nor recombined into the host chromosome. What is the abortive configuration? Increased numbers of complete transductants can be obtained at the expense of abortive tansductants by treating the phage with UV to produce gaps in its DNA. Recombinational processes can be initiated at gaps, and the effect of UV on transduction suggests that, not surprisingly, abortively transduced DNA lacks gaps. However, recombination is most often initiated at the ends of linear molecules, and the DNA injected during phage infection is linear. Are the ends of abortively transduced DNA somehow protected? The answer appears to be a definite yes. If host DNA is made heavy by density-labeling before infection, and the label removed at the time phage are added, all heavy particles in the eventual lysate will contain transducing DNA; these can be separated from the light infective phages and used to infect cells. The heavy transducing DNA can be extracted from transduced cells soon after transduction and its physical state examined. This experiment was performed using P1 with startling results: Abortively transduced DNA appears to consist of circular molecules, the ends of which are held together by protein. Unfortunately the identity and origin of this protein have not been established. An intriguing suggestion made by Yarmolinsky and Sternberg is that the protecting protein is in fact the packaging endonuclease. They suggest that when bound to a genuine *pac* site, the protein is transferred along the concatemer, initiating headful cuts at intervals and finally being released. When it is initially bound to bacterial DNA, perhaps at an end, this does not occur and it is instead packaged bound to the end of the transducing DNA. In the recipient, it could attach to the other end, as well, creating abortive circles. In contrast, for P22, an injected internal head protein is implicated in abortive transduction. P22 mutants deficient in this protein yield an increased number of complete transductants at the expense of abortives.

C. Stable Transduction

Only about 2% of transduced DNA can be recovered from the recipient chromosome in continuous segments long enough (>500 bp) to suggest that they were introduced into the chromosome by recombination. DNA of this length is only found in the chromosomes of recombination-proficient recipients, consistent with the idea that it is a recombination product. Because abortive circles appear stable and do not recombine once they are formed, the recombining DNA molecules that are the source of stable transductants presumably do not achieve the abortive state. This may be because they lack the protective protein at one or both ends, because circularization fails to occur, or because the DNA molecule is internally damaged in a way that makes it a suitable substrate for recombination.

In the *E. coli*-P1 system, in which the packaging frequency reflects gene frequency in the donor population, the range in frequencies of stable transduction is about fivefold greater than the variation in packaging frequency. Thus, there are factors in the recipient cell that favor the integrative recombination of particular markers. It seems plausible that the success of transductional recombination is affected by the proximity of the "recombinator" sequence chi, a 9-bp sequence that stimulates recombination in its vicinity. The stimulation of transductional recombination by chi has been demonstrated in specific crosses, but it would be interesting to correlate the known locations of chi with measured transduction frequencies, in order to determine whether the observed variations in transduction can be attributed to the presence of the element.

The introduction of recombinogenic structures into transduced DNA can stimulate integrative recombination at the expense of abortive transduction. Lysates treated with UV, which should contain what will become gapped DNA, give a greater proportion of complete transductants, as do transduced fragments containing origins of replication. The initiation on replication on such fragments in the recipient cell would be expected to give increased numbers of recombinogenic DNA ends.

Stable transductants are haploid cells in which the recipient allele has been replaced with the donor allele. The initial product of recombination, however, will be a heterogenote, most likely because only a single strand of recipient DNA is replaced. Two cycles of DNA replication will thus be required before two identical donor-type chromosomes are available for segregation to daughter cells. There is therefore a lag before the number of transduced cells can begin to increase; the earliest divisions will produce one transduced and one parental-type cell. Because of this, colonies formed by transduced mixtures that have been plated immediately after phage addition can contain cells with both donor and recipient genotypes. Further purification of such colonies is required to score the cotransduction of unselected recessive markers.

VI. OTHER TRANSDUCING SYSTEMS

A variety of other coliphages are capable of carrying out generalized transduction. These include the virulent headful-packaging phages T1 and T4, for which special mutants must be used to avoid donor DNA degradation and excessive killing of recipients, and the temperate headful-packaging phage Mu. Certain mutants of λ are capable of generalized transduction, but the packaging pathway is not understood and the production of usable generalized transducing lysates require special manipulations. Of the aforementioned, only transduction by T4 has found a niche. Because T4 packages 172 kb, or 3.5% of the chromosome's DNA per particle, it can transduce fragments about 50% longer than those transduced by P1; this has been useful. It is also less fussy about its host and can package DNA from mutants, such as those deficient in recombination, which are poor hosts for P1.

Generalized transduction has been described for a great many species of bacteria, suggesting that a generalized transducing vector can probably be found for any species susceptible to bacteriophage infection. There are 11 main branches on the eubacterial evolutionary tree; generalized transducing phages have been described for organisms on at least three of these, the spirochetes, the gram-positive eubacteria, and the proteobacteria. These latter two branches include most of the familiar and extensively studied bacterial species, whereas three of the others consist exclusively of thermophiles or intracellular parasites for which phages might be expected to be rarer. Generalized transduction has been used to study both the high and low G + C gram-positive bacteria (*Bacilli* and *Actinomycetes*). Generalized transducing phages have been described for species in at least three of the five subdivisions of the proteobacteria, the γ subdivision, which includes the coliforms and a variety of pseudomonads, the δ subdivision (*Myxococcus* and *Desulfovibrio*), and the ε subdivision (*Campylobacter*). Gene transfer in the α subgroup species, *Rhodobacter capsulatus,* is mediated by small particles, resembling phage heads, that package and deliver 4.6-kb fragments of host DNA, but the production of the particles has not been associated with presence of a phage. Interestingly, bacteriophages have been described for halophilic *Archaeobacteria,* but there is, as yet, no report of a transductional system.

The discussion has so far has centered on the transduction of chromosomal markers. There have, however, been many reports of transduction of plasmids by a variety of phages. Although the integration of the plasmid into the chromosome or the introduction of a *pac* site into a plasmid greatly facilitates plasmid transduction in several experimental systems, transduction can also occur in the absence of these facilitating factors. In at least some cases, plasmid–concatemer formation seems to accompany infection, providing a mechanism for the formation of a long-enough substrate to constitute a phage headful. It is not clear whether a long substrate is required in all cases or whether single circular molecules can sometimes be transduced.

VII. LABORATORY USES OF TRANSDUCTION

The understanding of the processes and structures that constitute a cell can be sought either with a gene-to-phenotype, or a phenotype-to-gene approach. Both strategies require the generation and analysis of mutants. Although many genome sequences have now been completed and methods have been devised for readily producing the protein products of selected genes the understanding of the roles of particular gene products in cells will continue to require the phenotypic analysis of specific mutants. Conversely, generating and analyzing groups of mutants sharing a particular or related phenotype permits the establishment of functional connections between the products of different genes. The study of mutations requires their generation, mapping, and transfer between strains for specialized phenotypic analyses. Transduction continues to have an important role in all three of these stages in the functional analysis of genes.

A. Mutagenesis

Because a transducing phage contains a relatively small fragment of DNA, transduction–cotransduction can be used in one of several ways to make sure that a mutation generated by random chemical means is confined to a limited segment of the genome. If a mutation is sought that is linked to a selectable marker, either a donor strain (UV or radiomimetic agents are popular for this) or an already prepared transducing lysate (hydroxylamine can be used to mutate DNA already in phage capsids) can be treated with a mutagen and transductants inheriting the linked marker can be analyzed for their mutational status. Happily, collections of *E. coli* strains exist, each with a mapped transposon insertion, such that a transductionally linked transposon in available for any chromosomal gene.

If mutation is to be by transposon mutagenesis, transducing phages can be used to deliver the transposon. A group of specialized transducing λ phages, which cannot replicate except in special (permissive) hosts, have been engineered to contain minitransposons (which have the advantage of being unable, once transposed, to independently transpose again). If a nonpermissive strain is infected with these phages and a transposon encoded drug-resistance marker selected, surviving progeny will be the products of transposition. P1 can inject its DNA into, but not productively infect, a wide range of nonpermissive species, e.g., P1::Tn5 has been used many times for the mutagenesis of *Myxococcus xanthus.*

B. Mapping

Transduction has been used both for relatively rough and for fine-structure genetic mapping. Approximate gene distances can be calculated from cotransduction frequencies by using the curves in Fig. 3. These curves were derived by assuming that transducing DNA is packaged at random from donor DNA and that inheritence in the recipient follows a pair of recombinational exchanges equally likely to occur at any point on the transduced fragment. Each of these assumptions is only approximately true, as described earlier. For P1, the first is reasonably correct, but recombination is favored at certain sites more than others. For P22, packaging is not random, although high-transducing P22 mutants package less selectively. Because cotransduction frequency is only a good measure of distance if each of the markers is transduced with the same frequency, marker-dependent differences in transduction frequencies can limit its usefulness as a measure of distance. The availability of the set of transposon insertion strains described previously has made it possible to roughly map, relative to a transposon insertion site, any mutation whose location is approximately known.

Fine-structure genetic mapping using multifactorial transductional crosses was an important technique in its day, but is unlikely to be done again. The molecular technologies of cloning or PCR amplification and sequencing can order closely linked markers more straightforwardly (if not more cheaply). Fine-structure mapping required the tedious construction of multiply marked, reciprocally mutant, strains that could then be crossed. Markers were ordered on the assumption that because two crossovers are more common than four, rare recombinant classes are likely to have required multiple exchanges. Anomalies have occurred that led to incorrect conclusions.

The use of these techniques requires that the approximate location of the mutated gene of interest be known. If a new mutation has been isolated on the basis of its phenotypic consequences, its position on the chromosome may be entirely unknown. There are several choices for mapping such a mutation. One genetic approach is to use conjugational analysis (rough mapping) followed by transduction (fine mapping). An alternative for *Salmonella* is to use the Benson and Goldman (1992) set of P22 prophages, which are located at intervals on the chromosome, to directly map mutations by transduction. Lysates prepared on these lysogens transduce at such high frequency that they can be used to localize a mutation with spot tests. A fully molecular approach, such as finding a complementing clone in a genomic library and analyzing its DNA directly, can of course bypass genetic procedures. For *E. coli,* with its fully sequenced genome, the affected gene can be quickly identified in this way.

C. Strain Construction

Transduction really comes into its own as the simplest available method for moving genes between strains. Most sophisticated studies require either the combination of mutations in a strain to detail their interactions, the transfer of a mutation to a standard strain for comparability of behavior, or the transfer of a mutation to a specilized strain to determine its effect, for instance, on the expression of particular reporter genes. Linked drug-resistance markers, which can, if necessary, first be transduced to the mutant strain, can be used to facilitate the necessary transductions.

Transduction can also be used in multistep procedures to introduce mutations or deletions constructed *in vitro* onto the chromosome. Plasmids containing the desired constructions may, in a minority of cells, integrate into the chromosome by a single-crossover event. The unstable cointegrate thus formed can resolve with an exchange of alleles. If the mutation incorporates a selectable marker, transduction can be used to separate the chromosomal DNA with the desired allele from the plasmid, by transducing it to a new plasmid-free strain.

VIII. TRANSDUCTION OUTSIDE THE LABORATORY

In this article, I have discussed the mechanism of transduction and the uses to which it has been put in the laboratory. We know from the epidemic spread of drug-resistance plasmids that conjugation occurs in natural environments. Does transduction occur in nature as well? This question has not been extensively investigated, but a number of reports suggest that it very likely does. The principal organism for which there are reports of transduction in natural environments is *Pseudomonas aeruginosa. P. aeruginosa* transduction has been demonstrated on leaves and even between bacteria on neighboring plants. Both chromosomal and plasmid markers were transferred between strains in lake water in which phages are present. There is probably also marine transfer. The transductional transfer of drug resistance between *P. aeruginosa* strains in hospitals has also been recorded. Transduction has also been reported to occur between strains of *Streptococcus thermophilis* in its natural environment, yogurt! It thus appears likely that transduction does occur in the wild and might be responsible for a significant fraction of natural gene transfer. The ability of P1 to transfer DNA between *E. coli* and the distantly related *Myxococcus xanthus* suggests that transduction plays a role in horizontal gene transfer between distantly related species in addition to a role in the transfer of genes between more closely related species.

Acknowledgment

I thank David Donachie for the preparation of Figures 1, 2, and 4.

See Also the Following Articles

Bacteriophages • Conjugation, Bacterial • Mapping Bacterial Genomes • Mutagenesis • Transformation, Genetic

Bibliography

Benson, N. R., and Goldman, B. S. (1992). Rapid mapping in *Salmonella typhimurium* with Mud-P22 prophages *J. Bacterial.* **132**, 1673–1681.

Masters, M. (1996). Generalized transduction. *In "Escherichia coli* and *Salmonella* Cellular and Molecular Biology" (F. C. Neidhardt, *et al.* eds.), pp. 2421–2441. ASM Press, Washington, DC.

Miller, J. H.(ed.) (1991). Bacterial Genetic Systems. *Meth. in Enzymol.* **204.**

Weisberg, R. A. (1996). Specialized transduction. *In "Escherichia coli* and *Salmonella* Cellular and Molecular Biology" (F. C. Neidhardt, *et el.* eds.), pp. 2442-2448. ASM Press, Washington, DC.

Yarmolinsky, M. B., and Sternberg, N. (1988). Bacteriophage P1. *In* "The Bacteriophages" (R. Calendar, ed.), Vol. 1, pp. 291–438, Plenum Press, NY.

Zinder, N. D. (1992). Forty years ago: The discovery of bacterial transduction. *Genetics* **132**, 291–294.

Transformation, Genetic

Brian M. Wilkins and Peter A. Meacock
University of Leicester

I. History of Transformation and Diversity of Systems
II. Regulation of Competence Development
III. DNA Binding and Uptake
IV. Integration of Transforming DNA into the Recipient Genome
V. Ecological Implications of Bacterial Transformation
VI. Artificial Transformation Systems
VII. Transformation of Microbial Eukaryotes

GLOSSARY

competence Physiological state of cells allowing binding and uptake of exogenous DNA.

homologous recombination Physical exchange between DNA molecules having the same or a very similar nucleotide sequence.

heteroduplex DNA Duplex DNA in which the two strands have different genetic origins. A heteroduplex may contain mismatched bases.

pheromone A species-specific chemical produced by one organism to alter gene expression or behavior in a second organism.

transformation The uptake of naked DNA from the extracellular environment with production of genetically different progeny cells.

GENETIC TRANSFORMATION is the process by which bacteria and microbial eukaryotes take up fragments of naked DNA from the extracellular medium to produce genetically different progeny cells called transformants. The process is manifested by genetically and ecologically diverse organisms in which it is encoded by chromosomal genes. Natural transformation systems require the development of competence by the recipient cell as a specialized physiological state necessary for the binding and uptake of exogenous DNA as linear fragments. Considerable diversity exists among the transformation apparatuses of various bacteria, with some showing specificity for homologous DNA, but a unifying theme of natural systems is the uptake of random fragments of single-stranded (ss) DNA. Transforming DNA that is related at the nucleotide-sequence level to the chromosome of the recipient cell can be incorporated into the resident genome by homologous recombination to give a region of heteroduplex DNA. Mismatched bases may block formation of the heteroduplex or be removed from the recombinant structure by excision from the donor strand. Plasmid DNA is also taken up by natural systems as single-stranded fragments; these may interact to regenerate circular plasmid molecules, albeit with low efficiency. Natural transformation is viewed as a gene-exchange process that enhances the genetic diversity of a bacterial population, as well as maintaining the constancy of the bacterial genome by potentiating recombinational repair of damaged DNA. Artificial transformation procedures have been devised for many microbial species, allowing the uptake of duplex and circular forms of DNA. Such procedures provide a cornerstone of recombinant DNA technology.

I. HISTORY OF TRANSFORMATION AND DIVERSITY OF SYSTEMS

Transformation was discovered by Frederick Griffith in 1928 through work with *Streptococcus pneumoniae,* which causes pneumonia in humans and lethal infections in mice. His experimental system used pathogenic wild-type pneumococci, which had a smooth colony morphology due to the production of a polysaccharide capsule, and an avirulent rough mutant that lacked a capsule. Griffith found that the rough strain acquired virulence and the ability to produce a capsule when mixed with heat-killed wild-type cells and injected into the peritoneal cavity of mice. Some 16 years later, Oswald Avery, Colin MacLeod, and Maclyn McCarty reported a systematic chemical analysis of the transforming principle, showing it to possess the properties of DNA. This was a historic landmark not only in paving the way for the identification of DNA as the genetic material, but also in establishing transformation as the first genetic transfer process to be described for bacteria.

Natural transformation is encoded by genes in the bacterial chromosome and involves the uptake of naked DNA from the extracellular environment. Hence, there is no requirement for a living donor cell. The process supports the uptake of random small fragments of DNA. These are rendered single-stranded during their transport into the recipient cell and are recoverable as material of about 10^4 nucleotides in size. Transformation differs significantly from the two other quasisexual processes for gene transfer between bacteria, namely bacterial conjugation and transduction. Transduction is a bacteriophage-mediated process. It involves the accidental packaging of a double-stranded fragment of bacterial or plasmid DNA of the donor cell into a phage particle, which then acts as the transfer vector to the recipient cell. The size of a transduced DNA fragment is generally greater than that of transforming DNA; it may exceed 100 kb, depending on the DNA packaging capacity of the phage particle. Bacterial conjugation is the dedicated transfer process of self-transmissible plasmids and a class of mobile genetic elements called conjugative transposons. According to the general model of conjugation based on systems studied in *Escherichia coli*, the process requires a specialized contact between the donor and recipient bacterium to form the transport pore for a specific DNA strand. This is transferred progressively in the 5′ to 3′ direction from a genetically defined origin. As a secondary property, conjugative plasmids can cause the transfer of very large sectors of chromosomal DNA, which may exceed 2000 kb in size.

Many genetically and ecologically distinct bacteria are known to be naturally transformable. The list includes gram-positive and gram-negative (eu)bacteria, including cyanobacteria, and archaea. Mechanistic studies of transformation have focussed on the similar systems determined by *S. pneumoniae* and the soil bacterium *Bacillus subtilis*. These are viewed as gram-positive models. Gram-negative paradigms are the systems specified by *Haemophilus influenzae* and *Neisseria gonorrhoeae*—the aetiological agents of spinal meningitis and gonorrhoea, respectively. The *Haemophilus* and *Neisseria* systems differ significantly from each other and from the gram-positive models.

Transformation is likely to contribute importantly to gene flux between bacteria in natural environments. The process has also been used for mapping the relative positions of chromosomal genes of some naturally competent bacteria. Mapping involves the selection of transformants containing a specific donor gene and scoring the fraction containing one or more unselected donor markers. The derived values allow the ordering of genes on the basis that cotransformation frequency is inversely proportional to genetic distance. The usefulness of the technique is limited by the small fraction of the genome that is transferred on any one DNA fragment and by the fastidious growth requirements of a number of transformable bacteria, which make it difficult to isolate mutations in sufficient density per unit of chromosomal DNA. Such problems are bypassed by recombinant DNA technologies that allow mapping with reference to restriction fragment maps or, in the case of *B. subtilis* and *H. influenzae,* to the known nucleotide sequence of the entire chromosome.

The need in molecular biology to introduce DNA modified *in vitro* back into cells has led to the development of artificial procedures for the transformation of prokaryotic and eukaryotic microorganisms that do not develop natural competence. Methods used to promote the uptake of DNA include the treatment of whole cells with metal cations in conjunction with a heat shock, use of a brief high-voltage electrical discharge, and polyethylene glycol

(PEG) treatment of osmotically stabilized protoplasts.

II. REGULATION OF COMPETENCE DEVELOPMENT

Transformation systems require the acquisition of competence as a physiological state enabling the bacteria to take up high-molecular-weight fragments of exogenous DNA. In many systems, competence is a transient condition associated with the induced synthesis of a small number of specific proteins. Growth conditions favoring the development of competence by different bacteria vary considerably. For example, competence may occur throughout the exponential growth phase to decline at the stationary phase, as found for *Deinococcus radiodurans* and *Mycobacterium smegmatis,* whereas the process develops in the early- to mid-exponential phase in *S. pneumoniae* and at the entry to the stationary phase in *B. subtilis.* Competence is observed to be a constitutive property in *N. gonorrhoeae.* Yet another variation is apparent in the fraction of cells that acquires competence under optimal inducing conditions.

Competence development has been characterized most extensively for *B. subtilis.* Here, it is a postexponential response involving a complex network of signals that sense the nutritional state of the environment and cell density. Many of the regulatory proteins in the cascade participate in other postexponential-phase responses such as sporulation, motility, and the production of extracellular degradative enzymes and antibiotics. Competence optimally develops late in exponential growth in a glucose-minimal medium, or early in stationary phase, but occurs in only a subfraction of cells making up 10% or less of the population. The state is associated with a cell type that is more buoyant and relatively dormant in macromolecular synthesis. Twenty or more genes have been identified for the development of competence, which collectively function in an elaborate complex of regulatory networks involving cell density-sensing systems, protein phosphorylation, and protein–protein and protein–DNA interactions (Fig. 1). The various regulatory signals converge to control the expression and activation of the key competence

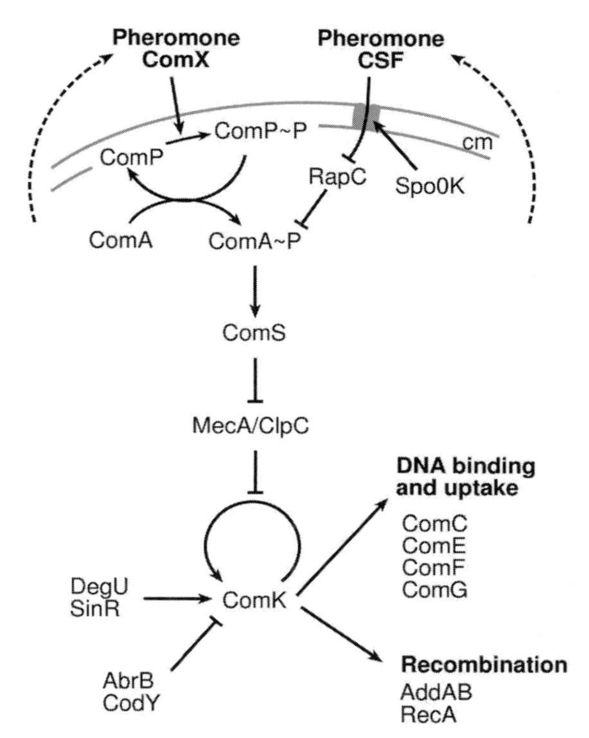

Fig. 1. Model for pheromone regulation in *B. subtilis,* showing control of the late competence genes, which function in DNA binding and uptake, and recombination genes. Blocked lines indicate negative action.

transcription factor, ComK. This is required for the expression of late genes in the *comC, comE, comF,* and *comG* loci, which determine the biosynthesis of the DNA-binding and -uptake apparatus.

The activity of ComK is negatively controlled by two proteins, ClpC and MecA. ClpC, in the presence of ATP, is thought to function as a molecular chaperone to increase the affinity of MecA for ComK, thereby forming a ternary complex that blocks the transcriptional function of ComK. The protein is released from the inhibitory complex via convergent pathways initiated by two cell-density-sensing pheromones. One pheromone, ComX, is a 9- or 10-amino-acid peptide that is processed posttranslationally and secreted from *B. subtilis* by a dedicated exporter system. ComX is detected by the ComP–ComA by two-component regulatory system; the pheromone is sensed by the membrane-associated ComP histidine protein kinase to trigger its autophosphorylation and the subsequent transfer of the phosphate group to the ComA response regulator. The phosphorylated form of the regulator, ComA~P, functions as a transcription factor to activate the expression of an operon specifying the ComS protein. In turn, ComS releases ComK from the ternary complex with MecA and ClpC.

The second pheromone, competence-stimulating factor (CSF), is a five-amino-acid peptide. It is probably taken up into the cell by the Spo0K oligopeptide permease transporter to modulate the phosphorylation level of ComA by inhibiting a phosphatase (RapC) active on ComA~P. The involvement of extracellular pheromones provides a quorum-sensing mechanism for monitoring cell density and triggering competence development at high cell concentrations. Under such circumstances, other cells of the same species might be releasing DNA through autolysis or, more intriguingly, by extrusion as an active donor function.

ComK protein is singularly required as a promoter-binding protein to stimulate the transcription of the late *com* loci, as well as for competence-related expression of recombination loci such as the *recA* and *add* genes. ComK protein in addition enhances transcription of its own gene via autoregulation. Transcription of *comK* is influenced by other DNA-binding proteins that integrate information on nutrient availability. Among these, AbrB and CodY are negative regulators, whereas DegU and SinR have stimulatory roles.

In contrast to the *B. subtilis* system, competence arises in *S. pneumoniae* as a transient state in the early- to mid-exponential phase of growth to affect nearly every cell in the culture. The state is marked by a shut-down of general protein synthesis and production of competence-specific proteins. As described for *B. subtilis,* competence development in *S. pneumoniae* depends on the accumulation of an extracellular pheromone. This is a 17-residue peptide, known as competence-stimulating peptide (CSP), which is sensed by a two-component signal transduction system to stimulate the expression of the competence genes. Interestingly, there are two promoters for the gene encoding the CSP precursor; one is constitutive and the other is autoinducible. The arrangement allows CSP that has accumulated to a basal level by expression of the constitutive promoter to trigger a cascade of pheromone production from the autoregulated promoter.

The regulation of competence in *H. influenzae* is quite different in that almost every cell can develop the state as a stable internally regulated condition. The same applies to a number of other gram-negative bacteria. The response is triggered by a transition into unbalanced growth following a shift-down from a rich to a nutrient-limited medium, with the cyclic AMP-catabolite activator receptor protein complex playing a regulatory role. *H. influenzae* also achieves low levels of competence in the late-exponential phase of growth.

III. DNA BINDING AND UPTAKE

A. Overview of DNA Transport Across Cell Membranes

A particularly challenging aspect of gene transfer is the nature of the specialized channels that allow DNA as a polyanion to pass through the hydrophobic environment of cell membranes and enter the cytoplasm of the recipient cell. Considerable variation exists in the protein complexes making up such channels. As discussed later, some of the DNA transporters show intriguing relationships to systems dedicated to protein secretion. These include systems for the secretion of exoenzymes and for biosynthesis of type IV pili. The latter are extracellular appendages that mediate the adherence of virulent bacteria to host eukaryotic cells, as well as a flagellum-independent form of translocation across surfaces called twitching motility.

A common feature of transformation systems that have been examined at the molecular level is that DNA enters the cytoplasm of the recipient cell in single-stranded form. Considerable diversity exists between systems with regard to the DNA uptake apparatus and whether or not it shows species specificity in using homologous DNA. In the *B. subtilis* and *S. pneumoniae* paradigms, DNA transport across the single-cell membrane of the cell envelope occurs through a surface-exposed complex of proteins. The transformation of gram-negative bacteria is complicated by the fact that the DNA must traverse an outer membrane, which characteristically contains an outer leaflet of lipopolysaccharide (LPS), as well as a cytoplasmic or inner membrane. In *H. influenzae,* DNA uptake proceeds through a specialized membrane vesicle, whereas passage of transforming DNA across the outer membrane of *N. gonorrhoeae* requires the type IV pilus.

B. *Bacillus subtilis* and *Streptococcus pneumoniae*

The inhibitory effects of nonhomologous competitor DNAs indicate that *B. subtilis* and *S. pneumoniae* take up DNA without sequence specificity. The first stage in DNA uptake involves the noncovalent binding of double-stranded (ds) DNA to about 50 competence-specific receptor sites on the surface of the cell. Duplex DNA is bound much more efficiently than ssDNA, ssRNA, or dsRNA. Binding is initially loose in that the DNA is released by washes of high ionic strength or by competitor DNA. Loose binding is followed by tight binding in which the DNA is resistant to release by washing but remains exposed to the external medium, as shown by its susceptibility to shear and exogenous DNase. The DNA then undergoes double-stranded cleavage by one or more surface-exposed endonucleases, which may be localized at the binding sites. The size distribution of the resulting fragments is in the range of ~5–15 kb.

The uptake of DNA, defined by its acquisition of resistance to exogenous DNase, commences 1–2 minutes after binding and occurs at a rate of ~100 nt/s at 28°C. Uptake culminates in the entry of ssDNA into the cytoplasm of the recipient cell. The uptake channel is thought to be an aqueous pore in a surface-exposed complex of proteins. The complex includes an entry nuclease that potentiates intracellular transport of ssDNA fragments with the associated degradation of the second strand and release of oligonucleotides into the medium. The entry nuclease of *S. pneumoniae* is the 30-kDa membrane-bound EndA endonuclease; the polarity of its activity is such that the non-hydrolyzed strand enters the cell with a leading 3′-terminus.

Further evidence for a single-stranded intermediate comes from the eclipse phase. This is a transient stage in the transformation process in which DNA that has been taken up into the recipient cell—as defined by its resistance to extracellular DNase—cannot be isolated in a form that transforms a second cell. The explanation is that entrant DNA is single-stranded in the eclipse phase and therefore an unsuitable substrate for uptake by a second cell. Another feature of the eclipse is the coating of the entrant strand with a ssDNA-binding protein. The eclipse protein is thought to protect the DNA from nucleases and aid the next step in the transformation process, namely homologous recombination with the resident genome.

Several models have been proposed to explain the bioenergetics of DNA transport. The classic proposal, made in 1962 by Sanford Lacks, that DNA uptake by *S. pneumoniae* is powered by the hydrolysis of the second strand by the membrane-localized EndA endonuclease remains controversial because the nuclease might be required to generate the single-stranded substrate of a transporter. Other mechanisms may include one or both components of the proton-motive force, an anion-exchange reaction dependent on proton cycling, an ATP-dependent system acting on ssDNA, and a transporter consisting of a complex of poly-β-hydroxybutyrate and calcium salts of inorganic polyphosphates, as discussed further in Section VI. Such a complex accumulates in the membrane fraction of competent bacteria and may serve to transport polyanionic salts, such as ssDNA, through the cytoplasmic membrane.

The DNA uptake apparatus of *B. subtilis,* as described by David Dubnau and his colleagues, is a complex of proteins encoded by some 15 genes belonging to the dispersed *comC, comE, comF,* and *comG* loci (Fig. 2). Homologs of some of these proteins have been described for other gram-positive and gram-negative bacterial-transformation systems. Similarities have also been detected at the predicted amino-acid-sequence level between ComG polypeptides and proteins involved in both the secretion of exoenzymes and the biosynthesis of type IV pili.

The seven ComG proteins of *B. subtilis* are membrane-associated and essential for DNA binding. ComGC, ComGD, ComGE, and ComGG have prepilin-like signal sequences, and together with ComGF, may be assembled in a complex on the outer face of the cell membrane. The processing of the

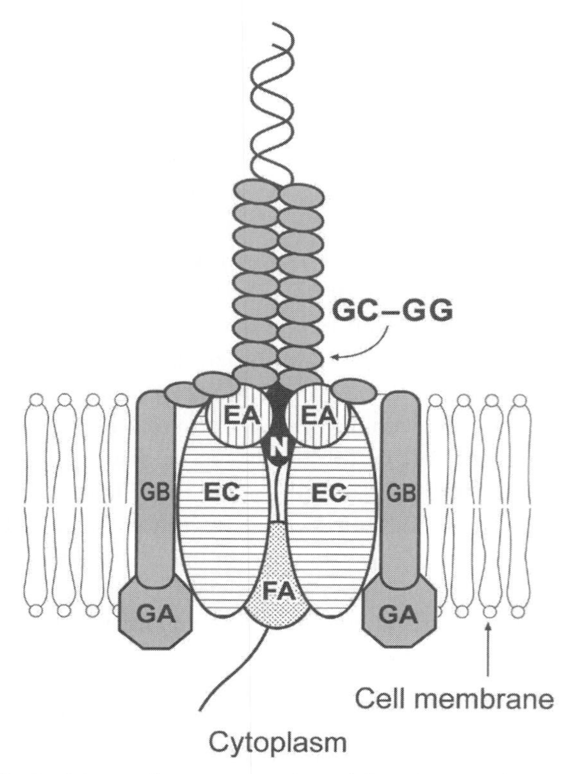

Fig. 2. Schematic representation of the DNA uptake apparatus of *B. subtilis.* Extracellular duplex DNA enters the protein complex at the top. Transport of one strand into the cytoplasm of the cell is associated with the degradation of the second strand by an entry nuclease (N). Other labeled structures in the complex are late Com proteins.

proteins is mediated by ComC, which resembles the prepilin protease family. ComGA and ComGB are located in the cell membrane and may play morphogenetic roles, as judged from properties of apparent equivalents active in type IV pilus biogenesis. The ComG apparatus might function to aid transfer of transforming DNA across the cell wall to receptors on the outer surface of the membrane or to remodel the cell wall to allow DNA access to the transmembrane transporter.

ComEA and ComEC are integral membrane proteins. Mutational studies suggest that ComEA function to couple DNA binding to uptake, whereas ComEC contributes to the DNA-uptake channel. ComFA is another membrane protein required for DNA uptake; it resembles by its amino acid sequence a family of ATP-dependent RNA–DNA helicases, in particular the subfamily containing the DEAD (Asp-Glu-Ala-Asp) motif. However, ComFA shows the highest similarity to *E. coli* replication protein PriA. This protein has the unusual property of translocating along ssDNA in the 3′ to 5′ direction via an ATP-dependent process. Thus, ComFA may use ATP hydrolysis to translocate ssDNA through an import channel consisting of ComEC subunits with ComEA localized at the entry port.

C. Gram-Negative Systems Showing Specificty of DNA Uptake

The transformation systems of *H. influenzae* and *N. gonorrhoeae* differ from the *B. subtilis* paradigm in showing specificity for the uptake of homologous DNA. Such selectivity is manifest through the lack of competitive inhibition by other types of DNA. Hamilton Smith and colleagues discovered in 1979 that the specificity of DNA uptake by *H. influenzae* is conferred by a specific oligonucleotide uptake sequence (US), subsequently identified as 5′-AAGTGCGGT-3′ and present at 1465 copies in the 1.8×10^3-kb genome. The US motif is widely distributed around the genome, but tends to be localized in inverted repeats downstream of genes. Here it might contribute to transcription termination. A different US, 5′-GCCGTCTGAA-3′, is responsible for the specificity of DNA uptake by *N. gonorrhoeae,* but again the sequence is found in transcriptional terminators. Such sequences may have acquired a role in transformation in response to their dispersal in the genome, which makes them suitable as tags for identifying diverse sectors of homologous DNA following its fragmentation.

DNA uptake by *H. influenzae* is effected by a membrane-protected structure called a transformasome. There are several of these structures per competent cell; each has a diameter of ~20 nm and consists of a membrane vesicle that emanates from a zone of adhesion of the inner and outer membranes to protrude beyond the cell surface. Duplex DNA is inferred to bind to a protein receptor on the outside of the transformasome and then to enter this structure irreversibly, where it is protected from exogenous DNase. Single-stranded linear DNA is released from the transformasome into the cytoplasm in a polar manner with a leading 3′-terminus.

In marked contrast, the uptake of transforming DNA by *N. gonorrhoeae* requires the type IV pilus of the organism (Fig. 3). The complete pilus is required, including PilC, which is associated with the cell surface and tip of the organelle. The pilus system might be required to present the US receptor at the cell surface or to translocate transforming DNA across the outer membrane. Transport might be achieved by the depolymerization of the pilus fiber, as is thought to occur in twitching motility. Passage of the transforming DNA across the remainder of the cell envelope involves competence proteins apparently unassociated with the pilus. Of these, ComL and Tpc are thought to cause localized restructuring of the polysaccharide–peptide polymers making up the peptidoglycan cell wall. Tpc appears to be loosely associated with the peptidoglycan, but ComL, a lipoprotein, is covalently bound to it. ComA as an integral membrane protein aids DNA transport across the inner membrane; interestingly, ComA is related at the amino-acid-sequence level to the ComEC DNA uptake protein of *B. subtilis*.

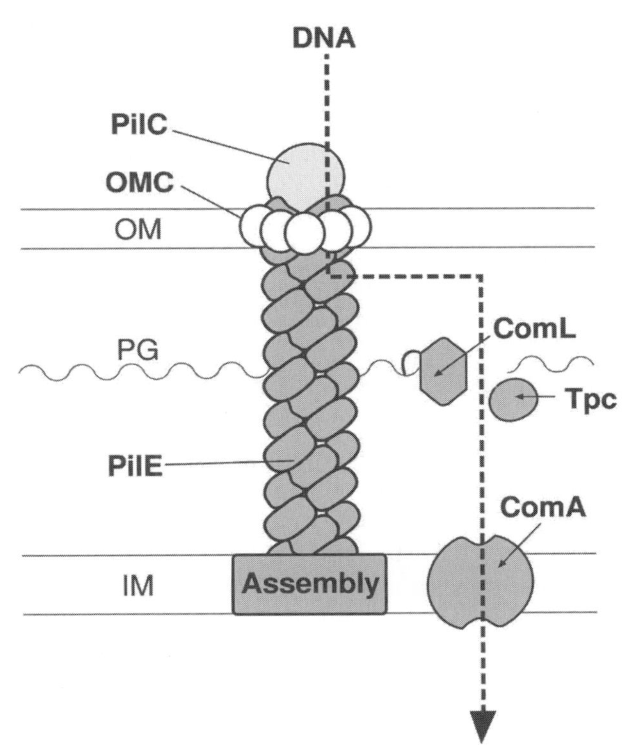

Fig. 3. Model of the route of DNA uptake through the cell envelope of *N. gonorrhoeae.* The structure to the left is the type IV pilus, consisting of the assembly center in the inner membrane, PilE subunits, and the terminal PilC protein. OMC is the multisubunit outer-membrance pore complex that is thought to act as the gated pore for export of the pilus. Cell envelope components are OM, outer membrane; PG, peptidoglycan; IM, inner membrane.

IV. INTEGRATION OF TRANSFORMING DNA INTO THE RECIPIENT GENOME

A. Homologous Recombination of Chromosomal Genes

Single-stranded fragments of chromosomal DNA entering the cell are integrated into the recipient genome by homologous recombination. The process has a central requirement for an equivalent of the *E. coli* K-12 *recA* gene product. This protein binds ssDNA to form long nucleoprotein filaments essential for strand pairing with the homologous duplex and subsequent exchange reactions. The expression of *recA*, as well as some other recombination genes, is induced as part of the competence response (Fig. 1). The product of strand exchange is a region of heteroduplex DNA (Fig. 4a). Such a structure may range in size from about 40 bp, which corresponds to the minimal segment that can be processed effectively, up to an average of ~12 kb. If heteroduplex DNA is formed from nonidentical DNA strands derived from genetically different organisms, the duplex will contain mismatched bases. These may influence recombination frequencies, as discussed later.

B. Processing of Plasmid DNA

As is the case for chromosomal DNA, plasmids are taken up as linear single-stranded molecules that are subject to extensive fragmentation. Plasmid DNA gives rise to very few transformants in natural systems, presumably due to the difficulty of forming circular unit-length molecules from ssDNA fragments. The frequency of transformation by plasmid monomers is dependent on the square of the DNA

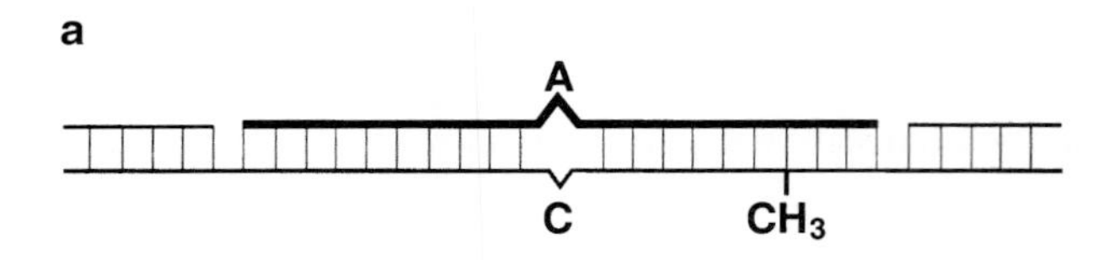

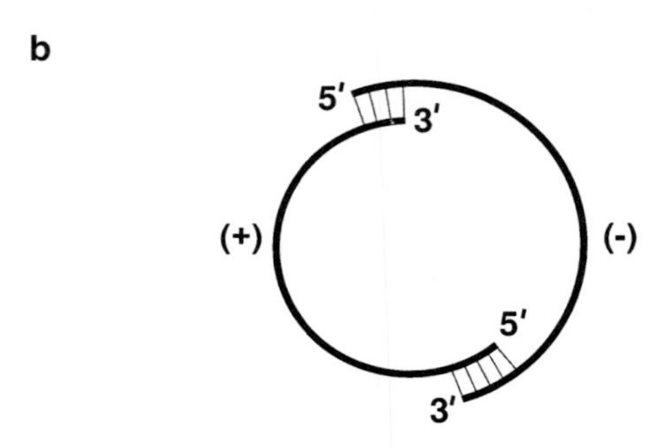

Fig. 4. Establishment of transforming DNA in the recipient cell. (a) Heteroduplex DNA formed by homologous recombination of an entrant strand of more than ~40 nt (bold line) and the complementary strand of the resident chromosome of the recipient cell (light lines). Gaps are sealed by DNA ligase. A and C represent mismatched nucleotides due to sequence divergence of the donor and recipient genomes. CH_3 indicates modification methylation of a restriction-enzyme recognition site in the recipient DNA; the entrant strand is unmodified. (b) Regeneration of a plasmid by the annealing of overlapping fragments of complementary (+) and (−) strands. A duplex circle is generated by DNA synthesis initiated at the 3′-OH termini.

concentration, suggesting that intact plasmids can be reassembled in the recipient cell from sections of different monomers taken up as separate events. The mechanism may involve the annealing of overlapping fragments of complementary (+) and (−) strands, followed by infilling of single-stranded regions by repair synthesis to give an intact duplex (Fig. 4b).

Plasmids also exist as multimers. Transformation frequencies by multimeric DNA show a linear dependence on DNA concentration, implying that a single multimer can give rise to a transformant. Monomeric circles might be generated by the annealing of overlapping parts of (+) and (−) strand fragments derived from one multimer. Another plausible explanation is that single fragments are sufficiently large to contain more than a complete set of plasmid genes; in effect one entrant (+) strand would possess a terminally redundant repeat. A unit-size circle would be produced if synthesis of the nascent (−) strand proceeded to copy one terminal (+) repeat, to be followed by the slippage of the newly synthesized end to pair with the other (+) terminus. The completion of the cycle of replication would give a circular (−) strand.

C. Sexual Isolation: Mismatch Repair and DNA Restriction

Nucleotide-sequence divergence is a barrier to the effective recombination of genes from closely related species, which is described as sexual isolation. The barrier is manifest by an exponential relationship between the reduction in transformation frequency and the sequence divergence of donor and recipient DNAs. One contributory factor to sexual isolation is the activity of mismatch-repair systems of the type that function to remove mispaired nucleotides inserted as replication errors during vegetative growth. An example is the Hex mismatch-repair system of *S. pneumoniae;* this operates preferentially to correct the donor contribution in a mismatched heteroduplex, thereby reducing the transformation efficiency of some markers by as much as 20-fold. Transition mismatches, such as A-C and G-T, are eliminated more effectively than transversions (e.g., A-G) and are inherited inefficiently as low-efficiency markers.

The Hex system is thought to recognize the mismatch in the newly formed heteroduplex and then initiate a bidirectional search for unsealed gaps flanking the donor-strand insertion. Thus, repair is targeted to the incoming strand, which may be removed entirely to be replaced by infilling through DNA synthesis. In this way, joints with imperfectly paired nucleotides are aborted. It is also possible that mismatch-repair enzymes inhibit the formation of mispaired heteroduplex joints in transformation. A precedent for this suggestion comes from properties of the *E. coli* counterpart of the Hex system, namely the Mut system. Purified Mut proteins have been shown to block RecA-mediated exchanges *in vitro* between nonidentical DNA molecules.

The mismatch-repair system is only partly responsible for impaired recombination of sequence diverged DNA in transformation. One line of evidence for this conclusion is the considerable sexual isolation detected when a repair-defective mutant is used as the transformation recipient; a second is the ready saturation of the Hex system by multiple mismatches in a single heteroduplex region. Thus, the main factor contributing to sexual isolation may be the physical obstacle that mismatches impose on RecA-mediated exchanges between nonidentical DNA strands.

DNA-restriction systems impose another barrier to productive gene exchange between different strains and species. Classic restriction endonucleases preferentially recognize dsDNA and cleave the DNA when the enzyme specificity sites lack modification methylation of specific bases in the target sequence. The prevalence and diversity of restriction-modification systems imply that restriction is a commonly encountered obstacle to productive gene exchange. Although restriction systems severely disrupt patterns of inheritance of unmodified chromosomal DNA transferred in conjugational and transductional crosses, they have little effect on the inheritance of unmodified chromosomal genes transferred by natural transformation systems. An explanation is that transforming DNA is refractory to cleavage because it is single-stranded and acquires protection following recombinational insertion into the recipient chromosome by the preexisting methylation of the recipient strand in the heteroduplex (Fig. 4a). Such hemimethylated DNA is the preferred substrate of modification enzymes, which primarily function as maintenance methyltransferases to effect the rapid conversion of hemimethylated DNA into the fully methylated state. In contrast to chromosomal DNA, unmodified plasmid DNA is sensitive to restriction during transformation. The explanation is that plasmids are regenerated from single-stranded fragments of entrant DNA and DNA synthesis on these strands gives a duplex molecule that is unmodified. Unmodified double-stranded plasmids transferred by artificial transformation procedures are likewise sensitive to restriction in the recipient cell.

V. ECOLOGICAL IMPLICATIONS OF BACTERIAL TRANSFORMATION

Transformation in natural environments has been examined for only a few organisms, but the elaborate nature of DNA uptake mechanisms and the ever-increasing list of species known to develop natural competence points to the conclusion that the process is important in the wild. Transformation is potentiated by the significant amounts of extracellular DNA—much being of microbial origin—that are present in a variety of environments, including marine water, freshwater, sediments, and soil. DNA concentrations in excess of 20 μg/liter water and up to 1 μg/g sediment have been recorded. In order to persist outside the cell, released DNA must acquire protection from ubiquitous extracellular DNases. Protection is conferred by the capacity of DNA to adsorb to the surfaces of clay and sand particles in soils and to humic acids. Binding to clay minerals is effected by H-bonds and electrostatic interactions between negative charges on DNA and positive charges on the edges of the clays. These interactions apparently change the conformation of the DNA in some way to confer protection from nucleases while, at the same time, allowing the DNA to interact with binding sites on competent cells. Evidence also exists that transformation occurs in marine water as well as in river epilithon—defined as the community of organisms in the slimy layer on the surfaces of stones in water—but not in sediments.

The notion that transformation is an evolutionary strategy for potentiating gene exchange raises the intriguing possibility that some cells in a population become genetic donors and release DNA under the conditions that favor competence development in recipients. In the well-studied *B. subtilis* system, DNA release has been detected at spore generation, competence development, and cell death. The release of DNA may reflect the activity of autolytic enzymes that hydrolyse moieties of the cell wall, as observed at the end of exponential growth of batch cultures. DNA may also be released by some active export process compatible with cellular survival. An interesting example of the coupling of donor and recipient activities is seen in the phenomenon of cell contact transformation. The process has been observed for several gram-positive and gram-negative organisms; it is bidirectional and requires cell-to-cell contact, yet is DNase sensitive and independent of detectable plasmids and phages.

In addition to promoting genetic diversity, gene transfer by bacterial transformation may contribute to maintaining the integrity of genomes by allowing recombinational repair of DNA damaged by various chemical and physical agents in the environment. DNA damage induces the expression of a network of dispersed genes, which collectively make up the SOS regulon. The repair scenario proposes that damaged sectors of the recipient genome are replaced by intact sequences acquired from other cells by transformation and processed by the recombination pathway of the SOS response. It should be pointed out that whereas classic SOS functions, including the derepression of the *recA* gene, are expressed spontaneously at competence development in *B. subtilis*, DNA-damaging treatments do not induce the organism to develop competence. The same applies to *H. influenzae*. Irradiation with ultraviolet light increases the rate of homologous recombination, but the response is apparently a consequence of the recombinogenic lesions that are generated in the recipient genome, rather than a reflection of stimulated DNA uptake.

A further idea is that transformation is a mechanism for scavenging the considerable amount of DNA present in natural habitats. According to this hypothesis, competence is an adaptation to starvation conditions that causes DNA uptake to potentiate alternate metabolic processes. Thus, nucleosides and bases salvaged from the DNA uptake process are channeled into the nucleotide pools or catabolized further to provide alternative sources of carbon, nitrogen, phosphate, and energy. The genetic complexity of the DNA uptake process, coupled with the specificity of some systems for homologous DNA, would seem to argue against the hypothesis that nutrient scavenging is the driving force in the evolution of transformation processes.

VI. ARTIFICIAL TRANSFORMATION SYSTEMS

The transformation of bacteria that do not develop natural competence was first reported in 1970 for the paradigm organism of bacterial genetics, the gram-negative *E. coli* strain K-12. During a study of bacteriophage transfection, M. Mandel and A. Higa discovered that *E. coli* could take up purified DNA of bacteriophage λ without the need for coinfection with helper virus if the cells were treated with high concentrations of $CaCl_2$ solution at 0°C; the entry of the transfecting DNA was achieved by exposing the mixture to a brief heat pulse after allowing time for adsorption of the molecules to the cell surface. Subsequently it was found that, once induced to the competent state by the ice-cold calcium treatment, amino acid-requiring mutants of *E. coli* could be transformed back to the wild-type state with linear fragments of dsDNA. However, such transformants, which arose by homologous recombination, could only be established in strains lacking exonuclease V, an enzyme that degrades linear DNA. Stanley Cohen and coworkers went on to report that both *E. coli* and its close relative *Salmonella typhimurium* could also take up plasmids as circular dsDNA molecules; their protocol used a 2 min heat pulse at 42°C followed by the incubation of the cells in growth medium to allow the phenotypic expression of the plasmid-borne drug-resistance genes prior to plating on antibiotic medium selective for the growth of transformants. This finding provided the mechanism whereby recombinant DNA molecules generated *in vitro* could be returned to bacterial cells for *in vivo* study, and this has since become a cornerstone of modern molecular biology. At present, this process is purely a laboratory phenomenon and whether *E. coli* uses something equivalent as a mechanism for genetic exchange in its natural habitat remains unresolved. From these initial observations of reported frequencies ranging across 10^3–10^6 transformants/μg input plasmid DNA, the method has been improved through the systematic manipulation of inducing conditions and strain genotypes, most notably by Douglas Hanahan, such that 100- to 1000-fold-higher frequencies can now be achieved. Using similar approaches, it is now possible to introduce DNA molecules into a diverse range of bacteria previously thought to be untransformable.

A number of parameters have been identified that contribute to the efficiency of artificial transformation of *E. coli*. Firstly, the genetic composition of the strain can have a marked effect, most likely accounted for by differences in the cell-surface lipopolysaccharide composition. Thus, variant rough strains with shortened, or completely lacking in, LPS O side chains are more easily transformed than wild strains. The peripheral long O side chains probably restrict access of the DNA to the uptake sites on the cell surface through either physical hindrance or ionic-charge interaction and are therefore dispensable, whereas the LPS core appears to be an essential component of the cell envelope for optimal uptake. Secondly, it is clear that although an artificially induced state, chemical competence is influenced by the physiological state of cells at the time of treatment. In general, the highest transformation frequencies are achieved with competent cells made from rapidly proliferating, late-exponential-phase cultures growing in rich medium. Moreover, growth at reduced temperatures (25–30°C) can also improve competence, and such cells can be induced to take up DNA by a less severe heat pulse. It has long been known that lipid composition is markedly affected by growth conditions, with membranes of low-temperature cells having a higher proportion of unsaturated fatty acids, which makes them more fluid; such membranes might allow easier transit of macromolecules. In addition, the growth of cells in the presence of Mg^{2+} ions improves competence, probably by modifying the structure of the LPS layer through the substitution of protein–LPS interactions with divalent-cation ionic bonds, thereby facilitating the removal or loosening of the LPS layer during the competence-inducing procedure.

Treatment with multivalent cations at 0°C is an absolute requirement for the induction of competence. However, other divalent ions can be substituted for calcium; barium, hexamine-cobalt, manganese, and rubidium have all been shown to work, and in some cases more effectively. Moreover, the addition of other chemicals, in particular dimethyl sulfoxide and the sulfydryl-reducing agent dithiothreitol, and certain physical procedures such as freeze–thaw can also be beneficial with particular strains. All of these probably influence cell-surface structure and membrane fluidity and so enable DNA molecules to access and transit the cell envelope. Thus simple and complex conditions have been developed that can be applied to all strains of *E. coli* with the expectation of one or other giving a high transformation yield; the reader is directed to the excellent review by Hanahan and Bloom (1996).

Chemical transformation may kill a large proportion of the cell population and, even under optimal conditions with excess DNA, only around 10% of the total viable cell population becomes transformed. The use of equimolar mixtures of two compatible and distinguishable plasmids has shown that co-transformation ocurs at a relatively high frequency; under conditions of DNA saturation, 70–90% of cells may receive both plasmid species. Thus, although only a small fraction of the cells may become induced to a competent state, those that do so are able to take up multiple DNA molecules very efficiently. However, when DNA is limiting, still less than 1% of the input plasmid molecules give rise to a transformant.

What then are the sites of DNA entry and what is the actual mechanism of DNA uptake in chemically induced cells? The cell envelope of a gram-negative bacterium like *E. coli* comprises outer and inner membranes separated by a peptidoglycan wall layer. The two membranes are fused to each other through holes in the rigid wall. It is thought that these zones of adhesion, estimated at 400/cell, constitute the channels through which DNA and other molecules are transported into the cell. Consistent with this, titration of transformation frequency against increasing amounts of transforming DNA suggests that there are probably a few hundred uptake sites per cell. Furthermore, there is a linear relationship between the increase in plasmid size and the decrease in transformation probability per plasmid molecule; and both supercoiled and relaxed DNAs transform with comparable frequencies across the size range of 2–66 kb. It seems most likely that uptake occurs by an active process; if it were simply passive diffusion through pores of a fixed size, there would be a dramatic drop in transformation efficiency with molecules larger than the pore size and compact supercoiled forms of large molecules would transform more efficiently than their relaxed isomers. There also is no evidence from competition experiments for the requirement of specific uptake sequences in the transforming DNA. Finally, the ability of competent *E. coli* cells to excise pyrimidine dimers introduced into transforming DNA molecules by prior UV irradiation indicates that it is dsDNA that is taken up, rather than ssDNA as found for natural transformation processes.

The uptake process itself can be divided into

stages. First, DNA molecules must bind to the cell surface at the sites of uptake. Chemical treatment probably exposes these sites via modification of the LPS structure, while the 0°C environment causes the membrances to adopt a suitable phase state for DNA association. Because both DNA and the phopholipid-based LPS are anionic polymers, it is likely that another function of the divalent cations is to shield and neutralize the phosphate backbone of the DNA, so overcoming repellent-charge interactions with the cell surface.

Two possibilities exist for the molecular channel that provides the route of DNA uptake. One suggestion is the cobalamin (vitamin B_{12}) transporter, which is known to be located at zones of adhesion; evidence comes from the observation that increasing amounts of cobalamin reduce transformation frequency. Thus the reason that inclusion of hexamine-cobalt salts during cell preparation boosts the transformation efficiency may be through their activation of the cobalamin transport system. Alternatively, it has been proposed by Rosetta Reusch and colleagues that the uptake channel is composed of poly-β-hydroxybutyrate (PHB) complexed with polyphosphate (polyP) and calcium ions. PHB has been isolated from the cytoplasmic membrances of many bacteria and eukaryotic cells, and found to accumulate to significant levels in *E. coli* during the induction of competence. Accumulation coincides with the appearance of a new high-temperature phase transition of the membrances, indicating that the lipid bilayer has undergone structural modification. The view is that the PHB–polyP–Ca^{2+} complex forms a cylindrical channel that spans the cytoplasmic membrane, with the PHB constituting the lipophilic exterior and the polyP–Ca^{2+} within the hydrophilic interior. After passage across the outer membrane, facilitated by disruption of the LPS layer by divalent cations, the introduced DNA displaces the polyP from the PHB pore and so enters the cell. Although transformants can be obtained at low frequencies without the heat pulse, this almost certainly increases membrance fluidity and so aids the release of DNA into the bacterial cytoplasm. Interestingly PHB accumulation has been correlated with the development of competence in the naturally transformable organisms *H. influenzae, B. subtilis,* and *Azotobacter vinlandii;* in the latter two cases an altered membrane phase transition has also been detected. Therefore, there may be similarities between natural and chemically induced competence at the level of the DNA-transport mechanism.

Electroporation is an alternative method for the introduction of naked DNA molecules into microbial cells and has been applied successfully to many organisms previously intractable to even chemical induction of competence. DNA and cells suspended in a solution of very low ionic strength are placed in a cuvette with two closely spaced electrodes and subjected to a brief high-voltage electrical pulse. The discharge causes a transient reversible depolarization and permeabilization of the cell membranes, so inducing the formation of pores, which allow entry of DNA molecules. The procedure can be used to introduce a variety of molecules including proteins and RNA, as well as DNA of various conformations. The yield of transformants is optimized by the alteration of the parameters of the electrical field discharged through the cells, and frequencies of 10^{10} transformants/μg plasmid DNA are achievable in *E. coli*. In contrast to the chemical method, electroporation can achieve the transformation of almost every viable cell if the DNA is in excess. Only 10% of the input molecules are taken up when the DNA is the limiting factor. Studies using mixtures of plasmid DNAs have shown that, as in the chemical method, uptake is efficient and multiple plasmid molecules can be transferred in a single process. Although this procedure has no requirement for treatment with divalent cations, *E. coli* mutant strains lacking LPS O side chains generally give higher yields of transformants, suggesting that, here again, accessibility of the DNA through the cell surface is important.

Once introduced by either chemical transformation or electroporation, covalently closed plasmids possessing a functional replication origin will be propagated as autonomous replicons. In contrast, genes carried on linear DNA fragments can only be established through homologous recombination. However, linear DNA fragments are subject to degradation by the exonuclease V component of RecBCD enzyme. Thus, efficient transformation with chromosomal genes can only be achieved in *E. coli* strains that lack exonuclease V, but remain recombination proficient. Appropriate strains are *recB recC* of *sbcB* triple mutants or the more robust *recD* mutants. The use of such strains in conjunction with linearized transforming DNA is a valuable procedure for targeting gene replacements and insertions to the bacterial chromosome and large plasmids.

The establishment of transforming duplex DNA is subject to the action of classic restriction systems that target molecules lacking modification methylation. There also exist methylation-dependent systems (Mcr) that recognize and cleave DNA methylated at certain cytosine residues. These systems may be particularly obstructive to the cloning of DNA from higher eukaryotes because this is often methylated extensively at CpG motifs. Although restriction systems reduce the overall transformation frequency, none constitutes an absolute barrier because escaping molecules acquire the modification status of the host cell and thus become immune to further attack.

A third method for the introduction of plasmid DNA, protoplast transformation, was developed for the antibiotic-producing streptomycetes by Mervyn Bibb, Judith Ward, and David Hopwood, and was subsequently applied by other workers to several other commercially important organisms, such as the corynebacteria. Osmotically stabilized protoplasts were treated with PEG, which precipitated the transforming DNA onto the exposed membrane surfaces, to promote uptake, and then were allowed to regenerate cell walls. Although the procedure was highly effective with plasmid and phage dsDNAs that underwent autonomous replication, giving frequencies approaching 10^7 transformants/μg input DNA, it worked only poorly with circularized and linear fragments of chromosomal dsDNA that required homologous recombination in order to be inherited. Subsequently, it was discovered that the latter type of transformation could be stimulated several hundredfold by the denaturation of the transforming DNA, which presumably generates ssDNA for strand invasion of the resident chromosome during homologous recombination. The procedure has significant practical application in streptomycetes because it facilitates targeted genetic manipulations of the chromosome such as gene disruption and replacement, mutational cloning, and complementation screening with ordered gene libraries. The stimulation of homologous recombination by transformation with denatured linear chromosomal DNA may also be applicable to genetic manipulation of other organisms for which transformation with dsDNA normally gives a high background of illegitimate integration events.

VII. TRANSFORMATION OF MICROBIAL EUKARYOTES

Although the initial claims for the transformation of yeast cells were recorded in 1960, these early studies failed to eliminate other explanations such as genetic reversion. Thus, the first convincing demonstration that a microbial eukaryote could be transformed with naked DNA was that of Albert Hinnen and colleagues who, in 1978, succeeded in converting a nonreverting *leu2* double mutant of bakers' yeast, *Saccharomyces cerevisiae,* to wild type using a cloned copy of the *LEU2* gene carried on a bacterial plasmid vector. The integration of the transforming DNA into the host chromosome was confirmed via Southern blot hybridization. Simultaneously, Jean Beggs showed that yeast could also be transformed with *in vitro* constructed recombinant plasmids capable of autonomous replication. In both cases, transformation involved the conversion of the yeast cells to spheroplasts by enzymatic degradation of the cell-wall β-glucan in the presence of sorbitol as an osmotic stabilizer. Polyethylene glycol, a known inducer of membrane fusion, and calcium were added to precipitate the DNA onto the spheroplasts, which were then plated in agar containing sorbitol in order to stabilize the transformants during their regeneration of a cell wall. Although technically difficult and time consuming, this procedure is capable of giving transformation frequencies up to 10^4/μg input DNA. Similar spheroplast transformation methods exist for several microbial eukaryotes, including the fission yeast *Schizosaccharomyces pombe,* and filamentous fungi, such as *Neurospora crassa, Aspergillus nidulans,* and *Trichoderma reesei.*

Most yeast researchers now use one of two procedures that allow the direct transformation of intact cells. The first, initially developed by H. Ito and colleagues and later optimized by R. H. Schiestl and R. D. Geitz, relies on the treatment of yeast cells with lithium ions. As with chemical transformation of bacteria, the lithium ions seem to permeabilize the yeast cell wall and facilitate the access of the DNA to the cytoplasmic membrane. To achieve maximal transformation frequencies, an excess of carrier DNA is added along with the transforming DNA and the mixture precipitated onto the cells with PEG at 30°C; DNA entry is effected by a 42°C heat pulse. Single-stranded DNA has a greater stimulatory effect as a carrier than dsDNA. Possibly ssDNA, which is thought to be a less-suitable substrate for uptake than dsDNA, saturates non-productive DNA binding sites. The net result is that ssDNA increases the effective concentration of transforming duplex DNA. The nature of the uptake channel and the molecular mechanism by which the DNA is taken into the cell nucleus are unknown. The second method is electroporation, which is performed in a manner very similar to that used for bacterial cells. In both methods, yeast cells are plated directly onto medium selective for growth of transformants and transformation frequencies as high as $10^7/\mu g$ DNA have been reported, although frequencies in the range 10^3–10^5 are more common. Again, related methods have been developed that can be applied to a range of yeasts, including several of commercial interest such as *Kluyveromyces spp., Hansenula polymorpha, Pichia pastoris,* and *Yarrowia lipolytica.*

All of these methods allow the introduction of both linear and circular dsDNA molecules and synthetic oligonucleotides. Because the frequency of mitotic recombination is high, recombinants of nuclear genes can be recovered easily. Indeed, homologous recombination between transforming DNA and the corresponding chromosomal region is stimulated several hundredfold if the DNA is linearized before introduction. Thus, targeted integration, gene replacement, and rescue of mutant chromosomal alleles into low-copy centromere-containing plasmids are all possible, as is the standard introduction of genes on extrachromosomal vectors. Such attributes make yeasts the most versatile eukaryotic organisms for molecular genetic studies.

A development for transforming microbial eukaryotes involves the use of biolistic devices. This process, originally developed for the transformation of plant material, literally involves firing microscopic tungsten or gold particles, previously loaded with DNA, at cells or tissues; the explosive force is generated either by the percussion of a gunpowder cartridge or rapid discharge of highly compressed gas, the net effect being to propel the microprojectiles at supersonic speeds into, and often through, the target material. This technology has been used to transform yeast cells not only with nuclear chromosomal genes, but also with mitochondrial DNA; thus, nonrespiring mitochondrial mutants could be transformed to respiratory proficiency and rho-zero (ρ^0) mutants lacking all mitochondrial DNA could be transformed to a stable rho-minus (ρ^-) state with a recombinant plasmid carrying a section of yeast mitochondrial DNA. The direct selection of mitochondrial transformants was not possible and they were recovered as cotransformants of nuclear gene transformants at a frequency of 10^{-3}–10^{-4}. However, despite this limitation, the discovery offers real possibilities for the molecular manipulation and study of eukaryotic organelle genomes.

See Also the Following Articles

Cell Membrane: Structure and Function • DNA Restriction and Modification • Quorum Sensing in Gram-Negative Bacteria • Plasmids, Bacterial • *recA* • Transduction: Host DNA Transfer by Bacteriophages

Bibliography

Bibb, M. J., Ward, J. M., and Hopwood, D. A. (1978). Transformation of plasmid DNA into *Streptomyces* at high frequency. *Nature* **274**, 398–400.

Dreiseikelmann, B. (1994). Translocation of DNA across bacterial membranes. *Microbiol. Rev.* **58**, 293–316.

Dubnau, D. (1993). Genetic exchange and homologous recombination. In "*Bacillus subtilis* and Other Gram-Positive Bacteria: Biochemistry, Physiology, and Molecular Genetics" (A. L. Sonenshein *et al.*, eds.), pp. 555–584. American Society for Microbiology, Washington, DC.

Dubnau, D. (1997). Binding and transport of transforming DNA by *Bacillus subtilis:* The role of type-IV pilin-like proteins—a review. *Gene* **192**, 191–198.

Fussenegger, M., Rudel, T., Barten, R., Ryll, R., and Meyer, T. F. (1997). Transformation competence and type-4 pilus biosynthesis in *Neisseria gonorrhoeae*—a review. *Gene* **192**, 125–134.

Gietz, R. D., Schiestl, R. H., Willems, A. R., and Woods, R. A. (1995). Studies on the transformation of intact yeast cells by the LiAc/ssDNA/PEG procedure. *Yeast* **11**, 355–360.

Grossman, A. D. (1995). Genetic networks controlling the

initation of sporulation and the development of genetic competence in *Bacillus subtilis. Annu. Rev. Genet.* **29**, 477–508.

Hanahan, D., and Bloom, F. R. (1996). Mechanisms of DNA transformation. *In* "*Escherichia coli* and *Salmonella:* Cellular and Molecular Biology" (F. C. Neidhardt *et al.*, eds.), 2nd ed., pp. 2449–2459. American Society for Microbiology, Washington, DC.

Kleerebezem, M., Quadri, L. E., Kuipers, O. P., and de Vos, W. M. (1997). Quorum sensing by peptide pheromones and two-component signal-transduction systems in Gram-positive bacteria. *Mol. Microbiol.* **24**, 895–904.

Lorenz, M. G., and Wackernagel, W. (1994). Bacterial gene transfer by natural genetic transformation in the environment. *Microbiol. Rev.* **58**, 563–602.

Majewski, J., and Cohan, F. M. (1998). The effect of mismatch repair and heteroduplex formation on sexual isolation in Bacillus. *Genetics* **148**, 13–18.

Se-Hoon, O., and Chater, K. F. (1997). Denaturation of circular or linear DNA facilitates targeted integrative transformation of *Streptomyces coelicolor* A3(2): possible relevance to other organisms. *J. Bacteriol.* **179**, 122–127.

Sherman, F. (1997). An introduction to the genetics and molecular biology of the yeast *Saccharomyces cerevisiae.* In "The Encyclopedia of Molecular Biology and Molecular Medicine" (R. A. Meyers, ed.), Vol. 6, pp. 302–325. VCH, Weinheim, Germany.

Yin, X., and Stotzky, G. (1997). Gene transfer among bacteria in natural environments. *Adv. Appl. Microbiol.* **45**, 153–212.

Transgenic Animal Technology

Simon M. Temperley, Alexander J. Kind, Angelika Schnieke, and Ian Garner
PPL Therapeutics, Ltd.

I. Generation of Transgenic Animals
II. Transgenic Livestock as Bioreactors
III. Xenotransplantation

GLOSSARY

construct An *in vitro*-generated fusion of DNA segments consisting of regulatory and coding regions designed to spacially or temporally express the protein of interest in a novel way.

DNA microinjection A technique for introducing nucleic acids into cells, such as the oocytes of *Xenopus*, mammalian cells in culture, or fertilized mammalian eggs, via a small-bore glass needle using a micromanipulator.

embryonic stem (ES) cells The pluripotent or totipotent cells of normal karyotype derived from the developing mammalian embryo. When introduced into other embryos of the same species, ES cells will contribute to the developing embryo in such a way that the resulting animal will be a chimaera of two cell types. If the germ cells consist of two populations of cells, whole animals derived from the original ES cells can be obtained by conventional breeding.

founder A transgenic animal resulting directly from the manipulation of a mammalian egg involving gene transfer. Such animals may be mosaic.

IVM/IVF The *in vitro* maturation of mammalian oocytes and subsequent *in vitro* fertilization of eggs.

lactation The period during which milk is produced following parturition. This varies from species to species and can be influenced by such factors as weaning times and machine milking. For example, the latter can extend lactation times dramatically. Typical lengths of lactation periods for the mouse, rabbit, pig, sheep, goat, and cow are 3, 4, 3, 8, 52, and 52 weeks, respectively.

line A family of animals derived from a founder animal and possessing the same germ-line modification.

mosaic A founder transgenic animal that possesses the desired germ-line modification in only a subset of its cells. Such animals are thought to arise following integration of the transgene after the first cell division of the embryo and may be incapable of transmitting the transgene to their progeny by virtue of an absence of the modification in their germ cells.

transgenic Referring to a plant or animal into which a gene from another species (the transgene) has been introduced. The term is also used more broadly to describe any organism whose genome has been altered by *in vitro* manipulation to induce, for example, specific gene knockout.

xenotransplantation The grafting of a whole organ or tissue from one animal species into another. Relates particularly to use of the domestic pig as an organ donor for humans.

zoonotic infection The cross-species transfer of a pathogen. For example, the risk of pig viruses entering the human population via xenotransplant patients and causing a pandermic is viewed as a very serious risk.

THE ABILITY TO GENETICALLY ALTER the chromosomal material of mammals has opened up exciting new possibilities for the use of farm animals. One such application is to modify the mammary gland in such a way that additional, commercially valuable proteins are secreted into milk during lactation. Such transgenic livestock effectively become walking bioreactors, producing large quantities of (usually human) recombinant proteins in their milk at a relatively low cost. Furthermore, not only has the gland an amazing capacity for protein synthesis, but it appears capable of performing posttranslational modifications that are frequently essential for full biological activity. Such animals represent an easily maintained and infinitely renewable source of disease-curing and nutritional recombinant proteins. The scope of transgenic technology is not limited merely to protein production. It plays a central role in the ever-growing possibility that animal tissues and whole organs, in particular those of the pig, may be used in human transplant surgery, circumventing the chronic shortage of human donor organs. Genetic modification of pigs offers us the chance to overcome the considerable immunological and physiological challenges faced by replacing human organs and tissues with those from another species.

I. GENERATION OF TRANSGENIC ANIMALS

In 1974, Rudolf Jaenisch and Beatrice Mintz produced the first viable transgenic animal by injecting viral DNA into the cavity of a mouse blastocyst-stage embryo (Jaenisch, 1974). Since that time, the study of genetically modified mice has made a huge contribution to biological understanding. This has been a direct result of the development of two broad methods of genetic manipulation. Direct DNA transfer into the embryo provides a straightforward and rapid means of adding exogenous DNA fragments at random locations in the host genome. Alternatively, genetic manipulation can be carried out in cultured cells that are capable of supporting or participating in the development of a whole animal. Such cell-mediated transgenesis permits greater control over the genotype of the resulting animal and facilitates the engineering of a wide range of precisely defined modifications at predetermined loci, termed gene targeting. Both of these approaches have allowed the elucidation of many mechanisms involved in both gene expression and fetal development. Although mice remain a fundamental tool of experimental biology, there is now a growing emphasis on the production of genetically modified livestock animals with direct medical or agricultural benefits. This article concentrates on livestock.

The single most important application of genetically modified livestock has been the production of therapeutically valuable proteins in milk. Several drugs produced in this way are close to being available for clinical use; these include alpha; 1-antitrypsin, fibrinogen, and antithrombin III. Also notable has been the modification of animal, principally porcine, organs for use as human transplants. The production of pharmaceutical proteins in the milk of transgenic animals and the possible use of transgenic pigs for xenotransplantation are discussed in more detail later. Work in both of these fields was based exclusively on animals produced by random transgene addition. The opportunity to carry out gene targeting would add significantly to these and other applications. For example, transgene expression could be enhanced by placement at a favourable chromosomal locus, bovine milk could be improved as a source of human-infant nutrition by the removal of antigenic proteins, and development of transplantable organs would be aided by the ablation of xenoantigens on porcine tissue. The recent development of a cell-mediated method of transgenesis has now brought the prospect of gene targeting in livestock significantly closer to realization.

A. DNA Transfer Directly into Embryos

1. *Pronuclear Microinjection*

Graham and Ruddle first reported that naked DNA can be introduced by microinjection into the pronuclei of fertilized mouse eggs in 1980. DNA introduced in this way can stably integrate into the host genome, becoming incorporated into the somatic and germ tissues and passed on to progeny in Mendelian fashion. It was subsequently shown that by inclusion of appropriate regulatory elements, the tissue-specific expression of the integrated transgene can be specified. Microinjection was successfully extended to livestock by Hammer in 1985. Since then, it has been the method of choice for the production of transgenic pigs, cows, and sheep.

Microinjection is a straightforward method that has been used to generate transgenic animals of many livestock species using a wide variety of transgenes. Providing precautions are taken to protect the DNA from shearing, large DNA fragments can also be successfully microinjected. A 0.5-Mb transgene introduced by the microinjection of DNA protected by the polyamines spermine and spermidine has been characterized as an intact integrant.

Microinjected DNA usually integrates as tandem repeats oriented head to tail at a single locus randomly located in the host genome. Because coinjected DNA fragments tend to cointegrate and often coexpress, the method lends itself to the production of animals with multiple transgenes. This has been used to generate mice and sheep that coexpress three distinct transgenes encoding the α-, β-, and γ-chains of human fibrinogen in their milk. The copy number and ratio of each transgene can to a large extent be controlled by varying the amount of DNA microinjected.

Microinjection is, however, inherently an inefficient process that presents producers of transgenic livestock with practical difficulties that arise largely as a result of the high costs incurred and the long generation intervals of large animals. Multiple copies of the transgene are randomly inserted at a single locus. Transgene expression is generally unpredictable. The delayed integration of exogenous DNA can lead to the production of mosaic animals in which the transgene is not present in all cells. Incomplete contribution to the germ line can therefore reduce the frequency of transmission to the offspring, thus increasing both cost and the timeline.

Typically 5–20% of mice born after microinjection are transgenic. However, current data from sheep, goats, pigs, and cattle indicate that a significantly smaller proportion (1–5%) of liveborn offspring of these species are transgenic. This may be due in part to the relative opacity of livestock oocytes and the consequent difficulty in visualizing the pronuclei. Improvements in the proportion of transgenic mice to 40–56% after microinjection has been claimed by consecutive introduction of DNA into both pronuclei of a fertilized oocyte; however, there are no reports that such a procedure increases the transgenic rate in large animals.

Because species such as sheep and cattle can gestate only a few embryos, a large number of animals are required as embryo-transfer recipients, most of which are effectively wasted gestating nontransgenic embryos. This is a major source of inefficiency and a considerable burden on large animal studies. For example, Eyestone (1994) has estimated that approximately 1200 microinjected bovine zygotes are required to produce a single transgenic calf. This requires approximately 300 donor cows to provide fertilized eggs and 600 recipient cows to gestate the microinjected embryos. There has, therefore, been a considerable incentive to reduce the number of animals required. The use of donor animals can be avoided by the use of oocytes extracted from ovaries excised from animals at slaughter. Such oocytes can be matured and fertilized *in vitro* and are capable of producing animals on transfer to foster mothers. Although the fertilized oocytes produced *in vitro* are less viable then those derived *in vivo,* this is offset by the greater number available. This technique was first developed in cattle and has been used to produce transgenic calves. Further refinements of oocyte maturation, fertilization, and culture conditions can be expected to improve the quality of such *in vitro*-derived embryos in the future.

The number of animals used as recipients could be reduced significantly if embryos carrying an integrated transgene could be identified prior to transfer. One approach to this has been to analyze biopsy samples from embryos cultured to blastocyst stage for the presence of transgene DNA, usually by the polymerase chain reaction (PCR). There have been several reports of transgenic cattle produced from such biopsied embryos. However, nonintegrated DNA molecules persist for several days after microinjection and this results in a high proportion of embryos falsely identified as transgenic by the PCR assay.

The best means of identifying embryos carrying an integrated transgene is to coinject the transgene with a construct that expresses the green fluorescent protein (GFP) of the jellyfish *Aequoria victoria.* GFP fluorescence in intact living embryos can be detected nondestructively by illumination with light of the excitation wavelength. Because coinjected DNA fragments usually cointegrate into the host genome, the presence of the GFP transgene can be used as a marker for the presence of the transgene. One published report demonstrated that 8 of 12 mice identified as GFP fluorescent at the blastocyst stage were transgenic. This approach has not been fully evaluated in large animals.

2. *Retroviral Gene Transfer*

Retroviruses infect susceptible mitotic cells with very high efficiency and result in the integration of a single copy of proviral vector DNA into the host genome. This has led to the development of replication-incompetent retroviral vectors capable of transducing non-viral genes into cells *in vitro* and *in vivo*, including zygotes and early embryos.

Transgenic mice were first produced by the coculture of early embryos with cells producing replication-competent retroviruses in 1975 by Jaenisch. Since then, there have been several reports of mice produced by retroviral infection and retrovirus-mediated gene transfer also facilitated the production of transgenic chickens, which because of the inaccessibility of the early avian embryos cannot easily be achieved by other methods.

Although retroviral gene transfer is potentially more efficient than pronuclear injection, it has not been widely used to produce transgenic large animals. Haskell and Bowen injected cells producing replication-defective virus into the perivitelline space of bovine zygotes and achieved 7% transgenic fetuses at day 90. This approach has been improved more recently, and viable cattle have been produced by the infection of mature oocytes using a Moloney murine leukemia virus-based vector carrying the envelope glycoprotein from the vesicular stomatitis virus, which promotes retroviral entry through the cell membrane. Infection at the mature-oocyte stage is argued to result in a higher frequency of retroviral integration than infection at the zygote or later stages. Retroviruses gain access to the chromatin when the nuclear envelope breaks down during mitosis. Therefore, the lack of a nuclear envelope in mature oocytes offers a greater opportunity for retroviral preintegration complexes to integrate into the chromatin than the short period occurring during mitosis in cycling cells.

Retroviruses do, however, suffer several disadvantages that severely limit their usefulness. The size of the transduced gene is limited by the capacity of the retroviral particle, which effectively restricts the choice of transgene construct to cDNAs, which are generally poorly expressed as transgenes. Delayed retroviral integration and the possibility of several different independent integrations leads to the frequent production of mosaic animals, which can fail to transmit the transgene through the germ line. Insertion of retroviral long terminal repeats that have powerful promoter activity can also cause the activation of genes adjacent to the integration site with possible deleterious effects. Perhaps the most serious problem with transgenic animals carrying retroviral vectors is the risk of producing replication-competent virus by recombination. Uncertainty regarding this possiblity excludes the use of animals containing retroviral transgenes for most human applications. Nevertheless they are used in gene therapy in humans.

3. *Sperm-Mediated DNA Transfer*

It has long been known that rabbit spermatozoa can associate with and take up DNA into the sperm head. A 1989 report by Lavitrano claimed that mouse spermatozoa exposed to exogenous DNA could be used as a vector to generate transgenic mice by artificial insemination. This report stimulated considerable interest because it seemed to offer a simple approach to the production of transgenic animals of many species. Subsequent research in this field has established that if seminal plasma is removed, the sperm cells of every species associate with exogenous DNA. However, despite considerable efforts from many laboratories around the world, the production of transgenic mice by this approach has not been repeated. Furthermore, results in other mammalian species have not been encouraging. The highest rates of transgenesis (23–37%) have come from fish, although such transgenic fish frequently failed to express the transgene. There have been two reports of transgenic calves and pigs, but in both cases the transgenes were apparently rearranged. This is consistent with findings that DNA that penetrates the sperm nucleus is subject to fragmentation and rearrangement. However, the details of the mechanism of exogenous DNA uptake by sperm cells and the fate of internalized DNA have yet to be properly elucidated.

Sperm-mediated gene transfer is now regarded with general scepticism. However, it remains possible that further development of the approach will result in a useful means of transgenesis.

B. Stem Cell-Mediated Transgenesis

1. *Embryonic Stem Cells*

Embryonic stem (ES) cells are totipotent cells derived from early embryos that can be cultured for extended periods in an undifferentiated state. When reintroduced into a host embryo, they are able to participate in development and can contribute to all tissues of the animal, including the germ line. In the mouse, the use of appropriate coat-color markers allow the host embryo and ES-derived components in the resulting chimeric animal to be distinguished.

ES cells provided the first means whereby genetic manipulation could be carried out *in vitro* and the manipulated genotype transferred to a whole animal. ES cells offer a direct alternative to microinjection if DNA microinjection is problematic, for example, if the transgene to be introduced is large, such as a yeast artificial chromosome. However, the most important advantage is that DNA transfer into a cell intermediate offers the opportunity to select, isolate, and analyze cells carrying desired genetic modifications before whole animals are produced. In contrast, direct transfer into embryos requires that detailed genetic analysis be carried out after the transgenic animals have been produced. The ability to carry out such analysis *in vitro* is a major advantage to transgenic large-animal programs in which there are strong ethical and financial incentives for reducing the number of animals used.

The most important application of ES cells in the past decade has been in the use of gene targeting, which has greatly extended the range of genetic modifications possible in mice. Gene targeting exploits the ability of ES cells to support recombination between exogenous DNA molecules and their cognate chromosomal sequences at regions of shared homology. Essentially, DNA constructs designed to induce genetic modification by homologous recombination are transfected into ES cells and selection strategies used to identify those transfectants that have undergone homologous recombination. There have been many demonstrations of gene targeting in mice, including inactivation of individual endogenous genes by insertion or deletion, replacement of whole genes, precise placement of transgenes in the host genome, subtle gene modifications by precise mutagenesis and deletion of megabase-size DNA fragments to render large chromosomal regions hemizygous. These have dramatically improved our understanding of mammalian development.

Although great efforts have been made to derive ES lines from other mammals, definitive ES cells capable of contributing to the germ line of a chimeric animal have not yet been isolated. However, there are numerous reports from other mammalian species based on the rather looser definition of ES cells as "cells capable of differentiating *in vitro* along at least three different embryonic lineages." These include hamster, mink, sheep, cattle, pig, rhesus monkey, and human. Although the production of pig and rat chimeras have been reported, in neither case has the ES contribution to the germ line been demonstrated. The functionality of human ES cells remains untestable for obvious ethical reasons. Therefore, the use of ES cells as a means of transferring genetic modifications from the culture dish to whole animals remains at present restricted to the mouse.

If and when large-animal ES cells do become available, their use in the production of chimeras in the same way as mouse ES cells will be time consuming. Farm animals have a long generation interval, and the production and breeding of chimeras can delay the analysis of the phenotype by several years. Chimera production can be avoided in mice by deriving animals entirely from ES cells (e.g., by the aggregation of ES cells with disadvantaged tetraploid embryos). However, this method is exquisitely sensitive to the particular ES cells used and the usefulness of this method for large-animal ES cells is unknown.

2. *Embryonic Germ Cells*

Embryonic germ (EG) cells are undifferentiated cells functionally equivalent to ES cells; that is, they can be cultured and transfected *in vitro* and then contribute to the somatic- and germ-cell lineages of a chimera. EG cells are derived by the culturing of primordial germ cells, the progenitors of the gametes, with a combination of growth factors—leukemia inhibitory factor, steel factor, and basic fibroblast growth factor. These factors promote the long-term growth of the primordial germ cells and their conversion to an undifferentiated cell type closely resembling ES cells.

In contrast with the small number of cells available from blastocyst-stage embryos, a relatively large number of primordial germ cells can be derived from somite-stage embryos. Thus, the isolation of EG cells from farm animals might be a viable alternative to ES-cell derivation.

EG lines have been isolated from rat, pig, and cattle. Blastocyst injection of cultured EG cells led to production of chimeric bovine embryos that survived to mid-gestation. Chimeric male piglets have been produced from both genetically manipulated and normal porcine EG cells. In both instances, the EG cell contribution to the testis was detected. However, unfortunately, the germ-line transmission could not be demonstrated, as one of the animals was stillborn and the other failed to thrive after birth and was sacrificed.

3. *Spermatogonial Stem Cells*

Spermatogonia are self-renewing stem cells of the adult testes that give rise to the male gametes, spermatozoa. Laboratory work by Ralph Brinster has shown that spermatogonia from one mouse can be injected into the testes of another, where they locate to the seminiferous tubules and produce functional spermatozoa. Most remarkably, rat spermatogonia transplanted into mouse testes were functional and could produce spermatozoa.

These findings indicate that spermatogonia may in the future provide a means of producing transgenic animals by direct manipulation of the germ line. Such a method, which circumvents embryo manipulation, could be of particular value to the producers of transgenic livestock. If transspecific spermatogonial transplantation can be extended to other species, it is conceivable that sperm from a valuable or difficult species could be produced in the testes of a more accommodating species.

Whether spermatogonia-mediated transgenesis becomes a reality will depend on the successful development of conditions for isolating and culturing spermatogonial cells to allow DNA transfer, and the selection or identification of transfected cells and their expansion to sufficient numbers to facilitate transplantation. Brinster's lab has cultured spermatogonia *in vitro* for up to 4 months while retaining their ability to repopulate a recipient testis. There have been no published reports of spermatogonial transfection *in vitro*.

C. Nuclear Transfer

1. *Background*

The replacement of the nucleus from an egg with that of another cell was first suggested by Spemann in 1938 as a means of determining whether nuclei of differentiated and undifferentiated cells have the equivalent development potential. In the 1950s, Briggs and King showed that nuclei from cells from blastocyst stage embryos of the frog *Rana pipiens* could be transplanted to enucleated eggs and were able to direct normal development to feeding-stage larva, whereas nuclei from the mesoderm or endoderm of the late-gastrulation stage were unable to do so. This led to the conclusion that the totipotency of somatic-cell nuclei is lost during development. This view gained support from the results of subsequent experiments using the frog *Xenopus laevis*, most notably by John Gurdon. The transfer of nuclei from early embryonic cells or larval intestinal cells to enucleated oocytes produced normal larva and some sexually mature frogs, whereas the use of terminally differentiated, adult foot-web skin cells as nuclear donors failed to produce animals that survived beyond metamorphosis. Whether these findings reflected real restrictions in the developmental capacity of adult cells or were due to technical aspects of the procedure is unclear.

During the 1980s, the culture conditions and micromanipulation techniques for mammalian embryos were significantly improved and nuclear transfer in mammals became a practical proposition. Somewhat unusually, the major developments and breakthroughs in the field of mammalian nuclear transfer were made using livestock rather than mouse embryos. Nuclear transfer in livestock is carried out by enucleating a zygote, or more commonly an unfertilized oocyte, by either microsurgery, chemical enucleation, or chromosome inactivation using UV or laser treatment. The donor nucleus is introduced into the cytoplast either by microinjection or by cell fusion using polyethylene glycol, inactivated Sendai virus, or electrofusion. Reconstructed embryos are then transferred to the oviduct of foster mothers to complete gestation.

In 1986, Willadsen reported the first viable mammal produced by nuclear transfer, a sheep produced by the transfer of a nucleus from an eight-cell embryo. For a decade after that time, successful nuclear transfer in mammals was limited to the use of cells obtained directly from early embryos or subjected to very short periods in culture. Nuclear transfer in mice proved to be particularly difficult. A common observation was that blastomeres from two- or four-cell-stage mouse embryos could support nuclear transfer, but nuclei from later-stage embryos, could not. It was thought that the difficulties associated with mice were related to the time at which the embryonic genome becomes transcriptionally active. Mouse embryos show a marked increase in transcription at the two-cell stage, pigs at the four-cell stage, cows and sheep at the eight-cell stage, and rabbits at 16-cell stage. It is conceivable that a delay before embryonic genome activation allows the transplanted nucleus to be reprogrammed by the oocyte cytoplasm. Alternatively the difficulty associated with mouse nuclear transfer may have been purely a technical problem. Mouse oocytes are very fragile, and techniques have been developed that achieve high survival rates of the manipulated oocyte; fertile mice were obtained by nuclear transplantation using cumulus donor nuclei derived from adult animals.

One feature that emerged from all the previous work was the importance of matching the cell-cycle stage of the donor nucleus and the oocyte cytoplasm. Oocytes of most mammalian species pause twice during meiosis, before the first and then at the second meiotic metaphase, at which stage, the oocyte is mature and can be fertilized. Oocyte maturation and arrest are induced by the activity of a regulatory factor known as either maturation or mitosis promoting factor (MPF). MPF is a two-subunit protein composed of cyclin B and protein kinase p34(cdc2), both of which are known to be important regulatory factors in the mitotic cell cycle. The high level of MPF in arrested oocytes is maintained by the multiprotein complex cytostatic factor (CSF), an essential component of which is the C-mos (protooncogene) protein. CSF activity is mediated by the mitogen-activated protein (MAP) kinase. When the oocyte is fertilized by a sperm, there is a large transient increase in cytoplasmic free Ca^{2+} ions. This causes the inactivation of CSF and proteolytic cleavage of the cyclin component of MPF. Reduction in the level of MPF breaks the arrested state and the fertilized oocyte completes meiotic division.

The level of MPF in the oocyte has a profound effect on the outcome of nuclear transfer. If a Nucleus is transferred into oocyte cytoplasm with high MPF, the nuclear envelope breaks down and chromatin undergoes chromosome condensation, followed by nuclear reformation and DNA replication. A nucleus from a cell in G1 phase will undergo normal DNA replication and can support normal development. However, where the donor nucleus is in S or G2 phase, aberrant rereplication of DNA occurs, causing aneuploidy or chromosomal damage and consequent failure of development. Nuclear transfer efficiency into unfertilized oocytes can therefore be improved by synchronizing donor nuclei in G1.

In contrast, if a nucleus is transferred to a fertilized oocyte in which MPF levels have declined, nuclear-envelope breakdown does not occur and the cell cycle of an incoming nucleus in either the G1, S, or G2 phase will be completed normally and normal development can occur. MPF can also be induced to decline by experimentally activating the oocyte by exposure to ionomycin, ethanol, or strontium, or by an electrical pulse and has been used to produce live offspring by the transfer of nuclei in S or G2 phase.

DNA replication occurs as a consequence of the breakdown of the nuclear envelope. It subverts the mechanism that restricts the duplication of genomic DNA to precisely once per cell cycle. DNA licensing factor is a protein complex that binds to replication origins, licensing them for replication in S phase. The access of licensing factor is strictly restricted by the nuclear envelope. At mitosis, the nuclear envelope breaks down and licensing factor in the cell cytoplasm gains access to chromatin where it binds to replication origins. At S phase, DNA replication irreversibly displaces licensing factor, relicensing and rereplication is prevented because the nuclear envelope prevents further access.

2. *Nuclear Transfer from Cultured Cells*

Embryonic blastomeres are unsuitable for cell-mediated transgenesis because they are available in very small numbers and cannot be cultured for a sufficient time to allow genetic manipulation. However in 1996, Wilmut and collaborators showed that sheep embryonic cells that had been maintained in culture for several weeks could be used as nuclear donors. It was proposed that the key to successful nuclear transfer was the induction of a quiescent state in the donor cell by culturing them in the presence of reduced amounts of serum, termed serum starvation. In the quiescent state, termed G0, cell-cycle activity is absent, protein synthesis is reduced, and changes occur to the chromatin that may render the nucleus more susceptible to reprogramming by the oocyte cytoplasm. This approach was subsequently extended to the use of ovine fetal fibroblasts and ovine adult mammary cells. The precise treatment of the nuclear donor cells to achieve nuclear transfer and the requirement for quiescence has, however, remained the subject of some debate.

Nuclear-transfer technology is being applied to many other species and cell types. The technique is in the early stages of development and several problems remain to be addressed. In particular, there is a higher than normal incidence of perinatal mortality among animals generated by nuclear transfer. It is possible that this is not a result of nuclear transfer as such, but of technical aspects of the procedure. Many types of manipulation of preimplantation embryos other than nuclear transfer have been reported as increasing fetal morbidity and mortality (e.g., *in vitro* culture, asynchronous embryo transfer, and progesterone treatment of the mother). Factors already known to affect late fetal development include the presence of serum in embryo culture medium and imprinting of the early embryonic genome.

An increased understanding of the factors that determine the success of nuclear transfer can only be gained by further experiments. In this respect, it is particularly noteworthy that mice have been produced from naturally quiescent cumulus cells using a modified method of nuclear transfer by Wakayama (1998). Mice are clearly more amenable to experiment than livestock and their use will significantly speed progress in the field.

3. *Nuclear Transfer and Transgenic Animals*

Although still in its infancy, the use of somatic-cell donors for nuclear transfer in livestock offers many advantages over pronuclear microinjection as a means of generating transgenic animals. Perhaps the most significant of these is a large reduction in the number of experimental animals required, which is desirable for both ethical and commercial reasons.

Primary fetal fibroblasts provide a convenient cell type for cell-mediated transgenesis in livestock species. A large number of them can be obtained and they can be cryopreserved to provide a uniform stock of early passage cells. They are readily transfectable and with care can undergo genetic manipulation and selection in culture within a short time and retain their ability to support successful nuclear transfer.

The demonstration of nuclear transfer as a practical means of producing transgenic large animals came with the production of cows by transfer of nuclei from primary fetal fibroblasts stably transfected with a selectable marker gene (neomycin) and by the production of sheep transgenic for the human clotting factor IX. Recombinant Factor IX protein expressed in the mammary gland of these animals provides a safe alternative to human blood-derived products for the treatment of hemophilia B. Furthermore, if nuclear transfer-competent cells support gene targeting by homologous recombination, this would provide the opportunity for precise positioning of the transgene, as well as the deletion, replacement, and mutation of an endogenous gene. (See Figs. 1 and 2.)

II. TRANSGENIC LIVESTOCK AS BIOREACTORS

A. Gene Expression in Transgenic Livestock

1. *Milk Proteins*

The major milk proteins are the caseins (e.g., as1, as2, β, and κ), which form large spherical complexes called micelles, and the whey proteins, (e.g., β-lactoglobulin, whey acidic protein, and α-lactalbumin). Apart from their obvious nutritional function, these proteins also have roles in maintaining the integrity of the milk itself. Between them, the two classes constitute greater than 90% of the protein in milk. Consequently, the genes for these proteins have been the main vehicles for transgenic studies.

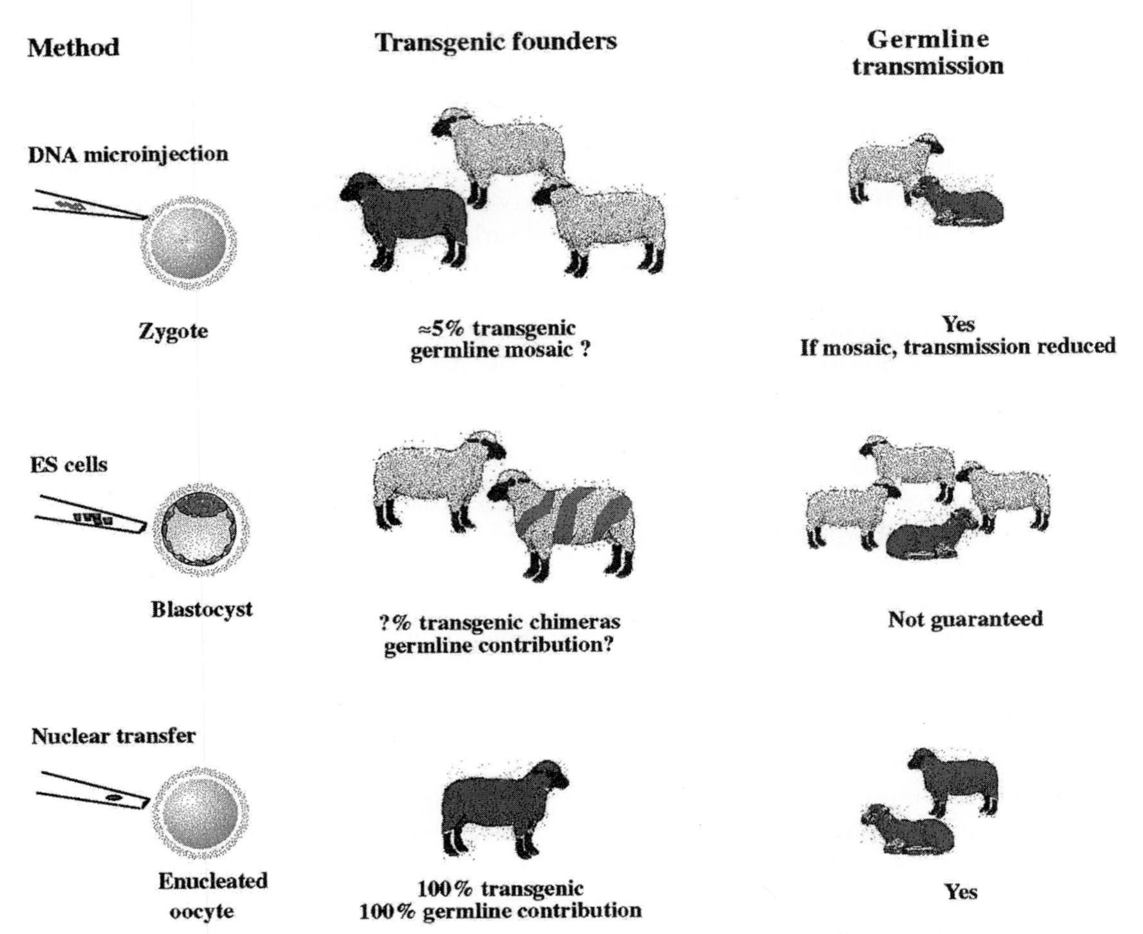

Fig. 1. Production of transgenic sheep. Schematic diagram of three methods for the production of transgenic animals—DNA microinjection, ES cells, and nuclear transfer of genetically modified donor nuclei. Expected germ-line transmission is also indicated.

Caseins are almost always the most abundant milk proteins. They are usually around 20,000 MW, relatively hydrophobic, and, under the ionic conditions in milk, associated into colloidal micelles. There are five casein genes known in the mouse and four major casein genes in cows and sheep. In all these animals, the genes are tightly linked on a single chromosome and fall within 400 kb of DNA. The proportion of casein differs considerably between humans (45%) and cows (80%), and this may be reflected in the fact that only two caseins are known in humans β and κ.

Of the whey proteins, α-lactalbumin (α-lac) has been detected in all milks that contain lactose and this is presumably a consequence of this protein's role in lactose synthesis. α-lactalbumin is a subunit of lactose synthetase and has a molecular mass of around 14,000. It plays a key role in the regulation of milk lactose content, contributing to milk osmotic pressure and volume. A deficiency in α-lac severely disrupts the ability of a mammal to lactate. β-lactoglobulin (β-LG) is a protein of 162 amino acids and is the most common whey protein in the milk of ruminants. Homology to retinoic acid-binding proteins has led to the suggestion that β-lactoglobulin may function in vitamin A transport. Whey Acidic Protein (WAP) is the major whey protein in rodent milk, constituting some 10–15% of the total polyA+ mRNA of mouse mammary gland. It is a member of the disulfide-core family of proteins. Initially thought to behave as a nutritional source, WAP may have a wider role, suggested by its structural similarity to mucous protease inhibitors (proteins that participate in tissue modeling).

Given the huge synthetic capacity of the mammary gland, it was immediately obvious to some that, if a heterologous gene could be driven by a milk gene promoter, it should be possible to secrete high levels of the gene product in milk. Assuming that high levels of expression can be achieved, there are additional advantages. The dairy industry, especially for cows but also for sheep and goats, is well established and scientifically advanced, providing the necessary infrastructure for the production and processing of transgenic milk; the production of milk is easy, cheap, renewable, and noninvasive; and milk has a very positive image in the mind of the consumer.

Genes for the major milk proteins have been the vehicles for both the study of milk proteins in transgenic animals and the elaboration of vector systems for the expression of heterologous proteins in the mammary gland. Due to the extended timescale and cost involved in work with livestock, most initial studies are performed using the mouse as a model system. A collection of published results for various species are shown in Table I. In the following two sections, we concentrate on some of the highlights of this work.

2. *Regulation of Transgene Expression*

a. Caseins

Investigating the regulation of casein gene expression, Rosen and co-workers generated lines of transgenic mice containing either the entire rat β-casein gene or the rat β-casein promoter driving expression of chloramphenicol acetyl-transferase (CAT). The levels of transgene mRNA expression were only 0.01–1% of the endogenous mouse β-casein. CAT activity was specific to the mammary gland, but levels were low. The removal of the first exon and most of the first intron from the gene completely abolished expression, suggesting a role for intragenic sequences in control of expression. The authors drew a parallel between the casein gene family and the globin gene family, which are coordinately expressed under the control of a locus control region (LCR). The whey protein genes, WAP, β-LG, and α-lactalbumin appear to be unlinked in the genome; however, the casein genes are arranged on a single chromosome and are tightly linked. This suggests these genes could be under coordinated, long-range control also, although this hypothesis awaits convincing evidence to back it up.

Mammary explant cultures derived from these transgenics have been used to examine hormonal control of β-casein. Endogenous and transgenic β-casein genes were appropriately stimulated by prolactin; however, β-casein–CAT constructs were not. Similar effects were seen with insulin and glucocorticoids, implying that intragenic sequences mediated hormonal regulation of β-casein. This type of control, however, is not entirely due to DNA downstream from the promoter because some stimulation can be detected. Work with hormone-responsive cell lines has indicated that induction of the rat β-casein gene may be by relief of transcription repression. An analysis of the mouse β-casein gene in pregnant and lactating mice identified a pregnancy-specific mammary nuclear factor that appears to serve as a repressor mediating the inhibitory action of progesterone on β-casein. Thus, the design of efficient constructs to drive heterologous protein expression to the transgenic mammary gland is often a complex task.

Experiments designed to express heterologous proteins from casein promoters have been both successful and innovative. Interleukin-2 (IL-2) has been expressed in rabbit milk from 2 kb of rabbit β-casein promoter. Of four transgenic females analyzed, the highest level of expression was 450 ng/ml. A massive 21 kb of bovine αs1-casein promoter was used by Meade and co-workers to drive the expression of the human urokinase gene in transgenic mice. Three founder animals were obtained, but only one transmitted the transgene. The female offspring of this mouse secreted active urokinase at concentrations of 1–2 mg/ml in the milk. In both these cases, the expressed heterologous gene included some or all of its natural introns, a feature that seems to be neces-

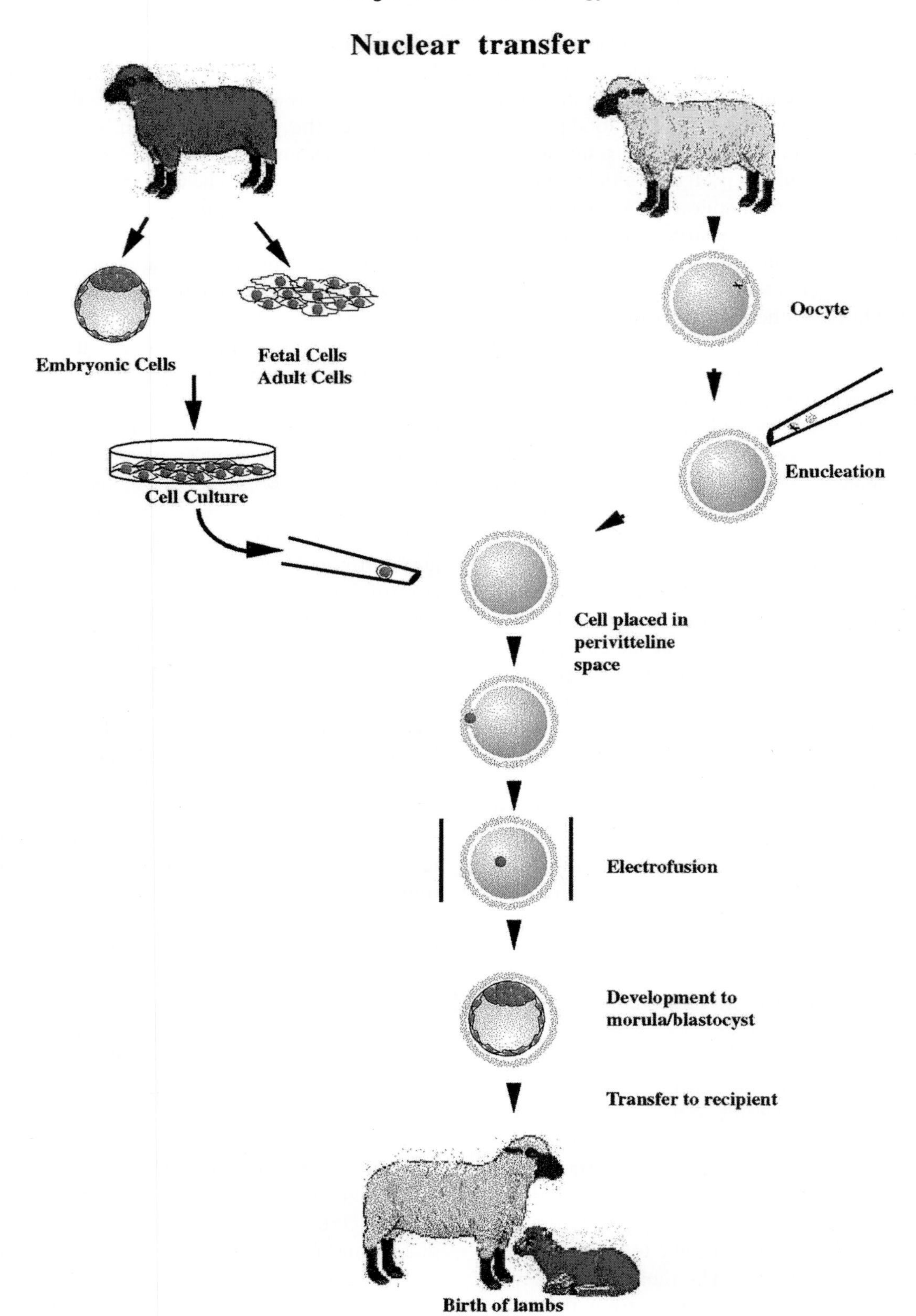
Nuclear transfer
Embryonic Cells
Fetal Cells
Adult Cells
Cell Culture
Oocyte
Enucleation
Cell placed in
perivitteline
space
Electrofusion
Development to
morula/blastocyst
Transfer to recipient
Birth of lambs

sary, but not to be the sole determining factor, for high expression.

Experiments conducted by Rosen's group first addressed the possibility of secreting both subunits of a heterodimeric protein into the milk of transgenic mice. Separate constructs placed cDNAs for both the α and β-subunits of follicle-stimulating hormone (FSH) under the control of the rat β-casein promoter. Mice harboring both transgenes were made either by single injection and mating or by coinjection. Bigenic mice expressed functional heterodimeric FSH at levels up to 15 μg/ml. Interestingly, mRNA for the α-subunit accumulated up to 7- to 17-fold higher levels than that for the β-subunit. One construct placed 408 bp of MMTV LTR (carrying four glucocorticoid response elements, (GRE) upstream of the β-casein promoter and this transgene gave more consistent expression and at higher levels than the non-MMTV counterpart. Since then, it has consistently been demonstrated that the mammary gland is capable of assembling a range of bioactive multisubunit transgenic proteins from antibodies to human fibrinogen. The latter is discussed in further detail later.

The caprine β-casein promoter has been used to express a normally membrane-associated protein—cystic fibrosis transmembrance conductance regulator (CFTR)—in mouse milk. Fifty microliters of milk contained the equivalent of 3×10^6 CHO cells of CFTR, and the protein appeared to be associated with the milk lipid globules. This opens up the possibility of secreting other membrane-associated receptors in milk, although the yields would appear to be limited.

b. Whey Proteins

Whey Acid Protein (WAP) expression has been particularly difficult to detect in mammary cell lines, leading to the conclusion that cell interactions and shape play an important role in its expression. The mouse and rat WAP genes have been cloned and reintroduced into mice by microinjection. Rosen and co-workers introduced a 4.3 kb genomic rat WAP fragment into mice. The WAP expression levels ranged from 1–95% (average 27%) of the endogenous mouse gene and the expression was mammary specific. However, the temporal expression of the transgene appeared to be slightly different, being switched on several days before the endogenous gene at around 7 days of pregnancy. When a different construct was used, having fewer 3′ flanking sequences, expression became uniformly high, but was still initiated earlier than the mouse gene. Hennighausen's group also noted this phenomenon. These workers reintroduced a 7.2-kb fragment of the mouse WAP gene, modified in such a way as to allow its discrimination from the endogenous mouse gene. Six of 13 lines expressed the new WAP mRNA during lactation and specifically in the mammary gland at levels from 3–54% of normal. Surprisingly, these transgenic mice also exhibited a peculiar phenotype, *milchlos,* in which female mice failed to lactate. This failure was associated with an arrest in mammary development and implied that the early onset of WAP expression in these mice was the cause. It is possible, then, that WAP plays a regulatory role in the terminal differentiation of mammary tissue. The *milchlos* phenotype was also seen in transgenic pigs expressing murine WAP.

A variety of heterologous genes have been hooked onto the WAP promoter and expressed in transgenic mice—tissue plasminogen activator (tPA), protein C, growth hormone, breast cancer protein PS2, and extracellular superoxide dismutase (EC-SOD). In general, levels of expression have been variable in mice (100 μg/ml tPA, 1.5 μg/ml PS2, 700 μg/ml

Fig. 2. Schematic outline of the nuclear transfer procedure. Oocytes derived from Scottish blackface (symbolized as a yellow sheep) are enucleated. The donor cell derived from a different sheep breed (symbolized as a red sheep) is placed under the zona pelucida into the perivitelline space. The cell nucleus is introduced into the cytoplast by electrofusion, which also activates the oocyte. The reconstructed embryo is then either cultured *in vitro* up to blastocyst stage or is transferred into a pseudopregnant intermediate-recipient ewe. At day 7, embryos are assessed for development. Late morulae and blastocysts are transferred into the final recipients. Pregnancies resulting from nuclear transfer are determined by ultrasound scan at about 60 days after estrus, and development is subsequently monitored at regular intervals. See color insert.

TABLE I
Published Protein-Expression Levels of Heterologous Proteins in Mammalian Milk

Milk gene promoter and species of origin	*Protein*[a]	*Gene configuration*	*Expression level*	*Transgenic species*
β-casein rabbit	h Interleukin 2	Genomic	430 ng/ml	Rabbit
WAP mouse	h tPA	cDNA	50 μg/ml	Mouse
WAP mouse	h growth hormone	Genomic	3.5 mg/ml	Mouse
αs1-casein bovine	h urokinase	Genomic	1–2 mg/ml	Mouse
αs1-casein bovine	h lactoferrin	Genomic	10.6 mg/ml	Mouse
β-casein goat	h longer-acting tPA	cDNA	2–3 mg/ml	Goat
WAP mouse	h PS2	cDNA	1.5 μg/ml	Mouse
WAP mouse	h tPA	cDNA	460 ng/ml	Mouse
β-LG sheep	h FIX	cDNA	25 ng/ml	Sheep
β-LG sheep	h AAT	cDNA	5 μg/ml	Sheep
WAP mouse	h PC	cDNA	1 mg/ml	Pig
α-Lac bovine	Bovine α-lactalbumin	cDNA	0.45 mg/ml	Mouse
α-Lac bovine	α-Lactalbumin	Genomic	2.4 mg/ml	Rat
α-Lac goat	Caprine α-lactalbumin	Genomic	1.2–3.7 mg/ml	Mouse
WAP mouse	WAP	Genomic	1–2 mg/ml	Pig
β-LG sheep	h AAT	Minigene	35 mg/ml	Sheep
β-casein goat	h CFTR	cDNA	1 μg/ml	Mouse
WAP rabbit	h AAT variant	Genomic	10 mg/ml	Mouse
WAP mouse	h longer-acting tPA	cDNA	3 μg/ml	Goat
β-LG sheep	h FIX	Minigene	15 μg/ml	Mouse
β-LG sheep	h PC	Genomic	560 μg/ml	Mouse
β-LG sheep	h PC	cDNA	108 μg/ml	Mouse
WAP mouse	h tPA	cDNA	250 μg/ml	Mouse
β-casein goat	β-casein	Genomic	24 mg/ml	Mouse
αs1-casein bovine	h tPA	Minigene	0.5 μg/ml	Mouse
α-Lac human	h EPO	Genomic	200 μg/ml	Mouse
WAP mouse	h PC	cDNA	3 μg/ml	Mouse
WAP mouse	h Fibrinogen	cDNA	60 μg/ml	Mouse
β LG sheep	h Fibrinogen	Genomic	1.6 mg/ml	Mouse
WAP mouse	h EC-SOD	cDNA	0.7 μg/ml	Mouse
αs1-casein bovine	h Lactoferrin	cDNA	36 μg/ml	Mouse
WAP rabbit	h GH	Genomic	22 mg/ml	Mouse
β-LG sheep	h serum albumin	Minigene	10 mg/ml	Mouse
WAP mouse	h PC	Genomic	0.7 mg/ml	Mouse
β-LG sheep	h AAT	cDNA	5 mg/ml	Mouse
β-LG sheep	h AAT	Genomic	21.3 mg/ml	Mouse
β-casein goat	h mAb	cDNA	10 mg/ml	Mouse
β-casein goat	h AAT	cDNA	4 mg/ml	Rabbit
β-casein goat	h ATIII	cDNA	1 mg/ml	Mouse
β-casein goat	h ATIII	Minigene	10 mg/ml	Mouse
β-casein goat	h longer-acting tPA	cDNA	6 mg/ml	Mouse
β-casein goat	h SA	cDNA	0.8 mg/ml	Mouse
β-casein goat	h AAT	Genomic	35 mg/ml	Mouse
α-Lac bovine	Ovine troph. interferon	cDNA	1 μg/ml	Mouse
WAP mouse	h PC	cDNA	30 ng/ml	Mouse

continues

Continued

Milk gene promoter and species of origin	*Protein*[a]	*Gene configuration*	*Expression level*	*Transgenic species*
β-casein goat	th ATIII	cDNA	6.0 mg/ml	Goat
as1-casein bovine	h insulin growth factor-1	cDNA	1 mg/ml	Rabbit
β-LG sheep	h AAT	cDNA	600 μg/ml	Mouse
β-LG sheep	h FIX	cDNA	60 μg/ml	Mouse
WAP rat	Rat WAP	Genomic	27% of endogenous	Mouse
WAP mouse	Mouse WAP	Genomic	100–500 μg/ml	Sheep
WAP mouse	h parathyroid hormone	cDNA	415 ng/ml	Mouse
β-casein goat	h LA tPA	cDNA	1–2 mg/ml	Goat
β-casein goat	h LA tPA	cDNA	0.85 mg/ml	Goat
β LG sheep	h Fibrinogen	Genomic	5 mg/ml	Sheep
β LG sheep	h interferon-γ	cDNA	20 ng/ml	Mouse
β LG sheep	h FIX	Genomic	580 μg/ml	Mouse
WAP mouse	h EC-SOD	cDNA	2.9 mg/ml	Rabbit
α-Lac bovine	Bovine α-lactalbumin	Genomic	1.5 mg/ml	Mouse
β-LG sheep	Sheep β-LG	Genomic	23 mg/ml	Mouse
β-LG sheep	h AAT	Genomic	7.3 mg/ml	Mouse
αs1-casein bovine	h lysozyme	cDNA	0.71 mg/ml	Mouse
β-casein rat	bovine FSH	cDNA	15 μg/ml	Mouse
αs1-casein bovine	h granulocyte macrophage stimulating factor	cDNA	0.2–4.6 mg/ml	Mouse
β-LG bovine	β-LG-h EPO fusion	cDNA	0.3 mg/ml	Mouse
β-LG bovine	β-LG-h EPO fusion	cDNA	0.5 mg/ml	Rabbit
αs1-casein bovine	h EPO	cDNA	205 ng/ml	Mouse
WAP mouse	bovine TAP (peptide)	cDNA	5 μg/ml	Mouse
WAP mouse	h longer-acting tPA	cDNA	6 mg/ml	Goat
WAP mouse	ATIII		14 mg/ml	Goat
WAP mouse	h AAT	Genomic	20 mg/ml	Goat
WAP mouse	anticancer mAb		10 mg/ml	Goat
WAP mouse	anti Lewis BR96 mAb		4 mg/ml	Mouse
WAP mouse	Single-chain antibody		1 mg/ml	Mouse
WAP mouse	Soluble receptor CD4		8 mg/ml	Mouse
WAP rat	Bovine GH	Genomic	16 mg/ml	Mouse
β-casein	h Transferrin receptor		2 mg/ml	Mouse
β-casein	hSA		35 mg/ml	Mouse
αs1-casein bovine	Bovine αs2-casein	Genomic	>20 mg/ml	Mouse
WAP mouse	h factor VIII	cDNA	2.7 μg/ml	Pig
WAP mouse	h Furin	cDNA	81–325 μg/ml	Mouse
WAP mouse	h EC-SOD	cDNA	3 mg/ml	Rabbit
β-casein bovine	h GH	Genomic	2.63 mg/ml	Mouse
WAP mouse	Anti-TGEV recombinant mAB	cDNA	5 mg/ml	Mouse
β-LG sheep	h Fibrinogen	Genomic	5 mg/ml	Sheep
WAP mouse	h Fibrinogen	cDNA	0.6 mg/ml	Mouse
β-LG sheep	h α-lactalbumin/saimon calcitonin	Fusion genomic	2.1 mg/ml	Rabbit
β-LG sheep	Anti-TGEV recombinant mAb	cDNA	6 mg/ml	Mouse
β-casein goat	h ATIII	cDNA	3.2 mg/ml	Goat
β-LG bovine	Bovine β-LG	Genomic	1–2 mg/ml	Mouse
αs1-casein bovine	h acid α-glucosidase	Genomic	2 mg/ml	Mouse

[a] ATIII, antithrombin III; CFTR, cystic fibrosis transmembrane conductance regulator; EPO, erythropoietin; FIX, factor IX; FSH, follicle-stimulating hormone; GH, growth hormone; mAb, monoclonal antibody; PC, protein C; +PA, tissue plasminogen activator.

protein C, and 700 μg/ml EC-SOD), probably reflecting the properties of the various proteins and constructs used. However Stromqvist and co-workers went on to generate rabbits transgenic for a WAP promoter-driven EC-SOD-expression construct that produced up to 3 mg/ml recombinant EC-SOD in the milk.

The sheep β-lactoglobulin (β-LG) gene was the first to be introduced into transgenic mice and has since been proven to drive very high levels of heterologous gene expression in mice, rabbits, and livestock. Rodents do not possess a β-LG gene and that allows the easy detection of β-LG expression in transgenic mice. Simons and co-workers took 16.2- or 10.5-kb genomic fragments of the ovine β-LG gene and introduced them into transgenic mice. These constructs differed in the number of 3′ flanking sequences, each containing 4.2 kb of 5′ flank and 4.9 kb of intron–exon sequence. From nine founder mice selected for further study, expression was generally very high; one line expressed β-LG protein at around 23 mg/ml, as judged by gel electrophoresis.

Developmental studies on these mice showed that the expression of ovine β-LG in mice, which have no endogenous β-LG gene, roughly paralleled that found in sheep. RNA analyses of a variety of tissues showed that expression of the transgene was restricted to the mammary gland. Temporal expression of β-LG was compared with the endogenous β-casein gene. Both genes were expressed at low levels in virgin gland. In 1–10 days of pregnancy, there was a small accumulation of transcript; in the period 10 days to parturition, there was a rise to between 65–80% of the final level. Following parturition, a further 1.5- to 2-fold increase occurred. The differences between sheep and mice were explained in terms of the relative state of differentiation of the gland in the two animals. Indices of differentiation show that, by parturition, mouse mammary cells are more differentiated than their ruminant counterparts.

The minimal β-LG promoter has been determined. Only 406 bp of sequences 5′ to the transcription start site are necessary for both a high-level and tissue-specific expression of the β-LG gene in transgenic mice. Several heterologous genes have been expressed from the β-LG promoter. Archibald *et al.* expressed human α_1-antitrypsin (AAT) in mice. In this case, the AAT gene used was a full genomic fragment, minus the first exon and intron, but including all the other intron sequences. This β-LG-AAT transgene (AATB) gave levels of fully functional AAT protein in the milk up to 7 mg/ml. PPL Therapeutics attained 21.3 mg/ml with the same construct. PPL Therapeutics have also expressed a wide range of other products, including human growth hormone, fibrinogen, protein C, and salmon calcitonin fusion peptide in transgenic mice at very high levels. The best expressing β-LG-GH mice gave 1.7 mg/ml, β-LG-Fibrinogen mice secrete in excess of 1 mg/ml, and the best β-LG-protein C mice secreted 0.5 mg/ml and β-LG-calcitonin as a fusion with α-lactalbumin at over 2 mg/ml. Substantially higher expression levels for human AAT and fibrinogen in transgenic sheep were achieved and are discussed in Section II.B. Published expression data for a range of proteins expressed in transgenic animals are summarized in Table I.

In general, native introns are required in order to maximize the levels of expression of transgenes. For example, if an AAT cDNA is used instead of a genomic fragment, expression levels of AAT in mammary gland fall to low or undetectable levels. The analysis of expression from a completely intronless FIX transgene showed no detectable RNA in mammary tissue. Our own results with human protein C show almost undetectable levels of protein in milk, in which there are no introns in the transgene. If one assigns the highest level attainable with full genomic protein C DNA constructs as 100%, then the maximum expression we have achieved with a cDNA has been about 1%, in a transgene containing the last β-LG intron only. These observations are consistent with the results obtained with other milk-protein systems and, as previously noted, indicate a fundamental feature of the control of gene expression in animals.

The α-lactalbumin system has also been studied in transgenics and used as a vector for the expression of heterologous proteins. In one experiment, the entire bovine α-lactalbumin gene was introduced into mice and the expression of protein and mRNA monitored. Two mice from six transgenics gave levels approaching those of the endogenous α-lactalbumin gene. The gene also shows developmental expression patterns similar to its endogenous counterpart. Very high expression levels of bovine α-lactalbumin have been reported in transgenic rats, reaching 2.4 mg/ml.

In general, the promoter sequences derived from milk-protein genes direct expression quite specifically to mammary tissue in transgenic mice. There are, however, cases in which apparently aberrant expression has been detected. A β-LG-AAT transgene showed human AAT mRNA in the salivary glands of some transgenic lines. There is also one report that a guinea pig α-lactalbumin gene introduced into transgenic mice showed expression in the basal layer of the sebaceous glands in the skin. It would appear that such examples are rare and this serves to confirm that the expression of heterologous proteins in the mammary gland is a good system for producing recombinant pharmaceuticals.

3. *Expression of cDNAs*

High-level expression of heterologous proteins in livestock milk is generally achieved using a fusion of the milk gene-promoter region to genomic sequences coding for the protein of interest. Analogous cDNA-based constructs are found to be generally less efficiently expressed in transgenic mice. Despite some exceptions, the poorer expression of cDNA-based constructs has been well documented by most workers in the field. Observed problems include an increase in the influence of negative chromosomal-position effects, which cause wide variation in transgene-expression levels in different lines transgenic for the same construct. In addition, there is also evidence from the work of John Clark's group suggesting that in some cases, cDNAs, as well as being transcriptionally shut down themselves, may create a domain whose effect extends outward, causing the silencing of adjacent transgenes. However, the genetic material encoding many candidate human proteins for production by the transgenic mammary gland is very often, due to lack of availability of a genomic clone or the size of the natural gene, limited to cDNAs. As such, the development of a technique giving more consistent expression from cDNA-based expression constructs is viewed as commercially important.

The addition of natural or heterologous introns to cDNA-based transgenic constructs has been found to improve transgene expression levels, but previously these have rarely approached those attained with constructs containing the entire gene sequence for an heterologous protein. However, in a number transgenic mouse and rabbit studies conducted by PPL Therapeutics, the expression levels from cDNA constructs based on a novel BLG-expression vector have approached, or even matched, that of the equivalent construct containing genomic sequences. Based on the observation that a number of highly expressed genes, whose expression is tissue specific, possess a noncoding exon and an intron upstream with respect to the translation start site, a heterologous 5′ intron from the bovine β-casein gene was positioned in this transgene expression vector. PPL Therapeutics have achieved expression levels of 6 mg/ml of a soluble receptor protein in transgenic rabbit milk using such a construct.

Another approach to the expression of cDNAs is "rescue" technology, a technique developed by John Clark and co-workers to overcome cDNA-related expression problems. It makes use of the group's observation that coinjection of the entire ovine BLG gene with an intronless construct allows the expression of that construct, whereas no expression is achieved when it is injected alone, Clark and colleagues have demonstrated the expression of up to 800 μg/ml AAT in the milk of mice transgenic for BLG and an intronless AAT expression construct. In mice transgenic for the latter construct alone, only one out of eight mice expressed and then only at a level of only 3.9 μg/ml. Similarly, using this technology, PPL Therapeutics has achieved an expression level of 200 μg/ml from a cDNA-based construct that did not give rise to expression in 15 lines of mice transgenic for the intronless construct alone. However, this represented only ~40% of the expression level obtained with an equivalent construct containing the entire gene. In most cases, coinjected genes integrate at the same chromosomal location. The cointegrated BLG gene "rescues" the expression of the intronless construct by an undefined mechanism. However, Clark and colleagues have proposed that this phenomenon probably results from an open chromatin conformation associated with the actively expressing BLG gene, which spreads to encompass adjacent intronless genes. The BLG gene may thus create a permissive domain that allows access to the intronless genes by the transcriptional machinery of the cell, perhaps via an enhancer-like sequence. In the absence of adjacent BLG genes, the intronless construct is likely to be inaccessible, possibly within condensed chromatin. Another possible explanation for this phenomenon is simply that the BLG gene, if flanking the cDNA construct, acts as a passive insulator for the intronless gene from the negative effects of adjacent chromatin.

It is likely that the reasons for poor cDNA transgene expression are complex and often construct specific, involving a number of genomic influences and interactions between transgene and host genome. However, the armory of techniques available has gone a significant way to overcoming the problem for a significant proportion of cDNA-based transgenic constructs.

B. Transgenic Products

A vast number of proteins have been expressed in the milk of transgenic animals. Many of these have been in the mouse, either as pilot studies to assess the feasibility of transgenic livestock as producer animals for pharmaceutical proteins or for use as research tools. A number of the former, expressed in livestock, have started proceeding toward or are in clinical development. One such example, that of human α-1-Antitrypsin is discussed in greater detail here. (See Table I.) There is potential for a growing number of transgenic products to meet a wide range of therapeutic, nutritional, and even surgical requirements when expressed at high volume in transgenic livestock.

1. *Human α-1-antitrypsin*

PPL Therapeutics has generated sheep containing a transgene encoding human α-1-antitrypsin (AAT). Five founder animals were produced, of which four were female. After mating and subsequent milking of the animals, it was found that the females all expressed human AAT in their milk at levels greater than 1 g/liter. One animal produced as much as 65 g/liter initially, which stabilized at approximately 35 g/liter and represented nearly 50% of the total milk protein throughout the lactation period.

After initial purification, the transgenic protein was shown to be more than 95% pure and of a similar size and immunosensitivity to human plasma-derived protein. Bioactivity and glycosylation comparisons were also very favorable. This comparison was extended to more-detailed analyses of glycosylation state, amino-terminal sequencing, pI value, and molecular weight determined by mass spectrometry. The transgenic protein was found to be extremely similar to human plasma-derived material, except for a slight difference observed in the transgenic protein when analysed by isoelectric focusing. This could be explained by minor differences in terminal sialylation. The similarities between the transgenic and human plasma-derived proteins merited the development of a production process for recombinant human AAT from these original founder sheep.

The male transgenic was used as the founder animal for a transgenic production flock. Transgenic offspring from this animal and subsequent generations have all stably inherited eight copies of the AAT transgene from the founder and the ewes all express from 13–16g/liter of AAT in their milk. The protein produced by each animal appears identical. These important results showed that stable transmission and stable expression levels can be obtained from a transgenic livestock animal. It also demonstrates that a transgenic production flock can be derived from one founder male or several half-brothers containing the same integrant. This is essential for livestock animals, for which the rapid expansion of a flock or herd would be greatly facilitated by using a male founder rather than employing modern breeding techniques, such as *in vitro* culture, embryo splitting, and transfer. The flock has now been expanded to over 3000 transgenic ewes. These animals provide the raw material for a pilot production plant currently producing upward of 1 kg purified clinical-grade AAT per week.

Human α-1-antitrypsin is the serine protease inhibitor primarily reponsible for the inhibition of neutrophil elastase. Excessive levels of lung neutrophil elastase are in large part responsible for the irreversible decline in lung function associated with the inherited disease cystic fibrosis (CF). CF is characterized by the systemic thickening of mucus, which in the lung seriously interferes with pathogen clearance, in turn triggering an excessive accumulation of phagocytic immune cells, including neutrophils. Elastase secreted by these cells attacks the gaseous-exchange surfaces, causing progressive scarring that ultimately reduces lung function to a level incompatable with life. By irreversibly inactivating neutrophil elastase, it is hoped, AAT could prove a useful intervention for both helping reduce the number of serious infections encountered and reducing the overall rate of decline in lung function.

Similarly, genetic deficiencies resulting in low circulating plasma levels of AAT are one of the most common lethal hereditary disorders affecting Caucasian males, being associated with an increased risk of developing life-threatening emphysema. Emphysema too is characterized by progressive accelerated irreversible lung damage resulting from excess activity of neutrophil elastase on gaseous-exchange surfaces. Some deficiencies have been successfully treated with AAT fractionated from human plasma, but the dose requirement (200 g/patient/year) means that this source is insufficient to meet the needs of the entire patient population. Attendant with this is the risk of transfer of human pathogens via human plasma-derived products.

Clearly both patient groups might benefit from the existence of a safe, effective, high-quality recombinant source of AAT. Steps to develop AAT purified from the milk provided by the transgenic producer flock as a human therapeutic are currently at an advanced stage. Results from a multicenter placebo-controlled double-blind study of the effects of inhaled nebulized AAT on cystic fibrosis patients indicated a promising trend toward the reduction of serious lung infections, and a reduction in the rate of lung-function decline in those patients receiving AAT. Extended clinical trials involving more patients and longer time intervals are scheduled to take place.

2. *Fibrinogen*

Fibrinogen is a complex plasma glycoprotein composed of two each of three polypeptide chains linked by 29 inter- and intrachain disulfide bonds. Synthesized in the liver, the assembled fibrinogen molecule is secreted into the bloodstream, where it functions as the final molecule in the blood coagulation cascade. During the coagulation process, enzymatic cleavage by thrombin converts fibrinogen to fibrin monomers, which polymerize into insoluble fibrin clots. Adjacent fibrin strands are then cross-linked by the transglutaminase activity of factor XIII, stabilizing the clot. It has been demonstrated that placing α-β- and γ-chains of humans under the control of the ovine β-lactoglobulin promoter brings about expression of the three chains in the mammary gland of transgenic mice and sheep. The three subunits are assembled in the mammary gland into mature hexameric fibrinogen that is functional in clotting reactions and can be cross-linked by factor XIII. Expression levels of up to 5 g/liter have been achieved for recombinant fibrinogen produced in this way in transgenic sheep. In contrast, attempts to express fibrinogen in mammalian cell cultures have been relatively unsuccessful, with levels of only 4 μg/ml culture medium being achieved. This example amply demonstrates the capacity of the transgenic ruminant mammary gland to produce highly complex biologically active molecules. A cheaper, readily available source of recombinant human fibrinogen would provide a valuable source of biological sealant for use in a wide range of surgical applications.

3. *Peptides*

Biologically active peptides are recognized for their potential as clinically useful therapeutics to treat a wide range of human diseases. Although usually chemically synthesized, the production costs for larger peptides (i.e., greater than 10 amino acids long) may be prohibitive. The modifications required for full biological activity, such as carboxy-terminal amidation also potentially limit their practical utility. Although usually unstable in biological systems, a novel use of transgenic technology for expressing peptides was recently developed by McKee and coworkers at PPL Therapeutics, offering a potentially cheaper route for peptide production. The peptide sequence is expressed as a carboxy-terminal fusion with a carrier protein. Using this technology, it has been demonstrated that salmon calcitonin can be expressed in the mammary gland of transgenic rabbits as a fusion with the human milk protein α-lactalbumin. Calcitonin is a 32-amino-acid peptide produced in the C cells of the thyroid gland in mammals and the ultimobrachial gland of fish. It is essential for correct calcium metabolism and has been found to be useful in treatment of bone disorders, such as osteoporosis and Paget's disease. A cleavage site for the bacterial protease enterokinase engineered at the junction between α-lactalbumin and the calcitonin amino acid sequence allows the cleavage of the peptide from its fusion partner. Furthermore C-terminal amidation of the peptide, an important posttranslational modification for calcitonin biological activity, was achieved by adding a single C-terminal glycine to act as a substrate for endogenous mammary gland-amidating activity. The cleaved peptide was found to have a biological activity equivalent to that of chemically synthesized calcitonin. Notably, no bioactivity is detectable in the fusion because the correct presentation of the peptide to its receptor is prevented in such a configuration. As a number of peptides and proteins potentially exhibit high or disruptive biological activity in a biological system (e.g., the transgenic production of human growth hormone in the mammary gland has been observed to cause precocious gland development and fertility problems in transgenic mice), the masking of activity in the fusion is an important factor.

The production of peptides as fusions in the milk of transgenic animals has several advantages over chemical synthesis. The cost of production is not related to the length of the peptide and the scale of production is not in any way limited by considerations such as reactor size and reagent handling and disposal. Some therapeutic peptides, especially those for antimicrobial use, potentially will require production in guantities of hundreds of kilograms per year, so transgenic production lends itself ideally to meeting such a high potential demand. The mammary gland's ability to carry out a range of posttranslational modifications that are often essential to biological activity also makes it an attractive alternative for production of therapeutic peptides.

C. Product Efficacy, Safety, and Acceptability

More established production routes for therapeutic proteins have been available for some time, such as bacterial and yeast fermentation. However, most of the proteins of interest to the biotechnology and pharmaceutical sector are relatively complex in nature, requiring correct folding and substantial posttranslational modifications, such as glycosylation, for biological activity. Prokaryotic systems are incapable of performing most of such modifications. Yeast-based processes are more accomplished, but still fall short of the mark for many modifications. The mainstay of more complex recombinant protein production has been mammalian cell culture, which has produced a regular stream of effective, safe products. However, irrespective of which of the cell-based production routes is taken, the costs and low yields involved have proved prohibitive for the production of some proteins. The production of proteins via the transgenic mammary gland clearly offers a superior alternative in these instances. Although legitimate concerns do remain about whether the mammary gland is able to carry out appropriate posttranslational modifications to products, it must be borne in mind that other recombinant systems used to produce therapeutic proteins do not produce, glycosylation patterns, for example, identical to those found on protein extracted from human sources. Indeed, changing the cultured mammalian cells that are used to produce a protein, or even the conditions under which the same cells are grown, can substantially alter the glycosylation pattern of the product. Similarly, it is recognized that transgenic livestock do not always produce human-identical patterns of glycosylation on recombinant human proteins. What is important is that these modifications allow the product to be safe and efficacious.

Another primary safety concern relates to transfer of animal pathogens to humans via transgenic products. The first major issue here is prion infection. These agents promote neuropathies in animals, such as scrapie and bovine spongiform encephalopathy. PPL Therapeutics has adopted a two-pronged approach to combat this risk. First, the purification process is extensively validated for the removal of these particles. Second, all producer sheep flocks are of New Zealand origin and all cows are from the United States, as these countries are considered free of the relevant ovine and bovine prion diseases, respectively. The other major issue is viral infection of producer animals. Again PPL Therapeutics has identified viruses of specific concern, and the producer flocks and herds are monitored for infections on a regular basis. As a second line of defense, the purification processes used are validated for the removal or inactivation of these viruses even if infections are subclinical.

All areas of product safety and efficacy are extensively controlled by regulatory authorities before any product is allowed into the marketplace. The regulatory authorities of Europe and the United States both have formulated guidelines that extend to products from transgenic animals. All those involved in the field must comply with these regulations. In spite of the safeguards, one of the main challenges that transgenic technology has encountered since its inception has been that of public misperceptions but the resistance to the use of this production route is gradually being overcome by advances in the field. Two leading companies in the area have proteins from transgenic livestock in advanced clinical trials, PPL Therapeutics with human α-1-antitrypsin from sheep and Genzyme Transgenics with human antithrombin III from goats. Both have helped significantly in widening the exposure of the possibilities of transgenic technology to the public. With an increase in the number and diversity of transgenic products expected in the near future, it appears that transgenic technology is finally coming of age. It clearly offers not just a rival to existing technologies for recombinant therapeutic-protein production, but rather an alternative approach providing a wider choice to meet the challenges of human health care in the twenty-first century.

III. XENOTRANSPLANTATION

The massive shortfall in human-donor transplant organs, in particular kidneys, presents an acute problem for medicine. Because many of the complications historically associated with human-to-human organ transplantation (allotransplantation), such as loss of function and rejection, have been overcome by an ever-increasing knowledge of the physiological–molecular mechanisms underlying these problems, the demand for donor organs has risen as transplantation is judged to be an appropriate treatment for an ever-increasing list of serious medical conditions. Although the idea that such a shortage might be met by using animal organs is not a new one, it is not until relatively recently that it has presented itself as a possible solution. This has come from a significant increase in our understanding, at the molecular level, of the previously insurmountable barriers preventing interspecies whole-organ grafting. In combination with the advent of transgenic technology since the 1970s, xenotransplantation now at least presents a realistic hope for bridging the gap between the supply and demand for some tissues.

A. Choice of Donor Species

Some animal tissues have already proved their usefulness in humans. For example, the replacement of defective human heart valves with those from the pig heart has for some time been accepted as an effective and safe treatment. However, heart valves are composed of relatively inert, nonvascularized tissue and rarely elicit rejection. Clinical trials are underway for the transplantation of pig cells for controlling diseases not previously considered treatable by transplantation—pancreatic islet cells are being grafted for treating diabetes and fetal brain cells are being grafted for treating Parkinson's disease and Huntington's chorea. Whole organs, however, present a considerably greater challenge, both physiologically and immunologically, in large part because they are linked directly to the host vasculature. Our nearest animal relatives, Old World monkeys and great apes, would at first seem the most obvious source of organs for human transplant because they are physiologically similar and their tissues are more likely to be immunologically tolerated. The transplantation of tissues between two species as closely related as great apes and humans, termed a concordant xenograft, however, is not a realistic option. Concerns over transfer and pathogenicity of primate viruses to humans and ethical constraints over the farming of such animals effectively rule them out as candidate donors. In choosing an alternative source of tissues and organs, much work has been focused on the use of pigs as a donor animal. Although transplantation of tissue between distantly related species (discordant xenograft) presents additional difficulties, the pig has a number of advantages over apes and other domesticated animals. The widespread farming of swine for their meat presents no ethical difficulties. Their internal organs are very similar in size to those of humans, they produce large litters, and their young develop rapidly. However, there are numerous problems that must first be overcome before there is any hope that pig organs can survive and function properly in humans. The major issues and some of the current approaches to tackling them are discussed here.

B. Barriers to Xenotransplantation and the Role of Transgenic Technology

1. *Immunological*

The major barrier to discordant xenotransplantations of whole organs is hyperacute rejection (HAR). Characterized by the destruction of the vascular endothelium of the donor organ within minutes of exposure to the host's blood, it occurs through a complement-dependent mechanism activated following the binding of naturally occuring host antibodies to glycoproteins and glycolipids on the donor cells. The major xenoantigen recognised on the porcine endothelial cells is a terminal sugar residue, galactose α-1,3-galactose (αGal). Nearly all mammals, including pigs, express a 1,3-galacytosyl transferase, which places αGal on glyco-conjugates. However, humans, in common with apes and Old World monkeys, are natural knockouts for this gene and do not possess a homolog performing the same function. Because gut bacteria have αGal, all humans make a range of αGal antibodies that immediately bind to the endothelium of pig organs, setting in motion the complement cascade. Attention is therefore focused on various ways in which HAR can be circumvented. Theoretically, transgenic technology offers two broad approaches to this problem. One is to knock out the function of αGal transferase, which would prevent αGal from being present on the cell surfaces of donor tissue. This approach relies entirely on using gene-targeting technology to disable the αGal transferase gene in cultured porcine cells, and then using these as donors for nuclear transfer. Although theoretically the most effective solution, pig nuclear transfer technology is still some way behind that of other species such as sheep and cattle, so in the short term such animals are not available. Alternatively, an approach requiring only standard egg-microinjection technology is to engineer transgenic pigs that express human complement regulatory proteins. These proteins, CD55 (decay-accelerating factor, DAF), CD46 (membrane cofactor protein, MCP-1), and CD59 (protectin), although not preventing αGal antibodies from binding graft endothelia and activating complement, inhibit downstream steps in the complement cascade, which could prevent endothelial cell lysis in the graft. Indeed, transgenic pig herds expressing one or more of the human complement inhibitors have already been developed and it has been claimed that hearts transplanted from such pigs survive for much extended periods of time (up to 40 days), compared to normal pig hearts, in immunosuppressed monkeys. Still another approach is to generate transgenic pigs that overexpress fucosyltransferase, which competes for the same substrate as αGal transferase, so that a fucose group, a natural human blood group antigen, is substituted for αGal. An alternative that does not involve transgenic technology is to circulate human plasma over αGal immunoadsorbant columns to remove αGal antibodies, although the utility of this approach is unlikely to be effective in isolation because anti-αGal antibody titers in human plasma rapidly return to normal levels.

Acute vascular rejection, also known as delayed xenograft rejection, is also triggered in part by the binding of αGal antibodies to the graft. Characterized by an inflammatory response leading to platelet coagulation and leucocyte extravasation in the capillaries of the grafted organ, it severely affects the organ's vasculature about 3–5 days after transplantation. The strategies that prevent the binding of aGal antibodies to the graft are likely to be effective in inhibiting this mechanism. Approaches that specifically target some of the mechanisms elicited may also prove useful. For example, by generating an animal transgenic for coagulation inhibitor proteins under the control of an endothelial cell-specific promoter, it may be possible to ameliorate blood clotting in the locality of the graft.

Cell-mediated rejection mechanisms, which also play a major role in allograft rejection, pose an equal threat to xenotransplantations. Various cell-surface antigens in the donor tissue are recognized as foreign by the host immune response and come under attack by specific cytotoxic T cells, leading to rejection from 1–2 weeks after transplantation. Standard therapies using immunosuppressive drugs such as steroids and cyclosporin, already used in allograft surgery, are likely to be effective in helping suppress this response for xenografts as well. What has become clear is that it is only by the judicious use of a number of different approaches, involving preparation and modification of the donor animal using transgenic technologies, and by appropriate administration of therapeutic regimes to the patient, that the considerable challenges posed by the host immune system to graft survival will be overcome.

2. *Physiological*

If the considerable immunological issues relating to xenotransplantation are eventually overcome, there are still physiological problems to be addressed. Although chosen largely for their compatability in size, a pig-to-human xenotransplant organ must perform a complex range of functions and respond to host signalling pathways in an appropriate manner. Whether this is the case will largely depend on the tissue or organ in question. For example, because pig insulin has for many years proved an effective treatment for diabetes, it seems a reasonable expectation that pig islet of Langerhans cells, if successfully engrafted, should perform the desired function in a patient with insuling-dependent diabetes. However, it remains an open question whether a pig kidney, for example, will be able to respond to humans diuretic hormone signals. In addition, the kidney, as well as performing its urinary functions, synthesizes erythropoetin, a hormone required for the production and maturation of erythrocytes. It is known that pig erythropoetin is not recognized by the human receptor on erythrocyte precursors, so a human recipient of a porcine kidney might well require treatment with human EPO. There may be many discordant physiological signals that are uncharacterized. Of particular concern might be factors that influence the function and health of the liver, for example. More data from expererimental xenotransplantation of pig tissues into primates may yield much needed information in this area, as we simply do not have enough knowledge to confidently predict which organs and tissues will function appropriately when set in a different species. Should protein and peptide therapies prove to be a general requirement as an adjunct to xenotransplantation procedures, transgenic protein production could play a valuable role in meeting this need.

3. *Zoonosis*

A significant threat to human health is posed by the possibility of the transmission of animal viruses to human via xenotransplants, especially because many species-specific viruses go unrecognized because they do not cause a disease state in their normal host. Examples are already known. Herpes virus B of macaques elicits no more than cold sores in the moneky, but can cause a fatal encephalitis in humans. Convincing evidence that human immunodeficiency virus originated in chimpanzees provides further support for these concerns. Although it can be argued that pig viruses may pose less of a threat to humans because the two species are less closely related, it is a widely held belief that we ought to proceed with considerable caution. Despite arguments by some in the field that because pigs have long been domesticated by human we have been exposed to a full range of pig microbes, it must be borne in mind that xenotransplantation affords an easier passage for animal viruses to humans. Physical barriers are breached if porcine tissue is connected to the human vasculature and the immune suppression required to prevent graft rejection may help a virus to adapt to and propogate a new host. In addition, introduction of human genes into pigs to make them more suitable organ donors could facilitate the adaptation of pig viruses, allowing them to infect humans. In particular, cell-surface proteins frequently act as viral receptors; therefore, there is a real danger that human antigens transgenically expressed on the surface of pig cells with the intention of facilitating graft acceptence may provide a backdoor route for porcine viruses to infect human cells. For example, it is known that CD46 is a receptor for the measles virus, so it is possible that related morbilliviruses of animals, such as those causing distemper and rinderpest, could become adapted by exposure to human cell-surface proteins in transgenic pigs to infect human cells. The genetic modification of swine also might increase the threat of porcine-enveloped viruses transferring to humans. These viruses are surrounded by a lipid envelope derived from the host-cell membrane from which they bud. If the cell membrane of a pig organ is engineered specifically to evade human immune responses, then a primary natural defense mechanism preventing this type of virus moving between species may be lost because the viral envelope does not trigger human complement.

Possible solutions to this problem exist. For example, breeding programs designed to eliminate pathogens from transgenic herds to be used for xenotransplantation are envisioned. However, only previously known and characterized pathogens may be screened for, and the threat posed by unidentified infectious agents is one that cannot be disregarded. A significant challenge is also set by porcine endogenous retroviruses. Retroviral genomes, because they are inserted into the chromosomal DNA of the host, are inherited as Mendelian traits and cannot be readily removed. Activated retroviruses are known to cause serious diseases such as leukemia in a wide range of species. Weiss and co-workers have identified at least two recognized pig retroviruses that infect human cells in culture. It is as yet unknown whether these could cause disease in humans.

C. Prospects and Alternatives

Although undoubtedly holding out major promise for the future of medicine, it is generally accepted by all involved in the field, and by governmental advisory bodies internationally, that a high degree of caution must be exercised in the development of xenotransplantation technology. The possible threat to public health posed by the spread of zoonotic infections from a xenotransplant patient to the public at large is of particular concern, and it is now widely recognized that in addition to research programs to more fully evaluate the problem, long-term surveillance of the recipients will be required. This in itself may be problematic because it will touch on basic issues of personal liberty for patients.

In the longer term, advances in research into cellular differentiation may offer an alternative that will negate the need for animal cell and organ transplantation to humans altogether. It may one day be possible to induce cultured human somatic cells to differentiate into virtually any functional tissue, or even organ of choice. In this instance, the patient, him- or herself, could be the donor for autografts, thereby circumventing a whole range of the problems associated with immunological graft rejection and zoonotic infections. However, it is unlikely that developments in human cell therapy will provide a generally applicable solution before xenotransplantation becomes possible.

See Also the Following Article

GENETICALLY MODIFIED ORGANISMS: GUIDELINES AND REGULATIONS FOR RESEARCH

Bibliography

Auchincloss, H., and Sachs, D. H. (1998). Xenogeneic transplantation. *Ann. Rev. Immunol.* **16**, 433–470.

Carver, A. S., Dalrymple, M. A., Wright, G., Cottom, D. S., Reeves, D. B., Gibson, Y. H., Keenan, J. K., Barrass, D., Scott, A. R., Colman, A., and Garner, I. (1993), Transgenic livestock as bioreactors: Stable expression of human alpha-1-antitrypsin by a flock of sheep. *Bio/Technology* **11**, 1263–1270.

Ebert, K. M., Selgrath, J. P, Ditullio, P., Denman, J., Smith, T. E., Memon, M. A., Schindler, J. E., Monastersky, G. M., Vitale, J. A., and Gordon, K. (1991). Transgenic production of a variant of human tissue-type plasminogen activator in goat milk: Generation of transgenic goats and analysis of expression. *Bio/Technology* **9**, 835–838.

Gordon J. W., and Ruddle F. H. (1981). Integration and stable germline transmission of genes injected into mouse pronuclei. *Science* **214**, 1244–1246.

Gordon, J. W., and Ruddle, F. H. (1983). Gene transfer into mouse embryos: Production of transgenic mice by pronuclear injection. *Meth. Enzymol.* **101**, 411–433.

Gordon, J. W., and Ruddle, F. H. (1985). DNA-mediated genetic transformation of mouse embryos and bone marrow—a review. *Gene* **33**, 121–136.

Hammer, R. E., Pursel, V. G., Rexroad, C. E., Wall, R. J., Bolt, D. J., Ebert, K. M., Palmiter, R. D., and Brinster, R. L. (1985). Production of transgenic rabbits, sheep and pigs by microinjection. *Nature* **315**, 680–683.

Hansson, L., Edlund, M., Edlund, A., Johansson, T., Marklund, S. L., Fromm, S., Stromqvist, M., and Tornell, J. (1994). Expression and characterization of biologically active human extracellular superoxide dismutase in milk of transgenic mice. *J. Biol. Chem* **269**, 5358–5363.

Hogan, B., Constantini, F., and Lacy, E. (1994). "Manipulating the Mouse Embryo: A Laboratory Manual." Cold Spring Harbor Laboratory Press, New York.

Houdebine, L-M. (1994). Production of pharmaceutical proteins from transgenic animals. *J. Biotechnol.* **34**, 269–287.

McKee, C., Gibson, A., Dalrymple, M., Emslie, E., Garner, I., and Cottingham, I. (1998). Production of biologically active salmon calcitonin in the milk of transgenic rabbits. *Nat. Biotechnol.* **16**, 647–651.

Schnieke, A. E., Kind, A. J., Ritchie, W. A., Mycock, K., Scott, A. R., Ritchie, M., Wilmut, I., Colman, A., and Campbell, K. H. S. (1997). Factor IX transgenic sheep produced by transfer of nuclei from transfected fetal fibroblasts. *Science* **278**, 2130–2133.

Stromqvist, M., Houdebine, L-M., Andersson, J-O., Edlund, A., Johansson, T., Viglietta, C., Puissant, C., and Hansson, L. (1997). Recombinant human extracellular suproxide dismutase produced in milk of transgenic rabbits. *Transgenic Res.* **6**, 271–278.

Velander, W. H., Johnson, J. L., Page, R. L., Russell, A., Subramanian, A., Wilkins, T. D., Gwazdauskas, F. C., Pittius, C., and Drohan, W. N. (1992). High-level expression of heterologous protein in the milk of transgenic swine using the cDNA encoding human Protein C. *Proc. Natl. Acad. Sci. U.S.A.* **89**, 12003–12007.

Weiss, R. A., (1998). Xenotransplantation. *Br. Med. J.* **317**, 931–934.

Wright, G., Carver, A., Cottom, D., Reeves, D., Scott, A., Simons, P., Wilmut, I., Garner, I., and Colman, A. (1991). High level expression of active human alpha-1-antitrypsin in the milk of transgenic sheep. *Bio/Technology* **9**, 830–834.

Translational Control and Fidelity

Philip J. Farabaugh
University of Maryland, Baltimore County

I. Mechanism of Translation
II. Gene-Specific Translational Control of Gene Output
III. Global Control of Translation
IV. Translational Accuracy
V. Programmed Alternative Translation

GLOSSARY

codon A unit of translational coding. Each three-nucleotide codon specifies the insertion of one of 20 amino acids. Genes consisting of strings of codons encode proteins. Special codons indicate the sites of initiation and termination.

recoding Any of three types of event in which ribosomes perform noncanonical forms of protein synthesis. In programmed translational readthrough, the ribosome translates through termination codons. In programmed translational frameshifting, the ribosome spontaneously shifts into a different reading frame. In programmed hopping, the ribosome bypasses a large number of codons in the mRNA. Special sequences in mRNAs stimulate each of these events, although they occur rarely in genes.

ribosome A molecular machine that catalyzes the process of translation. It consists of several ribosomal RNA molecules (rRNAs) and a few dozen ribosomal proteins.

translation The process of synthesis of polypeptides directed by the nucleic acid sequence of a messenger RNA (mRNA) molecule.

translational attenuation A specialized form of translational repression in which a translating ribosome is caused to stop the continued synthesis of a nascent protein.

translational repression A process of specifically turning off the translation of a specific mRNA often by a RNA-binding protein that occludes an essential sequence required for recognition of the mRNA by the ribosome.

A MOLECULAR MACHINE governs the process, termed translation, that transfers genetic information from nucleic acids to proteins. This machine, the ribosome, is a complex organelle composed of several RNA molecules complexed with several dozen proteins. The ribosome choreographs the movements of a large assembly of other proteins and RNA molecules using an RNA template, the messenger RNA (mRNA), as a guide for the order of insertion of amino acids into a growing polypeptide chain. The processs is complex and living cells depend critically on its speed and accuracy. Control mechanisms have evolved that manipulate the process of translation to modulate the production of specific protein products or of generic classes of proteins.

There are two general categories of translational control of gene expression, regulating the amount of protein expressed from a particular gene and regulating the structure of the protein product. The first has to do with the speed of translation and the second with its accuracy. Gene output responds to changes either in the rate of recruitment of mRNAs to the ribosome (regulation of initiation) or the rate of completion of nascent polypeptide chains (regulation of elongation). Control of translational output may be global, for example, through specific inactivation of factors generally required for efficient translation; or specific, through stimulation or inhibition of translation of single mRNAs. Some genes have evolved processes that allow the expression of alternative protein products by programming a nonstandard translational event. For example, some genes specify readthrough of in-frame termination codons, whereas others allow the expression of alternative forms of the protein by reading a portion of the mRNA in alternative reading frames. These processes are not

100% efficient, so that a given mRNA can program the expression of one protein by the default or canonical translational pathway while encoding an alternative structural form of the protein using a noncanonical translation event.

I. MECHANISM OF TRANSLATION

Evolution has conserved the mechanism of translation throughout all living things, though it differs in various details among the three domains—Prokarya, Archaea, and Eukarya. In general, translation in prokaryotes is somewhat simpler than in eukaryotes. The ribosomes of prokaryotes include only three ribosomal RNAs (rRNAs), 23*S*, 16*S*, and 5*S*, whereas eukaryotes encode four, a 26–28*S*, a 17–18*S*, 5.8*S*, and 5*S*. The ribosome consists of two dissociable subunits; the second-largest rRNA is the only rRNA in the smaller subunit in all cells. Prokaryotes encode only 50–60 ribosomal proteins, whereas eukaryotes encode 75–90 proteins. As this implies, the ribosomes of eukaryotes are somewhat larger and more complicated. Part of the reason for this difference is the greater physiological complexity of eukaryotes in general, but, in addition, the process of translation in eukaryotes is somewhat more complicated. The understanding of translation in Archaea has lagged behind the other two domains, as expected, although genome sequences of archeons may allow swifter progress in the future. The genome sequences of *Archaeoglobus fulgidis* and *Methanococcus jannaschii* reveal the presence of 62–66 ribosomal proteins and three rRNAs, 23*S*, 16*S*, and 5*S*. In keeping with our understanding of the evolution of the three domains, the ribosomal proteins of the Archaea are more similar to their eukaryotic homologs than to prokaryotes. Because relatively little is known about translation in Archaea, the main focus here is on the other two domains.

A. Translation Initiation

Initiation is the step of translation most markedly different among the three domains. The mechanism used to specify the site of translation initiation has profound consequences for initiation-control mechanisms. Prokaryotes use a different mechanism of initiation than do eukaryotes and archaea. In prokaryotes, the ribosome binds directly to the site of translation initiation. In eukaryotes, the ribosome generally binds to the 5′-end of the mRNA and searches along the messenger in a 5′-to-3′ manner, called scanning, until it encounters the initiation codon. Given the very different mechanisms employed, it should not be surprising to find that regulation of translational initiation also differs markedly.

In general, ribosomes do not have an intrinsic ability to identify initiation codons, but rather require specific initiation factors. Prokaryotes have three initiation factors, IF-1, IF-2, and IF-3. The roles of IF-1 and IF-3 in initiation are controversial, but IF-2 has a clear biochemical role, forming a ternary complex with the initiator tRNA (f-Met-$tRNA_i^{Met}$) and GTP. The IF-3 ternary complex directs the accurate binding of $tRNA_i^{Met}$ to an initiation codon, most often either AUG or GUG, at the beginning of a structural gene. GTP hydrolysis probably performs an error-correction function, allowing discrimination between correct and incorrect sites of initiation. After GTP is hydrolyzed to GDP, the initiation factor dissociates and the larger ribosomal subunit joins to form a complete ribosome. This allows the initiation of the elongation of the protein by the recruitment of the second aminoacyl-tRNA (aa-tRNA).

No special codons specify only initiation. Instead, initiation codons also occur in the middle of structural genes. In addition, these codons may occur outside translated regions or out of the normal translational reading frame within genes. The problem, then, is to identify the small proportion of these codons at the beginning of structural genes and to ignore all of the others. Prokaryotic ribosomes have a direct role in selecting the start site of translation. The rRNA of the small subunit has a short sequence near its 3′-end that is complementary in sequence to a short sequence immediately upstream of true initiation codons. The upstream sequence is termed the Shine–Dalgarno box (SD-box), commemorating the pair of researchers who first identified it. A correct initiation complex is one in which $tRNA_f^{Met}$ binds to an initiation codon while the Shine–Dalgarno interaction occurs immediately upstream. As this would imply, the distance between the two sequences is critical. Increasing or decreasing the distance between the initiation codon and the SD-box can drastically reduce translational initiation. This interaction provides a potent target for translational regulation because anything that interferes with the formation of the SD interaction would reduce the efficiency of initiation.

The mechanism of translational initiation in prokaryotes allows a fundamentally prokaryotic form of gene control, the operon. Because ribosomes bind directly to initiation sites, they may be positioned anywhere in an mRNA, and in fact a given mRNA can have multiple sites of initiation. Because of this feature, prokaryotic genes tend to occur in polycistronic mRNAs, those including more than one structural gene. Eukaryotic ribosomes adopt a different strategy to identify the proper sites for translational initiation. In eukaryotes, nearly all initiation occurs at AUG codons. Most eukaryotic genes have evolved so that the initiation codon is the AUG codon nearest the 5′-end of the mRNA. Ribosomes can merely bind to the 5′-end of the mRNA and scan to the first AUG codon. Consequently, eukaryotic ribosomes do not need any further sequence information to identify initiation codons, so they make no SD interaction.

The critical step for eukaryotic ribosomes is to identify the 5′-ends of true mRNAs. All mRNAs have diagnostic structures at their 5′-ends, a 5′–5′ triphosphate-linked guanosine, termed the 5′ Cap, which is added posttranscriptionally. A trimeric elongation factor, eIF-4F, binds to the Cap and directs eukaryotic initiation complexes to bind to the mRNA and initiate scanning. The initiation complex consists of the small ribosomal subunit bound by a large number of protein factors. Among these factors is eIF-2, which, as does IF-2, forms a GTP-containing ternary complex with the initiator tRNA, called Met-$tRNA_i^{Met}$ in eukaryotes. The hydrolysis of GTP, stimulated by the factor eIF-5, occurs along with the dissociation of the bound initiation factors, recruitment of the large subunit, and the start of elongation. Another important initiation factor, eIF-3, is critical to efficient translational initiation. It facilitates the dissociation of the large and small ribosomal subunits, binding of the eIF-2 ternary complex to the small subunit, and binding of the small subunit to mRNAs.

B. Elongation and Termination

The elongation phase of translation shows remarkable universal conservation. Elongation involves the repeated addition of amino acids to a growing polypeptide chain. This addition occurs in the context of the three tRNA-binding sites of the ribosome. The P site is the binding site of the tRNA attached by an ester linkage to the nascent peptide. Each aminoacyl-tRNA is recruited to the ribosomal A site. After aa-tRNA binding, an intrinsic activity of the ribosome catalyzes the transfer of the nascent peptide chain to the amino terminus of the amino acid on the aminoacyl-tRNA, forming a peptide bond. After peptide transfer, the newly deacyl-tRNA (tRNA lacking any amino acid) moves to a third site, termed E for exit, along with movement of the new peptidyl-tRNA into the P site. The cycle then repeats with recruitment of an aminoacyl-tRNA to the 3-nt codon immediately downstream of the codon bound to the peptidyl-tRNA. It is important that mRNAs have no internal punctuation that enables the ribosome to identify the correct reading frame. The ribosome appears to be constrained to read only successive adjacent codons. Exactly how this is accomplished is, remarkably, unknown.

Termination is really a specialized version of a normal elongation cycle that occurs when a special termination codon (UAG, UAA, or UGA) occupies the A site. At that point, the nascent peptide dissociates from the ribosome, and the two ribosomal subunits dissociate from the mRNA.

The ribosome requires only two elongation factors, EF-1A and EF-2 in prokaryotes (previously EF-Tu and EF-G) or their homologs eEF-1A or eEF-2 in eukaryotes (formerly EF-1α and EF-2). Each of these factors, similarly to IF-2 and eIF-2, bind GTP and have an intrinsic GTPase activity. EF-1A facilitates the recruitment of aa-tRNA to the ribosome, and EF-2 catalyzes the translocation of the peptidyl-tRNA–mRNA complex by 3 nt after peptide transfer occurs. The aa-tRNA forms a ternary complex with EF-1A and GTP, which is the form recruited to the ribosomal A site. As do IF-2 and eIF-2, GTP hydrolysis provides a proofreading function that is thought to reduce translational errors. Error correction appears to use a kinetic proofreading mechanism in which a difference in the rate of dissociation of correct (cognate) and incorrect (noncognate) aa-tRNAs is used to eliminate errors. Error correction occurs in two steps. Cognate aa-tRNA binds to the A site and dissociates only extremely slowly, whereas noncognate tRNA readily dissociates. Therefore most noncognate tRNA dissociates before GTP hydrolysis. There is a short pause after GTP hydrolysis before EF-1A–GDP dissociates. During this pause, almost all of the cognate tRNA remains bound, but almost all of the noncognate dissociates. This two-step process assures the low frequency of translational errors. An additional source of error correction may come from an allosteric interaction between a deacyl-tRNA in the E site and the A site. This interaction is thought to interfere with binding of noncognate tRNAs to the A site.

The translocation of the peptidyl-tRNA–mRNA complex appears to be an intrinsic function of the ribosome because it occurs in the absence of EF-2. The rate of translocation is extremely slow, however. Precise kinetic studies suggest that the energy from hydrolysis of a GTP bound to EF-2 stimulates the movement of the peptidyl-tRNA rather than regulating the release of EF-2 from the ribosome. The accurate movement of the mRNA only 3 nt during each translocation step is essential.

Translational termination differs between prokaryotes and eukaryotes. Prokaryotes have two peptide release factors, RF-1 and RF-2. The factors are codon specific, so that RF-1 recognizes the codons UAG and UAA and RF-2 recognizes UAA and UGA. A third factor, RF-3, has no codon specificity. Again, RF-3 is a GTP-binding protein that stimulates the activity of the other two factors. In eukaryotes there are only two factors, eRF-1 (the cognate of RF-1 and RF-2) and eRF-3 (the cognate of RF-3). Release factors bind to a ribosomal A site occupied by a termination codon and stimulate the peptidyl transferase center of the ribosome to hydrolyze the bond between the nascent peptide and the tRNA, releasing the nascent chain from the ribosome. In prokaryotes, another factor, ribosome recycling factor (RRF), facilitates the dissociation of the ribosomal subunits from the mRNA.

II. GENE-SPECIFIC TRANSLATIONAL CONTROL OF GENE OUTPUT

Translation, a complex process, provides multiple steps at which the amount of protein expressed from a given gene can be regulated. Most obviously, the rate that ribosomes bind to an mRNA and initiate translation clearly should affect the amount of protein expressed. The rate of production of protein, in the absence of some other effect, should be directly proportional to the frequency that ribosomes begin translating. Consequently, many genes have evolved regulatory mechanisms that modulate the frequency or initiation. Other mechanisms can indirectly regulate gene output without directly regulating translational initiation. For example, blocking translational elongation would also reduce gene output. Few systems have evolved that use such a block, and, where they have, the elongation block affects gene output indirectly by either causing early termination of transcription (transcriptional attenuation) or blocking downstream translational initiation (translational attenuation).

A. Alternative Pathways of Translational Initiation in Prokaryotes

Normally, translational initiation in prokaryotes occurs by direct binding of the initiation complex to the site of initiation. The vast majority of prokaryotic genes use this system, but some genes have evolved a different mode of initiation, translational coupling. In translationally coupled systems, ribosomes initiate translation in an upstream gene and the same ribosome translates one or more downstream genes by reinitiating at initiation codons overlapping or immediately adjacent to translational termination codons. For example, the end of a gene may have the sequence AUGA, in which the termination codon of the upstream gene (UGA) overlaps the initiation codon of a downstream gene (AUG). A translating ribosome reaches the end of the upstream gene, recognizes the UGA codon, and terminates translation by releasing the nascent peptide, allowing the last elongator tRNA to dissociate, and releasing the large ribosomal subunit. The initiation region of the downstream gene, including the AUG and upstream SD site, lies immediately adjacent to the termination site. Unlike eukaryotic ribosomes, the small subunit of prokaryotic ribosomes can bind to an initiation site (ribosome binding site, RBS) in the absence of either initiator tRNA or any initiation factor. Thus, the small subunit can be trapped on the mRNA after termination. Only later, when initiation factors and f-met-$tRNA_i^{Met}$ join the ribosome, does translation begin. This process theoretically is 100% efficient and allows a single upstream gene to regulate the efficiency of translation of multiple downstream genes. To make the system entirely dependent on upstream translation, a secondary structure must sequester the RBS of the downstream gene so that free initiation complexes cannot directly recognize it.

An alternative form of this control occurs when a ribosome translating through an upstream gene disrupts a secondary structure that sequesters the RBS of a downstream gene, making it unavailable for initiation. The ribosome translating the upstream gene may pause during elongation, unwinding the structure so that the RBS becomes available. A second ribosome can then initiate the translation of the downstream gene. This system also allows translational coregulation of multiple genes, but without using the same small ribosomal subunit.

B. Alternative Pathways of Translational Initiation in Eukaryotes

In most eukaryotic mRNAs, translation begins at the first AUG downstream of the 5′-end. In a large minority of genes, the initiation site is not the first AUG, and, in some, there are large numbers of AUG codons in the 5′ nontranslated region. In these genes, translation initiation occurs by an alternative mechanism. There are two classes of alternative initiation mechanisms, internal initiation and ribosomal shunting. Internal initiation occurs by recruitment of an initiation complex to a site within the body of the mRNA. Because the complex does not bind to the 5′-end of the mRNA, this form of initiation does not require the 5′ Cap. In fact, eukaryotic viruses have evolved systems blocking the normal recognition of cap-containing mRNAs. Viral mRNAs using a Cap-independent mechanism in this situation compete much more efficiently than endogenous cellular mRNAs for the cellular translational machinery. The internal binding of the initiation complex depends on a highly structured region in the 5′ nontranslated region. The initiation complex probably recognizes proteins bound to the mRNA to dock with the mRNA. The identity of these proteins remains unknown, although research has identified several candidate proteins. The initiation complex probably does not bind directly to the initiation site, as in prokaryotes, but probably scans a short distance from the binding site to the initiation codon, the first downstream AUG codon. The only clear difference between this and the normal mechanism of initiation is the way the initiation complex binds to the mRNA. Subsequent steps may be quite similar.

Ribosome shunting occurs in a very small minority of mRNAs. The mechanism is distinct from internal initiation because the initiation complex binds to the 5′-end of the mRNA, recognizing the eIF-4F complex as in normal initiation. It then begins scanning down the mRNA, but bypasses a large highly structured segment of the 5′ nontranslated region (shunts) before reinitiating scanning and initiating at the first AUG it encounters. The first example of shunting came from the plant virus cauliflower mosaic virus, but other examples have been found since.

C. Cotranslational Control of Gene Transcription in Prokaryotes: Transcriptional and Translational Attenuation

Transcription and translation in prokaryotes are not physically separated into different subcellular compartments, as they are in eukaryotes. This provides the possibility of a unique form of genetic control in bacteria, in which transcription is coupled to translation of the nascent mRNA. Control of certain bacterial genes occurs at the level of transcriptional termination, in response to pausing by ribosomes translating the nascent transcript. This process, termed transcriptional attenuation, governs the expression of several amino acid biosynthetic genes, first recognized in the *trp* and *his* operons. Ribosomes bind to and begin translating the nascent transcripts of these genes. The ribosomes initiate on the first open reading frame (ORF) in the operon, which is a short upstream open reading frame (uORF). In the *trp* and *his* systems, the uORFs contain tryptophan or histidine codons. Ribosomes reading the uORF stall at the Trp or His codon because of the lack of their cognate amino acid. The stalled ribosome disrupts an mRNA secondary structure, allowing the mRNA to adopt an alternative structure that signals the transcribing polymerase to terminate transcription. Thus, the lack of continuing translation transduces a signal to discontinue transcription.

Other similar systems modulate expression using a uORF, but the regulation is at the level of translational initiation. A dramatic example comes from the *cat86* gene of *Bacillus subtilis*. This gene encodes an enzyme that acetylates the antibiotic chloramphenicol, a protein-synthesis inhibitor that blocks peptidyltransferase. The gene has a uORF that is translated cotranscriptionally. The nascent polypeptide is itself a peptidyltransferase inhibitor, so that ribosomes translating the uORF become paused because the nascent peptide blocks the further extension of the protein. This pause is only transient, but in the presence of chloramphenicol it becomes extremely prolonged. The paused ribosome disrupts a mRNA secondary structure that, when formed, sequesters the ribosome-binding site of the *cat86* gene. Pausing thus allows other ribosomes to initiate the translation of *cat86*, ultimately leading to the inactivation of chloramphenicol. In this case, a gene exquisitely sensitive to inhibition by chloramphenicol transduces the blockage of translation into the stimulation of translation.

D. Control of Translational Reinitiation

Not all eukaryotic mRNAs in which the initiator AUG is not the first in the message use these mechanisms to bypass scanning. On these mRNAs, the ribosome binds to the 5′-end, as normal, and initiates at the first AUG it encounters. This AUG begins a short uORF similar to those found in attenuation systems in bacteria. Of course, because transcription and translation are physically separated in eukaryotes, the mRNA is not recognized as a nascent transcript, but as a full-length cytoplasmic mRNA. Therefore, these genes are regulated translationally. When the ribosome finishes translating the uORF, it can reinitiate scanning, eventually recognizing and initiating translation at the initiator AUG of the encoded gene. Translational reinitiation occurs in a minority of mRNAs, including many cytokine genes in higher eukaryotes, and the *GCN4* gene in the yeast *S. cerevisiae*. In the cytokine genes, the uORF has the role of attenuating translation into the downstream gene. For some mRNAs, this blocking function is conditional, although in many it appears to be constitutive.

The *GCN4* gene, by contrast, uses translational reinitiation to regulate output from its mRNA. The *GCN4* mRNA has four uORFs. Ribosomes scanning from the 5′-end initiate at either the first or second uORF. After translating it, the ribosomes reinitiate scanning. During the scanning they must restock themselves with methionyl-tRNA^{Met}–eIF-2–GTP ternary complex in order to be able to reinitiate at a downstream AUG. If they become competent to reinitiate before scanning past the second pair of uORFs, the ribosomes will initiate on them; but if they become competent after scanning past the uORFs, they may initiate on the *GCN4* gene. When starved for amino acids, yeast cells have a lower concentration of eIF-2 ternary complex. The GCN4 protein is a transcriptional regulator of amino acid biosynthetic genes, so this system allows cells to turn on coordinately all of these genes during amino acid starvation.

E. Translational Repressors

Translational repression is a ubiquitous and common form of translational control. The best-known example comes from ribosomal proteins in bacteria. When present in excess, certain ribosomal proteins bind to their own mRNAs, inhibiting translation initiation, often of entire operons. This mechanism achieves feedback control of ribosomal proteins and contributes to homeostasis in ribosome biogenesis. In prokaryotes, the binding site for the proteins lie within the 5′ noncoding region and block access by ribosomes to the ribosome-binding site of the first controlled gene. By translational coupling, downregulation of the first gene can proceed down the mRNA to other genes located on the same mRNA. A second well-known example is the gene 32 protein of bacteriophage T4. This single-stranded DNA-binding protein can also bind a pseudoknot within the 5′ nontranslated region of its own mRNA. When the protein is in excess, it binds to a pseudoknot in the 5′ nontranslated region of the mRNA and blocks ribosome binding.

One eukaryotic repressor, IRE-BP, works by a mechanism very similar to the prokaryotic proteins. The protein is the cytosolic isozyme of the enzyme aconitase, but it has the additional function of regulating gene expression in response to ferric iron (Fe^{3+}). IRE-BP has an iron–sulfur cluster that loses a single Fe^{3+} when the iron concentration is low. Lacking the Fe^{3+}, the protein can bind to an iron response element (IRE) in the 5′ noncoding region of the ferritin mRNA. When bound, it blocks access by the initiation complex to the mRNA. When the Fe^{3+} concentration is high, another Fe^{3+} can bind to IRE-BP, reducing its affinity for the IRE, and releasing the repression of ferritin. Ferritin chelates intracellular Fe^{3+}, protecting the cell from oxidation. The translational regulation assures that enough ferritin will be available for that purpose.

Other eukaryotic translational repressors function differently. First, their binding sites are in the 3′ noncoding regions of mRNAs. This means that unlike IRE-BP, they can not sterically block access by the ribosome to the mRNA. Instead, they must affect a long-range structure essential to initiation. Work has shown that mRNAs in eukaryotes adopt a circular structure in which the 5′- and 3′-ends are held in close contact by specific RNA-binding proteins. The eIF-4G protein, a subunit of the eIF-4F complex bound to the 5′ Cap, interacts with the poly-A binding protein (PABP), which binds to the posttranscriptionally added runs of adenosines at the 3′-ends of mRNAs. The initiation complex may recognize this structure to bind the mRNA, and translational repressors bound to the 3′ noncoding region could block its availability by blocking the recognition of the eIF-4F–PABP complex.

F. Control of mRNA Degradation

Although at first the process of mRNA degradation would seem unrelated to the issue of translational control, much degradation occurs in the context of ongoing translation. In eukaryotes, whose mRNAs usually have much longer half-lives than those of prokaryotes, a change in mRNA stability can profoundly affect gene output.

The classic example of cotranslational degradation comes from mammalian tubulin genes. In addition to long-term transcriptional control, the amount of both α- and β-tubulin is controlled by regulated cotranslational mRNA degradation. The mechanism of this controlled is better understood for β-tubulin, although the regulation of α-tubulin appears to be quite similar. When the concentration of free β-tubulin subunits is high enough, the mRNA for β-tubulin is rapidly and specifically degraded. The signal for this degradation is, surprisingly, the first four amino acids of the nascent polypeptide, Met-Arg-Glu-Ile. As this peptide emerges from the ribosome, some cellular factor (perhaps β-tubulin, itself) recognizes it and activates an unknown endonuclease, which cleaves the mRNA, leading to its ultimate degradation. This system provides a very rapid response to the presence of excess subunits, as when microtubules depolymerize into tubulin monomers, or to adjust the relative expression of α- and β-subunits.

The second example of cotranslational control concerns the degradation of mRNAs containing premature termination codons. An mRNA of a nonsense mutant gene can be much less stable than the identical mRNA lacking the nonsense mutation. The mechanism governing this difference is not yet clear, but the general pathway of the degradation is. When a ribosome encounters a premature termination codon, there are a group of ribosome-associated proteins that sense the presence of the nonsense codon and signal the degradation of the mRNA. Exactly how that signal is sent is not clear. It may involve a specific mRNA sequence present distal of the nonsense site in many mRNAs. However, it is sensed or signaled, the result is that the mRNA is degraded rapidly. This system may have evolved to rid cells of improperly spliced mRNAs. A ribosome reading an mRNA in which a single intron has not been excised would rapidly encounter an in-frame termination codon, which could signal the destruction of the defective messenger.

For some mRNAs, sequence-specific RNA-binding

proteins recognize special sites to activate mRNA degradation. Interestingly, the same protein that regulates the translation of ferritin mRNA in response to Fe^{3+} also regulates transferrin, the protein responsible for importing iron into eukaryotic cells. Transferrin regulation is opposite to ferritin, which is most abundant in the presence of high concentrations of iron. When cells are deficient in iron, transferrin must be abundant to allow recruitment of the diminished supply. When iron is abundant, transferrin must be downregualted so that cells do not accumulate too much of the potentially damaging ion. When Fe^{3+} concentration is low, IRE-BP binds to multiple IREs in the 3′ noncoding region of the transferrin mRNA. The IREs are interspersed with instability elements, and bound IRE-BP interferes with their function and stabilizes the mRNA. In high Fe^{3+}, IRE-BP dissociates from the IREs, and the instability elements then promote rapid degradation of the mRNA. There are many examples of genes with instability elements in their 3′ noncoding region, but most of them function constitutively to maintain a short half-life for the mRNA. These elements allow for the rapid readjustment of mRNA pools when the rate of transcription either increases or decreases. The elements are common in mRNAs encoding regulatory proteins, such as transcription factors.

III. GLOBAL CONTROL OF TRANSLATION

In addition to the many forms of gene-specific regulation, translation is often regulated globally by mechanisms that modulate the activity of translation factors. Cells globally regulate translation efficiency by modulating the activity of translation factors. One purpose of this regulation is to diminish the ability of invading viruses to replicate by inactivating the translational machinery. Ironically, after infection, cytotoxic viruses commonly use a similar mechanism to take over the translational machinery of the infected cell, directing much of the synthetic capacity of the cell to the production of virus-encoded proteins. This shift usually involves the inhibition of cellular translation through the covalent modification (phosphorylation or proteolysis) of cellular translation factors. Viral genes have evolved a lower dependence on these factors for their expression so that they compete more effectively for the translational machinery under these circumstances than do cellular genes.

The phosphorylation of translation factors is the most common form of global translational control. The commonest targets are the elongation factors eIF-2 and eIF-4E. The phosphorylation of eIF-2 reduces its activity, whereas the phosphorylation of eIF-4E stimulates its activity.

Reticulocytes dedicate virtually their entire protein-synthetic machinery to the production of hemoglobin. The rate of synthesis of the apoprotein and synthesis of the heme cofactor must be carefully coregulated. Reticulocytes express an eIF-2-specific protein kinase, heme-regulated inhibitor (HRI), which phosphorylates the α-subunit of eIF-2 in conditions of low heme. When phosphorylated, eIF-2 binds much more tightly to its recycling factor, eIF-2B, reducing the ability to generate the ternary complex met-$tRNA_f^{Met}$–eIF-2–GTP. Translation initiation declines as the concentration of ternary complex falls, so that the ultimate effect of eIF-2 phosphorylation is a loss of translation. This reduction would tend to equalize the concentration of apo-hemoglobin and the heme cofactor.

A similar mechanism of phosphorylation inhibits protein synthesis in virus-infected cells. A homolog of HRI, the dsRNA-dependent protein kinase (PKR), phosphorylates eIF-2α in virus-infected cells. This reduces protein synthesis, which reduces the virus yield from the infected cells. A third homolog, the product of the *GCN2* gene in the yeast *S. cerevisiae,* is the sensor of amino acid starvation in the *GCN4* translational reinitiation scheme. Cells starved for amino acids accumulate deacylated tRNAs, which bind to Gcn2p, activating its eIF-2α kinase activity. This leads to a reduction in eIF-2 ternary complex. The reduction is not enough to affect bulk protein synthesis, as PKR and HRI do, but it is enough to reduce the efficiency that ribosomes scanning past the first of the *GCN4* uORFs regain competence to reinitiate translation. In starved cells, the ribosomes tend to become competent after moving farther down the mRNA, and so tend to initiate translation on the

GCN4 gene rather than on the other uORFs, causing an increase in translation of *GCN4*.

The phosphorylation of eIF-4E, the Cap-binding factor, is an example of the opposite form of global control. The phosphorylation of eIF-4E is correlated with increased, rather than diminished, translation. In higher eukaryotic systems, various mitogens stimulate phosphorylation. Cells stimulated to proliferate or initiate developmental programs require an increase in translation. Modulating eIF-4E activity is especially effective because of its very low abundance compared to other initiation factors and because it is critical to the first step in recruitment of ribosomes to the mRNA. A second subunit of the Cap-recognition complex, eIF-4G, also becomes highly phosphorylated in response to mitogens. The third subunit, eIF-4A, may be developmentally regulated by phosphorylation in *Drosophila*. Finally, eIF-4B, which along with eIF-4A prepares the mRNA for ribosomal scanning, is also highly phosphorylated in response to mitogens. All of these events may stimulate initiation by increasing the efficiency with which ribosomes bind to the mRNA.

The Cap-recognition complex is subject to the opposite type of control by specific degradation. During picornavirus infection, a virally encoded protease specifically cleaves eIF-4G. This cleavage reduces translation, although not completely. Because picornaviruses use a Cap-independent method of initiation, reducing efficiency of Cap recognition would tend to increase the ability of the picornavirus mRNA to compete for the limiting supply of ribosomes in the cell, thus increasing its relative efficiency. The point of this control may not be to suppress cellular mRNA translation completely, but only to maximize viral expression.

IV. TRANSLATIONAL ACCURACY

The control of the structure of the primary translation product is a second ubiquitous, but perhaps not as prevalent, form of translational control. This type of control has been called recoding to suggest that the rules of translational coding are changed to result in an unconventional or noncanonical event. In most cases, this type of control allows a single gene to specify two structurally related, yet distinct primary gene products. Termination codon readthrough, the simplest of these events, results in the insertion of an amino acid at a termination codon. Readthrough is a stochastic event, usually much less than 100% efficient. Consequently, readthrough allows the expression of two forms of a protein, the first resulting from normal termination and a second consisting of the first protein fused to the product of the reading frame downstream from the terminator. Other events call for the ribosome to shift its reading frame either by reading a codon overlapping one in the previous frame or by causing the ribosome to bypass a short stretch of codons. These types of events superficially resemble the kinds of errors that occur during translation, although the efficiency of recoding events is several orders of magnitude higher. Understanding recoding requires, first, a consideration of the mechanism of translational errors and of the mechanisms used by the ribosome to forestall them.

A. Types of Translational Errors

Missense errors, the simplest type of translational error, occur when one amino acid inserts in the place of another. Missense errors result either from incorrect tRNA aminoacylation or recruitment of an incorrect (noncognate) aminoacyl-tRNA to the ribosome. They are the most frequent type of translational error, occurring at about 5×10^{-4} per codon. Although this seems like a low frequency, it results in inaccurate expression of a significant proportion of proteins. For example, the protein β-galactosidase is 1000 amino acids in length. Errors occurring at 5×10^{-4} per codon would result in 39% of proteins having at least one missense error. Yet, missense errors tend to have a relatively low phenotypic effect because most errant proteins retain full activity, and few have seriously lowered activity.

The translational machinery has to compromise between the cost of translational inaccuracy and the advantage of increased translation speed. To the extent that translation rate is increased, the ribosome will tend to accept a greater proportion of erroneous noncognate tRNAs. To reduce translation errors below the observed frequency would require a significant reduction in translation elongation rates. The observed rate of errors probably optimizes these antagonistic negative and positive effects.

The mechanism underlying missense error correction remains controversial. One theory proposes that ribosomes distinguish cognate from noncognate tRNAs using a process of kinetic proofreading. The idea is that both cognate and noncognate aa-tRNA–EF-1A–GTP complexes can enter the ribosomal A site, but that noncognates rapidly dissociate whereas cognates remain bound. After GTP hydrolysis, the complex remains bound transiently. During this pause, any noncognate tRNA again tends to dissociate whereas cognates remain bound. This is a proofreading step because it consumes GTP. In this model, discrimination uses the greater rate of dissociation of noncognates. The second model for discrimination involves a putative allostery between the ribosomal A and E sites. In this model, deacyl-tRNA bound to the E site causes the A site to discriminate between cognate and noncognate tRNAs. In this model, the discrimination is at the step of initial binding and is an equilibrium rather than a kinetic effect. It is possible that these mechanisms operate in concert to preclude missense decoding.

The second general class of errors are those that cause the premature termination of translation. These are termed processivity errors because they interfere with the processive nature of translation. Processivity errors occur on average at approximately the frequency of missense errors. Again, that would imply a significant loss in terms of gene output because of these errors. At a rate of 5×10^{-4} terminations per codon, only about 60% of ribosomes that begin translating β-galactosidase would actually produce a full-length product. These errors generally have a much more serious effect on the activity of the gene product. The vast majority of C-terminally truncated products retain no enzymatic activity, and would be rapidly degraded. Usually only those proteins retaining all but the most C-terminal portions of the protein retain significant activity. It is perhaps surprising that the translational machinery has not evolved a more efficient correction mechanism for these errors given their seriousness. This failure implies that the compromise in terms of the rate of translation elongation would be too drastic to reduce these errors further.

The simplest kind of processivity failures are those in which the process of elongation is aborted prematurely. Surprisingly, few of these errors result from peptide release factor mistakenly acting at a sense codon, so these errors are not termination events per se. Instead, the vast majority occurs when peptidyl-tRNA spontaneously dissociates from the translating ribosome (ribosome drop-off). Cells have a special enzyme that removes nascent peptides from these tRNAs, peptidyl-tRNA hydrolase. Inactivating the enzyme causes cells to fill rapidly with peptidyl-tRNA, underscoring the frequency of the error. Treatments or genetic states that favor missense translation also stimulate this error, suggesting that it may be peptidyl-tRNAs bound to noncognate codons that dissociate from the ribosome.

Translational frameshifting is also a processivity error because a ribosome that changes reading frame cannot complete the intended translational product, and, in fact, should rapidly encounter a termination codon in the shifted frame. Perhaps significantly, frameshifting errors occur at least 10-fold less frequently than do either missense errors or ribosome drop-off. Because these errors should be no more serious than other processivity failures in terms of enzyme activity, their lower occurrence cannot reflect a greater selection against them. Rather, it may reflect a fundamental feature of the error mechanism.

Spontaneous frameshifting normally occurs by changes of single nucleotides, either in the 5′ or leftward direction (−1), or the 3′ or rightward direction (+1). Frameshifting can cause the phenotypic suppression of a 1-bp insertion or deletion mutations. Judged by the level of suppression, +1 frameshifting appears to occur much more frequently. Sequence contexts greatly affect the efficiency of frameshifting because suppression can vary strikingly among various frameshift mutations. This implies that specific signals can stimulate frameshifting. The analysis of programmed frameshift sites, sequences in genes that program highly efficient frameshifting (see later), clarified the nature of the signals.

B. Genetics of Translational Accuracy

The classic approach to genetic analysis of translational accuracy was to identify suppressor mutations that restored normal activity to missense or frameshift mutant forms of a reporter gene. For example, using a gene that had suffered a 1-bp insertion and therefore was nonfunctional, one could identify second-site suppressors that restored normal expression of the gene. Frameshift-suppressor mutations can affect tRNAs, ribosomal proteins, rRNA, and translation factors. Similarly, suppressors can promote the readthrough of premature termination codons or allow the bypass of missense mutations. These experiments have identified factors responsible for maintaining translational accuracy.

1. *tRNA Mutations*

The easiest of the suppressors to understand were the tRNA suppressors. Nonsense, frameshift, and missense suppressors all seem to work in the same manner. By changing the sequence of the anticodon, the decoding properties of a tRNA change. Nonsense suppressors have an anticodon complementary to one of the nonsense codons, allowing the ribosome to read it as sense. Most +1-frameshift suppressors have expanded anticodons that are thought to shift reading by pairing to an expanded anticodon. Similarly, missense suppressors have anticodons corresponding to the misread codon. Creating these suppressors is complicated by the fact that most aminoacyl-tRNA synthetases contact bases in anticodons to discriminate among possible substrates. Changing the anticodon can also change recognition by the synthetase. This is most clearly an issue for the missense suppressors, but inefficient aminoacylation would reduce the function of any of the suppressors. This effect limited the examples in each suppressor class to a small number of mutant tRNAs.

Work has shown that the accepted model for +1-frameshift suppressors is incorrect for a major class of mutations and might be invalid in general. The Quadruplet Translocation model proposed that the expanded anticodon of +1-frameshift suppressors induced EF-2 to catalyze a 4-nt translocation by making a 4-bp codon–anticodon interaction with the mRNA. This model cannot explain frameshifting by certain suppressors because posttranscriptional modification of their anticodon loops precludes the required 4-bp codon–anticodon interaction. For these tRNAs, and perhaps for all, it appears that the suppressing tRNA causes a normal 3-nt translocation, but that the peptidyl-tRNA slips +1 on the mRNA while in the ribosomal P site. This Peptidyl-tRNA Slippage model can explain suppression by all classes of frameshift-suppressing tRNAs, although direct evidence supporting the model for many of them is not available.

2. *Ribosomal Mutations*

Because translational errors occur in the context of the ribosomal decoding sites, one might expect that some suppressors alter ribosomal components. In fact, some of the earliest suppressor mutations affected the ribosomal proteins S4, S5, and S12 of *E. coli*. Mutations in S4 or S5, called *ram*, for *r*ibosomal *am*biguity, increase the error frequency, whereas those in S12 cause resistance to the antibiotic streptomycin concomitant with reduced error frequency. These data suggest that the three ribosomal proteins may form the core of an accuracy center on the ribosome. Significantly, the three are proteins of the small subunit of the ribosome, the site of codon–anticodon pairing, suggesting that they may directly modulate that interaction.

Mutations affecting the rRNA in both subunits of the ribosome can can also stimulate translational errors. Three regions of the small subunit rRNA can mutate to suppress either frameshift or nonsense mutations. Frameshift-suppressing mutations fall in the part of the rRNA in the ribosome's decoding center (the 1400 region of 16*S* rRNA in bacteria). Other mutations affect the 530 loop, a region that is proposed as modulating cognate recognition of the A site codon, or a region near position 912. Both regions seem to modulate recognition in the decoding site and both are involved in rearrangements of rRNA secondary or tertiary structure. This fact emphasizes the idea that the ribosome functions as a molecular machine (i.e., that it has moving parts) to accomplish error correction. Other frameshift-suppressing rRNA mutations affect the large subunit rRNA. These fall within the α-sarcin loop, named this because it is cleaved by the cytotoxic nuclease α-sarcin. Significantly, the ribosomal proteins S4, S5, and S12 all interact closely with the 530 and α-sarcin loops, suggesting that their suppressive effect may result from the effect they have on these critical rRNA structures.

Mutations of rRNA that cause UGA-specific nonsense suppression affect two regions of a domain of 16*S* rRNA in bacteria called helix 34. Within this helix is a sequence UCAUCA, two copies of a sequence that could potentially base-pair with UGA. It is unclear whether the UCA repeat actually functions by recognizing UGA codons through basepairing, but that is an attractive model to explain the phenotype of these suppressors.

3. *Translation Factor Mutations*

The final class of suppressors affect the translation elongation factors EF-1A and eEF-1A, and peptide-release factors. Mutant forms of bacterial EF-1A and yeast (*S. cerevisiae*) eEF-1A have the same effect of suppressing nonsense or frameshift mutations, or both. The mutations target two of the three domains of the molecule, the GTP-binding and the tRNA-binding domains. The exact mechanism of suppression is unclear. Those affecting the GTP-binding domain may affect the kinetics of GTP hydrolysis, which under the Kinetic Proofreading model should perturb discrimination between correct and incorrect tRNAs. Those in the tRNA-binding domain may change the interaction between the EF-1A ternary complex and the ribosome, perhaps stabilizing the interaction of noncognate tRNAs, which again under that model would tend to reduce discrimination.

The availability of peptide-release factors influence the frequency of nonsense codon readthrough both *in vivo* and *in vitro*. The reduction in the effective concentration of release factors tends to allow readthrough, to varying degrees, of all three types of nonsense codons. The amino acid inserted tends to be one encoded by a codon near in sequence to the nonsense codon. This implies that when the rate of recognition of a nonsense codon is limiting, the ribosome will accept noncognate tRNAs. This is consistent with the idea that ribosomal accuracy is kinetically controlled. In yeast, an epigenetic state can arise in which the peptide-release factor is limiting. The aggregation of the factor brings on the state, causing increased nonsense codon readthrough. This situation resembles a prion infection (e.g., those causing scrapie or bovine spongiform encephalopathy) because it involves the formation of protein aggregates *in vivo*, and the condition is heritable, although not genetically encoded. This problem is an active subject of investigation given its significance as a clinical model system.

V. PROGRAMMED ALTERNATIVE TRANSLATION

Genetic control by induced alteration of the rules of translation occurs ubiquitously, but infrequently. The majority of programmed alternative translation systems occur in viruses or virus-like genetic elements. A small but slowly growing list of cellular genes use this form of control. The expectation is that this form of control will continue to be rare. Most frequently, alternative translation provides the capacity to express a C-terminally extended form of a protein. This extended form is usually required for a morphogenetic process related to the life cycle of the virus or virus-like element. A minority of genes use this system to express alternative proteins with different enzymatic features, or as a system of autogenous or feedback control.

A. Programmed Readthrough of Termination Codons

Readthrough of termination codons is arguably the simplest form of alternative decoding. The sequence context surrounding the termination codon probably acts to reduce the efficiency of translational termination and thus increase the probability that a noncognate tRNA will recognize the nonsense codon. Some programmed readthrough sites are quite simple. In Sindbis virus, a positive-strand RNA animal virus, the expression of a full-length nonstructural polyprotein occurs by readthrough of a single UGA codon. Readthrough requires only the sequence UGA-C. Peptide-release factor itself recognizes true termination codons, using a tetranucleotide signal consisting of the codon and its 3′-neighbor nucleotide, and an inappropriate 3′ nucleotide can drastically reduce the rate of termination. This appears to be the mechanism of readthrough suppression in Sindbis virus; slow recognition of this very poor termination sequence allows the recognition of the UGA codon, probably by tryptophanyl-$tRNA^{Trp}$.

Other programmed readthrough sites are much more complicated. For example, in the retrovirus Moloney murine leukemia virus (MoMLV), the product of the *pol* gene (for *pol*ymerase), which encodes the enzymatic activities of reverse transcription and integration, is fused translationally to the structural proteins encoded by the upstream *gag* gene. The fusion occurs by readthrough of a single UAG codon separating the two reading frames. The sequence context following this codon determines the frequency of readthrough. Maximal readthrough requires a 54-nt downstream region that folds to form a pseudoknot (a pseudoknot is a complex secondary structure in which the loop portion of a stem–loop structure base-pairs with an immediately flanking sequence). Readthrough requires the secondary structure of the pseudoknot, but not most of its primary sequence. The pseudoknot causes the misreading of the UAG codon as glutamine, which would require a U–G wobble interaction at the first position of the codon. It remains unclear how the pseudoknot stimulates this event.

Perhaps the most bizarre form of programmed readthrough occurs during cotranslational insertion of the amino acid selenocysteine. Specialized UGA codons encode selenocysteine. An mRNA secondary structure promotes the recognition of the codon by a specialized selenocysteyl-$tRNA^{Sec}$. This tRNA has an anticodon complementary to the UGA codon. Because of its unusual structure, it cannot bind EF-1A, but instead binds to a specialized cognate factor, the SELB protein. SELB in turn binds to a specialized mRNA secondary structure called the SECIS (selenocysteine insertion sequence). When bound to the SECIS, SELB delivers selenocysteyl-$tRNA^{Sec}$ into the A site at the UGA codon. In bacteria, the SECIS is located immediately downstream of the UGA, although in eukaryotes it resides in the 3′ nontranslated region of the mRNA. In a sense, this system is an example of specialized nonsense suppression because it employs a dedicated tRNA cognate for the UGA codon.

B. Programmed Translational Frameshifting

The second form of programmed alternative decoding occurs when a sequence in a gene causes the translating ribosome to shift into a new reading frame while continuing elongation. The most common form of programmed translational frameshifting shifts the ribosome in the −1 direction. Frameshifting in the −1 direction occurs in many metazoan viruses, including retroviruses, in various plant viruses, lower-eukaryotic virus-like elements, and in bacterial insertion sequences. Only one cellular gene appears to use this mechanism, the *dnaX* gene of *E. coli,* which encodes alternative forms of a subunit of DNA polymerase III. Frameshifting occurs on a slippery heptameric sequence of the form X-XXY-YYZ, shown in codons of the unshifted frame, by simultaneous slippage of two tRNAs from XXY-YYZ to XXX-YYY. A downstream secondary structure stimulates slippage, a pseudoknot in most eukaryotic sites and a stem-loop in most prokaryotic sites. Ribosomes arriving at this site pause before progressing. During the pause, they can shift −1 on the slippery heptamer. We do not understand how the structure stimulates frameshifting, although it involves some second event in addition to ribosome stalling. In prokaryotes, frameshifting requires a second site upstream of the slippery heptamer. At this site the SD sequence of the 16*S* rRNA engages the mRNA and appears to literally pull the mRNA to stimulate slippage of the tRNAs.

Frameshifting in the +1 direction is much less common. It occurs in a yeast retrotransposon (a transposable genetic element related to retroviruses), and in cellular genes in *E. coli,* yeast, and mammalian cells. In each case, frameshifting occurs when a poorly recognized codon occupies the A site while a special codon occupies the P site. The poorly recognized codon again induces a translational pause, during which the shift occurs. In most cases, during the pause a tRNA occupies the P-site codon, which can slip +1 while still maintaining at least 2 bp with the mRNA. In some cases, however, frameshifting appears to occur without the need for slipping. In these cases, a special feature of the peptidyl-tRNA stimulates out-of-frame decoding in the A site, shifting reading in the +1 direction. In yeast, it is clear that the feature causing this behavior is the fact that the peptidyl-tRNA is a near-cognate tRNA, an isoacceptor that cannot form a normal wobble interaction with the codon.

C. Programmed Translational Hopping

Translational hops are the rarest form of programmed alternative decoding. We have evidence for only one programmed hop, which occurs in T4 bacteriophage gene *60* encoding topoisomerase. A 50-nt insertion interrupts the transcript of the gene. Despite the interruption, the ribosome is able to decode the gene, bypassing the 50-nt insertion entirely. Hopping depends on several stimulatory features—matching "take-off" and "landing" codons, an in-frame termination codon immediately after the take-off codon, a stem-loop structure encompassing these two codons, and a nascent peptide sequence encoded immediately upstream of the take-off codon. The matching take-off and landing codons reflect the need for the peptidyl-tRNA to dissociate from the mRNA and re-pair after the 50-nt interruption. The slow recognition of the next codon, the in-frame termination codon, probably allows sufficient time for the dissociation of the peptidyl-tRNA. The exact function of the other features is still unclear.

A second putative example of a translational hop in the *trpR* gene of *E. coli* appears to be erroneous. Despite this, it is possible that more such sites exist in other genes. Finding them requires an unusual diligence because in a database a gene encoded by hopping would look either like two adjoining genes or like a single gene including an intron. The generality of the gene *60* hop has recently been reinforced by a study showing that inefficient hopping can be stimulated merely by providing a sufficient translational pause at a poorly recognized codon. During the pause, peptidyl-tRNA can dissociate from the mRNA and scan in the 3′ direction in search of a matching codon. The efficiency of the hop depends on the distance to the next codon because peptidyl-tRNAs tend to dissociate spontaneously from the ribosome during the search.

See Also the Following Articles

Attenuation, Transcriptional • Ribosome Synthesis and Regulation • Transcriptional Regulation in Prokaryotes

Bibliography

Farabaugh, P. J. (1996). Programmed translational frameshifting. *Microbiol. Rev.* **60**, 103–134.

Kurland, C. G. (1992). Translational accuracy and the fitness of bacteria. *Annu. Rev. Genet.* **26**, 29–50.

Lindahl, L., and Hinnebusch, A. (1992). Diversity of mechanisms in the regulation of translation in prokaryotes and lower eukaryotes. *Curr. Opin. Genet. Dev.* **2**, 720–726.

Transposable Elements

Peter M. Bennett
University of Bristol

I. Introduction
II. IS Elements and Composite Transposons
III. Tn3 and Related Transposons
IV. Conjugative Transposons
V. Site-Specific Transposons
VI. Transposing Bacteriophages
VII. Yeast Transposons
VIII. Consequences of Transposition in Bacteria
IX. Conclusion

GLOSSARY

composite transposon A modular structure comprising an intrinsically nontransposing sequence flanked by copies of the same IS element.

conjugative transposon A DNA element that mediates both its own transposition and conjugal transfer from one bacterial cell to another.

insertion sequence A small cryptic DNA element that can migrate from one genetic location to another completely unrelated location.

inverted repeat A sequence that defines both ends of a transposable element and that is found as an inverted duplication.

target site A site at which a transposition event occurs (occurred).

transposing bacteriophage A bacteriophage that replicates using a form of transposition.

transposition immunity The inhibition of transposition of one copy of a transposable element by a second copy of the same element on the target DNA.

transposon A DNA element that can migrate from one genetic site to another unrelated site and that encodes functions other than that required for transposition.

TRANSPOSABLE ELEMENTS are discrete DNA sequences that can move from one location on a DNA molecule to another location on the same or on a different DNA molecule. The process of transposition is, *ipso facto,* a recombination event in that it involves breaking and reforming phosphodiester bonds. It requires no extended homology between the element and its sites of insertion; accordingly, it does not depend on the homologous recombination system of the host cell. In practice, this means that transposition is *recA*-independent. Several types of transposable elements are known and transposition can occur by one of several different mechanisms. Hence, although the term ''transposable element'' implies a defined genetic entity, the term ''transposition'' indicates simply that genetic rearrangement by DNA relocation has occurred, by any one of a number of mechanisms. Transposable elements have been discovered in gram-negative and gram-positive bacteria, in archaebacteria, and in yeasts and are probably responsible for much of the macromolecular rearrangements of microbial genomes.

I. INTRODUCTION

A. General Structure of a Transposable Element

Two points of caution should be stated at the outset. First, although transposition results in the insertion of one DNA sequence into another, in some instances, but not all, the transposed sequence is also retained at its original location (i.e. the event is not only recombinational but also replicative; see later). Second, not all transposition events involve the movement of a defined DNA element (e.g., IS*91*-

like elements can effect contransposition of varying lengths of DNA adjacent to one side of the element; see later).

Most reported transposition events in microbes involve transposable elements alone. Each element is a defined structure that is preserved precisely from one transposition event to the next. Most, but not all, elements terminate in short, perfect or near perfect inverted repeats (IRs), which function as recognition signals for the transposition enzymes. Terminal IR sequences are usually 15–40 bp long (but may be much longer, e.g., the imperfect IRs of Tn7 and the bacteriophage Mu) and differ from one transposable element to another. Related elements have related IRs. Each fully functional element encodes at least one protein that is needed for its own transposition, normally termed a transposase. In most cases, the transposase processes the ends of the element in readiness for recombination; in addition, functions unrelated to transposition may be encoded.

B. Insertion Sequences and Transposons

The smallest elements capable of self-promoted transposition are the prokaryotic insertion sequences (IS elements), which range in size from about 0.7–2 kb. Each encodes only one or possibly two functions necessary for the control and execution of its own transposition. Transposons are larger structures that, in addition to a transposition system, accommodate functions unrelated to transposition, which confer a predictable phenotype on the host cell (e.g., resistance to an antibiotic), unlike IS elements, which may alter the cell phenotype at random by mutation. Indeed, it was this mutational activity that first drew attention to the existence of transposable elements in bacteria in the mid-1960s. However, transposable DNA sequences were first discovered in maize by Nobel laureate Barbara McClintock in the early 1950s; she suggested, correctly, that they were the cause of the variations in pigmentation of maize kernels. This conclusion was based on a detailed and elegant genetic analysis. Physical evidence to support the genetic case for transposable elements was first obtained from studies of bacterial systems in the late 1960s, culminating in the descriptions of three bacterial insertion sequences, IS*1*, IS*2*, and IS*3*. Modern molecular genetics has revolutionized and considerably simplified the task of detecting, isolating and characterizing transposable elements in both prokaryotic and eukaryotic systems.

Transposons fall into two general categories, sometimes called composite transposons and complex transposons. However, a better description would be composite and noncomposite, for reasons that will become clear. A composite transposon has a modular structure in which a central unique sequence, accommodating genes not involved in transposition, is flanked by two copies of an IS element in direct or, more commonly, in inverted repeat. One or both of these terminal elements provide the unique functions necessary for the transposition of the composite structure and often retain the ability to transpose as independent IS elements (see later).

Complex or noncomposite transposons have no obvious modular structure. Functions unconcerned with transposition (i.e., those equivalent to the central unique section of composite elements) are encoded by genes that are integrated into the basic transposable element, alongside transposition functions. In most cases, the whole gene ensemble is flanked by short IRs, as seen for the majority of IS elements, and no component of a noncomposite transposon can transpose independently of the rest of the structure. In crude comparative terms, a noncomposite transposon is more analogous to an IS element than to a composite transposon. Indeed, some noncomposite transposons (e.g., Tn3-like elements), as do IS elements, can operate in pairs to transpose the DNA sequences flanked by them. This cooperative behavior seems to be a feature of many prokaryotic transposable elements and has been exploited in the use of mini-Mu, a truncated derivative of the transposing bacteriophage Mu, to clone chromosomal genes *in vivo*.

C. Transposing Bacteriophages

Transposing bacteriophages are typified by bacteriophage Mu. This element was identified by J. Taylor in the early 1960s in a culture of *Escherichia coli* and was the first bacterial transposable element to be discovered, although this was not fully appreciated at the time. The bacteriophage was so named because Taylor deduced, correctly, that the increased number of auxotrophic mutants he found among populations of *E. coli* lysogenized by this temperate phage had been generated by phage integration into genes encoding biosynthetic enzymes. Hence, the phage was described as a mutator phage, now abbreviated to Mu. He concluded that, unlike bacteriophage λ, Mu inserts at many different sites in the *E. coli* chromosome. Transposing bacteriophages replicate by transposition. They differ from other transposable elements in that they can exist independently of other DNA molecules and the host cell as bacteriophage particles.

D. Distribution of Transposable Elements

Transposable elements are widely distributed among eubacteria and archeae and are found in lower and higher eukaryotes. Many IS elements have been identified in both gram-positive and gram-negative bacteria and several have been found in Archaebacteria such as *Halobacterium* sp. and *Methanobrevibacter smithii*. Indeed, wherever a serious search has been made for such elements, they have usually been found. In addition, as the complete sequences of various bacterial genomes have been determined, so a number of putative, previously unknown elements have been revealed. However, the confirmation of the identities of these elements must await the demonstration that they can transpose, and do not simply represent antique transposition events of obsolete elements. The failure to detect a transposable element is more likely to indicate a deficiency in the detection system than the absence of transposable elements.

Transposons, both composite and complex, have been found in many gram-positive and gram-negative bacteria, often in bacteria of clinical or veterinary origin. These sources account for the preponderance of antibiotic resistance transposons; the relative paucity of other markers found on transposons probably reflects no more than the intense interest in aspects of bacteria that impinge directly on human welfare, rather than that antibiotic-resistance genes have a special relationship with transposons. Most bacterial transposons have been found on plasmids rather than on bacterial chromosomes, although there are a few notable exceptions. This finding probably indicates no more than that once a gene is located on a plasmid its ability to spread horizontally to other bacteria is markedly enhanced and, hence, the likelihood of detection is also significantly increased. It also seems that, for some transposons, transposition to a plasmid location occurs at a significantly higher frequency than to bacterial chromosomes, despite the greater capacities of the latter and despite the fact that insertions are not site-specific. The reason for this is unknown.

E. Nomenclature

The designations given to transposable elements are not usually assigned at random or at the whim of the researchers who discovered them. To avoid accidental assignment of the same code to two different elements, it has been widely accepted that numbers are allocated from a central directory. This resource is administered by Dr. Esther Lederberg at the Department of Medical Microbiology, Stanford University, California. Periodically, lists of new assignments are published. The convention originally adopted is simple. Insertion sequences are designed IS followed by one of a set of numbers allocated to the research worker or laboratory from the central directory, on request. The digits are written in italics. Likewise, transposons are designated Tn followed by an appropriate number (in italics). The designation is intended as no more than a simple, unique identifier. In some cases, additional letters are added to indicate the bacterial source of the element and the numbering system has been restarted (e.g., IS*Rm2*, an IS element from *Rhizobium meliloti*, and IS*M1*, an IS element from *Methanobrevibacter smithii*, are quite distinct from IS*2* and IS*1*, respectively, both of which originated in *E. coli*, but which are also found in other members of the Enterobacteriaceae). These simple rules have, however, not been followed when naming some conjugative transposons, particularly those found in *Bacteroides* sp., because they fail to indicate the nature of the element.

II. IS ELEMENTS AND COMPOSITE TRANSPOSONS

A. History

The first transposable elements to be recognized as such in bacteria were three small cryptic elements, appropriately designated IS*1*, IS2 and IS3, discovered as a consequence of the analysis of strongly polar mutations in the *gal* and *lac* operons of *E. coli* (i.e., mutations that substantially eliminated the expression not only of the gene in which the mutations were located but also the expression of the other genes in the operon, located promoter distal to the mutant gene). Genetically, the mutations were found to behave like point mutations, but, although they reverted spontaneously, the frequencies of reversion were unaffected by chemical mutagens known to induce base-substitution and frameshift mutations. Hence, it was suggested that the genetic lesions resulted from the insertion of additional DNA sequences at the sites of mutation. A series of elegant physical studies using, initially, λ.*gal*-transducing phages and then electron-microscope analysis of heteroduplex DNA structures provided physical evidence for mutation by DNA insertion.

B. Structure

Bacterial IS elements are, structurally, the simplest form of transposable element. To date, approximately 500 have been reported. Their sizes (Table I), rarely larger than 2 kb, preclude them encoding more than the one or two functions that are necessary for transposition. Of those that have been examined in sufficient detail, many have one large open reading frame (ORF) that uses the majority of the sequence. In a few cases, notably IS*1*, IS*3*, IS*10*, IS*50*, IS*903* and IS*911*, the products of these predicted genes have been shown to encode proteins necessary for transposition. For the remainder, it is assumed the presumptive genes encode transposition functions. Some elements have smaller ORFs, which overlap or are contained within the large ORF. Whether these are real genes, the products of which have a role in transposition of the element, is largely unknown, although in one or two cases (e.g., IS*10* and IS*50*), the products have been shown to have regulatory functions (i.e., to control the frequency of transposition).

TABLE I
Some Prokaryotic IS Elements[a]

Insertion element	*Family*	*Size (bp)*	*Terminal IRs (bp)*	*Target site duplication (bp)*
		Gram-negative bacteria		
IS*1A*	IS*1*	768	18/23	9
IS2	IS3	1331	32/41	5
IS3	IS3	1258	29/40	3,4
IS4	IS4	1426	16/18	11–13
IS5	IS5	1195	15/16	4
IS6	IS6	820	14/14	8
IS*10R*	IS4	1329	17/22	9
IS*21*	IS*21*	2131	30/42	4
IS*50R*	IS4	1534	8/9	8–10
IS*91*	IS*91*	1830	0	0
IS*150*	IS3	1443	22/31	3
IS492	IS*110*	1202	0	5
		Gram-positive bacteria		
IS*110*	IS*110*	1558	0	0
IS*231A*	IS4	1656	20	10–12
IS*431R*	IS6	790	17/20	NR
IS*904*	IS3	1241	31/39	4
IS*L1*	IS3	1257	21/40	3
IS*S1S*	IS6	808	18	8
		Archaebacteria		
IS*H1*	IS5	1118	8/9	8
IS*H2*	IS4	521	19	10–12
IS*H23*	IS*NCY*	1000	23/29	9
IS*H51-1*	IS4	1371	15/16	5
IS*M1*	IS*NCY*	1381	33/34	8

[a] NR, not recorded; IS*NCY*, family not assigned.

Most of the IS elements characterized to sequence level appear to belong to one of a relatively small number of extended families. Seventeen independent familial groups have been identified, and what has become increasingly clear is that members of most individual families can be distributed across many different eubacterial and archaebacterial genera. There are exceptions. Members of the IS*1* family appears to be restricted to the enterobacteria, whereas IS*66* and similar elements have only been found in bacteria of the rhizosphere. The largest group is the

IS3 family, which now comprises more than 40 members distributed among both the eubacteria and the archeae, suggestive, perhaps, of a somewhat ancient lineage. One current database contains sequence information on approximately 500 IS elements isolated from 73 bacterial genera representing 159 bacterial species (see Mahillion and Chandler, 1998).

One variation on gene arrangement is seen on IS*1*, one of the smallest transposable elements known. It is 768 bp long, delimited by near-perfect 23-bp IRs, and has two short adjacent ORFs, *insA* and *insB*, which have been shown by mutation analysis to be necessary for transposition. The two sequences, in different reading frames, are cotranscribed. However, *insB* is not a true gene, in the sense that it does not encode a specific protein. Rather, the ORF-designated *insB* encodes the second half (C terminus) of a fusion peptide, InsAB, that has the bulk of InsA as its N-terminal half. This apparent violation of the rules governing the termination of peptide synthesis (i.e., the failure to terminate protein synthesis at end of the *insA* transcript) is accomplished by a proportion of the ribosomes that initiate the translation of *insA* making a −1 frameshift toward the end of the *insA* section, before the termination codon is reached. The frameshift, which is directed by the nucleotide sequence, puts these ribosomes into the reading frame that accommodates *insB* and translation continues until the termination codon signalling the completion of InsB synthesis is encountered.

The InsAB protein is the IS*1* transposase, whereas the InsA peptide functions as a transposition regulator. Both proteins compete to bind to the IRs of IS*1*. However, whereas InsAB can mediate transposition, InsA cannot, so, by occluding InsAB from the IRs, InsA can inhibit IS*1* transposition. Interestingly, the transposition frequency is determined not by the absolute amount of either protein, but by their ratio. Elements of the IS3 group also encode fusion peptides that are transposases (see later).

Several well-studied IS elements were discovered as terminal repeats of composite-resistance transposons. Notable among these are IS*10*, IS*50*, and IS*903*. The first constitutes the flanking elements, in inverted repeat, of Tn*10*, which encodes resistance to tetracycline; the second forms the terminal IRs of Tn*5*, and element that confers resistance to kanamycin (Km), bleomycin (Bl), and streptomycin (Sm), although the last determinant is silent in some hosts, including *E. coli*. IS*903* forms the terminal IRs of Tn*903*, another transposon encoding resistance to kanamycin. Although IS*10* (Fig. 1) and IS*50* (Fig. 2) both belong to the IS*4* group of elements, it is interesting to note that they differ in the way that each controls its frequency of transposition. IS*903* controls its frequency of transposition in yet another

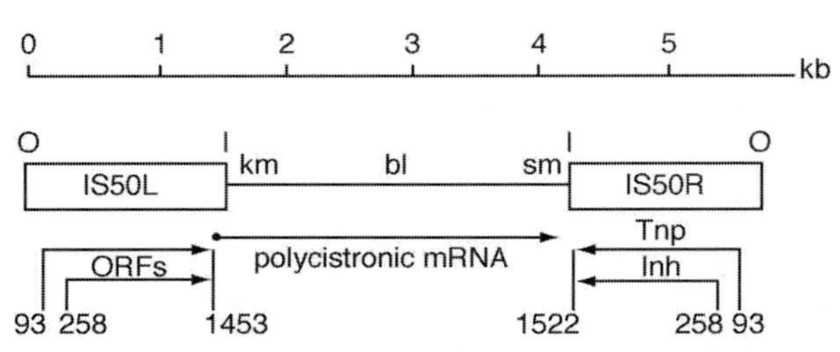

Fig. 1. Tn*5*. Tn*5* was detected on the resistance plasmid JR67, which originated in a strain of *Klebsiella pneumoniae,* when it transposed to phage λ. Tn*5* is 5.7 kb, with a central module of 2.6 kb encoding resistance to kanamycin (Km), bleomycin (Bl), and streptomycin (Sm), flanked by inverted copies of IS*50,* designated IS*50L,R.* The resistance genes are transcribed as a single operon from a promoter within IS*50L.* Deletion analysis has indicated that IS*50R* alone is responsible for Tn*5*–IS*50* transposition. IS*50R* is 1534 bp, has imperfect 19-bp IRs (seven mismatches) and encodes two peptides designated Inh and Tnp, of 421 and 476 amno acids, respectively. Tnp is the IS*50* transposase; Inh is an inhibitor of Tn*5*/IS*50* transposition. (It reduces the frequency of transposition after Tn*5*/IS*50* has established in a new host. On moving to another cell, the transposition of IS*50*/Tn*5* is subject to zygotic induction.) The Tnp and Inh proteins are virually identical; they differ in that Inh lacks the first 55 amino acids of Tnp. The shortening is not a result of posttranslational modification; rather, Inh is produced from its own transcript. IS*50R* and IS*50L* differ at only one position, 1453. The change has profound effects. It simultaneously creates the promoter for the resistance operon and a premature translation termination signal for the two IS*50* peptides, both of which are shortened by 23 amino acids at their C termini. The loss of these amino acids inactivates both proteins. I, junctions of IS*50* elements with central resistance module; mRNA, messenger RNA; O, junctions of Tn*5* and carrier molecule. Numbers indicate translation initiation and termination points (relative to nucleotide sequence). Arrows indicate direction of transcription and translation.

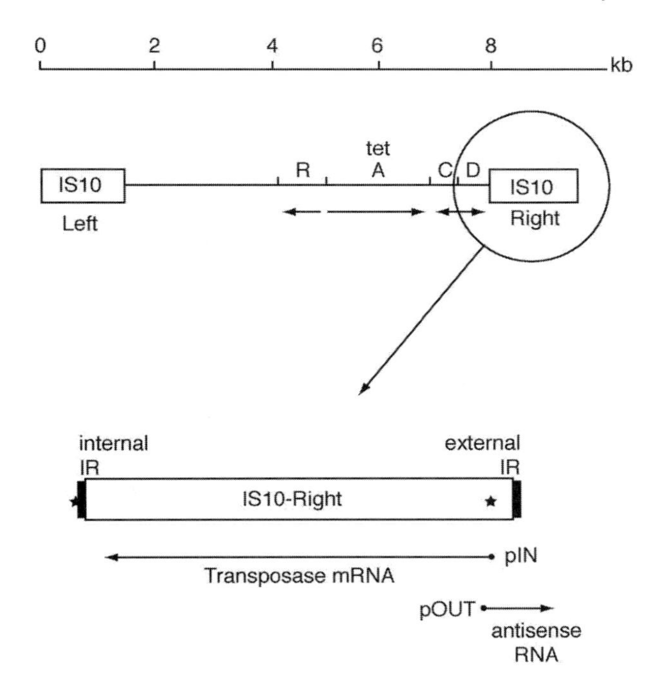

Fig. 2. Tn*10.* Tn*10* originated on the resistance plasmid 222 (otherwise called NR1 and R100). It is 6.7 kb with an approximate 4-kb central module encoding inducible tetracycline resistance, flanked by inverted copies of IS*10,* designated IS*10R,L.* Tetracycline resistance is conferred by the TetA protein, which forms a tetracycline antiport that pumps tetracycline out of the cell. Expression of *tetA* is repressed by TetR and induced by tetracycline. The expression of the divergently transcribed *tetC* and *tetD* (functions unknown) is also induced by tetracycline. The sequences of the two copies of IS*10* differ at several positions; IS*10R* is responsible for transposition. The IS*10* transposase is encoded by a 1206-nt open reading frame transcribed towards the *tet* genes from promoter pIN. A second promoter, pOUT, located 35 bp downstream from pIN, directs transcription in the opposite direction to pIN to produce a short RNA molecule (69 nt) that is partially complementary to the *tnp* mRNA. This small RNA molecule is practically stable, with a half-life of approximately 70 min and a level of about five copies per copy IS*10.* In contrast, the *tnp* transcript is unstable (half-life, 40 s) and of low abundance (approximately 0.25 copy per copy IS*10*). When the 5′-ends of these RNA molecules anneal, the translation of tnp mRNA is blocked (an example of antisense RNA translation inhibition). The transposition activity of IS*10* is also regulated by *dam* methylation of DNA. Two *dam* sites (*) are involved. One overlaps the −10 box of the *tnp* promoter (pIN); the second is at the other end of IS*10* where transposase binds (to mediate the transposition of IS*10*). Methylation of the pIN *dam* sequence reduces the efficiency of pIN, whereas methylation of the second site reduces the binding of transposase to that end of the element. Both effects act to damp down transposition activity of IS*10;* however, only methylation at pIN will affect the transposition of Tn*10,* a differential affect that enhances the coherence of the composite element.

fashion. IS*903* is an IS*5*-like element of 1057 bp with perfect 18-bp terminal IRs. It encodes a single peptide of 306 amino acids, the transposase, which is unstable with a half-life of approximately 3 min. The decay of transposase prevents a build-up of transposase activity in the cell and ensures a low transposition frequency.

C. Mechanisms of Transposition

What is known about the mechanisms of transposition of IS elements comes primarily from studies using IS*1,* IS*10,* IS*50,* IS*903,* and IS*911.* One early important finding was that, in most cases, IS sequences at new insertion sites are flanked by direct repeats (DRs) of short sequences (generally 2–12 bp) that are found as single copies at the insertion target sites. It was realized that this arrangement would result automatically if, in preparing the target site for insertion, both DNA strands at the target site are cut, with the individual cleavage sites on the different strands slightly staggered. Then, when the IS element is joined to the short single-strand extensions created by the staggered cuts and the single-strand gaps on both sides of the element are filled in (by gap-repair DNA synthesis), short DRs will be created at the junctions of the element and the target. This is essentially what happens with many transposable elements. Extensive sequence analysis has also shown that, although the sequences of the DRs vary from one insertion to another and the sizes of the DRs vary from one element to another, for any given element, the size of the DRs is usually constant, suggesting that target-site cleavage is element, rather than host, directed. These findings apply to most transposable elements, including phage Mu. Hence, if, during the course of determining the nucleotide sequence of a length of DNA, a structure is discovered that is delimited by IRs and flanked by short DRs, then it is reasonable to conclude that a transposable element has been discovered, even if the particular copy is no longer active. Examples of such serendipidous discoveries are readily found in the total genome sequences of a number of bacterial species. With one or two elements, transposition does not result in target-site duplications (i.e., following a transposition event the element is not flanked by short DRs). In these cases, insertion may be into flush-cut target sites.

The transposition mechanism of many IS elements, including IS*10* and IS*50*, is conservative; the element is disconnected from the donor molecule as a double-stranded sequence that is then moved to a new site (note that this is not true of all transpositions, as seen later). The simplest conservative model is that described as the cut-and-paste mechanism (Fig. 3). In this process, the two ends of the element are brought together (synapsed) and the cognate transposase cleaves both DNA strands at both its ends, disconnecting it completely from the donor DNA molecule. Formally, this generates a free form of the element, in the sense that the element is no longer covalently joined to a carrier DNA. However, free transposable elements are not normally seen in cells, although IS circles, in which the ends are held together by protein, have been reported for IS*10* when

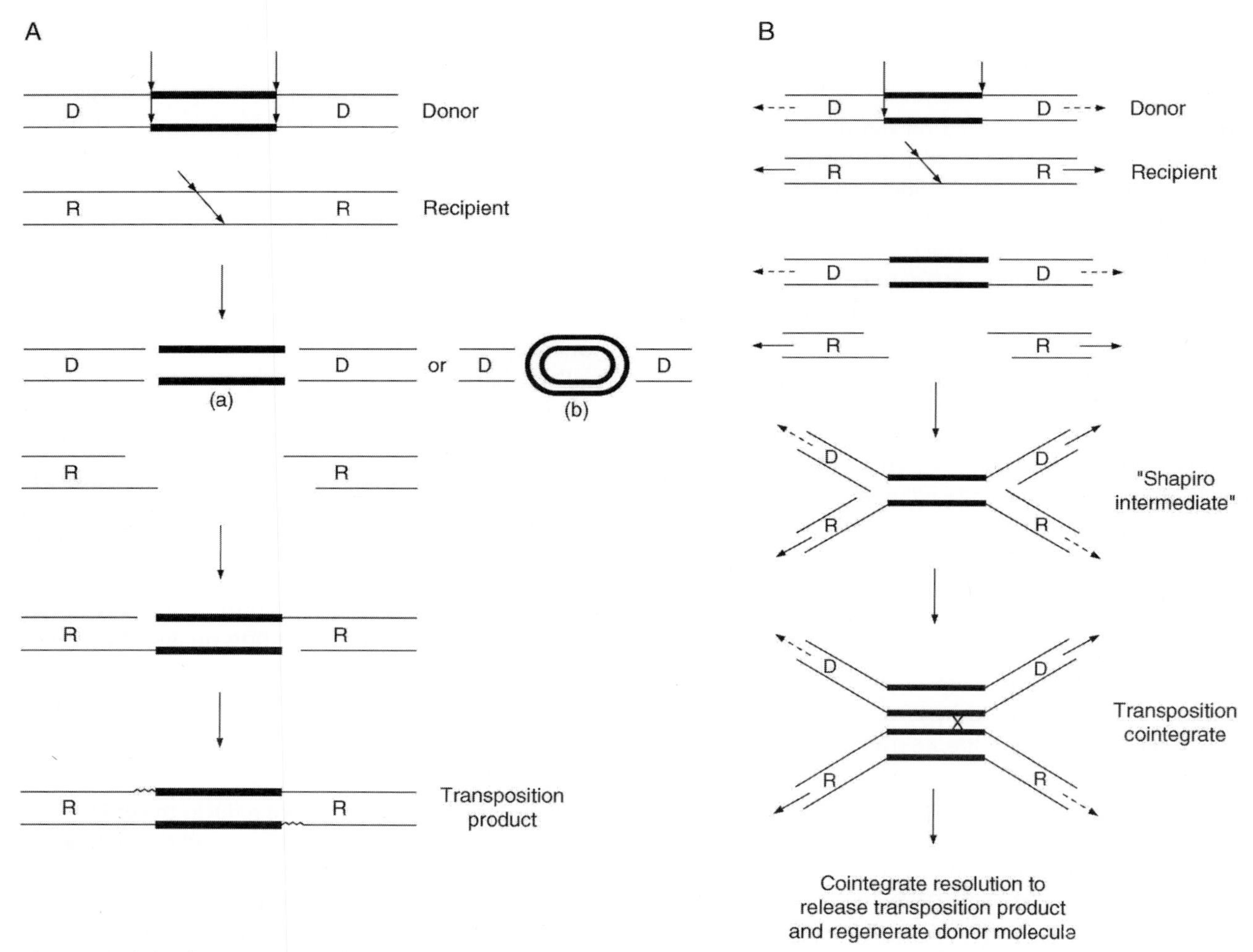

Fig. 3. Models of transpositon. (A) Conservative, cut-and-paste transposition with (a) linear DNA intermediate (although the ends may be held together by transposase), and (b) circular DNA intermediate; (B) replicative transposition.

the transposase is greatly overexpressed. Evidence from experiments with IS*10* and other elements suggests that these circular DNA–transposase protein complexes represent a transposition intermediate that is normally part of a transposition complex called a transpososome. The synapsing of the ends of IS*10*, prior to cleavage, is believed not to be a directed reaction but, rather, one that relies on the random collision of IR–transposase complexes that assemble at the ends of the element. The transpososome then attacks the target site directed by transposase. The free 3′-OH groups at both ends of the element act as nucleophiles, attacking 5′-phosphate groups in the DNA backbone at the target site to effect transesterification reactions. Synapsing the ends of the element holds them in the correct alignment to execute what is probably normally a coordinated, but slightly staggered attack by the element on the two strands of DNA. These reactions, which require no additional energy input, insert the transposon into its new site, in the process creating short single-strand gaps, one on each strand of the DNA, that flank the element. Gap-repair DNA synthesis completes the insertion by restoring strand continuities and generating the flanking DRs. The fate of the donor molecule, which has been cleaved at the original IS insertion site, is unknown, but it is thought normally to be lost (degraded).

Some IS elements not only transpose conservatively in this manner, but their transposition can also, on occasion, give rise to what are called transposition cointegrates. These structures result from transposition mechanisms that generate what is called a Shapiro intermediate, which allows the semiconservative replication of the element (Fig. 3; also, see later). Transposition cointegrates carry two copies of the element as direct repeats. IS*1* can generate transposition cointegrates as well as simple insertions. IS-generated cointegrates are relatively stable entities, particularly if generated and maintained in a *recA* background (i.e., in a host that is deficient for homologous recombination). Cointegrate formation has been exploited to identify and isolate new transposable elements.

The transposition of some IS elements involves the production of a true circular intermediate (i.e., the element is disconnected from its donor site and its ends are joined to produce a circular DNA molecule) (Fig. 3). IS3 and IS*911*, both of which are IS3-like, give rise to circular intermediary forms. The production of IS circles is dependent on two element-encoded proteins, OrfA and OrfAB. As in the case of IS*1*, the latter protein is a fusion protein, the product of *orfA* and *orfB*, two *orfs* that have relative reading frames 0 and −1 and that overlap slightly. As is IS*1* transposase, IS3 and IS*911* fusion proteins are the results of translational frameshifting. Both OrfA and OrfAB proteins bind to the terminal IRs of their respective elements. The increased production of OrfA and OrfAB stimulates the production of IS circles and both proteins are needed for high-frequency transposition. This is in contrast to IS*1*, in which the smaller protein InsA inhibits the activity of the fusion protein InsAB. IS*911* circles have been shown to be efficient substrates for intermolecular transposition *in vitro*, although it has not formally been demonstrated that they are true transposition intermediates *in vivo*. When the element is circularized, the terminal IRs are separated by 3 bp, which were originally linked to one end of the donor molecule. This spacing segment is lost when the element is inserted into the target site. One interesting consequence of circularizing both IS3 and IS*911* is that strong promoters are created that can drive the expression of *orfA* and *orfAB*. These new promoters are significantly stronger that the indigenous promoters on the linear structures. The −35 hexamer box of the new promoter is provided by one IR and the −10 hexamer box by the other, which are side by side in the circular form of the elements. The formation of these strong promoters by IS circularization may be a device to ensure that there are adequate levels of transposition proteins to drive the second stage of the transposition process (i.e., the insertion of the element into target sites) and to ensure that the transposase level is boosted only when there is a suitable substrate for it.

Although transposition of many IS elements is essentially a cut-and-paste exercise, some elements use a replication strategy. IS*91* and related elements are distinctive in that they have no IRs, insert into a specific tetranucleotide sequence, and do not generate target-site duplications.

The sequence analysis of IS*91*-like elements, each approximately 1.7 kb in size, has revealed that at one end of each element is a highly conserved sequence that resembles the leading-strand replication origins (ls-origin) of a number of gram-positive plasmids that replicate by rolling circle (RC) replication to generate single-stranded (ss) DNA copies, that are then converted to double-stranded DNA. The other end of each IS element is also highly conserved with a region of dyad symmetry followed by a tetranucleotide sequence that matches the specific insertion site. Each IS*91*-like element possesses a single large ORF that encodes a protein with striking similarities to the RC-plasmid replication proteins (REPs). These findings suggest that the transposition of IS*91* and similar elements involves RC replication and may also involve a ssDNA intermediate.

D. Transposition of Composite Transposons

Many composite transposons, originating in a broad spectrum of bacteria, including gram-positive and gram-negative organisms, encode resistance to antibiotics; a few encode functions other than drug-resistance (Table II). In general, these elements appear not to be phylogenetically related, although different transposons carrying essentially the same drug-resistance gene are known, as are transposons in which completely different central sequences are bracketed by copies of the same IS element (Table II). Composite structures transpose because each is treated simply as an extended version of the element that forms the terminal repeats. This is possible because each terminal element is delimited by short IRs, so the composite structure also is delimited by the same IRs. An interesting quirk of these systems is that, because all that is needed for transposition, apart from the cognate transposase, is a pair of IRs, what is called "inside-out" transposition can also occur, provided that the donor molecule is circular. In this case, the terminal elements form a composite structure with the carrier DNA molecule, abandoning the functions that characterized the original transposon. The new structure is delimited by the pair of IRs that were located at the inside junctions of the IS elements and their original central module, rather than the outside pair that delimit the original transposon.

TABLE II
Some Prokaryotic Composite Transposons[a]

Transposon	*Size (kb)*	*Terminal elements*	*Target duplication (bp)*	*Marker(s)*[b]
		Gram-negative bacteria		
Tn*5*	5.7	IS*50* (IR)	9	KmBlSm
Tn*9*	2.5	IS*1* (DR)	9	Cm
Tn*10*	9.3	IS*10* (IR)	9	Tc
Tn*903*	3.1	IS*903* (IR)	9	Km
Tn*1525*	4.4	IS*15* (DR)	8	Km
Tn*1681*	4.7	IS*1* (IR)	9	HST
Tn*2350*	10.4	IS*1* (DR)	9	Km
Tn*2680*	5.0	IS*26*[c] (DR)	8	Km
		Gram-positive bacteria		
Tn*3851*	5.2	NR	NR	GmTbKm
Tn*4001*	4.7	IS*256* (IR)	8	GmTbKm
Tn*4003*	3.6	IS*257* (DR)	8	Tm

[a] DR, direct repeat; HST, heat-stable enterotoxin; IR, inverted repeat; NR, not recorded.
[b] Resistance genes: B1, bleomycin; Cm, chloramphenicol; Gm, gentamicin; Km, kanamycin; Sm, streptomycin; Tb, tobramycin; Tc, tetracycline.
[c] IS*26* and IS*6* are synonyms.

III. Tn*3* AND RELATED TRANSPOSONS

A. Complex Transposons—General

Although it is convenient to consider composite transposons as a group because of their structural similarities, few appear to be phylogenetically related and individual elements may transpose by different mechanisms. Complex transposons show a more complicated genetic arrangement than either IS elements or composite transposons, in that genes that do not encode transposition functions have been recruited into and become part of the basic transposable element. However, just as the term "composite transposon" denotes only that such an element has a modular structure with terminal IS repeats, so the term "complex transposon" means no more than that the transposon does not possess a modular structure with terminal IS repeats and that it is not a bacteriophage.

B. Tn*3*

The archtype complex transposon is Tn*3*, originating on the Resistance (R) plasmid, R1. It is virtually identical to the first drug-resistance transposon discovered, Tn*1*, found on another R plasmid, RP4. Both elements were originally called TnA and the transposition functions are fully interchangable. The transposons encode TEM β-lactamases that confer resistance to ampicillin and carbenicillin and a few other β-lactam antibiotics. The two enzymes, designated TEM 1 and TEM 2 (encoded by Tn*3* and Tn*1*, respectively) differ by only one amino acid (a Gln-to-Lys substitution at position 37) and have essentially identical substrate specificities. Both Tn*1* and Tn*3* are widely distributed among gram-negative bacteria from clinical and veterinary sources, with Tn3 being somewhat more common. In recent years, these enzymes have evolved to create a large family of more than 70 members that are collectively referred to as extended-spectrum β-lactamases. These variants have expanded substrate spectra that incorporate some of the newer cephalosporins (the third-generation cephalosporins, e.g., ceftazidime and cefotaxime). A second set of variants are insensitive to β-lactamase inhibitors, such as clavulanic acid (inhibitor-resistant β-lactamases). Undoubtedly, the speed at which these mutants have arisen and have become established owes much to the widespread distribution of the parent genes on Tn*1* and Tn*3*.

C. Tn*3*-related Transposons

Tn*3* is typical of an extended family of phylogenetically related transposons, members of which have been found in both gram-negative and gram-positive bacteria. Between them they encode resistance to several antibiotics and to mercuric ions; a couple specify catabolic functions, whereas several are cryptic (Table III). The evolutionary relationships among these elements were first revealed when it was discovered that apparently unrelated transposons (Tn*3*, Tn*501*, Tn*551*, and Tn*1722*) have different, but clearly similar, short (35- to 40-bp) terminal IR sequences. When the elements were fully sequenced, it became apparent that the similarities extend to the transposition genes as well (Fig. 4).

D. Transposition Functions

Each Tn*3*-related element encodes a transposase of approximately 1000 amino acids. The genes, approximately 3 kb long and designated *tnpA*, are clearly ancestrally related. Most of these transposons also encode a second recombination enzyme, called a resolvase. These enzymes, encoded by genes designated *tnpR*, are site-specific recombinases. Each acts at a particular site, designated *res*, located on the transposon adjacent to *tnpR*. These enzymes mediate the second stage of a two-step transposition mechanism.

E. Family Branches

The Tn*3*-related transposons split naturally into two main branches of the family. On each of the main branches, the transposition functions of the elements are closely related. Tn*3* is the type element for one branch (which includes Tn*1000* and Tn*1331*) and Tn*21* for the other (Table III). Tn*21* is larger than Tn*3* (20 kb vs. 5 kb) and confers resistance to streptomycin, spectinomycin, and sulfonamide, as well as to mercuric ions. The antibiotic-resistance

TABLE III
Some Tn*3*-like Transposons[a]

Transposon	*Size (kb)*	*Terminal IRs (bp)*	*Target (bp)*	*Marker(s)*[b]
		Gram-negative bacteria		
Tn*1*	5	38/38	5	Ap
Tn*3*	4.957	38/38	5	Ap
Tn*21*	19.6	35/38	5	HgSmSu
Tn*501*	8.2	35/38	5	Hg
Tn*1000*	5.8	36/37	5	None
Tn*1721*[c]	11.4	35/38	5	Tc
Tn*1722*	5.6	35/38	5	None
Tn*2501*	6.3	45/48	5	None
Tn*3926*	7.8	36/38	5	Hg
Tn*4651*	56	32/38	5	xyl
		Gram-positive bacteria		
Tn*551*	5.3	35	5	Ery
Tn*917*	5.3	38	5	Ery
Tn*4430*	4.2	38	5	None
Tn*4451*	6.2	12	NR	Cm
Tn*4556*	6.8	38	5	None

[a] NR, not recorded.
[b] Ap, ampicillin; Cm, chloramphenicol; Ery, erythromycin; Hg, mercuric ions; Sm, streptomycin; Su, sulphonamide; Tc, tetracycline; xyl, xylose catabolism.
[c] Tn*1721* is a composite structure that uses Tn*1722* as its basis for transposition.

genes on Tn*21* and related elements have been acquired as gene cassettes inserted into integrons located on the elements (see later). Notwithstanding their different sizes, both Tn*3* and Tn*21* devote approximately the same genetic capacity, 4 kb, to transposition. In addition to the two main groups, there are one or two elements that are clearly Tn*3*-like, but that are not closely related to either Tn*3* or Tn*21* or, indeed, to each other.

The Tn*21* branch of the family contains many elements that differ principally in the number of resistance determinants carried (Table III). These elements have the same or almost the same transposition functions, which are functionally interchangable. Much of the diversification seen in the group is thought to be of relatively recent origin, reflecting the activities of integrons and the acquisition and loss of gene cassettes (see later). Some elements, such as Tn*501* and Tn*1721*, are less closely related to Tn*21* than others, such as Tn*2424* and Tn*2603*. Sequence analysis of the transposition functions of Tn*501* and Tn*1721* indicates significant degrees of divergence from the type element, Tn*21*, but they are much more closely related to Tn*21* than to Tn*3*. The Tn*21*-like transposons present what is possibly the most successful diversification of a single transposable element discovered. However, the identity of the parent element (i.e., the one that lacks accessory genes) is unknown.

The structures of Tn*3* and some related elements are depicted in Fig. 4, which illustrates the two arrangements of the transposition functions seen in the extended family. On Tn*21* and its close relatives, both transposition genes have the same orientation and the gene arrangement—*res, tnpR, tnpA*. On Tn*3* and closely related elements, *tnpR* and *tnpA* are adjacent but opposed. They are transcribed divergently from a short common promoter region, in which is also located the *res* site. One consequence of this gene arrangement is that TnpR acts not only as a site-specific recombinase, but also as a repressor for the transposase gene, *tnpA*. Knockout mutations of Tn*3* *tnpR* generate Tn*3* derivatives that transpose at elevated frequencies to produce transpo-

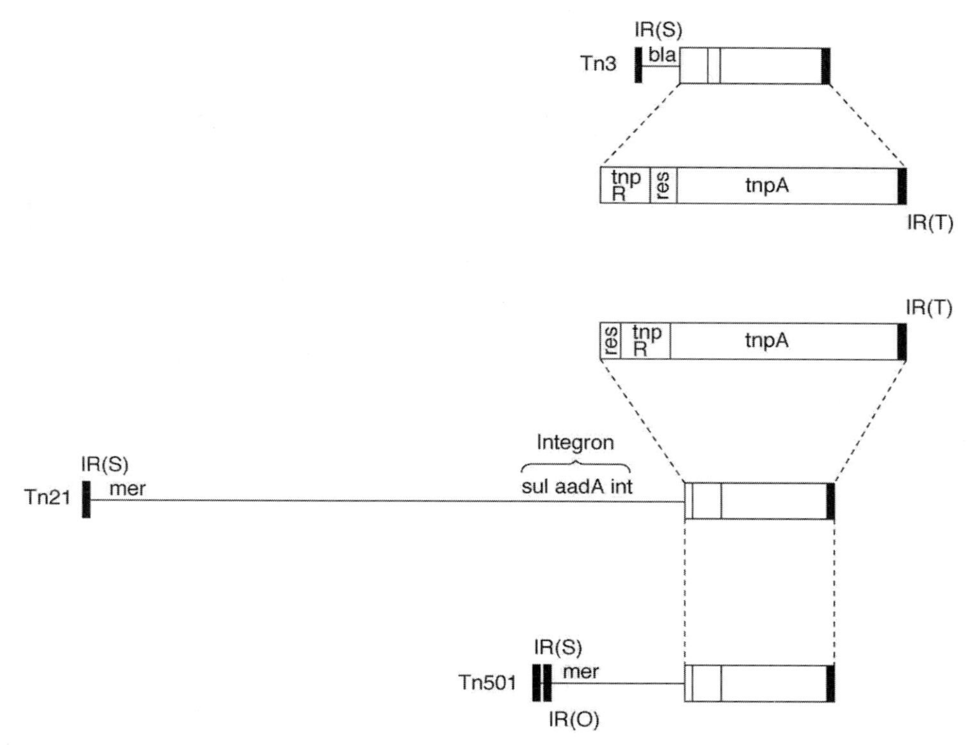

Fig. 4. Schematic representation of Tn*3* and some related elements. Genes depicted: *aadA,* aminoglycoside adenylyltransferase A; *bla,* TEM β-lactamase; *int,* integrase. IR(T), inverted repeat adjacent to *tnpA,* and IR(S), second IR, are represented by thick vertical bars; *mer,* operon conferring resistance to mercuric ions; *res,* resolution site; *sul,* sulfonamide-resistant dihydropteroate synthetase gene; *tnpA,* transposase gene; and *tnpR,* resolvase gene.

sition cointegrates that are easily recovered in a *recA* host.

F. Recruitment of Antibiotic Resistance Genes to Tn*21*-like Transposons: Integrons and Gene Cassettes

Molecular analysis of several Tn*21*-like elements revealed that many of the antibiotic resistance genes carried by these elements have not been inserted at random. Rather, they are located, as single genes, or sets of genes, at essentially the same place on the transposon backbone, specifically on one side or the other of the *aad* gene, if this is carried, or as a replacement for it if the *aad* gene is absent. This finding, given the variety of genes involved (approximately 40 resistance-gene cassettes have been reported, as well as a handful of cassettes with genes of unknown function) and the size of Tn*21* (20 kb, of which only 4 kb is needed for transposition), suggested that the diverse resistance genes had been captured by a site-specific recombination mechanism. This is now known to involve elements termed integrons and mobile DNA sequences called gene cassettes.

An integron is defined as a genetic element that has a site, *attI,* at which DNA, in the form of a gene cassette, can be integrated by site-specific recombination, and that encodes an enzyme, integrase (product of *int*), that mediates the site-specific recombinations. A number of integrons also accommodate a gene for sulfonamide resistance, *sul,* at one end. A gene cassette is a discrete genetic element that may exist as a free circular nonreplicating DNA molecule when moving from one site to another, but that is normally found in linear form as part of another DNA element, such as a transposon, plasmid, or bacterial chromosome. Gene cassettes normally accommodate only one gene and a short sequence called a 59-base element that serves as the specific cassette recombination site. Accordingly, gene cassettes are small, normally about 500–1000 bp. The genes on cassettes, in general, lack promoters and are expressed from a specific promoter on the integron located beside *attI*. In a few cases, cassettes carry two genes; these are likely to have arisen from the fusion of two gene cassettes which at one time were side by side. Because the expression of the genes on the majority of cassettes is from the integron promoter adjacent to *attI*, integration of cassettes is not only site-specific, but also orientation-specific. Several cassettes can be integrated sequentially at *attI*, which is regenerated with each insertion. When this occurs, the genes on the cassettes are expressed as an operon. Hence, a distinctive genetic array for an integron can be envisaged as *int attI* gene cassette(s) *sul* (Fig. 5).

Three distinctive classes of integrons have been reported. Class 1 accommodates the majority of integrons found, including those on Tn*21*-like transposons. The basic integron of class 1, designated In0, lacks gene cassettes; that is, it has the genetic array *int attI sul*. New integrons are generated by the insertion of one or more cassettes at *attI*, or the deletion of one or more cassettes from an existing integron. The order of cassettes, from *attI*, can also be changed by site-specific excision of a promoter-distal cassette and its reinsertion at *attI*.

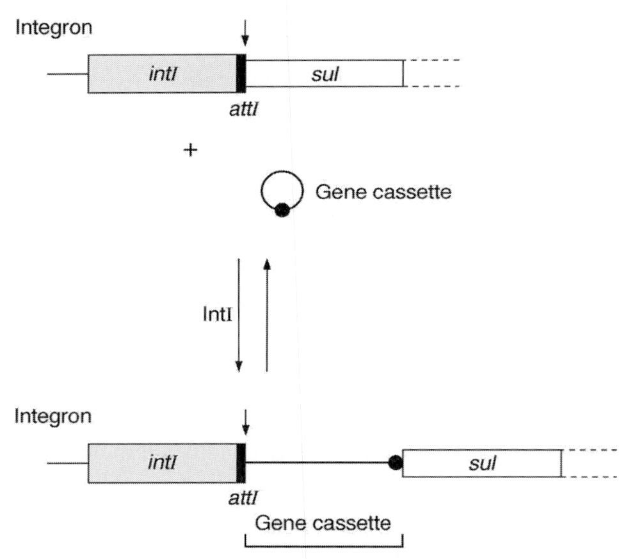

Fig. 5. Integrons and gene cassette integration and excision. The recipient integron shown represents the basic class 1 element, In0; that is, it has no gene cassette between the integrase gene *intI* and the sulfonamide resistance gene, *sul*. IntI first mediates the excision of a gene cassette from one integron array, releasing it as a free circular form comprising a single gene followed by a 59-base element (solid circle). The gene cassette is then inserted, by site-specific recombination mediated also by IntI, into *attI* (represented as the solid vertical bar) on the recipient integron.

The movement of gene cassettes in and out of integrons is a random process, similar to most other genetic rearrangements in bacteria. Which rearrangements survive and which are lost is a matter of circumstance and natural selection; if the gene combination is of benefit to the host, it will be selected and established in the population by clone amplification. Most of the genes discovered on cassettes are antibiotic- or disinfectant-resistance genes; however, this is probably not a true picture of the diversity of genes on these elements, but, again, only a reflection of the intense interest of medical microbiologists in resistance genes for the past two to three decades.

G. Mechanism of Transposition

The transposition of members of the Tn*3* family of transposons involves replication and recombination (Fig. 3). Each transposition event generates a target-site duplication, usually one of 5 bp and, as in the previously described mechanisms, the duplication arises because the cleavage of the target to allow transposon insertion involves slightly staggered cuts on the two DNA strands. The mechanism of transposition of Tn*3*-like elements differs from those described previously in that the element is not wholly disconnected from the donor site. Rather, single-strand cleavages are introduced by the cognate transposase at both ends of the transposon on opposite strands to expose 3′-OH groups. These are then used as nucleophiles to attack two 5′-phosphate groups at the target site, one on each DNA strand, separated by 5 bp. Two transesterification reactions join the transposon to the target DNA, generating what is termed a "Shapiro intermediate," after J. A. Shapiro, who first proposed its existence (the structure was also proposed independently, at about the same time, by A. Arthur and D. Sherratt). The transesterification reactions not only join the transposon to the target, but also generate free 3′-OH groups on both sides of the transposon that can be used to prime the replication of the element. Whether one or both is used *in vivo* is not known. The replication of the transposon is executed by host-cell enzymes.

When the element has been replicated and double-strand continuity has been restored, the product generated is a transposition cointegrate. This molecule contains the transposon donor and target replicons in their entireties, joined by directly repeated copies of the transposon, one at each replicon junction. The final product, a copy of the target replicon with a single copy of the transposon, is released by cointegrate resolution. Normally this is achieved by a site-specific recombination reaction, mediated by the *tnpR* gene product resolvase, using the two *res* sites present on the cointegrate (one on each copy of the transposon). Should either *tnpR* or *res* be damaged, then host-mediated *recA*-dependent recombination across the duplicate transposon sequences can also resolve the cointegrate to the final transposition product and a molecule that is indistinguishable from the original transposon donor (Fig. 3).

H. One-Ended Transposition

The replicative mechanism of transposition outlined in Section III.G explicitly involves both ends of the element in a nondiscriminatory manner (i.e., both ends are needed, but are treated identically). This explains the need for perfect or near-perfect terminal IRs because both ends must bind transposase. Surprisingly, however, when one IR sequence is deleted TnpA-mediated transposition still occurs, albeit at much-reduced frequency and necessarily modified. Such events are referred to as one-ended transpositions. A one-ended transposition is a replicative recombination mediated by some (Tn*3*, Tn*21*, and Tn*1721*), but not all Tn*3*-like elements. It involves one IR sequence (instead of two) and the cognate transposase. The products resemble replicon fusions in which the donor replicon, rather than a discrete part of it (i.e., the transposon), has been inserted into the recipient replicon. One end of the inserted DNA is always the solitary IR sequence, and the inserted sequence is flanked by 5-bp DRs, as for a normal transposition event, consistent with one-ended transpositions being true transpositions. Unlike normal transpositions, however, the products are not uniform in size and insertions both less than and greater than the unit length of the donor replicon have been reported. The different inserts form a nested set of DNA fragments with a copy of the IR sequence defining an anchored end. These data are not accommodated by the model described (Fig. 3B). However, a nested set of insertions, that all start from the same point on the donor molecule are indicative of a mechanism that involves RC replication. Whereas the two-ended mechanism precisely determines the sequence to be transposed, rolling circle transposition determines precisely only one end of the insert. How the other end is determined is not known.

I. Transposition Immunity

Tn*3* and some of its relatives, but not Tn*21*, display a characteristic termed transposition immunity. Elements that show this behavior will not normally transpose on to another DNA molecule that already carries a copy of the transposon. The effector for inhibition has been identified as the element's IR sequence and a single copy affords significant protection, but the mechanism is not understood. In the case of Tn*3*, transposition immunity can be largely overcome by increasing the level of transposase (by the use of *tnpR* knockout mutants). Transposition immunity has also been demonstrated for Tn*7* and bacteriophage Mu, in which the mechanism is somewhat better understood.

IV. CONJUGATIVE TRANSPOSONS

A. Origin and Properties

Conjugative transposons, first described in gram-positive bacteria, have also been found in gram-negative anaerobes such as *Bacteroides* sp. The type element Tn*916*, which encodes resistance to tetracycline, was discovered in *Enterococcus faecalis*. As the collective name for the elements implies, they not only transpose but also promote their own transfer from one cell to another. They differ from plasmids in that they do not replicate autonomously, but are replicated passively, as parts of the replicons into which they insert.

Tn*916* is 16.4 kb and indistinguishable from Tn*918*, Tn*919*, and Tn*925*, which are independent isolations of essentially the same element. A related conjugative transposon, Tn*1545*, is 23.5 kb and encodes resistance to kanamycin and erythromycin in addition to tetracycline. These elements promote their own conjugal transfer at frequencies of 10^{-9} to 10^{-5}.

B. Transposition

Tn*916* is typical of the group and is the best studied. These elements can mediate conjugal transfer followed by transposition into the chromosome of the recipient cell, chromosome-to-plasmid transposition, and transposon excision. The last of these activities is often tested in *E. coli* rather than in *Enterococcus faecalis* because the frequency of excision is considerably higher in the former microbe. More than 50% of the Tn*916* sequence encodes conjugation functions, located as a block of genes at one end of the element. These functions are not strictly necessary for transposition, although mutations in some interfere with chromosome-to-plasmid transposition. Mutations that block transposition map to the other end of the element. Such mutations also inhibit transposon excision. Two transposition genes, designated *xis-Tn* and *int-Tn,* have been identified on Tn*1545*.

The transposition of Tn*916* and related elements is a two-stage process. The first step is the excision of the element as a circular intermediate, similar to that involved in transposition of IS*911* (Fig. 3). Staggered double-strand cuts at both ends of the element generate a linear DNA molecule with 5′ single-stranded hexanucleotide overhangs. End-to-end joining creates a circular DNA molecule with a 6-bp heteroduplex spacer separating the ends of the element. The circular DNA species may then undergo one of two productive fates: (1) the insertion of the transposon into one of the DNA molecules in the cell, or (2) the conjugal transfer to another cell followed by insertion of the transposon into a suitable DNA molecule in the new host (e.g., the bacterial chromosome). The mechanism appears to be similar to the integration–excision systems of lambdoid phages, and involves functions related to the Int and Xis proteins that mediate integration and excision of bacteriophage λ into the *E. coli* chromosome.

V. SITE-SPECIFIC TRANSPOSONS

A. General Comments

Most transposable elements show little or no target-site specificity, although bias toward insertion into certain regions of DNA molecules has been reported for some elements (e.g., IS*1* has been reported to favor insertion in AT-rich tracts, as has Tn*3;* and although Tn*5* usually displays low target-site selectivity, a preference for some "hotspots" has been observed); but these biases are preferences, not requirements. In contrast, a few elements display a marked, if not exclusive, specificity of insertion.

B. Tn*7*

Tn*7* was first identified on the R plasmid R483. It carries two resistance genes, *dfrI* and *aadA,* the former encoding a trimethoprim-resistant dihydrofolate reductase, whereas the latter codes for an aminoglycoside adenylyltransferase, which mediates resistance to streptomycin and spectinomycin. Tn*7* is 14 kb long, generates 5-bp target-site duplications on transposition, and inserts in one orientation at high frequency into single sites on the chromosomes of a number of gram-negative bacteria, including *E. coli, Klebsiella pneumoniae, Pseudomonas* sp., and *Vibrio* sp. At a lower frequency (10^{-4}), it will insert, with little target-site selectivity, into many different DNA molecules (e.g., plasmids) that lack the specific insertion site. However, if the specific insertion site is engineered into a plasmid, Tn*7* will insert into that plasmid at high frequency in a site-specific orientation-specific manner.

C. Chromosomal Site of Tn*7* Insertion

The locus of Tn*7* insertion on the *E. coli* chromosome has been located at 84′, between *phoS* (encodes a periplasmic phosphate-binding protein) and *glmS* (encodes glucosamine phosphate isomerase), and designated *att*Tn7. The 5-bp target site is part of the *glmS* transcriptional terminator, which is located about 30 bp from the end of the translational reading frame. Unexpectedly, the target site (i.e., the site of insertion) and the attachment site (*att*Tn7, the sequence that determines the locus specificity) are distinct. The attachment site is a sequence of approximately 50 bp, located 12 bp from the 5-bp target site and extending into *glmS*. This relationship between *att*Tn7 and *glmS* is also seen in other bacteria, such as *Klebsiella pneumoniae* and *Serratia marcescens,* in which chromosomal insertion is specific. Although the various attachment sites in the different bacteria are homologous, the target sites differ completely, indicating that the sequence of the actual site of insertion of Tn7 has little or no role to play in the specificity, but is determined solely by its distance from *att*Tn7. Hence, Tn7 exploits sequence data held in a highly conserved host gene, *glmS,* to select an insertion site, but is then inserted beside rather than in *glmS,* ensuring that the insertion doesn't harm the host.

D. Transposition Functions of Tn*7*

Tn7 is unusual in the number of transposition functions it requires. Five have been identified, sequenced, and designated *tnsA,B,C,D,* and *E.* The genes encode peptides of 31, 78, 63, 59, and 61 kDa, respectively. The first three are required for all transpositions. High-frequency site-specific transposition requires, in addition, the product of *tnsD,* whereas low-frequency random-site transposition requires instead the *tnsE* product.

E. Mechanism of Transposition of Tn7

A Tn7 *in vitro* transposition system, with a plasmid carrying *att*Tn7 as target, has been developed. Using this system, it has been shown that the transposon is first disconnected from the donor molecule by two double-strand staggered cuts, each of which generates 5′ overhangs of three bases at the ends of the element. The DNA of these overhangs comes from the carrier DNA molecule, not Tn7. The uncoupled linear Tn7 sequence is then joined to the target (i.e. the mechanism is cut-and-paste) (Fig. 3). This is thought to involve nucleophilic attacks by the 3′-OH groups at the ends of the uncoupled transposon on phosphate groups at the target site. Two transesterification reactions, one targeted to each strand of the DNA duplex and separated by 5 bp, cleave the recipient DNA at the target site and fuse it to Tn7 (which is then joined at both ends to the target DNA by short ssDNA sequences). It is assumed that repair processes then remove the short ssDNA extensions at the 5′-ends of the transposon (i.e., the remnants of the previous insertion site, which would be mismatched with the short stretches of ssDNA to which the transposon is attached) and fill in the resulting single-strand gaps, generating the 5-bp DRs that flank the Tn7 insertions.

It has been found *in vitro* that the transposition of Tn7 proceeds via a DNA–protein complex that contains a transposon donor and target DNA molecules, four transposition proteins (TnsA, B, C, and D) and ATP. ATP may be directly involved in the reaction because TnsC is an ATP-binding protein, but it may also be needed to ensure that the DNA substrates have an appropriate degree of supercoiling, maintained by DNA gyrase.

The cell-free system faithfully reproduces several of the features of Tn7 transposition that are characteristic of the element. Transposition *in vitro* is site- and orientation-specific with respect to *att*Tn7, as it is *in vivo,* and shows transposition immunity: When the target molecule carries a copy of Tn7, transposition of a second copy of Tn7 into it is blocked.

F. IRs of Tn7

Tn7 is unusual in the length of the terminal sequences needed for transposition; 75 bp are needed at one end of the element and 150 bp at the other, but these sequences form IRs of only 30 bp. Artificial elements flanked by IRs of the shorter sequence will transpose when transposition functions are provided. In contrast, a DNA sequence flanked by IRs of the 150-bp sequence does not transpose. The basis for this differentiation has not been determined. Each terminal sequence contains several analogs of a 22-bp consensus sequence; these have been shown to bind TnsB and are thought to serve as nucleation centers for the assembly of the DNA–protein complex required for transposition.

G. Tn*554*

Tn*554* is a 6.7-kb transposon encoding resistance to erythromycin and spectinomycin in *Staphylococcus aureus,* where it transposes into the chromosome at high frequency, primarily into a single site, designated *att*Tn*554,* in one orientation only. From its sequence, six Tn*554* ORFs were identified, of which five have been demonstrated to be genes—the three-gene *tnsABC* transposition operon accounts for approximately half the transposon's coding capacity, whereas *ery* and *spc* code for resistance to erythromycin and spectinomycin, respectively. Tn*554* also displays transposition immunity.

The mechanism of transposition of Tn*554* differs from that of most other elements in the sense that not only is the element transposed, but so also are a few base pairs from the carrier molecule. This short additional sequence is located on one side of the element in the donor, but is transferred to the opposite side in the transposition products. Hence, the sequence changes with each sequential transposition. The transposition mechanism is not known, but current data suggest a cut-and-paste mechanism involving a circular intermediate.

H. Tn*502*

Tn*502* is a poorly characterized gram-negative transposon that encodes resistance to mercuric ions. It appears not to be related to Tn*21*, Tn*501*, or other Hg^r transposons of the Tn*3* family. It is notable because it displays site-specific insertion into a plasmid, rather than into a chromosome. Tn*502* is 9.6 kb and inserts at high frequency in one orientation into a single site on the IncP plasmid RP1 (RP4). If this site is deleted, insertion then occurs into many sites at a much lower frequency. In this respect, its behavior resembles that of Tn*7*, high-frequency site-specific transposition versus low-frequency random-site transposition. There is insufficient information to determine whether Tn*502* is related to other site-specific elements.

VI. TRANSPOSING BACTERIOPHAGES

A. Structure of Mu

Transposing bacteriophages use a transposition strategy to replicate. The type element of the group is bacteriophage Mu, discovered by L.Taylor in 1963. Taylor recognized that Mu was a temperate phage and concluded that because lysogenization often created auxotrophs, Mu must be able to integrate into the *E. coli* chromosome at many different sites, in some cases damaging genes responsible for biosynthetic functions. It is now well established that Mu integrates into the chromosome using a conservative (cut-and-paste) transposition strategy and that, subsequently, Mu replicates by multiple rounds of transposition via Shapiro intermediates (Fig. 5).

The linear genome of Mu is 37 kb, but each phage particle carries a DNA molecule of approximately 39 kb. This is constructed from the Mu genome with about 150 bp of host DNA on one side and 1–2 kb of host DNA on the other. These flanking sequences are different for individual copies of the phage genome and reflect the site of insertion of that copy of Mu, prior to the assembly of the phage particle. Genome packaging proceeds by a headful mechanism that disconnects the phage from the carrier molecule by first cutting the carrier DNA approximately 150 bp beyond one end of Mu (the end nearest the replication functions) and then again approximately 39 kb away, on the other side of the phage genome, so generating a linear copy of the genome flanked by host DNA.

In its genome structure, Mu is unremarkable. It has the usual arrays of genes needed for phage assembly, that is, head and tail production. It has two replication (transposition) genes, designated A and B, which are located close to one end of the genome (traditionally depicted on the left in linear maps) and it has a 3-kb segment that can invert by site-specific recombination. This region is homologous to an invertible section of the phage P1 genome; inversion of these regions changes the host specificity of the phages by promoting the expression of alternative tail fiber genes. Mu and P1 are not related in any other respect.

B. Mu Replication and Integration

More than a decade after its discovery, it was realized that Mu replicates by transposition and the study of Mu replication *in vitro* has been an invaluable model transposition system. Phage replication involves formation of a Shapiro intermediate, which establishes a replication origin(s) from which the replication of the phage genome proceeds. Because the transposition or replication is primarily intramolecular, at least initially, this results in gross molecular rearrangements (i.e., deletions and inversions) of the bacterial chromosome, which is also progressively fragmented into many small circular DNA molecules, each carrying one or two copies of the Mu genome (an inevitable consequence of rounds of intramolecular transposition). Unlike Tn3 and related transposon cointegrates, which are resolved by a site-specific resolution system (*res*/resolvase), Mu cointegrates are not specifically resolved. If resolution does occur, as it may, it is the result of host-mediated, *recA*-dependent recombination. The resolution of cointegrates is not important to the phage and lack of resolution does not inhibit further rounds of transposition or replication. Multiple cycles of Mu transposition do not require the presence of a second replicon in the cell; the chromosome suffices, although if a second replicon is present in the cell, it suffers the same fate as the chromsome.

In addition to replicating by transposition, Mu is established in a new host also by transposition, from the extended linear sequence carried in the phage particle into a replicon carried by the host, usually the bacterial chromosome. The transposition initially generates a Shapiro-type intermediate, but one in which the DNA molecule carrying the transposon (Mu) is linear instead of circular. Again, the intermediate is formed by transesterification reactions involving 3′-OH groups at the ends of the phage DNA, created by single-strand cleavage by the phage A protein. Then, instead of the inserted DNA (i.e., the phage genome) being replicated, a second pair of cleavages, this time at the 5′-ends of the genome, disconnects it completely from the terminal remnants of previous host DNA, which are discarded, and double-strand continuity is restored to seal the phage genome into a new site, flanked by 5-bp DRs.

That Mu lysogeny is established by a conservative integration was elegantly demonstrated as follows. Phage particles produced in a dam^+ host were used to infect a *dam* mutant of *E. coli* (which cannot methylate its DNA). The newly integrated phage DNA was found to be fully methylated, not hemimethylated, indicating that no replication had occurred prior to or as a consequence of insertion in the new host. The conservative transposition used to establish Mu in a new host and phage replication both require the A gene product, which is the Mu transposase. The role of the B gene product, which is also needed for efficient integration and replication, is largely one of target-site selection and facilitating the assembly of the transposition complex.

C. D108 and Other Transposing Bacteriophages

Mu was the first transposing bacteriophage to be discovered. Only one other transposing coliphage is known, D108, which is closely related to Mu. The two genomes display 90% sequence homology, with the main divergence being at the ends of the genomes. One such region includes the 5′-ends of the A genes, the consequence of which is that the two A gene products, although they do complement each other, do so only poorly. Transposing phages have also been identified in *Pseudomonas* sp. The sizes of their genomes (~37 kb) and the structures of the DNA packaged to form the phage particles are strikingly similar to those of Mu and D108, but no gross similarities in DNA sequence with Mu have been found. Possible transposing phages have also been found in *Vibrio* sp., identified on the basis that lysogenization can result in auxotrophy, which was how Mu was originally detected.

VII. YEAST TRANSPOSONS

A. Types

Finally, mention must be made of transposons found in yeast. Several types have been identified in

Saccharomyces cerevisiae. The majority are components of the nuclear DNA, but one, designated Ω, is found in mitochondrial DNA. No one element has been found in both compartments.

Several of the yeast transposons (e.g., Ty1, Ty2, Ty3) are what are known as retrotransposons because the way in which they transpose is clearly akin to the mechanism of retroviral replication. Ty1 has been shown to transpose by a process involving a reverse-transcriptase step. Ty2 and Ty3 closely resemble Ty1, and also require reverse transcriptase for transposition.

B. Structure of Ty Elements

The Ty elements are 5–6 kb long and typically have long-terminal direct repeats (LTRs), which can themselves transpose (called δ in the cases of Ty1 and Ty2, and σ for the LTRs of Ty3). The Ty elements and their LTRs have no IRs, as such, although, as do many retroviruses, the transposable sequences terminate 5′-TG . . . CA. In keeping with the view that these elements can be regarded as transposons is the finding that intact elements are nearly always flanked by 5-bp duplications of host sequence (i.e. target-site duplication). Both Ty1 and Ty2 accommodate two ORFs, designated TYA and TYB. The equivalent ORFs of the two elements indicate a considerable degree of similarity between the proteins of the two transposons. Several domains of the Ty1 TYB protein show significant similarities to retroviral *pol* functions.

C. Transposition of Yeast Transposons

Yeast cells engaged in high-frequency transposition of Ty1 or TY2 contain large numbers of virus-like particles that contain Ty–RNA, reverse transcriptase, and capsid proteins encoded by the Ty element. It is thought that these particles are transposition intermediates. Transposition requires the conversion of the RNA to DNA and integration of this into the nuclear DNA. That the mechanism of retrotransposon transposition was likely to be analogous to the replication of retroviruses was originally inferred from the structural analysis of the elements and their transcripts, which resemble the structures of retroviral proviruses and viral RNA, respectively.

In contrast to the retrotransposons, the mitochondrial element Ω appears to transpose directly as DNA. In this sense, it more closely resembles transposable elements in bacterial cells.

A common type of transposable element in eukaryotes are what are called LINE elements (long interspersed nucleotide elements), after the first-characterized L1 of mammals. Nucleotide-sequence analysis of the elements revealed ORFs with striking similarities to the *gag* and *pol* genes of retroviruses. These elements are also known as the non-LTR retrotransposons, to emphasize their structural difference from LTR retrotransposons, such Ty1, and to indicate that transposition involves an RNA intermediate. Although mainly identified in higher eukaryotes, a LINE-like element, TAD, has been identified in the fungus *Neurospora crassa.* These elements represent a second class of retrotransposon.

VIII. CONSEQUENCES OF TRANSPOSITION IN BACTERIA

A. Genome Rearrangements

The most obvious consequence of transposition is the insertion of one DNA sequence into another, an event that may disrupt a gene and cause a mutation. This property has been widely exploited in genetic analysis, and insertional inactivation by transposons is commonly used to locate genes of interest. Several ingenious "suicide systems" have been devised to deliver the transposon into the cell, where it becomes established only if it is transposed on to a resident replicon because the delivery vehicle cannot itself be replicated.

Insertional inactivation can also be exploited to capture transposable elements, including IS elements that encode no predictable phenotype. Vectors that express products that are lethal when the cells are exposed to particular culture conditions have been developed; insertions into the gene that encodes the lethal protein prevent the potentially fatal expression; that is, cells that contain a damaged copy of the gene survive and form colonies on the selective agar, whereas cells that contain an undamaged copy of the gene express it and die. The vector pKGR, carrying the *sacRB* cassette from *Bacillus subtilis,* is one such capture system. The expression of *sacB* is lethal to *E. coli* when cultured on medium containing sucrose, so carrying pKGR is lethal under these growth conditions; such cells fail to form colonies on sucrose-containing agar. In contrast, cells with pKGR derivatives with *sacB* inactivated readily form colonies. The system allows, therefore, positive selection for loss of function, hence, loss of expression of *sacB*.

Transposable elements mediate DNA rearrangements other than simple insertions. The sites at which transposable elements have been inserted are often hot spots for deletions and inversions. These rearrangements involve the sequences on one side or other of the element, but not the element itself, which remains in place and intact. Many of these events are the results of intramolecular transpositions; that is, the element and its prospective target site are on the same DNA molecule. For elements such as Tn*3* and Mu that transposase via Shapiro intermediates, whether intramolecular transposition results in a deletion or an inversion of the adjacent sequence depends only on how the 3′-OH ends of the element attack the target. If each attacks a phosphate group on the same DNA strand as itself, then the transposition event automatically results in DNA fragmentation. The two sections on the starting molecule separated by the element (at its original site) and the target-site end become separate circular DNA molecules, each with a copy of the transposable element. However, only one of the fragments will carry the origin of replication and so can be replicated and survive; the result is a deletion adjacent to the original copy of the transposon. Conversely, if each 3′-OH attacks a phosphate group on the opposite DNA strand, then the transposition will generate a duplicate but inverted copy of the element at the new site, and the two sections of the carrier DNA separated by the old and new copies of the element are inverted with respect to each other.

Many other transposable elements also promote the formation of adjacent deletions and inversions, including elements thought to transpose by conservative cut-and-paste mechanisms. Precisely how these rearrangements are achieved is largely unknown.

B. Replicon Fusion and Conduction

When plasmid-to-plasmid transposition gives rise to cointegrates, these may offer the opportunity for conduction, that is, conjugal transfer of a nonconjugative plasmid by a conjugative one fused to it. The plasmids are cotransferred to the recipient cell as a single cointegrate DNA molecule, after which resolution may occur to release the donor replicon carrying one copy of the element and the target replicon carring a second copy of the transposable element. The transposable element may initially be on either of the participating plasmids. Conduction systems have also been used to study transposable elements that are cryptic (e.g., IS elements) exploting the element's ability to create replicon fusions and using plasmid markers to follow the movement of the cointegrate.

IX. CONCLUSION

The ability of transposable elements to insert and to generate deletions and inversions accounts for much of the macromolecular rearrangement that is observed among related bacterial plasmids. That bacterial chromosomes are subject to the same array of mutational events can readily be demonstrated in the laboratory. Nonetheless, chromosomal rearrangement seems to occur less often than might be expected, given the sizes of the DNA molecules and the apparent abundance of transposable elements. Whether this is because many chromosomal rearrangements prove, immediately or in the longer term, to be deleterious to host survival, particularly in competitive situations, and so are never established in the population or because chromosomal rerrangements occur less frequently than we might expect remains to be seen.

See Also the Following Articles

Bacteriophages • Conjugation, Bacterial • Plasmids • Yeasts

Bibliography

Bennett, P. M. (1999). Integrons and gene cassettes: A genetic construction kit for bacteria. *J. Antimicrob. Chemother.* 43, 1–4.

Bennett, P. M., and Hawkey, P. M. (1991). The future contribution of transposition to antimicrobial resistance. *J. Hos. Infect.* **18**(Suppl. A), 211–221.

Berg, D. E., and Howe, M. M. (eds.) (1989). "Mobile DNA." American Society for Microbiology, Washington, DC.

Craig, N. L. (1995). Unity in transposition reactions. *Science* **270**, 253–254.

Craig, N. L. (1997). Target site selection in transposition. *Annu. Rev. Biochem.* **66**, 437–474.

Eickbush, T. H. (1992). Transposing without ends: The non-LTR retrotransposable elements. *New Biol.* **4**, 430–440.

Goryshin, I. Y., and Reznikoff, W. S. (1998). Tn*5 in vitro* transposition. *J. Biol. Chem.* **273**, 7367–7374.

Hallet, B., and Sherratt, D. J. (1997). Transposition and site-specific recombination: adapting DNA cut-and-paste mechanisms to a variety of genetic rearrangements. *FEMS Microbiol. Rev.* **21**, 157–178.

Kleckner, N., Chalmers, R. M., Kwon, D., Sakai, J., and Bolland, S. (1996) Tn*10* and IS*10* transposition and chromosome rearrangements; mechanism and regulation *in vivo* and *in vitro*. *Curr. Top. Microbiol. Immunol.* **204**, 49–82.

Mahillon, J., and Chandler, M. (1998). Insertion sequences. *Microbiol. Mol. Biol. Rev.* **62**, 725–774.

Mizuuchi, K. (1992). Transpositional recombination: Mechanistic insights from studies of Mu and other elements. *Annu. Rev. Biochem.* **61**, 1011–1051.

Mizuuchi, M., Baker, T. A., and Mizuuchi, K. (1995). Assembly of phage Mu transpososomes—cooperative transitions assisted by protein and DNA scaffolds. *Cell* **83**, 375–385.

Reznikoff, W. S. (1993). The Tn*5* transposon. *Annu. Rev. Microbiol.* **47**, 945–963.

Rice, L. B. (1998). Tn916 family conjugative transposons and dissemination of antimicrobial resistance determinants. *Antimicrob. Agents Chemother.* **42**, 1871–1877.

Salyers, A. A., Shoemaker, N. B., Stevens, A. M., and Li, L-Y. (1995). Conjugative transposons: an unusual and diverse set of integrated gene transfer elements. *Microbiol. Rev.* **59**, 579–590.

Sarnovsky, R. J., May, E. W., and Craig, N. L. (1996). The Tn7 transposase is a heteromeric complex in which DNA breakage and joining activities are distributed between different gene products. *EMBO J.* **15**, 6348–6361.

Ton-Hoang, B., Polard, P., and Chandler, M. (1998). Efficient transposition of IS*911* circles *in vitro*. *EMBO J.* **17**, 1169–1181.

Trypanosomes

Herbert B. Tanowitz
Murray Wittner
Craig Werner
Louis M. Weiss
Albert Einstein College of Medicine

Louis V. Kirchhoff
University of Iowa

Cyrus Bacchi
Haskins Laboratories, Pace University

I. *Trypanosoma cruzi*
II. *Trypanosoma brucei*

GLOSSARY

amastigote The multiplying intracellular form of *T. cruzi*.

***American trypanosomiasis* (*Chagas' disease*)** An illness endemic to most countries in Latin America, which is caused by the protozoan parasite *Trypanosoma cruzi* and is transmitted by triatomine insects.

***African trypanosomiasis* (*sleeping sickness*)** An illness endemic to large areas of Africa, which is transmitted by tsetse flies. *Trypanosoma brucei gambiense* is the cause of West African or gambiense trypanosomiasis, and *Trypanosoma brucei rhodesiense* is the cause of East African or rhodesiense trypanosomiasis.

blood-form trypomastigote The life-cycle stage of both American and African typanosomiase found in the blood, cerebrospinal fluid, and other extracellular spaces.

epimastigote The multiplying life-cycle stage of trypanosomes found in insect vectors.

metacyclic trypomastigote The life-cycle stage of American and African trypanosomes that develops in insect vectors and is infective for humans and other mammals.

trypanosomiasis A group of diverse diseases caused by members of the genus *Trypanosoma*, which are hemoflagellate protozoa.

THE GENUS *TRYPANOSOMA* consists of approximately 20 species. However, only *Trypanosoma cruzi*, the cause of American trypanosomiasis, and *Trypanosoma brucei gambiense* and *T.b. rhodesiense*, the causes of African trypanosomiasis have been linked to human disease. In Africa, there are a number of *Trypanosoma* species, including *T. b. rhodesiense*, that cause trypanosomiasis in wild and domesticated animals and are of immense economic importance.

I. *TRYPANOSOMA CRUZI*

Trypanosoma cruzi is the protozoan parasite that causes American trypanosomiasis or Chagas' disease. It is an important cause of chronic heart and digestive diseases and is endemic in many areas of Central and South America. Studies of autopsied Native American mummies suggest that Chagas' disease in humans dates back to the pre-Colombian era. Seroepidemiologic and clinical studies indicate that in the endemic areas of Latin America 16–18 million people are infected and 50,000 die annually as a result of this infection. Five autochthonous cases have been reported in the United States since 1955. Naturally infected dogs as well as a wide variety of wild mammals have also been described, and infected insect vectors have been found in areas of the southern and southwestern United States.

A. Transmission

In endemic areas transmission is most common among the rural poor whose dwellings become infested with infected vectors due to the proximity of sylvatic reservoirs. The natural transmission of *T. cruzi* is via a process known as posterior station, which indicates that the infective parasites are in the excreta of the vectors. During a blood meal, insect feces containing metacyclic trypomastigotes by chance find their way onto mucosal surfaces, the conjunctivas, or abraded skin. There they enter susceptible cells of the new host and establish infection. Control programs directed at decreasing contact with insect vectors through housing improvement, insecticide spraying, and education have met with considerable success in many endemic areas. The transmission of *T. cruzi* by blood transfusion is the second most important mode of spread of the parasite. This has been a major public health problem in some endemic regions, but its importance has been reduced in many areas by screening programs. Several cases of transfusion-associated transmission of *T. cruzi* have been reported in the United States and Canada, resulting from the transfusion of blood donated by asymptomatic *T. cruzi*-infected immigrants from Latin America. Other modes of transmission include congenital, laboratory accidents, organ transplantation, and breast-feeding.

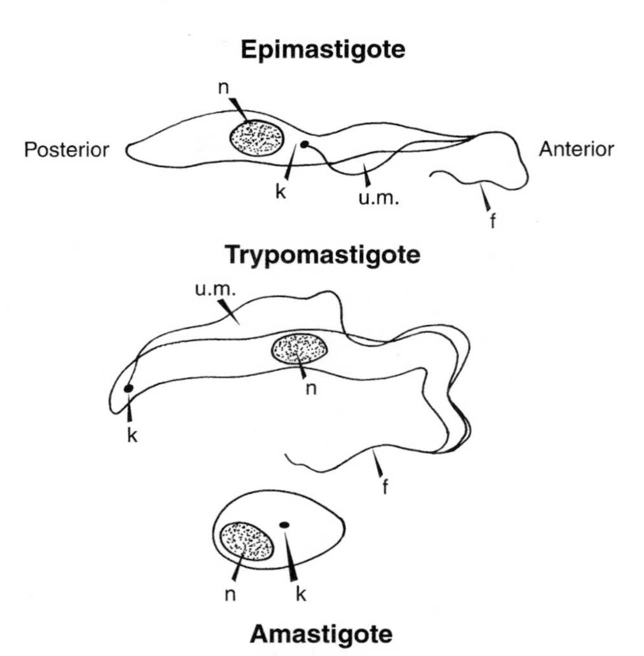

Fig. 1. Life-stage forms of *Trypanosoma cruzi.* Note the kinetoplasts; see Fig. 5 for the untrastructure of this organelle. The trypomastigotes measure 15–20 μm in length, the epimastigotes measure 20 μm in length, and the amastigotes measure 3–4 μm in diameter.

B. Organism and Life Cycle

T. cruzi has a complex life cycle in which four morphologically and biochemically distinct forms exist (Fig. 1). In mammalian hosts, there are two forms, extracellular nondividing trypomastigotes (blood forms) (Figs. 2–3), and intracellular dividing amastigotes (Figs. 4–5). During a blood meal, the insect vector ingests blood forms, which then transform into epimastigotes in the midgut. After 3–4 weeks, infective metacyclic trypomastigotes are present in the hindgut of the vector. These forms are then deposited with the excreta of the vector during subsequent blood meals. Transmission occurs to a new mammalian host when these infective parasites contaminate susceptible tissues, such as the oral or nasal mucosa. The infective trypomastigotes parasitize cells by direct penetration or phagocytosis and then transform into amastigotes. The mechanisms by which the trypomastigotes gain entry into host cells are not entirely understood. Studies have implicated the enzyme *trans*-silalidase in this process. Penetrin, another parasite molecule, binds to heparin sulfate receptors on nonphagocytic host cells, thereby promoting adhesion and subsequent parasite invasion. Furthermore, recent data indicate the transforming-growth-factor-β (TGF-β) receptor may be the host-cell receptor for this parasite. In the parasitophorous vacuole, amastigotes synthesize a hemolysin that lyses the vacuolar membrane, thus allowing the parasite to escape the cytocidal mechanism of the cell. Once in the cytoplasm of the host cells, amastigotes (Figs. 4–5) undergo binary fission and the parasitized cells eventually rupture, releasing amastigotes and trypomastigotes that infect adjacent unifected cells. In addition, trypomastigotes enter the bloodstream and infect distant cells of the reticuloendothelial system and muscle cells, as well as cells of the nervous system. Both cell necrosis and apoptosis have been reported to be associated with this parasite.

The chromosomal complement of *T. cruzi* is approximately 50 Mb or roughly 10 times the amount of DNA found in *E. coli*. Pulse-field gel electrophoresis reveals at least 20, and as many as 40, discrete chromosomal bands, ranging in size from 130–1600 Mb.

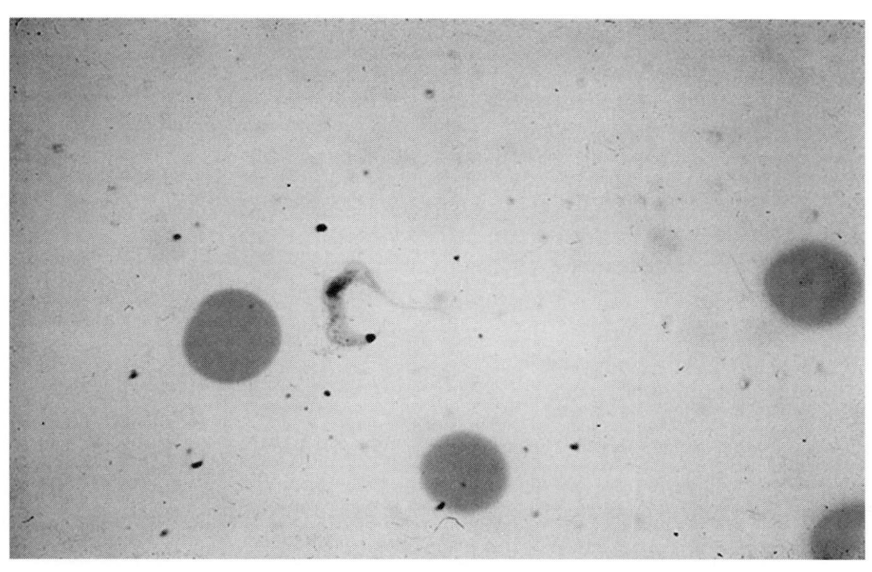

Fig. 2. Trypomastigote of *Trypanosoma cruzi* in the bloodstream. Note the "C" shape. From Zaiman, H. (ed.) (1986). "A "Pictorial Presentation of Parasites." Valley City, ND.

This is somewhat strain-dependent (i.e., the reference strain CL/Brener contains approximately 60–80 Mb DNA). Among characterized strains, there appears to be only one that contains minichromosomes, in sharp contrast to *T. brucei*. The genome of *T. cruzi* appears to be diploid in all stages. Homologous chromosomes, however, can differ markedly in size, and chromosomes containing conserved linkage groups can vary in size extensively among isolates.

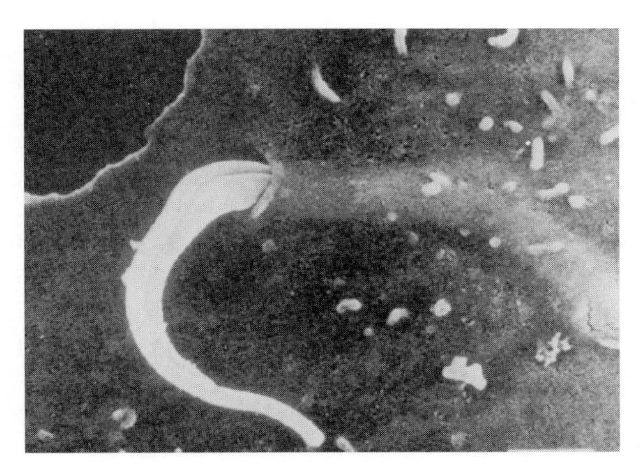

Fig. 3. Scanning electron microscopy of a trypomastigote invading a mammalian host cell.

T. cruzi exhibits significant polymorphism at a number of characterized alleles, far in excess of that seen in higher eukaryotes. Analysis of 121 stocks at 15 enzyme loci revealed that 14/15 were polymorphic, the number of alleles ranged from 2–12 with a mean of 5.14 per locus. There is a trend toward homozygosity in any given strain. The actual average frequency of heterozygous loci per strain is 0.059, below the 0.47 theoretical maximum calculated from the number given. There is also a significant linkage disequilibrium. The 121 stocks made up only 43 genotype patterns, 16 of which differed by only a single allele (out of 15), whereas combinatorial equilibrium (Hardy–Weinberg equilibrium) predicted so many possible genotypes (7×1015) that identical allozyme patterns would happen rarely, if at all. In addition, although there is a trend toward Hardy–Weinberg equilibrium among the alleles contained in strains coexisting in small isolated geographic ranges, no direct observation of exchange of genetic material has been observed in coinfection studies, in contrast to the demonstration of hybrid *T. brucei* following coinfection and passage through the tsetse fly vector. Thus, it appears that multiplication of *T. cruzi* is exclusively, or nearly exclusively, clonal without any evidence of a sexual stage.

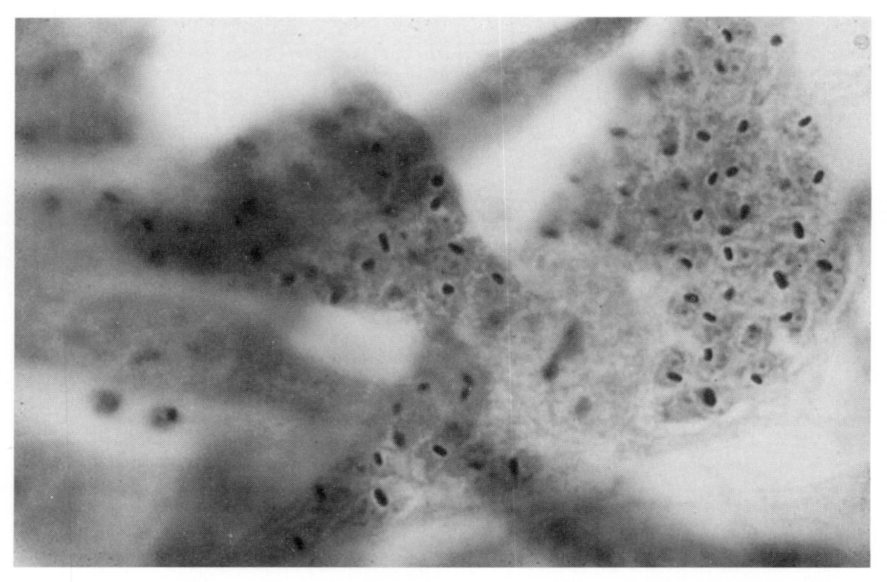

Fig. 4. Giemsa-stained endothelial cells infected with *T. cruzi*. Note the numerous amastigotes. See color insert.

Numerous genes of *T. cruzi* have been cloned and characterized. In a number of cases, the genes are arranged in tandem arrays of nearly identical copies, as is the case with gGAPDH, Histone H1, and heat-shock protein (Hsp70). Several libraries have been constructed that encompass the entire genome, and the mapping and sequencing of chromosome 3 is well underway. The sequence suggests that all *T. cruzi* genes are transcribed from a single strand outward from a central noncoding region toward the telomere. This would be similar to *T. brucei*, in which it has been demonstrated that large polycistronic pre-mRNAs are transcribed in a telomeric direction. Importantly, all *T. cruzi* genes sequenced lack introns and all mature mRNAs contain a highly conserved 5′ cap that is added to pre-mRNAs by *trans*-splicing. *T. cruzi* mRNAs contain poly-A tails, but there is no highly conserved poly-A addition-site consensus sequence, as is the case with other eukaryotes. The classification of *T. cruzi* strains has been accomplished by a variety of methods including isoenzyme differences (zymodemes), restriction endonuclease digest of kinetoplast DNA (schizodemes), and sequencing of noncoding segments of rRNA genes. There is considerable correlation among the groupings determined by the various approaches. These observations suggest that *T. cruzi* consists of many clonal lineages with little sexual recombination. The rRNA sequence data also suggest that the clonal lineages fall into two main branches with a very ancient split between them. If one assumes a constancy of the molecular clock, then the divergence may have occurred contemporaneously with

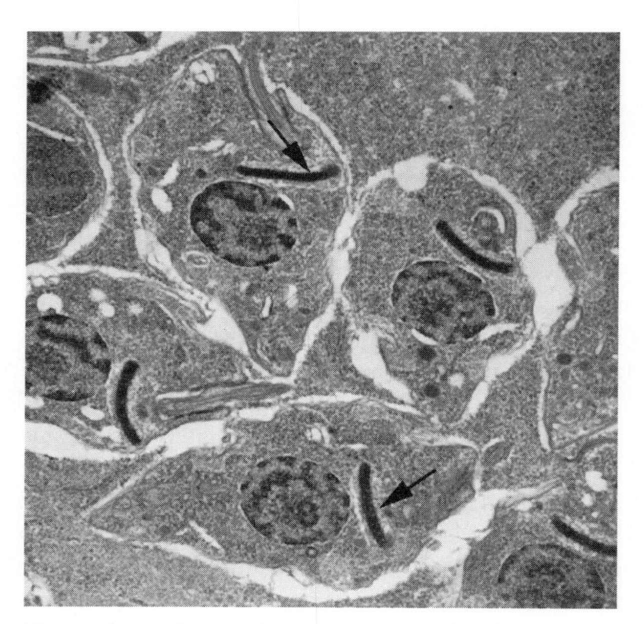

Fig. 5. Transmission electron micrograph of parasitized myoblast cell line. Note the amastigotes in the cytoplasm. The kinetoplasts are indicated with arrows.

the divergence of reptiles and amphibians, which predated the first mammals.

C. Epizootiology and Epidemiology

T. cruzi is found only in the Western hemisphere. Infection with this parasite is a zoonosis and the involvement of humans in the life cycle has little to do with the perpetuation of the parasite in nature. The parasite primarily infects both wild and domestic mammals and insects. The triatomine insect vectors of *T. cruzi* are distributed from the southern part of the United States to central Argentina. Transmission of the parasite typically occurs between infected vectors and nonhuman mammalian hosts in hollow logs, palm trees, burrows, and other animal shelters. A large number of insect vectors have also been found near houses in piles of roof tiles, wood, and vegetation.

T. cruzi has been found in more than 100 species of domestic and wild mammals, from the southern United States to central Argentina. Opossums, wood rats, armadillos, raccoons, dogs, and cats all are typical hosts, but *T. cruzi* is not a problem in livestock. Nontypical hosts, such as polar bears, can become infected when held in zoos in areas in which *T. cruzi* is enzootic. This lack of species-specificity, combined with the fact that infected mammalian hosts have life-long parasitemias, results in an enormous domestic and sylvatic reservoir in enzootic areas. Humans have become part of the cycle of *T. cruzi* transmission as farmers and ranchers open new lands in enzootic areas. When this development takes place, the vectors, known variously as the "kissing bug" or "assassin bug" (e.g., *Rhodnius prolixus, Triatoma infestans,* and *Panstrongylus megistus*) invade the nooks and crannies of the primitive wood, mud-walled, and stone houses that are typical of rural Latin America. In this manner, the vectors become domiciliary, establishing a cycle of transmission involving humans and peri-domestic mammals that is independent of the sylvatic cycle. Historically, Chagas' disease has been an infection of poor people living in rural environments. However, an enormous number of infected people have migrated to cities, thus urbanizing the disease and resulting in its frequent transmission by blood transfusion. The epidemiology of *T. cruzi* infection has been improving in several of the endemic countries, as vector and blood-bank control programs have met with considerable success. A major international eradication program in the "southern cone" countries of South America has provided the framework for this progress. Most of the obstacles hindering the elimination of *T. cruzi* transmission to humans are economic and political, and no technological breakthroughs are necessary for overall control of the problem.

Approximately 70% of people who harbor *T. cruzi* chronically will never develop associated cardiac or gastrointestinal symptoms. In the past, the high frequency of sudden death among young adults in some areas was attributed to rhythm disturbances due to chronic Chagas' disease and in one highly endemic area of Brazil, chagasic cardiac disease was found to be the most frequent cause of death in young adults. There is geographic variation in the relative prevalence of cardiac and gastrointestinal disease in patients with chronic Chagas' disease. It is not known if this variable clinical expression is caused by parasite strain or host factors. In this regard, studies have implicated host genetic factors as important variables in host response to infection.

The number of people in the United States with chronic *T. cruzi* infections has increased. Since the early 1970s, more than 1 million people have emigrated to the United States legally from Latin America and several million more may have entered illegally. A large number of these immigrants have come from Central America, where the prevalence of *T. cruzi* infection is high. A study among Nicaraguans and Salvadorans in Washington, DC, found a 5% prevalence rate of *T. cruzi* infection. Seroprevalence studies done in a Los Angeles hospital in which half of the donors are Hispanic have shown that between 1/1000 and 1/500 donors have *T. cruzi* infections. The presence of these immigrants creates a risk of transfusion-associated transmission *T. cruzi* here, and five such cases have been reported in the United States and Canada. All the reported cases occurred in immunocompromised patients in whom the diagnosis of *T. cruzi* was made because the transfusion recipients became severely ill. This suggests that additional cases have occurred in immunocompetent patients and have gone unnoticed because they were more able to control the acute infection. However, the risk may have been reduced by screening prospective blood donors with a questionnaire that focuses on residence in countries in which Chagas' disease is endemic.

D. Clinical Manifestations

1. *Acute Chagas' Disease*

Seroepidemiologic studies in endemic areas indicate that most seropositive patients have no recollection of having had acute Chagas' disease. Some infected patients may develop severe symptoms after an incubation period of 7–14 days. This has been documented with vector-borne infection as well as infections acquired by blood transfusions or laboratory accidents. Signs and symptoms include fever, chills, nausea, vomiting, rash, diarrhea, and meningeal irritation. A chagoma, a raised inflammatory lesion at the site of parasite entry, Romaña's sign (unilateral periorbital edema), conjunctivitis, lymphadenopathy, and hepatosplenomegaly have all been described in patients with acute Chagas' disease. Laboratory abnormalities include anemia, thrombocytopenia, and elevated liver and cardiac enzymes. Motile trypomastigotes can be found in peripheral blood and cerebrospinal fluid. Serologic tests are usually negative during this stage. During the acute phase of the illness, asynchronous cycles of parasite multiplication, cell destruction, and infection of new cells occur.

Myocarditis and cardiomegaly, sometimes associated with congestive heart failure, are present in some patients and are probably more common than was previously appreciated. The appearance of arrhythmias, heart block, or congestive heart failure in the setting of acute. *T. cruzi* infection are poor prognostic indicators. It is unclear if the severity of acute myocarditis is related to the development of chronic cardiomyopathy. A small percentage of acutely infected patients, often children, die of acute myocarditis or meningoencephalitis. As antibodies appear, the parasitemia wanes, and the majority of infected persons recover after 2–4 months, but remain infected for life. These patients then enter the indeterminate phase of the illness, which is characterized by a lack of symptoms, easily detectable antibodies to a variety of *T. cruzi* antigens, and low parasitemias. Congenital Chagas' disease is another example of acute infection. It occurs as often as 1/500 births in some highly endemic areas. These babies present with fever, jaundice, and occasionally seizures, and are indistinguishable clinically from infants with other congenital infections such as toxoplasmosis.

2. *Chronic Chagas' Disease*

a. *Chronic Cardiomyopathy*

Chronic cardiac Chagas' disease may present insidiously as congestive heart failure or abruptly with arrhythmias and thromboembolic events. Dilated congestive cardiomyopathy is an important manifestation of chronic Chagas' disease that typically occurs years after acute infection. An apical aneurysm of the left ventricle is one of the hallmarks of this disease. Chronic chagasic heart disease is associated with myonecrosis and myocytolysis. Contraction-band necrosis occurs after transient hypoperfusion followed by reperfusion, such as after local spasm of the coronary microvasculature. Focal and diffuse areas of myocellular hypertrophy are observed with or without inflammatory infiltrates (Fig. 6), and fibrosis replacing previously damaged myocardial tissue is evident. The destruction of conduction tissue results in atrioventricular and intraventricular conduction abnormalities. In areas where the disease is endemic, the presence of right-bundle branch block (RBBB), associated with an anterior fascicular block, is highly suggestive of chagasic cardiomyopathy. Conduction defects may necessitate the placement of a pacemaker.

b. *Chronic Gastrointestinal Manifestations*

Chagas' disease is associated with the loss of neurons within the myenteric plexus. Lesions of the autonomic nervous system produce the functional disorders in hollow visceral muscular organs. Initial damage to the nervous system occurs during acute *T. cruzi* infection and may reuslt from the increase in local cytokines and nitric oxide resulting from the induction of the inducible isoform of nitric oxide synthase (NOS2). Further neuronal loss occurs slowly during the chronic phase of the infection. The development of chronic gastrointestinal disease may

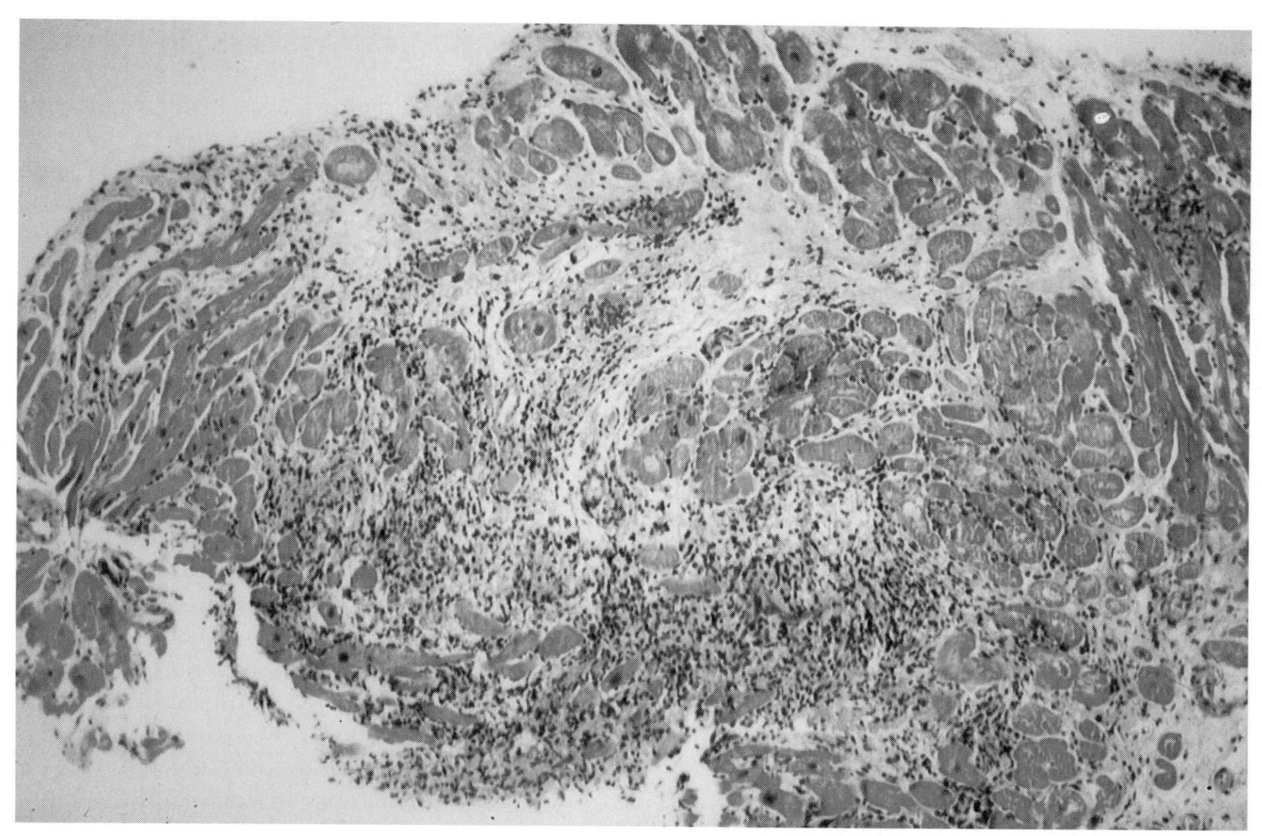

Fig. 6. Endomyocardial biopsy from a patient with Chagas' disease. Note the inflammation and fibrosis. See color insert. Courtesy of Dr. Alain C. Borczuk, North Shore University Hospital, Manhasset, NY.

take several decades, and the determinants of progression are unknown.

The colon is frequently involved in chronic Chagas' disease. Although the entire colon may be enlarged, usually only sigmoid dilation is found. Constipation is a common complaint. Symptomatic patients may have disorders of motility that antedate by years the radiological diagnosis of megacolon. In patients with chagasic megacolon, there is a reduction in ganglia throughout the colon. The esophagus is also frequently affected in patients with chronic *T. cruzi* infection. Patients with megaesophagus may have less than 5% of the normal number of ganglia and the loss of neurons is uniform along the length of the organ, including the grossly normal abdominal portion of the esophagus. Typical symptoms of megaesophagus include dysphagia, odynophagia, chest pain, cough, and regurgitation. Mega-gall bladder and megaureter have been reported in patients with chronic *T. cruzi* infections.

3. *Other Clinical Syndromes*

a. Transfusion-Associated T. cruzi Infection

The transmission of *T. cruzi* by transfusion of blood donated by asymptomatic persons who harbor the parasite is a major public health problem in some endemic areas. As noted, several million persons have emigrated from Latin America to the United States. Studies suggest that as many as 1/500 of these immigrants is infected with the parasite.

b. Immunosuppression and Transplantation

The incidence of reactivation of *T. cruzi* infection in immunosuppressed people is not known. In view of the chance of acute reactivation of *T. cruzi*, patients who are at risk of being infected and for whom immunosuppressive treatment is being planned, either as primary or post-transplanation therapy, should undergo serological testing and, if positive, should be monitored closely while immunosuppressed for evidence of reactivation. Several dozen HIV-infected patients with reactivation of chronic *T. cruzi* infections have been reported. Interestingly, these patients developed cerebral lesions similar to those observed in AIDS patients with cerebral toxoplasmosis, a process that does not occur in immunocompetent patients chronically infected with *T. cruzi*. Finally, more than three dozen patients with end-stage Chagas' heart disease have undergone cardiac transplantation in Brazil, and the procedure has been done in a dozen or so *T. cruzi*-infected people in the United States. Because of the postoperative immunosuppression, life-threatening reactivation of *T. cruzi* infection occurred in some of the Brazilian patients. However, lower doses of immunosuppressive drugs have been used in these patients and the incidence of reactivation of *T. cruzi* infection has become less of a problem.

D. Pathogenesis

Animal models have been used to evaluate the pathogenesis of chagasic heart disease. For example, alterations in cardiac choline acetyltransferase, acetylcholine, norepinephrine, and β-adrenergic adenylate cyclase complex have been described in experimental animals. The expression of cytokines and NOS2 in the myocardium has been suggested as a possible cause of myocardial dysfunction. Morphological changes during acute *T. cruzi* infection may be also be important in the pathogenesis of the myocardial lesions. For example, reduced numbers of autonomic ganglia is associated with cardiac as well as gastrointestinal Chagas' disease, and this may be a direct result of destructive processes that are part of the early phase of the illness.

The immunology of *T. cruzi* infection has been studied extensively in animal models and humans, and a large portion of these efforts have focused on a possible role for autoimmunity in the pathogenesis of chagasic lesions. There is evidence from studies in mice that both humoral and cell-mediated components of the immune system are important in host resistance, as is genetic background. The role of CD8+ T cells in the pathogenesis of *T. cruzi* infection has been investigated. Infected mice depleted of CD8+ cells by antibody treatment and β-microglobulin-deficient mice, which lack mature CD8+ T cells, develop high parasitemias and die early in the course of the acute infection. In addition, parasitized tissues in these mice lack an inflammatory response. The role of cytokines in the pathogenesis of *T. cruzi* infections have been the subject of intense investigation. Both humoral and cellular immunity have been suppressed in experimental infections and altered interleukin 2 (IL-2) concentrations are associated with this immunosuppressed state. Furthermore, studies in mice have suggested that IL-2 may participate in disease-related autoimmune phenomena. TNF-α, IL-5, and IFN-γ may be elevated during acute murine infection and may contribute to the pathogenesis of the disease. Increased levels of IL-1β, TNF-α, and IL-6, expressed in infected endothelial cells, result in leukocyte recruitment, coagulation, and smooth muscle cell proliferation. The role of cytokines and nitric oxide in the killing of the parasite has received increased attention. Nitric oxide, IFN-γ, TNF-α, and IL-12 all appear to be involved in intracellular killing.

Trypanosomes contain proteases, gelatinases, and collagenases capable of degrading native type I collagen, heat-denatured type I collagen (gelatin), and native type IV collagen. Proteolytic activities against laminin and fibronectin are also present. These enzymes may play important roles in the degradation of extracellular matrix and the subsequent tissue invasion by *T. cruzi*. It has been proposed that the degradation of the collagen matrix, evident in acute murine Chagas' disease, may result in chronic pathology such as apical thinning of the myocardium.

E. Diagnosis

The diagnosis of acute *T. cruzi* infection is generally made by the detection of parasites. Active trypomastigotes frequently can be seen by microscopic examination of fresh anticoagulated blood or buffy coat, and organisms can often be seen in Giemsa-stained thin and thick blood smears as well. If the organisms cannot be detected by these approaches, one may inoculate blood specimens or buffy coat into specialized liquid medium or intraperitoneally into mice. The disadvantages of these approaches are their lack of sensitivity and the fact that the parasites are usually not seen in positive cultures or infected mice for several weeks. Assays based on the polymerase chain reaction (PCR) have been described, and the results obtained suggest that this approach may be the most sensitive method for detecting acute *T. cruzi* infections. When acute Chagas' disease is suspected in an immunocompromised patient and these methods fail to demonstrate the presence of parasites, additional tissue specimens should be examined. These patients can pose a difficult diagnostic problem because they may present with fulminant clinical disease and low parasitemias that cannot be readily detected. Surprisingly, parasites can sometimes be seen in atypical sites, such as pericardial fluid, bone marrow, brain, skin, and lymph nodes, and thus, these tissues also should be investigated when indicated.

The diagnosis of chronic Chagas' disease is generally based on detecting specific antibodies that bind to *T. cruzi* antigens. A number of highly sensitive serological assays are used in Latin America for detecting antibodies, such as the complement fixation and indirect immunofluorescence tests and the enzyme-linked immunosorbent assay (ELISA). These and other conventional serologic assays are used widely for clinical diagnosis and screening donated blood, as well as in epidemiological studies. False-positive reactions, however, have been a persistent problem with these assays. Tests need to be developed that have the sensitivity of conventional assays, but are highly specific and easy to perform. An assay based on the immunoprecipitation of radiolabeled protein antigens followed by electrophoresis was found to be highly specific as well as sensitive when used in a clinically and geographically diverse group of infected people. Although this complex test has not been adapted for use on a large number of sera, it is presently available as a confirmatory assay. In addition, laboratories have performed DNA-sequence determinations on *T. cruzi* genes that encode antigenic proteins in an effort to produce recombinant proteins and synthetic peptides that could be used as targets in assays for anti-*T. cruzi* antibodies. A sizable number of DNA sequences encoding proteins that are specifically recognized by sera from infected patients have been published, and in many cases, these proteins have been evaluated in diagnostic assays. Encouraging initial results have been obtained with some of these recombinant-based serologic assays, but none is available commercially. The detection of chronic infection by testing for parasite antigens in blood and urine has been studied, but this approach has not achieved results comparable to those obtained by serologic methods. The usefulness of PCR-based assays for detecting chronic *T. cruzi* infections has not been established. It would appear that this approach would be particularly suited for the task of detecting the low number of parasites circulating in the blood of chronically infected patients, but parasitemias may in fact be intermittent and this may limit the sensitivity of the assays.

F. Treatment

The treatment of *T. cruzi* is unsatisfactory. Nifurtimox (Lampit, Bayer 2502) and benznidazole (Rochagan, Roche 7-1051), the two drugs available for treating this infection, lack efficacy, must be taken for extended periods, and may cause severe side effects. Both drugs reduce the severity of acute Chagas' disease, and it is thought that about 70% of patients treated with a full course of either nifurtimox or benznidazole are cured parasitologically. This cure rate decreases as a function of the time patients have been infected and is about 20% in persons who have harbored the parasite for many years. Evidence from studies of experimental animals and humans has accumulated indicating that the elimination of the parasites reduces the likelihood of the progression of cardiac and gastrointestinal lesions. In view of this, an international panel of experts has recommended that all infected patients be treated with one of the available drugs, regardless of the time elapsed since the acquisition of the infection. Allopurinol and several antifungal azoles have been shown to have some anti-*T. cruzi* activity in *in vitro* experiments and in animal studies, but none has a level of activity that would warrant its use in place of nifurtimox or benznidazole. The anti-*T. cruzi* activity of several inhibitors of bacterial topoisomerases *in vivo* have been studied. Marked inhibition of parasite proliferation and differentiation was caused by several drugs in this class, suggesting that *T. cruzi* has an enzyme similar to the topoisomerase II of bacteria, thus opening a new avenue in the search for an effective and nontoxic agent for treating *T. cruzi*. Subsequent work by some of the authors of the latter study resulted in the molecular cloning of a protein in *T. cruzi* that has an 80% amino acid identity with topoisomerase II of *T. brucei*. D0870, a bis-triazole derivative being developed as an antifungal agent, has been shown to eradicate *T. cruzi* in a mouse model of chronic Chagas' disease, but this agent is only in the early stages of development.

II. *TRYPANOSOMA BRUCEI*

Human African trypanosomiasis, or "sleeping sickness," is caused by two subspecies of hemoflagellate protozoa; *Trypanosoma brucei gambiense,* and *T. brucei rhodesiense.* The former causes the West African or gambiense form and the latter, the East African or rhodesiense form. Although these parasites produce similar diseases, the West African form usually evolves slowly, over many years and ends fatally if it is not treated. The East African form usually kills its host in weeks to months. These diseases exist in Africa wherever the various species of tsetse files (which belong to the genus *Glossina*) are found.

A. Organism and Life Cycle

Trypanosoma brucei gambiense and *T. brucei rhodesiense* are pleomorphic flagellates 15–30 μ in length by 1.5–3.5 μ in breadth. The two subspecies are morphologically indistinguishable. There are two forms of trypomastigotes that circulate in the bloodstream, long slender organisms that are capable of dividing, and short stumpy forms that are thought to be nondividing parasites that are infective for tsetse files. There are no intracellular forms. At various stages of the disease, trypomastigotes may be found in peripheral blood, lymphatics, lymph nodes, cerebrospinal fluid, and neural tissue (Fig. 7). Other than humans, there is no important reservoir host for *T. brucei gambiense,* whereas *T. b. rhodesiense* is primarily a parasite of wild game animals.

In the tsetse fly, trypomastigote forms ingested with a blood meal settle in the posterior midgut, where they multiply by binary fission for approximately, 7–10 days and then migrate anteriorly to the foregut, where they remain for the next 2–3 weeks. Finally, they enter the salivary glands, continue to

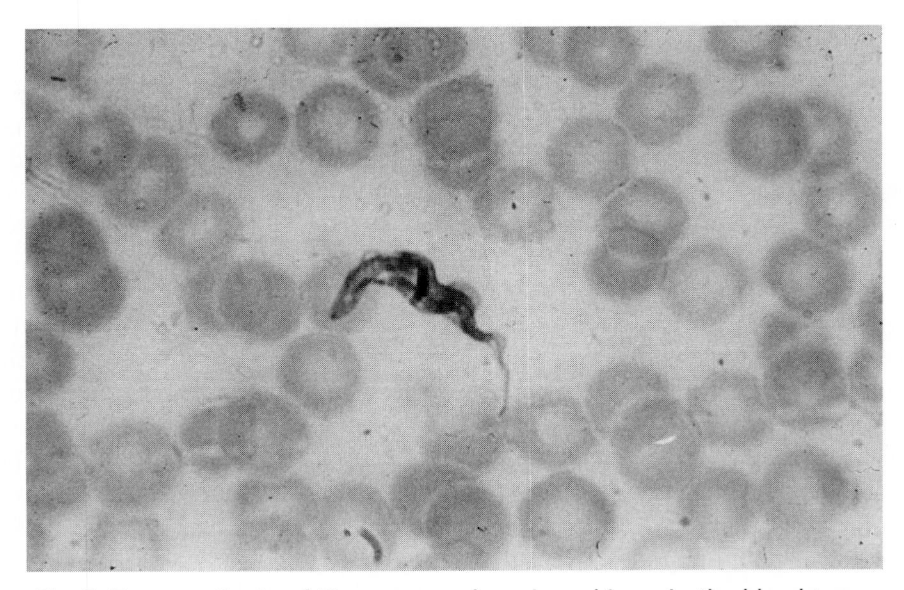

Fig. 7. Trypomastigote of *Trypanosoma brucei gambiense* in the bloodstream. From Zaiman, H. (ed.) (1986). "A Pictorial Presentation of Parasite." Valley City, ND.

replicate, and, after several cycles of division, transform into infective metacyclic trypomastigote forms. These organisms are inoculated the next time a mammalian host is bitten, and once in a human host, trypomastigotes multiply by binary fission in the blood, lymph, and other extracellular spaces. The central nervous system (CNS) eventually is invaded and multiplication continues there as well.

The haploid genome size of *T. brucei* spp. is approximately 40 Mb, although there is up to 14% variation in isolates of the same subspecies and up to 29% between the two subspecies. There is a minimum of seven resolvable chromosome pairs on pulse-field gel electrophoresis in the size range of 1.1–6 Mb. Homologous chromosomes, when probed in Southern blots, can differ in size by up to 20%. In addition to the large chromosome pairs, *T. brucei* contains approximately 100 linear minichromosomes ranging in size from 50–150 kb. Minichromosomes contain tandem arrays of a 177-bp repeat and contain transcriptionally silent copies of variant surface glycoprotein (*VSG*) genes proximal to their telomeres. *T. brucei* contains approximately 1000 genes capable of coding for *VSG* genes, which are switched at a rate of 10^2–10^6 switches per generation, which serves as the main mechanism of immune evasion for *T. brucei*. Only one *VSG* expression site is active at any given time. There are 15–20 expression sites (ESs) per genome, all at subtelomeric locations. In addition to the *VSG* genes, several upstream genes, called expression site-associated genes (ESAGs), are also transcribed. Three types of DNA rearrangements are associated with ES switching—duplicative transposition, telomere exchange, and telomere conversion. Transposition involves the 1.6-kb *VSG* and another 1.5-kb proximal sequence. Thus, a minimum of 8% of the genome is devoted to *VSG* coding and flanking sequences. As mentioned previously, a large fraction of the 25% of the genome found in the minichromosomes also contributes to *VSG* diversity. The modified DNA base "J" (β-D-glucosyl-hydroxymethyl-uracil), unique to organisms in the order Trypanosomatidae, replaces up to 1% of thymidines, and its presence is higher in repetitive DNA adjacent to transcriptionally silent telomeres, including the (GGGTTA)n telomeric hexamer.

T. brucei genes are transcribed as large polycistronic units that then undergo *trans*-splicing so that all mature mRNAs contain the same 39-nt sliced leader at their 5′-ends. In contrast, with the exception of an 11-nt intron in a tRNAtyr, *T. brucei* genes contain no introns and, hence, do not undergo *cis*-splicing. The spliced leader is also notable for the presence of a 7-methyl-guanosine 5′ cap, a promoter-like region consisting of four methylated nucleotides at the 5′-end, similar to that found upstream from ES regions and from genes that encode procyclic stage-specific coat protein (procyclic acid-rich protein, PARP; also called procyclin). Transcription from the PARP and VSG promoters, similar to that from the trypanosome rRNA promoter, is α-amanitin-resistant, suggesting transcription by RNA Polymerase I. Stage-specific gene expression is also influenced by 3′ untranslated portions of mRNAs, mediated through changes in mRNA stability and in efficiency of mRNA maturation. The sequence influencing stage specificity of *VSG* mRNA abundance is localized to a region 97 nt upstream from the polyadenylation site. Retroposon-like elements are also scattered throughout the genomes of these parasites. The best studied is a 5-kb sequence, designated *ingi* (Swahili for many), that is similar to reverse transcriptase genes in other organisms. There are approximately 400 copies of *ingi*, making up to 5% of the *T. brucei* genome.

Another interesting feature of trypanosome genetics, which is shared with other members of the order Kinetoplastida, is the phenomenon known as RNA editing. The kinetoplast DNA is organized as an interlocking and supercoiled network of approximately 50 maxicircle (20- to 30-kb) DNA molecules and many thousand minicircle DNA molecules (1.0 kb in *T. brucei* and 1.6 kb in *T. cruzi*). The maxicircle DNAs encode about a dozen mitochondrial proteins. The maxicircles and minicircles both encode small (50–100 nt) guide RNAs that serve as templates for the insertion, and less frequently deletion, of uridines in the primary RNA transcripts of the maxicircle mitochondrial genes. In some cases, nearly 50% of the mature mRNAs consist of Us inserted posttranscriptionally by the editing process. Studies comparing homologous nuclear genes between *T. brucei* and *T. cruzi* have demonstrated a large evolutionary divergence in codon use. A comparison of the nuclear small and large subunit rRNA gene sequences yields genetic distances comparable to those between plants and animals.

B. Epidemiology

Sleeping sickness (*Trypanosoma brucei gambiense* and *T. b. rhodesiense*) and veterinary trypanosomiasis caused by *Trypanosoma brucei* subgroup parasites continue to be responsible for much human suffering and economic loss (see Table 1). These agents are endemic and enzootic in an area of sub-Saharan African covering 10 million square kilometers. Approximately 50 million people are at risk for becoming infected with these parasites, and tens of thousands of new cases of human African trypanosomiasis occur each year. The exact numbers are not available because the acquisition of reliable health statistics is difficult in the developing countries where sleeping sickness is endemic. Human African trypanosomiasis has undergone a resurgence, and major epidemics have occurred in the Central African Republic, Ivory Coast, Chad, the Sudan, and several other endemic countries. Losses of cattle due to trypanosomiasis have had an enormous economic impact in many regions.

Human African trypanosomiasis is restricted to those areas south of the Sahara in which the annual rainfall exceeds 500 mm (i.e., 20 inches) because the larval stages of the tsetse fly are vulnerable to desiccation. Thus, the gambiense form occurs primarily in the western portion of tropical Africa and focal incursions eastward, north of Lake Victoria into

TABLE I
African Animal Trypanosomiasis[a]

Parasite	*Host*	*Occurrence*
T. vivax	Cattle	Common, mild
	Equines	Rare, mild
	Sheep and goats	Rare, severe
	Camels	Rare, mild
	Dogs	Not reported
	Pigs	Not reported
T. brucei brucei	Cattle	Common, but most cattle tolerate well
	Equines	Common, severe
	Sheep and goats	Rare
	Camels	Common, severe
	Dogs	Common, severe
	Pigs	Not reported
T. congolense	Cattle	Common, severe
	Equines	Rare, mild disease
	Sheep and goats	Rare
	Camels	Not reported
	Dogs	Not reported
T. evansi	Cattle	Mild disease
	Equines	Common, severe
	Sheep and goats	Not reported
	Camels	Common, severe

[a] Animal trypanosomiasis is of great economic importance in Africa and is associated with fever, anemia, thrombocytopenia, wasting, and death. These manifestations are mediated, in part, by cytokines.

Sudan. The rhodesiense form is found in the southwestern portion of Africa, north of South Africa. There is some overlap in the endemic ranges of these two forms of the disease.

Humans are the only important reservoir of *T. b. gambiense*. The cycle of gambiense disease is maintained only where there is a close relationship between humans and tseste flies (members of the *G. palpalis* group that preferentially feed on human blood). In contrast to gambiense disease, the cycle of *T. brucei rhodesiense* is maintained in wild mammals, and humans are only incidental hosts. The vectors of *T. b. rhodesiense* (which belong to the *G. morsitans* group) inhabit the relatively dry eastern African savannas and preferentially feed on wild ungulates, which are trypanotolerant and maintain infective parasitemias for long periods. Nonetheless, there have been instances in which East African trypanosomiasis has reached epidemic proportions. Human cases of rhodesiense trypanosomiasis typically occur in young adult men who have occupations in which they are exposed to tsetse flies. In epidemics, however, all age groups are infected, and mechanical transmission is said to occur. This occurs via the blood-filled proboscis of a fly, which may be interrupted while taking a blood meal from an infected individual. When the fly bites an uninfected host within 2–3 hours, the blood from the infected host becomes a parasite-bearing inoculum. In contrast to *T. cruzi*, congenital transmission of African trypanosomes is extremely rare. Laboratory-acquired transmission has been reported.

C. Pathology and Pathogenesis

Metacyclic infective trypomastigote forms are inoculated into the skin by the tsetse fly and multiply there. A characteristic hard and sometimes painful chancre is formed. By about day 10, long slender forms are found in the bloodstream and lymphatics, and for the next several days their numbers increase exponentially. Soon thereafter, the organisms nearly disappear from the bloodstream, only to reappear later. The interval between waves of parasitemia may vary from 1–8 days, with clinical symptoms accompanying each bout of parasitemia. Each successive wave of organisms represents a new crop of parasites expressing a *VSG* not previously expressed in that host, and it is through this process of sequential antigenic variation that the parasites stay one step ahead of the host's specific antibody responses.

In response to infection, a marked early humoral antibody response, consisting predominantly of IgM, is seen regularly. These macroglobulins consist not only of antitrypanosomal antibodies directed against parasite surface antigens, but also include a variety of other antibodies such as heterophile and rheumatoid factor. As a result of polyclonal B-cell activation, there also are many antibodies produced to a wide variety of antigens, including brain-specific autoantibodies. In addition, antibodies directed against myelin basic protein, gangliosides, and cerebrosides have been found in experimental models. Circulating immune complexes have been reported regularly and these may be responsible for the glomerulonephritis often accompanying acute and chronic disease. Cell-mediated immunity also is important in this disease and nitric oxide may be important in the depression of T-cell responsiveness and generalized immunosuppression.

Lymphadenopathy usually involves the posterior cervical, submaxillary, supraclavicular, and mesenteric lymph nodes. The advanced disease, often called stage II disease, involves the CNS. The microscopic examination of lymphatic tissues usually reveals generalized hyperplasia with diffuse proliferation of lymphocytes. Initially, affected lymph nodes are markedly hemorrhagic and contain a large number of trypomastigotes; later, the nodes may become small and fibrotic. A progressive, chronic leptomeningitis develops in stage II disease. The brain becomes edematous and there is prominent perivascular cuffing by glial cells, lymphocytes, and plasma cells. Morula cells, reactive astrocytes, and hyperplasia of microglial cells all have been described in brain specimens from patients with CNS disease, and demyelination occurs in chronic cases. Organisms can be found in the brain tissue near vessels and also may be present in the cerebrospinal fluid. There is a striking lymphocytosis in the cerebrospinal fluid and most of these lymphocytes are B cells. Glomerulonephritis, myocarditis, pericardial effusion, pulmonary edema, and hypoplastic bone marrow with an associated anemia may develop in some patients. The pathogenesis of the neuropsychiatric manifestations is poorly understood. Experimental models suggest that a variety of factors may be responsible (e.g., deposition of immune complexes and aberrant levels of brain neurotransmitters, prostaglandins, and cytokines).

D. Clinical Manifestations

Clinical manifestations of infection with *T. brucei gambiense* and *T. b. rhodesiense* are similar, except that the rhodesiense disease usually runs a much more fulminant course. Untreated patients with the latter typically die in weeks to months, whereas persons infected with *T. b. gambiense* may live for years and have long periods without symptoms. In both forms, the trypanosomal chancre may be evident several days after the bite of an infected tsetse fly. Within a week or two, the lesion becomes a large, red, painful nodule that may reach 5–10 cm in diameter. This nodule, which is reported to be more common in people of European descent, subsides spontaneously in a few weeks. In rhodesiense trypanosomiasis, the incubation period of the systemic disease typically is 2–3 weeks, whereas with the gambiense infection, the first symptoms may be noted weeks or months after the acquisition of the infection. The early systemic disease is characterized by intermittent fevers, chills, headache, and generalized lymphadenopathy. In the gambiense disease, the nodes in the posterior cervical triangle may become enlarged (i.e., Winterbottom sign) and, when present, strongly suggest the diagnosis in a patient with exposure to tsetse flies. Delayed deep hyperesthesia may occur over the tibia and moderate hepatosplenomegaly may be noted. Anemia and thrombocytopenia are frequent as well. Intermittent fevers may last for months to years with the gambiense infection. In Europeans, a circinate erythematous rash or erythema multiforme is sometimes present.

Untreated patients eventually develop signs of central nervous system invasion. Severe headaches, loss of nocturnal sleep, and a feeling of impending doom are typical. Following this, there may be progressive mental deterioration, with patients becoming incapable of caring for themselves. Tremors, especially of the tongue, hands, or feet, as well as generalized or focal convulsions may occur. Almost any neurologic and psychiatric manifestation can be seen with progressive mental deterioration until patients finally lapse into coma and die of intercurrent infections.

E. Diagnosis

A definitive diagnosis of African trypanosomiasis is made by detecting trypanosomes, and this can be done by looking for parasites in fresh and stained specimens of blood, bone marrow, lymph node aspirates, or, in late disease, the cerebrospinal fluid. If parasites are not seen in these specimens obtained from a patient whose history and clinical findings suggest African trypanosomiasis, efforts should be made to concentrate the organisms in blood. This can be done most simply by using commercially available quantitative buffy-coat analysis tubes. The parasites are separated from the blood components by centrifugation in these acridine orange-coated tubes, and are easily seen under light microscopy because of the stain. Alternatively, buffy coat obtained by centrifugation of 10–15 ml of anticoagulated blood can be examined microscopically as a wet preparation and after Giemsa-staining. All of these specimens also can be inoculated into specialized liquid-culture medium. Finally, a highly sensitive method for detecting *T. b. rhodesiense* is the inoculation of small volumes of specimens obtained from the patient into rodents. Patent parasitemias usually develop within a week or two in animals injected with specimens from people infected with *T. b. rhodesiense*. Unfortunately, host specificity precludes the use of this approach for diagnosing *T. b. gambiense*.

Several serologic assays are available to aid in the diagnosis of sleeping sickness, but the variable accuracy of these tests mandate that treatment decisions still be based on detection of the parasite. These assays are useful, none the less, in epidemiologic surveys. Simple direct agglutination tests for trypanosomes performed on cards (CATT and TrypTect CIATT) are available commercially. A role for PCR-based assays for detecting African trypanosomes has not been defined.

In advanced untreated sleeping sickness, the IgM level in the cerebrospinal fluid often is elevated, but it has no relationship to the presence of trypanosomes in the cerebrospinal fluid. After successful treatment, the IgM level declines gradually, disappearing after approximately 1 year, and, thus, a persistently elevated level or an abrupt rise in the IgM months after treatment may indicate a relapse. IgM levels, however, should not be used as the sole method of diagnosis or prognosis.

F. Treatment

The chemotherapy of these diseases, both human and veterinary, has lagged remarkably behind that of other tropical diseases. The main chemotherapeutic agents for human trypanosomiasis remain pentamidine and suramin for the early-stage disease, and melarsoprol (Mel B, Arsobal) for the late-stage (CNS) disease. Difluoromethylornithine (DFMO, Ornidyl) is the only new useful addition to this list since the early 1950s. With the exception of suramin, resistance to the established agents is growing, and the toxicity of these drugs continues to be a problem.

Pentamidine is a water-soluble aromatic diamidine that has been in use since the 1930s. It is effective against early-stage *T. b. gambiense* infection, but is less effective against *T. b. rhodesiense* infection, and is ineffective against late-stage disease. African trypanosomes have a nucleoside (adenine/adenosine: P2) transporter that takes up pentamidine, resulting in the concentration of the agent at levels many times that in plasma. Many studies have focused on the mechanism of pentamidine action; however, none appears to conclusively define the target. It is known to bind to the minor groove of kinetoplast (mitochondrial) DNA, and to promote the cleavage of kinetoplast minicircle DNA, eventually leading to the development of dyskinetoplastic cells. Despite its effects on kinetoplast DNA, pentamidine has no effect on nuclear DNA, and dyskinetoplastic forms can persist in the bloodstream of mammals. Pentamidine was also found to be a reversible inhibitor of S-adenosylmethionine (AdoMet) decarboxylase, an enzyme in the polyamine biosynthetic pathway. Although Ki values were in the 200-mM range, we now know that this internal concentration is achievable via uptake through the P2 nucleoside–pentamidine transporter. Other targets studied previously in trypanosomes include the inhibition of glycolysis and lipid metabolism, as well as the effects on amino acid transport and ion exchange. The facts that pentamidine does not kill trypanosomes outright and bloodstream forms persist after treatment argue for a sustained effect more consistent with interference with the nucleic acid metabolism.

Diminazene aceturate (Berenil) is an aromatic diamidine developed by Hoechst as treatment for bovine trypanosomiasis; however, its apparent low incidence of adverse reactions and significant therapeutic activity have led some physicians in endemic countries to use it extensively. It is effective against early-stage *T. b. gambiense* and *T. b. rhodesiense*. Diminazene has also been used in combination with melarsoprol for the late-stage disease. Mechanistically, like pentamidine, diminazene has also been linked to kinetoplast DNA binding at the minor groove and cleavage of minicircle DNA. As with pentamidine, diminazene may also interfere with RNA editing and *trans*-splicing. Diminazene is also a more effective and noncompetitive inhibitor of AdoMet decarboxylase in trypanosomes, resulting in the reduction of spermidine content and elevating putrescine in the parasite. As with pentamidine, diminazene uptake occurs via the P2 nucleoside transporter, which allows significant accumulation from the external environment. Although diminazene has been used for many years on thousands of sleeping sickness patients, there is little published on its toxicity. This may in part be due to physicians who are unwilling to document human studies with an agent licensed for veterinary use. However, personal accounts of those using diminazene in humans indicate it is well tolerated.

Suramin is a sulfonated naphthylamine, which has been used successfully against early-stage sleeping sickness caused chiefly by *T. b. rhodesiense*. It was first used in 1922, developed from the closely related azo dyes, trypan red and trypan blue. Suramin has an extremely long half-life in humans, 44–54 days, the result of avid binding to serum proteins. Suramin binds to many plasma proteins including LDL, which trypanosomes avidly bind and endocytose as a result of specific membrane receptors. LDL is a prime source of sterols for bloodstream trypanosomes. The uptake of suramin as a protein complex results in an internal concentration of 100 mM. Suramin has been shown to inhibit all of the glycolytic enzymes in *T. b. brucei* in the range of 10–100 mM, which in most cases is several-fold lower than for the corresponding mammalian enzyme. This specificity for trypanosomal enzymes was attributed to higher (basic) isoelectric points for the parasite enzyme than the mammalian enzymes, allowing the negatively charged suramin to bind preferentially to the parasite enzymes. In practice, because most trypanosome glycolytic enzymes are contained in a membrane-bound cytosolic organelle, the glycosome, it is not likely that rapid massive binding occurs. This would rapidly induce lysis in bloodstream forms that depend on glycolysis as the sole energy-generating source. Rather, animals that are heavily infected with trypanosomes and given suramin show a slow decrease in parasite numbers, indicating that enzyme inhibition occurs slowly. Suramin may be affecting newly synthesized enzyme molecules in the cytosol before they are imported into the glycosome. Beyond the inhibition of glycolytic enzymes, suramin has also been found to affect thymidine kinase and dihydrofolate reductase. It is likely that suramin's action may be attributable to the inhibition of several of these enzymes.

Melarsoprol (Mel B, Arsobal) is an arsenical resulting from the efforts of Ernst Freidheim in the late 1940s. His initial compound, melarsen oxide, *p*-(4,6-diamino-*s*-triazinyl-2-*yl*)aminophenylarsenoxide was complexed with dimercaptopropanol (British Anti-Lewisite) to form a less-toxic complex, merlarsoprol. Until 1990, this was the only agent available for curing the late-stage (CNS) disease both of East African and West African origin. It is usually given as two to four series of three daily injections. It is insoluble in water and must be dissolved in propylene glycol, and it must be given intravenously. Toxicity is an important concern with melarsoprol. This takes the form of reactive arsenical-induced encephalopathy, which is often followed by pulmonary edema and death in more than half the cases within 48 hr. Although the mechanism of melarsoprol action has been extensively studied, it still remains unclear. Parasites exposed to low (1–10 mM) levels rapidly lyse. Because the bloodstream forms are intensely glycolytic, any interruption of glycolysis should produce this effect. Thus a series of reports has detailed melarsoprol inhibition of trypanosome pyruvate kinase (ki, 100 mM), phosphofructokinase (ki, < 1 mM), and fructose-2,6-bisphosphate (ki, 2 mM). It is likely that the rapid inhibition of fructose 2,6,bis-phosphate production is a key factor in halting glycolysis through down-regulation of pyruvate kinase. Other studies indicated that melarsoprol and melarsen oxide formed adducts with trypanothione (*N*1,*N*8-bisglutathionyl spermidine), a metabolite unique to trypanosomes and believed to be responsible for the redox balance of the cell and detoxification of peroxides. The melarsen–trypanothionine adduct (Mel T) inhibits trypanothione reductase, which has been attributed to the mode of action. However, melarsoprol and related arsenicals may also bind to other sulfhydryl-containing agents in the cell, including dihydrolipoate and the closely adjacent cysteine residues of many proteins. Similar to pentamidine and diminazene, melarsoprol uptake into African trypanosomes has been attributed to entry through the P2 purine nucleoside transporter; thus, significant levels can be concentrated in the cell from a low external (plasma) concentration. Although most laboratory-generated melarsoprol-resistant strains have lost or modified the P2 transporter, clinical isolates appear to have retained uptake capacity.

Eflornithine DFMO (DL-*a*-difluoromethylornithine, Ornidyl) is the most recently developed agent for late-stage *T. b. gambiense* sleeping sickness. After initial testing in model infections, DFMO was studied extensively in human trials. The standard treatment regimen resulting from the trials indicate DFMO is >95% active when given intravenously. DFMO cured children, adults, and patients with melarsoprol-refractory strains, and patients with late-stage disease. The short plasma half-life of DFMO necessitates constant dosing when given as an i.v. drip. The most frequent toxic reaction was reversible bone marrow suppression, which was alleviated upon reduction of the doses. The major drawbacks with respect to DFMO are its cost, the duration of treatment, and its availability. DFMO rapidly and irreversibly binds to the catalytic site (cysteine 360 in mouse ODC), inactivating it. In culture, it blocks division of bloodstream trypanosomes, but it is not trypanocidal. In laboratory infections, DFMO cures when administered continuously in the drinking water as a 2% solution. Within 48 hr of administration, DFMO reduces putrescine levels to zero, and reduces spermidine levels by ~75%. Trypanothione levels are also significantly reduced. As noted, DFMO is not trypanocidal and depends on a functional immune system to rid the host of nondividing forms. Morphologically, trypanosomes with multiple kinetoplasts and nuclei are common, as are forms resembling "stumpy" blood forms. DFMO is curative for laboratory infections of *T. b. brucei* and *T. b. gambiense,* but not to all strains of *T. b. rhodesiense*. The reason for this selectivity is not completely evident, although it is not due to uptake of DFMO, because it enters by passive diffusion, not transport. Levels of AdoMet are highly elevated in susceptible strains, but less so in refractory isolates. The elevated levels of AdoMet are due to an AdoMet synthase insensitive to its product. DFMO treatment leads to intracellular concentrations of ~5mM, an increase of ~50-fold over untreated parasites. Trypanosome ODC is missing the PEST sequence in both procyclic and bloodstream trypanosomes, and this appears to be the major reason for the stability of the trypanosome enzyme. The remainder of the ODC molecule has >60% sequence identity with the mammalian enzyme, including a cysteine 360 residue at the demonstrated DFMO-binding site for the mammalian enzyme. Beyond this, trypanosomes lack a polyamine oxidase, which in mammalian cells converts spermine to the biologically active spermidine. Trypanosomes are also limited in their ability to transport putrescine and spermidine.

G. Prophylaxis and Prevention

Trypanosomes causes complex public health and epizootic problems in many developing countries in Africa. Control programs concentrating on the eradication of vectors and drug treatment of infected people and animals have been in operation in some areas for decades. Considerable progress has been made in a number of regions, but the lack of agreement on the best appraoch to solving the problem of African trypanosomiasis, combined with a paucity of resources, stands in the way of effective control. Individuals can reduce their risk of becoming infected with trypanosomes by avoiding tsetse fly-infested areas, by wearing clothing that reduces the biting of the flies, and by using insect repellants. Chemoprophylaxis with suramin or pentamidine can be effective, but it is not clear which populations should use this as a preventive measure. No vaccine is available to prevent the transmission of the parasites.

See Also the Following Articles

Cellular Immunity • Diagnostic Microbiology • Intestinal Protozoan Infections in Humans • Zoonoses

Bibliography

Anonymous. (1998). Drugs for parasitic infections. *Med. Lett. Drugs Ther.* **40**, 1–12.

Donelson, J. E., Hill, K. L., and El-Sayed, N. M. A. (1998). Multiple mechanisms of immune evasion by African trypanosomes. *Mol. Biochem. Parasitol.* **91**, 51–66.

Hagar, J. M., and Rahimtoola, S. H. (1995). Chagas' heart disease. *Curr. Probl. Cardiol.* **20**, 825–924.

Jordan, A. M. (1986). "Trypanosomiasis Control and African Rural Development." Longmans, London.

Kirchhoff, L. V. (1993). American trypanosomiasis (Chagas' disease)—A tropical disease now in the United States. *N. Engl. J. Med.* **329**, 639–644.

Kirchhoff, L. V. (1996). American trypanosomiasis (Chagas' disease). *Gastroenterol. Clinics N. Am.* **25**, 517–533.

Moncayo, A. (1997). Progress toward elimination of transmission of Chagas disease in Latin America. *World Health Stat. Q.* **50**, 195–198.

Pentreath, V. W. (1995). Trypanosomiasis and the nervous system. *Trans. R. Soc. Trop. Med. Hyg.* **89**, 9–15.

Pepin, J., and Milord, F. (1994). The treatment of African trypanosomiasis. *Adv. Parasitol.* **33**, 1–47.

Pepin, J., and Milord, F. (1995). The treatment of human African trypanosomiasis. *Adv. Paristol.* **33**, 1–47.

Reed, S. G. (1998). Immunology of *Trypanosoma cruzi* infections. *Chem. Immunol.* **70**, 124–143.

Schmunis, G. A. (1991). *Trypanosoma cruzi,* the etiologic agent of Chagas' disease: Status in the blood supply in endemic and nonendemic countries. *Transfusion* **31**, 547–557.

Schmunis, G. A., Zicker, F., and Moncayo, A. Interruption of Chagas' disease transmission through vector elimination. *Lancet* **348**, 1171–1172.

Sternberg, J. M. (1998). Immunobiology of African trypanosomiasis. *Chem. Immunol.* **70**, 186–199.

Villanueva, M. S. (1993). Trypanosomiasis of the central nervous system. *Sem. Neurol.* **13**, 209–218.

Wang, C. C. (1995). Molecular mechanisms and therapeutic appraoches to the treatment of African trypanosomiasis. *Ann. Rev. Pharmacol. Toxicol.* **35**, 93–127.

Two-Component Systems

Alexander J. Ninfa and Mariette R. Atkinson
University of Michigan Medical School

I. Overview
II. Structure and Function Relationships in the Transmitter, Receiver, and Phosphotransfer Domains
III. Incorporation of T, R, and PT Domains in Signal Transduction Systems
IV. Factors Affecting the Phosphorylation State of the Response Regulator
V. Occurrence of T, R, and PT Domains

GLOSSARY

autophosphatase reaction The catalysis of the dephosphorylation of a phosphorylated residue found on itself by an enzyme.

autophosphorylation reaction The catalysis of the phosphorylation of a residue found on itself by an enzyme.

cross-regulation The convergence of parallel signal transduction pathways permitting the stimulation of one pathway to affect the output of the parallel pathways. The phosphorylation of a receiver domain by multiple, independently regulated, transmitter domains is an example of cross-regulation.

phosphotransfer reaction Reaction in which a phosphoryl group is transferred from a site on a protein to a different site on the same or a different protein, or to a small molecule.

response regulator A protein containing a receiver domain, which brings about the final step in a signal trandsuction pathway resulting in regulation of the target of the system. For example, many response regulators are transcription factors that activate or repress gene expression on phosphorylation of their receiver domain.

THE TWO-COMPONENT REGULATORY SYSTEMS are a related family of signal transduction systems that use the transfer of phosphoryl groups to control gene transcription and enzyme activity in response to various stimuli.

I. OVERVIEW

The key components of these systems are two distinct protein domains, referred to as the transmitter domain (T) and the receiver domain (R). Each of these domains has enzymatic activities (Fig. 1). As examples, we shall use the NRI-NRII two-component system regulating nitrogen assimilation in *Escherichia coli* and the KinA-Spo0F-Spo0B-Spo0A two-component system regulating sporulation in *Bacillus subtilis*. In these systems, NRII and KinA are proteins containing the T domain. The common feature of the T domain, which is always dimeric, is that these proteins bind ATP and phosphorylate themselves on a conserved histidine residue (Fig. 1A). The R domain catalyzes the transfer of phosphoryl groups from the phosphorylated histidine of the T domain to a highly conserved aspartate residue on itself (Fig. 1A). In the NRI-NRII system, the NRI protein contains an R domain, which transfers phosphoryl groups from NRII~P to itself, whereas in the sporulation system, the Spo0F protein consists of an R domain that catalyzes the transfer of phosphoryl groups from KinA~P to itself. From the phosphorylated R domain, phosphoryl groups may be subsequently transferred to water, as in the NRI-NRII system, or in some cases to another histidine residue within a domain referred to as the phosphotransfer (PT) domain (Fig. 1B). In the sporulation system, phos-

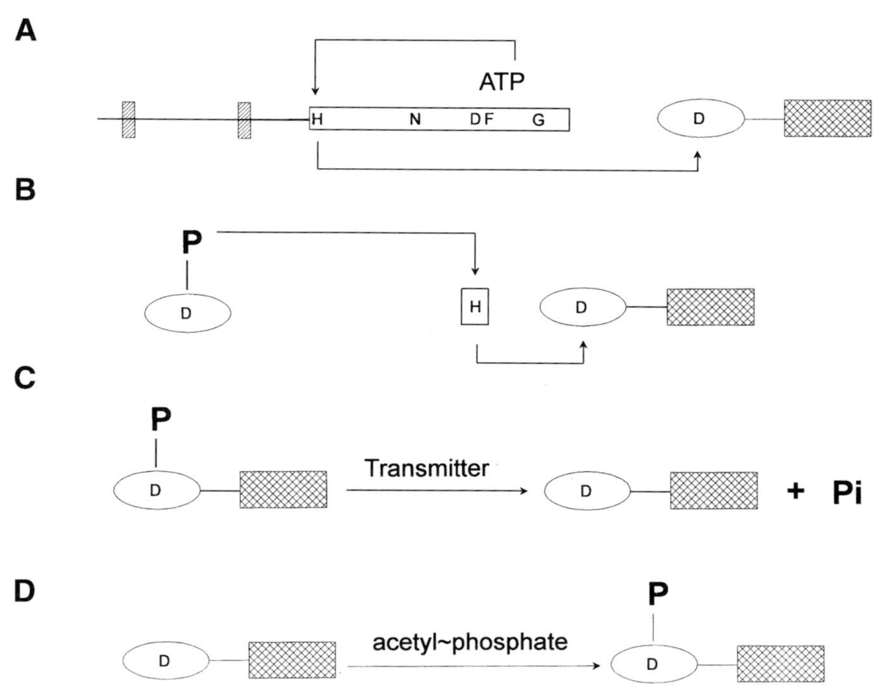

Fig. 1. Enzymatic activities of transmitter, receiver, and phosphotransfer domains. Transmitter domains are depicted as rectangles containing H, N, D, F, and G motifs. Receiver domains are depicted as ovals with the conserved aspartate represented with D. Phosphotransfer domains are depicted by small rectangles with the conserved histidine represented with H. (A) Flow of phosporyl groups to the response regulator in a typical two-component system. The transmitter protein is depicted as containing an N-terminal domain (thin line) with two transmembrane segments (depicted as hatched rectangles). The response regulator is depicted as containing an N-terminal receiver domain and a C-terminal transcriptional activation domain (crosshatched). (B) Flow of phosphoryl groups from a phosphorylated receiver domain to a phosphotransfer domain, to the receiver domain of a response regulator. (C) Dephosphorylation of a phosphorylated response regulator by a transmitter domain. (D) Phosphorylation of a response regulator receiver domain by acetyl phosphate. For details, see text.

phoryl groups are transferred from Spo0F~P to Spo0B. From the phosphorylated PT domain, phosphoryl groups may be transferred to another R domain (Fig. 1B). In the sporulation system, the Spo0A protein contains an R domain that catalyzes the transfer of phosphoryl groups from Spo0B to itself. Thus, in this latter class of systems, a three-site phosphorelay (T → R → PT) is used to deliver phosphoryl groups to an R domain.

For each two-component system, the phosphorylation of a key R domain is used to control the activities of various other protein domains, which may be either directly associated with the R domain or part of a distinct protein or macromolecular complex. The protein bearing this key R domain, which directly affects a cellular activity, has been referred to as the response regulator. For example, NRI and Spo0A, but not Spo0F, are considered to be response regulators. Phosphorylation of the R domain of the response regulator brings about a conformational change in the R domain that may be propagated to associated domains or alter the interactions of the R domain with other proteins. Information is transduced by mechanisms that alter the phosphorylation state of the R domain of the response regulator and by so doing alter the associated or interacting proteins.

Two-component systems form the core of a wide variety of different cellular signal-transduction systems, and not surprisingly the domains noted above are used in different ways in different systems. In many cases the T domain has another activity; it catalyzes the dephosphorylation of the phosphorylated R domain (Fig. 1C). This is true of the NRI-NRII system, where NRII catalyzes the dephosphorylation of NRI~P. In some signal transduction systems, multiple T and R domains, each with distinct functions, are employed. Finally, in almost all systems, additional regulatory proteins and/or protein domains are present. These permit the perception of stimuli, and they may affect the autokinase or phosphatase activities of the T domain in response to stimuli, or in cases where a phosphorelay is used, may affect the flow of phosphoryl groups through the relay or the phosphorylation state of the R domain of the response regulator at the end of the chain. In the *B. subtilis* sporulation system, distinct phosphatases act at each step to regulate the flow of phosphoryl groups to the response regulator Spo0A.

Yet another important feature of the R domain is that in many cases this domain may catalyze its own phosphorylation directly from small molecule phosphorylated metabolic intermediates, such as acetyl phosphate (Fig. 1D). Acetyl phosphate seems to be a key intracellular stress messenger; it is formed from acetyl~CoA and may signal a perturbation in one or more of the metabolic pathways using acetyl~CoA. When acetyl phosphate accumulates, the phosphorylation of the R domain of certain response regulators may participate in the adaptation to stress. Since a number of different response regulators are phosphorylated by acetyl phosphate, factors affecting this common source of phosphoryl groups may simultaneously affect many cellular processes. As will be discussed later in the article, the common use of acetyl phosphate may serve as a mechanism for communication between different two-component signal transduction systems.

II. STRUCTURE AND FUNCTION RELATIONSHIPS IN THE TRANSMITTER, RECEIVER, AND PHOSPHOTRANSFER DOMAINS

The distinguishing feature of the two-component signal transduction systems is the presence of the two protein domains corresponding to the T domain and the R domain. In addition, some two-component systems also contain the PT domain. The primary amino acid sequence of these domains is sufficiently conserved to permit the identification of homologous domains from the conceptual translation of DNA sequences.

A. Transmitter Domains

The T domain (Fig. 2A) is about 250 amino acids in length, and is less well conserved than the R domain. However, within this domain are several short segments that are highly conserved (Fig. 2A). Near the N terminus of the domain is a conserved segment known as the H-box, which contains the highly conserved histidine residue that is the site of autophosphorylation (see below). Approximately 100 amino acids away, another highly conserved segment known as the N-box is found, followed by the D-box, F-box, and G-box. Although more information will be required for a definitive functional assignment, some evidence suggests that the D-, F-, and G-boxes are involved in the binding of ATP and catalysis (see below). These segments of the T domain are shared with other classes of proteins that bind ATP, such as serine kinases (Fig. 3G), topoisomerases, and chaperones.

Essentially none of the conserved residues depicted in Fig. 2A are completely conserved. However, the D- and G- boxes are the best conserved parts of the T domain, and as noted above, these features are also found in other classes of proteins that bind ATP. The F-box is the most variable, and is not recognizable in several bona fide T domains. The H-box is invariant with the exception of the CheA and FrzE proteins, which are orthologous. CheA and FrzE are involved in the regulation of bacterial chemotaxis toward attractants and away from repellants. In these two proteins, an H-box is completely absent and is

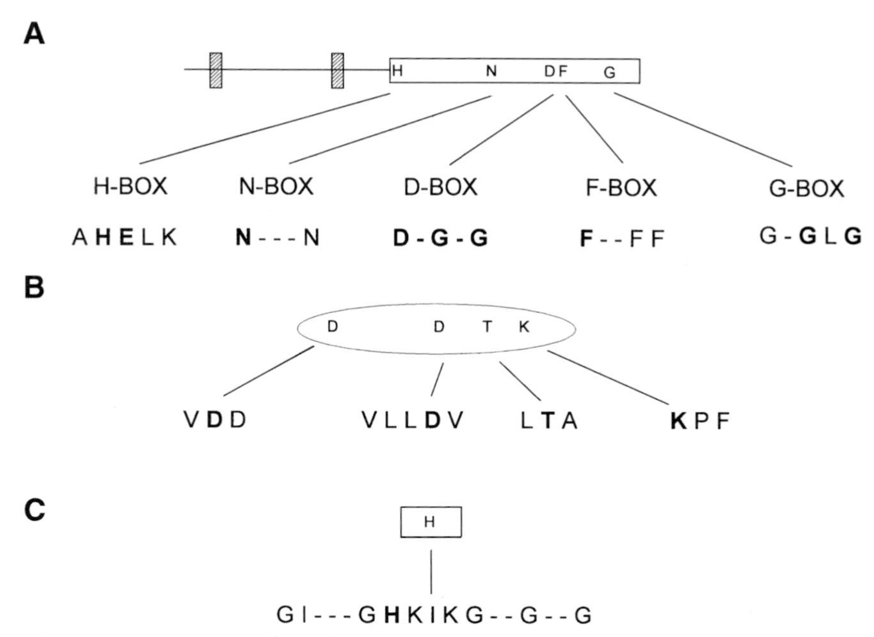

Fig. 2. Conserved motifs within transmitter, receiver, and phosphotransfer domains. Amino acids are depicted using the standard single-letter code; the most highly conserved residues are depicted in bold. (A) The transmitter domain. (B) The receiver domain. (C) The phosphotransfer domain.

replaced functionally by a PT domain found near the N-terminus of the protein (Fig. 3F).

To examine the role of these conserved regions, each of the highly conserved residues found within T domains was mutagenized by site-specific mutagenesis of the *E. coli glnL* gene encoding the paradigmatic transmitter, NRII (NtrB). NRII is involved in the regulation of nitrogen assimilation in response to signals of nitrogen and carbon status. As already noted, NRII is among those transmitter proteins that has both kinase and response regulator phosphatase activity. Alteration of the highly conserved residues affected the ability of NRII to become autophosphorylated, but did not diminish the phosphatase activity of NRII. Indeed, some of the mutations appeared to increase the phosphatase activity of NRII. These results indicate that the conserved sequences are directly involved in the kinase activity. Also, alteration of the conserved H-box histidine in NRII did not diminish the NRII phosphatase activity, indicating that this phosphatase activity is not the reversal of the phosphotransfer reaction by which the R domain becomes phosphorylated.

The site of NRII phosphorylation was directly determined by proteolysis and peptide mapping after labeling the protein with radioactive ^{32}P. This analysis directly confirmed that the H-box histidine is the site of NRII autophosphorylation. Similarly, the site of CheA autophosphorylation was mapped to the conserved histidine within the N-terminal PT domain of this protein.

The role of the G-box residues in NRII was also examined by characterization of altered proteins with conservative replacements in this region. This analysis suggested that the conserved glycine residues are important for the binding of ATP. Whereas the wild-type protein has very high affinity for ATP and can be readily cross-linked to ATP by UV irradiation, the proteins with G-box mutations were defective in autophosphorylation at low ATP concentrations, and could not be readily cross-linked to ATP by UV irradiation.

In addition to the conserved regions discussed so far, several recent studies suggest that some T domains may also share a conserved helical structural motif adjacent to the H-box, known as the X-box.

This helical segment forms part of the dimer interface in these proteins. In NRII this region is predicted to lie immediately downstream from the H-box, whereas in several other T domains, this region is predicted to map immediately upstream of the H-box. The H-box and X-box regions of transmitter proteins are similar to the region surrounding the site of phosphorylation in the Spo0B PT protein.

The autophosphorylation of the dimeric T domain occurs by a trans-intramolecular mechanism in which ATP bound to one subunit is used to phosphorylate the H-box of the other subunit within the dimer. This was demonstrated conclusively for the NRII T domain. Although this domain forms a stable dimer, the subunits may be reversibly dissociated and reassociated by treatment with low concentrations of urea, followed by dialysis. This has permitted the formation of heterodimers containing subunits with different mutations *in vitro*. Heterodimers containing one subunit lacking the histidine site of autophosphorylation and one subunit defective in the ATP-binding G-box are capable of autophosphorylation, because the intact ATP-binding portion of the H-box mutant can bring about the phosphorylation of the intact H-box site of the G-box mutant subunit.

Several lines of evidence have suggested that the T domain is actually two protein domains, not one. First, as already noted, a number of other ATP-binding proteins have the N-, D-, and G-boxes, but lack the H-box. Second, the CheA kinase has been cleaved into two pieces bearing the PT domain and the N-, D-, F-, and G-boxes. The latter of these peptides is able to bring about the phosphorylation of the disconnected PT domain. Finally, the T domain of the *E. coli* EnvZ protein (depicted in Fig. 3B) has been cleaved into two peptides by scission between the H-box and N-box. The peptide containing the N-, D-, F-, and G-boxes is able to phosphorylate the disconnected H-box. The EnvZ transmitter is involved in the regulation of porin gene expression in response to the osmolarity of the growth medium.

As already noted, some of the T domains catalyze not only the phosphorylation of the H-box histidine, but also the dephosphorylation of the phosphorylated response regulator R domain. Among this class of T domains are those found in the NRII and EnvZ proteins of *E. coli*. Mutations affecting the phosphatase activity of these proteins have been identified. Some of these mutations map in other domains of these proteins, and may affect signal perception (see below). However, some of the mutations mapped to the H-box (but not to the conserved histidine residue) and nearby flanking portions of the T-domain. One interpretation of these findings is that the non-phosphorylated form of the H-box must interact with the phosphorylated receiver domain to bring about its dephosphorylation. This conclusion is consistent with the observation, already noted, that mutations reducing the binding of ATP appear to increase the phosphatase activity of NRII.

How are stimuli perceived in two-component regulatory systems? One mechanism seems to involve control of the autophosphorylation and/or phosphatase activates of the T domain by associated sensory domains. The T domain is always found associated with other unrelated domains in proteins. The most common arrangement is shown in Fig. 3A and Fig. 3B; in these proteins the T domain is found at the C-terminal end of the protein, and the N-terminal end is involved in signal perception. In many cases, the N-terminal signal-perception domain contains two transmembrane segments, with the portion of the protein between these forming an extracellular domain. The NarX and NarQ transmitters of *E. coli* control respiratory gene expression in response to the presence of the alternative electron acceptors nitrate and nitrite. In these transmitters, the ligands nitrate and nitrite have been shown to bind to the extracellular sensory domain. This binding appears to reciprocally regulate the autophosphorylation and response regulator phosphatase activities of the transmitter domains. In some other *E. coli* systems, a transmembrane transmitter protein interacts not with small molecules, but with other membrane proteins that are apparently responsible for the detection of small molecule signals.

Another transmitter of interest is the FixL protein, which regulates nitrogen fixation in rhizobia. This protein (Fig. 3C) contains an extracellular sensory domain and an additional cytoplasmic sensory domain that is bound to heme and functions in oxygen sensation. Signals from the oxygen-sensing domain appear to reciprocally regulate the kinase and phosphatase activities of the transmitter domain.

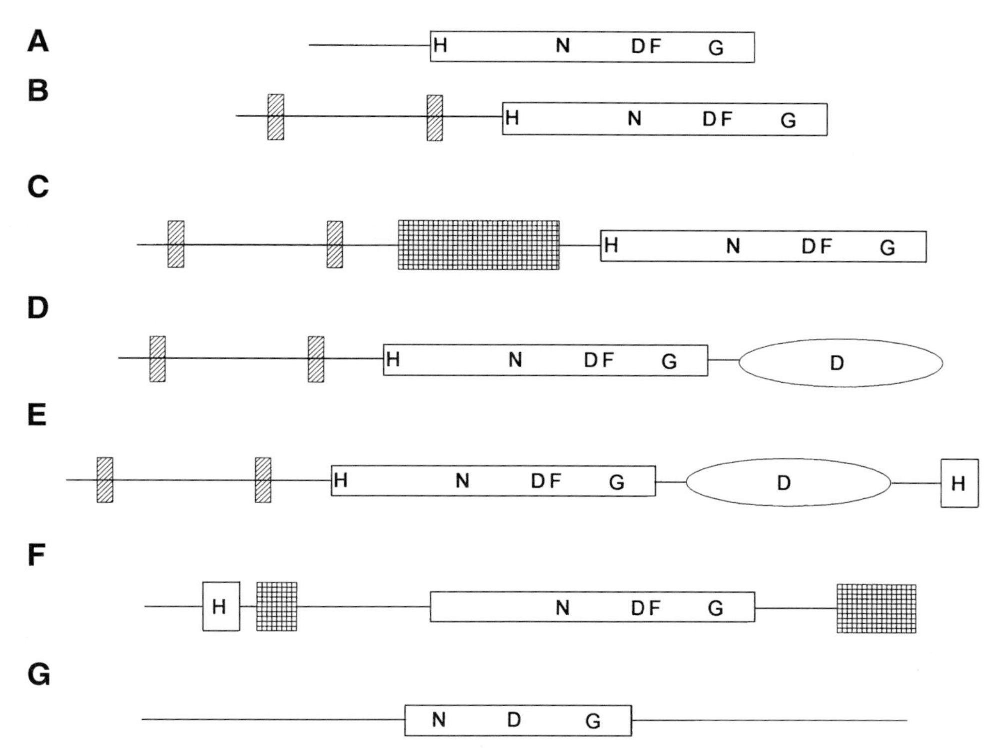

Fig. 3. Association of transmitter domains with other protein domains. Symbols are as in Figure 1, and as below. For C, an additional sensory domain is depicted with checkerboard hatching; for F, domains for the association with the receiver and transmembrane receptors are depicted with checkerboard hatching. For details, see text.

In the case of *E. coli* NRII (Fig. 3A), the transmitter protein is soluble, and information on the availability of carbon and nitrogen is transmitted by interaction of NRII with another signal-transduction protein called PII. When PII binds to NRII, the kinase activity of NRII is inhibited and the phosphatase activity of NRII is activated. Signals of carbon and nitrogen status regulate the ability of PII to bind to NRII.

The chemotaxis system of enteric bacteria is an example of the integration of a two-component regulatory system with a complex signal-perception mechanism. The CheA transmitter protein binds transmembrane receptor proteins that detect ligands, and this binding requires an adapter protein, CheW. CheA that is not coupled to the receptors has very low autophosphorylation activity, whereas CheA in the complex with receptors and CheW is very active. This activity is greatly reduced on interaction of the receptors with their ligands. Thus, in this system regulation appears to be mainly due to the regulation of the rate of transmitter autophosphorylation.

B. Receiver Domains

The receiver domain is about 125 amino acids in length, and the entire domain is generally highly conserved (Fig. 2). Within this domain there are four segments that are essentially invariant. Near the N-terminus, one or more acidic residues are found, which play a role in the chelation of Mg^{2+}. The phosphorylated aspartate residue occurs near position 55 in the R domain. A highly conserved threonine residue, which is occasionally replaced by serine, is found near position 87. Finally, a very highly conserved lysine residue is found at approximately position 109.

In some cases, the receiver domain alone comprises the whole protein (Fig. 4A). For two such

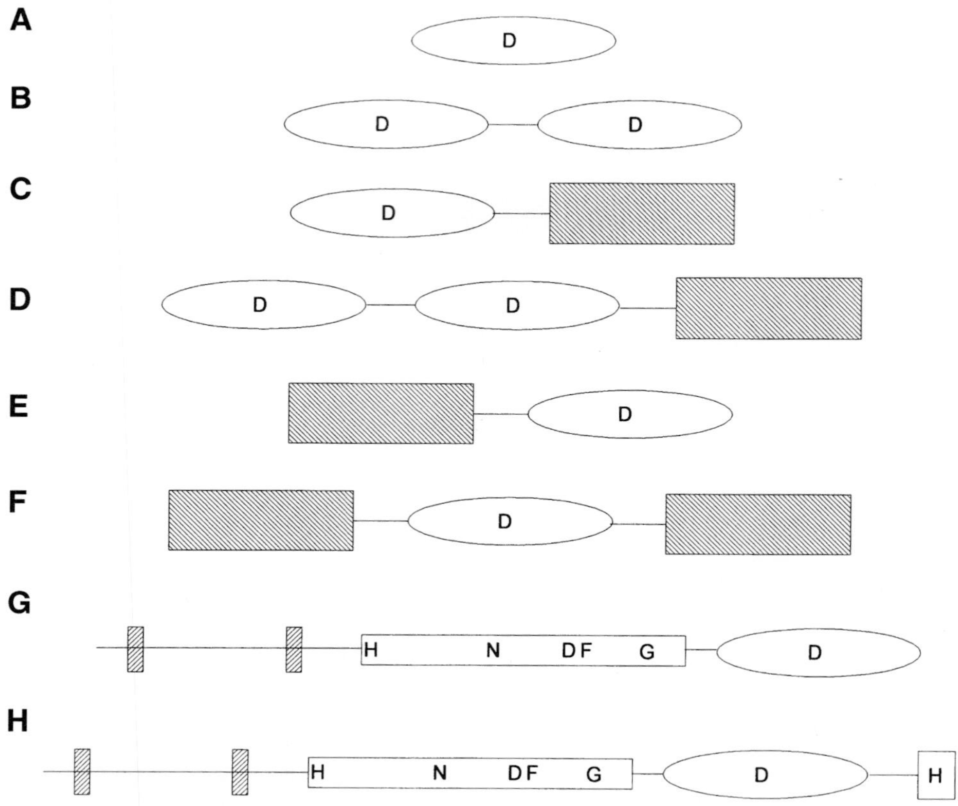

Fig. 4. Association of receiver domains with other protein domains. Symbols are as in Fig. 1. For details, see text.

cases, the Spo0F protein controlling sporulation in *Bacillus subtilis* and the CheY protein controlling chemotaxis in enteric gram-negative bacteria, the structure of the domain has been solved by nuclear magnetic resonance (NMR) or x-ray cyrstallographic methods. In the *E. coli* NRI (NtrC) protein, a receiver domain is found at the N terminus of the protein, and the associated domains are involved in transcriptional activation. The structure of the isolated N terminal receiver domain of NRI has been solved by NMR methods. Finally, the NarL transcription factor, like NRI, contains a receiver domain at its N terminus, and a transcriptional activation domain at its C terminus. The complete structure of NarL has been solved by x-ray diffraction methods. In all of these cases, the receiver domain consists of a five-stranded β-sheet surrounded by α-helices. At one end of the receiver domain, the highly conserved acidic residues form an acidic pocket. The aspartate at approximately position 55 is the site of phosphorylation, while the aspartate near position 10 plays a key role in chelation of a Mg^{2+} metal ion necessary for catalysis. Given the high sequence conservation among response regulator domains, they probably all share a similar structure.

The site of phosphorylation of the receiver domains of *E. coli* CheY and NRI were determined by reduction of the acyl phosphate with tritiated borohydride, followed by mapping of the site. These experiments confirmed the identity of the modified residue. The site of phosphorylation of a number of different receiver domains has been altered by site-specific mutagenesis, and either the resulting mutant proteins were not modified or the modification was vastly decreased. Among the latter group of proteins, the residual phosphorylation was observed to occur on a serine residue, or the modified protein had properties suggesting that a serine or threonine was modified. This modification may represent an artifact that is due to the absence of the usual site of phosphorylation. Interestingly, in some cases receiver domains have been mutated so that the aspartate at the site of phosphorylation is converted to glutamate. This mutation results in a protein with properties similar to the phosphorylated form of the receiver, suggesting that the larger glutamate residue at this position may mimic the effect of phosphorylating the natural aspartate residue.

A number of protein domains have been identified that are remarkably similar to the receiver domain but lack the aspartate residue that is phosphorylated in bona fide receivers. One of these proteins, the FlbD transcription factor of *Caulobacter crescentus*, is active when unphosphorylated and was not able to be phosphorylated using several different transmitter domains. Apparently, the pseudoreceiver domain of this protein and its relatives serves some other regulatory function that does not require phosphorylation.

Aspartyl phosphates in proteins are relatively unstable, especially at basic pH. At neutral pH, studies of model compounds have suggested an intrinsic stability of acyl phosphates of several hours in aqueous solutions. A similar stability is observed with most phosphorylated receiver proteins when they are denatured with sodium dodecyl sulfate (SDS). When proteins are not denatured, however, a wide range of stabilities is observed for the receiver domains, ranging from a few seconds to the predicted several hours. Thus, many of the phosphorylated receiver domains are considerably more unstable than predicted. Since denaturation of the phosphorylated proteins results in the expected phosphoryl-group stability, it is clear that in the native structure, the phosphorylated aspartate is rendered unstable by nearby structural features of the protein. One of the factors affecting the stability of the phosphorylated receiver is the conserved lysine found at position 109, since alteration of this residue has been observed in several cases to stabilize the phosphorylated form of the protein. It has been proposed that the receiver domain "catalyzes" its own dephosphorylation, and this activity has been referred to as the "autophosphatase" activity. The relationship between this autophosphatase activity and the phosphatase activities of transmitter domains is unclear at the present time. It has been proposed that the transmitter domains may not actually be phosphatases, but rather may be activators of the autophosphatase activity intrinsic to the receiver domain.

In addition to the receiver domain autophosphatase activity and phosphatase activity that may be associated with a transmitter domain, the phosphorylated receiver domains are frequently subject to dephosphorylation by additional, unrelated phosphatases. The mechanism of action of these phosphatases is currently not well understood. They may act by stimulating the intrinsic autophosphatase activity. Regulation of the appearance or activity of these unrelated phosphatases serves to permit additional stimuli to affect the phosphorylation state of the receiver module.

How does the phosphorylation of the receiver domain bring about signal transduction? The response regulator proteins contain receivers that regulate other protein domains, either in the same or different proteins. The phosphorylation of the receiver domain in these proteins apparently results in a conformational change that affects the target domains that are regulated. A typical response regulator organization is shown in Fig. 4C. In these proteins, an N-terminal receiver domain is associated with a C-terminal domain. Various types of C-terminal domains are found, such as transcriptional activation domains and domains with various enzyme activities. In the case of the Spo0A response regulator, which is a transcription factor regulating the initiation of sporulation in *Bacillus subtilis*, the unphosphorylated N-terminal receiver domain inhibits transcriptional activation activity. Phosphorylation of this domain results in an altered conformation in which the receiver no longer is able to inhibit the binding of DNA and the activation of transcription at the regulated promoter elements.

A similar mechanism is observed with the *E. coli* NRI (NtrC) response regulator protein, which is an enhancer-binding transcriptional activator that works in concert with a specialized form of RNA polymerase containing σ^{54} to activate genes in response to nitrogen starvation. Unphosphorylated NRI is a dimer of identical subunits, which is essentially unable to activate transcription. Phosphorylation of the receiver domain of NRI results in the formation of an oligomer, likely a tetramer or octamer of NRI subunits (i.e., a dimer or tetramer of dimers), that is able to activate transcription. Thus, the unphosphorylated receiver domain of NRI appears to act by preventing the oligomerization necessary for the formation of the active form of the protein.

The mechanism of signal transduction by the *E. coli* CheY chemotaxis regulator is apparently somewhat different. This protein is a response regulator that consists entirely of the receiver domain (Fig. 4A). CheY interacts with the switch proteins of the flagellar motor to bring about reversals or pauses in the rotation of the flagella. This in turn causes a jittery swimming motion known as "tumbling," which serves to randomize the direction of the bacteria. Phosphorylated CheY, but not unphosphorylated CheY, is able to interact with the motor switch proteins. Thus, activity of the motors is controlled by the extent of CheY phosphorylation. Both unphosphorylated CheY and CheY~P are monomeric, and the distinction made by the motor switch proteins must therefore be due to conformational differences between the two forms of CheY.

C. Phosphotransfer Domains

As the phosphotransfer domain was only recently discovered, much less is known about it. The domain is approximately 60 amino acids in length, and it contains a highly conserved histidine and a pattern of conserved glycine residues (Fig. 2C). The conserved histidine is apparently the site of phosphorylation, as deduced from the properties of the phosphorylated domain and the result of alteration of the histidine to a nonphosphorylatable residue.

As noted above, the CheA transmitter lacks an H-box and is instead directly phosphorylated within its N-terminal PT domain. This observation indicates that the PT domain may serve as a target for direct phosphorylation by the kinase domain of transmitters. However, it is clear from experiments with the ArcB transmitter of *E. coli* and several other transmitters that the PT domain usually requires phosphotransfer from an R domain to become phosphorylated.

Interestingly, the first protein observed to have the function of a PT domain is not related by homology to the PT domains found in other systems. This protein was the Spo0B protein of *B. subtilis,* which, as already noted, forms part of the phosphorelay that delivers phosphoryl groups to the receiver domain of the response regulator Spo0A. The sporulation phosphorelay consists of several transmitters (KinA, KinB, perhaps others) that phosphorylate the Spo0F receiver. Spo0F~P then transfers its phosphoryl groups to Spo0B, which in turn transfers its phosphoryl group to Spo0A. Currently, there are no known homologs of Spo0B in the GenBank database. However, since the function of Spo0B has been clearly demonstrated, it raises the possibility that there exist additional proteins with the functions of the PT domain, which are not related to either Spo0B or the PT domain by homology.

Recently, an *E. coli* protein has been identified that is a phosphatase specific for phosphorylated PT domains. This phosphatase, the SixA protein, contains a conserved arginine-histidine-glycine (RHG) motif found in many other phosphatases, and becomes transiently phosphorylated on the RHG histidine motif as part of the phosphatase catalytic mechanism.

III. INCORPORATION OF T, R, AND PT DOMAINS IN SIGNAL TRANSDUCTION SYSTEMS

The arrangements in which the T, R, and PT domains are found in signal transduction systems are summarized in Figs. 3 and 4. As already noted, the T domain is typically found associated with an N-terminal sensory domain (Fig. 3A,B) or multiple N-terminal sensory domains (Fig. 3C). In numerous cases, T and R domains are found in the same protein, termed hybrid transmitters (Fig. 3D). In yet other cases, T, R, and PT domains are contained within a single protein, which we call compound transmitters (Fig. 3E). The CheA and FrzE proteins represent a special class in that they lack the H-box and contain instead an N-terminal PT domain (Fig. 3F). Finally, as already noted, there are numerous cases of proteins that are not bona fide transmitters but contain the N-, D-, and G-boxes characteristic of the transmitter domain (Fig. 3G). This plethora of domain arrangements indicates the modular nature of the

T, R, and PT domains and suggests that additional arrangements are likely to be discovered.

Receiver domains may be found unassociated with other protein domains, as in the Spo0F and CheY proteins (Fig. 4A), or associated with another receiver domain (Fig. 4B). These proteins may serve as response regulators (CheY) or as part of phosphorelay systems. The typical response regulator contains a receiver domain associated with the domains it controls (Fig. 4C). Many different types of associated domains have been observed in this arrangement. The PleD protein of *C. crescentus* represents a variation in which two receiver domains are associated with a domain that is controlled (Fig. 4D). The receiver domain of response regulators is less often found downstream from associated domains under its control (Fig. 4E) or sandwiched between unrelated domains (Fig. 4F). Finally, as already noted, receiver domains may be found associated with T domains in hybrid transmitters (Fig. 4G) and with T and PT domains in compound transmitters (Fig. 4H). These proteins are almost certainly part of phosphorelay systems.

IV. FACTORS AFFECTING THE PHOSPHORYLATION STATE OF THE RESPONSE REGULATOR

A. Flow of Phosphoryl Groups

The flow of phosphoryl groups to the response regulator is depicted in Fig. 5 for some of the arrangements of T, R, and PT domains. In the typical bacterial two-component system, phosphoryl groups are transferred from ATP to the H-box histidine, followed by transfer to the aspartate residue of the R domain, from which they are transferred to water (Fig. 5A). In the various phosphorelay systems depicted in Fig. 5B–D, the flow of phosphoryl groups is from the H-box histidine to an R domain, then to the PT domain, and finally to the receiver domain of the response regulator. This basic theme seems to occur regardless of whether the various domains are on separate proteins (Fig. 5B,D) or are part of a compound transmitter (Fig. 5C).

B. Role of Acetyl Phosphate

As already noted, some receiver domains are efficiently phosphorylated by acetyl phosphate. In one case, the RssB response regulator of *E. coli* that controls the stability of σ^S, phosphorylation by acetyl phosphate seems to be the main mechanism for activating the response regulator (i.e., there is no cognate transmitter, it is a one-component system). In other cases, a transmitter and acetyl phosphate both contribute to response regulator phosphorylation. These systems usually have a transmitter protein with phosphatase activity toward the phosphorylated receiver. *In vivo,* this phosphatase activity is required to prevent the inappropriate activation of the system by acetyl phosphate. Indeed, these phosphatase activities have a role in regulating the intracellular concentration of acetyl phosphate, since the result of their activity is the conversion of acetyl phosphate to acetate and inorganic phosphate (P_i). Since multiple two-component systems are affected by acetyl phosphate, this serves as a means for efficient communication between regulatory systems.

Various roles have been proposed for acetyl phosphate activation of receiver domains. Phosphorylation of these domains by acetyl phosphate may provide a basal extent of phosphorylation that can be quickly increased above the regulatory threshold on activation of the transmitter kinase activity. Alternatively, acetyl phosphate may be required to achieve the full phosphorylation of the receiver domain, even when the transmitter kinase activity is fully stimulated.

C. Cross-Regulation

In numerous cases, multiple transmitter domains have been shown to bring about the phosphorylation of a receiver domain. When each receiver domain has an apparent "cognate" transmitter as well as the ability to be phosphorylated by other transmitters, this phenomenon is an example of cross-regulation. In every case where cross-regulation has been shown to occur, only one of the transmitter domains, the "cognate" transmitter, has the ability to bring about the dephosphorylation of the receiver domain. Cross-regulation permits the activation of receivers by stimuli affecting different transmitters, while the

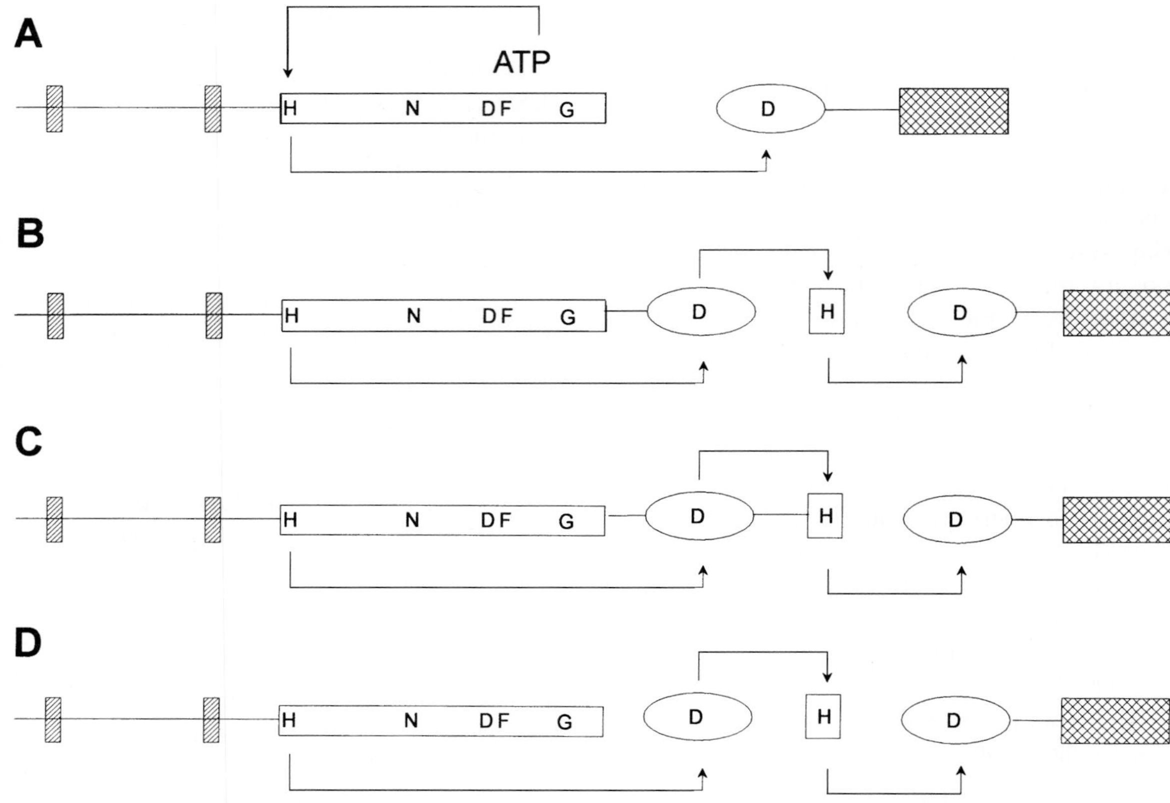

Fig. 5. Flow of phosporyl groups in a typical two-component system and in phosphorelay systems. Symbols are as in Fig. 1. (A) Flow of phosphoryl groups in a typical two-component system. (B, C, and D) Flow of phosphoryl groups in phosphorelay systems with differing domain associations. For details, see text.

phosphatase activity of the cognate transmitter may prevent inappropriate activation of the system. Cross-regulation may permit more rapid phosphorylation of the receiver under certain conditions than would be possible if only a single transmitter could phosphorylate the receiver.

D. Multiple, Independently Regulated Phosphatases

The obvious advantage of signal transduction phosphorelay systems is that each phosphorelay step may serve as a point for control of the system output. This has been shown to be true for the phosphorelay system regulating *B. subtilis* sporulation. The decision to initiate sporulation in this organism is a "consensus" decision affected by numerous stimuli, such as the availability of nutrients and the density of the bacterial culture. A multiplicity of phosphatases have been found, which affect the various steps of the phosphorelay. Each of these phosphatases imposes a requirement for the appropriate stimuli before the initiation of sporulation may proceed.

V. OCCURRENCE OF T, R, AND PT DOMAINS

The two-component systems constitute the most common type of signal transduction system found in bacteria, with over 100 different systems known. In many cases, a given system has been identified in numerous bacteria, such that about 300 examples of bacterial transmitter proteins are currently known. At the time this article was prepared (July 1998) new bacterial transmitter proteins were appearing in the

GenBank database at the rate of 15–20 per month, largely because of rapid progress in the sequencing of bacterial genomes. The biological functions of many of the newly discovered systems are not known.

In the course of preparing this article, a search was made of the completed *Escherichia coli* genome sequence for homologs of the paradigmatic transmitter protein, nitrogen regulator II (NRII or NtrB). Twenty-three clearly homologous sequences were identified, as well as a number of proteins that aligned to part of the conserved T domain. Since some transmitter proteins may have diverged from NRII sufficiently to remain unrecognized in the homology search (performed using advanced BLAST at the National Center for Biotechnology Information Web site: *http://www.ncbi.nlm.nih.gov/cgi-bin/BLAST*), this represents the minimal number of transmitter proteins in *E. coli*.

A somewhat higher number of receiver domains are known in bacteria, as might be expected since multiple receiver domains may be functionally linked to a single transmitter protein. An advanced BLAST search of GenBank using the paradigmatic *E. coli* CheY response regulator protein sequence as the query sequence revealed 506 sequence entries displaying high similarity; the vast majority of these were bacterial receiver proteins. A similar search of the *E. coli* genome sequence revealed 32 receiver domains highly homologous to CheY, as well as several proteins that shared homology to parts of CheY. Thus, *E. coli* has 32 receiver domains. At the time this article was prepared, the number of receiver domain sequences in GenBank was growing at the rate of about 20 entries per month.

Until recently, it was thought that the two-component systems were uniquely prokaryotic in distribution; however, it is now clear that this is not the case. The budding yeast *Saccharomyces cerevisiae* is known to contain a hybrid protein with both T and R domains, Sln1. In addition, *S. cerevisiae* contains a PT protein, Ypd1, and the response regulator SSK1, which work in concert with Sln1 to regulate gene expression in response to medium osmolarity, as well as another protein with a receiver domain, SKN7. During the preparation of this article, it was observed that ORF YLR006c (accession number e245767) from the yeast genome sequence is homologous to the receiver domain. The fission yeast *Schizosaccharomyces pombe* contains the receiver protein Mcs4 and two transmitter proteins, Mak1 and Mak2. Mcs4 is involved in cell cycle control. *Neurospora crassa* contains a hybrid transmitter, Nik1, and a second transmitter protein, Nik2. *Aspergillus nidulans* also contains a Nik1 protein similar to the *N. crassa* Nik1 protein. Finally, *Candida albicans* has been shown to contain orthologs of the Sln1 and Nik1 proteins. Thus, the yeasts contain a limited number of two-component systems. *Dictyostelium discoideum* contains at least three transmitter proteins, two of which are hybrid transmitters with associated R domains. In addition, *D. discoideum* contains a response regulator protein, RegA, which acts in concert with the DhkA transmitter protein. The *D. discoideum* two-component systems are involved in various aspects of sporulation and germination control.

The higher plant *Arabidopsis thaliana* contains three transmitter proteins, ETR1, ERS, and CKI1; of these the ETR1 and CKI1 proteins, regulating responses to ethylene and cytokines, respectively, are hybrids containing R domains as well. Tomato contains an ortholog of the ETR1 protein, NR (never-ripe). In addition, *A. thaliana* contains five additional proteins with receiver domains and four PT proteins, as deduced by PCR amplification using primers to the conserved regions of these domains. Thus, two-component systems are clearly present in plants.

In addition to the proteins listed above, a large number of proteins are present in GenBank that are clearly related to the transmitter proteins, yet lack the conserved H-box motif containing the phosphorylated histidine residue of bona fide transmitter proteins. These proteins are probably not histidine kinases, and in two cases, *B. subtilis* anti-sigma factor SpoIIAB protein and mitochondrial branched chain keto acid dehydrogenase kinase, the proteins have been shown to be serine kinases. The phytochromes of higher plants are also likely to be in this category. The ATP-binding site of the bona fide transmitter proteins is conserved in these proteins, as well as in a number of other proteins such as the HS90 family of chaperone proteins and type II DNA topoisomerases. Included among the latter group are many putative proteins found in human sequences from the high throughput genome sequencing projects, as well as the human disease gene associated with hereditary nonpolyposis colon cancer (accession U07418). During the course of preparing this article, many such putative proteins were identified within the human genome (e.g., accession numbers AF029367, AC002353, AC003035). However, bona fide transmitter proteins, with an intact H-box, seem to be lacking in the human sequences available so far (as of July 1998). A simple BLAST search of human genomic and expressed sequences using the NRII sequence as query readily identifies many such sequences, but further analysis revealed that they are the result of bacterial sequence contamination of the human sequence libraries. Similarly, the rice expressed sequence library within GenBank is extensively contaminated with bacterial sequences, and therefore is essentially useless for the identification of bona fide transmitter proteins in this organism.

See Also the Following Articles

PEP: Carbohydrate Phosphotransferase Systems • Transduction

Bibliography

Bourret, R. B., Borkavitch, K. A., and Simon, M. I. (1991). Signal transduction pathways involving protein phosphorylation in prokaryotes. *Annu. Rev. Biochem.* **60**, 401–441.

Hoch, J. A., and Silhavy, T. J. (eds.) (1995). "Two-Component Signal Transduction." American Society for Microbiology Press, Washington, D.C.

Loomis, W. F., Shaulsky, G., and Wang, N. (1997). Histidine kinases in signal transduction pathways of eukaryotes. *J. Cell Sci.* **110**, 1141–1145.

Ninfa, A. J. (1996). Regulation of gene expression by extracellular stimuli. *In* "*Escherichia coli* and *Salmonella typhimurium:* Cellular and Molecular Biology." (F. C. Neidhardt, (ed.), 2nd Ed., Chap. 39. American Society for Microbiology Press, Washington, D.C.

Parkinson, J. S. (1993). Signal transduction schemes of bacteria. *Cell* 73, 857–871.

Typhoid, Historical

William C. Summers
Yale University

I. History
II. Taxonomy and Classification
III. Prevention and Treatment
IV. Epidemiology

GLOSSARY

carrier An individual who harbors and may excrete virulent bacteria while remaining free of symptoms of the disease.

cyclogeny A theory, widely held in the 1920s and 1930s, that asserted that bacteria in cultures go through stages of a life cycle in which the bacterial cells may undergo changes in shape, size, staining, and biochemical properties.

endotoxin Molecules contained in the bacterial cell, but which when released by lysis of the cell can exert toxic effects on the host cells by a variety of biological mechanisms.

flagellum A flexible fiber-like structure on the surface of some bacteria. Flagellar movement is responsible for bacterial motility. The flagellum is driven by an internal cellular structure, the "motor," which is powered by a proton gradient across the cell membrane.

Peyer's patches Localized lymphoid tissue in the small intestine of mammals.

TYPHOID is the classic enteric fever that begins with malaise, anorexia, and headache, followed by high fever, abdominal tenderness, and distention. Rose spots appear on the skin. Patients may become delirious, the typhoid state for with the disease is named (typhus from the Greek τυφος, stupor), have shock from intestinal hemorrhage, and die. It often occurs in epidemics, especially under conditions of poor sanitation such as in military camps and after widespread natural disasters. Its symptoms and epidemiology overlap with that of typhus (camp fever), so typhoid has been confused with this rickettsial disease for much of its history. The name typhoid implies a ''typhus-like'' condition, and these two diseases have been clearly delineated only since the mid-nineteenth century. The typhoid organism infects only humans and chimpanzees, and the bacteria enters the body through the alimentary tract. In the sick individual, the organisms are usually found in the spleen, bone marrow, and lymphoid tissue associated with the gut (Peyer's patches), and almost always found in the gall bladder. Much of the pathology and the symptoms of typhoid are the result of endotoxin, which is released by the lysis of the bacteria. The typical incubation period is 7–14 days. Mortality in untreated cases is about 10%; 75% of these have intestinal hemorrhage or perforation. About 3% of clinically recovered patients still excrete the typhoid organism in the feces after 1 year and are designated as carriers.

I. HISTORY

In England, Huxham (1739) in his "Essay on Fevers" described the Plymouth epidemic of 1737, and he distinguished between putrid typhus (febris putrida) and slow nervous fever (febris nervosa lenta), or what is now recognized as typhoid. As disease nosology evolved, the concept of specific disease entities developed. By the nineteenth century, typhoid and typhus were recognized as diseases with distinct clinical features. The American physician and medical educator Nathan Smith provided the classic description of typhoid in 1824 in his essay "A Practical Essay on Typhous Fever," in which he recounted his observations of what was clearly typhoid in the

Connecticut River valley in New England. The great Parisian clinician Pierre-Charles-Alexandré Louis named the condition fièvre typhoïde in his major work on the disease in 1829. Schönlein (1839) in Germany distinguished Typhus exanthematicus (typhus) and Typhus abdominalis (typhoid), terms that became established in the German literature.

The English physician William Budd, who studied under Louis at the La Pitié hospital, published a detailed "natural history" of typhoid (what would now be called an epidemiological study) in 1856. Even before the rise of bacteriology, Budd concluded that typhoid was contagious and that in epidemic outbreaks the agent of contagion was spread by fecal contamination of water and milk. He also proposed that a convalescent patient could still be a source of contagion.

Carl Joseph Eberth described the typhoid organism in the tissues of patients in 1880, and Georg Gaffky isolated and grew this organism in 1884. Because the typhoid bacterium is relatively easy to grow in culture, soon many different bacterial isolates were found in typhoid-like cases, both in humans and in other species. For example, in 1888 Gaerntner isolated what is now designated *Salmonella enteritidis* from a patient who had eaten contaminated meat, and shortly thereafter Durham and de Noeble isolated *S. typhimurium* from gastroenteritis patients who had also eaten contaminated meat. The classification of these early isolates was based on the disease entity (*S. cholerasuis*), the geographic place of isolation (e.g., *S. montivideo* and *S. newport*), or the name of the investigator (*S. schottmülleri*). In 1896, Grünbaum and Widal independently observed that serum from infected or recovered patients could agglutinate dead typhoid bacteria and this test (Widal reaction) became the basis for a serological diagnosis of infection, as well as a tool to study *Salmonella* pathogenesis.

II. TAXONOMY AND CLASSIFICATION

The early nomenclature of the typhoid bacillus is as inconsistent and confusing as that of most bacteria. The typhoid organism was variously called *Bacillus typhosa, Bacterium typhosum, Eberthella tyhposa, Salmonella typhosa,* and most recently *Salmonella typhi.* The genus is named in honor of Daniel Elmer Salmon (1850–1914), the first director of the Bureau of Animal Industry of the U.S. Department of Agriculture, and one of the founders of bacteriology in America. Salmon investigated hog cholera and, along with his protege Theobald Smith, identified the hog cholera bacillus (*S. cholerasuis*), long believed to be the cause of hog cholera (later found to be caused by a filterable virus, however).

When it was found that individual isolates could be distinguished serologically from one another, the classification of the typhoid bacillus and its relatives was revised and based on the antigenic reactivities of a particular isolate. Some strains grew in more diffuse colonial form on surface cultures, and this growth form had come to be described as similar to the appearance of "breath" (perhaps the water condensate) on the agar surface. These strains shared antigenic properties termed H-antigens (German *Hauch,* breath). Later it was recognized that this diffuse colony morphology results from the motility of the individual organisms and that the H-antigens are associated with the flagella of these strains. The antigens of the strains without H-antigens were designated as O-antigens (German *ohne,* without) and are now recognized as somatic antigens, belonging to the body of the bacterium. A third category of serological reactions was believed to be related to the virulence of the organism in experimental infections and these antigens were designated as the Vi (virulence) antigens. The H-antigens were observed to vary with the culture conditions, and this two-state variation was interpreted in terms of the cyclogenic theories of the interwar period as characteristic of different growth phases of the culture, hence the concept "phase variation." Once the H-antigens were understood in terms of the flagellar antigens and structure, phase variation was reinterpreted as a problem in flagellar biosynthesis and regulation.

Fritz Kauffmann (1937) and Philip B. White (1926) studied the antigenic structures of the typhoid and typhoid-like bacteria in detail and the Kauffmann–White system of classification became standard. This system of classification of the genus *Salmonella* employs the H, O, and Vi antigenic specificities to identify a particular isolate. One species is recognized, sometimes referred to *enterica* (first proposed in 1987, but not officially adopted). The species includes over 2300 serotypes, many of which were formerly designated as different species (e.g., *S. cholerasuis, S. typhi, S. paratyphi* A, *S. typhimurium*). The variants are termed serovars (e.g., *S. enterica* serovar Typhimurium).

III. PREVENTION AND TREATMENT

Although vaccines for typhoid have been in use for a long time, it was only in the 1960s that definitive evidence of their effectiveness was obtained through studies on human volunteers. The best vaccine preparations give over 90% protection for at least 3 years against challenges with doses of typhoid similar to that expected in contaminated water supplies. Higher challenge doses, however, can overcome the vaccine-induced immunity and cause disease. Chloramphenicol (discovered in 1948) was the first effective drug to be used for typhoid, and it was followed by the newer penicillin derivatives as treatment for typhoid and other related *Salmonella* infections.

IV. EPIDEMIOLOGY

The epidemiology of typhoid has been of great interest, both for medical and sociological reasons. Based on epidemiological evidence, Robert Koch suggested that new cases of typhoid could come from convalescents long after the disease disappeared, and in support of this assertion Drigalski (1904) isolated typhoid bacteria from individuals in complete health. Observations such as these led to the concept of the healthy carrier of disease. Soon it was recognized that the gall bladder was a common site in which to find bacteria in these carriers, and ways were sought to eradicate the asymptomatic infections in order to eliminate the carrier state. Gall bladder removal (cholecystectomy) was a favored treatment. Less drastic approaches such as chemotherapy and various disinfectants were tried without success. Food contamination, usually by flies in summer, but sometimes by food handlers, led to the notion of the carrier as a danger to society. Quarantine and sometimes long-term detention of suspected or proven carriers were justified in the cause of protecting the public from contamination. Probably the most notorious and egregious case was that of Mary Mallon, an Irish immigrant domestic worker in New York, who seemed resistant to treatment and who was deemed uncooperative by the authorities. As "Typhoid Mary" she was *de facto* incarcerated for many years.

Typhoid and its agent played another role in the American legal system when the city of Chicago constructed a drainage canal to help channel Chicago sewage from the Chicago river into the Mississippi River watershed. The city of Saint Louis, Missouri, downstream was the potential recipient of the diverted sewage. Worried about typhoid outbreaks from Chicago, St. Louis went to court to block the Chicago plan. This case was the first in U.S. history in which the new science of bacteriology was accepted as expert testimony in court.

In circumstances of social disruption, such as wars, civil unrest, famine, or natural disasters that affect the water and sewage systems, typhoid is still a major health problem. However, immunization, antibiotics, and sanitary measures have reduced the U.S. mortality from typhoid from 26/100,000 in the period 1906–1910, to about 500 in the entire country in 1967, to a few sporadic cases.

See Also the Following Articles

Cholera • Plague • Typhus Fever and Other Rickettsial Diseases • Vaccines, Bacterial

Bibliography

Garrison, F. H. (1929). "An Introduction to the History of Medicine," 4th ed. Saunders, Phildelphia, PA.

Leavitt, J. W. (1996). "Typhoid Mary: Captive to the Public's Health." Beacon Press, Boston, MA.

Morgan, H. R. (1965). The enteric bacteria. *In* "Bacterial and Mycotic Infections of Man" (R. J. Dubos and J. G. Hirsch, eds.), 4th ed., pp. 610–648. Lippincott, Philadelphia, PA.

Topley, W. W. C., and Wilson, G. S. (1929). "The Principles of Bacteriology and Immunity." William Wood, New York.

Typhus Fevers and Other Rickettsial Diseases

Theodore E. Woodward

University of Maryland School of Medicine and Baltimore Veterans Administration Medical Center

I. History of the Rickettsial Diseases
II. Pathogenesis, Transmission, and Pathophysiology
III. Differential Diagnosis
IV. Laboratory Findings
V. Treatment
VI. Specific Rickettsial Diseases

GLOSSARY

animal reservoir A source in nature where an infectious agent can be maintained in enzootic or epizootic form for occasional or usual transmission to human beings by a vector.

arthropod A group of invertebrate animals including insects, mites, spiders, and ticks, relevant as carriers of infectious diseases between humans or between other animals and humans.

epidemiologist A person who studies the relationships of various factors that determine the frequency, causes, and distribution of diseases in a human community.

eschar A dark dry area on the skin where the skin is devitalized, as in a burn. A small area of tissue destruction surrounding a bite of an insect vector; a scab.

fulminant Describing a sudden severe condition occurring with great intensity.

immunofluorescence A technique for staining organisms or cells for microscopic examination in which a specific antibody molecule is chemically linked to a fluorescent dye to render the sites of antibody binding to cell or microbe structures visible when illuminated with light of the appropriate wavelength (direct immunofluorescence), or in which the binding of the specific antibody is detected by the reaction of a fluorescent-linked antibody directed against the first antibody molecule (indirect immunofluorescence). This assay can also be used with known target organisms to detect the presence or absence of specific antibodies in, for example, blood serum.

macular rash A skin discoloration characterized by circular or oval spots that are not raised above the surface of the skin. They may fade if mild pressure is applied, indicating their color is related to flowing capillary-blood supply. They may be fixed and persist with mild pressure, indicating that the capillary blood supply has congested with some clotting or vascular obstruction.

papular rash A skin discoloration characterized by circular or oval spots that are raised above the surface of the skin. A maculopapular rash is slightly raised.

Rickettsia Small, gram-negative bacteria characterized by obligate intracellular growth. The various species are separated into a typhus group, a spotted fever group, a tsutsugamushi (scrub typhus) group, and a miscellaneous group.

serologic Pertaining to the study of antigen–antibody reaction in blood serum and other tissue fluids.

vector An animal, often an arthropod, that carries an infectious agent from one host to another. Sometimes the infection is carried passively, but sometimes the vector is actually infected with the agent, which may undergo a complex life cycle in the vector organism distinct from that in the human or animal alternate host.

MOST AGENTS OF THE ORDER RICKETTSIALES are obligate intracellular pleomorphic coccobacilli. They are true bacteria with metabolic enzymes and cell walls; they use oxygen and are susceptible to antibiotics. Living cells are required for viability. Most *Rickettsia* that infect humans are maintained naturally by cycles involving animal reservoirs and an insect vector (usually arthropods). *Rickettsia* are localized in certain geographic areas; hence, knowledge of the habitat of the patient or recent travel information often helps in diagnosis.

I. HISTORY OF THE RICKETTSIAL DISEASES

While practicing medicine in Philadelphia in 1836, William Gerhard was the first to distinguish epidemic louse-borne typhus from typhoid fever, with which it was often confused. He described the differences of the febrile course, rapidity of development of the body rash, and absence of ulceration of the intestine. Epidemic typhus or European typhus, as it was traditionally known, was the foremost cause of human suffering and death. It ravaged Russia, east Poland, and the Balkan countries; from 1915–1922, 30 million persons were infected, with 3 million deaths in Russia and Poland alone.

The pioneer Howard T. Ricketts, working in Hamilton, Montana from 1906–1909, confirmed that ticks transmitted the agent of Rocky Mountain spotted fever to humans and established an animal model in rabbits and guinea pigs. The organism that he discovered was named *Rickettsia rickettsii.* Later in 1909, Charles Nicolle, while working in Tunisia, identified the body louse as the vector of typhus and reproduced the infection in primates using blood from patients. It was not until 1914 and 1916 that Von Prowazek and DaRocha Lima, respectively, stained and identified small bacteria-like microorganisms in the tissues of lice taken from patients.

During the years 1896–1910, Nathan Brill studied 221 febrile patients in New York City. Although many physicians regarded this illness as typhoid or pseudo-typhoid, Brill noted its resemblance to mild typhus fever in the lack of positive blood cultures, absence of serum agglutinins (Widal test), and the sporadic occurrence of these infections. He referred to it as a disease of unknown origin. For several decades, this mild type of typhus, which occurred in the southern United States and elsewhere, was called Brill's disease. It was not until 1934 that Hans Zinsser of Harvard, through his clinical and epidemiological observations, suggested that this disease was really a recurrence of typhus fever years after a prior initial attack in the Old World. He postulated that the illness occurred during a period of waning immunity. Snyder and Murray of the Harvard School of Public Health proved this hypothesis by the transmission of the causative rickettsia in animals from patients with Brill's disease. Lice were used in their transmission studies. It was also shown that Brill's disease patients showed an elevation of specific antibodies to epidemic typhus. This type of sporadic illness is now called Brill-Zinsser disease (see Table I).

A. Identification of Endemic (Murine) Typhus

In the United States, particularly in the southeastern areas, a typhus-like illness usually called Brill's disease, occurred, often in small outbreaks. Kenneth Maxcy, a wise epidemiologist, evaluated these minor outbreaks in the South in 1926, particularly the Carolinas, Virginia, Georgia, and Alabama. He noted that young adults were frequently affected, particularly those who worked in the vicinity of food depots and grainaries. The illnesses were short, usually nonfatal, except in the elderly, and were associated with about 2 weeks of fever, headache, and a rash. Blood from patients infected guinea pigs, which were later shown to be immune to epidemic typhus rickettsia. Based on his epidemiologic detective work, Maxcy surmised that these cases had a rodent reservoir and the agent was transmitted to humans probably by ticks or fleas. Dyer and his associates from the Public Health Service in Washington confirmed Maxcy's hypothesis in Baltimore. They isolated rickettsia from the brains of rats, as well as from their fleas, taken from areas where a few typhus cases had occurred. In 1930, Pincoffs and Krause of the University of Maryland had identified several cases of typhus in Baltimore, which alerted Dyer and his associates of this occurrence. Rats were trapped at the Lafayette Market, located on Pennsylvania Avenue, and the vicinity. Thus, in 1930 the disease became known as flea-borne endemic (murine) typhus, distinct from epidemic typhus or Brill's disease. The causative agent was later shown by Mooser to be a rickettsia, now known as *R. mooseri* or *R. typhi.*

II. PATHOGENESIS, TRANSMISSION, AND PATHOPHYSIOLOGY

Most agents of the order Rickettsiales are obligate, intracellular, pleomorphic coccobacilli. They are true

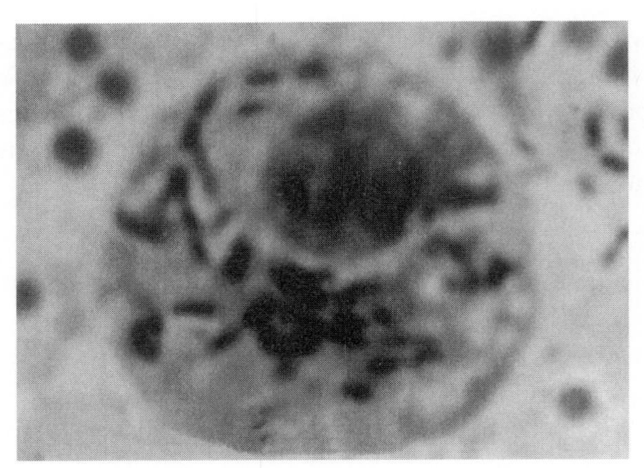

Fig. 1. Cell with intracellular *Rickettsia.* See color insert.

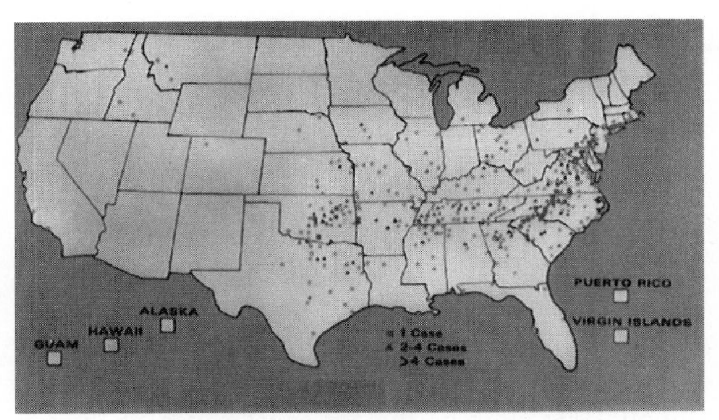

Fig. 2. Distribution of Rocky Mountain spotted fever in the United States around 1980. See color insert.

bacteria with metabolic enzymes and cell walls; they use oxygen and are susceptible to antibiotics. Living cells are required for viability (Fig. 1). Most *Rickettsia* that infect humans are maintained naturally by cycles involving animal reservoirs and an insect vector (usually arthropods). *Rickettsia* are localized in certain geographic areas; hence, knowledge of the habitat of the patient or recent travel information often helps in diagnosis (Fig. 2).

Many *Rickettsia* multiply at the arthropod attachment site and produce a local lesion (eschar) after penetrating the skin or mucus membranes. The *Rickettsia* of the contaminated feces of the arthropod enter and multiply in endothelial cells of small blood vessels. They cause a widespread blood vessel inflammation, consisting of swelling of the cells lining the vessels, and blood clotting. This type of blood vessel inflammation results in rash toxic manifestations, central nervous system signs, and gangrene of the skin and tissues. A short incubation period generally indicates a more serious infection.

Similar features typify the pathophysiological abnormalities of patients severely ill with the typhus fevers and Rocky Mountain Spotted Fever (RMSF). Peripheral vascular collapse during the later stages of illness often causes death in fulminating cases, particularly in epidemic typhus, scrub typhus, and RMSF. As inflammation and clotting develop in the small blood vessels, oxygen depletion occurs in these areas, resulting in cell death (necrosis) and increased capillary permeability. At this stage, there may be loss of fluid, electrolytes, proteins, and red blood cells into the extravascular space. This, in turn, leads to a decrease in the blood volume, expansion of the extracellular tissues, and swelling. Electrocardiograms may indicate signs of myocardial inflammation, swelling, and oxygen deprivation in the heart muscle. Liver function is impaired, and the high blood levels of ammonia in such severely ill patients results from failure of the liver to properly metabolize nitrogen waste products. Clinical manifestations resulting from the physiological abnormalities are low or absent urine output, high ammonia levels, anemia, low serum proteins, low blood sodium, tissue swelling, and coma. At these later stages, intravenous fluids should be restricted, in contrast to the earlier phase, when fluids can be given carefully to maintain blood volume. In severely ill patients with typhus and RMSF, with diffuse hemorrhagic and tissue lesions, and impaired blood clotting are manifested by platelet deficiency, low blood fibrinogen, and other coagulation abnormalities. These serious abnormali-

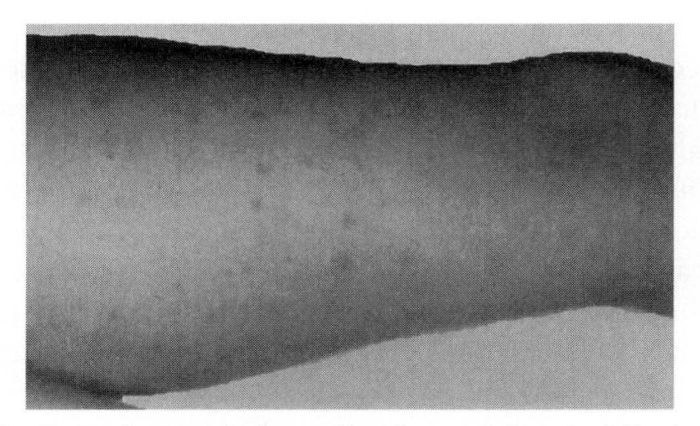

Fig. 3. Rocky Mountain spotted fever pink rash of the leg, day 4.

TABLE I
Rickettsial Diseases

	Agent	Geographic distribution	Vector	Mammal reservoir	Serologic reactions[a]		
					CF	MI	IFA
Typhus Group							
Epidemic	*R. prowazekii*	Worldwide	Body louse	Humans, squirrels	+	+	+
Brill-Zinsser disease (recurrent)	*R. prowazekii*	Worldwide	Recurrent, years after original attack of epidemic typhus	?	+(7s)	+	+
Endemic (murine)	*R. typhi(mooseri)*	Worldwide	Flea	Small rodents	+	+	+
Scrub	*R. tsutsugamushi*	Asia, Australia	Trombiculid	Wild rodents	+	+	+
Spotted-Fever Group							
Rocky Mountain spotted fever	*R. rickettsii*	Western Hemisphere	Ticks	Wild rodents, dogs	+	+	+
Tick typhus, eastern hemisphere	*R. conorii*	Africa, Europe, Middle East, Asia, India	Ticks	Wild rodents, dogs	+	+	+
Queensland tick typhus	*R. australis*	Australia	Ticks	Marsupials, wild rodents	+	+	+
North Asian tick-borne Rickettsiosis	*R. siberica*	Siberia, Mongolia	Ticks	Wild rodents	+	+	+
Rickettsial pox	*R. akari*	United States, Russia, Africa	Mite	House mouse, other rodents	+	+	+
Other Rickettsial Diseases							
Q-fever	*R. burnetti*	Worldwide	Tick, inhalation of dried infected material	Small mammals, cattle, sheep, goats	+	+	+

[a] CF, complement fixation; MI, microscopic agglutination; IFA, immunofluorescent antibody.

TABLE II
Salient Epidemiological, Clinical, Diagnostic, and Therapeutic Features of Typhus Fevers and RMSF

Clinical features	Epidemiological features		Tests	Diagnosis	Therapy
	Content	Season			
Usually high and continuous fever associated with headache and prostration with a macular rash: pink, discrete, macules beginning on second to sixth day of illness, later becomes maculopapular, petechial	Louse infestation in endemic area of typhus or contact with flying squirrels(US)	Any season (usually winter)	Serological tests (CF.IFA)	Epidemic typhus	Begin specific treatment
	Rat or flea contact	Any season	Serological tests (CF.IFA)	Murine typhus	Begin specific treatment
	Mite (chigger) bite or exposure in known endemic area (particularly Asia, southwest Pacific area)	Any season	Serological tests (*Proteus* Ox K, CF.IFA)	Scrub (tropical typhus, Tsutsugamushi disease)	Begin specific treatment
	Tick bite or tick contact in wooded-bushy setting in known endemic areas	Usually April to Sept.	Serological tests (CF.IFA)	Most likely RMSP	Specific treatment immediately

ties are absent or minimal in mild cases or in those given specific antibiotic treatment early.

III. DIFFERENTIAL DIAGNOSIS

The early stages of many serious infectious diseases are similar and associated with fever, headache, muscular pains, and toxic substances in the blood (toxemia). During the early phases of infection before a characteristic rash has appeared, the differentiation of the various acute infections is difficult. Hence, epidemiologic considerations such as the geographic area and habitat of the patient involved, season of the year, and daily activities of the ill person can be helpful.

Epidemic typhus is more prevalent during the colder months when louse infestation is more likely. In eastern areas of the United States, flying squirrels serve as a reservoir of infection, with transmission to humans by their ectoparasites. Generally, the reservoir of *R. prowazekii* infection (epidemic typhus) is in humans maintained by a cycle involving human–louse–human. In these same areas, recurrent typhus (Brill-Zinsser disease) occurs years after a primary infection. Typically, the rash of classic typhus occurs initially in the axillary folds and on the trunk; it later extends peripherally, rarely involving the soles, palms, and face. A pink flat rash appears on the fifth febrile day, or occasionally earlier, and subsequently becomes small bloody spots (petechial), and then fixed, confluent, and bruise-like. The eyes become bloodshot and the tongue reddened, with headache, malaise, and prostration pronounced.

Epidemic typhus is capable of causing all of the pronounced clinical, physiological, and anatomical alterations that occur in RMSF and seriously ill patients with scrub typhus; these include low blood pressure, peripheral vascular failure, pallor, skin necrosis, gangrene of the digits, renal failure with elevated ammonia, and neurological manifestations.

Murine typhus is a milder disease. The toxic signs are less severe; the rash is nonbloody, nonconfluent, and less extensive. It may be slightly raised. Contact of the patient with rats and their fleas in crowded areas, particularly in port cities, are helpful clues.

Scrub typhus fever is prevalent geographically in Asia, China, Southeast Asia, Malaysia, Indonesia, the Philippines, and Japan. In northern Thailand, scrub typhus is said to account for about 30% of patients with fever of unknown origin (FUO). Rubber plantations and wooded scrub areas are favorite sites for infected rodents and their mite vectors. Headache, sensitivity to light, bloodshot eyes, malaise, prostration, and muscle aches are all typical. A rash appears on about the third to fifth days, first pink and flat and later becoming bloody, diffuse, and then bruise-like. A primary lesion or eschar is often present, which is a quite helpful clue, although a small primary lesion may occur in RMSF and is common in Mediterranean spotted fever.

The history of a tick bite in a person who resides or walks in tick-infested endemic areas of RMSF is important diagnostic information. Illness usually occurs in the spring and summer, when ticks are active. The rash of RMSF occurs from the third to fifth days, particularly as pink flat lesions on the palms, soles, forearms, and legs (Fig. 3). A warm compress will accentuate the rash at this stage, which helps. The rash later spreads to the buttocks, trunk, and face. It then becomes bloody, purplish, and confluent, particularly over pressure areas such as the sacrum; these areas may become necrotic and ulcerative.

The rickettsial diseases such as endemic and mild epidemic typhus require differential serologic tests for confirmation. Classic typhus, RMSF, and scrub typhus may be clearly distinguished using serologic tests. Gene probes for specific antigens are diagnostic. Agent identification may be made with immune fluorescence techniques, using skin biopsy specimens. These clearly distinguish all of the rickettsial diseases (see Table II).

IV. LABORATORY FINDINGS

A. Routine Tests

In the severe rickettsial diseases, such as the typhus fevers and RMSF, routine laboratory procedures are not specifically diagnostic and merely reflect the severity of physiological and chemical abnormalities.

The white blood cell count is usually normal (6000 to 10,000 cu/mm). Low white cell counts occur occasionally and elevated white cell counts often signify complications, such as superimposed infections or widespread vascular inflammation. Anemia occurs in patients with widespread hemorrhagic lesions. There is usually elevated ammonia and low values of sodium, chloride, and serum albumin caused by widespread hemorrhagic necrosis, kidney involvement, and increased capillary permeability. These blood-plasma elements diffuse into the extravascular space. Appropriate laboratory tests are used to monitor these changes. Disseminated intravascular clotting with significant depression of platelets, fibrinogen, and associated coagulation factors deserves careful monitoring. There is no convincing evidence of the efficacy of heparin treatment to prevent clotting.

B. Specific Laboratory Procedures

1. Isolation and Identification of Rickettsia

The isolation of causative rickettsia in animals or tissue culture from blood or tissues is time-consuming, expensive, and hazardous to laboratory personnel. Such procedures are now useful for research purposes. Techniques using immunoflouresence are available and are suitable for indentifying the various types of rickettsia. They may be identified specifically in skin tissues taken from the macular lesions or from other tissues. Each *Rickettsia* has an identifiable morphology and staining property. Also, the immunofluorescent technique effectively identifies *Rickettsia* in formaldehyde-fixed paraffin-embedded tissues. Gene probes are now available for the specific identification of *Rickettsia* in blood, other tissues, or the insect vectors.

2. Specific Serologic Tests

The demonstration of specific antibodies in serum specimens is now the standard diagnostic procedure and includes the indirect fluoresent antibody and micro-immunofluorescent tests. These procedures use whole-tissue culture-propagated *Rickettsia* and are very sensitive. To be useful, they require two and preferably three serum samples taken early in the illness, during the first week, and later samples in the second, third, and even fourth weeks to demonstrate a rise in specific antibody titer during convalescence. Very early antibiotic treatment may delay the development of specific antibodies. Nevertheless, specific antibiotic therapy must be initiated early in the illness based on clinical suspicion rather than waiting for the results of laboratory values. The latex agglutination test using purified rickettsial antigens coated on latex beads is simple, rapidly performed, sensitive, and specific.

Older tests such as the Weil-Felix reaction (proteus OX-19 and OX-2) are no longer used. Some laboratories employ the complement fixation reaction using specific rickettsial antigens, which is very dependable. Here again, several serum specimens are necessary to demonstrate a rise in antibody titer.

V. TREATMENT

Therapeutic standards for patients who are seriously ill with the typhus–spotted fever groups are specific antibiotic treatment and supportive care. Each is essential for the patient encountered late in the course of illness. For the moderately ill, specific therapy usually suffices and energetic supportive care is less essential, in contrast to fulminant patients who require careful hospital management. Specific therapy is very effective when initiated early in the illness, coincident with the appearance of the rash. When the rashes become hemorrhagic and widespread, and associated with toxemia, the response is less dramatic.

The antibiotics of choice are the tetracycline–doxycycline group and chloramphenicol. These agents inhibit rickettsial growth rather than killing the *Rechettsia*. Ciprofloxacin is effective in patients with Mediterranean spotted fever (tick-borne) and probably in other rickettsial diseases. Specific treatment is continued for a day or two after the fever subsides. In about 24–36 hours, the patient looks and feels better. The patient's temperature reaches normal in 2–3 days. The favorable response for scrub typhus is a little earlier. In patients too ill to take medication orally, intravenous administration of the specific antibiotic is indicated, switching to the oral route when feasible.

A. Supportive Care

Frequent turning of the patient helps prevent choking on food and secretions, and the development of necrotic pressure lesions. Usually, food is well tolerated and the daily diet should provide 1–2 gm protein/kg normal body weight to help maintain the normal nitrogen balance. Mouth care, with swabbing of the oral cavity, will help avert gum and salivary gland infections. As mentioned previously, seriously ill patients, later in the illness, develop increased capillary permeability. Hence, intravenous fluids should be given cautiously to avoid fluid accumulation in the lungs and brain swelling. The seriously ill can recover, but obviously it takes more time.

VI. SPECIFIC RICKETTSIAL DISEASES

A. Epidemic Typhus Fever

Included in this category are European typhus, classic typhus, jail fever, ship fever, dermotypho (Italy), tifo exanthematico, tabardillo (Spain), and Fleckfieber (Germany).

1. *Clinical Manifestations*

Caused by *R. prowazeki,* typhus is transmitted from the body louse to humans after an incubation period of about 7 days. Severe headache and prostration occur early, and the patient appears "dusky" or "smoky". Light-sensitivity and bloodshot eyes are common. Rash is a distinguishing feature, apparent on about the fourth to fifth day, first appearing as small pink flat spots on the axillary folds and trunk. The rash progresses to red, bloody, and fixed flat spots. In several days, it becomes hemorrhagic, bloody, confluent, and occasionally necrotic at pressure sites. It spreads peripherally to the extremities, rarely involving the palms, soles, or face (in contrast to RMSF, in which the rash is initially peripheral).

The fever is continuously high for about 2 weeks with slight morning remissions. Neurological manifestations are a severe frontal headache, mental confusion, delirium, and coma; ringing in the ears and partial deafness are common. The pulse may be rapid and thready, with low blood pressure; the associated lowered blood flow results in diminished urine output and elevated blood ammonia levels. Extreme pallor and gangrene of the digits, hand, or feet may occur in late severe cases.

In uncontrolled epidemics, mortality rates have reached 60% (generally 10–20%). With specific antibiotic treatment and adequate supportive care, even under field conditions, case mortality is virtually eliminated and the course of illness is sharply reduced.

B. Brill-Zinsser Disease (Recurrent Typhus Fever)

Occasionally, there is recurrent infection, which is a milder illness many years after an initial attack of louse-borne typhus contracted in an endemic area, such as Europe or Asia. Latent *Rickettsia* apparently reactivate when host defenses falter. Recurrent typhus is sporadic and occurs in any season in the absence of infected lice. The usual setting is in larger cities in an immigrant population.

The clinical manifestations are milder than in the original illness. Headache and toxic signs are milder; the rash is pink, flat, and sparse; and a remittent-type febrile course lasts 7–10 days. Mortality is nil. Serum-specific antibodies appear early. The tetracycline antibiotics and chloramphenicol are effective.

C. Scrub Typhus

This is also known as tropical typhus or tsutsugamushi disease.

1. *Clinical Manifestations*

After a mite bite, the incubation period is 9–12 days. An eschar often appears as a small raised area, develops into a reddish swollen and hardened area, and forms a darkened scab with surrounding reddening. Regional lymph nodes are enlarged, often in the neck, axillae, and groin. Fever may suddenly develop (with the primary lesion present), associated with severe headache, general muscle pains, aches, light-sensitivity, pain behind the eyes, prostration, and lethargy. With the headache there is often ringing in the ears and moderate deafness. Bloodshot eyes are common. An irritative cough with moderate expectoration is indicative of infection of the bronchioles in the lungs. Diarrhea occasionally occurs.

A flat rash is readily observed involving the trunk on about the fifth febrile day, occasionally earlier. It is first pink and may spread peripherally, become slightly raised diffuse, and bruise-like, and persist until the end of the second week of illness.

The febrile course, as in most severe rickettsial diseases, is continuous in character, slightly remittent, and lasts for an average of 2 weeks in the untreated. It falls slowly. Mental confusion with disorientation may last until the fever subsides. The pulse, initially slow, may weaken and become rapid if decreased blood flow occurs associated with low blood pressure, decreased or absent urine output, and elevated blood ammonia. In some severe cases, the widespread, blood-vessel inflammation may lead to bronchopneumonia, inflammation of the heart, and even inflammation of the brain or its covering.

First, in 1948, chloramphenicol and later the tetracycline anibiotics led to rapid and dramatic cures. Severely ill, compromised patients recover with antibiotic and careful supportive treatment.

C. Murine (Endemic) Typhus

This is also known as rat flea typhus fever.

1. Clinical Manifestations

Rickettsia typhi (*mooseri*) is transmitted to humans by fleas. After an incubation period of 8–16 days (a mean of 10 days), the acute febrile illness is characterized by headache, backache, and joint pain as common signals of the disease, followed by chills and a high fever. Nausea, vomiting, prostration, and general weakness are present. A sustained febrile course, ranging from 102–104°F or higher usually lasts for about 12 days. Most patients develop a dull red flat rash on about the fifth febrile day, appearing on the abdomen, shoulders, chest, arms, and thighs. Initially, they fade when pressed and then become slightly raised and do not fade. Cardiovascular, respiratory and neurologic manifestations resemble epidemic typhus, although they are mild. Mortality from murine typhus is low; occasionally, death occurs in the elderly. The illness is mild in children. RMSF and murine typhus may occur in similar geographic areas, which makes their differentation difficult. Serologic diagnostic procedures are definitive and antibiotic as well as supportive measures lead to a rapid cure.

In the United States from 1990–1997, there were from 409–831 (average ~500) cases of murine typhus reported. After 1997, it ceased to be a reportable illness.

D. Rocky Mountain Spotted Fever

Rocky Mountain spotted fever (RMSF) is also known as spotted fever, tick fever, tick typhus. Not until the late 1920s was RMSF reported in the southeastern United States. It was thought to be unique to the Rocky Mountain area. The agent was identified by Ricketts in 1906, who confirmed the tick as the vector. *R. rickettsii* is transmitted to humans by Ixodid ticks, principally *Dermacentor andersoni* (wood tick) in the West and *Dermacentor variabilis* (dog tick) in the East. The tick serves as the actual reservoir of infection, principally because of transovorial transmission of *Rickettsia* in female ticks.

1. Clinical Manifestations

After an incubation period of 3–12 days (average 7), there is usually an abrupt onset with fever, headache, chills, prostration, and muscle pain. A short incubation usually signifies a more serious infection. In several days, the patient's temperature reaches 104–105°F and remains high, with slight morning remissions for as long as 12–20 days if untreated. A distinctive rash appears from the second to sixth days of illness, first as pink flat spots on the palms, soles, forearms, and legs, gradually spreading centrally to the body. It later becomes darker, red, fixed, and slightly raised; and eventually it becomes bloody and purplish, and coalesces with necrosis and ulceration over pressure sites. Characteristic neurologic manifestations are headache (usually frontal), confusion, restlessness, and coma, all indicative of an inflammation of the blood vessels in the brain. Rapid heart rates and low blood pressure in the later stages are expressions of blood vessel inflammation and low blood flow, often associated with low urine output and kidney failure. Lung and bronchial involvement and heart inflammation with various types of cardiac abnormalities mark the critically ill patient. The clinical course simulates epidemic typhus, with the exception that in typhus the rash is central on the trunk and sparse peripherally.

Mortality in adults, age 50 and older, could reach 50% or higher. Specific antibiotic treatment with good supportive care leads to remarkably favorable results.

Acknowledgment

Grateful appreciation is expressed to Mrs. Harriet Kerr for her typing and retyping of many copies of the manuscript. The author is indebted to her for this considerable contribution.

See Also the Following Articles

Lyme Disease • Rickettsiae • Typhoid, Historical

Bibliography

Brill, N. E. (1916). An acute febrile illness of unknown origin. *Am. J. Med. Sci.* **139**, 484–502.

Dyer, R. E., Rumreich, A., and Badger, L. F. (1931). A virus of the typhus type derived from fleas collected from wild rats. *Pub. Health Ref.* **46**, 334–338.

Greisman, S. E., and Wisseman, C. L., Jr. (1958). Studies of rickettsial toxins: Cardiovascular functional abnormalities induced by rickettsia in white rats. *J. Immun.* **81**, 345–354.

Hattwick, M. A., O'Brian, R. J., and Hanson, B. P. (1976). Rocky mountain spotted fever: epidemiology of an increasing problem. *Ann. Int. Med.* **84**, 732–739.

Lewthwaite, R., and Savoor, S. R. (1936). The typhus group of diseases in Mayala III: The relation of rural typhus to Tsutsugamushi disease (with special reference to cross-immunity tests). *B. J. Exp. Path.* **17**, 448.

Maxcy, K. F. (1926). An epidemiologic study of endemic typhus, (Brill's disease) in the south eastern United States with special reference to its mode of transmission. *Pub. Health Rep. USPHS* **41**, 267.

Murray, E. S., and Snyder, J. C. (1951). Brills disease II etiology. *Am. J. Hyg.* **53**, 22.

Smadel, J. E., Traub, R. Ley, H. L., Jr., Philip, C. B., Woodward, T. E., and Lewthwaite, R. (1949). Chloramphenicol (chloromycetin) in the chemo prophylaxis of scrub typhus (Tsutsugamushi disease). Results with volunteers exposed in hyperondomic areas of scrub typhus. *Am. J. Hyg.* **50**, 7591.

Sonenshine, D. E., Bozeman, F. M., Williams, M. S., Masiello, S. A., Chadwick, D. P., Stocks, N. I., Lauer, D. M., and Elisberg, B. L. (1978). Epizootiology of epidemic typhus (Rickettsia Prowazeki) in flying squirrels. *Am. J. Trop. Med.* **27**, 339–349.

Stuart, B. M., and Pullen, R. L. (1945). Endemic (Murine) typhus fever, clinical observations of one hundred and eighty cases. *Ann. Int. Med.* **23**, 520.

Walker, D. H., and Cain, B. G. (1978). A method for specific diagnosis of rocky mountain spotted fever on fixed, paraffin-embedded tissues by immunofluorescence. *J. Infect. Dis.* **137**, 206–209.

Woodward, T. E. (1999). Rickettsial Diseases (Rickettsioses) *In* "The Merck Manual," 16th ed.

Woodward, T. E. (1994). Rickettsial diseases. *In* "Harrison's Principles of Internal Medicine." 13th ed.

Woodward T. E., and Dumler J. S. (1998). Rocky mountain spotted fever. *In* "Bacterial Infections of Humans" (Evans and Bachman, eds.), 3rd ed. Plenum Medical, New York.

Woodward, T. E., Pederson, C. E., Jr., Oster, C. M., Bagley, L. R., Romberger, J., and Snyder, M. J. (1976). Prompt confirmation of rocky mountain spotted fever: Identification of rickettsiae in skin tissues. *J. Int. Dis.* **134**, 293–301.

Zinsser, H. (1934). Varieties of typhus fever and the epidemiology of the American form of European typhus fever. (Brills Discase). *Am. J. Hyg.* **20**, 513.

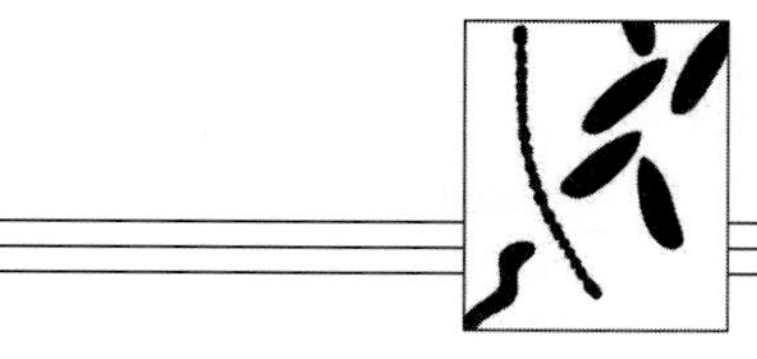

Vaccines, Bacterial

Susan K. Hoiseth
Wyeth–Lederle Vaccines

I. Brief History
II. Bacterial Pathogenesis: Toxins, Capsules, and Other Virulence Factors
III. Licensed Vaccines
IV. Concluding Remarks: Disease Eradication

VACCINATION is the practice of using modified (killed or attenuated) microorganisms, or portions thereof, to induce immunity to a particular disease without actually causing the disease. This may be through preventing infection or by limiting the effects of infection.

GLOSSARY

antibody A serum protein produced in response to an antigen and which has the property of combining specifically with that antigen. Different classes of antibody (lgG, lgM, lgA, etc.) have different functional properties and may be opsonic, bactericidal, or neutralizing, or interfere with the attachment of pathogens to the host cell.

antigen A molecule which is able to react specifically with antibody.

capsule A loose, gel-like structure on the surface of some bacteria. Capsules are usually, but not always, polysaccharide in nature.

cell-mediated immune response A specific immune response mediated by T cells and activated macrophages rather than by antibody.

endotoxin The lipopolysaccharide component of the outer membrane of gram-negative bacteria. The lipid A portion triggers a host inflammatory response via tumor necrosis factor and interleukin-1.

T-independent antigen An antigen capable of stimulating B cells to produce antibody directly without requiring T cell help.

toxin (exotoxin) A protein, present in cell-free extracts of pathogenic bacteria, capable of causing a toxic effect on the host.

toxoid A toxin which has been treated so as to inactivate its toxicity but which is still capable of inducing immunity to the active toxin.

I. BRIEF HISTORY

Jenner's work using live cowpox virus to vaccinate humans against smallpox dates back to 1796. The practice of variolation (the introduction of dried pus from smallpox pustules into healthy individuals) dates back to even earlier times. Variolation is known to have been practiced during the sixteenth century in India and China and may have originated in central Asia as early as the tenth or eleventh centuries. Although often effective, variolation was quite risky and resulted in death in 2 to 3% of those receiving the treatment. It is Jenner, however, who is generally credited with the first scientific study and the development of a true "vaccine" safe enough for widespread use. Jenner's vaccine utilized live cowpox virus, which caused only a mild disease in humans but which was sufficiently related to the human virus to confer cross-protection against smallpox. The first bacterial vaccines for human use were not developed until 100 years after Jenner's first viral vaccine. These were killed, whole cell vaccines for typhoid, cholera, and plague, developed in the late 1890s.

Vaccines are one of the greatest achievements of biomedical science and public health. Since the introduction of wide-spread vaccination for many bacterial diseases, there has been a dramatic decline in

TABLE I
Effect of Vaccination on Bacterial Diseases for Which Universal Infant Immunization Is Recommended (United States)[a]

Disease	*Baseline (prevaccine) annual cases*	*1998 Provisional cases*	*% Decrease*
Diphtheria	175,885[b]	1	100[c]
Pertussis	147,271[d]	6279	95.7
Tetanus	1,314[e]	34	97.4
Haemophilus influenzae type b	20,000[f]	54[g]	99.7

[a] Data from the Centers for Disease Control (1999).
[b] Average annual number of cases, 1920–1922.
[c] Rounded to nearest tenth.
[d] Average annual number of cases reported during 1922–1925.
[e] Estimated number of cases based on reported number of deaths during 1922–1926 assuming a case fatality rate of 90%.
[f] Estimated number of cases from population-based surveillance studies before vaccine licensure in 1985.
[g] In children younger than 5 years of age. Excludes 71 cases of *Haemophilus influenzae* disease of unknown serotype.

the morbidity and mortality associated with these diseases (Table I). A good example is diphtheria vaccine, which became available in 1923. In 1920, there were 147,991 cases of diphtheria reported in the United States, with 13,170 deaths; in 1998 there was only a single case reported. This decline is not seen in countries in which diphtheria vaccination rates are low. A striking example occurred in the newly independent states of the former Soviet Union following the breakup of the union. Vaccine coverage decreased, and the incidence increased dramatically starting in 1990. The incidence peaked in 1995, after which renewed efforts in vaccinating helped lower the incidence.

II. BACTERIAL PATHOGENESIS: TOXINS, CAPSULES, AND OTHER VIRULENCE FACTORS

Although the early killed whole cell typhoid, cholera, and plague vaccines were moderately effective, later vaccines relied on an understanding of some of the mechanisms by which various bacteria cause disease and the factors they elaborate in order to circumvent host defense mechanisms. The recognition of tetanus and diphtheria as toxin-mediated diseases led to the production of inactivated toxins (toxoids) as vaccines for these two diseases. Similarly, the recognition that bacterial capsules impart antiphagocytic properties to bacteria, and that antibodies to the capsule are often protective, led to the development of purified capsular polysaccharide or polysaccharide–protein conjugate vaccines against pneumococcal, meningococcal, and *Haemophilus influenzae* infections (see Sections III, B, C, and D).

Diseases caused by bacteria that are either facultative or obligate intracellular pathogens have been more difficult to vaccinate against. These organisms will require identification of novel antigens and/or vaccine delivery methods. One approach currently being tested is the use of DNA vaccines. This approach involves injecting nonreplicating plasmid DNA directly into an animal or human. The antigen of interest is expressed under the control of a mammalian promoter and is synthesized in the tissues of the vaccine recipient (usually muscle). DNA vaccines generally elicit a strong cell-mediated immune response, which is thought to be necessary to protect against intracellular pathogens. Other approaches currently being investigated include the use of novel adjuvants (immune-enhancing agents) and/or use of oral or intranasal delivery of vaccines for mucosal pathogens. These approaches are still mainly experimental (although oral administration is used for some existing vaccines). The reader is referred to Levine *et al.* (1997) and Plotkin and Orenstein (1999) for further discussion of new vaccines that are still being tested in animal models or which are in preliminary stages of testing in humans. The remainder of this article will focus on bacterial vaccines currently licensed in the United States for use in humans. These vaccines are listed in Table II, along with references for the current U.S. Advisory Committee for Immunization Practices (ACIP) recommendations for use of each vaccine.

TABLE II
Bacterial Vaccines Licensed for Use in the United States[a]

Vaccine	*Reference for ACIP recommendation*[b]
Diphtheria, tetanus, pertussis (DTP)	*MMWR* 40 (RR-10), 1991
Acellular pertussis (DTaP)	*MMWR* 46 (RR-7), 1997
Haemophilus influenzae type b	*MMWR* 40 (RR-1), 1991
Pneumococcal	*MMWR* 46 (RR-8), 1997
Meningococcal	*MMWR* 46 (RR-5), 1997
Typhoid	*MMWR* 43 (RR-14), 1994
Cholera	*MMWR* 37 (40), 1988
Anthrax	*MMWR* 40 (RR-12), 1991[c]
Plague	*MMWR* 45 (RR-14), 1996
Tuberculosis (BCG)	*MMWR* 45 (RR-4), 1996
Lyme	*MMWR* 48 (RR-7), 1999

[a] Licensed for use by the Food and Drug Administration, Center for Biologics Evaluation and Research.

[b] Recommendations for vaccine use are provided by the Advisory Committee on Immunization Practices, administered through the U.S. Centers for Disease Control (CDC). They are published as Recommendations and Reports (RR) in *Morbidity and Mortality Weekly Report* (MMWR) and are available on-line at *http://wonder.cdc.gov*. Other sources for vaccine recommendations include those provided by the Committee on Infectious Diseases of the American Academy of Pediatrics (published in their *Red Book*, with periodic updates published in the journal *Pediatrics*) and *The Guide for Adult Immunization* published by the American College of Physicians.

[c] *In General Adult Immunization Guidelines;* see also *MMRW* 48(4) (1999) for an updated statement from the CDC.

III. LICENSED VACCINES

A. DTP/DTaP (Diphtheria, Tetanus, and Pertussis/DT Plus Acellular Pertussis)

1. Tetanus

Tetanus is an acute, often fatal disease caused by release of a potent neurotoxin from the anaerobic bacterium, *Clostridium tetani*. The organism is acquired through environmental exposure, often contamination of a wound with soil. It is also found in animal feces. Neonatal tetanus, due to infection of the umbilical cord stump, is an enormous problem in developing countries, with an estimated 1.2 million deaths per year worldwide.

Vaccines for tetanus followed from von Behring and Kitasatos's purification of the toxin in 1890 and their finding that injection of minute amounts of the toxin into animals generated antibodies in the survivors that neutralized the toxin. This work led to the production of antitoxin in horses that was then used to treat humans. Active immunity in humans was not produced until the 1920s, when it was demonstrated that the toxin could be inactivated by treatment with formaldehyde but was still capable of generating neutralizing antibodies. The resulting tetanus "toxoids" became commercially available in the United States in 1938, but they were not widely used until the military began routine prophylactic use in 1941. Tetanus vaccination has been part of routine infant immunization in the United States since 1944 and has been highly successful in reducing the morbidity and mortality associated with the disease (Table I). Tetanus toxin is one of the more potent toxins known, and the disease has a high fatality rate (up to 90% without treatment; ~20% with antibiotics and good intensive care). By the time the patient is symptomatic, a significant amount of toxin has already been released. Hence, pre-existing immunity is crucial. Current U.S. guidelines call for routine immunization of infants at 2, 4, 6, and 15–18 months of age, followed by a booster dose prior to school entry. An additional dose is given at 11 or 12 years of age and then every 10 years thereafter. The first five doses are usually combined with diphtheria and pertussis vaccines. There are additional guidelines for wound prophylaxis, and for dirty wounds tetanus immune globulin may also be given.

2. Diphtheria

Diphtheria is a respiratory infection caused by the gram-positive bacterium, *Corynebacterium diphtheriae*. Like tetanus, diphtheria is a toxin-mediated disease. The hallmark of the disease is a membranous inflammation of the upper respiratory tract, with widespread damage to major organ systems from the toxin. The toxin is a protein synthesis inhibitor and causes damage mainly to the cardiovascular and nervous systems and sometimes the kidneys. The bacterium generally remains localized in the respiratory tract, and deep tissue invasion and bacteremia are extremely rare. The major manifestations of disease are due to systemic spread of the toxin.

Diphtheria antitoxin was first given in 1891 and was produced commercially in Germany in 1892. A combination of toxin/antitoxin was used in the United States beginning in 1914. By 1923, formalin-inactivated toxoids became available, and in 1948 diphtheria toxoid was combined with tetanus toxoid and pertussis as DTP. Current U.S. guidelines call for doses at 2, 4, 6, and 15–18 months of age, followed by a booster dose prior to school entry. The amount of diphtheria toxoid given to older children and adults is usually less than that given to infants in order to prevent local injection site reactogenicity in people who may already have significant antibody titers. This reduced dose is often given combined with a tetanus booster and is referred to as Td, with the lower, case "d" indicating the reduced amount of diphtheria toxoid. As indicated in Table I, the use of diphtheria toxoid has been highly effective in controlling disease in the United States.

3. *Pertussis*

Pertussis (whooping cough) is caused by the gram-negative bacterium, *Bordetella pertussis*. The hallmark of the disease is a protracted cough, often lasting many weeks. Pertussis is spread by the respiratory route, and the initial symptoms are indistinguishable from those of other upper respiratory infections. The disease is most severe in infants, in whom intense coughing leads to a forced inspiratory "whoop." Major complications include bronchopneumonia, encephalopathy, and prolonged vomiting (eating may trigger coughing episodes, followed by vomiting). Prior to the introduction of a pertussis vaccine, an average of 147,271 cases per year were reported in the United States (Table I), with 5000–10,000 deaths. Epidemic peaks of disease occur at 2- to 5-year intervals, and as many as 270,000 cases were reported in a single peak year. By the late 1930s, there was evidence from controlled field trials that several whole cell pertussis vaccines provided significant protection against disease. Following introduction of widespread vaccination in the 1940s, the incidence of pertussis in the United States decreased to a low of 1010 cases in 1976. The incidence has risen slightly since 1976, with 7796 cases reported in 1996 and 6279 in 1998. Some of this increase reflects increased disease in older children and adults, perhaps due to waning immunity.

The original pertussis vaccines were inactivated whole cell preparations (typically heat, thimerosal, or formalin inactivated). Like other gram-negative whole cell vaccines, they contained considerable amounts of endotoxin (lipopolysaccharide from the gram-negative outer membrane) and were commonly associated with local injection site pain and erythema, fever, drowsiness, and fretfulness. Other rare but more serious events (seizures and encephalopathy) have also been reported in temporal association with whole cell pertussis vaccines. Cause and effect of these rare events has been difficult to prove, however, because these vaccines are given to infants at an age when previously unrecognized underlying neurological and developmental disorders first become manifest. Because of concerns regarding the use of whole cell pertussis vaccines, however, several countries (Japan, Sweden, England, and Wales) curtailed or suspended pertussis vaccination, and epidemic pertussis recurred. They subsequently resumed routine childhood vaccination with either whole cell vaccine or acellular vaccines. Many studies have shown that if there was a relationship between whole cell pertussis vaccine and neurological problems, it was sufficiently rare that the benefits of vaccination far exceeded the risks. In the absence of vaccination, the risk of death or encephalopathy from disease was much greater than the risks (real or perceived) from vaccination. This issue should be largely put aside because new, more purified (and less pyrogenic) acellular pertussis vaccines have recently been licensed.

Although the correlates of protective immunity for pertussis are still not fully defined, several acellular pertussis vaccines have been tested in clinical efficacy studies and have been shown to be effective. Acellular vaccines were first licensed in Japan in 1981 and were tested extensively in large efficacy trials in Sweden, Italy, Germany, and Senegal in the late 1980s and early 1990s. They were licensed in the United States in 1991 for use as the fourth and fifth doses of the DTP series and for use as the primary series in infants in 1996. These vaccines contain inactivated pertussis toxin, either singly or with one or more of the following: filamentous hemagglutinin (FHA), pertactin (a 69-kDa nonfimbrial, outer membrane agglutinogen), and fimbrial agglutinogens (attachment factors that allow binding of the organism to ciliated epithelial cells of the upper respiratory tract). Pertussis toxin (previously known as lymphocytosis-promoting factor, histamine-sensitizing factor, or islet cell-activating factor) plays a major role in the systemic manifestations of disease and contributes to protective immunity. However, other toxins (adenylate cyclase, tracheal cytotoxin, and heat-labile toxin) and colonization factors may also contribute to pathogenesis and to protective immunity (based on animal models). The correlation of antibody titers to the various acellular vaccine components and clinical efficacy has not been clearcut. However, several different acellular vaccines gave overall efficacies approximately comparable to the whole cell vaccines, and local reactions, fever, and other systemic events occur significantly less often with acellular vaccines than with whole cell vaccines. Acellular vaccines, in combination with diphtheria and tetanus toxoids (DTaP), are currently recommended by the ACIP for all five doses.

B. *Haemophilus influenzae*

Haemophilus influenzae type b (Hib) causes serious invasive diseases (meningitis, septicemia, cellulitis, and epiglottitis) primarily in young children. Before the availability of vaccines, Hib was the most common cause of bacterial meningitis in children younger than 5 years of age. Other capsular serotypes of *H. influenzae* (a,c,d,e, and f) only rarely cause disease.

The first-generation Hib vaccine, licensed in 1985, consisted of purified type b capsular polysaccharide. The capsule is a major virulence determinant for the organism, allowing it to avoid phagocytosis. Antibodies directed to the capsule promote phagocytosis and clearance of the organism. The first-generation polysaccharide vaccine was licensed only for use in children 18 months of age and older. Infants posed a greater challenge and required new technologies for development of a successful vaccine. This is due to the fact that the immature infant immune system does not respond well to T-independent antigens. Whereas infants respond to T-dependent protein antigens such as diphtheria and tetanus toxoids, they do not respond well to T-independent polysaccharide antigens (they lack the subset of B cells capable of responding directly to antigens without T cell help). This problem was overcome by chemically conjugating the polysaccharide to a protein carrier and converting the response to a T-dependent response. The first Hib conjugate vaccine was licensed for use in infants in 1990. Prior to introduction of Hib vaccines, there were an estimated 20,000 cases of invasive Hib disease per year in the United States, with approximately 1000 deaths and considerable long-term sequelae in survivors (hearing loss, learning disabilities, or mental retardation). After less than 10 years of routine infant immunization, the number of cases in children younger than 5 years of age decreased to 54 in 1998 (excluding 71 cases of unknown serotype). The reduction in disease incidence has exceeded that predicted based on the number of completely immunized infants and suggests an element of "herd immunity" (through reduction in transmission of the organism from vaccinated individuals to unvaccinated infants).

The type b vaccine will not protect against unencapsulated ("nontypeable") *H. influenzae*. These latter strains are a frequent cause of otitis media but rarely cause systemic disease. Efforts are under way to develop protein-based vaccines to protect against otitis media caused by nontypeable strains.

C. Pneumococcal Vaccine

Infections with *Streptococcus pneumoniae* (pneumococci) are most common in the very young (<2 years of age) and in the elderly (>65 years of age) and in individuals with other underlying medical conditions. Predisposing conditions include immunocompromized individuals, those with functional or anatomic asplenia, chronic cardiovascular diseases (congestive heart failure or cardiomyopathy), chronic pulmonary diseases (emphysema and chronic obstructive pulmonary disease), and chronic liver disease. The organism colonizes the upper respiratory tract, and many people carry the organism without developing disease. Damage to respiratory mucosa by a viral respiratory infection, cigarette smoke, air pollutants, opiates, or aging may predispose to infection. Disease can occur with the following manifestations: (i) disseminated infection (bacteremia or meningitis), (ii) pneumonia or other lower respiratory tract infection, or (iii) upper respiratory tract infection, including otitis media and sinusitis. It is estimated that each year in the United States pneumococcal disease accounts for 3000 cases of meningitis, 50,000 cases of bacteremia, 500,000 cases of pneumonia, and 7 million cases of otitis media (*Morbid. Mortal. Weekly Rep.* **46**, RR-8, 1997).

More than 90 immunologically distinct serotypes of pneumococci have been described, varying in structure of the capsular polysaccharide. The capsule is a major virulence determinant for the pneumococcus, and antibodies to the capsule are protective. Protection, however, is type specific, necessitating the inclusion of multiple serotypes in the vaccine. Fortunately, some types are more frequent causes of disease than others, and vaccines are designed to include the most common types. Hexavalent pneumococcal polysaccharide vaccines were available for use in the United States starting in 1946 but were withdrawn after only a few years because of the apparent success of treating pneumococcal disease with penicillin. It was not until the 1960s that it became apparent that there was still considerable morbidity and mortality from pneumococcal disease, despite antimicrobial therapy. This is perhaps more true today, with the emergence of penicillin-resistant pneumococci. Extensive field trials were conducted in South African gold miners in the 1970s, and a 14-valent capsular polysaccharide vaccine was licensed in 1977, followed by a 23-valent vaccine in 1983. As was the case for *H. influenzae,* these polysaccharide vaccines do not work well in infants. However, unlike *Haemophilus,* for which the vast majority of disease occurs in infants and young children, pneumococcal disease is also a significant problem in the elderly. Pneumococcal polysaccharide vaccine is recommended for those age 65 or older as well as individuals 2 years of age or older who have certain chronic diseases, asplenia, or who are immunocompromised. The vaccine is currently underutilized in the elderly population, with only approximately 30% of those age 65 or older having received the vaccine. The ACIP issued a revised recommendation in 1997 (*Morbid. Mortal. Weekly Rep.* **46**, RR-08) that the vaccine be used more extensively. The serotypes in 23-valent vaccine cover at least 85–90% of the serotypes causing invasive disease in the United States.

Pneumococcal polysaccharide conjugate vaccines have recently been developed and should be available for infant use in the near future. A large-scale efficacy trial of a 7-valent conjugate vaccine in infants showed 97% efficacy against invasive pneumococcal disease caused by vaccine serotypes. Because the conjugates are considerably more complex to make than the polysaccharide vaccine, the conjugate vaccines will not contain all 23 serotypes currently available in the polysaccharide vaccine. Fortunately, the 7 most common serotypes account for ~80–90% of invasive disease in young children in the United States and Canada, and a 7-valent vaccine is therefore expected to have a significant impact on disease. Serotype prevalence varies in different parts of the world, and 9- or 11-valent conjugates are being developed for global use.

D. Meningococcal Vaccines

Neisseria meningitidis causes both endemic and epidemic disease, mainly in the form of meningitis or meningococcemia (fulminant sepsis). In the United States, the disease is mainly endemic, with small outbreaks occurring in localized settings. Epidemic disease occurs mainly in the developing world, especially sub-Saharan Africa and Asia. There are approximately 2600 cases of meningococcal disease per year in the United States with a case fatality rate of 12%. These fatalities occur despite appropriate antimicrobial therapy. Particularly for menigococcemia, disease progresses very rapidly, and 60% of patients have experienced symptoms for less than 24 hr prior to hospital admission; death may occur within hours. Death from meningococcemia is due to disseminated intravascular coagulation and hypovolemic shock (gram-negative endotoxic shock); purpura is frequently present, and if the patient survives, skin graft or amputation may be required. Neurological complications may occur in up to 20% of survivors of meningococcemia or meningitis. Although the incidence of meningococcal disease is not as high as that of pneumococcal disease (or *Haemophilus* before introduction of Hib conjugate vaccines), the severity of the disease and the rapidity with which it strikes previously healthy individuals make it a target for vaccination.

Like pneumococcal and *Haemophilus* infections, the ability of the meningococcus to survive in the bloodstream is related to the presence of a polysaccharide capsule, and antibodies to the capsule are protective. Although there are at least 13 different capsular serogroups of meningococci, the majority of disease is caused by groups A, B, and C, with small numbers of cases of Y and W-135. In the United States, most cases are B or C; the number of cases of Y has increased slightly during the past several years. Epidemics in the developing world are mainly serogroup A, especially in the "meningitis belt" of sub-Saharan Africa.

Previously, military recruits experienced high rates of meningococcal disease. Since introduction of routine vaccination with an A/C polysaccharide vaccine in 1971 (and later a quadrivalent A, C, Y, and W-135 vaccine), rates in recruits have decreased substantially. Routine vaccination of civilians is not recommended, however, because, like other polysaccharide vaccines, meningococcal polysaccharide vaccine is relatively ineffective in children younger than 2 years of age (among whom risk for disease is highest). The A and C vaccines have estimated efficacies of 85–100% in older children and adults and have been useful for controlling serogroup C outbreaks (provided a large segment of population at risk is vaccinated). In Quebec, Canada, 1.6 million doses of polysaccharide vaccine were administered to 6-month-olds through 20-year-olds during a serogroup C outbreak in 1993. The overall efficacy was estimated to be 79%, with greater efficacies in teenagers and lower efficacies in children younger than 5 years of age. In addition to use in outbreak settings, the A, C, Y, and W-135 polysaccharide vaccine licensed for use in the United States is recommended for use in people with terminal complement component deficiencies, people with asplenia, and for travelers to parts of the world where there is epidemic disease. Several conjugate vaccines are being developed, and these should result in better efficacies in the younger age groups. The conjugates are not yet licensed but should be available in the next few years.

Unfortunately, there is no serogroup B vaccine currently available in the United States. The serogroup B capsule contains polysialic acid and is poorly immunogenic in humans. This is likely related to the fact that human tissues also contain sialic acid determinants, and vaccine-induced cross-reactivity with human tissue may be undesirable from a safety standpoint. Efforts to develop a serogroup B vaccine have therefore focused mainly on surface proteins. This has presented a challenge since there is considerable strain to strain variability in the major surface proteins. A protein-based vaccine is available in Cuba and some Latin American countries, but it is specific for the predominant strain in Cuba. This vaccine can provide protection against epidemics caused by strains homologous to the vaccine but not against endemic disease caused by strains of heterologous subtypes. This vaccine would not be expected to have much of an impact in countries such as the United States or Canada, where endemic disease is caused by a diverse number of strains of differing surface protein subtypes. Efforts are ongoing to identify conserved, protective surface proteins and/or to combine several of the less conserved proteins into a multivalent vaccine.

E. Special-Use Vaccines: Typhoid, Cholera, Plague, and Anthrax

These are all diseases of very low incidence in the United States and for which vaccination is indicated for only limited numbers of individuals (travelers to parts of the world where disease is a problem, military personnel, etc.).

1. Typhoid

Typhoid fever is an enteric fever caused by *Salmonella typhi*. It is an acute, systemic illness resulting from ingestion of food or water contaminated with *S. typhi*. The organism causes a severe, generalized infection of the reticuloendothelial system, intestinal lymphoid tissue, and gallbladder and results in high fever, headache, malaise, and abdominal discomfort. Without treatment, typhoid fever has a case fatality rate of 10–20%; with appropriate antibiotic treatment, fatality rates decrease to <1%. Antibiotic resistance, however, is a continuing problem.

Prior to the introduction of water treatment at the beginning of the twentieth century, typhoid fever was a major problem in large cites in Europe and the United States. Once water filtration and chlorination were introduced, typhoid rates plummeted. There are currently only a few hundred cases per year reported in the United States, and the majority of these occur in people who have traveled to parts of the world where typhoid is still endemic or who are immigrants from these countries. Most other cases can be traced to contaminated food, handled by food service workers who can be classified into one of the two previous categories. Worldwide, typhoid remains an enormous problem, with an estimated 33 million cases and 500,000 deaths per year. These cases occur largely in countries that lack proper sewer and water systems.

There have been a variety of typhoid vaccines available since the late 1890s, and they have been evaluated in numerous efficacy trials. These were mainly heat-killed, phenol-preserved whole cell vaccines, or acetone-killed and dried whole cell vaccines. Efficacies are in the 70% range, with some higher and some lower. Efficacy is dependent on the number of organisms ingested, and this will vary from country to country depending on the level of contamination of the water supply, etc. The killed whole cell vaccines, although providing considerable protection, do have unpleasant side effects. These are due mainly to the endotoxin component (lipopolysaccharide from the gram-negative outer membrane). Side effects include fever, malaise, local injection site erythema, induration, and pain; some vaccine recipients experience general disability for 1 or 2 days. Considerable efforts have gone into developing less reactogenic typhoid vaccines, and two new vaccines have been licensed in the United States during the past 10 years. The live attenuated strain, Ty21a, was generated in the 1970s by chemical mutagenesis. It has a mutation in the *galE* (galactose epimerase) gene, and in the absence of exogenous galactose it is unable to synthesize full-length lipopolysaccharide (LPS); it is also Vi [virulence antigen (capsule)] negative. For many years it was believed that the *galE* and *via* (Vi) mutations were the major attenuating lesions in Ty21a. However, recent work has shown that a strain with defined deletions in *galE* and *via* is still capable of causing typhoid in humans. The precise mechanism of attenuation in Ty21a is still not understood, although a mutation in *rpoS* (stress-induced sigma factor) may also contribute. Despite the undefined nature of attenuation, Ty21a has been found to be extremely safe. Ty21a is given orally and was tested extensively in efficacy studies in Alexandria, Egypt, in the late 1970s and in Chile and Indonesia during the 1980s. It gave efficacies of 96% in Egypt with a liquid formulation, but gave lower efficacies for several different formulations in Chile and Indonesia. The liquid formulation used in Egypt was not suitable for mass production, and a variety of enteric (acid-resistant) and gelatin capsule formulations were tested in Chile along with a sachet formulation containing lyophilized organisms and buffer to be reconstituted with water just prior to drinking the vaccine. The enteric capsules were superior to the gelatin capsules plus bicarbonate and gave an efficacy of 67%. The sachet formulation provided 78% protection. Efficacy in Indonesia was 42% for three doses of enteric capsule and 53% for the sachet formulation. A four-dose regimen of the enteric capsule formulation was licensed for use in the United States in late 1989. Efficacy in persons from nonendemic areas who travel to endemic areas, however, has not been tested.

An alternative approach is the Vi capsule. Purified Vi capsular polysaccharide vaccines were tested in field trials in Nepal and South Africa in the 1980s and conferred protection rates of 55–72%. A Vi typhoid vaccine was licensed in the United States in 1994 and is given intramuscularly.

2. *Cholera*

Like typhoid, cholera is a disease related to poor sanitation. The causative agent, *Vibrio cholerae*, excretes a potent enterotoxin (cholera toxin) that leads to severe, dehydrating diarrhea, frequently with vomiting. Fluid loss can reach 15–20 liters per day, resulting in death from hypovolemic shock and/or electrolyte imbalance. Fluid and electrolyte replacement is essential, and such treatment dramatically reduces the fatality rates. Disease in the United States is rare (~10–20 cases per year), but it is endemic parts of Asia, Africa, South America, and the Middle East.

Parenteral, killed whole cell vaccines for cholera have been available in many parts of the world since the late 1800s. These vaccines confer only approximately 50% protection, and booster doses are needed every 6 months. They have many of the side effects of other whole cell gram-negative vaccines (fever, headache, and general malaise). Efforts have been under way to make an improved cholera vaccine, and an oral formulation consisting of killed whole cells plus the B subunit (non-toxic, binding subunit) of cholera toxin has been licensed in Sweden. The vaccine is less reactogenic when given orally and gave efficacies in the 60% range for vaccinees older than 5 years of age. Efforts have also focused on live attenuated strains, and a genetically engineered strain (CVD103 HgR) is licensed in several countries (but not in the United States). This strain is deleted for the A (active) subunit of cholera toxin but makes the B subunit; therefore, it is capable of generating antitoxic (B subunit) antibodies and antibody to the bacterial cell surface (LPS, etc.). Both types of antibodies are thought to contribute to protection. However, results from a recent efficacy trial of a single oral dose of CVD103 HgR in Indonesia are disappointing.

Current ACIP guidelines do not call for routine vaccination of U.S. residents traveling to parts of the world where cholera is endemic, but some countries may require proof of vaccination for entry. The vaccine available in the United States is the parenteral killed whole cell product.

3. *Anthrax*

Anthrax is extremely rare in the United States and in most parts of the world, with an average of 0.25 cases per year reported in the United States between 1988 and 1996. Disease is acquired by contact with *Bacillis anthracis* spores, either by cutaneous exposure or through inhalation. In the industrialized world, disease is associated with processing of animal materials such as wool, hair, hides, and bones. In agricultural settings in Africa and Asia, meat and animal carcasses may also be a source of infection. Although disease in the United States is rare, interest in anthrax vaccines was renewed following a 1979 epidemic in Sverdlovsk, Russia, that was associated with accidental release of spores from a military laboratory and because Iraq produced weapons containing anthrax spores during the 1991 Gulf War.

Anthrax produces two main toxins, lethal factor and edema factor, which are unusual in that they share the same binding (B) subunit, which is referred to as "protective antigen." The vaccine available in the United States is composed of a cell-free culture filtrate, adsorbed onto aluminum hydroxide, plus low concentrations of formaldehyde and benzethonium chloride as preservatives. Vaccine potency is tested by guinea pig challenge, and although the protective antigen is believed to contribute to immunity the importance of the various components is not completely defined. The vaccine is recommended for industrial workers who handle wool, goat hair, hides, and bones imported from countries in which animal anthrax occurs (mainly from Asia, Africa, and parts of South America and the Caribbean) and for laboratory researchers working with virulent strains. The U.S. military also routinely vaccinates against anthrax. The vaccine has not been tested for safety or efficacy in individuals younger than 18 years of age. A live, attenuated anthrax vaccine is available in the former Soviet Union, and a similar attenuated vaccine is used for vaccinating domestic animals in many countries throughout the world.

4. *Plague*

Plague is a natural (zoonotic) infection of rodents and their fleas. During the Middle Ages, plague killed approximately 25 million people in Europe and was known as the "Black Death." The causative organism, *Yersinia pestis,* is a gram-negative coccobacillus belonging to the Enterobacteriaceae. Although currently disease is rare, the potential for epidemics still exists. In the United States, an average of 13 cases per year were reported to the Centers for Disease Control during the years 1970–1995. Disease in the United States occurs mainly in the southwestern states (New Mexico, Arizona, and Colorado, with smaller numbers of cases reported from California and nine other western states). On a worldwide basis, an average of 1087 cases per year were reported to the World Health Organization during the years 1980–1994 (although disease may be under-reported from countries in which surveillance and laboratory capabilities are inadequate). Epidemics are most likely to occur in countries with poor sanitary conditions, where large populations of rats live in close proximity to humans.

The plague vaccine currently available in the United States is a formaldehyde-inactivated whole cell preparation. It is recommended for use in laboratory researchers working with virulent strains and for field-workers, agricultural consultants, and military personnel working in areas where they may be exposed to enzootic or epizootic disease.

F. Tuberculosis

Mycobacterium tuberculosis can cause disease in any organ of the body, but pulmonary tuberculosis is by far the most common form of disease. The organism is spread via the respiratory route, and infection at sites other than the lung is usually the result of dissemination from a primary lung lesion. The study of tuberculosis and tuberculosis vaccines is complicated by the fact that only a small percentage of those infected with the organism develop clinical disease; in those that do develop disease, the onset of symptoms may occur from several weeks to many decades following initial infection.

In the mid-1800s, mortality from tuberculosis in the larger eastern cities of the United States averaged 400 per 100,000 people per year. It had long been recognized that crowded living and working conditions favored the spread of tuberculosis, and improvement of these conditions reduced the incidence of tuberculosis, even before the availability of antimicrobial therapy or vaccination. In the United States, the mainstay of tuberculosis control has been early detection and treatment of patients with active disease and contact identification to provide preventive therapy for persons infected but not yet symptomatic. This strategy worked reasonably well until the 1980s, when the incidence of disease began to increase and outbreaks of multi-drug-resistant strains began to appear (mainly in HIV-positive individuals and their health care workers and in correctional facilities and among the homeless). Although a tuberculosis vaccine was developed in the 1920s, and has been given to billions of people throughout the world, it has never been widely used in the United States mainly because it converts people to skin test positive and thereby eliminates an important screening test used to identify infected but asymptomatic people. Additionally, vaccine efficacy estimates from different trials have varied widely from 0 to 80%.

The current vaccine is the live, attenuated bacille Calmette–Guerin (BCG) strain, developed by Calmette and Guerin at the Pasteur Institute by subculturing a strain of *Mycobacterium bovis* on artificial media every 3 weeks for 13 years. Unfortunately, different sublines of BCG have resulted from propagation and production of vaccine in different laboratories throughout the world, and this may partially account for some of the discrepancies in efficacy reported for the various trials. Despite the widespread use of BCG vaccine in most countries of the world, there are still an estimated 8 million cases of active disease, and 2 or 3 million deaths per year worldwide. Although overall efficacy may be less than ideal, BCG vaccination is believed to reduce the more serious meningeal and disseminated forms of tuberculosis in children.

In the United States, the risk of tuberculosis in the general population is relatively low, and routine BCG vaccination is not recommended. Vaccination should be considered for an infant or child continually exposed to an untreated or ineffectively treated patient who has pulmonary tuberculosis, where the child cannot be removed from the infectious patient or given long-term preventive antimicrobial therapy. Efforts are ongoing to develop a better tuberculosis vaccine.

G. Lyme Disease

The newest bacterial vaccine is that for Lyme disease. This vaccine was licensed in the United States in December 1998, and contains lipidated outer surface protein A (OspA) from *Borrelia burgdorferi,* produced recombinantly in *Escherichia coli.* Lyme disease is named for the community in Connecticut in which disease was first recognized to occur in epidemic fashion. Clinical descriptions matching that of Lyme disease had been described earlier, but it was not until 1975 in Lyme, Connecticut, that the infectious nature was recognized. The causative agent was first isolated in 1982 by Burgdorfer and colleagues.

Lyme disease is transmitted through the bite of an infected tick, and disease generally occurs in stages. In the initial stage there is usually (~85% of cases) a characteristic skin lesion (erythema migrans) emanating from the site of the tick bite, accompanied by fever, headache, malaise, or stiff neck. Approximately 5–15% of infected individuals develop neurologic or cardiac symptoms within a few months of the initial infection, and if untreated this frequently leads to late-stage infection and Lyme arthritis. Most patients respond to antibiotic therapy although refractory cases do occur. Treatment is more likely to be delayed in patients who do not develop the initial skin lesion.

The majority of Lyme disease in the United States (~88%) occurs in the northeast, the upper Midwest, and northern California. It is also prevalent in the temperate European countries, including Germany, Sweden, and Austria, and the central regions of the former Soviet Union. More than 12,000 cases of Lyme disease were reported in the United States in 1998. Control efforts have focused on tick avoidance, the wearing of clothing that covers the arms and legs when working or playing in tick-infested areas, and the use of insect repellents. The availability of the vaccine now provides an additional option for people who live, work, or recreate in areas of high to moderate risk. Routine use of the vaccine, even in areas with the highest incidence of disease, is not believed to be cost-effective. However, vaccination should be considered for individuals in areas of high or moderate risk who engage in recreational, property maintenance, or occupational activities that result in frequent or prolonged exposure to tick-infested habitats. It is not recommended for persons in high- to moderate-risk areas who have minimal or no exposure to tick-infested habitats or for persons who reside, work, or recreate in areas of low to no risk.

The mechanism of action of the Lyme disease vaccine is unlike that of other vaccines. The bacteria express OspA in the infected tick but downregulate OspA and express mainly OspC in humans. Antibodies to OspA are apparently ingested by the tick when it takes a blood meal on the human host, and these kill the *Borrelia* within the tick gut. This suggests that relatively high levels of antibody will need to be maintained over time. Prelicensure efficacy studies showed 76% efficacy 1 year after the third dose of vaccine, but the long-term efficacy has not been evaluated. Additional studies will be required to address the need (and timing) for booster doses.

IV. CONCLUDING REMARKS: DISEASE ERADICATION

Global eradication of smallpox through vaccination was a milestone in history. Polio eradication is on the horizon, and elimination of measles and rubella from the United States, Canada, and several European countries is also targeted. Whether bacterial diseases can be similarly eradicated will depend on several factors: (i) Is there an animal or environmental reservoir for the pathogen? (ii) Does immunity prevent actual infection/colonization (as opposed to protection against the systemic effects of the disease)? (iii) Is there a long-term carrier state? and (iv) Can sustained global vaccination efforts be implemented? *Haemophilus influenzae* type b vaccine has had the unanticipated benefit of helping reduce transmission of the organism to unvaccinated infants (a form of "herd immunity"). Although Hib disease is still a problem globally, the dramatic reduction in Hib disease in the United States following introduction of conjugate vaccines, combined with the fact that there is no animal reservoir for the organism, suggests that global eradication of Hib might be achievable through vaccination. Global eradication, however, requires tremendous effort and resources. Major bacterial diseases are more likely to be controlled by vaccination but may not be completely eradicated.

See Also the Following Articles

Exotoxins • *Haemophilus influenzae*, Genetics • Lipopolysaccharides • Mycobacterium • Plague • *Streptococcus pneumoniae* • Typhoid, Historical

Bibliography

Centers for Disease Control (1999). Achievements in public health, 1900–1999. Impact of vaccines universally recommended for children—United States, 1990–1998. *Morbid. Mortal. Weekly Rep.* 48 (12), 243–248.

Centers for Disease Control. *Morbid. Mortal. Weekly Rep.* Annual Summaries.

Centers for Disease Control. *Morbid. Mortal. Weekly Rep.* Recommendations and Reports. (Specific references for each vaccine are listed in Table II.)

Institute of Medicine. Vaccine-related reports available online at *www.search.nationalacademies.org/* (search "vaccines").

Levine, M. M., Woodrow, G. C., Kaper, J. B., and Cobon, G. S. (Eds.) (1997). "New Generation Vaccines," 2nd ed. Dekker, New York.

Plotkin, S. A., and Orenstein, W. A. (Eds.) (1999). "Vaccines," 3rd ed. Saunders, Philadelphia.

Vaccine supplement (1998, May). *Nature Med.* 4(5).

Vaccines, Viral

Ann M. Arvin
Stanford University School of Medicine

I. General Principles
II. Live Viral Vaccines
III. Noninfectious Vaccines
IV. Vaccine Immunology
V. Molecular Approaches to Viral Vaccine Design
VI. Public Health Impact

GLOSSARY

adaptive immunity B and T lymphocyte-mediated, memory immune responses that control viral infections through specific interactions with virus-infected cells and virions.

attenuation Genetic alteration of infectious viruses to reduce their potential to cause disease.

immunogenicity Capacity to elicit adaptive immunity to proteins of the virus.

protective efficacy Capacity to protect against disease usually caused by the virus.

tropism Pattern of infectivity for cells and organs that is characteristic of the viral pathogen.

THE FUNDAMENTAL OBJECTIVE OF VACCINATION against viral pathogens is to induce adaptive immunity in the naive host which protects from disease upon any subsequent exposures to the infectious agent. In the absence of vaccine-induced immunity, the initial control of viral infection depends on mechanisms that comprise the innate immune system, such as the production of interferon-α or lysis of virus-infected cells by natural killer cells. Innate immunity limits viral spread, but these defenses are often not sufficient to block symptoms of illness during the interval necessary to elicit adaptive immunity against the virus. In the extreme circumstance, life-threatening complications may result in the interim. Adaptive antiviral immunity consists of the clonal expansion of T lymphocytes and B lymphocytes that have the functional capacity to recognize specific viral proteins and to interfere with viral replication and transfer of virions from infected to uninfected cells within the host.

I. GENERAL PRINCIPLES

In order to induce adaptive immunity, viral proteins must be processed by dendritic cells or macrophages, which are specialized antigen-presenting cells that mediate the cell surface expression of viral peptides in combination with the class I or class II major histocompatibility complex (MHC) proteins. MHC-restricted antigen presentation creates populations of "memory". T lymphocytes within the CD4 and CD8 subsets that are primed to synthesize cytokines, such as interleukin-2 or interferon-γ, when exposed to the same viral peptide–MHC class I or class II protein complex. Cytokines modulate the inflammatory response, expanding and recruiting antigen-specific, cytotoxic T lymphocytes to the site of viral infection and inducing B lymphocytes to produce antibodies of the IgM, IgG, and IgA subclasses that can bind to proteins made by the pathogen or mediate antibody-dependent cellular cytotoxicity.

Adaptive immunity that protects against viral pathogens can be achieved by inoculation of the naive host with infectious virus which has been attenuated for its capacity to cause disease or by exposure of the host to viral proteins administered in a noninfectious formulation. Effective priming of adaptive T lymphocyte and B lymphocyte responses by a viral vaccine is expected to block most or all symptoms of infection when the host is exposed to the pathogen. The immunogencity of a vaccine is defined as its capacity to elicit adaptive immunity, whereas protective efficacy refers to the prevention of disease which is a consequence of the effective induction of virus-specific immunity. Vaccine-induced immunity may not prevent asymptomatic or abortive infection during these encounters, but memory or "recall" responses to the viral proteins should eliminate any serious morbidity or risk of mortality known to be associated with the infection in a susceptible, non-immunized individual. Antiviral responses elicited by vaccination provide "active," as distinguished from "passive," immunity. Passive antiviral immunity is provided by virus-specific lgG antibodies which may be acquired transplacentally by infants or by administration of immunoglobulins, such as rabies or varicella zoster immune globulin. Active immunity as elicited by effective vaccines mimics the memory immunity that follows natural infection and is persistent, whereas passively acquired antibodies are metabolized over a half-life of approximately 4 weeks and protection is transient.

The challenge of designing viral vaccines that elicit adaptive immunity that is sustained and protective can be addressed using several different strategies, which are often dicated by characteristics of the pathogen and the target population requiring protection. Historically, variolation against smallpox was the first attempt to induce active immunity against a virus by inoculation, as described in early texts from China. Nevertheless, variolation differs from vaccination because unaltered variola virus was given, with disease modification presumed to result from administering a low infectious inoculum by a cutaneous route. The first success of viral vaccination is attributed to Benjamin Jesty, an English farmer, who used cowpox to prevent smallpox in 1774, as recounted by Edward Jenner, who published his own experience with vaccination in *Variolae Vaccinae* (1798). The remarkable achievement of the global eradication of smallpox was accomplished 200 years later using vaccinia vaccine. Many viral diseases can be prevented by immunization, and efforts are under way to make new vaccines that will provide effective prophylaxis against many other human viral pathogens. In some cases, vaccines may be developed as "therapeutic" vaccines which are intended to control the progression of chronic viral infections, such as human immunodeficiency virus (HIV) (Table I). Vaccination is also used to control viral diseases in non-human species.

TABLE I
Live Viral Vaccines for Prevention of Human Disease

Current vaccine	*Vaccines under development*
Measles	Influenza A and B
Mumps	Respiratory syncytial virus
Rubella	Parainfluenza virus 1–3
Varicella	Herpes simplex viruses 1 and 2
Polioviruses 1–3	Cytomegalovirus
Rotavirus	
Yellow fever	
Adenovirus	
Vaccinia	

II. LIVE VIRAL VACCINES

Live attenuated viral vaccines are licensed in the United States and elsewhere for the prevention of measles, mumps, rubella, varicella, polioviruses 1–3, and rotavirus (Table II). When conditions of special risk for exposure exist, live attenuated yellow fever vaccine, live adenovirus, and vaccinia are given as prophylaxis. Live viral vaccines contain an infectious virus as the primary component, which has been

TABLE II
Noninfectious Vaccines for Prevention of Human Viral Diseases

Current vaccine	*Vaccines under development*
Polioviruses 1–3	Human papillomavirus
Influenza A and B	Human immunodeficiency virus
Hepatitis A	
Hepatitis B	Herpes simplex virus 1 and 2
Japanese encephalitis virus	Respiratory syncytial virus
Tick-borne encephalitis	Hepatitis C

attenuated in order to reduce or eliminate its potential to cause disease in the naive host. Vaccine strains are made from RNA viruses, including measles, mumps, rubella, poliovirus, rotavirus, and yellow fever, and from DNA viruses, such as varicella, adenovirus, and vaccinia. Attenuation of virulence is accomplished by laboratory manipulations of the naturally occurring, "wild-type" virus, which is referred to as the parental strain of the vaccine virus. The parental strains of live attenuated viral vaccines are obtained from an individual experiencing the typical disease caused by the virus. Alternatively, attenuation is achieved by taking advantage of host range differences in virulence between human and closely related animal viruses. Proteins made by the animal virus are similar enough to those encoded by the human pathogen to elicit protective, adaptive immune responses. When a suitable animal model is available, a reduction in the capacity of the vaccine virus to replicate is demonstrated and an alteration in the protential to cause disease may be documented. Although strain selection and characterization can be done *in vitro* and in animal models to predict safety, sequential evaluation of vaccine strains in individuals who have natural immunity, followed by gradual dose escalation studies in susceptible individuals, is required to prove safety for human use.

The attenuation of a live vaccine strain is defined clinically by its loss of the potential to cause disease. The attenuated virus should retain infectivity at the site of inoculation, which may be by subcutaneous injection or by oral or intranasal delivery to mucosal cells. In order to be attenuated, the tropisms of the parent virus that would otherwise allow it to produce damage to the host must be incapacitated. For example, the attenuation of polioviruses requires that the vaccine strains be incapable of infecting cells of the central nervous system. Some recipients may have reactions to live attenuated vaccines, including low fever, but the incidence of reactogenicity such as fever associated with the administration of effectively attenuated vaccine strains is low, and other manifestations such as rash are mild.

The attenuation of RNA and DNA viruses to produce vaccine strains used in most licensed vaccines is accomplished by traditional approaches in which the parent virus undergoes passage *in vitro,* using human or nonhuman cells, sequentially in both human and nonhuman cells, or by growth in chick or duck embryo cells in eggs. An additional strategy for achieving attenuation is to modify environmental conditions, such as adapting the virus to grow at low temperatures. Cold-adapted viruses are less able to replicate at human body temperature. Measles vaccine is made by passage in chick embryo cells, with further attenuation achieved by cold passage at 32°C. Rubella vaccine, RA27/3 strain, is derived by passage in human cells only, including growth at 30°C. Varicella vaccine was attenuated by passage in guinea pig embryo cells and cold passage. In the case of rotavirus, attenuation is achieved by reassortment of genes from strains infectious for humans with genes from related, nonhuman virus. As a consequence of these manipulations, the vaccine virus remains infectious, but its ability to replicate in the human host is limited and the cycles of viral replication that occur in the vaccine recipient do not result in a reversion to virulence.

By definition, the vaccine strain must retain genetic stability in order to preserve its attenuation. Sequence differences from the parent strain have been implicated in the attenuation of poliovirus strains 1–3 that are used to make live poliovirus vaccines. However, the genetic basis for the attenuation of most traditional vaccine strains is not known. The evidence that these strains are genetically stable is inferred from the preservation of the attenuation phenotype when the vaccine is given to susceptible individuals. Even when sequence information is available for vaccine strains, it is difficult to determine which sequence differences from the parent strain are most essential for the biologically observed modification of virulence. The definition of genetic markers of attenuation is complex because the traditional procedures for making live vaccines typically yield many variations in the genome sequence of the vaccine strain and genetic stability can be predicted to be multifactorial. In general, genetic stability is enhanced as the number of mutations in the vaccine strain increases. For example, the vaccine poliovirus type 1 has 56 mutations in 7441 nucleotides compared to only 10 of 7429 differences in the type 3 strain. Live attenuated vaccines also may contain mixed populations of the vaccine virus that have different genetic alterations, as has been described for rubella vaccine. In most cases, biological attenuation means that the vaccine virus also loses its transmissibility to other susceptible individuals who are in close contact with the vaccine recipient. However, when vaccine strains are transmissible, the genetic stability of the vaccine virus must also be preserved after replication in secondary contacts.

The immunogenicity of live attenuated viral vaccines depends on the selection of an appropriate infectious dose for inoculation, whether the vaccine is given by systemic or mucosal routes. The identification of the proper dosage regimen for administration is also important. Many live attenuated viral vaccines must be given as several doses in order to establish persistent adaptive immunity in the majority of naive recipients. Some live virus vaccines consist of mixtures of vaccine strains because protection must be conferred against disease caused by different subgroups of the wild-type virus, as illustrated by the trivalent oral poliovirus vaccine. In other cases, several live attenuated virus strains are combined to facilitate simultaneous immunization of the susceptible host against unrelated viruses, as exemplified by the measles–mumps–rubella vaccine. The challenge of designing these multivalent vaccines is to ensure that each attenuated vaccine strain is present in a high enough inoculum to allow it to replicate adequately at the site of inoculation in the presence of the other strains. A balance of the components must be achieved to prevent interference by more potent vaccine viruses which might impair the immunogenicity of the other vaccine strains. The establishment of adaptive immunity to all components may depend on a multiple-dose regimen, as is recommended for live attenuated polio vaccine. The timing between doses of live attenuated viral vaccines is also important because interference can occur when one live virus vaccine is given too soon after another. The required interval is usually at least 4 weeks to avoid reducing the infectivity of the second vaccine strain as a result of replication of the first vaccine strain or antiviral immune responses elicited by the first vaccine, such as interferon production.

Healthy young children constitute the primary target population for the live attenuated vaccines to prevent measles, mumps, rubella, varicella, polioviruses 1–3, and rotavirus. In contrast, the live yellow fever and adenovirus vaccines are used in individuals who are considered to have a particular risk of disease. Yellow fever vaccines are used to prevent disease in local populations and visitors to endemic areas. These vaccines are made from a strain first developed in the 1930s which was attenuated by passage in monkeys and then prolonged tissue culture passage. The vaccine strain causes a low level of viremia, which is also characteristic of infection with wild-type virus, but multiple sequence changes from the parent strain have been demonstrated and clinical experience demonstrates that its pathogenic potential to cause life-threatening dissemination is eliminated. Adenovirus vaccines against serotypes 4 and 7 have been used to control outbreaks among military recruits. Prevention of respiratory tract infection is achieved by the oral administration of live adenovirus in tablets which are coated to prevent acid inactivation in the upper gastrointestinal tract. In this instance, attenuation results from the route of inoculation without any molecular alteration of the viral genome.

III. NONINFECTIOUS VACCINES

Noninfectious vaccines are licensed for influenza, polio, hepatitis A, hepatitis B, rabies, Japanese encephalitis virus, and tick-borne encephalitis (Table II). The vaccines are made by inactivating infectious virus after growth in tissue culture or eggs, or by using protein components of the virus only. These vaccines are referred to as "killed," "inactivated," or "subunit" vaccines. This approach to vaccine design has the advantage of eliminating concerns about the infectious component of attenuated live viruses. Although attenuation of virulence is the major issue in making live viral vaccines, immunogenicity is the primary concern in designing inactivated vaccines. Alum is used as an adjuvant to provide the amplification of adaptive immunity that is achieved by viral replication in the case of live attenuated vaccines. The induction of a balanced host response against viral proteins is of critical importance in the production of inactivated vaccines, as illustrated by the occurrence of atypical measles disease in children who were immunized with a formalin-inactivated, alum-precipitated measles vaccine. Formalin-inactivated respiratory syncytial virus vaccine was associated with severe lower respiratory tract infection in immunized infants who were infected with the wild-type virus. Although formalin inactivation creates safe, inactivated vaccines for other viral pathogens, cross-linking by formaldehyde probably changed the conformation of measles and respiratory syncytial virus proteins, inducing antibodies against amino acid epitopes that were not elicited in the normal host response to viral proteins made during replication in host cells. This misdirection of the adaptive immune response resulted in immune-mediated disease, instead of protective immunity in some vaccine recipients. Since inactivated vaccines are not as immunogenic for inducing memory host responses as natural infection or live viral vaccines, most dose regimens for inactivated or subunit vaccines incorporate "booster" doses to ensure the long-term persistence of virus-specific immunity.

Inactivated influenza vaccine is used to protect individuals who are at risk for life-threatening disease during the annual epidemics of influenza A and B. The target populations for this vaccine are elderly adults, immunocompromised patients, and those with chronic pulmonary or cardiac diseases. Because influenza viruses undergo rapid antigenic changes, it is necessary to formulate the vaccine annually to contain the hemagglutinin and neuraminidase proteins from the two predominant circulating strains of influenza A and the major influenza B strain, which are identified through a global surveillance network. The component viruses are grown in embryonated eggs, inactivated by formalin, and combined in a trivalent vaccine. Subunit preparations of influenza vaccine are made by detergent treatment to increase relative concentrations of HA and NA proteins. In the case of influenza vaccine, the need for repeated immunization is dictated by the genetic capacity of influenza viruses to undergo antigenic drift and shift, requiring administration of the new vaccine to high-risk populations before each winter epidemic begins. Inactivated polio vaccine is also a trivalent vaccine made from formalin-inactivated strains of polioviruses 1–3. The manufacture of inactivated polio vaccine is complicated by the need to achieve complete inactivation of these nonattenuated viruses while maintaining immunogenicity that is protective against paralytic disease caused by each of the three polio serotypes. The current enhanced potency vaccine given as five doses beginning in infancy with later booster doses, is recommended as an alternative to live attenuated polio vaccine in developed countries. Regimens combining initial immunization with inactivated vaccine followed by doses of live attenuated vaccine are also effective. Inactivated polio vaccine must be given by injection, which is a practical limitation to its use in developing countries.

Whereas most viral vaccines are designed to prevent the disease caused by acute, primary infection, the benefit of hepatitis B vaccine results from preventing chronic active infection and the late sequelae of hepatic failure and hepatocellular carcinoma. In contrast to influenza and polio, hepatitis B virus does not replicate in tissue culture. The vaccine for hepatitis B consists of the surface antigen of the virus, which is a glycoprotein that forms the outer envelope of the virion. When expressed by introducing the gene sequence into yeast or mammalian cells, the hepatitis B surface antigen self-assembles into a particle structure, which contributes to its immunogenicity as a single viral protein and allows its use as an effective single protein subunit vaccine. Recombinant DNA vaccines have replaced vaccines in which surface antigen particles were extracted from plasma of chronic carriers. The success of hepatitis B vaccine depends in particular on the timely delivery of the vaccine to infants. Vaccination beginning at birth blocks the transmission of hepatitis B virus from carrier mothers to their infants, who are otherwise at high risk for chronic infection. Although both viruses cause hepatitis, hepatitis B is a DNA virus, whereas hepatitis A is an enterovirus belonging to the same family as polioviruses. New vaccines for hepatitis A have been licensed which contain formalin-inactivated virus grown in tissue culture and administered with alum or liposomal adjuvants. Hepatitis A vaccine is useful for susceptible adults who may be exposed by occupation, travel, and other risk factors as well as during community outbreaks. Hepatitis A vaccine may be recommended for universal administration to children in developed countries in which the cost of vaccination is acceptable even though the risk of serious disease in young children is low.

Inactivated vaccines are licensed to prevent three viral causes of central nervous system disease: rabies, Japanese encephalitis virus, and tick-borne encephalitis. In 1885, Louis Pasteur inoculated Joseph Meister with spinal cord material from infected rabbits that was inactivated by drying but contained some infectious virus. This work initiated a vaccine method based on use of nervous tissue from infected animals that continued to be used during the first half of the twentieth century, with later modifications made to improve viral inactivation. The current rabies vaccine is made from virus grown in human cells in tissue culture and inactivated with $-\beta$ propiolactone, which eliminates the risks of adverse reactions to myelinated animal tissues. The administration of the vaccine to exposed individuals is simplified to a 5 or 6 dose regimen compared to the 14–23 doses required for earlier rabies vaccines. Whereas most viral vaccines protect against infections acquired by human to human transmission, immunization of domestic animals with inactivated rabies vaccine is critical for disease prevention. Japanese encephalitis virus is a flavivirus, related to St. Louis encephalitis virus and other members of this family, which is maintained as a mosquito-borne pathogen in Asia. Although most infections are asymptomatic, some individuals develop encephalitis that is fatal or causes severe, permanent neurologic damage. The licensed vaccine for Japanese encephalitis is an inactivated preparation purified from infected mouse brain, although inactivated and live attenuated vaccines made in tissue culture are used in China. The need for immunization is restricted to populations in endemic areas and for travelers who are visiting rural areas in these countries during the peak season for transmission in summer and fall. Tick-borne encephalitis virus is also a flavivirus, with subgroups called Far Eastern and Western virus types. The distribution of endemic areas includes parts of Europe and Russia. The vaccine used in Europe is made from formalin-inactivated virus grown in chick embryo cells. Immunization is recommended for populations in endemic areas and for travelers to these areas who may have increased risk of exposure to ticks.

IV. VACCINE IMMUNOLOGY

Because the development of new viral vaccines takes years and is very costly, immunologic criteria are used to judge the probable efficacy of candidate vaccines. Laboratory assays for assessing vaccine immunogenicity measure the production of antibodies directed against viral proteins, including IgG and secretory IgA antibodies, as well as their functional capacity to neutralize the virus *in vitro* or to mediate antibody-dependent cellular cytotoxicity. Because of the importance of cell-mediated immunity for defense against viral infections, assays for cytokine production by T cells stimulated with viral antigens *in vitro* and for T cell-mediated cytotoxicity are useful measures of the establishment of virus-specific memory immunity. Establishing accurate correlates of protection must be done in field trails during which large cohorts of vaccinees are exposed to wild-type virus. In most instances, only simple labortory assays can be performed when many individuals must be tested. Serologic assays are used for this purpose even though protection is likely to require adequate T cell-mediated immunity. It is necessary to have a reliable expected attack rate for transmission of the viral pathogen, whereas under field conditions rates of transmission are affected by many variables, such as the proximity and duration of contact with the index case. Therefore, most clinical vaccine trials require large populations of subjects. Whether protective immunity is induced by viral vaccines is often proved conclusively only after widespread implementation of immunization programs, as illustrated by the impact of vaccines against childhood diseases such as measles, mumps, and rubella.

The specific goals of vaccine immunology are to demonstrate that vaccination induced adaptive immunity against relevant viral antigens in the naive host and, when possible, to identify immunologic responses that are associated particularly with protection of vaccine recipients against the usual consequences of infection with the wild-type virus. In practice, the definition of immunologic correlates of vaccine protection is rarely straightforward. The immunogenicity of vaccines is usually assessed by comparison with immune responses that follow natural infection with the same virus, but in most instances specific correlates of protection are not known for naturally acquired immunity. The redundancy of the mammalian immune system means that a broad range of adaptive immune responses to the pathogen can be measured in the healthy immune individual. For example, individuals who have antibodies to viral proteins can also be expected to have antigen-specific CD4+ and CD8+ T cells. Vaccine-induced antibodies, especially those with neutralizing activity against the virus, are considered the first line of defense against infection when the immunized host encounters the wild-type virus. These antibodies may limit initial replication at the site of viral inoculation. Primary vaccine failure is defined as a failure of the initial doses of the vaccine regimen to induce virus-specific antibodies. Effective vaccines are expected to elicit seroconversion in most vaccine recipients, which often requires the administration of several doses. Nevertheless, seroconversion is not invariably a predictable marker of protective immunity. Immunization usually induces a range of antigen-specific antibody responses in different individuals. In some cases, detection of any antibodies to viral proteins correlates with protection, whereas in other cases a "protective" titer is defined as greater than or equal to a particular concentration of antibodies. The appropriate laboratory marker of protection may also differ depending on the nature of the vaccine that is being evaluated. For example, inactivated vaccines often elicit high titers of virus-specific antibodies that correlate with protection, whereas live attenuated vaccines are more likely to induce cellular immunity and lower antibody titers. Despite these differences, the inactivated and live attenuated forms of vaccine may be equally effective against the same pathogen. Live attenuated viral vaccines often induce a more persistent cell-mediated immune response, which affords protection even when antibody titers fall below the threshold of detection in standard serologic assays. In the case of live viral vaccines, rates of vaccine virus shedding after the inoculation of naive subjects may be as reliable a marker of protection as immunologic assays. Depending on the assessment of risk, correlates of vaccine protection may be defined by deliberate, direct challenge of immunized volunteers with the wild-type virus.

Whether or not precise correlates of protection can be defined, immunologic assays are useful for demonstrating the effects of vaccine composition and host factors on the response to viral vaccines. These analyses provide valuable insights about the effect of age. For example, immunologic studies of live attenuated varicella vaccine revealed that adolescents and young adults require a two-dose regimen to achieve humoral and cell-mediated immune responses that are equivalent to those induced by a single dose in young children. The immunogenicity of vaccines given to young infants may be diminished by transplacentally acquired antibodies, as is observed in measles immunization. Immunologic assays are also useful to determine whether different viral vaccines are compatible when administered concurrently. Since the immunogenicity of viral vaccines in infants is influenced by nutritional status and factors such as the prevalence of intercurrent infections with gastrointestinal pathogens, vaccine formulations that are effective in developed countries may not be appropriate in other circumstances. A need to adjust vaccine dosage or regimen may be evident from comparative immunogenicity studies. Assessing the interval over which adaptive immune responses remain detectable is necessary because waning immunity may indicate the need for booster doses of the vaccine. In addition to primary vaccine failure in which the initial immunogenicity is inadequate to prevent disease caused by wild-type virus, secondary vaccine failures occur when immunity declines over time to non-protective levels. Finally, the control of vaccine-preventable disease often depends not just on immunogenicity in individual vaccinees but also on achieving adequate levels of herd immunity. Immunologic assays provide information necessary to perdict whether the local introduction of the virus is likely to be sustained through secondary transmissions and result in a community outbreak.

V. MOLECULAR APPROACHES TO VIRAL VACCINE DESIGN

The advances in molecular biology that have occurred during the past several decades have generated new opportunities for making viral vaccines.

Molecular approaches to the design of human viral vaccines will have a major impact in helping to address deficiencies of existing vaccines and allowing the invention of vaccines against infectious diseases that are currently not preventable by immunization (Table III). Molecular techniques are already being implemented to improve licensed vaccines, as illustrated by the use of cDNA clones to reduce the frequency of poliovirus mutations during vaccine manufacture and the use of reassortant methods to incorporate new influenza viral antigens into available virus strains that replicate to the levels required for production of the inactivated influenza vaccine. Hepatitis B vaccine is now made from recombinant surface antigen protein, and the live attenuated rotavirus vaccine contains genetic ressortants that incorporate genes from human and primate viruses.

Molecular approaches that are being developed to create new or improved vaccines include the genetic engineering of live attenuated viral vaccines, in which targeted mutations or deletions are made in genes that are determinants of virulence or tissue tropism, the synthesis of replication-defective viruses as vaccines, the expression of recombinant viral proteins and peptides from plasmid and in constitutively expressing mammalian cells, the synthesis of virus-like particles from viral proteins made in the absence of the viral genome, the administration of "naked DNA" corresponding to viral genome sequences that are immunogenic, and the use of attenuated human viruses or host range mutants as vectors for expressing genes from unrelated viruses. Progress is also being made in the creation of novel adjuvants such as cytokins or immunostimulatory DNA sequences that modulate the host response to enhance antiviral immunity.

TABLE III
Molecular Approaches to the Design of Human Viral Vaccines

Genetically engineered attenuation
Genome reassortants
Host range variants
Replication-defective viruses
Recombinant viral vectors
Recombinant proteins and peptides
Virus-like particles
DNA vaccines

VI. PUBLIC HEALTH IMPACT

The ultimate success of a viral vaccine is realized when the implementation of vaccine delivery programs results in a global reduction of the disease burden caused by the pathogen. The standard set by the smallpox vaccine campaign provides a challenge to eradicate other viral diseases that continue to cause serious disease and death. The worldwide control of measles and polio is the current priority for eradication initiatives. Even when effective vaccines are available, the need to vaccinate very high percentages of the susceptible population in order to block transmission is an obstacle to disease control. Viral vaccines are often labile unless frozen, necessitating an intact "cold chain" during transport to remote areas, and sterile needles and syringes must be available. Mass vaccine campaigns supported by funds from international agencies provide a practical response to these problems through "vaccination days," as was demonstrated by the successful administration of polio vaccine to millions of children in India in a single day.

The opportunity for reducing disease burden by immunization depends on the viral pathogen, how it is transmitted, and the pathogenic mechanisms by which it causes disease. Viral pathogens have evolved concurrently with human evolution to achieve persistence in human populations. Smallpox eradication succeeded by case identification and vaccination of close contacts, but polioviruses circulate by causing asymptomatic infection in most individuals. The cycle of transmission of these viruses may be broken by achieving high levels of vaccine immunity through several summer–fall seasons. Measles may be difficult to control because it is highly contagious, requiring only a few susceptibles in the population to cause an outbreak. Some viruses, most notably the herpesviruses and human immunodeficiency virus, cause a lifelong persistent infection, associated with intermittent or chronic viral shedding. Control of these viruses differs from that of those that cause acute infection because reintroduction of the virus into the population can occur readily. Human immunodeficiency virus presents the exceptionally difficult problem of marked antigenic diversity and rapid emergence of virus subpopulations that can escape adaptive immune responses.

Despite these obstacles, vaccine strategies are essential to reduce the impact of viral diseases because of the limited availability of effective antiviral drugs for most viruses, their short-term efficacy in many circumstances, and the relative cost of antiviral drugs compared with that of vaccines. The global impact of viral vaccines on public health is recognized as the most important intervention provided by modern medicine.

See Also the Following Articles

Global Burden of Infectious Diseases • Hepatitis Viruses • Influenza Viruses • Polio • T Lymphocytes

Bibliography

Ada. G., and Ramsay, A. (1997). "Vaccines, Vaccination and the Immune Response." Lippincott–Raven Press, Philadelphia.

Fields, B. N., Knipe, D. M., and Howley, P. M. (1996). "Fields Virology," 3rd ed. Lippincott–Raven Press, Philadelphia.

Long, S., and Prober, C. G. (1997). "Textbook of Pediatric Infectious Diseases." Lippincott–Raven Press, Philadelphia.

Plotkin, S. A., and Orenstein, W. A. (1999). "Vaccines," 3rd ed. Saunders, Philadelphia.

Verticillium

Jane Robb
University of Guelph

I. Culture of *Verticillium*
II. Life Cycle of Plant Pathogenic *Verticillium*
III. Vascular Wilt Disease
IV. The *Verticillium*–Host Interaction
V. Genetics of *Verticillium*
VI. Diagnostics
VII. Control

GLOSSARY

defense mechanism A physiological process that the host deploys in resisting pathogen invasion and colonization.

host tolerance A condition in which a pathogen colonizes a plant extensively but few or no symptoms develop.

pathogenicity factor An agent of a virulent pathogen which in some ways mediates the extent of damage to the host.

phytoalexins Antimicrobial toxic compounds, such as phenylpropanoids, isoprenoids, and acetylenes, synthesized in plants in response to pathogen infection.

tylose Outgrowth of xylem parenchyma cell which balloons into the xylem vessel lumen.

vascular wilt disease A plant disease usually involving foliar flaccidity, caused by a fungal or bacterial pathogen that colonizes the vascular system, usually the xylem.

vegetative compatibility The ability of individual fungal strains to undergo mutual hyphal anastamosis, which results in viable fused cells containing nuclei of both parental strains in a common cytoplasm (i.e., heterokaryon formation).

THE MICROBIOLOGY OF *VERTICILLIUM* encompasses the study of more than 40 diverse fungal species represented within the genus; some are nonpathogenic, whereas others infect a wide range of hosts including other fungi, insects, and plants.

The genus was erected in 1816 by Nees von Esenbeck, with the name *Verticillium* denoting the importance of the branched verticillate conidiophores (Fig. 1) which are a defining characteristic of all included members. Further taxonomic classification depends on other characteristics such as type and size of reproductive spores and survival structures and, where appropriate, type of host. By far the best studied are the following five species which are collectively responsible for wide ranging plant disease, called verticillium wilt, which affects a diversity of crops: *V. dahliae, V. albo-atrum, V. tricorpus, V. nigrescens,* and *V. nubilum,* all of which form characteristic resting structures. *Verticillium albo-atrum* has dark mycelium, *V. dahliae* has microsclerotia, and *V. tricorpus* produces both dark mycelium and microsclerotia along with chlamydospores. *Verticillium nigrescens* and *V. nubilium* produce only chlamydospores. This article emphasizes these plant pathogenic species, the diseases which they cause, and their control.

I. CULTURE OF *VERTICILLIUM*

Verticillium is readily cultured on agar plates or slants or in liquid solution. Popular media include potato dextrose, V8 juice and, Czapek's and there are also selective media for the various species. Grain cultures are preferred by some. Cultures can be maintained for short periods of time by subculturing; however, virulence may be rapidly lost and other undesirable physiological changes may occur. Therefore, storage on silica gel or in 25% aqueous glycerol

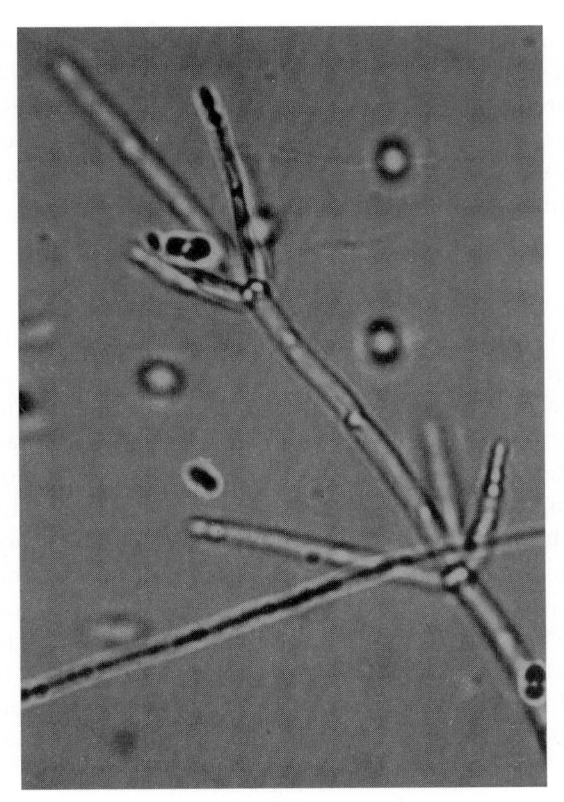

Fig. 1. Verticillate condiophores of *V. dahliae.*

solution at −80°C is usually preferable. Culture conditions, especially temperature, are important. *Verticillium albo-atrum* thrives at 21°C. *Verticillium dahliae* generally prefers warmer conditions (23–27°C), and the remaining plant pathogens prefer the temperature in between those of the other two. Temperatures higher than 28°C become detrimental, resulting in reduced growth rate, sporulation, and spore viability.

Cultures can be started by seeding antibiotic plates with microsclerotia sieved from soil. Alternatively, the fungus can be grown from surface-sterilized plant parts such as roots, stems, petioles, or leaves.

II. LIFE CYCLE OF PLANT PATHOGENIC *VERTICILLIUM*

A. Dormant Phase

Verticillium is a soil-borne pathogen. Various types of propagules serve as long-term resting structures either in laboratory cultures or in infested soils. For example, dark mycelia are the major resting structures of *V. albo-atrum,* and microsclerotia, mycelia, and clusters of hyaline cells all contribute to survival of *V. dahliae. Verticillium* has been found to survive in farm fields up to 15 years in the absence of a host.

The resting structures are stimulated to germinate by root exudates which contain a variety of nutrients as well as enzymes, alkaloids, nucleotides, and inorganic ions. However, the activity of plant exudates is very nonspecific and propagules will germinate equally well in the presence of host and nonhost plants. Germination gives rise to conidia and infection hyphae which invade the roots either via naturally occurring wounds or through the root cap.

B. Parasitic Phase

Once the initial invasion has been accomplished, hyphae grow both inter- and intracellularly. Mycelia that penetrate through the root cap and meristem encounter little resistance on their way to the vascular tissue. However, fungus that enters through wounds may encounter barriers such as lignified epidermal or cortical cell walls and suberized endodermis, and unless invasion is massive penetration of the xylem usually does not occur. A variety of studies involving different hosts suggests that it requires 2 or 3 days for *Verticillium* to grow into the vascular system and enter the xylem elements.

Once established in the xylem, hyphae proliferate and rapidly penetrate bordered pits to colonize adjacent vessels in a lateral direction. Eventually, conidiospores are produced and released to travel upwards in the transpiration stream until stopped by vessel end walls. These spore-trapping sites become new infection centers in which the spores germinate to produce germ tubes which again invade adjacent vessel elements and initiate a second round of hyphal growth and sporulation. In this way *Verticillium* progresses in stepwise fashion into the aerial portions of the plant, rapidly becoming fully systemic.

Unlike other fungal vascular pathogens, such as *Fusarium* and *Ceratocystis, Verticillium* apparently does not leave the xylem elements until the plant is moribund. In these final stages of the parasitic phase, the fungus penetrates into xylem parenchyma cells and surrounding tissues and forms resting structures which are returned to the soil when the host dies.

III. VASCULAR WILT DISEASE

A. Symptoms

All vascular wilt diseases, whether caused by *Verticillium* or other fungal or bacterial pathogens, result in similar symptomology. One of the main diagnostic features used by field pathologists is a brown discoloration of the vascular tissues which is easily visualized by cutting through a stem or petiole. The browning effect results from the deposition of phenolic compounds which act as antimicrobial agents. Other common symptoms include stunting, leaf epinasty and abscission, and the development of distinctive foliar symptoms such as flaccidity and chlorosis followed by necrosis (Fig. 2). If symptoms are severe enough the plant will die. Some *Verticillium* pathosystems produce two forms of the disease, one very virulent and the other milder (e.g., "progressive" and "fluctuating" wilt of hops and "defoliating" and nondefoliating wilt of cotton).

B. Crops Affected

Verticillium can infect virtually all agriculturally significant plant species other than some grasses, causing billions of dollars in yield loss annually. Important crops in which *Verticillium* is a widespread, often limiting factor to production include hops, cotton, olive, potato, tomato, and eggplant. Disease distribution is worldwide. However, *V. albo-atrum* prefers cooler climates and *V. dahliae* the warmer latitudes, reflecting their individual temperature preferences. Together these two *Verticillium* species account for the vast majority of plant disease. The other three pathogenic species occasionally produce mild symptoms in very specialized niches.

Fig. 2. *Verticillium* wilt disease of potato.

IV. THE *VERTICILLIUM*–HOST INTERACTION

A. Resistance, Tolerance, and Susceptibility

When the virulence of the pathogen is sufficiently great and the defenses of the host are sufficiently weak, a susceptible interaction occurs; the fungus proliferates in the host and disease symptoms become overt. In a resistant interaction *Verticillium* colonization is very restricted, often to the root and base of the stem, and "wilt" does not develop. In some pathosystems resistance is controlled by one or a few major genes with corresponding virulence genes in the pathogen, following Flor's well-established gene-for-gene concept. The dominant Ve gene of tomato is a classic example. However, in many cases, such as alfalfa, resistance seems to be polygenic.

"Tolerance" has been defined in various ways by plant pathologists, but the preferred usage is that suggested by D. D. Clarke in 1986. In a tolerant interaction *Verticillium* proliferates and spreads throughout the plant to a level normally associated with susceptibility, but the plant develops few if any symptoms. Tolerance generally has been associated with polygenic inheritance in the host.

B. Pathogenicity Factors

Some possible pathogenicity factors are summarized in Fig. 3 and listed, along with some additional factors, as follows:

1. Mycelial occlusion: Fungal hyphae are an obstruction which, in extensively colonized vessels, can block water flow and contribute directly to water deficit in the leaves.
2. Elicitors and suppressors: Fragments of fungal walls may induce or, in some cases, suppress plant defense responses.
3. Growth regulators: *Verticillium* produces indole acetic acid (IAA)-like compounds which are very similar to plant IAA. High levels of IAA have been demonstrated in *Verticillium*-infected hosts, but the contribution of fungal IAA to the elevated pool is unknown. IAA levels have profound, if complex, ef-

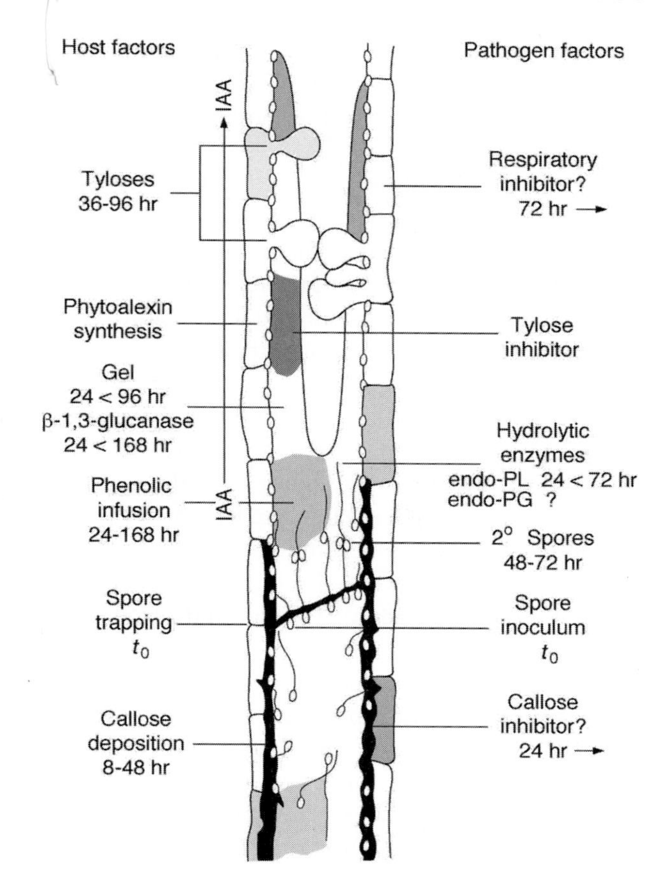

Fig. 3. Time–space model of host–parasite interactions in an infected vascular element and the surrounding contact parenchyma cells. The time frame is the period from t_0 through t_7 (the 7-day period after initial infection through a severed taproot), and the space frame corresponds to S_0 and S_1 where S_0 and S_1 are longitudinal segments of the vascular cylinder corresponding to the initial inoculum uptake distance (S_0) and the first spore passage distance (S_1). Shown are a spore-trapping site, which delays the advance of the pathogen and marks the end of S_0, and (left) components of the host defense system that subsequently serve to localize and neutralize the infection. (Right) Pathogen factors that may circumvent or overcome the host responses and permit systemic distribution of the parasite within the plant. Times noted are hours after t_0, the time of initial infection. The symbol $<$ indicates an increasing response during the period indicated; arrows indicate continuing responses. PL, pectin lyase; PG, polygalacturonase (reproduced with permission from Beckman, 1987, Fig. 12).

fects on many other factors of the interaction such as other hormones and antibiotic substances produced by the plant as well as both plant and fungal enzymes.

4. Fungal hydrolases: *Verticilliums* have been shown to secrete a variety of extracellular enzymes which target plant cell wall material (e.g., polygalacturonase, cellulase, and pectin methylesterase). Roles for these enzymes in initial invasion of the root, subsequent colonization of the vascular system, and the final parasitic stages, when the fungus spreads to other tissues in the stem and leaf, are obvious.

5. Fungal vivotoxins: Vivotoxin is defined as a substance produced in the infected host by the pathogen and/or its host, which functions in the production of disease but is not itself the initial inciting agent of disease. A variety of compounds from *Verticillium* culture filtrates have been shown to have toxic properties and even induce wilt-like symptoms; however, there is little proof any of these are produced naturally within infected plants. Thus, although toxins may be found to play a role in disease induction or symptom expression in the future, much more work is required.

C. Defense Mechanisms

Often, defense mechanisms are classified broadly as biochemical or structural, although these are obviously related; plant physiology is necessarily involved in mounting structural barriers. Some of the more thoroughly studied defenses are summarized in Fig. 3 and are listed, along with some additional factors, as follows:

1. Phytoalexins: Speed, magnitude, and site of accumulation of antimicrobial compounds are critical to resistance in numerous *Verticillium* pathosystems. For example, resistance in cotton appears to depend on timely amassing of terpenoid compounds in tylose-occluded vessels.

2. Plant hydrolases: Plant cells have been demonstrated to secrete hydrolytic enzymes such as β-1,3 glucanase, chitinase, and β-glycosidase into xylem vessel walls in response to *Verticillium* colonization. These enzymes target fungal cell walls and have been implicated in hyphal lysis.

3. Tylose formation: Tyloses, collectively, can block both water flow and fungal colonization. Tylosis is induced by *Verticillium* in vessel elements above conidial trapping sites.

4. Gelation: Gels form in vessels when fungal and plant hydrolases release cell wall fragments which subsequently accumulate in the lumina. Calcium cross-linking produces a gel-like substance that is further stabilized by phenolic infusion. Gels help to occlude vessels and act as phytoalexin-saturated barriers to *Verticillium*.

5. Wall modifications: Callose deposits form inside plant cell walls which are in close proximity to *Verticillium* as a very early defense against penetration of parenchyma cells during invasion of the root or xylem. Because of their complex chemical structure, plant lignins and suberins resist the activity of fungal hydrolases. Thus, lignification and suberization of cell walls is induced by *Verticillium* as an effective defense mechanism. Some of these barriers, such as the suberinized endodermal walls of the root, are preformed; others, such as suberin coatings of vessel walls and pit membranes, are an active defense induced by the pathogen.

D. The Interaction of Fungal Pathogenicity Factors and Plant Defense Mechanisms

The factors involved in producing wilt have been well studied but are so complex that particular roles in specific interactions remain unclear. However, there is general agreement that in any one pathosystem the same fungal and plant factors participate in both compatible and incompatible interactions, with the difference being in the amount, timing, and position of the response. In resistant plants the parenchyma cells immediately adjacent to spore trapping sites respond very rapidly, deploying an array of defense responses (see Section IV,C) which effectively block pathogen escape. In susceptible plants these responses are temporarily delayed or suppressed in the immediate vicinity of the pathogen allowing *Verticillium* to proliferate, penetrate new vessel elements, and spread systemically. The same responses then contribute to symptom expression.

V. GENETICS OF *VERTICILLIUM*

Most evidence suggests that whether *Verticillium* is growing on a plate or in a plant, the cells are normally haploid. Because of the very small size of the chromosomes the exact chromosome number is unclear. Several lines of evidence suggest that there are four; for example, in 1967 Hastie defined four linkage groups genetically. Although *Verticillium* has no sexual cycle, it does have a parasexual one—two hyphal cells meet and fuse to form binucleate heterokaryons. Synkaryons apparently arise occasionally as well, and mitotic recombination can be demonstrated both in culture and in the plant.

Variants occur readily in monokaryotic haploid culture. Most are stable and appear to be genuine genetic mutants. These may include morphological, nutritional, and pathogenicity mutants of various kinds. The ease with which such mutants can be obtained should facilitate future molecular studies.

Verticillium is ripe for molecular analysis. The molecular technologies now exist to probe into this fascinating arena and *Verticillium* has been transformed successfully but none of these tools have been used to advantage. The structure, regulation, and function of the avirulence genes and pathogenicity factor genes, which are very important to the outcome of *Verticillium*–host interactions, remain unstudied. This is a key area of focus in which future research is vital.

VI. DIAGNOSTICS

A. Traditional Diagnostics

The field of diagnostics encompasses the identification of a microorganism to the species or even the subspecies level as well as its quantification. Traditionally, identification of a *Verticillium* isolate required reisolation of the pathogen followed by light microscope examination and standard taxonomic classification. Classical techniques to quantify the amount of fungus include microscopic assessment of host tissue sections, maceration and plating of host tissues, or chemical estimation of fungal chitin. The amount of fungi in soil can be estimated by wet sieving, dilution plating, or the agar-film technique. Such methods are very laborious, time-consuming, and at best only semiquantitative.

B. Molecular Diagnostics

During the past 5 years there has been a rapid proliferation of molecular assays designed to improve *Verticillium* diagnostics. One approach utilizes ELISA techniques. A quantitative immunoassay established by using monoclonal antibodies can detect as little as 4 ng of fungal protein; unfortunately, cross-reactivity between *V. dahliae* and *V. albo-atrum* limits its application.

Polymerase chain reaction (PCR) technologies also offer very promising opportunities for further advancement. These methods depend on the development of DNA probes which identify nuclear or mitochondrial genes of *Verticillium;* the intervening transcribed spacer regions of the rRNA genes have been a favorite target. Primer sets which can distinguish all pathogenic species are now in routine use. Coupled with the use of internal control DNAs which serve as standards for quantification and improved simple methods for extracting nucleic acids from plant tissues and soils, these techniques are very powerful tools because they are fast, accurate, and extremely sensitive. Further classification of species into various subgroups can be carried out by studying restriction fragment length polymorphisms, another powerful tool for the determination of molecular variation at the DNA level. Also, randomly amplified polymorphic DNA PCR techniques, which use a wide range of decamers with different base compositions and GC content as primers, are capable of differentiating various strains of *Verticillium* and further extend our ability to subclassify. Of all the diagnostic methods available, those based on PCR technologies are probably the most promising in the long term for both pathogenic and nonpathogenic species. Because they are fast, cost-effective, and suitable for automation, they seem to be the most adaptable to the large-scale screening that would be required for large-scale taxomic screening or for agricultural use.

C. Vegetative Compatibility Groups

Vegetative compatibility can be demonstrated by indirect genetic complementation tests in which two isolates of varying genetic background are grown together on minimal medium plates. Each isolate is usually represented by recessive auxotrophic or pigmentation mutants which are complementary to each other. Wild-type growth can occur only if heterokaryons are formed.

However, not all strains are mutually compatible. It is generally recognized that there are approximately five vegetative compatibility groups (VCGs) occurring worldwide (VCG1–VCG5). Strains within a single VCG are compatible with each other but not with strains from other VCGs.

VCGs are important from a diagnostic perspective because certain VCGs can be associated with particular geographical or host ranges. For example, all potato strains of *V. dahliae* found in the United States appear to belong to VCG4. In a practical sense this means that a farm field found to be infested with VCG2 *V. dahliae* strains, resulting from prior tomato crops, would still be suitable for potato rotation. On the other hand, a field infested with other VCG4 strains clearly would not. The use of VCGs can be an additional diagnostic aid to supplement results obtained by either traditional diagnostic tools or molecular ones. Currently, several methods of diagnosis are probably better than one.

VII. CONTROL

In nature, plant health is the rule rather than the exception. This occurs, in large measure, because other organisms in a complex environment strike an ecological balance that helps hold potential pathogens in check. Vascular pathogens share a unique ecological niche in which they spend a large part of their life cycle sequestered within the xylem, in which they can successfully avoid many of these competitive influences. For this reason, control measures, particularly biocontrol techniques, generally target those stages of the life cycle outside the vascular system. Control techniques favored for vascular pathogens are usually classified into two categories: chemical control and biocontrol. The following are the most common techniques employed in control of *Verticillium* wilt:

1. Chemical application: Application of chemical fumigants is probably the most widely used and successful method of controlling *Verticillium* for agricultural purposes. Metham sodium products are very popular because they can be applied easily through sprinkler irrigation systems and are usually very effective.
2. Crop rotation: This is one of the oldest methods of control. Short-term rotations of 2 or 3 years with alternative crops that are not susceptible to *Verticillium* can be effective if the levels of pathogen are near threshold levels.
3. Soil solarization: Heating the soil targets microsclerotia and other long-term storage propagules, effectively reducing the initial inoculum in the soil. This method has been used with considerable success in Israel and some areas of the United States.
4. Balanced management strategies: Farming practices can be managed to reduce other stresses on the crop, promoting general vigor and increased host resistance. These can include balanced fertility programs based on soil and foliar analysis and the manipulation of moisture levels in the fields.
5. Resistant cultivars: Planting resistant cultivars is the most practical solution for many crops. Unfortunately, in some crops such as potato the most resistant cultivars are not the ones preferred for commercial use, and in others such as eggplant (i.e., universal susceptibility) resistance is limited. The vast majority of resistance breeding programs still use traditional breeding strategies based on Mendelian genetic principles. Modern genetic engineering approaches hold much promise for creating new cultivars with increased resistance to *Verticillium*. Currently, progress in this direction is hampered by lack of understanding of the structure, function, and regulation of the resistance (R) genes, and knowledge of the defense genes of the host is also in its infancy. Nevertheless, the tomato Ve gene has been trans-

ferred into potato, providing increased resistance to *V. dahliae* race 1. One advantage of genetic engineering is that genetic material can be incorporated from a wide range of biological sources. For example, in a recent Australian study *Verticillium*-susceptible cotton was transformed using a gene construct developed from a full-length tobacco chitinase gene. The resulting transgenic chitinase-expressing cotton line was more tolerant of the pathogen. This is clearly another key area of focus for future research.

Until recently, farming practice relied heavily on chemical control. However fumigants are costly in dollars and costly in terms of environmental hazard. The current trend among producers, government, and the general public is a gradual reduction in pesticide use countered by increasing reliance on integrated pest management strategies which focus on the use of biocontrol methods managed individually for the unique situation. The integrated approach tends to be slower to implement but more long-lasting in effect; for maximum effectiveness it requires constant monitoring of *Verticillium* in the environment. In this sense commercial production is the main beneficiary of the recent development of molecular diagnostics. Further developments in diagnostic techniques and improvements in integrated pest management strategies will go hand and hand in the future.

See Also the Following Articles

Plant Disease Resistance • Plant Pathogens • Soil Microbiology

Bibliography

Beckman, C. H. (1987). "The Nature of Wilt Diseases of Plants." American Phytopathological Society, St. Paul, MN.

Beckman, C. H., and Roberts, E. H. (1995). On the nature and genetic basis for resistance and tolerance to fungal wilt diseases of plants. *Adv. Bot. Res.* **21**, 35–77.

Mace, M. E., Bell, A. A., and Beckman, C. H. (1981). "Fungal Wilt Diseases of Plants." Academic Press, New York.

Robb, E. J., and Nazar, R. N. (1995). Factors contributing to successful PCR-based diagnostics for potato and other crop plants. *In* "Diagnostics in Crop Production" (G. Marshall, Ed.), pp. 75–84. Major Print Ltd., Nottingham, UK.

Tjamos, E. C., Heale, J., and Rowe, R. C. (1998). "Advances in *Verticillium* Research and Disease Management." American Phytopathological Society. St. Paul, MN.

Viruses

Sondra Schlesinger and Milton J. Schlesinger
Washington University School of Medicine

I. Structure, Composition, and Classification
II. Historical Development
III. Replication
IV. Genetics
V. Virus–Host Interactions
VI. Viruses as Vectors for Gene Expression

GLOSSARY

genome replication The synthesis of multiple copies of a virus genome transcribed from a complementary strand of RNA or DNA.

plaques Clear areas in a lawn of bacteria or in a monolayer of cultured cells caused by the cytopathic effects of a virus infection. The plaque arises from infection of a single cell. The spread of the virus from the original cell to neighboring cells is limited by the addition of a solid medium (such as agar) to the dish which confines the spread to a small area. The number of plaques is used as a measure of the number (titer) of infectious particles.

transformation A change in the properties of a cell usually associated with a loss in growth control. Transformed cells are more tumorigenic than their normal counterparts.

virion Extracellular virus particle. These particles may be only nucleoproteins or the nucleoprotein may be surrounded by an envelope consisting of lipid and protein.

virus replication Synthesis of new virus particles includes both replication of the genome and assembly of virus particles.

IN 1953, LURIA DEFINED VIRUSES as "submicroscopic entities, composed of nucleic acid and protein, capable of being introduced into specific living cells and of reproducing inside such cells only." Any definition of a virus, however, should emphasize the unique mechanism of their replication: Viral nucleic acid genomes and virus-encoded proteins are synthesized as separate components followed by their assembly into new particles within the host cell. Organized growth followed by binary division, characteristic of all other microorganisms and eukaryotic cells, does not occur during virus replication. A second distinctive characteristic of viruses is that they are obligate parasites and depend totally on metabolic activities of the host cell for progeny formation. Most viruses are much smaller in size than bacteria; this factor led to the recognition that they were distinct entities from bacteria as causes of transmissible diseases.

The term "virus" is derived from the Latin word for poison and stems from the original observation that a virus infection caused detectable damage to the host organism, often leading to lethality.

I. STRUCTURE, COMPOSITION, AND CLASSIFICATION

Virus particles are mostly in the shape of helical rods or icosahedral spheres in which organized arrays of protein subunits surround and protect the nucleic acid genome. Several viruses are pleomorphic in shape and others, such as some of the bacteriophages, have complex structures composed of a head which contains the packaged genome and a tail which functions in the attachment to the host cell and in the injection of the genome into the cell. In general, cell-free viruses, referred to as virions, are structured to be relatively stable to most environmental conditions but also capable of conversion to a labile form for rapid disassembly in their host cell.

Many viruses have highly uniform and symmetrical structures and several have been crystallized. X-ray diffraction patterns of these crystals have led to the determination of the virus structure at the atomic level. The first high-resolution structure based on X-ray diffraction was tomato bushy stunt virus, an icosahedral plant virus. Structures of the small icosahedral RNA animal viruses, poliovirus and rhinovirus, quickly followed and enhanced technologies soon led to structures of the small DNA viruses, polyoma and SV40. More than two dozen complete viral capsids have been solved by X-ray crystallography. These have revealed a common structure among the small icosahedral viruses consisting of a protein subunit core composed of a wedge-shaped, eight-stranded anti-parallel β-barrel motif. Atomic structures of several virus subcomponents, such as the hemagglutinin and neuraminidase of influenza virus, the hexon protein of adenovirus, and the protease and reverse transcriptase of human immunodeficiency virus (HIV), have been determined.

Electron microscopy is required for visualization of viruses and was utilized initially to also interpret structures of viruses. Isolated, purified virus particles and those associated with intracellular structures can be detected after fixation, dehydration, and staining of samples (Fig. 1). Technical advances in electron microscopes and computer analysis of images have allowed virions to be examined in vitreous ice (cryo-electron microscopy) and have led to a determination of virus structures with resolution on the order of 9–10 Å.

In addition to protein and nucleic acid, many viruses are enveloped by a lipid bilayer which contains transmembranal proteins encoded by virus genomes, but the lipids and lipid components are derived from the host cell. They are selectively incorporated into the virus particle during the stages of virus assembly and budding from host cell membranes. The proteins are arrayed as oligomeric spikes protruding outward from the membrane. Many viruses also contain enzymes, encoded by virus genomes, that are required for the early stages of virus replication.

The major groups of viruses are distinguished by their nucleic acid genome, which consists of either DNA or RNA. Except for the retroviruses, which are diploid, all viruses contain only a single copy of their genome. The nucleic acid can either be single or double stranded and arranged as a single molecule or in separate segments. The size of the nucleic acid ranges from 1000 or 2000 to 300,000 bases or base pairs, which is equivalent to 3 or 4 genes to more than 100 genes. The genome may be circular or linear and covalently linked or modified at its termini by various chemical structures.

Viruses with DNA as their genome are classified based on whether they are single or double stranded. Viruses with RNA as their genome are classified into three groups. One contains single-strand RNA that has been designated as positive sense because the genomic RNA serves as an mRNA and the isolated genomic nucleic acid is infectious by itself. The second group is composed of RNA genomes that have been designated as negative sense because the genomic RNA is complementary to the viral mRNAs. The third group contains double-stranded RNA. Negative-sense RNA and the double-strand RNA viruses require the initial transcription of the input genome RNA by virion-associated transcriptases upon entry to produce mRNAs and initiate the replication cycle. Additional viruses, i.e., the retroviruses and hepadnaviruses, are classified as RNA and DNA reverse-transcribing viruses.

Within each of the groups, the viruses are further subdivided based on the host used for their replication, i.e., bacteria, protozoa, fungi, plants, invertebrates, and vertebrates. Most viruses are restricted to a single species of host and within a multicellular host grow only in certain cells. Some viruses use more than one species as a host during their natural life cycle. One such group was originally called arboviruses (arthropod-borne viruses) because they are transmitted by mosquitoes or other arthropods to vertebrates, replicating in both hosts.

Early efforts to classify viruses were based on common pathogenesis, common organ tropism, or common ecological niches or modes of transmission. These attempts broke down when it became clear that the same virus could cause vastly different symptoms in different hosts. For example, the monkey B virus (a herpes virus) is asymptomatic in these animals but usually lethal in humans. In addition, different viruses may cause similar symptoms; for example, rhinoviruses, coronaviruses, and adenoviruses all can cause the common cold. As more information about isolated and purified viruses was obtained, it became more rational to group viruses based on common structures and composition. Table I illustrates some examples of the current classification system. A complete taxonomic list contains one order, 71 families, and 164 genera. Approximately 4000 different species of virus are placed in this classification. Virus orders are groupings of families that share common characteristics but are distinct from other virus families and orders. Most virus families have distinct particle morphology (Fig. 1); similar physicochemical properties of the genome, such as particle mass, density, and pH stability; similar genome structure with regard to type, size, and segmentation of the nucleic acid as well as strategies of replication; similar kinds of proteins and lipids; and antigenic properties. Subfamilies allow for separation of members that have more complex relations within the family. Genera group viruses that share common characteristics which distinguish them from viruses of other genera. A virus species has been defined by the International Committee on Taxonomy of Viruses as a polythetic class of viruses that constitutes a replicating lineage and occupies a particular ecological niche.

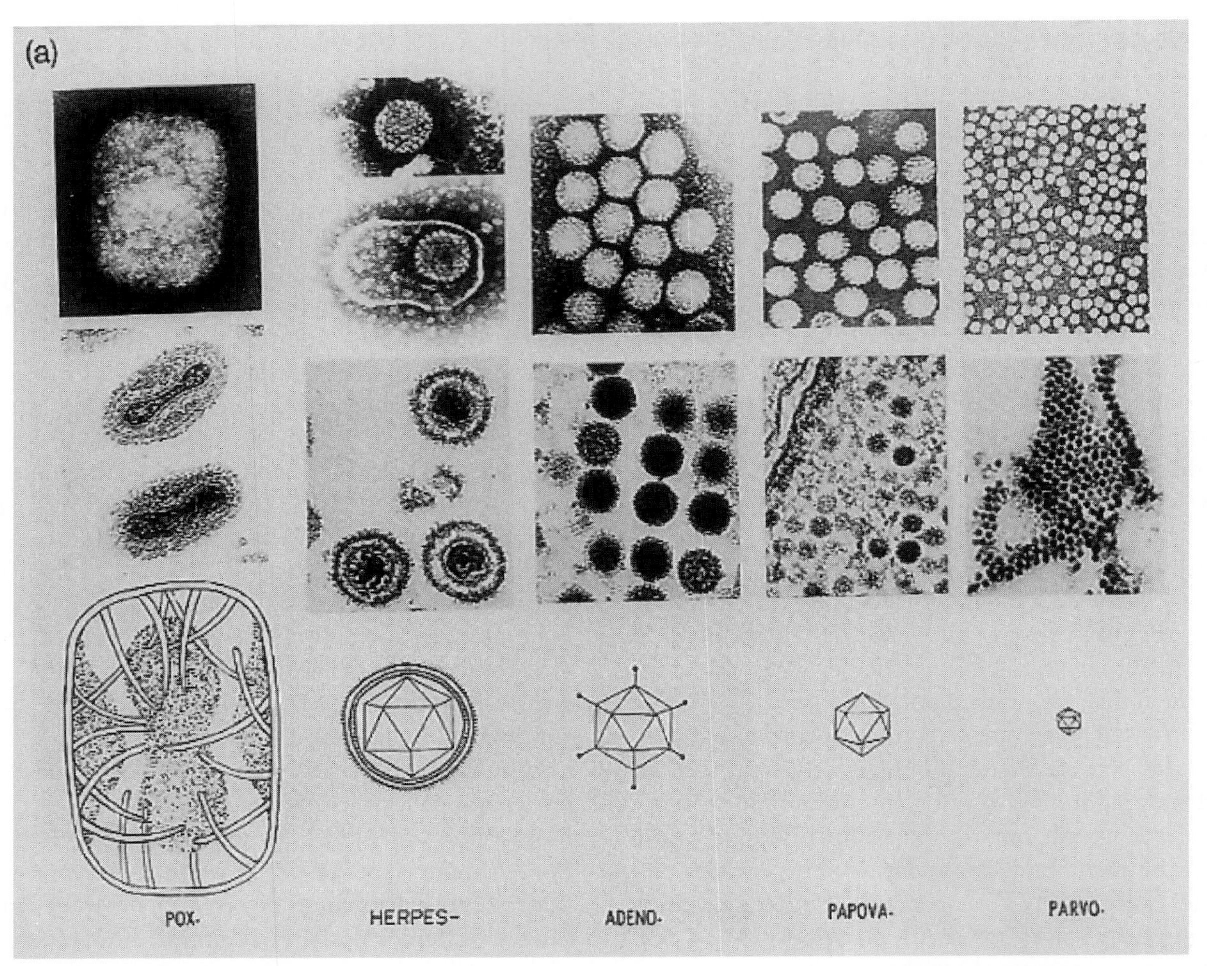

Fig. 1. Examples of several DNA viruses (a) and RNA viruses (b). The top rows show negatively stained preparations of extracellular virus; the middle rows show virus particles in thin-sectioned cells. Magnification × 50,000. The bottom row depicts the viruses and indicates relative sizes. Myxo refers to the Orthomyxoviridae family (reproduced with permission from Landry and Hsiung, 1996, copyright Yale University Press).

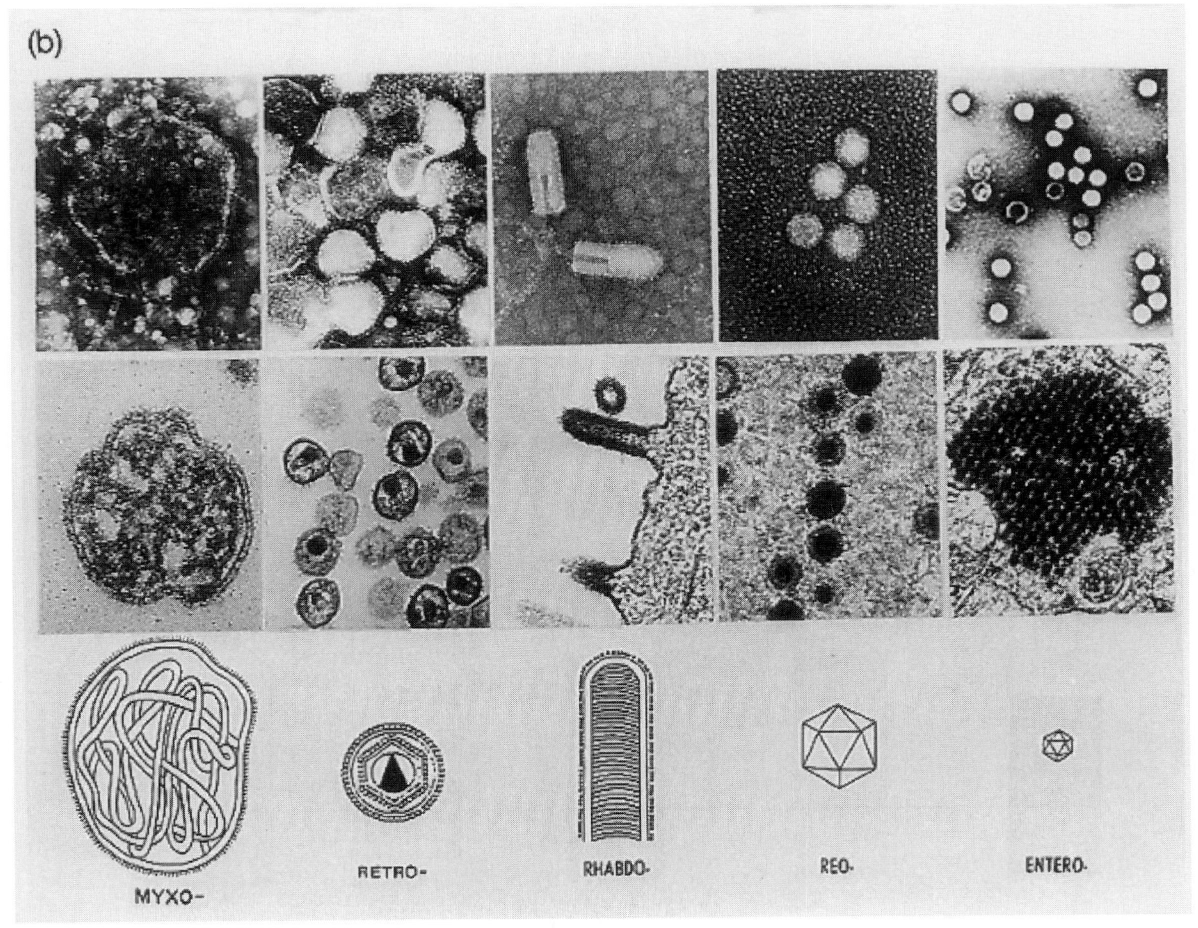

Fig. 1. continued.

Supergroups represent a new kind of virus classification which recognizes similarities in genome sequences among positive-sense RNA viruses that cut across traditional families. Most of these are in genes encoding enzymes involved with virus replication and this suggests that these viruses evolved from a common ancestor. One superfamily includes the enveloped positive strand RNA alphaviruses, a member of the Togaviridae family, the human hepatitis E virus, a nonenveloped RNA virus that has not been classified, and the plant virus family Bromoviridae.

Some viruses are referred to as satellites in that they require other closely related virus gene expression in order to complete their replication cycle. These include hepatitis delta virus, which grows only in the presence of coinfection with hepatitis B virus, and the adenoassociated virus, a member of the Parvoviridae family, which requires coinfection with adenovirus to replicate.

II. HISTORICAL DEVELOPMENT

The first virus discovered was tobacco mosaic virus (TMV), the agent responsible for a disease in tobacco plants. That this disease was transmissible by an agent produced in the sick plant was noted by Adolph Mayer in Holland in 1879. The identification of the agent as nonbacterial was made in 1892 in St. Petersburg by Ivanovsky, who recovered the agent in a filtrate from a filter that blocked passage of bacteria. The term virus had been used somewhat generally in the late nineteenth and early twentieth centuries to signify a toxic preparation of microorganisms

TABLE I
Examples of Virus Taxonomy[a]

Family	Genus	Host category type species
	Double-stranded DNA viruses	
Siphoviridae	Unnamed, the λ-like phages	Bacteria: coliphage λ
Baculoviridae	Nucleopolyhedrosis	Invertebrates: *Autographa californica* nuclear polyhedrosis virus
	Granulovirus	Invertebrates: *Plodia interpunctella* virus
Adenoviridae	Mastadenovirus	Vertebrates: human adenovirus 2
	DNA and RNA reverse transcribing viruses	
Retroviridae	"Unnamed mammalian type B retroviruses"	Vertebrates: mouse mammary tumor virus
	Lentiviruses	Vertebrates: human immunodeficiency virus
	Negative-sense, segmented RNA viruses	
Orthomyxoviridae	Influenza virus A, B	Vertebrates: influenza A A/PR/8/34(H1N1)
Bunyaviridae	Hantavirus	Vertebrates: Hantaan virus
	Positive-sense, single-stranded RNA viruses	
Picornaviridae	Enterovirus	Vertebrates: polio virus 1
	Rhinovirus	Vertebrates: human rhinovirus 1
Togaviridae	Alphavirus	Vertebrates: Sindbis virus

[a] From Murphy (1996), in *Fields Virology* (B. N. Fields, D. M. Knipe, and P. M. Howley, Eds.), copyright Lippincott–Raven press, 1996.

made by biological material but distinct from other known microorganisms. In 1898, Beijerinck repeated Ivanovsky's experiments but proposed that the transmissible agent was a fluid rather than a virus particle. The first animal virus reported was the filterable agent that caused foot and mouth disease in cattle which was discovered by Loffler and Frosch in 1898; the first human virus was identified as the filterable agent that caused yellow fever and was discovered by Walter Reed and associates of the U.S. Army Commission in 1900. Viruses causing tumors were first noted in chickens in 1911. They were discovered by Peyton Rous (at the Rockefeller Institute), who also worked on the rabbit papilloma virus identified by Shope in the late 1920s. In all these cases, it was the ability to cause disease in its host that actually defined and led to the name of the agent, and its designation as a virus was based on passage through filters that retained bacteria.

In 1915, F. W. Twort noted that viruses could kill bacteria. Also in 1915, F d'Herelle observed clear areas in lawns of shigella bacilli spread on agar culture dishes and referred to them as *taches vierges* or plaques. He showed that the killer agent was a filterable virus and called it bacteriophage in a 1917 publication that described the plaque assay, which involved a limiting dilution procedure of the toxic agent and led him to conclude that the agent was particulate. d'Herelle first described the binding of virus to a host cell as the initial stage of the infectious cycle, noting that the initial adsorption step was a major determinant of host cell specificity. He described cell lysis as the cytopathic effect of virus growth in the bacteria, which led to the release of newly replicated virus. Bacteriophages were discovered for many other bacteria and were initially tested as potential therapeutic agents for curing bacterial diseases.

The first evidence that viruses contained nucleic acids as well as proteins was obtained from experiments with bacteriophage by M. Schlesinger in Germany in the early 1930s. He also determined the mass for the phage particle based on diffusion studies, and his studies led to the hypothesis that viruses were essentially small, dense nucleoprotein particles.

The morphology of viruses had to await the development of the electron microscope because they were too small to be resolved by optical instruments. TMV was the first virus visualized in the electron microscope, which was developed in Germany in the late 1930s. A T-even bacteriophage was the first virus specimen observed with the electron microscope in the United States by Tom Anderson in 1942.

Crystals of TMV were obtained by Wendell Stanley in 1935, and their chemical analysis revealed that 5% of their mass was RNA. An X-ray diffraction pattern of these crystals was reported by Bernal in 1941 and showed the repeating unit structure of the virus rod. Experiments in the 1950s demonstrated that TMV could be reversibly dissociated into its component RNA and protein and that the RNA contained the genetic information for this virus. Prior to these studies, Hershey and Chase had reported in 1952 that the DNA of bacteriophage contained the genetic information for these viruses. This result provided one of the important landmarks in establishing that the genetic information was carried by nucleic acid.

A major concept which helped lead to an initial understanding of virus structure was the proposal by Watson and Crick in the early 1950s that the small size of viral genomes contained limited amounts of genetic information; thus, the virus capsid coats must be organized in repeating subunits of single or small numbers of protein species and a self-assembly mechanism should lead to their final structure. Caspar and Klug laid down the ground rules for this type of self-assembly that could lead to isometric structures in a classic paper published in 1962.

The report in 1926 that variants of TMV bred true and were revertible suggested that this virus contained mutable genetic information, but virus genetics did not advance significantly until studies of the bacteriophages in the 1940s. Emory Ellis and Max Delbruck were primarily responsible for initiating studies with the bacteriophages of *Escherichia coli*, and they elaborated the one-step growth curve experiment originally devised by d'Herelle and elaborated on by Burnet in the mid-1930s, who also noted distinctive phage variants in host range and in antigenic responsiveness. The ready isolation of mutants and their analysis by recombination led to major insights into the molecular nature of genetic events.

The breakthrough that led to advances in identifying and characterizing animal viruses can be attributed to the development of tissue culture systems in the 1950s. The various cells and culture conditions allowed for development of plaque assays for animal viruses and for analysis of animal virus growth under controlled and reproducible conditions. Identification of biochemical events in the replication of these viruses became tractable in these cell cultures. Importantly, such cultures could be used for growing virus in amounts useful for vaccine production, as exemplified by the work of Enders and colleagues at Harvard Medical School using monkey kidney cells to produce human poliovirus. Thus, in the 1960 and 1970s, molecular components of many different kinds of viruses were identified and viral genomes began to be mapped initially by recombination, then by DNA restriction enzymes, and, ultimately, by determination of the sequence of the nucleotides.

A surprise that came from these initial studies was the observation that some viruses contained, within the particle, enzymes that were essential for the initial stages of virus replication; the first identified was a DNA-dependent RNA polymerase carried in the poxvirus. Some RNA viruses were found to contain RNA-dependent RNA transcriptases tightly bound to the ribonucleoprotein structure. Another major breakthrough occurred in 1970 with the independent discoveries by Baltimore and Temin that a class of RNA-enveloped viruses responsible for cell transformation contained an enzyme that copied DNA from RNA. Temin proposed several years earlier the existence of a DNA copy of the Rous sarcoma RNA virus based on the requirement for cell replication to obtain progeny of the replicating virus and the inhibition of virus replication by a drug which blocked DNA transcription to RNA. In addition to the essential role of this enzyme in the life cycle of the RNA viruses, now known as retroviruses, it became possible to convert any RNA molecule into DNA. In the 1980s and 1990s, with the introduction of genetic engineering, the genomes of both DNA- and RNA-containing viruses could be cloned and modified. The ability to clone and amplify viral genes has played an essential role in identifying new viruses and developing diagnostic assays for viruses such as HIV and hepatitis C virus Undoubtedly, HIV is the most important virus currently circulating in the human population. This retrovirus, the etiologic agent for adult immunodeficiency syndrome (AIDS), was first recognized in the early 1980s. Transmission of this virus—primarily through sexual contact and, initially, blood transfusions—has led to an ongoing worldwide pandemic in which an estimated 30 million individuals have been infected.

III. REPLICATION

A. Attachment

The first step in the virus replication cycle is the attachment of the virus to its host cell. This interaction is relatively specific and, in general, accounts for the selectivity of a virus for its particular host and, within multicellular hosts, for a particular type of cell. There are unique domains within proteins on the outer surfaces of viruses which function as sites for attachment to host cell components, referred to as virus receptors. The latter are frequently molecules on the outer surface of the host cell which function as normal cellular components but have been coopted by a virus in order to promulgate its entry into the cell. The type of specificity that appears to play such an important role in the tropism of animal and bacterial viruses has not been observed for plant viruses.

Virus receptors are highly diverse and range from simple subcomponents of surface proteins, such as sialic acid, to more complex polypeptides, such as the members of the Ig superfamily of membrane-embedded surface proteins. Measles virus utilizes the complement receptor CD46. HIV recognizes CD4 on T cells. The bacteriophage lambda uses the bacteria surface protein that normally functions to bind maltose for utilization of this sugar by the bacteria. Closely related viruses may use quite distinct cell receptors; for example, the intercellular adhesion molecule-1 is the receptor for one group of rhinoviruses, whereas another group uses a receptor that recognizes the low-density lipoproteins. Different viruses can use similar receptors; for example, sialic acid is recognized by orthomyxo viruses, papovaviruses, reoviruses, and coronaviruses. Heparin sulfate is recognized by herpesviruses and by some alphaviruses. Many viruses use more than a single receptor for host-cell binding, which can extend their host range. For example, HIV binds to CD4 on T cells and also to the glycolipid, galactosyl ceramide, on neuronal cells and intestinal epithelium.

Binding affinities of viruses to cells are not exceptional, and effective attachment is the result of multiple interactions due to the number of repeating domains on the surface of a virus that can bind receptors. In several cases, binding may require a coreceptor which most likely functions in the second stage of virus entry. Another mechanism of virus binding is one in which the virus is complexed with an antibody, and the Fc region of the antibody interacts with cells, such as macrophages, that contain Fc receptors on their surface. This type of interaction can lead to enhanced infection in cases in which an individual has been previously exposed to the virus.

B. Entry and Uncoating

The binding of virus to a host cell is followed rapidly by uptake and delivery of the virus genome into the cytoplasm or nucleus of the cell. Two major pathways are known for accomplishing this step in animal cells. One is receptor-mediated endocytosis, a general mechanism for uptake which consists of the formation of a membrane pit coated with clathrin, a protein capable of forming a protein-coated lipid vesicle. The receptor-bound virus enters the cell incorporated in the clathrin-coated vesicle. The clathrin coat dissociates, leaving an endocytic vesicle that takes up protons via an ATP-driven pump which lowers the pH within the vesicle. Proteins on the surface of the virus respond to this lower pH by changes in their conformation such that they can form pores within the endosome membrane that act as entry channels for translocation of the virus genome sequestered within the particles.

For many viruses, particularly RNA viruses such as influenza virus, alphaviruses, and vesicular stomatitis virus, which contain a lipid envelope surrounding their nucleocapsid, channel formation requires a fusion of the virus with cell lipids. This fusion is driven by virus-encoded glycoproteins that localize to the surface of the virus particle, but their fusogenic potential is cryptic and nonfunctional in the virus particle outside the host cell. The low pH of the endosome provides one kind of signal for activation of the fusogenic structure. The conformation change in the influenza glycoprotein, the hemagglutinin, leads to an elongated, hydrophobic coiled coil structure.

The second entry pathway delivers virus genetic material directly from the cell surface and is generally believed to require coreceptors for pore formation. Their role would presumably be to trigger conformational changes in virus surface proteins that mimic the low pH-driven systems. For HIV, two specific chemokines have been identified as coreceptors.

For picornaviruses such as poliovirus and rhinovirus and some plant icosahedral viruses, crystallographic analyses coupled with cryo-EM provide information about how the binding of these viruses to host cell receptors may trigger them for disassembly. Rhinovirus has on its surface a crevice or canyon that contains sites for binding the host cell receptor and the floor of the canyon has a pocket occupied by a long-chain fatty acid or sphingosine. The latter stabilizes virus against heat and low pH. In a model for virus penetration into the host cell the binding of receptor releases the pocket factor and leads to conformational changes in protein capsid subunits VP1 and VP3 and movement of VP4 from an internal position in the capsid shell to an external location. This creates in the membrane a pore that allows the threading of the RNA from its compact form in the virus to an extended configuration in the cell cytoplasm.

C. Genome Expression and Replication

Uptake of a virus into a cell initiates the eclipse phase of replication, which is defined as that period in which infectious particles are not detected within the cell. This illustrates one of the unique characteristics of viruses: the synthesis of the different components followed by their assembly into infectious particles. Virus genome expression generally is initiated very soon after entry into the host cell. For DNA viruses that infect higher eukaryotic cells, with the exception of the Poxviridae, the genomic material must be transported to the nucleus. Poxviruses replicate entirely in the cytoplasm and their genes code for the enzymes required for their DNA replication and transcription. DNA viruses that replicate their genomes in the nucleus (and bacterial DNA viruses) may use host enzymes to carry out these reactions. For most DNA viruses, however, genomic replication includes reactions that are distinct from the ones carried out by the host cell and are often dissimilar for different viruses. Those activities that are unique to the replication of a specific DNA viral genome would require the introduction of new proteins encoded by the viral genome. Some enzymes that seem to duplicate host cell functions may also be encoded by a virus genome; this redundancy may serve as a means to increase or to differentially regulate some of the reactions required for viral genome replication.

Most RNA viruses, with some exceptions such as members of the Orthomyxoviridae and Retroviridae families, replicate entirely in the cytoplasm and no nuclear functions are required for their replication. Most retroviruses require insertion of their genome into the host cell chromosome for their expression, and for many such viruses cell division is required for the DNA copy of the virus to be integrated. Lentiviruses, which include HIV and are members of the Retroviridae family, possess nuclear localization signals in their nucleoproteins and other virus gene products that enable virus genome integration to occur in the absence of host cell division.

An essential and common feature for all viruses is their need to synthesize mRNAs that will be translated into the proteins required for replication and assembly. For positive-sense RNA viruses, the incoming genomic RNA is a messenger RNA and can be translated directly to produce proteins required for further transcription and replication of the RNA. Cells do not contain the enzymes for transcribing the genomes of negative-sense RNA viruses or the DNA of poxviruses into mRNA in their cytoplasm, and for these viruses, transcriptases carried within the particles transcribe the genome into mRNA as the first stage of intracellular replication. Viral mRNAs are translated using the host cell ribosomes and translation factors. For many positive-sense RNA viruses, the mRNA is translated from a single initiation site near the 5′ end of the genome. A polyprotein is produced which is proteolytically processed to smaller, individual polypeptides that function in genome replication and in assembly of progeny virus. Much of this proteolytic processing is carried out by proteases encoded in the virus genome. The genes of most DNA viruses and both the double-stranded and negative-sense RNA viruses are transcribed as mRNAs encoding single polypeptides in the manner similar to those of the host cell. Viruses can utilize a variety of means to expand the number of proteins encoded by their genome. These include translation of proteins from more than one initiation site in a gene, alternative splicing, frameshifting, and post-transcriptional editing which involves the insertion of additional nucleotides into mRNAs causing a change in the reading frame.

A separation of virus replication into early and late stages can be readily distinguished for the large DNA bacteriophages and for DNA viruses infecting animals. The early stages are those that occur before the genome is replicated and include transcription of the genome into mRNAs and, for the large DNA viruses, synthesis of proteins involved in regulation of transcription and enzymes required for viral DNA replication. Genome replication—the synthesis of many copies of the genome which will be incorporated into newly formed progeny—is required to initiate the late steps, which involve the synthesis of the structural proteins of the virus and the interaction of the genome with those proteins leading to assembly of infectious particles and, for many viruses, their release from the cell. Viral mutants blocked in early steps are unable to carry out the late steps, but mutants with defects in late steps can complete the early steps.

The division into early and late steps can also be made for some RNA viruses but not for all. Members of the Picornaviridae and Flaviviridae families are examples of viruses that initiate translation at a single site on the RNA genome. The viral proteins required for genome replication and virus structure are synthesized at essentially the same time. There is still a kinetic division, as there is for all viruses, into the eclipse phase in which the viral proteins are synthesized and the genome is replicated and a phase of assembly and release. However, the steps cannot be separated genetically as described previously. Replication of these RNA genomes requires the synthesis of complementary copies of the genome which then serve as templates for the synthesis of new copies of genomic RNA. The viral proteins essential for these steps have been termed the nonstructural proteins; they are encoded in the virus genome but are not part of the virus particle. Host proteins may also be involved in genome replication, but for most viruses the proteins and how they function have not been determined.

Some positive-strand RNA viruses, such as members of the Togaviridae, the Coronaviridae, and the plant viruses Bromoviridae, translate nonstructural proteins required for transcription and replication from genomic RNA and other proteins, including those required for virus structure and assembly from subgenomic mRNAs. This division permits different proteins to be produced in different amounts. Those proteins that function as enzymes in transcription and replication are needed at lower concentrations than structural proteins that are required in large amounts for assembly. Early steps are those involved in translation of the incoming genomic RNA to produce the proteins for transcription and replication followed by their interaction with genomic RNA to synthesize complementary strands. Later steps include the production of more genomic RNA and the subgenomic RNAs, translation of these RNAs, and assembly, but these are difficult to distinguish except in the cartoons shown in textbooks.

The first step in expression of the genome of negative-sense RNA viruses is transcription of the genome into mRNAs. This step can be distinguished from subsequent steps because it can occur in the absence of any viral protein synthesis and does not require the host cell. Host cell factors are required, however, for the subsequent steps to generate the positive-strand complement of the genome that serves as a template for the replication of more genomic RNA.

D. Assembly

The final stages in virus production involve selective interactions between the newly formed virus-encoded structural proteins and the newly replicated genomic nucleic acid to form nucleoprotein particles, often referred to as nucleocapsids. These structures may have helical or icosahedral symmetry. For some viruses, assembly initiates at a nucleation site in which specific regions of the virus nucleic acid bind the protein subunit followed by interactions with additional polypeptides. In general, it is the sequential addition of closely related or identical protein subunits that drives a self-assembly pathway leading to the formation of the nucleocapsids. Alternatively, in other viruses, oligomeric substructures of proteins form prior to binding nucleic acid, and for some of the larger DNA viruses the nucleic acid is incorporated into the particle after the capsid has been almost completely formed. For these more complex viruses, substructures form prior to the final assembly process, and additional scaffolding proteins and chaperones as well as enzymes that cleave protein and nucleic acids are required transiently to effect the assembly. There does not appear to be a specific mechanism by which nonenveloped viruses are released from cells. The cytopathic effects of infection can cause sufficient disintegration of the cells to allow for virus release and spread to neighboring cells.

The final stages of assembly and release of newly replicated viruses that are enveloped by lipid bilayers occurs by a process called budding. In this activity, the assembled nucleocapsids bind to the cytoplasmic face of a cellular membrane. In most cases, the clustering of a viral-encoded, membrane-associated protein acts as a nucleation site and binding is followed by additional protein–lipid interactions which drive the lipid bilayer to fold around the particle, ultimately fusing and pinching off the virus segment from the host cell but keeping the host cell sealed. Figure 2 shows electron micrographs illustrating the assembly of the alphavirus, Sindbis virus. The virus-encoded proteins at the budding sites are either transmembranal glycoproteins synthesized and transported to a specific membrane or proteins with lipophilic domains attracted to the cytoplasmic face of membranes. The different membranes of a eukaryotic cell (i.e., the endoplasmic reticulum, the Golgi stacks, the nuclear membrane, or the plasma membrane) can be utilized for this budding process, and the choice of membrane is determined by where the virus-encoded membrane proteins are localized. In the process, host cell membrane proteins are for the most part excluded from the virus membrane, but there is a selection or enrichment of specific lipids at the budding sites.

IV. GENETICS

A low level of mistakes in base pairing can occur during replication of the genome of a virus and may result in the formation of mutants detectable as phenotypic variants. For DNA-containing viruses, corrective repair or proofreading decreases the rate of mutation to about 10^{-6}–10^{-8}. However, with RNA-containing viruses or those utilizing RNA in genome replication (e.g., retroviruses), these values are much higher—on the order of 10^{-3}–10^{-5}. A stock of an RNA virus is essentially a population that almost always contains variants, and for this reason the population has been termed "quasi-species." Selective pressures during the replication cycle can lead to the emergence of different virus species. For example, antiviral drugs can give rise to new variants that have become drug resistant. Variants can also be isolated by specifically inserting nucleotide base changes into viral genomes by site-directed mutagenesis utilizing recombinant DNA methodologies.

A special set of deletion mutants may arise during continuing virus replication and act to limit virus growth. These mutants retain the set of sequences required for genome replication and for packaging of the genome into cell-free particles but otherwise have extensive deletions and rearrangements of the virus genome. They are defective and require the nondefective virus as helper to provide the missing enzymes and factors required for replication. These mutants are termed defective interfering or DI particles. The replication of their genomes competes effec-

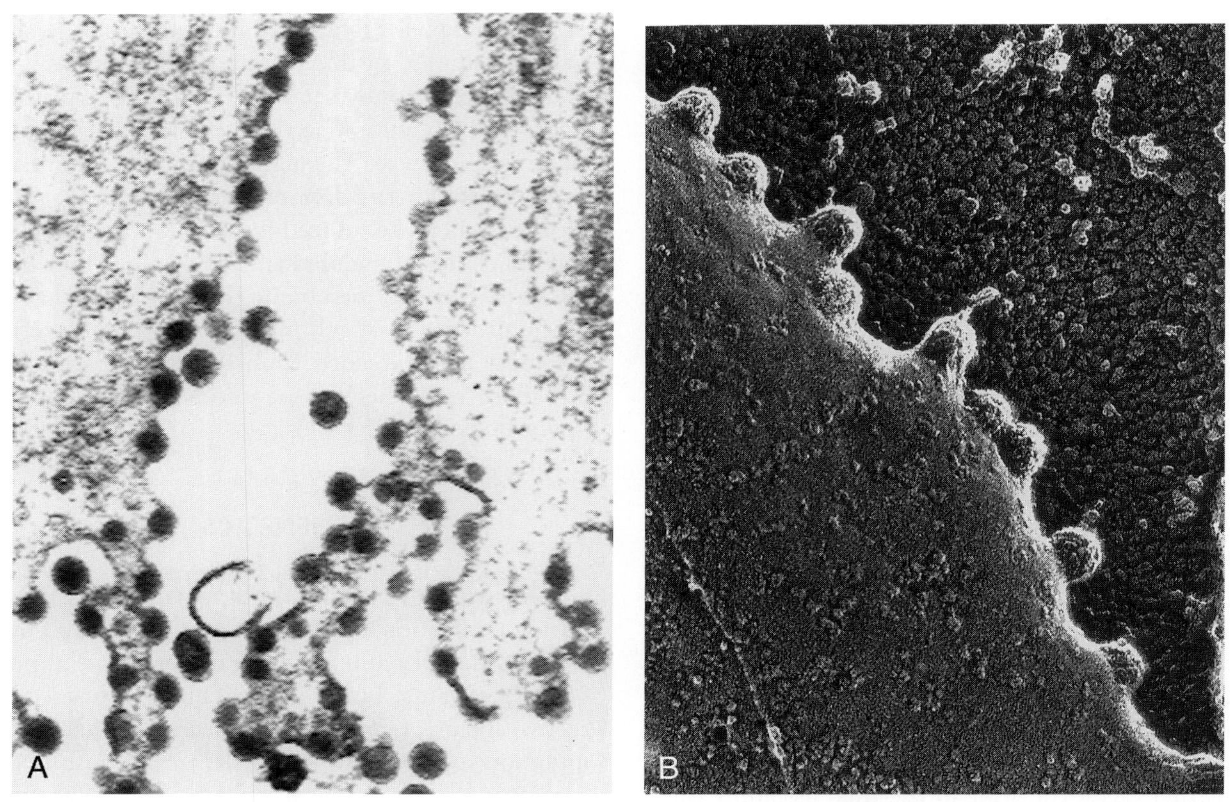

Fig. 2. Electron micrographs of Sindbis virus, a member of the Togaviridae family, budding from the plasma membrane of infected cells. (A) Fixed and stained thin sections. Darkly stained, icosahedral nucleocapsids of the virus are shown at the cytoplasmic face of the plasma membrane prior to release. Magnification × 100,000 (courtesy of M. Aach-Levy, Department of Cell Biology and Physiology, Washington University School of Medicine). (B) Samples were quick frozen, deep etched, and freeze fractured and surface replicas were prepared. Structures protruding from the cell surface represent membrane envelopment of virus nucleocapsids. Magnification × 150,000 (courtesy of Professor John Heuser, Department of Cell Biology and Physiology. Washington University School of Medicine).

tively with that of the standard virus genome and can damp down the severity of infection.

Genetic recombination between viral DNA genomes may occur at homologous regions and has been analyzed in detail, especially for several of the DNA bacteriophage. DNA viruses also recombine with and become incorporated into the host cell genome, leading to what has been termed a provirus or latent state. Recombination between DNA molecules that occurs by a mechanism of break and rejoin of molecules is a well-established phenomenon; recombination between nonsegmented RNA virus genomes is less well understood. It appears to occur by a copy choice or template switching mechanism during nucleic acid synthesis. RNA viruses with segmented genomes, such as the influenza and reoviruses, can participate in a reassortment of genetic information in cells that are multiply infected with different strains of the same virus. Some progeny will then contain gene segments from both parents. For reassortment to be detected, some of the gene segments in the different strains must differ genetically so that the progeny that had undergone reassortment is phenotypically different from the parents. The ability of different strains of influenza virus to undergo reassortment in nature has led to some of the pandemics caused by this virus. This type of reassortment can introduce new variants of the viral surface proteins into an influenza virus, producing a strain to which the human population has not been exposed and would not be immune.

V. VIRUS–HOST INTERACTIONS

A. Interactions between Viruses and the Host Cell

Viruses were first recognized by their ability to kill cells and cause disease, but interaction of a virus with a host cell does not always lead to death. Perhaps the most detailed study of how a cell survives viral infection is that of the bacteriophage lambda. Lambda is a DNA-containing bacteriophage that can follow one of two pathways when it infects its bacterial host, *E. coli*. In one pathway the virus synthesizes viral-specific proteins, replicates its DNA, assembles new particles, and kills the cell. Under some conditions, however, infection follows a different pathway. Replication of the viral DNA as an independent entity is repressed by the presence of a specific viral-encoded protein termed the repressor, the lambda genome becomes integrated into the chromosome of the bacteria, and the cell survives. This is known as the lysogenic state. Lambda DNA can be excised from the chromosome under conditions that are adverse for bacterial growth leading to a lytic state and the production of high levels of new bacteriophage. The molecular mechanism by which lambda bacteriophage becomes latent and is reactivated is very well understood. There are many examples of viral latency among animal DNA viruses, in particular the herpesviruses, but the mechanisms of latency and activation for these viruses are not as well understood.

In recent years, mechanisms by which viruses are able to kill cells have come under further scrutiny, particularly in studies with cultured mammalian cells. In many cases, infection causes the cell to undergo a programmed series of reactions, known as apoptosis, that result in death. Some viruses (e.g., adenovirus and Epstein–Barr virus) code for proteins that interfere with the steps leading to apoptosis, and cells infected with these viruses may survive. In these and in many other examples in which cells survive, virus replication is suppressed. Cells that are infected with adenovirus and survive may be altered in growth control (transformed) and are able to cause tumors in animals. Epstein–Barr virus infects human B lymphocytes and does not complete its replication cycle in most cells. Although the infection is aborted, it can lead to the immortalization of the cells, allowing them to survive in culture.

Several DNA viruses of animals are able to transform cells, making them tumorigenic, and the viruses can cause tumors in animals. These viruses encode proteins such as the T antigen of polyoma and SV40, the E6 and E7 gene products of papillomavirus, and the E1a and E1b proteins of adenovirus that interact with and alter tumor suppressor proteins of normal mammalian cells. Transformation is usually observed only in cells in which the complete virus replication cycle is blocked.

The only RNA viruses associated with transformation and tumorigenesis are members of the Retroviridae family. In the first decade of the twentieth century, viruses (later identified as retroviruses) were shown to cause leukemia in chickens and a different retrovirus designated as Rous sarcoma virus was determined to cause solid tumors in chickens. More than 60 years later, the ability of retroviruses to cause tumors could be explained by the incorporation of a modified host gene into the viral genome. Infection of normal cells by a virus carrying one of these modified host genes (such as the *src* gene or the *ras* gene, which code for proteins that are involved in normal growth control) can transform that cell into one that has become cancerous. In other cases, retrovirus sequences were integrated into a position in a host chromosome that altered the level of transcription of a particular regulatory gene.

Many RNA viruses are cytopathic to the cell that they infect. These viruses may usurp the translational machinery of the cell so that viral proteins are synthesized at the expense of host cell proteins. It has been possible to isolate variants of some of these cytopathic viruses that can establish persistent infections. In most cases that have been studied in detail, cells infected with these variants produce lower levels of viral products and this may be an important factor allowing survival of both the host cell and the virus. In this type of persistent infection essentially all the cells remain infected. In other cases, infected cells synthesize cytokines such as interferons, which are proteins that induce many cellular pathways that inhibit viral replication. Cells that are producing interferons remain for the most part refractory to virus infection, but a small fraction of the cells may remain permissive for a productive infection. In this type of persistent infection most of the cells remain uninfected.

B. Interaction of Viruses with the Host Organism

Virus infection of humans most often conjures up diseases such as polio, influenza, hepatitis, or AIDS. Mechanisms by which viruses cause disease are complex and depend on multiple factors, but for many viruses disease does not always follow infection. An important finding in the studies that led to the development of vaccines to prevent the disease of poliomyelitis was that infection with this virus was a common event and only a small fraction (approximately 1%) of those infected developed a paralytic disease. Poliovirus enters the body through the mouth and replicates in the intestine. It is only when poliovirus infects the neurons of the gray matter of the brain (medulla oblongata) and spinal cord (anterior horn cells) that the paralytic disease is likely to develop. In contrast to infection by poliovirus, infection by viruses such as measles, rabies, influenza, and HIV almost always lead to disease.

There are several different routes by which a virus can enter an individual to cause disease. In addition to the gastrointestinal route taken by enteric viruses such as poliovirus, viruses such as rhinoviruses and influenza virus enter through the respiratory tract and viruses such as HIV, papillomaviruses, and herpesvirus type 2 are most likely to enter via the genitourinary tract or the rectal mucosa, although they can also be introduced by direct inoculation (e.g., with a needle). Some viruses are usually not transmitted from person to person but are carried by insect vectors or infected animals. Viruses, particularly those in the Togaviridae and Flaviviridae families, replicate in insects and are then transmitted to humans by an insect bite. These viruses are among those originally grouped as arboviruses. A few of them, such as the virus that causes yellow fever, have been controlled in some parts of the world by elimination of their mosquito host. Both rabies virus, a member of the Rhabdoviridae family, and hantaviruses, a member of the Bunyaviridae family, are introduced into humans through infected animals. Humans are then considered "dead-end hosts" because the virus usually is not spread from the infected person to other people.

For some viruses, the site of entry is also the site of replication and disease manifestation. This is clearly illustrated by viruses such as rhinoviruses and influenza virus which cause symptoms in the respiratory tract. This is not always the case, however, as mentioned previously for poliovirus. Viruses such as measles virus and varicella virus (the cause of chicken pox) also enter through the respiratory tract, but the symptoms of the disease (rashes) appear when the virus replicates in the skin. The most common route for a virus to spread from one site to another is through the blood stream, but some viruses such as HIV, rubella virus, and Epstein–Barr virus infect lymphocytes and can be transported within those cells to other sites in the body.

Most virus infections of humans that have been studied in detail involve acute disease, meaning that the symptoms of disease appear within a limited and defined period after infection. Although viruses are thought to also be associated with chronic diseases, it has been much more difficult to establish this connection. AIDS, however, represents a clear example in which the disease symptoms may not appear until years after the initial infection. The onset of AIDS is attributed to the ability of HIV to destroy those T cells expressing CD4 that are required in the immune response. In most cases, this leads to an inability of the infected individual to mount an immune response against opportunistic infections. The reasons why the length of time between HIV infection and disease can vary so dramatically remain unclear. For most chronic diseases, a connection between the disease and a virus infection has been much more difficult to establish, and the evidence in support of an association is derived mainly from animal models. There are several animal models in which the disease, initiated by a specific virus infection, shows a striking similarity to a human chronic disease, such as multiple sclerosis or some of the arthritic diseases. In some of these cases, the disease can be attributed to an aberrant immune response. In particular, extensive studies with animals have demonstrated that virus infection can lead to autoimmunity. One explanation for this is that virus proteins may share antigenic determinants with host proteins (molecular mimicry) and the immune response against the virus proteins could subsequently lead to an autoimmune response.

C. Protection against Virus Infection

1. *Vaccines*

The first virus used to inoculate humans as a way of protecting them against infection was a poxvirus isolated from cows. In 1798, Jenner performed his classic experiment of inoculation with cowpox to protect against the dreaded disease of smallpox, and the term vaccination derives from that time (*vacca* is Latin for cow). The mechanism of protection was not understood at that time, but we now know that infection with an attenuated virus will induce an immune response that can protect against subsequent infections with more virulent strains of that or very closely related viruses. The use of an attenuated strain does pose some risks, particularly of reversion to virulence, and it has not always been possible to isolate a strain that is both nonvirulent and able to induce a protective immune response. Another means of vaccination has been to use inactivated virus as the immunizing agent. The most successful example was the inactivated polio virus developed by Jonas Salk in the 1950s. Immunization with killed virus or with purified components of a virus is safer because reversion is not a risk, but protection usually does not last as long. In the past few decades vaccination against major diseases, including measles, mumps, and hepatitis B, has had a major impact on decreasing the frequency of these diseases.

2. *Eradication of Virus Diseases*

One of the major advances in public health in the last half of the twentieth century has been the eradication of smallpox. The elimination of this disease was made possible by the very great effort put forth by the World Health Organization (WHO) in providing mass vaccinations accompanied by extensive surveillance and containment. As a result, the world was declared free of smallpox in 1979.

Many factors made eradication of smallpox a reality, but perhaps the most important was that humans are the only natural host. There are no pockets of this virus existing in animal populations and so there would not be a natural source for reintroducing the virus into the human population. Two other important factors were the availability of a stable vaccine and the ability to recognize all cases of smallpox by its clear symptoms.

WHO has targeted polio for eradication. The problems in carrying out this program are more complex than those involved with smallpox, in part because infection by this virus does not always produce recognizable symptoms. Due to the extensive work of the Pan American Health Organization, eradication of wild-type poliovirus was achieved in the Americas in 1991. The only cases reported in these areas have been those associated with the live vaccine strains. An important part of the polio eradication effort has been the use of sophisticated molecular techniques which made it possible to identify the source of any strain of this virus and to trace its origins.

VI. VIRUSES AS VECTORS FOR GENE EXPRESSION

Traditionally, viruses have been considered as agents of disease, but in the past decade recombinant DNA technology has provided a new impetus for using viruses in beneficial ways. The genomes of many DNA and RNA viruses have been cloned as cDNAs. These cloned molecules, when introduced into cells either as DNA plasmids or after transcription into RNA, can produce infectious virus. It is becoming possible to identify each viral gene and to determine which genes are essential and at which step in the replication cycle they function. Moreover, other viral or cellular genes can be inserted into a viral genome, either in addition to or in place of native viral genes.

There are a variety of ways in which virus vectors may be used. The most obvious is in the large-scale production of a particular cellular or viral protein.

Many viruses take over the protein synthetic machinery and cause the infected cell to synthesize large quantities of viral-encoded proteins. A heterologous gene inserted into a viral genome would be produced in amounts comparable to those of the virus proteins. A second area that holds much promise is that of vaccination. Attenuated viruses are now being used as vaccines. These vaccine strains could also contain genes of other viruses or other infectious agents for which there is no attenuated strain—the result being that a person would be vaccinated against several different infectious agents simultaneously.

The most valuable use of viruses as vectors would be in gene therapy. This type of therapy is being considered in the treatment of genetic diseases such as cystic fibrosis or hemophilia. Viruses provide an extremely effective means for introducing new genetic information into cells and several are currently being study for possible use in genetic diseases and also in the treatment of cancer and cardiovascular and neurological diseases. Those viruses being tested include the DNA-containing poxviruses and adenoviruses, the RNA alphaviruses, and lentiviruses such as HIV. HIV could have special advantages for use in gene therapy, particularly because of its ability to be incorporated into the genome and express its genes in nondividing cells. At the end of the twentieth century, HIV is considered one of the most devastating infectious agents that is rampant in the human population. A hope for the future is that we will learn how this virus causes disease and delete or disarm those genes involved in pathogenesis while harnessing other components that allow the virus to be used for curing diseases.

Bibliography

Baltimore, D. (1971). Expression of animal virus genomes. *Bacteriol. Rev.* 35, 235–241.

Caspar, D. L. D., and Klug, A. (1962). Physical principles in the construction of regular viruses. *Cold Spring Harbor Symp. Quant. Biol.* 27, 1–24.

Eigen, M., and Biebricher, C. K., (1988). Sequence space and quasispecies distribution. *In* "RNA Genetics" (E. Domingo, J. J. Holland, and P. Ahlquist, Eds.), Vol. 3, pp. 211–245. CRC Press, Boca Raton, FL.

Landry, M. L., and Hsiung, G. D. (1996). Diagnostic virology. *In* "Encyclopedia of Virology" (R. G. Webster and A. Granoff, Eds.). Academic Press, London.

Murphy, F. A. (1996). Virus taxonomy. *In* "Fields Virology" (B. N. Fields, D. M. Knipe, and P. M. Howley, Eds.), pp. 15–57. Lippincott–Raven, Philadelphia.

Nevins, J. R., and Vogt, P. K. (1996). Cell transformation by viruses. *In* "Fields Virology" (B. N. Fields, D. M. Knipe, and P. M. Howley, Eds.), pp. 301–343. Lippincott–Raven, Philadelphia.

Ptashne, M. (1992). "A Genetic Switch." Cell Press/Blackwell, Cambridge, MA.

Viruses, Emerging

Stephen S. Morse
Columbia University

I. Defining Emerging Viruses: "New" versus Newly Recognized
II. Examples of Potentially Emerging Viruses
III. What Are the Origins of New Viruses?
IV. Factors Precipitating Viral Emergence
V. What Restrains Viral Emergence?
VI. Prospects for Prediction, Control, and Eradication

GLOSSARY

arbovirus Arthropod-borne virus; it replicates in arthropods (e.g., mosquitoes, biting files, and ticks) and is transmitted by bite to a vertebrate host. Arbovirus is an ecological rather than a taxonomic definition; various arboviruses belong to a variety of virus families, including the Flavidiridae, the Bunyaviridae, and the genus *Alphavirus* of the family Togaviridae.

emerging viruses Viruses for which their incidence has recently increased and appears likely to continue increasing.

endemic Occurring naturally and constantly in a particular area (as opposed to epidemic).

epidemic The appearance of a disease in a population at a prevalence greater than expected.

hemorrhagic fever An infection manifested by acute onset of fever and hemorrhagic signs (blood vessel damage as indicated by petechiae on skin, internal bleeding, and, in severe cases, shock); typical of many zoonotic viruses.

incidence The number of people (in a given population) developing a specified disease or becoming infected at a specified period of time.

pandemic From Greek *pan* (" = all"); an epidemic so widespread that it covers virtually the entire world.

prevalence The frequency of a disease in a population. An epidemiologic measurement, it is similar to but not technically synonymous with incidence.

re-emerging infection An infection that had formerly been controlled but is now increasing again; usually an indication of a breakdown in public health measures.

vector An agent, usually animate, that serves to transmit an infection, e.g., mosquitoes.

xenotransplantation The transplantation of organs or cells from another species in order to replace a nonfunctioning organ (e.g., heart, liver, kidney, or pancreas) or cellular component (e.g., a bone marrow transplant).

zoonosis An infection or infectious disease of vertebrate animals that is transmissible under natural conditions to humans.

EMERGING VIRUSES are viruses that have recently increased their incidence and appear likely to continue to do so. In less formal terms, they are viruses that have newly appeared in the human population or are rapidly expanding their range, with a corresponding increase in cases of disease. In recent years, many viral diseases have been identified for the first time. Some, such as AIDS, have made their debut alarmingly and dramatically. Other viruses, such as influenza, have long been known for their tendency to reappear periodically to cause major epidemics or pandemics. The reasons for these sudden manifestations of new viral diseases have in general been poorly understood, making it difficult to determine whether anything can be done to anticipate and prevent disease emergence. In many cases, the causes of viral emergence, although complex, are often less random than they seem. Because of the large number of viruses, this article is selective, focusing on common features shared by emerging viruses; no attempt has been made to include all possible candidates. For additional information on specific viruses or particular aspects, the reader is referred to the appropriate articles in this encyclopedia and to the Bibliography.

I. DEFINING EMERGING VIRUSES: "NEW" VERSUS NEWLY RECOGNIZED

A. Categories of Emerging Viruses: New or Newly Recognized?

The most noticeable category of emerging viruses includes those that seem genuinely new in some important respect, such as a first appearance in an epidemic of dramatic disease, a sudden increase in distribution, or novel mechanisms of pathogenesis. Many of the emerging viruses that immediately come to mind, such as HIV, are classified into this group. Most of this article will consider viruses in this category and what is known about the causes of their emergence.

A second group of potentially emergent viruses consists of viruses that, although not new in the human population, are newly recognized. A recent example is human herpesvirus 6 (HHV-6). Although identified only in the mid-1980s, HHV-6 appears to be almost universal in distribution and has recently been implicated as the cause of roseola (exanthem subitum), a very common childhood disease. Since roseola has been known since at least 1910, HHV-6 is likely to have been widespread for at least decades if not longer. Suggested roles of HHV-6 in chronic fatigue syndrome or as a cofactor in AIDS are still under study. Other viruses discussed later, such as parvovirus B19, are also likely to be in this category.

B. The Newly Recognized

The significance of the newly recognized but common viruses is still debatable. Some have been implicated in various chronic diseases, although much of the evidence is inconclusive. However, as chronic diseases have become increasingly important in industrialized societies, the impact of agents responsible for chronic diseases could be considerable. Otherwise, epidemiologically, newly recognized but common viruses generally do not represent an apparent threat because they are already widespread and may have reached an equilibrium in the population. Recognition of the agent can even be advantageous, offering new promise of more accurate diagnosis and possibly control.

Conceivably, with any type of agent, a change in the agent or (more frequently) in host condition might result in a new or more serious disease. This is evident with "opportunistic pathogens," agents that generally have limited ability to cause human infections and disease but may do so in suitable circumstances. Immunosuppressed or immunocompromised individuals, such as people with AIDS, are particularly prone to such infections. With many pathogens, expression of disease may also be altered by such host factors as nutritional or immune status or age at first infection. The effects of nutritional status are often obvious, but there may be more subtle effects as well. Recent experiments have demonstrated that a virulent variant could occur during infection of selenium-deficient mice with a normally avirulent (mild) Coxsackievirus B3 isolate. The mechanism is unknown. The virulent virus that appeared closely resembled other known virulent genotypes of the same virus. The genotype was stable and could lethally infect healthy mice. Although no other examples of this remarkable phenomenon have been found, it would be suprising if this was an isolated case.

C. The Significance of New Technologies in Identifying New Viruses

The importance of technological advances in identifying New viruses should be mentioned. Viruses such as HHV-6 and hepatitis C became apparent because the means were developed to demonstrate their existence. The recognition of HIV was dependent on the previous development of methods for growing T lymphocytes in culture. In a more general sense, the introduction of tissue culture methods, in the 1940s, was a major breakthrough in the study and characterization of viruses. It can be expected that new tools for detection will uncover new viruses. In particular, many new avenues have been opened by the recent development of such techniques as the polymerase chain reaction (PCR), which is capable of detecting 1 HIV-infected cell in 100,000. Because PCR can detect and amplify DNA in minuscule amounts of sample and is comparatively undemanding, it is rapidly finding favor in many applications. PCR has great potential for disease archeology and the study of evolution. Using PCR, many otherwise intractable samples can now be tested, even those from mummified human bodies 7000 years old and files preserved in amber. Recently, PCR was used, with generic primers to relatively well-conserved influenza virus sequences, to obtain sequence data on the influenza virus that caused the devastating pandemic of 1918 and 1919 from nucleic acid extracted from autopsy specimens preserved from victims of the pandemic.

The recent discovery of the Kaposi's sarcoma-associated herpesvirus (human herpesvirus 8) demonstrates the power of molecular diagnostic technologies, guided by epidemiological evidence, for pathogen discovery. Kaposi's sarcoma has long been known as a slowly progressing cancer in elderly Mediterranean men (essentially a chronic disease), but the disease, in a rapidly progressive form, achieved prominence in the 1980s as one of the first AIDS-associated diseases. (Interestingly, since then Kaposi's sarcoma as an AIDS-associated disease appears to have declined in incidence for unknown reasons.) Despite suspicions that Kaposi's sarcoma might be of infectious etiology, the cause remained elusive. A recently developed molecular technique, representational difference analysis, has been used to compare the DNA in tissues from Kaposi's sarcoma patients and controls. The sequence data were used to develop additional specific probes and made it possible to identify a herpesvirus in the Kaposi's patients. The virus, which has since been more extensively characterized, appears to be a gamma-herpesvirus related to such primate viruses as herpesvirus saimiri and (more distantly) Epstein–Barr virus.

II. EXAMPLES OF POTENTIALLY EMERGING VIRUSES

Predicting the greatest threats is a difficult task which is made more difficult by significant gaps in our knowledge. Although specifics differ, many viruses seem likely to merit inclusion on a world list of emerging viruses (Table I). It must be cautioned that predicting the future is an endeavor notoriously fraught with pitfalls. HIV was arguably underappreciated when AIDS was first identified, and attempts to predict the next influenza pandemic are still largely guesswork. Unanticipated effects of changes in environmental or other conditions can bring an as yet undescribed or currently obscure zoonotic virus to world prominence, as occurred with HIV a few years ago or Lassa fever. Conversely, a prominent virus may become submerged by environmental changes or even driven to extinction, as was smallpox. Therefore, consideration of emerging viruses should emphasize the principles of viral emergence, and listings of specific viruses are offered only as examples.

Many viruses are currently prominent as emerging viruses or have become prominent in the recent past, including influenza; several members of the Bunyaviridae, including the hantaviruses (Hantaan, Seoul, and related viruses classified in the genus *Hantavirus*) and Rift Valley fever (in a separate genus, *Phlebovirus*, of the Bunyavirdae); yellow fever and dengue (in the Flaviviridae); possibly the arenavirus hemorrhagic fevers, including Junin (Argentine hemorrhagic fever) and Lassa fever, probably the Filoviridae (Marburg and Ebola); human retroviruses, especially HIV but also (human T-lymphotropic virus (HTLV); and various arthropod-borne encephalitides, including St. Louis encephalitis and Japanese encephalitis (both in the Flaviviridae) and Venezuelan equine encephalomyelitis, Eastern equine encephalomyelitis, Eastern equine encephalomyelitis, and (in Australia) Ross River (all three in the *Alphavirus* genus of the Togaviridae; the latter causes a dengue-like disease with arthritis as a frequent complication). These encephalitides are mosquito-borne but also have natural (non-human) vertebrate hosts.

Regarding the health of people in the United States, some viruses are of a greater concern than others. Of the viruses listed, and excluding viruses already established in the United States, influenza, HIV-2, dengue, and possibly some hantaviruses are of greatest immediate potential importance. Proximity makes dengue, which is spreading over the Caribbean basin, a special concern. Some native mosquito-borne encephalitides, such as LaCrosse and Califor-

TABLE I
Examples of Emerging Viruses[a]

Virus	Signs/symptoms	Distribution	Natural host	Transmission
Orthomyxoviridae (RNA, 8 segments)				
Influenza[b]	Respiratory	Worldwide (from China?)	Fowl, pigs	Respiratory
Bunyaviridae (RNA, 3 segments)				
Hantaan, Seoul	Hemorrhagic fever + renal syndrome	Asia, Europe, United States	Rodent (e.g., *Apodemus*)	Contact with infected secretions
Rift Valley fever[c]	Fever ± hemorrhage	Africa	Mosquito, ungulates	Vector: *Aedes* mosquitoes
Flaviviridae (RNA)				
Yellow fever[c]	Fever, jaundice	Africa, South America	Mosquito, monkey	Vector: *Aedes aegyptia* (ubran), other *Aedes* species (sylvan)
Dengue[b]	Fever ± hemorrhage	Asia, Africa, Caribbean	Mosquito, human/monkey	Vector: *Aedes aegyptia* (Asia: also *Aedes albopictus*)
Arenaviridae (RNA, 2 segments)				
Junin (Argen. HF)	Fever, hemorrhage	South America	Rodent (*Calomys musculinus*)	Contact with infected secretions
Machupo (Boliv.)	Fever, hemorrhage	South America	Rodent (*Calomys callosus*)	Contact with infected secretions
Lassa fever	Fever, hemorrhage	West Africa	Rodent (*Mastomys natalensis*)	Contact with infected secretions
Filoviridae (RNA)				
Marburg, Ebola	Fever, hemorrhage	Africa	Unknown	Contact; nosocomial through contaminated needles
Retroviridae (RNA + reverse transcriptase)				
HIV[b]	AIDS, etc.	Worldwide	?Primate	Blood transfusion; nosocomial through contaminated needles; sexual transmission

[a] Modified from Morse and Schluederberg (1990). © 1990 by The University of Chicago Press.
[b] Viruses of special concern for the near future.
[c] Transmitted by arthropod vector.

nia encephalitis (both in the California group of the Bunyaviridae), St. Louis encephalitis, and Eastern equine encephalomyelitis, are generally sporadic. Venezuelan equine encephalomyelitis has been almost entirely eliminated from the United States, but reintroduction is possible.

In considering emerging viruses, it is essential to take a global view. A virus emerging anywhere in the world could, under favorable conditions, reach any other part of the globe within days. Accelerating environmental change and increased immigration also expose new populations to microbes that were once buried in the depths of rain forests or confined to remote villages. For example, in 1989, a man in an Illinois hospital died of Lassa fever. The virus is normally endemic to west Africa; the patient contracted the virus while visiting family in Nigeria but fell ill only after returning home a few days later. The ability of a virus to disseminate throughout the world in a short time is clearly demonstrated by new influenza pandemics, which blanket the globe within a few months of their inception (which usually occurs in China). Additionally, many diseases, such as Korean hemorrhagic fever (caused by Hantaan, the prototype hantavirus, with an estimated 100,000–200,000 cases a year in China), may be serious causes of illness or death in certain regions of the world, even if they are not an imminent threat to the health of people in the United States. The following sections provide a brief description of some of these viruses. Additional information on specific viruses or viral families can be found under the appropriate cross-references in this encyclopedia or in the sources cited in the Bibliography.

A. Arenaviruses

Members of the Arenaviridae cause hemorrhagic fevers in humans; their natural hosts are rodents. Both Old World and New World representatives are known. New World representatives include Junin (Argentine hemorrhagic fever), whose natural hosts include the rodent *Calomys musculinus,* and Machupo (Bolivian hemorrhagic fever), whose host is the rodent *Calomys callosus.* Several additional New World arenaviruses have been identified throughout South America in the past few years. For example, one arenavirus, provisionally named Guaranito, was described in 1991 during a dengue epidemic in Venezuela. The major Old World arenavirus, Lassa fever of west Africa, has as its natural host the rodent *Mastomys natalensis.* Lassa became infamous for its high mortality rate in Western medical missionaries who first came in contact with the virus in the late 1960s and early 1970s. Infected rodent hosts usually shed virus asymptomatically in their urine, and primary infection in humans is generally by contact with infected excreta from the rodent, probably through inhalation of aerosolized virus in the excreta. Secondary cases of Lassa fever in health care providers or family members can occur through contact with patients' blood or infected secretions.

B. Filoviruses (Ebola and Marburg)

The Filoviruses (Ebola and Marburg) are among the least understood of all viruses. Their natural hosts are unknown; some believe they originated in primates, but recent evidence favors bats as the most likely natural hosts. Human disease is typically fever with hemorrhage, and mortality can be high. Ebola has caused several epidemics in Africa. An epidemic in Zaire (now Congo) in 1976 involved 278 known cases, almost all of which were hospital acquired through contaminated hypodemic equipment or contact with patients; mortality was approximately 90%. In the same year, a separate epidemic in the Sudan involved almost 300 cases, with more than 50% mortality. Outbreaks since then have included the Cote d'lvoire (1 case in a field naturalist, who survived) and Gabon, both associated with handling chimpanzee carcasses, and a well-publicized 1995 outbreak in Kikwit, Zaire (Congo), involving 315 known cases and 77% mortality. Filoviruses in imported Old World monkeys have also been of recent concern. Marburg virus was first identified in 1967 when 25 laboratory technicians in Marburg, Germany, and in Belgrade, Yugoslavia (Serbia), became sick after handling tissues from African green monkeys (7 died). Six medical workers and family contacts subsequently became infected but recovered. Since then, there have been 3 documented primary cases (all fatal) of Marburg which was acquired by travelers in Kenya or (in one instance) Zimbabwe; three individuals who became infected by contact with the primary cases all survived. In 1989 and 1990, monkeys (Asian macaques) in a facility in Reston, Virginia, died suddenly. A filovirus was isolated, although whether this was the primary cause of death in the monkeys is still not certain. The virus, now variously known as Reston filovirus or Reston strain of Ebola, appears less virulent than classic Ebola. Several animal handlers who apparently became infected did not develop acute disease. This outbreak was the subject of the best-selling book, *The Hot Zone,* by Richard Preston.

C. Flaviviruses (Dengue and Yellow Fever)

Yellow fever, a mosquito-borne disease characterized by fever and jaundice, remains a significant world health problem. Historically, the impact of yellow fever virus was tremendous. Yellow fever was so devastating to workers in the region that the Panama Canal could be completed only after yellow fever was controlled. Historians have documented the high mortality caused by yellow fever in European settlers coming to Africa in the nineteenth century. The development of a yellow fever vaccine was one of the first successes of the Rockefeller Foundation; the same vaccine is still in use today and remains effective. Despite the availability of effective vaccines, yellow fever virus is still unvanquished and is widespread in Afica and South America. Its origins as a human pathogen date back several hundred years at least, but human infection is only incidental for yellow fever. The natural cycle of infection is the sylvatic ("jungle") cycle in monkeys in tropical areas of Africa and South America (where it was probably introduced from Africa). Most human cases occur by incidental infection of people in areas where sylvatic yellow fever is well established. The virus is mosquito-borne, but different mosquito species are involved in different settings. In the sylvatic form, the virus is carried by local forest mosquitoes, which can also transmit infections among humans. The *Aedes aegypti* mosquito, a species well adapted to living within human habitations (the genus name is Latin for "house"), is generally the vector in "urban" yellow fever. It is generally believed that transport and movement of people in the slave trade disseminated both the yellow fever virus and the *A. aegypti* mosquito from Africa to other tropical areas. Although *A. aegypti* can be found in some portions of the southeastern United States, the last yellow fever epidemic in the United States was in New Orleans in 1905.

Dengue, another flavivirus carried by many of the same mosquito species as yellow fever, is in tropical areas worldwide (Africa, Asia, the South Pacific, South America, and the Caribbean) and continues to spread. Dengue is now widespread in the Caribbean basin. Cuba had more than 300,000 cases in a 1981 epidemic, and there was an epidemic in Venezuela in the winter of 1990 and in Brazil in 1991. Travelers returning to the United States from the tropics occasionally return with dengue: The federal Centers for Disease Control reported at least 27 confirmed cases of imported dengue (in 17 states) in the United States in 1988, which is a typical yearly rate. A more severe form, known as dengue hemorrhagic fever, occurs in many areas where dengue is hyperendemic and has been postulated to result from sequential infection with different dengue viruses that now overlap geographically in many tropical areas. In the New World, dengue hemorrhagic fever was first seen during the Cuban dengue epidemic of 1981. The frequency of dengue hemorrhagic fever is increasing as several types of dengue virus extend their range.

D. Hantaviruses

In the family Bunyaviridae, members of the Hantavius genus (Hantaan, Seoul, and related viruses such as Puumala and others in Europe) classically cause hemorrhagic fevers with renal syndrome (fever, bleeding, kidney damage, and shock). Hantavirus pulmonary syndrome (HPS), caused by some New World hantaviruses, was first recognized in 1993. Various hantaviruses are found in Asia, Europe, and the United States as naturally occurring viruses of rodents. The most prominent member of this family is Hantaan virus, the cause of Korean hemorrhagic fever. Named for a river in Korea, the disease first came to Western attention during the Korean War. At least 3000 U.S. and UN troops developed Korean hemorrhagic fever, more than 300 of which died. The disease has long been known in Asia, and it has been suggested that there is a description of Korean hemorrhagic fever in a Chinese medical text dating to the tenth century. Currently, approximately 100,000–200,000 cases of Hantaan are diagnosed annually in China compared with 471 in 1955. The major natural host of Hantaan virus in Asia is the striped field mouse, *Apodemus agrarius*. This rodent is not native to the United States; consequently, Hantaan virus is not likely to become established in the United States (of course, the evolution of a host range variant capable of infecting native rodents cannot be excluded as a theoretical possibility). There are many

New World hantaviruses with native rodent hosts that have not been clearly associated with human disease (as well as other New World hantaviruses that do cause disease). Because any new variant would probably have to compete with these existing viruses or would have to be geographically distinct or have a different host range, new introductions of related viruses may be limited (however, many hantaviruses are already present). This does not necessarily apply to the rodent hosts, and a new rodent-borne virus could be introduced into the United States if a suitable rodent host established itself or if the host were a cosmopolitan rodent such as the domestic rat. For example, another Hantavirus, Seoul virus, is found in rats. The virus was originally identified in rats in Korea and has since been identified in urban rats living in American cities. It has been suggested that rats carried on ships from Asia may have introduced Seoul virus into the United States. In Korea, Seoul virus has caused hemorrhagic fever with renal syndrome similar to Hantaan virus but usually considerably milder. In the United States, acute disease has not been identified, although seropositive individuals have been found in some inner-city areas. There is some evidence, although inconclusive, for a possible association with chronic renal disease, including renal hypertension.

Pathogenic hantaviruses native to the Americas were not recognized before 1993. Then, in the late spring and summer of 1993, several patients, mostly young and previously healthy adults, were admitted to hospitals in New Mexico, as well as Arizona and Colorado, with fever and acute respiratory distress; many subsequently died from pulmonary edema and respiratory failure. By the time the outbreak ended in late summer, there had been more than two dozen patients, with 60% mortality. Serology and detection of genetic sequences by PCR provided evidence that a previously unrecognized hantavirus was the cause of the outbreak, and the condition was subsequently called HPS. The major reservoir was identified as *Peromyscus maniculatus,* the deer mouse, which was also the rodent most commonly trapped near houses in the area. A high percentage (20–30%) of captured *P. maniculatus* proved positive by serology or PCR. Using PCR, the same sequences were identified in tissues taken at autopsy from several of the patients and in tissue samples from local rodents. As of January 1999, 205 cases of HPS (mostly sporadic; that is, as isolated individual cases) have been identified in 30 states in the United States and 30 cases in three western Canadian provinces. On the basis of identification of hantaviruses in stored samples and other evidence, it appears very likely that most of these viruses have long been present in their natural hosts but are newly recognized as a cause of human disease.

Many related hantaviruses have been identified throughout the Americas. In North America, other HPS-associated hantaviruses have been identified in the rodents *Peromyscus leucopus, Sigmodon hispidus* (the cotton rat), and *Oryzomys palustris.* Several hantaviruses have also been identified in South America in the past few years. One virus, dubbed Andes, caused cases of HPS in Argentina as well as several outbreaks in Chile (one in 1997 involved 25 cases). Like most other zoonotic viruses, most hantaviruses do not spread readily from person to person, but there was evidence of person-to-person transmission of Andes virus during the outbreak in Argentina.

E. Hendra and Nipah Viruses

In 1994, an outbreak of respiratory illness claimed the lives of 13 horses (of 20 that became ill during the outbreak) and a horse trainer (one other worker was affected but recovered) in Hendra (a suburb of Brisbane), Queensland, Australia. A virus was isolated and identified as a paramyxovirus by electron microscopy and partial gene sequencing. Because the virus appeared most closely related to the morbilliviruses (the genus within the Paramyxovirus family that includes measles and canine distemper viruses), it was originally named "equine morbillivirus." However, gene sequencing data support its classification as a member of the Paramyxovirus family that, although more closely related to the morbilliviruses than to other genera, is distinct and probably deserving of a separate genus. Other studies determined that the natural host of the virus appeared to be fruit bats (particularly *Pteropus* species). Since the virus was neither "equine" nor a true morbillivirus, it was proposed to rename the virus "Hendra." In another outbreak (in a different part of Queensland and believed to have occurred in August 1994), 2 horses died and one human subsequently developed encephalitis.

A related virus, now called Nipah, was identified in Malaysia in 1999 when, beginning in September 1998 and continuing into April 1999, cases of encephalitis appeared. The cases were originally ascribed to Japanese encephalitis, but laboratory results proved negative. Subsequent laboratory work demonstrated that a different virus, related to but distinct from Hendra, was responsible. Pigs were infected as well as humans, and most of the human cases were occupationally related. As of April 4, 1999, there were 229 cases reported, with 111 deaths. Several cases (11 cases with 1 death) also occurred among slaughterhouse workers in Singapore who worked with imported pigs. Control measures in Malaysia included the slaughter of more than 900,000 pigs by April 1999. Studies are under way to determine natural host (quite possibly bats, by analogy with Hendra) and occurrence in other domestic animals.

F. Hepatitis

Five viruses, all unrelated, are now known as possible causes of human viral hepatitis, and several others have been suggested based on molecular evidence. In addition to the familiar hepatitis A and B viruses, there are hepatitis C virus, the transfusion non-A non-B hepatitis Virus characterized in the late 1980s hepatitis E, a water-borne virus from Asia identified several years ago; and delta agent, or hepatitis D.

Hepatitis C is common in the United States and may be responsible for 98% of current post-transfusion hepatitis, especially since routine testing of blood has greatly reduced hepatitis B in transfusions. Its identification, led rapidly to development of serologic and molecular tests for the virus (which it is hoped will lead to its reduction or elimination in the blood supply) as well as to trials of therapeutic interferon. The agent is an RNA virus which appears "flavivirus-like" (resembling; but not closely related to yellow fever and dengue viruses). In addition to post-transfusion non-A non-B hepatitis, hepatitis C may be an important cause of community-acquired hepatitis. In a recent study, more than half of 59 patients with "community-acquired non-A non-B hepatitis" (no blood transfusion) in one U.S. county were seropositive, suggesting additional modes of transmission for hepatitis C. In the plast few years, other putative blood-borne hepatitis viruses (GB, TT, and others) have been identified by molecular methods, but their natural history and public health significance are still unclear. Hepatitis E was identified several years ago. It is a water-borne disease widespread in Asia and South America and is caused by an RNA virus that has not been fully characterized but appears to belong to a different viral family from that of the other known hepatitis viruses.

Delta hepatitis, discovered in 1977 in Italy, causes an acute fulminant hepatitis in hepatitis B carriers. Delta is a defective agent, consisting of a small RNA and a distinctive protein known as delta antigen, wrapped in a covering of hepatitis B surface antigen. Delta therefore requires hepatitis B as a helper virus, in essence parasitizing the parasite. Luckily, delta is still not common in the United States, except in some groups that are also at high risk for hepatitis B, but it is endemic in Italy and parts of South America. Also, delta is comparatively rare in Asia, where hepatitis B is very common. Because it "borrows" the hepatitis B surface antigen and requires co-infection with hepatitis B virus, increasing use of the vaccine for hepatitis B should reduce the occurrence of disease from delta. In many parts of the world, where hepatitis B immunization may not be widely practiced, delta remains a great potential menace. Delta is unique among known animal viruses due to its small size and the fact that portions of the agent appear to be related to viroids, which are very small infectious RNA agents of plants. These ultimate parasites are tiny pieces of nucleic acid with an independent agenda. Other small RNA agents similar to delta probably exist in animals but have not been identified.

G. Influenza

Influenza is one of the most familiar viruses, but it remains an important threat. Annual or biennial epidemics of influenza A are due to antigenic drift, a mutational change in the hemagglutin (H) Surface protein of the virus so that the host's immune system no longer recognizes the antigen and the new virus can reinfect until host immune responses are mounted to the new variant antigen. Pandemics, very large epidemics that occur periodically and involve virtually the entire world, are the result of antigenic shift, a reassortment of viral genes usually involving the acquisition by a mammalian influenza virus of a new hemagglutinin gene from an avian influenza virus. There are approximately 13 subtypes of the hemagglutin gene, although only a few H subtypes have been associated with human infection. These pandemic strains have generally originated in China. In June 1991, it was reported that an influenza virus with the H3 hemagglutinin but distinct from currently circulating varieties and believed to be of avian origin had appeared in horses in north-eastern China in 1989 and 1990. In 1997, several cases of human disease associated with H5 influenza were reported from Hong Kong, eventually involving 18 confirmed human cases with 6 deaths. Although H5 influenza has long been associated with highly lethal outbreaks in poultry, human infection with H5 influenza had not previously been noted. Control measures instituted by Hong Kong officials included the slaughter of all 1.6 million chickens in Hong Kong. The virus did not appear to spread beyond Hong Kong in this outbreak.

H. Parvovirus B19

Parvoviruses are the smallest DNA viruses of vertebrates (*parvo* is from the Latin word for small). The virus particle measures approximately 20–22 nm in diameter, and the viral DNA genome is approximately 5,000 bases long. Parvoviruses typically replicate in rapidly dividing cells, such as cells lining the intestine or blood cell precursors in bone marrow. Canine parvovirus is another member of this family. Another human parvovirus, adenoassociated virus, has not been associated with human disease. Parvovirus B19 was discovered fortuitously in sera from healthy blood donors with false-positive reactions for hepatitis B antigen. After the virus was characterized, additionally studies indicated that approximately 60% of adults are seropositive. The virus is worldwide in distribution. Evidence implicates B19 as the cause of erythema infection, or "fifth disease," which is a mild, self-limited febrile disease of childhood. B19 has also been associated with aplastic crises in chronic hemolytic anemias (the sudden disappearance of blood cell precursors in bone marrow). B19 and other parvoviruses have also been suggested as causes of joint disease, but this is not clearly resolved.

I. Retroviruses

Human retroviruses have become the subject of intense scientific interest largely because of HIV. The HIV pandemic has become one of the defining conditions of the late twentieth century. Two major types, HIV-1 and -2 are known. HIV-1, which has several subtypes, is the cause of the main AIDS pandemic. HIV-2 has not yet disseminated as far as HIV-1, although the potential exists. Although HIV-2 appears to cause a more slowly progressing disease than HIV-1, AIDS caused by HIV-2 has been documented. The origin of HIV-1 remains obscure, but recent evidence support the hypothesis that HIV was introduced from a subspecies of chimpanzee, *Pan troglodytes troglodytes,* in Africa. Molecular evidence from currently circulating HIV-1 strains suggests that there have been at least three successful introductions of HIV-1 viruses into the human population from this chimpanzee subspecies.

Considerable work has also been done recently on HTLV types I and II. HTLV is widespread in some populations in Asia, the Pacific, and parts of the Caribbean, and HTLV II has been reported from an isolated aboriginal group in South America, possibly suggesting relative antiquity as a human virus. HTLV II is also spreading rapidly within certain populations in the United States, such as intravenous drug users. The spectrum of disease due to HTLV is still being defined. HTLV was originally identified in adult T cell leukemia–lymphoma. In tropical areas, HTLV I has been associated with a neurological disease, tropical spastic paraparesis, and a related condition, HTLV I-associated myelopathy, has been defined. It is not known whether HTLV II is responsible for human disease, although some atypical T cell leukemias have been advanced as possibilities. Because of serologic cross-reactivity, older studies did not always distinguish the two viruses, and it is possible that some disease attributed to HTLV I might have been caused by HTLV II.

More speculatively, various researchers have recently suggested that other neurological diseases or autoimmune diseases may be caused by human retroviruses, either HTLV or currently unknown types. There is only limited evidence in human disease; however, the recent demonstration in mice of viral superantigens, homologs of cellular genes that are carried by a mouse retrovirus and that can induce abnormalities in T cell development and possibly autoimmunity, suggests that similar roles for human viruses are possible.

Speculation about endogenous retroviruses is possible, but there are no data to decide the question. Humans, like virtually all eukaryotes, contain endogenous retroviral elements (retroviral-like sequences) in cellular DNA. In certain circumstances, an endogenous retrovirus, silent for generations in the host DNA, can regain independent existence and become capable of infecting other hosts. Murine retroviruses have demonstrated this property in the past, and Robert Huebner suggested years ago that feline leukemia virus, a common and sometimes fatal infection of cats, probably originated in this way from an endogenous rodent retrovirus. Despite this potential, a similar event has not been identified in humans.

J. Rift Valley Fever

Rift Valley fever virus (RVF), a mosquito-borne virus found in Africa, was first recognized in 1931 as a livestock disease in European breed sheep and cattle introduced into Africa. The apparent recipe, as expressed by Karl M. Johnson, was "foreign animals, local virus, new disease." For unknown reasons, the severity of human disease has increased since the virus was first identified. The virus is found naturally in Africa in a number of ungulates (including camels), and RVF caused epizootics in several parts of Africa, with occassional disease in occupationally exposed animal handlers, veterinarians and butchers, but no human deaths. This changed in 1975 when South Africa experienced an epizootic with some human deaths. In 1977, RVF suddenly emerged in a dramatic zoonotic outbreak in Egypt that resulted in thousands of human cases and 598 reported human deaths. Rapid action (quarantine and immunization of livestock in Israel) prevented the virus from spreading over the Mediterranean basin. An outbreak in Mauritania in 1987 followed the Egyptian pattern but on a smaller scale, with an estimate of 1264 human cases and 224 deaths. The human infection is characterized by a fever, usually with hemorrhaging; retinitis is frequently seen.

K. Spongiform Encephalopathies

Bovine spongiform encephalopathy (BSE), currently a cause of great concern in the United Kingdom, is another recently emerged disease. BSE belongs to a family of diseases which also includes scrapie in sheep and whose human counterparts include Creutzfeldt–Jakob disease and kuru. All are uniformly fatal. The term spongiform encephalopathy refers to characteristic histopathological lesions in the brain. Attempts to implicate conventional viruses or viroids have been unsuccessful, and Stanley Prusiner, in work that earned him the 1997 Nobel prize in medicine or physiology, proposed that the spongiform encephalopathy agents are a novel form of infectious material which he terms "prions," infectious self-replicating proteins. Creutzfeldt–Jakob disease is a sporadic presenile dementia (occurring in people younger than the age expected for senility) with an estimated prevalence of approximately 1 case per 1 million population. Familial forms, due to mutation in a gene coding for a specific protein (the prion protein, also known as amyloid A), have recently been identified. Kuru was extensively studied by D. Carleton Gajdusek (a 1976 Nobel laureate in medicine or physiology), who determined the now well known epidemiology of kuru, which was sustained within an aboriginal New Guinea population by ritual cannibalism.

At least in crude extracts, strains of the spongiform encephalopathy agents are among the most heat-stable infectious materials known, BSE appears to be an example of interspecies transfer, in this case possibly scarpie from sheep introduced into a new host species. Incompletely rendered sheep byproducts fed to cattle have been suggested as the vehicle, although the possibility cannot be excluded that BSE existed earlier but was unrecognized. It has been proposed that changes in rendering processes in the late 1970s and early 1980s, using lower temperatures and (perhaps more critically) lesser amounts of solvents, may have permitted some infectious material to survive the process. A similar interspecies transfer seems to have previously occurred (in the 1940s), resulting in the disease now known as transmissible mink encephalopathy. The recent identification of spongiform encephalopathy in felines that were fed BSE-contaminated meat by-products seems to have a similar history. Although the risk to humans is unknown, the relative rarity of Creutzfeldt–Jakob disease, the human equivalent, suggests that the risk with scrapie may be comparatively low. Creutzfeldt–Jakob disease is not noticeably more prevalent among sheep or cattle farmers, even though scrapie has been known in sheep for at least two centuries and there have presumably been many opportunities for human exposure comparable to the acquisition of BSE by cattle. However, BSE may possibly be more transmissible. In the United Kingdom, many cases of "variant CJD" in people younger than would normally be expected to develop classic Creutzfeldt–Jacob disease have been described and are believed to be linked to BSE.

Interestingly, except for the spongiform encephalopathy agents and parvoviruses, most of these emerging viruses contain RNA genomes. Although there is no compelling reason for this, one might speculate that this might be related to the diversity and mutability of RNA viruses. At least superficially, RNA viruses represent a great diversity of viruses and replication strategies. This diversity may be due at least in part the high mutation rates shown by many RNA viruses. This in turn has been attributed to the error-prone nature of RNA replication and especially to the lack of a "proofreading" function in this process.

III. WHAT ARE THE ORIGINS OF NEW VIRUSES?

A. Possible Sources of New Viruses

Although it is not possible to attribute underlying causes or precipitating factors to all episodes of viral emergence, for many episodes at least some of the causes can be identified. Of course, we cannot be certain that these are the only causes, and the known examples probably show some unintentional ascertainment bias because the most explainable are the most likely to be included. Nevertheless, it is noteworthy that many episodes are explainable.

An examination of emerging viruses might appropriately question how a new pathogen might originate. If we assume the constraints of organic evolution, which essentially require that new organisms must descend from an existing ancestor (evolutionary constraints will be briefly considered later, there are fundamentally three sources (which are not necessarily mutually exclusive): (i) the evolution of a new viral variant (*de novo* evolution), (ii) the introduction of an existing virus from another species, and (iii) dissemination of an agent from a smaller human population in which the agent may have arisen or been introduced originally. The term "viral traffic" was coined to represent processes involving the access, introduction, or dissemination of viruses to their hosts, as distinct from *de novo evolution*.

B. Evolution of New Viral Variants

Considerable debate has centered around the relative importance of viral evolution versus transfer and dissemination of viruses to new host populations (viral traffic) in the emergence of "new" viral diseases. Evaluating the significance of *de novo* evolution is complicated by the difficulty of demonstrating that a new isolate is truly newly evolved and not merely a new introduction of an organism that has long existed in nature but was previously unrecognized. There appear to be a few documented examples of this "*de novo* evolution" in nature. Antigenic drift in influenza is probably the best known example of viral emergence due to the evolution of a new variant. Some additional examples of possibly or apparently newly evolved viruses are listed in Table II. The list is not exhaustive, although attempts have been made to make it as inclusive as possible. Most viruses on this list cause diseases typical of their viral families or similar to the parental virus from which the new variant evolved. In both humans and horses, the recombinant Western equine encephalomyelitis (WEE), for example, causes a disease similar to Eastern equine encephalomyelitis but somewhat milder (its apparent evolutionary advantage is that WEE generally has different insect and bird hosts). A polio-like syndrome described in China and South America in the early 1990s is not included because the respon-

TABLE II
Known or Suggested Newly Evolved Viruses

Virus	*Virus family*	*Remarks*	*Disease*
Rocio encephalitis (Brazil)	Flaviviridae	?Recombinant	H[b]
Western equine encephalomyelitis (United States)	*Alphavirus* genus (Togaviridae)	Recombinant	H
Influenza H5 mutant (chickens: Pennsylvania, 1983)	Myxoviridae	New variant	Severe respiratory infection in chickens
Influenza H7 (seals: United States, 1980)	Myxoviridae		H: Conjunctivitis
Enterovirus 70	Picornaviridae	?New strain	H: Conjunctivitis
Rev-T (strain of avian reticulo-endotheliosis virus)	Retroviridae	Avian	Fulminant lymphoma in fowl
Friend virus, spleen focus-forming strains	Retroviridae	Mouse	
Canine parvovirus 2	Parvoviridae	Dogs	Enteritis, cardiomyopathy (similar to parvoviruses infecting other species)

[a] From Morse (1994).
[b] H, associated with human disease.

sible virus is still not characterized and it is not clear whether this is newly evolved or the result of viral traffic.

In addition, there have been a few examples of variants with altered biological properties. Howard Temin offers the example of an avain retrovirus [reticuloendotheliosis virus strain T (REV T), which derived from the considerably less virulent REV-A strain] that became more virulent as a result of several accumulated genetic changes. Recent evidence demonstrates that human chronic hepatitis B infection was associated with a mutation in a viral gene for precore protein. Vaccine escape mutants of hepatitis B were also recently described in a few infected individuals, although some dispute the interpretation. Despite many demonstrations of this phenomenon *in vitro,* this is one of the few documented vaccine escape mutants isolated from natural infection in the field.

C. Role of Viral Traffic

The previously discussed examples notwithstanding, critical examination of known examples of viral emergence indicates that the overwhelming majority of such instances can be accounted for by viral traffic. At least over the admittedly limited time span of human history, most emerging pathogens have probably not been newly evolved. Rather, they are existing agents conquering new territory. The overwhelming majority probably already exist in nature and simply gain access to new host populations. The most novel of these emerging pathogens are zoonotic (naturally occurring agents of other animal species); rodents are among the particularly important natural reservoirs. Many instances of emergence can be attributed to precipitating factors which facilitate the introduction of viruses from the environment into human hosts or aid their dissemination or expansion from a smaller human population. The following section discusses some of these causes.

IV. FACTORS PRECIPITATING VIRAL EMERGENCE

A. Causes of Viral Emergence: The Role of Viral Traffic

The previous analysis suggests that viral emergence can be viewed as a two-step process, involving the introduction of a virus into a human population followed by dissemination. Emphasis should therefore be placed on understanding the conditions that affect each of these steps. Although this discussion concentrates on viruses, most of the considerations are also applicable to most other emerging pathogens.

Many of the known examples of viral emergence share common features. They are usually precipitated by environmental or social changes, often induced by human activities (Table III). The significance of the zoonotic pool is most apparent for viruses, which generally require a host in order to be maintained in nature. Considering that the total number and variety of viruses in animal species are probably very large, this offers a large pool of potential new virus introductions. In such cases, introduction of viruses into the human population is often the result of human activities, such as agriculture, that cause changes in natural environments. Often, these changes place humans in contact with previously inaccessible agents or increase the density of a natural host or vector, thereby increasing the chances of human infection. Examples among the viruses reviewed here include hantaviruses, Lassa fever, and Argentine hemorrhagic fever, all of which are natural infections of rodents, and probably Marburg, Ebola, and Rift Valley fever.

In the 1993 outbreak of HPS (the outbreak in which the disease first came to the attention of medical science), unusual climatic conditions in the winter and spring preceding the outbreak may have caused a massive increase in the rodent population, increasing human exposure to the virus. Reports from the Four Corners area suggested that the winter had been unusually wet, resulting in a large crop of nuts and other rodent food and, in turn, an exceptionally large rodent population and thereby offering more opportunities for people to come in contact with infected rodents (and hence the virus). This has apparently been confirmed by the ecological research station at Sevilleta, New Mexico, which documented a 10 fold increase in area rodent populations begin-

TABLE III
Probable Factors in the Emergence of Some Emerging Viruses[a]

Virus family, virus	*Probable factors in emergence*
Arenaviridae	
Junin (Argentine HF[b])	Changes in agriculture (maize, changed conditions favoring *Calomys musculinus,* rodent host for virus)
Lassa fever	Human settlement, favoring *Mastomys natalensis*
Bunyaviridae	
Hantaan	Agriculture (contact with mouse *Apodemus agrarius* during rice harvest)
Seoul	?Increasing population density of urban rats in contact with humans
Rift Valley Fever	Dams, irrigation
Oropouche	Agriculture (cacao hulls encourage breeding of *Culicoides* vector)
Filoviridae	
Marburg, Ebola	Unknown; in Europe and United States, importation of monkeys
Flaviviridae	
Dengue	Increasing population density in cities and other factors causing increased open-water storage, favoring increased population of mosquito vectors
Orthomyxoviridae	
Influenza (pandemic)	?Integrated pig–duck agriculture
Retroviridae	
HIV	Medical technology (transfusion, contaminated hypodermic needles); sexual transmission; other social factors
HTLV	Medical technology (transfusion, contaminated hypodermic needles); sexual transmission; other social factors

[a] Modified from Morse (1992).
[b] HF, hemorrhagic fever.

ning in the spring 1993. The abnormally high precipitation has been attributed to climatic events caused by the ocean current known as El Niño.

Although this outbreak was precipitated by natural environmental changes that favored an increased rodent population, the responsible environmental changes are often human induced. For example, Argentine hemorrhagic fever has spread as the pampas were cleared for maize planting. The natural host of this virus, the mouse *Calomys musculinus,* flourishes in this environment, propagating its virus in the process. Cases of Argentine hemorrhagic fever have increased proportionately. Hantaan, although unrelated to the Argentine hemorrhagic fever virus, is acquired by humans in a similar way: Increased rice planting encouraged the little field mouse *A. agrarius,* the natural host of Hantaan virus. Infected mice shed virus in secretions such as urine. Humans normally become infected during the rice harvest by contact with infected secretions in the rice fields.

There are numerous examples of this situation in which the emergence of the new virus or pathogen is associated with changing environmental conditions that favor contact of humans with a natural host for an existing virus. Although Lyme disease is bacterial rather than viral, it appears likely that similar environmental conditions are responsible for its recent emergence. Another example is monkey pox. The name seems misleading, because various arboreal mammals in the rain forest, mostly squirrels and probably not including monkeys, appear to be the natural reservoir hosts. Human monkey pox exposures appear most likely to orginate by contact with infected arboreal rodents as a result of hunting the animals for meat of exposure while foraging. It is still uncertain whether monkey pox is emerging. There is no indication that this is occurring, and deforestation in Africa is reducing human exposure to the virus.

Agriculture provides some other unexpected examples. Viroids, infectious agents of plants that consist entirely of small RNA without a protein coat, as far as is known are spread entirely by mechanical transmission on agricultural implements such as pruning knives and harvesters. The evolution of viroids could very likely have been shaped, unbeknownst to its Its human agents, by these human activities.

B. Pandemic Influenza

Remarkably, the same principles as those discussed previously also seem to apply in certain circumstances to viruses in which there is an essential role for viral evolution in the success of new viral variants. Influenza is perhaps the most interesting example. Robert Webster calls influenza the oldest emerging virus that is still emerging; influenza A virus is one of the few known examples (aside from some arguable cases, such as HIV, it may be the only example) of an emerging virus whose emergence (actually, for influenza, periodic reemergence) can clearly be ascribed to viral evolution. Approximately every 20 years, influenza A undergoes a major antigenic shift in one key protein, known as the hemagglutinin (H) protein, and a pandemic results. Although most changes in influenza virus H proteins occur by so-called antigenic drift involving the accumulation of random mutations (this drift can lead to the smaller, but still medically important, influenza epidemics seen every few years), new pandemic influenza viruses occur by a different route—that of major antigenic shifts. These invariably seem to involve a reassortment of viral genes belonging to different influenza strains. Thus, the important event in generating new pandemic influenza strains has not been mutational evolution but rather reshuffling of existing genes. Where do the genes come from? It has recently been found that most influenza genes are maintained in wildflow; every known subtype of the H protein can be found in waterfowl, such as ducks. Many virologists believe that pigs are an important mixing vessel allowing influenza virus to make a transition from birds to humans. Every known major influenza epidemic has originated in south China, where a traditional and unique form of integrated pig–duck farming has been long practiced. Christoph Scholtissek and Ernest Naylor suggested that this form of agriculture may facilitate the development of new influenza reassortants by placing ducks (the reservoir of a variety of influenza strains) and pigs, (thought to be "mixing vessels" for mammalian influenza strains) in close proximity. Agriculture may therefore play the leading role in emergence of this virus. Also viral traffic, reassortant viruses from the mixing of animal influenza strains and the transmission of the resulting virus to humans, appears more important than new viral evolution for human disease.

C. Arboviruses

Water is an essential factor for arthropod-borne viruses, which include many important diseases world wide, because many of the insect vectors breed in water. Japanese encephalitis accounts for almost 30,000 human cases annually in Asia, with approximately 7000 deaths (although immunization programs in several countries promise to control the disease in humans). The incidence of the virus is closely associated with flooding of fields for rice growing. In the outbreaks of RVF in Mauritania, the human cases occurred in villages near dams on the Senegal River.

Rapid urbanization has been blamed for the high prevalence of dengue in Asia. Profusion of water storage containers in cities, necessary to supply the dense and rapidly expanding human population, has caused a mosquito population boom, with a concomitant increase in the transmission of dengue.

Many types of human activities may also disseminate mosquito vectors or reservoir hosts for viruses. It was mentioned previously that both yellow fever virus and its principal vector, the *A. aegypti* mosquito, are believed to have been spread from Africa via the slave trade. The mosquitoes were apparently stowaways in the large open-water containers that were kept on the ships to provide water during the voyage. In a more modern repetition, an aggressive vector of dengue virus, *Aedes albopictus* (the Asiatic tiger mosquito), was recently introduced into the United States in shipments of used tires imported from Asia. From its entry in Houston, Texas, the mosquito has established itself in at least 17 states. In this century, many other mosquito species have been introduced to new areas in war material being returned from foreign theaters of action.

The recent rapid spread of raccoon rabies in the United States has a similar cause. In the past few years, rabies in raccoons has moved from the southeast, where it has been localized for some time, to the northeast, and it appear that raccoons will become a major wildlife source of rabies in the northeastern United States. After identifying its first rabid raccoon in October 1989, New Jersey reported 37 rabid raccoons during the first third of 1990. The Centers for Disease Control has implicated sport hunting as the main factor in this explosive spread of raccoon rabies. In order to ensure an adequate supply of raccoons for hunting, a group of hunters imported Florida raccoons, some of which were apparently rabid, to the area of western Virginia, from whence it spread further north.

D. Gateways for Viral Traffic: The Expansion of Human Viruses

Highways and human migration to cities, especially in tropical areas, can introduce remote viruses to a larger population. On a global, scale, similar opportunities are provided by rapid air travel. For example, HIV probably traveled along the Mombasa–Kinshasa highway and came to the United States presumably through travel. Health officials have also linked the movement of young men from villages to the cities, and resulting freedom from local restraints on behavior, with dissemination of HIV in Africa.

Once introduced into a human host, the success of a newly introduced pathogen depends on its ability to spread within the human population. A similar situation applies to agents already present in a limited or isolated human population because the agents best adapted to human transmission are likely to be those that already infect people. Here, too, human intervention is providing increasing opportunities for dissemination of previously localized viruses. The example of HIV demonstrates that human activities can be especially important in disseminating newly introduced pathogens that are not yet well adapted to the human host and do not spread efficiently from person to person.

Finally, as a highway for viral traffic, the possibilities for iatrogenic disease should also be mentioned. Cases of Lassa fever, Ebola, and Crimean–Congo hemorrhagic fever, in addition to more familiar viruses, have been acquired by health care workers caring for infected individuals or spread in hospitals. As demonstrated by well known examples such as HIV and hepatitis B, many viruses that might not otherwise transmit easily from person to person may be transmitted, through transfusions, organ transplants, or contaminated hypodermic needles, allowing the donor's viruses direct access to new hosts. For many viruses that were not able to spread efficiently from person to person, including HIV, this circumvents their lack of effective means of transmission. As these life-saving procedures become more widely used, and as the scarcity of donors forces medical centers to search farther afield, it is reasonable to expect more instances of disease. In the past few years, similar concerns have been raised about xenotransplantation as a possible route for introducing new zoonotic infections into the human population.

E. Animal Viruses as Models of Interspecies Transfer

In order to supplement our knowledge of interspecies transfer of viruses, examples of emerging viruses in animals might also be considered as useful models for interspecies transfer and mechanisms of viral emergence.

Two of the best studied examples are canine parvovirus and seal plague. Canine parvovirus [officially called canine parvovirus type 2 (CPV-2)] first emerged in approximately 1978 as an epidemic disease of dogs. By 1980, the virus spread globally in the dog population, and the virus is now endemic in every country that has been tested. Slight variants, designated as CPV-2a and CPV-2b, appear to have subsequently displaced the original CPV-2 in the dog population. The virus appears to be closely related to two other parvoviruses, mink enteritis virus and feline panleukopenia virus (the cause of feline distemper). Typical signs are enteritis and cardiomyopathy, often with leukopenia, in animals infected with these viruses. The ability of CPV-2 to infect domestic dogs may have been conferred by a mutation in the capsid (coat protein) gene. Consistent with this hypothesis, it is known that the host range for parvoviruses is at least partly determined by the capsid.

Seal plague, a newly recognized paramyxovirus, is related to measles and canine distemper viruses, but it is distinct. The viruses may have been transferred directly from another species of seal; it has also been speculated that an outbreak of canine distemper might have been related, although this is not clear and seems less probable. Outbreaks of canine distemper, probably from infected dogs, have been associated with die-offs of lions in the Serengeti park in Africa in recent years.

In addition to their possible utility as model systems, emerging viruses of other vertebrates are worth further consideration because serious diseases of livestock or of food plants can have great economic impact and, in the worst case, can cause widespread starvation. Occasionally, some might be potential zoonotic viruses, as was RVF.

V. WHAT RESTRAINS VIRAL EMERGENCE?

Understanding the factors restraining emergence is of great importance, but data are limited. This section, summarizing current knowledge, is necessarily speculative in nature. It is clear that we are still discovering many of the applicable rules, as indicated by the Hong Kong example of H5 influenza in humans—a surprise when it was identified because human infection with H5 had not been known previously.

Although many viruses have high mutation rates and thus potentially may be evolving rapidly, few have shown striking changes in pathogenesis. In order to survive, viruses must be maintained in nature in a living host. There are constraints imposed by the requirements for a means of transmission and the relatively few routes by which a virus can infect a host. Such requirements must impose strong selective pressures on a virus. Therefore, although variants are continually being generated, presumably a stabilizing influence is exerted by natural selection as the virus replicates in its natural hosts. Even if viral mutation is currently unpredictable, genotypic variation and phenotypic change are not equivalent, and there remain evolutionary constraints at the phenotypic level at least. This may be the reason that ecological and demographic factors appear to be at least as important in influencing viral emergence as viral mutation or evolution. Unfortunately, when changes have occurred, they have generally not been predictable. Occasionally, an apparent recombinant virus will emerge, such as Rocio encephalitis or WEE. Why most of them are unsuccessful is not known.

In addition to emerging, viruses may also disappear or be displaced by new variants for reasons that are often poorly understood. Influenza A H7N7, once frequent in horses, has almost disappeared, having apparently been displaced by the H3 subtype. Some have suggested that a similar fate may befall H2N2 influenza in humans. The original strain of CPV-2 is being displaced by a variant, CPV-2a, which in turn will be displaced by CPV-2b. However, although smallpox was eradicated, it is apparently not being replaced by monkey pox, despite the latter's ability to cause sporadic human cases. Several recent viral emergences, such as Rocio encephalitis, Lassa fever, and Marburg disease, have fortunately remained limited or (for Rocio) apparently have resubmerged.

As Edwin Kilbourne pointed out, many zoonotic introductions fail to become self-sustaining in the human population. The restricted range of disease, often severe acute hemorrhagic fevers, and the often limited ability of the virus to spread from person to person are evidence that most of these viruses are not well adapted to human infection. Viruses already adapted to humans—many of which may have been zoonotic introductions in the past that did evolve successfully—and that are present in an isolated human population are more likely than newly introduced zoonotic viruses to disseminate if conditions favor their transmission. HIV is a case in point.

Mathematical ecologists have suggested from mathematical analysis that self-sustaining infection involves a trade-off between transmissibility and virulence (Paul Ewald, taking this analysis a step further, suggested that rate of transmission can affect virulence so that factors that cause more rapid transmission of the virus favor increased virulence and, conversely, slowing down transmission can select for decreased virulence). As in the cause of smallpox, a highly virulent virus can sustain itself if it is also easily transmitted. Many zoonotic introductions cause severe primary disease, (hence they are highly virulent), but appear to have low transmissibility and therefore have not become established in the human population. This is not an evolutionary imperative for them because they survive in nature in their natural hosts, to which they are better adapted. Nevertheless, the impact of zoonotic introductions can be severe, as shown by the Ebola outbreaks in Africa. In addition, as discussed previously, some can now be transmitted inadvertently by artificial means, such as blood transfusions or contaminated needles, circumventing the requirement to evolve greater transmissibility.

Therefore, understanding restrictions on viral variation and emergence would appear to be of prime importance. Some of this will hinge on improved understanding of viral evolution, especially in the ecological context. The relatively low rate of emergence of viral disease may be due in part to the limited entry points whereby viruses gain access to new hosts.

In addition to the considerations mentioned previously, successful viruses also need to evade or subvert host defenses. The macrophage is likely to be one important cell in this process. Other factors restraining the emergence of new viruses, such as inability to spread from person to person, are still not understood at the molecular level. Moreover, an improved understanding of receptors, tissue-specific transcription factors, and tissue-specific mechanisms of cell killing will be essential for defining target cell specificity and host selective pressures restraining viral variation.

VI. PROSPECTS FOR PREDICTION, CONTROL, AND ERADICATION

A. Strategies for Anticipating Viral Emergence

Although many of the questions about emerging viruses may have scientific foundations, allocating resources and setting priorities are often more affected by social, economic, and political factors. This is nowhere more apparent than in considering strategies for anticipating and controlling emerging infections. Environmental and social factors can be major determinants of viral emergence. In principle, this means that emergence can be better anticipated and potentially controlled. In practice, this requires making political decisions such as defining appropriate and workable precautions for development programs, balancing programs so that both development needs and health protection requirements can be met, and mobilizing resources to accomplish these objectives on a world-wide basis. However, although molecular technologies such as sensitive immunoassays and PCR are providing powerful tools for identifying and tracking viruses, resources for global surveillance and control are currently inadequate. Research and clinical facilities are in dangerously short supply, and a critical dearth of trained researchers is expected by the next generation.

From historical experience, new viruses appear most likely to emerge from tropical areas undergoing agricultural and demographic changes and in the periphery of cities in these areas. Surveillance of such areas for the appearance of disease outbreaks or novel diseases would therefore seem advisable. In 1989, D.A. Henderson proposed an international network of surveillance centers located in tropical areas, especially on the edges of expanding tropical cities. Each center would include clinical facilities, diagnostic and research laboratories, an epidemiological unit which could include disease investigation and local response capability, and a professional training unit and would be linked to an international network for data analysis and instant response to emergencies.

One challenge has long been how to develop surveillance capabilities that could provide warning at the early stages of disease emergence, wherever this may occur, to complement the coverage provided by established centers in urban areas. The rise of the internet, and the global connectivity it offers, provides unprecedented opportunities for better disease surveillance worldwide. An example is the e-mail list ProMED-mail. In order to provide a vehicle for bringing together scientists and health officials from around the world to develop and promote coordinated plans for global disease surveillance, ProMED, the Program for Monitoring Emerging Diseases, was formed in 1993 under the auspices of the Federation of American Scientists. Because the need for effective communications was repeatedly mentioned in our discussions, in 1994 we sought the help of SatelLife, a nonprofit organization in Boston, to connect participants worldwide by e-mail. The e-mail list, dubbed ProMED-mail, rapidly evolved into a prototype system for real-time reporting of disease events and discussion of emerging diseases. The technology also makes it possible to cross-correlate similar events that may be occurring in different places. Open to all interested persons at no charge, ProMED-mail now has more than 1000 subscribers worldwide.

If we are often the engineers of viral traffic, we need better traffic engineering. Currently, analytic knowledge of viral traffic is not advanced enough to allow predictions and long-term advance planning based on its principles. However, it is conceivable that such knowledge could be made more systematic in the future, allowing better anticipation of viral and microbial traffic, and better predictive ability.

B. Existing Methods for Control of Viral Diseases

Human viruses that have been substantially controlled in industrialized countries by immunization include polio, rubella, and measles. In some populations (primarily American and European travelers from areas without yellow fever who are immunized before exposure if they are traveling to endemic areas), yellow fever is prevented by immunization. Potential for immunization exists for several other viruses, such as hepatitis B, mumps, and perhaps cytomegalovirus; vaccines of reasonable efficacy are available for these viruses. Rabies immunization in domestic animals has greatly reduced the occurrence of human rabies in the United States to one to three cases a year, although wildlife remain a source of potential exposure for both humans and domestic animals.

Public health measures have traditionally been directed at combating transmission or protecting potential susceptibles through immunization. In addition to immunization, traditional control measures include improved sanitation, mosquito control programs, health certification of travelers, and health inspection of imported livestock. Traditional public health programs have been instrumental in containing many potential threats but also have several drawbacks. Their success with the targeted diseases depends on vigilance and assiduity. Efforts may fall victim to their own success, being prematurely relaxed or abandoned, usually to save money, and allowing the conditions that precipitated the program in the first place to be reestablished. Many mosquito control programs have met with this fate after initial partial success.

In many cases, relatively simple solutions are possible, if there is sufficient global resolve to implement them. For example, rebuilding water supply systems in tropical cities to reduce or eliminate open-water storage could have a real impact on dengue. In other cases, there may be efficacious vaccines or other preventive measures, but problems in deployment allow old agents, placed under control by improved environmental conditions and public health measures, to regain a foothold or expand. Epidemics of yellow fever in some areas (Nigeria reported 600 deaths in a July 1991 outbreak) are an example. Adding yellow fever immunization to the World Health Organization (WHO) world-wide Expanded Program on Immunization, might well prevent further epidemics such as the one in Nigeria. Several U.S. cities experienced epidemics of childhood measles in 1990 and 1991. A recent government report attributes much of the increase in measles cases to cutbacks in childhood vaccination programs, with the result that some children were inadequately immunized or immunized too late. Most programs also cannot contain viruses that can spread efficiently from person to person, such as influenza. The current strategy used for influenza is to attempt to track emerging new strains and to immunize when feasible.

C. Prospects and Requirements for Eradication

Given that viruses that had been major scourges in the past have been controlled or (rarely) even eliminated by immunization or public health measures, it is logical to consider the prospects of eradication for some of the diseases discussed here. The greatest victory so far has been smallpox, now officially extinct. One of the most feared of all viral diseases, as well as reputedly one of the most easily transmissible, smallpox has had a long history. The historian William McNeill suggested that the Spanish conquest of Mexico may have been aided by the effects of smallpox, which the Europeans brought with them. Jenner's work with vaccination made smallpox one of the first examples of immunization; a form of immunization, using material from smallpox lesions, may also have been practiced earlier in China. In the twentieth century, universal childhood vaccination and health control at borders succeeding in bringing the virus under control in Western industrialized countries, except for occasional imported cases and their contacts. Smallpox was last seen in the United States in 1949. As control measures continued to reduce the geographic distribution of smallpox, eradication was adopted as a feasible goal by WHO. An intensive world-wide campaign of vaccination and surveillance in smallpox-endemic areas eventually succeeded in eliminating the disease. In 1979, with an official certification by WHO that was confirmed in May 1980 by the World Health Assembly, smallpox became the first virus to be declared extinct.

This success with a previously feared virus encouraged public health agencies to consider additional candidates for eradication. (Ironically, this success has led to the dilemma of whether to destroy the existing strain collections of smallpox—currently centralized at two laboratories—or to retain them for future research. Some have expressed concern that smallpox might be obtained by terrorists or by countries interested in developing biowarfare capabilities, and there is a lack of effective antiviral treatment. In May 1999, in part because of these concerns, the World Health Assembly agreed to defer destruction for at least another 3 years.) For successful eradication, self-sustaining infection within the human population must be eliminated, and there must be no natural non-human reservoir of infection from which the virus could be reintroduced. In practice, eliminating self-sustaining infection in humans usually involves immunizing a large proportion of the susceptible population and requires effective vaccines. These characteristics—the ability to prevent infection through immunization and the lack of an additional reservoir of the virus—made smallpox vulnerable to eradication. On the other hand, yellow fever does not meet these criteria. There is a suitable vaccine so transmission to people could theoretically be eliminated, but its maintenance in the sylvatic cycle would prevent eradication of yellow fever from the environment. Many of the other viruses discussed here, including by definition all of the zoonotic and arthropod-borne viruses, have natural reservoirs that would render eradication impracticable. Viruses that have no additional reservoirs but that have long periods of infection or transmission, such as human herpesviruses and HIV, would also be difficult to eradicate unless effective lifelong immunity could be established in susceptibles. On the other hand, measles and polio meet the criteria, and are likely targets for the future. They have no known reservoirs outside humans and effective vaccines are available. Polio eradication, a world-wide goal for the beginning of the twenty-first century, will probably not be achieved but natural polio has essentially been eliminated from the Western Hemisphere. For measles, there is currently no definitive timetable for worldwide eradication, but health agencies hope to eradicate it from the Western Hemisphere by late 2000. Thus, even when eradication is possible in principle, the goal may prove elusive or impracticable.

The successes have been encouraging, but the examples of viral emergence and reemergence or resurgence should caution us against complacency. Continuing improvement in nutrition and sanitation has been responsible for reducing the overall impact of infectious disease in industrialized countries. However, most of the world has still to benefit from these simple improvements, with predictable consequences. With respect to viral emergence, human intervention is providing increasing opportunities for dissemination of previously localized viruses, as in the case of HIV and dengue. Because human activities are a key factor in emergence, anticipating and limiting viral emergence is more feasible than previously believed but requires mobilizing effort and funds, especially on behalf of the Third World. One speculation is that episodes of disease emergence may become more frequent as environmental and demographic change accelerate. Evidence suggests that both the scientific and the social challenges of viral emergence are likely to continue for the foreseeable future.

Acknowledgments

I thank Joshua Lederberg, Mirko Grmek, Howard Temin, John Holland, Frank Fenner, Walter Fitch, Gerald Myers, Edwin D. Kilbourne, Robert E. Shope, Thomas P. Monath, Karl M. Johnson, Peter Palese, Hugh Robertson, Baruch Blumberg, D.A. Henderson, Seth Berkley, Barry Bloom, James M. Hughes, James LeDuc, Patrick S. Moore, and other colleagues for invaluable comments and discussions. Special acknowledgments for ProMED to Jack Woodall, Barbara Hatch Rosenberg, Dorothy Preslar and the Federation of American Scientists, and the members of the steering committee. In modified form, some of the views and interpretations in portions of this article are based on those expressed in earlier articles in other publications, including "The Origins of 'New' Viral Diseases" (*Environ. Carcinogen. Ecotoxicol. Rev.* **C9**(2), 1991).

I was supported by the National Institutes of Health, U.S. Department of Health and Human Services (RR 03121 and RR 01180) and by the Arts & Letters Foundation (Milford Gerton Memorial Fund). Work on emerging viruses was supported by the Division of Microbiology and Infectious Diseases, National Institute of Allergy and Infectious Diseases (NIAID), National Institutes of Health (NIH), and by the Fogarty International Center of NIH. I am grateful to Dr. John R. La Montagne, Deputy Director, NIAID, and Dr. Ann Schluederberg, former virology branch chief, for their support and encouragement.

See Also the Following Articles

ARBOVIRUSES • HEPATITIS VIRUSES • INFLUENZA VIRUSES • PRIONS • RETROVIRUSES

Bibliography

Anderson, R. M., and May, R. M. (1991). "Infectious Diseases of Humans. Dynamics and Control." Oxford Univ. Press, Oxford.

Benenson, A. S. (Ed.) (1995). "Control of Communicable Diseases Manual," 16th ed. American Public Health Association, Washington, DC.

Emerging Infectious Diseases [journal published by the Centers for Disease Control and Prevention (CDC)]: Current and past issues available on-line at *http://www.cdc.gov/EID/* or from the CDC homepage (*http://www.cdc.gov*), which also provides links to additional sources of information.

Fields, B. N., Knipe, D. M., and Howley, P. M. (Eds.) (1996). "Fields Virology," 3rd ed. Lippincott–Raven, Philadelphia.

LeDuc, J. W. (1989). Epidemiology of hemorrhagic fever viruses. *Rev. Infect. Dis.* **11**, (Suppl. 4), S730–S735.

McNeill, W. (1976). "Plagues and Peoples." Doubleday, New York.

Morse, S. S. (1991). Emerging Viruses: Defining the rules for viral traffic. *Perspect. Biol. Med.* **34**, 387–409.

Morse, S. S. (Ed.) (1992). "Emerging Viruses." Oxford Univ. Press, New York.

Morse, S. S. (1994). Toward an evolutionary biology of viruses. *In* "The Evolutionary Biology of Viruses" (S. S. Morse. Ed.). Lippincott–Raven, Philadelphia.

Morse, S. S. (1995). Factors in the emergence of infectious diseases. *Emerg. Infect. Dis.* **1**, 7–15. [Also available on the Internet from the CDC website]

Morse, S. S., and Schluederberg, A. (1990). Emerging viruses: The evolution of viruses and viral diseases. *J. Infect. Dis.* **162**, 1–7.

ProMED-mail (e-mail list for reporting and discussion of infectious disease events): Available on-line (from SatelLife website) at *http://www.healthnet.org/programs/promed.html* or (from Federation of American Scientists website) at *http://www.fas.org/promed/index.html/*.

Wilson, M. E. (1991). "A World Guide to Infections. Diseases, Distribution, Diagnosis." Oxford Univ. Press, New York.

Virus Infection

William C. Summers
Yale University

I. Virus Life Cycles
II. Modes of Entry and Transmission of Viruses
III. Responses to Viral Infection
IV. Outcomes of Viral Infections
V. Viral Virulence

GLOSSARY

attenuation The reduction in virulence or ability to cause disease or other consequences of infection by a microbe while it still remains viable.

burst The abrupt release of many progeny virus on the disintegration of the infected cell in which the virus has been growing.

eclipse The period between the entry of a virus into a cell when it loses its infectivity as an independent particle and the appearance inside of the cell of fully infectious progeny virus particles.

interferon Specific cellular proteins synthesized in response to viral infection which are secreted outside the infected cell and which render neighboring uninfected cells resistant to viral infection.

latent Describing or referring to a virus that has entered into a nonreplicative mode of existence in a cell but is propagated along with the cell by certain mechanisms which allow limited viral genome replication without full expression of virus functions and production of mature infectious virus particles.

lysogenic Referring to a mode of virus interaction with cells in which the virus enters the latent state and can be induced to enter the full replicative cycle with the subsequent disruption (lysis) of the cell and production of a full burst of progeny virus.

lytic Referring to a mode of virus interaction with cells in which the virus undergoes a complete replicative cycle with the production of many progeny virus particles and subsequent release of these virus on the disruption (lysis) of the cell.

persistence A mode of virus interaction with a population of cells in which a few cells are always in a lytic mode of infection but the majority of cells are uninfected but potentially susceptible to lytic virus infection.

tropism The specificity of a virus for infection of a specific cell type or specific species.

viremia The presence of virus in the bloodstream.

virulence The ability of a virus to cause more or less severe disease symptoms of a specific type.

VIRUSES are obligate intracellular parasites that can exist as potentially active but inert entities outside of cells. Although there are viruses that infect many animal, plant, and protist cells and which result in effects on the host ranging from inapparent infection to lethality, all virus infections have some features in common. These include an entry phase, an intracellular phase consisting of multiplication, integration, or latency formation, a virus release phase, and usually some sort of host responses to the presence of the virus. It is often these host responses that appear as signs and symptoms of virus infection.

I. VIRUS LIFE CYCLES

At the cellular level a virus first must have some way of entry into the cell, often by adsorption or attachment to some structure or specific molecule on the surface of the target host cell. The virus attachment site can often be a molecule or group of molecules that the cell uses for other purposes; for example, a protein in the maltose transport system is used by bacteriophage lambda for attachment to *Escherichia coli*, and one of the lymphocyte cell recognition molecules is used by the human immunodeficiency virus (HIV) as its cell attachment site. In all virus infections the genome of the virus enters the host cell. In some cases only the viral nucleic acid enters the cell, leaving the protein coat of the virus outside of the cell; in other cases, the entire virus is taken into the cell, and the genome is exposed after a process of intracellular "uncoating." In some instances, the viral nucleic acid enters the host cell with one or a very few genome-associated proteins while the bulk of the virus structural proteins remain on the outside of the host cell. On entering the cell, or soon thereafter, the infectious virus particles are disrupted, and even if the cell is artificially broken open no infectious viruses are found. This period between loss of infectivity and appearance of fully infectious progeny virus production is called the "eclipse" phase of the virus life cycle.

In the cases where some virus proteins enter along with the genome, the proteins play a necessary role in helping to express the viral genes or in bringing about the replication of the viral genome. In some instances, some of the imported viral proteins function to suppress the host gene expression to help the virus effectively shut down host functions as the virus subverts the cellular processes to its own program.

After the viral genome enters the cell, some or all of its genetic information is expressed. In the case of viruses that have evolved to be virulent and that will replicate in and kill the host cell, some of the genes of the virus are expressed immediately, and their translation into proteins results in the beginning of the intracellular replication phase of the virus. The usual genetic program of such viruses (e.g., bacteriophage T4, herpes simplex virus) is to direct the synthesis of viral DNA and, when there are many copies of the viral genome, to express the genes for the structural components of the virus, for example, the capsid proteins, and the envelope proteins, in the case of enveloped viruses. Once a large number of viral genomes have been replicated, and once a sufficiently large pool of virus structural proteins has accumulated, virus assembly is possible, and when a large number of mature virus particles have accumulated, the cell often bursts or is lysed from within by the presence of specific lysis enzymes. This process releases the progeny virus in a "burst" of hundreds to thousands of new infectious virus particles.

Some viruses, however, do not undergo this "lytic cycle" but instead have evolved to enter into a symbiotic relationship with the host cell by promoting the integration of the viral genome into the host cell chromosome in a "repressed" or latent state. Because these latent virus genomes can usually be induced by some conditions to full virus replication and gene expression with consequent virus production, burst, and cell lysis, they are often called lysogenic viruses (this terminology is most commonly used for bacterial viruses). The processes by which the infecting virus genome is integrated into the host chromosome is complex and differs for RNA- and DNA-containing viruses.

II. MODES OF ENTRY AND TRANSMISSION OF VIRUSES

At the organismal level, virus infection is related to the physiology of the particular organism. In plants, for example, virus entry and release are often promoted by cellular injury, and the virus is carried through the vascular system of the plant. In animals, there are many routes of entry, each exploited by different viruses. Common routes of infection are through the respiratory tract, the gastrointestinal tract, directly into the bloodstream, and by venereal contact.

Airborne viruses, such as the common cold virus, measles virus, and influenza virus, enter the body in small droplets (aerosols), and the viruses attach to and penetrate the cells lining the surface of the respiratory tract. These viruses often replicate in the cells of the respiratory tract and cause these cells to initiate a local inflammatory response which results in many of the symptoms of these viral diseases. Viruses present in respiratory secretions can be subsequently transmitted by coughing, sneezing, and other similar modes of spread to other susceptible individuals. Some viruses spread from these localized infections to the bloodstream (causing viremia, or virus in the blood) which allows for dissemination throughout the body.

Other viruses, such as polioviruses, enter the body

on ingested material (contaminated food and water) and, because of their structural features, are able to survive the digestive actions of the stomach and intestines and then to infect cells of the intestinal tract. These viruses may cause local inflammation, such as various enteric fevers (various diarrheas, for example), or they may replicate and then be shed into the bloodstream for dissemination to other parts of the body. Poliovirus, for example, initially replicates in the gut with few symptoms but is borne by the blood to the central nervous system, where it infects specific cells to cause its devastating effects. These viruses that replicate in the intestinal cells often are shed into the feces and are passed on to others by the fecal–oral route of transmission.

Some viruses are efficiently transmitted by direct inoculation into the bloodstream. In nature these viruses require insect vectors to effect this transmission. Well-known examples include the yellow fever virus, dengue fever virus, and the encephalitis viruses, all transmitted by blood-feeding arthropods such as mosquitos and ticks. Many viruses, although not transmitted by direct inoculation into the bloodstream in nature, can be transmitted by blood inoculation through medical procedures (transfusions, injections) or trauma.

The venereal route of transmission is also utilized by some viruses. Well-known examples include herpes simplex virus type 2, certain human papillomavirus strains, and human immunodeficiency virus (HIV).

After the local infection of susceptible cells with viral multiplication, the initial viremia (primary viremia) serves to transport the virus to specific target cells or tissues in the body where the virus may replicate further, giving rise to additional virus in the blood (secondary viremia). Often the immunological responses of the individual are provoked by the massive secondary viremia because the primary viremia may be inadequate in duration or intensity to do so.

Certain unusual modes of virus transmission have been observed. For example, in rabies, where the virus enters the tissues by trauma, often an animal bite, it enters the peripheral nerve cells and the virus migrates along the nerves to the central nervous system, where it then replicates and causes damage. The virus can find its way, perhaps by the bloodstream or by the nerves, to the salivary glands, where it can be excreted into the saliva and thereby be transmissible to another susceptible host.

III. RESPONSES TO VIRAL INFECTIONS

Most virus infections are asymptomatic or, at most, cause such common and inconsequential symptoms that the infection passes unnoticed. Analysis of the antibodies in normal human serum shows that we have many antibodies, which indicate a history of prior encounters, to viruses of which we have been unaware. For example, approximately 85% of adults in the U.S. harbor latent Epstein-Barr virus (a herpesvirus) in their lymphocytes. Likewise, many individuals who cannot recall having fever blisters carry herpes simplex virus type 1 in a latent form in their bodies. Infection with poliovirus in infancy often provokes only a mild, self-limited febrile illness, in contrast to the devastating infections of the central nervous system seen in infection of older children and adults.

The first cellular response to infection by many viruses seems to be the induction of interferon, a set of proteins that are secreted by the infected cells and function to render neighboring cells more resistant to viral replication. The interferon response appears to be aimed at producing local resistance to virus infection so as to limit the spread of the virus. This response is immediate and occurs within hours to days of the initial infection. Some side effects of the production of interferon include fever as well as the general malaise associated with many viral infections.

The viremic phase of virus infection allows the cells of the immune system to detect and respond to the presence of virus. If the virus is sufficiently immunogenic, the immune system produces a primary antibody response in about a week. This primary immune response results in the production of long-lasting memory B-cells, which can be activated later by subsequent exposure to the same virus to provide a more rapid and more intense secondary immune response. This immunological memory is the primary reason that we usually are immune for life once we have survived a particular virus infection.

The specific antibodies produced by the primary immune response can combine with the virus in the blood and result in circulating immune complexes that facilitate the destruction and clearance of the virus from the body, but also result in activation of some other processes such as the production of fever. Some viruses, such as herpes simplex virus, cause local immunological reactions of such intensity that much of the inflammation and pain at the site of the infection are the result of the action of the immune cells rather than the destruction of the infected cells by the virus alone. Another unusual immune reaction is observed in the case of infection by Epstein-Barr virus. This virus enters certain cells of the B-lymphocyte lineage and results in a growth transformation of these cells, providing them with the potential for unlimited cell division ("immortalization," the first step in the formation of a B-cell malignancy). The normal immune surveillance mechanism which involves the T-cell system is activated to respond to these transformed B-cells and kill them. The process of T-cell activation gives rise to a large population of unusual T-cells, and it was this population of activated T-cells (large cells with a large nucleus, initially thought to be unusual monocytes) which gave Epstein-Barr virus infection its name, infectious mononucleosis, or "mono." That is, it was named for the cells which reacted to the virus infection rather than for the virus-infected cells themselves.

Some viruses that enter into a latent or symbiotic state within the host cell can provoke the cell to behave in abnormal ways. Many such viruses carry extra genes that regulate cell division and can result in the malignant transformation of the cell to produce a cancer. These cancer-causing viruses (oncogenic viruses) are a special group of viruses that are of great current interest in terms of both their special biology and their practical importance.

IV. OUTCOMES OF VIRAL INFECTIONS

The usual outcome of a viral infection is recovery of the organism with long-lasting immunity. After the initial local multiplication, viremic phase, and immunological responses, the virus is eliminated from the body, and the memory cells of the immune system stand ready to guard against another infection. It is this sort of immunity that is produced by successful iatrogenic virus infection called vaccination. If, however, the immune system is compromised, the virus replication overwhelms the immune system, or the virus manages to get into cells or tissues that are hidden from the immune system, the virus may destroy critical tissues or organs and result in serious illness or death.

Some viruses, after the primary infection, may enter into a latent form and be asymptomatic until periodic reactivation at later times. The herpes group of viruses are especially prone to such latent infections. Initial infection, such as with the chicken pox virus, gives rise to the viremia and generalized skin rash. The virus then enters into a latent infection of the dorsal root ganglia of the spinal cord, and later, at times of lowered immunity, the virus may replicate and cause lesions in the skin along the local distribution of a particular spinal nerve, giving rise to the condition called "shingles." Both chicken pox and shingles are disease manifestations of the same virus, the varicella-zoster virus.

A few viruses are known that may be present in the body and replicate at such a low level and be relatively benign yet escape the immune system and thereby establish a true persistent infection. The early phase of HIV infection seems to be an example of this mode of virus–host interaction.

Various degrees of cell proliferation may result from latent virus infections. These outcomes can result in local, limited growths such as viral warts and the small skin lesions caused by the virus of molluscum contagiosum, or they can lead, in steps not yet fully understood, to malignant diseases such as Burkitt's lymphoma, nasopharyngeal carcinoma, Kaposi's sarcoma, and some types of cervical cancer.

V. VIRAL VIRULENCE

Viruses vary in their ability to infect and cause changes in their host cells. Even the specific cell type (tropism) may vary. These variations may be because of heritable properties (genetic mutations) or because of properties acquired from the most recent host, for example, viral envelope structures of cellular origin (pseudotype variations). Viruses are said to be virulent if they have a high propensity to cause disease or other evidence of infection in the specific test organism. Thus, a virus stock may be virulent for one species and avirulent for another. Repeated selection for virulence in one species may select for mutations that render the virus less virulent (attenuated) in another. This principle has been widely exploited to produce vaccine strains of viruses.

Some virulence may be related to the interaction of essential viral functions with related cellular functions. Other aspects of virulence may be simply a matter of virus interactions with the specific cell receptors for the virus. In certain cases, the genes of the virus that are known to be required for certain functions can be deleted or modified to make avirulent variants. Thus, nononcogenic forms of some retroviruses can be constructed by deletion of their specific viral oncogene.

Virulence is a concept reserved for the capacity for the virus to produce an effect, not for the ability of the virus to survive inactivation. Some viruses are especially sensitive to drying, for example, and others are sensitive to organic solvents. These viruses may be virulent, even though they are, in some sense, very fragile and easily killed.

See Also the Following Articles

Aerosol Infections • Interferons • Rabies • Vaccines, Viral • Water, Drinking

Bibliography

Dimmock, N. J., and Primrose, S. B. (1994). "Introduction to Modern Virology," 4th Ed. Blackwell, Oxford.

Fields, B. N, Knipe, D. M., and Howley, P. N. (eds.) (1996). "Fields' Virology," 3rd Ed. Lippincott-Raven, Philadelphia.

Levy, J. A., Fraenkel-Conrat, H., and Owens, R. A. (eds.) (1994). "Virology," 4th Ed. Prentice-Hall, Englewood Cliffs, New Jersey.

Vitamins and Related Biofactors, Microbial Production

S. De Baets, S. Vandedrinck, and E. J. Vandamme

University of Gent

I. Need and Use of Vitamins
II. Fat-Soluble Vitamins
III. Water-Soluble Vitamins
IV. Strategies for Improved Biotechnological Vitamin Production
V. Application Range
VI. Concluding Remarks

GLOSSARY

bioconversion The use of enzymes or living organisms to transform matter from one form to another. Also, **biotransformation.**

biotechnological production The application of advanced biological techniques in the manufacture of industrial products.

coenzyme An organic nonprotein molecule that is required for certain enzymatic reactions.

fermentation The process of culturing cells or other microorganisms in a container, bioreactor, or fermenter for experimental or commercial purposes.

vitamin Any of various compounds that occur in many foods in small amounts and are also produced synthetically; trace amounts are required for the normal physiologic and metabolic functioning of the body.

VITAMINS AND RELATED BIOFACTORS belong to those few chemicals with a direct positive appeal to people. There is indeed a need for extra vitamins, other than those derived from plant and animal sources, due to unbalanced food habits or processing, food shortage, or disease.

Added vitamins are now prepared either chemically or biotechnologically via fermentation or bioconversion processes. Several vitamins and related biofactors until recently were mainly produced either chemically [vitamin A, cholecalciferol (D_3), tocopherol (E), vitamin K_2, thiamin (B_1), niacin (PP or B), pantothenic acid (B_5), pyridoxine (B_6), biotin (H or B_8), and folic acid (B_9)] or via extraction processes (β-carotene, provitamin A, provitamin D_3, tocopherol, and the vitamin F group).

However, for most of these valuable compounds, microbiological production methods also exist or are rapidly emerging. Others are produced practically exclusively via fermentation [ergosterol or provitamin D_2, riboflavin (B_2), cyanocobalamin (B_{12}), orotic acid (B_{13}), the vitamin F group, ATP, nucleosides, coenzymes, etc.] or via microalgal culture (β-carotene, E, and F). Both chemical and microbiological processes are run industrially for vitamin B_2, whereas vitamin C (L-ascorbic acid) is produced via a combination of chemical reactions and fermentation processes. This article discusses the current state of vitamin production, with emphasis on developments and strategies for improved biotechnological production and its significance compared to existing chemical processes. The application range of vitamins in the food/feed sector, health and medical field, and as technical aid is also discussed. The screening or construction of vitamin hyperproducing microbial strains is a difficult task: pathway elucidation and metabolic (de)regulation need further study, r-DNA technology has only recently been introduced, improved fermentation processes and immobilized biocatalyst bioconversions for the synthesis of chiral vitamins and related compounds or intermediates or derivatives are gaining importance, and the recovery and purification of these vitamin compounds from their fermentation broths remain equally complex tasks.

I. NEED AND USE OF VITAMINS

Vitamins and related biofactors belong to those few chemicals with a positive appeal to most people: All of us indeed need our daily intake of vitamins, which should normally be provided via a balanced and varied diet. Principally, the staple food of man, including cereals, rice, potatoes, vegetables, fruits, milk, fish, meat, and eggs, form the basic source of vitamins and biofactors. Adequate nutrition should thus supply this daily vitamin need; however, the need increases with physical exercise, pregnancy, lactation, active growth, convalescence, drug abuse, stress, air pollution, etc. However, current food habits or preferences or food processing and preservation methods do not always provide a sufficient natural daily vitamin supply, even for a healthy human being; this is all the more true for stressed or sick individuals. Pathological situations (intestinal malabsorption; stressed intestinal flora; liver/gall diseases; drug, antibiotic, or hormone treatment; enzyme deficiencies; etc.) can also lead to vitamin shortages despite a sufficient intake. Malnourishment in many countries also requires direct medical attention, combined with diet and vitamin adjustment. Although modern society is seldom confronted with the notorious avitaminoses of the past, these do still occur frequently in overpopulated and poverty- and famine-struck regions in many parts of the world.

Apart from their *in vivo* nutritional–physiological roles as growth factors and/or coenzymes for man, animals, plants, and microorganisms, vitamin compounds are increasingly being introduced as food/feed additives, as medical–therapeutic agents, as health aids, and as cosmetic and technical aids. Indeed, today many processed foods, feeds, pharmaceuticals, cosmetics, and chemicals contain added vitamins or vitamin-related compounds, and single or multivitamin preparations are commonly taken or prescribed. Furthermore, vitamin-enriched and medicated feed is used worldwide to procure healthy livestock.

Thus, there is a need for vitamin supply other than that provided from plant and animal food resources. Most added vitamins are indeed now prepared chemically and/or biotechnologically via fermentation or bioconversion processes.

Similarly, other related biofactors, provitamins and pigments, vitamin-like compounds, special unsaturated fatty acids, coenzymes, etc., some of which are increasingly used in pharmacy, medicine, agriculture, food/feed production or processing, and as health aids, are equally important biotechnological products for which production via microbial fermentation or microalgal bioconversion is now applied industrially. Indeed, microbial biotechnology has been instrumental in procuring sufficient amounts of these valuable complex molecules via natural processes, although for certain vitamin products there

TABLE I
Industrial Production of Fat-Soluble Vitamins

Compound	*Organic chemistry*	*Extraction chemistry*	*Biotechnology*			*World production (tons/year)*
			Bacterial	*Fungal*	*Algal*	
β-Carotene	+	+		+	+	100
Vitamin A	+					2500
Provitamin D_2				+		
Provitamin D_3	+	+				5000
Vitamin E	+	+			(+)[a]	7000
Vitamin F		+		+	(+)[a]	100
Vitamin K_2	+					2

[a] Parentheses indicate pilot scale process.

TABLE II
Industrial Production of Water-Soluble Vitamins and Related Biofactors

Compound	Organic chemistry	Extraction chemistry	Biotechnology: Bacterial	Biotechnology: Fungal	Biotechnology: Algal	World production (tons/year)
Vitamin B_1	+		+			2,000
Vitamin B_2	+		+	+		>2,000
FAD			(+)[a]			
Niacin (B_3, PP)	+		+			>8,500
NAD, NADP		+		+		
Pantothenic acid (B_5)	+					4,000
Coenzyme A			+			
Vitamin B_6	+					1,600
Biotin (B_8)	+		(+)			30
Folic acid (B_9)	+					300
Vitamin B_{12}			+			10
Vitamin B_{13}			+			100
Vitamin C[b]	+		+			>70,000
ATP			+	+		10

[a] Parentheses indicate pilot scale process.
[b] Combination of organic chemistry and fermentation process.

is still fierce competition with organic chemical synthesis.

A survey of the current vitamin and biofactor industrial production methodology and its yearly tonnage is given in Tables I and II. The microorganisms involved in these processes are listed in Tables III and IV as well as the application range of vitamins in the food/feed sector and medical or technical fields.

II. FAT-SOLUBLE VITAMINS

A. β-Carotene

Provitamin A (β-carotene) (Fig. 1) is used as a natural vitamin, as an antioxidant, and as an orange/red pigment in food, feed, pharmaceuticals, and cosmetics. Large-scale trials have confirmed its

TABLE III
Microbial and Microalgal Synthesis of Fat-Soluble Vitamins

Compound	(Industrial) microorganism	Application[a]
β-Carotene	*Dunaliella salina, D. bardawil* (algal)[b]	F, M, T
	Blakeslea trispora (fungal)[b]	
Vitamin D_2	*Saccharomyces cerevisiae*[b] (via ergosterol $\xrightarrow{UV}$ D_2)	F, M
Vitamin E	*Euglena gracilis* (algal)	F, M, T
	Bioconversion reactions: *Pseudomonas, Clostridium*	
Vitamin F	*Mortierella isabellina* (GLA),[b] *M. alpina* (ARA, DHGLA, EPA, DHA)[b]	F, M
	Marine microalgae (EPA, DHA)	
	Bioconversion (linseed oil → EPA)	
Vitamin K_2	*Flavobacterium meningosepticum*	M

[a] Abbreviations used: F, food/feed; M, medical; T, technical.
[b] Commercial bioprocess.

TABLE IV
Microbial Synthesis of Water-Soluble Vitamins and Biofactors

Compound	*(Industrial) microorganism*	*Application*[a]
Vitamin B_1 (thiamin)	*Saccharomyces cerevisiae* (bioconversion)	F, M
Vitamin B_2 (riboflavin)	*Ashbiya gossypii:* from D-ribose (*Bacillus* sp.)[b]	F, M
Niacin (B_3, PP)	Bioconversion of 3-cyanopyridin: *Nocardia rhodochrous*	F, M
Pantothenic acid (B_5)	Bioconversion of (keto)pantoyllactone: *Rhodotorula minuta, Candida parapsilosis, Rhodococcus erythropolis*	F, M
Coenzyme A	Pantothenic acid + L-cystein + adenin: *Brevibacterium ammoniagenes*[b]	M, T
Vitamin B_6	*Flavobacterium*	F, M
	Pichia guilliermondii	
Biotin (H, B_8)	*Bacillus sphaericus*	F, M
Vitamin B_{12}	*Propionibacterium shermanii, Pseudomonas denitrificans*[b]	F, M
Vitamin B_{13} (orotic acid)	*Corynebacterium glutamicum*[b]	F, M
	Brevibacterium ammoniagenes	
	Bacillus	
Vitamin C	*Gluconobacter oxydans* (D-sorbitol → L-sorbose)[b]	F, M, T
ATP	Yeast[b]	M, T
	Brevibacterium ammoniagenes	
NAD	Yeast	T
	Brevibacterium ammoniagenes	
NADP	*Achromobacter aceris* (NAD + ATP)	T
Coenzyme Q	Bacteria, yeasts	M, T
Ado-Met	Yeast	M
L-Ado-HCy	*Pseudomonas putida* and *Alcaligenes faecalis* (adenine + DL-homocystein)	M

[a] Abbreviations used: F, food/feed; M, medical; T, technical.
[b] Commercial bioprocess.

"anticancer"/"antitumor" activity. Efficient commercial chemical synthesis and extraction processes have been used since the 1950s. A recently developed economic bioprocess is based on the culture of hypersaline green microalgae (*Dunaliella salina* and *D. bardawil*) in salt ponds or lakes. This bioprocess is used in countries in which suitable climatic conditions and pristine salt-lake environments prevail, such as Australia, the United States, Russia, Israel, Chile, Spain, and China.

β-Carotene contents of 14% on a dry cell weight (DCW) basis have been reported. By-products of the process include glycerol, which can comprise 30% of biomass dry weight, and high-quality protein meal left after β-carotene extraction.

A second bioprocess is the β-carotene fermentation with the fungus *Blakeslea trispora*. This process has been industrialized for years in Russia, yielding purified β-carotene and β-carotene-enriched mycelium as an animal feed additive. In this fermentation, a fungal

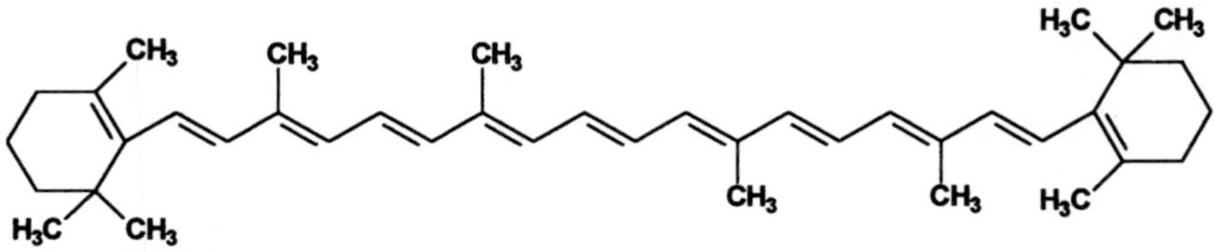

Fig. 1. Chemical structure of β-carotene (provitamin A).

mated culture is used with a preferred ratio of so-called minus and plus mating strains of *Blakeslea trispora*. Accumulation of β-carotene is strongly linked to sexual interaction between the two mating types. A hormone-like substance, the β factor, that is produced upon mating (of which the major component is trisporic acid) stimulates β-carotene formation.

Other chemicals that stimulate β-carotene production include β-ionon, retinol, kerosene, aromatics such as dimethylphtalate and veratrol, and nitrogenous heterocyclic compounds such as isoniazid and iproniazid. Mutants, obtained after UV, ethylmethyl sulfonate (EMS), or N-methyl-N-nitro-N-nitrosoguanidin (NTG) treatment, were selected based on resistance to lovastatin, acetoanilide, and β-ionon. Using such pooled mutants, up to 7 g/liter of β-carotene was produced in a 7-day fermentation process (approximately 200 mg of β-carotene per gram DCW).

Other *Mucorales* fungi (family *Choanephoraceae*), which are natural β-carotene producers, include *Choanephora, Mucor, Parasitella, Phycomyces,* and *Pilaria*. However, these are lower producers compared to *Blakeslea* species.

Recently, several Western companies have started similar fermentation processes for β-carotene. The food-grade yeast *Candida utilis* has recently been engineered to confer a novel biosynthetic pathway for the production of carotenoids such as provitamin A (β-carotene) and the pigments lycopene and astaxanthin. The endogenous carotenoid biosynthesis genes were derived from the epiphytic bacterium *Erwinia uredovora* and from the marine bacterium *Agrobacterium aurantiacum*. The resultant recombinant yeast strains accumulated 0.4 mg of β-carotene per gram DCW. Further optimization of this process could lead to a simple β-carotene fermentation process based on a food-grade yeast.

B. Vitamins D_2 and D_3

Vitamins D_2 and D_3 (Fig. 2) are used for antirachitic treatment; large amounts are also used for fortification of food and feed. Vitamin D_2 (ergocalciferol) is obtained by UV radiation (280–300 nm) of yeast ergosterol (provitamin D_2). Efficient fermentation processes for ergosterol accumulation in *Saccharomyces cerevisiae* and *S. uvarum* have been established, but most information is restricted to patent literature. 7-Dehydrocholesterol, extracted from mussels or chemically produced from cholesterol extracted from wool grease, is converted into cholecalciferol (or vitamin D_3). Efforts have been made to produce an ergosterol precursor, cholestatetraenol, useful for vitamin D_3 production by yeast fermentation. 1α,25-Dehydroxyvitamin D_3 is the most physiologically active form of vitamin D_3 and has been used clinically for chronic renal failure, hypoparathyroidism, osteoporosis, and psoriasis. Chemical synthesis requires a complicated procedure. Recently, bacteria have been found which can convert vitamin D_3 into its 1α,25-dehydroxy derivative, with *Amycolata autotrophica* being a promising strain.

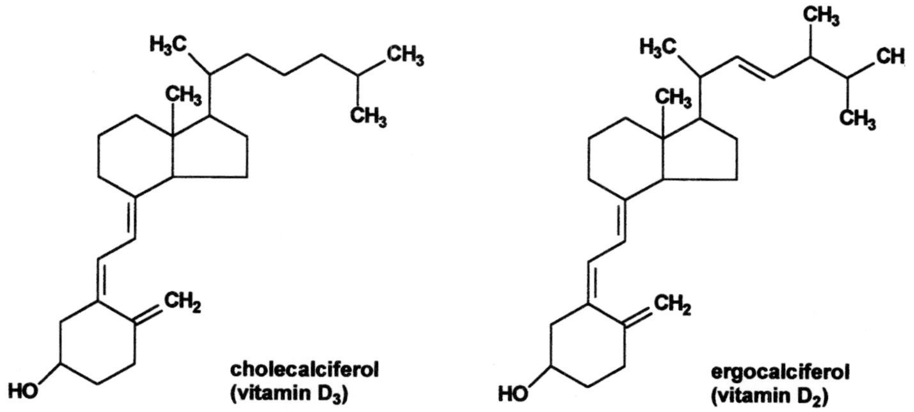

Fig. 2. Chemical structures of vitamin D_3 and D_2.

C. Vitamin E (Tocopherols)

Vitamin E (Fig. 3) is used as an antioxidant in clinical and nutritional applications. The chemically synthesized racemate of α-tocopherol (composed of a chromane ring and a C_{14} aliphatic side chain) or a mixture of tocopherols extracted from vegetable oil are currently the main sources, but there is much interest in applying microbial and microalgal-based processes.

The microalgal process has received much attention. *Euglena gracilis* seems to be the best producer of α-tocopherol when cultured photoheterotrophically in the presence of L-tyrosine, homogentisate, ethanol, and peptone. Since chemical synthesis of vitamin E yields a racemic mixture, attempts are now being made to synthesize the naturally optically active compound using biocatalysis. The formation of the optically active aliphatic side chain of α-tocopherol starts from a small chiral building block (C_4 or C_5 synthons), which is produced via microbial action. In this respect, $S(+)\beta$-hydroxyisobutyric acid, formed from isobutyric acid using a.o. *Pseudomonas putida* as a biocatalyst, has been used to produce the chiral C_{14}–vitamin E side chain.

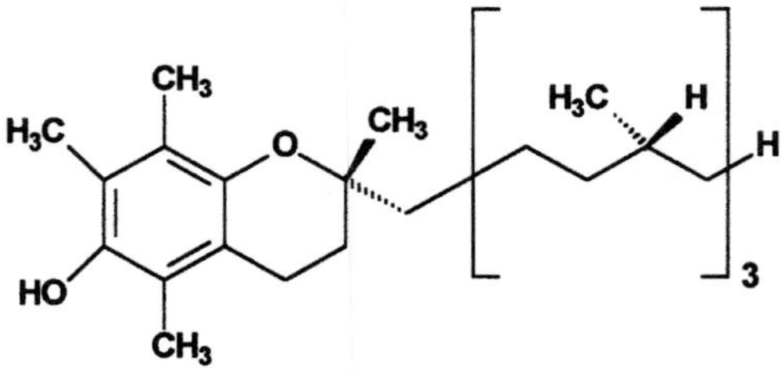

D. Polyunsaturated Fatty Acids or Vitamin F Group

Polyunsaturated fatty acids (PUFAs) (Fig. 4) are essential fatty acids with important applications in pharmacy, medicine, and human and animal nutrition. They are prostaglandin precursors and display important biological activity such as controlling cholesterol levels, thrombosis prevention, etc. Sources rich in C_{18} (γ-linolenic acid; or GLA) and C_{20} PUFAs are evening primrose seeds, animal tissue and fish oil, respectively. Certain fungi and algae are also rich sources of these compounds and fermentation technology has also recently become appropriate for their production.

Fungi such as *Mortierella isabellina* and *Mucor circinelloides* accumulate up to 5 g/liter of GLA in a medium based on glucose or molasses. This fermentation process has been commercialized in Japan and the United Kingdom. Related strains, such as *Mortierella alpina,* are good producers of C_{20} PUFAs such as dihomo-γ-linolenic acid (DHGLA) and arachidonic acid (ARA). It was also found that eicosapentanoic acid (EPA) accumulated when the strain was cultured at low temperature (at 12°C instead of 30°C) and that DHGLA could be overproduced by specific inhibition with sesamin, present in sesame oil, of the $\Delta5$-desaturase enzyme, which normally converts DHGLA into ARA.

Recently, it was also demonstrated that *Mortierella* strains can convert linseed oil, rich in α-linolenic acid (ALA) into EPA, indicating the possibility of upgrading cheap linseed oil using biotechnological processes.

Marine microalgae, such as *Nannochloropsis,* also provide a rich source of PUFAs, especially EPA and DHA. Attempts in Japan and the United States are now focussed on culturing these algae in fermentors either photosynthetically or heterotrophically on glucose.

E. Vitamin K

Vitamin K_2 (Fig. 5) is known as an antihemorrhagic factor and is now widely used in medicine for various types of bleeding symptoms. Menaquinone (vitamin K_2) has higher therapeutic activity than phylloquinone (vitamin K_1). Today, vitamin K is produced by chemical synthesis and only recently has work on its fermentative production been initiated.

An intensive screening program demonstrated that vitamin K_2 was only found in bacteria and not in fungi or yeasts. In particular, facultative anaerobic bacteria contained vitamin K_2 and its homologs. *Flavobacterium meningosepticum* was identified as a particularly high producer. A mutant of this organism was selected which was resistant to 1-hydroxy-2-naphtoate (HNA), a selective inhibitor of vitamin

Fig. 3. Chemical structure of vitamin E (tocopherol).

K_2 biosynthesis and which overproduced vitamin K_2 intracellularly (5.5 mg/g dry cell mass) on a medium containing L-tyrosine and isopentenol. These results indicate the feasibility of establishing a microbial fermentation process for specific production of vitamin K homologs.

III. WATER-SOLUBLE VITAMINS

A. Thiamin (Vitamin B_1)

Vitamin B_1 (Fig. 6) is currently produced chemically and is used in medicine and foods. Its deficiency is the cause of the disease beriberi. Yeasts and bacteria do not normally overproduce large amounts; also, the biosynthesis pathway has not been fully elucidated. Thiamin-excreting (up to 1 mg/liter) mutants of yeast (*Saccharomyces cerevisiae*) have been isolated. Another possible microbial process would involve the enzymatic coupling by *E. coli* and baker's yeast of preformed pyrimidine and thiazole moieties of thiamin (i.e., 2-methyl-4-amino-5-hydroxymethyl-pyrimidine and 4-methyl-5-β-(hydroxyethylthiazole, respectively).

B. Riboflavin (Vitamin B_2)

Riboflavin is used in human nutrition and therapy and in animal feed. It is produced both by synthetic and by fermentation processes, with the latter recently increasing in application. Although bacteria (*Clostridium* sp.) and yeasts (*Candida* sp.) are also good producers, currently two closely related ascomycete fungi, *Ashbya gossypii* and *Eremothecium ashbyii,* are considered to be the best riboflavin producers. Strain improvement related to riboflavin has been studied thoroughly, such that this fermentation process is well established, but further yield improvement remains possible.

The total biosynthesis pathway of riboflavin has only recently been resolved completely; the biosynthetic origin of its xylene ring has only been known since 1990 (Fig. 7).

The well-known direct precursor of riboflavin (compound 6), 6,7-dimethyl-8-ribityllumazine (compound 5), is formed by condensation of 5-amino-6-ribitylamino-2,4 (1H, 3H)-pyrimidindione (compound 2), originating from guanosine triphosphate (GTP; compound 1) with the novel carbohydrate 3,4-dihydroxy-2-butanone-4-phosphate (compound 4), which is produced from ribulose-5-phosphate (compound 3) by the novel enzyme 3,4-dihydroxy-2-butanone-4-phosphate synthase. The enzyme has been purified from the flavinogenic yeast *Candida guilliermondii* and from *E. coli.* Recently, the *E. coli* synthase gene has been cloned, sequenced, and expressed: the gene codes for a protein of 24 kDa, which is also the size of the yeast enzyme.

Fermentation solids such as residues from the acetone–butanol *Clostridium* fermentation contain low levels of riboflavin and may be recovered for use as an animal feed supplement. Another industrial process for riboflavin is based on microbial production of D-ribose which then serves as a starting material for further chemical synthesis of riboflavin. Asporogenous *Bacillus subtilis* mutants, deficient in transketolase, accumulate high levels of D-ribose ($>$70 g/liter). Riboflavin can also be coupled with adenine to form FAD by microbial processes using *Sarcina lutea* or yeast as a biocatalyst.

C. Niacin and Nicotinamide (Vitamin B_3 or PP)

Vitamin B_3 (Fig. 8) is used for vitamin enrichment of cereal products and as a meat additive. A deficiency in vitamin B_3 leads to the disease pellagra. The vitamin is also used as a supplement in animal feed. It has widespread medical use as an antihyperlipidemic agent and a peripheral vasodilator. Most nicotinic acid is produced chemically by liquid-phase oxidation of 2-methyl-5-ethylpyridine, which is synthesized from ethylene or acetaldehyde. An alternative process involves oxidative ammonolysis of an alkylpyridine such as 3-methylpyridine or 2-methyl-5-ethylpyridine to produce 3-cyanopyridine. Subsequent hydrolysis of the nitrile with aqueous ammonia yields either nicotinic acid or nicotinamide, depending on the reaction conditions. Recently, a microbial process was described for the bioconversion of 3-cyanopyridine into nicotinic acid and ammonia; the microorganism used is a *Rhodococcus rhodochrous*

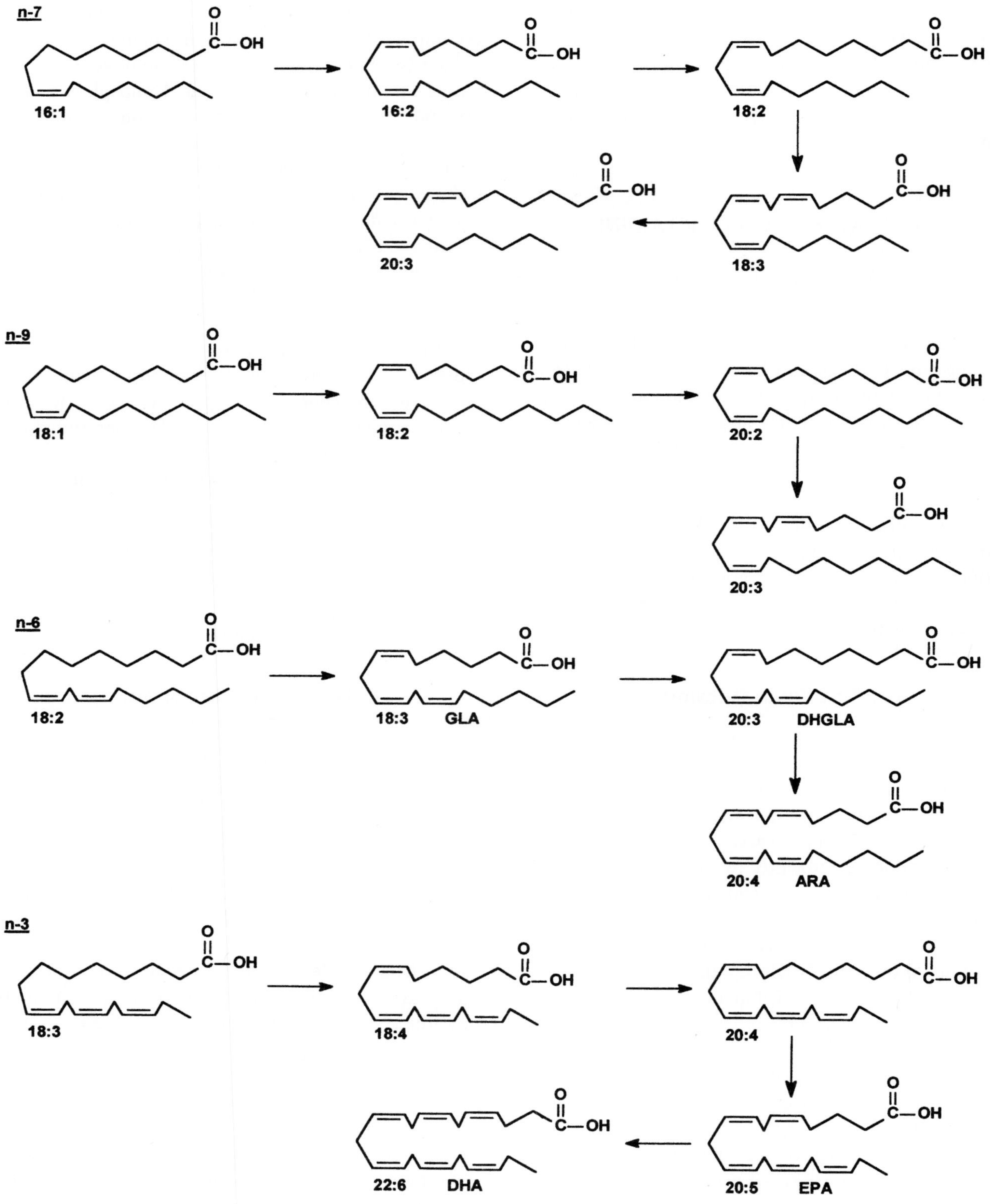

n-7

16:1 → 16:2 → 18:2 → 18:3 → 20:3

n-9

18:1 → 18:2 → 20:2 → 20:3

n-6

18:2 → 18:3 GLA → 20:3 DHGLA → 20:4 ARA

n-3

18:3 → 18:4 → 20:4 → 20:5 EPA → 22:6 DHA

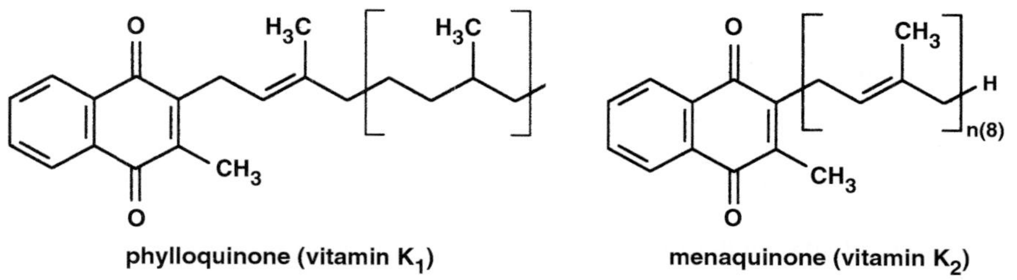

phylloquinone (vitamin K_1) menaquinone (vitamin K_2)

Fig. 5. Chemical structures of vitamin K_1 and K_2.

strain expressing 3-cyanopyridinase (a nitrilase) activity. When the enzyme was induced with benzonitrile, the liberated ammonia was utilized as a nitrogen source for growth, whereas the nicotinic acid was not further metabolized.

Nicotinamide can be converted by a *Brevibacterium* fermentation process into NAD^+; this can be phosphorylated to $NADP^+$ with *Achromobacter aceris.*

D. Pantothenic Acid (Vitamin B_5) and Coenzyme A (CoA-SH)

Pantothenic acid (Fig. 9) is mainly used as an animal feed additive and as a pharmaceutical. Commercial production of D(−)-pantothenate depends on chemical synthesis, which includes a difficult optical resolution step. To overcome these problems, microbial bioconversion steps have been introduced. Starting from ketopantoyl lactone, a stereospecific reduction into D(−)-pantoyllactone has been reported using washed cells of *Rhodotorula minuta* or *Candida parapsilosis.* The responsible enzyme is a novel carbonyl reductase. Condensation with β-alanine then gives pantothenic acid.

With racemic pantoyllactone as a substrate, *Nocardia asteroides* specifically oxidized the L-(+)-isomer to ketopantoyl lactone, which was then further converted as discussed previously. These two successive bioconversion steps could also be performed by a single strain, *Rhodococcus erythropolis,* with high yield.

In 1993, a direct high-yielding fermentation process for D-pantoic acid and/or D-pantothenic acid was disclosed by Takeda Chem. Ind. (Japan). Several *Enterobacteriaceae* genera (e.g., *Citrobacter, Klebsiella, Enterobacter,* and *Escherichia*), having resistance to salicylic acid and/or α-ketoisovaleric acid, α-ketobutyric acid, α-aminobutyric acid, β-hydroxyaspartic acid, and *O*-methylthreonine, were selected as high producer strains. *Escherichia coli* mutants and recombinant strains, transformed with plasmids, carrying genes involved in pantothenic acid biosynthesis yielded up to 65 g/liter D-pantothenic acid from glucose upon addition of β-alanine as a precursor.

The related compound coenzyme A (Fig. 10) can be synthesized chemically, but this is not practical due to the complexity of the process. Therefore, commercial production is performed using microorganisms. A successful process consists of the formation of coenzyme A from pantothenic acid, L-cysteine, ATP or AMP, and adenosine or adenine in the presence of surfactants using *Brevibacterium ammoniagenes,* growing on high levels of glucose. Higher yields (up to 23 g/liter) were obtained when 4-phos-

Fig. 4. Pathways for the biosynthesis of polyunsaturated fatty acids (PUFAs) of the *n*-7 (palmitoleic), *n*-9 (oleic), *n*-6 (linoleic), and *n*-3 (α-linolenic) families. PUFAs include the *n*-7, *n*-9, *n*-6, and *n*-3 families, which are defined by the position of the double bond closest to the methyl end of the fatty acid molecule. Desaturation occurs toward the carboxyl end of the molecule and chain elongation at the carboxyl end, leaving the methyl end unaltered. Usually, PUFAs of each family are not convertible. GLA, γ-linolenic acid; DHGLA, dihomo-γ-linolenic acid; ARA, arachidonic acid; EPA, eicopentaenoic acid; DHA, docosahexaenoic acid.

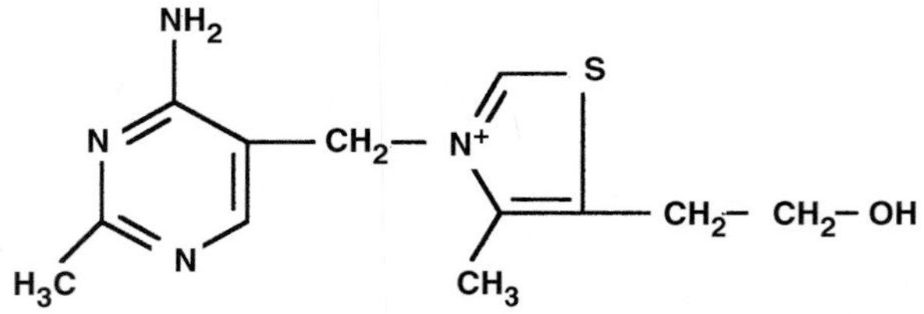

Fig. 6. Chemical structure of vitamin B_1 (thiamin).

phopantothenic acid was used instead of pantothenic acid; a further increase was obtained by using feedback resistant mutants.

E. Pyridoxine (Vitamin B_6)

Vitamin B_6 (Fig. 11) compounds, mainly pyridoxine and pyridoxalphosphate, are currently produced chemically. They have many applications in medicine and foodstuff. The complete biosynthetic pathway of vitamin B_6 is not known in detail. Extensive screening for vitamin B_6 producers among microorganisms has not yielded strains likely to be useful for industrial fermentation. The best producers are *Flavobacterium* strains, producing up to 20 mg/liter extracellularly, and the yeast *Pichia guilliermondii*, yielding up to 25 mg/liter.

F. Biotin (H, B_8)

Biotin (Fig. 12) is widely used in health care, food, and feed sectors. Industrial synthesis of biotin is performed via a chemical process, the so-called Goldberg and Sternbach synthesis. The biosynthesis process of biotin in microorganisms has also been studied in detail using rDNA techniques. Only the last step, the conversion of dethiobiotin (DTB) into biotin by the so-called "biotin synthetase," has not been enzymatically resolved, although *Bacillus sphaericus* could convert DTB into (+)-biotin. A strong feedback repressive action of biotin has been demonstrated on all biotin-synthesizing enzymes, which makes microbial overproduction difficult. Regulatory mutants have been selected with enhanced biotin formation capacity. Acidomycin (ACM) and 5-(2-thienyl)-*n*-valeric acid (TVA) are biotin analogs and ACM- and TVA-resistant mutants of *B. sphaericus* and *Serratia marcescens* excreted biotin at concentrations of up to 20 mg/liter on a medium with glucose and urea. Several biotin synthesis genes from *B. sphaericus* and *S. marcescens* have recently been cloned and expressed in *E. coli* and *B. subtilis.* In order to further increase the biotin productivity of the previously mentioned *S. marcescens* mutants, their biotin operons were cloned on several plasmids. These hybrid plasmids harbored the 7.2-kb DNA fragments, coding for the five biotin synthesis genes. Reintroduction into *S. marcescens* mutants led to the selection of a stable recombinant strain exhibiting a D-biotin production level of 200 mg/liter. For successful industrial application, the biotin yields in fermentation need to be increased further and the cost of precursors such as pimelic acid should be taken into consideration.

G. Vitamin B_{12}

Vitamin B_{12} (Fig. 13) is obtained exclusively by fermentation processes. The primary natural source of vitamin B_{12} is bacterial metabolic activity. An overall mechanism for vitamin B_{12} biosynthesis has only recently been presented and a few genes have already been cloned in *E. coli* a.o. urogen III synthetase. At first, vitamin B_{12} used for human therapy and as a food or feed supplement was obtained as a by-product of *Streptomyces* antibiotic (streptomycin, neomycin, and chlortetracycline) fermentations. Good producing strains were also isolated from feces, manure, and sewage sludge. For industrial purposes, *Propionibacterium shermanii* and *Pseudomonas denitrificans* were selected because of high yield and rapid growth. Mutagenic treatments have resulted in improved productivity, but in all cases cobalt ions and frequently 5,6-dimethylbenzimidazole (5,6-DBI) have to be added, in addition to precursors such as glycine, threonine, δ-aminolevulinic acid, and aminopropanol. Betaine, present in sugar beet molasses, and choline also have a strong stimulatory effect on bacterial vitamin B_{12} production. Porphyrineless mutants and catalase-negative mutants usually are vitamin B_{12} overproducers.

Microaerophilic propionibacteria produce cobalt-corrinoids in conventional media (e.g., molasses and corn steep liquor), supplemented with cobalt and without aeration. They can synthesize 5,6-DBI under

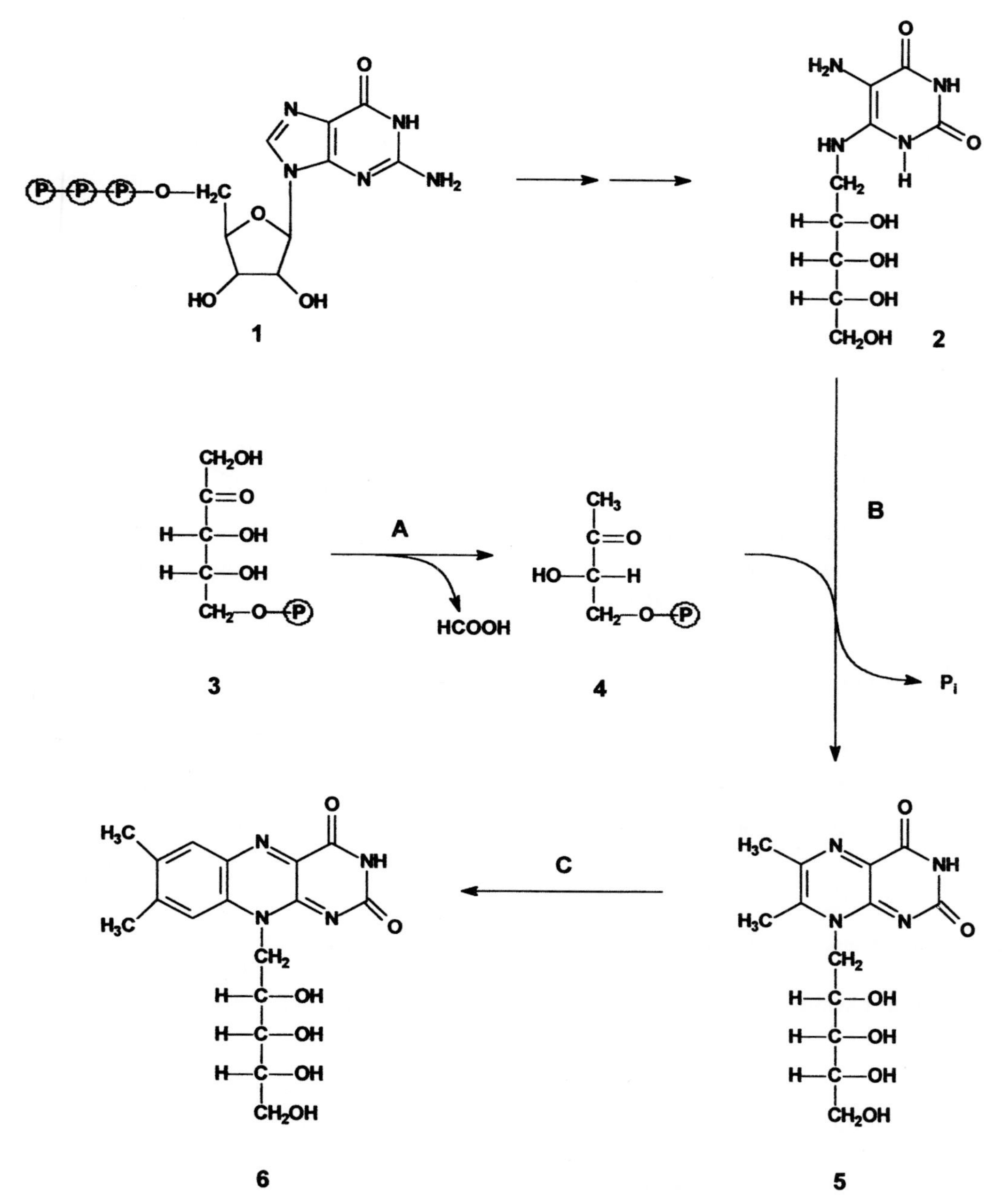

Fig. 7. Biosynthesis pathway of riboflavin. (1) GTP, (2) 5-amino-6-ribitylamino-2,4 (1H, 3H)-pyrimidindione, (3) ribulose-5-phosphate, (4) carbohydrate 3,4-dihydroxy-2-butanone-4-phosphate, (5) 6,7-dimethyl-8-ribityllumazine, and (6) riboflavin. (A) 3,4-dihydroxy-2-butanone 4-phosphate synthase, (B) 6,7-dimethyl-8-ribityllumazine synthase, and (C) riboflavin synthase.

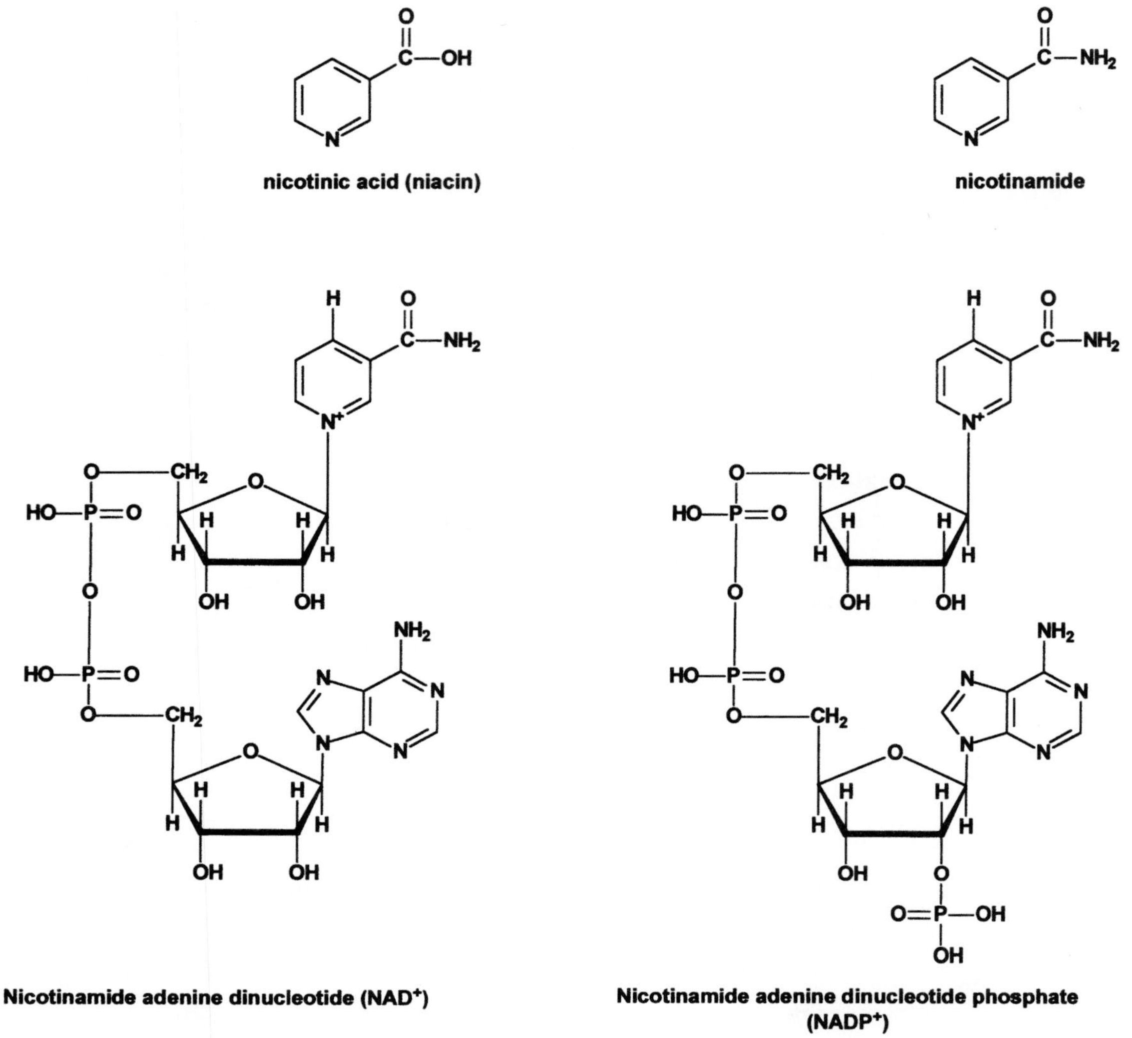

nicotinic acid (niacin)

nicotinamide

Nicotinamide adenine dinucleotide (NAD^+)

Nicotinamide adenine dinucleotide phosphate ($NADP^+$)

Fig. 8. Chemical structure of niacin, nicotinamide, and the derivatives NAD^+ and $NADP^+$.

aerated cultural conditions. This has led to a two-stage industrial fermentation: a first anaerobic stage promoting growth and cobamide biosynthesis and a subsequent shift to aeration of the culture that promotes DBI biosynthesis and conversion of cobamide to cobalamine. *Pseudomonas denitrificans* growth parallels cobalamin synthesis under aerobic conditions if the culture is directly supplemented with 5,6-DBI and cobalt salts. Maintaining a low dissolved oxygen level has a favorable effect. By mutation and selection, strains have been obtained that produce concentrations higher than 150 mg/liter. Several vitamin B_{12} derivatives can be produced either by direct fermentation (hydroxycobalamin and coenzyme B_{12}) or by chemical conversion of cyanocobalamin. The vitamin B_{12} fermentation can be considered a well-established industrial process, but one that is still capable of improvement.

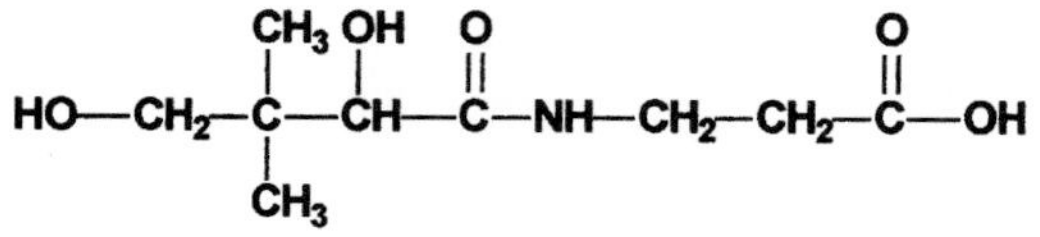

$$HO-CH_2-C(CH_3)_2-CH(OH)-C(=O)-NH-CH_2-CH_2-C(=O)-OH$$

Fig. 9. Chemical structure of vitamin B_5 (pantothenic acid).

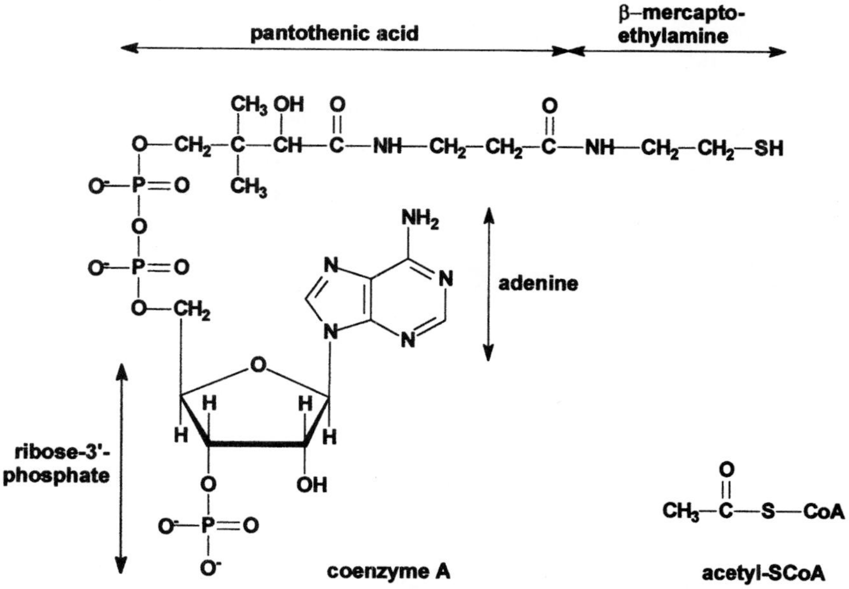

pantothenic acid
β-mercapto-ethylamine
adenine
ribose-3'-phosphate
coenzyme A
acetyl-SCoA

Fig. 10. Chemical structure of coenzyme A, which is related to pantothenic acid as shown.

H. Vitamin B_{13} (Orotic Acid)

Orotic acid (Fig. 14) is produced industrially exclusively by fermentation; it is used in medicine as a hepatic drug and in health food. Orotic acid is an intermediate in the biosynthesis of pyrimidine nucleotides. The biosynthetic pathway and regulation of pyrimidine nucleotides have been studied in detail, such that rational selection of overproducer mutants is possible. Most orotic acid-overproducing strains (*Corynebacterium glutamicum, C. ammoniagenes,* and *B. subtilis*) require uracil for growth. Suboptimal uracil feeding results in high orotic acid levels. Mutants resistant to 5-fluorouracil produced up to 50 g/liter. Orotic acid is a substrate in the microbial synthesis process of cytidine diphosphatecholine.

I. Vitamin C (L-Ascorbic Acid)

Total world production of vitamin C (Fig. 15) is estimated to be more than 70,000 tons/year. It is an important vitamin in human and animal nutrition and in medicine, and it has food technology applications as an antioxidant. Vitamin C is currently manufactured by a well-established five-step process, the Reichstein–Grüssner synthesis (Fig. 16), in which the conversion step of D-sorbitol into L-sorbose is a fermentation process. This fermentation step is carried out with *Gluconobacter oxydans* bacteria. D-Sorbitol is obtained from D-glucose by chemical hydro-

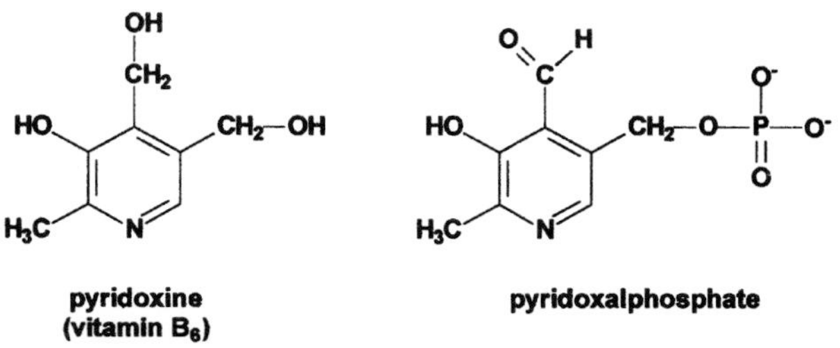

pyridoxine (vitamin B_6)
pyridoxalphosphate

Fig. 11. Chemical structures of vitamin B_6 (pyridoxine) and the derivative pyridoxalphosphate.

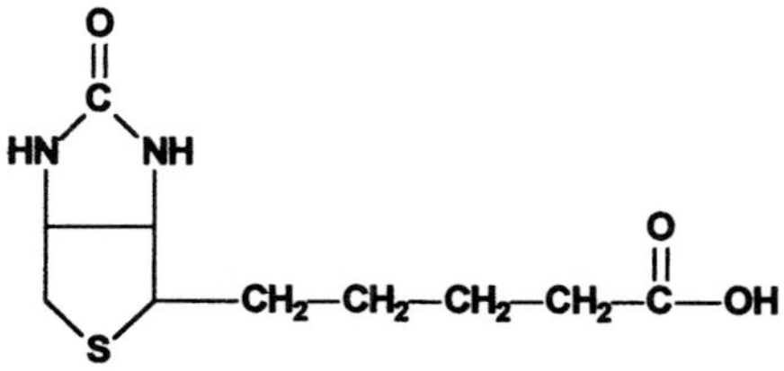

Fig. 12. Chemical structure of vitamin B_8 (biotin).

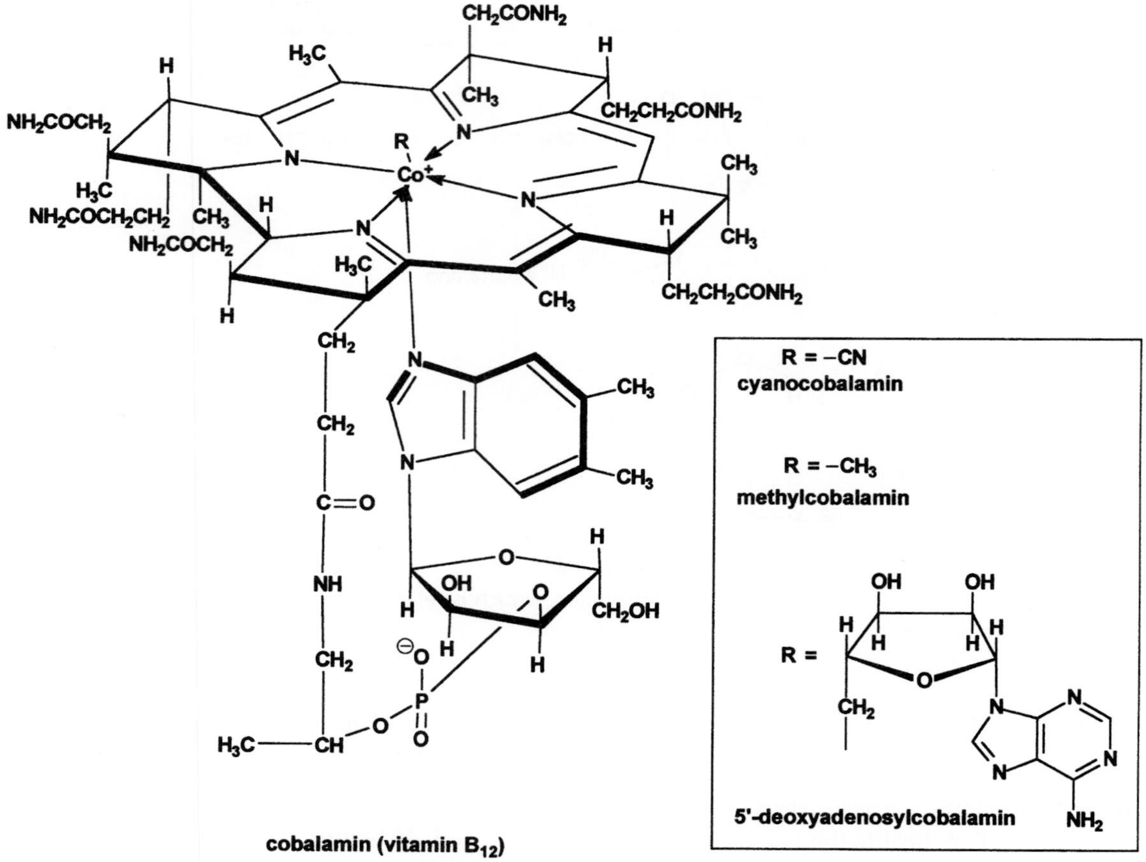

cobalamin (vitamin B_{12})

R = –CN cyanocobalamin

R = –CH_3 methylcobalamin

5'-deoxyadenosylcobalamin

Fig. 13. Chemical structure of vitamin B_{12}.

genation. The L-sorbose is then condensed with acetone to form sorbose diacetone which is oxidized to 2-keto-L-gulonic acid (2-KLG), which is then converted into L-ascorbic acid. Alternatively, the 2-keto-L-gulonic acid can be produced from glucose via 2,5-diketo-D-gluconic acid (2,5-DKG) by a tandem fermentation utilizing selected strains of *Erwinia herbicola* and *Corynebacterium*. The introduction of the gene encoding the enzyme 2,5-DKG reductase from *Corynebacterium* into *Erwinia* has enabled the rDNA *Erwinia* strain to produce 2-KLG directly, albeit at low levels. Another approach is to convert D-sorbitol or L-sorbose into 2-KLG via L-sorbosone. *Gluconobacter melanogenus* and *Gluconobacter oxydans* UV10 have the necessary enzymes: a membrane-located sorbose-FAD-dehydrogenase (SDH) and a cytoplasmic NAD(P)-L-sorbosone dehydrogenase (SNDH). The genes coding for SDH and SNDH have been cloned into 2-KLG producing *G. oxydans* strains. The recombinant strains were able to produce

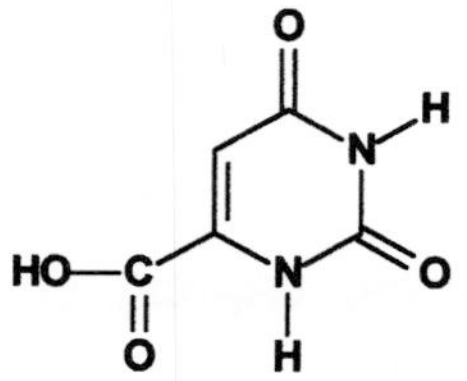

Fig. 14. Chemical structure of vitamin B_{13} (orotic acid).

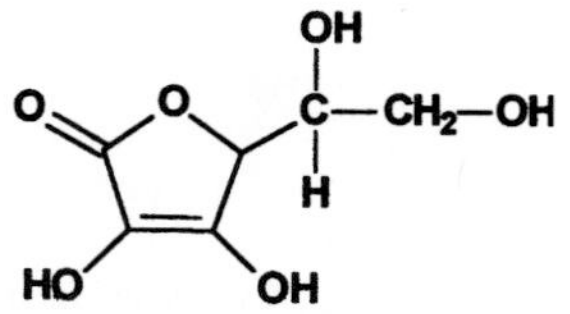

Fig. 15. Chemical structure of vitamin C (L-ascorbic acid).

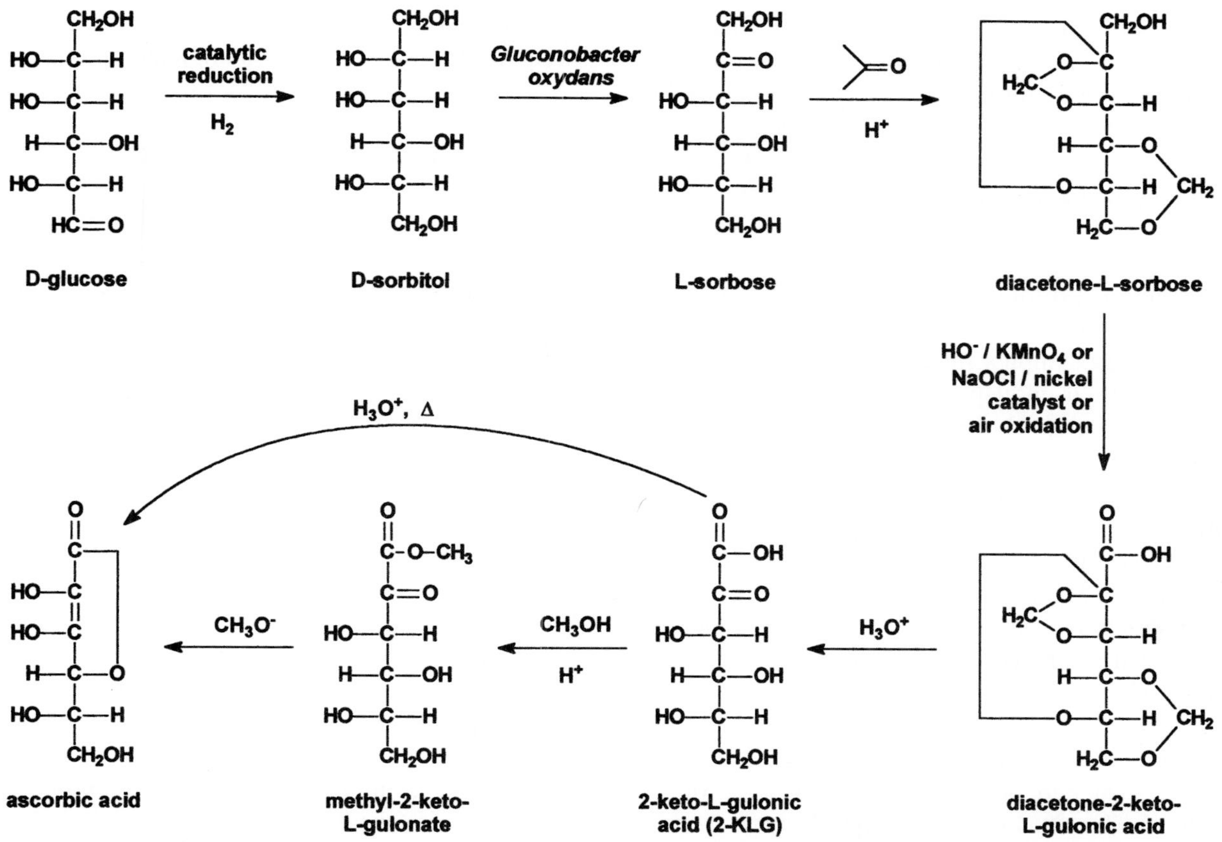

Fig. 16. The Reichstein–Grüssner synthesis of L-ascorbic acid. The oxidation of D-sorbitol to L-sorbose is accomplished by a *Gluconobacter suboxydans* fermentation; all the other steps are chemically catalyzed. The curved arrow from 2-keto-L-gulonic acid to L-ascorbic acid represents a nonaqueous acid-catalyzed cyclization. Currently, in the commercial process ascorbic acid is produced by acid cyclization of diacetone-2-keto-L-gulonic acid.

130 g/liter of 2-KLG from a 15% D-sorbitol-based medium.

Candida yeast strains have been reported that are able to directly produce D-erythroascorbic acid during cultivation on L-galactonic substrates.

J. Other Biofactors

Other biofactors—vitamin-like compounds and coenzymes such as ATP, NAD, FAD, coenzyme Q, *S*-adenosyl-methionine, folic acid, and lipoic acid—are also produced on a large scale by fermentation and are used as drugs against various diseases because of their unique pharmaceutical properties. They are also used as agents in diagnostic analyses and in biological and biochemical research.

IV. STRATEGIES FOR IMPROVED BIOTECHNOLOGICAL VITAMIN PRODUCTION

Extensive and ingenious screening (especially in Japan) has revealed microalgae and fungi to be excellent producers of most fat-soluble vitamins and pigments. In this respect, controlled microalgal culture in ponds or special bioreactors needs further attention. Conversely, vitamin K_2, for example, has only been found in bacteria, with *Flavobacterium meningosepticum* as a promising strain, and not in fungi or algae. Screening for thiamin (B_1) and pyridoxine (B_6) excreting strains should be intensified.

Since vitamin biosynthesis pathways are very complex (and in some cases not fully elucidated, i.e.,

B_1, B_6, biotin, B_{12}, etc.) and under strict regulatory control, mutation and selection for improved producer strains remains a challenging task. High-yielding laboratory mutants have recently been obtained for vitamin K_2 and biotin (B_8) synthesis. Hyperproducing industrial mutants have already been selected for large-scale production of riboflavin (and D-ribose), vitamin B_{12}, orotic acid, and L-sorbose, needed for vitamin C synthesis; obtaining a further increase in yield needs delicate fine-tuning of these bioprocesses. As complex indirect gene products, vitamins and their industrial production have so far gained only relatively little from rDNA technology; the cloning of genes, encoding for crucial/limiting enzymes in the pathways, appears promising however.

Only recently has interest focused on enzymatic reactions for bioconversion of chemically prepared precursors into desirable chiral vitamin compounds, intermediates, or derivatives. Examples include the following:

1. Synthesis of the chiral aliphatic side chain of α-tocopherol and of vitamin K_1 from a C_4 or C_5 synthon (i.e., $S(+)$-β-hydroxyisobutyric acid), formed with yeast or bacteria from isobutyric acid
2. Linseed oil (rich in α-linolenic acid) conversion by *Mortierella* fungi into EPA
3. Bioconversion of 3-cyanopyridine into niacin by *Rhodococcus rhodochrous* nitrilase
4. Stereospecific reduction with *Rhodotorula* of keto-pantoyl lactone into D-(−)-pantoyllactone, which upon condensation with β-alanine yields D-(−)-pantothenic acid
5. Regiospecific bioconversion of vitamin C into its 2-*O*-α-glucoside derivative

The application of immobilized biocatalyst technology (viable cells, multienzymes, etc.) for continuous vitamin (B_3, B_5, B_{12}, etc.), biofactor (CoA, ATP, NAD(P), FAD, etc.), or chiral intermediate (for E, F, K, B_5, C, etc.) synthesis should receive more interest.

Because high added-value biochemicals are often produced in low levels—even in optimized fermentations—the recovery and purification of these vitamin compounds from the cells or fermentation broths is an equally complex task, in which novel recovery techniques (e.g., extractive fermentation and membrane technology) should be implemented.

V. APPLICATION RANGE

A broad range of applications exist for these vitamins and related compounds in food, feed, cosmetics, and technical and pharmaceutical preparations or processes. The following are important applications:

Revitamination: Restoring the original vitamin level of a foodstuff

Standardization: Addition of a vitamin(s) to compensate for natural fluctuations

Vitamin enrichment: Addition of a vitamin(s) to a level higher than the natural level (health food, diet food industry, and medicated feed)

Vitamination: Addition of a vitamin(s) to products lacking them (feed and cosmetics)

Medical and pharmaceutical applications to alleviate hypo- or even avitaminoses or other physiological disorders: Considerable evidence indicates that vitamins and related biofactors have biochemical, medical, and nutritional functions far beyond their historical roles as coenzymes or for the prevention of deficiency symptoms; epidemiological and other data indicate that the risk of diseases such as cancer, liver and cardiovascular disease, cataracts, viral infections, and Parkinson's disease can be influenced by vitamin status

Use as technical additive (in food, feed, and cosmetics), e.g., β-carotene and astaxanthin as pigment: β-carotene and vitamins C and E as antioxidants; and vitamin C as acidulant

Use in diagnostics, agrobiological, medical, microbiological, biochemical, and genetic research

VI. CONCLUDING REMARKS

Compared to chemical synthesis processes (if available at all), there are several important advantages of microbial or microalgal processes for vitamin and biofactor production. Aside from obtaining these vitamins and biofactors via a natural way, fermentation or bioconversion reactions yield the desired enantiomeric compound (provitamin A, E, K, B_3, biotin, L-sorbose, etc.), whereas products from organic chemical synthesis are often racemic mixtures displaying a different biological activity. Furthermore, yields in fermentation broths can be very high—especially when using genetically improved strains—compared with the vitamin and biofactor levels in plants or animals. Also, fermentation processes and products generally have a positive environmental impact and a positive appeal to the consumer.

See Also the Following Articles

Biocatalysis for Synthesis of Chiral Pharmaceutical Intermediates • Biotransformations • Industrial Fermentation Processes • Pigments, Microbially Produced

Bibliography

Benemann, J. R. (1990). Microalgal products and production: An overview. *Dev. Ind. Microbiol.* **31**, 247–256.

Borowitzka, M. A., and Borowitzka, L. J. (Eds.) (1988). "Micro-algal Biotechnology." Cambridge Univ. Press, Cambridge, UK.

Boudrant, J. (1990). Microbial processes for ascorbic acid biosynthesis: A review. *Enzyme Microbial Technol.* **12**, 285–297.

Brodelius, P., and Vandamme, E. J. (1987). Immobilized cell systems. *In* "Biotechnology" (H. J. Rehm and G. Reed, Eds.; J. F. Kennedy, Series Ed.), Vol. 7a (Enzyme Technology), pp. 405–464. VCH, Weinheim.

De Wulf, P., and Vandamme, E. J. (1997). Microbial synthesis of D-ribose: Metabolic deregulation and fermentation process. *Adv. Appl. Microbiol.* **44**, 167–214.

Fusio, T., and Maruyama, A. (1997). Enzymatic production of pyrimidine nucleotides using *Corynebacterium ammoniagenes* cells and recombinant *E. coli* cells: Enzymatic production of CDP-choline from orotic acid and choline chloride. *Biosci. Biotechnol. Biochem.* **61**, 956–959.

Guiot, P., Ryan, M. A., and Scriven, E. F. V. (1996). Preparations and applications of nicotinic acid and nicotinamide. *Chim. Oggi/Chem. Today* **14**, 55–57.

Johnson, E. A., and An, G. H. (1991). Astaxanthin from microbial sources. *Crit. Rev. Biotechnol.* **11**, 297–326.

Kendrick, A. J., and Ratledge, C. (1992). Microbial polyunsaturated fatty acids of potential commercial interest. *SIM-Ind. Microbiol. News* **42**, 59–65.

Leuenberger, H. G. W. (1985). Microbiologically catalyzed reaction steps in the field of vitamin and carotenoid synthesis. *In* "Biocatalysis in Organic Synthesis" (J. Tramper, M. C. Van der Plas, and P. Linko, Eds.), pp. 99–118. Elsevier, Amsterdam.

Miura, Y., Kondo, K., Saito, T., Shimada, H., Fraser, P. D., and Misawa, N. (1998). Production of the carotenoids lycopene, β-carotene and astaxanthin in the food yeast *Candida utilis*. *Appl. Environ. Microbiol.* **64**, 1226–1229.

Ratledge, C. (1989). Biotechnology of oils and fats. *In* "Microbial Lipids" (C. Ratledge and S. G. Wilkinson, Eds.), Vol. 2, pp. 567–668. Academic Press, London.

Saito, Y., Ishii, Y., Hayashi, M., Imao, Y., Akashi, T., Yoshikawa, K., Noguchi, Y., Soeda, S., Yoshida, M., Niwa, M., Hosoda, J., and Shimomura, K. (1997). Cloning of genes coding for L-sorbose and L-sorbosone dehydrogenases from *Gluconobacter oxydans* and microbial production of 2-keto-L-gulonate, a precursor of L-ascorbic acid, in a recombinant *G. oxydans* strain. *Appl. Environ. Microbiol.* **63**, 454–460.

Sakurai, N., Imai, Y., Masuda, M., Komatsubara, S., and Tosa, T. (1993). Molecular breeding of a biotin-hyperproducing *Serratia marcescens* strain. *Appl. Environ. Microbiol.* **59**, 3225–3232.

Sasaki, J., Miyazaki, A., Saito, M., Adachi, T., Mizoue, K., Manada, K., and Omura, S. (1992). Transformation of vitamin D_3 using *Amycolata* strains. *Appl. Environ. Microbiol.* **38**, 152–157.

Shimizu, S., and Yamada, H. (1986). Coenzymes. *In* "Biotechnology" (H. J. Rehm and G. Reed, Eds.), Vol. 4, pp. 159–184. VCH-Verlag, Weinheim

Soetaert, W., Holemans, D., and Vandamme, E. J. (1989). Production of L-sorbose by *Gluconobacter oxydans* cells, immobilized on reticulated polyurethane foam. *Med. Fac. Landbouww. Rÿksuniv. Gent* **54** (46), 1511–1513.

Vandamme, E. J. (Ed.) (1989). "Biotechnology of Vitamins, Pigments and Growth Factors." Elsevier, London.

Vaughan, P. A., Knowles, C. J., and Cheetham, P. S. J. (1989). Conversion of 3-cyanopyridine to nicotinic acid by *Nocardia rhodochrous* LL100–21. *Enzyme Microbial Technol.* **1**, 815–823.

Volk, R., and Bacher, A. (1990). Studies on the 4-carbon precursor in the biosynthesis of riboflavin. *J. Biol. Chem.* **265**, 19479–19485.

Yamada, H., and Shimizu, S. (1987). Microbial and enzymatic processes for the production of biologically and chemically useful compounds. *Angew. Chem. Int. Eng. Ed.* **27**, 622–642.

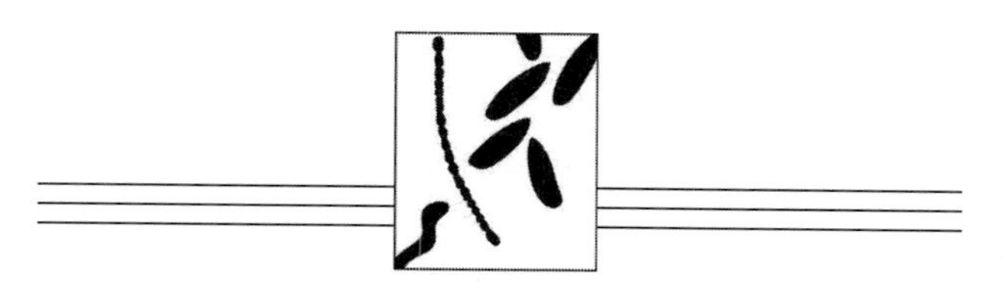

Wastewater Treatment, Industrial

Angela R. Bielefeldt
University of Colorado

H. David Stensel
University of Washington

I. Treatment System Design Parameters
II. Treatability Evaluations
III. Biological Treatment Processes
IV. Fluidized Bed Reactors
V. Anaerobic Bioreactors
VI. Use of Membranes in Bioreactors

GLOSSARY

activated sludge process An aerobic biological treatment process consisting of a mixed and aerated basin followed by liquid–solids separation and return of biomass to the aeration basin.

anaerobic treatment process A biological treatment process in which no oxygen is available. Electron acceptors may be organic substrates, sulfate, nitrate, or carbon dioxide. Carbon dioxide is more common for treatment of high-strength industrial wastewaters, with conversion of a large portion of the organic carbon to methane.

fluidized bed reactor A biological reactor containing bacteria attached to media continuously mixed and suspended by a high upflow liquid velocity.

lagoons A biological treatment process without separate liquid–solids separation and return of biomass.

membrane bioreactor A biological treatment reactor containing a selective material that acts as a barrier to prevent the transport of solids and allows transport of organic compounds and/or water.

respirometer A closed vessel in which the oxygen consumption due to biological activity can be measured.

solids retention time Average time biomass is maintained in a biological treatment process reactor.

BIOLOGICAL TREATMENT OF INDUSTRIAL WASTEWATER has been applied for most all types of industrial wastewater and for a wide range of compounds. Industrial wastewaters are generated from a wide variety of sources and have a broad diversity of chemical properties and constituents.

In the United States, wastewater standards have been developed for more than 50 industries and pollution control permits are in force for more than 57,000 facilities. The use of pretreatment standards that regulate the quality of wastewater discharged into public sewers has reduced the release of toxic compounds by an estimated 75% or 1 billion pounds during the past 25 years. To achieve the wastewater characteristics required by the standards, more industries are treating their wastewater on-site by a variety of processes. These processes are used to remove a mixture of organic compounds broadly measured by chemical oxygen demand (COD), biochemical oxygen demand (BOD), and total volatile suspended solids (VSS), or they are aimed at reducing the concentrations of specific chemicals to acceptable levels. Where it is possible to use biological treatment, it is generally the most cost-effective alternative. Biological treatment designs are continually evolving to provide lower effluent discharge concentrations and to reduce costs. Microbial methods to treat industrial wastewaters must be more robust than traditional municipal wastewater treatment processes designed to remove easily degradable organic compounds or nutrients such as nitrogen and phosphorous, due to the possible presence of toxic compounds, variations in pH, and shock loadings. The unique characteristics of industrial wastewaters frequently require more "engineered" designs that allow greater control over the conditions in the bioreactors. More innovative approaches are continually evolving to better meet the needs of biological industrial wastewater treatment.

I. TREATMENT SYSTEM DESIGN PARAMETERS

Industrial wastewater treatment generally requires reactor designs to successfully apply the degradative ability of the microorganisms. Basic design issues for a given reactor configuration are the biomass yield, oxygen or other electron donor supply, nutrient addition, pH control, and biological reaction rates or kinetics. The biomass yield is expressed as the quantity of biomass produced per unit of electron donor removed in the system. The greater the energy produced by the biological oxidation–reduction reaction, the greater the biomass yield. The yield value affects the amount of biomass that can be maintained in the biological reactor and the amount of solids that must be periodically wasted or removed from the system. For aerobic systems net yields are in the range of 0.20–0.40 g biomass as VSS/g COD removed compared to 0.03–0.05 g/g for anaerobic systems. The oxygen required in aerobic systems may range from 60 to 90% of the COD removed from the wastewater.

The system reactor design requires knowledge of the biotransformation rates or kinetics. The rate is generally dependent on a limiting substrate, which in many cases is the wastewater organic compounds or other waste constituents. These rates are also affected by the concentration of the electron donor. For aerobic systems a dissolved oxygen concentration of 2.0 mg/liter or more ensures that oxygen is not limiting the reaction rate.

Most biotransformation reactions can be described by Michaelis–Menten or Monod kinetics. The following Michaelis–Menten equation was first developed to describe the rate of enzyme transformation; however, since enzymes are the catalysts for degradation in bacteria, the equation is also applicable to many microbial degradation processes:

$$dS/dt = kSX/(K_s + S) \quad (1)$$

where S is the substrate concentration (mg/liter), dS/dt is the substrate removal rate (mg/liter/day), k is the maximum specific substrate degradation rate (g S/g VSS/day), X is the enzyme concentration or biomass concentration (mg/liter), and K_s is the half-saturation concentration (mg S/liter). This equation describes substrate utilization rates as a function of the reactor substrate concentration. It has also been used to describe cometabolic degradation rates.

Monod kinetics describe the bacteria specific growth rates as a function of the reactor substrate concentration:

$$\mu = \mu_{max} S/(K_s + S) \quad (2)$$

where μ is the specific growth rate (g VSS produced/g VSS present-day) and μ_{max} is the maximum specific growth rate (g/g-day).

Monod and Michaelis–Menten kinetics are related by the cell yield (Y) of the bacteria on the substrate, where

$$k = \mu_{max}/Y \quad (3)$$

For both Monod and Michaelis–Menten, at low substrate concentrations ($S < K_s$) the substrate degradation rate is approximately first order with respect to the substrate concentration. At very high substrate concentrations ($S \gg K_s$) the substrate-degradation rate is zero order with respect to substrate concentration (the rate does not change with substrate concentration). The substrate-degradation rate is always first order dependent on biomass concentration; the more bacteria present, the higher the transformation rate of the bacterial substrate.

For many compounds of interest in industrial wastewaters, high concentrations are toxic to bacteria (Table I). The most common expression used to account for substrate toxicity is the Haldane or Andrews kinetics:

$$dS/dt = kSX/(K_s + S + S^2/I) \quad (4)$$

where I is an inhibition constant. In this case, when $S > K_s$ and $S > I$, there is a decrease in the substrate degradation rate with increasing substrate concentrations. Examples of substrate inhibition by degradable substrates are BTEX compounds (toxicity begins to be evident above 50 mg/liter) and phenol (toxicity evident above 5 mg/liter phenol).

TABLE I
Most Significant Industrial Waste Compounds Based on Quantity and/or Toxicity

Chemical	TRI rank[a]	lbs to SW + POTWs (K)[b]	ATSDR rank[c]	Class[d]
Methanol	1	88,986	NR	VOC
Zinc compounds	2	1,476	65	Inorganic
Ammonia	3	22,905	153	Inorganic
Nitrate compounds	4	177,660	229	Inorganic
Toluene	5	669	61	VOC, AHC
Xylenes	6	528	83	VOC, AHC
Benzene		242	5	VOC, AHC
n-Hexane	8	311	NR	VOC
Methyl ethyl ketone	13	673	NR	VOC
Vinyl chloride	NR	1.1	4	VOC, CHC
Dichloromethane (methlene chloride)	15	650.4	70	VOC, CHC
Styrene	16	277	258	VOC, AHC
Trichloroethene	NR	87	15	VOC, CHC
Chloroform	NR	670	11	VOC, CHC
Lead compounds	20	110	2	Inorganic
Polychlorinated biphenyls	NR	0	6	SVOC, CHC
Polyaromatic hydrocarbons	NR	8.6	8, 9, 10	SVOC, AHC

Note. NR, not in top 20 TRI or top 250 ATSDR.
[a] 1996 TRI; rank by total lbs released via all media (air, water, and solids).
[b] 1996 TRI; thousands of lbs to surface water (SW) and publicly owned wastewater treatment works (POTWs).
[c] 1997 ATSDR.
[d] VOC, volatile organic compound; AHC, aromatic hydrocarbon; CHC, chlorinated hydrocarbon; SVOC, semivolatile organic compound.

The presence of other dissolved organic and inorganic compounds can influence the biotransformation rates. Other organics may be competitive or non-competitive inhibitors of biodegradation. In competitive inhibition more than one compound competes for the same bacteria enzyme. For competitive inhibition, the K_s term in the Michaelis–Menten equation is replaced by K_s (1 + conc. inhibitor/K_i), where K_i is generally the K_s value of the inhibitor when degraded alone. For noncompetitive effects, the term k in the Michaelis–Menten equation is replaced by k (1 + conc. inhibitor/inhibition constant). Inorganics generally exhibit noncompetitive effects on organic compound degradation. Bacterial sensitivity to inorganics varies, with anaerobic bacteria generally more susceptible to toxic effects from inorganics.

Biodegradation rates tend to increase with temperature in the bacterial survival range, and optimal temperature ranges exist in which the specific maximum growth rate is highest. Below the optimal range the reaction rate K can be approximated by the following equation:

$$KT = KT_o(\theta)(T - T_o) \quad (5)$$

where KT is the maximum specific degradation rate at temperature T (g/g-day), KT_o is the maximum specific degradation rate at optimal temperature T_o (g/g-day), and θ is the temperature correction coefficient (1.04–1.075).

An important design parameter for biological systems is the solids retention time (SRT), which is the average time that the biomass is in the system. Biokinetic design models can relate treatment performance to the reactor SRT. The SRT can be controlled during treatment by the amount of solids wasted from the system each day. Longer SRT systems have more biomass since the solids produced are in the reactor longer and thus produce lower effluent substrate concentrations. More difficult to degrade compounds require longer SRTs. Aerobic industrial wastewater treatment processes have SRTs ranging from 10 to 40 days, and SRTs more than 20 days are generally used for anaerobic processes.

II. TREATABILITY EVALUATIONS

Goals for treatability tests can be divided into two levels. The first is to simply test the wastewater for biodegradability and the second is to develop more specific process information that can be used to design a pilot plant or full-scale facility. A critical aspect at the first level of determining if a wastewater or particular chemical compound is biodegradable is the bacterial seed source. Unless the wastewater contains easily degradable compounds, special care must be taken to find and develop a culture. Comparison of biochemical oxygen demand (BOD) test results to COD data provides a first estimation of whether a wastewater is highly biodegradable. The closer the ultimate BOD concentration is to the COD concentration, the more biodegradable is the wastewater or chemical compound. For new xenobiotic compounds or more recalcitrant compounds, seed from many sources may be tried and may require long-term acclimation on the order of weeks to months. Seed sources include existing biological tratement facilities treating related industrial wastewater, sludge from municipal wastewater treatment facilities, sediment or soil with long-term exposure to the wastewater, and bacterial cultures known to degrade compounds of similar structure. Sample bottles or flasks are prepared with the wastewater, seed source, buffer, and nutrient media. The acclimation and degradation studies can be done in shake flasks with direct measurement of specific compounds or COD with time or in respirometer instruments. Various types of respirometers are available that provide continuous monitoring and data storage of oxygen consumption in sample bottles. Oxygen consumption data over time can indicate acclimation and can also be used to determine degradation rates and degradability. Units with automatic data acquisition tend to be more expensive but provide a large amount of data with minimal labor.

At the second level, tests are done to obtain bioreactor design parameters that include biokinetic degradation model coefficients, biomass yield and decay coefficients, and oxygen consumption ratios. A common approach is to operate an activated sludge reactor long term to observe treatment performance, to develop a culture for kinetic testing, and to perform mass balances to determine the biomass yield relative to the amount of substrate consumed and oxygen requirements. Similarly, anaerobic reactors can also be operated to develop performance data and to relate substrate consumption to biomass and methane production.

The Environmental Protection Agency has established a standardized biodegradation test method (method 304) for the operation of aerobic treatability reactors. A laboratory reactor that simulates the activated sludge aeration tank and clarifier is operated. The reactor may be closed or open to the atmosphere; if open, calculations are used to correct for compound volatilization. Adequate supply of nutrients, buffer, and oxygen must be provided to the system. Dissolved oxygen and pH are monitored during the system operation. The system is operated and controlled at a selected SRT. For slowly degraded compounds the SRT may be very long, on the order of 10–30 days versus 5–10 days for easily degraded compounds. Influent and effluent samples are collected and analyzed for organic compounds and biomass concentrations. An alternative approach is to use a sequence batch reactor (SBR) in which a single vessel serves as the aeration and clarifier steps. A programmable timer and pumps control the operation, and the sequence consists of fill, react, settling, and effluent withdrawal steps. The system is also operated at a selected SRT. The previously mentioned reactors can be operated at many SRTs to develop a relationship between SRT and degradation efficiency and to also develop biokinetic model coefficients. However, the multiple-SRT operation is time-consuming and labor intensive. A more common approach is to use the biomass developed at a selected SRT to obtain biokinetic coefficients by batch testing methods. Anaerobic degradation treatability tests are also easier to carry out using the SBR method.

Batch tests can be done in bottles or flasks in which samples are withdrawn with time after batch feeding to measure COD or specific compound concentrations. Through model-fitting techniques the biokinetic coefficient values can be determined. Consideration must be given to the batch testing conditions or the kinetic coefficient values obtained may not represent what occurs in actual operation. Wastewater treatment reactors are normally operated with low substrate concentrations—well below that commonly added in batch kinetic tests. In some cases, exposing the bacteria to higher substrate concentrations may induce a greater level of enzyme production and thus higher degradation rates. Thus, a lower initial batch feed concentration should be used. At such lower concentrations the substrate consumption rate may best be measured by oxygen consumption, assuming that the oxygen consumption rate is directly proportional to the substrate utilization rate. The oxygen consumption rate may be measured by relatively sensitive respirometers or simply by a dissolved oxygen probe in a fully submerged test bottle without any head space. A distinct advantage of this method is the ability to measure very low K_s values in the Michaelis–Menten model. For wastewaters with mixed compounds a greater amount of oxygen would be consumed, and respirometers have been used to study the kinetics. In this case, the oxygen consumption is related to a COD or BOD removal instead of an individual compound. All these methods can evaluate the effect of operating conditions such as toxicity, pH, nutrient additions, and temperature.

For any of the previously mentioned test methods it is important to carefully analyze the liquid effluent. Many bioprocesses only partially transform organic compounds. In some cases, the degradation products can be more toxic than the parent compound. For example, anaerobic bacteria reductively dechlorinate tetrachloroethene to trichloroethene, then to dichloroethene, and then to vinyl chloride. Degradation of vinyl chloride (VC) is usually slower than that of any of the previous compounds, and it can accumulate. Since VC is more toxic than the preceding chlorinated alkenes, a net increase in overall toxicity can occur. Care must be taken to ensure that the end products of biodegradation are indeed harmless.

III. BIOLOGICAL TREATMENT PROCESSES

There are many types of biological processes that can be selected for treatment of industrial wastewaters. The process selection depends on many factors, including the biodegradability and complexity of the wastewater, the wastewater strength, the variability of the wastewater flow and organic loading, the potential for inhibitory compounds, and the degree of treatment needed. In some applications the industrial wastewater may be discharged to municipal or publicly owned treatment works (POTWs) so that only pretreatment standards must be met. BOD removals for pretreatment can vary from 50 to 80%. Alternatively, >90% removal must be achieved to meet secondary treatment permit levels or the discharge must have very low concentrations of particular regulated compounds. A variety of suspended growth biological processes are used for moderate-strength (100–1000 mg/liter BOD) and high-strength industrial wastewaters (>1000 mg/liter BOD). For weaker wastewaters or for treatment of streams with only a few milligrams per liter of a specific priority pollutant, fixed film processes may be used to better maintain a sufficient biomass in the treatment reactor. Aerobic biological processes are commonly used for industrial wastewater treatment, but for higher strength wastewaters anaerobic treatment processes may be more advantageous. The common types of biological processes used can be classified as lagoons, activated sludge, and fixed film processes. In addition, fluidized bed reactors and membrane reactors are more current developments that are described later.

A. Lagoon Treatment

Lagoons are relatively long-detention time basins that do not employ solids recycle as in activated sludge processes. They can be anaerobic, facultative, or aerobic. Anaerobic lagoons are loaded high enough so that anaerobic conditions exist. The depth of anaerobic ponds is selected to provide a minimum surface area-to-volume ratio to minimize heat loss. They have been used for pretreatment (50–70% BOD removal) for meat and poultry processing, canning, sugar processing, wine and brewery, potato processing, and rendering wastewaters. BOD loadings have been in the range of 300 to 1000 lb/acre-day with detention times from 10 to 50 days. The ponds are generally not covered, but in many cases a scum layer accumulates to minimize oxygen transfer. Alternately, membrane liners have been used to cover anaerobic ponds and provide a means for methane gas collection. Potential odor problems exist for uncovered anaerobic ponds. Sequential lagoon system designs, consisting of an anaerobic lagoon followed by an aerobic lagoon, have been used to meet secondary effluent requirements.

Facultative ponds consist of an aerobic surface and anaerobic bottom. The surface is kept aerobic by oxygen transfer due to diffusion and wind action and by photosynthetic activity of algae. Solids at the bottom of the pond undergo anaerobic decomposition producing methane, volatile fatty acids, and other gases. The upper aerobic layer provides treatment of odorous substances and dissolved organic compounds. Depths of facultative ponds typically vary from 4 to 10 ft, and the loadings are lower than those for anaerobic ponds—in the range of 50–200 lb BOD/acre-day. BOD removals may range from 50 to 90%. In some cases, mechanical surface aerators are used to maintain the upper aerobic layer of a facultative pond. Eckenfelder (1989) reports a power input for mechanical aerators of 4 horsepower/million gallons (0.79 W/m^3) for pulp and paper industry applications.

In aerated lagoons mechanical or diffused aeration is used to provide oxygen and to maintain solids mixing and suspension. Aerated lagoons are considered flowthrough activated sludge processes without solids setting and recycle. The SRT equals the hydraulic retention time (HRT). The pond depths range from 8 to 16 ft and HRTs from 2 to 5 days. Aerated lagoons can provide 50–80% BOD removal, but it is possible to meet secondary treatment standards when they are followed by a facultative pond and settling pond.

Low loaded aerobic ponds or stabilization ponds with detention times of a few days to 2 or 3 weeks have been used for biological treatment or BOD removal after anaerobic or facultative lagoons. These depend on photosynthesis for oxygen supply, use lower loadings of 20–50 lb BOD/acre-day, and have shallower depths of 4–6 ft.

B. Activated Sludge Processes

The activated sludge process consists of an aeration basin followed by liquid–solids separation, generally in a clarifier by gravity separation and return of settled biomass to the activated sludge basin. Excess sludge is produced and removed for further processing and disposal or reuse. Various aeration-basin configurations have been used, including long, narrow tanks termed "plug flow" systems, basins in series, and single completely mixed basins. Completely mixed basins are most common due to their inherent ability to dilute and buffer shock loads. Aeration equipment provides both oxygen and sufficient mixing to maintain a uniform mixed liquor suspended solids (MLSS) concentration. MLSS concentrations generally vary from 1500 to 4000 mg/liter depending on the settling characteristics of the biomass, and HRTs vary from 5–24 hr to days depending on the wastewater strength and design SRT. For a given SRT and MLSS concentration, a wastewater with a 2000 mg/liter BOD would require about a 10 times longer HRT than that for a 200 mg/liter BOD wastewater. The design and operating SRT will vary from 5 to 50 days depending on the wastewater characteristics, temperature, and treatment level needed. Longer SRT systems provide a more stable operation, greater removal of more recalcitrant chemical compounds, and better handling of variable loads.

The longer SRT systems (20- to 50-day SRTs) result in longer aeration times and are generally termed extended aeration processes. These processes produce less sludge, handle variable loads well, and can produce high-quality effluents. A cost-effective aeration basin design for extended aeration processes is an oxidation ditch. In its classic sense, an oxidation ditch consists of a "race track" or loop channel reactor, in which aeration or mixing equipment provide a unidirectional flow down the channel. A complete recirculation occurs in 5–15 min. Brush rotor, jet aerators, and mechanical surface aerators have been used to provide channel flow. With the ditch basin configuration and method of mixing, long HRTs can be used with minimal energy needed for mixing. In most ditch designs the aeration horsepower is based on oxygen transfer and is not due to mixing energy needs.

The performance of the activated sludge processes depends on the ability of the clarifier to provide efficient clarification and thickening of the mixed liquor. Wastewaters with a significant amount of readily degradable soluble BOD encourage the growth of filamentous bacteria, which results in poor sludge settling characteristics in extended aeration and completely mixed activated sludge systems. The sludge volume index (SVI) may increase to 200–400 ml/g when bulking by filamentous bacteria occurs. The SVI is the volume per gram of sludge after 30 min of quiescent settling. SVI values in the range of 120–150 ml/g are preferred for activated sludge, and high values can lead to unacceptably high effluent suspended solids concentrations from the clarifier. Alternatively, at high SVIs the activated sludge system could be operated with a lower MLSS concentration, which reduces treatment capacity.

A modification to the activated sludge process, termed a selector design (Fig. 1), can prevent filamentous growth problems and lead to a stable activated sludge process for industrial wastewater treatment. The selector design consists of adding a contact basin before the activated sludge aeration basin. Three selector designs have been used: anaerobic, high food to mass (F : M) ratio, and anoxic. In the anaerobic selector return activated sludge is mixed with the influent wastewater in a contact tank with an HRT of approximately 30–60 min. The process requires excess phosphorus in the wastewater influent since it selects for bacteria with high phosphorus storage ability. These bacteria help to provide a dense activated sludge floc with SVI values in the range of 70–100 ml/g. A disadvantage of this type of selector for industrial wastewater treatment is that it may require the addition of much more phosphorus to support growth. The high F : M selector typically involves adding an aerated three-stage reactor of equal volumes for contacting influent and return sludge before the main activated sludge unit. Usually, coarse bubble aeration is used and dissolved oxygen concentration in the selector reactor is close to zero. The F : M ratio in the first stage is in the range of 4–8 lb BOD/lb MLSS-day. The soluble BOD is sorbed quickly in the selector stages and stored in intracellular organic products. SVI values from high F : M selectors have been in the range of 70–100 ml/g. The third type, the anoxic selector, has been used when the aerobic basin produces significant levels of nitrate due to biological nitrification of ammonia or the wastewater influent has significant amounts of nitrite or nitrate. The most common application is that in which the nitrate produced by nitrification in the aeration basin is fed to the selector anoxic tank by recycling aeration basin mixed liquor to the selector tank. The selector zone is mixed and not aerated, and bacteria use nitrate or nitrite to oxidize the influent-soluble BOD. The anoxic selector may have an HRT of 2–4 hr, depending on the amount of nitrate available and influent-soluble BOD concentration. Sufficient nitrate must be available to consume most of the influent-soluble BOD. SVI values in the range of 100–150 ml/g have been achieved with anoxic selectors.

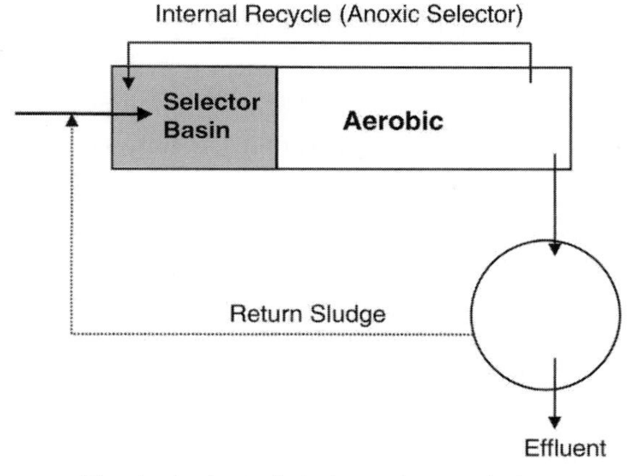

Fig. 1. Activated sludge selector design.

A sequence batch reactor (SBR) is very advantageous for highly variable or intermittent flows. In a SBR a single reactor is used for biodegradation reactions followed by quiescent settling of the biomass and then decanting of the treated wastewater. A typical SBR operating cycle is shown in Fig. 2. The cycle consist of a sequence of fill (feeding), react, settle, decant, and possibly idle times. The process provides excellent flow and load equalization and has been used where industrial discharges occur infrequently during the day. A large discharge can be taken at

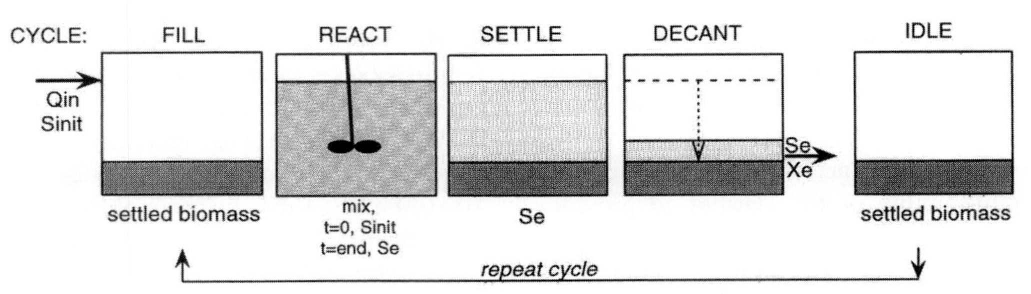

Fig. 2. Sequence batch reactor operating steps.

once during the tank fill period. Following aeration, settling, and decanting the reactor is ready for another wastewater charge, which can vary in intensity over time. The process can be operated as a selector to control SVI by having mixing only during the fill period. If nitrate is available during the mixed fill period, it may function as an anoxic selector or, if no nitrate is available, as an anaerobic selector. Other advantages of the SBR are a good-quality effluent due to its long SRT operation and quiescent settling period, flexibility due to the ability to vary the times of the sequential steps, and low cost for small facilities.

The performance of activated sludge processes can be enhanced by the addition of powdered activated carbon (PAC). The PAC can adsorb inhibiting chemicals or adsorb chemicals to buffer variable loads. PAC has proven very useful in industrial wastewater plants in which nitrification is required. Autotrophic bacteria that oxidize ammonia to nitrite and nitrate are much more sensitive than heterotrophic bacteria, and there are many organic compounds that can inhibit their activity. Addition of PAC at doses in the range of 10–50 mg/liter has helped to remove organic inhibitors and to encourage nitrification. The PAC addition also increases sludge settleability, clarifier performance, and mixed liquor concentrations.

C. Fixed Film Reactors

Fixed film reactors provide an alternative to suspended growth processes and are especially useful for treating dilute wastewater streams. In these reactors wastewater flows over a solid material or "media" covered with a biological film. Substrate, nutrients, and oxygen diffuse from the bulk liquid into the biofilm. The biofilm thickness may range from 100 to 10,000 mm and the biofilm density may range from 20 to 50 g/liter, depending on the system design and operation. For some designs, the resultant reactor biomass concentration, based on empty bed volume, has been in the range of 10,000–20,000 mg/liter, which is much higher than that found in activated sludge aeration basins. Fixed film reactors may be submerged or nonsubmerged, such as plastic or sand downflow packed bed reactors or downflow reactors containing plastic media as in the case of trickling filters. Downflow packed bed reactors are useful for treating more dilute wastewaters, in which it is important to capture and maintain the limited biomass produced. These reactors have used solid media in the size range of 1–4 mm with hydraulic application rates in the range of 1–2.5 gallons per minute per square foot of tank cross-sectional area (gpm/ft^2, 0.04–0.10 m^3/m^2-min). Periodic backwashing is normally required to remove excess solids.

Trickling filters contain a plastic packing material or 6- to 8- in. rocks, and the wastewater is distributed over the top of the media. An underdrain collects the wastewater before transport to a gravity settler for suspended solids removal. Air is supplied by natural draft or by a more preferred method of forced ventilation to ensure aerobic conditions. The plastic media has a specific surface area of approximately 27 ft^2/ft^3 (88.5 m^2/m^3) of bulk volume and the trickling filter towers may be 20–40 ft in height. Hydraulic application rates in the range of 1.5–4.0 gpm/ft^2 (0.06–0.16 m^3/m^2-min) have been used. Excess biomass and solids slough periodically and a clarifier provides liquid–solids separation of the tower effluent. Trickling filters have been used to meet both pretreatment and final discharge standards for industrial wastewaters. Organic loadings range from 100 to 400 lb BOD/ft^3 of reactor volume per day (1.6–6.4 kg/m^2-day) for pretreatment applications and from 10 to 40 lb BOD/ft^3-day (0.16–0.64 kg/m^2-day) for secondary treatment levels. The advantages of trickling filters are a low energy requirement and simple operation. Disadvantages include a lower effluent quality than that of activated sludge, longer start-up time, and higher capital cost.

Fluidized bed reactors use a media that is suspended and mixed and offers mass transfer advantages over other fixed film processes. Because of many advantages of fluidized bed reactors for aerobic treatment of dilute industrial wastewaters, this system will be described in more detail in the following section.

IV. FLUIDIZED BED REACTORS

A sketch of a fluidized bed reactor (FBR) is shown in Fig. 3. Due to the upflow of liquid into the reactor, the media is fluidized or suspended to create essentially completely mixed conditions. Above the level of the fluidized media there is a "free water" volume to prevent washout of the media with attached bio-

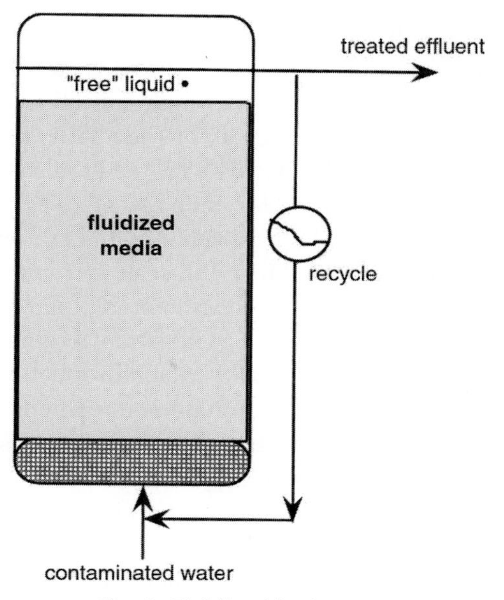

Fig. 3. Fluidized bed reactor.

mass from the reactor. Typical types of media include sand, plastic rings, activated carbon, and celite. The type and density of the chosen media will determine the water flow required for its fluidization; therefore, "lighter" materials have somewhat less energy costs associated with their use. The largest full-scale aerobic FBR treats a flow of 30 million gallons per day (30 Mgal/day, 1314 liters/s) for nitrification, with sand as the supporting media. Due to the high liquid flows needed to fluidize most media, FBRs usually have liquid recycle. For sand media, upflow liquid velocities to achieve fluidization are about 15 gpm/ft^2 (0.60 m^3/m^2-min). Typical HRTs reported for FBR operation are in the range of 10–20 min. In addition, the turbulent conditions in the reactor maintain a thinner biofilm on the particles of the suspended media than would be present on trickling filter media. This generally maintains bioactivity throughout the biofilm depth and avoids complete depletion of substrate and electron acceptor concentrations prior to penetration through the biofilm depth.

Biofilm accumulation on the reactor media is critical to the proper operation and treatment efficiency of the FBR. Biofilm accumulation on the media affects its size and apparent density, thereby changing the hydrodynamics of the bed. If the carrier material is not of uniform size, stratification of the bed can occur. In addition, biokinetic rates that are applicable to suspended growth cultures will be lower in the FBR, largely due to diffusional resistance in the biofilm.

The major advantages of the FBR are related to its efficient mass transfer between the biofilm and bulk liquid and its ability to maintain a long SRT and high volumetric biomass concentration. Even for treating dilute wastewaters biomass is able to accumulate to high concentrations on the media. Due to the high biomass concentration, slowly degradable compounds can be removed to very low bulk liquid concentrations. With the use of activated carbon media the FBR can also treat a variety of toxic and more recalcitrant compounds to low concentrations. For these reasons, FBRs have been well suited for the treatment of hazardous or industrial wastewaters with priority pollutants.

Disadvantages of FBRs include high energy and a more sophisticated design compared to activated sludge treatment. The high energy consumption is related mainly to the pumping needs for the feed and recycle flows. A more complex design is needed to provide aeration and influent flow distribution. The compact installation and no requirement for a conventional secondary clarification means that capital construction costs are approximately 50% those of activated sludge plants. However, annual operation and maintenance costs are higher.

FBR designs can be grouped into three main categories: three-phase (liquid, gas, and solid) aerobic FBRs with oxygen added as air, two-phase FBRs which would be anaerobic or aerobic with oxygen predissolved in recycled water, and FBRs with granular-activated carbon (GAC) as media. Specific advantages and disadvantages unique to each of these designs will be discussed and specific examples of their use for industrial wastewater treatment are provided.

Aerobic FBRs, which are simultaneously fed both air (for oxygen) and wastewater, have very turbulent conditions in the reactor that improves mass transfer rates of substances from the liquid to the biofilm and create high shear forces that minimize biofilm thickness. Disadvantages of this type of operation are the difficulty achieving homogenous flow distribution and control of bed expansion and higher effluent suspended solids concentration. Energy use for these systems is reported to be similar to that of activated sludge and less than that of FBR systems that predissolve oxygen in recycle flows. These types of reactors have been primarily used in lab and pilot studies, and there are little or no full-scale data. The University of Montpellier in France has tested the system with a clay-based media operated at loadings of 1500 lb COD/ft^3 of reactor volume-day (24 kg COD/m^3-day), resulting in a reactor biomass concentration of 2 or 3 g total suspended solids (TSS)/liter (based on empty-bed volume).

The most common type of aerobic FBR configuration involves predissolving oxygen in FBR effluent that is recycled and fed with the influent wastewater to the reactor. This type of process with a sand media was developed and patented by Ecolotrol in the United States. There are approximately 65 full-scale industrial plants in the United States and Europe that use this type of reactor for wastewater treatment. They have been shown to be efficient for achieving nitrification and treating wastewaters with low organic pollutant concentrations. Typical loadings to these systems are in the range of 37–50 lb BOD/ft^3-day (0.6–0.72 kg BOD/m^3-day) or 37–80 lb NH_4–N/ft^3-day (0.6–1.3 kg NH_4–N/m^3-day), with reactor biomass concentrations of 5.5–22 g TSS/liter (based on empty-bed volume).

FBRs have also been used for anaerobic treatment. Since oxygen is not required in these systems, operating costs are lower. An anaerobic, sand-packed FBR for denitrification was operated 30 months as a demonstration project; methanol was added as a supplemental carbon source, and nitrate loadings of 1.84–6.41 kg/m^3-day produced effluent nitrate concentrations of 2.5–4.1 mg NO_3–N/liter. A 70-liter pilot-scale anaerobic FBR with sand media was used to treat brewery waste at 35°C with inlet concentrations of up to 12,000 mg/liter COD. In comparisons with a lab-scale reactor to determine optimal scaling factors for design, it was found that a similar Reynold's number gave similar process efficiency and the Peclet number gave very similar methane production and substrate utilization. An anaerobic FBR treating wastewater from an industrial plant producing MSG removed more than 70% of the inlet COD of 5000–15,000 mg/liter at temperatures higher than 15°C. It has been found that methanogenic gas production in the anaerobic FBR has little effect on hydrodynamics in the bed.

Another advance in FBR technology is the use of GAC as a packing media. GAC provides a sorption capacity for toxic organics, which can allow faster start-up of the reactor and can provide better process stability for variable loads. GAC media also has a very high surface area for biomass growth and is lighter than sand. GAC has been used in both aerobic and anaerobic FBRs. Most full-scale aerobic GAC reactors are used for treatment of petroleum-contaminated groundwater, whereas anaerobic GAC reactors are used to treat pentachlorophenol and wood-treating chemicals contained in contaminated groundwater. The largest of these reactors are 7.3 m tall × 23.6 m^3, treating 6 Mgal/day of gasoline-contaminated groundwater. Pilot-scale systems have demonstrated treatment of chlorinated solvents such as dichloromethane (DCM), which was treated at a General Electric plant in Mt. Vernon, Indiana, at a

TABLE II
Biological Transformations of Important Industrial Wastewater Compounds

Compound	Aerobic	Aerobic cometabolic	Anaerobic	Comment
Ammonia	***			Autotrophs
Benzene	***		*	
Chloroform		*	**	
Dichloromethane	**	*	**	
Methanol			**	
Methyl ethyl ketone				
Nitrate			**	Electron acceptor, converted to nitrite and N_2 gas
Styrene	**		**	
Toluene	***		**	
Trichloroethene		**	**	Anaerobically dechlorinated to DCE
Vinyl chloride	**	*	*	
Xylenes	***	*	*	

flow of 30 gpm to a 20-in. diameter × 12-ft tall reactor with 50% bed fluidization. Inlet DCM concentrations of 1000–1200 mg/liter with spikes of 3000 and 4000 mg/liter were reduced to 50 μg/liter, after a start-up period of about 10 days. A fairly stable operation was maintained for 50 days.

FBRs have been used to treat wastewaters with volatile aromatic compounds (BTEX), alkanes, polyaromatic hydrocarbons, chlorinated phenol compounds, and chlorinated aliphatic compounds (Table II). One example is the use of a FBR for cometabolic treatment of TCE-contaminated groundwater. In a lab study, the 5-cm diameter × 100-cm tall reactor containing quartz filter sand media and with a biomass concentration of 21 g/liter treated 0.8 liter/min of 0.1 mg/liter TCE with 5.0 mg/liter phenol. During the 3-min empty bed contact time, phenol was removed to below detection limits and TCE was 70–80% removed.

V. ANAEROBIC BIOREACTORS

Anaerobic treatment is an economically attractive alternative for high-strength industrial wastewaters. It consists of three steps: (i) hydrolysis of particulate matter to soluble organic compounds, (ii) fermentation of the soluble organic compounds to produce volatile fatty acids (acetate, propionate, and butyrate) and hydrogen, and (iii) utilization of volatile fatty acids (VFA) and hydrogen by methanogenic bacteria to produce methane. The process primarily converts organic material to methane gas, a useful energy source. About 0.35 m^3 of methane is produced per kilogram of COD converted. The exit gas consists of approximately 65% methane and 35% carbon dioxide plus trace amounts of hydrogen sulfide and other compounds. Because of the high CO_2 content in the process it must be well buffered to maintain a pH near neutral. Reactor alkalinity in the range of 3000–5000 mg/liter as $CaCO_3$ is common. For industrial wastewaters, the addition of alkalinity is sometimes necessary. Due to the conversion of most of the organic material to methane, the process has a much lower biomass yield of about 0.05 g VSS/g COD used.

The anaerobic process performance depends on a balance between acid production by fermentation bacteria and VFA utilization by the methanogenic bacteria. Methanogenic bacteria are slower growing than fermentation bacteria and are very sensitive to low or high pH (6.8–7.4 is preferred), temperature changes, toxicity, high ammonia concentrations, and high dissolved solids. Anaerobic systems work best at higher mesophilic temperatures (approximately 30–35°C), but systems have been operated at 20°C and at thermophilic temperatures (50–60°C).

The operation of anaerobic systems requires a balance between fermentation and methanogenic bacteria activity. Significant increases in feeding rates can lead to a VFA production rate that is higher than the possible VFA consumption rate by methanogenic bacteria, which will increase the reactor VFA concentration and thus decrease the pH. The methanogenic bacteria activity will decrease as the pH is lowered so that the system becomes unstable and moves toward failure, unless a correction in feed rate or pH can be made.

Table III compares the advantages and disadvantages of anaerobic and aerobic activated sludge treatment. With anaerobic treatment, no aeration energy is required and a much lower sludge production results. The lower sludge production reduces solids disposal costs and requires less nutrient addition. For high-strength, high-temperature industrial wastewaters, anaerobic treatment can be very economical and can generate a net income if the methane produced can be used in the industrial facility. The process operation can be stopped for extended time periods, with anaerobic degradation activity returning upon substrate addition. If large amounts of alkalinity must be added to maintain the reactor pH, the economics of the process can be reduced considerably. The process usually has high effluent suspended solids concentrations and effluent BOD concentrations that cannot meet secondary treatment requirements. To meet secondary treatment discharge levels aerobic treatment after an anaerobic process is used.

TABLE III
Advantages and Disadvantages of Anaerobic Treatment

Advantages	*Disadvantages*
Low sludge production	May require significant amounts of alkalinity addition
Low nutrient requirements	Long start-up periods are needed
Energy savings since no aeration energy needed	Mainly limited to pretreatment applications
Produces a useful product—methane	Sensitive to variable loads and toxicity
Seasonal and intermittent operation is possible	Requires more operational skill
More than 90% conversion is possible	

Many process designs have been used for anaerobic treatment. Lagoons and fluidized bed reactor designs were described previously. Other designs include the anaerobic contact process, upflow packed bed reactors, and the upflow anaerobic sludge blanket (UASB) reactor. The anaerobic contact process is similar to an activated sludge process, in that a clarifier or other liquid–solids separation process follows the biological reactor with return of the thickened sludge. Anaerobic contact process reactors have been operated with industrial wastewater at applied organic loadings in the range of 125–187 lb COD/ft^3-day (2.0–3.0 kg COD/m^3-day) at 35°C. Upflow packed bed reactors using plastic media have been operated at loadings of 218–436 lb COD/ft^3-day (3.5–7.0 kg COD/m^3-day) at 35°C. The UASB reactor has been used to treat high-strength food processing wastewaters. A dense granular sludge forms when treating high-carbohydrate wastewaters. Solids concentrations in the range of 5–8% are maintained at upflow velocities of 1 m/hr due to the formation of dense granular bacteria floc particles. Applied loadings in the range of 624–1248 lb COD/ft^3-day (10.0–20.0 kg COD/m^3-day) at 35°C have been used with UASB reactors.

VI. USE OF MEMBRANES IN BIOREACTORS

The use of membranes in biological treatment processes for wastewater is a fairly new innovation. Uses of membranes can be classified as follows:

1. Using the membrane as a surface for attached biological growth, with permeation of oxygen into the biofilm through the membrane
2. Using the membrane as a selective barrier, allowing hydrophobic organic compounds from the industrial wastewater to permeate through the membrane into a bioreactor while excluding ions (such as toxic metals) from entry to the bioreactor
3. The use of the membrane in the biomass-separation step, in place of a clarifier, after a conventional

suspended-growth (activated sludge) biotreatment reactor

Each of these uses is discussed in detail in the following sections.

A. Membranes Permeable to Oxygen

Hollow-fiber gas-permeable membranes are being investigated for use in wastewater treatment. Reports of the use of hollow-fiber membranes in bioreactors first appeared in 1992, demonstrating oxygen transfer to reactor liquid without stripping out volatile organic compounds. The optimal materials used for oxygen-permeable membranes are microporous, hydrophobic polypropylene. These membranes have high oxygen permeability due to pores that stay dry and air-filled. Approximately 100% oxygen transfer efficiency can be achieved, making optimal use of oxygen provided. The fibers also have a large surface area for biofilm growth, allowing a high biomass density in a limited space. Typical hollow fibers have an internal diameter of 0.03–0.11 cm and a wall thickness of 25–400 mm. To avoid formation of bubbles on the biofilm side of the membrane, low gas pressure should be used. In these systems, the liquid film resistance of oxygen transfer is much greater than the resistance of the membrane or the gas film. The hollow polypropylene fibers can be operated in a flow through mode or as dead-end passages. By sealing one end of the fibers and introducing pure oxygen into the tubes, nearly 100% oxygen transfer efficiency can be achieved. By fixing the tube ends that are attached to the air supply and leaving the sealed end free, turbulence through the bundle of fibers can serve as a self-cleaning mechanism to keep biofilm accumulation on the surface of the fibers to a minimum.

Tests have also been conducted using nonporous silicone rubber as a membrane material. These membranes can be operated with a higher gas pressure without formation of bubbles; higher gas pressure allows a greater concentration gradient across the membrane and therefore a higher mass transfer potential. Here, the oxygen dissolves in the membrane and diffuses across it. A disadvantage of this membrane type is that the material is usually thicker than polypropylene membranes and has a significant resistance to mass transfer. In addition, this material is not available in a hollow fiber form with small diameter, so there is less surface area per volume for mass transfer. Finally, the material is more expensive than polypropylene.

The membranes can be used to transfer oxygen into a suspended bacterial culture or as a support media for fixed biofilm growth. Some biomass growth on the membrane is unavoidable. However, excess growth can lead to solids fouling or channeling of water through the membrane bundle. Backwashing with a more turbulent water flow may be an effective way to control excess biomass growth in these systems. Bubbleless transfer of pure oxygen was used in a pilot-scale MBR to treat a high-strength wastewater from a brewery (average COD 2250 mg/liter). This system was backwashed with a combination of air and water daily to maintain good operation.

B. Membranes as a Selective Barrier

The use of membranes as a selective barrier in biological treatment processes for industrial wastewater treatment has been termed "extractive membrane bioreactors" (EMB) by Livingston *et al.* (1996). The membrane allows transport of the organic compounds out of the industrial wastewater and through the membrane into a bioreactor operated with controlled, selected conditions (pH, nutrients, pure culture bacteria, etc.). The membrane does not allow transport of ions into the bioreactor. In this way, biodegradable organic compounds in the industrial wastewater can be selectively treated. An advantage of this process is the ability to treat industrial wastewaters with conditions that are inhibitory to direct biotreatment, such as extreme pH or high inorganic salt concentrations. It also allows the use of selected bacteria for treating more recalcitrant compounds.

The most common type of membrane material that has been used for this application is silicone rubber. It is permeable to hydrophobic organic compounds (including aromatics, chlorinated aromatics, and chlorinated aliphatics), but it is virtually nonpermeable to ionic species and water. It is also biocompatible, elastomeric, and fairly robust over a wide pH range. Industrial wastewaters with $pH < 1$ and pH > 12 have been treated with this membrane over periods of 25 days to 6 months without visible deterioration or measurable reduction in organic compound flux. This method has also been used to treat organic compounds in air in which both the organic compound and oxygen can diffuse through the membrane (silicone rubber and polysulfone have both been used). The most common configuration of the membrane is as small-diameter tubing, with the industrial wastewater running through the inside and the bioreactor on the outside. The inside diameter of the silicone tubing used has ranged from 0.5 to 2 mm, with a tubing wall thickness of 0.25–0.5 mm. Tests comparing membrane thickness found no effect of the thickness on the mass transfer of the organic compound into the bioreactor; however, liquid film resistance on the inside of the tubing was found to be critical to the overall mass transfer rate out of the industrial wastewater. Operating at a high Reynolds number gave a thinner liquid film and therefore better mass transfer. Biomass will tend to accumulate on the surface of the membrane that is in contact with the bioreactor. Under aggressive wastewater characteristics or high internal shear, biofilm growth inside the membrane is limited, which will help maintain optimal mass transfer across the membrane. Contact times for treating industrial wastewater with the membrane to achieve 90–99% removal of organics have been on the order of 20 min to 3 hr.

C. Membranes to Retain Biomass

Ultrafiltration or microfiltration membranes have been used in place of secondary clarifiers following activated sludge bioreactors at some biotreatment facilities. Activated sludge processes generally require a long SRT and thus long liquid contact time to achieve significant degradation of industrial chemicals that are difficult to degrade. For slow-growing bacterial populations, such as those needed to degrade toxic organics present in many industrial wastewaters, retention of the biomass is key to process stability and performance. The use of membranes allows a high biomass retention and long SRT with a comparatively short HRT.

Membrane separation is able to provide a high-quality effluent with little suspended solids present. Winnen *et al.* (1996) reported that a ceramic membrane, with a 30-kDa molecular weight cut-off, retained 100% of heterotrophic organisms and also retained the MS-2 virus (although the virus is smaller than the membrane pores). The ability of membranes to remove microbial particles results in an effluent microbial concentration that is similar to effluent microbial levels from conventional wastewater treatment processes with chlorine or ultraviolet disinfection. The membrane is not susceptible to "upsets" related to bulking or rising sludges, allowing better SRT control and a more reliable high-quality effluent. The membranes are more compact than a conventional clarification, and it is easy to implement automatic process control. An activated sludge process with membrane separation also tends to produce less sludge due to its long SRT, thereby reducing sludge processing costs. The disadvantages of the system are clear: high initial investment cost, higher operating costs associated with the energy needed to overcome pressure drops across the membrane, limited membrane lifespan, and a restricted effluent flux rate. Periodic cleaning is also required to remove biofouling and solids accumulations that further increases the pressure drop across the membrane. In the study by Winnen *et al.* (1996), their ceramic membrane was regenerated once per week using a warm alkaline solution.

Zaloum *et al.* (1994) reported on activated sludge treatment of oily waste from a metal transformation mill using an ultrafiltration membrane with molecular weight retention of 10 kDa. The wastewater contained toxic organics, such as chlorobenzenes, toluene, and phenols, in addition to having high oil and grease content. When compared to the existing ultrafiltration system used alone at the site, the coupled bioreactor plus membrane system achieved greater detoxification of the effluent water and less hazardous waste was produced (one-third the sludge compared to ultrafiltration alone). The additional maintenance required for the bioprocess was minimal compared to that for the ultrafiltration units.

An alternative configuration to placing a membrane process downstream of an activated sludge reactor is to place the membrane inside the activated sludge basin. This was tested at a plant treating pulping process wastewater. Hollow-fiber membranes with a 0.1-mm pore size and a total surface area of 1.7 m^2 were placed inside a 90-liter activated sludge basin. The two aerators inside the basin were used to provide oxygen and enough turbulence to minimize biogrowth on the surface of the membranes. At a 13.5-kPa vacuum on the inside of the membrane, a wastewater flux of 30 liters/m^2-hr was achieved. The membrane-containing reactor achieved better removal of COD and suspended solids, and it produced a lower effluent toxicity compared to a conventional activated sludge and clarifier system treating the same inlet water. The membrane reactor also retained lignins in the aeration tank to produce lower effluent lignin concentrations. The MLSS concentration in the membrane reactor averaged 24,200 mg/liter (twice the conventional reactor) with a 24-hr HRT and 15-day SRT. Some biofouling did occur, and so the membranes were backwashed daily with a chlorine solution inside the membranes in a closed circuit (operated at a low pressure so that the chlorine did not enter the bioreactor).

See Also the Following Articles

BIODEGRADATION • BIOFILMS AND BIOFOULING • BIOREACTORS • BIOTRANSFORMATIONS

Bibliography

Bielefeldt, A. R., and Stensel, H. D. (1999). Evaluation of biodegradation kinetic testing methods and long-term variability in biokinetics for BTEX metabolism., *Water Res. J.* **33**.

Brindle, K., Stephenson, T., and Semmens, M. J. (1998). Nitrification and oxygen utilization in a membrane aerated bioreactor. *J. Membrane Sci.* **144**, 197–209.

Brook, P. R., and Livingston, A. G. (1994). Biological detoxification of a 3-chloronitrobenzene manufacturing wastewater in an extractive membrane bioreactor. *Water Res.* **28**(6), 1347–1354.

Dufresne, R., LaVallee, H. C., Lebrun, R. E., and Lo, S. N. (1998). Comparison of performance between membrane bioreactor and activated sludge system for the treatment of pulping process wastewater., *TAPPI* **81**(4), 131–135.

Eckenfelder, W. W., Jr. (1989). "Industrial Water Pollution Control," pp. 189–210. McGraw-Hill, New York.

Ellis, T. G., Barbeau, D. S., Smets, B. F., and Grady, C. P. L. (1996). Respirometric technique for determination of extant kinetic parameters describing biodegradation. *Water Environ. Res.* **68**(5), 917–926.

Environmental Protection Agency (1998). 1996 TRI public data release. *www.epa.gov*.

Hickey, R. F., and Smith, G. (1996). "Biotechnology in Industrial Waste Treatment and Bioremediation." Lewis, Boca Raton, FL.

Lewandowski, G. A., and DeFilippi, L. J. (1998). "Biological Treatment of Hazardous Waste." Wiley, New York.

Livingston, A. G., Freitas dos Santos, L. M., Pavasant, P., Piskikopoulos, E. N., and Strachan, L. F. (1996). Detoxification of industrial wastewater in an extractive membrane bioreactor. *Water Sci. Technol.* **33**(3), 1–8.

Maloney, S. W., Engbert, E. G., Suidan, M. T., and Hickey, R. F. (1998). Anaerobic fluidized-bed treatment of propellant wastewater., *Water Environ. Res.* **70**, 52–59.

Sutton, P. M., and Mishra, P. N. (1996). The membrane biological reactor for industrial wastewater treatment and bioremediation. *In* "Biotechnology in Industrial Waste Treatment and Bioremediation" (R. F. Hickey and G. Smith, Eds.), pp. 175–191. Lewis, Boca Raton, FL.

Winnen, H., Suidan, M. T., Scarpino, P. V., Wrenn, B., Cicek, N., Urbain, V., and Manem, J. (1996). Effectiveness of the membrane bioreactor in the biodegradation of high molecular weight compounds., *Water Sci. Technol.* **34**(9), 197–203.

Zaloum, R., Lessard, S., Mourato, D., and Carriere J. (1994). Membrane bioreactor treatment of oily wastes from a metal transformation mill. *Water Sci. Technol.* **30**(9), 21–27.

Wastewater Treatment, Municipal

Ross E. McKinney
University of Kansas

I. Wastewater Characteristics
II. Important Microorganisms
III. Energy-Synthesis Metabolism
IV. Aerobic Treatment Systems
V. Anaerobic Treatment Systems
IV. Mixed-Treatment Systems
VII. Pathogenic Microorganisms
VIII. Regulatory Controls
IX. Future Biological Treatment Systems

GLOSSARY

activated sludge A flocculated mixture of bacteria and protozoa grown in an aerobic environment under carbon-limiting conditions.

hydraulic retention time The average time in hours that the wastewaters remain in the aeration tanks.

MLSS Suspended solids in the aeration tank, concentration in mg/liter or quantity in lb.

return activated sludge (RAS) Activated-sludge solids returned from the final sedimentation tanks to the aeration tanks on a continuous basis.

solids retention time The average time in days that the activated sludge remains in the aeration tanks.

waste activated sludge (WAS) Activated-sludge solids removed from the treatment system on a continuous or periodic basis, measured as pounds per day.

MUNICIPAL BIOLOGICAL WASTEWATER TREATMENT is an essential part of municipal operations in the United States and Europe. Slowly, it is becoming part of the operations in larger cities in the developing countries of the world. Municipal wastewater treatment has evolved over the years, as its importance in public health became established. Enteric diseases were recognized as being easily transmitted by water from both improperly treated sewage and poorly treated drinking water. Early research at the Lawrence Experiment Station in the United States starting in 1887 and in England in the 1890s put municipal biological wastewater treatment on a constantly changing path that continues to change and evolve, producing more efficient systems to treat municipal wastewater. It will not be long before continuous recycling of wastewater will be practical and economical, not only throughout the world, but also in space.

I. WASTEWATER CHARACTERISTICS

The design of wastewater treatment plants starts with the municipal-wastewater characteristics and the desired effluent quality established by the regulatory agency responsible for water quality below the treatment-plant discharge. In the United States, each wastewater treatment plant is issued a NPDES permit that indicates the anticipated influent characteristics and the required effluent characteristics. Normally, the NPDES permit provides the concentration units (mg/liter) and mass units (lb/day or kg/day) for all major pollutants. It also indicates the frequency for collecting samples for analyses and the frequency for reporting data to the regulatory agency. NPDES permits are issued for 5-year periods.

A. Biochemical Oxygen Demand (5-Day, 20°C)

One of the major reasons for municipal wastewater treatment is the excessive oxygen demand that occurs downstream from municipal wastewater discharges.

The 5-day, 20°C, biochemical oxygen demand (BOD) test is the primary method for analyzing the potential oxygen demand of untreated and treated municipal wastewater. The BOD5 test uses a mixed microbial population growing under aerobic conditions in a dilute organic solution at a constant temperature over a 5-day period. At 20°C, the oxygen of saturation at 1 atm pressure is 9.1 mg/liter. To produce valid data for the BOD5 test, the microorganisms must use at least 1.0 mg/liter dissolved oxygen (DO) and leave at least 1.0 mg/liter DO at the end of the 5-day incubation period. With a suitable seed of microorganisms, approximately two-thirds of the ultimate carbonaceous oxygen demand will be used in 5 days. It takes about 20 days to reach the ultimate carbonaceous oxygen demand. Because some dead cell mass remains unoxidized in the BOD bottles, only 87% of the theoretical carbonaceous oxygen demand is measured by the ultimate carbonaceous oxygen demand. If fresh municipal wastewaters are used as the source of the microbial seed, the BOD5 test will measure only the carbonaceous oxygen demand in 5 days. Although the nitrifying bacteria are also growing in the BOD bottles, they will not show sufficient growth to create a significant oxygen demand in 5 days. By the seventh day of incubation, the nitrifying bacteria will exert a measurable oxygen demand. If the microbial seed contains a high population of nitrifying bacteria, the nitrification oxygen demand will occur partially within the 5-day period. Because the BOD5 test is dependent on a suitable microbial seed for valid results, the chemical oxygen demand (COD) test has been used to measure the total oxygen demand in the wastewaters. The COD test is an acid-dichromate oxidation test that provides results in about 2–3 hours.

Unfortunately, the COD test measures the nonbiodegradable organics as well as the biodegradable organics. The only way to measure the biodegradable organics is with the BOD5 and its correlation with the theoretical COD. Data indicate that the biodegradable COD (BCOD) is about 1.7 times the carbonaceous BOD5. The nonbiodegradable COD (NBCOD) is simply the COD minus the BCOD. Municipal wastewater in the United States contains between 170 mg/liter and 200 mg/liter BOD5 and between 340 mg/liter and 440 mg/liter COD.

B. Suspended Solids

Municipal wastewaters contain suspended solids that must be removed if the receiving water quality is to be maintained at high levels. Data on suspended solids are among the simplest to determine. Total suspended solids are measured by filtering a wastewater sample through fiberglass filters having openings of 1.0 μm or smaller. Although small particles will pass through the filter, the quantity is not sufficient to affect the results. The fiberglass filter is dried at 103°C for 1 hour, cooled, and weighed. The change in weight from the original filter is the total suspended solids (TSS) in the filtered sample. The dried filter is placed into a muffle furnace at 550°C to burn off the organics. The filter is cooled and the change in weight is the ash residue, the inorganic suspended solids (NVSS). The organic suspended solids (VSS) fraction is determined by the difference between the TSS and the NVSS. Municipal wastewater contains TSS of the same magnitude as BOD5, with the VSS close to 80% of the TSS. Unfortunately, the VSS in municipal wastewater are not completely biodegradable. Data have shown that 30–40% of the VSS are nonbiodegradable.

C. Nitrogen

Nitrogen exists in municipal wastewater as organic nitrogen (Org-N) and ammonia nitrogen (NH_3-N). Nitrite nitrogen (NO_2-N) and nitrate nitrogen (NO_3-N) are not normally found in municipal wastewater. Anaerobic conditions in the sewage-collection system reduce any nitrite and nitrate in the wastewater before they reach the wastewater-treatment plant. Most of the Org-N is in the form of urea. Bacteria quickly hydrolyze the urea to form ammonia nitrogen. The remaining Org-N is primarily protein that hydrolyzes to form amino acids. The quantity of available nitrogen in municipal wastewater ranges from 30–40 mg/liter and exceeds the demand created by the biodegradable carbon, making carbon the limiting element in bacteria metabolism.

D. Phosphorus

Phosphorus, as is nitrogen, is an essential part of the microbial cell mass. Municipal wastewater contains orthophosphates and complex polyphosphates. Orthophosphates are metabolized by bacteria to create energy for cell synthesis. Most of the complex polyphosphates can be hydrolyzed by bacteria to form orthophosphates. In municipal wastewater, the orthophosphates will be a mixture of dibasic phosphates and monobasic phosphates. Orthophosphates react with divalent and trivalent metal ions to form colloidal precipitates. Changes in the formulation of synthetic detergents over the years have resulted in a significant decrease in phosphates in municipal wastewater, so that the excess of phosphorus to carbon results in only a small excess of phosphorus in biologically treated effluents. The concentration of phosphorus found in municipal wastewater is between 5 and 10 mg/liter as P.

E. Alkalinity and pH

Alkalinity measurements in municipal wastewater have taken on greater significance because more wastewater-treatment plants are concerned with nitrification and denitrification. Alkalinity measures the buffering capacity of the wastewater. It has two basic components, the alkalinity of the carriage water and the alkalinity created by the wastewater constituents. In soft-water areas of the United States, the carriage water has little alkalinity. The limited alkalinity will be from ammonium bicarbonate in the wastewater. In hard-water areas, the carriage water will contain calcium bicarbonate as the major source of alkalinity. A few areas of the United States have alkaline waters with sodium bicarbonate as the primary source of alkalinity. The alkalinity is important in keeping the pH in the proper range for good microbial growth. The optimum pH is between 6.5 and 8.5. Because organic solids react with the acid used in the alkalinity test, alkalinity measurements are best made on centrifuged samples.

II. IMPORTANT MICROORGANISMS

A. Bacteria

Bacteria are the most important microorganisms in municipal wastewater-treatment systems. They are responsible for the stabilization of the biodegradable pollutants in wastewater. Because each cell weighs only about 10^{-12} g and is about 80% water, it takes a large number of bacteria to stabilize the pollutants in municipal wastewater. The bacteria that metabolize the organics to carbon dioxide and water the fastest predominate in aerobic wastewater-treatment units and produce the highest quality effluent. Bacteria are single-cell organisms that metabolize the soluble pollutants in municipal wastewater first. The ability of bacteria to metabolize suspended organic pollutants depends on enzymes located on the surface of the bacteria. The surface enzymes hydrolyze the organics to small soluble compounds that can be taken through the cell wall. A few bacteria in aerobic wastewater-treatment units are autotrophic. Nitrifying bacteria and sulfur-oxidizing bacteria are the most common autotrophic bacteria in wastewater-treatment systems.

B. Fungi

Fungi compete with the bacteria for nutrients. The common fungi in municipal wastewater systems are aerobic, even though there are some anaerobic fungi. Most fungi are larger than bacteria and are filamentous. The smaller bacteria have a greater surface area-to-mass ratio than the fungi and will outgrow the fungi in most environments. Fungi have an advantage over bacteria at pH levels below 6.0 and in low-nitrogen environments. Fungi reproduce using spores that survive for long periods of time under adverse environments. Fungi also have the ability to metabolize complex organics, making them very useful in industrial-waste treatment. Terrestrial fungi can metabolize both lignin and the complex polysaccharides forming the residues of dead bacteria. Aqueous fungi cannot metabolize either lignin or the dead bacterial residues within a reasonable period of time.

C. Algae

Algae are the photosynthetic plants. They use light energy to create cell protoplasm from inorganic materials in the wastewaters. Essentially, algae produce organic materials from the stable end products of bacterial and fungal metabolism. More importantly, algae produce oxygen as an end product of cell synthesis. Excessive growths of algae in rivers and lakes can damage these water resources. Algae can grow as filaments and as individual cells. Studies have shown that the blue-green algae (cyanobacteria) are actually photosynthetic bacteria rather than algae. This change in classification has created some confusion in the literature. It illustrates the changing nature of microbial classification. The 16S rRNA genetic classification of microorganisms will create major classification changes into the twenty-first century.

D. Protozoa and Rotifers

Protozoa are single-cell animals that primarily exist by consuming bacteria and dispersed algae. They cannot metabolize the filamentous fungi or algae. Most of the protozoa are aerobic. They must consume a large number of bacteria in order to produce a single new cell. Protozoa range in size from around 10 μM to several hundred micrometers. The flagellated protozoa are not very efficient food gatherers, except in very high bacterial populations. The free-swimming ciliated protozoa are the most efficient food gatherers, but they require lots of energy and must consume a large number of bacteria and algae to survive. Stalked ciliated protozoa attach themselves to particles and feed on fewer bacteria than do the free-swimming ciliated protozoa. Their lower energy requirements allow the stalked ciliated protozoa to survive better than the free-swimming ciliated protozoa under food-limiting conditions. The crawling ciliated protozoa are similar in appearance to the free-swimming ciliated protozoa. Similar to the stalked ciliated protozoa, the crawling ciliated protozoa do not require as much energy as the free-swimming ciliated protozoa. They find their food by crawling over suspended organics and consuming the bacteria that are growing on particle surfaces. The protozoa survive adverse environments by forming cysts until they find a suitable environment for growth.

Rotifers are multicellular animals that are strict aerobes. They also consume bacteria and algae. Rotifers have forked tails to help them hold onto suspended particles and flexible bodies to allow them to bend around and consume bacteria on the surface of the particles. The rotifers reproduce by producing eggs in egg sacs. The presence of a large number of egg sacs indicates that a population explosion of rotifers is about to occur.

Because of the sizes of protozoa and rotifers, they are easy to see and identify under the optical microscope. They also can be used to identify the general environmental conditions in biological systems.

E. Worms and Crustaceans

Nematodes are the most common, microscopic worms found in municipal wastewater systems. Nematodes come from the soil and consume a large number of dispersed bacteria for their nutrients. The thrashing action of nematodes dislodges the bacteria from the surface of suspended particles. Similar to higher animals, the nematodes are aerobic organisms. Other worms include bristle worms, with their large transparent bodies, orange spots, and bristles projecting along the length of their bodies. There are even water mites and water bears in the proper environments.

Crustaceans are large complex animals that are found in a few treatment systems with clean environments. *Daphnia* is the largest microscopic crustacean. It metabolizes bacteria and algae in large numbers and is considered a polishing organism in complex wastewater-treatment systems.

III. ENERGY-SYNTHESIS METABOLISM

A. Heterotrophic Bacteria

The stabilization of biodegradable organic pollutants in municipal wastewaters is brought about by the growth of heterotrophic bacteria. Although the growth of bacteria is continuous, it is usually presented in two steps, the energy step and the synthesis step. The energy step begins with the bacteria oxidizing the organic pollutants to lower energy-containing end products. As quickly as the bacteria obtain energy, they use that energy to convert some of the organic pollutants into protoplasm. Ammonia nitrogen, phosphates, and trace metals are also incorporated into the new cell mass. In carbon-limiting environments, the bacteria that grow in municipal wastewater have the same general chemical formulation as the VSS that are produced, $C_5H_8O_{2.5}N$ or $CH_{1.6}O_{0.5}N_{0.2}$. Under aerobic conditions, the primary bacteria metabolize the energy fraction in the biodegradable organics to carbon dioxide, water, ammonia, and energy. Data indicate that one-third of the total biodegradable energy content of the organic pollutants is oxidized, with two-thirds converted to cell mass. It appears that 19.3 kJ of energy is tied up in each gram of new cell mass, with 9.6 kJ of energy required for the synthesis reactions. In an anaerobic environment the bacteria metabolize part of the substrate for energy and part for synthesis. Each gram of VSS cell mass, produced anaerobically, has the same energy content as aerobic bacteria and requires the same expenditure of energy per unit of cell mass formed. The difference between aerobic metabolism and anaerobic metabolism is the limited energy yield for anaerobic bacteria. Essentially, the anaerobic bacteria have to process more nutrients to obtain sufficient energy for cell synthesis than do the aerobic bacteria. The net effect is that anaerobic bacteria produce less cell mass per unit of substrate metabolized than do aerobic bacteria. The primary end products of anaerobic metabolism are short-chain organic acids, aldehydes, ketones, and alcohols. The methane-forming bacteria consume the short-chain organic compounds with the production of methane gas, which is relatively insoluble in water. The high energy content of the methane gas limits the production of cell mass of the methane-producing bacteria. Many faculative bacteria have the ability to use nitrite and nitrate as their electron acceptor under anaerobic conditions. Denitrification uses the energy stored in the nitrite and nitrate to produce new cell mass and N_2 gas. Specialized sulfate-reducing bacteria can use sulfate as their electron acceptor, producing cell mass and reduced sulfur compounds. If there is an excess of organics, the sulfate-reducing bacteria produce hydrogen sulfide as their end product in addition to new cell mass. In organic-limiting environments, the sulfate-reducing bacteria can produce a range of reduced sulfur compounds, even free sulfur. Some bacteria can use ferric iron as their electron acceptor with the production of ferrous iron as the primary end product in addition to cell mass.

B. Autotrophic Bacteria

The autotrophic bacteria use carbon dioxide, water, and ammonia to create their cell mass, and oxidize various inorganic compounds for energy synthesis. The nitrifying bacteria oxidize ammonia to nitrite and then oxidize the nitrite to nitrate. The two groups of nitrifying bacteria are aerobic bacteria, using DO for oxidation. The energy yield for the nitrifying bacteria is reduced in proportion to the energy tied up as nitrite and nitrate. The cell-mass yields of the nitrifying bacteria are low compared to heterotrophic metabolism. The sulfur-oxidizing bacteria use hydrogen sulfide or reduced sulfur compounds and DO to produce sulfate for their energy reactions. Similar to the nitrifying bacteria, the sulfur-oxidizing bacteria tie up energy in sulfate, reducing cell synthesis. The iron-oxidizing bacteria are an environmentally important group of autotrophic bacteria, oxidizing ferrous iron to ferric iron. Although there is a group of hydrogen-oxidizing bacteria, the methane-forming bacteria are more efficient in hydrogen metabolism, using the hydrogen to reduce carbon dioxide to methane and water. All of the energy-synthesis reactions are based on oxidation–reduction reactions. A portion of the pollutants in wastewater is oxidized to yield energy to reduce the remaining pollutants to cell mass.

C. Endogenous Respiration

Endogenous respiration is the third part of the metabolic reactions in microorganisms. Although endogenous respiration occurs continuously, it passes without notice until the microbial population increases to large masses and is retained in the bioreactor for long periods of time at low feed rates. Endogenous respiration is quite evident in aerobic biological wastewater-treatment systems. Chemically, endogenous respiration is the continuous degradation of cellular materials in direct proportion to the living mass of bacteria in the system. In aerobic biological wastewater-treatment systems, it is not possible to separate the endogenous respiration of the bacteria from the metabolism of bacteria by protozoa. For this reason, the two metabolic reactions are normally included in the overall endogenous-respiration rate. The apparent accumulation of dead microbial residue occurs as a function of the overall endogenous-respiration rate. About 20% of the cell mass undergoing endogenous respiration accumulates as dead cell mass. Endogenous respiration increases the oxygen demand and reduces the size of the microbial mass.

IV. AEROBIC TREATMENT SYSTEMS

Municipal wastewater treatment begins with preliminary treatment units, screens to remove the large trash and grit chambers to remove sand. The preliminary treatment units are followed by primary treatment units, large sedimentation tanks to remove the settleable suspended solids by gravity. The primary sludge must be continuously collected, treated in other units, and returned to the land environment. The primary effluent must be further treated in aerobic treatment units to remove the soluble biodegradable pollutants, and the remaining suspended solids that were not removed in the primary sedimentation tanks. The primary sedimentation tanks remove 60–70% suspended solids and 30–35% BOD5 from municipal wastewater.

The primary stabilization reactions in aerobic biological-treatment units are oxidation and synthesis, together with endogenous respiration. The microbial cells produced by synthesis are removed in secondary sedimentation tanks, following the aerobic treatment units. The settled microbial solids are collected continuously and recycled back to the aerobic treatment units to maintain an excess of microorganisms in the system at all times. The excess microbial solids, produced by metabolism, must be removed from the treatment system, stabilized, and returned to the land environment. If the wastewater effluent does not meet regulatory requirements after the secondary treatment, tertiary treatment units may be required. Tertiary treatment units are defined as any treatment units after secondary treatment units, including disinfection, sand filtration, and carbon absorption.

Activated-sludge systems are the primary aerobic treatment systems used in the United States for municipal wastewater-treatment plants having wastewater flows in excess of 2–3 million gallons/day (MGD). Although many smaller municipal wastewater-treatment plants use activated sludge, the small plants normally require simpler, less costly systems. Fundamentally, activated-sludge systems are aerobic dispersed-growth biological systems employing bacteria as the primary microorganisms and protozoa as the secondary microorganisms. The major bacteria are soil bacteria and include *Pseudomonas, Alcaligenes, Achromobacter, Flavobacterium,* and *Bacillus.* The protozoa range from *Amoeba* to flagellates, such as *Peranema, Bodo, and Monas;* to free-swimming ciliates, such as *Lionotus, Paramecium, Colpidium, Tetrahymena, Euplotes, Aspidiscus,* and *Stylonychia;* to stalked ciliates, such as *Vorticella, Epistylis, Opercularia,* and *Charchesium.* Higher animals will include rotifers and nematodes under special circumstances. The key for activated-sludge systems is the ability of bacteria to flocculate when the biodegradable pollutants have been stabilized and for the protozoa to consume the dispersed bacteria. Flocculation allows the removal of the nonbiodegradable suspended pollutants in the secondary sedimentation tanks and permits the accumulation of excess bacteria in the treatment system. The removal of dispersed bacteria by the protozoa produces the clarified effluent that is essential for a high-quality effluent. Activated-sludge plants can produce effluents containing from 10–30 mg/liter TSS and from 5–30 mg/liter BOD5.

A. Conventional Systems

Conventional activated-sludge systems, also called plug-flow systems, consist of long narrow aeration tanks followed by secondary sedimentation tanks, shown in Fig. 1. Return activated sludge (RAS) is continuously pumped back to the head end of the aeration tank to insure an excess of microbes in the system. A small fraction of the activated sludge is lost in the settled effluent. Most of the excess activated sludge is removed from the treatment system as waste activated sludge (WAS) on a daily basis. Aeration is supplied by either diffused aerators or mechanical surface aerators. Air diffusers are placed

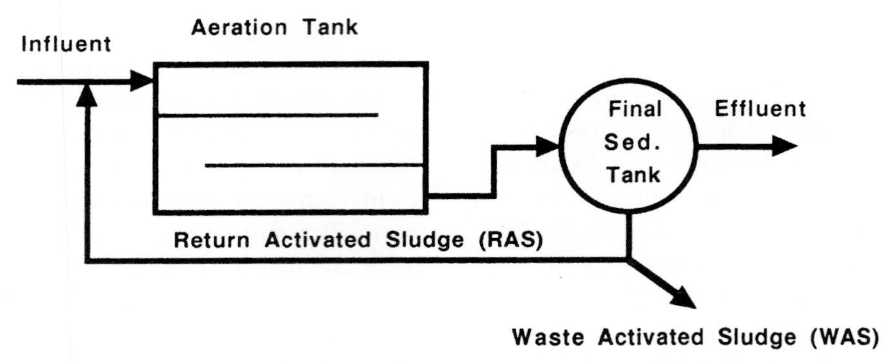

Fig. 1. Schematic diagram of a conventional activated-sludge system.

along one side of the aeration tank or spread over the entire tank bottom. Conventional systems tend to have insufficient oxygen at the head end of the aeration tank because of the very high oxygen demand created by the introduction of primary effluent with the large bacterial population from the RAS. Eventually, the oxygen transfer exceeds the oxygen demand and excess DO builds up at the end of the aeration tank. Good treatment depends on sufficient oxygen transfer to the microbes and good mixing to insure proper contact among the bacteria, the nutrients, and the oxygen.

B. Completely Mixed Systems

The examination of bacterial metabolism and oxygen-transfer characteristics led to the development of completely mixed activated-sludge (CMAS) systems. CMAS systems depend on the rapid mixing of the primary effluent and the RAS to disperse the nutrients and the bacteria uniformly throughout the aeration tanks. A schematic diagram of a typical CMAS system is shown in Fig. 2. It is easy to adjust the aeration to maintain excess DO in all parts of the aeration tanks, maximizing the biological reactions with the production of the highest quality effluent with the least amount of WAS. Because of the greater efficiency in oxygen transfer in CMAS systems over conventional systems, some engineers reduced the size of the aeration equipment. The smaller aerators reduced the mixing in the aeration tanks and resulted in a greater growth of filamentous bacteria, producing poorer settling sludge in the secondary sedimentation tanks. With proper mixing, filamentous bacteria growth will be minimal and will not create settling problems. The real value of CMAS systems is their ability to process potentially toxic industrial wastewater that is often combined with municipal wastewater in large industrial cities. CMAS design allows the bacteria to acclimate to minimum concentrations of potentially toxic pollutants.

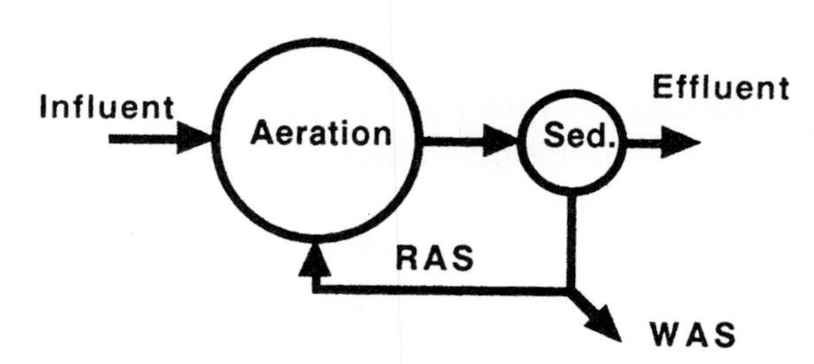

Fig. 2. Schematic diagram of a completely mixed activated-sludge system.

C. Contact Stabilization Systems

It was long recognized that activated sludge had the ability to quickly remove suspended solids from wastewaters. This reaction led to the development of the contact stabilization modification. Unsettled municipal wastewater is mixed with the RAS in a short-retention-time mixing tank and then separated in the secondary sedimentation tank. The RAS is then aerated to stabilize the adsorbed organics before being returned to the mixing tank. A schematic diagram of the contact stabilization system is shown in Fig. 3. By aerating RAS at a concentration of 8000–10,000 mg/liter rather than aerating MLSS at 2000 mg/liter, the aeration tank used can be smaller. If the aeration is complete, the bacteria will also absorb

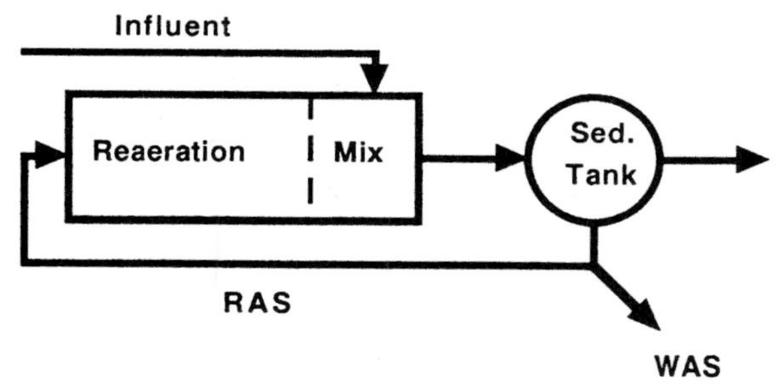

Fig. 3. Schematic diagram of a contact stabilization activated-sludge system.

a portion of the soluble organics during the mixing phase. By using aeration mixing, the bacteria are able to metabolize the soluble organics while adsorbing the suspended organics. Contact stabilization uses 30 min aeration mixing of RAS and raw wastewaters, with 2 hr reaeration of RAS to produce a high-quality effluent.

D. Extended Aeration Systems

Extended aeration was developed to treat a small volume of wastewater from isolated subdivisions, motels, and small industrial plants. It consists of an aeration tank having a raw waste volume of 24 hr of flow, followed by a secondary sedimentation tank. The large volume of the aeration tank minimizes the production of excess activated sludge by aerobic digestion and allows the activated sludge to be retained in the aeration tank for at least 30 days. The small amount of WAS does not need further treatment before being placed on the land.

E. Oxidation Ditch Systems

Oxidation ditches are simply one form of extended aeration system that employs a horizontal brush-type mechanical aerator to supply both oxygen and mixing. The oxidation ditches began with 48 hr of raw wastewater retention and 30–60 days, or more, solids retention time under aeration. A schematic diagram of a simple oxidation ditch system is shown in Fig. 4. The simplicity of the oxidation ditch system resulted in shifting the loads from extended aeration to conventional activated-sludge systems. Multiple rotor aerators were placed around the oxidation ditch and the depth was increased from 4–10 ft or deeper. Instead of single ditches, multiple ditches were constructed with common wall construction. The multiple-ditch systems are operated in series with the raw wastewater and RAS added to the outer ditch having the greatest volume. The simplicity of operation and the high effluent quality have made oxidation ditches popular.

F. Sequencing Batch Reactors

Sequencing batch reactors (SBR) are a return to the initial activated-sludge systems that were operatd on a fill-and-draw basis. In SBR systems, a single tank is used as both the aeration tank and the secondary sedimentation tank. The raw wastewater is added to the settled activated sludge in the initial fill cycle. When the liquid level reaches a given depth, the aeration equipment is turned on to produce an aerated fill until the tank reaches the normal operating depth. Then the influent flow is stopped and aeration is continued for a short time to stabilize the organics. The aeration equipment is turned off and the activated sludge is allowed to flocculate and settle. After a short settling period, the supernatant is decanted slowly while the activated sludge finishes settling and compacts on the tank bottom. The excess activated sludge is removed from the settled sludge, and the

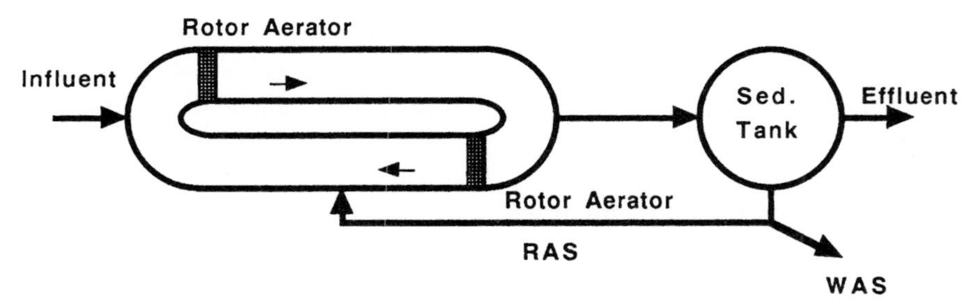

Fig. 4. Schematic diagram of an oxidation ditch activated-sludge system.

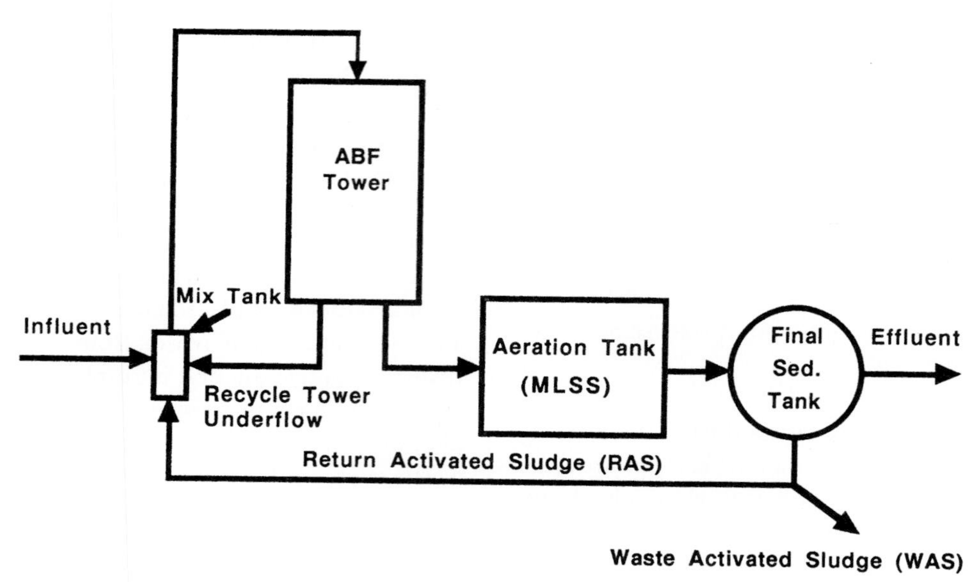

Fig. 5. Schematic diagram of an activated biofiltration activated-sludge system.

cycle is started again. The alternating feed–settle routine normally requires two tanks, with one tank being fed while the contents of the other tank settle. New designs allow continuous feeding and effluent removal from different sections of the same tank. SBR operations are microprocessor-controlled, based on the specific wastewater characteristics. The simplicity of SBR systems and automatic controls have made them quite popular.

G. Activated Biofiltration Systems

Activated biofiltration (ABF) units have been placed ahead of the aeration tanks to act as rapid synthesis units with a high rate of oxygen transfer. RAS from the secondary sedimentation tanks are recycled back to the influent sump, as shown in Fig. 5, and mixed with incoming wastewaters and tower underflow before being pumped at a fixed rate over the top of the plastic media tower. The RAS bacteria metabolize the soluble organic matter aerobically as the thin film moves over the plastic media with a rapid oxygen transfer from the air. The ABF tower effluent is discharged to the aeration tank, where the suspended organics are metabolized along with any residual soluble organics. The aeration tank does not have to supply as much oxygen as do other activated-sludge systems. Nitrification can also take place in the aeration tank if it is designed for the additional oxygen demand.

V. ANAEROBIC TREATMENT SYSTEMS

Anaerobic treatment units are used for concentrated organic wastes. Anaerobic digesters are used to treat primary sludge and WAS. Primary sludge is collected and pumped from the primary sedimentation tanks to the anaerobic digesters at 4.0% to 6.0% total solids (TS). Primary sludge is about 80% organic solids (VS), of which about 65% is biodegradable. This means that about half of the primary-sludge TS is inert and will not be changed in the anaerobic digester. The biodegradable fraction of the primary sludge is rapidly metabolized anaerobically by facultative soil bacteria to short-chain organic acids and new cell mass. The short-chain organic acids are metabolized by the methane-forming bacteria to methane and carbon dioxide. Ammonia is released from protein metabolism and reacts with the fatty acids to form neutralized fatty acids. The metabolism of the neutralized fatty acids releases the ammonia to react with carbon dioxide and water to form ammonium bicarbonate. The ammonium bicarbonate created in the anaerobic digester accumulates as alkalinity, keeping the pH above 6.5 for good methane production. The metabolism of complex hydrocarbons by β-oxidation results in the production of hydrogen, which can be metabolized by methane-forming and acetogenic bacteria. The methane-forming bacteria can use hydrogen and carbon dioxide to produce methane and water plus new cell mass. The acetogenic bacteria use hydrogen and carbon dioxide to produce acetic acid, water, and new cell mass. Based on energy, both the methane-forming bacteria and the acetogenic bacteria are competitive for nutrients.

WAS contains the inert fraction of primary suspended solids, dead cell mass, and active microbial mass. Only the active microbial mass is biodegradable. About 80% of the active mass VSS is biodegradable in the anaerobic digester. After partial aerobic digestion in the aeration tanks, the WAS is only about 25–35% biodegradable. WAS concentrations are between 0.6 and 1.2% in secondary sedimentation tanks. It is normally thickened to 3–4% by flotation thickeners, gravity belt thickeners, or centrifuges before being added to anaerobic digesters with the primary sludge.

Anaerobic digesters are normally operated at 37°C, with the heat supplied by burning the methane gas generated by the digestion process. A few municipal wastewater-treatment plants use thermophilic digesters operating at 50–54°C to destroy potentially pathogenic microorganisms. Thermophilic digestion is more difficult to operate than mesophilic digestion, limiting its use in most plants. Anaerobic digesters are operated as continuous-flow dispersed bioreactors, having retention times of 20–30 or more days. The large volume is required because the upper part of the digester has been designed as an active bioreactor and the lower part of the digester has been designed as a solids-settling tank. Limited mechanical mixing in the upper part of the digester does not produce adequate mixing for good biological activity. Anaerobic digesters have been designed in an egg shape, with the upper part of the digester narrowing down. It was found that the egg shape helped concentrate the gas in a smaller area, producing better mixing in the upper part of the digester. The heavy, digested sludge settles to the bottom of the digester for easy removal. In the 1990s, it was discovered that gas mixers placed over the floor of the anaerobic digester provided mixing throughout the entire digester. The digestion time with good mixing could be reduced to less than 10 days. A second storage digester is used for concentrating the digested sludge prior to dewatering and return to the land. Digested sludge is quite stable and can be placed on the land without creating odors or attracting flies. It consists primarily of inert suspended solids and residues of dead cells. Most of these materials will slowly decompose in soil.

VI. MIXED-TREATMENT SYSTEMS

More and more municipal biological-treatment systems consist of mixtures of aerobic and anaerobic metabolism. Each group of bacteria is selected to produce the optimum reaction in a given tank or portion of the tank. Together, the bacteria produce a greater reaction than they could produce by themselves. Knowledge of the biochemistry of the various bacteria has been the key to producing the new mixed biological treatment systems for municipal wastewaters.

A. Trickling Filters

Trickling filters were widely used for municipal wastewater treatment. Unfortunately, trickling filters could not produce the effluent quality that activated-sludge systems could produce. The simplicity of their operation made trickling filters popular. Initially, trickling filters employed rock media. After World War II, rock media was slowly replaced with plastic media. The microbial populations in trickling filters are quite diverse. Aerobic metabolism occurs in the upper layer of microbial growth attached to the media, and anaerobic metabolism occurs in the lower layers of microbial growth. The short fluid-retention time in contact with the microbes limits the effluent quality from trickling filters. Only lightly loaded trickling filters can meet the effluent criteria of 30 mg/liter TSS and 30 mg/liter BOD5.

B. Rotating Biological Contactor Systems

Rotating biological contactor (RBC) systems use the fixed growth of trickling filters with better con-

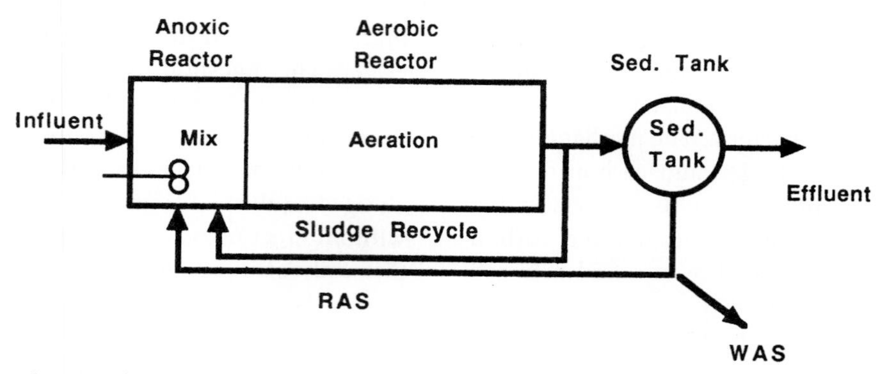

Fig. 6. Schematic diagram of a nitrification–denitrification activated-sludge system.

tact time in a submerged fluid. The fixed media in the RBC systems consist of a series of large-diameter plastic discs with perforations in a series of small tanks. The plastic discs are attached to a rotating shaft that turns slowly, moving through the liquid contained in the small tanks into the air above the liquid and back into the liquid. The bacteria become attached to the plastic media, assimilate nutrients as the discs pass through the liquid in the small tanks, and metabolize the nutrients aerobically as the discs pass through the air. Heavy microbial growths in the first tank can produce anaerobic conditions as well as aerobic metabolism. Microbial growths in the second and third tanks become less and less, as the nutrients are reduced. The growths in the second and third tanks tend to be aerobic. If there are enough tanks in the series, nitrification will occur. The excess bacterial growth is removed from the plastic discs by hydraulic shear created when the discs move through the liquid. The sheared growths flow through the small tanks into secondary sedimentation tanks, where they settle and are removed from the system. Very little power is required to turn the discs.

If odors are produced by the anaerobic growths, it is necessary to cover the system with plastic covers and to push air through the gas space between the liquid and the plastic cover to remove the volatile organics from the system. Treatment is further complicated by having to treat the contaminated air. Odors can be prevented by recycling the treated effluent to dilute the incoming wastewaters and distributing the load over more of the discs.

C. Nitrification–Denitrification Systems

Nitrification–denitrification systems are classic mixed-treatment systems that employ aerobic and anaerobic treatment to accomplish the desired treatment results. Nitrification–denitrification systems consist of an anoxic tank ahead of an aerobic tank, as shown in Fig. 6. The anoxic tank is the denitrifying unit and the aerobic tank is the nitrifying unit. The settled wastewater is added to the RAS in the anoxic tank together with some recycled mixed liquor, and mechanically mixed. The facultative bacteria in the RAS and recycled mixed liquor metabolize the soluble organics in the incoming wastewaters and use nitrate as their electron acceptor. The nitrate is reduced to nitrite and to N_2 gas in an excess of organics. The discharge from the anoxic tank goes to the aeration tank, where the suspended organics are stabilized and the excess nitrogen is oxidized to nitrate.

It is also possible to use a two-stage nitrification–denitrification system. The first stage is a nitrifying activated-sludge system. The second stage is an anoxic, mechanically mixed system fed with methanol to reduce the nitrate. The activated sludge formed from the methanol feed is separated in a conventional sedimentation tank and recycled back to the anoxic unit to maintain an acclimated microbial seed for rapid metabolism.

D. Biological Phosphorus-Removal Systems

The biological phosphorus-removal systems are the newest mixed-microbial-treatment systems. They are known as Bio-P systems. Bio-P systems have three bioreactors in series. The first bioreactor is an anaerobic unit with mechanical mixing. The second bioreactor is an anoxic denitrification unit. The third bioreactor is an aerobic nitrification–phosphorus-removal unit. The anaerobic tank receives the incoming wastewater and mixed liquor from the anoxic unit. The strongly anaerobic conditions allow the Bio-P bacteria to metabolize the soluble short-chain fatty acids for growth with the release of phosphate. The mixed liquor from the anaerobic tank flows to the anoxic tank that also receives RAS from the secondary sedimentation tank and mixed liquor from the aeration tank. The anoxic tank is mechanically mixed to allow anaerobic metabolism of the remaining soluble organics with a reduction of the nitrate to N_2 gas. The mixed liquor from the anoxic tank flows into the aeration tank, where the organic stabilization is completed, nitrification is completed, and the Bio-P bacteria take up most of the excess phosphorus. Phosphorus removal is accomplished by removing the WAS after the aeration tank and dewatering it before anaerobic conditions allow its release. Bio-P systems are the most complex biological wastewater-treatment systems in current use. Their advantage lies in the use of bacteria to remove phosphates instead of using chemical precipitation.

E. Wastewater Lagoons

Wastewater lagoons are the simplest wastewater systems capable of producing a high-quality effluent. Essentially, wastewater lagoons are a series of large shallow ponds. Raw wastewater is added directly at the bottom of the first lagoon. The lack of fluid motion allows the suspended solids to settle out around the inlet and undergo anaerobic metabolism. Over time, methane-forming bacteria will accumulate and produce methane gas, which rises to the pond surface and returns to the atmosphere. The soluble organic compounds are metabolized by bacteria under aerobic conditions to carbon dioxide, water, and new cell mass. The surface of the lagoon is exposed to sunlight during the daylight hours. The presence of carbon dioxide, ammonia, phosphate, trace metals, water, and sunlight allows algae to grow, with the production of oxygen and new cell mass. Wind action is important in keeping the lagoons aerobic and mixed. The algae and bacteria remain dispersed, slowly settling out to the bottom of the lagoons. Whereas the limited nutrients and long retention time in the lagoons allow the bacteria to die off, the algae die and recycle the nutrients. The regrowth of the algae allows them to become the predominant group of organisms in lagoon systems. The effluent quality from lagoons will be determined by the algae in the effluent. Multicell lagoons with submerged drawoffs are used to minimize the discharge of algae in the effluent. As long as there are adequate nutrients available and light energy, algae will grow in the lagoons. When the algae die, they will slowly settle to the bottom of the lagoons.

The overloading of wastewater lagoons has resulted in the development of aerated lagoons. Oxygen has been added to lagoons through plastic hoses and plastic pipe, as well as, by floating surface aerators. Experience has shown that the use of a 24-hr aerated cell with sufficient mixing and oxygen transfer will stabilize the incoming waste materials. The bacteria produced in the aerated cell will have to be removed by settling in subsequent lagoon cells. The growth of algae in these cells will determine the effluent quality.

VII. PATHOGENIC MICROORGANISMS

One of the primary purposes of municipal wastewater treatment is the removal of enteric pathogenic microorganisms from wastewater before the water is discharged back into the environment. Biological wastewater-treatment plants are able to reduce the number of all types of enteric pathogens. The competition for nutrients favors the nonpathogenic microorganisms over the pathogens. Predation is responsible for the destruction of a large number of pathogens. The adsorption of pathogens onto suspended solids results in most of the remaining pathogens being concentrated in the sludge. Sludge treatment can effectively reduce the adsorbed pathogens before the residual sludge is returned to the land. Long-retention-time digestion systems will result in pathogen die-off. The additional storage of sludge will produce even greater pathogen reduction. Increasing the temperature of the sludge to the thermophilic range during digestion, or afterward, will kill pathogens. The treated effluent contains sufficient indicators of possible pathogens to warrant disinfection by chlorination–dechlorination or ultraviolet light. Chlorination has been used for wastewater disinfection for about 85 years. The U.S. EPA has required greater chlorination, followed by dechlorination, to insure the destruction of the pathogens and the protection of the receiving stream.

Ultraviolet light is the newest form of disinfection. The popularity of UV light stems from the fact that pathogens are killed without the use of chemicals in the process. Ultraviolet light works best with low-concentration suspended-solids effluents. For most biological wastewater-treatment systems, close operational control is required to produce a low-concentration suspended-solids effluent. Often, a high-rate sand filter is required after the secondary sedimentation tanks to produce the low-concentration suspended-solids liquid required for good ultraviolet disinfection.

Ozone has also been used as a disinfectant in pure-oxygen plants. The efficiency of ozone generation is much higher with pure oxygen than with air.

Although municipal wastewater may contain enteric pathogens, personnel in wastewater treatment plants have not suffered from these pathogens. A healthy respect for good sanitation and a clean environment have helped prevent personnel in wastewater treatment plants from becoming infected with enteric pathogens. The spread of enteric pathogens has been greatly reduced in the United States and Europe with the increased application of municipal wastewater-treatment systems. Cholera and typhoid fever are still endemic in those parts of the world where municipal wastewater treatment has not been installed. Enteroviruses and the cysts of parasitic protozoa can also be found in improperly treated domestic wastewater. As the world's population increases and the potential for spreading enteric diseases rises, biological wastewater treatment will have an even greater role to play.

VIII. REGULATORY CONTROLS

Every country has ultimately recognized that regulatory control is required to insure the proper design, construction, and operation of municipal wastewater-treatment plants. Even the most advanced nations have demonstrated the inability of local governments to accept the primary responsibility for properly treating their own wastewater. The British were among the first to recognize the need for regulatory control to prevent stream pollution. They passed legislation prohibiting the discharge of municipal wastewater into receiving streams without adequate treatment. Unfortunately, the recognition of a problem is only one step in its solution. The ultimate solution for municipal wastewater-pollution problems lies in the development of a partnership between the municipalities and the regulatory agency. The municipalities must build and operate the wastewater-treatment plants to obtain the desired results. The regulatory agency's job is to examine the regional requirements for water quality and coordinate the municipalities in the region to achieve the desired results. The regulatory agency should also be prepared to offer technical assistance when problems arise. When government operates properly, the regulatory agency is simply the facilitator to help all water users arrive at reasonable criteria to protect the water resources. Everyone involved understands the problems and the solutions, making it easy to achieve the desired results at all levels. When government operates improperly, special interest groups persuade the legislature to pass unreasonable legislation requiring the regulatory agency to set water-quality standards that favor one group over the other groups involved. When the local municipalities are forced to install more complex treatment systems than necessary, problems are generated and economic resources are wasted. Education shows people how to do things correctly; but education cannot insure that the correct thing will be done. People ultimately decide what they will do and why they will do it. Regulatory agencies will always have a role to play in controlling water pollution, but society ultimately determines the role for both the regulatory agencies and the municipal wastewater-treatment plants.

IX. FUTURE BIOLOGICAL TREATMENT SYSTEMS

When faced with challenges, people tend to find real solutions and move forward. This has been true since the beginning of time. Progress seems to have been more rapid in recent years. Unfortunately, the volume of new information should not be used as a measure of the amount of progress that has been made. People have always made progress one step at a time. There are people who clamor for zero discharge of contaminants into the environment. If people could turn off their own personal production of waste materials, the concept of zero contamination might be realistic. Until that time, waste treatment will be a necessity for society.

Engineered biological wastewater-treatment systems have evolved over the past 100 years from crude intermittent sand filters to complex mixtures of aerobic and anaerobic biological treatment units. The chemical characteristics and the quantities of wastewater determine the optimum wastewater-treatment systems. Indications point to the development of simple photosynthetic systems that use bacteria and algae together in high-rate systems to remove the contaminants and produce organic fertilizers that can be used in agriculture to meet the food-production needs for future generations. As the need for water increases, biological treatment will be combined with chemical membranes to reduce the salts in the treated effluent, allowing the water to be used over again many times. The key to understanding biological wastewater-treatment systems has always been the application of basic concepts of microbiology and microbial biochemistry with engineering to achieve the desired results. From an engineering point of view, simplicity of design and operation will insure the success of future municipal wastewater-treatment systems. The important aspect of biological wastewater treatment is its use of a natural system that is self-generating while converting the wastewater pollutants to products that are easily assimilated into the natural environment for reuse.

See Also the Following Articles

Alkaline Environments • Autotrophic CO_2 Metabolism • Enteropathogenic Bacteria • Heterotrophic Microorganisms • Protozoan Predation

Bibliography

Battley, E. H. (1987). "Energetics of Microbial Growth." Wiley-Interscience, New York.

Hunter, G. L., O'Brien, W. J., Hulsey, R. A., Carns, K. E., and Ehrhard, R. (1998). *Water Environ. Technol.* **10**, 40–44.

Ingraham, J. L., Maaloe, O., and Neidhardt, F. C. (1983). "Growth of the Bacterial Cell." Sinauer, Sunderland, MA.

Speece, R. E. (1988). *Water Res.* **22**, 365–372.

Water-Deficient Environments

Bettina Kempf and Erhard Bremer
Philipps University Marburg

I. Flexible Adaptation of Microorganisms to Low-Water Activities
II. Microbial Strategies to Cope with High-Osmolality Environments
III. Characteristics of Compatible Solutes
IV. Biosynthesis of Compatible Solutes
V. Transport of Compatible Solutes
VI. Efflux of Compatible Solutes
VII. Future Prospects

GLOSSARY

compatible solute An organic solute that provides osmotic balance without interfering with the physiological activities of the cell.

osmolality The mole fraction of osmotically active particles of solute per kilogram of water. The osmolality of a particular solute depends on the degree of its dissociation in water.

osmoprotectant A compound that, when provided exogenously, stimulates bacterial growth in media of high osmotic strength. Osmoprotectants are taken up by the cells into the cytoplasm where they act as, or are converted to, compatible solutes.

osmosis The net movement of water across a partially permeable membrane from a region of lower solute concentration to a region of higher solute concentration.

semipermeable Allowing certain molecules (e.g., water) to cross membranes freely while blocking the passage of other molecules (e.g., ions).

turgor The pressure inside a cell resting on the cytoplasmic membrane.

water activity (a_w) An index of the amount of water that is free to react. It is expressed as a_w, which represents the ratio of the vapor pressure of the air in equilibrium with a solution (or a substance) to the vapor pressure of pure water at the same temperature. Pure distilled water has $a_w = 1$.

WATER is essential for the survival and functioning of both prokaryotic and eukaryotic cells. Its critical role for life on Earth depends on its function as a solvent, to which the structure of proteins, nucleic acids, and cell components have been evolutionarily optimized. It is also the foundation for the biochemistry of the cell, either by serving as a direct partner in chemical reactions or by providing the appropriate milieu for biochemical transformations.

Given the remarkable physiological and metabolic diversity of microorganisms, it is not surprising that in the course of evolution they effectively adapted not only to environments with an ample supply of water, but also to habitats subjected to either frequent fluctuations in their water content or permanent water deficits. Examples of such habitats are freshwaters, soils, and salt brines. The concentration of osmotically active compounds inside a bacterial cell generally exceeds that of its aqueous surroundings. Because the cytoplasmic membrane is semipermeable, changes in the external salinity or osmolality will immediately trigger fluxes of water along the osmotic gradient. Consequently, the water content of the cytoplasm must be sensitively adjusted throughout the entire cell cycle.

Water-deficient environments pose a considerable challenge to prokaryotic cells because the osmotically instigated fluxes of water result in a dehydration of the cytoplasm and finally in a collapse of turgor, an outward-directed hydrostatic pressure. To survive and grow in such environments, cells must use active countermeasures to retain a suitable level of cytoplasmic water. Because microorganisms do not possess active transport mechanisms for water, turgor is adjusted by controlling the pool of osmotically active solutes in their cytoplasm. Bacterial cells accomplish this either through the synthesis of organic osmolytes or through the uptake of ions and preformed organic, osmotically active solutes from the environment. Here, we focus on the physiological and molecular mechanisms that allow bacterial cells to survive osmotic stress and thrive in habitats with a low water content.

I. FLEXIBLE ADAPTATION OF MICROORGANISMS TO LOW-WATER ACTIVITIES

An important step in the development of self-reproducing cells was the development of the cell membrane, which formed a closed compartment in which biochemical transformations and the copying of the genetic material could take place. Because solutes and cell components are concentrated in this compartment, water is drawn across the semipermeable lipid bilayer into the cell from the more dilute environment. The resulting buildup of turgor presses the cytoplasmic membrane against the elastic cell wall, whose mechanical stability allows the bacteria to withstand a remarkable level of strain. Although turgor is quite difficult to quantify, values of $3–10 \times 10^5$ Pa (3–10 bar) for gram-negative bacteria and approximately 20×10^5 Pa (20 bar) for gram-positive microorganisms have been estimated. In the case of the gram-positive soil bacterium *Bacillus subtilis*, this is equivalent to 10 times the pressure present in a standard car tire. The surface-stress theory, originally advanced by A. Koch, proposes that the bacteria use turgor to stretch the cell wall, thereby permitting the elongation of the peptidoglycan chains for the purpose of cell expansion and finally cell division. Therefore, the maintenance of an outward-directed pressure by the cell within physiologically acceptable boundaries is a key determinant for the proliferation of microorganisms.

Water permeation across the cytoplasmic membrane proceeds by two distinct pathways.

1. Simple diffusion through the lipid bilayer is characterized by a high Arrhenius activation energy ($E_a > 10$ kcal/mol), which indicates that water movement is most effective at higher temperatures when the lipid mobilities are increased.
2. Channel-mediated water transport exhibits a low Arrhenius activation energy ($E_a < 5$ kcal/mol) and accounts for the more rapid transmembrane water movement; this process can frequently be reversibly inhibited by mercuric chloride. The channels are water selective and do not allow the passage of ions or metabolites. They are therefore called aquaporins. The aquaporins were first discovered in eukaryotic tissues characterized by a high water permeability. Their identification in microorganisms suggests that these specific water channels participate in turgor control in prokaryotes as well.

The water requirements of microorganisms are generally described in terms of water activity, a_w, an index of the amount of water that is free to react and thus is available for the microbial cell. This parameter is defined as the ratio of vapor pressure of the air in equilibrium with a solution (or a substance) to the vapor pressure of pure water at the same temperature, $a_w = p/p_0$, where p is the vapor pressure of the solution and p_0 is the vapor pressure of the solvent (usually water). Pure water has an a_w of 1.0, a 22% NaCl solution (w/v) has an a_w of 0.86, and a saturated solution of NaCl has an a_w of 0.75. Microbial growth is possible in the range of water activity between 0.998 and 0.6. Bacteria require higher values of a_w for growth than do fungi, and, in general, gram-negative microorganisms have higher requirements than gram-positive microorganisms. The lowest reported a_w value permitting the growth of bacteria is approximately 0.75 for true halophiles (salt-loving bacteria), whereas xerophilic (dry-loving) molds such as *Xeromyces bisporus* and osmotolerant yeasts such as *Saccharomyces rouxii* have been reported to grow at a_w values of approximately 0.6. The a_w of most fresh foods is above 0.99, and the demands of microorganisms for a certain humidity have long been exploited by humans for the conservation of food by desiccation and its preservation in the presence of high concentrations of salt or sugars. The general effect of lowering a_w below the optimum required for the efficient proliferation of a particular microorganism is to increase the length of the lag phase and to decrease the growth rate and growth yield. Habitats with a low water content not only result from desiccation on solid surfaces and from increases in the external solute concentration causing cellular dehydration via osmosis, but also from the removal of free water by freezing.

Despite the fact that low a_w values considerably restrict bacterial growth, microorganisms can colonize a wide variety of water-deficient environments, such as salt lakes and brines, saline soils, arctic salt-water sources, salted fish, candied fruits, and the phylloplane (leaf surface) of salt-excreting plants. Even the unusual and high-saline environment of the nasal cavities of desert iguanas has been colonized by a highly halotolerant *Bacillus* species. A commonly used classification scheme that was originally introduced by D. J. Kushner describes the salt tolerance and salt requirements of microorganisms. Nonhalophilic microorganisms can grow in up to 1.0 M salt, moderately halophilic bacteria grow at salt levels between 0.4 and 3.5 M, and extremely halophilic bacteria grow in media with salt concentrations between 2.5 and 5.2 M. The use of the term halophile is restricted to those microorganisms that actually require high salt concentrations for their growth. In contrast, organisms capable of growing over a range of salt concentrations, but with their growth rate optimal in the absence of salt, are referred to as halotolerant. The unusual group of microorganisms capable of growing over a very broad range of salt concentrations (from 0 M to saturated NaCl solutions) with their growth-rate optima in the presence of salt have been termed haloversatile. We note that the borderlines between the various categories overlap to a certain degree and that the salt tolerance of a given microorganism can vary widely depending on environmental conditions, such as temperature, the presence of oxygen, and the supply of nutrients.

II. MICROBIAL STRATEGIES TO COPE WITH HIGH-OSMOLALITY ENVIRONMENTS

The inability of microorganisms to actively pump water from the environment into the cell requires an active control over the intracellular solute pool to properly adjust turgor. To cope with dry and high-osmolality habitats, prokaryotic cells employ two very different schemes of adaptation that are frequently referred to as the salt-in and salt-out strategies.

1. Extremely halophilic *Archaea* and *Bacteria,* whose entire physiology has been adapted to a permanent life in high-osmolality environments, accumulate large amounts of ions in their cytoplasm (salt-in).
2. Microorganisms that are periodically subjected to conditions of low-water activity avoid high-ionic conditions in their cytoplasm (salt-out) and instead amass a large amount of a defined group of organic osmolytes. Because these osmoprotective compounds are highly congruous with the entire cellular physiology even when accumulated to molar concentrations, they are frequently referred to as "compatible solutes."

A. Truly Halophilic Organisms

A number of halophilic *Archaea* and *Bacteria* exploit hypersaline environments with salt concentrations ranging between 2 M and 5 M as their preferred habitats. In such hypersaline habitats, K^+, Mg^{2+}, and Ca^{2+} are generally the dominant cations, and Cl^-, SO_4^{2-}, and CO_3^{2-} serve as the major anions. Truly halophilic organisms such as *Halobacterium salinarum* and *H. halobium* balance their internal osmolality with the external osmolality by amassing a large quantity of ions, mainly K^+ and Cl^-, in their cytoplasms and usually actively extrude Na^+. The preference of K^+ over Na^+ might be connected with the observation that less water is needed in the hydration of a K^+ ion than of a Na^+ ion, and therefore a larger proportion of water is left free in the cytoplasm for biological purposes.

High concentrations of inorganic ions have severely negative effects on the structure and functioning of polypeptides and cell components. They induce the aggregation of macromolecules by enhancing hydrophobic interactions; induce charge shielding that interferes with electrostatic attraction and repulsion; and induce salt-ion hydration that restricts the availability of water for cellular processes. Microorganisms that use the salt-in osmoadaptation scheme had to evolutionarily adjust their entire cellular physiology such that molar concentrations of intracellular ions do not interfere with the normal functioning of the cell. A prominent structural modification observed in polypeptides from halophilic microorganisms is the incorporation of additional acidic (glutamic and aspartic acid) residues into proteins, which permits the formation of strong hydration shells, the organization of a hydrated salt-ion network, and the formation of additional salt bridges, thus providing further structural rigidity to polypeptides. Halophilic bacteria often lyse in solutions of low ionic strength and their proteins denature. Hence, their salt-in scheme of osmoadaptation limits their habitats to rather high-osmolality and high-ionic conditions and precludes the colonization of environments with more moderate salinities.

B. Moderately Halophilic and Halotolerant Organisms

In contrast to truly halophilic microorganisms, moderately halophilic and halotolerant bacteria do not grow optimally under high-ionic and high-osmolality conditions. However, these organisms can effectively withstand temporary increases in the osmolality of their habitats, and a number of them can grow in environments exhibiting a broad range of water activities. In general, this group of microorganisms avoids the high-ionic cytoplasm characteristic of the true halophiles. Instead, they actively amass via synthesis or take up large amounts of organic osmolytes that are highly compatible with cellular functions. This strategy allows the cells to evade the detrimental effects of inorganic ions and obviates the need to evolutionarily adjust their entire cellular physiology to low-water activities. It is not surprising that this versatile and flexible stress response is widespread among *Bacteria* and *Archaea.*

III. CHARACTERISTICS OF COMPATIBLE SOLUTES

Compatible solutes are operationally defined as compounds that do not disturb the functioning of the cell. In general, compatible solutes are polar, highly soluble molecules, and they usually do not carry a net charge at physiological pH. They serve a dual role in osmoregulating cells. First, because they are frequently accumulated by bacteria up to molar concentrations, compatible solutes lower the cytosolic osmotic potential, and, hence, they make major contributions to the restoration and maintenance of turgor under conditions of low-water activity. The free cytoplasmic water (unbound water, in contrast to water bound by macromolecules) is a key determinant for cell growth, and the high-level accumulation of compatible solutes thus increases the free-water content of the cytoplasm and hence its volume. Second, compatible solutes serve as stabilizers of proteins and cellular components against the denaturing effects of high ionic strength. This protective property is not fully understood, but is generally explained in terms of the "preferential exclusion" model.

Water present in the immediate vicinity of polypeptides is structurally different (more dense) than that located a bit further away from surfaces of biomolecules. Compatible solutes are strong water-structure formers. This biophysical property allows them to avoid the dense water fraction around polypeptides and to assemble in less dense water areas. As a consequence, compatible solutes are excluded from the immediate hydration shell of polypeptides, resulting in a preferential hydration of protein surfaces. This solvent distribution leads to a situation in which the disruption of water structure in the hydration shell of proteins by local or global unfolding of the polypeptide chain is energetically unfavorable, and hence the native conformations of proteins are stabilized. Remarkably, the accumulation of compatible solutes as an adaptive strategy to high osmolalities has been adopted not only in the microbial world but also by plant, animal, and even human cells. Furthermore, the types of compounds that serve as compatible solutes are the same across the kingdoms, reflecting fundamental constraints on the type of solutes that are congruous with macromolecular and cellular functions.

High-performance liquid chromatography (HPLC) and natural abundance ^{13}C-nuclear magnetic resonance (NMR) spectroscopy procedures have been the major tools used to assess compatible-solute production and accumulation in bacteria. The spectrum of compatible solutes found in microorganisms comprises only a limited number of compounds: sugars (e.g., trehalose and 2-sulfotrehalose), polyols (e.g., glycerol and glucosylglycerol), free amino acids (e.g., proline and glutamate) and derivatives thereof (e.g., proline betaine and ectoine), quaternary amines and their sulfonium analogs (e.g., glycine betaine, carnitine, and dimethylsulfonopropionate), sulfate esters

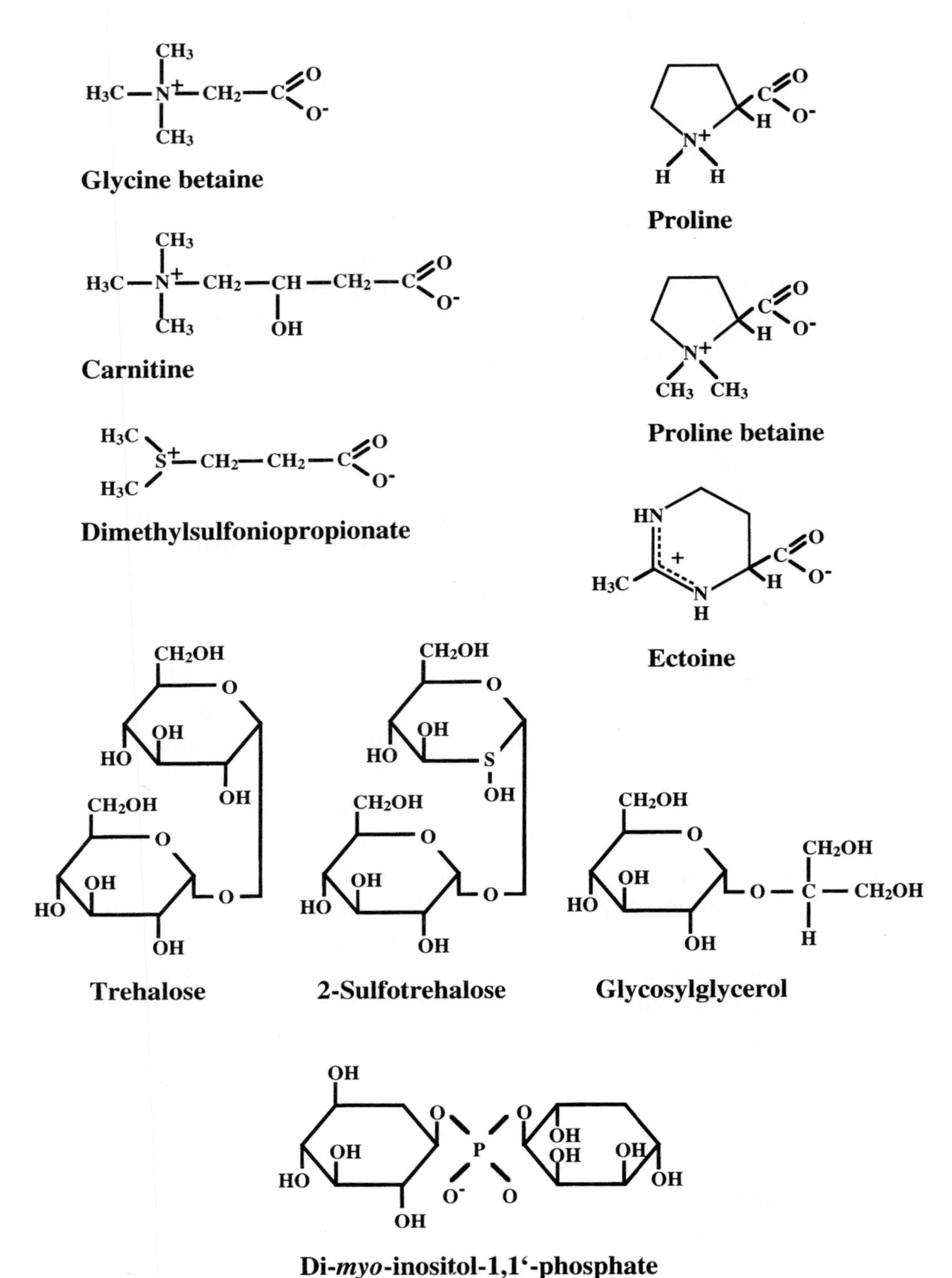

Fig. 1. Osmoprotectants in eubacteria and halophilic Archaea.

(e.g., choline-*O*-sulfate), *N*-acetylated diamino acids, and small peptides (e.g., *N*δ-acetylornithine and *N*-acetylglutaminylglutamine amide) (Fig. 1). A survey of compatible solutes synthesized in a wide spectrum of *Bacteria* and *Archaea* has demonstrated that proline, ectoine, and glycine betaine are frequently used as osmoprotectants. However, in the 1990s, more organisms (in particular, methanogens, thermophiles, and hyperthermophiles) have been examined for their compatible-solute content, and several new osmoprotectants (e.g., di-*myo*-inositol-1, 1′-phosphate; Fig. 1) have been identified. A given bacterium usually employs a spectrum of compatible solutes for osmoregulatory purposes, and the composition of its compatible-solute pool can vary in response to its growth phase and growth medium. For instance, cells of *Halomonas israelensis* contain trehalose as the predominant organic osmolyte when they are grown in media with less then 0.6 M salt, but ectoine is the major solute when the osmolality of the growth medium is raised.

The accumulation of compatible solutes not only allows microbial cells to withstand a given osmolality, but it also extends their ability to colonize habitats with low-water content that are otherwise strongly inhibitory for their proliferation. Depending on the type, compatible solutes can also protect microorganisms against stresses other than dehydration. An example is the increased cold tolerance conferred on *Listeria monocytogenes* by the accumulation of the compatible solutes glycine betaine and carnitine from food sources. Thus, the accumulation of compatible solutes has direct consequences for the safety of food. The finding that compatible solutes exhibit a general stabilizing effect on macromolecules by preventing their unfolding under unfavorable conditions (e.g., heating, freezing, and drying) has intensified the biotechnological interest in this class of compounds and has fostered the search for microbial producers of new compatible solutes and of compatible solutes that are very difficult to synthesize by classic chemical procedures, such as the tetrahydropyrimidine ectoine (Fig. 1). Large-scale biotechnological production of ectoine is being carried out by high-cell-density fermentation of *H. elongata* in high-osmolality media and the "milking" of this osmoprotectant from the microbial producer by severe osmotic downshock.

IV. BIOSYNTHESIS OF COMPATIBLE SOLUTES

Although microorganisms are known to produce a considerable variety of compatible solutes, the complete biosynthetic pathways have been elucidated for only a few of them. Examples are the cyclic amino acid derivative ectoine, the trimethylammonium compound glycine betaine, and the sugar trehalose.

Ectoine (Fig. 1) was originally discovered as a compatible solute produced by the extremely halophilic phototrophic eubacterium *Ectothiorhodospira halochloris*. Ectoine is a highly effective compatible solute, and its production has now been detected in a wide variety of microorganisms under high-osmolality growth conditions. Among the known ectoine producers are anoxygenic phototrophic bacteria (e.g., *Rhodospirillum salinarum*), aerobic chemoheterotrophic proteobacteria (e.g., *H. elongata*), and aerobic chemoheterotrophic gram-positive microorganisms (e.g., *Brevibacterium lines, Streptomyces parvulus, B. pasteurii,* and *Marinococcus halophilus*). In addition, the ability to recover preformed ectoine from exogenous sources is widespread in nature; effective ectoine-transport systems have been detected in the gram-negative enterobacterium *Escherichia coli* and the gram-positive soil bacterium *Corynebacterium glutamicum.* The pathway for ectoine biosynthesis was elucidated in the moderately halophilic eubacteria *M. halophilus* and *H. elongata* (Fig. 2). L-Aspartate β-semialdehyde serves as the precursor for ectoine production, and the consecutive action of three biosynthetic enzymes is required. Their structural genes are genetically organized into an operon (*ectABC*) whose expression is stimulated in response to increases in medium osmolality. Some microorganisms can modify ectoine by hydroxylation, and in low-water environments they accumulate a mixture of ectoine and hydroxyectoine. Other species, such as the *Marinococcus* isolate M52, convert ectoine entirely into hydroxyectoine when the cells enter stationary phase.

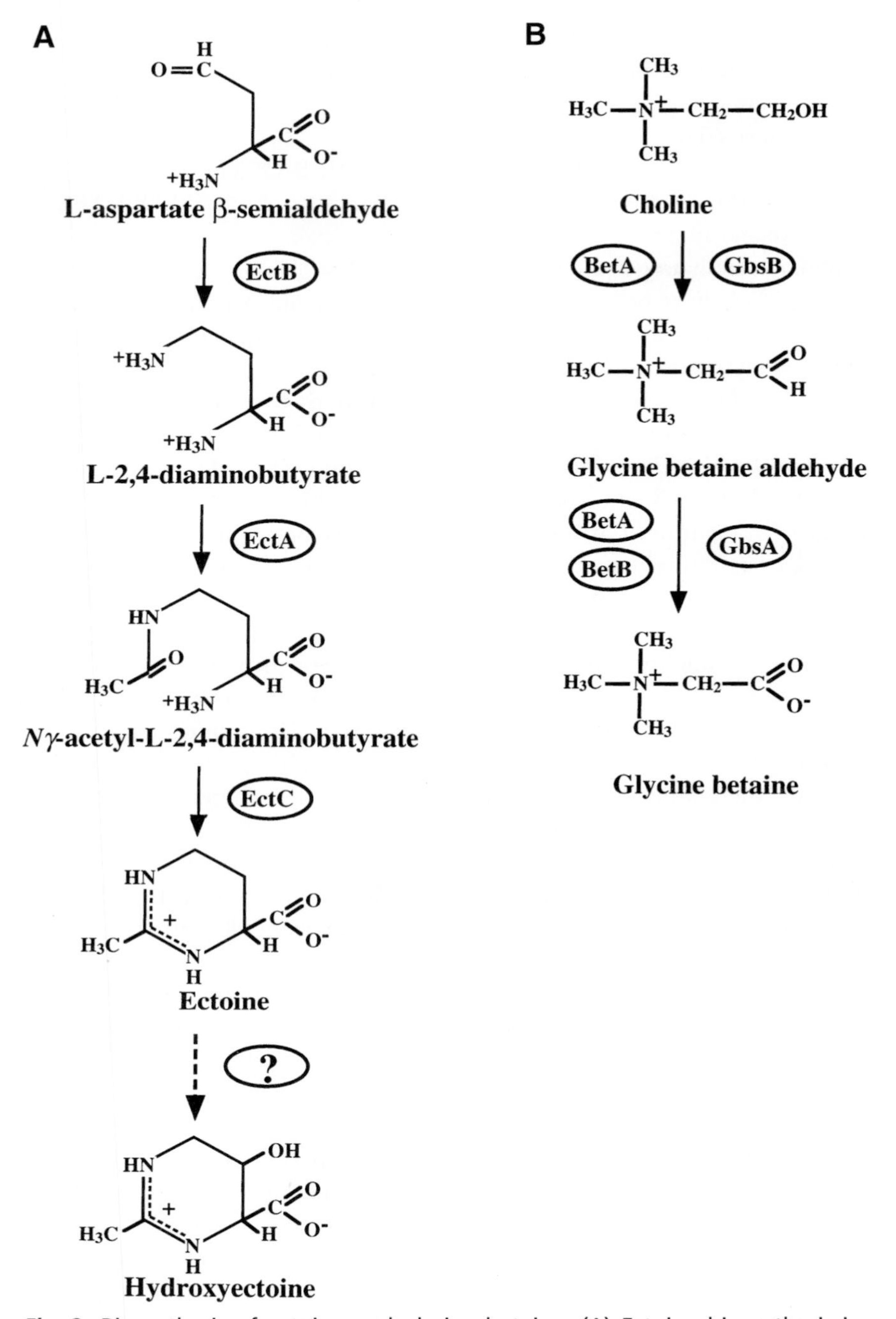

Fig. 2. Biosynthesis of ectoine and glycine betaine. (A) Ectoine biosynthesis in *M. halophilus* and *H. elongata.* EctB, L-2,4-diaminobutyrate aminotransferase; EctA, L-2,4-diaminobutyrate acetyltransferase; EctC, *N*-acetyldiaminobutyrate dehydratase (ectoine synthetase). (B) Choline-to-glycine betaine biosynthetic pathway in *E. coli* (left side) and *B. subtilis* (right side). BetA, choline dehydrogenase; BetB, glycine betaine aldehyde dehydrogenase; GbsA, glycine betaine aldehyde dehydrogenase; GbsB, alcohol dehydrogenase (type III).

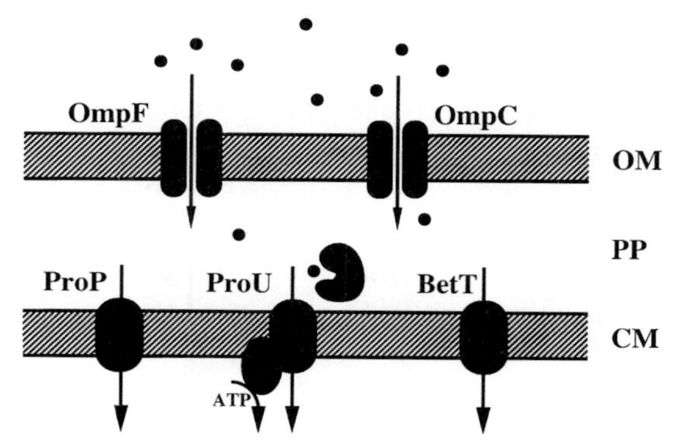

Fig. 3. Uptake systems for osmoprotectants in *E. coli*. The ProP and ProU transporters serve for the uptake of a variety of osmoprotectants, whereas BetT is a specific transport system for choline, the biosynthetic precursor for glycine betaine. OM, PP, and CM denote outer membrane, periplasm, and cytoplasmic membrane, respectively. The dots represent compatible solutes or their biosynthetic precursors.

The most widespread osmoprotectant used by prokaryotic and eukaryotic cells is glycine betaine (Fig. 1). Two routes for its production have been detected in the microbial world. Some microorganisms have the ability to synthesize it *de novo* by a stepwise methylation of the amino acid glycine, involving sarcosine and dimethylglycine as the intermediates and *S*-adenosyl methionine as the methyl donor. However, the molecular and biochemical details of this process have not been fully elucidated. Among the producers of glycine betaine *de novo* are the halophilic methanogen *Methanohalophilus portucalensis*, the extremely haloalcalophilic sulfur bacterium *E. halochloris*, the salt-tolerant cyanobacterium *Aphanothece halophytica*, and the moderately halophilic actinomycete *Actinopolyspora halophila*.

Glycine betaine production via the second biosynthetic pathway, the enzymatic oxidation of the precursor choline to glycine betaine via the intermediate glycine betaine aldehyde, has been studied in considerable detail in a number of organisms, and the complete osmoregulatory choline-to-glycine betaine pathway has been elucidated for the gram-negative and gram-positive model organisms, *E. coli* and *B. subtilis* (Fig. 2). *S. typhimurium* lacks the ability to synthesize glycine betaine from the precursor choline.

The precursor for glycine betaine production is usually acquired by the cells via uptake from exogenous sources because most microorganisms do not synthesize choline. Although the overall reaction pathway of glycine betaine synthesis from choline appears to be conserved in many organisms, there is considerable variation with respect to the number and types of choline transporters and enzymes involved. A single-component choline transporter (BetT) is found in *E. coli* (Fig. 3), whereas two multicomponent, binding-protein-dependent ATP binding cassette (ABC) transporters (OpuB and OpuC) are present in *B. subtilis* (Fig. 4). Each of these transporters is well suited for its physiological task because BetT, OpuB, and OpuC exhibit a high uptake velocity and recognize choline effectively, with K_m values in the low micromolar range. In both organisms, an evolutionarily well-conserved and highly salt-tolerant glycine betaine aldehyde dehydrogenase (BetB and GbsA) is involved in the synthesis path-

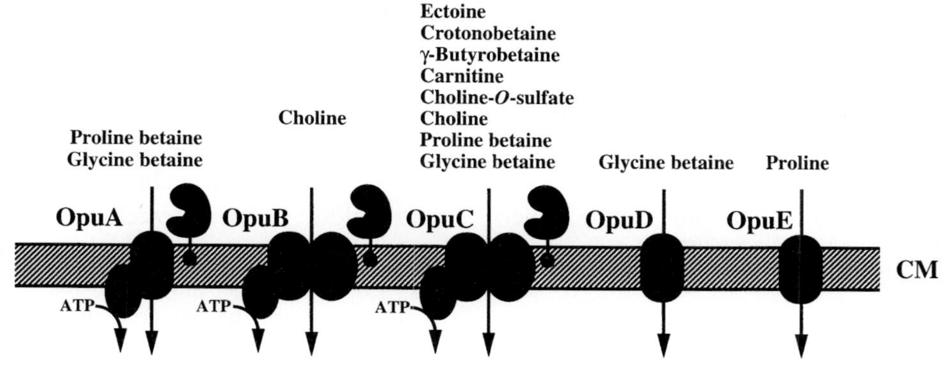

Fig. 4. Systems for osmoprotectant transport in *B. subtilis*. CM, cytoplasmic membrane.

way; however, the first step in glycine betaine synthesis is performed by two distinct types of enzymes in *E. coli* and *B. subtilis*. A soluble, metal-containing, type III alcohol dehydrogenase (GbsB) functions in *B. subtilis* to convert choline into glycine betaine aldehyde, whereas this reaction is catalyzed in *E. coli* by a flavin adenine dinucleotide (FAD)-containing, membrane-bound choline dehydrogenase (BetA), which can also oxidize glycine betaine aldehyde to glycine betaine at the same rate (Fig. 2). *B. subtilis* and *E. coli* synthesize glycine betaine as a metabolically inert stress compound, but in *Sinorhizobium meliloti* glycine betaine and choline have both osmoregulatory and nutritional roles. Additional genetic and cellular control mechanisms are therefore required to curb the consumption of glycine betaine under conditions of high osmolality stress.

The disaccharide trehalose (Fig. 1) is an important stress compound in both prokaryotic and eukaryotic organisms. *E. coli* and *Salmonella typhimurium* accumulate it via synthesis *de novo* as their predominant endogenous compatible solute. Two enzymes encoded by the *otsBA* operon determine this osmoregulatory trehalose synthesis. OtsA, the trehalose-6-phosphate synthase, catalyzes the enzymatic condensation of the precursors glucose-6-phosphate and uridine diphosphate (UDP)-glucose; free trehalose is then generated from this intermediate by the *otsB*-encoded trehalose-6-phosphate phosphatase. Osmotic stress induces the expression of the *otsBA* operon, which is entirely dependent on RpoS, an alternative transcription factor controlling or contributing to gene expression under a variety of stress conditions (e.g., osmotic stress) and in stationary phase. Hence, elevated levels of trehalose are also found in stationary-phase cells of *E. coli* not subjected to elevated osmolality, reflecting the function of this sugar as a general stress protectant.

In contrast to *E. coli* and *S. typhimurium, B. subtilis* does not produce trehalose under high-osmolality conditions and instead synthesizes large amounts of proline as its endogenous compatible solute. The formation of a high level of proline for osmoprotective purposes has been detected in various microorganisms (e.g., *Streptomyces griseus* and *Planococcus citreus*) and also occurs widely in plants. In contrast to glycine betaine and ectoine, which are accumulated by most microorganisms only as metabolically inert stress compounds, proline is also required for the production of proteins, and frequently serves as nitrogen and carbon source. It is generally assumed, but has not been proven, that the biosynthetic pathway for proline as an osmoprotectant in microorganisms is the same as that for proteogenic proline, which usually is synthesized from glutamate. These multiple roles of the amino acid proline in cellular physiology require dedicated regulatory circuits to avoid a wasteful futile cycle of energy-costly proline biosynthesis and degradation in high-osmolality stressed cells.

V. TRANSPORT OF COMPATIBLE SOLUTES

A. Features of Transporters for Osmoprotectants

In addition to accumulating compatible solutes by endogenous synthesis, a wide variety of *Bacteria* and *Archaea* have developed the ability to acquire preformed osmoprotectants from exogenous sources. These compounds are released into ecosystems by primary microbial producers through dilution stress; by decaying microbial, plant, and animal cells; and by mammals in their excretion fluids (e.g., urine). In general, the accumulation of compatible solutes from exogenous sources strongly decreases, at least over a certain range of osmolalities, the synthesis of the endogenously produced osmoprotectants. This observation implies a dedicated control over the intracellular pool of compatible solutes and attests to the ability of microorganisms to effectively coordinate the synthesis and uptake of osmoprotectants and to integrate both processes into a finely tuned homeostatic cellular system. In this way, precious energy sources are preserved, which would otherwise be wasted by the unnecessary buildup of the intracellular compatible solute pool via synthesis, and environmental resources are effectively used. We note that the standard rich media used in the laboratory frequently contain substantial amounts of compatible solutes (e.g., glycine betaine and trehalose in yeast extract and hydroxyproline in tryptone), which are accumulated by the cells after an increase in osmolality in the medium and, hence, influence the pattern of osmotically controlled gene expression in bacteria.

Transporters for osmoprotectants have evolved to meet the special demands imposed by their physiological tasks. In natural ecosystems, the supply of compatible solutes and their biosynthetic precursors is varying and generally very low; they usually occur in concentrations in the nanomolar to micromolar range. Therefore, osmoprotectant transporters frequently exhibit a very high affinity for their substrates (K_m values in the low micromolar range), and their capacity (V_{max}) is geared to permit the intracellular accumulation of compatible solutes to molar concentrations. In addition, they also function most effectively at high osmolality and at high ionic strength, conditions that otherwise inhibit uptake systems for nutrients. Furthermore, their level of activity and the expression of the structural genes for osmoprotectant uptake systems are frequently stimulated in response to increases in the external osmolality, permitting the adjustment of compatible solute transport to the actual demand for these stress compounds. Microorganisms frequently possess several transport systems for osmoprotectants, many of which often exhibit a broad substrate specificity so that the cell can take maximal advantage of a spectrum of compatible solutes that might be present in their environment. The presence of compatible solutes in natural habitats has far-reaching ecological consequences because these compounds will provide a selective advantage under unfavorable osmotic conditions to those bacteria that can effectively scavenge them from exogenous sources.

B. Molecular Analysis of Osmoprotectant-Uptake Systems

E. coli (along with its closely related cousin *S. typhimurium*) and *B. subtilis* have long served as prototypes for the gram-negative and gram-positive groups of microorganisms. They are amenable to sophisticated genetic, physiological, and biochemical investigative approaches, and their role as model organisms has been reinforced by the determination of the entire genome sequence of *E. coli* and *B. subtilis*. The ease of genetic manipulation of *E. coli*, *S. typhimurium*, and *B. subtilis* has facilitated studies of their stress responses to water-deficient environments, with particular emphasis on the molecular details of the systems for compatible-solute transport.

1. *Escherichia coli* and *Salmonella typhimurium*

A sudden increase in the external osmolality causes a reduction in turgor, which in turn triggers several cellular events that are aimed at restoring the cellular water balance and turgor, and eventually aimed at resuming growth under the new environmental conditions. The net outflow of water from the cell leads to a strongly increased influx of potassium via activation of the Trk K^+ uptake system(s), a transporter with a low affinity but a high transport rate. Under conditions in which K^+ is limiting or when a reduction in turgor persists after an osmotic upshock, the cells transiently induce the high-affinity K^+-uptake system Kdp. Genetic control of the *kdpFABC* operon is mediated by the KdpDE two-component regulatory system. It is thought that the membrane-embedded sensor kinase KdpD perceives a reduction in turgor and relates this information via phosphorylation to the transcriptional activator KdpE. The induction of *kdpFABC* expression occurs only transiently because the influx of K^+ restores turgor, and this in turn is sensed by the KdpD protein. To counterbalance the increase in positive charges, the cell synthesizes large amounts of glutamate; at high external osmolalities, intracellular K^+-glutamate concentrations can reach approximately 0.8 M in *E. coli* and *S. typhimurium*. This primary stress response allows the cells to withstand a sudden osmotic upshock by limiting water loss and the restoration of turgor. However, high concentrations of K^+ cannot be tolerated for prolonged periods; therefore, the cell initiates a series of secondary responses that are aimed at reducing the intracellular K^+ levels by exchanging this inorganic ion with more compatible organic osmolytes. This is accomplished either via the synthesis of trehalose and glycine betaine or the uptake of osmoprotectants from environmental sources.

Two transport systems, ProP and ProU, are responsible for osmoprotectant uptake both in *E. coli* and

S. typhimurium. ProP is a single-component transporter located in the cytoplasmic membrane and is driven by cation symport. In contrast, ProU is a multicomponent system and is a member of the ABC superfamily of transporters. It consists of a periplasmic substrate binding protein (ProX), which recognizes glycine betaine and proline betaine with high affinity and delivers them to the integral inner-membrane component ProW. The ProW-mediated substrate translocation across the cytoplasmic membrane depends on the hydrolysis of ATP by the inner-membrane-associated ATPase ProV (Fig. 3). Two molecules of ATP are hydrolyzed per molecule of substrate transported via ProU, and hence the high-level intracellular accumulation of glycine betaine requires a substantial supply of energy. Sudden osmotic upshocks result in a rapid increase in *proP* and *proU* expression to a level that is proportionally linked to the osmolality of the growth medium. Expression is kept at elevated levels for as long as the osmotic stimulus exists, thus permitting the adjustment of the number of the osmoprotectant-uptake systems to the degree of osmotic stress. Thus, the osmotic regulation of *proP* and *proU* loci differs fundamentally from that of the transient induction exhibited by the *kdpFABC* operon subsequent to osmotic upshifts.

The access of osmoprotectants to the periplasmic space is provided by the OmpC and OmpF porins (Fig. 3). These proteins form nonspecific channels in the outer membrane and allow the passive diffusion of a wide variety of compounds with a molecular mass up to approximately 600 Da. The channel formed by OmpC is of particular importance for compatible solute acquisition because its synthesis is induced in hypertonic environments. In contrast, the synthesis of OmpF predominates over that of OmpC in low-osmolality environments. A two-component regulatory system consisting of a membrane-bound sensor kinase (EnvZ) and a cytoplasmic response regulator (OmpR) serves as a molecular device to detect changes in environmental osmolality and to regulate the expression of the *ompC* and *ompF* genes in a reciprocal fashion in response to this environmental stimulus. EnvZ and OmpR are members of a large family of homologous proteins that are widely employed in the bacterial world to sense and respond to a large array of environmental parameters. EnvZ is responsible for monitoring the environmental osmolality, transducing this information across the cytoplasmic membrane, and relating it by either phosphorylation or dephosphorylation reactions to the soluble transcription factor OmpR. The degree of phosphorylation of this regulatory protein is critical for its DNA interactions with both the *ompC* and *ompF* regulatory regions.

Despite intensive efforts, it has not been possible to unambiguously decipher the molecular and cellular events allowing *E. coli* and *S. typhimurium* to sense changes in the environmental osmolality and to adjust the level of transcription of those genes that are essential for its osmostress response. In particular, it is still uncertain which physiological and biophysical parameters are actually sensed by cells when subjected to sudden osmotic increases or grown for prolonged periods in high-osmolality environments. It is also unknown whether there is a globally acting osmosensor that coordinates the cellular responses to hypertonic conditions. The two-component regulatory system EnvZ and OmpR is clearly not the central osmosensing device because it is not involved in the osmoregulation of the compatible solute-uptake systems ProP and ProU, the synthesis of the osmoprotectant trehalose, or the genetic control of K^+ uptake in *E. coli* and *S. typhimurium*. As outlined, the cell's initial response to a rise in the environmental osmolality is a rapid amassing of K^+-glutamate to cytoplasmic levels suitable for the restoration of turgor. It is thought that the increase in K^+-glutamate serves as a second messenger for the initiation of secondary defense reactions (e.g, the synthesis of trehalose and uptake of osmoprotectants) that eventually allows the cells to adjust effectively to high-osmolality environments and to resume growth under unfavorable conditions.

2. *Bacillus subtilis*

There has been a long-standing focus on osmoadaptation in gram-negative bacteria, but only in the 1990s has the osmostress response in gram-positive bacteria attracted wider attention. The pathogens *Staphylococcus aureus* and *L. monocytogenes*, the soil microorganisms *B. subtilis* and *C. glutamicum*, and certain bacteria used in dairy industry (*Lactobacillus plantarum* and *B. linens*) are currently intensively studied by physiological and genetic approaches.

B. subtilis is a facultative anaerobic endospore-forming rod-shaped bacterium that is wide-spread in nature and belongs to the group of gram-positive bacteria with a low G-C content. It is a ubiquitous inhabitant of the upper layers of the soil, where frequent fluctuations in the availability of water often cause severe alterations in the osmolality of this habitat. It is also exposed to lateral transport into both freshwater and marine environments, thus imposing considerable strains on the water balance of the cell. Sudden osmotic changes trigger a behavioral response (osmotaxis), such that the *B. subtilis* cells are repelled by both high and low osmolality. The key physiological role of compatible-solute accumulation as an important cellular-stress response has been firmly established, and the transport systems for osmoprotectant uptake have been studied in considerable detail at the molecular level. *B. subtilis* responds to a sudden increase in the external osmolality by an initial rapid uptake of K^+, followed by the accumulation of large amounts of the compatible solute proline through synthesis *de novo*. The influx of K^+ is essential for the recovery of turgor, increased proline biosynthesis, and the resumption of growth subsequent to an osmotic challenge. The nature of the counterion for K^+ in *B. subtilis* is unclear because, in contrast to *E. coli,* its glutamate levels increase only slightly after osmotic upshock. The number of transporters found for the acquisition of osmoprotectants is larger in *B. subtilis* (Fig. 4) than in *E. coli* (Fig. 3), and both single-component and multicomponent systems are employed. The growth of *B. subtilis* in high-osmolality environments induces each of these transport systems at the level of transcription, reflecting the increased demand for compatible solutes accumulation under these conditions. The glycine betaine transporter OpuD (Opu denotes osmoprotectant uptake) and the proline transporter OpuE are secondary uptake systems; in contrast, OpuA, OpuB, and OpuC are members of the ABC superfamily of transporters and each possesses an extracellular substrate-binding protein tethered with a lipid modification at its amino-terminal end to the cytoplasmic membrane (Fig. 4). *B. subtilis* can effectively use a wide spectrum of osmoprotectants to proliferate under highly saline conditions, and the OpuC transporter plays a particular important role in scavenging them from environmental sources (Fig. 4). The presence of five high-affinity transporters for the acquisition of osmoprotectants or their biosynthetic precursors in *B. subtilis* attests to the physiological importance of osmoprotectant uptake in this soil bacterium. Likewise, several (BetP, EctP, and ProP) high-affinity transporters for compatible solutes have been characterized at the molecular level in the soil bacterium *C. glutamicum*. There is considerable variation in the number, substrate specificity, and type of osmoprotectant transport systems present in various bacteria, probably reflecting their adaptation to various habitats with different compatible-solute content.

An important difference exists between gram-negative and gram-positive bacteria with respect to the accumulation of compatible solutes in osmotically nonstressed cells. Gram-negative bacteria do not amass these compounds unless they are subjected to hyperosmotic conditions, whereas gram-positive microorganisms tend to accumulate compatible solutes in standard rich and minimal laboratory media. This is likely to reflect the difference in turgor between gram-negative and gram-positive bacteria. Compatible solutes might be accumulated by nonstressed gram-positive cells to assist in maintaining high turgor in preference to ionic osmolytes, which are deleterious at high concentrations.

VI. EFFLUX OF COMPATIBLE SOLUTES

The growth of microorganisms in high-osmolality environments leads to the massive intracellular accumulation of compatible solutes. In their natural ecosystems, bacteria are likely to experience hypoosmotic shocks caused by rain, flooding, and washout into freshwater sources. Such conditions lead to a massive influx of water into the cell, requiring the bacteria to quickly reduce their intracellular solute pool. Under such conditions, the metabolism of compatible solutes can not make a substantial contribution to the required reduction in the intracellular osmoprotectant concentration because the very rapid increase in turgor must be counteracted quickly to avoid cell lysis. For a variety of gram-negative and gram-positive microorganisms, there is increasing experimental evidence for the presence of osmoprotectant efflux systems that operate independently of their transporters. Within seconds of hypotonic shocks, *E. coli, S. typhimurium, L. plantarum, C. glutamicum,* and *L. monocytogenes* exhibit a massive efflux of osmoprotectants, thus implicating stretch-activated channels in the fast release of compatible solutes from the cell. The opening of the mechanosensitive efflux channels is linked to the extent of the osmotic downshock, and in certain microorganisms, these channels also appear to possess a certain degree of substrate specificity. Hence, a finely tuned release of compounds that are preferentially accumulated by the bacterial cells under hyperosmotic conditions is possible, and a new steady-state level in intracellular solute content can be achieved within a very short time. Carrier-like systems also seem to contribute to the discharge of compatible solutes because in some microorganisms (e.g., *L. plantarum*) the initial rapid efflux of compatible solutes is followed by a slow release with kinetic parameters different from those characteristic of channel-like proteins. The release of compatible solutes into the environment is of importance in natural habitats because this process not only supplies osmoprotectants for other microorganisms, but their breakdown also provides additional resources for gaining energy and nutrients. For instance, glycine betaine can be degraded aerobically by *S. meliloti* via sequential demethylation reactions. Microbial degradation of glycine betaine can also occur anaerobically and can proceed in a number of ways—by fermentation, by reduction using external electron donors from certain amino acids, or by oxidation with sulfate or elemental sulfur as electron acceptors. Anaerobic metabolism of glycine betaine occurs in *Clostridium sporogenes, Eubacterium limosum,* and in the sulfur-reducing bacterium *Desulforomonas*.

VII. FUTURE PROSPECTS

The ability to adapt to highly-saline habitats and to fluctuations in environmental osmolality is essential for the survival and growth of many prokaryotic and eukaryotic cells. Understanding the underlying genetic, biochemical, and physiological mechanisms of this adaptive response is important not only as basic scientific knowledge, but also because of its application in agriculture and biotechnology. The common response of plant and microbial cells to high osmolality by synthesizing compatible solutes (e.g., glycine betaine and proline) has fostered interest in using bacterial systems for osmoprotectant synthesis as resources for the metabolic engineering of stress-tolerance in plants. For instance, bacterial genes encoding glycine betaine biosynthetic enzymes, such as the choline oxidase *codA* gene from *Arthrobacter globiformis* or the *betB* glycine betaine aldehyde dehydrogenase gene from *E. coli,* have been successfully expressed in plant cells. Thus, it might be possible in the future to generate desiccation-resistant varieties of commercially important crops such as rice, tomatoes, and potatoes, which are not natural glycine betaine producers.

See Also the Following Articles

Bacillus subtilis, Genetics • *Escherichia coli* and *Salmonella*, Genetics • Extremophiles • Freshwater Microbiology • Soil Microbiology

Bibliography

Blomberg, A. (1997). Osmoresponsive proteins and functional assessment strategies in *Saccharomyces cerevisiae. Electrophoresis* **18**, 1429–1440.

Booth, I. R., and Louis, P. (1999). Managing hypoosmotic stress: Aquaporins and mechanosensitive channels in *Escherichia coli. Curr. Opin. Microbiol.* **2**, 166–169.

Burg, M., Kwon, E., and Kultz, D. (1997). Regulation of gene expression by hypertonicity. *Annu. Rev. Physiol.* **59**, 437–455.

da Costa, M., Santos, H., and Galinski, E. (1998). An overview on the role and diversity of compatible solutes in *Bacteria* and *Archaea. Adv. Biochem. Eng. Biotechnol.* **61**, 117–153.

Csonka, L. N., and Epstein, W. (1996). Osmoregulation. *In* "*Escherichia coli* and *Salmonella.* Cellular and molecular biology" (F. C. Neidhard *et al.*, eds.), pp. 1210–1223. ASM Press, Washington, DC.

Galinski, E. A., and Trüper, H. G. (1994). Microbial behaviour in salt-stressed ecosystems. *FEMS Microbiol. Rev.* **15**, 95–108.

Hecker, M., and Völker, U. (1998). Non-specific, general and multiple stress resistance of growth-restricted *Bacillus su-*

brilis cells by the expression of the σ^B regulon. *Mol. Microbiol.* **29**, 1129–1136.

Hengge-Aronis, R. (1996). Back to log phase: σ^S as a global regulator in the osmotic control of gene expression in *Escherichia coli. Mol. Microbiol.* **21**, 887–893.

Kempf, B., and Bremer, E. (1998). Uptake and synthesis of compatible solutes as microbial stress responses to high osmolality environments. *Arch. Microbiol.* **170**, 319–330.

Koch, A. (1983). The surface stress theory of microbial morphogenesis. *Adv. Microb. Physiol.* **24**, 301–366.

Miller, K. J., and Wood, J. M. (1996). Osmoadaptation by rhizosphere bacteria. *Annu. Rev. Microbiol.* **50**, 101–136.

Nevoigt, E., and Stahl, U. (1997). Osmoregulation and glycerol metabolism in the yeast *Saccharomyces cerevisiae. FEMS Microbiol. Rev.* **21**, 231–241.

Oren, A. (1990). Formation and breakdown of glycine betaine and trimethylamine in hypersaline environments. *Antonie van Leeuwenhoek* **58**, 291–298.

Poolman, B., and Glaasker, E. (1998). Regulation of compatible solute accumulation in bacteria. *Mol. Microbiol.* **29**, 397–407.

Potts, M. (1994). Desiccation tolerance of prokaryotes. *Microbiol. Rev.* **58**, 755–805.

Pratt, L. A., Hsing, W., Gibson, K. E., and Silhavy, T. J. (1996). From acids to *osmZ:* Multiple factors influence synthesis of the OmpF and OmpC porins in *Escherichia coli. Mol Microbiol.* **20**, 911–917.

Record, M. T., Jr., Courtenay, E. S., Cayley, D. S., and Guttman, H. J. (1998). Responses of *E. coli* to osmotic stress: Large changes in amounts of cytoplasmic solutes and water. *Trends Biochem. Sci.* **23**, 143–148.

Rhodes, D., and Hanson, A. D. (1993). Quaternary ammonium and tertiary sulfonium compounds in higher plants. *Annu. Rev. Plant Physiol. Plant Mol. Biol.* **44**, 357–384.

Strøm, A. R., and I. Kaasen. (1993). Trehalose metabolism in *Escherichia coli:* Stress protection and stress regulation of gene expression. *Mol. Microbiol.* **8**, 205–210.

Ventosa, A., Nieto, J. J., and Oren, A. (1998). Biology of moderately halophilic aerobic bacteria. *Microbiol. Mol. Biol. Rev.* **62**, 504–544.

Wood, J. M. (1999). Osmosensing in bacteria: Signals and membrane-based sensors. *Microbiol. Mol. Biol. Rev.* **63**, 230–262.

Water, Drinking

Paul S. Berger, Robert M. Clark, and Donald J. Reasoner

U.S. Environmental Protection Agency

I. Introduction
II. Indicator Organisms in Microbial Monitoring
III. EPA Drinking Water Regulations
IV. Sources, Treatment, and Distribution
V. Alternative Water Sources
VI. Conclusion

GLOSSARY

biofilm A biologically active layer on surfaces exposed to water, consisting of adsorbed inorganic and organic chemicals and microorganisms held together by a matrix of organic polymers produced and excreted by the microorganisms.

disinfection The chemical or physical process used to destroy or otherwise inactivate pathogenic microorganisms

indicator microorganism An organism that is used to monitor the microbial quality of drinking water.

THE PRIMARY OBJECT OF THE MICROBIOLOGY OF DRINKING WATER is to prevent waterborne disease. A drinking-water system can minimize waterborne disease by employing proper treatment and control practices, and by monitoring the effectiveness of these practices. Here, these issues are addressed, and the regulations in the United States that attempt to protect the public from pathogens in drinking water are described.

I. INTRODUCTION

Safe drinking water is one of the oldest public health concerns known. Ancient civilizations practiced water treatment, as evidenced by Egyptian inscriptions and Sanskrit writings. Sand filtration was in use by some cities as early as the sixteenth century. Chlorination of drinking water, so widely used, was not introduced until the first decade of the twentieth century. These two treatment practices dramatically decreased the incidence of waterborne disease, although waterborne disease is still a problem in the United States and elsewhere.

The control of waterborne disease depends on the presence of natural and artificial barriers. Natural barriers may include soil and bedrock that inactivate, remove, or otherwise prevent the passage of fecal pathogens into well water. Systems using surface water may control human activities on the watershed supplying the system, thereby creating another barrier. Artificial barriers include the treatment of the raw source water and programs for protecting the integrity of the underground pipe network that transports the treated water to consumers. A water system must employ at least one, and usually several, of these barriers to control waterborne disease. Regardless of the nature of the barrier system used, it must be adequately monitored to insure its continued effectiveness.

This article discusses how water is treated and monitored to insure its microbiological safety for human consumption. The primary focus is on practices in the United States, although those of other countries will be mentioned. A brief description of the drinking-water regulations for microbiology in the United States is also included.

II. INDICATOR ORGANISMS IN MICROBIAL MONITORING

Ideally, the specific detection of the various pathogenic agents of waterborne disease would be the most direct approach in determining their presence and the need for control. Unfortunately, in actual practice, this is quite impractical. The variety of potential waterborne pathogens—including various species of bacteria, viruses, and protozoa—makes such a search, especially on a routine basis, extremely difficult, time-consuming, and expensive. Moreover, the efficiency of techniques for recovering and detecting known waterborne pathogens in drinking water is often very low.

Although techniques for many bacterial pathogens are available, some are nonselective, thus allowing nontarget organisms to proliferate in numbers that overgrow the pathogen. Viral pathogens are fastidious in their growth requirements and grow only in special tissue cultures that are expensive and often difficult to maintain. Some viruses cannot be cultivated in the laboratory. The recovery and assay methods for pathogenic protozoa are inefficient. Extended delays can be involved in carrying out the specific identification procedures for pathogens, and only certain laboratories with specially trained technologists may have the expertise to carry out some of these procedures. Moreover, if pathogens were present in drinking water, their concentrations would usually be sufficiently small to require the analysis of large-volume samples. Finally, extended delays would usually be involved in carrying out identification procedures, under the extremely doubtful assumption that laboratories involved in water testing would have such resources.

Because of the problems associated with trying to detect specific enteric pathogens, indicator organisms are used as surrogates. Such indicator organisms are used to assess the microbiological quality of drinking water. The ideal drinking-water indicator has the following attributes.

1. It is suitable for all types of drinking water.
2. It is present in sewage and polluted waters at much higher densities than fecal pathogens.
3. Its survival time in water is at least that of waterborne pathogens.
4. It is at least as resistant to disinfection as the waterborne pathogens.
5. It is easily detected by simple, inexpensive laboratory tests in the shortest time with accurate results.
6. It is stable and nonpathogenic.
7. It is generally not present in waters uncontaminated by mammalian feces.

The indicators generally employed worldwide are total coliforms, fecal coliforms, and *Escherichia coli*.

Total coliforms are a group of closely related bacteria in the family Enterobacteriaceae that are usually not pathogenic and are widespread in ambient water. They are usually, but not necessarily, associated with sewage. Total coliforms are not defined in precise taxonomic terms, but instead by whether they can produce β-galactosidase, which hydrolyzes lactose to yield acid and gas. These media vary by country. Total coliforms include most species of the genera *Enterobacter, Klebsiella, Citrobacter,* and *Escherichia,* although some species of *Serratia* and other genera are also often included. Total coliforms are used to determine the effectiveness of water treatment in removing enteric microorganisms, to monitor the integrity of the underground pipe network (called the distribution system), and as a screen for fresh fecal contamination. Treatment and other water-system practices that provide coliform-free water should also reduce pathogens to minimal levels. A major shortcoming of using total coliforms as an indicator is that they are only marginally adequate for predicting the potential presence of some pathogenic protozoa and viruses because total coliforms may be more susceptible to disinfection than are these other organisms.

Fecal coliforms are a subset of the total coliform group, primarily including *E. coli* and a few thermotolerant strains of *Klebsiella.* Fecal coliforms and *E. coli* are more suitable indicators than total coliforms of fresh fecal contamination, and most waterborne pathogens are associated with fecal contamination. However, total coliforms are a more suitable indicator than fecal coliforms or *E. coli* for determining the vulnerability of a system to fecal contamination, especially in the absence of fecally contaminated samples at the times and locations of the sample collection. Total coliforms are usually present at much higher concentrations in source waters than these other indicators, and they are relatively more resistant to chlorine and other environmental stresses. Fecal coliform or *E. coli* monitoring may be preferable in countries where monitoring for total coliforms is impractical due to consistently high densities of these organisms in the drinking water. Most countries that have formal drinking-water rules, including the United States, monitor for total coliforms. Many of these countries, including the United States, also require fecal coliform or *E. coli* monitoring.

In addition to the indicators described, other drinking-water indicators also have been used or suggested, usually as a supplement to the initial indicators. These supplemental indicators include fecal streptococci, enterococci, coliphage, *Clostridium perfringens, Bacteroides,* heterotrophic bacteria (as measured by plate-counting techniques such as the heterotrophic plate count, HPC), and other organisms. Several chemicals have also been mentioned, for example, fecal sterols, caffeine, assimilable organic carbon (AOC), and disinfectant residuals. For monitoring ambient waters, the U.S. Environmental Protection Agency (EPA) suggests the use of enterococci or *E. coli* as indicators.

III. EPA DRINKING-WATER REGULATIONS

A. General

The EPA publishes enforceable regulations under the Safe Drinking Water Act, which was passed by the U.S. Congress in 1974 and subsequently amended several times. Regulations under this Act apply to all public water systems, that is, to those systems that regularly serve 25 or more people at least 60 days out of the year, or that have 15 or more service connections. The number of public water systems in the United States is about 173,000, of which about 15,000 are surface water systems and the rest are groundwater systems. There are about 8000 systems that serve 3300 people or more, and these 8000 systems serve a total of approximately 222 million people.

The EPA has published two regulations to protect the public against waterborne pathogens, the Total Coliform Rule and the Surface Water Treatment Rule. These two rules are summarized here, along with a brief mention of the Ground Water Rule that is under development to control waterborne pathogens in groundwater. The Code of Federal Regulations (40 CFR 141 and 142) provides detailed requirements about the two existing rules. The *Federal Register* (FR) also provides detailed requirements, along with the rationale for these requirements.

B. Total Coliform Rule

The Total Coliform Rule (54 FR 27544; June 29, 1989) was revised in June 1989, and became effective on December 31, 1990. This regulation sets a maximum contaminant level (MCL) for total coliforms as follows. For systems that collect 40 or more samples per month, no more than 5.0% may have any coliforms; for systems that collect fewer than 40 samples per month, no more than one sample may be total coliform-positive. If a system exceeds the MCL, it must notify the public using mandatory language developed by the EPA. The required monitoring frequency for a system depends on the size of the population served. This frequency ranges from 480 samples per month for the largest systems to once annually for certain of the smallest systems. The regulation also requires all systems to have a written plan identifying where samples are to be collected.

If a system has a total coliform-positive sample, it must take three (for small systems, four) repeat samples within 24 hours of being notified of the positive sample. In addition, systems that collect fewer than five samples per month must, with some exceptions, collect at least five routine samples the next month of operation. Both routine and repeat samples count toward calculating compliance with the MCL.

If any sample is total coliform-positive, the system must also test the positive culture for the presence of either fecal coliforms or *E. coli*. Any positive fecal coliform or *E. coli* test must be reported to the state. If two consecutive samples at a site are total coliform-positive and one is also fecal coliform- or *E. coli*-positive, the system is in violation of the MCL and must notify the public using more urgent mandatory language than that used for the presence of total coliforms alone.

The Total Coliform Rule also requires each system that collects fewer than five samples per month to have the system inspected every 5 years (10 years for certain types of systems using only protected and disinfected groundwater). This on-site inspection (referred to as a sanitary survey) must be performed by the state or by an agent approved by the state.

The regulation also defines when a state may invalidate a total coliform-positive sample, and when a laboratory must invalidate a total coliform-negative sample. It also specifies which analytical methods are approved for compliance samples. The sample volume must be 100 ml, regardless of the method used.

C. Surface Water Treatment Requirements

The Surface Water Treatment Rule (SWTR) (54 FR 27486; June 29, 1989) requires all systems using surface water or groundwater under the direct influence of surface water to disinfect. It also requires all such systems to filter their water, unless the system has an effective watershed control program; it uses source water that meets EPA-specified limits on the level of total or fecal coliforms, and turbidity (opaqueness); and it meets stringent disinfection conditions.

The SWTR also controls *Giardia lamblia,* viruses, and *Legionella* in drinking water by requiring systems to insure that water treatment removes or inactivates at least 99.9% of the *Giardia lamblia* cysts and at least 99.99% of the enteric viruses in the surface water. The regulation and associated EPA guidance assist the system in meeting these levels by identifying pertinent CT values for disinfection inactivation (C, disinfectant concentration in mg/liter; T, time of disinfectant contact with the water in min). CT values are provided for *Giardia* and enteric viruses by disinfectant type (e.g., chlorine, chloramines, ozone, or chlorine dioxide), water pH, and temperature. The regulation also requires a system using surface water to have at least a 0.2 mg/liter disinfectant residual continually entering the distribution system, at least a detectable disinfectant residual in the distribution system, and an annual on-site inspection of the treatment facility. The regulation specifies the monitoring frequency for determining these disinfection residuals and (for unfiltered systems) testing source water quality.

In 1998, the EPA strengthened the SWTR, primarily by tightening the filtration process, to control *Cryptosporidium* and to ensure that new EPA regulations to minimize the risk of disinfectants and toxic disinfectant byproducts, which are formed by the reaction of disinfectants with organic material in the source water, do not undermine protection against waterborne pathogens. Under these strengthened provisions (called the Interim Enhanced SWTR) (63 FR 69478; December 16, 1998), systems must remove 99% of *Cryptosporidium* oocysts by filtration. The filtration process must remove sufficient turbidity such that the filtered water never exceeds one nephelometric turbidity unit (NTU), and averages no more than 0.3 NTU in 95% of the measurements during a month. The system must also conduct additional monitoring of the system operation, including the measurement of turbidity levels leaving each individual filter.

D. Ground Water Rule

The Ground Water Rule (GWR), which is under development, applies to all public water systems that use groundwater; exceptions will be those systems that are under the direct influence of surface water and, consequently, must comply with the SWTR. Under the GWR, the state will determine whether a system is fecally contaminated or is vulnerable to fecal contamination. This determination may be made by a variety of means, including direct monitoring of the well, an on-site sanitary survey by a trained inspector, and an examination of the site's hydrology. A threatened site would be required to disinfect or take other corrective action. The regulation may specify when, where, and how often a system must monitor, the frequency of sanitary surveys, the minimum disinfectant requirements, and other provisions.

E. Other Rules

The EPA has published a list of pathogens and chemical contaminants that are not specifically regulated, but that occur, or are likely to occur, in water systems, pose a health risk, and may need to be regulated (63 FR 10274; March 2, 1998). The list includes microsporidia, adenoviruses, caliciviruses, coxsackieviruses, echoviruses, *Mycobacterium avium intracellulare, Aeromonas hydrophila,* and cyanobacteria and other algae and their toxins. The EPA will conduct research on each of these pathogens and determine whether new regulations are needed to control them. Some of these organisms are opportunistic pathogens; the EPA is required to pay special attention to people who are immunocompromised.

In developing future regulations, EPA will assess the risk of waterborne disease associated with specific pathogens. This assessment will facilitate the development of a risk model(s) that would specify, for example, the level of pathogen inactivation or removal needed in a groundwater system, and, thus, whether that system would be required to disinfect. A satisfactory assessment of risk depends on the evaluation of a number of factors, including the infectious dose, virulence, ratio of infection to disease, identification of susceptible populations, the extent to which a pathogen is present in water and the percentage that are viable, and level of water treatment. The development of risk assessment and modeling for waterborne pathogens is in its infancy.

IV. SOURCES, TREATMENT, AND DISTRIBUTION

A. Sources of Water

Drinking water sources can be divided into two categories, surface water and groundwater; treatment varies with the source. Microbiologically, groundwaters tend to be of better quality than surface waters and, consequently, require less-intensive treatment. Most groundwater supplies used for drinking water are pumped wells (dug, drilled, or driven), artesian wells, or springs.

Surface water sources include streams, rivers, ponds, lakes, and manmade impoundments and reservoirs. These sources represent precipitation runoff that is not lost to the atmosphere by evaporation and does not enter the ground via infiltration and percolation. Surface waters become contaminated by human pathogens through direct and indirect inputs of municipal sewage and other sources of human and animal excreta. The removal or inactivation of these microorganisms to provide safe drinking water involves a multibarrier concept, using a variety of treatment processes in addition to measures aimed at controlling or reducing source-water pollution (source protection).

Bacterial concentrations in groundwater, as measured by the HPC, are usually below 100 colony-forming units (CFU)/ml; coliforms are normally absent. Bacterial concentrations in surface waters depend on the degree of contamination with human and animal fecal material. Pristine and relatively uncontaminated surface waters commonly contain bacterial concentrations of $10\text{–}10^4$ CFU/ml, whereas contaminated surface waters may contain more than 10^6 CFU/ml. Coliform bacteria concentrations in surface waters range from $<1/100$ ml to more than $10^6/100$ ml, and fecal coliforms range from $<1/100$ ml to more than $10^5/100$ ml. Raw water containing more than 2000 fecal coliforms/100 ml should not be used as a drinking water source if at all possible.

B. Water Treatment

The major drinking-water treatment processes for controlling microbiological occurrence in surface waters include coagulation and flocculation, sedimentation, filtration, and disinfection. Most groundwaters are only disinfected, if they are treated at all. Figure 1 is a schematic of the unit processes in a conventional drinking-water treatment system used for surface waters. Several additional treatment processes are widely used, for example, softening, fluoridation, and iron removal, but these are not addressed in this article.

1. Coagulation, Flocculation, and Sedimentation

In the first step, raw water is pumped into a rapid mix unit, where chemicals are added to destabilize particles in the water electrostatically (i.e., make them "sticky"). This step is referred to as coagulation. Then water enters a flocculation basin, which is often a series of chambers with slow moving paddles. During flocculation, the destabilized particles are brought into contact with each other so that aggregation into larger particles, or flocs, can occur. Microorganisms become trapped by, or attached to, the flocs. The optimal coagulation practice varies with water

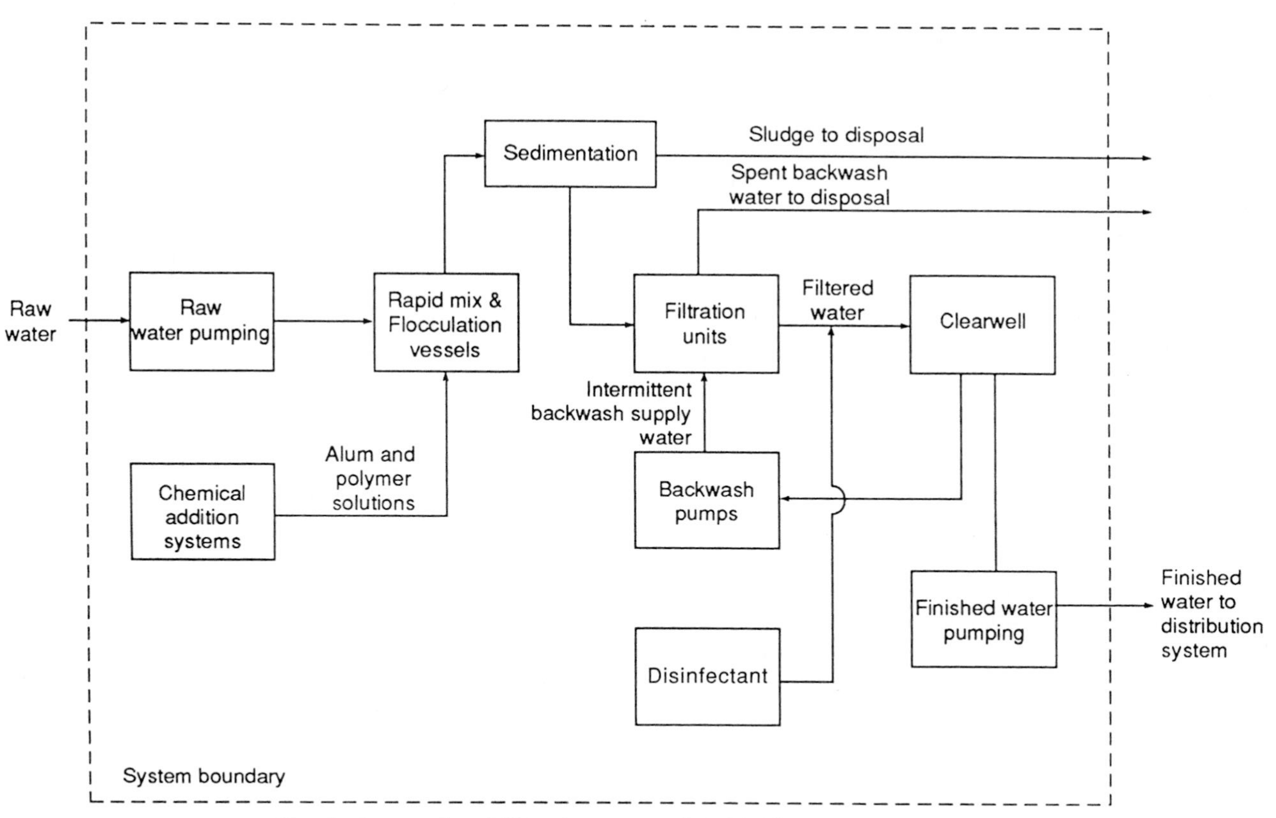

Fig. 1. Conventional filtration system for drinking water treatment.

pH and temperature, raw water turbidity, and type of coagulant(s) used. Commonly used coagulants include aluminum sulfate (alum), calcium oxide (lime), ferrous sulfate, and ferric chloride. Often, coagulant aids such as activated silica, bentonite clay, and polyelectrolytes (synthetic polymers of varying charge) are also used to reduce the concentration of primary coagulants employed.

After flocculation, water enters the sedimentation basin(s), where flocs are given time to settle by gravity. Some systems omit this step and feed the flocculated water directly to filters (direct filtration). If the source waters are highly turbid, systems may have an additional sedimentation step before coagulation (presedimentation). Sludge in the sedimentation basins is removed and discharged to a municipal sewer, lagooned, or dewatered and hauled to a landfill. The sedimentation basins are generally equipped with sludge-removal mechanisms; if not, they usually have a sloping bottom so that most of the sludge flows out with the water when the basin is drained for cleaning.

2. *Filtration*

The next treatment step is filtration, which removes suspended and colloidal material that has not settled. The filtration unit consists of steel or concrete vessels containing granular materials such as graded sand, anthracite, or gravel. Three types of filters are commonly used—rapid granular filters, slow sand filters, and diatomaceous earth (DE) filters.

a. Rapid Granular Filters

The most commonly used filter in the United States is the rapid granular filter. These filters may consist of silica sand, anthracite coal, or other materials. Filters are usually set up to provide gradation in filter media, with the coarsest particles on top and the smallest on the bottom. In a rapid granular filter, the rate at which water is applied is at least 2 gal/min/ft^2 surface area.

Rapid granular filters gradually accumulate a large amount of particles, which impedes water flow; thus, the filters must be periodically cleaned. In this process, water flow through the filter is reversed in a process termed backwashing. Backwashing expands the filter media, thereby releasing trapped particles into the water. Jets of water, air injection, or mechanical agitation at the surface may improve the process. The filter backwash water should be disposed of in an environmentally acceptable manner.

b. Slow Sand Filters

Slow sand filters are similar to the rapid granular filters, except that water flows through the filter at a much slower rate, 0.05–0.15 gal/min/ft^2 surface area, and pretreatment (coagulation, flocculation, and sedimentation) is often omitted. Removal occurs primarily in the upper portion of the sand, by straining, sedimentation, adsorption, and chemical and microbiological action. As contaminants are removed, a layer of deposited material called a schmutzdecke forms. A schmutzdecke contains a large number of bacteria that break down organic material in the water. When the schmutzdecke clogs the filter, it must be removed by scraping the top layer of sand to improve water flow. Normally, slow sand filters are used by systems with relatively clean source water (turbidity of 10 NTU or less, and no undesirable chemical contaminants). They require far less maintenance than rapid granular filters.

c. Diatomaceous Earth Filters

For diatomaceous earth filtration, solids are removed by passing water through a thin filter consisting of a layer of diatomaceous earth supported on a rigid base (septum). Diatomaceous earth is composed of crushed siliceous shells of diatoms (microscopic algae). To maintain adequate water flow, additional diatomaceous earth, called body feed, is continually added to the raw water during operation. As in slow sand filters, systems that use diatomaceous earth are usually small with relatively clean source water, and often pretreatment is omitted.

d. Granular Activated Carbon (GAC) Filters

GAC filters are sometimes used in conjunction with rapid granular filters to control taste and odor problems. In Europe, many systems ozonate the water before applying it to the GAC filters. This promotes the growth of a high density of heterotrophic bacteria in the filter; the resulting high metabolic rate significantly reduces the level of organic substances in the water as it passes through the filter.

e. Membrane Processes

Membrane technology has become increasingly popular in potable water treatment. It has been used for desalinization, removal of dissolved inorganic chemicals, water softening, and removal of solids. The simplest way to describe membrane technology is that water is forced through a porous membrane under pressure while suspended solids, larger molecules, or ions are held back or rejected.

Membrane processes may be pressure-driven or electric-driven. The pressure-driven membrane processes are microfiltration, ultrafiltration, nanofiltration, and reverse osmosis. In microfiltration (MF), water is forced through a porous membrane with a pore size of 0.45 μm, which is relatively large compared to other membrane processes. It is probably best used for small particles such as bacteria or protozoa. Ultrafiltration (UF) uses a membrane size smaller than 0.1 μm, which removes colloids and other high-molecular-weight materials. It is effective for the removal of most organic compounds. Nanofiltration (NF) rejects even smaller molecules than UF, and has been used for water softening and the removal of total dissolved solids and viruses. Reverse osmosis (RO) has the highest rejection rate of all of the membrane processes. It has been used primarily for the desalinization of seawater. The electric-driven processes are electrodialysis and electrodialysis reversal. Electrodialysis (ED) transfers ions though a membrane as a result of direct electric current. Electrodialysis reversal is similar to ED except that the polarity of the direct-current process is periodically reversed. These type of systems are used primarily for the treatment of brackish water. Of these processes, NF seems to have the greatest potential increased use in water treatment.

3. Disinfection

Disinfection is the primary means for inactivating pathogenic microorganisms in water. For systems using groundwater, disinfection is generally the only treatment practiced. Several disinfectants are available, including chlorine, chloramines (chlorine combined with ammonia or organic amines), ozone, chlorine dioxide, and ultraviolet light. All are oxidizing agents.

Chlorination is, by far, the most common disinfection technique practiced in the United States. The dose of chlorine usually applied is sufficiently high to meet chlorine demand (i.e., the tendency of organic substances and ammonia to react with chlorine) and still leave a sufficient concentration (chlorine residual) to inactivate microorganisms throughout the distribution system. A major shortcoming is that chlorine combines with organic substances in the water to produce chlorinated by-products, some of which are toxic (e.g., trihalomethanes such as chloroform). Unlike chlorine, chloramines do not result in any significant trihalomethane formation, but their microbial inactivation rates are substantially slower than that of chlorine. Chlorine dioxide also does not generate significant by-products, but its intermediates (chlorite and chlorate) are toxic. Like chlorine, chloramines and chlorine dioxide are useful because they provide disinfectant residuals in the water throughout the distribution system.

Ozone is the most effective disinfectant generally available and is widely used in water treatment in Europe. Fewer toxic by-products have been identified for ozone than for chlorine, but ozone is more expensive and does not leave a significant residual. There is increased interest in the United States in using ozone as a predisinfectant, followed by chlorine or a chloramine. This process would allow a system to control the formation of toxic by-products, yet maintain a disinfectant residual in the distribution system.

Ultraviolet light is sometimes used by smaller systems and individual domestic systems, especially those using groundwater. The optimum wavelength for biocidal effectiveness is 254 nm. Dosage is expressed as the product of radiation intensity (μW) and the time (s) per unit area (cm^2). Ultraviolet light does not produce disinfectant by-products in water, but leaves no disinfectant residual. Its effectiveness is reduced by high turbidity, air bubbles, some dissolved chemicals that block light penetration, lack of reliable methods or meters to measure dosage, and equipment maintenance and reliability considerations.

C. Microbiology of Treatment Processes

The removal of microorganisms by the treatment processes through the filtration step is variable due to a variety of factors, including fluctuating source-water quality, coagulant dose, pH, water temperature, depth of filter, filter-medium particle size, and filtration rate.

Table I shows drinking-water-treatment processes and the approximate percentage removal or inactivation of microorganisms achieved by each. The removal percentages represent only removal percentages for that stage, not the cumulative percentage. It is difficult to establish the cumulative removal percentages from one step to the next because of the considerable variation in the data reported from different studies. These differences reflect different source waters and qualities, different treatment processes, and different operating conditions. Thus, direct comparison of study results is difficult, but sufficient studies have been conducted to permit a general characterization of the effectiveness of water-treatment processes for removing microorganisms.

1. *Slow Sand Filters*

The efficiency of microorganism removal by slow sand filters is influenced by several factors, including the particle size of the filter medium and the extent to which the scum layer (schmutzdecke) has developed. A filter with a smaller particle size is more efficient in microbe removal, but results in shortened filter runs; that is, more frequent filter cleaning is necessary to maintain a suitable water flow rate. In addition, new or cleaned slow sand filters require some conditioning (ripening) before they are effective in removing microorganisms. The ripening period allows a biologically active schmutzdecke to build up on the particles and in the filter bed. This layer then assists in the filtration of other particles and colloids from the water. The schmutzdecke is important in microbe removal, particularly for bacteria and viruses. Virus removal by sterile sand is negli-

TABLE I
Microorganism Removal Efficiency by Water Treatment Processes[a]

Unit process	Bacteria	Viruses	Protozoa	Helminths
Storage reservoirs	80–90	80–90	—	—
Aeration	—	—	—	—
Pretreatment[b]	90–99	90–99	>90	>90
Hardness reduction				
high lime	90–99.9	99–99.9	—	—
low lime	90–99	90–99	—	—
Slow sand filtration				
without pretreatment[b]	35–99.5	10–99.9	59–94	—
with pretreatment[b]	90–99.9	90–99.9	59–99.98	—
Rapid granular filtration				
without pretreatment[b]	0–90	0–50	0–90	—
with pretreatment[c]	90–99	90–99	90–99	—
with pretreatment[b]	90–99.9	90–99	90–99.9	—
Diatomaceous earth filtration	90–99.9	99–99.96	99–99.999	—
Activated carbon	—	10–99	—	—
Disinfection	99–99.99	99	27–78	—
Full conventional treatment[e]	99–99.9999	99.9 → 99.99	99.9–99.98	—

[a] Values in percent. Dashes indicate value not known. Reprinted in part from Amiratharaja, A. (1986). *J. Amer. Water Works Assoc.* 78(3), 34–49, by permission. Copyright © 1986, American Water Works Association.
[b] Pretreatment includes coagulation, flocculation, and sedimentation.
[c] Pretreatment without sedimentation.
[d] With pretreatment and precoating of filter.
[e] Pretreatment, filtration, and disinfection.

gible, and removal by clean nonsterile sand is variable but poor.

The removal of coliform bacteria by slow sand filters ranges from about 83% with new filter sand to nearly 100% for sand with an established schmutzdecke. Removal of poliovirus type 1 by slow sand filters ranges from 22–96% with clean sand to ≥99.9% for sand with an established biological population. The removal of *Giardia* (protozoan) cysts by slow sand filtration ranges from 59–99.98%.

Cold water temperatures generally decrease microorganism removals by slow sand filtration. At water temperatures ≤5°C, the removal of heterotrophic bacteria (measured by HPC) and coliforms decreases by about 2% and 2–10%, respectively, compared to removals at temperatures greater than 15°C. Virus removals decrease by about 0.5% and protozoan cyst removals decline by 0–6.2%. Increased water flow through the filter (filtration rate) also results in decreased microorganism removals.

2. *Rapid Granular Filters*

Systems that use rapid granular filtration must first pretreat the raw source water (by coagulation, flocculation, and sedimentation) for effective microbial removal, unless the water is clear (turbidity less than 10 NTU). Effective pretreatment should reduce the turbidity of muddy surface water to a level well below 1 NTU. Effective pretreatment and filtration collectively should reduce the turbidity to 0.1 NTU or less.

Microorganism removals achieved by pretreatment and rapid granular filtration are high. Bacterial removals (heterotrophic bacteria and coliforms) range from 86–98.8%, and virus removals range from 90 to >99.99%. The coagulation and filtration processes generally can achieve removals of protozoan cysts ranging from 83–99.99%. The efficient removal of *Giardia* and other protozoan cysts by filtration is dependent on an adequate coagulant dose. A change in alum dose from 5–10 mg/liter can increase the removal of *Giardia* cysts from 96 to >99%.

Factors that adversely affect removal efficiency include the interruption of chemical feed (coagulants and polymers), poor filter efficiency at the beginning of a filter run, sudden increases in water flow, and turbidity breakthrough that can occur with higher filter head loss (resistance to water passage through the filter) at the end of a filter run. Any of these factors can seriously degrade the microbiological quality of the filtered water. In addition, some source waters are so highly polluted that full treatment (pretreatment, filtration, and disinfection) may not achieve a suitable level of microbiological reduction.

3. *Diatomaceous Earth*

Bacterial removals by diatomaceous earth (DE) filtration are affected by the grade of DE used. When fine DE is used, yielding a median pore size of 1.5 μm in the filter cake, bacterial removals of nearly 100% can be achieved. Lower percentage removals occur when coarser DE is used that provides an increased median pore size. However, by chemical conditioning of the coarser grades of DE, good bacterial removals can be obtained.

DE filtration can satisfactorily remove viruses from water but, to be most effective, the raw water must be pretreated with a coagulant aid (polymer) or the DE filter cake must be chemically conditioned to enhance virus attachment to the filter material during filtration. Overall, DE filtration is most effective for the removal of microorganisms in the size range of *Giardia* or *Entamoeba histolytica* cysts. *Giardia* cyst removals of 99% or greater can be achieved by proper DE filter operation. The use of DE to remove smaller organisms (bacteria and viruses) can be improved by coating it with aluminum hydroxide precipitate. This coating gives the DE a positive surface electrical charge. Bacteria and viruses with a negative electrical charge are then removed by surface attachment.

4. *Disinfection*

a. General Considerations

The final treatment process for drinking water is chemical or physical disinfection intended to inactivate any coliforms and pathogenic microorganisms that penetrate the filter. The effectiveness of disinfection is a function of the types of organisms to be inactivated, the quality of the water, the type and concentration of the disinfectant, the exposure or contact time, and the temperature of the water.

As stated previously, CT values (explained in Section II.B) are used to identify the level of inactivation provided by a given disinfectant for an organism under a specific environmental condition. These values are useful for comparing biocidal efficiency. Table II provides CT values for several organisms. Most of the available CT data for microorganisms of health concern were developed from laboratory studies that might not be indicative of field operations.

TABLE II
CT Value Ranges for Inactivation of Microorganisms by Disinfectants[a]

Microorganism	*Free chlorine, pH 6–7*	*Preformed chloramine, pH 8–9*	*Chlorine dioxide, pH 6–7*	*Ozone, pH–7*
E. coli	0.034–0.05	95–180	0.4–0.75	0.02
Polio virus-1	1.1–2.5	768–3740	0.2–6.7	0.1–0.2
Rotavirus	0.01–0.05	3806–6476	0.2–2.1	0.006–0.06
Phage f_2	0.08–0.18	ND	ND	ND
G. lamblia cysts	47–150	2200[b]	26[b]	0.5–0.6
G. muris cysts	30–630	1400	7.2–18.5	1.8–2.0
Cryptosporidium parvum	7200[c]	7200[d]	78[d]	5–10[c]

[a] All CT values are for 99% inactivation at 5°C except where noted. ND indicates no data.
[b] Values for 99.9% inactivation at pH 6–9.
[c] Values for 99% inactivation at pH 7 and 25°C.
[d] Values for 90% inactivation at pH 7 and 25°C.

Water temperature can affect disinfection rates (and thus CT values). Microorganism inactivation rates decrease as water temperature decreases. Water pH can also affect disinfection rates. In most water systems, the pH is kept in the range of 7–9. Water pH, for example, determines the proportions of the most important chlorine species, hypochlorous acid (HOCl) and hypochlorite (OCl^-). Lower pH values (pH 6–7) result in the formation of HOCl, which is favorable for rapid inactivation, whereas higher pH values (pH 8–10) result in formation of OCl^-, which results in slower inactivation. For chlorine dioxide (ClO_2), which does not dissociate, inactivation is more rapid at higher pH values (pH 9) than at lower pH values (pH 7). Ozone disinfection efficacy does not appear to be affected by pH.

An important factor in bacterial inactivation is the phenomenon of cell injury. Disinfection and other environmental stresses may cause nonlethal physiological injury to waterborne bacteria. This phenomenon causes problems for monitoring water quality and calculating CT values, because injured bacteria may not grow on selective media normally used to detect or enumerate the bacteria. Thus, the actual number of viable cells may be underestimated. In some cases, injured pathogens remain infective. Problems with detecting injured cells can be mitigated by the use of media and procedures that remain selective, yet permit the injured cells to repair metabolic damage.

b. Microorganism Inactivation

In general, the order of microbial disinfectant efficiency is $O_3 > ClO_2 > HOCl > OCl^- > NH_2Cl > NHCl_2 >$ RNHCl (organic chloramines). However, for technical reasons, practical handling considerations, cost, and effectiveness, the frequency of use of disinfectants by utilities in the United States is generally chlorine $\gg$ chloramines $> O_3 > ClO_2$.

1. Chlorine—Enteric viruses (represented by poliovirus type 1) are more resistant to inactivation by chlorine than are bacteria (represented by *E. coli*), and protozoan cysts are nearly two orders of magnitude more resistant than the enteric viruses.

2. Chloramines—Comparison of chloramines with chlorine for disinfection of microorganisms shows that, in general, for all types of microorganisms, CT values for chloramines are higher than CT values for free chlorine species. CT values for *Giardia lamblia* cysts are lower, in contrast to the results for free chlorine.

3. Chlorine dioxide—Chlorine dioxide CT values show that at pH 7.0, ClO_2 is not as strong a bactericide and virucide as HOCl. However, as the pH is increased, the efficiency of ClO_2 for the inactivation of viruses increases. CT data for protozoan cyst inactivation are not available.

4. Ozone—Overall, the comparison of CT values for ozone with those for chlorine and ClO_2 indicates that ozone is a much more effective biocide than the other disinfectants. *E. coli* is about 10-fold more sensitive to ozone than is poliovirus type 1. *Giardia muris* cysts are about 10-fold more resistant to ozone than poliovirus type 1. Because ozone is a powerful oxidant, it reacts rapidly with both microorganisms and organic solutes and is very useful as a primary disinfectant.

5. Ultraviolet light—The sensitivity of the various microbial groups to ultraviolet light is similar to that for chemical disinfectants. Enteric bacteria are most sensitive, followed by enteric viruses; protozoan cysts are least sensitive. Organisms that are sublethally injured by UV light exposure may, under appropriate conditions, be able to repair the damage (i.e., photoreactivation or dark repair). Ranges of UV dosages required for 99.9% inactivation of microorganisms of concern in drinking water are bacteria, 1400–12,000 μW-s/cm^2; viruses, 21,000–46,800 μW-s/cm^2; and protozoan cysts, 105,000–300,000 μW-s/cm^2. In general, the UV disinfection values given for protozoan cysts are not practical with current UV technology used for water treatment. However, recent data indicate that *Cryptosporidium parvum* oocysts can be inactivated by a UV dose as low as 20 μW-s/cm^2.

D. Distribution Systems

1. Description

Water-transmission and -distribution systems are needed to deliver water to the consumers. Distribution systems represent the major investment of a municipal water works and consist of large mains that carry water from the source or treatment plant, service lines that carry water from the mains to the buildings or properties being served, and storage reservoirs that provide water storage to meet demand fluctuations, for firefighting use, and to stabilize water pressure. The branch and loop (or grid) are the two basic configurations for most water-distribution systems.

The layout of a branch system is similar to that of a tree branch, with smaller pipes branching off from larger pipes throughout the area served. This system, or a derivative of it, is normally used to supply rural areas where water demand is relatively low and long distances must be covered. The disadvantages of this configuration are the possibility that a large number of customers will be without service should a main break occur, and the potential water-quality problems in parts of the system resulting from the presence of stagnant water. System flushing should be done at regular intervals to reduce the possibility of water-quality problems.

The loop configuration is the most widely used distribution system design. Good design practices for smaller systems call for feeder mains to form a loop approximately 1 mile (1600 m) in radius around the center of the town, with additional feeder loops according to the particular layout and geography of the area to be served. The area inside and immediately surrounding the feeder loops should be gridded with connecting water mains on every street.

The most commonly used pipes for water mains are ductile iron, prestressed concrete cylinders, polyvinyl chloride (PVC), reinforced plastic, steel, and asbestos cement.

2. *Microbiology*

Microbiologically, water-distribution systems are interesting bacterial ecosystems that present a real challenge to the water utilities in terms of maintaining good water quality with low bacterial densities. The construction characteristics, operation, and maintenance of a water-distribution system provide ample opportunities for microbial recontamination of the treated water during distribution. Pipe joints, valves, elbows, tees, and other fittings as well as the vast amount of pipe surface provide both changing water movement and stagnant areas where bacteria can attach and colonize. Water-distribution systems are susceptible to cross-connections that may allow the entry of pathogens into the system. (A cross-connection is any direct connection between the drinking-water-distribution system and any nondrinkable fluid or substance.)

a. Biofilms in Water-Distribution Systems

Bacteria found in water-distribution systems can be classified as indigenous (autochthonous) and exogenous (allochthonous) populations. The indigenous organisms are well-adapted biofilm-forming bacteria that represent a stable ecosystem that is difficult to eradicate. The exogenous bacteria are contaminants that are transported into the system by a variety of mechanisms.

The development of a permanent biofilm in the distribution system occurs because the bacteria find physical and chemical conditions conducive to colonization and growth at the solid surface–water interface. These conditions include an ample supply of nutrients (assimilable organic carbon, AOC) for growth, a relatively stable temperature, and some degree of protection from exposure to harmful chemicals, such as the disinfectant(s) used to treat the water.

When an adequate disinfectant residual is maintained in the water throughout a distribution system, the growth of bacteria is usually well controlled and the density of bacteria in the water will remain low—in the range of <10 to several hundred CFU/ml (HPC). The disinfectant residual concentration needed to control the growth of the bacteria varies from one water system to another. The choice of the medium and method used to determine the bacterial density may result in low or high counts of heterotrophic bacteria. In general, rich culture media and incubation at 35°C will yield lower counts than dilute nutrient media and incubation at 20–28°C. Factors that appear to be critical are pH, temperature, dissolved organic carbon (DOC) concentration, AOC concentration, and the type of disinfectant used. The disinfectant residual will also reduce or suppress extensive biofilm growth if the residual is maintained throughout the system.

Disinfectant residuals should be at least 0.2 mg/liter for chlorine and 0.4 mg/liter for monochloramine (NH_2Cl). Higher concentrations of disinfectant residual may be applied and maintained, but if the water contains high levels of DOC, it may be difficult to maintain an adequate disinfectant residual to control bacterial growth and still have water that is aesthetically acceptable to the consumers.

Bacterial concentrations in distribution water vary from <1 CFU/ml in the water leaving the treatment plant to as high as 10^5–10^6CFU/ml in water from slow-flowing or stagnant areas of the distribution system. The concentrations of bacteria in the water and on the pipe surfaces vary spatially and temporally in the distribution system. Bacterial densities in the pipe-wall biofilm and in sediments may reach 10^7 CFU/cm^2. The biofilm contributes bacteria to the flowing water through shear loss (erosion) and by migration of actively motile bacterial cells into the water. Table III lists some of the bacteria commonly found in drinking water, sediments, and biofilm. Many of these bacteria are found in both the water and the biofilm, indicating the influence of the biofilm on the bacterial quality of distribution system water.

TABLE III
Microorganisms Found in Treated Distribution Water, Sediment, and Distribution System Biofilm

Microorganism	*Distribution water*	*Sediment*	*Biofilm*
Pseudomonas spp.	X	X	X
Alcaligenes spp.	X		X
Acinetobacter spp.	X		X
Moraxella spp.	X	X	X
Arthrobacter spp.	X	X	X
Corynebacterium spp.			X
Bacillus spp.	X	X	X
Enterobacter spp.	X	X	X
Micrococus spp.	X		
Flavobacterium spp.	X	X	X
Klebsiella spp.	X	X	X
Mycobacterium spp.	X		X
Iron and sulfur bacteria	X	X	X
Nitrifying bacteria	X	X	X
Yeasts and fungi	X	X	X
Invertebrates and protozoa	X	X	X

The biofilm has been shown to be the major contributing source of bacteria found in the bulk water in the distribution water because the biofilm is where the most significant bacterial growth occurs. Iron and sulfur bacteria are nuisance organisms that cause taste and odor problems and are often associated with groundwater sources. Nitrifying bacteria and fungi cause problems in chloraminated drinking-water systems, including the depletion of total chlorine residual, conversion of reduced nitrogen compounds such as ammonia to nitrite and nitrate, and high levels of heterotrophic bacteria. Invertebrates and protozoa are nuisance organisms associated with aesthetic complaints about water quality. Invertebrates and protozoa may also harbor pathogenic and nonpathogenic bacteria, either internally or attached to their surfaces, and this association provides the bacteria with some degree of protection from the inactivation by disinfectant residuals.

Information about the development and control of biofilms in drinking water is sparse. In addition, there are few data on other issues related to biofilms, including the role of iron and sulfur bacteria in biofilm development, the role of biofilms in the corrosion of pipe material, the role of biofilms in nitrification for systems that use chloramines as a secondary disinfectant, and the effect of added corrosion inhibitors on bacterial populations in biofilms.

The presence of chlorine or chloramine retards the development and affects the spatial distribution of biofilm. The type of disinfectant residual provided selects for bacteria that are more tolerant to the specific disinfectant used. The biofilm environment provides protection for the cells by diffusional resistance and neutralization of the disinfectant. Therefore, biofilm organisms are less inhibited by the disinfectant residual than are planktonic cells. The differential effectiveness of chlorine and monochloramine for control of biofilm growth has been shown. Monochloramine, because it is less reactive and therefore more persistent, apparently can penetrate the biofilm and is more effective than chlorine in controlling biofilm growth.

In some cases, coliforms that gain entry into the distribution system may attach to pipes or pipe sediments and proliferate, thus becoming a biofilm constituent. The intermittent, sporadic, or persistent sloughing of coliform bacteria from biofilms into the water of the distribution system may cause systems to repeatedly violate standards for total coliforms. This problem is most frequently associated with utilities that use a surface water supply where the water temperature is 15°C or higher. Public safety dictates that all coliforms detected by testing be regarded as representing system vulnerability, unless strong evidence suggests otherwise.

The utility should review its treatment operations and increase its monitoring of the quality of water entering the distribution system to be sure that inadequate or failed treatment is not responsible for the total coliform occurrences and that *E. coli* is not present in the water. The water utility should also review the operation and management of the water-distribution system to insure that no *E. coli* are present, a disinfectant residual is present, and an adequate cross-connection control program is in effect. Finally, the water utility should insure that large volumes of water held in storage in reservoirs, standpipes, or above-ground tanks are not the source of the total coliform problem.

During warm-water periods, it is more difficult to maintain a disinfectant residual in the water in all parts of the distribution system. With increasing temperature, the disinfectant reacts more rapidly with dissolved organic chemicals in the water and in the biofilm, and bacteria grow faster. Increased reaction rates and bacterial growth are often compounded by a lack of knowledge of system hydraulics and of the actual water movement throughout the system. In many cases, there are large areas where minimal movement of the water occurs. Many water utilities may create microbial or chemical contaminant problems by their failure to understand how their systems work; better system management can reduce those problems.

b. Biofilm in Household and Building Plumbing Systems

Biofilms develop in household and building plumbing systems and can degrade water quality more rapidly and to a greater extent than in the distribution mains. In such systems, pipe materials, long residence time, stagnation or low flow conditions, residual disinfectant concentration, and water temperature may differ considerably from those found in the main distribution piping. Increased disinfectant demand in household and building networks may be related to the smaller diameters of service-line piping, which results in a greater surface-to-volume ratio than in the pipes of the main distribution system. The depletion of the disinfectant residual due to water stagnation, organic deposits on the pipe surfaces, and increased water temperature result in increased biofilm bacterial growth.

In general, there is little evidence that high levels of heterotrophic bacteria in drinking water from plumbing systems have adverse health effects. However, for specific organisms such as *Legionella* spp. and nontuberculous *Mycobacterium* spp., which have been found in the biofilm of plumbing systems, there is evidence of adverse health effects. Plumbing-system components such as hot-water storage tanks and showerheads may permit the amplification of the bacteria to significant levels. For example, outbreaks of Legionnaire's disease in health care facilities have been related to the amplification of *Legionella pneumophila* in hot-water storage tanks and showerheads.

In the case of nontuberculous mycobacteria, some evidence suggests that the type of residual disinfectant used in the drinking-water treatment process may select for colonization of biofilms by these organisms, and species found in treated drinking water have been linked to human infections in immunocompromised patients.

V. ALTERNATIVE WATER SOURCES

Under some circumstances, for example in flooding emergencies, other sources of drinking water may be necessary.

A. Bottled Water

Bottled water is one alternative source, but is very expensive and is therefore not viewed as a long-term solution. The bacterial quality of noncarbonated bottled water varies both among brands and from lot to lot within a brand. The chemical quality varies as well. Bottled water in the United States is usually treated by reverse osmosis, ozone, and/or ultraviolet light. Bacterial densities in noncarbonated bottled water, determined by plate count methods, vary from 0 to $>1.0 \times 10^5$ CFU/ml. Fresh noncarbonated bottled waters often have low bacterial densities, but during storage prior to sale the bacteria multiply, often by several orders of magnitude. Carbonated (sparkling) bottled waters generally have low or no detectable bacteria because of the low pH. In most bottled waters examined, coliform bacteria have been absent. Bottled-water quality is regulated in the United States by the U.S. Food and Drug Administration, but must comply with EPA drinking water regulations to the extent that it is possible to do so.

B. Emergency Drinking Water

Treatment of water for drinking in an emergency generally involves either boiling the water or chemically disinfecting the water using liquid chlorine laundry bleach, tincture of iodine, or iodine or chlorine tablets. Water to be disinfected should be strained through clean cloth to remove particles and floating matter.

Boiling is the most effective means of inactivating pathogens. For heat disinfection, the strained water should be boiled vigorously for 1 full minute and allowed to cool before use. For chemical disinfection, add liquid chlorine bleach (4–6% chlorine) at a rate of 2 drops (clean clear water) or 4 drops (cloudy water) per quart, mix thoroughly, and allow to stand for 30 min. A slight chlorine odor should be detectable in the water; if not, repeat the dosage and let stand for an additional 15 min before using. If tincture of iodine is used, add 5 drops (clean clear water) or 10 drops (cloudy water) of 2% tincture of iodine solution per quart of water. Allow the water to stand for 30 min before use. For chlorine or iodine tablets (obtained from drug or sporting goods stores), follow the instructions on the package. Keep all purified water in clean, closed containers. Chlorine and iodine may not inactivate the protozoan pathogens, especially *Cryptosporidium*.

C. Point-of-Use/Point-of-Entry Treatment

Point-of-use/point-of-entry treatment (POU/POE) devices (or filters) are small package units designed to treat water at a designated tap in a household or to treat all of the water entering the household. These devices are becoming increasingly sophisticated and are growing in popularity. POU devices are generally placed under a sink and POE devices might be located in a basement or a garage. Currently POU/POE treatment is used to reduce the levels of a wide variety of contaminants in drinking water, including organic material, turbidity, fluoride, iron, radium, chlorine, arsenic, nitrate, ammonia, and pathogens (including cysts and oocysts). Taste, odor, and color can also be improved with POU/POE treatment.

The types of technology that may be used in POU/POE treatment devices include adsorption, ion exchange, reverse osmosis, filtration, chemical oxidation, distillation, aeration, and disinfection such as chlorination, iodination, ozonation, and ultraviolet light. Activated carbon filtration, a type of adsorption device, is the most widely used point-of-use system for home water treatment. The performance of an individual unit depends on a combination of factors, such as unit design, type and amount of activated carbon, and contact time of the water with the carbon. Most units use granular activated carbon (GAC) in their design, although other types of carbon such as pressed block, briquettes of powdered carbon, and powdered carbon are available.

Bacterial growth in activated carbon units may be a problem. The organic material adsorbed onto the carbon provides nutrients for bacterial growth. Stagnation periods between operations allows time for bacteria to grow. To help prevent this problem, some units contain GAC impregnated with silver or some other bacteriostatic agent. Silver, although useful against some bacteria (e.g., coliforms and related enteric bacteria) as a bacteriostatic agent, does not effectively inhibit the growth of all bacteria in GAC, POU, or POE filters. Bacterial levels in water leaving GAC filters, with or without silver, range from 10^2–10^5 CFU/ml, but it may take longer for bacterial levels in a GAC filter with silver to reach the higher concentrations. When a GAC filter is used as the key treatment technique and the incoming water is not microbiologically safe, water from the unit should be disinfected either chemically or by UV irradiation to reduce the bacterial levels in the water and to inactivate opportunistic pathogens that may colonize the carbon. Some carbon block units do not contain silver but claim bactericidal effects, probably due to straining through small carbon pore size (0.5 μm). Some POU/POE systems use a disinfectant such as iodinated resins, ozone or UV irradiation to control bacteria.

VI. CONCLUSION

Currently, technology is capable of providing microbiologically safe drinking water for all consumers. In practice, however, many systems are still vulnerable to waterborne disease. The reasons for this include the inability to fund the installation and use of adequate technology, ignorance of the need for adequate technology, inadequate or improper monitoring to insure that barriers against waterborne disease remain intact, lack of real-time monitoring data, and lack of adequate data on the effectiveness of various treatment practices against certain pathogens. In addition, with increased population pressure, source-water quality may decline over time, especially in areas where water is scarce; thus the technology used will become inadequate. These problems are common on a global basis. The global challenge is to highlight the imperative of safe drinking water and to find solutions to these problems.

See Also the Following Articles

Biofilms and Biofouling • Biomonitors of Environmental Contamination • Heterotrophic Microorganisms • Wastewater Treatment, Municipal

Bibliography

American Public Health Association (1998). "Standard Methods for the Examination of Water and Wastewater," 20th ed. American Public Health Association, Washington, D.C.

Amirtharajah, A. (1986). *J. Am. Water Works Assoc.* **78**, 34–49.

Bukhari, Z., Hargy, T. M., Bolton, J. R., Dussert, B., and Clancy, J. L. (1999). Medium-pressure UV for oocyst inactivation. *J. Am. Water Works Assoc.* **91**(3), 86–94.

Clark, R. M., and Tippen, D. L. (1990). Water supply. *In* "Standard Handbook of Environmental Engineering" (R. Corbitt, ed.), pp. 5.1–5.225. McGraw-Hill, New York.

Craun, G. F. (ed.) (1986). "Waterborne Diseases in the United States." CRC Press, Boca Raton, FL.

Geldreich, E. E. (1996). "Microbial Quality of Water Supply in Distribution Systems." Lewis Publishers, New York.

McFeters, G. A. (ed.) (1989). "Drinking Water Microbiology." Springer-Verlag, New York.

Montgomery, J. M. (1985). "Water Treatment Principles and Design." John Wiley & Sons, New York.

O'Melia, C. R. (1985). Coagulation. *In* "Water Treatment Plant Design" (R. L. Sanks, ed.), pp. 65–81. Ann Arbor Science, Ann. Arbor, MI.

U.S. Environmental Protection Agency (1991). "Manual of Small Public Water Supply Systems." Report No. 570/9-91-003. USEPA, Washington, DC.

Wine

Keith H. Steinkraus
Cornell University

I. Antiquity of Wine Fermentation
II. Alcoholic Fermentation
III. Yeasts
IV. Malolactic Fermentation
V. Nobel Mold
VI. Microorganisms Causing Spoilage in Wines
VII. Microbiology of Sherry Wines
VIII. Wine Grapes
IX. Wine Classification
X. Steps in Manufacture of Wine
XI. Sparkling Wines
XII. Fortified Wines
XIII. Flavor and Acceptability

GLOSSARY

dry Of wine, lacking sweetness.
enology The science of wine-making.
must The expressed grape juice or the mashed grape before and during fermentation.
vinification The technology of wine-making.
viticulture The growing of grapes.

WINE is an alcoholic beverage made by the fermentation of grape juice or other sugar-containing substrates including honey, sugar cane, fruit juices, and other plant juices containing sugars such as palm tree sap, floral extracts, and *Agave*, the century cactus plant. Wine is distinguished from beer, in which the fermentable sugars are derived principally from starches by means of amylases produced in cereal grains, particularly barley, by germination or malting.

The term "wine" probably should not be applied to beverages that involve the hydrolysis of starch in cereals through ptyalin in saliva introduced by chewing, the hydrolysis of starch to sugars via fungal or bacterial amylases, or the application of microorganisms that hydrolyze the starch to sugars and then ferment the sugars to ethanol. Thus, South American chicha, made by chewing maize and subsequently fermenting it, should probably be called a maize beer. Rice wine, such as Japanese sake, made by overgrowing boiled rice with *Aspergillus oryzae* as a source of amylases and then fermenting with yeast, should probably more correctly be termed rice beer; however, accepted usage may preclude this.

This article will restrict the term "wine" to products made by the fermentation of sugar-rich substrates in which starch hydrolysis is not required.

I. ANTIQUITY OF WINE FERMENTATION

Wine fermentation is one of the oldest and perhaps the oldest fermentation known to humans. Honey diluted with rainwater ferments spontaneously with yeasts in the environment. Fruit juices ferment spontaneously with yeasts on the fruit. Mexican pulque, one of the most ancient beverages of the Mayans and Aztecs, required that the core of the *Agave* cactus plant be harvested and mashed, and then undergo spontaneous fermentation by yeasts and other microorganisms in the environment.

Wine was truly a gift of God (or nature) to primitive humans. Fruit juices ferment spontaneously, yielding a product that preserves and, in fact, enhances the fruit's nutritive value and retains its essential flavors. The conversion of the sugar to ethanol and ethanol's striking effect on the psyche of the consumer led naturally to the conclusion that the "spirit" of the fruit had been extracted for the benefit of the consumer.

Wine or alcoholic fermentation was present on Earth many millions of years before humans evolved or were created. Fossil microorganisms have been found in rocks more than 3 billion years old. Plants, the basis for human food, likely evolved about a billion years later. Thus, the microorganism required for fermentation and the plants providing the substrates were well established on Earth at least a billion years before humans.

Microorganisms have the primary task of recycling organic matter, including sugars; yeasts in the environment ferment the sugars to alcohol as fruits become overripe and fall to the ground. Alcohol is one of the products of recycling reactions. It, in turn, is further fermented to acetic acid or respired to CO_2 and H_2O in continuing recycling reactions.

There are many references to wine in the Bible. Genesis IX: 20 and 21 states: "And Noah began to be a husbandman, and he planted a vineyard. And he drank of the wine and was drunken." Some anthropologists have suggested that it was the alcoholic fermentation that caused humankind to change from hunter–gatherer subsistence to agriculture to satisfy humans' desire for alcohol.

From the beginnings of civilization, wines have been associated with culture, art, and religion because early humans considered wine a miraculous conversion. Alcoholic beverages have always been closely related to politics, and even today governments use alcoholic fermentations as a major method of raising taxes. In very primitive times in South America, emperors could hold office only as long as they provided the people with a sufficient supply of chicha. From ancient times, wines have played a vital role in religious ceremonies, and they continue to do so today. From the Middle Ages on, religious orders and monasteries had a major role in the manufacture and supply of wines for religious and other cultural activities.

II. ALCOHOLIC FERMENTATION

The biochemistry of the alcoholic fermentation is among the most thoroughly studied enzyme–substrate sequences ever investigated. Black, in the eighteenth century, reported that ethyl alcohol and carbon dioxide were the products of sugar fermentation. Lavoisier, in the same century, reported that 95.9 parts of sugar yielded 57.7% ethyl alcohol, 33.3% carbon dioxide, and 2.5% acetic acid. Gay-Lussac was the first to show that glucose fermentation yielded approximately equal weights of ethanol and carbon dioxide (45 parts glucose yielded 23 parts alcohol and 22 parts carbon dioxide). Pasteur found that 100 parts sucrose yielded 105.4 parts invert sugar, which was fermented to 51.1 parts ethyl alcohol, 49.4 parts carbon dioxide, 3.2 parts glycerol, and 1.7 parts succinic acid. In Neuberg's normal fermentation (Scheme 1), 1 mole glucose yields 2 moles pyruvic acid, which, in turn, yield 2 moles carbon dioxide and 2 moles acetaldehyde, which, in turn, are reduced to 2 moles ethanol. Detailed research has shown that there are actually 12 intermediate steps between glucose and ethyl alcohol (the Embden–Meyerhof scheme), which account for the production of 2 moles carbon dioxide, 2 moles ethanol, and smaller quantities of glycerol from 1 mole glucose.

Buchner in the nineteenth century produced ethyl alcohol from sugar using cell-free juice, made by grinding yeast cells with sand under pressure and filtering to remove the whole cells. This demonstrated that it was enzymes that were required for fermentation and not the whole cells themselves.

III. YEASTS

The principal microorganisms in the alcoholic fermentation of wines are yeasts belonging to genus *Saccharomyces*, in particular *S. cerevisiae*, *S. cerevisiae* var. *ellipsoideus*, *S. bayanus*, and *S. oviformis*. Although yeasts belonging to genus *Saccharomyces* are generally present on the grapes or other fruits and they are capable of fermenting the juice to wine, it is generally accepted procedure, particularly in wine companies, to inoculate pure cultures of selected wine yeasts. These may be selected for flavor, production of little or no foam, ability to ferment at low temperatures, or floculation–sedimentation behavior. They are more often selected for the absence of any undesirable flavors in wines.

To ensure that the selected wine yeast predominates over wild yeasts and other microorganisms that may be present, it is also common practice to add 20–50 mg sulfur dioxide/liter to the juice, usually in the form of bisulfite, before inoculating the selected yeast, which is generally tolerant or acclimated to these concentrations of sulfite.

Saccharomyces cerevisiae and other wines yeasts have the ability to grow and metabolize sugars, such as glucose, fructose, and sucrose, aerobically or anaerobically. Aerobic growth yields far more energy and the yeasts multiply vigorously, but little if any ethanol is produced. Anaerobically, yeasts dissimilate glucose according to the Emden–Meyerhof scheme. Theoretically 1 mole glucose (180 g) yields 2 moles (92 g) ethanol and 2 moles (88 g) carbon dioxide. Actually, yields are slightly less because of the production of glycerol and other minor compounds; nevertheless, it is practical to consider that 2 g sugar yields slightly less than 1 g ethanol. Anaerobic dissimilation of 1 mole glucose yields 56 kcal of energy to the yeast, part of which is used in the fermentation itself and a portion in multiplication of the yeast.

Yeasts multiply by budding, and the percentage of yeast cells showing buds when examined microscopically in culture or in a fermenting wine showing buds is a measure of the vigor of the culture.

Calculated on an overall fermentation basis, an individual wine yeast cell produces about 30 million molecules of ethanol/cell/s. In the earlier stages of fermentation, a yeast cell may produce 100 million ethanol molecules/cell/s. As the ethanol content reaches 12% v/v, the rate of ethanol production may fall to 10 million molecules/cell/s. To maintain high levels of ethanol production, it is necessary to introduce small amounts of oxygen (e.g., 13% oxygen saturation) into the must. Some wine producers, wanting to achieve high concentrations of ethanol, expose the must to air during the early stages of fermentation.

IV. MALOLACTIC FERMENTATION

Another group of microorganisms accompanying the yeast in wine fermentations is the lactic acid bacteria. Lactic acid bacteria–yeast interactions are rather common in food and beverage fermentations. Grape juice contains mainly tartaric acid and malic acid. If the acidity is too high for the best flavor, it may be desirable for lactic acid bacteria to ferment the malic to lactic acid, which decreases the total acidity and raises the pH of the wine. It also modifies the aroma, body, and aftertaste (Henick-Kling). If the grape juice is low acid, a malolactic fermentation may be undesirable because it will decrease the acidity still further, making the wine taste flat. Malolactic acid fermentation should occur during the alcoholic fermentation or during holding before the final clarification and fining because it increases the microbiological stability, decreases fermentable carbohydrates, and produces inhibitory bacteriocins in the wine (Henick-Kling). If malolactic fermentation occurs in the bottle, it may result in cloudy wine or carbonation of a still wine.

V. NOBEL MOLD

Under high-humidity conditions, unripened grapes may become infected with *Botrytis cinerea*, the nobel rot. If the humidity decreases, the grapes lose moisture and, in fact, become shriveled. Such grapes yield a juice with a higher sugar content, the grape flavor may also be concentrated, and the mold may introduce a delicate flavor. Under ideal conditions, very fine flavors can be produced from grapes infected with the nobel mold.

VI. MICROORGANISMS CAUSING SPOILAGE IN WINES

The major organisms causing spoilage in wines are those belonging to genus *Gluconobacter* and genus *Acetobacter*. *Acetobacter* includes the organisms producing acetic acid or vinegar from ethyl alcohol. *Acetobacter* is highly aerobic and, in fact, grows on the surface of the must or wine if oxygen is present. The acetic acid bacteria are not generally a problem as long as the fermenting must and the wines are kept totally anaerobic. During active fermentation, carbon dioxide is being rapidly produced, and it blankets the must, keeping it anaerobic.

Other wine-spoilage organisms include film-forming yeasts such as genera *Hansenula* and *Pichia*. *Hansenula* is an esterifying yeast that produces fruity aromas such as ethyl acetate. Osmotolerant yeasts such as *Schizosaccharomyces pombe* may produce haze in stored wines and bottles. *Brettanomyces* produces off-flavors if it develops during storage.

VII. MICROBIOLOGY OF SHERRY WINES

Sherry wines have an oxidized flavor, color, or aroma and are highly acceptable among many wine consumers. The oxidized flavor arises by conversion of a portion of the ethanol to acetaldehyde and other complex reactions. A type of sherry flavor can be introduced by baking a still wine. The still wine is baked (i.e., held at elevated temperatures with air being injected into the wine) to produce the sherry flavor or aroma. In a true fermented sherry, the still wine is filled into half-filled barrels that are incubated in the sun, often on the roof of the winery. A type of baking occurs over a long period of time, and the surface of the wine is exposed to air. The sherry flavor is produced by film-forming yeasts such as *Saccharomyces oviformis* and *S. bayanus* or related yeasts. In the absence of these yeasts, *Acetobacter* could invade and spoil the wine.

VIII. WINE GRAPES

Grape wines are made mainly from species *Vitis vinifera*. This is particularly true in Europe, where the major producers are France and Italy. The native grapes in the United States belong to the species *Vitis labrusca* and several other species that have a characteristic fruity, floral ("foxy") aroma, distinctly different from *vinifera* grapes. Hybrids of *V. vinifera and V. labrusca* have been developed to yield grape varieties that are able to withstand a colder climate and yield wines more similar to the European types. *Vitis rotundifolia* is an important species in the southeastern United States. Most American wine is produced in California, based on the cultivation of *V. vinifera*. *Vitis vinifera* wines are becoming increasingly more important in the northeast United States.

IX. WINE CLASSIFICATION

Wines are classified according to color (white, red, or rosè, pink) and how much ethanol they contain. Wines with 7–14% ethanol are described as table wines; more than 14% ethanol designates fortified wines, such as port and sherry. Wines are dry (nonsweet), semidry, semisweet, or sweet. Even a dry (nonsweet) wine contains a small quantity of sugar.

Still wines given a secondary fermentation in the bottle or in a tank to produce carbon dioxide under pressure are called "sparkling." Outside the United States, the term "champagne" is legally reserved for use only with sparkling wines made in the Champagne region of France. Wines are also designated, at times, by the type of grape from which they have been made.

There are thousands of cultivars of *Vitis* species. Many yield wines with flavors and aromas reflecting their distinct type. The reader is referred to the references at the end of this article for further information. A very wide variety of flavors, aromas, and colors of wines can be produced. The climate and the soil on which the grape variety grows and the particular growing year can have a marked effect on the flavor of the resulting wine. This leads to "vintage" years, in which a particular variety yields unusually high-quality wine.

X. STEPS IN MANUFACTURE OF WINE

1. *Harvesting:* Grapes should be harvested when they are at their peak of maturity and desired flavor with a high sugar content and desirable acidity. They should be as intact as possible and be processed quickly.

2. For white wines, the stems are removed and the grapes are crushed. The skins are retained for red wines. Sulfite may be added to inhibit the oxidation and growth of the natural microbial flora.

3. Pressing is the separation of the juice from the skins, seeds, and pulp—the pomace. Pectinase may

be added to facilitate juice extraction. Free-run juice, the juice running from the press before pressure is applied, is considered to be best for the highest quality wines.

4. *Fermentation:* Fermentation can proceed either through the action of the naturally present yeast or by the inoculation of a selected pure yeast culture. In industry, the yeast inoculum may be produced in the company's own laboratory or it may be purchased as an active dry yeast from companies specializing in the production of industrial cultures. Fermentation temperature is generally 10–15°C for white wines and somewhat higher for red wines. The higher the temperature, up to about 30°C, the more rapid the fermentation; however, the slower, lower temperature fermentations are considered to lead to higher-quality wines because less volatile aroma is lost. As the fermentation proceeds, the actively metabolizing yeasts produce heat and the temperature of the must tends to rise. If not controlled, the temperature can reach a level at which yeast cells are killed or the ethanol becomes so inhibitory that the maximum ethanol content is decreased.

The sugar content of the juice must be sufficient to yield a wine with the desired alcohol content. This requires a sugar content of about 13% w/v to yield a 7% v/v alcohol wine or about 22% to yield a 12% v/v alcohol wine. The sugar content of the juice may be raised before or during fermentation to adjust the final ethanol content. Water may be added to lower the acidity where the law allows amelioration.

A weight of carbon dioxide gas almost equivalent to the weight of ethanol produced is released during fermentation. This causes the fermentation to "boil" during its most active phases.

The fermentation gradually ends as the fermentable sugar is consumed. With red wines, the pomace, skins, seeds, and other insoluble matter must be removed. This is usually performed when the color and tannin extraction have reached the maximum toward the end of fermentation or several days after completion of the fermentation. The separation of the pomace is accomplished by draining and pressing the must. With white wines, the pomace has already been removed during the pressing before fermentation.

5. *Racking:* As the fermentation nears completion, the yeast cells tend to settle, and after fermentation has been completed, the wine can be decanted or siphoned (racked) as the first step in clarification of the wine. Racking is repeated periodically until most or all the yeast cells (the lees, tartrate crystals, and other insolubles) have been separated from the wine. This may be combined with cold stabilization, which is accomplished by storing the wine at about −2°C. Cold stabilization removes excess tartrates and other materials that might cause cloudiness in the bottle later.

6. *Clarification:* Most wine consumers like a crystal-clear wine with no haze or sediment. To accomplish this objective, gelatin, isinglass, or egg white may be added to the wine. These ingredients combine with tannin and form a fine precipitate that removes some of the materials that might cause turbidity. White wines may not contain sufficient tannin. In that case, some added tannin sufficient to react with the added protein may be necessary to produce the precipitate and stabilize the wine. Bentonite, a clay, is also widely used for clarification. Ion exchange may be used to change potassium to sodium tartrate, thus increasing the solubility and reducing the likelihood of its precipitation later.

7. *Bottling:* Wines are generally bottled in dark green or brown bottles to decrease the deleterious effects on quality that can be caused by sunlight. Small amounts of sulfite may be added to inhibit oxidation. Sweet wines may be fortified with ethanol (18–20%) or the wines may be pasteurized or sterile filtered (the choice for high-quality wines) to prevent the growth of contaminants in the bottle.

8. *Aging:* Aging occurs after fermentation. It can occur in tanks, in barrels, or in the bottle. The flavor may be improved through esterification reactions in which small portions of ethanol combine with organic acids in the wine. Aging improves some wines, but has little effect on others.

The great enemy of stability in wine is oxygen and oxidation. Barrels, tanks, or bottles holding the wine must be kept free of oxygen. This is done principally by keeping the containers full or replacing the atmosphere in the container with nitrogen or carbon dioxide. In the case of sherries, however, there is a deliberate oxidation to modify the flavor, as discussed earlier.

XI. SPARKLING WINES

Sparkling wines require unique processing, at least in the secondary fermentation that occurs in the bottles or in tanks under pressure (Charmat process). The initial sugar content of the grapes should yield a dry still wine with about 10% v/v ethanol. The base wine, cuvee, is stabilized, fined, and filtered. It is often a blend of several wines to obtain the acidity and alcohol content desired with a clean flavor—no off-flavors or off-odors. Special wine yeasts that tend to agglomerate and thus can be more easily removed after the secondary fermentation are generally used. These same yeasts may also be used for the primary fermentation if desired. It is particularly important that the yeasts do not produce any off-odors or off-flavors, as part of the sparkling-wine flavor is derived from the yeast cells as they autolyze after the secondary fermentation. Sugar addition for the secondary fermentation is carefully controlled, as too little added will lead to a flat sparkling wine and too much sugar can lead to dangerous pressures and even explosions of the bottles. A sugar concentration of 2.4% w/v is enough to produce 7 atm pressure in the sparkling wine bottles, which generally will withstand 8 atm pressure.

The secondary fermentation can occur either in the bottle or in special pressurized tanks designed for the purpose (Charmat process). The tanks offer the most efficient, lowest cost method of manufacturing sparkling wines; however, fermentation in the bottle with the longer contact of the wine with the yeast, particularly during autolysis, even though less efficient and more costly, generally yields the highest quality sparkling wines. If fermentation occurs in the bottle, the yeast cells and any precipitates must be removed after the secondary fermentation. This involves the riddling process, in which bottles are placed neck down in racks and turned a quarter turn every day or so until the yeast and all sediment in the bottle is in the neck against the cork or a crown seal. The bottles are then cooled to close to freezing, and the neck of the bottle containing the yeasts and precipitates is frozen. The bottle is then opened and held at such angle that the ice plug is blown out of the bottle. The volume lost is immediately replaced with a small quantity of sugar in wine (sometimes brandy), called "dosage." A dry sparkling wine is called brut and contains from 0.5–1.5% sugar; sec indicates 2.5–4.5% sugar; doux contains 10% sugar. The bottles are sealed with a final cork and stored on their sides to age or labeled and sent to the distributor.

XII. FORTIFIED WINES

Although it is possible to ferment wines to ethanol levels of 18% v/v or higher if sufficient sugar is present and a lengthy fermentation is acceptable, wines containing higher concentrations of ethanol (i.e., 17–21%) have been generally fortified to the higher ethanol concentration by the addition of wine spirits produced by distillation of wine. Wine spirits contain about 95% ethanol v/v and are generally manufactured by the distillation of lower-quality wines.

Sweet dessert wines such as port, muscatel, or sherry, with sugar contents of 12%, generally start with grapes containing more than 25% sugar. Such wines are fortified by the addition of wine spirits to an ethanol content of 17–21% v/v. *Acetobacter* is inhibited at this ethanol concentration, and such wines are quite stable at room temperature even in partially filled bottles.

XIII. FLAVOR AND ACCEPTABILITY

Wine preferences are strictly personal. No other fermented beverages offer so much variety in flavor, aroma, color, sweetness, and carbonation. Consumers should drink what they like within their economic means. There are many excellent values at reasonable cost on the market. Many new wine consumers prefer the sweet wines initially, but, over time, come to prefer dry wines, which provide more of the subtle flavors and aromas characteristic of individual grape varieties and which vary from country to country.

See Also the Following Articles

BACTERIOCINS • FOOD SPOILAGE AND PRESERVATION • INDUSTRIAL FERMENTATION PROCESSES • YEASTS

Bibliography

Amerine, M. A. (1980). "The Technology of Wine Making," 4th ed. Avi, Westport, CT.

Amerine, M. A., and Roessler, E. B. (1983). "Wines. Their Sensory Evaluation." W. H. Freeman & Co., San Francisco.

Foy, J. J. (1994). Use and manufacturing of active dry wine yeast cultures. *In* "Proceedings of the 23rd Annual New York Wine Industry Workshop" (T. Henick-Kling, ed.), pp. 21–28. New York State Agricultural Experimental Station, Cornell University, Geneva, NY.

Henick-Kling, T. (1994). Considerations for the use of yeast and bacterial starter cultures: SO_2 and timing of inoculation. *Am. J. Enol. Vitic.* 45, 464–469.

Henick-Kling, T. (1995). Control of malo-lactic fermentation in wine: Energetics, flavour modification and methods of starter culture preparation. *J. Appl. Bacteriology Symp.* Suppl. 79, 29S–37S.

Henick-Kling, T., and T. E. Acree. (1998). Modification of wine flavor by malolactic fermentation. *In* "Proceedings of the 27th Annual New York Wine Industry Workshop" (T. Henick-Kling, ed.), pp. 67–74. New York State Agricultural Experimental Station, Cornell University, Geneva, NY.

Johnson, H. (1977/1985). "The World Atlas of Wine." Simon and Schuster, New York.

Kunkee, R. (1994). The making of wine: An overview about the microorganisms in wine. *In* "Proceedings of the 23rd Annual New York Wine Industry Workshop" (T. Henick-Kling, ed.), pp. 1–6. New York State Agricultural Experimental Station, Cornell University, Geneva, NY.

Lafon-Lafourcade, S. (1983). Wine and brandy. *In* "Biotechnology, vol. 5. Food and Feed Production with Microorganisms" (H. J. Rehm and G. Reed, eds.), VCH, New York.

Pool, R., and Henick-Kling, T. (1991). "Production Methods in Champagne." New York State Agricultural Experiment Station, Cornell University, Geneva, NY.

Reed, G., and Nagodawithana, T. W. (1991). "Yeast Technology," 2nd ed. Van Nostrand Reinhold, New York.

Sponholz, W. R. (1994). Identification of wine aroma defects caused by yeast and bacteria. *In* "Proceedings of the 23rd Annual N.Y. Wine Industry Workshop" (T. Henick-Kling, ed.), pp. 78–95. New York State Agricultural Experimental Station, Cornell University, Geneva, NY.

Troost, R. (1988). Technologie des Weines. "Handbuch der Lebensmitteltechnologie." Ulmer (Stuttgart).

Vine, R. P. (1981). "Commercial Winemaking." Avi, Westport, CT.

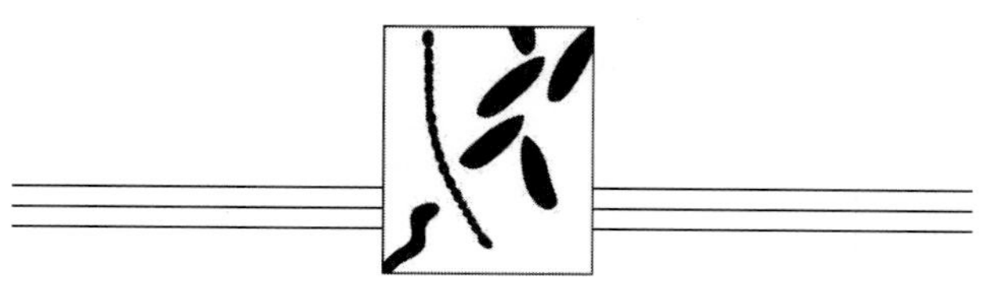

Xanthomonas

Twng Wah Mew
International Rice Research Institute, Philippines

Jean Swings
Laboratory of Microbiology, Gent, Belgium

I. The Bacteria
II. Diseases Caused by *Xanthomonas*
III. By-Products for Industry

GLOSSARY

aerobic Providing adequate oxygen for microorganisms to grow.

aerosol Fine solid or liquid particles that through the operation of harvesting of a crop dispense in the air.

avirulent Weakly or less pathogenic.

epitope An antigenic determinant of defined structure (e.g., an identified oligosaccharide).

gene-for-gene relationship The correspondence of each gene determining resistance in the host variety to the related gene determining virulence in the pathogen.

genotype The genetic makeup of individual crop varieties, often an indication of their parental relationship.

pathogenicity The ability of an organism (microorganism) to cause disease on a plant in nature. The plant or plant species is the host of the organism and the organism is a pathogen.

pathovar **(*pv.*)** A subdivision of bacterial taxonomy within a named species, characterized by a pathogenic reaction of the bacteria in one or more hosts.

phenotype The morphological or other properties of individual crop varieties, often a particular trait, plant height, type, or yield.

phylogenetics The genetics between closely related organisms, in a species or population, indicating an evolutionary relationship.

polyphasic Applying information from different disciplines or areas of research in bacterial taxonomy.

seed health The health status of seeds or seed lots, often in relation to the physical appearance, discoloration due to diseases or injury or other abnormal physiology, or contamination with other undesirable objects, such as weed seeds, soil particles or plant debris.

virulent Strongly pathogenic.

ALL SPECIES OF *XANTHOMONAS* (after the exclusion of *X. maltophilia* from the genus) are traditionally known to be plant pathogens. However, there is growing evidence that many bacterial populations, living in close association with plants but causing no apparent disease symptoms on the host from which they were isolated, also belong to genus *Xanthomonas.* The nonpathogenic forms of *Xanthomonas* differing from avirulent strains of the pathogens appear to be distributed widely in plants, especially among agricultural and horticultural crops. They are found in postharvest soft rot of vegetables and fruits. Many of the pectolytic bacteria from tomato and pepper transplants, in plant debris, weeds from soils under citrus trees, and from rice seeds and olive leaves are also identified to be *Xanthomonas,* even though it has not been established that they cause any disease on these plants. It is known, however, that they do not cause diseases when artificially inoculated into the host plants from which they were isolated. From an economic point of view, these bacteria appear to have no major importance. As they are frequently isolated and as they occur often in large number, their ecological and evolutionary role in the microbial community needs to be further understood. In addition to being plant pathogens, the bacterial cells of *Xanthomonas* produce a large amount of extracellular polysaccharide, a unique product used in oil drilling and the food industries. Although these microorganisms are plant pathogens and the pathogenic activity poses a threat to our food, vegetable, fruit, and fiber supply annually, the by-products produced contribute significantly to our daily life.

I. THE BACTERIA

A. Taxonomy

Xanthomonas is a strictly aerobic gram-negative rod, with motile bipolar flagella. The majority of the strains form mucoid colonies with characteristic yellow water-soluble pigments (i.e., xanthomonadins) that have a mono- or di-bromoaryl polyene structure typical of the genus *Xanthomonas*. The high phenotypic resemblance among the various species and strains within the genus is very striking.

For glucose metabolism, *Xanthomonas* uses the Entner–Doudoroff pathway in conjunction with the tricarboxylic acid cycle (80–90%); a small portion (10–20%) is routed via the pentose phosphate pathway. A number of important enzymatic activities involved in pathogenesis have been demonstrated (e.g., starch hydrolysis; lignin breakdown; and proteolytic, cellulolytic, and pectinolytic activities).

A polyphasic approach to the taxonomy of *Xanthomonas* has allowed its recent reclassification. In the former classification system, only six species were recognized, whereas the new classification recognizes 20 species—*X. albilineans, X. alboricola, X. axonopodis, X. bromi, X. campestris, X. cassavae, X. codiaei, X. cucurbitae, X. fragariae, X. hortorum, X. hyacinthi, X. melonis, X. oryzae, X. pisi, X. populi, X. sacchari, X. theicola, X. translucens, X. vasicola, and X. vesicatoria.*

Some species have narrow host specificity, whereas others have a broad host range; for example, *X. arboricola, X. axonopodis, X. hortorum, X. translucens,* and *X. vasicola,* contain more than three pathovars. The pathovar system, based on host specificity and pathogenicity to variety of host species, should be treated with some caution as it has been shown that some pathovars do not represent real biological entities inasmuch as they are composed of different genotypes that can be detected by fine genotypic fingerprinting methods.

A phylogenetic study based on 16*S* rDNA sequences of all the species of *Xanthomonas* has revealed their high mutual relatedness and also their relatedness to *Stenotrophomonas* (its nearest neighbor) and to *Xylella,* both of which are typical plant-associated bacteria.

At one time, *X. maltophilia* was considered a species under *Xanthomonas,* the only species in the genus that is not a plant pathogen. *X. maltophilia* appears to be a heterogeneous population of many opportunistic bacteria that are, to some extent, similar but not related to *Xanthomonas.* They have limited nutritional requirements. The colonies are smooth and glistening, with the entire margin white, grayish, or pale yellow. The pigments of the pale yellow *X. maltophilia* are not xanthomonadins. Later, a new genus, *Stenotrophomonas,* was established and all the bacteria in *X. maltophilia* were included in this genus as *Stenotrophomonas maltophilia* (the only species listed in this genus). There may be some overlap in natural habitats of soil and plant materials with *Xanthomonas,* but *Stenotrophomonas* bacteria are more frequently isolated from clinical materials, causing human infection.

B. Genetics

Research during the 1990s has shown that *Xanthomonas* is amenable to genetic analysis through mutagenesis and genetic exchange. In the beginning, the major constraint to genetic analysis was the difficulty of isolating genes and introducing them into hosts, but since then mobilizable broad-host-range vectors and various molecular tools have been described. This allows the genetics of *Xanthomonas* as a plant pathogen to be studied, which has provided a better understanding of microbial population biology as well as a wider view of the host–parasite interaction. In the past few years, research achievements in the genetics of *Xanthomonas* include the cloning of avirulence genes and the development of techniques to detect the pathogen and to assess the genetic diversity of bacterial populations in different environments.

1. Genetics of Host Recognition and Response

In the gene-for-gene hypothesis of host–pathogen interaction, an elicitor molecule that is directly or indirectly a product of an avirulence gene is recognized by a plant receptor, and this results in the induction of a cascade of reactions in the host that leads to a hypersensitive response. Pathogen resistance is often correlated with hypersensitive response, which manifests as localized cell death at the site of infection.

In host–parasite coevolution, avirulence genes may have been significant in the adaptation of *Xanthomonas* pathovars to host cultivars. Apparently there are two types of avirulence genes; one is involved in the recognition of related pathogens in nonhost plants, whereas the other has the ability to elicit resistance in a cultivar-specific manner. The first of such genes to be identified was *avr*Rxv from the pepper pathogen *X. campestris* pv. *vesicatoria.*

The development of various molecular techniques paved the way for the efficient cloning of genes involved in pathogenesis, that is, avirulence genes from a wide range of bacteria and resistance genes from various plant species. Over 30 bacterial avirulence genes have been cloned and characterized. Interestingly, avirulence genes that were isolated from *Xanthomonas* pathovars, such as pv. *vesicatoria,* pv. *malvacearum,* pv. *citri,* pv. *raphani,* and *Xanthomonas oryzae* pv. oryzae, have been shown to share sequence similarity and belong to one gene family, the *avrBs3.* There are also other *avr* genes from *Xanthomonas,* which show no apparent similarity to *avrBs3.* As in other bacterial systems, the physiological function of these *avr* genes in eliciting host response has not been completely understood. Only a few are known and *avr*D in *Pseudomonas syringae* pv. *syringae* is one of them (for details, see Wang and Leung, 1998).

Another set of genes, the *hrp* (hypersensitive response and pathogenicity) genes, also functions to elicit the hypersensitive response. In most if not all cases, *avr* gene function has been shown to be dependent on the *hrp* genes. Research in these areas will continue to provide insights into the mechanism of host–pathogen interaction.

2. *Resistance Genes*

Resistant variety has been one of the most effective, ecologically sound, and economically feasible method of disease control. There are many resistance genes that have been identified against many *Xanthomonas* pathogens in many crop species. In rice, there are over 20 *Xa* genes identified and two of them, *Xa1* and *Xa21,* have been cloned. Interestingly, all the resistance genes cloned so far encode one or more conserved motifs such as leucine-rich repeats (LRR), nucleotide-binding sites (NBS), and kinase motifs. For instance, *Xa21* encodes a protein containing LRR and kinase motifs, whereas *Xa1* encodes a product with LRR and NBS. The implication of these cloned R genes is that the genes can be introduced directly into popular cultivars through tissue transformation to develop transgenic lines. Cloned *Xa21* has been used to develop transgenic lines derived from a tissue culture of IR 72, a popular cultivar in rice production in tropical Asia. The significance of this technology advancement is that in conventional breeding, it will take 6–7 years to develop a new crop variety with a new gene. Using genetic engineering through transformation, the time needed to develop a variety with a targeted gene will be shortened to between 6 months and 1 year. The constraint is that not all crop varieties can be easily transformed. Also, the evaluation of such transgenic lines in the field is guided by a set of biosafety guidelines available in many countries before they are released to farmers.

3. *Genetic Diversity of Pathogen Population*

In their natural habitat, bacterial pathogens survive as a heterogeneous population. The challenge is to develop appropriate methods to identify and characterize them in order to effectively manage the diseases they cause. Populations of *Xanthomonas* pathogens have been differentiated and described in many ways.

To define the virulence of the pathogen strains and to understand the variability of the pathogen, plant pathologists rely conventionally on host cultivars or lines with different resistance genes to test a range of pathogen strains. The differential interactions are confined to specific cultivar-strain combinations and thus demonstrate a degree of pathogenic specialization. This differential system used to separate bacterial populations is important in plant pathology and resistance breeding. It has offered a simple means of locating resistance genes for developing resistant cultivars and to classify populations to reveal pathogen variability. The phenomenon is common in many *Xanthomonas*–host plant systems, such as *X. oryzae* pv. *oryzae*–rice, *X. campestris* pv. *malvacearum*–cotton, and *X. campestris* pv. *vesicatoria*–tomato or sweet pepper. In rice, near-isogenic lines containing a series of resistance genes from *Xa1* to *Xa21* in the genetic background of the recurrent parent IR24 have been developed. These lines are used for race diagnosis in many Asian countries to establish the different populations of the bacteria. The variability of the bacterial pathogen was observed in response to the planting of host cultivars with resistance to the disease, and, thus, the population of the bacterial pathogen demonstrated a shift in virulence. This is a phenomenon closely associated with host–parasite coevolution.

Molecular markers have been identified that could differentiate strains of *Xanthomonas* pathovars. Among these are the 16*S* + 23*S* rRNA genes, the *avr* genes, the *hrp* gene cluster, and the repetitive elements such as the IS *1112* from the genome of *Xanthomonas oryzae* pv. *oryzae* and the REP, BOX, and ERIC sequences from other bacterial genomes. These techniques provide a means of assessing the diversity of the pathogen population and its relation to the host and geographical origin in determining the phylogeny and in monitoring changes in the pathogen population structure in relation to the resistance genes in crop cultivars.

Xanthomonas campestris pathovars were characterized by rRNA gene-restriction patterns, which allow differentiation at the genus, species, and pathovar levels. Intrapathovar variability was not shown for pv. *begoniae,* but it could be related to ecotypes for pv. *manihotis,* and to both the host plant and pathogenicity for pv. *diffenbachiae.* In some pathovars such as the pv. *malvacearum,* however, the variability observed was not correlated with pathogenicity or with geographical origin. The genetic analysis of *hrp*-related DNA sequences of *X. campestris* strains causing diseases in citrus indicated that strains in group A, B, and C, which cause citrus canker A, B, and C, respectively, produced characteristic restriction-banding patterns of amplified *hrp* fragments. The restriction fragment length polymorphism (RFLP) was evident among strains of the moderately and weakly aggressive groups.

Molecular markers are also available for tracking the evolution of *Xanthomonas* in relation to the resistance of the host plant in the field. Polymerase chain reaction (PCR)-based assays including RAPD (random amplified polymorphic DNA), LMPCR (ligase-mediated PCR), and rep-PCR have been useful for this purpose. Using a combination of molecular markers together with virulence typing (inoculation of bacterial strains on a set of differential cultivars), the differentiation of the *X. oryzae* pv. *oryzae* population was revealed on various spatial scales in farmers' fields in Asia. The information provided insights into the evolution of the rice *Xanthomonas* pathogen in relation to the rice cultivars grown by the farmers in the different countries.

4. Detection

Molecular techniques have been developed and used to detect the bacteria in soil, water, and plants. The application of molecular tools in the routine detection of the pathogen often relied on the speed and sensitivity of the polymerase chain reaction. Various primers based on bacterial repetitive sequences, plasmid, RAPD fragments, and rDNA sequences have been designed to be specific for a species or pathovar. Among these are the primers for detecting *X. c.* pv. *vesicatoria*, *X. c.* pv. *phaseoli* and var. *fuscans*, *X.* albilineans, *X. c.* pv. *pelargonii*, *X. c. translucens*, *X. axonopodis* pv. *citri*, and *X. o.* pv. *oryzae*.

5. Other Areas

Advances in *Xanthomonas* genetics have also been made in other areas. A brief summary is provided here. It is likely that more progress will appear in the literature in the coming years.

1. Extracellular enzyme production (e.g., amylase, proteases, pectinases, and lipases). Both pectinolytic and nonpectinolytic activities have been demonstrated in *Xanthomonas* (*in vitro* and *in planta*), but the significance of pectinolysis in pathogenicity is not understood. By cloning protease genes and Tn5 mutagenesis, it was shown that the proteases are not critically important for plant pathogenicity.
2. Extracellular polysaccharide biosynthesis. A linear sequence of 12 genes responsible for xanthan production is known. The genes are designated as *gum* B to *gum* M, comprising transferase, acetylases, ketalase, and polymerase functions.

3. Regulatory genes that regulate the expression of other genes in response to environmental signals are also characterized. The positive regulatory gene *rpf*C allows the bacteria to adjust their biosynthetic activities to the changes in their environment, the plant.

C. Ecology

Prior to the advances of molecular markers, serological markers were widely used to identify strains and species of *Xanthomonas* for ecological and epidemiological studies. In serological techniques, both polyclonal and monoclonal antibodies (PAb, and MAb) are used to distinguish strains within species or subspecies. In general, PAbs are specific at species, subspecies, or pathovar level, whereas MAbs are specific at the strain level.

Several monoclonal antibodies of *X. oryzae* pv. *oryzae* (Xoo) have been produced for identifying Xoo strains and that have become useful for ecological studies. For instance, MAb Xco-1 reacts with strains of Xoo from different geographical locations, but does not react with strains of other species of *Xanthomonas* or with bacterial genera of plant or animal origin. The epitope detected by this antibody appears to be a unique marker characteristic of *X. oryzae* pv. *oryzae* strains. There are also MAbs that appear useful for identifying subgrouping or races of *Xanthomonas* species. For instance, Xco-2 reacts with most but not all Xoo strains, whereas Xco-5 reacts with strains of limited geographical location.

Monoclonal antibodies that are capable of differentiating strains of other pathovars of *X. campestris* have also been developed. These were useful in separating strains of pv. *dieffenbachiae* into groups that corresponded loosely with the host of origin, and in *X. albilineans.*

The molecular techniques discussed in Section I. B are also applicable to ecological studies and have provided very useful information, especially in relation to the genetic diversity of the bacterial population in spatial and temporal distributions.

1. *Functional Relationship*

Xanthomonas has been found to be in close association with plants. All the species in the genus have been reported to be plant pathogens. However, many nonpathogenic *Xanthomonas* are also reported. Nonpathogenic *Xanthomonas* differ from the avirulent forms of the pathogenic strains. They do not cause disease in the host plant from which they are isolated. Whether this is due to loss of virulence is not certain. Whether they are pathogens of other plant species is not known. In principle, these forms can be tested against many plant species to determine if they are pathogenic to any of the plants from which they are not isolated. In reality, it is not feasible to test all plant species.

The nonpathogenic forms do impose a problem for diagnosis in relation to seed-health testing for the purposes of plant-quarantine regulation. Because the nonpathogenic forms are distributed widely and abundantly on plant surfaces and seeds, their detection on seeds or plant products intended for international trade could cause a disadvantage for the senders. It is important, therefore, that the nonpathogenic *Xanthomonas* be distinguished accurately and rapidly from the pathogenic forms. The wide occurrence of nonpathogenic seed-borne or saprophytic *Xanthomonas* associated with seed lots of rice, cereal, and legumes has complicated plant quarantine and disease treatment. If nonpathogenic *Xanthomonas* cannot be differentiated from the pathogenic forms, seed lots may be unnecessarily rejected or subjected to expensive control treatment.

The actual identity of nonpathogenic *Xanthomonas* is an issue of concern. The pathovar systems of *X. campestris* does not allow them to be classified by host pathogenicity or specificity. Because they are not pathogens, there is no clear host range for this group of *Xanthomonas.* None of the nonpathogenic strains of *Xanthomonas* have been identified as belonging to the pathovars of the host plant from which they were isolated. In most cases, fatty acid analysis and SDS-PAGE protein profile analysis allow the identification of nonpathogenic *Xanthomonas* within the genus. Although most of the strains investigated seem to belong to the *Xanthomonas* species, in particular *X. arboricola* and *X. axonopodis,* they usually cannot be unequivocally identified as known pathovars within these species. Nonpathogenic *Xanthomonas* strains clearly form a heterogeneous population. Their relationship to characteristic pathogenic strains has not been sufficiently clarified and this area of study deserves more attention. A better understanding of the taxonomy and ecology of nonpathogenic *Xanthomonas* is expected to contribute to the detection and management of plant diseases associated with *Xanthomonas* taxa.

In general, each host species may have unique species or pathovars of *Xanthomonas*. The nonpathogenic forms may be present together with the pathogenic forms. However, the relationship between the two is not clear. Because all the classified species of *Xanthomonas* are plant pathogens, it is evident that each species has a unique relationship to the host-plant species or related species. Thus, *X. oryzae* is associated with plant species in the genus *Oryza,* whereas *X. fragariae* is only associated with species in the genus *Fragaria*. Because *X. campestris* consists of many pathovars, each pathovar has a unique association with the host-plant species or related species. Thus *X. c.* pv. *cassavae* attacks only *Manihot* spp. In rice, *X. o.* pv. *oryzae* and *X. o.* pv. *oryzicola* may infect the same rice plant and produce distinct lesions. Both bacteria have been detected from the same plant tissues or seeds. In the leaf tissue, however, pv. *oryzae* only infects the xylem tissues and pv. *oryzicola* attacks the parenchyma tissue.

2. *Survival in Natural Habitat*

Bacteria in the genus *Xanthomonas* do not survive well outside of their primary host because of desiccation and solar radiation. *Xanthomonas* pathogens survive in several ways—in seed, in plant residues, in perennial hosts as epiphytes on nonhost or related host plants, or as saprophytes in soil and insects. The life cycle of *Xanthomonas* may be distinguished into parasitic, epiphytic, and saprophytic phases. Although not all *Xanthomonas* have these three phases, when the conditions require it, the *Xanthomonas* may survive in any one of these phases to sustain its life cycle. The parasitic phase is when the bacterial cells enter the host tissues and begin to multiply and eventually induce disease symptoms. Bacteria may survive as epiphytes on the plant surface, including seeds before entering the host tissues. Epiphytic survival of *Xanthomonas* is well established (e.g., *X. c.* pv. *manihotis, X. c.* pv. *glycines, X. c.* pv. *pelargonii,* and *X. c.* pv. *pruni*). Some important seed-borne *Xanthomonas* include *X. c.* pv. *campestris, X. vesicatoria, X. translucens, X. c.* pv. *glycines, X. c.* pv. *phaseoli, X. c.* pv. *malvacearum, X. c.* pv. *manihotis, X. c.* pv. *carotae, X. c.* pv. *pelargonii,* and *X. c.* pv. *zinniae*. The saprophytic survival ability of most *Xanthomonas* is very poor. Saprophytic survival in soil and plant debris is usually short-lived because of the presence of other antagonistic bacteria. *X. oryzae* pv. *oryzae* survive in paddy water for several months, but not in sterile distilled water at room temperature.

Bacteria may survive parasitically on secondary or volunteer host plants, either as epiphytes or as pathogens in diseased tissues. For instance, *X. vesicatoria,* the bacterial spot pathogen of tomato and pepper, may survive between seasons in lesions on volunteer plants and on seeds harvested from diseased plants. Its survival on weeds and solanaceous plants, either as epiphytes or in restricted lesions, may occur in some areas of the world. However, the saprophytic survival of the bacterium is very poor. It does not compete well with soil bacteria and disappears from soils as the vegetative materials containing the bacteria decompose. These examples illustrate the survival of xanthomonad pathogens:

1. *X. c.* pv. *citri* can survive in trees from year to year. The bacterial cells survive in soils and on the surfaces of weeds for short periods.
2. *X. c.* pv. *campestris,* causal agent of black rot in cabbage, survives in the seeds of crucifers and crop residues, which then serve as a major source of inoculum and also as a medium for survival away from host plants. In addition, the bacterium attacks many weed hosts and also survives as epiphytes on symptomless leaves.
3. *Xanthomonas* may survive in plant debris or on soil as long as the host tissue is not decomposed. Thus, *X. c.* pv. *campestris* survives in cabbage stalks in soil for as long as 2 years. Similar observations have been reported for *X. c.* pv. *malvacearum* in dry plant tissues and for *X. c.* pv. *phaseoli* and *X. o.* pv. *oryzae* in infected or infested straws.

II. DISEASES CAUSED BY *XANTHOMONAS*

Bacteria in genus *Xanthomonas* are important because of their ability to cause diseases in agricultural, horticultural, and fiber crops. Species of *Xanthomonas* are known to be plant pathogens. The damage caused by these pathogens often leads to epidemics, resulting in serious food and fiber shortage. Acting singly or in combination with one another, plant pathogens are agents that cause diseases.

A. Host Range

Xanthomonas has infected at least 124 monocotyledonous and 268 dicotyledonous plant species. Among monocotyledonous species, the range of hosts extends across 11 families, comprising at least 70 genera. The bacteria have a host range in 57 dicotyledonous families comprising more than 170 genera. From the view point of agriculture and horticulture, the hosts of plant species consist of those that are major food crops (e.g., rice, wheat, maize, and cassava), horticultural crops such as vegetables (e.g., cabbage, beans, tomato, and pepper) and fruits (e.g., citrus and stone fruits), oil crops (e.g., soybean), fiber crops (e.g., cotton), important forage crops (e.g., alfalfa), and industry crops (e.g., sugar beet, sugar cane, and poplar). The host specificity of *Xanthomonas* species or pathovars is generally limited to one genus or some closely related plant genera. In many cases, the name of the pathogen was derived from the genus or species name of the host plant (e.g., *X. oryzae* from *Oryza sativa,* rice, *X. c.* pv. *manihotis* from *Manihot* spp, cassava, and *X. c.* pv. *citri,* citrus). The host range of the bacterial pathogen is ideally based on natural hosts. Negative results from the artificial inoculation of a bacterial pathogen in a range of plant species do not necessarily prove that the plant species is not a host because conditions and the susceptibility of the host-plant species vary. Likewise, if the results are positive, the symptom(s) or lesions induced by the pathogen must resemble the symptoms found in natural disease situations.

B. Invasion of the Host Plant

Xanthomonas causes many types of diseases that are often associated with various visible changes or symptoms in the plant.

1. Invasion and the Infection Process

Xanthomonas, as in other pathogenic bacteria, have no mechanism of active penetration into host tissues. *Xanthomonas* infects and enters the host plant via several venues—natural openings such as stomata, water pores of hydathodes, lenticels, nectaries, or wounds.

When the main infection takes place in vascular tissues, the primary portals of entry are the water pores of the hydathodes. For instance, *X. c.* pv. *campestris* and *X. o.* pv. *oryzae* enter cabbage or rice plants via the water pores. They may, by accident, be pushed into the stomata; the multiplication of the bacterial cells in these "unnatural" sites is marginalized and the resulting lesions are also localized. However, in the "natural" sites of entry (perhaps related to the tissue specificity of infection), as in the case of *X. o.* pv. *oryzae,* within a few hours after inoculation virulent bacteria gain entrance through the water pores. The bacteria then multiply in the epithem and invade the vessels through the vascular pass, linked directly to the vascular tissue.

For bacteria that cause the main infection in parenchyma tissues, entry through stomata is common. The resulting symptoms are normally the necrotic type. A few examples are the angular leaf spot of bacterial blight of cotton caused by *X. c.* pv. *malvacearum,* lesions (leaf spots) of common bacterial blight of bean caused by *X. c.* pv. *phaseoli,* and circular lesions of bacterial spot of tomato and pepper caused by *X. c.* pv. *vesicatoria.*

2. Long-Distance Dissemination

The transmission of bacterial cells from sites, organs, or tissues to new host plants is critical in the life cycle of a plant pathogen. The means of transmission determines whether this is long-distance dissemination. In the absence of a primary host, the duration of survival determines the efficiency of the pathogen. The mechanism of pathogen dissemination affects the type of epidemic that *Xanthomonas* may cause. Generally, *Xanthomonas* are disseminated through seed, planting materials, and farm tools. Wind-driven rain or sprinkler irrigation is important for the dispersal of bacterial pathogens.

Seed is a common means of long-distance dissemination of most bacteria in the genus. Seeds are known to carry bacterial pathogens harvested from fields with severe infection. Seeds are either infected or contaminated during harvest.

Movement or introduction of infected planting or propagating materials is another means of disseminating *Xanthomonas*. In the bacterial blight of cassava, the distribution of infected vegetative planting material from diseased plantations has been the means of disseminating the pathogenic bacterium, *X. c.* pv. *manihotis,* over long distances from Latin America to Africa and Asia. It is also the source of primary infection in newly established plantations. This is due mainly to the lack of visible symptoms on the stems and to the ability of the pathogen to survive in the invaded tissue for a very long time. Long-distance spread of *X. c.* pv. *citri* is usually through the introduction of diseased propagating material such as budwood, root stocks, and budded trees. *X. c.* pv. *pruni* can be transmitted to plum and apricot nursery trees by budding.

Transmission of *Xanthomonas* also occurs through plantation operations. For instance, *X. c.* pv. *pruni* is transmitted from diseased to healthy trees during pruning. The spread of *X. c.* pv. *hyacinthi,* the causal pathogen of yellow disease in *Hyacinthus,* is mainly due to infested tools, knives used for flower cuttings, and inexperienced workers.

All *Xanthomonas* pathogens can be dispersed by wind-driven rain for short distances. DNA typing of *X. o.* pv. *oryzae* in rice fields during the rainy season indicates that most of the inoculum originates from neighboring fields. The distance appears to be closely related to wind speed during the rainstorms or typhoons.

Fruits may harbor the pathogen for potential long-distance dissemination. Bacterial cells have been detected in diseased citrus or other stone fruits in trade, but the establishment of disease due to inoculum carried by these fruits has not been reported.

Aerosol particles are often naturally generated when water droplets hit wet bacteria-covered leaf surfaces. Aerosols remain suspended in the air for varying periods of time, depending on particle size, rain, and wind. Aerosol dispersal from cruciferous weeds was an important source of primary inoculum for short-distance dispersal

C. Damage

Xanthomonas causes varying kinds of damage to host plants. One direct effect is interference with the normal physiological processes of the plant. Growth and productivity of the grains and fruits or weight of the fibers that the host plants normally produce are affected. The indirect effect has something to do with the quality of the product produced by the host plant. For instance, in rice, cotton, and cassava, the severe damage caused by their pathogens in terms of yield losses ranged from 10–50% or more, depending on locality, planting time, resistance of host varieties, plant growth stage when infection took place, and environment (both physical and nutritional conditions of the host plant, as well as the site where the host plant is grown). The quality losses are reflected in the marketability of the product. For instance, not only will the market value of citrus with citrus canker or mango with bacterial black spot decrease but potentially, the fruit can be targets of international trade restrictions. Similarly, the market value of beans, and tomato and pepper may be affected by common bacterial blight caused by *X. c.* pv. *phaseoli* and bacterial spot caused by *X. vesicatoria,* respectively. In general, quantitative losses especially of cereals, fibers, and horticultural crops are a serious concern.

III. BY-PRODUCTS FOR INDUSTRY

A. Xanthan Production

Xanthomonas typically produces slimy colonies on solid media and viscous liquid cultures. Hydrolysis of xanthan results in D-glucose, D-mannose, and D-gluconic acid in the molar ratio of 2:2:1. Trisaccharide side chains are attached to the main cellulose chain at the C-3 position on alternate D-glycosyl residues. Varying amounts of *O*-acetyl groups and pyruvate are also present. A linear sequence of the genes responsible for xanthan production has been determined.

In the laboratory, xanthan can be produced by batch and continuous cultures; for commercial uses, batch production is used. Laboratory synthesis is performed in a medium containing 2–3% glucose and 0.1–0.2% ammonium salt and requires oxygen. The increasing viscosity of the medium rapidly leads to oxygen limitation and necessitates vigorous agitation. In industrial xanthan production, molasses and starch may be used as a carbon source and yeast hydrolysates or distiller's solubles as the nitrogen source.

Xanthan production continues during the growth and stationary phases. The recovery of xanthan is preceded by pasteurization in order to kill the bacteria, followed by the precipitation of xanthan by using ethanol or isopropanol.

B. Properties and Uses of Xanthan

All industrial uses of xanthan are derived from its physical properties. Aqeuous solutions of xanthan at high temperatures, in various electrolyte concentrations, and in a broad pH range are important for industrial applications. These applications include agrochemicals, explosives, laundry chemicals, and food applications such as in salad dressings, cheese, ice cream, and beer. Unfortunately, xanthan commands a very high price.

Acknowledgment

The senior author acknowledges the assistance of Ms. Ellen Regalado in reviewing the manuscript.

See Also the Following Articles

Aerosol Infections • Biopolymers, Production and Uses of • Extracellular Enzymes

Bibliography

Goto, M. (1992). Fundamentals of bacterial plant pathology. Academic Press, San Diego, CA.

George, M. L. C., Bustamam, M., Cruz, W. T., Leach, J. L., and Nelson, R. J., (1997). Movement of Xanthomonas oryzae pv. oryzae in southeast Asia detected using PCR-based DNA fingerprinting. *Phytopathology* **87**, 302–309.

Leyne, F., De Cleene, M., Swings, J. G., and De Ley J. (1984). The host range of the genus *Xanthomonas*. *Botan. Rev.* **50**, 308–356.

Leach, J. E., Leung, H., Nelson, R. J., and Mew, T. W. (1995). Population biology of Xanthomonas oryzae pv. oryzae and approaches to its control. *Curr. Opin. Biotechnol.* **6**, 298–304.

Leach, J. E. and White, F. F. (1996). Bacterial avirulence genes. *Annu. Rev. Phytopathol.* **34**, 153–179.

Palleroni, N. J., and Bradbury, J. F. (1993). Stenotrophomonas, a new bacterial genus for xanthomonas maltophilia (Hugh 1980) (Swings *et al.*, 1983). *Int. J. Sys. Bacteriol.* **43**, 606–609.

Swings, J. G., and Civerolo, E. L. (ed.) (1993). "Xanthomonas." Chapman & Hall, London.

Vauterin, L., Yang P., Alvarez, A. Takikawa, Y., Roth, D. A., Vidaver, A. K., Stall, R. E., Kersters, K., and Swings, J. (1996). Identification of non-pathogenic Xanthomonas strains associated with plants. *Sys. Appl. Microbiol.* **19**, 96–105.

Vivian A., and Gibbnon, M. J. (1997). Avirulence genes in plant-pathogen bacteria: Signal or weapons. *Microbiology* **144**, 693–704.

Wang, G. L., and Leung, H. (1999). Molecular biology of host-pathogen interactions in rice diseases. *In* "Molecular Biology of Rice" (K. Shimamotoa, ed.), pp. 201–232. Springer-Verlag, Tokyo.

Xylanases

Pratima Bajpai
Thapar Corporate Research and Development Centre, India

I. Structure of Xylanase
II. Properties of Xylanases
III. Production of Xylanases
IV. Purification of Xylanases
V. Mode of Action of Xylanases
VI. Application of Xylanases
VII. Conclusion

GLOSSARY

cellulases A family of enzymes that hydrolyze β-1, 4-glucosidic bonds in native cellulose and derived substrates.

constitutive enzyme An enzyme whose formation is not dependent on the presence of a specific substrate.

induced enzyme An enzyme produced by an organism in response to the presence of its substrate or a related substance.

mutant An organism with a changed or new gene.

recombinant A cell or clone of cells resulting from recombination.

β-xylosidase An enzyme that hydrolyzes xylooligosaccharides to D-xylose.

XYLANASES have received a great deal of attention in the last 10 years, mainly due to their potential application in the pulp and paper, food, and chemical industries. Endo xylanase (1,4-β-D-xylan xylanohydrolase EC 3.2.1.8) is a crucial enzyme for general xylan depolymerization. It catalyzes the random hydrolysis of 1,4,β-D-xylosidic linkages in xylans. Biotechnology is not the only field with such a great interest in xylanases. Evidence suggests that xylanases play important physiological roles in plant tissues. There is an indication that they may be involved in fruit softening, seed generation, and plant defense mechanisms. Striking results were obtained in studies of the interaction of microbial xylanases with plant tissues. Xylanases from *Trichoderma viride* were found to be able to induce, in tobacco, the biosynthesis of ethylene and two of three classes of pathogenesis-related proteins including, chitinases. Therefore, it appears that certain xylanases can elicit defense mechanisms in plants. This action may be mediated by specific signal oligosaccharides, collectively known as oligosaccharins, which were apparently not produced by all xylanases from *Trichoderma viride*. However, the possibility that the enzymes themselves or their fragments serve as the elicitors of the biosynthesis has not been ruled out.

I. STRUCTURE OF XYLANASE

Xylans are among the most abundant biopolymers (after cellulose) synthesized in the biosphere. They are the most represented hemicellulosic polysaccharides found in the cell walls of land plants, in which they may constitute more than 30% of the dry weight. Xylan structure is variable, ranging from linear 1,4-β-linked polyxylose chains to highly branched heteropolysaccharides. The prefix "hetero" denotes the presence of sugars other than D-xylose. Some major structural features are presented in Fig. 1. The main chain of xylan is analogous to that of cellulose, but is composed of D-xylose instead of D-glucose. Branches consist of L-arabinofuranose linked to the 0–3 positions of D-xylose residues and of D-glucuronic acid or 4-0-methyl-D-glucuronic acid linked to the 0–2 positions. Both side-chain sugars are linked α-glycosidically. The degree of branching varies depending on the source. The xylans of several wood species, particularly of hardwoods, are acetylated; for exam-

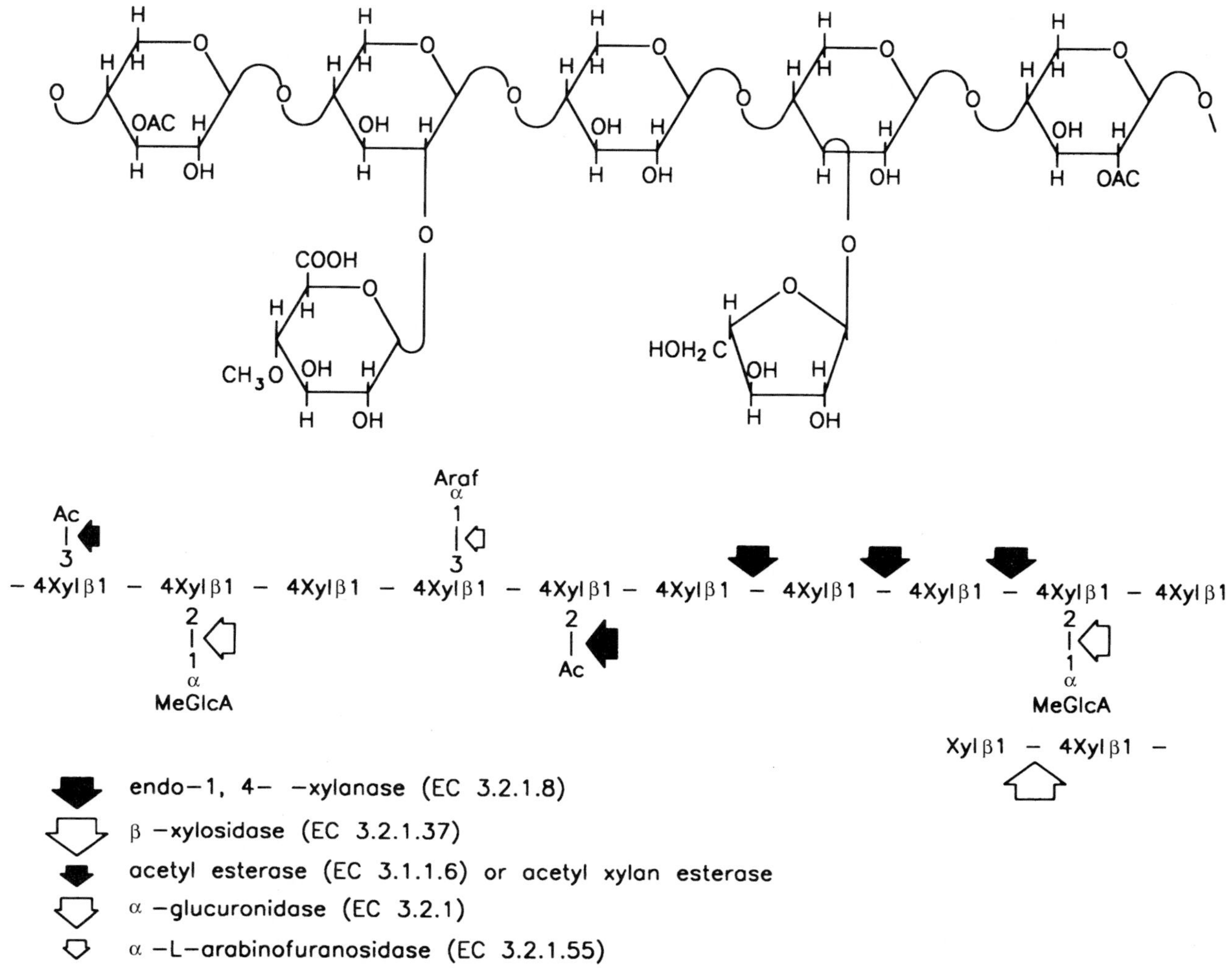

endo-1, 4- -xylanase (EC 3.2.1.8)
β -xylosidase (EC 3.2.1.37)
acetyl esterase (EC 3.1.1.6) or acetyl xylan esterase
α -glucuronidase (EC 3.2.1)
α -L-arabinofuranosidase (EC 3.2.1.55)

Fig. 1. A hypothetical plant xylan and the sites of its attack by microbial xylanolytic enzymes. (Reproduced from Bieley, P. (1985). Microbial xylanolytic system. *Trends Biotechnol.* **3**(11), Copyright (1985), with permission from Elsevier Science.)

ple birch xylan contains >1 mol acetic acid/2 mol D-xylose. Acetylation occurs more frequently at the 0–3 than the 0–2 positions and double acetylation of a D-xylose unit has been also reported. There is a relationship between the chemical structure of xylans and their botanical origin and also their cytological localization. Xylans interconnect other cell-wall components via covalent linkages and also secondary forces. The interactions are at the level of the cellulose microfibril, as shown by several electron microscopy studies using specific markers of heteroxylans.

II. PROPERTIES OF XYLANASES

Fungal xylanases of *Trichoderma* and *Aspergillus* species, as well as bacterial xylanases of *Bacillus, Streptomyces,* and *Clostridium* species have been intensively studied. Several of the xylanases that have been purified are rather small (molecular weight 20 kDa) monomeric proteins with basic isoelectric points (pl 8–9.5). They also exhibit great homology at the molecular level and belong to the family G (or 11) of glycosyl hydrolases. The other xylanases with high molecular mass and lower pl values belong to the other identified endoxylanase family, F (or 10). The optimum pH for xylan hydrolysis is around five for most fungal xylanases, and they are normally stable between pH values of 2, and 9. The pH optima of bacterial xylanases are generally slightly higher than those of fungal xylanases. Alkalophilic *Bacillus* species and alkalophilic actinomycetes produce xylanases with high activity at alkaline pH values. Most of the fungi and bacteria produce xylanases that tolerate temperatures below 40–50°C. Xylanases have also been characterized from thermophilic organisms. The most thermostable xylanase described is that from an extremely thermophilic species of *Thermotoga* with a half-life of 20 min at 105°C. Xylanases with half-lives from a few minutes up to 90 min at 80°C are produced by *Thermoascus aurantiacus, Bacillus stearothermophilus, Caldocellum saccharolyticum, Clostridium stercorarium,* and *Thermomonospora* spp.

Xylanases show the highest activity against polymeric xylan; the rate of the hydrolysis reaction normally decreases with the decreasing chain length of oligomeric substrates. They do not hydrolyze xylobiose, and the hydrolysis of xylotriose is in most cases negligible or at least limited. The main products formed from the hydrolysis of xylan are xylobiose, xylotriose, and substituted oligomers of two to four xylosyl residues. The length and type of the substituted products depend on the mode of action of the individual xylanases. Most of the enzymes studied cleave the xylan backbone, leaving the substituent on the nonreducing end of the xylosyl chain of the oligosaccharide. Some xylanases leave the substituent on the nonreducing end and in the middle of the oligosaccharide chain of the end products; xylotriose has been reported to inhibit the action of xylanases. In addition to hydrolytic activity, transferase activity has been detected in several xylanases.

Of the xylanases produced by *Trichoderma* species, two main groups can be identified. Both types have low molecular weights, but the isoelectric points are different. Whereas xylanases with high pl have been quite extensively studied, the other type of xylanase (with pl near to pH 5) has been purified and characterized only from *T. lignorum* and *T. reesei*. Basic xylanases have been isolated from *T. harzianum, T. koningii,* and *T. longibrachiatum*. However, only one of these enzymes made a major contribution to the total xylanase activity in the culture filtrate. The pl 9.0 and pl 5.5 xylanases of *T. reesei* have been shown to be different gene products and were both classified as belonging to the family G. Multiple xylanases have also been purified from culture filtrates of *Aspergillus niger, A. oryzae, A. kawachii, A. awamori,* and other fungi, as well as from bacteria. Xylanases with high pH and temperature optima have been isolated and tested for improving the bleachability of kraft pulps. Table I shows the properties of some thermostable xylanases. Several alkali-tolerant strains of *Bacillus* have been used for the production of xylanases with pH optima of around 9.0. *Thermomyces lanuginosus* has been found to be an excellent producer of thermostable β-xylanase. A wild-type strain has been found to produce about 60,000 n Kat/ml xylanase activity in 6 days. The most thermophilic xylanases described are produced by an extremely thermophilic bacterium, *Thermotoga* sp. *T. maritima* produces at least two hyperthermophilic xylanases; one has its temperature optimum at 90°C and the other at 105°C. However, microorganisms living in extreme conditions are often difficult to grow in the laboratory and the productivity of xylanases is usually low. The amino acid homology of xylanases has opened up the possibility of using novel genetic techniques to screen for better xylanases for industrial applications. Thus, several xylanase genes encoding proteins active at temperatures from 75°C up to 95°C (pH 6–8) have been isolated from the extremely thermophilic bacteria *Thermotoga* and *Dictyoglomus* without the laborious production and purification of the enzymes.

TABLE I
Xylanases Active at High Temperature

Organism	*Activity half-life*
Thermotoga sp.	90 min at 95°C
Thermomonospora fusca	30–60 min at 80°C
Thermoascus auranticus	54 min at 80°C
Dictyoglomus sp.	80 min at 90°C
Caldocellum saccharolyticum	2–3 min at 80°C
Bacillus sp.	40–50 min at 80°C
Bacillus stearothermophillus	30 min at 80°C

III. PRODUCTION OF XYLANASES

The cost of enzymes is one of the factors determining the economics of a biocatalytic process and can be reduced by finding optimum conditions for their production by the isolation of hyperproducing mutants and (possibly) by the construction of efficient producers using genetic engineering. A rational approach to these goals requires knowledge of the regulatory mechanisms governing enzyme production. Studies of the regulation of xylanolytic enzymes have largely focused on the induction of enzyme activities under various conditions rather than on gene regulation. In addition, the studies have mainly focused on xylanase and xylosidase. Xylanases appear to be inducible. They are produced in high amounts during growth on xylan and the synthesis of the enzyme is catabolite repressed by easily metabolized carbon sources such as glucose or xylose. Xylan can not enter the cells, so the signal for accelerated synthesis of xylanolytic enzymes must involve low-molecular-weight fragments, mainly xylobiose and xylotriose. The oligosaccharides are formed by the hydrolysis of xylan in the medium by small amounts of enzymes produced constitutively. Induction can also be achieved by various synthetic alkyl and aryl β-D-xylosides in a *Streptomyces* sp. and by methyl β-*D* xyloside in yeast. These compounds enable the production of xylanolytic enzymes in the absence of xylan and xylooligosaccharides. In the yeast *C. albidus,* only methyl β-D-xyloside induced the xylanolytic system. Other alkyl and aryl β-D-xylosides were unable to do so; they could induce a nonspecific β-glucosidase that hydrolyzed aryl β-D-xylosides but not xylooligosaccharides. Xylanolytic systems of yeast can also be induced by positional isomers of xylobiose. Induction with 1,2-β-xylobiose is analogous to the induction of cellulases in filamentous fungi by sophorose. However, the slow response of cells to 1,2-β-xylobiose compared to 1,4-β-xylobiose, as well as the evidence for its transformation into 1,4-β-xylobiose, the natural inducer, suggested that, in yeast, the isomeric diasaccharide is a precursor of the natural inducer. The nature of the regulation in filamentous fungi has not been established. Generalization will perhaps never be possible due to the diversity of cellular control mechanisms. Fungal xylanases appear to be inducible or under derepression control, which includes enzyme production on carbon sources that are used slowly. Regulatory studies in fungi are often complicated by the concurrent production of xylanase and cellulase and by substrate cross-specificity of cellulases and xylanases. There are different types of cellulases and xylanases, the substrate specificities of which range from absolute for one polymer to about equal affinity for both polymers. The xylanolytic and cellulolytic systems in some filamentous fungus are likely to be under separate regulatory control. During growth on xylan, several species produce specific xylanases with little or no cellulase. However, when grown on cellulose, cellulases are produced together with xylanases. The reason for production on cellulase of specific xylanases is unclear. Perhaps, it results from the presence of xylan remnants in cellulose or depression on cellulose, a carbon source that is used slowly. Experiments with defined low-molecular-weight inducers in *T. reesei* afforded similar results. Sophorose induced both specific and nonspecific endo-1,4-β-glucanases, cellobiohydrolase I, and very little xylanase. Induction with xylobiose produced only specific xylanases. Therefore, the strategy for xylanolytic systems free of cellulases might be simply to grow cells on xylan uncontaminated by cellulose. However, this strategy could not be applied to all fungi. In *Schizophyllum commune,* high xylanase production is linked strictly to cellulase production. The fungus grows poorly on xylan in the absence of cellulose. The possibility of producing xylanolytic systems free of cellulase should be clarified in strains of *Aspergillus* because they belong to the best xylanase and xylosidase producers. An alternative and promising approach to the production of xylanolytic systems free of cellulases is the isolation of cellulase-deficient mutants. Another possibility is the construction of appropriate recombinants by genetic engineering.

Recombinant DNA techniques offer opportunities for the construction of microbial strains with selected enzyme machinery. In xylan bioconversion, the main objectives for recombinant DNA technology are the construction of producers of xylanolytic systems free of cellulolytic enzymes and the improvement of the fermentation characteristics of industrially important xylose-fermenting organisms by introducing genes for xylanase and xylosidase so that the direct fermentation of xylan is possible.

All the published cloning work has been restricted to bacterial genes. Three groups have reported the isolation of xylanase genes from *Bacillus* sp and their expression in *Escherichia coli.* In only one case does the expressed enzyme appear to be secreted from the host cells. Further biochemical studies of xylanase-secreting and nonsecreting transformants could lead to a better understanding of the secretary process and to the development of cloning strategies that would guarantee the secretion of the desired products. Further difficulties with cloning genes from eukaryotic microorganisms can be expected. In addition to permitting the introduction of novel genes, cloning techniques could enable the amplification of the expression genes already present. For instance, the production of xylanase in *B. subtilis* was enhanced successfully using a plasmid vector carrying the *B. pumilus* genes. The transformant produced approximately three times more extracellular xylanase than the donor strain. Moreover, the enzyme was produced constitutively, suggesting that regulatory elements of the donor organism were absent in the vector used for the transformation.

IV. PURIFICATION OF XYLANASES

Xylanase purification schemes have generally used standard column chromatographic techniques, mainly ion exchange and gel filtration, but also hydrophobic interaction. The low molecular weight of certain xylanases has also enabled their separation from other proteins using ultrafiltration. Furthermore, adsorptive interactions of certain low-molecular-weight xylanases (~20 kDa) with gel-filtration resin matrices cause these enzymes to elute as proteins smaller than 12 kDa. This phenomenon has probably facilitated xylanase purification in certain cases, and it is probably due to the affinity of dextran gel matrices for the aromatic group of certain acids. The various purification schemes have successfully isolated a number of apparent xylanases from *Trichoderma* spp. to electrophoretic homogeneity. There is, however, some evidence that one of the low-molecular-weight xylanases from *T. viride* can be nonhomogenous under denaturing conditions, showing smaller peptides under SDS-PAGE or urea-SDS-PAGE. These fragments may be due to the cleavage of certain peptide linkages by proteases excreted by the fungus. For application on pulp fibers, the purification of xylanases does not have to be complete as long as cellulase activity is eliminated. Various strategies for preparing cellulase-free xylanases have been reviewed by Biely (1985). Ultrafiltration can be used in conjunction with solvent exchange through ion exchangers to produce high yields of cellulase-free xylanase from the culture filtrate of *T. harzianum.* Alternatively, selective adsorption of cellulases to a cellulose column may be used to enrich xylanases; however, the column would also retain certain xylanases that contain cellulose-binding domains.

V. MODE OF ACTION OF XYLANASES

Endoxylanases show the highest activity against polymeric xylan and the rate of the hydrolysis reaction normally decreases with decreasing chain length of the oliogomeric substrates. They do not hydrolyze xylobiose and the hydrolysis of xylotriose is in most cases negligible or at least limited. The main products formed from the hydrolysis of xylan are xylobiose, xylotriose, and substituted oligomers containing two to four xylosyl residues. The chain length and the type of substituted products depend on the mode of action of the individual xylanases. Dekker (1985) has compared the structures of some substituted xylooligosaccharides that are produced during the hydrolysis of arabinoxylans and glucuronoxylans by various xylanases. Most enzymes (*Ceratocystis paradoxa, Sporotrichium dimorphosphorum, Streptomyces* sp., *Trichoderma viride, Aspergillus niger,* and *Trametes hirsuta* xylanases) cleaved the xylan backbone, leaving the substituent on the nonreducing end of the xylosyl chain of the oligosaccharide. For example, the presence of glucopyranosyl uronic groups sterically hinders the hydrolysis of the third and second xylosylic linkages to the right of the branch point by the xylanases of *S. dimorphosporum* and *T. hirsuta*, respectively. *Aspergillus niger* and *C. sacchari* xylanases were reported to leave the substituent on the reducing end and in the middle of the oligosaccharide chain, respectively. Some of the end products, particularly xylotrose, have been reported to inhibit the action of xylanases. In addition to hydrolytic activity, transferase activity has also been detected in some xylanases from *Cryptococcus, Aspergillus,* and several species of *Trichoderma.* The three basic xylanases of *T. harzianum* have been reported to act synergistically. No synergism was detected between the two *T. reesei* xylanases with pl values 5.5 and 9.0.

VI. APPLICATION OF XYLANASES

A. Pulp and Papermaking

The most promising application of xylanase is in the prebleaching of kraft pulps. Other applications proposed for xylanases include debarking, modification of fiber properties for improving drainage and beatability, preparation of dissolving pulp, shive removal, and de-inking.

1. *Prebleaching of Kraft Pulps*

Faced with market, environmental, and legislative pressures, the pulp and paper industry is modifying its pulping, bleaching, and effluent-treatment technologies to reduce the environmental impact of mill effluents. The incorporation of xylanase prebleaching is being considered because it permits the use of lower chlorine charges during the bleaching of kraft pulps, the bleach-boosting effect being associated with reduced chloroorganic discharges. Xylanase can also contribute to bleaching sequences in which chlorine is completely replaced with chlorine dioxide and in which chlorine-containing chemicals have been replaced by peroxide or ozone.

The results from laboratory studies and mill trials show about a 35–41% reduction in active chlorine at the chlorination stage for hardwoods and 10–20% for softwoods, whereas the savings in total active chlorine were found to be 20–25% for hardwoods and 10–15% for softwoods. In the elementary chlorine-free bleaching sequences, the use of xylanase increases the productivity of the bleaching plant when the production capacity of ClO_2 is a limiting factor. This is often the case when the use of chlorine gas has been abandoned. In totally chlorine-free (TCF) bleaching sequences, the addition of xylanase increases the final brightness value, which is a key parameter in the marketing of chlorine-free pulps. In addition, the savings in TCF bleaching are important with respect both to costs and to the strength properties of the pulp. Xylanase pretreatment leads to reductions in effluent AOX and dioxin concentrations due to reduced amount of chlorine required to achieve a given brightness. The level of AOX in effluent is significantly lower for xylanase-pretreated pulps as compared to conventionally bleached control pulps. The xylanase-treated pulps show unchanged or improved strength properties. Also, these pulps are easier to refine than the control pulps. The viscosity of the pulp is improved as a result of xylanase treatment. However, the viscosity of the pulp is adversely affected when cellulase activity is present. Therefore, the presence of cellulase activity in the enzyme preparation is not desirable. A significant number of Scandinavian and North American mills are bleaching entirely with xylanases. Various paper products including magazine paper and tissue papers manufactured from enzymatically treated pulps have been successfully introduced to the market.

2. *Debarking*

The removal of tree bark is the first step in all processing of wood. This step consumes a substantial amount of energy. Extensive debarking is needed for high-quality mechanical and chemical pulps because even a small amount of bark residue causes the darkening of the product. In addition to its high energy demand, complete debarking leads to losses of raw material due to prolonged treatment in the mechanical drums. The border between the wood and bark is the cambium, which consists of only one layer of cells. This living cell layer produces xylem cells toward the inside of the stem and phloem cells toward the outside. In all the wood species studied, the common characteristics of the cambium include a high content of pectins and the absence or low content of lignin. The content of pectins in cambium cells varies among the wood species, but may be as high as 40% dry weight. The content of pectic and hemicellulosic compounds is very high in the phloem. Pectinases are found to be key enzymes in the process, but xylanases may also play a role because of the high hemicellulose content in the phloem of the cambium.

3. *Fiber Modification*

The enzymatic modification of fibers aims at the decreased energy consumption in the production of thermomechanical pulps and the increased beatability of chemical pulps or improvement of fiber properties. Xylanase action has been found to produce pulp fibers with properties similar to those of slightly beaten pulps. The energy required for pulp beating is reduced by 20–30%. The water-retention value, which describes fiber swelling, is also considerably increased.

Water removal on the paper machine has been shown to improve as a result of limited hydrolysis of the fibers in recycled paper. A mixture of xylanase and cellulase enzymes at low concentrations has been found to markedly increase the freeness of recycled fibers without reducing the yield. The lower the initial freeness, the greater the gain following treatment. Freeness shows a rapid initial increase with over half of the observed effect occurring in the first 30 minutes. A relatively small amount of enzyme is required. Although the initial effects are largely beneficial, extending the reaction time with large concentrations of enzyme is detrimental. Several commercial enzymes are available that have been specially designed to improve the drainability of pulp.

4. *Production of Dissolving Pulp*

Dissolving pulps are used to produce cellulosic materials such as acetate, cellophanes, and rayons. Their manufacture is characterized by the derivatization and thus solubilization of highly purified cellulose. Hemicellulosic contaminants lead to colour and haze in the product, as well as to insolubles that hamper the manufacturing process. Their extraction from the pulps requires the use of high caustic loadings and appropriate pulping conditions, the latter restricted to sulfite pulping and acid-pretreated kraft pulping. The use of xylanase for purifying cellulose has been tried. The complete enzymatic hydrolysis of the hemicellulose in the pulp is difficult to achieve. Even with very high enzyme loading, only a relatively small amount of xylan could be removed. The inaccessibility of a large portion of the xylan in pulps, however, has limited the potential of this application. Nevertheless, xylanase treatment may reduce the chemical loading required during caustic extraction or facilitate xylan extraction from kraft pulps.

5. *Removal of Shives*

Xylanases are also effective in removing shives, the small bundles of fibers that have not been separated into individual fibers during the pulping process. Shives appear as splinters that are darker than the pulp. One of the most important quality criteria for bleached kraft pulp is shive count. By treating the brownstock with xyanases, mills can substantially increase the degree of shive removal in the subsequent bleaching.

6. *De-inking*

Xylanase enzymes in combination with cellulases are being used for the de-inking of waste paper. This method has proven effective and economical on both a laboratory and industrial scale. The side benefits include higher freeness, greater paper strength, higher brightness, reduced chemical use, lower bleaching costs, and simplified waste disposal.

7. *Retting of Flax Fibers*

Xylanase enzymes have been also used in processing plant-fiber sources such as flax and hemp. At present, fiber liberation is affected by retting, that is, the removal of binding material present in plant tissues using enzymes produced *in situ* by microorganisms. Pectinases are believed to play the main role in this process, but xylanases may also be involved. The replacement of slow natural retting by treatment with artificial mixtures of enzymes could become a new fiber-liberation technology.

B. Other Applications

The use of xylanases has also been proposed for clarifying juices and wines; for extracting coffee, plant oils, and starch; for improving the nutritional properties of agriculture silage and grain feed; for macreating plant cell walls; for producing food thickeners; and for providing different textures in bakery products. A particular endo-1,4-β xylanase produced by *Aspergillus niger* has been identified that is very effective in increasing the specific volume of breads without giving rise to a negative side effect on dough handling, similar to that observed with xylanases derived from other bacterial or fungal sources. The effect of xylanase on bread-volume improvement can be ascribed to the redistribution of water from the pentosan phase to the gluten phase. The increase in gluten volume fraction gives the gluten more extensibility, which eventually results in better ovenspring. Xylanase-based enzyme products are now widely used throughout the world to supplement pig and poultry diets based on wheat, triticale, and rye. Other enzyme activities in such products normally include β-glucanase, pectinase, amylase, and protease. The addition of these products can improve feed use and live weight gain primarily by improving the digestion of such nutrients as starch, proteins, and fats in the intestines. It would appear that enzymes reduce variation in live weight, essentially eliminate litter problems, and help aileviate digestive disorders.

There has also been an interest in the production of xylose, xylobiose, and xylooligomers. Such sugars can be prepared by enzymatic hydrolysis of xylan, whereas other sugar residues can be added using the transglucosylation activity of such enzymes as β-xylosidase. These xylose-containing sugars may be useful for research as well as for their rheological properties.

With the exception of xylose conversion to xylitol, the bioconversion of lignocellulosic materials to fermentable sugars does not appear to be an economic prospect because of other more competitive sources of sugars such as starch and sucrose. However, the massive accumulation of agricultural, forestry, and municipal solid-waste residues are creating a large volume of low-value feedstock. Alternative technologies are desirable for dealing with all these materials, even if they are from the perspective of waste management. One alternative being considered is bioconversion to produce fuel ethanol, single-cell protein, xylanases, and other chemicals from xylan-rich material.

VII. CONCLUSION

The new major large-scale application area of xylanases is clearly in the pulp and paper industry to increasing the bleachability of kraft pulps. The improved bleachability is based mainly on the action of endo-β-xylanases, which can be produced efficiently on an industrial scale. The partial hydrolysis of xylan facilitates the extraction of lignin from the pulp in higher amounts and with high molecular masses. The enzymatic pretreatment method is applicable to any traditional or modern bleaching sequence at existing plants without significant investment.

The potential for the use of xylanases in the food and feed industries is also high. The main aim in the application of xylanolytic enzymes has been the hydrolysis of hemicellulosic substrates for the production of fermentable sugars. The knowledge gathered about the hydrolysis mechanism of the hemicelluloses, especially xylans, has greatly promoted the rapid application of these enzymes in new areas.

See Also the Following Articles

Cellulases • Lignocellulose, Lignin, Ligninases • Pulp and Paper

Bibliography

Bajpai, P. (1997). Bleaching pulp. *In* "Enzymes in Pulp and Paper Processing" pp. 1–37. Miller Freeman Books, San Francisco, CA.

Biely, P. (1985). Microbial xylanolytic systems. *Trends Biotechnol.* **311**, 286–290.

Dekker, R. F. H. (1985).*In* "Biosynthesis and Biodegradation of Wood Components," pp. 503–533. Academic Press, Orlando, FL.

Paice, M. G., and Jurasek, L. (1984). Removing hemicellulose from pulps by specific enzyme hydrolysis. *J. Wood Chem. Technol.* 4(2), 187–198.

Suurnakki, A., Tenkanen, M., Buchert, J., and Viikari, L. (1997). Hemicellulases in the Bleaching of Chemical Pulps.

In "Advances in Biochemical Engineering and Biotechnology" (T. Scheper, ed.), Vol. 57, pp. 261–287. Springer-Verlag, Berlin.

Viikari, L., Kantelinen, A., Sundquist, J, and Linko, M. (1994). Xylanases in bleaching: From an idea to industry. *FEMS Microb. Rev.* **13**, 335–350.

Viikari, L., Ranua, M., Kantelinen, A., Sundquist, J., and Linko M. (1986). Bleaching with enzymes. *In* "Proceedings of Third International Conference on Biotechnology in Pulp and Paper Industry," pp. 67–69. Stockholm, Sweden.

Wong, K. K. Y., and Saddler, J. N. (1992). *Trichoderma* xylanases, their properties and application. *Crit. Rev. Biotechnol.* **12** (5/6), 413–435.

Woodword, J. (1984). Xylanases: Functions, properties and applications. *Topics Enzyme Ferment. Biotechnol.* **8**, 9–30.

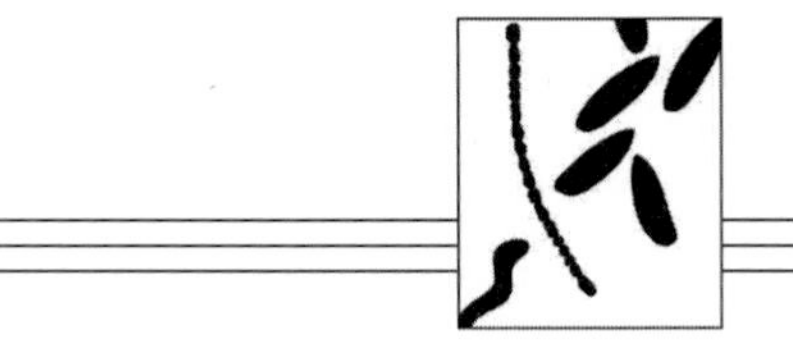

Yeasts

Graeme M. Walker
University of Abertay Dundee, Scotland

I. Definition and Classification of Yeasts
II. Yeast Ecology
III. Yeast Cell Structure
IV. Nutrition, Metabolism, and Growth of Yeasts
V. Yeast Genetics
VI. Industrial, Agricultural, and Medical Importance of Yeasts

GLOSSARY

bioethanol Ethyl alcohol (ethanol) produced by yeast fermentation for use as a fuel or industrial commodity.

birth scar Concave indentations that remain on the surface of daughter cells following budding.

budding A mode of vegetative reproduction in many yeast species in which a small outgrowth, the daughter bud, appears and grows from the surface of a mother cell and eventually separates to form a new cell at cell division.

bud scar The chitin-rich, convex, ringed protrusions that remain on the mother-cell surface of budding yeasts following the birth of daughter cells.

Crabtree effect The suppression of yeast respiration by high levels of glucose. This phenomenon is found in *Saccharomyces cerevisiae* cells which continue to ferment irrespective of oxygen availability due to glucose repressing or inactivating the respiratory enzymes or due to the inherent limited capacity of cells to respire.

fission A mode of vegetative reproduction found in the yeast genus *Schizosaccharomyces*. Fission yeasts grow lengthwise and divide by forming a cell septum that contricts mother cells into two equal-size daughters.

Pasteur effect The activation of yeast sugar-consumption rate by oxygen; alternatively, under anaerobic conditions glycolysis proceeds faster than it does under aerobic conditions. In *Saccharomyces cerevisiae,* the Pasteur effect is only observable when the glucose concentration is low (below around 5 mM) or in resting or starved cells.

respirofermentation The yeast fermentative metabolism in the presence of oxygen.

Saccharomyces cerevisiae Baker's or brewer's yeast species, which is used widely in the food and fermentation industries and is also being exploited in modern biotechnology (e.g., in the production of recombinant proteins) and as a model eukaryotic cell in fundamental biological research.

sporulation The production of haploid spores when sexually reproductive yeasts conjugate and undergo meiosis.

YEASTS are eukaryotic unicellular microfungi that are widely distributed in the natural environment. Around 800 yeast species are known, but this represents only a fraction of yeast biodiversity on Earth. The fermentative activities of yeasts have been exploited by humans for millennia in the production of beer, wine, and bread. The most widely exploited and studied yeast species is *Saccharomyces cerevisiae,* commonly referred to as "baker's yeast." This species reproduces asexually by budding and sexually following the conjugation of cells of the opposite mating type. Other yeasts reproduce by fission (e.g., *Schizosaccharomyces pombe*) and by formation of pseudohyphae as in dimorphic yeasts, such as the opportunistic human pathogen *Candida albicans.* In addition to their wide exploitation in the production of foods, beverages, and pharmaceuticals, yeasts play significant roles as model eukaryotic cells in furthering our knowledge in the biological and biomedical sciences. The complete genome of *S. cerevisiae* was sequenced in 1996 and research is underway to assign a physiological function to each of these sequenced genes. Work with yeasts is not only leading to insights into how a simple eukaryote works, but also insights into human genetics and an understanding of human heritable disorders.

I. DEFINITION AND CLASSIFICATION OF YEASTS

A. Definition and Characterization of Yeasts

Yeasts are recognized as unicellular fungi that reproduce primarily by budding, and occasionally by fission, and that do not form their sexual states in or on a fruiting body. Yeast species may be identified and characterized according to various criteria based on cell morphology (e.g., mode of cell division and spore shape), physiology (e.g., sugar fermentation tests), immunology (e.g., antibody agglutination and immunofluorescence), and molecular biology (e.g., ribosomal DNA phylogeny, DNA base composition and hybridization, karyotyping, and random amplification of polymorphic DNA). Molecular-sequence analyses are being increasingly used by yeast taxonomists to categorize new species.

B. Yeast Taxonomy

The most commonly exploited yeast species, *Saccharomyces cerevisiae* (baker's yeast), belongs to the fungal kingdom subdivision Ascomycotina. Other yeast genera are represented in Basidiomycotina (e.g., *Cryptococcus* spp.) and *Rhodotorula* spp. and Deuteromycotina (e.g., *Candida* spp. and *Brettanomyces* spp.). There are around 100 recognized yeast genera (see Table I).

C. Yeast Biodiversity

Around 800 species of yeast have been described, but new species are being characterized on a regular

TABLE I
Alphabetical Listing of 100 Yeast Genera

Aciculoconidium	*Dipodascopsis*	*Malassezia*	*Sporobolomyces*
Agaricostilbium	*Dipodascus*	*Metschnikowia*	*Sporopachydermia*
Ambrosiozyma	*Endomyces*	*Moniliella*	*Stephanoascus*
Arxiozyma	*Eremothecium*	*Mrakia*	*Sterigmatomyces*
Arxula	*Erythrobasidium*	*Myxozyma*	*Sterigmatosporidium*
Ascoidea	*Fellomyces*	*Nadsonia*	*Sympodiomyces*
Aureobasidium	*Fibulobasidium*	*Oosporidium*	*Sympodiomycopsis*
Babjevia	*Filobasidiella*	*Pachysolen*	*Tilletiaria*
Besingtonia	*Filobasidium*	*Phaffia*	*Tilletiopsis*
Blastobotrys	*Galactomyces*	*Pichia*	*Torulaspora*
Botryozyma	*Geotrichum*	*Protomyces*	*Tremella*
Brettanomyces	*Guilliermondella*	*Prototheca*	*Trichosporon*
Bullera	*Hanseniaspora*	*Pseudozyma*	*Trichosporonoides*
Bulleromyces	*Hansenula*	*Reniforma*	*Trigonopsis*
Candida	*Holtermannia*	*Rhodosporidium*	*Trimorphomyces*
Cephaloascus	*Hyalodendron*	*Rhodotorula*	*Tsuchiyaea*
Chionosphaera	*Issatchenkia*	*Saccharomyces*	*Ustilago*
Citeromyces	*Itersonila*	*Saccharomycodes*	*Wickerhamia*
Clavispora	*Kloeckera*	*Saccharomycopsis*	*Wickerhamiella*
Coccidiascus	*Kluyveromyces*	*Saitoella*	*Williopsis*
Cryptococcus	*Kockovaella*	*Saturnispora*	*Xanthophyllomyces*
Cyniclomyces	*Kurtzmanomyces*	*Schizoblastosporion*	*Yarrowia*
Cystofilobasidium	*Leucosporidium*	*Schizosaccharomyces*	*Zygoascus*
Debaryomyces	*Lipomyces*	*Sirobasidium*	*Zygosaccharomyces*
Dekkera	*Lodderomyces*	*Sporidiobolus*	*Zygozyma*

TABLE II
Natural Yeast Habitats

Habitat	*Comments*
Soil	Soil may only be a reservoir for the long-term survival of many yeasts, rather than a habitat for growth. However, some genera are isolated exclusively from soil (e.g., *Lipomyces* and *Schwanniomyces*)
Water	*Debaryomyces hansenii* is a halotolerant yeast that can grow in nearly saturated brine solutions
Atmosphere	A few viable yeast cells may be expected per cubic meter of air. From layers above soil surfaces, *Cryptococcus, Rhodotorula, Sporobolomyces,* and *Debaryomyces* spp. are dispersed by air currents
Plants	The interface between soluble nutrients of plants (sugars) and the septic world are common niches for yeasts (e.g. the surface of grapes); the spread of yeasts on the phyllosphere is aided by insects (e.g., *Drosophila* spp.); a few yeasts are plant pathogens
Animals	Several nonpathogenic yeasts are associated with the intestinal tract and skin of warm-blood animals; several yeasts (e.g., *Candida albicans*) are opportunistically pathogenic toward humans and animals; numerous yeasts are commensally associated with insects, which act as important vectors in the natural distribution of yeasts
Built environment	Yeasts are fairly ubiquitous in buildings; for example, *Aureobasidium pullulans* (black yeast) is common on damp household wallpaper and *S. cerevisiae* is readily isolated from surfaces (pipework, vessels) in wineries

basis and there is considerable untapped yeast biodiversity on Earth. Several molecular biological techniques are used to assist in the detection of new yeast species in the natural environment, and together with input from cell physiologists, provide ways to conserve and exploit yeast biodiversity. *S. cerevisiae* is the most studied and exploited of all the yeasts, but the biotechnological potential of non-*Saccharomyces* yeasts is gradually being realized, particularly with regard to recombinant DNA technology.

II. YEAST ECOLOGY

A. Natural Habitats of Yeast Communities

Yeasts are not as ubiquitous as bacteria in the natural environment, but nevertheless yeasts can be isolated from terrestrial, aquatic, and aerial samples. Yeast communities are also found in association with plants, animals, and insects. Preferred yeast habitats are plant tissues, but a few species are found in commensal or parasitic relationships with animals. Some yeasts, most notably *Candida albicans,* are opportunistic human pathogens. Several species of yeast may be isolated from specialized or extreme environments, such as those with low water potential (i.e., high sugar or salt concentrations), low temperature (e.g., some psychrophilic yeasts have been isolated from polar regions), and low oxygen availability (e.g., intestinal tracts of animals). Table II summarizes the main yeast habitats.

B. Yeasts in the Food Chain

Yeasts play important roles in the food chain. Numerous insect species, notably *Drosophila* spp., feed on yeasts that colonize plant material. As insect foods, ascomycetous yeasts convert low-molecular-weight nitrogenous compounds into proteins beneficial to insect nutrition. In addition to providing a food source, yeasts may also affect the physiology and sexual reproduction of drosophilids. In marine environments, yeasts may serve as food for filter feeders.

C. Microbial Ecology of Yeasts

In microbial ecology, yeasts are not involved in biogeochemical cycling as much as bacteria or filamentous fungi. Nevertheless, yeasts can use a wide range of carbon sources and thus play an important role as saprophytes in the carbon cycle. In the cycling of nitrogen, some yeasts can reduce nitrate or ammonify nitrite, although most yeasts assimilate ammonium ions or amino acids into organic nitrogen. Most yeasts can reduce sulfate, although some are sulfur auxotrophs.

III. YEAST CELL STRUCTURE

A. General Cellular Characteristics

Yeasts are unicellular eukaryotes that portray the ultrastructural features of higher eukaryotic cells. This, together with their ease of growth, and amenability to biochemical, genetic, and molecular biological analyses, makes yeasts model organisms in studies of eukaryotic cell biology. Yeast-cell size can vary widely, depending on the species and conditions of growth. Some yeasts may be only 2–3 μm in length, whereas others may attain lengths of 20–50 μm. Cell width appears less variable, between 1–10 μm. Table III summarizes the diversity of yeast-cell shapes.

Several yeast species are pigmented and various colors may be visualized in surface-grown colonies, for example: cream (e.g., *S. cerevisiae*), white (e.g., *Geotrichum* spp.), black (e.g., *Aureobasidium pullulans*), pink (e.g., *Phaffia rhodozyma*), red (e.g., *Rhodotorula* spp.), orange (e.g., *Rhodosporidium* spp.), and yellow (e.g., *Bullera* spp.). Some pigmented yeasts have uses in biotechnology. For example, the astaxanthin pigments of *Phaffia rhodozyma* have applications as fish-feed colorants for farmed salmonids, which have no means of synthesizing these red compounds.

B. Methods in Yeast Cytology

By using various cytochemical and cytofluorescent dyes and phase-contrast microscopy, it is possible to visualize several subcellular structures in yeasts (e.g., cell walls; capsules, if present; nuclei; vacuoles; mitochondria; and several cytoplasmic inclusion bodies). The *GFP* gene from the jellyfish (*Aequorea victoria*) encodes the green fluorescent protein (which fluoresces in blue light) and can be used to follow the subcellular destiny of certain expressed proteins when GFP is fused with the genes of interest. Immunofluorescense can also be used to visualize yeast cellular features when dyes such as fluorescein isothiocyanate and rhodamine B are conjugated with monospecific antibodies raised against yeast structural proteins. Confocal scanning laser immunofluorescence microscopy can also be used to detect the intracellular localization of proteins within yeast cells and to give three-dimensional ultrastructural information. Fluorescence-activated cell sorting (FACS)

TABLE III
Diversity of Yeast-Cell Shapes

Gross cellular morphology	*Description*	*Typical genera*
Ellipsoidal	Oval- or 'rugby-ball'-shaped	*Saccharomyces*
Spherical	Complete spheres	*Debaryomyces*
Cylindrical	Cylinders with hemispherical ends	*Schizosaccharomyces*
Apiculate	Lemon-shaped	*Hanseniaspora, Saccharomycodes*
Ogival	Elongated cell rounded at one end and pointed at other	*Dekkera, Brettanomyces*
Bud-fission	Flask-shaped cells with septum at neck of bud	*Pityrosporum*
Filamentous	Pseudohyphal (chains of elongated budding cells) or hyphal (branched or unbranched filamentous cells that may show septa)	*Candida albicans, Yarrowia lipolytica*
Triangular	Buds are restricted to the three apices of triangular cells	*Trigonopsis*
Curved	Crescent-shaped	*Cryptococcus cereanus*
Stalked	Buds formed on short denticles or long stalks	*Sterigmatomyces*

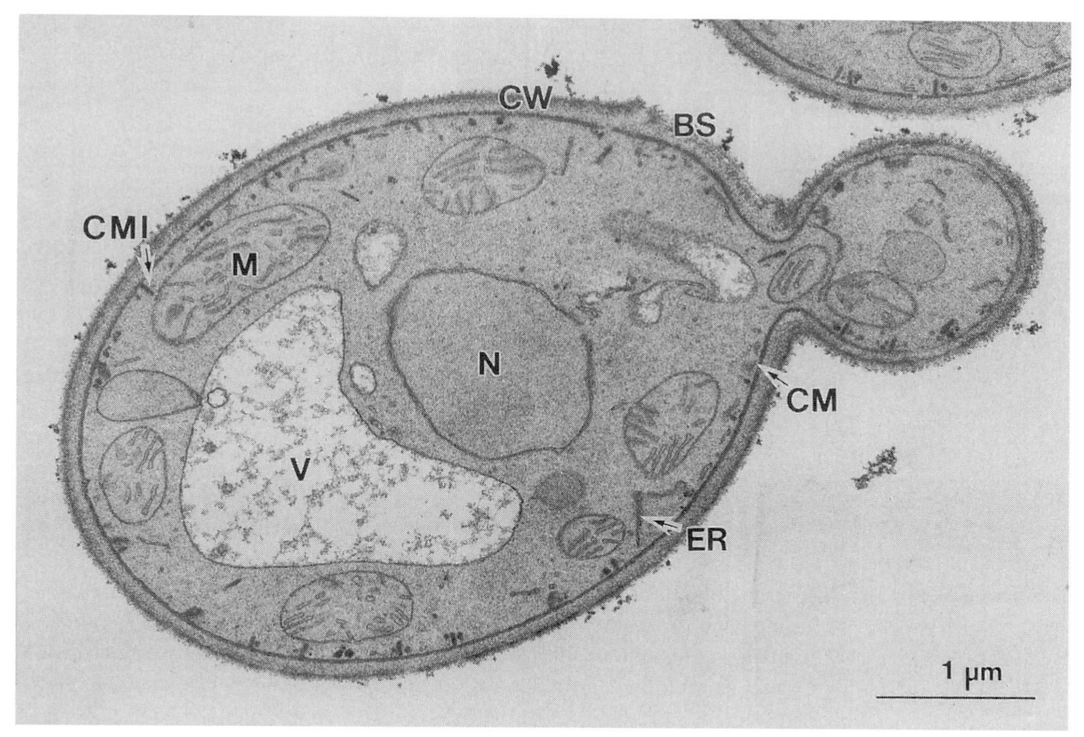

Fig. 1. Ultrastructural features of a yeast cell. The transmission electron micrograph is of a *Candida albicans* cell. BS, bud scar; CM, cell membrane; CMI, cell membrane invagination; CW, cell wall; ER, endoplasmic reticulum; M, mitochondrion; N, nucleus; V, vacuole. (Courtesy of M. Osumi, Japan Women's University, Tokyo.)

has proved very useful in studies of the yeast cell cycle and in monitoring changes in organelle (e.g., mitochondrial) biogenesis. Scanning-electron microscopy is useful in revealing the cell-surface topology of yeasts, as is atomic force microscopy, which has achieved high-contrast nanometer resolution of yeast-cell walls. Transmission-electron microscopy, however, is essential for visualizing the intracellular fine structure of ultrathin yeast-cell sections.

C. Subcellular Yeast Architecture and Function

Transmission-electron microscopy of a yeast cell would typically reveal the cell wall, nucleus, mitochondria, endoplasmic reticulum, Golgi aparatus, vacuoles, microbodies, and secretory vesicles. Figure 1 shows an electron micrograph of a yeast cell.

Several of these organelles are not completely independent of each other and derive from an extended intramembranous system. For example, the movement and positioning of organelles depends on the cytoskeleton and the trafficking of proteins in and out of cells relies on vesicular communication between the endoplasmic reticulum (ER), Golgi, vacuole, and plasma membrane. Yeast organelles can be readily isolated for further studies by physical, chemical, or enzymic disruption of the cell wall, and the purity of organelle preparations can be evaluated using specific marker enzyme assays.

In the yeast cytoplasm, ribosomes and occasionally plasmids are found and the structural organization of the cell is maintained by a cytoskeleton of microtubules and actin microfilaments. The yeast-cell envelope, which encases the cytoplasm, comprises (from the inside looking out) the plasma membrane, periplasm, cell wall, and, in certain yeasts, a capsule and a fibrillar layer. Spores encased in an ascus may be revealed in those yeasts that undergo differentiation following sexual conjugation and meiosis. Table

TABLE IV
Functional Components of an Idealized Yeast Cell

Organelle or subcellular compartment	Function
Plasma membrane	Primary barrier for the passage of hydrophilic molecules in and out of yeast cells; selective permeability is mediated by specialized membrane proteins (e.g., proton-pumping ATPase)
Periplasm	This cell wall-associated region external to the membrane and internal to the wall mainly is made up of secreted mannoproteins (e.g., invertase and acid phosphatase) that are unable to permeate the cell wall
Cell wall	Involved in cell protection, cell-shape maintenance, cell–cell interactions, signal reception, surface attachment, and specialized enzymatic activities
Fimbriae	Proteinaceous protrusions emanating from surface of several basidiomycetous and ascomycetous yeasts that are mainly involved in cell–cell interactions before sexual conjugation
Capsules	Slimy extramural layers, prevalent in basidiomycetous yeasts (e.g., *Cryptococcus*), which serve to protect cells from physical (e.g., dehydration) and biological (e.g., phagocytosis) stresses
Peroxisomes	Membrane-bound organelles found in some yeasts that serve oxidative functions (e.g., use of methanol by methylotrophic yeasts)
Nucleus	The nucleoplasm contains DNA, RNA, and proteins (e.g., protamines and histones). DNA–histone complexes (chromatin) are organized in chromosomes that pass genetic information to daughter cells at cell division
Nucleolus	Crescent-shaped region within the nucleus, which is the site of ribosomal RNA transcription and processing
ER, Golgi, secretory vesicles	Secretory system for import (via endocytosis) and export (via exocytosis) of proteins
Vacuole	Membrane-bound organelle involved in intracellular protein trafficking in yeasts; also responsible for nonspecific intracellular proteolysis and as a site for storage of basic amino acids, polyphosphate, and metal cations
Mitochondria	Respiratory metabolism (aerobic conditions); fatty acid, sterol, and amino acid metabolism (anaerobic conditions)

IV provides a summary of the physiological function of the various structural components to be found in yeast cells.

IV. NUTRITION, METABOLISM, AND GROWTH OF YEASTS

A. Nutritional and Physical Requirements for Yeast Growth

1. Yeast Nutritional Requirements

Yeast cells require macronutrients (sources of carbon, nitrogen, oxygen, sulfur, phosphorus, potassium, and magnesium) at the millimolar level in growth media, and they require trace elements (e.g., Ca, Cu, Fe, Mn, Zn) at the micromolar level. Most yeasts grow quite well in simple nutritional media, which supplies carbon and nitrogen-backbone compounds together with inorganic ions and a few growth factors. Growth factors are organic compounds required in very low concentrations for specific catalytic or structural roles in yeast, but are not used as energy sources. Yeast growth factors include vitamins, which serve vital functions as components of coenzymes; purines and pyrimidines; nucleosides and nucleotides; amino acids; fatty acids; sterols; and other miscellaneous compounds (e.g., polyamines and choline). Growth-factor requirements vary among yeasts, but when a yeast species is said to have a growth-factor requirement, this indicates that it cannot synthesize the particular factor, resulting in the curtailment of growth without its addition to the culture medium.

2. *Yeast Culture Media*

It is quite easy to grow yeasts in the laboratory on a variety of complex and synthetic media. Malt extract or yeast extract supplemented with peptone and glucose (as in YEPG) is commonly employed for the maintenance and growth of most yeasts. Yeast Nitrogen Base (YNB) is a commercially available chemically defined medium that contains ammonium sulfate and asparagine as nitrogen sources, together with mineral salts, vitamins, and trace elements. The carbon source of choice (e.g., glucose) is usually added to a final concentration of 1% (w/v). For the continuous cultivation of yeasts in chemostats, media are usually designed that ensure that all the nutrients for growth are present in excess except one (the growth-limiting nutrient). Chemostats can therefore facilitate studies into the influence of a single nutrient (e.g., glucose, in carbon-limited chemostats) on yeast-cell physiology, with all other factors being kept constant. In industry, yeasts are grown in a variety of fermentation feedstocks including malt wort, molasses, grape juice, cheese whey, glucose syrups, and sufite liquor.

3. *Physical Requirements for Yeast Growth*

Most species thrive in warm, dilute, sugary, acidic, and aerobic environments. Most laboratory and industrial yeasts (e.g., *S. cerevisiae* strains) grow best from 20–30°C. The lowest maximum temperature for growth of yeasts is around 20°C, whereas the highest is around 50°C.

Yeast need water in high concentration for growth and metabolism. Several food-spoilage yeasts (e.g., *Zygosaccharomyces* spp.) are able to withstand conditions of low water potential (i.e., high sugar or salt concentrations), and such yeasts are referred to as osmotolerant or xerotolerant.

Most yeasts grow very well between pH 4.5 and 6.5. Media acidified with organic acids (e.g., acetic and lactic) are more inhibitory to yeast growth than are media acidified with mineral acids (e.g., hydrochloric). This is because undissociated organic acids can lower intracellular pH following their translocation across the yeast-cell membrane. This forms the basis of action of weak-acid preservatives in inhibiting food-spoilage yeast growth. Actively growing yeasts acidify their growth environment through a combination of differential ion uptake, proton secretion during nutrient transport (see later), direct secretion of organic acids (e.g., succinate and acetate), and CO_2 evolution and dissolution. Intracellular pH is regulated within relatively narrow ranges in growing yeast cells (e.g., around pH 5 in *S. cerevisiae*), mainly through the action of the plasma-membrane proton-pumping ATPase.

Most yeasts are aerobes. Yeasts are generally unable to grow well under completely anaerobic conditions because, in addition to providing the terminal electron acceptor in respiration, oxygen is needed as a growth factor for membrane fatty acid (e.g., oleic acid) and sterol (e.g., ergosterol) biosynthesis. In fact, *S. cerevisiae* is auxotrophic for oleic acid and ergosterol under anaerobic conditions and this yeast is not, strictly speaking, a facultative anaerobe. Table V categorizes yeasts based on their fermentative properties and growth responses to oxygen availability.

B. Carbon Metabolism by Yeasts

1. *Carbon Sources for Yeast Growth*

As chemorganotrophic organisms, yeasts obtain carbon and energy in the form of organic compounds. Sugars are widely used by yeasts. *Saccharomyces cerevisiae* can grow well on glucose, fructose, mannose, galactose, sucrose, and maltose. These sugars are also readily fermented to ethanol and carbon dioxide by this yeast, but other carbon substrates such as ethanol, glycerol, and acetate can only be respired by *S. cerevisiae* in the presence of oxygen. Some yeasts (e.g., *Pichia stipitis* and *Candida shehatae*) can use five-carbon pentose sugars such as D-xylose and L-arabinose as growth and fermentation substrates. A few amylolytic yeasts exist (e.g., *Saccharomyces diastaticus*) that can use starch, and several oleaginous yeasts (e.g., *Candida tropicalis*) can grow on hydrocarbons such as straight-chain alkanes in the C_{10}–C_{20} range. Several methylotrophic yeasts (e.g., *Hansenula polymorpha* and *Pichia pastoris*) can grow very well on methanol as the sole carbon and energy source, and these yeasts have industrial po-

TABLE V
Classification of Yeasts Based on Fermentative Capacity

Class	Examples	Comments
Obligately fermentative	*Candida pintolopesis*	Naturally occurring respiratory-deficient yeasts; only ferment, even in presence of oxygen
Facultatively fermentative		
Crabtree positive	*Saccharomyces cerevisiae*	Predominantly ferment high sugar-containing media in the presence of oxygen
Crabtree negative	*Candida utilis*	Do not form ethanol under aerobic conditions and cannot grow anaerobically
Nonfermentative	*Rhodotorula rubra*	Do not produce ethanol, either in the presence or absence of oxygen

tential in the production of recombinant proteins for pharmaceutical use.

2. *Yeast Sugar Transport*

Sugars are transported into yeast cells across the plasma membrane by a variety of mechanisms from simple net diffusion (a passive or free mechanism), facilitated (catalyzed) diffusion, and active (energy-dependent) transport. The precise mode of sugar translocation will depend on the sugar, yeast species, and growth conditions. For example, *S. cerevisiae* takes up glucose by facilitated diffusion and maltose by active transport. Active transport means that the plasma-membrane ATPases act as directional proton pumps in accordance with chemiosmotic principles. The pH gradients thus drive nutrient transport either via proton symporters (as is the case with certain sugars and amino acids) or via proton antiporters (as is the case with potassium ions).

3. *Yeast Sugar Metabolism*

The principal metabolic fates of sugars in yeasts are the dissimilatory pathways of fermentation and respiration (summarized in Fig. 2), and the assimilatory pathways of gluconeogenesis and carbohydrate biosynthesis.

Yeasts described as fermentative are able to use organic substrates (sugars) anaerobically as electron donor, electron acceptor, and carbon source. During alcoholic fermentation of sugars, *S. cerevisiae* and other fermentative yeasts reoxidize the reduced coenzyme NADH to NAD (nicotinamide adenine dinucleotide) in terminal step reactions from pyruvate.

$$\text{Glucose} \xrightarrow{\nearrow 2\text{ATP}} 2\text{Pyruvate} \xrightarrow[\text{pyruvate decarboxylase}]{\nearrow 2CO_2}$$

$$2\text{ Acetaldehyde} \xrightarrow[\text{alcohol dehydrogenase}]{2\text{NADH} + 2\text{H}^+ \rightarrow 2\text{NAD}^+} 2\text{ Ethanol}$$

$$\text{Sum: Glucose} + 2\text{Pi} + 2\text{ADP} \longrightarrow 2\text{ Ethanol} + 2Co_2 + 2\text{ATP} + 2H_2O$$

In the first of these terminal reactions, catalyzed by pyruvate decarboxylase, pyruvate is decarboxylated to acetaldehyde, which is finally reduced by alcohol dehydrogenase to ethanol. The regeneration of NAD is necessary to maintain the redox balance and prevent the stalling of glycolysis. In alcoholic-beverage fermentations (e.g., of beer, wine, and distilled spirits), other fermentation metabolites, in addition to ethanol and carbon dioxide, are produced by yeast that are very important in the development of flavor. These metabolites include fusel alcohols (e.g., isoamyl alcohol), polyols (e.g., glycerol), esters (e.g., ethyl acetate), organic acids (e.g., succinate), vicinyl diketones (e.g., diacetyl), and aldehydes (e.g., acetaldehyde). The production of glycerol (an important industrial commodity) can be enhanced in yeast fermentations by the addition of sulfite, which chemically traps acetaldehyde.

$$\text{Glucose} + HSO_3^- \rightarrow \text{Glycerol} + \text{acetaldehyde-}HSO_3^- + CO_2$$

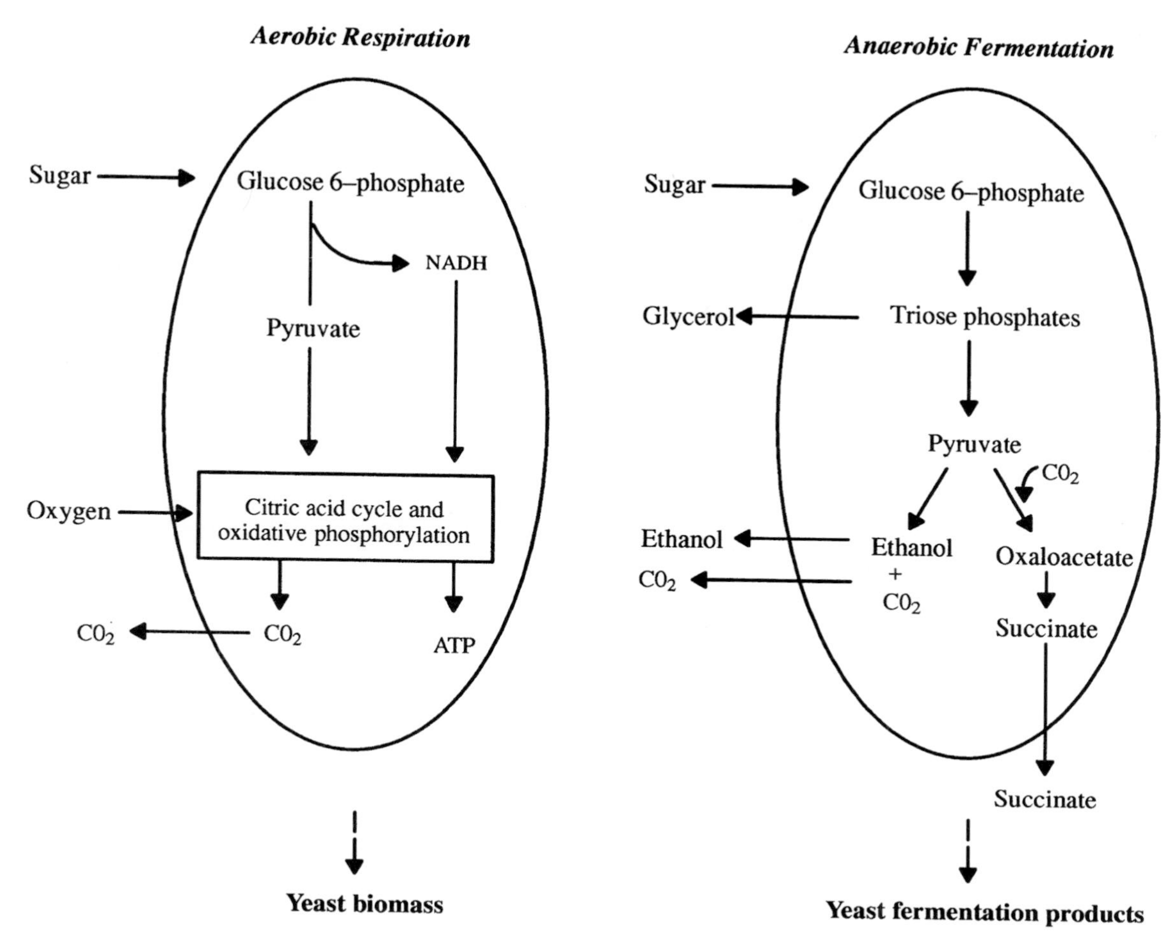

Fig. 2. Summary of major sugar catabolic pathways in yeast cells. From Walker (1998). "Yeast Physiology and Biotechnology." Copyright John Wiley & Sons Limited. Reproduced with permission.

Of the environmental factors that regulate respiration and fermentation in yeast cells, the availability by glucose and oxygen are the best understood and are linked to the expression of regulatory phenomena, referred to as the Pasteur effect and the Crabtree effect. A summary of the description of these phenomena is provided in Table VI.

C. Nitrogen Metabolism by Yeasts

1. *Nitrogen Sources for Yeast Growth*

Although yeasts cannot fix molecular nitrogen, simple inorganic nitrogen sources such as ammonium salts are widely used. Ammonium sulfate is a commonly used nutrient in yeast growth media because it provides a source of both assimilable nitrogen and sulfur. Some yeasts can also grow on nitrate as a source of nitrogen, and, if able to do so, may also use subtoxic concentrations of nitrite. A variety of organic nitrogen compounds (amino acids, peptides, purines, pyrimidines, and amines) can also provide the nitrogenous requirements of the yeast cell. Glutamine and aspartic acids are readily deaminated by yeasts and therefore act as good nitrogen sources.

2. *Yeast Transport of Nitrogenous Compounds*

Ammonium ions are transported is *S. cerevisiae* by both high-affinity and low-affinity carrier-mediated transport systems. Two classes of amino acid uptake systems operate in yeast cells. One is broadly

TABLE VI
Summary of Pasteur and Crabtree Effects in Yeast Sugar Metabolism

Regulatory phenomenon	*Description*	*Comments*
Pasteur effect	Activation of sugar consumption rate by anaerobiosis	Only observable in resting or nutrient-starved cells (e.g., *S. cerevisiae*)
Crabtree effect	Suppression of respiration by high glucose levels	Cells (e.g., *S. cerevisiae, Schiz. pombe*) continue to ferment irrespective of oxygen availability due to glucose repressing or inactivating respiratory enzymes or due to the inherent respiratory capacity of cells

specific, the general amino acid permease (GAP), and effects the uptake of all naturally occuring amino acids. The other system includes a variety of tranporters that display specificity for one or a small number of related amino acids. Both the general and the specific transport systems are energy dependent.

3. *Yeast Metabolism of Nitrogenous Compounds*

Yeasts can either incorporate ammonium ions or amino acids into cellular protein, or these nitrogen sources can be intracellularly catabolized to serve as nitrogen sources (see Fig. 3). Yeasts also store relatively large pools of endogenous amino acids in the vacuole, most notably arginine. Ammonium ions can be directly assimilated into glutamate and glutamine, which serve as precursors for the biosynthesis of other amino acids. The precise mode of ammonium assimilation adopted by yeasts will depend mainly on the concentration of available ammonium ions and the intracellular amino acid pools. Amino acids may be dissimilated (by decarboxylation, transamination, or fermentation) to yield ammonium and

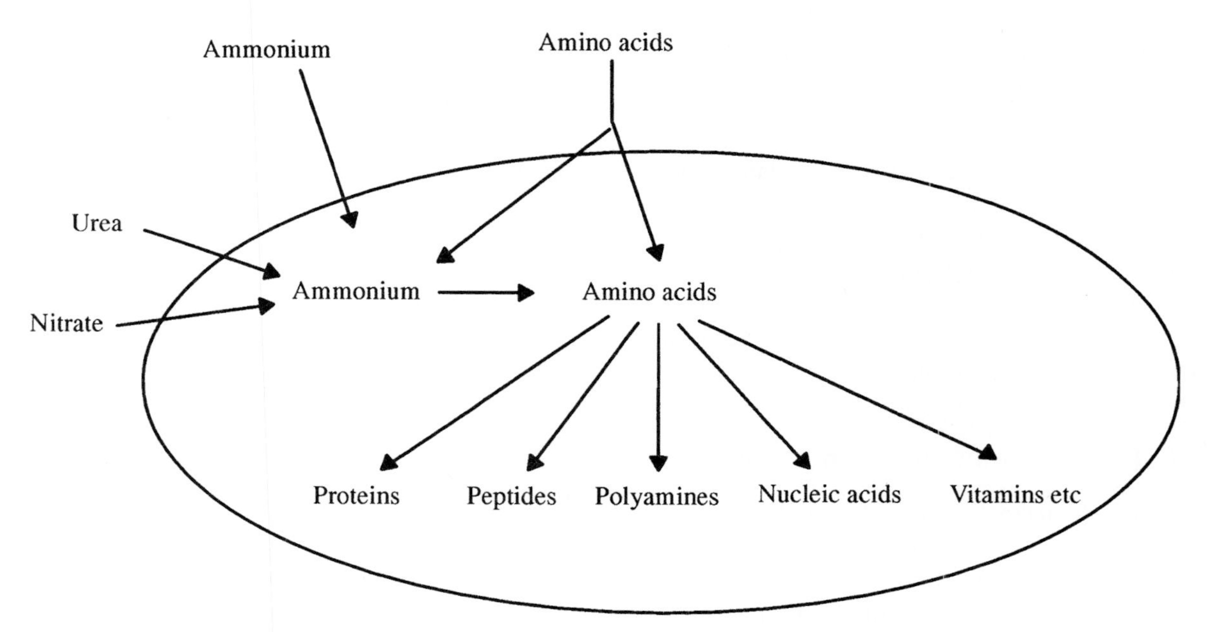

Fig. 3. Overview of nitrogen assimilation in yeasts. From Walker (1998). "Yeast Physiology and Biotechnology." Copyright John Wiley & Sons Limited. Reproduction with permission.

glutamate, or they may be directly assimilated into proteins.

D. Yeast Growth

The growth of yeasts is concerned with how cells transport and assimilate nutrients and then integrate numerous component functions in the cell in order to increase in mass and eventually divide. Yeasts have proved invaluable in unravelling the major control elements of the eukaryotic cell cycle and research with the budding yeast, *Saccharomyces cerevisiae,* and the fission yeast, *Schizosaccharomyces pombe,* has significantly advanced our understanding of cell-cycle regulation, which is particularly important in the field of human cancer.

1. *Vegetative Reproduction in Yeasts*

Budding in the most common mode of vegetative reproduction in yeasts and is typical in ascomycetous yeasts such as *S. cerevisiae.* Yeast buds (see Fig. 4) are initiated when mother cells attain a critical cell size at a time that coincides with the onset of DNA synthesis. This is followed by localized weakening of the cell wall and this, together with tension exerted by turgor pressure, allows the extrusion of the cytoplasm in an area bounded by new cell-wall material. The mother and daughter bud cell walls are contiguous during bud development. In *S. cerevisiae,* cell size at division is asymmetrical, with buds being smaller than mother cells when they separate. Scar tissue on the yeast cell wall, the bud and birth scars, remain on the daughter bud and mother cells, respectively (see Fig. 4).

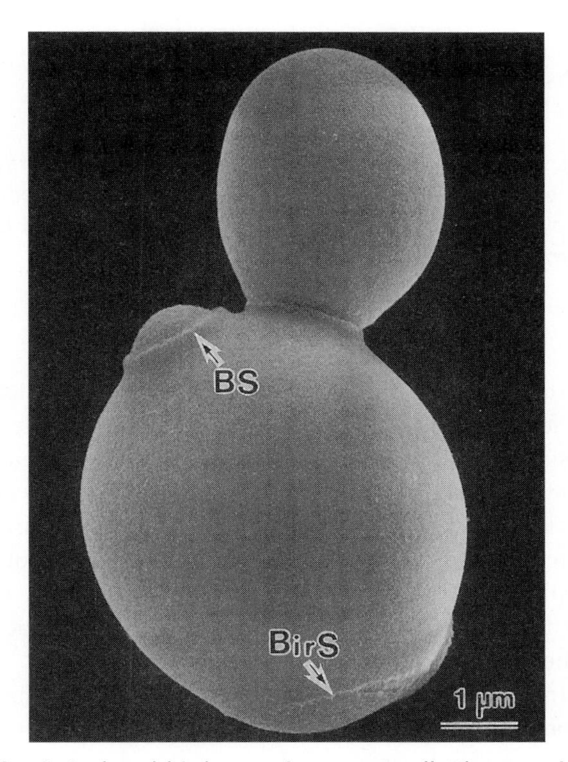

Fig. 4. Bud and birth scars in a yeast cell. The scanning electron micrograph shows a bud scar (BS) and a birth scar (BirS) on the surface of a *Saccharomyces cerevisiae* cell. (Courtesy of M. Osumi, Japan Women's University, Tokyo.)

Fission is a mode of vegetative reproduction typified by species of *Schizosaccharomyces,* which divide exclusively by forming a cell septum that constricts the cell into two equal-size daughters. In *Sch. pombe,* which has been used extensively in eukaryotic cell cycle studies, newly divided daughter cells grow lengthways in a monopolar fashion for about one-third of their new cell cycle. Cells then switch to bipolar growth for about three-quarters of the cell cycle until mitosis is initiated at a constant cell-length stage.

Filamentous growth occurs in numerous yeast species and may be regarded as a mode of vegetative growth alternative to budding or fission. Some yeasts exhibit a propensity to grow with true hyphae initiated from germ tubes (e.g., *Candida albicans*), but others (including *S. cerevisiae*) may grow in a pseudohyphal fashion when induced to do so by unfavorable conditions. Hyphal and pseudohyphal growth represent different developmental pathways in yeasts, but cells can revert to unicellular growth upon return to more conducive growth conditions. Filamentation may therefore represent an adaptation by yeasts to foraging when nutrients are scarce.

2. *Population Growth of Yeasts*

As in most microorganisms, when yeast cells are inoculated into a liquid nutrient medium and incubated under optimal physical growth conditions, a typical batch-growth curve will result when the viable cell population is plotted against time. This growth curve is made up of a lag phase (period of no growth, but physiological adaptation of cells to their new environment), exponential phase (limited period of logarithmic cell doublings), and stationary phase (resting period with zero growth rate).

Diauxic growth is characterized by two exponential phases and occurs when yeasts are exposed to two carbon growth substrates that are used sequentially. This occurs during aerobic growth of *S. cerevisiae* on glucose (the second substrate being ethanol formed from glucose fermentation).

In addition to batch cultivation of yeasts, cells can also be propagated in continuous culture in which exponential growth is prolonged without lag or stationary phases. Chemostats are continuous cultures that are based on the controlled feeding of a sole growth-limiting nutrient into an open culture vessel, which permits the outflow of cells and spent medium. The feeding rate is referred to as the dilution rate, which is employed to govern the yeast growth rate under the steady-state conditions that prevail in a chemostat.

Specialized yeast culture systems include immobilized bioreactors. Yeast cells can be readily immobilized or entrapped in a variety of natural and synthetic materials (e.g., calcium aliginate gel or microporous glass beads) and such systems have applications in the food and fermentation industries.

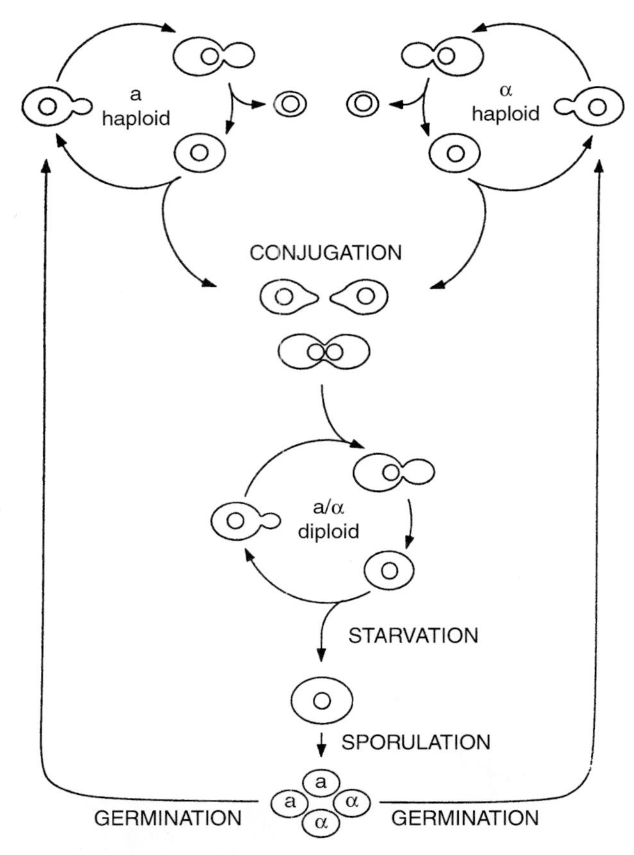

Fig. 5. Life cycle of *Saccharomyces cerevisae.* From Murray, A., and Hunt, T. (1993). "The Cell Cycle, an Introduction." Copyright © 1993 Oxford University Press, Inc. Used by permission of Oxford University Press, Inc.

V. YEAST GENETICS

A. Life Cycle of Yeasts

Many yeasts have the ability to reproduce sexually, but the processes involved are best understood in the budding yeast, *S. cerevisiae,* and the fission yeast, *Sch. pombe.* Both species have the ability to mate, undergo meiosis, and sporulate. The development of spores by yeasts represents a process of morphological, physiological, and biochemical differentiation of sexually reproductive cells.

Mating in *S. cerevisiae* involves the conjugation of two haploid cells of opposite mating types, designated **a** and α. These cells synchronize one another's cell cycles in response to peptide-mating pheromones, known as **a** factor and α factor (see Fig. 5). The conjugation of mating cells occurs by cell-wall surface contact followed by plasma-membrane fusion to form a common cytoplasm. Karyogamy (nuclear fusion) then follows, resulting in a diploid nucleus. The stable diploid zygote will continue mitotic cell cycles in rich growth media, but, if starved of nitrogen, the diploid cells will sporulate to yield four haploid spores. These germinate in rich media to form haploid budding cells that can mate with each other to restore the diploid state.

In *Sch. pombe,* haploid cells of the opposite mating types (designated $h+$ and $h-$) secrete mating pheromones and, when starved of nitrogen, undergo conjugation to form diploids. In *Sch. pombe,* however, such diploidization is transient under starvation conditions and cells soon enter meiosis and sporulate to produce four haploid spores.

B. Genetic Manipulation of Yeasts

There are several ways of genetically manipulating yeast cells including hybridization, mutation, rare mating, cytoduction, spheroplast fusion, single chromosome transfer, and transformation using recombinant DNA technology.

Classic genetic approaches in *S. cerevisiae* involve mating of haploids of opposite mating type. Subsequent meiosis and sporulation results in the production of a *tetrad ascus* with four spores, which can be isolated, propagated, and genetically analyzed (i.e., tetrad analysis). This process forms the basis of genetic breeding programs for laboratory-reference strains of *S. cerevisiae*. However, industrial (e.g., brewing) strains of this yeast are polyploid, are reticent to mate, and exhibit poor sporulation with low spore viability. It is, therefore, generally fruitless to perform tetrad analysis and breeding with brewer's yeasts. Genetic manipulation strategies for circumventing the sexual reproductive deficiencies of brewer's yeast include spheroplast fusion and recombinant DNA technology.

Intergeneric and intrageneric yeast hybrids may be obtained using the technique of spheroplast fusion. This involves the removal of yeast-cell walls using lytic enzymes (e.g., glucanases from snail gut juice or microbial sources), followed by the fusion of the resulting spheroplasts in the presence of polyethylene glycol and calcium ions.

Recombinant DNA technology (genetic engineering) of yeast is summarized in Fig. 6. Yeast cells possess particular attributes for expressing foreign genes and have now become the preferred hosts, over bacteria, for producing human proteins for pharmaceutical use. Although the majority of research and development into recombinant protein synthesis in yeasts has been conducted using *S. cerevisiae*, several non-*Saccharomyces* species are being studied and exploited in biotechnology. For example, *Hansenula polymorpha* and *Pichia pastoris* (both methylotrophic yeasts) exhibit particular advantages over *S. cerevisiae* in cloning technology.

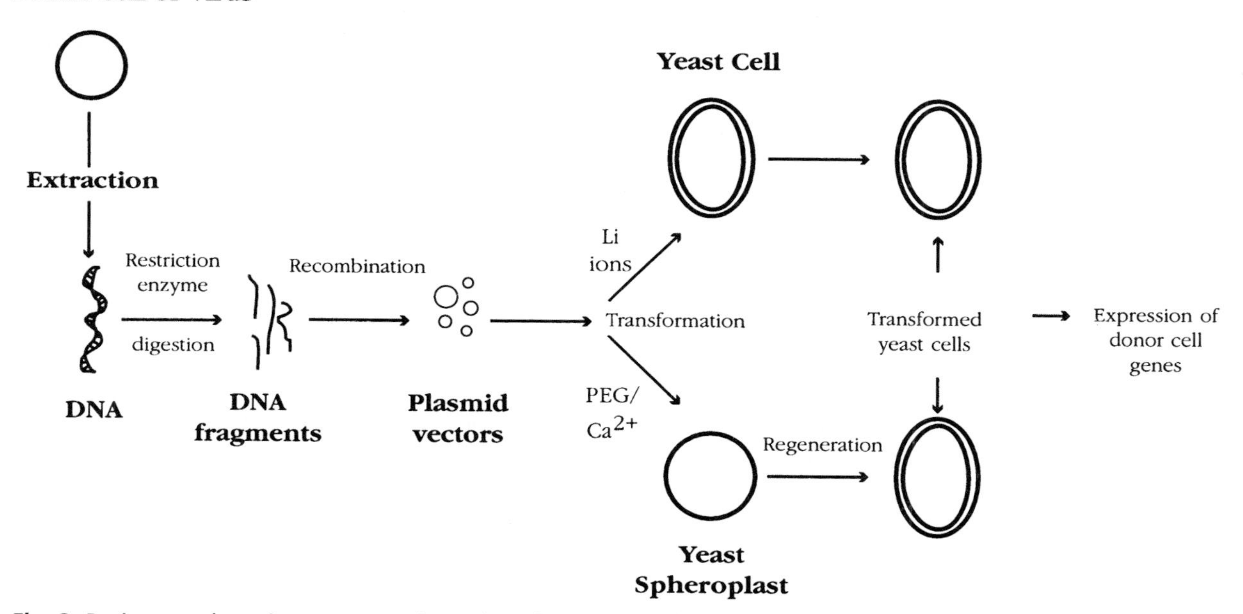

Fig. 6. Basic procedures in yeast genetic engineering. From Walker (1998). "Yeast Physiology and Biotechnology." Copyright John Wiley & Sons Limited. Reproduced with permission.

TABLE VII
Industrial Commodities Produced by Yeasts

Commodity	Examples
Beverages	Potable alcoholic beverages: beer, wine, cider, saké, distilled spirits (whisky, rum, gin, vodka, cognac)
Food and animal feed	Baker's yeast, yeast extracts, fodder yeast and livestock growth factor, feed pigments
Chemicals	Fuel ethanol, carbon dioxide, glycerol, citric acid vitamins; yeasts are also used as bioreductive catalysts in organic chemistry
Enzymes	Invertase, inulinase, pectinase, lactase, lipase
Recombinant proteins	Hormones (e.g., insulin), viral vaccines (e.g., hepatitis B vaccine), antibodies (e.g., IgE receptor), growth factors (e.g., tumor necrosis factor), interferons (e.g., leucocyte interferon-α), blood proteins (e.g., human serum albumin), enzymes (e.g., gastric lipase)

C. Yeast Genome and Proteome Projects

A landmark in biotechnology was reached in 1996 with completion of the sequencing of the entire genome of *S. cerevisiae.* The *Sch. pombe* genome project is ongoing. The functional analysis of the many orphan genes of *S. cerevisiae,* for which no function has yet been assigned, is underway through international research collaborations. Elucidation by cell physiologists of the biological function of all *S. cerevisiae* genes, that is, the complete analysis of the yeast proteome, will not only lead to an understanding of how a simple eukaryotic cell works, but will also provide insight into molecular biological aspects of heritable human disorders.

VI. INDUSTRIAL, AGRICULTURAL, AND MEDICAL IMPORTANCE OF YEASTS

A. Industrial Significance of Yeasts

Yeasts have been exploited for thousands of years in traditional fermentation processes to produce beer, wine, and bread. The products of modern yeast biotechnologies impinge on many commercially important sectors, including food, beverages, chemicals, industrial enzymes, pharmaceuticals, agriculture, and the environment. Table VII lists some of the principal industrial commodities from yeasts. *S. cerevisiae* is the most exploited microorganism known and is the yeast responsible for producing potable and industrial ethanol, which is the world's premier biotechnological commodity.

Some yeasts play detrimental roles in industry, particularly as spoilage yeasts in food and beverage production (see Table VIII).

B. Yeasts of Environmental and Agricultural Significance

Although a few yeast species are known to be plant pathogens (e.g., *Ophiostoma ulmi,* the causative agent of Dutch Elm disease), several have been shown to be beneficial to plants in preventing fungal disease. For example, *S. cerevisiae* has potential as a phytoalexin elicitor in stimulating cereal plant defenses against fungal pathogens, and several yeasts (e.g.,

TABLE VIII
Some Food and Beverage-Spoilage Yeasts

Yeast species	Food spoiled
Cryptococcus laurentii, Candida zeylanoides	Frozen poultry carcasses
Zygosaccharomyces bailii, Z. rouxii	Fruits, fruit juices, vegetables
Kluyveromyces, Rhodotorula, and *Candida* spp.	Milk, yogurt, cheeses
Torulopsis, Pichia, Candida, Hansenula spp., and 'wild' species of *Saccharomyces*	Beer
Zygosaccharomyces bailii and many other yeasts	Wine

Debaryomyces hansenii and *Metschnikowia pulcherrima*) may be used in the biocontrol of fungal fruit diseases. Other environmental benefits of yeasts are to be found in aspects of pollution control. For example, yeasts can effectively biosorb heavy metals and detoxify chemical pollutants from industrial effluents. Some yeasts (e.g., *Candida utilis*) can effectively remove carbon and nitrogen from organic wastewater.

In agriculture, live cultures of *S. cerevisiae* have been shown to stabilize the rumen environment of ruminant animals (e.g., cattle) and improve the nutrient availability to increase animal growth or milk yields. The yeasts may be acting to scavenge oxygen and prevent oxidative stress to rumen bacteria, or they may provide malic and other dicarboxylic acids to stimulate rumen bacterial growth.

C. Medical Significance of Yeasts

The vast majority of yeasts are beneficial to human life. However, some yeasts are opportunistically pathogenic towards humans. Mycoses caused by *Candida albicans*, collectively referred to as candidosis (candidiasis), are the most common opportunistic yeast infections. There are many predisposing factors to yeast infections, but immunocompromised individuals appear particularly susceptible to candidosis. *C. albicans* infections in AIDS patients are frequently life-threatening.

The beneficial medical aspects of yeasts are apparent in the provision of novel human therapeutic agents through yeast recombinant DNA technology (see Table VII). Yeasts are also extremely valuable as experimental models in biomedical research, particulalry in the fields of oncology, pharmacology, toxicology, virology, and human genetics.

See Also the Following Articles

Beer/Brewing • Carbohydrate Synthesis and Metabolism • Microscopy, Confocal • Rumen Microbiology • Sporulation

Bibliography

Johnston, J. R. (1994). "Molecular Genetics of Yeast, a Practical Approach." Oxford University Press, New York.

Kurtzman, C. P., and Fell, J. W. (1998) "The Yeasts. A Taxonomic Study." Elsevier Science, Amsterdam.

Kocková-Kratochvilová, A. (1990). "Yeasts and Yeast-like Organisms." VCH, New York.

Oliver, S. G., Winson, M. K., Kell, D. B., and Baganz, F. (1998). Systematic functional analysis of the yeast genome. *Trends Biotechnol.* **16**, 373–378.

Panchal, C. J. (1990) "Yeast Strain Selection." Marcel Dekker, New York.

Pringle, J. R., Broach, J. R., and Jones, E. W. (1997). "The Molecular and Cellular Biology of the Yeast Saccharomyces." Cold Spring Harbor Laboratory Press, Cold Spring Harbor, NY.

Rose, A. H., and Harrison, J. S. (1989–1995). "The Yeasts." Academic Press, London.

Spencer, J. F. T., and Spencer, D. M. (1997). "Yeasts in Natural and Artifical Habitats." Springer-Verlag, Berlin.

Walker, G. M. (1998). "Yeast Physiology and Biotechnology." John Wiley & Sons, Chichester, UK.

Wolf, K. (1996). "Nonconventional Yeasts in Biotechnology, A Handbook." Springer-Verlag, Berlin.

Zimmerman, F. K., and Entian, K-D. (1997). "Yeast Sugar Metabolism, Biochemistry, Genetics, Biotechnology and Applications." Technomic Publishing, Basel, Switzerland.

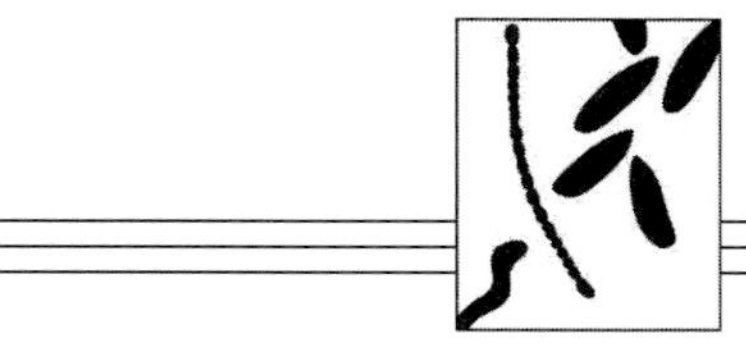

Zoonoses

Bruno B. Chomel
University of California, Davis

I. Classification and Importance
II. Orthozoonoses (Direct Zoonoses)
III. Cyclozoonoses
IV. Metazoonoses
V. Saprozoonoses
VI. New and Emerging Zoonoses
VII. Prevention and Control

GLOSSARY

endemic A disease that is restricted to and constantly present in a particular country or locality.

epidemic A disease that is prevalent and spreading rapidly among people.

infection The state of being infected, especially by the presence in the body of bacteria, protozoans, viruses, or other parasites.

lethality The percentage of people who die of a disease of all people sickened by that disease.

pandemic An epidemic with a worldwide distribution.

prodrome A warning symptom indicating the onset of a disease.

reservoir host A species that serves as an immune host for a parasite that can cause disease in another species. Member of a population in which certain infectious agents can perpetuate.

sporadic Occurring on an occasional basis.

vector An animal, such as an insect, that transmits a disease-producing organism from one host to another.

ZOONOSES are the "diseases and infections that are naturally transmitted between vertebrate animals and man," as defined in 1959 by the World Health Organization (WHO) Expert Committee on Zoonoses. The word zoonosis (*zoonoses*, plural) is the combination of two Greek words (*zoon,* animals; and *noson,* disease), and was coined at the end of the nineteenth century by Rudolph Virchow to designate human diseases caused by animals. Nevertheless, the term should also include vertebrate animal diseases caused by exposure to humans, such as measles in nonhuman primates, which is of major concern in any major primate center. The term "zoonosis" is also considered to be shorter and more convenient than "anthropozoonosis" (animal to humans) and "zooanthroponosis" (humans to animal), which are based on the prevailing direction of transmission between humans and lower vertebrates. The word anthropozoonosis (*anthropozoonoses*, plural) is much more in use in the medical community than in the veterinary community.

The well-accepted WHO definition of zoonoses requires some comments. The definition should include the world "infestations" to cover more properly parasitic diseases. Zoonoses are usually limited to agents that can replicate in the animal host; therefore it does not include diseases caused by inoculation of venom or toxins of reptile or fish origin or by allergies to vertebrates. It also excludes diseases transmitted by animals or food of animal origin, which are vehicles of human pathogens, such as ice cream contaminated with the hepatitis A virus or a poliovirus. The notion of transmissibility in natural conditions is also important for zoonoses and excludes all experimental infections of nonhuman vertebrates with human pathogens, such as human hepatitis B virus experimentally inoculated in chimpanzees. Transmissibility also distinguishes the true zoonoses, for which there must be epidemiological evidence of direct or indirect transmission from animal to humans, from the communicable diseases common to humans and animals such as tetanus, for which the contamination of humans and animals occurs from a common source without an epidemiological interrelation. However, animals may play a role in the amplification of the agent, such as when a horse sheds *Clostridium tetani* in the environment.

For the most ubiquitous zoonoses, humans and lower vertebrates are equally suitable reservoir hosts, and infection may be transmitted in either direction. The term "amphixenoses" applies to these infections, which include some of the staphylococcoses and streptococcoses. Zoonoses exclude those diseases that are vector-borne only from humans to humans, such as malaria or many arboviral diseases. Finally, the reciprocity of the infection from animals to humans or from humans to animals illustrates the intertransmissibility of zoonoses. However, humans represent, in most instances, an epiphenomenon in the natural cycle of infection between nonhuman vertebrates. Only in a few parasitic diseases, such as for *Taenia solium* or *T. saginata* infestations, are humans an obligatory host of the infection. Zoonoses are a type of animal contagious infections, in which humans become an occasional victim, sometimes able to reinfect animals or humans (extensive and reversible zoonosis), sometimes unable to spread the infection (restricted or limited zoonosis). In the latter case, humans can be considered an epidemiological dead end or cul-de-sac.

I. CLASSIFICATION AND IMPORTANCE

A. Classification

Zoonoses can be classified according to the etiologic agent—viral, bacterial, parasitic, or mycotic. However, it is the primary epidemiological classification based on the zoonosis maintenance cycle that is of major importance when considering alternatives for control measures. This classification divides the zoonoses into four categories.

1. Direct zoonoses (orthozoonoses) are transmitted from an infected to a susceptible vertebrate host by direct contact, by contact with a fomite, or by a mechanical vector. Direct zoonoses may be perpetuated in nature by a single vertebrate species. Examples are rabies or brucellosis (Fig. 1).

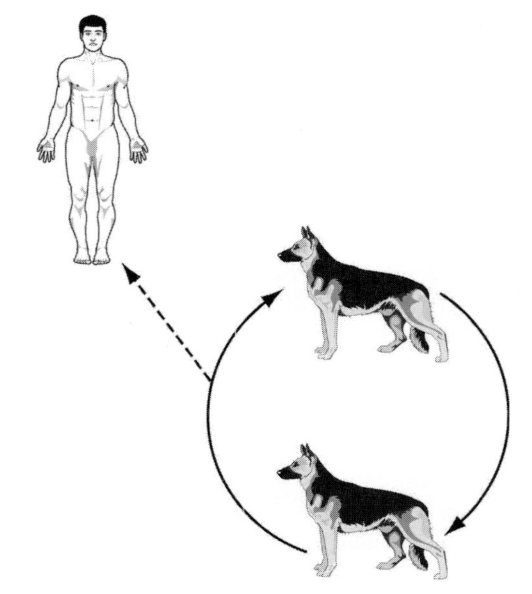

Fig. 1. Epidemiological maintenance cycle of rabies, a direct zoonosis (orthozoonosis).

2. Cyclozoonoses require more than one vertebrate species, but no invertebrate host, in order to complete the developmental cycle of the agent. Examples are human taeniases or pentastomid infections (Fig. 2). Most of the comparatively few cyclozoonoses are cestodiases.

3. Metazoonoses (also called pherozoonoses) are zoonoses that require both vertebrates and invertebrates for the completion of their infectious cycle (Fig. 3). In metazoonoses, the infectious agent multiplies (propagative or cyclopropagative transmission) or merely develops (developmental transmission) in the invertebrate; there is always an extrinsic incubation period in the invertebrate host before transmission to a vertebrate host is possible. Examples are arbovirus infections, plague, or rickettsial infections.

4. Saprozoonoses have both a vertebrate host and an inanimate developmental site or reservoir (Fig. 4). The developmental reservoir is considered nonanimal, such as organic matter, including food, soil, and plants. In this group of zoonoses, direct infection is usually rare or absent. Examples are histoplasmosis, *Erysipelothrix* infection, or listeriosis.

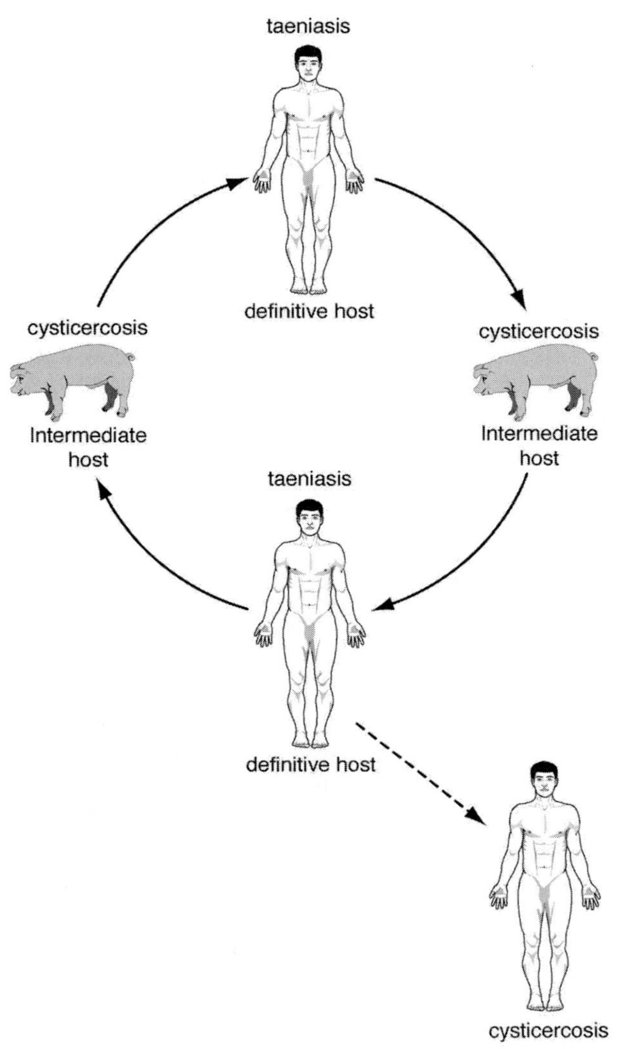

Fig. 2. Epidemiological maintenance cycle of taeniasis and cysticercosis (*Taenia solium*), a cyclozoonosis.

Other classifications of zoonoses may include a classification based on the categories of people at risk or relating to the type of human activity, such as occupational zoonoses (which occur when people are infected during their professional activity; e.g., brucellosis in farmers, veterinarians, or slaughterhouse employees, Lyme disease in foresters, rabies in wildlife trappers or taxidermists), zoonoses associated with recreational activities (e.g., plague, hantavirus infection, Lyme disease, tularemia, or larva migrans), domestic zoonoses (diseases acquired from pets), or accidental zoonoses (some very rare and peculiar circumstances of infection, as well as food-borne outbreaks).

Another aspect of zoonoses classification concerns their clinical manifestations and their diagnosis. Clinical diagnosis of zoonoses is not always easy, especially if the symptoms are different in animals and humans, or if they are present only in humans. If clinical signs are observed in animals and humans, zoonoses are designated as phanerozoonoses. If symptoms are similar in both animals and humans, they are considered isosymptomatic (rabies and tu-

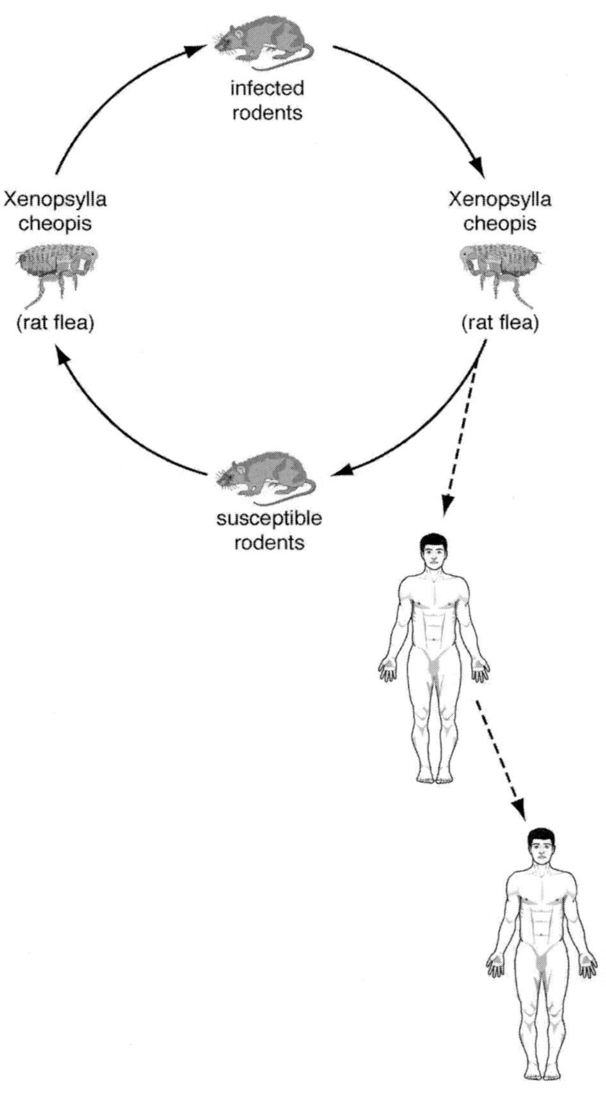

Fig. 3. Epidemiological maintenance cycle of plague, a metazoonosis.

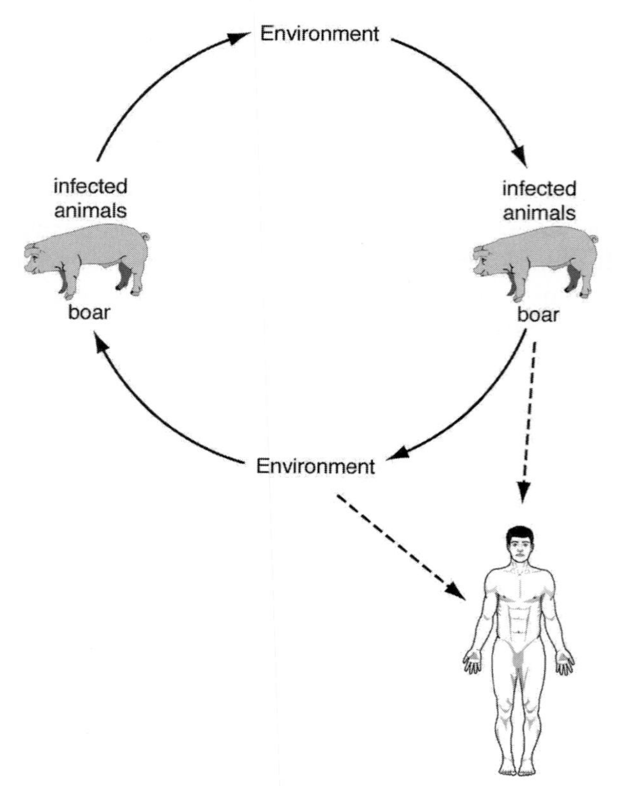

Fig. 4. Epidemiological maintenance cycle of *Erysipelothrix* infection, a saprozoonosis.

berculosis), whereas they are anisosymptomatic if the symptoms are different in humans and animals (anthrax, brucellosis, psittacosis, and Rift Valley fever). In some instances, subclinical infection is observed in animals and clinical illness in humans, or vice versa. In such cases, these zoonoses are designated as cryptozoonoses. Examples of animal infection and human disease are ornithosis pneumonia in humans and latent infection in pigeons and turkey, or leptospirosis menigitis in swinekeepers. The inverse can also occur, such as the viral Ebola-like (Reston virus) infection, which is deadly in nonhuman primates and leads only to seroconversion in infected humans. Finally, infection without any clinical symptoms may occur in both animals and humans, such as some arboviral infections, detected only serologically (Tahyna virus).

B. Medical and Economic Importance

Zoonoses are important to Public Health because of their number, their frequency, and their severity in relation to human health. There are more than 150 zoonoses according to the World Health Organization (WHO) Zoonoses Expert Committee. There are very few vertebrates that are not involved with one or more zoonoses. Human infection most often occurs when infection persists in animals, such as rabies, brucellosis, or tuberculosis. Zoonoses frequency varies for each disease and depends on the geographical distribution of reservoirs, agents, and population density, as well as efficiency of controlled measures. Some zoonoses are ubiquitous, such as salmonellosis and leptospirosis. Others are geographically restricted, such as plague, or various arboviral diseases, including yellow fever. Some zoonoses are very rare or restricted to very limited areas, such as Ebola or Reston virus hemorrhagic fevers. The high lethality rate of some zoonotic diseases, despite their rarity, classify them as major zoonoses, such as hantavirus respiratory syndrome, rabies (in developed countries), herpes B infection, or Marburg disease. Many zoonoses induce severe but usually nonlethal diseases in humans, such as brucellosis, and some are characterized by very mild symptoms in humans, such as Newcastle disease, pseudo-cowpox, or erysipeloid.

In order to prevent such risks for human health, most of the major zoonoses and those that are exotic but particularly life-threatening are reported to the Ministry of Health, or Human Health Services in the United States. Many of them are also reported to the Ministry or Department of Agriculture, especially if these zoonoses are of economic importance to livestock production, such as bovine brucellosis or tuberculosis. However, it may happen that some diseases are declared only to one department, underlining the need of complete cooperation between Human Health Services and Agriculture Services.

Of more than 50 million deaths worldwide in 1997, about one-third were due to infectious and parasitic diseases such as acute lower respiratory diseases, tuberculosis, diarrhea, HIV/AIDS, and malaria. Tuberculosis, mainly caused by *Mycobacterium tuberculosis,* killed about 3 million people and malaria about 2 million. In developing countries, zoonoses account for millions of cases of illness and deaths, underlining not only the loses of human life and suffering, but also their economic burden. Chagas' disease, a zoonosis found in Latin America, affects 17 million people in 21 countries, causing 45,000 deaths and 400,000 cases of heart and stomach disease annually. Even in developed countries, zoonoses affect rural populations. A study of zoonoses prevalence conducted from 1991–1996 in the United Kingdom (Thomas and Salmon, 1997) indicated that antibody prevalence to *Coxiella* (Q fever) (27.3%), *Chlamydia* (79.6%), and *Toxoplasma* (50.2%) were high in UK farmworkers, whereas the prevalence of antibodies to Lyme disease (0.3%), leptospirosis (0.2%), and brucellosis (0.7%) were low. Other zoonoses detected were hantavirus (seroprevalence, 4.7%), orthopoxvirus (0.7%), parapoxvirus (4.5%), *Bartonella* (2%), and *Echinococcus granulosus* (1.5%).

Zoonoses are considered to represent a tremendous economic burden to humans due to the loss of diseased animals and agricultural production, cost of prevention, and treatment, debilitation of and productivity losses to humans. It is quite difficult to evaluate such costs precisely, but some estimates have been published that illustrate the economic impact of zoonotic diseases. For instance, the total yearly cost of food-borne illnesses was estimated to be between $1.5 and $2.7 billion by U.S. Department of Agriculture economists for the five major food-borne diseases, with salmonellosis accounting for 36% and campylobacteriosis for 52% of the total. Productivity losses represented about 95% of the total cost. According to Steele, there are more than 1 billion cattle in the world; if the incidence of reactors can be estimated at 5% of the world population, the economic losses associated with tuberculosis are staggering, the magnitude being in the range of $1.5 billion.

Economic losses resulting from food-borne parasitic zoonoses are difficult to assess, as underlined by Murrell. In Mexico, for example, porcine cysticercosis was reported to be responsible for a loss of more than one-half of the national investment in swine production and for more than $17 million annually in hospitalization and treatment costs for humans with neurocysticercosis. For all of Latin America, porcine cysticercosis accounted for an economic loss of $164 million. In Africa, losses of 1–2 billion per year due to bovine cysticercosis have been reported. Human toxoplasmosis in the United States is estimated to be an annual economic and public health burden of more than $400 million.

Zoonoses also underline the important contribution of the veterinary profession and veterinary preventive medicine to human public health. The activities of veterinarians are to investigate zoonosis outbreaks, to establish surveillance systems in animal populations, to reduce disease prevalence in domestic animals by culling infected animals and restricting animal movements, and to monitor wildlife. Veterinarians are also a natural link between physicians, ecologists, and environment specialists.

II. ORTHOZOONOSES (DIRECT ZOONOSES)

The direct zoonoses are transmitted from an infected vertebrate to a susceptible vertebrate by direct contact, vehicle, or mechanical vector (Fig. 1). The infectious cycle requires only one vertebrate species to perpetuate. Most of these zoonoses are transmitted to humans through direct physical contact with an infected animal (e.g., rabies). The dog is certainly the main animal species associated with rabies worldwide and with its transmission to humans. Fortunately, rabies is a rare disease in humans in the United States. Since canine rabies has been controlled, the incidence of human rabies, excluding cases acquired abroad, has fallen from 43 (0.03/100,000) per year in 1945 to <1 (<0.001/100,000) per year in the 1980s. Although rabies in the United States is mainly a wildlife disease affecting raccoons, skunks, foxes, and bats, humans are probably at greater risk of acquiring rabies from a rabid dog, because dogs account for a large majority of bite incidents reported in the United States. In a review of dog rabies cases in 1988 in the United States, most dog rabies cases occurred in young dogs (57% in dogs ≤1 year old), dogs considered as pets (84% of the cases), and mostly from a rural environment (85%). More important, virtually all rabid dogs were never vaccinated, or their vaccination status was unknown. The number of rabid dogs decreased from 182 in 1992 to 126 in 1997. North Carolina reported the largest number of cases of rabies in dogs (18) for that year, followed by Puerto Rico (15) and Texas (11). Two of the Texas cases were associated with the ongoing epizootic of rabies in dogs and coyotes in southern Texas, down from a high of 42 rabid dogs reported in 1993.

When introduced to the victim's wound via a bite, the rabies virus will multiply locally at the wound site and will then invade the peripheral nerve(s) supplying that area. The virus migrates along the nerves to reach the central nervous system and radiates further through the nerves in various organs, including the salivary glands. The virus is shed in the saliva and can be transmitted to a new victim through a bite. The incubation of rabies varies from a few days to several months, and the virus can be shed in the saliva a few days prior to the appearance of any clinical neurological signs. This is the main reason why a dog that bites a human is quarantined for 10 days. If the dog is still healthy 10 days after the bite incident, one can assume that the dog was not shedding the virus at the time of the bite, and, consequently, that the bite victim does not need a rabies postexposure treatment (a series of five doses of 1.0 ml of rabies vaccine injected intramuscularly on days 0, 3, 7, 14, and 28).

Rabies is mainly characterized by neurological and behavioral disorders. During the initial prodromic phase, the dog is anxious, nervous, and suffers changes in personality. After 2 or 3 days, the furious or paralytic form will start. In the furious form, lasting 1–7 days, the dog shows irritability, aggression, hypersensitivity, disorientation, and sometimes grand mal seizures. In the paralytic or dumb form, lasting 1–10 days, paralysis will affect one or more limbs. Paralysis then progresses to affect the entire nervous system. Cranial nerve paralysis, especially laryngeal paralysis, is often the first recognized sign. Death occurs within 10 days.

Prevention of rabies in dogs is based on the vaccination of susceptible animals, confinement of animals, quarantine of biting dogs, and removal of stray animals. In the United States, the recommendations for state and local rabies vaccination and control measures are published yearly by the National Association of State Public Health Veterinarians. All states require rabies vaccinations for dogs. Interstate travel usually requires the possession of a health certificate and a proof of rabies vaccination. Eleven rabies vaccines are currently marketed for dogs in the United States and should be administered according to the manufacturer's instructions. Puppies should be vaccinated at 3 months old and revaccinated 1 year later at 15 months, after which they should receive annual or triennial boosters. If dogs with a current rabies immunization are exposed to a rabid animal, they should be revaccinated immediately, and then confined and observed for 90 days. Unvaccinated animals are usually euthanized or may be placed in strict isolation for 6 months and vaccinated 1 month before release.

III. CYCLOZOONOSES

Cyclozoonoses require more than one vertebrate species, but no invertebrate host for the completion of the agent's development cycle (Fig. 2). Most of these zoonoses are cestodiases. Infection with *Taenia solium* is an obligatory cyclozoonosis, as a human must be one of the vertebrate hosts. Cysticercosis, infection with the larval stage of the pork tapeworm *Taenia solium,* has been recognized as a cause of severe disease and occasional mortality in the United States, principally among immigrants from Latin America.

Gravid proglottids shed in feces of infected humans are ingested by pigs (in parts of the world where pigs are free-roaming and no sanitation system for eliminating human dejections is available). Humans acquire *T. solium* taeniasis by eating raw or undercooked pork containing cysticerci (*Cysticercus cellulosae*), which is very common in developing countries. Pigs usually do not suffer from the infestation. In humans, the scolex of the larva evaginates in the small intestine and attaches to the intestinal wall. In 2–3 months, the larva has developed into an adult taenia and the first proglotids are expelled in the feces, thus renewing the cycle. The public health significance of *T. solium* lies in the fact that humans can also be infected by the eggs and develop cysticerci in their tissues. Humans acquire cysticercosis by either ingestion of eggs with food (vegetables and fruits) and water contaminated by the fecal material of a taenia carrier, or ingestion of eggs introduced to the mouth by the contaminated hands of an individual with poor hygenic habits or self-autocontamination (anus–hand–mouth).

Taeniasis is usually subclinical, but anal pruritus can cause autoinfestation. Cysticercosis is a more severe disease, neurocysticercosis being most dramatic with ocular and periocular forms. Months after the ingestion of eggs, the patient may develop seizures, epilepsia, dementia, hydrocephalus, or meningitis.

Treatment is based on the use of albendazole or praziquantel. Steroids may be required to decrease the inflammatory sequelae after killing the parasite with antiparasitic agents. Hygenic measures and meat inspection at slaughterhouses are the basis for the prevention and control of human and swine infestation.

IV. METAZOONOSES

The metazoonoses require invertebrate vectors for their transmission to vertebrate hosts. Many vectors can be involved, such as fleas, ticks, mosquitoes for many bacterial and viral diseases. Many parasitic infestations require crustaceans or molluscs for the completion of the parasite cycle. We will use as examples three zoonoses, transmitted by fleas (plague), ticks (Lyme disease), and mosquitoes (equine meningo-encephalitides) (Fig. 3).

A. Plague

Plague is caused by the bacterium *Yersinia pestis,* a gram-negative coccobacillus. In the United States, human cases are uncommon, ranging from 2–17/year in the 1990s, resulting from infected flea, rodent, or cat contact. Sciurid rodents (rock squirrels, California ground squirrels, chipmunks, and prairie dogs) are the primary plague reservoir in the western United States. The oriental flea (*Xenopsylla cheopis*), the common vector of plague, is well established in the United States. Plague is a seasonal disease that occurs mainly in the late spring until early fall, when rodents and fleas are abundant.

Bubonic plague in humans is usually characterized by fever after an incubation period of 2–7 days and the development of a large, tender, and swollen lymph node called buboe. The infestation may result in severe pneumonic or systemic plague, especially in people exposed to infected cats suffering from pneumonic plague. Human infection is most often caused by an infected flea bite, but occasionally can result from exposure to infected materials via cuts or abrasions in the skin or via infected aerosols.

Streptomycin and gentamicin are the antibiotics of choice for plague treatment. Tetracyclines or sulfonamides are usually given to case contacts to prevent disease spread. Mortality can be high (up to 60%) in the absence of treatment, especially in cases of pneumonic plague. Rodent and flea control, especially in endemic areas, is an essential part of controlling the infection, as well as the closing of infected campgrounds.

B. Lyme Disease

Lyme disease, also called Lyme borreliosis, is caused by a spirochete, *Borrelia burgdorferi.* This disease accounts for more than 60% of all tick-borne human infections reported in the United States, with more than 10,000 human cases per year. In the northeastern United States, where more than 90% of the human cases have occurred, the reservoir is constituted by the white-footed mouse (*Peromyscus leucopus*) and a deer tick (*Ixodes scapularis*). In the western United States, the black-legged tick (*Ixodes pacificus*) is the main vector of infection. Small mammals (particularly rodents) are an important host of these ticks' larvae and nymphs and are critical for maintenance of *B. burgdorferi* in nature; deer are an important host for the adult tick stages. The considerable increase in deer population in the United States since the middle of the twentieth century has led to an unprecedented proliferation and extension in the distribution of the deer tick, which could explain the increase of this human health problem in peri-urban habitats. Most human infections with *B. burgdorferi* in the United States, occur during the months of May through August, when both *I. Scapularis* nymphal-stage activity and human outdoor activity are at their peak.

Acute Lyme disease is a flu-like illness that develops after a 7- to 10-day incubation period and is generally accompanied by a characteristic annular rash (erythema migrans) that appears where the infecting tick has attached. Neurologic lesions usually follow, including cranial neuritis, peripheral neuropathy, encephalopathy, arthralgia, carditis, or ocular lesions. Weeks or months later, chronic arthritic lesions may appear.

When detected early, antibiotics such as doxycycline or amoxicillin are very effective in preventing chronic forms of the disease. Prevention is largely based on wearing protective clothes and tick repellents when going outdoors in tick-infested areas.

C. Equine Encephalitides

Arboviral diseases have a worldwide impact, especially in tropical and subtropical countries. The first record of eastern equine encephalomyelitis (EEE) in the United States was in 1831 by a physician who described an episode of neurologic disease in horses in Massachusetts. Venezuelan equine encephalomyelitis (VEE) was first described in northern South America in the 1920s and western equine encephalomyelitis (WEE) in the western United States in 1847. Equine encephalomyelitis viruses have been isolated only in the Western Hemisphere.

In horses, clinical signs range from subclinical or inapparent infections to a mild or severe, and frequently fatal, clinical course of disease. For the first few days, nonspecific fever is observed followed by neurological signs, such as profound depression and stupor or hyperexcitability. Human beings can be infected by EEE virus, sylvatic and epizootic VEE virus subtypes and variants, and WEE virus. The clinical syndrome can vary from a mild influenza-like illness to a severe encephalitic disease. Deaths have been reported primarily in children and the elderly. Direct transmission from horses to humans usually does not occur; the vectors are various mosquito species. The human disease has been reported frequently during equine epizootics, but human infections generally follow equine infections by approximately 2–3 weeks.

Equids are the most important amplifiers and indicators of epizootic VEE virus activity. Usually, reservoirs are birds (EEE) or rodents (VEE) or both (WEE), and various species of mosquitoes (*Culex tarsalis* for WEE, *Culiseta melanura* and *Aedes sollicitans* for EEE, and several species for VEE). EEE epidemics occurred in recent years in Florida, Georgia, and South Carolina in horses, with few human cases, because of heavy spring rains leading to an exceptionally large population of *Culiseta melanura.* In 1994, health departments in 20 states reported 100 presumptive or confirmed human cases of arboviral disease to the Centers for Disease Control and Prevention. A major outbreak of VEE occurred in 1995 in Venezuela and Colombia with an estimated 13,000 and 45,000 human cases, respectively. Hundreds of horses died. Cases were also reported in humans and horses in El Salvador and Panama in the fall of 1995.

Effective prevention measures are based on destroying mosquito-breeding places; destroying adult mosquitoes; avoiding nighttime outdoor activity in affected areas, especially at dusk and dawn; applying mosquito repellents, and wearing long-sleeved shirts and long pants. Safe and effective monovalent or bivalent formalin-inactivated EEE and WEE virus vaccines are commercially available for horses and should be administered regularly in endemic areas.

V. SAPROZOONOSES

The saprozoonoses are zoonoses that require a nonanimal site to serve either as a true reservoir of infection or as a site for an essential phase of development. Human erysipeloid and animal erysipelas are examples of this type of zoonotic infection (Fig. 4). Swine erysipelas is a common infection in many parts of the world. Many species are susceptible to *Erysipelothrix insidiosa* (*rhusiopathiae*), but it is an important economic problem in the pig industry in absence of vaccination.

After an incubation of 1–7 days, pigs develop a very high fever, with anorexia, prostration, and usually the apparition of characteristic cutaneous lesions, presenting as red urticarial plaques (Diamond disease). The animals either die or recover rapidly. The chronic form is more insidious and characterized by arthritis and endocarditis. In humans, infection results from an accidental inoculation during a necropsy or during slaughtering of infected pigs, and is mainly a professional disease. In most of the cases, the infection is localized at the inoculation site (fingers, hands, or arms). After an incubation of 1–2 days, an erythematous and edematous skin lesion, usually very itchy, is observed. Arthritis can occur in some instances. Recovery occurs in 2–4 weeks. Septicemia is rare, but often fatal, death being caused by the endocarditis.

The treatment of choice in both humans and animals is penicillin. Wearing protective clothes, especially gloves, is an important measure to prevent human infection.

VI. NEW AND EMERGING ZOONOSES

The concept of emerging infectious diseases appeared in the late 1980s, when major outbreaks occurred around the globe and surprised many scientists who considered infectious diseases to be maladies of the past or limited to the underdeveloped world. The spectrum of infectious diseases is changing rapidly in conjunction with dramatic societal and environmental changes. Worldwide explosive population growth with expanding poverty and urban migration is occurring, international travel and trade is increasing, and technology is rapidly changing—all of which affect the risk of exposure to infectious agents. Disease emergence often follows ecological changes caused by human activities such as agriculture or agricultural change, migration, urbanization, deforestation, or dam building. Of these new diseases, surprisingly, most of the emergent viruses and many of the emergent bacteria are zoonotic. “Emerging zoonoses” are defined as zoonotic diseases caused either by apparently new agents, or by previously known microorganisms, appearing in places or in species in which the disease was previously unknown. Natural animal reservoirs represent a more frequent source of new agents of human disease than the sudden appearance of a completely new agent. Most emerging infections appear to be caused by pathogens already present in the environment, brought out of obscurity or given a selective advantage by changing conditions and afforded an opportunity to infect new host populations. The zoonotic pool, mentioned by Morse, is an important and potentially rich source of emerging diseases. Of the 22 major etiologic agents causing infectious diseases identified between 1973 and 1994, 14 (63%) are zoonotic agents. Since then, equine morbillivirus (EMV), South American hantaviruses, new *Ehrlichia* and *Babesia* species, and new rabies viruses (pteropid lyssavirus, also known as Lyssavirus genotype 7) have been identified as causes of human diseases or deaths.

Factors explaining the emergence of a zoonotic or potentially zoonotic disease are usually complex and related to the infectious agent itself, involving mechanisms at the molecular level such as genetic drift and shift enhancing the virulence (many viral infections) or the acquisition of multidrug resistance (bacteria); modifications of the immunological status of individuals or populations; environmental and social changes, including ecological changes caused by human activities (such as agriculture or agricultural change, migration, urbanization, deforestation, or dam building), and human demographic and behavioral changes; travel and trade; technology and industry; and the breakdown of public health measures.

Water irrigation is certainly one of the most important agricultural techniques used to expand agricultural land. It may also be associated with the emergence of mosquito-borne diseases. In Asia, the incidence of Japanese encephalitis, which causes almost 30,000 human cases and 7,000 deaths annually, is closely associated with rice-field irrigation, creating large areas of stagnant water. In Africa, the dramatic expansion of Rift Valley fever in Egypt in 1977 and in Mauritania in 1987, appears to be associated with dam construction and irrigation in areas either densely populated with humans and domestic animals (Egypt) or with large naive ruminant populations (both Egypt and Mauritania).

Human population movements or upheavals, caused by migration or war, are often important factors in disease emergence. Urbanization has led to mass movement of workers from rural areas to cities. Recent outbreaks of brucellosis in urban communities such as in Lima, Peru, or in some of the conflict areas of the Gaza strip and West Bank in Israel and Palestine are illustrations of the effects of uncontrolled urbanization. The continuous spread of human populations to new living areas is also a factor to consider in the emergence of new zoonoses. In South America, major outbreaks of vampire-bat rabies occurred in Peru and Brazil following the settlement of new agricultural communities in the remote jungle.

Farming of new animal species, such as ostriches, may lead to new outbreaks. An outbreak of Congo-Crimean hemorrhagic fever occurred in South Africa in a slaughterhouse specialized in slaughtering ostriches. In that outbreak, 16 human cases were suspected, with at least one death. Infected ticks may have been the source of human infection or exposure to the viremic birds. Similarly, major concern has been raised about deer farming and outbreaks of bovine tuberculosis occurring in these herds. Translocation of exotic animal species for farming can also lead to new diseases or new reservoirs of infection. Outbreaks of western equine encephalitis occurred in emus in Texas in 1992 and in California in 1994.

The need for new exotic pets can also lead to the emergence of new zoonotic diseases, such as the epidemic of chorio-lympho-meningitis in pet-hamster owners in the United States, and Germany in the early 1970s, and the present epidemic of iguana-related salmonellosis in the United States.

The spread of diseases has for centuries been associated with human migrations. From the plague epidemics of the sixth, fourteenth, and nineteenth centuries to the dispersion of yellow fever from Africa to the Americas or dog rabies from Europe to the Americas, zoonotic diseases have found new reservoirs to adapt to while humans were expanding their territories. One major modern concern is the large amount of human flux associated with the dramatic reduction in travel time over long distances, which has increased the possibility of the global transport of infectious agents in a short period of time. Transoceanic travel by steamer was considered responsible for the spread of the third plague pandemic at the end of the nineteenth century. Now, any location on the globe can be reached within 24 hours. The latest outbreak of Ebola fever clearly exemplifies the risk, as the first case of Ebola fever reported in South Africa was caused by a sick patient seeking care far from the initial outbreak.

The importation of monkeys for research was the source of the first Ebola-like outbreak investigated in the United States. In September 1989, numerous cynomolgus monkeys shipped from the Philippines died during their quarantine of suspected simian hemorrhagic fever, but in fact also of Reston virus. The raccoon rabies epidemic in the United States, was created by the importation for hunting purpose of infected animals incubating the disease to disease-free areas with a large susceptible population. Expanding human habitats to suburban areas cause us to be in contact with the wildlife reservoir of these expanding zoonoses. The dramatic increase of Lyme disease cases in the United States and the considerable number of postexposure rabies treatments required in the northeastern part of the United States are direct consequences of this expansion.

The increasing international trade in live animals and foodstuff has favored the spread of enteric zoonotic infections, especially salmonellosis. Foodborne zoonoses are increasing in both developed and developing countries. According to the Pan American Health Organization many of these diseases have increased as much as 100% during the 1990s. In some countries, the incidence is estimated to be as high as 10% of the population. Salmonella outbreaks caused by new serotypes or multidrug-resistant serotypes are reported more and more frequently. The identification of infection caused by *E. coli* (O:157;H:7) is an example of better tools for the diagnosis of infectious diseases, but also allows the assessment of a pathogen increasingly found in our food chain. It is suspected that modifications in the processing of rendering products led to the bovine spongiform encephalopathy (BSE) epidemic in Great Britain.

The recognition of zoonotic diseases may also be related to a better technology for investigating and identifying the pathogens. Many new infectious agents have been identified following carefully conducted epidemiological investigations, leading, for instance, to the suspicion of animal contacts (e.g., for bacillary angiomatosis and cat scratch disease, or hantavirus pulmonary syndrome) and the use of modern molecular techniques. Molecular epidemiology is one of the promising outcomes of the interaction between epidemiology and the laboratory. The agent of the hantavirus pulmonary syndrome was first identified by detecting genetic material in specimens collected from dead and sick patients with polymerase chain reaction (PCR). It was one of the first cases in which a disease of unknown origin was identified through molecular epidemiology. In other instances, molecular epidemiology was essential for the detection and identification of bacteria resistant to cultivation, such as the agent of bacillary angiomatosis and cat scratch disease, *Bartonella henselae*.

Among the various factors associated with the emergence of new zoonoses, microbial adaptation must be considered. Such an event appears more likely to occur with viruses than bacteria. Influenza viruses are examples of genetic shift and drift, which may lead to new variants and recombinant strains, causing worldwide pandemics. The few cases of equine morbillivirus in Australia clearly exemplify the risk of new zoonoses from an unknown reservoir. It is thought that western equine encephalitis virus arose from a recombination event that seems to have involved a Sindbis-like virus and eastern equine encephalomyelitis virus approximately 100–200 years ago. Strains of *Salmonella* resistant to various antibiotics are emerging in the industrial farming environment of the poultry, pig, and calf industries. The risk that these strains may be entering the food chain is of major concern.

The breakdown of public health measures and deficiencies in public health infrastructure must not be neglected among the various risk factors associated with disease emergence or re-emergence. The economic crisis of the late 1980s and early 1990s has had a major impact on the financial and human resources available to prevent and control infectious diseases. Disease prevention is funded by only a few percent (3%) of all the billion of dollars devoted to human health in the United States, and even less is devoted to public health surveillance and information systems. The increase in urban population, increase in poverty, increase in susceptibility to infectious agents by lack of immunization, reduction of public health support and privatization, and reduction of vaccination programs have led to an explosive situation in which infectious diseases can claim more ground.

VII. PREVENTION AND CONTROL

The prevention and control of zoonotic diseases has a three-tier action—the direct protection of humans, reduction or elimination of the infection in the animal reservoir, and antivector measures.

The direct protection of humans applies mainly to occupational diseases in the laboratory, the workplace, or the rural environment. Preventive measures include the wearing of protective clothing, including gloves and glasses or goggles, appropriate air filtration systems, regular disinfection, vector (e.g., insect or rodent) control, and water treatment. Health education, including safe dietary habits and proper food hygiene, is also a major component of zoonosis prevention. Specific vaccination against zoonotic pathogens may be appropriate, such as leptospirosis vaccination in sewage workers or rabies vaccination in abattoir workers.

Zoonoses control in animals is an imperative goal of veterinary public health activity. Priority should always be given to disease eradication by the culling of sick and infected animals. The treatment of sick animals should be reserved to very specific instances in which it does not compromise the screening of infected animals, such as cases of anthrax in cattle. Disease control is based on the quarantine of sick and infected animals, testing and segregation of infected flocks or herds, restriction of animal movements, and immunization of exposed animals whenever vaccines are safe and available. Hygienic management on the farm or at the slaughterhouse also must be emphasized.

The control of vectors and vehicles is essential to prevent the spread of infectious agents to noninfected animals or humans and to disease-free areas. The destruction of infected material or products is essential to control the spread of infection, especially from the food chain. The thorough disinfection of contaminated areas will reduce or stop the spread of zoonoses. Feed hygiene and the elimination of pests and vermin must be performed continuously to control all types of zoonoses, especially meta- and saprozoonoses. The harmonization of national and international rules and regulations; and interagency cooperation under the sponsorship of the U.N. World Health Organization, the U.N. Food and Agricultural Organization and the Office International des Epizooties are necessary to prevent and control the spread of zoonoses.

See Also the Following Articles

Arboviruses • Economic Consequences of Infectious Diseases • Emerging Infections • Lyme Disease • Mycobacteria • Plague • Rabies

Bibliography

Acha, P. N., and Szyfres, B. (1989). "Zoonoses and Communicable Diseases Common to Man and Animals," 2nd ed. Pan American Health Organization, Washington, DC.

Chomel, B. B. (1998). New emerging zoonoses: A challenge and an opportunity for the veterinary profession. *Comp. Immun. Microbiol. Infect. Dis.* **21**, 1–14.

Goret, P., and Joubert, L. (1966). Les zoonoses—maladies animales transmissibles à l'homme. *Gazette médicale de France,* **1-X**, 3551–3578.

Mantovani, A. (1992). Zoonoses and veterinary public health. *Rev. Sci. Tech. Off. Int. Epiz.* **11**, 205–218.

Meslin, F. X. (1992). Surveillance and control of emerging zoonoses. *World Health Statist. Quat.* **45**, 200–207.

Morse, S. S. (1995). Factors in the emergence of infectious diseases. *Emerg. Infect. Dis.* **1**, 7–15.

Murrell, K. D. (1991). Economic losses resulting from food-borne parasitic zoonoses. *Southeast Asian J. Trop. Med. Pub. Health* **22**, 377–381.

Schwabe, C. W. (1984). Veterinary medicine and human health, 3rd ed., pp. 194–251. Williams and Wilkins, Baltimore, MD.

Steele, J. H. (1985). The zoonoses. *Int. J. Zoon.* **12**, 87–97.

Thomas, D., and Salmon, R. L. (1997) Zoonotic illness in farm workers and their families: Clinical presentation and extent, a prospective collaborative study. Final report, April 1997. Health and Safety Executive Commissioned Research project number: 1/HPD/126/308/90 Public Health Laboratory System, U.K.

Contributors

David W. K. Acheson
Tufts–New England Medical Center
Boston, Massachusetts
Food-borne Illnesses

Hans-Wolfgang Ackermann
Laval University
Laval, Quebec, Canada
Bacteriophages

George N. Agrios
University of Florida
Gainesville, Florida
Plant Pathogens

Adriano Aguzzi
University of Zurich
Zurich, Switzerland
Prions

Shin-Ichi Aizawa
Teikyo University
Utsunomiya, Japan
Flagella

Michael F. Allen
University of California, Riverside
Riverside, California
Mycorrhizae

M. R. K. Alley
Imperial College of Science, Technology and Medicine
London, England, UK
Caulobacter, Genetics

Laurel S. Almer
University of Wisconsin
Madison, Wisconsin
Detection of Bacteria in Blood: Centrifugation and Filtration

Orna Amster-Choder
The Hebrew University Medical School
Jerusalem, Israel
Transcriptional Regulation in Prokaryotes

Zhiqiang An
Merck Research Laboratories
Rahway, New Jersey
Polyketide Antibiotics

Annaliesa S. Anderson
Merck Research Laboratories
Rahway, New Jersey
Polyketide Antibiotics

Thomas M. Anderson
Archer Daniels Midland BioProducts
Decatur, Illinois
Industrial Fermentation Processes

Adriane M. Antler
Pennie & Edmonds LLP
New York, New York
Patenting of Living Organisms and Natural Products

Jesús F. Aparicio
University of León
León, Spain
Secondary Metabolites

Eurico Arruda
University of São Paulo School of Medicine
São Paulo, Brazil
Rhinoviruses

Ann M. Arvin
Stanford University School of Medicine
Stanford, California
Vaccines, Viral

Mariette R. Atkinson
University of Michigan Medical School
Ann Arbor, Michigan
Carbon and Nitrogen Assimilation, Regulation of
Two-Component Systems

Jean-Paul Aubert (deceased)
Institut Pasteur
Paris, France
Cellulases

Cyrus Bacchi
Pace University
New York, New York
Trypanosomes

Farah K. Bahrani-Mougeot[1]
University of Maryland
Baltimore, Maryland
Enteropathogenic Bacteria

Pratima Bajpai
Thapar Corporate Research and Development Centre
Patiala, Punjab, India
Xylanases

Richard H. Baltz
CognoGen Biotechnology Consulting
Indianapolis, Indiana
Mutagenesis

Joseph T. Barbieri
Medical College of Wisconsin
Milwaukee, Wisconsin
Exotoxins

S. F. Barefoot
Clemson University
Clemson, South Carolina
Bacteriocins

Tamar Barkay
Rutgers University
New Brunswick, New Jersey
Mercury Cycle

Douglas H. Bartlett
University of California, San Diego
La Jolla, California
Osmotic Stress

Christopher F. Basler
Mount Sinai School of Medicine
New York, New York
Influenza Viruses

1. Present address: United Arab Emirates University, United Arab Emirates.

Carl E. Bauer
Indiana University
Bloomington, Indiana
Photosensory Behavior

Samuel I. Beale
Brown University
Providence, Rhode Island
Tetrapyrrole Biosynthesis in Bacteria

Pierre Béguin
Institut Pasteur
Paris, France
Cellulases

John T. Belisle
Colorado State University
Fort Collins, Colorado
Mycobacteria

Arnold J. Bendich
University of Washington
Seattle, Washington
Chromosome, Bacterial

Joan W. Bennett
Tulane University
New Orleans, Louisiana
Fungi, Filamentous

Peter M. Bennett
University of Bristol
Bristol, England, UK
Transposable Elements

Paul S. Berger
U.S. Environmental Protection Agency
Alexandria, Virginia
Water, Drinking

Ruth L. Berkelman
Centers for Disease Control and Prevention
Atlanta, Georgia
Surveillance of Infectious Diseases

Mary K. B. Berlyn
Yale University
New Haven, Connecticut
Stock Culture Collections and Their Databases

Mathias Bernhardt
University of Wisconsin
Madison, Wisconsin
Detection of Bacteria in Blood: Centrifugation and Filtration

Terrance J. Beveridge
University of Guelph
Guelph, Ontario, Canada
Microscopy, Electron

Angela Bielefeldt
University of Colorado
Boulder, Colorado
Wastewater Treatment, Industrial

Ronald H. Bishop
University of Ulster
Newtownabbey, Ulster, Northern Ireland
Education in Microbiology

Martin J. Blaser
Vanderbilt University School of Medicine
Nashville, Tennessee
Helicobacter pylori

August Böck
University of Munich
Munich, Germany
Fermentation

Jean-Marc Bollag
The Pennsylvania State University
University Park, Pennsylvania
Biodegradation

Wendy B. Bollag
The Pennsylvania State University
University Park, Pennsylvania
Biodegradation

John C. Boothroyd
Stanford University School of Medicine
Stanford, California
Toxoplasmosis

Rodney J. Bothast
U.S. Department of Agriculture
Peoria, Illinois
Enzymes in Biotechnology

Lucas A. Bouwman
Alterra Soil
Wageningen, The Netherlands
Protozoan Predation

George H. Bowden
University of Manitoba
Winnipeg, Manitoba, Canada
Oral Microbiology

Andrea Denise Branch
The Mount Sinai School of Medicine
New York, New York
Antisense RNAs

Allan M. Brandt
Harvard Medical School
Cambridge, Massachusetts
AIDS, Historical

Daniel K. Brannan
Abilene Christian University
Abilene, Texas
Cosmetic Microbiology

Erhard Bremer
Philipps University
Marburg, Germany
Water-Deficient Environments

Patrick J. Brennan
Colorado State University
Fort Collins, Colorado
Mycobacteria

James A. Brierley
Newmont Metallurgical Services
Englewood, Colorado
Metal Extraction and Ore Discovery
Ore Leaching by Microbes

Yves V. Brun
Indiana University
Bloomington, Indiana
Developmental Processes in Bacteria

Trevor N. Bryant
University of Southampton
Southampton, England, UK
Identification of Bacteria, Computerized

Peter Burnett
Agriculture and Agri-Food Canada Research Centre
Winnipeg, Manitoba, Canada
Smuts, Bunts, and Ergot

Mark P. Buttner
University of Nevada, Las Vegas
Las Vegas, Nevada
Airborne Microorganisms and Indoor Air Quality

Steven M. Callister
Gundersen Lutheran Medical Center
La Crosse, Wisconsin
Detection of Bacteria in Blood: Centrifugation and Filtration

Elisabeth Carniel
Institut Pasteur
Paris, France
Plague

Timothy L. W. Carver
Institute of Environmental and Grassland Research
Aberystwyth, Wales, UK
Powdery Mildews

Arturo Casadevall
Albert Einstein College of Medicine
Bronx, New York
Fungal Infections, Systemic

Michael Cashel
National Institute of Child Health and Human Development
Bethesda, Maryland
Stringent Response

Stan W. Casteel
University of Missouri
Columbia, Missouri
Mycotoxicoses

Ricardo Cavicchioli
University of New South Wales
Sydney, Australia
Extremophiles

Jane A. Cecil
The Johns Hopkins University
Baltimore, Maryland
Sexually Transmitted Diseases

Nora M. Chapman
University of Nebraska Medical Center
Omaha, Nebraska
Enteroviruses

Arun K. Chatterjee
University of Missouri
Columbia, Missouri
Erwinia: Genetics of Pathogenicity Factors

Asita Chatterjee
University of Missouri
Columbia, Missouri
Erwinia: Genetics of Pathogenicity Factors

Munir Cheryan
University of Illinois
Urbana, Illinois
Acetic Acid Production

Bruno B. Chomel
University of California, Davis
Davis, California
Zoonoses

Peter J. Christie
University of Texas Health Science Center
Houston, Texas
Agrobacterium and Plant Cell Transformation

Sudha Chugani
University of Illinois College of Medicine
Chicago, Illinois
Pseudomonas

Orio Ciferri
University of Pavia
Pavia, Italy
Conservation of Cultural Heritage

Jeremy S. C. Clark
University of the West of England
Bristol, England, UK
Downy Mildews

Robert M. Clark
U.S. Environmental Protection Agency
Alexandria, Virginia
Water, Drinking

Jenifer L. Coburn
Tufts-New England Medical Center
Boston, Massachusetts
Lyme Disease

Richard W. Compans
Emory University
Atlanta, Georgia
Paramyxoviruses

Laurie E. Comstock
Harvard Medical School
Boston, Massachusetts
Gram-Negative Anaerobic Pathogens

Ralf Conrad
Max Planck Institut für Terrestrische Mikrobiologie
Marburg, Germany
Microbes and the Atmosphere

M. W. Covert
University of California, San Diego
La Jolla, California
Genomic Engineering of Bacterial Metabolism

Marc A. Cubeta
North Carolina State University
Plymouth, North Carolina
Rhizoctonia

Cleora J. D'Arcy
University of Illinois
Urbana, Illinois
Luteoviridae

Sandra Da Re
Princeton University
Princeton, New Jersey
Chemotaxis

Robert B. Dadson
University of Maryland, Eastern Shore
Princess Anne, Maryland
Nitrogen Fixation

Amaresh Das
University of Georgia
Athens, Georgia
Acetogenesis and Acetogenic Bacteria

Gregory A. Dasch
Naval Medical Research Center
Bethesda, Maryland
Rickettsiae

Elie Dassa
Institut Pasteur
Paris, France
ABC Transport

Michael J. Davis
University of Florida
Homestead, Florida
Phloem-Limited Bacteria

Robert E. Davis
U.S. Department of Agriculture
Beltsville, Maryland
Phytoplasma

Ciro A. de Quadros
Pan American Health Organization
Washington, DC
Polio

Rutger de Wit
Universite Bordeaux I and CNRS
Bordeaux, France
Sulfide-Containing Environments

S. De Baets
University of Ghent
Ghent, Belgium
Vitamins and Related Biofactors, Microbial Production

Jerzy Dec
The Pennsylvania State University
University Park, Pennsylvania
Biodegradation

Bruce Demple
Harvard School of Public Health
Boston, Massachusetts
Oxidative Stress

Walter B. Dempsey
University of Texas Southwestern Medical Center
Dallas, Texas
Coenzyme and Prosthetic Group Biosynthesis

Gerald A. Denys
Clarian Health Methodist Hospital
Indianapolis, Indiana
Infectious Waste Management

Kevin M. Devine
Trinity College
Dublin, Ireland
Bacillus subtilis, Genetics

David J. Dickinson
Imperial College of Science, Technology and Medicine
London, England, UK
Timber and Forest Products

Anh Ngoc Dang Do
University of Wisconsin Medical School
Madison, Wisconsin
Oncogenic Viruses

Karen W. Dodson
Washington University School of Medicine
St. Louis, Missouri
Fimbriae, Pili

Leslie L. Domier
University of Illinois
Urbana, Illinois
Luteoviridae

Michael S. Donnenberg
University of Maryland
Baltimore, Maryland
Enteropathogenic Bacteria

Ralph Dornburg
Thomas Jefferson University
Philadelphia, Pennsylvania
Retroviruses

Brian A. Dougherty
Bristol-Myers Squibb Pharmaceutical Research Institute
Wallingford, Connecticut
DNA Sequencing and Genomics

Angela E. Douglas
University of York
York, England, UK
Symbiotic Microorganisms in Insects

Karl Drlica
Public Health Research Institute
New York, New York
Chromosome, Bacterial

C. Korsi Dumenyo
University of Missouri
Columbia, Missouri
Erwinia: Genetics of Pathogenicity Factors

Paul S. Dyer
University of Nottingham
Nottingham, England, UK
Eyespot

Daniel E. Dykhuizen
State University of New York, Stony Brook
Stony Brook, New York
Natural Selection, Bacterial

Paul J. Dyson
University of Wales
Swansea, Wales, UK
Streptomyces, Genetics

Charles F. Earhart
The University of Texas at Austin
Austin, Texas
Iron Metabolism

Jeremy S. Edwards
University of California, San Diego
La Jolla, California
Genomic Engineering of Bacterial Metabolism

Sharon Egan
Warwick University
Coventry, England, UK
Actinomycetes

Lothar Eggeling
Research Center Jülich
Jülich, Germany
Amino Acid Production

Thomas Egli
Swiss Federal Institute for Environmental Science and Technology
Dübendorf, Switzerland
Nutrition of Microorganisms

Gerald H. Elkan
North Carolina State University
Raleigh, North Carolina
Nitrogen Fixation

Karen T. Elvers
University of Wales Institute
Cardiff, Wales, UK
Biofilms and Biofouling

N. Cary Engleberg
University of Michigan Medical School
Ann Arbor, Michigan
Legionella

Marina E. Eremeeva
University of Maryland
Baltimore, Maryland
Rickettsiae

Larry E. Erickson
Kansas State University
Manhattan, Kansas
Bioreactors

Karl-Erik L. Eriksson
University of Georgia
Athens, Georgia
Lignocellulose, Lignin, Ligninases

Ana A. Espinel-Ingroff
Medical College of Virginia
Richmond, Virginia
Antifungal Agents

Martha Espinosa-Cantellano
Center for Research and Advanced Studies
Mexico City, Mexico
Intestinal Protozoan Infections in Humans

Philip J. Farabaugh
University of Maryland, Baltimore County
Baltimore, Maryland
Translational Control and Fidelity

Linda Farrell
University of Wisconsin Medical School
Madison, Wisconsin
Oncogenic Viruses

Brian Federici
University of California, Riverside
Riverside, California
Insecticides, Microbial

Stuart J. Ferguson
University of Oxford
Oxford, England
Energy Transduction Processes

David P. Fidler
Indiana University School of Law
Bloomington, Indiana
International Law and Infectious Disease

William R. Finnerty
Finnerty Enterprises, Inc.
Athens, Georgia
Biopolymers, Production and Uses of

Glen M. Ford
Dexall Biomedical Labs, Inc.
Gaithersburg, Maryland
Industrial Biotechnology, Overview

Terrence G. Frey
San Diego State University
San Diego, California
Microscopy, Confocal
Microscopy, Optical

Laura S. Frost
University of Alberta
Edmonton, Alberta, Canada
Conjugation, Bacterial

William E. Fry
Cornell University
Ithaca, New York
Phytophthora infestans

Daniel Y. C. Fung
Kansas State University
Manhattan, Kansas
Food Spoilage and Preservation

Clay Fuqua
Indiana University
San Antonio, Texas
Quorum Sensing in Gram-Negative Bacteria

Geoffrey M. Gadd
University of Dundee
Dundee, Scotland, UK
Heavy Metal Pollutants: Environmental and Biotechnological Aspects

Jorge Galan
Yale University
New Haven, Connecticut
Protein Secretion

Ferran Garcia-Pichel
Arizona State University
Tempe, Arizona
Cyanobacteria

Ian Garner
PPL Therapeutics, Ltd.
Roslyn, Midlothian, UK
Transgenic Animal Technology

Denis Gaudet
Agriculture and Agri-Food Canada Research Centre
Lethbridge, Alberta, Canada
Smuts, Bunts, and Ergot

Charles J. Gauntt
University of Texas Health Sciences Center
San Antonio, Texas
Enteroviruses

Simonida Gencic
Uniformed Services University of the Health Sciences
Bethesda, Maryland
Methane Biochemistry

Howard Gest
Indiana University
Indianapolis, Indiana
Photosensory Behavior

Janet L. Gibson
The Ohio State University
Columbus, Ohio
Autotrophic CO_2 Metabolism

Emil C. Gotschlich
The Rockefeller University
New York, New York
Gram-Negative Cocci, Pathogenic

Jan C. Gottschal
University of Groningen
Haren, The Netherlands
Continuous Culture

Peter H. Graham
University of Minnesota
St. Paul, Minnesota
Nodule Formation in Legumes

David A. Grahame
Uniformed Services University of the Health Sciences
Bethesda, Maryland
Methane Biochemistry

William D. Grant
University of Leicester
Leicester, England, UK
Alkaline Environments

Julia E. Grimwade
Florida Institute of Technology
Melbourne, Florida
Chromosome Replication and Segregation

D. A. Grinstead
Clemson University
Clemson, South Carolina
Bacteriocins

Carol A. Gross
University of California, San Francisco
San Francisco, California
Heat Stress

Lawrence Grossman
The Johns Hopkins University
Baltimore, Maryland
DNA Repair

Janine Guespin-Michel
Université de Rouen
Rouen, France
Mapping Bacterial Genomes

Robert P. Gunsalus
University of California, Los Angeles
Los Angeles, California
Anaerobic Respiration

Sarah J. Gurr
University of Oxford
Oxford, England, UK
Powdery Mildews

Santiago Gutiérrez
University of León
León, Spain
Secondary Metabolites

Alison A. Hall
University of Oxford
Oxford, England, UK
Powdery Mildews

Ian R. Hamilton
University of Manitoba
Winnipeg, Manitoba, Canada
Oral Microbiology

John Hammond
U.S. Department of Agriculture
Beltsville, Maryland
Potyviruses

Thomas E. Hanson
The Ohio State University
Columbus, Ohio
Autotrophic CO_2 Metabolism

K. M. Harmon
Clemson University
Clemson, South Carolina
Bacteriocins

Mark A. Harrison
University of Georgia
Athens, Georgia
Beer/Brewing

Claudia C. Häse
Harvard Medical School
Boston, Massachusetts
Cholera

Fawzy Hashem
University of Maryland, Eastern Shore
Princess Anne, Maryland
Nitrogen Fixation

J. Woodland Hastings
Harvard University
Cambridge, Massachusetts
Bioluminescence, Microbial

Anthony G. Hay
University of Tennessee
Knoxville, Tennessee
Plasmids, Catabolic

Alan Christopher Hayward
University of Queensland
St. Lucia, Queensland, Australia
Ralstonia solanacearum

Jack A. Heinemann
University of Canterbury
Christchurch, New Zealand
Horizontal Transfer of Genes between Microorganisms

James A. Hejna
Oregon Health Sciences University
Portland, Oregon
DNA Replication

Donald A. Henderson
The Johns Hopkins University School of Public Health
Baltimore, Maryland
Smallpox

Joan M. Henson
Montana State University
Bozeman, Montana
Gaeumannomyces graminis

Christophe Herman
University of California, San Francisco
San Francisco, California
Heat Stress

David L. Heymann
World Health Organization
Geneva, Switzerland
Emerging Infections

Susan Hill
John Innes Centre
Norwich, England, UK
Azotobacter

Bradley I. Hillman
Rutgers University, Cook College
New Brunswick, New Jersey
Plant Disease Resistance: Natural Mechanisms and Engineered Resistance

Bernd Hitzmann
University of Hannover
Hannover, Germany
Bioreactor Monitoring and Control

Philip M. Hoekstra
Buckman Laboratories International
Memphis, Tennessee
Pulp and Paper

Susan K. Hoiseth
Wyeth–Lederle Vaccines
West Henrietta, New York
Vaccines, Bacterial

Herbert L. Holland
Brock University
St. Catherine's, Ontario, Canada
Biotransformations

Joachim-Volker Höltje
Max-Planck-Institut für Entwicklungsbiologie
Tübingen, Germany
Cell Walls, Bacterial

Dallas G. Hoover
University of Delaware
Newark, Delaware
Bacteriocins

Paul J. J. Hooykaas
Leiden University
Leiden, The Netherlands
Agrobacterium

Ching T. Hou
U.S. Department of Agriculture
Peoria, Illinois
Lipases, Industrial Uses

Joseph B. Hughes
Rice University
Houston, Texas
Bioremediation

Roger Hull
John Innes Centre
Norwich, England, UK
Plant Virology, Overview

Scott J. Hultgren
Washington University School of Medicine
St. Louis, Missouri
Adhesion, Bacterial
Fimbriae, Pili

John E. Hyde
University of Manchester Institute of Science and Technology
Manchester, England
Plasmodium

John J. Iandolo
University of Oklahoma Health Sciences Center
Oklahoma City, Oklahoma
Staphylococcus

Mark A. Jackson
U.S. Department of Agriculture
Peoria, Illinois
Biopesticides, Microbial

Rosemary Jagus
University of Maryland Biotechnology Institute
Baltimore, Maryland
Protein Biosynthesis

Charles A. Janeway
Yale University School of Medicine
New Haven, Connecticut
T Lymphocytes

William J. Jewell
Cornell University
Ithaca, New York
Methane Production/Agricultural Waste Management

Ze-Yu Jiang
Indiana University
Indianapolis, Indiana
Photosensory Behavior

Allison R. Jilbert
Institute of Medical and Veterinary Science
Adelaide, Australia
Hepatitis Viruses

Rolf D. Joerger
University of Delaware
Newark, Delaware
Bacteriocins

Eric A. Johnson
University of Wisconsin
Madison, Wisconsin
Clostridia
Pigments, Microbially Produced

Brian E. Jones
Genencor International B.V.
Delft, The Netherlands
Alkaline Environments

David Shumway Jones
Harvard Medical School
Brookline, Massachusetts
AIDS, Historical
Syphilis, Historical

Ronald D. Jones
Florida International University
Miami, Florida
Phosphorus Cycle

Francoise Joset
Laboratoire de Chimie Bactérienne
Marseille, France
Mapping Bacterial Genomes

Bhavesh Joshi
University of Maryland
College Park, Maryland
Protein Biosynthesis

Nicholas Judson
Harvard Medical School
Boston, Massachusetts
Cholera

Robert J. Kadner
University of Virginia
Charlottesville, Virginia
Cell Membrane: Structure and Function

Robert A. Kalish
Tufts-New England Medical Center
Boston, Massachusetts
Lyme Disease

Shilpa Kamath
University of Illinois College of Medicine
Chicago, Illinois
Pseudomonas

Vinayak Kapatral
University of Illinois College of Medicine
Chicago, Illinois
Pseudomonas

Louis A. Kaplan
Stroud Water Research Center
Avondale, Pennsylvania
Freshwater Microbiology

Isao Karube
University of Tokyo
Tokyo, Japan
Biosensors

Dennis L. Kasper
Harvard Medical School
Boston, Massachusetts
Gram-Negative Anaerobic Pathogens

Stefan H. E. Kaufmann
Max Planck Institute for Infection Biology
Berlin, Germany
Cellular Immunity

Bettina Kempf
Philipps University
Marburg, Germany
Water-Deficient Environments

Kitai Kim
University of Wisconsin Medical School
Madison, Wisconsin
Oncogenic Viruses

Alexander J. Kind
PPL Therapeutics, Ltd.
Roslyn, Midlothian, UK
Transgenic Animal Technology

Louis V. Kirchhoff
University of Iowa
Iowa City, Iowa
Trypanosomes

Donald A. Klein
Colorado State University
Fort Collins, Colorado
Rhizosphere

Michael J. Klug
Michigan State University
East Lansing, Michigan
Ecology, Microbial

Roger Knowles
McGill University
Ste-Anne de-Bellevue, Quebec, Canada
Nitrogen Cycle

Renate Koenig
Institut für Pflanzenvirologie, Mikrobiologie und Biologische Sicherheit
Braunschweig, Germany
Beet Necrotic Yellow Vein Virus

Allan E. Konopka
Purdue University
West Lafayette, Indiana
Freshwater Microbiology

Denise Koo
Centers for Disease Control and Prevention
Atlanta, Georgia
Surveillance of Infectious Diseases

Naim Kosaric
University of Western Ontario
London, Ontario, Canada
Biosurfactants
Industrial Effluents: Sources, Properties, and Treatments

Susan F. Koval
University of Western Ontario
London, Ontario, Canada
Microscopy, Electron

Martin Krsek
Warwick University
Coventry, England, UK
Actinomycetes

L. David Kuykendall
U.S. Department of Agriculture
Beltsville, Maryland
Nitrogen Fixation

Paul F. Lambert
University of Wisconsin Medical School
Madison, Wisconsin
Oncogenic Viruses

David C. LaPorte
University of Minnesota
Minneapolis, Minnesota
Glyoxylate Bypass in Escherichia coli

Hilary M. Lappin-Scott
University of Exeter
Devon, England, UK
Biofilms and Biofouling

David S. Latchman
University College, London
London, England, UK
Transcription, Viral

Edward R. Leadbetter
University of Connecticut
Storrs, Connecticut
Soil Dynamics and Organic Matter, Decomposition

Ing-Ming Lee
U.S. Department of Agriculture
Beltsville, Maryland
Phytoplasma

Jacek Leman
Olsztyn University of Agriculture and Technology
Olsztyn, Poland
Lipids, Microbially Produced

Piet Lens
Wageningen Agricultural University
Wageningen, The Netherlands
Sulfur Cycle

Richard E. Lenski
Michigan State University
East Lansing, Michigan
Evolution, Theory and Experiments

Alan C. Leonard
Florida Institute of Technology
Melbourne, Florida
Chromosome Replication and Segregation

Dietrich-Eckhardt Lesemann
Institut für Pflanzenvirologie, Mikrobiologie und Biologische Sicherheit
Braunschweig, Germany
Beet Necrotic Yellow Vein Virus

John F. Levy
Imperial College of Science, Technology and Medicine
London, England, UK
Timber and Forest Products

Erik P. Lillehoj
Dexall Biomedical Labs, Inc.
Gaithersburg, Maryland
Industrial Biotechnology, Overview

John H. Litchfield
Batelle Memorial Institute
Columbus, Ohio
Lactic Acid, Microbially Produced

Haibin Liu
Virginia Commonwealth University
Richmond, Virginia
Antibiotic Biosynthesis

Yang Liu
University of Missouri
Columbia, Missouri
Erwinia: Genetics of Pathogenicity Factors

Lars G. Ljungdahl
University of Georgia
Athens, Georgia
Acetogenesis and Acetogenic Bacteria

William F. Loomis
University of California, San Diego
La Jolla, California
Origin of Life

Charles R. Lovell
University of South Carolina
Columbia, South Carolina
Diversity, Microbial

K. Brooks Low
Yale University
New Haven, Connecticut
Escherichia coli and Salmonella, Genetics

Karl Maramorosch
Rutgers University
New Brunswick, New Jersey
Plant Disease Resistance: Natural Mechanisms and Engineered Resistance

Lynn Margulis
University of Massachusetts
Amherst, Massachusetts
Spirochetes

Martin G. Marinus
University of Massachusetts Medical School
Worcester, Massachusetts
Methylation of Nucleic Acids and Proteins

Howard Markel
University of Michigan Medical School
Ann Arbor, Michigan
Cholera, Historical

Allen G. Marr
University of California, Davis
Davis, California
Growth Kinetics, Bacterial

Juan F. Martín
University of León
León, Spain
Secondary Metabolites

Adolfo Martínez-Palomo
Center for Research and Advanced Studies
Mexico City, Mexico
Intestinal Protozoan Infections in Humans

Esperanza Martínez-Romero
Universidad Nacional Autónoma de Mexico
Mexico City, Mexico
Biological Nitrogen Fixation

William S. Mason
Fox Chase Cancer Center
Philadelphia, Pennsylvania
Hepatitis Viruses

Millicent Masters
University of Edinburgh
Edinburgh, Scotland, UK
Transduction: Host DNA Transfer by Bacteriophages

Kostas D. Mathiopoulos
University of Patras
Patras, Greece
Malaria

A. Matin
Stanford University
Stanford, California
Starvation, Bacterial

Ian R. McDonald
University of Warwick
Coventry, England, UK
Methylotrophy

Ross E. McKinney
University of Kansas
Durham, North Carolina
Wastewater Treatment, Municipal

Peter A. Meacock
University of Leicester
Leicester, England, UK
Transformation, Genetic

Christopher D. Meehan
University of Michigan Medical School
Ann Arbor, Michigan
Cholera, Historical

John J. Mekalanos
Harvard Medical School
Boston, Massachusetts
Cholera

Peter C. Melby
University of Texas Health Science Center
San Antonio, Texas
Leishmania

Martin I. Meltzer
Centers for Disease Control and Prevention
Atlanta, Georgia
Economic Consequences of Infectious Diseases

Jim Menzies
Agriculture and Agri-Food Canada Research Centre
Winnipeg, Manitoba, Canada
Smuts, Bunts, and Ergot

Paul Messner
University of Agricultural Sciences
Vienna, Austria
Crystalline Bacterial Cell Surface Layers

Twng Wah Mew
International Rice Research Institute
Manila, The Philippines
Xanthomonas

Catherine M. Michaud
Harvard School of Public Health
Cambridge, Massachusetts
Global Burden of Infectious Diseases

Linda A. Miller
SmithKline Beecham Pharmaceuticals
Collegeville, Pennsylvania
Biological Warfare

Robert V. Miller
Oklahoma State University
Stillwater, Oklahoma
recA

Stephen P. Miller
University of Minnesota
Minneapolis, Minnesota
Glyoxylate Bypass in Escherichia coli

Saroj K. Mishra
NASA/Johnson Space Center
Houston, Texas
Space Flight, Effects on Microorganisms

Tapan K. Misra
University of Illinois at Chicago
Chicago, Illinois
Heavy Metals, Bacterial Resistances

S. Leslie Misrock
Pennie & Edmonds LLP
New York, New York
Patenting of Living Organisms and Natural Products

Hervé Moreau
Centre National de Recherche Scientifique
Banyuls-sur-Mer, France
Dinoflagellates

Richard Y. Morita
Oregon State University
Corvallis, Oregon
Low-Nutrient Environments
Low-Temperature Environments

Nathan Morris
Warwick University
Coventry, England, UK
Actinomycetes

Stephen S. Morse
Columbia University
Arlington, Virginia
Viruses, Emerging

Knud Mortensen
Agriculture and Agri-Food Canada Research Centre
Saskatoon, Saskatchewan, Canada
Biological Control of Weeds

Robb E. Moses
Oregon Health Sciences University
Portland, Oregon
DNA Replication

James W. Moyer
North Carolina State University
Raleigh, North Carolina
Tospoviruses

Matthew A. Mulvey
Washington University School of Medicine
St. Louis, Missouri
Adhesion, Bacterial
Fimbriae, Pili

Erik L. Munson
University of Wisconsin
Madison, Wisconsin
Detection of Bacteria in Blood: Centrifugation and Filtration

Noreen E. Murray
University of Edinburgh
Edinburgh, Scotland, UK
DNA Restriction and Modification

J. Colin Murrell
University of Warwick
Coventry, England, UK
Methylotrophy

Christine Musahl
University of Zurich
Zurich, Switzerland
Prions

T. G. Nagaraja
Kansas State University
Manhattan, Kansas
Gastrointestinal Microbiology

Nanne Nanninga
University of Amsterdam
Amsterdam, The Netherlands
Cell Division, Prokaryotes

Edward A. Nardell
Harvard Medical School
Cambridge, Massachusetts
Aerosol Infections

C. Nelson Neale
Rice University
Houston, Texas
Bioremediation

C. G. Nettles Cutter
Clemson University
Clemson, South Carolina
Bacteriocins

Dianne K. Newman
California Institute of Technology
Pasadena, California
Arsenic

Marie Lockstein Nguyen
University of Wisconsin Medical School
Madison, Wisconsin
Oncogenic Viruses

Jerry Nielsen
Oklahoma State University
Oklahoma City, Oklahoma
Meat and Meat Products

Alexander J. Ninfa
University of Michigan Medical School
Ann Arbor, Michigan
Carbon and Nitrogen Assimilation, Regulation of
Two-Component Systems

Yoko Nomura
University of Tokyo
Tokyo, Japan
Biosensors

Brian Nummer
University of Georgia
Athens, Georgia
Beer/Brewing

Per Nygaard
University of Copenhagen
Copenhagen, Denmark
Nucleotide Metabolism

David A. Odelson
Eco Soil Systems, Inc.
San Diego, California
Ecology, Microbial

Gary B. Ogden
St. Mary's University
San Antonio, Texas
Leishmania

Donald B. Oliver
Wesleyan University
Middletown, Connecticut
Protein Secretion

Ronald S. Oremland
U.S. Geological Survey
Menlo Park, California
Selenium

José-Julio Ortega-Calvo
Instituto de Recursos Naturales y Agrobiologia
Seville, Spain
Biodeterioration

Mary J. Osborn
University of Connecticut Health Center
Farmington, Connecticut
Outer Membrane, Gram-Negative Bacteria

Li-Tse Ou
University of Florida
Gainesville, Florida
Pesticide Biodegradation

Peter Palese
Mount Sinai School of Medicine
New York, NY
Influenza Viruses

Carol J. Palmer
Nova Southeastern University
Ft. Lauderdale, Florida
Polymerase Chain Reaction

Bernhard O. Palsson
University of California, San Diego
La Jolla, California
Genomic Engineering of Bacterial Metabolism

Sarad R. Parekh
Dow AgroSciences
Indianapolis, Indiana
Strain Improvement

Robert F. Park
University of Sydney
Camden, New South Wales, Australia
Rust Fungi

Christine Paszko-Kolva
Accelerated Technology Laboratories, Inc.
West End, North Carolina
Polymerase Chain Reaction

Ramesh N. Patel
Bristol-Myers Squibb Pharmaceutical Research Institute
New Brunswick, New Jersey
Biocatalysis for Synthesis of Chiral Pharmaceutical Intermediates

Etienne Pays
Free University of Brussels
Brussels, Belgium
Antigenic Variation

Guy A. Perkins
San Diego State University
San Diego, California
Microscopy, Confocal
Microscopy, Optical

David H. Persing
Corixa Corporation/Infectious Disease Research Institute
Rochester, Minnesota
Diagnostic Microbiology

Donald A. Phillips
University of California, Davis
Davis, California
Biological Nitrogen Fixation

Wolfgang Piepersberg
Bergische University
Wuppertal, Germany
Aminoglycosides, Bioactive Bacterial Metabolites

D. L. Pierson
NASA/Johnson Space Center
Houston, Texas
Space Flight, Effects on Microorganisms

Patrick J. Piggot
Temple University School of Medicine
Philadelphia, Pennsylvania
Sporulation

Robert W. Pinner
Centers for Disease Control and Prevention
Atlanta, Georgia
Surveillance of Infectious Diseases

Look Hulshoff Pol
Wageningen Agricultural University
Wageningen, The Netherlands
Sulfur Cycle

Roger J. Pomerantz
Thomas Jefferson University
Philadelphia, Pennsylvania
Retroviruses

Pablo Pomposiello
Harvard School of Public Health
Boston, Massachusetts
Oxidative Stress

Robert K. Poole
University of Sheffield
Sheffield, England, UK
Aerobic Respiration

Pieter Postma
University of Amsterdam
Amsterdam, The Netherlands
PEP: Carbohydrate Phosphotransferase Systems

James A. Poupard
SmithKline Beecham Pharmaceuticals
Collegeville, Pennsylvania
Biological Warfare

Jack Preiss
Michigan State University
East Lansing, Michigan
Glycogen Biosynthesis

Fergus G. Priest
Heriot Watt University
Edinburgh, Scotland, UK
Enzymes, Extracellular

Thomas C. Quinn
The Johns Hopkins University
Baltimore, Maryland
Sexually Transmitted Diseases

Jane E. Raulston
University of North Carolina
Chapel Hill, North Carolina
Chlamydia

Kenneth Reardon
Colorado State University
Fort Collins, Colorado
Bioreactor Monitoring and Control

Donald J. Reasoner
U.S. Environmental Protection Agency
Alexandria, Virginia
Water, Drinking

Rosemary J. Redfield
University of British Columbia
Vancouver, British Columbia, Canada
Haemophilus influenzae, Genetics

Alice G. Reinarz
University of Michigan
Ann Arbor, Michigan
Careers in Microbiology

Larry Reitzer
University of Texas at Dallas
Richardson, Texas
Amino Acid Function and Synthesis

Kevin A. Reynolds
Virginia Commonwealth University
Richmond, Virginia
Antibiotic Biosynthesis

Anna-Louise Reysenbach
Portland State University
Portland, Oregon
Archaea

Steven Ripp
University of Tennessee
Knoxville, Tennessee
Plasmids, Catabolic

Jane Robb
University of Guelph
Guelph, Ontario, Canada
Verticillium

Donald Roberts
Utah State University
Logan, Utah
Insecticides, Microbial

Mary F. Roberts
Boston College
Chestnut Hill, Massachusetts
Osmotic Stress

Charles O. Rock
St. Jude Children's Research Hospital
Memphis, Tennessee
Lipid Biosynthesis

Michael S. Rolph
Max Planck Institute for Infection Biology
Berlin, Germany
Cellular Immunity

Antonio H. Romano
University of Connecticut
Storrs, Connecticut
Carbohydrate Synthesis and Metabolism

George E. Rottinghaus
University of Missouri
Columbia, Missouri
Mycotoxicoses

Brenda G. Rushing
Indiana University
Indianapolis, Indiana
Photosensory Behavior

James B. Russell
Cornell University
Ithaca, New York
Rumen Fermentation

N. Jamie Ryding
University of Georgia
Athens, Georgia
Myxococcus, Genetics

Badal C. Saha
U.S. Department of Agriculture
Peoria, Illinois
Enzymes in Biotechnology

Hermann Sahm
Research Center Jülich
Jülich, Germany
Amino Acid Production

Joseph P. Salanitro
Shell Oil Company
Houston, Texas
Oil Pollution

Mary Ellen Sanders
Dairy and Food Culture Technologies
Littleton, Colorado
Dairy Products

Alexei Savchenko
Michigan State University
East Lansing, Michigan
Amylases, Microbial

Gary Sawers
John Innes Centre
Norwich, England, UK
Azotobacter

Hans Henrik Saxild
Technical University of Denmark
Lyngby, Denmark
Nucleotide Metabolism

Gary S. Sayler
University of Tennessee
Knoxville, Tennessee
Plasmids, Catabolic

Moselio Schaechter
San Diego State University
San Diego, California
Escherichia coli, General Biology

Kenneth F. Schaffner
George Washington University
Washington, DC
Method, Philosophy of

Ronald F. Schell
University of Wisconsin
Madison, Wisconsin
Detection of Bacteria in Blood: Centrifugation and Filtration

Thomas H. Scheper
University of Hannover
Hannover, Germany
Bioreactor Monitoring and Control

Christophe H. Schilling
University of California, San Diego
La Jolla, California
Genomic Engineering of Bacterial Metabolism

Milton J. Schlesinger
Washington University School of Medicine
St. Louis, Missouri
Viruses

Sondra Schlesinger
Washington University School of Medicine
St. Louis, Missouri
Viruses

Anne M. Schneiderman
Pennie & Edmonds LLP
New York, New York
Patenting of Living Organisms and Natural Products

Angelika Schnieke
PPL Therapeutics, Ltd.
Roslyn, Midlothian, UK
Transgenic Animal Technology

Kate M. Scow
University of California, Davis
Davis, California
Soil Microbiology

Leora A. Shelef
Wayne State University
Detroit, Michigan
Refrigerated Foods

Lawrence J. Shimkets
University of Georgia
Athens, Georgia
Myxococcus, Genetics

Robert E. Shope
University of Texas Medical Branch
Galveston, Texas
Arboviruses

David A. Shub
State University of New York, Albany
Albany, New York
RNA Splicing, Bacterial

Satinder K. Singh
University of Minnesota
Minneapolis, Minnesota
Glyoxylate Bypass in Escherichia coli

Uwe B. Sleytr
University of Agricultural Sciences
Vienna, Austria
Crystalline Bacterial Cell Surface Layers

Joan L. Slonczewski
Kenyon College
Gambier, Ohio
pH Stress

S. J. Smith
University of California, San Diego
La Jolla, California
Genomic Engineering of Bacterial Metabolism

Richard B. Smittle
Silliker Laboratories
Clear Brook, Virginia
Foods, Quality Control

Gerald Soffen
National Aeronautics and Space Administration
Greenbelt, Maryland
Exobiology

Peter G. Sohnle
Medical College of Wisconsin
Milwaukee, Wisconsin
Fungal Infections, Cutaneous

George A. Somkuti
U.S. Department of Agriculture
Wyndmoor, Pennsylvania
Lactic Acid Bacteria

Mohammad Sondossi
Weber State University
Ogden, Utah
Biocides

Terje Sørhaug
Agricultural University of Norway
Ås, Norway
Temperature Control

Gabriel E. Soto
Washington University School of Medicine
St. Louis, Missouri
Fimbriae, Pili

Hiroshi Souzu
Hokkaido University
Sapporo, Japan
Freeze-Drying of Microorganisms

Kevin R. Sowers
University of Maryland Biotechnology Institute
Baltimore, Maryland
Methanogenesis

Marie-Odile Soyer-Gobillard
Centre National de Recherche Scientifique
Banyuls-sur-Mer, France
Dinoflagellates

Peter T. N. Spencer-Phillips
University of the West of England
Bristol, England, UK
Downy Mildews

Catherine L. Squires
Tufts University
Boston, Massachusetts
Ribosome Synthesis and Regulation

James T. Staley
University of Washington
Seattle, Washington
Heterotrophic Microorganisms

Keith H. Steinkraus
Cornell University
Ithaca, New York
Fermented Foods
Wine

H. David Stensel
University of Washington
Seattle, Washington
Wastewater Treatment, Industrial

Linda D. Stetzenbach
University of Nevada, Las Vegas
Las Vegas, Nevada
Airborne Microorganisms and Indoor Air Quality

Jeff B. Stock
Princeton University
Princeton, New Jersey
Chemotaxis

James Strick
Arizona State University
Tempe, Arizona
Spontaneous Generation

William R. Strohl
Merck Research Laboratories
Rahway, New Jersey
Polyketide Antibiotics

Fred Stutzenberger
Clemson University
Clemson, South Carolina
Pectinases

Sebastian Suerbaum
University of Würzburg
Würzburg, Germany
Helicobacter pylori

William C. Summers
Yale University
New Haven, Connecticut
History of Microbiology
Typhoid, Historical
Virus Infection

Morton N. Swartz
Massachusetts General Hospital and Harvard Medical School
Boston, Massachusetts
Skin Microbiology

Jean Swings
Laboratory of Microbiology
Ghent, Belgium
Xanthomonas

F. Robert Tabita
The Ohio State University
Columbus, Ohio
Autotrophic CO_2 *Metabolism*

Yi-Wei Tang
Vanderbilt University School of Medicine
Nashville, Tennessee
Diagnostic Microbiology

Herbert B. Tanowitz
Albert Einstein College of Medicine
Bronx, New York
Trypanosomes

Simon M. Temperley
PPL Therapeutics, Ltd.
Roslyn, Midlothian, UK
Transgenic Animal Technology

Christopher M. Thomas
The University of Birmingham
Birmingham, England, UK
Plasmids, Bacterial

Torsten Thomas
University of New South Wales
Sydney, Australia
Extremophiles

Richard Tichy
Institute of Landscape Ecology
Ceske Budejovice, Czech Republic
Sulfur Cycle

Sue Tolin
Virginia Polytechnic Institute and State University
Blacksburg, Virginia
Genetically Modified Organisms: Guidelines and Regulations for Research

Alexander Tomasz
The Rockefeller University
New York, New York
Streptococcus pneumoniae

Suxiang Tong
Emory University
Atlanta, Georgia
Paramyxoviruses

Steven Tracy
University of Nebraska Medical Center
Omaha, Nebraska
Enteroviruses

Arthur O. Tzianabos
Harvard Medical School
Boston, Massachusetts
Gram-Negative Anaerobic Pathogens

Roland Ulber
University of Hannover
Hannover, Germany
Bioreactor Monitoring and Control

Erick J. Vandamme
University of Ghent
Ghent, Belgium
Vitamins and Related Biofactors, Microbial Production

S. Vandedrinck
University of Ghent
Ghent, Belgium
Vitamins and Related Biofactors, Microbial Production

Luc Vanhamme
Free University of Brussels
Brussels, Belgium
Antigenic Variation

Fazilet Vardar-Sukan
Ege University
Izmir, Turkey
Biosurfactants
Industrial Effluents: Sources, Properties, and Treatments

Costantino Vetriani
Rutgers University
New Brunswick, New Jersey
Archaea

Anne Vidaver
University of Nebraska
Lincoln, Nebraska
Genetically Modified Organisms: Guidelines and Regulations for Research

Claire Vieille
Michigan State University
East Lansing, Michigan
Amylases, Microbial

Rytas Vilgalys
Duke University
Durham, North Carolina
Rhizoctonia

Robert T. Vinopal
University of Connecticut
Storrs, Connecticut
Carbohydrate Synthesis and Metabolism

David K. Wagner
Medical College of Wisconsin
Milwaukee, Wisconsin
Fungal Infections, Cutaneous

Graeme M. Walker
University of Albertay
Dundee, Scotland, UK
Yeasts

C. H. Ward
Rice University
Houston, Texas
Bioremediation

Louis M. Weiss
Albert Einstein College of Medicine
Bronx, New York
Trypanosomes

Elizabeth M. H. Wellington
Warwick University
Coventry, England, UK
Actinomycetes

Craig Werner
Albert Einstein College of Medicine
Bronx, New York
Trypanosomes

David White
Indiana University
Bloomington, Indiana
Myxobacteria

Chris Whitfield
University of Guelph
Guelph, Ontario, Canada
Lipopolysaccharides

Richard J. Whitley
University of Alabama School of Medicine
Birmingham, Alabama
Antiviral Agents

Brian M. Wilkins
University of Leicester
Leicester, England, UK
Transformation, Genetic

Henry T. Wilkinson
University of Illinois
Urbana, Illinois
Gaeumannomyces graminis

Bryan R. G. Williams
Cleveland Clinic Foundation
Cleveland, Ohio
Interferons

Kevin W. Winterling
Emory and Henry College
Emory, Virginia
SOS Response

Murray Wittner
Albert Einstein College of Medicine
Bronx, New York
Trypanosomes

Theodore E. Woodward
University of Maryland School of Medicine
Baltimore, Maryland
Typhus Fevers and Other Rickettsial Diseases

Bernard S. Wostmann
University of Notre Dame
Granger, Indiana
Germfree Animal Techniques

William H. Wunner
University of Pennsylvania
Philadelphia, Pennsylvania
Rabies

Priscilla B. Wyrick
University of North Carolina
Chapel Hill, North Carolina
Chlamydia

Charles Yanofsky
Stanford University
Stanford, California
Attenuation, Transcriptional

Qizhi Yao
Emory University
Atlanta, Georgia
Paramyxoviruses

Marylynn V. Yates
University of California, Riverside
Riverside, California
Biomonitors of Environmental Contamination

A. Aristides Yayanos
University of California, San Diego
La Jolla, California
High-Pressure Habitats

Ki-Seok Yoon
The Ohio State University
Columbus, Ohio
Autotrophic CO_2 Metabolism

Allan A. Yousten
Virginia Polytechnic Institute and State University
Blacksburg, Virginia
Insecticides, Microbial

Anna Zago
University of Illinois College of Medicine
Chicago, Illinois
Pseudomonas

J. Gregory Zeikus
Michigan State University
East Lansing, Michigan
Amylases, Microbial

Ziguo Zhang
University of Oxford
Oxford, England, UK
Powdery Mildews

Ian M. Zitron
William Beaumont Hospital
Royal Oak, Michigan
Antibodies and B Cells

Judith W. Zyskind
San Diego State University
San Diego, California
Recombinant DNA, Basic Procedures

Glossary of Key Terms

A

ABC protein a protein that contains the widely conserved ATP-binding cassette, a motif that couples energy from ATP (adenosine triphosphate) binding and hydrolysis to various transport processes (**ABC transport**).

abiogenesis 1. a historic term used (1870) by Thomas H. Huxley to mean life arising from a combination of inorganic starting materials. 2. more generally, the origin of life from any nonliving matter.

abiotic not living; relating to the nonliving aspects of the environment.

acceptor see ELECTRON ACCEPTOR.

accessory hosts see COLLATERAL HOSTS.

acetogenesis the process of forming acetate by a linear combination of two molecules of CO_2 or CO.

acetogenic of microorganisms, carrying out or capable of acetogenesis.

acetone–butanol fermentation a bacterial fermentation carried out by certain sugar-metabolizing species of *Clostridium*. The typical endproducts include acetone, *n*-butanol, and acetic acid.

acid fermentation fermentation in which lactic acid, acetic acid, or other organic acids are produced, lowering the pH sufficiently to make the substrate resistant to growth of microorganisms that might otherwise spoil the food.

acidophile literally, acid lover; an organism whose optimal external pH for growth is in the acidic range, generally below pH 6. **Hyperacidophiles** may grow below pH 3.

acidophilic growing optimally in an acidic setting.

acid resistance the ability of a microbial strain to exist for an extended period at pH values too acidic for growth, retaining the ability to be cultured after the acidity is neutralized. Also, **acid survival** or **acid tolerance**.

acid shock a sudden decrease of the pH level of a growth medium to a level of acidity at or below the limit for growth.

acquired immunity see ADAPTIVE IMMUNITY.

Actinomycetes widespread gram-positive bacteria that range morphologically from cocci-like fragments to mycelium which may form complex spore structures.

activated sludge process an aerobic biological treatment process consisting of a mixed and aerated basin followed by liquid–solids separation and return of biomass to the aeration basin.

active surveillance a public health surveillance method in which health officials or other persons conducting surveillance take measures to ensure reporting, e.g., by collecting the data themselves or by reminding providers to report.

adaptation a change in response to some condition; e.g.: 1. a correlation between a particular feature of an organism and its environment, which results from natural selection. 2. a change in behavior in response to an altered environmental condition. 3. a change in the ability of a microbe to utilize a specific nutrient after exposure to that nutrient.

adaptive immunity the immune response that occurs following the recognition of specific antigen by B and T lymphocytes; develops later than innate immunity.

adenylate-forming enzymes a large family of enzymes that activate short chain fatty acids or amino acids as acyl adenylates (acyl-AMP).

adhesin a molecule, typically a protein, that mediates bacterial attachment by interacting with specific receptors.

adjuvant one of a heterogeneous group of substances that enhance the immune response to an antigen in a nonspecific manner, and therefore act as an important component of vaccines; several mechanisms may be involved, but prolonging exposure of the antigen to immunocompetent cells is thought to be a major factor.

adulteration the addition of any organism or substance that may make a consumer product harmful to users under usual conditions of use.

aerobe an organism able to use oxygen as an electron acceptor in metabolism.

aerobic 1. having to do with or occurring in the presence of oyygen. 2. using or requiring atmospheric oxygen for metabolism and growth.

aerobic process a physiological process that requires the presence of oxygen to proceed.

aerobic respiration the use of molecular oxygen to accept electrons derived from energy-conserving oxidation reactions by cells.

aerosol 1. a colloidal suspension of fine solid or liquid particles in air. 2. specifically, a suspension of true airborne particles; i.e., particles with a negligible settling tendency under average room conditions.

aerosol infection an infection spread by airborne particles, as distinct from droplet-borne respiratory infections, which spread via larger respiratory droplets.

African trypanosomiasis (sleeping sickness) an illness endemic in large areas of Africa that is transmitted by tsetse flies. *Trypanosoma brucei gambiense* is the cause of West African or gambiense trypanosomiasis, and *Trypanosoma brucei rhodesiense* is the cause of East African or rhodesiense trypanosomiasis.

aged residue a term for pesticide that over time diffuses into the internal soil matrix and is no longer available for microbial attack.

aglycon an organic molecule that can be glycosylated.

Agrobacterium a genus of gram-negative, aerobic bacteria of the family Rhizobiaceae that are noted as plant pathogens; e.g., *A. tumefaciens,* the inducer of crown gall tumors.

Airy disk a diffuse central spot of light with concentric rings of light of decreasing intensity, formed by a microscope imaging a point source of light.

alanine amino transferase an enzyme found in the liver and blood serum, the concentration of which is often elevated in cases of liver damage.

alate 1. having wings or winglike parts. 2. a winged individual of organisms that have both winged and wingless forms or phases.

alcoholic fermentation fermentation in which carbon dioxide flushes out residual oxygen, rendering the substrate anaerobic and producing ethanol at levels that make the substrate resistant to development of food poisoning and spoilage microorganims.

alginate a hydrated, viscous polysaccharide copolymer comprising variable amounts of β-D-mannuronic acid and α-L-guluronic acid. Alginate is excreted from the cells of *Azotobacter* species, giving them a slimy glutinous appearance.

alkaline fermentation fermentation in which the pH level rises to 7.5 or higher, allowing growth of the essential microorganisms but inhibiting growth of organisms that might otherwise spoil food.

alkaliphile literally, alkali lover; an organism whose optimal external pH for growth is in the alkaline range, generally above pH 8. **Obligate alkaliphiles** are incapable of growth at neutral pH; alkali-tolerant organisms are capable of growth at high pH but are also capable of growth under non-alkaline conditions.

alkaliphilic growing optimally in an alkaline setting.

alkanotrophic capable of assimilating propane, *n*-butane, and other *n*-alkane gases and liquid hydrocarbons as sole carbon and energy sources.

allochthonous not native; rising from outside a given system or habitat.

ALT alanine amino transferase, an enzyme liberated into the blood stream when hepatocytes are destroyed or damaged; serum levels are determined for evidence of hepatitis.

alternate host one of two possible hosts that a microbe may infect.

alveolar macrophages scavenging host-defense cells that reside in the alveoli (see below), engulfing inhaled particles, including microorganisms. They destroy some engulfed organisms directly, while others replicate within them, triggering the release of mediators of cellular immunity.

alveoli the numerous tiny air spaces of the lungs in which the exchange of oxygen and carbon dioxide takes place.

amastigote the aflagelleted stage of parasites such as *Leishmania*. Amastigotes are obligate intracellular parasites of mononuclear phagocytes of the vertebrate host.

***American trypanosomiasis* (*Chagas' disease*)** an illness endemic in most countries in Latin America, caused by the protozoan parasite *Trypanosoma cruzi* and transmitted by triatomine insects.

aminoacyl tRNA an activated form of amino acid used in protein biosynthesis. It consists of amino acid linked through an acid-ester linkage to the hydroxyl group at the 3′ terminus of tRNA.

aminocyclitol a cyclitol with one or more hydroxyls substituted by amino groups.

aminoglycoside a low-molecular-weight (pseudo-)saccharidic substance based mainly on aminated sugars. Aminoglycosides are a class of antimicrobial agents that inhibit protein synthesis; used widely against gram-negative bacteria.

amplification the tandem reiteration of a segment of DNA.

amplification of transmission the increased spread of infectious disease, occurring naturally or because of facilitating factors such as nonsterilized needles and syringes.

amplifying host a host that serves to increase the natural pool of an invading pathogenic agent.

amyloid plaques characteristic aggregated fibrils with high beta-sheet content, stainable by a dye and pH indicator called Congo Red (CR). CR intercalates with a universal structure shared by all amyloids, which consists of antiparallel beta sheet extensions arranged in a quasicrystalline fashion.

anabiosis return to life; the restoration of life processes in a dormant organism that did not have the external appearance of life.

anabolism the process of synthesis of cell components from a metabolic pool of precursor compounds.

anaerobe an organism that can survive and proliferate in the absence of molecular oxygen. **Obligate anaerobes** are those microbes that are able to grow only under conditions in which oxygen is completely excluded; they are usually killed by oxygen.

anaerobic **1.** having to do with or occurring in the absence of oyygen. **2.** using or requiring molecules or ions other than atmospheric oxygen for metabolism and growth.

anaerobic process a physiological process that requires the absence of oxygen to proceed.

anaerobic respiration the use of ions or molecules (usually inorganic) other than molecular oxygen to accept electrons derived from energy-conserving oxidation reactions by cells.

anaerobic treatment a biological treatment process in which no oxygen is available. Electron acceptors may be organic substrates, sulfate, nitrate, or carbon dioxide. The last is more common for treatment of high-strength industrial wastewaters with conversion of a large portion of the organic carbon to methane.

analog or ***analogue*** something that has very similar properties to another, but also is different in some fundamental way; for example: **1.** a gene that is structurally or functionally similar to another, but not related by descent. **2.** a naturally occurring or synthetic compound with structural similarity to natural nucleobases or nucleosides.

analogy the fact of being an analog.

anamorph the asexual stage(s) of a fungus.

animalcules a historic term used for the microscopic organisms first described by Leeuwenhoek.

animal reservoir a source in nature in which an infectious agent can be maintained for occasional or usual transmission to human beings by a vector.

anorexia lack or loss of the desire to eat; aversion to food.

anoxybiosis a condition of life in the absence of oxygen.

anoxygenic photosynthesis light-driven metabolic reduction of small carbon compounds (often CO_2) using electrons derived from a compound other than water; consequently, oxygen is not produced. Typical electron donors include H_2S, other reduced sulfur compounds, H_2, and small organic substrates. This type of photosynthesis has been found only in the Bacteria domain.

antibiotic a microbial product that kills or inhibits the growth of susceptible microorganisms.

antibiotic resistance the ability of bacteria to withstand the killing effect of antimicrobial agents.

antibody a serum protein that is produced in response to an antigen and that has the property of combining specifically with that antigen. Different classes of antibody (IgG, IgM, IgA, etc.) have different functional properties.

anti-contagionism a historic theory of public health holding that disease originates not from the transmission of germs from person to person but from interaction with an unsuitable environment, contact with rotting or deteriorating organic material, inhalation of polluted or contaminated air, or immoral behaviors. This premise had significant public support in the first half of the 19th century but was generally rejected thereafter through the studies of Pasteur, Koch, and others.

antifungal drug a drug used to treat a fungal infection. Antifungal agents can be fungicidal or fungistatic depending on their ability to kill or inhibit the growth of the fungal pathogen, respectively. Examples of currently used antifungal drugs include amphotericin B, fluconazole, ketoconazole, itraconazole, and 5-fluorocytosine.

antigen a substance or agent that is recognized as foreign or "non-self" and that thus provokes an immune response; a molecule that induces synthesis of an antibody.

antigenic relating to or acting as an antigen.

antigenic drift change of antigenic structure caused by amino-acid substitutions, which themselves result from the accumulation of random point mutations in the gene.

antigenic shift change of antigenic structure resulting from random genetic reassortment between different sequences of nucleic acid.

antigenic variation change of antigenic structure due to active specific mechanisms happening at higher frequencies than random mutation rates.

antigen-presenting cells (APCs) cells that process and present antigen to T cells along with the essential co-stimulatory molecules required for T cell activation.

anti-infective (drug) resistance the ability of a virus, bacterium, or parasite to defend itself against a drug which was previously effective. Drug resistance is currently occurring for bacterial infections such as tuberculosis and gonorrhoea, for parasitic infections such as malaria, and for the human immunodeficiency virus (HIV).

antimicrobial describing compounds that kill or inhibit the growth of microbes.

antiporter a membrane transporter protein catalyzing coupled, simultaneous counter transport across the membrane of two solutes in opposite directions.

antiretroviral agent any drug used in treating patients with human immunodeficiency virus (HIV) infection.

antisense a region of DNA or RNA sequence (usually short) that is complementary, is of opposite polarity, and hybridizes to a naturally functional (sense) target sequence, resulting in decreased function.

antisepsis the process of using agents to kill microbes during surgery or other procedures.

antitermination a process in which RNA polymerase is modified by cellular factors so that it makes transcripts at an increased elongation rate and it becomes resistant to transcription terminators.

antiviral resistance the developed resistance of a virus to a specific drug.

apnea an interruption or suspension of breathing.

apoptosis programmed cell death; a death pathway which involves the activation of cellular enzymes and results in the death of the cell by organellar disruption, while the plasma membrane remains intact, preventing the release of cellular contents. Morphologically, apoptotic cells show nuclear condensation, swollen mitochondria, and characteristic membrane blebs. An end-stage event is the degradation of nuclear DNA by an activated cellular endonuclease.

apothecia cup-shaped or saucer-shaped sexual reproductive structures formed by certain fungi.

appressorium in certain parasitic fungi, a flattened, but swollen, hyphal structure from which the infection peg emerges to penetrate the host.

apterous lacking wings.

arbovirus a virus that replicates in blood-sucking arthropods, e.g., mosquitoes, biting flies, or ticks, and that is transmitted by bite to a vertebrate host. Arbovirus is an ecological rather than a taxonomic definition; various arboviruses belong to a variety of virus families, including the Flavidiridae, the Bunyaviridae, and the genus *Alphavirus* of the family Togaviridae.

arbuscule a structure produced by a fungus, responsible for nutrient and carbon exchange between plant and fungus.

Archaea a group of prokaryotes that constitute a phylogenetically separate domain of organisms distinct from the other two domains, the Bacteria (Eubacteria) and Eukarya (Eukaryotes). Archaea are often (although not exclusively) found living in extreme terrestrial or aquatic environments such as hot springs, under-sea thermal vents, or under conditions of high alkalinity and/or concentrated saline.

archaeal relating to or designating an organism belonging to the Archaea (see above).

Archaebacteria another name for the Archaea, which is the newer and preferred term (see above).

archebiosis a term coined by Bastian in 1870 to describe life arising from strictly inorganic starting materials.

arthropods invertebrate animals such as insects, mites, spiders and ticks, relevant as carriers of infectious diseases between humans or between other animals and humans.

artificial RNA RNA molecules expressed from genes that have been introduced into cells (transgenes), or RNA molecules synthesized in cell-free systems.

ascigerous describing tissues in which asci are produced, or fungi having such tissue; generally refers to the sexual stage in Ascomycetes.

ascus *plural*, **asci**. a sac-like cell containing ascospores (usually 8 in number) formed after nuclear fusion (karyogamy) and meiosis.

asepsis surgery or other procedures carried out in the absence of microbes.

aseptic **1.** in medicine, referring to the absence of infectious organisms. **2.** in fermentation, referring to the absence of contaminating microorganisms.

assimilation in this context, the incorporation of inorganic compounds, such as ammonia or sulfate, into organic intermediates of cellular metabolism.

assimilatory reduction the reduction of a compound for the purpose of introducing its building elements into cellular material.

associative nitrogen fixation nitrogen fixation by free-living bacteria in the rhizosphere.

astaxanthin a reddish carotenoid pigment found in salmon, crustacean shells, and bird feathers (e.g., pink flamingos).

astrobiology the study of life in the universe; the chemistry, physics and adaptations that influence the origins, evolution, and destiny of life. A term used to describe a new field that will expand upon the limited studies embraced within the field of exobiology.

asymptomatic carrier a person harboring an infectious microbe without symptoms of the disease but with the capacity to transmit to other persons.

ATP adenosine triphosphate, the energy-rich molecule synthesized in energy-yielding reactions and consumed in biosynthetic processes.

ATP synthetase the membrane-bound enzyme that synthesizes ATP (adenosine triphosphate) from ADP (adenosine diphosphate) and inorganic phosphate by utilizing the energy stored as membrane potential.

attachment site a site on a phage (attP) or chromosome (attB or attλ, phi80, etc.) at which a site-specific integrase can join circular phage DNA and host chromosome.

attenuation a reduction in the ability of a still-viable microbe to cause disease or other consequences of infection by a microbe while it still remains viable.

attenuator a short DNA region that functions as a site of regulated transcription termination.

attractant a chemical that causes a positive chemotaxis response.

aufwuchs a collective term for a community of organisms firmly attached to, but not penetrating, a streambed substratum. (From German; literally, growing on or upon.)

authochthonous native; indigenous; rising from within a given system or habitat.

autoecious of a parasitic organism, requiring only one host species or group of closely related host species to complete the life cycle.

autoinducer a homoserine lactone produced by bacteria which, after accumulating in the medium to a critical concentration, initiates transcription of specific genes by a mechanism referred to as **autoinduction.**

autolysin endogenous murein hydrolases that cleave bonds in the peptidoglycan sacculus, that are critical for the mechanical strength of the structure and thus cause lysis (**autolysis**) of the bacterium.

autonomously replicating sequences (ARS) yeast chromosomal sequences that provide a replicator function to plasmids and produce high frequency transformation. ARSs are the most thoroughly studied replication origins in yeast.

autophagous capable of or carrying out a process of autophagy.

autophagy literally, self-eating; a process of self-digestion by cells that are starved of amino acids in which portions of the cytoplasm are circumscribed by endoplasmic reticulum and then catabolized as an internal supply of nutrients.

autophosphorylation reaction the catalysis of the phosphorylation of a residue found on itself by an enzyme.

autotroph literally, self-feeder; an organism that is able to utilize inorganic carbon (carbon dioxide) as the sole carbon source for growth; contrasted with heterotrophs (feeders on others) that must utilize organic materials for this purpose. Thus, **autotrophic.**

autotrophy the fact of being an autotroph; using CO_2 as a sole source of carbon.

auxostat a continuous culture system in which a growth-dependent parameter is held constant and all other parameters, including the specific growth rate, vary accordingly.

auxotroph a mutant that requires additional nutrients beyond what a "wild-type" strain needs for growth on minimal media. Thus, **auxotrophic.**

auxotrophy the inability of a mutant organism to produce an essential component or growth requirement.

avirulence a lack of virulence; the inability of a pathogen race or strain to cause disease on a particular strain or cultivar of of the host.

avirulence (avr) gene a gene in a plant pathogen that encodes a protein for which there is a corresponding dominant resistance (R) gene in a plant whose expression results in an active defense response.

avirulent not virulent; unable to cause disease due to resistance of the potential host.

axenic culture a known single species of microorganisms maintained in the laboratory.

B

bacillus 1. a rod-shaped bacterium. **2.** ***Bacillus.*** specifically, a diverse genus of gram-positive, endospore-forming rod-shaped bacteria; generally not disease agents, though two species, *B. anthracis* and *B. cereus,* are known to be pathogenic to humans. ***Bacillus subtilis*** produces enzymes that are widely used in the brewing, baking, and washing powder industries. It is also a model system for bacterial research.

bacteremia the presence of bacteria in the blood.

bacteria *singular*, **bacterium.** unicellular, prokaryotic, microscopic, generally heterotrophic organisms, present in great numbers in soil and in water; largely responsible for the decomposition of primary and secondary produced organic matter and for mineralization of its constituent elements, C, N, P, S, etc. The ***Bacteria*** (formerly called Eubacteria) are one of the three domains of life, the others being Archaea and Eukarya (eukaryotes). (From the Greek *bacterion* staff or rod, based on the characteristic shape of some forms; a term coined by the 19th century German naturalist Christian Ehrenberg.)

bacterial relating to, caused by, or involving bacteria.

bacterial artificial chromosomes (BACs) vectors based on plasmid F (from *E. coli*) origin of replication. Such vectors, which can carry 24 to 100 kbp, are stable in *E. coli*.

bacteriochlorin a porphyrin with two nonadjacent reduced pyrrole rings; the tetrapyrrole macrocycle of bacteriochlorophylls.

bacteriochlorophyll forms of chlorophyll present in photosynthetic bacteria.

bacteriocin an antibacterial compound or protein produced by some bacterial strains that is active against certain strains of the same or closely related species. Bacteriocin-producing strains usually carry gene(s) that confer resistance to the bacteriocin.

bacteriocyte an insect cell harboring symbiotic microorganisms.

bacteriophage literally, eater of bacteria; a virus that infects bacterial host cells, usually consisting of a nucleic acid molecule enclosed by a protein coat. Certain bacteriophages (lysogenic) can sometimes insert stably into the chromosome without replicating. Other bacteriophages (lytic) always replicate following infection, and usually kill their host bacterial cell as they produce a burst of progeny phage particles.

bacteriorhodopsin a protein of the cytoplasmic membrane of the halophilic archaebacterium *Halobacterium salinarum* (formerly *halobium*) that has a covalently attached retinal molecule. Absorption of light by the latter pigment results in proton translocation across the membrane.

bacterium see BACTERIA.

barophile literally, pressure lover; an organism that lives optimally at a pressure greater than atmospheric pressure; e.g., certain deep-sea bacteria. Thus, **barophilic.**

base resistance the ability of a microbial strain to exist for an extended period at pH values too alkaline for growth, retaining the ability to be cultured after the alkali is neutralized.

basic reproduction rate in public health, the estimated number of secondary infections potentially transmitted within a susceptible population from a single nonimmune individual.

basidium *plural*, **basidia.** an enlarged club-like cell that bears basidiospores (usually four in number) formed after nuclear fusion and meiosis.

batch process(ing) a technique in biotechnology in which all nutrients for the target microorganism are added to the bioreactor only once and the products are separated in their entirety when the fermentation process is completed.

B cell or ***B lymphocyte*** a cell that plays the central role in the humoral immune system; B cells are the only cells that can produce antibodies. Along with T cells, they are one of the two basic types of lymphocyte cell.

(Given the designation B from *bursa cell*; in birds the site of lymphocyte development is the bursa of Fabricius, a saclike projection connected to the cloaca.)

benthos 1. bottom-dwelling life forms; those organisms living on or within the sediments of a lake, sea, or other body of water. 2. the biogeographic region inhabited by such organisms; the soil–water interface at the bottom of a body of water. (From a Greek word meaning "depth.")

binary fission the division of a mother cell into two daughters or in some cases into a mother and daughter.

bioaerosol an aerosol of living matter; i.e., an air-borne suspension of microorganisms, microbial byproducts, and/or pollen.

bioattenuation the nonengineered, natural decomposition of organic contaminants in soil and ground water systems.

bioaugmentation the stimulation of microbial activity for production of a specific compound.

bioavailability in this context, the fact of being available in a viable form; e.g.: 1. the capacity of a drug to be absorbed and distributed within the body in a way that preserves its useful characteristics; i.e., it is not broken down, inactivated, or made insoluble. 2. the property of a chemical that is in the solution phase and available for microbial attack.

biocatalyst a biochemical compound (particularly an enzyme) that catalyzes chemical reactions.

biochemical marker in biotechnology, an indicator of biodegradation or biotransformation.

biochemical oxygen demand (BOD) an indicator of the amount of dissolved oxygen (DO) needed to biologically degrade the organic compounds in an aquatic environment.

biocidal capable of killing living organisms.

biocide 1. any agent that kills living organisms. 2. specifically, a chemical substance or composition used to kill microorganisms considered to be undesirable; i.e., pest organisms.

biocontrol 1. in agriculture, medicine, etc., the deliberate use of living organisms (e.g., an insect, nematode, fungus, bacterium, virus) to control other unwanted organisms (e.g., insect pests, weeds), as opposed to control via toxic chemicals. 2. also, **biocontrol agent.** an organism used for this purpose.

bioconversion see BIOTRANSFORMATION.

biodegradation the complete or partial breakdown of a substance by microorganisms or their enzymes, usually by microorganisms present in soils, sediments, biosludges, or marine and fresh waters.

biodeterioration any undesirable change in the properties of a material caused by the vital activities of microorganisms. It is more restrictive than the term BIODEGRADATION because the transformation implies the concept of a lowering in usefulness, quality, or value.

biodispersant a chemical additive that will inhibit the accumulation of a biofilm, or disperse a biofilm that has formed, but that is not in itself toxic to microorganisms at reasonable treatment levels.

bioenrichment enrichment of the protein, essential amino acid, or vitamin content of food through growth and synthetic activities of microorganisms overgrowing the substrate.

bioethanol ethyl alcohol (ethanol) produced by yeast fermentation for use as a fuel or industrial commodity.

biofilm a biologically active layer on a surface or interface, consisting of a complex association of microorganisms, microbial products, water, and suspended and dissolved solids.

biofouling damage caused to a surface by microorganisms attached to the surface.

biogenesis a historic term used (1870) by Thomas H. Huxley to mean life arising only from other living things, to which he contrasted the opposing term ABIOGENESIS, life arising from inorganic matter.

biogenic produced or formed by organisms, such as the formation of some metal sulfide deposits.

bioinformatics the interdisciplinary application of principles and techniques of information technology to issues of biology; specifically, the computer-assisted analysis of genome sequence and experimental data.

bioleaching the solubilization of metal values from an ore by microbially mediated oxidation of the minerals.

biological control see BIOCONTROL.

biological nitrogen fixation (BNF) see NITROGEN FIXATION.

biological warfare (BW) the use of microorganisms, such as bacteria, fungi, viruses, and rickettsiae, to produce death or disease in humans, animals, or plants. The use of toxins to produce death or disease is often included as well.

biological weapon(s) living organisms, whatever their nature, that are intended to cause disease or death in man, animals or plants, and which depend for their effects on their ability to multiply in the person, animal, or plant attacked.

bioluminescence the emission of visible light by living organisms.

biomagnification a process by which an increase in the concentration of a substance in biological tissues occurs when organisms are consumed by others at higher trophic levels; a term applied especially to toxic substances such as mercury.

biooxidation pretreatment the microbial oxidation of minerals in order to release a metal value contained within the mineral.

biopesticide a living microbial pathogen that selectively infects and kills its insect or weed host, or that is a microbial antagonist of a plant pathogen.

bioreactor a closed vessel or system used for fermentation or enzyme reaction.

bioremediation the use of microorganisms or microbial processes to control and detoxify environmental pollutants.

biosensor an instrument that combines a biological recognition unit with a physical transducer.

bioserotype see SEROTYPE.

bioslurry treatment accelerated biodegradation of contaminants by the suspension of contaminated soil or sediment in water through mixing energy.

biosorption the removal of metal or metalloid species, compounds and particulates, radionuclides, and organometalloid compounds from solution by physicochemical interactions with microbial or other biological material.

biostatic describing an agent that inhibits or halts growth and multiplication of organisms.

biosurfactant a biodegradable surfactant (surface-modifying agent) produced by microorganisms or enzymes.

biosynthesis the synthesis of cellular components by organisms.

biotechnology the use of living organisms, or their processes or products, for commerical or industrial purposes.

biotope a specific, defined site or environment containing a biological community.

biotransformation the alteration of the structure of a compound by an organism, especially a microorganism.

biotroph literally, eater of life; an organism that requires nutrients from a living host in order to grow and reproduce. Thus, **biotrophic.**

biotrophy the fact or process of feeding by extracting nutrients from living host tissues.

biotype see BIOVAR.

biovar an infrasubspecific grouping (taxon) of isolates of a species differing in biochemical or physiological properties.

bioventing accelerated biodegradation of contaminants in contaminated subsurface materials by forcing and/or drawing air through the unsaturated zone.

blackleg a bacterial disease of potato caused by *Erwinia carotovora* subspecies *atroseptica,* characterized by black shriveled lower stems and rotting of the tuber.

blue fluorescent proteins accessory proteins in the bioluminescence system in some bacteria, carrying lumazine chromophores and serving as secondary emitters under some conditions.

B lymphocyte see B CELL.

bottom-fermenting describing the yeasts used to produce the type of beer known as lager, as opposed to ale which is top-fermented.

botulism a rare paralytic disease of humans and other animals, caused by a potent neurotoxin (**botulin**) produced by the bacterium *Clostridium botulinum.*

bradyzoite in *Toxoplasma,* the slowly dividing, infectious, asexual stage found within tissue cysts in the intermediate host. (From Greek *brady* "slow" and *zoite* "little animal.")

budding a mode of vegetative reproduction in many yeast species whereby a small outgrowth, the daughter bud, appears and grows from the surface of a mother cell and eventually separates to form a new cell at cell division.

bud scar the chitin-rich, convex ringed protrusions that remain on the mother cell surface of budding yeasts following birth of daughter cells.

building-related illness a specific adverse health reaction resulting from indoor exposure to a known pollutant or contaminant.

bulk soil a term for the portion of the soil not under the influence of plant roots.

burst a term for the abrupt release of many progeny viruses upon the disintegration of the infected cell in which the virus has been growing.

C

candidiasis infection caused by fungal organisms of the genus *Candida.*

capsid the protein coat surrounding the nucleic acid of a virus.

capsule in some species of bacteria, a loose, gel-like structure external to the outer membrane; usually, but not always, polysaccharide in nature.

carotenoids a group of yellow, orange, or red oil-soluble polyene pigments, occurring in plants and certain microorganisms.

carrier in pathology, a term for an individual who harbors and who may excrete virulent bacteria, while remaining free of symptoms of the disease.

catabolism the breakdown of nutrients to precursor compounds for anabolism or for dissimilation.

catabolite repression a process in which glucose and certain other sugars prevent the uptake or utilization of lactose, glycerol, and other sugars.

CD4, CD8 two cell surface glycoproteins that are expressed on mutually exclusive subsets of mature T lymphocytes (T cells); CD4 is described as the coreceptor on T cells specific for MHC (major histocompatiblity complex) Class II molecules and CD8 as the coreceptor on T cells specific for MHC class I molecules.

cDNA complementary DNA copy transcribed from a messenger RNA template rather than a DNA template.

cell envelope in bacterial cells, a multilayered structure that engulfs the cytoplasm. The innermost layer, the cytoplasmic membrane, is stabilized by an exoskeleton of peptidoglycan (murein). Gram-negative bacteria have a second lipid bilayer, called the outer membrane, making the envelope of gram-negative bacteria less permeable as compared to gram-positive bacteria.

cell-mediated immune response specific immune response mediated by T cells and activated macrophages, rather than by antibody.

cell sorter an instrument that uses optical or mechanical technologies to allow the separation of cells on the basis of size or cellular properties.

cellular immunity a T cell-dependent host response to an antigen.

cellulases a family of enzymes that hydrolyze β-1,4-glucosidic bonds in native cellulose and derived substrates.

cellulitis an acute edematous, suppurative, spreading inflammation of the deep subcutaneous tissues producing an area of erythema and tenderness with indistinct margins. The most common etiologic agents are group A streptococci and *Staphylococcus aureus.*

cellulosome a high-molecular-weight, multienzyme complex, in which cellulases and hemicellulases are noncovalently bound to a multivalent scaffolding protein.

cell wall one of the outer layers of the bacterial cell that protects the cell from osmotic perturbations and provides mechanical protection to the fragile cellular membrane.

central metabolic pathway a pathway used by an organism for the production of energy and biosynthetic precursors regardless of the source of carbon and energy being utilized.

centromere cis-acting sequences that control movement and segregation of chromosomes. In eukaryotic microbes, centromeres are attached to microtubules of the mitotic spindle apparatus. In prokaryotic microbes, the centromere and its attachment structure are less well defined.

Cer regulon the set of genes directly regulated by the complex of cAMP (cylic adenosine monophosphate) and its receptor, the CAP protein.

Chagas' disease see AMERICAN TRYPANOSOMIASIS.

chaperones proteins that are required for the correct folding of newly synthesized proteins. They also prevent protein denaturation during stresses, and can renature damaged proteins.

chemiosmosis an electron transport process that pumps protons across the inner membrane to the outer aqueous phase, generating an H^+ gradient across the inner membrane. The osmotic energy in this gradient supplies the energy for ATP (adenosine triphosphate) synthesis. Examples of such membranes are the cytoplasmic membranes of bacteria, the inner mitochondrial membranes of eukaryotes, and the thylakoid membranes of algae.

chemiosmotic relating to or involved in chemiosmosis.

chemoautotroph a microorganism that can grow through the utilization of inorganic carbon (CO_2) and inorganic substrates as sources of carbon and energy, respectively. Thus, **chemoautotrophic.**

chemoautotrophy a type of biosynthetic metabolism in which cellular carbon is derived from inorganic carbon (CO_2), with the oxidation of inorganic electron donors providing the source of energy.

chemoheterotroph an organism that uses organic carbon both as a source of carbon and as a source of energy. Thus, **chemoheterotrophic.**

chemoheterotrophy a type of metabolism in which cellular carbon is derived from organic carbon compounds, with energy derived from its oxidation.

chemokine a subset of the cytokine family distinguished by a characteristic molecular structure, and mostly functioning to attract immune cells to the site of infection.

chemolithotroph see CHEMOAUTOTROPH.

chemolithotrophy see CHEMOAUTOTROPHY.

chemoorganotroph see CHEMOHETEROTROPH.

chemoorganotrophy see CHEMOHETEROTROPHY.

chemoprophylaxis preventive treatment with chemical agents such as drugs.

chemostat 1. a type of continuous culture system. Fresh medium, with one essential nutrient limited, is continuously added to and mixed with a bacterial culture, and an equivalent volume of the culture is withdrawn. In time the culture will reach a constant population that is thereafter maintained. 2. the apparatus used in this process.

chemotaxin a factor that induces a process of chemotaxis (see below).

chemotaxis 1. movement of a cell or organism towards or away from a chemical substance. 2. specifically, the directed migration of cells towards an inflammatory site.

chemotherapy treatment for disease with specific drugs of known composition, usually a natural or synthetic organic compound, sometimes including the toxic, antimicrobial natural products of other organisms.

chiral having the property of chirality; describing a molecule that cannot be superimposed on its mirror image.

chirality the property of a molecule with carbon atoms to which four different ligands are bonded tetrahedrally. This gives a pair of stereoisomers per chiral carbon atom that exhibit mirror symmetry, but cannot be superimposed on each other.

chlorin a porphyrin with one reduced pyrrole ring; the tetrapyrrole macrocycle of chlorophylls and heme *d*.

chlorophyll the generic name of any of several oil-soluble tetrapyrrole pigments that function as photoreceptors of light energy in photosynthesis.

chromosome a package of genes representing part or all of the inherited information of an organism.

chronic active gastritis infiltration of the gastric mucosa with lymphocytes, plasma cells and granulocytes. In effect all persons colonized with *Helicobacter pylori* develop chronic active gastritis, but the intensity varies.

chronic obstructive pulmonary disease any chronic condition which results in the blockage of the outflow of air from the lungs, for example, emphysema.

cistron the basic unit of genetic function. The term usually refers to a gene or coding region for a protein.

classical biocontrol the use of co-evolved, native predators or pathogens to control plants or insects which have become pests in a foreign environment. Control of the pest in the new environment is achieved through the introduction of the native predators or pathogens. The goal of this control method is the establishment of an ongoing host–pathogen relationship that significantly reduces pest populations.

classification the orderly arrangement of individuals into units (taxa) on the basis of similarity; each unit (taxon) should be homogenous and different from all others.

cleistothecium the closed, fruiting body of an Ascomycete fungus which contains sexually produced ascospores.

clone 1. a population of cells descended from a single parental cell. 2. a group of genetically identical cells or organisms. 3. in popular use, an organism that is technologically developed as a genetic copy of another organism.

cloning vector a DNA molecule suitable for transporting new genetic material into host bacteria.

Clostridium a genus of obligate anaerobic, endospore-forming eubacteria that occur widely in soils and in the intestines of humans and other animals.

coat protein-mediated resistance a phenomenon whereby the expression of a viral coat protein gene confers resistance to plant infection by the same or a closely related virus (with $>60\%$ amino acid homology in the coat protein).

codon a triplet of three consecutive nucleotide components in the genetic code in DNA or messenger RNA; each three-nucleotide codon designates a specific amino acid in the linear sequence of a protein molecule.

coelome the general cavity of coelomate animals (vertebrates and invertebrates).

coenzyme an organic nonprotein molecule that is required for certain enzymatic reactions.

coiling phagocytosis an asymmetrical mode of particle uptake by phagocytic cells in which pseudopods appear to form coils around the object being ingested.

cold see COMMON COLD.

cold shock response a rapid change in gene expression that occurs when there is a temperature shift to a low temperature.

colicin a bacteriocin produced by *Escherichia coli* or closely related enteric species.

coliform a member of a group of bacteria that has traditionally been used as an indicator of the presence of pathogenic microorganisms in water.

coliphage a virus that infects *E. coli* bacteria host cells.

colitis inflammation of the large intestine (colon).

collateral hosts a group of closely related plants parasitized by a species of rust fungus.

colony **1.** a group of individual organisms physically connected or associated with each other in some manner. **2.** specifically, a localized population of microrganisms derived from a single cell in culture.

colony-forming unit a colony (see above) formed from a group of microorganisms rather than from one.

combinatorial joining the process by which a relatively small number of gene segments can be recombined to give a large number of full-length, functional genes. This is the mechanism by which diversity is generated in the variable region genes for both immunoglobulins and T cell receptors. The term is also applied to the independent assortment of full length heavy and light chains, such that the variable regions of any heavy chain and any light chain can associate to form a combining site.

cometabolism the biotransformation of a compound by a microorganism that is incapable of utilizing the compound as a source of energy or growth.

commercial sterility a condition that results from the processing (usually retort processing) of food to eliminate all pathogenic and spoilage microorganisms that can contribute to food spoilage under normal storage conditions. A commercially sterile product is not necessarily entirely sterile. The only viable microbes, if any, remaining in a commercially sterile product are extremely heat resistant bacterial spores, which can cause spoilage of products stored at unusually high storage temperatures.

common cold a widespread clinical syndrome of acute upper respiratory symptoms: sore throat, nasal discharge, nasal obstruction, sneezing, cough, resulting from infection by a number of respiratory viral pathogens (**common cold viruses**), most of which are in the genus *Rhinovirus.*

community a collection of microbial populations growing together in a defined habitat.

compatible solutes low-molecular-weight organic compounds that provide osmotic balance without interfering with the physiological activities of the cell; they enhance the ability of organisms to survive in environments where the extracellular solute concentration exceeds that of the cell cytoplasm.

competitive exclusion a term for the protective function of the normal flora of the gut to prevent entry and colonization of pathogens.

compost **1.** a material used in or produced by composting (see below). **2.** to carry out a process of composting.

composting a process in which organic compounds are biodegraded, biotransformed, or otherwise stabilized by mesophilic and thermophilic bacteria. The addition of other readily biodegradable materials, bulking agents, moisture and possibly nutrients to contaminated soils enhances or stimulates the composting process.

concatemer a single molecule of DNA consisting of more than two phage genomes joined head to tail.

confinement a term for procedures used to keep genetically-modified organisms within bounds or limits, usually in the environment with the result of preventing widespread dissemination.

confocal microscopy a contemporary technique of microsocopy that excludes light from parts other than the specimen or point of interest, thus eliminating the high influence of background which can seriously degrade the contrast and sharpness of the image in conventional microscopy.

conidium an asexual fungal spore formed from a generative cell known as a conidiophore.

conjugal involving or taking part in conjugation (see below).

conjugal pilus an extracellular filament encoded by a conjugative plasmid involved in establishing contact between plasmid-carrying donor cells and recipient cells.

conjugation the fact of being joined together; specifically, a process of DNA transfer between bacterial cells involving cell-to-cell contact, and requiring the functions of fertility factor genes in the donor cell.

conjugative or ***conjugational*** involving or taking part in conjugation (see above).

conjugative (conjugational) transfer see CONJUGATION.

conjugative transposon a DNA element that mediates both its own transposition and conjugal transfer from one bacterial cell to another.

consortia a spatial grouping of bacterial cells within a biofilm in which different species are physiologically coordinated with each other, often to produce phenomenally efficient chemical transformations.

constitutive relating to or describing continuously expressed or unregulated genes in the absence of an inducer.

constitutive enzyme an enzyme whose formation is not dependent upon the presence of a specific substrate.

constitutive expression gene expression from which RNA synthesis occurs continuously without the requirement of an inducing molecule or complex.

constitutive gene a gene whose transcription is dependent on RNA polymerase only, and no other regulatory factor.

constriction a mode of cell envelope invagination during division. During constriction all envelope layers move inwards simultaneously and the daughter cells move gradually apart.

constructed wetlands the intentional creation of wetland areas with aquatic plants and their root-associated microbes.

contagion the transmission of an illness from one individual to another by contact or by indirect transfer of germs.

contagionism the principle that disease is transmitted by direct or indirect contact between living organisms, especially humans, in the form of microscopic living entities, e.g., germs.

containment in this context, conditions or procedures that limit the dissemination of genetically modified organisms and their exposure to humans and the environment.

contigs a term for adjacent or partially overlapping clones.

continuous fever an elevated fever with a fluctuation of not more than 1°C. Contrasted with fevers that are remittent (temperatures fluctuating more than 1°C) or intermittent (temperatures occasionally returning to normal).

coprotease a protein that acts to stimulate the autoproteolytic activity of another protein.

correlated species species of fungi that differ principally in their life cycles, but that are presumed to be derived from a common ancestor because of similarities in morphology and host range.

corrin a contracted macrocyclic tetrapyrrole in which two of the pyrrole rings are connected directly at their α positions instead of by a methylene or methene bridge; the tetrapyrrole macrocycle of cofactor B_{12}.

corrinoid coenzyme a coenzyme group that participates in methyl transfer reactions of negatively charged methyl groups, as in the methylation of Hg(II).

cosmid a plasmid cloning vector containing bacteriophage lambda *cos* (cohesive end) sites incorporated into the plasmid DNA.

cosuppression a type of post-transcriptional gene silencing in which transcripts of both an endogenous gene and an homologous transgene are synthesized and then degraded.

cotransducible relating to or involved in cotransduction (see below).

cotransduction the delivery of two genetic markers to a recipient cell by a single phage; cotransducible chromosomal markers must normally be close enough to be included on a single piece of DNA small enough to fit in a phage head.

CPE see CYTOPATHIC EFFECTS.

Crabtree effect the suppression of yeast respiration by high levels of glucose. This phenomenon is found in *Saccharomyces cerevisiae* cells which continue to ferment irrespective of oxygen availability, due to glucose repressing or inactivating respiratory enzymes, or due to the inherent limited capacity of cells to respire.

creatinine an end-product of energy metabolism found in the blood in uniform concentration, excreted by the kidney at a constant rate. Alterations of this rate are taken as an indication of kidney malfunction.

Crenarchaeota one of two kingdoms of organisms of the domain *Archaea*, the other being *Euryarchaeota*. Includes sulfur-metabolizing, extreme thermophiles. (From Greek *crene-* "spring, fountain," for the resemblance of these organisms to the ancestor of the *Archaea*, and *archaios* "ancient.")

crenulations characteristic ruffles in the outer cell wall layers characteristic of spirochetes of the genus *Pillotina*.

cross inoculation the ability of bacterial strains from two or more different legumes to each produce nodules on the other's host(s).

cross protection the phenomenon in which plant tissues infected with one strain of a virus are protected from infection by other strains of the same virus; protection can thus be conferred by the use of a related mild strain.

cross-regulation the convergence of parallel signal transduction pathways permitting the stimulation of one pathway to affect the output of the parallel pathways. The phosphorylation of a receiver domain by multiple, independently regulated, transmitter domains is an example of this.

crown gall a tumorous plant disease induced by *Agrobacterium tumefaciens*.

cry genes genes found in *Bacillus thuringiensis* and a few other bacteria that encode proteins toxic to insects; the genes are frequently located on large plasmids that may be transmissible by conjugation between bacteria.

cryobiosis 1. the existence of life at very low temperatures. 2. the absence of metabolic activity in a still-extant organism due to freezing.

cryptobiosis hidden life; the absence of any evidence of metabolic activity in a dormant organism.

curli a class of thin, irregular, and highly aggregated adhesive surface fibers expressed by *E. coli* and *Salmonella enteritidis*.

cuticle a thin layer, principally composed of wax and cutin, external to the wall of plant epidermal cells.

Cyanobacteria a structurally diverse group of photosynthetic prokaryotes, some species of which can fix nitrogen as both free-living and symbiotic cells; found symbiotically with a broad range of lower plant taxa.

cyclogeny a theory, widely held of the 1920s and 1930s, asserting that bacteria in cultures go through stages of a life cycle in which the bacterial cells may undergo changes in shape, size, staining, and biochemical properties.

cyst in this context, a metabolically dormant vegetative *Azotobacter* cell that has differentiated to a form resistant to desiccation and damage by chemical agents.

cytochrome a hemoprotein in which one or more hemes is alternately oxidized and reduced in electron transfer processes.

cytocidal causing or capable of causing death to cells.

cytokine any of various members of a group of low molecular weight proteins secreted from cells of the immune system that help regulate immune responses by exerting a range of effects on other cells of the system; the term includes the monokines, lymphokines and interleukins.

cytokinesis in prokaryote organisms, the process of cell division. In eukaryotes, cell division also includes mitosis.

cytopathic causing or capable of causing disease in cells.

cytopathic effects (CPE) the existence of tissue deterioration caused by viruses. CPE is widely used for the identification of virus isolates in the diagnostic virology laboratory.

cytostome in dinoflagellates, the buccal aperture (mouth), permanently open but able to dilate. It allows the prey to be transported into the cell.

cytotoxic having a toxic effect on cells.

cytotoxic T cells T cells that kill other cells; generally these are CD8 T cells that recognize peptides generated in cytosol and that are presented on the cell surface by MHC class I molecules.

cytotoxin a substance or agent that damages cells.

D

darkfield microscope a microscope that has a device to scatter light from the illuminator so that the specimen appears white against a black background. It is widely used for direct examination of spirochete microorganisms.

dcw cluster the cluster of genes involved in division (*d*) and cell wall (*cw*) synthesis. In many prokaryotes this cluster is evolutionary conserved.

dead-end host a term for a host that can be infected by a pathogenic agent but that is not part of the cycle of transmission of the agent to other organisms.

decomposition the breakdown of a compound into simpler compounds, often by microorganisms.

dedicated pathway a term for the entire series of biosynthetic steps, beginning with the first committed step, that converts a common metabolite to a specific metabolite.

defense mechanism a physiological process that a host deploys in resisting pathogen invasion and colonization.

degassing a natural process by which Hg^0 is transported from soils and waters to the atmosphere due to its volatility.

degradative pathway specific enzyme-catalyzed chemical reactions for the metabolism of a particular substrate.

delayed-type hypersensitivity reaction a cell-mediated immune reaction in the skin in response to foreign antigens; it is indicative of prior sensitization to those antigens.

deletion the loss of a segment of DNA from the genome. Spontaneous deletions can range in size from several nucleotides to multiple contiguous genes. Deletions can also be generated by recombinant DNA methods.

delta endotoxin a protein toxin produced by the bacterium *Bacillus thuringiensis* that is lethal when ingested by insects of a specific order.

demethylation a microbial process which removes the methyl groups from volatile forms of selenium, such as dimethylselenide or methane selenol.

denitrification the biochemical reduction of nitrate or nitrite to nitrogen gas (N_2) and nitrous oxide (N_2O) during microbial respiration without oxygen present.

dental plaque microbial biofilms on teeth.

depth of field in microscopy, the slice of the specimen that contributes to the in-focus image at a particular focus level.

derived sugars sugars in which the aldehyde group is oxidized or reduced, or a hydroxyl group is substituted or derivatized.

dermatophytosis a condition resulting from infection of keratinized structures, including hair, nails, and stratum corneum of the skin, by three genera of fungi termed the **dermatophytes.**

dermis a thick layer of skin tissue below the epidermis, consisting of loose connective tissue and containing blood and lymph vessels, nerves, sweat and sebaceous glands, and hair follicles.

desorption the non-destructive elution and recovery of, for example, metal or metalloid species, radionuclides, and organometal(loid) compounds from loaded biological material by physicochemical treatment(s).

detergent a molecule with polar and nonpolar portions that can disrupt membranes by stabilizing the dispersion of hydrophobic lipids and proteins in water.

deteriogenic describing organisms causing or capable of causing biodeterioration.

detoxification a reduction in the hazardous nature of a compound.

dextrose equivalent (DE) an indication of total reducing sugars as glucose percentage. Unhydrolyzed starch has a DE of zero and glucose has a DE of 100.

diastereoselectivity the ability of a catalyst or reagent to distinguish between stereoisomers other than enantiomers.

diazotroph an organism capable of biological nitrogen fixation.Thus, **diazotrophic.**

diazotrophy the ability of a microorganism to carry out biological nitrogen fixation.

diffraction the spatial distribution of waves that are scattered from an object and recombine interfering with one another.

dimorphism dual form; the ability of a fungus to grow as a yeast or a filament; sometimes also used to describe fungi that produce two kinds of motile spores.

din genes genes regulated as part of the DNA inducible network of gene regulation; a term usually only applied to genes of unknown function; when function is identified, the "*din*" designation is then replaced by a more appropriate function-related name.

dinoflagellates a phylum (Dinoflagellata) of unicellular eukaryotic microorganisms among the Protoctista, showing great diversity and widely distributed throughout all seas and freshwaters around the world. They play a prominent role in the trophic chains.

dinokaryon the nucleus of dinoflagellates, characterized by the presence of a permanent nuclear envelope and chromosomes quasi-permanently condensed during the entire cell cycle.

dinomitosis mitosis of the dinokaryon (see above), characterized by the presence of an extranuclear mitotic spindle crossing the nucleus via cytoplasmic channels.

Disability-Adjusted Life Year (DALY) a measurement of life span using as the basic unit one year of healthy life, as opposed to life while afflicted with a disability.

disease association a circumstance existing when the pathogenicity of a microorganism has not been established by rigorous standards that eliminate all other possible causes of disease, but the possibility that the microorganism causes the disease is strongly supported by circumstantial evidence, such as a constant association with diseased, but not healthy, hosts both in nature and following experimental transmission of the disease, and by remission of symptoms following treatment of diseased hosts with antibiotics.

disease cycle the chain of events involved in disease development, including the stages of development of the pathogen and the effects of the disease on the host.

disinfect to carry out a process of disinfection; destroy or otherwise inactivate pathogenic microorganisms.

disinfection any of various chemical or physical processes used to destroy or otherwise inactivate pathogenic microorganisms.

disproportionation a mechanism whereby two molecules of an intermediate redox status interact, resulting in the production of a more reduced and a more oxidized molecule, as in the case of 2Hg(I) producing Hg^0 and Hg(II).

dissimilation see DISSIMILATORY REDUCTION.

dissimilatory reduction the reduction of a compound in an energy-yielding reaction. The reduced product does not become incorporated into cellular material.

dissociation the phenomenon by which bacteria with different properties, usually related to virulence, arise from a homogeneous population of bacteria. Now attributed to gene mutations.

division the largest phylogenetic grouping within a domain. This grouping can be considered analogous to the kingdom and includes one or more phyla.

divisome in prokaryote cell division, the macromolecular complex that carries out division at the cell center.

DNA deoxyribonucleic acid, the substance that constitutes the genetic material of cellular organisms and certain viruses.

DNA-directed RNA polymerase an enzyme that synthesizes RNA using a DNA template.

DNA hybridization the process by which two complimentary strands of DNA bind to each other in a mixture of DNA strands.

DNA microarray a miniaturized array containing thousands of DNA fragments representing genes in an area of a few square centimeters. These "chips" are used mostly to monitor the expression of genomes at the level of messenger RNA.

DNA microinjection a technique for introducing nucleic acids into cells such as the oocytes of *Xenopus*, mammalian cells in culture or fertilized mammalian eggs via a small bore glass needle using a micromanipulator.

DNA polymerase an enzyme that catalyzes the synthesis of DNA from a DNA template.

DNA recombination see RECOMBINATION.

DNA supercoiling a phenomenon occurring in circular, covalently closed, duplex DNA molecules when the number of turns differs from the number found in DNA molecules of the same length but containing an end that can rotate. Supercoiling creates strain in closed DNA molecules. A deficiency of duplex turns generates **negative supercoiling**; a surplus generates positive **supercoiling.**

DNA topoisomerase see TOPOISOMERASE.

DNA unwinding element (DUE) a region of double-stranded DNA where the helix is easily destabilized to achieve a single-stranded state. The DUE is associated with replication origins and is rich in A-T base pairings.

domain **1.** the largest phylogenetically coherent grouping of organisms. The three domains of living organisms are defined as the Bacteria, the Archaea, and the Eukarya. **2.** a discrete amino acid sequence in a protein that can be equated with a particular function.

donor see ELECTRON DONOR.

downstream processing a series of unit operations or processes to recover, isolate and purify the desired component from a fermentation mixture.

droplet nuclei dried residua of larger respiratory droplets which may carry infectious microorganisms. Droplet nuclei of clinical importance are in the 1 to 5 μm diameter range.

dysentery an inflammatory disease of the intestines characterized by severe abdominal cramps, rectal urgency and pain during stool passage, and the presence of blood, pus and mucus in stool; the various causative agents of this condition include bacteria and protozoa.

E

EBV Epstein–Barr virus, a member of the herpesvirus family. EBV is associated with various forms of cancer.

EC effective concentration; the concentration of a drug that will produce a 50% effectiveness rate; e.g., in virus yield.

eclipse in virology, a term for the period between the entry of a virus into a cell when it loses it infectivity as an independent particle to the appearance inside the cell of fully infectious progeny virus particles.

E. coli see *ESCHERICHIA COLI.*

ecosystem **1.** generally, a definable community of organisms within a localized environment, described in terms of its interactions with this environment. **2.** in terms of microorganisms, the totality of biotic and abiotic interactions to which the organism is exposed.

ecotone an ecological interface; a zone of transition between two habitats.

ectopic expression the artificial expression of a gene under conditions where it would not normally be expressed.

ectotrophic obtaining nutrients from an external or surface source; e.g., certain parasitic plant fungi located on the surface of a root.

effectiveness in pathology, the reduction in incidence of a disease due to routinely applying an intervention to a population. The level of effectiveness is often less than the *efficacy* (see below).

efficacy in pathology, the maximum possible reduction in incidence of a disease attributable to an intervention designed to prevent and control the disease.

effluent **1.** flowing out or away. **2.** liquid waste resulting from sewage treatment or industrial processes, especially such matter released into waterways.

electrochemical gradient the sum of the electrical gradient, or membrane potential ($\Delta\varphi$), and ion concentration gradient across a membrane (the latter is often defined as ΔpH for protons).

electron acceptor a low molecular weight inorganic or organic species (compound or ion) that is reduced in the final step of an electron transfer process.

electron donor a low molecular weight inorganic or organic species (compound or ion) that is oxidized in the first step of an electron transfer process.

electron energy loss spectroscopy (EELS) a technique by which the change in energy and wavelength of a transmitted electron can be detected and related to an element for compositional analysis.

electron spectroscopic imaging (ESI) a technique similar to EELS (see above), but in which the electron energy lines are separated and used for imaging to produce a TEM (transmission electron microscopy) image of the distribution of a desired element.

electron transport the transfer of electrons from a donor molecule (or ion) to an acceptor molecule (or ion) via a series of components (i.e., a respiratory chain), each capable of undergoing alternate oxidation and reduction. The electron transfer can either be energetically downhill, in which case it is often called respiration, or energetically uphill, when it is called reversed electron transfer.

electron transport coupled phosphorylation the process by which ATP (adenosine triphosphate) is formed from ADP (adenosine diphosphate) and inorganic phosphate as electrons are transferred from electron donors to electron acceptors.

electroporation the transfer of DNA into a cell by the application of an electrical field.

elementary bodies (EB) a term for infectious, extracellular chlamydial forms that are metabolically inert and unable to grow and divide.

ELISA enzyme-linked immunosorbent assay, a highly sensitive serological technique that is widely used to detect and measure the levels of various specific antibodies and antigens.

embryonic stem cells (ES cells) pluripotent or totipotent cells of normal karyotype derived from the developing mammalian embryo. When introduced into other embryos of the same species, ES cells will contribute to the developing embryo in such a way that the resulting animal will be a chimera of two cell types. If the germ cells consist of two populations of cells, whole animals derived from the original ES cells can be obtained by conventional breeding.

emerging in public health, describing a condition or phenomenon that has increased in incidence in the recent past and appears likely to continue to increase in the near future; especially, such a condition that previously was unknown or severely restricted in range. Thus, **emerging disease**, **emerging infection**, and so on.

emerging virus a virus that has recently increased its incidence and appears likely to continue increasing. Generally, these are viruses that have newly appeared in the human population or are rapidly expanding their range, with a corresponding increase in cases of disease. In recent years, a number of viral diseases have been identified for the first time. Some, such as AIDS, have made their debut alarmingly and dramatically. Others, such as influenza, have long been known for their tendency to reappear periodically to cause major epidemics or pandemics.

enablement or ***enabling disclosure*** in patent law, a statutory requirement for patentability, mandating that the written description of an invention which is included in a patent specification shall enable a person who is "skilled in the art" to make and use the invention. (The underlying principle is that an invention which cannot be produced by anyone other than the inventor would not require patent protection, since no one else could duplicate it.)

enantiomer one of a pair of organic chemicals that are structural mirror images of each other.

enantioselective reaction a chemical reaction or synthesis that produces the two enantiomers of a chiral product in unequal amounts.

enantioselectivity the ability of a catalyst or reagent to distinguish between two mirror-image isomers.

encephalitis inflammation of the brain.

endemic 1. describing a disease or condition that occurs constantly over an extended period of time in a given population or geographic area. 2. describing a pathogen that is native to the environment in which its host occurs.

endocytosis in plant pathology, a process releasing rhizobia from the infection thread but surrounding them with plant-derived membrane material limiting host defense responses.

endodyogeny a process by which tachyzoites and bradyzoites divide involving creation of two daughter cells wholly within a mother cell.

endolithotrophic describing organisms living inside rocks, as some fungi.

endonucleases enzymes that have specific binding sites of varying complexity at which they sever phosphodiester bonds of DNA.

endoplasmic reticulum (ER) a system of cytoplasmic membrane compartments involved in transport of proteins and lipids.

endosomes small, bag-like structures in the cytoplasm of cells that are involved in the uptake of material from the exterior of the cell.

endospore a metabolically quiescent cell that is resistant to desiccation, ultraviolet light, and other environmental insults; formed by certain genera of bacteria including *Bacillus* and *Clostridium*.

endosymbiont an organism living inside a member of a different species.

endosymbiosis the condition of an organism living inside a member of a different species.

endotoxins molecules contained within the bacterial cell, but that when released by lysis of the cell can exert toxic effects on the host cells by a variety of biological mechanisms.

endotrophic obtaining nutrients from an internal source; e.g., certain parasitic plant fungi living inside the plant root.

energy dispersive X-ray spectroscopy (EDS) a technique in which X-rays that are emitted from the surface of a specimen are detected and separated into their unique energy lines so that compositional analysis can be performed.

energy spilling a term for the process microorganisms use to dissipate excess energy.

enhancer a region of DNA that, although incapable of directing transcription itself, can enhance the ability of a promoter to do so.

enteric found in or affecting the intestine.

enteric microorganism a microorganism that replicates in the intestinal tract of humans and other warm-blooded animals and is shed in fecal material.

enteritis inflammation of the intestine, especially the lining of the small intestine.

Enterobacteriaceae a family of the Gamma-Proteobacteria that includes *Escherichia coli* and related gram-negative bacteria.

enterohemorrhagic causing intestinal hemorrhage, especially hemorrhagic colitis.

enteropathogenic acting as an intestinal pathogen.

enterotoxic or ***enterotoxigenic*** describing an organism or substance that is harmful to the intestine.

enterotoxin an organism or substance substance able to induce damage to the intestine, generally leading to fluid accumulation.

enterovirus one of a group of viruses infecting the intestinal tract.

environmental scanning electron microscopy (ESEM) a contemporary form of scanning electron microscopy that allows the specimen to be viewed at partial pressure and humidity so that its natural topography can be seen.

enzyme a protein molecule that catalyzes specific chemical reactions by reducing the energy of activation.

enzyme-linked immunosorbent assay see ELISA.

epidemic 1. the incidence of a disease in a given population or geographic area at a greater rate (typically far greater) than would be expected based on past trends. 2. describing a disease prevalent and spreading rapidly in a given population or area.

epidemiology the scientific study of factors causing or affecting the frequency and distribution of disease in a given community. Thus, **epidemiologist**.

epidermis the outer layer of the skin, consisting of dividing and maturing epidermal cells with an outermost layer of dead, keratinized cells called the stratum corneum.

epigenetic describing gene activity that produces phenotypic but not genotypic change.

epi-illumination in microscopy, illumination produced when the specimen is illuminated from above by light emerging from the objective lens.

epilimnion the surface layer above the metalimnion in a lake or other body of water that thermally stratifies.

epimastigote the multiplying life cycle stage of trypanosomes found in insect vectors.

epimural describing bacteria attached to the epithelial cells lining the gut.

episome an extrachromosomal element in cells which can sometimes become associated with the chromosome.

epistatic describing differential genetic effect on phenotype in which a gene suppresses the activity of another, nonallelic gene.

epitope an individual small region of a molecule to which an antibody can bind.

Epstein–Barr virus see EBV.

equipartition the allocation of one from each pair of newly replicated chromosomes into each progeny cell.

eradication in public health, the complete absence of transmission of an infectious disease and the disappearance of the virus, bacterium, or parasite that caused that infection. The only significant infectious disease now described as eradicated is smallpox, which was so declared in 1980.

ergot a disease of cereals and grasses caused by the fungus *Claviceps purpurea.*

ergotism poisoning in humans and animals brought about by ingestion of ergot sclerotia.

eructation the process by which ruminants let fermentation gases escape from the rumen.

erysipelas an acute superficial form of cellulitis involving the dermal lymphatics, usually caused by group A streptococci, and characterized by a bright red, edematous, spreading process with a raised, indurated border.

ES cells see EMBRYONIC STEM CELLS.

eschar a small area of tissue destruction on the skin surrounding a bite of an insect vector.

Escherichia coli a gram-negative, facultative anaerobic, non-spore-forming, motile rod-shaped bacteria. The species belongs to the Enterobacteriaceae family of the Gamma Proteobacteria and includes certain strains that are common innocuous residents of the intestine of mammals, and others cause human and animal infections of the digestive and urinary tracts, blood, and central nervous system. The structure, biochemical functions, and genetics of this organism have been extremely well studied. (First isolated in 1884 by Theodor *Escherich,* German physician.)

Eubacteria another name for the domain *Bacteria,* which is now the preferred term (see BACTERIA).

Eukarya or ***Eucarya*** one of the three domains of life, the others being the Archaea and Bacteria. (From Greek *eu-* "good; true," and *karion* "nut," referring to the presence of a true nucleus in such organisms.)

eukaryote or ***eucaryote*** one of the Eukarya; an organism other than the Archaea and Bacteria.

eukaryotic generally, describing a cell in which DNA is housed within a nuclear envelope, or an organism having this cellular feature.

Euryarchaeota one of two kingdoms within the domain *Archaea,* the other being *Crenarchaeota.* Includes halophiles, methanogens, and some anaerobic, sulfur-metabolizing, extreme thermophiles. (From Greek *eurys* "broad, wide," for the relatively broad patterns of metabolism of these organisms, and *archaios* "ancient.")

eutroph an organism that grows well when the presence of nutrients is ample or high.

evaporite a residue of salts left behind after brines evaporate to dryness.

evolution change in the genetic properties of populations and species over generations, which requires the origin of variation (by mutation and mixis) as well as the subsequent spread or extinction of variants (by natural selection and genetic drift).

excision the removal of damaged nucleotides from incised nucleic acids.

exobiology a term originally used to refer to life that is not indigenous to the Earth, or extraterrestrial biology. Use of the term has now expanded to refer to studies dealing with the origin of terrestrial life, investigations to determine the possibility of life, or its precursor chemistry of non-terrestrial origin and related technical issues.

exons sequences that flank introns in precursor RNA molecules prior to splicing, and that are joined together in the mature RNA.

exonuclease an enzyme that degrades DNA from a terminus.

exotoxin a soluble protein produced by a microorganism, which can enter a host cell and damage the host cell.

extracellular matrix a complex network of proteins and polysaccharides secreted by eukaryotic cells. It functions as a structural element in tissues in addition to modulating tissue development and physiology.

extreme acidophile an organism with a pH optimum for growth at, or below, pH 3.

extremophile literally, extreme lover; an organism that is isolated from an extreme environment and that often requires this extreme condition for growth. *Extreme* in this sense is a subjective term referring to a condition markedly different from the optimum conditions for other, more well studied organisms; e.g., conditions of temperature or pressure for mammals.

extremophilic requiring, or growing optimally in, extreme environmental conditions; e.g., extremes of temperature or pressure.

exudates materials exuding from cells; e.g., compounds of low molecular weight that leak from plant cells either into the intercellular spaces or directly from the epidermal cell walls into the root environment.

F

facultative able to grow in the presence of a given environmental factor, but not restricted only to the presence of that factor; e.g., bacteria that can grow in the presence or absence of oxygen.

facultative autotroph a versatile organism that is capable of utilizing either inorganic (CO_2) or organic carbon for biosynthetic metabolism.

fatty acids long chain aliphatic carboxylic acids without (saturated) or with one (unsaturated) or more (polyunsaturated) double ($-CH=CH-$) bonds. Normally found esterified to glycerol (acylglycerols), but can be esterified to hydroxy compounds (waxes) or sterols.

fed batch processing in bioreactor technology, a process in which liquid media is fed to the reactor continuously.

feedback control a control mechanism in which the product of a transcription unit can negatively control its own expression.

feedback inhibition the inhibition of enzyme activity by a product of a reaction or a pathway. The first enzyme of a pathway is usually the only target of such control.

fermentation **1**. an anaerobic type of metabolism in which organic compounds are degraded in the absence of external electron acceptors. The organism produces a mixture of oxidized and reduced compounds. **2**. classically, the process of converting sugar into alcohol. **3**. more generally, any metabolic process whereby microbes gain energy from the breakdown and assimilation of organic and inorganic nutrients. **4**. in biotechnology, the process of culturing cells or other microorganisms in a container, bioreactor, or other vessel for experimental or commercial purposes.

fermentation balance the sum of the oxidized and the reduced compounds arising during a process of fermentation; this must equal the oxidation state of the substrate, which is calculated in arbitrary units.

fermenter or ***fermentor*** a large growth vessel used to culture microorganisms on a large scale for the production of commercially valuable products.

ferredoxin a low molecular weight protein containing one or more iron-sulfur clusters that functions to transfer electrons from one enzyme system to another.

ferritin (FTN) a large intracellular iron storage protein present in both prokaryotic and eukaryotic cells.

field resistance see HORIZONTAL RESISTANCE.

fimbriae hairlike proteinaceous appendages (100 to 1000 per cell) 2 to 8 nm in diameter, on the surface of bacteria. Often involved in adhesion to specific types of eukaryotic host cells. Also, **pili**.

FISH see FLUORESCENCE IN SITU HYBRIDIZATION.

fission a mode of vegetative reproduction found (for example) in the yeast genus *Schizosaccharomyces*. Fission yeasts grow lengthways and divide by forming a cell septum that contricts mother cells into two equal-sized daughters.

fitness the ability of an organism to pass on its genes to the next generation; the average reproductive success of a genotype in a particular environment, usually expressed relative to another genotype in that environment.

fixed nitrogen chemically reduced or biologically converted nitrogen in a chemical form that can be assimilated readily by most microbes or plants.

fixed specimen a mode of preservation intended to minimize swelling, shrinkage, evaporation, decay and all forms of distortion of cellular components. Such fixation is usually accomplished by chemical action to stabilize or fix the cell contents so that they may be subjected to further treatment that would otherwise damage them.

fixed virus a laboratory strain that grows in tissue culture at a predictable rate, with predictable biological properties, and that will regularly kill inoculated animals within a predictable period.

flagella *singular,* **flagellum.** a thin, flexible, fiber-like structure (5–10 per cell, about 15–20 nm in diameter) originating at random points on the surface of bacteria, which can rotate and provide the cell with the ability to move toward nutrients and away from toxic substances. Flagella are driven by an internal cellular structure, the "motor," which is powered by a proton gradient across the cell membrane.

flagellar having to do with or taking place by means of flagella.

flagellar basal body the major structure of the flagellar motor, consisting of ring structures and a rod.

flagellar motor a molecular machine that converts the energy of proton flow into the rotational force.

flagellum see FLAGELLA.

fluoresce to produce or exhibit fluorescence (see below).

fluorescence the property of a molecule that will absorb light and after a delay of less than one-millionth of a second re-emit light of a longer wavelength (lower energy).

fluorescence in situ hybridization (FISH) a microscopy technique used to locate specific genetic sites within an intact cell. DNA probes tagged with a fluorescent protein are hybridized to fixed cells and the position of the glowing spot is determined relative to intracellular structures or other genetic sites.

fluorescent having to do with, displaying, or employing the property of fluorescence (see above).

fluorescent microscope a microscope that uses an ultraviolet light source to illuminate specimens that will fluoresce.

fluorophore a molecule that will fluoresce when excited by light of a certain wavelength.

fomite a term for an inanimate object that harbors pathogenic microorganisms and that may serve as a vehicle for transmission of infection.

foodborne carried by food; resulting from the ingestion of a certain product in food.

food-borne microbe a living organism less than 0.01 mm in diameter (mainly bacteria, yeasts, and molds) that can grow in foods and beverages and cause undesirable changes (food spoilage and foodborne diseases) or desirable changes (food preservation and fermentation).

foregut fermentation microbial fermentation prior to gastric or peptic digestion.

forespore the cell compartment of a sporangium destined to become a spore.

fouling a term for biodeterioration caused by the presence of living organisms, without significant structural damage to the material; in the case of microorganisms, it usually consists of aesthetic changes that create a generally unacceptable appearance.

founder in gene technology, a transgenic animal resulting directly from the manipulation of a mammalian egg involving gene transfer.

frameshift mutation the deletion or addition of a small number of nucleotides (but not three or a multiple of three) in an open reading frame that shifts the translational register out of frame, causing the production of a protein containing an altered amino acid sequence downstream of the shift.

Frankia symbiotic, nitrogen-fixing actinomycetes that form root nodules on diverse shrubs and trees which are not legumes.

freeze-etching/fracturing a technique wherein cells are rapidly frozen, fractured while frozen, etched under vacuum, and a platinum-carbon replica is made. Especially useful for viewing the insides of membranes.

fruiting body a macroscopic, multicellular structure produced by myxobacteria in response to environmental stress.

functional genomics the study of gene function at the genome level.

fungal having to do with or belonging to the Fungi.

Fungi a kingdom of life forms that are eukaryotic, mycelial or yeast-like, heterotrophic, lacking in chlorophyll, sexually and/or asexually reproductive, and mostly aerobic.

fungicide a substance or agent used to kill fungi or to inhibit fungal growth.

fungus one of the Fungi (see above).

furuncle a painful erythematous nodule (boil) caused by *Staphylococcus aureus* and formed by localized inflammation of the dermis and subcutaneous tissue surrounding hair follicles.

fusogen a chemical that induces the fusion of protoplasts, resulting in genetic recombination.

G

gamete a sex-differentiated (male or female) reproductive cell.

gametocyte 1. an undifferentiated cell that subsequently develops into a gamete. 2. specifically, in this context, an erythrocytic form of the malaria parasite that has differentiated into either a male (**microgametocyte**) or female (**macrogametocyte**) cell, appearing after several rounds of blood-stage infection. Upon ingestion by the mosquito, the gametocytes mature into gametes, ready for fusion.

gas gangrene an acute, severe infection that results from the infection of a wound by certain anaerobic bacteria, particularly of the genus *Clostridium*.

gastritis inflammation of the stomach.

gastroenteritis inflammation of the intestinal wall, often as the result of exposure to a foodborne microbe or toxin.

gene the fundamental physical unit of heredity. The majority of genes are segments of DNA that determine the sequence of amino acids in specific polypeptides. Some others are regulatory genes that presumably control expression of the structural genes.

gene disruption the physical disruption of the integrity of a gene by insertion mutagenesis or by deletion, usually of an internal segment of the gene. Gene disruptions are commonly constructed by recombinant DNA methods to study gene function.

gene encyclopedia or ***library*** an array of cloned DNA fragments, ordered in a sequence reconstructing the gene order on the corresponding chromosome.

gene-for-gene relationship the relationship of many specialized plant pathogens with their hosts, in which there is a correspondence of each gene determining resistance in the host variety with a related gene determining virulence in the pathogen. A concept proposed by U.S. plant pathologist H. Flor, to account for the observation that such species continue to exist side by side in nature; if either the host evolved to unopposed resistance or the pathogen evolved to unopposed virulence, then the other species would no longer survive. Also known as **gene-for-gene concept, hypothesis, interaction**, etc.

gene fusion genetic manipulation to fuse two genes together. **Transcriptional fusions** are done in such a way that expression of the first gene could be monitored by the phenotype of the second. **Translational fusions** lead to the in-frame fusion of two coding sequences. The latter were used to establish the topological disposition of integral membrane proteins.

generalized transduction the transfer of random fragments of chromosomal DNA from one cell to another cell via a bacteriophage.

general secretion pathway a pathway by which proteins containing signal peptides or membrane signal-anchor domains are integrated into, or transported across, the cytoplasmic membrane of bacteria.

genes see GENE.

genetically modified organism (GMO) an organism that acquires heritable traits not found in the parent organism. While traditional scientific techniques such as mutation can result in a GMO, the term is most frequently used to refer to modified plants, animals, and microorganisms that result from deliberate insertion, deletion, or other manipulation of DNA. Also, **genetically engineered organism.**

genetic code the linear sequence of bases in DNA that ultimately determines the sequence of the amino acids into proteins.

genetic drift changes in gene frequency caused by the random sampling of genes during transmission across generations (rather than by natural selection).

genetic engineering a method by which the genomes of plants, animals, and microorganisms can be manipulated; this includes but is not limited to recombinant DNA technology.

genetic linkage an association of genes in inheritance due to their localization on the same chromosome or other nucleic acid molecule.

genetic nomenclature the designation of genes according to an abbreviated letter code that is italicized, for example *fabH* and *plsB*. The corresponding protein products of these genes are abbreviated as FabH and PlsB.

genetic recombination see RECOMBINATION.

genetics the scientific study of genes (hereditary units) and of the implications of gene activity in organisms.

genetic transduction see TRANSDUCTION.

genetic transformation see TRANSFORMATION.

genome the complete genetic material of an organism.

genome replication see REPLICATION.

genomics the study of the genome, or all of the genes of an organism.

genomic species a group of organisms that share a high (~70%) DNA-to-DNA similarity value.

genotype the particular genetic makeup of a cell, organism, or population.

germfree (GF) free of all known microorganisms, though not necessarily free of virus for all species.

gingival relating to or affecting the gums.

gingival margin the soft tissue (gums) next to the tooth.

glass state a highly viscous liquid that appears to be solid. Biopesticides dried in the presence of various carbohydrates can be stabilized when encapsulated in the glass state formed by the carbohydrate.

gliding motility spontaneous movement on solid surfaces in the absence of flagella.

global burden of disease the collective impact of disease on all nations of the modern world.

globalization in public health, the phenomenon that the processes of food importation, international trade, travel, tourism, etc., significantly affect the ability of a sovereign state to protect the health of its citizenry.

globin a hemoprotein that reacts with dioxygen, generally reversibly and without reduction of the released oxygen.

glucose oxidase (GOD) an enzyme that catalyzes the process of glucose oxidation. It is usually suitable for industrial use because of its high stability. One of the typical biological sensing elements; used in the first biosensor.

glycogenesis the synthesis of central metabolic 6-carbon sugars for biosynthetic precursors during growth on carbon sources other than these sugars.

glycolysis the process of sugar catabolism; commonly used as a synonym for the most widely known glycolytic pathway, the Embden–Meyerhof–Parnas path.

glycolytic pathways general central pathways for the metabolism of sugars for production of energy and precursors for biosynthesis.

glycoproteins a class of sugar proteins present as monomolecular arrays on surfaces of archaea and bacteria.

glycosylase an enzyme that hydrolyzes *N*-glycosyl bonds linking purines and pyrimidines to carbohydrate components of nucleic acids.

glycosylation the bonding of sugar residue to another organic compound.

glycosyl transferase an enzyme catalyzing the transfer of a sugar residue, generally from its activated nucleotide sugar derivative, to an acceptor molecule.

gnotobiote an environment or organism that has been specifically monitored or analyzed so that its entire complement of microorganisms has been identified.

gnotobiotic (GN) indicating the presence of well defined microbial forms.

gold standard a term for the best available approximation of the truth, generally indicating a test method currently accepted as reasonably, but not necessarily 100%, accurate.

Gram stain an iodine-gentian violet complex that is retained by some bacteria (called **gram-positive**) but is released from the envelope by acetone or ethanol from another group of bacteria (called **gram-negative**). Whereas gram-positive bacteria have a multilayered shell of peptidoglycan, gram-negative bacteria have a thin (monolayered) peptidoglycan and a second membrane, the outer membrane. (From *Gram's method,* the identification technique of staining bacteria in this manner; developed by Hans Christian Joachin *Gram,* Danish bacteriologist.)

granulocytopenia acquired or chemically induced immunosuppression caused by low blood white cells.

granulosis a type of enveloped, double-stranded DNA virus reported only from insects of the order Lepidoptera (moths, butterflies) that replicate initially in the nucleus and later in the cytoplasm of infected cells; virions are occluded individually in small occlusion bodies called granules.

GRAS Generally Regarded As Safe; an official designation for a microorganism, food additive, or the like not known to be pathogenic and thus rated as acceptable to be used in food production or preservation.

greenhouse gases long-lived gases such as CO_2, CH_4, and N_2O, which absorb infrared radiation in the lower atmosphere, leading to the **greenhouse effect**, a possible elevation in global surface temperatures.

green nonsulfur photosynthetic bacteria a group of filamentous anoxygenic photosynthetic bacteria that use the 3-hydroxypropionate pathway of CO_2 fixation.

green sulfur photosynthetic bacteria a metabolically limited group of anoxygenic photosynthetic bacteria that are obligately phototrophic and strictly anaerobic. Sulfide is the most common electron donor and CO_2 is assimilated via the reductive tricarboxylic acid cycle.

growth medium an aqueous solution containing all the nutrients necessary for microbial growth.

growth optimum the particular state of a variable environmental factor (e.g., temperature or salinity) at which an organism grows fastest.

Guillain–Barré syndrome an ascending paralysis associated with *Campylobacter jejuni.*

gyrase an enzyme that introduces supercoiling into the DNA duplex in an ATP-dependent reaction.

H

HAART highly active antiretroviral therapy; a combination of therapies, typically one protease inhibitor and two reverse transcriptase inhibitors, that have powerful inhibitory effects against viral reproduction.

habitat a location where a given organism occurs, especially a usual or characteristic location for this organism.

hairy root a tumorous plant disease provoked by *Agrobacterium rhizogenes*.

haloalkaliphile an organism that grows optimally in an alkaline setting (pH in excess of pH 8) and at high salt concentrations (above 15% w/v NaCl).

halophile literally, salt-lover; an organism that grows optimally at high salt concentrations.

halophilic growing optimally in a saline environment (above 15% w/v NaCl in the growth medium).

hantavirus pulmonary syndrome a pneumonia-like illness resulting from infection with **hantavirus**, a virus normally carried in rodents.

hapten a small, monovalent molecule that may be bound by specific antibody, but which will not form a precipitating complex. Hapten molecules are also capable of inhibiting precipitate formation when added to a mixture of antibody and bivalent antigen.

haptoglobin (Hp) a serum protein that scavenges hemoglobin.

hartig net fungal hyphae that weave between plant root cortical cells.

haustorium *plural,* **haustoria.** a structure formed by many obligate fungal pathogens upon penetration of the host cell wall. The haustorium forms within an invagination of the host plasma membrane and absorbs nutrients from the host.

hazardous waste any material or substance that poses a significant threat to human health or environment, requiring special handling, processing, or disposa.

headful packaging see PACKAGE.

heat shock response a rapid change in gene expression that occurs when there is a temperature shift to an elevated temperature.

heavy metals an ill-defined group of biologically essential and non-essential metallic elements, generally of density >5, exhibiting diverse physical, chemical, and biological properties with the potential to exert toxic effects on microorganisms and other life forms.

helicase an enzyme that unwinds duplex DNA and requires ATP (adenosine triphosphate).

Helicobacter pylori common bacterium residing in the human stomach, affecting a large portion of the world population and now associated with various gastric disturbances, including ulcer disease; also classified as a carcinogen.

helper T cells T cells that help B cells respond to antigen, or activate macrophages to become microbicidal; generally, these are CD4 T cells that recognize peptides generated in the vesicles of the cell and presented on the cell surface by MHC class II molecules.

hemagglutinins specific glycoprotein molecules on the surface of some viruses which have the property of binding to the surface of the red blood cells of some animal species. Because there are multiple binding sites, one virus can bind to two red cells causing them to clump (agglutinate).

heme (Hm) a metal ion, customarily iron in the FeII state, chelated in a tetrapyrrole ring.

hemolytic uremic syndrome a triad of renal failure, thrombocytopenia, and hemolytic anemia frequently associated with Shiga toxin-producing *E. coli*.

hemopexin a serum protein that strongly binds heme.

hemophore (Hbp) a secreted bacterial protein that shuttles environmental heme to the bacterial surface.

hemorrhagic colitis bloody diarrhea caused by enterohemorrhagic (EHEC) *Escherichia coli*.

hemorrhagic fever an infection manifested by acute onset of fever and hemorrhagic signs; typical of many zoonotic viruses.

hepatitis inflammatory reaction in the liver, with infiltration of leukocytes generally leading to liver cell damage or destruction.

hepatocellular carcinoma a cancer originating in the liver, probably from neoplastic transformation of hepatocytes.

hepatocyte a major cell type of the liver, comprising ~70% of liver cells; a specific target of infection by viruses that cause hepatitis.

hepatoxicity a toxic condition of the liver.

herbicide a substance or agent used to kill plants, especially weeds.

heritability the proportion of the variation among individuals in a phenotype that is attributable to differences in genotype and not to environmental effects.

herpes labialis infection of the lips with **herpesvirus**, commonly called cold sores or fever blisters.

heteroduplex DNA duplex DNA in which the two strands have different genetic origins. A heteroduplex may contain mismatched bases.

heteroecious requiring two different host species for completion of the life cycle.

heteroencapsidation the partial or complete encapsidation of the RNA of one virus in the coat protein of another isolate or distinct virus. The virion may have a mosaic of two different coat proteins.

heterofermentation **1.** a process of fermentation with more than a single end product. **2.** specifically, fermentation producing lactic acid, together with other organic acids, alcohols, aldehydes, ketones, and carbon dioxide. Thus, **heterofermentative.**

heterogenesis a nineteenth-century term used to mean the origin of living things from organic materials, e.g., from infusions of plant or animal matter, or from decaying tissue in a diseased or dying organism. Heterogenesis was thought to be the source of tumors, parasitic worms, and microorganisms from putrefaction, as these organized themselves from the smallest microscopically visible particles ("molecules") of organic tissues. Supporters of heterogenesis might not support the more extreme position of archebiosis/abiogenesis (life arising from inorganic matter).

heterokaryon genetically dissimilar nuclei found in a common cytoplasm, as in certain fungal cells.

heterologous derived from a different source or species; not native to the host.

heterothallism a condition of fungi that produce compatible male and female gametes on two physiologically distinct mycelia. In smuts and bunts, two anatomically similar sporidia, each possessing a different mating types, fuse to form the dikaryotic infection hyphae which is the parasitic stage of the fungus in the host plan.

heterotroph literally, feeder on others; an organism dependent upon organic material from an external source to provide carbon for growth; contrasted with autotrophs (self feeders) that can utilize inorganic materials for this purpose. Thus, **heterotrophic.**

heterotrophy the fact of being a heterotroph; obtaining carbon from some external organic source.

hexaflora a stable, six-member microflora often introduced to normalize specific anomalies of germfree animals such as the enlarged cecum. They most often consist of *Lactobacillus brevis, Streptococcus fecalis, Staphylococcus epidermidis, Bacteroides fragilis.* var. *vulgatus, Enterobacter aerogenes,* and a *Fusibacterium* sp.

high-efficiency particle air (HEPA) filters a designation for air filters that remove 99.97% of particles 0.3 μm or greater in diameter. HEPA filters are commonly used in industry to remove airborne particulates, and in laboratories and health care facilities to sterilize air.

HIV human immunodeficiency virus, an RNA retrovirus that has been identified as the causative agent of AIDS. It is related to a series of simian immunodeficiency viruses (SIVs).

homeostasis a state of steadiness or balance; the maintenance by an organism or cell of appropriate levels of fluid content, temperature, or the like. Homeostasis can be disturbed by various inputs from the external environment.

homofermentatative describing or referring to microorganisms that produce lactic acid as the sole product of carbohydrate metabolism.

homologous **1.** very similar or the same, as in form, structure, position, appearance, etc. **2.** specifically, having a shared common ancestry with another.

homologous recombination a process by which two nearly identical genetic elements are brought together for the purpose of combining them or separating them.

homolog or ***homologue*** **1.** an individual organism, part, etc., that is the same as or very similar to another. **2.** an individual having the same ancestry as another, often determined by structural comparison.

homology the fact of being homologous; being the same as or very similar to another.

homology-dependent viral resistance a form of post-transcriptional gene silencing in which viral RNAs are degraded in transgenic plants expressing RNAs homologous to viral RNAs, resulting in inhibition of viral replication and attenuation of virus symptoms.

hops the dried flowers of the female *Humulus lupulus* plant that contribute flavor and antibacterial compounds to beer.

horizontal gene transfer (HGT) the introduction of new genetic material into a cell's genome as a result of uptake from an outside source such as by mating (conjugation), transduction, or transformation; i.e., not simply by mutation and vertical inheritance into vegetative descendants. Usually used to refer to gene transfer between somewhat unrelated species.

horizontal gene transmission the reappearance of genetic material received by HGT in the offspring of the original recipient.

horizontally mobile element (HME) a collection of genetic material or genes reproducing by horizontal reproduction and that may also reproduce by vertical reproduction.

horizontal reproduction an increase in the amount of genetic material (e.g., DNA), genes, or sets of genes (e.g., viruses and plasmids) by transfer between organisms rather than through organismal reproduction (vertical reproduction).

horizontal resistance resistance conferred by the combined action of several genes; can vary from low to high susceptibility.

horizontal transmission see HORIZONTAL GENE TRANSMISSION.

host 1. generally, any living organism that provides the site where another organism of a different species resides and obtains nutrition. 2. an organism in which a pathogen can replicate and potentially cause disease. 3. an organism or cell culture in which a virus or group of viruses may replicate.

host range 1. the number and nature of organisms in which a given virus or group of viruses replicate. 2. the particular group of species in which genes transferred from another particular organism can replicate.

human immunodeficiency virus see HIV.

human rhinovirus (HRV) see RHINOVIRUS.

humus the organic fraction of soil, characteristically black-brown in color; organic by-products consisting of microbial cell walls and other resistant molecules formed from free-radical reactions of sugars, amino acids, and products of lignin decomposition.

hurdles technology in the temperature control of microbial activity, a calculated use of combinations of different preservation factors or techniques ("hurdles") in order to achieve multi-target, mild but reliable preservation effects.

hybridization the mating of two genetically distinct individuals.

hybridoma a somatic cell hybrid formed between a normal B cell and a B cell tumor called a myeloma. The normal B cell provides the gentic information to produce a particular antibody, while the myeloma cell provides immortality. These characteristics combine to yield a cell that can be grown long-term in tissue culture, or as a tumor in animals. In each case, the result is a system which produces large amounts of homogeneous antibody.

hydrocarbons organic compounds present in crude oils, natural gases and fuel oil products, composed primarily of carbon and hydrogen atoms.

hydrogenase an enzyme capable of redox activation of molecular hydrogen.

hydrogenosomes cytoplasmic organelles present in anaerobic protozoa and fungi, containing enzymes that produce hydrogen from the oxidation of reduced cofactors.

hydrogenotrophic methanogen A methanogen that generates methane by the reduction of CO_2 with H_2.

hydroperoxidase an enzyme whose major function is the decomposition of peroxides and that is involved in the oxidation of Hg^0 to Hg(II) by some bacteria.

hydrophobic literally, water-fearing; describing molecules or regions of molecules that cannot form hydrogen bonds or other polar interactions with water.

hydrophobic effect the tendency of hydrophobic regions of molecules to avoid contact with water.

hydrophobin a small cysteine-rich protein found associated with the aerial hyphae of filamentous fungi.

hyperacidophile see ACIDOPHILE.

hyperbilirubinemia abnormally high levels of the liver compound bilirubin in the blood; associated with the use of ribavirin, an antiviral agent.

hypermutation see SOMATIC MUTATION.

hyperosmotic relating to or causing an increase in osmotic pressure or osmolality.

hyperpiezophile an organism whose maximum rate of growth over all possible growth temperatures occurs at a pressure greater than 50 MPa. To date, only **hyperpiezopsychrophiles** have been isolated (organisms growing maximally at high pressure and low temperature, 0–20°C).

hyperplasia 1. excessive growth; an abnormal increase in the size of an organ or tissue due to the proliferation of component cells. 2. excessive abnormal growth in a plant due increased cell division.

hypersensitive reaction or ***response (HR)*** a localized and rapid physiological reaction in non-host plants in response to bacterial, fungal, or viral plant pathogens or their elicitors. The manifestation of the HR is due to the death of the plant cells that come in contact with the pathogen or the elicitor.

hypersensitivity see HYPERSENSITIVE REACTION.

hypertension abnormally elevated blood pressure.

hyperthermophile literally, heat-lover; an organism having an optimal temperature for growth of 80°C or higher.

hyperthermophilic relating to or describing organisms that grow optimally at temperatures above 80°C.

hypertrophy excessive, abnormal size; e.g., in a plant due to cell enlargement.

hypha *plural*, **hyphae.** the thread-like tubular cell that constitutes the basic growth unit of filamentous fungi.

hyphomycetes a grouping of asexually reproducing fungi ("fungi imperfecti") in which the asexual spores (conidia) are produced on loose, cottony hyphae.

hyphopodium *plural*, **hyphopodia.** in certain fungi, an attachment and infection structure that develops from short hyphal branches.

hypnozoite a resting life stage of certain parasites.

hypocalcemia an abnormally low level of calcium in the blood.

hypolimnion the bottom layer below the metalimnion in a lake or other body of water that thermally stratifies.

hypoplasia 1. reduced growth; an abnormal decrease in the size of an organ or tissue due to a reduction in the number of component cells. 2. abnormal reduction in cell production in a plant due to disease.

hyporheic zone the ecotone between a streambed and the groundwater, characterized by a mixture of surface water and groundwater.

hypotension abnormally low blood pressure.

hypovirulence the fact of having low virulence.

hypovirulent describing a pathogen with low virulence; e.g., fungi that are reduced in virulence because of infection with a transmissible virus, and may therefore be used for biological control of the pathogen.

hysteresis the ability of viscous solutions to recover instantaneously to initial viscosity values after removal of shear stress forces.

I

icosahedral symmetry a type of cubic symmetry found in isometric viruses. The basic form is 12 pentameric subunits forming the vertices of the particle and giving twofold, threefold, and fivefold rotational symmetry axes. For larger icosahedra, the number of subunits is increased in a regular manner by the addition of hexameric subunits in defined numbers and places.

image contrast in microscopy, the ability to distinguish various components of the structure of the object by different intensity levels in the image.

immobilization the entrapment of a biocatalyst in an inert three-dimensional matrix, or the attachment of a biocatalyst onto a solid surface, by chemical or physical means.

immobilized cell a cell attached to a solid support over which substrate is passed and is converted into product.

immobilized cell bioreactor a device in which microbial cells are entrapped in polymers or covalently bonded to or adsorbed onto inert materials in the bioreactor, which is operated in a continuous flow mode.

immortalization the capacity of a cell in tissue culture to grow *ad infinitum*. Normally somatic cells have a finite lifespan at the end of which the cells senesce. Immortalization requires that the cell overide the mechanisms that normally cause senescence.

immunity the ability to prevent infection, which results in no development of disease and no detectable level of pathogen multiplication, i.e., total resistance.

immunoassay the use of antibodies in the detection or assay of compounds of interest.

immunocompetence the ability or capacity to develop an immune response (i.e., antibody production and/or cell-mediated immunity) following antigenic challenge. Thus, **immunocompetent.**

immunocompromised having lost the capacity to produce an immune response, because of either a disease (i.e., AIDS) or immunosuppressive therapy (i.e., cancer treatment).

immunodominant epitope the major antigenic determinant of a particular antigen in its native state, stimulating the strongest antigen-specific immune response of the immunized animal.

immunofluorescence a technique for staining organisms or cells for microscopic examination in which a specific antibody molecule is chemically linked to a fluorescent dye to render the sites of antibody binding to cell or microbe structures visible when illuminated with light of the appropriate wavelength (**direct immunofluorescence**), or in which the binding of the specific antibody is detected by the reaction of a fluorescent-linked antibody directed against the first antibody molecule (**indirect immunofluorescence**).

immunogen a molecule that is capable of eliciting an immune response when introduced into an immunologically naive animal.

immunogenicity the capacity to elicit adaptive immunity.

immunosuppressant an agent capable of suppressing immune responses.

impaired immunity a defect in host defense; a lack of immune response.

impetigo a contagious pyoderma caused by direct inoculation of group A streptococci or *Staphylococcus aureus* into superficial abrasions of the skin. The lesion is confined to the epidermis and initially consists of a fragile vesico-pustule with an erythematous halo progressing to a yellow-brown crust.

inapparent infection the presence of infection in a host without recognizable clinical signs or symptoms. It can only be identified through laboratory tests.

incidence a measure of the rate of new cases of a disease; the number of people in a given population developing a specified disease or becoming infected within a specified period of time.

incision an endonucleolytic break in damaged nucleic acids.

inclusion the unique intracellular, membrane-enclosed organelle that supports the growth of chlamydia.

incubation the period of time between exposure to a pathogen and the time at which symptoms begin.

index of refraction see REFRACTIVE INDEX.

indicator something used as the sign of a larger trend; for example: 1. a microorganism that on detection gives a quantitative representation of a group of organisms, often of contamination or pathogens in foods or drinking water. 2. a plant that responds with conspicuous symptoms to infection with a particular virus. 3. a target organism sensitive to bacteriocin.

indigenous native or natural in a given site or region; not exotic.

indigenous flora microorganisms that constitute the normal or resident flora of a system, especially those that colonize animals or humans without causing known disease.

induced enzyme an enzyme that is produced by an organism in response to the presence of its substrate or a related substance (the **inducer**).

inducible describing genes that are transcribed when an appropriate signal is present.

inducible expression the increased transcription of specific genes in the presence of an appropriate molecule or complex.

infection the successful establishment of a pathogen in a host; the presence in a human, animal, or plant of bacteria, protozoa, viruses, or other disease-causing microorganisms (though not necessarily the manisfestation of the disease).

infection thread a plant-derived tube through which rhizobia move down the root hair, or between cells in the root cortex.

infectious agent any bacterium, virus, protozoan, prion, or other such form that can cause infection or disease in a human, animal, or other host.

infectious dose the number of microbes needed to cause symptoms.

infectious waste any material capable of producing an infectious disease; a subset of medical waste.

infusion a suspension resulting from boiling material in water; for example, tea.

initiation factor a protein factor that promotes the correct association of ribosomes with mRNA; required for the initiation step of protein biosynthesis.

innate immunity the early stage of an immune response, involving cells such as natural killer cells, polymorphonuclear granulocytes and macrophages, and factors such as complement, in which recognition of specific antigen is not involved.

inoculation 1. generally, any deliberate transfer of material from one source into another to bring about some altered condition. 2. the introduction into the skin of a mild form of a disease agent, e.g., pustular or scab material containing smallpox virus, in order to induce immunity to the virulent form of the disease. 3. the transfer of microorganism to a sterile medium in order to initiate a laboratory culture. 4. the application of artificially-cultured rhizobia to legume seed or soil in order to improve nodulation.

inoculative biocontrol see CLASSICAL BIOCONTROL.

insertional mutagenesis introduction of an in-frame addition of one or more codons, resulting in altered local or global properties of the protein. Used to disrupt and thereby identify gene functions.

insertion mutation a mutation caused by the insertion of a segment of DNA into another segment of DNA. Spontaneous insertions are often caused by the transposition of IS elements or transposons. Insertions can be generated in the laboratory by transposons or by the use of recombinant DNA.

insertion sequence element (IS element) a discrete nucleotide sequence that is involved in the transfer and integration of pieces of DNA to various foreign sites.

in silico literally, in silicon; describing biological studies carried out with computer simulations.

in situ in the site; in the natural or normal location.

integrase a viral enzyme mediating the integration of viral DNA into host DNA.

integrated pest management a practice for managing insect pest populations in which chemical, biological, and cultural techniques are integrated to achieve effective, environmentally safe pest control; emphasis is on the reduction or elimination of synthetic chemical insecticides.

integrating plasmid a plasmid that cannot replicate autonomously in a host bacterium, but can establish itself by integration into the chromosome through recombination between homologous plasmid and chromosomal sequences.

interspecies transfer the mutually beneficial transfer of H_2 from microorganisms producing H_2 to microorganisms comsuming H_2 in an anaerobic ecosystem.

interesterification the interchange of esters between two triglycerides.

interference contest in patent law, an administrative proceeding conducted by the U.S. Patent and Trademark Office to determine who is entitled to a patent on an invention when more than one inventor (or group of inventors) claim the same invention.

interferons specific cellular proteins synthesized in response to virus infection which are secreted outside the infected cell and which render neighboring uninfected cells resistant to virus infection.

intermittent fever a condition of fever with temperature returning to normal one or more times during a 24-hour period.

intra-abdominal sepsis a disease process in humans that occurs following spillage of colonic contents within the peritoneal cavity.

intramuscular in this context, referring to the injection of a drug directly into the muscle for fast absorption.

intrinsic incubation In the vertebrate animal, the time from infection of the animal until viremia occurs at a level sufficient to infect an arthropod.

intrinsic parameters a term for naturally occurring attributes of foods which may influence food spoilage, preservation and fermentation: pH, oxidation/reduction potential, moisture contents, nutrient contents, antimicrobial agents, and biological structures.

introns intervening sequences in the coding portions of genes that are transcribed into RNA and are removed during RNA splicing.

inundative biocontrol a process in which large quantities of biocontrol agents are applied to a target weed in a similar manner to chemical herbicides. Mainly indigenous fungi (mycoherbicides) have been used in this manner for control of weeds in crops.

invasin an adhesin that can mediate bacterial invasion into host eukaryotic cells.

inversion the physical breakage and rejoining of a segment of DNA that places the segment in opposite orientation relative to its normal orientation.

inverted repeat a sequence that defines both ends of a transposable element and that is found as an inverted duplication.

in vitro literally, in glass; describing studies carried out in the test tube, or, more broadly, in any setting outside the living organism.

in vivo in life; describing studies carried out in the living organism.

ionophore one of a class of antibiotics that alter ruminal fermentation end-products.

IS element see INSERTION SEQUENCE ELEMENT.

isoreceptors eukaryotic cell membrane components which contain identical receptor determinants recognized by a bacterial adhesin.

K

keratin an epidermal cell protein that makes up the hair, nails, and stratum corneum of the skin.

keratitis inflammation of the surface of the eye.

knockout mice genetically engineered mice lacking the ability to express a specific protein because the gene of interest was ablated ("knocked out").

Koch's postulates 1. fundamental criteria for investigation of the role of a specific microbe in the causation of a specific disease, established in a landmark paper (1882) on the etiology of tuberculosis, by the German physician and bacteriologist Robert *Koch* (1843–1910). 2. more generally, a set of fundamental questions needed to address the function of a microbial population within a given habitat.

Köhler illumination illumination used with light sources of irregular form or brightness, hence with most conventional optical microscopes. The image of the light filament is made large enough to fill the aperture iris and is focused on the condenser lens, which is focused so that the image of the aperture iris is in focus with the specimen. (Introduced in 1893 by August *Köhler*.)

L

lactation in mammals, the production and secretion of milk from the mammary glands, or the period following parturition when this process occurs.

lactic acid an organic acid that is produced by the reduction of pyruvic acid, e.g., through the anaerobic respiration of microorganisms.

lactic acid bacteria a diverse group of bacteria, composed chiefly of genera that produce lactic acid as the primary metabolic endproduct from the fermentation of carbohydrates; members of the genera *Lactococcus*, *Lactobacillus*, *Leuconostoc*, *Pediococcus*, and *Enterococcus*.

lactoferrin (LF) a glycoprotein present in the mucosal secretions and phagocytic cells of vertebrates that binds and transports iron.

land treatment the accelerated biodegradation of contaminated media through application to surface soils, to enhance aeration and in some cases to allow for nutrient amendment.

late blight the disease of potatoes caused by *Phytopthora infestans* that led to the disastrous Irish Potato Famine of the mid-19th century. The term also applies to infections of *P. infestans* on tomatoes.

latent infection a symptomless or inapparent infection that is chronic and in which a certain host-parasite relationship is established.

latent virus a virus that has entered into a non-replicative mode of existence in a cell but is propagated along with the cell by certain mechanisms which allow limited viral genome replication without full expression of virus functions and production of mature infectious virus particles.

latex a colorless or milky sap that exudes from wounds of some plants, such as milkweed.

laticifers cells or a series of cells forming ducts containing latex and occurring in some vascular plants. Elements of the laticiferous system are found in the phloem of some plant species.

lawn a term for an extended growth of bacterial cells covering a surface.

lentic describing still or standing waters, as in a lake or pond.

leprosy a wasting disease in humans characterized by lesions in the extremities; the causative agent is the bacillus *Mycobacterium leprae*.

lethality in pathology, a measure of the potency of a disease, defined as the percentage of people who die of a disease among the total of people sickened by that disease.

leukocytosis an abnormal increase in circulating white blood cells.

leukopenia an abnormal decrease in circulating white blood cells.

life cycle the sum of successive stages in the growth and development of an organism that occur between the appearance and reappearance of the same stage (e.g., spore) of the organism.

ligase an enzyme that covalently joins strands of DNA.

lignin a highly stable polymer of mostly methoxylated phenyl-propanoic residues, synthesized as part of the cell wall of vascular plants; the second most abundant organic polymer on earth after cellulose.

lignocellulose a collective term for plants (trees) whose main components are cellulose, lignin, and heme-celluloses. The proportions between these three main components vary considerably in different plants (trees).

line a term for a family of animals derived from a founder animal and possessing the same germline modification.

linear plasmid a mobile genetic element that exists as a linear molecule, found in many *Streptomyces* spp.

lipid a molecule containing aliphatic or aromatic hydrocarbon. These can be categorized as **polar lipids** (with charged group, e.g. phospholipids, glycolipids, sulfolipids) and **neutral lipids** (with no charged group, e.g. acylglycerols, hydrocarbons, carotenoids, sterols) or as **storage lipids** (acylglycerols, wax esters) that are laid down as reserves of depot fat and source of energy when required, and **structural lipids** (phospholipids, lipoglycans) that structure the membranes of cells.

lipid A an acylated and phosphorylated di- or monosaccharide that forms the hydrophobic part of LPS (lipopolysaccharide).

lipid bilayer the fundamental structural basis of biological membranes, in which two layers of lipids are apposed, with hydrophilic groups of each facing the aqueous medium and hydrophobic fatty acyl chains forming the interior.

lipooligosaccharide (LOS) a form of LPS (lipopolysaccharide) produced by many mucosal pathogens, including members of the genera *Neisseria, Haemophilus, Bordetella,* and others.

lipopolysaccharide (LPS) an amphiphilic glycolipid found exclusively in gram-negative bacteria. LPS forms the outer leaflet of the outer membrane in the majority of gram-negative bacteria.

lipoprotein one of a class of proteins carrying at their amino terminus a cysteine to which glycerol is linked that has its hydroxyl groups substituted by fatty acids.

lithotroph literally, stone eater; an organism deriving its energy from the oxidation of inorganic compounds; i.e., an autotroph Thus, **lithotrophic.**

lithotrophy the fact of being a lithotroph; deriving energy from the oxidation of inorganic compounds.

local lesion a symptom found in a plant leaf inoculated with a virus, caused by the death of, or changes in, cells around the original point of entry.

long-pass filter an optical filter that blocks wavelengths shorter than a characteristic cutoff wavelength and transmits longer wavelengths.

long terminal repeat (LTR) an element found at either end of the retroviral genome that contains the promoter/enhancer element driving viral transcription.

lotic having to do with swiftly flowing waters, as in a stream or river.

luciferase the generic name for enzymes that catalyze bioluminescent reactions. Luciferases from different major groups of organisms are not homologous; thus firefly and jellyfish luciferases are unrelated to bacterial luciferase, so the organism must be specified in referring to a specific luciferase.

luciferin literally, light bearing; the generic name for a substrate that is oxidized to give light in a bioluminescent reaction, identified as a flavin in bacteria and a tetrapyrrole in dinoflagellates.

lymphangitis inflammation of lymphatic vessels. Acute lymphangitis is evidenced by painful subcutaneous red streaks along the course of lymphatics.

lyophilization the process of simultaneously freezing and drying materials under a vacuum.

lysis the breaking down or destruction of a cell, with the release of its contents.

lysogen a bacterial or other cell carrying a viral prophage and having the potential of producing and releasing bacteriophage under proper conditions.

lysogenic relating to or having the property of lysogeny; capable of producing a full burst of progeny virus.

lysogenic conversion a change in bacterial phenotype due to either phage infection or stable establishment (lysogeny) of a bacteriophage in the genome, whereby new (bacteriophage) genes are expressed, such as genes for toxin production or a new cell surface antigen.

lysogeny the heritable, stable potentiality of bacteria to produce and release bacteriophages due to the presence of a prophage (viral genome) within the bacterial cell.

lysozyme β-1,4-*N*-acetylmuramidase that hydrolyzes the β-1,4-glycosidic bond between *N*-acetylmuramic acid and *N*-acetylglucosamine in peptidoglycan. It is found in many bacteria, where it is involved in the growth processes of the cell wall. It is also present in various tissues and secretions of higher organisms (e.g., hen egg white lysozyme), where it functions as a powerful antibacterial agent.

lytic having to do with or involved in lysis (cell disruption).

lytic phase a mode of virus interaction with cells in which the virus undergoes a complete replicative cycle with the production of many progeny virus particles and subsequent release of these virus upon the disruption (lysis) of the cell.

M

macrobe a (usually) eukaryotic organism, either single-celled or multicellular, that is visible to the unaided eye.

macrocycle the overall ring structure formed by four pyrrole rings linked one to another through their α positions, usually via methylene or methene bridges.

macrolactone a lactone ring that typically contains 14 or more atoms.

macromolecule a (relatively) large molecule, often containing thousands of atoms.

macrophage literally, large eater; a large cell that ingests and destroys particles (e.g., bacteria) by an engulfing process known as phagocytosis; macrophages are an active component of the immune response.

macrophagy the process of feeding on or ingesting relatively large amounts of material.

macular rash a skin discoloration characterized by circular or oval spots which are not raised above the skin surface.

maculopapular rash a skin rash characterized by slightly elevated, colored spots.

Magic Bullet the name given to Paul Ehrlich's Salvarsan, the first specific antibiotic, this term being chosen to represent the hope that use of a specific drug to kill the specific bacterial cause of syphilis would control the disease.

maintenance therapy drug treatment given for an extended time to maintain its effect after the condition has been controlled or to prevent recurrence.

major histocompatibility complex (MHC) a gene complex found in all mammals, the products of which are intimately involved in antigen presentation to T cells. The principal components of the MHC are divided into MHC class I and MHC class II proteins. These are recognized by T cells bearing two different coreceptor proteins, called CD8 and CD4 respectively. MHC class I and MHC class II molecules are encoded in genes that are more polymorphic than any others in the mammalian genome.

malt the major raw material used in brewing that provides the appropriate substrate and enzymes needed to yield wort.

MALT mucosa-associated lymphoid tissue. A "germ-free" stomach does not have MALT, but MALT is frequently acquired in the course of *H. pylori* gastritis. The presence of gastric MALT can give rise to malignant B-cell lymphomas.

mass transfer the movement of a given amount of fluid or the material carried by the fluid. In bioreactors, it refers to the dispersal and solubilization of nutrients and gases in the fermentation medium.

master genes a term for flagellar genes regulating the expression of all the other flagellar genes, sitting at the top of the hierarchy of flagellar regulons.

mating type 1. the characteristic of an oomycete that determines its sexual response to another organism. 2. generally, any subpopulation of organisms that is able to mate only with another subpopulation of the same species that is genetically distinct.

maturases proteins sometimes encoded within self-splicing introns, that assist in the splicing reaction.

mechanical transmission the transmission of a virus to a plant by mechanical contact. The transmission rate is greatly increased when sap from the infected plant is rubbed on the leaves of the not yet infected one.

mechanism of action the specific reaction that an exotoxin catalyzes to covalently modify a host cell component, such as the ADP-ribosylation of elongation factor-2 by diphtheria toxin. Each exotoxin catalyzes the modification of a host cell component by a specific mechanism of action.

medical waste a general term for material generated as a result of patient care, diagnosis, treatment, or immunization. It may or may not be hazardous.

megaplasmid an extremely large ($\geq$500 kb) extrachromosomal circle of DNA present in some cells in addition to the chromosome.

melanins high molecular weight polymeric indole quinone pigments having brown or black colors, occurring predominately in vertebrates and insects, and occasionally in microorganisms.

membrane bioreactor a biological treatment reactor containing a selective material that acts as a barrier to prevent the transport of solids and allows transport of organic compounds and/or water.

membrane-permeant weak acid a weak acid, usually an organic acid, that can permeate the cell membrane in the hydrophobic protonated form, then dissociate in the cytoplasm, producing hydronium ions and depressing intracellular pH.

membrane-permeant weak base a weak base, usually an organic base, that can permeate the cell membrane in the hydrophobic unprotonated form, then become protonated in the cytoplasm, producing hydroxyl ions and increasing intracellular pH.

membrane potential the pH and electrical gradients generated by the electron transport chains.

memory in immunology, the ability of the immune system to respond faster and more vigorously to the second (or third) exposure to an antigen, compared to its response to the first, or primary, exposure. This phenomenon is due to the generation of populations of **memory cells** at the time of primary immunization.

meningococcemia the infection of the blood stream by meningococci in the absence of meningitis.

meningococcus *plural*, **meningococci.** an organism of the bacterial species *Neisseria meningitidis*, a major causative agent of meningitis.

meninigitis inflammation of the meninges (membranes covering the brain and the spinal cord), resulting from a bacterial or viral infection.

meromixis a condition in which stratification of the water mass in a lake is maintained during the whole year, often due to a solute concentration gradient.

mer operon a structural organization of genes that encode the various functions needed for the reduction of Hg (II) and dissimilation of organomercury compounds. Often carried on plasmids and transposons.

merozoite an invasive form of the malaria parasite; first released from liver cells after the maturation of the hepatic schizonts, and then from erythrocytes in the cycles of blood-stage infection.

mesophile literally, middle lover; an organism that grows optimally at temperatures considered to be in a moderate range; i.e., between 20 and 50°C.

mesophilic growing optimally at temperatures between 20 and 50°C.

meso selectivity the ability of a catalyst or reagent to distinguish between two chemically identical but stereochemically different parts of a single molecule that contains chiral centers and a symmetry plane.

metabolic relating to, involved in, or produced by metabolism.

metabolic engineering the technique of manipulating metabolic processes through the use of recombinant DNA technology, with the goal of improving the properties or activity of a cell.

metabolic flux the rate at which material is converted via metabolic reactions and pathways.

metabolic genotype the complete set of metabolic genes identified in a genome. The utilization of the metabolic genotype under a given condition characterizes the **metabolic phenotype.**

metabolism a collective term for any and all chemical and physical processes occurring within a living organism, especially those that are involved in maintaining normal life status and function.

metabolite any substance produced by metabolism.

metacyclic trypomastigote the life cycle stage of American and African trypanosomes that develops in insect vectors and that is infective for humans and other mammals.

metacyclogenesis the developmental and biochemical transformation, occurring within the sand fly midgut, of the noninfectious procyclic promastigote into the highly infectious metacyclic promastigote.

metalimnion the water layer in a thermally stratified lake between the epilimnion layer above and the hypolimnion layer below, characterized by steep vertical gradients of water temperature and dissolved oxygen.

metalloprotease an enzyme that breaks down other proteins by cleavage of the peptide bonds, and in which heavy metal ions are bound directly to some of the amino acid side chains.

metallothionein a low-molecular-weight cysteine-rich protein capable of binding essential metals (e.g., Cu and Zn), as well as non-essential metals (e.g., Cd).

metal resistance the ability of a microorganism to survive toxic effects of heavy metal exposure by means of a detoxification mechanism usually produced in response to the metal species concerned.

metal tolerance the ability of a microorganism to survive toxic effects of heavy metal exposure because of intrinsic properties and/or environmental modification of toxicity.

methane monooxygenase the key enzyme for the oxidation of methane to methanol in bacteria.

methanofuran a coenzyme that serves as a carbon carrier during the initial reduction of CO_2 in the methanogenic pathway.

methanogen a microorganism that produces methane as its metabolic end-product. Methanogens constitute a large group within the *Archaea* domain. They are strict anaerobes that utilize and are limited to growth on a relatively narrow range of simple one-carbon compounds and acetate.

methanogenesis the process of methane formation by methanogens.

methanogenic producing methane as the primary metabolic end-product.

methanotroph literally, methane eater; an aerobic bacterium that uses methane as the sole source of carbon and energy. Thus, **methanotrophic.**

methanotrophy the fact or process of utilizing methane as the sole carbon and energy source, found in certain bacteria.

methylation the biological mechanism by which inorganic forms of selenium are reduced to the Se (-II) state and receive methyl groups from a carrier protein, resulting in the formation of volatiles such as dimethylselenide.

methyl-CoM reductase the primary enzyme responsible for methane formation in all methanogens.

methylotroph an organism that grows aerobically with one-carbon compounds as its carbon and energy source.

methylotrophic capable of utilizing a variety of simple methylated compounds as sole source of carbon and energy.

methylotrophy the ability of microorganisms to utilize one-carbon compounds more reduced than CO_2 as sole energy sources and to assimilate carbon into cell biomass at the oxidation level of formaldehyde. Methylotrophic organisms must synthesize all cellular constituents from methylotrophic compounds such as methane, methanol, methylated amines, halogenated methanes and methylated sulfur species. A diverse range of both aerobic and anaerobic prokaryotes and eukaryotes can utilize methanotrophic substrates for growth.

methyltransferase an enzyme methylating particular residues in a macromolecule.

MHC see MAJOR HISTOCOMPATIBILITY COMPLEX.

miasma 1. literally, foul or bad air arising from decaying animal or vegetable matter. 2. a historic concept that disease could be caused by the emanation of such "bad airs."

microbe 1. a generic term for microorganisms; living organisms, prokaryotic and eukaryotic, too small to be seen by the unaided eye. 2. in nontechnical use, those microorganisms that cause disease in humans and others.

microbial relating to or involving microbes.

microbial ecology the scientific study of the distribution, abundance, and interactions of microorganisms in nature, and the impact of microorganisms on their environment.

microbial infallibility a term for the premise that any substance can be biodegraded under the appropriate conditions by the appropriate microorganisms.

microbial mat a sediment ecosystem with very high population densities of microorganisms. Often the top millimeters comprise a clearly laminated structure: the top layer is dominated by oxygen-producing phototrophs, especially cyanobacteria; the anoxic bottom layer is rich in sulfide as a result of the degradation of organic matter by fermenting and sulfate-reducing bacteria, and phototrophic and/or chemotrophic bacteria sandwich in between the top and bottom layers, forming a fine lamina at the oxygen-sulfide interface. Microbial mats represent some of the oldest ecosystems on Earth and their presence in the fossil record accounts for some of the most ancient evidence of life on the planet.

microbicide a chemical additive used to restrict the growth of microorganisms.

microbiology the study of organisms so small that their visualization requires a microscope rather than the unaided eye; e.g., bacteria.

microcolony a microscopic aggregate of bacterial cells.

microcosm 1. any highly limited system or environment that can be used as a model for the larger whole of which it is a part. 2. specifically, a highly controlled laboratory-scale apparatus used to model the fate or transport of compounds under the conditions of the natural environment.

micronemes small, thin, anterior organelles that secrete their contents on a parasite's surface at the start of the invasion process.

micropore a tiny pore; e.g., a small, single invagination anterior to the nucleus with an apparent clathrin coat presumed but not yet demonstrated to be involved in phagocytosis.

microsomes small bodies derived from the protein synthesis apparatus of the cell when the cell is rather violently disrupted, for example by sonic shock waves or grinding with abrasives.

mineralization the complete degradation of an organic compound to simple inorganic constituents, such as carbon dioxide, water, ammonia, chloride and sulfate.

Mip macrophage infectivity potentiator, a *Legionella* protein with activity as a *cis-trans* peptidyl prolyl isomerase that is needed for full expression of bacterial virulence.

mistranslation a misreading of the genetic code on the ribosome due to improper codon–anticodon pairing.

mixis the production of a new genotype by recombination of genes from two sources.

mixotroph literally, mixed eater; an organism able to assimilate organic compounds as carbon sources while using inorganic compounds as electron donors. Thus, **mixotrophic**.

mixotrophy the fact of obtaining nutrients through a combination of organic and inorganic sources.

mobilizable plasmid a conjugal plasmid that carries an origin of transfer (*oriT*) but lacks genes coding for its own transfer across the bacterial envelope.

molecular relating to or occurring at the level of the molecule.

molecular biology the study of biological phenomena at the level of the molecule, including techniques with which genes can be purified, sequenced, changed, and introduced into cells. It provides an integrated experimental approach to problems in genetics, biochemistry, and prokaryotic and eukaryotic cell biology.

molecule 1. the smallest entity of a substance or system retaining all the essential chemical and physical properties of that substance or system. **2.** historically, a term used from the writings of Buffon (c.1750) onward to indicate the tiniest (bacteria-sized) microscopically visible particles of organic matter in blood, tissue, infusions, and so on. (Taken from a Latin phrase meaning "a tiny mass.")

monoclonal antibodies homogeneous antibodies produced (usually in cell cultures) by specialized white blood cells derived from a single cell (a clonal population) and hence of a single molecular form.

monocyte a leukocyte subset derived from bone marrow precursor cells which subsequently migrate into the blood, where they are identified as monocytes, and then to tissue sites where they may differentiate to become macrophages.

monomorphism literally, single form; one component of a 19th-century debate over the morphology of bacteria. It held that bacteria are of a fixed form and that the various different forms indicate distinct bacterial species. The contrasting position was pleomorphism, which held that bacteria did not exist in distinct, stable species but rather that a given organism could have various forms depending on environmental conditions.

monophyletic referring to a group of organisms descending from a common ancestor having the significant traits that define the group.

monotherapy treatment with a single drug, contrasted with combination therapies using more that one drug at the same time.

morbidity rate the incidence of nonfatal cases of a given disease in the total population at risk during a specified period of time.

morbus gallicus literally, the French disease; the most common name given to a new disease that appeared in Europe in the late 15th century; many physicians and historians now identify morbus gallicus as syphilis.

morphometrics analysis of structure by measurement of the physical features of cells and organisms; Also, **morphometric analysis.**

mortality rate the incidence of fatal cases of a given disease in a particular population during a specified period of time.

mosaic a founder transgenic animal that possesses the desired germ line modification in only a subset of its cells. Such animals are thought to arise following integration of the transgene after the first cell division of the embryo and may be incapable of transmitting the transgene to their progeny by virtue of an absence of the modification in their germ cells.

mother cell the compartment of the sporangium, which engulfs the forespore, synthesizes spore coat proteins and lyses when the mature endospore is formed.

motile capable of spontaneous movement.

motility the capacity for spontaneous movement; i.e., movement not induced by wind, water currents, or other phenomena of the surrounding environment.

mucigel gelatinous material on a root surface, derived from plant sources (mucilages), as well as microbial cells, soil colloids, and soil organic materials.

mucilage gelatinous organic materials released by a plant in the root cap region, derived from the Golgi apparatus, polysaccharide hydrolysis, and epidermal materials.

mucocutaneous referring to an area of both exterior skin and mucus membranes, as at the borders of the mouth.

multihazardous waste waste materials that are infectious and chemically hazardous and/or radioactive.

muramic acid a compound in bacterial cell walls, consisting of lactic acid and glucosamine joined by an ether linkage.

murein peptidoglycan, an outer layer of the bacterial cell wall; a biopolymer crosslinked by peptide bridges that are linked to the lactyl group of the muramic acid residues.

murein sacculus the bacterial exoskeleton that forms a bag-shaped macromolecule completely enclosing the cell. It endows the cell with mechanical strength and confers the specific shape to the bacterium.

must in winemaking, expressed grape juice or the mashed grape before and during fermentation

mutagen a physical or chemical agent that induces mutations.

mutagenesis the generation of mutations; any alteration or aberration in the native sequence of DNA that results in permanently inheritable changes in the organism.

mutant an organism that carries a mutation; i.e., an altered or new genetic sequence that is inheritable.

mutation an abrupt, inheritable change in the nucleotide sequence of an organism. Mutations include base pair substitutions, deletions, duplications, transpositions/(translocations), and inversions. Mutations can arise spontaneously and can also be induced by certain chemicals and by radiation. The rates of spontaneous mutation in viruses and bacteria are controlled by DNA replication and repair enzymes. Induced mutation rates are also under genetic control, and are often mediated by repair enzymes.

mutualism a living association of two organisms of different species that is beneficial to both.

mycelium *plural*, **mycelia.** the vegetative body of a fungus, composed of a cumulative mass of hyphae.

mycetoma a chronic cutaneous and subcutaneous infection resulting from direct implantation of actinomycetes or true fungi.

Mycobacterium a bacterial genus containing at least 30 different species, including *M. tuberculosis* and *M. leprae*, the causative agents of tuberculosis and leprosy, respectively. Members of this genus are also opportunistic pathogens and are common causes of infection in AIDS patients. Although this group of bacteria is positive by gram stain, the bacteria are not true gram-positive organisms, but are classified as acid-fast bacilli, a staining characteristic that is attributed to the unique cell wall of mycobacteria.

mycology the scientific study of fungi.

mycoplasmalike organisms (MLOs) a descriptive name for a group of nonhelical, cell wall-less, prokaryotic microbes that resemble mycoplasmas in morphology and ultrastructure, found in phloem tissues of infected plants and in the bodies of insects; now usually identified by the newer term phytoplasmas.

mycoplasmas members of the class Mollicutes, a group of prokaryotic microorganisms resembling bacteria but lacking a cell wall and thus of variable morphology.

mycorrhiza a mutualistic association co-evolved between a host plant and a fungus localized in the root or root-like structure of the plant; this aids the plants in obtaining phosphorus and other elements.

mycorrhizal relating to or taking part in a relationship of mycorrhiza.

mycorrhizosphere the region around a mycorrhizal fungus that derives its carbon from the plant. Materials released from the fungus increase the microbial populations and their activities around the fungal hyphae.

mycosis *plural*, **mycoses.** a disease caused by the active invasion of living tissue by actively growing fungi. Also, **mycotic infection.**

mycotoxicosis *plural*, **mycotoxicoses.** a disease induced in humans and animals by secondary mold metabolites.

mycotoxin a toxin associated with a fungus; any of a group of structurally diverse, mold-elaborated compounds that induce diseases known as mycotoxicoses in humans and animals. Mycotoxins contaminate animal feed and human food ingredients in the absence of intact fungal elements; an estimated 25% of the world's food crops are thought to be contaminated with mycotoxins.

myelosuppression suppression of the production of the blood cells from the bone marrow.

myoclonus rhythmic, rapid contractions of a muscle in response to stretching.

myoneme a fibrillar striated contractile structure located in the cytoplasm of some protozoa (ciliates, sporozoa, dinoflagellates). They are partially responsible for cell contraction.

myxobacteria aerobic, free-living, rod-shaped gliding gram-negative bacteria that grow primarily in soil, on tree bark, on herbivore animal dung, and on decaying vegetation. Their distinguishing characteristic is the formation of multicellular structures called fruiting bodies.

myxospore an asexual, dormant cell produced within the fruiting body.

N

natural killer cells (NK cells) large granular white blood cells that mediate innate, nonspecific immunity.

natural selection 1. changes in gene frequency caused by specific detrimental or beneficial effects of those genes. 2. variation in the capacity of individuals of the same species to survive and reproduce, based on differentiation in their ability to adapt to environmental conditions.

NBD (nucleotide binding domain) part of the ABC protein transporter involved in energy coupling. This part is hydrophilic, highly conserved and is involved in ATP binding and hydrolysis.

near-obligate parasite an organism that survives mainly as a parasite in its natural habitat.

necrotizing fasciitis a fulminating form of cellulitis that spreads to involve the superficial and deep fascia, causing thrombosis of subcutaneous vessels and gangrene of overlying tissues. Type I necrotizing fasciitis is commonly due to a mixture of one or more anaerobes with one or more facultative species; Type II is due to group A streptococci.

negative sense genome a viral RNA genome that codes for proteins in the viral complementary sense. Transcription of viral complementary mRNA is required for translation of viral proteins.

negative staining a technique in which a particulate sample is surrounded by a solution of a heavy metal salt and allowed to dry; especially useful for small particles such as viruses.

Negri body a cytoplasmic matrix of rabies viral/nucleocapsid material. (Named for Adelchi *Negri*, 1876–1912, Italian physician.)

nematode any member of the phylum Nematoda; invertebrate animals, worms of microscopic size, millions found per square meter in terrestrial soils and benthic sediments, feeding there on bacteria, fungi, algae, plant roots, and on each other; comprising also parasitic forms that occupy all moist ecological niches with the appropriate food in plants, animals, and humans, and inside almost all organs.

nephrolithiasis the presence of kidney stones.

nephrotoxicity damage to the kidney cells.

nested PCRs sequential PCRs (polymerase chain reactions), in which each successive reaction is primed by an oligonucleotide pair whose annealing positions on template DNA are located (nested) within the annealing positions of oligonucleotides used to prime the preceeding reaction.

neuraminidase an enzyme, present on the surface of some viruses, which catalyzes the cleavage of a sugar derivative called **neuraminic acid.**

neuritis inflammation of the nerves.

neuropathy pathological changes in the nervous system.

neutropenia deficiency in circulating white blood cells of the neutrophil type.

neutrophil a large leukocyte containing a lobed nucleus and abundant cytoplasmic granules, that stains with neutral dyes.

neutrophile literally, middle lover; an organism that grows optimally in the neutral range of external pH, generally within pH 6–8.

neutrophilic growing optimally in the neutral range of external pH.

nif genes bacterial genes required for nitrogen fixation, which encode nitrogenase proteins, regulate synthesis of nitrogenase, or are involved in processing nitrogenase.

nisin a bacteriocin produced by *Lactococcus lactis* subsp. *lactis*; has applications in food preservation.

nitrification the biochemical oxidation of ammonium to nitrite and nitrate by microorganisms.

nitrogenase highly conserved enzyme complex that catalyzes the ATP-dependent reduction of dinitrogen to ammonia; nitrogenase consists of two components, dinitrogenase reductase (an iron protein) and dinitrogenase (an iron molybdenum protein).

nitrogen fixation the property of some taxa of prokaryotic organisms (Bacteria or Archaea) to enzymatically reduce atmospheric N_2 to ammonia. The ammonia produced can then be incorporated by means of other enzymes into cellular protoplasm. Nitrogen fixation only occurs in these prokaryotes and not in higher taxa.

nitrogen limitation the slower nitrogen assimilation that results from utilization of a single nitrogen source other than ammonia.

nitrogen recycling metabolism of insect nitrogenous waste products (e.g., uric acid, ammonia) by symbiotic microorganisms to nitrogenous compounds (e.g., essential amino acids) that are transferred to and utilized by the insect.

NK cells see NATURAL KILLER CELLS.

Nod factors symbiotic signal molecules synthesized in rhizobia by proteins encoded by nodulation genes; all share a similar lipo-chitin structure and have additional chemical substituents that allow the molecule to regulate root nodule formation in particular host legumes at very low concentrations.

nod genes rhizobial genes that play a role in root nodule formation. Some *nod* genes encode proteins required for synthesizing Nod factors; others have regulatory functions; many rhizobia carry *nod* genes on plasmids. The large number of nodulation genes has led to the use of the abbreviations *nol* and *noe* for recently discovered genes.

nodule a gall-like structure developed on the root or stem of legumes following infection by compatible rhizobia.

nodulins gene products expressed in host tissue during nodulation and N_2 fixation.

nonrandom segregation oriented distribution of chromosomes into progeny cells; cosegregation of chromosomes and conserved cell structures.

nonsense codon see TERMINATION CODON.

normal flora a term for the population of microbes that normally reside in a host and for the most part live in harmony.

Ntr regulon the set of genes directly regulated by the phosphorylated form of the transcription factor NRI (NtrC).

nuclear polyhedrosis virus a type of enveloped, double-stranded DNA virus reported from insects and other invertebrate organisms; replicates in nuclei of infected cells and is occluded in large protein crystals known as polyhedra.

nuclease an enzyme that hydrolyzes in the internucleotide phosphodiester bonds in nucleic acids.

nucleic acid probe a piece of labeled single-stranded nucleotide used to detect complementary DNA in clinical specimens or a culture and thus to specifically identify the presence of an organism identical to that used to make the probe.

nucleocapsid (core) protein virus protein(s) associating with viral nucleic acids to form the internal structure of an enveloped virus.

nucleoid the bacterial chromosome when it is in a compact configuration, either inside a cell or as an isolated structure.

nucleoside a molecule composed of a heterocyclic nitrogenous base (a nucleobase) joined by an *N*-glycosidic bond to ribose (ribonucleosides) or 2′-deoxyribose (deoxyribonucleosides).

nucleosome a fundamental subunit present in the eukaryotic genome (except dinoflagellates), composed by octamer of histones and DNA.

nucleotide a substance composed of a nucleoside with one or more phosphoroyl groups joined by phosphodiester bonds to the pentose moiety. Nucleotides are synthesized *de novo* from small precursor molecules such as carbon dioxide, formate, amino acids and 5-phosphoribosyl–1-pyrophosphate (PRPP) or, through the salvage pathway from PRPP and bases and from nucleosides.

numerical aperture (N.A.) in microscopy, a measure of the resolving power of the objective lens, equal to the product of the refractive index of the medium in front of the lens and the sine of the angle between the outermost ray entering the lens and the optical axis.

nutrient any organic or inorganic compound utilized by an organism in a process of growth, maintenance, or repair of the living body.

nutrition the fact or process of supplying calories, energy, protein, essential amino acids/peptides, essential fatty acids, vitamins, and mineral requirements to satisfy the metabolic needs of the consuming organism.

nutritional category the classification of organisms based on their principal carbon and energy sources; microorganisms are classified into four nutritional categories: photoautotrophs, photoheterotrophs, chemoautotrophs, and chemoheterotrophs (see entries).

O

objective lens the lens system closest to the object that forms an image of the object.

obligate describing an organism that is restricted to a single type of environmental condition, behavioral mode, or the like.

obligate anaerobe see ANAEROBE.

obligate biotroph an organism that is dependent on a living host for its survival in nature.

obligate parasite an organism that has a parasitic phase as an essential part of its life cycle.

obligate syntroph a microorganism that must interact mutualistically with one or more other microorganisms to metabolize a specific substrate.

occlusal referring to the surface of the tooth that is used for chewing.

ocular lens the eyepiece (lens) of an optical microscope.

oligogalacturonate lyase one of a group of diverse, often cytoplasmic, enzymes that catalyze the eliminative removal of unsaturated terminal residues from D-galacturonate oligosaccharides.

oligotroph literally, small eater; an organism that is able to grow well in conditions of limited nutrients. Thus, **oligotrophic.**

oligotrophy the fact of growing well in nutrient-imited conditions.

omasum in ruminant mammals, a chamber that traps large feed particles as digesta passes from the rumen to the abomasum.

oncogene a gene whose expression is associated with the development and proliferation of tumor cells.

oocyst a life stage in certain parasitic mircoorganisms when the parasite rapidly grows into a spherical cyst enclosed by an elastic membrane.

ookinete a life stage in certain parasitic mircoorganisms which forms a few hours after fertilization; the zygote resulting from fusion becomes a mobile, invasive form.

oomycete one of a group of filamentous organisms, closely related to some algae, that may appear superficially like fungi. Cells of these organisms have many diploid nuclei in the vegetative stage.

oospore the spore that results from fertilization of an oogonium (female sex cell). This spore form in *Phytophthora infestans* is thick-walled and is the stage that is best able to survive adverse conditions.

open reading frame (ORF) a nucleic acid sequence containing a series of codons which is uninterrupted by stop codons, and which potentially encodes a protein.

operon a collection of contiguous genes that are transcribed together into a single mRNA molecule, together with any *cis*-acting regulatory sequences. The gene products are thus controlled coordinately.

opportunistic describing an organism that is not generally pathogenic but that can cause disease in a compromised host.

opportunistic infection an infection with an agent that is not normally pathogenic but that can become so under certain conditions of a particular host. Many fungal infections have been so described because they are more likely to cause disease in patients with impaired immunity.

opportunity costs in public health, the total value of the resources lost due to infectious diseases, valued in terms of foregone alternative uses. The use of opportunity costs is a core concept of economic analyses of disease burden.

opsonin a substance coating an antigen either by antibody alone, or antibody plus complement components. This greatly increases phagocytosis by cells of the reticuloendothelial system.

opsonization the process by which foreign particles are coated with factors known as opsonins, such as antibody and complement, facilitating their uptake by phagocytes. Also, **opsonic action.**

optical path the product of the distance traveled by a ray of light and the index of refraction of the various materials it passes through.

optical sectioning in microscopy, the process of imaging sections of the specimen by stepping the focal plane through a specimen and recording two-dimensional images on each focal plane.

oral bioavailability the rate of, and extent to which, an active drug or metabolite enters the circulation by the way of the gastrointestinal tract.

ORF see OPEN READING FRAME.

organic 1. describing or involving any carbon substance bearing molecules other than free carbon, carbon monoxide, or carbon dioxide. 2. describing or involving an organism or organisms.

organism a living or formerly living being; an entity in which life processes are or were present.

organometallic describing any compound containing at least one metal-carbon bond, often exhibiting enhanced microbial toxicity. When such compounds contain "metalloid" elements (e.g., Ge, As, Se, and Te), the term **organometalloid** may be used.

organotroph literally, organic eater; an organism that derives its energy from organic sources; i.e., a heterotroph. Thus, **organotrophic.**

organotrophy the fact of being an organotroph; deriving energy from organic compounds.

origin of replication a chromosomal site that interacts with specific proteins to produce unwound DNA regions required to start polymerization of new DNA from existing template.

orthologous describing genes that are related by descent from a common ancestor but that are divergent since organismal speciation.

orthologs or ***orthologues*** genes that began diverging from each other after ancestral speciation; contrasted with paralog, genes that began diverging from each other after an ancestral gene duplication.

osmolality the mole fraction of osmotically active particles of solute per kilogram of water (or another such solvent). The osmolality of a particular solute depends upon the degree of its dissociation in water.

osmolarity The tendency of water (or another such solvent) to flow across a membrane in the direction of the more concentrated solution.

osmophile literally, lover of osmotic pressure; an organism that grows optimally in conditions of high osmotic pressure.

osmophilic growing optimally in conditions of high osmotic pressure.

osmoprotectant a compound that, when provided exogenously, stimulates bacterial growth in media of high osmotic strength. Osmoprotectants are taken up by the cells into the cytoplasm where they act as, or are converted to, compatible solutes.

osmosis the net movement of water or another such solvent across a partially permeable membrane, from a region of lower solute concentration to a region of higher solute concentration. This process tends to equalize the concentration on either side of the membrane.

osmotic relating to or involved in osmosis.

oversight the application of appropriate laws, regulations, guidelines, or accepted standards of practice to control the use of a microorganism, based on the degree of risk or uncertainty associated with that organism.

oxidase a hemoprotein that binds and reduces oxygen, generally to water.

oxidation the the withdrawal (or donation) of electrons from a chemical compound. An oxidation reaction is always coupled to a reduction of a chemical compound.

oxidative demethylation the degradation of $CH_3Hg(I)$ by anaerobic bacteria that results in the production of CO_2 rather than CH_4.

oxidative phosphorylation ATP (adenosine triphosphate) synthesis coupled to a proton or sodium electrochemical gradient, generated by electron transport, across an energy-transducing membrane.

oxidative stress the sum of the deleterious intracellular events associated with excess production or insufficient disposal of intracellular oxidants.

oxygenic photosynthesis a type of metabolism based on the coordinated action of two photosystems by means of which radiant energy is converted to chemical energy in the form of ATP (adenosine triphosphate), and reduction equivalents are obtained in the form of NADPH from the (photo)oxidation of water to free molecular oxygen.

P

package to actively incorporate DNA into a maturing bacteriophage particle; the initiation of phage DNA packaging is site specific and starts from a *pac* (P1, P22) or *cos* (λ) site. In **headful packaging** a single site-specific cut initiates the entry of DNA into the phage head; when the head is filled the DNA is again cut to create a chromosome with a terminal repeat longer than a genome. In **site-specific packaging** the length of DNA packaged is determined solely by the distance between specific cutting sites, each of which is cut.

paleopathology the technique of examining ancient skeletal remains to identify the diseases suffered by historical populations.

palindrome an inverted repeated sequence; in double-stranded DNA the ends of a palindrome are self-complementary, and can melt and reanneal to form hairpin-like structures.

pancreatitis inflammation of the pancreas.

pandemic an epidemic so widespread that it affects virtually the entire world.

papilla 1. a small elevation or projection above a body surface. 2. a localized deposition of host material lying on the inner side of a plant cell wall subtending the site of fungal attack.

papillomavirus one of a group of viruses causing common warts of various kinds. Certain types are also associated with various cancers; e.g., cervical carcimona.

papular rash a skin discoloration characterized by circular or oval spots that are raised above the surface of the skin.

parainfluenza virus (PIV) an important respiratory tract pathogen of young children, which also is responsible for many acute respiratory tract infections in older children and adults.

paralogous describing genes that are related by descent from a common ancestor but that are divergent since an ancestral gene duplication. Their products generally serve similar functions in different organisms.

paralogs or ***paralogues*** genes that began diverging from each other after an ancestral gene duplication; distinguished from orthologs, which began diverging after an ancestral speciation.

parasite an organism that lives upon or within another organism at whose expense it obtains some advantage.

parasitism an association between two organisms of different species that provides benefit to one and some detriment to the other.

parasitophorous vacuole (PV) the vacuole within which a malaria parasite resides after invasion of the red blood cell; the vacuole is formed initially by the internal membrane of the erythrocyte and grows as the parasite expands to fill the cell. The PV membrane is actively involved in the transport of nutrients to, and waste products from, the parasite.

parasporal inclusion body an aggregate of one or more proteins formed inside a bacterial cell at the time of sporulation; may be toxic to insects that ingest them.

parenchyma plant tissue composed of thin-walled cells that synthesize or store foodstuffs and usually leave intercellular spaces between them.

paronychia an infection involving the folds of tissue surrounding the nail.

passive surveillance in public health, a surveillance method that relies on reporting of cases from health care providers (e.g., physicians, clinical laboratories), based on a published list of conditions.

Pasteur effect the activation of yeast sugar consumption rate by oxygen. Alternatively, this phenomenon states that under anaerobic conditions, glycolysis proceeds faster than it does under aerobic conditions. In *Saccharomyces cerevisiae*, the Pasteur effect is only observable when glucose concentrations are low (below around 5 mM) or in resting or starved cells.

pasteurization the process of food preservation developed by Louis Pasteur, in which food is subjected to a (relatively) mild heat treatment for a specified period of time, in order to render harmless any resident pathogens without consequently harming the taste or quality of the food.

pathogen an organism that causes disease; an infectious agent.

pathogen-derived resistance resistance derived by expression of a gene or nucleic acid from the pathogen in a transgenic host, that interferes with the ability of the pathogen to complete its normal life cycle.

pathogenesis 1. the process of development of a disease. 2. see PATHOLOGY.

pathogenic harmful to living organisms; causing disease.

pathogenicity the ability of an organism to infect a host and cause disease.

pathogenicity factor an agent of a virulent pathogen that in some ways mediates the extent of damage to the host.

pathogenicity island (PAI) a large fragment of DNA in the genome of a pathogenic microorganism, which contains virulence-related genes and which appears to have been acquired by horizontal gene transfer. PAIs are found in numerous bacterial species.

pathology the study of disease, including its causes, progression, and effects.

pathotype a pathogen phenotype distinguished by differential reactions on hosts carrying different resistance genes.

pathovar a subdivision of bacterial taxonomy within a named species characterized by a pathogenic reaction of the bacteria in one or more hosts.

PCR polymerase chain reaction, an *in vitro* method for the enzymatic synthesis of specific DNA sequences, using two oligonucleotide primers that hybridize to opposite strands and flank the region of interest in the target DNA. Following a series of repetitive cycles that involve template denaturation, primer annealing, and the extension of the annealed primers by *Taq* polymerase, there is an exponential amplification of the target DNA. PCR is a widely used contemporary technique since its development in the 1980s and was cited in the Nobel Prize in Chemistry for 1993.

pelagic describing or referring to animals living in marine waters, especially those living within the ocean rather than on the surface or on the sea bottom.

peplomer surface projections of homo (or hetero-) polymers of viral glycoprotein molecules on virus particles.

peptic ulcer a breach in the epithelium of stomach or duodenum, caused by an imbalance between aggressive factors (acid and pepsin) and mucosal protection mechanisms. Ulcers have a strong tendency to relapse, and can progress to the potentially fatal complications of bleeding and perforation. The removal of *H. pylori* colonization is found to ameliorate ulcer disease.

peptidase see PEPTIDE HYDROLASE.

peptide any member of a class of compounds of low molecular weight that yield two or more amino acids on hydrolysis, and that form the constituent part of proteins.

peptide bond linkage between two amino acids.

peptide hydrolase any of a group of enzymes that catalyze the hydrolysis of peptide bonds.

peptide synthetase any of a group of complex enzymes that catalyze a large number of enzymatic reactions (activation, condensation, elongation, etc.), leading to the formation of non-ribosomal peptides.

peptidoglycan see MUREIN.

peptidomimetic describing a molecule having properties similar to a peptide, or short protein.

peribacteroid membrane a host-derived membrane that surrounds rhizobial cells following their release into cells of the host.

peridinin a carotenoid pigment specific to most autotrophic dinoflagellates.

periodontal relating to or involving the area around the teeth.

periplasm the space beteen the cytoplasmic and outer membranes in gram-negative bacteria and between the cytoplasmic membrane and cell wall in gram-positive bacteria. Also, **periplasmic space**.

periplasmic chaperones a class of proteins localized within the periplasm of gram-negative bacteria which facilitate the folding and assembly of pilus subunits, but which are not components of the final pilus structure.

peristalsis the waves of muscle contraction followed by relaxation that propel digested matter down the gastrointestinal tract.

perithecium *plural*, **perithecia.** a flask-like fruiting body, usually dark in color and thick-walled, containing ascospores within asci.

peritrichous describing flagella that are uniformly distributed over the bacterial body.

permafrost soil, subsoil, or other ground deposit that remains in a continually frozen condition, defined as a temperature below 0°C for more than two years.

permeant acid see MEMBRANE-PERMEANT WEAK ACID.

permeant base see MEMBRANE-PERMEANT WEAK BASE.

persistence a term for a mode of virus interaction with a population of cells in which a few cells are always in a lytic mode of infection, while the majority of cells are uninfected but potentially susceptible to lytic virus infection.

persistent transmission a term for vectored transmission by a more intimate association of virions with the vector. Virions may accumulate within salivary glands or other organs, from which they can later be passed back into plant tissue in saliva or regurgitant. Infectivity is retained for much longer periods than with nonpersistently transmitted viruses, and in some cases for the life of the vector.

Peyer's patches localized lymphoid tissue in the small intestine of mammals. (From Johann Conrad *Peyer,* 1653–1712, Swiss anatomist.)

phage short for BACTERIOPHAGE, a bacterial virus.

phage lambda a bacteriophage that on infection of a bacterial cell may follow one of two alternative pathways; either the lytic pathway in which the bacterium is sacrificed and progeny phages are produced, or the temperate (lysogenic) pathway in which the phage genome is repressed and, if it integrates into the host chromosome, will be stably maintained in the progeny of the surviving bacterium. Phage lambda was isolated from *E. coli* K-12 in which it resided in its temperate (prophage) state.

phagocyte literally, cell eater; a designation given to cells such as macrophages (large eaters) that can engulf and destroy particles, such as bacteria.

phagocytic relating to or having the property of phagocytosis; able to engulf and destroy foreign particles.

phagocytosis the process in which large material such as microorganisms is engulfed by cells.

phagolysosome a host cell vacuole that forms after fusion of a phagosome (see below) with lysosomal vesicles.

phagosome a host cell vacuole that forms upon bacterial entry into a host cell; the host cell environment of ehrlichiae.

pharmacokinetics the rate and efficiency of uptake, distribution, and disposition of a drug in the body.

phase variation reversible switching on and off of the expression of a bacterial gene, resulting in different modifications of the cell surface; e.g., the phenomenon in several strains of *E. coli* in which the bacteria alternate between the fimbriated and nonfimbriated conditions.

phenazines compounds based on the trycyclid phenzaine ring system; secreted by certain species of bacteria.

phenotype the sum of observable characteristics of an organism, based on its cellular or colonial morphology, biochemical structure, growth characteristics, and/or serological specificity, and resulting from the interaction of its genetic constitution with its environment.

phenotypic relating to the phenotype of an organism.

pheromone a specific chemical produced by one organism to alter gene expression, or affect behavior, in a second organism of the same species.

pH homeostasis the maintenance of intracellular pH within a narrow range, during growth over a wider range of extracellular pH.

phialide a cell that develops one or more open ends from which conidia (**phialospores**) are formed.

phloem tube-like cells of the plant conductive system that carry sugars and other organic substances from leaves to other parts of the plant. Certain elements of this system provide passageways for the spread of phloem-limited bacteria within plants.

phosphate a compound chemically identified as PO_4^{3-} and in its soluble reactive forms (SRP) able to be readily utilized by organisms. The principal form of phosphorus found in the earth's crust.

phospholipids the major components of biological membranes; any lipid containing a phospho (H_2PO_3) group, especially lipids based on 1,2-diacylglycero-3-phosphate (phosphatidic acid).

phosphorelay a signal transduction pathway in which a phosphate group is passed along a series of proteins.

phosphotransfer a process in which a phosphoryl group is transferred from a site on a protein to a different site on the same or a different protein, or to a small molecule.

phosphotransferase system (PTS) a system of proteins that catalyzes the transport and phosphorylation of certain sugars. This system also participates in catabolite repression and inducer exclusion.

photoautotroph an organism that uses inorganic material as a source of carbon for growth, and light as an energy source. Thus, **photoautotrophic.**

photoautotrophy a type of metabolism in which cellular carbon is derived from inorganic carbon (CO_2), with photochemical light reactions providing the source of energy.

photoheterotroph an organism that uses organic material as a source of carbon for growth, and light as an energy source. Thus, **photoheterotrophic.**

photoheterotrophy a type of metabolism in which cellular carbon is predominantly provided by organic carbon, with photochemical light reactions providing the source of energy.

photolithotroph see PHOTOAUTOTROPH.

photolithotrophy see PHOTOAUTOTROPHY.

photophobia literally, fear of light; a movement away from light.

photophobic responding to light (or certain light) by movement away from the light source.

photophosphorylation ATP (adenosine triphosphate) synthesis coupled to a proton or sodium electrochemical gradient generated by light-driven electron transport, which is often cyclic in bacteria.

photopigment a pigment molecule that is sensitive to light; types of photopigments include bacteriochlorophyll, carotenoid, flavin, retinal, or phycobilin.

photoreceptor a protein with an attached photopigment molecule that can absorb light energy.

photoresponse a general term for the reaction of a cell to changes in the direction, quality, or intensity of light.

photosynthesis a series of processes in which light or other electromagnetic energy is converted to chemical energy to be used for the biosynthesis of organic cell materials.

phototaxis the oriented movement of a motile organism with respect to the direction of light. Phototaxis can either be toward (**positive phototaxis**) or away from (**negative phototaxis**) the light.

phototroph literally, light eater; an organism that uses light as an energy source. Thus, **phototrophic.**

phototrophy the fact or process of deriving energy from light.

phototropism growth or orientation in response to the location of light.

phycobilin a photosynthetic pigment found in cyanobacteria.

phycobiliproteins subunits of phycobilisomes, each with different spectral properties, that collect light energy from a variety of wavelengths and transfer it to the phycobilisomes. Phycobiliproteins are polypeptides containing covalently bound open tetrapyrrole chromophores, the phycobilins.

phycobilisomes macromolecular aggregates serving as antenna systems for the capture of light in photosynthesis, typical of cyanobacteria. They are composed of multimers of different phycobiliproteins and of linker polypeptides.

phyllocoenosis the association of a microorganism with the stem or leaves of a plant, which results in a benefit to one or both of the partners.

phylloplane a collective term for the aboveground exposed surfaces of plants that are available for the colonization of microorganisms.

phyllosphere a collective term for the leaves and stem of a plant that can be colonized by microorganisms.

phylogenetic relating to or involving phylogeny; describing an analysis that groups organisms according to evolutionary relationships.

phylogeny evolutionary history; the ordering of species into higher taxa and the construction of evolutionary trees based on evolutionary relationships rather than general resemblances.

physiologic race a group of isolates that share common pathogenic attributes. Isolates classified as a single race may or may not be genetically identical.

phytoalexin an antimicrobial toxic compound, such as phenylpropanoids, isoprenoids and acetylenes, synthesized in plants in response to pathogen infection.

Phytophthora a genus containing many important plant pathogens (notably the type species *P. infestans*; see below), most of which are near-obligate parasites.

Phytophthora infestans a filamentous plant pathogen that causes the late blight disease of potatoes (and tomatoes). The devastation caused by this disease, beginning in the fall of 1845, led to the Irish potato famine. The disease remains a major concern in potato production, with a recent resurgence in importance. Although *P. infestans* appears superficially to resemble true fungi, it is phylogenetically unrelated, though its ecology is similar.

phytoplankton the portion of the plankton community defined as plant life.

phytoplasmas minute, cell wall-less bacteria that are found in the phloem tissues of plants and in the bodies of insect vectors that transmit them from plant to plant. Formerly designated as mycoplasmalike organisms (MLOs) in reference to their resemblance to the mycoplasmas.

phytoremediation the use of plants and their associated microorganisms to facilitate degradation, concentration and/or extraction of soil-borne contaminants. This process may involve uptake and modification by the plant itself, microbial modification followed by plant uptake, or direct modification by the rhizosphere microbes, possibly involving cometabolism.

picornavirus a group of viruses with small RNA genomes, such as poliovirus.

piezophile literally, pressure lover; an organism whose maximum rate of growth, over all possible growth temperatures, occurs at a pressure greater than atmospheric pressure and less than 50 MPa.

piezophilic growing optimally at a pressure greater than atmospheric pressure and less than 50 MPa.

pigment any coloring matter in microbial, plant, or animal cells.

pilin the individual protein subunit of a pilus organelle. Immature pilins, containing leader signal sequences that direct the transport of pilins across the inner membrane of gram-negative bacteria, are called **propilins** or **prepilins.**

pilosebaceous relating to or involving hair follicles and sebaceous glands.

pilus *plural*, **pili.** a short, filamentous appendage extending from the surface of certain bacterial cells, composed mainly of the protein pilin.

placebo an agent without the specific effects of a drug under test, used to determine to what extent any observed effects of the drug are due to psychological influences or expectations. It is usually given to some patients while the test drug is given to others, with neither group knowing which agent it has received, the so-called "blind" design. (From a Latin phrase meaning "I will please," referring to a practice of giving patients a placebo drug to satisfy their wish to receive actual medication that is not deemed suitable to administer.)

planetary protection the sum of provisions taken to prevent the contamination of any planet by biological organisms from another planet due to spacecraft missions. This also includes the possible contamination of Earth from non-terrestrial samples from another planet.

plankton a term for a wide variety of organisms living suspended in open waters, both marine and freshwater, able to float, drift, or swim to some extent. Plankton are classified according to size; many types are microscopic.

planktonic referring to organisms living as plankton or in the manner of plankton; i.e., free-floating in a water environment.

plant cell culture the production of plant cells in a bioreactor to produce useful products.

plaque **1.** an identifiable clear area in a growth layer of bacteria or in a monolayer of cultured cells, caused by the cytopathic effects of a virus infection. The plaque arises from infection of a single cell. **2.** see DENTAL PLAQUE.

plasmid **1.** a nonessential extrachromosomal genetic element, usually circular, that is capable of autonomous replication via a segment of the plasmid called the replicon. A self-transferable plasmid possesses the machinery required for its transfer from one bacterial cell to another bacterial cell in close contact. **2. plasmids.** a term encompassing the study of the constituent parts of these genetic elements, the phenotypes they confer on their hosts, and the genetic processes they promote within the bacteria carrying them. (Coined by J. Lederberg, from *-plasm,* "living tissue," as in *cytoplasm,* and -id, "diminutive," by analogy with *chromatic* and *plastic.*)

plasmid incompatibility the inability of a plasmid to be maintained in the same host as another different plasmid.

plasmodesmata cytoplasmic connections between plant cells.

Plasmodium the genus to which the malaria parasites belong, notably *P. falciparum* and *P. vivax.* These organisms are obligate intracellular parasites for most of their life cycle and have two hosts; the vertebrate, in which two phases of asexual reproduction occur, and a blood-sucking insect, in which fertilization between male and female forms of the parasite occurs, followed by a further phase of asexual reproduction.

plasmolyzed describing cells incubated in a hypertonic medium to retract the inner membrane and expand the periplasmic space.

pleomorphic **1.** relating to or displaying more than one distinct form in the same organism. **2.** referring to the ability of fungi to produce more than one type of spore, and to grow in more than one form. **3.** having to do with the doctrine of pleomorphism (see below).

pleomorphism literally, different form; one component of a 19th-century debate over the morphology of bacteria. It held that the various different forms observed in bacteria did not reflect distinct species but rather were the response of the organism to differing environmental conditions. The contrasting position was monomorphism, the doctrine that microorganisms do exist in distinct, stable Linnean species. More extreme versions of pleomorphism held that bacteria, yeasts, molds, and algae are all interconvertible stages in the life cycle of a single organism.

plume a term for dissolved contaminants emanating from a source region due to ground water transport processes.

poikilothermic describing a "cold-blooded" animal; i.e., an organism having a body temperature that approximately varies with that of the surrounding environment, rather than being internally regulated to a relatively constant temeperature (homeothermic). Thus, **poikilotherm.**

poikilothermy the fact of having a body temperature that approximately follows that of the surroundings.

point spread function (PSF) the relationship between a point object and the image of it after passing through the objective of a microscope; i.e., the image-forming characteristics of an objective lens.

Poisson distribution a statistical function that predicts the probability of occurrence of all-or-none rare events, such as mutations, under the assumption that their occurrence is random. (Formulated by Simon Denis *Poisson,* 1781–1840, French mathematician and physicist.)

polarized chromosomal transfer the sequential transfer of all or part of a bacterial chromosome from a donor to a recipient cell, mediated by a conjugative plasmid.

polarized light the direction of an electric field, which is in a plane for plane-polarized light and rotates in a circle for circularly polarized light, as opposed to ordinary light whose waves vibrate in all directions.

polar organelle a proteinaceous vesicular structure approximately 200 nm in diameter, just below the poles of spirochetes and other motile bacteria; in some cases shown to be composed of ATPase, cytochrome oxidase, and stalked donut-shaped particles 5–6 nm in diameter arranged in a hexagonal array.

polyketide a carbon chain containing multiple keto groups on alternate carbon atoms, derived from acetate, propionate, or butyrate units activated in the form of acetyl-CoA (or malonyl-CoA), methylmalonyl-CoA, and ethylmalonyl-CoA, by the action of polyketide synthases.

polyketide synthase an enzyme involved in the activation and condensation of precursor units to form polyketides. Type I polyketide synthases (PKSs) are large multienzyme proteins with many catalytic centers encoded by a single large gene. Type II polyketide synthases are complexes of several polypeptides encoded by specific small genes.

polymerase an enzyme that synthesizes a nucleic acid polymer.

polymerase chain reaction see PCR.

polymorphonuclear granulocyte one of a leukocyte group, comprising neutrophils, eosinophils, and basophils.

polymyositis inflammation of the muscles, involving multiple muscles.

polyploid an organism having more than two sets of chromosomes.

polyploidy the fact of having multiple sets of chromosomes.

polyprotein a multicistronic protein translated from a single open reading frame. The polyprotein is subsequently cleaved into individual mature gene products.

population a group of individuals belonging to the same species and living within a defined area or space, typically so that individuals may potentially recombine, compete for limiting resources, or otherwise interact.

pore-forming toxin a soluble protein secreted by bacteria which bind to the surface of the host cell and oligomerize to form a pore to release soluble components from the host cell.

porin a protein molecule found in outer membranes of gram-negative bacteria serving as channels for the diffusion of water and small molecular weight solutes.

porphin the parent compound of the porphyrins, a cyclic tetrapyrrole structure in which the pyrrole rings are connected by one-carbon methine groups.

porphyrin a macrocyclic tetrapyrrole that has a conjugated double bond system conferring aromatic characteristics.

porphyrinogen a macrocyclic tetrapyrrole that is not aromatic because at least one of the links between adjacent pyrrole rings is saturated and the double bond conjugation system is interrupted.

portal of entry the site at which a pathogen enters the host and establishes the primary infection. Most fungal infections are contained at the portal of entry, but dissemination can occur, especially in patients with impaired immunity.

positive strand polarity an RNA molecule with a sequence able to be translated by the cell's translational machinery, similar to a cellular mRNA.

postreplication DNA repair a DNA repair mechanism using homologous recombination to insert complementary DNA from a sister chromosome into gaps associated with inhibition of DNA replication at or near DNA damage-induced lesions.

posttranscriptional gene silencing a process through which specific RNAs are degraded posttranscriptionally, resulting in loss of expression of associated genes.

posttranslational modification covalent modification to a cellular component that occurs subsequent to its synthesis.

potentiator a nonmicrobicidal substance that enhances the effect of a microbicide, with the effect of reducing the overall toxicity of the control program. Potentiators can work synergistically with the microbicide.

powdery mildews an array of plant diseases, caused by many species of Ascomycete fungi that are grouped into seven main genera. Infection with powdery mildew produces a white lawn of fungal mycelium that covers the plant surface while chains of aerial spores give the characteristic powdery appearance. Taken collectively, powdery mildews cause greater losses in terms of crop yield than any other single type of plant disease.

prebiotic literally, before life; a nondigestive food ingredient that beneficially affects a host by selectively stimulating the growth and/or activity of one or a limited number of bacterial species already in the colon.

prenyltransferases a family of enzymes from different organisms involved in the synthesis of isoprenoid molecules by condensation of isoprene units.

preprotein a term for a protein bearing a signal peptide prior to maturation by signal peptidase.

preservative a chemical agent used to prevent microbial growth in finished products. It prevents their multiplication or kills them to prevent spoilage or contamination of the product with pathogens.

prespore see FORESPORE.

pressure the ratio of the force acting on a surface divided by the area of the surface on which the force acts.

prevalence the proportion of a defined population that has a given disease or condition at a specific point in time, or over a defined time period.

primary metabolite simple molecules and precursor compounds such as amino acids and organic acids that are involved in the pathways essential for life processes and the reproduction of cells.

primase an enzyme that synthesizes an RNA primer.

prion the infectious agent that causes transmissible spongiform encephalopathies (TSEs). A large body of evidence indicates that prions do not contain informational nucleic acids. In 1996 Stanley B. Prusiner was awarded the Nobel Prize for Medicine for the hypothesis that prions consist of a modified form of the normal cellular protein called PrP^{C}. Although the human forms of prion diseases are rare, the recent epidemic of bovine spongiform encephalopathy (BSE) in the UK has dramatically raised the issue of transmissibility of these diseases from affected animals to humans.

prior art in patent law, the previous developments in the field of an invention against which the novelty and non-obviousness of the invention are assessed, including prior developments that have been described, published, or otherwise disclosed publicly.

probabilistic identification a determination of the likelihood that the observed pattern of results of tests carried out on an unknown bacterium can be attributed to the results of a known taxon within an identification matrix.

probe 1. a chemical or molecular technique that allows the detection or quantitation of a population or activity of microorganisms. 2. specifically, a labeled nucleic acid fragment useful in the detection of homologous nucleotide sequences.

probe amplification A nucleic acid amplification procedure in which many copies of the probe that hybridizes the target nucleic acid are made.

probiotic 1. promoting life; beneficial to a living organism. 2. a living microorganism that if ingested in certain numbers will exert health benefits beyond inherent basic nutrition. 3. specifically, a blend of live microbes exerting a beneficial effect on a host by improving the properties of the native gastrointestinal microflora.

prochiral selectivity the ability of a catalyst or reagent to distinguish between two chemically identical but stereochemically different parts of a single molecule that does not contain a chiral center.

prodrome a warning symptom indicating the onset of a disease.

prodrug a drug that is given in an inactive form and that thus has to be metabolized in the body to attain its active form.

proenzyme the inactive precursor of an enzyme; an enzyme that must be processed in order to exhibit catalytic activity.

prokaryote a cell or organism lacking a nucleus and other membrane-enclosed organelles, usually having its DNA in a single circular molecule. A term used (loosely) to describe archaeal and bacterial organisms in contrast to eukaryotic organisms.

prokaryotic generally, describing a cell whose DNA is not enclosed in a nuclear membrane, or an organism having this cellular feature.

promastigote a flagellated insect stage of *Leishmania*. Promastigotes exist in the sand fly vector as either the replicative procyclic form (noninfectious) or the highly infectious (and nonreplicative) metacyclic form.

promiscuity in this context, the ability of some rhizobia to nodulate with many legumes.

promoter a sequence of DNA whose function is to be recognized by RNA polymerase in order to initiate the transcription of a particular gene. A typical *E. coli* promoter contains two conserved elements.

prophage a bacteriophage that has had its DNA incorporated into the bacterial chromosome in a stable, heritable manner. Expression of the viral genes is repressed but, in many cases, may be induced by exposure of the host to various stress-producing agents such as DNA-damaging ultraviolet light.

prophylaxis the fact or process of preventing disease; efforts to prevent the spread of disease.

protease an enzyme that catalyzes the cleavage of proteins. In the case of HIV, for example, a virus-specific protease is needed to cleave some of the virus coat proteins into their final, active form.

protease inhibitor a substance that inhibits the action of enzymes.

protective substance in freeze-drying procedures, smaller or larger molecular weight compounds that are dispersed into a suspension to protect biological materials from injury arising from freezing or drying.

protein engineering the design, development, and production of new protein products having properties of commercial value.

protein export or ***secretion*** the transport of proteins to extracytoplasmic compartments, including the exterior of the cell or to other cells.

Proteobacteria a kingdom of bacteria divided into five groups based on their 16S ribosomal RNA sequences; a group of bacteria diverse in their morphology, physiology, and life patterns.

proteoliposomes membrane vesicles made *in vitro* by incorporating detergent-solubilized membrane proteins into artificial bilayer lipid vesicles liposomes or liposomes.

proteome a collective term for the entire protein content contained in an organism.

proteomics the study of the protein complement, or proteome, of an organism.

protist an organism of the Protoctista (Protista).

Protista an earlier (and still used) term for the Protoctista.

protoctist one of the Protoctista; a eukaryotic single-celled microorganism classified separately from the animals, plants, and fungi.

Protoctista a group of eukaryotic organisms distinct from the prokaryotes and from the other three main groups of eukaryotes (animals, plants, and fungi). Protoctists are single-celled, aquatic organisms with a nucleus, and include protozoa, algae, and slime-molds, among others.

protonmotive force (pmf) the electrochemical measure of the transmembrane gradient of protons, consisting of an electrical potential due to separation of charge and the pH gradient due to different concentration of protons.

protonolysis the mechanism by which a C-Hg bond is cleaved by the enzyme organomercurial lyase to produce Hg(II) and a reduced organic moiety.

protooncogene a normal cellular gene that is the precursor to an oncogene (tumor-inducing gene); i.e., a gene that can evolve into an oncogene through mutation, deletion, or overexpression.

protoplasm literally, first form or first material; the organic and inorganic compounds of which cells are composed.

protoplasmic relating to or consisting of protoplasm.

Protozoa literally, first animals; a taxonomic grouping of nonphotosynthetic, single-celled organisms with a nucleus, about 5–50 nm in size, living in aquatic environments (swimming) and in soil pore water (swimming, creeping) in densities of millions per liter of water and up to 10^6 per gram of soil, feeding mainly on bacteria. The group also includes forms parasitic to plants, animals, and humans.

protozoan *plural,* **protozoa** or **protozoans.** one of the Protozoa, free-living, unicellular, eukaryotic organisms that possess motility.

provirus the double-stranded DNA replicative intermediate of a retroviral genome found integrated into the host chromosome.

PrP^C cellular, normal host prion protein.

PrP^{Sc} scrapie-associated, pathological prion protein.

PrP protein prion cellular protein, a protein of unknown function, primarily expressed on cells of the central and peripheral nervous system as well as on lymphocytes. PrP is expressed in at least two different isoforms, one of which is thought to be the infectious agent of transmissible spongiform encephalopathies (TSEs).

pseudocyst see TISSUE CYST.

pseudoplasticity a property of change that occurs in viscous solutions as a result of increases in stress forces. The polymer molecules tend to orient in the direction of solution flow as the force (shear stress) is increased. The resistance to flow (viscosity) is thereby decreased. When lesser forces are applied, the solution viscosity remains high due to random orientation of nonaligned molecules effecting increased resistance to flow.

psychrophile literally, cold lover; an organism able to grow over the temperature range of below 20°C down to below 0°C, with an optimum range of 10–15°C.

psychrophilic describing an organism having a growth temperature range of below 0°C to 20°C, and optimally at 10–15°C.

psychrotroph an organism that can grow at 7°C (or less) irrespective of its optimum growth temperature.

psychrotrophic describing a microorganism that can grow at 7°C (or less).

PTS phosphoenolpyruvate: carbohydrate phosphotransferase system, a major bacterial transport system that catalyzes the transport and phosphorylation of a number of carbohydrates in gram-negative and gram-positive bacteria.

public health the branch of medicine concerned with measuring, safeguarding, and improving the health of a community at large.

public health surveillance see SURVEILLANCE.

pulsed field gel electrophoresis (PFGE) an electrophoretic device allowing the separation of very long DNA molecules.

purple nonsulfur photosynthetic bacteria a large metabolically diverse group of bacteria that carry out anoxygenic photosynthesis. Organisms within this group may grow via photoautotrophic or photoheterotrophic metabolism, and also via a type of anaerobic fermentation in the absence of light. They are all chemoheterotrophic and some can grow under chemoautotrophic conditions.

putrefaction the breaking down, decomposition, or rotting of organic matter; similar to but often contrasted with fermentation.

pyoderma a term for any purulent skin disease.

Q

Quality-Adjusted Life Year (QALY) a measurement of life span that adjusts pure chronological age by incorporating judgments about the value of time spent in different health states (e.g., years lived in normal health versus years lived with a disability).

quarantine 1. the isolation of an infected or exposed individual in order to limit the spread of a disease. 2. specifically, the historic practice of detaining and inspecting all ships, cargos, and passengers entering port for evidence of contagious diseases. (From an Italian phrase meaning “forty days,” originally referring to the period of time for which ships entering the port of Venice were required to remain in isolation before any disembarkment.)

quasispecies a near species or seeming species; e.g., viral strains within the viral population in a single host, derived by mutation during viral replication.

quinone a lipid-soluble hydrogen (i.e., proton plus electron) carrier that mediates electron transfer between respiratory chain components.

quorum sensing a phenomenon by which bacteria monitor their population by the levels of signal molecules (e.g., acyl homoserine lactones) present in the culture. The signal molecules are usually diffusible and their production is cell density-dependent. The accumulation of such molecules to the threshold concentration signals the bacteria that their density has reached a "quorum." The bacteria then begin expressing genes or phenotypes that are controlled in a cell density-dependent manner.

R

racemate an equimolar mixture of enantiomers, which therefore no longer rotate plane-polarized light.

race-specific describing plant resistance to pathogen invasion conferred by a single gene held by a particular cultivar of a plant, corresponding to an avirulence gene held by a particular race of the pathogen.

RAG genes recombination activating genes; the protein products of these genes catalyze the DNA recombinations that give rise to the variable domains of immunoglobulin chains and, therefore, form the specific binding sites on antibody molecules. These genes are expressed only in cells of the lymphocytic lineage, during their early development. They also perform a similar function in developing T cells, where they are responsible for DNA rearrangements leading to the expression of the T cell receptors.

rare-cutters a term for endonucleases (mostly Class II restriction enzymes) that have only a limited number of recognition/cutting sites on a whole chromosome.

RAS return activated sludge; activated sludge solids returned from the final sedimentation tanks to the aeration tanks on a continuous basis.

reaction center a membrane-spanning complex composed of pigments and proteins. Electromagnetic energy is converted to chemical energy within the reaction center.

reactive arthropathy a painful joint inflammation that can follow infection with a number of foodborne pathogens.

reactive gliosis the strong proliferation of reactive astrocytes as seen by staining with antibodies against glial fibrillary acidic protein (GFAP). While not specific for TSEs (transmissible spongiform encephalopathies), gliosis is extremely prominent in Creutzfeldt–Jakob disease and scrapie.

reactive oxygen species (ROS) partially reduced oxygen derivatives, such as superoxide and hydrogen peroxide.

reactogenicity the undesired symptoms produced by a vaccine strain.

reading frame the phase in which nucleotides are read in sets of three (triplets) to encode a protein.

reagin a substance made in response to a treponemal infection, characterized by its ability to combine with lipids.

reassortant a virus containing segments derived from two different influenza viruses which resulted from the co-infection of a single cell.

recalcitrance the resistance of a compound to biodegradation, resulting in its persistence in the environment.

receptor a cell surface protein that specifically interacts with a particular stimulus to generate a signal leading to a cellular response.

recoding any of several types of events in which ribosomes perform noncanonical forms of protein synthesis.

recombinant relating to or resulting from recombination.

recombinant DNA technology a range of techniques in which DNA, usually from different sources, is combined *in vitro* and then transferred to a living organism (e.g., *E. coli*) to assess its properties.

recombination 1. generally, any process in which new combinations of DNA sequences are generated. The general recombination process relies on enzymes that use DNA sequence homology for the recognition of the recombining partner. 2. a laboratory method in which DNA segments from different sources are combined into a single unit and manipulated to create a new sequence of DNA.

recovery resistance the ability of plant tissues to grow away from an infection to produce a healthy shoot.

recrudescence a recurrence of symptoms in a patient whose bloodstream infection previously was at such a low level as not to cause symptoms.

recycling the reuse of waste materials as valuable raw materials for the production of new industrial products.

redox short for oxidation–reduction.

redox-coupled reaction the combination of an oxidation reaction coupled to a reduction reaction that results in the release (or consumption) of free energy.

redox potential a measure of the thermodynamic tendency of an ion or molecule to accept or donate one or more electrons. By convention, the more negative the redox potential, the greater the propensity for donating electrons, and vice versa.

reduction the acceptance of electrons by a chemical compound. An oxidation reaction is always coupled to a reduction of a chemical compound.

reduction to practice in patent law, the inventor's act of physically carrying out his or her invention after initially conceiving of it, or the inventor's act of filing a patent application that satisfies certain statutory requirements.

re-emerging infection a known infectious disease that had previously fallen to such low prevalence or incidence that it was no longer considered a public health problem, but that is presently increasing in prevalence and/or incidence. Re-emerging infections include tuberculosis, which has increased worldwide since the early 1980s, dengue in tropical regions, and diphtheria in eastern Europe.

refraction the change in direction of propagation of a wave when it passes from one material to another in which the index of refraction is different.

refractive index the effect of a material object on the incident wave defined as the ratio of the velocity of light in a vacuum to that in the object.

refractory a metallurgical term for the "locking in" of precious metals, usually in sulfide minerals, preventing their recovery by conventional processes. Extraction of such metals can be accomplished by certain microorganisms.

regioselectivity the ability of a catalyst or reagent to direct reactivity towards a particular portion or region of a molecule.

regulon a group of genes whose expression is controlled by a common regulatory factor or signal.

relapse a recurrence of a disease that takes place after complete initial clearing of the original infection, and that implies reinvasion of the bloodstream by new pathogens.

release factor a protein factor that promotes the termination of protein biosynthesis and release of the mRNA and newly synthesized polypeptide from the ribosome.

remineralization the conversion of an element from an organic form to an inorganic state as a result of microbial activity.

remittent fever an elevated fever with a fluctuation of more than 1°C.

repellent any chemical that causes a negative chemotaxis response.

repertoire a term for the total number of distinct binding sites that a individual organism can generate either in antibody or T cell receptor combining sites.

replacement culture the resuspension of a microorganism in a medium other than its growth medium.

replicase the major virus-coded component of the viral replication complex; the RNA-dependent RNA polymerase.

replication 1. the synthesis of a copy of a DNA molecule, using the original as a template. 2. the synthesis of multiple copies of a virus genome transcribed from a complementary strand of RNA or DNA.

replication cycle the series of steps that a virus or cell goes through to multiply itself.

replication fork the point at which duplex DNA separates into two single strands during the process of DNA replication.

replicon a region of DNA that contains the origin for initiation of replication.

replisome the ensemble of proteins at the replication fork engaged in DNA replication.

reporter a gene that is fused to another gene, so that the expression of that gene may be assayed. The product of the reporter gene is typically more stable and easier to detect than the gene to which it is fused.

repression the inhibition of synthesis of an enzyme or other gene product often brought about by an end-product or other molecule that makes the enzyme reaction or the pathway in which it occurs unnecessary.

reservoir a host species or inanimate source that supports the long-term growth of a pathogen in nature, but that is not necessarily susceptible to disease caused by the pathogen.

resistance the ability of a host organism to restrict or mitigate the damaging effects of a pathogen, following infection.

resolution in microscopy, the ability to distinguish objects that are close together, defined as the minimum distance between points or parts of the object that appear as distinct points in the image.

respiration the sum of electron transfer reactions resulting in the reduction of oxygen (aerobically) or other electron acceptor (anaerobically) and the generation of protonmotive force.

respiratory chain a set of electron transfer components, thought of as arranged in a linear or branched fashion, that mediate electron transfer from a donor to an acceptor in aerobic or anaerobic respiration.

respiratory droplet a droplet produced by high velocity airflow over the wet respiratory mucosa, generated by coughs, sneezes, and other forced expiratory maneuvers.

respiratory protection a term for the process in which the highly active respiratory chain of *Azotobacter* species intercepts oxygen, thus preventing it from irreversibly inactivating nitrogenase.

respiratory syncytial virus (RSV) the most important cause of viral lower respiratory tract infections in infants and children. Related viruses are found in sheep and cattle.

respirofermentation yeast fermentative metabolism in the presence of oxygen.

respirometer a closed vessel in which oxygen consumption due to biological activity can be measured.

response regulator a protein containing a receiver domain, which brings about the final step in a signal transduction pathway resulting in regulation of the target of the system. For example, many response regulators are transcription factors that activate or repress gene expression upon phosphorylation of their receiver domain.

resting cell a quiescent, nongrowing cell.

restriction the inactivation of DNA by nucleolytic cleavage.

restriction endonuclease an enzyme that protects organisms by recognizing a specific foreign DNA sequence and degrading (restricting) the DNA strand at this site.

resuscitation any procedure that restores the ability of an organism to grow and develop after sublethal injury.

resynthesis the polymerization of nucleotides into excised regions of damaged nucleic acids.

reticulate bodies (RBs) noninfectious, intracellular chlamydial forms that grow and divide by binary fission.

reticulocytosis an increase in the number of immature forms of red blood cells (reticulocytes) in the bloodstream.

reticulum in ruminant mammals, a small pouch of that extends from anterior of the rumen. It collects either large feed particles for rumination or small particles for passage to the lower gut.

retinitis inflammation of the retina of the eye.

reversed electron transport the transfer of electrons energetically uphill toward the components of an electron transfer chain that have the more negative redox potentials. Such electron transfer can be regarded as the opposite of respiration and is driven by the protonmotive force.

reverse genetics a term for techniques that allow the introduction of specific mutations into a gene.

reverse transcriptase an RNA-dependent DNA polymerase that catalyzes the synthesis of DNA from an RNA template.

reverse transcription a process in which RNA is converted to DNA by the reverse transcriptase enzyme.

RFLP restriction fragment length polymorphism; a process in which restriction enzymes cleave DNA at very specific base sequences (restriction sites) into fragments of different length from different sources.

R-gene the plant gene that in response to a message from a specific pathogen attribute initiates a signal cascade that results typically in death of the host cell and cessation of further pathogen growth. Most R-genes are dominant.

Rhinovirus the genus of the common cold virus, human rhinovirus (HRV), constituting the largest genus in the family Picornaviridae, which includes small, nonenveloped, single-stranded (+)RNA viruses. HRV is the single most frequent causative agent of common colds, and probably the single most frequent etiologic agent of infectious disease in humans.

rhizobacteria a group of root-colonizing bacteria.

rhizobia a diverse group of nitrogen-fixing bacteria that form symbiotic relationships with certain legume plants; development of the symbiosis requires a series of molecular signals between the plant and bacteria, which results in formation of a plant nodule occupied by the bacteria.

rhizobiophages specific bacterial viruses (bacteriophages) virulent for *Rhizobium* and related genera.

rhizocoenosis the association of a microoganism with the root system of a plant, which results in a benefit to one or both of the partners.

rhizodeposition the release of materials (gaseous, soluble, particulate) from roots.

rhizoplane a root surface that can be colonized by microorganisms.

rhizosphere a region of soil around the plant root where materials released from the root modify microbial populations and their activities.

rhoptries club-shaped, apical organelles of certain parasitic protozoa that release their contents into the nascent parasitophorous vacuole.

riboflavin vitamin B_2, a water-soluble, yellow orange fluorescent flavin-based pigment that is essential to human nutrition.

ribosomal RNA (rRNA) a form of RNA that comprises part (approximately 60%) of the structure of a ribosome and participates in the synthesis of proteins.

ribosome a molecular machine that catalyzes the process of translation. It consists of several RNA molecules (ribosomal RNAs or rRNAs) and a few dozen ribosomal proteins.

ribotype a specific profile obtained after hybridization of digested bacterial DNA with a ribosomal RNA or DNA probe.

ribozyme an RNA molecule that can catalyze a chemical reaction.

Rickettsia any of various small, gram-negative bacteria characterized by obligate intracellular growth, associated with several diseases, including typhus and Rocky Mountain spotted fever. The various species are separated into a typhus group, a spotted fever group, a tsutsugamushi (Scrub typhus) group, and a miscellaneous group. (Named for Howard Taylor *Ricketts*, 1871–1910, U.S. pathologist who first established the etiology of Rocky Mountain spotted fever.)

rinderpest a morbillivirus that causes severe gastroenteritis in cattle, with inflammation and necrosis.

riparian relating to or occuring in the terrestrial area along river or stream banks.

RI plasmid root-inducing plasmid, a 200-kbp plasmid present in virulent *A. rhizogenes* strains, which confers virulence on the bacterium.

RNA ribonucleic acid, the nucleic acid material found in all living cells that transfers DNA information to direct protein synthesis. In certain viruses RNA serves as genetic material.

RNA-dependent see RNA-DIRECTED (below).

RNA-directed DNA polymerase a virally encoded reverse transcriptase that synthesizes complimentary strands of DNA using an RNA template. Also active on DNA templates.

RNA-directed RNA polymerase a virally encoded enzyme that synthesizes complimentary strands of RNA using an RNA template.

RNA hairpin structure a base-paired stem and loop structure that has sufficient stability to remain in the base-paired, hairpin configuration.

RNA polymerase one of several enzymes that synthesize RNA using a DNA template.

RNA polymerase II the specific enzyme that catalyzes the synthesis of RNA complementary to a DNA template. RNA polymerase II transcribes all genes except those encoding small RNAs such as transfer RNAs or ribosomal RNAs.

RNA splicing an RNA processing reaction in which intervening sequences (introns) in the primary RNA gene transcript are removed and the flanking sequences (exons) are joined together.

RNP a ribonucleoprotein complex consisting of the viral RNA genome segment, N protein, and a small number of RdRp molecules.

rosetting the adhesion of uninfected erythrocytes to an infected erythrocyte, particularly common in *P. falciparum* infections, though apparently strain dependent. The resulting clustering of cells is thought to exacerbate the blockage of blood capillaries in vital organs, particularly the brain.

rRNA see RIBOSOMAL RNA.

RSV see RESPIRATORY SYNCYTIAL VIRUS.

RT-PCR reverse transcriptase polymerase chain reaction, a process similar to conventional PCR, but with the starting material being RNA rather than DNA. Because of this, a DNA copy must first be made from the RNA, utilizing an enzyme known as reverse transcriptase. Once the copy of the DNA is made, the PCR proceeds as normal.

rumen the largest of the four compartments of the stomach of certain animals, harboring numerous microorganisms that metabolize ingested food and provide metabolic intermediates to the animal which serve as energy and biochemical precursors.

ruminant a herbivorous mammal that utilizes ruminal fermentation as a method of feed digestion.

rumination cud-chewing; the process by which certain mammals force large feed particles from the rumen up the esophagus to the mouth to be chewed again.

rust the disease that results from the interaction of a pathogenic rust fungus with a host plant. Although not strictly valid, the term is sometimes used to refer to the fungus itself.

rust fungus any of a group of phytopathogenic microfungi that comprise the order Uredinales of the phylum Basidiomycota. Rust fungi are cosmopolitan in distribution and parasitize a wide range of plants, including ferns and conifers and most families of dicotyledon and monocotyledon angiosperms. The term "rust" derives from the characteristic rust-colored spores produced on plants by many species.

S

saccharification a process in which a substrate is converted into low molecular weight saccharides by enzymes.

Saccharomyces cerevisiae baker's or brewer's yeast; a species which is used widely in the food and fermentation industries and is also being exploited in modern biotechnology (e.g., in the production of recombinant proteins) and as a model eukaryotic cell organism in fundamental biological research.

salvage pathways enzymatic reactions that allow an organism to utilize a partially preformed coenzyme or prosthetic group that originates either from the external environment (growth medium) or from intracellular degradation. A common example of "salvage" would be phosphorylation of a coenzyme that had become dephosphorylated.

sanitarian a term for a public health official who works to eradicate disease by disinfecting the environment and imposing sanitary restrictions on citizens.

sanitizer a chemical or physical agent used to eliminate or control microoorganisms.

sapwood the outer part of the trunk or branch of a tree between the bark and the heartwood.

scanning electron microscope (SEM) an electron microscope that is used to discern the topography of a sample through the emission of secondary electrons once the specimen has been energized by an electron beam. Thus, **scanning electron microscopy.**

scanning transmission electron microscope (STEM) an electron microscope that combines many of the operational principles of both an SEM and a TEM (transmission electron microscope).

schizont a multinucleate form of the malaria parasite; hepatic schizonts develop in the liver and contain many thousands of maturing merozoites, which are released into the bloodstream when the host cell ruptures; erythrocytic schizonts develop in the red blood cells and, depending upon the species of *Plasmodium*, each gives rise to 8–24 merozoites.

scintillons bioluminescent organelles unique to dinoflagellates which emit brief bright flashes of light following stimulation.

sclerotium *plural*, **sclerotia.** a survival structure consisting of a hard, frequently rounded or irregular mass of woven hyphal cells.

scotophobia literally, fear of darkness; a reversal or stoppage of flagellar rotation when a motile cell that is illuminated with attractant light begins to cross a light/dark boundary or is subjected to a decrease in light intensity.

scotophobic reacting to an absence of, or decrease in, light with a stoppage or reversal of motion.

Sec machinery components that promote protein export via the general secretion pathway.

secondary metabolite a molecule or substance usually not required for energy generation and growth, whose accumulation occurs at the later stages of growth and is regulated differently from primary metabolism.

Sec protein a protein involved in the secretory apparatus of an organism.

selected area electron diffraction (SAED) a microscopy technique in which a TEM (transmission electron microscope) is converted into an electron diffractometer so that the lattice spacings of small crystalline objects can be seen.

selection in this context, the laboratory technique of placing organisms under conditions where the growth of those with a particular genotype will be favored.

selective toxicity the ability of a chemotherapeutic agent (antibiotic, etc.) to kill or inhibit a microbial pathogen with minimal damage to the host at the concentrations used.

selenium a nonmetallic chemical element having the symbol Se; a micronutrient required by all life forms, as well as a potent toxicant when it is present at high concentrations in ecosystems.

selenium cycle the flux of selenium between components of the biosphere which includes the atmosphere, hydrosphere, and lithosphere. The chemical forms and oxidation states of selenium found in nature govern its hydrologic mobility, volatility, bioavailability, and toxicity. Important aspects of the environmental chemistry of this element are controlled by biologically-mediated redox reactions, and microorganisms play a significant role in these biogeochemical processes.

self-splicing describing splicing pathways where the determination of the splicing boundaries and catalysis are properties of the intron RNA.

SEM see SCANNING ELECTRON MICROSCOPE.

semi-permeable the property of a membrane that allows certain molecules (e.g., water) to freely cross the membrane boundary while blocking the passage of others (e.g., ions).

semi-starvation a state in which bacteria grow at a rate less than their maximal potential due to the limitation of an essential nutrient. Existence under conditions of starvation or semi-starvation is the norm for bacteria in nature.

semisynthetic describing compounds modified from natural products by chemical modifications.

Sendai virus (SN virus) a parainfluenza virus that is indigenous to mice.

sentinel surveillance in public health, the monitoring of key or "sentinel" health events in the general population or in special populations, which may be an early warning of a condition.

septation a mode of cell envelope invagination during division; it involves the ingrowth of the cell envelope forming a T-like structure.

septicemia the presence of pathogenic organisms or their toxins in the blood along with signs of clinical infection.

septum *plural*, **septa.** a partition that separates a cell or other body into two compartments.

ser see SEROTYPE.

serodiagnosis a diagnosis based on the presence (or absence) of antigen–antibody reactions in the blood serum.

serogroup bacteria of the same species with different antigenic determinants on the cell surface.

serologic or ***serological*** relating to antigen–antibody reaction in blood serum and other tissue fluids.

serology the study of antigen–antibody reaction in blood serum and other tissue fluids.

serotype 1. the process of classifying an infectious agent according to its antigenic constitution. 2. organisms grouped in this manner; bacteria with different proportions of the same antigenic determinants on the cell surface.

serovar see SEROTYPE.

sessile describing bacteria living within a biofilm.

sharps a collective laboratory term for hypodermic needles, syringes, disposable pipettes, capillary tubes, microscope slides, coverslips, and broken glass.

shear a process in which parallel bodies, layers, or surfaces move against each other in such a way as to remain parallel, but with deformation or displacement in a direction parallel to themselves.

shear rate a measure of the speed at which intermediate fluid layers move with respect to each other. The unit of measure is the reciprocal second.

shear stress a measure of the force per unit area required to produce shearing action. The unit of measure is dynes/cm^2.

sheath a structurally well defined, usually laminated, extracellular polysaccharide covering, escpecially such a layer in filamentous cyanobacteria.

shell vial culture a technique combining cell culture and immunofluorescence assay for the rapid detection of virus organisms.

Shine–Dalgarno sequence a region in 5′ UTR of eubacterial and archaeal mRNAs complementary to the region of 16S rRNA.

shingles an eruptive rash, usually in a girdle-like distribution on the trunk of the body, resulting from infection with varicella-zoster virus.

short-pass filter an optical filter that blocks wavelengths longer than a characteristic cutoff wavelength and transmits shorter wavelengths.

shotgun sequencing a random approach to whole genome sequencing, based on shearing of genomic DNA, followed by cloning, sequencing, and assembly of the entire genome, and ultimately leading to a finished, annotated genomic sequence. Sequencing of entire genomes can also be achieved by a sequential shotgun sequencing of a set of overlapping, large insert clones.

sick building syndrome (SBS) random symptoms of illness (e.g., respiratory ailments, headache, nausea) reported by building occupants when no causative agent has been identified; a modern diagnosis associated especially with workers in mechanically ventilated office buildings or factories.

siderophore a small non-protein molecule having a high affinity for iron, secreted by different microorganisms to scavenge ferric iron bound to eukaryotic proteins. Siderophores are separated into different biochemical groups.

sieve tube a series of cells, called sieve cells or elements, joined end to end, forming a tube through which nutrients are conducted.

sigma (σ) factor a transcription factor that recognizes specific DNA sequences and directs RNA polymerase to initiate transcription at these sites.

signal amplification a nucleic acid amplification procedure in which a signal or reporter molecule attached to the probe is detected, and the signal is amplified enormously.

signal peptide a region of a secreted protein that contains information for interaction with the secretion machinery.

sillon a longitudinally aligned groove on the spirochete surface where the outer membrane contacts the protoplasmic cylinder.

simian relating to, affecting, or involving apes or monkeys.

simian virus (SV) any of various viruses first identified in apes or monkeys, or isolated from their cell cultures; e.g., SV40, a contaminant of polio vaccines, or SV5, a virus related to human parainfluenza viruses that is a natural cause of respiratory disease in dogs.

sister chromosomes the two products of chromosome duplication, each containing a newly-polymerized DNA strand and a template strand formed during an earlier round of DNA replication.

site-specific packaging see PACKAGE.

S layer a crystalline surface layer, also called the microcapsule layer, which contains the species-specific surface protein antigen(s) in *Rickettsia*.

sleeping sickness see AFRICAN TRYPANOSOMIASIS.

snurps small nuclear ribonucleoprotein particles; particles of proteins and RNA that provide the specificity for splice boundaries and the catalytic center for the splicing of mRNA in the eukaryotic nucleus.

soda lake an alkaline lake in which the alkalinity is due to large amounts of Na_2CO_3 (or complexes of this salt).

soil the dynamic natural body comprising the surface layer of the earth, composed of mineral and organic materials and living organisms.

soil food web the intimately connected complex of organisms living within the soil pore labyrinth, with bacteria and fungi feeding primarily on organic matter and protozoa and nematodes feeding on microbes and on each other.

solids retention time (SRT) a measurement of the average time that biomass is maintained in a biological treatment process reactor.

soma the whole body of an organism excluding reproductive cells.

somaclone a clone produced from single cells grown in tissue culture, differing in resistance or other characteristics from parental plants.

somatic relating to or describing the body or cells of an organism other than reproductive cells.

somatic incompatibility the inability of hyphae from two fungi to recognize each other, fuse, and establish a continuous network of mycelium.

somatic mutation a process in which mutations are introduced into the DNA encoding the variable domains of antibody chains. The mutations change the specificity and binding strength of the antibody molecules. This, coupled with a selection process, produces a population of B cells, and therefore antibody molecules, that bind the antigen more tightly, resulting in the more rapid and efficient clearance of pathogens or their products. Since the mutation rate observed in the variable domains is several orders of magnitude higher than that generally seen in mammalian DNA, it is often referred to as *hypermutation*.

SOS regulatory network a group of at least 15 *E. coli* operons, located at different sites in the genome, that are induced in response to exposure of the cell to DNA-damaging agents; SOS-like regulons have been identified in several other bacterial species.

SOS response DNA damage-induced expression of a set of genes, the SOS genes, involved in the repair of DNA damage.

Southern blot(ting) a very sensitive method for detecting the presence among restriction fragments of DNA sequences complementary to a radiolabeled DNA or RNA sequence. Restriction fragments are separated by agarose gel electrophoresis, denatured to form single-stranded chains and then trapped in a cellulose nitrate filter on to which a probe suspension is poured. Hybridized fragments are detected by autoradiography, after washing off excess probe.

Southern transfer the transfer of denatured DNA from a gel to a solid matrix, such as a nitrocellulose filter, within which the denatured DNA can be maintained and hybridized to labeled probes (single-stranded DNA or RNA molecules).

speciation the fact of organisms being divided into different species, or the processes by which this takes place.

species the smallest functional unit of organismal diversity, consisting of a lineage evolving separately from others and with its own specific ecological niche.

specific immunity the ability of the system to discriminate between different antigens. The antigens may be significantly different, e.g., influenza and poliomyelitis viruses; or they may be very similar, e.g. human blood group substances A and B. Also, **specificity.**

spherical aberration aberration arising from rays that enter a lens at different distances from the optical axis and are focused at different distances along the axis when they are refracted by the lens.

spitzenkorper a refractive region near the hyphal tip in certain fungi, filled with many small vesicles.

spliceosome a large assembly of pre-mRNA, snurps, and other factors that constitute the mRNA splicing machinery in nuclei.

splicing the production of a mature messenger RNA molecule from the initial RNA transcript by the removal of intervening sequences (introns).

spongiform describing diseases marked by the formation of vacuoles (microscopic holes within cells) in the gray matter of the brain, giving it a sponge-like appearance. A highly characteristic feature of most TSEs (transmissible spongiform encephalopathies).

spontaneous generation historically, a proposed theory to account for the existence of microorganisms; the doctrine that living organisms arise from non-living matter. The matter may be inorganic (abiogenesis) or organic matter from previously living organisms (heterogenesis).

sporadic describing a disease that occurs on an occasional basis.

sporangiophore a filamentous cell bearing sporangia.

sporangiospore a spore produced within a sporangium.

sporangium *plural,* **sporangia.** a case, capsule, or other such structure in which spores are produced.

spore a specialized form of an organism that aids in survival or dispersion. Spores are a dormant form of bacteria. They are resistant to a variety of environmental stresses that would kill the vegetative (growing) form of the bacteria, including heat, desiccation, irradiation, and chemicals such as ethanol and chloroform.

sporotrichosis a chronic infection of the skin, subcutaneous tissue, and sometimes deep tissues with the fungus *Sporothrix schenckii.*

sporozoite the worm-like invasive form of the malaria parasite that develops in its thousands in the oocyst, and migrates to the salivary glands of the mosquito after the oocyst bursts. When the female mosquito takes a blood meal from the vertebrate host, the sporozoites are injected along with the saliva, and rapidly invade the liver cells.

sporulation the process by which quiescent spores are formed from vegetatively growing bacteria, a response to nutrient depletion or stress.

starch an α-glucan-based polymer, containing a mixture of amylose and amylopectin (amylose accounts for 17 to 25% of the total, depending on the source).

starter culture a microbial strain or a mixture of strains, species, or genera, used to effect a fermentation and bring about functional changes in food that lead to desirable characteristics in the fermented product.

startpoint the position on DNA that corresponds to the first base transcribed into RNA.

starvation an environmental condition in which bacteria do not grow at all due to the lack of an essential nutrient.

starvation promoters promoters that are selectively switched on during starvation or semi-starvation conditions. Their sequences are recognized by starvation-specific RNA polymerases.

starvation protein a protein whose level goes up in starving bacteria or that is uniquely synthesized in starving bacteria.

steady state the condition of a continuous culture in which changes in density and physiological state of the cells are no longer detectable.

stereoisomers molecules which are identical in all compositional matters (e.g., number, types of atoms and chemical bonds) but differing from each other in spacial orientation and in some physical property.

sterigma *plural,* **sterigmata.** a finger-like structure associated with a basidium on which basidiospores are borne.

sterilization the removal of all microorganisms from a material or space. Standard conditions for steam sterilization are 121°C, 15 psig for 15 minutes.

stop codon see TERMINATION CODON.

street rabies virus a term for a rabies virus isolated from natural infection that has not been serially passaged in animals or tissue culture.

stress 1. generally, any factor, external or internal, that disturbs the normal physiological balance (homeostasis) of an organism. 2. more specifically, an external condition that negatively affects the ability of an organism to grow and reproduce.

stromal keratitis inflammation of the deep layers of the cornea of the eye.

stylets the interlocking mandibles and maxillae of aphids; the mouthparts used to penetrate plant tissue and suck up sap.

subclinical describing an infection with no overt clinical signs of disease.

subcutaneous 1. occurring or located beneath the skin. 2. specifically, referring to the injection of a drug just under the skin, but not into the underlying muscle; a site for rather slow absorption of drug.

subdivision the largest phylogenetic grouping within a division. This grouping can be considered analogous to a phylum, and includes one or more orders.

subgingival occurring or located below the gingival margin between the the teeth and gingival tissue.

sublethal of an injury or condition, significantly damaging but not fatal.

sublethal injury the temporary loss in a microorganism of tolerance for specific conditions.

subliminal infection an infection in which the virus multiplies in the originally infected cell but is unable to spread to adjacent cells.

substrate 1. a structural layer beneath the exteranl surface layer. 2. a growth medium on which microorganisms feed.

substrate-level phosphorylation ATP (adenosine triphosphate) production in which an organic phosphate intermediate compound with a high-energy bond is formed, which then drives the phosphorylation of ADP (adenosine diphosphate) to ATP.

sulfate-reducing bacteria a group of bacteria belonging to a diversity of genera that gain their metabolic energy from the reduction of sulfate into sulfide.

sulfide-oxidizing bacteria a group of bacteria belonging to a diversity of genera that gain their metabolic energy from the oxidation of sulfide into sulfate. Under certain conditions, the oxidation is incomplete and stops at elemental sulfur, thiosulfate, or sulfite.

sulfur cycling a natural environmental process in which sequential transformation reactions convert the sulfur atom in its different valence states ranging from -2 to $+6$. The reactions of the sulfur cycle alter the chemical, physical and biological status of sulfur and its compounds so that the cycling can occur. Many of the reactions of the sulfur cycle are mediated by microorganisms.

sulfuretum *plural*, **sulfureta.** a habitat with a complete sulfur cycle.

sulfur-reducing bacteria a group of bacteria belonging to a diversity of genera that gain metabolic energy from the reduction of elemental sulfur into sulfide.

superantigen any of a number of soluble proteins that are secreted by bacteria and bind to the major histocompatibility complex of T lymphocytes. This stimulates a massive antigen-independent proliferation of T lymphocytes.

superinfection the subsequent infection of a virus-infected host by a second virus.

supragingival occurring or located on the tooth surface above the gingival margin.

surfactant a substance that reduces the surface tension of a liquid.

surveillance the ongoing, systematic collection, analysis, and interpretation of data essential to the planning, implementation, and evaluation of public health practice.

susceptibility the absence of protection against an antigen; the tendency to allow pathogens to invade and cause disease.

swarmer cell a motile cell that is unable to replicate its chromosome, and must differentiate into a stalked cell to divide.

symbiosis 1. an association between two organisms of different species from which each derives some benefit. 2. more generally, any living association of two dissimilar organisms, regardless of the effect on either.

symport the transport of a molecule up its chemical or electrochemical gradient with the concomitant movement, in the same direction but down its electrochemical gradient, of one or more protons or sodium ions.

symporter a membrane transporter protein catalyzing movement of the coupling ion in the same direction as the substrate molecule.

synapsis the portion of the recombinational process that positions homologous molecules of DNA to bring them into alignment (register).

synthase any of a broad class of enzymes that catalyze biosynthetic reactions. Synthetases require ATP to perform the condensation whereas synthases do not require ATP (they use pre-activated units, usually CoA thioesters).

synthetase see SYNTHASE.

syntrophic providing nutritional requirements for each other.

syntrophy an ecological relationship in which different organisms (or populations) provide nourishment for each other.

systemic involving the system; extending to or affecting internal organs or tissues.

systemic infection an infection that pervades the host tissue and sometimes the whole organism.

T

tachycardia a condition of abnormally rapid heartbeat.

tachyzoite the rapidly dividing, asexual life stage of certain protozoa. (From Greek *tachos* "fast" and *zoite* "little animal."

tannin a water-soluble phenolic compound having the property of combining with proteins, cellulose, gelatin, and pectin to form an insoluble complex.

Taq polymerase a thermostable DNA polymerase that was isolated from *Thermus aquaticus*. This enzyme allowed PCR to be easily automated. Due to the heat stability of the enzyme, fresh enzyme no longer had to be added after each amplification cycle.

target amplification a nucleic acid amplification procedure in which many copies of the nucleic acid target are made.

target site a site at which a transposition event occurs or has occurred.

taxon *plural*, **taxa.** a group of organisms classified as distinct from other groups on some basis, especially on the basis of sharing a common ancestor.

taxonomic relating to or based on principles of taxonomy.

taxonomy 1. the scientific principles and techniques that are involved in classifying organisms, especially in grouping organisms on the basis of common ancestry. 2. the classification of a particular organism in this manner.

TCA cycle the cyclic pathway by which the 2-carbon acetyl groups of acetyl-CoA are oxidized to carbon dioxide and water.

T cell or ***T lymphocyte*** a cell that plays the central role in the cellular immune system; one of the two basic types of lymphocyte cells along with B cells. Subtypes of T cells include the **cytotoxic T cell (CTL)**, which is directly cytolytic to the recognized cell, and the **T cell helper (Th)**, which augments the immune response by the production of cytokine in response to antigen-specific activation. (Given the designation *T* because they differentiate in the *thymus*.)

T cell receptor (TCR) the antigen receptor on T cells that is encoded in a set of rearranging gene segments in the genome.

T-DNA a segment of the *Agrobacterium* genome transferred to plant cells.

teichoic acid a cell wall-associated polymer in gram-positive bacteria; consisting of sugar alcohols connected by phosphodiester bonds.

teleomorph the sexual stage(s) of a fungus.

teliospore a globose, resting-spore stage of smuts and bunts, surviving for up to 25 years in storage.

telomere the specialized end region of a linear chromosome that carries multiple copies of a short DNA sequence, and which protects the chromosome from enzymic degradation and inappropriate joining to other chromosomes.

TEM see TRANSMISSION ELECTRON MICROSCOPE.

temperate in this context, a bacteriophage that can either infect and lyse a host with the production of further phages in an infectious cycle, or enter the lysogenic state and be stably inherited as part of the host genome.

temporal regulation the derepression of enzyme synthesis that occurs as a batch culture enters the stationary phase.

teratogenesis the production of fetal abnormalities by some agent or process.

terminal redundancy a DNA sequence repeated at the ends of a phage chromosome; generally indicative of headful packaging.

termination codon a group of three nucleotides (can be UAA, UGA, or UAG) on an mRNA molecule that do not encode an amino acid and therefore signal the end of protein synthesis.

terminator a DNA sequence that causes RNA polymerase to terminate transcription and to dissociate from the DNA template.

terpenoid lipids lipids that are based on isopentenyl group [$CH_2{=}C(CH_3)CH{=}CH_2$] giving rise to sterols, carotenoids, polyprenols and the side chains of chlorophylls and quinons.

tetanus an extremely painful spastic disease (also popularly known as **lockjaw**) affecting humans and animals, produced by the bacterium *Clostridium tetani.*

tetrahydromethanopterin a coenzyme that serves as the carbon carrier during sequential reduction of bound CO_2 to CH_3.

tetrapyrrole a molecule containing four pyrrole rings linked one to another through their α positions, usually via methylene or methene bridges; the tetrapyrrole can have a closed (macrocycle) or open (bilin) structure.

T-even phages a group of virulent bacteriophages (T2, T4, and T6) sharing the unusual characteristic that their DNA includes hydroxymethylcytosine rather than cytosine.

theca a protective case or covering; e.g., the total cell wall composed of cellulose plates in dinoflagellates.

therapeutic index the numerical ratio of the concentration of a substance that will achieve a desired effect in 50% of the patients, and the concentration that produces unacceptable toxicity in 50% of the patients.

thermal stratification the zonal separation of a body of water maintained by density gradients based on water temperature.

thermization the procedure of exposing microrganisms to a relatively mild thermal process (e.g. 63°C, 10–20 s).

thermoacidophile an organism thriving on heat and acidity; one of several archaean species, usually sulfur oxidizers, that grow optimally in extreme heat and extreme acid.

thermoacidophilic growing optimally in conditions of both extreme heat and extreme acid.

thermocline in the stratification of warm surface water over colder, deeper water, the transition zone of rapid temperature decline between the two layers.

thermoduric describing the ability of a microorganism to survive pasteurization.

thermophile literally, heat lover; an organism that grows optimally at temperatures between 50 and 80°C.

thermophilic growing optimally at temperatures between 50 and 80°C.

thioester a chemical entity formed by the esterification of an organic acid with a thiol.

thiol any of a group of organic compounds resembling alcohols, but with the oxygen of the hydroxl group replaced by sulfur.

thiotemplate mechanism a mechanism of non-ribosomal peptide biosynthesis in which the amino acids are activated as thioesters and transferred from one domain to another of the peptide synthetases by using a pantetheine arm.

thrombocytopenia a deficiency of platelets, the blood-clotting agents, in the blood.

thymus the central lymphoid organ in which T lymphocytes rearrange their receptor genes and undergo selection for self MHC (major histocompatibility complex) specificity and for self tolerance.

T-independent antigen an antigen capable of stimulating B cells to produce antibody directly without requiring T cell help.

tinea (pityriasis) versicolor an infection of the stratum corneum of the skin with the yeast-like fungal organism *Malassezia furfur.*

Ti plasmid tumor-inducing plasmid, a 200-kbp plasmid present in virulent *A.tumefaciens* strains, which confers virulence on the bacterium.

tissue cyst in certain protozoa, a protective structure containing the asexual bradyzoites in the intermediate host.

tissue engineering the design, development, and production of tissue cells (biomaterials) for use on or in humans.

tissue print immunoblotting a sensitive serological test in which the freshly cut surface of parts of a plant is pressed first on an adsorbing filter membrane. The virus particles that are bound to the membrane are detected by means of enzyme-labeled antibodies. The enzyme converts an unstained substrate into a dark-colored insoluble reaction product.

titer a measure of the concentration of specific antibodies in a serum sample. This is frequently expressed as the reciprocal of the highest dilution of the serum at which specific antibodies can be detected.

TK thymidine kinase, an enzyme that catalyzes the transfer of a phosphoryl group from a donor such as ATP (adenosine triphosphate) to the sugar (deoxyribose) component of the thymidine molecule, a building block of DNA.

T lymphocyte see T CELL.

TMD (transmembrane domain) the intramembranous part of an ABC transporter. It is formed of several hydrophobic helices that cross the lipid bilayer of the membrane. It contains the substrate-recognition sites.

tolerance a condition in which a pathogen colonizes a host extensively but few or no symptoms develop.

tolerant able to endure infection by a particular pathogen without showing disease symptoms, and therefore prone to be associated with latent infections.

top-fermenting describing the yeasts used to produce the type of beer known as ale, as opposed to lager beer which is bottom-fermented.

topical referring to the application of a drug directly onto the affected area, usually the skin; e.g., in ointment form.

topoisomerase an enzyme that alters the topology of DNA, either one strand at a time (Type I) or two strands at a time (Type II). Topoisomerases introduce and remove supercoils, tie and untie knots, and catenate and decatenate circular DNA molecules.

total/viable count estimates of the number of bacterial cells (individuals) in a population. After suitable dilution, total counts are made by microscopic observation or electronically. Viable counts are made by determining the number of colonies formed on a suitable medium. In general, total counts exceed viable counts.

toxicity the fact of being toxic (harmful to living organisms).

toxicology the study of organisms, substances, and agents that can induce disease or otherwise damage biological systems.

toxigenic or ***toxicogenic*** capable of producing toxins.

toxin 1. any substance or agent capable of harming living organisms. 2. specifically, a microbial organism, product, or component that causes injury or disease in multicellular organisms, including humans and animals.

toxin weapon (TW) any poisonous substance, whatever its origin or method of production, which can be produced by a living organism, or any poisonous isomer, homolog, or derivative of such a substance

toxoid a toxin which has been treated so as to inactivate its toxicity, but which is still capable of inducing immunity to the active toxin.

toxoiding the process of utlizing a detoxified form of a toxin for immunization.

trans-acting able to act on DNA after diffusing through the cytoplasm into the nucleus.

transconjugant a cell that has received a plasmid from a donor cell as a result of conjugation.

transcribe to synthesize RNA using the coding strand of a DNA.

transcription the synthesis of RNA using DNA as a template.

transcriptional attenuation a mechanism used to regulate continuation or termination of transcription.

transcription factors regulatory proteins that bind short specific DNA sequences which control the transcription of genes to which they are linked.

transcription pause a temporary pause or delay in RNA polymerase movement on its DNA template.

transcription pause structure an RNA hairpin that causes RNA polymerase to pause or stall during transcription.

transcription termination the cessation of RNA synthesis and release of transcript and DNA template from RNA polymerase.

transcription unit the DNA sequence expressed from a single promoter/control region.

transducer a device that provides an electronic signal generated by changes in a specific property in its microenvironment (e.g., for temperature, piezoelectric effects, light density effect).

transductant a bacterium that has received DNA material via the proces of transduction.

transduction the transfer, from a donor to a recipient cell, of nonviral genetic material within a viral coat. Transduction is one of the three ways by which genetic material can be moved from one bacterium to another. The others are transformation and conjugation. In **specialized transduction** only specific genes can be transferred, while in **generalized transduction** many genetic elements (plasmids or chromosomes) can be transferred between bacterial cells.

transesterification an interchange of esters.

transfectant carrying genetic material introduced by experimental genetics techniques.

transfection the introduction of genetic material into a cell by experimental procedures.

transfer intermediate a nucleoprotein particle composed of a single strand of the DNA destined for export and one or more proteins that facilitate DNA delivery to recipient cells.

transfer range the particular group of species to which genes can be transferred from another organism by a particular mechanism, e.g., conjugation.

transferrin (TF) an iron-binding and transport protein found in the serum and lymph of vertebrates.

transformation the process in which bacteria and microbial eukaryotes take up fragments of naked DNA from the extracellular medium to produce genetically different progeny cells called **transformants.** Natural transformation systems require the development of competence by the recipient cell as a specialized physiological state necessary for binding and uptake of exogenous DNA as linear fragments.

transgenic 1. describing a plant or animal into which a gene from another species (the **transgene**) has been introduced. 2. more generally, describing any organism whose genome has been altered by *in vitro* manipulation.

transgenic mice genetically modified mouse strains, expressing at least one foreign or altered gene; used as laboratory models; e.g., as cancer models.

transition a base pair substitution mutation that substitutes a purine for a purine or a pyrimidine for a pyrimidine.

translation the process of synthesis of polypeptides directed by the nucleic acid sequence of an mRNA (messenger RNA) molecule.

translational attenuation a specialized form of translational repression in which a translating ribosome is caused to cease the continued synthesis of a nascent protein.

translational repression a process of specifically turning off the translation of a specific mRNA, often by an RNA-binding protein that occludes an essential sequence required for recognition of the mRNA by the ribosome.

translocation see TRANSPOSITION.

transmissible spongiform encephalopathies (TSEs) a group of transmissible neurodegenerative diseases of animals and humans, characterized by distinctive spongy cortical formations, neuronal loss, and an accumulation of prions. Included are Creutzfeldt–Jakob disease (CJD), Gerstmann–Sträussler–Scheinker syndrome (GSS), fatal familial insomnia (FFI), bovine spongiform encephalopathy (BSE) and scrapie.

transmission electron microscope (TEM) an electron microscope producing an image from electrons that have passed through a (thin) specimen. The image depends on the electrons which are scattered by the specimen. Thus, **transmission electron microscopy.**

transovarial transmission in arthropods, the passage of an infectious agent (e.g., rickettsiae) from the ovary of an infected female to its eggs and ensuing offspring.

transposable element 1. generally, any DNA sequence that can move from one one location on a DNA molecule to another. 2. see TRANSPOSON.

transpose to move DNA material to a different site in the process of transposition.

transposition genetic rearrangement brought about by the relocation of a segment of DNA from one location to another. Transpositions are often mediated by transposons.

transposition immunity the inhibition of transposition of one copy of a transposable element by a second copy of the same element on the target DNA.

transposon a defined segment of DNA that can move (transpose) from one site to another in plasmids, chromosomes, or viruses via an event called transposition, in some cases leaving a gap at the original site but in other cases replicating the element so that one copy remains at the original site. In addition to genes involved in transposition, transposons carry other genes conferring selectable phenotypes such as antibiotic resistance.

transversion a type of base pair substitution mutation that substitutes a purine for a pyrimidine or a pyrimidine for a purine.

T region part of the Ti plasmid which is bracketed by a 24-bp direct repeat and which is transferred by *Agrobacterium* to plant cells. Similarly, the region in a binary cloning vector which is surrounded by this repeat and which is to be delivered into plant cells.

Treponema pallidum the causative agent of syphilis. It is extremely similar, possibly identical, to the organisms thatcause endemic syphilis, yaws, and pinta.

triacylglycerols a class of lipids in which the three alcohol groups of glycerol ($CH_2OH.CHOH.CH_2OH$) are esterified with a fatty acid. These, together with phospholipids, are the predominant lipid types in most eukaryotic cells.

trichocyst in dinoflagellates, an extrusome capable of sudden discharge outside of the cell. The exact role of trichocysts in dinoflagellates is not yet known.

trichome a row of vegetative cells in filamentous cyanobacteria, excluding the extracellular polysaccharide structures (sheaths). The term is complemented by "filament," which includes both the trichome and the sheaths.

triglyceride a substance composed of three fatty acids attached to a glycerol molecule.

tRNA transfer RNA, an adaptor molecule that translates information encoded in mRNA into an appropriate amino acid sequence.

trophozoite the feeding forms of the malaria parasite; hepatic trophozoites absorb nutrients from the liver cell, growing rapidly, and then divide internally to give multinucleate hepatic schizonts; erythrocytic trophozoites ingest hemoglobin from the red cell and break it down to provide free amino acids, as well as importing nutrients from the vertebrate plasma, before differentiating into erythrocytic schizonts.

tropism 1. generally, any movement or response of an organism to an external stimulus; e.g., light or heat. 2. the specific pattern of infectivity of a virus for the infection of a specific cell type or specific species.

trypanosomiasis a group of diverse diseases caused by members of the genus *Trypanosoma*, which are hemoflagellate protozoan parasites.

TSE see TRANSMISSIBLE SPONGIFORM ENCEPHALOPATHIES.

tuberculosis a chronic, widely distributed infectious condition affecting the lungs and other organs; an airborne disease caused by *Mycobacterium tuberculosis*.

tubular necrosis cell death in the tubule cells in the kidney.

tumor suppressor gene a gene whose product is normally involved in the negative regulation of cellular growth and division, thus inhibiting cancerous growth. The loss of function of a tumor suppressor gene, usually due to deletion or mutation, contributes to abnormal growth and transformation.

tumor necrosis factor (TNF) one of a family of membrane-anchored cytokines produced by cells of the immune system that can orchestrate innate and acquired host defenses against infectious and malignant diseases, through a corresponding group of specific cell surface receptors.

turgor the pressure exerted by the contents of a cell against the cell membrane or cell wall.

two-component system a signal transduction system composed of a sensor kinase and a response regulator. The kinase is activated when it senses some environmental or nutritional parameter. It then activates the response regulator, which alters gene expression in a manner that allows the bacterium to respond to the prevailing conditions.

two-dimensional protein crystals regular arrays of (glyco)proteins present as the outermost envelope component in many prokaryotic organisms.

tylose an abnormal outgrowth of xylem parenchyma cells that balloons into the xylem vessel lumen, restricting the flow of water.

U

ultraviolet germicidal irradiation (UVGI) a method of air disinfection using ultraviolet irradiation predominantly at the 254-nm wavelength (UV-C). UVGI has been used in ventilation ducts and in free-standing room air handling units, and more efficiently, in the upper room area to disinfect room air of airborne pathogens such as measles, influenza and tuberculosis.

umuD a genetic operon found on the *E. coli* chromosome that alters DNA polymerase III in such a way that fidelity is relaxed and synthesis can proceed past certain types of DNA damage.

uncoupler a term for a chemical that can disrupt linkage between phosphorylation of ADP and electron transport.

uniport the transport of an ionic species in direct response to the membrane potential across a membrane.

universal precaution the concept that all blood and body fluids should be considered potentially infectious, and appropriate procedures and barrier precautions (skin and mucus membranes) should be used to prevent exposure.

unsaturated zone in a body of water, the region that spans the area located just beneath the surface and directly above the water table.

UP element a sequence in the DNA of the promoter region of some transcription units that interacts directly with RNA polymerase to increase its activity at that promoter by up to 30-fold.

upland soils soils that, in contrast to wetland soils, are usually well aerated and generally oxic, but may contain anoxic microniches, especially when the water content is higher than field capacity, i.e., the water content that can be retained in soil against gravity.

uptake signal sequence (USS) the DNA sequence AAGTGCGGT or its complement, which causes fragments containing it to be preferentially taken up by competent cells.

upwelling the transport of water from the deep ocean to the surface, replacing the surface water that has moved offshore.

urease an enzyme that hydrolyzes urea; abundantly produced by *H. pylori* and essential for gastric colonization. Detection of urease activity is used to diagnose *H. pylori*.

usher a term for oligomeric outer membrane proteins that serve as assembly platforms for some types of pili. Usher proteins can also form channels through which nascent pili are extruded from bacteria.

uveitis inflammation of the iris or related structures in the eye.

V

vaccination the administration of an immunogen (toxoid) to stimulate an immune response that protects the host from infection by the microorganism that produces the immunogen. (From the Latin word for "cow"; the historically first vaccinations performed by Edward Jenner introduced cowpox matter in order to induce immunity against smallpox.)

vaccine an antigenic preparation (made in various ways) from an infectious organism, administered to humans or animals so as to induce protective immunity against a pathogen, but not produce full infection.

vacuolating cytotoxin (VacA) a novel protein that affects epithelial cell function. Multiple alleles exist that vary in VacA production *in vitro*. Colonization with particular genotypes affects the risk of disease development.

varicella-zoster virus (VZV) a member of the herpesvirus family that causes the disease chickenpox on initial infection and the condition zoster (shingles) on reactivation.

variola another term for smallpox, deriving from the Latin word for "pock" or "pustule."

variolation the introduction into the skin of pustular or scab material containing smallpox virus.

vascular system a network in vascular plants composed of two tissues, the phloem and the xylem, involved in the transport of water and nutrients.

vascular wilt disease a plant disease usually involving foliar flaccidity, caused by a fungal or bacterial pathogen that colonizes the vascular system, usually the xylem.

vector 1. an agent, usually animate, that serves to transmit a pathogen from one host to another, e.g., mosquitoes. 2. a genetically engineered DNA construct used to introduce and express a gene in a target cell.

vectorial capacity a mathematical form that expresses the malaria transmission risk, or, in other words, the 'receptivity' to malaria of a defined area.

vegetative cell a nonstarved actively growing bacterial cell.

vegetative compatibility the ability of individual fungal strains to undergo mutual hyphal anastamosis, which results in viable fused cells containing nuclei of both parental strains in a common cytoplasm (i.e., heterokaryon formation).

vegetative growth exponential growth that usually occurs by simple binary cell division and produces two identical progeny cells.

vertical reproduction an increase in the amount of genetic material, genes, or sets of genes through organismal reproduction rather than by transfer between organisms.

vertical resistance resistance conferred by the action of a single gene exerting a major effect expressed as immunity or hypersensitivity.

vertical transmission the transmission of microorganisms from a parent insect to its offspring.

viable count see TOTAL/VIABLE COUNT.

viniculture the growing of grapes for winemaking.

vinification the process or technology of winemaking.

viremia the presence of virus in the blood.

Vir gene a gene which is located in the virulence region of the Ti plasmid and which participates in the system with which the T region is transferred to plants.

virion a complete infectious virus particle. These particles may be composed only of nucleoproteins, or the nucleoprotein may be surrounded by an envelope consisting of lipid and protein.

viroid the smallest known plant pathogens, consisting of covalently closed circular RNA molecules of low molecular weight that do not encode proteins, but replicate using host plant enzymes.

viropexis the entry of virus particles into cells by engulfment, followed by the fusion of membrane-bound vesicles with lysosomes.

virulence the capacity of a specific organism to cause disease when introduced into a susceptible host.

virulence factor a protein or other molecule, or an inherent property, that contributes to the ability of a pathogen to cause disease.

virulent having the property of virulence; capable of causing disease when introduced into a susceptible host.

virus a noncellular microbial entity that is the smallest known biological system, consisting of a core of either DNA or RNA (but not both) enclosed in an outer protein coat, and, in some forms, an outer membrane. Viruses are not capable of independent metabolic and reproductive activity and remain in an inert state unless inside susceptible host cells (bacteria, plants, or animals). (From the Latin word for "poison," because of the association of viruses with disease.)

virus receptor a protein on the outer surface of a cell that serves as a site for adsorption of the virus to the cell and facilitates entry into the cytoplasm.

virus replication the synthesis of new virus particles, including both replication of the genome and assembly of virus particles.

viscosity a measure of the internal friction of a fluid in motion; the resistance to flow of a fluid subjected to a deforming force.

vitalism the historic concept that all living organisms possess a "life-force" or "vital principle" which accounts for the distinct properties of life, beyond the chemical and physical organization of the organism.

vitamin any of various compounds that occur in many foods in small amounts and are also produced synthetically; trace amounts are required for the normal physiologic and metabolic functioning of the body.

volatile fatty acids short-chain fatty acids that are major products of microbial fermentation in the gut.

VPg a virus-coded genome-linked protein, covalently attached at the 5′ end of the genomic RNA.

VZV see VARICELLA-ZOSTER VIRUS.

W

Walker motifs short-sequence motifs found by J. E. Walker in ATPases, myosin, kinases and other ATP-requiring enzymes. These motifs are involved in the recognition of nucleotides. They are also found in the nucleotide binding domains of ABC ATPases.

WAS waste activated sludge, activated sludge solids removed from the treatment system on a continuous or periodic basis, measured as pounds per day.

water activity (a_w) an index of the amount of water that is free to react. It is expressed as a_w, representing the ratio of the vapor pressure of the air in equilibrium with a solution (or a substance) to the vapor pressure of pure water at the same temperature. Pure distilled water has a water activity (a_w) of 1.

weathering a process that results from physical, chemical, and biological forces, leading to the destruction of original stone, and finally to soil formation.

Western blotting an immunological technique for the identification and characterization of protein antigen or antibody; a process in which proteins are first separated electrophoretically in a polyacrylamide gel on the basis of their molecular masses and then transferred, usually by means of an electric field, to a protein-binding membrane where they are detected by means of enzyme-labeled antibodies. The enzyme converts an unstained substrate into an dark-colored insoluble reaction product indicating the location of the virus.

wild yeasts a term for yeasts that are present in the brewing process that were not introduced purposely nor tolerated for a specific purpose during brewing.

wort the liquid that remains after mash is strained, containing soluble fermentable compounds.

X

xenobiotic foreign to life; describing a synthetic product not formed by natural biosynthetic processes; i.e., a manufactured substance.

xenotransplantation the transplantation into a human (or animal) of organs or cells from another species, in order to replace a nonfunctioning organ (such as heart, liver, kidney, or pancreas) or cellular component (such as a bone marrow transplant); e.g., the use of the domestic pig as an organ donor for humans.

xylem tube-like cells of the plant conductive tissue that carry water and minerals from the roots to other parts of the plant and also provide structural strength.

Y

Years Lived with Disability (YLDs) in public health, a measurement of the number of years lived with a disability of known severity and duration.

Years of Life Lost (YLLs) in public health, a measurement of the number of years of life lost due to premature death. Defined as a death occurring before the age to which the dying person could have expected to survive if he/she were a member of a standardized model population with a life expectancy at birth equal to that of one of the world's longest surviving populations, Japan (82.5 years for females and 80 years for males).

yeast a single-celled fungus that reproduces by budding or fission.

yeast artificial chromosomes (YACs) shuttle yeast–*E. coli* vectors, possessing all requirements (telomeric and centromere regions) allowing their reproduction and segregation in *S. cerevisiae*.

yellow fluorescent proteins accessory proteins in the bioluminescence system in some bacteria, carrying flavin chromophores and serving as secondary emitters under some conditions.

Yersinia pestis the agent of plague, a small gram-negative bacillus belonging to the genus *Yersinia*. This genus is part of the family Enterobacteriaceae but displays several characteristics that differentiate it from the other members of the family.

Z

zinc metalloprotease a proteolytic enzyme that requires a molecule of zinc at the active site.

zoonosis 1. an infection or infectious disease of vertebrate animals (e.g., birds, pigs, monkeys, cattle) that is transmissible under natural conditions to humans. 2. more generally, any disease of other organisms that may be transmitted to humans. 3. the process in which such disease transmission takes place.

zoonotic describing a human disease that derives from a nonhuman infection, especially a disease transmitted to people from vertebrate animals.

zooplankton the portion of the plankton community defined as animal life.

zoospore a free-living and flagellated reproductive structure of fungi.

zoster infection with varicella-zoster virus which leads to skin lesions on the trunk (usually) following the distribution of the sensory nerves; commonly called shingles.

zoster ophthalmicus eye infection with varicella-zoster virus.

Subject Index

Volume numbers are boldfaced, separated from the first page reference with a colon. Subsequent references to the same material are separated by commas.

A

Abacavir, antiretroviral therapy, **1**:305
ABC transport, *see* ATP-binding cassette transport
Abiogenesis, definition, **4**:364
Abiontic soil enzyme, definition, **3**:256
Abomasum, definition, **4**:185
Abscess, definition, **2**:562
Abzyme
  biocatalysts, **1**:640
  definition, **1**:636
ACDS, *see* Acetyl-CoA decarbonylase/synthase
Ace, cholera virulence factor, **1**:794
ACE, *see* Angiotensin-converting enzyme
aceBAK operon
  glyoxylate bypass, coarse control, **2**:556
  IclR regulation, **2**:558–559
  integrated host factor regulation, **2**:558–559
  transcriptional regulation, **2**:557–558
Acetic acid, cleavage to carbon dioxide and methane, **3**:195–197, 219
Acetic acid production
  Acetobacter aceti production, **1**:14
  aerobic processes, **1**:14
  anaerobic processes, **1**:14–15
  downstream processing, **1**:17
  fermentor designs
    batch fermenters, **1**:16
    cell-recycle fermenters, **1**:16–17
    draw-and-fill bioreactor, **1**:16
    evolution, **1**:15–16
  history, **1**:13–14
  uses, **1**:13–14
Acetogenesis
  autotrophic acetyl-coenzyme A pathway
    bacteria distribution, **1**:23
    carbon dioxide fixation, **1**:19–20
    carbon monoxide dehydrogenase/acetyl-CoA synthase, **1**:19–20, 26
    formate dehydrogenase, **1**:20
    free energy changes, **1**:24
    overview, **1**:18–19
  conservation of energy by acetogens
    acetyl-coenzyme A pathway, **1**:23–26
    electron receptor utilization other than carbon dioxide, **1**:26
  definition, **1**:18
  ecological impact, **1**:23
  general properties, **1**:18
  history of study, **1**:18
  species, **1**:20
Acetyl-CoA carboxylase, fatty acid biosynthesis, **3**:56–57
Acetyl-CoA decarbonylase/synthase (ACDS)
  acetate cleavage to carbon dioxide and methane, **3**:196–197
  definition, **3**:188
Acetyl phosphate, role in carbon and nitrate assimilation, **1**:681
Acid fermentation, definition, **2**:350–351, 356
Acid resistance, definition, **3**:625
Acid shock, definition, **3**:625
Acidicin B, features, **1**:391
Acidophile, *see also* pH stress; Sulfide-containing environments
  definition, **3**:625, **4**:495
  extreme
    adaptation biochemistry and physiology, **2**:323
    definition, **2**:317
    habitats
      artificial, **2**:322–323
      natural, **2**:322
    species, **2**:322–323
Acinus, definition, **4**:15
Acne, definition, **4**:271
ACP, *see* Acyl carrier protein
Acquired immunodeficiency syndrome (AIDS), *see also* Human immunodeficiency virus
  bath houses in transmission, **1**:109–110
  clinical course, **1**:105
  definition, **4**:81
  early history, **1**:108–112
  economic analysis, **2**:138–141
  emergence and recognition, **1**:106–108
  epidemiology, **1**:107–108, 112, 114–115
  government response, **1**:110, 112
  Koch's postulates and virology, **3**:233–234
  overview, **1**:104
  prevention, **1**:112, 114
  stigmatization, **1**:109
  surveillance, **4**:520–521
  transmission, **1**:105, 110
Acquired pellicle, definition, **3**:466
Acr, efflux system, **1**:724

Actinomycetes
antibiotic production
applications, 1:39–40
genes, 1:38
types, 1:39
biodegradation uses, 1:40
differentiation, 1:38
ecology
aquatic environment, 1:35
composts and moldy fodders, 1:35
miscellaneous habitats, 1:35–36
overview, 1:33–34
rhizosphere, 1:34–35
soil, 1:34
enzyme production, 1:40
genetic exchange, 1:37–38
genome, 1:37
morphology, 1:29–30
overview, 1:28
pathogenicity, 1:36–37
phages, 1:38
phylogenetic tree, 1:29
spores, 1:29–30
taxonomy
Actinomycineae, 1:33
cell wall chemotypes, 1:30
Corynebacterineae, 1:33
Frankineae, 1:32–33
Micromonosporinae, 1:31–32
Pseudonocardineae, 1:33
ribosomal DNA sequence analysis, 1:30–32
Streptomycineae, 1:33
table, 1:32
whole organism sugar patterns, 1:30
Actinorhodin, biosynthesis, 1:191
Action spectrum, definition, 3:618
Activated biofiltration system, municipal wastewater treatment, 4:878
Activated sludge, definition, 4:870
Activated sludge process
definition, 4:855
industrial effluent treatment, 2:751–752
industrial wastewater treatment, 4:860–862
municipal wastewater treatment
completely mixed activated sludge systems, 4:876
contact stabilization systems, 4:876–877
conventional activated sludge systems, 4:875–876
Active learning, definition, 2:156, 164
Active surveillance, definition, 4:506
Acyclovir
adverse effects, 1:292
formulations, 1:289
indications
genital herpes, 1:289–291
herpes labialis, 1:291
herpes simplex encephalitis, 1:291
herpes zoster, 1:292
immunocompromised host, 1:291
neonatal herpes simplex virus infection, 1:291
varicella, 1:291
mechanism of antiviral activity, 1:289
resistance, 1:292
valaciclovir prodrug, 1:289
Acylated homoserine lactone
definition, 4:1
quorum sensing signal
biofilms, 4:11
discovery, 4:3–4
ecology and functions, 4:10–12
host responses, 4:12
integration into cellular physiology, 4:10
interspecies communication, 4:11
mechanisms, 4:4
modulation of responses, 4:8
overview, 4:1–2
receptors
DNA binding and transcription activation, 4:7–8
interactions and cell entry, 4:6
repressor activity, 4:8
structure, 4:6–7
synthases, 4:5–6
Acyl carrier protein (ACP)
definition, 3:55
fatty acid biosynthesis, 3:56–57
Ada, DNA repair, 2:75–76
Adaptation, *see also* Extremophile; High-pressure habitats
definition and overview, 1:772–774, 2:283–284, 677
history of study, 2:693
natural selection, 2:287–288
protein methylation in adaptation
enzymology of methylation, 1:778
methionine requirement for chemotaxis, 1:776
methylaccepting chemotaxis proteins
structure and organization of sensing and signaling receptor domains, 1:777–778
types, 1:776–777
Adaptive immunity, definition, 1:729, 4:779
Adenovirus, transcription, 4:632–633
AdhE, fermentation regulation, 2:348
Adhesion, bacterial
adhesins
definition, 1:42
Escherichia coli, 1:43, 45–46, 50
examples and receptors, 1:44
Haemophilus influenzae, 1:48–49
Neisseria, 1:46, 48, 50
overview, 1:42–43
targeting for antimicrobial therapy, 1:50–51
consequences, 1:49–50
extracellular matrix binding, 1:49
Helicobacter pylori, 2:631–632
interaction forces, 1:42
skin bacteria, 4:274
Adjuvant
biotechnology industry, 2:736
definition, 3:745
ADP-glucose pyrophosphorylase
classification by activators, 2:544
cloning enzymes with altered allosteric properties, 2:546–547
ligand-binding sites, 2:546
regulation, 2:544–546
structure, 2:544
Adulteration, definition, 1:887
Aerated lagoon, industrial effluent treatment, 2:752, 754
Aerobe, definition, 4:65
Aerobic process, definition, 1:13
Aerobic respiration
definition, 1:53–54, 2:177, 4:316
Escherichia coli, 1:54
globins
flavohemoglobins, 1:67
single-domain myoglobin-like globins, 1:67
heterotrophs, 2:660–661
overview, 1:54
oxidases
cytochrome a_1 identity, 1:64
cytochrome *bd*, 1:63–65

cytochrome *bo'*, 1:63, 65
cytochrome cd_1, 1:64
cytochrome *c* oxidase
overview, 1:60–61
proton pumping, 1:62–63
steps in reaction, 1:61–62
subunits, 1:61
multiplicity and functions, 1:64–65
nomenclature, 1:60
non-respiratory functions, 1:65
overview, 1:60
synthesis, regulation in *Escherichia coli*, 1:64
taxonomic applications, 1:65
oxygen as electron acceptor
radicals and toxicity, 1:56–57
solubility, 1:56
thermodynamics, 1:54–56
oxygen-reactive proteins, classification, 1:59–60
respiratory chains
architecture, 1:57
assembly, 1:57
Helicobacter pylori, 1:58
mitochondrial paradigm, 1:57–58
organization in bacteria, 1:58–59
Paracoccus denitrificans, 1:58
Aeromonas
enteric disease, 2:199
food-borne illness, 2:403
Aeromonas hydrophila
refrigerated food pathogens, 4:70
skin microbiology, 4:284–286
Aerosol infection, *see also* Measles; Tuberculosis
comparison with droplet-borne infections, 1:77
definition, 1:69–70, 3:18, 4:921
disinfection of air, 1:74–76
environmental sources, 1:74
history of study, 1:70–72
human sources, 1:72–74
outbreak modeling, 1:73
prevention, 1:70
transport, 1:74
types, 1:70
AFA, adhesins, 1:46
A-factor, definition, 1:162
Aflatoxin
animal disease, 3:344
food poisoning, 2:410
hepatocarcinogenic activity, 3:340–341
human disease, 3:340–341
sources, 3:340
tolerances for foods, 3:340
types, 3:340
Aged residue, definition, 3:594
Agglutination reaction, diagnostics, 2:35–36
Aglycon, definition, 1:189
Agrobacterium
agrocin84 susceptibility, 1:80–81
autoinducer, 1:89
chromosomally-encoded virulence genes
attachment to plant cells, genes, 1:92
regulators of *vir* gene expression, 1:90–92
crown gall disease, 1:78, 86
genetic engineering of plants
applications, 1:100
barriers to transformation, 1:100
insertion site control, 1:100–101
monocot transformation, 1:99
overview, 1:84–86, 99–100
hairy root disease, 1:78, 86
host range, 1:98–99
infection cycle, 1:86–87
Ri plasmid, 1:79–80
T-DNA
definition, 1:78, 86
genes, 1:81, 87–89
integration, 1:81
processing
co-transported protein roles in transfer and plasmid conjugation, 1:93–95
movement and integration, 1:95
overview, 1:82–84
piloting and protection, 1:93–95
transfer intermediate formation, 1:92
VirE2 role, 1:92–93
transport system
components, 1:95
type IV transporters, 1:95–97, 101
functional similarities, 1:97
architecture, 1:97–98
T-pilus, 1:97
VirB ATPases, 1:97–98
VirB stimulation of IncQ plasmid uptake, 1:98
taxonomy, 1:78–79
VirB6 lipoprotein, 1:98
Ti plasmid
conjugation, 1:89, 859–860
genes, 1:79–81, 87–89
overview, 1:79–81, 87
virulence genes, *see vir* genes
yeast and fungi transformations, 1:99
AHL, regulation of *Erwinia* gene expression, 2:242–243
AIDS, *see* Acquired immunodeficiency syndrome
Airborne microorganisms, *see also* Aerosol infection; Ventilation
assays
biochemical assay, 1:124
chemical assay, 1:124
immunoassay, 1:124
polymerase chain reaction, 1:124
culture
algae, 1:123
bacteria, 1:123
fungi, 1:123
parasites, 1:123
viruses, 1:123
microbial volatile organic compounds, 1:119
microscopy, 1:123–124
overview, 1:116
sampling
forced air-flow samplers
filtration samplers, 1:122
impactor samplers, 1:121–122
impinger samplers, 1:122
passive/depositional sampling, 1:122–123
sources
building reservoirs, 1:120–121
occupants, 1:120
outdoor sources, 1:119–120
types
algae, 1:119
bacteria
gram-negative bacilli and endotoxin, 1:118
gram-positive bacilli, 1:117–118
gram-positive cocci, 1:117
fungi and mycotoxin, 1:118–119
parasites, 1:119
thermophilic actinomycetes, 1:119
viruses, 1:117
Airlift reactor, definition, 1:579
Airy disk, definition, 3:264

Akinetes, cyanobacteria, 1:916
Alanine
  biosynthesis, 1:143
  functions, 1:143
Alanine aminotransferase
  definition, 2:635
  liver function monitoring, 1:286, 2:637
Alate, definition, 3:99
Alcohol
  fermentation, *see* Beer brewing
  production from starch, 2:225–226
Alcoholic fermentation, *see also* Beer brewing; Wine
  definition, 2:350–351
  foods and organisms, 2:356–357
  techniques, 2:355–356
Algae
  diversity of species, 2:65–66
  lipid features, 3:66
Alginate
  Azobacter production, 1:366, 369–371
  definition, 1:359
  Pseudomonas virulence factor, 3:879
Alkaline environment, *see also* Soda desert; Soda lake
  genesis, 1:127–129
  overview, 1:126–127
Alkaline fermentation
  definition, 2:350–351
  foods and organisms, 2:356–357
Alkaline protease, *Pseudomonas* virulence factor, 3:882
Alkaliphile, *see also* pH stress
  adaptation biochemistry and physiology
    bioenergetics, 2:332
    pH homeostasis, 2:331–332
  definition, 1:126, 2:317, 3:625
  energy transduction, 2:185
  habitats, 2:331
  species, 2:331
Alkanotrophic, definition, 4:404
AlkS, transcriptional regulation of plasmids, 3:742
Allergy, overview, 1:228
Allochthonous, definition, 2:438, 485
Alpha helix, definition, 3:490
Alternative host, definition, 4:195
Alveolar bone, definition, 3:466
Alveolar macrophage, definition, 1:69, 3:18
Amantadine
  adverse effects, 1:299
  indications, 1:298–299
  influenza virus treatment, 2:808–809
  mechanism of antiviral activity, 1:298
  resistance, 1:299
Amastigote, definition, 3:27, 4:725
Ambisense genome, definition, 4:592
Amebiasis, *see Entamoeba histolytica*
Amino acid
  antifungal analogs, 1:253
  biosynthesis
    Lrp-controlled genes, 1:151
    pathways, *see specific amino acids and enzymes*
  commercial uses, 1:152
  composition of proteins, 1:136
  enzymatic production processes
    aspartate, 1:158–159
    prospects, 1:160–161
    tryptophan, 1:159–160
  fermentation, 2:346
  metabolic relationships, 1:134–135
  microbial production processes
    development of overproducing strains
      amino acid analogs for selection, 1:153
      genetic engineering, 1:153
      mutagenesis, 1:153
      overview, 1:152–153
    glutamate, 1:153–154
    lysine, 1:154–157
    prospects, 1:160–161
    threonine, 1:157–158
  secondary metabolites
    precursors, 4:216
    types of amino acids in secondary metabolites, 4:218
  stringent response, 1:150–151
  transport systems in *Escherichia coli*, 2:264
Aminoacyl transfer RNA, definition, 3:824, 828
Aminocyclitol, definition, 1:162
Aminoglycoside
  biosynthesis
    C_7-aminocyclitol-containing compounds, 1:168
    cyclitol pathways, 1:167
    2-deoxystreptamine-containing compounds, 1:168
    fortimicin group, 1:168
    streptomycins, 1:167–168
  clinical uses, 1:165, 168–169
  definition, 1:162, 2:9
  glycosidase inhibitors, 1:166–167
  nomenclature, 1:162
  resistance, 1:168–169
  structural classes and sources, 1:162, 164
  translational inhibitors, 1:165–166
6-Aminohexanoate, biodegradation pathway, 3:738
δ-Aminolevulinic acid, biosynthesis from metabolites
  glutamate, 4:559
  glycine and succinate, 4:559
  regulation, 4:568–569
Ammonia
  atmosphere cycling by microbes, 3:261–262
  biosynthesis, *see* Nitrogen fixation
  nitrate ammonification, 3:391
  oxidation, 3:384–385
Ammonia toxicity
  definition, 3:199
  methane production systems, 3:202
Amodiaquine, malaria treatment, 3:139
Amorolfine, antifungal activity, 1:248–249
ampC operon, *Escherichia coli*, 1:342–343
A motility gene, definition, 3:349, 360
Amphotericin B
  antifungal activity, 1:233, 237
  lipid formulations, 1:237–239
  mechanism of action, 3:778
  pharmacokinetics, 1:233
  resistance, 1:233
  structure, 1:233
Amplification, DNA, 3:307, 310
Amplification of transmission, definition, 2:170
Amylase
  α-amylase features and applications, 2:211, 223
  β-amylase features and applications, 2:212, 223
  catalytic mechanism, 1:175–176
  commercial uses
    detergents, 1:178
    products, 1:171, 177–178
    starch processing, 1:177–178
  families, 1:172–173
  microbiological sources

archaea, 1:173–174
bacteria, 1:173–174
eukaryotes, 1:173
selected species, 1:174
nomenclature and classification, 1:171–172
structures, 1:174–175
substrate specificity, 1:176–177
Amyloid plaque, definition, 3:809
Anabiosis, definition, 3:86
Anabolism, definition, 3:431
Anaerobe, definition, 1:834, 2:485, 562, 3:188, 4:65
Anaerobic bioreactors
definition, 4:855
industrial wastewater treatment, 4:865–866
Anaerobic contact reactor, industrial effluent treatment, 2:755
Anaerobic filter, industrial effluent treatment, 2:755–756
Anaerobic pond, industrial effluent treatment, 2:757
Anaerobic process, definition, 1:13
Anaerobic respiration
arsenate reduction, 1:336–337
components, 1:180–181
definition, 1:53, 180, 2:177, 4:316
diversity of electron acceptors and donors, 1:182
electron transport chain, 1:181
examples
carbon dioxide reduction, 1:187
dimethylsulfoxide reduction, 1:186–187
fumarate reduction, 1:186
heavy metal reduction, 1:187
nitrate reduction
denitrification pathway, 1:185
nitrate ammonification pathway, 1:185
overview, 1:184
sulfate reduction, 1:185–186
triethylamine-*N*-oxide reduction, 1:186
habitats of microbes, 1:181
heterotrophs, 2:661
regulation, 1:187–188
thermodynamics
coupling electron transport to proton motive force, 1:183–184
free energy calculations, 1:182–183
redox potentials, 1:183
Analog, definition, 3:418
Anamorph, definition, 2:338, 4:109
Ancillary factor, definition, 4:394
Angiotensin-converting enzyme (ACE), chiral biocatalysis of inhibitors, 1:431–433
Animalcule, definition, 2:677, 679
Animal feed, enzymatic processing, 2:228–229
Anorexia, definition, 1:286
Anoxygenic photosynthesis, definition, 4:478
Anthrax, *see Bacillus anthracis*; Biological warfare
Anti-contagionism
cholera, 1:803
definition, 1:801
Antibiotic, *see also specific compounds*
Actinomycetes production
applications, 1:39–40
genes, 1:38
types, 1:39
aminoglycosides, *see* Aminoglycoside
biosynthesis, *see also* Nonribosomal peptide synthesis
aromatic polyketides
act apo-acyl carrier protein structure, 1:196
actinorhodin, 1:191
gene cloning and analysis, 1:191, 193
structure, 1:191
synthases, genetic manipulation, 1:193, 195–196
β-lactams
gene cloning and analysis, 1:202–204
genetic manipulation of genes, 1:204
precursor tripeptide ACV synthesis, 1:202, 205
complex polyketides
erythromycin, 1:196–197
gene cloning and analysis, 1:197, 199
structure, 1:196
DEBS modular polyketide synthase, genetic manipulation
β-carbonyl-processing functions, 1:201
extender units, 1:201
polyketide chain length, 1:201
starter units, 1:201
structure studies, 1:202
systems for manipulation, 1:199
nonribosomal peptide antibiotics
gene cloning and analysis, 1:205–206
pathway for biosynthesis, 1:205
structure, 1:205
synthetases, genetic manipulation and structure, 1:206
overview, 1:189, 191
definition, 1:28, 2:722
enzymatic production, 2:233–234
Erwinia production, 2:256–257
immunogenicity, 1:392
β-lactams, *see* β-Lactams
plant disease resistance engineering, 3:674
polyketides, *see* Polyketides
Streptomyces biosynthesis organization and regulation of genes, 4:461, 463
Antibiotic resistance
definition, 2:562
gram-negative anaerobic pathogens
Bacteroides genes, 2:567–568
conjugative elements
clindamycin resistance, 2:569–570
5-nitroimidazole resistance, 2:570
plasmids, 2:569
tetracycline resistance, 2:569
transposons, 2:568–569
Pseudomonas, 3:882–883
Streptococcus pneumoniae, 4:447–448
Antibody, *see also* B cell
allelic exclusion, 1:215
binding
affinity of immunoglobulin classes, 1:221
forces, 1:213
biotechnology industry, 2:730–731
definition, 1:209, 211, 2:722, 4:767
diagnosis of disease, 1:230
function, 1:211
gene rearrangements, 1:214–216
monoclonal antibody, production and applications, 1:229–230
Neisseria immunity, 2:581–582
phage display of chains, 1:230
plant disease resistance engineering, 3:674
sources in mouth, 3:470

Antibody (*continued*)
structure
binding sites, 1:213
chains, 1:212
variable regions, 1:213, 222
Anticodon, definition, 3:824
Antifungal agents, *see also specific agents*
allylamines, 1:248
amino acid analogs, 1:253
azoles, 1:240–248
benanomycins, 1:251–252
benzimidazoles, 1:248
benzylamines, 1:248
cationic peptides, 1:252–253
cycloheximide, 1:239
definition, 1:232
echinocandins, 1:249
griseofulvin, 1:239
licensed agents
structures, 1:236
types, 1:232–233
morpholines, 1:248–249
nikkomycins, 1:252
papulocandins, 1:249
phthalimides, 1:252
pneumocandins, 1:251
polyenes, 1:233, 237–239
polyoxins, 1:252
pradimicins, 1:251
pyridines, 1:249
pyrimidines, synthetic, 1:240
pyrrolnitrin and derivatives, 1:240
sordarins, 1:252
summary of types, mechanisms and uses, 1:234
Antigen
definition, 1:208, 729, 3:1, 109, 4:81, 767
processing, *see* Major histocompatibility complex
Antigenic drift, definition, 1:254, 2:797
Antigenic shift, definition, 1:254, 2:797
Antigenic variation
Borrelia, 1:265
Candida albicans, 1:262–263
definition and overview, 1:254
functions, 1:266
Giardia lamblia, 1:260–261
Haemophilus influenzae, 1:263–264
human immunodeficiency virus, 1:266
influenza virus, 1:266
Neisseria, 1:264–265
Paramecium tetraurelia, 1:261–262
Plasmodium, 1:261
Pneumocystis carinii, 1:262
Salmonella, 1:265–266
Streptococcus, 1:265
telomeric location of expression sites, 1:266
Trypanosoma brucei
expression site-associated genes, variation of encoded receptors, 1:259
expression site, *in situ* activation, 1:257
invariant antigen protection, 1:259–260
life cycle, 1:255
procyclin, 1:255–256
variant surface glycoprotein
developmental control of expression, 1:256–257
repertoire programming and evolution, 1:258–259
replacement of gene, 1:257–258
stage-specific expression, 1:256
Antigen-presenting cell, definition, 4:583
Antimicrobial, definition, 1:887
Antimony, *see* Heavy metals
Antiport, definition, 2:177, 181
Antiporter, definition, 3:188
Antipsychotic agents, chiral biocatalysis, 1:441–443
Antisense
definition, 3:662, 4:81
plant disease resistance engineering, 3:671–672
Antisense RNA
definition, 1:268
naturally occuring RNAs, 1:269–270
overview, 1:268–269
posttranscriptional gene silencing
Caenorhabditis elegans, 1:282–283
homology-dependent virus resistance in transgenic plants, 1:280–281
overview, 1:280, 283
RNA duplexes in co-suppression of nuclear genes, 1:281–282
principles of antisense effects, 1:270–271
prokaryotic systems
artificial RNA design, 1:274
ColE1 copy number regulation by RNA I, 1:271–272
RNase III mediation of antisense activity, 1:274–276
stepwise binding of RNAs, 1:274
structure of antisense RNAs, 1:272, 274
protein mediation of effects, 1:270–271, 274–276
virus-infected mammalian cells
double-stranded RNA-binding proteins in interferon response
double-stranded RNA adenosine deaminase, 1:279
2′,5′-oligo(A) synthetase, 1:278–279
overview, 1:277–278
RNA-dependentprotein kinase, 1:278
evidence for viral sources of double-stranded RNA, 1:277, 279–280
immune signaling by double-stranded RNA, 1:271, 276
Antisepsis, definition, 2:677
Antitermination
definition, 4:127
overview, 4:625–626
ribosomal RNA synthesis regulation, 4:135–138
Antiterminator, definition, 1:339–340
α-1–Antitrypsin, expression in transgenic animal milk, 4:682–683
Antiviral agents, *see also specific agents*
chiral biocatalysis, 1:444
definition and overview, 1:288
enterovirus, 1:302
hepatitis
interferons, 1:300–301
prospects, 1:301–302
herpesvirus
acyclovir, 1:289–292
cidofovir, 1:292–293
famciclovir, 1:296–297
foscarnet, 1:293–294
ganciclovir, 1:294–295
idoxuridine, 1:295
prospects, 1:297–298
trifluorothymidine, 1:295
vidarabine, 1:297

paramyxoviruses, 3:545
respiratory viruses
amantadine, 1:298–299
prospects, 1:300
ribavirin, 1:299–300
rimantadine, 1:298–299
retrovirus
protease inhibitors
indinavir, 1:307–308
nelfinavir, 1:308–309
ritonavir, 1:308
saquinavir, 1:307
VX-478, 1:309
reverse transcriptase inhibitors
abacavir, 1:305
azidothymidine, 1:302–303
delavirdine, 1:306
didanosine, 1:303
efavirenz, 1:306–307
lamivudine, 1:305
nevirapine, 1:305–306
stavudine, 1:304–305
zalcitabine, 1:303–304
Antiviral state, definition, 2:826
Aperture iris, definition, 3:288
Apicoplast, definition, 3:745, 752–753, 4:598
Apnea, definition, 1:286
Apoptosis, definition, 1:208, 2:826, 3:456
Apothecia, definition, 2:338
Appressorium, definition, 3:801
Apterous, definition, 3:99
AqpZ, water transport in hypoosmotic stress, 3:513
Aqueous humor, definition, 1:286
Arabanase, features, 3:573
Arabinofuranosidase, features, 3:573
Arbovirus
definition, 1:311, 4:811
diseases
arthritis, 1:316–317
Colorado tick fever, 1:316
dengue, 1:313–314
equine encephalitis, 1:315
hemorhhagic fevers, 1:317
history and overview, 1:311–312
LaCrosse encephalitis, 1:316
St. Louis encephalitis, 1:316
tick-borne encephalitis, 1:316
undifferentiated fevers, 1:317
yellow fever, 1:311–312, 314–315
emerging infection, 4:825
focality and transport, 1:313
overwintering, 1:313
prevention and control, 1:317
transmission and pathogenesis, 1:312–313
Arbuscule, definition, 3:328
Archaea, *see also* Methanogen
cell membrane structure, 1:321–322
cell wall structure, 1:320–321
definition, 1:319, 2:317, 3:188
diversity of species, 2:59–61, 63
DNA replication, 1:329
Eubacteria comparison, 2:656
evolution, 1:330
extreme thermophiles
habitats, 1:324–325
molecular adaptations, 1:329–330
physiology, 1:325, 327
extremophiles, 2:318
genetic organization, 1:327–328
global distribution, 1:327
glycolysis, 1:655–656, 659
halophiles
habitats, 1:322
physiology, 1:322–323
heterotrophs, 2:656–658
lipid features, 3:64
methanogens
habitats, 1:323–324
physiology, 1:324
phylogenetic placement, 2:655–656
phylogenetic tree, 2:62
scotophobia, 3:622–623
taxonomy and phylogeny, 1:320
transcription and translation, 1:328–329
Archebiosis, definition, 4:364
Arenaviruses, emerging infection, 4:815
Arginine
biosynthesis
pathway, 1:139–140
regulation, 1:140
functions, 1:139
ARS, *see* Autonomously replicating sequence
Arsenic, *see also* Heavy metals
exposure, 1:335
forms, 1:332–333
fresh water cycling, 1:334
geochemical cycle, 1:332
microbial transformations
bioremediation applications, 1:338
chemolithotrophic metabolism, 1:332
dissimilatory reduction, 1:332
methylation, 1:337–338
oxidation, 1:335
reduction, 1:336–337
reservoirs, 1:333–334
sea water speciation, 1:334–335
toxicity, 1:333
uses, 1:333
Arthritis
arboviral, 1:316–317
definition, 3:109
Arthropod, definition, 4:758
Ascaris, food-borne illness, 2:408
Ascigerous tissue, definition, 4:297
Ascorbic acid, microbial production processes, 4:849–851
Ascus, definition, 2:468
Asepsis, definition, 2:677
A-site, ribosome, 3:824
Asparagine
biosynthesis, 1:140
functions, 1:140
Aspartame, enzymatic production, 2:233
Aspartate
biosynthesis, 1:140
enzymatic production process, 1:158–159
functions, 1:140
Aspergillosis, systemic infection, 2:463
Assimilation, definition, 1:134, 3:431
Assimilatory reduction, definition, 4:238, 495
Astaxanthin, definition, 3:647
Astrobiology, definition, 2:299, 305
Asymptomatic carrier, definition, 2:852
Atmosphere
composition, 3:257
oxygen evolution, 3:256–257
trace gases
consumption reactions
ammonia, 3:262
carbon monoxide, 3:261
carbonyl sulfide, 3:261
dimethylsulfide, 3:261–262
hydrogen, 3:261
methane, 3:261
nitric oxide, 3:262
nitrous oxide, 3:262
importance, 3:257–258
microbial processes in cycling, 3:258–259

Atmosphere (*continued*)
production reactions
ammonia, 3:261
carbon monoxide, 3:260
carbonyl sulfide, 3:260
dimethylsulfide, 3:260
hydrogen, 3:259–260
methane, 3:260
nitric oxide, 3:260–262
nitrous oxide, 3:260
sources, 3:257–258
types, 3:257
ATP, definition, 3:379
ATP-binding cassette (ABC) transport
binding protein dependence, 1:1–3
clustering patterns of subunits, 1:5–6
hydrophobic cytoplasmic membrane protein components, 1:4–5
iron uptake role, 2:861–862
mechanism
conformational changes, 1:9–10
domain interactions, 1:10–11
fusion protein studies, 1:9
models, 1:11–12
mutant analysis, 1:6, 9
reconstitution studies, 1:11
nucleotide-binding component, 1:5
outer membrane components of gram-negative bacteria, 1:3
periplasmic or extracytoplasmic substrate-binding components, 1:3–4
protein secretion pathway
energetics, 3:861–862
exporter structure, 3:860
overview, 3:860
signals, 3:860
translocation mechanism, 3:861
sequence alignments
conserved EAA region, 1:7
nucleotide-binding domains, 1:9
siderophore transport, 1:3
structure, 1:1, 12
subunit sequences and evolutionary implications, 1:5–6
ATP synthase
acetogenic bacteria, 1:24–26
definition, 1:180
F0F1, 2:177, 179
Atrazine, biodegradation pathway, 3:598, 731, 733
Attenuation
translation, *see* Protein biosynthesis
viruses, 2:677, 4:779–781, 832
Attenuation, transcription, *see also* Transcription
antiterminator, 1:340
binding protein-mediated attenuation
bgl operon, *Escherichia coli*, 1:343–344
pyr operon, *Bacillus subtilis*, 1:345
S10 operon, *Escherichia coli*, 1:345–346
trp operon, *Bacillus subtilis*, 1:344–345
definition, 1:134, 339
intrinsic terminator, 1:340
N protein-mediated antitermination in phage λ, 1:347
objectives and features, 1:339–340, 348
ribosome-mediated attenuation
ampC operon, *Escherichia coli*, 1:342–343
pyrBI operon, *Escherichia coli*, 1:342
rationale, 1:341
trp operon, *Escherichia coli*, 1:341–342
translation-mediated antitermination in *tna* operon, 1:347
tRNA-mediated attenuation, *Bacillus subtilis*, 1:346
Attenuator, definition, 1:339
Aufwuchs, definition, 2:438
Autochthonous, definition, 2:438, 485
Autoecious organism, definition, 4:195
Autoimmune disease, overview, 1:228
Autoinducer
Agrobacterium, 1:86, 89
definition, 1:520
Autolysin, definition, 1:759
Autonomously replicating sequence (ARS), definition, 1:822
Autophagy, definition, 3:18
Autoradiography, definition, 4:55
Autotroph, definition, 1:18, 2:317, 651
Autotrophic carbon dioxide metabolism
definition, 1:349
ecological importance, 1:349
fixation pathways
Calvin–Benson–Bassham pathway
molecular regulation, 1:352–353, 355
reactions, 1:351–352
rubisco, 1:351–352
distribution in nature, 1:351
3-hydroxypropionate pathway, 1:357
overview, 1:349–351
reductive tricarboxylic acid cycle
enzyme distribution, 1:356–357
molecular regulation, 1:356
reactions, 1:355–356
Auxostat, definition, 1:873
Auxotroph, definition, 1:692
Auxotrophy, definition, 2:767
Avermectin, mechanism of action, 3:776–777
Avirulence, definition, 4:297
Avirulence gene, definition, 3:662
Avirulent, definition, 3:801, 4:921
Axenic culture, definition, 1:461
Axial fibril, definition, 4:353
Azidothymidine (AZT)
adverse effects, 1:303
development, 1:112
indications, 1:302–303
mechanism of antiviral activity, 1:302
resistance, 1:303
Azobacter
alginate production, 1:366, 369–371
assimilation of fixed nitrogen, 1:367
carbon metabolism, 1:366–367
chromosome number, 1:364
commercial applications, 1:370–371
DNA content, 1:363
ecology
plant associations, 1:361–362
soil distribution, 1:361
life cycle
encystment, 1:362–363
germination, 1:363
molybdenum regulation, 1:366
nif
gene clusters, 1:364
NifL control, 1:365
regulation of genes, 1:364–366

nitrogenase
protection from oxygen
conformational protection, 1:368
respiratory protection, 1:368–369
reaction, 1:367
structure, 1:367–368
overview, 1:359–360
polyhydroxybutyrate production, 1:366
taxonomy, 1:360–361
terminal oxidases, 1:369
AZT, *see* Azidothymidine

B

Bacillus anthracis
skin microbiology, 4:283
vaccine, 4:775
Bacillus cereus
food-borne illness, 2:394
meat products as source, 3:166
noninflammatory diarrhea, 2:196
refrigerated food pathogens, 4:72–73
toxins, 2:196, 394
Bacillus sphaericus, insecticide application
strains, 2:815
toxins, 2:815–816
Bacillus subtilis, *see also* Sporulation
antibiotic production, 1:375
compatible solute transporters, 4:894–895
competence, 1:373–374
endospore formation, 2:16–17
enzyme production, 1:374
flux balance analysis, 2:518–519
genetic analysis
competence development, 1:378–379
enzyme production, 1:379
integrating plasmids, 1:377
sporulation
endospore development, 1:380–382
initiation, 1:379–380
two-component signal transduction systems, 1:378
genome
gene composition, 1:375
gene identity, 1:375–377
organization, 1:375
habitat, 1:374
heat shock response
CIRCE regulation, 2:603–605
HrcA regulation, 2:605
overview, 2:603–605
sigma factor B regulation, 2:605
species distribution of heat shock response mechanism, 2:605
operons, *see specific operons*
overview, 1:373–374
SOS response, 4:341–343
sporulation, 1:375
stringent response, 4:474
taxonomy, 1:374
transformation, 4:655–656
Bacillus thuringiensis
insecticide application
delta endotoxins and genes, 2:814–815
mechanism of action, 2:815
production and application, 2:815
strains, 2:814
toxins, overview, 1:544
plant disease resistance engineering with toxins, 3:674
Bacteremia
definition, 2:9
gram-negative anaerobic pathogens, 2:564
Neisseria, 2:573
Bacteria, *see also* Eubacteria
definition, 1:319
identification, computerized, *see* Identification of bacteria, computerized
Bacterial adhesion, *see* Adhesion, bacterial
Bacterial artificial chromosome
definition, 3:151
genome mapping, 3:158
Bacterial starvation, *see* Starvation, bacteria
Bacterial wilt, *see Ralstonia solanacaerum*
Bacteriochlorin, definition, 4:558
Bacteriochlorophyll, definition, 3:647
Bacteriocin
applications
alcoholic beverage production, 1:395
Ambicin, 1:395
cost-effectiveness, 1:394, 396
food preservatives, 1:392–396
genetic markers, 1:395–396
nisin, 1:392–394
classification, 1:384
cloacin DF13, 1:387–388
colicins, 1:387–388
definition, 1:383, 405, 2:236, 3:1, 163
detection assays, 1:384–385
gram-positive bacteria
class I bacteriocins, 1:388–390
class II bacteriocins, 1:390–391
class III bacteriocins, 1:391–392
history of study, 1:383–384
lactic acid bacteria production, 3:4–5
meat preservation, 3:170
microcins, 1:388
processing, 1:387
production, 1:385–386, 2:257
purification, 1:386–387
Bacteriocyte, definition, 4:526
Bacterioferritin
definition, 2:860
iron storage, 2:866–867
Bacteriophage, *see also* Transduction
applications
bacteria identification and classification, 1:409
disease therapy and prophylaxis, 1:409
genetic engineering, 1:409–410
miscellaneous applications, 1:410
biomonitoring application, 1:538
capsid, 4:637
concentration and purification, 1:399
dairy fermentations, effects and control, 2:6–7
definition and overview, 1:398, 530, 2:1, 270, 767, 3:1, 240, 363, 711, 4:451
geographical distribution, 1:406
gut, 2:496
habitats, 1:406
Haemophilus influenzae, 2:593
history of study, 2:690–691, 693
host range, 1:404, 406
identification, 1:399–400
isolation, 1:399
lytic cycle
adsorption, 1:407

Bacteriophage (*continued*)
infection, 1:407
multiplication, 1:407–408
release, 1:408
Mycobacterium, 3:322–323
origin and evolution, 1:405–406
persistence in environment, 1:407
pests and control, 1:410
pilus attachement, 1:853
plasmid comparison, 1:405
population sizes in nature, 1:406–407
propagation, 1:398–399
rhizobiophages, 3:405
Staphylococcus, 4:389–390
storage, 1:399
Streptomyces, 4:459–460
taxonomy
comparative biological properties, 1:402
cubic phages
DNA phages, 1:404
RNA phages, 1:404–405
dimensions and physiochemical properties, 1:402
family properties and frequency, 1:401, 403
filamentous phages, 1:405
overview, 1:400
pleomorphic phages, 1:405
tailed phages, 1:400–401, 404
temperate cycle
lysogeny, 1:408
pseudolysogeny and steady-state infections, 1:408–409
transduction and conversion, 1:409
transposing bacteriophages
D108, 4:721
Mu
replication and integration, 4:721
structure, 4:720–721
overview, 4:705–706
Bacteriophage λ
antirestriction systems, 2:101
cloning vectors, 4:59
features, 2:91
Bacteriophage T-even
features, 2:91
glucosylation of DNA, 2:91
Bacteriorhodopsin
definition, 2:177
electrochemical gradient generation, 2:184
Bacteroid, definition, 3:407
Bacteroides fragilis
antibiotic resistance, 2:567–568
capsular polysaccharide in abscess induction, 2:565
enteric disease, 2:199
BAL, *see* Bronchial alveolar lavage
Barophile, *see also* High-pressure habitats
adaptation biochemistry and physiology, 2:326–327
definition, 2:317, 664
habitats, 2:326
species, 2:326
Bartonella, *see also* Rickettsiae
cell wall, antigens, and virulence factors, 4:173–174
cultivation, 4:157, 160
genome, 4:166
growth environment, 4:153–155
identification, 4:175–176
species and disease, 4:148–149
Base excision repair, overview, 2:76–77
Base resistance, definition, 3:625
Basic nuclear protein, definition, 2:42
Basidium, definition, 2:468, 4:109
Basophil, immune response, 1:731
Batch process
definition, 3:9
lactic acid production, 3:13
B cell, *see also* Antibody
clonal selection hypothesis, 1:210
definition, 1:209
development
antigen induction
CD4+ T cell role, 1:222–224
class switching, 1:221, 224
CpG-containing DNA as antigen, 1:221–222
phosphorylative signal transduction, 1:225
plasma cell differentiation, 1:225–226
primary response, 1:220–221
proliferation induction, 1:219, 224–225
whole animal observations, 1:220–222
antigen-independent development
B1 cells, 1:219
bone marrow development, 1:216
mature virgin cells, 1:218
recombinant activating genes, 1:217–218
history of study, 1:216
overview, 1:211
termination of humoral response, 1:226
diseases
allergy, 1:228
autoimmune disease, 1:228
Bruton's X-linked agammaglobulinemia, 1:227–228
immunoglobulin A deficiency, 1:228
lymphomas, 1:229
immortalization by Epstein–Barr virus, 3:462
immunization, 1:230
prion pathogenesis role, 3:822
tolerance
comparison with antigen induction, 1:220
definition, 1:226–227
onset during development, 1:227
transgenic mouse studies, 1:227
Bdellovibrio, predatory lifestyle, 2:27–28
Beer brewing
African beers, 1:420–421
aging, 1:416
classification of beers, 1:412–413
fermentation, 1:415–416
finishing, 1:416
history, 1:413
home brewing, 1:421
ingredients, 1:413–414
malting, 1:414
mashing, 1:414–415
properties of beer, 1:416–417
spoilage problems
acetic acid bacteria, 1:418
anaerobic gram-negative bacteria, 1:419–420
control, 1:420
enterobacteriaceae, 1:418
killer yeasts, 1:420
lactic acid bacteria, 1:418
miscellaneous bacteria, 1:420
molds, 1:420
overview, 1:418
spore-forming bacteria, 1:419
wild yeasts, 1:420
wort processing, 1:415
yeasts
factors affecting growth, 1:417
Saccharomyces characteristics, 1:417

strain development, 1:417–418
Beet necrotic yellow vein virus (BNYVV)
cytopathic effects, 1:426
genome properties, 1:424–426
host range, 1:423
overview, 1:422–423
particles, properties and composition, 1:424
resistance genes, 1:423–424
serology, 1:426
strains, 1:426
taxonomy, 1:428
transmission, 1:423
Benanomycin A, antifungal activity, 1:251–252
Benthos, definition, 1:907, 2:438, 4:478
Benzene, toluene, ethylbenzene, and xylene isomers (BTEX), bioremediation, 1:589–591, 596, 598
Beta barrel, definition, 3:517
Beta blockers, synthesis with lipases, 3:53
Beta strand, definition, 3:490
bgl operon, *Escherichia coli*, 1:343–344
BglG, antitermination, 4:626
Bifonazole, antifungal activity, 1:241
Bilin
biosynthesis, 4:565
definition, 4:558
Bilirubin, definition, 1:286
Binary fission, definition, 2:584
Binary vector, definition, 1:78
Binding protein, definition, 1:1
Bioaerosol, definition, 1:116
Bioattenuation, *see* Bioremediation
Bioaugmentation, definition, 2:722, 724
Bioavailability
biodegradation of xenobiotics, 1:466–468
definition, 1:286, 461, 3:594
pesticides in soils, 3:605
Biocatalysis, *see* Biotransformation; Chiral pharmaceutical intermediates, biocatalysis
Biocatalyst, definition, 1:834
Biochemical oxygen demand (BOD)
biosensor monitoring, 1:615–616
definition, 1:611, 2:740
Biocide, industrial
applications, 1:448
classification, 1:449
combination biocides, 1:456–457
definition, 1:445
evaluation, 1:449
examples
overview, 1:448–449
table, 1:450–454
formaldehyde–adduct biocides, 1:456
historical perspective, 1:447–448
market, 1:445, 447
mechanisms of action, 1:449, 455–456
problems associated with use, 1:458–457
regulatory restrictions, 1:457–458
terminology related to control of microbial growth, 1:446
Biocontrol of weeds, *see* Weeds, biocontrol
Bioconversion, definition, 2:722, 724, 4:837
Biodegradation, *see also* Bioremediation
applications, 1:468–470
bioavailability factor, 1:466–468
definition, 1:445, 461, 3:449, 730
ecology of microorganisms, 1:462–463
enzymes, 1:463–464
mechanisms, 1:464–465
recalcitrance of xenobiotics, 1:465–466
wood, *see* Lignocellulose; Timber and forest products
Biodeterioration, *see also* Conservation, cultural heritage
architecture
algae, 1:474–475
bacteria, 1:473–474
cyanobacteria, 1:474–475
fungi, 1:474
significance of biodeterioration, 1:475
cultural heritage in hypogeal sites, 1:476
definition, 1:445, 472
paintings and sculptures, 1:476
wood rot, 1:472–473
Biodispersant, definition, 3:893
Bioenrichment
definition, 2:350
fermented foods
amino acids, 2:358
protein, 2:358
rationale, 2:357–358
vitamins, 2:358–359
Bioethanol, definition, 4:939
Biofilm, *see also* Oral microbiology
analytical techniques, 1:481
definition, 1:478, 2:131, 3:379, 466, 893, 4:1, 898
drinking water distribution system, 4:909–911
ecology, 2:133
exopolysaccharide production, 1:479–480
formation and detachment, 1:478–479
glycocalyx, 1:479–480
industrial systems and biofouling, 1:484
medicine
categories of biofilms, 1:481–482
dental plaque, 1:483
immune response, 1:483–484
prosthetic substrata, 1:482
urinary tract infection, 1:482–483
mixed-species biofilms, 1:480–481
paper machine biofilms, 3:894–896
quorum sensing, 4:11
structure, 1:480
Biogas, definition, 3:199
Biogenesis, definition, 4:364
Bioinformatics
biotechnology industry, 2:735
definition, 2:106
genome databases, 2:111–112
Bioleaching
definition, 3:482
ore leaching, *see* Ore leaching
Biological control, definition, 2:813
Biological nitrogen fixation, *see* Nitrogen fixation
Biological oxygen demand (BOD), definition, 2:767
Biological warfare
definition, 1:506
epidemiology models, 1:515–516
history
contemporary developments, 1:509–510
overview, 1:506–507
post-World War I to 1990, 1:508–509
pre-1925, 1:507–508
low-level conflict, 1:516
organisms regulated by the Centers for Disease control, 1:513

Biological warfare (*continued*)
research programs
Biological Defense Research Program, 1:509, 512–514
genetic engineering, 1:514–515
international programs, 1:514
offensive versus defensive research, 1:516–517
secrecy, 1:517–518
smallpox, 4:296
terrorism, 1:516
toxins, 1:506
treaties
biological warfare convention of 1972, 1:510–512
Geneva protocol of 1925, 1:510
review conferences, 1:512
United States legislation, 1:512
verification protocols, 1:518
Bioluminescence
bacteria
functions, 1:521–522
habitats, 1:520–522
luciferase
fluorescent proteins as emitters, 1:523
lux operon, 1:524
reaction, 1:522–523
structure, 1:522–523
regulation
autoinduction and quorum sensing, 1:524–525
dark variants, 1:525
glucose, 1:525
iron, 1:525
oxygen, 1:525
species, 1:521–522
definition, 1:520, 4:1
dinoflagellates
cell biology, 1:526
circadian control, 1:527–528
flashing, 1:527
luciferase
luciferin release, 1:526–527
structure, 1:526
species and habitats, 1:526
fungi
functions, 1:529
source of luminescence, 1:529
species and habitats, 1:528–529
Vibrio quorum sensing
autoinduction of *lux* genes, 4:2–3
model for cell density sensing, 4:3
regulation, 4:2
Vibrio fisheri, 4:9–10
Vibrio harveyi, 4:9
Biomagnification, definition, 3:171
Biomonitor, *see also* Biosensor
anaerobic bacteria, 1:537
applications, 1:536
bacteriophages, 1:538
coliform bacteria
fecal coliform, 1:536–537
overview, 1:535–536
sensitivity as indicators, 1:537–538
standards for drinking or recreational waters, 1:538–539
total coliform, 1:536
criteria for ideal indicator microorganism, 1:536
definition, 1:530
enteric microorganism
diseases, 1:533
features, 1:531–532
risks of microorganisms in water, 1:533–535
fecal streptococci and enterococci, 1:537
heterotrophic plate count bacteria, 1:537
regulation of water contamination, 1:538–539
waterborne diseases
agents, 1:532
outbreaks, 1:530–531
sources of microbial contamination, 1:532–533
Biopesticide, *see also* Insecticides, microbial; Weeds, biocontrol
bacteria, 1:544–545
commercialization, 1:553
definition, 1:541
fungi, 1:546
overview, 1:541–542
production methods
biphasic spore production, 1:550
liquid culture fermentation, 1:548–550
living host, 1:548
solid substrate fermentation, 1:550–552
selection, 1:547–548
stabilization
drying microbes, 1:552–553
freezing, 1:553
storage environment factors, 1:553–554
strategies
insecticidal toxins, 1:544
living microbes, 1:542–544
viruses, 1:546
Biopolymers, *see* Gums; Polyhydroxyalkanoates
Bioreactor
analytical strategies
invasive versus noninvasive analysis, 1:569, 575
ex situ analysis, 1:568–569
in situ analysis, 1:568
in-time analysis, 1:569
classification by reactants or products
gas phase, 1:580–581
liquid phase, 1:581–582
solid phase, 1:582–583
concentration of microorganisms, 1:583
control systems, 1:576–577
definition, 1:567, 579, 2:767, 3:9
design principles and analysis, 1:583–584
flow injection analysis, 1:570–571, 611–612
genetic stability, 1:585
historical perspective, 1:579–580
large-scale fermentation operations, 2:774, 777
media, *see* Nutrition, microbial culture
metabolic engineering, 1:584–585
modeling, 1:567–568, 575–576
photobioreactors, 1:583
protein engineering, 1:585
sampling, 1:569–570
sensors
biomass concentration, 1:574–575
biosensors, 1:572–573
definition, 1:567
foam, 1:571
gas volume, 1:572
metabolic state sensors, 1:574–575
optical sensors, 1:573–574
overview, 1:584
partial pressure for oxygen, 1:572
pH, 1:572
pressure, 1:571
software sensors, 1:576
temperature, 1:571
velocity, 1:571

viscosity, 1:572
volume, 1:571
sequential injection analysis, 1:570–571
sterilization, 1:585
Bioremediation, *see also* Biodegradation; Heavy metals; Oil pollution; Pesticide biodegradation
benzene, toluene, ethylbenzene, and xylene isomers, 1:589–591, 596, 598
bioaugmentation, 1:589
chlorinated hydrocarbons, 1:592, 599, 603, 606, 608
definition and overview, 1:587–588, 608, 834, 2:722, 724, 3:449, 594, 4:238
enhanced *ex situ* bioremediation
bioslurry treatment, 1:587, 607–608
composting, 1:605–607
land treatment, 1:604–605
overview, 1:603, 609
enhanced *in situ* bioremediation
biostimulation, 1:597–599
bioventing, 1:587, 600–602
electron donor delivery, 1:599–600
overview, 1:597–603, 608–609
permeable reactive barriers, 1:602–603
growth requirements, 1:589, 593–596
metabolic processes, 1:588–589
methanotrophs, 3:246–247
natural bioattenuation
demonstration, 1:596–597
overview, 1:593–596, 608
nitroaromatic compounds, 1:592–593, 606
oxygen demand, 1:595–596
phenols, 1:591–592
plasmid-encoded degradation enzymes, *see* Plasmid
polynuclear aromatic hydrocarbons, 1:591
Pseudomonas biodegradation of pollutants
alkanes, 3:889
benzoates, 3:887
chlorinated aromatic hydrocarbons, 3:886–887
cycloalkanes, 3:889
naphthalene, 3:887
nylon, 3:889
overview, 3:885–886
polychlorinated biphenyls, 3:887, 889
trichloroethane, 3:889–890
selection of technique by contaminant, 1:594
site characterization, 1:593
Biosensor, *see also* Biomonitor
advantages, 1:614
applications
bioreactors, 1:572–573
environmental monitoring, 1:615–616
food analysis, 1:615
medical diagnosis, 1:615
calibration, 1:613–614
configurations, 1:612
definition, 1:611, 2:722
enzymes, 2:234
historical perspective, 1:611
sensitivity, 1:614
stability, 1:616
transducers, 1:612
Bioslurry treatment, *see* Bioremediation
Biosorption
binding metals to microbial surfaces, 3:184–185
definition, 2:607, 611, 3:182–183
heavy metals
cell walls, 2:611
desorption, 2:611–614
free and immobilized biomass, 2:611
intracellular accumulation and complexation, 3:185–186
mechanisms, 3:183
precipitation, 3:184
Biostatic agent, definition, 1:445
Biosurfactant, *see also* Lipid production
advantages, 1:618–619
applications, 1:632–633, 635
classification schemes
enzyme-synthesized biosurfactants, 1:626–628
microbial producing source, 1:620, 622–626
structure of product, 1:619–620
comparison of types, 1:629
definition, 1:618, 3:62
enzyme synthesis, 1:626–628
exopolysaccharide bioemulsifiers, 1:624
glucose lipids, 1:623
mannosylerythrotol lipids, 1:626
microorganisms for production, table, 1:621
ornithin-containing lipids, 1:623–624
pentasaccharide lipids, 1:624
production from microorganisms
fermentation, 1:631, 633–634
media formulations, 1:628–631
overview of methods, 1:630
recovery, 1:634–635
rhamnose lipids, 1:620, 622–623
sophorose lipids, 1:625–626
sucrose lipids, 1:623
surfactin, 1:624
trehalose lipids, 1:623
Biosynthesis, definition, 4:316
Biotechnology, *see also specific applications*
agriculture industry
animal engineering, 2:728–729
plant engineering, 2:727–728
prospects, 2:729
definition, 2:222, 722
economics of industry, 2:722–723
fungus applications, 2:475–476
legal issues, 2:737
medical industry
adjuvants, 2:736
antibody production, 2:730–731
biochips, 2:735
bioinformatrics, 2:735
biosensors, 2:733–734
enzyme-linked immunosorbent assay, 2:731–732
genomics, 2:734
immunodiagnostics, 2:730
molecular diagnostics, 2:734
molecular imprinting, 2:734
proteomics, 2:734–735
rapid immunoassay, 2:732–733
therapeutics, 2:735–736
vaccination, needle-free, 2:736
microbial industry
bioaugmentation, 2:722, 724–725
bioconversion, 2:722, 724–725
bioremediation, 2:722, 724–725
enzymes, 2:723–724
metabolites, 2:723
recombinant DNA industry
complementary DNA and genomic libraries, 2:725
expression vectors, 2:725–727

Biotechnology (*continued*)
overview, 2:725
recombinant protein production, 2:725–727
social issues and ethics, 2:736–737
strain improvement, *see* Strain improvement
Biotin
biosynthesis, 1:841, 843
microbial production processes, 4:846
Biotope, definition, 2:438
Biotransformation, *see also* Chiral pharmaceutical intermediates, biocatalysis; Vitamin
applications
biodegradation intermediates, 1:642
drug metabolites, 1:642
green chemistry, 1:641–642
large-scale applications, 1:642–643
range of reaction, 1:640
stereochemistry, 1:640–641
biocatalysts
abzymes, 1:640
enzymes, 1:638–639, 643–644
microorganisms, 1:639–640, 644
overview of types, 1:638
plant cells, 1:640
definition, 1:461, 587, 636
history and development
enzyme catalysis, 1:637
utility in modern era, 1:637–638
whole-cell catalysis, 1:636–637
immobilization of biocatalysts, 1:644–645
mechanisms, 1:464–465
solvent selection for catalysis, 1:645
Biotroph, definition, 3:801
Biotrophic parasite, definition, 4:297
Biotrophy, definition, 2:117
Biotype
definition, 1:789
Vibrio cholerae, 1:790–792
Biovar, definition, 4:32
Bioventing, *see* Bioremediation
Birefringent structures, definition, 3:288
Birth scar, definition, 4:939
Black Death, economic analysis, 2:138–139
Blackleg, definition, 2:236
Blastomycosis, systemic infection, 2:463
bldA, sporulation role, 2:17
Blood culture
background, 2:9
devices, 2:9–10
Blood, microbe detection, *see* Blood culture; Diagnostic microbiology
Blue fluorescent protein, definition, 1:520, 523
BMS-181184, antifungal activity, 1:251–252
BMS-186318, chiral biocatalysis, 1:444
BMS-207147, antifungal activity, 1:246–247
BMY 14802, chiral biocatalysis, 1:441–443
BNYVV, *see* Beet necrotic yellow vein virus
BOD, *see* Biochemical oxygen demand
Border repeat, definition, 1:78, 86, 87
Borrelia burgdorferi, *see also* Lyme disease
adherence to host cells
adhesins, 3:123–124
laboratory culture, 3:124
receptors, 3:124–126
specificity for cell types, 3:124
animal reservoir, 3:112
antigenic variation, 1:265
arthropod vector, 3:112–113
cultivation, 3:111
genome structure, 3:111–112, 4:363
immune response, 3:115
intoxication of host cells, 3:127
invasion of host tissues, 3:126–127
Osp protein expression, 3:113
persistence of infection, 3:127–128
structure, 3:110–111
systems development for study, 3:129
vaccination, 3:128–129, 4:776–777
virulence mechanisms, overview, 3:123–124
Botulism
definition, 1:834
toxin, *see* Clostridia
Bovine spongiform encephalopathy (BSE)
emerging infection, 4:820–821
features, 3:819–820
Bradyzoite, definition, 4:598
Brewing, *see* Beer brewing
Bright-field microscopy, *see* Optical microscopy
Broad-spectrum antibacterial, definition, 3:773
Bromoxynil, biodegradation pathway, 3:733
Bronchial alveolar lavage (BAL), definition, 1:286
Bruton's X-linked agammaglobulinemia, overview, 1:227–228
BSE, *see* Bovine spongiform encephalopathy
BTEX, *see* Benzene, toluene, ethylbenzene, and xylene isomers
Bubble reactor, definition, 1:579
Bubonic plague, *see* Plague
Budding, definition, 4:939
Bud scar, definition, 4:939
Buffered charcoal yeast extract agar, definition, 3:18
Building, *see also* Sick building syndrome
building-related illness, definition, 1:116
deterioration, *see* Biodeterioration
sanitary design for cosmetic production, 1:891–892
ventilation, *see* ventilation
Bulk soil, definition, 4:321
Bunsen solubility coefficient, definition, 3:256
Bunts
control, 4:304–307, 314
economic impact, 4:302–303
history, 4:298–299
host genetic resistance, 4:307–308
host–parasite interactions, 4:304
infection cycle, 4:303–304
morphology, 4:299–300
overview, 4:297–298
taxonomy, 4:300–302
Burst, viral, 4:832
Butanediol operon, regulation in *Klebsiella*, 2:349
Buthiobate, antifungal activity, 1:249

C

Cadmium, *see* Heavy metals
Caenorhabditis elegans, double-stranded RNA-induced homology-dependent posttranscriptional gene silencing, 1:282–283
CagA, *Helicobacter pylori* virulence factor, 2:632

Calcitonin, expression in transgenic animal milk, 4:683–684
Calcium, growth requirements and functions, 3:438–439
Calcium channel blockers, chiral biocatalysis, 1:440
Calcium chloride transfection, overview, 4:660–661
Calvin–Benson–Bassham pathway
  molecular regulation, 1:352–353, 355
  reactions, 1:351–352
  rubisco, 1:351–352
cAMP, *see* Cyclic AMP
Camptosa, synthesis with lipases, 3:53
Campylobacter
  food-borne illness, 2:399–400
  growth conditions, 2:399
  inflammatory diarrhea, 2:191–192
  toxins, 2:192
Campylobacter jejuni, meat products as source, 3:167
CamR, transcriptional regulation of plasmids, 3:742
Candicidin, antifungal activity, 1:239
Candida albicans, antigenic variation, 1:262–263
Candidiasis
  cutaneous candidiasis, 2:453–455
  definition, 2:451
  systemic infection, 2:463–464
CAP, *see* Catabolite activator protein
Capsid, definition, 1:398
Capsular polysaccharide, *see* Extracellular polyaccharide
Capsule
  definition, 2:591, 4:387, 767
  Staphylococcus, 4:388
  Streptococcus pneumoniae, 4:444–445
Captopril
  chiral biocatalysis, 1:431–432
  synthesis with lipases, 3:53
Carbanion, definition, 3:171
Carbofuran, biodegradation, 3:598–599
Carbohydrate
  biosynthesis
    cell constituents, 1:662–663
    derived sugars, 1:648, 660, 662
    glycogenesis, 1:647, 659–660
  cell wall structure, 1:661–662
  central metabolism, 1:650
  definition and overview, 1:647–648
  degradation, *see also* Glycolysis
    monosaccharide, 1:649–650
    oligosaccharide, 1:649–650
    polysaccharide, 1:648–649
    recovery and salvage pathways, 1:663–665
  dry weight of microorganisms, 1:660
  export, 1:724
  genomic sequence studies, 1:665–667
  glycogen, *see* Glycogen
  origin and occurrence, 1:648
Carbon, growth requirements and functions, 3:436
Carbon assimilation, regulation with nitrogen assimilation
  acetyl phosphate role, 1:681
  carbon source selection, 1:673–674
  Escherichia coli, metabolism in log-phase aerobic cultures, 1:670, 672–673
  glnA regulation, 1:678–681
  glutamine synthetase regulation, 1:672, 676–678
  ntr genes, regulation, 1:678–679
  overview, 1:669–670
Carbon dioxide, *see also* Autotrophic carbon dioxide metabolism
  fixation
    acetogenesis, 1:19–20
    definition, 1:18
    methanogens, 3:220–221
  reduction in anaerobic respiration, 1:187
Carbon monoxide, atmosphere cycling by microbes, 3:260–262
Carbon monoxide dehydrogenase, definition, 3:188
Carbon monoxide dehydrogenase/acetyl-CoA synthase, acetogenesis, 1:19–20, 26
Carbonyl sulfide, atmosphere cycling by microbes, 3:260–262
Carboxymethyl cellulose, definition, 1:744
Carbuncle, definition, 4:271
Cardiolipin, biosynthesis, 3:58
Cardiomyopathy, definition, 1:286
Careers, microbiology, *see also* Education, microbiology
  academic career, 1:686–687
  clinical careers, 1:688
  demand, 1:684
  disciplines, definitions, 1:683
  education requirements, 1:684–686
  government careers, 1:687–688
  industry careers, 1:687
  management, 1:689
  medical, dental, or veterinary professions, 1:688
  professional affiliations, 1:689–690
  recruitment, 1:684
  research institutes, 1:687
  resources, 1:689–690
  salaries, 1:689
  teaching, 1:689
  technicians, 1:688
β-Carotene, microbial production processes, 4:839–841
Carotenoid
  commercial uses, 3:652
  definition, 3:647
  functions, 3:652
Carrier, definition, 4:755
Case definition, public health surveillance, 4:506, 509–510
Case-fatality rate, definition, 4:289
Caspofungin, antifungal activity, 1:251
Catabolism, definition, 3:431
Catabolite-activator protein (CAP)
  pilus synthesis regulation, 2:377
  regulation of gene expression, 1:674–675
Catabolite repression
  carbon source selection, 1:673
  definition, 1:669, 3:580
  extracellular enzymes, 2:217
  phosphotransferase system role, 3:586–590
Catalase, antioxidant defense, 3:529
Catechol, enzymatic production, 2:234
Caulobacter
  cell cycle
    Caulobacter crescentus
      cell division and DNA replication, regulation, 2:23, 25–26
      flagellum synthesis regulation, 2:22–23
      overview, 2:21–22
    overview, 1:692–693
  chemotaxis gene identification, 1:699–701
  counter-selectable markers, 1:697–699
  essential genes and inducible promoters, 1:699

Caulobacter (*continued*)
flagellum
gene identification, 1:699–701
synthesis and assembly studies, 1:702–703
generalized transduction and gene mapping, 1:695–696
genome sequencing, 1:703
mutagens, 1:694
plasmids
conjugation, 1:694
electroporation, 1:694
selection, 1:693–694
types, 1:693
promoters
analysis with reporter genes, 1:696–697
fusion, 1:701
transposon mutagenesis, 1:694, 701
Cause, definition, 3:227
CBE, *see* Cocoa butter equivalent
CCD, *see* Charge-coupled device
CD antigen, *see* Cluster of differentiation antigen
cDNA, *see* Complementary DNA
CEA, *see* Cost-effectiveness analysis
Cecropin, antifungal activity, 1:252
Cecum, definition, 2:485
Cell division, prokaryotes
chromosomal gene organization for division, 1:706–707
comparison to eukaryotes, 1:704
compartmentalization of division proteins, 1:707–708
divisisome assembly and fission, 1:708–709
Fts protein roles, 1:706–708, 725
model organisms, 1:704–705
patterns of cell groupings after division, 1:705
regulation, 1:709
Cell envelope, *see* Cell wall, bacterial; Cytoplasmic membrane; Outer membrane, gram-negative bacteria
Cell-mediated immunity, *see also* T cell
cell types in response
basophil, 1:731
dendritic cell, 1:733
eosinophil, 1:730–731
macrophage, 1:731–732, 742
natural killer cell, 1:731
neutrophil, 1:730
cytokines, *see specific cytokines*
definition, 2:562, 4:767
evasion by pathogens, 1:743
granulomatous response, 1:742–743
immunopathology, 1:743
innate immunity, 1:741
intracellular killing, 1:736–737
overview, 1:729–730
phagocytosis, 1:735–736
T cells
activation, 1:738–739
antigen processing
class I antigens, 1:737–738
class II antigens, 1:738
CD1-restricted T cells, 1:733
CD4+ T cells, 1:732
CD8+ T cells, 1:732
classification, overview, 1:732
cytotoxic T lymphocytes
target cell killing, 1:740–741
viral defense, 1:742
$\gamma\delta$ T cells, 1:732–733
helper cell differentiation, 1:739–740
memory, 1:741
unconventional T cells, 1:732
Cell membrane, *see* Cytoplasmic membrane
Cell sorter, definition, 2:131
Cell wall, bacteria
classification of cell envelopes, 1:899–900
crystalline bacterial cell surface layer, *see* Surface layer
definition, 4:387
enzyme secretion, 2:210–211, 219
Escherichia coli, 2:263
murein
biosynthesis
cytoplasmic reaction steps, 1:764
insertion into cell wall, 1:765–768
membrane translocation of precursors, 1:764–765
penicillin inhibition of transpeptidases, 1:767
lipoprotein, 1:763
metabolizing enzymes, 1:768–769
pseudomurein, 1:764
sacculus
barrier function, 1:768
definition, 1:759
growth and division, 1:769–771
structure, 1:768
structure, 1:761
Mycobacterium
arabinan galactan-linker unit biosynthesis, 3:317–318
architecture, 3:316–317
core structure, 3:317
lipoarabinomannan biosynthesis, 3:319
mycolic acids, structure and biosynthesis, 3:318–319
overview of cell envelope, 1:759–760
Staphylococcus, 4:387–388
Streptococcus pneumoniae, 4:445–446
structure, 1:661, 760
teichoic acid, 1:763–764
Cellobiose, definition, 1:744
Cellulase, *see also* Xylanase
applications
features and applications, 2:214–215
fuel and chemical production, 1:756–757
lignocellulosic material modification, 1:757–758
cell surface association, 1:749–750, 754–755
cellulosomes
anchoring to cell surface, 1:754–755
definition, 1:744, 749
quaternary structure, 1:753–754
classification, 1:750
definition, 1:744, 4:930
domains
catalytic domains, 1:750–752
cellulose-binding domains, 1:752–753
cohesin domains, 1:753
dockerin domains, 1:753
lignocellulose structure, 1:744–745
microorganisms
attachment to cellulose, 1:749–750
distribution and diversity, 1:745–747
regulation of expression, 1:755
substrate physical properties, 1:747–748
unassociated or transiently-associated cellulases, 1:748–749
Cellulitis, definition, 4:271, 282

Cellulose, *see also* Lignocellulose
degradation by insect symbiotic microorganisms, 4:532–533
enzymatic biomass conversions, 2:226
Centromere, definition, 1:822
Cephalosporin, *see* β-Lactams
Cephamycin, *see* β-Lactams
Ceranopril, chiral biocatalysis, 1:431, 433
Cer regulon, definition, 1:669
Chagas' disease, *see* *Trypanosoma cruzi*
Chaperone, *see also specific chaperones*
definition, 2:210, 236, 598, 4:394
heat stress response, 2:599–600
starvation proteins, 4:397–398
Charge-coupled device (CCD), definition, 3:264
Charged tRNA, definition, 1:339
Che proteins
chemotaxis role, 1:775–776, 778–780
phosphotransferase system interactions, 3:592
scotophobia role, 3:623
two-component system in chemotaxis, 4:747, 750, 4:753
Cheese, *see* Dairy products
Chemical oxygen demand (COD), definition, 2:740
Chemilithotroph, definition, 3:256
Chemiosmosis, definition, 1:53, 2:177, 618, 3:188
Chemoautotroph, definition, 1:18, definition, 1:349
Chemoheterotroph, definition, 2:651–652
Chemoheterotrophy, definition, 1:349
Chemokine
definition, 1:729
functions, 1:735
Chemolithotroph, definition, 2:317, 3:482
Chemolithotrophy, definition, 3:431, 4:316
Chemoorganotroph, definition, 2:317
Chemoorganotrophy, definition, 3:431
Chemostat, definition, 1:873, 2:584, 4:394
Chemotaxis
Caulobacter gene identification, 1:699–701
cyanobacteria, 1:923
definition, 1:729, 772, 873, 2:15, 131, 3:580
genetic analysis
Escherichia coli chemoreceptors, 1:775–776
mutants, 1:774–775
protein methylation in adaptation
enzymology of methylation, 1:778
methionine requirement for chemotaxis, 1:776
methylaccepting chemotaxis proteins
structure and organization of sensing and signaling receptor domains, 1:777–778
types, 1:776–777
response strategy
biased random walk, 1:772–773
excitation and adaptation, 1:773–774
temporal sensing and memory, 1:773
signal transduction
motor regulation and feedback control, 1:779
receptor–CheW–CheA signaling complexes, 1:778
Chemotherapy, definition, 2:677
Chemotrophic sulfur bacteria
competition with phototrophic bacteria, 4:493–494
ecological niches, 4:491–493
metabolism, 4:482, 498, 500
taxonomy, 4:491–492, 499–500
CheR, methylation, 3:243–244
Chill-proofing, beer brewing, 1:412
Chiral pharmaceutical intermediates, biocatalysis, *see also* Biotransformation
advantages, 1:430–431
angiotensin-converting enzyme inhibitors, 1:431–433
anti-arrhythmic agents, 1:441
antipsychotics, 1:441–443
antiviral agents, 1:444
calcium channel blockers, 1:440
cholesterol-lowering drugs, 1:437–438
paclitaxel semi-synthesis, 1:433–436
potassium channel openers, 1:440–441
thromboxane A2 antagonist, 1:436–437
Tigemonam, 1:443–444
Chirality, definition, 1:430, 2:222
Chlamydia
Chlamydia trachomatis features and pathogenesis, 4:261–263
clinical manifestations of infection
men, 4:259–260
miscellaneous infections, 4:260
neonates, 4:260
overview, 4:258–259
women, 4:259
diagnosis, 4:260–261
economic impact, 1:787
epidemiology, 4:257–258
growth, 1:782
immune response, 1:787
inclusion, 1:782
molecular biology, 1:784
morphology, 1:782
pathogenesis
associated disease syndromes, 1:785
diagnosis and treatment, 1:785
diversity, 1:782–784
genital infection, 1:784–785
non-human hosts, 1:787
ocular infection, 1:784
overview, 1:781–782
respiratory infection, 1:785
taxonomy, 1:781–782
treatment, 4:261
Chlorguanidine, malaria treatment, 3:139
Chlorin, definition, 4:558
Chlorinated hydrocarbons, bioremediation, 1:592, 599, 603, 606, 608
1-Chlorobenzyl-2–methylbenzimidazole, antifungal activity, 1:241
Chlorophyll
biosynthesis
bacteriochlorophylls, 4:567–568
chlorophyll *a*, 4:567
chlorophyll *b*, 4:567
chlorophyll *d*, 4:567
chlorosome chlorophylls, 4:568
isocyclic ring formation, 4:566
magnesium chelation, 4:566
photosynthesis reaction center pigments, 4:568
reduction to chlorin macrocycle, 4:566–567
vinyl reduction to ethyl group, 4:567

Chlorophyll (*continued*)
definition, 3:647, 4:558
Chloroquine, malaria treatment, 3:139
Cholera, *see also* Enteropathogenic bacteria; *Vibrio cholerae*
clinical presentation, 1:789, 802
diagnosis, 1:796–797
epidemiology, 1:798, 806
etymology, 1:801–802
history
anti-contagionism, 1:803
control and treatment, 1:790, 797–798, 806–807
geographical considerations, 1:806
miasmic theories, 1:802
overview, 1:789–791, 801
pandemics
fifth, 1:805
first, 1:790–791, 803–804
fourth, 1:805
second, 1:791, 804
seventh, 1:791, 805–806
sixth, 1:791, 805
third, 1:791, 804–805
immunity, 1:790, 798–799
pathogenesis, 1:789, 793–796, 802
societal impacts, 1:807
transmission, 1:792–793
treatment, 1:797–798
vaccination, 1:790, 798–799
Vibrio cholerae
biotypes, 1:790–792
features, 1:791, 799
serogroups, 1:790–792
virulence factors
Ace, 1:794
cholera toxin, 1:793–794, 799
hemagglutinins, 1:794–795
mutant screening, 1:795
regulation of expression, 1:795–796, 799
toxin-coregulated pili, 1:794, 799
Zot, 1:794
Cholesterol-lowering drugs, chiral biocatalysis, 1:437–438
Chromatic aberration, definition, 3:264
Chromium, *see* Heavy metals
Chromoblastomycosis, features, 2:457
Chromophore, definition, 3:647
Chromosome, bacterial
comparison with eukaryotes, 1:820
composition, 1:808
DNA bending, 1:815–816
DNA supercoiling, 1:808–809, 814–815
folding domains, 1:815
forms
circular, 1:810–811
linear, 1:811
gene arrangements and mapping, 1:811–813
Haemophilus influenzae, 2:592
historical perspective, 1:809–810
inactivation, 1:816
mobilization, 1:850–851
polyploidy, 1:811
recombination, 1:813–814
replication, *see also* DNA replication, *Escherichia coli*
chromosome packaging dynamics, 1:818–820
history of study, 1:816–817
origin of replication
binding sites for primary initiation proteins, 1:825–826
DNA methylation, 1:827
DNA unwinding elements, 1:826–827
initiation of replication, 1:817–818, 820, 823–824
nucleoprotein complex assembly/disassembly, 1:827–830
spacer sequences, 1:827
structure, 1:824–825
overview, 1:808, 822
replication fork, 1:817
tangle resolution, 1:818, 820
segregation, 1:832–833
Streptomyces
architecture, 4:452–453, 455
genetic instability, 4:455–456
Chromosome, dinoflagellates
segregation, 2:50
structure, 2:46
transcription and replication, 2:49
Chromosome, yeast
replication origin
binding sites for primary initiation proteins, 1:825–826
nucleoprotein complex assembly/disassembly, 1:829–830
overview, 1:824
segregation, 1:830–832
Chronic obstructive pulmonary disease (COPD), definition, 1:286
Chv proteins, functions, 1:92
Cidofovir
adverse effects, 1:293
indications, 1:293
mechanism of antiviral activity, 1:292–293
resistance, 1:293
Ciguatera, poisoning, 2:409
Cilofungin, antifungal activity, 1:249
CIRCE, heat shock response regulation, 2:603–605
Cistron, definition, 3:697
CJD, *see* Creutzfeldt–Jacob disease
Class II recall, definition, 1:887
Classification, definition, 2:709
Claviceps, *see* Ergot
ClcR, transcriptional regulation of plasmids, 3:742
Cleistothecum, definition, 3:801
Cloacin DF13, features, 1:387–388
Clone, definition, 3:466
Cloning
definition, 4:55, 428
direct selection by complementation, 4:56–57
Escherichia coli host selection, 4:58
library screening, 4:57–58
protein product characterization, 4:58–59
reverse genetics, 4:58
vectors
bacteriophage λ, 4:59
definition, 3:1
plasmids, 4:59
Clostridia, *see also* Enteropathogenic bacteria
acetic acid production, 1:14–15, 18
applications
enzyme production, 1:838
solvent production, 1:838
toxins as pharmaceuticals, 1:839
definition, 1:834
endospores, 1:836
food poisoning, 1:838
genome features, 1:836–837
history of study, 1:834–835
metabolic properties, 1:835–836
phages, 1:837
plasmids, 1:837
taxonomy, 1:835
toxins, 1:837–838
Clostridium acetobutylicum, butyrate–butanol fermentation regulation, 2:349

Clostridium botulinum
enteric disease, 2:199–200
food-borne illness, 2:392–393
inflammatory diarrhea, 2:192
meat products as source, 3:166
refrigerated food pathogens, 4:73
toxins, 2:200, 192, 392–393
treatment of botulism, 2:393
Clostridium perfringens
food-borne illness, 2:396
meat products as source, 3:166
noninflammatory diarrhea, 2:196–197
skin microbiology, 4:283
toxins, 2:197, 396
Clotrimazole, antifungal activity, 1:241
ClpXP, sigma factor S degradation, 4:399–400, 403
Cluster of differentiation (CD) antigen, definition, 1:208
Cnidocyst, definition, 2:42
CoA, *see* Coenzyme A
Coat protein-mediated resistance
definition, 3:662
plant disease resistance engineering, 3:668, 673
Cobalamin, *see* Vitamin B_{12}
Cobalt, growth requirements and functions, 3:439
Coccidioidomycosis, systemic infection, 2:464
Cocoa butter equivalent (CBE), lipid production, 3:66–68
COD, *see* Chemical oxygen demand
Codon, definition, 1:286, 3:824, 4:451, 690
Coelome, definition, 2:42
Coenzyme, *see also* Vitamin
biosynthesis, *see specific compounds*
compounds with unknown synthetic pathways, 1:841, 846
definition, 1:840, 4:837
vitamin relationships, 1:840–841
Coenzyme A (CoA)
biosynthesis, 1:843
microbial production processes, 4:845–846
Coenzyme M (CoM)
carbon dioxide reduction, 3:189, 193–194
definition, 3:188, 204
Cohesive domain, definition, 1:744, 753
Coiling phagocytosis, definition, 3:18
Cointegration, definition, 3:730
Cold shock response, definition, 4:545
ColE1, copy number regulation by RNA I, 1:271–272
Colicins
definition, 1:383
features, 1:387–388
Coliform
definition, 1:530
fecal coliform, 1:536–537
overview, 1:535–536
sensitivity as indicators, 1:537–538
standards for drinking or recreational waters, 1:538–539
total coliform, 1:536
Coliphage, definition, 1:530
Colitis, definition, 2:187
Collateral host, definition, 4:195
Colon, definition, 2:485
Colony-forming unit, definition, 2:29
Color, definition, 3:647
Colorado tick fever, overview, 1:316
CoM, *see* Coenzyme M
Combinatorial joining, definition, 1:208
Cometabolism
definition, 1:461, 587, 3:594, 4:117
pesticide biodegradation, 3:599–601
ComK, regulation of competence development, 4:653–654
Commensalism, definition, 4:344
Commercial sterility, definition, 2:1
Common cold
clinical illness, 4:104–105
definition, 4:97
diagnosis
isolation of virus in cell culture, 4:105
polymerase chain reaction, 4:105–106
serology, 4:105
immune responses, 4:104
incidence and prevalence, 4:101–102
pathogenesis of symptoms, 4:102–104
prevention of infection, 4:106
seasonality, 4:102
transmission routes, 4:102
treatment
antiviral therapy, 4:106–107
symptomatic remedies, 4:107
Community, definition, 2:131, 3:466
Compatible solute, *see also* Osmotic stress; *specific compounds*
amino acids and derivatives, 3:508, 511
biosynthesis, 4:889, 891–892
characteristics, 4:887–889
commercial applications, 4:896
definition, 3:502, 4:884
function, 3:505
metabolism, 4:895–896
methylamines, 3:509, 512
miscellaneous osmolytes, 3:510
sugars and polyols, 3:506–507, 509–511
transporters
Bacillus subtilis, 4:894–895
efflux systems, 4:895–896
Escherichia coli, 4:893–894
features, 4:892–893
Salmonella typhimurium, 4:893–894
Competence
definition, 1:373, 4:651
regulation of development, 4:653–654
Competitive exclusion, definition, 2:485
Complement, definition, 1:208
Complementary DNA (cDNA), definition, 4:55, 81
Complementarity, oligonucleotides, 1:268
α-Complementation, definition, 4:55
Complement fixation assay, diagnostics, 2:35–36
Composite transposon, definition, 4:704
Composting
bioremediation, 1:605–607
definition, 1:587
sludge treatment, 2:761–762
Computerized identification of bacteria, *see* Identification of bacteria, computerized
Concatemer, definition, 4:637
Condenser lens, definition, 3:288
Condyloma acuminatum
definition, 1:286
interferon therapy, 1:301
Confinement, definition, 2:499
Confocal microscopy
aberrations, 3:271
applications, 3:265
comparison with conventional microscopy, 3:265–266

Confocal microscopy (*continued*)
fluorescence recovery after photobleaching, 3:299–300
instrumentation, 3:268–269
multiwavelength experiments, 3:267–268
optical sectioning, 3:266–267
4–Pi microscopy, 3:275
resolution, spatial and temporal, 3:269–270, 274
specimen preparation, 3:271–272
tandem scanning microscope, 3:270–271
Theta confocal microscopy, 3:275
two-photon excitation, 3:273–275
Conformational protection
definition, 1:359
nitrogenase, 1:368
Conidium
definition, 3:801
downy mildew, 2:120
Conjugal pilus, definition, 1:86
Conjugation, *see also* Horizontal gene transfer
chromosome mobilization, 1:850–851
definition, 1:86, 692, 847, 2:91, 270, 699–700, 3:711, 4:43
efficiency, physiological factors, 1:849
evolutionary relationships of systems, 1:861
examples in nature, 1:861
functions, 1:847–848
gram-negative conjugation
DNA transfer mechanism, 1:855
fertility inhibition, 1:856
leading region expression, 1:855–856
mating pair formation, 1:854
oriT organization, 1:854–855
overview of systems, 1:851, 2:280
pilus
conjugation role, 1:853–854
phage attachment, 1:853
structure, 1:852–853
regulation of genes, 1:856
gram-positive conjugation
elements, 1:856–857
Enterococcus faecalis, 1:857–858
Streptomyces, 1:858
history of study, 1:847
induction, 1:860
lactic acid bacteria, 3:6
liquid versus solid support transfer, 1:849
overview of process, 1:848–849
plant transfer, *see Agrobacterium*
plasmids
conjugative types, 1:849–850
mobilization, 1:858–859
transfer, 3:724–725, 727
restriction and modification, 2:92–93
strain improvement, 4:439
transposons, 1:851
Conjugative transposon, definition, 4:704
Conservation, cultural heritage, *see also* Biodeterioration
aims, 1:863
evaluation of damage, 1:867–868
flora types by substrate
inorganic substrates, 1:865–866
organic substrates, 1:866–867
overview, 1:864–865
mechanisms oof microbial damage, 1:867
overview, 1:863
prevention and control of colonization
chemicals
fumigation, 1:871
overview, 1:870–871
spraying, 1:871–872
types, 1:870
humidity control, 1:869
irradiation, 1:869–870
mechanical removal, 1:869
object type dependence, 1:868–869
substrates, 1:864
Consortia, biofilms, 1:478
Constitutive enzyme, definition, 4:930
Constitutive expression, definition, 3:730, 4:336
Constitutive gene, definition, 2:618
Constriction, cell division, 1:704–705
Construct, definition, 4:666
Constructed wetland, definition, 4:117
Contact stabilization system, municipal wastewater treatment, 4:876–877
Contagion
definition, 2:677
discovery, 2:682
Contagionism, definition, 1:801
Containment, definition, 2:499
Contig, definition, 3:151
Continuous culture
advantages, 1:873, 886
applications
enrichment and selection, 1:884–885
mixed culture studies, 1:883–884
pure culture studies, 1:881–883
definition, 1:873
equipment, 1:880–881
growth kinetics, 2:587–588
mixed and multiple substrate-limited growth, principles, 1:876–877
non-substrate-limited growth, principles, 1:879–880
overview of types, 1:873
single substrate-limited growth, principles, 1:874–876
substrate-limited growth in mixed cultures, principles, 1:878–879
Continuous fever, definition, 3:131
Continuous process
definition, 3:9
lactic acid production, 3:13–15
Coo proteins, Cs1 pilus assembly, 2:369–370
COPD, *see* Chronic obstructive pulmonary disease
Copiotroph, definition, 3:86
Coporphyrinogen III, oxidative decarboxylation, 4:564
Copper
growth requirements and functions, 3:439
ore bioleaching, 3:483, 486–487
Coproporphyrin III, biosynthesis, 4:564
Coprotease
definition, 4:43
RecA activity, 4:46
Core oligosaccharide
adaptive responses and lipid A modifications, 3:84–85
biosynthesis, 3:77, 79–80
definition, 3:71
features, 3:74
Correlated species
definition, 4:195
rust fungi, 4:199–200
Corrin, definition, 3:188, 4:558
Corrinoid coenzyme

definition, 3:171
methylation of mercury, 3:179–180
Cortex, spore, 4:377
Corynebacterium diphtheriae
skin microbiology, 4:283
vaccine, 4:769–770
Corynebacterium glutamicum
flux balance analysis, 2:518
glutamate microbial production process, 1:153–154
lysine microbial production process, 1:155–157
Coryneform bacteria, skin resident species
Corynebacterium, 4:277
miscellaneous genera, 4:277–278
pathogens, 4:278–279
Cosmetic microbiology, *see also* Skin microbiology
background and importance, 1:887–888
microbial content test, 1:897–898
preservation, 1:895–896
preservative challenge test, 1:896–897
regulations and history, 1:888–889
sanitary manufacture
employee commitment, 1:893
good manufacturing practice, 1:889
personal hygiene, 1:891
raw materials, 1:890–891
sanitary design
buildings, 1:891–892
equipment, 1:892–893
water, 1:889
sanitization
cleaning, 1:894
sanitizing, 1:894
water system, 1:895
Cosmid, definition, 3:363
cos site, definition, 4:55
Cost–benefit analysis, definition, 2:137, 148–149
Cost-effectiveness analysis (CEA), definition, 2:137, 149
Cost–utility analysis (CUA), definition, 2:137, 149–151
Cotransduction, definition, 4:637
Coupled translation, definition, 4:127
Coxiella burnetii, *see also* Rickettsiae
cell wall, antigens, and virulence factors, 4:173
cultivation, 4:156–157
genome, 4:166
identification, 4:175
Q fever, 4:147
CPE, *see* Cytopathic effect
Cpx, regulation of pilus assembly, 2:366
Crabtree effect, definition, 4:939
CRE1, regulation of cellulase expression, 1:755
Creatinine, definition, 1:286
Crenarchaeota, definition, 1:319
Crenulation, definition, 4:353
Creutzfeldt–Jacob disease (CJD), features, 3:820–821
Cross-inoculation, definition, 3:407, 414
Cross-protection, definition, 3:99, 662
Cross-regulation, definition, 4:742, 751–752
Crown gall, definition, 1:78, 86
crr, definition, 3:580, 587
Crude oil, definition, 3:449
Cryobiosis, definition, 3:93
Cryptococcosis, systemic infection, 2:464
Cryptosporidium
clinical features, 2:858
diagnosis, 2:858
epidemiology, 2:857–858
life cycle, 2:858
pathogenesis and pathology, 2:858
prevention and control, 2:859
treatment, 2:858–859
Cryptosporidium parvum, food-borne illness, 2:406
Crystalline bacterial cell surface layer, *see* Surface layer
Csg proteins, extracellular nucleation–precipitation pilus assembly pathway, 2:375–376
csrA, regulation of glycogen biosynthesis, 2:552–553
CtrA, cell division control, 2:25–26
CTX, *Pseudomonas* virulence factor, 3:881
CUA, *see* Cost–utility analysis
Cucumovirus, features of infection, 3:702, 705
Cultural heritage
conservation, *see* Conservation, cultural heritage
definition, 1:863
Curdlan
production, 1:563
properties, 1:562
structure, 1:562
uses, 1:563
Curing, replicon elimination, 2:270
Curli, definition, 2:361
Curriculum, definition, 2:156
Cuticle, definition, 3:801
Cyanobacteria
commercial applications, 1:928–929
cytology, 1:911
dark metabolism, 1:919
definition and overview, 1:349, 492, 907–908, 3:392
fossil record and evolutionary history, 1:927–928
gene expression, 1:923–924
gene transfer, 1:923
genome, 1:923–924
growth and division
filamentous types, 1:915
rates, 1:913
unicellular and colonial types, 1:913
habitats
extreme environments, 1:924–925
freshwater plankton, 1:925–926
marine plankton, 1:925
sulfidogenic environments, 1:926
terrestrial environments, 1:926
heterocyst differentiation, 2:27
iron assimilation, 1:921
molecular genetic studies, 1:924
motility, 1:922–923
multicellularity and cell differentiation
akinetes, 1:916
heterocysts, 1:915–916
hormogonia, 1:915
necridic cells, 1:916
overview, 1:915
terminal hairs, 1:916
nitrogen assimilation, 1:921–922
photosynthesis
dark reactions, 1:918–919
light harvesting, 1:917–918
light reactions, 1:918
overview, 1:916–917
phototaxis, 3:620
phylogeny, 1:909–910
regulation of metabolism, 1:922
secondary metabolism, 1:919, 921
sulfur assimilation, 1:922
symbioses, 1:926–927

Cyanobacteria (*continued*)
taxis, 1:923
taxonomy, 1:908–909
ultrastructure, 1:911, 913
Cyclic AMP (cAMP), regulation of gene expression, 1:674–675
Cyclin-dependent kinases, replication initiation role, 1:829–830
Cyclodextrin, production from starch, 2:225
Cyclogeny, definition, 2:677, 692, 4:755
Cycloheximide, antifungal activity, 1:239
Cyclospora cayetanensis, food-borne illness, 2:407
Cylindrical inclusion body, definition, 3:792
Cyprodinil, antifungal activity, 1:240
Cyst
Azobacter, 1:362–363
definition, 1:359
Cysteine
biosynthesis
pathway, 1:146
regulation, 1:147
functions, 1:146
Cystic fibrosis, *Pseudomonas* infection, 3:883–884
Cystosome, definition, 2:42
Cytochrome
cytochrome a_1 identity, 1:64
cytochrome *bd*, 1:63–65
cytochrome *bo'*, 1:63, 65
cytochrome cd_1, 1:64
definition, 1:53, 59, 2:177
Cytochrome *c* oxidase
overview, 1:60–61
proton pumping, 1:62–63
steps in reaction, 1:61–62
subunits, 1:61
Cytokines, *see also specific cytokines*
definition, 1:729, 2:826, 3:745
functional overview, 1:733, 741
T cell secretion assay, 4:586–587
Cytokinesis, definition, 1:704
Cytopathic effect (CPE)
definition, 2:29
diagnostics, 2:33–34
Cytopathic virus, definition, 2:635
Cytoplasmic membrane, *see also* Outer membrane, gram-negative bacteria
assembly and control of composition, 1:727–728
carbohydrate export, 1:724
energy generation
coupled proceses, 1:719
gradients, 1:716–717
overview of mechanisms, 1:717
photosynthesis, 1:717–719
proton motive force, 1:717
respiration, 1:719
Escherichia coli, 2:263–264
isolation, 1:726–727
lipids
bilayer, 1:714–715
composition and structure, 1:712–713
hydrophobic effect, 1:713–714
phase behavior, 1:715
overview in bacteria, 1:710
protein secretion, 1:724–725
proteins
anchoring, 1:715–716
structure, 1:715–716
topology, 1:715–716
roles
cell boundary, 1:712
growth and cell division, 1:712, 725
motility, 1:726
osmotic barrier, 1:711–712
signal transduction, 1:725–726
transport regulation, 1:712
transport systems, *see also* ATP-binding cassette transporter
ATP-driven active transport, 1:720–722
efflux systems, 1:723–724
ion gradient-coupled transporters, 1:722–723
phosphotransferase system, 1:724
rationale, 1:719–720
types, 1:720
ultrastructure, 1:710–711
Cytosine, deamination, 3:308–309
Cytotoxin
definition, 2:187
type III secretion, 2:307

D

D 0870, antifungal activity, 1:247
Dairy products, *see also* Lactic acid bacteria
enzyme biotechnology, 2:230–231
fermentation
bacteriophage effects and control, 2:6–7
function of microbes, 2:5–6
genetics of starter cultures, 2:6
microbe types, 2:4–5
industrial effluents and treatment, *see* Industrial effluents
microbial spoilage
canned milk, 2:3
cheese, 2:2–3
dried milk products, 2:3
fluid milk, 2:2
psychotrophs, 2:2
yogurt, 2:3
natural flora of milk, 2:2
overview of microbiology, 2:1
pathogens, 2:3–4
probiotics, 2:7–8
refrigeration
butter, 4:76–77
cheese, 4:77
milk, 4:76
thermal handling of milk
enzymes in spoilage, 4:549–550
pasteurization, 4:548–549, 556
recontamination following pasteurization, 4:549
refrigeration, 4:548
spore-forming bacteria, 4:549
DALY, *see* Disability-adjusted life year
Dam, *see* DNA methylation
Dark-field microscopy, *see* Optical microscopy
Daunorubicin, mechanism of action, 3:780
Dcm, *see* DNA methylation
dcw cluster
definition, 1:704
organization, 1:706
DEBS modular polyketide synthase, genetic manipulation
β-carbonyl-processing functions, 1:201
extender units, 1:201
polyketide chain length, 1:201
starter units, 1:201
structure studies, 1:202
systems for manipulation, 1:199
Decimal reduction time, definition, 4:545, 555
Decomposition, definition, 3:866, 4:321
Deconvolution, definition, 3:264

Deconvolution microscopy, *see* Optical microscopy
Dedicated pathway, definition, 1:840
Deep learning, definition, 2:156, 162–163
Degassing, definition, 3:171
Deg proteins, enzyme production regulation in *Bacillus subtilis*, 1:379
Degradative pathway, definition, 3:730
Delavirdine
  adverse effects, 1:307
  indications, 1:307
  mechanism of antiviral activity, 1:307
  resistance, 1:307
Delayed-type hypersensitivity, definition, 3:27
Deletion, DNA, 3:307, 310
Demethylation, definition, 4:238
Demicyclic, definition, 4:195
Dendritic cell, immune response, 1:733
Dengue
  emerging infection, 4:816
  overview, 1:313–314
  polymerase chain reaction diagnostics, 3:789
Denitrification, definition, 4:117, 321
Dental plaque, definition, 3:466
Dermatophytosis
  definition, 2:451
  features, 2:455
Dermis, definition, 2:451
Desorption, definition, 2:607
Detergent
  definition, 1:710
  enzyme biotechnology, 2:231
Detoxification
  biodegradation, 1:468–470
  definition, 1:461, 3:171
Development, bacteria, *see also specific examples*
  functions, 2:15–16
  overview, 2:15
Dextran
  production, 1:561–562
  structure, 1:561
  uses, 1:562
Dextrose equivalent, definition, 2:222
DFMO, *see* Eflornithine
Diagnostic microbiology, *see also specific pathogens and techniques*
  antibody response measurement
    agglutination reactions, 2:35–36
    complement fixation assay, 2:35–36
    enzyme immunoassay, 2:35
    immunofluorescence, 2:35
    immunoglobulin classes, 2:35
    Western blot, 2:35
  blood, detection of bacteria
    centrifugation plus filtration
      background, 2:11
      clinical trials, 2:14
      concerns, 2:13
      Ficoll–Hypaque centrifugation, 2:11
      recovery of bacteria, 2:11–13
      refinement, 2:13–14
    culture
      background, 2:9
      devices, 2:9–10
    lysis of erythrocytes and filtration, 2:10–11
  culture
    artificial media, 2:33
    egg culture, 2:34
    living cells, 2:33–34
    phenotypic identification, 2:34
  direct identification
    antigen testing, 2:31–32
    microscopy, 2:31
    nucleic acid probe hybridization, 2:32–33
    stains, 2:32
  Escherichia coli, 2:268
  historical perspectives, 2:30–31
  ligase chain reaction, 2:37
  overview, 2:29–30
  performance assessment
    accuracy, 2:38, 40
    costs, 2:41
    precision, 2:40
    turnaround time, 2:40–41
  polymerase chain reaction
    overview, 2:31, 36–37
    probe amplification systems, 2:37
    signal amplification systems, 2:37–38
    species-specific probes, 2:38
    target amplification systems, 2:36–37
  techniques, table, 2:30
Diarrhea
  definition, 2:187
  inflammatory diarrhea pathogens
    Campylobacter, 2:191–192
    Clostridium difficile, 2:192
    Escherichia coli, 2:190
    Salmonella, nontyphoidal, 2:190–191
    Shigella, 2:189–190
  noninflammatory diarrhea pathogens
    Bacillus cereus, 2:196
    Clostridium perfringens, 2:196–197
    Escherichia coli
      enteroaggregative, 2:195
      enterohemorrhagic, 2:194–195
      enteropathogenic, 2:193–194
      enterotoxogenic, 2:194
    Staphylococcus aureus, 2:196
    Vibrio cholerae, 2:193
Diastereoselectivity, definition, 1:636
Diatomaceous earth filtration, drinking water treatment, 4:904, 907
Diazotroph, definition, 3:379, 392
Diazotrophy, definition, 1:359
Dichloroethane, biodegradation pathway, 3:733, 735
2,4-Dichlorophenoxy acetate, biodegradation, 3:597–598, 601–602, 735, 742
Didanosine
  adverse effects, 1:303
  indications, 1:303
  mechanism of antiviral activity, 1:303
  resistance, 1:303
Diffraction, definition, 3:288
Digitalis, toxicity, 1:286
Dilthiazem, chiral biocatalysis, 1:440
Dilution rate, continuous culture, 1:873–874
Dimethomorph, antifungal activity, 1:252
Dimethylsulfide, atmosphere cycling by microbes, 3:260–262
Dimethylsulfoxide (DMSO), reduction in anaerobic respiration, 1:186–187
Diminazene aceturate, sleeping sickness treatment, 4:739
Dimorphism, fungi, 2:468, 473
Dinoflagellates
  chromosome
    segregation, 2:50
    structure, 2:46
    transcription and replication, 2:49
  DNA content, 2:46

Dinoflagellates (*continued*)
evolution, 2:50–52
features, overview, 2:42–43
life cycle, 2:43, 45–46
microtubules and mitotic spindle, 2:50
nuclear envelope and lamins, 2:49–50
nucleolus, 2:49
plastid origins, 2:52
toxic species, 2:52–53
Dinokaryon, definition, 2:42
Dinomitosis, definition, 2:42
Diphtheria, *see Corynebacterium diphtheriae*
Diphyllobothrium latum, food-borne illness, 2:408
Disability-adjusted life year (DALY), definition, 2:529
Discounting, definition, 2:137
Disease association, definition, 3:607
Disease cycle, definition, 3:676
Disinfect, definition, 1:887
Disinfection, definition, 4:898
Disproportionation, definition, 3:171
Dissimilation, definition, 3:431
Dissimilatory reduction, definition, 4:238, 495
Divergicin A, features, 1:391
Diversity, microbial
algae, 2:65–66
approaches for phylogenetic analysis, 2:56–58
Archaea, 2:59–61, 63
Eubacteria, 2:58–59
functional survey, 2:68–69
fungi, 2:63–65
protozoa, 2:66–67
significance, 2:55–56
species number estimates, 2:58
viruses, 2:67–68
World Wide Web resources, 2:69
Divisisome
assembly and fission, 1:708–709
definition, 1:704
DMSO, *see* Dimethylsulfoxide
DnaA, initiation of replication, 1:825, 828–829, 2:84–85
DnaB, replication role, 2:85
DnaC, replication role, 2:85
DNA damage
endogenous damage
deamination, 2:72
depurination, 2:72
metabolic damage, 2:72
mismatced bases, 2:72
oxidation, 2:72
rates, 2:71–72
exogenous damage
alkylation, 2:73
bulky adducts, 2:73
ionizing radiation, 2:73
overview, 2:72–73
ultraviolet irradiation, 2:73–74
repair, *see* DNA repair
targets, 2:71
DnaG, primase activity, 2:85
DnaKJ, chaperone activity, 2:599–600
DNA methylation
S-adenosylmethionine as donor, 3:240
Dam methylation
bacteriophage DNA packaging, control, 3:243
chromosome replication initiation, 3:242
mismatch repair, 3:242
regulation of gene expression, 3:242
site specificity, 3:242
Dcm methylation, 3:243
definition, 2:826
DNA methyltransferase mechanism, 3:240–241
restriction and modification, *see* Restriction and modification
DNA methyltransferase, *see* DNA repair; Restriction and modification
DNA recombination, definition, 4:428
DNA repair
base excision repair, 2:76–77
damage, *see* DNA damage
methyl transferases, 2:74–76
mismatch correction, 2:81
mutagenic protein processing in *Escherichia coli*, 4:48–49
nucleotide excision repair, 2:77
phenotypes associated with gene mutations, 4:49
photolyases, 2:74
postreplication, 4:43
radiation-resistant microorganisms, 2:334–337
RecA role, 4:46, 48
structure, 4:49–51
transcription-coupled nucleotide excision repair
overview, 2:77–78
pyrimidine dimer and RNA polymerase effects, 2:78, 80
transcription repair coupling factor, 2:80–81
transformation, 4:658–659
DNA replication
definition, 4:451
plasmids
control, 3:720
DNA polymerase I-dependent replicons, 3:718
iteron-activated replicons, 3:718–720
overview, 3:716
rolling circle replication, 3:716, 718
viruses, *see* Virus
DNA replication, *Escherichia coli*, *see also* Chromosome, bacteria
lysate replication systems, 2:83
macroinitiation, 2:84–85
microinitiation, 2:85–86
mutants, 2:82–83
polymerases
DNA polymerase I, 2:87
DNA polymerase II, 2:87
DNA polymerase III, 2:87, 89
types, overview, 2:83, 86–87
proteins, table, 2:88
termination, 2:89–90
DNA restriction and modification, *see* Restriction and modification
DNA sequencing
automation, 2:107–108, 115
bioinformatics and genome databases, 2:111–112
chemical degradation method, 4:62–63
data presentation, 2:106
dideoxynucleotide chain-termination method, 4:62–63
genome sequencing approaches, 3:159, 161
high-throughput sequencing, 2:106–108
horizontal gene transfer, gene archaeology
caveats, 2:700
homology compared to sequence, 2:703–704
large rare events versus small common events, 2:703
mutation versus recombination, 2:702–703

orthologs versus paralogs, 2:702
whole genome shotgun sequencing
closure, 2:110–111
editing and annotation, 2:111
library construction, 2:109–110
overview, 2:106, 108–109
random sequencing, 2:110
DNA unwinding element (DUE)
definition, 1:822
functions, 1:826–827
Dockerin domain, definition, 1:744, 753
Domain
definition, 3:773
phylogeny, 2:55
dot/icm, *Legionella* virulence factor, 3:23–24
Double-stranded RNA adenosine deaminase (dsRAD), antiviral effects, 1:279
Double-stranded RNA-dependent protein kinase (PKR)
antiviral effects, 1:278
interferon induction and antiviral activity, 2:838
Downstream processing
acetic acid production, 1:17
definition, 1:13
Downy mildew
compatible interaction, 2:122
control
fungicides, 2:125
integrated control
biological control, 2:127
cultural control, 2:126–127
nonspecific resistance, 2:127–128
overview, 2:126
race-specific resistance and fungicide insensitivity, 2:125–126
culture, 2:123–124
disease assessment, 2:124
economic impact, 2:118
geographic distribution and damage, 2:117–118
host range, 2:119
incompatible interaction
hypersensitive response, 2:122–123
resistance markers, 2:123
infection cycle
conidia, 2:120
endophytic mycelium, 2:121–122
oospores, 2:119–120
sporangia, 2:120
symptoms, 2:120–121
pathotype screening, 2:124–125
prospects for research, 2:128–129
taxonomy
class Peronosporomycetes, 2:118–119
kingdom Straminipila, 2:118
species, 2:119
Doxorubicin, mechanism of action, 3:780
Drinking water, *see* Water, drinking
Droplet nuclei, definition, 1:69, 72
Drug resistance
definition, 2:170
emerging infection, 2:173–174
dsRAD, *see* Double-stranded RNA adenosine deaminase
DtxR, iron regulation, 2:866
DUE, *see* DNA unwinding element
Duplication, DNA, 3:307
Dysentary, definition, 2:187

E

EB, *see* Elementary body
Ebola, characteristics, 4:815
EBV, *see* Epstein–Barr virus
EC_{50}, definition, 1:286
ECA, *see* Enterobacterial common antigen
Echinocandins, antifungal activity, 1:249
Eclipse, definition, 4:832–833
Ecology, microbial
approaches for study, 2:134–135
communities, 2:132–134
evolution as a discipline, 2:131–132
prospects, 2:135–136
Econazole, antifungal activity, 1:241
Economic consequences, infectious disease
acquired immunodeficiency syndrome, 2:138–141
annual costs by disease, 2:142
assessment
cost-of-illness method, 2:143
infection versus clinical disease, 2:142
opportunity costs, 2:142–143
overview, 2:138
perspective of analysis, 2:143
willingness-to-pay method, 2:143–144
benefits versus harms from intervention, 2:147–148
Black Death, 2:138–139
cost categorization
data
limitations, 2:145–146
sources, 2:145
fixed cost, 2:144–145
intangible cost, 2:145
steps in analysis, 2:150–152
variable cost, 2:144–145
cost–benefit analysis, 2:137, 148–149
cost-effectiveness analysis, 2:137, 149
costs versus benefits of intervention, 2:147
cost–utility analysis, 2:137, 149–150
decision tree
building outline, 2:152–153
construction rules, 2:153
sensitivity analysis, 2:153
discounting costs over time
discount rate selection, 2:146
formula, 2:146
nonmonetary costs and benefits, 2:146–147
rationale, 2:146
efficacy versus effectiveness of interventions, 2:147
historical impacts, 2:138
incidence, 2:141
Markov models, 2:154
prevalence, 2:141
resource allocation, 2:154–155
time-related elements, 2:139–141
utility
direct valuation, 2:150
indirect valuation, 2:150
Ecosystem, definition, 2:131
Ecotone, definition, 2:438
Ecthyma, definition, 4:271, 281
Ectoine, biosynthesis, 4:889
Ectopic expression
definition, 3:363
Myxococcus xanthus, 3:369
Ectotrophic
definition, 2:479

Ectotrophic (*continued*)
Gaeumannomyces graminis colonization, 2:482
EDS, *see* Energy dispersive x-ray spectroscopy
Education, microbiology, *see also* Careers, microbiology
educating of educators, 2:168–169
evaluation of students, 2:167–168
graduate student supervision, 2:168
learning approaches, 2:162–163
reasons for education
personal, 2:157
political, 2:157
scientific, 2:156–157
resources, 2:168
subject matter
core content, 2:159
customized content
advanced study, 2:159–160
college nonmajors, 2:160
noncollege education, 2:160–162
introductory course content, 2:160
introductory laboratory content, 2:161
selection elements, 2:158–159
targets of people, 2:157–158
teaching techniques
independent study, 2:167
laboratory work
aims, 2:165
open investigations, 2:166
overview, 2:164–165
simulation and data-handling exercises, 2:166
structured exercises, 2:165–166
lecture, 2:163–164
small group sessions, 2:164
work experience, 2:166–167
EELS, *see* Electron energy loss spectroscopy
Efavirenz
adverse effects, 1:308
indications, 1:308
mechanism of antiviral activity, 1:307–308
resistance, 1:308
Effectiveness, definition, 2:137
Efficacy, definition, 2:137
Effluent, *see* Industrial effluents
Efflux, definition, 2:618
Eflornithine (DFMO), sleeping sickness treatment, 4:740–741
EF-Tu, *see* Elongation factor-Tu
Ehrlichia, *see also* Rickettsiae
cell wall, antigens, and virulence factors, 4:168–173
cultivation, 4:157
identification, 4:175
species and disease
groups and features, 4:147–148
table, 4:146
EIA, *see* Enzyme immunoassay
Elastase, *Pseudomonas* virulence factor, 3:878
Electrochemical gradient
definition, 1:53, 2:177, 3:188
generation
electron transport, 2:181–184
ATP hydrolysis, 2:184
bacteriorhodopsin, 2:184
decarboxylation linked to ion translocation, 2:184
metabolite ion exchange, 2:185
methyl group transfer, 2:184
Electron acceptor
definition, 1:53, 2:177, 4:478
hierarchy, 3:204
respiratory electron-transport linked ATP synthesis in bacteria, 2:181–184
Electron donor
definition, 1:53, 2:177, 4:478
respiratory electron-transport linked ATP synthesis in bacteria, 2:181–184
Electron energy loss spectroscopy (EELS), definition, 3:276, 285–286
Electron microscopy, *see also* Scanning electron microscopy; Transmission electron microscopy
applications, 3:276–277
electron energy loss spectroscopy, 3:276, 285–286
electron spectroscopic imaging, 3:276, 285–286
energy dispersive x-ray spectroscopy, 3:276, 285–286
scanning transmission electron microscopy, 3:276, 285
selected area electron diffraction, 3:276, 286
Electron spectroscopic imaging (ESI), definition, 3:276, 285–286
Electron transport
definition, 1:53, 2:177
photosynthesis, 2:185
respiratory electron-transport linked ATP synthesis in bacteria, 2:181–184
Electron transport chain
anaerobic respiration, 1:181
definition, 1:180
Electron transport-coupled phosphorylation, definition, 1:180, 647
Electroporation
definition, 1:692, 3:363, 4:55
lactic acid bacteria, 3:6
overview, 4:662–663
Elementary body (EB), definition, 1:781–782
ELISA, *see* Enzyme-linked immunosorbent assay
Elongation factor, definition, 3:824
Elongation factor-Tu (EF-Tu), methylation, 3:244
Embden–Meyerhof–Parnas pathway, features and distribution, 1:653–654
Embryonic stem cell
definition, 4:666
transgenesis, 4:669–670
Emerging infection, *see also specific diseases*
bovine spongiform encephalopathy, 4:820–821
drug resistance, 2:173–174
International Health Regulations, 2:170, 175
international law revision, 2:849–850
international travel, 2:174, 4:822–826
perspective of last 20 years, 2:170–172
principal newly identified organisms, 2:171
re-emerging infection, 2:170, 4:811
viruses
anticipation of emergence, 4:827–828
arenaviruses, 4:815
causes of emerging infection
animal viruses as interspecies transfer models, 4:826
arboviral diseases, 4:825
influenza pandemic, 4:824–825
overview, 4:823
trafficking, 4:822–826

control of disease, 4:828–829
eradication, prospects and requirements, 4:829–830
evolution of variants, 4:821–822
examples, overview, 4:813–815
filoviruses, 4:815
flaviviruses, 4:816
hantavirus, 4:816–817
Hendra virus, 4:817–818
hepatitis, 4:818
influenza virus, 4:818–819
molecular diagnostics, 4:812–813
newly recognized viruses, 4:812
Nipah virus, 4:818
overview, 4:811–812
parvovirus B19, 4:819
restraints on emergence, 4:826–827
retroviruses, 4:819–820
Rift Valley fever virus, 4:820
sources of new viruses, 4:821
weaknesses in public health infrastructure, 2:172–173
zoonosis, 4:963–965
Enablement, patent law, 3:546, 553–554, 557–559
Enalapril, chiral biocatalysis, 1:431
Enantiomer, definition, 1:286, 430
Enantiomeric excess, definition, 1:636
Enantioresolution, definition, 3:49
Enantioselectivity, definition, 1:636, 2:222
Encephalitis, definition, 1:286
Endemic, definition, 3:131, 4:811, 955
Endocytosis, definition, 3:407
Endodyogeny, definition, 4:598
Endogenous, definition, 4:344
Endolithotrophic organism, definition, 3:93
Endonuclease, definition, 2:71, 4:55, 387
Endoplasmic reticulum (ER), definition, 2:635
Endopolygalacturonase
definition, 3:562
features, 3:568–569
Endopolygalacturonate lyase, features, 3:570–571
Endopolymethylgalacturonase, features, 3:570
Endopolymethylgalacturonate lyase, features, 3:572–573
Endosome, definition, 1:286
Endospore
Bacillus subtilis, regulation of development
features, 1:381–382
SigmaE activation in mother cell, 1:380–381
SigmaF activation in forespore, 1:380
SigmaG activation in forespore, 1:381
SigmaK activation in mother cell, 1:381
clostridia, 1:836
definition, 1:373, 834
formation in *Bacillus subtilis*, 2:16–17
Endosymbiosis, definition, 2:42
Endotoxin
definition, 1:116, 3:71, 4:755, 767
airborne, 1:118
Endotrophic
definition, 2:479
Gaeumannomyces graminis colonization, 2:482–483
End-product repression, definition, 1:134
Energy dispersive x-ray spectroscopy (EDS), definition, 3:276, 285–286
Energy spilling, definition, 4:185, 189
Energy transduction, *see also specific processes*
alkaliphiles, 2:185
bacterial energetics, 2:180–181
electrochemical gradient generation
ATP hydrolysis, 2:184
bacteriorhodopsin, 2:184
decarboxylation linked to ion translocation, 2:184
electron transport, 2:181–184
metabolite ion exchange, 2:185
methyl group transfer, 2:184
membrane features, 2:178
mitochondrial energetics, 2:179–180
respiratory electron-transport linked ATP synthesis in bacteria, 2:181–184
Engulfment, sporulation, 4:377, 382
Enhancer, definition, 4:81, 628
Enilconazole, antifungal activity, 1:243
Enology, definition, 4:914
Enrichment, microorganisms, 1:461
Entamoeba histolytica
amebiasis
clinical features
intestinal disease, 2:853–854
liver abscess, 2:854
complications, 2:854–846
diagnosis
intestinal disease, 2:846
liver abscess, 2:846
epidemiology, 2:852–853
pathogenesis, 2:854
prevention and control, 2:846–856
treatment, 2:846
food-borne illness, 2:406–407
life cycle, 2:853
Enteric microorganism, definition, 1:530–531
Enteric virus, definition, 2:635
Enteritis, definition, 2:187
Enterobacterial common antigen (ECA)
biosynthesis, 3:524
structure, 3:519
Enterococcus faecalis, conjugation, 1:857–858
Enteropathogenic bacteria, *see also specific bacteria and diseases*
enteric fever pathogens
Salmonella typhi, 2:197
Yersinia, 2:197–198
epidemiology and clinical characteristics of pathogens, 2:188
gastritis and *Helicobacter pylori*, 2:198
inflammatory diarrhea pathogens
Campylobacter, 2:191–192
Clostridium difficile, 2:192
Escherichia coli, 2:190
Salmonella, non-typhi, 2:190–191
Shigella, 2:189–190
miscellaneous pathogens
Aeromonas, 2:199
Bacteroides fragilis, 2:199
Clostridium botulinum, 2:199–200
Listeria monocytogenes, 2:199
Plesiomonas shigelloides, 2:199
Vibrio, non-cholerae, 2:198–199
noninflammatory diarrhea pathogens
Bacillus cereus, 2:196

Enteropathogenic bacteria (*continued*)
Clostridium perfringens, 2:196–197
Escherichia coli
enteroaggregative, 2:195
enterohemorrhagic, 2:194–195
enteropathogenic, 2:193–194
enterotoxogenic, 2:194
Staphylococcus aureus, 2:196
Vibrio cholerae, 2:193
overview of diseases, 2:187
skin microbiology, 4:284
Enterotoxin, definition, 2:187, 4:65
Enterovirus
classification, 2:201–202
control
anti-viral drugs, 2:209
environment, 2:208
vaccination, 2:208–209
definition and overview, 1:286, 2:201
diseases, 2:201–202, 205–206
epidemiology, 2:205–206
identification, 2:202–203
isolation, 2:202
pathogenesis, 2:206–207
propagation, 2:202
receptors, 2:203–204
structure, 2:203
transcription, 2:205
translation, 2:204
virulence factors, 2:207–208
Entner–Doudoroff pathway, features and distribution, 1:654–655
Envelope
definition, 4:592
virus, 1:398
Environmental scanning electron microscopy (ESEM), definition, 3:276
EnvZ, osmosensor, 3:513–514
Enzyme I, structure, 3:581
Enzyme II, structure, 3:582–583
Enzyme immobilization, overview 2:723–724
Enzyme immunoassay (EIA), diagnostics, 2:35
Enzyme-linked immunosorbent assay (ELISA)
beet necrotic yellow vein virus, 1:426
definition, 1:104, 422, 3:654
principles, 2:731–732
rhinovirus diagnostics, 4:105
Enzymes, *see also specific enzymes*
biotechnology applications, *see also* Biosensor; Biotransformation; Chiral pharmaceutical intermediates, biocatalysis
analytical applications, 2:234
animal feed processing, 2:228–229
dairy industry, 2:230–231
detergents, 2:231
fat and oil modification, 2:228
fine chemical production, 2:233–234
fruit juice processing, 2:229–230
leather industry, 2:231
lignocellulosic biomass conversions
cellulose, 2:226
hemicellulose, 2:226–227
lignin, 2:227
overview, 2:226
meat and fish processing, 2:230
overview, 2:222–223
pulp and paper, 2:229
starch conversion
alcohol production, 2:225–226
cyclodextrin production, 2:225
enzyme overview, 2:223–224
glucose production, 2:224
high-fructose corn syrup production, 2:213, 224
high-maltose conversion syrup production, 2:224–225
classification by reaction, 2:223
Erwinia, *see Erwinia*
extracellular enzymes
catabolite repression, 2:217
cellulase, 2:214–215
definition, 2:210
β-glucanase, 2:215
induction, 2:215–217
industrial production, 2:219–220
lipase, 2:215, 228
pectinase, 2:215
proteases, 2:213–214
secretion, 2:210–211, 218–219
sources and applications, 2:216, 220, 232–233
starch-hydrolyzing enzymes and applications, 2:211–213
temporal regulation, 2:217–218
Eosinophil, immune response, 1:730–731
Epidemic, definition, 3:131, 4:811, 955
Epidemiology, definition, 2:338, 4:248, 758
Epidermis, definition, 2:451
Epigenetic, definition, 3:240
Epigrowth, definition, 3:866
Epi-illumination, definition, 3:264, 288
Epilimnion, definition, 2:438
Epimastigote, definition, 4:725
Epimural bacteria, definition, 2:485
Episome, definition, 1:405, 2:677
Epistatic, definition, 3:662
Epitope, definition, 1:208, 4:15, 921
Epoxiconazole, antifungal activity, 1:243
EPS, *see* Extracellular polyaccharide
Epstein–Barr virus (EBV)
Burkitt's lymphoma association, 3:461–462
definition, 1:286
diagnosis, 2:41
Equine encephalomyelitis
overview, 1:315
zoonosis, 4:962
Equipartition, definition, 1:822
ER, *see* Endoplasmic reticulum
Eradication
bacterial diseases through vaccination, 4:777
definition, 2:170, 3:762
polio
global situation, 3:771
goals, 3:762
impact of program, 3:770–771
program strategy, 3:769–770
smallpox, 2:172, 4:289, 293–295
viral disease, prospects and requirements, 4:809, 829–830
Ergot
alkaloids
sources, 3:339
structure, 3:339
tolerances for grain foods, 3:339
animal disease
cattle, 3:343–344
forms, 3:343
horses, 3:343
sows, 3:344
control, 4:312–313
economic impact, 4:310–311
history, 4:308–310
host range, 4:311
human disease

epidemics, 3:340
forms, 3:339
infection cycle, 4:311–312
overview, 4:297–298
poisoning, 4:313–314
Eructation, definition, 4:185
Erwinia
antibiotic production, 2:256–257
bacteriocin production, 2:257
cytokinin synthesis, 2:253
diseases, 2:236–237, 258
exoenzymes
cellulase, 2:240
pectate lyase, 2:238–239
pectin acetyl esterase, 2:240–241
pectin lyase, 2:240, 246–247
pectin methyl esterase, 2:240
polygalacturonidase, 2:239–240
protease, 2:238
regulation of gene expression
calcium, 2:243
classification of enzymes, 2:241–242
culture conditions, 2:243–244
DNA damage response, 2:246–247
negative regulators, 2:244–245
positive regulators, 2:245–246
posttranscriptional regulation, 2:246
quorum sensing, 2:242–243
secretion systems, 2:241
soft rot species, 2:237, 239
extracellular polysaccharide
biosynthesis
genes, 2:247–248
regulation, 2:248–250
composition, 2:247
pathogenesis role, 2:247
flavohemoglobin in pathogenesis, 2:255
hypersensitive reaction
hairpins
overview, 2:250–251
type III secretion pathway, 2:251
hrp genes
organization, 2:250
regulation, 2:251–252
virulence loci associated with *hrp* regulon
dsp, 2:252–253
hsvG, 2:253
indole-3–acetic acid synthesis, 2:253–254
iron-acquisition systems, 2:254–255
methionine sulfoxide reductase, 2:254
pathogenicity islands, 2:255–256
pigment synthesis, 2:257
prospects for research, 2:258
Sap functions, 2:254
taxonomy, 2:237
Erysipelas, definition, 4:271, 281
Erysipelothrix rhusiopathiae, skin microbiology, 4:283
Erysiphe graminis, *see* Powdery mildew
Erythromycin
biosynthesis, 1:196–197
mechanism of action, 3:776
Eschar, definition, 4:758
Escherichia coli
adhesins, 1:43, 45–46, 50
aerobic respiration
overview, 1:54
oxidase synthesis, regulation, 1:64
amino acid transport systems, 2:264
anaerobic respiration regulation, 1:187–188
capsule, 2:263
carbon and nitrogen metabolism in log-phase aerobic cultures, 1:670, 672–673
cell wall, 2:263
chemoreceptors, 1:775–776
Che proteins in chemotaxis, 1:775–776, 778–780
chromosomal map of K12, 3:161–162
cloning, host selection, 4:58
compatible solute transporters, 4:893–894
conjugation, 2:280
cyclic AMP regulation of gene expression, 1:674–675
cytoplasm, 2:264
cytoplasmic membrane, 2:263–264
diagnostics, 2:268
diversity, 2:271
DNA replication, *see* Chromosome, bacteria; DNA replication, *Escherichia coli*
ecology, 2:261
enteroinvasive form and inflammatory diarrhea, 2:190, 268
extrachromosomal elements, 2:275
fimbriae, 2:261–262
flagella, 2:262
food-borne illness
enteroaggregative, 2:404
enteroinvasive, 2:402–403
enteropathogenic, 2:403–404
enterotoxigenic, 2:397
Shiga toxin production, 2:396–397
fueling reactions, 2:265
genitourinary tract infection, 2:268
genome
chromosomes, 2:273, 275
K-12 features, 2:276
overview, 2:264–265
glycogen synthesis, genetic regulation
csrA gene regulation, 2:552–553
culture condition responses, 2:548
cyclic AMP regulation, 2:551
gene cluster, 2:548–549
glgCAP transcription, 2:550–551
guanosine diphosphate regulation, 2:552
integrated model, 2:553–554
kat F regulation, 2:550
loci affecting enzyme levels, 2:550
glyoxylate bypass, *see* Glyoxylate bypass, *Escherichia coli*
growth, 2:266
heat shock response
extracytoplasmic heat shock response and sigma factor E regulation, 2:605–606
molecular thermometers, 2:602–603
overview, 2:600–601
sigma factors, regulation, 2:601–603
species distribution of heat shock response mechanism, 2:603
lipid biosynthesis, *see* Lipid biosynthesis, *Escherichia coli*
macromolecular composition, 2:262
metabolism, 2:265–266
mutation, 2:275, 277
noninflammatory diarrhea
enteroaggregative, 2:195, 268
enterohemorrhagic, 2:194–195, 260, 267–268
enteropathogenic, 2:193–194, 260, 267

Escherichia coli (*continued*)
enterotoxogenic, 2:194, 260, 267
nucleoid, 2:265
nutrition, 2:266
operons, *see specific operons*
outer membrane, *see* Outer membrane, gram-negative bacteria
periplasm, 2:263
phase variation, 2:260, 270
pilus role in pathogenesis, 2:377–378
recombination
homologous, 2:277–278
site-specific, 2:278
transpositional, 2:278–280
refrigerated food pathogens, 4:73
taxonomy, 2:260–261, 271–273
transduction, 2:280, 282
transformation, 2:282, 4:661–662
ESEM, *see* Environmental scanning electron microscopy
ESI, *see* Electron spectroscopic imaging
E-site, ribosome, 3:824
Esophagitis, definition, 1:286
Esophagus, bacteria, 2:489
Essential chemical element, definition, 3:614
EST, *see* Expressed sequence tag
Eubacteria
diversity of species, 2:58–59
heterotrophs, 2:656
lipid features, 3:64–65
phylogenetic placement, 2:655–656
Eukarya, definition, 1:319
Eukaryote
heterotrophs, 2:658–660
lipid features, 3:65–66
phylogenetic placement, 2:655–656
Euryarchaeota, definition, 1:319
Eutroph, definition, 2:651
Evaporite, definition, 1:126
Evolution, *see also* Adaptation; Natural selection, bacterial
adaptation, 2:284
coevolution of interacting genomes
exploitive interactions, 2:293–294
mutualistic interactions, 2:294
overview, 2:293
definition, 2:283
divergence and speciation, 2:284
experimental tests
adaptation by natural selection, 2:287–288
random mutation
fluctuation test, 2:286–287
replica-plating experiment, 2:287
fitness, factors affecting
genetic background, 2:290
metabolic activity variation, 2:289–290
mutations, 2:288–289
possession of unused functions, 2:289
genetic drift, 2:283, 286
genetic systems evolution
directed mutations, 2:297
mutator gene effects, 2:296–297
overview, 2:295–296
sexuality and mixis, 2:296
genetic variation
frequency-dependent selection
nontransitive interactions, 2:292–293
stable equilibria, 2:291–292
unstable equilibria, 2:292
mixis, 2:285, 296
mutation, 2:284–285
overview, 2:284
selective neutrality, 2:291
transient polymorphisms, 2:291
hyperthermophiles, 2:321–322
metabolic functions
acquisition by gene transfer, 2:295
changes in regulatory and structural genes, 2:295
reactivation of cryptic genes, 2:295
mutagenesis role, 3:308
natural selection, 2:283, 285–286
oxidative stress adaptation, 3:531
patterns, 2:283–284
ExbB
ATP-binding cassette transport role, 1:3
iron uptake role, 2:861
ExbD
ATP-binding cassette transport role, 1:3
iron uptake role, 2:861
Excision, definition, 2:71
Exobiology, *see also* Space flight, effects on microorganisms
astrobiology, 2:299, 305
definition, 2:299
experiments on origin of life, 2:300–301, 304
extraterrestrial intelligence search, 2:305–306
extraterrestrial life detection efforts, 2:301–304
history, 2:300
planetary protection, 2:299, 304–305
Exoenzyme S, *Pseudomonas* virulence factor, 3:878–879
Exogenous, definition, 4:344
Exon, definition, 3:490, 4:181
Exonuclease, definition, 2:71
Exopolygalactouronase
definition, 3:562
features, 3:569–570
Exopolygalacturonate lyase, features, 3:571–572
Exopolysaccharide, *see* Extracellular polyaccharide
Exotoxin
AB structure-function, 2:307–308, 311–312
classification, 2:308–309
covalent modification of host cell components, 2:312–313
definition, 2:307, 4:767
genetic organization, 2:309–310
heat-stable enterotoxins, 2:307, 316
host entry, 2:312
molecular and structural properties, 2:313–314
pore-forming toxins, 2:315
proenzyme processing, 2:310–311
properties of major types, 2:309
secretion, 2:310, 315–316
therapeutic applications, 2:314–315
toxoid conversion, 2:314
Exotoxin A, *Pseudomonas* virulence factor, 3:878
Expanded bed reactor, industrial effluent treatment, 2:756
Exponential growth, definition, 2:584

Expressed sequence tag (EST), patentability, 3:554, 556–557
Extended aeration system, municipal wastewater treatment, 4:877
Extracellular matrix
 adhesin binding, 1:49
 definition, 1:42
Extracellular polyaccharide (EPS), *see also* capsule
 Bacteroides fragilis capsular polysaccharide in abscess induction, 2:565
 definition, 1:647
 Erwinia
  biosynthesis
   genes, 2:247–248
   regulation, 2:248–250
  composition, 2:247
  pathogenesis role, 2:247
 features, 1:624
 Neisseria, 2:576, 582–583
 Pseudomonas, 3:885
 structure, 1:661–662
Extremophile, *see also* Acidophile, extreme; Alkaliphile; Barophile; Halophile; Hyperthermophile; Low-temperature environments; Oligotroph; pH stress; Psychrophile; Radiation-resistant microorganisms
 biotechnological applications
  detergents, 2:337
  molecular biology, 2:337
  recombinant organisms, 2:337
  table, 2:336
 definition and overview, 2:317–318
 methanogens, 2:337
 temperature optima classifications, 2:318
 toxitolerants, 2:337
 xerophiles, 2:337
Extrinsic incubation, definition, 1:311
Exudate, definition, 4:117
Eyespot
 control
  crop management, 2:340–341
  fungicides, 2:341
  prospects, 2:341
  resistant plants, 2:341
 definition, 2:338
 epidemiology, 2:340
 pathogens, 2:338
 symptoms, 2:338

F

FA-2, antifungal activity, 1:251
FabA, fatty acid biosynthesis, 3:57
Factor F_{430}, biosynthesis, 4:563
Factor for inversion stimulation (FIS)
 DNA bending, 1:816
 replication origin binding, 1:826, 828–829
Facultative, definition, 3:676, 4:65
Facultative autotroph, definition, 1:349
Facultative pond, industrial effluent treatment, 2:757
Famciclovir
 adverse effects, 1:297
 indications
  genital herpes, 1:297
  herpes labialis, 1:297
  herpes zoster, 1:296
 mechanism of antiviral activity, 1:296
 prodrug of penciclovir, 1:296
 resistance, 1:297
Fatty acid
 biosynthesis, *see* Lipid biosynthesis, *Escherichia coli*
 definition, 3:55, 62, 773
 β-oxidation, 3:60
 production, *see* Lipid production
FBA, *see* Flux balance analysis
Feasible set, definition, 2:510
Fed batch, definition, 1:579
Fed-batch process, definition, 3:9
Feedback control, definition, 4:127
Feedback inhibition, definition, 1:134
Fenarimol, antifungal activity, 1:240
Fenpiclonil, antifungal activity, 1:240
Fenpropimorph, antifungal activity, 1:249
Fermentation
 agricultural waste, *see* Methanogen
 alcoholic beverages, *see* Beer brewing; Wine
 amino acids, 2:346
 biosurfactant production, 1:631, 634
 butanediol operon regulation in *Klebsiella*, 2:349
 butyrate–butanol fermentation regulation in *Clostridium acetobutylicum*, 2:349
 carbohydrates, 2:344
 cultures
  selection, 2:770
  development, 2:770–771
 dairy products, *see* Dairy products
 definition, 1:647, 834, 2:343, 677, 4:316, 428, 837
 energy conservation reactions, 2:347
 fermentation balance, 2:343–344
 foods, *see* Fermented foods
 heterotrophs, 2:660
 history of study, 2:683
 history of use, 2:768–770
 industrial production
  advantages, 2:767
  limitations, 2:767–768
  media development and optimization, 2:773–774
  process development, 2:771
  products, overview, 2:768
  scale-up, 2:771–773
 lactic acid production, *see* Lactic acid production
 large-scale operations
  batch mode, 2:777
  bioreactors, 2:774, 777
  continuous mode, 2:777–778
  downstream processing, 2:781
  fed batch mode, 2:777
  inoculum production, 2:774
  monitoring and control, 2:778
  sterilization and contamination control, 2:778–780
  utility requirements, 2:780–781
 mixed-acid fermentation, regulation of downstream branches, 2:348
 nucleic acids, 2:346
 organic acids, 2:344–346
 pH stress
  acidic fermentation, 3:630
  alkaline fermentation, 3:630–631
 pyruvate cleavage, regulation, 2:347–348
 ruminants, *see* Rumen fermentation
 solvent production with clostridia, 1:838
 strain improvement, *see* Strain improvement
 vitamin production, *see* Vitamin
Fermented foods, *see also* Dairy products
 bioenrichment
  amino acids, 2:358

Fermented foods (*continued*)
protein, 2:358
rationale, 2:357–358
vitamins, 2:358–359
classification, 2:350–351
cooking and fuel requirement minimization, 2:359
fermentation types
acid fermentation, 2:356
alcoholic fermentation, techniques, 2:355–356
alkaline fermentation, foods and organisms, 2:356–357
bread fermentation, 2:356
high-salt savory-flavored amino acid–peptide sauces and pastes, 2:357
lactic acid fermentation, *see also* Lactic acid bacteria
bacteria, 2:353
balao balao, 2:354
cereal–legume foods, 2:354–355
foods, overview, 2:353–354
pit fermentation, 2:354
yogurt–cereal mixtures, 2:354
overview, 2:351
meat products, 3:164
safety, 2:351–353
toxin reduction, 2:359
Fermenter
acetic acid production, 1:15–17
definition, 1:13
Ferredoxin, definition, 3:188
Ferritin
definition, 2:860
iron storage, 2:866–867
FhlA, regulation of formate hydrogen–lyase system, 2:348
FIA, *see* Flow injection analysis
Fibrinogen, expression in transgenic animal milk, 4:683
Ficoll–Hypaque centrifugation, *see* Diagnostic microbiology
Field inversion gel electrophoresis (FIGE), definition, 4:55
FIGE, *see* Field inversion gel electrophoresis
Filamentous fungus, *see* Fungus, filamentous
Filovirus, emerging infection, 4:815
Fimbriae, *see also* Pilus
definition, 2:187, 270
Escherichia coli, 2:261–262
type IV, 2:187
Yersinia pestis, 3:661
FimH
adhesin, 1:45, 50–51
sequence variation, 3:377
Fingerprint assembly, definition, 3:151
FIS, *see* Factor for inversion stimulation
Fish oil, fatty acid improvements with lipases, 3:52–53
FISH, *see* Fluorescence *in situ* hybridization
Fission, definition, 4:939
Fitness
definition, 2:283, 3:373
factors affecting
genetic background, 2:290
metabolic activity variation, 2:289–290
mutations, 2:288–289
possession of unused functions, 2:289
Fixation, microscopy specimens, 3:264, 288
Fixed cost, definition, 2:137
Fixed film reactor, industrial wastewater treatment, 4:862–863
FixL, functions, 1:67
FK506, mechanism of action, 3:777–778
Flaccid paralysis, definition, 3:762
Flagellin, methylation, 3:244
Flagellum
assembly
cytoplasm, 2:387–388
kinetics
filament growth, 2:388–389
hook growth, 2:389
outside of cell, 2:388
periplasmic space, 2:388
basal structure
basal body, 2:380, 383
C ring, 2:384
export apparatus, 2:384
LP-ring complex, 2:383
MS-ring complex, 2:383–384
rod, 2:384
Caulobacter
cell cycle regulation of synthesis in *Caulobacter crescentus*, 2:22–23
gene identification, 1:699–701
synthesis and assembly studies, 1:702–703
definition, 2:187, 4:755
energy source, 2:385
Escherichia coli, 2:262
filament
cap proteins, 2:382
definition, 2:380
flagellin, 2:381–382
shape, 2:381
gene classification
che, 2:386
fla, 2:385
mot, 2:385–386
gene clusters, 2:386
gene expression, transcriptional regulation, 2:386–387
hook
associated proteins, 2:382–383
FlgD scaffolding, 2:382
protein structure, 2:382
shape, 2:382
master genes, 2:380
motor, 2:380
number per cell, 2:380–381
peritrichous, 2:236
polymorphic transition, 2:380
Pseudomonas, 3:877
rotation and cell motility, 1:726
rotational direction, switching, 2:385
spirochetes, 4:354
torque
definition, 2:384
rotational direction, 2:385
rotational speed, 2:385
type III export, 2:380, 388
Flavin adenine nucleotides, *see* Riboflavin
Flavivirus, emerging infection, 4:816
Flavohemoglobin, functions, 1:67
FlgD, scaffolding of flagellar hook, 2:382
flhDC operon
flagellum gene expression, 2:386–387
swarming differentiation control, 2:19–21
Flotation concentrate, definition, 3:482
Flow injection analysis (FIA), bioreactors, 1:570–571, 611–612
Flower-breaking, definition, 3:792
Fluazinam, antifungal activity, 1:252
Fluconazole, antifungal activity, 1:243–244
Fludioxinil, antifungal activity, 1:240

Fluid milk, definition, 2:1
Fluidized bed reactor, industrial wastewater treatment, 4:855, 863–865
Fluidized bed reactor, industrial effluent treatment, 2:756
Fluorescence, definition, 3:264, 288
Fluorescence microscope, definition, 2:29
Fluorescence microscopy, *see* Confocal microscopy; Optical microscopy
Fluorescence *in situ* hybridization (FISH), definition, 1:822
5-Fluorocytosine, antifungal activity, 1:240
Fluorophore, definition, 3:264
Fluquinconazole, antifungal activity, 1:243
Flux balance analysis (FBA)
  definition, 2:510
  metabolic engineering
    applications
      Bacillus subtilis, 2:518–519
      Corynebacterium glutamicum, 2:518
      gene deletions, 2:516–517
      Penicilliuum chrysogenium, 2:517–518
    overview, 2:514–516
FNR, regulation of oxidase synthesis in *Escherichia coli*, 1:64
Folic acid, biosynthesis, 1:844
Fomite, definition, 4:97
Food-borne illness, *see also specific diseases and pathogens*
  Aeromonas, 2:403
  Ascaris, 2:408
  Bacillus cereus, 2:394
  Campylobacter, 2:399–400
  Clostridium botulinum, 2:392–393
  Clostridium perfringens, 2:396
  definition, 2:390
  Diphyllobothrium latum, 2:408
  epidemiology, 2:390–392, 411
  Escherichia coli
    enteroaggregative, 2:404
    enteroinvasive, 2:402–403
    enteropathogenic, 2:403–404
    enterotoxigenic, 2:397
    Shiga toxin production, 2:396–397
  hepatitis A virus, 2:404–405
  hepatitis E virus, 2:405
  Hymenolepis nana, 2:408
  Listeria monocytogenes, 2:401–402
  Norwalk virus, 2:405
  organism testing in quality control of foods, 2:428–429
  pathogen types by mechanism, 2:392
  Plesiomonas shigelloides, 2:403
  prevention, 2:411
  protozoa
    Cryptosporidium parvum, 2:406
    Cyclospora cayetanensis, 2:407
    Entamoeba histolytica, 2:406–407
    Giardia lamblia, 2:406
    miscellaneous organisms, 2:407
  refrigerated foods, *see* Refrigeration, foods
  rotavirus, 2:405
  Salmonella, 2:398–399
  Shigella, 2:402
  Staphylococcus aureus, 2:393–394
  surveillance
    Escherichia coli O157:H7, 4:523
    FoodNet, 2:391–392, 4:513–514
    PulseNet, 4:523–524
  Taenia saginata, 2:407
  Taenia solium, 2:407–408
  toxins, natural
    aflatoxin, 2:410
    ciguatera poisoning, 2:409
    miscellaneous toxins, 2:410
    mushroom toxins, 2:410
    scrombroid poisoning, 2:409
    shellfish poisoning, 2:409–410
    tetrodotoxin poisoning, 2:410
  Trichinella spiralis, 2:408–409
  Trichuris trichiura, 2:408
  Vibrio species
    miscellaneous species, 2:396
    Vibrio cholerae, 2:394–395
    Vibrio parahemolyticus, 2:395
    Vibrio vulnificus, 2:395
  Yersinia, 2:400–401
Food fermentation, *see also* Fermented foods
  definition, 2:412
  preservation of foods, 2:420
Food poisoning, definition, 2:390
Food preservation, *see also* Refrigeration, foods
  chemical preservatives, 2:419–420
  definition, 2:412
  drying, 2:415
  extrinsic parameters of food
    gaseous environment, 2:414
    physical stress, 2:415
    relative humidity, 2:414
    storage time, 2:415
    temperature, 2:414
  fermentation, 2:420
  freeze-drying, 2:416
  freezing, 4:548
  gas modification techniques, 2:419
  high temperature preservation
    pasteurization, 2:416–417
    sterilization, 2:417–418
  intrinsic parameters of food
    antimicrobial agents, 2:414
    biological structure, 2:414
    moisture content, 2:414
    nutrient content, 2:414
    pH, 2:413–414
    redox potential, 2:414
  low temperature preservation, 2:415–416
  radiation
    ionizing irradiation, 2:418
    nonionizing irradiation, 2:418–419
Food, quality control
  Good Manufacturing Practices, 2:426
  hazard analysis and critical control point, 2:429–430
  health hazard testing, 2:428–429
  indicator microorganisms, 2:427–428
  intrinsic and extrinsic parameters of food, 2:423
  microbiological criteria, 2:422
  overview, 2:421–422
  reference methods for microbiological criteria, 2:425
  sampling plan, 2:424–425
  sanitation, 2:425–427
  spoilage organism testing, 2:428
  statistical sampling, 2:423–424
Food spoilage, *see also* Meat products
  aesthetic differences, 2:412
  definition, 2:412
  economic significance, 2:412–413
  health hazard, 2:413
  microbial activities, 2:413
  organism testing, 2:428
Food spoilage, *see also* Meat products
Food web, definition, 2:131
Foregut fermentation, definition, 2:485
Forespore
  definition, 1:373
  SigmaF activation, 1:380

Forespore (*continued*)
SigmaG activation, 1:381
Formate dehydrogenase, acetogenesis, 1:20
Foscarnet
adverse effects, 1:293–294
indications, 1:293
mechanism of antiviral activity, 1:293
resistance, 1:293
Fouling, definition, 1:472
F-prime, definition, 2:270
Frameshift mutation, definition, 3:307
Frankia
definition, 3:392
host specificity, 3:398
nitrogen fixation, 3:397–398
Freeze-drying
foods, 2:416
microorganisms
cooling procedure, 2:432
drying
secondary drying, 2:431, 432–434
sublimation and isothermal desorption, 2:432–434
termination, 2:434
drying stage, 2:431
growth environment and cell tolerance, 2:436
operational processes affecting viability, 2:435
optimization, 2:436
preliminary preparation, 2:431–432
protective additives, 2:435–436
reconstitution, 2:434
storage, 2:434
Freeze-etching
definition, 3:276
transmission electron microscopy specimens, 3:281–282
Freeze-substitution
definition, 3:276
transmission electron microscopy specimens, 3:282
Freshwater microbiology
importance of microbes in lakes and streams, 2:439–440
lakes
spatial heterogeneities
inorganic nitrogen and phosphorous, 2:443
light, 2:442–443
light–hydrogen sulfide gradient, 2:443
light–macronutrient gradient, 2:443
oxygen, 2:443–444
reduced substances, 2:443
thermal stratification, 2:441–442
temporal heterogeneities, 2:444–445
metabolic types of microbes, 2:440
microbial loop, 2:449–450
microbial processes, 2:440–441
microbial productivity, 2:439–440
streams
spatial heterogeneities
habitat scale, 2:447–448
stream reach scale, 2:447
watershed scale, 2:445–447
temporal heterogeneities
decades, 2:449
diel or shorter, 2:449
seasonal, 2:448–449
substrates for growth, 2:441
water bodies, overview, 2:438–439
Fruiting body
definition, 3:349, 363
formation in myxobacteria, 2:26–27, 3:353, 355
genetic analysis of behavior in *Myxococcus xanthus*, 3:372
morphogenetic movements during formation, 3:360
survival advantage, 3:358
Fruit juice, enzymatic processing, 2:229–230
FruR, phosphotransferase system interactions, 3:592
Fts proteins
cell division role, 1:706–708, 725
definition, 1:704
FtsZ, cell division control, 2:25–26
Fulminant disease, definition, 4:758
Fumarate, reduction in anaerobic respiration, 1:186
Fumonsin
animal disease, 3:346–347
human disease, 3:342
sources, 3:342
Fungal infection, cutaneous
antifungal agents, *see* Antifungal agents
cutaneous host defenses
antifungal substances, 2:452
cutaneous immune system, 2:453
inflammatory response, 2:452–453
keratinization and epidermal proliferation, 2:452
skin structure, 2:451–452
transferrin, 2:452
overview, 2:451
subcutaneous mycoses
chromoblastomycosis, 2:457
mycotic mycetoma, 2:457–458
sporotrichosis, 2:456–457
superficial infections
cutaneous candidiasis, 2:453–455
dermatophytosis, 2:455
miscellaneous infections, 2:456
tinea versicolor, 2:455–456
systemic disease with cutaneous manifestations, 2:458
Fungal infection, plants, *see also* Bunts; Downy mildew; Ergot; Powdery mildew; *Rhizoctonia*; Rust fungi; Smuts; *Verticillium*
epidemiology, 3:685
management, 3:685
overview, 3:678–679
pathogens
classification, 3:681–683
dissemination, 3:684
ecology, 3:683–684
genetics of disease, 3:684–685
infection, 3:684
morphology, 3:679, 681
reproduction, 3:681
Fungal infection, systemic
acquisition and infection, 2:462
antifungal agents, *see* Antifungal agents
aspergillosis, 2:463
blastomycosis, 2:463
candidiasis, 2:463–464
coccidioidomycosis, 2:464
cryptococcosis, 2:464
definition and overview, 2:460–462
histoplasmosis, 2:464
host defenses, 2:462–463
miscellaneous infections, 2:464–466
opportunistic infection, 2:460–462
portal of entry, 2:460
species, 2:461
susceptibility to infection, 2:462–463
treatment and prevention, 2:466
Fungicide, definition, 2:338
Fungus
definition, 3:773

diversity of species, 2:63–65
gut, 2:494–495
insecticide application
conservation or environmental manipulation, 2:823–824
diversity of species, 2:821
history of use, 2:821
inoculative augmentation, 2:823
integrated pest control, 2:824
limitations, 2:824
mass production and formulation, 2:824
pathogenicity, 2:822
permanent introduction, 2:822–823
prospects, 2:824
opportunistic pathogens, 1:232
skin resident species, 4:279–280
toxins, *see* Mycotoxin
Fungus, filamentous, *see also specific fungi*
biotechnology applications, 2:475–476
culture collections and resources, 2:476–477
definition and overview, 2:468–469, 477
dimorphism, 2:468, 473
diseases
cutaneous infection, *see* Fungal infection, cutaneous
overview, 2:473–474
plant infection, *see* Fungal infection, plants
systemic infection, *see* Fungal infection, systemic
hypha features, 2:469–470
physiology and ecology, 2:474–475
reproduction, 2:470–471
spore types, 2:471
taxonomy, 2:471–473
Fur
iron regulation, 2:865–866
osmotic stress regulation, 3:531
Furuncle, definition, 4:271
Fusogen, definition, 3:1

G

Gaeumannomyces graminis
disease management, 2:483
ectotrophic colonization, 2:482
endotrophic colonization, 2:482–483
epidemiology
geographic distribution, 2:483
host range, 2:483
inoculation, 2:482
overview, 2:479
penetration of host, 2:482
survival, 2:482
taxonomy, 2:479–482
Gametocyte, definition, 3:745
Ganciclovir
adverse effects, 1:295
immunocompromised patient indications
human immunodeficiency virus, 1:294
transplant recipients, 1:295
mechanism of antiviral activity, 1:294
resistance, 1:295
val-anciclovir prodrug, 1:294
Gas gangrene
definition, 1:834, 4:271
toxin, *see* Clostridia
Gastric carcinom, *Helicobacter pylori* role, 2:633
Gastritis
definition, 2:187
Helicobacter pylori role, 2:198, 632–633
Gastroenteritis, definition, 1:530, 2:390
Gastroesophageal reflux disease (GERD), *Helicobacter pylori* role, 2:633
Gastrointestinal microbiology
animal–flora interactions
competition, 2:486
cooperation, 2:486, 488
cooperation and competition, 2:488
bacteria
esophagus, 2:489
intestine
chicken, 2:492
human, 2:490–492
pig, 2:492
population types, 2:488–489
rumen, 2:489–490
stomach, 2:489
bacteriophages, 2:496
ecosystems, 2:486
fermentation sites, 2:487
fungi, 2:494–495
gastrointestinal tract features, 2:485
host nutrition role, 2:496
normal flora
definition, 2:485
infection prevention, 2:496–497
pathogenesis, 2:497
population types, 2:488
protozoa, 2:493–494
GC, *see* Germinal center
GDH, *see* Glutamate dehydrogenase
Gellan, features, 1:560
Gene, definition, 4:428
Gene conversion, definition, 1:254
Gene disruption, definition, 3:307
Gene fusion, definition, 1:1
Gene mapping, *see* Mapping bacterial genomes
Generalized transduction, definition, 1:692
General secretion pathway, *see* Secretion, proteins
Gene silencing, posttranscriptional, 1:268, 280–283
Gene therapy, retroviral vectors, 4:95–96
Genetically modified organism (GMO), *see also* Transgenic animal
biohazard classification by risk group, 2:505
comparison with exotic species, 2:500–501
concerns
agriculture, 2:501
environment, 2:501
safety, 2:500
definition, 2:499
guidelines and regulations
appropriateness
contained research, 2:506–507
needs and options, 2:507–508
research involving planned introduction into the environment, 2:507
contained research under National Institutes of Health guidelines, 2:504–506
history, 2:501–503
research outside of containment, 2:506
oversight mechanisms
guidelines and directives, 2:503–504
regulations, 2:504
standards of practice, 2:503
risk assessment and management, 2:508–509
Genetic code, definition, 4:428

Genetic drift
definition, 2:283, 286, 3:373
evolution contribution, 3:373–374
Genetic engineering, definition, 2:722
Genetic linkage, definition, 4:43
Genetic locus, definition, 4:297
Genetic recombination, definition, 4:428
Genome
Borrelia burgdorferi, 3:111–112
carbohydrate metabolism studies, 1:665–667
Caulobacter, 1:703
cyanobacteria, 1:923–924
definition, 1:398, 2:270, 698, 722, 4:451, 628
Escherichia coli, 2:264–265
Haemophilus influenzae sequencing, 2:108
Helicobacter pylori, 2:629–630
mapping, *see* Mapping bacterial genomes
methanogens
genomics and gene function analysis, 3:224–225
structure, 3:221–222
Mycobacterium leprae, 3:319, 323
Mycobacterium tuberculosis, 3:319, 321
sequencing, *see also* DNA sequencing
approaches, 3:159, 161
published sequences, 2:107
Staphylococcus, 4:388–389
Streptomyces
composition, 4:451–452
sequencing, 4:451, 455
Genomic engineering, bacterial metabolism
applications, 2:512–513
flux balance analysis
applications
Bacillus subtilis, 2:518–519
Corynebacterium glutamicum, 2:518
gene deletions, 2:516–517
Penicilliuum chrysogenium, 2:517–518
overview, 2:514–516
genomics, 2:513–514
historical perspective, 2:510–512
linear programming, 2:516
metabolic flux control, 2:512
rationale, 2:510, 512
Genomics
biotechnology industry, 2:734
definition, 2:106
functional genomics
comparative genomics, 2:112–113
differential gene expression, 2:113–114
differential protein expression, 2:114
miscellaneous studies, 2:115
mutagenesis, 2:114–115
metabolic engineering, 2:513–514
World Wide Web resources, 2:112
Genotype, definition, 2:270, 4:32, 921
GERD, *see* Gastroesophageal reflux disease
Germfree animal
definition, 2:521
diets, 2:524–526
gnotobiote applications, 2:527–528
isolators
history, 2:521, 523
uses, 2:523, 528
production, 2:523–524
Germinal center (GC), definition, 1:208
Giardia lamblia
antigenic variation, 1:260–261
clinical features, 2:856–857
diagnosis, 2:857
epidemiology, 2:856
food-borne illness, 2:406
life cycle, 2:856
pathogenesis and pathology, 2:857
prevention and control, 2:857
treatment, 2:857
Gingival margin, definition, 3:466
Glabrous skin, definition, 4:271
Glass state
definition, 1:541
stability of biopesticides, 1:553
Gliding motility
definition, 3:363
genetic analysis of behavior in *Myxococcus xanthus*, 3:372
myxobacteria, 3:351, 360
glnA, regulation, 1:678–681
GlnN, functions, 1:67
Global burden, infectious disease
burden attributable to risk factors for infectious disease, 2:535, 539
causes of death, 2:531
deaths by region, 2:531
disability-adjusted life years
distribution by cause, 2:531–534
ranking by cause
developed regions, 2:538
developing regions, 2:537
global results, 2:536
summary, 2:534
risk factors for other diseases, 2:534–535, 539
Global Burden of Disease Study
classification of diseases and injuries, 2:531
objectives, 2:530
overview, 2:530
policy implications, 2:540
regions, 2:530–531
overview, 2:529–530
projections, 2:535, 540
sexually transmitted diseases, 4:249–250
years lived with disability, 2:529, 532–531
years of life lost, 2:529, 531–532
Global regulator, definition, 2:236
Globin
definition, 1:53
flavohemoglobins, 1:67
single-domain myoglobin-like globins, 1:67
gltBD operon, regulation, 1:672
β-Glucanase, features and applications, 2:215, 226
Glucoamylase, features and applications, 2:212, 223
Glucose, production from starch, 2:224
Glucose lipids, biosurfactant production and features, 1:623
Glucose oxidase, biosensor applications, 1:611–612
β-Glucosidase
catalytic domains, 1:750–752
definition, 1:744, 748
features and applications , 2:226–227, 230
regulation of expression, 1:755
Glutamate
microbial production process, 1:153–154
uses, 1:153
Glutamate dehydrogenase (GDH)
features, 1:138
nitrogen assimilation, 1:135–136
Glutamate synthase

features, 1:138
nitrogen assimilation, 1:135
Glutamine:oxoglutarate amidotransferase (GOGAT), operon regulation, 1:672
Glutamine synthetase (GS)
nitrogen assimilation, 1:135–136
regulation
adenylylation, 1:137, 676–678
feedback inhibition, 1:136–137, 676
Ntr response, 1:137–138
PII protein, 1:677–678
structure, 1:676
Glutaredoxin, antioxidant defense, 3:528
Glutathione, antioxidant defense, 3:528
Glycine
biosynthesis
pathway, 1:145–146
regulation, 1:146
functions, 1:145
Glycine betaine, biosynthesis, 4:889, 891–892
Glycogen, *see also* Carbohydrate
biosynthesis
ADP-glucose pyrophosphorylase
classification by activators, 2:544
cloning enzymes with altered allosteric properties, 2:546–547
ligand-binding sites, 2:546
regulation, 2:544–546
structure, 2:544
branching enzymes, 2:547–548
disaccharides as starting material, 2:543–544
genetic regulation in *Escherichia coli*
csrA gene regulation, 2:552–553
culture condition responses, 2:548
cyclic AMP regulation, 2:551
gene cluster, 2:548–549
glgCAP transcription, 2:550–551
guanosine diphosphate regulation, 2:552
integrated model, 2:553–554
kat F regulation, 2:550
loci affecting enzyme levels, 2:550
glycogen synthase, 2:547
sugar nucleotide pathway, 2:543
criteria for energy storage compound, 2:541–542
functions in bacteria, 2:542
occurrence in bacteria, 2:542
structure, 2:542–543
Glycogenesis
definition, 1:647
reactions, 1:659–660
Glycolysis
definition, 1:647
distribution of pathways
Embden–Meyerhof–Parnas pathway, 1:653–654
Entner–Doudoroff pathway, 1:654–655
hexose monophosphate pathway, 1:653–654
pentose phosphoketolase pathway, 1:654
fates of pyruvate and nicotinic adenine dinucleotides
aerobic or facultative bacteria under oxidative conditions, 1:656–657
anaerobic bacteria, 1:658–659
archaea, 1:659
facultative bacteria under anaerobic conditions, 1:657–658
overview of pathways, 1:651, 653
reactions, 1:652
variations in pathways
archaea, 1:655–656
methylglyoxal bypass, 1:655
Glycosidase inhibitor
aminoglycosides, 1:166–167
definition, 1:162
Glycosylase, definition, 2:71
Glycosyl transferase, definition, 3:517
Glyoxylate bypass, *Escherichia coli*
aceBAK operon
coarse control of bypass, 2:556
IclR regulation, 2:558–559
integrated host factor regulation, 2:558–559
transcriptional regulation, 2:557–558
control of metabolism
growth on acetate, 2:560
growth on glucose, 2:560
metabolic transitions, 2:560–561
function, 2:556
isocitrate dehydrogenase
flux control, 1:672, 674, 2:556–557, 559–561
kinase/phosphatase
regulation, 2:559–560
structure and mechanism, 2:559
GMO, *see* Genetically modified organism
Gnotobiotic
definition, 2:521
gnotobiote applications, 2:527–528
GOGAT, *see* Glutamine:oxoglutarate amidotransferase
Gold, biooxidation pretreatment of ores, 3:487–488
Gold standard, definition, 2:29
Gonorrhea
clinical manifestations
disseminated gonococcal infection, 4:254
men, 4:253
miscellaneous infections, 4:253–254
women, 4:252–253
diagnosis, 4:254–255
epidemiology, 4:251–252
pathogen, *see Neisseria gonorrhoeae*
pathogenesis, 4:255–257
treatment, 4:255
Gram-negative anaerobic pathogens, *see also specific bacteria*
antibiotic resistance
Bacteroides genes, 2:567–568
conjugative elements
clindamycin resistance, 2:569–570
5-nitroimidazole resistance, 2:570
plasmids, 2:569
tetracycline resistance, 2:569
transposons, 2:568–569
genome features, 2:567
immune response, 2:566–567
infection
bacteremia, 2:564
bone and joint, 2:564
central nervous system, 2:564
head, neck, and mouth, 2:563
intraabdominal infection, 2:564
pelvis, 2:564
pleuropulmonary infections, 2:563–564
skin and soft tissue, 2:564
normal flora and epidemiology, 2:562–563
pathogenesis

Gram-negative anaerobic pathogens (*continued*)
Bacteroides fragilis capsular polysaccharide in abscess induction, 2:565
synergy, 2:564–565
T cells, abscess formation role, 2:565
virulence factors, 2:565–566
Gram-negative cocci pathogens, *see also Neisseria*
overview, 2:571
species, 2:572
Gram stain, definition, 1:759
Granulocytopenia, definition, 1:232
Granulosis virus, definition, 2:813
Greenhouse gas, definition, 4:117
Gribble, definition, 4:571
Griseofulvin, antifungal activity, 1:239
GroELS, chaperone activity, 2:600, 605
Growth kinetics, bacteria
balanced growth, 2:586
binary fission, 2:584
continuous culture, 2:587–588
counting of populations, 2:585
crop yield, 2:586–587
definition, 2:584
growth curve, 2:586
nutrient concentration effects, 2:587
temperature effects, 2:588–589
unrestricted growth modeling, 2:585–586
Growth medium
definition, 3:431
design, *see* Nutrition, microbial culture
Growth optimum, definition, 2:317
GS, *see* Glutamine synthetase
Guanosine pentaphosphate, *see* Stringent response
Guillain–Barr/ syndrome, definition, 2:390
Gums
curdlan
production, 1:563
properties, 1:562
structure, 1:562
uses, 1:563
definition, 1:556
dextran
production, 1:561–562
structure, 1:561
uses, 1:562
gellan, 1:560
overview, 1:556–557
pullulan
production, 1:563
properties, 1:563
uses, 1:563–564
rhamsan, 1:561
scleroglucan, 1:564
sources and uses, 1:558
welan, 1:560–561
xanthan
physical properties, 1:557–559
production, 1:559
structure, 1:557
uses, 1:559–560

H

HAART, *see* Highly active anti-retroviral therapy
Haber–Bosch process, comparison with biological nitrogen fixation, 1:493
Habitat, definition, 2:131
HACCP, *see* Hazard analysis and critical control point
Haemophilus influenzae
adhesins, 1:48–49
antigenic variation, 1:263–264
chromosome, 2:592
diseases, 2:591
genes
cloning, 2:594
types and gene families, 2:592
genetic regulatory mechanisms, 2:594
genetic variation processes, 2:593–594
genome sequencing, 2:108
mutagenesis, 2:594–595
phages, 2:593
phylogenetic relationships, 2:591–592
plasmids, 2:593
repeated sequences, 2:592–593
restriction systems, 2:596
selectable markers, 2:595
skin microbiology, 4:284
surveillance, 4:517
transformation, 2:595–596, 4:656–657
uptake signal sequence, 2:591–593
vaccine, 4:771
Hairy root, definition, 1:78, 86
Half-life, definition, 1:288
Half-saturation constant for growth, continuous culture, 1:873, 875–876
Haloalkaliphile, definition, 1:126
Halofantrine, malaria treatment, 3:139
Halophile, *see also* Osmotic stress; Water-deficient environments
adaptation biochemistry and physiology
compatible-solute adaptation, 2:330–331
salt-in-cytoplasm addaptation, 2:329–330
Archaea
habitats, 1:322
physiology, 1:322–323
definition, 1:126, 319, 2:317, 3:502
habitats, 2:327–329
ion concentration of brines, 2:328
osmotolerance, 3:503
species, 2:327–329
Hantavirus
emerging infection, 4:816–817
pulmonary syndrome, definition, 1:287
surveillance, 4:519–520
Hap, adhesin, 1:48–49
Hapten, definition, 1:208–209
Haptoglobin, definition, 2:860
Hardwood, definition, 4:571
Hartig net, definition, 3:328
Haustoria, definition, 2:117, 3:676, 801
Hazard Analysis and Critical Control Points (HACCP)
food quality control, 2:429–430
meat safety, 3:168
Hazardous waste, definition, 2:782
Headful packaging, definition, 4:637
Heartwood, definition, 4:571
Heat stress response, *see also* Temperature control
Bacillus subtilis heat shock response
CIRCE regulation, 2:603–605
HrcA regulation, 2:605
overview, 2:603–605
sigma factor B regulation, 2:605
species distribution of heat shock response mechanism, 2:605

Escherichia coli heat shock response
extracytoplasmic heat shock response and sigma factor E regulation, 2:605–606
molecular thermometers, 2:602–603
overview, 2:600–601
sigma factors, regulation, 2:601–603
species distribution of heat shock response mechanism, 2:603
inputs, 2:598–599
outputs
chaperones, 2:599–600
peptidyl prolyl isomerase, 2:600
proteases, 2:600
protein disulfide isomerase, 2:600
overview, 2:598
Heavy metals
accumulation levels in species, 2:612, 614
binding molecules, 2:614
biosorption
cell walls, 2:611
desorption, 2:611–614
free and immobilized biomass, 2:611
definition, 2:607, 618
detoxification mechanisms of microbes, 2:609–610
environmental distribution, 2:607–608
environmental modification of toxicity, 2:608–609
microbial population impact, 2:608
precipitation
high gradient magnetic separation, 2:615
metal-reducing bacteria, 2:614–615
phosphatase-mediated precipitation, 2:615
sulfate-reducing bacteria, 2:615
protozoa vulnerability, 3:875
Pseudomonas resistance, 3:890
removal and recovery processes with microbes, 2:610–611
resistance
antimony, 2:619–620
approaches, 2:619
arsenic, 2:619–620
cadmium, 2:620–621
chromium, 2:621–622
definition, 2:607–608
mercury
gene organization in bacteria, 2:623–624
mercuric reductase features, 2:622–623
resistance operon regulation, 2:625–626
sources in environment, 2:622
spectrum of resistance, 2:622
transport into cells, 2:624–625
miscellaneous metals, 2:626
solubilization
autotrophic leaching, 2:611, 616
heterotrophic leaching, 2:611
tolerance, 2:607–608
transformation of metals and compounds
mercury and organometallic compounds, 2:616
metalloids, 2:616
methylation, 2:616
reduction, 2:616
transport systems, 2:614, 618
Helicobacter pylori
colonization and virulence factors
adherence, 2:631–632
CagA, 2:632
motility, 2:631
urease, 2:631
vacuolating cytotoxin, 2:632
diagnosis, 2:633–634
diseases
gastric carcinoma, 2:633
gastric lymphomas, 2:633
gastritis, 2:632–633
gastroesophageal reflux disease, 2:633
ulcer, 2:628, 633
epidemiology, 2:632
gastritis, 2:198
genome sequence, 2:629–630
historical perspective, 2:628–629
population genetics and evolution, 2:630–631
respiratory chain, 1:58
subpopulations in stomach, 2:631
taxonomy, 2:629
transmission, 2:198
treatment, 2:634
vaccine development, 2:634
virulence factors, 2:198
Helveticin J, features, 1:391–392
Hemagglutinin
definition, 1:287, 3:533
influenza virus
features, 2:800–802, 805
inhibitors, 2:810
Hematopoietic toxicity, definition, 1:287
Heme
biosynthesis
heme *a*, 4:565
heme *c*, 4:564–565
heme *d*, 4:565
heme d_1, 4:563
heme *o*, 4:565
protoheme, 4:564
definition, 2:860, 4:558
iron uptake, 2:864–865
Hemicellulose, enzymatic biomass conversions, 2:226–227
Hemolytic uremic syndrome (HUS)
definition, 2:390
features, 2:396
pathogen, 2:195, 267–268, 3:167
polymerase chain reaction diagnostics, 3:791
surveillance, 4:523–524
Hemopexin, definition, 2:860
Hemophore, definition, 2:860
Hemorrhagic fever, definition, 4:811
Hendra virus, emerging infection, 4:817–818
HEPA filter, *see* High-efficiency particle air filter
Hepatitis viruses
clinical features and course, 2:636–639
definition, 2:635
emerging infection, 4:818
fulminant hepatitis, 2:636
hepatitis A virus
epidemiology, 2:639
food-borne illness, 2:404–405
replication, 2:639–640
unique clinical features, 2:640–641
hepatitis B virus
epidemiology, 2:643
hepatocellular carcinoma association, 3:459–461
interferon therapy, 1:301
replication, 2:644–645
surface antigen, 2:643
transmission, 2:643–644
treatment, 2:646
unique clinical features, 2:645–646

Hepatitis viruses (*continued*)
vaccination, 2:643
hepatitis C virus
historical perspective, 2:647–648
interferon therapy, 1:301
replication, 2:648–649
surveillance, 4:517, 519
treatment, 2:649
unique clinical features, 2:649
hepatitis D virus
historical perspective, 2:646
replication, 2:646–647
treatment, 2:647
unique clinical features, 2:647
hepatitis E virus
epidemiology, 2:641
food-borne illness, 2:405
replication, 2:641–642
unique clinical features, 2:642–643
host specificity, 2:636
transmission, 2:636, 639, 643
types, overview, 2:635–636
Hepatocellular carcinoma, definition, 2:635
Hepatocyte, definition, 2:635
Herbicide, definition, 2:722
Herpes labialis, definition, 1:287
Herpes simplex virus (HSV)
transcription, 4:631–632, 636
treatment
acyclovir, 1:289–291
famciclovir, 1:297
Heterocyst, cyanobacteria, 1:915–916, 2:27
Heteroduplex DNA, definition, 4:651
Heteroecious organism, definition, 4:195
Heteroencapsulation, definition, 3:792
Heterofermentative microorganisms, definition, 3:9
Heterogenesis, definition, 4:364
Heterokaryon, definition, 2:468
Heterologous gene, definition, 3:1
Heterothallism, definition, 4:297
Heterotroph
Archaea, 2:656–658
chemoheterotrophs, 2:652
definition, 1:18, 579, 2:651, 3:379, 482
ecological considerations
eutrophs, 2:661
microbial loop, 2:661
oligotrophs, 2:661–662
substrates for growth, 2:662
Eubacteria, 2:656
eukaryotes, 2:658–660
fermentation, 2:660
oxygen requirements for growth, 2:653
photoheterotrophs, 2:652
prokaryotic versus eukaryotic microorganisms, 2:653–654
respiration
aerobic, 2:660–661
anaerobic, 2:661
role in carbon cycle, 2:651
Hexaflora, definition, 2:521
Hexose monophosphate pathway, features and distribution, 1:653–654
Hex system, transformation, 4:658
HFCS, *see* High-fructose corn syrup
HGMS, *see* High gradient magnetic separation
HGT, *see* Horizontal gene transfer
Hia, adhesin, 1:48
High-efficiency particle air (HEPA) filter
air disinfection, 1:75
definition, 1:69
High-fructose corn syrup (HFCS), production, 2:213, 224
High gradient magnetic separation (HGMS), metal-containing bacteria, 2:615
Highly active anti-retroviral therapy (HAART), effectiveness against AIDS, 1:113
High-maltose conversion syrup, production from starch, 2:224–225
High-pressure habitats, *see also* Barophile
apparatus for laboratory studies, 2:669–671
examples, 2:665–668
extraterrestrial environments, 2:666
history of study, 2:664–665
microorganism properties
adaptation overview, 2:671
fatty acid composition, 2:674–675
growth rates, 2:672
light sensitivity, 2:675
protein composition, 2:674
PTk diagrams, 2:664, 672
pressure, definition, 2:664–665
pressure-based classification of microorganisms, 2:672–674
species distribution, 2:671
survey instruments, 2:668–669
Hindgut fermentation, definition, 2:485
Histidine
biosynthesis, 1:150–151
functions, 1:149
Histoplasmosis, systemic infection, 2:464
History, microbiology *see also* Spontaneous generation
antimicrobial therapies, 2:687–688
classification of microbes, 2:680–682
clinical microbiology, 2:688
contagion theory, 2:682
culture, 2:684
genetics, 2:691–694
germ theories and applications, 2:682–687
Koch's postulates, 2:684–685
microscopy
electron, 2:691
light, 2:678–679
physiological genetics, 2:696–697
physiology of microbes, 2:691
public health, 2:688
spontaneous generation theories, 2:679–680
transformation, 2:694–696
vaccination, 2:686–687
virology, 2:688–691
HIV, *see* Human immunodeficiency virus
HME, *see* Horizontally mobile element
HMW, adhesins, 1:48
H-NS, DNA bending, 1:816
Homeophasic adaptation, definition, 3:93
Homofermentative microorganisms, definition, 3:9
Homolog, definition, 4:1
Homologous recombination
definition, 3:363, 4:651
tranformation, 4:657
Homology, definition, 2:698
Hops, beer brewing, 1:412, 414
Horizontal gene transfer (HGT), *see also* Conjugation; Transduction; Transformation
contribution to gene evolution, 2:699

definition, 2:270, 698
effects of transfer versus transmission, 2:705–706
frequency estimation, 2:705
gene archaeology
genome structure, 2:704
sequence comparisons
caveats, 2:700
homology compared to sequence, 2:703–704
large rare events versus small common events, 2:703
mutation versus recombination, 2:702–703
orthologs versus paralogs, 2:702
host range, 2:700
mechanisms, overview, 2:699–700
overview, 2:698–699
potentiating conditions, 2:706
Horizontal reproduction, definition, 2:698
Horizontal resistance
definition, 3:662
plant disease resistance, 3:664–665
Horizontally mobile element (HME), definition, 2:698
Hormogonia, cyanobacteria, 1:915
Host, definition, 1:311, 2:201, 3:109
Host range
definition, 1:398, 2:698
horizontal gene transfer, 2:700
HPr
phosphorylation, 3:590–592
structure, 3:581–582
HPV, *see* Human papilloma virus
HrcA, heat shock response regulation regulation, 2:605
Hsf, adhesin, 1:48
HSV, *see* Herpes simplex virus
HTLV-1, *see* Human T cell leuekotropic virus type 1
HU, DNA bending, 1:815
Human immunodeficiency virus (HIV), *see also* Acquired immunodeficiency syndrome
antigenic variation, 1:266
antiviral therapy, 4:91–92
azidothymidine therapy, 1:112
blood screening for antibodies, 1:107, 110–111, 114
definition, 1:104
discovery, 1:105
emerging infection, 4:819
highly active anti-retroviral therapy, 1:113
immunodeficiency, 4:90–91
levels in body fluids, 4:91
origins, 1:105
pathogenesis, 4:89
polymerase chain reaction diagnostics, 3:788–789
protease inhibitors
indinavir, 1:307–308
nelfinavir, 1:308–309
ritonavir, 1:308
saquinavir, 1:307
VX-478, 1:309
proteins and functions, 4:89
receptors, 1:105, 4:88
replication, 1:104–105
reverse transcriptase inhibitors
abacavir, 1:305
azidothymidine, 1:302–303
delavirdine, 1:306
didanosine, 1:303
efavirenz, 1:306–307
lamivudine, 1:305
nevirapine, 1:305–306
stavudine, 1:304–305
zalcitabine, 1:303–304
sexually transmitted disease, 4:250
simian immunodeficiency virus comparison, 1:106
structure, 1:104
transmission, 4:90
types, 4:88
vaccine development, 1:114
Human papilloma virus (HPV), *see also* Papillomaviruses
cervical cancer association, 3:462–464
transcription, 4:630
Human T cell leuekotropic virus type 1 (HTLV-1), *see also* Retrovirus
emerging infection, 4:819–820
T cell cancer association, 3:464
Humoral immunity, *see also* Antibody; B cell
comparison with cell-mediated immunity, 1:210
evolution, 1:211
Humus, definition, 4:316, 321, 331
Hungry codon, definition, 4:467
Hurdle technology, definition, 4:545, 550–551
HUS, *see* Hemolytic uremic syndrome
Hybridization, definition, 4:297
Hybridoma
definition, 1:209
production and applications, 1:229–230
Hydraulic retention time, definition, 4:870
Hydrogen, growth requirements and functions, 3:436–437
Hydrogen gas, atmosphere cycling by microbes, 3:259–261
Hydrogenase, definition, 3:188
Hydrogenosome, definition, 2:485
Hydrophobic effect
cytoplasmic membrane, 1:713–714
definition, 1:710
Hydrophobin, definition, 2:468
3-Hydroxypropionate pathway, autotrophic carbon dioxide metabolism, 1:357
Hymenium, definition, 4:109
Hymenolepis nana, food-borne illness, 2:408
Hyperbilirubinemia, definition, 1:287
Hyperemia, definition, 1:287
Hyperpiezophile, definition, 2:664
Hyperplasia, definition, 3:697, 4:297
Hypersensitive reaction
definition, 2:117, 236, 3:662, 801, 876, 4:32
downy mildew, 2:122–123
hairpins
overview, 2:250–251
type III secretion pathway, 2:251
hrp genes
organization, 2:250
regulation, 2:251–252
plant disease resistance, 3:665
Pseudomonas, 3:884–885
virulence loci associated with *hrp* regulon
dsp, 2:252–253
hsvG, 2:253
Hyperthermophile
adaptation, 2:320–321
Archaea
habitats, 1:324–325
molecular adaptations, 1:329–330
physiology, 1:325, 327
cell envelopes, 2:320–321
definition, 1:319, 2:317–318
evolution, 2:321–322

Hyperthermophile (*continued*)
habitats, 2:318–319
nucleic acids, 2:321
proteins and solutes, 2:321
species, 2:319–320
temperature classification of microorganisms, 2:672
Hypertrophy, definition, 4:297
Hypha
definition, 2:15, 338, 468, 4:109
features, 2:469–470
Hyphomycetes, definition, 4:404
Hypnozoite, definition, 3:131
Hypocalcemia, definition, 1:287
Hypolimnion, definition, 2:438
Hypophodium, definition, 2:479
Hyporheic zone, definition, 2:438
Hypotension, definition, 1:287
Hypovirulent pathogen, definition, 3:662
Hysteresis, definition, 1:556

I

IAA, *see* Indole-3–acetic acid
IclR, glyoxylate bypass regulation, 2:558–559
Icosahedral symmetry, definition, 3:697, 4:97
Identification, definition, 2:709
Identification of bacteria, computerized
breathprint analysis, 2:719–720
identification matrix
evaluation
practical, 2:715, 717
software, 2:717–718
theoretical, 2:715
generation
character selection, 2:714–715
cluster analysis, 2:713
grouping known strains, 2:713–714
literature data collection, 2:714
published matrices, 2:716–718
lipids, gas chromatogram analysis, 2:720–721
nucleic acid fingerprint analysis, 2:721
numerical codes, 2:710
overview, 2:709
principle, 2:709–710
probabilistic identification
attribute space, 2:711
Bayes theorem, 2:711
definition, 2:709
identification scores, 2:711–712
likelihood of unknown, calculation, 2:712
maximum possible likelihoods, 2:713
model likelihood fraction, 2:713
normalized likelihoods, 2:712
outcomes, 2:712
taxonomic distance, 2:713
proteins, gel electrophoresis analysis, 2:721
software, 2:718
IDH, *see* Isocitrate dehydrogenase
Idoxuridine
adverse effects, 1:295
indications, 1:295
mechanism of antiviral activity, 1:295
resistance, 1:295
IHF, *see* Integration host factor
IL-1, *see* Interleukin-1
IL-2, *see* Interleukin-2
IL-4, *see* Interleukin-4
IL-6, *see* Interleukin-6
IL-10, *see* Interleukin-10
IL-12, *see* Interleukin-12
IL-18, *see* Interleukin-18
Image contrast, definition, 3:264, 288
Immobilized cell bioreactor
aspartate production, 1:158–159
definition, 1:152, 3:9
lactic acid production, 3:15
Immortalization, definition, 3:456
Immunity
definition, 3:762
plants, 3:662
Immunoassay, definition, 2:722
Immunocompetence, definition, 2:852
Immunocompromised, definition, 2:852
Immunodominant epitope, definition, 3:792
Immunofluorescence, definition, 4:758
Immunogen, definition, 1:209
Immunogenicity, definition, 4:779
Immunoglobulin A deficiency, overview, 1:228
Immunogold labeling electron microscopy
beet necrotic yellow vein virus, 1:426
definition, 1:422
Immunosuppressant, definition, 3:773
Impaired immunity, definition, 2:460
Impetigo, definition, 4:271, 281
Inapparent infection, definition, 3:762
Incidence
cumulative incidence, 2:141
definition, 2:137, 141, 3:131, 4:248, 811
Incision, definition, 2:71
Inclusion, *Chlamydia*, 1:781–782
Incubation period, definition, 2:390
Independent study, definition, 2:156, 167
Indicator microorganism, *see also* Biomonitor
definition, 4:898
detection, 4:898–899
drinking water
ideal characteristics, 4:899
species, 4:899–900
Indicator organism, definition, 4:545
Indicator plant, definition, 1:422
Indigenous flora, definition, 3:466
Indigo
enzymatic production, 2:233
microbial synthesis, 3:651
Indinavir
adverse effects, 1:309
indications, 1:309
mechanism of antiviral activity, 1:309
resistance, 1:309
Indole-3-acetic acid (IAA), *Erwinia* synthesis, 2:253–254
Indolicidin, antifungal activity, 1:252
Induced enzyme, definition, 4:930
Inducer exclusion, definition, 1:669
Inducible expression, definition, 3:730
Induction
definition, 4:637
extracellular enzymes, 2:215–217
Industrial biocide, *see* Biocide, industrial
Industrial biotechnology, *see* Biotechnology
Industrial effluents, *see also* Industrial wastewater
definition, 2:738
food industry

canning industry, 2:747
dairy industry, 2:749
effluent properties, 2:746–747
meat industry, 2:748
methane fermentation of agricultural wastes
background, 3:199–200
historical overview, 3:200
potential, 3:200
wastewater treatment, 3:202
microbial treatment
aerobic treatment
activated sludge processes, 2:751–752
aerated lagoons, 2:752
aeration theory, 2:750–751
anaerobic treatment
anaerobic contact reactor, 2:755
anaerobic filter, 2:755–756
anaerobic ponds, 2:757
facultative ponds, 2:757
fluidized or expanded bed reactor, 2:756
overview, 2:754–755
upflow anaerobic sludge blanket reactor, 2:756
overview, 2:749–750
parameters in evaluation, 2:740
petroleum industry
processes and effluent properties, 2:741–745
sour condensate sources, 2:741
treatment processes, 2:745–746
wastewater usage, 2:740–741
pollutants, 2:739
prospects for treatment, 2:762, 766
pulp and paper industry, 2:746
sludge treatment
aerobic digestion, 2:759–760
anaerobic digestion, 2:760–761
composition, 2:758
composting, 2:761–762
objectives, 2:759
overview, 2:757–759
summary of origins, characteristics, and treatment, 2:763–765
wastes by industry, 2:740
water usage by industry, 2:739
Industrial fermentation, *see* Fermentation
Industrial wastewater
biological treatment processes
activated sludge processes, 4:860–862
anaerobic bioreactors, 4:865–866
fixed film reactors, 4:862–863
fluidized bed reactors, 4:863–865
lagoon treatment, 4:859–860
selection, 4:859
design parameters for treatment systems
competitive inhibitors, 4:857
Haldane kinetics for substrate toxicity, 4:856
Michaelis–Menten kinetics, 4:856
Monod kinetics, 4:856
solids retention time, 4:857–858
temperature, 4:857
yields, 4:856
membranes in treatment bioreactors
biomass-retaining membranes, 4:868
classification of uses, 4:866–867
oxygen-permeable membranes, 4:867
selective barrier membranes, 4:867–868
standards, 4:855
toxic compounds, 4:857
treatability evaluations, 4:858–859
treatment goals, 4:855
Infection, definition, 4:297, 955
Infection thread, definition, 3:407
Infectious disease, *see also specific diseases*
economic consequences, *see* Economic consequences, infectious disease
emerging infection, *see* Emerging infection
global burden, *see* Global burden, infectious disease
historical impacts, 2:138
international law, *see* International law, infectious disease
surveillance, *see* Surveillance, infectious disease
Infectious dose, definition, 2:390
Infectious waste
definitions, 2:782–783
management plan
components, 2:793–795
emergency planning, 2:794
implementation, 2:795
rationale, 2:782
recordkeeping, 2:795
regulatory compliance, 2:792–793
training, 2:794–795
minimization techniques, 2:792
public health risk, 2:784–785
separation at source, 2:784
sources, 2:783
treatment and disposal methods
autoclave, 2:785
chemical/mechanical treatment, 2:789
comparison of methods, 2:786
electrothermal deactivation, 2:789
incineration, 2:785
microwave/mechanical treatment, 2:789
oxidation, 2:791
sanitary landfill, 2:785–788
sanitary sewer system, 2:788
selection considerations, 2:791–792
steam/compaction treatment, 2:789
steam/mechanical treatment, 2:788–789
types, 2:783–784
Influenza A, treatment, 1:298–299
Influenza virus
antigenic variation, 1:266
antiviral therapy
amantidine, 2:808–809
hemagglutinin inhibitors, 2:810
neuraminidase inhibitors, 2:809–810
ribavirin, 2:809
rimantidine, 2:808–809
clinical features and complications, 2:804–805
emergence
infection, 4:818–819
pandemic strains, 2:806–807
epidemics and pandemics, 2:806
genes and proteins, 2:799–801
genetic manipulation
reassortment, 2:810–811
reverse genetics, 2:811–812
hemagluttinin, 2:800–802, 805
immune response, 2:805–806
neuraminidase, 2:801
overview, 2:797
pandemic causes, 4:824–825
polymerase, 2:800
replication cycle

Influenza virus (*continued*)
assembly, 2:803–804
entry, 2:801–802
release, 2:804
ribonucleoprotein, nuclear import, 2:803
RNA transcription and replication, 2:803
species specificity, 2:806–807
structure, 2:798
taxonomy, 2:798–799
vaccination, 2:807–808
Infusion, definition, 2:677
Initiation codon, definition, 3:824
Initiation factor, definition, 3:824
Innate immunity, definition and response, 1:729, 741
Inoculation, definition, 2:677, 3:407, 4:289
Insect, symbiotic microorganisms
density determinants
gut microorganisms, 4:533
intracellular microorganisms, 4:533–534
distribution in insects
abundance in species, 4:526
cells and tissues, 4:529–530
gut, 4:527–529
distribution of microorganism species, 4:526–527
functions
cellulose degradation, 4:532–533
nitrogen nutrition, 4:530–531
novel metabolic capabilities, 4:530
sterol synthesis, 4:532
vitamin synthesis, 4:531–532
pest management, 4:537
transmission
gut microbiota, 4:534–535
transovarial transmission, 4:535
vertical transmission, 4:535–537
Insecticides, microbial, *see also* Biopesticide
advantages and disadvantages, 2:813–814
Bacillus sphaericus
strains, 2:815
toxins, 2:815–816
Bacillus thuringiensis
delta endotoxins and genes, 2:814–815
mechanism of action, 2:815
production and application, 2:815
strains, 2:814
bacteria in development, 2:817
fungi
conservation or environmental manipulation, 2:823–824
diversity of species, 2:821
history of use, 2:821
inoculative augmentation, 2:823
integrated pest control, 2:824
limitations, 2:824
mass production and formulation, 2:824
pathogenicity, 2:822
permanent introduction, 2:822–823
prospects, 2:824
Paenibacillus lentimorbus, 2:816
Paenibacillus popilliae, 2:816–817
Serratia entomophilia, 2:817
symbiotic organisms, 4:537
viruses
granulosis virus, 2:818–820
host specificity, 2:818
limitations, 2:820
nuclear polyhedrosis virus, 2:818–820
overview, 2:817–818
pest control examples, 2:819–820
production and use, 2:820
recombinant viral insecticides, 2:820–821
types of viruses, 2:818–819
Insertional mutagenesis, definition, 3:792
Insertion mutation, definition, 3:307
Insertion sequence, *see also* Transposable element
composite transposons, 4:712
definition, 3:594, 4:451, 704
history, 4:707
structure, 4:707–709
transposition mechanisms, 4:709–712
Insertion sequence element
definition, 3:1, 307, 310
Mycobacterium, 3:321
Integrase, definition, 1:398
Integrated host factor
definition, 2:556
glyoxylate bypass regulation, 2:558–559
Integrated pest management, definition, 1:541, 2:813
Integrating plasmid, definition, 1:373, 377
Integration host factor (IHF)
DNA bending, 1:815–816
replication origin binding, 1:826, 828–829
Interference contest, patent law, 3:546
Interference, definition, 3:662
Interferon
adverse effects, 1:299, 301
antiviral pathway induction
2–5A synthetase/RNase L system, 2:838–839
double-stranded RNA-dependent protein kinase, 2:838
miscellaneous pathways, 2:839
Mx proteins, 2:837
overview of pathways, 2:837
cell growth inhibition, 2:839
clinical applications, 2:840–841
definition, 2:826, 4:832
discovery, 2:826
functions, 1:735
immune system effects, 2:839–840
indications
condyloma acuminatum, 1:301
hepatitis B, 1:301
hepatitis C, 1:301
induction of synthesis
double-stranded RNA, 2:828
inducers, 2:828
interferon-β, 2:828–829
interferon-γ, 2:829–830
interferon-γ
features, 2:827–828
Legionella defense, 3:23–25
mechanism of antiviral activity, 1:300–301
protein induction response, 2:836–837
receptors
Janus kinase in signal transduction, 2:832–833
Stat proteins in signal transduction, 2:832–833, 836
type I
signal transduction, 2:832
structure, 2:830
type II
signal transduction, 2:832–833
structure, 2:830–832
resistance, 1:299, 301
type I features, 2:827
Interleukin-1 (IL-1), functions, 1:733–734

Interleukin-2 (IL-2), functions, 1:734
Interleukin-4 (IL-4), functions, 1:734
Interleukin-6 (IL-6), functions, 1:734
Interleukin-10 (IL-10), functions, 1:734
Interleukin-12 (IL-12), functions, 1:734
Interleukin-18 (IL-18), functions, 1:735
Intermittent fever, definition, 3:131
Internal ribosome entry site (IRES), definition, 2:635
International law, infectious disease, *see also* Biological warfare
  arms control law, 2:849
  customary law, 2:842
  emerging disease and revisions, 2:849–850
  environmental law, 2:849
  globalization of public health, 2:842–843
  harmonization, 2:842
  historical perspective
    creation of permanent organizations, 2:845
    protection against foreign disease, 2:843
    regulation of national quarantines, 2:843–844
    surveillance system establishment, 2:844–845
    treaties, 2:844
  humanitarian law, 2:848
  human rights law, 2:848
  International Health Regulations, 2:170, 175, 845–846, 849
  legal regimes
    animal disease, 2:846
    definition, 2:842, 845
    human disease, 2:845–846
    plant diseases, 2:846
  sources of law, 2:843
  system of international law, 2:842–843
  trade law, 2:846–848
  World Health Organization, growing interest, 2:850
International Plant Protection Convention (IPCC), functions, 2:846
Interproximal, definition, 3:466
Interspecies hydrogen transfer, definition, 2:131
Interspecific protoplast fusion, definition, 4:428
Interstitial nephritis, definition, 1:286
Intestinal bacteria
  chicken, 2:492
  human, 2:490–492
  pig, 2:492
Intestinal protozoan infection, *see Cryptosporidium*; *Entamoeba histolytica*; *Giardia lamblia*
Intrinsic incubation, definition, 1:311
Intron
  definition, 3:490, 4:181
  self-splicing, 3:498–499
Invasin, definition, 1:42
Inversion, definition, 3:307
Inverted repeat, definition, 4:704
Ionophore
  definition, 4:185
  effects on rumen microorganisms, 4:190
IPCC, *see* International Plant Protection Convention
IRE, *see* Iron response element
IRES, *see* Internal ribosome entry site
Irish potato famine, *see Phytophthora infestans*
Iron
  assimilation by cyanobacteria, 1:921
  availability for growth, 2:860
  energy generation
    electron acceptor, 2:867
    oxidation, 2:867
  gene regulation in bacteria, 2:865–866
  growth requirements and functions, 3:438
  pathogenicity role, 2:867–868, 3:23
  storage proteins, 2:866–867
  uptake systems in bacteria
    ferric iron uptake, 2:864
    ferrous iron uptake, 2:864
    heme iron uptake, 2:864–865
    lactoferrin iron uptake, 2:865
    low-affinity transport, 2:865
    overview, 2:861
    receptors, 2:861–862
    regulation, 2:865–866
    siderophore-mediated systems, 2:862
    transferrin iron uptake, 2:865
Iron–molybdenum cofactor, definition, 1:359
Iron ore, bioleaching, 3:483–484
Iron response element (IRE)
  stabilization of messenger RNA, 4:697
  translational repressors, 4:696
Isobacteriochlorin, definition, 4:558
Isocitrate dehydrogenase (IDH)
  glyoxylate bypass flux control 1:672, 674, 2:556–557, 559–561
  kinase/phosphatase
    regulation, 2:559–560
    structure and mechanism, 2:559
Isoconazole, antifungal activity, 1:241
Isocyclic ring, definition, 4:558
Isoleucine
  biosynthesis
    pathway, 1:143
    regulation, 1:144–145
    substrate specificities of enzymes, 1:143–144
  functions, 1:143
Isoreceptor, definition, 1:42
Itraconazole, antifungal activity, 1:244

J

JAK, *see* Janus kinase
Janus kinase (JAK), interferon signal transduction, 2:832–833
JC virus, transcription, 4:630

K

Keratin, definition, 2:451
Keratitis, definition, 1:287
Ketoconazole, antifungal activity, 1:242–243
α-Ketoglutarate dehydrogenase
  regulation of tricarboxylic acid cycle, 1:674
  structure, 1:674, 672
Ketosynthase, definition, 1:189
KinA-Spo0F-Spo0B-Spo0A two-component system, sporulation regulation, 4:742–744, 753
Knockout mouse, definition, 3:809
Koch's postulates
  acquired immunodeficiency syndrome virology, 3:233–234
  definition, 2:131
  description, 3:233

Koch's postulates (*continued*)
development, 2:684–685
prions and transmissible spongiform encephalopathies, 3:234–236
Köhler illumination, definition, 3:288

L

L 693989, antifungal activity, 1:251
L 705589, antifungal activity, 1:251
L 731373, antifungal activity, 1:251
L 733560, antifungal activity, 1:251
LAB, *see* Lactic acid bacteria
Laccase, features and lignin degradation, 3:46–47
LaCrosse encephalitis, overview, 1:316
β-Lactam, biosynthesis of penicillins, cephalosporins, and cephamycins
cephalosporin, 4:227, 229
cephamycin, 4:227, 229
cyclization, 4:225, 227
isopenicillin N
conversion to penicillin G, 4:227
formation, 4:227
overview, 4:223
tripeptide formation, 4:223, 225
β-Lactamase, *Bacteroides fragilis* genes and antibiotic resistance, 2:568
Lactation, definition, 4:666
Lactic acid bacteria (LAB)
antimicrobial substance production
bacteriocins, 3:4–5
primary metabolites, 3:4
classification, 3:1–2
clinical applications, 3:7
culture types, 3:1
dairy product production
Lactobacillus, 3:3–4
Lactococcus, 3:3
Leuconostoc, 3:3
starter cultures, 3:2
Streptococcus, 3:3
definition, 1:383, 2:1, 3:1, 9, 163
genetic engineering
cloning vectors, 3:6–7
conjugation, 3:6
electroporation, 3:6
integration elements, 3:7
promoters for gene expression, 3:7
protoplast fusion and transformation, 3:6
purposes, 3:5
secretion elements, 3:7
transduction, 3:6
meat fermentation and preservation, 3:164, 169–170
metabolism, 3:4
milled grain food production, 3:4
plasmids, 3:5
prospects for use, 3:7
Lactic acid fermentation
bacteria, 2:353
balao balao, 2:354
cereal–legume foods, 2:354–355
foods, overview, 2:353–354
pit fermentation, 2:354
yogurt–cereal mixtures, 2:354
Lactic acid production
batch processes, 3:13
commercial applications and market, 3:16–17
continuous processes, 3:13–15
definition, 3:9
history, 3:9
inoculum development, 3:12–13
microorganisms for production
bacteria, 3:10–11
molds, 3:11
process control, 3:15–16
process systems, 3:12
product recovery and purifications, 3:16
raw materials, 3:11–12
stereochemistry, 3:9–10
Lactococcin A, features, 1:391
Lactoferrin
definition, 2:860
iron uptake, 2:865
Lagoon
definition, 4:855
industrial wastewater treatment, 4:859–860
municipal wastewater treatment, 4:881
Lake, *see* Freshwater microbiology
Lamin, dinoflagellates, 2:50
Lamivudine
adverse effects, 1:305
chiral biocatalysis, 1:444
indications, 1:305
mechanism of antiviral activity, 1:305
resistance, 1:305
Land treatment
bioremediation, 1:603–605
definition, 1:587
LasA protease, *Pseudomonas* virulence factor, 3:878
Laser-scanning confocal microscopy, *see* Confocal microscopy
Late blight
definition, 3:633
pathogen, *see Phytophthora infestans*
Latent infection, definition, 4:32
Latent virus, definition, 4:832
Latex, definition, 3:607
Laticifer
definition, 3:607
phloem-limited bacteria, 3:612
LCR, *see* Ligase chain reaction
Leader peptide, definition, 1:339
Leather, enzymes in industry, 2:231
Lectin
definition, 1:42
Pseudomonas virulence factor, 3:881
Legionaminic acid, definition, 3:18
Legionella
coiling phagocytosis, 3:22
history of outbreaks, 3:19
immune response, 3:24–25
interferon-γ in defense, 3:23–25
life cycle of *Legionella pneumophila*, 3:21–23
metabolism, 3:20
natural habitats, 3:20–21
overview, 3:18
structure, 3:19–20
transmission, 3:21–23, 25
virulence factors
dot/icm, 3:23–24
iron acquisition genes, 3:23
Mip, 3:23
Leishmania
classification, 3:27–28
endoplasmic reticulum, 3:31
flagellum, 3:31
gene expression control, 3:32–33
genome organization, 3:31–32
Golgi apparatus, 3:31
immune rsponse, 3:33–35
leishmaniasis
control, 3:38
diagnosis, 3:37
diffuse cutaneous disease, 3:37
epidemiology, 3:35

localized cutaneous disease, 3:35–37
mucosal disease, 3:37
treatment, 3:37–38
visceral disease, 3:37
life cycle, 3:28–29
messenger RNA
editing, 3:32
trans-splicing, 3:32
mitochondrion and kinetoplast, 3:31
morphology, 3:28
nucleus, 3:31
overview, 3:27
pathogenesis, 3:33
plasma membrane
gp63, 3:31
lipophosphoglycan, 3:30–31
structure, 3:30
species and geographic distribution, 3:35
transfection and gene targeting, 3:33
transmission, 3:29–30
Lentic, definition, 2:438
Leprosy, *see also Mycobacterium*
borderline disease, 3:326
definition, 1:28
history and epidemiology, 3:314
immune response, 3:326–327
multibacillary form, 3:326
pathogenesis, 3:325–327
transmission, 3:325–326
tuberculoid leprosy, 3:326
Lethality, definition, 4:955
Leucine
biosynthesis
pathway, 1:143
regulation, 1:144–145
substrate specificities of enzymes, 1:143–144
functions, 1:143
Leucine-responsive protein (LRP)
DNA bending, 1:816
gltBD operon regulation, 1:672
pilus synthesis regulation, 2:377
Leukocytosis, definition, 2:9
Leukopenia, definition, 1:287
LexA
definition, 4:43
SOS response
cleavage, 4:340–341
homologs, 4:343
repressor activity, 4:337–340
Life, origin, *see* Origin of life
Life cycle, definition, 3:676
Ligase chain reaction (LCR), diagnostics, 2:37
Ligation, definition, 2:71, 4:55
Light, definition, 3:647
Light microscopy, *see* Optical microscopy
Lignin
definition, 3:39
enzymatic biomass conversions, 2:227
Pseudomonas degradation, 3:889
types and functions, 3:40
Lignocellulose, *see also* Xylanase
cellulase modification biotechnology, 1:757–758
definition, 3:39
degradation enzymes
laccase, 3:46–47
lignin peroxidase, 3:43–45
manganese peroxidase, 3:45–46
regulation in *Phanerochaete chrysosporium*, 3:42–43
enzymatic biomass conversions
cellulose, 2:226
hemicellulose, 2:226–227
lignin, 2:227
overview, 2:226
microorganisms in degradation, 3:41–42
structure, 1:744–745, 3:39–40
Linear plasmid, definition, 1:28
Linear programming
definition, 2:510
metabolic engineering, 2:516
Linking clone, definition, 3:151
Lipase
definition, 3:49
features and applications, 2:215, 228, 230–231
history of study, 3:49
industrial applications
advantages, 3:50
ester enantioresolution, 3:53
ester hydrolysis, 3:50–51
ester synthesis, 3:51
fish oil, fatty acid improvements, 3:52–53
structured lipid synthesis, 3:52
transesterification, 3:51–52
Pseudomonas virulence factor, 3:879
regioselectivity, 3:50
stereospecificity, 3:50
structure and mechanism, 3:49–50
substrate specificity, 3:50
Lipid
characteristics
algae, 3:66
Archaea, 3:64
eubacteria, 3:64–65
eukaryotes, 3:65–66
definition, 3:62
functions, 3:63–64
nomenclature and structure, 3:62–63, 68
Lipid A
biosynthesis, 3:76–77
definition, 3:71
features, 3:73–74, 84–85
Lipid bilayer, definition, 3:517
Lipid biosynthesis, *Escherichia coli*
comparison with other bacteria, 3:60
dry weight composition, 3:55
fatty acids
biosynthesis, 3:55–57
β-oxidation, 3:60
inhibitors, 3:60
lipopolysaccharides, *see* Lipopolysaccharide
membrane protein acylation, 3:59
overview, 3:63
phospholipids
biosynthesis, 3:57–58
turnover, 3:58–59
regulation, 3:59–60
Lipid production, *see also* Biosurfactant
cocoa butter equivalent, 3:66–68
oils rich in polyunsaturated fatty acids, 3:68
oleaginicity, biochemistry and physiology, 3:66
single-cell oils, 3:62, 66
Lipoic acid, biosynthesis, 1:844
Lipooligosaccharide
definition, 2:591, 3:71
features, 3:74
Lipopolysaccharide (LPS)
adaptive responses and lipid A modifications, 3:84–85
biosynthesis and assembly
core oligosaccharide, 3:77
lipid A, 3:59, 76–77
minimal structure for viability, 3:74–76
O-polysaccharide, 3:77, 79–80
terminal reactions, 3:80
biosynthesis, 3:522–524

Lipopolysaccharide (LPS) (*continued*)
core oligosaccharide, 3:74
definition, 1:647, 3:18, 71, 517
functions
outer membrane stabilization, 3:80–81
pathogenicity, 3:58, 82–84, 522
protective barrier, 3:81
lipid A, 3:73–74, 84–85
lipooligosaccharide, 3:74
Neisseria, 2:579–581
nodule formation, role in legumes, 3:410–411
O-polysaccharide, 3:71–72, 74, 81–82
pathogenicity role, 3:82–84, 522
Pseudomonas virulence factor, 3:879
rough, 3:71–72
sepsis role, 3:71
smooth, 3:71–72
structure, 3:71–73, 518–519
therapeutic targeting, 3:84
topology of biogenesis and export, 3:524
Lipoprotein, definition, 1:759
Liquefaction, definition, 2:222
Listeria monocytogenes
cell invasion, 2:199
enteric disease, 2:199
food-borne illness, 2:401–402
meat products as source, 3:167
refrigerated food pathogens, 4:70, 72
Lithotroph, definition, 3:379
LN-2, adhesin binding, 1:49
Long terminal repeat (LTR), definition, 4:81, 628
Long-pass filter, definition, 3:288
Lotic, definition, 2:438
Lovastatin, biosynthesis, 3:779
Low-nutrient environments, *see* Oligotrophic environments
Low-temperature environments, *see also* Psychrophile; Psychrotroph
atmosphere, 3:94
definition, 3:93
microbial activity, 3:97
mountains, 3:94
oceans, 3:94
polar regions, 3:93–94
LPS, *see* Lipopolysaccharide
Lrp, control of genes, 1:151
LRP, *see* Leucine-responsive protein
LTR, *see* Long terminal repeat
Luciferase
bacteria
fluorescent proteins as emitters, 1:523
lux operon, 1:524
reaction, 1:522–523
structure, 1:522–523
definition, 1:520
dinoflagellates
luciferin release, 1:526–527
structure, 1:526
Luciferin, definition, 1:520
Luminescence, *see* Bioluminescence
Luteoviridae
control, 3:108
diagnosis of disease, 3:107–108
epidemiology, 3:106–107
genome structure, 3:100
host interactions
effects on hosts, 3:103–104
host range, 3:103
immunogenicity, 3:106
noncoding sequences, 3:100, 102
open reading frames, 3:100
overview, 3:99
protein functions
ORF0, 3:102
ORF1, 3:102
ORF2, 3:102
ORF3, 3:102
ORF4, 3:103
ORF5, 3:102
ORF6, 3:103
ORF7, 3:103
RNA expression, 3:102
structure and composition, 3:99–100
taxonomy
enamovirus, 3:106
luteovirus, 3:105
overview, 3:104–105
polerovirus, 3:105–106
relationships with other viruses, 3:106
vector interactions, 3:104
LuxR
acylated homoserine lactone interactions and cell entry, 4:6
DNA binding and transcription activation, 4:7–8
repressor activity, 4:8
structure, 4:6–7
Lux-type box, definition, 4:1
LY 303366, antifungal activity, 1:249
Lyme disease, *see also Borrelia burgdorferi*
animal models
mouse
factors affecting severity, 3:120–121
histopathology, 3:121
immune response, 3:121–123
presentation of disease, 3:120
route of infection, 3:120
strains of mice, 3:121
overview, 3:119–120
rhesus monkey, 3:123
arthropod vector, 3:112–113
autoimmune component, 3:117
clinical manifestations
course, 3:116–118
stages, 3:115–117
diagnosis, 3:118–119
history, 3:109–110
immune response
acquired immunity, 3:114
innate immunity, 3:113–114
measurement, 3:114–115
pathogens, 3:110
treatment, 3:119
vaccination, 3:128–129
vectors, 4:358–361
zoonosis, 4:961–962
Lymphangitis, definition, 4:271, 282
Lymphoma, B cell, 1:229
Lyophilization, *see* Freeze-drying
Lysine
biosynthesis
pathway, 1:141
regulation, 1:141–143
exporter in *Corynebacterium glutamicum*, 1:156
functions, 1:141
microbial production processes, 1:155–157
uses, 1:154–155
Lysine decarboxylase, pH neutralization mechanisms, 3:629–630
Lysogen, definition, 3:876
Lysogenic conversion, definition, 2:270, 4:832
Lysogeny, definition, 4:43, 637
Lysozyme, definition, 1:759
LysR, transcriptional regulation of plasmids, 3:739, 741
Lytic, definition, 4:832
Lytic cycle, phage
adsorption, 1:407

infection, 1:407
multiplication, 1:407–408
release, 1:408
Lytic phase, definition, 4:336

M

Macrobe, definition, 4:316
Macrocycle, definition, 4:558
Macrocyclic, definition, 4:195
Macrolactone, definition, 1:189
Macrophage
definition, 1:729
immune response, 1:731–732, 742
Macular rash, definition, 4:758
Maculopapular, definition, 1:287
Magnesium, growth requirements and functions, 3:438
Major histocompatibility complex (MHC)
antigen processing
class I antigens, 1:737–738
class II antigens, 1:738
class II molecules
B cell development role, 1:222
definition, 1:208
definition, 1:729, 4:583
T cell recognition
binding assay, 4:589
class I molecules, 4:584
class II molecules, 4:584–585
foreign peptides bound to self-major histocompatibility complex molecules, 4:585
overview, 4:583–584
Malaria, *see also Plasmodium*
chemotherapy
amodiaquine, 3:139
chlorguanidine, 3:139
chloroquine, 3:139
halofantrine, 3:139
mefloquine, 3:139
primaquine, 3:139
pyrimethamine, 3:139
qinghaosu, 3:139–140
quinine, 3:138–139
tetracycline, 3:139
clinical features
falciparum malaria, 3:134
overview, 3:751
quartan malaria, 3:134–135
vivax malaria, 3:134
control
antigenic variation exploitation, 3:760
chemotherapy, 3:756–758
vaccination, 3:758–760
vector control, 3:756
diagnosis, 3:137
diagnostics, 3:760–761
emerging infection, 4:816
epidemiology, 3:131–132
eradication programs, 3:133–134
evolution of malaria–mosquito–human relationship, 3:149
historical perspective, 3:747
history, 3:132–133
immune response, 3:138
immunity to parasite infection, 3:751–752
mosquito vector
control
bednets and curtains treated with insecticides, 3:147
environmental management, 3:148
genetic manipulation, 3:148–149
insecticides, 3:146–147
larval and biological control, 3:147
ecology, 3:143–144
life cycle, 3:142–143
species in transmission, 3:747–748
vectorial capacity, 3:144–146
pathophysiology, 3:137–138
Plasmodium species causing malaria, 3:134, 149
vaccination
asexual blood-stage vaccines, 3:140–141
DNA vaccines, 3:142
overview, 3:140
pre-erythrocytic vaccines, 3:142
second generation vaccines, 3:142
transmission-blocking vaccines, 3:141–142
MALT, *see* Mucosa-associated lymphoid tissue
Malt, beer brewing, 1:412, 414
Maltoporin, ATP-binding cassette transport role, 1:3
Manganese, growth requirements and functions, 3:439
Manganese peroxidase, features and lignin degradation, 3:45–46
Mannosylerythrotol lipids, biosurfactant production and features, 1:626
Mapping bacterial genomes
Escherichia coli K12 chromosomal map, 3:161–162
fine mapping, 3:153
genome sequencing, 3:159, 161
gram-negative bacteria, 3:153–154
gram-positive bacteria, 3:154
historical perspective, 3:151, 153, 162
ordered cloned DNA library
assembly, 3:158
definition, 3:151
vector selection, 3:158
physical mapping
gene positioning, 3:159
macrorestriction map construction, 3:156, 158
pulsed-field gel electrophoresis, 3:155–156
rarely cutting endonucleases, 3:154–155
transduction, 4:648–649
Marburg virus, characteristics of infection, 4:815
Mass transfer, definition, 2:767
Mating type, *Phytophthora infestans*, 3:633, 637
Maturase, definition, 4:181
MCT, *see* Microbial content test
Measles
clinical features, 3:543
outbreak modeling, 1:73
Meat products
chemical composition, 3:163
control of microorganisms
bacteriocins, 3:170
drying, 3:169
gas modification, 3:169, 4:79
irradiation, 3:169
lactic acid bacteria, 3:169–170
preservatives, 3:169
prevention of contamination, 3:168
refrigeration, 3:168, 4:74–75
thermal processing, 4:550
fermentation, 3:164
Hazard Analysis and Critical Control Points, 3:168
microbial content, 3:163–164
pathogens
food-borne infection
gastroenteritis, 3:167

Meat products (*continued*)
systemic, 3:167–168
food-borne intoxication
gastroenteritis, 3:166
systemic, 3:166
sources, 3:165–166
spoilage microorganisms
effects on meat, 3:165
signs, 3:164–165
sources, 3:165
Media, *see* Nutrition, microbial culture
Medical waste, *see* Infectious waste
Mefloquine, malaria treatment, 3:139
Megaplasmid, definition, 3:379, 392
Melanin
definition, 3:647
features, 3:651
Melarsoprol, sleeping sickness treatment, 4:740
Melt water, definition, 3:93
Membrane, *see* Cytoplasmic membrane
Membrane bioreactor
biomass-retaining membranes, 4:868
classification of uses, 4:866–867
definition, 4:855
oxygen-permeable membranes, 4:867
selective barrier membranes, 4:867–868
Membrane potential, definition, 1:180
Memory
immune system, 1:209
T cells, 1:741
Meningitis, definition, 2:571
Meningococemia, definition, 2:571
Mercury, *see also* Heavy metals
cycle, overview, 3:171–174
methylmercury
abiological formation, 3:180–181
photodegradation, 3:179
microbial transformations
methylation of mercury
corrinoid-mediated methylation, 3:179–180
detoxification role, 3:180
significance in mercury cycle, 3:181
methylmercury degradation, 3:178–179
overview, 3:174
oxidation of elemental mercury, 3:181
reduction of Hg(II)
direct reduction by *mer*-encoded enzymes, 3:176–178
indirect reduction, 3:178
significance in mercury cycle, 3:178
physiochemical properties, 3:172
resistance
anaerobes, 3:176
broad-spectrum resistance, 3:175
narrow-spectrum resistance, 3:175
species of bacteria, 3:175
sources, 3:172
toxicity, 3:172
Meristematic tissue, definition, 4:297
Meromixis, definition, 4:478
Merozoite, definition, 3:745, 749
Meso selectivity, definition, 1:636
Mesophile, definition, 1:319
Messenger RNA (mRNA)
cap, 3:824, 838
degradation, cotranslational control, 4:696–697
poly(A) tail, 3:824, 838
Shine–Dalgarno sequence, 3:824, 837
splicing, *see* RNA splicing
structure, 3:836–838
transcription, *see* Transcription
translation, *see* Protein biosynthesis
untranslated regions, 3:824, 836
Metabolic engineering, *see also* Genomic engineering, bacterial metabolism; Strain improvement
definition, 1:152, 4:428
strain improvement, 4:442
Metabolic flux, definition, 2:510
Metabolic genotype, definition, 2:510
Metabolic phenotype, definition, 2:510
Metacyclogenesis, definition, 3:27
Metal extraction, *see* Ore leaching
Metalimnion, definition, 2:438
Metallothionein, definition, 2:607, 614
Metal pollutants, *see* Heavy metals
Methane
biosynthesis, *see* Methanogen
global warming, 3:225
production systems
ammonia toxicity, 3:202
anaerobic bioreactors, 3:211–213
efficiency, 3:200, 202
organic matter fat in anaerobic digestion, 3:200
prediction of production, 3:200–202
utilization for growth, *see* Methanotroph
Methane monooxygenase
definition, 3:245
methanotrophs, 3:249
Methanofuran, definition, 3:204
Methanogen
Archaea
habitats, 1:323–324
physiology, 1:324
bioreactors, 3:211–213
carbon dioxide fixation, 3:220–221
definition, 1:18, 319, 2:317, 3:188, 256
diversity, 3:206–210
DNA replication, repair, modification, and metabolism, 3:222–223
ecological role, 3:189, 204, 225
fermentation of agricultural wastes
background, 3:199–200
historical overview, 3:200
potential, 3:200
wastewater treatment, 3:202
gene structure and transcription, 3:223
genome
genomics and gene function analysis, 3:224–225
structure, 3:221–222
growth substrates, 3:197
habitats
freshwater sediments, 3:210–211
geohydrothermal outsources, 3:215
human gastrointestinal tract, 3:215
interspecies hydrogen transfer, 3:210
marine habitats, 3:213–214
protozoan endosymbionts, 3:215
ruminants, 3:214
salt lakes, 3:216
soil, 3:215
termites, 3:214–215
Methylotrophy, atmosphere cycling by microbes

history of study, 3:205
methanogenesis
acetate cleavage to carbon dioxide and methane, 3:195–197, 219
bioenergetics, 3:219–220
carbon dioxide reduction, 3:189, 193–194, 216
cofactors
structures, 3:191, 217
types and functions, 3:190
energy generation, 3:193
methylated substrate reduction, 3:194–195
overview, 3:188–189, 204, 216
phylogeny, 3:206–210
RNA structure and translation, 3:223–224
species, 3:208–209
temperature optima of species, 3:199
Methanotroph, *see also* Methylotrophy
definition, 3:245, 4:117
gene regulation, 3:249
phylogenetic analysis
ecological studies, 3:249–251
taxonomy, 3:252
physiology and biochemistry, 3:248–249
Methionine
biosynthesis
pathway, 1:141
regulation, 1:141–143
functions, 1:141
requirement for chemotaxis, 1:776
Methionine sulfoxide reductase, *Erwinia*, 2:254
Method philosophy, *see* Scientific method
Methylation, *see* DNA methylation; Protein methylation; RNA methylation
Methyl-CoM reductase
definition, 3:188
reaction, 3:193–194
Methylmercury, *see* Mercury
Methylotrophy, *see also* Methanotroph
aerobic bacteria
characteristics of species, 3:252
halomethane utilization, 3:254
methanol utilizers, 3:251, 253
methylated amine utilization, 3:253
methylated sulfur compound utilization, 3:254
anaerobic bacteria, 3:254
atmosphere cycling by microbes, 3:260–262
definition, 3:245, 256
habitats and ecology, 3:247
history of study, 3:245
significance
bioremediation, 3:246–247
biotechnology, 3:246
carbon cycle, 3:245–246
recombinant protein production in *Pichia pastoris*, 3:247
yeasts, 3:254
Methyltransferase, *see also* DNA methylation; Protein methylation; Restriction and modification; RNA methylation
MHC, *see* Major histocompatibility complex
Miasma
cholera, 1:802
definition, 1:801, 2:677
MIC, *see* Minimum inhibitory concentration
Miconazole, antifungal activity, 1:242
Microarray, DNA, 2:106
Microbe, definition, 2:677–678, 4:316
Microbial content test (MCT), cosmetics, 1:897–898
Microbial loop, definition, 2:661
Microbial mat, definition, 4:353, 478
Microbial volatile organic compounds
airborne contamination, 1:119
chemical assay, 1:124
definition, 1:116
Microbiocide, definition, 3:893
Microbiology careers, *see* Careers, microbiology
Microcins, features, 1:388
Micrococcus, skin resident species, 4:279
Microcolony, definition, 4:1
Microcosm, definition, 1:587
Microcyclic, definition, 4:195
Microelectrode, definition, 2:131
Microgravity, definition, 4:344
Microinjection, DNA, 4:666–668
Microneme, definition, 4:598
Micropore, definition, 4:598
Microscopy, *see* Confocal microscopy; Electron microscopy; Optical microscopy
Microsome, definition, 1:287
Migration, evolution contribution, 3:373–374
Mildew, *see* Downy mildew; Powdery mildew
Milk, *see also* Dairy products
expression of transgenic products in livestock
α-1–antitrypsin, 4:682–683
calcitonin and other peptides, 4:683–684
efficacy, safety, and acceptability, 4:684–685
fibrinogen, 4:683
protein types and expression levels, 4:678–679
milk protein expression in transgenic livestock
protein types, 4:673–675
regulation of transgene expression
caseins, 4:675, 677
complementary DNA expression, 4:681–682
whey proteins, 4:677, 680–681
min operon, minicell formation role, 1:709
Mineralization, definition, 1:461, 2:651, 3:614, 866, 4:321
Minimal media, *see* Nutrition, microbial culture
Minimum inhibitory concentration (MIC), definition, 1:445
Mining, *see* Ore discovery; Ore leaching
Mip, *Legionella* virulence factor, 3:23
Mismatch repair, overview, 2:81
Mistranslation, definition, 1:162
Mitochondria, energetics, 2:179–180
Mixing ratio, trace gases in atmosphere, 3:256, 258
Mixis
definition, 2:283, 285
evolution conribution, 3:374
functions, 2:296
Mixotroph, definition, 2:317
Mobilized plasmid, definition, 1:86
Model, definition, 3:227
Modification, definition, 3:240
Modified Robbins device (MRD), biofilm analysis, 1:481
Moisture content, wood, 4:571
Molecular imprinting, features, 2:734
Molecule, history of term usage, 4:364, 370
Molybdenum, growth requirements and functions, 3:439

Monascus pigment, definition, 3:647
Monensin A, commercial uses, 3:779
Monkeypox, smallpox similarity, 4:295–296
Monocyte, definition, 1:729, 3:18
Monomorphism, definition, 2:677
Monopril, chiral biocatalysis, 1:431, 433
Morbidity, definition, 3:131
Morphometrics, definition, 4:353
Mortality, definition, 3:131
Mosquito, *see* Malaria
Mother cell
  definition, 1:373, 4:377
  SigmaE activation, 1:380–381
  SigmaK activation, 1:381
Motilide, definition, 3:773
MRD, *see* Modified Robbins device
mRNA, *see* Messenger RNA
Mucigel, definition, 4:117
Mucilage, definition, 4:117
Mucosa-associated lymphoid tissue (MALT)
  definition, 2:628
  gastric lymphomas, 2:633–634
Muk proteins, chromosome segregation role, 1:832–833
Multihazardous waste, definition, 2:782
Multiple cloning site, definition, 4:55
Mumps, clinical features, 3:543
Municipal wastewater treatment
  aerobic treatment systems
    activated biofiltration systems, 4:878
    completely mixed activated sludge systems, 4:876
    contact stabilization systems, 4:876–877
    conventional activated sludge systems, 4:875–876
    extended aeration systems, 4:877
    overview, 4:875
    oxidation ditch systems, 4:877
    sequence batch reactors, 4:877–878
  anaerobic treatment systems, 4:878–879
  characteristics of wastewater
    algae, 4:872–873
    bacteria, 4:872
    biochemical oxygen demand, 4:870–871
    crustaceans, 4:873
    fungi, 4:872
    nitrogen forms, 4:871
    pH, 4:872
    phosphorous, 4:871–872
    protozoa, 4:873
    rotifers, 4:873
    suspended solids, 4:871
    worms, 4:873
  microbial metabolism in treatment
    autotrophic bacteria, 4:874
    endogenous respiration, 4:874–875
    heterotrophic bacteria, 4:873–874
  mixed treatment systems
    biological phosphorous removal systems, 4:881
    lagoon treatment, 4:881
    nitrification–denitrification systems, 4:880
    rotating biological contactor systems, 4:879–880
    trickling filters, 4:879
  pathogen removal, 4:881–882
  prospects, 4:870, 883
  regulation, 4:882–883
Murein
  biosynthesis
    cytoplasmic reaction steps, 1:764
    insertion into cell wall, 1:765–768
    membrane translocation of precursors, 1:764–765
    penicillin inhibition of transpeptidases, 1:767
  definition, 1:759
  lipoprotein, 1:763, 3:521
  metabolizing enzymes, 1:768–769
  pseudomurein, 1:764
  sacculus
    barrier function, 1:768
    definition, 1:759
    growth and division, 1:769–771
    structure, 1:768
  structure, 1:761
Mushroom toxins, food poisoning, 2:410
Must, definition, 4:914
Mutagen, definition, 3:307
Mutagenesis
  applications
    basic research, 3:310–311
    strain development, 3:311
  evolution role, 3:308
  induced mutation, 3:310
  overview, 3:307–308
  spontaneous mutation
    amplifications, 3:310
    cytosine deamination, 3:308–309
    deletions, 3:310
    depurination, 3:309
    genome size effects, 3:308
    oxidative damage, 3:309
    replication errors, 3:309–310
    transposable elements, 3:310
Mutant, definition, 4:930
Mutation
  definition, 2:270, 677, 3:307, 373, 4:428, 430–431
  evolution contribution, 3:373–374
  history of study, 2:693–694
  selective fate of new mutations
    advantageous mutations, 3:375–376
    neutral theory of molecular evolution, 3:374–375
  strain improvement, *see* Strain improvement
Mutualism, definition, 3:328
Mx proteins, interferon induction and antiviral activity, 2:837
Myalgia, definition, 1:287
Mycelium
  definition, 1:28, 2:15, 117, 338, 468, 4:109
  downy mildew, 2:121–122
Mycobacterium
  bacteriophages, 3:322–323
  cell wall
    arabinan galactan-linker unit biosynthesis, 3:317–318
    architecture, 3:316–317
    core structure, 3:317
    lipoarabinomannan biosynthesis, 3:319
    mycolic acids, structure and biosynthesis, 3:318–319
  culture, 3:325
  diseases, *see also* Leprosy; Tuberculosis
    atypical mycobacterioses, 3:315
    overview, 3:312–313
  genome
    Mycobacterium leprae, 3:319, 323
    Mycobacterium tuberculosis, 3:319, 321
  insertion sequences, 3:321
  metabolism, 3:315–316
  plasmids, 3:321–322
  species and features, 3:312–313
  vaccine, *Mycobacterium tuberculosis*, 4:776

Mycology, definition, 2:468
Mycoplasmalike organisms, *see* Phytoplasma
Mycorrhizae
adaptation under changing global conditions, 3:335
apparent parasitism, 3:331
definition, 3:328–329, 614
evolution, 3:332
functioning
community interactions, 3:334–335
environmental limits, 3:333
interface between plant and fungus, 3:333–334
overview, 3:332–333
soil–fungus–root–leaf continuum, 3:334
multiple partnerships, 3:331–332
phylogenic mycotrophy, 3:331
structure, 3:329–331
types, 3:329–331
Mycorrhizosphere, definition, 4:117
Mycosis, *see* Fungal infection
Mycotic mycetoma
definition, 2:451
features, 2:457–458
Mycotoxin
aflatoxin
animal disease, 3:344
hepatocarcinogenic activity, 3:340–341
human disease, 3:340–341
sources, 3:340
tolerances for foods, 3:340
types, 3:340
airborne, 1:119
animal health impact, 3:342–343
chemical assay, 1:124
control, 3:347
definition, 1:116, 3:337
diagnosis of mycotoxicoses, 3:347
ergot alkaloids
animal disease
cattle, 3:343–344
forms, 3:343
horses, 3:343
sows, 3:344
human disease
epidemics, 3:340
forms, 3:339
sources, 3:339
structure, 3:339
tolerances for grain foods, 3:339
factors affecting production, 3:337–338
fumonsin
animal disease, 3:346–347
human disease, 3:342
sources, 3:342
human health, history of mycotoxicoses, 3:338–339
mycotoxicosis comparison to mycosis, 3:337
ochratoxin
animal disease, 3:345–346
human disease, 3:342
types, 3:342
species in production, 3:337–338
trichothecenes
animal disease, 3:344–345
human disease, 3:341–342
types, 3:341–342
zearalenone, animal disease, 3:347
Myelosuppression, definition, 1:287
Myoclonus, definition, 1:287
Myoneme, definition, 2:42
Myxobacteria
classification, 3:358–359
fruiting body
features and formation, 2:26–27, 3:353, 355
morphogenetic movements during formation, 3:360
survival advantage, 3:358
gliding motility, 3:351, 360
habitats, 3:359
intercelluolar signaling
development, 3:361
motility, 3:360
isolation, 3:359
life cycle
fruiting stage, 3:358
overview, 3:355, 357
vegetative stage and cooperative feeding, 3:357–358
myxospore formation, 3:353
overview, 3:349
purification, 3:359
reasons for study, 3:349, 351, 359
swarming, 3:351, 353
Myxococcus xanthus
cloning methods, 3:366–367
ectopic expression, 3:369
gene replacement studies, 3:367, 369
genetic analysis of behavior, 3:372
genetic mapping
long-range mapping, 3:367
short-range mapping and strain construction, 3:367
genetic markers, 3:370–371
genus characteristics, 3:363
life cycle, 3:363–364
mutagenesis, 3:364–365
reporter genes, 3:365, 369
sequence homolog analysis, 3:365–366
stringent response, 4:474–475
Myxospore, definition, 3:349, 363

N

N, antitermination mediation in phage λ, 1:347, 4:625–626
NAD, *see* Nicotinamide adenine dinucleotide
Naftifine, antifungal activity, 1:248
Naphthalene, biodegradation pathway, 3:735–736, 738
Natural killer (NK) cell
definition, 2:826
features, 4:590–591
immune response, 1:731
target cell killing, 1:740–741
Natural selection, *see also* Adaptation; Evolution
adaptation by natural selection, 2:287–288
antibiotic resistance, 3:377
balancing selection, 3:376
definition, 3:373
diversifying selection, 3:376–377
escape selection, 3:376
frequency-dependent selection, 3:376
overview, 2:283, 285–286
selective fate of new mutations
advantageous mutations, 3:375–376
neutral theory of molecular evolution, 3:374–375
Near-field scanning optical microscopy (NSOM), *see* Optical microscopy
Necridic cells, cyanobacteria, 1:916
Necrotizing fasciitis, definition, 4:271, 282
Negative sense genome, definition, 4:592
Negative stain, electron microscopy, 3:276, 278–280
Negri body, definition, 4:15

Neisseria
  adhesins, 1:46, 48, 50
  antibody response, 2:581–582
  antigen features
    capsular polysaccharides, 2:576, 582–583
    lipopolysaccharide, 2:579–581
    Opa proteins, 2:578–579
    pili, 2:574–576
    porins, 2:577
    Rmp, 2:577–578
  antigenic variation, 1:264–265
  bacteremia, 2:573
  infection mechanisms, 2:573–574
  local infection
    extension, 2:572
    sites, 2:571
    transmission, 2:571–572
  pelvic inflammatory disease, 2:572
  prevention of infection, 2:582–583
  treatment, 2:573
Neisseria gonorrhoeae
  gonorrhea, *see* Gonorrhea
  metabolism, 4:255
  pathogenesis, 4:255–257
  transformation, 4:657
Neisseria meningitidis
  surveillance, 4:523
  vaccine, 4:772–773
Nelfinavir
  adverse effects, 1:309
  indications, 1:309
  mechanism of antiviral activity, 1:309
  resistance, 1:309
Nematode, definition, 3:866
Nematode infection, plants
  epidemiology, 3:694
  management, 3:694
  overview, 3:692
  pathogens
    classification, 3:693
    dissemination, 3:693
    ecology, 3:693
    infection, 3:693–694
    morphology, 3:692
    reproduction, 3:692–693
Nephrolithiasis, definition, 1:287
Neuraminidase
  definition, 1:287, 3:533
  influenza virus
    features, 2:801
    inhibitors, 2:809–810
Neuritis, definition, 1:287
Neuropathy, definition, 1:287
Neutropenia, definition, 1:232, 287
Neutrophil, immune response, 1:730
Neutrophile, definition, 3:625
Nevirapine
  adverse effects, 1:307
  indications, 1:307
  mechanism of antiviral activity, 1:306–307
  resistance, 1:307
Newcastle disease virus, clinical features of infection, 3:544
NhaA, pH neutralization mechanisms, 3:629–630
Niacin, microbial production processes, 4:843, 845
Nickel, growth requirements and functions, 3:439
Nicotinamide adenine dinucleotide (NAD)
  antioxidant defense, 3:528
  biosynthesis, 1:844
nif genes
  Azobacter
    gene clusters, 1:364
    NifL control, 1:365
    regulation of genes, 1:364–366
  definition, 3:392
  rhizobial nitrogen fixation, 1:502–503
  types, 3:394, 403
Nikkomycin Z, antifungal activity, 1:252
Nipah virus, emerging infection, 4:818
Nisin
  applications, 1:392–394
  definition, 1:383
  gene, 1:389
  mechanism of action, 1:389, 394
  physical properties, 1:393
  processing, 1:388–389
  spectrum of activity, 1:393–394
Nitrate reduction, anaerobic respiration
  denitrification pathway, 1:185
  nitrate ammonification pathway, 1:185
  overview, 3:387–388
Nitric oxide (NO)
  atmosphere cycling by microbes, 3:260–262
  intracellular killing by immune cells, 1:736–737
  oxidative stress, 3;527–528
Nitrification, definition, 4:321
Nitrite, oxidation, 3:385–386
Nitrogen, growth requirements and functions, 3:437
Nitrogenase, *see also nif* genes
  assays, 1:495
  definition, 1:359, 492, 3:392
  genes, 1:493
  hydrogen production, 1:495
  iron–molybdenum cofactor, 1:493
  oxygen
    inactivation, 3:380
    protection
      conformational protection, 1:368
      overview, 1:493–495, 3:394–395
      respiratory protection, 1:368–369
  reaction, 1:367
  sequence homology among rhizobia, 3:404
  structure, 1:367–368, 493, 3:394
  substrate specificity, 3:381
Nitrogen assimilation, cyanobacteria, 1:921–922
Nitrogen assimilation, regulation with carbon assimilation
  acetyl phosphate role, 1:681
  carbon source selection, 1:673–674
  Escherichia coli, metabolism in log-phase aerobic cultures, 1:670, 672–673
  glnA regulation, 1:678–681
  glutamine synthetase regulation, 1:672, 676–678
  ntr genes, regulation, 1:678–679
  overview, 1:669–670
Nitrogen cycle
  assimilation and ammonification, 3:383–384
  denitrification
    control
      nitrogen oxide limitations, 3:389–390
      oxygen limitation, 3:388–389
      reductant limitations, 3:390
    genes, 3:388
    intermediates, accumukation and effects on plants, 3:390–391
    nitrate
      ammonification, 3:391
      reduction, 3:387–388
    overview, 3:387
    species, 3:388, 390
  dinitrogen fixation, *see also* Nitrogen fixation

ecology, 3:382–383
nif genes, 3:381
nitrogenase complex, 3:380–381
reactions, 3:380
species, 3:381
global implications, 3:391
nitrification
ammonia oxidation, 3:384–385
ecology, 3:386–387
heterotrophoic nitrification, 3:387
nitrite oxidation, 3:385–386
overview, 3:384
species, 3:385
overview, 3:379
Nitrogen fixation, *see also nif* genes; Nitrogenase; Nodule formation, legumes; *Rhizobium*; Rhizosphere
agricultural demands, 3:407
associative bacteria, 1:496–497
definition, 1:359, 492, 3:392, 4:117
ecological significance, 1:492–493
energy yield, 3:394
free-living bacteria, 1:495–496
insect symbiotic microorganisms, 4:530–531
nodule formation, 1:503–504
outputs from various systems, 3:393
overview, 1:493–495
rhizobial genes
nif, 1:502–503
nod, 1:503–504
species
associative bacteria, 1:497, 3:396–397
cyanobacteria, 1:494, 3:394
eubacteria, 1:493, 3:393
free-living bacteria, 3:395–396
symbiotic bacteria
classification, 1:497–498
cyanobacteria, 1:498, 3:398–399
Frankia, 1:498–499, 3:397–398
Rhizobium–legume symbiosis, 1:499–501, 3:399–405
Nitrogen limitation, definition, 1:134
Nitrogen recycling, definition, 4:526
Nitrous oxide, atmosphere cycling by microbes, 3:260, 262
NK cell, *see* Natural killer cell
NMR, *see* Nuclear magnetic resonance
NO, *see* Nitric oxide
Nod factors
definition, 3:392
types, 1:492, 504
nod genes
definition, 1:492, 3:392
nodule formation, 1:503–504
types, 1:503–504, 3:404–405
Nodule formation, legumes
evolution, 3:408–409
factors affecting nodulation
environmental factors, 3:416
soil fertility status, 3:416
ineffective mutants, 3:411
infection
molecular basis for nodule formation, 3:410–411
nodulin-specific gene expression, 3:411–412
plant defense response and symbiosis, 3:412–413
types, 3:409
visible changes during root-hair infection, 3:409–410
lipopolysaccharide role, 3:410–411
nodule characteristics
number, 3:413–414
shape, 3:413
size, 3:413
overview, 3:404–405
specificity in nodulation, 3:414
Nodulin
definition, 3:407
expression, 3:411–412
Nomarski microscopy, *see* Optical microscopy
Nonrandom segregation, definition, 1:822
Nonribosomal peptide synthesis
aminoacylation, 4:223
carboxyl activation, 4:221
modification reactions, 4:223
overview, 4:221
peptide bond formation, 4:223
peptide synthetases, 4:221
protein comparison with ribosome-synthesized proteins, 4:220
termination reactions, 4:223
thiotemplate mechanism, 4:213
Nontypable bacteria, definition, 2:591
Norwalk virus, food-borne illness, 2:405
NRI-NRII two-component system, nitrogen assimilation regulation, 4:742, 744–750
NSOM, *see* Near-field scanning optical microscopy
Ntr regulon, definition, 1:669
ntr genes, regulation, 1:678–679
Nuclear magnetic resonance (NMR), intracellular pH measurement using phosphorous-31, 3:627–628
Nuclear polyhedrosis virus, definition, 2:813
Nuclear transfer
background, 4:671–672
cultured cells, 4:672–673
transgenic animal generation, 4:673
Nuclease, definition, 2:71
Nucleocapsid protein, definition, 2:635
Nucleoid
definition, 1:808
Escherichia coli, 2:265
Nucleoside
catabolism, 3:427
definition, 3:418
salvage, 3:426–427
Nucleosome, definition, 2:42
Nucleotide
biosynthesis, *de novo*
purine nucleotides
AMP, 3:420
gene organization, 3:420
GMP, 3:420
regulation, 3:420
pyrimidine nucleotides
CTP, 3:422
gene organization, 3:424
regulation, 3:422–424
UMP, 3:420–422
catabolism of nucleobases
purines, 3:427
pyrimidines, 3:427–429
definition, 3:418
deoxyribonucleotide synthesis
alternative routes, 3:425
dTTP, 3:425
ribonucleotide reductase, 3:424–425
mutants in biosynthesis
5-fluoropyrimidine resistance, 3:430
industrial uses, 3:429
selection, 3:429
trimethoprim resistance, 3:430
occurrence, 3:418
regulation of enzymes, 3:418
salvage pathways

Nucleotide (*continued*)
growth requirement for nucleobases, 3:425–426
purine salvage
bases, 3:426
interconversion reactions, 3:426
ribonucleosides, 3:426
toxicity, 3:426
pyrimidine salvage, 3:427
starvation response, 3:420
structures, 3:419
Nucleotide-binding domain
ATP-binding cassettes, 1:5–6
definition, 1:1
Nucleotide excision repair, overview, 2:77
Numerical aperture, definition, 3:264
NusA, transcription termination, 4:625
NusG, transcription termination, 4:625
Nutrient, definition, 3:431
Nutrition, microbial culture
element composition of microbial biomass
polymeric constituents, 3:435
stability, 3:433
whole cells, 3:434
element requirements and functions
calcium, 3:438–439
carbon, 3:436
cobalt, 3:439
copper, 3:439
hydrogen, 3:436–437
iron, 3:438
magnesium, 3:438
manganese, 3:439
molybdenum, 3:439
nickel, 3:439
nitrogen, 3:437
oxygen, 3:437
phosphorous, 3:437
potassium, 3:438
sodium, 3:439
sulfur, 3:437
zinc, 3:439
limiting nutrient, growth modeling and comparison to unrestricted growth, 3:439–441
minimal growth media
applications, 3:441
design
calculations, 3:442
excess factors, 3:442
growth yield factors, 3:442
maximum biomass concentration setting, 3:441–442
maximum growth rate, setting during unrestricted growth, 3:441
optimization, 3:442–444
growth-limiting nutrient, experimental identification, 3:444–446
preparation, 3:444
nutrient classes, 3:433
nutritional categories of microorganisms, 3:432–433
quality assessment of media, 3:446
Nystatin
antifungal activity, 1:237
lipid formulations, 1:239

O

Objective lens, definition, 3:264, 288
Obligate anaerobe, definition, 1:18
Obligate parasite, definition, 4:195
Obligate syntroph, definition, 3:204
Occlusal, definition, 3:466
Ochratoxin
animal disease, 3:345–346
human disease, 3:342
types, 3:342
Ocular lens, definition, 3:264, 288
Oil pollution
biormediation
land spills, 3:453–454
open water spills, 3:452–453
prospects, 3:454–455
definition and overview, 3:449
hydrocarbon degradation by microbes, 3:451–452
sources, 3:450–451
Okazaki fragment, synthesis, 2:86
2′,5′-Oligo(A) synthetase, antiviral effects, 1:278–279
Oligogalacturonate hydrolase, features, 3:570
Oligogalacturonate lyase
definition, 3:562
features, 3:572
Oligotroph
adaptation biochemistry and physiology, 2:334
definition, 2:317, 651, 3:87–88
habitats, 2:332–333
species, 2:332–333
Oligotrophic environments, *see also* Starvation, bacterial; Stringent response
copiotrophs, 3:88
definition, 3:86
microbial activity measurement, 3:87
organic matter levels in ecosystems, 3:86–87
starvation-survival mode of microbes
overview, 3:88
starvation process, 3:88, 91
stress protein expression, 3:92
viable but nonculturable cells, 3:91–92
Omasum, definition, 4:185
OmpA, functions, 3:521
OmpC, ATP-binding cassette transport role, 1:3
OmpF, ATP-binding cassette transport role, 1:3
OmpR, osmosensor, 3:513–514
Oncogene, definition, 3:456, 4:81
Oncogenic viruses
cancer induction mechanisms, diversity, 3:464–465
DNA tumor viruses, 3:458–459
Epstein–Barr virus and Burkitt's lymphoma, 3:461–462
hepatitis B virus and hepatocellular carcinoma, 3:459–461
human papilloma virus and cervical cancer, 3:462–464
human T cell leuekotropic virus type 1 and T cell cancers, 3:464
onset of cancer, 3:457
prevalence in cancer etiology, 3:456
RNA tumor viruses, 3:457–458
tumor suppressor gene inactivation, 3:458–459, 463
types, overview, 3:456–457
Oocyst, definition, 3:745, 4:598
Ookinete, definition, 3:745
Oomycetes, definition, 2:117, 3:633
Oospore
definition, 3:633
downy mildew, 2:119–120
Opa, *see* Opacity-associated protein
Opacity-associated protein (Opa)
adhesins, 1:46, 50
Neisseria, 2:578–579

Open reading frame (ORF), definition, 1:383, 3:99, 697, 4:181, 451
Operon, definition, 1:383, 2:270, 618, 677, 3:363, 730, 876, 4:1
O-polysaccharide
definition, 3:71–72
features, 3:74
protective barrier function, 3:81
Opportunistic infection, definition, 1:104, 232, 2:460–462
Opportunistic organism, definition, 3:876
Opportunity cost, definition, 2:137, 142–143
Opsonization, definition, 1:209, 729
Optical microscopy, *see also* Confocal microscopy
bright-field microscopy, principles, 3:289–291
dark-field microscopy, principles, 3:291
deconvolution microscopy, 3:300–302
fluorescence microscopy
applications, 3:295–296, 298
confocal microscopy, *see* Confocal microscopy
epi-illumination, 3:296
filters, 3:296–297
fluorescence recovery after photobleaching, 3:299–300
immunofluorescence, 3:297–298
labeling, 3:297
live cell imaging, 3:298–299
time-resolved fluorescence microscopy, 3:299
near-field scanning optical microscopy, 3:304–306
Nomarski microscopy, 3:294–295
optical video and digital microscopy, 3:302–304
phase-contrast microscopy, 3:292–294
polarization microscopy, 3:295
selection of technique, 3:289
Optical path, definition, 3:288
Optical sectioning, definition, 3:264, 288
Optical video and digital microscopy, *see* Optical microscopy
Oral microbiology
age of host, effects, 3:471
antibody sources, 3:470
antimicrobial agents, 3:472
definition, 3:466
destructive perodontitis, 3:466, 468
disease control and prevention, 3:480
ecology
habitat, 3:471
importance of study, 3:466–467
microbial interactions, 3:471
mineralized tissue, 3:467–468
saliva, 3:469
soft tissue, 3:468–469
flora
host species dependence, 3:472–474
normal human species, 3:474–475
pathogens
actinomycosis, 3:479
enamel caries, 3:477–478
endocarditis, 3:479
overview, 3:476–477
periodontal disease, 3:478–479
root surface caries, 3:478
soft tissues, 3:475–476
pathology, overview, 3:466, 478
plaque biofilm
classification, 3:468
flora, 3:476–477
formation, 3:467
growth rate, 3:469
saliva
bacterial concentration, 3:469
components, 3:469–470
flow rate, 3:470
site effects on flora, 3:472
Ordered cloned DNA library, *see* Mapping bacterial genomes
Ore discovery, microbial prospecing, 3:186
Ore leaching
acidity requirements of bacteria, 3:482–483
bioleaching applications, 3:183
biorecovery of metals from solution
binding metals to microbial surfaces, 3:184–185
intracellular accumulation and complexation, 3:185–186
mechanisms, 3:183
precipitation, 3:184
commercial application
copper, 3:486–487
gold, biooxidation pretreatment, 3:487–488
historical perspective, 3:486
overview, 3:488–489
uranium, 3:487
direct versus indirect bioleaching, 3:483
microbial solubilization reactions
copper, 3:182
iron, 3:182–183
manganese, 3:183
sulfur, 3:183
overview, 3:182, 482
reactions
copper, 3:483
iron, 3:483–484
sulfur, 3:484
species for bioleaching, 3:485–486
temperature requirements of bacteria, 3:483
ORF, *see* Open reading frame
Organic acid stress, definition, 3:629
Organometallic compound, definition, 2:607
Orientia, *see also* Rickettsiae
cell wall, antigens, and virulence factors, 4:166, 168
cultivation, 4:156–157
identification, 4:174
species and disease, 4:145, 147
Origin of life, *see also* Exobiology; Spontaneous generation
coupled systems
primitive metabolism, 3:495–496
protocells, 3:496–497
survival by replication, 3:497
earliest cells
common descent, sequence homology evidence, 3:497–498
introns, self-splicing, 3:498–499
regulation, 3:499–500
fossil record, 3:500–501
microbial diversification, 3:500
nature of life
cells, 3:491
energetics, 3:491–492
growth, 3:492
overview, 3:490–491
reproduction, 3:492–493
prebiotic soup
hypercycles, 3:494–495
peptide translation, 3:494
RNA catalysis, 3:493–494

Origin of life (*continued*)
spark flask experiment, 3:493
Origin of replication
definition, 1:808, 822
OriC
binding sites for primary initiation proteins, 1:825–826
DNA methylation, 1:827
DNA unwinding elements, 1:826–827
initiation of replication, 1:817–818, 820, 823–824
nucleoprotein complex assembly/disassembly, 1:827–830
spacer sequences, 1:827
structure, 1:824–825
yeast
binding sites for primary initiation proteins, 1:825–826
nucleoprotein complex assembly/disassembly, 1:829–830
overview, 1:824
OriT, organization, 1:854–855
Ornithin-containing lipids, biosurfactant production and features, 1:623–624
Orotic acid, microbial production processes, 4:849
Ortholog, definition, 1:1
Osmolality, definition, 3:502, 4:884
Osmolarity, definition, 1:710
Osmophile, definition, 3:502
Osmoprotectant, definition, 4:884
Osmosis, definition, 4:884
Osmotic stress, *see also* Halophile; Water-deficient environments
adaptation, 3:502–504
compatible solute
amino acids and derivatives, 3:508, 511
definition, 3:502
function, 3:505
methylamines, 3:509, 512
miscellaneous osmolytes, 3:510
sugars and polyols, 3:506–507, 509–511
definition, 3:502
model of osmoregulation, 3:515
osmosensors
EnvZ, 3:513–514
OmpR, 3:513–514
ToxR, 3:513–514
osmotaxis, 3:504
osmotic downshift, 3:512–513
potassium uptake, 3:504–505
protein, effects of increased osmolality, 3:504
turgor pressure in bacteria, 3:503
water activity, 3:503
OspC, sequence variation, 3:376
Outer membrane, gram-negative bacteria
assembly
enterobacterial common antigen biosynthesis, 3:524
export of proteins, 3:522
lipopolysaccharide
biosynthesis, 3:522–524
topology of biogenesis and export, 3:524
phospholipid export, 3:524
composition
lipids
enterobacterial common antigen, 3:519
lipopolysaccharide, 3:518–519
phospholipid composition, 3:519
proteins, 3:518
Escherichia coli, 2:263
functions
barrier function, 3:521–522
host–pathogen interactions, 3:522
overview, 3:517
isolation, 3:517–518
outer membrane proteins, *see also specific proteins*
lipoproteins, 3:521
OmpA, 3:521
porins
general pores, 3:520
specific diffusion channels, 3:520
stabilization by lipopolysaccharide, 3:80–81
structure
enteric paradigm, 3:519
zones of adhesion, 3:519–520
Oxiconazole, antifungal activity, 1:241
Oxidase
definition, 1:53, 2:177
respiration
cytochrome a_1 identity, 1:64
cytochrome *bd*, 1:63–65
cytochrome *bo'*, 1:63, 65
cytochrome *c* oxidase
overview, 1:60–61
proton pumping, 1:101–63
steps in reaction, 1:61–101
subunits, 1:61
cytochrome cd_1, 1:64
multiplicity and functions, 1:64–65
nomenclature, 1:60
non-respiratory functions, 1:65
overview, 1:60
synthesis, regulation in *Escherichia coli*, 1:64
taxonomic applications, 1:65
Oxidation ditch system, municipal wastewater treatment, 4:877
Oxidative phosphorylation, definition, 1:53, 2:177–178
Oxidative stress
adaptation responses
biochemical basis, 3:529
evolution, 3:531
Fur, 3:531
OxyR, 3:529–530
Sigma factor S, 3:530–531
SoxR, 3:530
SoxS, 3:530
definition, 1:53, 3:526
Pseudomonas response, 3:883
scavengers of reactive oxygen species
catalase, 3:529
glutaredoxin, 3:528
glutathione, 3:528
nicotinamine adenine dinucleotide, 3:528
peroxidase, 3:529
superoxide dismutase, 3:526, 528
thioredoxin, 3:528
sources
aerobic metabolism, 3:526–527
chemical and physical agents, 3:527
nitric oxide, 3:527–528
photosynthesis, 3:527
yeast responses, 3:531
Oxygen
evolution in atmosphere, 3:256–257
growth requirements and functions, 3:437
Oxygenic photosynthesis, definition, 1:907
OxyR, osmotic stress regulation, 3:529–530

P

PII, regulation of glutamine synthetase, 1:677–678

p48, interferon signal transduction, 2:836
p53
  adenovirus regulation, 3:459
  human papilloma virus inactivation, 3:463
  SV40 inactivation, 3:458
Packaging, DNA, 4:637, 643–645
Paclitaxel, semi-synthesis with chiral biocatalysis, 1:433–436
Paenibacillus lentimorbus, insecticide application, 2:816
Paenibacillus popilliae, insecticide application, 2:816–817
PAHs, *see* Polynuclear aromatic hydrocarbons
Paleopathology, definition, 4:538
Palindromic sequence, definition, 2:270
Pancreatitis, definition, 1:287
Pandemic, definition, 2:797, 4:811, 955
Panmictic population structure, definition, 2:628
Pantothenic acid, *see* Coenzyme A
Paper, *see* Pulp and paper
PapG, adhesin, 1:43, 45, 51
Papilla, definition, 3:801
Papillomavirus, *see also* Human papilloma virus
  definition, 1:287
  transcription, 4:630
Pap proteins, pilus assembly, 2:362, 364, 366–369
Papular rash, definition, 4:758
Papulocandin, antifungal activity, 1:249
Paracoccus denitrificans, respiratory chain, 1:58
Parainfluenza virus, *see also* Paramyxovirus
  antiviral agents, 3:545
  clinical features of infection, 3:542–543
  definition, 3:533
  structure, 3:534–535
Paralog, definition, 1:1, 2:591
Paramecium tetraurelia, antigenic variation, 1:261–262
Paramyxovirus, *see also specific viruses*
  antiviral agents, 3:545
  definition, 3:533
  genome organization, 3:535–536
  life cycle
    adsorption, 3:539–540
    genome replication, 3:541–542
    overview, 3:539
    penetration, 3:540
    transcription, 3:540–541
    translation, 3:541
    uncoating, 3:540
    virion assembly and release, 3:542
  proteins
    C, 3:536
    F, 3:539
    G, 3:537
    H, 3:537
    HN, 3:537
    L, 3:537
    M, 3:537
    NP, 3:536
    NS, 3:539
    SH, 3:539
    V, 3:537
  taxonomy, 3:533–534
  vaccination
    attenuated live virus, 3:544
    genetically-engineered virus, 3:545
    killed vaccine, 3:544
    subunit vaccines, 3:544–545
  virion structure, 3:534–535
Parasite, definition, 2:852
Parasitism, definition, 3:328
Parasitophorous vacuole, definition, 3:745
Parasporal inclusion body, definition, 2:813
Parathion, biodegradation pathway, 3:738–739
Parenchyma, definition, 3:676
Paronychia, definition, 4:271
Parvovirus B19, emerging infection, 4:819
Passive surveillance, definition, 4:506
Pasteur effect, definition, 4:939
Patent
  biotechnological inventions, criteria for patentability and case law
    composition of matter criteria, 3:550–551
    genes and gene products, 3:559–560
    new and useful criteria, 3:548–550
    plants and animals, 3:551–553
  definition, 3:547
  depository, 4:404
  novelty and nonobviousness requirements, 3:554–557
  patentable subject matter, 3:547–548, 560
  utility requirement, 3:553–554
  written description and enablement requirements, 3:557–559
Pathogen, definition, 2:117, 338, 3:109, 676, 4:65
Pathogen-derived resistance
  definition, 3:662, 792
  plant disease resistance, 3:666–667
Pathogenesis, definition, 3:109
Pathogenicity, definition, 3:762, 4:921
Pathogenicity factor, definition, 4:788
Pathogenicity island
  cag island in *Helicobacter pylori*, 2:632
  definition, 1:789, 2:187, 236, 270, 628, 3:654
  Erwinia, 2:255–256
  Yersinia pestis, 3:661
Pathotype, definition, 3:792
Pathovar, definition, 4:921
Pausing, definition, 1:339
PBP, *see* Penicillin-binding protein
PCR, *see* Polymerase chain reaction
PCT, *see* Preservative challenge test
Pectate lyase
  definition, 3:562
  Pseudomonas, 3:885
Pectin
  definition, 3:562
  functions, 3:562
  pectolytic enzymes, *see also specific enzymes*
    antigenicity, 3:578
    arabanase, 3:573
    arabinofuranosidase, 3:573
    classification, 3:565
    commercial applications, 3:575, 577–578
    endopolygalacturonase, 3:562, 568–569
    endopolygalacturonate lyase, 3:570–571
    endopolymethylgalacturonase, 3:570
    endopolymethylgalacturonate lyase, 3:572–573
    exopolygalactouronase, 3:562, 569–570
    exopolygalacturonate lyase, 3:571–572
    oligogalacturonate lyase, 3:562, 572
    oligogalcturonate hydrolase, 3:570

Pectin (*continued*)
plant pathogenicity role, 3:564–565
regulation of synthesis, 3:573–575
protopectin, 3:564
structure, 3:563–564
Pectin acetyl esterase, *Erwinia*, 2:240–241
Pectinase, features and applications, 2:215
Pectin lyase, *Erwinia*, 2:240, 246–247
Pectin methylesterase
definition, 3:562
Erwinia, 2:240
features, 3:566, 568
Pedicel, definition, 4:344
Pediocins, features, 1:390
Pelagic organisms, definition, 2:42
Penicillin
mechanism of action, 1:767
synthesis, *see* β-Lactam
Penicillin-binding protein (PBP)
definition, 1:759
gene archaeology, 2:702–703
identification, 1:767–768
Streptococcus pneumoniae, 4:446–447
Penicilliuum chrysogenium, flux balance analysis, 2:517–518
Pentamidine, sleeping sickness treatment, 4:739
Pentasaccharide lipids, biosurfactant production and features, 1:624
Pentose phosphoketolase pathway, features and distribution, 1:654
Peplomer, definition, 4:15
Peptide bond, definition, 3:824
Peptide synthetase, nonribosomal peptide synthesis, 4:221
Peptidoglycan, *see* Murein
Peptidyl prolyl isomerase, heat stress response, 2:600
Peptidyl transferase, definition, 3:824
Perennation, definition, 3:807
Peribacteroid membrane, definition, 3:407
Peridinin, definition, 2:42
Periodontal pocket, definition, 3:466
Periplasm
definition, 1:759, 2:210, 3:379
Escherichia coli, 2:263
Periplasmic chaperone, *see also* Chaperone
definition, 2:361
pilus assembly, 2:366–367
Periplasmic space, definition, 3:517
Peristalsis, definition, 2:485
Peristence, definition, 4:832
Perithecium, definition, 2:479
Permafrost, definition, 3:93
Permeable reactive barrier, bioremediation, 1:602–603
Permeant acid
definition, 3:625
internal pH measurement, 3:627
uptake avoidance, 3:628
Permeant base, definition, 3:625
Permineralization, definition, 1:126
Peroxidase, antioxidant defense, 3:529
Persistent transmission, definition, 3:792
Personal respiratory protection
definition, 1:69
tuberculosis, 1:76–77
Perspective, economic analysis, 2:137, 143
Pertussis, vaccine, 4:770–771
Pesticide, definition, 2:722
Pesticide biodegradation
atrazine, 3:598
bacterial evolution, 3:602
biochemistry
degradation pathways, 3:595–599
initial reactions, 3:595
mineralization and cometabolism, 3:599–601
carbofuran, 3:598–599
2,4-dichlorophenoxy acetate, 3:597–598, 601–602
microorganisms, 3:595
pesticides
types, 3:606
use, 3:594–595
plasmids
encoded degradation enzymes, *see* Plasmid
overview, 3:601–602
soil pesticide degradation
aging and bioavailability, 3:605
bound residues, 3:605
enhanced degradation, 3:604–605
rates, 3:603–604
uses, 3:605
Petechia, definition, 2:571
Petroleum industry, *see* Industrial effluents; Oil pollution
Pex proteins, starvation response, 4:398–399
Peyer's patch, definition, 4:755
PFGE, *see* Pulsed-field gel electrophoresis
Phage, *see* Bacteriophage
Phagocyte, definition, 1:209
Phagocytosis
definition, 1:729
mechanisms
killing, 1:736–737
uptake, 1:735–736
Phagolysosome, definition, 4:140
Phagosome, definition, 4:140
Phanerochaete chrysosporium, *see* Lignocellulose
PHAs, *see* Polyhydroxyalkanoates
Phase, definition, 3:264, 288
Phase III trial, definition, 1:287
Phase-contrast microscopy, *see* Optical microscopy
Phase variation, definition, 1:254, 2:260, 270, 361, 591
Phenazine, definition, 3:647
Phenols, bioremediation, 1:591–592
Phenotype
B cell, 1:209
definition, 2:270, 3:1, 711, 4:32, 921
Phenylalanine
biosynthesis
pathway, 1:147–148
regulation, 1:148–149
functions, 1:147
Pheromone
competence signaling, 4:653–654
definition, 3:349, 4:651
myxobacteria, 3:361
Phialide, definition, 2:479
Philosophy of method, *see* Scientific method
Phloem, definition, 3:607, 676
Phloem-limited bacteria
definition, 3:607
habitats, 3:611–612
identification, 3:609–610
morphology, 3:607–608
plant diseases, 3:610–611
transmission, 3:612–613
ultrastructure, 3:608–609
Phosphenolpyruvate:carbohydrate phosphotransferase system (PTS)
carbohydrate transport and phosphorylation, 3:584–585
carbon source selection, 1:673, 681

catabolite repression in enteric bacteria
adenylate cyclase, activity regulation, 3:587
inducer exclusion, 3:587
overview, 3:586–587
physiological consequences of regulation, 3:589–590
protein interactions with non-phosphotransferase system proteins, 3:589
Che protein interactions, 3:592
composition, 3:581
definition and overview, 1:669, 3:580–581
FruR, 3:592
functions, 1:724
genes, 3:585–586
HPr phosphorylation, 3:590–592
oxidative metabolism control in *Escherichia coli*, 3:592
phosphoryl transfer reactions, 3:583–584
protein structures
enzyme I, 3:581
enzyme II complexes, 3:582–583
HPr, 3:581–582
regulation in gram-positive bacteria, 3:590–592
ruminal microorganisms, 4:187
Phospholipase C, *Pseudomonas* virulence factors, 3:879, 881
Phospholipid
biosynthesis, *see* Lipid biosynthesis, *Escherichia coli*
definition, 3:55, 62
production, *see* Lipid production
Phosphorelay, definition, 2:15
Phosphorous
cycling
direct microbial contributions, 3:615–616
indirect microbial contributions, 3:616
overview, 3:615
elemental forms, 3:614
growth requirements and functions, 3:437
internal pH measurement using phosphorous-31 nuclear magnetic resonance, 3:627–628
phosphate, 3:614
Phosphotransferase system, *see* Phosphenolpyruvate:carbohydrate phosphotransferase system
Photoautotroph, definition, 1:579
Photoautotrophy, definition, 1:349
Photobleaching, definition, 3:264, 288
Photodegradation, definition, 3:171
Photoheterotroph, definition, 2:651–652
Photoheterotrophy, definition, 1:349, 3:431
Photolithotrophy, definition, 3:431
Photomultiplier tube (PMT), definition, 3:264
Photophobia, definition, 1:287, 3:618
Photophosphorylation, definition, 2:178
Photoreceptor, definition, 3:618
Photoresponse, definition, 3:618
Photosynthesis
cyanobacteria
dark reactions, 1:918–919
light harvesting, 1:917–918
light reactions, 1:918
overview, 1:916–917
definition, 3:618, 4:316
electron transport, 2:185
lake microbes, 2:442–443
overview, 1:717–719
oxidative stress, 3:527
Phototaxis, *see also* Scotophobia
anoxygenic photosynthetic bacteria, 3:620
definition, 3:618, 621
eukaryotic green algae, 3:619–620
filamentous cyanobacteria, 3:620
functions, 3:618–619
Phototroph, definition, 2:317
Phototrophic sulfur bacteria
competition with chemotrophic bacteria, 4:493–494
consortia-based sulfur cycling, 4:487
ecological niches, 4:491
growth kinetics, 4:490–491
light quality as ecological factor, 4:487, 489
metabolism, 4:479–482, 489–490, 498, 500
taxonomy, 4:486–487, 499–500
pH stress
acidic fermentation, 3:630
alkaline fermentation, 3:630–631
classification of species by pH range for growth, 3:626–627
culture considerations an buffers, 3:626–627
definition, 3:629
environmental role, 3:631–632
genetic responses
neutralization mechanisms, 3:629–630
virulence regulators, 3:630
homeostasis
external pH modification, 3:629
importance to cell, 3:628
mechanisms during growth and survival, 3:628
permeant acid uptake, avoidance, 3:628
importance of study, 3:625–626
internal pH measurement
phosphorous-31 nuclear magnetic resonance of phosphates, 3:627–628
radiolabeled permeant acids, 3:627
ruminal microorganisms, 4:189–190
Phycobilisome, definition, 1:907
Phyllocoenosis, definition, 1:359
Phylloplane, definition, 2:131
Phyllosphere, definition, 1:359
Phylogenetic analysis, definition, 3:773, 4:921
Phylogeny, definition, 2:317, 3:379, 4:32
Physiologic race, definition, 4:195
Phytochelatin, definition, 2:607
Phytopalexin, definition, 4:788
Phytophthora infestans
ecology, 3:637
late blight diseases
management, 3:639
types, 3:633
life history, 3:634–637
pathogenicity, 3:638–639
population genetics
initial migrations, 3:637–638
mating types, 3:637
twentieth century migrations, 3:638
taxonomy, 3:633–634
Phytoplasma
classification, 3:643–644
definition, 3:640
detection
polymerase chain reaction, 3:641–642
serology, 3:641
genome, 3:644–645
pathogens

Phytoplasma (*continued*)
classification, 3:686–687
dissemination, 3:687
ecology, 3:687
epidemiology, 3:688
genetics of disease, 3:687–688
infection, 3:687
management, 3:688
morphology, 3:685–686
reproduction, 3:686
phylogeny, 3:644
plant diseases
epidemiology, 3:645
types, 3:640–641
taxonomy, 3:644
Phytoremediation, definition, 4:117
Picornavirus, definition, 1:287, 4:97
Pigment, microbial, *see also specific pigments*; Tetrapyrroles
antibiotic activities, 3:649, 651
classification
carotenoids, 3:652
heterocyclic pigments, 3:649, 651
metalloproteins, 3:653
pyrroles and tetrapyrroles, 3:648–649
functions and importance, 3:647–648
importance of study, 3:653
Pseudomonas virulence factor, 3:882
quantification, 3:648
Pilin
definition, 2:361
methylation, 3:244
pilus assembly, 2:362
Pilosebaceous, definition, 4:271
Pilus, *see also* Fimbriae
assembly pathways, overview, 2:363–364
chaperone–usher assembly pathway
molecular architecture, 2:362, 364, 366
outer-membrane ushers, 2:369
periplasmic chaperones, 2:366–367
classification, 2:362
conjugation role, 1:853–854
conjugative pilus
assembly pathway, 2:373–375
classification, 2:373
F pilus, 2:373–374
Cs1 pilus assembly, 2:369–370
definition, 2:361
extracellular nucleation–precipitation assembly pathway, 2:375–376
pathogenesis role, 2:377–378
phage attachment, 1:853
plasmid transfer role, 3:725
Pseudomonas, 3:877–878
regulation of pilus biogenesis, 2:377
species distribution, 2:361
structure, 1:852–853
type 4 pilus
classification, 2:370–371
PilE structure, 2:371
species distribution, 2:370
type II secretion pathway for assembly, 2:371–373
type III secretion pathway, 2:376
Pimaricin, antifungal activity, 1:239
Piscirickettsia salmonis, *see* Rickettsiae
Pitch, definition, 2:222
PKR, *see* Double-stranded RNA-dependent protein kinase
Placebo, definition, 1:287
Plague
antibiotic therapy, 3:658–659
clinical disease
bubonic plague, 3:657–658
general symptoms, 3:657
miscellaneous forms, 3:658
pneumonic plague, 3:658
diagnosis, 3:659
epidemiology, 3:657
pandemics, 3:654–655
pathogen, *see Yersinia pestis*
prophylactic therapy, 3:658
twentieth century cases, 3:655
vaccination, 3:658, 4:775–776
zoonosis, 4:961
Planetary protection, definition, 2:299, 304–305
Plankton, *see also* Cyanobacteria
definition, 1:907, 2:438
freshwater plankton, 1:925–926
marine plankton, 1:925
Plant disease, *see also* Phloem-limited bacterial; *Phytophthora infestans*; Phytoplasma
bacteria and phytoplasma pathogens
classification, 3:686–687
dissemination, 3:687
ecology, 3:687
epidemiology, 3:688
genetics of disease, 3:687–688
infection, 3:687
management, 3:688
morphology, 3:685–686
reproduction, 3:686
economic impact, 3:676
fungal pathogens, *see also* Fungal infection, plants
classification, 3:681–683
dissemination, 3:684
ecology, 3:683–684
epidemiology, 3:685
genetics of disease, 3:684–685
infection, 3:684
management, 3:685
morphology, 3:679, 681
overview, 3:678–679
reproduction, 3:681
nematode pathogens
classification, 3:693
dissemination, 3:693
ecology, 3:693
epidemiology, 3:694
infection, 3:693–694
management, 3:694
morphology, 3:692
overview, 3:692
reproduction, 3:692–693
overview of pathogenic mechanisms, 3:676
parasitic green algae, 3:696
parasitic higher plants, 3:695
protozoa pathogens
classification, 3:695
dissemination, 3:695
ecology, 3:695
infection, 3:695
management, 3:695
morphology, 3:694
overview, 3:694
reproduction, 3:694–695
resistance, *see* Plant disease resistance
species specificity of pathogens, 3:696
viruses and viroids, *see also* Beet necrotic yellow vein virus; Luteoviridae; Potyviruses
classification, 3:690, 697–698
cytological effects of infection, 3:702–703
dissemination, 3:690–691, 703–704

DNA viruses, 3:707
epidemiology, 3:691
gene expression, 3:707–708
host specificity, 3:701–702
infection, 3:691
management, 3:691–692, 709
morphology, 3:688, 690
movement through plant, 3:704–705
overview, 3:688
particle structure, 3:705–706
properties, table, 3:700
replication
DNA viruses, 3:709
double-stranded RNA viruses, 3:708
overview, 3:690, 708
single-stranded minus-sense viruses, 3:708
single-stranded plus-sense viruses, 3:708
RNA genomes
distribution, 3:706
double-stranded, 3:707
single-stranded minus-sense, 3:706–707
single-stranded plus-sense, 3:706
symptoms of infection, 3:702–703
Plant disease resistance
definition, 3:663
engineered resistance
interfering molecules
antibiotics, 3:675
antisense and ribozymes, 3:671–672
Bacillus thuringiensis toxin, 3:674
coat protein-mediated resistance, 3:668, 673
fungal killer systems, 3:674
interfering nucleic acids, 3:670–671
mammalian antibodies, 3:674
movement proteins, 3:673
nonstructural proteins, 3:673
protease inhibitors, 3:675
ribosome inactivating proteins, 3:674–675
satellite RNA, 3:672–673
whole viral/viroid genome expression, 3:674
natural resistance genes
pathogenesis-related proteins, 3:668–669
single resistance genes, 3:669–670
overview, 3:668
mechanisms, 3:663
natural resistance
horizontal resistance, 3:664–665
hypersensitivity, 3:665
insect vector resistance, 3:667–668
pathogen-derived resistance, 3:666–667
promotion, 3:663
qualitative and quantitative resistance, 3:665–666
somaclones, 3:667
sources, 3:666
systemic acquired resistance, 3:665–667
tolerance, 3:667
vertical resistance, 3:664
pathogenicity types, 3:663
Plaque, definition, 4:796
Plasma membrane, *Streptococcus pneumoniae*, 4:446
Plasmid
antibiotic resistance transfer in gram-negative anaerobic pathogens, 2:569
applications, 3:711–712
catabolic plasmids for biodegradation
alkane degradation, 3:731
6-aminohexanoate degradation, 3:738
atrazine degradation, 3:731, 733
bromoxynil degradation, 3:733
compounds degraded by encoded enzymes, 3:732
dichloroethane degradation, 3:733, 735
2,4-dichlorophenoxy acetate degradation, 3:735, 742
naphthalene degradation, 3:735–736, 738
overview of field, 3:730–731
parathion degradation, 3:738–739
toluene degradation, 3:739
transcriptional regulation, 3:739, 741–742
transposon roles, 3:742–743
chromosome mobilization and gene acquisition, 3:728
classification, 3:714–716
cloning vectors, 4:59
conjugation, *see* Conjugation
copy number, 3:712, 716, 720–721
definition, 1:383, 405, 692, 847, 2:187, 270, 677, 3:1, 594, 654, 711–712, 730, 4:387, 428, 451
discovery, 3:711, 728
evolution, 3:712, 714
Haemophilus influenzae, 2:593
history of study, 2:694–695
horizontal gene pool, 3:716
incompatibility testing, 3:714–716
Mycobacterium, 3:321–322
pesticide biodegradation, 3:601–602
phenotypes conferred in bacteria, 3:713
Pseudomonas, 3:890–891
replication
control, 3:720
DNA polymerase I-dependent replicons, 3:718
iteron-activated replicons, 3:718–720
overview, 3:716
rolling circle replication, 3:716, 718
size range, 3:712, 716, 720
stable inheritance
active partitioning, 3:723
killing of plasmid-free segregants, 3:721–723
multimer resolution, 3:721
plasmid loss rate, 3:720–721
Staphylococcus, 4:389
Streptomyces
chromosome interactions, 4:458–459
fertility, 4:457–458
genetic manipulation, 4:466
replication, 4:458
topology, 4:458
transfer
conjugation, 3:724–725, 727
establishment in new host, 3:727–728
transduction, 3:724
transformation, 3:724
transformation processing, 4:657–658
Yersinia pestis, 3:660
Plasmid incompatibility, definition, 1:692
Plasmodesmata, definition, 3:662

Plasmodium, *see also* Malaria
antigenic variation, 1:261
host specificity, 3:746–747
life cycle, 3:747–749
merozoite, 3:745, 749
phylogenetic relationships, 3:754
sporozoite, 3:745, 748
taxonomy, 3:746
Plasmodium falciparum
amino acid metabolism, 3:755–756
antigenic variation, 3:760
apicoplast, 3:752–753
chromosomes, 3:752
culture, 3:747
DNA replication and synthesis, 3:752, 755
energy production, 3:754–755
life cycle, 3:747–749
nucleic acid base composition, 3:753
proteases, 3:756
red blood cell reorganization, 3:754
ribosome biogenesis, 3:753
Plasmolyzed cell, definition, 3:517
Plastid, origin in dinoflagellates, 2:52
Plating efficiency, definition, 2:91
Pleomorphic fungi, definition, 2:468
Pleomorphism, definition, 2:677, 4:364
Plesiomonas shigelloides
enteric disease, 2:199
food-borne illness, 2:403
refrigerated food pathogens, 4:73
Plume, contaminants, 1:587
PMT, *see* Photomultiplier tube
Pneumocystis carinii, antigenic variation, 1:262
Pneumonic plague, *see* Plague
Pneumonitis, definition, 1:287
Poikilotherm, definition, 4:545
Point spread function (PSF), definition, 3:264, 288
Poisson distribution, definition, 2:677
Polar organelle, definition, 4:353
Polarization microscopy, *see* Optical microscopy
Polarized light, definition, 3:264, 288
Polio
clinical features, 3:764
differential diagnosis, 3:764–765
eradication
global situation, 3:771
goals, 3:762
impact of program, 3:770–771
program strategy, 3:769–770
etiology, 3:764–765
historical perspective, 3:762–764
immunity, 3:766
incubation period, 3:766
laboratory diagnosis
polymerase chain reaction, 3:767
specimens, 3:766–767
virus isolation, 3:767
occurrence, 3:765–766
reservoir, 3:766
strain identification, 3:767
surveillance, 4:521
transmission, 3:766
vaccines
administration, 3:768
adverse events, 3:768
contraindications, 3:768
dosage, 3:768–769
efficacy, 3:768
schedule, 3:768
storage, 3:769
types and selection factors, 3:767–768
Pollution, *see* Biomonitor; Heavy metals; Industrial effluents; Infectious waste; Oil pollution
Polyhydroxyalkanoates (PHAs)
Azobacter production of polyhydroxybutyrate, 1:366, 370–371
definition, 1:359
types, 1:565
uses, 1:564–566
Polyketide
biosynthesis
overview, 4:229–230
precursors, 4:230–232
synthases
comparison to fatty acid synthases, 3:780
evolution and phylogenetic relationships, 3:783, 785
type I gene organization, 4:234–235
type I iterative enzyme products, 3:781
type I processive enzyme products, 3:781
type II gene clusters, 4:232, 234
type II iterative enzyme products, 3:781, 783
classification, 3:774–775
definition, 1:189, 3:654, 773, 4:213, 229
history of study, 3:773–774
hybrid compounds, 3:785
market, 3:774
microorganisms for production, 3:774
rational drug design, 3:785
structure and activity
ansamycins, 3:778
anthracyclines, 3:780
ascomycins, 3:777–778
macrolactams, 3:778
macrolides, 3:775–776
pentacyclic lactones, 3:776–777
polyenes, 3:778
polyethers, 3:778–779
statins, 3:779
tetracyclines, 3:780
Polymerase chain reaction (PCR)
carryover, 3:787
definition, 1:287, 373, 2:131, 3:1, 109, 787, 4:55
diagnostics of infectious disease
dengue fever, 3:789
hemolytic uremic syndrome, 3:791
human immunodeficiency virus, 3:788–789
overview, 2:31, 36–37, 3:788
Plasmodium falciparum, 3:760–761
poliovirus, 3:767
probe amplification systems, 2:37
prospects, 3:789
rhinovirus, 4:105–106
signal amplification systems, 2:37–38
species-specific probes, 2:38
target amplification systems, 2:36–37
tropical diseases, 3:789
Verticillium diagnostics, 4:793
ecology applications, 2:135
emerging virus identification, 4:812–813
environmental microbiology applications, 3:789–790
commercial kits, 3:790
lactic acid bacteria classification, 3:2
modification using 5′-nuclease assay, 3:790–791
mutagenesis, 4:63

nested, 3:640
phylogenetic analysis, 2:57
phytoplasma detection and classification, 3:641–644
reverse transcriptase technique for RNA analysis, 3:787, 789–790
sensitivity, 3:788
Taq polymerase in amplification, 3:787
technique, overview, 3:788
Polymorphonuclear granulocyte, definition, 1:729
Polymyositis, definition, 1:287
Polynuclear aromatic hydrocarbons (PAHs), bioremediation, 1:591
Polyoxin D, antifungal activity, 1:252
Polyploidy, definition, 1:359
Polyprotein, definition, 3:792
Polyunsaturated fatty acid, *see* Vitamin F
Population, definition, 2:131, 283, 3:466
Population-based surveillance, definition, 4:506
P/O ratio, definition, 2:178
Porin
definition, 2:571
Neisseria, 2:577
Porphin, definition, 3:188
Porphyrinogen, definition, 4:558
Positive strand polarity, definition, 2:201
Potassium, growth requirements and functions, 3:438
Potassium channel openers, chiral biocatalysis, 1:440–441
Potato virus X, features of infection, 3:703
Potentiator, definition, 3:893
Potyviruses
control, 3:798–799
diagnosis of infection, 3:797–798
economic importance, 3:798
genome organization and regulation, 3:794–796
host range, 3:793
replication and gene functions, 3:796
symptoms of infection, 3:797
taxonomy, 3:793
transmission, 3:796–797
vectors, 3:793
Pour tube
definition, 2:664
high-pressure studies, 2:670
Powdery mildew
appressorial differentiation, 3:802–803
conidial germination, 3:802–803
control
chemicals, 3:807–808
host resistance in control, 3:807
prospects, 3:808
definition and overview, 3:801–802
developmental regulation following germination, 3:805
environmental influences on infection, 3:806–807
host
penetration, 3:803, 805
response to *Erysiphe graminis*
compatible interactions, 3:805–806
incompatible interactions, 3:806
specificity, 3:801, 806
life cycle, 3:802
perennation, 3:807
Pradimicin A, antifungal activity, 1:251
Pravastatin, chiral biocatalysis, 1:438
Prebiotic, definition, 2:222
Predivisional cell, definition, 1:692
Prenyltransferase, definition, 4:213
Preprotein, definition, 2:210
Preservative, definition, 1:887
Preservative challenge test (PCT), cosmetics, 1:896–897
Prespore, definition, 4:377
Pressure, *see* High-pressure habitats
Prevalence, definition, 2:137, 141, 3:131, 4:248, 811
Primaquine, malaria treatment, 3:139
Primary metabolite, definition, 3:1, 4:428
Prion
application of Koch's postulates, 3:234–236
copper binding octarepeats, 3:810–811
definition, 3:809
diseases
bovine spongiform encephalopathy, 3:819–820
Creutzfeldt–Jacob disease, 3:820–821
genetic diseases, 3:820
historical perspective, 3:818
peripheral pathogenesis, 3:821–822
treatment, 3:822–823
functions of protein, 3:810–811
gene features, 3:809–810
glycosylation, 3:816
pathogenesis
conversion mechanisms to pathogenic form, 3:816–817
knockout mouse studies, 3:813
protein only hypothesis, 3:812–813
virino hypothesis, 3:813–814
pathology, 3:821
protein features, 3:810
protein–protein interactions, 3:811
secondary structure, 3:811–812
strains and species barrier, 3:814–816
yeast prions, 3:818
Prior art, patent law, 3:546
Probabilistic identification, *see* Identification of bacteria, computerized
Probe, definition, 2:131, 3:1
Probiotic
dairy product inclusion, 2:7–8
definition, 2:1, 3:1
Prochiral selectivity, definition, 1:636
Prochloraz, antifungal activity, 1:243
Prodrome, definition, 4:955
Prodrug, definition, 1:287
Productivity, definition, 1:13
Prokaryote, definition, 1:398, 2:317, 3:773
Proline
biosynthesis
feedback inhibition, 1:139
pathway, 1:138
compatible solute, 4:892
functions, 1:138
Proline isomerase, protein folding role, 3:855
Promastigote, definition, 3:27
Promoter, *see also* Transcription
definition, 1:383, 2:618, 3:1, 240, 4:81, 394, 610, 628
recognition
elements, 4:613–615
mechanism, 4:615
Prophage, definition, 4:43, 336
Prophylaxis, definition, 1:287
Prosthetic group
biosynthesis, *see specific compounds*
compounds with unknown synthetic pathways, 1:841, 846
definition, 1:840

Prosthetic group (*continued*)
vitamin relationships, 1:840–841
Protease
definition, 1:287, 2:598
heat stress response, 2:600
inhibitors
overview, 1:288
plant disease resistance engineering, 3:675
sources and applications
acid proteases, 2:214
cysteine proteases, 2:214
meat and fish processing, 2:230
metalloproteases, 2:214
serine proteases, 2:213–214
Protective efficacy, definition, 4:779
Protein biosynthesis, *see also* Nonribosomal peptide synthesis; Ribosome
accuracy of translation
error types, 4:698–699
genetic analysis
ribosomal mutations, 4:700–701
transfer RNA mutations, 4:700
translation factor mutations, 4:701
attenuation of translation, 4:690, 694–695
codons, 3:825–826
elongation
aminoacyl transfer RNA entry into A-site, 3:831, 833
overview, 4:692–693
peptide bond formation, 3:833–834
rates, 3:831
ribosomal sites of action, 3:831
translocation, 3:834–835
evolution of translation mechanisms, 4:691
global control of translation, 4:697–698
initiation of translation
Archaea, 3:845
bacteria, 3:838–840, 4:693–694
eukaryotes
factors, 3:841
overview, 3:840, 4:694
cap-dependent initiation, 3:840, 843–845
cap-independent initiation, 3:845
overview, 4:691–692
reinitiation control, 4:695
messenger RNA degradation, cotranslational control, 4:696–697
overview of translation, 3:824–825, 4:690–691
programmed alternative translation
frameshifting, 4:703
hopping, 4:703
termination codon, 4:701–703
reading frames, 3:826
repression of translation, 4:690, 695–696
ribosome
abundance, 3:828
components, 3:828–830
consensus structure, 3:830
polysome formation, 3:830–831
termination and polypeptide release, 3:835–836, 4:693
transfer RNA
aminoacyl transfer RNA synthetases, 3:828
function, 3:826
initiator transfer RNA, 3:827–828
structure, 3:826–827
Protein disulfide isomerase
heat stress response, 2:600
protein folding role, 3:855
Protein F, adhesin binding, 1:49
Protein methylation
S-adenosylmethionine as donor, 3:240
CheR, 3:243–244
elongation factor-Tu, 3:244
flagellin, 3:244
pilin, 3:244
Protein secretion, *see* Secretion, proteins
Proteobacteria, definition, 3:245
Proteoliposome, definition, 1:1
Proteome, definition, 2:722
Proteomics
biotechnology industry, 2:734–735
definition, 2:106
Protista, definition, 2:852
Protoctist, definition, 2:42
Proton motive force
ATP generation, 1:717
definition, 1:710
Protonmotive force
definition, 1:54, 2:178, 181, 3:625
flagellum energy source, 2:385
Protonolysis, definition, 3:171
Protooncogene, definition, 4:81
Protopectinase
definition, 3:562
features, 3:565–566
Protoplasm, definition, 4:316
Protoplasmic cylinder, definition, 4:353
Protoplast
fusion, strain improvement, 4:438
overview, 4:663
transformation
Protoporphyrin IX, biosynthesis, 4:564
Protoporphyrinogen IX, oxidation, 4:564
Protozoa
definition and overview, 2:852, 3:18, 456, 3:866, 869
diversity of species, 2:66–67
encystment, 3:874–875
food-borne illness
Cryptosporidium parvum, 2:406
Cyclospora cayetanensis, 2:407
Entamoeba histolytica, 2:406–407
Giardia lamblia, 2:406
miscellaneous organisms, 2:407
gut, 2:493–494
heavy metal vulnerability, 3:875
intestinal infection, *see Cryptosporidium*; *Entamoeba histolytica*; *Giardia lamblia*
plant pathogens
classification, 3:695
dissemination, 3:695
ecology, 3:695
infection, 3:695
management, 3:695
morphology, 3:694
overview, 3:694
reproduction, 3:694–695
predation
aquatic environments, 3:871–872
feeding biology variation, 3:869–871
protozoan occurrence in various biotopes, 3:869
soil
biomass dependence, 3:872
metabolic processes, 3:872
mineralization of carbon, nitrogen, and phosphorous, 3:872–873
organic matter distribution, 3:873–874
rhizosphere, 3:874

Pseudomonas
antibiotic resistance, 3:882–883
biodegradation of pollutants
alkanes, 3:889
benzoates, 3:887
chlorinated aromatic hydrocarbons, 3:886–887
cycloalkanes, 3:889
naphthalene, 3:887
nylon, 3:889
overview, 3:885–886
polychlorinated biphenyls, 3:887, 889
trichloroethane, 3:889–890
biotechnology applications, 3:891–892
classification, 3:876–877
cystic fibrosis patient infection, 3:883–884
flagella, 3:877
heavy metal resistance, 3:890
lignin degradation, 3:889
overview, 3:876
oxidative stress response, 3:883
pathogens, 3:876
pili, 3:877–878
plant pathogens
exopolysaccharides, 3:885
hosts, 3:884–885
hypersensitive response, 3:884–885
pectate lyase, 3:885
syringomycin secretion, 3:885
plasmids, 3:890–891
population sensing, 3:883
protein secretion systems, 3:884
Pseudomonas aeruginosa, skin microbiology, 4:284
virulence factors
alginate, 3:879
alkaline protease, 3:882
CTX, 3:881
elastase, 3:878
exoenzyme S, 3:878–879
exotoxin A, 3:878
LasA protease, 3:878
lectins, 3:881
lipase, 3:879
lipopolysaccharide, 3:879
phospholipase C, types, 3:879, 881
pigments, 3:882
pycocins, 3:881
rhamnolipids, 3:881
siderophores, 3:881–882
Pseudoplasticity, definition, 1:556
Pseuocyst, definition, 4:598
PSF, *see* Point spread function
P-site, ribosome, 3:824
Psychotroph
definition, 4:545
microbial spoilage, 2:2
Psychrophile, *see also* Low-temperature environments
adaptation biochemistry and physiology
cold shock and acclimation, 2:325–326
enzymes, 2:325
membranes, 2:325
biodiversity, 3:95
definition, 1:319, 2:317, 4:65–66
enzymes, 3:97
growth characteristics, 3:95–96
habitats
artificial, 2:324
natural, 2:324
membrane features, 3:96–97
species, 2:323–324, 3:94–95
Psychrosphere, *see* Low-temperature environments
Psychrotroph, *see also* Low-temperature environments
biodiversity, 3:95
definition, 3:163, 4:65–66
enzymes, 3:97
growth characteristics, 3:95–96
membrane features, 3:96–97
PTS, *see* Phosphenolpyruvate:carbohydrate phosphotransferase system
Pullulan
definition, 1:171
production, 1:563
properties, 1:563
uses, 1:563–564
Pulp and paper
biofilms in paper machine, 3:894–896
control of microbiological problems
chemical treatment, 3:905–908
cleaning, 3:905
economic benefits, 3:909
environmental monitoring, 3:905
safety of chemicals, 3:908–909
environmental factors affecting growth on paper machine
nutrients, 3:897
oxygen, 3:898–899
pH, 3:898
temperature, 3:897–898
water, 3:896–897
enzymatic processing, 2:229
industrial effluents and treatment, *see* Industrial effluents
negative impacts of microbial growth
corrosion, 3:903–904
fresh water contamination, 3:901
gas production
hydrogen, 3:903
hydrogen sulfide, 3:902–903
methane, 3:903
machine additive contamination, 3:901–902
miscellaneous problems, 3:905
overview, 3:893–894
paper machine wet end, 3:901
pulp slurries in storage, 3:899–900
water storage, 3:900
wood fiber biodeterioration, 3:899
papermaking process, 3:893
preservation of paper products
food paperboard, spore reduction, 3:904
market pulp, 3:904
soap wrappers, 3:904–905
xylanase applications
de-inking, 4:936
debarking, 4:935–936
dissolving pulp production, 4:936
fiber modification, 4:936
prebleaching of pulps, 4:935
retting of flax fibers, 4:936
shive removal, 4:936
Pulsed-field gel electrophoresis (PFGE)
definition, 3:151
gene mapping, 3:155–156
PulseNet surveillance of bacteria, 4:523–524
Purple nonsulfur photosynthetic bacteria, definition, 1:349
Putrefaction, definition, 2:677–678
Pycocin, *Pseudomonas* virulence factor, 3:881
Pyoderma, definition, 4:271
pyr operon, *Bacillus subtilis*, 1:345
pyrBI operon, *Escherichia coli*, 1:342
Pyridoxal phosphate, biosynthesis, 1:842, 844
Pyridoxine, microbial production processes, 4:846

Pyrifenox, antifungal activity, 1:249
Pyrimethamine, malaria treatment, 3:139
Pyrimethanil, antifungal activity, 1:240
Pyrrole, definition, 3:647
Pyrrolnitrin, antifungal activity, 1:240
Pyruvate
cleavage regulation in fermentation, 2:347–348
fates following glycolysis
aerobic or facultative bacteria under oxidative conditions, 1:656–657
anaerobic bacteria, 1:658–659
archaea, 1:659
facultative bacteria under anaerobic conditions, 1:657–658

Q

Q, antitermination, 4:626
QALY, *see* Quality-adjusted life year
Qinghaosu, malaria treatment, 3:139–140
Quality-adjusted life year (QALY), definition, 2:529
Quantum yield, bioluminescent, 1:520
Quarantine
cholera, 1:806
definition, 1:801
history, 2:175
Quasi-equivalence, definition, 3:99
Quasispecies, definition, 2:635
Quinine, malaria treatment, 3:138–139
Quinone, definition, 1:54, 2:178
Quorum sensing
acylated homoserine lactone signaling
biofilms, 4:11
discovery, 4:3–4
ecology and functions, 4:10–12
host responses, 4:12
integration into cellular physiology, 4:10
interspecies communication, 4:11
mechanisms, 4:4
modulation of responses, 4:8
overview, 4:1–2
receptors
DNA binding and transcription activation, 4:7–8
interactions and cell entry, 4:6
repressor activity, 4:8
structure, 4:6–7
synthases, 4:5–6
definition, 2:236, 4:1
regulation of *Erwinia* gene expression, 2:242–243
Vibrio species
autoinduction of *lux* genes, 4:2–3
bioluminescence regulation, 4:2
model for cell density sensing, 4:3
Vibrio fisheri, 4:9–10
Vibrio harveyi, 4:9

R

Rabies
animal disease epidemiology, 4:17–18
clinical signs, 4:28
control and prevention, 4:30
definition and overview, 4:15–16
historical perspective, 4:16
immune responses to infection and vaccination, 4:28–30
pathogenesis and histopathology, 4:25–28
vaccination, 4:16
viruses
detection, 4:16–17
genome structure, 4:21–22
G protein, 4:24–25
isolation, 4:16–17
L protein, 4:25
M protein, 4:23–24
N protein, 4:22–23
NS/P protein, 4:23
replication, 4:25
serotypes, 4:18–19
structure, 4:19–21
Race-specific resistance
definition, 2:117
downy mildew control, 2:125–126
Racemate, definition, 1:430, 2:222
Radiation-resistant microorganisms
adaptation biochemistry and physiology, 2:334–337
habitats, 2:333–334
species, 2:333–334
RAG, *see* Recombinant activating gene
Ralstonia solanacaerum
control, 4:39–41
economic importance, 4:33
epidemiology, 4:38–39
geographic distribution, 4:37
host range, 4:37
phenotype conversion, 4:38
subspecific classifications and uses, 4:33–37, 41–42
taxonomy, 4:33
virulence genes, 4:37–38
Rapid granular filter, drinking water treatment, 4:903–904, 906–907
Rapid plasma reagin (RPR) test, definition, 2:29
Rare-cutter, endonucleases, 3:151, 154–155
RB, *see* Reticulate body
Rb, *see* Retinoblastoma protein
Rcs proteins, regulation of extracellular polysaccharide synthesis, 2:248–249
Reaction center, definition, 3:618
Reactive arthropathy, definition, 2:390
Reactive gliosis, definition, 3:809
Reactive oxygen species (ROS)
definition, 3:526
intracellular killing by immune cells, 1:737
oxidative stress, 3:527
Reactogenicity, definition, 1:789
Reading frame, definition, 3:824
Reassortant, definition, 2:797
RecA
coprotease activity, 4:46
DNA repair role, 4:46, 48
evolutionary conservation
DNA sequence homology, 4:52
homology between species, 4:44, 51
phenotypes of mutants, 4:51–52
phylogenetic tree analysis, 4:53
protein sequence and structure homology, 4:52–53
SOS response induction, 4:337, 340
synaptase activity, 4:44–46
Recalcitrance
definition, 1:461
xenobiotics, 1:465–466
Receptor, definition, 4:81

Recoding, definition, 4:690
Recombinant activating gene (RAG)
B cell development role, 1:217–218
definition, 1:209, 213
Recombinant DNA
biotechnology industry
complementary DNA and genomic libraries, 2:725
expression vectors, 2:725
overview, 2:725
recombinant protein production, 2:725–727
definition, 2:222, 499, 722
techniques, *see* Cloning; DNA sequencing; Mapping bacterial genomes; Restriction mapping; Strain improvement
Recombination
bacterial chromosomes, 1:813–814
definition, 1:808, 2:270–271, 4:43, 451, 930
homologous, 2:277–278
site-specific, 2:278
transpositional, 2:278–280
Recombination pathway, definition, 2:91
Recovery resistance
definition, 2:117
downy mildew control, 2:128
Recrudescence, definition, 3:131
Redox-coupled reaction, definition, 1:180
Redox potential
definition, 1:54, 2:178, 3:614
respiratory processes, 2:181
Reduction to practice, patent law, 3:546
Reductive tricarboxylic acid cycle (RTCA)
enzyme distribution, 1:356–357
molecular regulation, 1:356
reactions, 1:355–356
Refraction, definition, 3:288
Refractive index, definition, 3:264, 288–289
Refractory metals, definition, 3:482
Refrigeration, foods, *see also* Temperature control
adaptation of microorganisms to low temperature, 4:66–67
factors affecting food preservation, 4:65–66
features of refrigerated foods
butter, 4:76–77
cheese, 4:77
composition, 4:89
convenience foods, 4:78
cooked meat and poultry, 4:76
eggs, 4:77
fish, 4:75
fruits, 4:78
meat, 4:74
milk, 4:76
poultry, 4:75
shellfish, 4:75–76
vegetables, 4:77–78
microorganism types
bacteria, 4:67–68
overview, 4:65–66
modified atmosphere packaging
convenience foods, 4:80
fish, 4:79
meat, 4:79
overview, 4:78–79
poultry, 4:79
vegetables, 4:79–80
pathogens
Aeromonas hydrophila, 4:70
Bacillus cereus, 4:72–73
Clostridium botulinum, 4:73
Escherichia coli, 4:73
fungi, 4:73–74
Listeria monocytogenes, 4:70, 72
overview, 4:68–70
Plesiomonas shigelloides, 4:73
Salmonella, 4:73
Vibrio parahaemolyticus, 4:72
Yersinia enterocolitica, 4:72
Regioselectivity, definition, 1:636, 3:49
Regulon, definition, 1:789, 2:15, 236, 4:1, 336
RelA, *see also* Stringent response
abundance, 4:469
ATP:GTP pyrophosphoryl transferase activity, 4:469–470
guanosine pentaphosphate synthesis, 4:468–469
Relapse, definition, 3:131
Relaxosome, plasmid transfer role, 3:727
Release factor, definition, 3:824
Remittent fever, definition, 3:131
Renewable energy, definition, 3:199
Repertoire, T-cell, 1:209
Replacement culture, definition, 1:636
Replica plating, definition, 4:55
Replicase, definition, 3:792
Replication cycle, definition, 1:288
Replication fork, *see also* Chromosome, bacterial
chromosome replication, 1:817
definition, 1:808
termination of DNA replication, 2:90
Replicon, definition, 3:730
Reporter, definition, 3:363
Reporter gene, fusion protein constructs, 4:63
Repression, definition, 3:379
Repressor, scientific method application in isolation, 3:236–238
Reservoir, definition, 3:109, 4:758, 955
Resistance, definition, 3:662, 4:297
Resolution
calculation for microscopes, 3:290–291
definition, 3:265, 289
Respiration
definition, 1:54, 2:178
respiratory electron-transport linked ATP synthesis in bacteria, 2:181–184
Respiratory chain
architecture, 1:57
assembly, 1:57
definition, 1:54, 2:178
Helicobacter pylori, 1:58
mitochondrial paradigm, 1:57–58
organization in bacteria, 1:58–59
Paracoccus denitrificans, 1:58
Respiratory droplet, definition, 1:69
Respiratory protection
definition, 1:359
nitrogenase, 1:368–369
Respiratory syncytial virus (RSV), *see also* Paramyxovirus
antiviral agents, 3:545
clinical features of infection, 3:543
definition, 1:288, 3:533
ribavarin therapy, 1:300
structure, 3:534–535
Respirofermentation, definition, 4:939
Respirometer, definition, 4:855
Restriction, definition, 3:240
Restriction endonuclease, definition, 1:398, 3:204
Restriction fragment length polymorphism (RFLP), definition, 1:422, 4:32

Restriction mapping
  definition, 4:55
  restriction enzyme classification, 4:60
  Southern blot analysis, 4:61–62
  type II enzymes in recombinant DNA techniques, 4:60–62
Restriction and modification (R–M)
  alleviation of restriction, 2:100
  antirestriction systems, 2:100–101
  applications and clinical relevance, 2:104–105
  biological significance, 2:103–104
  classification
    miscellaneous systems, 2:98
    modification-dependent systems, 2:97–98
    overview, 2:95
    type I systems, 2:96–97
    type II systems, 2:95–96
    type III systems, 2:97
  detection of restriction systems
    fragmentation assays, *in vitro*, 2:93–94
    gene transfer barrier, 2:93
    sequence-specific screens, 2:94–95
  distribution in species, 2:101–102
  diversity, 2:102
  DNA translocation prior to cleavage, 2:99
  endonucleases and cleavage sites, 2:94
  evolution, 2:102–103
  gene expression regulation, 2:99–100
  history of study, 2:92
  nomenclature, 2:95
  sequence recognition, 2:98–99
Resuscitation, definition, 4:545
Reticulate body (RB), definition, 1:781–782
Reticulocytosis, definition, 1:288
Reticulum, rumen, 4:185
Retinitis, definition, 1:288
Retinoblastoma protein (Rb)
  adenovirus regulation, 3:459
  SV40 inactivation, 3:458–459
Retrovirus, *see also specific viruses*
  emerging infection, 4:819–820
  endogenous retroviruses, 4:86, 88
  genome, 4:82
  life cycle, 4:85–86
  oncogenic retroviruses, 4:92–93
  overview, 4:81
  structure, 4:82
  taxonomy, 4:82
  transcription
    complex viruses, 4:634–636
    simple viruses, 4:634
  vectors
    animal experiments, 4:94–95
    design, 4:93
    helper cells, 4:93
    human gene therapy, 4:95–96
    tissue culture experiments, 4:93–94
Return activated sludge, definition, 4:870
Reversed electron transport, definition, 2:178
Reverse genetics
  cloning, 4:58
  definition, 2:797, 4:55
Reverse transcription, definition, 4:628
Reversion, oligosaccharide production, 2:222
RFLP. *see* Restriction fragment length polymorphism
R gene, definition, 3:633
Rhamnolipid, *Pseudomonas* virulence factor, 3:881
Rhamnose lipids, biosurfactant production and features, 1:620, 622–623
α-Rhamnosidase, features and applications, 2:230
Rhamsan, features, 1:561
Rhinovirus
  clinical illness, 4:104–105
  culture, 4:101
  definition, 1:288
  diagnosis
    isolation of virus in cell culture, 4:105
    polymerase chain reaction, 4:105–106
    serology, 4:105
  epidemiology of common cold
    incidence and prevalence, 4:101–102
    seasonality, 4:102
    transmission routes, 4:102
  history of study, 4:97
  immune responses, 4:104
  pathogenesis of symptoms, 4:102–104
  physical and biological characteristics, 4:98–99
  prevention of infection, 4:106
  purification, 4:101
  receptor groups, 4:99–100
  replication cycle, 4:100–101
  serotypes, 4:99
  structure, 4:97–98
  treatment
    antiviral therapy, 4:106–107
    symptomatic remedies, 4:107
Rhizobacteria
  biocontrol of weeds, 1:489
  definition, 1:486
Rhizobia
  definition, 1:492
  legume symbiosis, 1:499–501
  taxonomic classification, 1:501
Rhizobium, *see also* Nodule formation, legumes; Rhizosphere
  factors affecting nodulation, 3:416
  growth rates, 3:400–402
  history of study, 3:399–400
  legume symbiosis, species, 3:399
  nitrogen fixation quantity by legume, 3:400
  nitrogenase, *see* Nitrogenase
  plasmids and megaplasmids, 3:403
  promiscuity, 3:407
  rhizobiophages, 3:392, 405
  taxonomy, 3:400–403
Rhizocoenosis, definition, 1:359
Rhizoctonia
  history of study, 4:109–110
  management
    cultural practices, 4:115–116
    host plant resistance, 4:116
    philosophy, 4:114–115
    prospects, 4:116
    soil and seed treatment, 4:115
  mycelium formation, 4:112
  overview, 4:109
  pathology, 4:113–114
  sclerotia formation, 4:111–112
  taxonomy, 4:109–111
Rhizodeposition, definition, 4:117
Rhizoplane, definition, 4:117
Rhizosphere
  definition and overview, 1:359, 2:131, 4:117–118, 321
  ecological impact, 4:122
  materials released by plants, 4:119
  microbes
    effects on plants, 4:121–122
    features, 4:118–119
    management and biotechnological applications, 4:122–125

responses, 4:119–121
protozoan predation, 3:874
Rho, transcription termination, 4:624–625
Rhodopsin, scotophobia role, 3:623
Rhodospirillum centenum, scotophobia, 3:621–622
Rhoptries, definition, 4:598
Ri plasmid, *see* Root-inducing plasmid
Ribavirin
adverse effects, 1:300
formulations, 1:299
indications, 1:300
influenza virus treatment, 2:809
mechanism of antiviral activity, 1:299
resistance, 1:300
Riboflavin
biosynthesis, 1:843
definition, 3:647
microbial production processes, 4:843
structure, 3:649
Ribonucleoprotein (RNP), definition, 4:592
Ribonucleotide reductase, deoxyribonucleotide synthesis, 3:424–425
Ribosomal RNA, definition, 3:824
Ribosome, *see also* Protein biosynthesis
abundance in cell, 3:828
assembly, 4:130–131
cell composition, 4:127
components, 3:828–830
consensus structure, 3:830
definition, 3:824, 4:690
idling, 4:467
polysome formation, 3:830–831
processing, 4:129–130
regulation of synthesis, overview, 4:127–128, 138
ribosomal protein synthesis
transcriptional regulation, 4:131–132
translational regulation, 4:131
ribosomal RNA synthesis regulation
antitermination
model, 4:135
protein factors, 4:136
regulatory regions, 4:135–136
regulatory roles, 4:138
roles, 4:136–137
termination regions, 4:138
initiation
activation mechanisms, 4:132–133
control sites in promoter, 4:134–135
growth rate-dependent regulation, 4:133–134
P2 promoter role, 4:135
stringent regulation, 4:133
structure, 4:127
transcription unit organization, 4:128–129
Ribotype, definition, 3:654
Ribozyme
definition, 3:662
definition, 4:181
plant disease resistance engineering, 3:671–672
self-splicing
bacteriophages and bacteria, 4:183
group I and group II introns, 4:182–183
mechanisms, 4:183
origin of introns, 4:183–184
Rickettsiae
antibiotic sensitivity, 4:176–179
Bartonella
cell wall, antigens, and virulence factors, 4:173–174
cultivation, 4:157, 160
genome, 4:166
growth environment, 4:153–155
identification, 4:175–176
species and disease, 4:148–149
classification and ecology, overview, 4:140–141
Coxiella burnetii
cell wall, antigens, and virulence factors, 4:173
cultivation, 4:156–157
genome, 4:166
identification, 4:175
Q fever, 4:147
cultivation of established strains, overview, 4:155–156
definition, 4:140, 758
Ehrlichia
cell wall, antigens, and virulence factors, 4:168–173
cultivation, 4:157
identification, 4:175
species and disease
groups and features, 4:147–148
table, 4:146
growth
cytoplasm and nucleus of host cells, 4:150–152
overview, 4:149–150
phagolysosomal growth, 4:153
phagosomal growth, 4:153
history of study, 4:140–141
identification, overview, 4:174
isolation and safety conditions, 4:160–161, 163
Orientia
cell wall, antigens, and virulence factors, 4:166, 168
cultivation, 4:156–157
identification, 4:174
species and disease, 4:145, 147
physical stability and chemical inactivation, 4:176
Piscirickettsia salmonis, 4:147
Rickettsia diseases and species
Brill–Zinsser disease, 4:764
cell wall, antigens, and virulence factors, 4:166, 168
clinical manifestations, 4:762
cultivation, 4:156–157
differential diagnosis, 4:762
endemic disease, 4:759
epidemic typhus fever, 4:764
genome features, 4:163–164
history of disease, 4:759
identification, 4:174
laboratory testing
bacteria isolation and identification, 4:763
routine tests, 4:762–763
serologic tests, 4:763
murine typhus, 4:759, 765
Rickettsia conorii, 4:143
Rickettsia prowazekii
features, 4:141
genome sequence and phylogenetic analysis, 4:163
Rickettsia typhi, 4:143
Rocky Mountain spotted fever, 4:765–766
scrub typhus, 4:764–765
spotted fever group, 4:143, 145
tables of diseases, 4:144–145
transmission, 4:760–761
treatment
antibiotics, 4:763
supportive care, 4:764
Rickettsiella species and disease, 4:147

Rifamycin, mechanism of action, 3:778
Riffle, definition, 2:438
Rift Valley fever virus, emerging infection, 4:820
Rimantadine
  adverse effects, 1:299
  indications, 1:298–299
  influenza virus treatment, 2:808–809
  mechanism of antiviral activity, 1:298
  resistance, 1:299
Rinderpest virus, definition, 3:533
Riparian zone, definition, 2:438
Rippling, definition, 3:349
Ritonavir
  adverse effects, 1:309
  indications, 1:309
  mechanism of antiviral activity, 1:309
  resistance, 1:309
R–M, *see* Restriction and modification
Rmp, *Neisseria*, 2:577–578
RNA, *see also* Antisense RNA; Messenger RNA; Protein biosynthesis; Transfer RNA; Transcription
  catalysis and origins of life, 3:493–494, 498–499
  methylation
    S-adenosylmethionine as donor, 3:240
    enzymes and functions, 3:243
RNA-dependent protein kinase, *see* Double-stranded RNA-dependent protein kinase
RNA polymerase
  catalysis, 4:611
  definition, 4:610
  fidelity, 4:619
  RNA-dependent polymerase, 2:201, 635
  RNA polymerase II, 2:826
  sigma factor, role in initiation, 4:612–613
  subunits, 4:611–612
RNase III, mediation of antisense RNA activity, 1:274–276
RNase L, interferon induction and antiviral activity, 2:838–839
RNA splicing
  definition, 4:181, 628
  history of study, 4:182–183
  self-splicing
    bacteriophages and bacteria, 4:183
    group I and group II introns, 4:182–183
    mechanisms, 4:183
    origin of introns, 4:183–184
  spliceosomes, 4:182
  transfer RNA splicing, 4:182
RNP, *see* Ribonucleoprotein
Rocky Mountain spotted fever, *see* Rickettsiae
Root-inducing (Ri) plasmid
  definition, 1:78
  virulence, 1:79–80
ROS, *see* Reactive oxygen species
Rosetting, definition, 3:745
Rotary-shadowing, definition, 3:276
Rotating biological contactor system, municipal wastewater treatment, 4:879–880
Rotavirus, food-borne illness, 2:405
R-prime, definition, 2:271
RPR test, *see* Rapid plasma reagin test
Rse proteins, extracytoplasmic heat shock response regulation, 2:606
RSV, *see* Respiratory syncytial virus
RTCA, *see* Reductive tricarboxylic acid cycle
Rubisco, *see* Calvin–Benson–Bassham pathway
Rumen
  bacteria, 2:489–490
  definition, 2:131, 485, 4:185
Rumen fermentation
  bacteria, 4:191
  biotechnology, 4:190–191
  computer modeling, 4:193
  crossfeeding among microorganisms, 4:187
  definition, 4:185
  ecology, 4:191
  energy spilling, 4:189
  fungi, 4:193
  growth and maintenance of bacteria, 4:189
  habitat, 4:186–187
  history of study, 4:185–186
  ionophore effects on rumen microorganisms, 4:190
  pathways, 4:188–189
  pH effects, 4:189–190
  polymer degradation, 4:187
  protozoa, 4:191–193
  toxic compounds, 4:190
  transport systems, 4:187
Rumination, definition, 4:185
Rust fungi
  axenic culture, 4:206
  control
    biological control, 4:209
    chemical control, 4:209–210
    cultural control, 4:208–209
    genetic host resistance, 4:210
  correlated species, 4:199–200
  cytology and genetics
    double-stranded RNA, 4:201
    inheritance studies, 4:201–203
    nuclear cycle, 4:200–201
  definition, 4:195
  diseases, 4:196–198
  effects on host
    growth, 4:204, 206
    physiology, 4:204
  establishment of infections, 4:203–204
  history of study, 4:196
  hosts, 4:198
  intraspecific variability
    formae speciales, 4:206–207
    pathogenic variability, 4:207–208
  phylogeny and taxonomy, 4:200
  spore states
    aeciospore, 4:198
    basidiopore, 4:199
    pycniospore, 4:198
    teliospore, 4:199
    urediniospore, 4:198–199
Rustmicin, mechanism of action, 3:776

S

S10 operon, *Escherichia coli*, 1:345–346
Saccharification, definition, 2:222
Saccharomyces cerevisiae, *see also* Yeast
  acetic acid production, 1:14
  prions, 3:818
SAED, *see* Selected area electron diffraction
St. Louis encephalitis, overview, 1:316
Salmonella
  antigenic variation, 1:265–266

chromosomes, 2:273, 275
conjugation, 2:280
diversity, 2:271
extrachromosomal elements, 2:275
food-borne illness, 2:398–399
invasiveness, 2:191
meat products as source, 3:167
mutation, 2:275, 277
nontyphoidal species and inflammatory diarrhea, 2:190–191
recombination
homologous, 2:277–278
site-specific, 2:278
transpositional, 2:278–280
refrigerated food pathogens, 4:73
surveillance, 4:523
taxonomy, 2:271–273, 398
transduction, 2:280, 282
transformation, 2:282
Salmonella typhi, *see also* Typhoid
enteric fever, 2:197
vaccine, 4:773–774
Salmonella typhimurium, compatible solute transporters, 4:893–894
Salvage pathway, definition, 1:840
Sanitarian, definition, 1:801
Sanitization, cosmetics
cleaning, 1:894
sanitizing, 1:894
water system, 1:895
Sanitizer
definition, 1:887, 2:421
food quality control, 2:426–427
Saperconazole, antifungal activity, 1:247
Sap functions, *Erwinia*, 2:254
Sapwood, definition, 4:571
Saquinavir
adverse effects, 1:308
indications, 1:308
mechanism of antiviral activity, 1:308
resistance, 1:308
SAR, *see* Systemic acquired resistance
Satellite RNA, plant disease resistance engineering, 3:672
Scanning electron microscopy (SEM)
definition, 3:276
principles, 3:282–284
specimen preparation
drying, 3:284
metal film deposition, 3:285
Scanning transmission electron microscopy (STEM), definition, 3:276, 285
SCH 39304, antifungal activity, 1:245
SCH 51048, antifungal activity, 1:245
SCH 56592, antifungal activity, 1:245–246
Schizont, definition, 3:745
SCID, *see* Severe combined immunodeficiency
Scientific method
definition, 3:227
examples
Koch's postulates and acquired immunodeficiency syndrome virology, 3:233–234
prions and transmissible spongiform encephalopathies, 3:234–236
repressor isolation, 3:236–238
scope of microbiology, 3:232–233
historical perspective, 3:227–230
Mill's methods, 3:230–232
physics compared with biology, 3:228–230
strong inference, 3:232
truth and progress in science, 3:230
Scintillon, definition, 1:520
Scleroglucan, features, 1:564
Sclerotium, definition, 2:468, 4:109, 297
SCO, *see* Single-cell oil
Scotophobia
Archaea, 3:622–623
definition, 3:618, 621
photoreception, 3:621
response in anoxygenic microbes, 3:620–621
Rhodospirillum centenum, 3:621–622
Scrombroid poisoning, 2:409
SDZ 89–485, antifungal activity, 1:247
Secondary metabolite, *see also specific metabolites*
biosynthetic reactions
classification, 4:215
late modification reactions, 4:220
polymerization and condensation, 4:219–220
comparison with primary metabolites, 4:214–215
definition, 2:236, 3:1, 4:213, 428
isoprenoid biosynthesis
chain elongation, 4:236
genes, 4:236
isomerization and cyclization, 4:236
overview, 4:235
steroids, 4:236
triterpenes, 4:236
β-lactams, *see* β-Lactams
nonribosomal peptides, *see* Nonribosomal peptide synthesis
polyketides, *see* Polyketides
precursors
amino acids, 4:216
aromatic intermediates, 4:216
conversion of primary metabolites, 4:217, 219
isoprene units, 4:216
nucleobases, 4:217
one-carbon groups, 4:217
organic acids, 4:215
overview, 4:215
pathway-committing reactions, 4:219
sugars, 4:216–217
structural overview, 4:213–214
Sec proteins
definition, 2:210
extracellular enzyme secretion role, 2:219
pilus assembly role, 2:366
protein secretion role
SecA, 3:852–854
SecB, 3:850–851
Secretion, protein
ATP-binding cassette protein secretion pathway
energetics, 3:861–862
exporter structure, 3:860
overview, 3:860
signals, 3:860
translocation mechanism, 3:861
cell wall translocation, 2:210–211, 219
classification of secretion pathways, 3:849
Erwinia exoenzymes, 2:241
exotoxins, 2:310, 315–316
flagellum proteins, type III export, 2:380, 388
folding chaperones, 3:855
general secretion pathway
chaperones and targeting factors, 3:850–852
distal steps during export, 3:854–855
energetics and mechanism, 3:854

Secretion, protein (*continued*)
overview, 3:847
signal sequences, 3:847–848, 850
translocon in *Escherichia coli*, 3:852–854
pilus assembly pathways, *see* Pilus
proline isomerases, 3:855
Pseudomonas systems, 3:884
signal hypothesis, 2:218–219
signal peptidases, 3:854
twin-arginine secretion pathway, 3:859
type II secretion pathway
exporter systems, 3:857–858
overview, 3:855
protein structure in secretion, 3:855–857
secretins, 3:858
type III secretion pathway
chaperones, 3:864
overview, 3:862
secretion apparatus, 3:862–863
signals, 3:864
substrates, 3:863–864
Secretion element, definition, 3:1
Segregation, chromosomes
budding yeast, 1:830–832
components of system, 1:830
prokaryotes, 1:832–833
Selected area electron diffraction (SAED), definition, 3:276, 286
Selective toxicity, definition, 1:445
Selenium
abundance, 4:239
assimilation by microbes, 4:239–240
cycle
biogeochemical cycling, 4:244–246
definition, 4:213
global mass balances, 4:243–244
demethylation, 4:240–241
diseases
deficiency, 4:240
toxicity, 4:240
dissimilatory reduction of inorganic compounds, 4:241–243
environmental sources, 4:239
methylation, 4:240–241
oxidation, 4:243
proprties, 4:238–239
Self-splicing, definition, 4:181
SEM, *see* Scanning electron microscopy
Semipermeable, definition, 4:884
Semisynthetic compound, definition, 3:773
Sendai virus, definition, 3:533
Sentinel surveillance, definition, 4:506
Sepsis, intraabdominal, 2:562, 564
Septation, cell division, 1:704–705
Septicemia, definition, 2:9, 678, 3:876
Septum, definition, 2:15, 468
Sequence batch reactor, municipal wastewater treatment, 4:877–878
Sequential injection analysis (SIA), bioreactors, 1:570–571
Serine
biosynthesis
pathway, 1:145–146
regulation, 1:146
functions, 1:145
toxicity, 1:146
Serodiagnosis, definition, 2:29
Serogroup
definition, 1:789
Vibrio cholerae, 1:790–792
Serotype, definition, 1:789, 2:271, 3:762
Serovar, definition, 3:163
Serratia entomophilia, insecticide application, 2:817
Sesile bacteria, definition, 1:478
Severe combined immunodeficiency (SCID), isolation of humans, 2:528
Sexually transmitted disease (STD), *see also* Acquired immunodeficiency syndrome; Chlamydia; Gonorrhea; Syphilis
definition, 4:248
economic impact, 4:250
global burden, 4:249–250
pathogens, major types, 4:249
prevalence, 4:248
risk factors, 4:250
Shadow-casting
definition, 3:276
transmission electron microscopy specimens, 3:280
Sharps, definition, 2:782
Shear rate, definition, 1:556
Shear stress, definition, 1:556
Sheath
cyanobacteria, 1:907
definition, 1:907, 4:353
Shell vial culture, definition, 2:29
Shellfish poisoning, 2:409–410
Shiga toxin, production in *Escherichia coli*, 2:396–397
Shigella
food-borne illness, 2:402
inflammatory diarrhea, 2:189–190
invasiveness genes, 2:189
Shiga toxin, 2:190
Shine–Dalgarno sequence, definition, 3:824, 837
Shingles, definition, 1:288
Short-pass filter, definition, 3:289
SIA, *see* Sequential injection analysis
Sick building syndrome, *see also* Airborne microorganisms; Ventilation
definition, 1:116
diagnosis, 1:116–117
Siderophore
definition, 2:607, 860, 3:654, 4:117
Erwinia features, 2:254–255
iron uptake role, 2:862
Pseudomonas virulence factor, 3:881–882
structures, 2:861
Sieve tube
definition, 3:607
phloem-limited bacteria, 3:611–612
Sigma factor
Bacillus subtilis, regulation of endospore development
features, 1:381–382
SigmaE activation in mother cell, 1:380–381, 4:382–383
SigmaF activation in forespore, 1:380, 4:381–382
SigmaG activation in forespore, 1:381, 4:382–383
SigmaK activation in mother cell, 1:381, 4:383
definition, 1:373, 2:236, 598, 3:876, 4:377, 394, 451, 610
endospore formation role, 2:17
F factor, flagellum gene expression, 2:386–387
heat shock response regulation
Bacillus subtilis, 2:605
Escherichia coli, 2:601–603, 606
RNA polymerase, 2:15
S factor
osmotic stress regulation, 3:530–531
starvation response

promoter target features, 4:401–402
stability and ClpXP degradation, 4:399–400, 403
synthesis rates, 4:399
translational regulation, 4:400–401
transcription initiation role, 4:612–613
Signal amplification, definition, 2:29
Signal peptide
definition, 2:210, 3:847
extracellular enzymes, 2:218–219
general secretion pathway, 3:847–848, 850
peptidases, 3:854
Signal recognition particle (SRP), protein secretion role, 3:850, 852
Signal transduction, *see* Two-component system, signal transduction
in Silico biology, definition, 2:510
Sillon, definition, 4:353
Simian virus 5 (SV5), definition, 3:533
Simian virus 40 (SV40), transcription, 4:628–630
Single-cell oil (SCO), definition, 3:62, 66
Single-strand conformation polymorphism (SSCP), definition, 1:422
Sister chromosome
definition, 1:822
segregation, 1:831
Site-directed mutagenesis, strain improvement, 4:440
Skin, *see also* Cosmetic microbiology; Fungal infection, cutaneous
adherence of bacteria, ecological niche determination role, 4:274
Aeromonas hydrophila, 4:284–286
Bacillus anthracis, 4:283
Clostridium perfringens, 4:283
Corynebacterium diphtheriae, 4:283
enteric gram-negative bacilli, 4:284
Erysipelothrix rhusiopathiae, 4:283
fungal infection, *see* Fungal infection, cutaneous
Haemophilus influenzae, 4:284
host defenses
antifungal substances, 2:452
cutaneous immune system, 2:453
inflammatory response, 2:452–453
keratinization and epidermal proliferation, 2:452
skin structure, 2:451–452
transferrin, 2:452
host determinants of colonization and invasion
anatomic features
gross regional variations, 4:273
hair follicles, 4:273
stratum corneum, 4:273
biochemical defenses, 4:272
immunologic defenses, 4:272–273
structural integrity of skin, 4:272
hydration role in flora composition, 4:273–274
Pseudomonas aeruginosa, 4:284
resident flora
Coryneform bacteria
Corynebacterium, 4:277
miscellaneous genera, 4:277–278
pathogens, 4:278–279
fungi, 4:279–280
gram-negative bactria, 4:279
Micrococcus, 4:279
overview, 4:274–275
Staphylococcus species
coagulase negative, 4:275–276
coagulase positive, 4:276–277
Staphylococcus aureus pyodermas, 4:280–281
Streptococci, non-group A pyodermas
progressive bacterial synergistic gangrene, 4:282–283
Streptococcus iniae, 4:283
Streptococcus pyogenes
acute cellulitis, 4:282
acute lymphangitis, 4:282
ecthyma, 4:281
erysipelas, 4:281
impetigo, 4:281
necrotizing fasciitis, 4:282
transmission of infections
direct contact, 4:285
hematogenous dissemination, 4:287
insect or animal bites, 4:286
Vibrio, 4:286–287
S-layer, *see* Surface layer
Sleeping sickness, *see Trypanosoma brucei*
Slime, definition, 3:349, 893
Slime trail, definition, 3:349
Slow sand filter, drinking water treatment, 4:904–906
Sludge, treatment
aerobic digestion, 2:759–760
anaerobic digestion, 2:760–761
composition, 2:758
composting, 2:761–762
objectives, 2:759
overview, 2:757–759
Small nuclear ribonucleoprotein particle (snRNP), definition, 4:181
Smallpox
biological warfare, 4:296
eradication, 2:172, 4:289, 293–295
etiology, 4:291
historical perspective, 4:289–291
laboratory diagnosis, 4:293
monkeypox similarity, 4:295–296
pathogenesis, 4:291–292
symptoms, 4:292–293
vaccination, 4:767, 780
virus stock destruction, 4:296
S motility gene, definition, 3:349, 360
Smuts
control, 4:304–307, 314
economic impact, 4:302–303
history, 4:298–299
host genetic resistance, 4:307–308
host–parasite interactions, 4:304
infection cycle, 4:303–304
morphology, 4:299–300
overview, 4:297–298
taxonomy, 4:300–302
snRNP, *see* Small nuclear ribonucleoprotein particle
Societal perspective, definition, 2:137, 143
SOD, *see* Superoxide dismutase
Soda desert
Archaea halophiles
habitats, 1:322
physiology, 1:322–323
geographic distribution, 1:127
overview, 1:126–127
Soda lake
Archaea halophiles
habitats, 1:322
physiology, 1:322–323
chemical analysis, 1:129
definition, 1:126
element cycles, 1:132–133
fossil records, 1:133

Soda lake (*continued*)
genesis, 1:127–129
geographic distribution, 1:127
overview, 1:126–127
taxa
aerobes, 1:131
anaerobes, 1:131–132
overview, 1:129–130
phototrophs, 1:130–131
Sodium, growth requirements and functions, 3:439
Softwood, definition, 4:571
Soil, *see also* Rhizosphere
aerobic versus anaerobic respiration, 4:318
biomass of flora and fauna, 4:323
chemolithotrophs, 4:319
composition, 4:316–317
definition, 4:321
factors affecting microbial communities
moisture, 4:327–328
organic content, 4:329
osmotic potential, 4:328
oxygen, 4:328
pH, 4:329
temperature, 4:328–329
toxic compounds, 4:329
fermentation and soluble organic molecule production, 4:318
food web of organisms, 3:867–869
habitat and microorganism distribution by soil size scale
field and landscape scale, 4:327
microscale, 4:324–325
profile scale, 4:325–327
interactions among organisms, 4:332–335
microbial biomass degradation, 4:319
microbial processes
biodegradation and transformation of pollutants, 4:332
decomposition, 4:331–332
elemental cycling, 4:330–331
gas exchange, 4:329–330
overview, 3:866–867
microorganisms
algae, 4:324
bacteria, 4:322–323
classification, 4:321–322
cyanobacteria, 4:324
fungi, 4:323–324
viruses, 4:322
mineralization, 4:319
organic matter
decomposition, 4:317–318
origins, 4:317
resistance to decomposition, 4:319
protozoan predation
biomass dependence, 3:872
metabolic processes, 3:872
mineralization of carbon, nitrogen, and phosphorous, 3:872–873
organic matter distribution, 3:873–874
rhizosphere, 3:874
Soil food web, definition, 3:866
Solar particle event, definition, 4:344
Solids retention time, definition, 4:855, 857–858, 870
Somaclonal variation
definition, 3:662
plant disease resistance, 3:667
Somatic incompatibility, definition, 4:109
Somatic mutation, definition, 1:209, 213
Sophorose lipids, biosurfactant production and features, 1:625–626
SOS response
Bacillus subtilis system, 4:341–343
definition, 1:373, 383, 2:91–92, 4:43, 336
discovery, 4:336
mutagenesis, 4:341
regulation
LexA
cleavage, 4:340–341
homologs, 4:343
repressor activity, 4:337–340
RecA induction, 4:337, 340
Sotalol, chiral biocatalysis, 1:441
Southern blot
definition, 2:92, 4:32, 55
restriction mapping, 4:61–62
SoxR, osmotic stress regulation, 3:530
SoxS, osmotic stress regulation, 3:530
Space flight, effects on microorganisms, *see also* Exobiology
antibiotic sensitivity, 4:347, 348–349
bacteria, 4:345–348
bacteriophage induction, 4:350–351
controls, 4:348
fungi, 4:346–347
growth, 4:347–348
human–microbe interactions, 4:351–352
importance of study, 4:344–345
missions carrying microbiology experiments, 4:345
organisms studied, 4:346
radiation sensitivity, 4:347, 349–351
simulated microgravity effects, 4:351
SpaceLab, definition, 4:344
Species, phylogeny, 2:55
Specific growth rate, continuous culture, 1:873–874
Spherical aberration, definition, 3:265
Spirochete, *see also* Lyme disease; Syphilis
definition, 4:353
ecology, 4:362–363
flagella, 4:354
genera, 4:355
history of study, 4:355–359, 361
morphology, 4:353–354
morphometric criteria, 4:356
pathogens, 4:359
physiology, 4:361–362
taxonomy and diversity, 4:354–355
Spitzenkorper, definition, 2:468
Splenomegaly, definition, 3:131
Spliceosome
definition, 4:181
mechanisms, 4:182
Spongiform changes, definition, 3:809
Spongiform encephalopathy, *see* Bovine spongiform encephalopathy; Prion; Transmissible spongiform encephalopathy
Spontaneous generation, *see also* History, microbiology; Origin of life
definition, 2:678
disagreements on theory, 4:364–365
early history, 4:365–366
evolutionary theory incorporation, 4:370
germ theory, 4:372–374
molecules, 4:364, 370
Needham/Spallanzani controversy, 4:366–368
parasitic worms, 4:370
Pasteur's experiments, 4:370–372

twentieth century ideas, 4:373–374
Sporadic, definition, 4:955
Sporangiole, definition, 3:349
Sporangium
definition, 1:373, 3:633
downy mildew, 2:120
Spore, definition, 1:28
Sporotrichosis
definition, 2:451
features, 2:456–457
Sporozoite, definition, 3:745, 748, 4:598
Sporulation, *see also Bacillus subtilis*
Bacillus subtilis
overview, 1:375
regulation
endospore development, 1:380–382
initiation, 1:379–380
coat structure, 4:384
compartmentalization of gene expression, 4:381–382
core, 4:385
definition, 1:373, 4:939
division, 4:380–381
engulfment, 4:382
germination, 4:385–386
initiation
phosphorylation and actions of SpoOA, 4:379–380
signals, 4:379
KinA-SpoOF-SpoOB-SpoOA two-component system in regulation, 4:742–744, 753
postengulfment development, 4:382–383
resistances of spores, development, 4:383–385
sigma factor roles
E, 4:382–383
F, 4:381–382
G, 4:382–383
K, 4:383
Spo mutants and gene loci, 4:378–379
stages, 4:377–378
Streptomyces coelicolor, 2:17–19
SpoT, *see also* Stringent response
domains, 4:472
guanosine pentaphosphate phosphorylase, 4:471
guanosine pentaphosphate synthetase activity, 4:471–472
SRP, *see* Signal recognition particle
Ssb, DNA replication role, 2:86
SSY 726, antifungal activity, 1:248
Stains, microbial, 2:32
Stalked cell, definition, 1:692
Staphylococcus
bacteriophages, 4:389–390
capsule, 4:388
cell wall, 4:387–388
culture, 4:387
diseases
sites of infection, 4:392
skin, 4:391–392
treatment, 4:393
extracellular proteins
cell surface proteins, 4:391
gene expression, 4:391
secretion, 4:390
soluble proteins, 4:391
types, 4:390
genome, 4:388–389
identification, 4:392
immunity, 4:392
plasmids, 4:389
skin resident species
coagulase negative, 4:275–276
coagulase positive, 4:276–277
species, 4:387–388
Staphylococcus aureus
food-borne illness, 2:393–394
meat products as source, 3:166
noninflammatory diarrhea, 2:196
pyodermas, 4:280–281
toxins, 2:196, 393–394
Starch
commercial processing, 1:177–178
definition, 1:171
hydrolyzing enzymes
α-amylase, 2:211, 223
β-amylase, 2:212, 223
applications, 2:212–213
debranching enzymes, 2:212
glucoamylase, 2:212, 223
structure, 2:211
Starter culture
definition, 2:1, 3:1, 163
genetics, 2:6
Starvation, bacteria, *see also* Oligotrophic environments; Stringent response
definition, 4:394
ecological importance, 4:394–395
morphological changes, 4:395
sigma factor S response
promoter target features, 4:401–402
stability and ClpXP degradation, 4:399–400, 403
synthesis rates, 4:399
translational regulation, 4:400–401
starvation proteins
cell envelope protection, 4:397
DNA protection, 4:397
metabolic amplification, 4:397
Pex protein expression, 4:398–399
protein protection chaperones, 4:397–398
stress response, 4:395
virulence role, 4:402
Stat proteins, interferon signal transduction, 2:832–833, 836
Stavudine
adverse effects, 1:304–305
indications, 1:304
mechanism of antiviral activity, 1:304
resistance, 1:304
STD, *see* Sexually transmitted disease
Steady state, continuous culture, 1:873–874
STEM, *see* Scanning transmission electron microscopy
Stereoisomer, definition, 1:430
Stereospecificity, definition, 3:49
Sterigma, definition, 4:109
Sterilization, definition, 2:767
Stock culture collection
definition, 4:404
federations sponsoring databases, 4:404–406
genetic stock centers
bacteria, 4:409–410
fungus and algae, 4:410
information management, 4:410–411
history
genetic stock centers, 4:408
germplasm and type culture collections, 4:406–408
locations, descriptions, and access, 4:411–426
national collections and consortia with broad holdings, 4:411
organism-specific collections, 4:426
organizations promoting culture collections, 4:426
Stomach, bacteria, 2:489

Strain improvement
cloning and genetic engineering
recombinant DNA technology, 4:437–438
site-directed mutagenesis, 4:440
engineering optimization
influence on strain performance, 4:441–442
metabolic engineering, 4:442
nutrition optimization, 4:441
fermentation applications, 4:429, 442–443
impact and benefits, 4:430
mutation
definition, 4:428, 430–431
mutagens, 4:432
random selection, 4:431–435
rationalized selection, 4:435–437
spontaneous mutations, 4:431
types of mutations, 4:431
overview, 4:428–429
rationale, 4:429–430
recombination techniques
conjugation, 4:439
overview, 4:437–438
protoplast fusion, 4:438
transformation, 4:438–439
transduction, 4:649
Stream, *see* Freshwater microbiology
Streptococci
antigenic variation, 1:265
group B disease surveillance, 4:512
non-group A pyodermas
progressive bacterial synergistic gangrene, 4:282–283
Streptococcus iniae, 4:283
Streptococcus pyogenes pyodermas
acute cellulitis, 4:282
acute lymphangitis, 4:282
ecthyma, 4:281
erysipelas, 4:281
impetigo, 4:281
necrotizing fasciitis, 4:282
Streptococcus equisimilis, stringent response, 4:472
Streptococcus pneumoniae
antibiotic resistance, 4:447–448
capsule, 4:444–445
cell wall, 4:445–446
epidemiology, 4:444
history of study, 4:444
morphology, 4:444
penicillin-binding proteins, 4:446–447
plasma membrane, 4:446
resources, 4:448
surveillance, 4:511–512
transformation, 4:655
vaccine, 4:771–772
Streptomyces
antibiotic biosynthesis, organization and regulation of genes, 4:461, 463
bacteriophages, 4:459–460
chromosome
architecture, 4:452–453, 455
genetic instability, 4:455–456
conjugation, 1:858
development, pleiotropic control of physiology and morphology, 4:460–461, 463, 465
gene disruption, 4:466
genome
composition, 4:451–452
sequencing, 4:451, 455
plasmids
chromosome interactions, 4:458–459
fertility, 4:457–458
genetic manipulation, 4:466
replication, 4:458
topology, 4:458
transposable elements, 4:456–457
Streptomyces coelicolor
sporulation, 2:17–19
stringent response, 4:474
Streptomycin, *see* Aminoglycoside
Stress, definition, 4:545
Stringent response, *see also* Oligotrophic environments; Starvation, bacteria
amino acid metabolism response, 1:150–151
bacterial distribution of guanosine pentaphosphate systems
Bacillus subtilis, 4:474
Myxococcus xanthus, 4:474–475
Streptococcus equisimilis, 4:472
Streptomyces coelicolor, 4:474
components, 4:470–471
definition, 1:134, 4:467
guanosine pentaphosphate
cycle, 4:468
regulation mechanisms, 4:475–476
overview, 4:467–468
RelA
abundance, 4:469
ATP:GTP pyrophosphoryl transferase activity, 4:469–470
guanosine pentaphosphate synthesis, 4:468–469
SpoT
domains, 4:472
guanosine pentaphosphate phosphorylase, 4:471
guanosine pentaphosphate synthetase activity, 4:471–472
Stromal keratitis, definition, 1:288
Strong inference, definition, 3:227, 232
Stylet, definition, 3:792
Subclinical infection, definition, 3:27
Subdivision, phylogeny, 2:55
Subgingival plaque, definition, 3:466
Sublethal injury, definition, 4:545, 553
Subliminal infection, definition, 3:697
Substrate, definition, 4:571
Substrate-level phosphorylation, definition, 1:647, 2:343
Sucrose lipids, biosurfactant production and features, 1:623
Sulfate-reducing bacteria
anaerobic respiration, 1:185–186
definition, 1:18, 3:171
methylation of mercury, 3:179–180
Sulfide-containing environments
chemotrophic sulfur bacteria
competition with phototrophic bacteria, 4:493–494
ecological niches, 4:491–493
metabolism, 4:482, 498, 500
taxonomy, 4:491–492, 499–500
coastal and marine environments, 4:486
inland environments, 4:484–486
metabolism
assimilatory sulfate reduction, 4:478–479
formation by sulfate-reducing bacteria, 4:479
free energy of metabolic reactions, 4:479
microorganism types, 4:478
overview, 4:483–485
phototrophic sulfur bacteria
competition with chemotrophic bacteria, 4:493–494

consortia-based sulfur cycling, 4:487
ecological niches, 4:491
growth kinetics, 4:490–491
light quality as ecological factor, 4:487, 489
metabolism, 4:479–482, 489–490, 498, 500
taxonomy, 4:486–487, 499–500
toxicity, 4:482
Sulfur
assimilation by cyanobacteria, 1:922
growth requirements and functions, 3:437
ore bioleaching, 3:484
Sulfur cycle
biotechnology applications, 4:503–504
cycling within ecosystems, 4:500–501
distribution on Earth, 4:495–496
environmental consequences of disruption, 4:501, 503
gaseous intermediates, 4:496
organic sulfur compound transformations, 4:500
overview, 4:495–497
sulfate reducing bacteria, 4:495, 497–498
sulfide oxidizing bacteria, *see* Sulfide-containing environments
Sulfuretum, definition, 4:495
Superantigen, definition, 2:307, 4:387
Supercoiling, DNA, 1:808–809, 814–815
Superficial punctate keratopathy, definition, 1:288
Superinfection, definition, 1:398
Superoxide dismutase (SOD), antioxidant defense, 3:526, 528
Supragingival plaque, definition, 3:466
Suramin, sleeping sickness treatment, 4:739–740
Surface layer (S-layer)
applications, 1:905
assembly, 1:902–903
biosynthesis, 1:903–904
chemical composition, 1:902
definition and overview, 1:899, 4:140
electron microscopy, 1:900–901
functions
cell adhesion and surface recognition promotion, 1:905
cell shape determination, 1:904
molecular sieves, 1:905
pathogenicity, 1:904–905
isolation, 1:901
Surface learning, definition, 2:156, 162–163
Surfactant, *see* Biosurfactant
Surfactin, biosurfactant production and features, 1:624
Surveillance, infectious disease
acquired immunodeficiency syndrome, 4:520–521
case definitions, 4:509–510
data sources, 4:508–509
definition, 3:762
FoodNet surveillance, 4:513–514
global programs, 4:516–517
Haemophilus influenzae, 4:517
hantavirus, 4:519–520
hepatitis C, 4:517, 519
historical perspective and overview, 4:506–508
Neisseria meningitidis, 4:523
notifiable diseases surveillance
active surveillance, 4:511–514
passive surveillance, 4:511
sentinel surveillance, 4:514
technological innovations in communications and information technology, 4:514–515
United States system, 4:510–511
polio, 4:521
purposes and uses, 4:508
Salmonella, 4:523
streptococcal disease, group B, 4:512
Streptococcus pneumoniae, 4:511–512
subtyping and molecular epidemiology, 4:521, 523–524
unexplained deaths and critical illness, 4:513
Susceptibility, definition, 3:762
SV5, *see* Simian virus 5
SV40, *see* Simian virus 40
Swarm, definition, 3:349
Swarmer cell, definition, 1:692
Swarming differentiation
flhDC operon, 2:19–21
signals, 2:19
species, 2:19
Syllabus, definition, 2:156
Symbiosis
definition, 3:328, 392, 407, 4:526
insects and microbes, *see* Insect, symbiotic microorganisms
rhizobacteria and legumes, *see* Nitrogen fixation
Symport, definition, 2:178, 181
Symporter, definition, 1:710
Synapsis, definition, 4:43
Synthase, definition, 4:1, 213
Synthetase, definition, 4:213
Syntrophy, definition, 2:131, 3:86
Syphilis
clinical manifestations, 4:264–266
diagnosis, 4:266–267
epidemiology, 4:263–264
history
biomedicine limitations and modern history, 4:542–543
management, 4:540–542
origins, 4:538–539
overview, 4:538–540
parasite, *see Treponema pallidum*
pathogenesis, 4:268–269
treatment, 4:267–268
Syringomycin, *Pseudomonas* secretion, 3:885
Systemic acquired resistance (SAR)
definition, 2:117
downy mildew control, 2:128
overview, 3:665–667
Systemic infection, definition, 2:117

T

T-8581, antifungal activity, 1:247
Tachycardia, definition, 2:9
Tachyzoite, definition, 4:598
Taenia saginata, food-borne illness, 2:407
Taenia solium, food-borne illness, 2:407–408
Taeniocyst, definition, 2:42
TAK 187, antifungal activity, 1:248
Tannin, definition, 2:222
Tapesia, *see* Eyespot
Target amplification, definition, 2:29
Target site, definition, 4:704
Taxol, synthesis with lipases, 3:53
Taxon, definition, 2:709
TB, *see* Tuberculosis
T cell
abscess formation role, 2:565
activation, 1:738–739
antigen processing

T cell (*continued*)
class I antigens, 1:737–738
class II antigens, 1:738
antigen recognition
class I molecules, 4:584
class II molecules, 4:584–585
foreign peptides bound to self-major histocompatibility complex molecules, 4:585
overview, 4:583–584
assays
cytokine secretion, 4:586–587
cytotoxicity , 4:587, 589
peptide binding, 4:589
proliferation, 4:586
B cell development, CD4+ T cell role, 1:222–224
CD1–restricted T cells, 1:733
CD4+ T cells, 1:732
CD8+ T cells, 1:732
classification, overview, 1:732
cytotoxic T lymphocytes
target cell killing, 1:740–741
viral defense, 1:742
definition, 1:729
development
negative intrathymic selection, 4:590
overview, 4:589
positive intrathymic selection, 4:590
survival signals, 4:590
T-cell receptor gene rearrangement, 4:589–590
$\gamma\delta$ T cells, 1:732–733, 4:591
helper cell differentiation, 1:739–740
human T cell leuekotropic virus type 1 and T cell cancers, 3:464
memory, 1:741
natural killer cells, 4:590–591
unconventional T cells, 1:732
T-DNA
border repeats, 1:78, 86, 87
definition, 1:78, 86
genes, 1:81, 87–89
genetic engineering of plants
applications, 1:100
barriers to transformation, 1:100
insertion site control, 1:100–101
monocot transformation, 1:99
overview, 1:84–86, 99–100
integration, 1:81
opine catabolism genes, 1:89
processing
co-transported protein roles in transfer and plasmid conjugation, 1:93–95
movement and integration, 1:95
overview, 1:82–84
piloting and protection, 1:93–95
transfer intermediate formation, 1:92
VirE2 role, 1:92–93
transport system
architecture, 1:97–98
components, 1:95
functional similarities, 1:97
T-pilus, 1:97
type IV transporters, 1:95–97, 101
VirB ATPases, 1:97–98
VirB stimulation of IncQ plasmid uptake, 1:98
Teichoic acid
cell wall, 1:763–764
definition, 1:647, 759
Teichuronic acid, definition, 1:647
Teleomorph, definition, 4:109
Teliospore, definition, 4:297
Telomere, definition, 3:745
TEM, *see* Transmission electron microscopy
Temperate cycle, phage
lysogeny, 1:408
pseudolysogeny and steady-state infections, 1:408–409
transduction and conversion , 1:409
Temperate phage, definition, 4:637
Temperature control
adaptation of microorganisms to thermal shock, 4:551–553
growth curve characteristics, 4:545–546
growth optimum classification of microorganisms, 4:545, 547
hurdle technology, 4:550–551
maintenance of constant temperature, 4:547–548
mathematical modeling of growth versus temperature, 4:546–547
meat, poultry, and fish, 4:550
milk handling
enzymes in spoilage, 4:549–550
pasteurization, 4:548–549, 556
recontamination following pasteurization, 4:549
refrigeration, 4:548
spore-forming bacteria, 4:549
sublethal injury at temperature extremes
definition, 4:545, 553
growth response curve, 4:553
implications for applied microbiology and research, 4:554–555
vegetative cells, 4:553–554
thermal inactivation
decimal reduction time, 4:555
pH effects, 4:556
survival curves, 4:555–556
Temporal regulation, extracellular enzymes, 2:217–218
Ter, termination of DNA replication, 2:90
Teratogenesis, definition, 1:288
Terbinafine, antifungal activity, 1:248
Terconazole, antifungal activity, 1:244
Terminal hairs, cyanobacteria, 1:916
Terminal redundancy, definition, 4:637
Termination codon, definition, 3:824
Terminator, definition, 1:339, 4:610
Terpenoid lipid, definition, 3:62
Tetanus
definition, 1:834
toxin, *see* Clostridia
vaccine, 4:769
Tetracycline
applications, 3:780
malaria treatment, 3:139
Tetrahydromethanopterin, definition, 3:204
Tetrapyrroles, *see also specific compounds*
δ-aminolevulinic acid precursor formation from metabolites
glutamate, 4:559
glycine and succinate, 4:559
biosynthesis
bilins, 4:565
coenzyme B_{12}, 4:563
coproporphyrin III, 4:564
factor F_{430}, 4:563
heme *a*, 4:565
heme *c*, 4:564–565
heme *d*, 4:565
heme d_1, 4:563
heme *o*, 4:565
protoheme, 4:564
protoporphyrin IX, 4:564

siroheme, 4:563
tolyporphyrins, 4:564
uroporphyrin III, 4:564
chlorophyll synthesis
bacteriochlorophylls, 4:567–568
chlorophyll *a*, 4:567
chlorophyll *b*, 4:567
chlorophyll *d*, 4:567
chlorosome chlorophylls, 4:568
isocyclic ring formation, 4:566
magnesium chelation, 4:566
photosynthesis reaction center pigments, 4:568
reduction to chlorin macrocycle, 4:566–567
vinyl reduction to ethyl group, 4:567
coporphyrinogen III oxidative decarboxylation, 4:564
definition, 4:558
protoporphyrinogen IX oxidation, 4:564
regulation of biosynthesis
enzyme activity, 4:568–569
gene expression, 4:569
structures, 4:561
uropophyrinogen III
biosynthesis, 4:561
decarboxylation, 4:563
Tetrodotoxin, poisoning, 2:410
TfdD, transcriptional regulation of plasmids, 3:742
Theca, definition, 2:42
Theory, definition, 3:227
Therapeutic index, definition, 1:288
Thermal stratification
definition, 2:438
lakes, 2:441–442
Thermization, definition, 4:545
Thermoacidophile, definition, 3:625
Thermocline, definition, 3:93
Thermoduric organism, definition, 4:545
Thermophile, definition, 1:319, 2:317–318
Thermoplastic adaptation, definition, 3:93
Theta confocal microscopy, *see* Confocal microscopy
Thiabendazole, antifungal activity, 1:241
Thiamin, microbial production processes, 4:843
Thiamine pyrophosphate (TPP), biosynthesis, 1:846
Thioester, definition, 1:189
Thioredoxin, antioxidant defense, 3:528
Threonine
biosynthesis
pathway, 1:141
regulation, 1:141–143
functions, 1:141
microbial production process, 1:157–158
uses, 1:157
Thrombocytopenia, definition, 1:288
Thromboxane A2 antagonist, chiral biocatalysis, 1:436–437
Thymidine kinase, definition, 1:288
Thymus, definition, 4:583
Ti plasmid, *see* Tumor-inducing plasmid
Tick-borne encephalitis, overview, 1:316
Tigemonam, chiral biocatalysis, 1:443–444
Tiller, definition, 2:338
Timber and forest products, *see also* Lignocellulose; Pulp and paper
colonization of wood in service
bacteria, 4:578–579
brown-rot fungi, 4:580
overview, 4:577–578
primary mold, 4:579
sapstain fungi, 4:579
secondary mold, 4:580–581
soft-rot fungi, 4:579–580
white-rot fungi, 4:580
durability of timber
induced, 4:581
natural, 4:581
preservatives, 4:581–582
ecosystems in conversion processes
converted timber before and after seasoning, 4:575
felled log, 4:574–575
overview, 4:573
standing tree, 4:574
habitat and substrates for microbes, 4:573
hazard classes
aboveground, covered, 4:575
aboveground, covered, occasional wetting, 4:575–577
abovground, uncovered, 4:577
ground or fresh water contact, 4:577
overview, 4:575
salt water contact, 4:577
wood degradation
abiotic factors, 4:572–573
biological factors, 4:572
importance in nature, 4:571
types, 4:572
wood structure, 4:571–572
Tinea versicolor
definition, 2:451
features, 2:455–456
Tioconazole, antifungal activity, 1:241
Tissue engineering, definition, 1:579
Tissue print immunoblot, definition, 1:422
Titer, antibody, definition, 1:209
TMAO, *see* Triethylamine-*N*-oxide
TMV, *see* Tobacco mosaic virus
tna operon, translation-mediated antitermination, 1:347
TNF, *see* Tumor necrosis factor
Tobacco mosaic virus (TMV)
coat protein-mediated resistance, 3:668, 673
features of infection, 3:702–704, 707–708
historical perspective, 2:689
structure, 3:705, 4:801
TOC, *see* Total organic carbon
Tolerance
definition, 3:662, 4:788
plant disease resistance, 3:667
Toluene, biodegradation pathway, 3:739
Tolyporphyrins, biosynthesis, 4:564
TonB
ATP-binding cassette transport role, 1:3
iron uptake role, 2:861
Topoisomerase
definition, 1:808
discovery, 1:810
Tospovirus
control, 4:597
genome
organization and protein function, 4:594–595
structure, 4:594
host range, 4:595–596
morphology and cytopathology, 4:592–594
overview, 4:592
replication, 4:595
symptomatology, 4:596
taxonomy, 4:592
transmission, 4:596–597

Tospovirus (continued)
vectors, 4:596–597
Total organic carbon (TOC), definition, 2:740
Total suspended solids (TSS), definition, 2:740
Toxin, *see also* Exotoxin
clostridia, 1:837–839
definition, 1:834, 2:390
fermented foods, reduction, 2:359
natural in food-borne illness
aflatoxin, 2:410
ciguatera poisoning, 2:409
miscellaneous toxins, 2:410
mushroom toxins, 2:410
scrombroid poisoning, 2:409
shellfish poisoning, 2:409–410
tetrodotoxin poisoning, 2:410
Toxoid
definition, 2:307, 4:767
exotoxin conversion, 2:314
Toxoplasma gondii
classification, 4:599
genome and gene expression, 4:603
genotypes, 4:602
immune response of host, 4:607
life cycle, 4:599–600
lytic cycle
attachment, 4:605
egress, 4:606–607
homology with viruses, 4:604–605
invasion, 4:605
vacuole modification and parasite replication, 4:605–606
mutants, 4:603
organelles, 4:604
overview, 4:598–599
recombinant DNA tools, 4:603–604
strain-specific virulence, 4:602–603
toxoplasmosis
behavioral effects, 4:607–608
control, 4:601, 608
diagnosis, 4:600–601
symptoms, 4:600
treatment, 4:601, 608
ToxR, osmosensor, 3:513–514
TPP, *see* Thiamine pyrophosphate
Trace gas
consumption reactions
ammonia, 3:262
carbon monoxide, 3:261
carbonyl sulfide, 3:261
dimethylsulfide, 3:261–262
hydrogen, 3:261
methane, 3:261
nitric oxide, 3:262
nitrous oxide, 3:262
definition, 3:256
importance, 3:257–258
microbial processes in cycling, 3:258–259
mixing ratio, 3:256, 258
production reactions
ammonia, 3:261
carbon monoxide, 3:260
carbonyl sulfide, 3:260
dimethylsulfide, 3:260
hydrogen, 3:259–260
methane, 3:260
nitric oxide, 3:260–262
nitrous oxide, 3:260
sources, 3:257–258
types, 3:257
Trans-acting element, definition, 2:618
Transconjugant, definition, 1:86, 847
Transcription
definition, 2:71, 618, 4:628
elongation
cleavage of transcripts, 4:620
DNA replication implications, 4:622
DNA topology effects, 4:622
inchworm model, 4:620–621
pausing and arrest, 4:619–620
rate, 4:618–619
slippage, 4:621–622
initiation
activation, 4:617–618
DNA topology in regulation, 4:618
repression, 4:617
stages, 4:615–617
overview, 4:610–611
promoter recognition
elements, 4:613–615
mechanism, 4:615
RNA polymerase in bacteria, *see also* Sigma factor
catalysis, 4:611
fidelity, 4:619
sigma factor, role in initiation, 4:612–613
subunits, 4:611–612
termination
antitermination, 4:625–626
auxiliary termination factors, 4:625
intrinsic terminators, 4:623–624
overview, 4:623
rho-dependent termination, 4:624–625
viruses
adenovirus, 4:632–633
herpes simplex virus, 4:631–632, 636
JC virus, 4:630
papillomaviruses, 4:630
retroviruses
complex, 4:634–636
simple, 4:634
simian virus 40, 4:628–630
Transcription-coupled nucleotide excision repair
overview, 2:77–78
pyrimidine dimer and RNA polymerase effects, 2:78, 80
transcription repair coupling factor, 2:80–81
Transcription factor, definition, 2:826
Transcription unit
definition, 4:127, 610
organization, 4:128–129
Transducer, biosensors, 1:611–612
Transduction, *see also* Horizontal gene transfer
definition, 2:271, 699–700, 3:363, 711, 4:43, 637, 638
generalized transduction
abortive transduction, 4:637, 645–646
degradation of transduced DNA, 4:645
DNA packaging
host, 4:643–645
phage, 4:643
miscellaneous systems, 4:647
overview, 4:638, 642–643
stable transduction, 4:637, 646–647
gram-negative bacteria, 2:280, 282
laboratory uses
mapping, 4:648–649
mutagenesis, 4:648
strain improvement, 4:649
lactic acid bacteria, 3:6
natural environments, 4:649
plasmid transfer, 3:724
specialized transduction
bacteriophage λ, 4:640

comparison to generalized transduction, 4:638
specialized phages, 4:640, 642
Transesterification
definition, 3:49
lipase reactions, 3:51–52
Transfectant virus, definition, 2:797
Transfection, definition, 4:81
Transfer range, definition, 2:698
Transferrin
definition, 2:860
iron uptake, 2:865
Transfer RNA, *see also* Protein biosynthesis; RNA splicing
aminoacyl transfer RNA synthetases, 3:828
attenuation mediation, *Bacillus subtilis*, 1:346
function, 3:826
initiator transfer RNA, 3:827–828
structure, 3:826–827
Transformation, *see also* Horizontal gene transfer
artificial systems
calcium chloride transfection, 4:660–661
electroporation, 4:662–663
protoplast transformation, 4:663
competence, regulation of development, 4:653–654
definition, 2:92, 271, 591, 678, 699–700, 3:456, 711, 4:43, 651, 796
diversity of systems, 4:652
DNA binding and uptake
Bacillus subtilis, 4:655–656
Escherichia coli, 4:661–662
Haemophilus influenzae, 4:656–657
Neisseria gonorrhoeae, 4:657
overview, 4:654–655
Streptococcus pneumoniae, 4:655
ecological importance, 4:659–660
eukaryotic microbes, 4:663–664
gram-negative bacteria, 2:282
Haemophilus influenzae, 2:595–596
history of study, 4:651–652
integration of DNA
homologous recombination, 4:657
plasmid DNA processing, 4:657–658
sexual isolation, DNA repair and restriction, 4:658–659
lactic acid bacteria, 3:6
plasmid transfer, 3:724
strain improvement, 4:438
Transgenic animal, *see also* Genetically modified organism
applications, overview, 4:666–667
definition, 3:809
DNA transfer directly into embryos
pronuclear microinjection, 4:667–668
retroviral gene transfer, 4:668–669
sperm-mediated DNA transfer, 4:669
founder, 4:666
historical perspective, 4:667
milk expression of transgenic products
α-1–antitrypsin, 4:682–683
calcitonin and other peptides, 4:683–684
efficacy, safety, and acceptability, 4:684–685
fibrinogen, 4:683
protein types and expression levels, 4:678–679
milk protein expression in transgenic livestock
protein types, 4:673–675
regulation of transgene expression
caseins, 4:675, 677
complementary DNA expression, 4:681–682
whey proteins, 4:677, 680–681
mosaic, 4:666
nuclear transfer
background, 4:671–672
cultured cells, 4:672–673
transgenic animal generation, 4:673
stem cell-mediated transgenesis
embryonic germ cells, 4:670–671
embryonic stem cells, 4:669–670
spermatogonial stem cells, 4:671
xenotransplantation applications
barriers to conventional xenotransplantation
immune response, 4:686–687
physiological problems, 4:687
zoonosis, 4:687–688
demand, 4:685
donor species selection, 4:685–686
prospects, 4:688
Transgenic plant, definition, 3:662
Transition mutation, definition, 3:307
Transition state analog, definition, 1:636
Translation, *see* Protein biosynthesis; Ribosome
Translational inhibitor
aminoglycosides, 1:165–166
definition, 1:162
Translation factor, definition, 3:824
Transmembrane domain
ATP-binding cassettes, 1:6
definition, 1:1
Transmembrane pH difference, definition, 3:625
Transmissible spongiform encephalopathy (TSE)
application of Koch's postulates, 3:234–236
definition, 3:809
types, 3:819–821
Transmission electron microscopy (TEM)
definition, 3:276
principles, 3:277–278
specimen preparation
freeze-etching, 3:281–282
freeze substitution, 3:282
negative staining, 3:278–280
shadow-casting, 3:280
thin sectioning, 3:278
Transovarial transmission
definition, 1:311, 4:140, 526
symbiotic microorganisms, 4:535
Transposable element
bacteriophages
bacteriophage Mu
replication and integration, 4:721
structure, 4:720–721
D108, 4:721
overview, 4:705–706
complex transposons
antibiotic resistance genes, recruitment to Tn*21*-like transposons, 4:715–716
family branches, 4:713–715
one-ended transposition, 4:717
overview, 4:713
Tn*3*, 4:713
Tn*3*-related transposons, 4:713
transposition functions, 4:713

Transposable element (*continued*)
transposition immunity, 4:704, 717
transposition mechanisms, 4:716–717
conjugative transposons
origin, 4:717–718
properties, 4:718
transposition, 4:718
definition, 3:711, 4:704
distribution, 4:706
genome rearrangement, 4:722–723
insertion sequence elements
composite transposons, 4:712
history, 4:707
structure, 4:707–709
transposition mechanisms, 4:709–712
nomenclature, 4:706
replicon fusion and conduction, 4:723
site-specific transposons
Tn7
chromosomal site of insertion, 4:718–719
inverted repeats, 4:719–720
overview, 4:718
transposition functions and mechanism, 4:719
Tn*502*, 4:720
Tn*554*, 4:720
Streptomyces, 4:456–457
structure, 4:704–705
yeast transposons
transposition, 4:722
Ty element structure, 4:722
types, 4:721–722
Transposition, definition, 3:307, 4:451
Transposon, *see also* Transposable element
antibiotic resistance transfer in gram-negative anaerobic pathogens, 2:568–569
catabolic plasmids for biodegradation, 3:742–743
Caulobacter mutagenesis, 1:694, 701
complex transposons
antibiotic resistance genes, recruitment to Tn*21*-like transposons, 4:715–716
family branches, 4:713–715
one-ended transposition, 4:717
overview, 4:713
Tn3, 4:713
Tn3-related transposons, 4:713
transposition functions, 4:713
transposition immunity, 4:704, 717
transposition mechanisms, 4:716–717
conjugative transposons
origin, 4:717–718
properties, 4:718
transposition, 4:718
definition, 1:692, 847, 2:618, 3:307, 363, 594, 4:451, 704
Myxococcus xanthus mutagenesis, 3:365
recombination in gram-negative bacteria, 2:278–280
site-specific transposons
Tn7
chromosomal site of insertion, 4:718–719
inverted repeats, 4:719–720
overview, 4:718
transposition functions and mechanism, 4:719
Tn*502*, 4:720
Tn*554*, 4:720
yeast transposons
transposition, 4:722
Ty element structure, 4:722
types, 4:721–722
Transversion, definition, 3:307
Tranzchel's law, definition, 4:195
Tra proteins, pilus assembly role, 2:373–374
TraR, Ti plasmid conjugation role, 1:89
Trehalose, biosynthesis, 4:892
Trehalose lipids, biosurfactant production and features, 1:623
Treponema pallidum
discovery, 4:356
features, 4:268–269
genome, 4:363
pathogenesis, 4:269
syphilis, *see* Syphilis
Triacylglycerol, definition, 3:62
Triangulation number, definition, 3:697
Triarimol, antifungal activity, 1:240
Trichinella spiralis
food-borne illness, 2:408–409
meat products as source, 3:167–168
Trichocyst, definition, 2:42
Trichome, definition, 1:907
Trichothecenes
animal disease, 3:344–345
human disease, 3:341–342
types, 3:341–342
Trichuris trichiura, food-borne illness, 2:408
Trickling filter, municipal wastewater treatment, 4:879
Tridemorph, antifungal activity, 1:249
Triethylamine-*N*-oxide (TMAO), reduction in anaerobic respiration, 1:186
Trifluorothymidine
adverse effects, 1:295
indications, 1:295
mechanism of antiviral activity, 1:295
resistance, 1:295
Triglyceride, definition, 3:49
Triticonazole, antifungal activity, 1:243
tRNA, *see* Transfer RNA
Trophozoite, definition, 3:746
Tropism, definition, 4:779, 832
trp operon
Bacillus subtilis, 1:344–345
Escherichia coli, 1:341–342
Trypanosoma brucei
antigenic variation
expression site
associated genes, variation of encoded receptors, 1:259
in situ activation, 1:257
invariant antigen protection, 1:259–260
life cycle, 1:255
procyclin, 1:255–256
variant surface glycoprotein
developmental control of expression, 1:256–257
repertoire programming and evolution, 1:258–259
replacement of gene, 1:257–258
stage-specific expression, 1:256
genome, 4:735–736
life cycle, 4:734–735
RNA editing, 4:735–736
sleeping sickness
clinical manifestations, 4:738
diagnosis, 4:738–739
epidemiology, 4:736–737

forms, 4:734
pathogenesis, 4:737–738
prophylaxis and prevention, 4:741
treatment
diminazene aceturate, 4:739
eflornithine, 4:740–741
melarsoprol, 4:740
pentamidine, 4:739
suramin, 4:739–740
subspecies, 4:733–734
Trypanosoma cruzi
animal infection, 4:729
Chagas' disease
acute manifestations, 4:730
chronic manifestations, 4:730–731
diagnosis, 4:732–733
epidemiology, 4:729
history, 4:725
immunosuppressed patients, 4:731–732
pathogenesis, 4:732
transfusion-associated infection, 4:731
treatment, 4:733–734
chromosomes, 4:726
genome organization, 4:728
immune response, 4:732
life cycle, 4:726
transmission, 4:725–726
Tryptophan
biosynthesis
pathway, 1:147–148
regulation, 1:148–149
enzymatic production process, 1:159–160
functions, 1:147
uses, 1:159
TSE, *see* Transmissible spongiform encephalopathy
TSS, *see* Total suspended solids
Tuberculosis (TB), *see also Mycobacterium*
aerosolization, 1:73–74
bovine disease, 3:314–315
definition, 1:28, 69
health threat compared to AIDS, 1:115
history of study, 1:71–72
human disease
diagnosis, 3:324–325
epidemiology, 3:313–314, 323–324
history, 3:313
multidrug resistance, 3:323–325
pulmonary pathogenesis, 3:324
treatment, 3:325
infectious dose, 1:76
prevention, 1:76–77
propagation, 1:72
treatment, 1:73
Tubular necrosis, definition, 1:288
Tumor-inducing (Ti) plasmid, *see also* T-DNA
agrocin84 susceptibility, 1:80–81
conjugation, 1:89
definition, 1:78
genes, 1:79–81, 87–89
virulence, 1:79–80
Tumor necrosis factor (TNF), functions, 1:735
Tumor suppressor gene, definition, 3:456
Turgor pressure, definition, 3:502
Turgor, definition, 4:884
Turnover, thermally-stratified lakes, 2:438
Tus, termination of DNA replication, 2:90
Twin-arginine protein secretion pathway, *see* Secretion, proteins
Two-component system, signal transduction
domain incorporation into signal transduction systems, 4:750–751
occurrence of domains
bacteria, 4:752–753
homology searching, 4:752–753
plants, 4:753
yeast, 4:753
overview, 4:742–744
phosphorylation state of response regulator, factors affecting
acetyl phosphate role, 4:751
cross-regulation, 4:751–752
flow of phosphoryl group, 4:751
multiple, independently-regulated phosphatases, 4:752
phosphotransfer domains, 4:742–743, 750
receiver domains, 4:742–744, 747–750
transmitter domains, 4:742–747
Tylose, definition, 4:788
Typhoid
clinical features, 4:755
epidemiology, 4:757
history, 4:755–757
prevention and treatment, 4:757
taxonomy and classification of bacteria, 4:756–757
vaccine, 4:773–774
Typhus fever, *see* Rickettsiae

U

Ulcer
definition, 2:628
Helicobacter pylori role, 2:628, 633
Ultraviolet germicidal irradiation (UVGI)
air disinfection, 1:75
definition, 1:69
umuDC
definition, 4:43
SOS response role, 4:48
Uniport, definition, 2:178, 181
Universal precautions, definition, 2:782
Unsaturated zone, definition, 1:587
Updwelling, definition, 3:93
UP element, definition, 4:127
Upflow anaerobic sludge blanket reactor, industrial effluent treatment, 2:756
Upland soil, definition, 3:256
Uptake signal sequence (USS), definition, 2:591–593
UR-9746, antifungal activity, 1:247–248
UR-9751, antifungal activity, 1:247–248
Uranium, ore bioleaching, 3:487
Urease
definition, 2:628
Helicobacter pylori virulence factor, 2:631
Uropophyrinogen III
biosynthesis, 4:561
decarboxylation, 4:563
Uroporphyrin III, biosynthesis, 4:564
Usher proteins
chaperone–usher assembly pathway
molecular architecture, 2:362, 364, 2:366
outer-membrane ushers, 2:369
periplasmic chaperones, 2:366–3614
definition, 2:361
USS, *see* Uptake signal sequence
Uveitis, definition, 1:288

UVGI, *see* Ultraviolet germicidal irradiation

V

VacA, *see* Vacuolating cytotoxin
Vaccination
  adaptive immunity, 4:779–780
  attenuation of viruses, 4:780–782
  definition, 2:307, 4:767
  diseases, *see* Vaccines
  eradication of bacterial diseases, 4:777
  history, 2:686–687, 4:809
  immunology of viral vaccines, 4:784–785
  live viral vaccines, 4:780–782
  molecular approaches to viral vaccine design, 4:785–786
  objectives, 4:779
  public health impact
    bacterial vaccines, 4:767–768
    viral vaccines, 4:786–787
  virulence factor targets, 4:768
Vaccines
  anthrax, 4:775
  Borrelia burgdorferi, 3:128–129, 4:776–777
  definition, 1:834, 2:722, 678
  diptheria, 4:769–770
  enterovirus, 2:208–209
  Haemophilus influenzae, 4:771
  Helicobacter pylori, 2:634
  influenza virus, 2:807–808
  malaria, 3:758–760
  Mycobacterium tuberculosis, 4:776
  needle-free, 2:736
  Neisseria meningitidis, 4:772–773
  noninfectious viral vaccines, 4:782–784
  paramyxoviruses
    attenuated live virus, 3:544
    genetically-engineered virus, 3:545
    killed vaccine, 3:544
    subunit vaccines, 3:544–545
  pertussis, 4:770–771
  plague, 4:775–776
  polio
    administration, 3:768
    adverse events, 3:768
    contraindications, 3:768
    dosage, 3:768–769
    efficacy, 3:768
    schedule, 3:768
    storage, 3:769
    types and selection factors, 3:767–768
  rabies
    history, 4:16
    immune response, 4:28–30
  Salmonella typhi, 4:773–774
  smallpox, 4:767, 780, 809
  Streptococcus pneumoniae, 4:771–772
  tetanus, 4:769
  types
    bacterial vaccines, 4:769
    viral vaccines, 4:780
  typhoid, 4:757
  Vibrio cholerae, 4:774–775
  Yersinia pestis, 3:658
Vacuolating cytotoxin (VacA)
  definition, 2:628
  Helicobacter pylori virulence factor, 2:632
Valine
  biosynthesis
    pathway, 1:143
    regulation, 1:144–145
    substrate specificities of enzymes, 1:143–144
  functions, 1:143
Variable cost, definition, 2:137
Variant surface glycoprotein (VSG), *Trypanosoma brucei*
  developmental control of expression, 1:256–257
  repertoire programming and evolution, 1:258–259
  replacement of gene, 1:257–258
  stage-specific expression, 1:256
Varicella, definition, 1:288
Varicella-zoster virus, definition, 1:288
Variola
  definition, 4:289
  disease, *see* Smallpox
Vascular system, definition, 3:607
Vascular wilt disease
  control, 4:794–795
  crop types affected, 4:790
  definition, 4:788
  diagnostics, 4:793–794
  symptoms, 4:790
Vector
  definition, 3:109, 607, 662, 676, 4:81, 758, 811, 955
  phloem-limited bacteria, 3:612–613
  plant disease resistance to insects, 3:667–668
Vectorial capacity
  definition, 3:131
  malaria-carrying mosquito, 3:144–146
Vegetative cell, definition, 4:377, 394
Vegetative compatibility, definition, 4:788
Vegetative growth, definition, 2:15
Ventilation
  air disinfection, 1:74–75
  building design and operation
    evaporative cooling systems, 1:120
    heating, ventilation, and air-conditioning systems, 1:120
    portable systems, 1:120
  definition, 1:69–70
Vertical reproduction, definition, 2:698
Vertical resistance
  definition, 3:662
  plant disease resistance, 3:664
Vertical transmission
  definition, 4:526
  symbiotic microorganisms, 4:535–537
Verticillium
  culture, 4:788–789
  genetics, 4:793
  host defense mechanisms, 4:792
  interaction of pathogenicity factors with defense mechanisms, 4:792–793
  life cycle
    dormant phase, 4:789
    parasitic phase, 4:789–790
  pathogenicity factors, 4:791–792
  resistance, 4:791
  species overview, 4:788
  susceptibility, 4:791
  tolerance, 4:791
  vascular wilt disease
    control, 4:794–795
    crop types affected, 4:790
    diagnostics, 4:793–794
    symptoms, 4:790
  vegetative compatibility groups, 4:794
Vgb, functions, 1:67
Vibrio
  food-borne illness

miscellaneous species, 2:396
non-cholerae, enteric disease, 2:198–199
Vibrio cholerae, 2:394–395
Vibrio parahemolyticus, 2:395
Vibrio vulnificus, 2:395
quorum sensing
autoinduction of *lux* genes, 4:2–3
bioluminescence regulation, 4:2
model for cell density sensing, 4:3
Vibrio fisheri, 4:9–10
Vibrio harveyi, 4:9
skin microbiology, 4:286–287
Vibrio cholerae, *see also* Cholera
biotypes, 1:790–792
features, 1:791, 799
noninflammatory diarrhea, 2:193
serogroups, 1:790–792
toxins, 2:193
vaccine, 4:774–775
Vibrio parahaemolyticus, refrigerated food pathogens, 4:72
Vidarabine
adverse effects, 1:297
indications, 1:297
mechanism of antiviral activity, 1:297
resistance, 1:297
Viniculture, definition, 4:914
Vinification, definition, 4:914
Viral infection, plants, *see also* Beet necrotic yellow vein virus; Luteoviridae; Potyviruses
cytological effects of infection, 3:702–703
dissemination, 3:690–691, 703–704
management, 3:691–692, 709
pathogens
classification, 3:690, 697–698
DNA viruses, 3:707
epidemiology, 3:691
gene expression, 3:707–708
host specificity, 3:701–702
infection, 3:691
morphology, 3:688, 690
movement through plant, 3:704–705
overview, 3:688
particle structure, 3:705–706
properties, table, 3:700
RNA genomes
distribution, 3:706
double-stranded, 3:707
single-stranded minus-sense, 3:706–707
single-stranded plus-sense, 3:706
replication
DNA viruses, 3:709
double-stranded RNA viruses, 3:708
overview, 3:690, 708
single-stranded minus-sense viruses, 3:708
single-stranded plus-sense viruses, 3:708
symptoms of infection, 3:702–703
Viremia, definition, 1:288, 311, 4:832
vir genes
activation, 1:89–90
definition, 1:78, 81–82
regulators of expression, 1:90–92
T-DNA processing and transport, 1:82–84, 92–98
VirB properties, 1:97–98
Virion, definition, 1:288, 398, 4:592, 796
Viroid, definition, 3:663
Virology, historical perspective, 2:688–691
Viropexis, definition, 4:15
Virulence, definition, 2:678, 3:676, 801, 4:297, 394, 832
Virulence factor, definition, 1:789, 3:109, 4:32
Virulent, definition, 4:637, 921
Virus, *see also specific viruses*
cellular interactions, 4:807–808
classification, 4:797–799
composition, 4:797
definition, 4:796, 832
diversity, 2:67–68
emerging infection
anticipation of emergence, 4:827–828
arenaviruses, 4:815
causes of emerging infection
animal viruses as interspecies transfer models, 4:826
arboviral diseases, 4:825
influenza pandemic, 4:824–825
overview, 4:823
trafficking, 4:822–826
control of disease, 4:828–829
eradication, prospects and requirements, 4:829–830
evolution of variants, 4:821–822
examples, overview, 4:813–815
filoviruses, 4:815
flaviviruses, 4:816
hantavirus, 4:816–817
Hendra virus, 4:817–818
hepatitis, 4:818
influenza virus, 4:818–819
molecular diagnostics, 4:812–813
newly recognized viruses, 4:812
Nipah virus, 4:818
overview, 4:811–812
parvovirus B19, 4:819
restraints on emergence, 4:826–827
retroviruses, 4:819–820
Rift Valley fever virus, 4:820
sources of new viruses, 4:821
enterovirus receptor, 2:201
eradication of diseae, 4:809
families, 4:800
history of study, 4:799–802
horizontal gene transfer, frequency estimation, 2:705
host organism interactions, 4:809
immune response, 4:834–835
insecticide applications
granulosis virus, 2:818–820
host specificity, 2:818
limitations, 2:820
nuclear polyhedrosis virus, 2:818–820
overview, 2:817–818
pest control examples, 2:819–820
production and use, 2:820
recombinant viral insecticides, 2:820–821
types of viruses, 2:818–819
mutation, 4:805–806
outcomes of infection, 4:835
plant infection, *see* Viral infection, plants
recombination, 4:806–807
replication
assembly, 4:805
attachment, 4:802
eclipse phase, 4:833
entry and uncoating, 4:802–803
genome expression and replication, 4:803–805
overview, 4:832–833
structure, 4:796–797
transcription

Virus (*continued*)
adenovirus, 4:632–633
herpes simplex virus, 4:631–632, 636
JC virus, 4:630
papillomaviruses, 4:630
retroviruses
complex, 4:634–636
simple, 4:634
simian virus 40, 4:628–630
transmission, 4:833–834
vaccines, *see* Vaccination; Vaccines
vectors for gene expression, 4:809–810
virulence, 4:835–836
Viscosity
bioreactor monitoring, 1:572
definition, 1:556
Vitalism, definition, 2:678
Vitamin, *see also* Coenzyme; *specific vitamins*
definition, 4:837
needs and ues, 4:838–839
production processes
advantages of fermentation, 4:852
applications for products, 4:852
ascorbic acid, 4:849–851
biotin, 4:846
β-carotene, 4:839–841
coenzyme A, 4:845–846
improvement strategies, 4:851–852
niacin, 4:843, 845
orotic acid, 4:849
pantothenic acid, 4:845
pyridoxine, 4:846
riboflavin, 4:843
thiamin, 4:843
vitamin B_{12}, 4:846
vitamin D, 4:841
vitamin E, 4:842
vitamin F, 4:842
vitamin K, 4:842–843
Vitamin B_1, *see* Thiamin
Vitamin B_2, *see* Riboflavin
Vitamin B_3, *see* Niacin; Nicotinamide
Vitamin B_5, *see* Pantothenic acid
Vitamin B_6, *see* Pyridoxine
Vitamin B_8, *see* Biotin
Vitamin B_{12}
biosynthesis, 1:843, 4:563
definition, 3:647
microbial production processes, 3:649, 4:846
Vitamin B_{13}, *see* Orotic acid
Vitamin C, *see* Ascorbic acid
Vitamin D, microbial production processes, 4:841
Vitamin E, microbial production processes, 4:842
Vitamin F, microbial production processes, 4:842
Vitamin K, microbial production processes, 4:842–843
Volatile fatty acid, definition, 2:485
Volatile suspended solids (VSS), definition, 2:740
Voriconazole, antifungal activity, 1:245
VPg, definition, 3:792
VSG, *see* Variant surface glycoprotein
VSS, *see* Volatile suspended solids
VX-478
adverse effects, 1:309
indications, 1:309
mechanism of antiviral activity, 1:309
resistance, 1:309

W

Walker motif, definition, 1:1
Waste activated sludge, definition, 4:870
Wastewater, *see* Industrial effluents; Industrial wastewater; Municipal wastewater
Water, *see also* Osmotic stress
total stocks on Earth, 2:739
water activity, definition, 3:502–503, 4:884
Waterborne disease, *see* Biomonitor
Water-deficient environments, *see also* Halophile; Osmotic stress
adaptation of microorganisms, 4:885–886
compatible solutes
biosynthesis, 4:889, 891–892
characteristics, 4:887–889
commercial applications, 4:896
metabolism, 4:895–896
transporters
Bacillus subtilis, 4:894–895
efflux systems, 4:895–896
Escherichia coli, 4:893–894
features, 4:892–893
Salmonella typhimurium, 4:893–894
halophile coping strategies, 4:886–887
overview, 4:884–885
Water, drinking, *see also* Freshwater microbiology
bottled water, 4:911–912
disease control, overview, 4:898
distribution systems
biofilms, 4:909–911
overview, 4:908–909
emergency drinking water, 4:912
Environmental Protection Agency regulations
general, 4:900
ground water rule, 4:901
miscellaneous rules, 4:901–902
surface water treatment requirements, 4:901
total coliform rule, 4:900–901
indicator microorganisms
detection, 4:898–899
ideal characteristics, 4:899
species, 4:899–900
plumbing system biofilms, 4:911
point-of-use–point-of-entry treatment, 4:912–913
sources, 4:902
treatment
coagulation, flocculation, and sedimentation, 4:902–903
disinfection
efficiency, 4:907–908
techniques, 4:904–905
efficiency of removal
diatomaceous earth filtration, 4:907
rapid granular filters, 4:906–907
slow sand filters, 4:905–906
filtration
diatomaceous earth filtration, 4:904
membrane processes, 4:904
rapid granular filters, 4:903–904
slow sand filters, 4:904
Weathering, definition, 1:472
Weeds, biocontrol, *see also* Biopesticide
classic biocontrol, 1:486–487, 541
inundative biocontrol
constraints, 1:489–490
foliar and stem pathogens, 1:487–488

overview, 1:486–487, 541
soil microbials, 1:488–489
rationale, 1:486
Welan, features, 1:560–561
Western blot
definition, 1:422, 2:29
diagnostics, 2:35
whi genes, sporulation role, 2:19
WHO, *see* World Health Organization
Wilt, *see Ralstonia solanacaerum*
Wine
alcoholic fermentation, 4:915
classification, 4:917
definition, 4:914
flavor and acceptability, 4:919
fortified wines, 4:919
grapes, 4:917
historical perspective, 4:914–915
malolactic fermentation, 4:916
manufacturing steps, 4:917–919
nobel mold, 4:916
sherry wines, 4:917
sparkling wines, 4:919
spoilage microorganisms, 4:916–917
yeasts, 4:915–916
Wood, *see* Lignocellulose; Pulp and paper; Timber and forest products
Wood cell wall, definition, 4:571
Wood rot, organisms, 1:472–473
World Health Organization (WHO), *see* International law, infectious disease
World Wide Web, stock culture collection resources, 4:407
Wort, beer brewing, 1:412, 415

X

Xanthan
physical properties, 1:557–559
production, 1:559, 4:928–929
structure, 1:557
uses, 1:559–560, 4:929
Xanthomonas
biotechnology
prospects, 4:924–925
xanthan prouction, 4:928–929
detection, 4:924
ecology
epitope studies, 4:925
functional relationship with plants, 4:925–926
survival, 4:926
genetic diversity, 4:923–924
host recognition and response, 4:922–923
host resistance genes, 4:923
plant diseases
damage to plants, 4:928
dissemination, 4:927–928
host range, 4:927
invasion, 4:927
overview, 4:921–922, 926–927
taxonomy, 4:922
Xenobiotic
definition, 1:461, 3:594
recalcitrance, 1:465–466
Xenotransplantation
barriers to conventional xenotransplantation
immune response, 4:686–687
physiological problems, 4:687
zoonosis, 4:687–688
definition, 2:842, 4:666, 811
transgenic animal applications
demand, 4:685
donor species selection, 4:685–686
prospects, 4:688
X protein, carcinogenesis role with hepatitis, 2:645
Xylanase
classification, 4:932
definition, 4:930
food industry applications, 4:937
mode of action, 4:934–935
pH optimum, 4:932
physical properties, 4:931–932
production, 4:933–934
pulp and papermaking applications
debarking, 4:935–936
de-inking, 4:936
dissolving pulp production, 4:936
fiber modification, 4:936
prebleaching of pulps, 4:935
retting of flax fibers, 4:936
shive removal, 4:936
purification, 4:934
substrate specificity, 4:932, 934
temperature optima, 4:932
xylan structure, 4:930–931
Xylem, definition, 3:607, 676
β-Xylosidase, features and applications, 2:227
Xyl proteins, transcriptional regulation of plasmids, 3:741

Y

Yeast, *see also individual species*
beer brewing, 1:412, 417–418
biodiversity, 4:940–941
carbon metabolism
sources for growth, 4:945–946
sugar metabolism, 4:946–947
sugar transport, 4:946
chromosome
replication origin
binding sites for primary initiation proteins, 1:825–826
nucleoprotein complex assembly/disassembly, 1:829–830
overview, 1:824
segregation, 1:830–832
cytology methods, 4:942–943
definition, 2:468, 4:939–940
ecology, 4:941–942
food chain roles, 4:941
functional components, 4:944
genetic manipulation, 4:951
genome and proteome projects, 4:952
growth
population growth, 4:949–950
vegetative reproduction, 4:949
habitats, 4:941
life cycle, 4:950–951
nitrogen metabolism
metabolism of compounds, 4:948–949
sources for growth, 4:947
transport of compounds, 4:947–948
nutrition
culture media, 4:945
requirements, 4:944–945
physical requirements for growth, 4:945
shapes, 4:942
significance
agriculture, 4:952–953

Yeast (*continued*)
industry, 4:952
medicine, 4:953
taxonomy, 4:940
ultrastructure, 4:943–944
wine making, 4:915–916
Yeast artificial chromosome
definition, 3:151
genome mapping, 3:158
Yellow diseases, *see* Phytoplasma
Yellow fever, overview, 1:311–312, 314–315
Yellow fluorescent protein, definition, 1:520, 523
Yersinia
enteric fever, 2:197–198
food-borne illness, 2:400–401
virulence factors, 2:197–198
Yersinia enterocolitica, refrigerated food pathogens, 4:72
Yersinia pestis
antibiotic therapy, 3:658–659
chromosome features, 3:660–661
diagnostics, 3:659
diseases, *see* Plague
features, 3:655–657
fimbriae, 3:661
hemin storage locus, 3:661
pathogenicity island, 3:661
plasmids
pFra, 3:660
pPla, 3:660
pYV, 3:660
prophylactic therapy, 3:658
taxonomy, 3:655
vaccination, 3:658
virulence, 3:659–660
Yield coefficient, definition, 1:873–874
Yogurt, *see* Dairy products; Lactic acid fermentation

Z

Zalcitabine
adverse effects, 1:304
indications, 1:304
mechanism of antiviral activity, 1:304
resistance, 1:304
Zaragozic acid, mechanism of action, 3:779
Zearalenone, animal disease, 3:347
Zidovudine, *see* Azidothymidine
Zinc, growth requirements and functions, 3:439
Zofenopril, chiral biocatalysis, 1:431–432
Zoonosis, *see also specific diseases*
Chlamydia, 1:787
classification, 4:956–958
cyclozoonoses, 4:960–961
definition, 1:781, 2:852, 3:27, 4:598, 666, 811, 955–956
emerging zoonoses, 4:963–965
metazoonoses
equine encephalomyelitis, 4:962
Lyme disease, 4:961–962
plague, 4:961
orthozoonoses, 4:959–960
prevention and control, 4:965
public health and economic impact, 4:958–959
saprozoonoses, 4:962–963
xenotransplantation, 4:687–688
Zoonotic disease, definition, 1:311
Zoospore, definition, 2:485, 3:633, 792
Zoster
definition, 1:288
ophthalmicus, 1:288
treatment
acyclovir, 1:292
famciclovir, 1:296
Zot, cholera virulence factor, 1:794

ISBN 0-12-226804-0